14th International Symposium on Magnetic Bearings 2014

Linz, Austria
11-14 August 2014

ISBN: 978-1-5108-2260-3

Printed from e-media with permission by:

Curran Associates, Inc.
57 Morehouse Lane
Red Hook, NY 12571

Some format issues inherent in the e-media version may also appear in this print version.

Printed by Curran Associates, Inc. (2016)

For permission requests, please contact Linz Center of Mechatronics GmbH
at the address below.

Linz Center of Mechatronics GmbH
Altenberger Straße 69
4040 Linz
Osterreich

Phone: +43 732 2468-6002
Fax: +43 732 2468-6005

office@lcm.at

Additional copies of this publication are available from:

Curran Associates, Inc.
57 Morehouse Lane
Red Hook, NY 12571 USA
Phone: 845-758-0400
Fax: 845-758-2633
Email: curran@proceedings.com
Web: www.proceedings.com

ISMB 2014 Table of Contents

Keynote Speakers

Active Magnetic Bearings I

Active Magnetic Bearings II

Active Thrust Bearings

Advanced Control I

Advanced Control II

Backup/Touch Down-Bearings

Bearingless 3-Phases Systems

Bearingless Motors I

Bearingless Motors II

Bearingless Pumps

Bearingless Slice Motors I

Bearingless Slice Motors II

Bearingless SRM

Development Tools

Electrodynamic Bearings

Fault Detection and Diagnosis

Flexible Rotor Control

Flywheels I

Flywheels II

Kinetic Energy Storage Systems I

Kinetic Energy Storage Systems II

Linear Actuators

MagLev Technology for Artificial Hearts

Magnetic Bearings in Energy Technology

Magnetic Suspension Systems

Materials in Magnetic Bearing Technology

Modeling and Identification

Passive Magnetic Bearings I

Passive Magnetic Bearings II

Power Electronics for MB

Rotor Dynamics

Self-Sensing Techniques

Sensor Related Topics I

Sensor Related Topics II

Special Actuators

Unbalance Handling

Zero Power Control

Welcome to ISMB14

Welcome to the jubilee edition of the ISMB event series. Within the last 25 years, the International Symposium on Magnetic Bearings has brought together a high number of international experts coming from science and industry in a two years sequence. Meanwhile more than 1000 papers representing the state of the art of magnetic bearing technology have been published in the ISMB proceedings.

The first international meeting, allowing the exchange and presentation of ideas, development and research results was organized in Zurich, Switzerland, in June 1988 by Prof. Gerhard Schweitzer. Besides himself, also Prof. Paul Allaire (University of Virginia) and Prof. Toshio Higuchi (University of Tokyo), who organized the two follow-up symposia in 1990 and 1992, can be counted to the co-founders of ISMB. The first International Advisory Board was composed of only five persons: P. Allaire, H. Habermann, T. Higuchi, R. Holmes and G. Schweitzer. Their primary mission to bring together and establish connections between international research groups from science as well as from industry has proceeded very successfully. Without any doubt, today the ISMB events can be considered the most significant researcher platform in the fields of magnetic bearings worldwide.

With the 14th International Symposium on Magnetic Bearings (ISMB14), the series of ISMB meetings is held in Austria for the first time. ISMB14 is organized by LCM GmbH with the support of the Johannes Kepler University Linz. It will provide an excellent opportunity to share new scientific and technical results, to get to know about challenging problems, to discuss methods and approaches, to cultivate old friendships, to build up new connections, and to meet well-known experts from science and industry as well as young ambitious scientists at the beginning of their professional career.

The proceedings of ISMB14 contain all of the peer-reviewed papers presented at the symposium, covering and illustrating many fields of magnetic bearing technology. In particular we are grateful to our keynote speakers, to the organizers of the Mini-Symposia, and to the exhibitors, whose contributions enrich the symposium's program to a great extent.

We would like to extend our thanks also to our sponsors Cerobear GmbH, evico GmbH, Hanning Elektro-Werke GmbH & Co., Land Oberösterreich, Levitec GmbH, Levitronix GmbH, Tourismusverband Linz, Rieter CZ s.r.o., Schaeffler AG, Stadt Linz, Vacuumschmelze GmbH & Co. and Waukesha Magnetic Bearings. Their support has been a great help for financing and arranging the symposium.

Our special thanks go to the numerous colleagues and volunteers coming from the Institute of Electrical Drives and Power Electronics, JKU, from LCM GmbH and from external partners. In particular, this concerns Mrs. Kornelia Leitenberger and Mrs. Sonja Sloma, who managed the Symposium secretariat, Wilhelm Hopfinger and Günther Seiberl, who were responsible for technical installations and equipment, Dietmar Andessner supplying us with his photography services, Walter Bauer, Gerd Bramerdorfer, Peter Dirnberger, Edmund Marth, Gertraud Müller, Josef Passenbrunner, Doris Prohaska, Andreas Pröll, and Karlo Radman, Rudolf Scheidl, and Wolfgang Weiß, who unselfishly supported us in multidutinousorganizational matters. ISMB14 would not have been possible without their help.

This also includes the members of the International Advisory Board and the International Program Committee who spent hours of their valuable time to review the submitted papers to eventually get an average of almost three reviews per paper. The review process was finished with an amazingly high review feedback of 95%! Furthermore, Tom Wehner (Alliance Management Group, LLC) provided an excellent electronic platform for all paper handling and Melanie Grasböck together with Thomas Zauner (OÖ Tourismus GmbH) kindly handled the registration process as well as the organization of the sight-seeing tours.

In consent with the International Advisory Board meeting at ISMB13 in Arlington, 2012, we have established the web portal www.magneticbearings.org which introduces and presents the ISMB, related events, technology and news of magnetic bearings, and an archive of all ISMB symposia papers. We would like to invite you to visit this new portal from time to time to get current information about magnetic bearing issues.

Finally, we would like to thank you for your participation and your contributions to ISMB14. We very much hope that you will be able to fully enjoy this symposium and its events, that you can gather innovative ideas and valuable incitation from presentations and discussions, and that you will make use of the opportunity to establish promising personal contacts during your stay.

We warmly welcome you to our symposium in Linz!

The ISMB14 Organization Committee

Wolfgang Amrhein	Wolfgang Gruber	Gerald Jungmayr	Hubert Mitterhofer	Siegfried Silber
(General Chair ISMB14)	(Co-Chairs)			

Committees

International Advisory Committee

Paul E.	Allaire	University of Virginia	USA
Wolfgang	Amrhein	Johannes Kepler University Linz	Austria
Hannes	Bleuler	EPF Lausanne	Switzerland
Giancarlo	Genta	Politecnico di Torino	Italy
Shawn	Gibson	SKF	Canada
Larry	Hawkins	Calnetix Technologies	USA
Raoul	Herzog	HEIG-VD	Switzerland
Toshiro	Higuchi	University of Tokyo	Japan
Yefa	Hu	Wuhan University of Technology	China
Mochimitsu	Komori	Kyushu Institute of Technology	Japan
Lichuan	Li	Xi'an Jiaotong University	China
Zongli	Lin	University of Virginia	USA
Toru	Masuzawa	Ibaraki University	Japan
Takeshi	Mizuno	Saitama University	Japan
Brian	Murphy	University of Texas	USA
Myounggyu	Noh	Chungnam National University	Korea
Kenzo	Nonami	Chiba University	Japan
Yohji	Okada	Ibaraki University	Japan
Alan	Palazzolo	Texas A&M University	USA
Frank	Pinckney	Dresser Rand	USA
Jerzy	Sawicki	Cleveland State University	USA
Mike	Swann	Waukesha Bearings	USA
Gerhard	Schweitzer	ETH Zurich	Switzerland
Richard	Stephan	UFRJ Rio de Janeiro	Brazil
Hirochika	Ueyama	MUTECS Incorporated	Japan
Guiseppi	Vannini	GE Oil and Gas	Italy
Lie	Yu	Xi'an Jiaotong University	China
Suyuan	Yu	Tsinghua University	China

Local Organization Committee

Wolfgang	Amrhein	Johannes Kepler University Linz	Austria
Wolfgang	Gruber	Johannes Kepler University Linz	Austria
Gerald	Jungmayr	Johannes Kepler University Linz	Austria
Hubert	Mitterhofer	Linz Center of Mechatronics GmbH	Austria
Siegfried	Silber	Linz Center of Mechatronics GmbH	Austria

International Program Committee

Johan	Abrahamsson	Uppsala university	Sweden
Nicola	Amati	Politecnico di Torino	Italy
Dietmar	Andessner	Johannes Kepler University Linz	Austria
Junichi	Asama	Shizuoka University	Japan
Walter	Bauer	Johannes Kepler University Linz	Austria
Andreas	Binder	Technical University Darmstadt	Germany
Gerd	Bramerdorfer	Johannes Kepler University Linz	Austria
Katia	Cavalca Dedini	UNICAMP	Brazil
Akira	Chiba	Tokyo Institute of Technology	Japan
Peter	Dirnberger	Linz Center of Mechatronics GmbH	Austria
Stephan	Düsterhaupt	HS Zittau/Görlitz	Germany
Alexei V.	Filatov	Calnetix Technologies	USA
Erik	Fleischer	Technical University Dresden	Germany
Rainer	Gausmann	Siemens AG	Germany
Afonso C. D. N.	Gomes	UFRJ-PEE	Brazil
Herbert	Grabner	Hietzinger GmbH	Austria
Wolfgang	Gruber	Johannes Kepler University Linz	Austria
Christoph	Grund	Siemens AG	Germany
Clay	Hearn	University of Texas	USA
Raoul	Herzog	HEIG-VD	Switzerland
Thomas	Hinterdorfer	Technical University Vienna	Austria
Wilfried	Hofmann	Technical University Dresden	Germany
Jan	Janse van Rensburg	North-West University	South Africa
Gerald	Jungmayr	Johannes Kepler University Linz	Austria
Nobuyuki	Kurita	Gunma University	Japan
Matthias	Lang	Siemens AG	Germany
Jun-Ho	Lee	Korean Railroad Institute	Korea
Torbjörn	Lembke	Lembke Innovation	Sweden
Li	Li	HS Zittau/Görlitz	Germany
Kun	Liu	National University of Defense Technology	China
Urban	Lundin	Uppsala university	Sweden
Edmund	Marth	Johannes Kepler University Linz	Austria
Eric H.	Maslen	James Madison University	USA
Hubert	Mitterhofer	Linz Center of Mechatronics GmbH	Austria
Mimpei	Morishita	Kogakuin University	Japan
Koichi	Oka	Kochi University of Technology	Japan
Masahide	Ooshima	Tokyo University of Science	Japan
Josef	Passenbrunner	Johannes Kepler University Linz	Austria
Fernando C.	Pinto	UFRJ-PEM	Brazil
Florian	Poltschak	Johannes Kepler University Linz	Austria
Andreas	Pröll	Johannes Kepler University Linz	Austria
Lukas	Quurck	Technical University Darmstadt	Germany
Manes	Recheis	Technical University Graz	Austria
Christian	Redemann	Levitec GmbH	Germany
Michael	Richter	Technical University Darmstadt	Germany
Andres O.	Salazar	UFRJ Rio de Janeiro	Brazil
Jan	Sandtner	EPF Lausanne	Switzerland
José A.	Santisteban	Fluminense Federal University	Brazil
Hendrik	Schaede	Technical University Darmstadt	Germany
Joachim	Schmied	Delta JS AG	Switzerland

Maximilian	Schneider	Technical University Darmstadt	Germany
Manfred	Schrödl	Technical University Vienna	Austria
Tadahiko	Shinshi	Tokyo Institute of Technology	Japan
Peter	Siebke	Schaeffler	Germany
Siegfried	Silber	Linz Center of Mechatronics GmbH	Austria
Ryouichi	Takahata	JTEKT Corporation	Japan
Gang	Tao	University of Virginia	USA
Andrea	Tonoli	Politecnico di Torino	Italy
David L.	Trumper	Massachusetts Institute of Technology	USA
Satoshi	Ueno	Ritsumeikan University	Japan
Heinz	Ulbrich	Technical University Munich	Germany
Johann	Wassermann	Technical University Vienna	Austria
Ralf	Werner	Technical University Chemnitz	Germany
Frank	Worlitz	HS Zittau/Görlitz	Germany
Longxiang	Xu	Nanjing University of Aeronautics and Astronautics	China
Pablo	Yoon	University of Virginia	USA
Jin	Zhou	Nanjing University of Aeronautics and Astronautics	China

Keynote Speakers

Prof. Paul Allaire

Chief Technical Officer, Rotor Bearing Solutions International, USA
Also, Mac Wade Professor, University of Virginia, Retired

Mu Synthesis Control of High Speed Flexible Rotor on Active Magnetic Bearings Simulating High Pressure Compressors with Aerodynamic Cross Coupling

Paul Allaire, Simon Mushi, Zongli Lin

Many industrial companies are adopting magnetic bearings for high speed compressors at this time. A flexible rotor test rig of approximately 1.2 m in length was developed with four active magnetic radial bearings. It was employed to evaluate the control of aerodynamic cross coupling stiffness in the center of the machine. The two end AMBs were used for support, one center AMB used to apply the cross coupled stiffness. The fourth AMB was employed as a system exciter. Unfortunately for the compressor manufacturer and the end user, often the magnitude of the cross coupled stiffness changes with various compressor operating conditions. Mu synthesis control was used to achieve very nice control and low vibration response with the selection of useful uncertainty weighting functions when the controller does not have advance notice of the cross coupling amplitude. The shaft operated up to about 18,000 rpm, through the first critical speed which was at approximately 14,000 rpm. Details of the application of the mu synthesis controller implementation are presented.

Prof. Hannes Bleuler

Director Robotic Systems Lab, EPFL Lausanne, Switzerland

Overview of the various types of magnetic bearings and their respective potential for applications

We all know many very different types of AMBs such as "active", "passive", "electromagnetic", "electrodynamic" etc. In this keynote, I propose to briefly review the known basic configurations and then discuss why certain types are very common, while others are either nearly forgotten or seem uninteresting for practical applications. More importantly, I will try to identify a few types which might have as yet unexploited potential for future applications.

Matthias Kroll

CEO LEViTEC GmbH, Germany

Electric Drives Systems with Active Magnetic Bearings in Industrial Applications: Possibilities and Challenges

High speed drive systems will be playing an even more important role in industrial applications. They are used in the chemical and food industry, as well as in power generation and in machine tools. The drive systems in the mentioned applications have to meet high standards. Furthermore, there is an increasing demand for reliability and productivity of machines and plants, which also includes high power density and efficiency. Due to the high requirements, active magnetic bearings are increasingly used. To achieve the best technical and economical solution for the customer, it is necessary to consider the strong interaction between the system components as well as manufacturing aspects. This provides the best solution with significant benefits for the end user.

Florian Lösch

CEO MECOS AG

Industrialisation of Active Magnetic Bearing Technology

This keynote describes the development of Magnetic Bearings from their origin as a highly innovative concept to an industrial product, a path that was strongly influenced by technological innovation as well as non-technological boundary constraints. Starting from the pilot applications, key scientific and technological advancements that paved the way for large scale application in industrial products are highlighted. It is shown that besides the technical aspects also complex economic factors must be taken into consideration if one is to understand the technology's industrial evolution to date and to make predictions for the future.

Jiri Sloupensky

Vice President of Product Research, Rieter CZ s.r.o., Czech Republic

Magnetic Bearings: Application to High Speed Textile Processing

This keynote describes the history and actual status of a successful mechatronic industry- academia project, related to the rotor spinning textile process. This textile technology was invented in Czech Republic in 1960's and brought a revolution to the yarn spinning industry. The key factor for the productivity of rotor spinning machines is the speed of the spinning rotor. Conventional machines are equipped with different types of mechanical bearings with limited lifetime due to the heavy duty operation. The target of this project was to develop new concept of single-drive rotor spinning unit magnetic bearing and at the same time use as many synergies as possible, given by this new concept, also for improving the textile process.

Michael K. Swann

General Manager, Waukesha Magnetic Bearings, USA

Industrialisation Trends for Active Magnetic Bearings in the Turbomachinery Industry

As the adoption rate of active magnetic bearings in the turbomachinery industry continues to increase, leading magnetic bearing suppliers have been meeting customer expectations by transforming what was once a university laboratory development into a well accepted industrial product. In turbomachinery applications for oil & gas and power generation this has meant the globalisation of not only sales and services functions, but also key project engineering, product development and manufacturing functions as well. This talk briefly examines the history of these trends up through the present day and what one supplier sees as the future of this industry.

Comparative Study of Axial/Radial Magnetic Bearing Arrangements for Turbocompressor Applications

Alexei Filatov, Larry Hawkins

Calnetix Technologies, 16323 Shoemaker Av., Cerritos, CA 90703
afilatov@calnetix.com lhawkins@calnetix.com

***Abstract*—This article presents results of a comparative study of different radial/axial actuator solutions for an Active Magnetic Bearing (AMB) system in a real-life turbocompressor application.**

I. INTRODUCTION

Electromagnetic actuators are the largest components of Active Magnetic Bearing systems inside a machine and their design have significant impact on the machine size and performance. In this paper we analyze three types of actuator arrangements commonly used to control two radial and one axial degrees of freedom of a rotor using a real-life gas compressor example. These actuator arrangements include:

1. Heteropolar electrically-biased radial actuator and conventional electrically-biased axial actuator [1];
2. Heteropolar electrically-biased radial actuator and electrically-biased axial actuator with low target OD [1];
3. 'Side-By-Side' (SBS) homopolar permanent-magnet-biased combination axial/radial actuator [2].

We will investigate impact of the choice of the actuator solution on the following performance characteristics of an example gas compressor:

1. Rotor length;
2. Separation margin from the first forward-whirl bending mode frequency;
3. Axial actuator gain and load capacity loss with frequency;
4. Increase of the phase lag between the axial force and control current with frequency;
5. Aerodynamic drag;
6. Actuator axial negative stiffness;
7. Actuator radial negative stiffness;
8. Radial actuator performance loss with speed.

II. APPLICATION EXAMPLE

A machine that we use as an example for this study is a real life 300kW power range high-speed (≈40kRPM) turbocompressor equipped with a Surface Mounted Permanent Magnet (SMPM) synchronous motor. Figure 1 shows all the major machine components using the first actuator arrangement (heteropolar electrically-biased radial actuator and conventional electrically-biased axial actuator) as an example. The static load capacity requirements for the bearing arrangement on the impeller side are 3000N axially and 1200N radially. The impeller weighs approximately 4kg and is coupled to the shaft through a face tooth coupling (such as Hirth or Curvic) and Tie-Rod arrangement.

For the purpose of this comparative study we did not change any other component of the machine in the other configurations except for the radial/axial actuator arrangement located on the impeller side. Fig. 2 shows all three

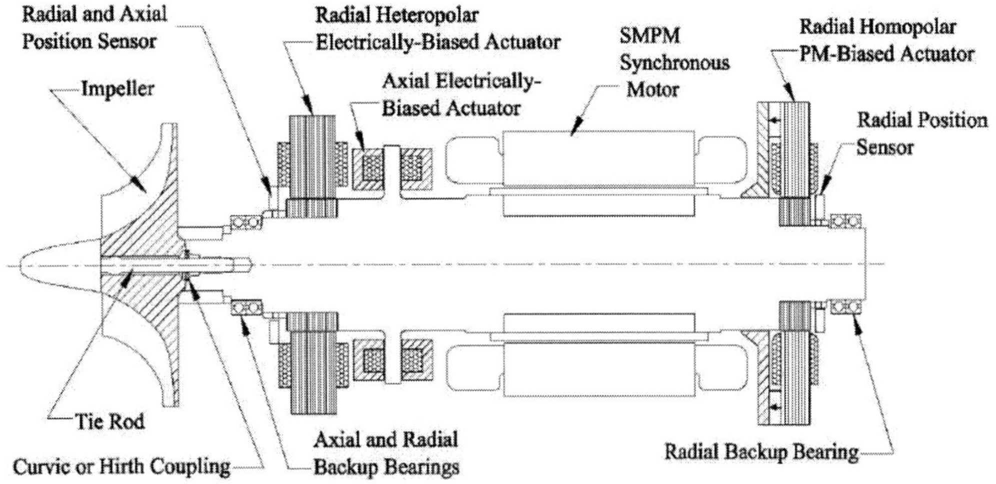

Figure 1. A turbocompressor example used to study effects of the radial/axial actuator arrangement. Shown with the heteropolar electrically-biased radial actuator and conventional electrically-biased axial actuator on the impeller side.

configurations side by side. In particular, the non-impeller side bearing is a homopolar permanent-magnet biased radial bearing in all three configuration. This is an effective solution for all configurations, but using a non-impeller side radial heteropolar bearing would not affect the results of the comparative study.

The comparison was based on the results of the numeric calculations, however, the configuration shown in Fig. 2c has

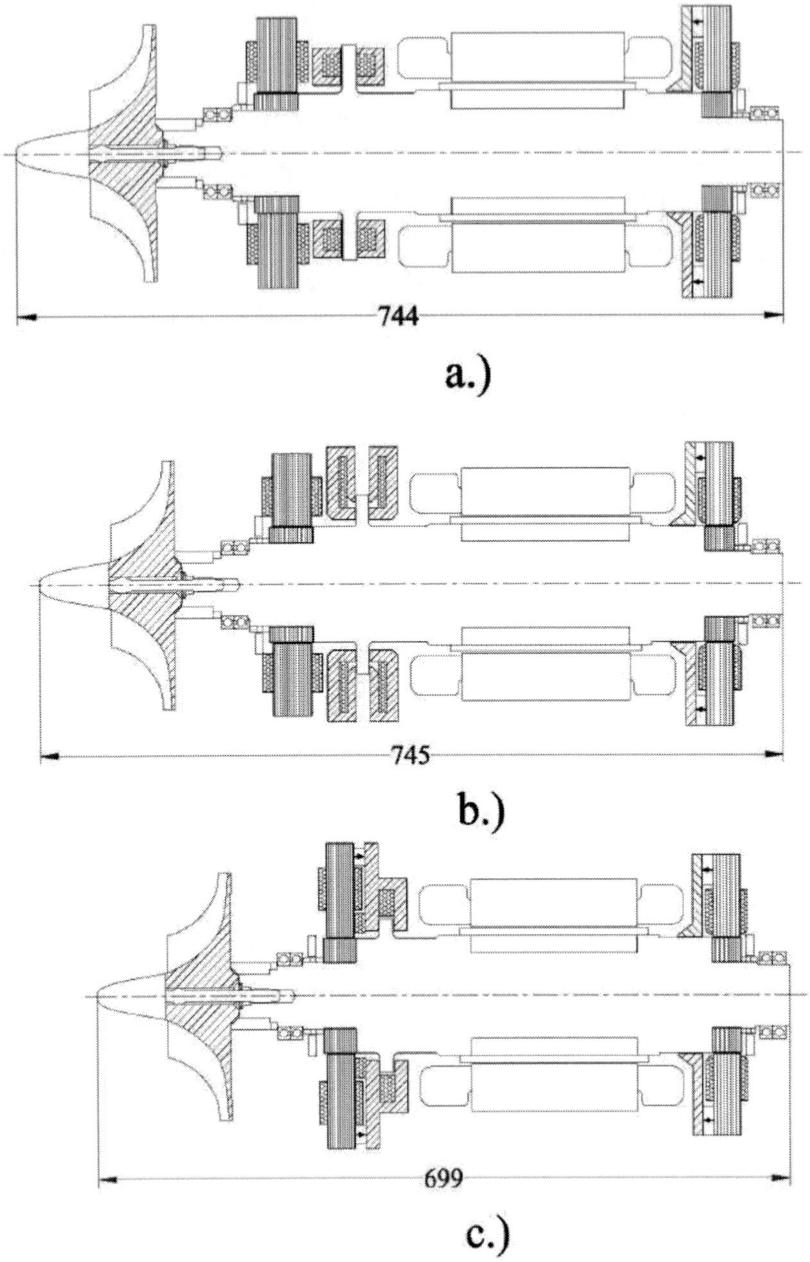

Figure 2. Turbocompressor configurations featuring different radial/axial actuator arrangements on the impeller side.

a) Heteropolar electrically-biased radial actuator and conventional electrically-biased axial actuator;
b) Heteropolar electrically-biased radial actuator and electrically-biased axial actuator with low target OD;
c) 'Side-By-Side' homopolar permanent-magnet-biased combination axial/radial actuator.

been built and tested. The measured performance characteristics in this case were found to be in a good agreement with theoretical predictions. Calnetix has also built several machines utilizing configuration similar to 2a with an exception of the radial bearing on the impeller side where a homopolar PM-biased radial magnetic bearing was used. In all of those machines the experimental results also matched well the analytical estimates.

III. COMPARISON METRICS

The following performance metrics were used to quantitatively characterize the machine performance for comparison purposes.

1. *Rotor length.*
The shorter rotor the more compact the machine would be.

2. *Rotordynamic margin at 37kRPM.*
This parameter defines separation (as a percentage of a rotational speed) between the rotor speed and the first forward-whirl bending mode frequency. In order to simplify controls it is preferable that the rotational speed stays below the first forward-whirl bending mode frequency and the separation is as large as possible.

3. *Axial actuator normalized gain at 40Hz.*
Bearing components defining the magnetic flux path in axial actuators are typically impractical to laminate. Therefore, there are significant eddy currents induced in those components whenever the magnetic flux changes in time. This parameter (dynamic load capacity as a percentage of the static load capacity) serves to compare actuator gain and load capacity loss with frequency between different designs.

4. *Axial actuator phase lag at 40Hz.*
In addition to the actuator gain loss, eddy currents also cause a phase lag between the forces and control currents in the axial actuators, which makes control design more difficult and at the end negatively impacts the machine performance. It is desirable to keep this phase lag to the minimum.

5. *Windage loss in actuator at 37kRPM.*
This is a portion of the total windage loss due to the axial/radial actuator arrangements shown in Fig. 3 only. Differences in overall windage losses between arrangements 2a, 2b and 2c show in Fig. 2 are caused exclusively by the differences between actuators 3a, 3b and 3c in Fig. 3.

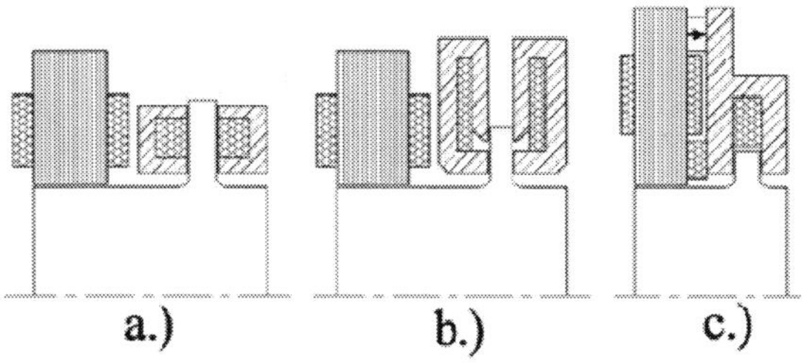

Figure 3. Portions of the systems 2a.), 2b.) and 2c.) shown in Fig. 2 in a respective order used for comparative study of windage losses. The rotor is assumed to be spinning in the air at 37kRPM.

6. *Axial negative stiffness.*
Negative stiffness effectively translates into a reduction of an actuator load capacity whenever the rotor position is off magnetic center due to manufacturing inaccuracies or a dynamic response to external loading. When comparing two magnetic bearings with otherwise similar performance metrics, the smaller this parameter - the better.

7. *Radial negative stiffness.*
Similar to the axial negative stiffness: when comparing two magnetic bearings with otherwise similar performance metrics, the smaller this parameter - the better.

8. *Skin-depth speed.*
Even though the radial actuator target is always made laminated in practical AMBs, at sufficiently high speeds this does not guarantee that there will be no expulsion of a magnetic flux from the laminations due to the skin effect, resulting in a loss of the bearing radial load capacity at speed, cross-coupling between the axes, and other (mainly negative) consequences. A numerical estimate of the effects of the eddy-currents in the rotor is a difficult task outside the scope of this paper. As a simple numerical parameter that can be used to carry out comparison between designs we used the rotor speed at which the skin depth associated with the fundamental electrical frequency seen by a spinning rotor becomes less than the lamination thickness. We considered two smallest standard off-the-shelf values of the silicon steel lamination thickness - 0.005in (0.127mm) and 0.007in (0.178mm).

TABLE I. Comparison of different axial/radial actuator arrangements on the impeller side.

	Heteropolar E-Biased Radial Conventional E-Biased Axial	Heteropolar E-Biased Radial Low Target OD E-Biased Axial	'Side-By-Side' Homopolar PM-Biased Combination
Rotor Length (mm)	744	745	699
Rotordynamic Margin at 37kRPM	42%	42%	42%
Normalized Axial Gain @ 40Hz	63%	38%	56%
Axial Phase Lag @ 40Hz (deg)	20	38	22
Windage Loss in Actuator at 37kRPM* (W)	3730	2050	1000
Axial Negative Stiffness (N/mm)	3100	3700	1800
Radial Negative Stiffness (N/mm)	9100	9100	7600
0.127mm Skin-Depth Rotor Speed (RPM)	42,000	42,000	84,000
0.178mm Skin-Depth Rotor Speed (RPM)	21,000	21,000	43,000

*In the air. See Fig.3 for definitions of the system portions to which these windage losses are attributed.

IV. COMPARISON RESULTS AND DISCUSSION

The results of the comparative study are summarized in Table 1. Each magnetic bearing arrangement analyzed here has its own advantages and disadvantages. In order to explain the origins of these advantages and disadvantages, Figs. 4, 5 and 6 illustrate structure, operating principles and key dimensions of a conventional heteropolar radial magnetic actuator, electrically biased thrust actuators and a homopolar permanent-biased 'Side-By-Side' combination radial/axial magnetic actuator shown in Fig. 2. More details of their operation can be found in [1] and [2].

As it can be seen from Fig.2 and Table 1, using 'Side-By-Side' combination actuator results in a shortest rotor because several components (particularly the permanent magnet) are used in both radial and axial parts of the actuator (see Fig. 6).

Another reason that the combo actuator is significantly more compact then alternatives is that a size of a modern rare-earth magnet needed to produce a certain magnetic field

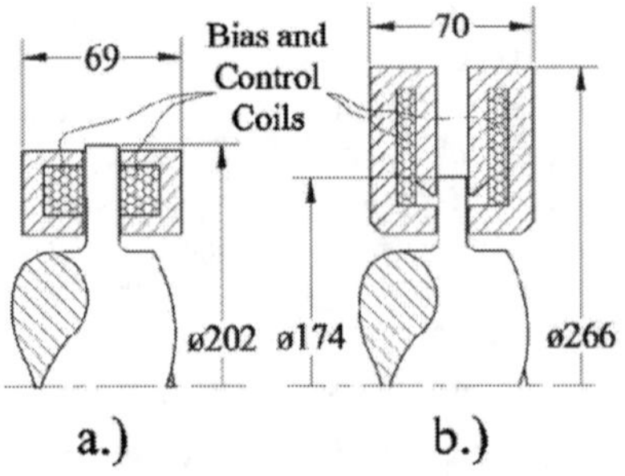

Figure 5. Structure and envelope dimensions of electrically-biased thrust bearings used in configurations shown in Figs. 2a and 2b respectively.

is only a small fraction of a size of a coil that would be needed to produce the same field. This is because reduction of a size of a coil with a given number of Ampere-turns inevitably leads to the increase of a total resistive power

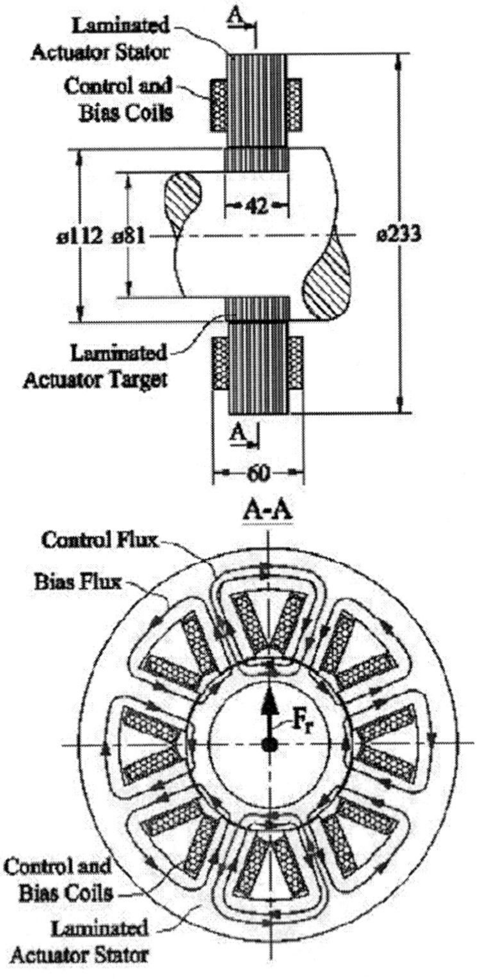

Figure 4. Structure, operating principle and key dimensions of a heteropolar radial magnetic actuator shown in Figs. 2a and 2b.

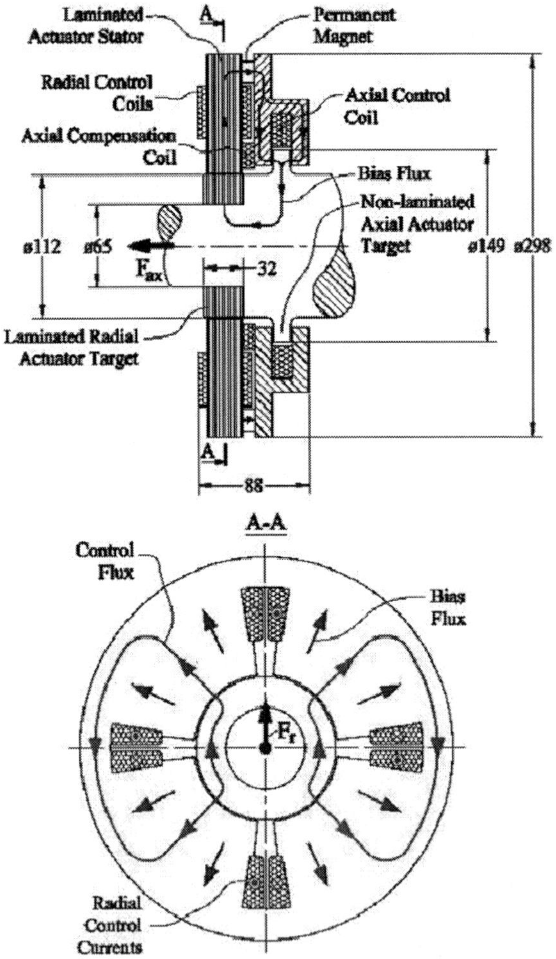

Figure 6. Structure, operating principle of a radial channel and key dimensions of a 'Side-By-Side' combination magnetic actuator shown in Fig. 2c.

dissipation in the coil, and even more to the increase of a resistive power dissipation in a unit volume of a coil, which may lead to the coil burn out.

A shorter rotor length in this example, however, does not translate into a rotordynamic advantage because heteropolar radial bearings used in the other arrangements require shallower lamination stacks on the rotor, leading to higher diameter of the rotor section under the stacks (compare Figs. 4 and 6). This diameter increase in arrangements 2a and 2b almost exactly compensates for the increases in rotor lengths in these arrangements (in this machine example) compared to arrangement 2c using a more compact combination bearing. See Fig. 7 for comparison of the first forward-whirl bending modes shapes and frequencies between variants 2a and 2c.

The reason for shallower actuator target laminations in heteropolar design is that both the bias and control fluxes are separated into numerous portions running short distances between neighboring poles (Fig. 4), whereas the conventional homopolar design utilizes four large poles with the entire control flux having to travel between the poles through the target laminations when the load is applied between the poles (Fig. 6). The price of this advantage of the heteropolar design is higher electrical frequency seen by the spinning rotor, leading to higher rotational losses, loss of the radial load capacity at speed and other complications which will be briefly discussed later herein. There is a known solution that

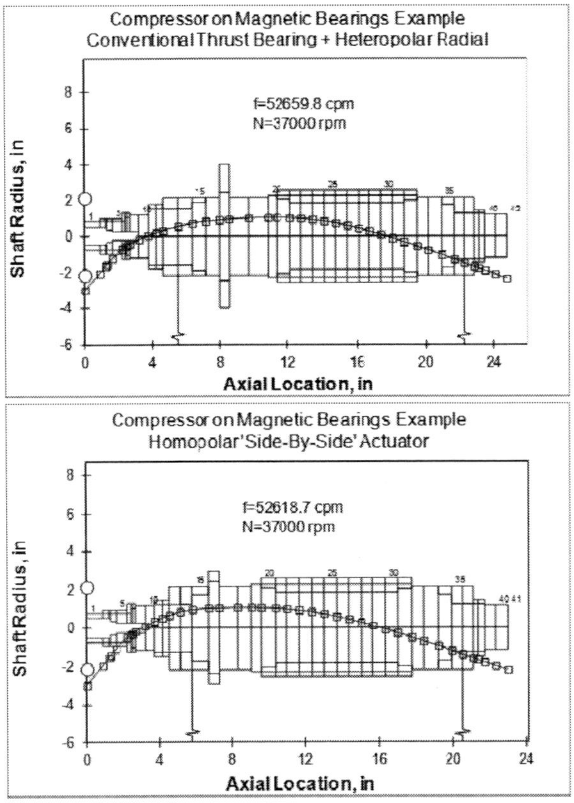

Figure 7. First forward-whirl bending mode shapes and frequencies at 37kRPM for the turbocompressor configurations shown in Fig. 2a (top) and 2c (bottom).

allows using shallow laminations in homopolar bearings as well [4], but it is more complicated to assemble and is not considered here.

The "classical" electrically-biased axial actuator shown in Fig. 2a and 5a offers the smallest losses of the axial gain and phase with frequency due to the shortest distance the control magnetic flux has to travel in a non-laminated iron from all the configurations. However, this actuator also requires a largest diameter target disk, resulting in much larger windage losses (windage losses for a spinning disk are increasing approximately as the fifth power of the disk diameter [3]). A modified version of this actuator shown in Fig. 2b needs a smaller disk diameter, resulting in almost two-fold reduction of the rotational loss (Table 1), but the gain and phase losses become much bigger due to a significant increase of the length of the control flux path in non-laminated iron.

The 'Side-By-Side' combination actuator offers the smallest target diameter from all the options (rotational losses are reduced by almost a factor of four compared to Fig. 2a) combined with a frequency response almost as good as in the version 2a.

FEA-calculated control magnetic flux distributions in different thrust bearings at 40Hz are shown in Fig. 8 and complete transfer functions are shown in Fig. 9. All calculations were done using FEMM [5].

The 'Side-By-Side' combination actuator also offers the lowest values of both radial and axial negative stiffness due to the bias flux being generated by a permanent magnet rather than a coil with a current. This is because permanent magnets have internal reluctances limiting bias fluxes to lower values than those produced by currents when either radial or axial targets deviate from the magnetic equilibriums.

The Skin-Depth Rotor Speed parameter comparison in Table 1 in effect simply illustrates that a better performance can be expected from homopolar radial bearings than heteropolar bearings at high speeds because of lower frequency of flux reversals seen by the rotor. However, Table 1 does not provide numerical estimates of the final parameters of interest affected by the eddy currents in the rotor such as loss of load capacity, amount of cross-coupling between the axes and rotational drag. The latter estimates are difficult to make and are not covered in this paper. It should be clarified that in case of the 8-pole heteropolar bearing (Fig. 4) we have chosen for the arrangements shown in Figs. 2a and 2b, the assumption was made that this bearing is used in NN-SS-NN-SS configuration to minimize electrical frequency seen by the rotor. Furthermore, we only considered the fundamental electrical frequency, which in this case is equal twice the rotational frequency of the rotor. There will be higher order frequencies caused by the gaps between the magnetic poles, which will get expelled from the rotor even faster. Therefore, the skin-depth frequency estimate for the heteropolar bearings listed in Table 1 is on the optimistic side.

In case of a homopolar bearing (Fig. 6) used in the machine arrangement shown in Figs. 2c, the fundamental frequency of the bias field will be zero in the absence of the radial loading and it won't be expelled from the rotor at any

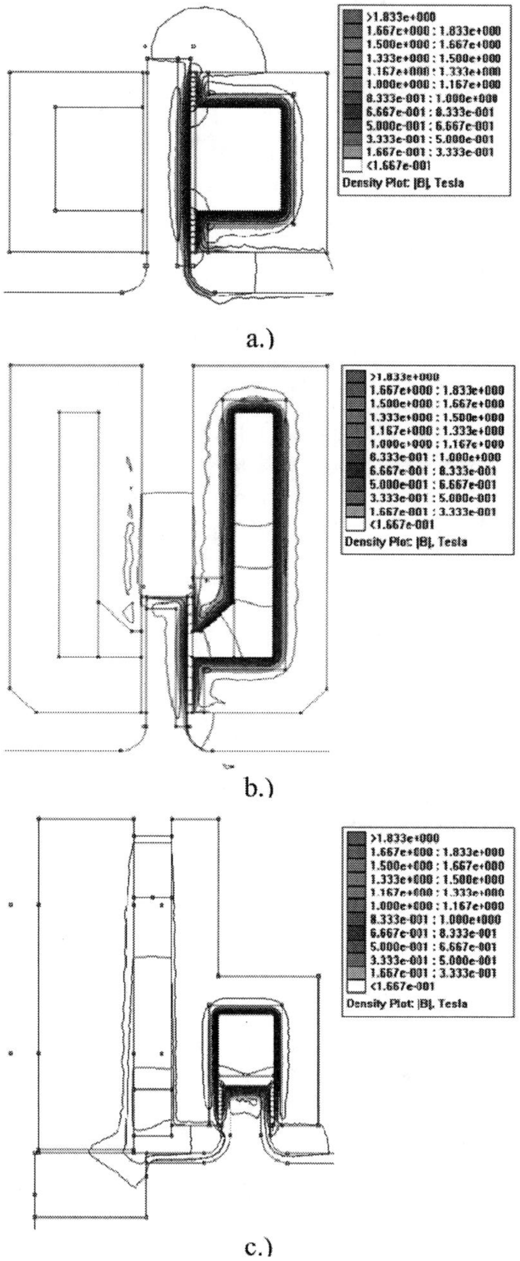

Figure 8. Axial control magnetic flux distributions in different configurations of electromagnetic actuators shown in Fig.2. The distribution shown in Fig. 8a corresponds to Fig. 2a, 8b corresponds to 2b, etc.

speed (there will be some field harmonics caused by having gaps between the poles which will be expelled similar to the heteropolar design). However, when a radial load is applied as illustrated in Fig. 6, a control magnetic field generated in response to that load will be non-uniform around the bearing resulting in a spinning rotor seen a time-varying magnetic

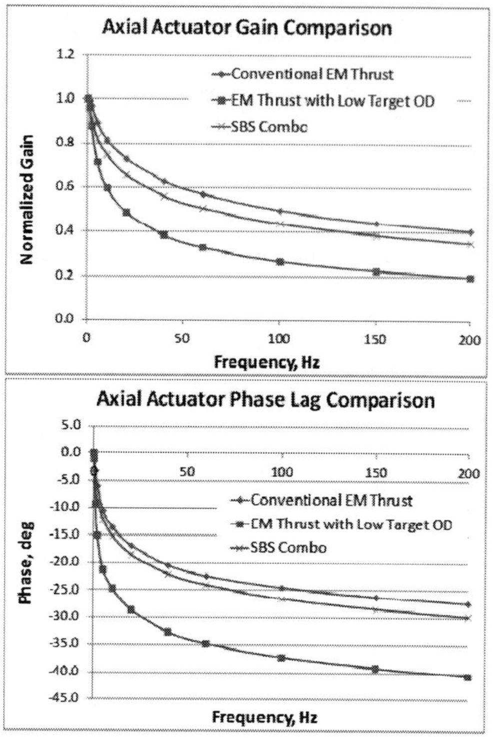

Figure 9. Comparison of the axial transfer functions for the different turbocompressor arrangements shown in Fig.2.

field with a fundamental frequency equal to the spinning frequency.

Based on a technical performance metric comparison in Table 1 it appears that the 'Side-By-Side' Homopolar Permanent would be a better choice for this particular application, but it has two additional drawbacks not reflected in the Table 1: higher cost and complexity, which may even the playfield in some applications.

V. CONCLUSIONS

A comparative study of three common radial/axial actuator arrangements for a real-life turbomachinery application has been carried out. While the answer to the question of which actuator would be the best clearly would depend on specifics of the application in hands, the results of the present study illustrate the importance of the actuator choice for the machine performance and typical trade-off associated with the most common actuator solutions.

REFERENCES

[1] G. H. Schweitzer, E. H. Maslen, "Magnetic Bearings: Theory, Design and Application to Rotating Machinery", 2009, Springer-Verlag, Berlin, Heidelberg.

[2] A. Filatov, L. Hawkins, "Novel Combination Radial/Axial Homopolar Active Magnetic Bearing", *Proceedings of the 1st Brazilian Workshop on Magnetic Bearings*, Rio-De-Janeiro, Brazil, 2013.

[3] F. J. Bayley, J. M. Owen, "Flow between a rotating and a stationary disk", Aeronotical Quarterly 20, pp. 333-354, 1969.

[4] A. Filatov, US Patent No. 8,169,118, "High-Aspect-Ratio Homopolar Magnetic Actuator", May 1, 2012.

[5] http://www.femm.info/wiki/HomePage

Field Experience with Automated Tools in Both Remote and Local Commissioning of Active Magnetic Bearing Systems.

Richard Jayawant[a], Nigel Davies[a]

[a] Waukesha Magnetic Bearings, Worthing, UK. Email rjayawant@waukbearing.com

***Abstract*—This paper describes field experience with 3rd generation AMB automated commissioning tools. This covers experience on "first of class" machinery on sites where both AMB specialists were present using the automated commissioning tools and other installations where no AMB specialist was present and all commissioning was conducted by remote guidance of local non-specialist staff and remote use of automated commissioning tools.**

I. Introduction

A. Automated Commissioning Tools

Automated commissioning tools, also known as 3rd generation magnetic bearing functionality [1], offer the potential for increased industrial acceptance of magnetic bearing systems due to a number of significant advantages:

- They lower the level of skill and training required to execute the commissioning process;
- They transfer commissioning activities to OEM and end user non-specialist engineers;
- They reduce the time required for commissioning;
- They improve the consistency and repeatability of commissioning measurements and record keeping.

Automated commissioning covers not only the processes traditionally associated with "automatic tuning" (ie configuration of the dynamic parameters of the AMB controller), but also the automation of the more "nuts and bolts" aspects of the commissioning process including such items as verification of mechanical build.

The automated commissioning tools, and in particular, automated tuning, use a network connection to link the "signal processing" functions integrated within the AMB controller with more capable analysis software running on an external computer. This external software can change parameters in the bearing controller to modify controller configuration, set up measurements, control data collection and apply results.

The tools discussed in this paper are written in MATLAB which provides advanced data processing and display within a general purpose programming environment including networking. For distribution of the tools, the MATLAB code is compiled to an easily distributed executable program. The architecture of the automated commissioning tools is described in [2]. Figure 1 shows the architecture of the system.

Figure 1 – Automated commissioning framework

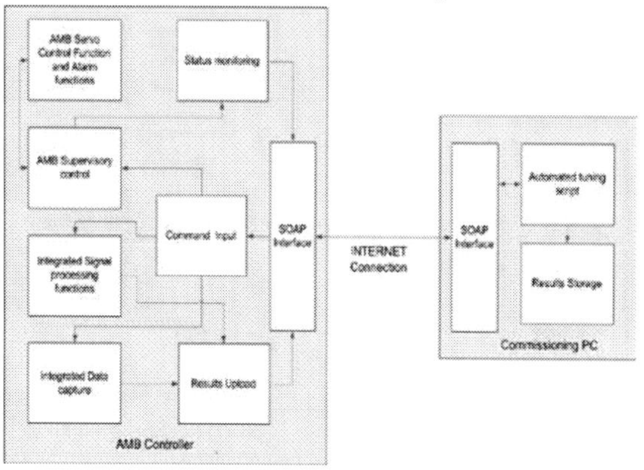

Automated commissioning works in parallel with the existing browser based AMB display system. The display system remains connected over the same network connection as the automatic commissioning tools and provides additional opportunities to monitor the commissioning progress. For remote commissioning it is occasionally helpful to exchange digital photographs of the bearing installation or arrange a video conference with a mobile camera to examine more details.

B. Test Installations

Results of automated commissioning on two active magnetic bearing (AMB) installations are presented in this paper. Both installations involved "first of class" applications and included both:

- Applications where commissioning was conducted on site by an AMB specialist using the automated commissioning tools.
- Applications which were commissioned remotely (in a separate continent) where no AMB trained specialist travelled to the OEM site and all on site support was provided by general electrical and mechanical technical staff.

The first machine, located in the USA, is a magnetic bearing supported test rig used for testing of fluid film and squeeze film damper bearings. The AMB in the test rig was commissioned under the remote supervision of Waukesha in the UK (WMB) with local support by engineers.

The technical specification of the fluid film test rig is shown in Table 1.

TABLE I. SPECIFICATION OF THE FLUID FILM TEST RIG

Parameter	Value
Rotor Mass	82 Kg
Bearing Span	524 mm
Speed range	0 to 25,000rpm
AMB configuration	2 radial bearings
Controller	Elephanta
Radial bearing load Capacity	18,000 N

The controller used in this application is an Elephanta cabinet customized to control only four axes. There is no AMB axial bearing. The test rig is shown in Figure 2. The magnetic bearing supported rotor is in the grey assembly on the left side of the photo.

Figure 2 – Fluid film bearing test rig

The second installation is a retrofit industrial turbo machine. The unit was pre-assembled with the magnetic bearings (supplied by a different vendor) prior to the arrival on site of the WMB commissioning specialists. For retrofit applications of this type where one is reliant on documentation from a separate vendor, the verification of the mechanical build and system check-out is of critical importance. The motivation for using the automated commissioning tools in this case was to expedite this verification of the mechanical system integrity.

The technical specification of the turbo-machine AMB's is as shown in table II.

TABLE II. SPECIFICATION OF THE TURBO-MACHINE

Parameter	Value
Rotor Mass	110 Kg
Bearing Span	400 mm
Speed range	0 to 22,000 rpm
AMB configuration	2 radial / 1 axial
Controller	Zephyr
Radial Bearing load capacity	5700 N
Axial bearing load Capacity	22,000 N

The controller used in this application is a custom Zephyr unit. This was fitted with an up-rated amplifier for the axial axis to allow a maximum current of 60A rather than the normal 27A. This current rating is at the standard Zephyr DC link voltage of 390V giving a per amplifier VA rating of 23.4 kVA. This was necessary to match the controller to the pre-existing magnets and sensors. In order to achieve the higher rating, the controller was modified so that rather than the standard natural convection cooling it was cooled by a forced air flow. The process of up-rating the controller is described in [3]. The controller is shown in Figure 3.

Figure 3 – AMB controller for turbo-machine

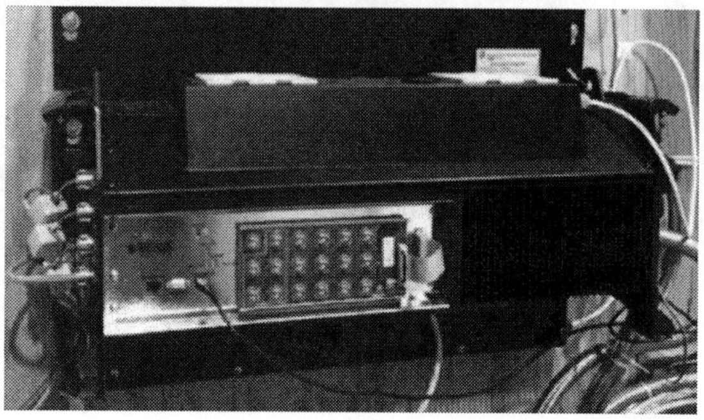

II. AUTOMATED-COMMISSIONING TOOLS

A. Build Verification Tools

These tools are concerned with the integrity of the mechanical build of the equipment and the correctness of the inter-connections between the AMB controller and the equipment. There are simple checks to verify this such as:

- When we attempt to drive current in one magnet, do we get current in the correct magnet?
- When we drive current in one magnet, are we able to move the rotor?
- When we drive current in one magnet, do we see motion in the expected sensor signal?
- When we pull the rotor with the magnets, do we see the expected magnitude of motion?
- Optimization of sensor phase for synchronous sensor demodulation [4]

These are all simple tests, but executing a full set of verification measurements manually is time consuming. For an application where we are retrofitting onto pre-existing AMB hardware from a competing vendor, the potential for misconnection is higher than normal (due to differing terminology / labeling and possible mis-interpretation of drawings). In this case the use of the build verification procedures is necessary and the use of the automated commissioning tools is essential to expedite this process.

The build verification tools used during the commissioning of the 2 units were:

- System check-out scripts
- The de-levitated sensor calibration tools.
- Levitation scripts

- The partially levitated sensor calibration tools.
- The fully levitated scripted clearance check.

The system check-out scripts are used to drive current in each of the magnets in turn, to ensure that it is possible to drive current and that it is applied to the correct winding. This eliminates the manual parameter manipulation (and the possibility of errors) that would normally be performed.

The de-levitated sensor calibration tools are used on the Zephyr controller to set up the phasing of the sensor phase sensitive synchronous de-modulation as described in references [2] and [4]. These tools are also used on both controller types to obtain a rough calibration of the sensor system, including setting up gains and offsets to allow levitation. This calibration can be either against the pre-defined (or measured) clearance in the auxiliary bearings, or against some other external measurement device such as dial gauges.

The levitation scripts support levitating each axis in turn followed by levitating each bearing in turn prior to levitation of the complete rotor. This eliminates the manual parameter manipulation (and the possibility of errors) that would normally be performed to achieve this process. The initial suspension is typically performed using a low-stiffness, low-bandwidth tuning configuration.

The partially levitated sensor calibration tools allow the rotor to be levitated on all axes (or planes) except one, and then the currents in the non-levitated axis (or plane) are manipulated to pull the rotor around within the auxiliary bearing clearance. This allows rapid evaluation of correct centering and calibration of the sensor system. The complexity of this parameter manipulation means that it is very difficult do this manually. Automated tools enable this capability.

The fully levitated scripted clearance check, replicates the functionality of the traditional AMB clearance check, where manipulation of the reference position is used to traverse the rotor within the auxiliary bearing clearance and current inflexion points and divergence between the actual and reference positions are used to determine contact points. This type of automatic functionality has been available within the Elephanta controller for many years, but bringing it into the external tools allows for several improvements:

- The functionality is available on all controllers with a consistent interface
- Faster measurement time (see below)
- Better graphical display and data recording

Due to the configuration of the integral loops within the MIMO controller architecture, the traditional clearance check is quite slow, with each measurement point requiring the integral loop to settle before making the associated measurements. With the external automated commissioning tools, rather than manipulating the reference positions, it is feasible to manipulate the offset in the sensor calibration block, with the automated tools keeping track of the "effective reference position". This eliminates the settling time associated with the integral loops and leads to faster measurements.

The use of the external tools also allow for more easy customization of such items as the step sizes, number of measurements at each position (which are then averaged) and measurements required to confirm an inflexion – all of which allows for faster (optimal) measurements.

Details of several of the algorithms are given in reference [2].

B. *Dynamic tuning tools*

These tools are concerned with configuring the dynamics of the AMB controller. The typical processes required in setting up the dynamics of the system includes:

- Configuration of flux estimators
- Amplifier bandwidth optimization
- Open loop transfer function tuning.
- Spectral measurements

The use of automated commissioning tools (for both dynamic tuning and system check-out), results in faster measurement times and greater consistency and repeatability of measurements, and reduces potential for human error in the set up of measurements.

III. Results

A. *Fluid Film Bearing Test Rig*

The fluid film bearing test rig is located in the USA. It was commissioned remotely from England by magnetic bearing commissioning engineers. Local staff, none of whom have magnetic bearing experience, installed the magnet bearings, made all the electrical connections and took measurements as required by for the automated commissioning tools. The tools used for remote commission of this AMB are:

- System check-out scripts
- Amplifier bandwidth optimization
- De-levitated sensor calibration tools
- Levitation scripts
- Spectral measurements
- Partially levitated sensor calibration tools
- The fully levitated scripted clearance check
- Open loop transfer function tuning

The system check-out scripts and initial calibration were used by the remote engineers with a network connection to the bearing controller and a local non-specialist engineer on hand to take current and dial gauge measurements. The current checks took 50 minutes. This time was dominated by the time for the levitation and de-levitation sequence. We discuss below how this delay can be eliminated with a special commissioning state for system check-out. The position checks and initial calibration took a similar time. Traditionally, these checks would have required a visit to site by a magnetic bearing commissioning engineer, but this project has demonstrated that these checks can be conducted remotely.

Having confirmed that the power and magnet connections are correct, the next step in the commissioning process is to tune the bandwidth of the amplifier and magnet circuit to a target determined by rotor dynamic analysis. The automated commissioning tools measure and tune this bandwidth using the transfer function capability built into the Elephanta controller. For the fluid film test rig, this was done remotely in 30 minutes for all axes, without needing local support. Without automated commissioning, this activity takes several hours. The automated commissioning tools also tune the parameters for a flux estimator used to control flux rather than current for better dynamic performance. Figure 4 shows the tuned amplifier bandwidth for both magnets of one axis of the test rig.

Figure 4 – Amplifier bandwidth, -45° deg @ 600Hz

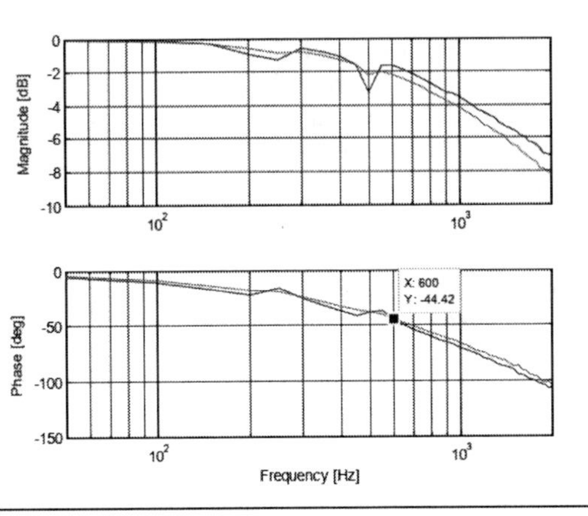

Next, the levitation script was used by the remote engineer to levitate and fine tune each axis in turn using initial calibration and loop shaping schemes from rotor dynamic analysis. The goal of this step is to achieve a soft suspension to allow better centering of the rotor. For this machine, we used remote manual adjustment of the controller terms to stabilize the suspension, but as discussed below, we plan to automate the heuristics used manually to achieve a stable suspension.

When the initial suspension was achieved, it was clear both audibly (though a telephone link to the bearing site) and from a remote display of bearing vibration level, that the suspension was very noisy. An exchange of digital photographs and a video review identified corrections to cable screening and to cable routing (to separate magnet cables from the same ducting as flux feedback sense wires). After the local engineers completed this work, the levitation tool was run remotely again to achieve a much quieter suspension.

The spectral measurement tool was used to measure the noise levels with no current flowing, with only bias current in each magnet and when suspended. This complete set of measurements takes approximately 20 minutes using the spectral measurement tool, compared with 2 hours required to take the measurements manually, due to all the changes to parameter settings required. The tool also ensures that all results are captured and saved in a consistent manner.

With a soft suspension, the partly levitated calibration tool was used to improve the position sensor calibration. The remote engineer ran the tool and the local engineers used a dial gauge to measure movement of the rotor. It took 11 minutes to calibrate all four axes using planar movement. This calibration procedure also revealed that one sensor was using a distinctly different portion of its electrical range to all the others, suggesting a possible build issue with the machine.

With a soft suspension and improved sensor gain calibration, the scripted clearance check was used to verify the design clearances and set the sensor offsets to center the rotor. Figure 5 shows the display generated during the J1-J4 parallel traverse clearance check. The top two traces show current against offset. The middle traces show current against measurement time and the bottom two traces show position against time. It is clear from the traces that J1 is better centered than J4. The 12 minutes required for this scripted clearance check is at least five times faster than the existing clearance check algorithm built into the controller.

Figure 5 – Scripted clearance check

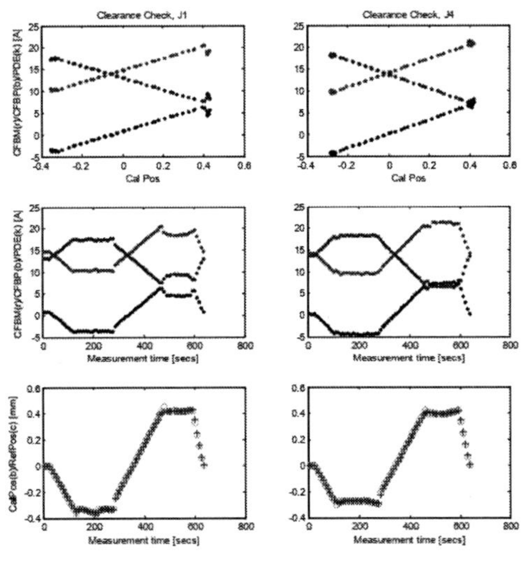

With an initial levitation and well centered rotor, it is possible to use the open loop transfer function tuning tool to measure and improve the stiffness and check the stability of suspension across the modes of the rotor. The Elephanta controller has a built in transfer function measurement capability, but this provides limited control over disturbance level, frequency step size and averaging of the measurement data. The open loop transfer function tool sequences a number of built in measurements then combines the results into a single transfer function measurement (together with a coherence estimate for the overall measurement). Combining measurements in this way allows us to use a reduced disturbance at low and critical frequencies to avoid destabilizing the suspension. It also allows closer frequency spacing near rotor modes and increased averaging where the response is low to improve the coherence of the measurement.

Figure 6 shows an intermediate result of one such transfer function measurement as a Nichols plot. The red rings demark the sensitivity zones defined in ISO-14389 [5]. It is clear that the system shown in the figure needs further tuning to stay in zone A at all frequencies (i.e. outside the outer red ring). This was resolved by adjustment of controller gain.

Figure 6 – Open loop transfer function

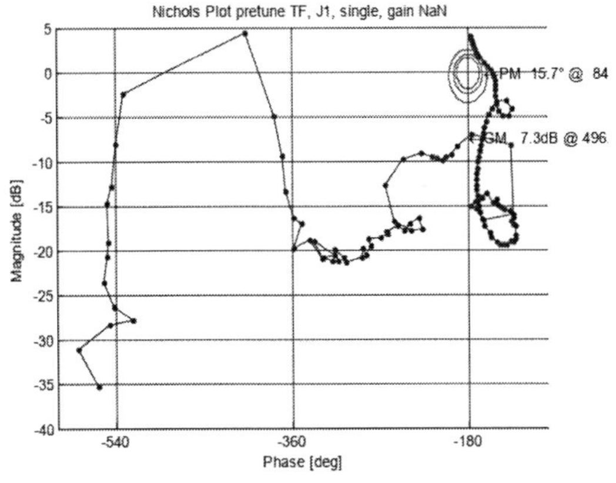

B. *Industrial Turbo-machine*

The turbo-machine was commissioned with the original AMB mechanical hardware together with a new Zephyr controller. The commissioning was conducted using an on-site commissioning presence. The Automated Commissioning tools used during this process were:

- The System check-out scripts
- The de-levitated sensor calibration tools.
- Levitation scripts
- The partially levitated sensor calibration tools.
- The fully levitated clearance check and calibration tools.

The system check-out scripts were used to drive current in each of the magnets in turn (remembering that these were the windings of the original bearing). Total time for execution of this script across all axes was 28 minutes, with no abnormal issues being identified. This compares favorably with the 120 minutes which this exercise would take when executed manually by a commissioning engineer.

A modification to the state machine of the AMB supervisory control system was identified which would allow this to be further speeded up. This change involves the creation of a commissioning state which would allow the direct enabling of individual axes (and/or amplifiers) without the need to undergo the normal full levitation sequence. This change is described in the review of further work.

The de-levitated sensor calibration tools were used to set up the phasing of the sensors (remembering that these were the sensors of the original bearing). These tools were also used to obtain a rough calibration of the sensor system. Time for execution of the sensor phasing on all 5 axes was 55 minutes compared to a typical time to execute this manually of 90 minutes. The phasing plots for one axis are shown in Figure 7.

Figure 7 – Sensor phasing

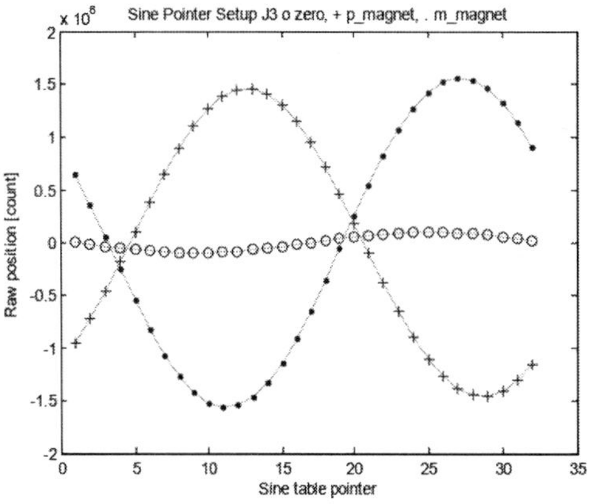

This figure also shows that the sensor has good sensitivity without risk of saturation. Time for the de-levitated sensor calibration (which was performed using dial gauges) was 24 minutes for the 5 individual axes and 9 minutes for the 2 radial bearing planes.

The initial levitation tools are a framework to guide the levitation process before the full verification of the build integrity has been completed. With this application being a retro-fit using another vendor's mechanical hardware, some manual manipulation of the AMB controller gains was necessary to establish levitation. Consequently the time taken to execute the levitation script was 14 minutes. The majority of this time was spent in adjustment of the controller gains as described above. One of the items for further work is to fully automate the process of initial levitation, capturing the heuristic algorithms employed by the expert commissioning engineer. This will be important in future retro-fit applications where there may be uncertainty around the mechanical configuration of the parts inside the machine.

The partially levitated sensor calibration tools were used to verify the clearances and sensor calibration. Time for the partially levitated sensor calibration (which was performed using dial gauges) was 17 minutes for the 2 radial bearing planes.

The initial configuration of the fully levitated clearance check tools required 17 minutes per axis to execute. It was identified that some of the configuration parameters were set very conservatively. Modifying the parameters (as identified in Table III) reduced the time required per axis to 3 minutes and the time required for one plane to 5.5 minutes.

TABLE III. PARAMETER CHANGES TO SPEED UP CLEARANCE CHECK

Parameter	Original	Revised	Notes
current noise margin	0.1A	0.03A	Controller averaging to DC current gives good results faster
number of intermediate steps	8	1	No need to suffer the time penalty of lots of little steps

Parameter	Original	Revised	Notes
samples to detect coarse-step inflexion	2	1	Stopping sooner avoids integral windup
samples to detect fine-step inflexion	2	3	More little steps after inflexion eliminates current noise

It was identified that when running a clearance check whilst simultaneously supporting multiple browser windows that the SOAP server response time was being degraded by limitations in the TCP/IP stack of the web server module associated with the Zephyr controller. An alternative web server module with superior processing capabilities has been identified and this is currently being evaluated. It is noted that no such performance limitations have been observed in the TCP/IP stack associated with the Chinook or Elephanta controllers. It is expected that once resolved the speed of the clearance check would improve significantly.

Following the system check-out the machine was tuned using conventional tuning tools. At the time this exercise was conducted, the Zephyr controller did not have the full integrated measurement capability; however, this is currently being added to the controller functionality.

The tuning exercise identified issues with the system:

- The screening of the sensor cables
- Incorrect impeller weights vs. that modeled

The screens on the sensor cables were not connected in the original assembly (identified through increased noise levels when driving current in the associated magnets). This was rectified by connecting the screens in the skid mounted junction box.

The impeller weights differed significantly from the values previously advised and consequently the rotor natural frequencies differed significantly from the values expected. This necessitated a revised tuning structure compatible with the actual rotor-dynamics.

Following resolution of these issues the machine ran to full speed.

IV. Conclusions and Further Work

Automated commissioning has proven to be an effective tool allowing both remote commissioning and making a commissioning engineer more effective when on-site.

To further improve automated commissioning, additional enhancements are under evaluation or being progressed:

- Adding a commissioning state to the AMB controller state machine. In this commissioning state amplifiers (and associated power supplies) are directly enabled, without needing a full de-levitation and levitation sequence. This capability is now being added to all controller types following the testing described here.

- In order to replicate the process undertaken by a skilled commissioning engineer when establishing a soft suspension on a new design of machine, a heuristic algorithm for establishing soft suspension is being investigated.

- Applying more checks to the measured data taken from the controller to warn about excessive noise level, limited use of the working range of the sensor, asymmetry between axes in a journal bearing etc.

- Incorporation of integrated transfer function measurement capability into the Zephyr controller is nearing completion and the deployment of the associated automated commissioning tools will follow.

- Adding a schedule of operations for an engineer to complete prior to the system check-out script including visual and photographic checks to guide a no specialist through the preparatory checks normally conducted by a magnetic bearing commissioning engineer.

- Evaluation of alternate web server modules on the Zephyr controller which offer a more robust TCP/IP stack performance has been completed subsequent to the testing described here. Faster response times will improve measurement like clearance check which control and track movement of the rotor in real time while avoiding integrator wind up. Alternate web server modules will be introduced in the next few months.

References

[1] M. Swann, "Recent Progress in AMB Technology and Manufacturing"; *ISMB13 Keynote Address; Arlington, VA; August 6-9, 2012*

[2] R. Jayawant and N. Davies, "Integration of Signal Processing Capability in an AMB Controller to Support Remote and Automated Commissioning." *1st Brazilian Workshop on Magnetic Bearings., Rio de Janeiro, Oct 2013,*

[3] P. Kumar and R. Jayawant, "Design Synthesis and Experimental Validation of an AMB Controller with a Power Dense Amplifier for Retrofit Applications." *Draft paper for ISMB 2014, Linz, August 2014,*

[4] R. Jayawant, S.Dabbs, "A Novel Digital Circuit for Position Sensor De-modulation Using Advanced FPGA, Including Field Experience with a 1.45MW Motor Driven Compressor.", *1st Brazilian Symposium on Magnetic Bearings, Rio de Janeiro, 2013*

[5] International Standard ISO 14389-3 "Mechanical vibration – Vibration of rotating machinery equiped with active magnetic bearings – Part 3: Evaluation of stability margin.

Linear and Nonlinear Control of a Three Pole Combined Active Magnetic Bearing - a Comparison

Erik Fleischer[a], Wilfried Hofmann[a]

[a] Technische Universiät Dresden, Elektrotechnisches Institut
Helmholtzstraße 9, 01062 Dresden, Germany, erik.fleischer@tu-dresden.de

***Abstract*—This paper analyses a linear and a nonlinear control strategy for a 5-axis magnetic suspension system built with two three pole combined bearings. In these soft magnetic composites (SMC) are used as iron core material to carry the required three-dimensional flux distribution.**

While the linear approach is designed around the CLARKE-Transformation to stabilize the bearing, the nonlinear one is based on feedback linearization to achieve a stable system even in the case of minimal bias flux.

The paper outlines the control strategies and the controller design. Both schemes are compared regarding tracking behavior, decoupling performance and stability margin based on experimental results in the time and frequency domain.

I. Introduction

Active magnetic bearings have found quite a few applications in vacuum technology especially in high speed machines like turbo molecular pumps. These require bearings which can operate in vacuum. Magnetic bearings are a highly attractive choice in such cases as they do not need any lubricants. Due to the changing magnetization losses are generated in the rotor, which are difficult to dissipate in vacuum.

The three pole combined magnetic bearing has been specifically designed with minimal rotor losses in mind: the pole count has been minimized to three and an axial disc is avoided. Thus it is possible to use mechanically weaker materials like soft magnetic composites (SMC) in the rotor.

Because of the structure there are no distinctive flux path' which can be mapped to the bearing axes like it is typical for conventional bearings. Therefore a different approach to the position control is necessary.

Therefore in this paper two control strategies are outlined and analyzed for a five axis magnetic suspension build using two three pole combined bearings.

Three pole configurations attracted significant attention in the magnetic bearing literature concerning position control strategies. In [1] the CLARKE-Transformation is proposed to map the orthogonal position axes onto the three coil currents and the rotor weight is used to compensate the net force generated by the bias flux.

In [2] an extension to this bearing structure is proposed which adds a homopolar flux path' to generate bias flux with permanent magnets and axial forces. These can be controlled using the zero component in the CLARKE transformation.

Operation without bias usually requires a nonlinear control law. In [3] a surprisingly simple equation is proposed which can be used to calculated the coil currents from a force command in the complex plane. Since such an approach is based on solving for a quadratic root, a systematic algorithm is required to select the best solution [4].

Experimental results report the presence of limiting cycles for both linear and nonlinear approaches in the three pole heteropolar bearing [5]. In [6] a control scheme based on feedback linearization is applied to such a combined bearing with seemingly good results.

II. Structure of the Combined Three Pole Bearing

The main part of the stator consists of three radial poles. Each is wound with one coil as shown in the left part of Fig. 1. This structure is extended by an axial flux path carrying the flux component Φ_a, which is depicted in the cross section view in right part of Fig. 1. Since this flux flows through the rotor frontal area it causes an axial force pulling the rotor along the negative z-direction.

In total there are three control currents, which are sufficient to set three forces in the x-, y- and z-directions. The only limitation exists in the form of a minimum z-axis force, which depends on the actual flux distribution and the forces in x- and y-directions.

A complete five axis suspension requires a second combined bearing flipped around by 180°, which is positioned at the other end of the rotor. Its axial force pulls the rotor in the positive z-direction. In such a structure six currents in total are used to control five degrees of freedom. This means there is

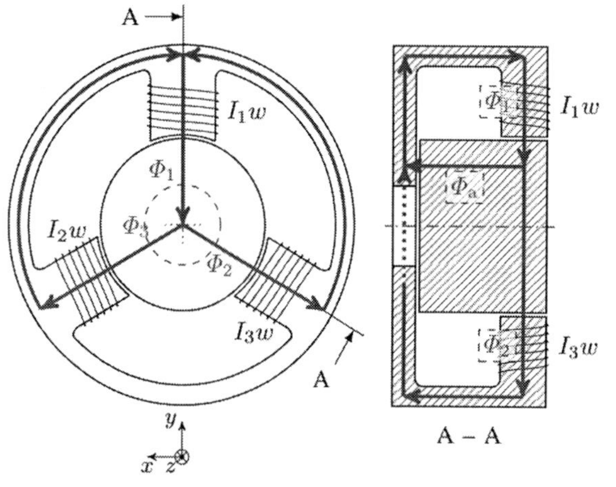

Figure 1. Front and cross section view of the combined bearing with the flux components driven by the current I_1

one current more than necessary, which can be used to either set a bias flux or to minimize the radial flux densities and therefore the rotor losses.

Introducing the axial flux path, one can write for all flux components

$$\Phi_a = \Phi_1 + \Phi_2 + \Phi_3. \quad (1)$$

Thus sum of the radial flux components Φ_i does not need to be zero. The resulting axial flux component Φ_a can be used for setting the bias flux or for an operation with minimal flux.

Similar to radial flux components the sum of the coil currents is no longer zero as well. A motor inverter consisting of three half bridges is therefore not suitable for operating the bearing.

This structure allows both the rotor and the stator to be build using soft magnetic composites as iron core material. As it can carry a magnetic flux in all directions it is used for all parts. A combination of massive iron parts and laminations can thus be avoided.

III. Control Design

A. Linear Position Control

Based on existing studies two approaches have been chosen for the position control loop. The first one is based on a linear transformation which maps the controller outputs onto the actual coil currents. As the bearing structure employs a pole pitch of 120°, the inverse Clarke transformation can be used as a starting point.

For a complete five axes suspension with two combined three pole bearings this leads to

$$\begin{bmatrix} I_1 \\ I_2 \\ I_3 \\ I_4 \\ I_5 \\ I_6 \end{bmatrix} = \begin{bmatrix} 0 & 1 & 0 & 0 & -1 & 1 \\ \frac{\sqrt{3}}{2} & -\frac{1}{2} & 0 & 0 & -1 & 1 \\ -\frac{\sqrt{3}}{2} & -\frac{1}{2} & 0 & 0 & -1 & 1 \\ 0 & 0 & 0 & 1 & 1 & 1 \\ 0 & 0 & -\frac{\sqrt{3}}{2} & -\frac{1}{2} & 1 & 1 \\ 0 & 0 & \frac{\sqrt{3}}{2} & -\frac{1}{2} & 1 & 1 \end{bmatrix} \begin{bmatrix} I_{x1} \\ I_{y1} \\ I_{x2} \\ I_{y2} \\ I_z \\ I_0 \end{bmatrix} \quad (2)$$

$$\boldsymbol{i}_c = \boldsymbol{T}_i \, \boldsymbol{i}_{xyz0} \quad (3)$$

in matrix formulation. The currents I_1 to I_3 represent the coil currents in the bearing on the left end of the rotor, while I_4 to I_6 belong to the bearing on the right end.

The vector $\boldsymbol{i}_{xyz0}$ contains one control current for each of the five control axes and the bias current I_0. This formulation leads to a quadratic transformation matrix $\boldsymbol{T}_i$ which can be inverted.

Because of the axial flux path the zero component of the Clarke transformation can be used for axial control and magnetic bias.

Based on previously conducted force measurements [7] it is possible to calculate the force current curves resulting from in the control currents in $\boldsymbol{i}_{xyz0}$. The results for the radial forces are shown in Fig. 2. The left plot shows the forces in x- and y-directions as a function of I_x. The relationship between I_x and F_x is nearly linear. But there is significant cross coupling resulting in a nonzero $F_y(I_x)$.

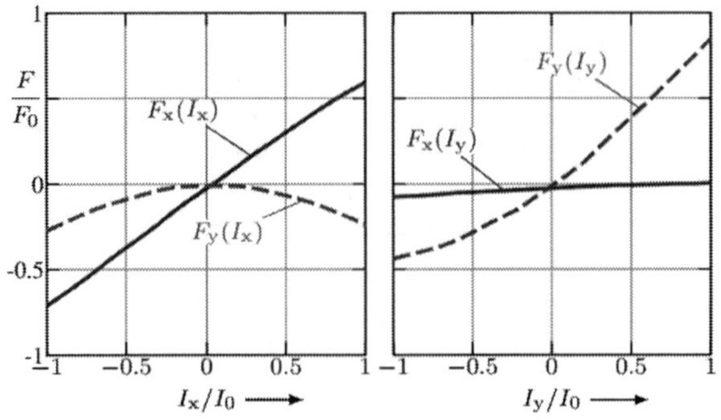

Figure 2. Measured force-current relationship. $I_0 = 2.5\,\text{A}$ and $F_0 = 80.5\,\text{N}$.

The relationship between I_y and F_y in Fig. 2(right) shows some nonlinearity, because the bearing is not symmetric about the x-axis.

The slight cross coupling between I_y and F_x should not occur in theory as there is a symmetry about the y-axis. This distortion is caused by a small excentricity of the rotor during the measurements.

The curves in Fig. 2 show that the control currents can directly be connected to the outputs of five decentral PID controllers. This leads to a reasonable simple control scheme for a magnetic suspension consisting of two combined three pole bearings.

B. Feedback Linearization Control

To deal with the nonlinearity caused by running the bearing with minimal bias flux, an approach based on feedback linearization has been designed [7].

It is based on force commands F_x, F_y and F_z from the position controllers. First F_x and F_y are transformed with

$$F_1 = \begin{cases} 0 & \text{if} \quad F_y < \dfrac{|F_x|}{\sqrt{3}} \\ F_y + \dfrac{|F_x|}{\sqrt{3}} & \text{else} \end{cases} \quad (4)$$

$$F_2 = F_1 - F_y + \frac{F_x}{\sqrt{3}} \quad (5)$$

$$F_3 = F_1 - F_y - \frac{F_x}{\sqrt{3}} \quad (6)$$

into the forces F_1, F_2 and F_3 for each pole [8]. From these the radial flux components and thus the currents can be calculated using a magnetic circuit model. Depending on the radial forces there is always an implied force $F_{4\text{imp}}$ along the z-axis in each combined bearing forming a five axis suspension.

The smaller one of these implied forces has to be increased in order to fulfill the force command F_z. This can be done by iteratively solving

$$0 = \pm\sqrt{F_1 + F_0} \pm \sqrt{F_2 + F_0} \pm \sqrt{F_3 + F_0} \pm \sqrt{F_4}k \quad (7)$$

for F_0 where k is the square root of the ratio between the axial and radial air gap cross sections and F_4 the axial force. This equation was derived from (1) by substituting the forces for the flux components and introducing the offset force F_0.

The offset force F_0 is then added to the previously calculated forces F_1, F_2 and F_3 and the result is used to calculate the flux components and the coil currents for the bearing, where the z-axis force has to be increased in order to fulfill the force command F_z.

C. Magnetic Offset Determination

While calculating the currents from the force commands the feedback linearization scheme takes the actual air gap length' into account. These can be calculated from the measured rotor position.

Due to manufacturing tolerances there is usually an offset between the magnetic center of the bearings and the geometrical center of the auxiliary bearings. As in most cases the measurement system is calibrated using the mechanical clearance in the auxiliary bearings, there is an offset in the air gap length' calculated from the measured rotor position.

A significant error can be observed between the force commands necessary for compensating gravity and the expected ones calculated from the rotor weight distribution. Thus the bearing behaves differently from the model used for the controller design. This reduces the stability of the closed control loops and the effectiveness of the decoupling between the x- and y-axes in one bearing.

As these offsets are caused by mechanical tolerances, they have to be determined after the complete assembly and for each unit separately in a series production. For this the observed error between the expected and referenced forces can be utilized. If one measures the currents needed for levitation at multiple rotor positions this error can be calculated. It can be minimized by introducing offsets into the calculation of the air gap length from the measured rotor position. Minimizing this error is a nonlinear optimization problem, which can for example be solved by Powells algorithm [9].

This procedure does not rely on any modifications in the control scheme and can therefore be easily carried out during commissioning of the system after the measurement system has been calibrated.

D. Controller Parameters

The use of Decentral $\mathrm{PIDT_1}$ controllers provides the reference commands for both control strategies with the transfer function

$$G_C = K_p \left(1 + \frac{1}{T_i s} + \frac{T_d}{T_1 s + 1}\right). \quad (8)$$

The used parameters can be found in Table I. For both strategies the same time constants have been used for comparability. The gain has been set for each controller and each axis separately based on the open loop frequency response in such a way that the phase shift is minimal at the cross over frequency, which was set to 50 Hz in both schemes. This results in a similar closed loop bandwidth in all cases.

The integral part has been set just strong enough to center the rotor with respect to the control axes. The resulting controller gains K_p differ for both control schemes as of the controller output is interpreted differently. Its assumed to be a current in the linear control scheme and a force for the feedback linearization.

Table I
PARAMETERS FOR LINEAR (A) AND FEEDBACK LINEARIZATION (B) CONTROLLERS

	$K_{p,x}$	$K_{p,y}$	T_i	T_d	T_1
(a)	$0.007\,\frac{A}{\mu m}$	$0.007\,\frac{A}{\mu m}$	0.3 s	3.2 ms	0.8 ms
(b)	$0.060\,\frac{N}{\mu m}$	$0.076\,\frac{N}{\mu m}$	0.3 s	3.2 ms	0.8 ms

E. Test Rig

A test rig has been built in which a five axis magnetic suspension using two three pole combined bearings has been realized.

The stators have been constructed using Somaloy Prototyping Material [10] and the active rotor parts using Somaloy 500 LB 1 [11]. Further details on the manufacturing of the bearings can be found in [12]. The main technical specifications have been summarized in Table II.

The rotor has been designed around a steel shaft onto which several rings for an induction machine, the position measurement system and backup bearings have been pressed. In the final manufacturing step massive SMC parts have been glued onto the shaft front faces using Delo Mono Pox 1197 (a single component epoxy glue). The hardening was carried out at 130° for 75 min.

The test rig features a horizontal shaft with the y-axis pointing up. Therefore the higher load capacity of the three pole bearings in the direction of a pole can be utilized for compensating gravity.

IV. Experimental Comparison

A. Overview

A stable operation of the magnetic suspension has been achieved with the two designed control schemes both with and without taking the magnetic offset described in section III-C into account.

The required currents are summarized in Table III. The results show that with feedback linearization the bearing runs with lower flux densities compared to the linear control scheme. But they are not as low as one would expect from the forces needed to levitate the rotor. Still, the power requirements for the bearing can be reduced to about 30 %.

Table II
TECHNICAL DATA OF THE BEARINGS IN THE TEST RIG

bore diameter	45 mm
outer diameter	90 mm
axial length	40 mm
nominal flux density	0.8 T
nominal current	5 A
nominal force	74.2 N

Table III
MEASURED CURRENTS AND RELATIVE TOTAL OHMIC LOSS P^*_{Cu} IN THE COILS OF ONE BEARING.

Control Scheme	I_1/A	I_2/A	I_3/A	P^*_{Cu}/%
Linear	2.7	2.4	2.5	100
Feedback Linearization	1.9	0.9	0.8	26

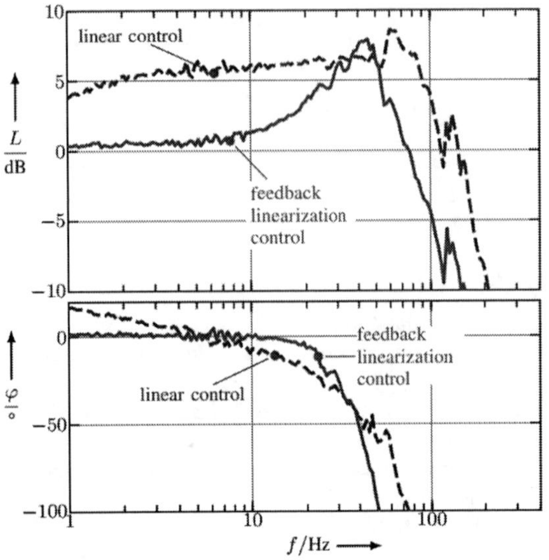

Figure 3. Comparison of closed loop frequency responses of the x_1-axis

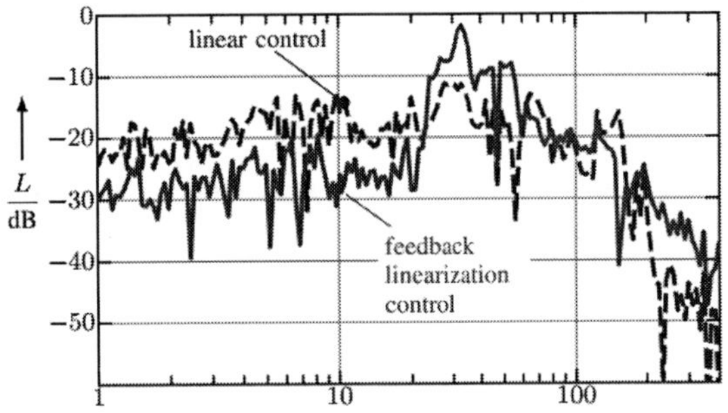

Figure 4. Comparison of cross coupling frequency responses $G = \frac{X_1}{Y_{1\mathrm{ref}}}$

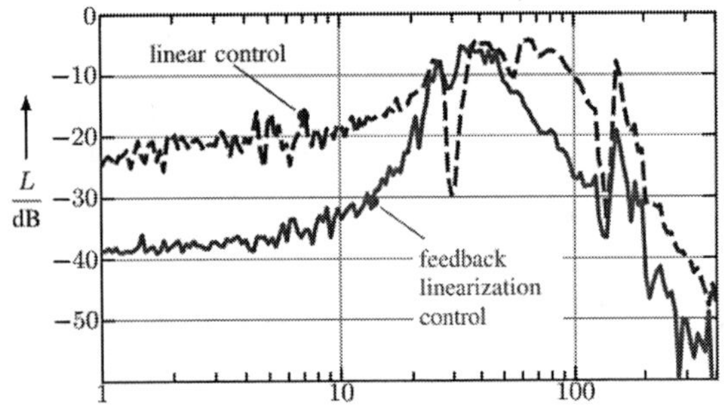

Figure 5. Comparison of cross coupling frequency responses $G = \frac{Y_1}{X_{1\mathrm{ref}}}$

As can be seen from the frequency responses in Fig. 3 both control schemes result in a similar bandwidth with a slightly higher one for the linear scheme. This is mainly due to the biasing, which allows the bearing force to be changed faster.

All measurement have been carried out with the same set of controller parameters found in Table I.

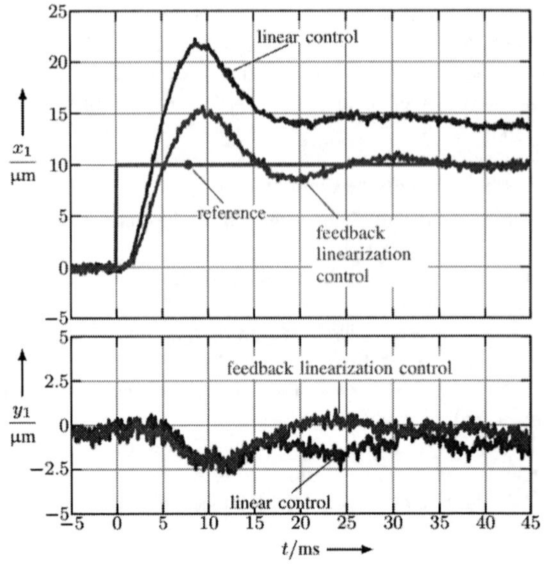

Figure 6. Comparison of step responses along the x_1-axis between linear and feedback linearization control

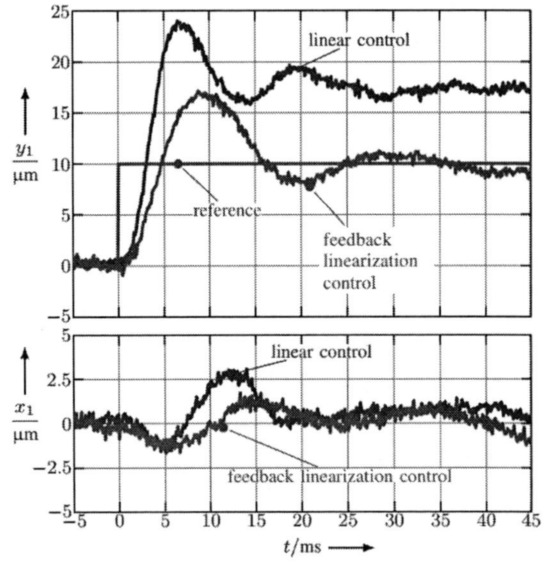

Figure 7. Comparison of step responses along the y_1-axis between linear and feedback linearization control

B. Impact of the Nonlinear Force-Current Relationship

Despite the introduction of a bias current in (3) a significant nonlinearity remains in the force-current relationship of the y-axis as shown in Fig. 2 (right). In conjunction with the inherent negative stiffness of the bearing this causes a 4 dB overshoot at frequencies up to 20 Hz as shown in the frequency response in Fig. 3. This indicates a poor tracking behavior, which is confirmed by the step response shown in Fig. 7. It shows a tracking error of 70 % after the initial oscillations have been attenuated. If one compares this result to the step response of the x-axis in Fig. 6, one can see that the overshoot and the tracking error (40 %) are smaller compared to the y-axis due to the better linearity of the force-current relationship along the x-axis.

As the feedback linearization takes the actual rotor position into account, this effect can be compensated fairly well. The frequency response in Fig. 3 shows a good tracking behavior up to 10 Hz despite the integral part of the PID controller being set to a corner frequency of 0.5 Hz. The step response in Fig. 7 confirms this. The feedback linearization control achieves nearly zero tracking error after the initial overshoot, which is smaller as well.

These results clearly demonstrate that inner loop formed by the feedback linearization can compensate both the static nonlinear behavior and the inherent negative stiffness of the bearings. This leads to good tracking behavior without the need of a strong integral part in the PID controller, which would reduce the stability margin of the control loop.

C. Analysis of Magnetic Cross Coupling Effects

The structure of the three pole bearing magnetically couples all control axes. The following section analyses how well the two schemes are able to decouple the position control loops.

Along the frequency responses of the closed loops the cross coupling has been measured as well. Fig. 4 shows the response of the x_1-axis to changes in the reference value for the y_1-axis in the frequency domain. The analysis in the opposite direction can be found in Fig. 5. Due to the improved tracking behavior the feedback linearization can achieve a better decoupling of the two axes at low frequencies up to 20 Hz. At higher frequencies decoupling behavior becomes similar for both schemes. The impact on the tracking behavior becomes obvious in the step responses.

The lower part of Fig. 6 shows the disturbance of the y-axis while the bearing performs a step on the x-axis on the same plane indicating that the decoupling of the control axes is not complete.

In general there are two possible mechanisms which can cause this effect. The first is that a cross coupling in the feed forward branch remains. Which means a control command along one axis causes a bearing force along another axis as well.

The second mechanism is caused by the position dependence of the magnetic force. That means a movement along one axis requires a change of the control command on another axis in order to keep the magnetic force constant.

Looking at the step response of the x-axis in Fig. 6 one can see that the disturbance on the y-axis is delayed compared to the response of the x-axis which indicates that the disturbance is caused by the position dependence of the magnetic force. As this is not taken into account in the linear control scheme, a tracking error remains.

Until the 15 ms mark the system behaves quite similar with feedback linearization control. After that mark the system actually returns to the reference position. This indicates that the feedback linearization scheme is able to compensate the position dependence of the magnetic force in the steady state. During dynamic processes this compensation does not work as well.

Note that both control schemes employ a very weak integral part in the PID in order to make these effects visible.

The step response of the y-axis shows a different behavior. The disturbance of the x-axis is not delayed indicating a cross coupling in the feed forward control. With both control schemes there is no remaining tracking error after initial oscillation. The main difference is the maximum displacement which is lower with the feedback linearization.

D. Stability Margin

The stability margin is a measure for the damping of oscillations in the system and of the robustness against changes in operating and equipment conditions. One method for evaluating the stability margin, which is also used in ISO 14839-3, is the sensitivity function.

For this analysis the closed loop frequency responses G_c of all control axes of the plant have been measured. From these the open loop frequency responses G_0 can be calculated.

$$G_0 = \frac{G_c}{1 - G_c} \tag{9}$$

The sensitivity function is then defined as in the inverse distance from the critical point (-1, 0).

$$G_S = \frac{1}{1 + G_0} \tag{10}$$

The smallest distance is a measure for the stability margin, which in turn is represented by the maximum of the sensitivity function and is the stability index defined in ISO 14839-3.

The sensitivity function for the y_1-axis has been plotted in Fig. 8. It indicates higher system robustness over wide range of frequencies when operating with feedback linearization control.

For all five control axes the sensitivity function have been calculated and the maximum of each can be found in Table IV. For the two x-axes the values show only small differences.

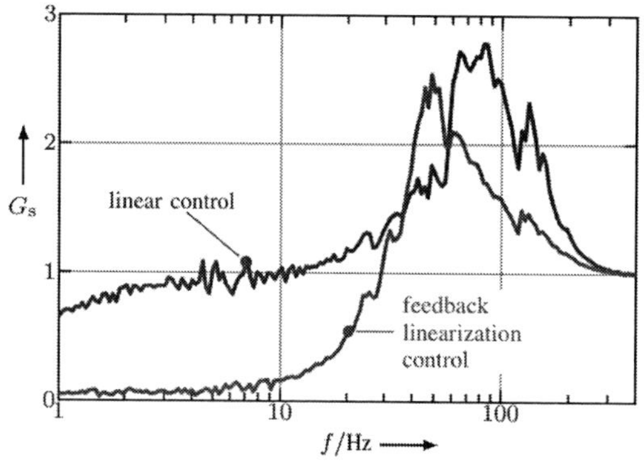

Figure 8. Impact of the control scheme on the sensitivity function of the y_1-axis

Table IV
STABILITY INDICES OF ALL FIVE CONTROL AXES

Axis	x_1	y_1	x_2	y_2	z
Linear Control	2.4	2.8	2.3	2.6	2.6
Feedback Linearization Control	2.4	2.5	2.4	2.5	2.2

This can be attributed to the good linearity of force-current relationship as was shown in Fig. 2. Therefore a similar performance can be expected with both control schemes.

In contrast the curves in Fig. 2 for the y-axes show a remaining nonlinearity in the force-current relationship, which reduces the stability margin of these axes compared to others and feedback linearization scheme. This is especially true for the y_1-axis.

Regarding the z-axis the linear control suffers from the coupling between radial and axial axes as the axial control current I_z acts similar to the bias current I_0 and a change in radial forces always implies an additional axial force in the same bearing.

With the feedback linearization scheme a greater stability margin can therefore be achieved as these cross coupling effects can at least partly be compensated.

It is therefore possible to achieve a similar or slightly better robustness of system with feedback linearization control despite the operation with minimal bias.

V. Conclusion and Outlook

It has been shown that both an approach based on a linear transformation and one based on feedback linearization can stabilize a magnetic suspension which consists of two combined three pole bearings. The stability analysis demonstrated good system robustness with both control schemes.

The comparison has shown that similar or better control performance can be achieved with feedback linearization while operating the bearing at minimal flux densities. Although there were no big gains in decoupling the axes, the tracking behavior can be significantly improved with feedback linearization. The decoupling performances is mainly limited by model accuracy both in terms of the mathematical formulation and the parameters.

Through the operation at minimal currents and flux densities the three pole combined bearing becomes an interesting alternative to permanent magnet bias homopolar bearings. While the copper losses are still somewhat larger, much lower rotor losses are possible.

One of the remaining problems of the feedback linearization scheme is the high computational cost which is mainly due to the ten square root operations for each sampling period of the controller.

Further work on the test rig will concentrate on measuring the rotor losses with different control schemes and comparing the results with conventional bearings.

Acknowledgement

The authors would like to thank the DFG (Deutsche Forschungsgemeinschaft) for funding this project.

References

[1] W. Hofmann, "Behaviour and control of an inverter-fed three-pole active radial magnetic bearing", in *Industrial Electronics, 2003. ISIE '03. 2003 IEEE International Symposium on*, vol. 2, 2003, 974–979 vol. 2.

[2] M. Reisinger, H. Grabner, S. Silber, W. Amrhein, C. Redemann, and P. Jenckel, "A novel design of a five axes active magnetic bearing system", in *12th Intern. Symp. on Mag. Bear.*, 2010.

[3] D. C. Meeker and E. H. Maslen, "Analysis and control of a three pole radial magnetic bearing", in *10th Intern. Symp. on Mag. Bear.*, Citeseer, 2006.

[4] S. Eckhardt and J. Rudolph, "Trajectory tracking for a magnetically levitated shaft with three-phase ambs in y-connection", in *10th Intern. Symp. on Mag. Bear.*, 2006.

[5] S.-L. Chen, Chia-Yi, S.-H. Chen, and S.-T. Yan, "Experimental validation of a current-controlled three-pole magnetic rotor-bearing system", *Magnetics, IEEE Transactions on*, vol. 41, no. 1, pp. 99–112, 2005.

[6] H. Grabner, M. Reisinger, S. Silber, W. Amrhein, C. Redemann, and P. Jenckel, "Nonlinear feedback control of a five axes active magnetic bearing", in *12th Intern. Symp. on Mag. Bear.*, 2010.

[7] E. Fleischer, S. Tröger, and W. Hofmann, "Control of a novel integrated radial-axial magnetic bearing", in *13th Intern. Symp. on Mag. Bear.*, 2012.

[8] S. Eckhardt and J. Rudolph, "High precision synchronous tool path tracking with an amb machine tool spindle", in *9th Intern. Symp. on Mag. Bear.*, 2004.

[9] M. J. D. Powell, "An efficient method for finding the minimum of a function of several variables without calculating derivatives", *The Computer Journal*, vol. 7, no. 2, pp. 155–162, 1964.

[10] Höganäs, *Somaloy technology for electrical motors*, Datenblatt, 2009.

[11] Höganäs, *Somaloy metal powders for smc components*, 2007.

[12] E. Fleischer, S. Tröger, and W. Hofmann, "Development of an active magnetic bearing with a soft magnetic composite core", in *Proceedings of the International Conference Magnetism and Metallurgy*, Jun. 2012.

Research on Redundancy of Weak Coupling Radial Magnetic Bearing

Xiaoguang Wang[a], Qian Liu[a], Yefa Hu[a], Haohui Chen[a], Qingling Geng[a]

[a] School of Mechanical and Electrical Engineering, , Wuhan University of Technology, 122 Luoshi Road, 430070, Wuhan, Hubei, P.R.China, xgwang55@163.com

***Abstract*—The weak coupling radial magnetic bearing structure and force balance compensation has been proposed in this paper. The force balance compensation of weak coupling radial magnetic bearing is simulated and comparison with bias current linearization. The research shows the force balance compensation can provide bigger electromagnetic force, the process of solving compensation matrix is simple, and the compensation current is more uniform. Meanwhile, The experimental results verify the feasibility of weak coupling radial magnetic bearing with redundant structure and force balance compensation.**

I. INTRODUCTION

Since the 1990s, aimed at how to further improve the reliability of magnetic bearing, many scholars, at home or abroad have done some studies in terms of actuators, controllers and sensors. In 1995, for radial magnetic bearing, bias current linearization method and algorithm of current distribution matrix was proposed by Eric H. Maslen[1]. Aimed at this method and algorithm, many scholars have conducted the confirmatory studies.

Based on bias current linearization theory, Myounggyu D. Noh describe the design and implementation of a fault tolerant magnetic bearing system for a turbo-molecular vacuum pump, the system is fault-tolerant design for actuators and position sensors of radial magnetic bearings[2]. Aimed at radial magnetic bearing in flexible magnetic suspension rotor , Uhn Joo Na in Texas A&M University use bias current linearization to solve the current distribution matrix of this structure, and use matlab software to simulate, the results show the linearized control forces can be realized up to certain combination of 5 poles failed for the 8 pole magnetic bearing. But with the number of failure coils increased, the stiffness and load capacity of magnetic bearings will decrease[3][4]. Ming-Hsiu Li studied on redundant control of magnetic bearings with flux coupling , and generalize the bias current linearization theory to the magnetic bearing with composite structure[5][6]. Satish Chand designed a three-pole magnetic bearings fault-tolerant controller based on nonlinear fuzzy logic control, when any of the three coils is failure, the rotor can stably suspended, and can obtain a good suspension properties[7].

For radial magnetic bearings with the configuration of multi-pole, independently driven coils, Wu Buzhou use the method of generalized bias current linearization to linearize the electromagnetic force. Then a redundant scheme of controller reconfiguration is developed on the basis of this method[8]. According to a six-core radial active magnetic bearing, the fault tolerant control method of displacement sensor based on coordinate transformation is studied by Cui Donghui, and the feasibility of this method has been verified through experiment[9][10].

II. REDUNDANCY SCHEME FOR WEAK COUPLING RADIAL MAGNETIC BEARINGS

Figure 1 shows the radial magnetic bearing structure based on Eric H. Maslen's idea. It works on the magnetic coupling between the poles, therefore can be called redundant structure for strong coupling radial magnetic bearing.

Figure 2 shows redundancy scheme for the radial magnetic bearing proposed in this paper. In the scheme, pole as an independent unit and pole pairs completely separated from each pole of the structure. The arrangement of using NNSS can further weaken the magnetic coupling between the magnetic poles, so it is called weak coupling radial magnetic bearing redundant structure. As shown in the two structures, each pole has a group of independent coils. The remaining coil can be reconstituted a new magnetic bearing systems when failure of anyone coil.

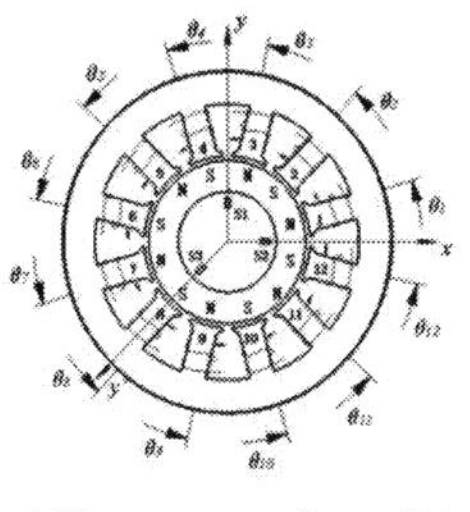

Figure 1.The strong coupling radial magnetic bearing structure

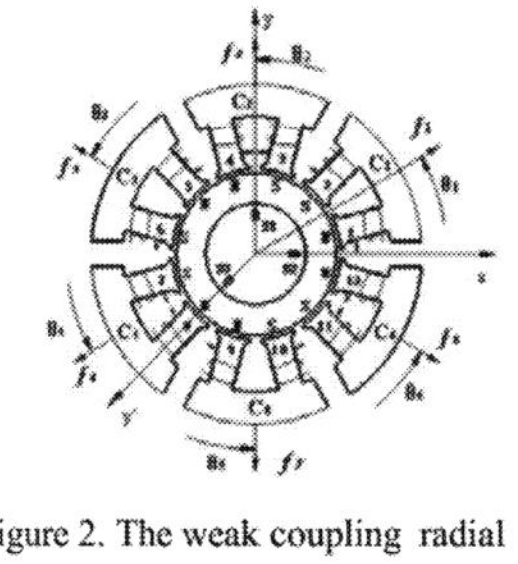

Figure 2. The weak coupling radial magnetic bearing structure

III. WORKING PRINCIPLE OF BIAS CURRENT LINEARIZATION AND FORCE BALANCE COMPENSATION

Eric H. Maslen and other scholars present the bias current linearization theory for strongly coupled radial magnetic bearing structure shown in Figure 1.

The theory is carried out by some reallocation of current control strategy[1], when part of the pole failure, which flux essentially redistribution and compensation.

A. Force balance compensation of weak coupling radial magnetic bearing

The magnetic poles coupling of the weak coupling magnetic radial magnetic bearings can be ignored, so the each poles pair can be used as an independent unit for analysis.

Define F_j as magnetic force of the jth electromagnetic pole, A as magnetic pole area, θ_j as orientation angle of the jth magnetic pole, δ_j as average value of air gap between the jth magnetic pole and rotor, δ_0 as air gap when rotor is the Equilibrium position, μ_0 as vacuum permeability, i_{2j-1} *and* i_{2j-1} as coil current of two poles for the j_{th} magnetic pole respectively, counter clock wise is order, N as the number of coil turns of the windings on a single pole.

Since each pole pair can be considered to be independent, thus it has the same impact for magnetic force by increasing the current of anyone pole coil in the same pole pair. It is assumed that the current in the two magnetic poles coil of the same pole pair is equal. that is $i_{j(1)}=i_{j(2)}$, make $i_{j(1)}=i_{j(2)}=i_j$, and record $i_{sj}=i_{j(1)}+i_{j(2)}$.

According to the electromagnetic force equation, the electromagnetic force of single magnetic pole pair is expressed as follow :

$$F_j = \mu_0 A N^2 \cdot \frac{(i_{j(1)}+i_{j(2)})^2}{4\delta_j^2} \cdot \cos\alpha = \mu_0 A N^2 \cdot \frac{i_{sj}^{\ 2}}{4\delta_j^2} \cdot \cos\alpha \qquad (1)$$

Where α represents half angle of adjacent poles, $\alpha=\pi/2n$, n represents total number of pole pairs of radial magnetic bearing,

$$\delta_j = \delta_0 - x\cos\theta_j - y\sin\theta_j \qquad (2)$$

When the rotor is at the equilibrium position, $\delta_j=\delta_0$.

If

$$k = \frac{\mu_0 A N^2 \cos\alpha}{4\delta_0^2} \qquad (3)$$

Then

$$F_j = k i_{sj}^{\ 2} \qquad (4)$$

The magnetic force on the rotor can be obtained as follows:

$$\begin{cases} F_x = \mathbf{F}\mathbf{D}_x \\ F_y = \mathbf{F}\mathbf{D}_y \\ \mathbf{F} = \begin{bmatrix} F_1 & F_2 & F_3 & F_4 & F_5 & F_6 \end{bmatrix} \\ \mathbf{D}_x = \begin{bmatrix} \cos\theta_1 & \cos\theta_2 & \cos\theta_3 & \cos\theta_4 & \cos\theta_5 & \cos\theta_6 \end{bmatrix}^T \\ \mathbf{D}_y = \begin{bmatrix} \sin\theta_1 & \sin\theta_2 & \sin\theta_3 & \sin\theta_4 & \sin\theta_5 & \sin\theta_6 \end{bmatrix}^T \end{cases} \qquad (5)$$

B. *Rule for force balance compensation*

According to failure modes of the weak coupling coil radial magnetic bearing, failure can be divided into two categories.

The first category failure: partial failure of a pole pair coil, that is, one coil is invalid.

The second category failure: a pole coil to fail completely, that is, two coils are invalid.

The electromagnetic force of magnetic pole pair can be achieved by increasing the current of the other coil when the first category failure to occur, that is, called the first class compensation.

The magnetic bearing will regain the same magnetic force as before coil failure or until the magnetic saturation by increasing the current of adjacent or near magnetic pole pair when the second category failure to occur, that is, called the second class compensation. The rule is the minimum energy and the minimum sum of compensation current for the method.

Define the current matrix without any coil failure as **I**.

Define a current compensation matrix as

$$\mathbf{M}_r = \begin{bmatrix} \cdots & 1 & 1 & 1 & 1 & 1 & 1 & \cdots \end{bmatrix}_{1\times 12}^T$$

Define the initial current of each magnetic pole coil as

$$\mathbf{I}_r = \begin{array}{c} \\ p_{(1)} \\ p_{(2)} \\ j_{(1)} \\ j_{(2)} \\ q_{(1)} \\ q_{(2)} \\ \\ \end{array} \begin{bmatrix} \cdots & & & & & & & \\ & i_p & & & & & & \\ & & i_p & & & & & \\ & & & i_j & & & & \\ & & & & i_j & & & \\ & & & & & i_q & & \\ & & & & & & i_q & \\ & & & & & & & \cdots \end{bmatrix}_{12\times 12}$$

Where $p=j-1$, $q=j+1$。

The current compensation matrix assigns compensated current to each coil,

$$\mathbf{I} = \mathbf{I}_r \mathbf{M}_r$$

(1)The first class compensation

The compensation would be achieved by increasing the current of the $j_{(2)}$ coil when the $j_{(1)}$ coil of the j_{th} magnetic pole pair invalid.

$$\mathbf{M}_r = \begin{bmatrix} \cdots & \underset{p_{(1)}}{1} & \underset{p_{(2)}}{1} & \underset{j_{(1)}}{0} & \underset{j_{(2)}}{2} & \underset{q_{(1)}}{1} & \underset{q_{(2)}}{1} & \cdots \end{bmatrix}_{1\times 12}^T$$

The compensation method is based on the maximum current in the coil $i_{max} \geqslant 2i_j$. It is necessary to adjust the current of adjacent magnetic pole pair coil when the maximum current in the coil $i_{max}<2i_j$.

(2) The second class compensation

Assumed the j_{th} magnetic pole pair failure completely, the compensation criteria are as follows: the force compensation for $j+1$ and $j-1$ should be provided by the nearest magnetic pole. The compensation matrix $\mathbf{M}_r$ follows as:

$$\mathbf{M}_r = \begin{bmatrix} \cdots & \underset{p_{(1)}}{\sqrt{\frac{F_p+F_j}{F_p}}} & \underset{p_{(2)}}{\sqrt{\frac{F_p+F_j}{F_p}}} & \underset{j_{(1)}}{0} & \underset{j_{(2)}}{0} & \underset{q_{(1)}}{\sqrt{\frac{F_q+F_j}{F_q}}} & \underset{q_{(2)}}{\sqrt{\frac{F_q+F_j}{F_q}}} & \cdots \end{bmatrix}_{1\times 12}^T$$

IV. SIMULATION OF BIAS CURRENT LINEARIZATION AND FORCE BALANCE COMPENSATION

Both magnetic flux density and magnetic force has been simulated using ANSYS on strong coupling and weak coupling radial magnetic bearing under bias current linearization and force balance compensation. The basic parameters of structure design are shown in Table I.

TABLE I. DESIGN PARAMETERS OF RADIAL MAGNETIC BEARING

	Parameter	Value
x_0	Single-side air gap	0.3 mm
I_m	Maximum current	6 A
N	coil turn of a single pole	56
A_a	Cross-section area of a magnetic pole	500 mm^2

A. *Simulation of bias current linearization on strong coupling structure*

Using bias current linearization, when there is no coil failure, the current distribution matrix of strong coupling radial magnetic bearing with 12 poles in figure 1 is obtained as $\mathbf{W}_q$ and when coil 1 and coil 2 are in failure, the corresponding current distribution matrix is obtained as $\mathbf{W}_{q(1,2)}$.

$$\mathbf{W}_q=\begin{bmatrix} 0.0631 & 0.2745 & 0.2328 \\ -0.1520 & -0.1532 & 0.1228 \\ -0.0257 & -0.1662 & 0.2235 \\ 0.1245 & -0.1998 & -0.1172 \\ 0.0484 & 0.0015 & 0.4503 \\ -0.1280 & 0.6745 & 0.2237 \\ -0.1859 & 0.3555 & 0.3560 \\ -0.1980 & 0.2277 & 0.3595 \\ -0.2206 & 0.1385 & 0.2710 \\ -0.1112 & -0.2600 & 0.2903 \\ 0.0947 & -0.2570 & -0.1892 \\ -0.1737 & 0.5632 & 0.1310 \end{bmatrix} \quad \mathbf{W}_{q(1,2)}=\begin{bmatrix} 0 & 0 & 0 \\ 0 & 0 & 0 \\ 0.1856 & 0.0743 & 0.2120 \\ -0.1318 & 0.2256 & 0.0690 \\ -0.1247 & -0.0519 & 0.1460 \\ 0.0412 & -0.0202 & 0.0180 \\ -0.0288 & 0.1127 & -0.0580 \\ -0.2575 & 0.1808 & 0.0203 \\ -0.1753 & 0.2168 & 0.1707 \\ -0.1279 & -0.0323 & 0.1464 \\ -0.0267 & 0.1115 & -0.0428 \\ 0.1423 & 0.2453 & -0.0030 \end{bmatrix}$$

According to the equation $\mathbf{I}=\mathbf{W}\mathbf{I}_c$, the current in each coil can be obtained. Then in the equilibrium state, there are relationships as $F_x=0$, $F_y=49$N.

According to the current distribution matrixes above, the magnetic flux density and magnetic force in the air gap of each magnetic pole can be obtained and then shown in figure 3 and 4. The magnetic forces in X and Y direction are shown in Table II.

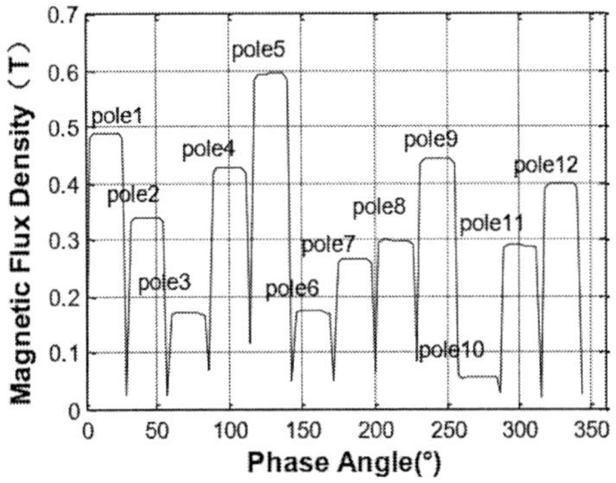

Figure 3. Magnetic flux density in the air gap before failure.

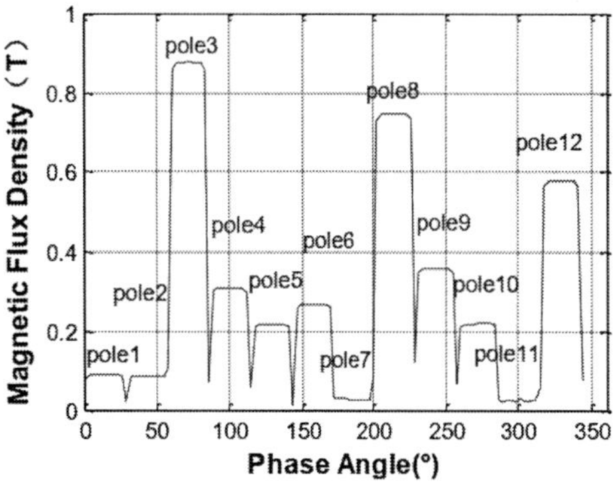

Figure 4. Magnetic flux density in the air gap after failure

TABLE II. MAGNETIC FORCE IN X、Y DIRECTION

Magnetic force direction \ *Working condition*		Magnetic force before failure(N)	Magnetic force when coil 1and coil 2 are in failure (N)
+Y, -X, +X, -Y	+X	105	109
	-X	101	110
	+Y	121	177
	-Y	74	129

B. *Simulation of bias current linearization on weak coupling structure*

Using bias current linearization, when there is no coil failure, the current distribution matrix of weak coupling radial magnetic bearing with 12 poles in figure 2 is obtained as $\mathbf{W}_r$ and when coil 1 and coil 2 are in failure, the corresponding current distribution matrix is obtained as $\mathbf{W}_{r(1,2)}$.

$$\mathbf{W}_r=\begin{bmatrix} 0.0810 & 0.1676 & 0.0109 \\ 0.0810 & 0.1676 & 0.0109 \\ -0.1170 & 0.1107 & -0.1145 \\ -0.1170 & 0.1107 & -0.1145 \\ 0.0662 & -0.1573 & 0.2611 \\ 0.0662 & -0.1573 & 0.2611 \\ 0.0872 & -0.1973 & 0.0343 \\ 0.0872 & -0.1973 & 0.0343 \\ 0.1125 & 0.0372 & -0.1099 \\ 0.1125 & 0.0372 & -0.1099 \\ 0.0737 & -0.1887 & 0.2631 \\ 0.0737 & -0.1887 & 0.2631 \end{bmatrix} \quad \mathbf{W}_{r(1,2)}=\begin{bmatrix} 0 & 0 & 0 \\ 0 & 0 & 0 \\ -0.0967 & -0.0713 & -0.1221 \\ -0.0967 & -0.0713 & -0.1221 \\ 0.1080 & -0.1258 & 0.0010 \\ 0.1080 & -0.1258 & 0.0010 \\ 0.0022 & -0.0204 & -0.0115 \\ 0.0022 & -0.0204 & -0.0115 \\ 0.0972 & -0.0692 & -0.1217 \\ 0.0972 & -0.0692 & -0.1217 \\ 0.1078 & 0.1270 & 0.0008 \\ 0.1078 & 0.1270 & 0.0008 \end{bmatrix}$$

With the same principle, the magnetic flux density and magnetic force in the air gap of each magnetic pole can be obtained and then shown in figure 5 and 6. The magnetic forces in X and Y direction are shown in Table III.

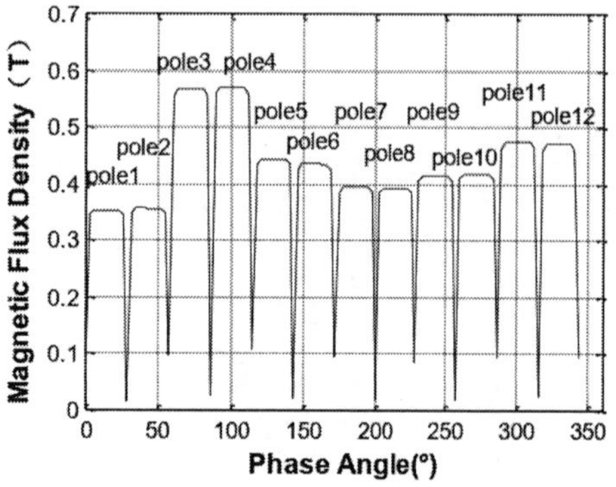

Figure 5. Magnetic flux density in the air gap before failure

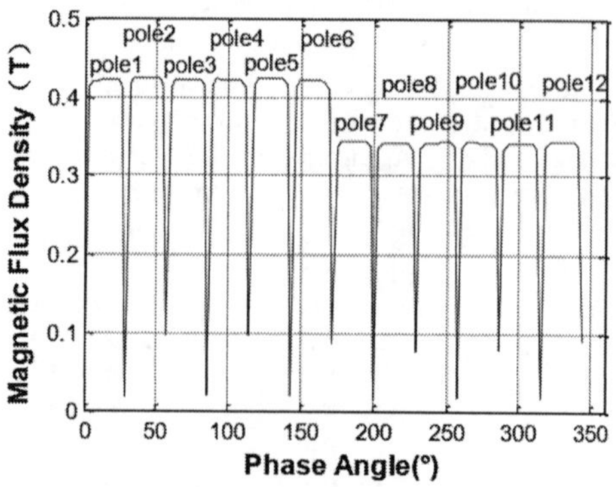

Figure 7 Magnetic flux density in the air gap before failure

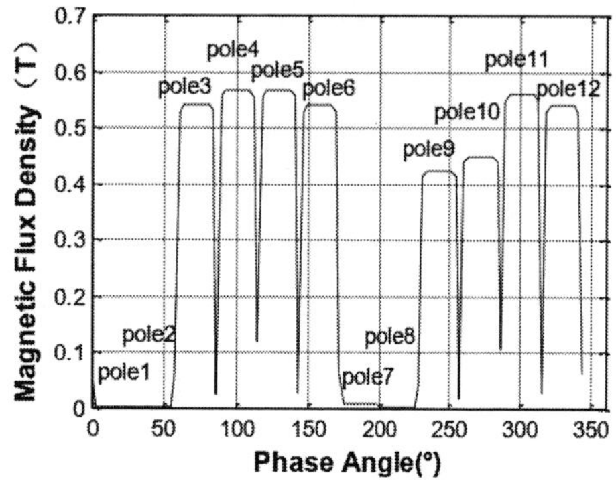

Figure 6. Magnetic flux density in the air gap after failure

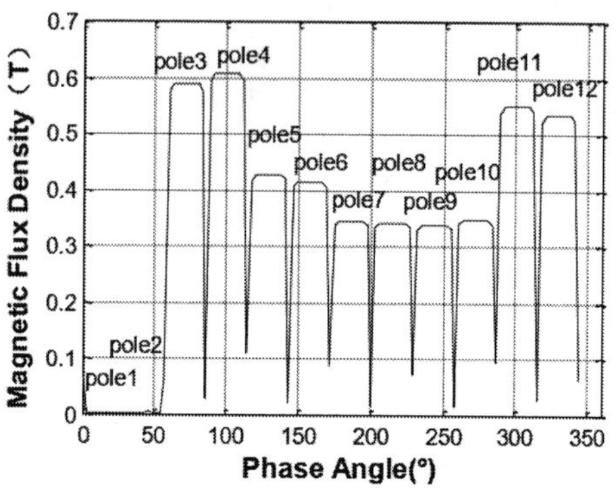

Figure 8 Magnetic flux density in the air gap after failure

TABLE III. MAGNETIC FORCE IN X、Y DIRECTION

Working condition / *Magnetic force direction*		Magnetic force before failure(N)	Magnetic force when coil 1and coil 2 are in failure (N)
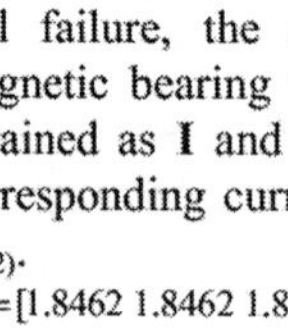	+X	99	94
	-X	99	94
	+Y	148	167
	-Y	99	119

TABLE IV. MAGNETIC FORCE IN X、Y DIRECTION

Working condition / *Magnetic force direction*		Magnetic force before failure(N)	Magnetic force when coil 1and coil 2 are in failure (N)
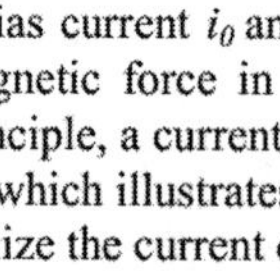	+X	104	104
	-X	104	104
	+Y	144	180
	-Y	95	131

C. Simulation of force balance compensation on weak coupling structure

Using force balance compensation, when there is no coil failure, the current matrix of weak coupling radial magnetic bearing with 12 poles in figure 2 can be directly obtained as **I** and when coil 1 and coil 2 are in failure, the corresponding current distribution matrix is also obtained as $\mathbf{I}_{(1,2)}$.

$$\mathbf{I}=[1.8462\ 1.8462\ 1.8462\ 1.8462\ 1.8462\ 1.8462\ 1.5\ 1.5\ 1.5\ 1.5\ 1.5\ 1.5]$$

$$\mathbf{I}_{(1,2)}=[0\ 0\ 2.6251\ 2.6251\ 1.8462\ 1.8462\ 1.5\ 1.5\ 1.5\ 1.5\ 2.3788\ 2.3788]$$

With the same principle, the magnetic flux density and magnetic force in the air gap of each magnetic pole can be obtained and then shown in figure 7 and 8. The magnetic forces in X and Y direction are shown in Table IV.

V. ANALYSIS AND COMPARISON BETWEEN TWO CONTROL METHODS

The basic principle of bias current linearization is using a bias current i_0 and two control currents i_x, i_y to express the magnetic force in X, Y direction as equation(4). With this principle, a current distribution matrix can then be deduced as **W** which illustrates the relationship between i_x, i_y and i_1~i_{12} to realize the current distribution.

The basic principle of force balance compensation is to analyse each pair of magnetic poles as an independent unit, ignoring the magnetic coupling between them. According to force balance compensation, the current in each coil can then be obtained and the current compensation is thus realized.

Based on the comparison of magnetic flux density and magnetic force in two structures when there is no coil failure and when coil 1 and coil 2 are in failure, conclusions can be made as followed.

- For bias current linearization, there are multiple solutions which means the outcome of compensation is not unique. Conversely, for force balance compensation, the solution is unique.
- For bias current linearization, the distribution of magnetic field is not uniform when there is no coil failure and when coil 1and coil 2 are in failure, so the magnetic force produced is relatively small. For force balance compensation, owing to the principle of minimum energy, the sum value of compensation current is then the least and the distribution of current and magnetic flux is relatively uniform, so under the unsaturation limit in magnetic material, magnetic force produced is larger.
- For bias current linearization, the calculation process is complicated. Different initial values will lead to a great deviation in the outcome when solving the current distribution matrix, so repeated calculation is needed for it. For force balance compensation, the calculation process of the matrix is simple. As long as the failure condition is known, corresponding compensation matrix can be obtained quickly according to the compensation principle, thus greatly simplifying the calculation process.

VI. FORCE BALANCE COMPENSATION EXPERIMENT OF WEAK COUPLING RADIAL MAGNETIC BEARING

To verify the feasibility of force balance compensation in weak coupling radial magnetic bearing, the bearing structure with 6 pole pairs is simplified to the basic form with one pole pair and the rotor is simplified as a armature which is shown in figure 9. The whole experimental device is shown in figure 10 and the design parameters are listed in Table V.

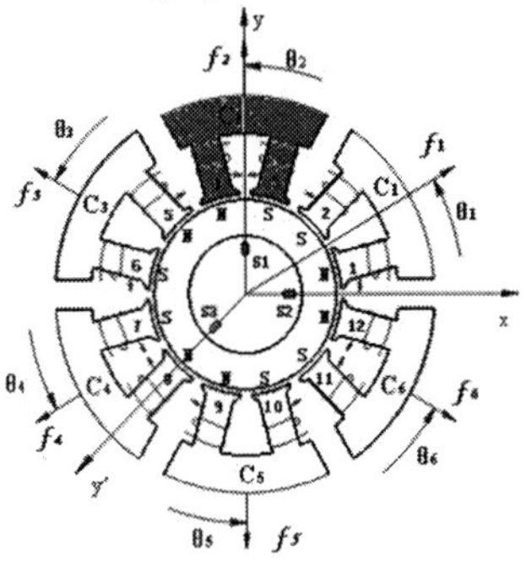

Figure 9 Experimental principle of force balance compensation

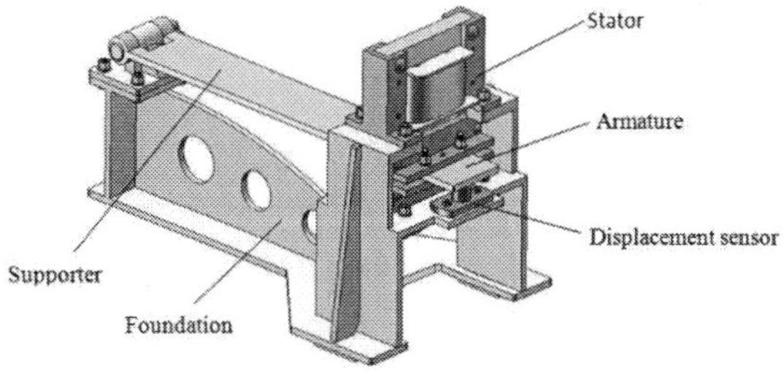

Figure 10 Experimental device of force balance compensation

TABLE V. DESIGN PARAMETERS OF MAGNETIC SUSPENDED REDUNDANT EXPERIMENTAL DEVICE

	Parameter	Value
A	Cross-section area of the armature	480 mm^2
x	Total air gap	1 mm
N	Turn of coil 1	400
N	Turn of coil 2	100
i_m	Maximum current	3A

A. *Control principle of experimental system*

Considering the generality, there are two independent coils with different turns winded onto the armature. Independent control can be realized on these two coils and the control principle is shown in figure11.

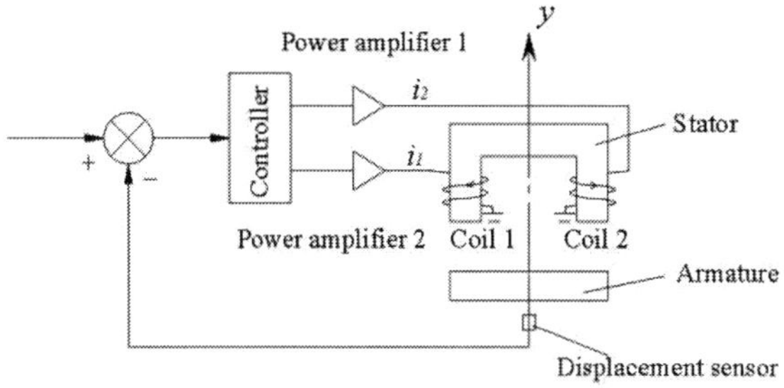

Figure 11 Control principle of experimental device

B. *Experimental result analysis*

Experimental research on the reconfiguration process is carried out on the experimental device of weak coupling radial magnetic bearing corresponding to coil 1 failure and coil 2 failure respectively. Firstly, the armature is suspended steadily and then a fault order is given to make a zero output of the failed coil, compensating the other coil at the same time. PID parameters are adjusted before the experiment to make sure the steady suspension of the armature.

Dynamic property curves of reconfiguration process under coil 1 failure and coil 2 failure are shown in figure 12 and 13 respectively.

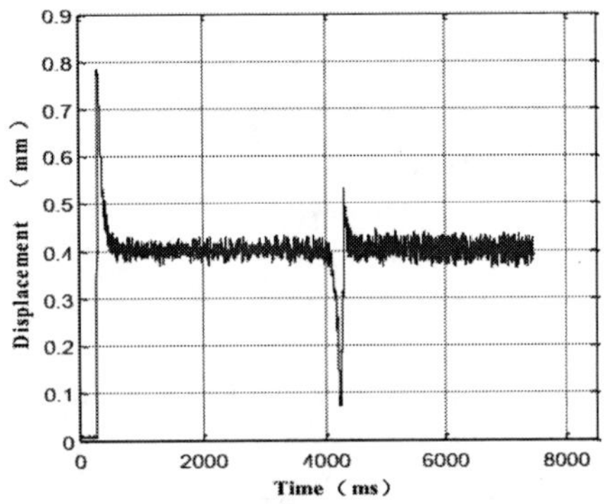

Figure 12 Reconfiguration process curve under coil 1 failure

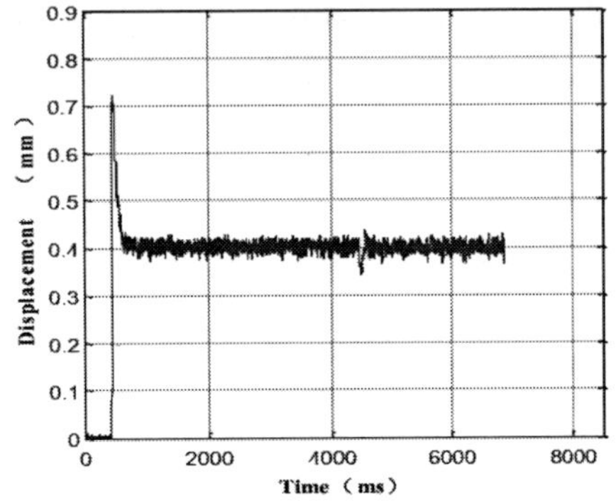

Figure 13 Reconfiguration process curve under coil 2 failure

From the two figures above, for coil 1 failure, the reconfiguration process needs a long time and the vibration of armature is quite large. For coil 2 failure, the reconfiguration process only needs a short time and the vibration of armature is really small. So coil 2 failure has a small influence on the steady-state performance of the system.

The experiment verifies the feasibility of force balance compensation in weak coupling radial magnetic bearing.

VII. CONCLUSIONS

The redundant structure of weak coupling radial magnetic bearing can realize the redundant reconfiguration. Compared to the strong coupling structure, the weak coupling structure has a couple of advantages. Bias current linearization is suitable for both strong and weak coupling structure. Compared to bias current linearization, force balance compensation has a couple of advantages.

VIII. ACKNOWLEDGEMENTS

The authors gratefully acknowledge the support of this research by the National Natural Science Foundation of China (no. 51205300, no.51147004, no. 51175390).

REFERENCES

[1] Eric H. Maslen and David C. Meeker, "Fault Tolerance of Magnetic Bearings by Generalized Bias Current Linearization," *IEEE Transactions on Magnetics*, vol. 313, no. 3, pp. 2304-2314, 1995.

[2] Myounggyu D.Noh, Seong-Rak Cho,et al, "Design and Implementation of a Fault-Tolerant Magnetic Bearing System for Turbo-Molecular Vacuum Pump," *IEEE/ASME Transactions on Mechatronics*, vol. 10, no. 6, pp. 626-631, 2005.

[3] Uhn Joo Na, "Fault-tolerant control of heteropolar magnetic bearings," *College Station: Texas A&M University*, 1999.

[4] Uhn Joo Na, Alan Palazzolo, "Optimized Realization of Fault-Tolerant Heteropolar Magnetic Bearings," *Journal of Vibration and Acoustics*, vol. 122, pp. 209-221, 2000.

[5] Ming-Hsiu Li, "Fault tolerant control of homopolar magnetic bearings and circular sensor arrays," *College Station: Texas A&M University*, 1999.

[6] Ming-Hsiu Li, Alan B.Palazzolo, Andrew Kenny, et al. Kascak, "Fault-Tolerant Homopolar Magnetic Bearings," *IEEE Transactions on Magnetics*, vol. 40, no. 5, pp. 3308 -3318, 2004.

[7] Praveen Kumar Agarwal, Satish Chand, "Fault-tolerant control of three-pole active magnetic bearing," *Expert Systems with Applications*, vol. 36, pp. 12592-12604, 2009.

[8] Wu Buzhou, Sun Yanhua, Wang Shihu,et al, "Coil Fault-tolerant Control for Radial Active Magnetic Bearings," *Chinese Journal of Mechanical Engineering*, vol. 41, no. 6, pp. 157-162, 2005.

[9] Cui Donghui, "Research on key technology of magnetic bearing with high reliability," *Nanjing University of Aeronautics and Astronautics*, 2010.

[10] Cui Donghui, XuLongxiang, "Fault tolerant control of radial active magnetic bearing based on coordinate transformation," *Control and Decision*, vol. 25, no.9, pp. 1420-1430, 2010.

Active Magnetic Bearings with Position and Flux Observer

Angelo Bonfitto [a], Lester D. Suarez [a], Andrea Tonoli [a]

[a] Laboratorio di Meccatronica – Politecnico di Torino,
Corso Duca degli Abruzzi 24, 10129 Torino, Italy, andrea.tonoli@polito.it

***Abstract* — Standard Active Magnetic Bearings are controlled using a current reference and the measured or observed current as state variable in the state space controller. Nevertheless since this approach selects the current as the state expressing the behavior of the magnetic bearing, several disadvantages occur. Indeed it is well known that this method leads to an inherent nonlinear relation between the generated force, the current and the air gap. This paper presents the modeling approach, the control design and the experimental validation of a strategy based on the combined observation of position and flux. The technique is designed and tested on a one-degree of freedom mechanical oscillator actuated by a couple of opposite electromagnets.**

I. INTRODUCTION

Contact-free levitation of rotors by means of active magnetic bearings has been a research topic for more than two decades. During this time, AMB evolved into an industry product that is used in many practical applications due to its numerous advantages over conventional bearings technology. The spread of this technology is mostly due to the growing demand of high accuracy and precision in industrial applications [1,2].

Currently the research effort is focused on many aspects of the whole suspension system i.e. power electronics, sensors, control layouts and actuators morphology, the aims being performance optimization and costs reduction.

Among these topics the definition of smart architectures allowing to optimize performance minimizing the use of sensors or getting rid of them at all is of great interest [3,4,5,6,7].

Many sensorless layouts have been developed and are currently used in industrial applications. The most spread architectures use AMB controlled with no gap sensors so to to save costs and size and improve performance thanks to the inherent collocation of sensing and actuators [8,9,10,11,12].

In this paper the modeling approach, the control design and the experimental validation of a control technique based on the combined observation of position and flux is illustrated. A one-degree of freedom mechanical oscillator actuated by a couple of electromagnets has been used to test the effectiveness of the proposed control strategy. The proposed layout exploits an observer to estimate the mechanical position, velocity and the flux on the electromagnets starting from current and position direct measurements. A cascaded control layout uses the flux observation in an inner flux loop whose reference is provided by the external position loop closed on the observed position.

As well known, the first advantage of a flux feedback approach is to minimize the effect of the displacement on the actuation force. Additionally, this layout allows to minimize the amount of noise introduced in the control loops. The model based observer approach permits to estimate the dynamics of interest, reducing noise issues and the effects of high order dynamics that are considered useless in terms of control performance.

Furthermore, the possibility of using the electromagnet either as actuator and sensor, permits to overcome actuators/sensors non-colocation problems, typical in rotors control layouts using direct position sensing as feedback in closed loops.

The overall system results to be more efficient than standard control layouts since it avoids measurements noise in the control loops while generating a system characterized by actuator/sensor colocation and using a linear formulation of the force generated by the electromagnets.

The paper starts on the description of modeling phases, details the steps followed for observer design and treats the selection of filters dynamics for both inner and outer control loops.

Afterwards the comparison of numerical and experimental results is presented aiming to validate the modeling approach of the mechanical and electrical subsystems and their interactions and to prove the effectiveness of the control action along with the performance and the efficiency of the resulting system.

The paper concludes providing an outlook on the next steps in the refinement of this scheme that will be mostly focused on the improvement of control architecture and of the model so to take into account hidden information such as magnetization or hysteresis of the core iron material, stray flux and eddy currents. The natural application of the presented technique is its transfer to the control of rotors suspended by means of active magnetic bearings.

II. MODELING AND CONTROL DESIGN

The considered system is a one degree of freedom mass-spring oscillator actuated by two opposite electromagnets, corresponding to an AMB configuration, as shown in Figure 1. The electromagnets are assumed to be identical, and the coupling between the two electromagnetic circuits is neglected. Also, each electromagnet can be considered as a two-port element (electrical and mechanical).

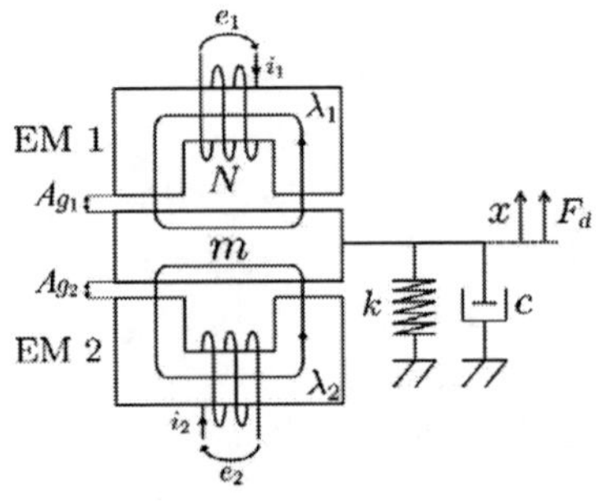

Figure 1. Model.

Owing to Newton's law in the mechanical domain and the Faraday and Kirchhoff laws in the electrical domain, the dynamics equations of the system are

$$m\ddot{x} + c\dot{x} + kx = F_j$$
$$\dot{\lambda}_j + Ri_j = e_j \pm k_m\dot{x} \quad ,j = 1,2 \tag{1}$$

where R is the coils resistance, c the viscous damping, k the stiffness, λ_j is the total flux and e_j, i_j and F_j are the voltage, current and force for each electromagnet j.

The force F_j can be derived from the energy store E_j in the electromagnet.

$$E_j(i_j, x_J) = \frac{1}{2}\frac{\lambda_j^2}{L(x)}$$
$$F_j(i_j, x_J) = \frac{\partial E_j(i_j, x_J)}{\partial x} \quad ,j = 1,2 \tag{2}$$

Thus, the generated force related to the control flux can be expressed as

$$F_j = \frac{\lambda_j^2}{\mu_0 Ag} \quad ,j = 1,2 \tag{3}$$

where Ag is the nominal airgap and μ_0 is the permeability.

The system dynamics is linearized around a working point corresponding to a bias flux λ_0 imposed to both electromagnets (Eq. 4)

$$\lambda_j = \lambda_0 \pm \lambda_c \quad ,j = 1,2 \tag{4}$$

$$m\ddot{x} + c\dot{x} + kx = \frac{1}{\mu_0 Ag}\left[\lambda_1^2 - \lambda_2^2\right]$$
$$m\ddot{x} + c\dot{x} + kx = \frac{1}{\mu_0 Ag}\left[(\lambda_0 + \lambda_c)^2 - (\lambda_0 - \lambda_c)^2\right] \tag{5}$$
$$m\ddot{x} + c\dot{x} + kx = \frac{4\lambda_0}{\mu_0 Ag}\lambda_c$$

where λ_c is the control flux.

Finally, considering a dynamic equilibrium, the mathematical model is expressed as follow

$$m\ddot{x} + c\dot{x} + kx = \frac{4\lambda_0}{\mu_0 Ag}\lambda_c$$
$$\dot{\lambda}_j + \frac{R}{L}\lambda_j = e_j \pm k_m\dot{x} \quad ,j = 1,2 \tag{6}$$

A three-state-space model is used to study system dynamics described in Eq. 7. The resulting linearized state-space model can be represented as follows:

$$\dot{\mathbf{X}} = \mathbf{AX} + \mathbf{B}u$$
$$\mathbf{y} = \mathbf{CX} \tag{7}$$

where $\mathbf{A}$, $\mathbf{B}$, and $\mathbf{C}$ are the dynamic, action, and output matrices, respectively, defined as

$$A = \begin{bmatrix} 0 & 1 & 0 \\ -k/m & -c/m & 4\lambda_0/(m\mu_0 Ag) \\ 0 & -k_m & -R/L \end{bmatrix}$$
$$B = \begin{bmatrix} 0 \\ 0 \\ 1 \end{bmatrix}, \quad C = \begin{bmatrix} 1 & 0 & 0 \\ 0 & 0 & 1/L \end{bmatrix} \tag{8}$$

with the associated input and output state vectors $\mathbf{X} = \{x, \dot{x}, \lambda\}^T$, $u = e_0$ and $\mathbf{y} = \{x, \lambda\}^T$

TABLE I. MODEL PARAMETERS

	Parameter	Value	Unit
R	Coils resistant	0.356	Ω
L	Coils inductance	5.8e-3	H
Ag	Airgap	0.6e-3	mm
μ_0	permeability	1.25e-6	H/m
m	mass	3.41	Kg
k_m	Back-FEM	3.96	Vs/rad
c	Viscous damping	1.15	kg/s
k	Stiffness	46500	kg/s^2

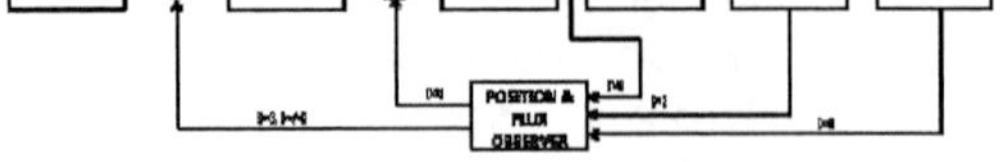

Figure 2. System Control Layout

The overall control scheme is illustrated in Figure 2. Here, the position and flux observer estimates the state variables based on the measurement of the output and control variables to generate the desired control vector.

From the system defined by Eq. 7, the dynamic model of the state observer can be obtained assuming that the state vector $\mathbf{X}$ is approximated by the observed state vector $\hat{\mathbf{X}}$

$$\dot{\hat{\mathbf{X}}} = \mathbf{A}\hat{\mathbf{X}} + \mathbf{B}\mathbf{u} + \mathbf{L}(\mathbf{y} - \mathbf{C}\hat{\mathbf{X}}) \tag{9}$$

$\mathbf{L}(\mathbf{y} - \mathbf{C}\hat{\mathbf{X}})$ is a correction term that involves the difference between the measured output $\mathbf{y}$ and the estimated output $\mathbf{C}\hat{\mathbf{X}}$ to minimize the difference between the dynamic model and the actual system. Thus, the problem of designing a state observer becomes that of determining the observer gain matrix $\mathbf{L}$ such that the error dynamic is asymptotically stable with sufficient speed of response. Furthermore, the asymptotically stability and the speed of response is determined by the eigenvalues of matrix $(\mathbf{A} - \mathbf{L}\mathbf{C})$

III. EXPERIMENTAL SETUP

The system used for the experimental investigation is a test rig available at the Mechatronics Laboratory of Politecnico di Torino presented in Figures 3 and 4. It is used for static characterization of radial magnetic bearings and for the analysis of different control strategies, sensors, and power amplifiers in a simple and well known experimental setup without the need of an expensive prototype of the whole suspension. This rig consists in a horizontal arm hinged at one extremity with a pivot and actuated with a single axis magnetic bearing. The electromagnets have 142 windings, an airgap area S=4.2x10-4 m^2, and a nominal airgap of approximately Ag =0.6 mm.

The length of the pendulum (320 mm) compared with the

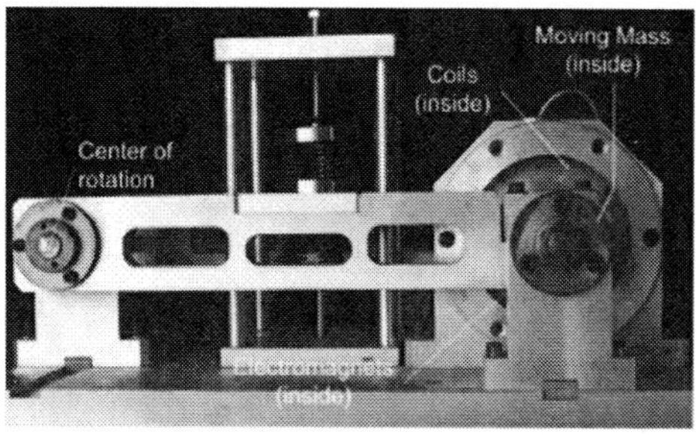

Figure 3. Picture of test rig

nominal airgap insures that the circular motion of the mass can be approximated with a linear translation (x). The base plate, connecting arms, and electromagnets' housing are made of aluminum, while the shafts are made of steel. The ferromagnetic circuit (stator and moving part) of the electromagnet is made of silicon iron sheets. Six springs in parallel are located at a distance of 160 mm from the hinge to provide a stabilizing stiffness to the system if necessary.

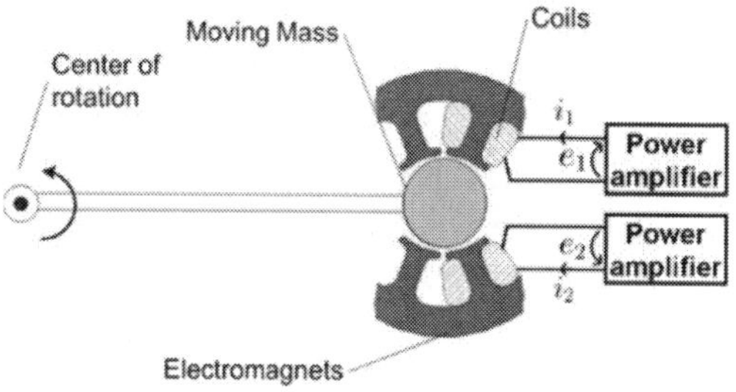

Figure 4. Picture of test rig

A DSP TI TMS320F28335 has been used for the control implementation. The analog-to-digital converters use 12 bit over a range of 3 V, with a sampling rate of 40 kHz.

The position was measured with a Bently Proximitor 3300XL eddy current sensor just for monitoring and validation, and not for feedback purposes. AMP25 Hall current sensors (range ±5A) were used for the current measurement in the two coils (lower than 100 mA, peak-to-peak, of noise).

The Vdc value of the power amplifier was set to 20 V, and the switching frequency was set to 40 kHz. Thus, as both the sampling rate and the switching frequency of the PWM were far beyond the mechanical frequency of the system (around 20 Hz), their dynamics was neglected during the study.

IV. EXPERIMENTAL RESULTS

The proposed technique has been validated experimentally so to verify that the proposed technique allows to avoid the direct effect of displacement on the actuation force typical of architecture with current feedback. To this end, after a preliminary phase of electromechanical parameters identification, an impact test with the external position loop disabled and the internal one (in current or in flux) enabled has been performed. To execute these tests, the plant has been mechanical centered by means of a set of springs (k in Table I). Results show that with electromagnets off, the system oscillates at a frequency of round 20 Hz (Figure 5.a). When the current loop is enabled with a bias of 0.5 A, the system oscillates at 15 Hz (Figure 5.b), while with the flux feedback the effects of negative stiffness disappears and the oscillation returns to be 20 Hz (Figure 5.c). The comparison between the three tests is reported in Figure 6.

The second phase of experimental tests is focused on the verification of system behaviour in closed loop configuration. The mechanical spring are uninstalled and the impact tests leads to results illustrated in Figure 7.

A further advantage of the proposed technique is the possibility of using observed position as feedback in the external loop. As a matter of fact this procedure acts as a filter against high frequency noise. Figure 9 shows the difference between the closed loop response obtained feed-backing measured position (a) and observed position (b). Apart from noise reduction, the use of a model based estimator could give benefits in terms of robustness in the presence of failure of one of the sensors.

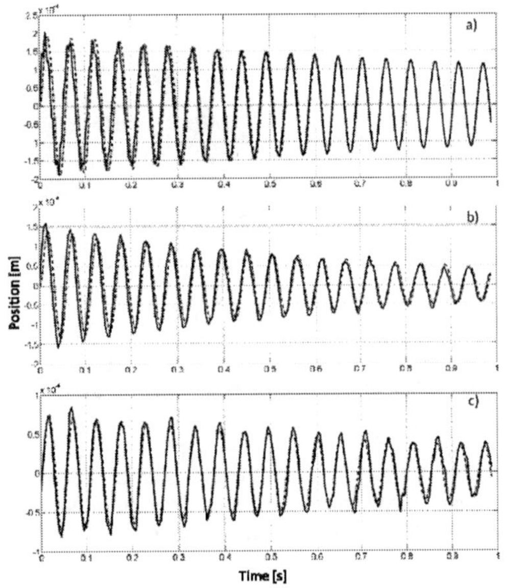

Figure 5. Tests with external position loop disabled and internal loop enabled. (Bias 0.5 A). Experimental (solid line) vs numerical results (dash-dotted lines). a) electromagnets off; b) internal current loop; c) internal flux loop.

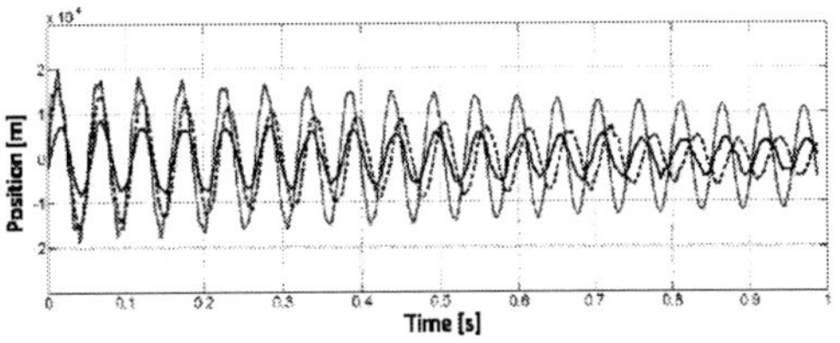

Figure 6. Tests with esternal loop disabled. Electromgantes off (solid thin line) vs Current feedback (dashed line) vs Flux feedback (solid bold line).

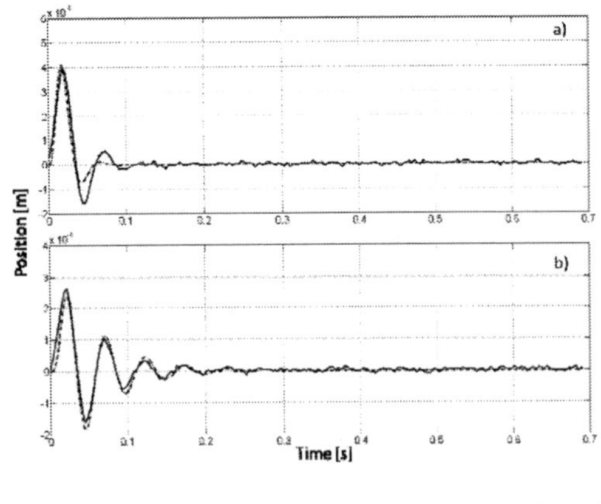

Figure 7. Test with external position loop enabled. Experimental (solid line) vs numerical results (dash-dotted lines). a) Current feedback; b) Flux feedback.

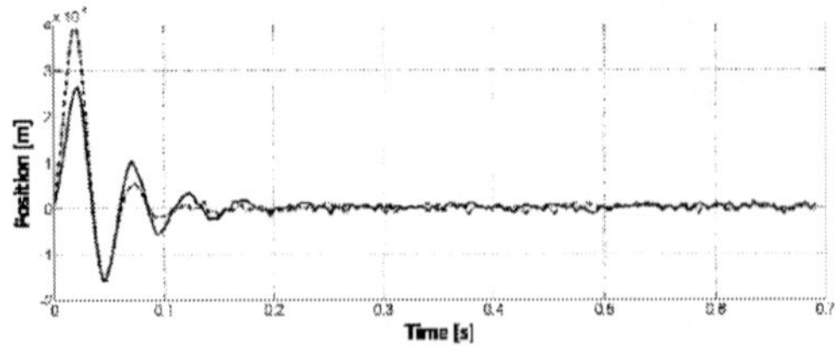

Figure 8. Test with external position loop enabled. Current feedback (dashed line) vs Flux feedback (solid line).

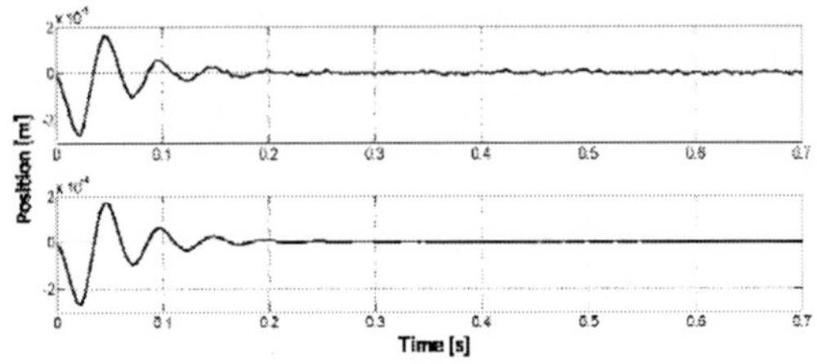

Figure 9. Closed position loop repsonse. a) Feedback with measured position; b) Feedback with observed position.

V. CONCLUSIONS

In this paper a technique of AMB control based on observed flux feedback has been presented. Starting from the measurement of mechanical displacement, current and voltage, the observer allows to estimate the flux and the position itself so to perform a cascaded control allowing to avoid the effect of negative stiffness on actuation force and to reduce noise by feed-backing observed position and flux.

Next steps will be focused on optimization of the control layout studying a feedback action based on pole placement technique and getting rid of cascaded architecture.

A further step will be the improvement of the model so to take into account hidden information such as magnetization or hysteresis of the core iron material, stray flux and eddy currents.

NOMENCLATURE

c = viscous damping coefficient
k = Stiffness
e = voltage applied to an electromagnet
Ag = Airgap
i = Current
k_m = back-electromotive force factor
m = mass
x = displacement of the mass
F = magnetic force
R = Resistance
L = Inductance
S = cross-sectional area of the magnetic circuit
λ = magnetic flux
μ_0 = magnetic permeability of the vacuum
E = magnetic energy

REFERENCES

[1] Bleuler H., Cole M., Keogh P., Larsonneur R., Maslen E., Nordmann R., OkadaY., Schweitzer G.,Traxler A.,"Magnetic Bearings Theory, Design, and Applicationto Rotating Machinery", Springer, Berlin Heidelberg, 2009

[2] Chiba A., Fukao T., Ichikawa O., Oshima M., Takemoto M., Dorrell D.G.,"Magnetic Bearings and Bearingless Drives" Elsevier, Oxford, 2005.

[3] Tonoli, A., Bonfitto, A., De L´epine, X., Silvagni, M., 2009, "Self-sensing Active Magnetic Dampers for vibration control", Journal of Dynamic Systems, Measurement and Control, 131, pp.

[4] Vischer, D., and Bleuler, H., 1990, "A new approach to sensorless and voltage controlled AMBs based on network theory concepts," Second International Symposium on Magnetic bearings, 12-14 July, Tokyo, Japan, pp. 301-306.

[5] Noh, M. D., and Maslen, E. H., 1997, "Self-sensing magnetic bearings using parameter estimation," IEEE Trans. Instr. Meas., 46(1), pp. 45-50.

[6] Schammass, A., Herzog, R., Buhler, P., and Bleuler, H., 2005, "New results for self-sensing active magnetic bearings using modulation approach," IEEE Trans. Contr. Syst. Tech., 13(4), pp. 509-516.

[7] Mizuno, T., Araki, and K., Bleuler, H., 1996, "Stability analysis of self-sensing magnetic bearing controllers," IEEE Trans. Contr. Syst. Tech., 4(5), pp. 572579.

[8] H. Bleuler, D. Vischer, Schweitzer T., Traxler A., ZLatnik D. "New Concepts for cost-effective Magnetic Bearings Control" Automatica, vol. 30 no. 5, pp.871-876, 1994.

[9] Mizuno, T., Namiki, H., and Araki, K., 1996, "Self-sensing operations of frequency-feedback magnetic bearings," Fifth International Symposium on Magnetic Bearings, 29-30 Aug., Kanazawa, Japan, pp. 119-123.

[10] Okada, Y., Matsuda, K., and Nagai,B., 1992, "Sensorless magnetic levitation control by measuring the PWM carrier frequency component," Third International Symposium on Magnetic Bearings, 21-31 July, Radisson Hotel, Alexandria, VA.

[11] Mizuno, T., Ishii, T., and Araki, K., 1998, "Self-sensing magnetic suspension using hysteresis amplifier," Cont. Eng. Pract., 6, pp. 1133-1140.

Vischer, D., and Bleuler, H., 1993, "Self-sensing magnetic levitation," IEEE Trans. on Magn., 29(2), pp. 1276-1281.

Interconnected Four Poles Magnetic Bearings Simulations and Testing

D. F. B. David[a], J. A. Santisteban[a], A .C. Del Nero Gomes[b]

[a] Universidade Federal Fluminense, Rua Passo da Pátria 156, 20420-240 Niteroi, Brazil, domingos@vm.uff.br
[b] Universidade Federal do Rio de Janeiro, Centro de Tecnologia, Rio de Janeiro, Brazil

***Abstract*—This article proposes and analyses a new idea to be used in magnetic bearings. It is an adaptation of a structure that has already worked successfully in split-winding self-bearing motors. Preliminary theoretical results predict that a greater equivalent stiffness could be achieved when compared with the values associated with the traditional active magnetic bearing conception in the literature.**

I. Introduction

The conventional structure of active magnetic bearings (AMBs) used with rotating machines [1][2][3], here called Type A, is shown in figure 1. It can be seen as four "U-shaped electromagnets", two of them in the x or horizontal direction and the other in the y or vertical direction, resulting in four independent magnetic flux loops .

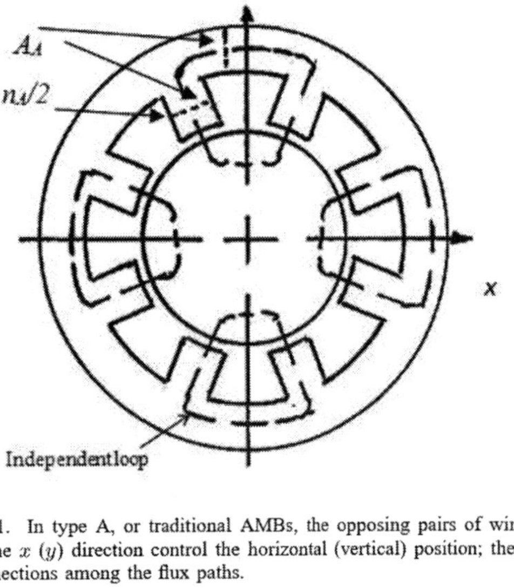

Figure 1. In type A, or traditional AMBs, the opposing pairs of windings along the x (y) direction control the horizontal (vertical) position; there are no connections among the flux paths.

The windings in the x and y direction are fed with currents $I_0 \pm i_x(t)$ and $I_0 \pm i_y(t)$; the constant DC current I_0 is the base, or bias, and the differential currents i_x and i_y will control the rotor position. Using basic reluctance concepts, the resultant forces f_x and f_y can be expressed in terms of these currents, the air magnetic permeability μ_0, the total number of coils n_a, the cross section area in the static ferromagnetic material A_a and the nominal length h of the air gaps. After a standard linearization procedure [1] around the operating point $x = y = i_x = i_y = 0$, the forces generated by the type A structure are shown in (1). Notice that the non connected nature of the magnetic fluxes leads to uncoupled forces.

$$\left.\begin{array}{l} f_x = k_p^a x + k_i^a i_x \\ f_y = k_p^a y + k_i^a i_y \end{array}\right\} \text{where} \left\{\begin{array}{l} k_p^a = \mu_0 A_a n_a^2 I_0^2 / h^3 \\ k_i^a = \mu_0 A_a n_a^2 I_0 / h^2 \end{array}\right. \quad (1)$$

An alternative structure for magnetic bearings, here named Type B, is possible; it contains four windings but now with interconnected magnetic loops, as depicted in figure 2. This structure is found in the split-windings self-bearing motors researched in Brazil [4],[5]. In that approach, to provide a simultaneous torque, alternate currents are injected in the windings; in AMBs, continuous currents are considered.

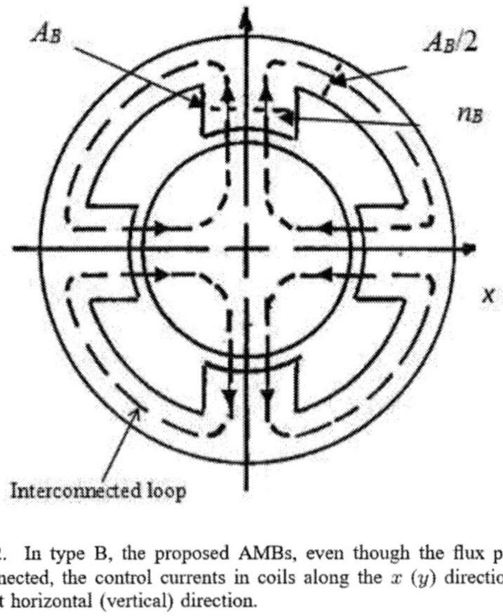

Figure 2. In type B, the proposed AMBs, even though the flux paths are interconnected, the control currents in coils along the x (y) direction affect only that horizontal (vertical) direction.

The generation of reluctance forces in Type B is presented in section II. The resulting linearized expressions also show a decoupled nature, and the position and current constants, k_p^b and k_i^b, have higher values than in the A case. Section III presents analytical results and simulations on how increasing $k_{p,i}^b$ affects dynamic and control aspects of AMBs. Discussions about tests, final comments and considerations on what remains to be done are made in sections IV and V.

The authors wish to thank FAPERJ for the financial support

Although other results are known with the Type B bearing concept [6], the authors of this article did not identify, up to the present time, the association of its interconnected structure with uncoupled equations for radial restoring forces or with higher values of the magnetic constants.

II. Reluctance Forces in Type B Bearings

A detailed study of the force generation was presented in [4] and in [7]; the main points are repeated below. The x and y components of a radial displacement of the rotor change the nominal gap width h as shown in figure 3.

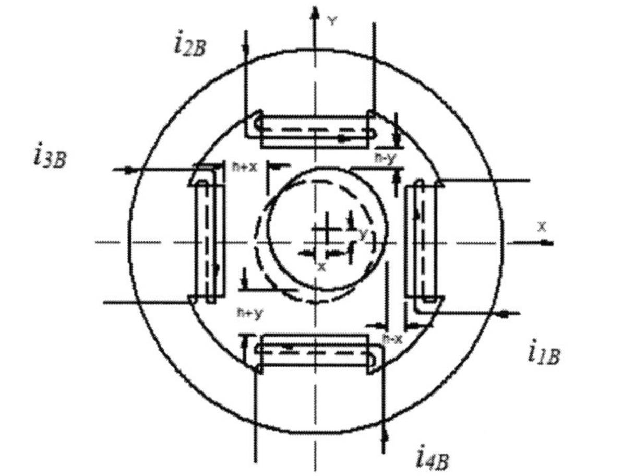

Figure 3. When the rotor moves x and y in the horizontal and vertical positions, the air gap widths change to $h - x$ in the right pole, $h + x$ (left pole), $h - y$ (upper pole) and $h + y$ (lower pole).

The usual procedure to compensate the displacements is the application of differential currents [1] to the pairs of windings: the differential, or control, currents $i_x(t)$, for the x or horizontal direction, and $i_y(t)$, for the vertical direction, are added an subtracted to a base, or bias, current I_0, a constant DC level. The total currents, neglecting the index $_B$ appearing in figure 3, imposed at each winding are

$$i_1(t) = I_0 + i_x(t),\ i_3(t) = I_0 - i_x(t) \text{ for the } x \text{ direction} \tag{2}$$

$$i_2(t) = I_0 + i_y(t),\ i_4(t) = I_0 - i_y(t) \text{ for the } y \text{ direction} \tag{3}$$

Dotted lines in figure 2 represent the magnetic flux distribution caused by these currents. The reluctance forces depend on the magnetic fluxes ϕ_k, $k = 1, 2, 3, 4$, in the four air gaps with cross section A_b:

$$f_x = \frac{\phi_1^2 - \phi_3^2}{2\mu_0 A_b} \quad \text{and} \quad f_y = \frac{\phi_2^2 - \phi_4^2}{2\mu_0 A_b} \tag{4}$$

Due to the ferromagnetic connections in type B, a current injected in any winding will cause fluxes in all four air gaps; figure 4 illustrates the effects of i_1 in all four "poles". If ϕ_{jk} denotes the flux in air gap j caused by a current in winding k, the total magnetic flux ϕ_1 in "pole" 1 is a function of the fluxes $\phi_{11}, \phi_{12}, \phi_{13}, \phi_{14}$. Assuming no air or ferromagnetic losses, and positive signs for fluxes headed to the rotating center, the total magnetic fluxes in the poles are:

$$\phi_1 = \phi_{11} + \phi_{12} - \phi_{13} + \phi_{14}. \tag{5}$$

$$\phi_2 = -\phi_{21} - \phi_{22} - \phi_{13} + \phi_{24} \tag{6}$$

$$\phi_3 = -\phi_{31} + \phi_{32} - \phi_{13} + \phi_{34} \tag{7}$$

$$\phi_4 = -\phi_{14} + \phi_{42} - \phi_{13} - \phi_{44}. \tag{8}$$

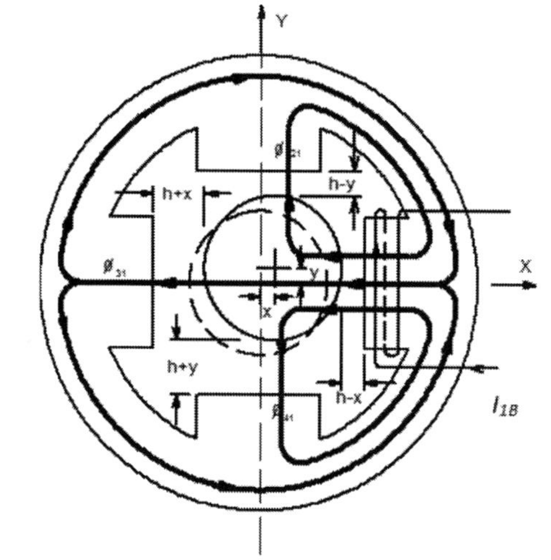

Figure 4. Magnetic flux distribution associated to i_1 in type B magnetic bearing; current injected only in winding 1 causes fluxes in all air gaps.

For the determination of the sixteen values of ϕ_{jk}, let the magneto-motive force generated by current i_1 be denoted by $\mathcal{F}_1$, and the reluctances of the air gaps in the four poles in figure 3 by $\mathcal{R}_1$, $\mathcal{R}_2$, $\mathcal{R}_3$ and $\mathcal{R}_4$. Figure 5 shows the equivalent magnetic circuit.

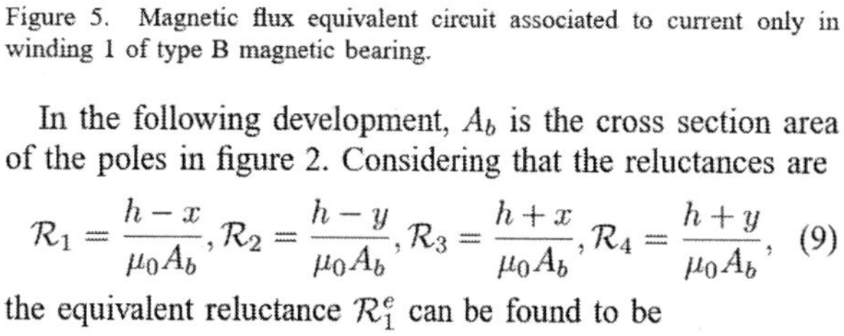

Figure 5. Magnetic flux equivalent circuit associated to current only in winding 1 of type B magnetic bearing.

In the following development, A_b is the cross section area of the poles in figure 2. Considering that the reluctances are

$$\mathcal{R}_1 = \frac{h - x}{\mu_0 A_b}, \mathcal{R}_2 = \frac{h - y}{\mu_0 A_b}, \mathcal{R}_3 = \frac{h + x}{\mu_0 A_b}, \mathcal{R}_4 = \frac{h + y}{\mu_0 A_b}, \tag{9}$$

the equivalent reluctance $\mathcal{R}_1^e$ can be found to be

$$\mathcal{R}_1^e = \frac{\mathcal{R}_1\mathcal{R}_2\mathcal{R}_3 + \mathcal{R}_1\mathcal{R}_2\mathcal{R}_4 + \mathcal{R}_1\mathcal{R}_3\mathcal{R}_4 + \mathcal{R}_2\mathcal{R}_3\mathcal{R}_4}{\mathcal{R}_2\mathcal{R}_3 + \mathcal{R}_2\mathcal{R}_4 + \mathcal{R}_3\mathcal{R}_4}. \tag{10}$$

To avoid cumbersome formulas, the auxiliary variables are defined

$$N = \mathcal{R}_1\mathcal{R}_2\mathcal{R}_3 + \mathcal{R}_1\mathcal{R}_2\mathcal{R}_4 + \mathcal{R}_1\mathcal{R}_3\mathcal{R}_4 + \mathcal{R}_2\mathcal{R}_3\mathcal{R}_4, \tag{11}$$

$$D_1 = \mathcal{R}_2\mathcal{R}_3 + \mathcal{R}_2\mathcal{R}_4 + \mathcal{R}_3\mathcal{R}_4, \quad (12)$$

$$D_2 = \mathcal{R}_1\mathcal{R}_3 + \mathcal{R}_1\mathcal{R}_4 + \mathcal{R}_3\mathcal{R}_4, \quad (13)$$

$$D_4 = \mathcal{R}_1\mathcal{R}_2 + \mathcal{R}_1\mathcal{R}_4 + \mathcal{R}_2\mathcal{R}_4, \quad (14)$$

$$D_4 = \mathcal{R}_1\mathcal{R}_2 + \mathcal{R}_1\mathcal{R}_3 + \mathcal{R}_2\mathcal{R}_3. \quad (15)$$

Since $\mathcal{F}_1 = n_b i_1$, algebraic operations lead to the following expressions for the magnetic fluxes associated to current $i_1 = I_0 + i_x$ imposed to the winding in pole 1 of figure 3:

$$\phi_{11} = \frac{\mathcal{F}_1}{\mathcal{R}_1^e} = n_b(I_0 + i_x)\frac{D_1}{N}, \quad (16)$$

$$\phi_{21} = n_b(I_0 + i_x)\frac{\mathcal{R}_3\mathcal{R}_4}{N}, \quad (17)$$

$$\phi_{31} = n_b(I_0 + i_x)\frac{\mathcal{R}_2\mathcal{R}_4}{N}, \quad (18)$$

$$\phi_{41} = n_b(I_0 + i_x)\frac{\mathcal{R}_2\mathcal{R}_3}{N}. \quad (19)$$

The same procedure, repeated for currents i_2, i_3, i_4 imposed at the windings in poles 2, 3 and 4 in figure 3, results in

$$\phi_{12} = n_b(I_0 + i_y)\frac{\mathcal{R}_3\mathcal{R}_4}{N} \qquad \phi_{22} = n_b(I_0 + i_y)\frac{D_2}{N}$$

$$\phi_{32} = n_b(I_0 + i_y)\frac{\mathcal{R}_1\mathcal{R}_4}{N} \qquad \phi_{42} = n_b(I_0 + i_y)\frac{\mathcal{R}_1\mathcal{R}_3}{N}$$

$$\phi_{13} = n_b(I_0 - i_x)\frac{\mathcal{R}_2\mathcal{R}_4}{N} \qquad \phi_{23} = n_b(I_0 - i_x)\frac{\mathcal{R}_1\mathcal{R}_4}{N}$$

$$\phi_{33} = n_b(I_0 - i_x)\frac{D_3}{N} \qquad \phi_{43} = n_b(I_0 - i_x)\frac{\mathcal{R}_1\mathcal{R}_2}{N}$$

$$\phi_{14} = n_b(I_0 - i_y)\frac{\mathcal{R}_2\mathcal{R}_3}{N} \qquad \phi_{24} = n_b(I_0 - i_y)\frac{\mathcal{R}_1\mathcal{R}_3}{N}$$

$$\phi_{34} = n_b(I_0 - i_y)\frac{\mathcal{R}_1\mathcal{R}_2}{N} \qquad \phi_{44} = n_b(I_0 - i_y)\frac{D_4}{N}$$

The total fluxes ϕ_k for $k = 1, 2, 3, 4$ can be determined by substituting the previous values of the partial fluxes ϕ_{jk} in equations (5) to (8). Then, with the help of (4), the total reluctance forces generated in a type B magnetic bearing can be expressed as

$$f_x = \frac{\mu_0 A_b n_b^2}{2} q_x(h, x, y, I_0, i_x, i_y) \quad (20)$$

$$f_y = \frac{\mu_0 A_b n_b^2}{2} q_y(h, x, y, I_0, i_x, i_y) \quad (21)$$

where $q_{x,y}$ are complicated functions of their arguments:

$$q_x(h, x, y, I_0, i_x, i_y) = \frac{N_1^2 - N_2^2}{\Delta^2} \quad (22)$$

$$q_y(h, x, y, I_0, i_x, i_y) = \frac{N_3^2 - N_4^2}{\Delta^2} \quad (23)$$

with

$$N_1 = i_1\Delta_1 + i_2\Delta_2 - i_3\Delta_3 + i_4\Delta_4$$

$$N_2 = i_3\Delta_5 - i_1\Delta_3 + i_2\Delta_6 + i_4\Delta_7$$

$$N_3 = -i_2\Delta_1 - i_3\Delta_2 - i_1\Delta_3 + i_4\Delta_8$$

$$N_4 = -i_4\Delta_5 + i_1\Delta_3 + i_2\Delta_8 + i_3\Delta_7.$$

The currents i_k are defined in equations (2) and (3); if the distances $h \pm x$ and $h \pm y$ are denoted by $\delta_x^\pm$ and $\delta_y^\pm$, the Δs above are

$$\Delta_1 = \Delta_2 + \Delta_3 + \Delta_4, \ \Delta_2 = \delta_x^+\delta_y^+, \ \Delta_3 = \delta_y^+\delta_y^-, \ \Delta_4 = \delta_x^+\delta_y^-$$

$$\Delta_5 = \Delta_3 + \Delta_6 + \Delta_7, \ \Delta_6 = \delta_x^-\delta_y^+, \ \Delta_7 = \delta_x^-\delta_y^-, \ \Delta_8 = \delta_x^+\delta_x^-$$

$$\Delta = \delta_x^-\delta_y^-\delta_x^+ + \delta_x^-\delta_y^-\delta_y^+ + \delta_x^-\delta_x^+\delta_y^+ + \delta_y^-\delta_x^+\delta_y^+$$

The complexity of the above formulas makes the linearization of (20) and (21) a hard task. Considering that the AMB operates around a point $P_0 = (x, y, i_x, i_y)_0 = (0, 0, 0, 0)$, the use of symbolical computation allows the calculation of the partial derivatives:

$$\left.\frac{\partial q_x}{\partial x}\right|_{P_0} = \frac{4I_0^2}{h^3} \qquad \left.\frac{\partial q_x}{\partial y}\right|_{P_0} = 0 \quad (24)$$

$$\left.\frac{\partial q_x}{\partial i_x}\right|_{P_0} = \frac{4I_0}{h^2} \qquad \left.\frac{\partial q_x}{\partial i_y}\right|_{P_0} = 0. \quad (25)$$

If a similar procedure is made for q_y, the combined results lead to the linear expressions for the type B structure forces:

$$\left.\begin{array}{l} f_x = k_p^b x + k_i^b i_x \\ f_y = k_p^b y + k_i^b i_y \end{array}\right\} \text{where} \left\{\begin{array}{l} k_p^b = 2\mu_0 A_b n_b^2 I_0^2/h^3 \\ k_i^b = 2\mu_0 A_b n_b^2 I_0/h^2 \end{array}\right. \quad (26)$$

Two remarkable aspects are to be noted: (a) even though the fluxes are interconnected in a type B structure, the forces are decoupled, exactly as they were in type A; (b) a factor 2 appears in the formulas above.

III. Theoretical Evaluations and Simulations

Assuming the same outside diameter of the stator, the following advantages can be identified for the Type B active magnetic bearing when it is compared with Type A:

1) The position and current constants k_p^b and k_i^b in (26) are two times bigger than their counterparts k_p^a and k_i^a in equation (1);
2) the cross section area A_b can be chosen greater than A_a; it is reasonable to have $A_b \approx 2A_a$;
3) the number of coils n_b can, possibly, be larger than n_a.

The net conclusion is: the position (k_p) and current (k_i) constants for Type B AMBs have values 2 times higher than in case A. Depending on design aspects (A_b and n_b), even higher rates can be achieved. How much can these constants be increased? The magnetic saturation seems to be the limit.

To evaluate the effects of k_b and k_i in an AMB performance, consider the problem, illustrated in figure 6, of positioning a particle of mass m that moves without friction in a horizontal, rectilinear path.

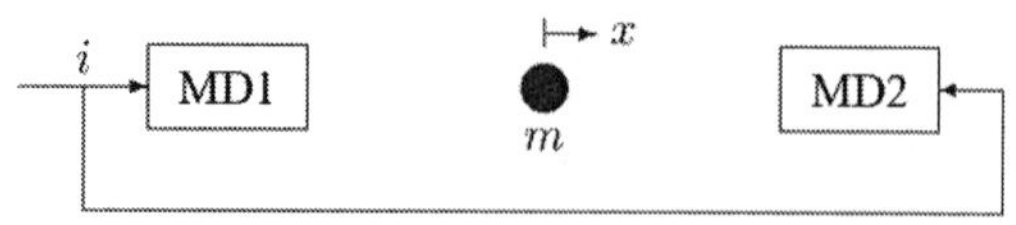

Figure 6. The mass position $x(t)$ is to be controlled by the magnetic devices.

The magnetic devices MD1 and MD2 apply a (horizontal) force $f(t) = k_p x(t) + k_i i(t)$ on the sphere, where i is a control current and x measures the displacement with respect to the reference position. A controller is desired, capable of driving the sphere position $x(t)$ to 0 for all possible initial conditions $x(0)$, and in the eventual presence of constant, horizontal disturbance forces d. This is a simple, but meaningful, problem: many theorectical aspects of the real life operation and control of AMBs are present in it.

A mathematical model is found, after a simple application of Newton's laws: $f(t)+d(t) = m\ddot{x}(t)$. The linear nature of f leads to $m\ddot{x}(t)-k_p x(t) = k_i i(t)+d(t)$, or $(ms^2-k_p)X(s) = k_i I(s) + D(s)$. A block diagram is

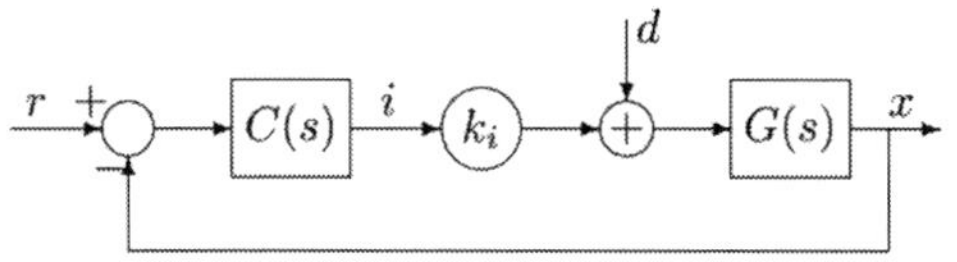

$$G(s) = \frac{1/m}{s^2-k_p/m}$$

Figure 7. Block diagram showing the plant dynamics and a closed loop control scheme with a controller $C(s)$.

A stabilizing PD controller $C(s) = \alpha s+\beta$ can be designed; it guarantees that initial displacements $x(0) \neq 0$ are corrected, when $d = 0$. The speed of convergence of $x(t)$ depends on the adjustable location of the closed loop poles, which is a function of the controller parameters α and β.

The effects of extra forces d on the sphere position can be evaluated by the expression $X(s) = T_d(s)D(s)$, where the disturbance transfer function $T_d(s)$ can be obtained from the above diagram, assuming $r = 0$:

$$T_d(s) = \frac{G(s)}{1+k_i G(s)C(s)} = \frac{1/m}{s^2+a_1 s+a_0}$$

where the characteristic polynomial coefficients are

$$a_1 = \frac{\alpha k_i}{m} \qquad \text{and} \qquad a_0 = \frac{\beta k_i - k_p}{m}.$$

The steady state influence of disturbances in the sphere position can be measured by $\rho = \lim x(t)$ when $t \to \infty$, when $r = 0$. Assuming closed loop stability, the Final Value Theorem can be used, and

$$\rho = \lim_{t\to\infty} x(t) = \lim_{s\to 0} sX(s) = \lim_{s\to 0} sT_d(s)D(s)$$

But, for constant disturbances, $D(s) = d_0/s$ leading to

$$\rho = \lim_{s\to 0} sT_d(s)\frac{d_0}{s} = d_0 T_d(0) = \frac{d_0}{\beta k_i - k_p} \qquad (27)$$

The well known fact that PD controllers do not completely reject ($\rho = 0$) constant disturbances becomes aparent. But equation (27) tells more: for a fixed, stabilizing controller, ρ decreases when k_p and k_i increase by the same factor.

In other words, if the position and current coefficients in a magnetic force generation law are both increased by the same amount, the resulting PD control is less sensitive to constant disturbances, and this characterizes a better, stiffer suspension.

Since a full, steady state rejection of constant disturbances can be achieved by PID controllers, consider now $C(s) = \alpha s+\beta+\gamma/s$. The new disturbance transfer function $T_d(s)$ is

$$T_d(s) = \frac{s/m}{s^3+a_2 s^2+a_1 s+a_0} \quad \text{with} \quad \begin{cases} a_2 = \alpha k_i/m \\ a_1 = (\beta k_i - k_p)/m \\ a_0 = \gamma k_i/m \end{cases}$$

Because each coefficient depends on a single controller parameter, the closed loop poles can be arbitrarily assigned, thus ensuring stability. A simple calculation would show that complete rejection ($\rho = 0$) of step disturbances is indeed achieved by the PID controller.

In order to feel the performance details of the situation, a simulation was performed. The numerical values, in the SI system, $m = 4$, $k_i^0 = 200$ and $k_p^0 = 2000000$ were used. A PID controller that assigns all the closed loop poles at -10 was calculated: $\alpha = 3/5$, $\beta = 10006$ and $\gamma = 20$.

An initial displacement of 1cm was imposed to the sphere, and in less than 1 second it returned to the desired rest position $x = 0$. After stabilization, a constant disturbance ($d_0 = 40$N) was applied and successfully rejected. Figure 8 shows the curves for different values of the k_i and k_p constants: (k_i^0, k_p^0), $(2k_i^0, 2k_p^0)$, $(4k_i^0, 4k_p^0)$ and $(8k_i^0, 8k_p^0)$.

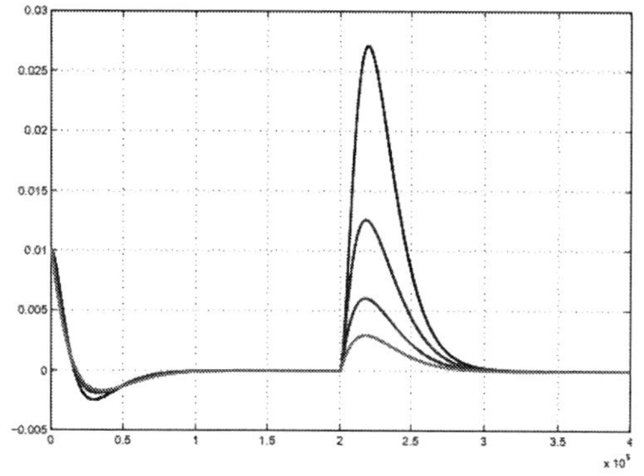

Figure 8. Sphere positioning with PID control for several values of the magnetic constants k_p and k_i. The highest rising, dark blue curve refers to the nominal values (NV) k_p^0 and k_i^0; twice the NV leads to the dark green (second highest rising) curve; four and eight times the NV generate the next curves, red and light blue. Total simulation time: 4s, displacements in meters.

All curves are very similar in the first 2 seconds, implying that high or low values for the magnetic constants are not crucial in the stabilizing stage. But when constant disturbance rejection is needed, better transient behaviours are a direct consequence of higher values in k_p and k_i.

It is important to stress, again, that this is a simple example, but the conclusions allowed by it are valid in much more general situations, involving real world applications of practical interest. And these conclusions are: increasing the values of the magnetic force constants k_p and k_i is a highly desirable goal in the AMB field.

IV. Prototype Testing

Section II final conclusions are that the interconnected fluxes in the type B structure increase the values of the magnetic force constants k_p and k_i. How sure can one be about the theoretical tools used in those developments? Are the symbolical algebra calculations safe?

It is our understanding that the idea of the Type B structure has already been tested. In fact, in the prototype used at UFRJ [4], a vertical rotor is radially positioned by a self-bearing motor based on the interconnected fluxes of the type B structure. Such a situation is more complicated, because the windings are fed with AC currents, to achieve the dual capabilities: torque generation and radial positioning. And the device has worked!

The best possible way to give definite answers to the seminal questions posted in the beginning of this section is by constructing prototypes and testing them in an exhaustive way. Only after this important stage, will the ideas proposed here be validated. Or not.

Some steps have already been walked in that direction: a pair of prototypes, one for type A and the other for type B, are under construction. Figures 9 and 10 show their geometries.

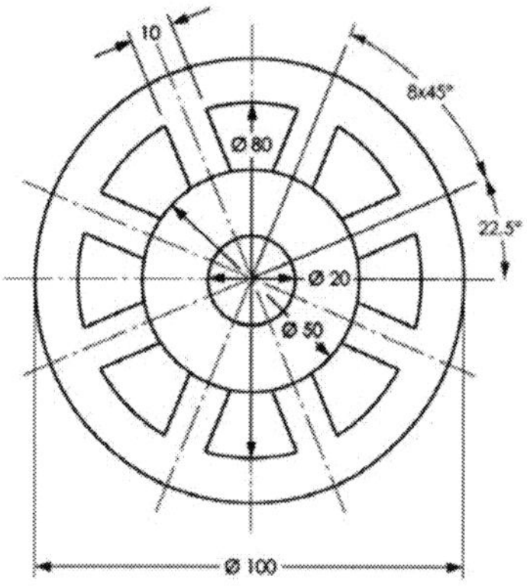

Figure 9. Type A magnetic bearing geometry.

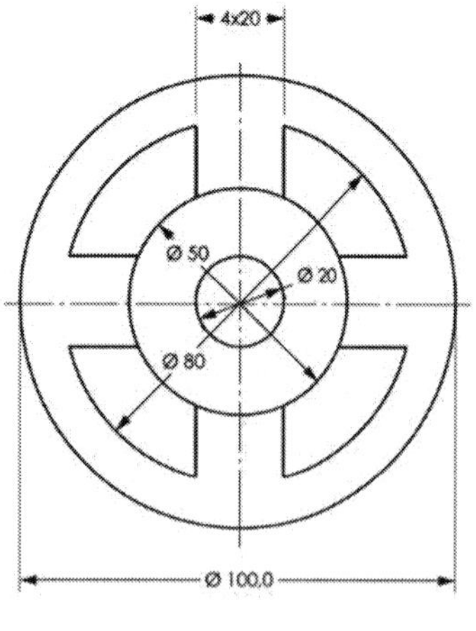

Figure 10. Type B magnetic bearing geometry.

They will be applied to a vertical rotor driven by a two-phase induction motor; the lower support will be provided by a mechanical bearing. Figure 11 shows these ideas.

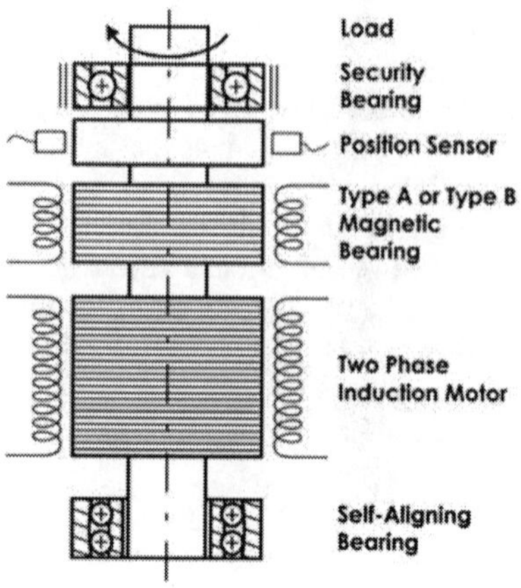

Figure 11. Testing prototype concept.

A finite element simulation of the magnetic characteristics of type A and type B structures was done, and some results are shown below.

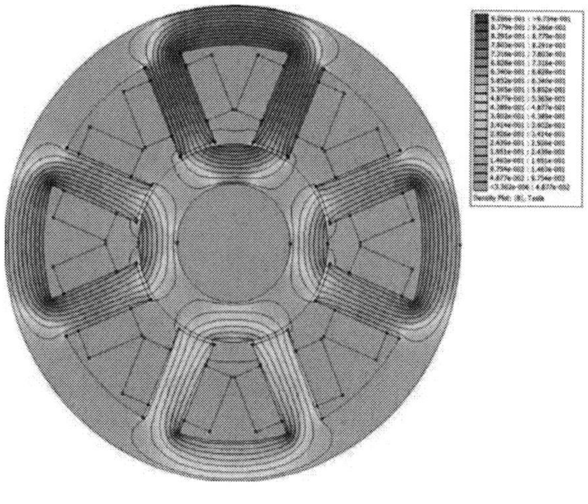

Figure 12. Flux distribution in a type A structure with balanced currents in the x axis and a differential current in the y direction.

Figure 13 shows clearly that a differential current in one direction does not affect the flux distribution in the other direction, thus confirming equation (25).

V. Final Comments and Conclusions

The authors have great expectations that the here called type B concept will be a valid contribution for the active magnetic bearings field, because of the possibility of increasing their equivalent mechanical stiffness.

References

[1] G. Schweitzer, H. Bleuler, and A. Traxler, *Active Magnetic Bearings*. Hochshulverlag AG an der ETH Zürich, 1994.

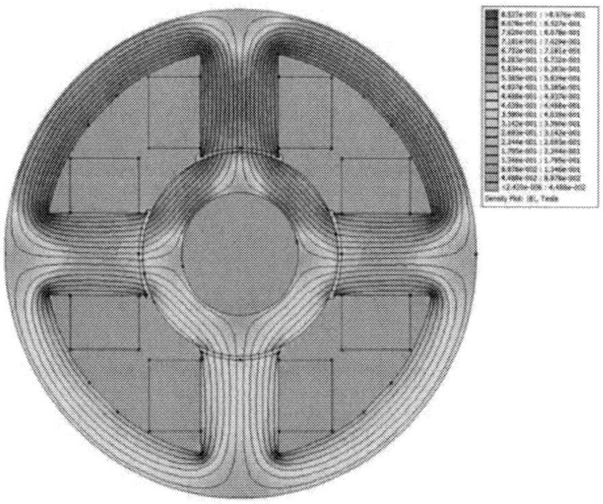

Figure 13. Flux distribution in a type B structure with balanced currents in the x axis and a differential current in the y direction.

[2] A. Chiba, T. Fukao, O. Ichikawa, M. Oshima, M. Takemoto, and D. Dorrell, *Magnetic Bearings and Bearingless Drives.* Newnes-Elsevier, 2005.

[3] G. Schweitzer, E. Maslen, H. Bleuler, M. Cole, P. Keogh, R. Larsonneur, R. Nordmann, and Y. Ogada, *Magnetic Bearings: Theory, Design and Applications to Rotating Machinery.* Springer-Verlag, 2009.

[4] D. F. B. David, "Levitação de rotor por mancais-motores radiais magnéticos e mancal axial sc auto-estável," D. Sc. tese, COPPE–UFRJ, 2000.

[5] E. F. Rodriguez and J. A. Santisteban, "An improved control system for a split winding bearingless induction motor," *IEEE Transactions on Industrial Electronics*, vol. 58, no. 8, pp. 3401–3408, 2011.

[6] L. Santos and K. Kjolhed, "Experimental contribution to high-precision characterization of magnetic forces in active magnetic bearings," *Journal of Engineering for Gas Turbines and Power*, vol. 129, pp. 503–, 2007.

[7] D. F. B. David, J. A. Santisteban, and A. C. D. N. Gomes, "Interconnected four poles magnetic bearings," in *Proceedings of the 1st Brazilian Workshop on Magnetic Bearings*, www.magneticbearings2103.com.br, october 2013.

Segmented Magnetic Thrust Bearings: Analytic Models and Predictions of Significant Improvement in Dynamic Performance

Zackary W. Whitlow[a], Roger L. Fittro[a], Carl R. Knospe[a]

[a] University of Virgina, Department of Mechanical and Aerospace Engineering, 122 Engineer's Way Charlottesville, VA 22904, USA, fittro@virginia.edu

***Abstract*—Non-laminated magnetic thrust bearings typically exhibit reduced dynamic performance due to eddy currents compared to laminated bearings. Segmented stators have been introduced to increase actuator performance by disrupting eddy current paths in the same manner as laminations in radial magnetic bearings. However, due to manufacturing limitations thrust bearing stators cannot be easily segmented to the extent that they would be considered fully laminated. Therefore, eddy currents continue to affect the dynamic performance significantly. Since analytic models of segmented stators do not presently exist the bearing design process requires the use of finite element analysis (FEA) and experimentation. Here we show, using finite element analysis, that a small number of cuts dramatically improves bandwidth of non-laminated magnetic actuators, validating the segmented design strategy. In addition, existing analytic models of non-laminated magnetic bearings are adapted to a segmented geometry and validated via FEA. Close agreement shows that the analytic model, rather than time consuming FEA, may be used in the design process.**

I. Introduction

Dynamic load capabilities for thrust bearings are critical in many magnetic bearing applications such as machine tool spindles and compressors. High speed milling machine tools may encounter high frequency vibrations due to the cutting process; it is critical that spindle actuators have adequate dynamic capabilities to dampen the response. Thrust bearing dynamic capability is also crucial to compressor performance. Non-laminated thrust bearings leave compressors susceptible to instability during surge and trip events. However, due to manufacturing and material constraints thrust bearings are typically non-laminated, which leads to reduced dynamic capabilities compared to their laminated radial bearing counterparts. Low bandwidth and poor dynamic response of non-laminated thrust bearings often makes them a performance limiting component in both of these applications. Increased dynamic capabilities of thrust bearings would enable a more robust closed loop system, with improved performance [1].

Powder metals, or soft magnetic composites, are one option for reducing eddy current effects. However, powder metalurgy introduces another set of deficiencies. Poor mechanical strength of powder metals precludes its use in thrust disks. Reduced permeability and low saturation flux density compared to the base material should also be considered. Powder metals may be a good option for some applications but will not be considered further here. Instead the focus will be on segmented stator geometry and a corresponding analytic model.

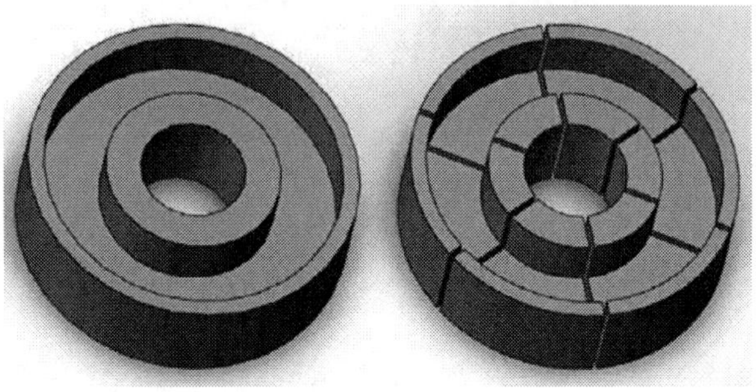

Figure 1: Depiction of a traditional thrust stator and a segmented stator (right).

Segmented axial stators, Fig. 1 right, are a promising method to mitigate problems associated with eddy currents in non-laminated bearings. Stator segmentation serves the same purpose as laminations and could be considered a coarse lamination strategy. By impeding the eddy current loop around the stator periphery with a single cut, or further breaking up the eddy current paths by introducing more stator segments, the geometry approaches that of a laminated thrust bearing. However, the size of these segments is constrained such that eddy current effects cannot be entirely ignored in segmented

Table I: Thrust bearing parameters

Parameter	Definition	Value
r_0	inner radius of inner pole	51.7 mm
r_1	outer radius of inner pole	68.3 mm
r_2	inner radius of outer pole	90.9 mm
r_3	outer radius of outer pole	101.3 mm
d_1	back iron thickness	16.5 mm
d_2	pole height	26.3 mm
d_3	floater thickness	15.2 mm
g	nominal air gap	0.5 mm
σ	conductivity	2 MS/m
μ_r	relative permeability	1000
N	coil turns	166

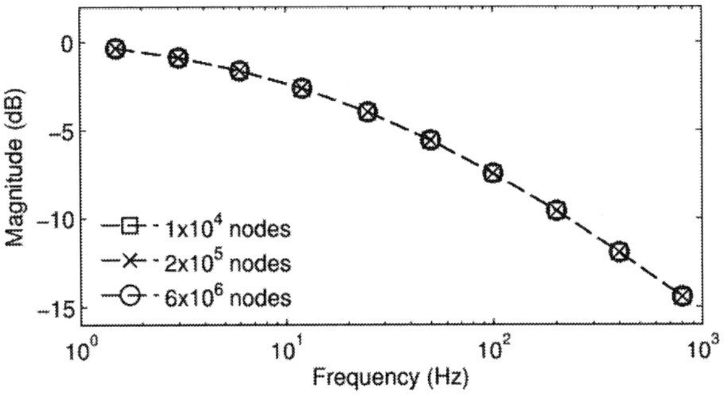

(a) Magnitude: 2-D mesh analysis

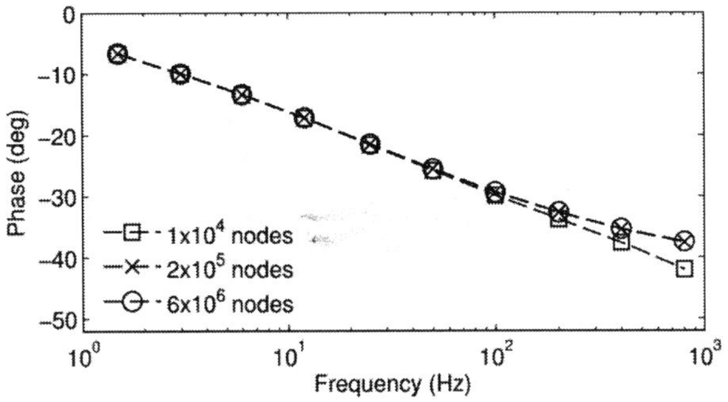

(b) Phase: 2-D mesh analysis

Figure 2: Two dimensional finite element modeling shows magnitude and phase response of the flux density for three mesh resolutions.

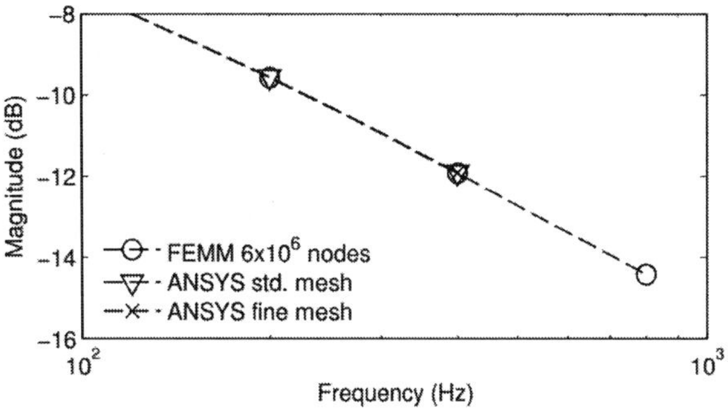

(a) Magnitude: 3-D mesh analysis

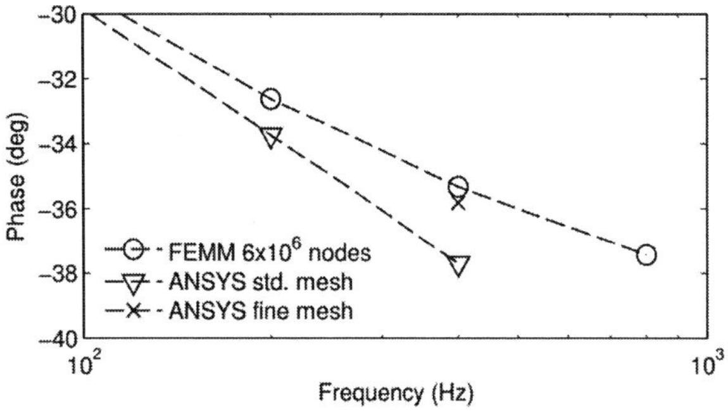

(b) Phase: 3-D mesh analysis

Figure 3: Finite element modeling with perturbation current input and flux output shows magnitude and phase with increasing mesh resolution using FEMM (a) and ANSYS compared to FEMM (b).

thrust bearing models as they can be for fully laminated radial bearings. The thrust disk, or floater, will remain solid to maintain its structural integrity. While segmented designs are currently in use, the authors are not aware of any systematic study of the effect on dynamic performance or of an analytical model for this type of geometry. Without an analytical model it is difficult to determine whether segmentation will yield adequate improvements for a given application.

Analytic models for non-laminated magnetic actuators have been developed by several groups. Feeley developed a two-dimensional (2-D) model for C-type magnetic actuators [2] and Kucera et al. developed a 2-D model for non-laminated axisymmetric geometries [3]. Both of these models account for the impedance of flux due to eddy currents but fail to account for reduced air gap flux. These models were further advanced by Zhu et al. who included the reduced air gap flux due to eddy currents to develop 2-D models for axisymmetric [4] and C-type [5] magnetic actuators. Zhu demonstrated with these 2-D analytical models that it is possible to accurately predict performance of C-type and solid axisymmetric non-laminated magnetic actuators. However, segmented stator geometries were not considered. The purpose of this work is to extend Zhu's concepts to segmented thrust bearings and develop a simple analytic model. Preliminary validation of the model for a test case of an existing thrust bearing geometry shows close agreement between analytic and finite element analysis.

II. Theory

Zhu et al. developed two models for non-laminated magnetic actuators [4], [5], one for an axisymmetric geometry and another for a C-type geometry. For each of these models, the geometry is divided into regions based on flux path and an effective reluctance is calculated for each region. In the axisymmetric case, the 2-D geometry is divided into six regions [4]. Regions 1 and 3 include parts of the floater (i.e. thrust disk), the air gaps, and the portion of the stator that borders the air gaps. Region 2 is the central portion of the floater. Regions 4, 5, and 6 make up the remaining parts of the stator. For the C-type actuator there are three regions; the floater, the air gaps, and the stator [5]. Using magnetic circuit theory and partial differential equations to model the skin effect produced by eddy currents, Zhu *et al.* developed analytic models of effective reluctance, R_1 ... R_6 for regions of an axisymmetric geometry and R_f, R_g, and R_s for the floater, air gap, and stator regions of a C-type geometry [4], [5]. This is the starting point for our development of an analytic model for segmented thrust bearing geometries.

Zhu's axisymmetric model assumed a geometry without a center hole which is not typical of thrust bearings. To accommodate geometries with a center hole the effective reluctance of Zhu's actuator regions 1 and 6 [4] which border the central axis of the actuator were modified. The derivation

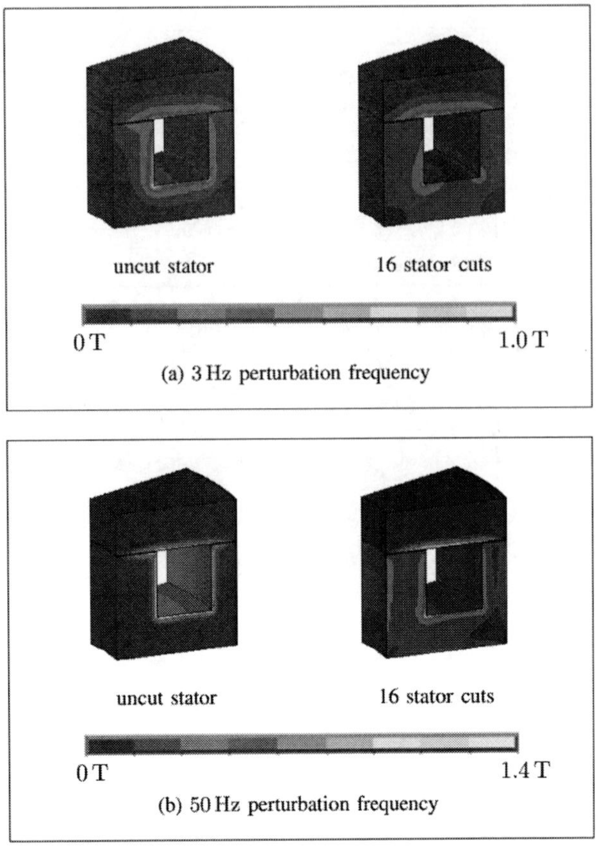

(a) 3 Hz perturbation frequency

(b) 50 Hz perturbation frequency

Figure 4: Magnetic flux density magnitude for a stator with 0 or 16 cuts. Floater is not cut.

results (details to be published) are as follows

$$R_1(s) = \frac{g\alpha_1}{r_1 2\pi\mu_0} \frac{I_1(\alpha_1 r_0)K_0(\alpha_1 r_1) + I_0(\alpha_1 r_1)K_1(\alpha_1 r_0)}{I_1(\alpha_1 r_1)K_1(\alpha_1 r_0) - I_1(\alpha_1 r_0)K_1(\alpha_1 r_1)} \tag{1}$$

$$R_6(s) = \frac{d_2\alpha}{r_1 2\pi\mu_0\mu_r} \frac{I_0(\alpha r_0)K_1(\alpha r_1) + I_1(\alpha r_1)K_0(\alpha_1 r_0)}{I_1(\alpha r_1)K_1(\alpha r_0) - I_1(\alpha r_0)K_0(\alpha r_1)} \tag{2}$$

$$\alpha_1 = \sqrt{\frac{2\alpha}{\mu_r g}} \qquad \alpha = \sqrt{s\sigma\mu_0\mu_r},$$

where $I_n(\cdot)$ is the nth-order modified Bessel function of the first kind, $K_n(\cdot)$ is the nth-order modified Bessel function of the second kind, σ is the conductivity, r_0 is the inner radius of the inner pole, r_1 is the outer radius of the inner pole, g is the air gap distance, d_2 is the pole height, μ_0 is the magnetic permeability of free space, and μ_r is the relative permeability.

The analytic model was then modified to accommodate a segmented stator. An axisymmetric thrust bearing cut like a pie has segments that resemble individual C-type actuators that are curved and fit together. Therefore, Zhu's C-type model for the segmented geometry was used as a starting point. If the floater and stator were cut Zhu's C-type model could be used directly on each segment and the pieces could be assembled as a parallel magnetic circuit to determine the total

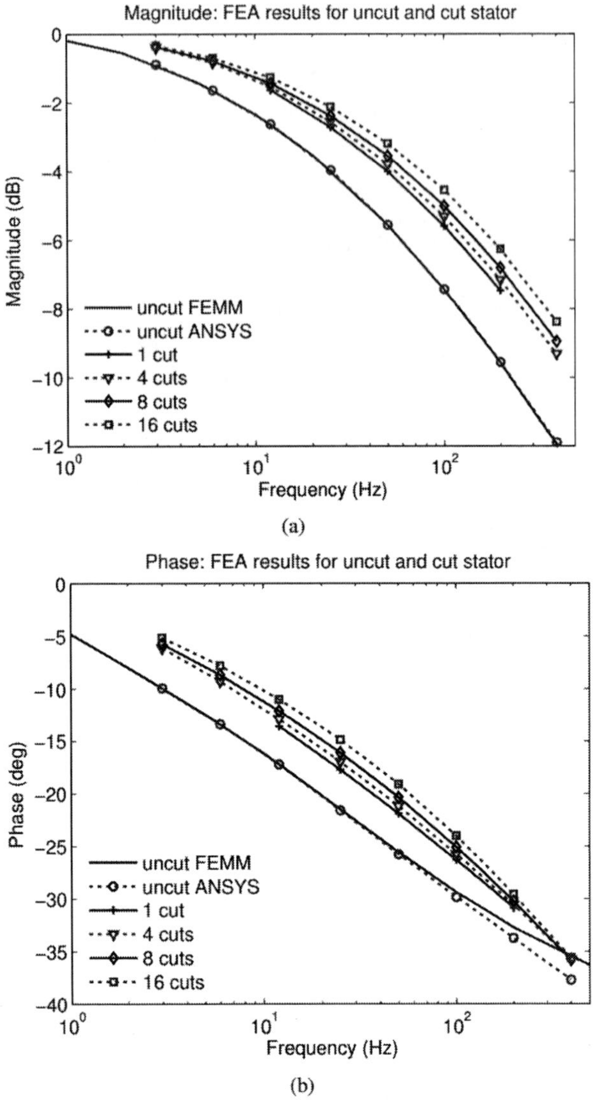

(a)

(b)

Figure 5: FEA Magnitude and phase from of thrust bearing with unsegmented stator or cut stator.

reluctance. However, a segmented floater is not practical in rotating machinery due to mechanical strength requirements. Since the C-type model has separate reluctance terms for the floater and stator the floater disk can be modeled as a c-type floater by neglecting the curvature and using the average circumference of the inner and outer poles as the length parameter. This strategy assumes the floater has a single cut since a C-type floater has two unconnected ends. However, if the length of a C-type actuator, pole circumference in this case, is much greater than the pole width then the electromagnetic field interactions at the ends of the actuator contribute little to the dynamics of the entire actuator [6]. Another difference between the C-type geometry and a segmented thrust bearing is the different widths of the inner and outer poles. A simple solution is to use the average length and width of the inner

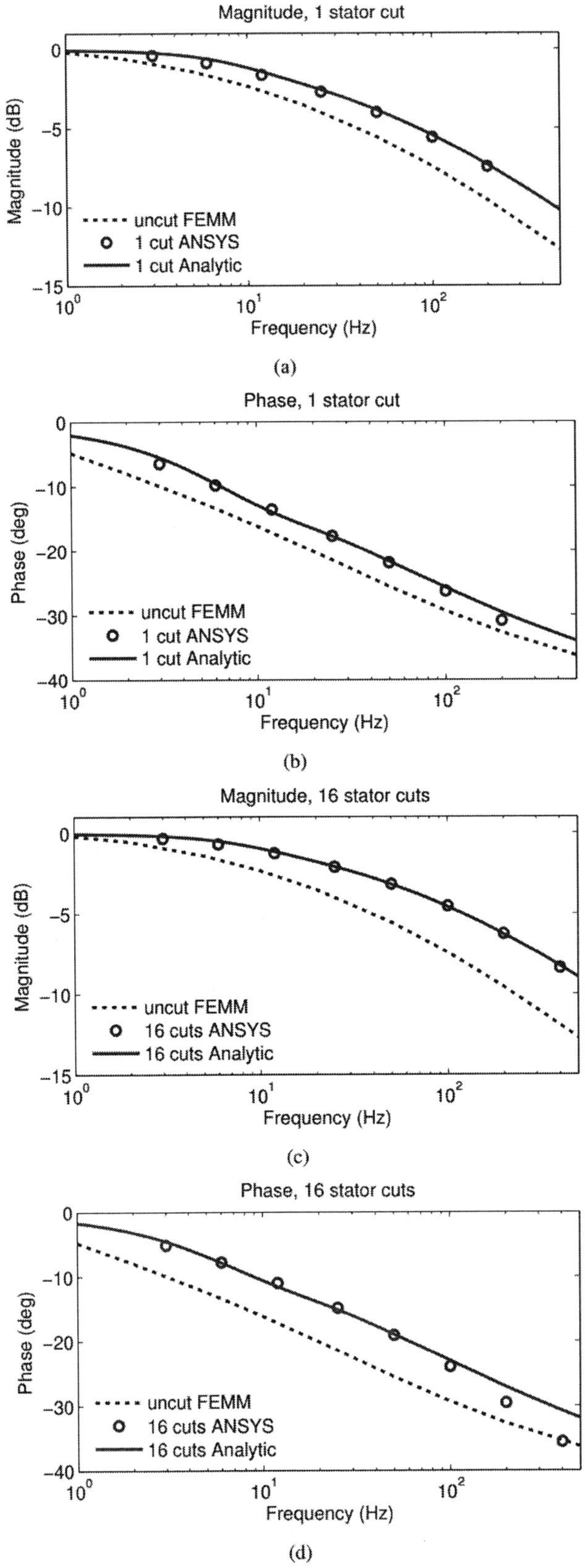

Figure 6: Magnitude and phase from FEA of thrust bearing with unsegmented stator or cut stator.

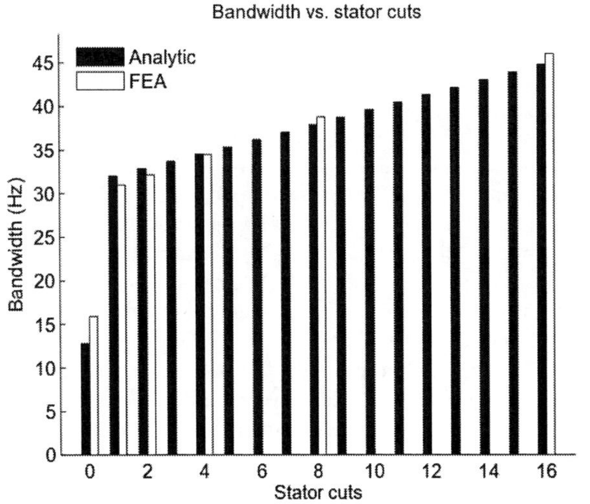

Figure 7: Actuator bandwidth predicted using the analytic model or FEA (0, 1, 2, 4, 8, and 16 stator cuts).

and outer poles which was the approach chosen here. Using Zhu's full analytic model of reluctance for floater, air gap and stator regions where the reluctance is the product of a static term, R^0, and dynamic term, R^d, as follows

$$R_i = R_i^0 R_i^d(s) \tag{3}$$

The total reluctance of an actuator with m stator segments is then

$$R = \frac{R_s^0 R_s^d(s)}{m} + R_f^0 R_f^d(s) + 2R_g^0 R_g^d(s) \tag{4}$$

The transfer function from perturbation current, i_p, to perturbation flux, ϕ_p, then becomes

$$\phi_p(s) = \frac{N}{\frac{R_s^0 R_s^d(s)}{m} + R_f^0 R_f^d(s) + 2R_g^0 R_g^d(s)} i_p(s) \tag{5}$$

The perturbation force F_p is then determined using the Maxwell stress tensor [4], [7].

$$\begin{aligned} f(t) &= \frac{1}{2\mu_0} \oint_S \mathbf{B}^2(t,r)\mathrm{d}S \\ &= \frac{1}{2\mu_0} \oint_S [\mathbf{B_b}^2 + 2\mathbf{B_b}\mathbf{B_p}(t,r) + \mathbf{B_p}^2(t,r)]\mathrm{d}S \end{aligned} \tag{6}$$

Such that for a small perturbation current we have the following relationship.

$$\mathbf{B_p}^2(t,r) \approx 0$$

$$\begin{aligned} F_p(s) &= \frac{1}{\mu_0}\left[\int_{r_0}^{r_1} \mathbf{B_b}\mathbf{B_p}(s,r)\cdot 2\pi r\mathrm{d}r + \int_{r_2}^{r_3} \mathbf{B_b}\mathbf{B_p}(s,r)\cdot 2\pi r\mathrm{d}r\right] \\ &= \frac{\phi_b}{\mu_0}\left(\frac{1}{A_1} + \frac{1}{A_2}\right)\phi_p(s) \\ &= \frac{\phi_b}{\mu_0}\left(\frac{1}{A_1} + \frac{1}{A_2}\right)\left[\frac{N}{\frac{R_s^0 R_s^d(s)}{m} + R_f^0 R_f^d(s) + 2R_g^0 R_g^d(s)}\right] i_p(s) \end{aligned} \tag{7}$$

III. Results of Finite Element Analysis

To support the validity and accuracy of the analytic model for segmented thrust bearings developed in section II, a sample test case was chosen to analyze and compare with FEA results. The test case chosen is an existing thrust bearing on the compressor surge test rig developed at the University of Virginia [8], [9]. The geometric and material parameters are summarized in Table I. Finite element analysis was performed using two different codes, Finite Element Method Magnetics (FEMM) [10] and ANSYS Mechanical, over a range of current excitation frequencies. Flux magnitude and phase were plotted to determine the minimal mesh resolutions required, Figs. 2,3.

Flux magnitudes are similar for each of the mesh resolutions tested in FEMM, Fig. 2a. However, for the lowest resolution mesh, the phase diverges from that of the finer meshes for frequencies above 100 Hz, Fig. 2b. At higher frequencies the eddy-current induced skin effect restricts the flux path to the surface region adjacent to the coil, Fig. 4. With the flux concentrated in fewer elements near the surface and a steep flux gradient at the edge of this region a finer mesh resolution is required in order to obtain an accurate result. The highest mesh resolution (6 million nodes) was used for subsequent FEMM-related analyses.

To analyze the effect of stator cuts, 3-D FEA is required. For benchmarking the 3-D ANSYS model the uncut geometry with 64-fold symmetry was used for comparison of mesh resolutions. Stator segmentation disrupts the axisymmetric characteristic of the actuator requiring that larger wedges be modeled. The worst case, from a computational perspective, being a 2-fold symmetry model for a stator with 1 or 2 cuts. Stator cuts are modeled by an air gap or by using a current parallel boundary condition.

3-D FEA showed the same phase inaccuracies at higher frequencies that were observed with 2-D FEA when standard mesh resolution was used, Fig. 3. The flux magnitudes predicted using the 3-D model (ANSYS) matched the 2-D results (FEMM) well. Using a finer 3-D mesh the phase results at 400 Hz are very close to matching the 2-D results. However, due to excessively large times required for computation, up to 72 hr for the 64-fold symmetry model, the standard mesh size was used for all subsequent 3-D FEA.

Plots of flux distribution in ANSYS clearly show the effect of stator segmentation on flux distribution, Fig. 4. Actuators with an uncut stator and a stator with 16 cuts were driven at frequencies of 3 Hz and 50 Hz. The magnitude of the predicted flux distributions were plotted and are shown in Fig. 4. With an uncut stator, eddy currents are more severe which leads to a dramatic skin effect restricting the flux path close to the iron surface. However, with a segmented stator, the eddy current loops circumscribing the stator are broken, restricting eddy currents to smaller regions. Smaller eddy current loops reduce the effective reluctance in the iron which leads to a more uniform flux distribution. The effect this has on dynamic performance is significant.

Flux magnitude and phase were predicted by FEA for perturbation frequencies between 3 and 400 Hz, Fig. 5. With a single stator cut a significant shift in magnitude can be seen. With smaller shifts for additional cuts, Fig. 5a. FEA also predicts a phase improvement with increasing number of stator cuts, Fig. 5b. These results show that non-laminated thrust bearings with a segmented stator can significantly improve dynamic performance. For frequencies above 100 Hz the phase lag predicted by 3-D FEA is an overestimate. This is obvious when comparing ANSYS and FEMM results shown in Fig. 5b for the uncut geometry. However, by comparing only 3-D FEA results it may be concluded, in spite of the FEA overestimate, that a significant reduction in phase lag will occur Fig. 5b. Focus will now be directed towards an analytic model for non-laminated thrust bearings, which will make the process of designing and modeling segmented or unsegmented thrust bearings much faster by eliminating the reliance on FEA for accurate results.

IV. Results of Analytic Analysis

FEA shows that the effective reluctance in the iron which increases with excitation frequency is reduced by cutting the stator into segments. The analytic model presented here closely matched the results of FEA, Fig. 6. For a stator with 1 cut the analytic model closely matched the normalized flux magnitude predicted by FEA, Fig. 6a, where FEA (FEMM) results from the uncut geometry are also shown as a reference. Fig. 6b shows close agreement for the phase predicted by the analytic model and FEA model for frequencies below 200 Hz. It is likely that the analytic model is more accurate than FEA at frequencies above 100 Hz since the analytic model does not suffer from the effects of discretization. The analytic results for a stator with 16 cuts also agree well with FEA, Figs. 6c and 6d, but with a similar discrepancy in phase at higher frequencies due to the standard mesh density used for FEA.

Plotting the bandwidth as a function of frequency shows once again that a single cut in the stator has the greatest impact on performance, Fig. 7. The relationship between bandwidth and cuts appears to be affine for 1 through 16 cuts, Fig. 7, with cuts beyond the first resulting in smaller bandwidth improvement. With 8 cuts in the stator, well within manufacturing and material constraints, a 2.5-fold increase in bandwidth is predicted. These results show that a segmented geometry will improve dynamic performance of non-laminated thrust bearing.

V. Conclusions

Existing analytic models for non-laminated C-type and axisymmetric magnetic actuators have been further developed for application to segmented thrust bearings. This analytic model was applied to a test case which predicted significant improvements in dynamic performance with segmentation and closely matches predictions from FEA. Enhanced dynamic performance was predicted with modifications that are well within practical manufacturing constraints, validating the segmentation strategy. Enhanced dynamic performance due to segmentation may not be as significant for all size bearings and geometries. Future studies will address the effect of geometric parameters on performance enhancement achieved via segmentation. Analysis of other geometries will also further assess the validity of the analytic model.

References

[1] C. R. Knospe and L. Zhu, "Performance limitations of non-laminated magnetic suspension systems," *IEEE Transactions on Control Systems Technology*, vol. 19, no. 2, pp. 327–336, Mar. 2011. [Online]. Available: http://ieeexplore.ieee.org/lpdocs/epic03/wrapper.htm?arnumber=5463016

[2] J. J. Feeley, "A simple dynamic model for eddy currents in a magnetic actuator," *IEEE TRANSACTIONS ON MAGNETICS*, vol. 32, no. 2, pp. 453–458, March 1996.

[3] L. K. M. Ahrens, "A model for axial magnetic bearings including eddy currents," *Proc. Third Int. Symp. Magnetic Suspension Technology*, vol. 2, pp. 421–438, Dec. 1995.

[4] L. Zhu, C. Knospe, and E. Maslen, "Analytic model for a nonlaminated cylindrical magnetic actuator including eddy currents," *IEEE Transactions on Magnetics*, vol. 41, no. 4, pp. 1248–1258, Apr. 2005. [Online]. Available: http://ieeexplore.ieee.org/lpdocs/epic03/wrapper.htm?arnumber=1420680

[5] L. Zhu and C. Knospe, "Modeling of nonlaminated electromagnetic suspension systems," *IEEE/ASME Transactions on Mechatronics*, vol. 15, no. 1, pp. 59–69, Feb. 2010. [Online]. Available: http://ieeexplore.ieee.org/lpdocs/epic03/wrapper.htm?arnumber=4838650

[6] L. Zhu, "Non-laminated magnetic actuators: Modeling and performance limitations," Ph.D. dissertation, University of Virginia, May 2005.

[7] D. J. Griffiths, *Introduction to Electrodynamics*. Prentice-Hall, 1989.

[8] S. Y. Yoon, "Surge control of active magnetic bearing suspended centrifugal compressors," Ph.D. dissertation, University of Virginia, December 2011.

[9] L. Z. A. P. E. Yoon, Se Young, *Control of Surge in Centrifugal Compressors by Active Magnetic Bearings*. Springer, 2013.

[10] D. C. Meeker. Finite element method magnetics. Version 4.2 (11Apr2012). [Online]. Available: http://www.femm.info/wiki/HomePage

Design of a Highly Reliable Fan with Magnetic Bearings

Gerald Jungmayr[a], Hans-Jörg Berroth[b], Martin Panholzer[a], Edmund Marth[a], Wolfgang Amrhein[a], Frank Jeske[b] and Martin Reisinger[c]

[a] Johannes Kepler University Linz, Institute of Electrical Drives and Power Electronics, Altenbergerstr. 69, 4040 Linz, Austria, gerald.jungmayr@jku.at
[b]ebm-papst St. Georgen GmbH&Co KG, 78112 St. Georgen, Germany
[c]Linz Center of Mechatronics GmbH, Altenbergerstr. 69, 4040 Linz, Austria

***Abstract*—Long term analysis at ebm-papst St. Georgen GmbH & Co. KG showed that bearing failures are responsible for 90% of compact fan breakdowns. A magnetically levitated impeller has no contact with the stator and therefore the life expectancy can be essentially increased. This paper presents the design of a magnetically levitated fan. To meet the low-cost requirement, passive magnetic bearings (PMBs) are used to stabilize the radial and tilt deflections of the rotor. By applying an optimized viscoelastic support to the stator, sufficient damping is introduced to the passively stabilized degrees of freedom. The optimization of the stiffness and damping is discussed and the design of the key components, namely the PMB, the active magnetic bearing (AMB) and the passive damping device, is described. Finally the built prototype and measurements are presented.**

I. Introduction

A. Reliability

The reliability of electronic systems essentially depends on the operating temperature. Fans are the common thermal management solution for cooling of electronic systems. However, wear in the bearings reduces the life time of conventional fans. A study by Tian [1] identified bearing defects as major failure cause of fans.

To estimate the rating life of roller bearings, ISO 281 specifies methods for calculation of the basic dynamic load rating [2]. As fans usually have a very light load, the use of the ISO 281 standard is not suitable for life estimation [3]. The reliability of the roller bearings instead depends mainly on the retention of its lubricant (grease or oil). Grease life equations can be found in [4].

Fan manufacturers usually specify the lifetime when ten percent of the population fails under the specified test conditions (e.g L_{10} at 40°C). However, care must be taken when the lifetimes L_{10} of different manufacturers are compared. The differences are briefly discussed below.

To cut down the test time, accelerated life tests (ALT) are used to get the life expectancy. As the bearing temperature is the most important acceleration factor for the roller bearing life, ALT are done at the highest possible operation temperature. The industry standard IPC-9591 was developed in 2006 in order to standardize fan's performance parameters, including reliability [5]. The norm suggests an acceleration factor of 1.5 for 10 Kelvin temperature rise ($AF|_{IPC9591} = 1.5/10\,\mathrm{K}$). Nevertheless, often unacceptably high acceleration factors are used by fan manufacturers [6]. The acceleration factor used at ebm-papst St. Georgen GmbH & Co. KG is even below the value suggested by the norm ($AF|_{ebmpapst} = 0.7/10\,\mathrm{K}$), leading to more conservative life time predictions [7].

Also the test strategy of the reliability test should be considered if the fan life is predicted. The *zero-failure test strategy* determines the test time for each fan dependent on the sample size n. The test time where the n samples must not fail depends on the estimated parameters of the Weibull distribution. The *test with failure strategy* leads to longer test times, as it stops when a predefined number of defects has been reached. As it is not necessary to assume the Weibull distribution parameters this strategy can be considered more reliable.

B. Measures for Improved Reliability

1) Nonrotating Air Moving Devices: One approach to improve the reliability of air moving devices is to use an oscillating beam instead a rotating fan blades. The vibration can be realized by a cantilever beam, which is actuated by a piezoelectric element [8] or by the magnetic forces of coils [9]. The beam is excited at the first or second eigenfrequency. However, conventional fans with comparable installation space achieve considerably higher pressures and flow rates. Nevertheless, a vibrating beam might be an alternative for applications demanding rather low flow rates.

2) Magnetic bearings: A magnetically levitated fan could be the solution to reach a durable air cooling device with high cooling power. This approach will be investigated in this paper.

II. Magnetic Bearing Concept

As magnetic bearings are employed primarily in high-tech applications, their realization in a compact fan poses a great challenge. Commercial success requires concepts that are only marginally more expensive to manufacture than traditional fans. The magnetically levitated fan uses two simple passive magnetic bearings (PMBs) to meet this low-cost requirement, see Fig. 1. The upper and lower PMB stabilize four degrees of freedom: the radial and tilt motions of the rotor. Due to Earnshaw's Theorem [10], the axial direction is unstable and therefore controlled by an active magnetic bearing (AMB). The remaining degree of freedom (rotation of the impeller) is actuated by a permanent magnet synchronous motor (PMSM). The cross section of a magnetically levitated fan that was jointly developed by ebm-papst St. Georgen, the Institute of Electrical Drives and Power Electronics at the Johannes Kepler University Linz and the Linz Center of Mechatronics (LCM) is shown in Fig. 1.

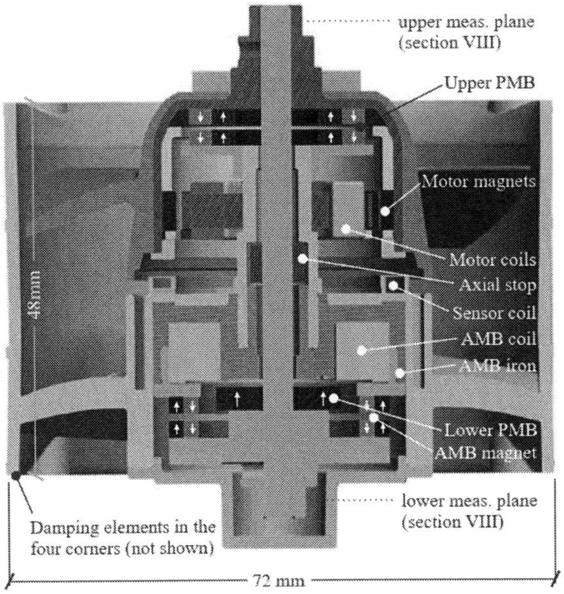

Figure 1. Cross-section of magnetically levitated fan. The radial and tilt directions are passively stabilized by two permanent magnetic bearings (PMB). Stabilization of the axial degree of freedom is achieved by an active magnetic bearing (AMB).

Each PMB consists of two axially magnetized permanent magnets on the stator and the rotor. The shown alignment of the magnets causes radially stable magnetic forces. All PMB magnets are mounted on a back iron which increases the radial stiffness. However, the value of the destabilizing axial stiffness increases disproportionately [11], [12]. The back iron ensures that the magnetic field of the PMB is kept inside the fan. The stacking of the magnets allows to achieve a higher radial stiffness per volume ratio [13], [14].

In Fig. 1 the rotor is shown in the levitating, axially force free position. Here, the sum of the axial forces of the upper and lower PMB and the axial force of the currentless active magnetic bearing (AMB) is zero. Only small positive and negative currents are therefore necessary to keep the rotor levitating. The axial position of the impeller is measured by an eddy current sensor.

Fig. 2 depicts the tool which was built to assemble the magnetically levitated fan. It is equipped with a load cell and two position sensors. This device enables to mount the rotor PMB parts so that the lower and the upper air gap of the axial stop are of same size in the force free position. This has proven to be a good method to compensate for the force tolerances due to variations in the PMB parts (magnet material, magnetization, magnet size, magnet position, magnet shape, etc.).

Figure 2. Assembly tool for magnetically levitated fan.

III. Passive Magnetic Bearing (PMB)

Although the permanent magnets of the upper PMB are mounted on a back iron, the axial force and the axial and radial stiffnesses can be calculated analytically. A novel calculation method based on the method of images for the stiffnesses is described in [12].

IV. Active Magnetic Bearing (AMB)

AMBs with one degree of freedom using permanent magnetic bias flux can be distinguished in two main groups.

A linear force per current relationship is achieved by placing the coil in the magnetic field of the permanent magnet. This arrangement corresponds to a voice coil except that the moving part in the magnetic bearing is the magnet (and the attached iron) and not the coil. These types of actuators have no negative stiffness, which makes them well suited for rotors with passive permanent magnetic radial bearings [15]. An actuator providing both axial forces and torque is shown in [16].

In the second AMB concept, which is used in the presented magnetically levitated fan, the coil is embedded in iron. The magnetically levitated fan uses a space-saving configuration where the magnetic circuits of the AMB and the lower PMB are combined. Fig. 3(a) shows the general geometry of this active/passive magnetic bearing. A drawback of the design is the rather high reluctance of the AMB coil. However, the design was chosen due to the simple construction and the easy assembly of rotor and stator.

In the available space this arrangement achieves a higher axial force per ampere winding ratio k_Θ compared to the voice coil type actuator. The voice coil actuator causes no magnetic forces if the AMB coil is not energized. Contrarily, the AMB concept of Fig. 3(a) has a negative axial stiffness, a negative tilt stiffness and a possibly negative radial stiffness[1]. Furthermore, all these stiffnesses change with the current of the AMB coil. As the AMB iron of the prototype fan was not laminated or built from soft magnetic composite (SMC), there is additionally a strong frequency dependence of the force per ampere winding ratio k_Θ due to eddy currents. The effect on the closed loop stability was investigated by Panholzer et al. [17].

[1]The radial stiffness with the lower PMB is always positive. However, the AMB alone can be radially stable or unstable depending on the actual geometrical dimensions of the AMB.

A proper design of the AMB involves the fulfillment of several criterias:

- low copper losses at rotor start
- high stability or at least low instability in radial direction
- low decrease of the radial stiffness when the coil is energized
- low overall height of the AMB
- low magnet volume

A multi-objective optimization using genetic algorithms was carried out to find a good compromise between these conflicting objectives using the software MagOpt [18]. All geometric parameters of the AMB have been used as design variables. For the optimization process a short calculation time of all stiffness values is crucial. Since the original rotational symmetry is lost in case of a radially deflected or tilted rotor, the radial stiffness and the tilt stiffness are usually calculated with 3D-FE software. But these time-consuming calculations would prevent fast optimization results. Therefore, fast approximate calculations of the radial and tilt stiffnesses which need only 2D finite element analyses were used. This method is described in [19]. The 2D finite element calculations were done in FEMM [20].

Fig. 3(b) shows the selected Pareto-optimal design which was realized in the magnetically levitated fan. The field lines are drawn for zero magnetomotive force and the rotor is shown in the force free rotor position, i.e. the sum of the axial forces of the AMB and the two PMBs is zero. The insulation thickness between the AMB and PMB is zero in this design.

To start from the upper mechanical stop, the axial force of the AMB has to be weakened by the current, shown in Fig. 3(c). Contrarily, to start from the lower mechanical stop the flux density in the air gap above the AMB magnet is increased by the coil, shown in Fig. 3(d).

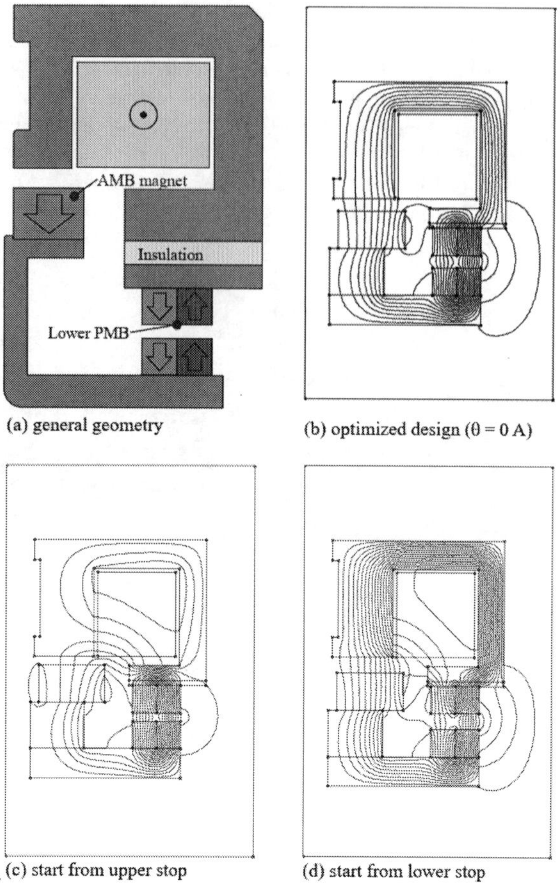

(a) general geometry
(b) optimized design (θ = 0 A)
1. (c) start from upper stop
(d) start from lower stop

Figure 3. Cross section of the active magnetic bearing (AMB) and the lower passive magnetic bearing (PMB) of the magnetically levitated fan.

V. Electrical Drive

The cross section of the electrical drive was optimized with MagOpt [18] and is shown in Fig. 4. The desired speed was 16300 rpm, where the load torque of the impeller is circa 13 mNm. The motor is operated with block voltage commutation without an angle sensor by detecting the zero-crossing of the back-emf. This operation mode was implemented in a motor simulation within the MagOpt project using the modeling strategy as suggested by Weidenholzer et al. [21]. A multi-objective optimization with the following objectives was carried out:

- low instability in radial direction (k_{drive})
- minimum losses at load point
- low motor height (including the end windings)
- low torque ripple (at load point, gained from block voltage commutation simulation)
- low magnet volume

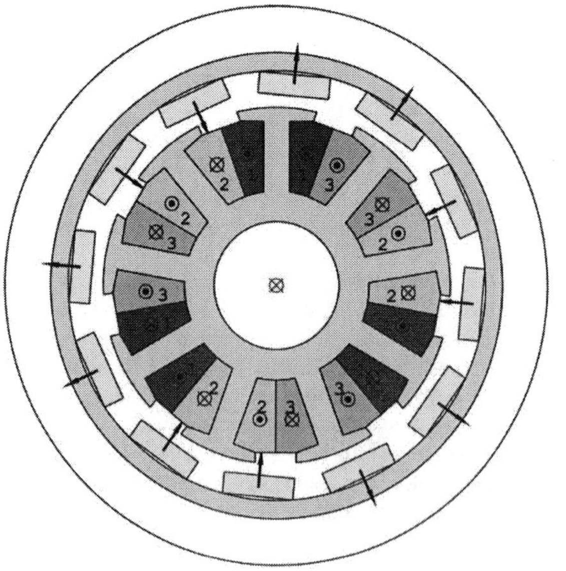

Figure 4. Cross section of the electrical drive of the magnetically levitated fan (9 slots, 6 poles, 2 magnets per pole)

VI. Optimization of PMBs and Damping Elements

A. Mathematical Model

For magnetic bearing systems using passively stabilized degrees of freedom, the optimization of the stiffnesses of the

PMBs and viscoelastic damping elements (VDE) is essential in the design process. Marth et al. showed that an analytical model is well suited to calculate the rotordynamic behavior of a passive magnetic bearing system including viscoelastic damping elements [22]. All stiffness and damping coefficients of the PMBs, the AMB, the motor and the viscoelastic dampers have to be considered.

The equations of motion were derived for the system illustrated in Fig. 5, where k_{B1} and $k_{\varphi B1}$ denote the radial and tilt stiffness of the upper PMB, k_{B2} and $k_{\varphi B2}$ denote the radial and tilt stiffness of the lower PMB and the AMB when it is powered off. Each PMB by itself is unstable against tilting ($k_{\varphi B1} < 0$, $k_{\varphi B2} < 0$). The positive tilt stiffness of the rotor is achieved by the positive radial stiffnesses ($k_{B1} > 0$, $k_{B2} > 0$) and the distance between the PMBs. The destabilizing effect caused by the radial reluctance force between motor magnets and the motor lamination stack is considered using k_{drive} ($k_{drive} < 0$). The rotor and the stator can move in the radial directions and incline about both diametral axes. The axial movement of the stator and the rotor as well as the rotation of the stator were neglected. Thus, eight degrees of freedom remain to describe the position of the rigid rotor and the rigid stator. The nonlinear equations of motion were linearized around the rest position for a fixed angular speed Ω.

B. Viscoelastic Damping Elements (VDE)

In order to damp the vibrations of the rotor, the stator was mounted on four VDEs placed in the corners of the fan housing. The dynamic behavior of the VDEs can be calculated as product of the VDE static stiffness with the dynamics of the viscoelastic material [23]. The oscillation frequency and the temperature have the biggest influence on the material properties[2]. Frequency dependent stiffness and damping values $[k_{VDE}(\omega), d_{VDE}(\omega), k_{\varphi VDE}(\omega), d_{\varphi VDE}(\omega)]$ can be used to describe the dynamics of the VDEs at one temperature, where k_{VDE} and d_{VDE} denote the radial stiffness and radial damping, $k_{\varphi VDE}$ and $d_{\varphi VDE}$ denote the tilt stiffness and tilt damping and ω is the excitation frequency. However, no transient simulation is possible with this model. Therefore, the frequency-dependence was implemented by using the generalized Maxwell or Wiechert model [24], which introduces additional state variables to the linear system in order to model the frequency-dependence.

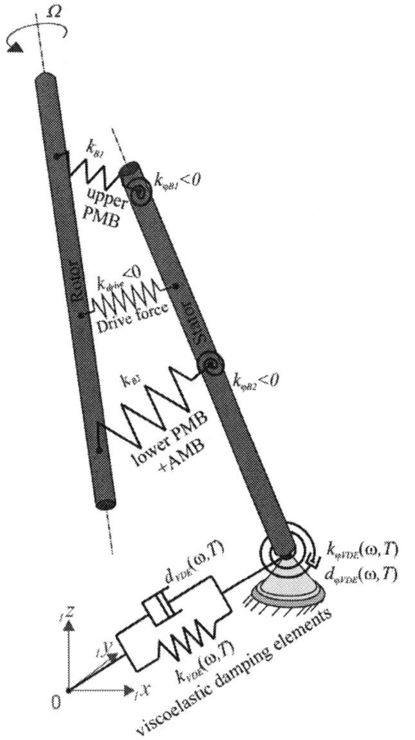

Figure 5. Schematic figure of the dynamic model. The rotor and stator can move in radial direction and incline about both diametral axes of inertia.

Table I
CONSIDERED DISTURBANCE FORCES AND OBJECTIVES FOR THE OPTIMIZATION OF THE PASSIVE MAGNETIC BEARING AND DAMPING ELEMENT STIFFNESSES

	Force application	Force type	Objective
1	stator	external acceleration	fulfillment of the vibration norm IEC 60721-3-4 Class 4M5
2	rotor	force step	high transient damping ratio
3	rotor	unbalance force	low deflection in the planes of the PMBs even if the rotor speed is equal to the frequency of a rigid body mode

C. Optimization with MagOpt

The mathematical model was implemented and optimized in MagOpt [18]. For a constant angular speed Ω the system is linear and has around 70 states[3].

The deflection of the rotor relative to the stator is the main criteria for the stability of a magnetically levitated fan. Table I lists the considered disturbance forces and the corresponding objectives. Each objective was calculated at $-10\,°C$ and $+60\,°C$ damping element temperature. The objectives of the optimization were set to the worst case value. In this paper the deflections due to unbalance will be discussed in more detail. A comparison with measurements is carried out in section VIII.

Since the axial force of the upper PMB has to compensate the axial force of the lower PMB and the axial force of the currentless AMB, the PMBs and the AMB have to be adjusted appropriately. Therefore feasible upper and lower bearing combinations were identified before starting the optimization. The upper PMB was calculated analytically [12], while the AMB and the lower PMB were calculated with FEMM (as described in section IV). The stiffnesses of every suitable design were saved to a table. Every table row represented

[2]Static pre-load or nonlinear material behavior due to large strain amplitudes were not considered in the model.

[3]The exact number depends on the number of state variables used to describe the viscoelastic material

a design with different radial stiffnesses but where the axial forces add up to zero. In the optimization the bearings were varied by using these precalculated table values. Furthermore, the geometry and material of the VDEs were altered during the optimization process.

VII. PROTOTYPE

Fig. 6 depicts the built magnetically levitated fan prototype and its basic data. The power electronics of the prototype were assembled in an external housing, but integration within the fan housing is possible. For levitation of the rotor (speed 0 rpm), the power consumption of the whole system is 1.7W.

Nominal data	
Operating voltage	48 V
Nominal speed	16.300 rpm
Power consumption	
...for levitation (standstill)	1.7 W
...at nominal speed	30 W

Figure 6. Prototype of magnetically levitated fan.

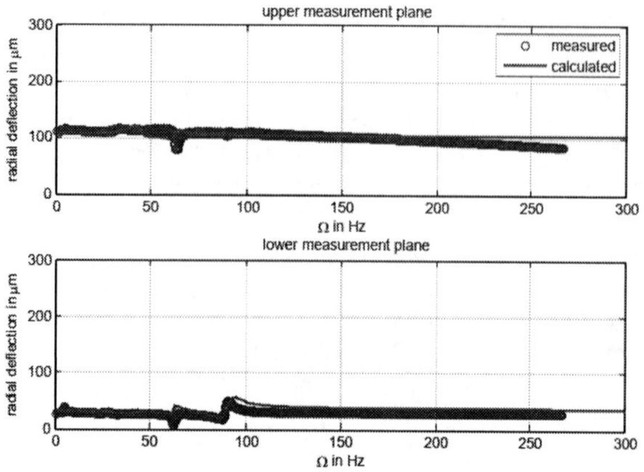

Figure 7. Measured and calculated rotor deflections of the balanced rotor.

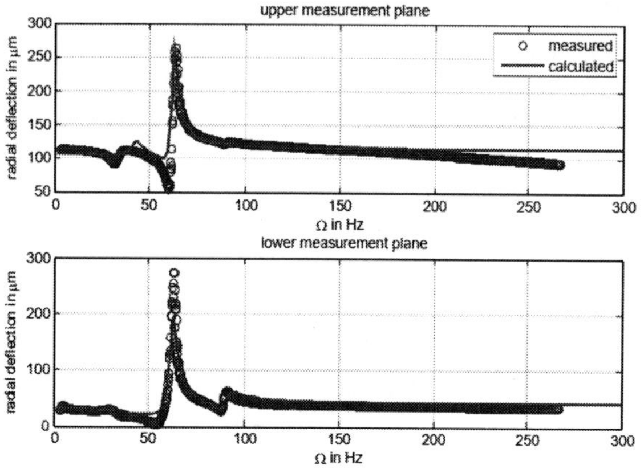

Figure 8. Measured and calculated rotor deflections. An additional weight causing an unbalance of 0.5 gmm was added to the rotor top.

VIII. MEASUREMENTS

In this section the measured rotor deflections due to unbalance and magnetic tolerances are compared to calculated values. Magnetic tolerance means that the magnetic axis, which is given by the position where the sum of the magnetic forces is zero, does not coincide with the symmetry axis of the rotor. The magnetic axis further depends on the rotor angle. Magnetic tolerances as well as unbalance are the major internal excitations in radial and tilt direction [25].

In a first step the rotor was balanced as good as possible. This was done by measuring the forces on the stator corresponding to the method described in [26, pages 24-28]. The radial rotor deflections caused by the remaining unbalance and by the magnetic tolerances were measured by laser distance sensors, see Fig. 7. The two measuring planes are shown in Fig. 1. Since the deflections at low speeds ($\Omega \approx 0\,rpm$) are caused by magnetic tolerances they cannot be compensated by balancing weights. As shown in Fig. 7, the balancing worked well since there are no distinct resonances.

The rotor deflections were also calculated by the mathematical model described in section VI-A. However, the remaining unbalance and magnetic tolerance of the rotor are initially unknown. Therefore these excitations were identified by using an optimization with genetic algorithm. The calculated rotor deflections are also shown in Fig. 7. It can be seen that the absolute value and phase of the unbalance and magnetic tolerance could be found and that the mathematical model is able to describe the behavior of the rotor.

In a second step a weight was added to the rotor top which caused an unbalance of $0.5\,$gmm. This unbalance mass was also considered in the mathematical model. In Fig. 8 the measured rotor deflections were compared to the calculated ones. Both Fig. 7 and Fig. 8 show that the measured deflections decline at higher rotor speeds, especially in the upper measurement plane. This effect, which might be caused by aerodynamic forces, was not considered in the mathematical model.

IX. FURTHER WORK

The focus of the development is to enhance the life expectancy of compact high speed fans under adverse environmental conditions. The goal is to achieve less than 2.5 percent failures in 10 years. Experiments must be made to prove if this ambitious goal can be met. In particular, the longtime stability of the viscoelastic damping elements and the electronics have to be tested. Before the tests can be started the damping elements must be integrated in the housing. To keep dirt and dust away from the air gaps of the passive magnetic bearings and the electrical drive further design changes might be necessary.

X. Conclusion

This paper describes the concept and design process of a magnetically levitated fan and its components. It was shown that two radial passive magnetic bearings and an axial active magnetic bearing can be successfully implemented in the available space of a 72 mm x 72 mm x 48 mm compact fan.

The necessary damping injection for the passively stabilized degrees of freedom was obtained by a viscoelastic support of the stator. In order to achieve the required damping, an optimization of the stiffnesses of the passive magnetic bearings and damping elements was absolutely necessary.

The calculated rotor deflections due to unbalance and magnetic tolerances were compared to measurements and it was shown that the mathematical model is sufficiently precise to predict the deflections of the rotor and the viscoelastically mounted stator.

Acknowledgment

This work has been carried out at LCM GmbH as part of a K2 project. K2 projects are financed using funding from the Austrian COMET-K2 program. The COMET-K2 projects at LCM are supported by the Austrian federal government, the federal state of Upper Austria, the Johannes Kepler University and all partners which form part of the COMET-K2 consortium.

References

[1] X. Tian, "Cooling fan reliability: failure criteria, accelerated life testing, modeling and qualification," in *Reliability and Maintainability Symposium, 2006. RAMS '06. Annual*, 2006, pp. 380–384.

[2] *Rolling bearings – Dynamic load ratings and rating life*, ISO 281 Std., 2007.

[3] H. Oh, M. Azarian, D. Das, and M. Pecht, "A critique of the ipc-9591 standard: Performance parameters for air moving devices," *IEEE Transactions on Device and Materials Reliability*, vol. 13, no. 1, pp. 146–155, 2013.

[4] E. R. Booser, *Tribology Data Handbook*. CRC Press, 1997.

[5] *Performance Parameters (Mechanical, Electrical, Environmental and Quality/Reliability) for Air Moving Devices*, IPC-9591 Std., 2006.

[6] X. Jin, E. Ma, T. Chow, and M. Pecht, "An investigation into fan reliability," in *Prognostics and System Health Management (PHM), 2012 IEEE Conference on*, 2012, pp. 1–7.

[7] L. Ramonat and A. Zeiff, "Im Dauerlauf, Verschiedene Berechnungen strategischer Ausfälle bei Lüftern," *QZ Qualität und Zuverlässigkeit*, vol. 1, 2012.

[8] M. Kimber, K. Suzuki, N. Kitsunai, K. Seki, and S. V. Garimella, "Pressure and flow rate performance of piezoelectric fans," *IEEE Transactions on Components and Packaging Technologies*, vol. 32, no. 4, pp. 766–775, 2009.

[9] R. Schacht, A. Hausdorf, B. Wunderle, and B. Michel, "Miniaturized frictionless fan concept for thermal management of electronic," in *18th International Workshop on Thermal Investigations of ICs and Systems (THERMINIC)*, 2012.

[10] S. Earnshaw, "On the nature of the molecular forces which regulate the constitution of the luminiferous ether," *Transactions of the Cambridge Philosophical Society*, vol. 7, London 1848.

[11] W. Braunbek, "Freischwebende Körper im elektrischen und magnetischen Feld," *Zeitschrift fuer Physik*, vol. 112, pp. 753–763, 1939.

[12] G. Jungmayr, E. Marth, W. Amrhein, H. Berroth, and F. Jeske, "Analytical stiffness calculation for permanent magnetic bearings with soft magnetic materials," *IEEE Transactions on Magnetics*, 2014.

[13] J. P. Yonnet, G. Lemarquand, S. Hemmerlin, and E. Olivier Rulliere, "Stacked structures of passive magnetic bearings," *Journal of Applied Physics*, vol. 70, no. 10, pp. 6633–6635, 1991.

[14] M. Lang, "Berechnung und Optimierung von passiven permanentmagnetischen Lagern für rotierende Maschinen," Ph.D. dissertation, TU Berlin, 2003.

[15] G. Jungmayr, W. Amrhein, W. Angelis, S. Silber, H. Grabner, and D. Andessner, "Design of a hybrid magnetic bearing," in *The 8th International Symposium on Magnetic Suspension Technology, Dresden, Germany*, Sep. 2005.

[16] W. Bauer and W. Amrhein, "Electrical design and winding selection for a bearingless axial-force/torque motor," in *2012 International Symposium on Power Electronics, Electrical Drives, Automation and Motion (SPEEDAM)*, 2012, pp. 1224–1229.

[17] M. Panholzer, G. Jungmayr, and E. Marth, "Performance and dynamic properties of an active axial bearing," Aug. 2012.

[18] S. Silber, W. Koppelstätter, G. Weidenholzer, and G. Bramerdorfer, "MagOpt - optimization tool for mechatronic components," in *14th International Symposium on Magnetic Bearings (ISMB14)*, 2014. [Online]. Available: http://www.magneticbearings.org/

[19] G. Jungmayr, E. Marth, M. Panholzer, W. Amrhein, and G. Weidenholzer, "Fast approximate calculation of the radial and tilt stiffness of magnetic bearings using magnetostatic 2d finite element analysis," Aug. 2012, p. 9. [Online]. Available: http://pages.shanti.virginia.edu/ismb13/

[20] D. Meeker, "Finite element method magnetics," *User's manual, Version 4.2*, accessed 12-June-2013. [Online]. Available: http://www.femm.info/

[21] G. Weidenholzer, S. Silber, G. Jungmayr, G. Bramerdorfer, H. Grabner, and W. Amrhein, "A flux-based pmsm motor model using rbf interpolation for time-stepping simulations," in *IEEE International Electric Machines Drives Conference (IEMDC)*, May 2013, pp. 1418–1423.

[22] E. Marth, G. Jungmayr, and W. Amrhein, "Influence of viscoelastic elements on magnetically passive stabilized degrees of freedom," in *12th International Symposium on Magnetic Bearings (ISMB12)*, Aug. 2010. [Online]. Available: http://www.magneticbearings.org/

[23] A. Bormann, "Elastomerringe zur Schwingungsberuhigung in der Rotordynamik," Ph.D. dissertation, Technische Universität Berlin, 2005.

[24] N. W. Tschoegl, *The Phenomenological Theory of Linear Viscoelastic Behavior*. Springer-Verlag, 1989.

[25] G. Jungmayr, "Der magnetisch gelagerte Lüfter," Ph.D. dissertation, Johannes Kepler University Linz, Institute for Electrical Drives and Power Electronics, 2008.

[26] R. Gasch, R. Nordmann, and H. Pfützner, *Rotordynamik*. Springer-Verlag, 2002.

Novel Permanent-Magnet-Biased Axial Hybrid Magnetic Bearings without a Thrust Disc

Pengfei Zhang, Zhiheng Wang, Guang Xi

School of Energy and Power Engineering, Xi'an Jiaotong University
28 Xianning West Road, 710049, Xi'an, P.R. China
pfzhang1234@gmail.com, wangzhiheng@mail.xjtu.edu.cn, xiguang@mail.xjtu.edu.cn

Abstract—**The paper proposes a pair of novel permanent-magnet-biased axial hybrid magnetic bearings, which are assembled at ends of a compressor spindle. The outer diameter of rotor is minimized to reduce the wind loss, which is regarded as the main loss in a typical axial magnetic bearing for a compressor. The bias magnetic flux, provided by an axial polarized permanent magnet ring, diminishes the loss further. The paper firstly introduces the structure and the working principle of the new axial hybrid magnetic bearing system. The cooling holes are aimed to improving the cooling capability without degrading the load capacity. Then, the axial magnetic attraction forces of axial magnetic bearings in pairs, under the condition without the cooling holes, are calculated by 2-D Finite Element Method (FEM) and 3-D FEM respectively. Additionally, another 3-D model is set up to investigate the influence of the designed cooling holes on the axial force. The coupling of the magnetic force and the flux density limit of iron cores are analyzed. The numerical results indicate that the proposed structure of the axial magnetic bearing system is feasible.**

I. INTRODUCTION

Magnetic bearings have been applied to high-speed turbomachinery progressively for their advantages such as noncontact, no lubrication requirement, lower power loss and less maintenance [1-3]. Compared with active magnetic bearings, permanent-magnet-biased hybrid magnetic bearings further reduce the power loss by generating the bias magnetic field instead of the current in coils [4].

Although much attention has been focused on the new design of permanent-magnet-biased radial and radial-axial magnetic bearings, quite few studies have been performed on the new structures of permanent-magnet-biased axial hybrid magnetic bearings. Khoo et al. [5] proposed a high load capacity bearing that generates the force from a number of parallel stator and rotor discs. Fang et al. [6] suggested a novel 3-DOF axial hybrid magnetic bearing, which is used in reaction flywheel system. The iron loss of magnetic bearing is reduced through a parallel-slot structure. A new structure of permanent-magnet-biased axial hybrid magnetic bearings was described in [7]. The power loss is diminished by separating the path of bias flux and control flux. However, all of those axial magnetic bearings mentioned above have an obvious thrust disc, which leads to a maximum material strain increase and larger rotor unbalances [8]. Moreover, in case of a rotor running in the air or other gases, the gas friction loss increases sharply with the increase of the outer diameter of a thrust disc [9]. Due to the high pressure condition in turbomachinery, the gas friction loss caused by fluid turbulence is the main power loss of magnetic bearings. A tubular linear actuator without a thrust disc was employed as a permanent-magnet-biased axial hybrid magnetic bearing in literature [8]. However, the axial load capacity is limited.

To overcome these drawbacks of the existing axial magnetic bearings, especially the high gas friction loss, a new structure of permanent-magnet-biased axial hybrid magnetic bearing system is proposed in this paper. Firstly its working principle is introduced. The characteristics of the magnetic force are analyzed based on 2-D FEM and 3-D FEM respectively. A 3-D Finite Element Analysis (FEA) is conducted to study the feasibility of the cooling holes design. In addition, the coupling of the force between axial direction and radial direction and saturation flux density limit are also analyzed by the 3-D FEM.

II. PRINCIPLE OF PERMANENT-MAGNET-BIASED AXIAL MAGNETIC BEARING SYSTEM

The configuration of the new axial magnetic bearing system is shown in Figure 1, which is assembled at both ends of a compressor spindle. The new system consists of the rotor, the stator, the working gaps and the cooling holes. The rotor includes an upper end plate, a spindle and a bottom end plate. The spindle is along the Z-axis of cylindrical coordinates. The stator has two parts, i.e., the upper stator and the bottom stator, each of which has an inner housing, a spacer, coils, a hood, a permanent magnet ring and an outer housing. The working gaps are named according to their location, that is, the working gap between the inner housing and the end plate is

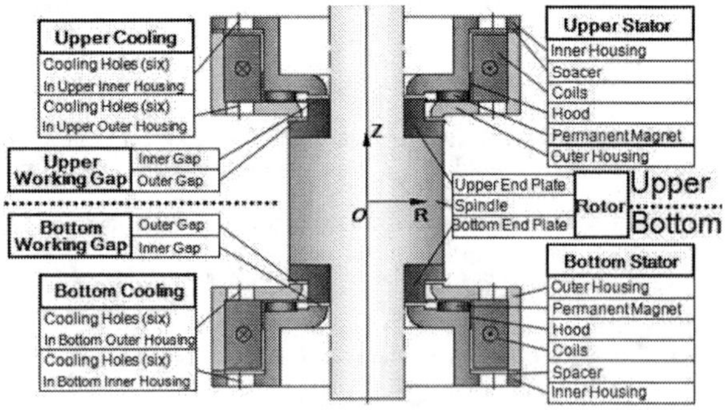

Figure 1. Permanent-magnet-biased axial hybrid magnetic bearing system.

named as the inner gap, while that between the outer housing and the end plate is called by the outer gap. Both the upper and bottom axial magnetic bearings have an inner gap and an outer gap, which together are named as upper working gap and bottom working gap respectively.

Figure 2 shows the magnetic flux schematic diagram in the half section of the bottom bearing. The magnetic flux is classified into the bias magnetic flux and the control magnetic flux. The dotted lines with the hollow arrows denote the flux paths generated by the permanent magnet and the solid lines with the solid arrows denote the flux paths generated by the coil. The flux provided by the permanent magnet is separated into two paths: One is the bias flux flowing across the inner housing, the inner gap, the end plate, the outer gap, the outer housing and the hood, and the other is the flux through the inner housing, the spacer, the outer housing and the hood. The control flux generated from the coil with the current flows across the inner housing, the inner gap, the end plate and the outer housing. Based on the working gap, the attractive magnetic force on the rotor is generated by the bias magnetic flux and control magnetic flux flowing through them. The spacer is inserted to guarantee that enough magnetic flux goes through the working gap, and the permanent magnet is thick enough to ensure the main control flux goes through the working gap other than through the permanent magnet. Therefore, the control flux circuits are of low reluctance, and the new axial magnetic bearing system is of low loss.

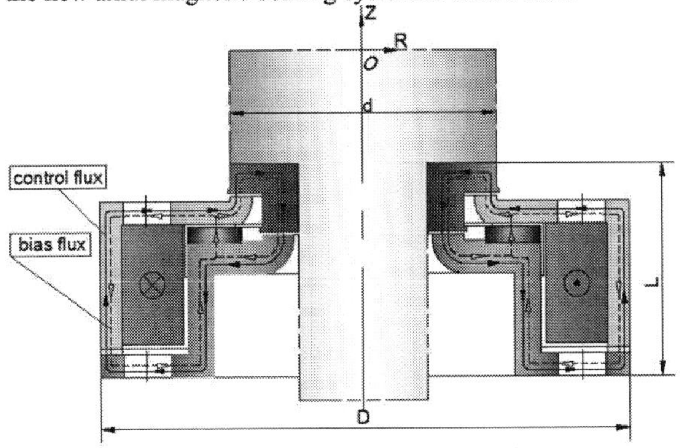

Figure 2. Schematic diagram of the bottom bearing system.

Assuming that the rotor is disturbed along positive Z-axis direction without loss of generality, it will move to positive Z-axis direction under the attractive magnetic force. As a result of the motion, the bottom working gap of the axial magnetic bearing gets larger, while the upper working gap gets smaller. The displacement sensors sense this motion and transmit it to the control system, and then the control system sends an amplified control current through the corresponding coils. The process results in the additional control flux of the bottom bearing denoted by the solid lines with the solid arrows in Figure 2, which is controlled by the controller to pull the rotor back to the equilibrium position. If the direction of current in coils is shown as Figure 1 and the magnetization of the upper magnetic ring is opposite to that of the bottom magnetic ring, the control flux direction of the upper bearing is the same as that of the bottom one. However, the bias flux direction is opposite between them. Thus, the bias flux in the upper working gap is reduced by the corresponding control flux, while the bias flux in the bottom working gap is increased by the corresponding control flux. By adding the flux in the bottom (i.e., larger) working gap and subtracting the flux in the upper (i.e., smaller) working gap, the axial magnetic bearing system generates an axial restoring force to stable the rotor.

The cooling holes on the inner housing and the outer housing are designed to improve the cooling capability without the degradation of magnetic force. It is guaranteed by the sectional area that the flux goes across is no less than that of magnetic poles, especially where cooling holes locate. The R coordinates of cylindrical holes' center here are the same, so cooling fluid could flow through the cavity between the inner and outer housing directly.

III. Finite Element Analysis

A. Models and Parameters

The main parameters of the proposed axial magnetic bearing system are listed in Table I. The working gap length, δ_i and δ_0 , is 0.3mm recommended by ISO standard [10], while the pole area A_i and A_o is nearly equivalent to increase the axial load capability as much as possible. The values of d and D are set based on the compressor. L, shown in Figure 2, is designed for enough volume of coils. The parameters, δ_h and δ_s, are set to adjust the bias flux in working gaps without reducing the control flux too much. Cooling holes' diameter d_h is maximized to improve the cooling capability with no impact on the magnetic force characteristics.

To verify the feasibility of the new structure, the proposed axial magnetic bearing system is analyzed by FEM with ANSYS Maxwell Software. The 2-D FEM and 3-D FEM analysis models were built as shown in Figures 3 and 4 respectively. As shown in Figure 3(a), a symmetrical configuration of bearings in pairs was built for the simulation. Since the axial bearing system is axisymmetric, the 2-D FEM can solve the steady problem with suitable boundary conditions. Figure 3(b) shows the whole mesh of the model above. In Figure 3(c), it's obvious that finite element refinement has been conducted in the working gap. For 2-D FEM analysis, the model is axisymmetric about Z-axis and is without cooling holes. However, for 3-D FEM analysis, the rotor in Figure 4(a) may be misaligned in order to study the force coupling between Z direction and R direction. Figure 4(b) shows the 3-D model with cooling holes. It was built to validate that the cooling holes have little influence on the force characteristics of this magnetic bearing system.

TABLE I. Parameters of the Axial Magnetic Bearing System

	Parameter	Value
d	Outer rotor diameter	50 mm
D	Outer stator diameter	100 mm
L	Axial length	38.5 mm
δ_i	Inner gap length	0.3 mm
δ_o	Outer gap length	0.3 mm
δ_s	Thickness of spacer	1.1 mm
δ_h	Thickness of hood bottom	0.7 mm
A_i	Pole area of inner housing	591.28 mm^2
A_o	Pole area of outer housing	591.22 mm^2

	Parameter	Value
h_{pm}	Thickness of permanent magnet	3 mm
N	Turns of coils	300
d_h	Cooling holes' diameter	9.6 mm
B_s	Saturation flux density of iron core	1.5 T

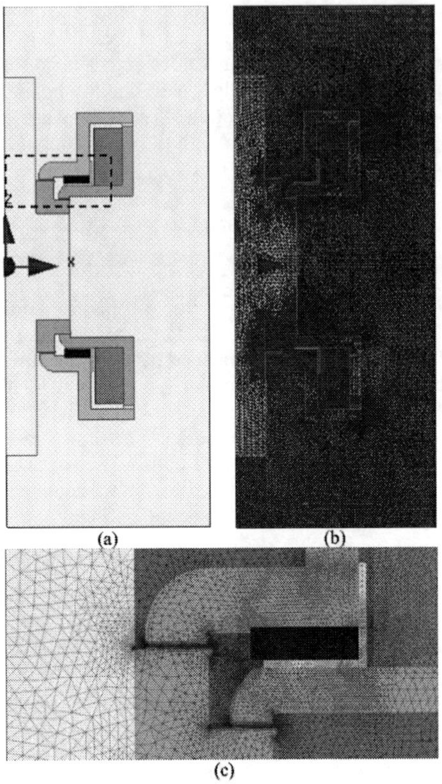

Figure 3. The 2-D FEM model and final mesh of the axial magnetic bearing system. (a) The 2-D FEM model of the axial magnetic bearing system. (b) The whole mesh of 2-D FEM model. (c) The local mesh in box of (a).

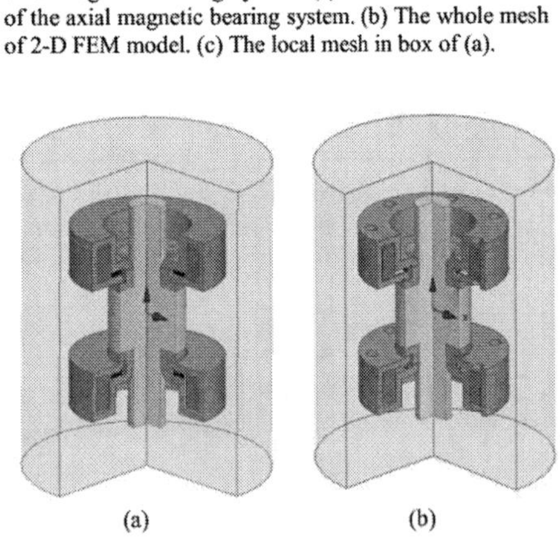

Figure 4. The 3-D FEM model of the axial magnetic bearing system. (a) The 3-D FEM model without cooling holes. (b) The 3-D FEM model with cooling holes.

B. Boundary and Excitation

The external boundaries are set as "balloon". The nonlinearity of iron cores is concerned and the magnetization direction of permanent magnet at the top is opposite to that at the bottom which pointing to +Z direction. For the upper coils and the bottom coils, their current sources are equal and share the same direction. As shown in Figure 1, the current in coils is assumed to be positive, and it will generate the negative force, pointing to –Z direction.

IV. Simulation Results and Analysis

A. Basic Characteristics

The force-displacement and force-current relationships are the main features of magnetic bearing. By 2-D FEM and 3-D FEM, they are displayed in Figure 5.The models are built without the cooling holes. The simulation results show that the results by 2-D FEM are generally consistent with the results by 3-D FEM. Figure 5(a) shows the proposed permanent-magnet-bias axial magnetic bearing system has a nearly linear force-current characteristic at whole range of current in coils, especially in 3-D FEM analysis. In addition, from Figure 5(b), the force-current characteristic is nearly linear when the displacement of rotor is less than 10^{-4} m in Z direction. Except the condition that the rotor displacement is zero, the force-displacement factor from 2-D FEA is larger than that from 3-D FEA. The force error between them increases if the rotor departures its equilibrium position.

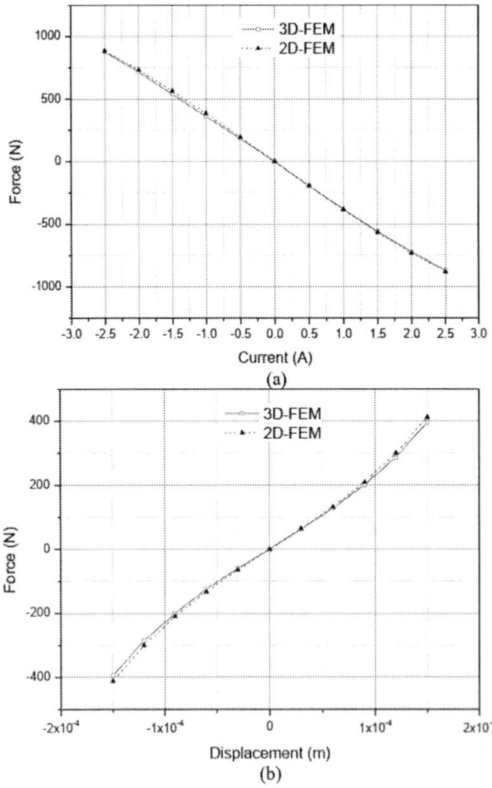

Figure 5. Force-displacement and force-current characteristic of the axial magnetic bearing system. (a) Axial force variation with current. (b) Axial force variation with displacement.

B. *Cooling Holes' Influnence*

The force performances of the presented magnetic bearings with and without cooling holes are compared in Figure 6, which illustrates that their deviations are very small at all the range of current and displacement. That indicates the cooling holes have little influence on the force-displacement and force-current relationships. However, the holes in the bearings can greatly contribute to heat dissipation of coils in working gas.

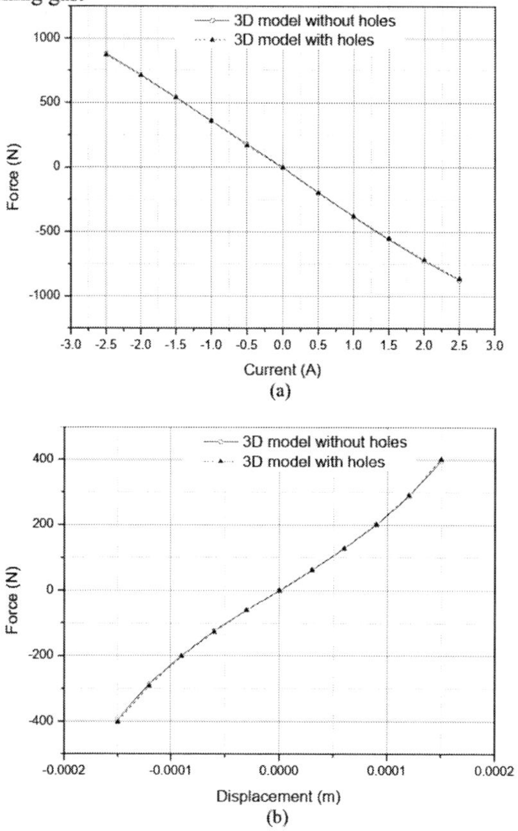

Figure 6. Influence of cooling holes on mechanical property. (a) Axial force variation with current. (b) Axial force variation with displacement.

C. *Coupling with R Direction*

Since the translation of the rotor along R direction is limited to 0.15mm, i.e., the nominal gap of radial back-up bearings will be employed as 0.15mm, the axial force variation with the displacement and the current is calculated by 3-D FEM with the assumption that the rotor translates 0.15mm along positive X-axis. In this section, the model analyzed is with the cooling holes. In Figure 7, it can be found that the force-current curves, as well as the force-displacement curves, are almost overlapped. Therefore, the motion effect on the axial force is small.

Figures 8 (a) & (b) depict the force variation along X-axis (i.e. F_x). In Figure 8(a), without the rotor motion, the forces deviate from zero due to the numerical simulation error. However, with the rotor motion, the forces reduce and are negative, except the condition that the current is zero. The larger the current is, the more the corresponding force reduction will be. But it should be further studied that whether it is a restoring force which may stable the rotor, or it is just caused by the numerical error. According to the results when x = 0.0mm, the numerical error is about 0.1N. Therefore, it seems that it could be a restoring force except the situation that the current is 0A. Here, the maximum value of F_x is less than 0.7N, so the influence of the rotor motion along X-axis can be ignored or be regarded as a small disturbance.

It can be seen from Figure 8(b) that, without rotor motion, the maximum force deviation from zero is about 0.1N, while with the rotor motion, it is about 0.15N. Both of them are quite small. So, the variation of force caused by rotor motion is very small, and the force variation can also be ignored.

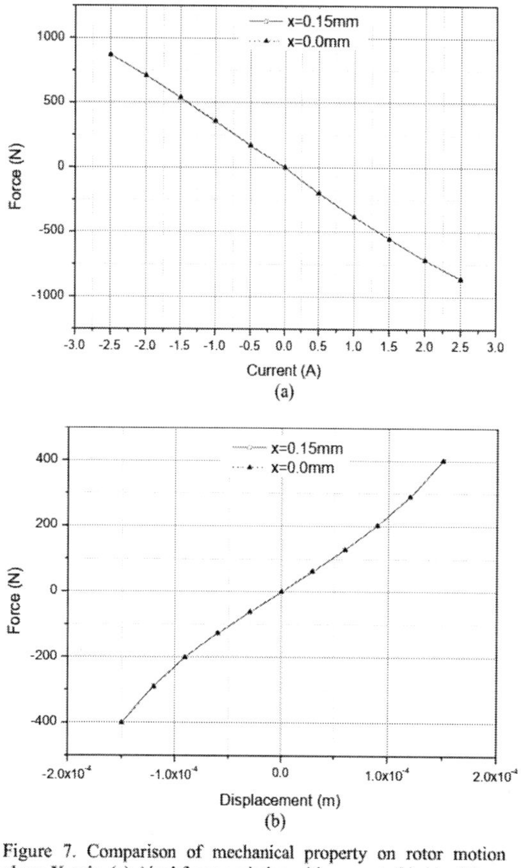

Figure 7. Comparison of mechanical property on rotor motion along X-axis. (a) Aixal force variation with current. (b) Aixal force variation with displacement.

D. *Flux Density Limit of Iron Core*

The maximum load capability is expressed as Equation (1):

$$F = \frac{B_s^2 A}{\mu_0} \tag{1}$$

where B_s is the saturation flux density, A denotes the total area of magnetic poles, and μ_0 is the permeability of free space. Figure 9 shows the flux density distribution at the axial sectional view, which only displays that in the components (i.e., the background distribution doesn't display here.)

The rotor is at the equilibrium position, and the current in coils is 2.0A in Figure 8(a) and 2.5A in Figure 8(b). The flux density in parts of inner housing and outer housing is more than saturation flux density of iron cores, 1.5T, as shown in circles of Figures 8(a) & (b). The area where flux density is no less than 1.5T gets larger, if the current in coils gets bigger. Nevertheless, the force-current factor is almost constant, which is plotted in Figure 6(a) and Figure 7(a).

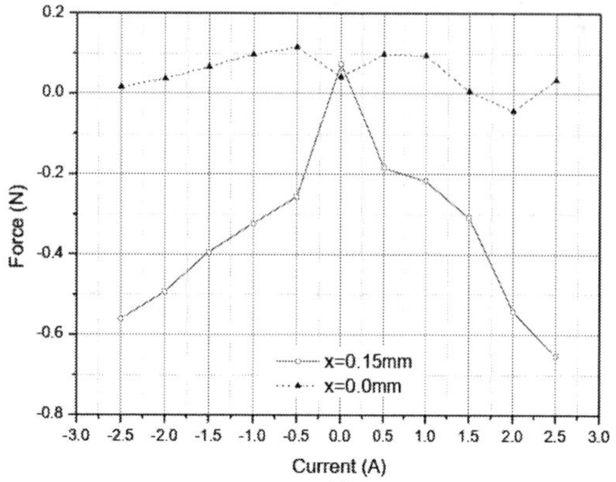

(a)

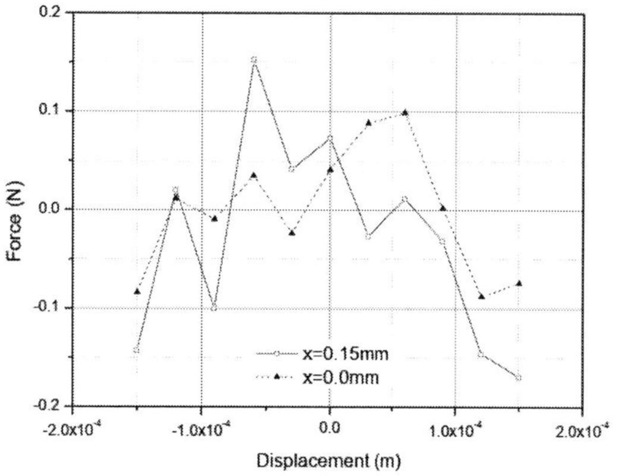

(b)

Figure 8. Force along X-axis under the influence of rotor motion. (a) F_x variation with current. (b) F_x variation with displacement.

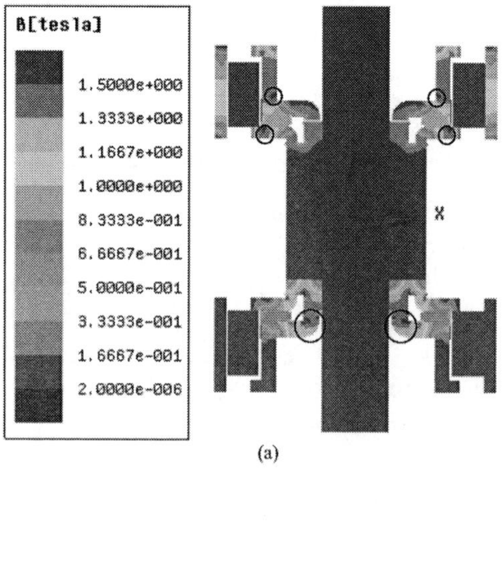

(a)

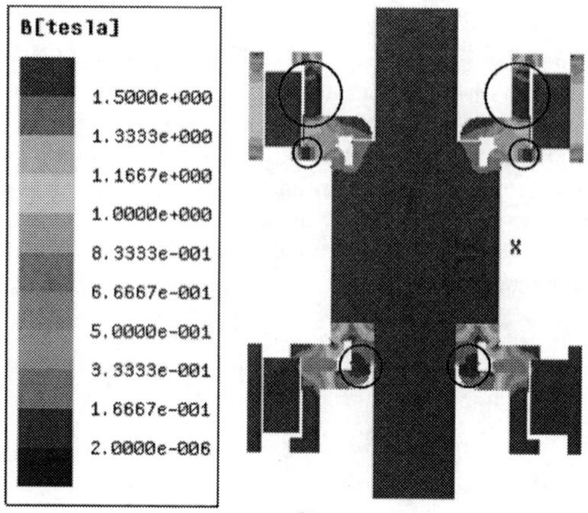

(b)

Figure 9. Flux density distribution at the axial sectional view. (a) Current =2.0A. (b) Current =2.5A.

V. CONCLUSIONS

A new structure of a permanent-magnet-biased axial magnetic bearing system without a thrust disc has been proposed to reduce the power loss, especially the gas friction loss, and to improve the cooling capability without the degradation of load capability. According to the results by 2-D FEM and 3-D FEM, we could make conclusions as follows:

- The proposed axial magnetic bearing system has a linear force-current factor and a force-displacement factor if the current in coils is no more than 2.5A and the displacement of rotor is less than 10^{-4} m.
- The designed cooling holes have little influence on the force characteristics, but it will contribute to improve the heat dissipation by convection.
- The effect of rotor motion along R direction on axial force is small, and the effect on radial force can also be ignored. It indicates that the axial force produced by the magnetic bearing system decouples with the radial force.
- Under the large current conditions, the flux density is saturate in some area. In spite of the flux density limit of iron cores, the axial force is still proportional to the current in coils.

It can be concluded that the presented structure of the axial magnetic bearing is feasible. It is interesting that the radial force by the current in coils seems to stable the rotor in corresponding radial direction, even though it's quite small. In future, the phenomenon will be further investigated, and the experiments will be conducted.

ACKNOWLEDGMENT

This present work is financially supported by the National Nature Science Foundation of China under grant No. 51236006.

REFERENCES

[1] J. Schmied. "Experience with magnetic bearings supporting a pipeline compressor," *Proceedings of 2nd Int. Symp. On Magnetic Bearings*, pp. 47–55, 1990.

[2] W. Canders, N. Ueffing, U. Schrader-Hausmann, R. Larsonneur, "MTG400:A magnetically levitated 400 KW turbo generator system for

natural gas expansion," *Proceedings of 4th Int. Symp. on Magnetic Bearings*, pp. 435–440, 1994.
[3] S. Yu, G. Yang, L. Shi, Y. Xu, "Application and research of the active magnetic bearing in the nuclear power plant of high temperature reactor", *Proceedings of 10th Int. Symp. On Magnetic Bearings*, pp. 1–7,2006.
[4] A. C. Lee, F. Z. Hsiao, D. Ko, "Analysis and testing of magnetic bearing with permanent magnets for bias," *JSME International Journal*, vol. 37, no. 3, pp. 774–782, 1994.
[5] W. K. S. Khoo, K. Kalita, S. D. Garvey, R. J. Hill-Cottingham, J. F. Eastham, "Active axial-magnetomotive force parallel-airgap serial flux magnetic bearings," *IEEE Transactions on Magnetics*, vol. 46, no. 7, pp. 2596–2602, 2010.
[6] J. Fang, J. Sun, H. Liu, J. Tang, "A novel 3-DOF axial hybrid magnetic bearing," *IEEE Transactions on Magnetics*, vol. 46, no. 12, pp. 4034-4045, 2010.
[7] J. Fang, J. Sun, Y. Xu, X. Wang, "A new structure for permanent-magnet-biased axial hybrid magnetic bearings," *IEEE Transactions on Magnetics*, vol. 45, no. 12, pp. 5319-5325, 2009.
[8] C. Weißbacher, H. Stelzer, K. Hameyer, "Application of a tubular linear actuator as an axial magnetic bearing," *IEEE/ASME Transactions on Mechatronics*, vol. 15, no. 4, pp. 615–622, 2010.
[9] Mack, M., "Luftreibungsverluste bei elektrischen Maschinen kleiner Baugröße," Diss. TH Stuttgart, 1967.
[10] ISO Standard 14839-4, Mechanical vibration - Vibrations of rotating machinery equipped with active magnetic bearings - Part 4: Technical guidelines, 2012.

LQG Control of an Active Magnetic Bearing with a Special Method to consider the Gyroscopic Effect

Markus Hutterer, Matthias Hofer, Thomas Nenning, Manfred Schrödl

Institute of Energy Systems and Electrical Drives, TU Wien, Gusshausstraße 25-27, 1040 Vienna, Austria,
markus.hutterer@tuwien.ac.at

***Abstract*—It could be very difficult to find an acceptable feedback path for coupled differential equations like a FIVE degree of freedom (DOF) AMB system. This paper describes a linear quadratic Gaussian (LQG) approach on this problem. Due to the time variant gyroscopic part of the AMB system equation the controller could only be optimized for a specific angular velocity. For this reason a time variant state feedback is developed and added to the LQG controller in a way that the system is nearly equal for all angular velocities. The feedback matrix which is the result of the LQG problem uses all feedback terms. Compared to a decentralized controller which only uses the diagonal terms, the LQG method also uses the cross feedback terms which are able to provide instability. After developing the LQG method was compared to a decentralized PID controller.**

I. INTRODUCTION

Magnetic Bearings are used in many industrial applications, because of their big advantages compared to other bearing types. The main advantages are that they operate frictionless and wearless, they need no lubricant and are maintenance free. In the last years also a few methods for sensorless control of AMB systems were developed, like the INFORM method which is described in [1] and [2]. The stable levitation is provided by magnetic forces exerted by electromagnets in a closed loop system. A FIVE DOF AMB system is described as a coupled MIMO structure, where it could be difficult to design a controller with a good performance. The most straightforward method is a decentralized control method, where each sensor is fed back to the actuator of the measured degree of freedom, which is used in [3]. A drawback of this method is that only the diagonal terms of the feedback matrix are used. To eliminate that drawback a central MIMO controller can be designed. For such a coupled MIMO control problem a LQR controller is a good solution. A way to solve that problem is described in [4]. But it has to be considered that the AMB system is a time variant system, because of the gyroscopic term. So the controller would only be optimal for one operating point and this could lead to instability for other rotational speeds [5]. In [6] a time variant cross coupling for a decentralized controller is developed, which canceled the time variant term. The aim of this paper is to combine the LQR controller with the time variant feedback path to design a robust control structure for a wide angular velocity range. For the LQR method alone all states have to be known. If the states cannot be measured, the states could be estimated with an observer. When a LQR controller is combined with a Kalman observer it is called LQG controller. In this paper the experimental results of this new control structure are compared to the results of a decentralized controller.

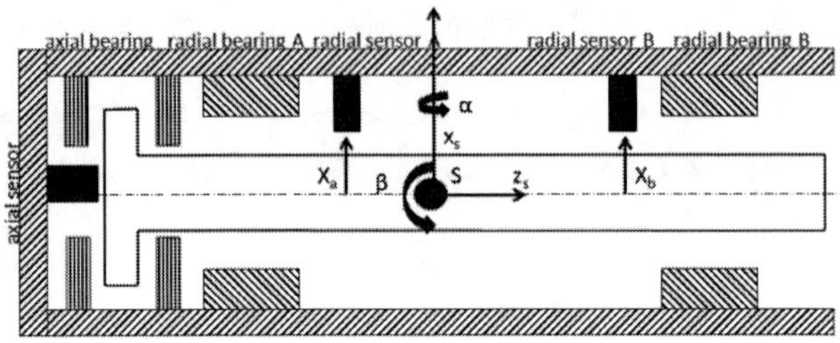

Figure 1. FIVE-DOF AMB SYSTEM

II. FIVE DOF AMB SYSTEM

A FIVE DOF AMB system consists of 2 radial bearings and one axial bearing as shown in Fig.1. Every DOF needs a position sensor for the feedback path. Under certain assumptions the radial and the axial bearings can be treated seperatly from each other. Here the seperated case is used. The equation to describe the radial part of a linear AMB-System is mathematically derived in [7] and is given by:

$$\mathbf{M}\ddot{\mathbf{x}} + \mathbf{G}(\Omega)\dot{\mathbf{x}} + \mathbf{B}\mathbf{K}_s\mathbf{B}^T\mathbf{x} = \mathbf{B}\mathbf{K}_i\mathbf{i} \quad (1)$$

$$\mathbf{y} = \mathbf{C}\mathbf{x} \quad (2)$$

with

$$\mathbf{M} = \begin{bmatrix} I_x & 0 & 0 & 0 \\ 0 & m & 0 & 0 \\ 0 & 0 & I_x & 0 \\ 0 & 0 & 0 & m \end{bmatrix} \qquad \mathbf{C} = \begin{bmatrix} c & 1 & 0 & 0 \\ d & 1 & 0 & 0 \\ 0 & 0 & c & 1 \\ 0 & 0 & d & 1 \end{bmatrix}$$

$$\mathbf{K}_s = \begin{bmatrix} k_s & 0 & 0 & 0 \\ 0 & k_s & 0 & 0 \\ 0 & 0 & k_s & 0 \\ 0 & 0 & 0 & k_s \end{bmatrix} \qquad \mathbf{B} = \begin{bmatrix} a & b & 0 & 0 \\ 1 & 1 & 0 & 0 \\ 0 & 0 & a & b \\ 0 & 0 & 1 & 1 \end{bmatrix}$$

$$\mathbf{K}_i = \begin{bmatrix} k_i & 0 & 0 & 0 \\ 0 & k_i & 0 & 0 \\ 0 & 0 & k_i & 0 \\ 0 & 0 & 0 & k_i \end{bmatrix} \qquad \mathbf{i} = \begin{bmatrix} i_{xA} \\ i_{xB} \\ i_{yA} \\ i_{yB} \end{bmatrix} \qquad \mathbf{x} = \begin{bmatrix} \beta \\ x_S \\ -\alpha \\ y_S \end{bmatrix}$$

$$\mathbf{G} = \begin{bmatrix} 0 & 0 & I_p\Omega & 0 \\ 0 & 0 & 0 & 0 \\ -I_p\Omega & 0 & 0 & 0 \\ 0 & 0 & 0 & 0 \end{bmatrix} \quad \mathbf{y} = \begin{bmatrix} x_{seA} \\ x_{seB} \\ y_{seA} \\ y_{seB} \end{bmatrix}$$

Where k_i is the linearized force current factor, k_s is the linearized force/displacement factor, m is the mass of the rotor, I_x is the equatorial moment of inertia, $\mathbf{G}$ is the time variant gyroscopic term, I_p is the polar moment of inertia, Ω is the angular velocity, and $\mathbf{i}$ is the current from the current controller. The equation of an AMB system is usually more complicated and nonlinear, but with k_i and k_s the nonlinear electrical equation is linearized to a linear one. The index S means that the axes are in the center of gravity and seA or seB that the axes are at the position sensors. The mechanical system has to fulfill a few assumptions that the linear equation (1) can be used [5]:

- The rotor is modeled as rigid body
- Deviations from the reference position are small compared to the rotor dimension
- Non-diagonal terms of the matrix of inertia are small compared to the diagonal terms
- The cross sectional area of the rotor is symetric ($I_x = I_y$)
- The bearing forces in two directions vertical to each other are independent
- The angular velocity around the longitudinal axis is constant or is changing slowly compared to the dynamic of the AMB system

III. Design of the Controller

A. Time variant feedback path

The result of the LQR method is an optimal controller only for a linear time invariant system. From this context it can be seen that the time variant path of equation (1) needs to be eliminated. If the control law

$$\mathbf{i} = (\mathbf{BK}_i)^{-1}\mathbf{G}(\Omega)\dot{\mathbf{x}} + \mathbf{v} \tag{3}$$

is inserted to equation (1)

$$\mathbf{M}\ddot{\mathbf{x}} + \mathbf{G}(\Omega)\dot{\mathbf{x}} + \mathbf{BK}_s\mathbf{B}^T\mathbf{x} = \mathbf{BK}_i(\mathbf{BK}_i)^{-1}\mathbf{G}(\Omega)\dot{\mathbf{x}} + \mathbf{BK}_i\mathbf{v} \tag{4}$$

the result is a time variant system in the form with a new input $\mathbf{v}$:

$$\mathbf{M}\ddot{\mathbf{x}} + \mathbf{BK}_s\mathbf{B}^T\mathbf{x} = \mathbf{BK}_i\mathbf{v} \tag{5}$$

In [6] it is suggested that a complete elimination of the gyroscopic term is not very robust, and therefore a factor C_{att} is introduced. After that step the control law has the following form:

$$i = (\mathbf{BK}_i)^{-1}C_{att}\mathbf{G}(\Omega)\dot{\mathbf{x}} + \mathbf{v} \tag{6}$$

Because a constant or a slowly changed angular velocity is assumed, this feedback term could be treated as a linear time invariant system for stability analyses. With this feedback law it is possible to find an optimal controller for a wide angular velocity range. The angular velocity can be measured with a sensor or is able to estimated with a special unbalance observer.

B. LQR Controller

The LQR method is a powerful technique for designing controllers for complex systems, for which it is hard to find a good solution with classical control theory. Compared to other control design techniques the LQR seeks to find an optimum that minimizes a cost function. In [8] this theory is described clearly and understandable. Because the controller is usually implemented in a Digital Processor, the cost function of the discrete system has to be minimized. The first step is to convert equation (5) in a time continuous state space representation.

$$\begin{aligned} \dot{\mathbf{z}} &= \mathbf{Az} + \mathbf{Bv} \\ \mathbf{y} &= \mathbf{Cz} \end{aligned} \tag{7}$$

with

$$\mathbf{A} = \begin{bmatrix} \mathbf{0} & \mathbf{E} \\ -\mathbf{M}^{-1}\mathbf{BK}_s\mathbf{B}^T & \mathbf{0} \end{bmatrix} \mathbf{B} = \begin{bmatrix} \mathbf{0} \\ \mathbf{M}^{-1}\mathbf{BK}_i \end{bmatrix} \tag{8}$$

$$\mathbf{C} = \begin{bmatrix} \mathbf{C} & \mathbf{0} \end{bmatrix} \mathbf{z} = \begin{bmatrix} \mathbf{x} \\ \dot{\mathbf{x}} \end{bmatrix} \tag{9}$$

Where $\mathbf{E}$ is the identity matrix and $\mathbf{0}$ is the zero matrix. This continuous state space system is transformed to a discrete system with the following Form:

$$\begin{aligned} \mathbf{z}_{k+1} &= \mathbf{\Phi z}_k + \mathbf{\Gamma v}_k \\ \mathbf{y}_k &= \mathbf{Cz}_k \end{aligned} \tag{10}$$

To have a controller which is able to eliminate constant disturbances like the weight force the system is extended with an integration action.

$$\mathbf{z}_{Ik+1} = \mathbf{z}_{Ik} + (\mathbf{r}_k - \mathbf{Cz}_k) \tag{11}$$

Herein $\mathbf{r}_k$ is the reference signal. This integrator can be considered in the discrete state space representation.

$$\begin{aligned} \begin{bmatrix} \mathbf{z}_{k+1} \\ \mathbf{z}_{Ik+1} \end{bmatrix} &= \begin{bmatrix} \mathbf{\Phi} & \mathbf{0} \\ -\mathbf{C} & \mathbf{E} \end{bmatrix} \begin{bmatrix} \mathbf{z}_k \\ \mathbf{z}_{Ik} \end{bmatrix} \\ &+ \begin{bmatrix} \mathbf{\Gamma} \\ \mathbf{0} \end{bmatrix} \mathbf{v}_k + \begin{bmatrix} \mathbf{0} \\ \mathbf{E} \end{bmatrix} \mathbf{r}_k \\ \mathbf{y}_k &= \mathbf{Cz}_k \end{aligned} \tag{12}$$

With this equation the cost function

$$J(z_0) = \sum_{k=0}^{N-1} \left(\mathbf{z}_k^T\mathbf{Qz}_k + \mathbf{u}_k^T\mathbf{Ru}_k\right) + \mathbf{z}_k^T\mathbf{Sz}_k \tag{13}$$

with the weighting matrices $\mathbf{Q}$, $\mathbf{R}$ and $\mathbf{S}$ can be minimized. From this equation it can be seen that $\mathbf{Q}$ weigths the states. When a diagonal matrix is used, the state which has the biggest entry in the matrix will decay very fast to zero compared to the other states. The $\mathbf{R}$ matrix weights the manipulated variables. When $\mathbf{R}$ increases the manipulated variables will decrease. The result of this optimization problem is:

$$\mathbf{u}_k = \mathbf{K}_k\mathbf{x}_k \tag{14}$$

$$\mathbf{K}_k = -\left(\mathbf{R} + \mathbf{\Gamma}^T\mathbf{P}_{k+1}\mathbf{\Gamma}\right)^{-1}\left(\mathbf{\Gamma}^T\mathbf{P}_{k+1}\mathbf{\Phi}\right) \tag{15}$$

$$\mathbf{P}_k = \left(\mathbf{Q} + \mathbf{\Phi}^T \mathbf{P}_{k+1} \mathbf{\Phi}\right) - \left(\mathbf{\Gamma}^T \mathbf{P}_{k+1} \mathbf{\Phi}\right)^T \left(\mathbf{R} + \mathbf{\Gamma}^T \mathbf{P}_{k+1} \mathbf{\Gamma}\right)^{-1} \left(\mathbf{\Gamma}^T \mathbf{P}_{k+1} \mathbf{\Phi}\right) \quad (16)$$

Equation (16) is a discrete Riccati equation, which operates backwards. With this constellation the feedback matrix is time variant and the end time has to be known. To overcome this issue the end time is set to infinite. Now the Riccati equation is an algebraic equation and the feedback matrix is time invariant. To get a stable system the matrices of (16) have to fulfill two assumptions:

- The eigenvalues ot the pair $(\mathbf{\Phi}, \mathbf{\Gamma})$ outside the unity circle need to be controllable.
- The eigenvalues ot the pair $(C, \mathbf{\Phi})$ outside the unity circle need to be observable

In [6] it is stated that a LQR controller is not very robust. It is possible to get a more robust system, if the system equation is modified in a specific way. The LQR method calculates a system, which has all poles of the system matrix $(\mathbf{\Phi} + \mathbf{\Gamma}_s)$ inside the unit circle. To increase robustness it is possible to get a system, which has the poles inside a circle with a radius of $r < 1$. The new system equation is

$$\mathbf{z}_{k+1} = \tilde{\mathbf{\Phi}} \mathbf{z}_k + \tilde{\mathbf{\Gamma}} \mathbf{v}_k \quad (17)$$

with

$$\tilde{\mathbf{\Phi}} = \frac{1}{r} \mathbf{\Phi} \quad \tilde{\mathbf{\Gamma}} = \frac{1}{r} \mathbf{\Gamma} \quad (18)$$

C. Kalman observer

A disadvantage of the LQR method is that all states have to be known. Generally this is not the case, so an observer is needed to estimate the states. For a MIMO system it could be hard to find an acceptable observer with the method of pole placement. A powerful approach for this problem is a Kalman observer, which is an optimal observer in the sense of control sciences. For filter design a discrete, time invariant and linear system

$$\begin{aligned} \mathbf{z}_{k+1} &= \mathbf{\Phi} \mathbf{z}_k + \mathbf{\Gamma} \mathbf{v}_k + \mathbf{G} \mathbf{w}_k \\ \mathbf{y}_k &= \mathbf{C} \mathbf{z}_k + \mathbf{H} \mathbf{w}_k + \mathbf{v}_k \end{aligned} \quad (19)$$

with the state vector $\mathbf{z}_k$, the output vector $\mathbf{y}_k$, the disturbances $\mathbf{w}_k$ and the measuring noise $\mathbf{v}_k$ is used. The Kalman observer has a few assumptions:

$$E(\mathbf{v}_k) = 0 \quad E(\mathbf{w}_k) = 0 \quad E(\mathbf{v}_k \mathbf{v}_j^T) = \mathbf{R}\delta_{kj}$$
$$E(\mathbf{w}_k \mathbf{w}_j^T) = \mathbf{Q}\delta_{kj} \quad E(\mathbf{w}_k \mathbf{v}_j^T) = 0$$
$$E(\mathbf{w}_k \mathbf{x}_j^T) = 0 \quad \text{for k} \geq \text{j}$$
$$E(\mathbf{v}_l \mathbf{x}_j^T) = 0 \quad \text{for all l, j}$$

The equation of the Kalman filter is derived in [8] and is

$$\hat{\mathbf{z}}_{k+1} = \mathbf{\Phi} \hat{\mathbf{z}}_k + \mathbf{\Gamma} \mathbf{v}_k + \hat{\mathbf{K}}_k \left(\mathbf{y}_k - \mathbf{C} \hat{\mathbf{z}}_k - \mathbf{D} \mathbf{v}_k\right) \quad (20)$$

$$\hat{\mathbf{K}}_k = \mathbf{\Phi} \mathbf{P}_k \mathbf{C}^T \left(\mathbf{C} \mathbf{P}_k \mathbf{C}^T + \mathbf{H} \mathbf{Q} \mathbf{H}^T + \mathbf{R}\right)^{-1} \quad (21)$$

$$\mathbf{P}_{k+1} = \mathbf{\Phi} \mathbf{P}_k \mathbf{\Phi}^T + \mathbf{G} \mathbf{Q} \mathbf{G}^T - \mathbf{\Phi} \mathbf{P}_k \mathbf{C}^T \left(\mathbf{C} \mathbf{P}_k \mathbf{C}^T + \mathbf{H} \mathbf{Q} \mathbf{H}^T + \mathbf{R}\right)^{-1} \mathbf{C} \mathbf{P}_k \mathbf{\Phi}^T \quad (22)$$

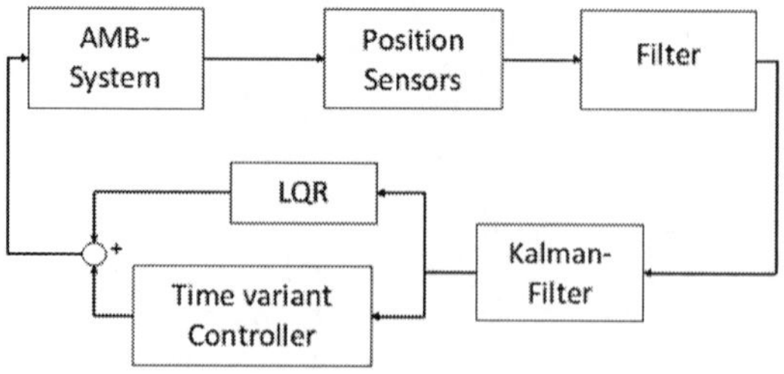

Figure 2. Control System of the AMB

Equation (22) is also a discrete Riccati equation, but compared to the LQR problem it runs forward. The feedback matrix is also time variant. If a time invariant feedback matrix is required, the end time could be set to infinite. The result of this method is a time invariant feedback matrix $\hat{\mathbf{K}}_k$. This feedback loop stabilizes the system under the following assumptions:

- The eigenvalues ot the pair $(C, \mathbf{\Phi})$ outside the unity circle need to be observable.
- The eigenvalues ot the pair $(\mathbf{\Phi}, \mathbf{G}\mathbf{Q}\mathbf{G}^T)$ outside the unity circle need to be controllable.
- The marix $\mathbf{H}\mathbf{Q}\mathbf{H}^T + \mathbf{R}$ needs to be positive definite.

Compared to the pole placement method the Kalman filter can affect the error system by the covariance matrix of the disturbances $\mathbf{Q}$ and the covariance matrix of the measuring noise $\mathbf{R}$. Where $\mathbf{R}$ considers the reliability of the sensors.

The total control system can be seen in Fig.2. Where LQR, time variant controller and Kalman filter are described in the sections above. The filter is a a second order IIR filter, which is used to stabilize bending modes at high frequencies.

D. Stabilization of the first bending mode

The rotor system has the first bending mode at 840 Hz. If the damping of the first bending modes are higher then zero the system is stable. Fig.3 shows the areas where the controller provides stability. The dead time of the sampling process would destabilize the system in the range of the first bending mode.

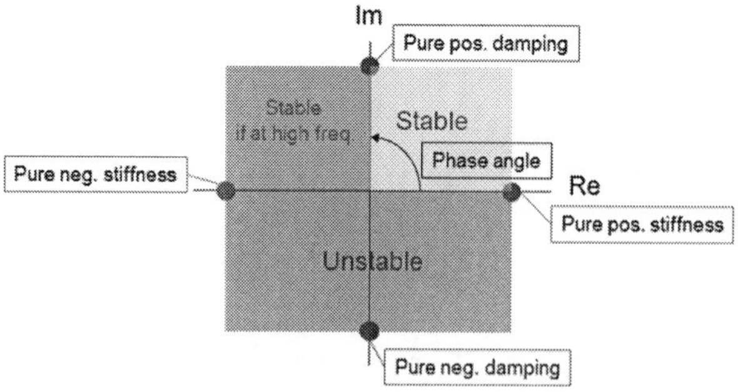

Figure 3. Requirement on the controller to provide stability [9]

Because no model for the flexible rotor was derived which could be used in the LQR design, a filter was designed to overcome this problem. This filter consists of an second order lowpass filter multiplied with a lead lag filter. The transfer function has the following form:

$$T(s) = \frac{1}{\frac{s^2}{(2\pi f_f)^2} + \frac{2D}{2\pi f_f}s + 1} \cdot \frac{\frac{s^2}{(2\pi f_n)^2} + \frac{2D_n}{2\pi f_n}s + 1}{\frac{s^2}{(2\pi f_d)^2} + \frac{2D_d}{2\pi f_d}s + 1} \quad (23)$$

With this method the phase is shifted in the stable area at high frequencies, which can be seen from the Bode plot (Fig.4).

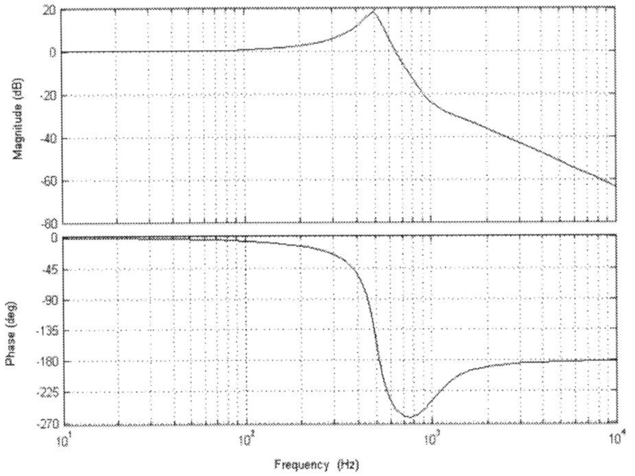

Figure 4. Bode plot of the filter to stabilize the first bending mode

A disadvantage of this method is that this filter limit the dynamic of the system and could lead to instability if this filter is not considered in the design of the PID or LQR Controller.

IV. Simulation Results

For the simulation the rigid body model from section 2 is used and performed on MATLAB/SIMULINK. The controller and observer was implemented with a Matlab function block, where the digitization is considered. A white noise with an amplitude of 10 µm is applied on the measured signals. Fig.5 shows the response of the position signals controlled with the LQR and the decentralized PID controller after a disturbance force step of $40N$ at a time of $0.2s$ with speed of $3000rpm$. The LQR method shows a better performance than the PID and the states decrease almost equally fast to zero with the LQR. With the PID controller two states perform much faster than the other two. By tuning the value of r the robustness of the system is able to be changed. Another effect of decreasing r is the increasing stiffness of the system. To test the robustness of the system the dynamic matrix was changed to simulate a wrong model. The system has an acceptable performance from 0.1 times the dynamic matrix to 2.8 times the dynamic matrix. From these values it can be seen, that the AMB system is robust against model inaccuracies.

Fig.7 shows the Campbell diagram of the decentralized PID controller and Fig.6 for the LQG controller with the compensation of the gyroscopic effect. The natural frequencies of the decentralized PID controller are drifting away from each other, because of the gyroscopic effect. This can decrease the performance of the system and is able to destabilize the control structure. The time variant feedback path in combination with the LQG method on the other hand has constant natural frequencies, because of the compensation of the gyroscopic effect. This means that the system is the same for all angular velocities.

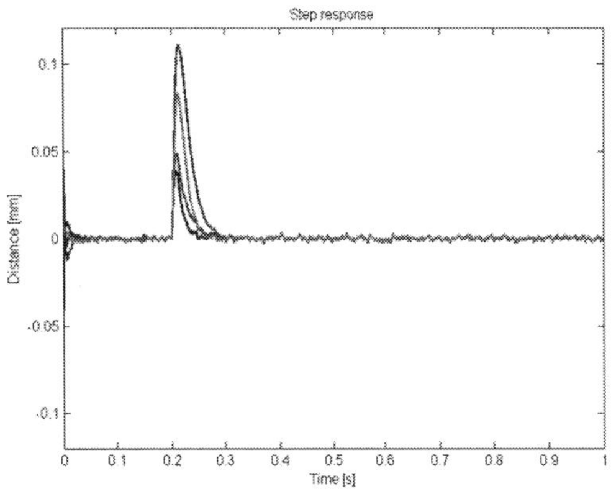

LQG

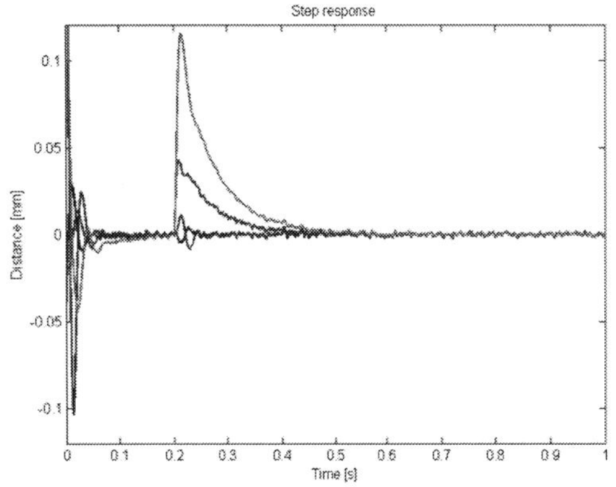

Decentralized PID-Controller

Figure 5. Disturbance-step-responses

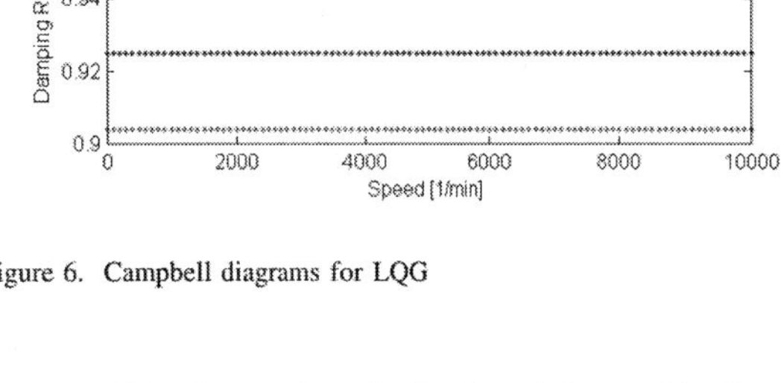

Figure 6. Campbell diagrams for LQG

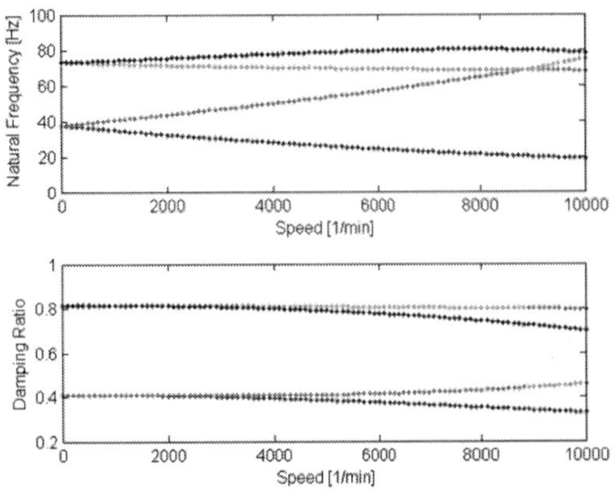

Figure 7. Campbell diagrams for decentralized PID-Controller

V. EXPERIMANTAL RESULTS

A. Behavior after a disturbance step

In this experiment a defined disturbance step of $26,5N$ at the upper bearings and $12N$ at the lower bearing was applied in both directions and the positions were measured. The LQR shows a better performance than the PID, because the step is lower and the settling time is faster. The reason is that the LQR uses the diagonal and the non-diagonal terms, the PID only uses the diagonal ones. The measurements are shown in Fig.5. The both higher signals are from the bottom bearing and the lower signals from the upper bearing. To plot all positions in one figure, offsets were used. In reality all signals starts at 0µm.

B. Comparison of the behavior at critical speed

In this section both systems operate at the first critical speed. To compare them the amplitude of the positions were measured. The operation speed of the LQG controlled system was $3600rpm$ and of the PID controlled system $3900rpm$. With the PID controller the amplitude of the upper bearing is very low compared to the LQG controlled system (Fig.9). In the lower bearing occurs the opposite case. The ratio between the amplitude of lower and upper bearing is smaller with the LQG controller. As consequence the amplitude of both bearings is able to increase to higher values before the rotor touches its mechanical boundaries in one of the bearings.

C. System behavior at standstill and at high rotational speed

Here the system behavior at standstill and at high rotational speed is evaluated. With an exact compensation, the behavior is expected to be angular velocity independent. Both controller operates at standstill and at $6000rpm$. For comparison of the systems behavior the transfer function at standstill and $6000rpm$ was measured. For measuring the system is excited with a sweep sine at the position reference value and the positions from the sensors were measured. After the measurement in the time domain the signal was transformed in the frequency domain using the Fast Fourier Transformation. With

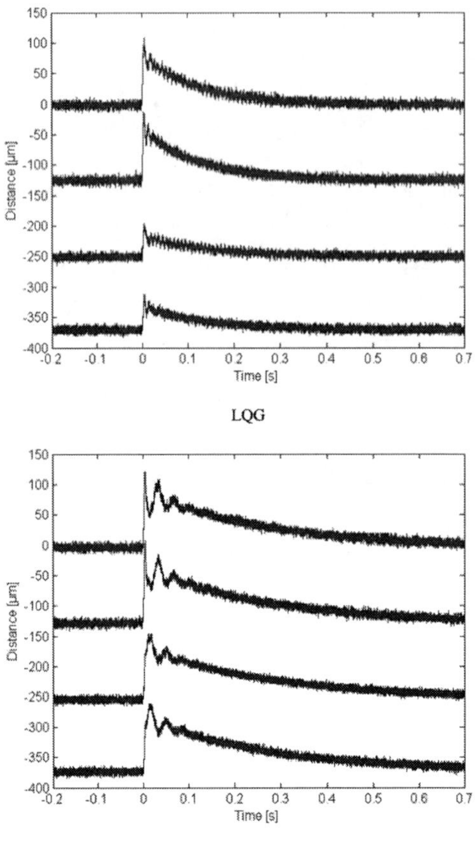

Figure 8. Disturbance step responce, x-y from lower bearing are the both higher signals, x-y from upper bearing are the both lower signals

the signals in the frequency domain the transfer functions was calculated (Fig.10). The LQG controlled system has nearly the same transfer functions for both angular velocities as expected. The peak in the range of $100Hz$ is not a resonance, its only a disturbance caused from the unbalance. In contrast the decentralized PID controller has different shapes for both operating points. These measurments established the simulation results of the compensation of the gyroscopic effect (Fig.6).

VI. CONCLUSIONS

In this paper a LQG controller and a time variant feedback path has been designed. The resulting system of the combination of these two structures is a linear time invariant optimal system for a high angular velocity range. It is verified that the transfer functions of the AMB system are nearly the same for all angular velocities up to $6000rpm$. In the experiments a damping effect of the time variant feedback path was measured at the critical speeds. The combination of the filter to stabilize the first bending mode and the LQG controller is not ideal. If the feedback matrix of the Kalman observer is to high the system with the filter for the first bending mode gets unstable. Another solution would be to stabilize the first bending mode with the LQR controller by using a

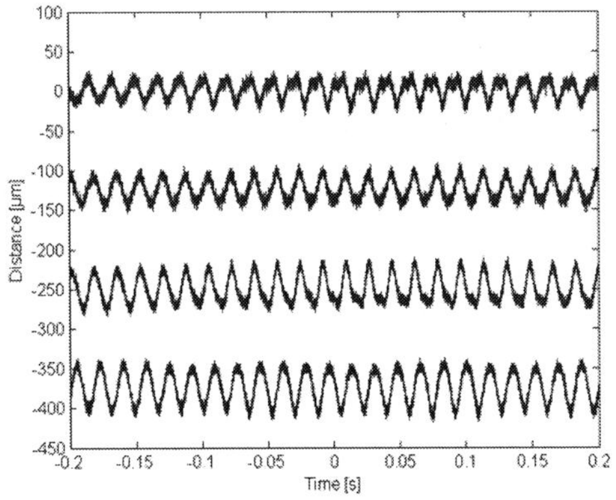

LQG

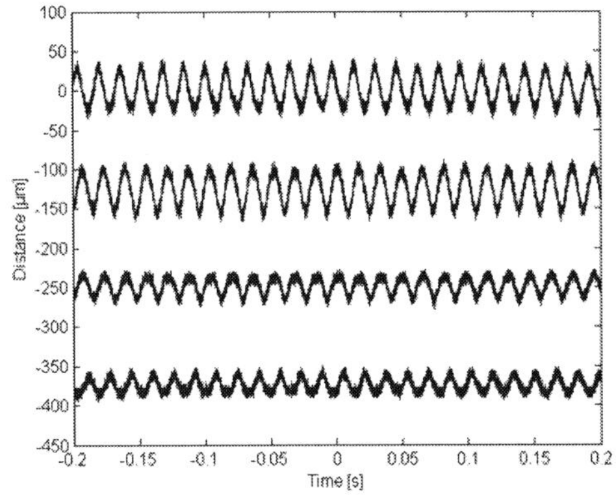

Decentralized PID-Controller

Figure 9. Behavior at critical speed, , x-y from lower bearing are the both higher signals, x-y from upper bearing are the both lower signals

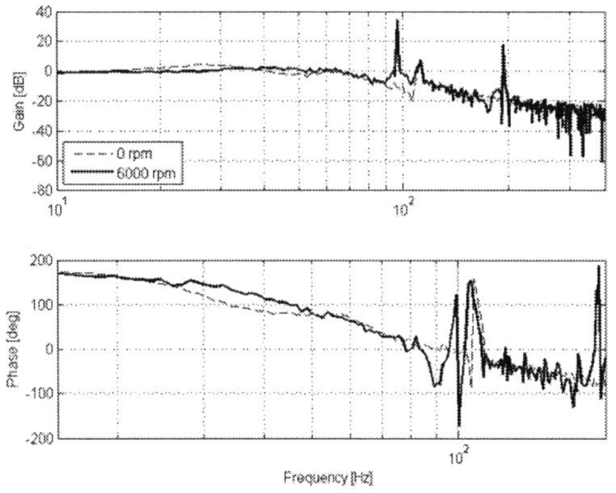

LQG

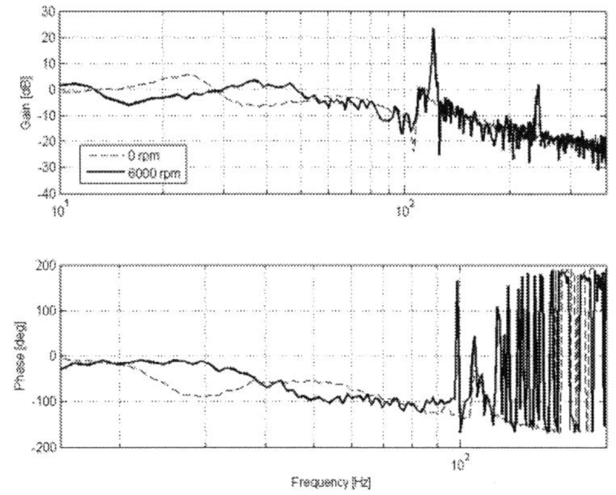

Decentralized PID-Controller

Figure 10. Transfer functions at standstill and 6000 rpm

flexible body model. If a Kalman observer is used for both, the LQR and to compensate the gyroscopic effect, a contradiction occurs, because the compensation of the gyroscopic effect for high frequencies depends on the dynamic of the Kalman filter. In summary, it can be stated, that it is easy to find a stable solution with the LQG method. If a system is required with a defiend stiffness and damping it is not so easy to adapt the solution with the LQG method compared to the decentralized PID controller. An advantage of the PID in this case is that the stiffness and damping can be changed directly by the control parameter. With the LQG controller only weighing matrices can be changed, which only changes the settling time of the states and not directly the stiffness and damping.

VII. Acknowledgments

This work is conducted within the project P21631-N22 of the Vienna University of Technology with the Austrian Science Fund (FWF). The authors thank the Austrian Science Fund for supporting this project.

References

[1] M. Hofer, "Design and Sensorless Position Control of a Permanent Magnet Biased Radial Active Magnetic Bearing," PhD thesis, Technical University Vienna, 2013.

[2] M. Hofer, E. Schmidt, and M. Schrödl, "Design of a Three Phase Permanent Magnet Biased Radial Active Magnetic Bearing Regarding a Position Sensorless Control," *IEEE*, 2009.

[3] H. Bleuler, "Decentralized Control of Magnetic Rotor Bearing Systems," PhD thesis, Federal Institute of Technology Zurich, 1984.

[4] Y. N. Zhuravlyov, "On LQ-control of Magnetic Bearing," *IEEE Transaction on control systems technology*, vol. 8, pp. 344–350, 2000.

[5] H. Ulbrich, "Entwurf und Lagerung einer berührungsfreihen Magnetlagerung für ein Rotorsystem," PhD thesis, Technical University Munich, 1979.

[6] M. Ahrens and L. Kucera, "Cross feedback control of a magnetic bearing system controller design considering gyroscopic effects," *Proceedings of the Third International Symposium on Magnetic Bearings*, pp. 177–191, 1996.

[7] G. Schweitzer, "Magnetic Bearings Theory, Design, and Applications to Rotating Machinery," *Springer Berlin*, 2009.

[8] K. J. Aström, "Computer Controlled Systems," *Prentice Hall*, 1997.

[9] J. Schmied and A. Kosenkov, "Practical controller design for rotors on magnetic bearings by means of an efficient simulation tool," *10th International Conference on Vibrations in Rotating Machines*, 2000.

Feedback Compensation of Tool Deflection in a High-Speed Machining AMB Spindle

Alexander Smirnov[a], Alexander Pesch[b], Adam C. Wroblewski[b], Olli Pyrhönen[a], Jerzy T. Sawicki[b]

[a] Lappeenranta University of Technology, PO Box 20, 53851, Lappeenranta, Finland, alexander.smirnov@lut.fi
[b] Center for Rotating Machinery Dynamics and Control Cleveland State University, Cleveland, OH 44115-2214 USA

***Abstract*—This paper describes a method for compensation of tool tip deflection in a high-speed, magnetically levitated, machining spindle. The proposed method utilizes measurements from the AMB sensors to estimate forces and calculate the resulting tool deflections. The deflections are then compensated online with a feedback signal that changes the loaded tool tip's position. The modeled results show significant performance improvements, which are experimentally validated.**

I. Introduction

High-speed machining is an emerging technology in industry, with advantages such as faster material removal rate and less tool wear. With increased rotational speeds, traditional rolling element bearings inherently impose limitations and require constant maintenance. To overcome these issues, active magnetic bearings (AMBs) are employed.

The AMB is a device that allows for levitation of a ferromagnetic body using electromagnetic forces. In this way, mechanical friction is absent and there is no wear of components. The system is also lubrication free and suitable for clean environments. In addition, for machining application, AMBs allow for a larger rotor diameter, thus increasing the rotor's stiffness.

The AMB system is naturally unstable as electromagnets can provide only attractive forces. To stabilize it, active feedback control is used. Actively controlled systems provide additional benefits such as programmable stiffness and damping characteristics for specific frequencies. In that way, lower vibration can be achieved at the required cutting speeds.

One of the possible machining applications is a boring operation. Boring requires a long and slender rotor for holding the tool bit inside the workpiece cavity. Therefore, a boring setup is a flexible system. When boring, the radial force acting on the tool causes its deflection and deviation from the desired tool path. Increasing the cutting speed to the high-speed region reduces the force without sacrificing the material removal rate but cannot totally eliminate the error.

The issue of compensating tool deflection has received significant attention in the literature. For example, it was considered for miniature ball mills in [1] in which the authors used an experimentally tuned model for offline compensation of the deflection. The offline compensation approach with trajectory correction for the flexible tools is also presented in [2, 3] in which force measurements are used to calibrate and estimate tool forces for the specific cutting conditions. This requires additional hardware and extensive experiments before actual machining. It was demonstrated by Auchet et al. [4] that it is possible to measure cutting forces with magnetic bearings.

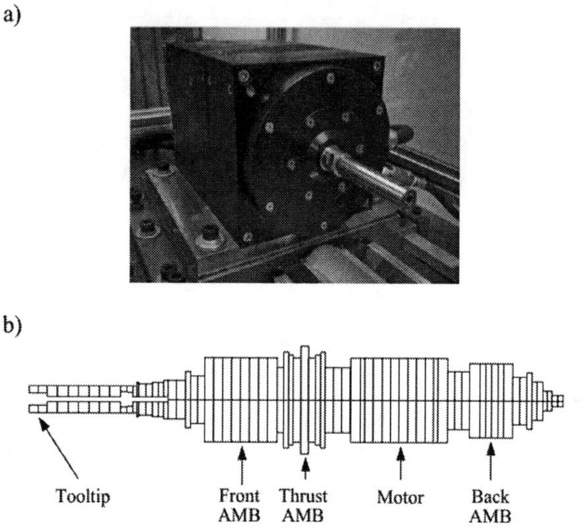

Figure 1. AMB Spindle a) photograph and b) finite element discretization.

Using AMBs, no additional hardware is required and measurements are available in real time.

This article presents a method to compensate tool deflection based on available current and rotor position measurements from the AMBs. To eliminate error due to tool tip deflection, the tool is shifted according to the estimated deflection. The shifting is achieved by using the bearings' airgap with various methods such as the one developed by Eckhardt and Rudolf [5].

II. System Description

The system used for experimental validation of the tool compensation method is presented in Fig. 1. The photograph is shown at the top and the corresponding rotor finite element model (FEM) is shown at the bottom. This system is an industrial grade high-speed machining spindle for single point boring, levitated with AMBs. The maximum speed of 50,000 rpm is provided by a 10 kW induction electrical motor whose rotor is fully levitated by two radial and one axial magnetic bearing. The static capacity for the front, back and axial bearings is 1400 N, 600 N, and 500 N, respectively.

For research, the industrial system is augmented to be controlled by a dSPACE rapid control prototyper. It includes DS1005 PPC processor board, DS2001 A/D and DS2101 D/A

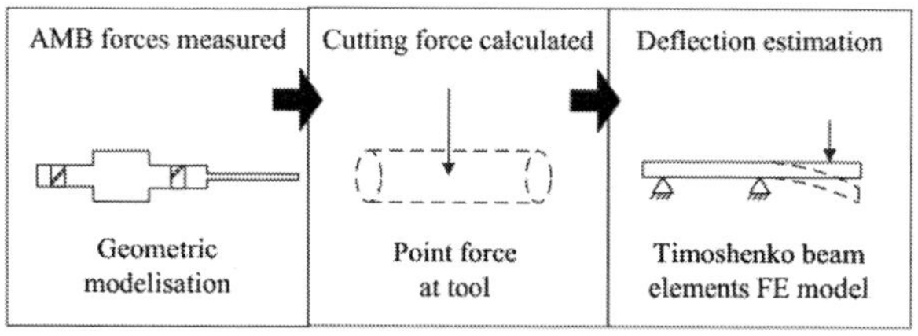

Figure 2. Estimation of the tool deflection

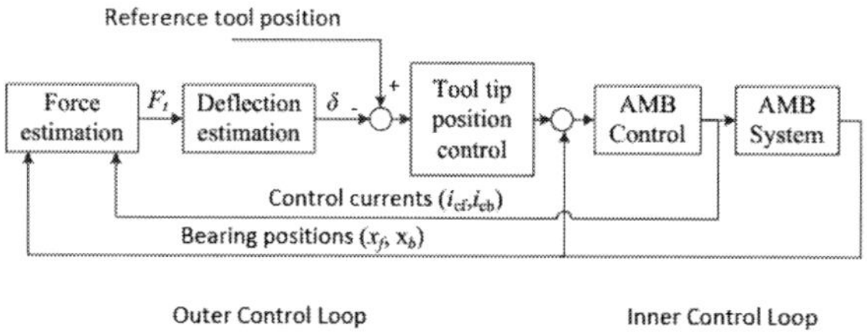

Figure 3. Compensation of the tool deflection

converters. The levitation and compensation control studied is implemented with sampling time 12.5 kHz.

The linear model of the actuator is obtained with bias current linearization with radial AMB force f characterized as

$$f = k_x x + k_i i, \tag{1}$$

where k_x is position stiffness, k_i is current stiffness, i is control current, and x is displacement from the center point. The stiffness values are experimentally identified by levitating the system and applying a set of static loads.

The rotor is discretized and modeled with the FEM. The obtained model is truncated to include only the first three flexural modes and tuned to the measured frequency response [6]. The frequency response of the open-loop system is obtained by exciting it with a sinusoidal current signal, one axis at a time, and measuring positions at all AMB sensors. Finally, the full response is obtained as described by [7]. The obtained rotor model accurately represents the system, with the first natural frequency of 1070 Hz. This frequency is higher than the maximum rotational frequency, thus the system is subcritical.

III. Deflection Compensation

In the previously discussed works [1- 3] the tool deflection is compensated by changing the tool path before the cutting process. To get an accurate estimate of the forces acting on the rotor, preliminary experimental measurements were used with specific conditions and particular material. With magnetic bearings, the forces can be estimated online [4] and that information is used to compensate for the tool deflection.

This work assumes that the reduced rotor model accurately represents the system. Thus, the contribution of the flexible modes above the third one is negligible. Another assumption is that the only external disturbance acting on the rotor is the cutting force which is applied to the tool tip.

The deflection is estimated using several modeling steps as proposed by Dépincé and Hascoët [2]. First, the geometric model of the system is used to find the relationship from tool load to reaction force at the bearing. Then, the corresponding cutting force is calculated based on the bearing forces. Finally,

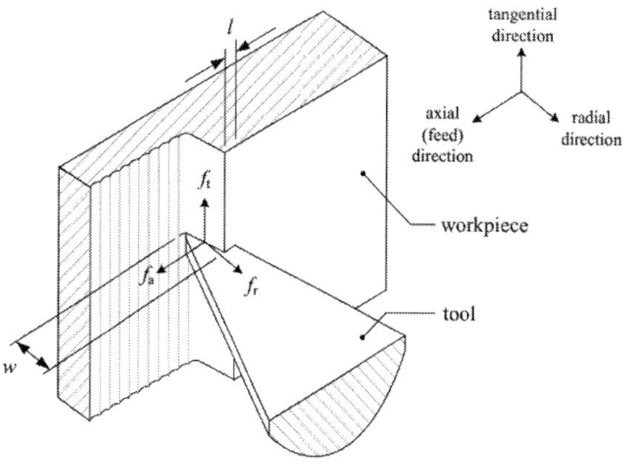

Figure 4. Diagram of the cutting tool imbedded in the workpiece while machining

the deflection is estimated using the FE model of the flexible rotor. The general order is presented in Fig. 2. At the second step, the point force acting on the tool is estimated based on AMB control currents and position measurements with the use of Eq. (1).

With the knowledge of tool deflection, the rotor position is corrected. The correction is done based on the rigid body motion of the rotor by shifting the AMB setpoints. That correction causes an increase in the acting force, which is sensed through control currents and bearing positions. In that way, closed-loop deflection compensation is achieved. The scheme describing the process is presented in Fig. 3. The inner control loop consists of a typical AMB stabilizing control. The outer control loop performs the tool deflection compensation.

A. Rotor model

The flexible rotor model derived from the FEM is described as follows

$$\mathbf{M}\ddot{\mathbf{x}} + (\mathbf{D} + \Omega\mathbf{G})\mathbf{x} + \mathbf{K}\mathbf{x} = \mathbf{f}, \tag{2}$$

where $\mathbf{M}$, $\mathbf{D}$, $\mathbf{G}$, and $\mathbf{K}$ are the mass, damping, gyroscopic and stiffness matrices, $\mathbf{f}$ is a vector of external forces acting on the rotor, Ω is the rotational speed, and $\mathbf{x}$ is a vector of modal coordinates. Equation (2) is transformed to the state space form of an unconstrained, or free-free ($_{\text{ff}}$) rotor [6]

$$\begin{aligned} \dot{\mathbf{x}} &= \mathbf{A}_{\text{ff}}\mathbf{x} + \mathbf{B}_{\text{ff}}\mathbf{u}, \\ \mathbf{y} &= \mathbf{C}_{\text{ff}}\mathbf{x}. \end{aligned} \tag{3}$$

This model includes two rigid rotor modes and the first three flexible modes. When the initial conditions and external forces acting on the system are known, the model represented by Eq. (3) allows for estimation of the position of the center of mass and any other point of the rotor such as the tool location. The inclusion of flexible modes takes into account the deformation under load. Thus, the vector $\mathbf{u}$ contains the forces acting on the rotor and the corresponding points of force application are described by matrix $\mathbf{B}_{\text{ff}}$. The tool tip displacement is provided by applying transformation to the state vector with matrix $\mathbf{C}_{\text{ff}}$. To estimate rotor deflection, an accurate force description should be provided.

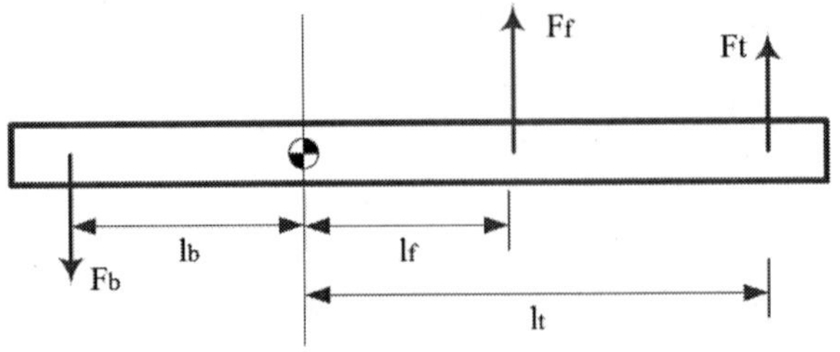

Figure 5. Scheme of forces acting on the rotor

B. Forces acting on the rotor

The rotor is supported by two bearings that counteract gravitational force and any disturbance force acting on the system. The gravitational force can be estimated with high precision by weighing the rotor before the assembly and then updating the value when a cutting tool and tool holder are added. The estimation of the center of mass can be done both experimentally and by FEM modelling of the rotor.

To estimate the forces of the magnetic bearings, the simple linearized representation (1) is used. This is a widely used model obtained from a nonlinear bearing description [8]. The displacement from the center point is provided by sensors that are present in most AMB systems and the control current is known from the control system.

As the bearing force model is linearized under the assumption of small displacements from the bearing magnetic center, it is important to keep the rotor near the bearing magnetic center during operation. In that way inaccuracies are minimized. There are several procedures described in the literature to position the rotor in the magnetic center, for example an iterative method [9].

The main issue of the article is cause by the radial cutting force at the tool tip. The system under consideration is a single point boring spindle, thus the machining force is decomposed into components in the directions axial, radial, and tangential to the rotor. The process of cutting and corresponding forces is illustrated in Fig. 4.

The axial component of the cutting force is addressed by the relatively simple dynamics of the thrust bearing. The tangential component of the cutting force is overcome by the motor. The radial component however may cause significant tool deflections perpendicular to the workpiece surface.

Without loss of generality only one plane is considered; the results are readily applied to the second plane by substituting the corresponding measurements (displacements and control currents). With the previously stated assumptions, the system under steady state conditions corresponds to the one described in Fig. 5. The balance of moments gives

$$\begin{aligned} &Fl_{ff}+F_b l_b+F_t l_t=0\\ &F_t=\frac{-F_b l_b-F_f l_f}{l_t}, \end{aligned} \tag{4}$$

where F_f, F_b, and F_t are the forces of front bearing, back bearing, and tool tiprespectively , l_f, l_b, and l_t are distance from the center of mass to the front bearings, back bearing, and the tool tip respectively.

The mass of the rotor can be taken into account in two ways. The first is by including it as a term in Eq. (4). The second is by subtracting the gravitational component from the bearing forces. The latter approach is more useful for practical implementation when this component is readily available after initial levitation of the system. In addition, this approach simplifies the equations.

Considering that only control currents and sensor signals are available, the full force vector acting on the system is obtained by combining Eqs. (1) and (4)

$$\mathbf{u}=\begin{bmatrix} f_f\\ f_b\\ f_t\end{bmatrix}=\begin{bmatrix} k_{if} & 0 & k_{xf} & 0\\ 0 & k_{ib} & 0 & k_{xb}\\ -\frac{l_f}{l_t}k_{if} & \frac{l_b}{l_t}k_{ib} & -\frac{l_f}{l_t}k_{xf} & \frac{l_b}{l_t}k_{xb}\end{bmatrix}\begin{bmatrix} i_{cf}\\ i_{cb}\\ x_f\\ x_b\end{bmatrix}, \tag{5}$$

where i_{cf} and i_{cb} are currents and x_f and x_b are displacements in front and back bearings respectively. The obtained force vector in Eq. (5) is directly substituted into Eq. (3) yielding the tool deflection as

$$\begin{aligned} &\dot{\mathbf{x}}=(\mathbf{A}_{ff}+\mathbf{B}_{act}\mathbf{K}_x\mathbf{C}_{act})\mathbf{x}+\mathbf{B}_{act}\mathbf{K}_i\mathbf{I}_c+\mathbf{B}_t F_t,\\ &\delta=\mathbf{C}_t\mathbf{x}, \end{aligned} \tag{6}$$

where $\mathbf{B}_{act}$, $\mathbf{C}_{act}$, $\mathbf{B}_t$, and $\mathbf{C}_t$ are the components of the $\mathbf{B}_{ff}$ and $\mathbf{C}_{ff}$ matrices which correspond to the magnetic actuator ($_{act}$) and tool ($_t$) locations and δ is the tool position. The rest of the matrices are defined as

$$\mathbf{K}_x=\begin{bmatrix} k_{xf} & 0\\ 0 & k_{xb}\end{bmatrix},\quad \mathbf{K}_i=\begin{bmatrix} k_{if} & 0\\ 0 & k_{ib}\end{bmatrix},\quad \mathbf{I}_c=\begin{bmatrix} i_{cf}\\ i_{cb}\end{bmatrix}.$$

The tool force F_t is obtained by substituting Eq. (1) into Eq. (4) resulting in

$$F_t=\left(-\frac{l_f}{l_t}k_{if}\right)i_{cf}+\left(\frac{l_b}{l_t}k_{ib}\right)i_{cb}+\left(-\frac{l_f}{l_t}k_{xf}\right)x_f+\left(\frac{l_b}{l_t}k_{xb}\right)x_b.$$

As dynamics is neglected the estimation is done at steady state as follows

$$\delta=\mathbf{C}_t(s\mathbf{I}-\mathbf{A}_{ff}-\mathbf{B}_{act}\mathbf{K}_x\mathbf{C}_{act})^{-1}(\mathbf{B}_{act}\mathbf{K}_i\mathbf{I}_c+\mathbf{B}_t F_t)|_{s\to 0}, \tag{7}$$

where s is a Laplace variable, and $\mathbf{I}$ is a unit matrix. Equation (7) yields to the transformation matrix, which is based on the assumptions discussed above, estimates the deflection of the tool under load.

There are several other sources of external forces acting on the rotor in high-speed systems. One is a magnetic pull from an electrical motor that appears when the active part of the rotor is shifted from the center point. Assuming that under small displacements, the resulting lateral motor force is small compared to the machining force and should not introduce significant error.

Another source of excitation is unbalance, which has a great importance in such applications. Although balancing requirements for machining spindles are high, the presence of residual unbalance cannot be avoided. However, unbalance has attracted a great amount of attention from the research community and there are techniques for estimation of these forces [10, 11]. Thus, they can be separated from the tool tip forces.

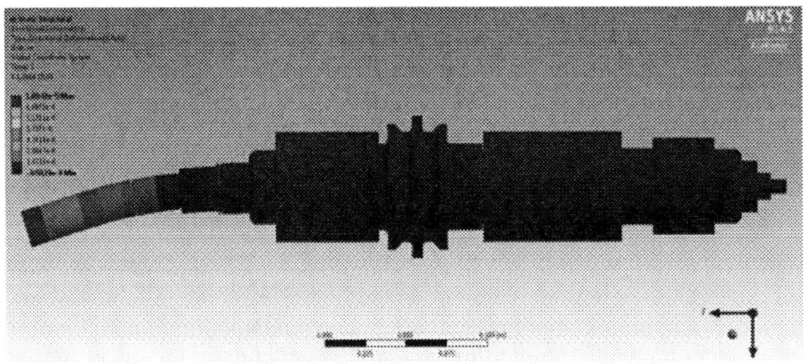

Figure 6. AMB spindle deformation under applied force

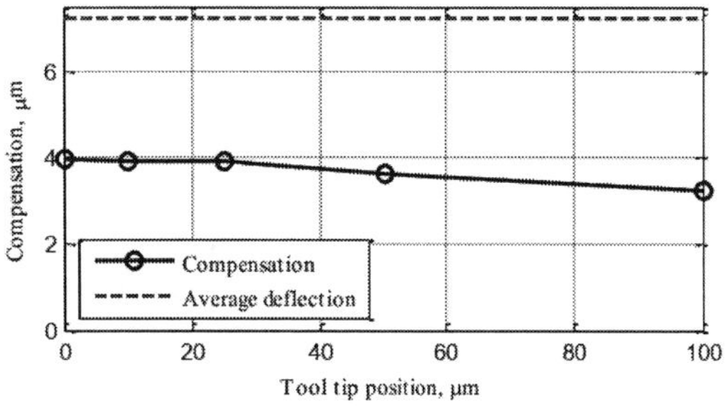

Figure 7. Compensation with respect to tool tip position

TABLE I
TOOL TIP POSITION MEASUREMENTS

Tool reference, µm	Load	Compensation	Tool tip position, µm
0	NO	NO	0
	YES	NO	-7.037
	YES	YES	-3.077
	NO	YES	0.276
10	NO	NO	0
	YES	NO	-7.472
	YES	YES	-3.542
	NO	YES	-0.369
25	NO	NO	0
	YES	NO	-7.725
	YES	YES	-3.822
	NO	YES	0.077
50	NO	NO	0
	YES	NO	-6.809
	YES	YES	-3.168
	NO	YES	-0.490
100	NO	NO	0
	YES	NO	-7.05
	YES	YES	-3.825
	NO	YES	-0.668

C. Tool tip control

The solution provided by Eqs. (3) and (5) gives an estimation of the tool tip position with the given assumptions. These equations correspond to the blocks "Force estimation" and "Deflection estimation" in Fig. 3. To eliminate the difference between the actual and desired tool positions, a control block should be applied that is in addition to the typical AMB control system.

The main priority of the bearing control is to keep the system stable. Thus, to change the tool tip position, several approaches can be applied.

The first one is to offset the rotor equally in both bearings according to the estimated deflection. In that way, a rigid (translational) mode is used. The main benefit of this approach is that it is inherently built-in in AMB systems and requires only supplying commands to the existing control system. The drawback is that the rotor may be moved out from the linearization center point. Thus, the error is increased and stability of the system is decreased.

The second approach is to apply a control system designed to control the position of the tool tip [12]. In that way the drawbacks of the previous method are alleviated but some effort on applying another control system is necessary.

The benefit of the proposed tool tip deflection compensation approach is that it is linear and can be included into any linear control system that is usually used for AMB control. The solution of Eq. (7), with some state space algebra according to scheme in Fig. 3, is combined with an existing controller into a single state space representation. Thus, the implementation of the proposed technique does not require any changes in control structure but only supplying new controller matrices.

IV. EXPERIMENTAL RESULTS

To experimentally validate the results of deflection compensation, the outer feedback loop is implemented in dSPACE for the industrial high-speed spindle. To attenuate noise the signals for the tool deflection blocks are filtered with a low pass filter. In addition to the AMB sensors, two additional eddy current sensors are mounted at the tool tip location. The measurements from the additional sensors are only logged and they are not connected to the feedback control system.

To achieve constant force, an external load is applied to the rotor at the tool tip location. A load of 3.2 kg freely hanging weight is attached with a wire. This weight provides a static force with magnitude greater than the force expected under cutting conditions. The direction of the load corresponds to that of a control axis of the AMB system for convenient measurements.

The same experiment is modeled with ANSYS® FEM software. The system is simply supported at the bearing locations and a force is applied at the tool tip position. The results of FEM analysis are presented in Fig. 6, with a tool deflection value of 8.93 µm.

The experiment is performed in the following order. At first, the load is applied to the system and the position of the tool tip is measured. Next, the compensation for the deflection is applied and the tool tip position is again measured. Then, the load is removed, compensation is disabled, and the results are recorded. Finally, the compensation is enabled without the applied load and the results are measured. This type of experiment is repeated for several tool tip reference positions. The data is presented in Table I with normalization to the case without load and without compensation.

The total tool deflection can be estimated as the difference between cases without load and with load. For all experiments the average value is 7.219 µm. The error when compared to the ANSYS® results is about 25%, which is explained by the complicated nature of the rotor with shrink fits and AMB laminations. These complex elements and nonlinearities are not accounted for in the ANSYS FEM model. An additional source of error is that the proposed method neglects sensor/actuator non-collocation.

With compensation enabled the tool deflection is reduced on average by 3.732 μm. This value depends on the tool tip reference position as presented in Fig. 7. It is seen that with the greater deviation from the center point, the compensation degrades. This is the result of electromagnetic model linearization. The linearization with bias current is valid for small deviation and control currents less than bias current. As the rotor moves further from this point, the force estimations become less accurate. In general the difference between estimated and actual tool deflection comes from the rotor model simplification and measurement noise. The measurements are available only at the bearings locations and small deviations propagate significantly to the tool tip position.

There is an almost indistinguishable difference between the experiments without load. When compensation is turned on it tries to compensate some small amount, less than one micrometer. It is the result of deviation from the linearization point and gravitational force that bends the rotor.

V. Conclusions and Further Work

In this article a method for compensation of tool deflection is presented for spindles supported on AMBs. The approach utilizes force estimation from magnetic bearings and calculates deflection with the reduced-order rotor model. The deflection is compensated for by correcting the rotor position at the bearing locations.

The experimental results demonstrated improvement in tool tip position under load by 52%. In addition, experiments showed a dependence on the model linearization which limits the accuracy at large displacements.

The proposed approach is simple to implement and can be done through appropriate replacement of the control law. Thus, no hardware and no software changes are necessary for the system. However, to obtain reasonable accuracy for industrial implementation, a full set of cutting experiments is necessary. In addition, the unbalance force should be separated from the cutting force, which requires on-line unbalance identification before cutting.

The future work will concentrate on implementation of nonlinear estimator allowing better accuracy for the full operating range. Another direction is toward increasing the bandwidth of the compensation and repeating experiments at some rotational speed with actual cutting.

References

[1] T. A. Dow, E. L. Miller, and K. Garrard, "Tool force and deflection compensation for small milling tools," *Precis. Eng.*, vol. 28, no. 1, pp. 31–45, Jan. 2004.

[2] P. Dépincé and J.-Y. Hascoët, "Active integration of tool deflection effects in end milling. Part 1. Prediction of milled surfaces," *Int. J. Mach. Tools Manuf.*, vol. 46, no. 9, pp. 937–944, Jul. 2006.

[3] P. Dépincé and J.-Y. Hascoët, "Active integration of tool deflection effects in end milling. Part 2. Compensation of tool deflection," *Int. J. Mach. Tools Manuf.*, vol. 46, no. 9, pp. 945–956, Jul. 2006.

[4] S. Auchet, P. Chevrier, M. Lacour, and P. Lipinski, "A new method of cutting force measurement based on command voltages of active electro-magnetic bearings," *Int. J. Mach. Tools Manuf.*, vol. 44, no. 14, pp. 1441–1449, Nov. 2004.

[5] S. Eckhardt and J. Rudolph, "High Precision Syncronous Tool Path Tracking with an AMB Machine Tool Spindle," in *Ninth International Symposium on Magnetic Bearings*, 2004.

[6] A. C. Wroblewski, J. T. Sawicki, and A. H. Pesch, "Rotor Model Updating and Validation for an Active Magnetic Bearing Based High-Speed Machining Spindle," *J. Eng. Gas Turbines Power*, vol. 134, no. 12, p. 6, 2012.

[7] J. T. Sawicki, E. H. Maslen, and K. R. Bischof, "Modeling and performance evaluation of machining spindle with active magnetic bearings," *J. Mech. Sci. Technol.*, vol. 21, no. 6, pp. 847–850, Jun. 2007.

[8] G. Schweitzer and E. H. Maslen, *Magnetic Bearings*. Berlin, Heidelberg: Springer Berlin Heidelberg, 2009.

[9] R. J. Prins, M. E. F. Kasarda, and S. C. B. Prins, "A System Identification Technique Using Bias Current Perturbation for Determining the Effective Rotor Origin of Active Magnetic Bearings," *J. Vib. Acoust.*, vol. 129, no. 3, p. 317, 2007.

[10] T.-J. Park, Y. Kanemitsu, S. Kijimoto, K. Matsuda, and Y. Koba, "Identification Of Unbalance And High Order Sensor Runout On Rigid Rotor Supported By Magnetic Bearings," in *Eighth International Symposium on Magnetic Bearings*, 2004.

[11] J. D. Setiawan, R. Mukherjee, and E. H. Maslen, "Synchronous Sensor Runout and Unbalance Compensation in Active Magnetic Bearings Using Bias Current Excitation," *J. Dyn. Syst. Meas. Control*, vol. 124, no. 1, p. 14, 2002.

[12] A. H. Pesch, A. Smirnov, O. Pyrhönen, J. T. Sawicki, "Magnetic Bearing Spindle Tool Tracking through μ-synthesis Robust Control," *Mechatronics, [under review].*

Feedback control of a magnetic bearing using fusion of rotor acceleration and position measurements

Matthew O T Cole [a], Samuel Jimenez [a], Patrick S Keogh [a]

[a] Department of Mechanical Engineering, University of Bath, Bath BA2 7AY, UK, m.o.t.cole@bath.ac.uk

***Abstract*—This paper considers controlling a magnetic bearing using combined feedback of rotor acceleration and position measurements. A controller design based on sensor fusion via complementary filtering is described. The control method can maintain stability while allowing the cross-over frequency for the acceleration and position feedback loops to be set arbitrarily. In this way, position measurements can be utilized primarily for disturbance cancellation over a low frequency band while acceleration feedback maintains bearing stability and disturbance rejection over a higher frequency band. The approach has the advantage that it gives enhanced robustness to poor quality position measurements. For flexible rotors, stabilization with large axial separation of position sensors and bearing locations can be easily achieved by utilizing collocated rotor-embedded accelerometers. The approach also offers improved robustness to measurement noise, position sensor run-out error and other forms of physical contamination or damage. Through analysis and simulation, important physical effects on vibration control performance are identified and further considered within the design methodology.**

I. Introduction

The use of rotor-embedded sensors in combination with high-speed wireless transmission holds significant promise for application to condition monitoring and vibration control of rotating machines. This paper considers the utilization of acceleration sensors fixed within a rotor in combination with external position sensing as a means to stabilize and control a magnetic bearing. A potential advantage of using acceleration sensors is a low vulnerability to physical damage, debris contamination and run-out errors that can adversely affect the operation of noncontact displacement sensors such as eddy current probes. The ability to obtain vibration measurements at locations that are inaccessible to externally mounted probes may also be important. The possibility of using additional accelerometer-based feedback to stabilize a system with very low bandwidth position sensors provides further motivation for this study. In this respect, the technique has relevance to, and may be useful in combination with, self-sensing magnetic bearings [1-4].

Sensor data-fusion is important for various motion control/tracking applications where there is no single reliable means to obtain absolute position information [5-8].

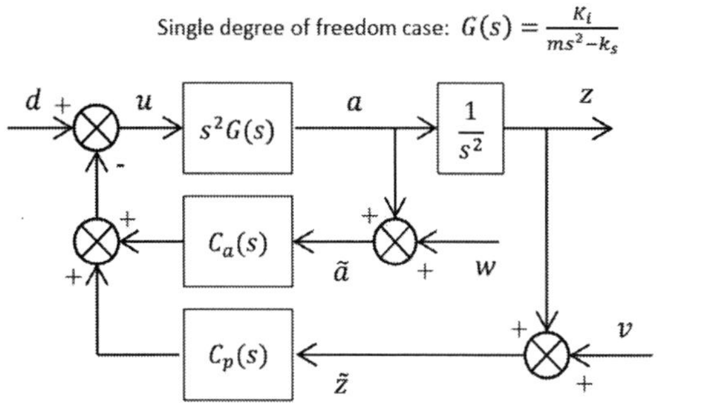

Figure 1. Combined acceleration-position feedback control of a rotor-AMB system.

Controllers for large-gap magnetic levitation systems have also been developed via this approach [9]. The successful utilization of rotor-mounted accelerometers, however, presents some unique challenges. This preliminary work shows that a given controller designed for position feedback stabilization can be used within a control strategy based on complementary filtering of acceleration and position measurements. In this way direct forcing disturbance rejection characteristics can be set according to a prototype feedback controller design. Sensitivity to measurement error, noise or other disturbances can be independently adjusted through a frequency domain weighting of sensor signals, chosen according to anticipated spectral characteristics of disturbances.

II. Hybrid Acceleration-Position Feedback Control

A. Complementary filter design

A parallel feedback of acceleration and position measurements to control a rotor-AMB system may be considered as shown in Fig. 1. The transfer function $G(s)$ represents the open loop dynamics from AMB control currents to rotor displacements z. Measurement of acceleration a is subject to an additive disturbance w, while the disturbance v acts additively on the displacement measurements. For illustration purposes, a single-degree-of-freedom (1-DOF) lumped-mass model may be considered (with G as in Fig. 1).

Formulation of a sensor-fusion strategy can be based on an arbitrary pair of complementary filters with transfer functions $F_1(s)$ and $F_2(s)$ satisfying

$$F_1 + F_2 = 1 \quad (1)$$

For the control structure shown in Fig. 1, a prototype stabilizing controller H may be included in the feedback path transfer functions according to

$$C_p = F_1 H, \quad C_a = \frac{F_2 H}{s^2} \quad (2)$$

The closed loop transfer functions relating the rotor displacement to each disturbance signal are then given by

$$T_{zd} = \frac{G}{1+GH}, T_{zv} = \frac{F_1 GH}{1+GH}, T_{zw} = \frac{F_2 GH}{s^2(1+GH)} \quad (3)$$

Stabilization and direct disturbance rejection characteristics, as determined by T_{zd}, can be assigned through standard design procedures for the controller H. Separately, the characteristics of T_{zv} and T_{zw}, which dictate sensor noise/disturbance sensitivity, can be balanced through an appropriate choice of F_1 (and F_2). Clearly, for position feedback only, $F_1 = 1$ and $F_2 = 0$. With combined feedback, the complementary filters can be chosen so that C_p and C_a both have low-pass characteristics, thereby achieving insensitivity to noise and other forms of zero-mean measurement error. Care must be taken within the design process to consider offset errors, particularly due to gravitational effects, that might affect acceleration measurements and result in a large or unbounded displacement response. Such instability is associated with zero-valued poles of T_{zw} and, in general, can be overcome by ensuring F_2 has a sufficient number of zero-valued zeros.

For illustration, consider a prototype design for H based on a standard PID control law:

$$H = K_I/s + K_P + K_D s \quad (4)$$

Complementary filters could be synthesized according to optimal estimation methods, typically in the form of an extended Kalman filter [6,7]. However, this approach requires stochastic information on expected measurement disturbances. In the absence of such information, a generic low-pass/high-pass filter pair may be considered:

$$F_1 = \frac{(s+\omega_c)^N - s^N}{(s+\omega_c)^N}, \quad F_2 = \frac{s^N}{(s+\omega_c)^N} \quad (5)$$

This leads to

$$C_p = \frac{((s+\omega_c)^N - s^N)(K_I + K_P s + K_D s^2)}{s(s+\omega_c)^N} \quad (6)$$

$$C_a = \frac{K_I + K_P s + K_D s^2}{s^{3-N}(s+\omega_c)^N} \quad (7)$$

Note that the cross-over frequency $\omega_c > 0$ directly determines the high frequency gain of C_p (as given by $\omega_c K_D$) and can be chosen arbitrarily small.

It follows from (3) that

$$T_{zw} = \frac{K_I + K_P s + K_D s^2}{(s+\omega_c)^N} \cdot \frac{G s^{N-2}}{(s + G(K_I + K_P s + K_D s^2))} \quad (8)$$

Examining the multiplicity of zero-valued zeros of T_{yw} indicates that, to eliminate steady-state displacement error subject to constant offset in the acceleration measurement, requires $N \geq 3$. A further conclusion here is that the use of acceleration signals alone ($N = 0$) will not achieve stabilization.

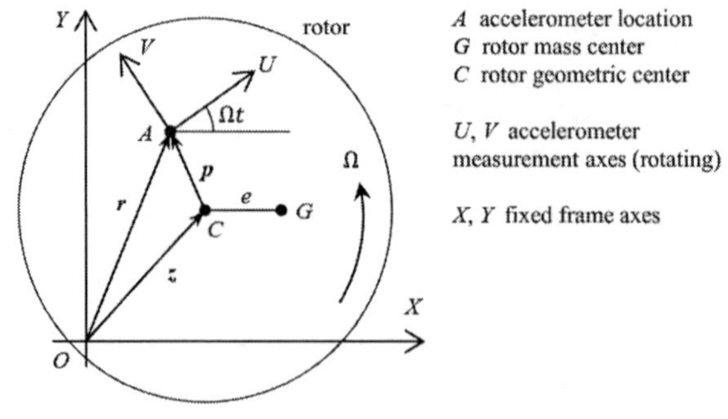

Figure 2. Geometry of rotor acceleration measurements

B. Rotational effects

For control of a spinning rotor using measurements of acceleration at points fixed within the rotor, consideration must be given to:

- The effect of accelerometer rotation and positioning eccentricity with respect to geometric and inertial centres of the rotor.
- The effect of gravity on measurements by the accelerometer.
- Possible non-collocation of position and acceleration sensors.

Measurements from accelerometers will relate to fixed-frame motion of the rotor according to the geometry shown in Fig. 2. Suppose that a two-axis accelerometer is positioned internally to the rotor at point A and has orthogonal sensing axes (U, V) lying within the plane of rotation. The position of point A relative to the origin of the fixed frame is

$$\boldsymbol{r} = \boldsymbol{z} + \boldsymbol{p} \quad (9)$$

where $\boldsymbol{z}$ is the position vector for the rotor center, as would be measured by rotor displacement sensors. For conciseness, complex notation is adopted: $\boldsymbol{r} = r_x + jr_y$. The corresponding vector in the rotating frame is $\overline{\boldsymbol{r}} = \boldsymbol{r}e^{-j\Omega t} = r_u + jr_v$. Accordingly, $\boldsymbol{p} = \overline{\boldsymbol{p}}e^{j\Omega t}$ where $\overline{\boldsymbol{p}}$ is a constant vector defining the accelerometer position relative to the rotor center. Therefore, the accelerometer eccentricity can be denoted $|\boldsymbol{p}|$. It follows that,

$$\ddot{\boldsymbol{r}} = \ddot{\boldsymbol{z}} - \Omega^2 \overline{\boldsymbol{p}} e^{j\Omega t} \quad (10)$$

An acceleration measurement in the non-rotating frame (including gravity effects) at point A would be

$$\tilde{\boldsymbol{a}} = \ddot{\boldsymbol{z}} - \Omega^2 \overline{\boldsymbol{p}} e^{j\Omega t} + jg \quad (11)$$

where the last two terms are considered as error terms. The actual measurements by the accelerometer a_u, a_v are given by

$$\begin{aligned} a_u + ja_v &= \tilde{\boldsymbol{a}} e^{-j\Omega t} \\ &= \ddot{\boldsymbol{z}} e^{-j\Omega t} - \Omega^2 \overline{\boldsymbol{p}} + jge^{-j\Omega t} \end{aligned} \quad (12)$$

The error component from centrifugal acceleration $-\Omega^2\overline{\boldsymbol{p}}$ acts in an unchanging direction but increases in magnitude with rotational speed. Although this component could be estimated and compensated for within a control implementation, care must still be taken to ensure it does not cause saturation problems.

The acceleration measurements are also contaminated by a synchronously rotating gravity component. Although various

TABLE I. PARAMETERS FOR ROTOR-AMB SIMULATION MODEL

	Parameter	Value
m	Rotor mass	10 Kg
R	Rotor diameter (in sensor plane)	0.03 m
k_s	AMB negative stiffness	0.3 x 10^6 N/m
K_i	AMB current gain	97.5 N/A
$K_I K_i$	Net integral gain	100 Nm/s
$K_P K_i$	Net proportional gain	1 x 10^6 N/m
$K_D K_i$	Net derivative gain	1000 Ns/m

methods to compensate for this component could be envisaged, inexact cancellation will lead to stability/drift problems, as discussed in Section II.A. Addressing this issue within the complementary filter design is therefore desirable.

To use the measured accelerations for control, a counter-rotating transformation can be applied to recover the fixed-frame acceleration signal prior to feedback. Complementary filtering can then be applied as described in Section II.A. Filtering could alternatively be applied in the co-rotating (accelerometer) frame rather than the stationary frame. A combination of filtering in both frames may also be possible and could help to eliminate both static and synchronous disturbance components from measurement/ response signals.

III. RESPONSE BEHAVIOUR

In this section of the paper the influence of filtering on vibration response and error sensitivity are investigated through time-step simulation of a linearized 2-DOF rotor-AMB model with the parameters detailed in Table I. The results presented in Fig. 4 involve a supercritical rotational speed of 400 rad/s. The three cases simulated correspond to three different complementary filter designs. In all cases, the underlying controller is PID type with parameter values given in Table I. All three simulation runs involve the same set of disturbance conditions. Initially, the rotor is well-balanced and vibration is induced only by the effect of centrifugal acceleration at the accelerometer location, which acts through the control feedback path. The accelerometer eccentricity is $|\boldsymbol{p}| = 0.1$ mm. After 0.3 seconds, a rotor unbalance condition is instantaneously introduced resulting in an increase in vibration in all three cases. The eccentricity of the rotor center of mass is $e = 0.1$ mm. After 0.6 seconds a run-out error is further introduced affecting the position measurement according to the geometry shown in Fig. 3.

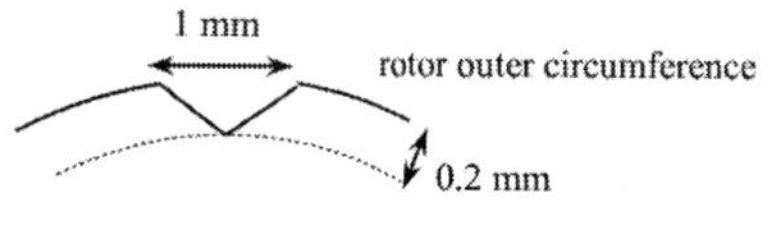

Figure 3. Geometry of run-out error used for simulations. This example case could represent circumferencial damage to the rotor at the displacement sensor location, e.g. due to impact

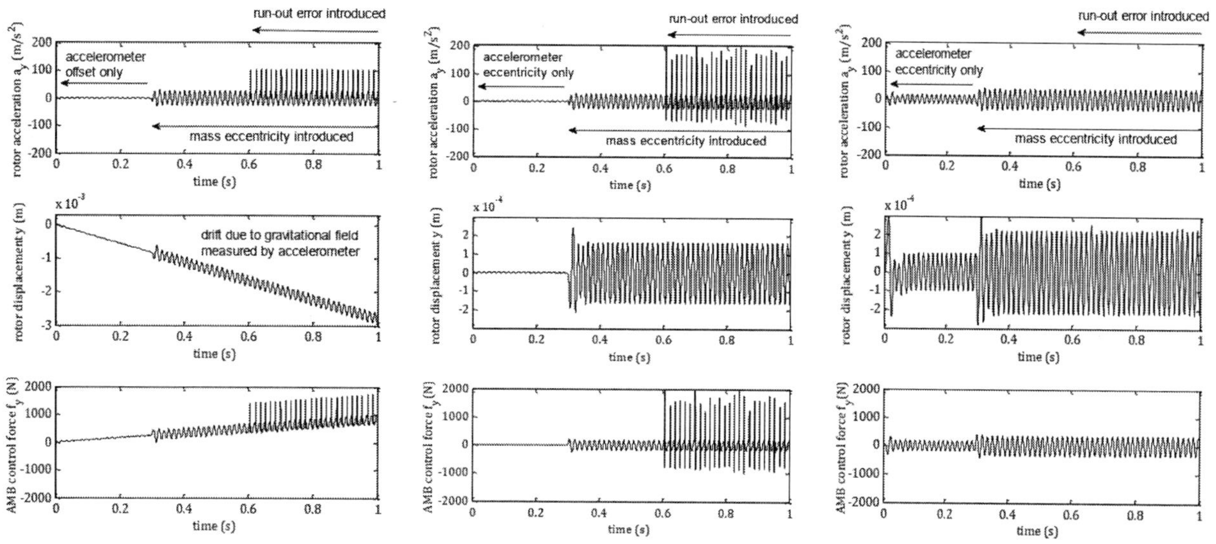

(a) *Case 1: N* = 1, ω_c = 5000 rad/s (b) *Case 2: N* = 3, ω_c = 5000 rad/s (c) *Case 3: N* = 3, ω_c = 100 rad/s

Figure 4. Simulated rotor-AMB behaviour under combined acceleration-position feedback. System model and controller parameters are given in Table I. Rotational speed Ω = 400 rad/s. Rotor unbalance (mass-eccentricity of 1x10^{-3} kg-m) is introduced after 0.3 seconds and run-out error introduced after 0.6 seconds. Results are shown for three different complementary filters designs.

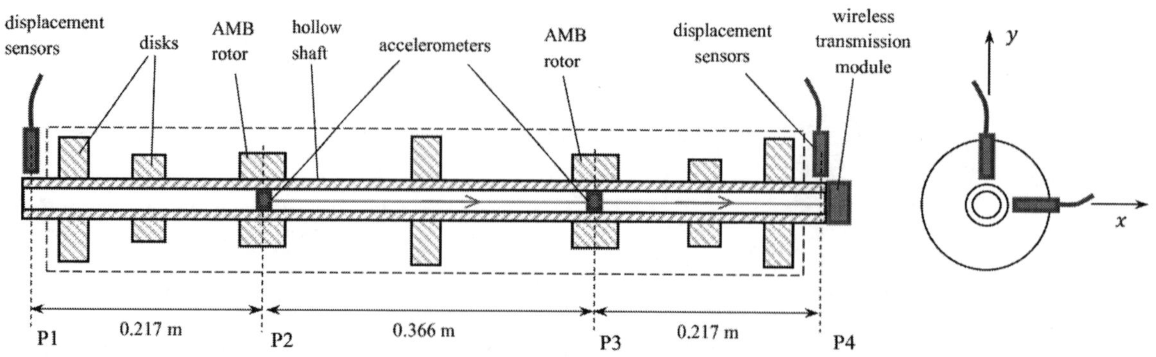

Figure 5. Layout of hollow-shaft flexible rotor system with position sensing in planes P1, P4 and acceleration sensing in planes P2, P3 collocated with AMBs

Some important features of the response behaviour can be identified from the simulation results. For Case 1, the complementary filter is a first order design ($N = 1$). Although the rotor acceleration seems well-controlled, the displacement of the rotor drifts in an unstable manner. This is caused by offset in the fixed-frame acceleration signal due to gravity, as discussed in Section II. Use of a third order complementary filter design ($N = 3$) overcomes this problem, as seen in Case 2. Rotor acceleration measurements are similar to the first case but unstable drift of rotor position is prevented. For Cases 1 and 2, the cross-over frequency for the complementary filter design is ω_c= 5000 rad/s and so the rotor is predominantly position-controlled over the frequency range of the disturbances. Introducing the run-out error shown in Fig. 3 causes large spikes in the rotor acceleration due to the action of the feedback control. Correspondingly, large spikes can be seen in the AMB control signal. Associated issues with vibration/noise transmission and force saturation would be expected. Case 3 involves the same 3rd order complementary filter design but with a lower cross-over frequency of ω_c=100 rad/s. In this case, position errors introduced by the run-out damage are effectively filtered by the position feedback loop such that control forces and rotor accelerations are unaffected. As a further observation, the rotor vibration response due to unbalance is the same in all three cases. This can be explained by the fact that the transfer function T_{zd} is independent of the filter transfer functions F_1 and F_2.

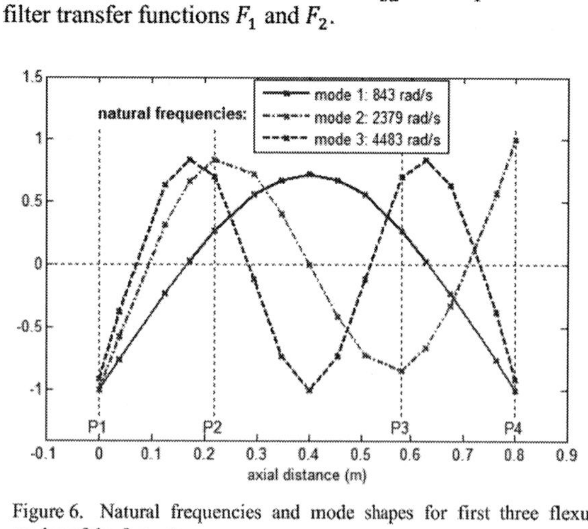

Figure 6. Natural frequencies and mode shapes for first three flexural modes of the free rotor

IV. FLEXIBLE ROTOR TEST SYSTEM

A numerical study on the application of the proposed approach to a fully levitated flexible rotor test system (Fig. 5) has been undertaken. The system has a hollow shaft rotor that is 0.8 m in length and 0.03 m in diameter. A number of disks are mounted along the rotor at the positions indicated in Fig. 5. The natural frequencies and mode shapes of the first three flexural modes of the free rotor are shown in Fig. 6. The nominal rotational speed range of 0-12000 rpm (0-1260 rad/s) covers the critical speed associated with the first flexural mode only. The characteristics of the AMBs match those given in Table I for the 1-DOF system.

The main focus of this study is the effect of non-collocation of position sensors and AMBs. A hybrid sensor configuration, as shown in Fig. 5, is of practical interest as it would separate all sensors from working sections of the rotor, thus protecting sensors from unfavorable environmental conditions such as high temperatures, direct contact with pumped/working fluids or contaminants. It could also lower the risk of debris or impact related damage and allow easy servicing and replacement. Nonetheless, for the sensor configuration shown in Fig. 5, use of either displacement or acceleration sensors alone would present difficulties in achieving stable levitation. For position-based feedback alone, the large axial separation of the displacement sensors and AMBs would necessitate model-based controller design to stabilize flexural modes, with stringent requirements on robustness to model-error. For acceleration-feedback control, collocation of accelerometers and magnetic bearings can be exploited by a passivity-based controller design (including modified PD controllers). However, stable drift-free levitation is unachievable with acceleration feedback alone. By combining sensor information using a MIMO version of the complementary filtering approach described in Section II, stable levitation and satisfactory vibration attenuation performance can be achieved with model-free controller designs.

To analyze achievable control performance, a rotordynamic model based on finite element methods has been constructed. Details on such methods and the conversion to standard first order state-space form can be found elsewhere [10, 11]. To construct an appropriate model of the complete system, the relation between lateral force components $f_m = [f_x, f_y]$ applied to the rotor in plane m and the rotor displacement response $z_n = [x, y]$ in plane n is considered in the standard form

$$\dot{x} = Ax + B_m f_m \quad (13)$$
$$z_n = C_n x \quad (14)$$

The rotor acceleration in plane n is

$$a_n = \ddot{z}_n = C_n \ddot{x} = C_n A^2 x + C_n A B_m f_m \quad (15)$$

The MIMO transfer function for the open loop system, combining four displacement measurements in planes 1 and 4 and four acceleration measurements in planes 2 and 3, is

$$C(Is - A)^{-1}B + D =$$
$$\begin{bmatrix} C_{1,4} \\ C_{2,3}A^2 \end{bmatrix}(Is - A)^{-1}B_{2,3} + \begin{bmatrix} 0 \\ C_{2,3}AB_{2,3} \end{bmatrix} \quad (16)$$

The controller transfer function is assigned as

$$C_c(Is - A_c)^{-1}B_c + D_c = \quad (17)$$
$$\begin{bmatrix} F_1H & 0 & 0 & 0 & F_2H/s^2 & 0 & 0 & 0 \\ 0 & F_1H & 0 & 0 & 0 & F_2H/s^2 & 0 & 0 \\ 0 & 0 & F_1H & 0 & 0 & 0 & F_2H/s^2 & 0 \\ 0 & 0 & 0 & F_1H & 0 & 0 & 0 & F_2H/s^2 \end{bmatrix}$$

A prototype PID feedback controller H was selected with the gain values given in Table 1. With standard collocated position-based feedback, the closed loop sensitivity function $T_{\bar{z}v}$ has the form shown in Fig. 7, which is within ISO recommended limits for AMB control [12]. The complementary filters were selected according to (5) with $N = 3$. The closed loop transfer functions under hybrid feedback defined by (16) and (17) can be computed as

$$\begin{bmatrix} T_{zd}(s) \\ T_{zw}(s) \end{bmatrix} = \begin{bmatrix} C_z(Is - A_{cl})^{-1}B_d \\ C_z(Is - A_{cl})^{-1}B_w \end{bmatrix} \quad (18)$$

where

$$A_{cl} = \begin{bmatrix} A + BD_cC & BC_c \\ CB_c & A_c + B_cDC_c \end{bmatrix},$$

$$C_z = [C_{1,4}\ 0], \quad B_d = \begin{bmatrix} B_{2,3} \\ B_cD \end{bmatrix}, \quad B_w = \begin{bmatrix} B_{2,3}D_c \\ B_c \end{bmatrix} \times \begin{bmatrix} 0 \\ I \end{bmatrix}$$

From the closed loop model (18) a parametric study on the influence of filter cross-over frequency ω_c on control performance may be undertaken. Figure 7 showing a map of how the natural frequencies and stability properties of closed loop eigenmodes (as calculated from A_{cl}) vary with ω_c. For large ω_c, the AMBs are predominantly positioned-controlled over the frequency range covering the first three flexural modes. Due to the non-collocation of planes P1 and P4 with planes P2 and P3 and phasing associated with the mode shapes (Fig. 6), all three modes are unstable. As ω_c is decreased the flexural modes are sequentially stabilized, until for $\omega_c < 120$ rad/s the closed loop system is stabilized.

Selection of an appropriate value of ω_c should also take into account the frequency domain properties of T_{zd} and T_{zw}, i.e. the rotor displacement response due to direct forcing and accelerometer measurement noise, respectively. For illustration, the maximum singular values of $T_{zd}(j\omega)$ and $T_{zw}(j\omega)$ are shown in Fig. 9 for three different values of ω_c. For $\omega_c \to 0$, the dynamic properties of T_{zd} tend to replicate those for the prototype (collocated position-based feedback) design. However, the down side of this is an increased sensitivity to low frequency noise affecting accelerometer measurements. As low frequency noise and drift can be a significant problem for conventional accelerometer designs, there is clearly a compromise to be made in this respect. Further theoretical work is recommended to fully investigate the performance trade-offs inherent in the proposed feedback control approach. It is anticipated that consideration of stochastic (or deterministic) information for expected sensor error/noise and disturbances should provide a systematic approach to shape the stability and response characteristics under combined acceleration and position feedback control. Comparisons with model-based optimal controller design methods would also be worthwhile.

V. WIRELESS SENSING IMPLEMENTATION

For practical implementation of the proposed rotor acceleration feedback, rotor-embedded sensors must satisfy requirements of compact size, high resolution and bandwidth, low noise and straightforward integration with wireless technology. Practical work has focused on combining MEMS digital accelerometers with a microcontroller and wireless radio module for data transmission (Arduino Mini with ATmega328 processor and XBee 802.15.4 wireless module). The accelerometer is a 3-axis device with ±16g measurement range and 13 bit resolution, giving a sensitivity of 4mg/LSB. It can provide up to 3.2 kHz sampling rate, measures 3 x 3 x 1 mm and can withstand accelerations up to 10,000 g. These devices work with low supply voltages (2.6V) and have low power consumption so that battery supply or inductive transmission of power are both viable. The microcontroller (with clock speed of 2 MHz) is sufficiently powerful to manage data transfer between multiple accelerometers and the wireless module. Based on four measurement channels, each with sampling frequency of 3.2 kHz, a radio transmission rate >1 Mbits/s is considered adequate. For

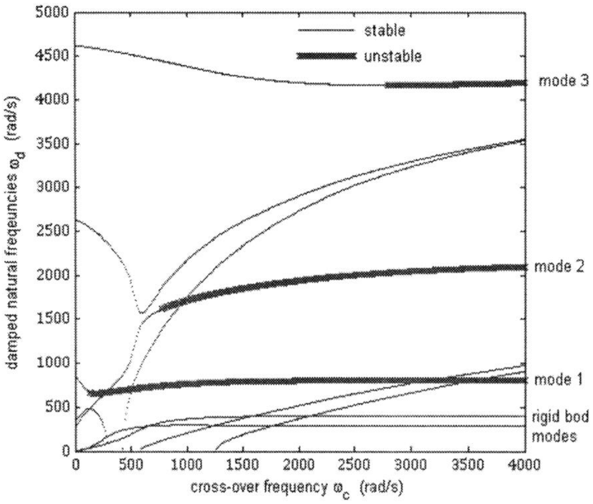

Figure 7. Map showing how the natural frequencies and stability properties of closed loop eigenmodes vary with cross-over frequency for complementary filtering

short range transmission, Wireless USB or Bluetooth 3.0 are suitable protocols. Testing of open-loop data acquisition under rotating conditions has confirmed viability of the approach and further work is in progress to realize full multi-sensor closed loop control.

VI. CONCLUSIONS

An investigation on the control of a rotor-AMB system using combined feedback of rotor acceleration and position measurements has been undertaken. A complementary filtering has been described that can maintain stability while allowing the cross-over frequency for the position and acceleration feedback loops to be freely chosen. The potential advantage of this approach in respect of achieving improved robustness to position sensor run-out error was demonstrated in simulation. Further investigations on the application of the method to model-free control of a multi-mode flexible rotor system have been undertaken. The potential to maintain control stability and vibration attenuation performance with large axial separation of position sensors and AMBs has been shown. Practical realizations will open up new possibilities for AMB-rotor design topologies, free of the limitations imposed by conventional proximity-based collocated sensing of rotor outer surfaces.

REFERENCES

[1] D. Atsumo, T. Yoshida, K. Ohniwa, "Sensorless control of axial magnetic bearings", *IEEJ Transactions on Industry Applications*, vol. 130, no. 6, pp. 826-827, 2010.

[2] R. Herzog, S. Vuilloud, R. Amstad, G. Galdos, P. Muellhaupt, R. Longchamp "Self–sensing of non–laminated axial magnetic bearings: modelling and validation", *Proceedings of 11th Int. Symp. Magnetic Bearings*, pp.575-581, pp. 2008

[3] E. H. Maslen, D. T. Montie, T. Iwasaki, "Robustness limitations in self-sensing magnetic bearings", *Journal of Dynamic Systems, Measurement and Control, Transactions of the ASME*, vol. 128, no. 2, pp. 197-203, 2006.

[4] G. Van Schoor, A.C. Niemann, C.P. Du Rand, "Evaluation of demodulation algorithms for robust self-sensing active magnetic bearings", *Sensors and Actuators, A: Physical*, vol. 189, pp. 441-450, 2013.

[5] D. Gruyer, A. Lambert, M. Perrollaz, D. Gingras, "Experimental comparison of Bayesian positioning methods based on multi-sensor data fusion", *International Journal of Vehicle Autonomous Systems*, vol. 12, no. 1, pp. 24-43, 2014.

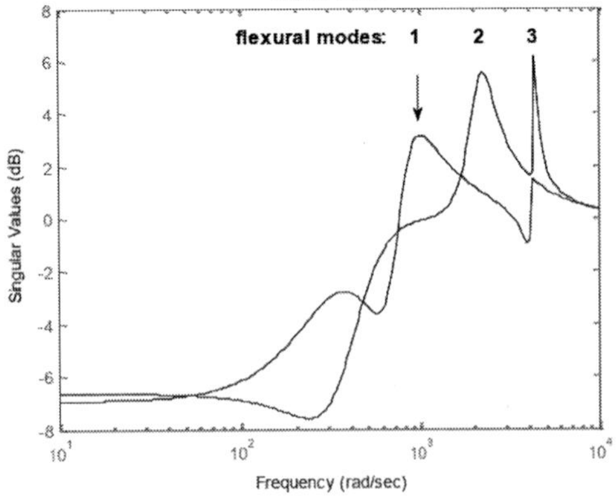

Figure 8. Sensitivity function T_{zv} for prototype feedback controller H designed for feedback of collocated rotor displacement measurements

[6] H. Zhao, Z. Wang, "Motion measurement using inertial sensors, ultrasonic sensors, and magnetometers with extended Kalman filter for data fusion", *IEEE Sensors Journal*, vol. 12, no. 5, pp. 943-953, 2012.

[7] I. A. Mahmood, S.O.R. Moheimani, K. Liu, "Tracking control of a nanopositioner using complementary sensors", *IEEE Transactions on Nanotechnology*, vol. 8, no. 1, pp. 55-65, 2009.

[8] K. K. Tan, P.V. Er, R. Yang, C. S. Teo, "Selective precision motion control using weighted sensor fusion approach", *Proceedings of IEEE International Conference on Mechatronics and Automation*, pp. 179-184, 2013.

[9] P. J. Berkelman, R. L. Hollis, S.E. Salcudean, "Interacting with virtual environments using a magnetic levitation haptic interface" *Proceedings of IEEE International Conference on Intelligent Robots and Systems*, vol. 1, pp. 117-122, 1995.

[10] Zorzi, E.S. and Nelson, H.D., Finite-element simulation of rotor-bearing systems with internal damping. Journal of Engineering for Power-Transactions of the Asme. vol. 99. No. 1, pp. 71-76, 1977.

[11] G. Schweitzer, E. Maslen (Eds.), *Magnetic Bearings - Theory, Design, and Application to Rotating Machinery*, Springer, ISBN 978-3-642-00496-4.

[12] ISO TC108/SC2 Working Group 7. *Active magnetic bearings - evaluation of stability margin*. ISO-14839-3, September 2006.

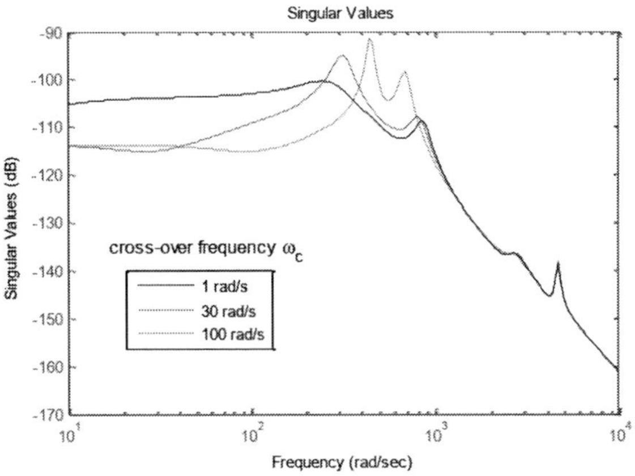

(a) Direct forcing disturbance sensitivity T_{zd}

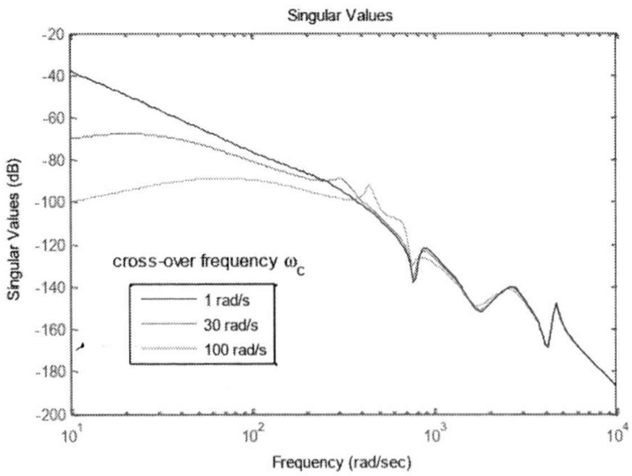

(b) Accelerometer noise sensitivity T_{zw}

Figure 9. Frequency response plots showing influence of filter cross-over frequency on disturbance rejection characteristics

Anti-windup Control of Active Magnetic Bearings Subject to Voltage Saturation

Jinxiang Zhou[a], Jiancheng Fang[a], Yin Zhang[a], Yinxiao Jiang[a]

[a] BeiHang University, XueYuan Road No.37, HaiDian District, Beijing, China, zhoujinxiang99@gmail.com

***Abstract*—Active magnetic bearings (AMBs) have been widely applied in industrial rotating machinery. Power amplifier as an essential actuator of the AMB faces the limitation of DC bus voltage. A simple approach is to increase the voltage to a high value, but it has to deal with large current ripples and a low resolution. Other than the ordinary method, this paper addresses to the anti-windup control of AMB that aims to eliminate the adverse effect of the voltage saturation. The anti-windup algorithm is an addition to the nominal linear controller. And only when the saturation occurs is it activated. The stability is guaranteed by the procedure of calculating the linear matrix inequality (LMI) form of the regional Lyapunov stability. Simulation test is done to verify the proposed controller. Finally in the motor drive experiment, the axial AMB equipped in the prototype supports the thrust force. With the load changes, the performance of the anti-windup controller is demonstrated.**

***Keywords*—AMB, voltage saturation, anti-windup control**

I. Introduction

Active magnetic bearings (AMBs) have so many advantages such as non-contact, no friction, no wear, long life, low noise, small vibration, no lubrication requirement, and adjustable dynamic performance [1, 2]. Due to the character of its natural unstable pole, each AMB needs a control loop. A typical AMB system consists of sensors, signal processers, power amplifiers, electromagnets (EMs) and ferromagnetic rotor components. Power amplifier converts the control signals to the changes of the current in the EM's windings, so that it can provide the magnetic force to suspend the rotor stably.

Nowadays Pulse Width Modulation (PWM) switch technic is quite common adopted in power converters, since it is simple, convenient, and efficient. This kind of power electronics together with EM's coils can be classified into the Class-D amplifier, which is generally modeled as a controlled voltage source [3]. As to the classification of the AMB control mode, there are current control mode, voltage control mode, and flux control mode reviewed in [1, 4]. No matter in which mode, the physical realization is the same way that the voltage across the coil is changed. So the DC bus voltage is the inherent restriction of the AMB PWM amplifier. And the voltage saturation was once discussed in the Annex C of the standard ISO 14839-2:2004.

In most controller design procedures, based on the small signal linear model, we can get a preliminary result. Then through the regulation to enhance the robustness, and maybe making some compromise, the effect of the voltage saturation is not significant. However, when the AMB is applied in some special industrial cases, or even in the rotor levitating moments, the windup phenomenon caused by the voltage saturation is observed, and may lead to a failure. Some AMB equipped in power motor products designed by famous manufactures such as Danfoss Turbocor, SKF and Synchrony, have employed a high power DC voltage of 250V or higher.

Since the high voltage across the switch converters causes more current ripples, which are the noise to the current sensors. And to the same PWM carrier frequency, the resolution of the amplifiers is reduced as the range of the output is widened. Besides, a low voltage is adequate for the most time of the AMB regular operation. So there are some works that have been done to seek solutions in other ways. For the low- and zero-bias AMB, a nonlinear controller based on Teel's small gain saturation theory, was reported in [5] to solve the voltage saturation. Lee, R. M. [6] and Haiping, Du [7] proposed fuzzy control methods subject to their special disturbance models, that is to solve the saturation problem in the AMB of milling machine.

This paper is to apply the anti-windup (AW) control synthesis to the AMB controller design in the 100KW 32000r/min magnetic levitation motor. The AMB power bus voltage 90V is selected. In the motor drive experiment, two rotors are attached to each other by a rope coupling. There is an additional thrust force upon the axial AMBs when the motor drives the generator. This changing load often causes the voltage saturation of the axial AMB amplifier, leading to a large-scale long-period oscillation.

In both fields of academy and engineering, the anti-windup problem has been studied for a quite long history. Until the last two decades, uniform control structure and the linear matrix inequality (LMI) approach have become the consensus [8]. The general anti-windup control structure is shown as Figure 1.

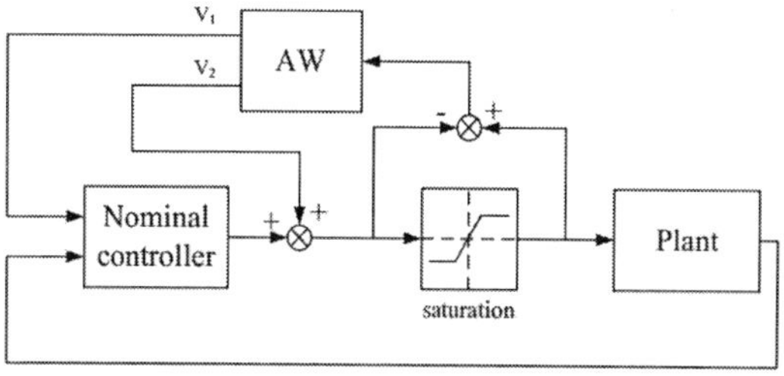

Figure 1. General anti-windup control structure

The AW refers to the anti-windup block, which takes the saturation information as inputs, and outputs two vectors. The design procedure is divided into the following two steps:

- Design the nominal controller without the consideration of the saturation.
- Transform the saturation function to the sector condition, and then use the LMI to search an AW matrix to minimize the optimization goal.

As to the unstable plant, a local case region of stability domain is guaranteed under the constraints of the actuator's output [9]. The anti-windup synthesis proposed by da Silva, J. M. G. etc. [10], takes the maximum stability region as the optimization goal. The methods based on L_2 norm performance are given in [11, 12]. Due to the simple procedure and few parameters, the former static compensator [10] is easier to calculate and to put into engineering practice. And this method is further clarified and promoted in [13]. So based on this theory, we takes the anti-windup algorithm into the axial AMB control. Although this work is done to a one degree of freedom (DOF) AMB, it could be easily applied to the control of radial AMBs if it is wanted.

The next section presents the structure of the axial AMB and derives its mathematical model. To be universal, section 3 gives control system configuration and the nominal controller design, which includes the PID form of the displacement signal and a current feedback loop. Then the anti-windup compensator is constructed and the LMIs are solved to get the AW matrix. Besides the simulation, the experiments using the AMB prototype with load cases are present in the following part. At last, the conclusion is made that the negative influence of the voltage saturation can be improved.

II. Structure of the Axial AMB

The axial magnetic bearing of the prototype employs the structure that is proposed in [14], but differs in dimensions and parameters. Figure 2 shows the profile of the thrust magnetic bearing that is tested in this work. Although it is named permanent-magnet-biased hybrid magnetic bearing in [14], there is little difference in the linear model for control consideration, as we will see in the Eq.(3). So in the opinion of [1], this kind hybrid magnetic bearing is still classified into the AMB.

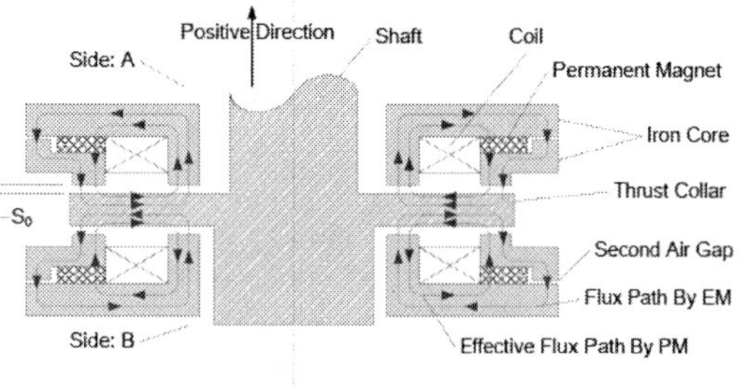

Figure 2. Configuration of axial magnetic bearing

The control flux is generated by the EMs, and flows across the iron cores, thrust collar, air gaps of the poles, and a second air gap. Two axially magnetized ring magnets act as permanent magnets (PMs). The flux circuit of PM has two branches. One flows across the second air gap, the other flows across the pole air gap providing the bias flux.

The cross sections in each pole air gap of the inner ring and the outer have the equal area A (m^2). The coils of the upper EM and the below are connected in series. Assuming the shaft has a positive displacement x (m) to the nominal position S_0 (m) and there is a current i (A) in the coil, the total magnetic force $F(x,i)$ (N) upon the collar can be calculated as follows:

$$F(x,i) = \frac{\phi_a^2}{2\mu_0 A} + \frac{\phi_a^2}{2\mu_0 A} - \frac{\phi_b^2}{2\mu_0 A} - \frac{\phi_b^2}{2\mu_0 A} \tag{1}$$

$$\begin{cases} \phi_a = \phi_{pa} + \phi_{ca} = \dfrac{F_{pm}\mu_0 A}{2(S_0 - x)} + \dfrac{Ni\mu_0 A}{2(S_0 - x) + S_2 A / A_2} \\ \phi_b = \phi_{pb} - \phi_{cb} = \dfrac{F_{pm}\mu_0 A}{2(S_0 + x)} - \dfrac{Ni\mu_0 A}{2(S_0 + x) + S_2 A / A_2} \end{cases} \tag{2}$$

where ϕ denotes the magnetic flux, and the subscript a, b, p, c refers to the side a, b, the bias flux by the permanent magnet, the control flux, respectively. μ_0 denotes the permeability of air. F_{pm} is the magnetomotive force (MMF) of the PM. S_2 and A_2 denote the width and cross section area of the second air gap.

The function of $F(x,i)$ can be expanded in Taylor's series at the equilibrium point. It can be predicted that the high order terms contain the same order of μ_0, which are quite small values. So a familiar linear term can be expressed in Eq.(3):

$$F(x,i) = K_x \cdot x + K_i \cdot i \tag{3}$$

where

$$\begin{cases} K_x = \left.\dfrac{\partial f}{\partial x}\right|_{\substack{x=0 \\ i=0}} = \dfrac{\mu_0 F_{pm}^2 A}{S_0^3} \\ K_i = \left.\dfrac{\partial f}{\partial i}\right|_{\substack{x=0 \\ i=0}} = \dfrac{2\mu_0 N F_{pm} A}{S_0(2S_0 + S_2 A / A_2)} \end{cases} \tag{4}$$

K_x and K_i denote the displacement stiffness and current stiffness.

III. System Model and the Nominal Controller

Figure 11 is the general schematic diagram of the AMB control system. A chip of DSP is used to realize the control algorithm. The PWM drive signals are produced in the FPGA. The amplifier circuits apply the H bridge topology, which consists of four power MOSFETs. The control signal is modulated in the varying voltage across the EM coil.

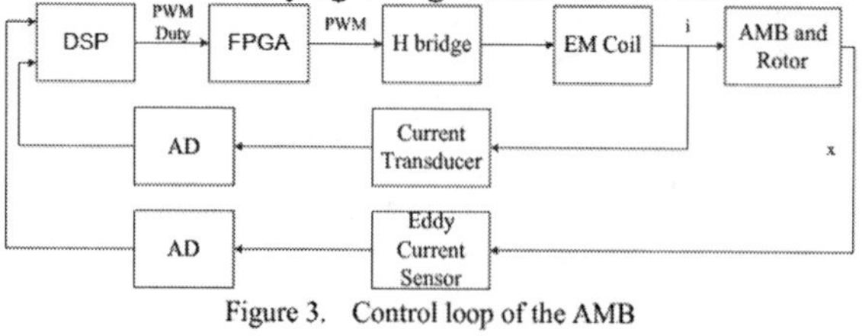

Figure 3. Control loop of the AMB

As the coils of two sides connected in series, the voltage can be calculated by the Eq.(5).

$$u = (R_a + R_b) \cdot i + N\frac{d\phi_a}{dt} + N\frac{d\phi_b}{dt} \tag{5}$$

Applying (2) to (5), and combining like terms, a simplified equation is obtained.

$$u=(R_a+R_b)\cdot i+\frac{2N^2\mu_0 A}{2S_0+S_2A/A_2}\cdot\frac{di}{dt}\triangleq R\cdot i+L\cdot\frac{di}{dt} \quad (6)$$

As to one DOF, the state vector X is defined as follows:

$$X=\begin{bmatrix}x & \dot{x} & i\end{bmatrix}^T \quad (7)$$

The system model in state space form can be easily obtained.

$$\begin{aligned}\dot{X}(t)&=A\cdot X(t)+B\cdot u(t)\\ Y(t)&=C\cdot X(t)\end{aligned} \quad (8)$$

The matrices, A, B, C are given as:

$$A=\begin{bmatrix}0 & 1 & 0\\ K_s/m & 0 & K_i/m\\ 0 & 0 & -R/L\end{bmatrix}$$

$$B=\begin{bmatrix}0 & 0 & V_c\end{bmatrix}^T \quad (9)$$

$$C=-diag\begin{pmatrix}S_x & S_x & S_i\end{pmatrix}$$

where V_c denotes the DC bus voltage, and S_x, S_i represent the position sensor's and current transducer's proportional factors, including the scale factor of AD chips.

It should be noticed about the meaning of the controller output u_c. In the actual system, it is realized by the PWM duty δ, which is bounded in the section of 0 to 1. Because of the H bridge topology, the full positive DC bus voltage is upon the coil when δ is 1, and the full negative voltage is when δ is 0. And a proper scale transform is needed. And in the abstract model, the plant input u and the controller output u_c have the relationship as:

$$u=sat(u_c)=\begin{cases}1 & u_c>1\\ u_c & -1\le u_c\le 1\\ -1 & u_c<-1\end{cases} \quad (10)$$

Although numerous modern control strategies applied in the AMB have been reported in academics, the PID control is still the primary choice of the industry. The reason that we take the PID as the original controller is its concise form and universality. So the parts of the anti-windup study will be prominent.

Specifically, the PID progress of the position signal contains a first-order derivative filter. A proportional feedback of the current signal is added to the operation result. The state form of the nominal controller is expressed as:

$$\begin{aligned}\begin{bmatrix}\dot{\eta}_1\\ \dot{\eta}_2\end{bmatrix}&=\begin{bmatrix}0 & 0\\ 0 & -1/T_f\end{bmatrix}\bullet\begin{bmatrix}\eta_1\\ \eta_2\end{bmatrix}+\begin{bmatrix}I & 0 & 0\\ 0 & D/T_f & 0\end{bmatrix}\bullet Y\\ u_c&=K_{amp}\bullet\begin{bmatrix}1 & 1\end{bmatrix}\bullet\begin{bmatrix}\eta_1\\ \eta_2\end{bmatrix}+K_{amp}\bullet\begin{bmatrix}P & 0 & I_{co}\end{bmatrix}\bullet Y\end{aligned} \quad (11)$$

where I_{co} denotes the coefficient of current feedback, K_{amp} denotes the factor that regulars the output's scale, P, I, D, and T_f denote the coefficients of the proportion, integration, differentiation, and derivative filter respectively.

Without consideration of the constraint (10), the stability and performance of the control system are checked by the Nyquist diagram (Figure 4) and magnitude plot of sensitivity function (Figure 5). Obviously, the AMB system is well stabilized. And the peak magnitude of the sensitivity function is 1.8, satisfying the performance requirements of most applications. The specific values of the prototype are given in TABLE I.

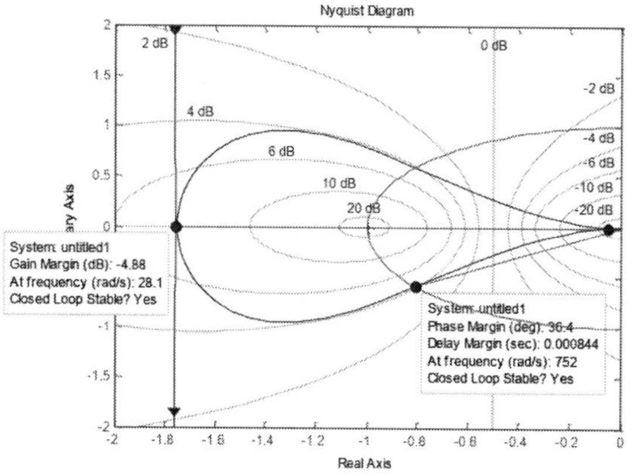

Figure 4. Nyquist diagam of the control system

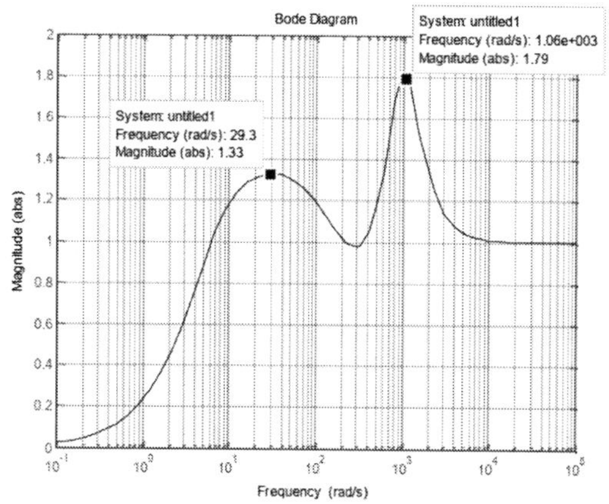

Figure 5. Magnitude plot of sensitivity function

TABLE I. PARAMETER VALUES OF THE AMB SYSTEM

Parameter	Value	Parameter	Value
K_s	3200 N/mm	P	9.6
K_i	960 N/A	I	23.6
L	0.34 H	D	0.038
R	50.8 Ω	T_f	0.038/1000
m	20.6 Kg	I_{co}	8.6
V_c	90 V	K_{amp}	0.64/1000

IV. ANTI-WINDUP COMPENSATOR DESIGN

In the last section, it has been mentioned that the controller output is restricted in Eq.(10). This is a typical saturation of the actuator output. When the control signal reaches or overflows the boundary, there is ineluctable performance degradation of the control system. The missing information ($sat(u_c)-u_c$) can be easily picked up in the digital processor. So an anti-windup term $E_c(sat(u_c)-u_c)$ is added to the controller to mitigate the undesirable influence of windup. Here E_c is a constant matrix of appropriate order, so this is called the static anti-windup compensator [12].

Considering the anti-windup strategy, rewrite the dynamic controller (11) and the closed-loop system (8), we gets:

$$\begin{aligned}
&\dot{X}(t) = AX(t) + Bsat(u_c(t)) \\
&Y(t) = CX(t) \\
&\dot{X}_c(t) = A_c X_c(t) + B_c Y(t) + \begin{bmatrix} I_2 & 0 \end{bmatrix} E_c \varphi(u_c) \\
&u_c(t) = C_c X_c(t) + D_c Y(t) + \begin{bmatrix} 0 & I_1 \end{bmatrix} E_c \varphi(u_c) \\
&\varphi(u_c) = sat(u_c(t)) - u_c(t)
\end{aligned} \tag{12}$$

where X_c denotes the state vector of the controller (11), A_c, B_c, C_c, D_c are abbreviations of the matrixes in (11).

The benefit of the (12) is obvious that when there is no saturation the system is the original one with no difference, when saturation occurs the controller's input and output get modified.

Now define an extended state vector:

$$\xi = \begin{bmatrix} X & X_c \end{bmatrix}^T = \begin{bmatrix} x & \dot{x} & i & \eta_1 & \eta_2 \end{bmatrix}^T \tag{13}$$

then the closed-loop system reads:

$$\begin{aligned}
&\dot{\xi} = \alpha\xi + (\beta + \gamma_1 E_c)\varphi(u_c) \\
&u_c = \lambda\xi + \gamma_2 E_c \varphi(u_c)
\end{aligned} \tag{14}$$

where

$$\begin{aligned}
&\alpha = \begin{bmatrix} A + BD_c C & BC_c \\ B_c C & A_c \end{bmatrix}, \beta = \begin{bmatrix} B \\ 0 \end{bmatrix} \\
&\gamma_1 = \begin{bmatrix} B\begin{bmatrix} 0 & I_1 \end{bmatrix} \\ I_2 \quad 0 \end{bmatrix}, \gamma_2 = \begin{bmatrix} 0 & I_1 \end{bmatrix} \\
&\lambda = \begin{bmatrix} D_c C & C_c \end{bmatrix}
\end{aligned} \tag{15}$$

The Theorem 1 in [13] gives the sufficient condition to calculate the asymptotic stability region for this problem. The proofs to it could be explored in the similar procedures of [10]. Here we rewrite the theorem in the form special to system (14) as follows:

Theorem 1: If there exist a symmetric positive definite matrix $W \in \mathbb{R}^{5\times5}$, a vector $Y \in \mathbb{R}^{1\times5}$, a vector $Z \in \mathbb{R}^{3\times1}$, and a positive scalar S satisfying

$$\begin{bmatrix} W\alpha^T + \alpha W & \beta S + \gamma_1 Z - W\lambda^T - Y^T \\ * & -2S - \gamma_2 Z - Z^T \gamma_2^T \end{bmatrix} < 0 \tag{16}$$

$$\begin{bmatrix} W & Y^T \\ * & 1 \end{bmatrix} \geq 0 \tag{17}$$

then the gain matrix $E_c = Z \cdot S^{-1}$ is such that the ellipsoid $\varepsilon(W) = \left\{ \xi \in \mathbb{R}^{5\times1} \middle| \xi^T W^{-1} \xi \leq 1 \right\}$ is an asymptotic stability region for system (14).

The LMI (16) is obtained from the Lyapunov candidate function:

$$V(\xi) = \xi^T(t) \cdot P \cdot \xi(t), P = P^T = W^{-1} > 0, \, and \, \dot{V}(\xi) < 0 \,. \tag{18}$$

A sector condition $\varphi^T(u_c) \cdot S^{-1} \cdot (sat(u_c) + \omega) \leq 0$ where ω satisfying $|\omega| \leq 1$ is applied in (18). And LMI (17) implies ω is constructed from the set $\varepsilon(W)$ and Y.

Based on the theorem 1, the direct objective to search the matrix E_c is to maximize the volume of $\varepsilon(W)$, which could be quantified by the determinant of the matrix W^{1}. So the solution to the anti-windup design is transformed to an optimization problem in the LMI form. The optimization goal should be set as

minimize: $J = -log(det(W^{-1})) = log(det(W))$

Unfortunately, we could not access a proper solver to this nonlinear goal. To be simplified, the $\varepsilon(W)$ is searched to maximize the size along two directions of the controller states. A linear objective is obtained by:

$$\begin{aligned}
&(\varepsilon v_1)^T \cdot W^{-1} \cdot (\varepsilon v_1) \leq 1 \\
&(\varepsilon v_2)^T \cdot W^{-1} \cdot (\varepsilon v_2) \leq 1 \\
&v_1 = \begin{bmatrix} 0 & 0 & 0 & 1 & 0 \end{bmatrix}^T \\
&v_1 = \begin{bmatrix} 0 & 0 & 0 & 0 & 1 \end{bmatrix}^T \\
&\max \varepsilon \triangleq \frac{1}{\sqrt{\mu}}
\end{aligned} \tag{19}$$

The inequalities (19) could be further written in matrix form (20). Besides, from the forth equation in (12), it can be found that $E_c = \begin{bmatrix} * & * & -1 \end{bmatrix}$ is an undesirable singular point, making the controller output u_c meaningless. So constraint (21) is added to avoid converging to that point. The final optimization problem reads as follows:

minimize: μ

subject to: $W > 0, S > 0$

$LMIs$ (16), and (17)

$$\begin{bmatrix} \mu & v_1^T \\ v_1 & W \end{bmatrix} \geq 0, \begin{bmatrix} \mu & v_2^T \\ v_2 & W \end{bmatrix} \geq 0 \tag{20}$$

$$\left| \begin{bmatrix} 0 & 0 & 1 \end{bmatrix} \cdot Z + S \right| > 0.1 \cdot S \tag{21}$$

The toolbox YALMIP in MATLAB is used in the numerical calculation. To avoid large conditions of matrixes during the computation, the state vector (13) is normalized by:

$$\bar{\xi} = T \cdot \xi = \left[\xi_i / \max(\xi_i) \right]^T \tag{22}$$

where $\max(\xi_i)$ denotes the max value or estimated max value of the respective element. The matrixes in (15) are transformed by T and T^{1} accordingly. The detailed process is simple and omitted here.

Adopting the data listed in TABLE I, the results of the optimization is given as:

$\mu = 0.348, E_c = \begin{bmatrix} 37.5 & -263.3 & -0.6 \end{bmatrix}^T$.

Since the E_c is obtained, there is a reverse procedure to put it back to the actual control system. All scale factors should be paid special attentions to. Otherwise it may lead to mistakes according to our experience. The next section will demonstrate the simulation and experiments results.

V. SIMULATION AND EXPERIMENT

The first occasion that the windup phenomenon in AMB caused our interest is the levitating moments of the prototype rotor. The controller parameters first designed by the system model without consideration of the saturation have a poor performance in the actual system.

The auxiliary thrust bearing restricts the position of the rotor to -0.5mm to 0.5mm. When the AMB control board is powered on, the first results of controller must reach the limits. On the other hand, the amplifier circuit is powered on after the drive signals are working properly. During this delay time of 0.1 seconds, the integration register of the controller may be overflowing. All these conditions result in a long time of the output saturation. The simulation of the AMB under this condition is shown in Figure 6. The Figure 7 is the plot of the calculated voltage values according to PWM duty during this time. Although the PWM duty in actual system cannot be exact 0 or 1 because of the security setting of dead-times. It could be still seen that the voltage has reached the limitation of DC bus in the starting time. And correspondingly there is a large overstrike in the displacement of the shaft. The black dash line simulates the condition that the voltage is unlimited. After applying the anti-windup strategy, the times of overstrike and settle are reduced.

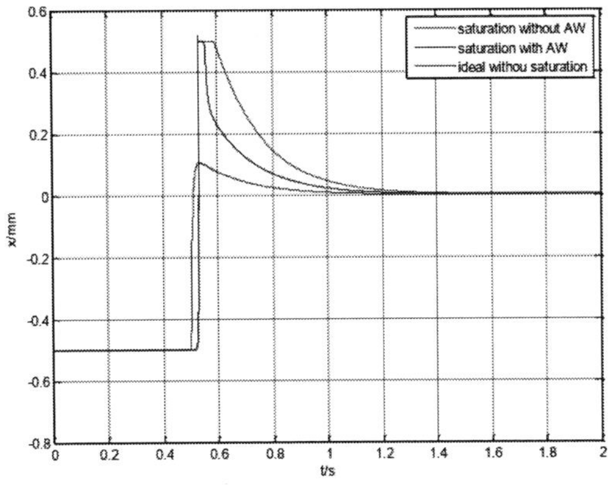

Figure 6. The simulation plot of rotor position during levitating

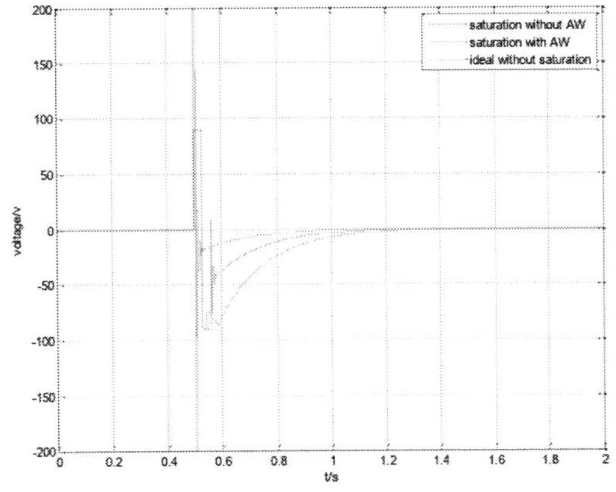

Figure 7. The computed voltage values during levitating

Figure 8 and Figure 9 simulate the occasions the thrust magnetic bearing carrying loads. The shaft is working at the rotation speed of 12000rpm. So a 200Hz displace sinusoid disturbance is added to the simulation. At the time of 0.5s, a first step 300N axial force is upon the shaft. Without controller saturation occurs, there is no difference in the three. At the time of 2.5s, another step 500N axial force is added. The total force is up to 800N and the output voltage reaches to the limits 90V. In the ideal conditions, the magnetic bearing still works well except for a little large oscillation. But when the voltage saturation is considered, the magnetic bearing could not hold the rotor properly. The shaft touches down on the auxiliary bearing and causes serious collisions. As to the robustness of the controller, the system recovered after about 0.25s without AW control. As to the controller with AW, the touchdown behavior is still not avoided, but the system recovers faster.

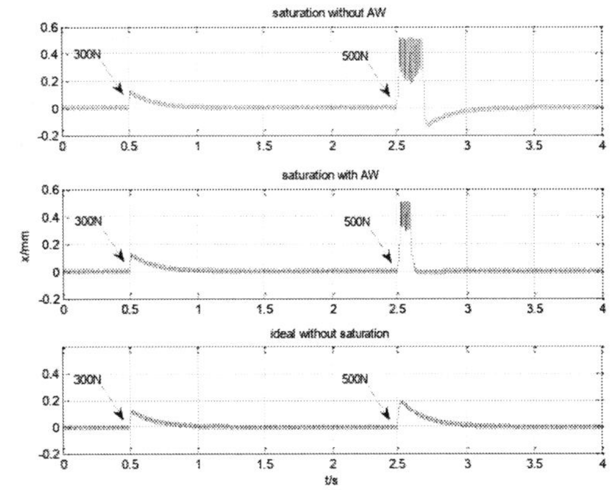

Figure 8. The simulation plot of rotor position when carrying loads

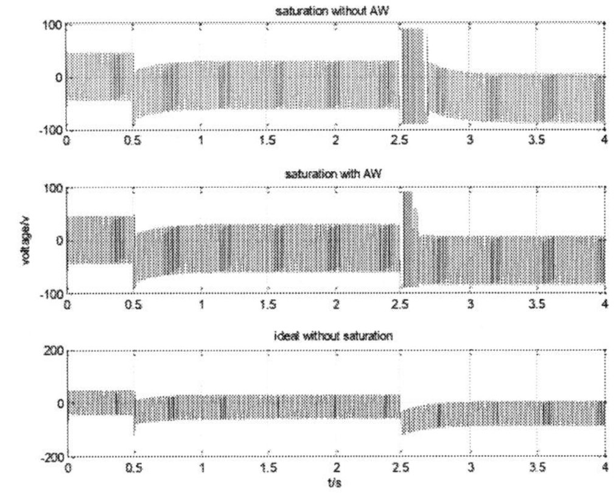

Figure 9. The computed voltage values when carrying loads

In the experiments of real AMB system, the comparison results of the AW control are given in Figure 10. The data are taken from the signals of the displacement sensor. It outputs voltage signals of 0.25V to 2.75V, corresponding to the shaft displacement -0.5mm to 0.5mm. It could be seen that the controller with anti-windup strategy have the advantage of a fast recovery speed when the saturation occurs.

Figure 11 gives the photo of the site of the motor drive experiment. In the picture, two prototypes are both equipped with AMBs. Two shafts are connected by a rope coupling. The generator armature is connected with resistances to produce resisting moment. When the fiber ropes conducts the motor's torque, an axial tensile force is generated simultaneously. The torque is determined by the rotation speed and the resistance's value. So when we switch the resistance's value to test the motor power, the axial force is changed suddenly. This axial impulse upon the magnetic bearing may lead to the voltage saturation, and caused a long time recovery. Due to the flexible connection of the rope coupling, it is hard to model and quantify the change of the axial force. So the controller must be strong robust. Figure 12 is on the circumstance that the rotor is operating at the speed of 12000rpm, and the generator's resistance is switched on to 1.2Ω from the off state. Although there is no significant difference between the two plots, the anti-windup effect still could be found in the period of the oscillations.

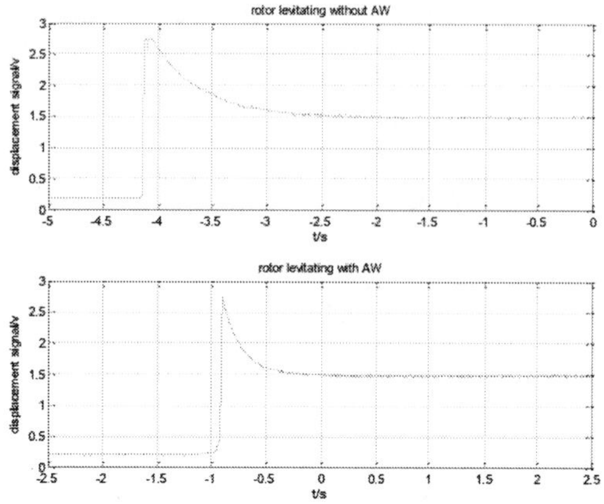

Figure 10. The signal of the real shaft displacement during levitating

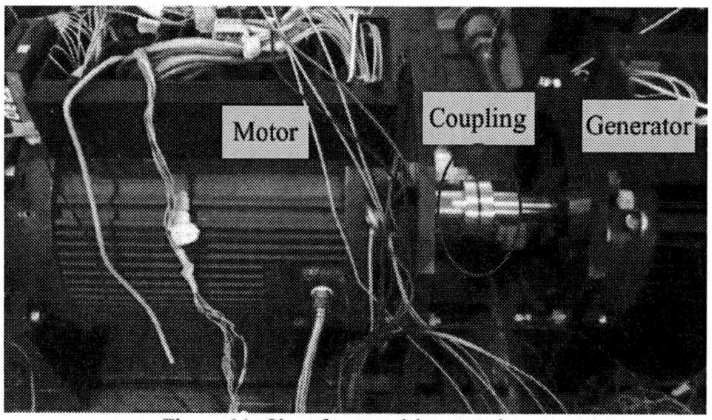

Figure 11. Site of motor drive experiment

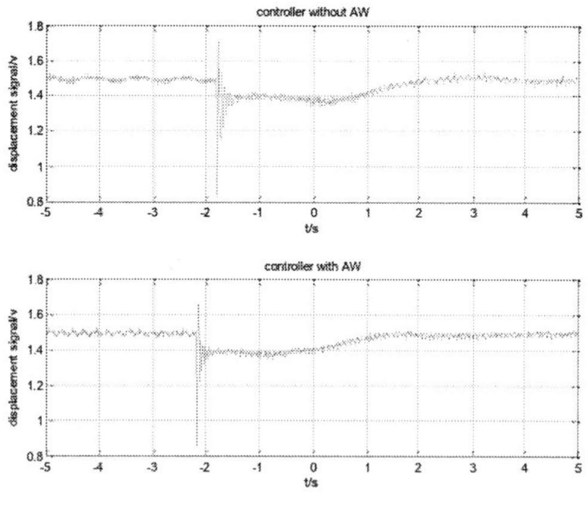

Figure 12. The displacement of the rotor with axial load changes

VI. CONCLUSION

Since the anti-windup control has been a hot topic in control theorem, and this paper aims to examine the application in the AMB control. An actual axial AMB is taken into the anti-windup controller design. One advantage of the strategy is that there is no change of the nominal controller when the voltage saturation does not occur. The other is that it improves the recovery performance of the system when there is saturation. Through the simulation and prototype experiments, the anti-windup control is proven to be a choice to mitigate the negative influence of the AMB voltage saturation.

REFERENCES

[1] Schweitzer, G. and E.H. Maslen, *Magnetic bearings: theory, design and application to rotating machinery* 2009.

[2] Schweitzer, G., *Applications and Research Topics for Active Magnetic Bearings.* Iutam Symposium on Emerging Trends in Rotor Dynamics, ed. K. Gupta. Vol. 25. 2011, New York: Springer. 263-273.

[3] Deslauriers, I., N. Avdiu, and B.T. Ooi, *Naturally sampled triangle carrier PWM bandwidth limit and output spectrum.* Ieee Transactions on Power Electronics, 2005. **20**(1): p. 100-106.

[4] Tsiotras, P. and B.C. Wilson, *Zero-and low-bias control designs for active magnetic bearings.* Control Systems Technology, IEEE Transactions on, 2003. **11**(6): p. 889-904.

[5] Tsiotras, P. and M. Arcak, *Low-bias control of AMB subject to voltage saturation: State-feedback and observer designs.* Control Systems Technology, IEEE Transactions on, 2005. **13**(2): p. 262-273.

[6] Lee, R.M. and N.C. Tsai, *Fuzzy compensator design against windup of a magnetic actuator during milling operation.* Proceedings of the Institution of Mechanical Engineers Part B-Journal of Engineering Manufacture, 2012. **226**(B9): p. 1454-1464.

[7] Haiping, D., et al., *Robust Fuzzy Control of an Active Magnetic Bearing Subject to Voltage Saturation.* Control Systems Technology, IEEE Transactions on, 2010. **18**(1): p. 164-169.

[8] Zaccarian, L. and A.R. Teel, *A common framework for anti-windup, bumpless transfer and reliable designs.* Automatica, 2002. **38**(10): p. 1735-1744.

[9] Teel, A.R., *Anti - windup for exponentially unstable linear systems.* International Journal of Robust and Nonlinear Control, 1999. **9**(10): p. 701-716.

[10] da Silva, J.M.G. and S. Tarbouriech, *Antiwindup design with guaranteed regions of stability: An LMI-based approach.* Ieee Transactions on Automatic Control, 2005. **50**(1): p. 106-111.

[11] Hu, T., A.R. Teel, and L. Zaccarian, *Anti-windup synthesis for linear control systems with input saturation: Achieving regional, nonlinear performance.* Automatica, 2008. **44**(2): p. 512-519.

[12] Hu, T., A.R. Teel, and L. Zaccarian, *Stability and performance for saturated systems via quadratic and nonquadratic Lyapunov functions.* Ieee Transactions on Automatic Control, 2006. **51**(11): p. 1770-1786.

[13] Tarbouriech, S. and M. Turner, *Anti-windup design: an overview of some recent advances and open problems.* Iet Control Theory and Applications, 2009. **3**(1): p. 1-19.

[14] Fang, J., et al., *A New Structure for Permanent-Magnet-Biased Axial Hybrid Magnetic Bearings.* Magnetics, IEEE Transactions on, 2009. **45**(12): p. 5319-5325.

Advanced Controller of a Hybrid Foil-magnetic Bearing

Ye Tian[a], Yanhua Sun[b], Lie Yu[c]

State Key Laboratory for Strength and Vibration of Mechanical Structures, Xi'an Jiaotong University, No.28 Xianning West Road, Xi'an, 710049, Shaanxi Province, China,
[a]a.g.o.u2@stu.xjtu.edu.cn
[b]sunyanhua@mail.xjtu.edu.cn
[c]yulie@mail.xjtu.edn.cn

Abstract—A hybrid foil-magnetic bearing (HFMB) is combination of a gas foil bearing (GFB) and a active magnetic bearing (AMB), which takes advantages of both bearings while compensating each other's weaknesses. It is a solution of friction and wear of the GFBs at low speeds and limited load capacity of the AMBs. Furthermore, load sharing and control of dynamics can be achieved in a hybrid foil-magnetic bearing. However, in order to achieve the ideal performances of the HFMBs, an advanced controller must be designed based on the static and dynamic performances of the HFMBs. Since the AMBs are 'smart' devices that the working condition can be measured and actively controlled, it is quite natural to design an advanced controller for the HFMBs to use the AMBs to identify the characteristics of the GFBs and with the function of load sharing between the GFBs and the AMBs. In this paper, a 4-DOF model of bearing-rotor system supported by two HFMBs was established firstly, and then an advanced controller with the function of identification and load sharing was presented. Finally, experiments were done on a test rig whose rotor is supported by two journal HFMBs and a thrust AMB. The test results show that when change the load shared by the GFBs and the AMBs, advanced controller can find the new proper working position of the HFMBs rapidly and smoothly. More importantly, HFMBs with advanced controller based on the identified characteristics of GFBs can work steadily at the speed range of 0~2,2000r/min.

I. Introduction

AMBs and GFBs are two typical bearings for the high speed rotating machines. Each bearing has its own advantages and disadvantages. The AMB generates magnetic force to support rotors without contact. There is no need of lubrication or sealing, and the wear is completely eliminated. Furthermore the working condition of AMBs can also be measured and actively controlled. However, because of magnetic saturation and thermal effect, the load capacity of AMBs is limited [1-2]. The GFB provides another non-contact way to support the high speed rotor. GFB is self-acting hydrodynamic bearing that uses ambient air or any process gas as the lubricating fluid. A hydrodynamic film pressure builds up within the small gap between the rotating shaft and the smooth foil. It has simple structure and favorable adaptability, and can work in high temperature circumstance. Its weakness is the contact and friction between the bearing and the journal at low speeds, which reduce the life of GFBs greatly. Moreover, it also has low load capacity at low speeds [3-7]. A HFMB is a combination of a GFB and an AMB and can share the load between them. The HFMB has the advantages of both bearings. It can improve not only the load carrying capacity but also the dynamic performance of the rotor-bearing system [8-11]. The HFMB is an ideal manner of the high speed rotating machines.

The control system is very important in operation of a HFMB. Using a conventional PID control algorithm, the controller would attempt to force the journal to a referred position at all times regardless of loads. However, in a loaded foil bearing, the journal runs at some eccentricity and attitude angle when supporting a load. Heshmat, Chen, and Walton did a great job in solving the contradiction, they used a supervisory control system to ensure the functionality of the hybrid bearings [9,10,12]. However, the supervisory control system strongly depends on the characteristics of GFBs obtained from numerical calculation or independent bearing experiments. If different GFBs are used or the working conditions are changed, numerical calculation or independent bearing experiments must be done again, and the control system should be adjusted as well. It causes repetitive work and takes a long time. Since the AMBs are 'smart' devices that the working condition can be measured and actively controlled, it is quite natural to design an advanced controller for HFMBs to use AMBs identify the characteristics of GFBsand with the function of load sharing between GFBs and AMBs[13, 14].

In this paper, a 4-DOF model of bearing-rotor system supported by two journal HFMBs was established firstly. Then an advanced controller with was presented. Dynamic loads were applied by two radial AMBs, and a frequency domain identification method was presented to identify the stiffness and damping coefficients of the GFBs. After that, a searching algorithm was presented to determine the static working position of the HFMBs with different sharing load. The test results show the advanced controller can rapidly find the new working position of HFMBs by using a searching algorithm when the load shared by GFBs and AMBs are changed. Furthermore, HFMBs with advanced controller work steadily in the transient operation tests.

A. Hybrid Foil-magnetic Bearings

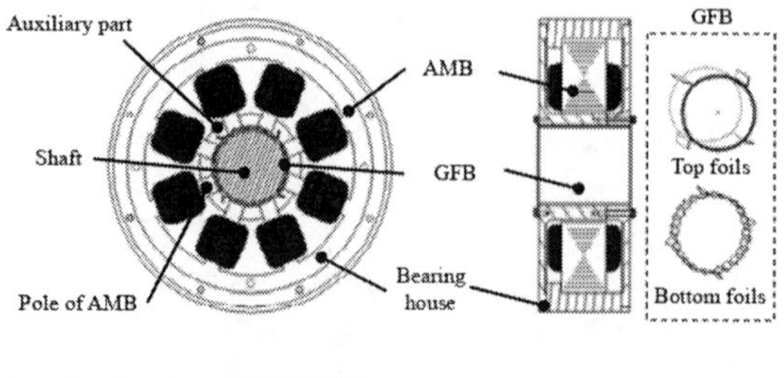

Figure 1. Structure of the HFMB

The structure of the HFMB studied in this paper is shown in Fig.1. The GFB consists of four top foil each supported by a bottom bump foils. All these foils are located in the air gap between the AMB poles and the journal of the rotor. The foils are made of non-magnetic materials so that they will not affect work of the AMB. The applied forces on the shaft generated by the AMB can be calculated using the following formula [15]:

$$f_n = ek_{th}\left(\frac{i_1^2}{G_1^2}-\frac{i_2^2}{G_2^2}\right),\ (n=x,y) \tag{1}$$

where k_{th} is a constant related to the AMB, i_1 and i_2 are the currents in the opposing coils, G_1 and G_2 are the upper and lower air gaps, and e is a derating factor. A series calibrations were conducted by applying deadweight to the AMB in multiple orientations beforehand witch makes the mean error of static forces within 2%.

B. Dynamic Model of 4-DOF Bearing-rotor System

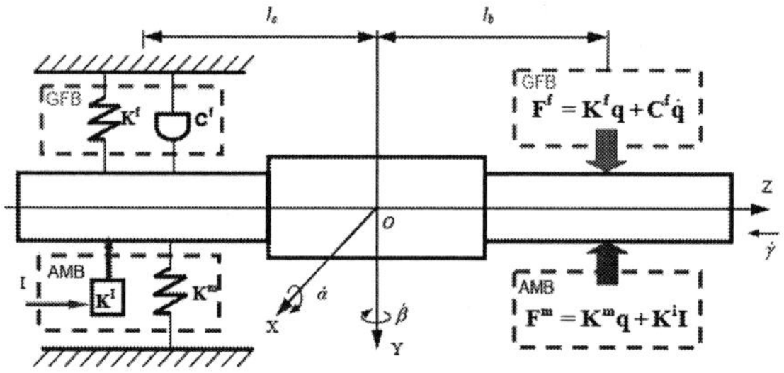

Figure 2. Schematic of the dynamic model

When the ratio of the rotor length to the diameter of the thrust disc is relatively large, influences of the thrust bearing to radial motion of the rotor can be neglected [2]. Therefore, the axial motion of the rotor could be studied separately. A rigid rotor supported by two radial bearings is shown in Fig.2. The distant between bearing locations is l and O is the mass center of the rotor. Define the journal displacements at the left journal bearing as x_a and y_a, and the displacements at the right journal bearing as x_b and y_b. The distant between mass center to the left and right HFMBs are l_a and l_b respectively. The rotation angle in the XZ plane is $\alpha=(x_b - x_a)/l$ and the rotation angle in the Y-Z plane is $\beta=(y_b - y_a)/l$. Both the GFB and the AMB at the left side are the same as the right ones; hence the stiffness and damping coefficients at left side and right side use the same expressions.

The generalized displacements are defined as $\mathbf{q_m}=[x,y,\alpha,\beta]^{\mathrm{T}}$, where x and y are the displacements of its center of mass O The radial displacements of the journal at bearing positions are defined as: $\mathbf{q}=[x_a,x_b,y_a,y_b]^{\mathrm{T}}$. The equations of motion of the rotor are:

$$\begin{cases}(1-a)m\ddot{x}+\Delta F_{xa}+\Delta F_{xb}=f_x\\ am\ddot{y}+\Delta F_{ya}+\Delta F_{yb}=f_y\\ J_y\ddot{\beta}+\dot{\gamma}J_z\dot{\alpha}=al\Delta F_{xa}-(1-a)l\Delta F_{xb}\\ J_x\ddot{\alpha}-\dot{\gamma}J_z\dot{\beta}=al\Delta F_{ya}-(1-a)l\Delta F_{yb}\end{cases} \tag{2}$$

where, m is the mass of the rotor, $\dot{\gamma}$ is the rotational speed, f_x and f_y are the external forces, J_x, J_y and J_z are moment of inertia of rotor at rotary point O. $\Delta F_{xa},\Delta F_{xb},\Delta F_{ya}$ and ΔF_{yb} are bearing forces in the X and Y directions.

$$\begin{cases}\Delta F_{xa}=k_x^m x_a+k_x^i i_{xa}+k_{xx}^f x_a+k_{yx}^f y_a+c_{xx}^f \dot{x}_a+c_{yx}^f \dot{y}_a\\ \Delta F_{xb}=k_x^m x_b+k_x^i i_{xb}+k_{xx}^f x_b+k_{yx}^f y_b+c_{xx}^f \dot{x}_b+c_{yx}^f \dot{y}_b\\ \Delta F_{ya}=k_y^m y_a+k_y^i i_{ya}+k_{yy}^f y_a+k_{xy}^f x_a+c_{yy}^f \dot{y}_a+c_{xy}^f \dot{x}_a\\ \Delta F_{yb}=k_y^m y_b+k_y^i i_{yb}+k_{yy}^f y_b+k_{xy}^f x_b+c_{yy}^f \dot{y}_b+c_{xy}^f \dot{x}_b\end{cases} \tag{3}$$

where i_{xa}, i_{xb}, i_{ya} and i_{ya} are the control currents. k_x^m, k_y^m are the force/displacement stiffness and k_x^i, k_y^i are force/current stiffness of the AMB. $k_{xx}^f, k_{xy}^f, k_{yx}^f, k_{yy}^f$ are the stiffness coefficients and $c_{xx}^f, c_{xy}^f, c_{yx}^f, c_{yy}^f$ are the damping coefficients of GFB. The equations of motion can also be written as:

$$\mathbf{M}\ddot{\mathbf{q}}+\mathbf{C^f}\dot{\mathbf{q}}+(\mathbf{K^f}-\mathbf{K^m})\mathbf{q}=\mathbf{K^i I} \tag{4}$$

$$\mathbf{M}=\begin{bmatrix}(1-a)m & am & 0 & 0\\ 0 & 0 & (1-a)m & am\\ -J_y/l & J_y/l & 0 & 0\\ 0 & 0 & -J_x/l & J_x/l\end{bmatrix},$$

$$\mathbf{C^f}=\begin{bmatrix}c_{xx}^f & c_{xx}^f & c_{xy}^f & c_{xy}^f\\ c_{yx}^f & c_{yx}^f & c_{yy}^f & c_{yy}^f\\ -alc_{xx}^f & (1-a)lc_{xx}^f & -alc_{xy}^f-\dot{\gamma}J_z/l & (1-a)lc_{xy}^f+\dot{\gamma}J_z/l\\ -alc_{yx}^f+\dot{\gamma}J_z/l & (1-a)c_{yx}^f-\dot{\gamma}J_z/l & -alc_{yy}^f & (1-a)lc_{yy}^f\end{bmatrix},$$

$$\mathbf{K^f}=\begin{bmatrix}k_{xx}^f & k_{xx}^f & k_{xy}^f & k_{xy}^f\\ k_{yx}^f & k_{yx}^f & k_{yy}^f & k_{yy}^f\\ -alk_{xx}^f & (1-a)lk_{xx}^f & -alk_{xy}^f & (1-a)lk_{xy}^f\\ -alk_{yx}^f & (1-a)k_{yx}^f & -alk_{yy}^f & (1-a)lk_{yy}^f\end{bmatrix}, \tag{5}$$

$$\mathbf{K^m}=\begin{bmatrix}k_x^m & k_x^m & 0 & 0\\ 0 & 0 & k_y^m & k_y^m\\ -alk_x^m & (1-a)lk_x^m & 0 & 0\\ 0 & 0 & -alk_y^m & (1-a)lk_y^m\end{bmatrix},$$

$$\mathbf{K^i}=\begin{bmatrix}k_x^i & k_x^i & 0 & 0\\ 0 & 0 & k_y^i & k_y^i\\ -alk_x^i & (1-a)lk_x^i & 0 & 0\\ 0 & 0 & -alk_y^i & (1-a)lk_y^i\end{bmatrix},\mathbf{I}=\begin{bmatrix}i_{xa}\\ i_{xb}\\ i_{ya}\\ i_{yb}\end{bmatrix}$$

M is the mass matrix, $\mathbf{K^m}$ and $\mathbf{K^i}$ are the force/displacement matrix and force/current matrix of the AMB, $\mathbf{K^f}$ and $\mathbf{C^f}$ are the stiffness and the damping matrices of the GFB, **I** is the control currents vector. The elements of matrix $\mathbf{K^m}$ and $\mathbf{K^i}$ can be obtained by linearizing Eq. (1) for small rotor displacements around the operating point. Take X direction for instance, linearized AMB force around operating point (x_0, i_{x0}) can be written as

$$f_x^m = \frac{\partial f_x}{\partial x}|_{x=x_0} x + \frac{\partial f_x}{\partial i}|_{i=i_{x0}} i_x = k_x^m x + k_x^i i_x \tag{6}$$

The matrices $\mathbf{K^m}$ and $\mathbf{K^i}$ of the GFB are unknown which need to be identified.

C. Advanced Control System

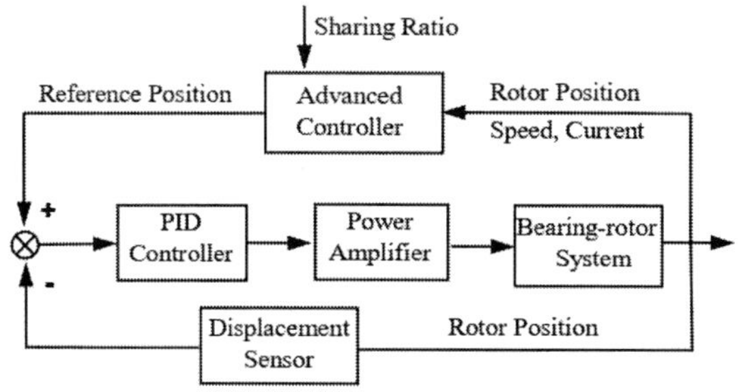

Figure 3. Schematic control system

The control system is designed as shown in Fig.3. It consists of a transient feedback control loop of the AMBs and an advanced controller which has the function of identification and load sharing. In the transient control loop, decoupling control is employed. At each DOF, a PID controller is used to ensure the shaft working at the reference position determined by the advanced controller. For identification of the characteristics of GFBs, the advanced controller output harmonic signals which lead to harmonic excitation of magnetic forces. With the function of sharing load between the GFBs and the AMBs, the advanced controller output constant signals based on the load sharing ratio λ, rotating speed ω, control current I and shaft position q.

Identify the dynamic characteristics of GFBs

For the harmonic excitation with an arbitrary excitation frequency ω, Eq. (4) can be rewritten as

$$\begin{bmatrix} D_{11} & D_{12} & D_{13} & D_{14} \\ D_{21} & D_{22} & D_{23} & D_{24} \\ D_{31} & D_{32} & D_{33} & D_{34} \\ D_{41} & D_{42} & D_{43} & D_{44} \end{bmatrix} \begin{Bmatrix} x_a(\omega) \\ x_b(\omega) \\ y_a(\omega) \\ y_b(\omega) \end{Bmatrix} = \begin{Bmatrix} i_{xa}(\omega) \\ i_{xb}(\omega) \\ i_{ya}(\omega) \\ i_{yb}(\omega) \end{Bmatrix} \tag{7}$$

where

$$\begin{aligned}
&D_{11} = \left\{k_{xx}^f - k_x^m - \omega^2\left[J_y / l^2 + (1-a)^2 m\right] + j\omega c_{xx}^f\right\} / k_x^i, D_{14} = 0 \\
&D_{12} = \omega^2\left[J_y / l^2 - (1-a)am\right] / k_x^i, D_{13} = \left(k_{xy}^f + j\omega c_{xy}^f\right) / k_x^i; \\
&D_{21} = \omega^2\left[J_y / l^2 - (1-a)am\right] / k_x^i, D_{24} = \left(k_{xy}^f + j\omega c_{xy}^f\right) / k_x^i, \\
&D_{22} = \left[k_{xx}^f - k_x^m - \omega^2\left(J_y / l^2 + a^2 m\right) + j\omega c_{xx}^f\right] / k_x^i, D_{23} = 0; \\
&D_{31} = \left(k_{yx}^f + j\omega c_{yx}^f\right) / k_x^i, D_{34} = \omega^2\left[J_x / l^2 - (1-a)am\right] / k_x^i, \\
&D_{33} = \left\{k_{yy}^f - k_y^m - \omega^2\left[J_x / l^2 + (1-a)^2 m\right] + j\omega c_{yy}^f\right\} / k_x^i, D_{32} = 0; \\
&D_{41} = 0, D_{44} = \left[k_{yy}^f - k_y^m - \omega^2\left(J_x / l^2 + a^2 m\right) + j\omega c_{yy}^f\right] / k_x^i, \\
&D_{43} = \omega^2\left[J_x / l^2 - (1-a)am\right] / k_x^i, D_{42} = \left(k_{yx}^f + j\omega c_{yx}^f\right) / k_x^i,
\end{aligned} \tag{8}$$

The dynamic coefficients of the GFB can be obtained experimentally by separating the real and imaginary part from the measured dynamic stiffness D_{mn} (m, n =1, 2, 3, 4). See ref. [14] for more details of identification of the GFBs parameters by using the AMBs as exciters.

Load sharing between the GFB and the AMB

The total bearing force F^B of the HFMB includes two parts, the force of the GFB F^f and the force of the AMB F^m. Define the load sharing ratio of F^f to F^B as λ. The total bearing force F^B equals to the rotor gravity.

$$F^B = F^m + F^f = G = \lambda G + (1-\lambda)G \tag{9}$$

Since both the forces of the GFB and the AMB are functions of the working place q [2,6], sharing the load between the GFB and the AMB can be achieved by finding a proper working position where the forces satisfy the following condition, where e_0 is the coefficient of precision.

$$\frac{\left|F^f(q_B) - (1-\lambda)G\right|}{\left|(1-\lambda)G\right|} \le e_0 \text{ or } \frac{\left|F^m(q_B) - \lambda G\right|}{\left|\lambda G\right|} \le e_0 \tag{10}$$

Use equilateral triangles to divide the working plane of the HFMB. And the working position can be obtained through continuous searching and dividing of the journal working area. Fig.4 (a) shows the initial working areas of the GFB, the AMB and the HFMB. Fig.4 (b) shows the initial area Δ123. The vertexes of Δ123 are the midpoints of the sides of the inscribed triangle of the bearing boundary circle.

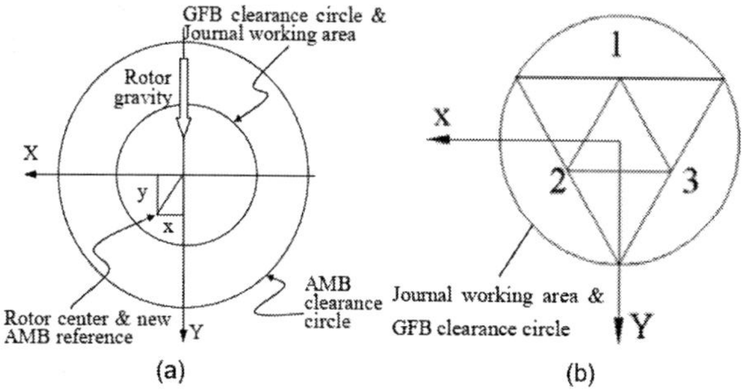

Fig. 4 Journal working area and initial Area

Define the vectors of AMB forces at vertices as $F^m(k) = [F_x^m(k), F_y^m(k)]^T$, k=1,2,3. The predetermined sharing load of the AMB are $F_s^m(k) = [F_{xs}^m(k), F_{ys}^m(k)]^T$, the difference vectors are $F_e^m(k) = F^m - F_s^m = [F_{xe}^m(k), F_{ye}^m(k)]^T$. So the total differences of forces are $F_e(k) = |F_e^m(k)|$, k=1,2,3. The steps of searching the working position are as follows

Step1. Move the journal center to the vertices of the initial triangle area calculate the forces of AMB and store.

Step2. Decide the next triangle area to be moved in based on the stored data. If the next area is in the present area, go to step 4, else go to step 3.

Step3. Move the journal center to the vertices of the next triangle area calculate the forces of AMB and store.

Step4. Divide the present area; the vertexes of the next triangle area are the midpoints of the sides of the present one. End searching if Eq. (10) is satisfied, else go to step 3.

The next triangle area can be decided as follows.

Define the vector of symbol coefficients $coe(k)=[coe_x(k),coe_y(k)]^{\mathrm{T}}$ for every side of Δ123

$$\text{if}\begin{cases}F_{ie}^{m}(k)\times F_{ie}^{m}(k+1)>0,(k=1,2)\\ F_{ie}^{m}(k)\times F_{ie}^{m}(k-2)>0,(k=3)\end{cases},(i=x,y)$$

then $coe_i(k)=1$, else $coe_i(k)=0$

Thus the four possible conditions of $coe(k)$ at each side of Δ123 are shown in Fig.5. (Permutation of vertices 1,2,3 do not affect the determination, and "01" equals to "10").

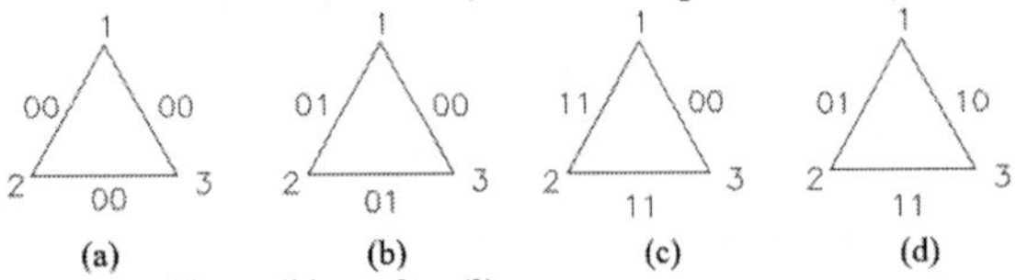

Fig. 5 Possible conditions of $coe(k)$

Case (a): Plus $F_e(k)$ of every side of Δ123 and find the minimum: min($F_e(1)+F_e(2)$, $F_e(2)+F_e(3)$, $F_e(3)+F_e(1)$).The side with the minimum value is the common side of the next triangle area and the present triangle area.

Case (b): Plus $F_e(k)$ of the side of Δ123 where the symbol coefficient is 01 and find the minimum. Take Fig.10 (b) for example, find min($F_e(1)+F_e(2)$, $F_e(2)+F_e(3)$). The side with the minimum value is the common side of the next triangle area and the present triangle area.

Case (c) or (d): The vertexes of the next triangle area are the midpoints of the sides of the present one.

D. Experiments

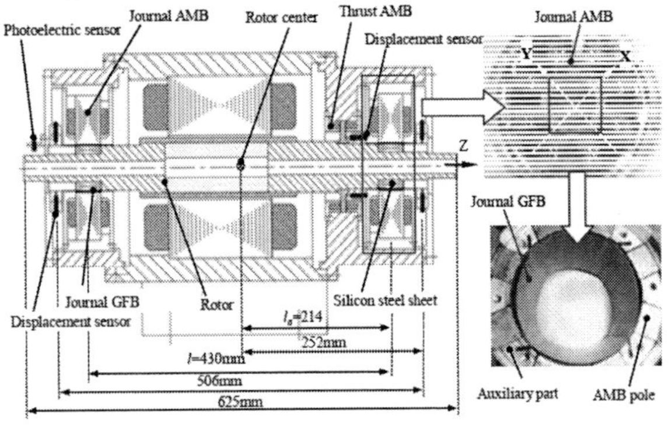

Figure 6. Schematic diagram of the test rig

Figure 6 displays the schematic diagram of the test rig and the photograph of the journal HFMB. There are ten eddy current displacement sensors, eight of them fixed beside the journal bearings measuring the displacements in X and Y direction in differential manner and the rest two fixed beside the thrust bearing measuring the axial displacements. A photoelectric sensor is used to measure the rotating speed, and the currents of the AMB coils are measured by current sensors which are not shown in Fig. 6. All these sensor signals are saved into a PC equipped with a commercial I/O boards. Parameters of the HFMB, the rotor and the control system are shown in Table 1. The parameters of PID controller are determined using a LQG method mentioned in reference [16].

TABLE I. PARAMETERS OF THE HFMB, THE ROTOR AND THE CONTROL SYSTEM

Parameters and dimensions	Values
Bearing housing inner diameter, D_h	61.4 mm
Axial bearing length, L	60 mm
Rotor diameter, d	59.8 mm
Arc length of the top foil, L_t	90 mm
Arc length of the bump foil, L_b	46.5 mm
Number of the bumps per bump foil, num	16
Top foil thickness, t_t	0.18 mm
Bump foil thickness, t_b	0.1 mm
Nominal clearance, C	0.14mm
Bump pitch, s_0	3.1 mm
Bump length, l_0	1.8 mm
Bump height, h_b	0.3 mm
Poisson's ratio, v	0.29
Modulus of elasticity, E	213.73 MPa
Ambient pressure, P_a	103.4 kPa
Lubricant viscosity, μ	2.896×10^{-5} Pa·s
Rotor mass, m	15.5 kg
Bias current of AMB, I_0	4 A
Number of turns of the coil windings, N	97
Air permeability, μ_0	$4\pi\times10^{-7}$N/A^2
Projection area of the magnetic pole, A	4.54×10^{-4}m^2
Gain of the power amplifier, A_a	50 A/V
Decay time of the power amplifier, T_a	3×10^{-5} s
Gain of the displacement sensor, A_s	10V/m
Decay time of the displacement sensor, T_s	3×10^{-5} s
Proportional coefficient of the PID controller, K_{pr}	15
Integral coefficient of the the PID controller, K_{ir}	300 s^{-1}
Differential coefficient of the PID controller, K_{dr}	0.03
Differential time constant of PID controller, T_{dr}	0.009 s
Position precision coefficient, e_0	0.01

Load sharing tests

Load sharing tests were conducted under conditions of no shaft rotation. Set the value of load sharing ratio λ to 0, 0.5, and 1. The track of the journal canter and the difference value F_{ex}^{m}, F_{ey}^{m} are shown in Fig.7, Fig.8 and Fig.9 respectively. Results shown here are correspond to the right journal HFMB. Similar results are found for the left ones.

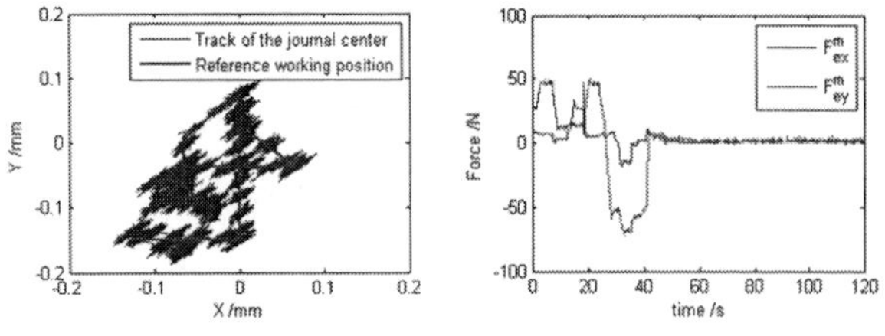

Figure 7. Schematic control system

In the case of λ=0 in Fig. 7, the GFBs take all the rotor gravity. The final position of the journal is (0.10mm,

0.09mm)^T. The difference value between AMB forces and sharing load F^m_{ex}=-1.5N, F^m_{ey}=0.5N.

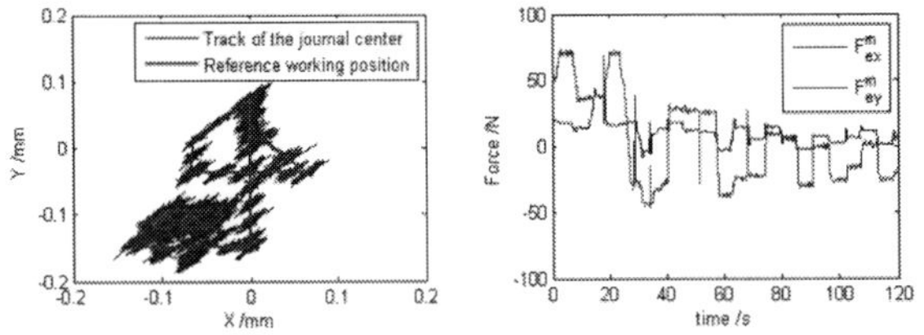

Figure 8. Schematic control system

In the case of λ=0.5 in Fig. 8, the AMBs take half of the rotor gravity. It shows that the final position of the journal is (0.08mm, 0.07mm)^T. The difference value between AMB forces and sharing load are F^m_{ex}=0.8N, F^m_{ey}=-0.2N.

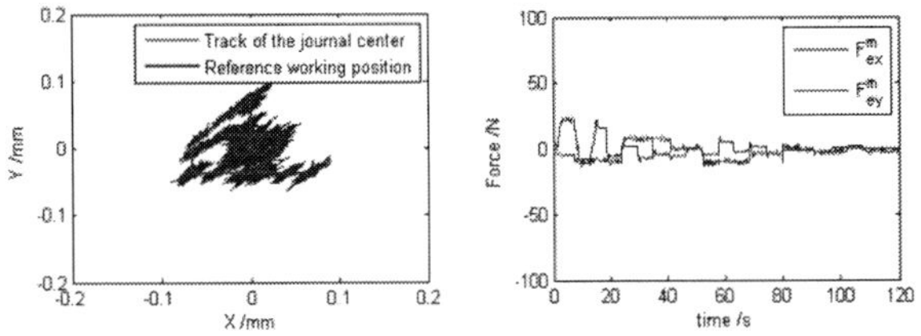

Figure 9. Schematic control system

In the case of λ=1 in Fig. 9, the AMBs take all of the rotor gravity. It shows that the final position of the journal is (0.01mm,-0.01mm)^T and the difference value between AMB forces and sharing load F^m_{ex}=-0.3N, F^m_{ey}=2.2N.

The results of load sharing tests show that the searching algorithm is able to locate the working position of the HFMB within a short time in different situations. Please note that the searching process only needs to be operated once before the uses of the HFMB and the position data are stored for future use.

Transient operation tests

Tests are conducted on the test rig to demonstrate the ability of the HFMB to support the rotor during transient events. The rotor was levitated by AMBs at zero speed and started to run up in the HFMB mode in order to evaluate the operation at increasing speeds, when the speed was above the GFB lift-off speed, the AMBs were tuned off to let rotor drop down on the GFBs in the GFB mode alone and the operation was evaluated. Finally the shaft was drove to slow down by the motor. Tests were successfully conducted at speeds up to 22,000rpm where AMB failures were simulated. Fig. 10 shows results of the shaft speed and the torque of the motor. The forces of the AMB and the rotor orbit are shown in Fig. 11 and Fig. 12 respectively.

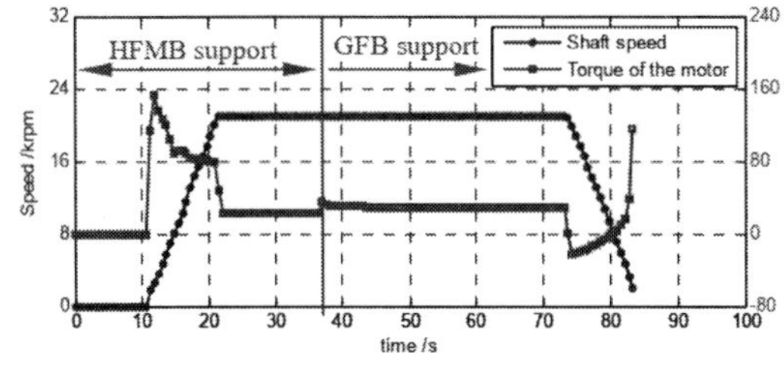

Figure 10. Shaft speed and motor torque vs. time

As shown in Fig. 10 there are peak signals of the motor torque caused by the friction between the shaft and GFBs during run-up and coast-down (Note that the negative value of the torque stands for the negative work generated by the motor when decelerate the shaft). When the speed was above the GFB lift-off speed, the torque quickly turns to a low value, a peak signal was caused by switching off the AMBs.

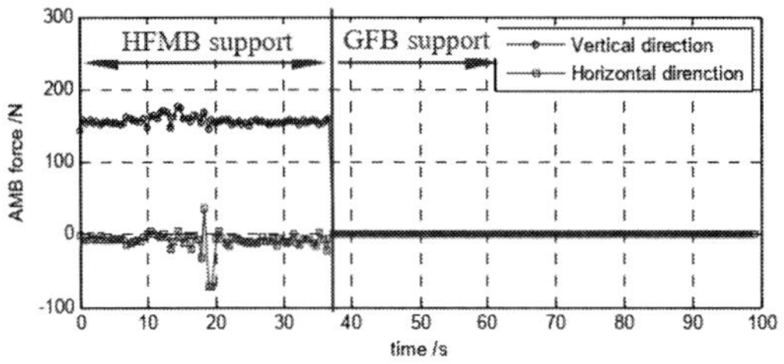

Figure 11. AMB forces vs. time

As shown in Fig. 11, since the rotor was levitated by AMBs at the center of the bearing housing at the beginning, the vertical force of AMBs equals to the gravity of rotor, the horizontal force of AMBs is almost zero. At the time when the AMBs were switched off, both the vertical and the horizontal forces of AMBs turn to zero.

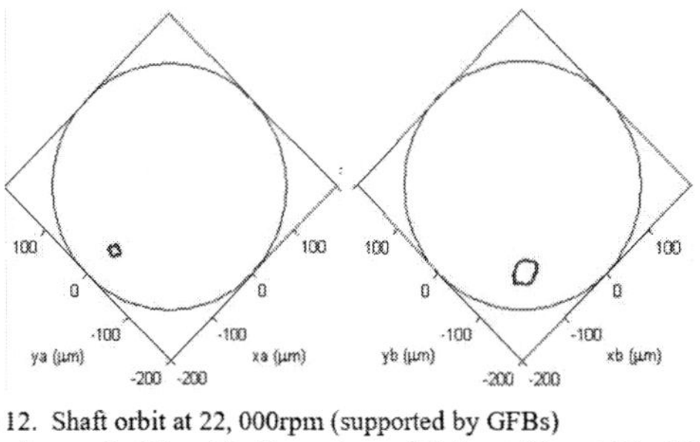

Figure 12. Shaft orbit at 22, 000rpm (supported by GFBs)

As shown in Fig. 12, the rotor orbit is stable at 22, 000rpm when the rotor is supported solely by the GFBs. Moreover, the final position of the journal is (-0.1mm, -0.1mm)^T, which is similar as the results shown in Fig. 7 with no shaft rotation. That means the working position of GFB is determined by structure deformation more than the gas films. Therefore, the results of load sharing tests of HFMBs under conditions of no shaft rotation can be used in a rotating bearing-rotor system.

E. Conclutions

An advanced controller of the HFMB is designed based on the 4-DOF bearing-rotor system in this paper. The characteristics of the GFB are identified using the AMBs as exciters and a frequency domain identification method was presented to identify the stiffness and damping coefficients. A searching algorithm is proposed to locate the working position of the HFMB. The test results show that when change the load shared by the GFBs and the AMBs, advanced controller can find the new proper working position of HFMBs rapidly and smoothly. More importantly, HFMBs with advanced controller can work steadily at the speed range of 0~2,2000r/min.

REFERENCES

[1] Schweitzer G, Traxler A, and Bleuler H, 1993, "Mageticlager," Springer-Verlag Press, Berlin, pp. 14-16, 48-50, 197-202.

[2] Schweitzer, G., and Maslen, E. H., 2009, "Magnetic bearings: theory, design, and application to rotating machinery," Springer, Verlag Berlin Heidelberg, pp. 167-174.

[3] Wang Yunfei, 1999, "Gas Lubrication Theory and Gas Bearings Design," Beijing Mechanical Industry Press, Beijing, pp. 1-143.

[4] Ehrich F F, Jacobson S A, 2003, "Development of high-speed gas bearings for high-power density microdevices," *Journal of Engineering for Gas Turbines and Power*, 125(1), pp. 141-148.

[5] S. Jahanmir, H. Heshmat, and C. Heshmat, 2009, "Evaluation of DLC coatings for high-temperature foil bearing applications," *Journal of Tribology*, 131, pp. 011301.

[6] Geng Haipeng, Qi Shemiao,. YU Lie. "Analysis of the multileaf foil journal bearings with large preload effect," *Journal of Aerospace Power*, 2006, 21(3), pp. 569-574

[7] Liu L X, Teo C J, Epstein A H, 2005, "Hydrostatic gas journal bearings for micro-turbomachinery," *Journal of Vibration and Acoustics*, Transactions of the ASME, 127(2), pp. 157-164.

[8] G. Foshage and E. C. Lovelace, 2007 "An integrated magnetic/foil bearing and methods for supporting a shaft journal using the same," WO Patent WO/2007/011,366.

[9] H. Heshmat, H. Ming Chen, J. F. Walton II, January, 2000, "On the performance of Hybird Foil-Magnetic Bearings," *Journal of Engineering for Gas Turbines and Power*, 122, pp. 73-81.

[10] E. E. Swanson, H. Heshmat, April 2002, "Performance of a Hybrid Foil-Magnetic Bearing," *Journal of Engineering for Gas Turbines and Power*, 124, pp. 375-381.

[11] K. J. Eakman, T. L. Coons, M. Andres, and L. F. Miller, 1998, "Backup bearing for magnetic bearings," US Patents 5,714,818.

[12] H. Heshmat, 2005, "Hybrid Foil-magnetic Bearing with Improved Load Sharing," US Patent 6,965,181.

[13] Ye Tian, Yanhua Sun, Lie Yu, 2012, "Steady-State Control of Hybrid Foil-Magnetic Bbearings," *Proceedings of ASME Turbo Expo 2012*, June 11-15, 2012, Copenhagen, Denmark, GT2012-68394.

[14] Ye Tian, Yanhua Sun, Lie Yu, 2013, "Structural Stiffness and Damping Coefficients of a Multi-leaf Foil Bearing with Bump Foils Underneath," ASME J Eng Gas Turbines Power, 136(4), pp. 044501.

[15] Kasarda, M. E., Marshall, J., and Prins, R., 2007, "Active Magnetic Bearing Based Force Measurement Using the Multi-Point Technique," *Mechanics Research Communications*, 34(1), pp. 44-53.

[16] Ye Tian, Yanhua Sun, Lie Yu, 2010, "LQG Control of Hybrid Foil - Magnetic Bearing," *The twelfth international symposium on magnetic bearings*, Wuhan, China.

A Notch Filter for Magnetically Suspended Rotors Based on Rotating Coordinate Transformation

Shiqiang Zheng[a], Rui Feng[b]

[a] School of Instrumentation Science & Opto–electronics Engineering, Beijing University of Aeronautics and Astronautics, No.37 Xueyuan Road, Beijing 100191, China, zhengshiqiang@buaa.edu.cn
[b] Huawei Technologies Co. Ltd, No.156, Beiqing Road, Beijing 100095, China

***Abstract*—Synchronous vibration are obstacles to the performance improvement of a rotating shaft supported by active magnetic bearings (AMBs). Based on the synchronous rotating coordinate transformation, this paper proposed a novel notch filter for the autobalancing of magnetically suspended rotors. After the analysis of the frequency characteristics of the novel notch filters, the calculating method of the convergence rate and comparison with the traditional notch filter are further investigated. Finally, the corresponding simulations and experiments are conducted. The results demonstrate that the novel notch filter has a good significant effect on the synchronous vibration of weakly gyroscopic rotors.**

I. INTRODCUTION

The basic idea of autobalancing control is introduced by Burrows et al, and a great number of the different techniques have been proposed to achieve this task, which can be divided into two categories. One approach is to estimate position of the rotor's principal axis of inertia and force the magnetically suspended rotor to rotate about the principal axis of inertia, named zero displacement. However, this approach depends on the accuracy of identification results of the system model, and will result in the housing vibrations. The other one, called zero magnetic force, is to eliminate synchronous force generated by magnetic bearings, which is based on the self-alignment theory. Different approaches, such the notch filter (Hui et al., 2010), adaptive feedforward (Knospe et al., 1996), and robust control (Kuseyri, 2012), have been proposed to eliminate the synchronous components in the currents of magnetic bearings. The two-modulation-step implementation of the generalized notch filters was introduced by Raoul Herzog (1996) and sensitivity function was used to analyze the stability of the closed-loop system with notch filters (Raoul Herzog et al., 1996), which is wildly considered as an effective approach.

Based on the relationships between two synchronous sinusoidal vibration forces and signals along the radial axes, which are caused by residue mass unbalance, this paper proposes a novel kind of notch filter using synchronous rotating coordinator transform. The main idea of this paper is that: (1) After being transformed by the synchronous rotating coordinator, the phases of synchronous spatial vectors (force, displacement, etc) become static, that is, the synchronous vectors become zero-frequency vectors; (2) The zero-frequency vectors can be extracted by passing through the low-pass filters; (3) The synchronous vectors can be retrieved by the extracted zero-frequency vectors and inverse synchronous rotating coordinator transform.

II. NOVEL NOTCH FILTER BASED ON SYNCHRONOUS ROTATING COORDINATE TRANSFORM

A. Principle

Assuming that the cross section of the rotor shaft is an ideal round and the generalized controlled channels adopted in the four degree of radial magnetic bearings are identical, the vibration signals of X-channel and Y-channel have the same amplitude and frequency, but with a phase difference of 90 degrees.

As shown in Figure 1, Let M and C denote the geometric centre and mass centre of the rotor, respectively. The fixed coordinate system CX_sY_s denotes the position sensor frame. CX_rY_r denotes the rotating coordinate system with a spin rate of Ω. When the rotor rotates stably around its inertial axis of the rotor, the trajectory of the geometric center is a circle with the center C and the eccentricity ε. Let $(x_s, y_s)^{\mathrm{T}}$ and $(x_r, y_r)^{\mathrm{T}}$ denote the motion of M in the fixed coordinate system and rotating coordinate system, respectively, we can get

$$\begin{pmatrix} x_r \\ y_r \end{pmatrix} = \boldsymbol{T}(\Omega t)\begin{pmatrix} x_s \\ y_s \end{pmatrix} = \begin{pmatrix} \cos(\Omega t+\theta) & \sin(\Omega t+\theta) \\ -\sin(\Omega t+\theta) & \cos(\Omega t+\theta) \end{pmatrix}\begin{pmatrix} x_s \\ y_s \end{pmatrix} \quad (1)$$

where $\boldsymbol{T}(\Omega t)$ is the the synchronous rotating coordinate transformation matrix; θ is the phase compensation angle. In fact, M is the constant vector in the rotating coordinate system. So $(x_r, y_r)^{\mathrm{T}}$ is frequency zero (DC) displacement vector, and can be filtered out by a low-pass filter. Through a multiplication by the inverse synchronous rotating coordinate transformation matrix $\hat{\boldsymbol{T}}(\Omega t)$, namely,

$$\begin{bmatrix} x_c \\ y_c \end{bmatrix} = \hat{\boldsymbol{T}}(\Omega t)\begin{bmatrix} x_r \\ y_r \end{bmatrix} = \begin{bmatrix} \cos(\Omega t) & -\sin(\Omega t) \\ \sin(\Omega t) & \cos(\Omega t) \end{bmatrix}\begin{bmatrix} x_r \\ y_r \end{bmatrix} \quad (2)$$

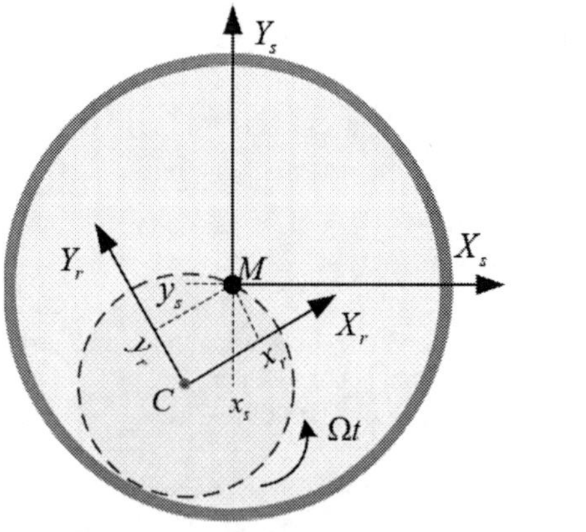

Figure 1. Schematic diagram of coordinate transformation

The DC displacement vector is shifted back to synchronous vibration signals, which need to be identified. This two-modulation procedure is shown in Figure 2.

Here we use the complex coefficient method to represent the displacement vector $(x,y)^{\mathrm{T}}$, transformation matrix $\boldsymbol{T}(\Omega t)$ and $\hat{\boldsymbol{T}}(\Omega t)$ by

$$\begin{aligned} d &= x + jy \\ \boldsymbol{T}(\Omega t) &= e^{-j(\Omega t+\theta)} \\ \hat{\boldsymbol{T}}(\Omega t) &= e^{j\Omega t} \end{aligned} \tag{3}$$

Set $\tilde{d}(j\omega)$, $\tilde{\boldsymbol{T}}(j\omega,\Omega)$ and $\tilde{\boldsymbol{T}}_{inv}(j\omega,\Omega)$ as the Fourier transformations of $d(t)$, $\boldsymbol{T}(\Omega t)$, $\boldsymbol{T}_{inv}(\Omega t)$, respectively. It can be derived that

$$\begin{cases} \tilde{\boldsymbol{T}}(j\omega,\Omega) = \delta(j\omega + j\Omega)e^{-j\theta} \\ \tilde{\boldsymbol{T}}_{inv}(j\omega,\Omega) = \delta(j\omega - j\Omega) \end{cases} \tag{4}$$

Meanwhile, it can also be achieved that

$$\begin{cases} F(d(t) * \boldsymbol{T}(\Omega t)) = \tilde{\boldsymbol{T}}(j\omega,\Omega) \otimes \tilde{d}(j\omega) = \tilde{d}(j\omega + j\Omega)e^{-j\theta} \\ F(d(t) * \boldsymbol{T}_{inv}(\Omega t)) = \tilde{\boldsymbol{T}}_{inv}(j\omega,\Omega) \otimes \tilde{d}(j\omega) = \tilde{d}(j\omega - j\Omega) \end{cases} \tag{5}$$

where $F(\cdot)$ and $\otimes$ denote the Fourier transformation and revolution operator, respectively. Assume the transfer function of the low-pass filter as $g_f(s)$, and combining with (1), (2) and (5), the transfer function of the band-pass filter shown in Figure 2 is given by

$$\begin{aligned} \hat{g}_n(s) &= \frac{\tilde{d}_{out}(s)}{\tilde{d}_{in}(s)} = \frac{\tilde{x}_{out}(s) + j\tilde{y}_{out}(s)}{\tilde{x}_{in}(s) + j\tilde{y}_{in}(s)} \\ &= \frac{\tilde{\boldsymbol{T}}_{inv}(j\omega,\Omega) \otimes (g_f(s) * \tilde{\boldsymbol{T}}(j\omega,\Omega) \otimes \tilde{d}_{in}(s))}{d_{in}(s)} \\ &= g_f(s - j\Omega)e^{-j\theta} \end{aligned} \tag{6}$$

Therefore, the transfer function of this novel notch filter can be achieved as

$$g_n(s) = \frac{1}{1 + \hat{g}_n(s)} = \frac{1}{1 + g_f(s - j\Omega)e^{-j\theta}} \tag{7}$$

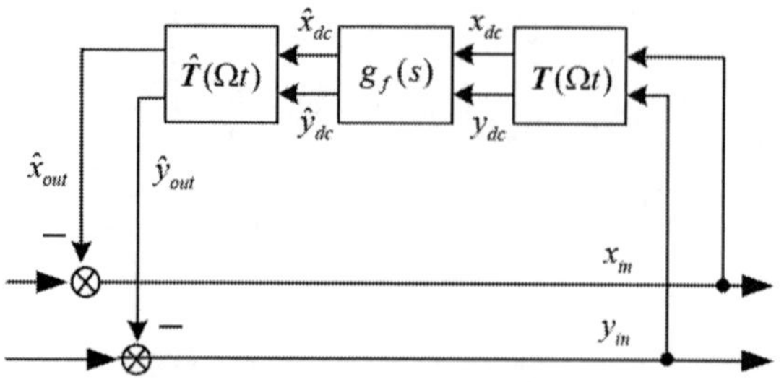

Figure 2. The implemetation principle of novel notch filter

B. Performance Comparisons

Usually, the classic notch filter has the following form

$$g_r(s) = \frac{s^2 + \varepsilon_1 s + \Omega^2}{s^2 + \varepsilon_2 s + \Omega^2} \tag{8}$$

When the first order low-pass filter is adopted, the novel notch filter is given as Eq. (8). The frequency response at the center frequency $j\Omega$ is $g_n(j\Omega) = \varepsilon / (\varepsilon + \kappa e^{-j\theta})$, and as for $j\omega = j(\sigma + \Omega)$ where $\sigma << \Omega$, it can resolved that

$$g_n(j(\Omega+\sigma)) = \frac{j(\Omega+\sigma) + \varepsilon - j\Omega}{j(\Omega+\sigma) + \varepsilon + \kappa e^{-j\theta} - j\Omega} = \frac{j\sigma + \varepsilon}{j\sigma + \varepsilon + \kappa e^{-j\theta}} \tag{9}$$

Thereby, it can be concluded that bandwidth and depth of the notch filters are decided by ε and κ, the influence of which are similar to the ε_1 and ε_2 in Eq. (8), respectively (as shown in Figure 3). While compared with the classic notch filter, $e^{-j\theta}$ in Eq. (8) not only leads to the free pole assignment of $g_n(s)$, but also affects the phase-frequency response in the frequencies adjacent to the center frequency Ω, which is essential to the system stability. The proposed notch filter has the same advantage as the two-modulation-step notch filter proposed by Herzog (1996), which can output the steady unbalance state in rotating coordinate without additional computation.

As for the signals $x = \gamma_1 \sin(\Omega t + \varphi_1)$ and $y = \gamma_2 \sin(\Omega t + \varphi_2)$, where $\gamma_1 \neq \gamma_1$ and $\varphi_1 - \varphi_1 \neq \pi/2$, we have that

$$\begin{aligned} x + jy &= \gamma_1 \frac{e^{j(\Omega t+\varphi_1)} - e^{-j(\Omega t+\varphi_1)}}{2j} + \gamma_2 \frac{e^{j(\Omega t+\varphi_2)} - e^{-j(\Omega t+\varphi_2)}}{2j} \\ &= \hat{\gamma}_1 e^{j\Omega t} + \hat{\gamma}_2 e^{-j\Omega t} \end{aligned} \tag{10}$$

where $\hat{\gamma}_1 = -j(\gamma_1 e^{j\varphi_1} + \gamma_2 e^{j\varphi_2})/2$ and $\hat{\gamma}_2 = -j(\gamma_1 e^{-j\varphi_1} + \gamma_2 e^{-j\varphi_2})/2$.

Based on the frequency characteristics of $g_n(s)$, the power of $\hat{\gamma}_1 e^{j\Omega t}$ will be greatly attenuated and that of $\hat{\gamma}_2 e^{-j\Omega t}$ will remain invariant when $x + jy$ passes though $g_n(s)$, while both of $\hat{\gamma}_1 e^{j\Omega t}$ and $\hat{\gamma}_2 e^{-j\Omega t}$ will be attenuated by passing through $g_r(s)$.

TABLE I. THE SIMULATION PAREMETER OF NOTCH FILTER

Parameters	Values
ε	8π
κ	36π
θ	0
ε_1	42
ε_2	210
Ω	500Hz

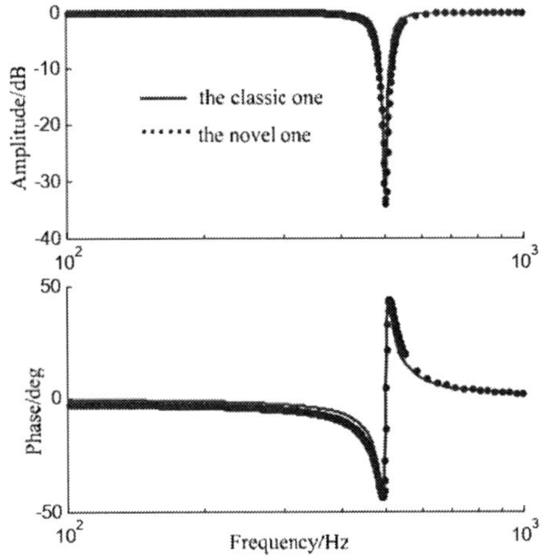

Figure 3. The Bode diagram of notch filter

III. IMPLEMENTATION

A. Control Scheme

To evaluate the effectiveness of the proposed method, experiments were carried out on an AMB based compressor test rig, which is fixed vertically on a metal base (as shown in Figure 4). The design parameters of this test rig are listed in Table 1.

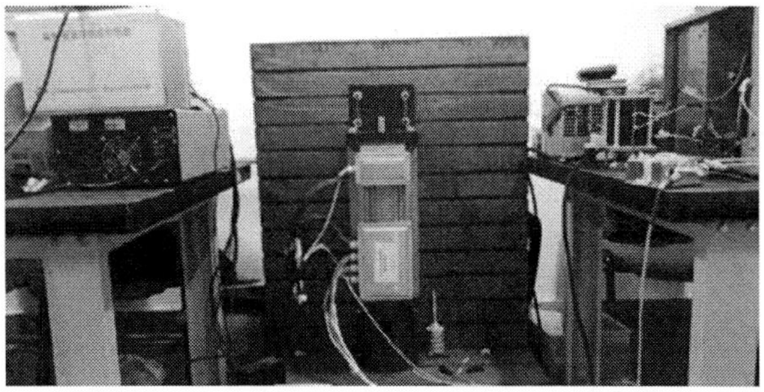

Figure 4. AMB test rig

TABLE II. DESIGN PARAMETERS OF AMB TEST RIG

Name	Value	Unit
Length of the rotor	554	mm
Mass of the rotor	6.85	kg
Power	4	kW
The first bending frequency	650	Hz
The transverse moment of inertia	0.1147	kg m^2
The polar moment of inertia	0.002529	kg m^2
Bias current	1.2	A
Current stiffness	43	N A^{-1}
Negative position stiffness	-0.21×10^6	N m^{-1}
Radial protective clearance	0.2	mm

In this paper, the inverse power amplifier model in feedforward channel is also introduced to compensate the displacement stiffness force to improve the accuracy of the synchronous displacement stiffness force compensation. The schematic diagram of the AMB control system combined with the novel notch filter is shown in Figure 5.

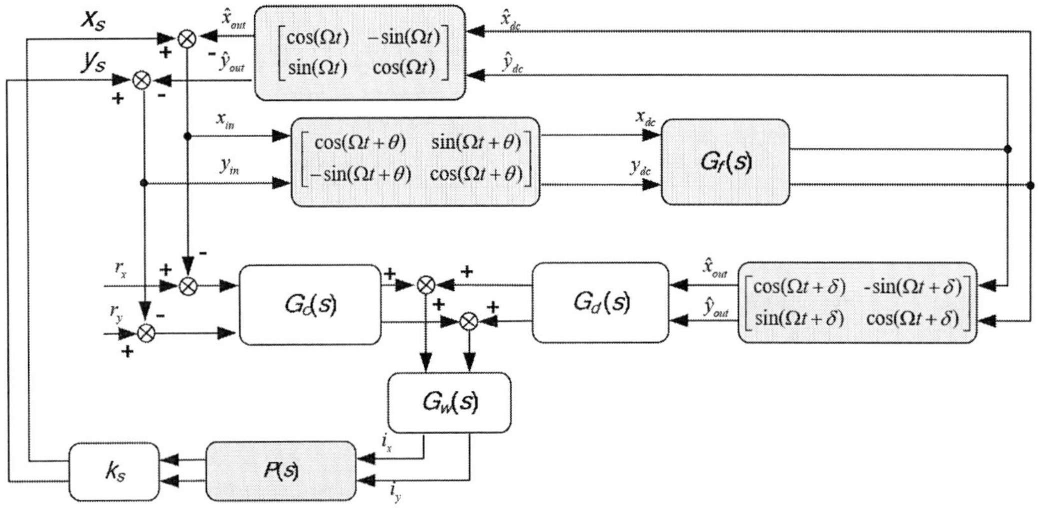

Figure 5. Vibration control scheme used for this paper

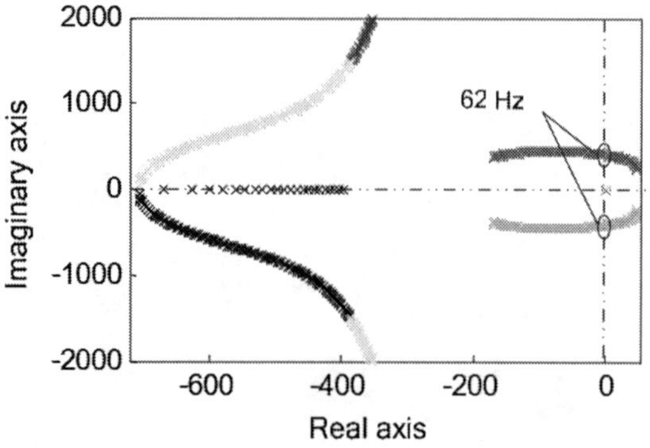

Figure 6. Stability analysis: dominant root locus

Stability is the primary concern, among the other specifications due to the energy and speed of the high-speed rotating machinery. For the linearized high-speed AMB-rotor system, the closed-loop root locus is widely employed in the controller design for its simplicity. Figure 6 reveals the dominant root locus when Ω varies from 20Hz to 800Hz with the control scheme shown in Figure 5. It can be seen that the system is stable if Ω >62Hz. Since the AMB rotor system runs at a speed of 30000r/min in this paper, stability of the system can meet the operation requirements.

B. *Experiments*

In order to evaluate the proposed compensation scheme, an AMB test rig as shown in Figure 4 is used to test compensation scheme performance on the synchronous vibration reduction. Vibration signals are obtained by an acceleration sensor mounted on the motor base.

Vibration acceleration signal spectrums are measured and analyzed in the two cases: with the proposed method and without the proposed method. Figure 7 shows the vibration acceleration signal spectrums of the motor housing without any vibration suppression measures. We can find that the synchronous component of the vibration acceleration is 0.0560 *g* (*g* is the acceleration of gravity). Figure 8 shows the vibration acceleration signal spectrums of the motor housing which is removing synchronous current by novel notch filter based on coordinate transformation. The synchronous component of the vibration acceleration is 0.0262 *g*. We see from these results that there is greater than 50% reduction in synchronous vibration. Therefore, experimental results clearly show that the proposed method significantly reduces the unbalance vibration of the AMB test rig.

IV. Conclusion

This paper presents a notch filter for the autocentering of the magnetically suspended rotors based on rotating coordinate transformation. Simulation and experimental results show that the synchronous vibration force is significantly reduced. However, we can find there is low-frequency vibration on the motor housing which is caused by the air flow when turbine blades is rotating, but we haven't found a suitable way to accurately analyze its mechanics. It still needs further research in the future.

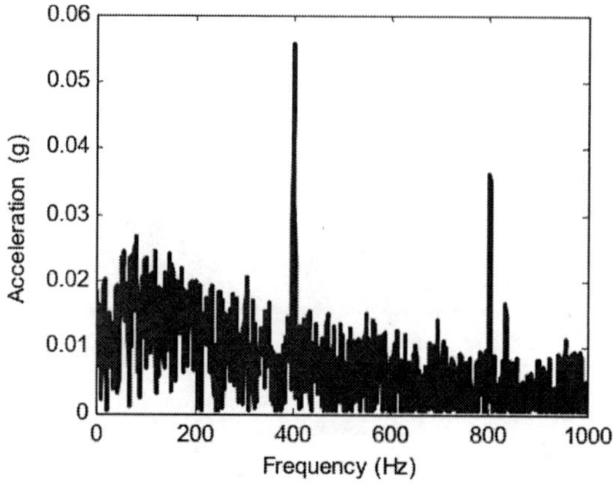

Figure 7. Vibration signal spectrum of the motor base without the proposed method

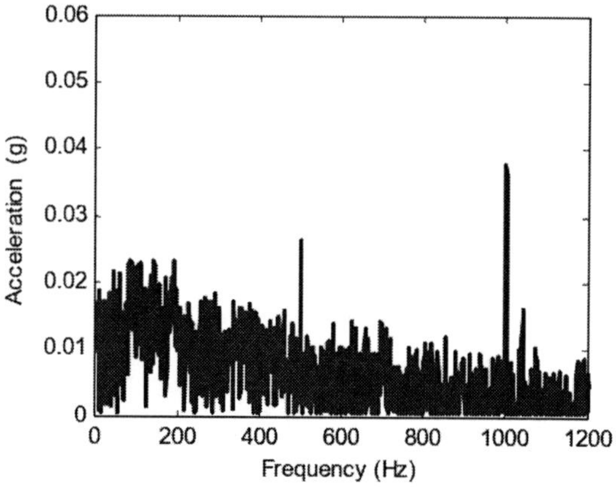

Figure 8. Vibration signal spectrum of the motor base with the proposed method

References

[1] C. Zhu, D.A. Robb, D.J. Ewins, The dynamics of a cracked rotor with an active magnetic bearing, Journal of Sound and Vibration 265 (2003) 469–487.

[2] G. Mania, D. D. Quinna, M. Kasardab, Active health monitoring in a rotating cracked shaft using active magnetic bearings as force actuators, Journal of sound and vibration 294 (2006) 454–465.

[3] R. Herzog, P. Buhler, C. Gahler, R. Larsonneur, Unbalance compensation using generalized notch filters in the multivariable feedback of magnetic bearings, IEEE Transactions on Control System Technology 4(5) (1996) 580–586.

[4] C. Bi, D. Wu, Q. Jiang and Z. Liu, Automatic learning control for unbalance compensation in active magnetic bearings, IEEE Transactions on Magnetics 41(7) (2005) 2270–2280.

[5] Y. Maruyama, T. Mizuno, M. Takasaki, Y. Ishino, H. Kameno, and A. Kubo, Application of rotor unbalance compensation to AMB-based gyroscopic sensor, Proceedings of the 11th International Symposium on Magnetic Bearings, Nara, Japan, August 2008, pp. 206-211.

[6] K. Jiang, C. Zhu, M. Tang, A uniform control method for imbalance compensation and automation balancing in active magnetic bearing-rotor systems, ASME Journal of Dynamic Systems, Measurement, and Control 134(2) (2012) DOI: 10.1115/1.4005279, in press.

[7] L. Li, T. Shinshi, C. Iijima, X. Zhang, A. Shimokohbe, Compensation of rotor imbalance for precision rotation of a planar magnetic bearing rotor, Precision Engineering 27(2) (2003) 140–150.

Identification of magnetic bearing system using a novel subspace identification method

Sun Zhe[a,b], Zhao Jingjing[a,b], Shi Zhengang[a,b]

[a] Institute of Nuclear and New Energy Technology, Tsinghua University, Beijing, 100084, PR China, {sun_zhe, zhao-jj, shizg}@tsinghua.edu.cn
[b] The key laboratory of advanced reactor engineering and safety, Ministry of Education
Beijing, 100084, PR China

***Abstract*—Active magnetic bearings (AMBs) are widely applied to support high-speed rotating machineries, especially in special environment. Identification is a useful method to obtain the mathematical model of an AMB system. Frequency domain subspace identification method (FDSIM) is an attractive identification technique. However, in practice one usually expects that the identified model possesses some properties, for example stability. FDSIM cannot ensure these expected properties. This paper firstly proposes a novel subspace identification method, FDSIM-λ. This method can guarantee that the resultant model possesses predefined poles. Moreover, an optimization procedure is introduced to obtain the optimal poles based on FDSIM-λ, the initial estimation of system poles and constraints on the system poles. The proposed method is then applied to identify an AMB system and validated by the experiments on a real AMB system.**

I. Introduction

Compared with conventional bearings, active magnetic bearings (AMBs)[1], [2] possess several attractive advantages, such as no friction, no need of lubrication, and the ability of long-term high speed running. Therefore AMBs are widely applied to support high-speed rotating machineries, especially in special environment. Nevertheless, for each application case, the AMB system should be specifically designed, installed and adjusted. This procedure is quite costly and time-consuming. The mathematical model of an AMB system plays an important role in designing and adjusting the whole system. The models of parts in an AMB system can be computed numerically [1], [2], [3]. However, the practical parts are usually so complex that the computational models of flexible rotors and AMB systems are not precise enough.

Identification is the procedure of estimating the model of a system based on measured input-response data of this system. It is an useful method to obtain the system model. From the viewpoint of control system analysis, an AMB system with a flexible rotor possesses several special features. First, it is well-known that an AMB system is open-loop unstable. This obstructs the application of most well-developed identification methods where the stability of system is assumed. Second, the flexible rotor lacks of internal damping, which results in a series of numerical problems in computation. Based on these reasons, frequency domain test and identification are often utilized to attain the model of an AMB system.

Frequency domain subspace identification method (FDSIM)[4], [5], [6] is an attractive identification technique and an improved version of FDSIM proposed in [5] and the so-called w-operator is introduced in this method to improve computational aspects. In this method, the continuous-time state space description $(\mathbf{A}, \mathbf{B}, \mathbf{C}, \mathbf{D})$ of a system is utilized and the system model is then transferred to a w-domain description $(\mathbf{A}_w, \mathbf{B}_w, \mathbf{C}_w, \mathbf{D}_w)$. The column space of the observation matrix $\begin{bmatrix}\mathbf{C}_w^\top & (\mathbf{C}_w\mathbf{A}_w)^\top & \cdots & (\mathbf{C}_w\mathbf{A}_w^{i-1})^\top\end{bmatrix}^\top, i \in \mathcal{N}$ is estimated based on frequency domain response (FDR) data and estimation of the model $(\mathbf{A}_w, \mathbf{B}_w, \mathbf{C}_w, \mathbf{D}_w)$ can be attained. The FDSIM possesses excellent numerical stability and generality, therefore it is applied to many identification problems, including the region of AMB systems[7].

However, the FDSIM possesses a serious defect, namely no prior knowledge and assumption is involved in the identification. In other words, the stability and any other properties of the resultant model cannot be guaranteed by this method. To overcome this defect, efforts on subspace identification method with guaranteed stability are reported [8], [9]. Nevertheless, as mentioned, an AMB system is open-loop unstable and the damping ratios of the bending modes are relatively small, these characteristics obstruct the direct application of these modified methods. It is important to notice that the poles of an AMB system can be roughly estimated by the numerical computations and the FDR measurements. In this paper, a novel FDSIM method is proposed, the proposed method ensures that the resultant model possesses predefined poles. Then, an optimization procedure is introduced to modify the estimated poles. The initial estimation of the system poles and constraints on the system poles will be introduced in this optimization procedure. In this way, the properties of the poles of the resultant system can be ensured. The proposed method is then applied to identify an AMB system and validated by the experiments on a real AMB system.

In this paper, the notation $\dagger$ is used to denote the Moore-Penrose generalized inverse of a matrix, $\|\circ\|$ denotes the Frobinius norm of a matrix. The notation $R_{(m_1,m_2)}(\mathbf{U})$ reflects the matrix composed of the m_1-th to m_2-th rows of matrix $\mathbf{U}$, $\Leftrightarrow$ denotes that two optimization problems are equivalent. For matrixes $(\mathbf{A}, \mathbf{C})$ and some integer i, the generalized observation matrix is defined as

$$\mathbf{G}_o(\mathbf{C},\mathbf{A},i)=\begin{bmatrix}\mathbf{C}\\ \mathbf{CA}\\ \vdots\\ \mathbf{CA}^{i-1}\end{bmatrix}. \quad (1)$$

Similarly, for $(\mathbf{A},\mathbf{V})$ and i, the generalized input matrix is defined as

$$\mathbf{H}_i(\mathbf{V},\mathbf{A},i)=\begin{bmatrix}\mathbf{V} & \mathbf{AV} & \cdots & \mathbf{A}^{i-1}\mathbf{V}\end{bmatrix}. \quad (2)$$

II. Subspace identification with predefined poles

A. Frequency domain subspace identification method

For a system $(\mathbf{A},\mathbf{B},\mathbf{C},\mathbf{D})$ with $\mathbf{A}\in\mathcal{R}^{n\times n},\mathbf{B}\in\mathcal{R}^{n\times h},\mathbf{C}\in\mathcal{R}^{m\times n},\mathbf{D}\in\mathcal{R}^{m\times h}$, for some $i\geq n$, the generalized observation matrix $\mathbf{G}_o(\mathbf{C},\mathbf{A},i)$ plays the center role in the subspace identification methods [4], [5], [6]. More specifically, in the subspace identification methods, the column space of $\mathbf{G}_o(\mathbf{C},\mathbf{A},i)$ can be estimated based on the time domain or frequency domain measurement of system inputs and outputs. Then the system matrixes $(\mathbf{A},\mathbf{B},\mathbf{C},\mathbf{D})$ are evaluated based on the estimation of $\mathbf{G}_o(\mathbf{C},\mathbf{A},i)$. As an improvement of the standard FDSIM, [5] introduced the so-called w-operator to prevent numerical illness or ill-condition. This paper concentrates on the w-FDSIM (namely FDSIM with w-operator) and this paper mainly focuses on the evaluation of w-domain system matrixes $(\mathbf{A}_w,\mathbf{C}_w)$ based on the estimation of column space of $\mathbf{G}_o(\mathbf{C}_w,\mathbf{A}_w,i)$. The estimation of the column space of $\mathbf{G}_o(\mathbf{C}_w,\mathbf{A}_w,i)$ and evaluation of $(\mathbf{B}_w,\mathbf{D}_w)$ are realized by the algorithm proposed in [5].

B. FDSIM with predefined poles

In the standard w-FDSIM, the column space of the observation matrix $\mathbf{G}_o(\mathbf{C}_w,\mathbf{A}_w,i)$ is estimated as a matrix $\hat{\mathbf{U}}_n$, which is constructed based on FDR measurement. The estimations $\hat{\mathbf{C}}_w$ and $\hat{\mathbf{A}}_w$ are evaluated directly by

$$\hat{\mathbf{C}}_w=R_{(1,m)}\left(\hat{\mathbf{U}}_n\right), \quad (3a)$$

$$\hat{\mathbf{A}}_w=R_{(1,m(i-1))}\left(\hat{\mathbf{U}}_n\right)^{\dagger}R_{(m+1,mi)}\left(\hat{\mathbf{U}}_n\right). \quad (3b)$$

However, no prior knowledge and assumptions about $\hat{\mathbf{C}}_w$ and $\hat{\mathbf{A}}_w$ are involved in (3). An obvious problem is that the estimated $\hat{\mathbf{A}}_w$ may be unstable, no matter $\mathbf{A}_w$ is stable or not. Some modified methods with guaranteed stability are reported [4], [8], [9]. Nevertheless, an AMB system is open-loop unstable and these methods cannot be applied directly.

To overcome these defects, we propose a novel method to estimate $\hat{\mathbf{C}}_w$ and $\hat{\mathbf{A}}_w$ based on $\hat{\mathbf{U}}_n$. Since a state space model is invariant under similar transformation, we can define

$$\hat{\mathbf{A}}_w=\mathbf{U}\bar{\mathbf{A}}_w\mathbf{V},\hat{\mathbf{C}}_w=\bar{\mathbf{C}}_w\mathbf{V}, \quad (4)$$

where $\bar{\mathbf{A}}_w$ is a real block diagonal matrix, whose eigenvalues equal the predefined poles $\boldsymbol{\lambda}=\{\lambda_k\}$ and $\mathbf{V}=\mathbf{U}^{-1}$ is a matrix to be optimized. In this paper, for given $\boldsymbol{\lambda}$, the following optimization problem is solved to obtain $\mathbf{V}$ and $\bar{\mathbf{C}}_w$

$$\min_{\mathbf{V},\bar{\mathbf{C}}_\mathbf{w}}\left\|\mathbf{G}_o\left(\hat{\mathbf{C}}_w,\hat{\mathbf{A}}_w,i\right)-\hat{\mathbf{U}}_n\right\|^2. \quad (5)$$

The optimization problem (5) can be further reduced. In fact, suppose that

$$\hat{\mathbf{U}}_n=\begin{bmatrix}\mathbf{U}_1\\ \vdots\\ \mathbf{U}_i\end{bmatrix}, \quad (6)$$

with $\mathbf{U}_k\in\mathcal{R}^{m\times n},k=1,\cdots,i$. Define the block rearrangement of $\hat{\mathbf{U}}_n$ as

$$\hat{\mathbf{U}}_n'=\begin{bmatrix}\mathbf{U}_1 & \cdots & \mathbf{U}_i\end{bmatrix}, \quad (7)$$

then it can be verified that

$$\min_{\mathbf{V},\bar{\mathbf{C}}_\mathbf{w}}\left\|\mathbf{G}_o\left(\hat{\mathbf{C}}_w,\hat{\mathbf{A}}_w,i\right)-\hat{\mathbf{U}}_n\right\|^2\Leftrightarrow\min_{\mathbf{V},\bar{\mathbf{C}}_\mathbf{w}}\left\|\begin{bmatrix}\bar{\mathbf{C}}_w\mathbf{V}\\ \bar{\mathbf{C}}_w\bar{\mathbf{A}}_w\mathbf{V}\\ \vdots\\ \bar{\mathbf{C}}_w\bar{\mathbf{A}}_w^{i-1}\mathbf{V}\end{bmatrix}-\hat{\mathbf{U}}_n\right\|^2$$

$$\overset{\mathbf{H}=\mathbf{H}_i(\mathbf{V},\bar{\mathbf{A}}_w,i)}{\Longleftrightarrow}\min_{\mathbf{V},\bar{\mathbf{C}}_\mathbf{w}}\left\|\bar{\mathbf{C}}_w\mathbf{H}-\hat{\mathbf{U}}_n'\right\|^2$$

$$\overset{\bar{\mathbf{C}}_w=\hat{\mathbf{U}}_n'\mathbf{H}^{\dagger}}{\Longleftrightarrow}\min_{\mathbf{V}}\left\|\hat{\mathbf{U}}_n'\mathbf{H}^{\dagger}\mathbf{H}-\hat{\mathbf{U}}_n'\right\|^2. \quad (8)$$

The last optimization problem in (8) is a nonlinear optimization problem associated with the invertible matrix $\mathbf{V}$. This optimization problem is not easy to solve in practice. In this paper the random optimization technique is utilized to obtain a "good enough" solution of $\hat{\mathbf{V}}$. That is, for a large enough number N, N invertible candidate matrixes $\{\mathbf{V}_k\}_{k=1}^{N}$ are generated from some probability distribution, then the "good enough" estimation $\hat{\mathbf{V}}$ is given by

$$\hat{\mathbf{V}}=\left\{\mathbf{V}_\kappa:\nu(\mathbf{V}_\kappa)=\min_{1\leq k\leq N}\nu(\mathbf{V}_k)\right\}, \quad (9)$$

where

$$\nu(\mathbf{V})=\left\|\hat{\mathbf{U}}_n'\mathbf{H}^{\dagger}\mathbf{H}-\hat{\mathbf{U}}_n'\right\|^2,\mathbf{H}=\mathbf{H}_i\left(\mathbf{V},\bar{\mathbf{A}}_w,i\right). \quad (10)$$

Finally,

$$\bar{\mathbf{C}}_w=\hat{\mathbf{U}}_n'\mathbf{H}_i\left(\hat{\mathbf{V}},\bar{\mathbf{A}}_w,i\right)^{\dagger}, \quad (11)$$

$$\hat{\mathbf{A}}_w=\hat{\mathbf{V}}^{-1}\bar{\mathbf{A}}_w\hat{\mathbf{V}},\hat{\mathbf{C}}_w=\bar{\mathbf{C}}_w\hat{\mathbf{V}}. \quad (12)$$

In this paper, the proposed FDSIM with predefined poles will be abbreviated as "FDSIM-$\boldsymbol{\lambda}$".

III. Identification based on FDSIM-$\boldsymbol{\lambda}$ with initial estimation of system poles

For a system, suppose the FDR data

$$\mathbf{G}_m(\mathrm{j}\omega_k),k=1,2,\cdots,M \quad (13)$$

is obtained by the swept-sine measurement. Moreover, suppose the poles of the subsystem can be roughly estimated as $\boldsymbol{\lambda}_0$.

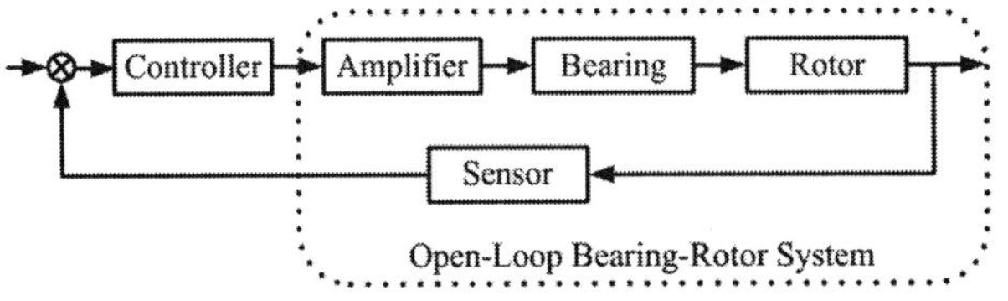

Figure 1: Scheme of an AMB system

For given pole set $\boldsymbol{\lambda}$, the identified model by FDSIM-$\boldsymbol{\lambda}$ is $\hat{\Psi}(\boldsymbol{\lambda}) = (\mathbf{A}(\boldsymbol{\lambda}), \mathbf{B}(\boldsymbol{\lambda}), \mathbf{C}(\boldsymbol{\lambda}), \mathbf{D}(\boldsymbol{\lambda}))$ and the corresponding frequency responses are

$$\hat{\mathbf{G}}_i(\boldsymbol{\lambda}, \mathrm{j}\omega_k), k = 1, 2, \cdots, M. \tag{14}$$

Then the pole set is estimated by solving the following optimization problem:

$$\hat{\boldsymbol{\lambda}} = \underset{\boldsymbol{\lambda}}{\operatorname{argmin}} \sum_{k=1}^{M} \left\| \hat{\mathbf{G}}_i(\boldsymbol{\lambda}, \mathrm{j}\omega_k) - \mathbf{G}_m(\mathrm{j}\omega_k) \right\|^2. \tag{15}$$

The initial estimation of poles $\boldsymbol{\lambda}_0$ is used as the initial value in this optimization problem. Some constraints can be incorporated into this optimization problem to ensure the resultant optimal poles obey some known properties. This optimization problem can be solved by standard nonlinear optimization algorithm and the details of solving is ignored in this paper.

IV. Experiment: identification of an AMB system

In this section, the proposed method is applied to identify an AMB system.

A. System model and problem description

This paper concentrates on AMB systems with fully suspended rotors. This paper supposes that the axial and radial dynamic characteristics of the rotor can be decoupled. Moreover, this paper considers static identification, that is, during the identification experiments the rotor does not rotate. Thus gyroscopic effects [3] are not involved in the experiments and the dynamic features in two orthogonal radial planes can also be decoupled. In this paper we only consider the dynamic model of the rotor in one radial plane.

Since AMB systems are open-loop unstable, only closed-loop experiments can be performed. However, the model of the controller is known, thus it is not difficult to compute the open-loop model by block transformations. The system scheme of a typical AMB system is shown in Fig. 1. As shown in Fig. 1, this paper mainly considers the open-loop system consisting of the power amplifier, the magnetic bearing, the flexible rotor and the sensor. In the following part of this paper, we use the term "Open-Loop Bearing-Rotor System (OLBRS)" to refer to this system. In a radial plane, the OLBRS is a (2×2)-dimensional system.

B. Experiment setup and data preprocessing

The magnetic bearing system used in the experiments is shown as Fig. 2. A detailed description of this system can be found in [10]. This system is characterized by an about

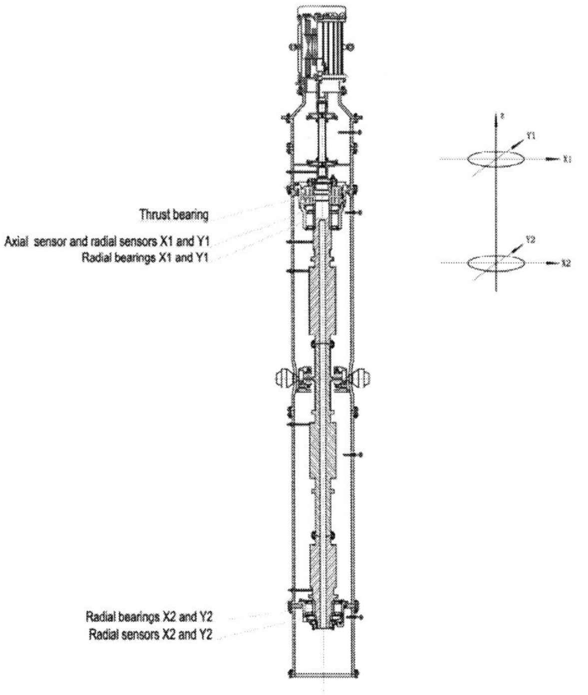

Figure 2: Magnetic bearing-rotor system

3.5m long and 630kg heavy rotor. Four radial electromagnetic bearings and an axial one are mounted to suspend the rotor. The nominal maximum displacement of rotor, restricted by the clearance of auxiliary bearing, is 150μm. The definition of coordinate system in this bearing system is illustrated in Fig. 2. The nature frequencies of the first three free-free bending modes of the rotor are approximately 45Hz, 122Hz and 239Hz, respectively. A maximum rotational speed of 230Hz is achieved on this system, namely beyond the second bending mode.

Closed-loop swept-sine measurements are performed on this AMB system and the FDR model of OLBRS is then calculated. For such a large-size AMB system, the system delay will significantly affect the FDR. However, the system delay model is not involved in a FDSIM model and will bring negative effect to identification, thus it should be eliminated. In this paper, the method proposed in [11] is applied to estimate and eliminate the system delay.

C. Experiment results

The experiment result is shown in Fig. 3. Here Bode plot is applied to express the magnitude of the frequency response gain and the phase shift. Since the OLBRMs discussed in this paper are 2×2 systems, four subplots are contained in the plot. The arrangement of the subplots corresponds to the arrangement of the transfer function matrix. In this Bode plot, blue dots denote the measured FDR and red lines denote the frequency response of identified model.

As shown in Fig. 3, the Bode plots of identified model fit the measured FDR precisely. This fact validates the proposed

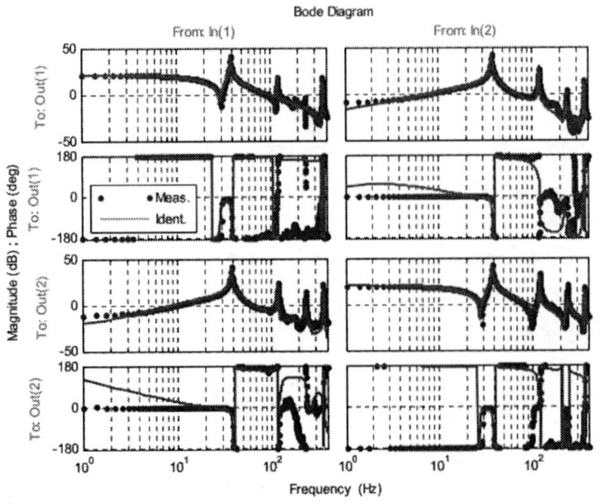

Figure 3: Experiment result

identification method. However, some antiresonances (e.g. the antiresonance of the second bending mode) are not correctly identified. In our opinion, the main reason is that FDSIM is essentially a least-square type method, the estimation error near antiresonances will be submerged by the error near resonances, as discussed in [11].

V. Conclusions and Discussions

In this paper a novel FDSIM is proposed. The proposed method FDSIM-λ can ensure that the resultant model possesses predefined poles. Based on the proposed method, an initial estimation of the system poles and constraints on the poles, the system can be identified and it can be guaranteed that the resultant system possesses some predefined properties. The proposed method is applied to identify AMB systems. Experiments on a large-size AMB system validate the proposed method.

The future works of this paper include:

1) In this paper, a random optimization strategy is applied to find a "good enough" transform matrix $\hat{\mathbf{V}}$. However, this is quite inefficient and the exact optimum cannot be achieved. Analytical or efficient numerical solution of Eq. (8) should be developed.

2) As mentioned, the antiresonances are not precisely fitted by the proposed method. On the other hand, as thoroughly discussed in [12], [13], [11], the antiresonance frequencies play an important role in defining the rotor's flexible behavior, therefore one expects precise agreement of the identified model with the measured data near antiresonance frequencies. A simple solution to this problem is adding a cost term associated to the antiresonances to Eq. (15). However, this method will not affect the least-square nature of FDSIM. Researches on this problem should be performed.

3) In this paper, the parameters in FDSIM are chosen empirically. Moreover, for a stable system, the numerical stability can be ensured by the w-FDSIM, however, an AMB system is essentially instable, so that numerical illness or ill-condition may arise while applied w-FDSIM-based methods and the paramter α in w-FDSIM (see [5]) should be selected carefully. In future works, we expect to make detailed researches on the choice of these parameters and develop novel identification method to deal with instable systems.

VI. Acknowledgements

This paper is financially supported by the National Science and Technology Major Project of China (2011ZX069) and Project 61305065 supported by NSFC.

References

[1] G. Schweitzer, A. Traxler, and H. Bleuler, *Magnetlager: Grundlagen, Eigenshaften und Anwendungen berührungsfreier elektromagnetischer Lager*. Berlin: Springer Verlag, 1992.

[2] G. Schweitzer and E. H. Malsen, *Magnetic Bearings: Theory, design and application to rotating machinery*. Berlin: Springer Verlag, 2009.

[3] G. Genta, *Dynamics of rotating systems*. New York, NY: Springer Science, Business Media Inc., 2005.

[4] P. van Overschee and B. D. Moor, *Subspace identification for linear systems: theory, implementation, applications*. Norwell, Massachusetts, USA: Kluwer Academic Publishers, 1996.

[5] Z. Yang and S. Sanada, "Frequency domain subspace identification with the aid of the w-operator," *Electrical Engineering in Japan*, vol. 132, no. 1, pp. 46–56, 2000.

[6] R. Mohd-Mokhtar and L. Wang, "System identification of mimo magnetic bearing via continuous time and frequency response data," in *Mechatronics, 2005. ICM'05. IEEE International Conference on*. IEEE, 2005, pp. 191–196.

[7] R. Mohd-Mokhtar, L. Wang, L. Qin, and T. Barry, "Continuous time system identification of magnetic bearing systems using frequency response data," in *Control Conference, 2004. 5th Asian*, vol. 3. IEEE, 2004, pp. 2066–2072.

[8] L. L. Seth and D. S. Bernstein, "Subspace identification with guaranteed stability using constrained optimization," *IEEE Transactions on Automatic Control*, vol. 48, no. 7, pp. 1259–1263, 2003.

[9] T. V. Gestel, J. A. K. Suykens, P. V. Dooren, and B. D. Moor, "Identification of stable models in subspace identification by using regularization," *IEEE Transactions on Automatic Control*, vol. 4, no. 9, pp. 1416–1420, 2001.

[10] L. Shi, S. Yu, G. Yang, Z. Shi, and Y. Xu, "Technical design and principle test of active magnetic bearings for the turbine compressor of htr-10gt," *Nuclear Engineering and Design*, vol. 251, pp. 38–46, 2012.

[11] Z. Sun, Y. He, J. Zhao, Z. Shi, L. Zhao, and S. Yu, "Identification of active magnetic bearing system with a flexible rotor," *Mechanical Systems and Signal Processing*, 2014, dOI: 10.1016/j.ymssp.2014.05.004.

[12] A. C. Wroblewski, J. T. Sawicki, and A. H. Pesch, "Rotor model updating and validation for an active magnetic bearing based high-speed machining spindle," *Journal of Engineering for Gas Turbines and Power*, vol. 134, no. 12, p. 122509, 2012.

[13] R. J. Madden and J. T. Sawicki, "Rotor model validation for an active magnetic bearing machining spindle using mu-synthesis approach," *Journal of Engineering for Gas Turbines and Power*, vol. 134, no. 9, p. 092501, 2012.

Adaptive Kalman filter for active magnetic bearings using softcomputing

Li Li, Wolfgang Kästner, Frank Worlitz,

Institut für Prozeßtechnik, Prozeßautomatisierung und Meßtechnik, Hochschule Zittau/Görlitz, Theodor-Körner-Allee 16, 02763, Zittau, Germany, f.worlitz@hszg.de

Abstract—**Magnetic bearings can not only solve the bearing wear and life problems but also reduce the loss and noise of bearing. However, the strong disturbance and noise from the system affects the control behavior. Based on the Kalman filter the influence of noise will be reduced. But the strong nonlinear and uncertainty of parameter of the magnetic bearings make it difficult to establish the estimation / prediction equation in Kalman filter. This paper presents a design method of system estimation / prediction for Kalman filter with using soft computing. Firstly, linear local model for axial magnetic bearing overall system will be deduced. Then a few system parameters, which is relative with nonlinear and uncertainty, will be obtained by a intelligence function, which uses soft computing algorithm as system identification. Finally, the identified system parameters will be used in state equation in Kalman filter. It aims at better filter performance and state estimation than the conventional linear Kalman filter.**

A. *Introduction*

Many works have shown, that AMBs have tremendous potential for many high speed industrial applications.

The new approach aims to get a better filter performance and system observer by introducing a Kalman filter. Because of the intense nonlinear performance of the magnetically suspended bearings, the traditional linear system model is difficult to guarantee the accuracy in comparison with the actual system, when the system is far from the working point. As a result the filter function will be reduced. Thus a new approach for a precise system prediction for Kalman filter will be searched.

The soft computing consists of fuzzy logic, artificial neural network, and neural fuzzy logic. In the method, the relationship between the large amount of input and output are established. As the widely used intelligent model, it is essentially one nonlinear model and is easy to describe a complicated dynamic system. It has been proved that soft computing model can identify arbitrary nonlinear system and system parameters with a high precision [2].

This paper demonstrates a soft computing variation of the system equation in Kalman filter, with experimental validation on a simulated active axial magnetic bearing. Despite its highly nonlinear and uncertain nature, the dynamics of AMB system are represented using an adaptive linear model with parameters that are identified by system identification with soft computing. In paragraph B a linear Kalman for this system will be introduced. In paragraph C the artificial neural networks are an important tool of the identification for the system parameters. In paragraph D a expert system, which is based on fuzzy rule, using a radial basis function and the result from identification, will be designed for a adaptive Kalman filter. Lastly the simulation's result will be showed.

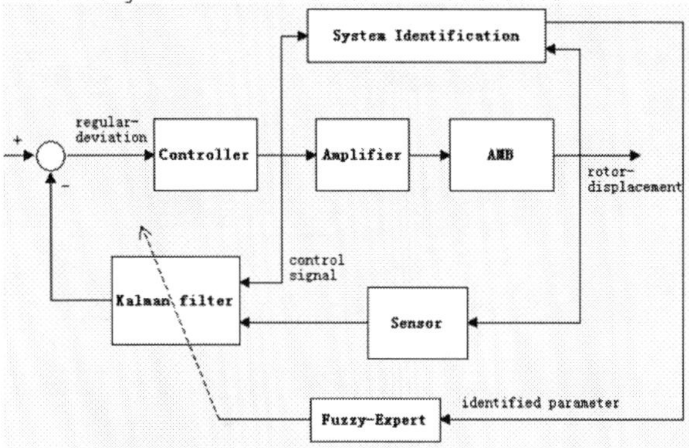

Fig 1: concept of the adaptive Kalman filtering in closed loop

B. *Traditional linear Kalman filter for magnet bearing*

The first step of Kalman filter is to design the system state equation. This paragraph proposes linear discrete equation model on the basis of the force analysis of rotors and linearization of the magnetic force. The Kalman filter embodies the process - and measurement noise. We design a Kalman filter with the constant noise. Detailed analysis is listed as follows.

Force Analysis of the Single Degree Axial Magnetic Bearing

Im magnetic bearing system, single degree magnet poles are usually assembled symmetrically as showed in Fig 2 a pair of electromagnetic forces opposite in direction is created simultaneously be adopting a pair of symmetric power amplification circuits and driving the electromagnet in differential mode. When the rotor is at the geometric center of the bearing, the distances between rotor and air gaps are s_0.

There is equal current i_0, which is also termed as magnetic biasing current between the upper and lower magnet poles to set up magnetic field. An any working state, if the rotor bias is x, then the air gap between rotor and lower magnet is $s_0 + x$. Accordingly, the air gap between rotor and lower magnet is $s_0 - x$. So the resultant force generated by this pair of magnet poles is: [3]

$$F_{Magnet} = k_{Magnet} \cdot \left[\left(\frac{i_0 + i_u}{s_0 - x} \right)^2 - \left(\frac{i_0 - i_u}{s_0 + x} \right)^2 \right] = k_{Magnet} \cdot \frac{i_u^{\ 2}}{x^2} \quad (1)$$

With using magnet parameter: $k_{Magnet} = \dfrac{\mu_0 \cdot A \cdot N^2}{4}$

μ_0 states for vacuum magnetic permeability, and A is the cross sectional area between core and air gap, N is turns of coil, i_u is control current.

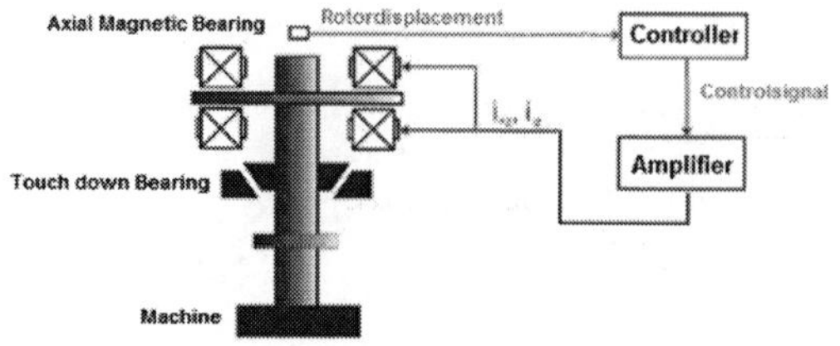

Fig 2: mode of single degreed magnetic bearing

Plant physical parameters	
Mass of rotor	0.518 kg
Force const k_{Magnet}	$N \cdot m^2 / A^2$
Total air gap	5 mm
Nominal air gap s_0	2 mm
Nominal current i_0	1.5 A

Table 1: parameter and value about experiment system

According to *Newton Second Law*, the motion differential equation of single degree magnetic bearing is:

$$F_{Magnet} - m_{Rotor} \cdot g + F_{St} = m_{Rotor} \cdot \ddot{x} \quad (2)$$

F_{St} is the disturbance force.

Linearization of the magnetic force

By taking the Taylor's series expansion of equation (1) for arbitrary operating points (x^* , $i_u^{\ *}$), the nonlinear equations of the magnetic force can be represented by the following equation:

$$F_{Magnet} = k_{Magnet} \cdot \frac{i_u^{\ 2}}{x^2} = k_i \,|_{(s_0, i_0)} \cdot i_u + k_s \,|_{(s_0, i_0)} \cdot x \quad (3)$$

With linear approximation with respect to an equilibrium point (s_0, i_0), position stiffness k_s and current stiffness k_i are given by the following:

$$k_i \,|_{(s_0, i_0)} = \frac{\partial F_{Magnet}}{\partial i} = 2 \cdot k_{Magnet} \cdot \frac{i_0}{s_0^{\ 2}} \quad (4)$$

$$k_s \,|_{(s_0, i_0)} = \frac{\partial F_{Magnet}}{\partial s} = -2 \cdot k_{Magnet} \cdot \frac{i_0^{\ 2}}{s_0^{\ 3}} \quad (5)$$

Because of the intense nonlinear performance of the magnet force, the traditional linearization of the force with the constant stiffness k_i and k_s is difficult to guarantee the long-term accuracy, which is showed by the work [4].

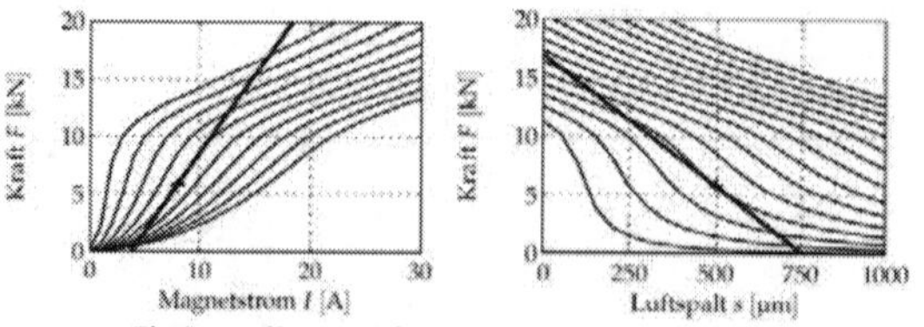

Fig 3: nonlinear performance of the magnet force [4]

According to this reason, the linear magnet force is not equal to the real magnet force, when the work point is far from the equilibrium point.

$$F_{Magnet} = k_{Magnet} \cdot \frac{i_u^{\ 2}}{x^2} \neq k_i \,|_{(s_0, i_0)} \cdot i_u + k_s \,|_{(s_0, i_0)} \cdot x \quad (6)$$

Review of Kalman filter

Im many cases, particularly in industrial practice, PID regulators are implemented for this purpose [5]. Considering the fact that measured rotor position signal typically contain amount of high-frequency measurement noise, a reliable filter with prediction function is desired, Kalman filter goes without differentiation.

The process of designing an optimal state estimator for a magnetic levitation system can be summed up in four main steps-choosing an appropriate filter algorithm, determining the discrete-time state-space model and noise variance. Those steps are outlined subsequently. [5]

In general, a discrete-time process model with stochastic disturbance will be given as:

System state equation:

$$X(k+1) = f_k(X(k), u(k)) + \upsilon(k) \quad (7)$$

Measurement equation:

$$Y(k+1) = h_k(X(k)) + n(k) \quad (8)$$

With the state vector $X(k)$, the input value $u(k)$, the output vector (measurement) $Y(k)$ at the time step k. $\upsilon(k)$ and $n(k)$ are the uncorrelated process noise and measurement noise. A Kalman filter consists of two parts, prediction step and correction step.

As the prediction step:

System state estimation:

$$X(k \mid k-1) = \bar{A} \cdot X(k-1 \mid k-1) + \bar{B} \cdot u(k) \quad (9)$$

Covariance matrix estimation:

$$P(k \mid k-1) = \bar{A} \cdot P(k-1 \mid k-1) \cdot \bar{A}^T + Q \quad (10)$$

As the correction step:

Calculation for Kalman gain:

$$K(k) = P(k \mid k-1) \cdot \bar{H}^T \cdot \left(\bar{H} \cdot P(k \mid k-1) \cdot \bar{H}^T + R \right)^{-1} \quad (11)$$

Optimal estimation for next step:

$$X(k \mid k) = X(k \mid k-1) + K(k) \cdot \left(Z(k) - \bar{H} \cdot X(k \mid k-1) \right) \quad (12)$$

Optimal covariance matrix:

$$P(k \mid k) = (\bar{I} - K_K \cdot \bar{H}) \cdot P(k \mid k-1) \quad (13)$$

The notation k|k-1 denotes an estimation at the time step k with measurement information from time step k-1. The notation k-1|k-1 denotes an optimal measurement at the time step k-1. A is system matrix, B is input matrix, H is measurement vector, P is covariance matrix of the estimation error, Q and R are system noise covariance matrix and measurement noise covariance matrix.

Kalman filter of magnetic bearing system

The active magnetic bearing treated in this paper is an axial bearing (see Fig 2). The electromagnetic force acting on the rotor in bearing axis, which is a function of axial rotor position x and control current i, can be calculated by

$$F_{Magnet} = k_i \,|_{(s_0,i_0)} \cdot i_u + k_s \,|_{(s_0,i_0)} \cdot x \tag{14}$$

The rotor position is linked to the magnetic force and disturbance force. It is connected to the magnet force, disturbance force, "delayed" rotor position and velocity by the discrete equation:

$$x(k+1) = x(k) + T \cdot v(k) + \frac{1}{2} \cdot \left(F_{Magnet} + F_{St}\right) \cdot \frac{T^2}{m_{Rotor}} \tag{15}$$

The rotor velocity can be described by:

$$\dot{x}(k+1) = v(k) + \left(F_{Magnet} + F_{St}\right) \cdot \frac{T}{m_{Rotor}} \tag{16}$$

The magnet force will be described by (3). The control current of the coil is generated by the control signal u from controller, and the magnetic force can be described by:

$$F_{Magnet} = k_i \,|_{(s_0,i_0)} \cdot k_{ui} \cdot u + k_s \,|_{(s_0,i_0)} \cdot x \tag{17}$$

For generating the measuring signal, the state signal will be delayed. The sensor is modeled as a first-order lag element with the equivalent time constant T_{Sen} and proportional gain K_{Sen}. The "delayed" rotor position x_M is connected to the undelayed rotor position x by the differential equation:

$$x_M = \frac{K_{Sen}}{1 + p \cdot T_{Sen}} \cdot x \tag{18}$$

In discrete form it will be described as:

$$x_M(k+1) = K_{Sen} \cdot \left(1 - e^{-T/T_{Sen}}\right) \cdot x(k) + e^{-T/T_{Sen}} \cdot x_M(k) \tag{19}$$

Active magnetic bearings are well known as nonlinear systems. But in general, active magnetic bearings are regarded as linear system. Thus, an estimation algorithm will be used, which regards the system as a linear system using a linear state space model. The equation (14)-(19) describe the 4 state variables of the system. They can be combined to the time discrete state space model:

$$\begin{bmatrix} x(k+1) \\ v(k+1) \\ F_{St}(k+1) \\ X_M(k+1) \end{bmatrix} = \begin{bmatrix} 1 + \frac{k_s \cdot T^2}{2 \cdot m_{Rotor}} & T & \frac{T^2}{2 \cdot m_{Rotor}} & 0 \\ \frac{T \cdot k_s}{m_{Rotor}} & 1 & \frac{T}{m} & 0 \\ 0 & 0 & 1 & 0 \\ K_{Sen} \cdot \left(1 - e^{-T/T_{Sen}}\right) & 0 & 0 & e^{-T/T_{Sen}} \end{bmatrix} \cdot \begin{bmatrix} x(k) \\ v(k) \\ F_{St}(k) \\ x_M(k) \end{bmatrix} + \begin{bmatrix} \frac{k_i \cdot k_{ui} \cdot T^2}{2 \cdot m_{Rotor}} \\ \frac{T \cdot k_i \cdot k_{ui}}{m_{Rotor}} \\ 0 \\ 0 \end{bmatrix} \cdot u(k) \tag{20}$$

The observation equation is:

$$y(k) = \begin{bmatrix} 0 & 0 & 0 & 1 \end{bmatrix} \cdot \begin{bmatrix} x(k) \\ v(k) \\ F_{St}(k) \\ X_M(k) \end{bmatrix} \tag{21}$$

Choice of measurement noise variance R is a quite simple problem, if a sequence of the Gaussian signal noise can be measured . For the experiments, R is the noise variance matrix of sensor $R = diag(\mathrm{var}_{Sensor})$. The process variance matrix Q will be calculated as: $Q = \overline{B} \cdot \mathrm{var}_{Process} \cdot \overline{B}^T$. For the element in matrix Q, which is relative with estimated disturbance, will be experimentally initialized.

Besides the undisturbed rotor position signal, the Kalman filter yields the remaining state variables, rotor velocity, disturbance force and measurement signal. The combination of a Kalman filter as a state estimator in process is depicted in Fig4.

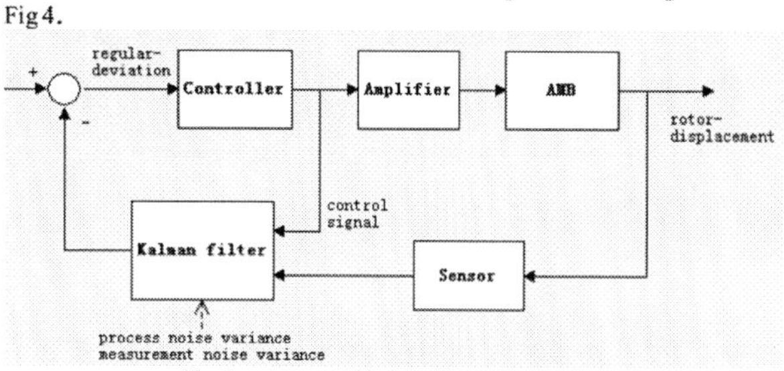

Fig 4: structure of kalman filter in process

C. *System identification with using softcomputing*

It is well known that active magnetic bearings are inherently unstable systems due to the indirect proportionality between the attractive force and the length of the air gap [5]. In addition, the uncertainty of the magnetic force, which is caused by the changed magnetic field, has effect an the calculating of the magnetic force. Thus a identification of the parameter k_{Magnet}, furthermore of k_i and k_s, are necessary.

Identification the system parameter

A neural network is a massively parallel distributed processor made up of simple processing units, called neurons, which has the natural propensity for storing experiential knowledge and making it available for use. And system identification, in general, is to determine the model structure and parameters for a dynamic system based on the measurable input and output data of the system. [7] Model structure in its classic form is multi input single output (MISO) linear system. The mathematical description of the used structure and adaptation law of the ADALINE (Adaptive Linear Element Neural Network) is presented in this section.

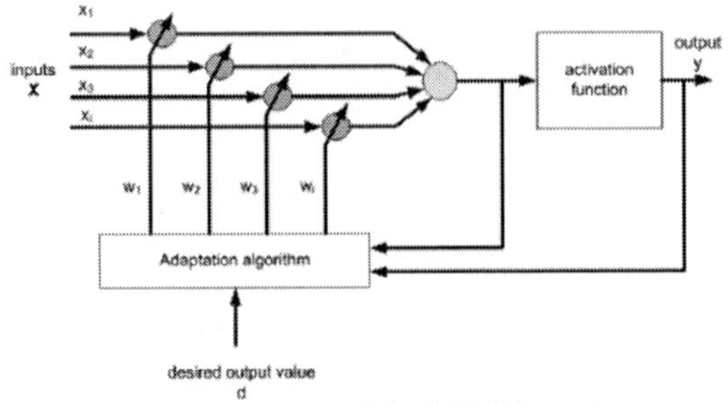

Fig 5: structure of ADALINE [6]

Output signal y(k) for ADALINE model is described by equations (without bias value):

$$y(k) = f(u(k)) \tag{22}$$

$$u(k) = \sum_{i=1}^{N} x_i(k) \cdot w_i(k) \tag{23}$$

Where f – activation function, w_i - weights coefficients, x_i - input signals, u – argument of the function. The online weight adaptation algorithm is realized in each iteration. All weight coefficients w_i are calculated according to the following equation: [6]

$w_i(k+1)=w_i(k)+\Delta w_i(k)$, with k=0,1,2...

Detail of the net training can be found in [6] and [7]. In this work, a ADALINE will be used for the online identification of linear time varying systems, which can be described by a discrete time model.

For axial magnetic supported rotor, the system (rotor dynamic - magnet force) can be presented as a MISO system:

$$x(k+1) = \overline{x} \cdot \overline{w}^T = x(k) + T \cdot v(k) + \frac{1}{2} \cdot \frac{T^2}{m_{Rotor}} \cdot F_{St}(k) + \frac{1}{2} \cdot \frac{T^2}{m_{Rotor}} \cdot F_{Magnet}(k) \tag{24}$$

In such systems, the system input vector

$$\overline{x} = \left[1, \quad T, \quad \frac{1}{2} \cdot \frac{T^2}{m_{Rotor}}, \quad \frac{1}{2} \cdot \frac{T^2}{m_{Rotor}} \right]$$

Weight vector:

$$\overline{w} = \left[x(k), \quad v(k), \quad F_{St}(k), \quad F_{Magnet}(k) \right]^T$$

Single output:

$[x(k+1)]$

In this example the weight parameter (magnet force) $w_4 = F_{Magnet}(k)$ will be adapted and identified.

For magnet force, the system (magnet force – current and air gap) can be presented as a SISO system:

$$F_{Magnet}(k) = \overline{x} \cdot \overline{w}^T = \left[\left(\frac{i_0 + i_u(k)}{s_0 - x(k)} \right)^2 - \left(\frac{i_0 - i_u(k)}{s_0 + x(k)} \right)^2 \right] \cdot k_{Magnet}(k) \tag{25}$$

In such systems, the system input vector is just with one element:

$$\overline{x} = \left[\left(\frac{i_0 + i_u(k)}{s_0 - x(k)} \right)^2 - \left(\frac{i_0 - i_u(k)}{s_0 + x(k)} \right)^2 \right]$$

Weight vector:

$\overline{w} = [k_{Magnet}(k)]^T$

Single output:

$[F_{Magnet}(k)]$

In this example the weight parameter (magnet parameter) $w = k_{Magnet}(k)$ will be adapted and identified.

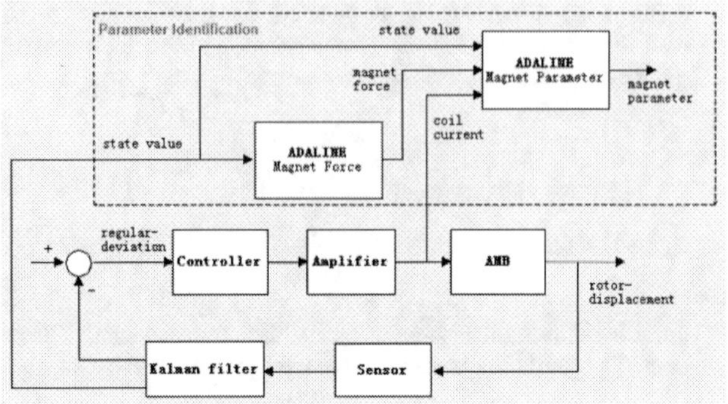

Fig 6: ADALINE used in system identification

Identification of the parameter for working point

The active magnetic bearing system is uncertain, nonlinear and unstable. The most useful approach of dealing with such a system is to linearize it about a single nominal equilibrium point. In this work, we apply least square method (LSM) to obtain the stiffness parameter of the different work point. In order to present the overall system, the work point shall be chosen from the F-(i, s)-map.

The nodes of control current will be chosen as:

control current i_u	
$i_{u1} = -1.5A$	$i_{u2} = -0.75A$
$i_{u3} = 0A$	$i_{u4} = 0.75A$
$i_{u5} = 1.5A$	

Table 2: chosen value of contril current

The nodes of rotor displacement will be chosen as:

rotor displacement x	
$x_1 = -0.5mm$	$x_2 = -0.25mm$
$x_3 = 0mm$	$x_4 = 0.25mm$
$x_5 = 0.5mm$	

Table 3: chose value of rotor displacement

To obtain a linear model of every work point $(i_{u|m}, x_n)$, with $m \in [1,5]$ and $n \in [1,5]$, which means the number of the input – fuzzy set. The magnet force must be linearized. The equation will be used:

$$F_{Magnet}\,|_{n,m} = k_i\,|_{(x_n, i_{u|m})} \cdot i_u + k_s\,|_{(x_n, i_{u|m})} \cdot x \tag{26}$$

The parameters of this linearized model will not vary with rotor position, when the work point is not far from the set point. For these work points, the process will be presented as:

$$F_{Magnet}\,|_{n,m} = k_i\,|_{(x_n, i_{u|m})} \cdot (i_u + \Delta i_u) + k_s\,|_{(x_n, i_{u|m})} \cdot (x + \Delta x) \tag{27}$$

With $|\Delta i_u| \leq 0.01A$, and $|\Delta x| \leq 0.01mm$. The work point series can be written as 9 equations:

$$F_{Magnet-1}|_{n,m} = k_i|_{(x_n, i_{u|m})} \cdot (i_u - 0.01) + k_s|_{(x_n, i_{u|m})} \cdot (x - 0.01)$$
$$F_{Magnet-2}|_{n,m} = k_i|_{(x_n, i_{u|m})} \cdot (i_u - 0.01) + k_s|_{(x_n, i_{u|m})} \cdot (x)$$
$$F_{Magnet-3}|_{n,m} = k_i|_{(x_n, i_{u|m})} \cdot (i_u - 0.01) + k_s|_{(x_n, i_{u|m})} \cdot (x + 0.01)$$
$$\vdots$$
$$F_{Magnet-8}|_{n,m} = k_i|_{(x_n, i_{u|m})} \cdot (i_u + 0.01) + k_s|_{(x_n, i_{u|m})} \cdot (x)$$
$$F_{Magnet-9}|_{n,m} = k_i|_{(x_n, i_{u|m})} \cdot (i_u + 0.01) + k_s|_{(x_n, i_{u|m})} \cdot (x + 0.01)$$

A weight vector does there exist with two constand values:

$$\theta = \left[k_i|_{(x_n, i_{u|m})} \quad k_s|_{(x_n, i_{u|m})} \right]$$

The input matrix:

$$\overline{A} = \begin{bmatrix} i_u - 0.01 & x - 0.01 \\ i_u - 0.01 & x \\ \vdots & \vdots \\ i_u + 0.01 & x + 0.01 \end{bmatrix}$$

The output vector:

$$Y = \begin{bmatrix} F_{Magnet-1}|_{n,m} \\ F_{Magnet-2}|_{n,m} \\ \vdots \\ F_{Magnet-9}|_{n,m} \end{bmatrix}$$

Consider the given set of samples θ, $\overline{A}$ and Y, with least square method (LSM), the relation underlying the data set is represented as a function of the following form:

$$Y = \overline{A} \cdot \theta^T + e \quad (28)$$

Here e is the error for every equation. For the LSM regression, the error variables for the fitting problem will be introduced as follows:

$$e = Y - \overline{A} \cdot \theta^T \quad (29)$$

And for the given data those weights θ will be searched, in which the summed quadratic error E of the training samples is smallest. The quadratic error is given as:

$$E(\theta) = \sum_{i=1}^{m} (y_i - a_i \cdot \theta)^2 = e^T \cdot e = \left(Y - \overline{A} \cdot \theta\right)^T \cdot \left(Y - \overline{A} \cdot \theta\right) \quad (30)$$

When the quadratic error is minimal, $E(\theta) \to 0$, the weights θ shall be:

$$\theta = \left(A^T \cdot A\right)^{-1} \cdot A^T \cdot Y \quad (31)$$

The Fig 7 shows a schematic of the parameter identification process for stiffness parameter, when the identified magnet parameter is identified as:

$$k_{Magnet} \approx 6.7 \frac{N \cdot mm^2}{A^2}.$$

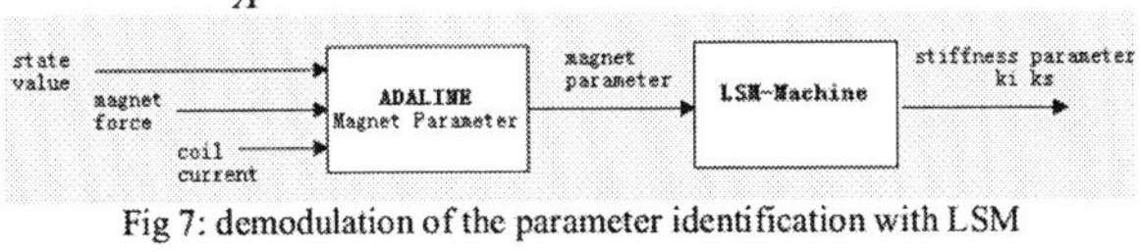

Fig 7: demodulation of the parameter identification with LSM

The table illustrates the experimental result of the stiffness parameter for every nominal work point:

Parameter identification k_i		i_u (A)				
		-1.5	**-0.75**	**0**	**0.75**	**1.5**
x_{Rotor} (mm)	**-0.5**	7.14	8.47	12.15	9.35	8.74
	-0.25	9.47	9.58	10.53	9.69	9.65
	0	10.05	10.05	10.05	10.05	10.05
	0.25	9.66	9.69	10.53	9.58	9.48
	0.5	8.75	9.35	12.15	8.45	7.14

Table 4: identified parameter ki for every sub space

Parameter identification k_s		i_u (A)				
		-1.5	**-0.75**	**0**	**0.75**	**1.5**
x_{Rotor} (mm)	**-0.5**	32.16	16.23	8.57	6.53	6.94
	-0.25	21.89	12.57	7.78	7.21	10.30
	0	15.08	9.42	7.53	9.42	15.08
	0.25	10.30	7.21	7.78	12.57	21.89
	0.5	6.95	6.53	8.57	16.23	32.16

Table 5: identified parameter ks for every sub space

D. Kalman filter with Expert system

In this paragraph, we apply a expert system, which is based on fuzzy basis function neural network (FBFNN), to obtain the actual stiffness parameter over the entire clearance and control current area in state equation in Kalman filter.

Generation of expert system with using FBFNN

Fuzzy logic system (FLS) is a logic in form of many-valued-rules. The FLS can be expressed mathematically as a linear combination of the input and output, such as TSK-fuzzy-logic, as a nonlinear combination it can be a Mamdani-fuzzy-logic. Detail for the fuzzy theory can be found in [8].

The expert system uses the control current i_u and rotor displacement x_{Rotor} as input, the stiffness parameter k_i and k_s as output. As shown in Table 2 and Table 3, the input i_u and x_{Rotor} will be divided into 5 levels. The conclusion value will be captured from LSM-machine. The rule base is constructed with Mamdani-structure, and presented as follows:

If <control current i_u is ..., rotor displacement x_{Rotor} is ...>

Then < stiffness parameter k_i and k_s are ...>

A convenient way to list all rules, is to use a tabular representation from Table 4 and Table 5.

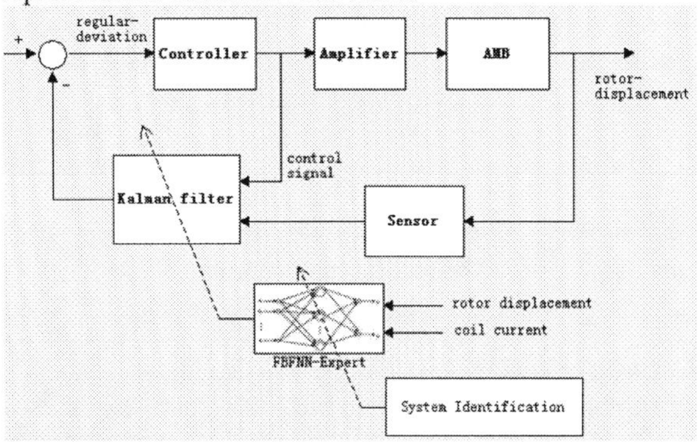

Fig 8: structure of adaptive kalman filter with using fuzzy expert system

Training of expert system with using FBFNN

RBFN is an adaptive network functionally equivalent to a fuzzy inference system. The fuzzy basis function has if-then of membership functions in the neural network. As a combination from both, the FBFNN has Input layer, hidden layer and output layer. The input layer receives input data into the network. Hidden layer is a fuzzy basis function which transfers input data into a membership function by using the Gaussian function. The output layer is linear combination for the hidden layer output.

We use two input layer node, 25 hidden nodes and one output layer nodes. The input value $x_{Rotor}(k)$, $i_u(k)$ as work point are measurable in experiment. The value k_i and k_s for the work point are generated by LSM - Machine. For every logic rule, the Gaussian membership function shall be determined by the training function. Fig 9 is the FBFNN structure. The function of the nodes in each of the layers will be described.

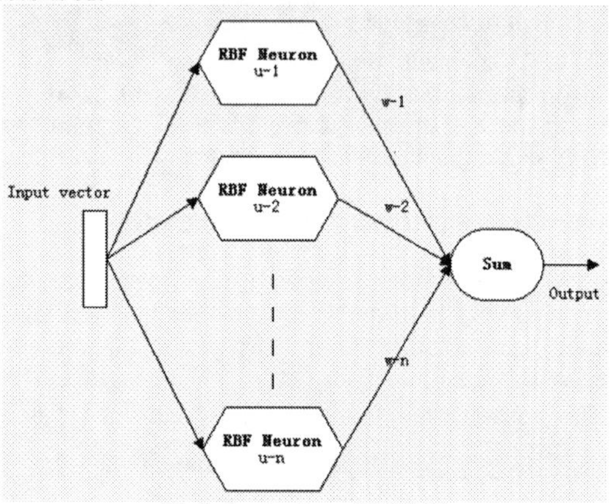

Fig 9: the structure of FRBFNN

Layer 1: Input layer for input vector.

Layer 2: the nonlinear activation function of hidden layer is a Gaussian function which expressed as:

$$u_i = e^{\frac{-(\|\bar{x}-\bar{\tau}_i\|)^2}{2\cdot\sigma_i^2}} \text{ ,with } i \in [1,n], \qquad (32)$$

Where τ_i is the Gauss function vertex for the i-th sub space (rule), which expresses the value i_u and x_{Rotor}. Parameter σ_i is the width (or variance) of the Gauss function.

Layer 3: the output of hidden layer is expressed as:

$$y = \sum_{i=1}^{m} w_i \cdot u_i \qquad (33)$$

Where w_i is the i-th weight between hidden layer nodes and output layer. The consequent parameters w_i for each sub soace are from the process of parameter identification.

FBFNN algorithmus uses the steepest gradient methode to learn the weights in the network. The steepest gradient methode is supervised leaning which reduces the error between output and input of neural network. Detail will be found in [13].

As a result in this work, the width of the Gaussian membership function σ_i of every rule (sub space) will be optimal adapted as shown in Fig 10, the training result is shown in Table 6.

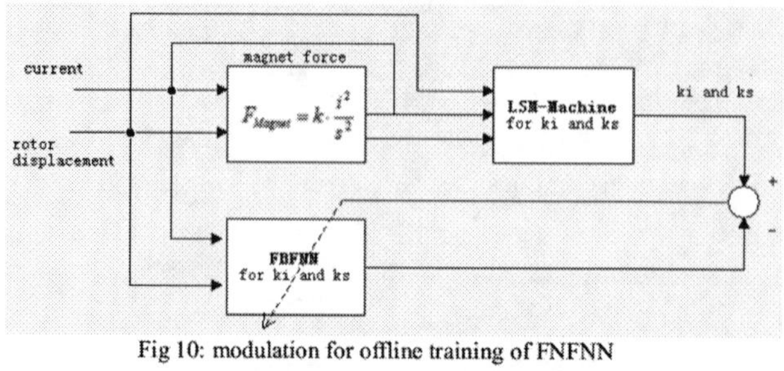

Fig 10: modulation for offline training of FNFNN

Gaussian width of the vector for the sub-space	Initial data	Training Result for ki	Training Result for ks
σ_1 (-1.5 A, -0.5 mm)	0.4	0.2621	0.2594
σ_2 (-0.75 A, -0.5 mm)	0.4	0.2982	0.301
σ_3 (0 A, -0.5 mm)	0.4	0.06776	0.08365
σ_4 (0.75 A, -0.5 mm)	0.4	0.2881	0.3599
σ_5 (1.5 A, -0.5 mm)	0.4	0.1182	0.2257
σ_6 (-1.5 A, -0.25 mm)	0.4	0.00654	0.007588
σ_7 (-0.75 A, -0.25 mm)	0.4	0.04799	0.04451
σ_8 (0 A, -0.25 mm)	0.4	0.2558	0.02858
σ_9 (0.75 A, -0.25 mm)	0.4	0.03555	0.05053
σ_{10} (1.5 A, -0.25 mm)	0.4	0.02587	0.04686
σ_{11} (-1.5 A, 0 mm)	0.4	0.00451	0.1859
σ_{12} (-0.75 A, 0 mm)	0.4	0.1372	0.08701
σ_{13} (0 A, 0 mm)	0.4	0.001998	0.02898
σ_{14} (0.75 A, 0 mm)	0.4	0.1399	0.09562
σ_{15} (1.5 A, 0 mm)	0.4	0.3374	0.01122
σ_{16} (-1.5 A, 0.25 mm)	0.4	0.4036	0.0000128
σ_{17} (-0.75 A, 0.25 mm)	0.4	0.000607	0.01736
σ_{18} (0 A, 0.25 mm)	0.4	0.09378	0.3585
σ_{19} (0.75 A, 0.25 mm)	0.4	0.006809	0.02371
σ_{20} (1.5 A, 0.25 mm)	0.4	0.006899	0.001145
σ_{21} (-1.5 A, 0.5 mm)	0.4	0.03358	0.2894
σ_{22} (-0.75 A, 0.5 mm)	0.4	0.2517	0.2875
σ_{23} (0 A, 0.5 mm)	0.4	0.2608	0.3154
σ_{24} (0.75 A, 0.5 mm)	0.4	0.3845	0.1358
σ_{25} (1.5 A, 0.5 mm)	0.4	0.02863	0.2194

Table 6: Gaussian width value for fuzzy set

Adaptive Kalman filter

In the majority of cases, a Kalman filter requires many system information for the prediction function. So a interesting problem is the possibility for the parameter adaptation in system state equation for a linear Kalman filter by using of the expert system, which is relative with system identification. Its aim is to enhance the prediction ability of the Kalman filter.

In this system, besides the constant parameter in state equation of the Kalman filter, the parameter k_i and k_s will be adapted. The combination of the kalman filter and the expert system is known as "adaptive Kalman filter with using softcomputing". The system structure and diagram of the work are depicted in Fig 8.

E. Simulation and result

In this section, the concept made in this work shall be verified by means of simulation and measurement. All relevant test rig parameters can be found in table.

Process noise variance	$8 \cdot 10^{-5}$ V
Measurement noise variance	$8.3 \cdot 10^{-5}$ V/mm
Rotor rotation	0 /min

Result of system identification

The identification of the magnet force and magnet parameter will be verified by means of a rotor position step response without force disturbance.

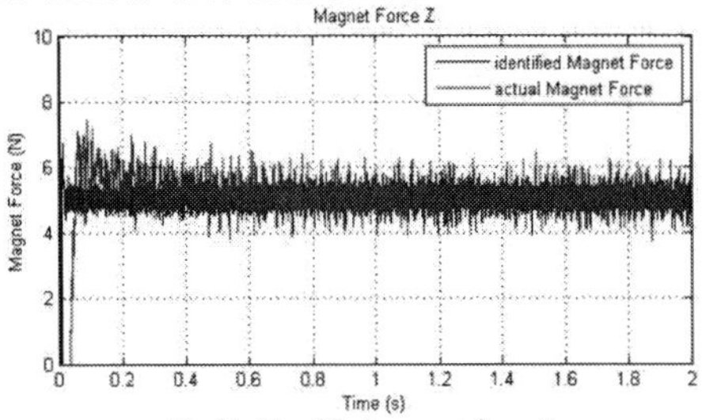

Fig 11: identified magnet force F

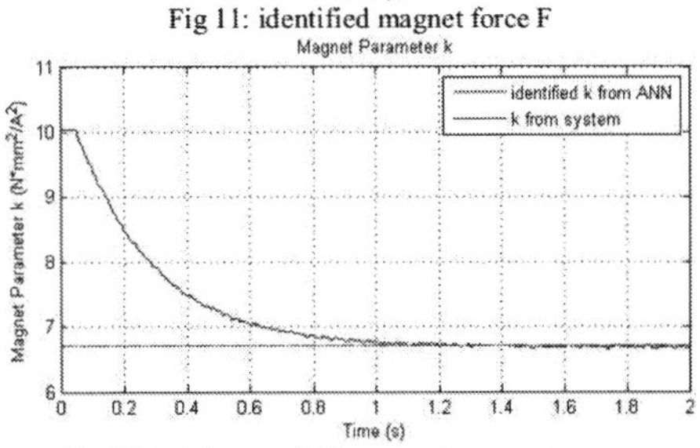

Fig 12: training result for magnet parameter k_{Magnet}

After the identification of magnet parameter and the calculation of stiffness parameter by LSM-machine, inclusive offline training, the RBFN expert systems are generated. In this case, the calculated stiffness parameter from expert system will be compared with the constant stiffness parameter from traditional linearization. The difference between the two values will be illustrated.

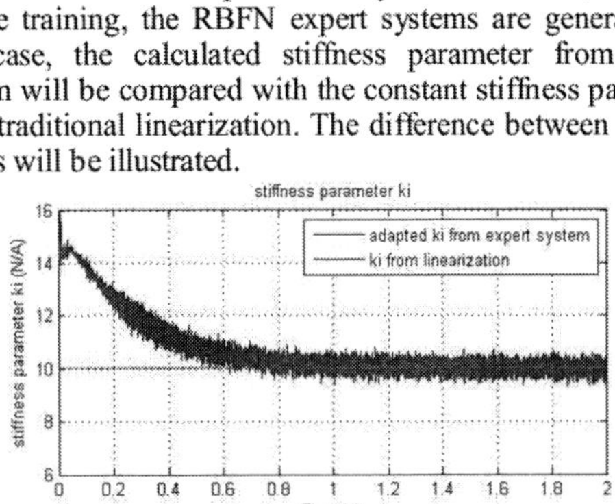

Fig 13: adapted k_i from fuzzy expert system

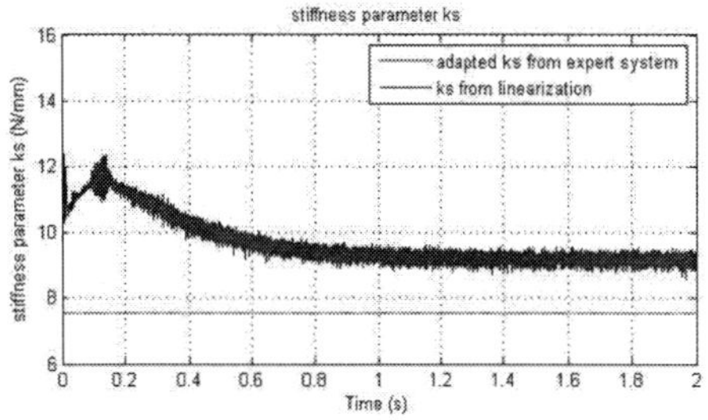

Fig 14: adapted k_s from fuzzy expert system

Result of rotor displacement control

As a position control, the rotor displacement in step response is shown. In this case, the measured rotor displacement with and without measure noise, the estimated rotor displacement from Kalman filter will be illustrated, where a high correlation between the actual and the estimated position signal can be observed.

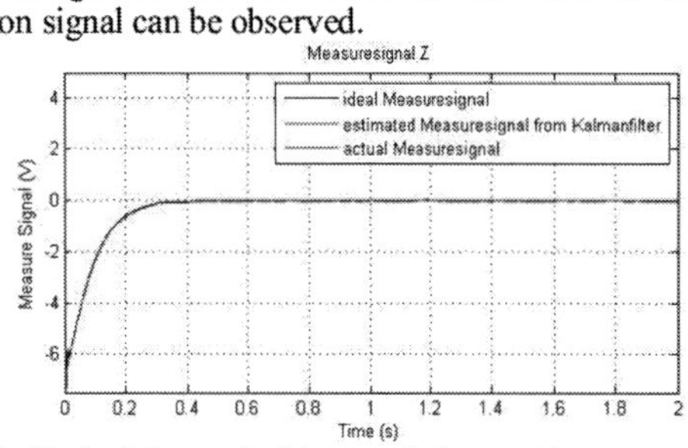

Fig 15: simulation result of the rotor displacement in step response

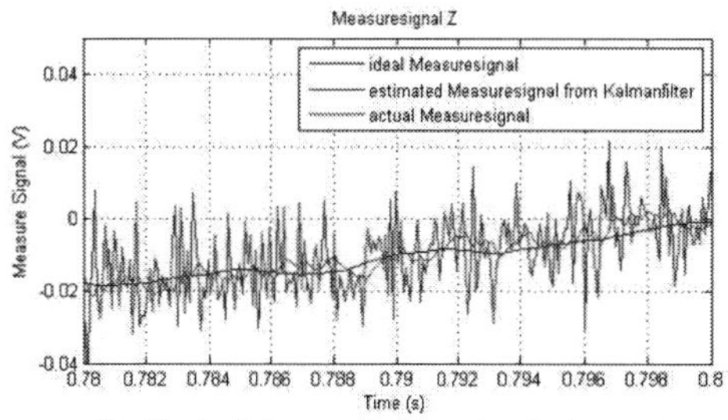

Fig 16: simulation result of the rotor displacement

F. Conclusion

A new concept of Kalman filtering for AMB system has been introduced. To cope with the strong sensor noise and process noise, the Kalman filter is suggested to AMB system. It is based upon system identification which are realized by artificial neural network. Thus the system parameter and magnetic force can be identified online. Experiments have proved that the trained fuzzy expert system can give the actual system parameter. The strong uncertainty and nonlinear of the system to the linear state equation in Kalman filter are reduced.

By estimating the magnet force, the ADALINE is used. By calculation of every sub-space, the LSM is suggested. All the calculation result will be saved in fuzzy system, which is optimal adjusted by RBFN. Results have shown, that the adapted linear state equation in Kalman filter is more reasonable.

An outlook has been given on more parameter estimation for state equation in Kalman filter, including noise variance. On principle, it is possible to reduce the estimation errors, and to realize a better state observation.

REFERENCES

[1]: H.Neumann, F.Worlitz, "Modellierung und Simulation eines Kalman-Filters für aktive Magnetlager", 8. Workshop Magnetlagerungstechnik Zittau-Chemnitz, Zittau, 2011

[2]: Dapeng Wang, Fengxiang Wang, "design of PDC Controller based on T-S Fuzzy Model for Magnetic Bearing of Highspeed Motors", school of electric engineering shenyang university of technology, IEEE, 2010

[3]: Z.C.Yu, D.Wen, H.Y,Zhang, "the identification model of magnetic bearing supporting system, IEEE,2008"

[4]: Dipl. Martin Ruskowski, „ Aufbau und Regelung aktiver Magnetführungen, Fachbebiet Maschinenbau, Universität Hannover", Dissertation, von 2004

[5]: Thomas Schuhmann, Wilfried Hoffmann, Ralf Werner, "improving operational performance of active magnetic bearings using Kalman filter and state feedback control", IEEE transaction on industrial electronics, 2011,

[6]: Marcin Kaminski, Teresa Orlowska-Kowalska, "Adaline-based Speed Controller of the Drive System with Elastic Joint", IEEE, 2012

[7]: Wenle Zhang, "A Generalized ADALINE Neural Network for System Identification", IEEE International Conference on Control and Automation,2007

[8]: Praveen Kumar Agarwal and Satish Chand, "Fuzzy logic control of four-pole Active Magnetic Bearing system",Processdings of the 2010 International Conference on Modelling, Identification and Control, Okayama, Japan

[9]: Safanah M.Raafat, Rini Akmeliawati, "intelligent estimation of uncertainty bounds of an active magnetic bearings using ANFIS", IEEE,2011

[10]: Thomas Schuhmann, Wilfried Hoffmann, Ralf Werner, "sensor integration and state estimation on magnetically levitated rotors", SPEEDAM 2006, IEEE, 2006

[11]: M Mendel, "fuzzy logic systems for engineering: a tutorial", proceedings of IEEE, 1995

[12]: Sung-Kyung Hong and Reza Langari, "robust fuzzy control of a magnetic bearing system subject to harmonic disturbances", IEEE transactions on control systems technology, 2000

[13]: Huann-Keng Chiang, Chao-Ting Chu and Yong-Tang Jhou, "Fuzzy Control with fuzzy basis function neural network in magnetic bearing system", IEEE

Dynamics of Magnetic Suspended Rotor Drop on Auto-Eliminating Clearance Auxiliary Bearings

Chaowu Jin[a], Qiuping Fan[a], Yili Zhu[b], Chengtao Yu[a], Xianxi Xia[a]

[a] College of Mechanical and Electrical Engineering, Nanjing University of Aeronautics and Astronautics, Yudao St 29, B218 Nanjing, China, jinchaowu@mail.com
[b]College of Electronic Information and Electrical Engineering, Changzhou Institute of Technology, Tongjiang Road No. 299, Changzhou, China, zhuyil@czu.cn

***Abstract*—In an active magnetic bearing (AMB) system, the auxiliary bearings are indispensable to protect the stators from damage in case the AMB system fails or overloads. This paper introduces a new type auxiliary bearing. An auto-eliminating clearance auxiliary bearing (ACAB) for AMB has been developed against the shortcomings of traditional auxiliary bearings recently. The operating principle of ACAB outline and the nonlinear contact force model of the rotor and the support are established. Detailed simulation models including containing rigid rotor model, ball bearing model and contact model between outer race and support after rotor drop are established using the multi-body dynamics simulation software MSC.ADAMS. The dynamic responses of rotor drop on ACABs are simulated using those established models. The rotor orbits and contact forces after AMB failure are analyzed with different parameters. ACAB design guides based on the simulation results are presented. An experimental rig is developed to demonstrate the feasibility of the ACAB after AMB system failure.**

1. Introduction

Active magnetic bearings (AMBs) have many advantages over conventional mechanical bearings. In addition to supporting the high-speed rotor without any mechanical friction and lubrication, AMBs enable the rotor position and induced vibration to be monitored and controlled by adjusting supporting stiffness and damping. The auxiliary bearing is an indispensable part to protect the AMBs from direct contact with the rotor. They prevent damages during the maintenance and destruction of the system after a possible AMBs failure or overloads.

Most conventional auxiliary bearings are rolling element or bushing type bearings with fixed clearance. Most of the researches focused on the rotor dynamic responses [1-4], heating [5] and experiments [6] after rotor drop. Some new type auxiliary bearings have been developed. Literature [5, 7] fixed an elastic ring on the auxiliary to buffer the series of impacts. Literature [8, 9, 10] proposed a double-decker auxiliary bearing. Literature [11 12] proposed a type of zero clearance auxiliary bearing and presented its performance over conventional auxiliary bearings. A hydrodynamic foil gas bearing was also advised to be used as the auxiliary bearing [13 14]. Wu [15] proposed a device with active electromagnetic as auxiliary bearing, which could eliminate the clearance between the rotor and itself by releasing the device under the power-down conditions. This paper introduces the structure and operating mechanism of a new type of auto-eliminating clearance auxiliary bearing. The dynamic responses of rotor drop on ACABs are simulated. Following that, the performances of the ACAB are experimentally analyzed.

2. Simulation

2.1. Structure of ACABs

In order to eliminate the possibility of backward whirl of the rotor and the clearance between the outer race of the ball bearing and the tilting supports after the AMBs fail, a ball bearing is mounted on a rotor and spin with the rotor. The outer race of this ball bearing is surrounded by several supports. There exists an initial protective clearance between the outer race of the ball bearing and the tilting supports when AMBs are operating. The bearing mounted rotor drops onto the supports after the AMBs fail. The tangential friction force generated between the outer race of the ball bearing and the tilting support will overcome resistance and rotate this support. The protective clearance between the outer race and all supports will be eliminated simultaneously. Meanwhile, the rotor will continue spinning stably within the inner race of the ball bearing around its original center sine all the supports pushes it back to the center. The concept sketch of ACAB is shown in Figure 1.

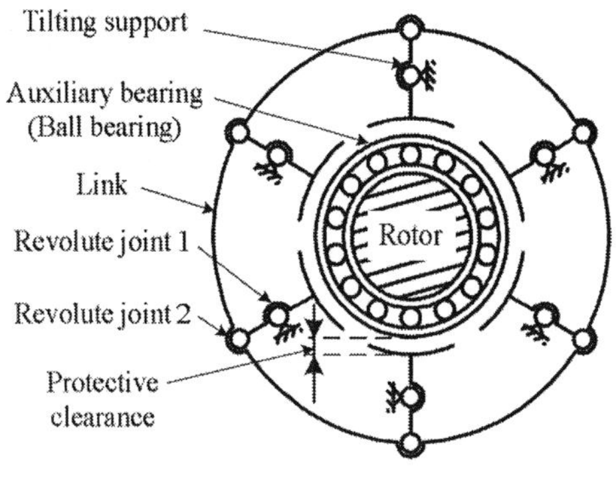

Figure 1 The concept sketch of auto-eliminating clearance auxiliary bearing

2.2. MSC.ADAMS model

The dynamic responses of rotor drop onto ACAB are simulated using multi-body dynamics simulation software MSC.ADAMS. According to the rotor parameters, the relevant parameters of rotor are listed in table 1. A ball bearing (61804) is mounted on a rotor and spin with the rotor.

Table 1 relevant parameters of rotor

Parameter	Value	Unit
Rotor radius	10	mm
Polar MOI	3.8E+2	kg.mm^2
Transverse MOI	1.612E+4	kg.mm^2
Protective clearance	0.15	mm
Mass	2.14	kg

Impact function and recovery coefficient are the two main methods to define the contact in ADAMS. Usually, the use of impact function is wider and more accurate. Contact stiffness k and damping c are used to calculate the crash force in this rotor drop model.

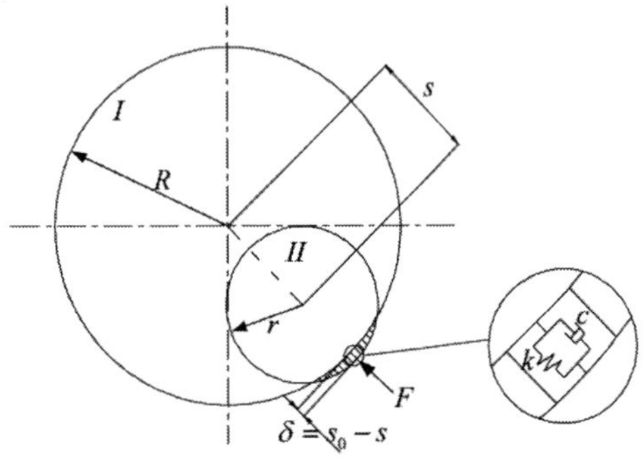

Figure 2 Contact model

Fig. 2 shows the contact model between tilting support and outer race, and the impact function can be expressed as a subsection function:

$$F = \begin{cases} 0 & s \le s_0 \\ k(s-s_0)^e + c_m \dot{s} \times STEP(s, s_0 - d_m, 1, s_0, 0) & s > s_0 \end{cases} \quad (1)$$

where k is the contact stiffness between tilting support and outer race.

$$k = 7.85 \times 10^{10} L^{8/9} \quad (2)$$

where L is the contact length; e is contact coefficient. For point contact e=3/2, line contact e=10/9.

The *STEP* function is used to prevent the contact damping not continuous. Here s is the relative displacement between outer race and titling support. s_0=R-r; here R is the bore radius of titling and r is the radius of outer race . Here c_m is the maximum damping coefficient; d_m is the distance when the damping ratio reached the maximum, d_m=0.01mm. s is variables of the *STEP* function; s_0-d_m is the initial value of variables; s_0 is the final value of variables.

Stiffness-damping contact model is also adopted to define the catcher bearing internal contacts. Based on Hertz theory [16], the contact stiffness between ball and race can be calculated using the bearing geometrical and material parameters.

$$k_b = e_\delta^{-\frac{3}{2}} (\sum \rho)^{-\frac{1}{2}} \quad (3)$$

where e_δ is the Hertz contact coefficient associated with the contact surface curvature, which can be calculated by the literature [17] interpolation; ρ is the contact surface curvature.

In addition to the normal contact force, there is tangential contact force and that is friction. Based on Coulomb, the friction can be calculated as equation 4 in ADAMS.

$$F_T = \mu F_N \quad (4)$$

Considering the effect of relative speed on the friction coefficient, the friction coefficient can be expressed as follow

$$\mu = \begin{cases} \mu_d & v \le -v_d \\ \mu_d - (\mu_d - \mu_s)\cdot \Delta_1^3 \cdot (10 - 15\Delta_1 + 6\Delta_1^2) & -v_d \le v \le -v_s \\ \mu_s - 2\mu_s \cdot \Delta_2^3 \cdot (10 - 15\Delta_2 + 6\Delta_2^2) & -v_s \le v \le v_s \\ -\mu_s - (\mu_d - \mu_s)\cdot \Delta_3^3 \cdot (10 - 15\Delta_3 + 6\Delta_3^2) & v_s \le v \le v_d \\ -\mu_d & v > v_d \end{cases}$$

(5)

where $\Delta_1 = \frac{v + v_d}{v_d - v_s}, \Delta_2 = \frac{v + v_s}{v_d + v_s}, \Delta_3 = \frac{v - v_s}{v_d - v_s}$. μ_d is the dynamic friction coefficient; μ_s is static friction coefficient; v_s is static friction critical speed (m/s) and v_d is dynamic friction critical speed (m/s).

Fig. 3 shows the relationship between the friction coefficient and relative speed.

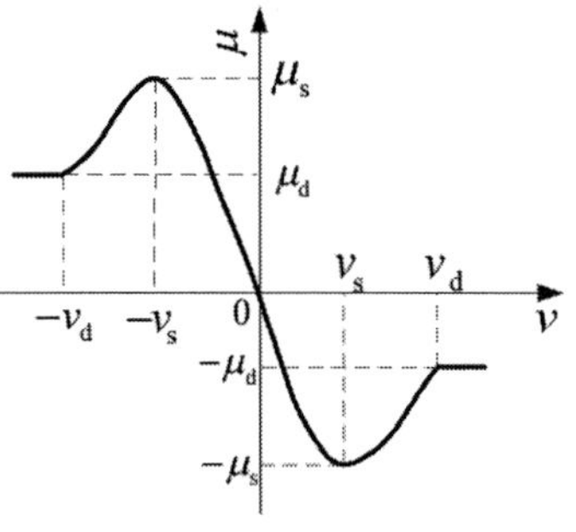

Figure 3 Coulomb friction model

As there is no reliable theoretical method to calculate the bearing damping, the collision contact damping between outer race and tilting support, the collision contact damping between tilting support and revolute joint, empirical values are chosen in this paper. According to the literature [18], the bearing damping take 200 N·s/m. According to the literature [19], the collision contact damping takes 1000 N·s/m. Figure 4 shows ADAMS model of ACAB. The relative displacement between outer race and titling support is 0.15mm.

For the whole ACAB model, it can be divided into two parts, respectively, adding constraints, namely the rotor component model (rotor and rolling bearings) and eliminate the gap model of the device (titling support, revolute joint 1, revolute joint 2, link and base). Firstly, constraint relationship between each part of the rotor component is divided into three kinds: (1) the rotor and the inner race were set as a fixed side. (2) Assume that rotor do cylindrical sports, and don't consider the rotor axial force, the ground and the rotor were set as a planar pair. (3) The bearing rolling body, the inner channel, the outer channel and the cage were set as contact pair. Secondly, constraint relationship between each part of the eliminating the gap device is divided into five kinds: (1) the base and the ground were set as a fixed side. (2) The base and the revolute joint 1 were set as a fixed side. (3) The tilting support and the revolute joint 1 were set as revolute pair. (4) The tilting support and the revolute joint 2 were set as a fixed side. (5) The revolute joint 2 and the link were set as a high pair. Finally, the constraint between rotor component and the eliminating the gap device was set as contact pair.

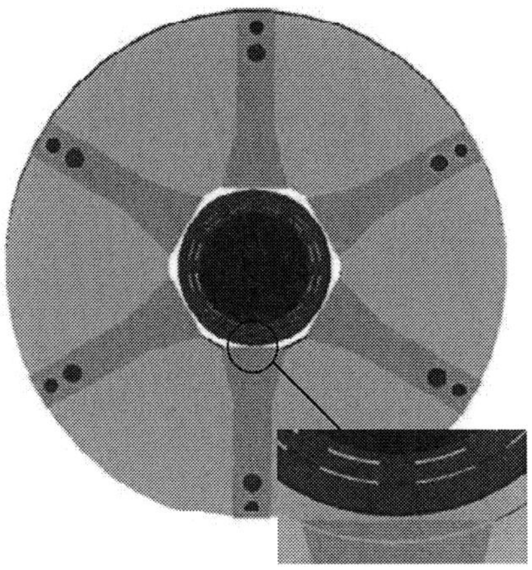

Figure 6 The ADAMS model of auto-eliminating clearance auxiliary bearing

2.3. *Simulation results*

After the AMBs failure, the rotor component drops on some tilting support. The tangential friction force generated between the outer race of the ball bearing and the tilting support will overcome resistance and rotate this support. The friction on the surface of the outer race of the ball bearing and the tilting support is the only driving force mechanism motion. The ACAB eliminating clearance speed and effect greatly depends on the size of the friction. So, the tilting support surface friction coefficient influence on the rotor dynamic behavior was simulated after the rotor drop. The other parameters were set: the unbalance of the rotor component was set to 2μm. The radial clearance of the ball bearing was set to 12μm. The moment of inertia of tilting support was set to 0.1 kg·mm^2. The friction coefficient between the revolute joint and tilting support was set to 0.4.

Figure 5 shows simulated orbits of the rotor component from 0 to 0.2 seconds after rotor drop at 18 000rpm. The result shows that the rotor is quickly pushed back to the center after a few times of collisions.

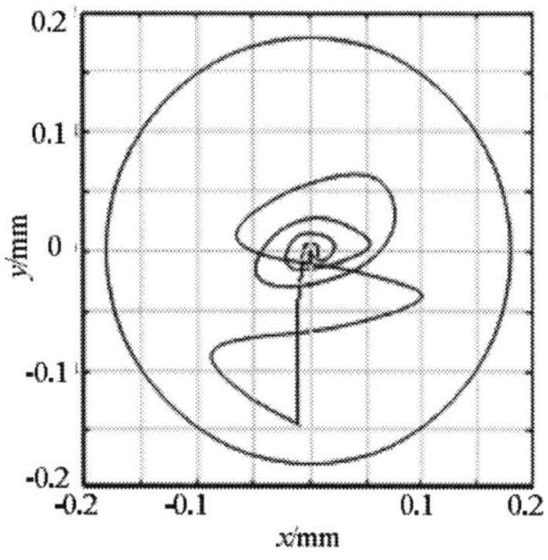

Figure 5 The orbits of the rotor obtained by simulations speed

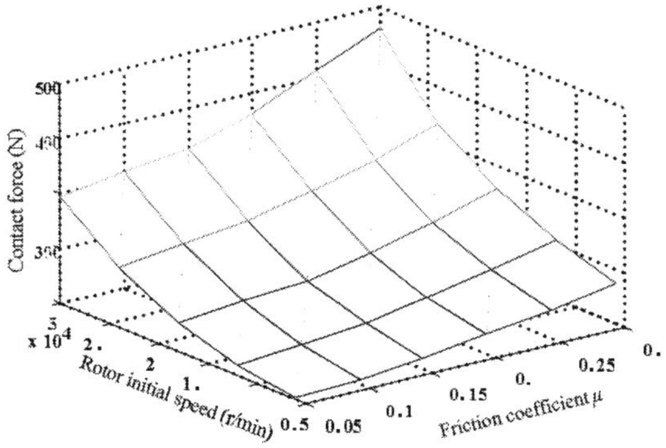

Figure 4 Contact force map between the outer race and the tilting support for various friction coefficient and the rotor

Figure 6 shows the maximum contact force between the outer race of ball bearing and the tilting support at various friction coefficients and rotor speed. With the increasing of surface friction coefficient of tilting support, the contact force between the outer race and the tilting support is a linear increasing trend, but the increase rate is weak. In order to reduce the impact, it can reduce the friction coefficient appropriately. However, the tilting support can not be driven and may not be able to eliminate the clearance if the friction coefficient is too small, which would not achieve the elimination of vibration and shock.

3. *Experiments*

Figure 7 shows the experimental facilities. The system adopts eddy current sensor to measure rotor vibration displacement, which have to be known for controlling the AMBs anyway. Three National Instruments boards NI-9215 are used to acquire the experiment data and the measurements are analyzed by the MATLAB software.

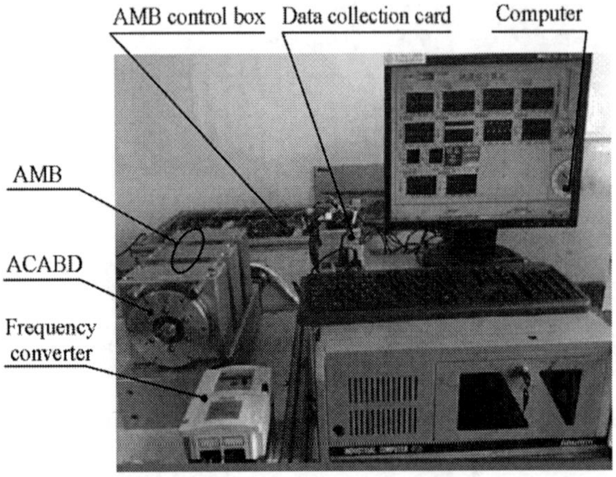

Figure 8 The concept sketch of auto-eliminating clearance auxiliary bearing

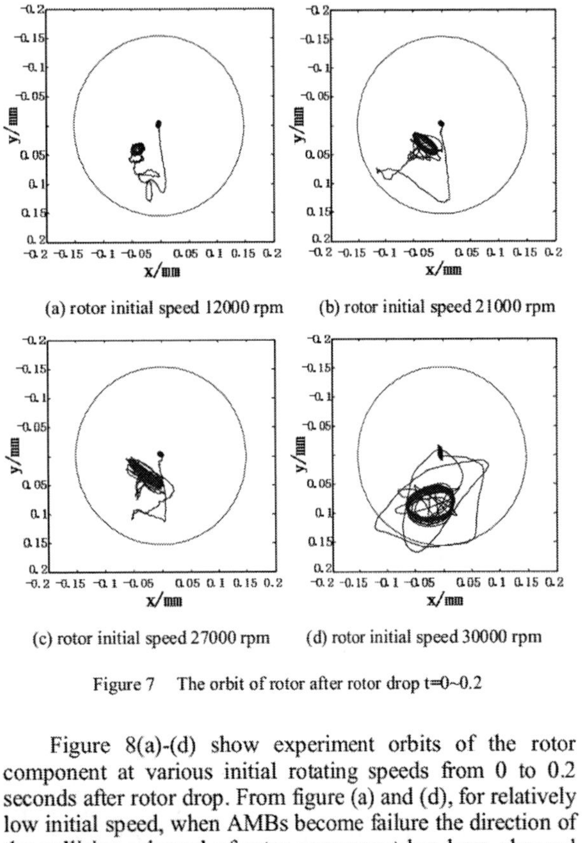

(a) rotor initial speed 12000 rpm (b) rotor initial speed 21000 rpm

(c) rotor initial speed 27000 rpm (d) rotor initial speed 30000 rpm

Figure 7 The orbit of rotor after rotor drop t=0~0.2

Figure 8(a)-(d) show experiment orbits of the rotor component at various initial rotating speeds from 0 to 0.2 seconds after rotor drop. From figure (a) and (d), for relatively low initial speed, when AMBs become failure the direction of the collision rebound of rotor component has been changed after the first collision. The rotor is quickly pushed back to nearly the center of AMBs after a few times of collisions. But the stable rotation center was not coincidence with the initial balance rotation center and there were still some rotor vibration. The following reasons can explain this phenomenon: (1) the radial clearance of the ball bearing provides one part of that vibration. (2) there existed some machining errors and assembly errors, which would make the clearance between the outer race and each tilting support can not be accurately and completely eliminate, and the remaining clearance offers the another part of the vibration. Greater the rotor initial speed was, higher the vibration was. From figure (c) and (d), for relatively higher initial speed, when AMBs become failure the collision motion of rotor component drop on the tilting support was more and more intense. When the rotor initial speed up to 27000 rpm was still able to eliminate the clearance, the vibration of the device is bigger. When the rotor initial speed reached 30000 rpm, the ACAB has completely no effect.

4. *Conclusion*

In this paper, ACAB is proposed to be used as auxiliary bearing in AMB system. The dynamic response of the rotor drops on ACAB is simulated on the platform of ADAMS. Finally, rotor drop experiments are carried out to validate the simulation results. The simulation and experiment results show that the ACAB could eliminate the clearance between the outer race of the ball bearing and the titling supports. Meanwhile, the rotor will continue spinning stably within the inner race of the ball bearing near to its original center sine all the supports pushes it back to the center. In the next research, optimization of structure parameters will be done to make the ACAB can also eliminate the clearance under high rotation speed, and the vibration of the rotor can fall.

Acknowledgment

This research is supported by National Nature Science Foundation of China (51205186)

REFERENCES

[1] Ishii I and Kirk R G, "Transient response technique applied to active magnetic bearing machinery during rotor drop, " *ASME Conference Proceedings-Rotating Mach Veh Dyn*, vol 35, pp. 191-199, 1991

[2] Krik R G, ISHII I, "Transient rotor drop analysis of rotors following magnetic bearing power outage, " *Proc of Mag'93 Magnetic Bearing, Magnetic Drives and Dry Gas seals Conference and Exhibition*, 1993

[3] Fumagalli M, Varadi P, Schweitzer G., "Impact dynamics of high speed rotors in retainer bearings and measurement concepts," *Proc of the Fourth International Symposium on Magnetic Bearing*, 1994.

[4] ZHU Changsheng, "Nonlinear dynamics of rotor drop on rolling element backup bearings after active magnetic bearing failure," *Journal of Mechanical Engineering*, vol 42, no 7, pp. 196-202, 2006.

[5] Sun Guanyoung, "Rotor drop and following thermal growth simulations using detailed auxiliary bearing and damper models," *Journal of Sound and Vibration*, vol 289, pp. 334-359, 2006.

[6] M.Helfert, "Rotorabstüze in wälzlager-experimentelle untersuchungen des rotor-fanglager -kontakts," PhD thesis, TU Darmstadt, Fachgebiet Mechatronik im Maschinenbau, 2008.

[7] Odd W. Reitsma, "Development of long-life auxiliary bearings for critical service turbo-machinery and high-speed motors," *Proceedings of the 8th International Symposium on Magnetic Bearings*, pp. 507-514, 2002.

[8] Zhu Yili, Xu Longxiang, "The radial stiffness and application of double-decker ball bearing," *Key Engneering Materials*, vol. 450, pp. 353–356, 2011.

[9] Zhu Yili, Xu Longxiang, "Mechanical research of double-decker ball bearing ," *Journal of Aerospace Power*, vol 26, no 8, pp. 1914–1920, 2011.

[10] Zhu Yili, Jin Chaowu and Xu Longxiang, "Dynamic response of rotor drops onto double-decker catcher bearing," *Chinese Journal of Mechanical Engineering*, vol 26, no 1, pp. 104–113, 2013.

[11] Chen H.M., Walton J and Heshmat H., "Zero clearance auxiliary bearing for magneitc bearing," *International Gas Turbine and Aeroengine Congress and Exhibition*, 1997.

[12] Salehi M., and Heshmat H., "On the dynamic and thermal performance of a zero clearanc auxiliary bearing for a magnetic bearing system," *Tribol. T*, vol 43, pp. 435-440, 2008.
[13] Yu L. et al, "Compliant foil bearings used as emergence bearings," *Proceedings of the 9th International Symposium on Magnetic Bearings*, 2004.
[14] Swanson E. E., Heshmat, H. and Walton J., "Performace of a foil-magnetic hybrid bearing," *J. Eng. Gas Turbines Power*, 2000.
[15] Wu L. S., et al., "Design of a new-style assistant bearing of magnetic beaing," *Machinery Design and Manufacture*, vol 3, pp. 32-34, 2010.
[16] T.A. Harris. "Rolling Bearing Analysis" [M], 4th Edition. John Wiley & Sons, Inc. New York, 2001
[17] Okamoto Junnami. "The design of ball bearing calculation" [M], Beijing: Mechanical industry press, 2003.
[18] Li zenggang. "ADAMS introductory explanation and the examlle" [M], Beijing: National defense industry press, 2009.
[19] Xue zheng. "Dynamic Analysis of Deep Groove Ball and Cylindrical Bearings" [D] zhejiang University, Hangzhou 2008.

Research and Experiment of Auxiliary Bearings with No Lubrication for Helium Blower of HTR-10

Guojun YANG[a], Zhengang SHI[b], Xingnan LIU[b], Jingjing ZHAO[b]

[a]Institute of Nuclear and New Energy Technology of Tsinghua University, the Key Laboratory of Advanced Reactor Engineering and Safety, Ministry of Education, Beijing, 100084, China. yanggj@tsinghua.edu.cn.
[b]Institute of Nuclear and New Energy Technology of Tsinghua University, the Key Laboratory of Advanced Reactor Engineering and Safety, Ministry of Education, Beijing, 100084, China.

***Abstract*— The auxiliary bearing is one of the important parts for the active magnetic bearing (AMB). A new auxiliary bearing design with no lubrication is proposed in this paper. This auxiliary bearing is composed of angular contact ceramic ball bearing. Two pairs of bearing will be distributed on both ends of the rotor. Each pair of bearing arrangement is face to face. Oil or grease will not be applied in this auxiliary bearing. But dry lubrication with MoS_2 will be used in this bearing. This auxiliary bearing will be applied in the helium blower of 10MW high-temperature gas-cooled reactor (HTR-10). Oil or grease can't be used in the primary loop of the HTR-10. The auxiliary bearing is used to support the blower rotor when the AMB fails to work. It must support the dropping rotor and bear the great impact force and friction heat. It is difficult to analyze the falling course of the rotor. There are two difficult problems to solve for the design of auxiliary bearing, including great impact force and friction heat. The preliminary analysis of the rotor was done. The impact force of the auxiliary bearing was computed for the axial and radial impact load by the finite element method. The temperature field will be analyzed and studied also in this paper. The auxiliary bearing's life will be evaluated. The scheme of auxiliary bearing and the simulation results offer the important theoretical base for this new auxiliary bearing style, and offer the protector design of the helium blower with AMB for HTR-10.**

A. *Introduction*

A 10MW high-temperature gas-cooled test reactor (HTR-10) was constructed by the Institute of Nuclear and New Energy Technology (INET) at Tsinghua University of China. It has been operated to generate electricity on January, 2003. The helium blower is the key equipment in the primary loop of THR-10. It makes the helium flow around the HTR-10. The rolling element bearing with grease lubrication was selected to support the rotor of the helium blower.

However, this bearing can't be used in the commercial reactor for replacement and lubrication. The replacement is difficult for the radiation, and the lubricant will pollute the reactor. So the Active Magnetic Bearing (AMB) can overcome these shortcomings, and the AMB instead of mechanical bearing will become the new supporting system in the helium blower. The auxiliary bearing will be applied in the AMB system as the backup protector[1].

The auxiliary bearing located at the outer-bound of the AMB is also called a catching bearing or a back-up bearing. Being assembled on the stator, the auxiliary bearing does not work during the normal operation of the magnetic bearing. The clearance of the auxiliary bearing is smaller than that of the magnetic bearing. Typically, fifty percent of the magnetic bearing clearance is used for the auxiliary bearing clearance.

The use of rolling element bearing with grease lubrication as auxiliary bearing is widespread in rotating machinery incorporates magnetic bearing. The function of the auxiliary bearing is to prevent rotor/stator contact, for which the inner race can experience a high impact force and rapid angular acceleration. Rapid deterioration of the auxiliary bearing can result from rotor impacts and high-speed touchdowns. It is therefore important to ascertain the influence of auxiliary bearing design parameters on the number of touchdowns that can be tolerated before replacement is required. A prerequisite is to understand the dynamic behavior of the system during a touchdown event, and this is also a necessary step before attempting to predict any thermal transients within the auxiliary bearing[2-3].

A new auxiliary bearing design with no lubrication will be introduced in this paper. The angular contact ceramic ball bearing will be used as auxiliary bearing. Two pairs of angle contact ball bearing will be distributed on both ends of the rotor. Each pair of bearing arrangement is face to face. Oil or grease will not be applied in this auxiliary bearing. But dry lubrication with MoS_2 will be used in this bearing.

This auxiliary bearing will be applied in the helium blower of 10MW high-temperature gas-cooled reactor (HTR-10). But, so far, this bearing with no lubricate applied in the high temperature reactor is a very difficult problem.

Some dropping tests for rotor have already been done, the impact force of the auxiliary bearing will be analyzed, and the friction heat of this bearing will also be discussed in this paper. Results will offer the basis for the application of this auxiliary bearing in special environment.

B. THE HELIUM BLOWER RIG WITH AMB

The AMB has been designed for helium blower. The rig has been built for research. The figure 1 shows the structure and picture of the helium blower rig. The parameters are listed in Table 1.

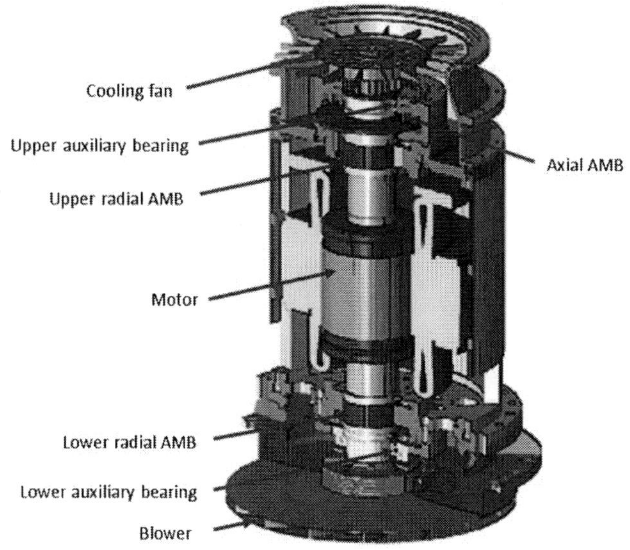

Figure 1. The structure and picture of the helium blower rig

Table 1: Parameters of helium blower

Parameter	Value
Mass of rotor	450kg.
Length of rotor	1518mm
Distance between two radial AMB	800mm
Axial moment of inertia	7.9kg•m^2
Transverse moment of inertia	78kg•m^2
Radial gap between rotor and auxiliary bearing	0.18mm
Axial gap between rotor and auxiliary bearing	0.52mm
Axial load	180kg
Rotating speed	5000r/min

The angle contact ball bearing is applied as the auxiliary bearing for helium blower with AMB. The model is 71926ACD. It is the ceramic bearing with dry lubrication instead of oil or grease. The outer ring and the inner ring are designed by special way.

Two pairs of angle contact ball bearing will be distributed on both ends of the rotor. Each pair of bearing arrangement is face to face. The upper auxiliary bearing carries the radial and axial load, and the lower auxiliary bearing carries the radial load.

The material of bearing rings is bearing steel, and the rolling element is ceramic ball. The material of the cage is ployimide.

The parameters of bearing rings are as followed. Young's modulus is 1.99 × 105MPa, Poisson's ratio is 0.269, the yielding limit is 1600MPa[4]. The parameters of ceramic ball are as followed. Young's modulus is 3.15×105MPa, Poisson's ratio is 0.3[5], the strength limit is 3500 MPa[4].

C. IMPACT LOAD OF DROPPING ROTOR

The normal rotating speed of the rotor is 5000r/min. If the AMB fails to work under this speed, the rotor will fall off suddenly to the auxiliary bearing. The auxiliary bearing must support the dropping rotor, and can bear the great impact load and friction heat. So the auxiliary bearing will fail to work because the impact load and friction heat.

The dropping test is very dangerous for the rotor and AMB system, so theoretical analysis may be done at first. And then the dropping test will be carried out from lower rotating speed to higher rotating speed. The maximum impact force and friction will be produced at the maximum dropping rotating speed (5000 r/min). But the theoretical analysis is difficult for the complex of the dropping course. The test results will be true. So the test results will be introduced in this section.

Figure 2 and 3 are the axial displacement curve and axial impact force curve at the maximum dropping rotating.

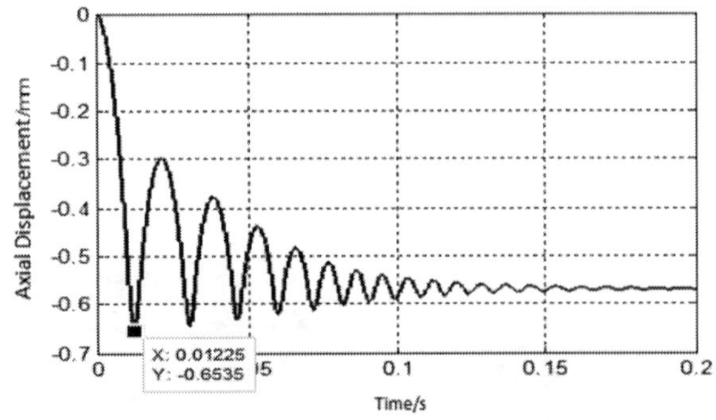

Figure 2. Axial displacement curve of the dropping rotor

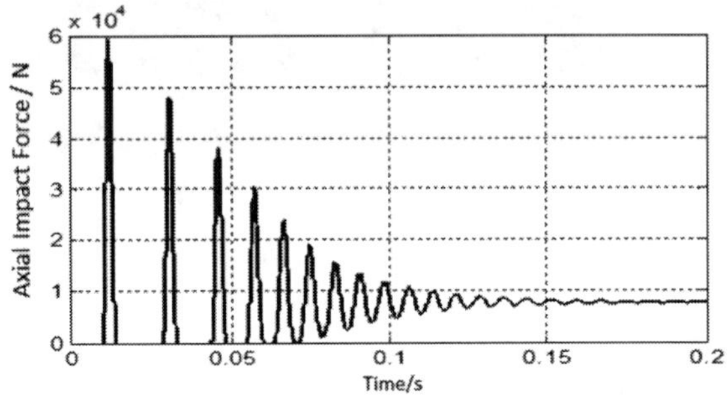

Figure 3. Axial impact force curve of the dropping rotor

The maximum displacement of the auxiliary bearing is 0.134mm from Figure 2, and the maximum axial impact force is 60 kN from Figure 3.

Figure 4 and 5 are the angle velocity curve of the rotor and radial impact force curve at the maximum dropping rotating speed.

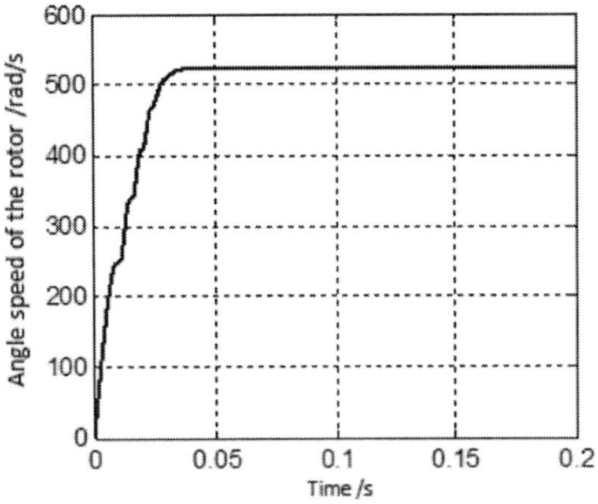

Figure 4. Radial displacement curve of the dropping rotor

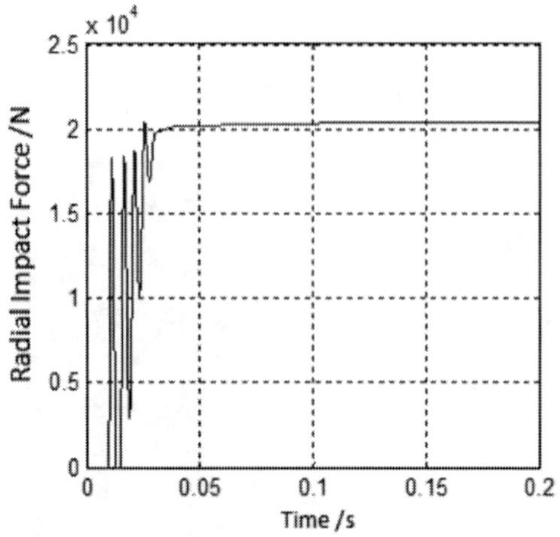

Figure 5. Radial impact force curve of the dropping rotor

The maximum angle velocity of the rotor is 521rad/s from Figure 4, and the maximum radial impact force is 20 kN from Figure 5.

The radial equivalent static load is:

$$P_r=F_r+0.76*F_a=20+0.76*60=65.6 \text{ kN}$$

The basic static load rating P_{or} of the auxiliary bearing is 110.9 kN. P_{or} is bigger than the P_r. So this auxiliary bearing can bear the impact force of the dropping rotor.

D. *FRICTION HEAT OF AUXILIARY BEARING*

When the rotor with high rotating speed falls to the auxiliary bearing, the bearing will produce great heat energy in a short time. The auxiliary bearing may be burned by the heat energy. So the analysis of friction heat is very important.

However the friction is very complicated. A simple simulation was completed by the finite element method in this paper. The initial temperature is supposed to be 20℃. But the time is very short for the first impact time, and the friction heat is not clearly. The temperature of the auxiliary bearing improves only 6.9℃. The temperature field is showed in figure 6. In fact, the friction heat may be very high in a short period of time because of the high rotating speed, full load and enclosed environment. But the temperature rises very little in this paper. The reason is on the light load or good heat dissipation.

Figure 6. The temperature field analysis of the inner ring

E. DROPPING TEST OF DIFFERENCE ROTATING SPEED

It has different influence for auxiliary bearing at different dropping speed. So some dropping tests have been done to study rotor dropping dynamics for different dripping speed.

Figure 7, 8, 9 and 10 give the result of axis trajectory and spectrum of radial displacement at 60Hz, 40Hz, 35Hz and 25Hz.

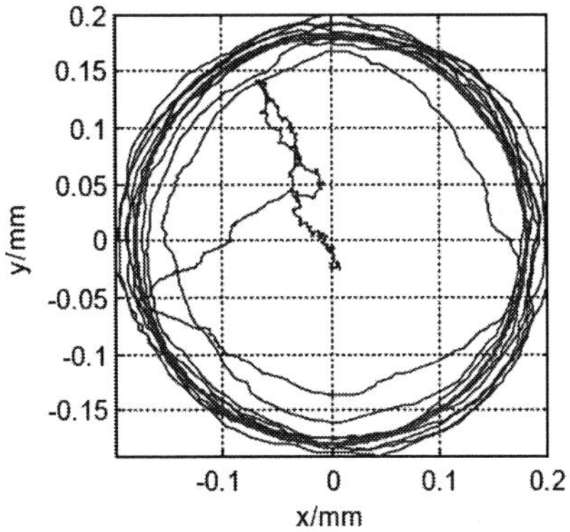

a) axis trajectory

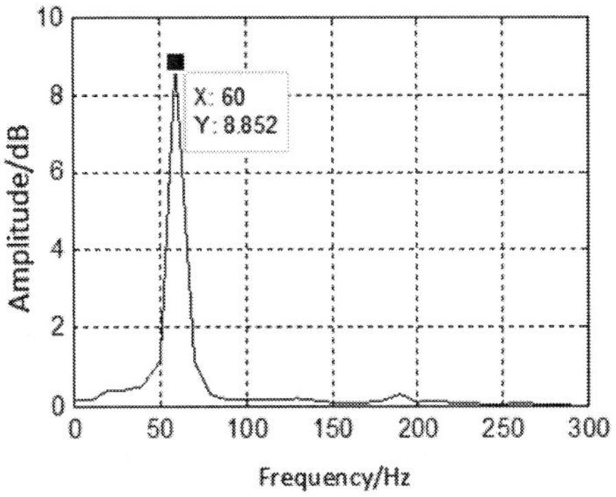

b) spectrum of radial displacement

Figure 7. Axis trajectory and spectrum at 60Hz

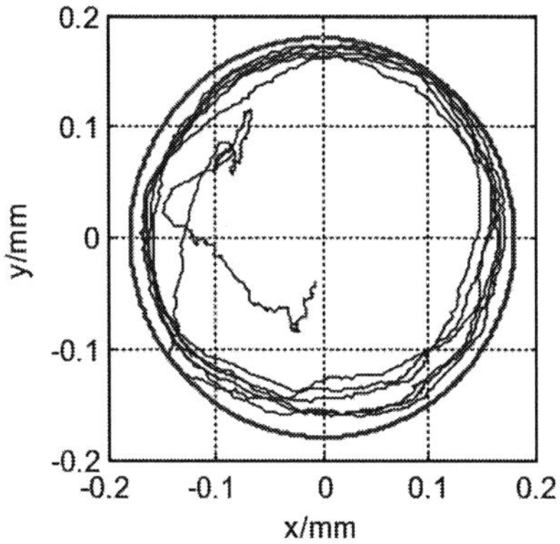

a) axis trajectory

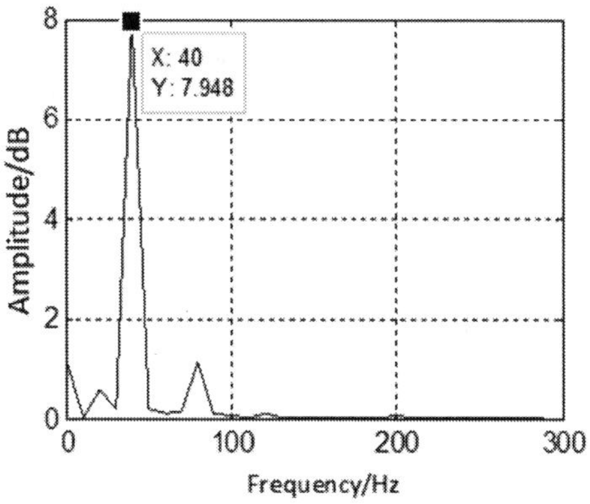

b) spectrum of radial displacement

Figure 8. Axis trajectory and spectrum at 40Hz

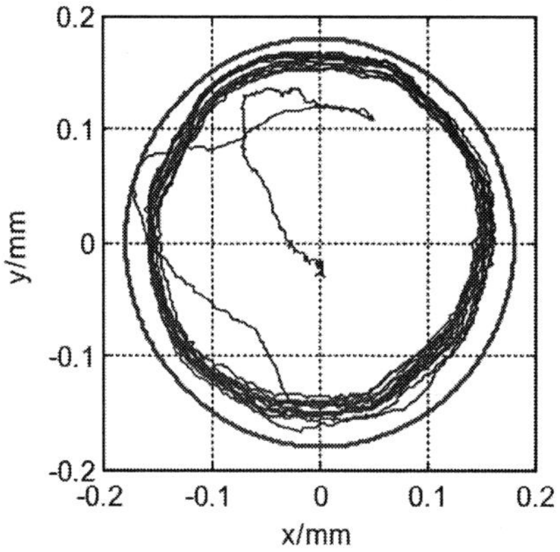

a) axis trajectory

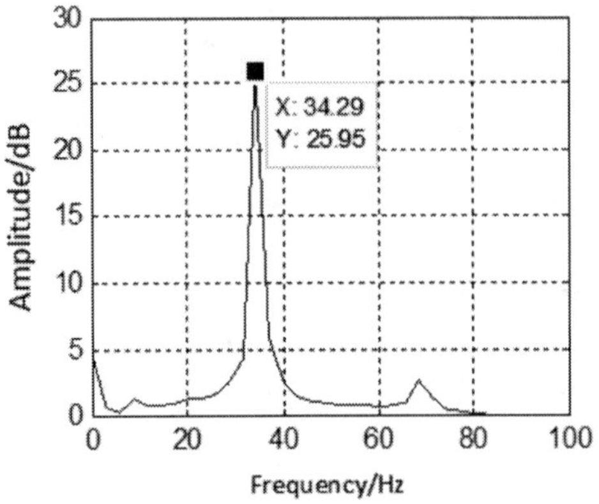

b) spectrum of radial displacement

Figure 9. Axis trajectory and spectrum at 35Hz

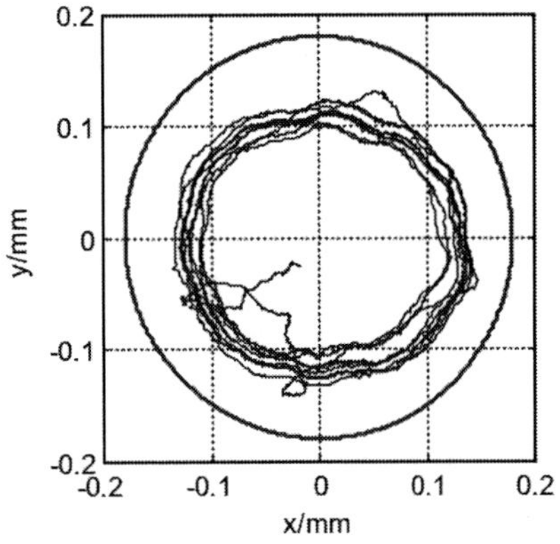

a) axis trajectory

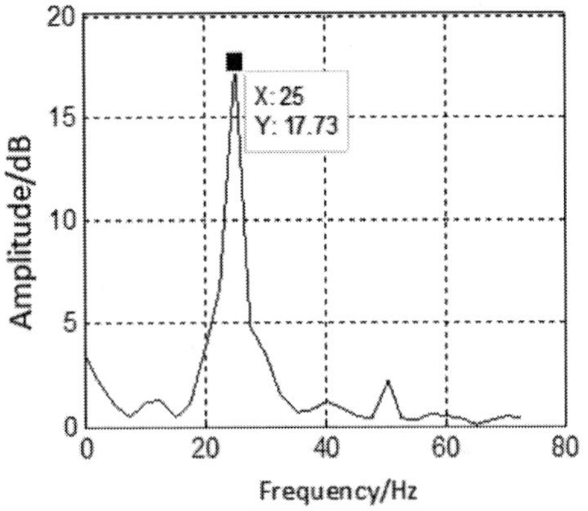

b) spectrum of radial displacement

Figure 10. Axis trajectory and spectrum at 25Hz

From the above figures, we can know that the rotor rotates around the axis with the reduction of the rotating speed, and the rotor don't touch the inner ring of the auxiliary bearing or the radial impact force will reduce little by little.

That is to say, the axial impact force will turn small with the reduction of the rotating speed. The axial impact force can influence the radial motion of the rotor because the generating of the cross-couple stiffness is mainly by the friction force at the thrust interface of the auxiliary bearing. So the reduction of the axial force may reduce the friction force. Then it may have positive effect on the decrease the forward whirl of the rotor. However, because such process is complicated, the relationship between the axial force and the cross-couple stiffness still need further research.

F. Conclusion

A new auxiliary bearing style is proposed in this paper. It is applied in the helium blower of HTR-10. The experiments of AMB and auxiliary bearing are implementing. The impact load and temperature field have been analyzed by the finite element method in this paper.

The drop process can be divided into 3 steps. The first step is the free fall process. The second step is the bounces and backward whirl period. The third part is the forward whirl period. The preliminary analysis has been done. The impact force is about 60kN (instant peak). The friction heat is not so high.

But this is the preliminary results. Further experiments and analysis have been designed. It will be done in the future time. For example, the dropping test in the helium and auxiliary bearing evaluation standard analysis, etc. These results will be introduced in the future paper.

ACKNOWLEDGMENTS

This paper is supported by the National Natural Science Foundation of China (No. 51275261) and National S&T Major Project (No. ZX069).

REFERENCES

[1] Yang Guojun, Xu Yang, Shi Zhengang, Gu Huidong. Characteristic analysis of rotor dynamics and experiments of active magnetic bearing for HTR-10GT. Nuclear Engineering and Design. 2007(237): 1363-1371.

[2] Yang Guojun, Shi Zhengang, Zhao Jingxiong, Yu Suyuan. RELIABILITY ANALYSIS OF HELIUM BLOWER AUXILIARY BEARINGS FOR HTR-10. San Francisco, American. 22nd International Structural Mechanics in Reactor Technology (SMiRT-22). August 18-23, 2013: 2912-2916

[3] Gerhard Schweitzer, etc. (1994) "Active Magnetic Bearings". Vdf Hochschulverlag AG an der Eth Zürich.

[4] Liu Yuanxin. Marc-based FEA (Finite Element Analysis) of Hybrid Ceramic Ball Bearings and Its' Experiment Confirmation. Master's thesis. Tianjin University, 2005. 1: 57-60

[5] Jiang Yiping. Si3N4: Most Promising Bearing Material. Ceramics Engineering. 2001.10: 23-25

High Speed Backup Bearings for Outer-Rotor-Type Flywheels – Proposed Test Rig Design

Lukas Quurck[a], Hendrik Schaede, Michael Richter, Stephan Rinderknecht

[a] Institute for Mechatronic Systems in Mechanical Engineering (IMS), TU Darmstadt, Otto-Berndt-Str. 2, 64287 Darmstadt, Germany, quurck@ims.tu-darmstadt.de

Abstract — **Magnetic levitated rotors for high speed applications need backup bearing (BB) for safe spin down or relevitation after magnetic bearing failure or overload situations. Especially in flywheel energy storage systems long spin down times due to the high kinetic energy content and the vacuum environment complicate the use of conventional bearings as BBs. Depending on the geometric dimension, the rotational speed of the rotor and associated rotordynamic effects today's available bearings (usually rolling element bearings) cannot fulfill their task with the desired lifetime. This paper shows mechanical and thermal problems as well as approaches to mitigate them. For a built and tested Outer-Rotor-Type flywheel its BB system is discussed. For an increase of the surface velocity limit and the extension of the BB lifetime several design improvements are suggested. Properties of a test rig and its special rotor design for BB testing are described and proposed for further experimental investigation. A design of the test-BB is proposed and the data acquisition as well as the dynamic behavior of the rotor is described.**

I. Introduction

A. General problems

Rotors for high speed applications such as turbo compressors, grinding spindles and advanced flywheels are often contact-free suspended with magnetic bearings in order to reduce wear, friction losses and to provide rotordynamic properties. In order to increase availability and to spin down safely after bearing failure or overload, BBs need to be installed. At high operation speeds, current bearing systems like rolling element bearings are utilized highly and often have to be replaced after only a few drop-down events. Common reasons are mechanical bearing failures induced by overload situations. Mechanical seize of rolling elements as a result of transient thermal growth and high accelerations have also to be brought into concern. For BBs non-cage and full complement bearing design is often recommended.

B. Flywheel related problems

Considering the high kinetic energy content and the vacuum environment under which advanced flywheels are often operated, conventional bearings may not withstand a single spin down without mechanical failures. This is caused by the insufficient cooling condition of the bearing and the challenging lubrication. Both might be worsened by rotordynamic events caused by a nonlinear contact situation and be influenced by the gyroscopic properties of the rotor. It is assumed that thermal degradation of material properties including strength and variable friction coefficients are the major reasons for failure. Due to the long spin down of high inertia rotors the BB systems are exposed to these unfavorable conditions for a longer time than thinner rotors in other applications are.

C. Outer-Rotor-Type related problems

Outer-Rotor-Type flywheels are built as a hollow cylinder in order to place the rotor mass as far as possible from the rotational axis. The desired increase of inertia has the side effect of increased surface speed. This surface speed can easily exceed 100 m/s, which is very high compared to most of BB systems known from literature. One reason why ball bearings are difficult to use at such high speeds is explained at the following equation which calculates the tangential hoop stress σ_{hoop} in a thin walled ring, which might be the rotating ring of a bearing.

$$\sigma_{hoop} = \rho \cdot \Omega^2 \cdot r^2 = \rho \cdot v^2 \qquad (1)$$

In Equation (1) ρ is the material density in kg/m³, Ω is the rotational speed in rad/s, r is the radius of the ring in meters and v is the velocity of the ring. Bearings are often classified according to the DN-number representing surface speed in mm·min^{-1}. The rotor described in the next passage has a DN-value of 7,2e6 mm min^{-1} which exceeds the known systems and leads to a hoop stress above 1000 MPa, which is in the range of material strengths of bearing steel. External bearing loads transferred through Hertzian stress and friction were not included in this calculation. Because of the high stresses inside the rotating bearing race and the expected high temperatures, the mechanical integrity of the bearing is obscure even without external loads.

D. Friction induced whirl movement of Outer-Rotor-Types

A typical behavior of Inner-Rotor-Types during a touchdown inside BBs is called the backward whirl. Thereby the rotors center of gravity performs an orbital movement inside the BB inner race against its own direction of rotation. This movement is induced by high friction forced between stator. Very high and destructive whirling frequencies can be reached. Vertical oriented rotors, like most flywheels, tend to slower and less destructive whirl movements in the direction of its rotation. [6] Because the rotor mass is not supported by radial BBs, the friction forces are lower than that of horizontally oriented rotors. Therefore the backward whirl was not reported in literature about vertical rotors.

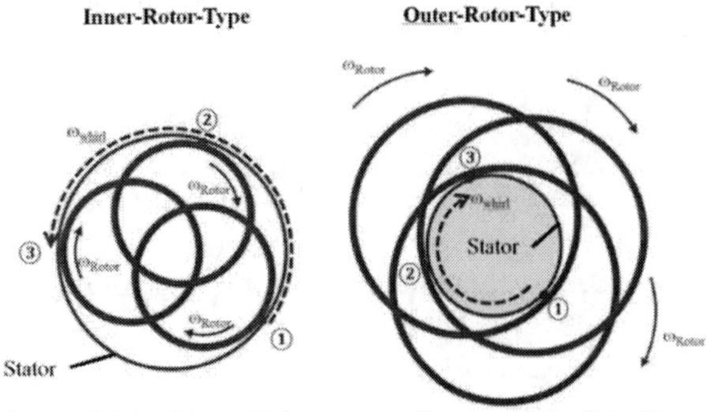

Figure 1. Friction driven whirl movements of Inner- and Outer-Rotor-Types. Numbers represent chronology of the rotor movement inside the BB.

With Outer-Rotor-Types, the kinematic description of the whirl movements "forward" and "backward" needs to be adapted. The directions of friction induced whirls are inverted at Outer-Rotor-Types. Friction driven whirl movements appear here as forward whirls due to the different geometric situation. Therefore the rotors center of gravity rotates in the same direction as the rotor. Fig. 1 shows a simplified 2-dimensional illustration of the differences between Inner- and Outer-Rotor-Types when performing friction driven whirls.

The forward whirl frequencies here tend to very high frequencies and to have a high destructive potential to the whole system like the "backward whirl" of Inner-Rotor-Types. BB failures or rigid contact between rotor and stator may cause high friction forces and drive these whirl movements. However, backward whirls are still expected to happen as long as the remaining friction forces stay under a certain level. This non-friction-driven whirls happen at a certain frequency which could be identified as the coupled elastic eigenfrequency of the rotor contacting the stator. [6]

II. First Demonsrtator Rotor and BB System

A. Description of the existing flywheel rotor

An Outer-Rotor-Type flywheel has been built at the IMS for feasibility proof and component testing. The Flywheel is built as a hollow cylinder from fiber reinforced plastics (FRP). The components of the electric drive and the magnetic bearings as well as the contact area of the BB are fully integrated at the inner diameter and are mechanically supported by the FRP. Some technical data of the rotor is given in Table I. Fig. 2 and 3 show some views and details of the system. The two radial active magnetic bearing plains are marked as well as the axial permanent magnetic bearing (PMB) can be seen there.

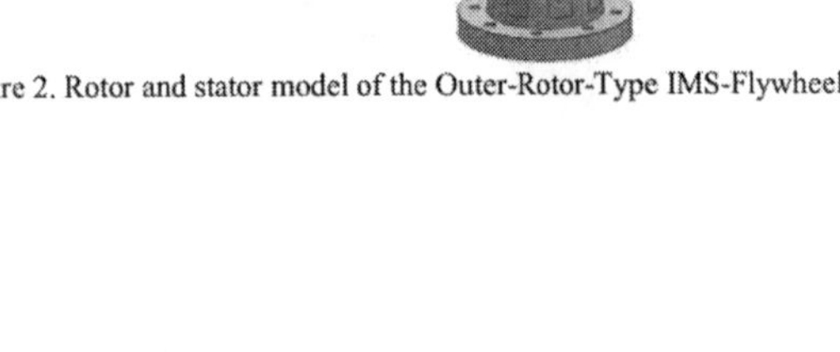

Figure 2. Rotor and stator model of the Outer-Rotor-Type IMS-Flywheel

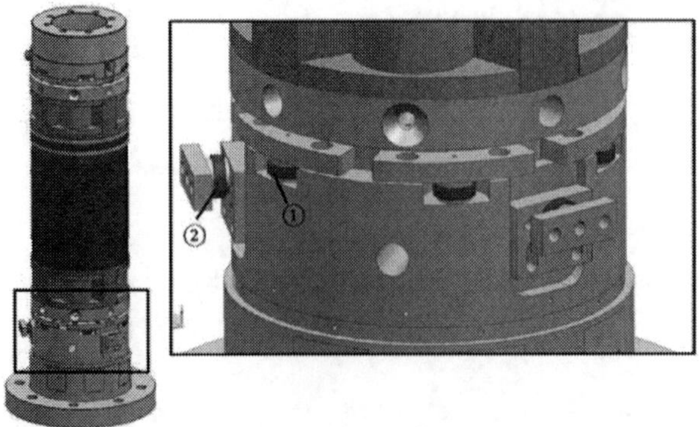

Figure 3. Detail view of the lower BB plain. ①: Radial rolling element bearing; ②: Axial rolling element bearing

The rotor can be described as rigid as numerical calculation and experimental modal analysis showed. First measured elastic eigenfrequency is above 1160 Hz, while the rotational speed of 40.000 results in a harmonic frequency of 667 Hz. It can be assumed that elastic eigenmodes are not heavily exited by the AMB which has a cut off frequency at around 1 kHz. Experimental tests confirm this assumption.

TABLE I. Technical data of the IMS Flywheel with its mechanical design limits

	Parameter	Value
d_o	outer-rotor diameter	300 mm
d_i	inner-rotor diameter	180 mm
h	rotor height	500 mm
N_{rotor}	max. revolutions per minute	40.000 rpm
v_{surf}	max. surface speed @ BB	376 m/s
I	rotor inertia	0,7 kg m²

The system was put into operation in Dec. 2013 and reached its preliminary max. speed of 21.000 rpm. For several technical reasons higher speed was not achieved yet, exemplary the first BB system was not completely supposed to withstand a highspeed drop safe without mechanical failures. The reached energy content at 21.000 rpm is 0,48 kWh. Mechanical challenges and uncertainties which lead to the reduced max. speed of the first BB system are discussed in the next paragaph II.B. An improved BB System with higher load capacity and higher predicted speed limitations is described and discussed at paragaph II.C and II.D.

B. Description of the first BB System

As no manufacturer can provide centrally mounted bearings with an outer diameter of 180 mm, that can withstand the occurring rotational speeds and accelerations, alternative concepts are considered. Instead of one single BB per BB plain, six smaller bearings are uniformly distributed around the circumference of the stator. Similar designs were proposed and tested in [1] and [2]. Clearance elimination at BB contact mentioned in [1] and [2] was not implemented. While being levitated the rotor does not contact the bearings. After being dropped into the bearing the rotor does contact one or two bearings in each plain at the same time. Because of the play between the inner diameter of the rotor and the bearings the rotor is able to change its position inside the BB system. Because of the different sizes of the rotor inner diameter and the outer bearing race diameter, a kinematic translation leads to different angular speeds which can be calculated using Equation (2). The surface velocity still stays the same as it is the boundary condition at contact area when slipping is disregarded.

$$N_{BB} = N_{rotor} \cdot d_i / d_{o,BB} \qquad (2)$$

Data of the chosen bearings are listed in Table II.

TABLE II. TECHNICAL DATA OF THE IMS FLYWHEEL

	Parameter	Value
$d_{o,BB}$	Outer bearing diameter	19 mm
$d_{i,BB}$	Inner bearing diameter	10 mm
$N_{rotor,prel.}$	Preliminary revolutions per minute	21.000 rpm
N_{BB}	Max. bearing revolutions per minute	189.000 rpm
n_{BB}	Number of bearings per plain	6

Because of the high mechanical stress in the outer races of the bearing the maximum angular speed of the rotor was reduced to the half of the projected speed of 40.000 rpm until the dynamic system behavior in BB is better investigated or design improvements mentioned in the next section are implemented.

C. *Discussion of the first BB system*

Fortunately the planetary arrangement of multiple smaller bearings brings some advantages. First of all the selected ball bearings provide dry lubrication for long lifetime and low friction coefficients in high speed and high temperature operation. Dry lubrication also does not suffer from evaporation under high vacuum conditions. According to the manufacturer's information these bearings are used in turbo molecular pumps as BB at around 90.000 rpm withstanding several dropping events without failure. A second application of this bearing type is in micro turbines as thrust bearings. They withstand more than 200.000 rpm at high temperature for several minutes. Because of these operational experiences the selected system was assumed to withstand at least one rotor drop from 20.000 rpm to safe standstill. Hoop stress stays under 280 MPa so that additional bearing loads can be tolerated.

The second advantage of the chosen concept is the polygon shaped orbit of the whole BB system. It is assumed that whirling movements of the rotor might be suppressed, which was tested in former works [3]. In addition, low BB inertia is assessed to be good during touchdown events because synchronization time between rotor and BB can be shortened. The shorter these synchronization times are, the is lower the probability of a friction driven whirl movement. The first BB system has several magnitude orders lower inertia compared to a theoretical possible central bearing configuration. Furthermore specific compliancy of the bearing can be achieved by design parameters in order to support rotordynamic behavior in BB. Some disadvantages still have to be taken into account with this design approach. The hoop stress stays the same following (1) and (2), as angular speed of the BB is inversely proportional to its outer diameter $d_{o,BB}$. Additionally extra load from the balls being pressed against the outer bearing ring occurs. In general outer ring rotation is not the best load case for ball bearings. Additional axial forces and tilt moments are possible loads during dropping events. When bearing tilting exceeds a certain level, the ball filling grooves of the bearing can cause seizing when roller elements roll over these grooves.

Regarding all drawbacks and uncertainties of rotordynamics and thermodynamics no high speed BB testing was made with the first BB system. More investigation and design improvements mentioned in next section are necessary to increase lifetime and to enhance performance.

D. *Load case optimization*

To reduce mechanical and thermal loads of the BB system some design changes are implemented in an improved stator design of the flywheel. The main aim for this improvement is to reduce angular BB speed without reducing rotor speed. This can be done by increasing the outer diameter of bearing interface. The bearing speed is reduced linearly with the increasing bearing interface shown in Equation (2). Adding roller elements at the bearings outer diameter is the easiest way to achieve this as shown in Fig. 4 on the left side. Major drawback of this configuration is the different strain of the roller element and the bearing which can result in a lift off at high speed. Depending on geometric parameters, high preload and elastic elements between the outer ring and the roller element can solve this problem. Calculations were made for the introduced flywheel and this solution was abandoned because lift of could not be prevented without seizing the bearing due to the necessary high preloads. Elastic interlayers were also discussed but calculations failed because of unknown thermal behavior of the BB.

A promising configuration can be found when the inner race of the bearing is chosen for rotation. Forces are transmitted with roller elements called "rpm-reduction-disc" (RRD). RRD are mounted at the inner bore by one or more ball bearings. By adding a second bearing the load capacity can be increased simultaneously and tilting of the rotor as well as axial forces can be supported. Thermal junction to the stator and general load case are improved at the same time. The RRD design is illustrated in Fig. 4 on the right side. In the next paragraph the improved BB system for the IMS Flywheel is described in more detail.

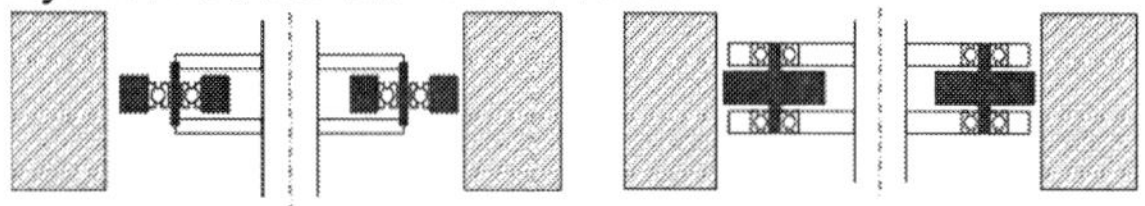

Figure 4. Left: Roller elements mounted on the outer bearing race; right: RRD mounted inbetween a set of two ball bearings.

E. *Improved BB data*

RRD are made with an outer diameter of 25 mm and turned out of tempering steel. Due to the full cylinder design and the reduced bearing speed, the mechanical stress induced by centrifugal forces stays under 460 MPa when the RRDs are accelerated to the surface speed of the flywheel rotating at 40.000 rpm. The maximum stress in the BB system is reduced to less than half compared to the first BB System without the RRDs at full speed. The bearing size stays the same as in the first system but a second bearing was added to heighten load capacity. Fig. 5 shows a cross-section of the lower BB plain for the IMS Flywheel. Experimental testing remains to be done. For efficient testing another experimental environment has to be set up. A special test rig is assumed to be the best environment for investigation as it is described in the following chapter.

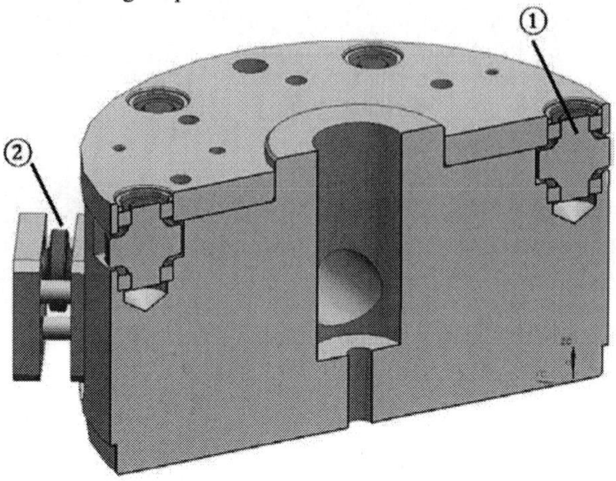

Figure 5. Cross-section of the lower BB plain of improved bearing BB system for the IMS Flywheel. ①: RRD with rolling element bearing for radial support; ②: RRD with rolling element bearing for axial support

III. PROPOSED TEST RIG

A. *Nesessity of experimental testing*

Because of complex dynamics, unknown thermal junctions and a lack of knowledge about mechanical limits like thermal induced degradation of material properties or lubrication, the experimental investigation of the BB system behavior is reasonable. Experimental investigation on the described flywheel would be desirable but brings the high risk of irreparable damages. Therefore an alternative test rig for intense and safe experiments is considered as the best way for observation and testing of design improvements. Furthermore the experimental investigations may deliver important information about parameters for numerical investigation and optimization. In the following section a design of an active magnetic levitated substitute rotor and its components are discussed.

B. *Properties of test rig*

The suggested substitute rotor needs to be magneticly levitated and accelerated to comparable surface speeds. After reaching the desired speed, the rotor has to be dropped into the BB for example by cutting the power supply of the AMB. To generate similar operation environment as for the flywheel the housing needs to be evacuateable.

In addition to the required mechanical strength and robustness of stator and rotor, modularity is one main feature. The test rig has to withstand numerous dropping tests. Increasing robustness and thermal stability can be achieved by using steel instead of FRP. For high speed testing full disc design leads to lower material stresses compared to hollow disc rotors. Inner-Rotor-Type design is proposed therefore. Except of the different direction of the friction driven whirl the Inner-Rotor-Type is supposed to behave like Outer-Rotor-Type. Moreover standard components and machinery can be used. Planetary bearing arrangement can still be provided and high surface speeds are achieved by the diameters at the BB plains. The AMBs and the drive system are restricted to lower diameters and therefore surface speeds due to the mechanical integrity of the interference fitted sheets and magnets. Vertical orientation of the rotational axis is important for the investigation especially because the majority of prior works deal with horizontal rotors. In [5] and [6] two different vertical oriented flywheel rotors were dropped into BB and only forward whirl movement was observed. The observed whirl frequencies seem to be limited by the rigid body eigenfrequency of the rotor contacting the BB. The backward whirl movement was not observed, which can be explained by the absence of a strong radial force like weight force. Such strong radial forces induce tangential friction forces and drive whirling. For a rotordynamic comparability of the test rig, the gyroscopic properties must match to that of the original flywheel rotor. This can be achieved by setting the ratio of axial and polar moments of inertia as described by Equation (5) near to the original rotors ratio.

$$K_{inertia}=I_p/I_a \qquad (5)$$

The ratio of the introduced flywheel is near 0.5. The higher the ratio of a rotor is, the bigger the gyroscopic behavior of the system is. Thin and long rotors have low ratios while flywheels tend to have bigger ratios. Formulas for axial and polar moment of inertia of a full cylinder are given in Equations (6 & 7).

$$I_p=1/2\ m\ r_o^2 \qquad (6)$$

$$I_a=1/4\ m\ r_o^2 + 1/12\ m\ l^2 \qquad (7)$$

m represents the cylinder mass, r_o is the outer radius and l is the length. $K_{inertia}$ is a major feature for a stable control of the rotordynamics in active magnetic bearings [4], as it represents the gyroscopic properties of the rotor. Another important dynamic characteristic is the elasticity of the substitute rotor. The first elastic bending eigenfrequency should be widely above the rotation frequency of the rotor as it is given for the original rotor made from FRP.

C. *Substitute rotor for BB testing*

Regarding the dynamic behavior of the substitute rotor a design has been done. Two discs act as the contact area to the stator bound planetary BB sets and provide high gyroscopic behavior. For assembly at least one disc has to be removable from the rest of the rotor. The rotor shown in Fig. 6 has an inertia ratio of $K_{inertia}$ = 0,5 and its first bending eigenfrequency is 1806 Hz. Turning at 20.000 rpm, FE-Analysis shows maximum mechanical equivalent stress around 300 MPa at the BB discs. The maximum surface velocity is at 209 m/s, which is assumed to be sufficient for BB testing. Polar moment of inertia is around 0.05 kg m² leading to a kinetic energy content of max. 0.03 kWh. The weight of the rotor is 17 kg. For fast breaking capability the max. power of the electric drive is around 15 kW. Electric emergency braking takes under 50 seconds from full speed to standstill, disregarding friction inside the BB system. One important feature of the rotor is the possibility of using a secondary BB plain which can be mounted at both ends of the rotor. The clearance at the secondary BB should be slightly larger compared to the primary BB. Thereby it can be used as a emergency BB when the primary BB has a severe mechanical damage. Moreover investigations on dynamic behavior of the rotor being dropped inside the conventional central mounted BB without the planetary BB can be done and may deliver important information about general behavior of both approaches.

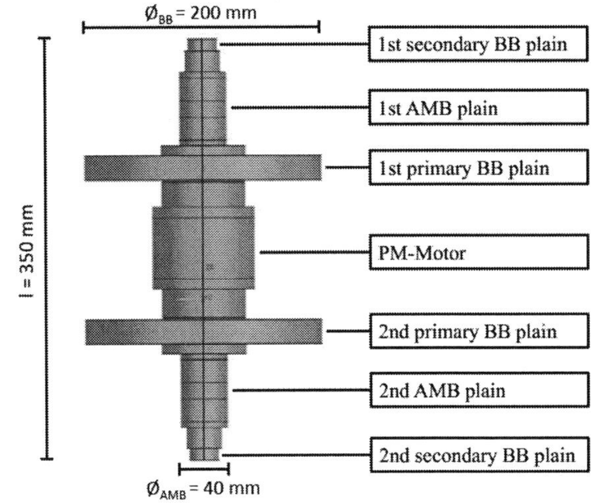

Figure 6: Model of the substitute rotor for BB testing

D. Test Rig Equipment

The substitute rotor is completely levitated by AMBs. The active axial bearing compensates the weight of the rotor and is capable to center the rotor during radial touchdown events in order to prevent axial contact. This is necessary if the influence of an axial BB contact is unwished. The positioning sensors of the AMB are used to record the orbits during testing with high sampling rates up to 20 kHz. Because of the high stiffness of the rotor no additional range sensor is needed in the BB system. Optical temperature sensing by pyrometers is used to calculate heat generation at the BB contact zones. Temperature probes can record the temperature at the non-rotating parts of the BB system. Thereby heat transfer is observable.

A synchronous drive system is used to accelerate the rotor to the desired dropping speed. After dropping the rotor into BB the drive is able to brake with up to 12 Nm. Idle coasting of the dropped rotor is another opportunity for simulating an unassisted spin down e.g. when the drive system has a serious failure and cannot actively decelerate rotation. For drive control and dynamic observation of the rotor speed, a rotational encoder for angular position is necessary test rig equipment.

E. Design of the advanced BB system

In Fig. 7 a design of a BB system with planetary arrangement of multiple RRDs is shown. The primary BB design for the test rig differs from the original flywheel system in the geometric dimensions and the inverse design of the Inner-Type-Rotor. A further design difference is the variable air gap between the rotor and the RRDs. Every single RRD is beared inside a radial slideable mount. Therefore the influence of the air gap size can be investigated. The air gap is not varied during a test in this design approach. Advanced approaches with variable actuated air gap control may be investigated in later works as well as predefined compliancy. Positive influences on bearing loads and lifetime is expected from certain dimensioning.

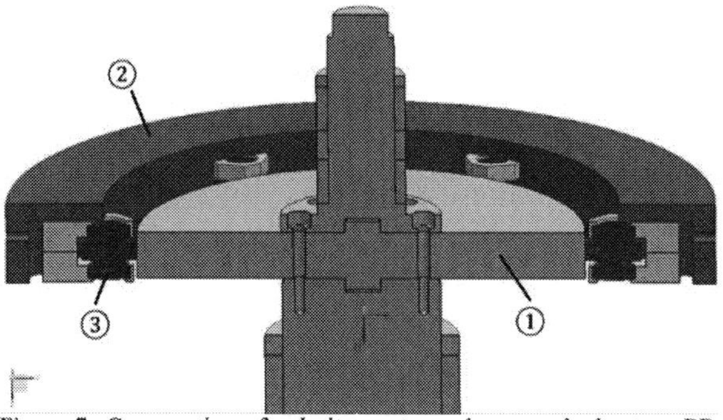

Figure 7: Cross-section of substitute rotor and proposed planetary BB. ①: Rotor; ②: Stator of planetary BB set; ③: RRD with rolling element Bearing inside a slidable mount

I. SUMMARY

This paper introduces a novel high speed Outer-Rotor-Type Flywheel with an experimentally achieved energy content of 0.47 kWh at 21.000 rpm. Full speed of 40.000 rpm was not achieved yet. One reason are the uncertainties of the first BBs load capacity. The risk of a BB failure at high speed operation was assumed to be too high. For heightening load capacity and speed range of the BB, a design improvement was made and introduced. For efficient testing a special test rig with a substitute rotor is suggested. The substitute rotor for this test rig is designed as Inner-Rotor-Type with comparable dynamic properties as the flywheel rotor. The modular steel design of the fully active levitated substitute rotor provides the needed thermal and mechanical robustness for multiple BB dropping tests and reduces risk and cost of future investigation. Properties of the data recording of the test rig are described and an exemplary BB system with planetary arrangement of multiple single bearing sets with adjustable air gaps is introduced.

II. OUTLOOK

Primary aim of the proposed test rig is to investigate a high speed BB system for Outer-Rotor-Type flywheels. The current configuration with 6 planetary bearings in each plain is not assessed to withstand multiple dropping events from high speed without failure. Because of difficulties in mechanical and thermal calculation and unknown thermal degradation of material properties inside the bearings, experimental testing is obligatory to learn more about limiting loads, temperatures and wear of the bearing. Influences of ambient pressure on friction, thermal behavior and wear are expected to deliver important information for further BB designs. When important parameters are identified and determined numerically simulation can achieve better results and further design improvements will be achieved more efficiently.

The use of roller elements called "rpm-reduction-disc" (RRD) as an improvement of the current flywheel BB may improve performance and lifetime of the system. Increasing the limiting speeds of the BB enables the possibility to increase the kinetic energy content of the energy storage under safe and economic circumstances. This is an important step towards industrial utilization and commercial operation of Outer-Rotor-Type flywheels.

REFERENCES

[1] Mohawk Innovative Technology, Inc.; "Auxiliary Bearing Developed gelagertfor 140 mm Shaft with 1000 Pound Load at 18,000 RPM!"; MiTi Developments, Vol 8, 1999

[2] Chengtao, Y. et al; "Auto-eliminating Clearance Auxiliary Bearings for Active Magnetic Bearing Systems"; ISMB 13, 2012

[3] Helfert, M.; "Rotorabstürze in Wälzlager"; Dissertation TU Darmstadt, 2008

[4] Gasch, R; Nordmann, R.; Pfützer, H; „Rotordynamik"; Springer, 2002

[5] Caprio, M.T.; Murphy, B.T.; Herbst, J.D.; "Spin commissioning and drop tests of a 130 kW-hr composite flywheel," in Proc. 9th Int. Symp. Magnetic Bearings, Paper, vol. 65, 2004

[6] Hawkins, L.; McMullen, P.; "Long Term Backup Bearing Testing Results"; ISMB 13, 2012

Active Touchdown Bearing Control for Recovery of Contact-Free Rotor Levitation in AMB Systems

Peichao Li[a], Necip Sahinkaya[a,b], Patrick Keogh[a]

[a] Department of Mechanical Engineering, University of Bath, Bath BA2 7AY, UK, p.s.keogh@bath.ac.uk
[b] Department of Mechanical and Automotive Engineering, Kingston University, London KT1 2EE, UK

***Abstract*—Control strategies for rotor displacements and transmitted forces in active magnetic bearing (AMB) systems are generally well-developed when the rotor is in contact-free levitation. In such a condition, the rotor/AMB system, the plant to be controlled, is well-defined to use for controller design. However, transient overload conditions or large scale disturbances may cause the rotor to make contact with one or more touchdown bearings (TDBs). The ensuing dynamic response is an effective change to the contact-free plant, which may cause an AMB controller to apply inappropriate forces, causing persistent contact, or even inducing system instability. It is therefore beneficial to restore contact-free levitation as quickly as possible in order that normal operation may resume. This would minimise outage time and also minimise high levels of potentially damaging contact that may be experienced by the TDBs. In this paper, the use of an active TDB is considered for the restoration of contact-free levitation. The contact problem is demonstrated in simulation and also through appropriate control applied in an experimental system. In this system, TDB motion is activated by piezoelectric stack actuators.**

I. INTRODUCTION

The study of the dynamics of a spinning rotor making contact with a stator component has been the subject of numerous publications. There are many ways in which the rotor may interact, including forward whirl rubbing, backward whirl rubbing, combined bouncing and rubbing, and chaotically induced contact. Direct contact will also involve heat generation, which may cause dynamic rotor thermal bending. These effects are possible in any type of rotating machinery.

Specific to rotor/AMB systems, the problem of rotor contact is a critical factor affecting machine operation. In order to cope with the existence of a clearance gap in AMBs, a TDB may prevent excessive rotor excursions. However, the step-like changes in contact force may be large and may also result in significant excitation of rotor dynamics.

A TDB must be able to cope with two basic cases: (a) Rotor drop in which the AMBs no longer function; and (b) Transient overload cases when the AMBs are still functional. In Case (a), a number of authors have addressed the problem, including the assessment of rotor dynamics through experimental tests [1-8]. Although Case (b) has received less attention, it is important to recognise that high TDB contact forces may occur at potentially damaging levels. Also, when contact forces are experienced for any significant period of time, the rotor lateral motion is resisted by an effective high stiffness compared to contact-free operation. Hence, the still functioning AMB controller may react inappropriately to measured rotor displacements. It is easy to envisage that notch filters designed over certain contact-free rotor flexural frequencies may be incorrectly positioned when persistent contact is experienced. This arises due to the changes of rotor dynamic modal frequencies.

Figure 1. Experimental rotor/AMB/TDB system. The TDBs are adjacent to the AMBs on the outboard sides. The right hand side TDB is actived by piezoelectric actuators.

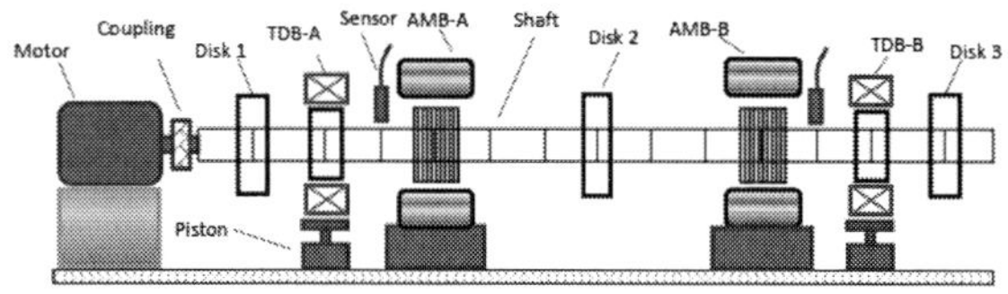

Figure 2. Rotor finite element discretisation for dynamic simulations.

A number of authors have recognised the benefits of minimising rotor contact by actively controlling the position of a TDB. In [9, 10], electodynamic actuators are used, though the AMB is not functional. In [11], piezoelectric actuators are used to activate a TDB. That paper focused on the system design to achieve active control of a TDB in orthogonal axes. There is a need to avoid side loads on piezoelectric stack actuators and the use of hydraulic couplings for this purpose is explained in [11], though active control was not applied.

In this paper, the control issues associated with a TDB while the AMBs remain operational are considered. The operating condition of the rotor is taken to be bi-stable with either contact-free levitation or persistent rotor/TDB contact possible. This paper addresses the problem of how to move from the contact to contact-free operation. The bi-stable rotor dynamics are shown experimentally, followed by control back to levitation.

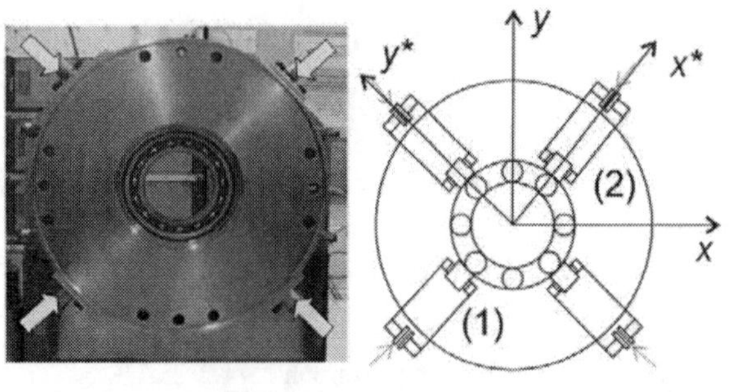

Figure 3. Active TDB system. The TDB is actived by piezoelectric actuators through closed hydraulic lines.

II. EXPERIMENTAL SYSTEM

The rotor/AMB/TDB system used in this study is shown in Fig. 1. For the purposes of dynamic modelling, the rotor is discretised into standard beam elements as shown in Fig. 2. Figure 2 also shows the positions of sensors, AMBs, TDBs, disks and the drive motor. The rotor shaft is 800 mm long. Each AMB has a clearance gap of 0.8 mm. The rolling element TDB shown in Fig. 3 has a radial clearance of 0.4 mm. However, for investigative purposes, TDB bushings were also fabricated with a reduced radial clearance of 0.29 mm. Each AMB has a force capacity of 690 N and break frequency of 250 Hz. The right hand side, non-driven end TDB (TDB-B in Fig. 2) is activated by four 10 kN piezoelectric stack actuators as shown in Fig. 3. The actuators are contained within a manifold (Fig. 4). They are coupled to the TDB through closed hydraulic lines, which prevent damaging side loads from being applied to the actuators.

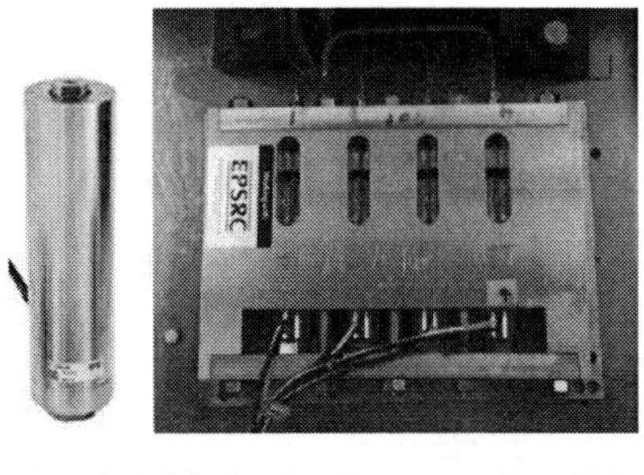

Figure 4. TDB piezoelectric actuator and manifold.

III. SIMULATIONS

A rotor dynamic model of the system was established using the finite element discretisation of Fig. 2. For this section, both TDBs were considered to be in a passive configuration, effectively with the piezoelectric actuators in an 'off' state. The matrix-vector rotor dynamic equation of motion has the form

$$\mathbf{M}\ddot{\mathbf{q}}_r + \mathbf{C}\dot{\mathbf{q}}_r + \mathbf{K}\mathbf{q}_r = \mathbf{f}_m + \mathbf{f}_t + \mathbf{f}_d + \mathbf{f}_u \tag{1}$$

where $\mathbf{q}_r$ is the rotor displacement vector, $\mathbf{M}$, $\mathbf{C}$ and $\mathbf{K}$ are the mass, damping/gyroscopic and stiffness matrices, $\mathbf{f}_m$, $\mathbf{f}_t$ and $\mathbf{f}_d$ are the AMB, TDB and disturbance force vectors, respectively. $\mathbf{f}_u$ is the unbalance force vector.

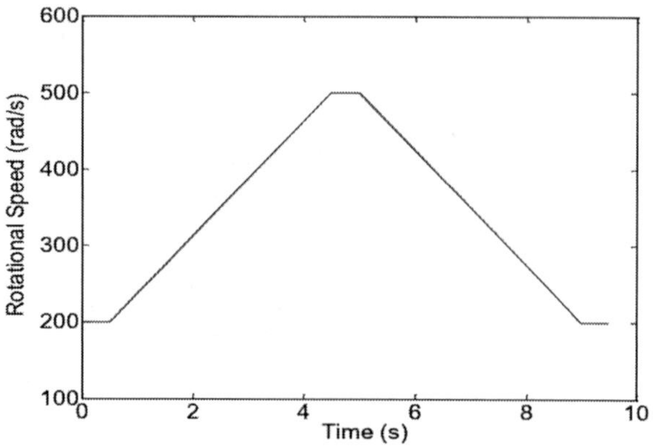

Figure 5. Simulated speed variation with time.

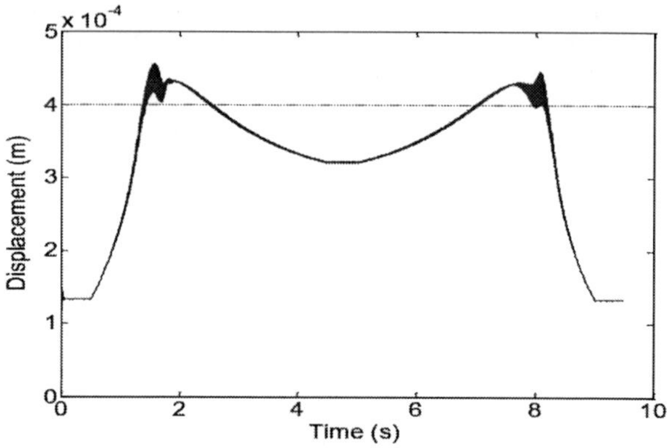

Figure 6. Radial displacement of the rotor at touchdown bearing position assuming that no TDBs are present.

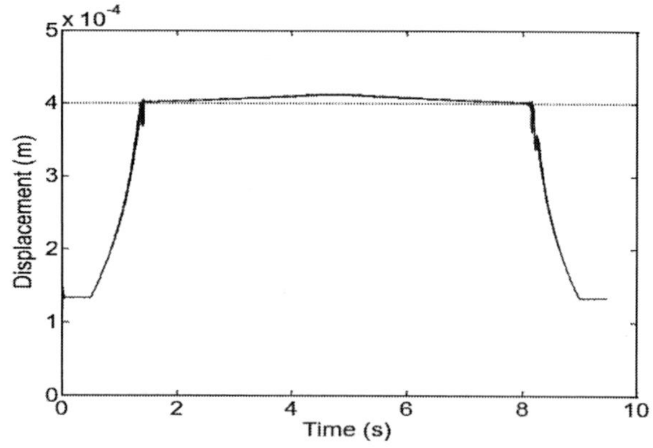

Figure 7. Radial displacement of the rotor at touchdown bearing position. The rotor remains in contact with the TDB for speeds above 270 rad/s.

The TDB force vector consists of normal and tangential (friction) components as derived from Hertzian contact theory. A coefficient of friction equal to 0.15 was used to evaluate the friction force. For the simulations, the nominal

TDB radial clearance of 0.4 mm was used. Since the TDBs are also moveable, another equation of motion is also formed:

$$\mathbf{M}_t\ddot{\mathbf{q}}_t + \mathbf{C}_t\dot{\mathbf{q}}_t + \mathbf{K}_t\mathbf{q}_t = -\mathbf{f}_t \qquad (2)$$

where $\mathbf{q}_t$ is the TDB displacement vector. $\mathbf{M}_t$, $\mathbf{C}_t$ and $\mathbf{K}_t$ are the TDB mass, damping and stiffness matrices that are tailored to match the displacement vectors.

In a simulation with an appropriate unbalance on the end disks (1 and 3 in Fig. 2), the rotor speed was varied according to that shown in Fig. 5. Without consideration of TDBs ($\mathbf{f}_t = \mathbf{0}$), Fig. 6 shows the rotor radial displacement to increase-decrease-increase-decrease as appropriate to passing through a rigid body critical frequency. It is noted that the radial displacement exceeds 0.4 mm in this transition. When the TDBs are included ($\mathbf{f}_t \neq \mathbf{0}$) in the simulation, Fig. 7 shows that rotor contact persists at speeds above 270 rad/s. It may be inferred from Figs 6 and 7 that bi-stable contact and contact-free rotor motions are possible at speeds between 320 rad/s and 500 rad/s.

IV. INDUCED ROTOR/TDB CONTACT TESTS

Since rotor/TDB contact may induce severe vibration, it was decided in the first instance to undertake low speed tests in which variable frequency rotating forces were applied to the rotor through the AMBs. In terms of rotor dynamic response, the gyroscopic moments are negligible, but this may be acceptable if disk conical motions are small. Hence equal in-phase rotating forces were applied through each AMB. Although the rotor is symmetric, the motor coupling caused the orbits at the motor end to be slightly smaller than those at the non-driven end over the range of frequencies applied (up to 450 rad/s). Furthermore, rotor/TDB contact tended to occur at TDB-B.

Initial tests with the rolling element bearing shown in Fig. 3 were problematic with large amplitude rotor vibration experienced. This was attributed to steel-steel friction and excessive inertia in the TDB. It was therefore decided upon to replace the rolling element bearings with bronze bushings. Furthermore, the radial clearance of TDB-B was reduced to 0.29 mm to guarantee that it was the only one to experience contact.

Figure 8 shows the measured maximum rotor radial displacement in the plane of the sensor pair associated with control of AMB-B. This pair is also relatively close to TDB-B (Fig. 2). The rotating forces applied through the AMBs simulate unbalance through speed-frequency association. The frequency was slowly increased from 200 rad/s to 450 rad/s, and then reduced at a similar rate. The regions in which the radial displacements overlap correspond with a non-contacting linear response either side of a resonance (~ 320 rad/s). With frequency increase, contact persists from 310 rad/s to 405 rad/s, before it is lost. As the frequency reduces, the rotor is contact-free up to 360 rad/s, then contact persists down to 300 rad/s after which the rotor becomes contact-free. The hysteresis effect clearly shows that bi-stable rotor dynamics are evident from 300 rad/s to 405 rad/s.

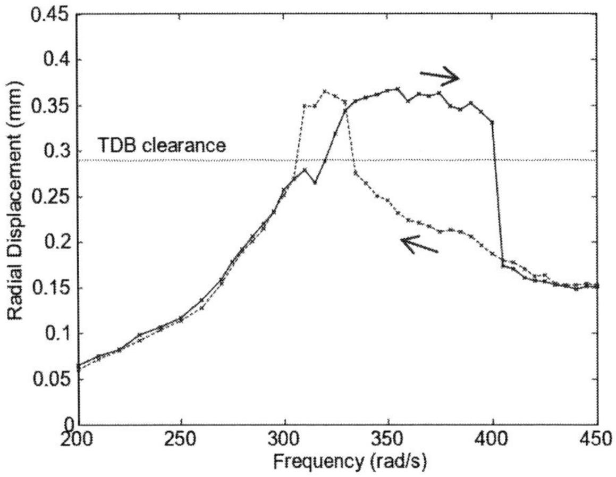

Figure 8. Rotor maximum radial displacement at AMB-B with respect to gradual run up and run down of the rotating force frequency. The rotating forces were equivalent to 500 g.mm and applied in-phase to the rotor through each magnetic bearing.

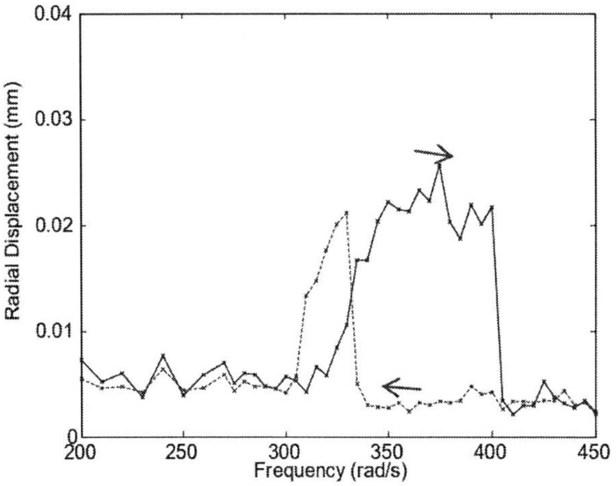

Figure 9. TDB-B maximum radial displacement with respect to gradual run up and run down of the rotating force frequency. The rotating forces were equivalent to 500 g.mm and applied in-phase to the rotor through each magnetic bearing.

A sensor pair was also available to measure the displacement of TDB-B during the frequency tests. Figure 9 also shows the effect of the bi-stable rotor response between 300 rad/s and 405 rad/s. The lower values of maximum radial TDB displacement (around 0.005 mm) correspond with contact-free motion, and are attributable to measurement noise and the eddy current displacement sensor sensitivity.

Figures 10 and 11 show the orbits at the sensor planes associated with the AMB control axes at 335 rad/s and 400 rad/s, respectively. It is noted that the orbits at AMB-B are directionally biased, which is likely due to slight radial misalignment of TDB-B relative to the centre of AMB-B.

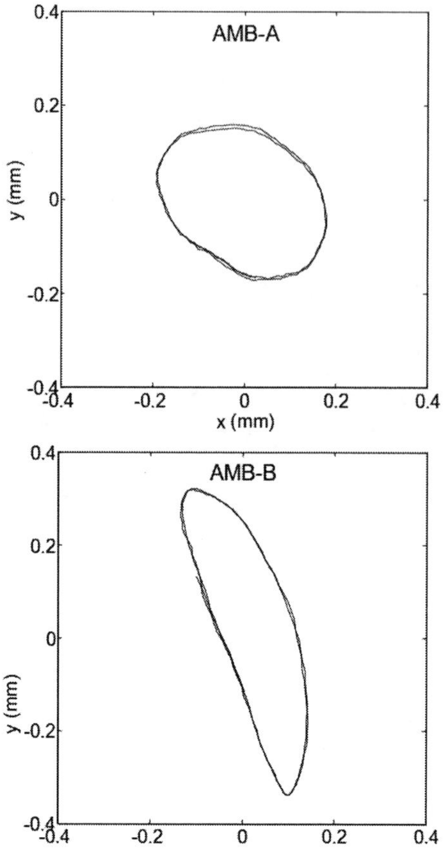

Figure 10. AMB Orbits at 335 rad/s

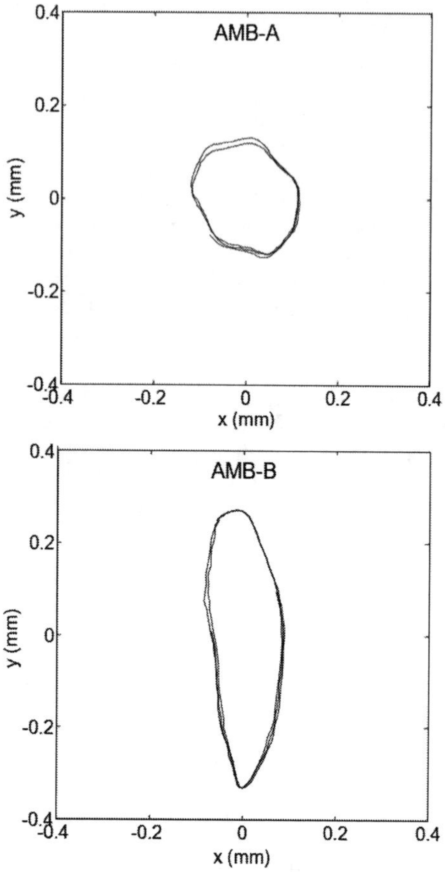

Figure 11. AMB Orbits at 400 rad/s

V. ROTATIONAL TESTS

Frequency based tests have a benefit of being independent of any inherent unbalance in the test rotor. Nonetheless, it is important to assess whether the low speed/frequency variation procedure is acceptable. The rotor was therefore balanced and Disks 1 and 3 were then unbalanced through the addition of in-phase masses (11 g each), equivalent to 500 g.mm. The rotor was then run up slowly in speed to 450 rad/s and then run down.

Figure 12 shows the measured rotor maximum radial displacement at AMB-B. The results correlate well with those of Fig. 8 with similar ranges of bi-stable frequencies and speeds. There is some sensitivity associated with residual unbalance arising from the balancing procedure. However, the results were deemed to be acceptable and indicate that low speed frequency based testing of contact may be a safer alternative to rotational tests, particularly during initial trials.

The corresponding TDB-B maximum radial displacements are shown in Fig. 13. The correlation with Fig. 9 is acceptable in terms of the bi-stable speed and frequency ranges. However, there are differences in profiles. The most notable is the TDB-B maximum radial displacement during

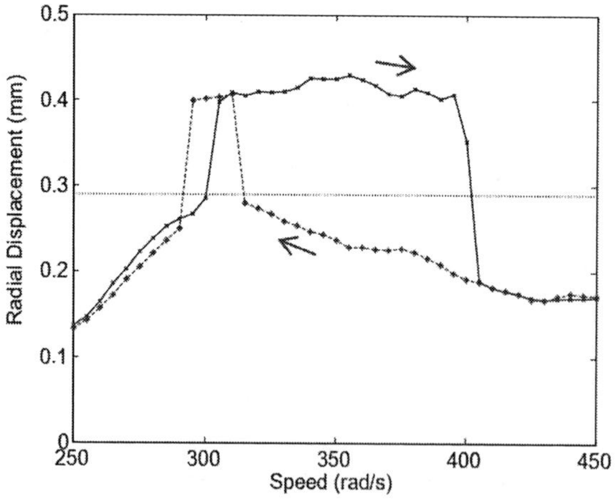

Figure 12. Rotor maximum radial displacement at AMB-B with respect to gradual run up and run down of the rotating speed. The in-phase unbalances of Disks 1 and 3 were 500 g.mm.

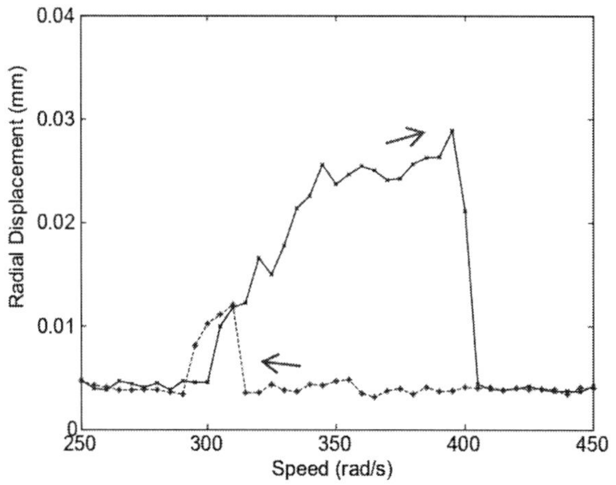

Figure 13. TDB-B maximum radial displacement with respect to gradual run up and run down of the rotating speed. The in-phase unbalances of Disks 1 and 3 were 500 g.mm.

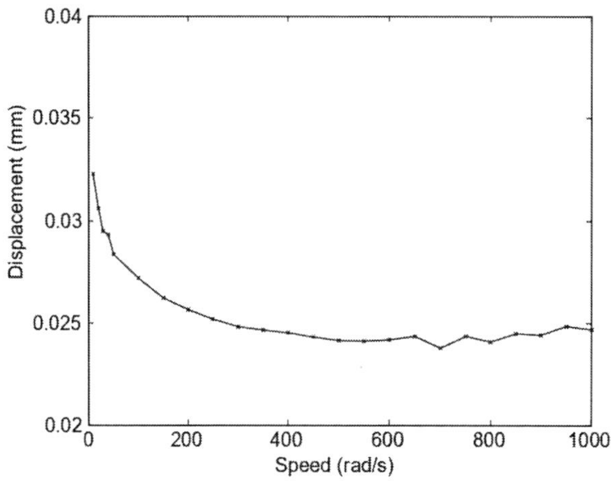

Figure 14. TDB-B average radial displacement when circular orbit demand of radius 0.05 mm is applied.

speed run down attaining 0.012 mm compared with a frequency run down value of 0.021 mm.

VI. RE-LEVITATION AFTER CONTACT

The derivation of a suitable TDB control strategy to move the rotor from a persistent contact condition to a bi-stable contact-free condition is a difficult problem since the plant dynamics depend on frequency/speed. Moreover, the plant dynamics change between the bi-stable states of the rotor. It is also a requirement that the TDB actuation system must be capable of reacting against the induced contact forces and it must not have excessive radial displacement so as to compromise the AMB air gap.

When tested in isolation from the rotor, Fig. 14 shows the effective frequency response of TDB-B under a 0.05 mm radius circular orbit demand. The decay in the response may be due to fluid compressibility in the hydraulic lines together

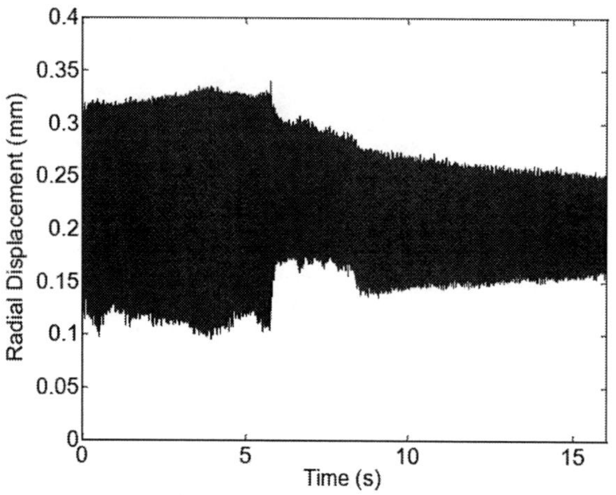

Figure 15. Rotor radial displacement at AMB-B. TDB-B control at frequency 380 rad/s is applied at 6 s to bring rotor out of contact.

with resistance from the pressurised seals. Notwithstanding this attenuation, this response would be capable of matching the passive displacements experienced by TDB-B in Figs 9 and 13. TDB-B would also maintain an AMB-B airgap.

The control strategy employed was assessed with respect to the rotor dynamic mode at 320 rad/s. The bi-stable modes are dominantly of a rigid body cylindrical nature. It was decided upon to attempt contact-free re-levitation at a frequency of 380 rad/s. An assessment of the nonlinear response indicated that it would be beneficial to move TDB-B in the same sense as the rotor motion in order to reduce contact forces. At other frequencies this may not be beneficial and some phase difference between the rotor and TDB induced motion may be more appropriate. Figure 15 shows the transient response from persistent contact to a contact-free state. The corresponding steady state orbits at AMB-B are shown in Fig. 16,

VII. CONCLUSIONS

It has been demonstrated that bi-stable rotor/TDB contact and contact-free orbits may co-exist under particular operating conditions within a rotor/AMB system. It is shown that for conditions in which gyroscopic moments are small, unbalance response tests may be replaced by low speed tests that utilise rotating forces from AMBs. Similar ranges of bi-stable modes were identified through speed (for unbalance) or frequency (for AMB rotating forces). The low speed tests have a greater margin of safety. They allow the user to prevent potentially destructive contact forces by simply switching off the rotating forces, rather than having to endure speed run-down.

It is always desirable to operate a rotor/AMB/TDB system in a contact-free condition. However, certain bi-stable contact modes may be persistent. This paper shows that relatively small TDB motions, synchronously applied with the rotational frequency/speed and appropriately phased relative to the rotor motion, may be effective in restoring contact-free rotor operation.

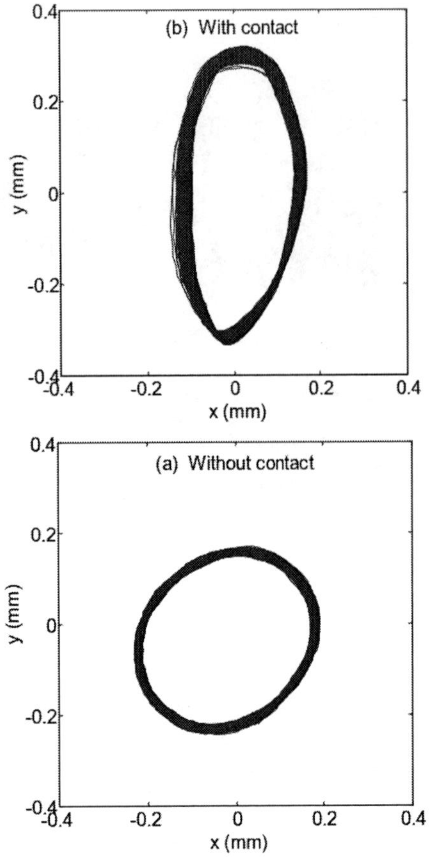

Figure 16. Rotor orbit at AMB-B before (a) and after (b) TDB control was applied at 380 rad/s.

VIII. ACKNOWLEDGEMENT

The authors are grateful for the support of the EPSRC under grant EP/D031389 for the development of the experimental facility. Peichao Li was also supported by a University of Bath research studentship for his PhD study.

REFERENCES

[1] J. Schmied, and J.C. Pradetto, "Behavior of a one ton rotor being dropped into auxiliary bearings," *Proceedings 3rd Int. Symp. Magnetic Bearings*, Alexandria, VA, pp. 145-156, 1992.

[2] R.G. Kirk, and T. Ishii, "Transient response drop analysis of rotors following magnetic bearing power outage," *Proceeding MAG'93*, Alexandria, VA, pp. 53-61, 1993.

[3] R.G. Kirk, E.E. Swanson, F.H. Kavarana, and X. Wang, "Rotor drop test stand for AMB rotating machinery, Part 1: Description of test stand and initial results," *Proceedings 4th Int. Symp. Magnetic Bearings*, ETH Zurich, pp. 207-212, 1994.

[4] E.E. Swanson, R.G. Kirk, and J. Wang, "AMB rotor drop initial transient on ball and solid bearings," *Proceedings MAG'95*, Alexandria, VA, pp. 207-216, 1995.

[5] M. Fumagalli, P. Varadi, and G. Schweitzer, "Impact dynamics of high speed rotors in retainer bearings and measurement concepts," *Proceedings 4th Int. Symp. Magnetic Bearings*, ETH Zurich, pp. 239-244, 1994.

[6] M. Fumagalli, and G. Schweitzer, "Measurements on a rotor contacting its housing," Paper C500/085/96, *Proceedings 6th Int. Conf. Vibrations in Rotating Machinery*, Oxford, UK, pp. 779-788, 1996.

[7] A.R Bartha, "Dry friction induced backward whirl: theory and experiment," *Proceedings 5th IFToMM Conf. Rotor Dynamics*, Darmstadt, pp. 756-767, 1988.

[8] L. Hawkins, A. Filatov, S. Imani, and D. Prosser, "Test results and analytical predictions for rotor drop testing of an active magnetic bearing expander/generator," *ASME Journal of Engineering for Gas Turbines and Power*, Vol.129, pp. 522-529, 2007.

[9] H. Ulbrich, A. Chavez, and R. Dhima, "Minimization of contact forces in case of rotor rubbing using an actively controlled auxiliary bearing," *Proceedings 10th Int. Symp. Transport Phenomena and Dynamics of Rotating Machinery*, Honolulu, Hawaii, pp. 1-10, 2004.

[10] H. Ulbrich, and L. Ginzinger, "Stabilization of a rubbing rotor using a robust feedback control," Paper-ID: 306, *Proceedings 7th IFToMM Conf. Rotor Dynamics*, Vienna, Austria, 2006.

[11] I.S. Cade, M.N. Sahinkaya, C.R. Burrows, and P.S. Keogh, "On the design of an active auxiliary bearing for rotor/magnetic bearing systems," *Proceedings 11th Int. Symp. Magnetic Bearings*, Nara, Japan, 2008.

Numerical and Experimental Results of Auxiliary Bearings Testing on a High Speed Test Rig

Tim Collins[a], Andrea Masala[a], Richard Shultz[a], Zenglin Guo[a]

[a] Waukesha Magnetic Bearings, Unit K, Lyons Way, Worthing - UK,
tcollins@waukbearing.com, amasala@waukbearing.com.com, rshultz@waukbearing.com, zguo@waukbearing.com

Abstract—A test stand including a rotor and active magnetic bearing system, located at Waukesha Magnetic Bearing premises was utilized in 2013 to validate a new auxiliary bearing design. This paper provides an overview of the preparatory activities and an insight on the numerical and experimental results obtained during the testing campaign.

I. INTRODUCTION

Auxiliary bearing systems are used on active magnetic bearing supported rotors to protect the rotor and stator part of the machine in case of overload of the active magnetic bearings (AMB) or failure of the AMB system resulting in a rotor drop or landing event. The auxiliary bearing system always includes a radial bearing system, two per rotor, and most often also an axial auxiliary bearing system. Different auxiliary bearing designs have been developed and tested over the years by Waukesha Magnetic Bearings (WMB) to meet specific customer and project requirements dictated by gas environment properties, rotordynamic performances and auxiliary bearing lives.

With reference to the radial auxiliary bearings, these can be classified in three main categories: bushing type, ball bearing type and hybrid type. The bushing type auxiliary bearing system (RDS) typically consists of a fixed stator part made of bearing pads mounted in a supporting housing by means of a compliant mounting system, and landing sleeve forming the rotor part.

A new type of auxiliary bearing landing sleeve was developed by Waukesha Magnetic Bearings in partnership with a major turbomachinery OEM using a new material and manufacturing process. The new auxiliary bearing design was intended to provide enhanced mechanical properties to enable the extension of this type of application to high speed rotors, and guarantee stability of performance over time.

The design and validation of the mechanical properties of the new auxiliary bearing sleeve was conducted in 2012 and was finalized in 2013 with an extensive test campaign on a test facility, prior to being approved for testing on the full scale machine.

The testing campaign, which included a repetition of a set of full speed rotor drops, was performed on a large size rotor test stand located at WMB premises and with capability to rotate a 1500 kg rotor up to 8000 rpm. The landing test and subsequent visual inspection and non destructive testing (NDT) confirmed successful results of the auxiliary bearing performance and good agreement between the predicted and measured results in terms of rotor drop dynamics and coefficient of friction at the rotor-stator interface.

Figure 1. RDS type radial auxiliary bearings

A summary of the test stand features, conditions and results are discussed here.

II. ENHANCED RADIAL AUXILIARY BEARINGS

RDS auxiliary bearings typically include separate radial and axial auxiliary bearing systems.

Radial RDS auxiliary bearings include a stator part consisting of a housing structure with compliantly mounted pads lined with a dry-self lubricating material, and a rotor part consisting of the landing sleeve which represents the journal part of the bearing system and provides the sliding contact interface with the lining of the stator part during a landing event or momentary contact. A schematic of the radial RDS auxiliary bearing is represented in Figure 2.

During normal operation of the AMB the rotor and stator parts of the auxiliary bearings are not in contact and a nominal gap, normally ranging between 0.2 and 0.35 mm will be present.

During a landing event, the rotor will move away from its nominal centered position, resulting with a contact with the stator parts. The RDS type auxiliary bearing provides a frictional sliding bearing surface to allow the rotor to coast down safely to rest. This process results in high thermal energy input to the auxiliary bearing landing sleeve and wear to the stator pad material. The endurance of the landing sleeve and stator pads depends on the capacity of the material to dissipate the thermal energy generated during the landing event.

The original auxiliary bearing sleeve design was based on a bronze/copper alloy characterized by high thermal conductivity and diffusivity properties which resulted in a relatively low temperature rise of the auxiliary bearing system during a landing event.

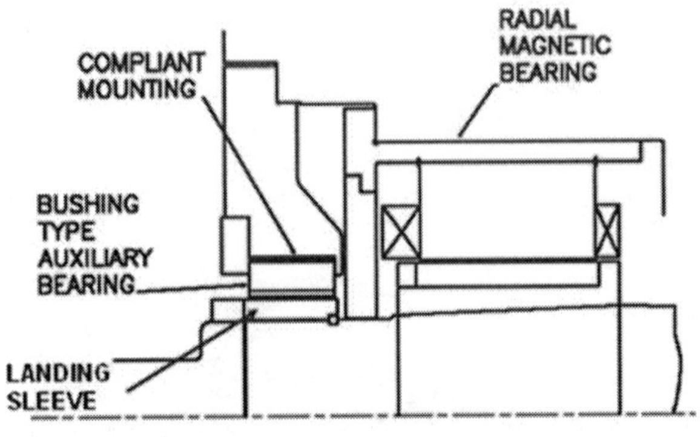

Figure 2. RDS type radial auxiliary bearings

The requirement to keep the landing sleeves in contact with the shaft when the machine is rotating at maximum speed with the worst case temperature gradient coupled with consideration of manufacturing tolerances sets the requirement for a robust interference fit between the landing sleeve and shaft. The resulting hoop stresses generated during the assembly process or during rotational operation may limit the application of RDS type auxiliary bearing systems for high speed machines.

To overcome this limitation, a new landing sleeve design was developed in cooperation with a major turbomachinery OEM. The new sleeve design (Figure 3) consists of a bi-metallic sleeve and is composed of an outer landing sleeve, which utilises a material with high thermal conductivity and diffusivity properties, bonded to an inner carrier sleeve, which utilises a material with higher strength characteristics, where higher stress levels are present.

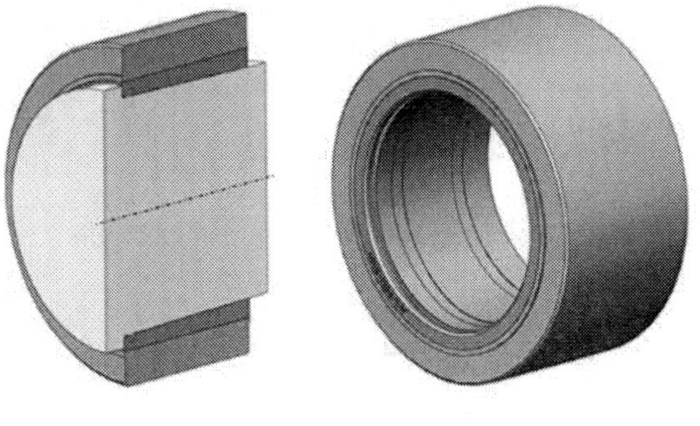

Figure 3. The new sleeve design

The design of the landing sleeve was verified with detailed Finite Element Analysis, able to predict the temperature profile of the sleeve and rotor by means of transient and non-linear analysis during a landing event from full speed.

III. TEST RIG DESCRIPTION

To validate the new landing sleeve design an experimental activity, including repetition of landing tests representative of a real high speed / heavy weight rotor were performed. The landing tests were performed on an existing stand, located on a confined and protected area at WMB premises. The test stand, known as the 'Red Rig' because of the characteristic color of the shielding structure is pictured in Figure 4. The Red Rig was originally developed in the late '90s by the company with the purpose of verifying AMB and auxiliary bearing performance when subject to shock loads.

Figure 4. The Red-Rig test stand

A schematic of the Red-Rig is represented in Figure 5. The red rig has a horizontal rotor supported by two radial AMBs and two split axial AMBs (with a thrust disc on each end of the rotor). Besides each radial AMB are located the two RDS type radial auxiliary bearings. The axial auxiliary bearing is located at the NDE of the rotor. The Red Rig includes an external shaker system to excite the rotor, when at standstill, at different position, and several emergency braking systems to absorb rotational energy from the rotor in case of a failure of the AMB or auxiliary bearing systems during the test.

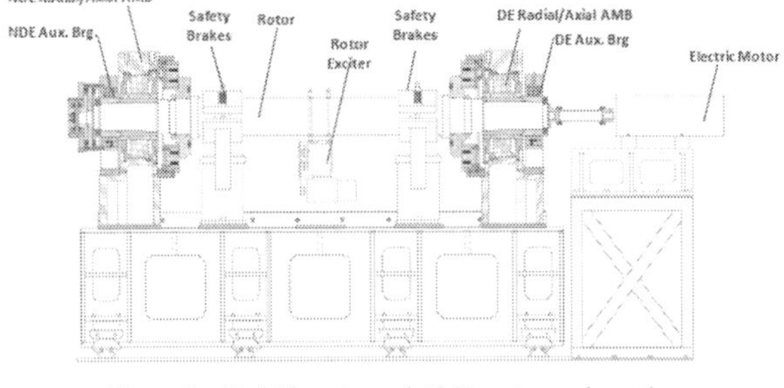

Figure 5. Red Rig rotor and AMB system schematic

Featuring a 1500 kg supercritical rotor, running up to 8000rpm, and able to accommodate some of the latest electronic and software capabilities developed by WMB over the years, such as auto-clearance check, remote commissioning and more recently auto-tuning capabilities, the Red-Rig was used over the years as a permanent laboratory able to reproduce some of the critical aspects that a rotor or AMB system will experience during real operation.

To perform the landing test repetitions required to validate the new landing sleeves design, while being able to guarantee the required level of safety for the equipment and personnel involved in the test, red rig was re-commissioned and operated under strict safety rules and verification procedures. These were based on a well developed pre-existing risk assessment and safety justification.

The commissioning included rotor balancing and AMB tuning operations to keep residual or desired unbalance levels and vibrations under strict control. Fine measurements of residual clearances and monitoring of landing sleeves temperature during operation and following a landing event were made.

IV. NUMERICAL SIMULATIONS

Prior to performing the testing activity on the new landing sleeve design, an extensive numerical activity was performed to predict the rotor drop results and determine the test conditions to have similitude between the test rig and the actual machine.

The simulations accounted for different phenomena related to the rotor auxiliary bearing performances during a landing event, including rotor drop dynamics and loads, transient thermal effects and stress analysis on the rotor sleeves.

The transient rotor drop dynamic simulation was performed by means of WMB proprietary numerical tools, which combine the rotor dynamics with the auxiliary bearing characteristics, including compliance system stiffness effects and, friction and contact models between rotor and auxiliary bearings.

A finite element model of the red rig rotor is represented in Figure 6. The model includes AMBs and auxiliary bearing systems and allows the determination of the rotor vibration amplitudes on selected rotor stations, and loads on the AMBs and auxiliary bearings to be transferred to the thermal and stress analysis.

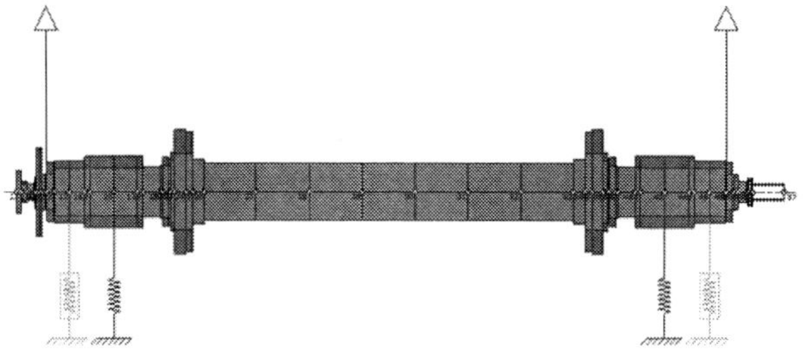

Figure 6. FEM model of the Red-Rig rotor

Different unbalance configurations, with a total unbalance equivalent to ISO G 2.5 were considered on the simulation runs. The transient responses of the system during a landing event with unbalances located on the two opposite shaft ends and in phase are shown in Figure 7 and Figure 8 where the bearing clearance and compliance circles are marked in green and red respectively.

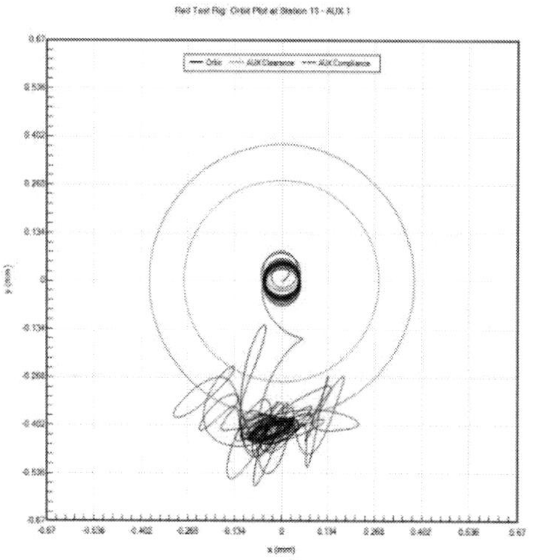

Figure 7. Rotor orbit at NDE

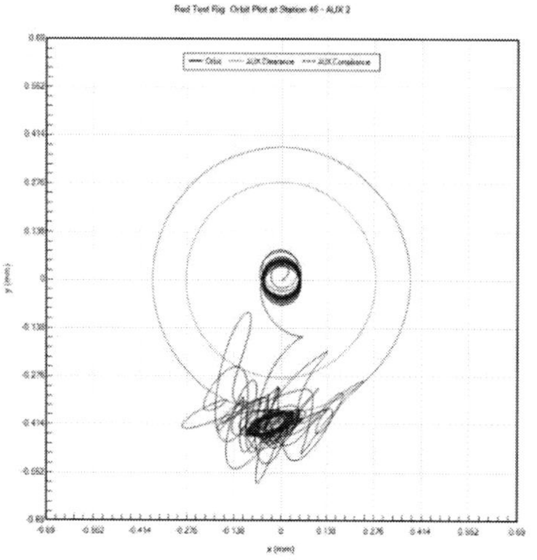

Figure 8. Rotor orbit at DE

Detailed thermal and structural FE analyses were necessary to investigate the performance of the new sleeve design during the normal operating cycles of the machine and during the worst case landing event from the maximum operating speed.

The analysis of normal operating cycles considered variations in operating speed and temperature conditions along with the interference boundary condition and utilized a non-linear material model to establish plastic strain cycles at the bonding interface between landing and carrier sleeves.

The landing event analysis utilised a combined thermal transient and structural FE model. The temperature distribution in the sleeves resulting from thermal energy input due to the frictional sliding interface between rotor and stator parts of the auxiliary bearing system were simulated during the landing transient to ensure that peak operating temperatures remained within allowable limits. Figure 9 shows a typical temperature distribution during a landing event,

Temperature distributions from the thermal transient analysis were mapped onto a structural FE mesh to study the resultant stress state within the sleeves along with interference fit and centrifugal loading during the landing event. Resulting contact pressures at the interference fits and thermal expansions at the sleeve ODs provided key justifications for determining the safe conditions to run the test.

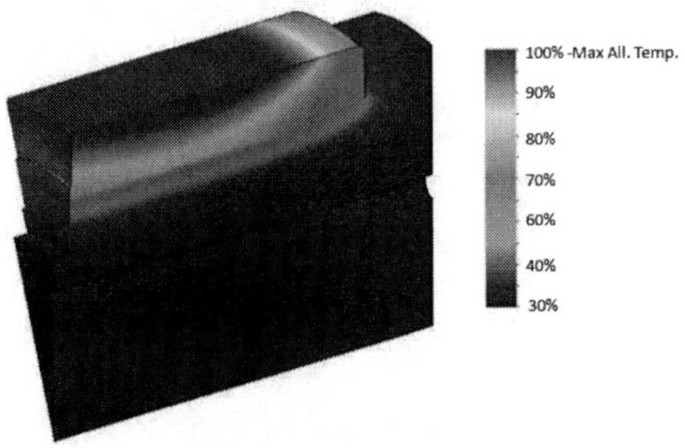

Figure 9. Typical temperature distribution during a landing event.

V. THE TEST PROGRAM

The commissioning and test programs were completed in a step wise manner where landing tests were completed from increasingly higher speeds. The braking systems and dynamic behavior of the RDS auxiliary bearing system were validated at each stage against the predicted behavior.

Test speeds were determined using an energy equivalency principle where the thermal energy input due to friction per unit volume of the landing sleeves was compared to that of the full sized machine.

To further guarantee the equivalency between the test and a landing event on the full sized machine it was necessary to apply an external braking torque to the rotor during the landing tests, This braking torque was provided by a combination of regenerative braking from the electric motor drive and a frictional braking torque developed by the axial auxiliary bearing. To this aim the axial AMB actuator was maintained active and able to develop a known amount of force and braking torque, during the landing test. In total, two levels of braking torque were used during the test program in order to study the coefficient of friction at the rotor-stator interface.

Using the energy equivalency principle it was determined that a landing test from 6000 rpm on the red rig with medium braking torque is equivalent to a landing event from maximum operating speed on the full sized machine.

Non contacting infrared temperature sensors were targeted at the OD surface of each auxiliary bearing sleeve. Sensor outputs were logged to give an indication of temperature rise during rotational operation and landing tests.

The diagnostic outputs from the AMB controller were logged to capture the rundown speed curves and rotor positions during the landing tests. This data was used to calculate the coefficient of friction in the radial auxiliary bearings and review radial stator pad wear behavior.

VI. TEST RESULTS

The landing test speeds, durations and energy inputs per unit volume of landing sleeve material used during the test program are given in Figure 10.

Speed (rpm)	Breaking Torque Level	Duration (s)	Energy Input (J mm^-3)
3200	Medium	27	0.172
5000	Medium	32	0.295
6000	Medium	34	0.369
6000	High	19	0.208
7500	High	17	0.275
7500	High	15	0.185
7600	High	17	0.237

Figure 10. Landing test speeds, durations and energy inputs.

Knowledge of the torque generated by the braking systems along with the speed-time curves for each landing test was used to calculcate the coefficient of friction at the rotor-stator interface. Figure 11 shows typical plot of coefficient of friction against speed for a landing test from 5000 rpm. The plot shows that coefficient of friction was in the range of 0.055 to 0.080 for the region where the rotor runs down from 4600 to 2000 rpm. This result is in good agreement with previous landing tests and data from the field.

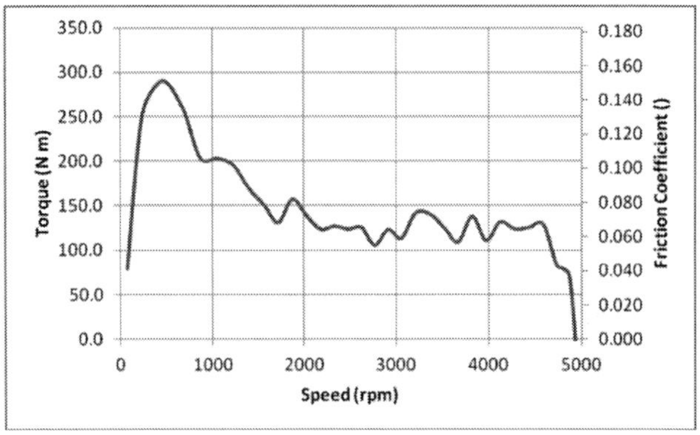

Figure 11. Friction Coefficient for 5000 rpm Landing Test

The rotor orbits were measured with fast dynamic signal and recorded for each landing event. The rotor drop orbits corresponding to the drop conditions measured during the landing event at 7600 rpm is represented in Figure 12. The rotor orbit showed a typical pendulum motion after some initial bouncing. Importantly no insurgence of backward or forward whirl was identified. The measured results were generally in good agreement in terms of vibration amplitudes with the simulation results represented in Figure 13.

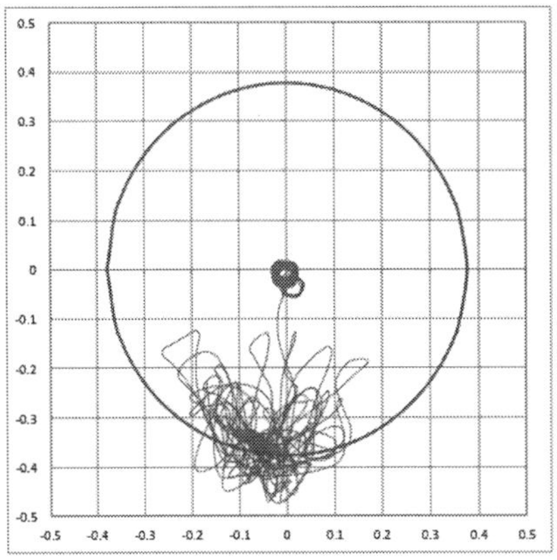

Figure 12. Measured orbit at the NDE Sensor

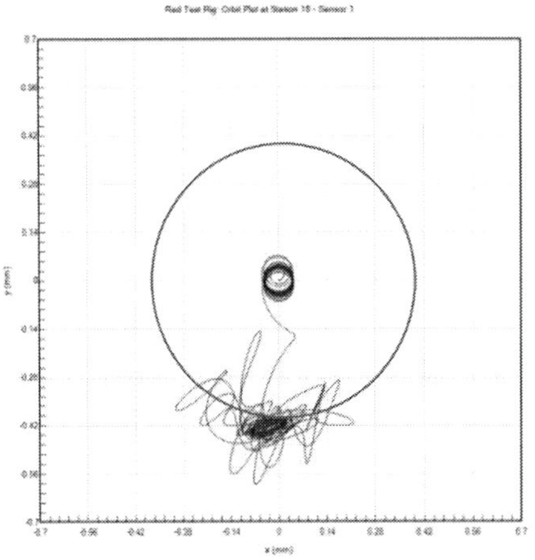

Figure 13. Simulated orbit at the NDE Sensor

VII. VISUAL INSPECTION AND NDT RESULTS

The auxiliary bearing stator pads were measured before and after each landing test to establish the degree of wear that occurred. Figure 14 shows the cumulative wear for lower pads of the auxiliary bearing stators after three successive landing tests from high speed.

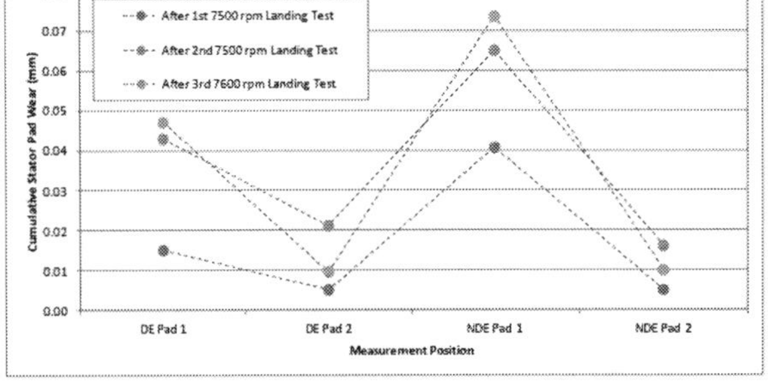

Figure 14. Cumulative pad wear

The figure shows that the wear rates were relatively low with the majority of wear being limited to within 0.08 mm across the high speed landings. The phenomenon whereby the stator bore reduces after landing tests/events has been witnessed previously during testing and in the field. This is a result of material transfer from other positions around the stator and from deposits on the rotor from previous landing tests. In general, the wear rates and behavior experienced during the landing tests on the red rig were consistent with experience of previous landing tests and landing events in the field.

The stator pad build up on the radial auxiliary bearing sleeve OD surfaces was measured at critical points during the test program. In general, the number of deposit patches increased with each landing test. Patches which existed before a landing test tended to increase in size and thickness during subsequent landing tests. After several high speed landing tests the majority of patched were found to be in the thickness range of 0.025 to 0.050 mm with the thickest being 0.080 mm.

Figure 15. Stator pad deposits on landing sleeve after 7600 rpm landing test.

Following the completion of the test program the radial auxiliary bearing sleeves were subjected to a detailed inspection to ensure that no defects had developed during the testing which would compromise their structural integrity for continued operational use. A 100% dye penetrant inspection was performed to examine for surface breaking flaws on the sleeve OD. An image of the NDE landing sleeve is represented in Figure 16. A 100% ultrasonic inspection was performed to check the integrity of the bonding between the carrier and landing sleeves portions of the sleeve. No referable indications of damage or flaws were observed during either of the inspections.

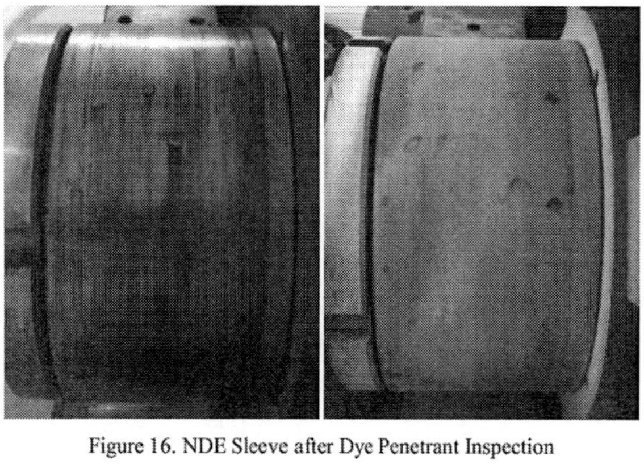

Figure 16. NDE Sleeve after Dye Penetrant Inspection

VIII. SUMMARY

Analysis of the diagnostic outputs from the AMB controller during the landing events show that the dynamic behavior of the auxiliary bearing system was as predicted and consistent with experience of previous landing tests and landing events in the field

In summary, the testing of the scale model sleeves on the red rig was successful and the sleeves proved fit for purpose. The new auxiliary bearing sleeve design was qualified for testing on the full sized machine on the basis of this successful testing.

REFERENCES

[1] S. Yates, R.R. Shultz, Landing, "Test Results for a 6 MW Motor Compressor and Auxiliary Bearing Systems for Corrosive Environments", *Proceedings of 13th International Symposium on Magnetic Bearings*, 2012.

[2] R.R. Shultz, E. Lucchetta, "Time transient simulation model and full scale experimental verification for high speed rotor delevitation events with a 1.5 ton supercritical rotor supported by dry lubricated bushing type auxiliary bearings" - *Proceedings of International Conference on Compressors and Their Systems*, 2005.

Reduction of Rotor Vibration in Passively Stabilized Direction in Single-Drive Bearingless Motor

Hiroya Sugimoto [a], Seiyu Tanaka [a], and Akira Chiba [a]

[a] Tokyo Institute of Technology, 2-12-1 Ookayama, Meguro, Tokyo, Japan, sugimoto@belm.ee.titech.ac.jp

***Abstract*— In this paper, an improvement of the rotor vibration in one-axis actively positioned single-drive bearingless motor (SDBM) with repulsive passive magnetic bearings is presented. The SDBM has only one set of three-phase windings. It generates both torque and active axial force independently driven by only one three-phase inverter and one displacement sensor. Therefore, the single-drive bearingless motors have the advantages of low cost and small size. Only axial z-axis is actively positioned. The other axes, radial movements x and y, and tilting movements θ_x and θ_y, are passively stable. Therefore, in the one degree-of-freedom (1DOF) bearingless motor, vibrations in the radial and tilting directions are serious problem. In this paper, an improvement method of the radial vibration in the 1DOF SDBM is presented. The principle of the radial vibration generation is shown. In the experiment, it is found that a correction of the rotor rotational angular position is effective for the radial vibration reduction.**

I. Introduction

Bearingless motors have advantages of no wear, no lubricant, non-pollution and maintenance-free because a magnetic bearing function is magnetically integrated in a single motor. Therefore, the bearingless motors have been applied in centrifugal pumps, contamination-free ventricular assist devices, high purity pharmaceutical mixing devices, rotating stages and flywheels.

In general magnetic suspension, five degree-of-freedom is actively positioned, however, for cost reduction, two degree-of-freedom (translational directions x, y) are actively positioned in the bearingless motor. On the other hands, in recent years, one degree-of-freedom (1DOF) bearingless motors have been studied because of the further cost reduction and down-sizing [1]-[7]. In the 1DOF bearingless motor, only the axial direction is actively positioned, thus, the cost is extremely low because only one displacement sensor and one or two inverters are necessary. The other radial and tilting directions are passively stabilized by passive magnetic bearings (PMB). The PMB is classified to attractive and repulsive types. In case of the attractive type PMBs in [1]-[3], the PMB is composed of the rotor permanent magnets and stator core, thus, the PMB functions are sometimes included in the bearingless motor structure. In case of the repulsive type in [4]-[7], additional two passive magnetic bearings are installed at both shaft ends of the bearingless motor in tandem. Improvement of the passive stiffness of the PMB is important because of the vibration reductions in the radial and tilting directions.

In the literatures, the rotor vibrations in the passively stable directions are reported [5], [7] and [8]. The vibration is increased at the critical speed, and in the worst case, the rotor is touch down. Most of the critical speed in the 1DOF bearingless motor is several thousand r/min. The authors have proposed a 1DOF single-drive bearingless motor (SDBM) with the repulsive passive magnetic bearing (RPMB) [9]. The critical speeds in the radial and tilting directions are improved because of high stiffness. However, the radial vibration is amplified around several thousand r/min.

In this paper, reduction of the radial vibration in a proposed 1DOF SDBM is presented. Two factors of the radial vibration generation are shown. The rotor angular position error results in the radial vibration because of interference between the axial and radial directions. A harmonic component in flux density in the air-gap also causes the radial vibration. Two rotors are fabricated and compared with respect to the flux density and radial vibration. It is found that a correction of the rotor rotational angular position and reduction of the harmonic component in the air-gap flux density are important for reducing the radial vibration.

II. Structure and Principle of Radial Vibration

A. Structure of Proposed Single-Drive Bearingless Motor

Fig. 1 illustrates xz cross-sectional view of one of the representative single-drive bearingless motors. The single-drive bearingless motor (SDBM) is installed in the center, and the repulsive passive magnetic bearings (RPMBs) are installed at both ends of the SDBM. Each RPMB is constructed with the two permanent magnets. Four ring-shaped permanent-magnets are faced so that the radial stiffness is improved. The displacement sensor and the hall sensors are installed in the right and left end of the shaft, respectively. In the design, the radial air-gap length is 1 mm. In the SDBM part, only axial

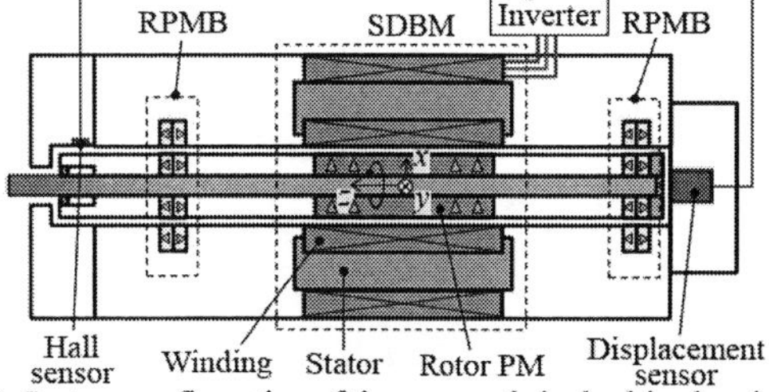

Fig. 1. System configuration of the proposed single-drive bearingless motor.

direction is actively positioned and the rotational torque is regulated, thus, only one displacement sensor and one three-phase inverter are necessary. Significant cost down can be realized.

Fig. 2 shows the detail of the winding structure of the proposed SDBM. Only one set of 3-phase 2-pole winding is installed. Fig. 2 shows the xy cross-sectional view. The windings are wound around the six stator cores. The black arrows in the windings indicate the positive current direction. Only the active axial force is generated by the x-y windings. To generate the rotating torque around z-axis, the current component in the z-axis is necessary. The torque by the Lorentz force is generated by the current component perpendicular to the current direction as the circle and cross marks shown in Fig. 2. The winding arrangement is similar to two-pole concentrated winding structure by a proposed V-shaped winding presented in [9]. Therefore, the rotating torque can be generated by the q-axis current i_q.

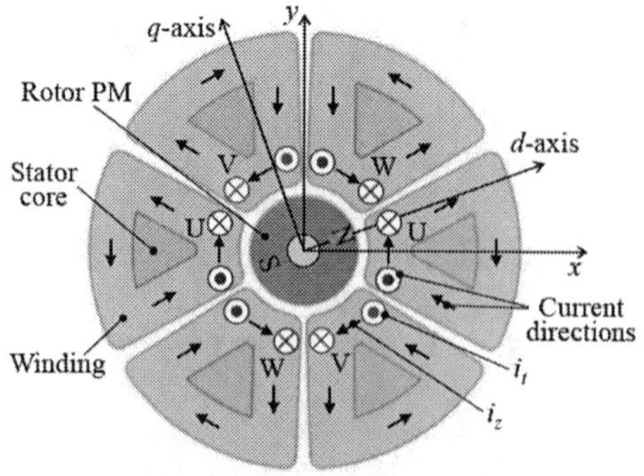

Fig. 2. xy cross-sectional view of the proposed SDBM.

B. Principle of Active Axial Force Generation

Figs. 3(a) and 3(b) show principle of the active axial force generation. Fig. 3(a) shows the principle of the Maxwell force generation. The stator teeth are constructed to face to the air-gap. In addition, the stator core is composed of the six bars so that the permanent magnet fluxes at both rotor ends are circulated in the air-gap as shown by red arrows in Fig. 3(a). The suspension flux mainly flows in the air-gap in the z-direction. When the suspension flux is superimposed on the permanent magnet flux as shown in Fig. 3(a), the flux strengthening and weakening occur in the air-gaps at both axial ends. Therefore, the axial suspension force is generated in the positive z-direction. Fig. 3(b) shows the principle of the Lorentz force generation. At the rotor center in the z-direction, the permanent magnet fluxes go across the inside conductors confronting the air-gap. When the d-axis current is provided in the direction as indicated in Fig. 3(b), the Lorentz force is generated in the negative z-direction in the conductors. Thus, the active axial force is generated in the rotor in the positive z-direction as the counteractive force. The active axial force is a sum of the Maxwell and the Lorentz forces. On the other hand, when there is an error of the rotor rotational angular position, undesirable vibration occurs in the passive stabilized directions.

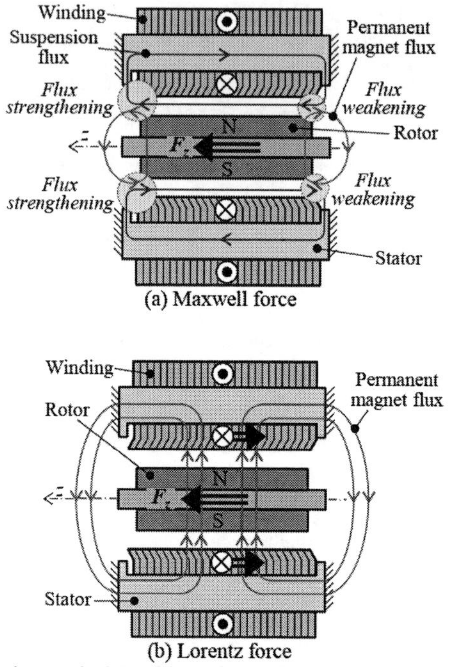

Fig. 3. Principle of the active axial force generation.

C. Principle of Radial Vibration Generation

Fig. 4 shows one of the factors in the axial vibration generation. In theoretical calculation, the axial force is not generated even if the q-axis current is excited. However, when there is an error between the actual and calculated rotor rotational angular positions, unfortunately, undesired axial force is generated because the d-axis current component is generated. In Fig. 4, the error angle is defined as θ_e. d_e-axis is shifted from d-axis by angle of θ_e. When i_{de} and i_{qe} are defined as currents on d_e- and q_e-axis, d- and q-axis currents i_d and i_q are given as

$$\begin{bmatrix} i_d \\ i_q \end{bmatrix} = \begin{bmatrix} cos\theta_e & sin\theta_e \\ -sin\theta_e & cos\theta_e \end{bmatrix} \begin{bmatrix} i_{de} \\ i_{qe} \end{bmatrix}. \tag{1}$$

Therefore, even when i_{de} is zero, i_d is not zero as shown in the following expression.

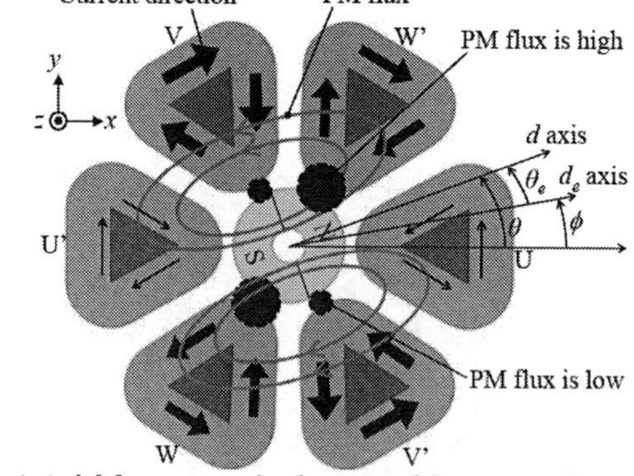

Fig. 4. Axial force generation because of the error angle.

$$\begin{bmatrix} i_d \\ i_q \end{bmatrix} = \begin{bmatrix} cos\theta_e & sin\theta_e \\ -sin\theta_e & cos\theta_e \end{bmatrix} \begin{bmatrix} 0 \\ i_{qe} \end{bmatrix} = \begin{bmatrix} i_{qe} sin\theta_e \\ i_{qe} cos\theta_e \end{bmatrix}. \tag{2}$$

The axial force is proportional to the d-axis current. When the error angle is equal to zero, the current-force coefficient is k_{zi}. The coefficient k_{zi} is varied with the error angle as sinusoidal waveform. Therefore, the undesired axial force F_{zd} is expressed as

$$F_{zd} = k_{zi} cos\theta_e i_d. \tag{3}$$

By substituting (2) into (3), F_{zd} is derived as follows:

$$F_{zd} = \frac{1}{2} k_{zi} i_{qe} sin 2\theta_e. \tag{4}$$

In addition, the angular position error is varied with respect to the rotor rotational angular position. Therefore, the variation of the angular position error results in a disturbance, thus, the rotor is vibrated in the axial direction.

Fig. 5 shows the radial stiffness with respect to the rotor axial position. In the proposed structure, the positive radial stiffness is generated in the repulsive passive magnetic bearings. The radial stiffness is the highest at the aligned position between the rotor and stator permanent magnets. On the other hand, when the rotor is displaced in the positive or negative axial direction, the radial stiffness is reduced. Therefore, the radial stiffness is varied with twice frequencies with respect to the axial vibration. The variation of the radial stiffness results in movement of the radial position, thus, the radial vibration is generated. Thus, when the rotor is set in the aligned position as shown in Fig. 5, the second-order harmonic radial vibration occurs because of first-order axial vibration. The resonance between the second-order harmonic vibration and the natural frequency in the radial direction is serious problem. Thus, the axial vibration reduction by minimization of the angular position error is very important.

III. Test Results

Figs. 6(a) and 6(b) show rotor 1 and rotor 2, respectively. Rotor permanent magnet installed at center is magnetized in two-pole parallel direction. Two repulsive passive magnetic bearings are magnetized in thrust direction. The shaft diameters of rotor 1 and rotor 2 are 3 mm and 5 mm, respectively. Two sensor targets for monitoring radial direction are slightly different. The diameter in the rotor 2 is decreased for reduction of the rotor mass. However, the rotor outer diameters are both 10 mm. In addition, dimensions of the rotor PM and RPMBs are exactly identical. The PM and shaft materials are Nd-Fe-B and stainless, respectively. The rest spaces at both side of the rotor PM are necessary to avoid interference between the rotor PM and the RPMBs.

Fig. 7 shows spectrum of the radial displacement when an impulse disturbance is applied in the rotor shaft end in the radial direction. The rotor radial movement is measured by two displacement sensors. The rotor is vibrated by the impulse disturbance. The radial natural frequencies of rotor 1 and rotor 2 are 85.4 Hz and 79.4 Hz, respectively. Therefore, critical speeds may be 5124 r/min and 4758 r/min. In the impulse disturbance test, amplitude of the radial displacement is not important, but rather the natural frequency is significant result for calculating a radial stiffness.

Table I shows radial stiffness of rotor 1 and rotor 2. When the rotor mass is m and the natural frequency is f_n, the radial stiffness k_r is given by

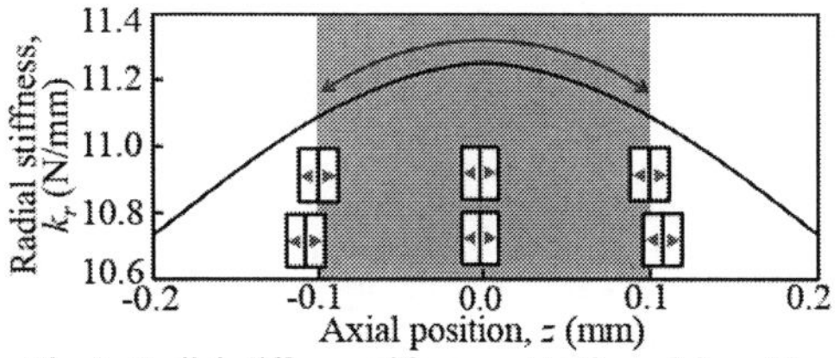

Fig. 5. Radial stiffness with respect to the axial position.

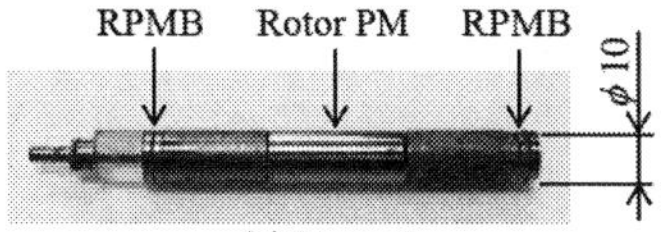

(a) Rotor 1.

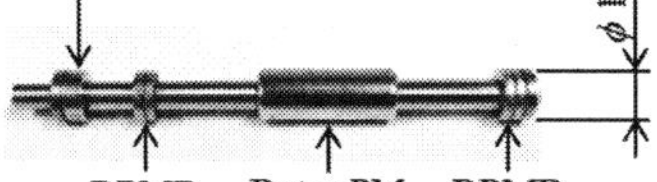

(b) Rotor 2

Fig. 6. Fabricated rotors.

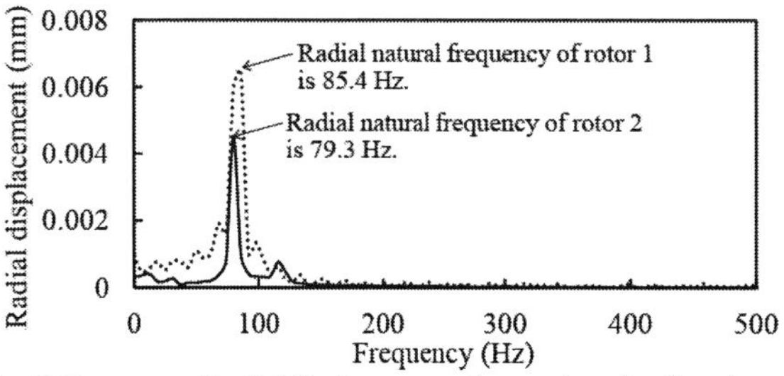

Fig. 7. Spectrum of radial displacement when an impulse disturbance is applied in rotor shaft end.

Table I Radial stiffness of two rotors.

	Rotor mass	Radial natural frequency	Radial stiffness
Rotor 1	0.039 kg	85.4 Hz	11.2 N/mm
Rotor 2	0.038 kg	79.3 Hz	9.4 N/mm

$$k_r = m(2\pi f_n)^2. \tag{5}$$

From expression (5), the radial stiffness of rotor 1 and rotor 2 are 11.2 N/mm and 9.4 N/mm, respectively. The radial stiffness of rotor 2 is low compared with that of rotor 1 because flux density in the air-gap is decreased. In the case of rotor 2, the shaft diameter is extended for enhancement of the shaft stiffness. Therefore, radial thickness of rotor permanent magnets is decreased so that permanent magnet flux is reduced.

Fig. 8 shows the angular position error with respect to the rotor rotational angular position. The angular position is calculated by linearized equation every 90 degrees. Before correction of the angular position in rotor 1, the maximum error is 7.1 degrees. On the other hand, after the gain and offset in the linearized equation are improved, the maximum error is reduced by 1.9 degrees. In the case of rotor 2, maximum error angle is 1.7 degrees. Therefore, it is expected that axial and radial vibrations in rotor 1 with angular position correction and rotor 2 are reduced compared with rotor 1 without angular position correction.

Fig. 9(a) shows harmonic order components in the error angle with respect to rotor 1 without angular position correction. Second-order component is the highest. When the rotor is rotated at constant angular speed ω, the second-order component in the error angle causes axial vibration of second-order 2ω. As a result, the second-order axial vibration generates radial vibration of fourth-order 4ω. After angular position correction, the second-order component is reduced by 89 % as shown in Fig. 9(b). Error angle in rotor 2 is also corrected so that the second-order component is reduced as shown in Fig. 9(c). However, the first- and second-order components are slightly increased compared with Fig. 9(b).

Fig. 10(a) and 10(b) show ratio of harmonic order component in surface flux density in the axial direction with respect to RPMBs in rotor 1 and rotor 2. The surface flux densities in the axial direction at positions a, b, c and d are measured. The measured four positions are indicated in Fig. 10(a) and 10(b). Ideally, the surface flux density is constant. However, in the fabricated RPMBs, harmonic components are included because of magnetization error. In the surface flux density of RPMBs in rotor 1, second-order harmonic component is high compared with first-order component. Therefore, in the case of rotor 1, second-order radial vibration may be increased. On the other hand, in the case of rotor 2, first-order component is high compared with second-order harmonic component. Thus, radial vibration synchronized a rotational speed may be increased.

Fig. 11 shows the radial displacement of rotor 1 before/after the correction of the rotor rotational angular position and that of rotor 2. The touch down length in the radial direction is 0.3 mm. Before the correction, the rotor is touch down in the radial direction at around 1600 r/min. On the other hand, after the angular position correction, the rotor radial vibration is reduced, thus, the rotational speed is improved up to 2200 r/min. However, when the rotational speed is around 2200 r/min, the rotor radial vibration is increased, and then, the rotor is touch down. In the case of rotor 2, radial vibration is higher than rotor 1 with the angular position correction at low speed because the radial stiffness is lower than rotor 1. However, rotor 2 can rotate without touch down at 2300 r/min because second-order harmonic components in the error angle and flux density in the air-gap are decreased. In the experiments, two valuable results are confirmed. Firstly, when radial stiffness is increased, the radial vibration is decreased. Secondly, it is found that reductions of the second-order harmonic component in the error angle and in the flux density generated by the RPMBs are important for reducing radial vibration.

ACKNOWLEDGMENT

This work was supported by JSPS KAKENHI Grant Number 24246046.

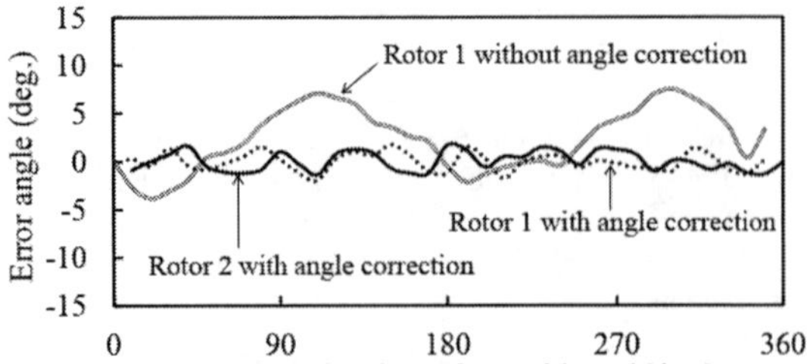

Fig. 8 Angular position error with respect to the rotor rotational angular position.

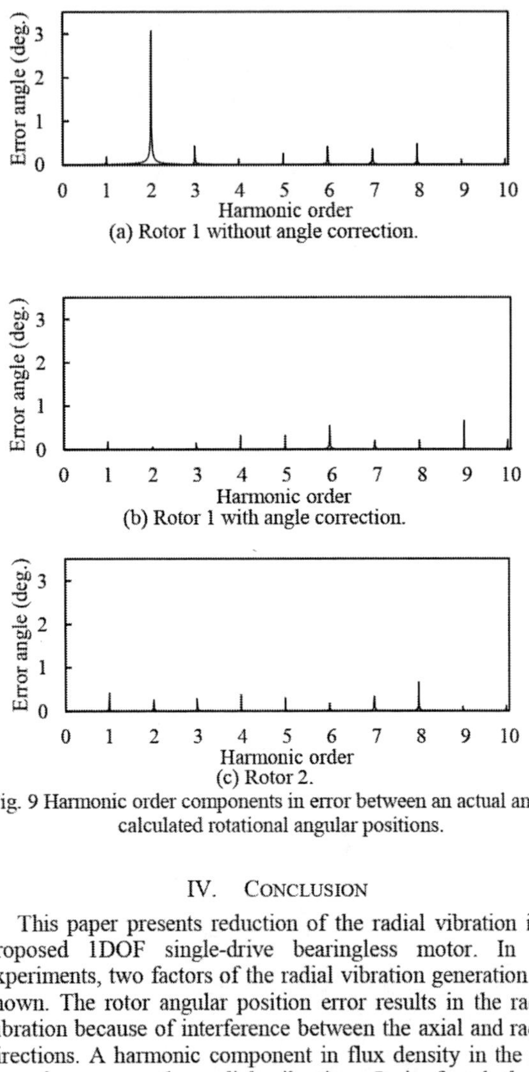

Fig. 9 Harmonic order components in error between an actual and calculated rotational angular positions.

IV. CONCLUSION

This paper presents reduction of the radial vibration in a proposed 1DOF single-drive bearingless motor. In the experiments, two factors of the radial vibration generation are shown. The rotor angular position error results in the radial vibration because of interference between the axial and radial directions. A harmonic component in flux density in the air-gap also causes the radial vibration. It is found that a correction of the rotor rotational angular position and reduction of the harmonic component in the air-gap flux density are important for reducing radial vibration. Moreover, when the radial stiffness is increased, the radial vibration is decreased.

References

[1] I. D. Silva and O. Horikawa, "An attraction-type magnetic bearing with control in a single direction," *IEEE Trans. Ind. Appl.*, vol. 36, no. 4, pp. 1138–1142, Jul./Aug. 2000.

[2] J. Kuroki, T. Shinshi, L. Li, and A. Shimokohbe, "Miniaturization of a one-axis-controlled magnetic bearing," *Precis. Eng.*, vol. 29, no. 2, pp. 208–218, Apr. 2005.

[3] J. Asama, Y. Hamasaki, T. Oiwa, and A. Chiba, "A Novel Concept of a Single-Drive Bearingless Motor", *IEEE Trans. Industrial Electronics,* vol. 60, no. 1, pp. 129-138, Jan. 2013.

[4] Q. D. Nguen and S. Ueno, "Modeling and control of salient-pole permanent magnet axial-gap self-bearing motor," *IEEE/ASME Trans. Mechatronics,* vol. 16, no. 3, pp. 518–526, Jun. 2011.

[5] S. Yang and M. Huang, "Design and Implementation of a Magnetically Levitated Single-Axis Controlled Axial Blood Pump", *IEEE Trans. Industrial Electronics,* vol. 56, no. 6, pp. 2213-2219, Jun. 2009.

[6] W. Bauer and W. Amrhein, "Electrical design and winding selection for a bearingless Axial-Force/Torque Motor", *in Proc. International Symposium on Power Electronics, Electrical Drives, Automation and Motion (SPEEDAM)*, pp. 1224-1229, 2012.

[7] T. Ohji, Y. Katsuda, K. Amei and M. Sakui, "Structure of One-Axis Controlled Repulsive Type Magnetic Bearing System With Surface Permanent Magnets Installed and Its Levitation and Rotation Tests," *IEEE Trans. Magn.*, vol. 47, no. 12, pp. 4734-4739, Dec. 2011.

[8] J. Asama, Y. Hamasaki, T. Oiwa, and A. Chiba, "Performance Improvement of a One-DOF Actively Positioned Bearingless Motor", in *Proc. 13th Int. Symp. Magnetic Bearings*, 2012.

[9] H. Sugimoto, S. Tanaka, A. Chiba, and J. Asama, "Design and Test Result of Novel Single-Drive Bearingless Motor with Cylindrical Radial Gap", *in Proc. Energy Convers. Congr. and Expo. (ECCE)*, pp. 2466-2473, 2013.

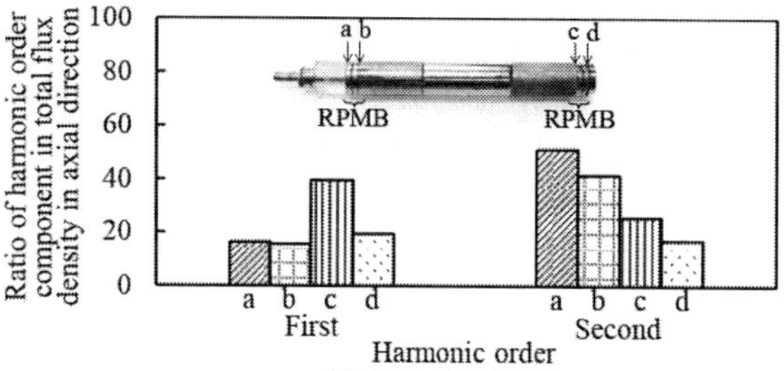

(a) Rotor 1.

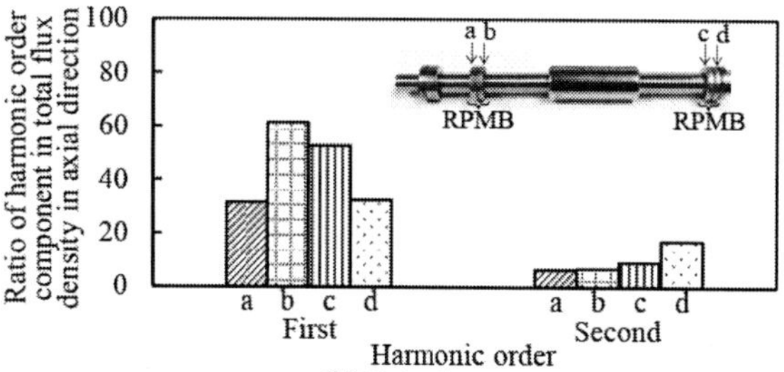

(b) Rotor 2.

Fig. 10 Harmonic order components in surface flux density on repulsive passive magnetic bearings.

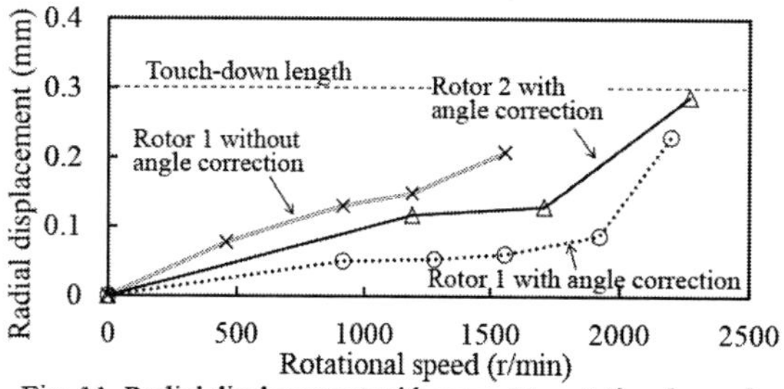

Fig. 11. Radial displacement with respect to rotational speed.

Experimental Characterization of a Bearingless Rotating Field Axial-Force/Torque Motor

Walter Bauer[a], Paul Freudenthaler[b] and Wolfgang Amrhein[c]

Institute of Electrical Drives and Power Electronics, Johannes Kepler University, 4040 Linz, AUSTRIA, http://www.eal.jku.at
[a]walter.bauer@jku.at, [b]paul.freudenthaler@students.jku.at, [c]wolfgang.amrhein@jku.at

***Abstract*—The Bearingless Axial-Force/Torque Motor is an air-core coil motor, that can generate axial forces and drive torques concurrently using a common stator winding system. The radial suspension and the tilting stabilization of the bearingless drive is realized by passive repulsive permanent magnetic ring bearings. The end-windings of the concentric, non-overlapping stator coils create axial directed levitation forces, provided that a non-vanishing common-mode current component is present. In contrast to the existing alternating field motor topology, the proposed redesign features a rotating field motor structure for constant, rotor-angle independent drive torque generation. The current investigation concentrates on the verification of the drive parameters via simple and robust measurements. A presentation of the functional prototype concludes the article.**

I. Introduction

In order to control the motion of a rigid rotor in three dimensional space, the dynamic behaviour of six degrees of freedom (DOFs) must be stabilized either passively by constructive measures or by active closed-loop position control.

A bearingless drive is a motor with an integrated magnetic bearing (levitation) functionality, which allows the reduction of the number of necessary (additional) bearings. The challenge in bearingless drive design is the stabilization of all the DOFs with the lowest possible constructive complexity.

In terms of control theory, the Bearingless Axial-Force/Torque Motors (AFTMs) are underactuated mechanical systems, which means, they possess fewer control inputs than DOFs. The passive dynamics[1] and the related passive DOFs must be asymptotically stable to ensure a proper overall drive functionality.

The passive permanent magnet ring bearing (PMB) unit is a basic component of the AFTM. A single repulsive PMB unit consists of two nested components, a stator ring and a rotor ring, both axially magnetized. The two interacting bearing rings provide stabilizing resilience forces in the radial directions, while destabilizing axial thrust and tilting torques are generated simultaneously. In order to eliminate those detrimental tilting torques, at least a second PMB unit, mounted on the same shaft axis with an axial displacement, is required. Thus, the lateral (radial) and the tilting directions (4 DOFs) of the AFTM are passively stabilized by the application of PMB technology avoiding any electrical energy consumption during drive operation.

[1]dynamical system behaviour, when not drawing energy from a supply

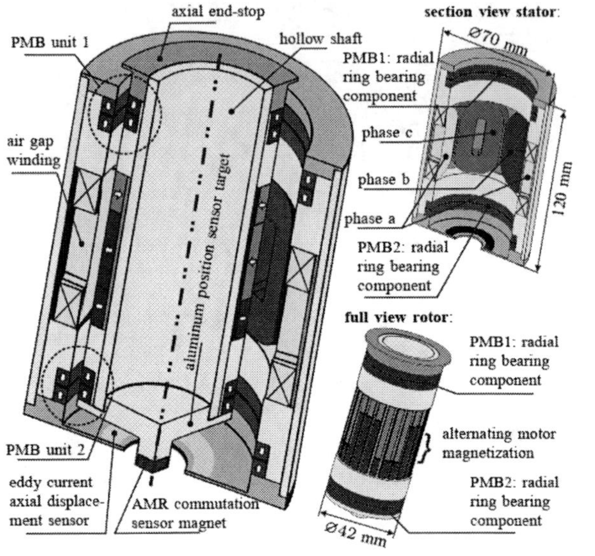

Figure 1. Schematic view of the Bearingless Rotating Field Axial-Force/Torque Motor in its operating position. The passive magnetic radial bearing (PMB) units are axially aligned. Inside rotor motor design with $p_r = 4$ segmented rotor pole pairs and $m = 6$ air-core coils.

The remaining two DOFs are the axial and angular rotor position. The axial position dynamics is unstable due to the introduced axial thrust caused by the PMB units and the gravitational forces.

The actuator part of the AFTM, basically a permanent magnet synchronous motor structure with an air gap stator winding, is responsible for active generation of axial compensation forces and drive torques. The air gap winding of an AFTM consists of m non-overlapping concentric air-core stator coils. The circumferential directed conductor sections of the bent stator coils are designated as the upper and lower end-windings. Together with a very specific magnetization pattern of the rotor magnet system, the energized coil ends can produce non-vanishing axial Lorentz forces. The impression of common-mode current components into the stator winding system result in rotor-angle independent active levitation forces.

The axial directed conductor sections of the stator coils provoke tangential Lorentz forces in combination with the centrally arranged alternating pole motor magnetization pattern of the permanent magnet excitation system. The application of rotor-angle dependent, harmonic stator coil currents yield to constant drive torques.

The combined generation of active axial bearing forces and drive torques using just a single stator winding system with superimposed bearing and motor coil current components is a key feature of the AFTM.

A. *State of the Art*

The first prototype [1], [2] consists of $m = 4$ non-overlapping stator air-core coils and $p_r = 2$ rotor pole pairs. The outcome of this topology selection is an alternating field motor, showing a rather detrimental single-phase motor characteristics. In addition to the non-zero mean drive torque value, a drive torque component of equal amplitude, pulsating with the double synchronous rotor frequency, is present during motor operation.

On the other hand, the choice of $m = 4$ stator coils reliably avoids the appearance of rotor tilting during drive torque generation. The off-centric axial rotor forces, provoked by the energized coil ends, induce tilting torques referring to the rotors center of mass. Diagonally arranged stator coils, that share the same phase/coil current in amplitude and phase, mutually compensate their destabilizing tilting behaviour caused by the coil end-windings.

The bell-shaped rotor design of the Single-Phase AFTM offers various technical advantages over conventional drive concepts. Due to the both rotating back-irons, no eddy current and hysteresis losses can occur.

The extremly compact bearingless motor design is integrated into an electronic ventilator housing.

The high radial drive stiffness values of the PMBs facilitate the realization of a moderate drive tilting stiffness even with a small axial distance between the two PMB units.

Even though a single-phase motor characteristics is fully acceptable for the target application, the pulsating drive torque excitation raises questions about the dynamical behaviour of the rotor. Moreover, the geometric shape of the bell-rotor, which is in between a slice and slim, elongated spindle, is controversial in terms of its rotor dynamics.

The characteristic properties of a Bearingless AFTM, whether favorable or detrimental, strongly depend on the number of motor pole pairs p_r and the number of stator coils m. Certain combinations provoke disturbing radial forces during motor operation, others create drive torque ripple in case of common-mode stator current excitation. Furthermore, the circumferentially directed conductors (end-windings) of the coil system might promote the rotor tilting during drive torque generation in particular configurations.

The investigations in [3] have shown, that all rotating-field motor topology comes along with a rotor-angle dependent torque ripple during levitation force generation via common-mode currents. For the alternating-field motor toplogy, no torque ripple is present at all.

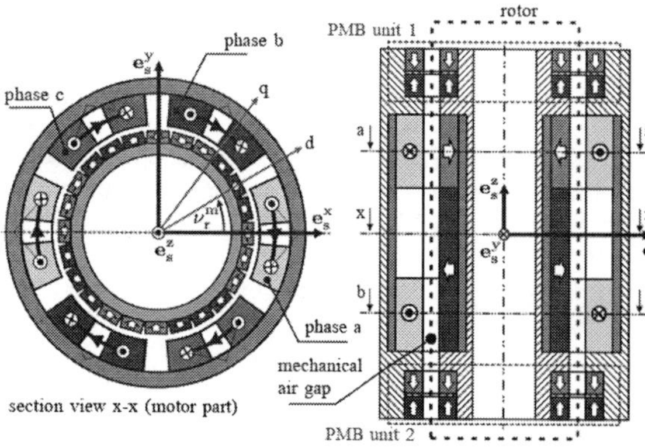

Figure 2. Full section view (2D) of the AFTM topology. Two diametrical opposing air core coils establish one electrical phase. (connection in series)

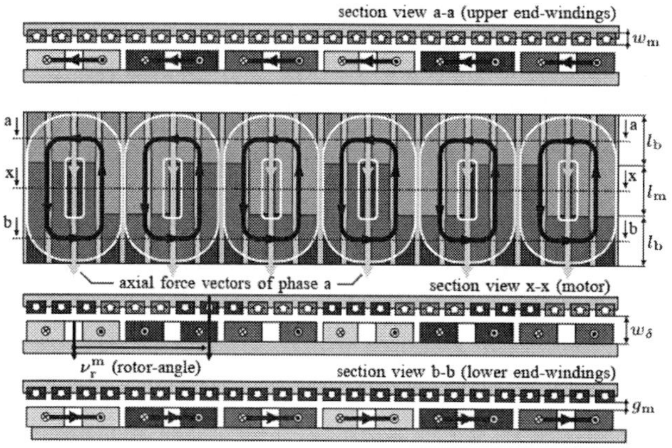

Figure 3. Developed view of the Rotating Field AFTM with segmented rotor magnet system.

B. *Novel Actuator Topology*

Referring to the first AFTM prototype, several modifcations have been undertaken at the current prototype design. The $m = 6$ non-overlapping air-core stator coils are grouped into three electrical phases. Together with $p_r = 4$ rotor pole-pairs realized by the segmented rotor permanent magnet system of Fig. 3, a rotor-angle independent, constant drive torque behaviour can be obtained. The expense for this favorable motor behaviour is the appearance of a small drive torque ripple during active levitation force generation via common-mode coil currents.

Moreover, the axial mounting distance of the radial PMB units has been augmented for higher resulting drive tilting stiffness values. While the chosen inside rotor drive topology benefits of one single mechanical air gap, the attainable drive speed is limited simultaneously due to its surface mounted rotor magnets.

C. *Objectives of the Proposed Investigation*

The current research is focused on the experimental verification of the system parameters for the new prototype drive using simple and robust state of the art measurement procedures. A brief summary of the advanced mathematical description framework for Rotating-field AFTM topologies should highlight its distinctive features comparing to the existing Single-Phase formulation presented in [4]. Special emphasis is placed on the complete characterization of the Rotating Field AFTM for comprehensive dynamic system simulation and control system design. A comparison between measurement data and electromagnetic FEA calculations proves the validity of the numerical simulation methods and highlights the (small) discrepancies. Last but not least, a prototype presentation of the novel rotating field AFTM structure concludes the current article.

II. Model Equations

Given is the state-space description of the three electrical phases of a Rotating field synchronous machine with surface-mounted permanent magnets (SPM) in the stator reference frame, e.g. [5]. The three electrical winding phases do no have a common star point, i. e. they are galvanically isolated. The application of the Blondel-Park transformation

$$\mathbf{T} = \frac{2}{3}\begin{bmatrix} 1/2 & 1/2 & 1/2 \\ \cos\left(\nu_r^e\right) & \cos\left(\nu_r^e - \frac{2\pi}{3}\right) & \cos\left(\nu_r^e + \frac{2\pi}{3}\right) \\ -\sin\left(\nu_r^e\right) & -\sin\left(\nu_r^e - \frac{2\pi}{3}\right) & -\sin\left(\nu_r^e + \frac{2\pi}{3}\right) \end{bmatrix} \tag{1}$$

to the input phase voltage vector $\mathbf{u}$ and the phase current $\mathbf{i}$ state vector

$$\mathbf{u}^t = [u_a,\, u_b,\, u_c] \quad \text{and} \quad \mathbf{i}^t = [i_a,\, i_b,\, i_c] \tag{2}$$

using the input and state transformations

$$\mathbf{u}_e = \mathbf{T}\mathbf{u} \quad \text{and} \quad \mathbf{i}_e = \mathbf{T}\mathbf{i} \tag{3}$$

yields the affine, time-invariant dynamical system description

$$\mathbf{u}_e = \left(\mathbf{R} + \omega_s L_{ph} \mathbf{G}\right)\mathbf{i}_e + \mathbf{L}\frac{\mathrm{d}}{\mathrm{d}t}\mathbf{i}_e + \mathbf{u}_{p,e} \tag{4}$$

in the synchronous rotor reference frame $\{dq\}$. The synchronous, electrical stator frequency

$$\omega_s = p_r \Omega_z \tag{5}$$

is a function of the number of motor pole-pairs p_r and the mechanical rotor speed Ω_z. The rotation of the reference frame is covered by the permutation matrix

$$\mathbf{G} = \begin{bmatrix} 0 & 0 & 0 \\ 0 & 0 & -1 \\ 0 & 1 & 0 \end{bmatrix} \tag{6}$$

and ν_r^e denotes the electrical rotor angle. The transformed input voltages and currents in the rotor reference frame are

$$\mathbf{u}_e^t = [u_0,\, u_d,\, u_q] \quad \text{and} \quad \mathbf{i}_e^t = [i_0,\, i_d,\, i_q]\,, \tag{7}$$

where the variables u_0 and i_0 designate the so called zero-sequence components. They are used for description of the common-mode voltage and current relations, which determine the dynamic behaviour of the active axial levitation forces of the AFTM. Due to the non-overlapping air-core coil winding, no mutual inductance is present between the electrical phases. The magnetic coupling for adjacent air core coils is below 3 % and therefore neglected for system modeling. The constant phase resistance $\mathbf{R}$ and phase inductance $\mathbf{L}$ matrices are

$$\mathbf{R} = \mathrm{diag}\{R_{ph}, R_{ph}, R_{ph}\} \text{ and} \tag{8}$$

$$\mathbf{L} = \mathrm{diag}\{L_{ph}, L_{ph}, L_{ph}\}\,, \tag{9}$$

where the phase inductance L_{ph} is determined by the leakage inductance only. As already mentioned, no magnetic coupling between the different winding phases is present. The transformed back-emf phase voltage vector of the SPM may be written as

$$\mathbf{u}_{p,e}\left(\Omega_z, \nu_r^e\right) = \Omega_z \begin{bmatrix} -k_{\Omega,3}\sin 3\nu_r^e \\ 0 \\ k_{\Omega,1} \end{bmatrix} \tag{10}$$

with the mechanical rotor speed $\Omega_z = \dot{\nu}_r^e / p_r$ and the speed constant of the harmonic order v

$$k_{\Omega,v} = p_r v \hat{\psi}_v\,. \tag{11}$$

$\hat{\psi}_v$ is the phase flux linkage amplitude of the electrical order number v. All higher harmonics above third order are neglected in (10). The mechanical output variables

$$\begin{bmatrix} t_r^z\left(\nu_r^e\right) \\ f_r^z \end{bmatrix} = \begin{bmatrix} -3k_{\Omega,3}\sin 3\nu_r^e & 0 & \frac{3}{2}k_{\Omega,1} \\ -k_f & 0 & 0 \end{bmatrix}\mathbf{i}_e \tag{12}$$

are linear functions of the transformed current vector $\mathbf{i}_e$. The active axial bearing forces f_r^z, realized by the coil end-windings, are controlled via the common-mode coil current component i_0, while zero-mean drive torque ripple is being generated simultaneously due to the presence of a third harmonic component in the back-emf phase voltage spectrum.

III. Measurements

Due to the complex actuator geometry, numerical field calculations are indispensable for dimensioning of the drive components to fulfill the predefined design specifications. After prototype manufacturing, the device characterization using simple and robust measurement procedures is essential not only for design validation, but also for dynamical system simulation and control system design.

A. Radial Drive Stiffness

The total radial drive stiffness is composed of the stabilizing forces f_r^x of the passive PMB system and the destabilizing radial reluctance forces of the actuator part. The results of the strain gauge radial force measurements and the numerical calculations are depicted in Fig. 4. In case of a centered and undeflected rotor, no radial resilience forces occur. According to the measurement results, the stabilizing radial magnetic forces of the passive ring bearing units (PMBs) are reduced by the (relatively small) reluctance forces of the stator back-iron. The radial motor reluctance forces of the present prototype are

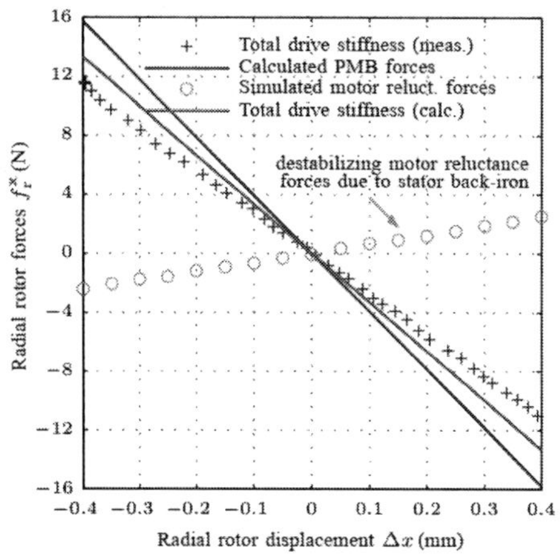

Figure 4. Radial drive stiffness measurements. Comparison between the semianalytic PMB radial stiffness calculations and the measured total radial drive forces. The radial force measurements include destabilizing radial reluctance forces of the actuator.

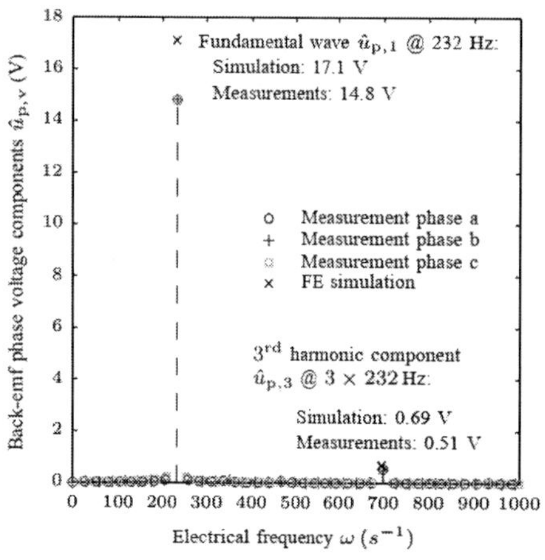

Figure 5. Spectral analysis of the back-emf voltage of phase a at Ω_z = 3480 rpm.

estimated using numerical 3D-fea simulations. The calculated destabilizing motor stiffness value is

$$s_{\mathrm{mot}}^{\mathrm{sim}} \approx -7\,\mathrm{N/mm}\,. \tag{13}$$

The radial force characteristic of the ironless, passive magnetic ring bearings (PMB) is calculated by analytical methods. The inclusion of the numerical calculated motor stiffness value reduces the gap between measurement and calculation.

$$\frac{s_{\mathrm{r}}^{\mathrm{m}}}{s_{\mathrm{r}}^{\mathrm{sim}}} = \frac{30\,\mathrm{N/mm}}{(39-7)\,\mathrm{N/mm}} = 0.93 \tag{14}$$

B. Back-emf (No-load) Phase Voltage

A simple back-emf phase voltage measurement can provide valuable information about the torque capability and the drive torque ripple. The spectral analysis of the back-emf phase voltage at a constant rotor speed is illustrated in Fig. 5. The fundamental wave and third harmonic voltage amplitudes can be clearly identified. The flux linkage of phase a with the permanent magnet rotor excitation system

$$\psi_{\mathrm{a}}\left(\nu_{\mathrm{r}}^{\mathrm{e}}\right) \approx \hat{\psi}_1 \cos \nu_{\mathrm{r}}^{\mathrm{e}} + \hat{\psi}_3 \cos 3\nu_{\mathrm{r}}^{\mathrm{e}} \tag{15}$$

is useful for the deduction of the back-emf voltage expression. The total time-derivative of (15) yields

$$\begin{aligned} u_{\mathrm{p,a}}\left(\nu_{\mathrm{r}}^{\mathrm{e}}\right) &= \frac{\mathrm{d}}{\mathrm{d}t}\psi_{\mathrm{a}} = p_{\mathrm{r}}\Omega_{\mathrm{z}}\frac{\partial}{\partial \nu_{\mathrm{r}}^{\mathrm{e}}}\psi_{\mathrm{a}} \\ &= -\,\Omega_{\mathrm{z}} p_{\mathrm{r}} \hat{\psi}_1 \sin \nu_{\mathrm{r}}^{\mathrm{e}} - \Omega_{\mathrm{z}} p_{\mathrm{r}} 3\hat{\psi}_3 \sin 3\nu_{\mathrm{r}}^{\mathrm{e}} \\ &= -\,\hat{u}_{\mathrm{p},1} \sin \nu_{\mathrm{r}}^{\mathrm{e}} - \hat{u}_{\mathrm{p},3} \sin 3\nu_{\mathrm{r}}^{\mathrm{e}} \end{aligned} \tag{16}$$

with the back-emf voltage amplitude of the order v

$$\hat{u}_{\mathrm{p,v}} = \Omega_{\mathrm{z}} k_{\Omega,v} \tag{17}$$

as a linear function of the mechanical rotor speed Ω_z. A comparison of the measured fundamental wave and third harmonic wave back-emf phase voltage components, taken at different mechanical speeds, is depicted in Fig. 6. The related speed constants are determined by calculation of the regression lines.

$$\frac{k_{\Omega,1}^{\mathrm{m}}}{k_{\Omega,1}^{\mathrm{sim}}} = \frac{4.242\,\mathrm{mV/rpm}}{4.915\,\mathrm{mV/rpm}} = 0.86 \tag{18}$$

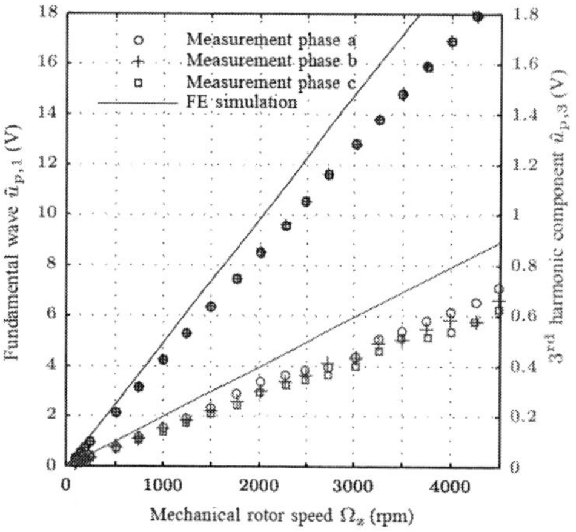

Figure 6. Comparison of the back-emf voltage measurements of three phases and determination of the back-emf speed constants $k_{\Omega,1}$ and $k_{\Omega,3}$.

C. Drive Torques

The second method for determination of the torque constant is the direct drive torque measurement via strain gauge torque sensor at constant phase current excitation.

1) Strain Gauge Measurement of Drive Torques: The drive torque expression of phase a at a constant phase current

$$\begin{aligned} t_{\mathrm{r,a}}^{\mathrm{z}}\left(\nu_{\mathrm{r}}^{\mathrm{e}}\right) &= -i_{\mathrm{a}}\left(k_{\Omega,1}\sin\nu_{\mathrm{r}}^{\mathrm{e}} + k_{\Omega,3}\sin 3\nu_{\mathrm{r}}^{\mathrm{e}}\right) \\ &= t_{\mathrm{r,a1}}^{\mathrm{z}}\left(\nu_{\mathrm{r}}^{\mathrm{e}}\right) + t_{\mathrm{r,a3}}^{\mathrm{z}}\left(\nu_{\mathrm{r}}^{\mathrm{e}}\right) \end{aligned} \tag{19}$$

is composed of a fundamental wave and a third harmonic drive torque component. All other harmonics are neglected in (19), which is justified in view of the following measurement results. A spectral analysis of the single phase drive torque measurements for a constant phase current is depicted in Fig. 7.

$$\frac{k^{\mathrm{m}}_{\Omega,1}}{k^{\mathrm{sim}}_{\Omega,1}} = \frac{41\,\mathrm{mN.m/A}}{46.5\,\mathrm{mNm/A}} = 0.88 \tag{20}$$

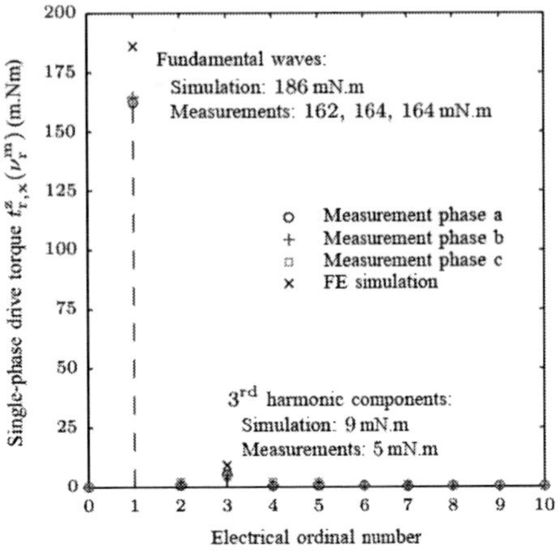

Figure 7. Single phase drive torque measurements via strain gauge torque sensor at $i_x = 4$ A with $x \in \{a, b, c\}$. The fundamental wave (dashed lines) and the third harmonic (plain lines) spectral decomposition of the measured torques have been added into the diagram.

2) Torque Ripple Measurement: For measuring the torque ripple provoked by the interacting third harmonic back-emf voltages and the common-mode stator currents, all electrical phases are excited with constant currents corresponding to

$$i_0 = i_a = i_b = i_c = 4\,\mathrm{A}. \tag{21}$$

Two independent measurements of the drive torque ripple have been carried out. The results are depicted in Fig. 8 together with the superponed single phase torque measurements of Fig. 7. The drive torque ripple as a result of a zero-sequence current component i_0 in absence of a mean torque producing current component i_q is

$$t^z_r(\nu^e_r) = -3k_{\Omega,3}i_0 \sin 3\nu^e_r, \tag{22}$$

referring to (12). The three phase-shifted fundamental harmonic torque components of the prototype machine do not compensate one another exactly, which can be seen easily in Fig. 7 . Due to a slight geometric unsymmetry of the stator winding geometry, the measured torque ripple in Fig. 8 contains a subharmonic component, that is not represented by (22).

D. Active Axial Bearing Forces

The measurement of the axial force as a function of the current i_0 has been effectuated using a strain gauge force/torque

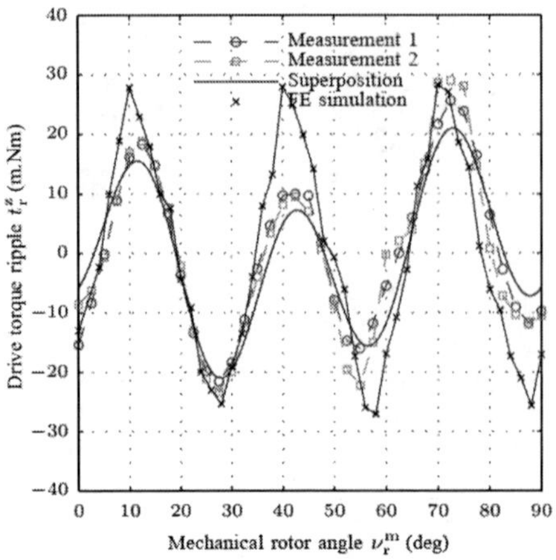

Figure 8. Direct measurement of the drive torque ripple at $i_0 = 4$ A. Because of the small amplitudes, two independent measurements have been effectuated and plotted into a single diagram. The black line represents the superposition of the single phase torque measurements of Fig. 7.

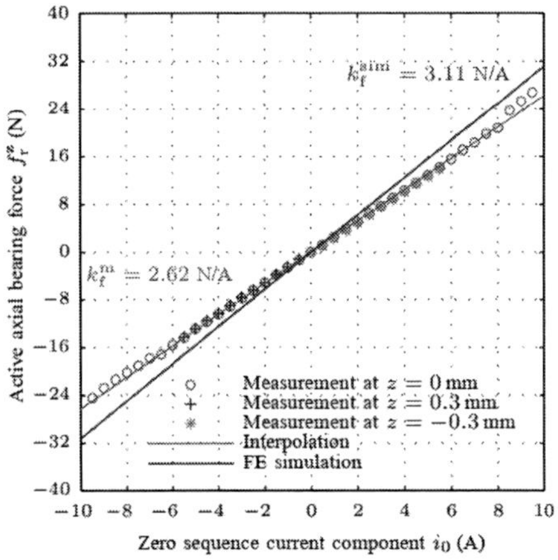

Figure 9. Determination of the axial force/current constant. Axial force measurements for two different axial deflections around the equilibrum position $z = 0$.

sensor. A comparison between measurement data and numerical force simulation is shown in Fig. 9. The discrepancy between the measured and simulated force/current constant

$$\frac{k^{\mathrm{m}}_{\mathrm{f}}}{k^{\mathrm{sim}}_{\mathrm{f}}} = \frac{2.62\,\mathrm{N/A}}{3.11\,\mathrm{N/A}} = 0.84 \tag{23}$$

is within the usual bandwidth of numerical 3d fea simulations.

IV. Prototype Presentation

The stator air gap winding is depicted in Fig. 10 and the rotor of the Bearingless AFTM is shown in Fig. 11.

V. Conclusion

The dynamic behaviour of the novel polyphase AFTM is determined by a few system parameters only, which can be

Figure 10. Experimental setup stator part. Air gap winding (expoxy resin bonded) with stator back iron.

Figure 11. Experimental setup rotor.

determined either by time-consuming numerical 3d simulations or by more or less complicated and expensive device measurement methods. Both approaches yield comparable results.

Table I
PROTOTYPE SPECIFICATIONS OF THE ROTATING-FIELD AXIAL-FORCE/TORQUE MOTOR.

Description	Symbol	Value	Unit
Number of stator coils	m	6	
Number of pole pairs	p_r	4	
Average air gap radius	R_δ	24	mm
Motor magnet height	l_m	10.5	mm
Active bearing magnet height	l_b	9.5	mm
Magnet width	w_m	2	mm
Electromagnetic air gap width	w_δ	6	mm
Mechanical air gap width	g_m	1	mm
Permanent magnet volume		6.4	cm^3
Copper volume		2.1	cm^3
Dc-link voltage	$U_{dc\text{-}link}$	48	V
Nb. of turns per coil	N_c	160	
Max. admissible drive speed	Ω_z	7900	rpm
Nb. of electrical phases		3	
Nb. of required half-bridges		6	
Max. output motor power	P_{mech}	122	W
Admissible RMS current density	S_t^{cu}	10	$A.mm^{-2}$
Short-time current density	S_f^{cu}	50	$A.mm^{-2}$
Admissible avg. drive torque	$\langle t_r^z \rangle$	147	mN.m
Active axial force at S_f^{cu}	$f_{s,50}^z$	21	N
Rotor mass	m_R	0.330	kg
Mass moment of inertia	J_z	1813	$g.cm^2$

Table II
COMPARISON OF MEASUREMENTS AND SIMULATION.

Description	Symbol	Calc.	Meas.	Unit
Radial PMB stiffness	s_r	39.4		N/mm
Tilting PMB stiffness	$s_{\varphi\varphi}$	25.2		N.m/rad
Coupling stiffness	$s_{\varphi r}$	−288		N
Radial motor stiffness	s_{mot}	−7		N/mm
Total radial drive stiffness	s_r	32	30	N/mm
Axial drive stiffness (unstable)	s_z	−78.8	−80.4	N/mm
Speed constant (fund. wave)	$k_{\Omega,1}$	4.915	4.242	mV/rpm
Torque constant	k_t	69.8	61.5	mN.m/A
Speed constant (3rd harmonic)	$k_{\Omega,3}$	198.2	137.7	uV/rpm
Axial force constant	k_f	3.1	2.6	N/A
Phase resistance	R_{ph}	2.78	2.72	V/A
Phase inductance	L_{ph}	2.36	2.16	mH

ACKNOWLEDGMENT

The presented research work has been supported by the Linz Center of Mechatronics (LCM GmbH), which is part of the COMET/K2 program of the Federal Ministry of Transport, Innovation and Technology and the Federal Ministry of Economics and Labour, Austria. The author would like to thank the Austrian Government, the Upper Austrian Government and the Johannes Kepler University for their support in making this work possible.

REFERENCES

[1] W. Bauer and W. Amhrein, "Design and sizing relations for a novel bearingless motor concept," *ICEMS: International Conference on Electrical Machines and Systems, Beijing, China*, Aug. 20-23, 2011, DOI: 10.1109/ICEMS.2011.6073470.

[2] ——, "Electrical design and winding selection for a bearingless axial-force/torque motor," *SPEEDAM 2012: International Symposium on Power Electronics, Electrical Drives, Automation and Motion, Sorrento, Italy*, Sep. 19-20, 2013, DOI: 10.1109/SPEEDAM.2012.6264458.

[3] ——, "Design study of air-core coil winding topologies for bearingless axial-force/torque motors," *ISMB13: The 13th International Symposium on Magnetic Bearings, Arlington, VA, USA*, Aug. 6-9, 2012, http://www.magneticbearings.org.

[4] ——, "Performance analysis of a bearingless axial-force/torque motor," *9. GMM/ETG Symposium 2013: Innovative Small Drives and Micro-Motor Systems, Nuernberg, Germany*, Sep. 19-20, 2013, ISBN: 978-3-8007-3537-2.

[5] P. Vas, *Electric Machines and Drives - A space-vector theory approach.* Oxford University Press, 1992.

Realization of Magnetic Suspension and Motor Operation with One Three-phase Voltage Source Inverter Using a Zero-Phase Current

Junichi Asama[a], Yusuke Fujii[a], Masato Fukuyo[a], Takaaki Oiwa[a], and Akira Chiba[b]

[a] Shizuoka University, 3-5-1 Johoku, Hamamatsu, Japan, tjasama@ipc.shizuoka.ac.jp
[b] Tokyo Institute of Technology, 2-12-1 Ookayama, Meguro, Tokyo, Japan, chiba@ee.titech.ac.jp

***Abstract*—This paper focuses on realization of both the single-axis active magnetic suspension and rotation of a three-phase permanent magnet motor using only one three-phase voltage source inverter. The suspension winding of the electromagnet for the single-axis magnetic suspension is connected between the neutral point of the Y-connected three-phase winding and the middle point of the voltage sources, and the resulting zero-phase current is utilized to control the suspension force. The control method of the zero-phase current is theoretically derived to avoid interference with the motor performance. Experimental apparatus was built and tested, which consists of an iron ball magnetic suspension system and a three-phase interior permanent magnet motor. The experimental results demonstrated stable magnetic suspension and motor rotation at 1200 rpm.**

I. Introduction

Conventional five-axis actively regulated bearingless motors and magnetic bearings require a number of electromagnets, displacement sensors, and inverters [1, 2]. To reduce the size, cost, and power consumption, one-axis actively positioned magnetic bearings have been previously developed [3, 4]. These machines generally combine both a thrust magnetic bearing for axial suspension and a motor for rotation. The remaining two-translational and two-tilting motions of the rotor are passively stabilized. In these machines, at least two inverters are needed to regulate the suspension and motor currents independently.

A one-axis actively positioned bearingless motor has been previously proposed [5]. This bearingless motor has unique rotor structure and winding configuration using four coreless coils to generate both axial suspension force and torque. In this bearingless motor, two independent currents are regulated to control the torque and axial suspension force. The radial and tilting motions of the rotor are passively stabilized by the repulsive type permanent magnet (PM) pairs. The authors have previously developed one-axis actively positioned bearingless motor using only one three-phase inverter, dubbed a single-drive bearingless motor [6]. Magnetic couplings between the rotor and the stator stabilize the radial and tilting motions of the rotor passively. The axial position and torque are stabilized by regulating the *d*- and *q*-axis currents, respectively. However, only two independent *d*- and *q*-axis currents are actively regulated by one three-phase inverter. If three-axis currents are independently regulated by only one three-phase inverter, further cost reduction, downsizing, and energy saving for the magnetically suspended machines can be achieved.

This paper focuses on realization of both the single-axis active magnetic suspension and rotation of a PM motor using only one three-phase voltage source inverter. The final goal of this study is to apply the proposed control method to a one-axis actively regulated bearingless machine, which consists of, for instance, a PM motor, repulsive type passive magnetic bearings for non-contact shaft suspension in the radial and tilting directions, and a thrust magnetic bearing. In this paper, we propose a novel concept that the suspension winding is connected between a neutral point of the Y-connected three-phase winding and a middle point of the voltage sources, and resulting zero-phase current generates the suspension force. To our knowledge, application of the zero-phase current to active actuator control is new to the field of magnetic bearings, bearingless motors, and electric machines and drives.

In section II, the proposed control method to avoid interference of the zero-phase current with motor performance is theoretically derived. Section III presents results of circuit simulation using a commercially available circuit simulator. To verify the proposed control method, we fabricated the experimental apparatus which consists of a three-phase interior permanent magnet (IPM) motor and a single active magnetic suspension system with an iron ball. The experimental results with this test machine are shown in Section IV.

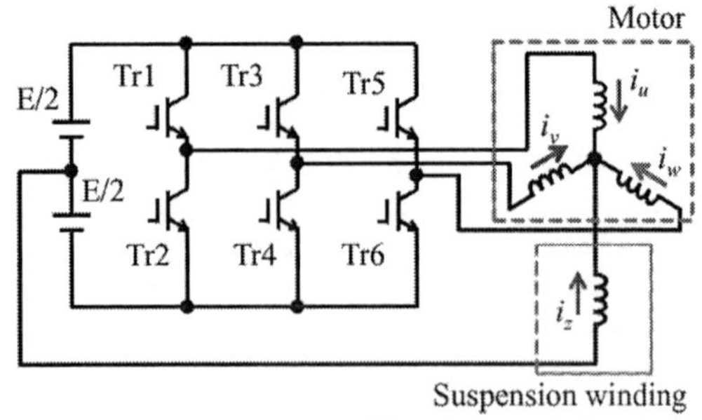

Figure 1. Circuit configuration of the proposed control method.

II. Control Method of Zero-Phase Current

Figure 1 shows a circuit configuration of the proposed control method. The suspension winding is connected between a neutral point of the Y-connected three-phase winding and a middle point of the voltage sources, and resulting zero-phase current i_z generates the suspension force. For example, when the all lower transistors, Tr2, Tr4, and Tr6, are turned off and Tr1, Tr3, and/or Tr5 are turned on, a zero-phase current i_z flows through the suspension winding from the neutral point to the voltage middle point. In this section, the control method of this zero-phase current to avoid influence on motor performance is theoretically introduced.

The transformation of the three-phase (I_u, I_v, I_w) frame into the zero-direct-quadrature (I_0, I_d, I_q) frame is given by

$$\begin{bmatrix} i_0 \\ i_d \\ i_q \end{bmatrix} = \sqrt{\frac{2}{3}} \begin{bmatrix} \frac{1}{\sqrt{2}} & \frac{1}{\sqrt{2}} & \frac{1}{\sqrt{2}} \\ \cos\omega t & \cos(\omega t - \frac{2\pi}{3}) & \cos(\omega t - \frac{4\pi}{3}) \\ -\sin\omega t & \sin(\omega t + \frac{\pi}{3}) & \sin(\omega t - \frac{\pi}{3}) \end{bmatrix} \begin{bmatrix} i_u \\ i_v \\ i_w \end{bmatrix} \quad (1)$$

where ω ant t are the angular frequency and time, respectively. In a general motor, three-phase current vector is given by

$$\begin{bmatrix} i_u \\ i_v \\ i_w \end{bmatrix} = \sqrt{2} I_e \begin{bmatrix} e^{j(\omega t+\theta)} \\ e^{j(\omega t+\theta-2\pi/3)} \\ e^{j(\omega t+\theta-4\pi/3)} \end{bmatrix} \quad (2)$$

Substituting (2) into (1) yields

$$\begin{bmatrix} i_0 \\ i_d \\ i_q \end{bmatrix} = \sqrt{3} I_e \begin{bmatrix} 0 \\ e^{j\theta} \\ -je^{j\theta} \end{bmatrix} \quad (3)$$

where I_e and θ are the root-mean-square value and phase of the current, respectively. It can be observed that the zero-phase component is always zero and d- and q- components are dc values and not a function of time.

On the other hand, to feed the zero-phase current actively into the suspension winding, dc components (I_{u0}, I_{v0}, I_{w0}) have to be added to each phase current, which can be written as,

$$\begin{bmatrix} i_u \\ i_v \\ i_w \end{bmatrix} = \sqrt{2} I_e \begin{bmatrix} e^{j(\omega t+\theta)} \\ e^{j(\omega t+\theta-2\pi/3)} \\ e^{j(\omega t+\theta-4\pi/3)} \end{bmatrix} + \begin{bmatrix} I_{u0} \\ I_{v0} \\ I_{w0} \end{bmatrix} \quad (4)$$

Substituting (4) into (1) yields

$$\begin{bmatrix} i_0 \\ i_d \\ i_q \end{bmatrix} = \begin{bmatrix} \frac{(I_{u0}+I_{v0}+I_{w0})}{\sqrt{3}} \\ \sqrt{3} I_e e^{j\theta} + \frac{(2I_{u0}-I_{v0}-I_{w0})\cos\omega t}{\sqrt{6}} + \frac{(I_{v0}-I_{w0})\sin\omega t}{\sqrt{2}} \\ -\sqrt{3} I_e j e^{j\theta} - \frac{(2I_{u0}-I_{v0}-I_{w0})\sin\omega t}{\sqrt{6}} + \frac{(I_{v0}-I_{w0})\cos\omega t}{\sqrt{2}} \end{bmatrix} \quad (5)$$

To obtain time-independent dc values of I_d, I_q, the following conditions should be satisfied, as,

$$\begin{aligned} 2I_{u0} - I_{v0} - I_{w0} &= 0 \\ I_{v0} - I_{w0} &= 0 \end{aligned} \quad (6)$$

This results in the following using a dc component, I_z, as,

$$I_{u0} = I_{v0} = I_{w0} = I_z \quad (7)$$

Substituting (7) into (5) yields

$$\begin{bmatrix} i_0 \\ i_d \\ i_q \end{bmatrix} = \sqrt{3} \begin{bmatrix} I_z \\ I_e e^{j\theta} \\ -I_e j e^{j\theta} \end{bmatrix} \quad (8)$$

Therefore, the equivalent dc components should be fed into the three-phase currents to have no influence on the motor performance.

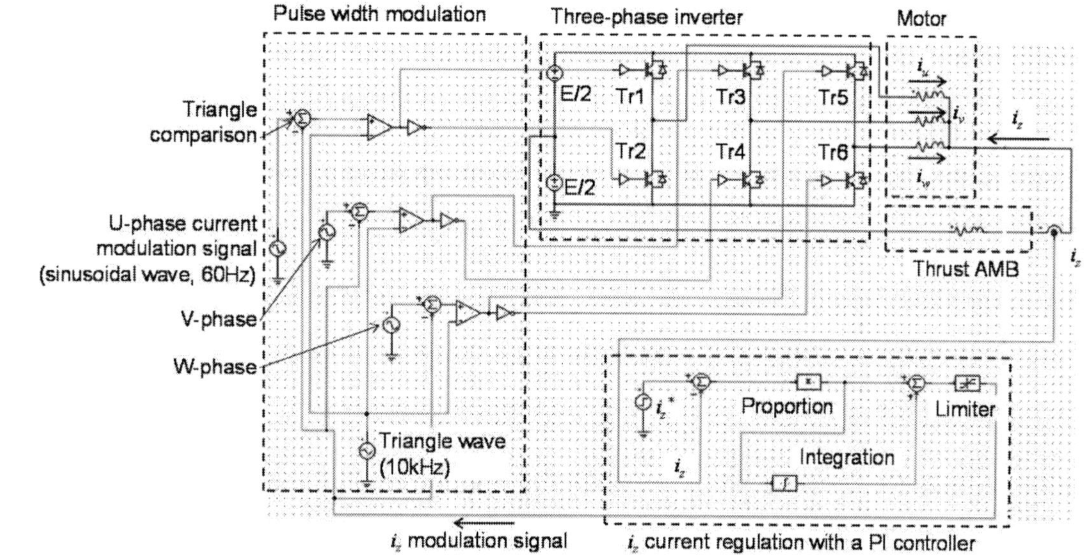

Figure 2. Circuit configuration of the proposed control method.

III. CIRCUIT SIMULATION

Circuit simulation is performed using a commercially available software (PSIM, Myway Plus Corp., Japan). Figure 2 shows simulation blocks using the pulse-width-modulation (PWM) voltage source inverter. A triangle wave generator with a carrier frequency of 10 kHz and a comparator generate PWM signals. Three-phase open-loop alternating currents with a frequency of 60 Hz are provided to the motor windings. To simplify the control circuit, all windings of the three-phase motor and magnetic bearing consist of resistance and inductance. The induced voltage of the motor is not considered in this simulation. A zero-phase current is measured and regulated by proportional-integral (PI) controller. This PI controller outputs a modulation component, which is then equally added to each phase.

Figure 3 shows the simulation result of the zero-phase current control. Three-phase currents and zero-phase current are shown in top and bottom graphs, respectively. The zero-phase current control is activated after 0.1 s to feed 3 A into the suspension winding. A dc component of 1 A is fed into each phase of the motor winding. As can be seen, the zero-phase current rises within 10 ms. The simulation results demonstrated that the zero-phase current was controlled by adding a desired dc component to the PWM modulation signal of the three-phase inverter.

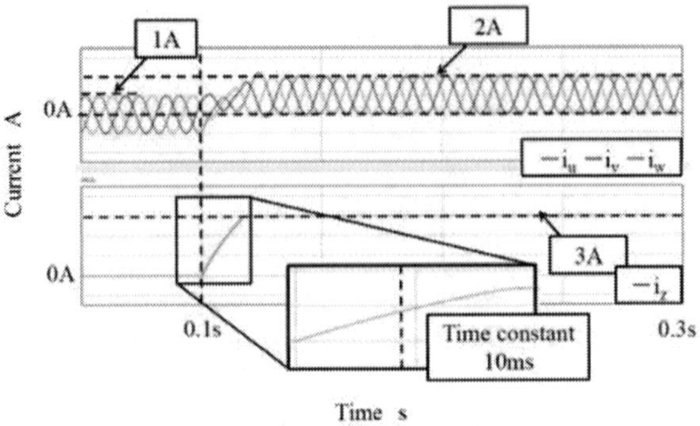

Figure 3. Simulation results of the zero-phase current control.

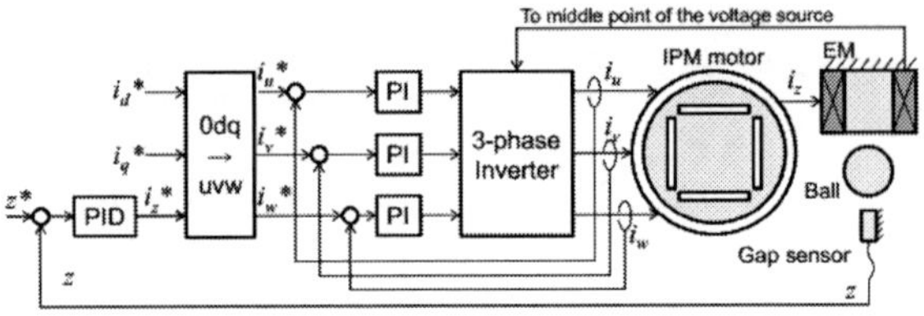

Figure 4. Schematic of the control diagram.

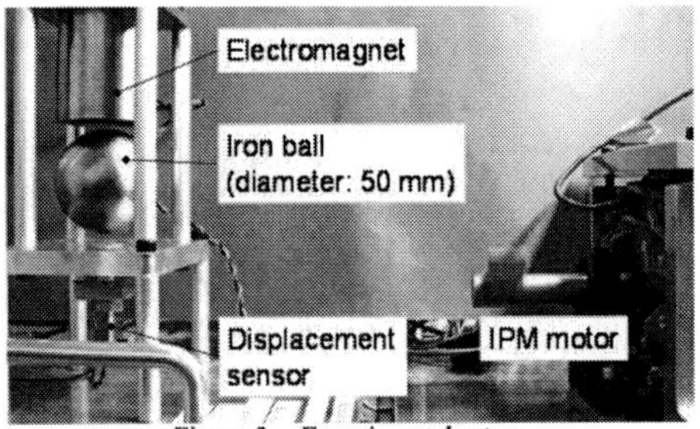

Figure 5. Experimental setup.

IV. EXPERIMENT

To verify the proposed method, we built an experimental apparatus which consists of a three-phase IPM motor and a single magnetic suspension system with an iron ball. Figure 4 shows a schematic of the control diagram. The suspension winding of the electromagnet is connected between the neutral point of the Y-connected three-phase motor winding and the middle point of the voltage source. An axial motion of the iron ball, z, is detected by an eddy current type displacement sensor. A proportional-integral-derivative (PID) controller outputs the zero-phase current reference, i_z*, and stables motion of the iron ball. Based on (1), the references of d-axis, q-axis, and zero-phase currents, i_d*, i_q*, and i_z* are transformed into the three-phase current references, i_u*, i_v*, and i_w*. Because of the Y-connected winding configuration, the three-phase current, i_u, i_v, and i_w, and zero-phase current, i_z, have the following relationship due to Kirchhoff's current law, as,

$$I_u + I_v + I_w + I_z = 0 \tag{9}$$

Three of those currents are independent variables. Therefore, three-phase currents are detected and actively regulated by the PI controller.

Figure 5 shows a picture of the experimental apparatus. The electromagnet for the single-axis magnetic suspension does not include any PMs. A hollow iron ball with a diameter of 50 mm and a weight of 160 g is used. A bias current of the electromagnet for magnetic suspension of the iron ball is set to be 1 A. A three-phase four-pole/six-slot IPM motor with a rated power of 1.5 kW is driven under no load condition. The d-axis current reference, i_d*, is set to 0 A. Two power supplies (ZX-400L, Takasago Ltd., Japan) with a dc voltage of 30 V each are used and thus the dc link voltage of the inverter is 60 V. A custom-made inverter equipped with an intelligent power module (PS21353-N, Mitsubishi Electric Corp., Japan) and a real-time digital controller (Nc-RIO-9022, National Instruments Corp., USA) is used. A carrier frequency of the PWM is 10 kHz.

The PID controller parameters for magnetic suspension were independently regulated and then the iron ball was successfully levitated. Figure 6 shows the U-phase current, i_u, the zero-phase current, i_z, and the axial displacement of the iron ball, z, when the magnetic suspension control is activated at a rotational speed of 1200 rpm. The zero-phase current increased rapidly to pull up the iron ball although undesirable current fluctuation occurred partially. After 0.2 s, the axial displacement, z, reached reference value of 5 mm. Figure 7 shows the three-phase currents, the zero-phase current, and the axial displacement in steady-state at 1200 rpm. The iron ball was stably suspended with the zero-phase current regulation. The vibration amplitude of the iron ball in the axial direction was 60 μm (3σ). Although the three-phase ac currents are flowing, the dc component in W-phase does not correspond with those of U-phase and V-phase. In future works, we will address an evaluation of the motor performance with magnetic suspension condition, a further investigation of interference between the magnetic suspension and motor drive, and a non-contact IPM motor drive using only one three-phase inverter.

V. Conclusion

This paper describes a novel concept that two independent motor currents, i_d and i_q, and one suspension current, i_z, are independently controlled using only one three-phase voltage source inverter. The zero-phase current that corresponds to the suspension current, i_z, is utilized to actively regulate the suspension force. A control method of this zero-phase current was theoretically derived to avoid an influence on the motor performance. Experimental results demonstrated active regulation of the zero-phase current and successful magnetic suspension during motor rotation at 1200 rpm using only one three-phase inverter.

Acknowledgment

This work was supported in part by a Grant-in-Aid for Scientific Research of the Japan Society for the Promotion of Science (26289064).

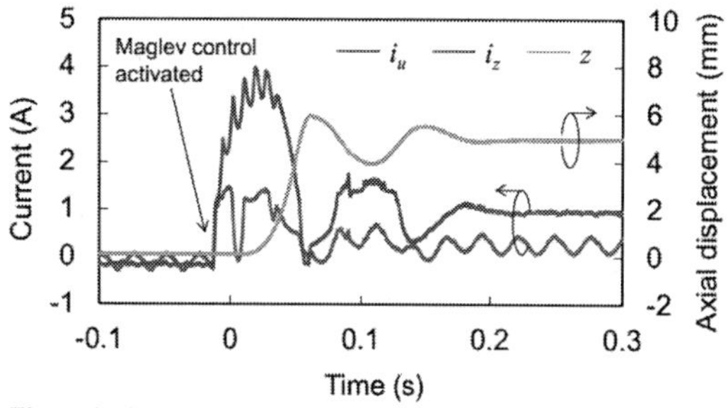

Figure 6. Measured U-phase current, zero-phase current, and axial displacement of the iron ball when magnetic suspension is activated at a motor speed of 1200 rpm.

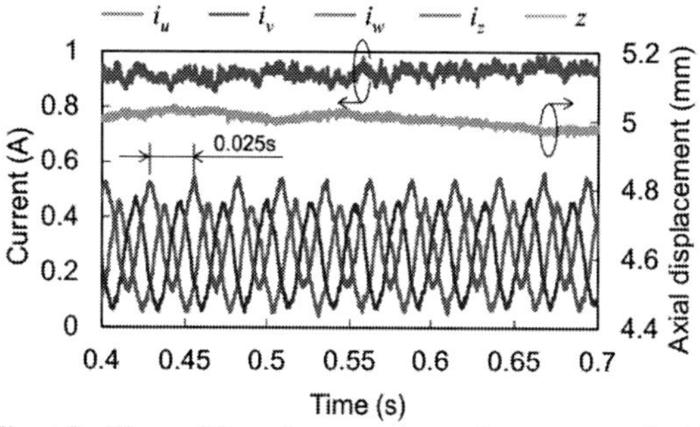

Figure 7. Measured three-phase current, zero-phase current, and axial displacement of the iron ball in steady-state at 1200 rpm.

References

[1] R. Siegwart, R. Larsonneur, and A. Traxler, "Design and Performance of a High Speed Milling Spindle in Digitally Controlled Active Magnetic Bearing", *Proceedings of the 2nd International Symposium on Magnetic Bearing*, pp. 197-204, July 12-14, 1990, Tokyo, Japan.

[2] N. Yamamoto, M. Takemoto, S. Ogasawara, and M. Hiragushi, "Experimental Estimation of a 5-axis Active Control Type Bearingless Canned Motor Pump", *Proceedings of the IEEE International Electric Machines and Drives Conference*, pp. 148-153, May 15-18, Niagara Falls, ON, Canada, 2011.

[3] I. D. Silva and O. Horikawa, "An attraction-type magnetic bearing with control in a single direction," *IEEE Transactions on Industry Applications*, vol. 36, no. 4, pp. 1138-1142, 2000.

[4] J. Kuroki, T. Shinshi, L. Li, and A. Shimokohbe, "Miniaturization of one-axis-controlled magnetic bearing," *Precision Engineering*, Vol. 29, No. 2, pp. 208-218, Apr. 2005.

[5] W. Bauer and W. Amrhein, "Electrical Design Considerations for a Bearingless Axial-Force/Torque Motor," *IEEE Transactions on Industry Applications*, available online, 2014.

[6] J. Asama, Y. Hamasaki, T. Oiwa, and A. Chiba, "Proposal and Analysis of a Novel Single-Drive Bearingless Motor," *IEEE Transactions on Industrial Electronics*, vol. 60, no. 1, pp. 129-138, 2013.

Study on Bearingless Motor with Rectified Circuit Coil

Kunihikoko Tachibana[a], Koichi Oka[a]

[a] Kochi University of Technology, Miyanoguchi 185, Tosa Yamada Cho, Kami City, Kochi 782-8502, Japan, tachibana.kunihiko@kochi-tech.ac.jp

***Abstract*—This paper represents the research finding of levitation and rotation control on bearingless motor with rectified circuit coil. Rectified circuit coil is functioned as transmitting and receiving coil of wireless electric power transmission, and supplies direct current to rotor coils. Rotor coils are used as rotor magnets not using permanent magnets. Even if magnetic force of magnet is strong, normal force will not be generated if magnets are horizontally configured on straight line. So as to generate vertical axial force, the tip form of rotor and stator is made to incline and levitation force (vertical axial force) is acquired using this inclination.**

I. INTRODUCTION

The permanent magnet is used for the stator or the rotor of bearingless motor in plenty of researches [1-5]. A permanent magnet has problem which cannot be used under high temperature environment and is demagnetized by the external magnetic field. In recent years, there are several magnets which can be used exceeding 100 centigrade degrees [6-8], however other problem has not been solved.

If the number of turns of coil is the same, the magnetic force of an electromagnet is proportional to the current of coil. In order to be stabilized levitating and rotating rotor, powerful magnetic force is required. Therefore, it is needed to feed much electric power for rotor coil. If a magnet size is small, the magnetic force of an electromagnet is small than a permanent magnet. Therefore, in order to levitate a rotor by small magnetic force, it has to generate magnetic support force efficiently. On that, it is effective that Z-axis inclination exists in the tooth edge face.

The authors have previously reported a bearingless motor with rectified circuit coil of half wave rectification type[9-11]. In this type, to acquire large magnetic force of rotor, magnetic saturation of rotor and stator core will occur. In order to solve this problem, we propose a method of wireless power transmission for rotor by full wave rectification type and an inclination of teeth of stator and rotor.

In this paper, generating levitation force by inclination of teeth and control method are proposed; basic principle of generating force and the calculated results of the three-dimensional finite element method (3D-FEM) and the experiment result of power transmission by full wave rectification type are presented. We assume that the motor is used as pump for material transfer used in high environment temperature, such as food and medicine.

II. PRINCIPLE OF RECTIFIED CIRCUIT COIL

A rectified circuit coil of half wave rectification type is shown in Figure 1, and that of full wave rectification type is shown in Figure 2. The rectified circuit coil transforms the electromotive force by electromagnetic induction into a direct current using diode, and fixes the generated magnetic pole in a coil. A difference between half wave rectification type and full wave rectification type is existence of power transmission coil. In half wave rectification type, the rotor coils receive electric power directly from the stators. In other hand, in full wave rectification type, the rotor coils receive electric power from power transmission coil. However, in both of the types, a direct current flows in rotor coils, and the fixed magnetic pole appears in the both sides of a coil. Therefore, it means attaching a permanent magnet in a rotor.

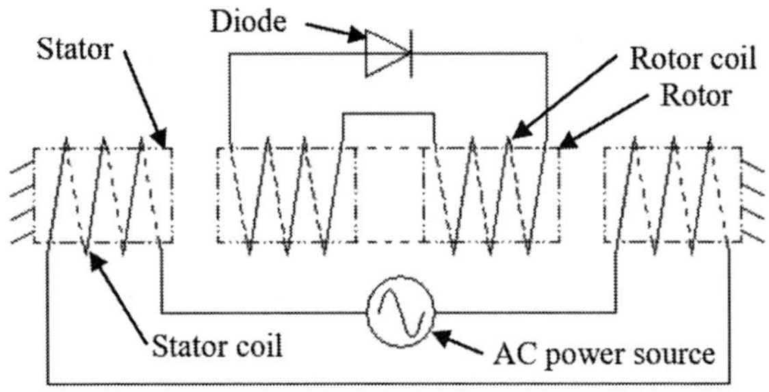

Figure 1. Half wave rectification type.

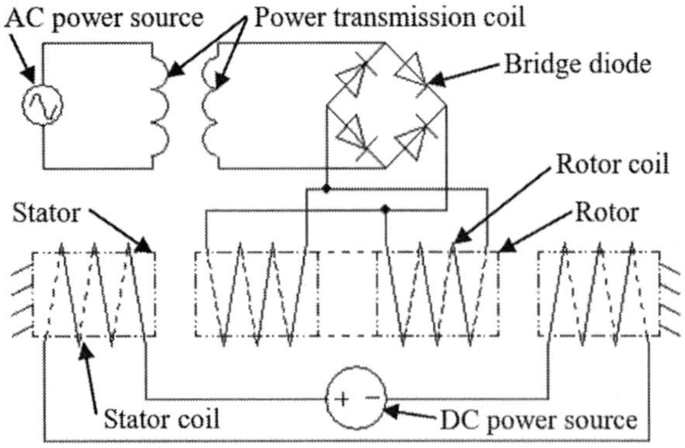

Figure 2. Full wave rectification type.

III. BEARINGLESS MOTOR WITH RECTIFIED CIRCUIT COIL

A. Structure

A bearing loess motor is classified into one axis to five axis control type according to the number of axes controlled actively. In case of two axis control type motor, the structure and control can be simplify, therefore we selected it. Figure 3 shows a structure of the proposed bearingless motor with two layers flat-shaped rotors. If a coil on one tooth has a small winding number and the coil current is small, it cannot generate strong magnetic force. In order to solve this problem, the tooth edge face is inclined for generating strong levitation force. Furthermore, to supply much electric power to rotor coils, it was considered as the full wave rectification type. In one layer, the number of the stator tooth is 24 pieces and the rotor tooth is 32 pieces. Suspension coils and motor coils are wound in stator core, and rotor coils are wound in rotor core. An electric power transmission coil is configured to space of the center of the rotor.

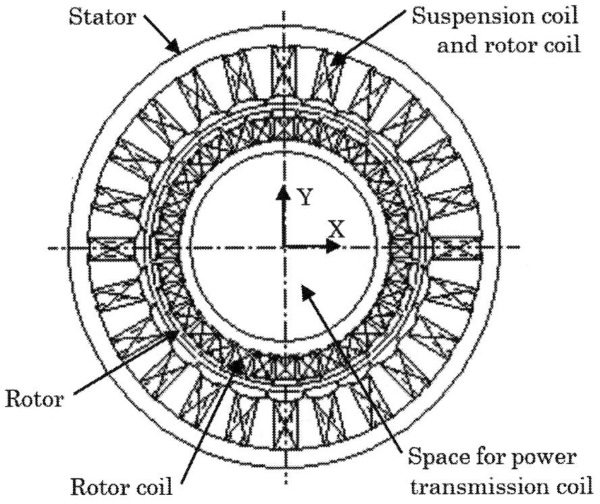

(a) Top view

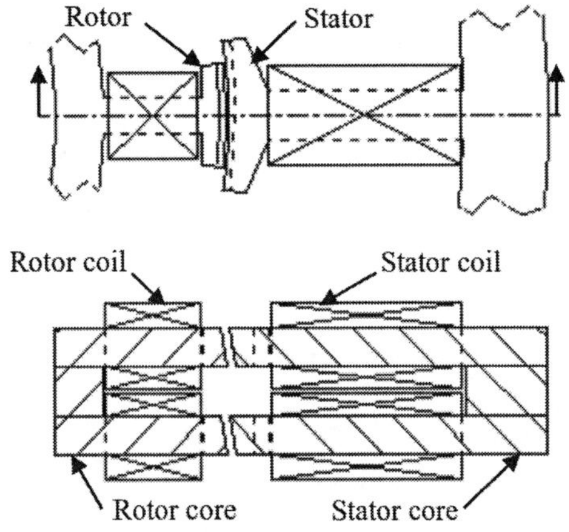

(b) Detail of teeth

Figure 3. Structure of proposed bearingless motor.

B. Principle of generating magnetic force

The rotor coils are connected to bridge diode for the monopolar of N pole or S pole in one layer. The stator is fixed as a different magnetic pole from the rotor. The attractive force occurs between the stator and the rotor, and the force is used for levitation of the rotor. If the same current is supplied to the stator coils, the rotor received force from the stator in X-axis and Y-axis is zero when it integrates the vector of magnetic force over all circumferences, therefore, the rotor axis is located at the center of the stator axis.

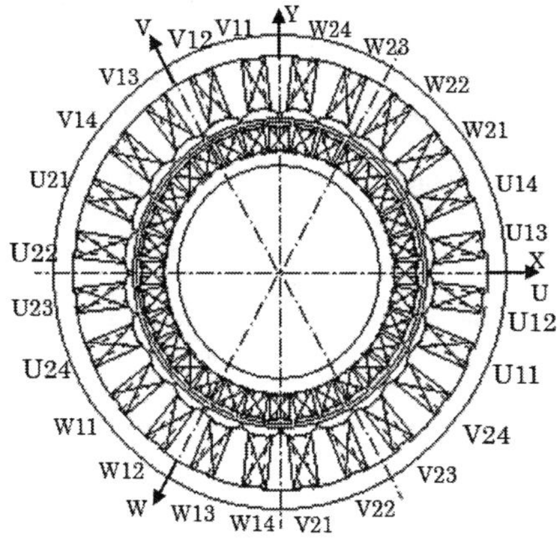

Figure 4. Principle of radial suspension force generation.

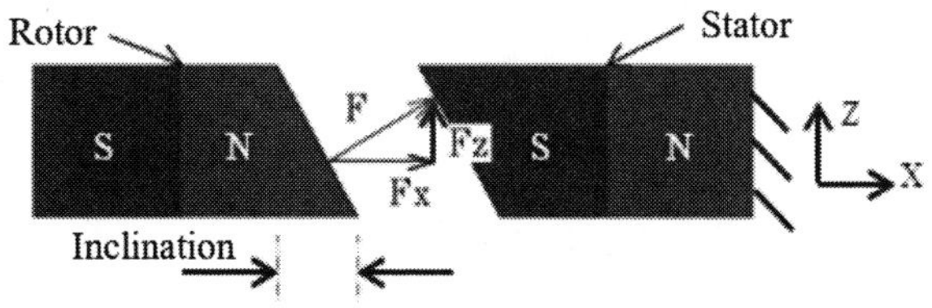

Figure 5. Principle of axial suspension force generation.

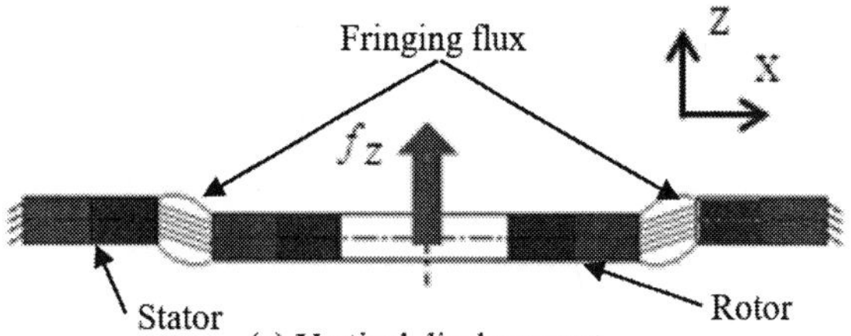

(a) Vertical displacement

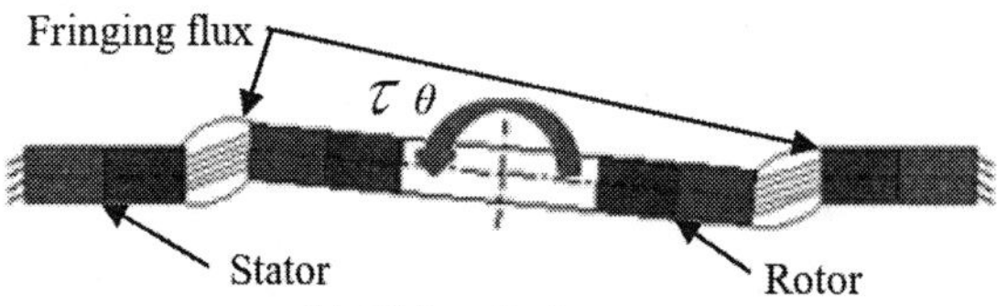

(b) Tilting displacement

Figure 6. Principle of axial and tilting restitutive force generation.

The suspension coils of the stator are configured on 3 axis of U, V and W axis, as shown in Figure 4. The position of XY coordinates can be displaced to the position of UVW coordinates. A displacement in XY axis of the rotor is displaced into UVW axis by 2-phase/3-phase conversion, and the magnetic force of UVW axis of the stator is controlled; magnetic force can be generated in radial by integral of the vector of applied force to the rotor. Radial position control of the rotor is attained by the disequilibrium of this force.

Principle of generating Z-axis magnetic suspension force is shown in Figure 5. Generally, it is difficult to acquire sufficient suspension force with short electromagnet of shaft orientation. Then, it is considered as the edge form of tooth which has inclination angle to the Z-axis, and the levitation force of Z-axis is acquired. Direction of the magnetic-flux line of the air-gap is perpendicular to the edge surface of the direction of magnetization. Even if the position of the rotor and the stator is the same on the Z-axis, Z-axis force will occur.

Figure 6 shows the principle of restitutive force of rotor when a displacement of the rotor occurs in shaft orientation and tilting. Magnetic couplings between the rotor and the stator provide stability in the axial and tilting directions.

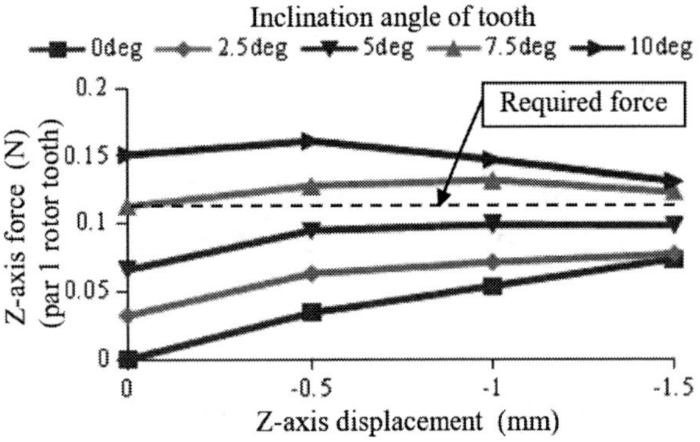

Figure 7. Compare of force by teeth inclination angle.

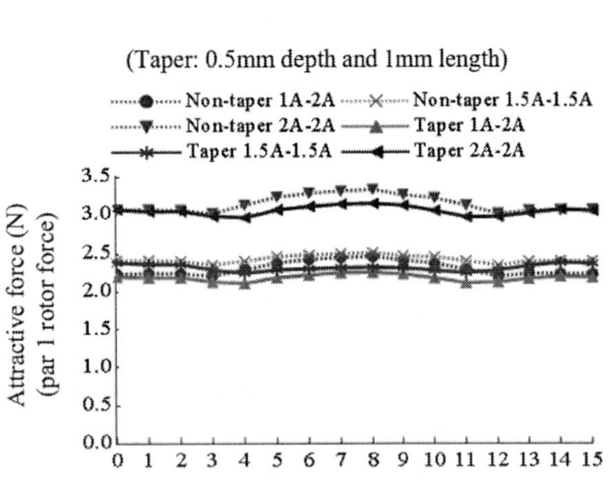

Figure 8. Compare of force by teeth inclination angle.

IV. Design of Bearingless Motor

A. Levitation force

Figure 7 shows the relation between inclination angle of teeth and generated Z-axis force (levitation force) with stator and rotor is calculated by 3D-FEM with 75% in design current. If inclination angle is large, generated force is also large. However, if the displacement of Z-axis is greater than 0.5mm, the force will become low rate of increase or will decrease. The Z-axis force for levitating the rotor in design is approximate 0.11N per the rotor tooth. Therefore, the inclination angle of the rotor and the stator is decided to be approximate 7.5 degrees.

B. Form of tooth edge

A taper of tooth edge is applied to reduce a change of the magnetic force by angle of rotation of the rotor. Calculations by 3D-FEM ware carried out using design dimension. Figure 8 shows a comparison of attractive force in case of 0.5mm depth and 1mm length of taper and in case of without taper, and indicates that the taper is functioning, and change of magnetic force is decrease. The variation of force becomes small with the increase in current, therefore the taper is acting effectively.

C. Power transmission to rotor

The coil for electric power transmission shown in Figure 9 is attached in the center of rotor not as the method of exciting rotor coil directly but as the method of transmitting electric power to rotor coil indirectly. In order that an electric power transmission coil may increase transmission efficiency, Spider winding applies a coil winding method, and uses resonance by

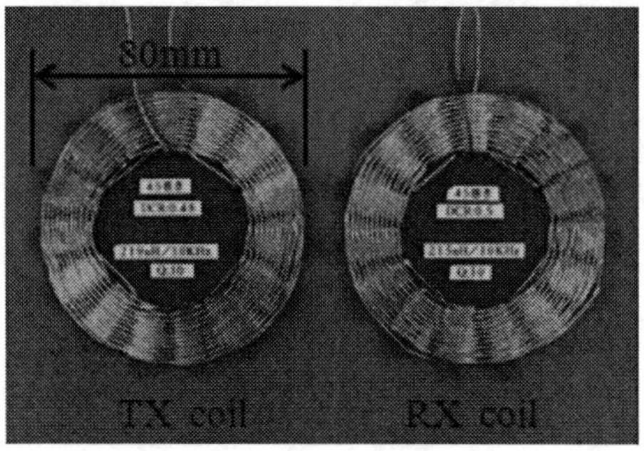

Figure 9. Wireless power transmission coil.

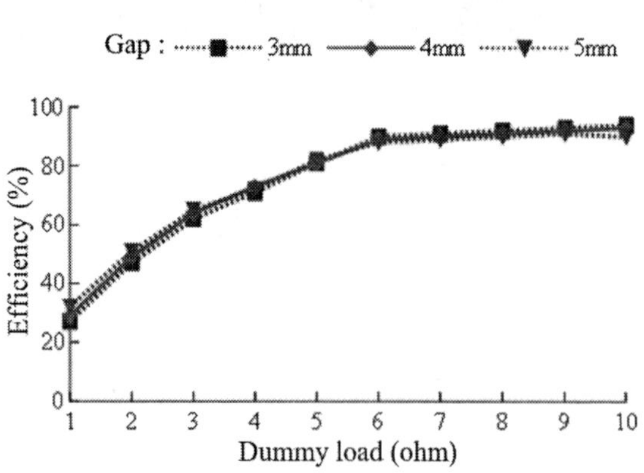

Figure 10. Power transmission efficiency.

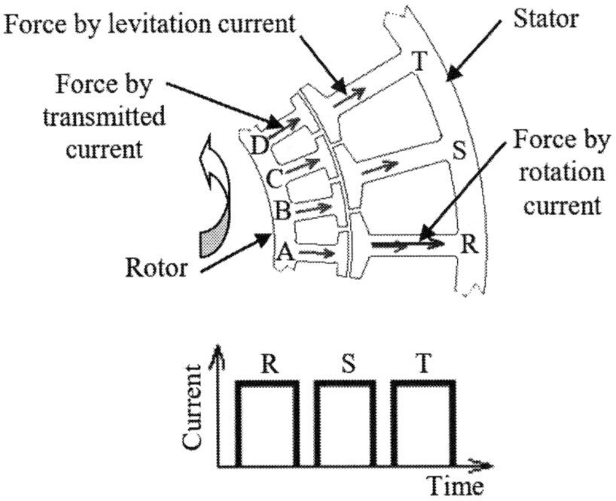

Figure 11. Principle of rotation.

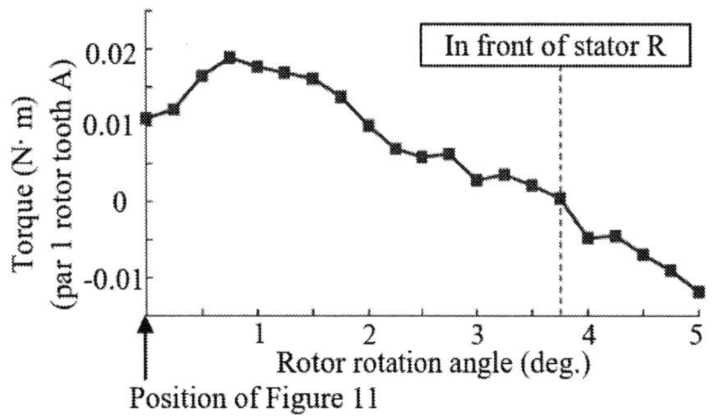

Figure 12. Calculated rotation force (1/8 model).

a capacitor for primary side (fixed side). The result of having measured the electric power transmission efficiency in which load resistance and a coil gap were changed is shown in Figure 10. The estimate value of rotor coil resistance is about 5.5ohms, and transmission efficiency of 5.5 ohms is 80% or above showed in Figure 10.

D. Levitation and XY position control

If large displacement of Z-axis rotor position does not occur, since the rotor position restores to an initial position passively, the levitating position of Z-axis of the rotor is controlled to the initial position. On the other hand, the displacement of XY-axis of the rotor is controlled actively. XY-axis position of the rotor is measured by displacement sensors, and the positions are controlled using the measured value. Specifically, if a X-axis displacement of the rotor occurs, the current of U2x coils in Figure 4 is fixed and the current of U1x coils in Figure 4 is changed. The relative force of X-axis between the rotor and the stator is changed, accordingly the rotor moves by this action of the force. In case of Y-axis, it carries out by the same method with V1x, V2x, W1x and W2x coils in Figure 4.

E. Rotation control

The magnetic pole of the stator on one layer which opposes the rotor is fixed N pole or S pole. By moving the tooth of N pole or S pole with the strong magnetic force of the stator, a force works between stator and rotor then the rotor rotates by difference of force on the rotor teeth shown in Figure 11. In Figure 11, the generated force between tooth A of the rotor and tooth R of the stator is largest, therefore the tooth A of the rotor is rotated to front of the tooth R of the stator. Rotation control coils of the stator are arranged to rotate the rotor, and resupplied rectangle wave current one by one as shown in Figure 11. Figure 12 shows the rotation force by calculation with 3D-FEM. The rotate angle by one pulse is 3.75 degrees, therefore 96 pulses are required for one rotation.

V. Conclusion

The tow layer flat monopolar type bearing loess motor using full wave rectification circuit which does not use permanent magnet was proposed. It was shown from the experiment and the simulation result that levitation force can be efficiently acquired by that it is possible to send electric power efficiently with an electric power transmission coil and using inclination of the tooth of the opposite side of the rotor and the stator. We think it is possible to realize the proposed bearing loess motor from these results. After this, a proto type motor will make for experiment, and the details of these proposal will verify.

References

[1] K. Matsuda, T. Kita, T. Masuzawa and Y. Okada, "Passive Stability Analysis of a Radial Type Self-Bearing Motor for Nonpulsatile Type Artifical Heart, " *JSME D&D 2000*, CD-ROM, 2000, (in Japanease).

[2] H. Kanebara, H. Konishi and Y. Okada "Development of Lorentz Force Type Radial Self-Bearing Motor," *JSME D&D 2000*, CD-ROM, 2000, (in Japanease).

[3] K. Oi, D. Matsuhashi and M. Nomura, "Characteristics Measurement of Bearingless Motor with Wide Air Gap Structure,., " *Meidensha Corp. Meidengiho*, no. 311, pp. 35-40, 2011, (in Japanease).

[4] P. Karutz, T. Nussbaumer, W. Gruber and J. W. Kolar, "Novel Magneticaly Levitated Tow-Level Motor," *IEEE Transactions on Mechatronics*, vol. 13, no. 6, pp. 658–668, 2008.

[5] H. Sugimoto, J. Asama and A. Chiba, "Suspension Characteristics of Mult-Consequent-Pole Bearingless Motor with Toroidal Windings," *IEEJTrans.IA*, vol. 132, no. 12, pp. 1112-1120, 2012, (in Japanease).

[6] http://www.neomag.jp/.

[7] http://www.magfine.co.jp/.

[8] http://www.26magnet.co.jp/.

[9] K. Oka, "Bearingless Motor with Rectifier Circuit, " *JoProc. Of the 8th Int. Symposium on Magnetic Bearings*, pp. 271-276, 2002.

[10] H.Aratani, L. Chen and K. Oka, "Development of Bearingless Motor with Rectified Circuit Coil –Study of Rotation Mechanism, " *2004 Shikoku-Section Joint ConventionRecord of the Institutes of Electrical and Related Engineer*, pp. 493-496, 2003, (in Japanease).

[11] K. Tachibana and K. Oka, "Levitaion of the Rotor of Bearingless Motor with Rectified Circuit Coil, " *JSME D&D 2012 CD-ROM*, 2012, (in Japanease).

Robust Speed Controller of a Bearingless AC Motor type Divided Winding, Based on a Conventional Squirrel Cage Induction Motor

Luciano P. dos Santos[a], José S. B. Lopes[a], Valcí F. Victor[b], André L. Maitelli[c] and Andrés O. Salazar[c]

[a] Federal Institute of Education, Science and Technology of Rio G. do Norte, Santa Cruz, RN, Brazil, luciano.junior@ifrn.edu.br
[b] Federal Institute of Tocantins - IFTO - Campus Palmas - 77.015-200 - Palmas - TO – Brazil.
[c] Federal University of Rio Grande do Norte, Natal, RN, Brazil.

Abstract— **This paper shows a new approach to control the speed of rotor field oriented for a bearingless AC Motor type divided winding for a conventional squirrel cage Induction Motor of 3.75 kW, 380V, 4-poles, 60Hz, 1.04 Nm. To implement the control strategies, was chosen the rotor flux referential (R-FOC). This referential is based on decomposition of the stator current into two components done in a decoupled manner: flux controlled by stator direct-axis current and torque controlled by quadrature-axis current. To implement the control algorithm, was chosen the rotor flux referential, which is being estimated through the stator winding currents and the rotor speed. The vectorial speed controller operates in synchronism with the radial positioning controller, but these controllers should work with low mutual influence.**

I. INTRODUÇÃO

The advance of technology has leveraged new challenges increasing order to produce goods and consumer products with a high degree of control, security, robustness, and competitive prices and environmentally correct. In the field of engineering of electrical machine has softly improved the quality of these machines, as well as, control methods, searching optimal operation.

In recent decades, due to the great development of microelectronics and computer engineering and control, rotating electrical machines of alternating current began to have greater demand into industrial market, especially the machines with rotors in cage, due to several factors: robustness, dimensions versus-power, reduced maintenance, low cost and simplicity of construction.

However, one of the major limitations to the expansion of the application of induction machines is the number of maintenance of mechanical parts, such as bearings and roll-tires, especially when access to these parts is difficult [1]. Thus, in order to minimize mechanical wear and, consequently, the number of maintenance, an area of research that has progressed is the engine-bearings of induction. It has been developed works involving contact control and radial-position control using different radial windings disposed in the stator (called double winding). Started in the mid-80s, works were developed involving the control strategies of the machine and its stator winding acting as magnetic bearings in alternating current engine.

In 1990s, was carried out a study about the effect of application of tensional load on the radial positioning of a bearing engine, in 2002, the proposal was the machine without bearing three-phase winding divided vertically for operation. In 2004, there were optimizations system radial positioning of the engine shaft with the same running horizontally, and control is accomplished through the DSP and in 2006 was conducted study that dealt with the mathematical modeling of engine-bearing phase based the vector model of the conventional three-phase induction machine. In 2007, was implemented a system of vector control rotational speed DSP via the engine-bearing three-phase winding divided, using two feedforward multilayer neural networks to estimate the magnitude of the magnetizing current and speed of the rotor flux [2].

The main objective of this paper is to present a robust controller for speed control for a Bearingless AC Motor type Divided Winding, Based on a Conventional Squirrel Cage Induction Motor.

In this work, was chosen the rotor flux reference to run the vectorial speed control. To implement the speed control is necessary to do the rotor flux position and magnitude estimation due to this variable presents a difficult measurement, mainly in this machines type [3], [4], [7], [8].

II. MACHINE DESCRIPTION AND MODELING APROACH

The model of bearingless induction machine (*BIM*) is based on the conventional induction machine model. This fact help the analysis and the attainment of the vectorial model, however the currents generated by the positioning control and the unbalancing influence become the machine modeling more complex.

A. Squirrel-cage machine equivalent circuit

For a balanced three-phase, the analysis of a *BIM* can be done by analyzing only one of the phases. Where $ia1$ is stator current and $i'a1$ is the rotor current refer to stator.

The steady state circuit model for *BIM* is give Fig. 1.

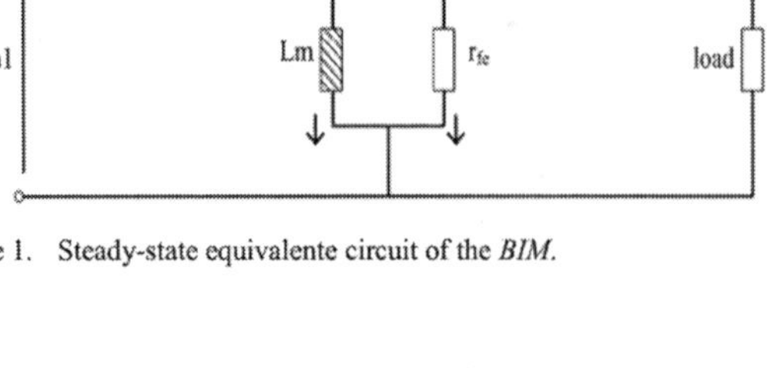

Figure 1. Steady-state equivalente circuit of the *BIM*.

III. THE CURRENT MODEL

To avoid the use the of additional sensors or measuring coils in the *BIM*, method rotor field-oriented-control (*R-FOC*). For generating torque on this machine, is necessary to feed the machine phases with phase currents given by

$$i_{a1} = I_m \cos(\omega_s t) \tag{1}$$

$$i_{b1} = I_m \cos(\omega_s t - 2\pi/3) \tag{2}$$

$$i_{c1} = I_m \cos(\omega_s t + 2\pi/3) \tag{3}$$

$$i_{a2} = I_m \cos(\omega_s t + \pi) \tag{4}$$

$$i_{b2} = I_m \cos(\omega_s t - \pi/3) \tag{5}$$

$$i_{c1} = I_m \cos(\omega_s t + \pi/3) \tag{6}$$

For the rotor flux estimation, is being used the vectorial model of conventional induction machine [3], [4], [7], [8]. For decoupling control, it is desirable that,

$$\Psi_{rd} = \Psi_r \tag{15}$$

$$\Psi_{rq} = 0 \tag{16}$$

Taking into account the previous result, the following equations are defined [3], [4], [7], [8]:

$$\frac{R_r}{L_r}\frac{d\Psi}{dt} + \psi_r \tag{9}$$

$$R_s i_{sd} + \sigma L_s \frac{d}{dt} i_{sd} - \omega_{mR}\sigma L_s i_{sq} - \frac{R_r M^2}{L_r^2}(i_{sd} - i_{mR}) = u_{sd} \tag{10}$$

$$R_s i_{sq} + \sigma L_s \frac{d}{dt} i_{sq} - \omega_{mR}\sigma L_s i_{sd} + \omega_{mR}\frac{M^2}{L_r^2} i_{mR} = u_{sq} \tag{11}$$

$$\frac{d}{dt} i_{mR} = \frac{R_r}{L_r}(i_{sd} - i_{mR}) \tag{12}$$

$$\frac{d\rho}{dt} = \omega_{mR} = p\omega_{mec} + \frac{R_r}{L_r}\frac{i_{sq}}{i_{mR}} \tag{13}$$

$$mM = \frac{2}{3}(1-\sigma)L_s i_{mR} i_{sq} \tag{14}$$

where

σ	dispersion factor
p	number of pole pairs
i_{sd}	stator current *d* component (field current)
i_{sq}	stator current *q* component (current torque)
L_r	rotor inductance
L_s	stator inductance
M	mutual inductance
ρ	rotor flux angle
ω_{mR}	rotor flux frequency
ω_{mec}	rotor speed
mM	Electromagnetic torque
u_{sd}	*d*-axis stator voltage
u_{sq}	*q*-axis stator voltage
i_{mR}	flux magnetizing current
Ψ_r	rotor flux linkage
Ψ_{rd}	*d*-axis rotor flux linkage
Ψ_{rq}	*q*-axis rotor flux linkage

IV. SPEED CONTROL

This paper presents a robust speed controller for an Bearingless AC Motor type Divided Winding, Based on a Conventional Squirrel Cage Induction Motor.

For designing the speed controller, the nonlinear model is linearizing around an operating point (op). This control will be realized by state feedback obtained through linearized model.

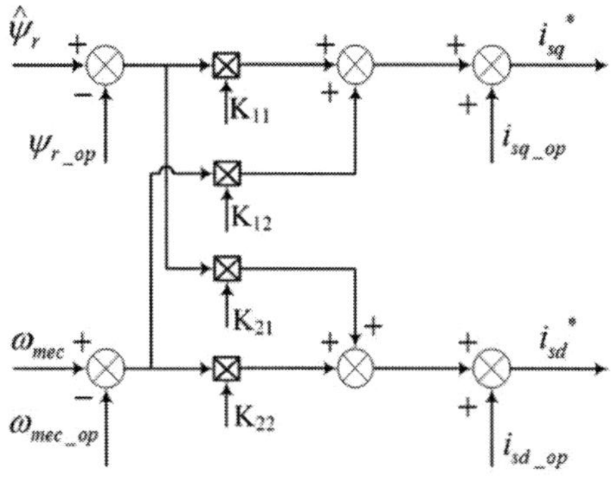

Figure 2. Block diagram of the synchronous PI current controller.

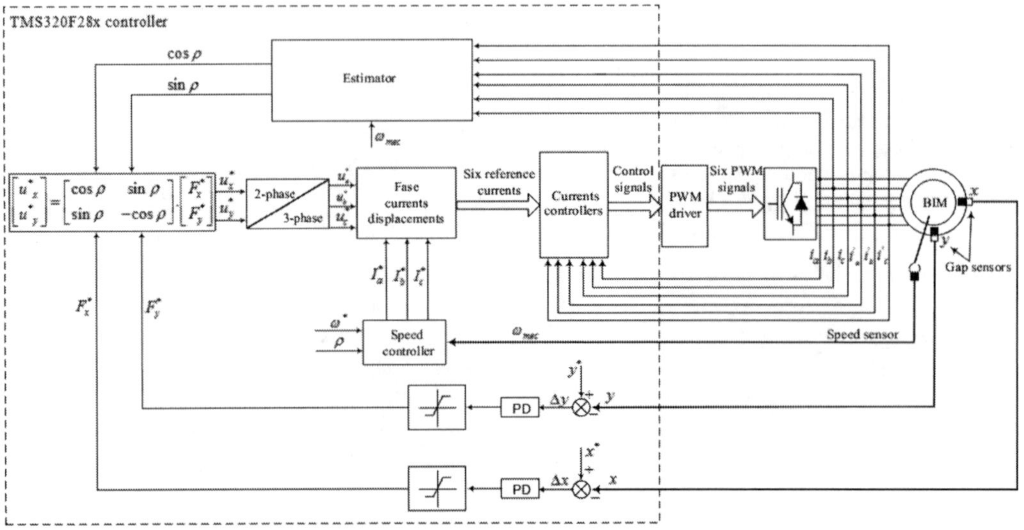

Figure 3. A digital control block diagram of BIM.

The schematically shown controller in Fig. 2 and the architecture of the overall control system is depicted in Fig. 3.

Bearingless motor receives six command currents for motor stator coils. These currents are driven using two parallel IGBTs (Insulated Gate Bipolar Transistors) inverters.

The control of the *BIM* divides in three parts: the current control, the radial positioning control and the speed control. The current controllers are of the type PI and they do the applied currents to the six bearingless machine windings.

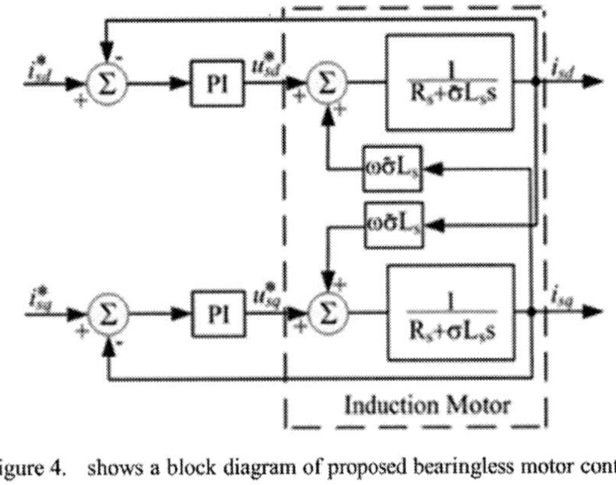

Figure 4. shows a block diagram of proposed bearingless motor control system.

The structure of the PI controller is presented in Fig. 4. To design these controllers is studied a mathematical model based on the impedance of the stator coils, since it is assumed that the currents flowing therein have no significant influence each other [3].

The transfer function of the PI controller is given by

$$G_{PI}(s) = k_{p+}\frac{k_i}{s} \tag{17}$$

The k_p and k_i controller parameters are tuned by (18) and (19) where a value of $\zeta = 0.707$ is chosen and $\omega_n = 631.7$ rad/s.

$$k_p = 2\zeta\omega_n L_s \tag{18}$$

$$k_i = \omega_n^2 L_s \tag{19}$$

The radial position of the rotor shaft is regulated by two digital PD controllers, one for the *x*-axis and another for *y*-axis. Both position controllers are implemented on synchronous reference frame [3], [4], [5], [6].

The ideal transfer function of a PD controller is shown in

$$G_c(s) = k_p + sT_d \tag{20}$$

where, K_p is constant gain of the proportional controller and T_d is time constant of the derivative.

The fig. 5 shows the control system in *x*-direction for this bearingless motor [3], [4].

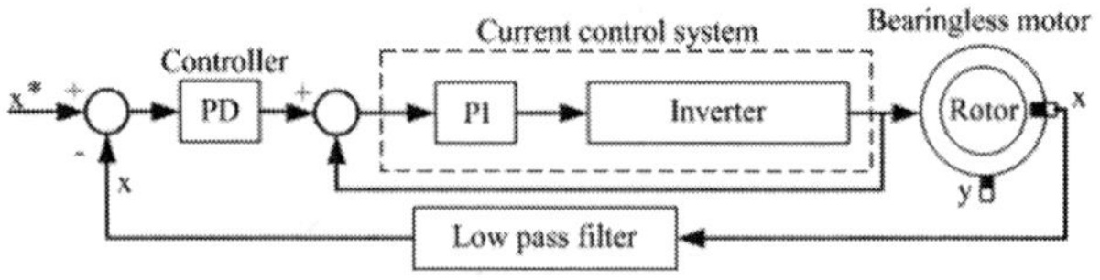

Figure 5. Dynamic model of position.

V. Simulation and Experimental Results

A robust controller is proposed for speed control for a Bearingless AC Motor, due some classical controllers find some difficulties in dealing with the tuning problem.

In order to verify the validity of the proposed controller, the Fig. 6 show the speed response of the robust control.

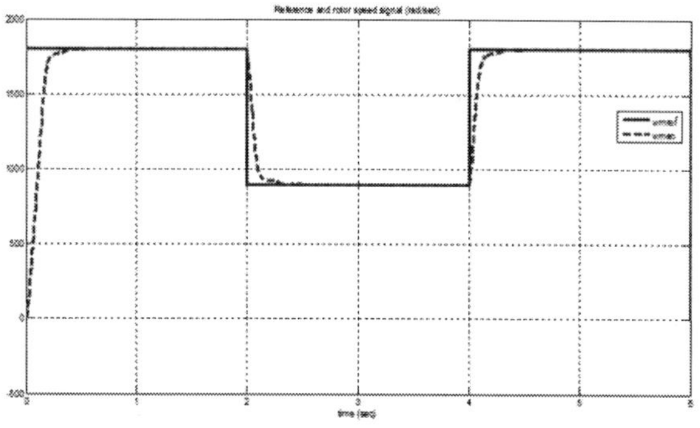

Figure 6. Steady state response.

Fig. 7(a)-(b) present the preliminary experimental results for the proposed BIM motor drive system.

To validate the current control of induction motor drive with PI controller, the following results confirms the satisfactory performance for this controller.

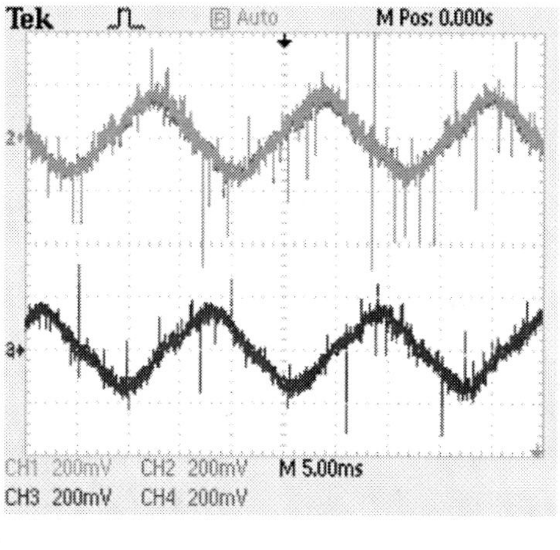

Figure 7(a).Two windings currents of different phases, at 60 Hz.

The Fig.7 (a) shows the currents waveforms in two of the six stator windings corresponding to two different phases, both operating under the nominal frequency of 60 Hz. Fig.7 (b) shows the waveforms the rotor x- and y-axis direction displacements whit frequency equal to 60 Hz.

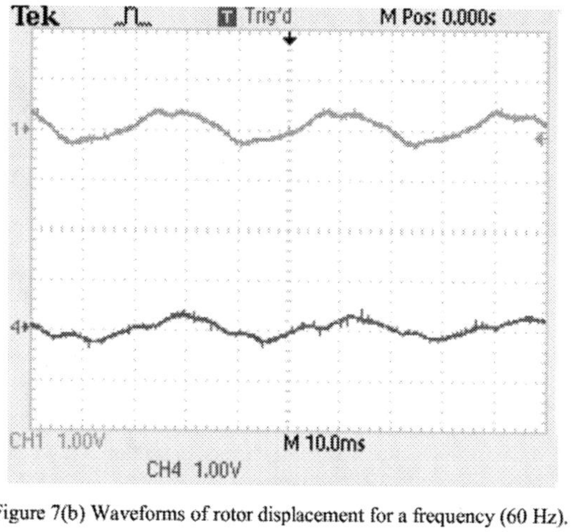

Figure 7(b) Waveforms of rotor displacement for a frequency (60 Hz).

VI. Conclusion

This result shows that with the proposed control strategy keeps the machine rotor in almost the center of the possible radial displacement.

According to obtained results, it was possible to confirm the high performance of the system.

An important result is the independent control of the radial forces in relation to the electric torque of the machine. This specific control was obtained through the generation of minimum induced currents.

In order to reduce the implementation costs, a speed observer been developed. This observer is based on the conventional methods of the state estimation.

Experimental results show that the radial forces are sufficient to control the rotor position on the rotating magnetic field.

References

[1] V. F. Vitor et. al. "Analysis and Study of a Bearingless AC Motor type Divided Winding, Based on a Conventional Squirrel Cage Induction Motor". IEEE Transactions on Magnetics, v. 48, p. 3571-3574, 2012.

[2] A.O. Salazar, R.M. Stephan, "A Bearingless method for induction machine", IEEE Trans.on Magn. Vol.29, N°6, pp.2965-2967, Nov. 1993

[3] A. Chiba, T. Fukao, O. Ichikawa, M. Oshima, M. Takemoto, D. G. Dorrell, 2005 Magnetic Bearings and Bearingless drives Elsevier Oxfor.

[4] W. Leonhard, "Control of electrical drives", Springer-Verlag, Third Edition, Berlin Heidelberg New York, Germany, 2001

[5] J.M.S Ferreira,. F. E. Castro, J. A. Paiva, A. O. Salazar, "DSP utilization in radial positioning control of bearingless machine", ISIE - IEEE International Symposium on Industrial Electronics, 1 , 2003 pp:312 – 317.

[6] A Chiba, K. Chida, and T. Fukao, "Principle and characteristics of a reluctance motor with windings of magnetic bearing", IPEC, Japan, pp.919-926, Apr. 1990

[7] Basak, S.; Ravi Teja, A.V.; Chakraborty, C.; Hori, Y., "A new model reference adaptive formulation to estimate stator resistance in field oriented induction motor drive," *Industrial Electronics Society, IECON 2013 - 39th Annual Conference of the IEEE* , vol., no., pp.8470,8475, 10-13 Nov. 2013.

[8] Fnaiech, M.A.; Khadraoui, S.; Nounou, H.N.; Nounou, M.N.; Guzinski, J.; Abu-Rub, H.; Datta, A.; Bhattacharyya, S.P., "A Measurement-Based Approach for Speed Control of Induction Machines," Emerging and Selected Topics in Power Electronics, IEEE Journal of , vol.2, no.2, pp.308,318, June 2014.

Rotor Flux Observer based Neuro-Fuzzy Techniques to Speed Vector Control of a Bearingless AC Motor

José S. B. Lopes[a], Luciano P. dos Santos[a], José Á. de Paiva[a], Valcí F. Victor[b], André L. Maitelli[c] and Andrés O. Salazar[c]

[a] Federal Institute of Education, Science and Technology of Rio G. do Norte, Natal, RN, Brazil, jose.soares@ifrn.edu.br
[b] Federal Institute of Tocantis - IFTO - Campus Palmas - 77.015-200 - Palmas - TO – Brazil.
[c] Federal University of Rio Grande do Norte, Natal, RN, Brazil.

Abstract— **This study presents the problem of rotor flux orientation control of bearingless induction motor. The key of this solution is the estimation of rotor flux. This work applied an inference system using fuzzy logic and the neural networks with the MATLAB®. The Adaptive Neuro-Fuzzy Inference System (ANFIS) is based in an input-output model. ANFIS is used to tune the membership functions in fuzzy system and utilize the backpropagation learning algorithm. This system was applied to estimate the rotor flux and the magnetization current of the purpose of identifying bearingless induction motor angular speed. The hybrid estimator aims at compensating possible parametric variations of the machine caused by agents such as temperature or nucleus saturation. The simulated results showed good performance. The results demonstrated the feasibility of the proposed technique. The ANFIS estimator proposed will be embedded in the DSP TMS 3208F28335 in future works.**

I. INTRODUCTION

In recent years, there is an increasing interest in bearingless motors around the world. Principally in some industry sectors, such as oil extraction industry, the common problems of maintenance, reliability, efficiency and longtime life issues in conventional electrical machines are leading to the use of bearingless motors [1-2].

Bearingless induction motor is an integration of magnetic bearing with the function of a motor [3]. The Bearingless motors were designed to solve the reducing volume issue of conventional Induction Motor (IM) [5]. In these motors, rotor positioning and torque generation are realized through magnetic forces provoked by stator current control of the motor. So, in this sense, bearingless motors behave as a conventional IM [4].

Research on control of induction motors in recent decades have focused on the improvement of control schemes for field-oriented or vector control, to solve the problems presented by the use of sensors attached to or installed near the rotor [6]. The Flux estimation is an important part in induction machine control system. The flux information is needed in induction machine control for the purpose of synchronous angle and synchronous speed estimation, flux regulation and torque regulation.

Neural Networks (NN) and Fuzzy Logic (FL) are widely used in the areas of modelling, identification, diagnostics and control [9]. In [10] use ANN to reject the influence of speed detection on system stability and precision for a bearingless induction motor. In order to compensate the disadvantages of one system with the advantages of another system, several researchers tried to combine fuzzy systems with Neural Networks [7]. A Neuro-Fuzzy system is simply a Fuzzy inference system trained by a Neural Network-learning algorithm. The Adaptive Neuro-Fuzzy Inference System (ANFIS) [8] combines Fuzzy Logic and Artificial Neural Networks.

The vector control technique needs flux sensors to determine the exact value of the magnitude and of the position of the rotating flow [7]. This limitation was resolved with the flux estimator based on vector control of motor model using as reference the rotor flux vector requiring only the stator phase currents and speed mechanics [11].

ANFIS simultaneously estimates flux speed and magnetization current that defines the rotor flux position (after integrating flux speed) and rotor flux magnitude. The results simulated in the MATLAB® showed satisfactory to estimation of flux with ANFIS due the excellent stabilization of variation parameters of the motor.

II. MODELING OF A BEARINGLESS INDUCTION MOTOR

A three phase, four poles, conventional squirrel cage IM, has its stator winding lodged in adjacent slots. The stator of the induction machine used has divided windings, as shown in Fig. 1.

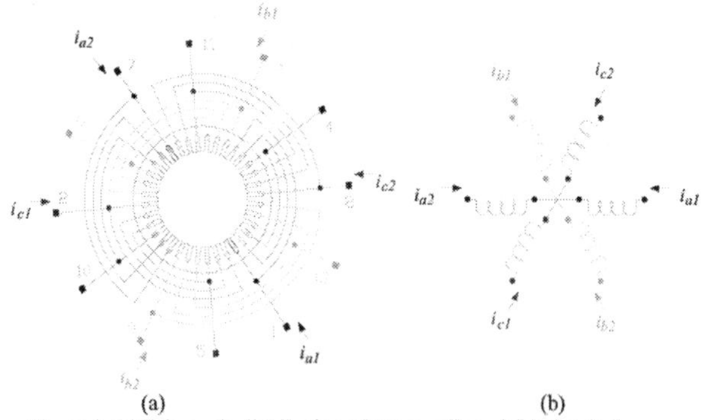

(a) (b)

Figure 1. (a) Schematic distribution of stator coils and (b) circuit diagram.

A. The Flux Model

For the flux estimation or conventional estimator, is being used the vectorial model of conventional induction machine [13]. The applied model is described for the following equations:

$$\frac{di_{mR}(t)}{dt} = \frac{i_{sd}(t)}{T_R} - \frac{i_{mR}(t)}{T_R} \quad (1)$$

$$i_{mR}(t+h) = i_{mR}(t) + h.\frac{di_{mR}(t)}{dt} \quad (2)$$

$$\frac{d\rho(t)}{dt} = \omega_{mec}(t) + \frac{i_{sq}(t)}{T_R i_{mR}(t)} \quad (3)$$

$$\rho(t+h) = \rho(t) + h.\frac{d\rho(t)}{dt} \quad (4)$$

$$m_M(t) = k.i_{mR}(t)i_{sq}(t) \quad (5)$$

$$k = \frac{2}{3}(1-\sigma)L_S \quad (6)$$

where h is the integrative step, $i_{mR}(t)$ is the magnetizing current, $i_{Sd}(t)$ *and* $i_{Sq}(t)$ are the Park currents, $\rho(t)$ is the rotor flux position, *mM(t)* is the electric torque, $\omega_{mec}(t)$ is the rotor mechanical speed, T_R is the rotor time constant, L_s is the stator inductance and σ is the leakage factor.

The use of conventional vector model of induction machines in the study of the bearingless induction machine with divided windings was only possible due to the similarity between their stators structures. Such similarities are equivalent on both models [5] allowing the implementation of the speed and radial positioning controllers.

B. Behavior of the Flux density

It was used a simulator based on the finite element method for acquiring flux density for a model of IM. Fig. 2 shows flux density distribution of the rotor to different current of each phase group.

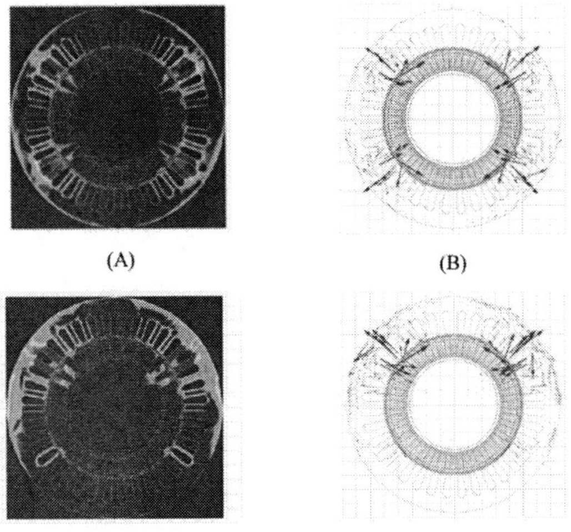

Figure 2. Behavior of the Flux density distribution of the Rotor.

The Fig. 2 (A) and (B) presents the rotor flux vector behavior to the centralized rotor positioning. Rotor magnetic field analysis was realized by means of Maxwell tensor method [12]. Fig. 2 (C) and (D) presents the rotor flux vector behavior to the decentralized rotor positioning. Current were unbalanced to fix the position of the rotor.

III. ANFIS

The chosen Neuro-Fuzzy system is used ANFIS (Adaptive Network-based Fuzzy Inference System) [8]. It adapts the values of inputs and outputs from the base of rules that establish all the input and output connections. Thus, it generates a robust base of rules that creates an Inference System Fuzzy in which contemplates all the possible inputs.

The Neuro-Fuzzy system can be analyzed as Fuzzy Inference, implemented under the architecture of the neural network. ANFIS structure based on the model Takagi-Sugeno first-class. A typical architecture of an ANFIS is as shown in Fig. 3.

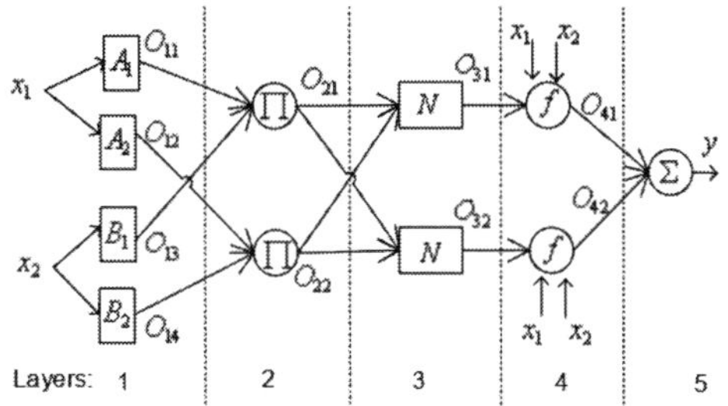

Figure 3. Basic structure of ANFIS.

Layers 1 – Every node i in this layer is a square node with a node function as:

$$O_{1,i} = \mu_{Ai}(x_1), i = 1,2. \quad O_{2,i} = \mu_{Bi}(x_2), i = 1,2. \quad (7)$$

where x is the input to node i, A_i or (B_i) is a linguistic label (such as "*smal*" or "*large*") associated with this node and μ_{Ai} and μ_{Bi} can adopt any fuzzy membership function(MF). Parameters in this layer are referred to as "premise parameters".

Layers 2 – Every node in this layer is a fixed node labeled as ∏, whose output in the product of all incoming signals:

$$O_{2,i} = w_i = \mu_{Ai}(x_1)\mu_{Bi}(x_2), i = 1,2. \quad (8)$$

Layers 3 – Every node in this layer is a fixed node labeled as N. The i node calculated the ratio of the rule's firing strength to sum of all rules firing strengths:

$$O_{3,i} = \overline{w}_i = w_i / (w_1 + w_2), i = 1,2. \quad (9)$$

The output represents the weight of the decision rule.

Layers 4 - The output of the neurons are calculated by the product of the consequences of the rules. Parameters in this layer are referred to as "consequent parameters".

Layers 5 – The single node in this layer is a fixed model labeled Σ that compute the overall output as the summation of all incoming signals:

$$y = O_{5,i} = \sum_i \overline{w}_i f_i \quad (10)$$

This structure can be trained by any learning mechanism used in the Neural Networks [8].

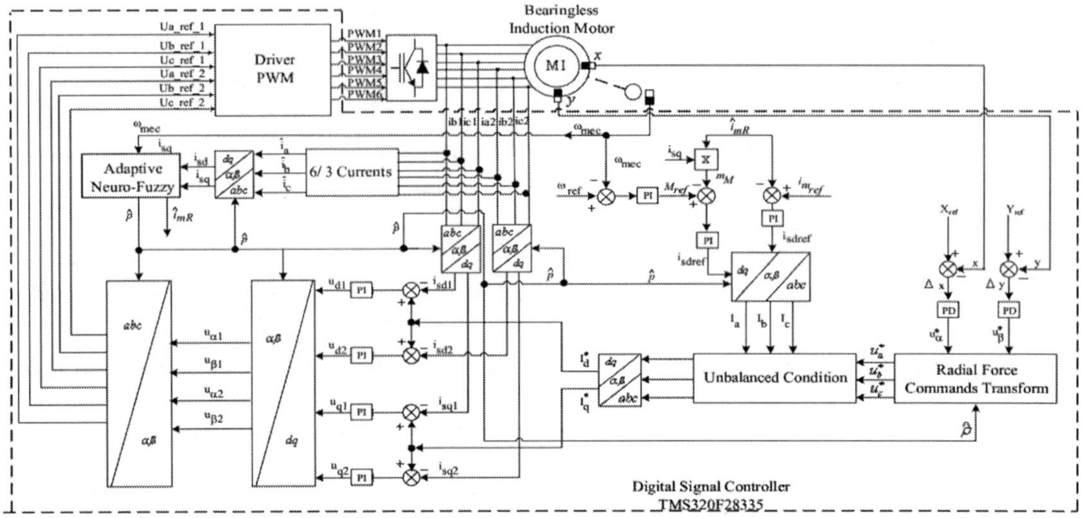

Figure 5. Block diagram of controller with implements DSP TMS 3208F28335.

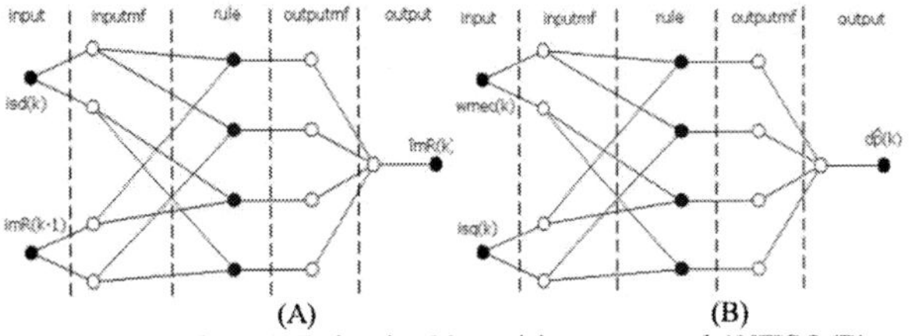

Figure 4. ANFIS 1 (A) -Estimation Magnetizing current and ANFIS 2 (B) - Estimation flux rotor.

The estimator of rotor flux used two ANFIS structures. The first structure estimates the magnetizing current and the second structure estimates was the rotor flux, Fig. 4.

It is possible observes in Fig. 4 that each node of the structure represents a layer ANFIS, Fig. 3. Although we do not have basic knowledge of the parameters adjustments, the initial parameter settings are designed by ANFIS. This system manages to achieve quick adjustments without the need knowledge the induction machine underlying dynamics.

The ANFIS training was performed from the conventional model presented in [6], using the nominal parameters of the induction machine without bearings with split winding which are given in Appendix A, in the condition of the rotor centered. The data of Tables I and II were obtained with ANFIS.

TABLE I. REGARDING TRAINING AND TESTING OF THE ANFIS 1.

MF Shape	*MF per entry*	*Rules*	*Epochs*	*Train.MSE*	*Test.MSE*
trimp	2	4	300	0.000105349	3.46523e-005
trapmf	2	4	300	9.97182e-005	2.79477e-005
gbellmf	3	9	500	7.71389e-005	1.26352e-005
gbellmf	**5**	**25**	**500**	**2.23735e-005**	**1.87042e-005**

TABLE II. REGARDING TRAINING AND TESTING OF THE ANFIS 2.

MF Shape	*MF per entry*	*Rules*	*Epochs*	*Train.MSE*	*Test.MSE*
trimp	2	4	300	0.0040629	0.00489893
trapmf	2	4	300	0.0039423	0.00482686
gbellmf	3	9	500	0.00170528	0.00241691
gbellmf	**5**	**25**	**500**	**0.00119589**	**0.00165979**

IV. SYSTEM DESCRIPTION COMPLETE

The stage of vector speed control was developed with the radial positioning controller and the current controllers. The Fig. 5 shows a block diagram of the proposed system. The radial positioning controller is compose by two PD controllers (Proportional-Derivative). These unbalanced currents try to compensate the position displacements. The current controllers are PI (Proportional-Integrative). The stage of control the rotor position was implemented in the paper [11]. One of the many contributions of this paper is continuity of the work [12] adding speed control based on Neuro-Fuzzy estimator.

V. SIMULATION RESULTS

The software environment used for this simulation was MATLAB® R2010a. The model of the induction machine adopted the balanced three-phase currents and despised the viscous friction of the bearings. These considerations were realized with the goal of approximation the motor conventional model with the model bearingless motor as in [6]. Fig. 6 shows the results of the ANFIS with different spe[illegible]

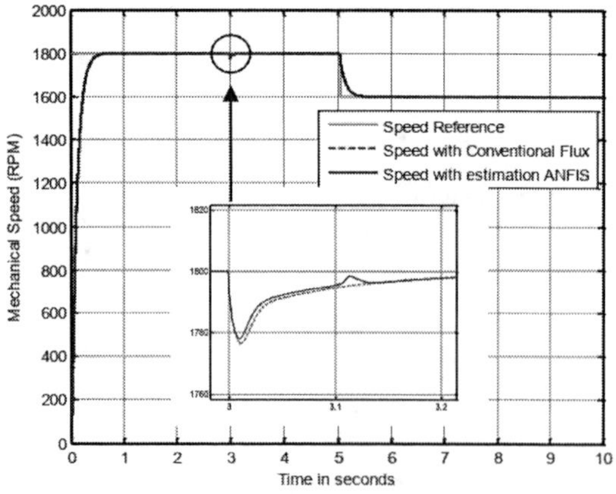

Figure 6. Behavior of speed with conventional flux and anfis.

The range of simulation time was 10 seconds. Fig. 6 show a better zoomed view for analysis of load variation: 0.028 N.m at the time of 3 seconds. Fig. 7 shows the performance of speed control by signal error obtained by conventional model,

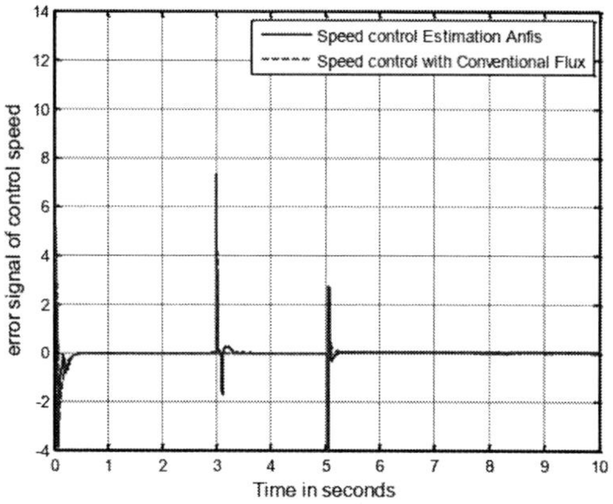

Figure 7. Behavior of error signal of speed control.

Fig. 7 referring to the amplitudes of the errors generated by the control time of 3 seconds while the load was applied in the instant process and 5 seconds in the speed change. Fig. 8

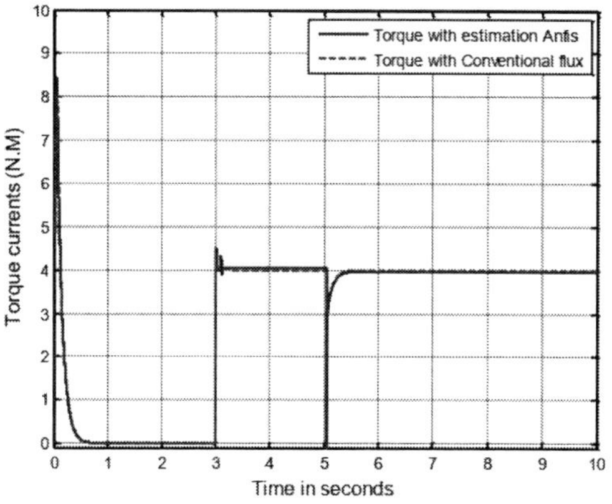

Figure 8. Behavior of torque currents.

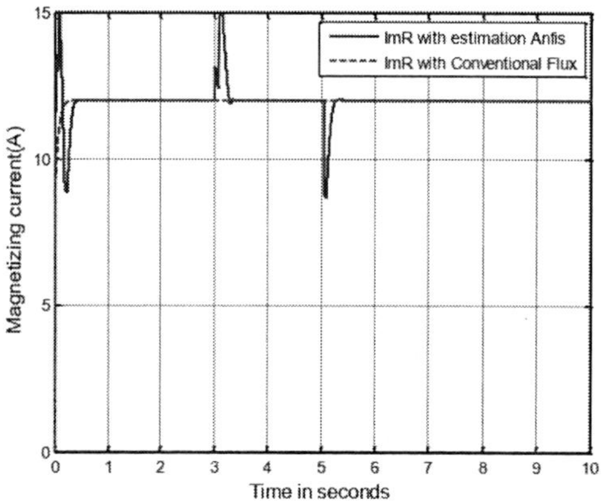

Figure 9. Behavior of magnetizing current.

Fig. 10 and 11 shows the current signal behavior $i_{sd}(t)$ and $i_{sq}(t)$. The signals currents tend to continuous and constant

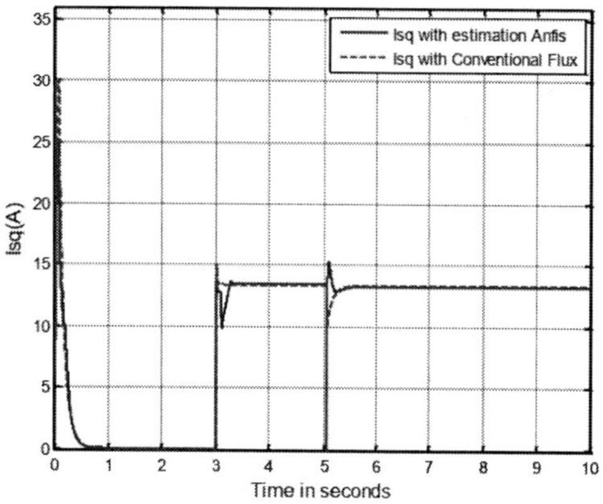

Figure 10. Behavior of quadrature component of stator current.

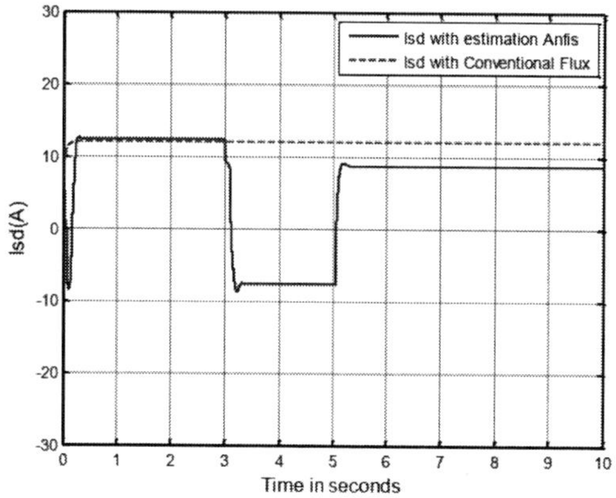

Figure 11. Behavior of direct component of stator current.

The Figure 12 shows the angular position in a zoom range from 0 to 0.4 seconds. Figure 13 shows rotor flux.

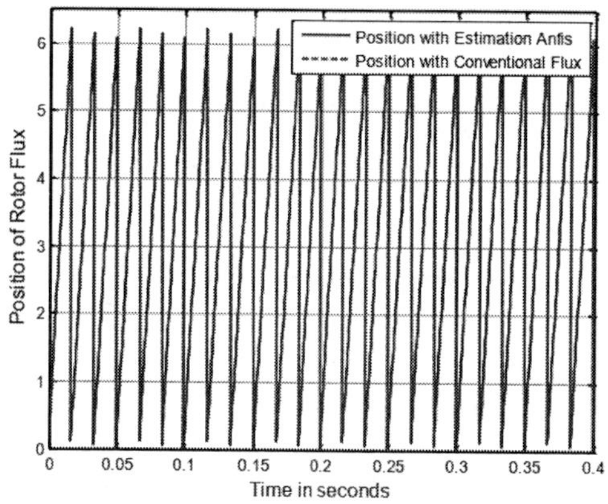

Figure 12. Behavior of position of rotor flux.

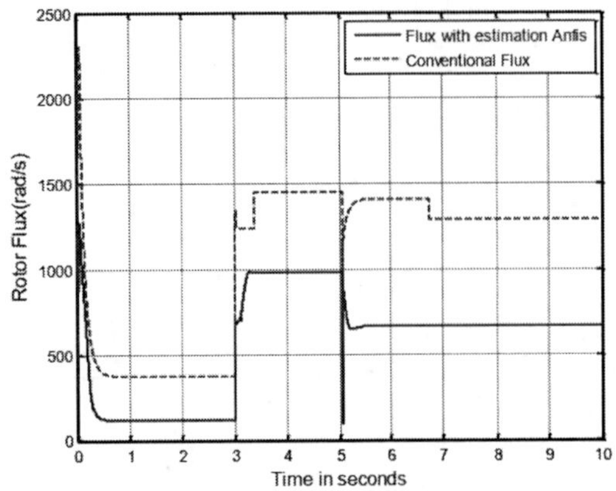

Figure 13. Behavior of rotor flux.

VI. Conclusion

This paper shows the implementation of an inference mechanism system – Neuro-Fuzzy what allows the representation of the knowledge structuralized in rules. It composes the Takagi-Sugeno model. Proposed systems have been simulated using MATLAB. This system used a set of data obtained by simulations.

The considered inference system proved itself capable of assisting or even replacing a human operator during real time process. The learning mechanism revealed itself efficient. The estimator ANFIS implemented was evaluated in close system loops. Results demonstrated the effectiveness of Neuro-Fuzzy estimator. Forthcoming researches will investigate estimator ANFIS in an experimental prototype in the laboratory worked in [12].

Acknowledgment

The authors would like to thank CAPES and CNPq for financial support.

References

[1] V. F. Vitor et. al. "Analysis and Study of a Bearingless AC Motor type Divided Winding, Based on a Conventional Squirrel Cage Induction Motor". IEEE Transactions on Magnetics, v. 48, p. 3571-3574, 2012.

[2] X. Sun et al, "Speed-Sensorless Vector Control of a Bearingless Induction Motor With Artificial Neural Network Inverse Speed Observer", *Mechatronics, IEEE/ASME Transactions on* , vol.18, no.4, pp.1357,1366, Aug. 2013.

[3] Liu Xianxing; Zhu Bin; Zheng Jingwen; Wang Zhengqi, "Air-gap-flux oriented optimized control of bearingless induction motor," Control Conference (CCC), 2011 30th Chinese , vol., no., pp.3549,3553, 22-24 July 2011.

[4] A.O. Salazar, R.M. Stephan, "A Bearingless method for induction machine", IEEE Trans.on Magn. Vol.29, N°6, pp.2965-2967, Nov. 1993.

[5] J. M. S. Ferreira et al. "Analyses of Bearingless Machine with divide windings", IEEE Transactions on Magnetics, vol. 41, No. 10, pp. 3931-3933, Oct. 2005.

[6] J. Á. Paiva,. A. O. Salazar and A. L. Maitelli, "Performance Analysis of a Neural Flux Observer for a Bearingless Induction Machine With Divided Windings". Eletrônica de Potência - Soabraep, Vol 15, pp. 107-114, 2010.

[7] J.S.B. Lopes et. al. "Adaptive Neuro-Fuzzy Inference System for Estimation of Rotor Flux of a Bearingless Induction Motor applied to Speed Control". 1st Brazilian Workshop on Magnetic Bearings, Oct, RJ, Brazil, 2013.

[8] T. S. R. Jang, C. T. Sun, and E. Mizutani, Neuro-Fuzzy and soft computing - A computational approach to learning and machine intelligence. Prentice Hall, 1997.

[9] D. Kukolj, F. Kulic, E. Levi, "Design of the speed controller for sensorless electric drives based on AI techniques: a comparative study", Artificial Intelligence in Engineering, Vol. 14, 165-174, Apr 2000.

[10] Xiaodong Sun; Long Chen; Zebin Yang; Huangqiu Zhu, "Speed-Sensorless Vector Control of a Bearingless Induction Motor With Artificial Neural Network Inverse Speed Observer," Mechatronics, IEEE/ASME Transactions on , vol.18, no.4, pp.1357,1366, Aug. 2013.

[11] E.F Rodriguez and J. A. Santisteban, "An Improved Control System for a Split Winding Bearingless Induction Motor", Industrial Electronics, IEEE Transactions on , vol.58, no.8, pp.3401,3408, 2011.

[12] V. F. Victor et al. "Analysis and Study of a Bearingless AC Motor type Divided Winding, Based on a Conventional Squirrel Cage Induction Motor". In: the 13th International Symposium on Magnetic Bearings (ISMB13), Virginia, 2012.

[13] W. Leonhard, *Control of electrical drives*, Springer-Verlag, Third Edition, Berlin Heidelberg New York, Germany, 2001.

Appendix A

R1	*Stator resistance*	*1,18 Ω*
R2	*Rotor resistance*	*1,42 Ω*
J	*Inertia moment*	*0,00995 kg.m2*
Ls	*Stator Inductance*	*6,56 mH*
Lr	*Rotor inductance*	*6,56 mH*
Lm	*Magnetizating inductance*	*0,14 H*

Ω = ohm, m =milli, H = Henry, mm = millimeter.

Performance Assessment of a Self-bearing Motor: an Application of ISO 14839

José Dalvio Ghirello Garcia[a], Afonso Celso del Nero Gomes[b] and Richard Magdalena Stephan[b]

[a]Instituto Federal de Educação, Ciência e Tecnologia do Rio de Janeiro(IFRJ)
Rua Lúcio Tavares, 1045, Nilópolis/RJ, Brazil, jose.garcia@ifrj.edu.br
[b]Programa de Engenharia Elétrica, COPPE/EE/UFRJ
Caixa Postal 68504, Rio de Janeiro/RJ, Brazil, nero@coep.ufrj.br, richard@coe.ufrj.br

***Abstract* — Two different controllers have been applied for the stabilization of a self-bearing motor prototype existing at LASUP/UFRJ (Laboratory of Applied Superconductivity – Federal University of Rio de Janeiro): a PID controller and a LQR controller. This paper compares the performance of those controllers based on criteria described on the ISO 14839 standard, which establishes requirements for AMB (Active Magnetic Bearing) applied to turbo machinery.**

I. INTRODUCTION

A. *Performance assessment for AMB controllers*

Controllers are essential for the operation of active magnetic bearings, due to their intrinsic instability. As many different control strategies have been applied, some articles have addressed the question on which controller is the best. Various performance criteria have been used, like: analysis of phase plane trajectories [1], sensitivity analisys[2], step response [3], etc.

Standard ISO 14839 has been proposed as a performance evaluation procedure for turbo-machines, aiming to be an answer to the lack of consensus on how to evaluate magnetic bearings applied to turbo machines [4]. This paper addresses this technique, applied to a self-bearing motor.

B. *LASUP self-bearing prototype*

LASUP self-bearing motor has been assembled as shown in Figure 1 below, having two 4-poles 2 phase motors. Although structured to have an upper and a lower radial bearing, only the upper magnetic bearing was used for the experiments, being the lower radial bearing, as well as the axial bearing, implemented by a ball bearing. Phase A of the motor is used for positioning, though generating some torque, while phase B is just applied for torque generation.

C. *Control structure*

For the stabilization of this self-bearing motor, two different controllers were designed and implemented: a PID controller [5] and a LQR controller [6]. Both controllers were realized using a DSP (Digital Signal Processor).

The question posed is "which controller is the best ?". In order to address this point, various performance criteria may be proposed, as mentioned before, provided the objective of the AMB is defined. For the purpose of this paper, the performance evaluation criterium used was the criterium proposed by ISO 14839 [4].

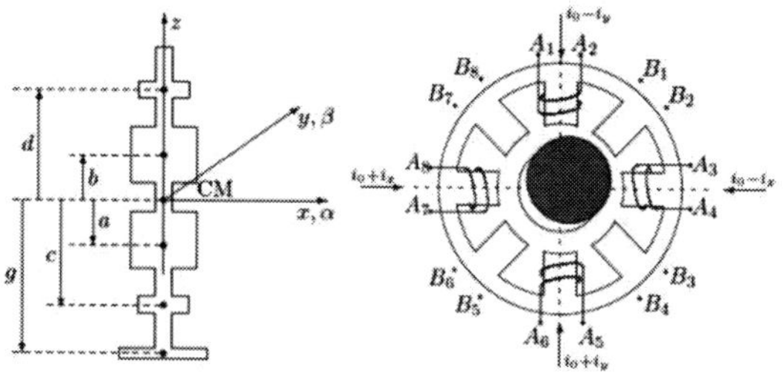

Figure 1. Self-bearing prototype geometry and fase coil arrangement.

II. ISO 14839 PERFORMANCE EVALUATION APPROACH

ISO 14839 Part II proposes a procedure to evaluate the vibration level that naturally arises in the turbo machine operation, defining acceptable displacements, calculated in terms of the minimum clearance established for the considered equipment. For that purpose ISO 14839 uses the definitions of ISO 7919-1 related to vibration zone guidelines for oil-film bearings. The definitions of each zone are as follows.

- ***Zone A:*** *The vibratory displacement of newly commissioned machines would normally fall within this zone.*
- ***Zone B:*** *Machines with vibratory displacement within this zone are normally considered acceptable for unrestricted long-term operation.*
- ***Zone C:*** *Machines with vibratory displacement within this zone are normally considered unsatisfactory for longterm continuous operation. Generally, the machine may be operated for a limited period in this condition until a suitable opportunity arises for remedial action.*
- ***Zone D:*** *Vibratory displacement within this zone is normally considered to be sufficiently severe to cause damage to the machine.*

ISO 14839 sets the maximum displacements in terms of C_{min}, minimum radial clearance, as presented in table I.

ISO 14839 Part III describes the steps for the evaluation of the stability margin, proposing the usage of random signal to be injected in the closed loop, to measure the plant sensitivity transfer function (1). Then, the standard proposes stability indexes based on the sensitivity transfer function, as shown in table II.

TABLE I. DISPLACEMENTS ZONE LIMITS

Zone	Displacement D_{max}
A/B	< 0.3 C_{min}
B/C	< 0.4 C_{min}
C/D	< 0.5 C_{min}

$$G_S(s) = \frac{G_p(s)G_c(s)}{1 + G_p(s)G_c(s)} \quad (1)$$

TABLE II. STABILITY ZONE LIMITS

Zone	Sensitivity peak
A/B	3 (9.5 dB)
B/C	4 (12 dB)
C/D	5 (14 dB)

III. EXPERIMENTAL DATA

A. Data gathering

For the purpose of performance evaluation, a pseudo-random binary signal (PRBS) was generated in the DSP, and made available at a DSP output. This signal was added to the self-bearing motor position measurement, as shown in figure 2.

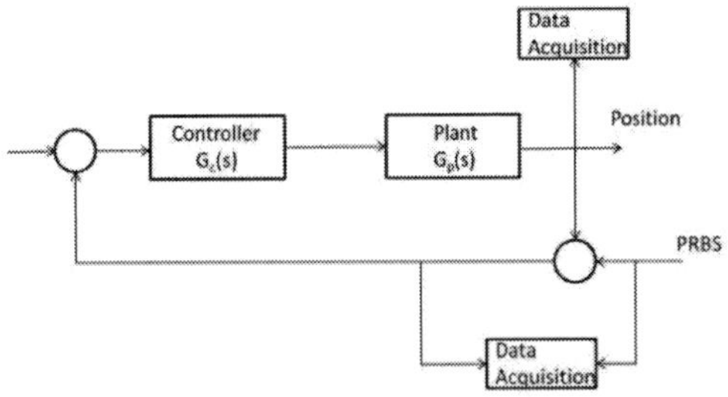

Figure 2. Data acquisiton arrangement

The series of position measurements were recorded, as well as the PRBS testing input and the PRBS added to position signal, in order to generate the performance figures refered by ISO 14839, namelly, the maximum displacement of the rotor axis from the central position, and the sensitivity. Data was recorded on a digital osciloscope, as may be seen in figures 3a and 3b below, showing experimental results for operation at 1800 rpm, both for X axis and Y axis displacements, for sample 1 of three experiments.

B. Performance analysis

In order to apply ISO 14839, experiments were conducted using the LASUP prototype, and time series were acquired for the different experiments. An example of such time series plot, showing its maximum value, is presented in figure 4.

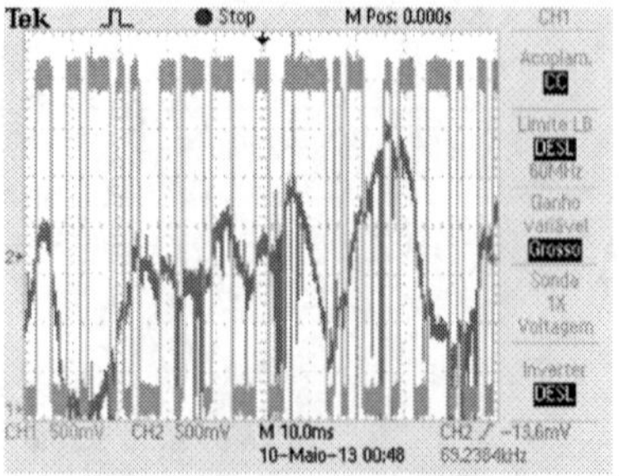

Figure 3a. Time series for X, 1800 rpm. Yellow – PRBS signal. Blue – Output signal

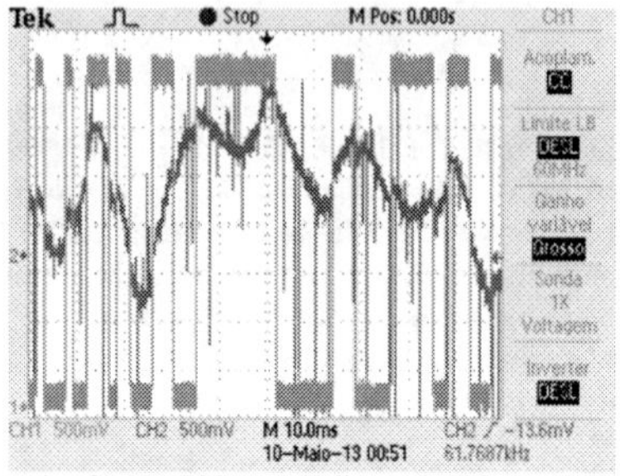

Figure 3b. Time series for Y, 1800 rpm. Yellow – PRBS signal. Blue – Output signal

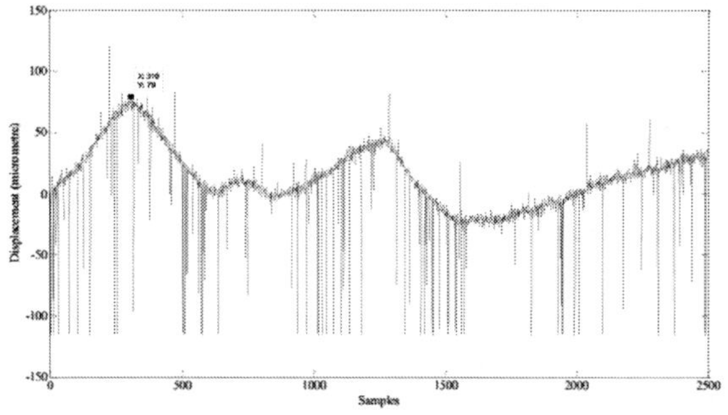

Figure 4. Time series for Y, 1800 rpm

Three experiments were conducted with the self-bearing motor operating at 1800 rpm, and times series were collected for X and Y positions, as shown in figure 2, in order to calculate de maximum displacement, and also series for the PRBS and position added to PRBS were recorded, so that sensitivity transfer function data could be calculated.

IV. PERFORMANCE ASSESSMENT

The analysis of experimental data, according to ISO 14839, was conducted in two steps: first, generate and analise data for the maximum displacement calculation, and then process data for the sensitivity evaluation.

A. Displacement analisys

Experimental data for displacements in the X and Y directions was submitted to MATLAB in order to calculate total displacement as

$$D(k) = \sqrt{(X(k)^2 + Y(k)^2)} \quad (2)$$

Plotting such series, as shown in figure 5, its maximum could be determined for each of the series. This data is shown on table I, for the three experiments using each controller.

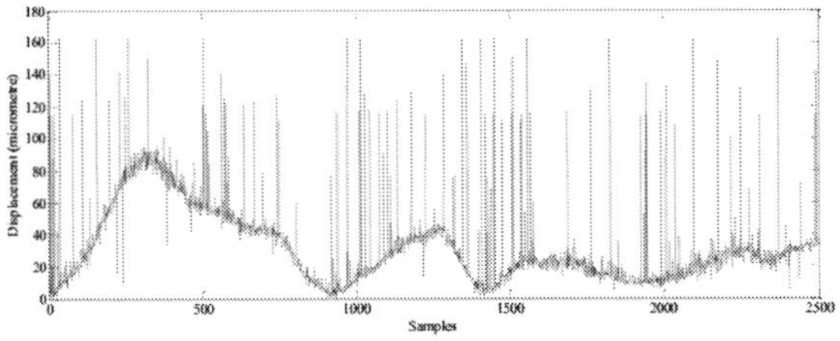

Figure 5 Displacement D(k) for LQR control at 1800 rpm

TABLE III. DISPLACEMENT FIGURES

Experiment	Maximum displacement (μm)
LQR 1800/1	74
PID 1800/1	94
LQR 1800/2	86
PID 1800/2	90
LQR 1800/3	68
PID 1800/3	87

This result can be graphically displayed, as in figure 6 below, showing that, regarding this figure of merit, the LQR controller has a better performance.

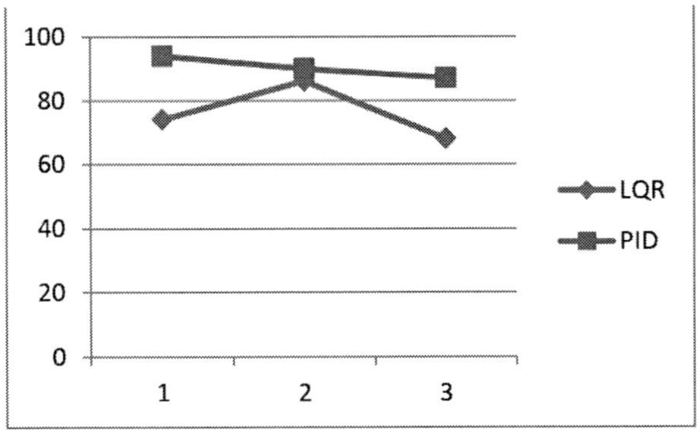

Figure 6 Maximum displacement, at 1800 rpm

B. Sensitivity analisys

Plant sensitivity, as defined by equation (1), was obtained by submitting experimental time series, for the three experiments, to MATLAB *etfe()* function in order to develop sensitivity Bode diagrams. Such Bode diagrams, for both controllers, are presented in figure 7. In the same figure, zone limits, defined in the standard ISO 14839, are shown as colored straight lines.

Those Bode diagrams show that LQR controller satisfies the gain limit for C/D zone, while PID controller does not satisfy the stability margin for none of the zones.

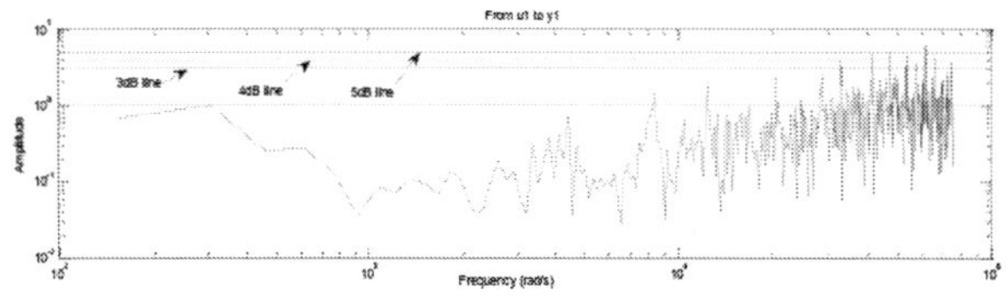

Figure 7a. Sensitivity for LQR serie, showing stability zone limits

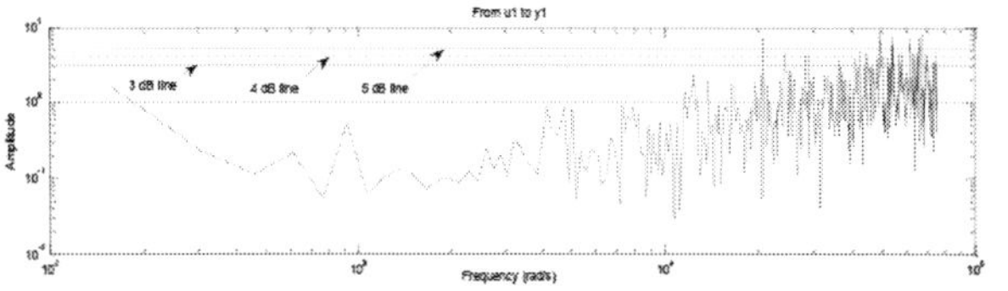

Figure 7b. Sensitivity for PID serie, showing stability zone limits

V. PERFORMANCE CRITERIA AS TUNNING CRITERIA

Previous paragraphs have led to the conclusion that the PID controller has, regarding the maximum displacement value, presented a performance poorer than the performance obtained with the LQR controller. A natural question arises: could the PID controller by retuned to display a better performance?

With that in mind, the self-bearing motor, controlled by the PID algorithm, was operated with different values for the PID proportional gain, beginning with the original gain, and the experimental data was analyzed regarding the maximum displacement.

Table III shows the displacement figures, and figure 8 displays graphically the performance obtained for each tested proportional gain.

TABLE IV. DISPLACEMENT FIGURES

Proportional gain	Maximum displacement (μm)
50	73
80	71
100	68
120	65

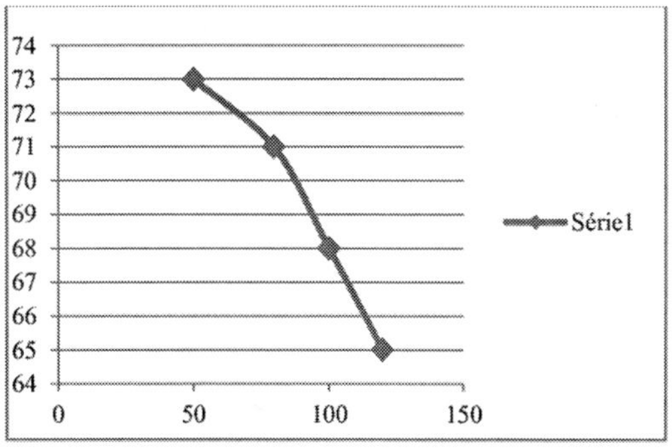

Figure 8 Maximum displacements for various PID proportional gains

As may be observed, increasing the proportional gain, the displacement is reduced, thus displaying a better performance for the plant. Nevertheless, a question should be posed regarding the stability of the new controller parameter: is the system still stable? Has the stability margin worsened with the new gain?

The ISO 14839 criteria address this sensible question by analyzing not only the error, but also the system stability. For the larger gain, the sensitivity Bode diagram was also obtained, and is shown in the figure 9.

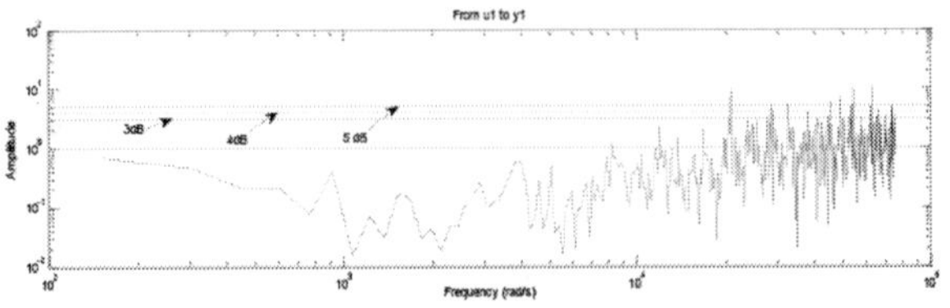

Figure 9a Sensitivity Bode diagram for original controller, showing stability zone limits

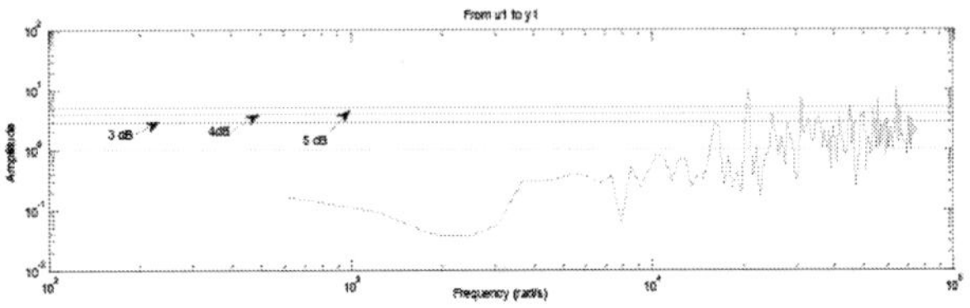

Figure 9b Sensitivity Bode diagram for optimized controller, showing stability zone limits

As can be observed, the stability margin has been reduced but not drastically, the system remaining stable for this point of operation.

VI. CONCLUSION

For the LASUP prototype, with position control applied to the upper rotor, running at 1800 rpm, the LQR controller shows a better performance than the PID controller, when considering the requirements of the ISO standard 14839. Frequency domain data shows that both controllers should be retuned in order to attain a better stability margin, according to the mentioned ISO norm.

A retuning of the PID controller, done in order to obtain better results in the sense of a reduction of maximum displacements, has shown to be feasible. Nevertheless, caution should be taken, as stability may be affected by this procedure. Ideally, one should consider a continuous monitoring of plant performance, calling the attention of operators, should the controller require retuning.

It should be stressed that those requirements are set for AMB´s applied to turbo machinery, meaning that different applications may require different performance requirements, as stressed by the very ISO 14839 standard.

REFERENCES

[1] Ahn, H-J., Han, D_C.: *System Modeling amd Robust Control of an AMB Spindle: Part II A Robust Controller Design and its Implementation*" In KSME International Journal, V. 17, N.12, pp 1855-1866, 2003.

[2] Jastrzbski,R.P.,Pollanen, R.: *Centralized optimal position control for active magnetic bearings: comparison with decentralized control* In Electr. Eng. 91, pp101-114, 2009.

[3] Jastrzbski,R.P., Hynynen, K.M.,Smirnov, A.: *Uncertainty Set, Design and Performance Evaluation of Centralized Controllers for AMB* Systems In: XII International Symposium on Magnetic Bearings, 2010, Wuhan.

[4] Matsushita, O., Kanemitsu, Y., Itou, M. et al: " *ISO Standardization for Active Magnetic Bearing Technology". Available at* http://www.nedo.go.jp/content/100084729.pdf

[5] Gomes, R.R., Stephan, R. M., Santisteban, J. A.: *Self-Bearing Motor with DSP Based Control System* In: X International Symposium on Magnetic Bearings, 2006, Martigny.

[6] Kauss, W. L., Gomes, A. C. N., Stephan, R. M., David, D.:*LRQ Control of a Bearingless Machine Implemented with a DSP* In: XI International Symposium on Magnetic Bearings, 2008, Nara.

[7] Mushi, S. E., Lin, Z., Allaire, P.E.: Stability Analysis for Flexible Rotor on Active Magnetic Bearings subject to Aerodynamic Loads. In XII International Symposium on Magnetic Bearings, 2010, Wuhan.

Vibration Suppression of A Middle-Point Current-Injection Type Bearingless Motor at 50000r/min

Yusuke Uemura, Takayuki Nakahata, Akira Chiba*

Tokyo Institute of Technology, 2-12-1 Ookayama Meguro, Tokyo, Japan

Abstract— **With the developments of power electronics technology and increments of rotational speed of motors, a critical speed is one of the problems. Vibration of rotational shaft has been reported to be suppressed by magnetic force of bearingless motors. In general bearingless motors, motor windings and suspension windings are separately enclosed in the stator slot. For only vibration suppression at a critical speed, however, suspension windings are not used except the critical speed. In this paper, vibration suppression of a rotor shaft by a middle-point current-injection type bearingless motor at 50000 r/min is presented. This motor can generate the torque and the magnetic force by only one set of three-phase windings. The test results of the vibration suppression are presented.**

I. Introduction

In recent years, the developments of power electronics technology and the appearances of strong permanent magnets and iron materials with low iron losses make possible to increase a motor rotational speed. High speed motors and generators are applied to turbines, superchargers and compressors. The critical speed is one of the serious problems in the high speed motor. The vibration of a rotor shaft at the critical speed is caused by resonance of the shaft natural frequency. In the worst case, the rotor shaft may be broken. Thus, various solutions for the vibration suppression have been studied.

To solve the problems, bearingless motors have been proposed. The bearingless motors combine a magnetic bearing and a motor function [1-14]. The advantages are low cost and small size compared with magnetic bearing motors. The Finland group and the authors reported the vibration suppression and operation over the critical speed with a bearingless motor [15-16]. In these bearingless motors, two sets of windings are installed in the stator slots. One set is the motor winding for the torque generation, and another set is the suspension winding for the radial magnetic force generation. However, in the vibration suppression, the suspension winding is not excited except the critical speed. Furthermore, the requirement of the additional suspension conductors causes torque reduction, or increased stator dimensions.

In the literatures, there are some bearingless motors which generate the torque and the magnetic force by only one set of three-phase winding. The bridge configured winding type bearingless motor [17] is one of the novel structures. The winding coils are separated and radial force currents are injected in the bridge circuit structure. This structure needs three units of a single-phase inverter for x- and y-axis force generation. Therefore, the cost of the bearingless motor driver is high. A simple homopolar winding structure for an induction motor bearingless motor has been proposed [18]. This winding structure is simple, however, an additional homopolar winding is necessary in addition to a three-phase winding. The authors have proposed a middle-point current-injection type bearingless motor [19]. The proposed bearingless motor has simple structure and does not require an additional winding because vibration suppression inverter is connected to the middle-points in the motor winding. In addition, the motor driver and a controller do not need modification. Suppression vibration of rotor shaft and operation over the critical speed (7200 r/min) was reported. The principle of this bearingless motor applied at a rotational speed up to 10000r/min. In [19-20], a controller with a DSP is applied, however, the calculation was not fast enough for the target speed of 100,000r/min.

In this paper, the principle of a middle-point current-injection type bearingless motor is applied to a high-speed prototype machine. The controllers are based on analog circuits and a FPGA as well as a microcomputer. The vibration suppression is confirmed at a critical speed of 50000r/min.

II. Winding Structure

Fig. 1 shows an example of the stator winding structure for two-pole SPM rotor in the proposed middle-point current-injection type bearingless motor. Fig. 2 shows the connection between the coils and the three-phase inverters. In the middle-point current-injection type bearingless motor, only one set of three-phase winding is installed in the stator slots. The U, V and W phase windings are installed by every 120 degrees so that 2-pole three-phase windings are built. The U, V and W phase motor currents i_{MU}, i_{MV} and i_{MW} generate 2-pole magnetic flux in the air gap to generate torque. On the other hands, the suspension currents i_{SU}, i_{SW} and i_{SV} are injected to the middle-points of each phase in order to generate 4-pole magnetic flux in the air gap to generate radial magnetic force. Consequently, both the motor and suspension currents are provided in the windings of U_2, V_2 and W_2.

*corresponding author

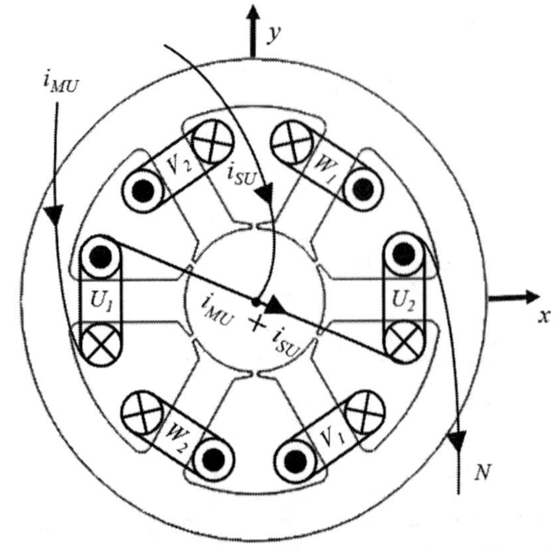

Fig. 1. Stator structure of the proposed middle-point current-injection type bearingless motor.

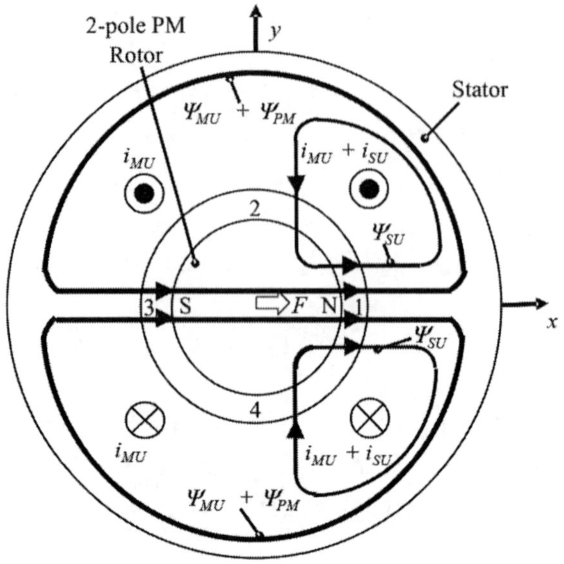

Fig. 3. Principle of radial force generation with the proposed middle-point current-injection type bearingless motor.

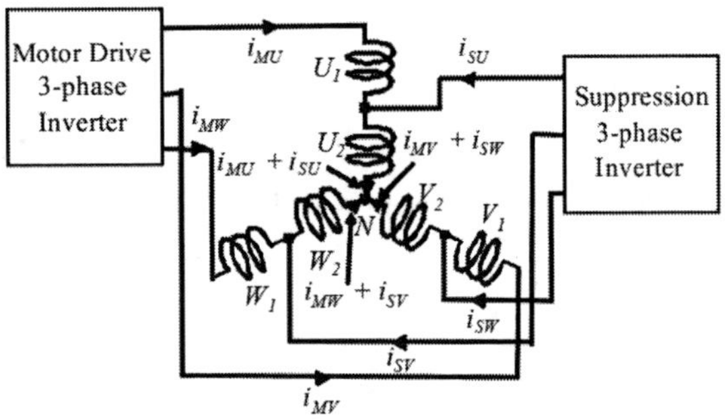

Fig. 2. Connection of windings and inverters in the proposed middle-point current-injection type bearingless motor.

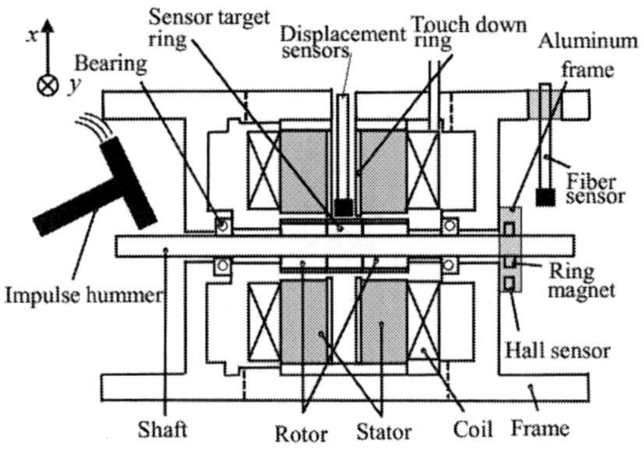

Fig. 4. Structure of the high speed test machine.

III. Principle of Radial Magnetic Force Generation

Fig. 3 shows principle of radial force generation with the proposed middle-point current-injection type bearingless motor. Only U-phase winding is illustrated. In the left-side conductors U_1, only the motor current i_{MU} is provided. On the contrary, in the right-side the suspension current i_{SU} is added to the motor current i_{MU}. As the result, the 2-pole flux $\Psi_{MU}+\Psi_{PM}$ is generated by the 2-pole motor current and the rotor PM. In addition, 4-pole flux Ψ_{SU} is generated in the right half side by the suspension current. The $\Psi_{MU}+\Psi_{PM}$ and Ψ_{SU} are the same direction in the air gap 1, and thus, air gap magnetic flux density is increased. Therefore, the radial suspension force is generated in the positive x-direction because the flux density in the air gap 1 is high compared with that in the air gap 3. When the i_{SU} is negative, the radial suspension force is generated in the negative x-direction because the flux density in the air gap 1 is low compared with that in the air gap 3.

IV. Experimental System

A. Prototype Machine Structure

Fig. 4 shows the cross-sectional view of a test machine. The both ends of the rotational shaft is suspended by the ball bearings. The stator consists of two iron cores with only two coil ends. Three displacement sensors are installed between the stator cores. A touch-down ring is installed in the stator to protect the eddy current type sensors. The touch-down gap length between the touch-down ring and the sensor target is 500 μm. Two hall sensors are installed to aluminum frame

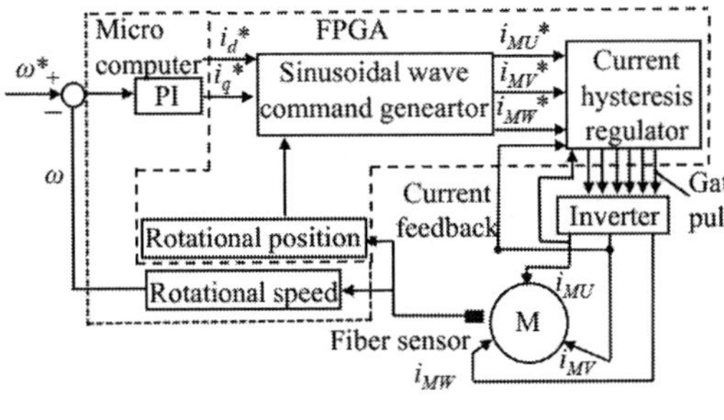

Fig.5. Block diagram of the motor drive system.

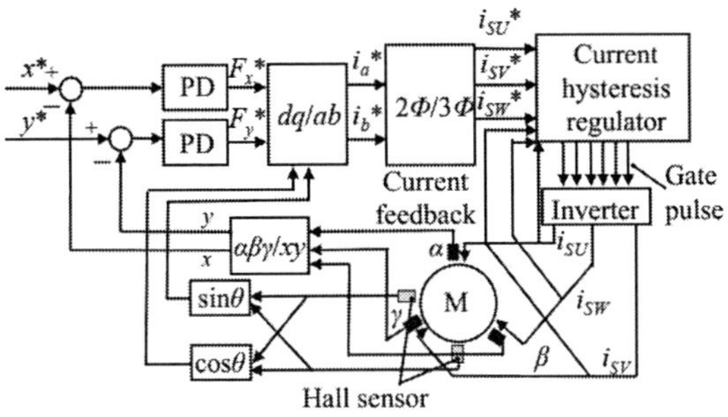

Fig.6. Block diagram of the vibration suppression system.

fixed to the right end. A fiber sensor for motor speed regulation is also installed, however, this is not necessary.

B. Motor Drive And Vibration Suppression System

Fig. 5 shows a block diagram of the motor drive system. The PI regulator and rotational speed calculation are realized in a microcomputer. A sinusoidal wave command generation, a current hysteresis regulator and a rotational position calculation are realized in the FPGA. The rotational speed signal ω is compared with the speed reference ω^*. The error signal is amplified by a proportional-integral (PI) regulator, and torque current reference i^*_q is generated. Based on the rotor rotational angular position, the current references i^*_q and i^*_d are transformed into sinusoidal current references i^*_{MU}, i^*_{MV} and i^*_{MW}. In the current hysteresis regulator, the current references i^*_{MU}, i^*_{MV} and i^*_{MW} are compared with the detected currents. Gate pulse signals are generated in the current hysteresis regulator. The PWM signals are given to the motor inverter, and the three-phase motor currents are regulated.

Fig. 6 shows the block diagram of the vibration suppression system. Analog circuits are adapted in the vibration suppression regulation system. Three displacement sensors are installed every 120 degrees in the stator. The detected signals are transformed into two perpendicular displacements x and y. The position references x^* and y^* are zero. The error signals between the references and the detected feedback signals are input in the proportional-

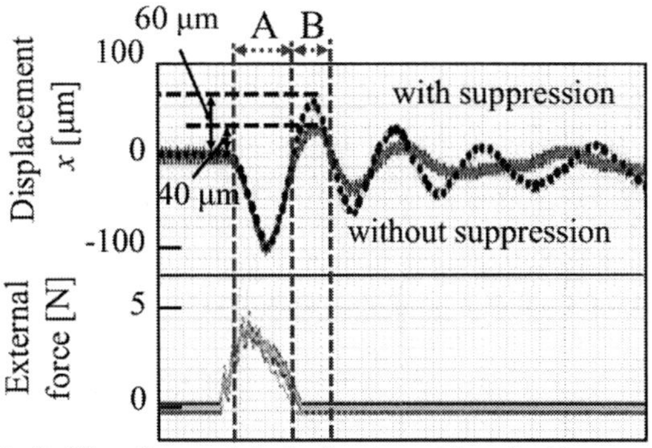

Fig.7. Waveforms of shaft vibration with and without suppression regulation.

derivative regulators, and then, radial magnetic force references F^*_x and F^*_y are generated. The radial magnetic force references are transformed into three-phase current references. The three-phase current references i^*_{SU}, i^*_{SV} and i^*_{SW} are given to the current hysteresis regulator. Thus, suspension currents are injected at the middle point in the motor main windings.

V. Experimental Results

Fig. 7 shows waveforms of the shaft vibration and external force with and without suppression feedback regulation in static condition. The vibration is given by an impulse hummer, which provides force at the end of shaft. The impulse hummer has output terminals to observe the applied external force. The vibration is measured by displacement sensors. In this experiment, the external force is about 5 N. The effect of vibration suppression regulation does not appear in the section A because the impulse hammer is touching to the rotor shaft. The maximum displacement without the suppression regulation in the section B is 60 μm. On the other hands, the maximum displacement with the suppression regulation in the section B is 40 μm. The vibration in the section B is decrease by 33 %.

Figs. 8 (a) and (b) show the waveforms of the vibration with and without the suppression regulation, respectively, at the rotational speed of 50000 r/min. The vibration peak to peak amplitudes without suppression regulation are 400 μm in the x- and y-axes. When suppression regulation is applied, the vibration peak to peak amplitudes are reduced to 320 μm in the x- and y-axes. Thus, the vibration is decreased by 20 %, respectively. The effectiveness is limited, however, better mechanical balancing is needed for future experiments.

VI. Conclusion

In this paper, development and experimental results of vibration suppression of 50000r/min test motor have been described. The vibration is suppressed by the current injection at the middle points of motor windings. The injected current make the magnetic field unbalanced, thus, radial force can be generated in two perpendicular axes. By detecting the rotor

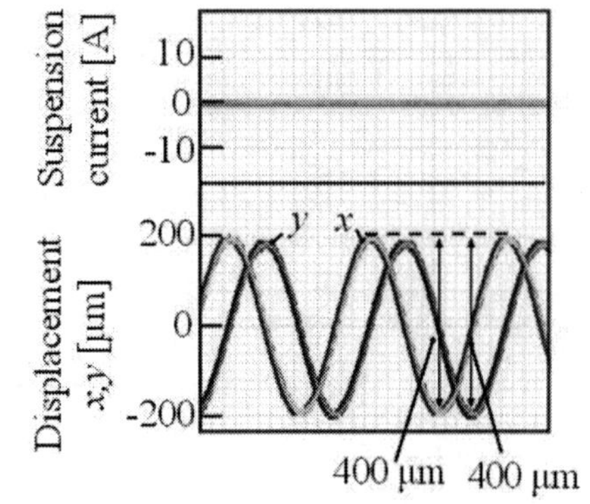

(a)Without suppression.

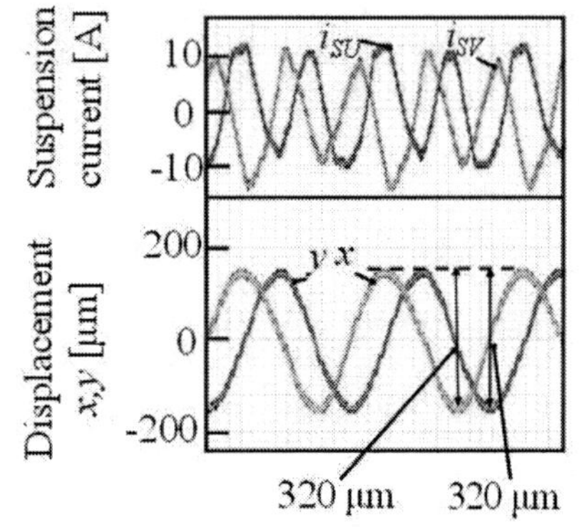

(b)With suppression.

Fig.8. Waveforms of shaft vibration with and without suppression regulation at rotational speed of 50000 r/min.

vibration, feedback controllers have been developed. The effectiveness has been verified in principle.

REFERENCES

[1] Ribeiro, R.L.A.; Castro, F.E.F.; Salazar, A.O.; Maitelli, A.L.; , "A Suitable Current Control Strategy for Split-Phase Bearingless Three-Phase Induction Machine," Power Electronics Specialists Conference, 2005. PESC '05. IEEE 36th , vol., no., pp.701-706, 16-16 June 2005

[2] Gruber, W.; Amrhein, W.; Haslmayr, M.; , "Bearingless Segment Motor With Five Stator Elements—Design and Optimization," Industry Applications, IEEE Transactions on , vol.45, no.4, pp.1301-1308, July-aug. 2009

[3] Gruber, W.; Nussbaumer, T.; Grabner, H.; Amrhein, W.; , "Wide Air Gap and Large-Scale Bearingless Segment Motor With Six Stator Elements," Magnetics, IEEE Transactions on , vol.46, no.6, pp.2438-2441, June 2010

[4] J. Asama, R. Kawata, T. Tamura, T. Oiwa, and A. Chiba, "Reduction of Force Interference and Performance Improvement of a Consequent-Pole Bearingless Motor", Precision Engineering, Vol.36, Issue 1, pp. 10-18, January 2012.

[5] Akira Chiba, Tadashi Fukao, Osamu Ichikawa, Masahide Oshima, Masatsugu Takemoto and David G. Dorrell, "Magnetic Bearings and Bearingless Drives", Newnes Elsevier, ISBN 0 7506 5727 8, 2005

[6] Akira Chiba, Tazumi Deido, Tadashi Fukao and M.A.Rahman, "An Analysis of Bearingless ac Motors", IEEE Transactions on Energy Conversion, vol.9, no.1, pp.61-68, 1994

[7] J. Asama, Y. Hamasaki, T. Oiwa, and A. Chiba, "A Novel Concept of a Single-Drive Bearingless Motor", Proceedings of the IEEE International Electric Machines and Drives Conference (IEMDC) 2011, Niagara Falls, 15 –18 May, 2011.

[8] S. Silber, et al., "Design aspects of bearingless slice motors", IEEE/ASME Trans. Mechatronics, vol. 10, no. 6, pp. 611 – 617, Dec. 2005.

[9] Z. Ren and L. S. Stephens : "Closed-Loop Performance of a Six Degree-of-Freedom Precision Magnetic Actuator", IEEE/ASME Trans. Mechatronics, Vol.10, No.6 pp-666-674 (2005)

[10] H. Kanebako, Y. Okada, "New design of hybrid-type self-bearing motor for small, high-speed spindle," IEEE Trans. Mechatronics, vol.8, no.1, pp. 111-119, Mar. 2003.

[11] J. Amemiya, A. Chiba, D. G. Dorrell and T. Fukao, "Basic Characteristics of a Consequent-Pole-Type Bearingless Motor," IEEE Trans. Magn., vol. 41, no. 1, pp.82-89, Jan. 2005.

[12] T. Schneeberger, T. Nussbaumer and J. W. Kolar, "Magnetically Levitated Homopolar Hollow-Shaft Motor," IEEE/ASME Trans. Mechatronics, vol. 15, no. 1, pp. 97 – 107, Feb. 2010.

[13] J. Asama, T. Asami, T. Imakawa, A. Chiba, A. Nakajima and M. A. Rahman, "Effects of Permanent-Magnet Passive Magnetic Bearing on a Two-Axis Actively Regulated Low-Speed Bearingless Motor," IEEE Trans. Energy Conversion, vol. 26, no. 1, pp. 46 – 54, Mar. 2011.

[14] N. Watanabe, H. Sugimoto, A. Chiba, T. Fukao, "Basic Characteristic of the Multi-Consequent-pole Bearingless Motor," in Proc. PCC Nagoya, pp.1565-1570, 2-5 Apr. 2007.

[15] Antti Laiho, Anssi Sinervo, Juha Orivuori, Kari Tammi, Antero Arkkio and Kai Zenger : "Attenuation of harmonic rotor vibration in a cage rotor induction machine by a self- bearing force actuator", *IEEE Trans. Magnetics 2009*, vol.45, No.12, pp.1-2 (2009)

[16] Akira Chiba, Tadashi Fukao, M. Azizur Rahman : "Vibration Suppression of a Flexible Shaft With a Simplified Bearingless Induction Motor Drive", *IEEE Trans. Industry Applications*, vol.44, Issue.4, pp.745-752 (2008).

[17] W. K. S. Khoo, Kalita K, Garvey S.D. : "Practical Implementation of the Bridge Configured Winding for Producing Controllable Transverse Forces in Electrical Machines", *IEEE Trans. Industry Magnetics*, vol.47, No.6, pp.1712-1718 (2011)

[18] A. Sinervo, T. Jokela, A. Arkkio,"Controlling Rotor Vibrations of a Two-Pole Induction Machine With Unipolar Actuator", IEEE Transactions on Magnetics Vol.48, no.7, 2012, pp.2205 - 2210

[19] A. Chiba, K.Sotome, Y. Iiyama, M. A. Rahman : "A Novel Middle-Point-Current-Injection-Type Bearingless PM Synchronous Motor for Vibration Suppression", *IEEE Trans. Industry Applications*, vol.47, Issue.4, pp.1700-1706 (2008)

[20] Ryohei Oishi, Satoshi Horima, Hiroya Sugimoto and Akira Chiba, "A Novel Parallel Motor Winding Structure for Bearingless Motors", IEEE Transactions on Magnetics, vol.49, no.5, May 2013, pp. 2287-2290

Characteristics of Torque and Suspension Force in a d-q Axis Current Control Bearingless Motor

Masahide Ooshima[a], Yuto Gomi[a]

[a] Science University of Tokyo, Suwa, 5000-1, Toyohira, Chino, Nagano, JAPAN, moshima@rs.suwa.tus.ac.jp

***Abstract*—This paper presents the characteristic of torque and suspension force in a d-q axis current control bearingless motor. It is derived based on Finite Element (FE) analysis using a simulation software, and furthermore compared with the computed results of the conventional bearingless motors, which two kinds of windings for motor drive and magnetic suspension are wound in a stator core. Three types of the conventional bearingless motor with n-pole motor winding and (n ± 2)-pole suspension winding are taken up and their computed results of the torque and suspension force are discussed in detail. As a result, the superiority of the proposed d-q axis current control bearingless motor is found.**

I. Introduction

Recently, the bearingless motors (BELMs) are attracting attention in variable speed drives due to some of its advantageous features as compared to the conventional motor drives [1]. In the BELM as there are no bearings, it is maintenance free and hence the longevity of the motor is increasing. The functions of motor and magnetic rotor levitation are successfully integrated so that the rotor shaft can be made shorter and at the same time it requires less number of inverters, controller and electric wires as compared to a motor with magnetic bearings. Thus, the overall size and cost of the bearingless machine is considerably reduced as compared to the machine with magnetic bearing. Furthermore, as the rotor shaft is shorter, there is no fear to decrease the critical speed of the BELM by the rotational axis bend.

The d-q axis current control BELM, which has been proposed by the authors, is one of the BELMs with the integrated winding [4]-[6]. The advantageous features of the proposed BELM are as follows. 1)The motor structure is just the same as the conventional brushless dc motors. The stator winding is short-pitched and hence, its distribution is quite simple. The induced voltage at the winding terminals is relatively low compared with that of the conventional BELMs with the distributed windings. From the difference of their inverter voltage ratings, the d-q axis current control BELM is competitive on the inverter cost nevertheless the number of inverters is much. In addition, the end coil length is also shorter and hence, it is superior as the rotor shaft length can be short, particularly in the two-unit tandem configuration of BELM. 2)The control method is similar to that of the interior permanent magnet synchronous motor (IPMSM), i.e., in the d-q axis current control BELM, the rotational torque is controlled by the q-axis current and the suspension force is controlled by the d-axis current (the field-weakening or field-strengthening controls). Thus, the control method is also quite simple and the general-use 3-phase inverter can be employed to control the torque and suspension force.

The authors have proposed the d-q axis current control BELM and introduced the motor structure, the principle of torque and suspension force generation and the control method. Furthermore, the validity of the proposed control method has been confirmed by the experimental test results using a prototype machine. However, the performance of motor drive and magnetic levitation in the proposed d-q axis current control BELM has not yet made clear.

In this paper, the characteristic of torque and suspension force is derived based on Finite Element (FE) analysis using a simulation software. It is compared with the computed results of the conventional BELMs, which two kinds of windings for motor drive and magnetic suspension are wound in a stator core. Three types of the conventional BELMs with n-pole motor winding and (n ± 2)-pole suspension winding are taken up; the first one is the combination of 2-pole motor drive and 4-pole magnetic suspension, the second one is the combination of 4-pole motor drive and 2-pole magnetic suspension and, the final one is the combination of 6-pole motor drive and 4-pole magnetic suspension. The characteristic of torque and suspension force in each type BELM is discussed in detail based on the computed results. From the comparison among their BELM types, it is found that the d-q axis current control BELM is absolutely superior regarding the characteristics of torque and suspension force.

II. Structure and Principle of d-q Axis Current Control Bearingless Motor

A. *Motor Structure*

Figure 1 shows the cross section of the proposed d-q axis current control BELM. As there is only one stator winding wound on one stator core, the structure of the d-q axis current control BELM is simple and also the winding arrangement looks like conventional blushless dc motor [4]-[6]. The stator core is classified into three sections as the section-α, the section-β and the section-γ. In the section-α, the three-phase three-wire windings $N_{u\alpha}$, $N_{v\alpha}$ and $N_{w\alpha}$ are wound; in the section-β, $N_{u\beta}$, $N_{v\beta}$ and $N_{w\beta}$ are wound; in the section-γ, $N_{u\gamma}$, $N_{v\gamma}$ and $N_{w\gamma}$ are wound. All these windings are short pitch and simple. Each section is independently controlled by separate general-use 3-phase inverters. Totally, three 3-phase inverters are needed to drive the proposed BELM. The number of inverters is much than that of the conventional BELMs. However, the capacity per an inverter may be decreased to 1/3 times that for the motor drive in the conventional BELM, and

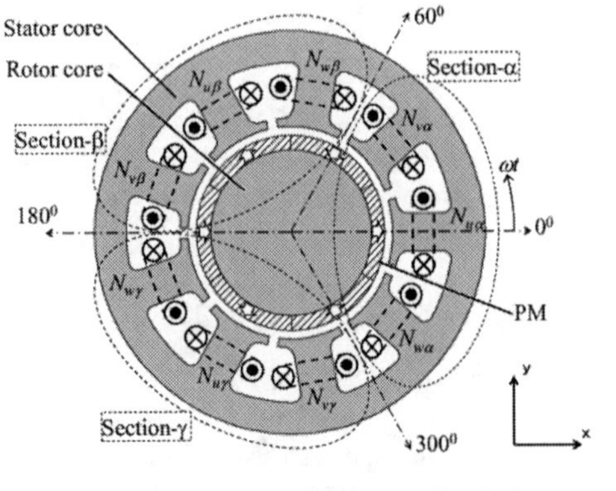

Figure 1. Cross section of a d-q axis current control bearingless motor.

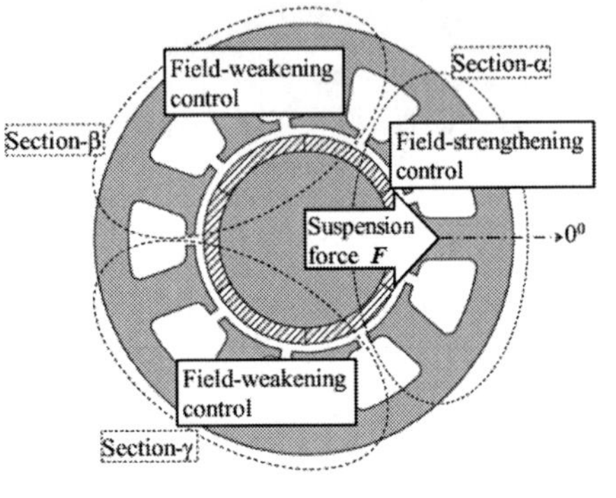

Figure 2. Principle of suspension force generation

particularly the voltage rating of inverter is lower. Thus, the d-q axis current control BELM is competitive on the inverter cost.

B. Principle of Suspension Force Generation

The suspension force is regulated to suspend the rotor of BELM. Figure 2 shows the principle of the suspension force generation in the proposed d-q axis current control BELM. The suspension force is generated by unbalanced flux density in air-gap with controlled d-axis currents $i_{d\alpha}$, $i_{d\beta}$ and $i_{d\gamma}$ in each section. For example, in section-α the field-strengthening control is done and then the air-gap flux density is increased; in the section-β and section-γ the field-weakening control is done and then the air-gap flux density is decreased. By the net vector sum in three sections, thus the suspension force is generated in the x-positive direction. By these controlled d-axis currents, the suspension force can be successfully generated in the arbitrary radial direction.

C. Control System Configuration

Figure 3 shows the control system configuration of the d-q axis current control BELM. In the motor controller, the rotor speed control method is the same as the field oriented control of the conventional ac motors. The difference $\Delta\omega$ between the detected rotor speed ω and command speed ω^* is input to the proportional-integral (PI) controller, and q-axis current command i_q^* is output of the PI speed controller which is the input to the current controllers.

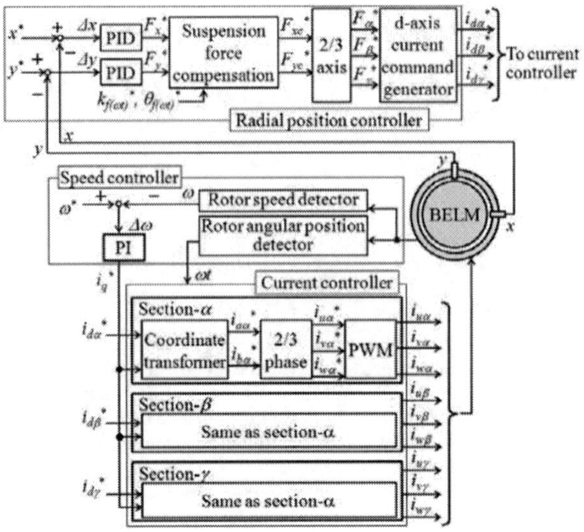

Figure 3. Control system of d-q axis current control BELM.

In the radial position controller, the detected rotor positions x and y on the x- and y- axes with eddy-current type gap sensors are input. The differences Δx and Δy between the detected rotor positions x, y and the commands x^*, y^* are input to the proportional-integral-derivative (PID) controller, and the suspension force commands F_x^* and F_y^* in the x- and y-axes coordinate are determined. In the block of suspension force compensation, F_x^* and F_y^* are significantly compensated. Because the magnitude and direction of suspension force are undesirably varied depending on the rotor angular position due to the rotation of d-axis direction. Hence, the authors have proposed the compensation method of the suspension force to stably support the rotor shaft [4]-[6]. The proposed compensation method can be explained using Figure 4. Figure 4 shows the relation between the actual suspension force $\boldsymbol{F}_{(\omega t)}$ and command force $\boldsymbol{F}_{(\omega t)}^*$. Figure 4 (a) shows the suspension forces without the compensation method. The magnitude and direction of $\boldsymbol{F}_{(\omega t)}$ does not agree with those of $\boldsymbol{F}_{(\omega t)}^*$. The ratio of the magnitude of $\boldsymbol{F}_{(\omega t)}$ to that of $\boldsymbol{F}_{(\omega t)}^*$ is defined as $k_{f(\omega t)}$, i.e., $k_{f(\omega t)}$ is shown as

$$k_{f(\omega t)} = \frac{F_{(\omega t)}}{F_{(\omega t)}^*} \tag{1}$$

Furthermore, the lead angle of $\boldsymbol{F}_{(\omega t)}$ as the reference of $\boldsymbol{F}_{(\omega t)}^*$ is defined as $\theta_{f(\omega t)}$. The $k_{f(\omega t)}$ and $\theta_{f(\omega t)}$ are defined as magnitude compensation coefficient and direction compensation coefficient, respectively.

The rotation matrix $\boldsymbol{R}_{(\omega t)}$ is defined using the $\theta_{f(\omega t)}$ as

$$\boldsymbol{R}_{(\omega t)} = \begin{bmatrix} \cos\theta_{f(\omega t)} & -\sin\theta_{f(\omega t)} \\ \sin\theta_{f(\omega t)} & \cos\theta_{f(\omega t)} \end{bmatrix} \tag{2}$$

Thus $\boldsymbol{F}_{(\omega t)}$ is represented using $k_{f(\omega t)}$, $\boldsymbol{R}_{(\omega t)}$ and $\boldsymbol{F}_{(\omega t)}^*$ as

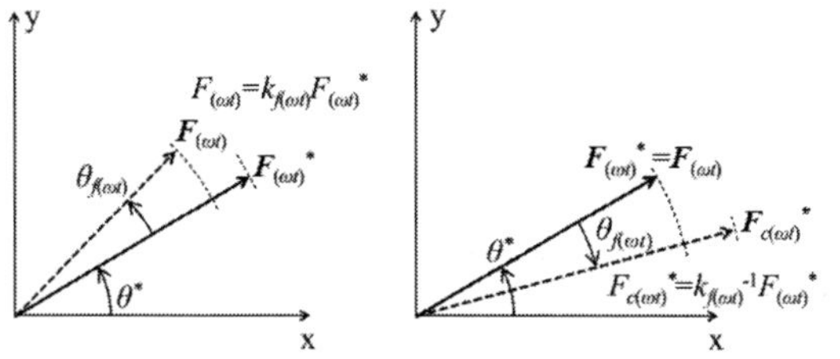

(a) Without compensation. (b) With compensation.

Figure 4. The relation between suspension force and its command.

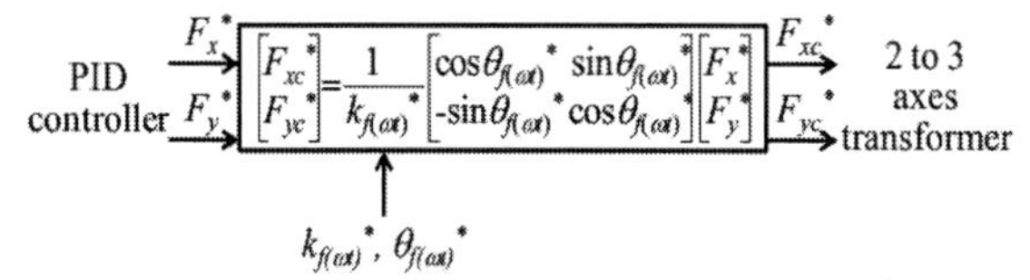

Figure 5. Block diagram of suspension force compensation.

$$F_{(\omega t)} = k_{f(\omega t)} R_{(\omega t)} F_{(\omega t)}^* \tag{3}$$

Figure 4 (b) shows the compensated suspension force command $F_{c(\omega t)}^*$. It is derived by solving for $F_{(\omega t)}^*$ in the equation (3), i.e., the compensated suspension force $F_{c(\omega t)}^*$ is defined as

$$F_{c(\omega t)}^* = \frac{1}{k_{f(\omega t)}} R_{(\omega t)}^{-1} F_{(\omega t)} \tag{4}$$

The x- and y- axes components of $F_{c(\omega t)}^*$ are defined as F_{xc}^* and F_{yc}^*, respectively, so that F_{xc}^* and F_{yc}^* are shown as

$$\begin{bmatrix} F_{xc}^* \\ F_{yc}^* \end{bmatrix} = \frac{1}{k_{f(\omega t)}} \begin{bmatrix} \cos\theta_{f(\omega t)} & \sin\theta_{f(\omega t)} \\ -\sin\theta_{f(\omega t)} & \cos\theta_{f(\omega t)} \end{bmatrix} \begin{bmatrix} F_x^* \\ F_y^* \end{bmatrix} \tag{5}$$

By the above compensation method, $F_{(\omega t)}$ is absolutely equal to $F_{(\omega t)}^*$ as shown in Figure 4 (b). The compensated suspension force command is installed in the radial position controller so that the stable magnetic suspension is realized.

Figure 5 shows the block diagram of the suspension force compensation. The $k_{f(\omega t)}$ and $\theta_{f(\omega t)}$ are estimated from the computed results by FE analysis. They can be also done from the detected suspension force during the practical operation. The estimated $k_{f(\omega t)}$ and $\theta_{f(\omega t)}$ are input as the references $k_{f(\omega t)}^*$ and $\theta_{f(\omega t)}^*$ in the radial position controller, respectively.

These commands F_{xc}^* and F_{yc}^* are transformed into F_α^*, F_β^* and F_γ^* in the α-, β- and γ- axes coordinate. Then, the d-axis current commands $i_{d\alpha}^*$, $i_{d\beta}^*$ and $i_{d\gamma}^*$ are determined to be proportional to the suspension force commands, respectively.

In the current controller, the current is independently controlled in each section. In the section-α, for example, the current commands i_q^* and $i_{d\alpha}^*$ are transformed into 2-phase the current commands $i_{a\alpha}^*$ and $i_{b\alpha}^*$ in the stationary coordinate. Then $i_{a\alpha}^*$ and $i_{b\alpha}^*$ are transformed into $i_{u\alpha}^*$, $i_{v\alpha}^*$ and $i_{w\alpha}^*$. The winding currents $i_{u\alpha}$, $i_{v\alpha}$ and $i_{w\alpha}$ are regulated to follow the current commands in the Pulse Width Modulation (PWM) block. The current controllers of the sections -β and -γ are the same as that of the section-α.

The validation of the proposed control system has been verified by the simulated results using an FE software and the experimental test results using a prototype machine [4]-[6].

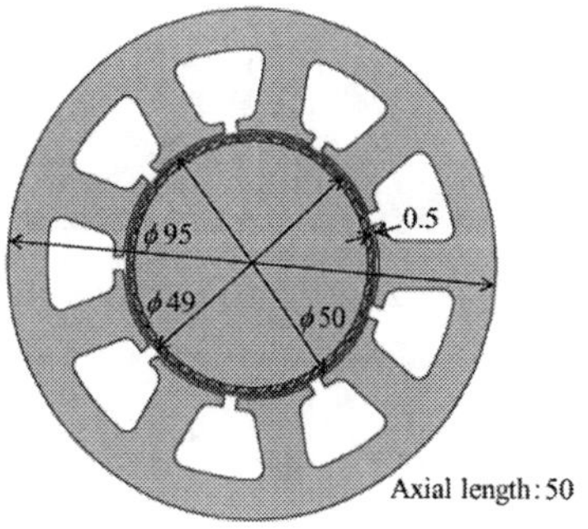

Figure 6. Dimension of FE model d-q axis current control BELM.

TABLE I. SPECIFICATION OF FE MODEL D-Q AXIS CURRENT CONTROL BELM

Rotor core, Stator core	Silicon steel
PM	Nd-Fe-B (thickness : 1 mm)
Stator winding	146 turns/tooth, ϕ0.6 mm
Current rating	1.7 A

III. CHARACTERISTICS OF TORQUE AND SUSPENSION FORCE

In this section, the torque and suspension force in a d-q axis current control BELM (d-q BELM) are computed by an FE analysis simulation software (JMAG Designer, Ver.11.1, 2-dimension) using a machine model, respectively, and the characteristic of them is obviously shown. In addition, the torque and suspension force in the conventional BELMs, which two kinds of windings for the motor drive and magnetic suspension are wound in a stator, are also computed by FE analysis. Based on the computed results, the characteristics of torque and suspension force are competitively compared and the feature of each BELM is comprehensively discussed. The combination of motor and magnetic suspension poles of the conventional BELMs is taken up to compare with the d-q axis current control BELM in this section is as follows,

- 2-pole motor and 4-pole magnetic suspension (2-4 BELM)
- 4-pole motor and 2-pole magnetic suspension (4-2 BELM)
- 6-pole motor and 4-pole magnetic suspension (6-4 BELM)

A. Specification of FE Models

Figure 6 shows the dimension of FE model of the d-q axis current control BELM. The rotor is an interior permanent magnet type (IPM), in which the small permanent magnets are buried just below the rotor core surface. The saliency ratio is relatively low and hence, the motor performance is almost the same as that of surface-mounted permanent magnet

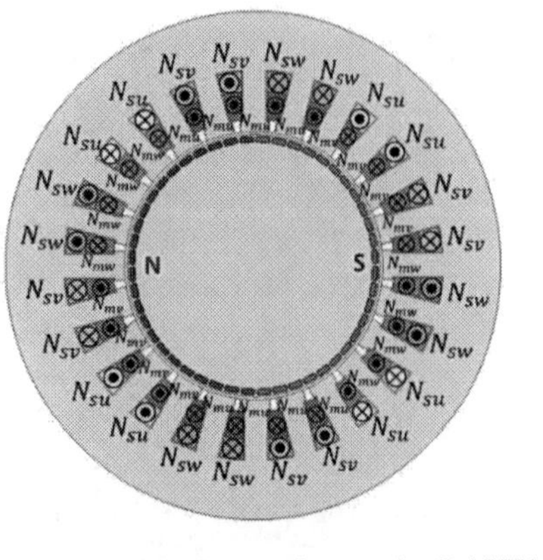

(a) 2-pole motor and 4-pole magnetic suspension (2-4 BELM).

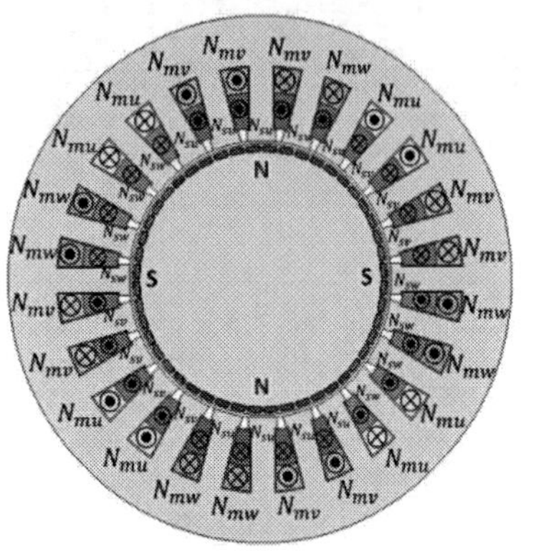

(b) 4-pole motor and 2-pole magnetic suspension (4-2 BELM).

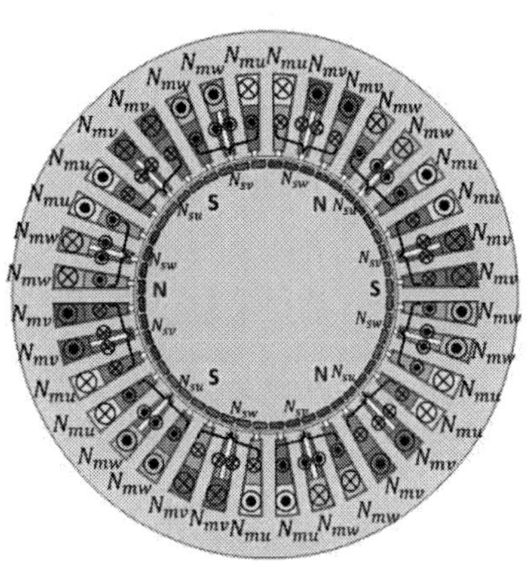

(c) 6-pole motor and 4-pole magnetic suspension (6-4 BELM).

Figure 7. Stator and rotor geometry and winding arrangement in the conventional BELMs.

synchronous motors (SPMSMs). The d-q axis current control BELM in Figure 6 is defined as a reference model to compare the characteristics of the torque and suspension force between the BELMs. Table 1 shows the specification of FE model of the d-q axis current control BELM.

Figures 7 (a)-(c) show the stator and rotor geometry and winding arrangement of the above conventional types of BELM. Figure 7 (a) shows that of the combination of 2-pole motor and 4-pole magnetic suspension. The inside windings N_{mu}, N_{mv} and N_{mw} in the stator slot are the 2-pole motor winding and the outside windings N_{su}, N_{sv} and N_{sw} are the 4-pole suspension winding. Two kinds of windings are wound in a stator core as described in Figure 7 (a). The rotor is 2-pole and its structure is basically the same as that of the reference model of d-q axis current control BELM in Figure 6.

Figure 7 (b) shows the cross section of the combination of 4-pole motor and 2-pole magnetic suspension. The outside windings N_{mu}, N_{mv} and N_{mw} are the 4-pole motor winding and the inside windings N_{su}, N_{sv} and N_{sw} are the 2-pole suspension winding. The rotor configurates 4-pole by the buried small permanent magnets.

Finally, Figure 7 (c) shows the cross section of the combination of 6-pole motor and 4-pole magnetic suspension. The outside windings N_{mu}, N_{mv} and N_{mw} are the 6-pole motor winding and the inside windings N_{su}, N_{sv} and N_{sw} are the 4-pole suspension winding. The rotor configurates 6-pole by the buried small permanent magnets.

The winding arrangement in all the above conventional BELM is a distributed winding. On the other hand, it is short-pitched in the d-q BELM. As mentioned above, there are significant differences between the d-q BELM and the conventional BELMs at the view points of the induced voltage at the winding terminals and the end coil length.

The FE models of the conventional BELMs are designed under the following conditions due to the comparison between the BELMs.

(1) In all the FE models, the outer diameter of the stator is the same as that of the model of d-q axis current control BELM. The rotor geometry is quite the same as that of the d-q BELM model except for the axial length.

(2) The maximum torque in the model of d-q axis current control BELM, i.e., the torque when the d-axis current is zero, is set as the reference torque. The conventional BELMs are designed by adjusting the axial length of the stator and rotor cores so that the maximum torque, i.e., the torque under the condition which there are no suspension windings and only motor winding occupies in the stator slots, is equal to the maximum torque of the d-q axis current control BELM. Therefore, the axial length of each FE model is different. Figure 8 shows the ratio of them.

(3) The winding current ratings are equal in all the FE models. In the d-q axis current control BELM, the ratio of the suspension force to the torque is regulated by changing the ratio of the d-axis current to the q-axis current. On the other hand, it is changed by adjusting the ratio of the number of suspension winding turns to that of the motor winding turns in the conventional BELMs. The characteristics of torque and suspension force are compared under the different conditions by changing the ratio of the torque and suspension force.

(4) The flux density in the stator yoke and teeth is less than 1.6 T. Hence, the relationship between the torque and the motor current and the relationship between the suspension force and suspension current are roughly linear under this condition, respectively.

B. *Analyzed Results of Torque and Suspension Force*

Figure 9 shows the relation between the computed torque

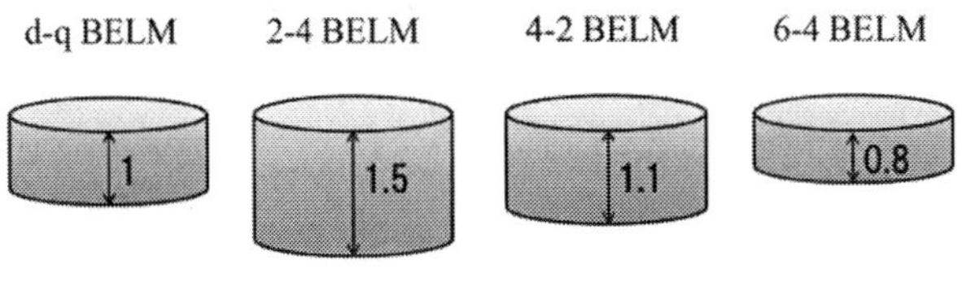

Figure 8. Ratio of axial length in FE models.

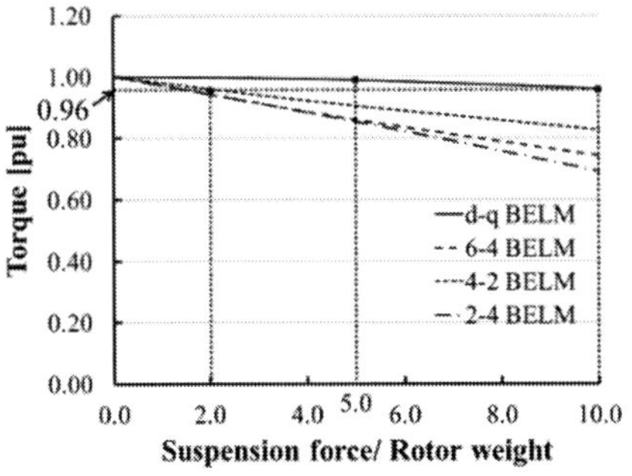

Figure 9. Characteristic of torque and suspension force .

and suspension force in the d-q axis current control BELM and the conventional BELMs. The vertical axis is the torque [p.u.] and the torque of each FE model is normalized by that when no suspension force is generated. In the d-q axis current control BELM, it is normalized by the torque when the d-axis current is set to zero, and in the other conventional BELMs, it is normalized by the torque of the FE model which there are no suspension windings in the stator core. The horizontal axis is the ratio of the suspension force to the rotor weight.

It is seen in Figure 9 that in the conventional BELMs, the torque is almost linearly decreased as the suspension force is increased by increasing the number of suspension winding turns. Because the number of motor winding turns should be decreased as the slot fill factor is remained constant. In the d-q axis current control BELM, on the other hand, the d-axis current is increased to generate the suspension force. The q-axis current should be decreased so that the winding current is within its rating. Therefore, the torque is decreased as described in Figure 9 in accordance with the increase in the suspension force. However, the decreasing ratio of the torque is less than those of the conventional BELMs. It means that the larger torque in the d-q axis current control BELM can be generated in comparison with the other BELMs when the suspension force per rotor weight is equal. In Figure 9, for example, when the suspension force per rotor weight is 5.0, the torque is decreased to less than 90% of its maximum in the conventional BELMs. However, the decreasing ratio in the d-q BELM is slightly a few percentages and the torque of d-q BELM is larger than those of the conventional BELMs.

It should be noted that the axial length of the model of 6-4 BELM is shorter than that of the d-q BELM as described in Figure 8. Hence, it is possible that its torque will be larger than that of the d-q BELM if the axial length of stator and rotor cores will be lengthened so as to be equal to that of the d-q BELM. As mentioned above, however, total axial length of the 6-4 BELM model including the end coil is significantly longer than that of d-q BELM. Because the winding arrangement of 6-4 BELM is a distributed winding. On the other hand, the winding arrangement of d-q BELM is short-pitched. Thus, it is difficult to lengthen the stator and rotor cores in the axial direction in the 6-4 BELM model.

On the contrary, the suspension force per rotor weight of the d-q BELM is larger than those of the conventional BELMs if the torque is equal. In Figure 9, for example, when the torque is 0.96 pu, the suspension force per rotor weight is 10.0 in the d-q BELM. However, it is only 2.0 even in the 4-2 BELM model, in which the suspension force per rotor weight is maximum in the conventional BELMs. The reason why the larger suspension forces than those of the other conventional BELMs is generated in the d-q BELM can be explained by the principle of d-q BELM. In the conventional BELMs, the air-gap flux densities are unbalanced by superimposing the suspension flux on the permanent magnet magnetic field. The flux density is increased in an air-gap and it is decreased in the air-gap in the opposite side across the rotor. As a result, the suspension force is successfully generated. In the d-q BELM, on the other hand, the flux densities in the air-gaps in three sections are unbalanced by the d-axis current in the sections. Hence, the suspension force in the d-q BELM is larger than those of the conventional BELMs.

From these results, the d-q axis current control BELM is absolutely superior to the conventional BELMs at the view point of the characteristics of the torque and suspension force.

IV. Conclosion

The characteristic of torque and suspension force in a d-q axis current control BELM has been found by the FE model calculation. Compared with those of the conventional BELMs, it is found that the proposed d-q axis current control BELM is absolutely superior. In the next step, the characteristic of torque and suspension force of the d-q axis current control BELM will be compared with several types of the conventional BELMs except for 3 conventional BELMs mentioned in this paper. Furthermore, it will be compared under the other conditions, for example, under the constant stack length of the stator and rotor cores. Based on the analyzed torque and suspension force, it will be discussed in detail which types of the BELM are more superior.

References

[1] A. Chiba, T. Fukao, O. Ichikawa, M. Ooshima, M. Takemoto and David G. Dorrell, "Magnetic Bearings and Bearingless Drives", Newnes Publishers, ISBN 0-7506-5727-8, March 2005.

[2] H. Grabner, S. Silber and W. Amrhein, "Feedback control of a novel bearingless torque motor using an extended FOC method for PMSMs", 2013 IEEE International Conference on Industrial Technology (ICIT), pp.325-330, 2013.

[3] S. Silber, W. Amrhein, P. Bosch, R. Schob and N. Barletta, "Design aspects of bearingless slice motor", IEEE/ASME Transactions on Mechatronics, Volume 10, Issue 6, pp. 611-617, 2005.

[4] Masahide Ooshima, Syunsuke Kobayashi and M. Nasir Uddin, "Magnetic Levitation Tests of a Bearingless Motor Based on d-q Axis Current Control", *IEEE Industry Application Society 2012 Annual Meeting*, 2012-IACC-212, CDROM, @Las Vegas, 2012.

[5] Masahide Ooshima, Toshiki Karasawa and M. Nasir Uddin, "Stabilized Control Strategy Under Loaded Conditions in a d-q Axis Current Control", *IEEE Industry Application Society 2013 Annual Meeting*, 2013-IACC-315, CDROM, @Orlando, 2013

[6] Syunsuke Kobayashi, Masahide Ooshima and M. Nasir Uddin, "A Radial Position Control Method of Bearingless Motor Based on d-q Axis Current Control", *IEEE Transactions on Industry Applications*, vol.49, No.4, pp.1827-1835, 2013.

Thin Maglev Ventricular Assist Device with Radial Type Self-bearing Motor

Michiko Murakami[a], Toru Masuzawa[a], Shoichi Yoshida[a], Hiroyuki Onuma[b], Takashi Nishimura[c], Syunei Kyo[c]

[a] College of Engineering, Ibaraki University, 4-12-1 Nakanarusawa, Hitachi, Ibaraki 316-8511, Japan, 12nd214y@hcs.ibaraki.ac.jp
[b] Department of Electronics and Control Engineering, Ibaraki National College of Technology, 866 Nakane, Hitachinaka, Ibaraki 312-8508, Japan,
[c] Therapeutic Strategy for Heart Failure, the University of Tokyo, 7-3-1 Hongo, Bunkyo, Tokyo,113-8655

***Abstract*— A small, thin, maglev ventricular assist device, which can be implanted under the thorax muscle layer like a pacemaker, has been developed. A cascade pump, which has inlet and outlet ports in the same plane and a radial type self-bearing motor are adopted to reduce the axial thickness of the device. The radial self-bearing motor is using the passive stability to restrict the impeller movement in the axial direction and its inclination. We evaluated the fluid force acted on the levitated impeller by using the computational fluid dynamics analysis and restoring force and torque by using the magnetic field analysis. Also, an axial hydrodynamic bearing is developed to avoid the impeller contact on the casing when the levitated impeller received a large disturbance. Finally, a prototype pump is developed and evaluated its levitation and pump performance. The prototype pump was levitated and rotated up to a rotating speed of 2,600 rpm successfully. The target performance of a flow rate of 2 L/min against a head pressure of 100 mmHg is achieved with a rotating speed of 2,000 rpm. The maximum oscillation amplitude and the total power consumption with the target performance were 17 μm and 4.5 watts.**

I. Introduction

Ventricular assist device (VAD) is applied for severe heart failure patient [1]. But, intended patients are limited the most severe type because of its high cost. We have been developing a thin maglev VAD with a high reliability, a good cost performance and a long-term usage with periodical replacement. Proposed VAD can be implanted under the thorax muscle layer like a pacemaker by adopting a magnetically levitated cascade blood pump [2] with a radial type self-bearing motor which has an advantage it can be made thinner than axially suspension motor [3]. The cascade pump which has inlet and outlet ports in the same plane is able to reduce the axial thickness of the device. The radial self-bearing motor is using the passive stability to restrict the impeller movement in the axial direction and its inclination to reduce complexity of the magnetic suspension system. In this paper, we evaluated the fluid force acted on the levitated impeller by using the computational fluid dynamics (CFD) analysis and restoring force and torque by using the magnetic field analysis. Also, an axial hydrodynamic bearing was developed to avoid the impeller contact on the casing when the impeller received a large disturbance. Finally, a prototype pump was developed. The basic levitation and pump performance of the prototype pump was evaluated.

II. Method

A. *Magnetically levitated cascade blood pump and Self-bearing motor*

Fig. 1 shows a schematic of the magnetically levitated cascade blood pump. Magnetic suspension techniques have been used to eliminate the mechanical contact parts for improvement of device durability. The self-bearing motor is set at the center of the device and the ring-shaped levitated impeller is set at the circumferential space. There is a cylindrical sensor target for the radial position detection on the outer periphery of the levitated impeller. The hydrodynamic bearing is formed on the rotor yoke's end face of the levitated impeller. The target pump performance is a flow rate of 2 L/min against a head pressure of 100 mmHg to assist a part of the cardiac function of patients with mild heart failure. Fig. 2 shows an oblique perspective figure of the radial type self-bearing motor. The outer rotor structure, that a rotor is set around a stator, is adopted to miniaturize the self-bearing motor. Eight thin permanent magnets are set on the inner circumferential surface of the rotor which is a yoke itself. The stator has twelve radial poles. Rotation coils to produce 3-phase 8-pole magnetic field and levitation coils to produce 2-phase 6-pole magnetic field are wound separately on the centered stator yoke. The plus minus two-pole algorithm is adopted to levitate and rotate the rotor. Two radial degrees of freedom of the rotor and rotation of the rotor are controlled actively by electromagnets constructed on the stator. The axial movement and the inclination of the levitated impeller are restricted with the passive stability to reduce complexity of the magnetic suspension system.

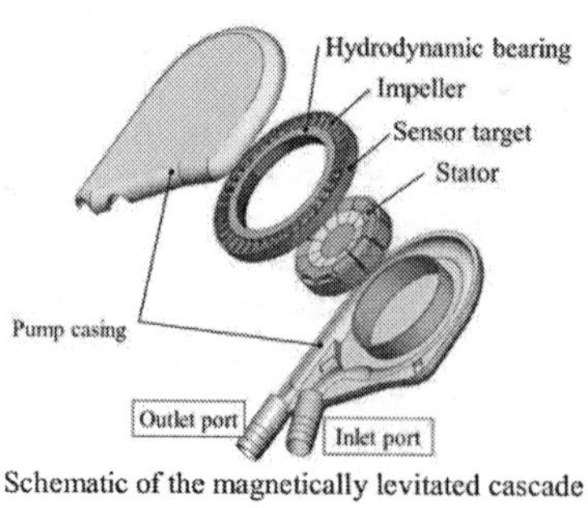

Figure 1. Schematic of the magnetically levitated cascade pump.

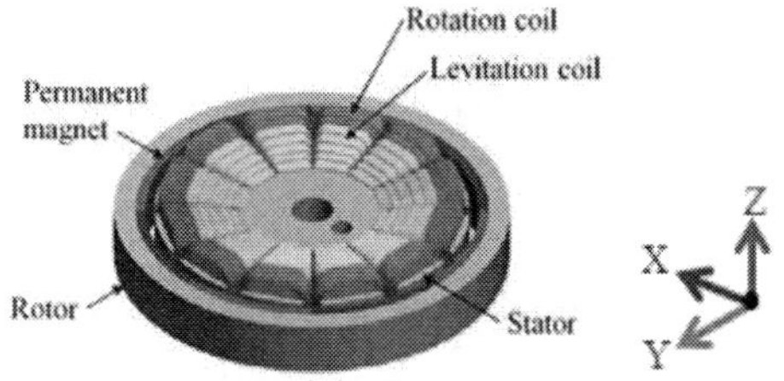

Figure 2. Oblique perspective figure of radial type self-bearing motor.

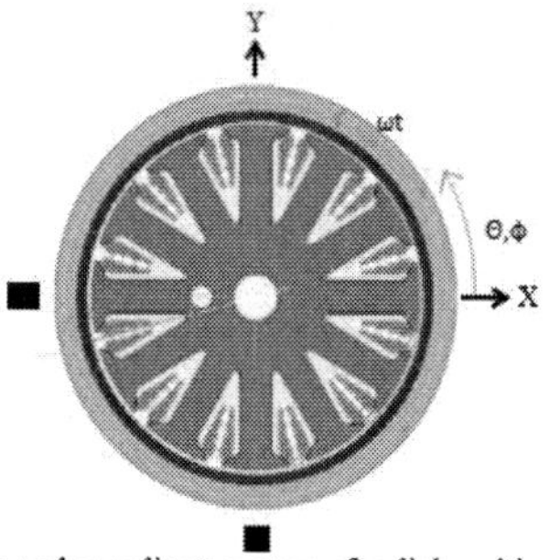

Figure 3. Schematic and coordinate system of radial position control's principle.

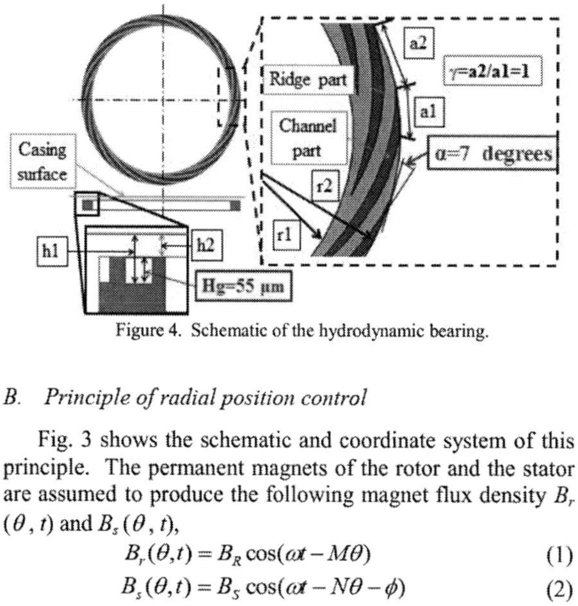

Figure 4. Schematic of the hydrodynamic bearing.

B. *Principle of radial position control*

Fig. 3 shows the schematic and coordinate system of this principle. The permanent magnets of the rotor and the stator are assumed to produce the following magnet flux density $B_r(\theta, t)$ and $B_s(\theta, t)$,

$$B_r(\theta,t) = B_R \cos(\omega t - M\theta) \tag{1}$$

$$B_s(\theta,t) = B_S \cos(\omega t - N\theta - \phi) \tag{2}$$

where B_R, B_S are peak density of magnetic flux produced by the rotor permanent magnets and the stator electromagnets, M is pole pair number of the rotor, θ is angular coordinate, and ϕ is an arbitrary angle. The magnet flux density of the air gap B_{rs} is sum of the flux density B_r and B_s.

$$\begin{aligned} B_{rs} &= B_r(\theta,t) + B_s(\theta,t) \\ &= B_R \cos(\omega t - M\theta) + B_S \cos(\omega t - N\theta - \phi) \end{aligned} \tag{3}$$

The control force can be generated by controlling imbalance of the magnetic flux distribution. The attractive force is expressed as the gradient of the magnetic energy. The magnetic flux energy in the air gap is given by equation (4),

$$W = \int_V \frac{B_{rs} H_{rs}}{2} dV = \int_0^{2\pi} \frac{B_{rs}^{\ 2}}{2\mu_0} zrRd\theta \tag{4}$$

where H_{rs} is magnetic field intensity in the air gap, z and r are axial height and radial thickness of the air gap, R is length between the coordinate center and the center of the air gap. Therefore, the attractive force inside of the stator and rotor which is the gradient of the magnetic energy in the normal direction is given by equation (5).

$$F_r = \frac{dW}{dr} = \int_0^{2\pi} \frac{B_{rs}^{\ 2}}{2\mu_0} zRd\theta \tag{5}$$

The x-directional force F_x is calculated by integrating the x component of the attractive force.

$$\begin{aligned} F_x &= \frac{zR}{2\mu_0} \int_0^{2\pi} B_{rs}^{\ 2} \cos(\theta) d\theta \\ &= \frac{zR}{2\mu_0} \int_0^{2\pi} [B_R \cos(\omega t - M\theta) \\ &\quad + B_S \cos(\omega t - N\theta - \phi)]^2 \cos(\theta) d\theta \\ &= \frac{B_R B_S zR}{4\mu_0} \int_0^{2\pi} \{\cos[(N - M + 1)\theta + \phi] \\ &\quad + \cos[(N - M - 1)\theta + \phi]\} d\theta \end{aligned} \tag{6}$$

This calculation produces a constant force when $N - M = \pm 1$.

$$F_x = \frac{B_R B_S zR\pi}{2\mu_0} \cos(\phi) \tag{7}$$

The y-directional force F_y when $N - M = \pm 1$ is similarly obtained as follows,

$$F_y = \frac{B_R B_S zR\pi}{2\mu_0} \sin(\phi) \tag{8}$$

The control force F_r is expressed the square root of sum of squares of F_x and F_y and is given by equation (9).

$$F_r = \sqrt{F_x^{\ 2} + F_y^{\ 2}} \tag{9}$$

Hence, two dimensional radial position of the rotor can be controlled by changing the B_S and ϕ.

C. *Hydrodynamic bearing*

The hydrodynamic bearing was designed to avoid impeller touchdown when the levitated impeller received a large disturbance. Fig. 4 shows a schematic of the hydrodynamic bearing. A load carrying capacity Q was calculated to determine dimensions of the hydrodynamic bearing [4] by equation (10).

$$Q = \frac{3\pi}{2} g_1 \left(\frac{r_{eff}}{r_2}\right)^4 \left\{ \left[1 - \left(\frac{r_2}{r_{2eff}}\right)^2\right]^2 - 2\left[1 - \left(\frac{r_2}{r_{2eff}}\right)^2\right] \frac{A}{\left(In \frac{r_b}{r_1}\right) g_5 - \left(In \frac{r_2}{r_{2eff}}\right) g_3} \right\} \frac{\mu\omega r_2^{\ 4}}{h_2^{\ 2}} \tag{10}$$

where, r_1 is the inner radius of the grooved part of the hydrodynamic bearing [m], r_2 is the outer radius of the grooved part of the hydrodynamic bearing [m], r_{2eff} is the effective outer radius [m], r_b is the radius in the middle of the contacting surfaces at which spiral grooving ends [m], μ is viscosity of a working fluid [Pa/s], ω is angular velocity of the hydrodynamic bearing [rad/s]. g_1, g_3, g_5, A and r_{2eff} in equation (10) are as follows,

$$g_1 = \frac{\gamma\left(\frac{h_2}{h_1}\right)^2\left[1-\left(\frac{h_2}{h_1}\right)\right]\left[1-\left(\frac{h_2}{h_1}\right)^3\right]\cot\alpha}{\left[1+\gamma\left(\frac{h_2}{h_1}\right)^3\right]\left[\gamma+\left(\frac{h_2}{h_1}\right)^3\right]+\left(\frac{h_2}{h_1}\right)^3(1+\gamma)^2\cot^2\alpha} \tag{11}$$

$$g_3 = \left(\frac{h_2}{h_1}\right)^3(1+\gamma)(1+\cot^2\alpha)(1+\gamma)^2 \tag{12}$$

$$g_5 = \left[1+\gamma\left(\frac{h_2}{h_1}\right)^3\right]\left[\gamma+\left(\frac{h_2}{h_1}\right)^3\right]+\left(\frac{h_2}{h_1}\right)^3(1+\gamma)^2\cot^2\alpha \tag{13}$$

$$A = \left[\frac{r_2}{r_{2eff}} In\frac{r_2}{r_{2eff}}+\frac{1}{2}-\frac{1}{2}\left(\frac{r_2}{r_{2eff}}\right)^2\right]g_3 - \left[\frac{r_2}{r_{2eff}} In\left(\frac{r_b}{r_1}\right)-\frac{1}{2}\left(\frac{r_2}{r_{2eff}}\right)^2+\frac{1}{2}\frac{r_1^{\,2}}{r_{2eff}^{\,2}}\right]g_5 - \frac{\pi}{2k}\left(1-\frac{\alpha}{90}\right)(\tan\alpha)\frac{2}{1+\gamma}\frac{1-\left(\frac{h_2}{h_1}\right)^3}{1+\left(\frac{h_2}{h_1}\right)^3}F_{r2} \tag{14}$$

$$r_{2eff} = r_2 e \tag{15}$$

with

$$s_{r2} = \frac{1}{2}\left[1-\left(\frac{r_b}{r_1}\right)^2\right]\frac{\gamma\left(\frac{h_2}{h_1}\right)^3(\cot\alpha)(1+\gamma)\left(1-\frac{h_2}{h_1}\right)\left[1-\left(\frac{h_2}{h_1}\right)^3\right]}{In\left(\frac{r_b}{r_1}\right)g_5-\left[In\left(\frac{r_b}{r_1}\right)\right]g_3} \tag{16}$$

$$F_{r2} = \left\langle 1+\frac{\gamma\left(\frac{h_2}{h_1}\right)^3(\cot\alpha)(1+\gamma)\left(1-\frac{h_2}{h_1}\right)+S_{r2}\left\{\left[\gamma+\left(\frac{h_2}{h_1}\right)^3\right]+\left(\frac{h_2}{h_1}\right)^3(\cot^2\alpha)(1+\gamma)\right\}}{-\gamma(\cot\alpha)\left[1-\left(\frac{h_2}{h_1}\right)\right]\left[1+\gamma\left(\frac{h_2}{h_1}\right)^3\right]+S_{r2}\gamma(\cot^2\alpha)\left[1-\left(\frac{h_2}{h_1}\right)^3\right]}\right\rangle^{-1} \tag{17}$$

where, h_1 is groove depth $H_g + h_2$ [m], h_2 is film height above the ridges [m], γ is ratio of ridge width a_2 divided by groove width a_1, α is groove angle.

The maximum load carrying capacity was calculated as 2.2 N with a bearing clearance of 10 μm, when a rotational speed of the impeller is 2,000 rpm and a viscosity of a fluid is 1 mPa·s. Groove depth, groove angle and ratio of ridge width to groove width were determined as 55 μm, 7 degrees and 1.0.

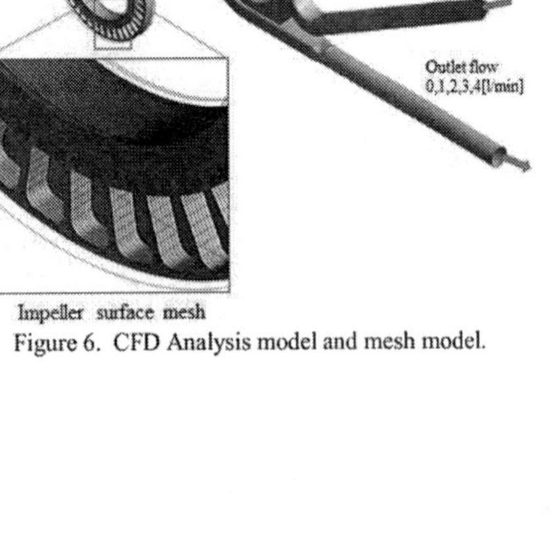

Figure 6. CFD Analysis model and mesh model.

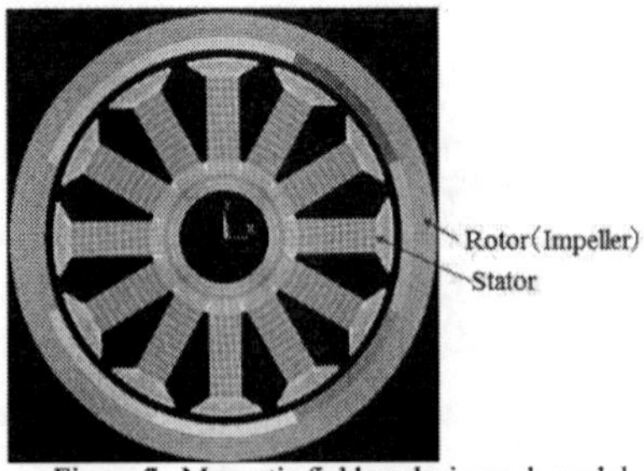

Figure 7. Magnetic field analysis mesh model.

III. Experiments

A. CFD analysis

The CFD analysis was conducted to estimate the hydrodynamic force acting on the levitated impeller in order to determine the passive stability performance required to suppress the axial movement of the levitated impeller caused by the fluid force. Fig. 6 shows CFD analysis model and impeller surface mesh model. The analysis was carried out in the steady-state analysis by using the frozen rotor method with ANSYS CFX 14.0. Shear stress transport model has been adopted as a turbulent flow model. The working fluid is water which has density ρ = 997 kg/m^3, viscosity μ = 0.8899 × 10^{-3} Pa·s. Fluid forces with different mass flow rate condition from 0 L/min to 4 L/min were calculated with a rotating speed of 2,000 rpm. Fluid forces were calculated by integrating the force caused by the pressure and the wall shear stress of each mesh. Number of elements in the CFD analysis model is about 10 million.

B. Magnetic field analysis

The magnetic field analysis was performed to evaluate the restoring torque generated on the impeller with passive stability. The restoring force in the axial direction generated by the axial impeller displacement and the restoring torque generated by the inclination of the levitated impeller were calculated with the magnetic field analysis with ANSYS 14.0. Fig.7 shows the magnetic field analysis mesh model. The material of the permanent magnets is the Samarium-Cobalt (magnetic coercive force is 748 kA/m, remanent magnetic flux density is 1.02 T). The material of the impeller yoke and the stator yoke are magnetic steel (SUY-1) and laminated silicon steel (35A440), respectively. The position in the axial direction of the impeller was changed from 0 mm at the center position to 0.4 mm. The impeller inclination around Y axis was changed from 0 degree to 1.2 degrees. Number of elements in the magnetic field analysis model is about 1.2 million.

C. Hydrodynamic bearing performance

Fig. 8 shows the developed hydrodynamic bearing. The hydrodynamic bearing is formed on the rotor yoke's end face of the levitated impeller. The performance evaluation equipment of the hydrodynamic bearing was shown in Fig. 9 and 10. The stator of the self-bearing motor was set at the center of the equipment. A ring shaped permanent magnet which gives a preload of 2.1 N in the axial direction on the levitated impeller is set on the axially movable stage under the rotated impeller. Glycerin-water solutions with a viscosity of 0.1 Pa・s was filled in the channel and the impeller was set in the channel and was immersed in the glycerin-water solution. The radial impeller position was controlled at the center and rotated with an axial preload by the self-bearing motor. At first, the impeller is set in the channel on the bottom. The rotational speed of the impeller increased gradually, the axial impeller position which is lifted off with hydrodynamic force was measured by using four eddy current sensors.

D. *Prototype pump and pump performance*

Fig. 11 shows the prototype pump of the maglev cascade VAD. The outer diameter and the axial thickness of the maglev VAD are 59 mm and 22 mm, respectively. The minimum clearances between the casing and the impeller in the radial and axial direction are 0.25 mm and 0.4 mm, respectively. The maximum tilt angle of the rotor is 1.2 degrees. Fig. 12 shows schematic of the radial type self-bearing motor's control system. Levitation and rotation of the rotor were controlled with a digital PID controller using a digital signal processor. Two eddy current sensors and three Hall elements are used to measure the rotor radial position and the rotation speed, respectively. Fig. 13 shows experimental set up for pump performance test. The pump performance, which is a relationship between the pump head and the flow rate, was evaluated. The prototype pump was connected to a closed mock circuit filled with saline. The closed mock circuit is composed of two pressure gauge sensors, an electromagnetic flow meter, and a reservoir tank. The pump head is the difference pressure between the static pressure of the inlet port and the static pressure of the outlet port. The maximum oscillation amplitude and power consumption during pumping were measured by using eddy current sensors and the power meter.

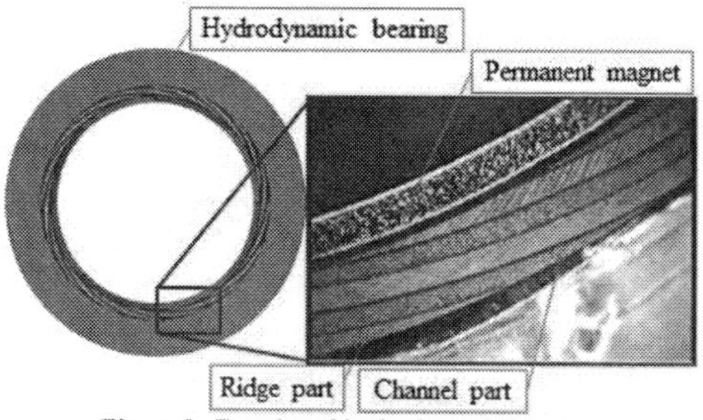

Figure 8. Developed hydrodynamic bearing.

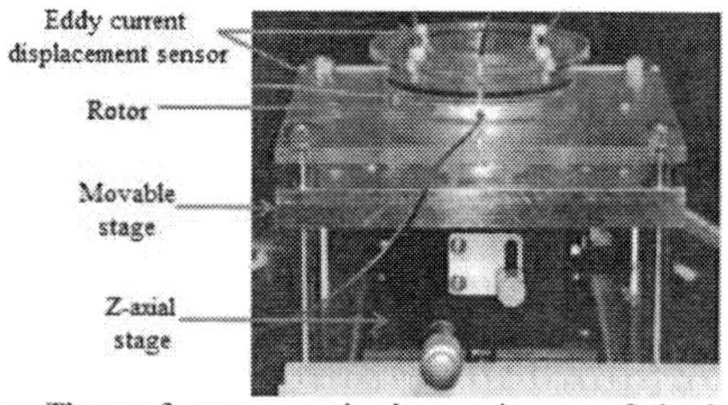

Figure 9. The performance evaluation equipment of the hydrodynamic bearing.

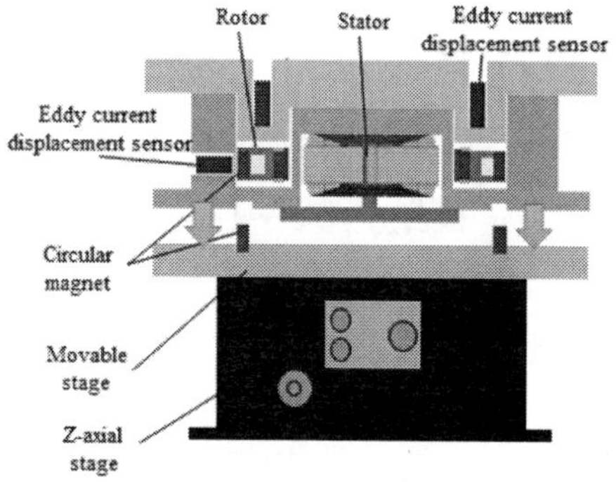

Figure 10. Schematic of the performance evaluation equipment of the hydrodynamic bearing.

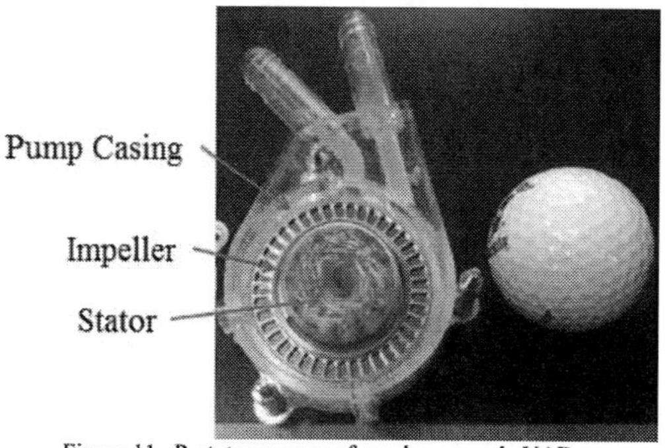

Figure 11. Prototype pump of maglev cascade VAD.

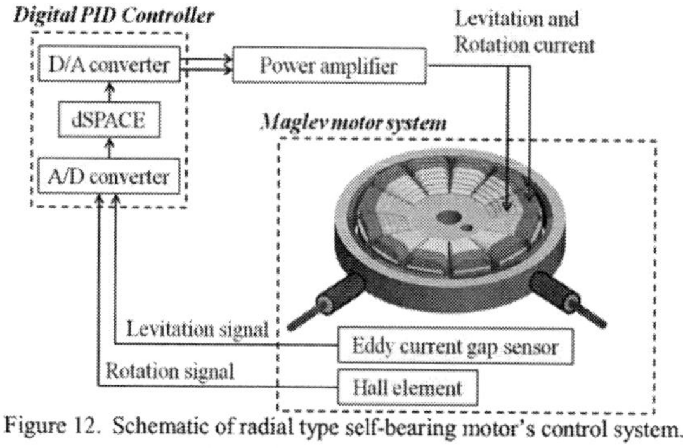

Figure 12. Schematic of radial type self-bearing motor's control system.

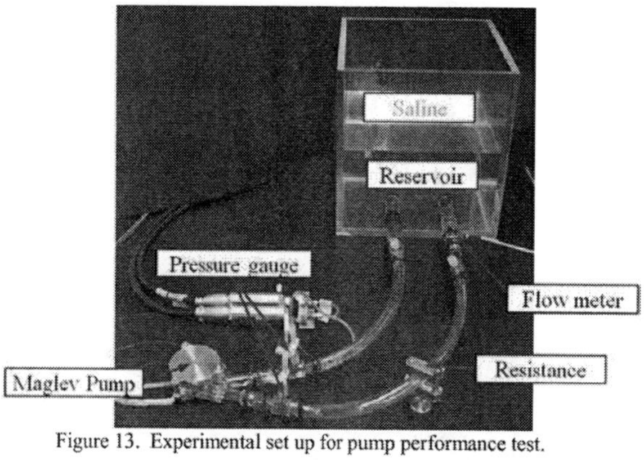

Figure 13. Experimental set up for pump performance test.

IV. RESULT

A. CFD analysis

Fig. 14 and 15 show hydrodynamic force in the axial direction and torque around Y axis corresponding to flow rate with a rotating speed of 2,000 rpm. Both of the axial hydrodynamic force and the torque around radial axis decreased with an increase in flow rate. The maximum axial hydrodynamic force was 0.16 N and the maximum torque around Y axis was 2.5 mNm at shutoff operation.

B. Magnetic field analysis

Fig. 16 and 17 show simulated restoring force in the axial direction and restoring torque around Y axis generated by the passive stability. The maximum restoring force in the axial direction was 1 N with a displacement of 0.4 mm. The restoring torque was 2.3 mNm with an inclination of 1.2 degrees.

C. Hydrodynamic bearing performance

Fig. 18 shows the relationship between the impeller axial position change and the rotating speed. The impeller was lifted off with the hydrodynamic effect when the rotating speed is over 1,000 rpm and the maximum levitated distance was 5 μm with a rotating speed of 3,000 rpm.

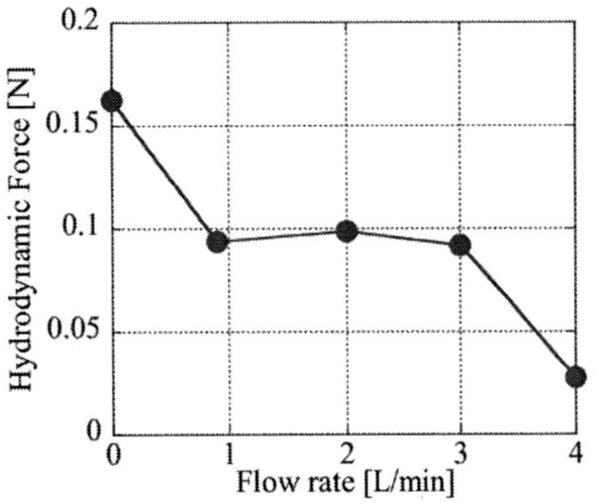

Figure 14. The relationship between the hydrodynamic force in the axial direction and flow rate calculated by the CFD analysis.

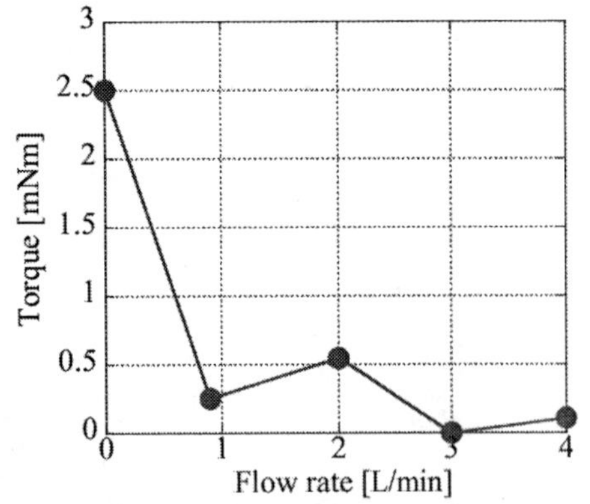

Figure 15. The relationship between the torque around the Y axis and flow rate calculated by the CFD analysis.

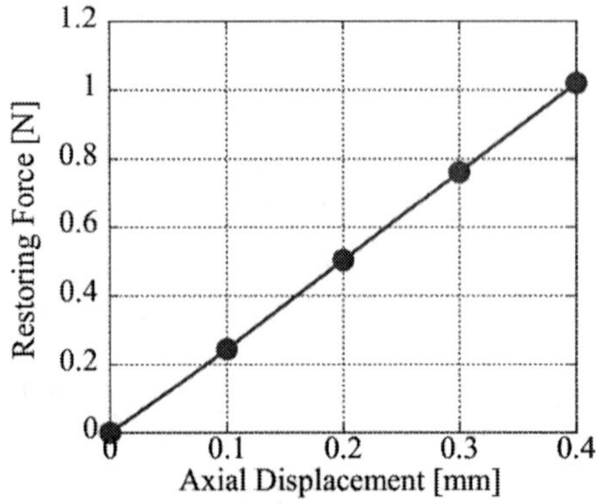

Figure 16. The relationship between the restoring force in the axial direction and the axial displacement of the rotor simulated by magnetic field analysis.

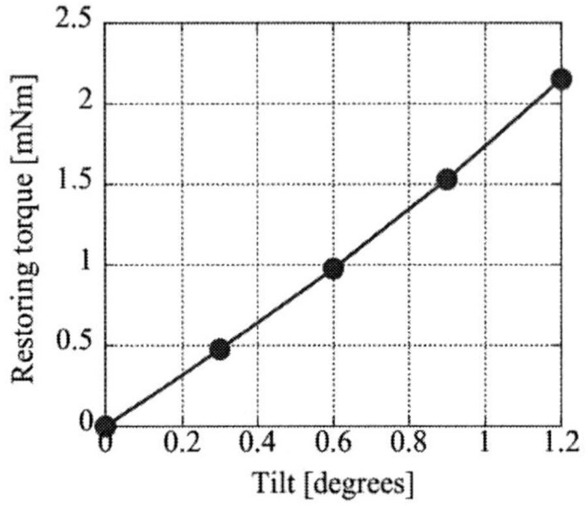

Figure 17. The relationship between the Y-axis restoring torque and the inclination of the rotor simulated by magnetic field analysis.

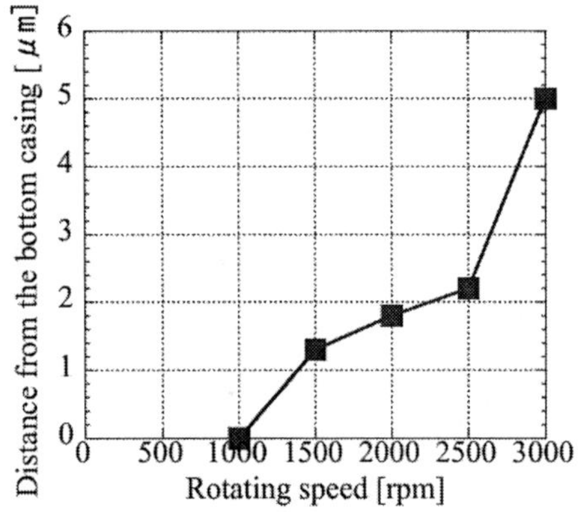

Figure 18. The relationship between the distance and the rotating speed of hydrodynamic bearing.

D. Pump performance

Fig. 19 shows the relationship between the pump head and the flow rate. Fig. 20 shows the relationship between maximum oscillation amplitude and the flow rate with different rotating speeds. The pump was levitated and rotated up to a rotating speed of 2,600 rpm successfully. The maximum head pressure and the maximum flow rate are 300 mmHg and 6.3 L/min.

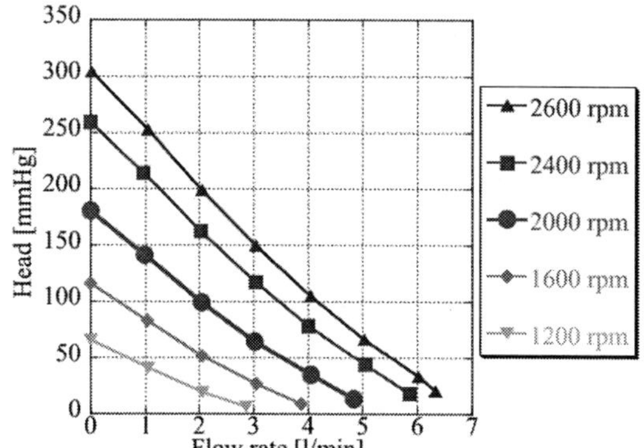

Figure 19. The relationship between the pump head and the flow rate.

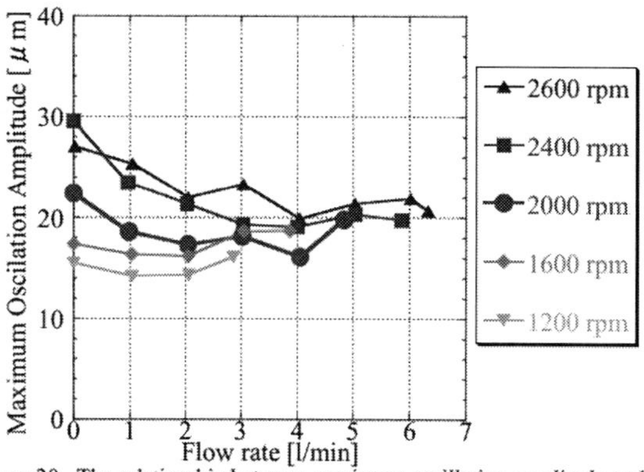

Figure 20. The relationship between maximum oscillation amplitude and the flow rate.

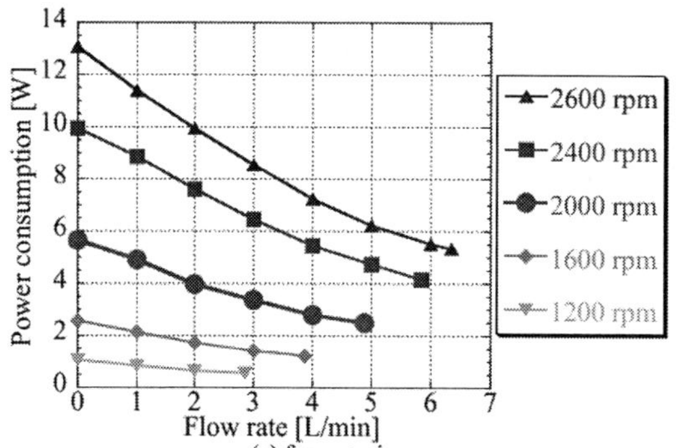

(a) for motoring

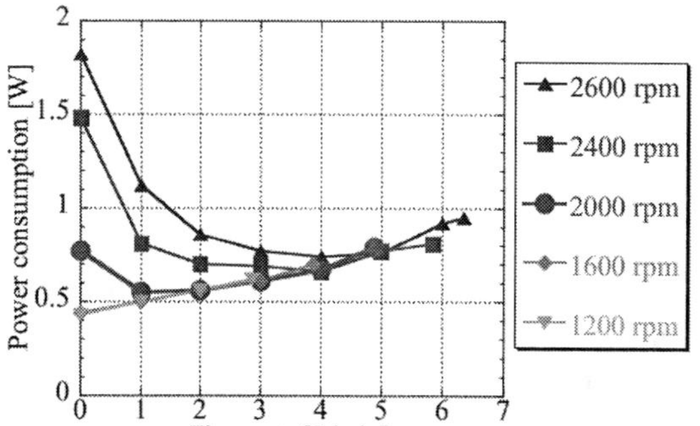

(b) for levitation

Figure 21. The relationship between the power consumption and the flow rate.

Fig. 21 shows the relationship between the power consumption and the flow rate. The maximum motor and levitation power consumption were 13 W and 1.8 W with a rotational speed of 2,600 rpm. The maximum oscillation amplitude of the levitated impeller is 30 μm with a rotating speed of 2,400 rpm. The target performance which is a flow rate of 2 L/min against a head pressure of 100 mmHg is achieved with a rotating speed of 2,000 rpm. The total power consumption and the maximum oscillation amplitude with the target performance were 4.5 W and 17 μm.

V. Discussion

The estimated axial fluid force by the CFD analysis was 0.01 N with a flow rate of 2 L/min and is smaller than an axial restoring force due to the passive stability that is 1 N estimated with the magnetic field analysis. The torque around the Y axis produced by the fluid force was 0.5 mNm and is also smaller than the maximum restoring torque by the passive stability which is 2.3 mNm. Therefore, the radial position and inclination of the levitated impeller is restricted by passive stability characteristics and it is also confirmed with the pumping experiment.

In the experiments of the hydrodynamic bearing, the levitated distance was 2 μm despite using glycerin-water solutions which has a viscosity of 0.1 Pa · s. The hydrodynamic bearing could produce lifting force to push the impeller away from the touchdown even the viscosity of the experimental working fluid is 10 times higher than the blood. The hydrodynamic bearing should be redesigned to enhance the lifting performance.

VI. Conclusion

A small, thin, maglev VAD, which can be implanted under the thorax muscle layer like a pacemaker, has been developed. A radial type self-bearing motor and a cascade pump are adopted to reduce the axial thickness of the device. The restoring force and toque produce by the passive stability in the uncontrolled axis were calculated by the magnetic field analysis and these are larger enough to restrict the levitated impeller posture compared with the fluid force which acts on the levitated impeller and was estimated by the CFD analysis. A developed hydrodynamic bearing to avoid touchdown of the impeller indicated the lift-off performance with higher viscous fluid. A prototype maglev pump indicated sufficient levitation performance and pump performance as a VAD.

References

[1] Eric A.Rose , "Long-term use of a left ventricure assist device for end-stage heart failure" , *The New England Journal of Medicine* , Vol.345,2001.11.

[2] K. Ukita, T. Masuzawa, H. Onuma, T. Nishimura, and S. Kyo, "A Radial Type Self-Bearing Motor for Small Maglev Regenerative Blood Pump ", Journal of the Japan Society of Applied Electromagnetics and Mechanics,Vol. 20, No. 2, pp. 312-318, 2012.

[3] H. Onuma, T. Masuzawa, K. Matsuda, and Y. Okada, "Magnetically levitated centrifugal blood pump with radially suspend self-bearing motor", Proc. 8th International Symposium on Magnetic Bearings, pp. 3-8, 2002.

[4] E.A.Muijderman, "SPAIRAL GROOVE BEARINGS", *Philips technical*, 1996.

Evaluation of Magnetic Suspension Characteristics and Levitation Performance of A Centrifugal Blood Pump Using Radial Type Self-Bearing Motor

Hiroyuki Onuma[a], Toru Masuzawa[b]

[a] Department of Electronics and Control Engineering, Ibaraki National College of Technology, 866 Nakane, Hitachinaka, Ibaraki 312-8508, Japan, onuma@ss.ibaraki-ct.ac.jp

[b] College of Engineering, Ibaraki University, 4-12-1 Nakanarusawa, Hitachi, Ibaraki 316-8511, Japan

***Abstract*—Magnetic levitation technology is applied for durability enhancement of artificial hearts. A centrifugal blood pump using a self-bearing motor was previously developed for use as an implantable artificial heart. Rotors of three types which are different of magnetic suspension characteristics were fabricated. The magnetic suspension characteristics, the levitation and the pump performances of the Maglev pump were examined. If the negative spring force in the radial direction was weak, the restoring force in the axial direction and the restoring torque of the inclination were not always weak. It is possible to improve the motor performance by increasing the flux density of permanent magnet, even if the air gap is increased. It revealed that the levitation performance does not influence the pump performance. And then, it is possible to improve the levitation performance by reducing the negative spring force and improving the passive stability ability.**

I. Introduction

Magnetic levitation technology is applied for durability enhancement of an artificial hearts [1-8]. The non-contacting behavior of the artificial hearts using the magnetic levitation technology is good for high durability, lower hemolytic properties, and anti-thrombogenesity. A centrifugal blood pump using a self-bearing motor (Maglev pump) was previously developed for use as an implantable artificial heart [1-4]. In order to realize an implantable artificial heart for a small patient, miniaturization and high efficiency of the Maglev pump is necessary. To find ways of miniaturization and high efficiency, it is necessary to verify in greater detail the magnetic suspension characteristics, and to demonstrate the influence of the magnetic suspension characteristics on the levitation performance. In this study, rotors of three types which are different of magnetic suspension characteristics were fabricated, and the magnetic suspension characteristics of self-bearing motor of three type's rotor were measured. The levitation and the pump performances of the Maglev pump were examined.

II. Methods

A. Centrifugal Blood Pump Using Self-Bearing Motor

Fig. 1 shows the centrifugal blood pump using the self-bearing motor. The radial position and rotation of the rotor-impeller of the Maglev pump were actively controlled through control of the magnetic field of the self-bearing motor. The axial position and inclination of the rotor-impeller were passively suspended by the strong magnetic attractive force of permanent magnets positioned on the inner surface of the rotor-impeller. The Maglev pump's outer diameter and thickness were 78.5 mm and 41.5 mm, respectively. Fig. 2 shows the self-bearing motor. The outer rotor structure, that a rotor is set around a stator, is adopted to miniaturize the Maglev pump. The rotor, which is a yoke itself, has eight permanent magnets on inner circumferential surface. The stator set at the center of the pump has twelve poles. The plus minus two-pole algorithm [9] is adopted to levitate and rotate the rotor. The rotation coil to produce 3-phase 8-pole magnetic field and the levitation coil to produce 2-phase 6-pole magnetic field are separately constructed in the stator. Two eddy current sensors and three hall elements are used to measure rotor radial position and rotation speed respectively.

B. Design of Rotor

Rotors of three types which are different of magnetic suspension characteristics were fabricated. Parameters of created rotors are shown in Table I. Based on previous studies [1], rotors were designed. Type-A and Type-B are the same shape. And, Type-A was employed the permanent magnet's material with good ability than that of Type-B. The permanent magnet's material of Type-B and Type-C are same.

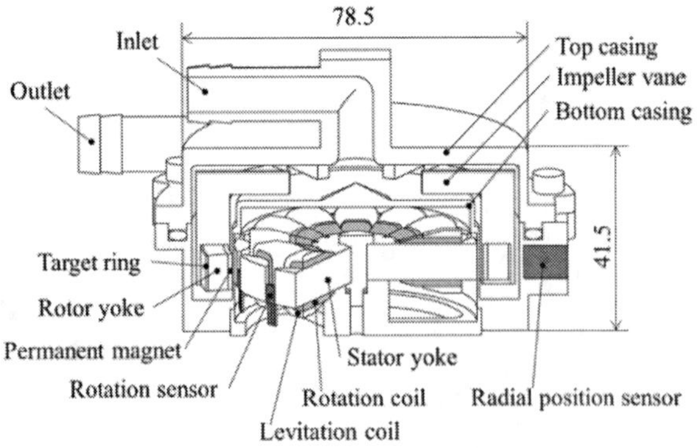

Figure 1. Schematic illustration of Maglev pump.

And, the thickness of the permanent magnet and the air gap length of Type-C are thicker and longer than those of Type-B. Therefore, the strength of the magnetic suspension force in the radial direction is Type-A, Type-B and Type-C, in that order.

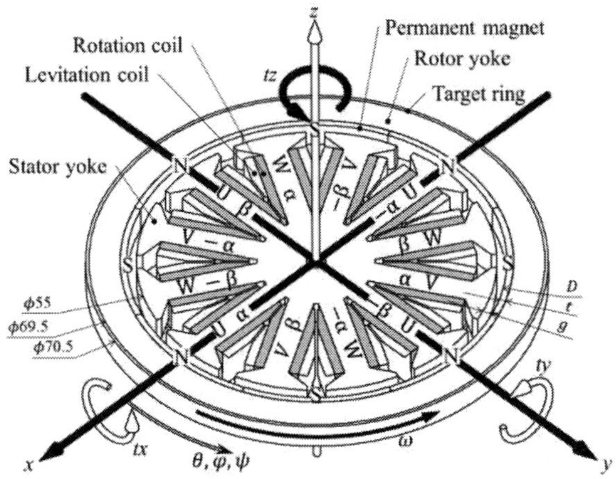

Figure 2. Cross-sectional perspective view of self-bearing motor.

TABLE I. PARAMETER OF DESIGNED ROTOR

Parameter		Type-A	Type-B	Type-C
Inner diameter of permanent magnet	D	58 mm	58 mm	59 mm
Air gap length between stator and rotor	g	1.5 mm	1.5 mm	2 mm
Thickness of permanent magnet	t	1 mm	1 mm	1.5 mm
Coercive force of permanent magnet		756 kA/m	692 kA/m	692 kA/m
Residual flux density of permanent magnet		1.03 T	0.95 T	0.95 T
Magnetic flux density in air gap generated by permanent magnets		0.392 T	0.360 T	0.387 T

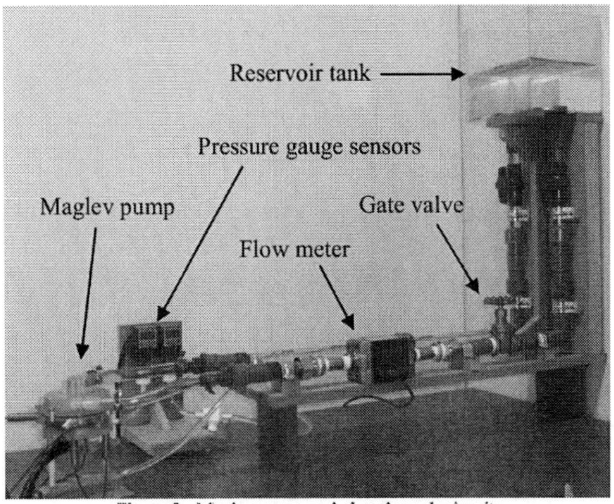

Figure 3. Maglev pump and closed mock circuit.

III. EXPERIMENTS

A. Measurement of Magnetic Suspension Characteristics

(1) Measurement of Attractive Force in Radial Direction

The attractive force in the radial direction was measured with the load cell to evaluate the magnetic suspension force and negative spring force in the radial direction. The rotor was fixed to the center position of Fig.1. To evaluate the magnetic suspension force, the excitation current was changed from 0 A to 3.0 A. To evaluate the negative spring force, the displacement of the rotor from the center position was changed from 0 mm to 0.5 mm.

(2) Measurement of Restoring Force in Axial Direction

The restoring force in the axial direction was measured with the load cell to evaluate the passive stability performance in the axial direction. The axial displacement of the rotor was changed from 0 mm to 8.0 mm.

(3) Measurement of Restoring Torque of Inclination

The restoring torque of the inclination was measured with the irrotational torque meter to evaluate the passive stability performance of the inclination. The inclination angle of the rotor was changed from 0 degrees to 5 degrees, at the axial position of 0 mm and 1 mm.

B. Measurement of Motor Performance

The motor power consumption to the load torque applied to the motor was measured to evaluate of the motor performance. And then, the motor efficiency was calculated from the shaft output of the motor and the motor power consumption. The rotor was attached to a shaft of the torque meter. The torque and electrical input power to the motor was measured with the torque meter and a power meter. The load torque to the motor was changed from 3 mNm to 50 mNm.

C. Measurement of Levitation and Pump Performances

The Maglev pump was connected to a closed mock circuit filled with water, and pump and levitation performances were examined. The closed mock circuit is composed of two pressure gauge sensors, a flow meter, a reservoir tank, and a gate valve. The Maglev pump and the closed mock circuit are shown in Fig. 3. The oscillation amplitude of the rotor-impeller was measured to evaluate the levitation performance during pumping. Then, the electrical input power to the maglev pump was measured with the power meter. In addition to two position sensors using to control the radial position of the rotor-impeller, four position sensors were placed on the bottom face of the Maglev pump to measure the axial displacement and the inclination of the rotor-impeller. Rotation and levitation of the rotor-impeller were controlled with a digital PID controller using a digital signal processor. The parameters of the PID controller were Kp = 5 A/mm, Ki = 5 A/mm/s, Kd = 0.02 A · s/mm respectively. The PID parameter of Type-A, the Type-B and Type-C is the same value. Rotational speed was set to 1500 rpm to achieve a flow rate of 5 L/min against a head pressure of 100 mmHg that is the pump condition of a ventricular assist device.

IV. Result

A. *Measurement of Magnetic Suspension Characteristics*

(1) *Measurement of Attractive Force in Radial Direction*

The measured magnetic suspension force in the radial direction is shown in Fig. 4. The magnetic attractive force of Type-A, Type-B and Type-C generated by a current of 1 A were 6.8 N, 6.2 N and 5.2 N, respectively. As designed, it was verified that the strength of the magnetic suspension force is Type-A, Type-B, and Type-C, in that order.

The measured negative spring force in the radial direction is shown in Fig. 5. The magnetic attractive force of Type-A, Type-B and Type-C generated by displacement of 0.1 mm were 2.1 N, 2.0 N and 1.2 N, respectively. Negative spring force was strong in the same order as the measurement of the magnetic suspension force.

(2) *Measurement of Restoring Force in Axial Direction*

The measured restoring force in the axial direction is shown in Fig. 6. The magnetic attractive force of Type-A, Type-B and Type-C generated by displacement of 1 mm were 3.3 N, 2.8 N and 3.0 N, respectively. The maximum of the magnetic attractive force of Type-A, Type-B, and Type-C were 13.5 N, 10.8 N and 12.6 N at 6 mm displacement, respectively. The strength of the restoring force in the axial direction is Type-A, Type-C, and Type-B, in that order.

(3) *Measurement of Restoring Torque of Inclination*

The measured restoring torque of inclination is shown in Fig. 7. The restoring torque of Type-A, Type-B and Type-C generated by inclination of 1 degree and axial displacement of 0 mm were 0.009 Nm, 0.006 Nm and 0.008 Nm, respectively. The restoring torque of Type-A, Type-B and Type-C generated by inclination of 1 degree and axial displacement of 1 mm were 0.013 Nm, 0.009 Nm and 0.011 Nm, respectively. The strength of the restoring torque of inclination is Type-A, Type-C, and Type-B, in that order. When the rotor 1mm displaced in the axial direction and the inclination angle of the rotor is the range of up to 3 degrees, the restoring torque is stronger than that in the case of no axial displacement. When the inclination angle of the rotor reaches 4 degrees or more, the restoring torque is weaker than that in the case of no axial displacement.

B. *Measurement of Motor Performance*

Fig. 8 shows the motor power consumption against each load torque at a rotational speed of 1400 rpm and 1600 rpm. When the load torque was 40 mNm at a rotational speed of 1400 rpm, motor power consumption of Type-A, Type-B and Type-C were 8.0 W, 8.6 W and 8.0 W respectively. When the load torque was 40 mNm at a rotational speed of 1600 rpm, motor power consumption of Type-A, Type-B and Type-C

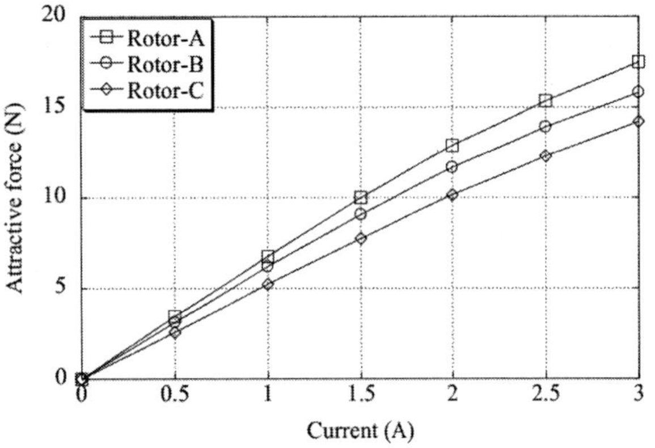

Figure 4. Magnetic suspension force in the radial direction.

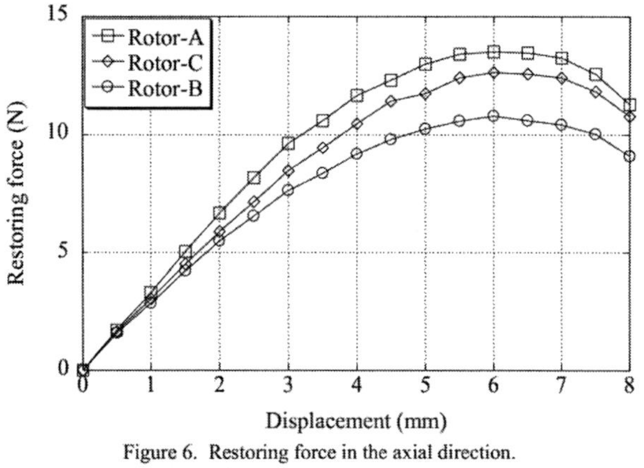

Figure 6. Restoring force in the axial direction.

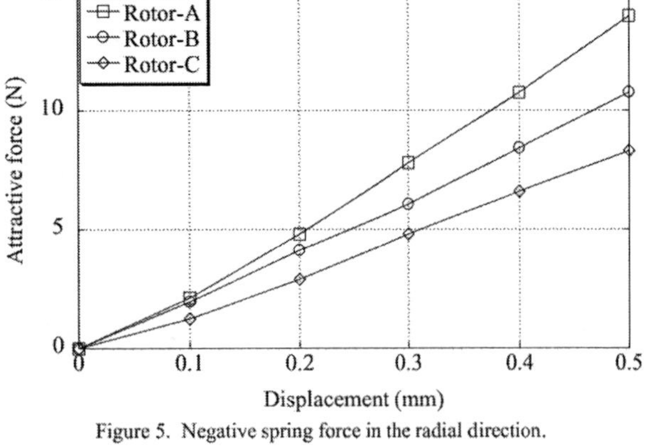

Figure 5. Negative spring force in the radial direction.

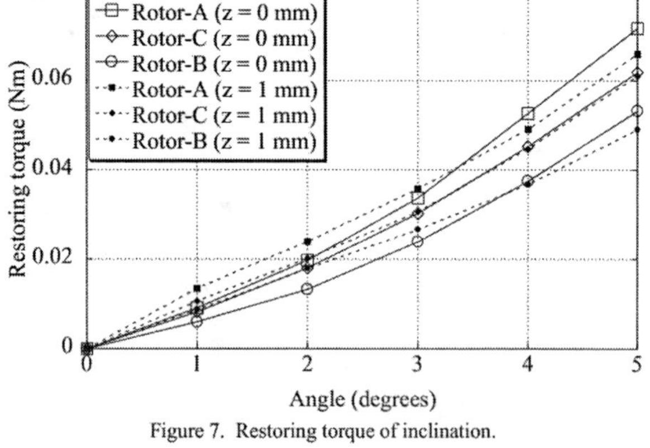

Figure 7. Restoring torque of inclination.

were 9.2 W, 9.7 W and 9.2 W respectively. Motor power consumption of Type-A and Type-C were approximately equal. And, these were lower than that of Type-B.

Fig. 9 shows the motor efficiency against each load torque at a rotational speed of 1400 rpm and 1600 rpm. At a load torque of 40 mNm, the motor efficiency was about maximum value. When the load torque was 40 mNm at a rotational speed of 1400 rpm, motor efficiency of Type-A, Type-B and Type-C were 73.7 %, 68.3 % and 73.1 % respectively. When the load torque was 40 mNm at a rotational speed of 1600 rpm, motor efficiency of Type-A, Type-B and Type-C were 72.7 %, 68.8 % and 73.0 % respectively. There is almost no difference between the motor efficiency of Type-A and Type-C at a torque load of 40 mNm. At a load torque is 30 mNm or less, the motor efficiency of Type-C was good compared to the other.

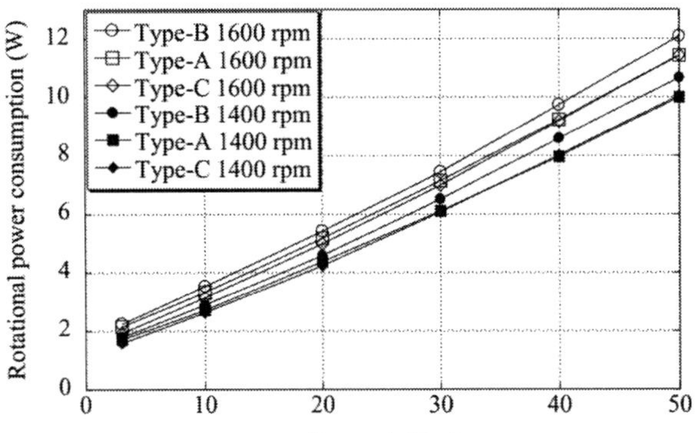

Figure 8. Rotational power consumption vs Motor torque.

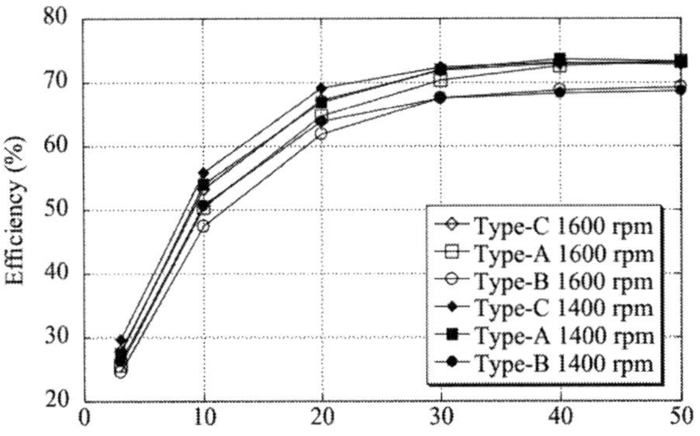

Figure 9. Motor efficiency vs Motor torque.

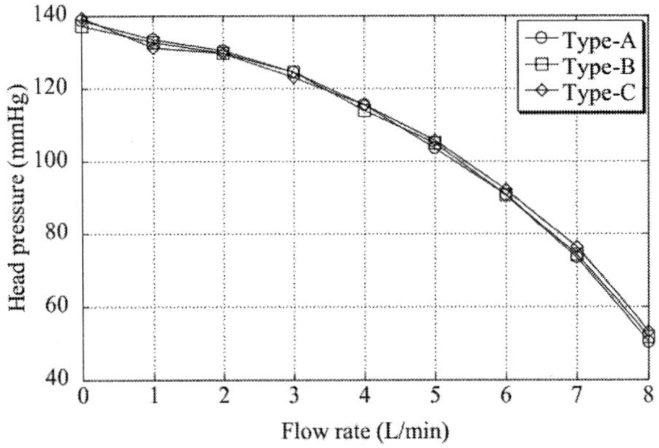

Figure 10. HQ characteristics.

C. Measurement of Levitation and Pump Performances

Fig. 10 shows HQ characteristics of Type-A, Type-B and Type-C with a rotational speed of 1500 rpm. At a flow rate of 5 L/min, head pressure of Type-A, Type-B and Type-C were 104 mmHg, 105 mmHg and 106 mmHg respectively. There was almost no difference in HQ characteristics. Maximum head pressure and flow rate were approximately 140 mm Hg and 8 L/min, respectively.

Fig. 11 shows oscillation amplitude in the radial direction (x, y) and axial direction (z) of the rotor-impeller against each flow rate at a rotational speed of 1500 rpm. At a flow rate of 5 L/min, oscillation amplitude in x, y and z direction of Type-A were 0.13 mm, 0.12 mm and 0.058 mm respectively. At a flow rate of 5 L/min, oscillation amplitude in x, y and z direction of Type-B were 0.15 mm, 0.13 mm and 0.036 mm respectively. At a flow rate of 5 L/min, oscillation amplitude in x, y and z direction of Type-C were 0.076 mm, 0.061 mm and 0.047 mm respectively. Fig. 12 shows oscillation amplitude in the inclination (tx, ty) of the rotor-impeller against each flow rate at a rotational speed of 1500 rpm. At a flow rate of 5 L/min, oscillation amplitude in tx and ty direction of Type-A were 0.65 degree and 0.73 degree, respectively. At a flow rate of 5 L/min, oscillation amplitude in tx and ty direction of Type-B were 0.33 degree and 0.38 degree, respectively. At a flow rate of 5 L/min, oscillation amplitude in tx and ty direction of Type-C were 0.35 degree and 0.39 degree, respectively. In all types, the rotor-impeller did not touch the casing during the experiment. The rotor-impeller of Type-C is most stable. Flow channel width in the radial direction and axial direction of Maglev pump are 0.5 mm and 1.5 mm, respectively. And, the rotor-impeller can be tilted geometrically 2 degrees in flow channel.

Fig. 13 shows the rotation, levitation and total power consumption against each flow rate with a rotational speed of 1500 rpm. Total power consumption is the sum of the rotation and levitation power consumption. At a flow rate of 5 L/min, rotation power consumption of Type-A, Type-B and Type-C were 8.6 W, 8.9 W and 8.6 W respectively. At a flow rate of 5 L/min, levitation power consumption of Type-A, Type-B and Type-C were 2.3 W, 4.9 W and 1.4 W respectively. At a flow rate of 5 L/min, total power consumption of Type-A, Type-B and Type-C were 10.9 W, 13.9 W and 10.0 W respectively. Levitation power consumption was substantially constant even if the flow rate changes. Levitation power consumption of Type-C was lower than that of the other. Levitation power consumption of the Type-B was higher most. Rotation power consumption was increased with an increase in the flow rate. There was no large difference in rotation power consumption of Type-A, Type-B and Type-C. Rotation power consumption of the Type-A and Type-C were approximately equal. And, these were slightly lower than that of Type-B. Therefore, total power consumption of Type-C was lower most.

V. Discussion

Magnetic suspension characteristics depend on factors such as the magnetic flux density in air gap generated by permanent magnets and the stator coil, and the air gap length.

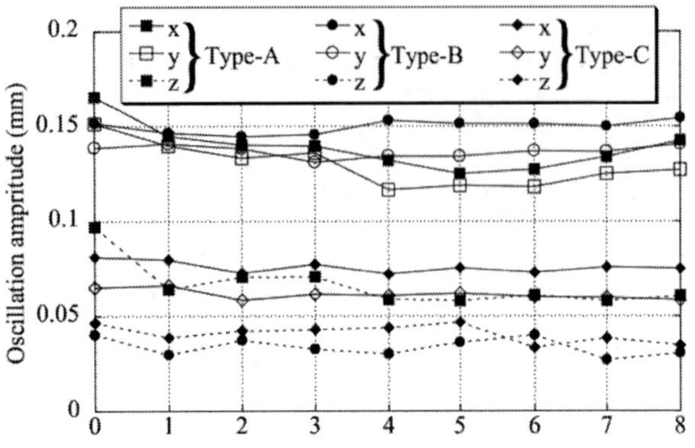

Figure 11. Oscillation amplitude of rotor-impeller in x, y, z direction.

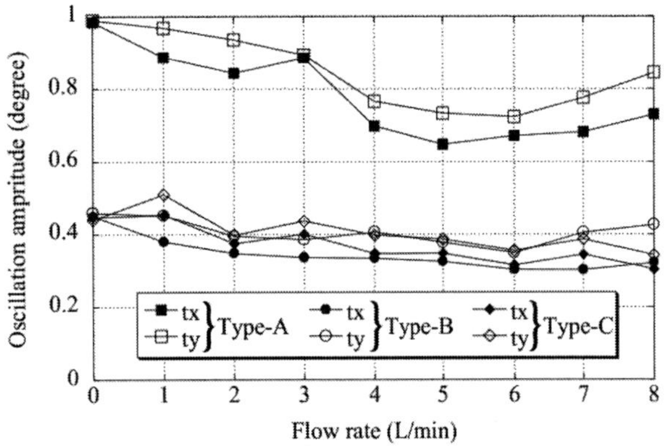

Figure 12. Oscillation amplitude of rotor-impeller in tx, ty direction.

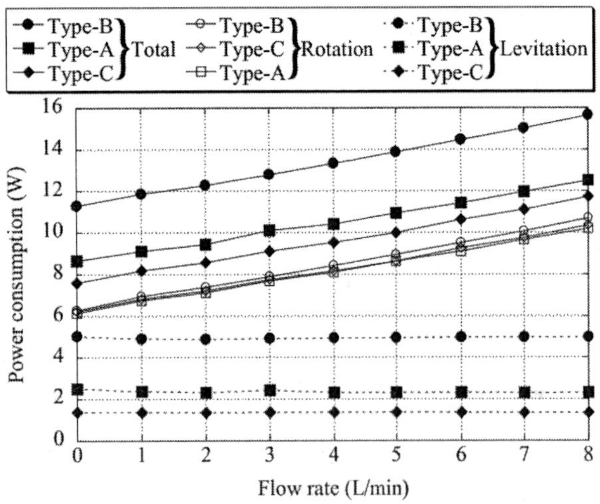

Figure 13. Power consumption in pumping.

The height of the magnetic flux density in air gap generated by permanent magnets is Type-A, Type-C and Type-B, in that order. However, Type-A and Type-C are similar in the height of the magnetic flux density in air gap generated by permanent magnets. The air gap length of Type-C is larger than the air gap of Type-A and Type-B. In the self-bearing motor of Type-A, Type-B and Type-C, the same stator was used. Therefore, it is conceivable that the air gap length have a large impact on determining the strength of the attractive force in the radial direction, whereas the air gap length has a small impact on determining the strength of the restoring force in the axial direction and the restoring torque of the inclination.

There was no difference of the motor performance of Type-C and Type-A at 40 mNm torque load. In load torque of 30 mNm or less, motor performance of Type-C was good compared to the other. The air gap length of Type-C is larger than the air gap of Type-A. But, the height of the magnetic flux density in air gap generated by permanent magnets of Type-A and Type-C are almost similar. Type-C has the most magnetic energy stored in the air gap. Therefore, motor performance of type C was good. The difference between the motor performance of Type-C and Type-A was reduced, probably due to the influence increase of magnetic saturation with an increase in the load torque. It is possible to improve the motor performance by increasing the flux density of permanent magnet, even if the air gap is increased.

There was almost no difference in HQ characteristics of Type-A, Type-B and Type-C. Therefore, it revealed that the levitation performance does not influence the pump performance.

Oscillation amplitude in the radial direction of the Type-C was lower most. It seems that this is because negative spring force was weaker most. Oscillation amplitude in the radial direction of the Type-B and Type-A were nearly equal value. However, oscillation amplitude in the inclination of the Type-A was large. In previous studies, it is clear that the force in the axial direction and the torque in the inclination are generated by generating a magnetic suspension force in the radial direction, in the case of the rotor-impeller is decentered [10]. If negative spring force is strong, it is possible that the interferential effect is stronger. The negative spring force of Type-A was the strongest. Consequently, the levitation performance of Type-C was the best. Total power consumption of Type-C that is the smallest oscillation amplitude in the radial direction was low. Levitation power consumption of Type-C that is the smallest oscillation amplitude in the radial direction was lower most. Oscillation amplitude have exert little influence on levitation power consumption. Consequently, it is possible to improve the levitation performance by reducing the negative spring force and improving the passive stability ability.

VI. Concludion

Rotors of three types which are different of magnetic suspension characteristics were fabricated. The magnetic suspension characteristics of self-bearing motor of three type's rotor were measured. If the negative spring force in the radial direction was weak, the restoring force in the axial direction and the restoring torque of the inclination were not always weak. Magnetic suspension characteristics depend on factors such as the magnetic flux density in air gap generated by permanent magnets and the stator coil, and the air gap length. It is conceivable that the air gap length have a large impact on determining the strength of the attractive force in the radial direction, whereas the air gap length has a small impact on determining the strength of the restoring force in the axial direction and the restoring torque of the inclination. As a result of measurement of motor performance, it is possible to improve the motor performance by increasing the flux density of permanent magnet, even if the air gap is increased. As a result of measurement of levitation and pump performance, it revealed that the levitation performance does not influence the pump performance. And then, it is possible to improve the levitation performance by reducing the negative spring force and improving the passive stability ability.

ACKNOWLEDGMENT

This work was supported by JSPS KAKENHI Grant Number 25820077.

REFERENCES

[1] H. Onuma and T. Masuzawa, " Estimation of suspension characteristics of a radial type self-bearing motor using 12 slot stator," *Proc. 22th MAGDA Conference inMiyazaki*, pp.79-84, 2013.(in Japanese)

[2] H. Onuma, K. Ukita and T. Masuzawa, " Optimum pole number of a radial type self-bearing motor using 12 slots stator for artificial heart," *Journal of JSAEM*, vol. 20, no. 1, pp.59-65, 2012. (in Japanese)

[3] H. Onuma, T. Masuzawa, K. Matsuda and Y.Okada, " Magnetically levitated centrifugal blood pump with radially suspend self-bearing motor," *Proc. 8th International Symposium on Magnetic Bearings*, pp. 3-8, 2002.

[4] T. Masuzawa, H. Onuma,SJ Kim, Y. Okada, " Magnetically suspended centrifugal blood pump with a self-bearing motor," *ASAIO Journal*, vol. 48, no. 4, pp.437-442, 2002.

[5] K. Ukita, T. Masuzawa, H. Onuma, T. Nishimura and S. Kyo, " A Radial Type Self-Bearing Motor for Small Maglev Regenerative Blood Pump," *Journal of JSAEM*, vol. 20, no. 2, pp.312-318, 2012.

[6] T. Nishinaka, H. Schima, W. Roethy, A. Rajek, C. Nojiri, E. Wolner, and G M Wieselthaler," The DuraHeart VAD, a Magnetically Levitated Centrifugal pump: The University of Vienna Bridge-to-transplant Experience", *Circulation journal*, vol. 70, no. 11, pp.1421-1425, 2006.

[7] H. Onuma, M. Murakami and T. Masuzawa, " Novel Maglev pump with a combined magnetic bearing", *ASAIO Journal*, vol. 51, no. 1, pp.50-55, 2005.

[8] T. Merkel, A. Arndt, J. Hoffmann, P. Nusser, K. Graichen, W. Neumann and J. Muller, " Magnetic bearing in INCOR axial blood pump acts as multifunctional sensor", *Proc. 9th International Symposium on Magnetic Bearings*, #33 on CDROM, 2004.

[9] Y. Okada, T. Ohishi and K. Dejima, " General Solution of Levitation Control of a Permanent Magnet (PM) –Type Rotating Motor," *JSME International Journal Series C*, vol. 38, no. 3, pp.538-542, 1995.

[10] H. Onuma, K. Ukita and T. Masuzawa, " Effect of rotor displacement on a magnetically levitated motor for artificial heart", *The 54th Japan joint automatic control conference*, pp.1481-1482, 2011. (in Japanese)

Development of Highly Efficient Magnetic Bearing and Application to Ultra-Low Temperature Fluid Pump

Yohji Okada[a], Hironari Suzuki[a], Ken-ichi Matsuda[a], Ryou Kondo[a], and Masato Enokizono[b]

[a] Ibaraki University, Nakanarusawa 4-12-1, Hitachi, Ibaraki Pref. 316-8511 Japan, y.okada@mx.ibaraki.ac.jp
[b] Oita University, Dannoharu 700, Oita, Oita Pref. 870-1192, Japan

***Abstract*— Highly efficient hybrid type magnetic bearing is proposed in this paper. It is intended to apply the proposed magnetic bearing to ultra low temperature fluid pump. Traditional low temperature pump is supported by a ceramic ball bearing which is damaged by the low temperature liquid. Non-contact support is highly requested. However, the standard electro-magnetic bearing consumes high electric power causing heat to the liquid. This paper proposes highly efficient hybrid type radial magnetic bearing which can operate with very low power consumption. Also the pump is canned by the FRP pipe which has thick wall to separate the stator from the liquid. The designed magnetic bearing has 3 mm wide air gap with high force factor.**

I. Introduction

Traditional ultra-low temperature pump uses seramic ball bearing in the liquid. Hence there are several problems of short bearing life time, polution of powdered ceramic and heart transfer to the liquid.

The standard electro-magnetic bearing consumes high electric power and is not applicable [1]. In this paper highly efficient magnetic bearing is proposed for supporting the pump in ultra-low tenperature liquid. A centrifugal pump is supported by a hybrid (HB) type radial magnetic bearing and the shaft is set in vertical position as shown in Fig. 1. The top of the shaft is supported by a standard ball bearing and motor, since the low temperature liquid is under this level. The stator of HB magnetic bearing is better to be separated from the fluid. A FRP pipe can use in such a low temperature which has thick wall of 2 [mm] thickness. The designed magnetic bearing has 3 mm wide air gap with high force factor [2]-[8]. Highly efficient magnetic bearing is realized by using strong bias flux produced by wide area neodymium permanent magnet (PM) and high cupper ratio by using a pressed winding coils made by Selco Co., Ltd [9].

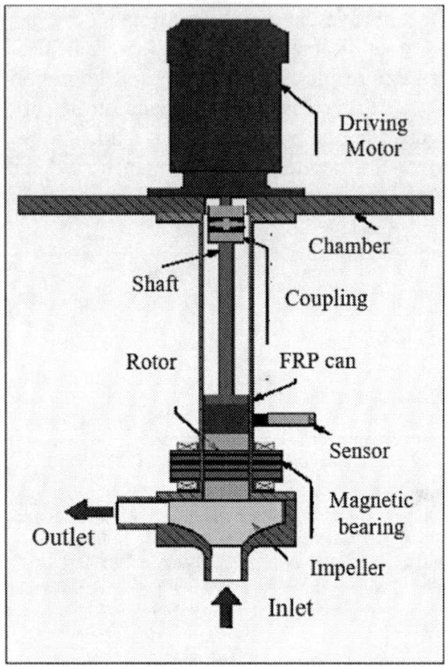

Figure 1. Ultra-low temperature fluid pump

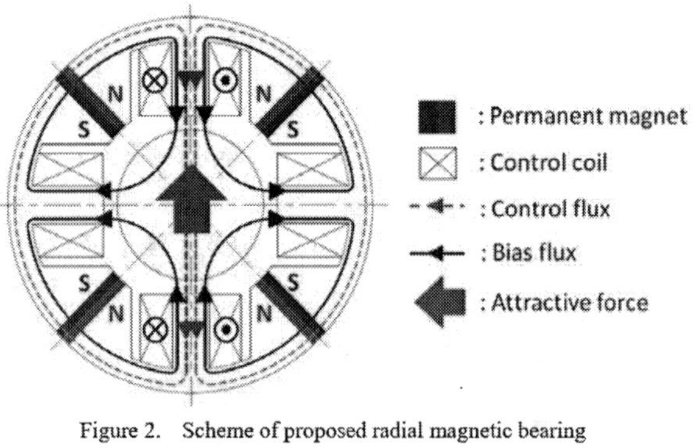

Figure 2. Scheme of proposed radial magnetic bearing

II. Highly Efficient Magnetic Bearing

Schematic of the proposed magnetic bearing is shown in Fig. 2. The stator has four main poles with pressed control windings. Between them sub-poles are arranged which has strong and wide area neodymium PMs to produce high bias flux as shown by the solid black lines. The upper and lower control windings are connected series. To produce upward force the current of two coils are flown as shown by the arrow symbols which produce the dotted red control flux lines. This increases the upper air gap flux while decreases the lower air gap flux. Similar control is applied to the left and right coil windings to control the left and right direction. Hence the radial two directions are actively controlled.

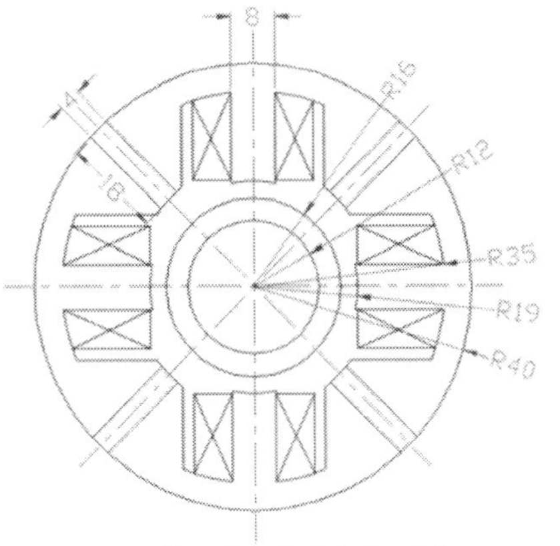

Figure 3. Analytical model

TABLE I Physical parameters

Item	Value, Material
Airgap	3 [mm]
Permanent magnet	Neodymium (NMX-48CH) ≒1000 [kA/m]
Coil turns	400 [turns/slot]
Core material	Electromagnetic silicon sheet 35A300

III. FEM Analysis and Design

The propsed magnetic bearing is designed through EFM analysis. Analytical model and the parameters are shown in Fig. 3 and Table 1, respectively. Analyzed air gap flux density, control force and negative force characteristics are shown by the dotted lines in Figs. 4, 5 and 6, respectively. The maximun flux of 0.48 [T], the control force factor of 26.1 [N/A] and the negaitve force factor of 25.4 [N/m] are obtained. These values are considered enough high compared with the wide gap of 3 [mm]. These graphs also include the experimental results mentioned later.

IV. Experimental Setup and Static Characteristics

To confirm the capability of the proposed magnetic bearing the designed magnetic bearing is fablicated and tested. The flux density is measured by a gauss meter (Lake Shore 421, MNT-4E02-VH) as shown in Fig. 7. The results are shown by the solid line in Fig. 4. The maximum flux density is about 0.36 [T] which is about 75 % of the calculated value. Also the control and the negative forces are measured as shown by the solid lines in Figs. 5 and 6, respectively. The force factor is about 21.1 [N/A] which is 88 % of the calculated one. The negative force factor is 16.4 [kN/m] which is 68 % of the calculated one. Even the measured values are lower than the calculated ones they are still enough high to control the pump.

V. Applying the Test Pump

To confirm the developed magnetic bearing a simple pump test is carried out. The developed pump is planned to apply to the liqid nitrogen. However, water pumping is tested due to the university laboratory limitation as shown in Fig. 8.

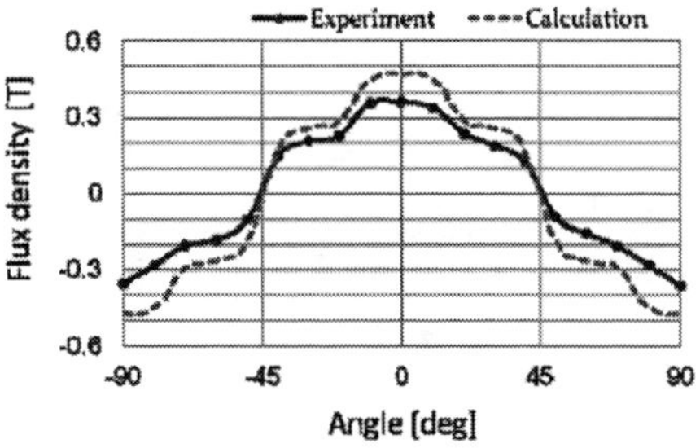

Figure 4. Flux densityies in air gap

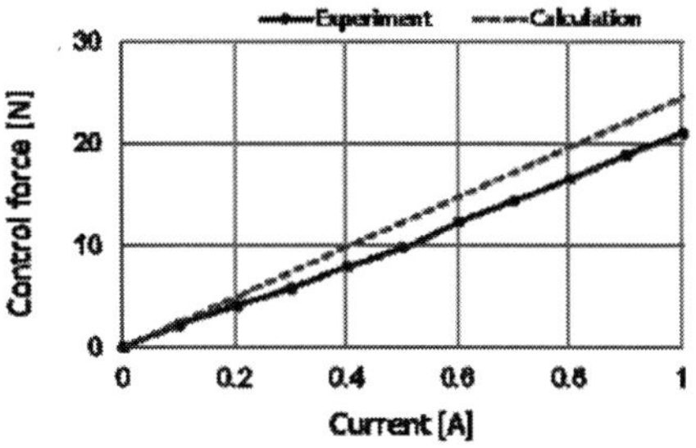

Figure 5. Control force characteristics

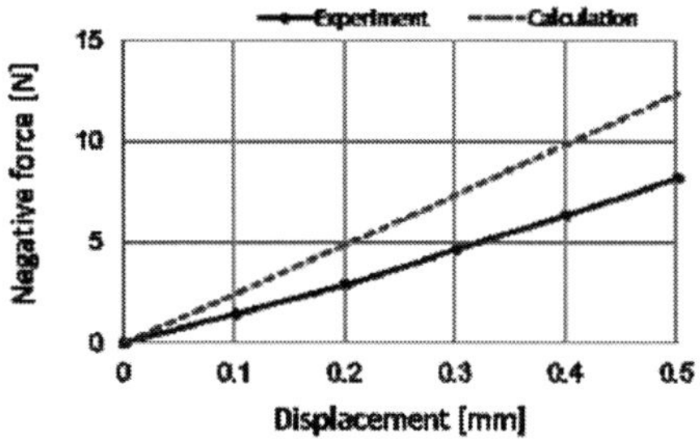

Figure 6. Negative force characteristics

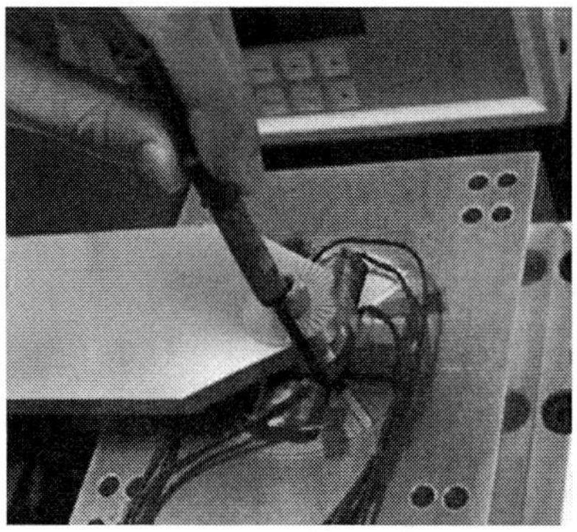

Figure 7. Fabricated magnetic bearing and flux measurement

As the same limitation the pumping pressure head is limited to 0.7 [m] while the actual pump is requested to be 8.7 [m]. Even the the tested pressure is low the levitation of the pump is stable. The controller used is the standard PID controller installed in dSPACE (DS-1104) the schematic of which is shown in Fig. 9. The controller gains are; $K_P = 10$ [A/m],

Figure 8. Pump test with water

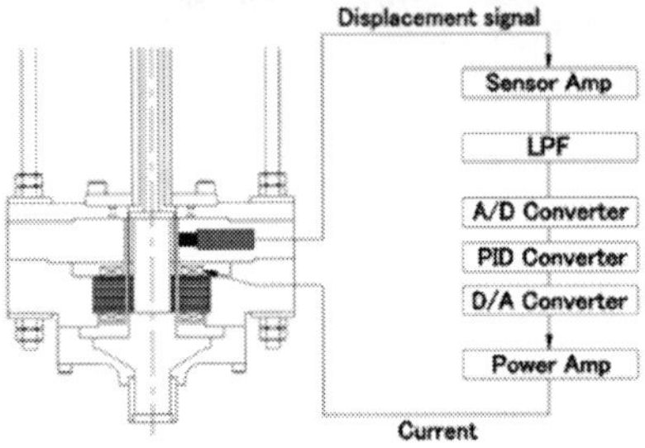

Figure 9. The control system

$K_D = 0.03$ [A/sm] with the derivative time constant $T_D = 0.8$ [ms], and $K_I = 0.0001$ [As/m]. The sampling interval used is $\tau = 0.1$ [ms].

VI. Results

Stable levitation test is realized during the pumping test. The flow rate versusu rotating speed is shown in Fig. 10 while the x- and y-directional power consumptions of magnetic bearing are shown in Fig. 11. Vibration level is under 50μm up to 5500 rpm. This means the high capability of the developed magnetic bearing to apply to the ultra-low temperature pump.

VII. Concluding Remarks

In this paper highly efficient magnetic bearing is developed which is planned to apply to ultra-low temperature pump. The develop magnetic bearing is successfully support the pump impeller up to 5500 rpm over the rated operating speed of 4800 rpm. Further research work is continuing to apply to high pressure output and actual fluid such as liquid nitrogen or liquid helium.

Acknowledgement

This work is supported by Oita Prefecture Collaboration of Regional Entities for Advancement of Technology Excellence, JST.

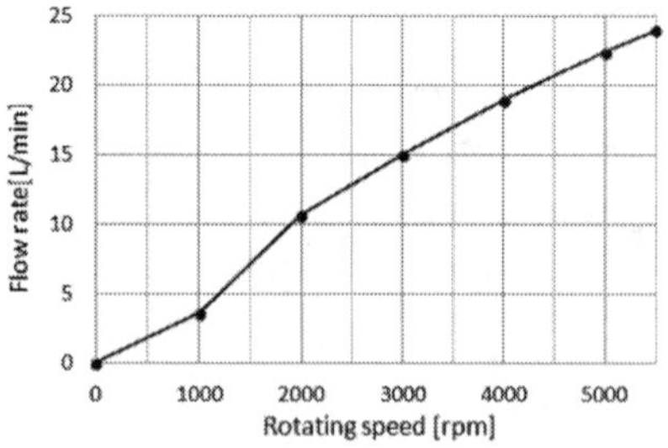

Figure 10 Flow rate versus rotating speed

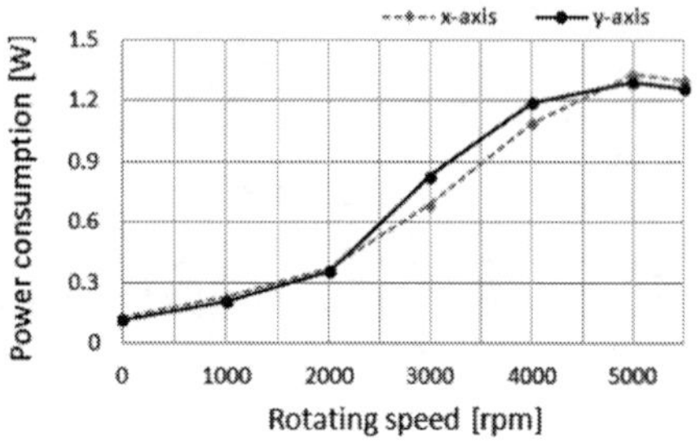

Figure 11 Power consumption of magnetic bearings versus rotating speed

References

[1] G. Schweitzer and E. H. Maslen, ed., "Magnetic Bearings", Springer, 2009.

[2] K. Kakihara, Y. Okada, and H. Koyanagi, "Development of IPM type Hybrid Magnetic Bearing", *Proc. of 6th Korea-Japan Symposium of Frontiers in Vibration Science and Technology*, Hitachi JAPAN, pp. 32-35, 2005.

[3] S. Kodama, et. al., "Development of 5 DOF IPM Type Magnetic Bearing", *CD-ROM Proc. of 8th Int. Conf. on Motion and Vibration Control*, pp. 1-6, 2006.

[4] R. Martin, et. al., "Development of A Low Cost Permanent Magnet Biased Bearing", *CD-ROM Proc. of 9th Int. Symp. on Magnetic Bearings*, Lexington, Kentucky, USA, pp.1-6, 2004.

[5] K. Kakihara, Y. Otsuka, N. Kurita, and Y. Okada, "Development of Wide-Gap Hybrid Magnetic Bearing", *CD-ROM Proc. of 8th Int. Conf. on Motion and Vibration Control*, KAIST, Daejeon, Korea, pp. 1-6, 2006.

[6] Y. Okada, K. Sagawa, E. Suzuki, and R. Kondo, "Development and Application of Parallel PM type Hybrid Magnetic Bearings", *JSME Journal of System Design and Dynamics*, pp.1-10, 2009.

[7] H. Miyazawa, Y. Okada, R. Kondo, and M. Enokizono, "Development of a Flux Concentrated PM Type Magnetic Bearings", *Proc. of First Japan-Korea Joint Symposium on Dynamics and Control*, Sapporo JAPAN, pp. 183-186, 2009.

[8] Y. Okada, H. Miyazawa, R. Kondo, and M. Enokizono, "Proposal of Flux Concentrated Radial and Axial Magnetic Bearings", *Applied Electromagnetic Engineering, Edited by A. G. Mamalis, M. Enokizono,* and A. Kladas, Trans Tech Publications LTD, pp. 435-446, (2011.

[9] http://selco-coil.com/

A Permanent Magnet Free Bearingless Motor For Disposable Centrifugal Blood Pump

Jun Rao[a], Wataru Hijikata[b] and Tadahiko Shinshi[b]

[a] Department of Mechano-Micro Engineering, Tokyo Institute of Technology, Yokohama, Japan, shinshi@pi.titech.ac.jp
[b] Precision and Intelligence Laboratory, Tokyo Institute of Technology, Yokohama, Japan

***Abstract*—The disadvantages of disposable centrifugal blood pumps (DCBPs) are the high cost of the disposable pump head due to the use of rare earth magnets and the complex-shaped secondary flow channel due to the space allocated for the magnetic coupling. In order to overcome these problems a permanent magnet free bearingless motor for DCBPs with a smooth secondary flow channel has been developed. The radial position and rotational speed of the rotor are actively controlled, and the axial position and tilt angle of the rotor are passively stabilized by a magnetic reluctance force. Results of magnetic field FEM analysis showed that the proposed bearingless motor has an axial stiffness of 3.50 N/mm and a tilt stiffness of 0.87 Nm/rad. In an experiment, the power consumption of the prototype was less than 30 W without load at a rotational speed of 3000 rpm.**

I. INTRODUCTION

To support blood circulation in heart failure patients, various types of blood pump have been developed and used at the clinical stage. In recent years, rotary blood pumps such as centrifugal and axial pumps, which generate continuous flow by rotating an impeller, have made progress in research and practical use [1].

Up till now, disposable centrifugal blood pumps (DCBPs) whose single-use pump heads can easily be attached to and detached from a reusable unit, have been used for circulation support in cardiopulmonary bypass operations, for extracorporeal membrane oxygenation, and for emergency circulatory resuscitation [2][3]. Not only the pressure-flow characteristics, but also prolonged durability, minimal destruction of blood cells (hemolysis), and no thrombus formation are important requirements for DCBPs. Furthermore, cost-effective disposable pump heads are also required for DCBPs.

Contactless support of the impeller by a magnetic bearing has been identified as a promising method for reducing blood damage and enhancing durability [4]-[7]. Our group has developed a DCBP employing a two-degree-of-freedom magnetic bearing [4]. Nevertheless, the shape of the secondary flow channel is very complicated due to the space allocated for the magnetic bearing and the contactless torque transmission mechanism as shown in Fig. 1. The complex secondary channel leads to small tolerances in manufacturing and a difficult fabrication process. A DCBP utilizing bearingless motor technology has also been proposed [5]. The disposable pump head of this DCBP utilizes a rare earth magnet. However, rare earth magnets increase the cost of the disposable pump head as well as causing serious damage to the environment during the mining and fabrication processes of the rare earth elements.

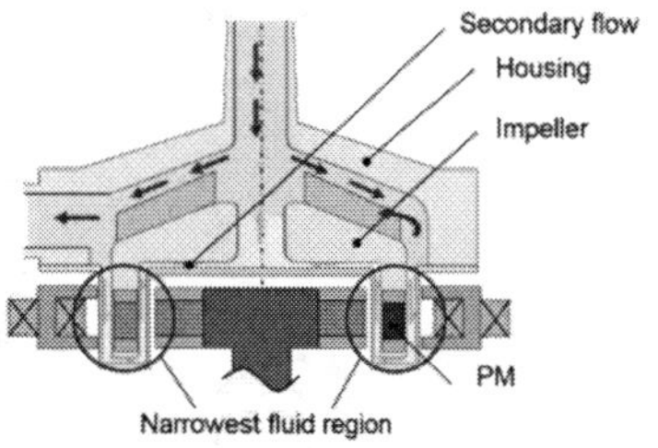

Figure 1 Disposable pump head structure of a previously proposed DCBP

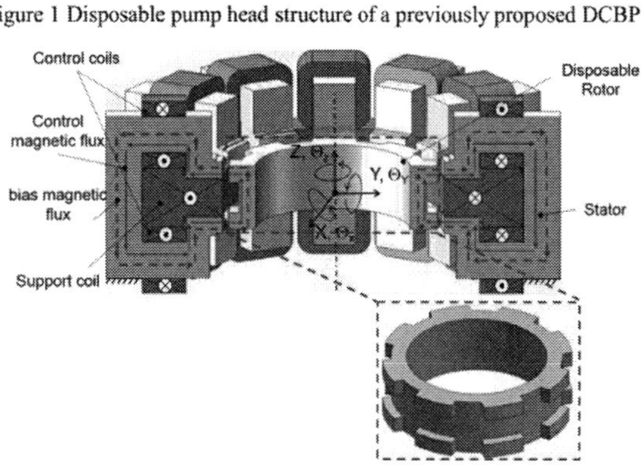

Figure 2 Structure of proposed bearingless motor

In this paper, a prototype bearingless motor for potential application in a DCBP is proposed. The proposed bearingless motor has a permanent magnet free rotor and a permanent magnet free magnetic bearing structure which is arranged around the rotor. Thus, a cost-effective pump head with a simple-shaped secondary flow channel can be achieved.

II. BEARINGLESS MOTOR FOR DCBP

A. Structure

Figure 2 shows the configuration of the proposed bearingless motor which is a hybrid of a two-degree-of-freedom controlled magnetic bearing and a three-phase twelve-eight-pole switched reluctance motor. The stator core consists of twelve C-shaped cores made of soft magnetic material allocated around the rotor at even intervals. Each core has a pair of square-shaped control coils placed symmetrically

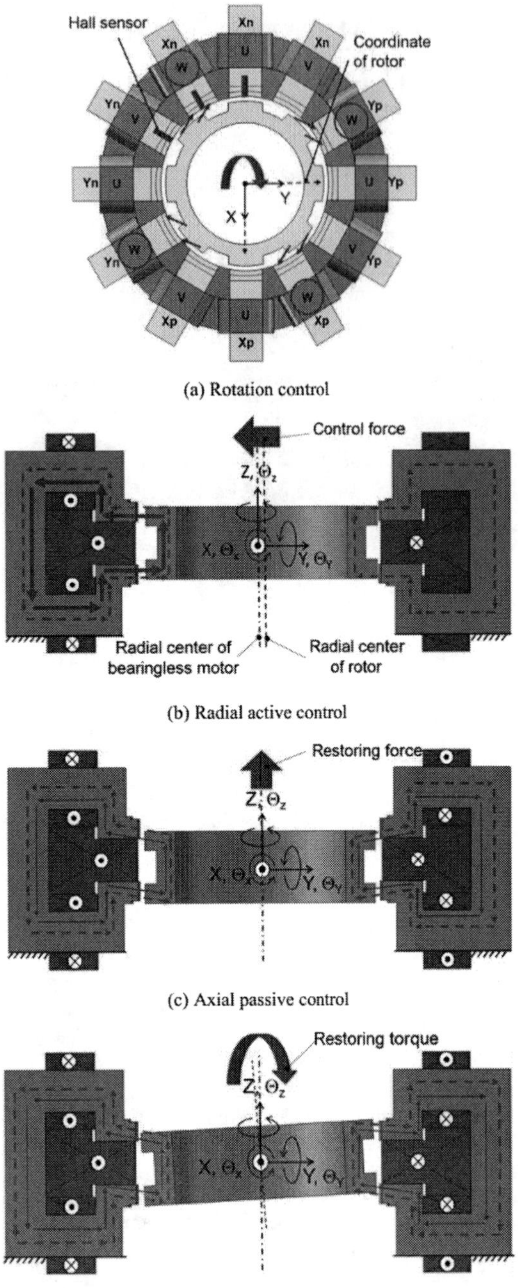

Figure 3 Control principle of proposed bearingless motor

in the axial direction. To rotate the rotor, a control magnetic flux is generated between the C-shaped stator cores and the rotor by applying current to the control coils. There is a large ring-shaped support coil wound in the stator core concentric with the rotor so that a bias magnetic flux is generated between the C-shaped stator cores and the rotor when a constant current is applied. This unique structure makes the proposed bearingless motor permanent magnet free and compact.

B. Rotation

The twelve stator cores are divided into four groups (Xp, Xn, Yp and Yn) for radial positioning control and three phase groups (U, V and W) for rotational control as shown in Fig. 3(a). Thus, during rotation of the rotor, the control current in each control coil is the sum of the positioning current and the rotational current. When the rotational angle of the rotor is at the position shown in Fig. 3(a), rotational current is applied to the four pairs of control coils in the W phase group to generate a rotational torque in the clockwise direction. The three Hall element sensors are used to measure the phase and its rate of change and calculate the rotational speed of the rotor. Thus, the proposed bearingless motor works as a traditional three-phase twelve-eight-pole switched reluctance motor.

C. Radial Suspension Force Generation

Figure 3(b) shows the magnetic circuit and the principle by which the positioning force is generated in the radial direction. The solid red and dashed blue lines indicate the control and bias magnetic fluxes, respectively. When the rotor leaves the radial center of the motor as shown in Fig. 3(b), for example, the displacement sensor will detect the error. The positioning current in the square-shaped coils wound on the C-shaped stator cores which are located in the same direction as the rotor displacement (positive Y direction, as shown in Fig.3(b); Yp stator group, as shown in Fig.3(a)) will be zero. There will be positioning current in the square-shaped coils only on the opposite side (negative Y direction, as shown in Fig.3(b); Yn stator group, as shown in Fig.3(a)). This current difference will generate a force in the Y direction. As a result, the rotor will be drawn back and kept in the radial center of the bearingless motor. This positioning method is different from the conventional one [8]. In the conventional positioning method, a bias magnetic flux on the same side as the rotor displacement is weakened by supplying a "reverse positioning current" to the control coil, which leads to lower control efficiency. Additionally, this reverse positioning current will weaken the axial and tilt stiffness of the rotor. On the other hand, in the proposed method no current is used to substitute for this reverse positioning current.

D. Passive Stabilization

Figures 3(c) and (d) show schematics of the passive stabilization of the rotor in the axial and tilt directions. The axial and tilt motions of the rotor are passively stabilized by a reluctance force generated from the bias and control magnetic fluxes. The axial and tilt stiffnesses are enhanced during rotation of the rotor because of the rise in control current leading to an increase in the control magnetic flux. Thus, due to the increase in stiffness, the rotor can be levitated more stably.

E. *Bearingless Motor Prototype*

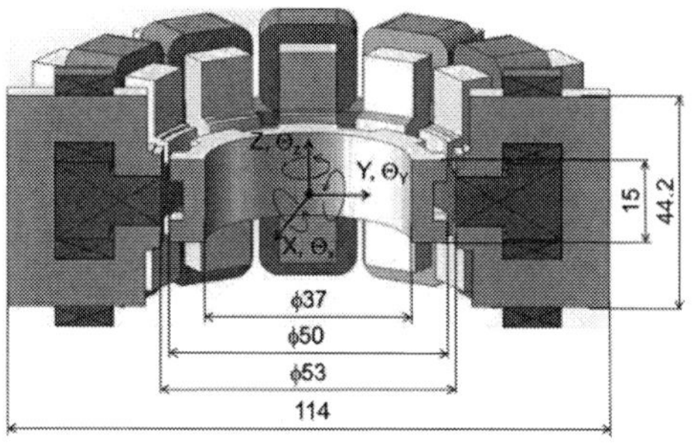

Figure 4 Design of the proposed bearingless motor

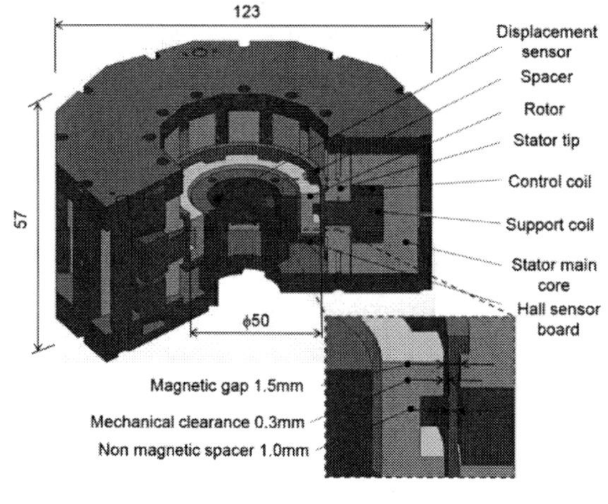

Figure 5 Mechanical structure

DCBPs are required to yield a flow rate of 5 L/min against a head pressure of more than 250 mmHg (33.3 kPa) for extracorporeal circulation support with the use of an artificial lung. According to our previously proposed magnetically-levitated DCBP with a 50 mm diameter impeller [8], a rotational speed of over 3000 rpm is necessary. Following the design aspects of our previous research, we employed a rotor with a diameter of 50 mm with a target rotational speed of 3000 rpm.

Figure 4 shows the dimensions of the proposed bearingless motor. The origin of the system is at the geometric center of the rotor. The height of the rotor is 15 mm. As shown in Fig. 5, the total height and diameter of the proposed motor is 57 mm and 123 mm, respectively. The mass of the rotor is 0.074 kg which was directly measured on precision scales. The magnetic gap between the rotor, which is made of soft iron, and the stator cores is 1.5 mm. Additionally, for future DCBPs, the mechanical clearance between the rotor and the housing wall of the pump head is 0.3 mm which is large enough to avoid substantial destruction of blood cells (hemolysis). A spacer with a thickness of 1.0 mm is placed between the stator and the rotor to generate a mechanical clearance of 0.3 mm in the prototype. In order to decrease the eddy current loss, laminated steel with a lamination thickness of 0.35 mm is used to fabricate parts of the C-shaped stator core. Due to the two-dimensional slice structure of laminated steel and the complex three-dimensional geometry of the C-shaped stator cores, each

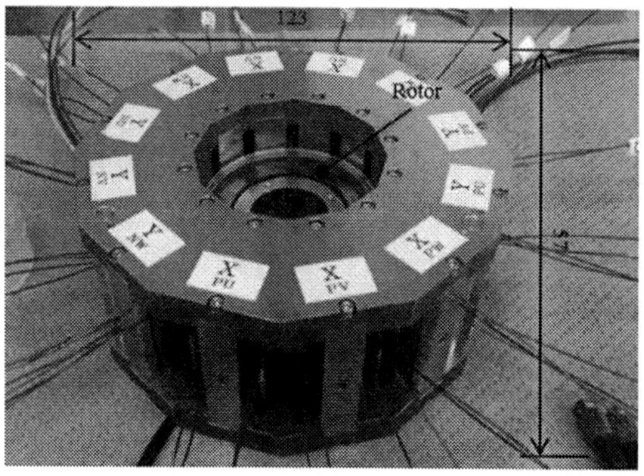

Figure 6 Fabricated bearingless motor

C-shaped stator core is divided into three parts. The main part of the stator core is made of laminated steel and the two tip cores are made of soft iron. An overview of the prototype is shown in Fig. 6. The control and support coils have 75 and 196 turns, respectively. The diameters of the wires in the control and support coils are 0.6 mm and 0.95 mm, respectively. For convenience, the three eddy current displacement sensors (PU-05A, AEC Corp., Japan) are allocated on the inner side of the rotor in this prototype design. For future DCBPs, the displacement sensors will be put on the outer side of the rotor.

III. Simulation And Experiment

A. *Stiffness Simulation*

In order to achieve continuous work, the maximum thermal balance temperature of the support coil is set to 65℃. During thermal testing of the support coil, a bias current of 3.6A achieved a thermal balance temperature of less than 65℃ after over half an hour of continuous work at room temperature of 25 ℃. Figure 7 shows the simulation performance of the proposed bearingless motor under stop mode (0 rpm) and rotation mode with a phase current of 1A. The bias current in the support coil is set to 3.6A in order to generate a stable and sufficient bias magnetic flux between the rotor and the stator and to achieve continuous work. Both the axial stiffness and the tilt stiffness increase by 15% and 13%, respectively, during rotation of the motor, because the rise in control current leads to a rise in the control magnetic flux. This enhancement in stiffness during rotation provides the possibility of us being able to reduce the bias current in future research.

B. *Motor Performance Measurement*

The rotational accuracy and power consumption of the prototype were evaluated experimentally. Large vibration amplitudes of the impeller and heat generated by the coils cause hemolysis. In order to stabilize the magnetic levitation and reduce heat generation, a pole-placement controller and a zero-power controller [9], respectively, were adopted for the control system. The viability of the zero-power controller was evaluated under conditions in which the rotor rotates without a load at room temperature (25℃).

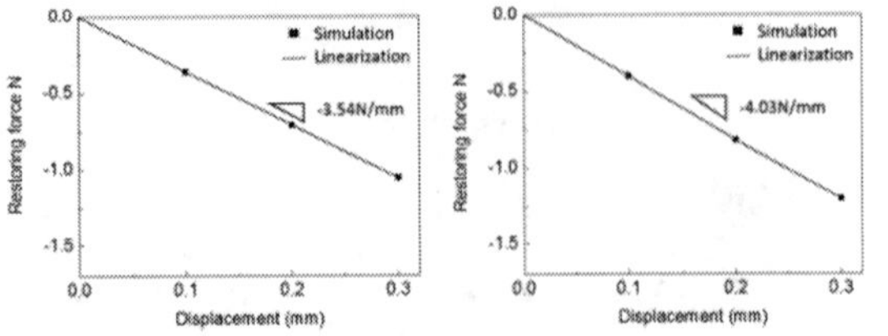

(a) Axial stiffness without phase current (b) Axial stiffness with phase current

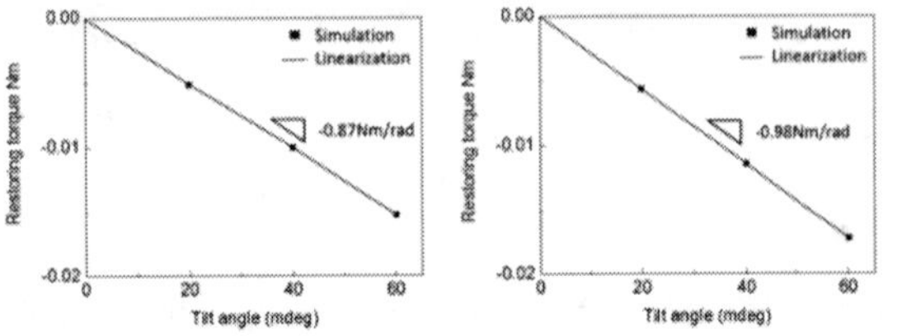

(c) Tilt stiffness without phase current (d) Tilt stiffness with phase current

Figure 7 Stiffness simulation results of the proposed bearingless motor

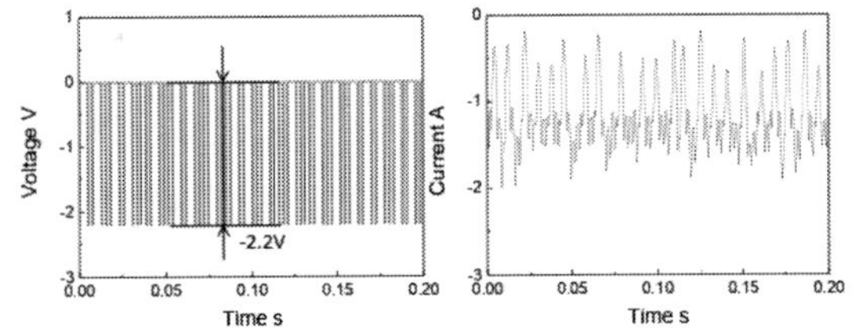

(a) Phase U voltage (b) Phase U current

Figure 8 Phase U voltage and current at 3,000rpm

In measurements of the performance, the motor was rotated without load under a bias current of 3.6 A which flows through the support coil. The power consumption of the motor was calculated using the feedback signal provided by the current sensor. During the measurement, the rotational speed of the rotor was under closed-loop control. The measured vibration amplitude of the rotor in the X direction at 3000 rpm without a load was less than ±25μm, which is sufficiently smaller than the mechanical clearance of 0.3 mm.

Figures 8 (a) and (b) show the phase U voltage and current of the fabricated motor at a target speed of 3000 rpm. The power consumption of one pair of control coils is 0.68 W at a rotational speed of 3000 rpm, and the power consumption of the support coil with a resistance of 1.66 Ω is 21.5 W at a bias current of 3.6 A. The total power consumption of the motor is 29.7 W. Over 70% of the power is consumed by the support coil. Thus, the power consumption of the support coil needs to be reduced.

Adjusting the bias current based on the control current is a solution for reducing the total power consumption without decreasing the passive stiffness.

Figures 9 (a) and (b) show the maximum rotational speed and phase U current of the motor. The maximum speed is over 3200 rpm. The average phase U current has increased

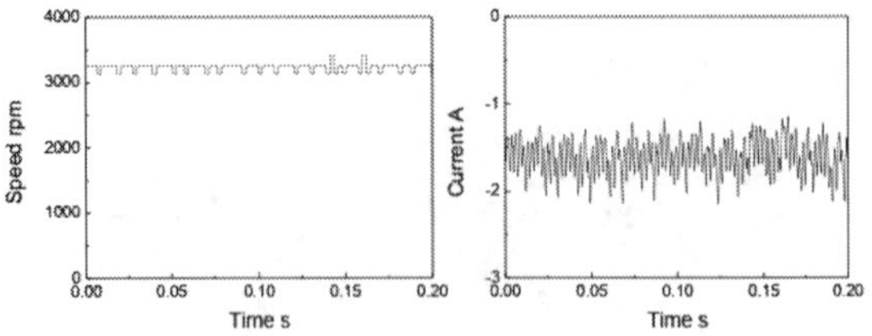

(a) Speed feedback (b) Phase U current

Figure 9 Maximum speed test

significantly compared with that shown in Figure 8 (b). This is because of the magnetic saturation of the stator and the eddy current generated during the high speed phase transformation.

IV. CONCLUSIONS

A bearingless motor for a DCBP with a permanent magnet free structure was designed and fabricated. The positioning and rotational performance of the motor was evaluated. The rotor can be levitated and rotated at a maximum speed of over 3200 rpm with a rotational error of less than ±25 μm. The power consumption of the proposed motor is less than 30 W at a rotational speed of 3000 rpm.

Future work is to reduce the bias current in the support coil and fabricate a DCBP using this proposed bearingless motor.

ACKNOWLEDGEMENT

This research is partly supported by JSPS KAKENHI Grant Number 24360059.

REFERENCES

[1] Takatani S, "Progress of rotary blood pumps," *Artificial Organs*, vol. 30, no.5, pp.317-321, 2006.

[2] George PN, Javier AL, and Suellen I. "Acute and temporary ventricular support with bioMedicus centrifugal pump." *Ann Thorac Surg* vol. 68, pp. 650-4, 1999.

[3] U.Mehlhorn, et al, "LIFEBRIDEGE: a portable, modular, rapidly available "plug-and-play" mechanical circulatory support system." *Ann Thorac Surg*, Vol. 80, no. 5, pp. 1887-1892, 2005

[4] Hijikata.W, Shinshi T, Asama J, et al. A magnetically levitated centrifugal blood pump with a simple-structured disposable pump head. *Artificial Organs 2008*;32:531-40.

[5] Schob R, Barletta N. Principle and application of a bearingless slice motor. *JSME Int J Ser C 1997*;40:593-8.

[6] Someya T, Kobayashi M, Waguri S, et al. Development of a disposable Mag-lev centrifugal blood pump(MedTech Dispo) intended for one-month support in bridge-to-bridge applications: in vitro and initial in vivo evaluation. *Artificial Organs 2009*;33:704-13.

[7] Hijikata W, Mamiya T, Shinshi T, Takatani S. A cost-effective extracorporeal magnetically-levitated centrifugal blood pump employing a disposable magnet-free impeller. *Proc. IMechE, J.Engineering in Medicine*, Vol.225, pp.1149-1157, Dec. 2011.

[8] Hijikata W, Sobajima H, Shinshi T, et al, "Disposable MagLev centrifugal blood pump utilizing a cone-shaped impeller." *Artificial Organs* 2010, vol. 34, pp. 669-677.

[9] Mizuno T, Takasaki M, Suzuki H, Ishino Y, "Development of a three-axis active vibration isolation system using zero-power magnetic suspension." *Proceedings of the 42nd IEEE Conference on Decision and Control*, pp. 4493-4498, 2003.

Radial Position Active Control of Double Stator Axial Gap Self-bearing Motor for Paediatric VAD

Masahiro Osa[a], Toru Masuzawa[a], Naoki Omori[a], Eisuke Tatsumi[b]

[a] Ibaraki University, 4-12-1, Hitachi, Japan, osa@mx.ibaraki.ac.jp
[b] National Cerebral and Cardiovascular Center, 5-7-1 Fujishirodai, Suita, Japan

***Abstract*—A novel axial gap double stator self-bearing motor which can control five-degrees of freedom (5-DOF) of rotor postures have been developed for paediatric ventricular assist device (VAD). The motor has a top stator, a bottom stator and a levitated rotor driven as a synchronous permanent magnet motor. The rotor is axially suspended between the stators which have an identical structure. A double stator mechanism enhances a higher torque production and realizes radial position active control. This paper proposes a concept of radial position active control and magnetic suspension ability of developed self-bearing motor. The rotor is successfully levitated and rotated up to 6400 rpm without any physical contact with 5-DOF active control in air. The radial oscillation amplitudes are actively suppressed with the proposed radial position control concept. The developed 5-DOF controlled self-bearing motor which is suitably miniature as an actuator for paediatric VAD indicates sufficient capability of the magnetic levitation and non-contact rotation.**

I. Introduction

Mechanical circulatory supports have been clinically applied to adult heart failure patients due to the high rate of heart disease and the shortage of donor hearts. The successful use of ventricular assist devices (VADs) in adult patients over the last decade has led to increased interest in this technology for paediatric patients [1]-[4]. However, the majority of VADs for paediatric heart disease have been applied to patients older than 10 years due to limitations of device size and structural functionalities. Thus, demands for paediatric VADs specifically designed for both infants and small children have been recently promoted widely. The paediatric VADs has to satisfy following design requirements: 1) high durability for long term support of patients, 2) minimal blood damage and no thrombosis formation in the pump cavity, 3) miniature device size in order to be fully implanted in a small peritoneal cavity from birth to more than two years of age, and 4) wide operational range from 2000 rpm to 5000 rpm in order to regulate the pump flow rate according to growth of paediatric patients. Due to significant advantages of eliminating mechanical wear components compared to traditional contact bearing systems, application of magnetically suspended motors in VADs have been main focus of the research [5]-[8]. Magnetically suspended rotary VADs offers advantages of contactless motor drive such as high speed rotation, no material wear, zero friction, less heat generation, better durability and better blood compatibility of the devices. However, the magnetic air-gap of magnetically suspended motors applied for rotary VADs is quite larger than in case of typical industrial magnetic bearing applications in order to avoid blood cell destruction due to the high share stress. These circumstances grossly reduce a magnetic suspension force and torque capacity. This reduction of magnetic suspension force and torque causes both a deterioration of magnetic suspension stability and a reduction of motor energy efficiency. In this study, a novel 5-degrees of freedom (DOF) controlled axial gap double stator self-bearing motor was proposed to overcome magnetic performance drawbacks due to the large air-gap length. The objectives of this paper were to demonstrate the radial position control capability and examine the feasibility of miniaturized 5-DOF controlled motor for paediatric VAD.

II. Materials and Methods

A. *Structure and control principle of 5-DOF controlled self-bearing motor*

The proposed 5-DOF controlled self-bearing motor shown in Fig. 1 is a permanent magnet synchronous motor which has a top stator, a bottom stator and a levitated rotor. The levitated rotor is axially sandwiched by both stators that have an identical structure. A double stator mechanism enhances a rotating torque production and also realizes the proposed 5-DOF active control of levitated rotor postures. Two types of concentrated windings are independently wound on each stator tooth. An axial position (z) and rotating speed (ω_z) of the rotor are regulated by using a vector control algorithm which can generate axial suspension force and rotating torque with a single rotating magnetic field [9]-[12]. The axial position of levitated rotor is actively regulated by field strengthening and field weakening as shown in Fig. 2.

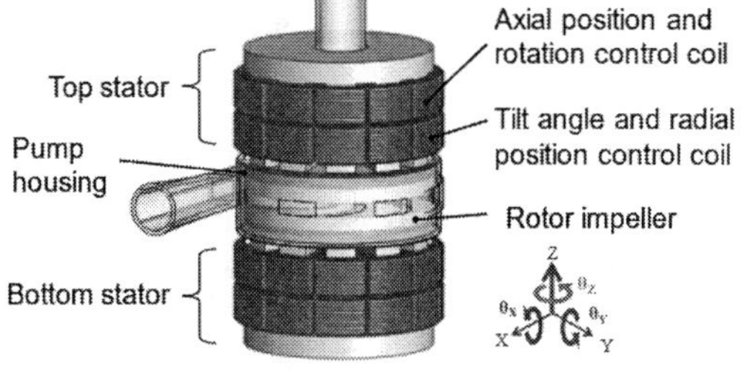

Figure 1. Schematic of proposed 5-DOF controlled axial gap double stator self-bearing motor for paediatric VAD

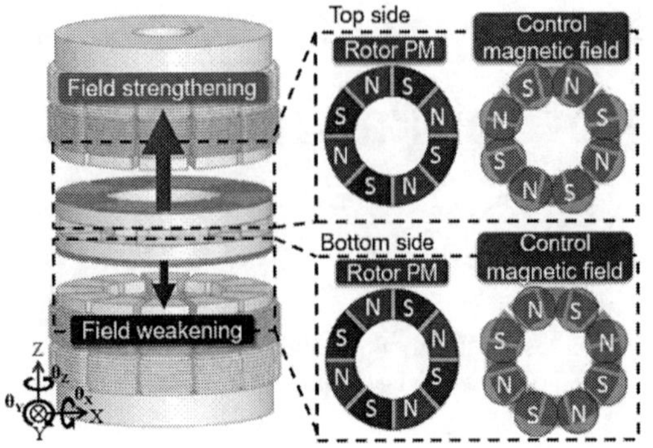

Figure 2. Principle of axial force production by utilizing field strengthning and field weakening

An axial attractive force F_z and a rotating torque $T_{\theta z}$ produced by the double stator are described by equation (1) and (2) using magnetic flux density produced by rotor permanent magnets B_p, inner and outer radius of rotor r_1 and r_2, pole pair number of rotor permanent magnet M, number of turns in windings N, air-gap length of z, d-axis current i_d and q-axis current i_q.

$$F_z = \sqrt{\frac{2}{3}} \frac{(r_2^2 - r_1^2)\pi B_p}{z} N i_d \quad (1)$$

$$T_{\theta z} = \sqrt{\frac{2}{3}} (r_2^2 - r_1^2)\pi B_p N i_q \quad (2)$$

From equation (1) and (2), axial attractive force and rotating torque are controlled independently with d-axis and q-axis current.

As shown in Fig. 3, a single rotating control flux density based on P±2 pole algorithm produces a restoring torque and a radial suspension force simultaneously [13]. The restoring torque and the radial suspension force can be produced independently with double stator mechanism. Fig. 4 shows the principles of inclination control around y-axis and radial position control in x direction of levitated rotor. Both stators produces restoring torque in the same direction; the radial forces produced by stators cancel each other, and the restoring torque produced as shown in Fig. 4(a). In contrast, the restoring torque produced by top stator and bottom stator is regulated in opposite direction, then the radial suspension force can be generated as shown in Fig. 4(b). In a similar manner, the inclination control around x-axis and the radial position control in y direction are available. Consequently, the inclination angles (θ_x, θ_y) and the radial positions (x, y) of levitated rotor can be controlled by regulating the magnitude and the direction of restoring torque and radial suspension force according to excitation currents fed into top stator and bottom stator.

When restoring torque and radial suspension force are proportional to excitation current, these are given by equation (3) and (4) using a restoring torque constant $k_{T\theta y}$ and a radial suspension force constant k_{Fx}.

$$T_{\theta y} = k_{T\theta y} i \quad (3)$$

$$F_x = k_{Fx} i \quad (4)$$

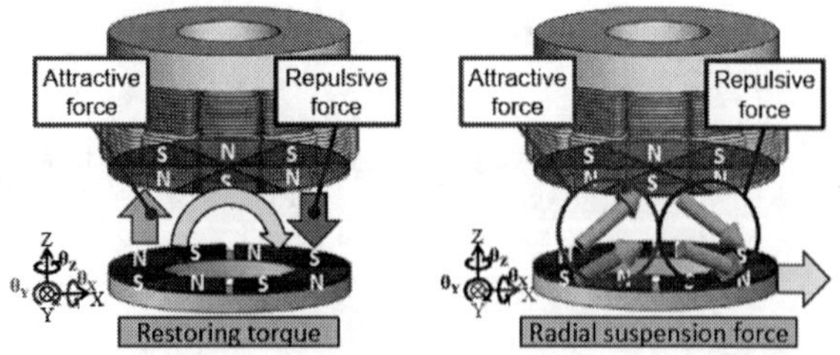

Figure 3. Principle of restoring torque and radial suspension force production by using single magnetic field based on P±2 pole algorithm

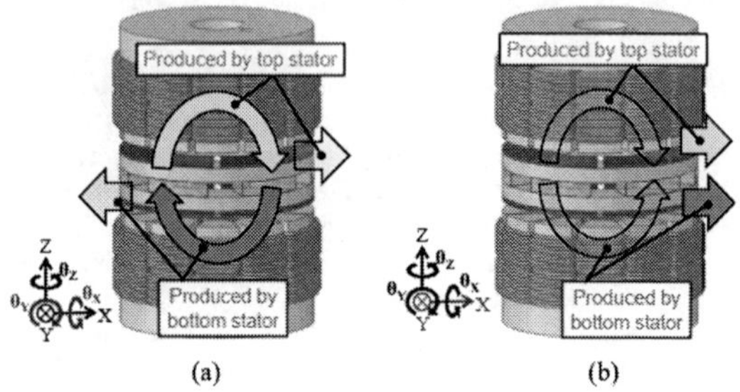

Figure 4. Independently produced restoring torque and radial suspension force with double stator mechanism: (a) Restoring torque production, (b) Radial suspension force production

A total restoring torque and a radial suspension force acting on levitated rotor with double stator mechanism are given by equation (5) and (6) when excitation current of top stator and bottom stator are defined as i_{top} and i_{bottom} respectively.

$$T_{\theta y} = k_{T\theta y}(i_{top} + i_{bottom}) \quad (5)$$

$$F_x = k_{Fx}(i_{top} - i_{bottom}) \quad (6)$$

When the both excitation current i_{top} and i_{bottom} are equal, only the restoring torque can be produced. In that case, the motor is operated as 3-DOF controlled self-bearing motor.

The double stator mechanism with both vector control algorithm and P±2 pole algorithm offers 5-DOF control of the levitated rotor postures without any additional active or passive magnetic bearings. Thus, the proposed 5-DOF controlled self-bearing motor has advantages of compactness and simplification of device structure compare to conventional 5-DOF controlled maglev motors.

B. FEM analysis and fabrication of 5-DOF controlled self-bearing motor

The principle of radial suspension force production was confirmed by using three dimensional magnetic field FEM analysis. A pole pair combination and detailed geometries of proposed 5-DOF controlled self-bearing motor were determined in FEM analysis. A double stator axial gap self-bearing motor which has twelve motor stator slots and eight rotor poles shown in Fig. 5 were developed based on the results of FEM analysis. Fig. 6 shows the geometric parameters of developed motor stator and levitated rotor. The developed motor has an outer diameter of 28 mm, a height of 41 mm and a magnetic air-gap length of 1.5 mm. The weight of the levitated rotor is 25 g. The thickness of rotor permanent magnets are 0.7 mm. A core material of stator and rotor is

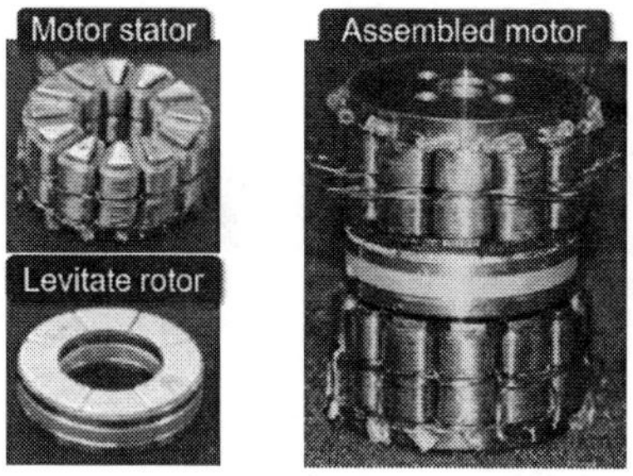

Figure 5. Photographs of developed 5-DOF controlled self-bearing motor

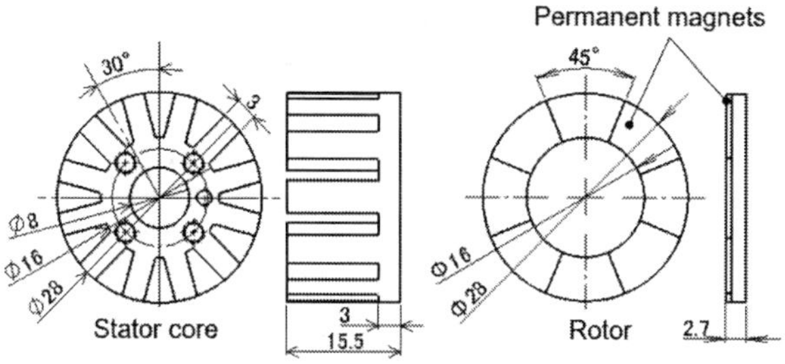

Figure 6. Geometric parameters of developed motor stator core and levitated rotor including permanent magnets

bulk soft iron (SUY-1). The permanent magnets are made of Nd-Fe-B which has a coercivity of 907 kA/m and a residual flux density of 1.36 T. The number of turns in concentrated windings wound on each stator tooth are 58. The diameter of isolated copper wire is 0.3 mm.

C. 5-DOF control system

A diagram of 5-DOF control system is shown in Fig. 7. Three eddy current sensors (PU-03A, Applied Electronics Corporation) are set onto inner side of stator tooth to measure axial position and inclination angles around x and y axes of the levitated rotor. Other two eddy current sensors are set on x-axis and y-axis to measure radial positions of the levitated rotor. Three Hall Effect sensors (Asahi KASEI Corporation) are set at stator slots to detect the rotating angles of the levitated rotor with a sensitivity of 30° electrical angle. The rotating speed is determined by calculating time derivative of rotating angles. 5-DOF control and rotating speed regulation are carried out with digital PID controllers that are implemented on a micro-processor board DS1104 (dSPACE GmbH, Paderborn Germany) with MATLAB/Simlink. Power amplifier (PA12A, Apex Microtechnology Corporation) supplies the calculated command excitation current to control windings of both stators. Sampling and control frequency is 10 kHz.

A block diagram for axial position and rotation control is shown in Fig. 8. The axial position of the levitated rotor is stabilized by regulating d-axis current i_d with a PID feedback loop. The PID controller calculate i_d from the error between detected axial position z and target axial position z^*. The target position z^* is the center of the motor, and it is set to zero. Positive and negative d-axis currents which are for field strengthening and field weakening generate an unbalanced magnetic attractive force in rotor axial direction. The rotating

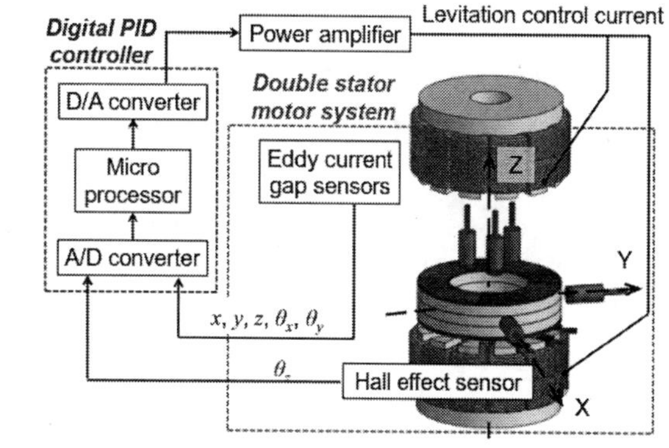

Figure 7. Feedback control system of proposed 5-DOF controlled self-bearing motor

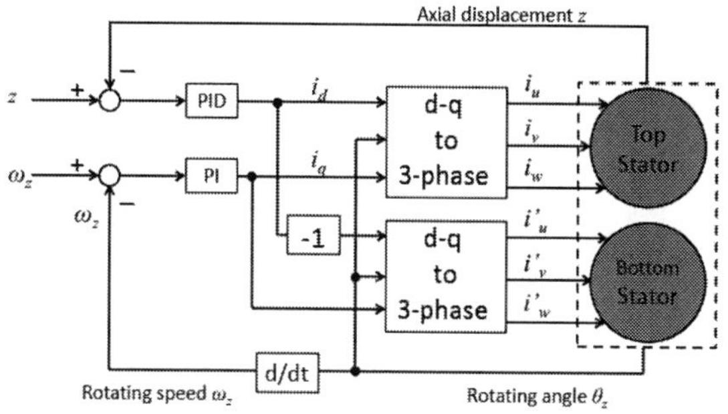

Figure 8. Block diagram for axial position and rotation control system with vector control algorithm

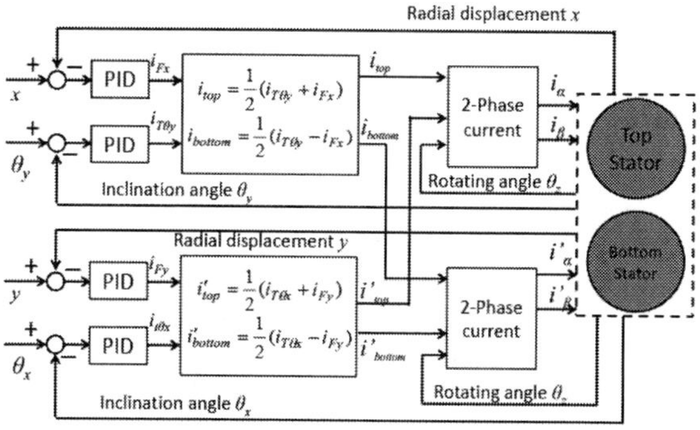

Figure 9. Block diagram for inclination and radial position control system with P±2 pole algorithm

speed is controlled by regulating q-axis current i_q with the PI feedback loop. The detected rotating speed ω_z is compared with a target rotating speed ω_z*. The calculated q-axis current for both stator windings are equal to i_q. The d-axis current i_d and the q-axis current i_q are transformed into three-phase current i_u, i_v and i_w. Then these currents supplied to axial position and rotation control windings of both stators to generate eight pole rotating magnetic field.

A diagram for inclination angle and radial position control is shown in Fig. 9. The inclination angle θ_y control corresponds to radial position x control, and the inclination angle θ_x control corresponds to the radial position y control.

The inclination angles and the radial positions of the levitated rotor are controlled by regulating the amplitude, the direction of the restoring torque and the radial suspension force. The required excitation current for generating the restoring torque and the radial suspension force, $i_{T\theta y}$ and i_{Fx}, are calculated with PID feedback loop. The excitation current for top stator and bottom stator, i_{top} and i_{bottom}, are determined from $i_{T\theta y}$ and i_{Fx} respectively. In a similar manner, excitation current i'_{top} and i'_{bottom} that are to control inclination angle θ_x and radial position y are calculated with another PID feedback loop. The amplitude and phase angle of the two-phase current i_a, i_b, i'_a and i'_b are determined from i_{top}, i_{bottom}, i'_{top} and i'_{bottom} respectively. Consequently, these excitation currents are supplied to the inclination angle and radial position control windings to generate six poles rotating magnetic field.

D. Radial suspension force and stiffness measurement

The primary objective of the study was to determine the radial suspension capability of proposed 5-DOF controlled self-bearing motor. Initially, the radial suspension forces with varying excitation currents and radial displacements were measured. A radial force measurement system is shown in Fig. 10. The rotor is fixed to a linear slider connected with a load cell in order to restrict the rotor movement in the axial direction z and the rotational direction θ_x, θ_y. At first, both centers of the rotor and stator are aligned. The radial suspension forces were evaluated by changing the excitation current from -2 A to 2 A. The radial restoring forces were then measured by displacing the rotor radially from the center position up to 1.0 mm with micrometer at the excitation of 0 A.

E. Frequency response of radial position control

The 5-DOF controlled self-bearing motor was assembled as shown in Fig. 11, and combined with the digital control system. The rotor was levitated without rotation. The initial parameters for the PID controller were determined based on the limit sensitivity method. The optimized gains were then tuned manually as shown in Table I. The frequency response of radial position control system was measured with a frequency sweeping method using a FFT analyzer (DS-2100, DS-0264, Ono Sokki Corporation). A sinusoidal disturbance was added to the measured radial position and the disturbance frequency was varied from 1 Hz to 1 kHz.

F. Comparison of 5-DOF control and 3-DOF control

The magnetic suspension characteristics in the radial directions (x, y) were evaluated under 3-DOF control and 5-DOF control in order to demonstrate the advantages of proposed 5-DOF control. The movable range of levitated rotor in radial and axial directions was ± 0.5 mm and ± 0.3 mm, respectively. The inclination range of the levitated rotor was $\pm 1°$. The PID gains of the 5-DOF control were same as given in Table I. In case of the 3-DOF control, PID gains of the radial position controller were zero. The P gain and I gain of a rotating speed PI controller were 0.00075 A/rpm and 0.007 A/(sec • rpm). As a first step, the radial vibration waveforms at a rotating speed of 0 rpm were evaluated. Then, the levitated rotor was accelerated in rotational direction θ_z, and maximum oscillation amplitudes in the radial direction were measured. The maximum oscillation amplitude was defined as half of the peak-to-peak value of rotor vibration.

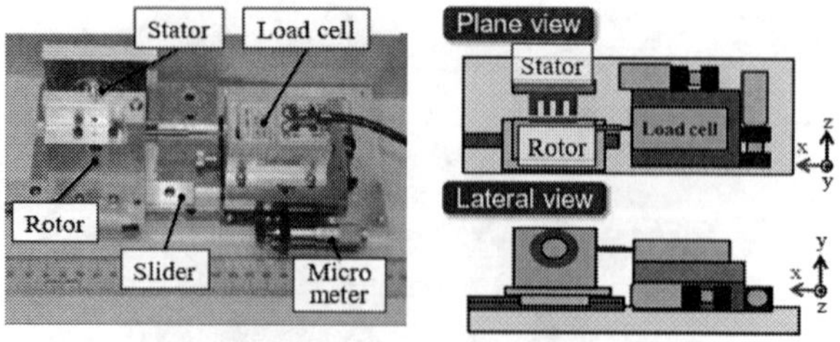

Figure 10. Experimental setup to measure the radial suspension force characteristics

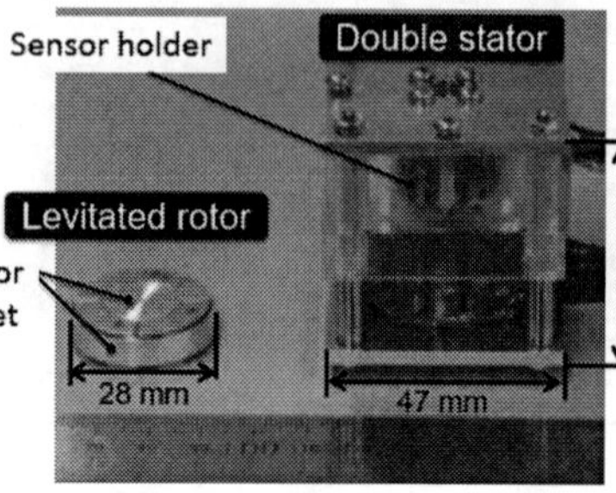

Figure 11. Photograph of assembled 5-DOF controlled self-bearing motor

TABLE I. PID CONTROLLER GAINS FOR 5-DOF CONTROLLER

Gain	Radial position	Axial position	Inclination angle
P	2.73 [A/mm]	10 [A/mm]	2.0 [A/deg]
I	0 [A/sec mm]	0.01 [A/sec mm]	0.024 [A/sec deg]
D	0.045 [A sec/mm]	0.01 [A sec/mm]	0.003 [A sec/deg]

III. EXPERIMENTAL RESULTS

A. Radial stiffness measurement

The developed motor produced radial suspension force according to proposed radial position control concept. The relationship between the radial suspension force and the excitation current is displayed in Fig. 12 (a). The developed 5-DOF controlled self-bearing motor produced radial suspension force of $\pm$ 0.8 N by varying excitation current from -2 A to 2 A. The radial suspension force to excitation current gradient of developed motor was 0.4 N/A. An acceleration coefficient defined as a mass specific force index to indicate the dynamic suspension ability was 16 m/s^{2}A. The radial restoring force with different rotor positions are shown in Fig. 12 (b). The restoring force is proportional to the radial displacement, and that gradient of the least square approximation showed a radial stiffness of 0.4 N/mm.

B. Frequency response of radial position control

Fig. 13 shows the bode diagrams of radial position control system. The results displayed similar characteristics in x direction and y direction. The resonant peaks in x and y directions are observed at 24 Hz and 30 Hz. The frequency bandwidths are 125 Hz in x direction and 112 Hz in y direction.

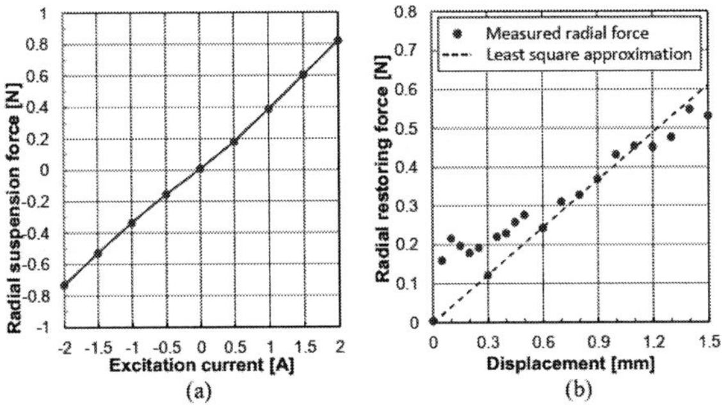

Figure 12. Measured radial suspension force: (a) Force and excitation current characteristics, (b) Force and rotor radial displacement characteristics

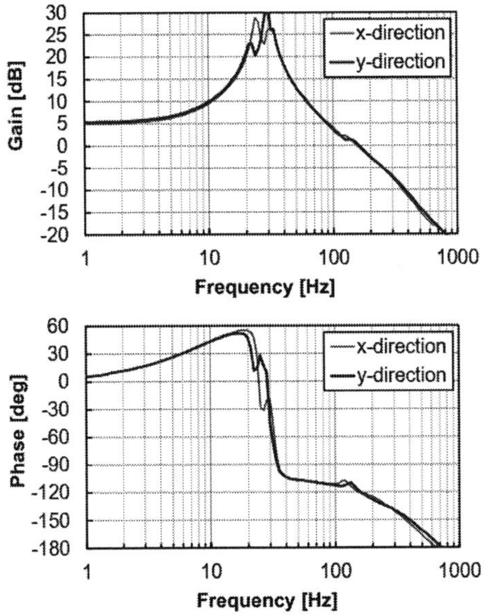

Figure 13. Bode diagrams obtaind from the results of frequency respons of radial position control system

C. *Comparison of 5-DOF control and 3-DOF control*

Fig. 14 shows the vibration waveforms of levitated rotor in radial directions at rotating speed of 0 rpm. The vibration amplitudes were less than 10 μm in both the 3-DOF control and the 5-DOF control. However, the levitated rotor was periodically fluctuated in the radial direction under the 3-DOF control as shown in Fig. 14 (a). The periodical radial fluctuation of the levitated rotor was sufficiently suppressed with 5-DOF control as shown in Fig. 14 (b).

The rotor was successfully levitated and rotated without any physical contact. The radial oscillation amplitudes of the levitated rotor with the 3-DOF control and the 5-DOF control are shown in Fig. 15. The maximum rotating speed of developed motor was expanded from 5400 rpm to 6400 rpm by implementing the 5-DOF control. Under the 3-DOF control, the double peaks of radial oscillation amplitude were observed with the rotating speeds under 2000 rpm. Except peak oscillations, the radial oscillation amplitudes were around 100 μm. By contrast, under the 5-DOF control, with rotating speeds under 2000 rpm, the resonances were

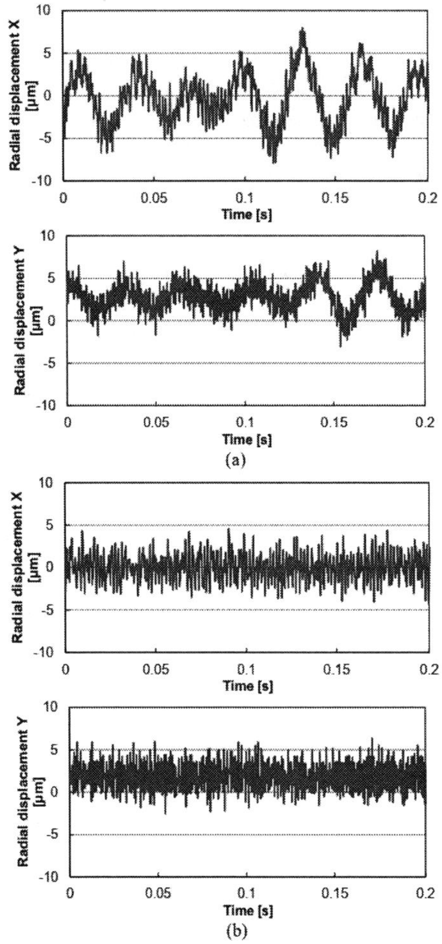

Figure 14. Radial displacement waveforms of levitated rotor at a rotating speed of 0 rpm: (a) 3-DOF of levitated rotor postures were controlled, (b) 5-DOF of levitated rotor postures were controlled

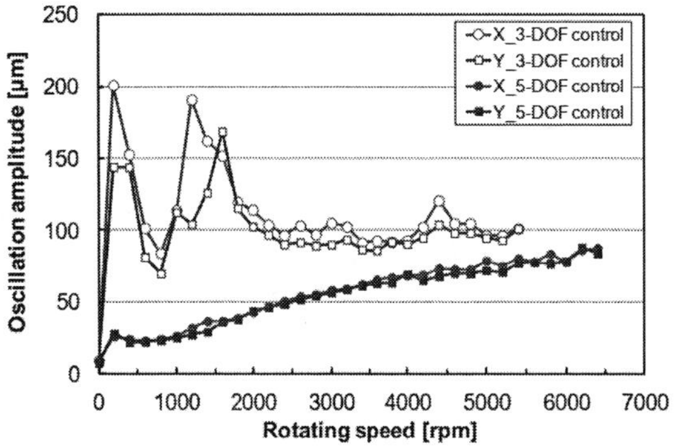

Figure 15. Maximum oscillation amplitudes in X and Y direcitions when the rotor is levitated and rotated.

successfully suppressed with the radial position active control. The oscillation amplitudes were less than 80 μm over every rotating speeds. These results demonstrate the potential of proposed radial position active control.

IV. DISCUSSION

Magnetic suspension technique is a strong candidate for VADs to improve device durability and reduce blood distraction. In the implantable paediatric VAD, size of magnetic suspension system is severely limited and magnetic air-gap is relatively large. The newly proposed 5-DOF controlled double stator axial gap self-bearing motor is one of the appropriate concepts to miniaturize the magnetically levitated system with large air-gap and guarantee non-contact stable levitation.

The results of developed radial suspension force shown in Fig.12 (a) indicate that the amplitude and direction of the radial suspension force are able to be regulated by varying excitation current. The radial position control system is further stabilized by utilizing the radial restoring force according to the rotor radial displacement. By assuming viscoelasticity of the air negligible, the radial position control system can be described as single DOF spring-mass system. A natural frequency of the system was calculated to be 28 Hz by using double stator radial stiffness of 0.8 N/mm and the levitated rotor weight of 25 g. Effects on radial fluid forces developed in the pump cavity and dumping force derived from the viscoelasticity of blood must be considered in a blood pump operation.

The frequency response tends to the vibration characteristics of the single DOF spring-mass system. The resonant frequencies in x and y directions are approximately equal to calculated natural frequency of 28 Hz. The low suppression of resonant frequency comes from weak performance of the radial suspension force production. Higher suppression performance would be achieved by optimizing the magnetic circuit geometries and combination of the pole number.

At the rotating speed of 0 rpm, the magnetically levitated rotor was fluctuated periodically under the 3-DOF control. The most dominant frequency components of the rotor fluctuation in x and y directions were analyzed with Fast Fourier transform, and these were calculated as 25 Hz in the x direction and 28 Hz in the y direction. The dominant frequencies were approximately equal to both calculated natural frequencies and the resonant frequencies observed from the bode diagrams.

The magnetic suspension performance of the developed motor was sufficient to maintain non-contact suspension and rotation at required rotating speeds for the paediatric VAD. At every rotating speed, the maximum radial oscillation amplitudes were much smaller than initial clearance of 500 μm with both the 3-DOF control and the 5-DOF control. However, under the 3-DOF control, the maximum oscillation amplitude increased at rotating speeds of 300 rpm, 1600 rpm and 1800 rpm. The first increase of radial oscillation was due to unstable characteristics at lower rotating speeds. The second increase of the radial oscillation at rotating speeds of 1600 rpm and 1800 rpm was caused by resonance when the angular frequencies come close to the resonant frequencies. By contrast, the instability at lower rotating speeds and the resonant were suppressed with the radial position control, and the proposed 5-DOF control concept demonstrated a potential of the stable levitation and rotation.

Future tasks are to develop a magnetically suspended paediatric VAD with developed 5-DOF controlled self-bearing motor and to evaluate the pump performance as well as the magnetic suspension ability and the rotor dynamics during pumping. Future design of the levitation system will focus on considering the effect of hydraulic force and viscoelastic characteristics.

V. CONCLUSIONS

The novel concept which can actively control 5-DOF of rotor postures by utilizing two types of magnetic fields based on the vector control algorithm and the P$\pm$2 pole algorithm was proposed for paediatric VAD. The proposed motor produced the sufficient radial suspension force to suspend the levitated rotor radially. The developed 5-DOF controlled self-bearing motor was successful with stable levitation and rotation. The resonances in the rotor radial directions were significantly suppressed with radial position active control. The results verified the sufficient radial position control capability and structure of 5-DOF controlled self-bearing motor.

REFERENCES

[1] J. Timothy Baldwin, Harvey S. Borovets, Brian W. Duncan, Mark J. Gartner, Robert K. Jarvik, William J. Weiss and Tracey R. Hoke, The National Heart, Lung, and Blood Institute Pediatric Circulatory Support, Journal of the American heart association, PP. 147-155, 2006.

[2] Marc Gibber, Zhongjun J. Wu, Won-Bae Chang, Giacomo Bianchi, Jingping Hu, Jose Garcia, Robert Jarvik, Bartley P. Griffith, In Vivo Experience of the Child-Size Pediatric Jarvik 2000 Heart: Update, ASAIO Journal, Vol. 56, No. 4, 2010.

[3] Noh, M.D., Antaki, J.F., Ricci, M.Gardiner, MagneticLevitation Design for the PediaFlow Ventricular Assist Device, Proc. 2005 IEEE/ASME International Conference on Advanced Intelligent Mechatronics, pp. 1077-1082, 2005.

[4] Setsuo Takatani, Hideo Hoshi, Kennichi Tajima, Katsuhiro Ohuchi, Makoto Nakamura, Tadahiko Shinshi, Masaharu Yoshikawa, Feasibility of a Miniature Centrifugal Rotary Blood Pump for Low-Flow Circulation in Children and Infants, ASAIO Journal, Vol. 51, No. 5, pp. 557-562, 2005.

[5] Daniel L. Timms, Nobuyuki Kurita, Nicholas Greatrex, Toru Masuzawa, BiVACOR A Magnetically Levitated Biventricualr Artificial Heart, Proc. of MAGDA conference in Pacific Asia, pp.482-487, 2011.

[6] Hideo Hoshi, Tadahiko Shinshi, Setsuo Takatani, Third-generation Blood Pumps With Mechanical Noncontact Magnetic Bearings, Artificial Organs, Vol. 30, No. 5, pp. 324-338, 2006.

[7] David J. Farrar, Kevin Bourque, Charles P. Dague, Christopher J. Cotter, Victor L. Poirier, Design Features, Developmental Status, and Experimental Results With the Heartmate III Centrifugal Left Ventricular Assist System with a Magnetically Levitated Rotor, ASAIO Journal, Vol. 53, No. 3, pp. 310-315, 2007.

[8] Atsushi Yumoto, Tadahiko Shinshi, Xiayou Zhang, Hiroyuki Tachikawa, Akira Shimokohbe, A One-DOF Controlled Magnetic Bearing for Compact Centrifugal Blood Pumps, Motion and Vibration Control, Springer Science+Business Media B.V. , pp. 357-366, 2009.

[9] Masahiro Osa, Toru Masuzawa, Eisuke Tatsumi, Miniaturized axial gap maglev motor with vector control for pediatric artificial heart, Journal of JSAEM, Vol. 20, No. 2, pp. 397-403, 2012.

[10] Junichi Asama, Yuki Hamasaki, Takaaki Oiwa, Akira Chiba, Proposal and Analysis of a Novel Single-Drive Bearingless Motor, IEEE Transactions on Industrial Electronics, Vol. 60, No. 1, pp. 129-138, 2013.

[11] Quang Dich Nguyen, Satoshi Ueno, Analysis and Control of Nonsalient Permanent Magnet Axial Gap Self-Bearing Motor, IEEE Transactions on Industrial Electronics, Vol. 58, No. 7, pp. 2644-2652, 2011.

[12] Satoshi Ueno, Yoji Okada, Characteristics and Control of a Bidirectional Axial Gap Combined Motor-Bearing, IEEE/ASME Transactions on Mechatronics, Vol. 5, No. 3, pp. 310-318, 2000.

[13] Masahiro Osa, Toru Masuzawa, Eisuke Tatsumi, 5-DOF control double stator motor for paediatric ventricular assist device, Proceedings of ISMB13, paper41, 2012

Influence of fault-tolerant stator structure on performance of permanent magnet bearingless slice machines

Qiang DING[a,b], Guyu WU[a] , Zhiquan DENG[a] , Xiaolin WANG[a]
[a]Nanjing University of Aeronautics and Astronautics, Yudao street 29, 20016 Nanjing, China, dzq@nuaa.edu.cn
[b]Nanjing Institute of Industry Technology, North Yangshan road 1, 210023 Nanjing, China

***Abstract*—Fault-tolerant capability is a vital requirement for bearingless slice machine which is used in safety critical application field. This paper compares two bearingless slice machines, conventional permanent magnet bearingless slice machine and fault-tolerant permanent magnet bearingless slice machine, from fault-tolerant capability, suspension and drive performance points of view. Due to introduction of fault-tolerant stator structure, a optimization method for reduction of total harmonic distortion(THD) of back-EMF and electromagnetic coupling between different phases is proposed. Then, some relevant performance comparison, such as fault-tolerant performance, suspension performance as well as drive torque performance, is analyzed by finite element analysis. Finally, several design considerations are presented as guideline for design of bearingless slice machines with fault-tolerant stator structure.**

Keywords-Machine design, permanent-magnet bearingless slice machine(PMBSM), fault-tolerant PMBSM, structure optimization

I. INTRODUCTION

During the last few years, a magnetic suspension machine technique has been successfully used in various industry field, especially for some safety critical applications, such as hazardous chemical material process, blood pumps for artificial heart assistance device and ultra-pure fluid pumps in semiconductor industry [1-2]. A bearingless machine combines a electric machine function and a magnetic bearing function [3-6]. Due to absence of mechanical bearing, it offers the advantages of no friction, no wear, lubricant- and maintenance-free operation and much longer service interval [7-8].

Stable suspension of five degrees of freedom must be achieved for conventional bearingless machine, which results in complex control system and increase of hardware cost. Thus, it blocks the development of bearingless technique in the application of limited space. In order to simplify control system and minimize hardware cost, a bearingless slice machine, which features a disk-shaped rotor, have been developed [9]. Only two radial degrees of freedom are needed active control, while one axial degree of freedom and two tilting degrees of freedom are passively stabilized by means of reluctance force.

Under safety critical application field of bearingless slice machine, two fundamental requirements, i.e., system durability and reliability, need to be considered. Due to magnetic suspension, the structure durability of machine can be enhanced drastically because bearing failure accounts for high possibility of machine failure. As for the control system reliability, only two radial positions need to be actively controlled because of disk-shaped rotor design. Thus, the number of power electronic devices and computational burden of digital signal processor are decreased. Certainly, the control system reliability is enhanced from control point of view. However, the system reliability also depends on the machine structure. Once it appears failure during operation, the machine must have capability (so-called fault-tolerant capability) to prevent further development of the failure. Meanwhile, it must have little impact on healthy components and exists sufficient healthy phase windings to keep stable operation for a period of time. Thus, the system reliability can not be enhanced if no specific fault-tolerant design is performed. However, to the best knowledge of the authors, there is a little available literature about fault-tolerant design of bearingless slice machine. In the preceding study of our research team in [7], the authors studied the fault-tolerant controllability of bearingless slice motor and gave theoretical analysis with emphasis on control strategy. However, the motor topology in [7] has no fault-tolerant capability. In [8-13], several topologies of bearingless slice motor were proposed and one of them, which has E-shape stator structure, has possibility of fault-tolerant operation. Nevertheless, the fault-tolerant operation can not be achieved because it is a two-phase motor. There are insufficient phase windings to keep operation when appearing failure in windings. In [14-16], a series of bearingless slice segment motors with four-, five- and six stator elements and surface mounted rotor magnets were proposed. In [17], a bearingless slice segment motor with four stator elements and halbach magnet ring rotor was designed. In [18], a bearingless slice segment motor with four stator elements and consequent pole rotor was delivered. In [19], a bearingless slice segment motor with five stator elements and axial magnetized rotor magnets was achieved. All these bearingless slice segment motors features no mutual inductance between different phases and the number of stator elements is larger than three, thus, this type of bearingless slice motor possesses potential fault-tolerant capability. However, due to stator segment it is difficult obtain analytical model between control currents and suspension force as well as torque, that is to say, some key parameters must be calculated by 2D or 3D finite element simulation for closed-loop control. Thus, the control algorithm loses its generality.

In this paper, a fault-tolerant stator structure is utilized for PMBSM to increase fault-tolerant capability and its influence on the machine performance is also studied and analyzed. Except for the stator structure, the rotor structure, main dimensions and machine materials of the fault-tolerant PMBSM in this paper are the same as our preceding work in [7], the aim of which is to keep the comparison as fair as possible. The fault-tolerant PMBSM consists of six-phase combined concentrated windings, which generates drive torque and bearing force simultaneously. The optimized stator shape with unequal tooth width between armature teeth and fault-tolerant teeth is for increase of fault-tolerant capacity. However, when the pole-pair number of rotor is one, the introduction of fault-tolerant teeth result in distortion of no-load back-EMF. Thus, a optimization method is proposed for reduction of total harmonic distortion of no-load back-EMF, which is helpful for decreasing torque ripple and increasing control performance. The detailed optimization process will be presented. Finally, drive and levitation performance comparison of fault-tolerant PMBSM and PMBSM is performed to show the influence of the fault-tolerant stator structure.

II. STRUCTURE AND ADVANTAGES

The fault-tolerant PMBSM proposed in this paper is depicted in Fig.1. There are twelve stator teeth with unequal tooth width, where the narrow teeth are fault-tolerant tooth and the wide teeth are armature tooth. Because the fault-tolerant teeth have little contribution to suspension force and torque, smaller tooth width than armature teeth is beneficial to increase slot area. It can be seen from Fig.1 that the shape of armature teeth is similar to duck flipper and the distances from inner edge of armature teeth and fault-tolerant teeth to centre point are different. All these design considerations are to reduce harmonic content of back-EMF.

If the fault-tolerant PMBSM is driven by six H-bridge inverters, it leads to three advantages from fault tolerance point of view. To begin with, it could achieve thermal isolation under short circuit condition because adjacent phase windings are in different stator slot. In addition, the it features magnetic isolation capability due to existence of fault-tolerant teeth and adoption of single-layer non-overlapping concentrated windings. That is to say, the healthy phases are difficult to be infect with the faulty phases. Third, the utilization of H-bridge inverter results in the electrical isolation between different phases.

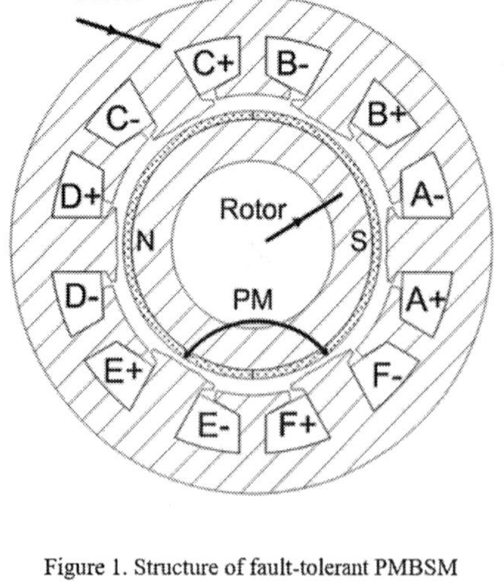

Figure 1. Structure of fault-tolerant PMBSM

III. OPERTIMIZATION DESIGN AND PERFORMANCE ANALYSIS

A. Stator Structure optimization

To reduce mutual inductance of inter-phase is a key issue with respect to the magnetic isolation. It can be achieved by introduction of fault tooth between adjacent armature teeth. However, the back-EMF deviates from sinusoidal waveform compared with its original stator structure. Thus, the maximum ratio of mutual inductance over self inductance (signed as max(M/L)) and the total harmonic distortion of back-EMF (signed as THD) are two variables needed to be considered and the target values of them are set as max(M/L)<=5% and THD<=3%.

In order to achieve optimization design, three structure variables of stator, W_f , L_g and θ shown in Fig.2, which are defined as width of fault-tolerant tooth, difference between fault-tolerant tooth radius and armature tooth radius, slot opening angle between two adjacent armature teeth respectively, are investigated by FEM. The results of this investigation provide the guideline for fault-tolerant PMBSM design.

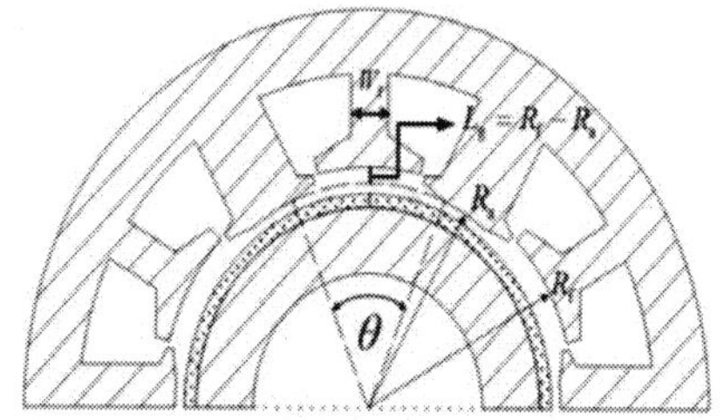

Figure 2. Definition of structure parameters for optimization

The detailed optimization process is as follows:

1. To adjust W_f and remain L_g , θ unchanged

$$\begin{cases} THD_{E_0} = f_1(W_f, L_g, \theta)\big|\theta = const,\ L_g = const \\ \max(M/L) = \tau_1(W_f, L_g, \theta)\big|\theta = const,\ L_g = const \end{cases} \quad (1)$$

where M and L are values of mutual inductance and self-inductance respectively.

It can be observed in Fig.3 that to change W_f has insignificant variations with respect to max(M/L)) and THD when W_f is smaller than eight millimeters. Hence, what should be considered to determine W_f is its influence on the slot area.

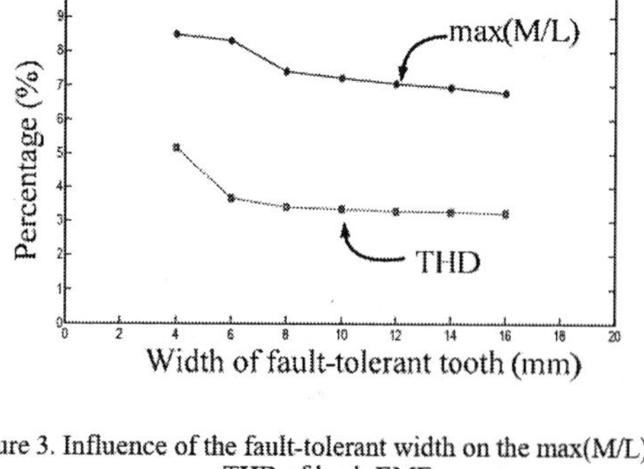

Figure 3. Influence of the fault-tolerant width on the max(M/L) and THD of back-EMF

2. To adjust L_g and remain W_f, θ unchanged

$$\begin{cases} THD_{E_0} = f_2(L_g, W_f, \theta)\big|W_f = const, \theta = const \\ \max(M/L) = \tau_2(L_g, W_f, \theta)\big|W_f = const, \theta = const \end{cases} \quad (2)$$

In this step, five samples of L_g, which is adjusted from 1mm to 5mm, are taken into account. Fig.4 shows the variation tendency of max(M/L) and THD. It can be observed that the variation tendency of THD is to decrease while max(M/L) is to increase with respect to the variation of L_g. Thus, there is a trade-off when to determine L_g.

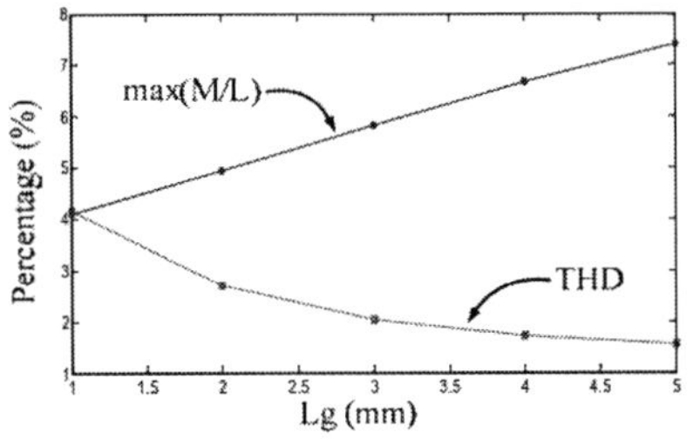

Figure 4. Influence of the eccentric distance of fault-tolerant tooth on the max(M/L) and THD of back-EMF

3. To adjust θ and remain W_f, L_g unchanged

$$\begin{cases} THD_{E_0} = f_3(\theta, L_g, W_f)\big|L_g = const, W_f = const \\ \max(M/L) = \tau_3(\theta, L_g, W_f)\big|L_g = const, W_f = const \end{cases} \quad (3)$$

It can be observed in Fig.5 that the variation tendency of max(M/L) and THD is to decrease simultaneously when the variable θ is to increase. Hence, to adjust θ is a effective method to achieve optimization objectives. However, it can be seen from Fig.6 that there is a trade-off because the adjustment of θ lead to the decrease of root-mean-square value of back-EMF.

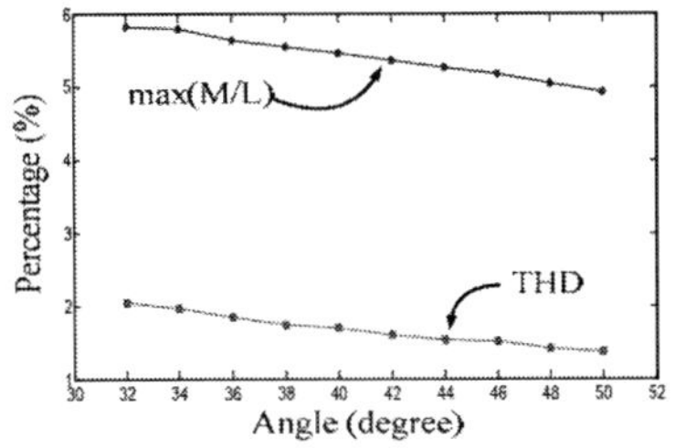

Figure 5. Influence of the angle θ on the max(M/L) and THD of back-EMF

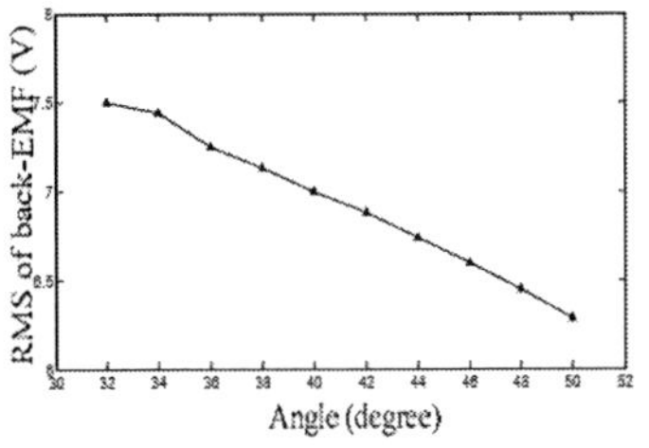

Figure 6. Influence of the angle θ on RMS value of back-EMF

B. Fault-tolerant capability analysis

Based on above-mentioned three optimization steps, the specifications of fault-tolerant PMBSM are listed in Table I. In order to show comparison between before and after optimization, the specifications of PMBSM in [7] are also listed.

Table I Specifications of fault tolerant PMBSM and PMBSM

Items	Bearingless Machine
Stator outside diameter	149mm
Stator inner diameter	84mm
Rotor inner diameter	50mm
Axial length	10mm
Number of stator slots	6 (PMBSM)
	12 (fault-tolerant PMBSM)
Number of PMs	2
Air gap length	2mm (PMBSM)
	2mm to 5mm (fault-tolerant PMBSM)
Width of armature tooth	16mm
W_f	8mm(fault-tolerant PMBSM)
L_g	3mm(fault-tolerant PMBSM)
θ	46 degree(fault-tolerant PMBSM)

Fig.7 shows the flux distribution of the two machines, in which only phase-A is excited and the magnets are un-magnetized for clarity. It can be observed that the large quantity of flux lines pass through the fault-tolerant tooth and small quantity of flux lines couple with other phases in fault-tolerant PMBSM, whereas the number of flux lines which couple with the other phases in PMBSM are larger than the number in fault-tolerant PMBSM. This means that fault-tolerant PMBSM has weak coupling between phases, whereas PMBSM has strong coupling between phases. The significant inter-phase coupling indicates that once a fault occurs in one phase, it would have an undesirable impact on other healthy phases.

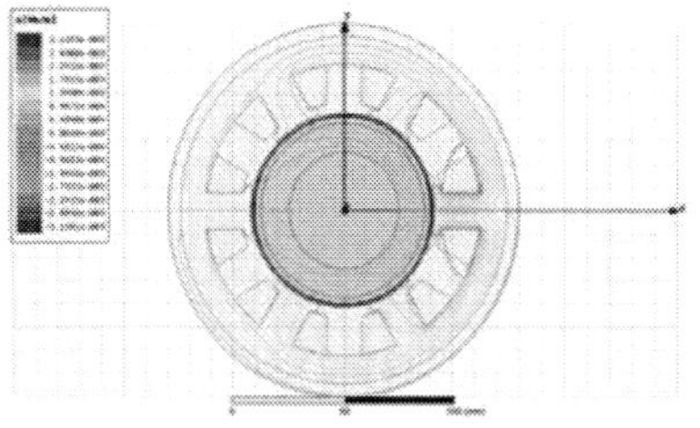

(a) PMBSM

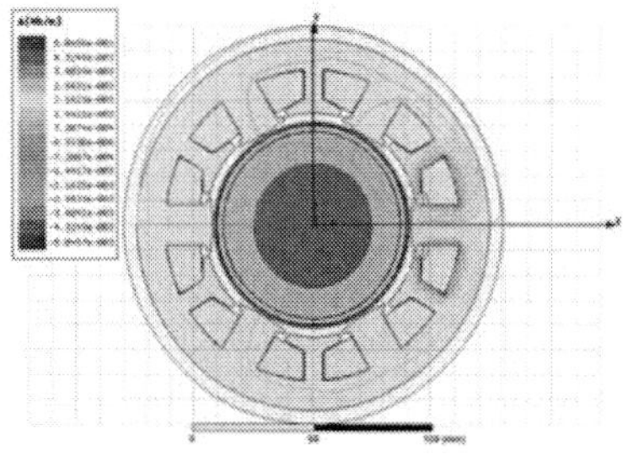

(b) Fault-tolerant PMBSM

Figure 7. Flux distribution of PMBSM and fault-tolerant PMBSM

Table II compares self- and mutual inductance of phase-A between PMBSM and fault-tolerant PMBSM. The max(M/L) values of PMBSM and fault-tolerant PMBSM are 17.71% and 4.96% respectively, which demonstrates that electromagnetic coupling in PMBSM is much more serious than that in fault-tolerant PMBSM and its value of fault-tolerant PMBSM satisfies the target value(<=5%). Furthermore, the self-inductance in fault-tolerant PMBSM is larger than that in PMBSM, it means that fault-tolerant PMBSM possesses the better short-circuit current restraint capability than PMBSM under winding short-circuit failure condition.

Table II Self- and mutual inductance in phase-A

	PMBSM		fault-tolerant PMBSM	
	Inductance(mH)	Percentage	Inductance(mH)	Percentage
L(A,A)	15.70	100%	25.20	100%
M(A,B)	2.78	17.71%	1.25	4.96%
M(A,C)	2.46	15.67%	1.16	4.60%
M(A,D)	2.42	15.41%	1.14	4.52%
M(A,E)	2.46	15.67%	1.16	4.60%
M(A,F)	2.78	17.71%	1.25	4.96%

Fig.8(a) illustrates the back-EMF waveform of phase A in fault-tolerant PMBSM at 2000rpm. Fig.8(b) shows its harmonic analysis. The THD value of phase A back-EMF equals to 1.51%. Thus, the high sinusoid degree of back-EMF can be achieved and satisfy its target value(<=3%) based on the optimization method above mentioned.

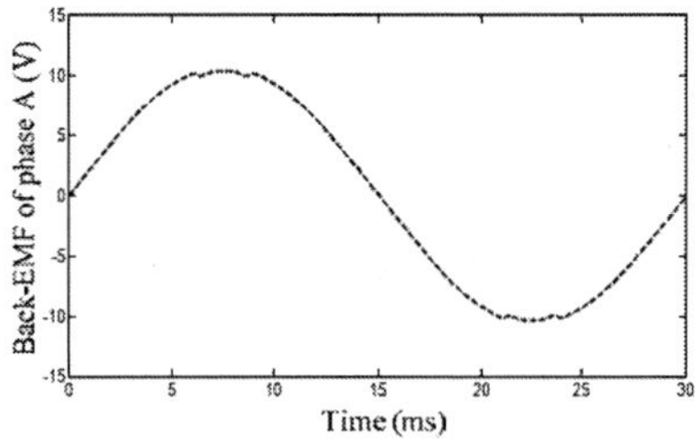

(a) Back-EMF of Phase A

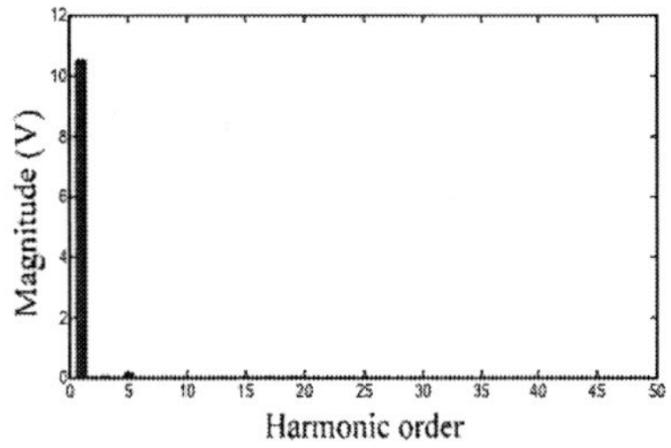

(b) Harmonics analysis of back-EMF in phase-A

Figure 8. Waveforms of back-EMF and its harmonics analysis

C. *Suspension and drive performance analysis*

In this section, the suspension force and drive torque performance of PMBSM and fault-tolerant PMBSM will be compared.

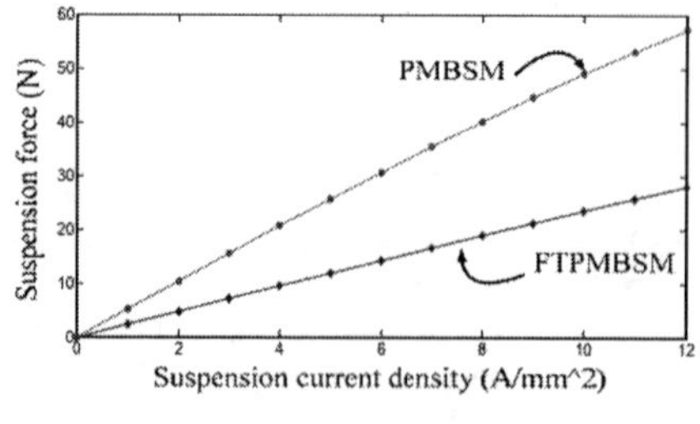

Figure 9. Comparison of suspension force

Fig. 9 demonstrates that suspension force generation capability at different suspension current density. It can be calculated from Fig.9 that the suspension force per unit levitation current density are 4.77N and 2.34N in PMBSM and fault-tolerant PMBSM respectively. Thus, there is 51% loss of suspension force in fault-tolerant BSM compared with PMBSM.

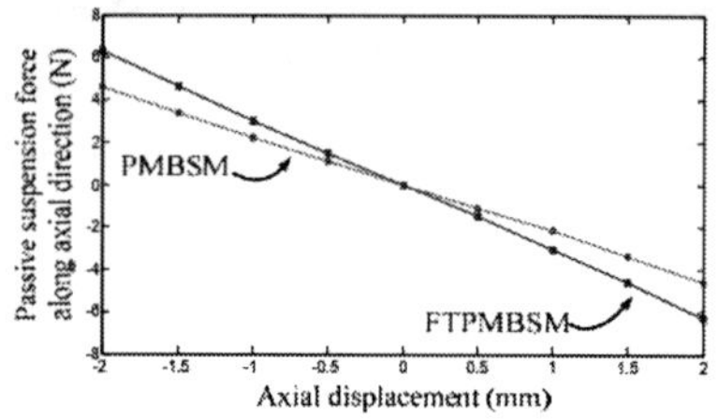

Figure 10. Comparison of passive suspension force in axial direction

Fig. 10 shows that passive levitation capability in axial direction when the rotor is in the centre of stator bore. It can be calculated from Fig.10 that the axial passive suspension force per unit axial displacement are -3.15N and -2.30N in PMBSM and fault-tolerant PMBSM respectively. Thus, there is 27% loss of axial passive levitation force in fault-tolerant PMBSM compared with PMBSM.

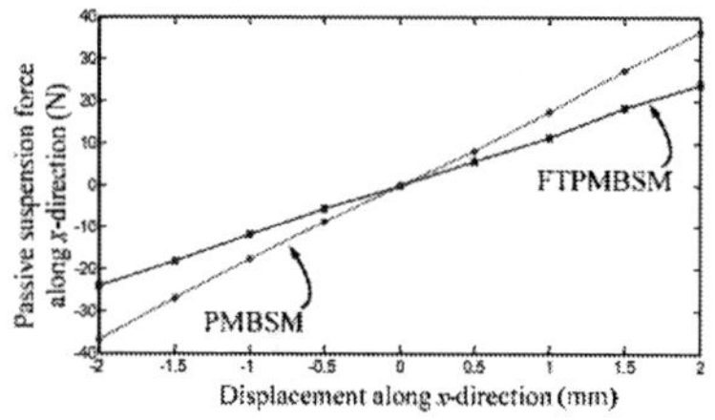

Figure 11. Comparison of passive suspension force in x direction

Fig. 11 shows that passive levitation capability in x-direction when the rotor at 0 degree position which is the rotor position shown in Fig.1. It can be calculated from Fig.11 that the passive suspension force per unit displacement are 11.98N and 18.33N in PMBSM and fault-tolerant PMBSM respectively. Thus, there is 35% reduction of horizontal passive suspension force in fault-tolerant PMBSM compared with PMBSM.

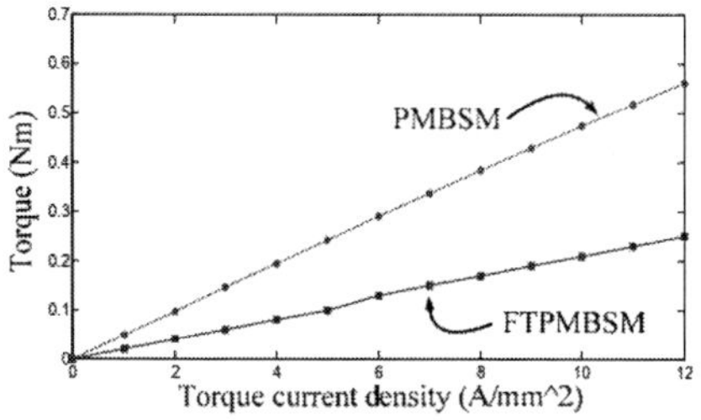

Figure 12. Comparison of torque

Fig. 12 demonstrates that torque generation capability at different torque current density. It can be calculated from Fig.12 that the torque per unit torque current density are 0.047Nm and 0.021Nm in PMBSM and fault-tolerant PMBSM respectively. Thus, there is 55% loss of torque generation capability in fault-tolerant PMBSM compared with PMBSM.

D. Design considerations

In this paper, performance comparison of two bearingless slice machines, PMBSM and fault-tolerant PMBSM, are investigated with respect to their fault-tolerant capability, suspension and drive performance. Based on performance comparison, there are some design considerations needed to be presented.

Firstly, the fault-tolerant stator structure and its optimization method can enhance fault-tolerant capability of this machine.

However, secondly, due to introduction of fault-tolerant stator structure and its optimization method, the active suspension force and drive torque performance decrease drastically. That is to say, enhancement of fault-tolerant capability is at the expense of loss of active suspension force and drive torque. There are two reasons for this result: one is the reduction of effective stator slot area due to the existence of fault-tolerant teeth; another is the increase of effective air gap length due to the optimization method.

Thirdly, for increase of active suspension force and drive torque performance, the following measures can be used based on the optimization process. The first is to decrease of main mechanical air gap length, which is defined as the difference between armature tooth inner radius and rotor radius. The second is to reduce parameter L_g, however, it should consider the increase of THD of back-EMF. The third is to decrease parameter θ, but it will lead to increase electromagnetic coupling between inter-phase.

IV. CONCLUSION

Bearingless slice machines with and without fault-tolerant stator are investigated in this paper. Three stator structure variables are defined for the stator shape optimization in fault-tolerant PMBSM. According to the proposed three optimization steps, the maximum ratio of mutual inductance over self inductance and the total harmonic distortion of back-EMF can be optimized and satisfied the target values. Then, the performance, such as fault-tolerant capability, suspension force and drive torque generation, of PMBSM and fault-tolerant PMBSM are compared and analyzed. The results of comparison and analysis show that fault-tolerant PMBSM have better fault-tolerant capability than PMBSM, nevertheless its suspension and drive performance is weaker than PMBSM due to small stator area and large air gap length. Thus, three design considerations are presented for fault-tolerant PMBSM to enhance suspension and drive performance.

REFERENCES

[1] M. Neff, N. Barletta, and R. Schöb, "Magnetically levitated centrifugal pump for highly pure and aggressive chemicals," in Proc. PCIM Conf., Nuremberg, Germany, Jun. 6–8, 2000.

[2] N. Barletta and R. Schöb, "Design of a bearingless blood pump," in Proc. 3rd Int. Symp. Magn. Suspension Technol., Tallahassee, FL, 1995, pp. 265–274.

[3] F. Zürcher, T. Nussbaumer, and J.W. Kolar, "Motor torque and magnetic levitation force generation in bearingless brushless multipole motors," IEEE/ASME Trans. Mechatron., vol. 17, no. 6, pp. 1088–1097, Dec. 2012.

[4] J. Asama, R.Kawata,T.Tamura, T. Oiwa, andA. Chiba, "Reduction of force interference and performance improvement of a consequent-pole bearingless motor," Precis. Eng., vol. 36, no. 1, pp. 10–18, Jan. 2012.

[5] T. Reichert, T. Nussbaumer, and J. W. Kolar, "Bearingless 300-W PMSM for bioreactor mixing," IEEE Trans. Ind. Electron., vol. 59, no.3, pp. 1376–1388, Mar. 2012.

[6] W. Gruber, T. Nussbaumer, H. Grabner, and W. Amrhein, "Wide air gap and large-scale bearingless segment motor with six stator elements," IEEE Trans. Magn., vol. 46, no. 6, pp. 2438–2441, Jun. 2010.

[7] X.L.Wang, Q.C. Zhong, Z.Q. Deng, S.Z. Yue, "Current-Controlled Multiphase Slice Permanent Magnetic Bearingless Motors With Open-Circuited Phases: Fault-Tolerant Controllability and Its Verification," IEEE Trans. Ind. Electron., vol.59, no.5, pp. 2059-2072, May, 2012.

[8] M.T.Bartholet, T.Nussbaumer, S. Silber, J.W. Kolar, "Comparative Evaluation of Polyphase Bearingless Slice Motors for Fluid-Handling Applications," IEEE Trans. Ind. Appl., vol.45, no.5, pp. 1821-1830, Sep. 2009.

[9] R. Schoeb and N. Barletta, , "Principle and application of a bearingless slice motor," JSME Int. J. Ser. C, vol.40, pp.593-598, 1997.

[10] M. T. Bartholet, T. Nussbaumer, D. Krahenbuhl, F. Zurcher, and J. W. Kolar, "Modulation concepts for the control of a two-phase bearingless slice motor utilizing three-phase power modules," IEEE Trans. Ind. Appl., vol. 46, no. 2, pp. 831–840, Mar. 2010.

[11] M. T. Bartholet, T. Nussbaumer, and J.W. Kolar, "Comparison of voltage source inverter topologies for two-phase bearingless slice motors," IEEE Trans. Ind. Electron., vol. 58, no. 5, pp. 1921–1925, May 2011.

[12] S. Silber, W. Amrhein, P. Bösch, R. Schöb, N. Barletta, "Design aspects of bearingless slice motors," IEEE Trans. Mecha., vol.10, no.6, pp. 611-617, Dec. 2005.

[13] K. Raggl, T. Nussbaumer, and J.W. Kolar, "A comparison of separated and combined winding concepts for bearingless centrifugal pumps," Journal of Power Electronics, Vol.9, no.2, pp. 243-258, Mar. 2009.

[14] W. Gruber, W. Amrhein, "Design of a bearingless motor," Proc. 10th Int. Sym. Magnetic Bearing, pp. 252-257, Aug. 2006

[15] W. Gruber, W. Amrhein, M. Haslmayr, "Bearingless segment motor with five stator elements—design and optimization," IEEE Trans. Ind. Appl., Vol.45, no.4, pp. 1301-1308, Jau. 2009.

[16] W. Gruber, T. Nussbaumer, H. Grabner and W. Amrhein, "Wide air gap and large-scale bearingless motor with six stator elements," IEEE Trans. Magnetics, Vol.46, no.6, pp. 2438-2441, Jun. 2010.

[17] G. Bramerdorfer, G. Jungmayr, W. Amrhein, W. Gruber, E. Marth and M. Reisinger, "Bearingless segment motor with halbach magnet," Proc. Int. Sym. Power Electronics, Electrical Drives, Automation and Motion, pp.1466-1471, 2010.

[18] T. Stallinger, W. Gruber, W. Amrhein, "Bearingless segment motor with a consequent pole rotor," IEEE Int. Electrical Machines and Drives Conference, pp. 1374-1380, May 2009.

[19] W. Gruber, J. Passenbrunner, G. Bramerdorfer and W. Amerhein, "Novel bearingless segment motor design with axial magnetized rotor magnets," Proc. 8th Int. Power Electronics-ECCE Asia, pp. 2225-2232,May,2011

Comparison of Homopolar and Heteropolar Bearingless Reluctance Slice Motor Prototypes

Wolfgang Gruber[a], Walter Bauer[a], Karlo Radman[b]

[a] Johannes Kepler University Linz, Altenberger Str. 69, 4040 Linz, Austria, wolfgang.gruber@jku.at
[b] University of Rijeka, Faculty of Engineering, Vukovarska 58, 51000 Rijeka, Croatia

***Abstract*—This paper focuses on the operational behavior of two bearingless reluctance slice motor prototypes. After a brief introduction of the two systems, experiences, obtained from both novel bearingless drives, are presented. On this basis, communalities and differences, in various areas, e. g. construction, geometric constraints, winding systems, force to torque ratio, power electronic utilization and efficiency, will be outlined. This will help to point out the applicability of the reluctance slice motor concepts for industrial utilization.**

I. Introduction

A. Bearingless Reluctance Slice Motors

Contactless operation and therefore, no lubrication, no mechanical wear and loss, extreme cleanness and purity, superior tightness as well as no need for seals are advantages of magnetically levitated systems. However, due to the relatively high cost magnetic levitation is normally used in very specialized areas and has not been implemented in large scale mass production. A design that has the capability to reduce the mechanical complexity and therefore production cost, are bearingless motors. These magnetically levitated drives are very compact and combine the generation of motor torque and suspension force in one common unit. The passive stabilization of various degrees of freedom by a permanent magnetic flux in the air gap further reduces the complexity. In bearingless slice motors the axial and tilting directions are passively stabilized and the radial suspension forces and the motor torque are created actively, featuring a fully magnetically levitated system of low mechanical complexity [1]. Pumps [2] and stirrers [3] are the main field of application for this topology by now.

However, the passive air gap field is normally created by permanent magnets located in the rotor. Thus, the rotor of bearingless motors resembles rotors of synchronous machines, with surface mounted, interior or inset permanent magnets [4]. Only recently, first theoretical works have been published on bearingless slice motors that have no permanent magnets in the rotor, but in the stator. These designs are called bearingless reluctance slice motors. No permanent magnets lead to very cheap and robust rotors. This is important especially in disposables and high temperature applications. In previous works the authors introduced theoretical considerations about homopolar and heteropolar bearingless reluctance slice motors [5]. Homopolar drives feature a unipolar permanent magnetic field in the air gap, whereas heteropolar drives possess alternating permanent magnetic air gap field. Based on that work, two prototype drives have been optimized, fabricated and put into operation recently [6], [8]. This paper now focuses on the experiences obtained from the manufactured heteropolar and homopolar bearingless motor and points out the differences in their characteristic behaviour based on measurement results.

B. Homopolar Design

The setup of the homopolar bearingless reluctance slice motor is depicted in Fig. 1. The rotor is an iron ring featuring four salient poles on the outer side and a circular surface on the inside. The stator composition resembles a temple motor design [1], where the six stator teeth are bent in axial direction and are closed over a back yoke plate on the bottom of the system. Hence, the winding axes of the stator coils are parallel to the rotor axis. However, common temple motor designs do not have a stator permanent magnet and the stator flux collector plate. It is also visible that the illustrated design features two radial air gaps between rotor and stator. There is one gap between the stator teeth and the rotor and an additional one between the rotor and the stator flux collector plate. The cylindrical stator permanent magnet is axially magnetized and creates a homopolar bias flux in both air gaps. The electromagnetic flux created by the stator coils passes

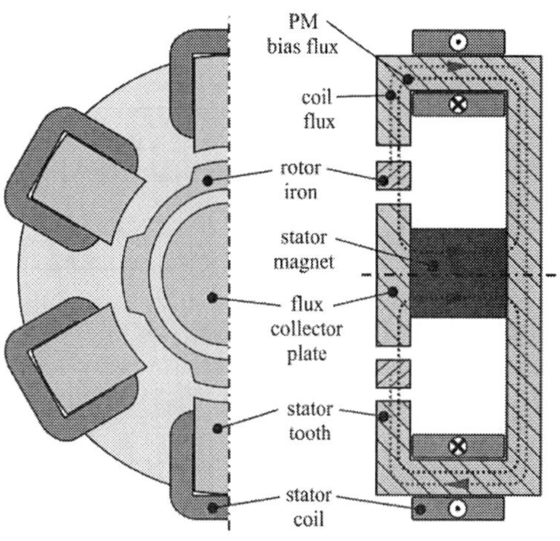

Figure 1. Composition of an 6/4 bearingless homoploar reluctance slice motor (featuring six stator teeth and four rotor teeth).

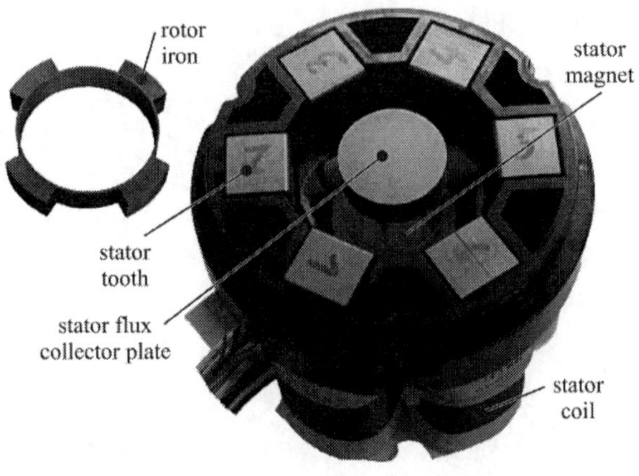

Figure 2. Rotor and stator of the bearingless homopolar reluctance slice motor prototype.

over the stator teeth and the rotor. Thus, the flux collector plate and the stator permanent magnet are not permeated by the coil flux path. However, each one of the six stator coils is capable to create coupled radial bearing forces and drive torque in dependence on the rotor angle.

Figure 2 depicts a photo of the manufactured prototype [6]. The six concentric coils are connected in two double star systems. All phases are capable to create both radial suspension forces and motor torque in dependence on the rotor angle.

C. Heteropolar Design

The main and striking constructional characteristic of the flux-switching motor, depicted in Fig. 3 and 4, is the placement of the permanent magnets. They are located in the midst of the stator teeth and separate them electromagnetically. The flux lines of each permanent magnet close mainly over the neighbouring stator teeth and, thus, do not penetrate the whole air gap, but only influence the air gap region close to the permanent magnet, creating two opposing magnetic poles in the air gap and therefore a heteropolar magnetic air gap field around the circumference. Unfortunately, there is also a considerable amount of fringing flux closing over the outer side of the stator. To keep this kind of stray flux low, a small air gap length (compared to the magnet width) is favourable and necessary.

However, the air gap flux density in flux-switching machines is often higher than in comparable other electric machines, leading to an increased torque capacity [9]. This is due to the flux concentration capability of the construction. The surfaces of the stator teeth normal to the magnetization direction collect permanent magnetic flux and concentrate it towards the much smaller area of the stator teeth adjoining the air gap.

In this novel kind of bearingless motor, two separated three phase winding systems are placed. One phase consists of two opposing coils. In dependence of their referring coil turn direction, either suspension force or drive torque is created by the considered phase. Thus, in contrast to the considered homopolar reluctance drive, radial suspension force and torque generation are decoupled in this bearingless drive.

The prototype features twelve stator coils and a rotor with ten poles [8]. Figure 4 depicts a picture of the manufactured bearingless flux-switching motor. The stator iron is held in an aluminium frame to strengthen the structure. The permanent magnets are glued in pockets that separate the stator elements. The externally wound stator coils are easily slipped on the stator teeth and connected accordingly.

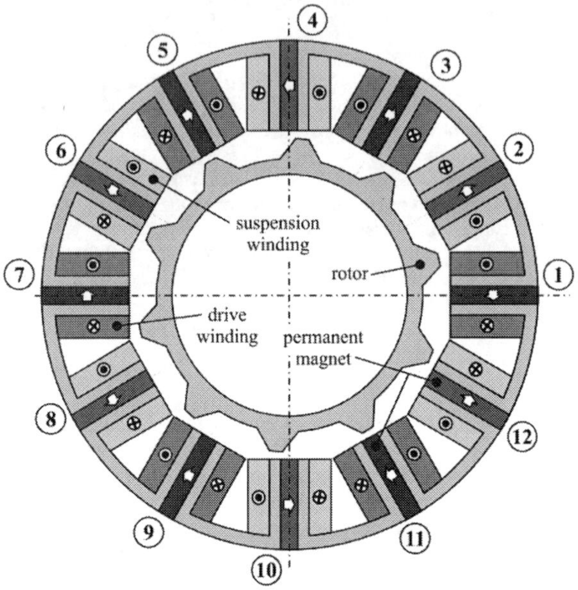

Figure 3. Principle cross section of the built flux-switching motor with twelve stator and ten rotor teeth. The arrows indicate the direction of the magnetization.

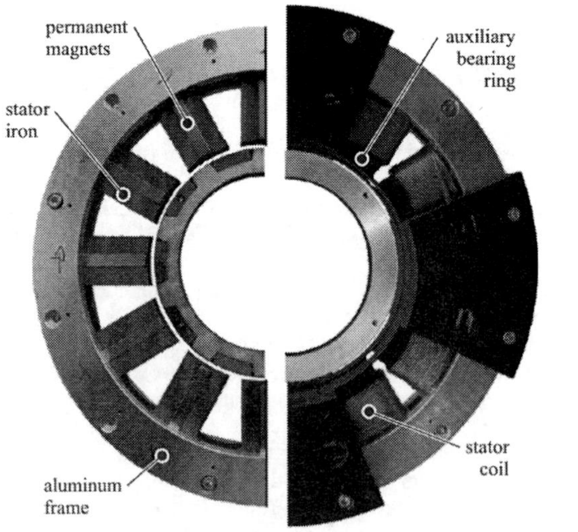

Figure 4. Stator of the bearingless flux-switching slice motor prototype without coils (left) and fully assembled (right).

II. Comparison of the Systems

Both systems feature very different geometric sizes but are driven by the same power electronics, featuring six half bridges with a DC voltage link of 300V and a maximum effective current of 15A. Thus, the electrical maximum input power is the same. In the following sections certain characteristics of the two prototypes are considered and discussed.

A. Geometric System Values

The two considered bearingless reluctance slice motor prototypes feature a quite different geometry. Table I summarizes the main values. Comparing the gap factor

$$g = \frac{\delta}{d_{ro}} \tag{1}$$

with $g_{hom} = 0.074$ and $g_{het} = 0.02$ it becomes obvious that the heteropolar flux-switching motor seems not very suitable for large air gaps. This originates from the distinct permanent magnetic fringing flux paths (on the outer side of the stator), that rise significantly with increasing air gap. Thus, the performance of the flux-switching designs decreases strongly for larger air gaps. This is not the case for the homopolar design.

B. Force and Torque Generation

The generation of force and torque works very different in the two considered systems. As indicated in [6] for the homopolar motor, each single phase creates both bearing forces and motor torque. Nevertheless it is feasible to use two separated 3-phase systems. Concerning the heteropolar drive there is are separated 3-phases system; one for torque generation and another for force generation. The phase characteristic of both systems is shown in Fig. 5 and 6. From these phases characteristic performance factors can be derived that describe the overall behaviour of the drive in comparison to the phase characteristic [5], [7]. Due to the fact that these factors are related to the phase characteristic, they cannot be compared for different design principles, but they show that the bearingless motor operation of both reluctance motors is limited to only a few design variants. Concerning six phases only the 6-4 stator-rotor teeth design is reasonable for homopolar drives. For the six phases heteropolar flux-switching motor a bearingless operation is only feasible for the 12-10 and 12-16 topologies (when single phase torque characteristics are not under consideration).

TABLE I. Geometric and Material Parameterss

Variable	Description	Value		Unit
		hom.	het.	
h_{rz}	axial rotor height	17	10	mm
d_{ro}	outer rotor diameter	65	150	mm
δ	mechanical air gap	4.8	3	mm
d_{so}	system outer diameter	130	280	mm
h_{sz}	system axial length	130	35	mm
n_r	number of rotor teeth	4	10	
n_s	number of stator teeth	6	12	
stator and rotor material		M330-50A		
stator permanent magnet material		N42	N38	

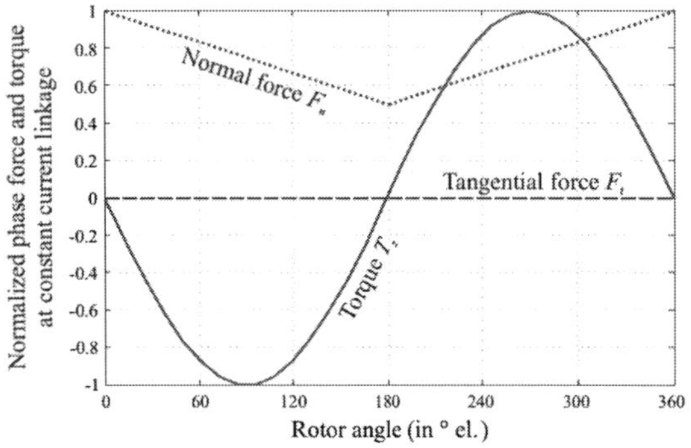

Figure 5. Resulting force and torque of one phase over the rotor angle at constant current linkage for the homopolar slice motor prototype.

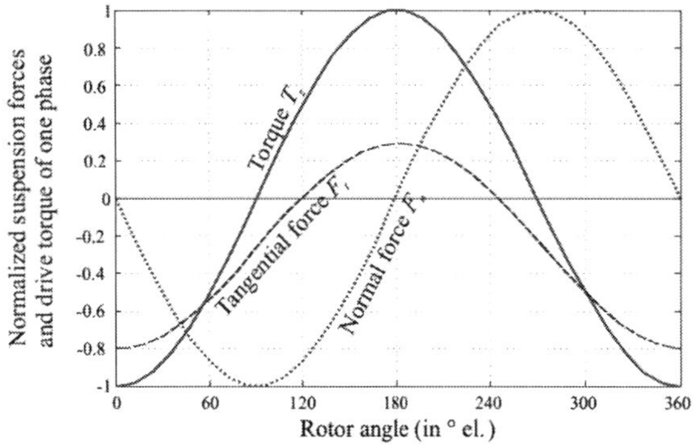

Figure 6. Bearing force (of a bearing phase) and motor torque (of a torque phase) over the rotor angle at constant current linkage for the heteropolar slice motor prototype.

However, the following analysis enables a comparison of homopolar and heteropolar torque generation capability. Assuming sinusoidal flux linkage in the coils, the drive torque

$$T(\varphi) = \frac{\partial \Psi_{PM}(\varphi)}{\partial \varphi} i_s = N_s A_{tooth} \hat{B} \cos(\varphi) \tag{2}$$

holds true. Considering non-saturated machines, for heteropolar drives the magnetic field strength in a stator tooth fluctuates in the range

$$-B_{sat} < B_{het}(\varphi) < B_{sat} . \tag{3}$$

In contrast to that, for homopolar drives the field in the tooth yields

$$0 < B_{hom}(\varphi) < B_{sat} . \tag{4}$$

Thus, it becomes clear, that for the machines of the same size with the same material utilization the homopolar and heteropolar mean drive torque can be estimated by the relation

$$\overline{T}_{het}(\varphi) \approx 2 \cdot \overline{T}_{hom}(\varphi) . \tag{5}$$

This impliess that for well utilized machines the torque density of homopolar drives will be superior to that of heteropolar motors. However, this consideration does not hold true for force generation. About the same force densities are possible with both machines.

C. Winding Scheme and Outline

As mentioned before the used winding schemes for the two considered drives are quite different, although both systems feature a double 3-phase star connected winding configuration. Figure 7 illustrates both of these winding systems. Due to the double 3-phase star connections only four currents needs to be measured for the nested current controllers in both prototypes.

The setting of the coil turn numbers can be considered easier in the separated winding system of the heteropolar flux-switching motor, because the suspension and drive coils can be designed independently and, thus, differently. In the combined winding system of the bearingless homopolar drive the set coil turn number influences both suspension and torque.

A remarkable issue of the two prototype drives is the poor power factor because of the large stator inductivities due to the small magnetic air gap [11].

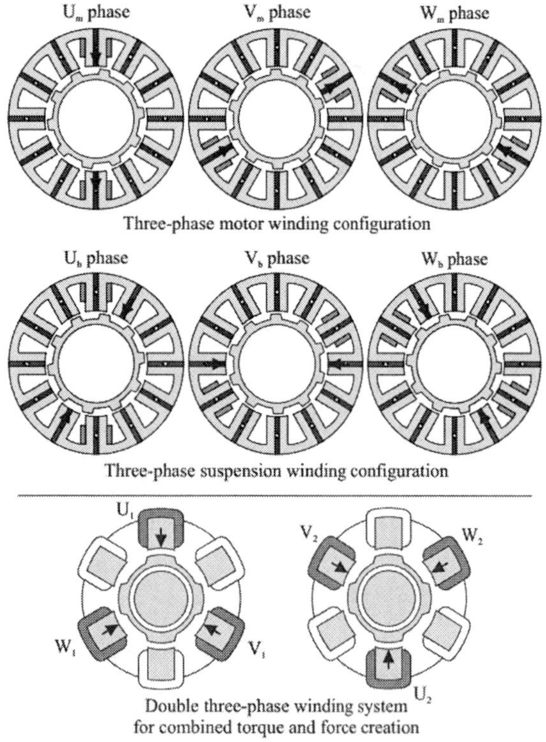

Figure 7. Winding schemes for the heteropolar (top) and the homopolar (bottom) prototype machine.

TABLE II. KEY PARAMETERS OF THE PROTOTYPEW

Variable	Description	Value		Unit
		hom.	het.	
k_r	radial stiffness	-60	-41.6	N/mm
k_z	axial stiffness	3.4	7.3	N/mm
k_φ	tilt stiffness	1	21.4	Nm/rad
$F_{r,p}$	force capacity/phase	20	18	mN/Aturn
$T_{r,p}$	torque capacity/phase	0.1	0.45	mNm/Aturn
N_s	windings per coil	250	80	turns
n_p	coils per phase	1	2	
n_r	rated speed	1000	3000	rpm
T_r	rated torque	0.65	1.6	Nm

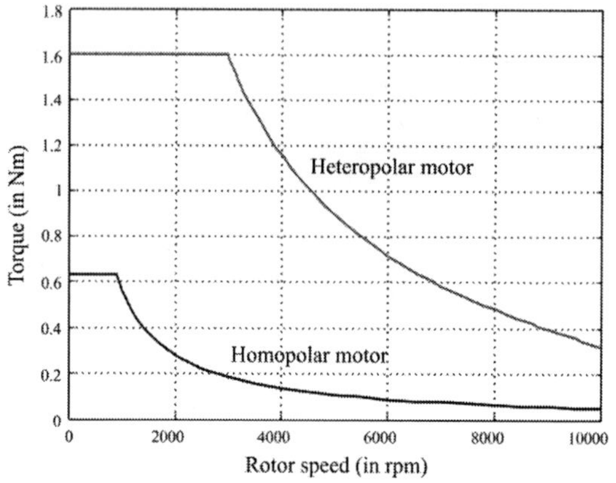

Figure 8. Torque-speed-characteristics of the two bearingless reluctance slice motors.

D. Operational Characteristic

The principal characteristics of both bearingless reluctance drives is the same as in permanent magnet excited bearingless synchronous machines. Even though the rotor has salient poles, the force and torque dependency from the stator currents is linear, due to the high permanent magnetic bias flux in the air gap. The torque created by the variance of the inductivity over the rotor angle is negligible. Force and torque measurements can be found in [6] and [7]. Figure 8 shows the torque-speed curve of both drives. It is clearly visible that the heteropolar flux-switching motor features higher output power. This is mainly because of its much larger rotor diameter and smaller air gap.

Concerning the motor losses and their efficiency measurements are currently conducted. First results are presented in [10]. Regarding the idle losses it is expected that homopolar drives are superior due to the reduced magnetic field variation.

After putting both systems into operation successfully, it can be claimed that their operational, dynamic and control characteristic and behaviour is very similar to common rotor permanent magnet excited slice motors. Thus, it has been shown that these new drive concepts represent feasible alternatives to state-of-the-art bearingless slice motors, especially in disposable systems and high temperature applications.

Conclusion

This paper comprises a comparison of the two bearingless reluctance slice motor prototypes. The characteristic data and main behaviour is discussed. Both drives feature a relatively low power factor due to the high inductivities of the drives. Normally, this results in a bad utilization of the power electronics. It also can be seen that the force and torque capacity per phase and ampere-turn are very similar. However, in the other characteristic parameter, like for instance the siffnesses, high differences do exist.

The homopolar motor features reduced iron losses, a capability for higher speed and larger air gaps, but lacks of torque density. On the other hand the heteropolar motor seems not suitable for large air gaps. The high number of rotor teeth increases the electrical speed and, therefore, prohibits high mechanical speeds. The generation of force and torque capability is decoupled and can be designed independently.

Thus, the bearingless homopolar reluctance slice motor seems more suitable for applications like pumps, blowers or fans (characterized by relatively high rotational speed with low torque), whereas the bearingless heteropolar flux-switching slice motor rather works for mixers and stirrers (demanding lower speeds but higher torque).

Acknowledgement

This work was supported by the Linz Centre of Competence (LCM) GmbH, a K2-centre of the COMET program of the Austrian Government. The authors thank the Austrian and Upper Austrian Government for their support.

References

[1] R. Schöb, N. Barletta, "Principle and application of a bearingless slice motor," *Proceedings on 5th Int. Symp. on Magnetic Bearings (ISMB)*, pp. 333-338, 1996.

[2] T. Nussbaumer, K. Raggl, P. Boesch, J. W. Kolar, "Trends in integration for magnetically levitated pump systems," *Proceedings on Power Conversion Conf. (PCC)*, pp. 1551-1558, 2007.

[3] B. Warberger, R. Kaelin, T. Nussbaumer, J.W. Kolar, "50Nm/2500W Bearingless Motor for High-Purity Pharmaceutical Mixing," *Proceedings on IEEE Trans. on Industrial Electronics*, vol. 59, no. 5, pp. 2236-2247, 2012

[4] T. Nussbaumer, P. Karutz, F. Zürcher and J. W. Kolar, "Magnetically levitated slice motors—an overview," *IEEE Transactions on Industry Applications*, vol. 47, no. 2, pp. 754-766, March-April 2011.

[5] W. Gruber, W. Briewasser, M. Rothböck, W. Amrhein and R. Schöb, "Bearingless reluctance slice motors," *Proceedings on 13th International Symposium on Magnetic Bearings (ISMB)*, Aug. 2012.

[6] W. Gruber, M. Rothböck, R. Schöb, "Design of a novel homopolar bearingless slice motor with reluctance rotor," *Proceedings on 5th IEEE Energy Conversion Congress and Exposition (ECCE)*, 2013.

[7] W. Gruber, W. Bauer, K. Radman, W. Amrhein, R. Schöb, "Considerations Regarding Bearingless Flux-Switching Slice Motors," *Proceedings on 1st Brazilian Workshop on Magnetic Bearings*, Oct. 2013

[8] W. Gruber, K. Radman, "Design of a Bearingless Flux-Switching Slice Motor," *Proceedings on 7th Int. Power Electronics Conference, (IPEC-ECCE Asia 2014)*, May 2014.

[9] Z. Q Zhu, Y. Pang, J. Chen, Z. P. Xia, D. Howe, "Influence of design parameters on output torque of flux-switching permanent magnet machines," *Proceedings on IEEE Vehicle Power and Propulsion Conference (VPPC)*, pp. 1-6, 2008.

[10] K. Radman, N. Bulic , W. Gruber, "Loss Analysis of Bearingless Flux-Switching Slice Motor," *Proceedings on 14th Int. Symp. on Magnetic Bearings (ISMB)*, accepted for publication, August 2014.

[11] K. Radman, N. Bulic, W. Gruber, "ECCE, Performance Evaluation of a Bearingless Flux-Switching Slice Motor," *Proceedings on 6th IEEE Energy Conversion Congress and Exposition (ECCE)*, accepted for publication, Sept. 2014

Two mathematical models of five-phase high speed permanent magnet bearingless slice motor

Tuocheng NI[a], Xiaolin WANG[a], Qiang DING[a],Guyu WU[a]

[a]Nanjing University of Aeronautics and Astronautics, Yudao street 29, 20016 Nanjing, China, wangxl@nuaa.edu.cn

Abstract—Most bearingless motors run at a relatively low speed because of the high iron loss problem. Considering that a slotless structure is beneficial to high speed, this paper deduces two mathematical models based on this existent structure. The first model is for traditional control and the second one is for fault-tolerant control. In the process of derivation, the concept of superposition is introduced to calculate the flux density in the air gap. At last, finite element analysis is done to verify the model.
Keywords-mathematical models, permanent magnet bearingless slice motor (PMBSM), slotless structure, fault-tolerant

I. Introduction

Due to its simple and compact structure, the permanent magnet bearingless slice motor (PMBSM) has been used in various industry fields such as hazardous chemical material process, blood pumps for artificial heart assistance device and ultra-pure fluid pumps in semiconductor industry[1]. However, these applications mainly focus on low speed field. In [2], a five-phase PMBSM with slotless stator structure is presented for verification of high speed capability about PMBSM, as shown in Fig.1. In [3], winding configuration, loss analysis as well as control system implementation are reported. Nevertheless, the mathematical model of radial levitation force is not given by the authors. It is well known that the mathematical model of radial levitation force is the foundation of control strategy and control system design[4], thus this paper is to establish two kinds of mathematical levitation force models, the first one is for traditional high speed field and the second one is for high reliability application where the PMBSM must possess fault-tolerant capability[5][6]. All of these two mathematical models are deduced from the distribution of air gap magnetic flux density and verified by finite element method.

II. Mathematical Model of Radial Levitation Force and Torque

We refer the slotless structure in [3] and deduce the mathematical model. Fig.1 shows the structure of this kind of bearingless motor. Two coils of each phase reverse series. The rotor is made up of rotor core and the permanent magnet (PM) . In order to reduce the complexity, the PM has only a pair of poles. To simplify the analysis, a few assumptions are made as follows:

a. Ignore the magnetic saturation and leakage flux;
b. Conductors of windings are evenly distributed along stator circumference;
c. Ignore magnetic motive force drop in iron.

A. Five-Phase Constraint Model

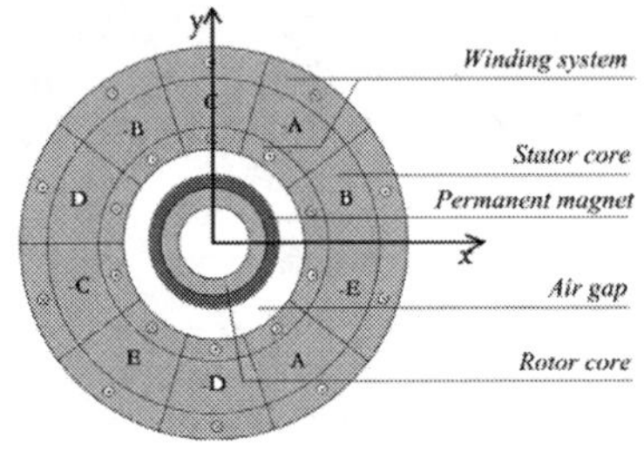

Figure 1. The slotless topology of the bearingless motor

According to working principle of PMBSM, the current in each phase can be spilt into two components, which are called torque current component I_t and levitation current component I_s. Torque current generates one pole-pair magnetic field for drive generation and levitation current generates two pole-pairs magnetic field for levitation force generation. Both of the two components satisfy five-phase constraint relationship, which is described as equation (1).

$$i_k = I_t \cos[\theta_t - \frac{2(k-1)\pi}{5}] + I_s \cos[\theta_t - \frac{4(k-1)\pi}{5}] \quad (1)$$

Where i_k (k=1,2,3,4,5) is the current in each phase. We know that magnetic motive force (MMF) changes linearly under each coil. Change of MMF is directly proportional to the value of the current.

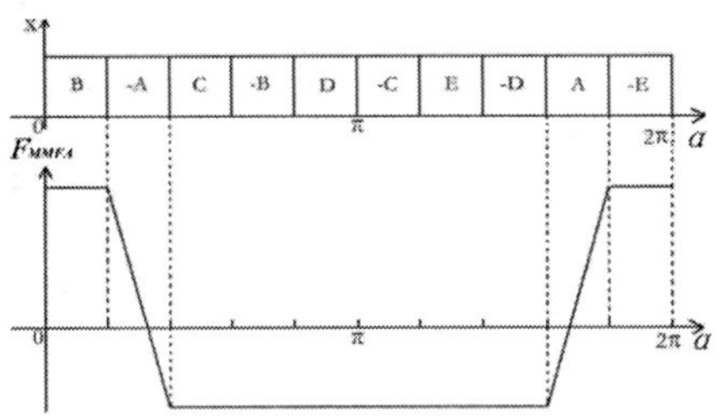

Figure 2. Schematic diagram of magnetic motive force generated by A phase

Take phase-A for example, its distribution of magnetic motive force can be shown in Fig.2 and be expressed as equation (2):

$$F_{MMFA}=\begin{cases}\frac{7\pi}{50}\cdot N_c i_1 \ldots\ldots\ldots\ldots 0\le\alpha\le\frac{\pi}{5}\\ (-\alpha+\frac{17\pi}{50})\cdot N_c i_1 \ldots\ldots \frac{\pi}{5}\le\alpha\le\frac{2\pi}{5}\\ -\frac{3\pi}{50}\cdot N_c i_1 \ldots\ldots\ldots \frac{2\pi}{5}\le\alpha\le\frac{8\pi}{5}\\ (\alpha-\frac{83\pi}{50})\cdot N_c i_1 \ldots\ldots \frac{8\pi}{5}\le\alpha\le\frac{9\pi}{5}\\ \frac{7\pi}{50}\cdot N_c i_1 \ldots\ldots\ldots \frac{9\pi}{5}\le\alpha\le 2\pi\end{cases} \quad (2)$$

Therefore, the expressions of the MMF generated by all five phases are in the form of piecewise function:

$$F_{MMF}=\begin{cases}[\frac{7\pi}{50}i_1+(\alpha-\frac{3\pi}{50})i_2-\frac{3\pi}{50}i_3-\frac{3\pi}{50}i_4-\frac{3\pi}{50}i_5]N_c \ldots\ldots 0\le\alpha\le\frac{\pi}{5}\\ [(-\alpha+\frac{17\pi}{50})i_1+\frac{7\pi}{50}i_2-\frac{3\pi}{50}i_3-\frac{3\pi}{50}i_4-\frac{3\pi}{50}i_5]N_c \ldots \frac{\pi}{5}\le\alpha\le\frac{2\pi}{5}\\ [-\frac{3\pi}{50}i_1+\frac{7\pi}{50}i_2+(\alpha-\frac{23\pi}{50})i_3-\frac{3\pi}{50}i_4-\frac{3\pi}{50}i_5]N_c \ldots \frac{2\pi}{5}\le\alpha\le\frac{3\pi}{5}\\ [-\frac{3\pi}{50}i_1+(-\alpha+\frac{37\pi}{50})i_2+\frac{7\pi}{50}i_3-\frac{3\pi}{50}i_4-\frac{3\pi}{50}i_5]N_c \ldots \frac{3\pi}{5}\le\alpha\le\frac{4\pi}{5}\\ [-\frac{3\pi}{50}i_1-\frac{3\pi}{50}i_2+\frac{7\pi}{50}i_3+(\alpha-\frac{43\pi}{50})i_4-\frac{3\pi}{50}i_5]N_c \ldots \frac{4\pi}{5}\le\alpha\le\pi\\ [-\frac{3\pi}{50}i_1-\frac{3\pi}{50}i_2+(-\alpha+\frac{57\pi}{50})i_3+\frac{7\pi}{50}i_4-\frac{3\pi}{50}i_5]N_c \ldots \pi\le\alpha\le\frac{6\pi}{5}\\ [-\frac{3\pi}{50}i_1-\frac{3\pi}{50}i_2-\frac{3\pi}{50}i_3+\frac{7\pi}{50}i_4+(\alpha-\frac{63\pi}{50})i_5]N_c \ldots \frac{6\pi}{5}\le\alpha\le\frac{7\pi}{5}\\ [-\frac{3\pi}{50}i_1-\frac{3\pi}{50}i_2-\frac{3\pi}{50}i_3+(-\alpha+\frac{77\pi}{50})i_4+\frac{7\pi}{50}i_5]N_c \ldots \frac{7\pi}{5}\le\alpha\le\frac{8\pi}{5}\\ [(\alpha-\frac{83\pi}{50})i_1-\frac{3\pi}{50}i_2-\frac{3\pi}{50}i_3-\frac{3\pi}{50}i_4+\frac{7\pi}{50}i_5]N_c \ldots\ldots \frac{8\pi}{5}\le\alpha\le\frac{9\pi}{5}\\ [\frac{7\pi}{50}i_1+\frac{3\pi}{50}i_2-\frac{3\pi}{50}i_3-\frac{3\pi}{50}i_4+(-\alpha+\frac{97\pi}{50})i_5]N_c \ldots\ldots \frac{9\pi}{5}\le\alpha\le 2\pi\end{cases} \quad (3)$$

Put the torque current component and levitation current component into (3) and we'll get the expressions of MMF distribution generated by these two current components respectively. Plot them and they'll be as shown in Fig.3 and Fig.4.

Fig.3 shows that the torque magnetic motive force distribution along air gap circumference when phase angle of torque current component (θ_t) equals to zero.

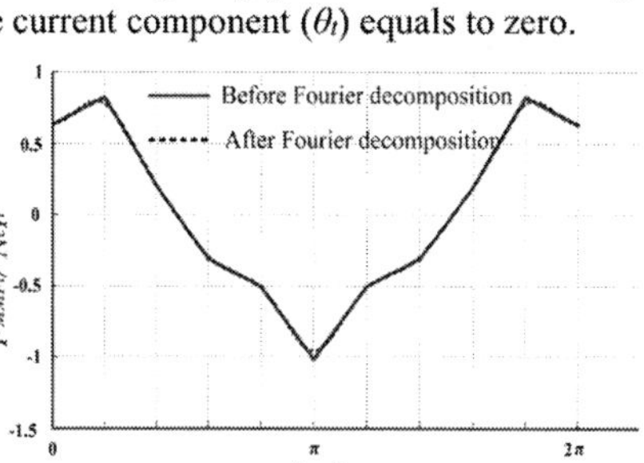

Figure 3. Torque magnet motive force distribution when θ_t=0

Fig.4 shows that the levitation magnetic motive force distribution along air gap circumference when phase angle of levitation current component (θ_s) equals to zero.

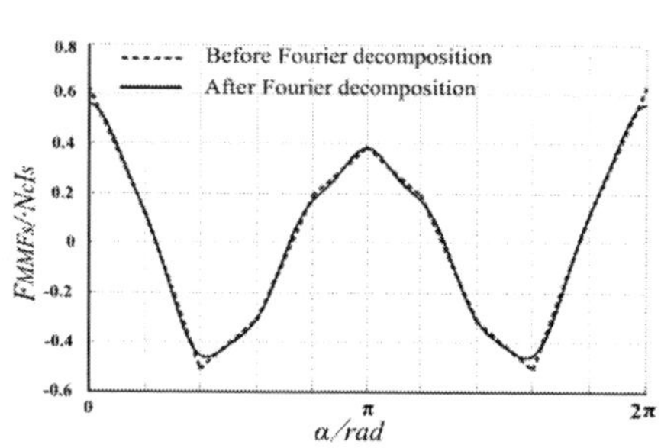

Figure 4. Levitation magnet motive force distribution when θ_s=0

Considering the MMF of air gap is a piecewise function, thus Fourier decomposition is used to obtain a unified expression. It can be seen from Fig.3 and Fig.4 that the fitting curves has good agreement with original curves. Due to the MMF has a linear relationship with flux dentisy when neglecting the rotor eccentricity and the magnetic saturation, the expressions of air gap flux density caused by torque current component and levitation current component can be shown as follows:

$$B_t(\alpha)|_{\theta_t=0}=[a_1\cos\alpha+a_2\cos 4\alpha+a_3\cos 6\alpha]N_c I_t G_\mu \quad (4)$$

$$B_s(\alpha)|_{\theta_s=0}=[a_4\cos 2\alpha+a_5\cos 3\alpha+a_6\cos 8\alpha]N_c I_s G_\mu \quad (5)$$

Where a_i (i=1,2...6) is Fourier decomposition coefficient; N_c is the number of winding conductor in unit radian; G_μ is air gap permeance.

Because the equation (4) and (5) are established at two special positions (θ_t =0 and θ_s =0), the expressions of air gap flux density at general positions can be expressed as:

$$B_t(\theta_t,\alpha)=[a_1\cos(\alpha-\theta_t)+a_2\cos(4\alpha+\theta_t)+a_3\cos(6\alpha-\theta_t)]N_c I_t G_\mu \quad (6)$$

$$B_s(\theta_s,\alpha)=[a_4\cos(2\alpha-\theta_s)+a_5\cos(3\alpha+\theta_s)+a_6\cos(8\alpha+\theta_s)]N_c I_s G_\mu \quad (7)$$

According to the Maxwell force formula, the radial forces can be deduced and decomposed into x and y directions:

$$\begin{cases}F_{Mx}=\frac{rh}{2\mu_0}\int_0^{2\pi}(B_t+B_s+B_{PM})^2\cos\alpha d\alpha\\ F_{My}=\frac{rh}{2\mu_0}\int_0^{2\pi}(B_t+B_s+B_{PM})^2\sin\alpha d\alpha\end{cases} \quad (8)$$

Where B_{PM} is flux density caused by rotor permanent magnets. The torque T can be obtained by virtual work method:

$$T=\frac{dE_g}{d\theta_r}=\frac{d}{d\theta_r}\int_0^{2\pi}\frac{(B_t+B_s+B_{PM})^2 l_m rh}{2\mu_0}d\alpha \quad (9)$$

Where E_g and θ_r stand for the stored energy of air gap and the rotor rotation angle; l_m is equivalent air gap length; r is the radius of rotor; h is the effective axial length. The model can be shown as:

$$F_{Mx}=\frac{N_c I_s h\pi r\mu_0}{2l_m^2}[a_4a_5N_cI_s\cos 2\theta_s+(a_1a_4+a_2a_5)N_cI_t\cos(\theta_s-\theta_t)+a_4\cos(\theta_s-\theta_r)A_{PM}] \quad (10)$$

$$F_{My}=\frac{N_c I_s h\pi r\mu_0}{2l_m^{\ 2}}[-a_4a_5A_fI_s\sin 2\theta_s+(a_1a_4+a_2a_5)N_cI_t\sin(\theta_s-\theta_t)$$
$$+a_4\sin(\theta_s-\theta_r)A_{PM}] \quad (11)$$

$$T=\frac{a_1N_ch\pi r\mu_0A_{PM}}{l_m}I_t\sin(\theta_t-\theta_r) \quad (12)$$

Where A_{PM} denotes the amplitude of magnetic motive force produced by permanent magnet. Considering that in (10) and (11), the values of first two terms are much smaller than the third one which is relevant to the PM magnetic motive. We neglect the first two terms and the model is simplified as:

$$\begin{cases} F_{Mx}=K_M\cos(\theta_s-\theta_r)I_s \\ F_{My}=K_M\sin(\theta_s-\theta_r)I_s \\ T=K_t\sin(\theta_t-\theta_r)I_t \end{cases} \quad (13)$$

Where K_M and K_t are only relevant to the structure of the motor and their values are:

$$\begin{cases} K_M=\dfrac{N_ch\pi r\mu A_{PM}a_4}{2l_m^{\ 2}} \\ K_t=\dfrac{N_ch\pi r\mu A_{PM}a_1}{l_m} \end{cases} \quad (14)$$

B. General Model

In order to achieve fault-tolerant operation, a general fault-tolerant mathematical model of PMBSM is necessary. In this mathematical model, the currents in five phases are independent with each other.

If we caculate the flux dentisy by using (3) directly and according to the Maxwell force formula, the forces will be expressed as:

$$\begin{cases} F_{Mx}=\dfrac{rh}{2\mu_0}\sum\limits_{i=1}^{10}\int_{\frac{i-1}{5}\pi}^{\frac{i}{5}\pi}(F_{MMF}G_\mu+B_{PM})^2\cos\alpha d\alpha \\ F_{My}=\dfrac{rh}{2\mu_0}\sum\limits_{i=1}^{10}\int_{\frac{i-1}{5}\pi}^{\frac{i}{5}\pi}(F_{MMF}G_\mu+B_{PM})^2\sin\alpha d\alpha \end{cases} \quad (15)$$

Simplified (15) and Maxwell forces are as the following:

$$\begin{cases} F_{Mx}=k_M\{[0.588(i_2-i_5)-0.951(i_3-i_4)]\sin\theta_r+ \\ \qquad [i_1-0.809(i_2+i_5)+0.309(i_3-i_4)]\cos\theta_r\} \\ F_{My}=k_M\{[0.588(i_2-i_5)-0.951(i_3-i_4)]\cos\theta_r+ \\ \qquad [-i_1+0.809(i_2+i_5)-0.309(i_3+i_4)]\sin\theta_r\} \end{cases} \quad (16)$$

Where $k_M=\dfrac{\sqrt{5}N_chr\mu_0A_{PM}}{8l_m^{\ 2}}$. The torque will also be obtained in a similar method as (9). At last, the torque model is expressed as:

$$T=k_t\{[0.951(i_2-i_5)+0.588(i_3-i_4)]\cos\theta_r$$
$$+[-i_1-0.309(i_2+i_5)+0.809(i_3+i_4)]\sin\theta_r\} \quad (17)$$

Where $k_t=\dfrac{N_chr\mu_0A_{PM}}{l_m}$. In fact, equtation (16) and (17) can be simplified in the maxtrix form as

$$[P]=[M]\cdot[i] \quad (18)$$

Where $[P]=[F_{Mx}\ F_{My}\ T]^T$,$[i]=[i_1\ i_2\ i_3\ i_4\ i_5]^T$ and $[M]$ is as the following:

$$\begin{bmatrix} k_M\cos\theta_r & k_M\cos(\theta_r+\frac{6\pi}{5}) & k_M\cos(\theta_r+\frac{2\pi}{5}) & k_M\cos(\theta_r+\frac{8\pi}{5}) & k_M\cos(\theta_r+\frac{4\pi}{5}) \\ k_M\sin(\theta_r+\pi) & k_M\sin(\theta_r+\frac{\pi}{5}) & k_M\sin(\theta_r-\frac{3\pi}{5}) & k_M\sin(\theta_r+\frac{3\pi}{5}) & k_M\sin(\theta_r-\frac{\pi}{5}) \\ k_t\sin(\theta_r+\pi) & k_t\sin(\theta_r+\frac{3\pi}{5}) & k_t\sin(\theta_r+\frac{\pi}{5}) & k_t\sin(\theta_r-\frac{\pi}{5}) & k_t\sin(\theta_r-\frac{3\pi}{5}) \end{bmatrix} \quad (19)$$

III. VERIFICATION BY FINITE ELEMENT ANALYSIS

In this section, finite element analysis (FEA) is done for detailed verification of the model. Table I shows the motor parameters in Ansoft.

TABLE I. PAPAMETEROF THE MOTOR

	Parameter	Value
r	Rotor diameter	16mm
W	Number of winding turns	12
h	Axial rotor length	10mm
l_m	Equivalent length of air gap	2.9mm
l_{PM}	Thickness of the PM	1.8mm

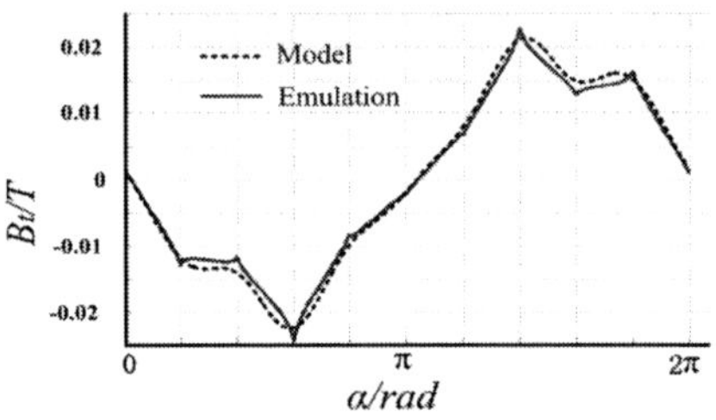

Figure 5. Comparison of the theoretical and emulational torque flux density when θ_t=56°

Fig.5 and Fig.6 show the comparison of torque air gap flux density and levitation air gap flux density between analytical result and finite element result, which are both only caused by stator currents. It can be seen from these two figures that the analytical results and finite element results agree well.

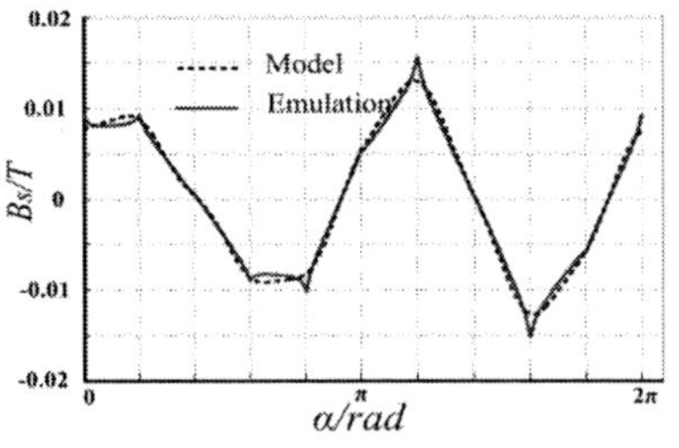

Figure 6. Comparison of the theoretical and emulational levitation flux density when θ_s=275°

To verify the exactitude of levitation, we do a preliminary validation in Ansoft. The rotor is kept still. Rotate the magnetic field produced by levitation currents and observe the levitation forces. The curves of them are shown in Fig.7.

In Fig.7, it is obvious that the levitation forces in FEA are much larger than the theoretical value. The error is high up to 30%. We have to conjecture that another kind of force may increase the levitation force.

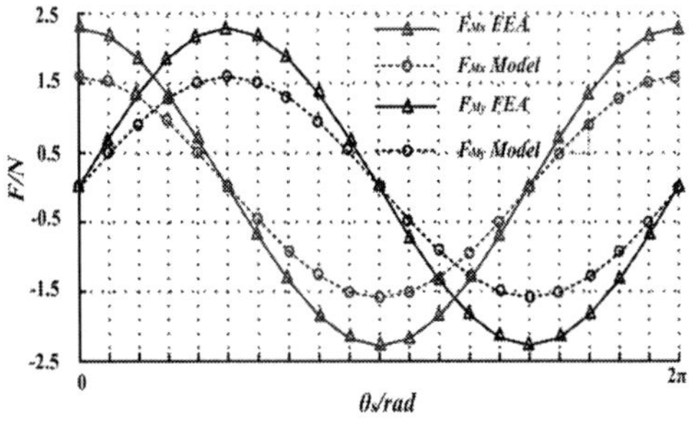

Figure 7. Comparison of the radial forces in Ansoft and the Maxwell forces

As we all know, there are two kinds of electromagnetic forces exsiting in the motor. One is the Maxwell force and its main effect is producing the levitation force as we deduced in last section. The other is Lorentz force, whose main effect is exerting a torque. In traditonal slice motor with stator teeth, the Lorentz force only has a minor effect on the levitation force. Usually, the Lorentz component in levitation force is small enough to be ignored. But it is still need to be verified whether the Lorentz force will cause such a big error in this slotless structure.

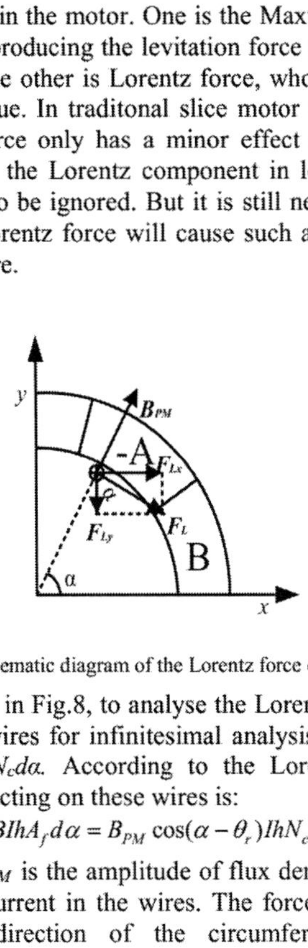

Figure 8. Schematic diagram of the Lorentz force on the stator wires

As shown in Fig.8, to analyse the Lorentz force, we take a part of the wires for infinitesimal analysis. The number of the wires is $N_c d\alpha$. According to the Lorentz fomula, the Lorentz force acting on these wires is:

$$dF_L = BIhA_f d\alpha = B_{PM}\cos(\alpha-\theta_r)IhN_c d\alpha \tag{20}$$

Where B_{PM} is the amplitude of flux density produced by PM, I is the current in the wires. The force is always along the tangent direction of the circumference. Then we decompose the force to x and y directions as (21):

$$\begin{cases} dF_{Lx} = B_{PM}IhN_c\cos(\alpha-\theta_r)\sin\alpha d\alpha \\ dF_{Ly} = -B_{PM}IhN_c\cos(\alpha-\theta_r)\cos\alpha d\alpha \end{cases} \tag{21}$$

Integrate the force F_{Lx} under each coil like fomula (22):

$$\begin{aligned} F_{Lx} &= \int_0^{2\pi} B_{PM}IhN_c\cos(\alpha-\theta_r)\sin\alpha d\alpha \\ &= B_{PM}hN_c\sum_{k=1}^{5}\Big[\int_{(2k-4)\pi/5}^{(2k-3)\pi/5} i_k\cos(\alpha-\theta_r)\sin\alpha d\alpha \\ &\quad + \int_{(2k-1)\pi/5}^{2k\pi/5} -i_k\cos(\alpha-\theta_r)\sin\alpha d\alpha\Big] \end{aligned} \tag{22}$$

If the constraint relationship of the currents (1) is put into (22) and the force in x direction is as the following:

$$F_{Lx} = -\frac{5\sqrt{5}}{8}N_c B_{PM}hI_s\cos(\theta_s-\theta_r) \tag{23}$$

Caculate the F_{Ly} in the same method and the result will be achieved:

$$F_{Ly} = \frac{5\sqrt{5}}{8}N_c A_{PM}\frac{\mu_0}{l_m}hI_s\sin(\theta_r-\theta_s) \tag{24}$$

Considering the force acting on the rotor is the opposite reaction of the force on the wires, we rewrite the forces as (25):

$$\begin{cases} F_{Lx} = K_L\cos(\theta_s-\theta_r)I_s \\ F_{Ly} = K_L\sin(\theta_s-\theta_r)I_s \end{cases} \tag{25}$$

Where $K_L = \frac{5\sqrt{5}N_c A_{PM}\mu_0 h}{8l_m}$. Compare the Lorentz forces (25) with Maxwell forces and we'll find that these two forces have no differences in the form except the coefficient. Hence, the radial levitaiton forces in five-phase constraint model, which is made up of Maxwell component and Lorentz component, are corrected as (19):

$$\begin{cases} F_x = (K_L+K_M)\cos(\theta_s-\theta_r)I_s = K_s\cos(\theta_s-\theta_r)I_s \\ F_y = (K_L+K_M)\sin(\theta_s-\theta_r)I_s = K_s\cos(\theta_s-\theta_r)I_s \end{cases} \tag{26}$$

Where K_s=K_L+K_M. If the current relationship is not put into (22), we'll get the general form of the Lonrentz forces as :

$$\begin{cases} F_{Lx} = k_L\{[0.588(i_2-i_5)-0.951(i_3-i_4)]\sin\theta_r + \\ \qquad [i_1-0.809(i_2+i_5)+0.309(i_3-i_4)]\cos\theta_r\} \\ F_{Ly} = k_L\{[0.588(i_2-i_5)-0.951(i_3-i_4)]\cos\theta_r + \\ \qquad [-i_1+0.809(i_2+i_5)-0.309(i_3+i_4)]\sin\theta_r\} \end{cases} \tag{27}$$

Where $k_L = \frac{\sqrt{5}N_c h\mu_0 A_{PM}}{4l_m}$. Similarly, correct the general model as

$$\begin{bmatrix} F_x \\ F_y \\ T \end{bmatrix} = \begin{bmatrix} k_s\cos(\theta_r) & k_s\cos(\theta_r+\frac{6\pi}{5}) & k_s\cos(\theta_r+\frac{2\pi}{5}) & k_s\cos(\theta_r+\frac{8\pi}{5}) & k_s\cos(\theta_r+\frac{4\pi}{5}) \\ k_s\sin(\theta_r+\pi) & k_s\sin(\theta_r+\frac{\pi}{5}) & k_s\sin(\theta_r-\frac{3\pi}{5}) & k_s\sin(\theta_r+\frac{3\pi}{5}) & k_s\sin(\theta_r-\frac{\pi}{5}) \\ k_t\sin(\theta_r+\pi) & k_t\sin(\theta_r+\frac{3\pi}{5}) & k_t\sin(\theta_r+\frac{\pi}{5}) & k_t\sin(\theta_r-\frac{\pi}{5}) & k_t\sin(\theta_r-\frac{3\pi}{5}) \end{bmatrix} \begin{bmatrix} i_1 \\ i_2 \\ i_3 \\ i_4 \\ i_5 \end{bmatrix} \tag{28}$$

Where k_s=k_L+k_M.

Replot the levitation forces as Fig.9. This time the force in model is greater than that in FEA on the contray, but their values are very close and the error is only about 5%. If the exsitence of flux leakage, magnetic saturation magnetic and motive force drop in iron are taken into account, this result is relatively reasonable.

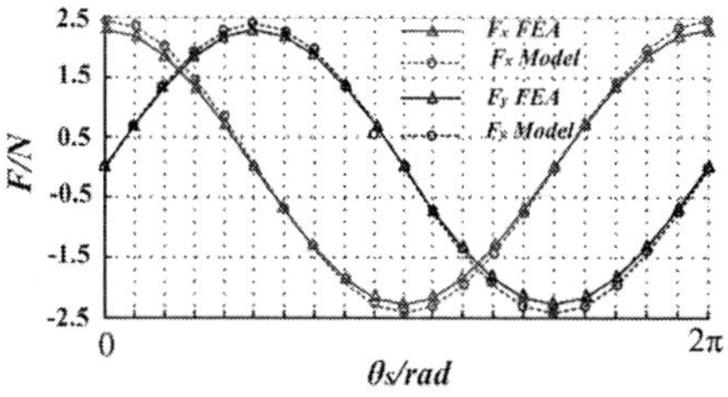

Figure 9. Comparison of the levitation forces

Next, levitation force and torque characteristics will be prensented. Here, we set I_t=3A, I_s=2A, $\theta_s=\theta_r+45°$,$\theta_t=\theta_r+90°$ and change the rotor angle θ_r from 0 to 2π. In Fig.10, it can be seen that the torque in FEA maintains stable, though a small error, which is no more than 4%, is observed. The causes of it are no other than the assumption *a* and *c*. These two assumptions both made the value of the flux density in our calculation greater.

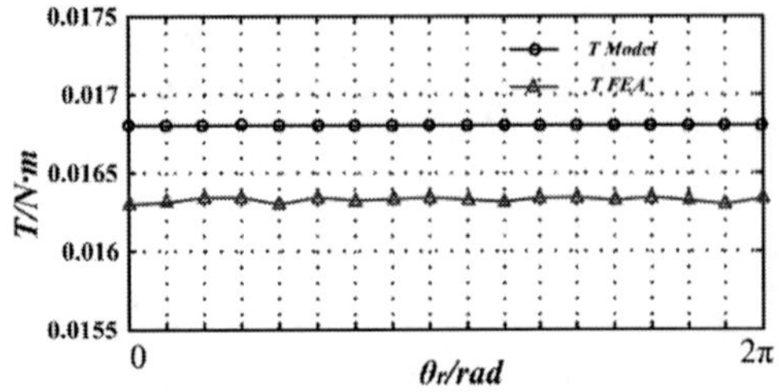

Figure 10. Curve of the torque changing with θ_r

Fig.11 illustrates that the levitation force is stable basically though there is pulsation in it. From (10) and (11), we can easily know the pulsation is caused by the first term. In addition, the force in *x* and *y* direction should be equal theoretically when $\theta_s=\theta_r+45°$, but these two values in FEA deviate a little. This is the effect of the second term, which is relevant to the torque current. Although these two problems affect the accuracy of the model, they are not so obvious and still in the acceptable range when the current is not so heavy.

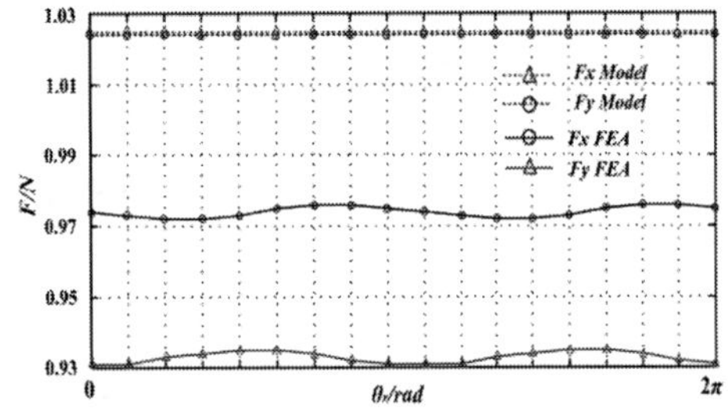

Figure 11 Curve of the levitation forces changing with θ_r

If we keep θ_r=0°, $\theta_s=\theta_r+45°$, $\theta_t=\theta_r+90°$ and change the value of I_s and I_t, curves of torque and levitation forces changing with amplitude of currents will be achieved as Fig.12 and Fig.13

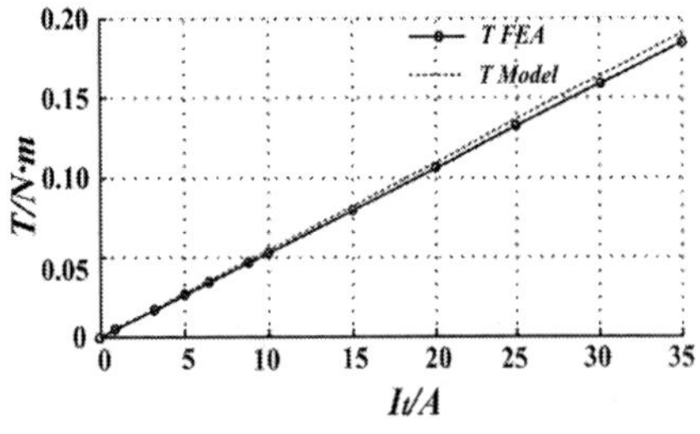

Figure 12. Curve of torque changing with I_t

It can be seen that there is high linearity between torque and I_t, though the torque in FEA is sightly less than that in theory, the error is about just 3%.

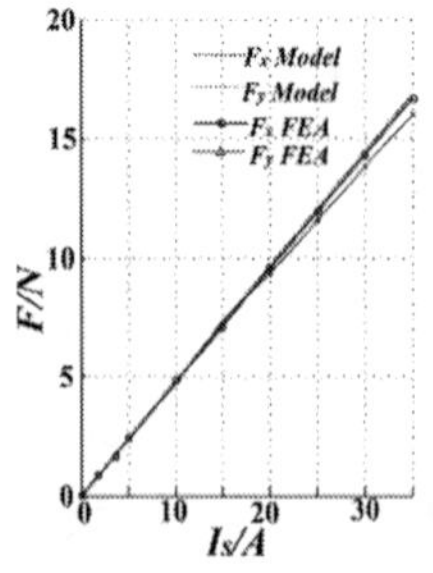

Figure 13. Curve of torque changing with I_s when I_t=0

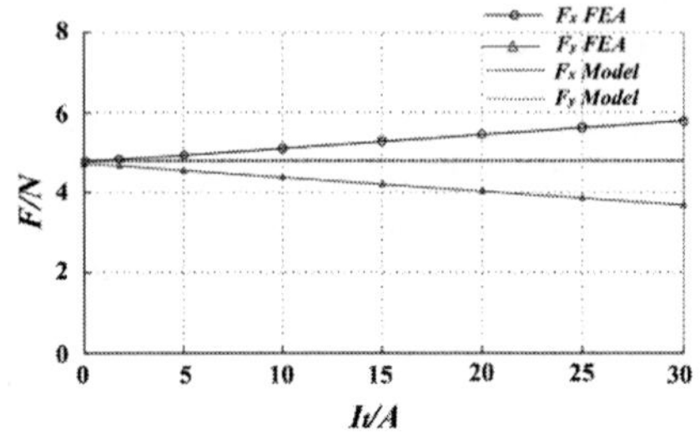

Figure 14. Curve of the levitation forces changing with I_t when I_s=10A

Fig.13 shows that the levitation forces in FEA has a linear relationship with I_s when there are no torque current component. But from Fig(14), it can be seen that with the increase of I_t, the levitation forces are becoming skewed. That's the result of the coupling between levitation force and I_t . From (10) and (11), we can kown that the effect of coupling will be obvious when I_t is large enough. Thus, in actual situation, I_t will affect the peformance of radial disposition control in a certain extent.

In order to make model more precise, two compensation coefficients σ_t and σ_s are introduced to fix the expressions. In five-phase constraint model, K_s and K_t are updated as:

$$\begin{cases} K_s = \sigma_s(K_L + K_M) \\ K_t = \sigma_t \dfrac{N_c h r \mu_0 A_{PM} a_1}{l_m} \end{cases} \tag{29}$$

Similarly, in the general model k_s and k_t are updated as :

$$\begin{cases} k_s = \sigma_s(k_L + k_M) \\ k_t = \sigma_t \dfrac{N_c h r \mu_0 A_{PM}}{l_m} \end{cases} \tag{30}$$

Where σt is 0.97 and σs is 0.95.

IV. Conclusion

Five-phase constraint model and general model of the slotless PMBSM are deduced in this paper. The first model is relatively simple in the form, though it is not as applicable as the general one. The general model is the theoretical basis of the research on fault-tolerant capability. Because the flux density in the air gap is the key to gain the torque and the levitation force, a verification about it is done. After that, levitaiton force and torque characteristics are investigated by FEA. The result shows both of the models are accurate enough to describe the slotless PMBSM in the case that current is not too heavy.

REFERENCES

[1] Steinert, Daniel, Thomas Nussbaumer, and Johann W. Kolar. "Concept of a 150 krpm bearingless slotless disc drive with combined windings." Electric Machines & Drives Conference (IEMDC), 2013 IEEE International. IEEE, 2013.

[2] H. Mitterhofer, D. Andessner, and W. Amrhein, "Analytical and experimental loss examination of a high speed bearingless drive," in Proc. International Symposium on Power Electronics Electrical DrivesAutomation and Motion, 2012.

[3] Hubert Mitterhofer, Wolfgang Gruber, and Wolfgang Amrhein, " On the High Speed Capacity of Bearingless Drives," in IEEE Transactions onIndustrial Electronics, 2013.

[4] Zhu, Huangqiu, et al. "Mathematical model and control technology of bearingless PMSM." Control and Decision Conference (CCDC), 2010 Chinese. IEEE, 2010.

[5] Xiao-lin, Wang, et al. "Current-controlled multiphase slice permanent magnetic bearingless motors with open-circuited phases: Fault-tolerant controllability and its verification." Industrial Electronics, IEEE Transactions on 59.5 (2012): 2059-2072.

[6] Wang, X. L., et al. "Fault-tolerant control of multi-phase permanent magnetic bearingless motors." Electrical Machines (ICEM), 2010 XIX International Conference on. IEEE, 2010.

Loss Analysis of a Bearingless Flux-Switching Slice Motor

Karlo Radman [a], Wolfgang Gruber [b], Neven Bulić [a]

[a] University of Rijeka, Vukovarska 58, 51000 Rijeka, Croatia, karlo.radman@riteh.hr, neven.bulic@riteh.hr
[b] Johannes Kepler University Linz, Altenberger Str. 69, 4040 Linz, Austria, wolfgang.gruber@jku.at

***Abstract*—The bearingless flux-switching slice motor is a recently developed motor concept combining the magnetic rotor suspension of a bearingless slice drive and the topology of a flux-switching drive. In comparison to state of the art bearingless slice motors this concept introduces a magnet free rotor and relatively high torque generation. This offers advantages in industrial and medical applications which require regular rotor replacements and high temperature working conditions. This work will describe the working principles, prototype design, implemented control theory and measurement to determine the losses and evaluate the efficiency.**

I. Introduction

Bearingless motors refer to electrical drives in which torque and the bearing forces are generated by one integrated motor element. Stator coils create both motor torque and radial bearing forces. The remaining degrees of freedom, of the rotor, can be passively stabilized by including a permanent magnetic bias flux in the air gap and a disc shaped rotor, as depicted in Fig. 1. With the additional actively controlled radial deflection, a fully magnetically suspended rotor is achieved. Such integrated levitating drives were developed in the mid 90's and are known as "bearingless slice motors" [1]. Bearingless slice motors are used in medical, pharmaceutical and semiconductor industry due to low vibrations and the ability to hermetically seal the rotor [2], [3].

In chemical and medical applications where the rotor is regularly replaced because of contamination, permanent magnets (PM) embedded in the rotor significantly increase the disposable rotor price. The working temperature conditions also limit the use of PM, and for such applications a magnet free rotor topology is preferable. The flux-switching permanent magnet (FSPM) motor topology solves that issue by placing the PM on the stator iron. The rotor is therefore reduced to a salient laminated steel ring. Two variants of the bearingless reluctance slice drives, without PM on the rotor, were recently proposed [4]. The first variant has a constant directional homopolar PM field, while the bearingless flux-switching drive represents the heteropolar version with an alternating PM field in the air gap.

This paper describes the working principle, control scheme and design of a bearingless flux-switching slice motor. A finished prototype was put into operation and the characteristics were measured to determine the eddy current losses and the overall efficiency of the motor. Three-dimensional finite element (3D FEM) simulations were compared to measurement results and show an acceptable congruence.

II. Bearingless Flux-Switching Motor

A. Flux-Switching Permanent Magnet Motor

The topology of a FSPM motor is shown in Fig. 2. The tangentially polarized permanent magnets are situated in each stator teeth and adjacent magnets have opposing directions (depicted in red and green in Fig. 2). This configuration separates the stator electromagnetically, in such a manner that the magnetic flux from each tooth is closing only over the two neighboring ones. With an even rotor and stator teeth number a symmetric layout is achieved with opposing coils having in phase induced voltages.

Due to the rotor saliency, an alternating angle depended linked flux is observed in the stator coils. A sinusoidal back-electromotive force is induced and linear PM torque generation is possible. Figure 3 illustrates the maximum PM flux entering the rotor, at an electrical rotor angle of 90° and 270° teeth. The PM flux is changing its direction over the electrical rotor angle and the direction switch occurs at 0° and 180° electrical.

B. Torque and Force Generation

The stator coil field superimposed on the PM field leads to local field weakening and strengthening which in turn exerts force on the rotor, as shown in Fig 3. Each coil is capable of generating both force and torque and for controlled magnetically suspended rotor operation, force and torque generation need to be decoupled. Additionally, radial bearing force needs to be linearly current dependent for the position control. Because of the motor symmetry, the opposing rotor coils can be connected in series, shown in Fig 4. If the fields

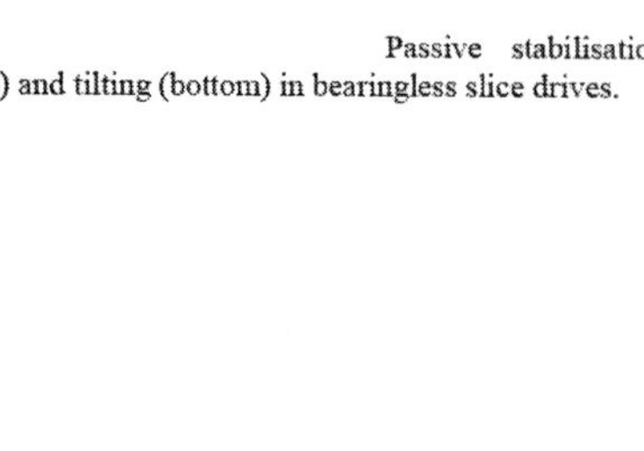

Figure 1. Passive stabilisation of axial deflection (top) and tilting (bottom) in bearingless slice drives.

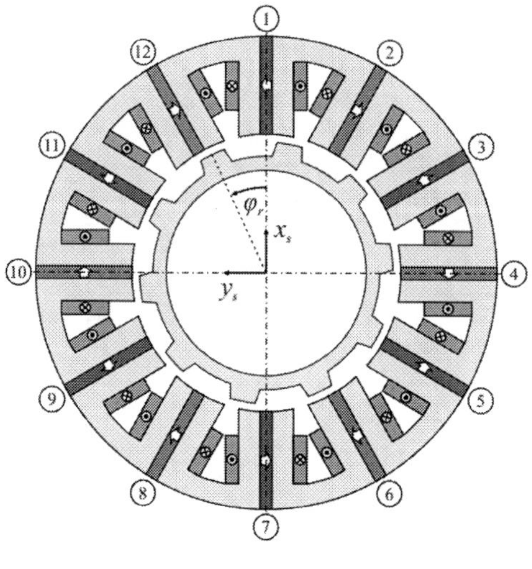

Figure 3. Topology of a FSPM drive

of both coils are oriented in the same direction, torque and all quadratic force components cancel out and linear force generation is achieved. Vice versa, with the coil fields pointing in opposing direction the force generation cancels out. This way, with different connections of the opposing coils a phase with decoupled linear torque or force generation is possible.

Figure 5 depicts a characteristic torque and force generation of a single phase energized with constant current over the electrical rotor angle. Torque is generated by the phase with concordantly connected coils, while radial force with inversely connected opposing coils. Maximal force normal to the phase is generated at a rotor electrical angle of 90° and 270° when the maximal PM field is closing over the rotor teeth. Similarly maximum torque is achievable at 0° and 180° degrees electrical. Tangential force angle dependency is in phase with the torque generation. A force phase is

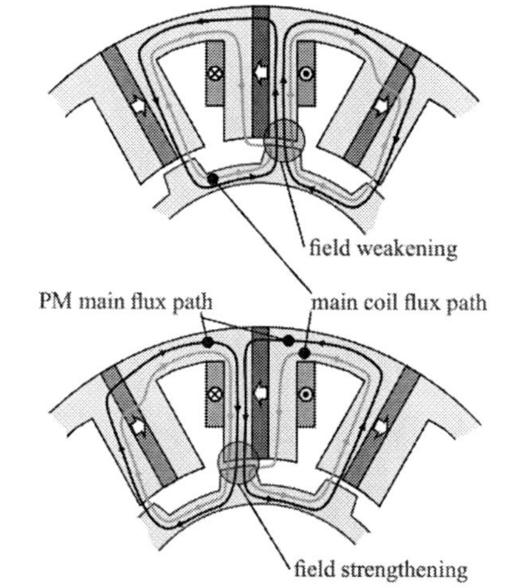

Figure 2. Basic principle of the switching flux path of a motor segment. additional field strengthening (top), and weakening (bottom) applied with a constant current and varying electrical rotor angle (-90°el. top, +90°bottom)

generating negligible torque while the torque phase is not creating any significant force on the rotor.

C. Passive stabilization

The PM field from each stator segment exerts reluctance forces to minimize the air gap. Reluctance forces in radial direction pull the rotor away from the center operating position, and are therefore destabilizing. The reluctance forces, in axial direction, strive to achieve position with minimal air gap length by aligning the rotor in plane with the stator, thus, stabilizing the axial displacement, as shown in Fig. 1 (top). Figure 1 (bottom) depicts the same principle that stabilizes the x- and y-axis tilting.

If the rotor is constructed with a relatively small height in regard to the diameter, a bearingless slice motor operation is viable, with three degrees of freedom passively stabilized [5]. At any rotor angle, with a high rotor/stator teeth number, there is always some rotor teeth over which the PM flux is closing, and therefore the axial stabilization force and x- and y-axis tilting stabilizing torque have a mostly constant term over the rotor angle, Figure 6 shows the 3D FEM simulated passive forces and torque for a full electrical rotor angle.

III. Prototype Design and Control

A. Mathematical model

Each stator coil is capable of creating drive torque T_z, and normal and tangential bearing force relative to the coil. The overall suspension forces F_x and F_y, in a stator fixed coordinate system and the motor torque T_z for the whole motor can be expressed over the phase current vector $\boldsymbol{i}_s$ as [5]

$$\begin{pmatrix} F_x(\varphi_r) \\ F_y(\varphi_r) \\ T_z(\varphi_r) \end{pmatrix} = \begin{pmatrix} i_s^{\mathrm{T}} & 0 & 0 \\ 0 & i_s^{\mathrm{T}} & 0 \\ 0 & 0 & i_s^{\mathrm{T}} \end{pmatrix} T_Q(\varphi_r) i_s + T_L(\varphi_r) i_s + T_C(\varphi_r) \quad (1)$$

with φ_r being the mechanical rotor angle. Centered rotor position is assumed. The mathematical model is valid only for linear material behaviour. Saturation effects are not considered and the motor is therefore expected to work below saturation levels in the laminated steel. The term $\boldsymbol{T}_C$, describing cogging torque and purely PM reluctance forces, is relatively small in FSPM drives they work on the principle of PM synchronous machines, opposed to the reluctance machines usually

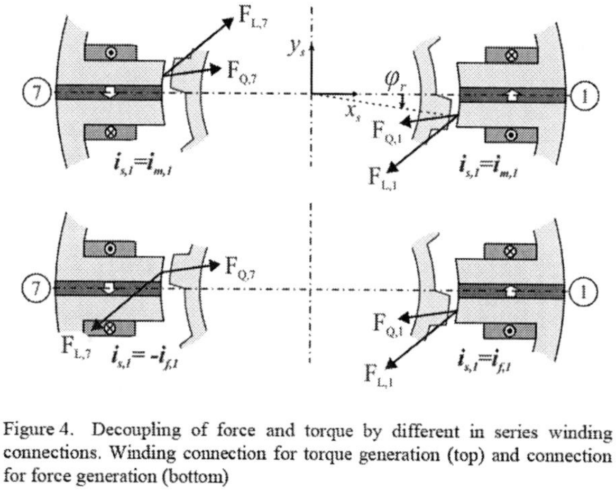

Figure 4. Decoupling of force and torque by different in series winding connections. Winding connection for torque generation (top) and connection for force generation (bottom)

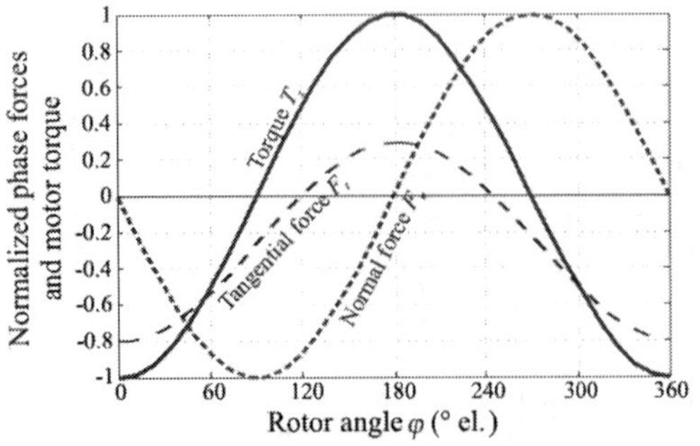

Figure 5. Normalized values generated by single DC energized stator phases over the electrical rotor angle. Torque generated by a torque phase and forces normal and tangential to a force phase.

connected to magnet-free rotors. With 12 stator and 10 rotor teeth and optimum rotor teeth width, the reluctance torque and higher harmonics influences can be minimized, as research shows for conventional FSPM drives [6], [7]. $\boldsymbol{T_Q}$ is the quadratic term influenced by the stator currents alone, without the PM. It is the dominant element in classic magnetic bearings and reluctance machines. The linearly current depended vector $\boldsymbol{T_L}$ represents the forces and torque that are created due to the interaction between the PM field and stator current field. This part is dominant in PM machines and is the only desirable element in the mathematical model (1), since a decoupling of torque and force generation has to be obtained based on linear superposition.

With the already mentioned series connection of opposing coils the quadratic current dependent terms $\boldsymbol{T_Q}$ are superimposed and the linear force and torque model of a bearingless flux-switching motor is

$$\begin{pmatrix} F_x(\varphi_r) \\ F_y(\varphi_r) \\ T_z(\varphi_r) \end{pmatrix} = \boldsymbol{T}_L(\varphi_r)\boldsymbol{i}_s. \qquad (2)$$

For a stator with 12 segment, which make six phases, the phase current vector can be defined as

$$\boldsymbol{i}_s = (i_1 \quad i_2 \quad i_3 \quad i_4 \quad i_5 \quad i_6)^{\mathrm{T}}. \qquad (3)$$

The optimal wiring topology and distribution of torque and force phases has been researched [8]. A three-phase system is used for the torque generation, as found in consecutive pole wound flux-switching machines [9]. The remaining three phases, shifted 120°, generate the radial suspending forces. The phase topology is presented in Table I and follows the winding directions marked in Fig. 2. The force and torque orbit of a 2x3-phase system is shown in Fig. 7. The orbits are calculated from the waveforms in Fig 5.

TABLE I. PROTOTYPE PHASE TOPOLOGY

Phase No.	Generation of	First Coil No.	First Coil Current	Second coil No.	Second coil Current
1	Force	1	i_1	7	$-i_1$
2	Torque	2	i_2	8	i_2
3	Force	5	i_3	11	$-i_3$
4	Torque	6	i_4	12	i_4
5	Force	9	i_5	3	$-i_5$
6	Torque	10	i_6	4	i_6

B. Control system

Equation (2) is used to calculate the pseudo inverse control matrix $\boldsymbol{K}_m$ [10]

$$\boldsymbol{K}_m(\varphi_r) = \boldsymbol{T}_L(\varphi_r)^{\mathrm{T}} \left(\boldsymbol{T}_L(\varphi_r)\, \boldsymbol{T}_L(\varphi_r)^{\mathrm{T}}\right)^{-1}. \qquad (4)$$

The nonlinear control matrix $\boldsymbol{K}_m$ gives the relation between suspension forces, drive torque, and therefore needed phase currents with the lowest possible resistive copper losses. The phase currents can be calculated from

$$\boldsymbol{i}_s = \boldsymbol{K}_m(\varphi_r)(F_x \quad F_y \quad T_z)^T \qquad (4)$$

for given reference suspension forces and drive torque. $\boldsymbol{K}_m$ can be implemented in the control system with feedback control loops for position and torque. Matrix $\boldsymbol{T}_L$ needed for (4) is calculated from the characteristic waveforms in Fig 5, which can be obtained from FEM simulations.

Due to symmetrical attraction from all sides, passive destabilizing radial reluctance forces are minimal in the central position. When the magnetic air gap is the same from all sides, only little force is needed to maintain the rotor levitating. An additional control loop for finding the reference central position is implemented. A relatively slow integration of the x- and y-position controller output can be fed back instead of a constant reference value. This zero-force control feedback loop enables the self calibration of the position

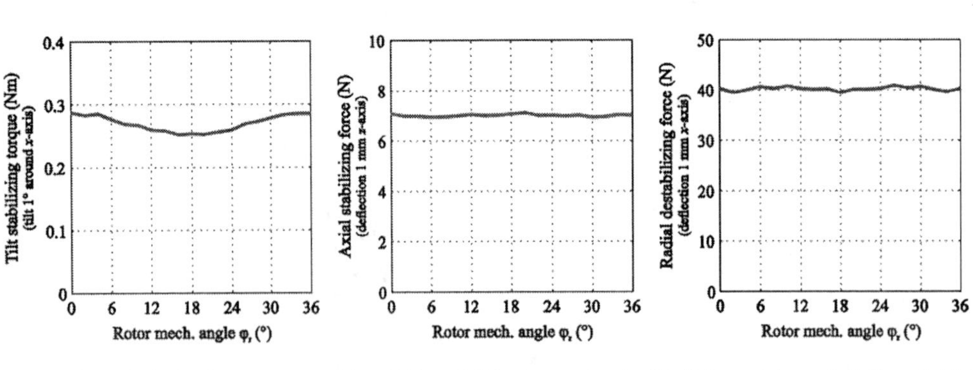

Figure 6. Prototype 3D FEM simulated passive rotor stiffness. Tilting stabilizing torque at a rotor tilt of 1° around the x-axis (left), axial stabilizing force at 1 mm axial rotor displacement (center), and radial destabilizing force at an radial x-axis displacement of 1 mm (right)

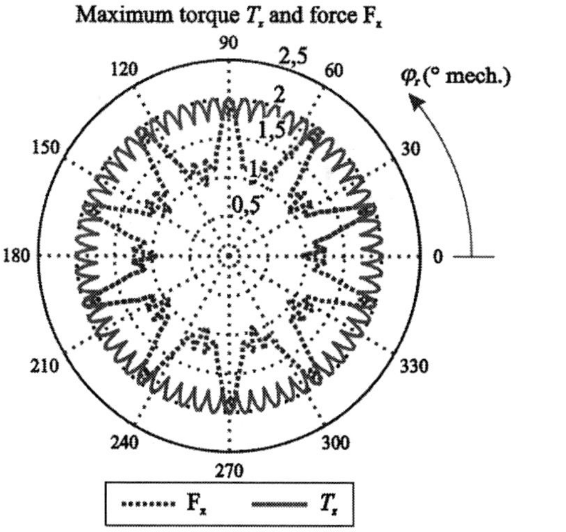

Figure 7. Orbit plots for maximum producible motor torque and bearing forces. Values are relative to the maximum per phase producible torque/force, as seen Fig. 5. Orbits are created by separate three-phase torque and force systems.

references for bearingless operation. The control scheme is depicted in Fig. 8.

Position is measured using four eddy currents sensors around the stator circumference. The sensor coil field is influenced by the induced currents in a thin aluminum layer around the rotor. Position in x- and y-axis is measured differentially with two opposing sensor coils. Rotor angle measurement is made with a three-phase Hall sensor arrangement between the stator segments, that measure the sinusoidal PM flux change over the rotor angle.

C. Prototype

A bearingless flux-switching slice motor prototype was optimized and built to validate the simulated model [11]. Optimizations were made using 3D FEM simulations. Criterions included adequate axial and tilting stiffness and motor operating point below iron saturation limits. Additional parameters regarding pole pair number, pitch ratio and the motor split ratio were chosen based on research of conventional flux-switching machine with bearings [6], [7]. The prototype in two assembly stages is shown in Fig. 9, and the main parameters are given in Table II.

The electronic part includes a 2 KW power inverter, with two three-phase half bridges. The control was implemented on a TI DSP using the X2C toolbox developed by the Linz Center of Mechatronic (LCM) [12]. Force and torque phase coils have the same number of windings, which is 80 per coil, and thus 160 per phase. Phase resistance equals 0.86 Ω.

TABLE II. PROTOTYPE PARAMETERS

Parameter	Value	Unit
stator iron outer diameter	266	mm
rotor outer diameter	150	mm
rotor and stator height	10	mm
magnetic air gap	3	mm
magnet width	8,5	mm
magnet length	54	mm
laminated iron material	M330-35A	
PM material	NdFeB N38	

IV. MEASUREMENTS

The assembled prototype was put into operation and static measurements were carried out. Forces in radial and axial direction, and motor torque were measured on a test bench. Static measurements were made with a force and torque load cell for a full electrical angle. Linearity of the force generation for a single phase can be seen in Fig. 10. Saturation affects the force generation above phase current of 10A, which is a magnetomotive force of 1600 Aturns. Figure 11 shows the force orbit of a single force phase energized with constant 9.5 A. The static simulated and measured data can be seen in Table III. Passive and actively generated electromagnetic characteristics show deviation around 10% from the simulation data, which can be contributed to the nonideal PM.

Dynamic measurement where conducted with the rotor mechanically attached to the torque sensor and test load. Suspension control was turned off, and only the torque phases were used. As the torque phases are separated by force phases in between them, cross coupling between torque phases is not present. This is due to the coil flux closing only over adjacent stator segments, as depicted for two rotor angles in Fig. 3.

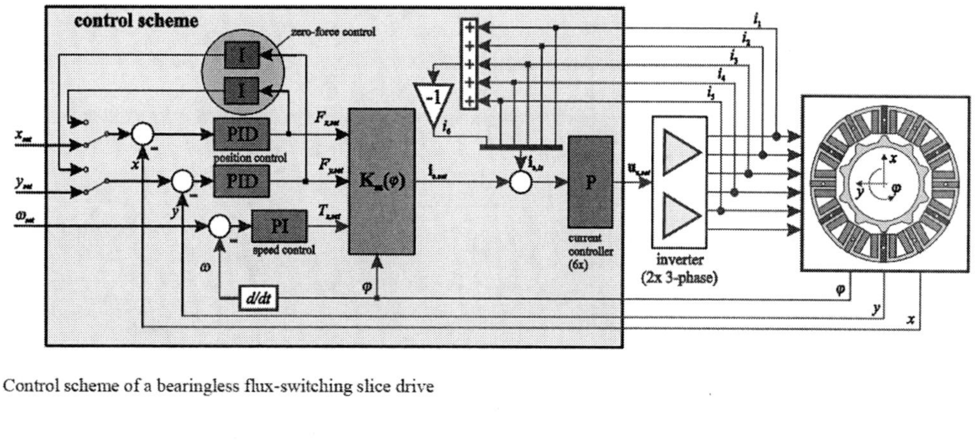

Figure 8. Control scheme of a bearingless flux-switching slice drive

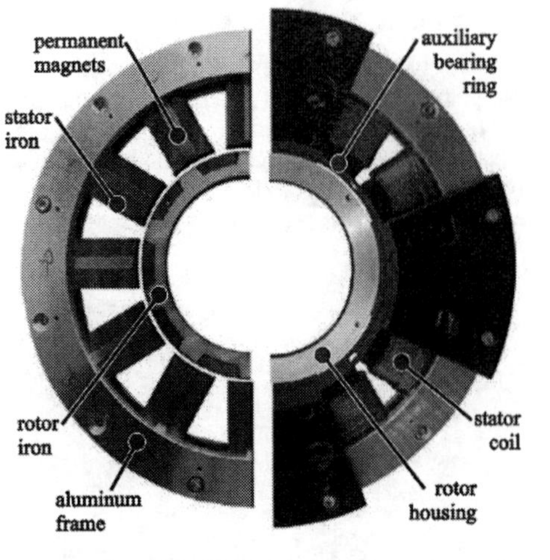

Figure 9. Half (left) and fully (right) assembled prototype.

TABLE III. SIMULATED AND MEASURED ELECTROMAGNETIC CHARACTERISTICS

Parameter	Simulated	Measured	Unit
Radial stifness	20.8	18	N/(0.5mm)
Axial stiffness	7.25	7.04	N/mm
Tilt stifness	21.4	NA	Nm/rad
Bearing force (phase, peak, at 9.5 A)	26.4	26	N
Drive torque (phase, peak, at 9.5 A)	0.708	0.66	Nm
Induced phase voltage constant (peak)	0.08	0.0757	V/(rad/s)
Max. current density	6		A/mm^2

When the motor is running in horizontal position, the rotor is axially displaced do to its weight. The prototype rotor weights 700 g and is displaced approximately 1 mm. The RMS induced phase voltage in reference to axial displacement, at a rotor speed of 500 rpm, is shown in Fig 12. With increasing axial displacement the motor torque is decreasing.

Cogging torque and lo load was measured with the test load spinning the rotor and the result are shown in Fig. 13.

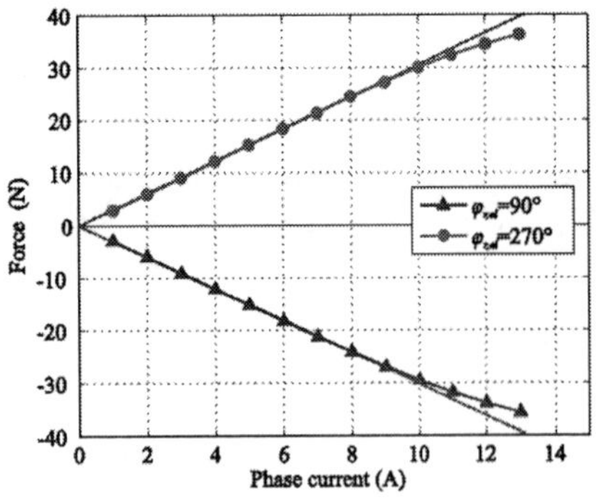

Figure 10. Force generation of a DC energized force phase.

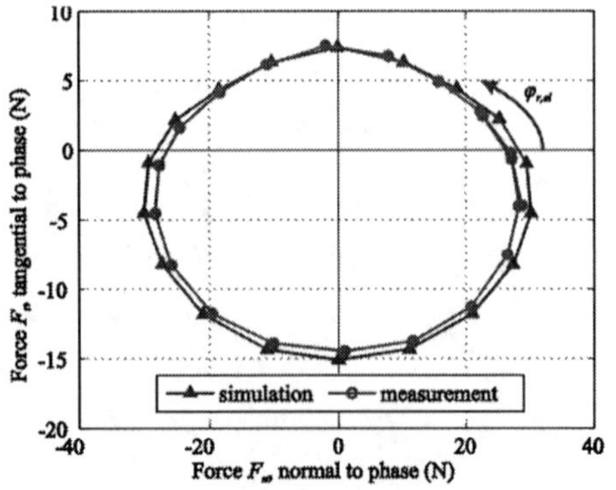

Figure 11. Measured and simulated force orbit of a single DC energized force phase for a full electrical rotor angle. Constant magneto-motive force of 1520 Aturns.

The aluminum 0.5 mm thick rotor housing needed for position measurements has been proven to create enormous eddy current losses. At a rotor speed of 1000 rpm the eddy current losses reach 180 W, which significantly increase the rotor temperature. For that reason, all measurements were done with a nonconductive rotor housing, as the position measurement was not needed with a mechanically suspended rotor. Measured cogging torque at lower speeds is about 12 mNm and the no load free-wheel losses at 1000 rpm are 3.12 W. These measurements prove the used optimization and that the bearingless flux-switching slice drive, although having a rotor similar to reluctance machines, is lacking the characteristic cogging torque.

Figure 14 shows the measurement of real, complex and motor output power at 1000 rpm. Due to the high stator coil inductivity the power factor is around 0.22 at the load of 1.55 Nm. Efficiency at different speeds is shown in Fig 15. Linearly and quadratic frequency dependent losses are not dominant and most losses are resistive heating. Therefore as the speed increases, the motor output power is rising in reference to the real electrical power, and the efficiency also increases.

V. CONCLUSION

The bearingless flux-switching slice motor concept was successfully realized into a working prototype and the first measurements satisfactorily coincide with the expected 3D

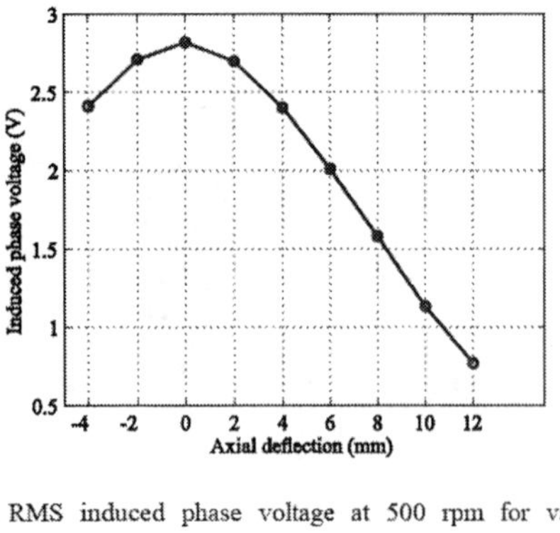

Figure 12. RMS induced phase voltage at 500 rpm for varying axial deflection.

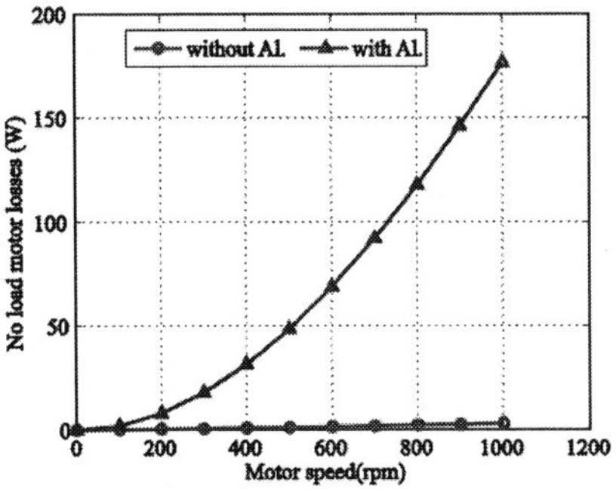

Figure 13. No load motor losses with and without the aluminum rotor housing. Eddy current losses in are quadratic dependent with the frequency.

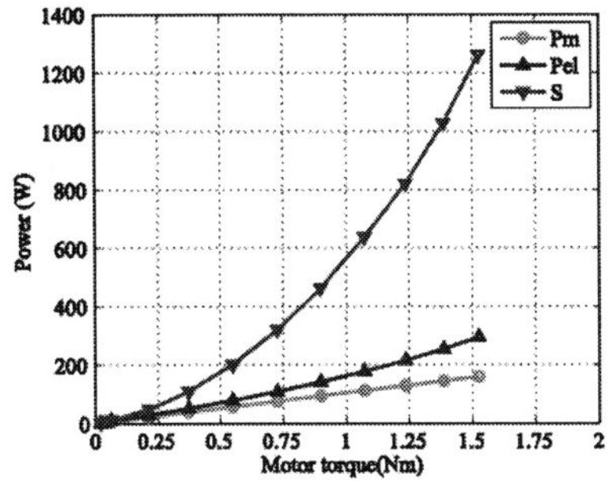

Figure 14. Measured motor, real and complex power at 1000 rpm. Due to high inductivity the power factor at maximum load is 0.22.

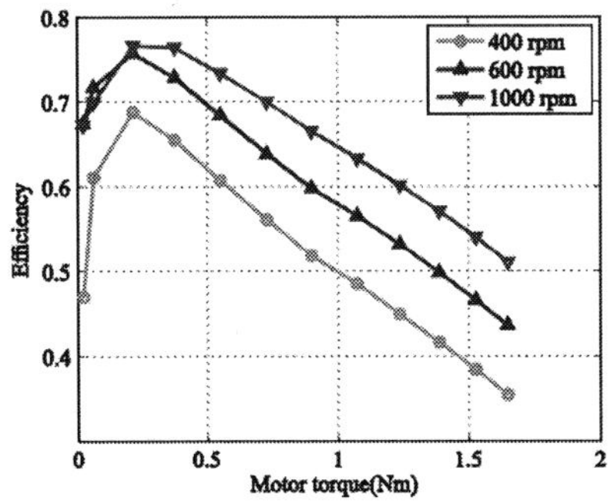

Figure 15. Motor efficiency at different speeds. The efficiency is increasing with motor speed.

FEM simulation results. The high losses of the position measurement aluminum housing have to be removed by finding new solutions for position measuring sensors in the next prototype. With a new position sensor solution, measurements of bearingless operation under varying load and speed would be measurable. Saturation levels in iron where found at expected levels at the rated motor current of 10A. The cogging torque was proven to be minimal, and neither the torque or force orbit has a single phase characteristic. Further step are taken to increase the motor stability and drive torque generation by increasing the motor height.

VI. Acknowledgment

Parts of this work were supported by the Linz Center of Mechatronics (LCM) GmbH, a K2-centre of the COMET program of the Austrian Government. The authors thank the Austrian and Upper Austrian Government for their support.

References

[1] Schöb, R., Barletta, N., "Principle and application of a bearingless slice motor," Proc. 5th Int. Symp. on Magnetic Bearings (ISMB), pp. 333-338, 1996

[2] Salazar, A. Chiba, and T. Fukao. "A review of developments in bearingless motors," In Proceedings of ISMB7, 2000

[3] Warberger, B.; Kaelin, R.; Nussbaumer, T.; Kolar, J.W., "50 Nm/2500-W Bearingless Motor for High-Purity Pharmaceutical Mixing," Industrial Electronics, IEEE Transactions on, vol.59, no.5, pp. 2236-2247, May 2012

[4] Gruber, W.; Briewasser, W.; Rothbock, M.; Schob, R.T., "Bearingless slice motor concepts without permanent magnets in the rotor," Industrial Technology (ICIT), 2013 IEEE International Conference on, pp. 259-265, 25-28 Feb. 2013

[5] Schöb, R., Barletta, N., "Principle and Application of a Bearingless Slice Motor," Proc. 5th Int. Symp. on Magnetic Bearings (ISMB), pp. 333-338, 1996

[6] Chen, J.T.; Zhu, Z.Q., "Winding Configurations and Optimal Stator and Rotor Pole Combination of Flux-Switching PM Brushless AC Machines," Energy Conversion, IEEE Transactions on , vol.25, no.2, pp. 293-302, June 2010

[7] Wei Hua; Ming Cheng; Zhu, Z.Q.; Howe, D., "Analysis and Optimization of Back-EMF Waveform of a Novel Flux-Switching Permanent Magnet Motor", Electric Machines & Drives Conference, 2007. IEMDC '07. IEEE International, vol.2, pp.1025-1030, 3-5 May 2007

[8] Gruber, W., Bauer, W., Radman, K., Amrhein, W. and Schöb, T., "Considerations Regarding Bearingless Flux-Switching Slice Motors", Proceeding on 1st Brazilian Workshop on Magnetic Bearings, 2013

[9] Owen, R.L.; Zhu, Z.Q.; Thomas, A.S.; Jewell, G.W.; Howe, D., "Alternate Poles Wound Flux-Switching Permanent-Magnet Brushless AC Machines," Industry Applications, IEEE Transactions on , vol.46, no.2, pp. 790-797, March-April 2010

[10] Silber, S., Amrhein, W., "Power Optimal Current Control Scheme for Bearingless PM Motors", Proc. 7 th Int. Symp. on Magnetic Bearings (ISMB), pp. 401-406, 2000

[11] Gruber, W, Radman, K., Schöb, R., "Design of a Bearingless Flux-Switching Slice Motor", Proc. Int. Power Elec. Conf. (IPEC), 2014, in press

[12] Radman, K.., Gruber, W., Bulić, N., "Control Design of a Bearingless Flux-Switching Slice Drive", European Embedded Design in Education and Research Conference (EDERC), 2014, in press

A 5-DOF Active Controlled Disk Type PM Motor with Cylindrical Flux Paths

Satoshi Ueno[a], Takuya Fukuura[a], Tran Van Toan[a]

[a] Ritsumeikan University, 1-1-1 Noji-higashi, Kusatsu, Shiga, Japan, sueno@se.ritsumei.ac.jp

***Abstract*—This paper proposes a novel disk type permanent magnet motor, which actively controls 5 degrees of freedom of a flat rotor without any contact. The proposed motor consists of an Axial Self-bearing Motor (ASBM) and additional cylindrical flux paths. The axial force and motor torque are controlled by the rotating flux with the same number of poles of the rotor, and the tilt control torque is produced by the rotating flux with the two plus or two minus poles of the rotor. The radial force is produced by supplying direct current to the stator windings. The direct current produces the flux through the stator and the cylindrical flux path, then the radial force can be generated by the interaction of the rotor poles and the edge of the cylindrical flux path. This paper introduces the structure and principle of the proposed motor, and shows the results of the magnetic forces calculated by the numerical electromagnet analysis.**

I. Introduction

An axial self-bearing motor (ASBM) is a combination of a disc motor and a thrust magnetic bearing, and can simultaneously produce rotating torque and control the axial position of the rotor[1]. Since the single rotating magnetic flux controls both the axial force and motor torque, the structure of the stator and control system is simple compared to other self-bearing motors. We are developing a miniaturized canned pump using the ASBM for a long lifetime[2].

However, the ASBM requires other radial magnetic bearings, such as permanent magnet repulsive type passive bearings and superconducting magnetic bearings[3] to support the radial directions of the rotor, then the overall structure tends to be complex and a lot of design effort is needed to meet specifications. Moreover, it is difficult to rotate over critical speed due to lack of the damping force by often-used passive magnetic bearings. In order to solve these problems, it is required to use the active control of 5 degrees of freedom with similar structure of the original ASBM.

This paper introduces a novel self-bearing motor, which actively controls 5-DOF of the flat rotor. Since only two stators can support the rotor, other radial bearings are unnecessary, and then the shaft of the rotor can be removed. This provides the flexibility of the overall design of the system such as canned pumps, vacuum pumps and so on.

In this paper, the structure of the proposed motor is introduced and the principles of the generation of the motor torque and bearing forces are shown. And the feasibility of the proposed motor is confirmed by finite element electromagnet analysis.

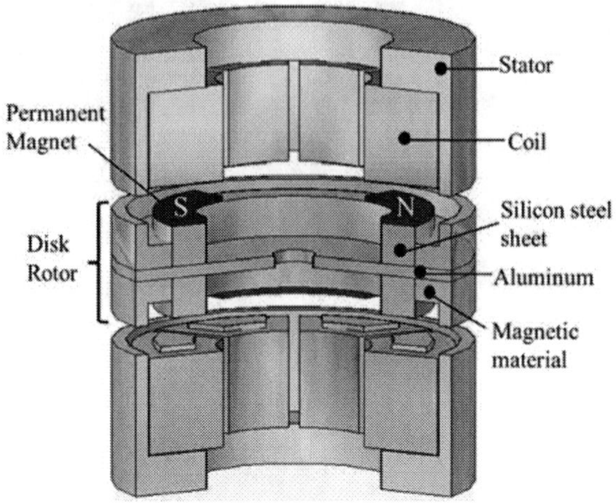

Figure 1. Structure of the disk motor.

II. Structure and Principle

A. Structure

The schematic structure of the proposed motor is shown in Fig. 1. The motor consists of one disc rotor and two stators placed on upper and lower side of the rotor. The cylindrical magnetic paths are attached on the outside of the stator and the rotor in order to generate uniform flux on the stator. The cylindrical flux paths are connected to the stator and rotor through back yoke. One stator has 6 concentrated coils and each coil is driven by a separate power amplifier. Two permanent magnets are attached on one side of the rotor surface to make 2-pole magnetic flux. The position of permanent magnet poles are shifted 90 degrees between upper and lower side of the rotor.

B. Motor Torque

The motor torque produced by the 2-pole rotating magnetic flux in upper and lower stator is out of phase with the rotor by 90 degrees. These fluxes do not flow in the cylindrical flux paths, and the motor works as a conventional one. The torque is controlled by the amplitude of the rotating fluxes. The rotating fluxes for the motor torque also produces the axial force. However, by using the same amplitude in both stators, the resultant axial force becomes zero.

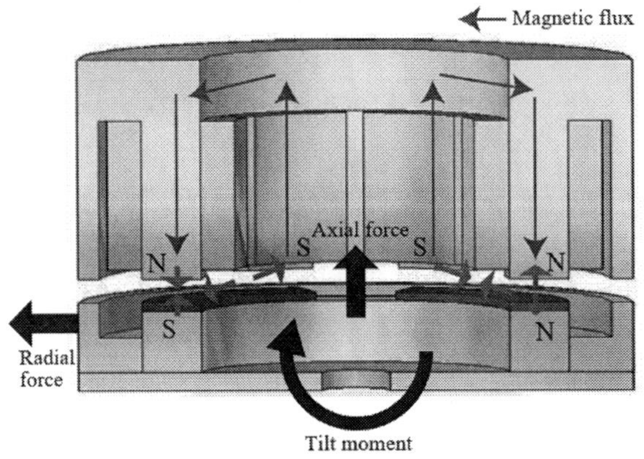

Figure 2. Tilt moment and radial force generation by 4-pole current.

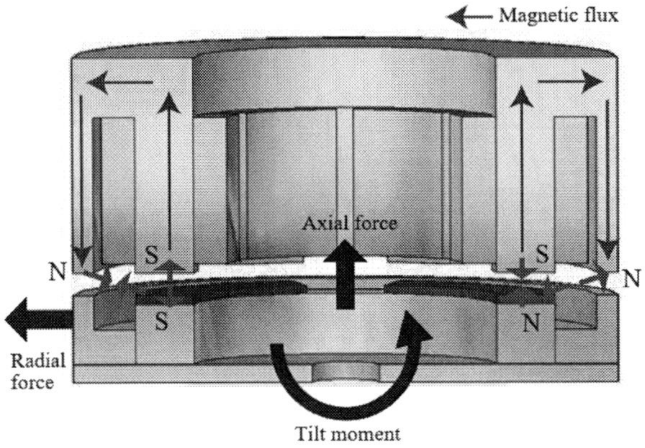

Figure 3. Tilt moment and radial force generation by direct current.

C. Axial Force

The 2-pole rotating magnetic flux with the same phase as the rotor, control the axial force. The axial force is generated by using the opposite amplitude in the upper and lower fluxes. The motor torque is not produced by these fluxes, and the axial force can control the amplitude of these fluxes.

D. Tilt Moment and Radial Force

The tilt moment and radial force are generated by 4-pole rotating magnetic flux and uniform flux. Fig. 2 shows the generation of tilt moment and radial force by 4-pole flux. The 4-pole flux unbalance the air gap flux density at right pole and left pole, and produce the difference of the attractive force. This imbalance produces the tilt moment around radial axis. The 4-pole flux also generate radial force due to south-poles of the stator and north- and south-poles of the rotor.

Fig. 3 shows the generation of tilt moment and radial force by uniform flux. Supplying direct current to all coils of the stator generates the magnetic flux flowing in the cylindrical magnetic path and makes uniform magnetic pole at the edge of the cylinder. The poles of the rotor and the cylinder edge

Table I. PARAMETERS OF THE ROTOR AND THE STATOR

Rotor and Stator	
Material	Silicon Steel
Rotor	$\phi 50 \times \phi 36 \times 7$ mm
Stator	$\phi 50 \times \phi 36 \times 25$ mm
Slot	10 wide × 18 deep mm
Coil Turns	200
Outer Cylinder	$\phi 65 \times \phi 60$ mm
Air Gap	2.7 mm
Rotor PM	
Material	NdFeB
Magnetization	Perpendicular
Shape	Segment
Size	r25 × r18 × 1 × 150° mm

generate the attractive and repulsive force, and then the radial force can be obtained. In case of Fig. 3, north-pole is generated at the edge of stator by the direct current, and the attractive force acts on left side of the rotor and repulsive force acts on right side, then the left directional radial force is generated. At the permanent magnets on the rotor, the attractive force at the north-pole increases, and the south-pole force decreases, then the tilt moment is generated. The direction of the tilt moment is opposite to the moment by the 4-pole current. Hence, the tilt moment and radial force along the permanent magnet can be controlled by combining these two kinds of current.

By using two pairs of the stator and rotor, which has 90 degrees shifted permanent magnet, we can obtain the tilt moments and radial forces along two directions.

III. FEM ANALYSIS

In order to confirm the generation of tilt moment and radial force, three-dimensional finite element analyses was carried out by using JMAG. The radial force and tilt moment of a pair of 2-pole rotor and 6-slot stator are calculated by static magnetic analysis. The geometries of the rotor and stator are shown in table I.

A coordinate system is defined as shown in Fig. 4. The direction of magnetization and the orthogonal direction of it are defined as a and b, respectively. The angle between x and a is defined as ψ. The rotation around each axis are represented by $\theta_z, \theta_x, \theta_y, \theta_a$ and θ_b.

The current of the stator coils is the summation of 2-pole current, 4-pole current and direct current;

$$i_k = i_{2pk} + i_{4pk} + i_{dc} \quad (1)$$

$$i_{2pk} = i_z \cos\left\{\psi - \frac{(k-1)\pi}{3}\right\} + i_m \sin\left\{\psi - \frac{(k-1)\pi}{3}\right\} \quad (2)$$

$$i_{4pk} = i_{tx} \sin\left\{\psi - \frac{2(k-1)\pi}{3}\right\} + i_{ty} \cos\left\{\psi - \frac{2(k-1)\pi}{3}\right\} \quad (3)$$

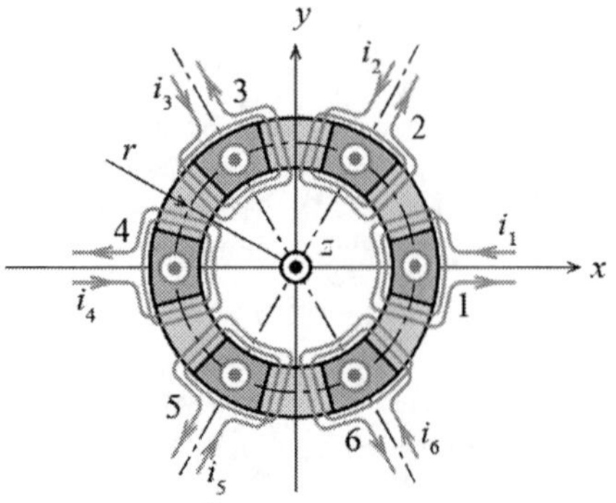

(a) Top view of stator

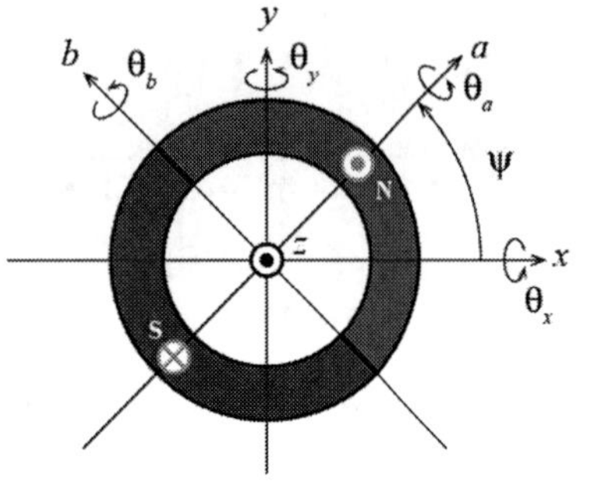

(b) Top view of rotor

Figure 4. Coordinate System.

where $k = 1 \sim 6$ is a coil number. i_z is direct axis current and controls axial force. i_m is quadrature axis current and controls motor torque. i_{tx} and i_{ty} control tilt moment around x and y, respectively.

Fig. 5 shows the results of the radial force and tilt moment. Only the direct current is supplied to the stator, and the radial forces along a and b and tilt moment around a and b are calculated at $\psi = 0$. The radial force F_a and tilt moment M_b decrease in proportion to the direct current. F_b and M_a are not affected by the direct current.

Fig. 6 shows the radial force and tilt moment for 4-pole current at $\psi = 0$. The 4-pole current is supplied to produce only the moment around b-axis. That is

$$i_{tx} = -i_{tb} \sin \psi \tag{4}$$

$$i_{ty} = i_{tb} \cos \psi \tag{5}$$

F_a and M_b are linear with respect to i_{tb}. In this case, the slope of F_a is positive.

Fig. 7 and 8 shows the radial force and tilt moment with constant tilt current, $i_{tb} = 1$A and direct current, $i_{dc} = 1$A, respectively. In case of Fig. 7, the radial force produced by i_{tb} is canceled by i_{dc}. In case of Fig. 8, the tilt moment produced by i_{dc} is canceled by i_{tb}. These results confirm that the undesirable force and moment can be canceled by using both i_{dc} and i_{tb}, and just radial force or tilt moment can be obtained.

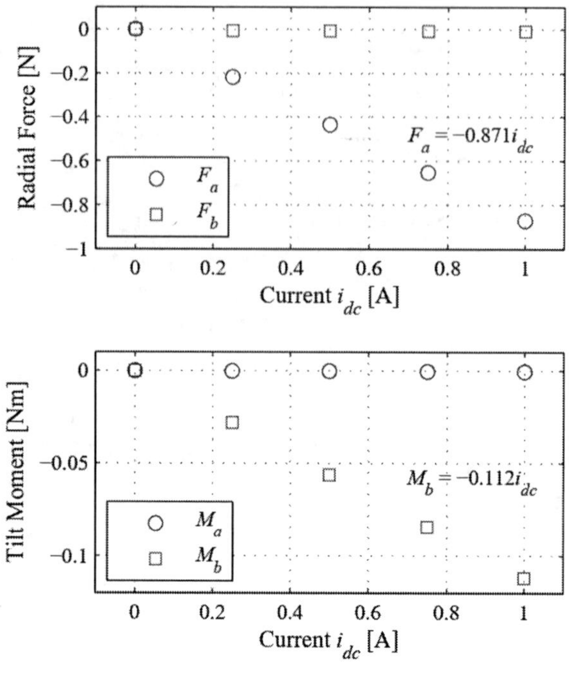

Figure 5. Radial force and tilt moment for direct current at $\psi = 0$.

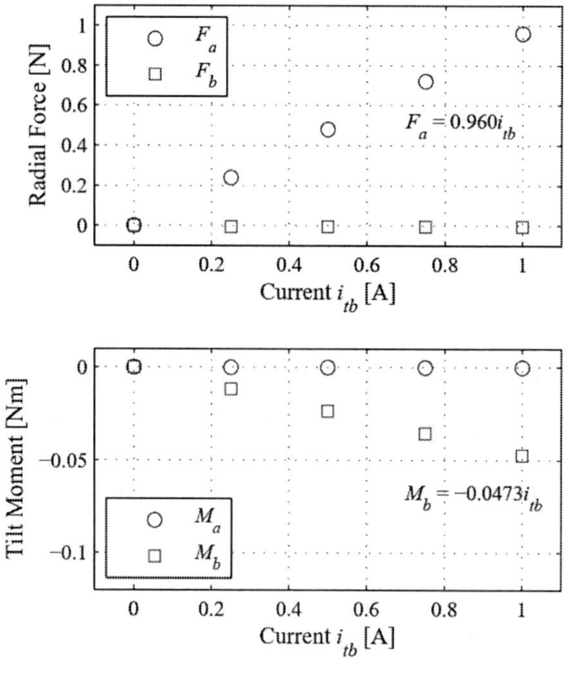

Figure 6. Radial force and tilt moment for 4-pole current at $\psi = 0$.

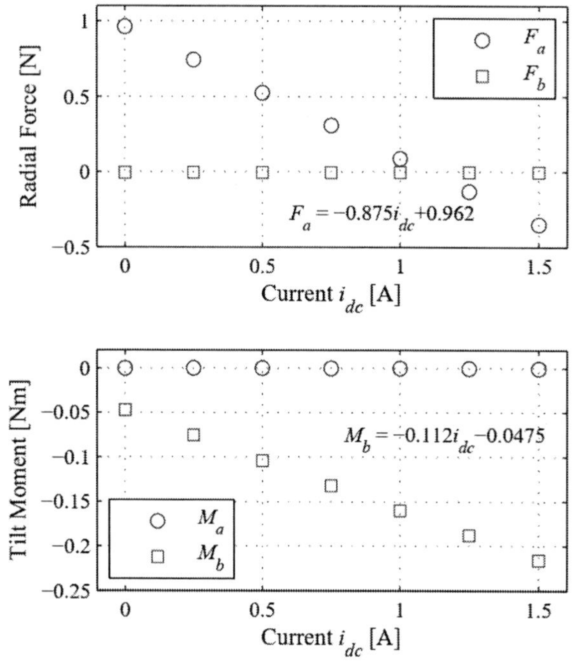

Figure 7. Radial force and tilt moment with $i_{tb} = 1$A at $\psi = 0$.

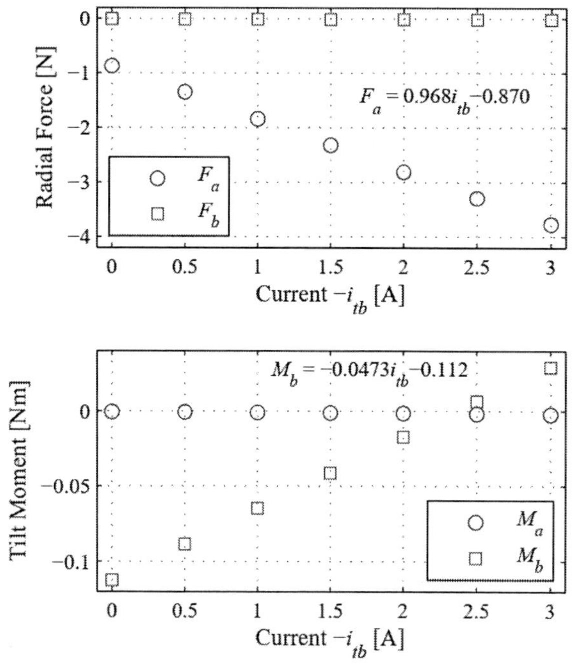

Figure 8. Radial force and tilt moment with $i_{dc} = 1$A at $\psi = 0$.

Moreover, the force gain and moment gain are almost same as the previous cases. Therefore, the radial force and tilt moment can be represented by

$$\begin{bmatrix} F_a \\ M_b \end{bmatrix} = - \begin{bmatrix} K_{fdc} & -K_{ftb} \\ K_{mdc} & K_{mtb} \end{bmatrix} \begin{bmatrix} i_{dc} \\ i_{tb} \end{bmatrix} \quad (6)$$

By solving the above equation, we have

$$\begin{bmatrix} i_{dc} \\ i_{tb} \end{bmatrix} = - \frac{1}{K_{fdc}K_{mtb} + K_{ftb}K_{mdc}} \begin{bmatrix} K_{mtb} & K_{ftb} \\ -K_{mdc} & K_{fdc} \end{bmatrix} \begin{bmatrix} F_a \\ M_b \end{bmatrix} \quad (7)$$

These current achieve decoupling control of radial force and tilt moment.

IV. Summary

This paper introduces a novel axial gap self-bearing motor, which has a capability of the 5-DOF active position control. The results of the FEM analysis show that the radial force and tilt moment can be controlled by combination of dc current and 4-pole current. We are making a prototype and the experimental results will be shown.

V. Acknowledgment

This work was supported by JSPS KAKENHI Grant Number 25420228.

References

[1] Q. D. Nguyen and S. Ueno, "Analysis and control of non-salient permanent magnet axial-gap self-bearing motor," *IEEE Transactions on Industrial Electronics*, vol. 58, no. 7, pp. 2644–2652, 2011.

[2] S. Ueno, T. Okamura, M. Sakagami, and S. Tanaka, "Development of miniaturized regenerative pump using axial self-bearing motor," *Proc. of the 12th International Symposium on Magnetic Bearing, Wuhan, China*, pp. 180–188, 2010.

[3] A. Matsuoka, M. Sumino, S. Ueno, and T. Takeda, "Development of high-speed motor at extremely low temperatures with axial self-bering motor and superconducting magnetic bearing," *International Symposium on Applied Electromagnetics and Mechanics, Quebec City, Canada*, 2013.

A Novel Stabilization Technique for Magnetically Levitated Disk Drives

Hubert Mitterhofer[a], Gerald Jungmayr[a], Wolfgang Amrhein[a]

[a] Johannes Kepler University Linz, Altenberger Strasse 69, 4040 Linz, Austria, hubert.mitterhofer@jku.at

Abstract—**Magnetically levitated disk drives mostly dispose of active stabilization in radial direction and passive stabilization in the axial and tilt degrees of freedom. This may create problems when external torques act on the rotor, tilting it and provoking gyroscopic movements which can destabilize or even destroy the drive. In a very simple fashion, an additional winding can damp this critical movement. The novel winding can furthermore be used to create active tilt torques when energized with the according currents. In addition, a geometric modification to the winding system provides active axial force and enhanced active tilt torque creation capacity without hampering the aforementioned stabilization properties.**

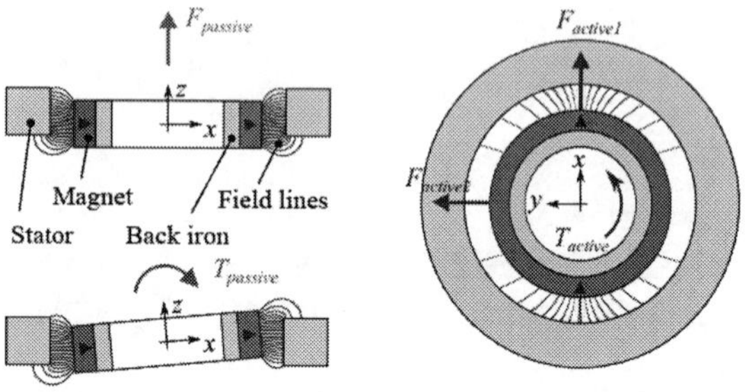

Figure 1. Active and passive stabilization of a bearingless or magnetically levitated disk drive

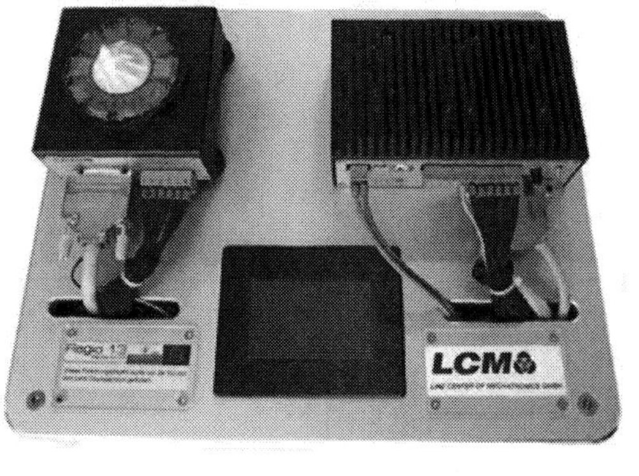

Figure 2. Bearingless disk drive on demonstration system board

I. Introduction

Disk drives are well researched and have covered certain application types for a long time. With the invention and increased research of magnetically levitated drives, disk-shaped rotors have again gained popularity [1]-[7] as they offer certain advantages in rotor dynamics since the axis of rotation is equal to the main axis of inertia. Additionally, they allow a mechanically extremely simplified drive type - the bearingless disk drive with combined bearing- and motor-windings [8]-[10]. One of the central ideas of this drive topology, which is also the reason for its mechanical compactness, is the distribution of passively and actively stabilized degrees of freedom. The axial and the tilt deflections are stabilized passively via the occurring reluctance forces of the drive. Figure 1 illustrates that only the two radial degrees of freedom need to be stabilized actively - a fact that allows concentrating this bearing function into the combined bearing-motor winding system[11]. One of the key flaws of this concept, however, also flows from this distribution since the passive magnetic bearings provide little to no damping and cannot be controlled in any way. What sounds annoying when looking at the axial degree of freedom may virtually become a knock-out criterion for many applications when the tilt deflections are considered as they can easily excite gyroscopic movements of the rotor. This is especially true for applications with high rotational speeds, non-static motor suspensions introducing disturbance forces and torques in all degrees of freedom, or a combination of both. Typical examples are e.g. magnetically levitated turbo chargers, pumps or compressors for vehicles.

To the knowledge of the authors, there is no existing solution to this tilting problem in literature except for adding external damping via fluids surrounding the rotor. Such kind of external damping, however, would seriously hamper many possible applications where the rotor is supposed to work in a low-viscous medium.

This work presents a very simple approach for stabilizing the tilt degree of freedom and proposes a relatively simple approach for providing additional active influence on the rotor movement.

II. Demonstration system

The principles for tilt stabilization and additional active degrees of freedom will be demonstrated at a bearingless disk drive design for high rotational speeds which was presented in earlier works. The drive disposes of a slotless stator core which carries a toroid litz winding system. The rotor consists of a diametrically magnetized permanent magnet with an inner back iron and an outer support sleeve for enhanced centrifugal force resistance. Figure 1 shows the respective motor parts.

The actual prototype drive shown in Figure 2 was successfully tested up to a high speed of 115 000 rpm until a destructive mechanical rotor failure occurred [12]. During these tests, the drive was mounted on a stable support without external forces or torques acting on the stator. However, it was observed that external tilting torques on the stator or rotor would be extremely destabilizing and dangerous for the motor operation due to the on-setting rotor precession. The following section describes an additional winding which introduces additional stabilization for this tilting degree of freedom and, additionally, provides a means of active force and tilting torque generation.

III. Passive tilt damping

A. Problem description

Moving the axis of rotation of a spinning body is problematic if this movement does not lead to a parallel axis deflection. A common example for such a non-parallel movement is the tipping of a spinning top, shown in Figure 3. The top has an angular momentum vector **L** pointing in direction of the axis of rotation. If the top is tipping over - here in negative y-direction - there is a gravitational force $\mathbf{F_g}$, acting on its center of mass. If the top is not completely upright in this moment, this force vector does not go through the point of contact with the ground (root point) and thus creates an external torque **T**, defined as

$$\mathbf{T} = \mathbf{r} \times \mathbf{F_g} \tag{1}$$

with **r** being the horizontal distance vector to the axis through the top's root point. For a constant spinning speed, this external torque defines the change of the angular momentum since

$$\frac{d\mathbf{L}}{dt} = \mathbf{I}\frac{d\omega}{dt} \tag{2}$$

with **I** and ω standing for the momentum of inertia matrix and the angular velocity vector, respectively. This means that the entire top tilts in counterclockwise direction (out of the drawing plane), thus also rotating the external torque vector **T** and, consequently, the vector $\frac{d\mathbf{L}}{dt}$. With the top in Figure 3 spinning in counterclockwise direction, the described tilting equally results in a counterclockwise, and thus, forward precession movement.

The same principle can be applied to the magnetically levitated disk rotor with passive tilt stabilization, depicted in Figure 4. The main difference, however, is that for a negative tilt around a certain horizontal axis, the "external" reaction torque **T** due to the reluctance forces acts in the positive direction of that very axis. Consequently, the vector $\frac{d\mathbf{L}}{dt}$ also points in that negative direction which means that, other than in the case of a top under the influence of an external torque due to gravity, the precession occurs in backward direction.

Unfortunately, the passive tilt stabilization of a bearingless disk drive usually disposes of very little inherent damping, which means that the precession movement can fully unfold. The precessing rotor can either contact a housing part or lead to chaotic movement which causes the rotor to fall into the touch down bearings, potentially damaging the drive.

B. Additional damping

Only few methods can be considered for adding damping to a magnetically passively stabilized degree of freedom.

One of them is to provide an appropriate support of the stator, e.g. in form of visco-elastic damping elements.

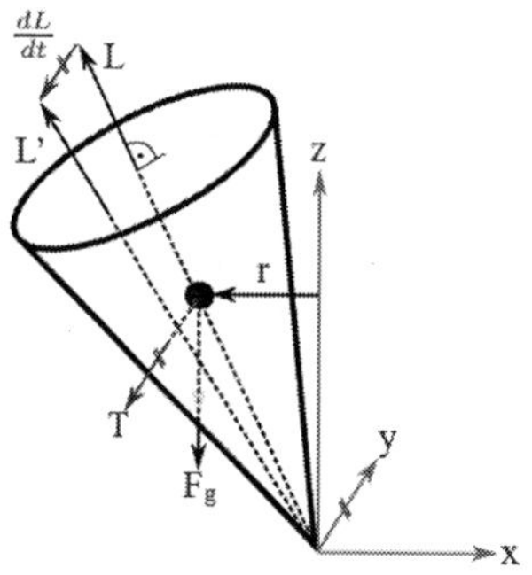

Figure 3. Gyroscopic situation of a spinning top due to an external torque **T** due to gravity. The external torque provokes a precession movement in the sense of rotation (forward whirl).

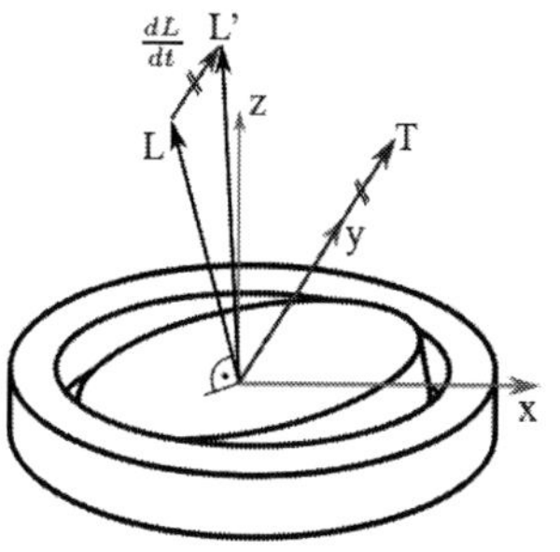

Figure 4. Gyroscopic situation of a spinning magnet rotor disk due to an external torque**T** due to passive stabilization. The stabilization torque provokes a precession movement opposite to the sense of rotation (backward whirl).

The authors of [13], for instance, have demonstrated the effectiveness of this method for suppressing vibrations of a magnetically levitated shaft. This work, however, focused on rotor translations with notable amplitudes and frequencies, while the precession movement in the present disk drive exerts torques which may be small in amplitude and frequency but yet cause big problems. Additionally, a visco-elastic stator support is very demanding for the used materials concerning life time, precision and stiffness and may not be suitable in many applications.

A second source of damping is a medium, surrounding the rotor, which disposes of a significantly higher viscosity than air. This is relevant for most magnetically levitated pump applications as e.g. in the disk drives used as blood pumps [6], [14] or for blenders and mixers for the pharmaceutical industry [15], [16] which operate in liquids. However, this sort of damping can usually not be added to a drive - dependent on the application it is either present or it is not.

Apart from the mentioned methods, an electrodynamic damper can be applied. The principle of electrodynamic force generation is based on Lenz' law. Changing magnetic flux in an electric conductor loop will induce a voltage which, in turn, will offset currents in this conductor. These currents will produce a magnetic field which oppose the change in the previous magnetic field, in other words, they tend to maintain that field constant. The principle of electrodynamic dampers appeared in the 1970's, usually connected to the support of railway vehicles [17][1] and surfaced again after the turn of the millennium, now focused on the stabilization of rotors [18]-[20]. Two very extensive works [21], [22] describe many different forms and design criteria for such electrodynamic bearing and damping elements. However, they mostly focus on bearings or dampers with homopolar permanent magnetic fields and rotor-bound electric conductors where stabilization currents are induced. This creates the problem that the *non-rotating* damping necessary for the stability of the electrodynamic levitation must be added externally.

C. Tilt stabilization coil (TSC)

The proposed stabilization uses the rotating 2-pole permanent magnet field of the bearingless disk drive described in [10] and just requires an additional short circuited winding in the stator, placed as shown in Figure 5. In the centered and non-tilted rotor position, the newly introduced tilt stabilization coil (TSC) does not experience an induced voltage. If the diametrically magnetized rotor is tilted, the flux at the magnet poles is not equally distributed vertically over the inner stator surface but is more concentrated in the upper or lower half, respectively. This is schematically demonstrated in Figure 5. Following the flux path, we see the a part of the flux leaving the magnet, crossing the air gap, entering the stator (here: in the upper half of the stator), flowing through the stator to the opposite side, leaving the stator (here: from the lower stator half), again crossing the air gap and entering back into the magnet. This means that there is a net flux closing through the TSC. This is illustrated in an equivalent flux path in the lower part of that figure.

The described flux path changes when the non-rotating rotor tilts back and forth around one axis but the frequency of this movement is relatively low. A low frequency of the flux linked to the coil would also result in low induced voltage. However, the presented arrangement is also sensitive to the high-frequency rotation when it is linked to the tilting. As the rotor rotates with a quasi-static tilt inclination of the axis of rotation, the pole orientation changes and so does the direction of the flux through the TSC. The quasi-static consideration seems legitimate when the rotational frequency is far above the tilting frequency. The mentioned flux change, of course, induces a voltage which now depends on both, the rotational-speed and the tilt angle. Since the coil is short circuited, that voltage will drive a current through the coil. Due to the high frequency of rotation compared to the frequency of pure tilting movement, the stabilization effect is significantly larger in this case. The resulting effect on the precession movement was confirmed by introducing a TSC, similar to the one shown here, into an existing disk drive.

The described effect is based on three assumptions:

- The rotor does not rotate at the exact same speed while precessing in the same direction (forward whirl). For this special case, the vertical positions of the rotor poles would not change and no voltage would be induced in the TSC.
- The center of the rotor is axially aligned with the center of the stator.
- Eddy current effects in the TSC have to be minimized by using thinly stranded wire. Otherwise the TSC would rather resemble an eddy current brake which operates independent of the rotor eccentricity or its tilt angle.

[1]The Japanese Railway Company only recently announced the planned construction of a 286 km long track, linking the cities of Tokyo and Nagoya, for magnetically levitated trains using an electrodynamic suspension principle. The track is supposed to allow travel at a speed of above $500\,\mathrm{km/h}$ and is the first part of a longer railway link from Tokyo to Osaka.

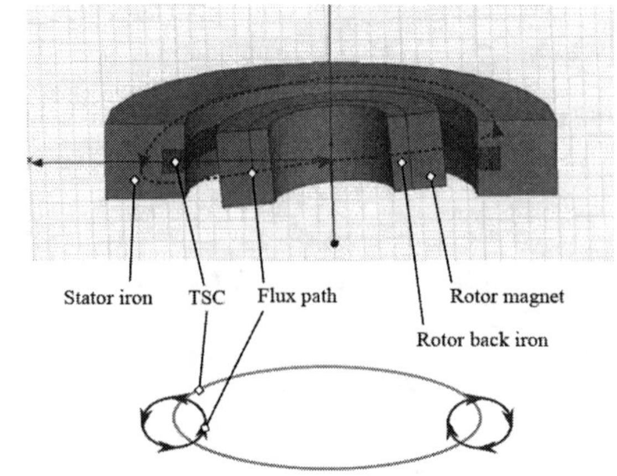

Figure 5. Flux path in motor cross section with TSC when rotor is tilted (top) with equivalent flux path (bottom)

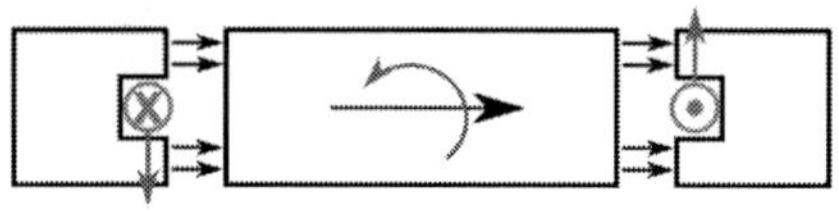

Figure 6. Flux path in motor cross section with resulting force on stator and reaction torque on rotor when coil is actively energized

IV. Active influence

A. Active tilting - 1 coil

The TSC mentioned in the above section was short-circuited for the passive stabilization. When connecting the coil endings to a power electronics circuit, it can also be used to exert active tilting torque. The effect is shown in Figure 6 and can be understood by either considering the Lorentz forces on the winding or by examining the Maxwell forces on the radial slot walls. The axis about which the torque is produced is always perpendicular to the axis of magnetization. Only the amplitude of this torque $T^{\perp}$ and its sense (positive or negative) can be modulated by the amplitude and direction of the coil current.

B. Active axial force - 2 coils

So far, the difference in the presented effects on the rotor was due to the either passive or active use of the TSC. For the next active method of influencing the rotor, namely by creating axial forces, the coil geometry has to be modified, forming two electrically insulated and geometrically separated, "C"-shaped coils, as shown in Figure 7.

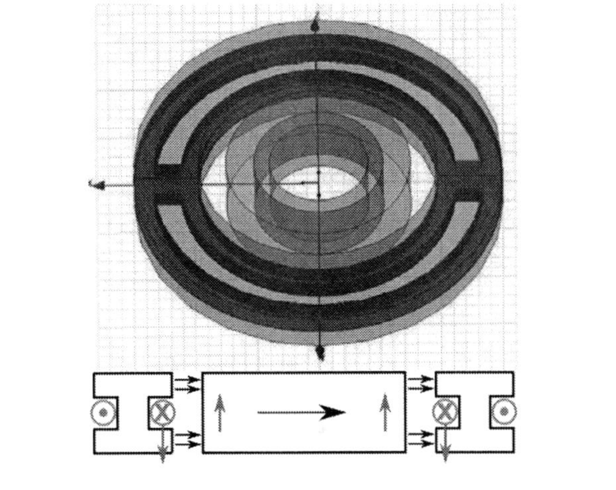

Figure 7. Horizontal cross section through the center of the drive showing the modified coil shape (top) and vertical motor cross section with flux path and resulting force on stator and rotor when the "C"-shaped coils are energized in opposite sense (bottom)

This modification to the stator reduces the effective iron volume but otherwise, has similar characteristics as the ring-shaped TSC. Both previously presented effects, the passive tilt stabilization and the active tilting torque generation can be achieved with this winding set by either short-circuiting the "C"-shaped coils (for passive stabilization) or by energizing them in identical sense (for active tilting torque), respectively.

If the two coils, however, are energized with currents of opposing sense, an axial force is actively created. Since the geometric setup is not rotationally symmetric any more, the achievable axial force F_z depends on the rotor angle. If the rotor angle φ_r is defined to be zero when the poles are aligned with the plane which separates the two coils, the achievable force can be expressed as

$$F_z(\varphi_r) = q_z i_c \cdot \sin(\varphi_r) \tag{3}$$

with i_c being the coil current and and q_z representing a force coefficient which relates the current to the maximum achievable force amplitude.

C. Active tilting - 4 coils

A last potential modification shall not stay unmentioned. When segmenting the coil further into four electrically insulated coils as shown in Figure 9, a large variety of active force and torque components can be created. The passive stabilization characteristic can be provided by accordingly short-circuiting the 4 coils. Energizing all coils in the same current sense creates a situation equivalent to Figure 6, describing the creation of $T^{\perp}$. If two adjacent coils are picked for one current direction and the other two for the opposite direction, the scenario is similar to Figure 7 with one main difference; The current direction can be changed every 90 deg, thus avoiding the configuration where zero F_z could be created. The resulting force capacity is displayed in Figure 8.

Only if the current configuration is chosen in such a way that the geometrically opposite coils are energized in the same

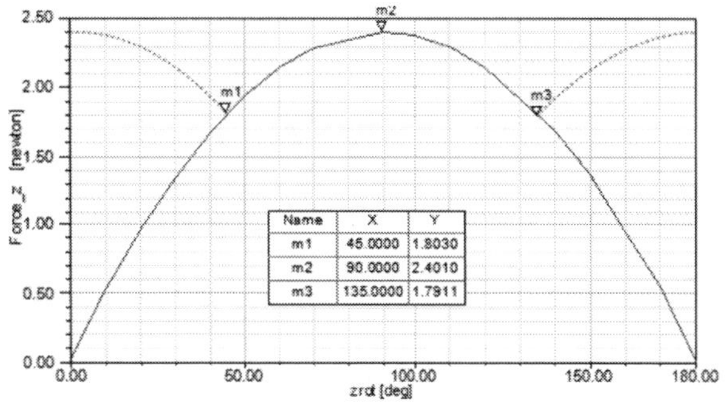

Figure 8. Simulated maximum axial force in the 4-coil configuration over the rotor angle

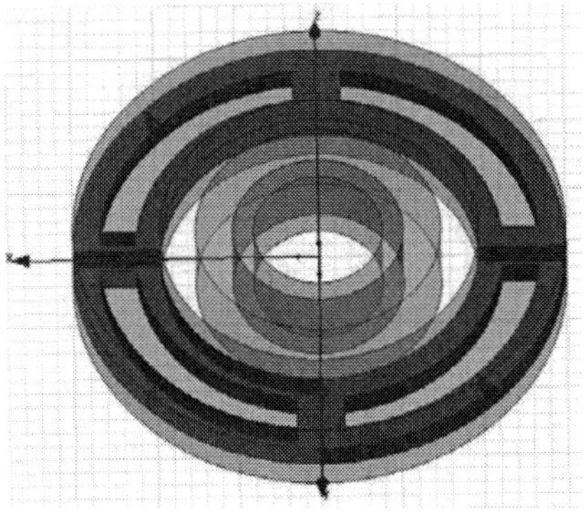

Figure 9. Transparent 3D view of the drive showing the modified coil shape for four coils

current direction, the rotor can additionally be tilted actively about the axis parallel to its magnetization axis by producing a torque $T^{\parallel}$. However, there is a main difference between to the generation of $T^{\perp}$ and $T^{\parallel}$. The torque component about the axis perpendicular to the magnetization axis due to constant coil currents turns out to be a rotating vector of constant amplitude. The new active torque component about the magnetization axis is sinusoidally dependent on the rotor angle just like the axial force component. In other words, constant coil currents will produce a rotating vector which also oscillates in amplitude. As a consequence, there are certain angular positions, at which the rotor cannot be tilted about its magnetization axis.

Apart from this deficit, the necessary effort for this solution is large, considering an even larger loss of effective iron volume compared to all previous variations and the necessary four independent coils with four independent power electronic circuits energizing them.

V. Simulation Results

A concluding 3D-simulation of the proposed concepts has been conducted. For the determination of the active force and torque components, the stator coils have been designed to have a cross section of 3 mmx3 mm. Apart from these new coils, the geometric dimensions and material parameters used in the simulation are identical to the referenced system described in [10]. For the verification of the passive stabilization, the TSC

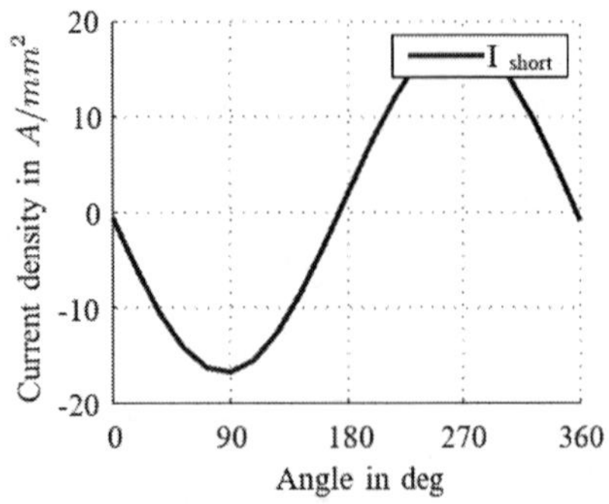

Figure 10. Short circuit current in TSC

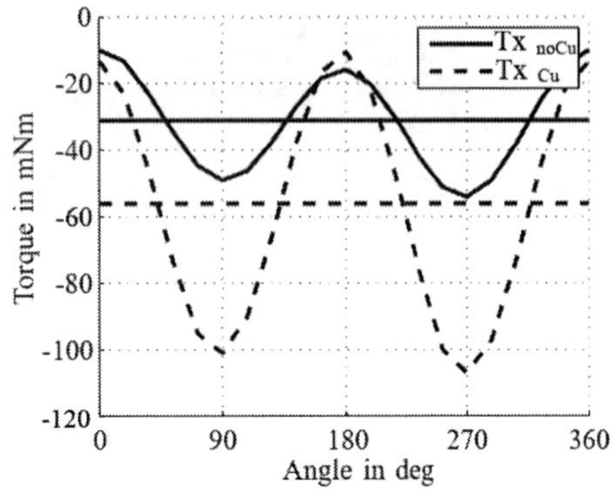

Figure 11. Torque curve and mean value in x-direction with and without the copper coil (TSC)

was set to have 10 winding turns and the rotor was tilted by 5° about the x-axis while rotating at a speed of 60 000 rpm. Figure 10 shows the short-circuit current in the TSC due to the induced voltage of the rotating magnetic field. Figures 11 and 12 show the occurring torque which acts on the rotor for both situations - with and without the TSC. It is clearly shown that even though the mean value of the torque does not change significantly for the non tilted y-axis, it does go to significantly lower values for the x-axis which was tilted in positive sense, thus forcing the rotor back to an upright position.

For the actively created torque and force values, the current density in the coils was set to constant value of $10\,\mathrm{A/mm^2}$ and their direction in the individual coils was set according to the requirements of the respective active component. Along with the achievable amplitudes of the active components, the passive tilt stiffness k_β is given in Table I in order to give a comparison value and also, to show the negligible influence of the present modification on this value.

VI. Conclusion and Outlook

This work discusses a novel winding for passively stabilizing the tilting movement in disk shaped rotors. Due to the weak damping in this passively stabilized degree of freedom, such movement may otherwise lead to gyroscopic movements such as precession which can seriously threaten successful operation of the drive. The presented additional winding may

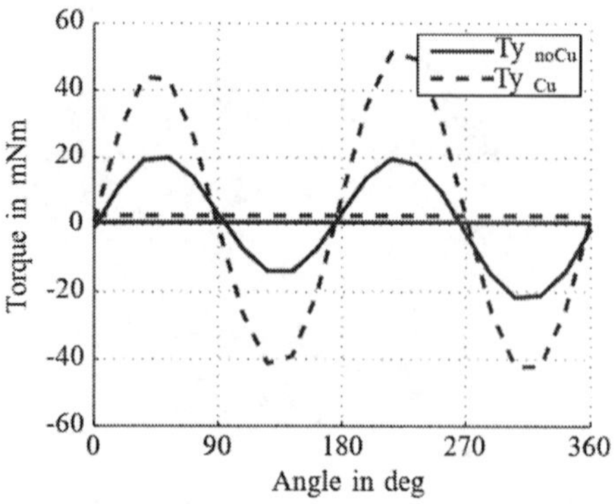

Figure 12. Torque curve and mean value in y-direction with and without the copper coil (TSC)

Table I
USE TABLES FOR A CLEAR OVERVIEW OF YOUR DATA

	Original setup	1-coil setup	2-coil setup	4-coil setup
k_β	$0.81\,\mathrm{Nm/rad}$	$0.78\,\mathrm{Nm/rad}$	$0.76\,\mathrm{Nm/rad}$	$0.76\,\mathrm{Nm/rad}$
$\hat{T}^{\perp}$	-	31.6 mN m	30.2 mN m	30.1 mN m
$\hat{F}_z$	-	-	2.4 N	2.4 N
$\hat{T}^{\parallel}$	-	-	-	$11.1\,\mathrm{Nm/rad}$

be used passively for the mentioned stabilization but can also serve to provide active influence in the form of active tilting torque and active axial forces. The necessary modifications are shown and some resulting side effects are mentioned. Finally, the effect of the additional windings is shown by simulation for both, the passive stabilization and the active force and torque generation. Even though the passive stabilization coil was observed to effectively stabilize any on-setting precession movements when an external torque was applied to a prototype system at speeds of up to 40 000 rpm, a more reliable prototype needs to be constructed in the future in order to measure the predicted consequences.

References

[1] R. Schob and N. Barletta, "Principle and application of a bearingless slice motor," *Proc. 5th International Symposium on Magnetic Bearings*, 1996.

[2] S. Ueno, C. Chen, T. Ohishi, K. Matsuda, Y. Okada, Y. Tanaka, and T. Masuzawa, "Design of a self-bearing slice motor for a centrifugal blood pump," *Proc. 6th International Symposium on Magnetic Bearings*, 1998.

[3] A. Salazar, A. Chiba, and T. Fukao, "A review of developments in bearingless motors," *Proc. 7th International Symposium on Magnetic Bearings*, 2000.

[4] H. Grabner, H. Bremer, W. Amrhein, and S. Silber, "Radial vibration analysis of bearingless slice motors," *Proc. 9th International Symposium on Magnetic Bearings*, 2004.

[5] S. Silber, W. Amrhein, P. Bosch, R. Schob, and N. Barletta, "Design aspects of bearingless slice motors," *IEEE/ASME Transactions on Mechatronics*, vol. 10, no. 6, pp. 611–617, 2005.

[6] T. Nussbaumer, P. Karutz, F. Zuercher, and J. Kolar, "Magnetically levitated slice motors—an overview," *IEEE Transactions on Industry Applications*, vol. 47, no. 2, pp. 754–766, 2011.

[7] J. Asama, Y. Hamasaki, T. Oiwa, and A. Chiba, "Proposal and analysis of a novel single-drive bearingless motor," *IEEE Transactions on Industrial Electronics*, 2013.

[8] W. Gruber, T. Nussbaumer, H. Grabner, and W. Amrhein, "Wide air gap and large-scale bearingless segment motor with six stator elements," *IEEE Transactions on Magnetics*, vol. 46, no. 6, pp. 2438–2441, 2010.

[9] H. Grabner, W. Amrhein, S. Silber, and W. Gruber, "Nonlinear feedback control of a bearingless brushless DC motor," *IEEE/ASME Transactions on Mechatronics*, vol. 15, no. 1, pp. 40–47, 2010.

[10] H. Mitterhofer, W. Gruber, and W. Amrhein, "On the high speed capacity of bearingless drives," *IEEE Transactions on Industrial Electronics*, vol. 61, no. 6, pp. 3119–3126, 2014.

[11] H. Mitterhofer and W. Amrhein, "Motion control strategy and operational behaviour of a high speed bearingless disc drive," *Proc. IET Power Electronics, Machines and Drives Conference*, 2012.

[12] ——, "Design aspects and test results of a high speed bearingless drive," *Proc. 9th Power Electronics and Drive Systems Conference*, 2011.

[13] E. Marth, G. Jungmayr, and W. Amrhein, "Influence of viscoelastic elements on magnetically passive stabilized degrees of freedom," *Proc. 13th International Symposium on Magnetic Bearings*, 2010.

[14] K. Raggl, B. Warberger, T. Nussbaumer, S. Burger, and J. W. Kolar, "Robust angle-sensorless control of a PMSM bearingless pump," *IEEE Transactions on Industrial Electronics*, vol. 56, pp. 2076–2085, 2009.

[15] B. Warberger, R. Kaelin, T. Nussbaumer, and J. Kolar, "50Nm / 2500W bearingless motor for high-purity pharmaceutical mixing," *IEEE Transactions on Industrial Electronics*, vol. 59, pp. 2236–2247, 2012.

[16] T. Reichert, T. Nussbaumer, and J. W. Kolar, "Bearingless 300-W PMSM for bioreactor mixing," *IEEE Transactions on Industrial Electronics*, vol. 59, no. 3, pp. 1376–1388, 2012.

[17] D. Atherton and A. R. Eastham, "Flat guidance schemes for magnetically levitated highspeed guided ground transport," *Journal of Applied Physics*, vol. 45, 1973.

[18] K. Davey, A. Filatov, and R. Thompson, "Design and analysis of passive homopolar null flux bearings," *IEEE Transactions on Magnetics*, vol. 41, 2005.

[19] J. G. Detoni, F. Impinna, N. Amati, and A. Tonoli, "Rotordynamic stabilization of rotors on electrodynamic bearings," *INTECH Open Publishing*, vol. Advances in Vibration Engineering and Structural Dynamics, pp. 51–66, 2012.

[20] V. Kluyskens and B. Dehez, "Dynamical electromechanical model for magnetic bearings subject to eddy currents," *IEEE Transactions on Magnetics*, vol. 49, pp. 1444–1453, 2013.

[21] T. Lembke, "Design and analysis of a novel low loss homopolar electrodynamic bearing," Ph.D. dissertation, KTH Stockholm, 2005.

[22] F. Impinna, "Electrodynamic bearings," Ph.D. dissertation, Politecnico di Torino, 2010.

Decoupling Control for Bearingless Synchronous Reluctance Motor Based on SVM with Ant Colony Optimization Algorithm

Huangqiu Zhu, Xiaoyan Diao, Enxiang Xu, Li Cao

School of Electrical and Information Engineering, Jiangsu University, 212013 Zhenjiang, China, zhuhuangqiu@ujs.edu.cn

***Abstract*—A decoupling control based on support vector machine (SVM) with ant colony optimization algorithm is proposed for a bearingless synchronous reluctance motor (BSRM). The α-order inverse model approximated by SVM using ant colony parameter optimization is cascaded with the original system to get the composite pseudo-linear system. The control system of the BSRM is simulated with Matlab software. The simulation and analysis have been carried out. The simulation results show that the dynamic, static performances and robustness of the control system are excellent.**

I . INTRODUCTION

A bearingless synchronous reluctance motor (BSRM) has many advantages such as high speed, high precision, low loss, long operating life, simple structure, reliable operation, low cost, and etc.. The BSRM is a great breakthrough in electric drive field, and is more applicable to high speed and high precision special applications. However, the BSRM is a nonlinear, strongly coupled and multi-input multi-output (MIMO) system, so it is not to meet the needs of its high speed and high precision operation with the distributed control method obviously. The key problem of control system is the couplings among the electromagnetic torque and the two radial suspension forces [1-5].

At present, there are the following kinds of control methods for the BSRM. A decoupling control strategy by using series feed-forward compensator was proposed in reference [4, 5]. Because parameters of this control method are influenced by the magnetic saturation easily, the control method is difficulty to realize dynamic decoupling control. L. Hertle proposed an improvement of mathematical model in reference [6]. By using approximation, online look-up table and detection of real-time parameters methods, it can reduce the magnetic saturation and the interaction of electromagnetic torque and radial suspension forces. This model has reached good high precision, but it can't realize the completely dynamic decoupling control. Moreover, inverse system method need know the mathematic model of the controlled device and also get its analytical solutions. Neural networks method has lots of inherent limitations such as low learning speed, long time expenditure, converging to a local minimum point and over-fitting learning problem effecting system generalization capability, and so on. Therefore, in the paper, a new control method, which is control method based support vector machine (SVM) is proposed. The SVM is a powerful machine learning technique based on statistical learning theory (SLT). Based on the structure risk minimization principle, SVM can solve the small sample, partial minimization, over-fitting effectively, high dimensionality and has good generality capacity. Appropriate parameters are very crucial to support vector machines learning results and generalization ability, so it is very important to select parameters. The ant colony algorithm is a heuristic optimization method, which can get feasible solutions of problems by using the path of artificial ants, and it has many merits such as strong optimization ability, fast convergence rate.

In the paper, the identification model of inverse system for the original nonlinear is gotten, an idea of optimum parameter of SVM by the ant colony algorithm is put forward, and the control strategy based on support vector machine can realize the completely dynamic decoupling control for the BSRM.

II . OPERATION PRINCIPLE OF BSRM AND ANALYSIS OF REVERSIBILITY FOR ORIGINAL SYSTEM

A. Operation Principle of BSRM

Fig. 1 shows the principle of radial suspension force production under no-load condition for the BSRM. The BSRM combines 4-pole motor windings N_a and 2-pole radial force windings N_x and N_y. Illustrated in Fig. 1, when currents pass 4-pole motor windings N_a and 2-pole radial force windings, the 4-pole motor windings N_a generate 4-pole magnetic field ψ_a, and the 2-pole radial force windings N_x and N_y generate separately 2-pole magnetic fluxes ψ_x and ψ_y. When the 2-pole winding currents are applied, the fluxes density in area 2 increases while the fluxes density in area 4 decreases. So the unbalanced revolving field results in the production of the radial force F_x in the positive x-direction. Similarly, it can produce the radial force F_y in the positive y-direction. So we can control the size and the direction of the radial force F by changing currents of radial force windings N_x and N_y, respectively, the rotor will achieve a stable suspension.

Because the BSRM has salient-pole rotor, the air gap variation caused by the rotor saliency is only considered in practical calculation of the radial forces [4-5]. The pole arc is assumed to be 30°, the number of pole pair of torque windings P_M=2, the number of pole pair of suspension windings P_B=1, and the magnetic saturation is neglected. Mathematical models of torque and radial suspension forces are deduced using Maxwell stress tensor method [2-6].

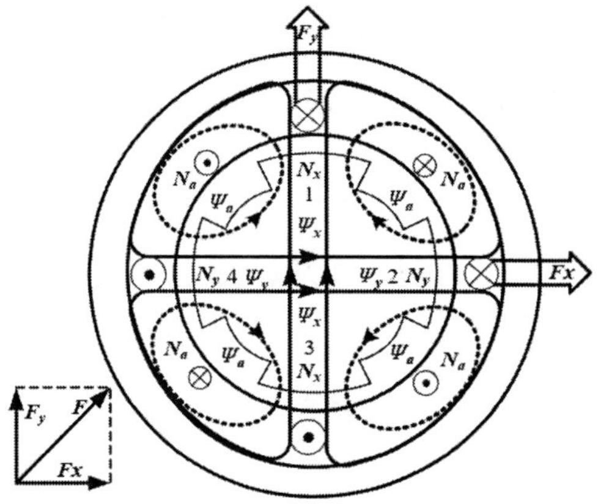

Figure 1. Principle of radial force generation for BSRM

In the *d-q* rotating reference frame, the stored energies in the machine inductance is calculated as follows

$$W_m = \frac{1}{2}\begin{bmatrix} i_d & i_q & i_x & i_y \end{bmatrix}\begin{bmatrix} L_d & L_{dq} & L_{dx} & L_{dy} \\ L_{qd} & L_q & L_{qx} & L_{qy} \\ L_{xd} & L_{xq} & L_x & L_{xy} \\ L_{yd} & L_{yq} & L_{yx} & L_y \end{bmatrix}\begin{bmatrix} i_d \\ i_q \\ i_x \\ i_y \end{bmatrix} \tag{1}$$

where i_d and i_q are the current components of the motor windings, respectively, i_x and i_y are the current components of radial force windings, respectively, L_d and L_q are the *d*-axis and *q*-axis inductances of the motor windings, respectively, L_{dx}, L_{qx}, L_{dy} and L_{qy} are the mutual inductances between the motor windings and the radial force windings, respectively.

Because two windings of the BSRM are piled in upper-under layers and arranged symmetrically under some rules, $L_x=L_y=L_2$, $L_{dq}=L_{qd}=0$, $L_{xy}=L_{yx}=0$. So Eq. (1) can be simplified as follows

$$W_m = \frac{1}{2}\begin{bmatrix} i_d & i_q & i_x & i_y \end{bmatrix}\begin{bmatrix} L_d & 0 & K_{m1}x & -K_{m1}y \\ 0 & L_q & K_{m2}y & K_{m2}x \\ K_{m1}x & K_{m2}y & L_2 & 0 \\ -K_{m1}y & K_{m2}x & 0 & L_2 \end{bmatrix}\begin{bmatrix} i_d \\ i_q \\ i_x \\ i_y \end{bmatrix} \tag{2}$$

where $K_{m1} = \frac{\mu_0 lrN_2N_4}{48\delta_0^2}(2\pi + 3\sqrt{3})$, $K_{m2} = \frac{\mu_0 lrN_2N_4}{48\delta_0^2}(2\pi - 3\sqrt{3})$, l is the stack length, μ_0 is the vacuum permeability, r is the rotor radius, N_2 and N_4 are the number of per-phase effective turns in series of the motor windings and radial force windings, respectively.

The radial forces acting on the rotor can be derived from the differential of the magnetic energy and can be obtained as follows

$$\begin{bmatrix} F_x \\ F_y \end{bmatrix} = \begin{bmatrix} \frac{\partial W_m}{\partial x} \\ \frac{\partial W_m}{\partial y} \end{bmatrix} = \begin{bmatrix} K_{m1}i_d & K_{m2}i_q \\ K_{m2}i_q & -K_{m1}i_d \end{bmatrix}\begin{bmatrix} i_x \\ i_y \end{bmatrix} \tag{3}$$

Supposed that $C_0 = \begin{bmatrix} \cos\theta & -\sin\theta \\ \sin\theta & \cos\theta \end{bmatrix}$, where θ is mechanical rotor angle and the above equations are transformed in the stationary coordinate system as follows

$$\begin{bmatrix} F_x \\ F_y \end{bmatrix} = C_0\begin{bmatrix} K_{m1}i_d & K_{m2}i_q \\ K_{m2}i_q & -K_{m1}i_d \end{bmatrix} C_0^{-1}\begin{bmatrix} i_\alpha \\ i_\beta \end{bmatrix} \tag{4}$$

where i_α and i_β are the currents of the radial force windings in the stationary coordinate system.

In addition, when the rotor offsets from the stator center, the rotor produce Maxwell stress which is direct proportional to the eccentric displacement, as followings

$$\begin{bmatrix} F_{sx} \\ F_{sy} \end{bmatrix} = k\frac{\pi rlB^2}{\mu_0\delta_0}\begin{bmatrix} x \\ y \end{bmatrix} = K_s\begin{bmatrix} x \\ y \end{bmatrix} \tag{5}$$

where k is the factor related to the motor structure, B is the amplitude of flux density in air gap.

F_{zx} and F_{zy} are assumed to the interference forces applied on the rotor in the *x*- and the *y*-direction, and the gravity is included. So the motion equations of the rotor are described as follows

$$\begin{cases} -F_x + F_{sx} + F_{zx} = m\ddot{x} \\ -F_y + F_{sy} + F_{zy} = m\ddot{y} \end{cases} \tag{6}$$

B. Equation of Torque

The mathematical model of the torque subsystem is totally constructed with voltage, flux linkage and torque equations. In the rotor oriented coordinate system, the equation of motor stator voltage as followings

$$\begin{cases} u_d = R_{s1}i_d + L_d\frac{di_d}{dt} - \omega L_q i_q \\ u_q = R_{s1}i_q + L_q\frac{di_q}{dt} + \omega L_d i_d \end{cases} \tag{7}$$

The equation of stator flux linkage is

$$\begin{cases} \Psi_d = L_d i_d \\ \Psi_q = L_q i_q \end{cases} \tag{8}$$

The equation of electromagnetic torque is

$$T_e = \frac{3}{2}P_M(L_d - L_q)i_d i_q \tag{9}$$

The equation of the motion is

$$T_e - T_L = \frac{J}{P_M}\frac{d\omega}{dt} \tag{10}$$

where u_d and u_q are the *d-q* axis stator voltage components, respectively, ψ_d and ψ_q are the *d-q* axis flux linkage components of the stator, L_d and L_q are the *d-q* axis inductance of the torque windings, R_{s1} is per-phase resistance of the stator, ω is the synchronous angular speed, P_M is the number of pole pairs of the rotor, J is the moment of inertia, T_e is the electromagnetic torque, T_L is the load torque.

C. Analysis of Reversibility for Original System

Based on mathematical model above, it is known the BSRM is MIMO nonlinear, time-delay system. The system state variables are defined as

$$\boldsymbol{x} = \begin{bmatrix} x_1 & x_2 & x_3 & x_4 & x_5 \end{bmatrix}^T = \begin{bmatrix} x & y & \dot{x} & \dot{y} & \omega \end{bmatrix}^T \tag{11}$$

The input variable is given as

$$\boldsymbol{u} = \begin{bmatrix} u_1 & u_2 & u_3 \end{bmatrix}^T = \begin{bmatrix} i_q & i_x & i_y \end{bmatrix}^T \tag{12}$$

The output variable is given as

$$y=\begin{bmatrix} y_1 & y_2 & y_3 \end{bmatrix}^{\mathrm{T}} = \begin{bmatrix} x & y & \omega \end{bmatrix}^{\mathrm{T}} \tag{13}$$

According to the mathematical model, the state equation of the BSRM can be derived as the followings

$$\dot{x} = f(x,u) = \begin{cases} \dot{x}_1 = x_3 \\ \dot{x}_2 = x_4 \\ \dot{x}_3 = \dfrac{1}{m}\left(-K_{m1} i_d u_2 - K_{m2} u_1 u_3 + k_s x_1 + F_{zx}\right) \\ \dot{x}_4 = \dfrac{1}{m}\left(-K_{m2} u_1 u_2 + K_{m1} i_d u_3 + k_s x_2 + F_{zy}\right) \\ \dot{x}_5 = \dfrac{3P_M^2}{2J}\left(L_d - L_q\right) i_d u_1 - \dfrac{P_M}{J} T_L \end{cases} \tag{14}$$

Based on Interactor algorithm, the reversibility of system is analyzed. Continuously differentiate the output function $y = h(x) = \begin{bmatrix} x_1 & x_2 & x_5 \end{bmatrix}^{\mathrm{T}}$ until it includes the input variable u. From Eq. (14), it can be obtained as the followings

$$A = \begin{bmatrix} -\dfrac{1}{m} K_{m2} u_3 & -\dfrac{1}{m} K_{m1} i_d & -\dfrac{1}{m} K_{m2} u_1 \\ -\dfrac{1}{m} K_{m2} u_2 & -\dfrac{1}{m} K_{m2} u_1 & \dfrac{1}{m} K_{m1} i_d \\ \dfrac{3P_M^2}{2J}\left(L_d - L_q\right) i_d & 0 & 0 \end{bmatrix} \tag{15}$$

The determinant of the Jacobi matrix A is

$$\det A = -\frac{3P_M^2}{2m^2 J}\left(L_d - L_q\right) i_d \left(K_{m1}^2 i_d^2 + K_{m2}^2 u_1^2\right) \tag{16}$$

Under normal condition, $i_d \neq 0$, $L_d \neq L_q$, and K_{m1} and K_{m2} are nonzero constants, so $\det A \neq 0$. So the system is reversible.

III. DECOUPLING CONTROL FOR BSRM BASED ON SVM WITH ANT COLONY OPTIMIZATION

A. SVM Theory and Ant Colony Optimization

SVM is deduced based on statistic studying theory and construction risk minimization principle, and it can solve the small sample, partial minimization etc. the basic theory of SVM is that training data are mapped into a high-dimensional feature space. For the undivided linear sample space, using constructive kernel covering lead data covering maps the awaiting sort samples of original space to a high dimensional feature space, makes the samples linear separable. Estimating linear regression function, as followings

$$f(x) = w^{\mathrm{T}} \varphi(x) + b \tag{17}$$

where w^{T} is the wight vector, b is offset, $\varphi(x)$ is the nonlinear function.

For k pairs of training sample set, $f(x)$ (function of optimization approximation in feature space) must make function J is minimum based on the principle of construction risk minimization, where

$$J = C\sum_{i=1}^{k} L\left(f(x_i), y_i\right) + \frac{w^{\mathrm{T}} \cdot w^{\mathrm{T}}}{2} \tag{18}$$

where C is normalization parameters, $L(f(x), y) = \max(0, |f(x) - y| - \varepsilon)$ the penalty function.

Through duality principle, Lagrangian's multiplier method, technology of kernel function, the Eq. (18) will be transformed as the problem of quadratic program. As followings

$$\min_{w,b,L} J = \frac{1}{2}\sum_{i,j=1}^{k} (\alpha_i^* - \alpha_i)(\alpha_j^* - \alpha_j) k\left(x_i, x_j\right) - \sum_{i=1}^{k} y_i \left(\alpha_i^* - \alpha_i\right) + \varepsilon \sum_{i=1}^{k} \alpha_i^* \tag{19}$$

$$s.t. \left\{ \sum_{i=1}^{k} \left(\alpha_i^* - \alpha_i\right) = 0, \quad 0 \leq \alpha_i^*, \alpha_i \leq C, \quad i = 1, \cdots, k \right.$$

where kernel function $k\left(x_i, x_j\right) = \varphi(x_i)\varphi(x_j)$ is symmetric function satisfying mercer condition.

α_i^*, α_i can be gotten based on Eq. (19), then b is gotten by combining KKT condition. Input samples are support vectors when $\left(\alpha_i^* - \alpha_i\right) \neq 0$. So output of SVM is equation as followings

$$f(x) = \sum_{i=1}^{k} \left(\alpha_i^* - \alpha_i\right) K\left(x, x_i\right) + b \tag{20}$$

There are many kinds of kernel functions, and different kernel functions can become different algorithms in SVM algorithm. In this paper, Gaussian kernel function is chosen as kernel function. Equation of Gaussian kernel function is as following: $K\left(x, x_k\right) = \exp(-\|x - x_k\|^2 / 2\sigma^2)$.

SVM has perfect theory basis, but it still exists some problems. Typical problem is the determination of parameter such as penalty factor C and parameter of kernel function σ^2, which are very important to the result of the classification. So ant colony algorithm is adopted to optimize the SVM parameters.

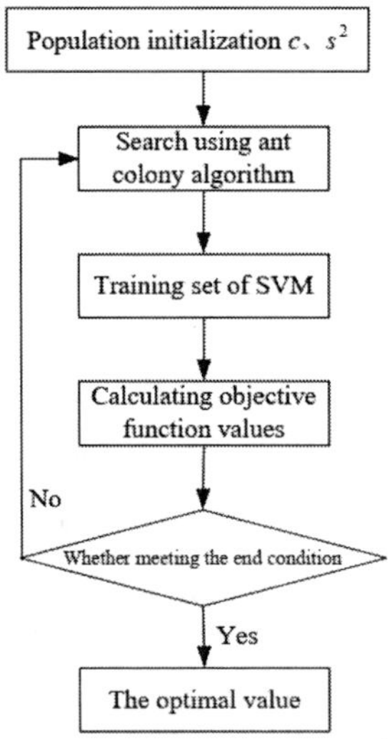

Figure 2. The flow diagram SVM parameters basted on ant colony algorithm.

Ant colony algorithm is a new heuristic optimization method. This paper gets the optimal solution of C and σ^2 in continuous space domain. Its algorithm flowchart is given as following Fig. 2 [9-10].

B. Decoupling Control of BSRM

Because the BSRM is MIMO system, the change of one

input signal affects multiple output signals. Under such circumstance, the decoupling control for system is of particular importance.

By applying uncorrelated white noises at input ports, we can get 100 groups of input/output data. These data become training sample sets.

Assumed fitted factors as followings

$$\begin{cases} X_1 = [\dot{\omega} \quad n] \\ X_2 = [\ddot{x} \quad \dot{x} \quad i_{q1} \quad i_{y2} \quad y] \\ X_3 = [\ddot{y} \quad \dot{y} \quad i_{q1} \quad i_{x2} \quad x] \end{cases} \tag{21}$$

By training input data based on inverse system of SVM decoupling theory, corresponding factor and threshold of input vector can be gotten. Inverse system of BSRM is identified by current input. The expression of the inverse system is as followings

$$\begin{cases} u_1 = \sum_{i=1}^{k} \left(a_i^{1^*} - a_i^1\right) K(x_i^1, x_1) + b_1 \\ u_2 = \sum_{i=1}^{k} \left(a_i^{2^*} - a_i^2\right) K(x_i^2, x_2) + b_2 \\ u_3 = \sum_{i=1}^{k} \left(a_i^{3^*} - a_i^3\right) K(x_i^3, x_3) + b_3 \end{cases} \tag{22}$$

whereu_1, u_2 and u_3 are input of BSRM, a^1, a^2, a^3, b_1, $b_2$and b_3 are respectively corresponding factor and threshold of u_1,u_2 and u_3.

The optimal solution of C and σ^2 are obtained by adopting ant colony algorithm, where C=4500, σ^2=36.

Affected by various factors in practice, the pseudo-linear compound system, which is formed by cascading with the inverse model (that is nonlinear system identification model based on SVM) before the original system, is not an ideal linear system. So using the pseudo-linear as accusing of the object, this paper designs an additional controller to compose a feedback control system. The closed-loop structure of the whole decoupled system is shown in Fig. 3.

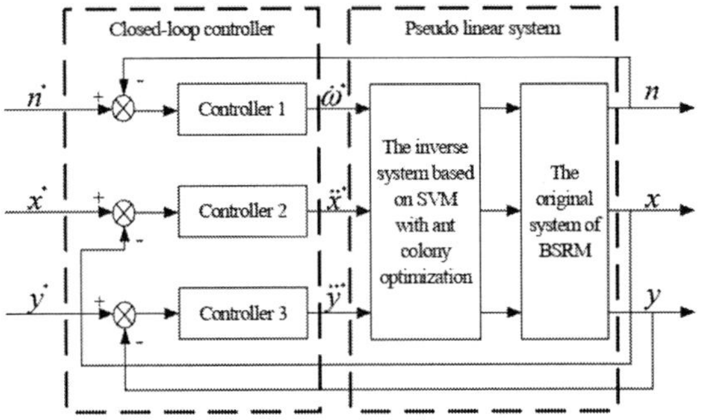

Figure 3. Structure diagram of control system

C. *Simulation Results*

In order to verify the effectiveness of the proposed decoupling control method in this paper, the simulation model is constructed based on Matlab/Simulink. The motor parameters are given as the followings: the motor power is 300W, the motor speed is 1500r/m, the rotor mass m=1kg, the rotor moment of inertia J = 0.002kg · m^2, air gap length δ_0= 0.25mm, the stator resistance R_{s1}=0.25 Ω, the number of the torque winding pole pair P_M=2, L_d=0.035H, L_q=0.0042H, the number of the suspension windings pole pair P_B=1, L_x=L_y=0.02H, the stator resistance R_{s2}=0.15Ω.

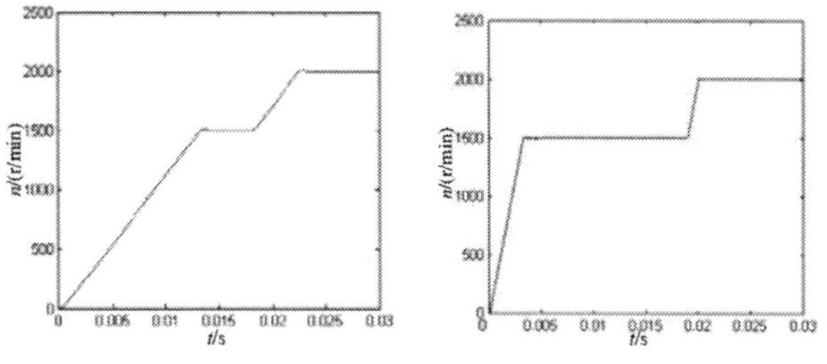

(a) Curve based on the inverse system (b) Curve based on the SVM

Figure 4. the speed tracking response curves.

To reveal its outstanding characters, this control method is used to compare to the inverse system method. The speed given tracking response curves is shown in Fig. 4. Fig. 4(a) show the simulation result of the inverse system control, and Fig. 4(b) show the simulation results of the SVM algorithm. As shown Fig. 4(a), the adjustment time of speed is about 0.013s when the motor run smoothly from 0 to 1500r/min and the overshoot is approximately 1.5%. When the time is 0.0018s, the speed accelerates suddenly from 1500r/min to 2000r/min. It can be seen from the adjustment time is 0.005s, and the overshoot is about 1%. Under the same condition, Fig. 4(b) show the adjustment time of speed is 0.004s when the motor starts to run smoothly, and the overshoot is approximately 0. In addition, when the motor accelerates suddenly from 1500r/min to 2000r/min, the adjustment of time is 0.002s, and the overshoot is also 0. Obviously adopting the method proposed in the paper, the system has good dynamic and static characters, quick response and high track precision.

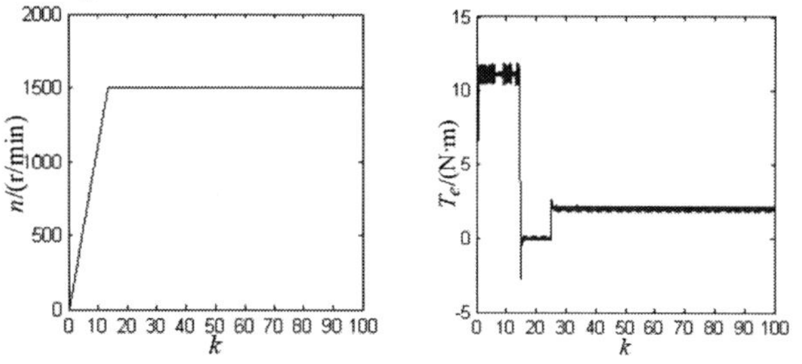

(a) Curve of speed (b) Curve of torque

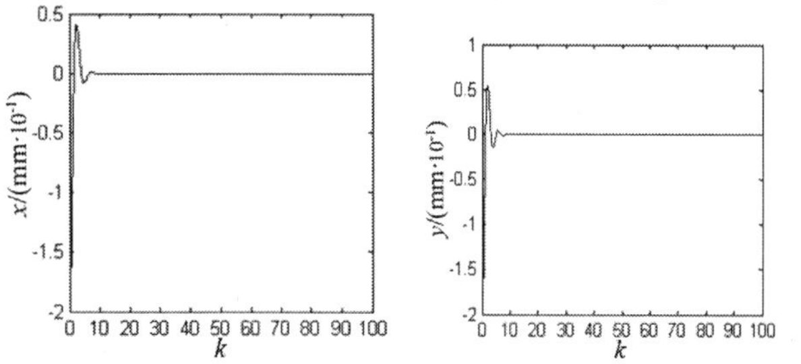

(c) Curve in the x-direction (d) Curve in the y-direction

Figure 5. Simulation results of decoupling control on BSRM.

The speed step response curve is shown in Fig. 5(a). As shown, the adjustment time of speed is small and the speed response is rapid. It shows the system has good speed performance. The torque response curve is shown in Fig. 5(b). As shown, when the motor starts in an unloaded state, the torque jumps to 0 after the motor running smoothly. And the torque T_e=2N·m is given when k=25, in the event of transient small fluctuations, the set value of torque can be quickly achieved. It can be seen that the torque subsystem has a good dynamic and static performance.

Fig. 5(c) and Fig. 5(d) show the displacement curves in the x- and y-direction when the rotor is suspended, respectively. The initial values of the rotor position are supposed that x_0 is -0.2 mm and y_0 is -0.2 mm. And the given reference value of the rotor position are supposed that x^* is 0 mm and y^* is 0 mm. As shown, the rotor reaches quickly given reference from starting-up and realizes stable suspension in central position. The overshoot and the steady-state error are approximately zero.

Fig. 6 shows Simulation results with the external disturbances and nonlinear parameters changing. When the step k is 55, 75 and 85, the strong disturbances n_d=300r/min, x_d=0.26mm, y_d=0.4mm are added respectively. Meanwhile, the original system parameters change suddenly into R_{s1}=10Ω, L_d=0.07H, L_q=0.008H. As shown in Fig. 6, when the speed, the radial displacements in the x- and y-direction is disturbed respectively, system gets back to the given value quickly. Moreover, the speed, the radial displacements in the x- and y-direction are almost not affected each other. It is shows that this control system has the good ability of restraint step disturbance and parameters disturbance, and the electromagnetic torque and the radial forces are decoupled completely.

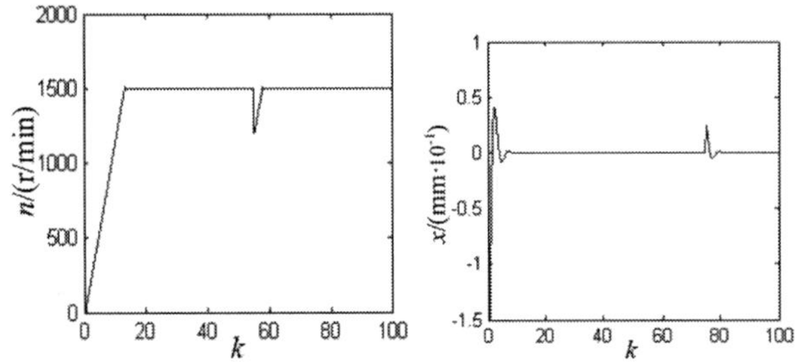

(a) Response curve of speed (b) Displacement curve in the x-direction

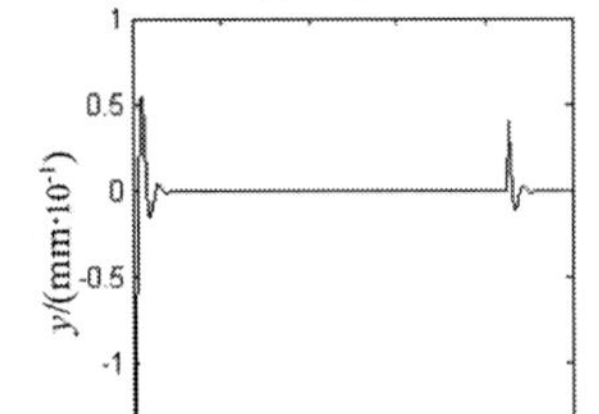

(c) Displacement curve in the y-direction

Figure 6. Simulation results with outside strong disturb and nonlinear parameter change.

The curves in Fig. 7 show simulation results with modeling errors. If the inverse system without the SVM training is connected with the original system in series, we can obtain a pseudo-linear system with modeling errors. Then this pseudo-linear system is decoupled control. The simulations are shown in Fig. 7, and it can be seen that the system existing modeling errors also has good control performance, rapid response and small steady-state errors.

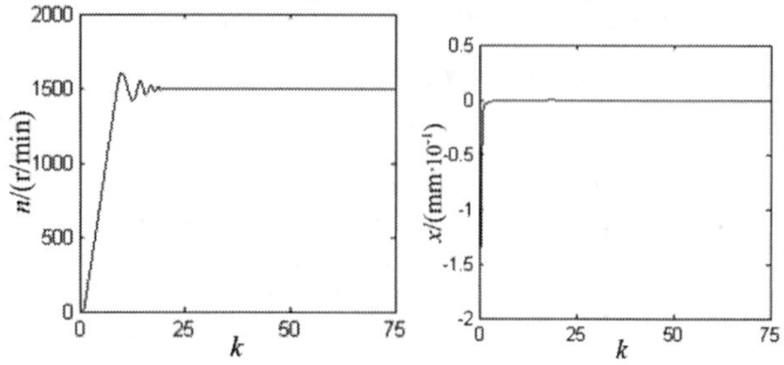

(a) Response curve of speed (b) Displacement curve in the x-direction

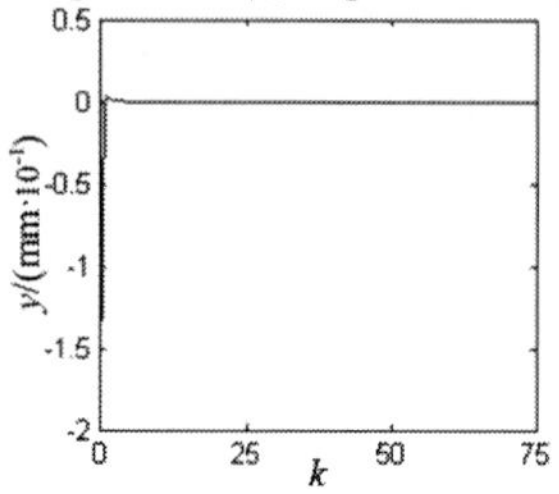

(c) Displacement curve in the y-direction

Figure 7. Simulation results with modeling errors

Ⅳ. CONCLUSIONS

Based on introducing principle and analyzing of reversibility of the BSRM system, the SVM algorithm with ant colony optimization is applied to realize decoupling control for the BSRM. The simulation results show that:

(1) Compared with inverse system control method, the control strategy based on support vector machine with ant colony optimization gets faster response and higher tracking precision to the given speed.

(2) When the external disturbances are in existence, simulation results demonstrate that the control strategy has strong ability of resisting disturbance, and realize dynamic decoupling control among the torque subsystem and the radial suspension force subsystem.

(3) For the original nonlinear system with parametric variation, the control method has good robustness.

(4) While the modeling errors exist, the simulation results show that the control system also possesses strong robusticity, and has a good control performance.

ACKNOWLEDGMENTS

This work is sponsored by National Natural Science Foundation of China (60974053), Natural Science Foundation of Jiangsu Province (BK2012707), Jiangsu University Achievements in Scientific Research Industrial Production Advancement Project(JHB2012-39), by Qing Lan Project and the Priority Academic Program Development of Jiangsu Higher Education Institutions.

REFERENCES

[1] A. O. Salazar, A. Chiba, and T. Fukao, "A review of developments in bearingless motors," *Proceedings of the 7th International Symposium on Magnetic Bearings*, Zurich, Switzerland, 2000.

[2] A. Chiba, T. Fukao, and O. Ichikawa, "*Magnetic Bearings and Bearingless Drives,*" Butterworth-Heinemann, London, 2005.

[3] L. Hertle, and W. Hofmann, "Theory and test results of a high speed bearingless reluctance machine," *Proceeding of Power Conversion Intelligent Motion*, Chemnitz, Germany, 1999.

[4] C. Michilka, T. Sakamoto, O. Ichikawa, and et al, "A decoupling control method of reluctance-type bearingless motors considering magnetic saturation," *IEEE Trans. Industry Application*, vol. 32, no. 5, pp. 1204-1210, 1996.

[5] H. N. Zhang, H. Q. Zhu, Z. B. Zhang, and et al, "Design and simulation of control system for bearingless synchronous reluctance motor, " *Proceedings of 8th International Conference on Electrical Machines and Systems*, Nanjing, China, 2005.

[6] L. Hertle, and W. Hofmann, "Magnetic couplings in a bearingless reluctance machine," *Proceeding of the International Conference on Electrical Machines*, Helsinki, Finland, 2000.

[7] F. H. Song, "*Study on support vector machine inverse system method and its application,*" PhD thesis, Zhejiang University, 2006.

[8] V. Vapnik, *The nature of statistical learning theory,*" New York, America, 1995.

[9] C. Cortes, and V. Vapnik, "Support vector networks", *Machine Learning,* No. 20, pp. 273-297, 1995.

[10] P. L. Zhang, L. F. Qian, J. J. Cao, and et al, "Parameter optimization of support vector machine based on ant colony optimization algorithm," *Journal of Nanjing University of Science and Technology (Natural Science)*, vol. 33, no. 4, pp. 464-468, 1999.

[11] H. B. Duan, "*The Principle and Application of the Ant Colony Algorithm,*" Beijing, Science Press, 2005.

Mass Eccentricity Comepensation Research of Bearingless Switched Reluctance Motor

Jie Chen[a], Zhiquan Deng[a], Yan Yang[b], Xin Cao[a], Qiang Ding[a]

[a] Department of Electrical Engineering, Nanjing University of Aeronautics and Astronautics, Nanjing 210016, China
[b] Department of Electrical Engineering, Nanjing University of Posts and Telecommunications, Nanjing 210013, China
dzq@nuaa.edu.com

***Abstract*—Due to the mechanical imbalance, the problem common to mechanical system, including the bearingless switched reluctance motor (BSRM), is the vibration caused by mass eccentricity. The mass imbalance leads to the additional unbalanced radial magnetic force and the flexural vibration. The vibration will be transmitted to the stator and housing case, which produces noise, influences the operation performance and limits the improving of speed. Because of the unbalance radial force, the suspension windings current and the load of the power devices increases, affecting the performance of the whole system. In this paper, how the mass eccentricity influences the suspension in bearingless motor is analyzed, and a compensation method based on the coordination transformation with one-order filter is designed to eliminate the vibration between stator and rotor. However, affection of analog PID in the experimental platform should be considered correctly. The simulation and experiment results have validated the effectiveness and the advantages of the method.**

I. Introduction

Bearingless switched reluctance motors (BSRMs) not only have superior performance of switched reluctance motors (SRM) due to the inherent advantageous features as simple construction, convenient maintenance and low cost, but also have good features of bearingless motors as high speed and low loss. The usage of the bearingless technology could broaden application of SRM. Because of its doubly salient structure, a BSRM produces the attraction magnetic force between stator and rotor poles. The force can be divided into tangential and radial components. The tangential force component compels the rotor to rotate, and the radial force component suspends the rotor without contact with bearing. However, the radial force also brings along the problem of noise and vibration [1]-[5].

In practical application, the machinery problems exist inevitable, such as the uneven distribution of the rotor lamination quality, radial displacement sensor measurement error, rotor shaft bending deformation and assembling processing. The problems and conditions lead to the rotor mass eccentricity and the principle axis of inertia is not coincided with the axis of geometry, producing the centrifugal force which is proportional to the square of the angular speed. Therefore, even very small eccentricity would produce large centrifugal force in high speed condition. When such rotor rotates about its geometric axis, the unbalanced radial force caused by acceleration of the inertial center is much larger. The unbalanced radial force would be transmitted to the stator or housing of the machine, causing the vibration of rotor.

For more than 20 years, the compensation techniques has experienced too much, from notch filter-based designs, model-based observer designs and off-line least square approximation designs to the modern adaptive method as LMS arithmetic. According to the objects that needed to be compensated, the strategies could be divided into two different kinds, compensating the unbalanced force and compensating the unbalanced displacement [6]-[12]. In order to reduce the generation of the vibration, we employ the compensation method to make the rotor of BSRM spin around its inertial axis and eliminate the radial, centrifugal force eventually.

The coordination transformation method is proposed in this paper to eliminate the synchronous signal with same frequency as speed and make the rotor revolve around the inertial axis with the features including easy implementation, negligible effect of the speed changing and compensation in a minimum energy state. As long as the air-gap of the auxiliary bearing is sufficiently large, the rotor can spin with little vibration transmitted to the stator or housing of the motor, reducing the vibration and noise. The compensation method is verified on Matlab/Simulink and the experiment platform to elaborate the features. Comparing to the former adaptive arithmetic, the result shows that the method can be less affected by the system parameters and cannot be limited by speed.

II. Unbalanced Vibration Analysis in BSRM

A. The Principle of BSRM

Fig.1 shows an A-phase stator winding configuration and the principle of radial force production in a double winding BSRM of 12/8 structure. In double winding BSRM, a set of suspension winding is added to the torque winding in the motor stator. The magnetic field generated by the windings is used to produce the radial force for suspension. The windings of each phase conduct in turns and the system is controlled by the rotor displacement closed-loop to ensure that the motor rotates and levitates accurately [5].

Considering that the magnetic attraction forces are similar in each phase during whole period, the theoretical formula of the forces of the A-phase are shown as following when considering rotor displacement [10].

$$\begin{aligned} F_\alpha &= k_1 i_{ma}{}^2 \alpha + k_2 i_{sa1}{}^2 \alpha \\ F_\beta &= k_1 i_{ma}{}^2 \beta + k_2 i_{sa2}{}^2 \beta \end{aligned} \tag{1}$$

This work was supported by the National Natural Science Foundation of China (51207073) (51107056).

Where, F_α, F_β are the magnetic attraction forces produced by i_{ma}, i_{sa1}, i_{sa2} in the α- and β-direction, respectively. k_1 and k_2 are the proportional coefficient of the magnetic attraction force. k_1, k_2 can be written as,

$$k_1 = N_m^2 \mu_0 hr(\pi - 12|\theta|)/6l_0^2 \quad (2)$$

$$k_2 = N_s^2 \mu_0 hr(\pi - 12|\theta|)(\pi + 12|\theta|)/12\pi l_0^3 \quad (3)$$

Where, N_m is the number of turns of the motor main winding, N_s is the number of the number of radial force winding. Similarly, the magnetic attraction forces of the B- and C-phases can also be derived.

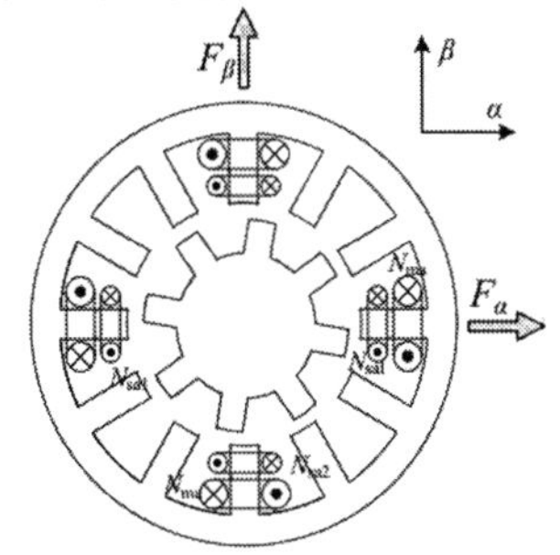

Figure 1. Winding strcuture of A phase in BSRM

B. Study of Rotor Dynamics

When the eccentric mass is ignored and the winding conducts, the center of the rotor is coincided with gravity center. There is no unbalanced magnetic force between stator and rotor when the rotor is stable. So the rotor is controlled to suspend stably in the geometric center O. However, the eccentricity exists exactly between the mass center C and geometric center O, as shown in fig.2, and the rotor will deviate from the geometric center along the direction of offset during the spinning. This makes the distance between rotor and stator instable, and unbalanced magnetic force is generated. Therefore the eccentricity of the rotor becomes larger and suspension effects become poor.

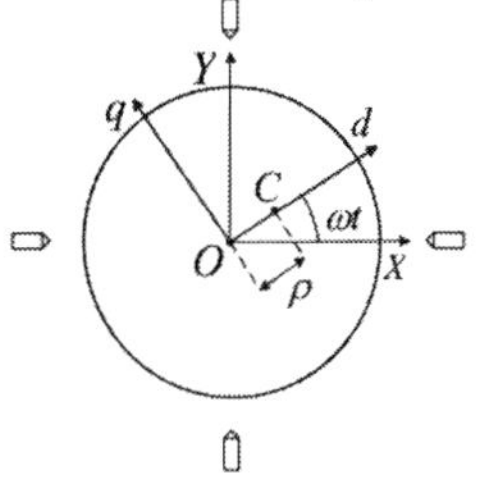

Figure 2. Mass eccentricity of rotor in BSRM

During the whole period, the dynamic equations of BSRM with gravity load are shown as following when the swirl effect is not considered.

$$m\ddot{\alpha} + c_\alpha \dot{\alpha} + k_\alpha \alpha = m\rho\omega^2 \cos(\omega t + \varphi) + F_\alpha \quad (4)$$

$$m\ddot{\beta} + c_\beta \dot{\beta} + k_\beta \beta = m\rho\omega^2 \sin(\omega t + \varphi) + F_\beta - mg \quad (5)$$

Where, (α, β) is the coordination of geometric center axis in α- and β-direction. c_α and c_β are the coupling coefficient, k_α and k_β are rotor rigidity coefficient. m is the mass of rotor.

The unbalanced magnetic force is related to both the displacement of the rotor eccentricity and the winding current. On the other hand, the centrifugal force generated by eccentricity is proportional to the square of the speed and the force is larger when the speed goes up. According to the equation (1)-(5), the solution of the steady state response is shown as,

$$\begin{aligned} \alpha &= \alpha_0 \cos(\omega t - \chi_\alpha) + \Delta\alpha \\ \beta &= \beta_0 \sin(\omega t - \chi_\beta) - m/k_\beta + \Delta\beta \end{aligned} \quad (6)$$

Where, χ_α, χ_β and k_β are related to the stiffness coefficient of rotor, and $\Delta\alpha$, $\Delta\beta$ are related to the levitation force F_α, F_β. The response is the displacement of rotor.

C. Production Mechanism of Vibration on Stator

Fig.3 shows the displacement close-loop control system and the rotor would rotate around the geometric center under the feedback regulation when the reference displacement x^* and y^* are set as zero.

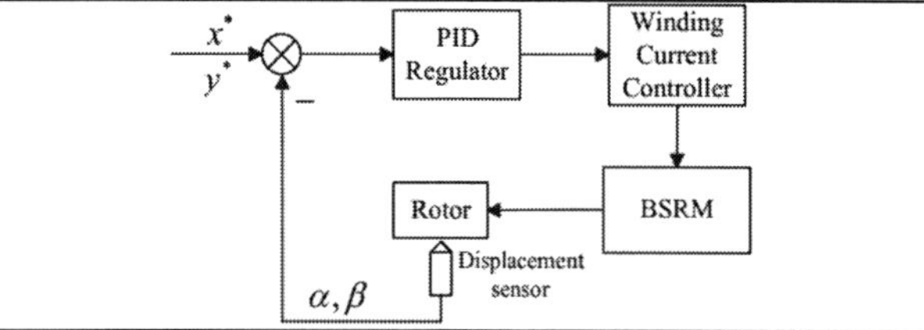

Figure 3. Dispalcement close-loop control diagram

Due to the mass eccentricity, the rotor has the tendency to deviate from the geometric center under the action of centrifugal force. However, the displacement closed-loop in the system makes the controller take the unbalanced displacement under control. So the rotor rotates around the geometric center and the displacement of rotor will be controlled very small. As a result, the force generated by the controller increased and it will be passed to the stator. Then the force aggravates vibration and noise inevitably. In order to reduce the vibration and noise, the coordination transform filter is needed to compensate the unbalanced force.

III. MASS ECCENTRICITY COMPENSATION BASED ON COORDINATION TRANSFORMATION

A. The Principle of Mass Eccentricity Compensation

The radial displacement closed-loop control block of BSRM is shown in fig.3, and the coordination transform filter is added to the block. Where, the displacement signal x_s, y_s contains random displacement and unbalanced displacement that with same frequency as speed. The unbalanced vibration displacement is transformed to direct component by matrix T, which could be found in x_{r1} and y_{r1}. The random displacement can be removed by the filter and the direct component reserves. After the transformation matrix T^{-1}, the direct component is transformed to synchronous sinusoidal signal, and the signal is used to compensate the unbalanced displacement. The given rotor displacement x^* and y^* is usually set as zero to ensure suspension stability. So there is no sinusoidal signal with same frequency as speed in the input of the controller.

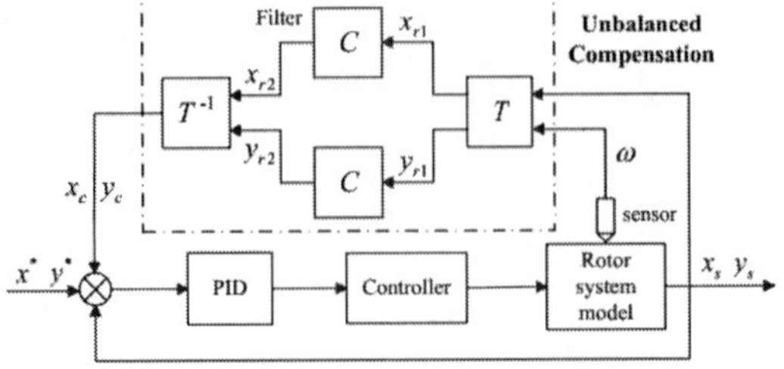

Figure 4. Coordination transformation control compensation diagram

The coordination transformation matrix T is shown as,

$$T = \begin{bmatrix} \cos \omega t & \sin \omega t \\ -\sin \omega t & \cos \omega t \end{bmatrix} \tag{7}$$

The transfer function of the one-order low pass filter C could be designed as,

$$G(s) = 1/\left[(1/2\pi\Delta f)s + 1\right] \tag{8}$$

Where, the cut-off frequency Δf is related to the frequency of the signal that needed to remove.

After the unbalanced compensation, the synchronous component would not be in the reference force generated by the PID regulator. So, the centrifugal force that worsens the vibration and noise of stator won't be controlled and the rotor would rotate around the axis of inertia under the action of centrifugal force. Therefore, the vibration on stator would be reduced significantly.

B. Influence of PID to the Compensation Method

According to the traditional experiment platform, the digital PID is not very effective due to the difficulties in adjusting parameters of PID. So the analog PID is needed for adjusting in time. The process of compensating the displacement directly may occupy more resources of the digital controller and affects the efficiency of the system. So the influence of PID to the compensation method should be considered without changing overall design of the platform.

According to the equation (6), the synchronous sinusoidal signal can be set as following without considering the influence of the random displacement.

$$x_s = \rho \cos \omega t \qquad y_s = \rho \sin \omega t \tag{9}$$

Due to the influence of the analog PID, phase-lead or phase-hysteresis, we need to analyze the influence of the PID. After the regulation of the PID, the displacement of α- and β direction is shown as,

$$x_{r1} = k_p \rho \cos \omega t + (k_i/\omega - k_d \omega)\rho \sin \omega t \tag{10}$$

$$y_{r1} = k_p \rho \sin \omega t - (k_i/\omega - k_d \omega)\rho \cos \omega t + k_i \rho/\omega \tag{11}$$

Where, k_p, k_i and k_d is the value of parameters in PID.

After the transformation, the compensation component needed is obtained as,

$$x_c = k_p \rho \cos \omega t - (k_d \omega - k_i/\omega)\rho \sin \omega t + b \sin \gamma \tag{12}$$

$$y_c = k_p \rho \sin \omega t + (k_d \omega - k_i/\omega)\rho \cos \omega t + b \cos \gamma + k_i \rho/\omega \tag{13}$$

So the displacement added compensation in the signal is shown as following.

$$x_0 = x_s - x_c = -b \sin \gamma \qquad y_0 = y_s - y_c = k_i \rho/\omega - b \cos \gamma \tag{15}$$

According to the formula (14) and (15), the offset exists in the displacement signal after compensation. And the bias is only related to the parameters b, k_i and γ. So the bias can be removed by setting the parameters. The parameters of the filter and PID should be chosen appropriately to obtain the result as, $x_0 = x_s - x_c = 0$, $y_0 = y_s - y_c = 0$. So the sinusoidal interference caused by mass eccentricity can be removed basically. As a result, the rotor rotates around the principle axis, reducing vibration and noise in stator and frame.

In order to realize the target, $b \approx 0$ and $\cos\gamma \approx 0$, Δf in the transfer function should be regulated. Due to the low speed, the frequency of random signal needed to remove is low. So the parameter Δf can be set as $1Hz$.

During the experiment and regulating the parameters of PID, we can find that the parameter of integration part k_i has less affection on suspension of the rotor. So the parameter k_i can be set to a very small value.

The compensation method is less affected by PID regulator. Considering that the design of the parameters is no related to the speed, the coordination transformation compensation model is not sensitive to the changing of speed and there is no need to adjust the parameters to the speed. Thus, the model has very good adaptability.

C. Simulation Analysis

In this section, the compensation control system is redesigned. The system model of BSRM is built in Matlab/Simulink and the model contains the block of the motor, levitation force control block, torque control block, the main winding current and suspension winding current control block, the three sets of power inverters and other auxiliary circuit related. The simulation parameters of the system are given in Table Ⅰ. The unbalanced compensation control diagram is shown in fig.3.

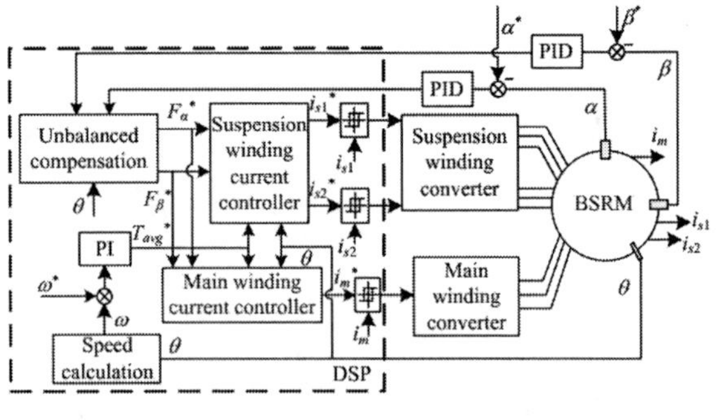

Figure 5. Block diagram of control system

Considering that the compensation of the displacement functions after the analog PID, what the coordination transformation filter dealing with is the unbalanced parts in the reference levitation force. And the synchronous sinusoidal signal that we need to compensate is added to the output of the displacement sensor to simulate the vibration caused by mass eccentricity.

The reference levitation force is the radial force that given to the rotor from stator by changing the wingding current. So, the amplitude of the reference levitation force given to the controller could respect the radial force between stator and rotor and compensation effect straightly. And the synchronous signal as the same frequency of speed in the reference force is caused by centrifugal force.

Simulation results are presented in fig.5 to demonstrate the effect of the compensation control on BSRM. For simulation, the nonlinear plant model in equation (1) and (2) are used. When in the conditions of different speed, the synchronous sinusoidal component caused by centrifugal force is removed and the amplitude of the reference force decreases. Hence, the force passed to the stator and frame reduces and the vibration and noise decreases actually.

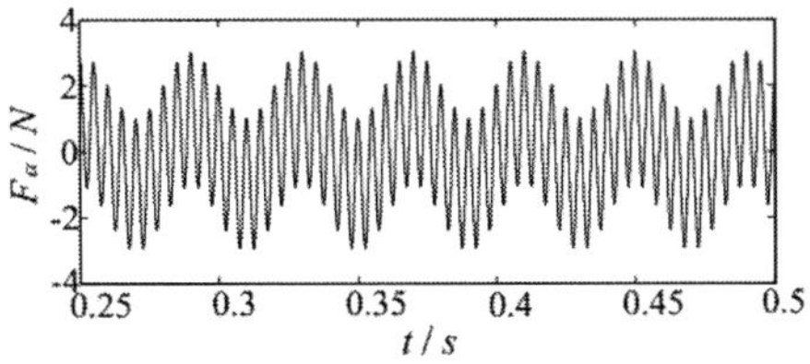

(a) n=1500rpm (Before compensation)

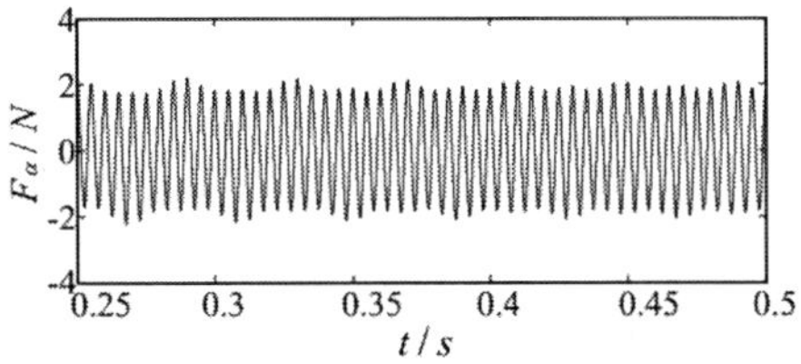

(b) n=1500rpm (After compensation)

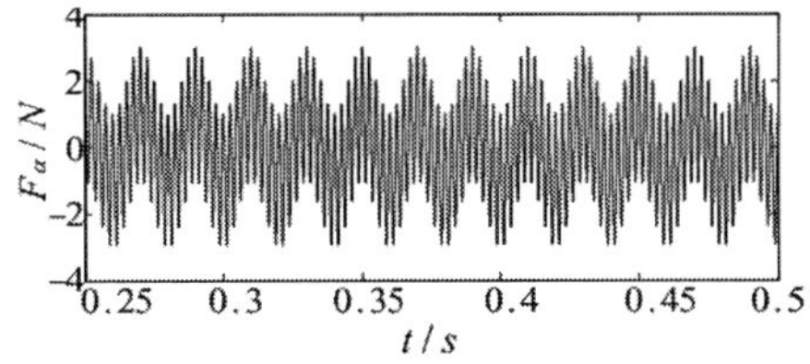

(c) n=3000rpm (Before compensation)

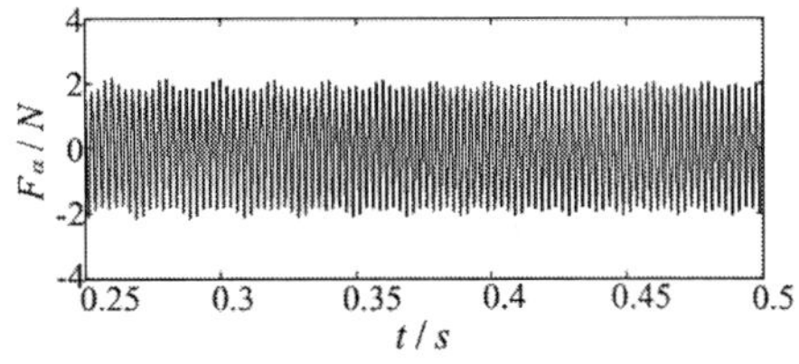

(d) n=3000rpm (After compensation)

Figure 6. Comparision of radial force in α direction before and after compensation

IV. EXPERIMENTS ANALYSIS

The results of above theory and simulation analysis were verified on the 12/8 doubling winding BSRM. The distance of the eccentricity is unknown. The dimensions of the test BSRM researched in this paper are shown in Table Ⅰ and the control system diagram is shown as fig.4. The signal of speed is achieved by photoelectric sensor. The motor is placed horizontally to apply the rotor's weight as an external force in β-axis. As the air gap length between suspension rotor and auxiliary protect bearing is only 0.2mm, the unbalanced vibration displacement amplitude should be less than 0.2mm.

TABLE I. DIMENSIONS OF THE MOTOR

Parameter	Value
Number of turns of main winding, N_m	14 turns
Number of turns of suspension winding, N_b	17 turns
Average air gap length between stator and rotor, l_0	0.25 mm
Air gap length between rotor and shaft	0.2 mm
Inner diameter of rotor	30 mm
Radius of rotor pole, r	30 mm
Stack length lamination, h	75 mm
Rated output power	6.5KW
Rated main winding current	28A
Rated suspension winding current	15A
Weight of rotor	5kg

Fig.6 shows the waveforms of rotor displacement and the reference levitation force before and after the unbalanced compensation. In fig.6 (a) and fig.6 (b), the role of the proposed method is shown effectually. And fig.6 (c) and fig.6 (d) are demonstrated to see the figures clearly. The peak-to-peak value of the reference levitation force F_α in fig.6 (a) and fig.6 (c) is 3.2V. Though, the peak-to-peak value of the force F_α in fig.6 (b) and fig.6 (d) is 1V. So making comparison with the performance of BSRM without compensation, the reference levitation force decreased significantly when the rotor was rotating around the axis of inertia. We can see that the sinusoidal component caused by centrifugal force is almost removed from the reference levitation force. So the unbalanced force caused by mass eccentricity is not controlled in the system. As a result, the force passed to the stator decreases and the vibration and noise on the stator may become smaller.

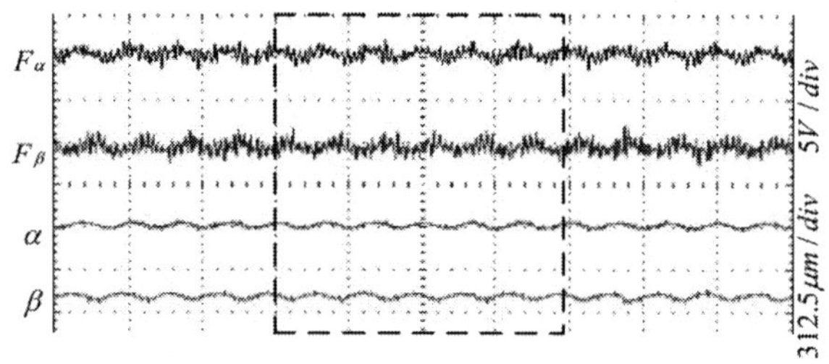

(a) before the compensation (100ms/div)

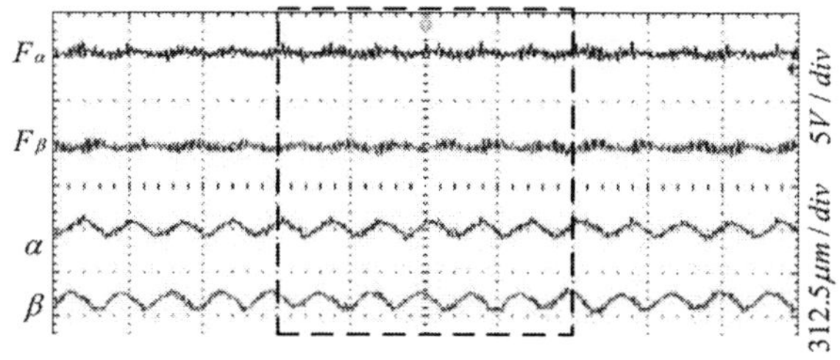

(b) after the compensation (100ms/div)

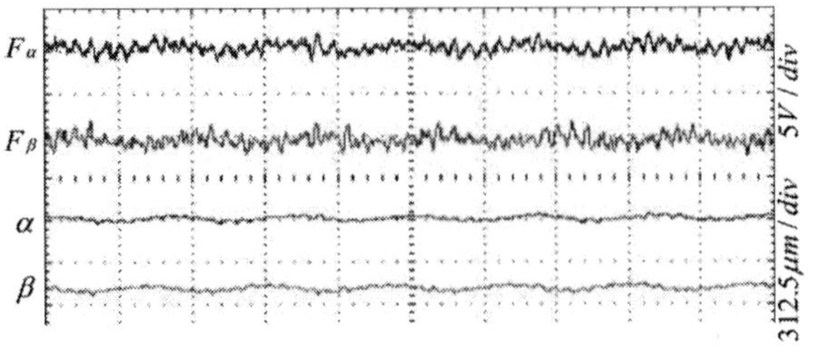

(c) before the compensation (40*ms/div*)

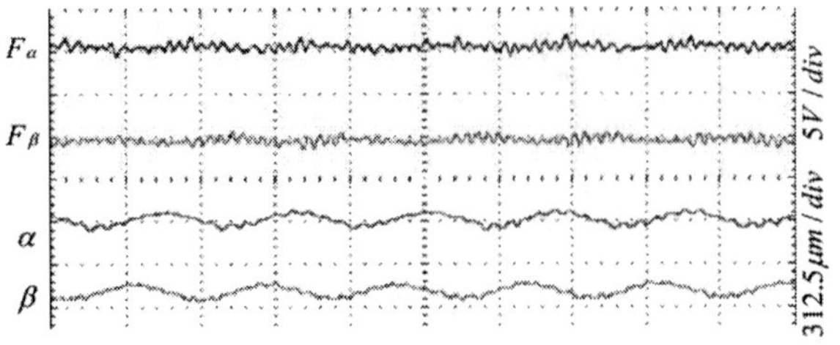

(d) after the compensation (40*ms/div*)

n=900rpm

Figure 7. waveforms of the reference levitation force and displacement

V. Conclusion

In this paper, a mass eccentricity compensation method is proposed practically. The compensation method based on the coordination transformation filter calculates the sinusoidal component caused by centrifugal force. And the sinusoidal signal is removed from the winding current controller and the centrifugal force won't be controlled in the system. As a result, the rotor of BSRM rotates around the axis of inertia after compensation. The method used in this paper is simple and less resource-intensive. The algorithm is adaptive to the speed and parameters of system. Therefore, the unbalanced force caused by mass eccentricity of BSRM rotor is compensated effectually. The peak-to-peak value of the reference levitation force from the suspension winding is decreased. Therefore, the force passed to the stator and frame decreases. The utility of the proposed method has been verified by the experimental results applied on the platform. The compensation method can eliminate the unbalanced vibration in the condition of different speed and different eccentricity.

References

[1] Takemoto M, Shimada K, Chiba A, et al. A design and characteristics of switched reluctance type bearingless motor [C]. Fourth International Symposium on Magnetic Suspension Technology, Gifu, Japan, 1998:49-63

[2] Takeshi Mizuno. An unified approach to controls for unbalance compensation in active magnetic bearings [C]. The 1998 IEEE International Conference on Control Applications. Trieste, Italy. 1998:1063-1067.

[3] Amrhein W, Silber S, Nenninger K. Developments on bearingless drive technology [C]. 8th International Symposium on Magnetic Bearings. Mito, Japan, 2002:229-234.

[4] Choi B B, Siebert M. A bearingless switched reluctance motor for high specific power applications [C]. 42nd AIAA/ASME/SAE/ASEE Joint Propulsion Conference and Exhibit. Sacramento, CA: AIAA, 2006:1-14.

[5] Cao X, Deng Z, Yang G, et al. Independent control of average torque and radial force in bearingless switched reluctance motors with hybrid excitations [J]. IEEE Transactions on Power Electronics, 2009, 24(5):1376-1385.

[6] S. Beale, B. Shafai, et al. Adaptive force balancing for magnetic bearing control systems [C]. 31st Conference on Decision and Control. Tucson, Arizona, 1992: 3535-3539.

[7] Abdelfatah M. Mohamed, Ilene Busch-Vishniac. Imbalance compensation and automatic balancing in magnetic bearing systems using the Q-parameterization theory [C]. The American Control Conference. Baltimore, Maryland. 1994:2952:1957.

[8] Raoul Herzog, Philipp Buhler, et al. Unbalanced compensation using generalized notch filters in the multivariable feedback of magnetic bearings [J]. IEEE Transaction on Control Systems Technology, vol. 4, No. 5, 1996:580-587.

[9] Takemoto M, Chiba A, Fukao T. A feed-forward compensator for vibration reduction considering magnetic attraction force in bearingless switched reluctance motors [C]. 7th International Symposium on Magnetic Bearings. ETH Zurich, 2000:395-400.

[10] Tao Zhang. Adaptive vibration compensation control research on rotor in active magnetic system [J]. International conference on measuring technology and mechatronics automation, 2010: 428-431.

[11] Wu C Y, Pollock C. Analysis and reduction of vibration and acoustic noise in the switched reluctance drive [J]. IEEE Trans. on Industry Applications, 1995, 31(1):91-98.

[12] Vijayakumar K, Karthikeyan R, Rajkumar S, et al. An investigation into vibration in high speed switched reluctance motor with soft magnetic composite material[C]. IEEE Region 10 Colloquium and the Third ICIIS, Kharagpur, India, 2008.

Eccentricity Analysis and Efficiency Estimation of 12/14 Hybrid Stator Pole Bearingless SRM

Zhenyao Xu, Jin-Woo Ahn

Kyungsung University, Busan, 608-736, Korea, zhenyao87@163.com

***Abstract*—The radial force to suspend the rotor in a constant air-gap for bearingless switched reluctance motor (BLSRM) is dependent on air-gap and force current. Due to the non-linear characteristic of inductance, the radial force is much affected by the rotor eccentricity. In order to keep constant air-gap of rotor, the nonlinear characteristics of eccentricity should be analyzed. Further, the auxiliary bearing is always used to restrict the maximum eccentricity of the rotor in industry. In this paper, the rotor eccentricity effect in the proposed 12/14 BLSRM is analyzed. Meanwhile, the efficiency of the 8/10 and proposed BLSRMs is estimated. The analysis results further verify the validity of the proposed BLSRM.**

I. INTRODUCTION

Switched reluctance motors (SRMs) have simple structure and inherent mechanical strength without rotor winding and permanent magnet, which make SRMs suitable for operating in harsh environments, e.g. high temperature and high speed applications [1-2]. Moreover, in SRM, a significant amount of magnetic attraction is generated in the radial direction, because switched reluctance motors generally have short air gap lengths in order to effectively produce rotational reluctance torque. It is quite possible to take advantage of this inherent large magnetic attraction for rotor shaft magnetic suspension. Therefore, SRMs could be a strong candidate for beringless motors.

Recently, several structures of bearingless switched reluctance motor (BLSRM) have been proposed [3-9]. But all of them are based on general SRM structure, the torque control is coupled with the suspending force control. A BLSRM with hybrid stator poles is proposed [1]. Compared with conventional BLSRM, the suspending force performance is improved, but in this structure, only half of the stator poles are used for the torque, output power density is very low. Moreover, in the motor long flux paths are present and flux reversal exists in the stator core, which increases the magneto motive force (MMF) requirements and leads to higher core losses.

To improve the performances of BLSRM, a novel 12/14 hybrid stator pole type BLSRM is proposed [10]. The proposed structure has separated torque and suspending force poles, the torque control can be decoupled from the suspending force control. Moreover, due to the short flux path without any reversal flux in the stator core, compared to the 8/10 hybrid stator pole type, the output torque is significantly improved and the air-gap is easier to control. But in this paper the effect of rotor eccentricity is not considered yet.

As is generally known, the rotor of the proposed BLSRM cannot be supported by mechanical bearing. So, the rotor can suffer displacement and eccentricity effects caused by a disturbance load and force. Rotor eccentricity, which changes rotor position in air gap, greatly influences the distribution of magnetic flux in air gap and causes a great variation in torque and force. Furthermore, the variations of torque and force may influence the suspension performance of BLSRM. In this paper, the eccentricity effect of proposed BLSRM is analyzed. Meanwhile, the efficiency of the proposed BLSRM is estimated. The validity of the proposed structure is verified by the analysis results.

II. A NOVEL 12/14 HYBRID STATOR POLE TYPE BLSRM

A. Structure of 12/14 BLSRM

A novel 12/14 hybrid stator pole type BLSRM with short flux paths and no flux reversal in the stator is proposed in Fig. 1. The proposed structure has separated torque and suspending force poles. Windings on the torque poles P_{A1}, P_{A2}, P_{A3} and P_{A4} are connected in series to construct phase A, and windings on the torque poles P_{B1}, P_{B2}, P_{B3} and P_{B4} are connected in series to construct phase B. The x-direction suspending force is generated by currents flowed in the suspending force poles P_{xp} and P_{xn}. Such as, the current i_{xp} in the stator pole P_{xp} generates positive x-direction suspending force, and the current i_{xn} in the stator pole P_{xn} can generate the negative x-direction suspending force. Similarly, the suspending forces for the y-direction can be generated by the

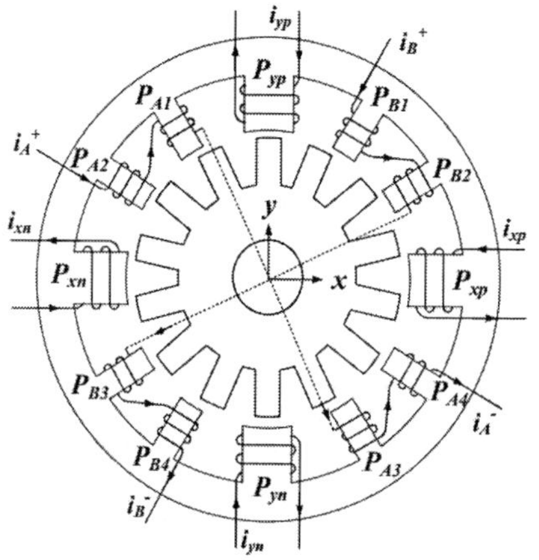

Figure 1. Basic structure of 12/14 BLSRM.

currents i_{yp} and i_{yn} which flow in suspending force poles P_{yp} and P_{yn}, respectively. Meanwhile, in order to get the continuous suspending force, the suspending force pole arc is selected not to be less than one-rotor pole pitch.

B. *Torque and Suspending Force Model*

Fig. 2 shows the attraction force produced by one stator and rotor poles. Because the magnetic flux passes through the overlapped area as well as the non-overlapped area due to fringing, the inductance of this pair of poles was modeled in [5] as,

$$L = \frac{\mu_0 N^2 L_{stk} r}{g}\left(\theta_o + K_f\right) \tag{1}$$

where, μ_0 is the permeability of the air, N is the number of coil turns, L_{stk} is the motor stack length, r is the rotor radius, g is the air gap length, K_f is a constant for the fringing inductance, θ_o and θ_{uo} are the overlapping and the non-overlapping angle, respectively.

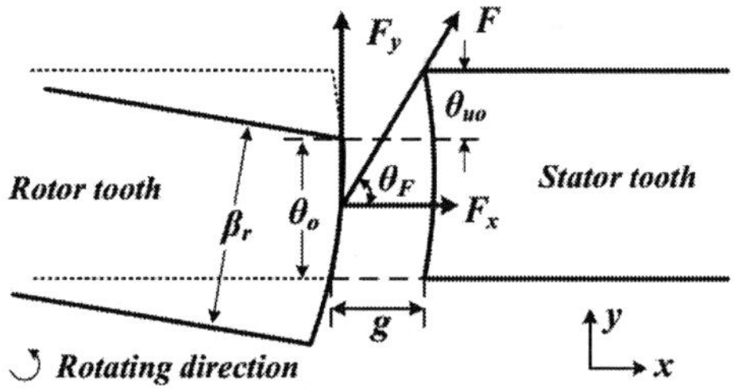

Figure 2. Attraction force produced by one stator and rotor poles.

From (1), the torque and radial force of the pair of poles can be approximately calculated as,

$$T = \frac{1}{2} i^2 \frac{dL}{d\theta} = \left(\frac{\mu_0 N^2 L_{stk} r i^2}{2g}\right) \cdot \frac{d\theta_o}{d\theta} \tag{2}$$

$$F = \frac{1}{2} i^2 \frac{dL}{dg} = \left(\frac{\mu_0 N^2 L_{stk} r i^2}{4g^2}\right) \cdot \left(\theta_o + K_f\right) \tag{3}$$

From (2) and (3), it can be seen that the torque is inversely proportional to the air-gap length and the suspending force is inversely proportional to the square of the air-gap length. Thus, any decrease in air-gap results in an increase in the electromagnetic torque and suspending force, vice versa.

III. Eccentricity Analysis of 12/14 BLSRM

As is generally known, the rotor of the proposed BLSRM cannot be supported by mechanical bearings. So, the center position of the rotor can suffer displacement and eccentricity effects caused by a disturbance load and force. However, the maximum rotor eccentricity is restricted by the capture bearing at *g/2*. Fig. 3 shows the possible rotor eccentricity in the proposed BLSRM. The output torque and suspending forces of the torque and suspending force winding can be changed according to the air gap displacements.

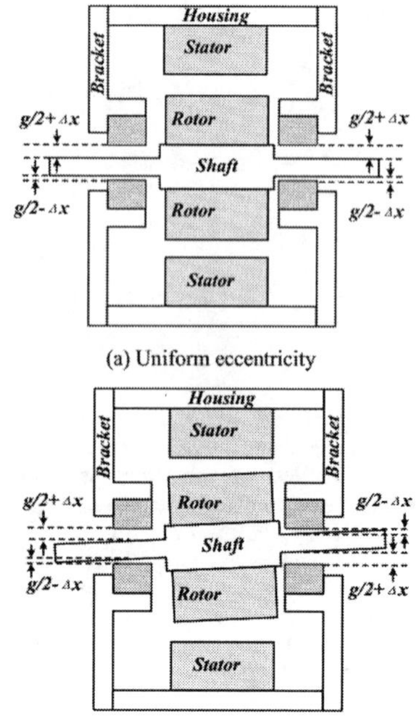

(a) Uniform eccentricity

(b) Nonuniform eccentricity

Figure 3. Rotor eccentricity in 12/14 BLSRM.

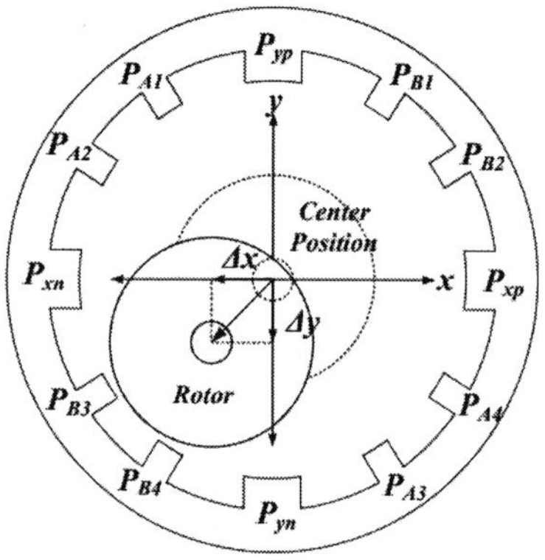

Figure 4. Uniform eccentricity of rotor in 12/14 BLSRM.

To get the eccentricity characteristics of the proposed structure, such as torque and suspending force, the proposed structure is analyzed by the software MATLAB combined with the FEA software, finite element method magnetics (FEMM).

A. *Torque Characteristics*

Fig. 4 shows the air-gap displacement according to the rotor eccentricity. When the rotor has eccentricity displacement in the negative x-direction, the output torques of the torque and suspending force winding with a constant value for various eccentric displacements are shown in Fig. 5. As shown in Fig. 5 (a), the peak value of the output torque increases slightly according to the rotor eccentricity. It is because the air gap distribution changes with the rotor eccentricity. That is, the air gap between stator poles P_{A1}, P_{A2} and the rotor is decreased, while the air gap between stator poles P_{A3}, P_{A4} and the rotor is increased. According to (2), it can be known that the torque generated by the windings on stator poles P_{A1} and P_{A2} increases and decreases under the stator pole P_{A3} and P_{A4}. However, the amount of increase in the torque generated by the windings on stator poles P_{A1}, P_{A2} is greater than that of decrease in the toque generated by the windings on stator poles P_{A3}, P_{A4}. Hence, the total torque will be increased.

For instance, when the rotor is in normal position, the air gap is g, the torque generated by the windings on stator poles P_{A1} and P_{A2} is equal to T and the torque generated by the windings on stator poles P_{A3} and P_{A4} is equal to T. Accordingly, the total torque is $2T$. When the rotor has a eccentricity of $g/2$ towards the center of the negative x-direction and the positive y-direction, the air-gap between stator poles P_{A1}, P_{A2} and the rotor is $g/2$, and the air-gap between stator poles P_{A3}, P_{A4} and rotor is the $3g/2$, so that if ignore the effect of saturation and fringing flux, according to (2), it can be found that the torque generated by the windings on stator poles P_{A1} and P_{A2} is $2T$, while the torque generated by the windings on stator poles P_{A3} and P_{A4} is $2T/3$. Accordingly, the total torque is $8T/3$, which is larger than $2T$. Therefore, the total torque is increased.

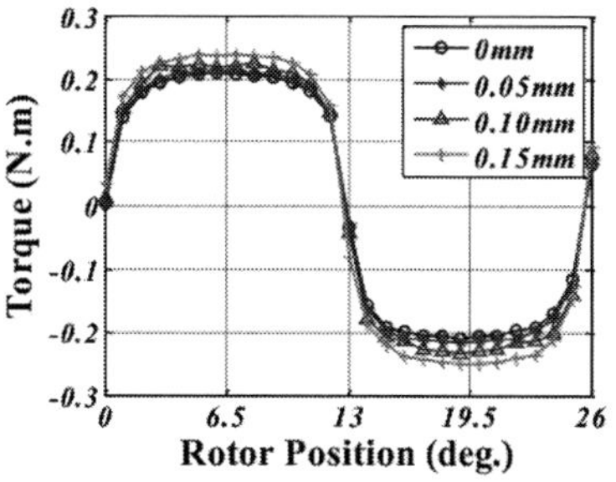

(a) Torque of torque winding at i_A=2A.

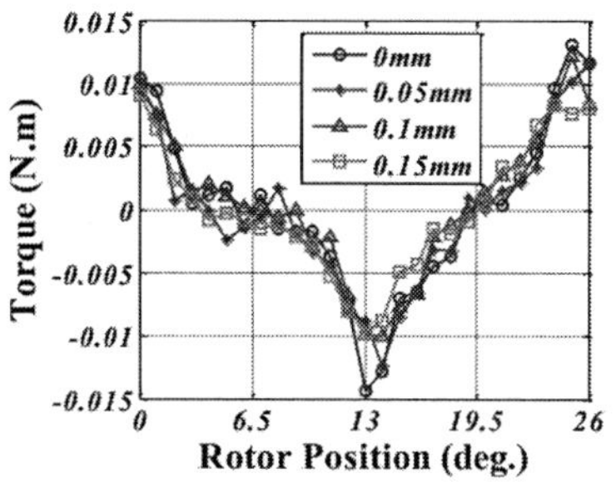

(b) Torque of suspending force winding at i_{xp}=2A.

Figure 5. Torque characteristics of 12/14 BLSRM for various rotor eccentric displacements (Δx).

Fig. 5 (b) shows the torque profiles with fixed suspending force current (i_{yp}=2A) at different eccentricities in the negative x-direction. As shown in Fig. 5 (b), the variation and peak value of the torque generated by the suspending force winding is much smaller than that generated by the torque winding. This is because the pole arc of the suspending force pole is not smaller than one rotor pole pitch, for the cause of the proposed BLSRM. Although the air-gap distribution changes with rotor eccentricity, the overlap area between suspending force pole and rotor pole is constant. Accordingly, the slope of the inductance with respect to rotor position is zero. Therefore, the torque control can be decoupled from suspending force control in the proposed structure.

B. *Suspending Force Characteristics*

Fig. 6 shows the suspending force generated by the torque winding with a constant current for different eccentricities in the negative x-direction. Unlike the rotor in the center position, in which the suspending force is almost zero, the suspending force generated by the torque winding is very large with different eccentricities. Moreover, with the eccentricity displacement increasing, the suspending force is also increased. It is because the air-gap between stator poles P_{A1}, P_{A2} and the rotor is decreased, while the air-gap between stator poles P_{A3}, P_{A4} and the rotor is increased. According to (3), it can be known that, compared with the suspending force generated in normal air-gap, the suspending force generated by the windings on stator poles P_{A1} and P_{A2} is increased due to the deceased air-gap, and the suspending force generated by the windings on stator poles P_{A3} and P_{A4} is decreased due to the increased air-gap. Therefore, an unbalance pull force will be generated by the torque windings along the direction towards the center of stator poles P_{A1} and P_{A2}. This force can be divided into two forces: one along the negative x-direction; the other along the positive y-direction.

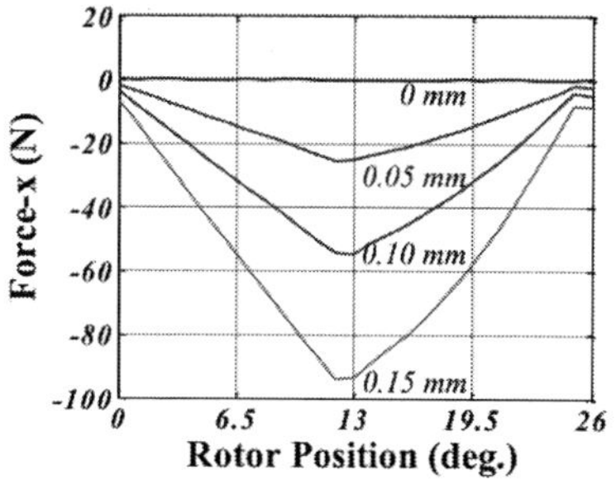

(a) Suspending force (F_x) of torque winding.

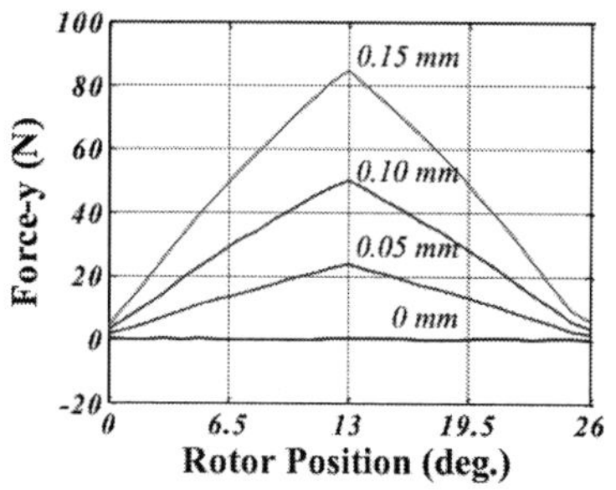

(b) Suspending force (F_y) of torque winding.

Figure 6. Suspending force chacracteristics of 12/14 BLSRM for various rotor eccentric displacements (Δx) with fixed torque winding current (i_A=2A).

Fig. 7 shows the suspending force generated by the suspending force pole P_{xp} with different eccentricities in the negative x-direction. It can be seen that the value of the suspending force in the x-direction decreases with increasing eccentricity, which can also be explained by (3). Compared with Fig. 6, it can be found that, although the value of the suspending force is decreased, the suspending force generated by the suspending force winding is still larger than the unbalance force generated by the torque windings when the rotor has eccentricity displacement 0.15mm. That is, the rotor can be pulled back to the center position with rotor eccentricity smaller than 0.15mm.

If the rotor eccentricity is larger than 0.15mm, the motor may cannot generate enough force to pull the rotor back to the center position, which makes the suspension failed. This is the reason why the capture bearing is used to limit the maximum eccentricity of the rotor at 0.15mm.

IV. SIMULATION ANALYSIS OF 12/14 BLSRM

According to the operation principle of the proposed BLSRM, a control scheme for proposed BLSRM is proposed in Fig. 8. As shown in Fig. 8, mathematical model is not used, only a PI type speed controller is adopted to regulate the motor speed, and two independent close-loop PID air-gap displacement controllers, one for x-direction and the other for y-direction, are used to generate the desired suspending force commands F_x^* and F_y^* to keep the rotor at the center position. Further, in the control scheme, the command currents, i_x^* and i_y^*, for which suspending force windings are determined by Table I according to the desired radial force. Meanwhile, the hysteresis current control method is used to regulate the current in each phase.

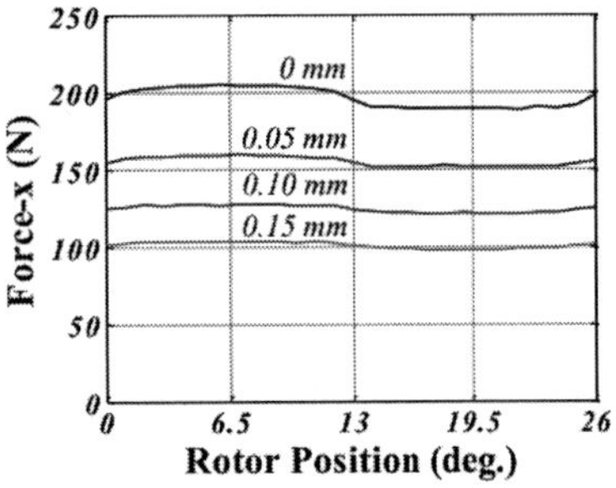

Figure 7. Suspending force chacracteristics with fixed suspending force current (i_{xp}=4A) at various eccentricities in negative x-direction.

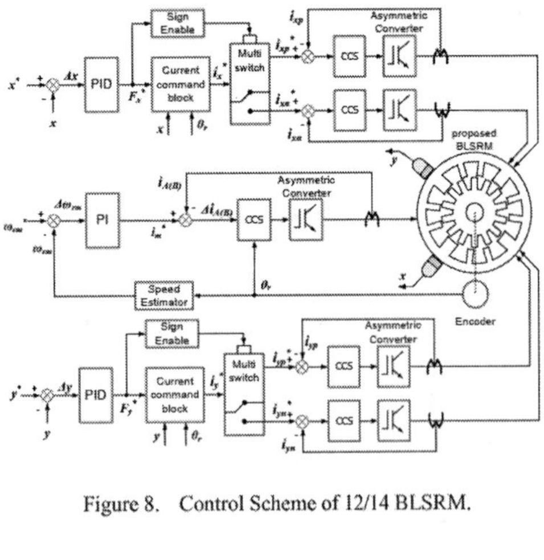

Figure 8. Control Scheme of 12/14 BLSRM.

TABLE I. SELECTION OF SUSPENDING FORCE CONTROL POLES

Desired force	Suspending force control pole
$F_x > 0$ & $F_y > 0$	P_{xp} & P_{yp}
$F_x > 0$ & $F_y < 0$	P_{xp} & P_{yn}
$F_x < 0$ & $F_y > 0$	P_{xn} & P_{yp}
$F_x < 0$ & $F_y < 0$	P_{xn} & P_{yn}

According to the control scheme of the proposed BLSRM, simulations are performed to analyze the dynamic characteristics of 12/14 BLSRM. The simulations are divided into three steps. Firstly, the rotor is locked and suspended to the center position. Secondly, while the rotor is being steadily suspended, a torque load is applied to the rotor shaft, and torque currents are injected to accelerate the rotor to the desired speed. Finally, while the rotor is being steadily suspended and steadily operates at the desired speed, loads of suspending force is suddenly changed and the characteristics of the proposed structure are observed.

Assume the following initial conditions: two 10*N* loads are applied to the rotor shaft in the x- and y-directions simultaneously. Meanwhile, in the x-direction, the initial rotor eccentricity displacement is -120μm, and in the y-direction, initial rotor eccentricity displacement is -90μm.

Fig. 9 shows the simulation results of 12/14 BLSRM. It can be seen from Fig. 9 that, when suspension control is applied, the suspending force windings on the suspending force poles P_{xp} and P_{yp} are excited according to Table I. Accordingly, the suspending forces are generated as shown in Fig. 9 (g) and (h). From Fig. 9 (c) and (d), it can be seen that, when the suspension control is applied, eccentric errors in the two directions can be rapidly reduced to zero, which means that the rotor can be kept in the center position.

While the rotor is being steadily suspended, a 0.25N.m torque load is applied to the rotor shaft at the time t=0.1 second, and then the torque currents are injected to accelerate the rotor to the desired speed, 1000rpm. As shown in Fig. 9, with the torque winding currents and motor speed increasing, the rotor can still be steadily suspended, and the suspending force winding currents and suspending force generated by the suspending force winding are almost same as before. That is, when the rotor in the center position, the torque current has almost no effect on the suspending force, which has a good match with the analysis results shown in [10].

When the rotor is being steadily suspended at 1000rpm, a suspending force load in the x-direction is suddenly increased from 10N to 30N at the time t=0.2 second. As shown in Fig. 9 (c), the rotor has a little eccentricity in x-direction when the suspending force load is increased from 10N to 30N, but it can be pulled back to the center position with suitable suspending force control. Meanwhile, from Fig. 9 (g) and (h), it can be found that, when the rotor has eccentricity, the torque winding will generate radial force. And the radial force is decreased with decreased eccentricity. To steadily suspend the rotor in the center, the suspending force currents have to be changed to generate the suspending force to compensate the radial force generated by the torque windings. Further, with the rotor eccentricity and change of suspending force currents, the speed and phase current are almost same as before, which has a good match with the analysis results in Section III.

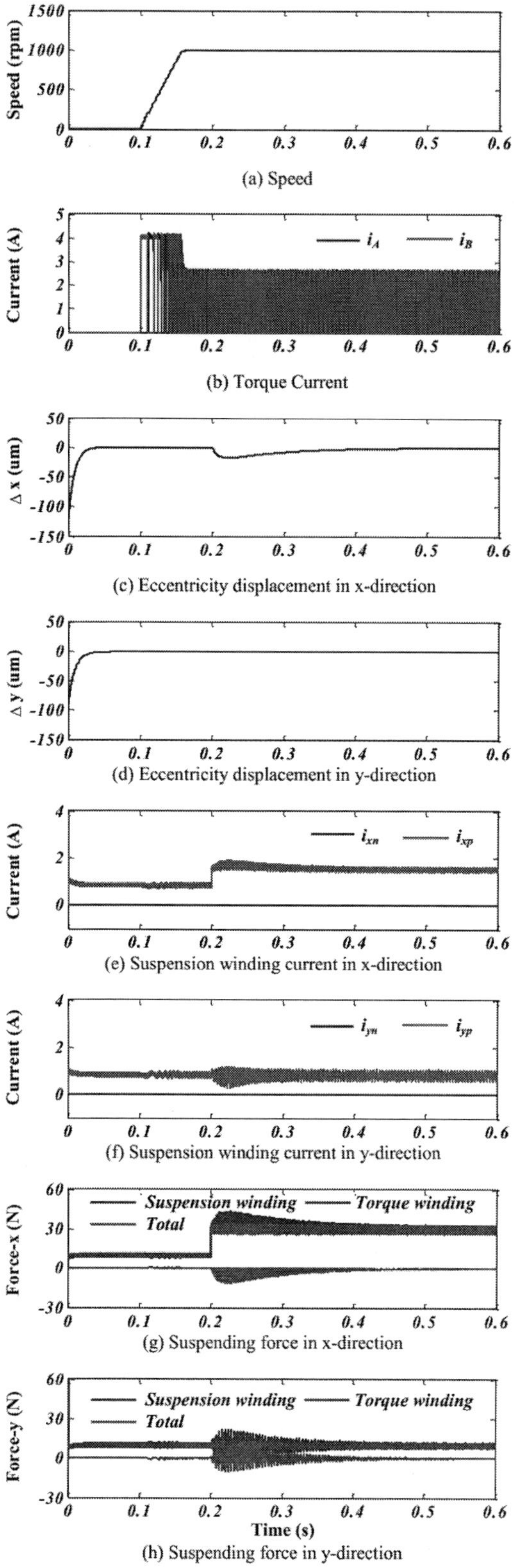

Figure 9. Simulation results of 12/14 BLSRM.

When the rotor is pulled back to the center position again, there is almost no radial force generated by the torque windings. However, compared with the radial force current before suspending force load change, the current in the y-direction has larger fluctuation than before. It is caused by the coupling effect of the radial force current in the x-direction.

The simulation results verify the correctness of the analysis results in [10] and Section III. It also verifies that the torque control can be decoupled from the suspending force control in 12/14 BLSRM.

V. EFFICIENCY ESTIMATION OF 12/14 BLSRM

To further verify the validity of the proposed structure, efficiency estimation is executed to the 8/10 and proposed BLSRM.

A. Copper Loss

The copper loss is proportional to the square of the phase RMS current and phase winding resistance. Hence, according to the current waveform, the copper loss can be calculated as,

$$P_{cu} = mI_{phrms}^2 R_{ph} \tag{4}$$

$$I_{phrms} = \sqrt{\frac{1}{T}\int_0^T I_m^2 dt} \tag{5}$$

where, m is the number of phase, I_{phrms} is the RMS current of phase, R_{ph} is the phase resistance, T is the time of one period, I_m is the phase current.

B. Core Loss

The core loss is composed of hysteresis loss, excessive loss and eddy-current loss. It can be calculated as,

$$P_{losses}(t) = P_h(t) + P_c(t) + P_e(t) \tag{6}$$

where,

$$P_h(t) = \left\{ \left| H_x \frac{dB_x}{dt} \right|^{\frac{2}{\beta}} + \left| H_y \frac{dB_y}{dt} \right|^{\frac{2}{\beta}} \right\} \tag{7}$$

$$P_e(t) = \frac{1}{C_e} k_e \left\{ \left| \frac{dB_x}{dt} \right|^2 + \left| \frac{dB_y}{dt} \right|^2 \right\}^{0.75} \tag{8}$$

$$P_c(t) = \frac{1}{2\pi^2} k_c \left\{ \left| \frac{dB_x}{dt} \right|^2 + \left| \frac{dB_y}{dt} \right|^2 \right\} \tag{9}$$

in which, P_h, P_c and P_e are the hysteresis loss, excessive loss, and eddy-current loss, respectively; k_h, k_e and k_c are the coefficient of hysteresis loss, excessive loss, and eddy-current loss, respectively. According to the material of the prototypes, the lamination steel 35PN440 is used as the stator and rotor material in the analysis. From the datasheet, k_h, k_e and k_c are given as 270.39764, 4.30046 and 0.30463 W/m^3, respectively.

C. Simulation Results

An external circuit is built and used to drive the two motors operating at rated condition. Meanwhile, hysteresis current control method is used to regulate the phase current. Figs. 10 and 11 show the transient phase current and core loss in the two motors. Compared with Figs. 10 (a) and 11 (a), it can be seen that the core losses are mainly produced at commutation region. Further, compared with core loss in 8/10 type, the core loss in 12/14 type is much smaller due to the short flux path and no flux reversal in this structure.

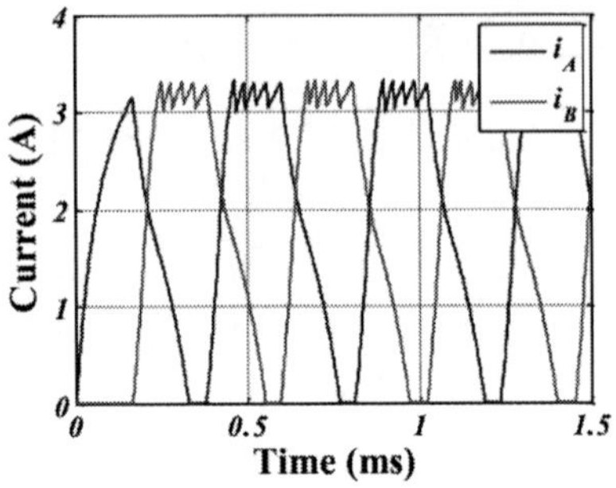

Figure 10. Phase current of 12/14 BLSRM operating at rated condition.

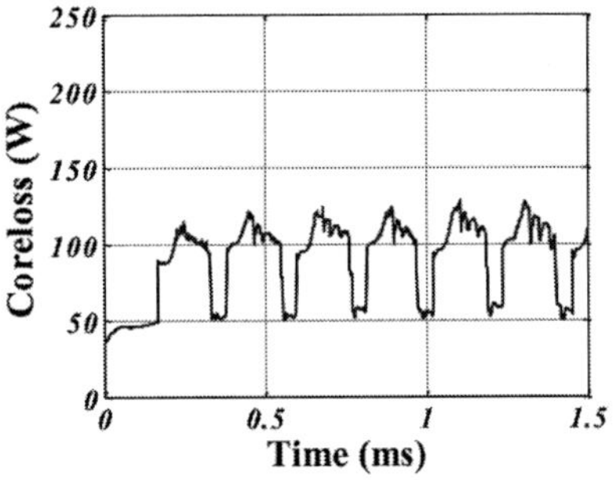

(a) 12/14 BLSRM

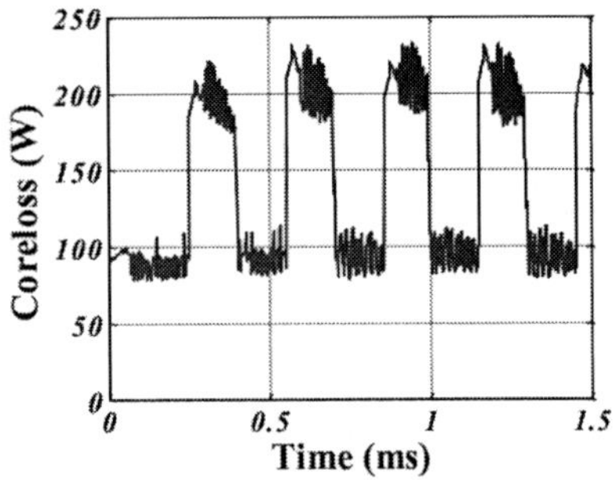

(b) 8/10 BLSRM

Figure 11. Core loss of 12/14 and 8/10 BLSRMs operating at rated condition

According to Section V.A and Fig. 10, the copper loss of 12/14 BLSRM can be calculated. Table II shows the estimated efficiency of 8/10 and 12/14 BLSRM. From Table II, it can be seen that the core loss of the 8/10 type is larger than the proposed type, it is because long flux and flux reversal exist in the flux path of the 8/10 type. Further, the output power and efficiency of the proposed type is larger than that in the 8/10 type, which further verifies the effectiveness of the 12/14 BLSRM.

TABLE II. ESTIMATED EFFICIENCY OF 8/10 AND PROPOSED BLSRM

	Parameter	Value	
		8/10 type	Proposed type
P_{cu_T}	Copper loss	15.54 W	20.5W
P_{Fe}	Core loss	148.93 W	95.63 W
P_{out}	Output power	355.70 W	415.47 W
η	Efficiency	68.38%	78.15%

VI. CONCLUSIONS

In this paper, the rotor eccentricity effect in the proposed BLSRM is analyzed. Based on the analysis results, the torque control can be decoupled from the suspending force control. Although the unbalance force generated by the torque winding is very large due to the rotor eccentricity, enough suspending force can be generated by the suspending force to pull the rotor back to the center positions. Moreover, due to short flux and no flux reversal in the stator, the efficiency of the proposed type is higher than 8/10 type.

ACKNOWLEGEMENT

This work was supported by the Ministry of Education (MOE) and National Research Foundation of Korea (NRF) though BK21 plus project.

REFERENCES

[1] D. H. Lee and J. W. Ahn, "Design and analysis of hybrid stator bearingless SRM," *Journal of Electrical Engineerig & Technology*, vol. 6, no. 1, pp. 94-103, 2011.

[2] D. H. Lee, T. H. Pham and J. W. Ahn, "Design and operation characteristics of four-two pole high-speed SRM for torque ripple reduction," *IEEE Transaction on Industrial Electronics*, vol. 60, no. 9, pp. 3637-3643, 2013.

[3] M. Takemoto, H. Suzuki, A. Chiba, T. Fukao, M. A. Rahman, "Improved analysis of a bearingless switched reluctance motor," *IEEE Transaction on Industrial Application*, vol. 37, no. 1, pp. 26-34, 2001.

[4] M. Takemoto, A. Chiba, H. Akagi, and T. Fukao, "Radial force and torque of a bearingless switched reluctance motor operating in a region of magnetic saturation," *in Conf. Record IEEE-IAS Annual Meeting*, 2002, pp. 35–42.

[5] M. Takemoto, A. Chiba, H. Akagi, and T. Fukao, "Radial force and torque of a bearingless switched reluctance motor operating in a region of magnetic saturation," *IEEE Transaction on Industry Applications*, vol. 40, no. 1, pp. 103-112, 2004.

[6] F. C. Lin and S. M. Yang, "Self-bearing control of a switched reluctance motor using sinusoidal currents," *IEEE Transaction on Power Electroics.*, vol. 22, no. 6, pp. 2518-2526, 2007.

[7] L. Chen and W. Hofmann, "Modeling and control of one bearingless 8-6 switched reluctance motor with single layer of winding structure," in *Proc. of the 2011-14th European Conference on Power Electronics and Applications (EPE 2011)*, Aug. 30-Sep. 1, 2011.

[8] L. Chen and W. Hofmann, "Speed regulation technique of one bearingless 8/6 switched reluctance motor with simpler single winding structure," *IEEE Transactions on Industrial Electronics*, vol. 59, no. 6, pp. 2592-2600, 2012.

[9] Carlos R. Morrison, Mark W. Siebert, and Eric J Ho, "Electromagnetic forces in a hybrid magnetic-bearing switched-reluctance motor," *IEEE Transactions on Magnetics*, vol.44, no.12, 2008.

[10] Z. Y. Xu, F. G. Zhang, and J. W. Ahn, "Design and analysis of a novel 12/14 hybrid pole type bearingless switched reluctance motor," *Proceedings on International Symposium on Industrial Electronics*, pp. 1922-1927, 2012.

MagOpt - Optimization Tool for Mechatronic Components

Siegfried Silber[a], Werner Koppelstätter[a], Günther Weidenholzer[a], Gerd Bramerdorfer[b]

[a] Linz Center of Mechatronics GmbH, Altenbergerstrasse 69, A-4040 Linz, Austria, siegfried.silber@lcm.at
[b] Johannes Kepler University Linz, Altenbergerstrasse 69, A-4040 Linz, Austria

***Abstract*—MagOpt is a modular software tool for the simulation and optimization of mechatronic components. It features a flexible structure for the storage of complex data and an open and modular interface to existing third party programs like CAD systems, finite element programs and other simulation software. Parametric design optimization can be carried out with various different optimization strategies like gradient-based methods or multi-objective evolutionary or genetic algorithms.**

I. Introduction

In the past the optimization of mechatronic components was done manually by varying key design parameters. With such an approach the number of different designs to be evaluated is limited from just a few up to several hundreds. To speed up the design process a more automated approach is demanded. Normally, calculating a mechatronic component is associated with analyzing a mathematical model that represents the characteristics of this component (in a specific domain). Characteristic quantities of an electric motor are, e.g., electrical properties, like voltages, currents, efficiency and many more properties of the geometry and the material parameters of the motor. Thus, parameter variation incorporates modifying any parameter that has an impact on the performance of the mechatronic component. This, of course, comprises geometric and material parameters. Simulation in different physical domains often necessitates solving nonlinear PDEs by means of finite element analysis (FEA). Usually it takes from several minutes up to hours to solve FEA problems. For a fully automated simulation of mechatronic components these FEA problems are the limiting factor concerning the throughput because of the high requirements in computation time. However, that bottleneck can be opened up by calculating the FEAs in parallel on a computer cluster.

For a systematic optimization approach the simulation and optimization tool MagOpt (short for Magnetic Optimization) has been developed. Although MagOpt was originally designed for magnetic problems it turned out that, due to the very general structure it can also be applied for analyses performed in different physical domains and thus for different mechatronic components. For instance electromagnetic actuators, motors, magnetic bearing systems, hydraulic actuators and many more can be evaluated using MagOpt. Various simulations and optimizations of electrical motors, magnetic bearings and bearingless motors have been carried out recently (see section IV).

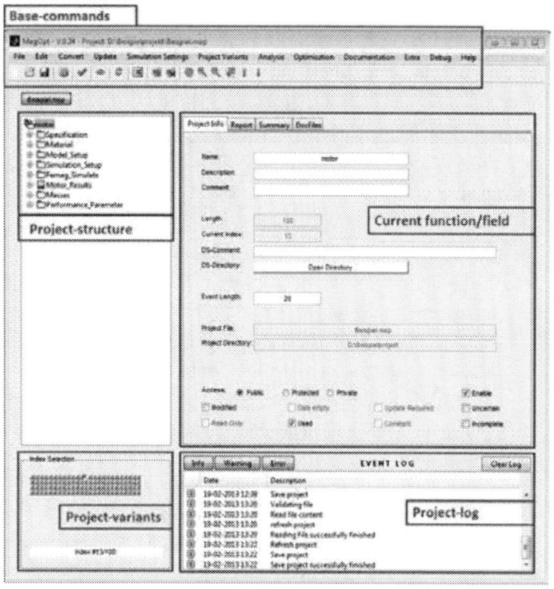

Figure 1. Screenshot of the MagOpt user interface.

II. Functionality of MagOpt

The simulation and optimization tool MagOpt has been developed at the Linz Center of Mechatronics and the Institute of Electrical Drives and Power Electronics of the Johannes Kepler University Linz.

Figure 1 shows a screenshot of the MagOpt main window. This graphical user interface (GUI) is partitioned into several panels where the key element "Project-structure" is used to define the project setup which is represented as a tree structure. Details of the selected item are summarized in the panel "Current function/field". A project may contain multiple variants differing by a set of altered parameter values based on the same project structure. Messages, warnings and errors which occur during the execution or simulation are displayed in the "Project-log" view.

A MagOpt project is organized by using a tree structure so that related parameters and functions can be arranged in separate branches. The following types of tree elements are currently supported:

- *Project*: Base element that specifies a MagOpt project. This is typically the root element in the parameter tree.

- *Container*: A node element for a branch and can be used for structuring of parameters. Any MagOpt element except a *Project* element can be inserted. The scope and visibility of the parameters inside such a *Container* can be specified (this means how the parameters are to be referenced inside and outside this *Container*).
- *Field*: Element for data storage. Various data types for numerical values, matrices, periodic signals, structs or character strings are supported.
- *Formula*: Simple algebraic equations. Any field element in the tree may be used as argument and the result is stored within the *Formula* element itself.
- *Function*: Interface to complex modules. This element is described in more detail in section *II-B*.

A. Scope and visibility of parameters

Scope and visibility refers to the availability of *Field* parameters for use in *Formulas* and *Functions*. There are three different levels to set the visibility of parameters within a *Container*:

- Public: Parameters in public containers are visible within the complete project tree unless a parameter with the same name exists at a higher level in the tree. These parameters can be referenced by the parameter name (e.g. `paramname`). If the scope of a parameter is hidden by a public parameter with the same name at a higher level in the tree (closer to the root) reference to that parameter can be achieved by applying a uniquely defined parameter identifier e.g., `container.paramname`.
- Protected: These parameters can be accessed inside a branch by using their parameter names. Outside this *Container* at least the name of the protected container has to be added with a dot-separator like `container.paramname`.
- Private: These parameters are only visible inside the container and sub-containers and cannot be referenced from outside.

B. Functions

The MagOpt element *Function* is intended to implement complex calculation modules and thus can be used as interface to external third party simulation tools. These *Functions* have to be coded either in Java or in script languages. Currently the script languages Matlab, Python, JavaScript and Lua are available. A *Function* element has the same properties as a *Container* so that associated parameters can be grouped. The interface of MagOpt parameters to internal Java or script variables is realized by import and export lists as shown in Fig. 2.

With these lists the internal Java or script variables are assigned to MagOpt parameters. If import parameters are not assigned default values are used instead, unless the particular variable is specified as required. In such a case an error message is generated. The interface also supports the specification of physical units. This means that the physical unit of the assigned parameter has to fit the expected unit in the *Function*. If feasible the values are automatically converted to

IMPORTS			EXPORTS		
Variable	Parameter	OK	Variable	Parameter	OK
N	Ns	✓	WdgData	WdgData	✓
p	pz	✓	Xi1	xi1	✓
m	m	✓	Nu	Nu	✓
SingleLayer	SingleLayer	✓	yn	yn	✓
dyn	dyn	✓	Xil		
PhaseConnection	PhaseConn	✓	Xim		
GroupConnection	GroupConn	✓	Xi		

Figure 2. Interface to MagOpt Functions. With the import and export list the internal Java or script variables are assigned to MagOpt parameters.

the specified unit in the *Function*, otherwise an error message is thrown.

C. Calculation and Simulation

Calculation or simulation means that the *Formulas* are evaluated and the *Functions* are executed. The order in which *Formulas* and *Functions* are processed is resolved by the import and export dependencies. The first *Function* that is evaluated must not have any input parameter that depends on export values of any *Function* or *Formula*. Subsequently, only *Functions* where all the input parameters are updated can be evaluated. This entails that circular dependencies cannot be resolved.

If parameters are modified only *Functions* with direct or indirect (via results of other *Functions*) dependencies of these parameters need to be evaluated. Thus, it is not necessary to evaluate all *Functions* of the project if just a few parameters are modified.

With *Functions* also external third party simulation tools can be executed. The processing time for these programs might become considerably long, especially when finite element programs are executed. To increase the throughput, MagOpt automatically distributes time consuming simulations to a computer cluster and collects the results upon completion of the calculation. As soon as the results are available the corresponding result parameters in the project are set.

D. Components

If MagOpt is employed for simulation and optimization of similar mechatronic components, the projects will also have a quite similar structure. For example, in the case of electric motors this means that any motor has a stator and a rotor and the simulation is done by using a magnetic FEA independent of the size of the motor. But the stators and rotors might be different in size and shape. To avoid that any project has to be set up manually from scratch *Components* are available in MagOpt. Such a *Component* is a collection of MagOpt parameters and can be inserted in the project as a sub-tree. *Components* could of course include *Functions*, *Formulas* and any other type of parameter. To ensure that a *Component* seamlessly fits a project, acceptance filters can be set. For example, if a motor with an internal rotor is simulated only corresponding stators for interior motors are accepted and stators for exterior rotors cannot be selected. For the simulation of various different mechatronic components template projects exist where just a few *Components* need to be added to completely setup the project.

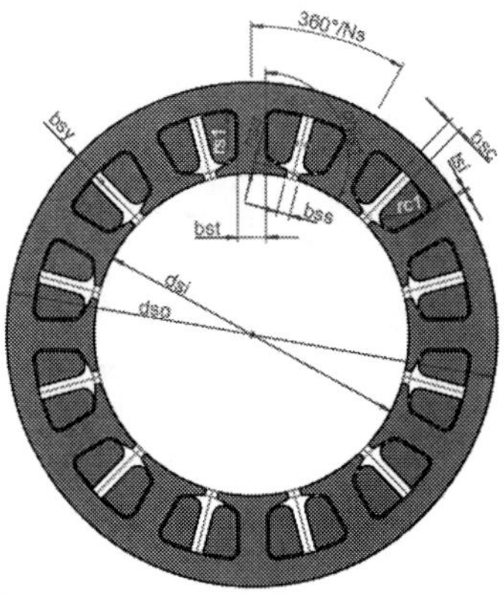

Figure 3. Parametric model of a stator of an electric motor available as a MagOpt *Component*.

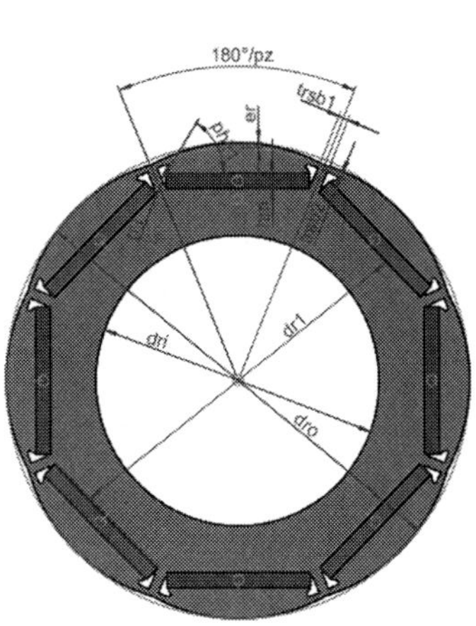

Figure 4. Parametric model of a rotor of an electric motor available as a MagOpt *Component*.

Components for various simulation tasks are currently available. For instance rotor and stator models for synchronous machines are available for different 3D CAD systems (PTC: Creo, Pro/Engineer; Dassault Systems: SolidWorks, Siemens: NX 9...). In Fig. 3 and Fig. 4 a parametric model of a stator and a rotor are shown respectively. All the geometric parameters are directly accessible via corresponding MagOpt parameters. If a MagOpt parameter is modified the CAD model and all other assigned mass properties and drawings are regenerated. The outputs of the CAD system are usually assigned to MagOpt parameters.

For magnetically levitated high speed drives an interface to the multi body simulation software HOTINT is available [1]. With this tool the eigenfrequencies of the shaft are calculated. Moreover, also the mechanical stress due to centrifugal forces in the material can be evaluated. An example of a high-speed rotor simulated with HOTINT is shown in Fig. 5.

III. Optimization algorithms

Optimization of a mechatronic component means to fulfilling the requirements in such a way that certain target parameters (objectives) reach their best or optimal values.

Figure 5. Example of a high-speed rotor where the first bending mode is simulated with HOTINT.

It is obvious that the requirements can be fulfilled with different embodiments. Nevertheless, those embodiments usually feature different characteristics for further performance parameters.

By means of optimization the objective parameters should reach an optimal (a minimal or maximal) value. In many cases there is not just one single objective to be optimized, but there are several simultaneously. This leads to multi-objective optimization problems. For a nontrivial multi-objective optimization problem, there exists no single individual that shows best behavior for each objective. In that case, the objective functions are said to be conflicting, and Pareto-optimal solutions exist. An individual is said to be Pareto-optimal if there are no other designs that better satisfy all the considered objectives. In other words, any improvement in one objective necessitates the worsening of at least one other objective [2].

In MagOpt well known and widely applied genetic algorithms for multi-objective optimization are implemented. Additionally we also implemented improved algorithms in terms of simulation speed. Especially for FEA problems an advanced algorithm was developed that improves convergence of the Pareto-front significantly [3]. Currently the following optimization algorithms are implemented in MagOpt:

- Grid calculation: Calculates any possible parameter combination and thus requires very high computational effort.
- Generational NSGA-II (Non-dominated Sorting Genetic Algorithm II) [4]
- Steady State Asynchronous NSGA-II
- Generational SPEA2 (Strength Pareto Evolutionary Algorithm 2) [5]
- Steady State Asynchronous SPEA2
- DECMO (Differential Evolution-based, Coevolutionary Multi-objective Optimization algorithm)

Evolutionary algorithms such as the Non-dominated Sorting Genetic Algorithm-II (NSGA-II) and Strength Pareto Evolutionary Algorithm 2 (SPEA2) are generation based approaches. This means that a new generation is created as soon as all individuals of the current generation have been evaluated. This, however, is a major disadvantage for optimization problems where FEAs are distributed on a computer cluster, especially when the simulation time varies much between the individuals. This is due to the fact that all the simulations have to be finished before new individuals can be submitted to the cluster. This leads to load imbalance on the computer cluster.

To overcome this problem steady-state asynchronous algorithms based on NSGA-II and SPEA2 have been developed and implemented in MagOpt [6]. In contrast to generation-based algorithms, using the steady-state algorithm it is not necessary to wait until each individual of a generation is evaluated until a new generation is created. This algorithm allows more generations to be evaluated in parallel even though it seems that this may lead to worse convergence. But for practical examples the convergence of the Pareto-front could even be improved as shown in [6].

A. DECMO

Modern multi-objective evolutionary algorithms use methods based on differential evolution (DE), because benchmark tests showed that with DE methods the design space can be explored far more efficiently. However, as regards some problems, state of the art algorithms like NSGA-II and SPEA2 still significantly outperform DE-based algorithms. These conventional algorithms also seem to be more robust with regard to their parameterization on a wide range of test problems with the disadvantage that the design space is not explored that comprehensively.

To improve the optimization algorithm we combined both algorithms as described above to gain advantage from the robustness of classic multi-objective algorithms and also from the very good performance exhibited by DE methods. This algorithm is called Differential Evolution-based, Coevolutionary Multi-objective Optimization algorithm (**DECMO**) and is described in detail in [3].

IV. Examples

In the last few years various optimization procedures have been performed on electric motors by using MagOpt. For instance, a wheel hub motor for an electric scooter was optimized [7]. Also the efficiency of industrial motors was increased whereas simultaneously the amount of rare earth magnets could be significantly reduced [8], [9], [10], [11]. MagOpt was also used for hardware in the loop optimization for a fast switching hydraulic valve. In that case the control parameters and the current waveform were optimized by controlling the system [12], [13]. Simulation of a passive permanent magnetic bearing and optimization of its stiffness further demonstrate the versatility of MagOpt [14].

Comparison of different embodiments of mechatronic components by their Pareto-fronts is also supported by MagOpt and was done in [15] for fast switching valves.

Acknowledgment

This work has been carried out at LCM GmbH as part of a K2 project. K2 projects are financed using funding from the Austrian COMET-K2 program. The COMET K2 projects at LCM are supported by the Austrian federal government, the federal state of Upper Austria, the Johannes Kepler University and all of the scientific partners which form part of the K2-COMET Consortium.

References

[1] D. Reischl, A. Dorninger, A. Fohler, J. Gerstmayr, W. Koppelstätter, S. Silber, and S. Weitzhofer, "Coupled mechanical and electromagnetic optimization of high speed rotors," in *Proc. 14th International Symposium on Magnetic Bearings*, 2014.

[2] R. G. Ajith Abraham, Lakhmi Jain, *Evolutionary Multiobjective Optimization*. Springer-Verlag London Limited, 2005.

[3] A.-C. Zăvoianu, E. Lughofer, W. Amrhein, and E. P. Klement, "Efficient multi-objective optimization using 2-population cooperative coevolution," 2013, pp. 251–258.

[4] K. Deb, A. Pratap, S. Agarwal, and T. Meyarivan, "A fast and elitist multiobjective genetic algorithm: Nsga-II," *IEEE Transactions on Evolutionary Computation*, vol. 6, no. 2, pp. 182–197, 2002.

[5] E. Zitzler, M. Laumanns, and L. Thiele, "SPEA2: Improving the Strength Pareto Evolutionary Algorithm for Multiobjective Optimization," in *Evolutionary Methods for Design, Optimisation and Control with Application to Industrial Problems (EUROGEN 2001)*, K. Giannakoglou *et al.*, Eds. International Center for Numerical Methods in Engineering (CIMNE), 2002, pp. 95–100.

[6] A.-C. Zăvoianu, E. Lughofer, W. Koppelstätter, G. Weidenholzer, W. Amrhein, and E. P. Klement, "On the performance of master-slave parallelization methods for multi-objective evolutionary algorithms," in *Artificial Intelligence and Soft Computing*, ser. Lecture Notes in Artificial Intelligence, L. R. et al., Ed. Springer Berlin Heidelberg, 2013, vol. 7895, pp. 122–134.

[7] W. Gruber, W. Back, and W. Amrhein, "Design and implementation of a wheel hub motor for an electric scooter," in *Vehicle Power and Propulsion Conference (VPPC), 2011 IEEE*, Sept 2011, pp. 1–6.

[8] G. Bramerdorfer, W. Amrhein, S. Silber, and R. Hagen, "Spectral-field design with respect to minimum cogging torque and maximum output power," in *IECON 2010 - 36th Annual Conference on IEEE Industrial Electronics Society*, Nov 2010, pp. 2264–2269.

[9] G. Bramerdorfer, S. Silber, G. Weidenholzer, and W. Amrhein, "Comprehensive cost optimization study of high-efficiency brushless synchronous machines," in *Electric Machines Drives Conference (IEMDC), 2013 IEEE International*, May 2013, pp. 1126–1131.

[10] R. Lohninger, H. Grabner, G. Weidenholzer, S. Silber, and W. Amrhein, "Modelling, simulation and design of a novel permanent magnetic reluctance machine," in *Power Electronics, Electrical Drives, Automation and Motion (SPEEDAM), 2012 International Symposium on*, June 2012, pp. 620–624.

[11] G. Weidenholzer, S. Silber, G. Jungmayr, G. Bramerdorfer, H. Grabner, and W. Amrhein, "A flux-based PMSM motor model using RBF interpolation for time-stepping simulations," in *Electric Machines Drives Conference (IEMDC), 2013 IEEE International*, May 2013, pp. 1418–1423.

[12] P. Foschum, A. Plöckinger, R. Scheidl, G. Weidenholzer, and B. Winkler, "Multi objective genetic optimization of fast switching valves," 2011, pp. 116–128.

[13] P. Foschum, A. Plöckinger, and R. Scheidl, "Hardware in the loop multi objective genetic optimization for efficient valve control," 2012, pp. 123–132.

[14] E. Marth, G. Jungmayr, J. Kobleder, M. Panholzer, and W. Amrhein, "Realization of a multi-pole permanent magnetic bearing with rotatin magnetization," in *Proc. 14th International Symposium on Magnetic Bearings*, 2014.

[15] P. Foschum, A. Plöckinger, G. Weidenholzer, and R. Scheidl, "Triz, design optimization and suh's 1st axion - a comparison on the example of fast switching valves," 2012, pp. 498–505.

Coupled Mechanical and Electromagnetic Optimization of High Speed Rotors

Daniel Reischl[a], Alexander Dorninger[a], Armin Fohler[a], Johannes Gerstmayr[a], Werner Koppelstätter[a], Siegfried Silber[a], Simon Weitzhofer[a]

[a] Linz Center of Mechatronics GmbH (LCM), Altenbergerstr. 69, 4040 Linz, Austria, daniel.reischl@lcm.at

***Abstract*—Optimizing mechatronic systems is of increasing importance, for example, to minimize the consumption of resources and energy during operation and production. The optimization objectives for rotordynamic systems are, for instance, critical resonances, durability and costs. In order to meet such objectives, good optimization algorithms as well as simulation tools for each physically relevant effect for the problem are required.**
In this paper, we present the coupling of the optimization software MagOpt and the mechatronic simulation tool HOTINT and an application of the system.

I. Introduction

For the coupled optimization of a magnetically levitated high speed rotor, it is necessary to consider vibrations and eigenfrequencies, mechanical strains, high efficiency and bearing design.
There exist specialized tools for specific physical effects, such as simulation programs for mechanics, electromagnetics, thermal or fluid problems. For simulating an electrical machine, the different specialized tools have to be coupled.
There already exist commercial solutions that can be used for multiphysics optimization, e.g., the multiphysics software ANSYS together with a third party optimization tool. In [1] the optimization of the acoustic behavior of electrical drives with these programs is described.
The main benefits of using a small open source program like HOTINT [2] are the flexibility of code that allows quick customization of the program for the specific problem. Furthermore small programs can be distributed over a cluster of computers for parallelized optimization, as done by MagOpt [3].

II. The multibody simulation software HOTINT

HOTINT is an open source engineering software based on a redundant multibody formulation. The work on this program started in 1997 and is continually developed further as joint project of LCM and Johannes Kepler University Linz. The software contains all features which are necessary to compute the quantities relevant for the optimization, such as the mechanical stresses due to centrifugal forces or critical resonances. Besides these necessary requirements HOTINT offers some key advantages over the available commercial programs. For instance, the program offers an easy-to-write scripting language and has an open architecture. Therefore, coupling the program with external software is especially easy, either through input parameters or through C++ code or the TCP/IP interface.
HOTINT provides a large variety of flexible and rigid bodies in 2D and 3D. Those can be connected via flexible and rigid joints in Lagrange or penalty formulation including nonlinear joints such as contact or friction. Additionally the user can include different sorts of loads and sensors into the model as depicted in Fig. 1. HOTINT is easy to extend, so if a particular object is not available yet, the comprehensive manual and the structured code enables the user to adapt existing objects or to implement new objects in short time.
After the optimization of the considered rotor is finished, HOTINT can also perform post-processing computations, e.g., model the controller with IO-elements in order to test its functionality, or do a parameter variation to generate a Campbell diagram. A fully dynamic run-up of the rotor can be simulated to verify the simulation results with measurements of real experiments.

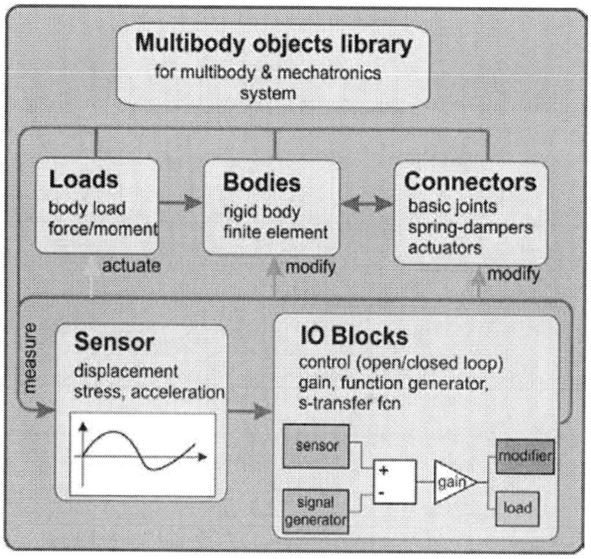

Figure 1. Structure of a HOTINT model.

III. The optimization tool MagOpt

MagOpt is an abbreviation for "Magnetic Optimization" and is a software package that does not only include magnetic optimization but also general simulation- and optimization-

tools in connection with mechatronic components and electrical machines in particular.
Starting in June 2008, the MagOpt project was continuously enhanced and contains modules for electromagnetic, thermal and circuit simulation today. A post processing toolbox as well as an ever-expanding material database is included for evaluating the optimal results.
Starting a project for a specific motor topology, MagOpt tries to find an optimal configuration for the given parameterized model considering the chosen objectives and constraints. During this optimization process (see Fig. 2), all needed simulation tools are automatically called by MagOpt. For electric machines, this can, e.g., range from the drawing of the motor geometry with a corresponding winding scheme to the particular electromagnetic simulation program.

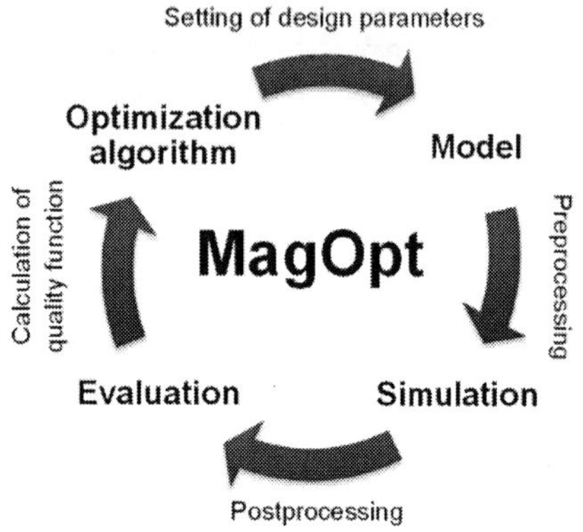

Figure 2. The MagOpt optimization process.

From the mechanical point of view, only rigid body calculations were carried out internally until now. Extensive mechanical simulations were conducted using external tools such as Pro/MECHANICA and the stand alone version of HOTINT.
The inclusion of a simulation program for rotor-dynamics is an inevitable next step for the efficient optimization of motor designs.

IV. Optimization of a rotor geometry

We consider the optimization of a rotor for several objectives such as energy efficiency, performance, durability and costs. In the optimization process, constraints regarding the size, weight, material types, maximum stresses and vibrational exitation have to be fulfilled for both applicability and safety regulations.
MagOpt uses genetic algorithms to acquire an optimal design starting from a user specified inital mechanical and electromagnetic setup of the rotor. In an optimization loop, MagOpt distributes numerous parallel subcalculations on a computer cluster.
HOTINT is one of these parallel executed tools, which receive a mechanical and electromagnetic configuration from the optimizer. It calculates relevant mechanical quantities necessary for the rating of the considered configuration.
In particular, HOTINT returns the maximum stresses due to centrifugal forces at operating speed.
Another characteristic indicator of the system is the eigenfrequency of the first bending mode. This frequency has to be considerably different from the operating frequency of the rotor. Ideally, this frequency is higher than the operating frequency to avoid resonance during run-up.
After an adequate number of iterations a set of optimal configurations is identified.

V. Coupling of MagOpt and HOTINT

The geometry of the rotor is defined in a parameterized model of the cross section which can be changed by MagOpt during the optimization process. This 2D geometry together with informations concerning the materials and connections of the individual parts of the rotor are set by MagOpt in the phase "setting of design parameters", compare Fig. 2. "Preprocessing" and the simulation itself are then performed by HOTINT, compare Fig. 3.

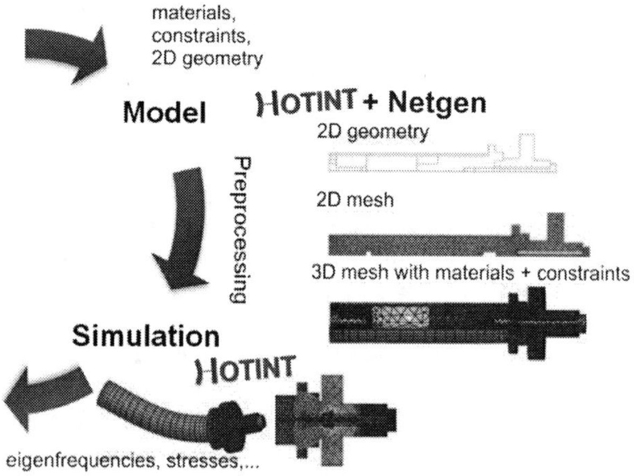

Figure 3. The coupling of MagOpt and HOTINT.

HOTINT is started from MagOpt with a single command including the path to an additional file in which a wide area of computational settings can be defined. These settings include the number of eigenfrequencies, coarsity of the mesh, path to the meshing tool, etc.

In order to avoid artificial numerical effects a symmetric mesh is required. Thus HOTINT generates a 3D mesh in a two step process from the provided cross section of the rotor. First HOTINT starts the open source meshing tool NETGEN [4] with appropriate settings to generate a 2D triangular mesh of the cross section.

The second step is to rotate this mesh in HOTINT to generate a symmetric 3D mesh: Depending on the number of points on the axis of rotation 3 different types of 3D finite elements are created when rotating a triangular mesh, see Fig. 4. If 2 points of the triangle are on the axis of rotation the resulting 3D object is a tetrahedral, 1 point leads to a pyramid and if no point is on the axis a prism is generated.

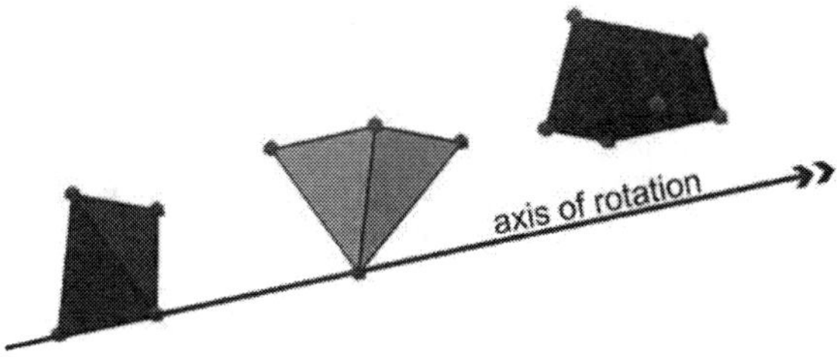

Figure 4. Rotation of a 2D triangular mesh leads to tetrahedrals, pyramids and prisms. Example for 3 triangles and a rotation angle of 45°.

The mesh is split into multiple meshes and the material data is set to represent the individual parts of the rotor. All joining surfaces between adjacent parts are cones, cylinders or discs with collocated nodes. Therefore joints with a node-node model are easy to implement and more complicated master-slave and node-surface models are not necessary if the relative motions are small. Corresponding to the specified connections of the individual parts of the rotor, the constraints are set automatically in HOTINT.

Depending on the settings either an eigenmode computation or a static FE computation is performed in the next step:

In the case of a static FE computation a centrifugal force due to the rotation of the rotor is applied to all elements of the mesh. The maximum value of the mises stress is then saved to an output file for MagOpt.

In the case of an eigenmode computation the number of free-free eigenmodes can be set as well as solver settings like the maximum number of iterations or an error bound. HOTINT uses the same mesh as described above to compute the eigenmodes and eigenfrequencies of the unsupported rotor. The lowest eigenfrequency corresponding to a bending eigenmode is the desired quantity for the rating of the current configuration in MagOpt. The eigenfrequencies are saved to an output file which is then read from MagOpt.

VI. Numerical Example of a Mechatronic System

The optimization of an electromagnetically levitated turbo charger is used in this work as example of a mechatronic system which shall be optimized. Different objectives for the optimization were defined by the involved branches of engineering:

- From the fluid mechanics point of view a large diameter of the wheel is advantageous because it leads to higher torques.
- Electro mechanics demand a large area and therefore length of the magnet in order to achieve a high efficiency.
- To avoid problems with the control, the first bending frequency of the rotor should be higher than the operating speed.
- Finally the stresses have to be sufficiently small to ensure a save and fatigue endurable operation.

With respect to additional specifications the following 2 variables were defined to be changed by MagOpt: The diameter d of the wheel is varied between 15.5 and 60.0 mm and the length l of the magnet can be chosen between 15.0 and 60.0 mm, compare Fig. 5.

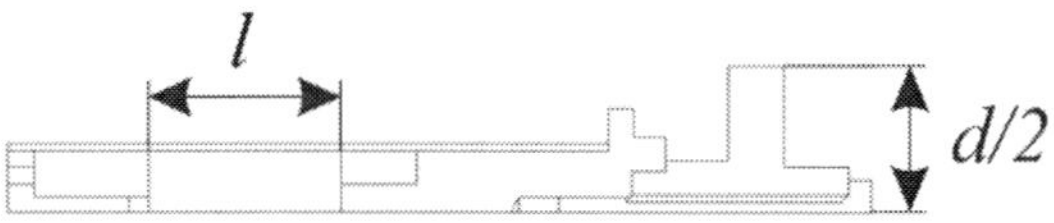

Figure 5. The diameter d of the wheel and the length l of the magnet are used as variables in the optimization.

Eigenfrequencies of the rotor and the stresses are obtained as results of the computations in HOTINT. As upper limit for the mises stress a value of 8e7 N/m^2 was chosen and the lowest eigenfrequency shall be maximized.

The result of the optimization with 4700 single computations is a pareto frontier as shown in Fig. 6.

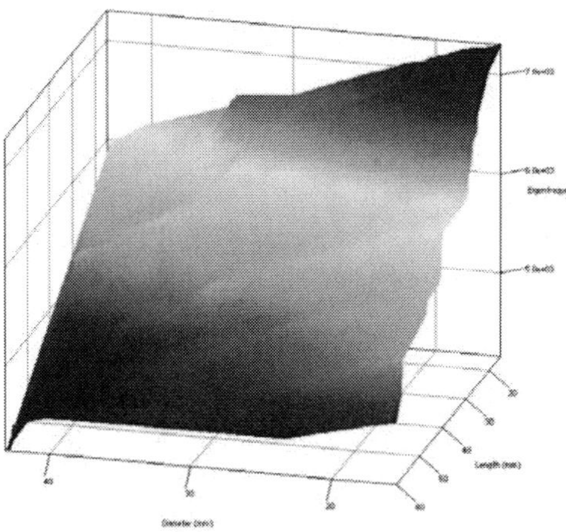

Figure 6. Pareto frontier as results of the optimization cycle: The lowest eigenfrequency as function of the diameter of the wheel and the length of the magnet.

VII. Validation of the Results

A prototype of the magnetically levitated high speed rotor was manufactured and the eigenfrequencies of this rotor have been measured.

Although strong simplifications concerning the interconnection of the individual parts of the rotor were made the results are very satisfying: In the manufactured prototype several parts are interconnected with force fits or glue. Parameters for these joints are hard to obtain and the evaluation of their influence on the eigenfrequencies was too time consuming for this first demonstrative work. All of these interconnections were modeled therefore as perfectly interconnected. The lowest bending frequency of the prototype obtained with HOTINT is 4620 Hz for a mesh with 5584 nodes. The measured eigenfrequency of the manufactured prototype is 3950 Hz.

Additionally the distribution of the mises stress due to the centrifugal force was computed with ANSYS with a much finer mesh with 46120 nodes. The results are shown in Fig. 7 for HOTINT and in Fig. 8 for ANSYS.

VIII. Verification by Additional Simulations

Before selecting one of the optimal configurations for producing a prototype additional simulations can be performed.

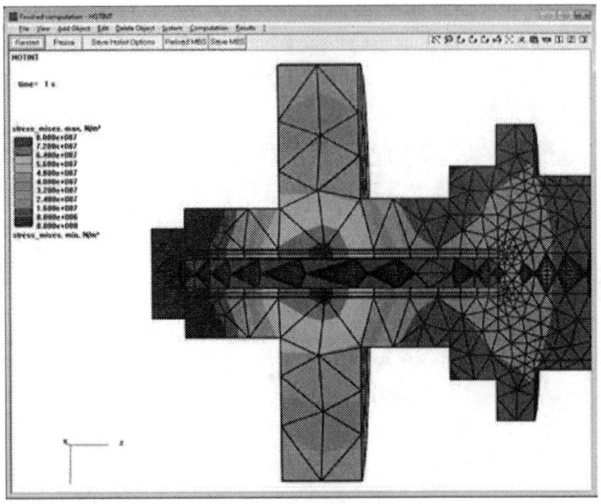

Figure 7. Results obtained with HOTINT: mises stress according to centrifugal forces due to a rotation with f = 1 kHz.

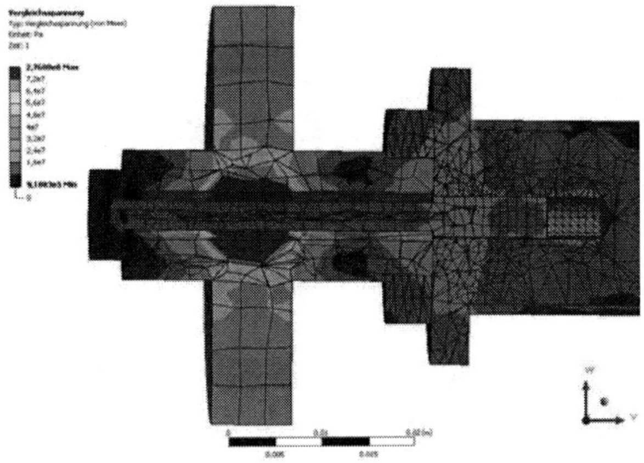

Figure 8. Results obtained with ANSYS: mises stress according to centrifugal forces due to a rotation with f = 1 kHz.

This is to verify the results of the optimization loop where certain assumptions have to be made to decrease computation time. The computation of the eigenfrequencies are made for the unsupported rotor as mentioned before.
Neither eigenfrequency computation nor computation of centrifugal forces can include sliding or contact phenomena. A dynamic run-up simulation identifies the following effects:

- rigid body oscillations in the bearing
- frequency shift in the bending eigenmodes due to shrink-fit connections in the rotor
- sliding of shrink-fit connections
- loosening of glued joints

Using IO-elements we can test a controller design for the specific rotor in a virtual environment. Additionally the required driving torque can be determined in order to get through resonances. Through this preliminary evaluations the number of required prototypes can be minimized.

IX. Acknowledgment

Support of the authors by the research project "Nachhaltig ressourcenschonende elektrische Antriebe durch höchste Energie- und Material-Effizienz" within the EU program "Regionale Wettbewerbsfähigkeit OÖ 2007-2013 (Regio 13)" by the ERDF and the Province of Upper Austria is gratefully acknowledged.

References

[1] D. B. Pinhal and M. Kellermeyer, "Understanding the acoustic behavior of electrical drives," in *9th Weimar Optimization and Stochastic Days 2012*, 2012.

[2] J. Gerstmayr, A. Dorninger, R. Eder, P. Gruber, D. Reischl, M. Saxinger, M. Schörgenhumer, A. Humer, K. Nachbagauer, A. Pechstein, and Y. Vetyukov, "Hotint - a script language based framework for the simulation of multibody dynamics systems," in *Proceedings of the ASME 2013 IDETC&CIE Conference, Portland, Oregon, USA*, 8 2013.

[3] S. Silber, W. Koppelstätter, G. Weidenholzer, and G. Bramerdorfer, "Magopt - optimization tool for mechatronic components," in *Proceedings of the ISMB14 - 14th International Symposium on Magnetic Bearings*, 2014.

[4] J. Schoeberl. Netgen mesh generator. [Online]. Available: http://sourceforge.net/projects/netgen-mesher

Code Agros2D and Possibilities of its Use for Modeling and Design of Magnetic Bearings

Pavel Karban, František Mach, Pavel Kůs, Lukáš Korous, Ivo Doležel

University of West Bohemia, Univerzitní 26, 30614 Pilsen, Czech Republic, {karban, fmach, pkus, korous, idolezel}@kte.zcu.cz

***Abstract*—A multi-platform software Agros2D is presented, whose purpose is to model complex 2D multiphysics problems in engineering practice. The code is intended for monolithic solving of systems of generally nonlinear and nonstationary partial differential equations describing particular physical fields. The numerical schemes implemented in the code are based on a fully adaptive higher-order finite element method. The paper discusses the unique features of the code and its internal structure. Its power is demonstrated on an example aimed at modeling of a passive axial magnetic bearing.**

I. Introduction

Numerous engineering problems are characterized by cooperation of several physical fields. In many cases one field (for example electromagnetic) produces another field (for example thermal) and these fields influence mutually one another.

Even when the physical essences and continuous mathematical models of many phenomena of this kind are well known, their numerical (or computer) modeling is still a challenging problem. Their continuous mathematical models consist of two or more generally nonlinear and nonstationary partial differential equations describing the distributions of involved physical fields. This system is often supplemented with a set of ordinary differential equations describing corresponding circuits (electric, magnetic, thermal, hydraulic, etc.). The coefficients of all these equations can be complicated functions of physical parameters of present materials and media that, however, depend on the state variables such as temperature or pressure.

For the solution of the indicated problems we can use nowadays two groups of codes codes. The first group are commercial professional codes developed by software companies and used both in industry and academy, the second group is represented by open source codes developed and used mostly in the academic sphere. The existing commercial codes are mostly based on lower-order finite element methods, work with elements of the first and second orders and miss advanced adaptive techniques. On the other hand, they are user-friendly due to very well elaborated preprocessors and postprocessors). In this group, just COMSOL Multiphysics works with elements up to the fifth order and a standard is h-adaptivity implemented in it.

The development of codes with advanced adaptive methods is intensively conducted mostly in the open-source academia groups. But even when they often contain strong algorithms and procedures, their user friendliness is rather low and some of them are yet not suitable for solving complicated technical problems. Nevertheless, even in this domain the progress is fast. Our group has been developing for years software Agros2D [1], [2] based on the library Hermes2D [3].

The aim of this paper is to show the structure of the code, its unique features, and an illustrative example aimed at modeling of a passive axial magnetic bearing.

II. Basic Characteristics and Unique Features of Agros2D

Agros2D is a multi-platform engineering software for solution of nonlinear coupled problems from a number of engineering disciplines (example of its main window being depicted in Fig. 1) The code representing a powerful preprocessor

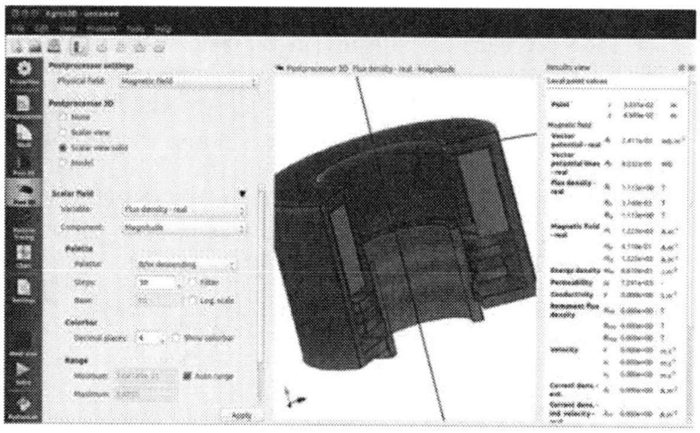

Figure 1. Main window of Agros2D

and postprocessor cooperates with the library Hermes that includes the most advanced numerical algorithms based on a fully adaptive higher-order finite element method and also discontinuous Galerkin method. Both packages written in C++ are freely distributable under the GNU General Public License.

The codes exhibit quite unique features, for example:

- Solution of multiphysics problems characterized by weak or hard coupling of the PDE system) with multi-mesh technology (every physical field can be calculated on a different mesh generally varying in time).
- Steady state, transient and harmonic analysis for physical fields (electric currents, electrostatics, magnetic, heat transfer, structural mechanics, radio frequency, acoustics, incompressible flow.

- Advanced implementations of the Newton and Picard solvers for the solutions of problems described by nonlinear PDEs.
- Full space (h, p and hp-adaptivity) [4] and also time adaptivity.
- Triangular, quadrilateral and curved elements with hanging nodes of any level.
- Advanced optimization techniques using the genetic algorithms (simulated annealing, NSGA II and several others).
- Higher-order methods of particle tracing, higher-order Eggshell Method for computation of forces acting in nonlinear magnetic fields.

Some of these features important for modeling and design of the magnetic bearings are described in more detail:

III. Solution of Nonlinear Magnetic Problems by Newton Method

The use of the Newton method is a well-established option for solving nonlinear PDEs, see, e.g., [5], [6]. The theory of its convergence is, however, rather complicated and the results are restricted to simple academic examples. The matter is even more complicated due to the necessity of damping, attempts to save computation time by using the same Jacobian for several successive steps, or other variants of the algorithm. In the case of solution of complicated problems arising in the engineering practice, one usually has to fine-tune the method for each problem individually in order to secure sufficiently fast convergence and avoid extensively long calculation.

The theory of the Newton method is widely discussed in the literature. Most of the results are, however, from the field of large nonlinear systems not coming from from the area of partial differential equations. There are generally several parameters of the Newton method, that may be changed to secure the maximum speed of the convergence process. We will first present the way Jacobian and residual are derived form the weak formulation. In the following, detailed implementation of the Newton solver will be described, together with all different variants and parameters.

A. Damping

By doing one step of the Newton method with damping coefficient $c_{\mathrm{D}} \leq 1$ we mean using equation

$$Y^{k+1} = Y^k + c_{\mathrm{D}} D^{k+1} . \tag{1}$$

The calculated direction D^{k+1} is used, but the step in that direction is made shorter. It makes the Newton method more secure. Damping is the key for solving difficult problems with strong nonlinearities. Without damping, some problems would not be solvable. On the other hand, using small fixed damping coefficient for all problems would seriously damage convergence of the method for simpler problems. The solution is the use of more variants of the method. Some of the parameters controlling damping in our algorithm follow:

- Selection of fixed or automatic damping.
- Damping coefficient c_{D} for fixed damping (if $c_{\mathrm{D}} = 1$, then no damping is used) and maximum acceptable damping coefficient for the automatic damping.

The following points only apply for the automatic damping.

- Maximum allowed ratio r_{D} between the actual and previous residual norm. If this ratio is exceeded, the step is refused, the damping factor is decreased and the step is repeated.
- Coefficient for decrease of the damping factor in the situation described in the previous item. For all calculations in this paper, the value 1/2 has been used.
- Number of steps, after which the damping factor may be increased again.

B. Jacobian reuse

The modified Newton method (using the same Jacobian for all iterations) has been proven useful by many authors. The advantage of not evaluating the Jacobian in each step is huge and justifies a somewhat slower convergence. For some problems, however, it is not applicable, since the method would not converge at all. In our algorithm, we introduced several parameters that control reuse of the Jacobian and allow benefit from both approaches.

- Jacobian reuse may be switched off.
- Maximum allowed ratio r_{J} between the actual and previous residual norms for Jacobian reuse.
- Maximum number m_{J} of the steps with the same Jacobian.

If the Jacobian reuse is allowed, after each full step with calculation of Jacobian, at most m_{J} steps with this Jacobian are performed. If, however, in some of those steps the residual is not reduced by the factor r_{J}, this step (with the reused Jacobian) is refused and the Jacobian is recalculated again. Again, it is not clear what are the optimal values of m_{J} for general nonlinear problems. Not to reuse Jacobian is the safest way, but, for some problems, the Jacobian reuse may speed up the algorithm significantly. On the other hand, using large values of r_{J} close to 1 (which means that even very small decrease of the residual norm is accepted) may be dangerous. In some cases such setting increases the possibility of getting in a wrong direction and ending up in a local minimum. Values $r_{\mathrm{J}} \geq 1$ are definitely not recommended.

IV. Force Calculation Methods

The evaluation of forces and torques acting on ferromagnetic elements in non-linear magnetic fields is still a challenge. The algorithms based on the Virtual Work and Maxwell Stress Tensor are slow, the convergence is rather poor and accuracy of the results is far from being high. Lower-order Eggshell Methods are generally better [7], but they often suffer from the same reasons Agros2D offers an efficient higher-order eggshell method whose algorithms are described below.

A. Computation of Forces Using Maxwell Stress Tensor

The Maxwell stress tensor $\boldsymbol{S}_{\mathrm{M}}$ can be written in the form

$$\boldsymbol{S}_{\mathrm{M}} = -\frac{1}{2\mu}(\boldsymbol{B}\cdot\boldsymbol{B})\boldsymbol{I} + \frac{1}{\mu}\boldsymbol{B}\otimes\boldsymbol{B}\,, \tag{2}$$

where $\boldsymbol{B}$ is the magnetic flux density vector, μ denotes the permeability, $\boldsymbol{I}$ is the unit diagonal matrix and the sign $\otimes$ represents the dyadic product. The force $\boldsymbol{F}_{\mathrm{M}}$ acting on the body of volume V and boundary S is now given by the integral

$$\boldsymbol{F}_{\mathrm{M}} = \int_S \boldsymbol{S}_{\mathrm{M}}\,\mathrm{d}\boldsymbol{S}\,, \tag{3}$$

where S denotes the outward unit normal vector to the boundary of the body. In the same way the torque acting on the same body is given by integral

$$\boldsymbol{T}_{\mathrm{M}} = \int_S \boldsymbol{r}\times(\boldsymbol{S}_{\mathrm{M}}\,\mathrm{d}\boldsymbol{S})\,, \tag{4}$$

where $\boldsymbol{r}$ is the position vector.

B. Eggshell method

The Eggshell method is based on introducing a thin shell over the surface of the body and a function satisfying the conditions $\gamma = 0$ along the external boundary of the shell and $\gamma = 1$ along its internal boundary. This function can be chosen in many ways. In our case, we used the solution of the Laplace equation in the eggshell with the prescribed boundary conditions. From (3) and definition of γ, the resultant force on the body can be expressed in the form

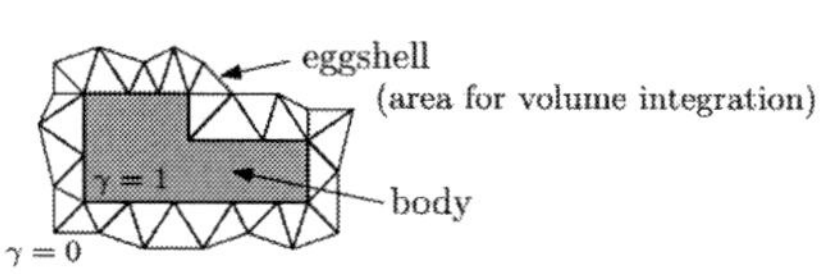

Figure 2. Principal scheme of Eggshell method

$$\boldsymbol{F}_{\mathrm{M}} = \int_V \boldsymbol{S}_{\mathrm{M}}\cdot\mathrm{grad}\,\gamma\,\mathrm{d}V\,, \tag{5}$$

where V is the volume of the eggshell. The torque can be determined similarly. In our implementation, we use the solution of the Laplace equation in the eggshell domain obtained using higher-order finite elements. These bring additional smoothness to the solution and we can choose their orders arbitrarily. Since the eggshell domain is covered with very few elements, the additional computational cost caused by increasing the element order is quite negligible.

V. Illustrative Example

The power of the code and also of the method itself is demonstrated on the computation of the axial force acting on the passive magnetic bearing for a novel design of a valve (see Fig. 3). The principal arrangement of the bearing together with the principal dimensions is depicted in Fig. 4.

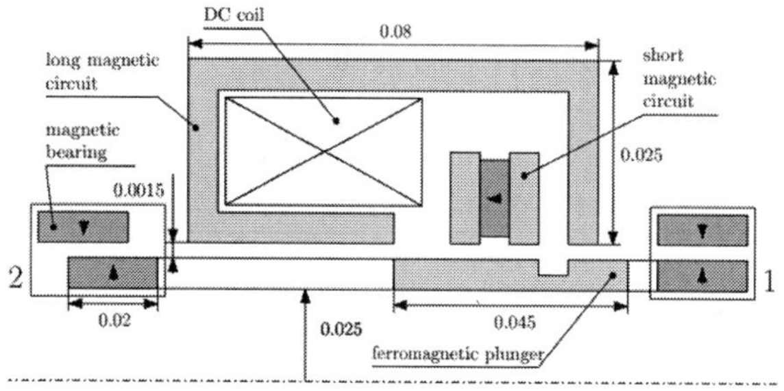

Figure 3. Magnetic circuit of valve with axial bearing with main dimensions

A. Mathematical model

The continuous mathematical model of the problem is represented by strongly nonlinear partial differential equations providing the distribution of static magnetic field in the system. The distribution of magnetic field is described in terms of magnetic vector potential $\boldsymbol{A}$ by the equation

$$\mathrm{curl}\left(\frac{1}{\mu}(\mathrm{curl}\,\boldsymbol{A} - \boldsymbol{B}_{\mathrm{r}})\right) = \boldsymbol{0}, \tag{6}$$

where μ stands for the magnetic permeability and $\boldsymbol{B}_{\mathrm{r}}$ is the remanent magnetization. A sufficiently distant artificial boundary is characterized by the Dirichlet condition $\boldsymbol{A} = \boldsymbol{0}$.

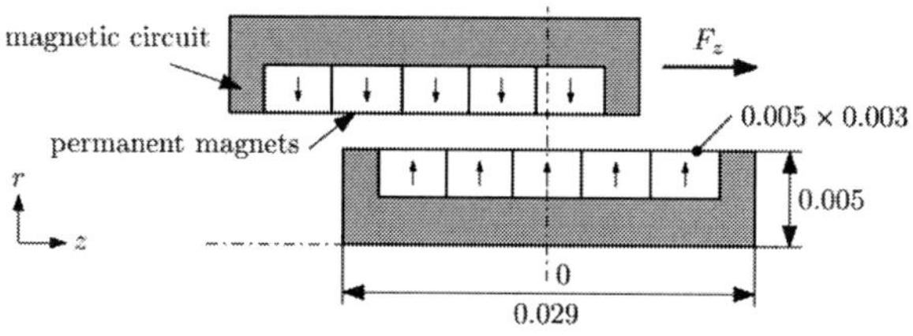

Figure 4. Detail of principal scheme of bearing

Magnetic parts are made of M-50 Bearing Steel whose saturation characteristic is shown in Fig. 5. The permanent

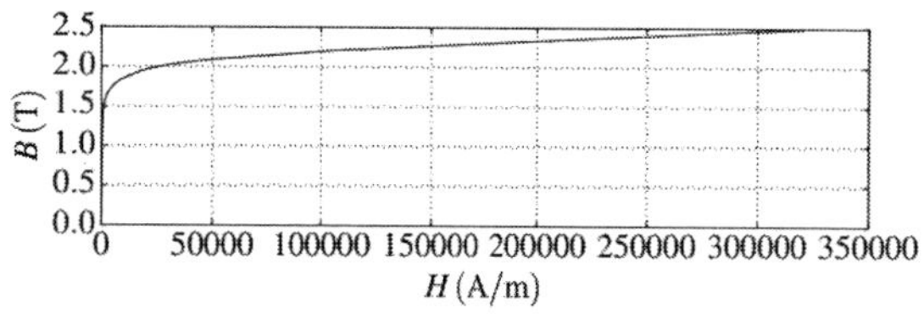

Figure 5. Saturation curve of bearing steel M-50

magnets VMM10 are of NdFeB type and their manufacturer provides the following parameters: remanence $B_{\mathrm{r}} = 1.4\,\mathrm{T}$, relative permeability in the second quadrant $\mu_{\mathrm{r}} = 1.2$ and maximum allowable temperature is 80 °C.

B. Computation

Table I contains the parameters of computation needed for determining the forces at one position of the axial bearing

($z = 0.01\,\mathrm{m}$). The number of DOFs was comparable, the prescribed residual norm was 10^{-4}.

Table I. COMPARISON OF PARAMETERS OF COMPUTATION USING AGROS2D, COMSOL AND FEMM (UNKNOWN RESIDUAL ACCURACY)

	Agros2D 3.2	COMSOL 4.4 Newton's	COMSOL 4.4 Dogleg	FEMM 4.2
DOFs	69037	69426	69426	79908
time (s)	41	74	54	(25)
iterations	31	54	43	(21)
force (N)	8.91	8.34	8.34	8.28

Fig. 6 depicts the behavior of the absolute norm of residual on the number of iterations. Fig. 7 shows the influence of

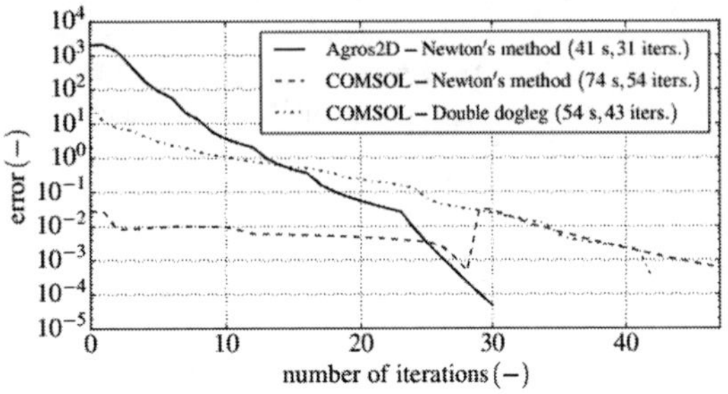

Figure 6. Comparison of absolute norm of residual on number of iterations

selection of the damping factor on the number of iterations. Fig. 8 depicts the mesh for the solution of the magnetic field

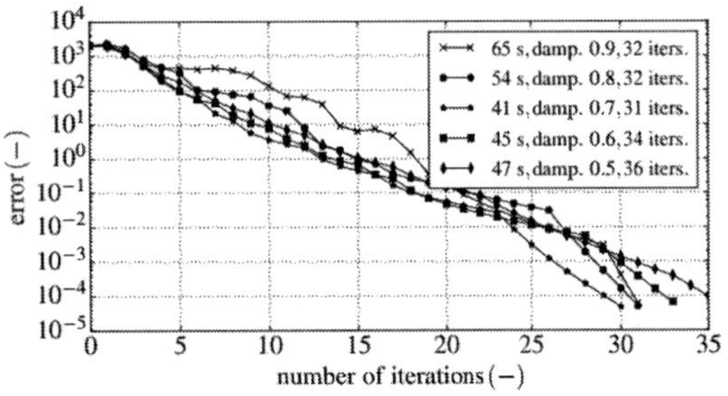

Figure 7. Influence of damping factor on curve error versus iterations

after the process of the automatic adaptivity. The number of DOFs is 69037 and solution of the system exhibits an error of 0.4 % compared to the reference solution.

C. Obtained results

Fig. 9 shows the distribution of magnetic field in the system. Fig. 10 shows the static characteristic computed by Agros2D and FEMM using the Maxwell tensor, Eggshell methods and virtual work principle. It is clear that even the results obtained by the Eggshell method are much better (the curve is smoother).

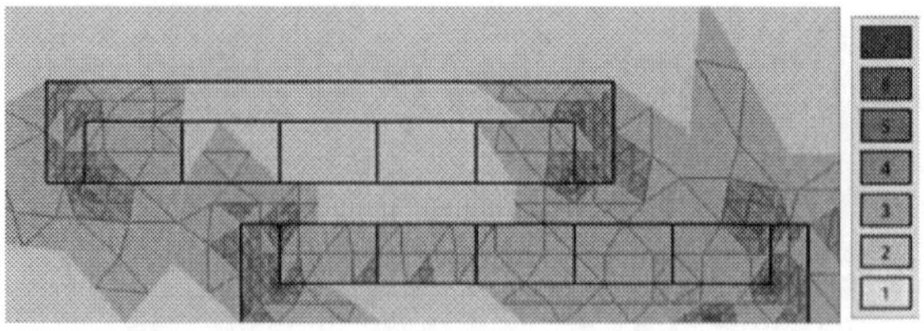

Figure 8. Discretization mesh and polynomial order used for computation

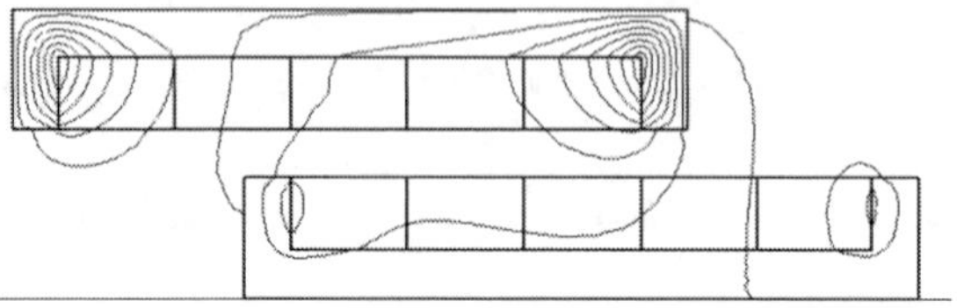

Figure 9. Magnetic force lines

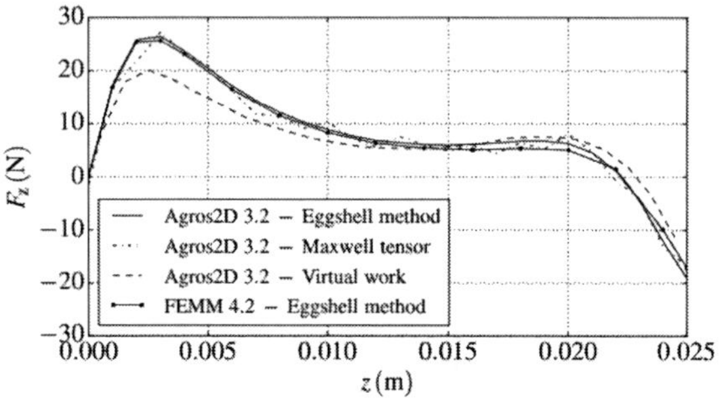

Figure 10. Comparison of Agros2D and FEMM for calculating force acting on the bearing

VI. TIME ADAPTIVITY

Time adaptivity is an important tool in improving efficiency of solution of time-dependent partial differential equations. Many problems exhibit rapid changes in solution in some parts of the time interval, while they do not change significantly otherwise. Using fixed time step for such problems would be inefficient. Rather than that, Agros2D allows the user to select adaptive time stepping. In such a case, optimal time step is determined automatically using an error estimate in the form of difference of two methods with different order.

VII. OPTIMIZATION

A continuous improvement of computers and their increasing computational power brings the possibility to solve completely new types of engineering problems. Not only larger and more complicated problems may be solved thank to the possibility of solving larger discrete problems, but also large sequence of calculations of the same model with different parameters may be performed. It can be done in parallel (thank to inherited parallelism of the considered problem), but also using ordinary PC, since the calculation time of each run may be short enough. It opens completely new areas, such as parametric analysis, material and shape optimization, inverse problems, etc., for an ordinary PC users. Agros2D brings several possibilities to perform such type of multiple calculations in a systematic way.

A. Scripting in Agros2D using Python language

The first possibility is the use of scripting. It is an integral part of the Agros2D software, which largely enhances its capabilities. Instead of developing a new scripting language, as it is usual in many commercial codes, we decided to use Python. This choice has several advantages. First of all, most users know Python, at least on an elementary level, which enables them to use scripting in Agros2D without the necessity to learn a new language. Another advantage is the possibility to use whole range of already developed scientific libraries for post-processing and visualization of results. This combination of advanced finite-element calculations with large amount of ready-to-use Python libraries is a very powerful tool.

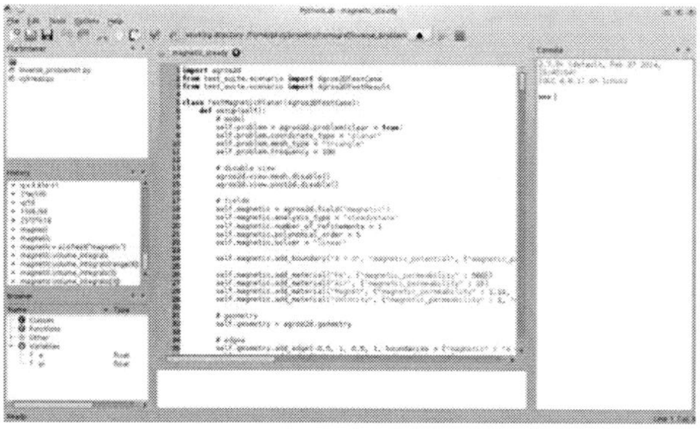

Figure 11. Main window of PythonLab

In Agros2D, almost every action or setting (including geometry definition, material specification, solver setting, evaluation of post-processor quantities, etc.), that can be done through graphical user interface, can be done from the script as well. This is a very important feature for more advanced analyses. It is also possible to use the GUI to create a basic model and convert it to an equivalent script, which can be used as a starting point.

B. Optimization framework OptiLab

The integration of a Python interpret into the application is a truly powerful tool that allows the user to perform whole range of different and specific analyses. Each user has different needs and no GUI, however complicated, can fulfill all such requirements. One of the most often used application of previously described functionality is a parametric analysis and optimization. Although it is perfectly possible to perform all the analysis (development of optimization algorithm, management of different designs, post-processing, graphs and figures generation) using our scripting tool PythonLab, it is convenient to approach those standard problems of engineering design in a more systematic way. A newly developed optimization framework OptiLab is intended to provide such service.

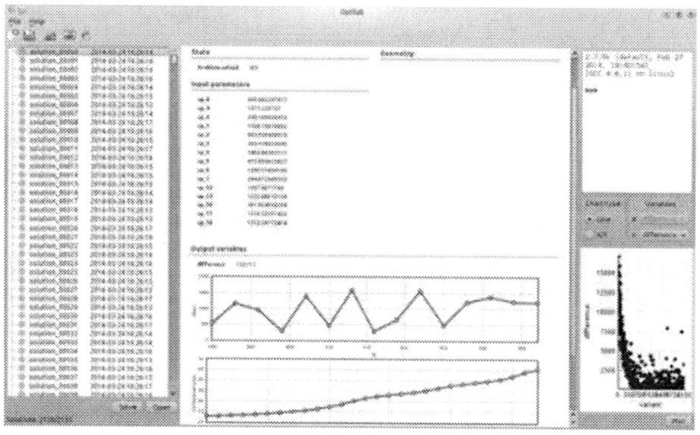

Figure 12. Main window of OptiLab

C. Parametric analysis

Parametric analysis can be viewed as a first part of the optimization process. It is very simple and productive using Agros2D and OptiLab. First an ordinary model is created with ease using Agros2D graphical user interface. After preliminary calculations are carried out and the model is verified, parametric analysis may start. Almost any number quantity defining shape, material properties or boundary conditions can be replaced by a variable defined in the startup script. This may be useful alone, but becomes extremely important after conversion of the model into Python script. All such defined variables may be controlled using the script itself (say, using for-cycle) or using diagnostic tools of OptiLab.

The OptiLab application is intended primarily as a management tool for results of multiple calculations analysis, no matter how it has been obtained, e.g. simple parameter sweep, optimization using gradient methods, genetic algorithms, multi-criteria optimization, etc. Its role is in managing sets of results described by design parameters such as geometry dimensions, material properties, etc. and calculated quantities, such as point values, integral values, etc. Entries corresponding to individual calculations may also contain images, graphs, convergence information and other more complex results, so that the analysis may be performed on those calculated data only without the necessity to perform time-consuming finite element calculations again.

D. Optimization algorithms

Optimization is a more advanced step in the engineering design, which automates the search of design parameters, which has been described in the previous section. The work-flow, however, does not differ much from the previously described case. One also has to create a parametric model and can use OptiLab for management of obtained parametric results and to create various graphs, statistics, etc. The only difference is the need of employment of some optimization algorithm. The user can create his own, use some of standard optimization algorithms implemented in some of Pythons scientific libraries or use one of the tools provided in our framework. The only thing that has to be done is to define optimization functionals

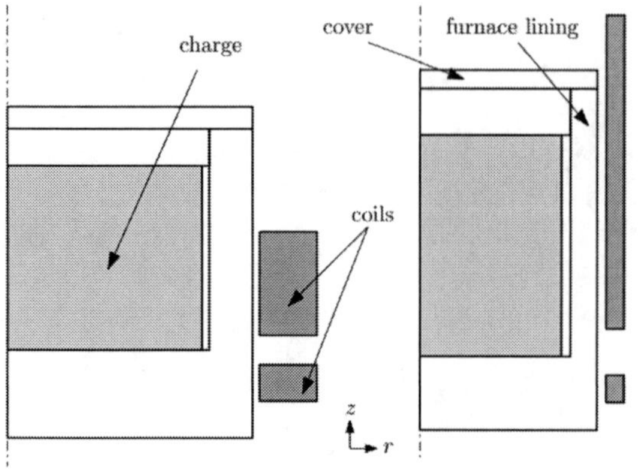

Figure 13. Two variants of geometry design of inductor

based on results of the calculation of the model. We have implemented some basic optimization algorithms, conjugated gradients and genetic algorithms.

E. Multi-objective optimization

In some situations, it is not possible to define one functional only, which should be maximized or minimized. In such cases, more functionals may be defined, see, e.g., [8]. Than it is not possible to find an optimal design, since the requirements usually go against each other (such as maximization of output power and minimization of energy consumption). Instead of that, a Pareto front (such as example shown in Fig. 14) is found. Designs on this Pareto front are all optimal in a sense that it is not possible to improve one functional without worsening the other one. Tools for a simple work with multi-objective optimization are also included in the OptiLab toolbox.

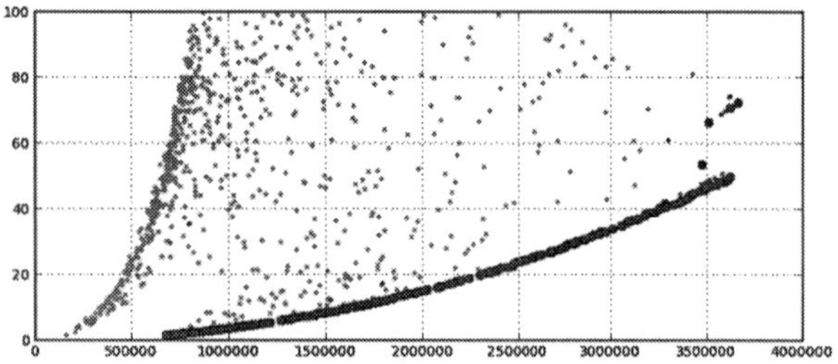

Figure 14. Pareto front as a result of optimization algorithm

The use of OptiLab is not restricted to multiple calculations with Agros2D software. It can be used as a stand alone application, provided that some other application would generate data in a required format. It is allowed by the fact that even though Agros2D, PythonLab and OptiLab are intended as collaborating pieces of software (and by this collaboration a maximal efficiency may be achieved), they can all work as individual applications. This loose interconnection brings large flexibility and allows to solve much more complicated and non-standard problems that one huge, overcomplicated application would permit.

F. Illustrative example

For the concrete examples of optimization of various engineering problems, that have been done using our software, we refer the reader to some of our previous works [8], [9], [10]. In order to show what type of results can be obtained using optimization tools we included sample results of one such example. The goal was to optimize shape of a furnace and an inductor used to heat up the metal. Two possible variants of such design can be seen in Fig. 13. The goal was to maximize heat generated inside the metal (x axis) and, at the same time, minimize differences in temperature of various areas of the metal (y axis in Fig. 14, respectively). Those two goals cannot be achieved at the same time, so a Pareto front is formed by optimal designs. Each point in the figure corresponds to one calculation with different parameters. Large points are those on the Pareto front. It is than a task for an expert to choose appropriate trade-off.

VIII. Acknowledgements

This work was supported by the European Regional Development Fund and Ministry of Education, Youth and Sports of the Czech Republic (project No. CZ.1.05/2.1.00/03.0094: Regional Innovation Center for Electrical Engineering - RICE) and by the project GACR P102/11/0498 (Grant Agency of the Czech Republic).

References

[1] Karban, P., Mach, F., Kus, P., Panek, D., Dolezel, I. *Numerical Solution of Coupled Problems Using Code Agros2D.* Computing 95 (2013), No. 1, Supplement, pp 381–408, DOI 10.1007/s00607-013-0294-4.

[2] Agros2D, http://www.agros2d.org

[3] Hermes, http://www.hpfem.org/hermes

[4] Solin, P., Cerveny, J, Dolezel, I. *Arbitrary-Level Hanging Nodes and Automatic Adaptivity in the hp-FEM.* Math. Comput. Simul., 2008, Volume 77, pp. 117–132.

[5] Wriggers, P., *Nonlinear Finite Element Methods*, Springer-Verlag, Berlin Heidelberg, 2010.

[6] Deuflhard, P. , *Newton Methods for Nonlinear Problems*, Springer Series in Computational Mathematics, Volume 35, 2004.

[7] Henrotte, F., Felden, M, vander Giet, M., Hameyer, K. *The Eggshell Approach for the Computation of Electromagnetic Forces in 2D and 3D.* COMPEL: The International Journal for Computation and Mathematics in Electrical and Electronic Engineering 23 (2004), No. 4, pp. 996–1005.

[8] Di Barba, P., Dolezel, I., Karban, P., Kus, P., Mach, F., Mognaschi, M.E., Savini, A. *Multiphysics field analysis and multiobjective design optimization: a benchmark problem.* Inverse Problems in Science and Engineering 22 (2014), No. 7, pp. 1214-1225.

[9] Mach, F., Kus, P., Karban, P., Dolezel, I. *Optimization of the system for induction heating of nonmagnetic cylindrical billets in rotating magnetic field produced by permanent magnets.* Computing 95 (2013), No. 1, Supplement, pp 537–552.

[10] Mach, F., Kus, P., Karban, P., Dolezel, I. *Optimized arrangement of device for electrostatic separation of plastic particles* ELEKTRO (2012), pp 431–434.

Review of Electrodynamic Bearings

Torbjörn A. Lembke [a]

[a] Lembke Elektromekaniska Verkstad LEV AB, Lärkträdsvägen 28, 740 11 Länna, Sweden,
torbjorn.lembke@lembke-elektromekaniska.se

Abstract—**When Earnshaw stated his theorem in 1839, it did not seem possible to levitate objects in static magnetic fields. 100 years later Braunbeck 1939 was able to show mathematically that levitation is actually possible in unvarying magnetic fields for diamagnetic materials. Today levitating toys based on weakly diamagnetic graphite and permanent magnets are available on the market. Some other materials like superconductors or moving conductors have properties that to a much higher extent resemble those of diamagnets, and these may also levitate in a static magnetic field. Levitating experimental full scale trains based on both of these effects exist today. In this paper the most important technical achievements regarding electrodynamic levitation will be reviewed, dating from 1922 and on, with particular focus on rotating bearings and demonstrated levitating prototypes.**

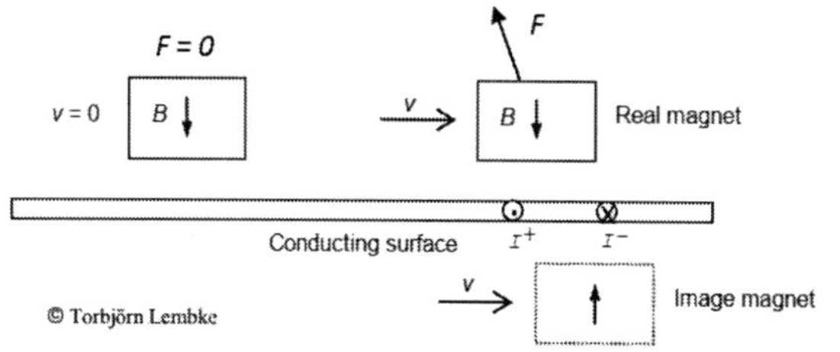

Figure 1. The electrodynamic repulsion principle. A magnet moves over a conducting surface, and forces act as if there is a mirror image below and slightly in front of the real magnet.

I. Early Work

A. Introduction

When Earnshaw [1] studied the "Luminiferous Ether" he found that electrically charged particles could not be stably levitated in a static electric field. The same equation is, at least mathematically, valid for magnetic "monopoles" in a static magnetic field. This theorem has later been extended to dipoles and magnets of arbitrary shapes.

1922 Anschütz-Kaempfe [2] patented an apparatus for stable levitation based on the electrodynamic effect from an alternating field, which not contradicts the Earnshaw theorem. However, in 1939 Braunbeck [3] was able to show that levitation is actually possible in unvarying magnetic fields for diamagnetic materials like graphite.

Superconductors with Meissner effect are often described as perfect diamagnets and are the best example of Braunbecks findings. Today diamagnetic bearings with either graphite or superconducting elements are available on the market.

Actually, the Meissner effect itself is not necessary for levitation. The zero resistance of superconductors is enough, and this is a special case of the electrodynamic effect, in which the induced currents do not decay. Thus permanent levitation is possible. Should the resistance increase, the currents have to be constantly renewed by an alternating field created by AC-coils or moving magnets.

B. The "Magnetic Mirror"

The electrodynamic levitation principle above is based on Lenz′ law. According to this law a change of magnetic flux through a conductor will induce a voltage in the conductor in such a way that a current is produced that tries to maintain the magnetic flux at the original level. In reality this means that any changing electromagnetic flux will be reflected on a conducting surface, thus giving rise to repulsive forces. The maximum force is generated at infinitely high speed. This force can be calculated as if there is a mirror image of the real magnet on the opposite side of the surface. This is very much like the superconducting effect described earlier, and the bearings are equally strong, at least for high frequencies. At low frequency an electromagnetic bearing has very low lift force. Thus at low speed electrodynamicly levitated vehicles will have to touch down on for instance retractable wheels, while rotating electrodynamic bearings will require auxiliary or integrated touch down bearings.

The "magnetic mirror" analogy brakes down at low speed. However, the author has found it quite useful from a pedagogic point of view to illustrate the image magnet as a weaker magnet slightly in front of the real one trying to push it upwards and backwards as in fig. 1. The explanation would be that the image magnet is induced, not by the field itself, but by the field gradients, and they of course come in front of the real magnet.

C. First Patents

In 1922 Anschütz and Kaempfe [3] patented an apparatus for stabilization based on the electrodynamic effect. They levitated a stationary conducting spherical gyro platform in a varying magnetic field. The electrodynamic effect was used to stabilize the gyro, but since the field was too weak to levitate the mass of the body, gravity was compensated for by letting it barely float in a liquid.

This product does not seem to have come to much use, but it is historically important from one point of view. It demonstrates that eddy currents alone are not enough to levitate an object over a long period, since the heat generation both in the coils and in the conductor is too high. One way of gravity compensation is necessary so that the eddy currents only need to be responsible for the stabilizing centering effect.

However, for some applications heat generation is advantageous, like the one patented by Okress and Wroughton et al in 1952 [4]. It combines the effects of levitating, melting and mixing pure metallurgic specimens, a method that is still in use.

D. The Null Flux Scheme

The first successful method of reducing eddy current losses is known to have been presented by Powell & Danby [5], [6] who patented their solution in 1969 for use in a ground transportation vehicle. Later, several authors have proposed means of reducing eddy current losses in linear electrodynamic bearings for magnetically levitated trains, maglevs, based on their results. Powell & Danby introduced what they called the "Null Flux Scheme" for maglev applications, where a track of conducting coils are placed between oppositely directed superconducting magnets.

Later, in 1974, Powell & Danby also introduced an unloading device of superconducting coils and an iron plate [8], which they integrated into the design. The unloader allows the stabilizing coils to operate in the null-flux region.

A variant of this principle, used for rotating bearings, was patented by Sacerdoti et al, [7] in 1971. They used a rotating conducting plate between permanent magnets. The concept is described by the author in fig. 2 left. The benefit with the null flux scheme is that the conducting plate operates in a null or at least reduced flux region. Thus the magnets are arranged in such a way that they do not induce currents until they are needed, that is when the conductor position is of center.

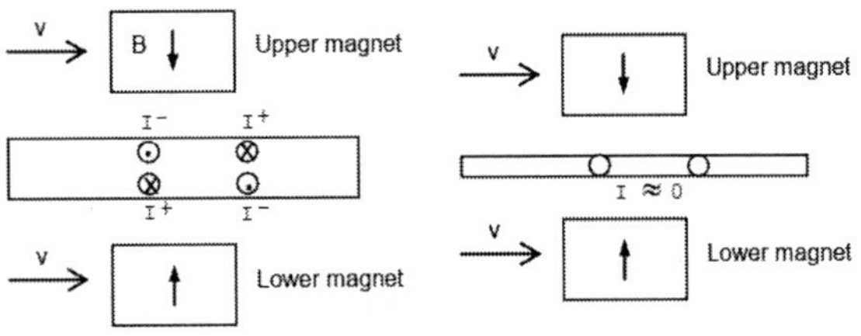

Figure 2. The null flux scheme. Magnets are placed over and under the plate oppositely. If the plate is thinner than one skin depth, most of the eddy currents cancel out.

For best utilization of the null-flux effect, the author [11] found that it is important to adjust the plate thickness to the skin depth. The plate to the left is more than two skin depths thick and there are surface currents counteracting each other but still producing losses. If the plate thickness is thinner than two skin depths the currents begin to cancel each other out, and when the thickness equals one skin depth or is slightly thinner, then the surface currents will almost cancel each other out, which is illustrated in fig. 2 right. This solution was analyzed by the author in 1989-1990 [11] and later patented for linear as well as rotating bearings in 1992 [12]. Figure 3 shows an axial and a radial bearing based on this concept.

To avoid confusion the author will refer to this concept as to the thin plate null flux scheme.

Most linear levitation principles above can easily be adopted to rotating bearings by simply "bending the track". Here there is the option to either rotate the magnets or to rotate the conductor. Since magnetic bearings are usually used for high speed applications, the solution with rotating magnets is not so attractive due to the need for proper bandages. Nevertheless, the majority of successfully demonstrated prototypes have rotating magnets, since this approach leads to less rotating losses and thus better stability.

The arrangement by Sacerdoti et al. was used as an axial bearing in a "Self-Centering Rotary Magnetic Suspension Device" which they patented [7]. It combined the electrodynamic axial bearing with rotating conductor with two permanent magnet radial bearings of reluctance type. These bearings were adjusted to compensate for the weight, so that the conductor could operate in the centered null flux region. It is not known to the author if this system ever levitated, but as described in the patent no damper was present, and thus stable levitation would not have been possible. On the other hand such details are often excluded in patent applications, so the author believes that the concept actually might have worked.

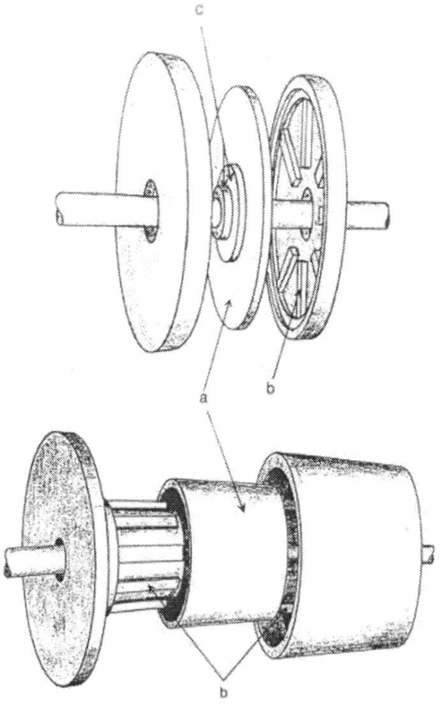

Figure 3. Axial eddy current bearing above, and radial bearing below, based on the thin plate null flux scheme. Both bearings are heteropolar.

One way to increase the lift capacity of these bearings is to use a halbach arrangement instead of the more simple south north arrangements shown. It is very well suited for the null flux scheme. Post et al. did lots of work in this field, both for linear guides and for rotating bearings for flywheels. [13], [16], [18]. Still, Halbach arrangements are seriously examined and investigated for linear transportation also without null flux configurations by for instance Cho et al. [31].

E. *Homopolar Bearings*

The bearings described above are all "heteropolar" and can be made as either linear or rotating bearings. However, for rotating bearings it is absolutely necessary to have some sort of loss reduction mechanism, otherways the conducting plate or coil will get overheated. A breakthrough for rotating bearings came with the invention of the "homopolar" design, shown in fig. 4 and 5. This concept is only possible with bearings with rotating conductors. The author filed a patent in 1997 [14], but the principle as such was described by Basore [9], [10] already in 1980.

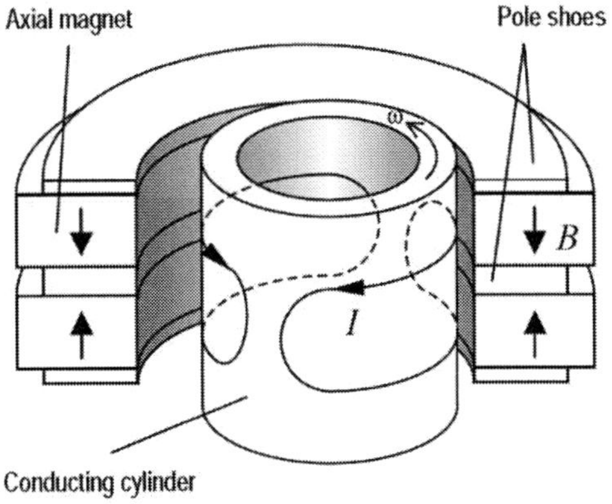

Figure 4. Working principle of homopolar bearing. Restoring eddy currents I are induced when the rotor is displaced.

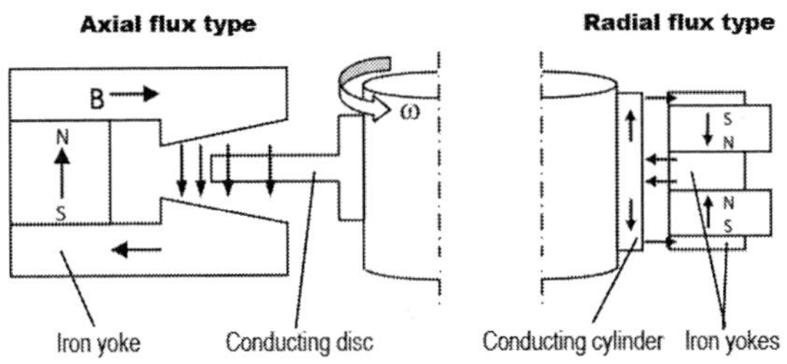

Figure 5. Two types of homopolar bearings. Axial flux type to the left and radial flux type to the right.

Bleuler et al. also did similar work at that time. [17] In 2000 they presented a levitating prototype, fig. 10. In a bearing with a homopolar field there are no induced voltages that need to be eliminated like in the multipole bearings above. Filatov [19] introduced the term "Null-E magnetic bearings" for this bearing concept, to distinguish it from the previously described "Null flux bearings". The circular design avoids the induction of any unnecessary eddy currents, as long as the rotor is centered in the bearing stator. The result is the same, but the homopolar concept is much simpler as no winding is required to avoid all the unnecessary eddy currents. Homopolar bearings can be of either of the axial flux or radial flux type, see figure 5. Filatov and Murakami et al. [15] focused on axial flux bearings and built as well a wire wound rotor and also proposed a solid rotor for a high speed satellite flywheel. The author [25], [26] focused on radial flux solid rotor bearings which have the advantage that they preferably are designed like conventional bearings, fig 11.

II. Demonstrated Prototypes

Many test rigs have been built by different research teams with the purpose to validate force measurements with calculated or FEM-simulated values. However, very few fully levitating prototypes are reported. The author designed one at SKF in 1990, fig. 6. It had one axial bearing, two radial bearings and one unloader of a push-pull magnet arrangement.

The shaft was connected to a gearbox via a flexibly mounted axial ball spline, so that the rotor could move freely both axially and radially, and it could also tilt.

The radial bearings were mounted in tuned rubber dampers, allowing the rotor to levitate stably from 8000 to 12000 RPM. The maximum tested speed was 14000 RPM, when the ball spline broke. A new axially rigid coupling was made, and the prototype still works after 24 years.

The temperature rise during long tests was interesting. The radial bearings got much warmer than the axial one. Obviously, the flat conductor in the axial bearing is more suited for null flux operation than the cylindrical conductor in the radial bearing, due to the geometrical difference between the inner and outer magnets.

Later the homopolar concept has also been demonstrated for full levitation. Bleulers and Sandtners prototype in fig. 10 was reported to levitate safely at 9000 RPM using an air turbine for propulsion. Much research is currently going on and several prototypes are about to be finnished in Europe based on different homopolar designs by Kluyskens et al. [28], [29] and Amati et al. [30], [32],[33].

Instead of solid conductors, it is possible to use wound coils. Filatov & Maslen [19], [20], [21] used rotating coils for the radial bearing and a repulsive PM-bearing for axial levitation and tilt control, fig 5. For improved stability they introduced a series connected inductor, something that was proposed already by Powell and Danby for linear applications. Filatov et al. made a testrig, and the prototype rotor is shown in fig. 7. It was demonstrated and safely levitated in the speed range of 1080 to 3000 RPM.

Figure 6. Prototype with heteropolar bearings designed by the author at SKF in 1990.

The rotor was powered by an external motor via a coupling which was disconnected for spin down tests, which could be performed down to the lower speed stability limit of 1080 RPM.

Figure 7. Electrodynamic bearing setup designed by Filatov et al.

Aspers [22] patented a system with rotating magnets between two wound coil systems forming an axial bearing. A null flux scheme is thus not required, since the induced coil voltage can be cancelled out by simply connecting the coils oppositely, fig. 8. This system, in a slightly simplified version, was analyzed by Sandtner & Bleuler [23], [24] and a fully operational prototype was demonstrated in 2004, fig. 9. Like Sacerdoti et al. they used permanent magnet radial bearings. They did mount them in dampers though. They reported a take-off speed of 4800 RPM and stable operation up to the maximum demonstrated speed of 6000 RPM.

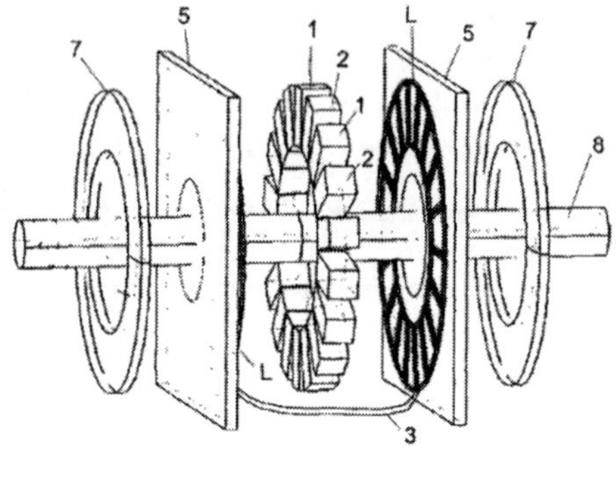

Figure 8. Aspers patented electrodynamic bearing

Figure 9. Aspers, Sandtners and Bleulers test rig.

The prototype also used the rotating magnets as motor magnets, by adding additional motor coils behind the stator windings.

As mentioned earlier, Bleulers and Sandtner also did work on homopolar bearings. Their prototype from year 2000 in fig. 10 was reported to levitate safely at 9000 RPM. The author has also tested this bearing, up to 90000 RPM in a vacuum pump [26], but it was not fully levitating since the pump had one ball bearing.

Damping is crucial for electrodynamic bearings [20], and particularly for homopolar designs, since the conductors need to be rotating. To understand the dynamics of electrodynamic bearings it is necessary take speed dependent rotating damping into account, and this has so far only been done by Filatov et al. [20], Kluyskens et al. [28] and Amati et al. [29]. To facilitate implementation of these bearings, some design methodologies have been developed by the author [11] and by Amati et al. [32].

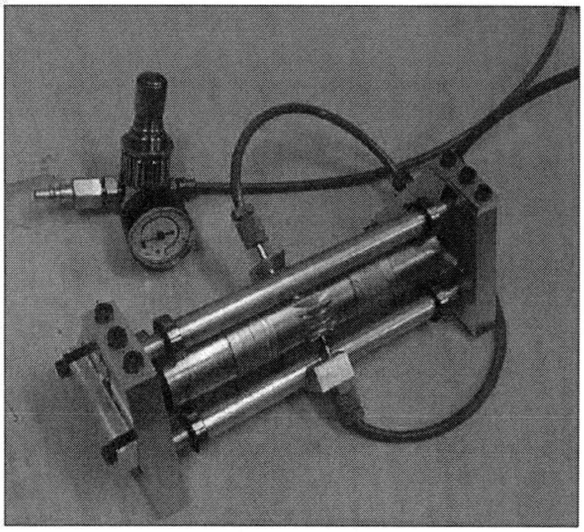

Fig. 10. Bleulers and Sandtners test rig with air turbine and homopolar bearings.

Figure 11. Homopolar bearings allow for simple and idustrially attractive designs.

III. Future Prospects

To prognosticate the future for electrodynamic bearings might be too early. However, by comparing the levitating prototypes presented in this paper, some predictions can be made:

- Heteropolar designs are stable and easy to use and will be used for low and medium high speed, but are not so well suited for high speed operations.
- Conventional permanent magnet radial bearings have been successfully demonstrated together with heteropolar axial bearings, so this combination will likely also find a market niche.
- Homopolar designs are much stiffer and have lower losses and are simpler to manufacture than heteropolar bearings, so they are good candidates for high speed applications.
- The choice of dampers plays an important role, and to understand system dynamics is crucial but not trivial. Thus damper design will be an important research topic.

It is also undisputable that homopolar bearings, particularly in combination with passive axial bearings, exhibits extremely low losses and might provide engineers with means to develop new products or toys inspired by the reachless Perpetum Mobile.

Hybrids between passive, active and electrodynamic bearings might also evolve, each optimized for its specific application. Figure 12 shows such an example with a 1-DOF bearing with electrodynamic/active damper currently being developed by the author, [30], [34].

Figure 12. Reluctance type bearing with electrodynamic damper.

ACKNOWLEDGEMENTS

The author would like to thank, Alexei Filatov, Virginie Kluyskens, J. José Perez Loya, Jan Sandtner and Andrea Tonoli for their contributions and pictures to this paper, a paper which could have been much longer if all who have contributed to this research field through the years should have been mentioned. Hopefully this can be compensated for in a coming review article.

REFERENCES

[1] Earnshaw, S., 1839. "On the Nature of the Moloecular Forces which Regulate the Constitution of the Luminiferous Ether", *Trans. Camb. Phil. Soc.*,v 7, p 97-112 (Published 1842).

[2] Anschütz-Kaempfe, U.S. *Patent no* 1,589,039, 1922.

[3] Braunbeck W.,. Free suspension of bodies in electric and magnetic fields. (Freischwebende Körper im elektrischen und magnetischen Feld.) *Zeitschrift für Physik*, v 112, n 11-12 (1939), p 753-763. 1939.

[4] Okress E.C., Wroughton D.M., Comentez G., Bracew P.H., Kelly J.C.R., "Electromagnetic levitation of solid and molten metals". *Journal of applied Physics*, v 23 n 5 (May, 1952) p 545-552; n 12 (Dec) p 1413, 1952.

[5] J. Powell and G. Danby, "Electromagnetic Inductive Suspension and Stabilization System for a Ground Vehicle", *U.S. Patent* 3 470 828, Oct.7, 1969.

[6] J. Powell and G. Danby, "Recent Advances in Engineering Science, Vol 5 *Gordon & Breach* 1970.

[7] Sacerdoti G., Catitti A., Soglia L-L, "Self-Centering Rotary Magnetic Suspension Device" *Italian patent* no 49943/71, 1971.

[8] Danby G.T., Jackson J.W. and Powell J.R. "Force Calculations for Hybrid (Ferro-Nullflux) Low-Drag Systems", *Toronto Intermag Conf.* 1974.

[9] Basore P. A., "Passive Stabilization of Flywheel Magnetic Bearings," Master's thesis, *Massachusetts Institute of Technology* (USA), 1980.

[10] Basore, P. A., "Self-Stabilizing Magnetic Bearings for Flywheels", PhD Thesis in Electrical Engineering and Computeer Science, *Massachusetts Institute of Technology* (USA), 1982.

[11] Torbjörn A. Lembke,"Design of Magnetic Induction Bearings", Master thesis, *Chalmers University of Technology*, Göteborg, Sweden, 1990.

[12] Lembke T. A., Eddy Induced Magnetic Bearing, *Swedish Patent Application* SE19920003090, 1992.

[13] Post, R. F., Fowler T. K. Post, S. F., "A High-Efficiency Electromechanical Battery", Proceedings of the IEEE, Vol 81 (1993), pp 462-474

[14] Lembke T. A., Electrodynamic Magnetic Bearing, *Patent Application* WO9832981 1998 (Priority date 1997)

[15] Murakami C. and Satoh I., "Experiments of a Very Simple Radial-Passive Magnetic Bearing Based on Eddy Currents", In *Proceedings of the 7th International Symposium on Magnetic Bearings*, March 2000.

[16] Bender D. and Post R. F., "Ambient Temperature Passive Magnetic Bearings for Flywheel Energy Storage Systems", In *Proceedings of the 7th International Symposium on Magnetic Bearings*, March 2000.

[17] Bermudez, J-L., Zanolli S., Sandtner J., Bleuler H., Benabderrahmane C., Preliminary Experiments on an Eddy Current Bearing, 7th *International Symposium on Magnetic Bearings*, 2000.

[18] Kratz R., Post R. F., "A Null-Current Electro-Dynamic Levitation System", *IEEE Trans on Appl Supercond* Vol 12, No 1 March 2002

[19] Filatov, A., "Null-E" Magnetic Bearings, PhD dissertation, *University of Virginia*, August 2002.

[20] Filatov, A. V., Maslen, E. H., and Gillies, G. T., "A Method of Suspension of Rotating Bodies Using Electromagnetic Forces", *Journal of Applied Physics*, Vol. 91 (2002) pp. 2355-2371.

[21] Filatov, A. V., Maslen, E. H., and Gillies, G. T., "Stability of an Electrodynamic Suspension" *Journal of Applied Physics*, Vol. 92 (2002), pp. 3345-3353.

[22] Asper H K, Passive, Dynamically Stabilizing Magnetic Beraing and Drive Unit, *Patent Application* WO 03/098064 2003.

[23] Sandtner J. and Bleuler H., "Electrodynamic Passive Magnetic Bearings with Planar Halbach Arrays", In *Proceedings of the 9th International Symposium on Magnetic Bearings*, August 2004.

[24] Sandtner J. and Bleuler H., "Passive Electrodynamic Magnetic Thrust Bearing Especially Designed for Constant Speed Applications ", In *Proceedings of the 9th International Symposium on Magnetic Bearings*, August 2004.

[25] Lembke T. A., 3D-FEM Analysis of a Low Loss Homopolar Induction Bearing, *Proceedings of the 9th International Symposium on Magnetic Bearings*, August 2004

[26] Lembke. T.A. "Design and Analysis of a Novel Low Loss Homopolar Electrodynamic Bearing", PhD Thesis in Electrical Machines and Drives, *Royal University of Technology*, Stockholm, Sweden 2005, ISBN 91-7178-032-7.

[27] Filatov A. V., McMullen P., Davey K., and Thompson R., "Flywheel Energy Storage System with Homopolar Electrodynamic Magnetic Bearing", In *Proceedings of the 10th International Symposium on Magnetic Bearings*, 2006

[28] Kluyskens V., Dehez B., "Dynamical electromechanical model for passive magnetic bearings", *IEEE Transactions on Magnetics*, 43, pp 3287-3292, 2007.

[29] Amati N., De Lépine X., and Tonoli A., "Modeling of electrodynamic Bearings", *ASME Journal of Vibration and Acoustics*, 130, 2008.

[30] Lembke T. A., Electrodynamic Actuator, *Patent Application* WO2010064978, 2008

[31] Cho, H-W.,Han, H-S., Bang, J-S., Sung, H-K., Kim, B-H. "Characteristic Analysis of Electrodynamic Suspension Device with Permanent Magnet Halbach Array", *Journal of Appl Physics* 105, 07A314, 2009.

[32] Amati, N., Tonoli, A., Zenerino, E., Detoni, J. G., Impinna, F., "Design Methodology of Electrodynamic Bearings", XXXVIII *Associazione Italiana per l'Analisi delle Solecitazioni*, Convegno Nazionale, No. 109, 2009.

[33] Kluyskens V., Dehez B., "Parameterized electromechanical model for magnetic bearings with induced currents", *Journal of System Design and Dynamics* - Special Issue on the *Eleventh International Symposium on Magnetic Bearings*, 2009.

[34] Lembke T. A., 1-DOF Bearing Arrangement with Passive Radial Bearings and Highly Efficient Integrated Electrodynamic Dampers, EDD, *Proceedings of ISMB13*, 2012

Test and theory of electrodynamic bearings coupled to active magnetic dampers

Fabrizio Impinna, Joaquim G. Detoni, Andrea Tonoli, Nicola Amati, Maria Pina Piccolo

Department of Mechanical and Aerospace Engineering, Mechatronics Laboratory,
Politecnico di Torino, Corso Duca degli Abruzzi, 24, 10129 Torino

Abstract— **Electrodynamic bearings (EDBs) are passive magnetic bearings that exploit the interaction between eddy currents developed in a rotating conductor and a static magnetic field to generate forces. Similar to other types of magnetic suspensions, EDBs provide contactless support, thus avoiding problems with lubrication, friction and wear. Electrodynamic bearings have also drawbacks such as the difficulty in insuring a stable levitation in a wide speed range. The paper presents a solution where the EDBs are coupled with active magnetic dampers (AMDs) to guarantee a stable levitation.**

I. Electrodynamic Levitation of Rotors

A. Introduction

Magnetic levitation of rotors has been an important research subject in the last decades. In addition to the developments on classical active magnetic bearings (AMB), recently, the research on more energy efficient systems associated to the developments on high-strength permanent magnet materials led the scientific community to focus on alternative contactless suspension systems. Usually this is achieved by combining the efforts of permanent magnet bearings (PMB), which are passive, with actively controlled electromagnets.

Examples of systems applying this concept are hybrid magnetic bearings (HMB) and self-bearing motor (SBM) devices. Among these systems, the only one that allows obtaining passive levitation is the PMB. However, since the forces arise from the interaction between magneto-static fields these bearings have a destabilizing stiffness contribution in at least one degree of freedom. An interesting alternative to generate stabilizing forces passively and that can be exploited in the field of magnetic bearings are electrodynamic bearings which have the unique characteristic of producing positive stiffness by passive means without introducing negative stiffness in any direction [1,2]. Despite the promising characteristics of this type of bearings, rotors supported by EDBs are affected by different types of instabilities that must be taken into account.

In the paper a combined EDBs – AMDs configuration is proposed as a possible solution to exploit the high reliability of EDBs overcoming the stability problem by means of AMDs. It requires studying the effect of the combination of electrodynamic and AMD forces both analytically and experimentally. This is aimed at developing an analytical model of the system taking into account the rotor and the actuators (EDBs and AMDs) characteristics. Furthermore a test rig was designed and built to test experimentally the suspension system.

The proposed hybrid configuration is attractive for high speed magnetically suspended systems because:

- it results in an increased global system reliability
- it allows downsizing the AMBs because the EDBs, for speed above the stability threshold, give a large force contribution for the levitation
- it is possible to discard the traditional ceramic landing bearings because, in case of AMBs failure, the EDBs are able to guarantee a stable levitation until a certain speed, considered safe to land, is reached.

The proposed solution can be considered as a suspension system for high speed applications such as turbomolecular pumps, oil and gas turbines and flywheels for energy storage.

B. Modeling

The paper presents the model of the complete levitating system composed of a 4 degrees of freedom model of the rotor supported by the combination of electrodynamic suspension with active magnetic dampers, shedding light onto the dynamic behavior of the system. The analytical model supports the design phase and the study of dynamic behavior and stability becomes a key factor for an aware design.

The main assumptions for the model validity are:

- 4 dof rotor, its inertia properties are represented by its mass, its moments of inertia J_p and J_t;
- conductor rigidly mounted on the rotor and its contribution is taken into account in the rotor inertia properties;
- electrical and electromechanical properties of the electrodynamic bearing are isotropic;
- stator part of the electrodynamic bearing rigidly fixed to the case of the machine;
- AMDs non-rotating damping introduced directly on the rotor [3], parameter c_n shown in Figure 1.

The analytical model of the 4 dof rotor on EDBs refers to the paper "Stability of a 4DOF rotor on electrodynamic passive magnetic bearings" published separately by the Authors in the ISMB14 conference [8]. A detailed description of EDBs analytical model is presented in [4] and the stability issue is deeply described in [5]. The guidelines to design and model the AMDs device can be found in [6].

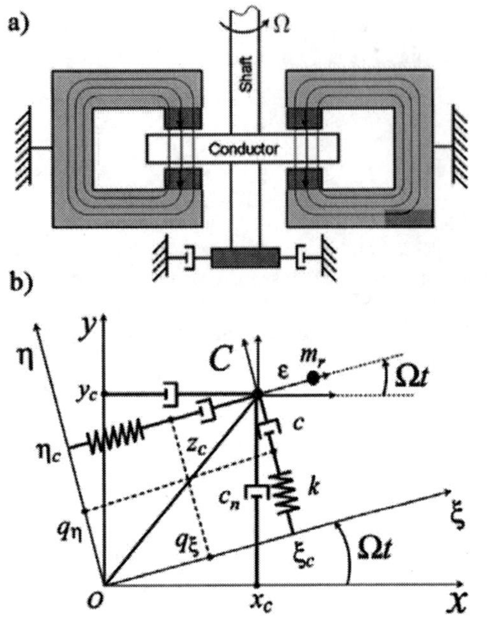

Figure 1: a) scheme of an electrodynamic bearing stabilized by introducing the damping between the rotating part of the bearing and the casing, b) schematization of the equivalent Jeffcott rotor model.

A scheme of a standard EDB suspension system with the introduction of the stabilizing non-rotating damping directly on the rotor is shown in Figure 1 a). Figure 1 b) shows the mechanical equivalent model of a single flux EDB supporting a Jeffcott rotor. The double flux EDB model can be obtained adding, in both directions η and ξ, a second branch with a damper-spring series in parallel to the one present in the scheme. Therefore the double flux EDB is characterized by the parameters k_1, c_1, k_2, c_2 instead of k and c.

The model of each single device considered separately from the others does not represent an innovative contribution to the magnetic levitation topic. By converse, the model which integrates the EDB double flux dynamic model, the AMDs system and the 4 dof model of the rotor gives new contribution to the subject since it allows describing the dynamic behavior of the whole system taking into account the cylindrical and conical instabilities representing a critical aspects of EDBs levitating systems.

Figure 2 shows the scheme of the complete model where the three main blocks represent the EDBs, the AMDs and the rotor. The scheme highlights the connections between the subsystems. Considering a reference frame where the z axis represents the axial coordinate and the x, y axes are the radial coordinates, the rotor block receives the forces in x and y directions by the EDBs, the AMDs and a generic external disturbance force contribution. The outputs of the rotor block are the x and y positions of the shaft referring to the EDBs and AMDs positions (Fig. 3 c). The EDB blocks receive the x and y positions and generate the force contribution in the same directions according to the rotating speed that is an input parameter to the whole model. No force in z direction is produced by the EDBs. The AMD blocks receive the x and y positions measured by the sensors and generate a force proportional to the first derivative of the position error to add non-rotating damping contribution to the system.
It must be taken into account that the rotor weight acts on the y axis.
Figure 3 shows the position of the EDB1, EDB2, AMD1, AMD2 with respect to the center of mass. The values of the parameters a_1, a_2, b_1, b_2 are listed in Table I.

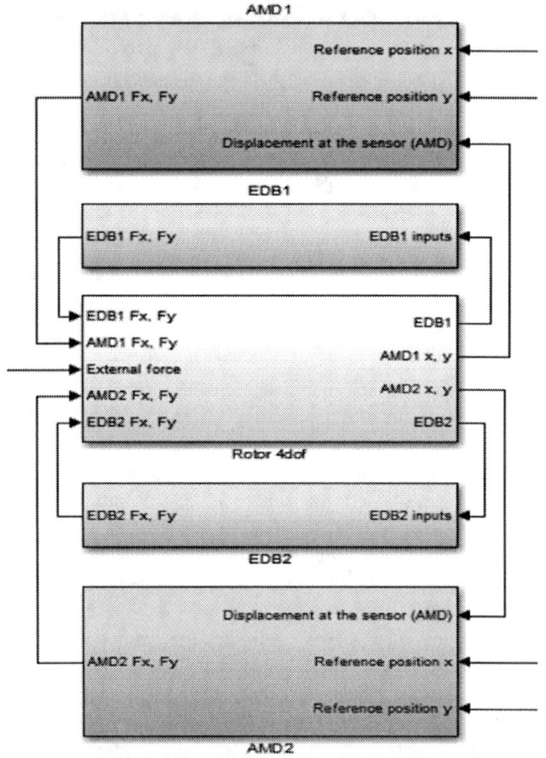

Figure 2: block diagram model of the Hybrid suspension system. AMD1 and EDB1 act on the left-hand side while AMD2 and EDB2 act on the right-hand side

C. Test Rig

A test rig was designed and built to demonstrate the feasibility of the proposed magnetic suspension and to validate the analytical model. The horizontal axis test rig is shown in Figure 3.

A standard EDB based suspension is characterized by the impossibility of providing levitation forces at zero spin speed and a stiff mechanical connection between the shaft and the basing must be ensured until the stabilization threshold speed is overcome. By converse, in the proposed solution the AMDs can be used as an active magnetic bearing to guarantee levitation for speed lower than the EDB stability threshold or for speed where the force contribution due to EDBs is not enough to levitate the rotor, and as pure damper for higher speed ensuring a stable levitation for a wide speed range. In this paragraph the mechanical layout of the test rig devised to fulfil the application requirements is presented.

The prototype is composed of four main parts: test rig structure, EDB's statoric magnetic circuit, AMD's stator coils and the rotating shaft. In Figure 3a) the main components of the test rig can be identified while Figure 3b) gives a clearer overview of the whole system. Figure 3c) highlights the rotor properties.

The structure of the test rig is composed by four stainless steel columns (1) connected to three aluminium layers that ensure a stiff construction. The connection of this structure to a seismic mass is done by means of two steel profiles (7). Two structural parts dedicated to the housing of radial position probes used for monitoring are attached to the end plates. These elements host also the supports for the statoric coils of the electromagnets (6) that works as AMBs for low rotating speed and as AMDs when the levitation is provided by EDBs. The stator of the EDB (5) contains four NdFeB N42 magnets oriented in attraction. The magnetic circuit closure is done using two plates of soft iron kept apart by an aluminium structure. The rotating part of the test rig contains a stainless steel shaft to which are attached the copper discs of the EDB (2), the motor's discs (3) and the rotating iron parts of the active actuator. A detailed view of the rotor is shown in Fig.3 c). The electric motor is a brushless axial flux motor and was designed exclusively for this application. The motor uses air wound coils in the stator to reduce the interactions between motor's disc (3) and stator (4), eliminating negative stiffness contributes due to reluctance forces. It will be used for the control of the z direction degree of freedom. The main parameters characterizing the test rig are summarized in Table I.

TABLE I. TEST BENCH PARAMETERS

	Parameter	Value
ω	Nominal speed	20000 rpm
m_r	Rotor mass	4.35 kg
l_r	Rotor length	305 mm
J_p	Polar moment of inertia	0.00572 kg m^2
J_t	Transversal moment of inertia	0.01995 kg m^2
g_r	Radial air gap	0.5 mm
a_1	EDB1 arm	67 mm
a_2	EDB2 arm	67 mm
b_1	AMD1 arm	125 mm
b_2	AMD2 arm	125 mm

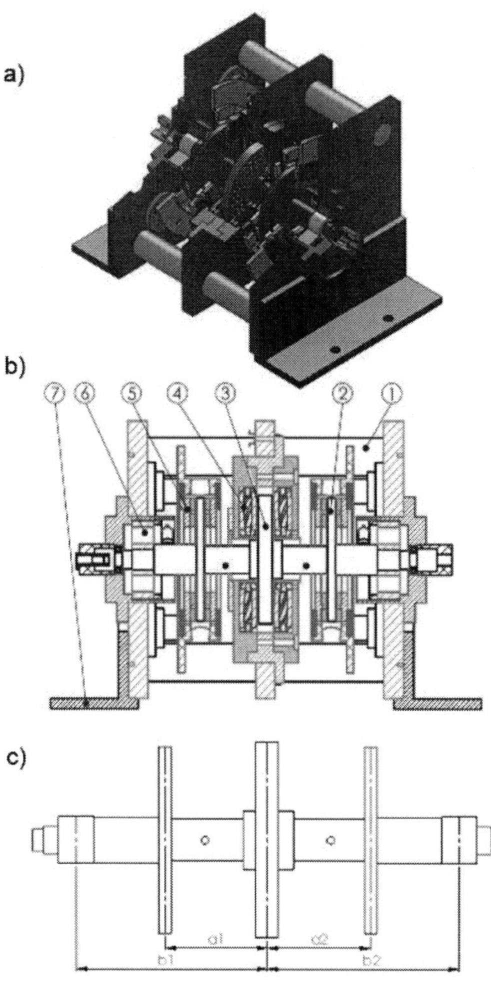

Figure 3 a): isometric section view of the test rig, b), Cross section view, c) rotor view with geometrical parameters.

It is worth to notice that $J_p < J_t$, therefore the rotor can be considered as a long rotor. This implies, according to what stated in [8], that both cylindrical and conical modes of a rotor supported by EDBs needs to be stabilized.

The double flux EDBs cross section view containing the main design parameters is shown in Figure 4. The geometry of the EDB has a very important role on its electromechanical properties and must be chosen to respect mechanical constraints and specifications that minimize the rotational speed at which the system becomes stable. To this end the analytical model presented in [4, 5] is used as design tool and the procedure described in [7] is applied. During the design phase the finite element (FE) analysis has a fundamental role to identify the EDB electromechanical performance. The design work flow is characterized by the following steps:

- perform a FE analysis and extract the parameters k and c
- estimate the total rotor's mass m_r and EDB's stator mass m_s

- calculate the minimum rotational speed ω for which all the poles of the system have negative real part considering a certain range of c_n
- change the geometric parameters and reiterate the previous points until a physically feasible solution that respect's the design constraints is found.

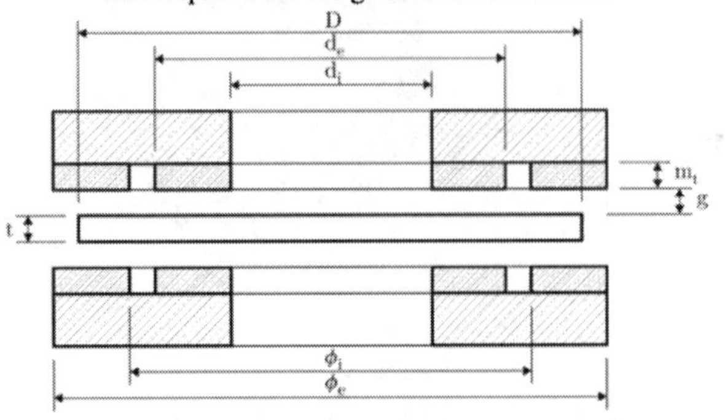

Figure 4: geometry of the axial flux EDB in the double flux configuration.

The list of the EDB parameters coming from the design procedure is presented in Table II.

TABLE II. DOUBLE FLUX EDBS PARAMETERS

	Parameter	Value
D	Conductor outer diameter	120 mm
d_e	Inner permanent magnet outer diameter	89 mm
d_i	Inner permanent magnet inner diameter	64 mm
m_t	Permanent magnet thickness	5.5 mm
t	Conductor disc thickness	8 mm
g	Axial air gap	0.75 mm
Φ_e	Outer permanent magnet outer diameter	120 mm
Φ_i	Outer permanent magnet outer diameter	95 mm
B_r	Residual induction	1.3 T
ω_{EDB}	EDB electric pole	440 rad/s
k_1	EDB equivalent stiffness1	129690 N/m
k_2	EDB equivalent stiffness2	123979 N/m
c_1	EDB equivalent damping1	72.2 Ns/m
c_2	EDB equivalent damping2	295.2 Ns/m
c_n	Non-rotating damping	500 Ns/m

The C_n parameter represents the smallest value to get the stability even at zero rotating speed.

The active actuator, devised to levitate the rotor at low speed and stabilize the EDB system, is characterized by the traditional eight poles configuration shown in Fig.5. The actuator working as AMB is designed to support the rotor weight and as AMD to provide the non-rotating damping needed to stabilize the EDB levitation.

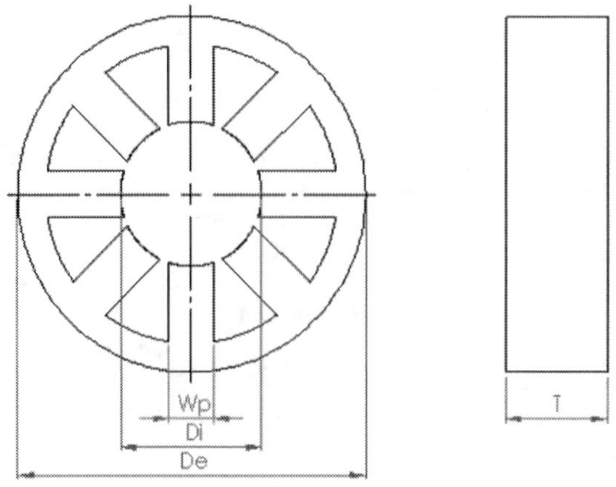

Figure 5: geometry of the AMD actuator.

Table III resumes the main characteristics of the active actuators.

TABLE III. AMDS PARAMETERS

	Parameter	Value
De	Outer diameter	152 mm
Di	Inner diameter	31 mm
Wp	Pole width	10 mm
T	Actuator thickness	22.5 mm
N	Number of turns	165
Wd	Wire diameter	0.63 mm
F	Actuator force	71 N

The AMD functionality is exploited for low rotating speed, where the EDBs, even if already stable, are not able to generate enough force to levitate the rotor.

D. Simulation Results

Simulations have been performed to evaluate the performance of the test rig. Most of the analyses were done considering the operating condition where the levitation is provided by the EDBs and the AMDs give the stabilizing contribution. This is the most critical condition.

The operating condition where the levitation is provided by AMBs is not described since it does not represent an innovative issue.

Figure 6 shows the root loci plot as function of the rotating speed for two values of non-rotating damping c_n. It is worth to notice that the right choice of c_n results in a system that is always stable, also for really low rotating speed. This condition is represented by Fig. 6 a) where the roots of the system are on the left hand side of the diagram for the whole range of the operating rotating speed. By converse an inappropriate design of the c_n parameter brings to an unstable system as shown by Fig. 6 b) where there is at least one pole on the right hand side that never crosses the imaginary axis.

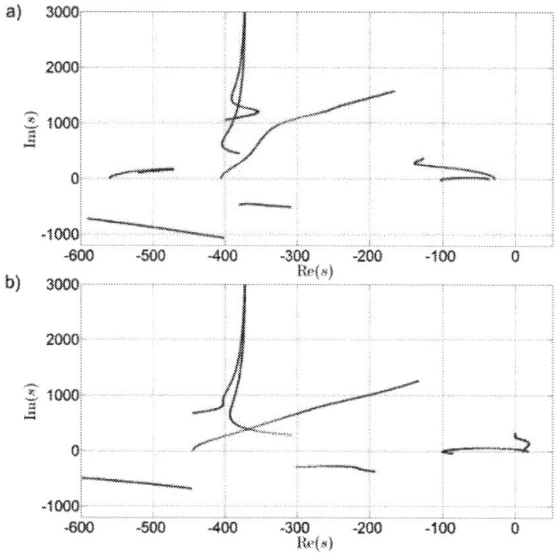

Figure 6: root loci diagram as function of the rotating speed. a) $c_n = 585$ Ns/m, stable system; b) $c_n = 455$ Ns/m, unstable system.

The simulations in the time domain are performed considering that the rotor is kept in levitation by means of the AMBs up to a certain rotating speed. Once this speed is reached, the active actuators stop operating as AMBs and start working as AMDs to make possible the electrodynamic levitation. The AMBs levitate the rotor in the equilibrium point x = 0, y = 0 that represents the initial condition for the electrodynamic levitation. The equilibrium point of the rotor supported by EDBs is not fixed and depends on the rotating speed. The main results of the simulations in the time domain are summarized in Fig. 7 and Fig. 8.

Fig. 7 shows rotor radial displacements (x blue, y red) at the left EDB level. Fig. 7 a) and b) refer to a rotational speed ω = 2000 rpm with non - rotating damping $c_n = 455$ Ns/m, and $c_n = 585$ Ns/m respectively in both the AMDs. The results in Fig. 7 c) varies from that in Fig. 7 b for the higher rotational speed (ω = 8000 rpm).

According to the root loci plot of Fig. 6 b), the time response of Figure 7 a) shows an unstable behavior that can be observed by the exponential increment of the radial displacements with respect to the time.

The result in Figure 7 b) and c), highlight a stable behavior as the x and y positions reach a constant value. These results are consistent with the root loci plot in Fig. 6b. It must be highlighted that in both cases the position oscillations are contained within the radial air gap and therefore it can be concluded that the EDBs are able to provide enough force to keep the rotor in levitation without any contact with the statoric parts starting from 2000 rpm. Another interesting consideration concerns the different behavior of the system as function of the rotating speed. From the graphs of Figure 7 b) an c) it is possible to note that for rotating speed lower than the EDBs electric pole ω_{EDB} (6 b), the x and y positions are larger than the one obtained for higher rotating speed (6 c). This is due to the fact that, according to the quasi-static characteristic of EDB [4], for rotating speed lower than ω_{EDB} the orthogonal force component is larger than the force parallel to the eccentricity, while above ω_{EDB} the higher the rotating speed the larger the parallel and the lower the orthogonal component.

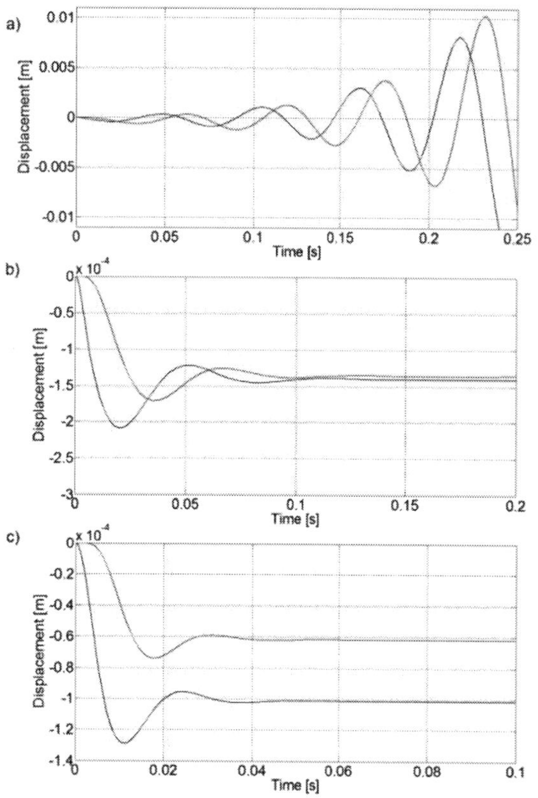

Figure 7: time response of the electrodynamic levitation system stabilized by AMDs. a) $c_n = 455$ Ns/m, unstable system; b) $c_n = 585$ Ns/m, ω = 2000 rpm, stable system; c) $c_n = 585$ Ns/m, ω = 8000 rpm, stable system.

Figure 8 describes the rotor orbit under different operating conditions. Graphs a) and b) are obtained with the same C_n = 585 Ns/m that stabilizes the system and describe the behavior at different rotating speed. Also in this graph it can be pointed out that the higher the rotating speed, the smaller the x, y coordinates of the equilibrium point. Figure 8 c) describes the orbit of the system with a c_n not enough to stabilize the system.

Experimental tests will be performed to validate the analytical model and to get experimental evidence of the advantages of the proposed solution.

E. Conclusions

The magnetic suspension proposed in the present work is an integrated EDB/AMD solution which exploits the high reliability of EDB with the controllability advantage of the AMD thus eliminating most of the problems due to both technologies.

The proposed solution is attractive for high speed applications because it increases the global system reliability,

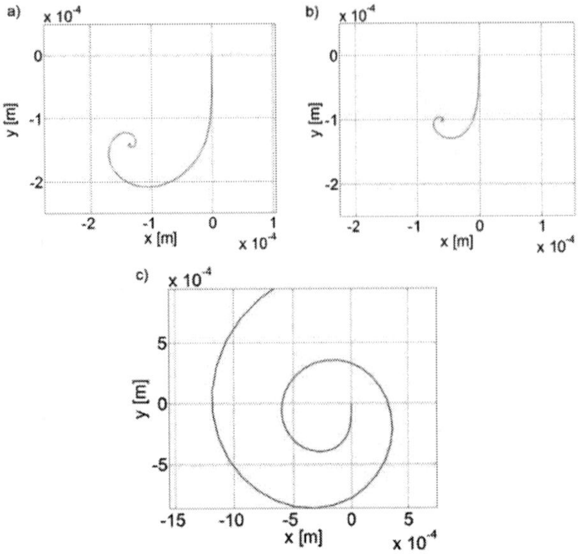

Figure 6 x-y orbit diagram of the electrodynamic levitation system stabilized by AMDs. a) $c_n = 585$ Ns/m, $\omega = 2000$ rpm, stable system; b) $c_n = 585$Ns/m, $\omega = 8000$ rpm, stable system; c) $c_n = 445$ Ns/m, unstable system.

makes possible an AMB downsizing and represents an alternative to the landing bearing problem.

The analytical model of the suspension system with a 4 degrees of freedom rotor was developed and used to design a test rig.

The simulation results shed light onto the promising characteristics of the hybrid EDB/AMD suspension.

Experimental tests will be performed to validate experimentally what observed in simulation.

REFERENCES

[1] P. A. Basore. Passive stabilization of flywheel magnetic bearings. Master's Thesis, Massachusetts Institute of Technology, Cambridge, USA, 1980.

[2] J. G. Detoni, "Progress on electrodynamic passive magnetic bearings for rotor levitation," Proceedings of the Institution of Mechanical Engineers, Part C: Journal of Mechanical Engineering Science, 2013, doi:101177/0954406213511798.

[3] A., Filatov, 2006, "Flywheel Energy Storage System With Homopolar Elec-trodynamic Magnetic Bearing," Proceedings of the Tenth International Sym-posium on Magnetic Bearings, Martigny, Switzerland.

[4] Amati, N., De Lépine, X., and Tonoli, A., 2008, "Modeling of Electrodynamic Bearing," ASME J. Vibr. Acoust., 130, p. 061007.

[5] A. Tonoli, N. Amati, F. Impinna, and J. G. Detoni, "A solution for the stabilization of electrodynamic bearings: Modeling and experimental validation," Journal of Vibrations and Acousticts, vol. 133, no. 021004, 2011.

[6] G. Schweitzer, H. Bleuler and A. Traxler. "Active Magnetic Bearings: Basics, Properties and Applications of Active Magnetic Bearings" vdf Hochschulverlag an der ETHZ, Zurich, 1994.

[7] N. Amati, J.G. Detoni, F. Impinna, A. Tonoli, "Model Based Design of a Rotor Supported by Radial Electrodynamic Bearings". 10th International Conference on Vibrations in Rotating Machinery, Londra, 11-13 September, 2012. pp. 123-132

[8] J. D. Detoni, F. Impinna, N. Amati, A. Tonoli, M.P. Piccolo, "Stability of a 4DOF rotor on electrodynamic passive magnetic bearings", 14th International Symposium on Magnetic Bearings, Linz, 11-14 August, 2014.

Stability of a 4 degree of freedom rotor on electrodynamic passive magnetic bearings

Joaquim G. Detoni[a], Fabrizio Impinna[a], Nicola Amati[a], Andrea Tonoli[a], Maria Pina Piccolo[a] ,Giancarlo Genta[a]

[a] Department of Mechanical and Aerospace Engineering, Politecnico di Torino, Corso Duca degli Abruzzi 24, 10129 Turin, Italy, joaquim.detoni@polito.it, fabrizio.impinna@polito.it, nicola.amati@polito.it, andrea.tonoli@polito.it, mp.piccolo@outlook.it, giancarlo.genta@polito.it

***Abstract*—Electrodynamic bearings exploit repulsive forces due to eddy currents to produce positive stiffness by passive means without violating the Earnshaw stability criterion. This remarkable characteristic makes this type of bearing a suitable alternative to active magnetic bearings in fields such as kinetic energy storage flywheels, turbo pumps, high speed compressors, among others. However, the suspension can become unstable due to rotating damping.**

To obtain deeper understanding of this instability phenomenon this paper presents the analysis of stability of a four degree of freedom (4dof) rotor supported by electrodynamic bearings. The 4dof rotor model is coupled to the dynamic model of the eddy current forces generated by the electrodynamic bearing and the stability of the complete system is analyzed. This model is used to study the stability of both cylindrical and conical whirling motion of the rotor. In addition to the well known cylindrical whirl instability the possible occurrence of conical instability is demonstrated. Finally the effects of two stabilization strategies are analyzed.

I. Introduction

Magnetic bearings provide contact-less support for rotating machines. They usually exploit reluctance or Lorentz forces to keep a rotor in stable levitation. Reluctance forces can be used to levitate a rotor on active magnetic bearing (AMB) and passive permanent magnet bearing (PMB) configurations. Magnetic bearings exploiting Lorentz forces are called electrodynamic bearings (EDB) and although some layouts can be actively controlled [1], in general they are passive [2].

Different from AMBs and PMBs, that have already been employed in many industrial applications, EDBs have not reached industry yet, remaining at the academic research stage. Despite this fact, the EDB unique characteristic of producing positive stiffness by passive means without violating the Earnshaw criterion has attracted some industrial interest in the last decades. One of the most interesting features of EDBs is the possibility of obtaining stable levitation using standard conductive materials at room temperature, in absence of power electronics and sensors. This makes them an interesting alternative to AMBs for high speed rotors in applications such as energy storage flywheels and turbo pumps, among others.

Nowadays the main challenge precluding industrial application of EDBs remains understanding the stability of rotors supported by them, and how to solve the remaining instability problems. This led most of the research groups working with EDBs to perform detailed studies about rotordynamic stability [3]–[6]. In these works different electromechanical models of EDBs were created and coupled to rotordynamic models. A common feature of these studies was the use of Jeffcott rotor models which, although insightful, neglects gyroscopic effects. Hence, only the instability of cylindrical whirling motion was addressed.

In the present paper a more detailed stability analysis is presented and discussed. To this end a four degree of freedom rotor (4dof) model is used allowing studying the stability of both cylindrical and conical whirling motion. Initially the models of rotor and EDB are described and the dynamic matrix of the complete system is obtained. This model is then used to study the stability of the rotor on EDBs by means of root locus analysis. The root locus analysis evidences the possible occurrence of instability of both cylindrical and conical whirling modes. Finally, two methods for the stabilization of the rotor which were previously presented in the literature are studied and their effectiveness in stabilizing each unstable mode is addressed.

II. Modeling of a rotor supported by EDBs

Due to the nature of the phenomena, studying the dynamics of a rotor on magnetic bearings requires one to consider that the rotor moves relative to the stator. This must be taken into account both in the rotor and bearing models which must then be coupled together to describe the dynamics of the whole system. In this section the models of a gyroscopic rotor and that of an EDB are presented, and the passages to perform the coupling are described.

A. Rotor model

We consider a rotor such as that shown in Fig. 1. The rotor is composed of a gyroscopic rigid body described by its mass m, transversal moment of inertia J_t, and polar moment of inertia J_p. This rigid body is clamped to a massless rigid shaft. This assumption is justified as in general magnetic bearings produce a stiffness which is much lower compared to the stiffness of the shaft.

Two reaction forces generated by the magnetic bearings are represented by two forces F_{b1} and F_{b2} positioned at distance a and b from the rotor center of mass as shown in the figure. The complete equations of motion of the described rotor have six degrees of freedom. Under the hypothesis of small transverse

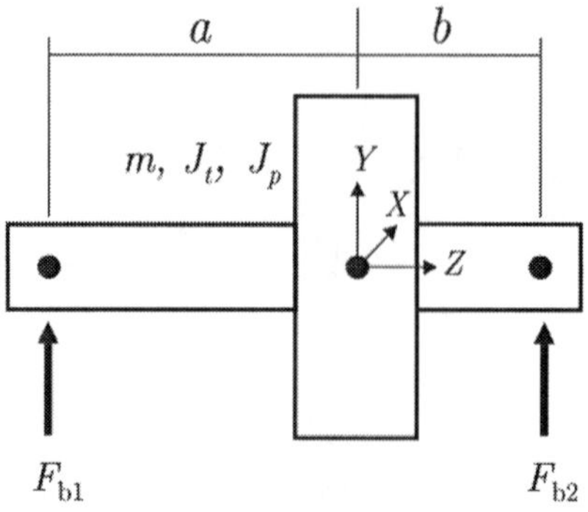

Figure 1. Scheme of the rotor model.

and angular displacements, and restricting our study to the constant rotational speed Ω condition, it can be demonstrated that a model with four degrees of freedom, or two complex degrees of freedom, is enough to study the flexural behavior of the rotor. The complete derivation of the rotor equations is not shown here. An extensive development is given in [7]. The equations describing the dynamic behavior of the four degrees of freedom rotor are the following:

$$\mathbf{M}\begin{Bmatrix}\ddot{q}\\ \ddot{\phi}\end{Bmatrix} - j\Omega\mathbf{G}\begin{Bmatrix}\dot{q}\\ \dot{\phi}\end{Bmatrix} + \begin{Bmatrix}F_b\\ M_b\end{Bmatrix} = \begin{Bmatrix}F_{ext}\\ M_{ext}\end{Bmatrix} \quad (1)$$

where **M** and **G** are the mass and gyroscopic matrices respectively. These are written as:

$$\mathbf{M} = \begin{bmatrix} m & 0\\ 0 & J_t\end{bmatrix} \qquad \mathbf{G} = \begin{bmatrix} 0 & 0\\ 0 & J_p\end{bmatrix}$$

These two equations are written in terms of two complex coordinates q and ϕ that represent the transverse and angular displacements respectively, which are written as:

$$\begin{aligned} q &= X + jY \\ \phi &= \phi_y - j\phi_{X'} \end{aligned} \quad (2)$$

B. Electrodynamic bearing

Electrodynamic bearings exploit repulsive forces due to eddy currents. Many different types of EDBs exist [8], however, here we restrict the analysis to homopolar EDBs which exploit relative motion between a conductor and a magnetic field. As for the rotor model, the complete derivation of the dynamic model is not described here. Only the fundamental equations are given. The extensive discussion of the model used in the present analysis is given in [9].

The dynamic equation of the EDB force can be written as:

$$\dot{F}_{EDB} = k\dot{q} - j\Omega kq - \left(\frac{k}{c} - j\Omega\right) F_{EDB} \quad (3)$$

The coefficients k and c give the stiffness and damping associated with the eddy current force model [10]. These two coefficients can be identified both by means of finite element simulations [11] or experimentally [6]. The notation used in

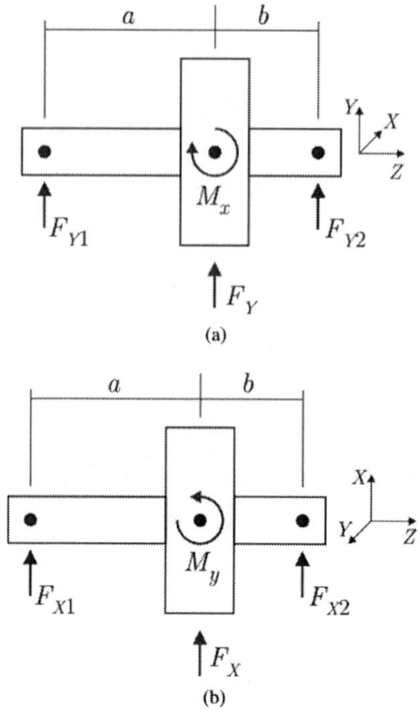

Figure 2. Transformation of reactions from bearing coordinates into center of mass coordinates. The black arrows represent the forces at the bearing and the red arrows represent the forces and moments at the center of mass.

Eq. 3 is in agreement with Eq. 2, hence, the coupling between rotor and EDB dynamic models is straightforward.

C. Rotor and EDB coupling

In Eq. 1 the reaction forces and moments generated by the two bearings are represented by F_b and M_b whereas the excitation forces and moments are F_{ext} and M_{ext}. Both contributions are considered as acting on the rotor center of mass. Since the bearing's reaction forces are applied away from the center of mass by a and b, it is simple to obtain a relation between reaction forces at the bearing position and force and moment at the center of mass. To do so the schemes in Fig. 2are considered.

Analyzing the scheme of Fig. 2 the forces and moments calculated at the rotor center of mass are equal to:

$$\begin{aligned} F_X &= F_{X1} + F_{X2} \\ M_y &= -aF_{X1} + bF_{X2} \\ F_Y &= F_{Y1} + F_{Y2} \\ M_x &= aF_{Y1} - bF_{Y2} \end{aligned} \quad (4)$$

Positive moments are oriented according to the right hand rule. In matrix format Eq. 4 becomes:

$$\begin{Bmatrix}F_X\\ M_y\\ F_Y\\ M_x\end{Bmatrix} = \mathbf{T}\begin{Bmatrix}F_{X1}\\ F_{Y1}\\ F_{X2}\\ F_{Y2}\end{Bmatrix} \quad (5)$$

where

$$\mathbf{T} = \begin{bmatrix} 1 & 0 & 1 & 0 \\ -a & 0 & b & 0 \\ 0 & 1 & 0 & 1 \\ 0 & a & 0 & -b \end{bmatrix} \tag{6}$$

Using the complex notation for the forces and moments one obtain:

$$\begin{aligned} F_b &= F_X + \mathrm{j}F_Y \\ M_b &= M_y - \mathrm{j}M_x \end{aligned} \tag{7}$$

Equation 6, written using the complex notation, reduces to:

$$\begin{Bmatrix} F_b \\ M_b \end{Bmatrix} = \mathbf{T} \begin{Bmatrix} F_{b1} \\ F_{b2} \end{Bmatrix} \tag{8}$$

where

$$\mathbf{T} = \begin{bmatrix} 1 & 1 \\ -a & b \end{bmatrix} \tag{9}$$

A similar procedure is used to obtain the relation between the displacements at the center of mass and those at the bearing position. The expression is:

$$\begin{Bmatrix} q_1 \\ q_2 \end{Bmatrix} = \mathbf{T}^{\mathrm{T}} \begin{Bmatrix} q \\ \phi \end{Bmatrix} \tag{10}$$

where $\mathbf{T}^{\mathrm{T}}$ is the transpose of $\mathbf{T}$, and q_1 and q_2 are the transverse displacements at the bearings in complex notation as given by Eq. 2.

At this stage it is possible to write a state space model where the model states are given by Eq. 1 and Eq. 3. The state space model of the complete system can be written as:

$$\begin{Bmatrix} \ddot{q} \\ \ddot{\phi} \\ \dot{q} \\ \dot{\phi} \\ \dot{F}_{b1} \\ \dot{F}_{b2} \end{Bmatrix} = \mathbf{A} \begin{Bmatrix} \dot{q} \\ \dot{\phi} \\ q \\ \phi \\ F_{b1} \\ F_{b2} \end{Bmatrix} + \mathbf{B} \begin{Bmatrix} F_{\mathrm{ext}} \\ M_{\mathrm{ext}} \end{Bmatrix}. \tag{11}$$

For the stability study only the dynamic matrix $\mathbf{A}$ is required and the poles of the system can be obtained as the eigenvalues of $\mathbf{A}$. The $\mathbf{A}$ matrix is written as:

$$\mathbf{A} = \begin{bmatrix} \mathrm{j}\Omega\mathbf{M}^{-1}\mathbf{G} & \mathbf{0}_{2\times2} & -\mathbf{M}^{-1}\mathbf{T} \\ \mathbf{I}_{2\times2} & \mathbf{0}_{2\times2} & \mathbf{0}_{2\times2} \\ \mathbf{K}_{\mathrm{EDB}}\mathbf{T}^{\mathrm{T}} & -\mathrm{j}\Omega\mathbf{K}_{\mathrm{EDB}}\mathbf{T}^{\mathrm{T}} & -\mathbf{C}_{\mathrm{EDB}}^{-1}\mathbf{K}_{\mathrm{EDB}} + \mathrm{j}\Omega \end{bmatrix} \tag{12}$$

The matrices associated with the dynamic behavior of the EDB contain the parameters of the two bearings. They are written as:

$$\mathbf{K}_{\mathrm{EDB}} = \begin{bmatrix} k_1 & 0 \\ 0 & k_2 \end{bmatrix} \qquad \mathbf{C}_{\mathrm{EDB}} = \begin{bmatrix} c_1 & 0 \\ 0 & c_2 \end{bmatrix}$$

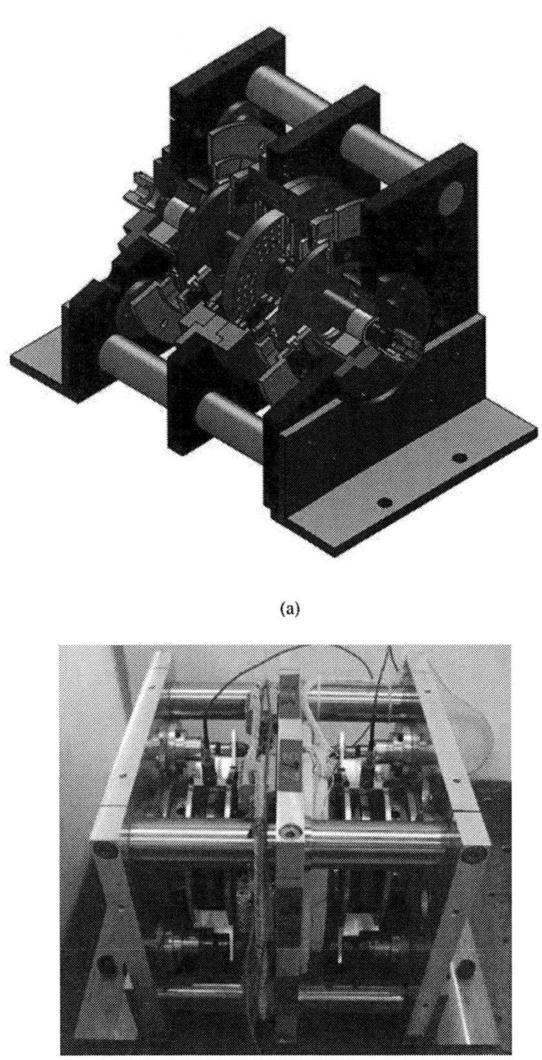

(a)

(b)

Figure 3. Test rig of a rotor supported by EDBs. The isometric cross section evidences how the rotor is positioned inside the machine.

where the indexes denote the properties of the EDBs positioned at ends 1 and 2 according to the scheme of Fig. 1.

III. Stability analysis

The present section studies the stability of the 4dof rotor model supported by two EDBs positioned at the two ends of the shaft. This analysis considers the parameters of a real laboratory rotor. The system studied is shown in Fig. 3 and is described in detail in the paper by Impinna et al. published separately in the present conference [12]. The two EDBs have equal properties and are positioned symmetrically with respect to the center of mass. The values of the parameters that characterize the whole suspension model in Eq. 12 are given in Tab. I.

The data from the table is introduced into Eq. 12 and the stability is studied analyzing the eigenvalues calculated for

Table I
PARAMETERS DESCRIBING THE DYNAMICS OF THE 4DOF ROTOR ON EDBS. SOME OF THE DATA REFFERS TO FIG. 6.

Parameter	Symbol	Value	Unit
Rotor mass	m	4.342	kg
Transverse moment of inertia	J_t	0.01982	kgm^2
Polar moment of inertia	J_p	0.00572	kgm^2
EDB 1 distance from center of mass	a	0.067	m
EDB 2 distance from center of mass	b	0.067	m
Damper 1 from center of mass	c	0.12425	m
Damper 2 from center of mass	d	0.12425	m
EDB eddy current stiffness	k_{1-2}	253051.3	Nm^{-1}
EDB eddy current damping	c_{1-2}	363.1	Nsm^{-1}

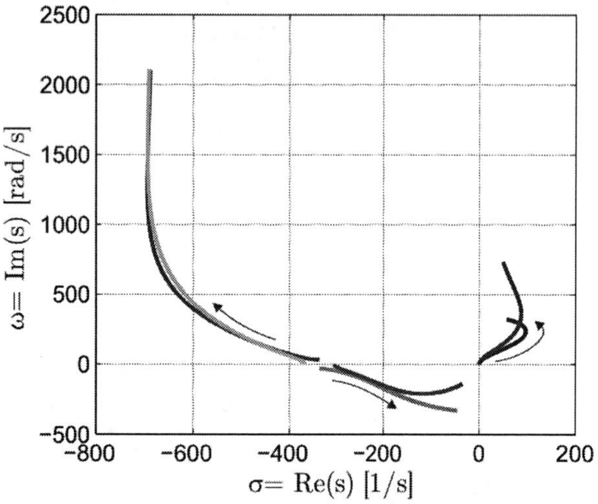

Figure 4. Root locus plot of the 4dof rotor supported by two EDBs. The arrows indicate the direction of the evolution of the poles with increasing rotation speed.

different values of Ω. The root locus plot is obtained plotting the real and imaginary parts of the calculated eigenvalues in the complex plane. Figure 4 shows the root locus of the rotor. Instability is evidenced by the presence of two poles with positive real part. These two unstable poles, illustrated by the blue and green lines, are linked to the cylindrical and conical forward whirl rotor modes.

In absence of external non rotating damping these two modes are unstable for any rotation speed different from zero. This is linked to the dynamic behavior of the EDB and is due to two main aspects. The first important aspect is the fact that the EDBs introduce rotating damping without any contribution of non rotating damping. The second is that EDB stiffness depends on the rotational speed; it does not provide any stiffness at zero rotational speed.

Intuitively one can think that the instability arises from the fact that the system operates always in supercritical regime. It is then clear that rotating damping causes instability since it is known to be destabilizing in these conditions. It is also clear that in case of disc rotors, where $J_p > J_t$, the conical instability does not exist since this mode is subcritical.

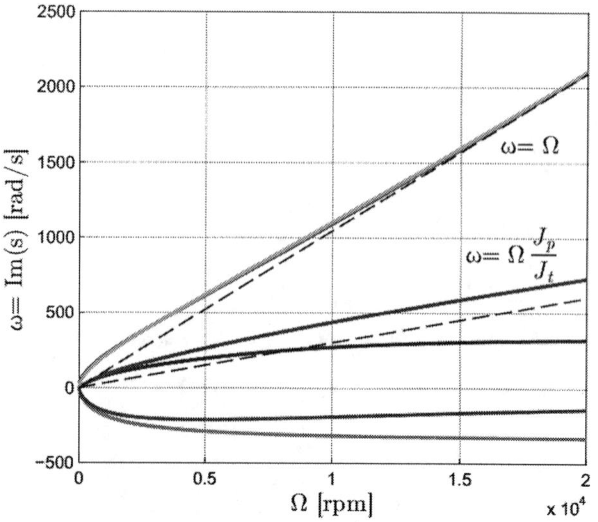

Figure 5. Campbell diagram of the rotor on EDBs. The lines show the evolution with speed of the imaginary part of the poles shown in Fig. 4.

The aforementioned behavior can be better understood observing the Campbell diagram of the system shown in Fig. 5. It can be noticed how the two lines showing the evolution of the two unstable modes are always below the line $\omega = \Omega$, thus these two modes are always supercritical.

IV. STABILIZATION OF ROTORS ON EDBS

In the previous section it was shown that the undamped behavior of a 4dof rotor on EDBs shows two unstable rotor modes. The instability is related to the forward cylindrical and conical whirl modes. In the past, several authors investigated the stability of the cylindrical whirl and two different stabilization methods were proposed [3], [6], [13]. Figure 6 shows the schemes of the two proposed methods.

The first scheme, in Fig. 6a, illustrates the most common stabilization method consisting in the introduction of non rotating damping by electromagnetic means between rotor and casing. In the second method, illustrated in Fig. 6b, non rotating damping is introduced by means of dissipative elements placed between the EDB stator and casing [6]. However, the effectiveness of each method to stabilize the conical mode has not been assessed. In this section the rotor and EDB models presented previously are used to understand the effects of the two stabilization methods upon the stability of the forward conical whirl mode.

A. Stabilization by rotor-casing interaction

The first stabilization method is the most common and straightforward. It consists in placing a permanent magnet attached to the rotor which interacts with a conductive element attached to the casing. Such method is well known and has been studied and employed in the field of passive magnetic bearings even at industrial level [14].

From the modeling point of view this stabilization system consists simply in introducing a damping element associated

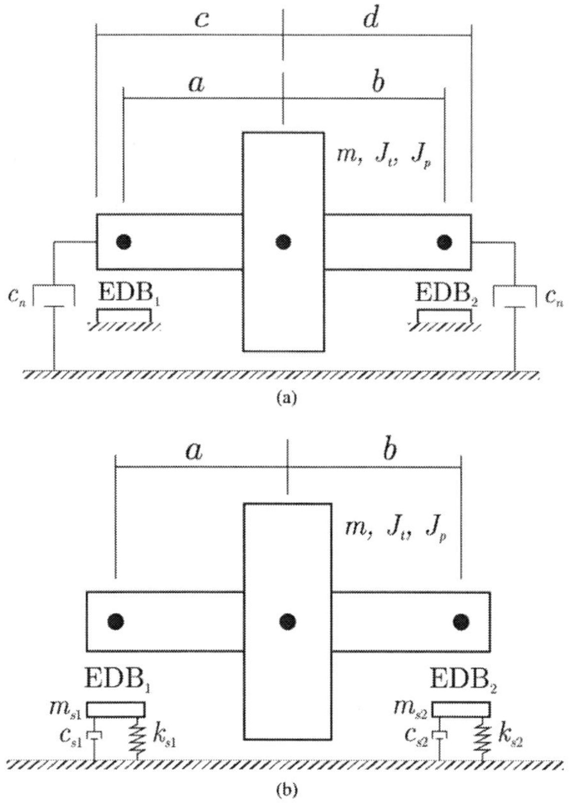

Figure 6. Scheme of the rotor on supported by EDBs with non rotating damping introduced between EDB stator and casing.

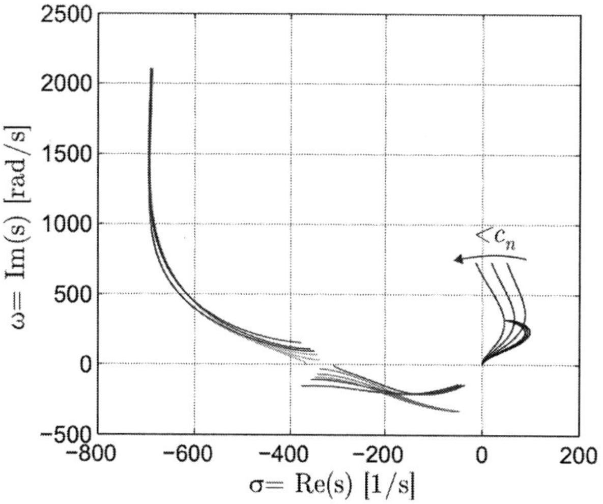

Figure 7. Root locus of the suspension calculated with different values of c_n. The arrow indicates the direction of the motion of the unstable poles when c_n is increased.

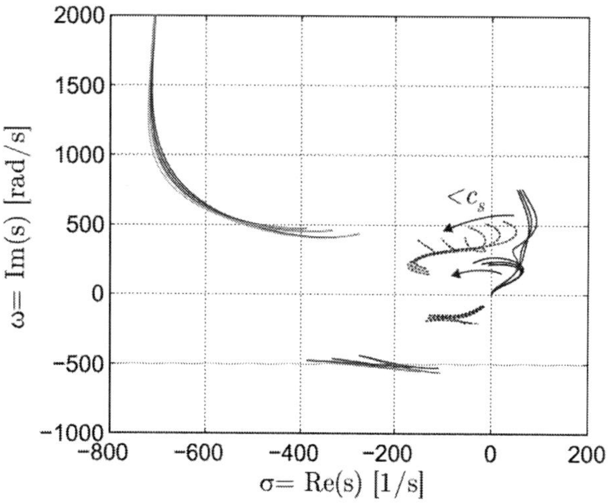

Figure 8. Root locus of the suspension calculated with constant values of m_s and k_s, and different values of c_s. The arrow indicates the direction of the motion of the unstable poles when c_s is increased.

with the translation speed of the rotor at the point where the damper is positioned. For the sake of generality, in the scheme of Fig. 6a two dampers are introduced at distances c and d from the center of mass. The damping matrix can be easily obtained using the procedure described in paragraph II-C.

A brief sensitivity analysis can be used to verify the influence of the non rotating damping parameter c_n upon the stability. The results are plotted in Fig. 7. It can be seen that c_n is mainly influential upon the behavior of the forward conical whirl. The poles associated with this mode tend to move towards the left half plane, eventually reaching a condition of stability. On the other hand, the cylindrical mode is much less sensitive to this parameter. Consequently, increasing the non rotating damping introduced between rotor and casing is particularly beneficial for the stability of the conical mode. This is specially related to the fact that a relativelly low value of c_n applied far from the rotor center of mass generates a large damping contribution with respect to the rotation degree of freedom ϕ.

B. Stabilization by stator-casing interaction

A different stabilization strategy for rotors on EDBs was proposed by Tonoli et al. [6]. The proposed technique consists in introducing the damping between the non rotating part of the bearing and the casing of the machine. In that work the authors assessed the effectiveness of their technique upon the stabilization of the cylindrical forward whirl. However, no mention to the influence upon the conical mode could be made.

Differently from the previous case, instead of introducing the damping on the rotor by means of contactless devices, here the underlying idea is to dissipate energy through a non rotating element. This opens to the possibility of using common damping techniques such as viscoelastic elements or fluid film dampers. Similar to the previous case, also this strategy to introduce non rotating damping is well known. It is widely used with rolling bearings and has been studied for passive [15], [16] and active magnetic bearings [17].

From the modeling point of view this solution is more complex than the previous. Two additional complex degrees of freedom associated with the transverse motion of the EDB stators must be included. The dynamics of the stators is described by simple mass-spring-damper systems.

To understand the influence of the stator damping parameter c_s a procedure similar to that presented in section IV-A is performed. The damping is modified, and the root locus of the whole system is plotted. The results are plotted in Fig. 8. As expected, the graph contains four more poles due to the new stator degrees of freedom. The lines describing the evolution of the stator poles are dashed whereas the rotor and EDB modes are shown in continuous lines. In accordance with the previous graphs, the forward cylindrical whirl is shown in blue and the forward conical is in green. In addition to the unstable rotor modes this new system may present an instability of the stator. Such behavior was verified experimentally by Tonoli et al. [18]. Even if this mode may become unstable and its occurrence should be verified at the design phase, its behavior can be stabilized by the addition of damping and is not of great concern.

For what concerns the rotor stability, it can be observed in the figure how increasing the stator damping is effective for the stabilization of the forward cylindrical mode; however it is not very effective in stabilizing the conical mode.

V. Conclusions

The present paper presented the modeling and stability analysis of a 4 dof rotor supported by electrodynamic bearings. The modeling of the two main subsystems, namely, rotor and EDB, was presented and a method to perform the coupling between the equations was described. Compared to formerly presented studies the present one employed a more complex rotor model. This upgraded rotor model allowed demonstrating that in addition to the well known cylindrical whirl instability, a rotor supported by EDBs may present an unstable conical whirl mode. The occurrence of the conical instability is directly linked to the ratio J_p/J_t. Disc rotors $(J_p/J_t > 1)$ do not present conical instability whereas long rotors $(J_p/J_t < 1)$ do.

The present study also analyzed the effectiveness of two known stabilization methods upon the conical mode stability. The stability analyses were conducted using the root locus method. It was observed that the stabilization method where damping is introduced directly between rotor and casing is particularly effective to stabilize the conical mode as it allows taking advantage of the damper position relative to the rotor center of mass. On the other hand this method is not so effective in stabilizing the forward cylindrical mode. The second stabilization method, based on the introduction of damping between the stator of the bearing and the casing showed to be less effective to stabilize the conical mode, however it is more effective in stabilizing the cylindrical mode. From this analysis it seems evident that the combination of the two methods may prove to be the best solution to guarantee stability of both cylindrical and conical modes for long rotors.

References

[1] W. K. S. Khoo, K. Kalita, S. D. Garvey, R. J. Hill-Cottingham, D. Rodger, and J. F. Eastham, "Active axial-magnetomotive force parallel-airgap serial flux magnetic bearings," *IEEE Transactions on Magnetics*, vol. 46, no. 7, pp. 2596–2602, 2010.

[2] G. Schweitzer and E. H. Maslen, *Magnetic bearings. Theory, design, and application to rotating machinery*. Berlin: Springer-Verlag, 2009.

[3] R. F. Post, "Stability issues in ambient-temperature passive magnetic bearing systems," *NASA STI/Recon Technical Report N*, vol. 3, Feb. 2000.

[4] A. V. Filatov, E. H. Maslen, and G. T. Gillies, "Stability of an electrodynamic suspension," *Journal of Applied Physics*, vol. 92, no. 6, pp. 3345–3353, 2002.

[5] V. Kluyskens and B. Dehez, "Dynamical electromechanical model for magnetic bearings subject to eddy currents," *IEEE Transactions on Magnetics*, vol. 49, no. 4, pp. 1444–1452, 2013.

[6] A. Tonoli, N. Amati, F. Impinna, and J. G. Detoni, "A solution for the stabilization of electrodynamic bearings: Modeling and experimental validation," *Journal of Vibrations and Acousticts*, vol. 133, no. 021004, 2011.

[7] G. Genta, *Dynamics of Rotating Systems*. New York: Springer, 2005.

[8] J. G. Detoni, "Progress on electrodynamic passive magnetic bearings for rotor levitation," *Proceedings of the Institution of Mechanical Engineers, Part C: Journal of Mechanical Engineering Science*, 2013, doi:10.1177/0954406213511798.

[9] J. G. Detoni, F. Impinna, A. Tonoli, and N. Amati, "Unified modelling of passive homopolar and heteropolar electrodynamic bearings," *Journal of Sound and Vibrations*, vol. 331, no. 19, pp. 4219–4232, 2012.

[10] N. Amati, X. De Lepine, and A. Tonoli, "Modeling of electrodynamic bearings," *Journal of Vibration and Acoustics*, vol. 130, no. 061007, 2008.

[11] G. Genta, X. De Lepine, F. Impinna, J. G. Detoni, N. Amati, and A. Tonoli, "Sensitivity analysis of the design parameters in electrodynamic bearings," in *IUTAM Symposium on Emerging Trends in Rotor Dynamics*, New Delhi, India, 2009, pp. 287 – 296.

[12] F. Impinna, J. G. Detoni, A. Tonoli, and N. Amati, "Test and theory of electrodynamic bearings coupled to active magnetic dampers," in *Proceedings of the 14th International Symposium on Magnetic Bearings*, Linz, Austria, Aug. 2014.

[13] A. V. Filatov and E. H. Maslen, "Passive magnetic bearing for flywheel energy storage systems," *IEEE Transactions on Magnetics*, vol. 37, no. 6, pp. 3913–3924, 2001.

[14] W. Becker, "Pump with magnetic bearings," US. Patent 4,541,772, Sep., 17, 1985.

[15] K. P. Lijesh and H. Hirani, "Stiffness and damping coefficients for rubber mounted hybrid bearing," *Lubrication Science*, 2014, doi:10.1002/ls.1252.

[16] E. Marth, G. Jungmayr, and W. Amrhein, "Influence of viscoelastic elements on magnetically passive stabilized degrees of freedom," in *Proceedings of the 12th International Symposium on Magnetic Bearings*, Aug. 2012, pp. 134–139.

[17] G. Jungmayr, E. Marth, W. Amrhein, and G. Weidenholzer, "Influence of viscoelastic elements on magnetically passive stabilized degrees of freedom," in *Proceedings of the 12th International Symposium on Magnetic Bearings*, Aug. 2012, pp. 148–153.

[18] A. Tonoli, N. Amati, G. Genta, F. Impinna, and J. G. Detoni, "A new stabilization technique for electrodynamic bearings," in *Proceedings of the 12th International Symposium on Magnetic Bearings*, Wuhan, China, Aug. 2010, pp. 551–560.

Impact of the Yoke Material on the Performance of Wounded Electrodynamic Bearings

Corentin Dumont, Virginie Kluyskens, Bruno Dehez

Center for Research in Energy and Mechatronics (CEREM), Université catholique de Louvain (UCL), Place du Levant 2, 1348 Louvain-la-Neuve, Belgium, bruno.dehez@uclouvain.be

***Abstract*—This paper concerns null-flux, centering electrodynamic bearings with an internal rotor made of permanent magnets creating a multipole radial magnetic field and with an external airgap wounded conductor attached to a yoke. The inclusion of a ferromagnetic yoke in front of the permanent magnets has the two following effects. On the one hand, it increases the magnetic flux density in the airgap and the magnitude of the centering Lorentz forces between the rotor and the stator of the device. This is positive since the stiffness associated with the centering force of the bearing increases too. On the other hand, it also induces a negative stiffness due to the reluctant force between the yoke and the magnets usually referred to as the unbalanced magnetic force. The goal of this paper is to investigate about the gain in stiffness and performance associated with the presence of a ferromagnetic yoke in multipole electrodynamic bearings. To this purpose, a method is exposed to evaluate the stiffness and stability of the bearing objectively using a root-loci plot. The results are based on a 2-D analytical model of electrodynamic bearings which is briefly presented in the paper. It is shown that in some cases, the magnetic permeability of the yoke has a very low impact on the bearing performance.**

I. Introduction

Magnetic bearings ensure a contactless guiding function. This makes them more suitable than mechanical bearings for high-speed and/or lubrication-free applications, among others [1]. Nowadays, the magnetic bearings used in the industry are controlled actively. This requires the use of sensors, controllers and power electronics. However, the complexity, cost or bulk associated with this control system can become prohibitive, especially for small rated power applications [2].

A way to overcome these disadvantages could be to use magnetic bearings which do not require external control means i.e., passive bearings. They include supraconducting and electrodynamic bearings. Unlike supraconducting bearings, electrodynamic bearings can operate at room temperature and therefore, they do not require a bulky cooling system. This makes them more compact and simpler [3] and more attractive for stationary applications. Some well-known examples of electrodynamic bearings were studied in [4], [5], [6].

In [7], an analytical study of a new centering electrodynamic bearing with a radial magnetic field was presented. The proposed topology is shown in Fig. 1. It has an internal rotor comprising a ferromagnetic shaft and permanent magnets generating a multipole radial magnetic field in the airgap. The stator is external and comprises a winding and a yoke. This yoke can be made of a ferromagnetic material or not. The topology of this bearing is very simple and allows to build a device using existing components developed for electrical motors, thereby reducing production costs.

In such a bearing with a radial magnetic field, the material of the winding yoke impacts the bearing performance. The goal of this paper is to investigate about the advantages associated with the presence of a ferromagnetic yoke.

In section II, the bearing topology is presented. In section III, a 2-D analytical model of multipole electrodynamic bearing is summarized. This model gives the forces between the rotor and the stator of the bearing, and the root-loci plot of the centering bearing is derived from it. The way to evaluate the bearing stiffness and stability on the basis of the root-loci plots is exposed in section IV and a comparison is made in section V.

II. Bearing Topology

The internal rotor of the proposed bearing comprises a ferromagnetic shaft surrounded by permanent magnets. These magnets induce a radial magnetic field with p pole pairs in the airgap. The external stator of the bearing comprises an airgap axial winding with $q = p+1$ pole pairs, k phases and a yoke. The $p = 1, q = 2$ case with three phases is illustrated in Fig. 1. The geometrical parameters of the bearing are given in Fig. 2. As demonstrated in [7], verifying identity $q = p+1$ ensures that the winding is null-flux and that it can interact with the permanent magnets to produce a radial centering force.

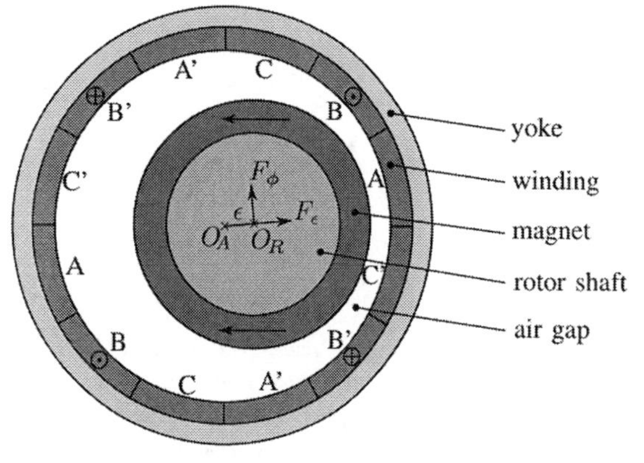

Figure 1. Bearing topology with $p = 1, q = 2$ and $k = 3$. The phases are denoted A, B and C. The airgap width is exaggerated.

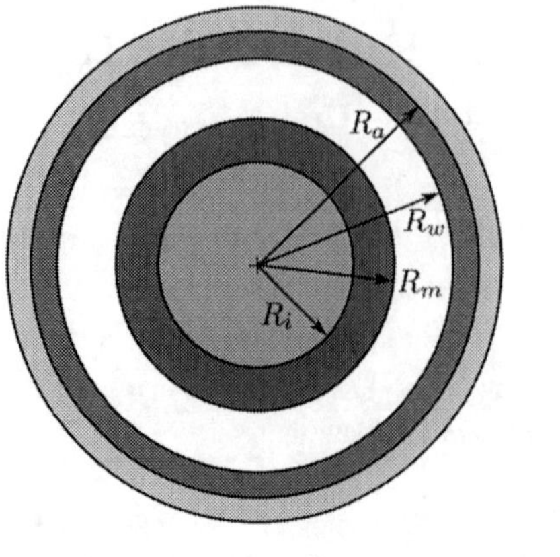

Figure 2. Geometrical parameters of the bearing.

III. FORCES

A 2-D analytical model was built to calculate the electromagnetic forces between the rotor permanent magnets and the windings of the bearing when the rotor is off-centered. This model is similar to the one used in [8], and only the main steps are given in this section.

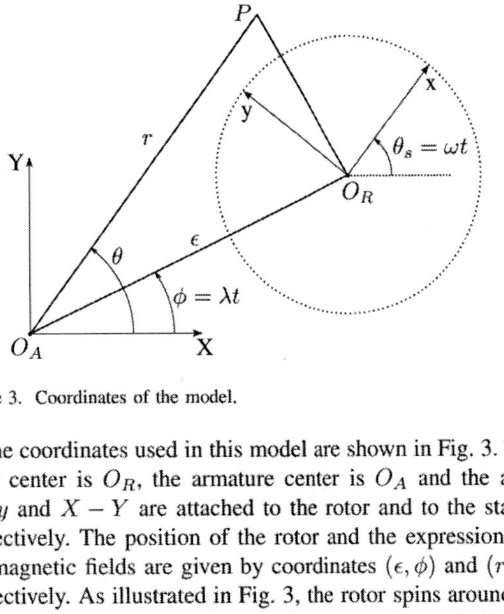

Figure 3. Coordinates of the model.

The coordinates used in this model are shown in Fig. 3. The rotor center is O_R, the armature center is O_A and the axes $x - y$ and $X - Y$ are attached to the rotor and to the stator, respectively. The position of the rotor and the expressions of the magnetic fields are given by coordinates (ϵ, ϕ) and (r, θ), respectively. As illustrated in Fig. 3, the rotor spins around its axis at speed ω and whirls around the axis of the winding at speed λ. To build the model, the following assumptions are made:

1) the end effects are not considered and the problem is assumed to be two-dimensional;
2) the permanent magnets have a linear magnetic characteristic, the rotor shaft is made of a ferromagnetic material with infinite magnetic permeability, and the winding and the permanent magnets have the same magnetic permeability as air. The magnetic permeability of the yoke is either infinite or the same as air, depending on whether the bearing has a ferromagnetic yoke or not;
3) Eddy current losses in the magnets and ferromagnetic parts are neglected;
4) the amplitude of the off-centering ϵ is small compared to the nominal airgap of the bearing;
5) ω and λ are constants and the electrical variables are in a steady-state regime;
6) when calculating the electromotive force in the winding, the effects of the displacement in the radial direction $\frac{d\epsilon}{dt}$ are assumed to be much smaller than ω and λ.

The goal of the model is to obtain the forces between the rotor and the stator of the bearing. These forces are calculated using the Maxwell stress tensor. For this purpose, the magnetic field created by the permanent magnets and by the currents induced in the winding should be calculated. This requires to calculate the induced currents which in turn depend on the electromotive force (emf), the phase resistance and the cyclic inductance of the winding.

First, the analytical expression of the vector potential from the permanent magnets was obtained from [9] and [7]. Because of hypothesis 1 from the previous list, it only has an axial component, and because of hypothesis 3 only the first order terms in ϵ were kept in the Taylor developments of the magnetic vector potential:

$$\begin{aligned} A_{Mz}(r,\theta) = \sum_{n,odd}^{\infty} \Big\{ & C_n(r)\sin\left(np(\theta-\theta_s)\right) \\ & + \epsilon\hat{C}_n(r)\sin\left((np+1)\,\theta - np\theta_s - \phi\right) \\ & + \epsilon\check{C}_n(r)\sin\left((np-1)\,\theta - np\theta_s + \phi\right) \Big\} \end{aligned} \tag{1}$$

with:

$$\begin{aligned} C_n(r) &= K_{1n}r^{-np} + K_{2n}r^{np} \\ \hat{C}_n(r) &= K_{1n}r^{-np-1}np + K_{5n}r^{np+1} + K_{6n}r^{-np-1} \\ \check{C}_n(r) &= -K_{2n}r^{np-1}np + K_{3n}r^{np-1} + K_{4n}r^{-np+1} \end{aligned} \tag{2}$$

In (2), coefficients K_{1n}, ..., K_{6n} are constants which depend on the geometry and on the magnetic properties of the rotor permanent magnets. The magnetic field fundamental harmonic has p pole pairs and the higher-order harmonics have np pole pairs with n being an odd number only. When the rotor is centered, $\epsilon = 0$ and only the terms with C_n remain in (1). When the rotor is off-centered, $\epsilon \neq 0$ and other harmonics appear. They correspond to the terms with $\hat{C}_n$ and $\check{C}_n$ in (1) and their amplitudes are proportional to ϵ.

The magnetic flux created by the permanent magnets in the winding phase i is denoted Φ_{Mi} and can be calculated as follows:

$$\Phi_{Mi} = L\int_{R_w}^{R_a}\int_0^{2\pi} A_{Mz}(r,\theta)\frac{J_i(r,\theta)}{I_i} r\,d\theta\,dr, \tag{3}$$

where L is the axial length of the bearing, and $J_i(r,\theta)$ and I_i are the current density and the current amplitude in phase i, respectively. For phase i in an axial winding with k phases:

$$\begin{aligned} &\frac{J_i(r,\theta)}{I_i} = \frac{2kqN(-1)^v}{\pi|R_a^2 - R_w^2|} \text{ if } r \in [R_w, R_a] \text{ and} \\ &\theta \in \left[\delta_i + \frac{\pi(k-1)}{2kq} + \frac{\pi v}{q}, \delta_i + \frac{\pi(k+1)}{2kq} + \frac{\pi v}{q}\right], \end{aligned} \tag{4}$$

and $\frac{J_i(r,\theta)}{I_i} = 0$ otherwise. In this expression, $\delta_i = \frac{\pi i}{qk}$ is the angular position of phase i and $v \in \{0; 2q-1\}$. The Fourier development of this function is:

$$\frac{J_i(r,\theta)}{I_i} = \sum_{m,odd}^{\infty} \frac{8Nkq}{\pi^2 m|R_a^2 - R_w^2|} \sin\left(\frac{m\pi}{2}\right) \sin\left(\frac{m\pi}{2k}\right) \sin\left(mq\left(\theta - \delta_i\right)\right). \quad (5)$$

Because $q = p + 1$, only the terms with coefficients $\hat{C}_n$ and $\check{C}_n$ in (1) can give a non-zero contribution to Φ_{Mi}, as exposed in [7]. The amplitude of Φ_{Mi} is thus proportional to ϵ too, as well as the corresponding emf induced in a phase i of the winding. Given hypothesis 5, it can be calculated as:

$$\begin{aligned} emf &= -\frac{d\Phi_{Mi}}{dt} = -\frac{\partial \Phi_{Mi}}{\partial \theta_s}\omega - \frac{\partial \Phi_{Mi}}{\partial \phi}\lambda \\ &= \sum_{m,odd}^{\infty} \Big\{ \epsilon \hat{E}_m \sin\left((mq-1)\theta_s - mq\delta_i + \phi\right) \\ &\quad + \epsilon \check{E}_m \sin\left((mq+1)\theta_s - mq\delta_i - \phi\right) \Big\}. \end{aligned} \quad (6)$$

In this expression, $\hat{E}_m$ and $\check{E}_m$ are constants depending on the geometry and magnetic properties of the winding, as well as the operating speed and pole pair numbers of the rotor and winding. These constants may cancel for some m.

To illustrate this, let us consider the first vector potential harmonic with coefficient $\hat{C}_{n=1}$ in (1). It has $(p+1)$ pole pairs which is the same as the winding. Therefore, it induces a non-zero time-varying magnetic flux in the winding and $\hat{E}_1 \neq 0$.

Let us continue with the model. Knowing the induced emf_i from (6) allows to calculate the current I_i in phase i through:

$$\begin{aligned} I_i = \sum_{m,odd}^{\infty} \Big\{ &\epsilon \hat{E}_m \frac{\sin\left((mq-1)\theta_s - mq\delta_i + \phi - \hat{\varphi}_m\right)}{\sqrt{R^2 + ((mq-1)\omega + \lambda)^2 L_c^2}} \\ &+ \epsilon \check{E}_m \frac{\sin\left((mq+1)\theta_s - mq\delta_i - \phi - \check{\varphi}_m\right)}{\sqrt{R^2 + ((mq+1)\omega + \lambda)^2 L_c^2}} \Big\} \end{aligned} \quad (7)$$

with:

$$\hat{\varphi}_m = atan\left(\frac{((mq-1)\omega + \lambda)L_c}{R}\right)$$

$$\check{\varphi}_m = atan\left(\frac{((mq+1)\omega + \lambda)L_c}{R}\right)$$

where R and L_c are the winding phase resistance and cyclic inductance, respectively. The phase difference between emf_i and I_i is denoted φ_m. The previous reasoning can be repeated for every winding phase. Once the current in each phase is known, the magnetic field created by this current in the airgap is found by solving Maxwell's equation for the magnetic field in the airgap. This was done in [10] for the bearing with a ferromagnetic yoke and a similar approach was used in the yokeless case. The magnetic field B_I in the airgap due to the induced current flowing in the entire winding is obtained by summing the contributions of each phase. Because of hypothesis 2 at the beginning of this section, the total magnetic field B_{tot} in the airgap is the sum of the magnetic field from the winding and from the rotor permanent magnets:

$$\vec{B}_{tot} = \vec{B}_M + \vec{B}_I, \quad (8)$$

with $\vec{B}_M = \nabla \times \vec{A}_{Mz}$. Finally, the force between the rotor and the winding was obtained by integrating Maxwell's stress tensor in the airgap. This allows to obtain the force component acting on the rotor in the direction of the off-centering and in the direction perpendicular to it F_ϵ and F_ϕ, see Fig. 1:

$$\begin{aligned} F_\epsilon &= \int_0^{2\pi} \Big\{ \frac{1}{2\mu_0}\left(B_{tot,r}^2 - B_{tot,\theta}^2\right)\cos(\theta - \phi) \\ &\quad - \frac{B_{tot,r}B_{tot,\theta}}{\mu_0}\sin(\theta - \phi) \Big\} Lrd\theta; \\ F_\phi &= \int_0^{2\pi} \Big\{ \frac{1}{2\mu_0}\left(B_{tot,r}^2 - B_{tot,\theta}^2\right)\sin(\theta - \phi) \\ &\quad + \frac{B_{tot,r}B_{tot,\theta}}{\mu_0}\cos(\theta - \phi) \Big\} Lrd\theta; \end{aligned} \quad (9)$$

where $B_{tot,r}$ and $B_{tot,\theta}$ are the radial and azimuthal components of the magnetic field, respectively. As will be shown in the next sections, these forces may be pulsating. The time averages of these forces are obtained by removing the time-dependent terms in (9) and are denoted $\bar{F}_\epsilon$ and $\bar{F}_\phi$. In the next sections, the pulsating component is neglected compared to the time average forces on the bearing rotor. Indeed, pulsating forces can be reduced by increasing the number of phases [7].

IV. Performance Evaluation

In this section, the effects of $\bar{F}_\epsilon$ and $\bar{F}_\phi$ on the stability of the bearing is first exposed. Then, a way to compare two bearings with and without ferromagnetic yoke using the model developed in the previous section is presented.

Due to the assumptions exposed in section II, the forces predicted by the model are directly proportional to ϵ. Therefore, the effects of $\bar{F}_\epsilon$ and $\bar{F}_\phi$ can be associated with stiffnesses K_ϵ and K_ϕ through $\bar{F}_\epsilon = -K_\epsilon \epsilon$ and $\bar{F}_\phi = -K_\phi \epsilon$.

When $K_\epsilon > 0$, it has a positive effect since it is acting on the rotor toward the axis of the winding, see Fig. 1. On the contrary, K_ϕ has a negative effect on the stability of the bearing whatever its sign. It induces a whirling motion of the rotor around the axis of the winding, resulting in centrifugal forces acting on the rotor. These centrifugal forces grow and finally overcome K_ϵ, thereby preventing the rotor from coming back toward the center of the winding. Finally, it appears that such a bearing is not stable without providing additional damping to cancel the effect of K_ϕ. This is consistent with [5], [11], [12] and with the results from the next sections. Since it is widely acknowledged that adding contactless damping in the system by passive means is not easy, the amplitude of $\bar{F}_\phi$ should therefore be minimized.

In order to compare or optimize electrodynamic bearings, it is necessary to evaluate their performance objectively. For this purpose, looking at $\bar{F}_\epsilon$ and $\bar{F}_\phi$ is not sufficient since it does not give the amount of damping which must be added to have a stable bearing e.g..

Let us show how a root loci plot can be used to evaluate the performance of a bearing objectively. This has been used in previous works on the stability of a system comprising electrodynamic bearings. In [13], a parametric model was derived. It was validated by identification of the parameters with experimental data and by showing that the predicted stability threshold spin speed corresponds with experimental results. Compared with the model exposed in the previous section, the model from [13] can be used with homopolar and heteropolar bearings and neither requires to derive the current amplitude nor to make assumptions 1, 5 and 6 from the previous section. However, this model alone cannot be used in a design optimization process because the parameters depend on the bearing geometry and must be identified first.

Let us derive the root loci plot using the 2-D model exposed previously. Using complex numbers for vector quantities and noting the position of the rotor $z = X + jY = \epsilon e^{j\phi}$, the motion equation of the rotor is:

$$m\ddot{z} + c\dot{z} + K_\epsilon z + jK_\phi z = 0, \tag{10}$$

where $j = \sqrt{-1}$, m is the rotor mass and c is the external damping added in the system. The stiffnesses K_ϵ and K_ϕ are calculated for a fixed spin speed ω and a zero whirl speed λ because $\lambda << \omega$. This will be validated later in this section. The governing set of equations of the associated state-space model are:

$$\begin{bmatrix} \ddot{z} \\ \dot{z} \end{bmatrix} = \mathbf{A} \begin{bmatrix} \dot{z} \\ z \end{bmatrix} + \frac{1}{m}\mathbf{B}, \tag{11}$$

with:

$$\mathbf{A} = \begin{bmatrix} \frac{-c}{m} & \frac{-K_\epsilon - jK_\phi}{m} \\ 1 & 0 \end{bmatrix}. \tag{12}$$

In (11), matrix $\mathbf{B}$ is the external input force matrix and $\mathbf{A}$ is the dynamic matrix whose eigenvalues give the root loci plot. The two eigenvalues are:

$$\rho_{1,2} = \frac{-c}{2m} \pm \sqrt{\left(\frac{c}{2m}\right)^2 - \frac{K_\epsilon + jK_\phi}{m}}. \tag{13}$$

The root loci of a system comprising a bearing without additional damping is shown in Fig. 4 for different speeds ω and yoke material. This was obtained for a bearing with the dimensions from Table I and with a rotor mass $m = 50[kg]$. The permanent magnet at the rotor has a parallel magnetization pattern with $p = 1$ and a remanent magnetic field density $B_{rem} = 1.32[T]$. The winding has two pole pairs and three phases.

Table I
BEARING DIMENSIONS $[mm]$.

R_i	R_m	R_w	R_a	L
2	18	19	21	70

As can be seen in Fig. 4, there is a rotational symmetry with respect to the origin so that whatever the material of the yoke, one of the two roots of the system is always on the right-hand side of the plane i.e., its real part is positive. This indicates that

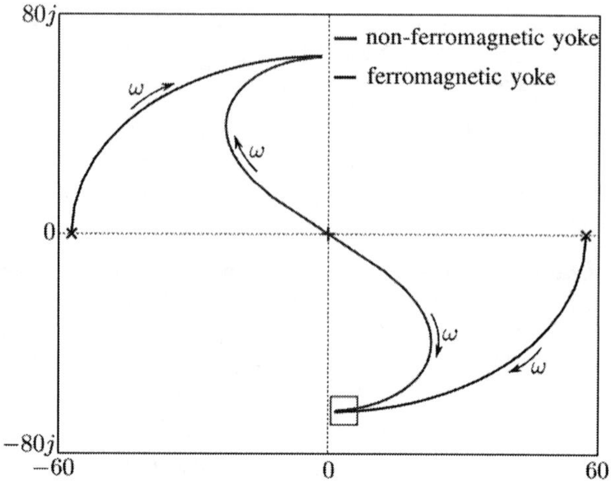

Figure 4. Root loci for $\omega \in [0, 2\pi 5000]$ and $c = 0$. The arrows show the direction of increasing ω.

the bearing is always unstable when no damping is added in the system and it corresponds to expectations. As ω increases, the winding becomes more inductive, F_ϵ (and K_ϵ) reach a maximum value while F_ϕ (and K_ϕ) decrease toward zero. The unstable root approaches the stability region on the left-hand side of the plane and $|\Re\{\rho\}| << |\Im\{\rho\}|$, see the region with a red square in Fig. 4. Let us focus on the behaviour of the unstable roots in this region. The root-loci is drawn for $c = 0$, so in this case (13) becomes:

$$\rho_{1,2} = \pm j\sqrt{\frac{K_\epsilon + jK_\phi}{m}}. \tag{14}$$

If the unstable root of the root loci with $c = 0$ is such that $|\Re\{\rho\}| << |\Im\{\rho\}|$, two observations can be made. Firstly, $K_\phi << K_\epsilon$ and therefore $\Im\{\rho_{1,2}\}$ can be associated with the stiffness K_ϵ toward the center of the stator through:

$$\Im\{\rho_{1,2}\} \approx \pm\sqrt{\frac{K_\epsilon}{m}}. \tag{15}$$

Secondly, if a reasonable amount of damping c is introduced to make the unstable root reach the imaginary axis toward the stability region, it can be derived from (13) that $\Re\{\rho\}$ moves $-\frac{c}{2m}$ to the left.

From these two observations, it appears that the stability and the stiffness of a bearing can be evaluated separately. Therefore, if two bearings with a same mass m and damping $c = 0$ are operating in a region such that $|\Re\{\rho\}| << |\Im\{\rho\}|$, looking at their real and imaginary parts allows to know which one has the greatest stiffness and/or needs less damping to be stable.

The assumption $\lambda << \omega$ can be validated as follows. When no external damping is added to the bearing, it was observed that the rotor trajectory is an outward spiral and that the value of λ quickly reaches a constant value given by $|\Im\{\rho\}|$. The evolution of this value with ω is given in Fig. 5, showing that the assumption $\lambda << \omega$ is valid whatever the value of ω.

Finally, let us add that these results are limited to bearings operating in a region such that $|\Re\{\rho\}| << |\Im\{\rho\}|$. However,

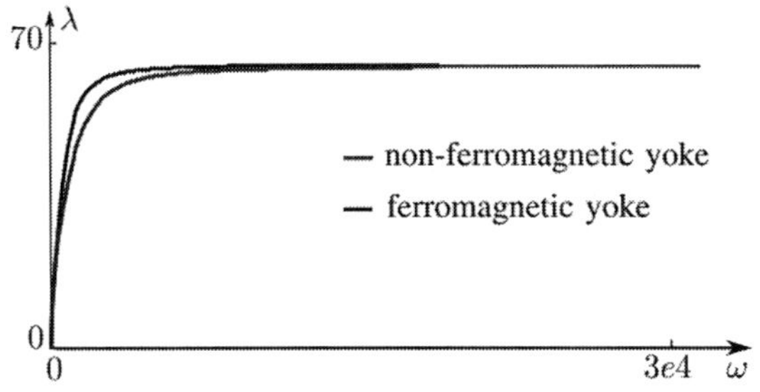

Figure 5. Evolution of the whirl speed λ with ω.

this assumption is not very constraining as this is the operating region that will be preferred when considering the difficulty of adding external damping in the system.

V. Bearing comparison

Using the results from the previous section, let us show an example of comparison between two bearings with and without ferromagnetic yoke. The study is based on the following assumptions. Both bearings have the same rotor and the same air gap thickness. Therefore, the bearings' rotors have the same mass m and the only remaining geometrical variable is R_a, see Fig. 2. The default geometrical and magnetic parameters are the same as in the previous section.

The selection process for the best bearing is the following. The bearings have to provide a minimum stiffness when operating above a given threshold speed. The best bearing will be the one meeting these requirements and requiring the less damping to reach stability.

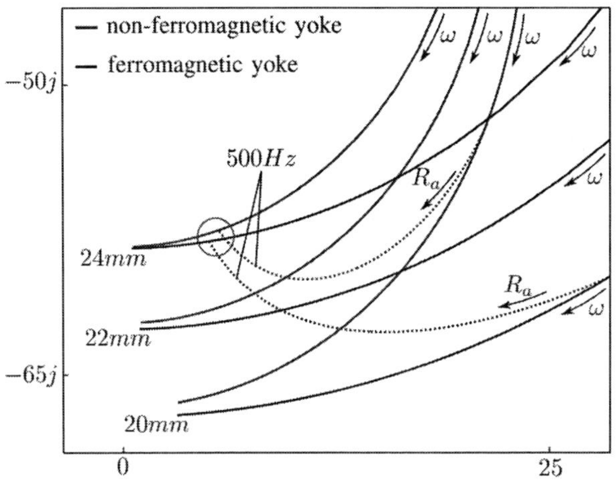

Figure 6. Unstable poles in the area of interest of the root loci for $c = 0$. The dotted (solid) lines correspond to a fixed ω (a fixed Ra).

Fig. 6 shows a zoom on the area of interest in the root loci plot. The dotted lines were obtained for a fixed speed ω and an increasing radius R_a. They give the minimum operating speed of the two bearings so that the area of interest with $\omega > 2\pi 500$ is below these curves. The solid lines were obtained for the fixed radii $R_a = 20, 22, 24[mm]$ and an increasing speed ω. The evolution of the bearing properties with speed can thus be observed by following these solid lines in the region above the threshold speed. The performance of a bearing for a given set of parameters (R_a, ω) is obtained by looking at the crossing points between the corresponding dotted and solid lines. E.g., two such points for $(R_a = 24[mm], \omega = 500[Hz])$ are in the red circle of Fig. 6. For this set of parameters, the bearing with a ferromagnetic yoke has a better stiffness and better stability properties. Looking at Fig. 6, it is clear that the bearing performance improve with speed.

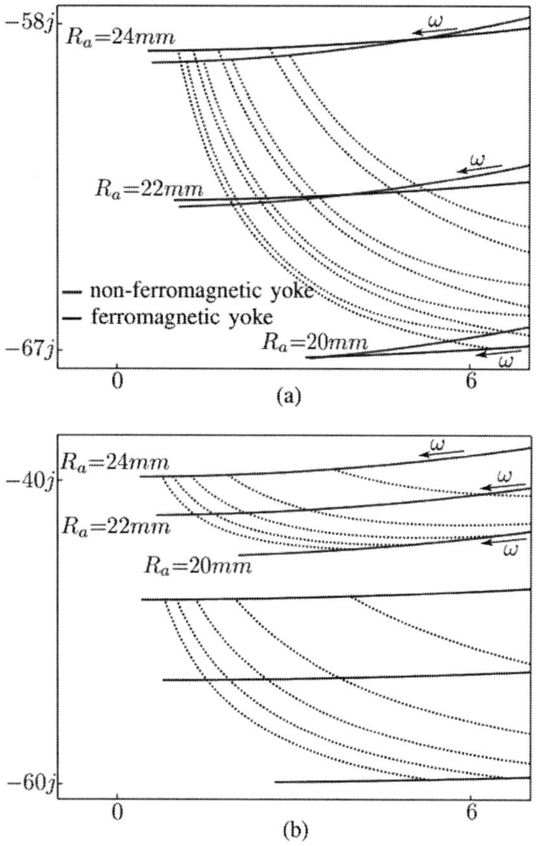

Figure 7. Root loci of the bearings with no damping, $p = 1$ and $R_i = 0[mm]$ (a) or $R_i = 10[mm]$ (b).

In order to see the evolution of the previous result with ω and R_a, more data is needed, as shown in Fig. 7. This was obtained for different rotor shaft radii $R_i = 0[mm]$ and $10[mm]$. From Fig. 7 (a), it appears that the yoke material does not have a significant influence on the bearing stiffness and stability in the area of interest when $R_i = 0[mm]$. E.g. for $R_a = 24[mm]$, a bearing with a ferromagnetic yoke has a better stiffness but is less stable than a bearing with no ferromagnetic yoke, but the differences are not very significant. However, these differences increase with R_i as shown in Fig. 7 (b). Comparing the corresponding points for each kind of yoke, it appears that the bearing with a ferromagnetic yoke has a significantly greater stiffness but is less stable. Finally, the stiffness of the two kinds of bearing decrease with R_i, which is consistent with intuition since the amount of permanent magnet decreases with R_i.

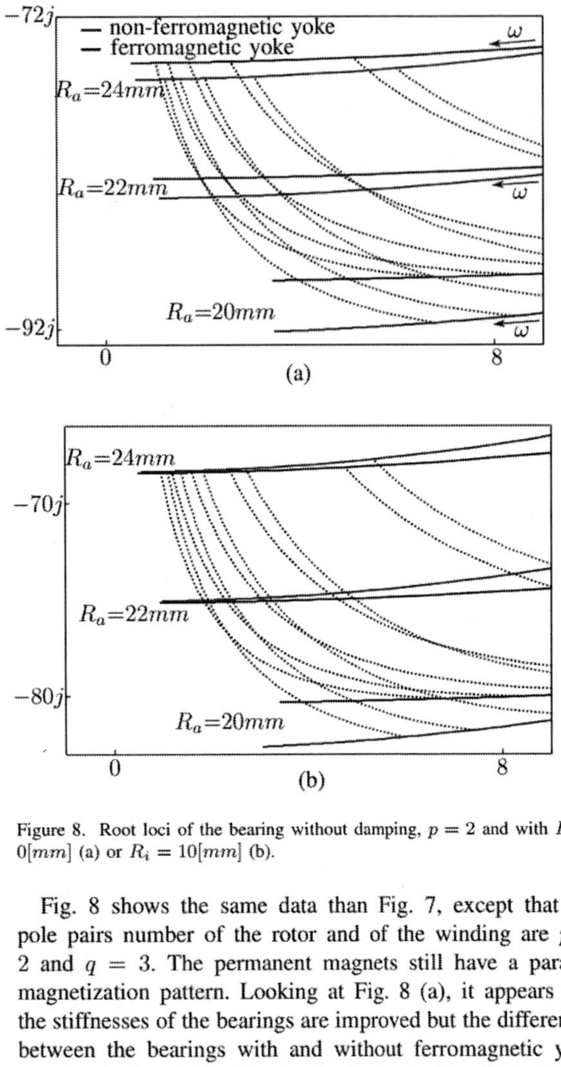

Figure 8. Root loci of the bearing without damping, $p = 2$ and with $R_i = 0[mm]$ (a) or $R_i = 10[mm]$ (b).

Fig. 8 shows the same data than Fig. 7, except that the pole pairs number of the rotor and of the winding are $p = 2$ and $q = 3$. The permanent magnets still have a parallel magnetization pattern. Looking at Fig. 8 (a), it appears that the stiffnesses of the bearings are improved but the differences between the bearings with and without ferromagnetic yoke remain small and the situation is similar to Fig. 7 (a). Fig. 8 (b) shows that increasing the shaft radius R_i is less favourable for the bearing without ferromagnetic yoke, but the difference between Fig. 8 (a) and (b) is much smaller than the difference between Fig. 7 (a) and (b). Last but not least, the bearing with a ferromagnetic yoke has a lower stiffness the the bearing without ferromagnetic yoke when $R_i = 20[mm]$ i.e., when the yoke is close to the permanent magnets. This is due to the higher order harmonics in the permanent magnet field. They contribute to the negative stiffness associated with the reluctant force between the permanent magnets and the ferromagnetic yoke, but they do not contribute to the centering Lorentz force between the permanent magnet and the winding. This effect cancels as R_a increases and was not observed with $p = 1$ because there was no higher order harmonics in the permanent magnet field in this case.

VI. Conclusion

In this paper, the influence of the magnetic permeability of the yoke on the behaviour of electrodynamic bearings was studied. To this end, a 2-D analytical model was presented and used to draw the root loci plot of the bearings without external damping. Considering the difficulty of adding the external damping to the bearing which is necessary to achieve stability, the area of interest to operate an electrodynamic bearing was identified in the root loci plot. Then, a way to use this plot to evaluate the stiffness and stability properties of the bearings independently was presented. Finally, this model was used to compare bearings with and without a ferromagnetic yoke.

It was shown that the yoke material does not have a significant influence on the performance of the bearing for a small rotor shaft radius. As this radius increases, the stiffness of the bearing decreases and this effect is more significant for a bearing without ferromagnetic yoke. Also, increasing the pole pairs number of the bearing has a positive effect on the bearing stiffness but is less advantageous for a bearing with a ferromagnetic yoke when the yoke is close to the permanent magnets. Finally, this paper shows that choosing the yoke material is not a trivial issue and requires a more comprehensive optimization process to be carried out.

References

[1] G. Schweitzer and E. H. Maslen, "Magnetic bearings: theory, design and application to rotating machinery," *Springer*, 2009.

[2] A. Looser and J. Kolar, "An active magnetic damper concept for stabilization of gas bearings in high-speed permanent-magnet machines," *IEEE Trans. Ind. Electron., vol.61, No 6*, pp. 3089–3098, 2014.

[3] R. F. Post and D. D. Ryutov, "Ambient-temperature passive magnetic bearings: Theory and design equations," *Proceedings of the Sixth International Symposium on Magnetic Bearings*, 1998.

[4] R. F. Post and D. A. Bender, "Ambient-temperature passive magnetic bearings for flywheel energy storage systems," *Proceedings of the Seventh International Symposium on Magnetic Bearings*, 2000.

[5] A. V. Filatov and E. H. Maslen, "Passive magnetic bearing for flywheel energy storage systems," *IEEE Trans. Magn., vol.37, No 6*, pp. 3913–3924, 2001.

[6] T. A. Lembke, "Design and analysis of a novel low loss homopolar electrodynamic bearing," *Ph.D. dissertation, Royal Inst. Tech., Sweden*, 2005.

[7] C. Dumont, V. Kluyskens, and B. Dehez, "Null flux radial electrodynamic bearing," *accepted for publication.*

[8] C. Dumont, V. Kluyskens, and B.Dehez, "Yokeless radial electrodynamic bearing," *11th International Conference on Modeling and Simulation of Electric Machines, Converters and Systems 'Electrimacs'*, 2014.

[9] A. Rahideh and T. Korakianitis, "Analytical open-circuit magnetic field distribution of slotless brushless permanent-magnet machines with rotor eccentricity," *IEEE Trans. Magn., vol.47, No 12*, pp. 4791–4808, 2011.

[10] K. Atallah, Z. Zhu, D. Howe, and T. Birch, "Armature reaction field and winding inductances of slotless permanent-magnet brushless machines," *IEEE Trans. Magn, vol.34, No 5*, pp. 3737–3744, 1998.

[11] V. Kluyskens and B. Dehez, "Dynamical electromechanical model for magnetic bearings subject to eddy currents," *IEEE Trans. Magn., vol.49, No 4*, pp. 1444–1452, 2013.

[12] A. Tonoli, N. Amati, F. Impinna, and J. G. Detoni, "A solution for the stabilization of electrodynamic bearings: Modeling and experimental validation," *J. Vib. Acoust., vol.133, No 2*, 2011.

[13] J. Detoni, F. Impinna, A. Tonoli, and N. Amati, "Unified modelling of passive homopolar and heteropolar electrodynamic bearings," *J. Sound Vibration, vol.331, No 19*, pp. 4219–4232, 2012.

Optimization of a new radial electrodynamic damper

Virginie Kluyskens[a], Bruno Dehez[a]

[a] Center for Research in Energy and Mechatronics, UCL, Bâtiment Stevin, bte L5.04.02 Place du Levant 2,B-1348 Louvain-la-Neuve, Belgium, virginie.kluyskens@uclouvain.be

Abstract—**The study of electrodynamic dampers is an important point when working with electrodynamic bearings. These dampers may also be used in combination with reluctant passive magnetic bearings. This paper proposes a new design of a robust and inherently stable radial electrodynamic damper with a homopolar axial magnetic flux. An optimization, based on finite element models, of this new bearing topology is made. For this to be possible, a 2-D representation of the damper is developed equivalent to the 3-D damper. The results of the optimizations are presented and a discussion of the influence of some of the parameters is done.**

I. Introduction

High speed electrical drives are a promising technology [1]. Fast spinning and high efficiency electrical drives may be obtained using for instance permanent magnet drives, and thus avoiding excitation losses. These high speed electrical drives need adapted bearings. One remaining challenge is to design high speed bearings with a high efficiency, a long lifetime and simple to use. Developped solutions concern magnetic bearings and gas bearings. However, gas bearings do present poor dynamic stability and need additional damping [2]. On the other hand active magnetic bearings lead to more complex, less compact, and possibly more energy consuming designs. Passive magnetic bearings could be a solution but they also suffer from rotor dynamic stability issues: passive magnetic bearings based on the reluctant principle have almost no intrinsic damping [3] and passive magnetic bearings based on the electrodynamic principle may also present a completely unstable behavior when the damping present in the system is not well managed [4]. Another solution for stabilizing electrodynamic bearings has been investigated in [5] and consists in adding dissipative elements between the statoric part of the bearing and the ground. In this context, when wishing to work with passive or electrodynamic bearings, the study of dampers is an important part of the conception of the bearing.

When designing electrodynamic dampers, attention has to be paid to stability issues, just as for electrodynamic bearings. Indeed, [6] observes that electrodynamic dampers using conductors spinning with the rotor show an unstable behavior at high spin speeds. They offer a first analytical approach to understand those phenomena and make experimental evaluations of it. In [7], a general parameterized electromechanical model, based on a simple analogy between the resistive-inductive dynamics of eddy currents and the spring-damper in series dynamics, is described. [8] presents a general parameterized electromechanical model for the radial behavior of any kind of magnetic system subject to eddy currents. It is shown that for stability issues, when the intended effect is to produce damping, this damping should be generated by the rotor whirling motion and not the spinning motion.

Electrodynamic dampers are based on the interaction between a magnetic field generated by a permanent magnet or an electromagnet and currents induced by the variation of this magnetic field. When the variation of the magnetic field can be generated by a time-variation of the magnetic field, one talks about a transformer e.m.f. on the conductor, and when the variation of the magnetic field is generated by a space-variation of the latter, induced for instance by the motion of the conductor relatively to the magnetic field, one talks about a motional e.m.f. A theoretical comparison between dampers based on each type of e.m.f. is carried out in [9], and it is shown that for a given magnetic field, a given volume of conductor, and assuming that there is no inductive behavior, dampers based on a transformer e.m.f. may lead to a better damping efficiency. However, it is also said that in practical cases, dampers based on a motional e.m.f. are often better suited.
Electrodynamic dampers have been developped for damping axial, radial or torsional rotor motions. They can be based on a radial or an axial magnetic flux. [10] gives an overview of all these dampers, and gives analytical tools to predict the damping, but with a focus on the case of a radial damper with a radial magnetic flux. In [11], the electrodynamic bearing is combined with a radial electrodynamic damper with axial magnetic flux. Torsional electrodynamic dampers are investigated in [12] and [13].
An example of an inherently stable electrodynamic damper, acting between the rotor and the stator, combined with a reluctant passive magnetic bearing is given in [14].

This papers proposes a radial electrodynamic damper with a homopolar axial magnetic flux, which could be used with a reluctant passive magnetic bearing, but also with an electrodynamic magnetic bearing. This damper is inherently stable thanks to its configuration, and is robust because it does not need to insert a permanent magnet on a fast spinning rotor. The effect of the insertion of teeth in the magnetic circuit of the damper is investigated. Parametric optimizations, based on finite element models, of this new bearing topology are made. For this to be possible, an original equivalent 2-D representation of the damper is developped equivalent to the

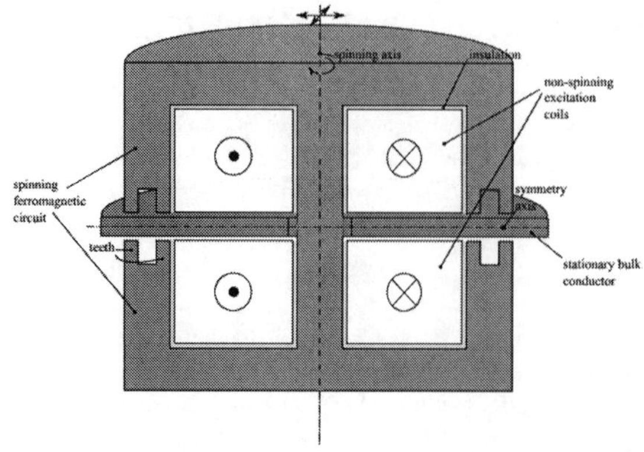

Figure 1. Schematic view of the studied electrodynamic damper.

3-D damper. The results of the optimizations are presented and discussed.

II. Structure and Principles of proposed electrodynamic damper

The electrodynamic bearing studied in this paper is a radial damper with a homopolar axial magnetic flux, and is presented on Fig. 1. The conductor is a bulk copper plate linked to the stator, and is thus stationary. The axial homopolar magnetic flux is generated by an excitation winding, fixed in the airgap, and fed by a DC-current. This fixed coil magnetizes a spinning steel magnetic circuit, which allows on one side to have a robust design, as there are no brushes, and no permanent magnets to be inserted in potentially high spin speed pieces, and on another side, the possibility exists to feed the coil by a current adapted to the needed damping, so as not to generate damping, and losses, when unnecessary.
This damper topology, with a conductor fixed to the stator, is inherently stable. Indeed, when the rotor is spinning in a fixed centered or out-centered position, no forces are produced. However, when moving radially, damping will be produced.

The magnetic path is constituted of one central leg, an external leg, and an airgap. The conductor is a copper annular disk. The external leg of the magnetic circuit is terminated by teeth, uniformly distributed, each of equal width. The width of those teeth are characterized by a parameter called x_{teeth} which is worth 0 when the teeth are so large that there is no air anymore between the teeth, and worth 1 when the teeth are infinitely thin. As these teeth are only present on the rotor, they do not produce any reluctant force, but they allow for another magnetic flux distribution in the air gap, which could increase the damping.
The damper geometry is characterized by the parameters summarized in Table I, and the parameters to be optimized are represented on Fig.2.

Table I
Geometric parameters of the damper

Parameter	Description	Value
L_1	half of damper axial length	to be optimized
L_2	damper radius	to be optimized
l_1	external leg width	to be optimized
l_2	yoke width	to be optimized
l_3	internal leg diameter	to be optimized
h_{cu}	conductor thickness	to be optimized
n_{teeth}	number of teeth	fixed
x_{teeth}	teeth proportion	to be optimized
l_{teeth}	teeth width	$(1 - x_{teeth}) * l_1/n_{teeth}$
h_{teeth}	teeth height	$g + h_{cu}$
g	airgap between conductor and ferromagnetic circuit	$0.5\,\mathrm{mm}$
e_{insh1}	thickness of insulation between central leg and coil	$3.8\,\mathrm{mm}$
e_{insh2}	thickness of insulation between external leg and coil	$2\,\mathrm{mm}$
e_{insv}	thickness of insulation between yoke and coil	$3\,\mathrm{mm}$

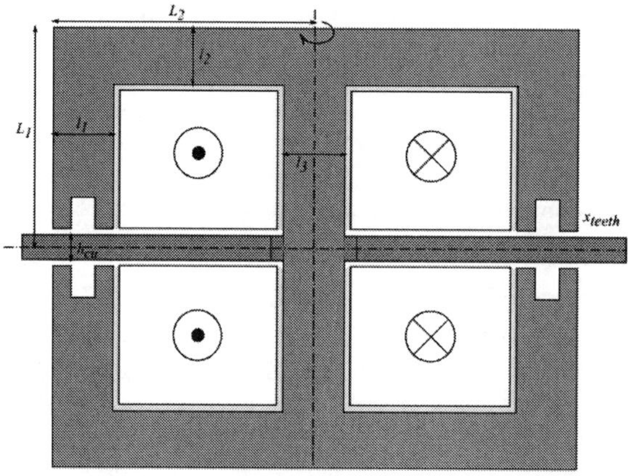

Figure 2. Geometric parameters to be optimized.

III. Optimization approach

A. Objective function and optimization tool

The optimization is based on a genetic algorithm. The evaluation of each individual of the population is made by a finite element model, as explained below. The objective function is to optimize the damping coefficient for a given excitation speed, per damper volume unit. This can be expressed as:

$$\max_{\{L_1,L_2,l_1,l_2,l_3,h_{cu},x_{teeth}\}} \frac{\iiint \bar{J}_{induced} \times \bar{B}\ dV_{conductor}}{\bar{v}\ V_{damper}}, \quad (1)$$

where $\bar{J}_{induced}$ is the current induced in the conductor, $\bar{B}$ is the magnetic flux density, $\bar{v}$ is the excitation speed, and V_{damper} is the damper volume.

B. Equivalent representation for 2-D FE model

A way to model the damper is to make a 3-D FE model of the damper, but this is not realistic to be used for an

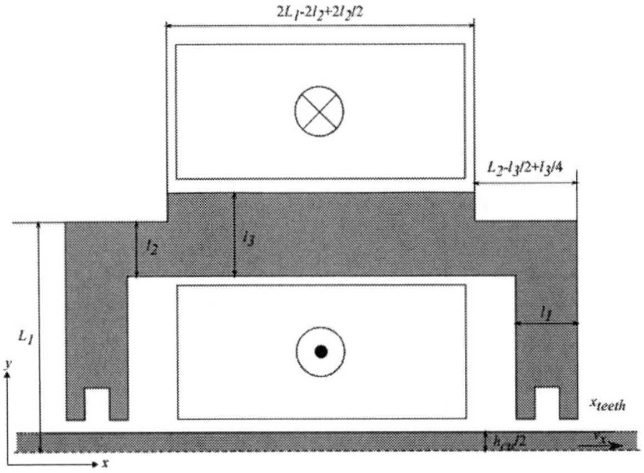

Figure 3. Equivalent 2D representation of the damper.

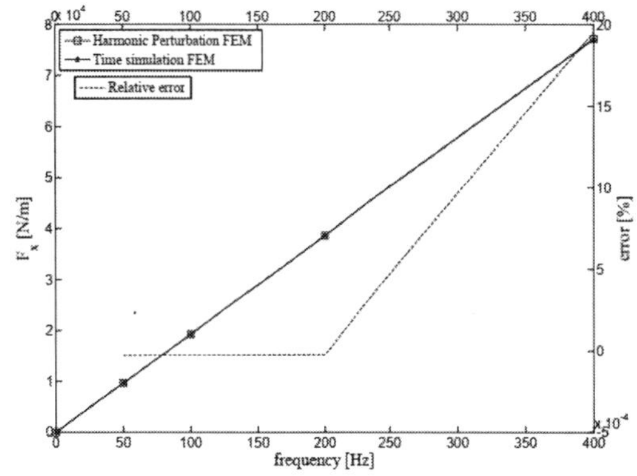

Figure 4. Evolution of the damping force as a function of the excitation frequency, for quasi-static time-simulations and for harmonic perturbation simulations.

optimization as it is too much time consuming. The damper has an axisymmetric geometry, however the excitation speed is not axisymmetric. This is why an equivalent 2-D model of the damper was made. It is obtained by unfolding the parts of the magnetic circuit above and underneath the conductor, as illustrated in Fig. 3. The geometric dimensions of this equivalent model depend on the initial geometry, and the corner reluctances are taken into account by a lengthening of the adjacent legs, based on the same idea as explained in [15]. The excitation speed is applied on the conductor.
In this model, the objective function (1) simplifies to:

$$\max_{\{L_1,L_2,l_1,l_2,l_3,h_{cu},x_{teeth}\}} \frac{\iint J_{induced,z} B_y \, dS_{conductor}}{v_x \, V_{damper}}, \quad (2)$$

As in the real system, the moving part of the damper will have a magnitude-limited motion, we considered a sine signal of amplitude v_x for the excitation speed in the optimization. The maximum displacement is worth $0.5\,\mathrm{mm}$, and the highest considered frequency is $400\,\mathrm{Hz}$. This gives a maximum amplitude of speed worth $v_x = 1.26\,\mathrm{m/s}$.

Therefore, in the first instance, quasi-static time simulations with a sinusoidal excitation speed have been conducted. For these simulations, it is assumed that the electric time constant is much smaller than the mechanical time constant. These quasi-static time simulations are then compared to harmonic perturbation simulations, where the problem is linearized around the stationary solution with no speed. The results for both simulations are shown on Fig. 4. The difference for the highest considered frequencies comes from the fact that for those speeds amplitudes, we are not in the linear part of the force vs speed characteristics anymore, as can be observed on Fig. 5. However, for the smaller frequencies, we are effectively working in the linear part of this characteristic, and even for the higher frequencies, the error remains small: less than 0.2%. This very small difference validates the harmonic perturbation approach.

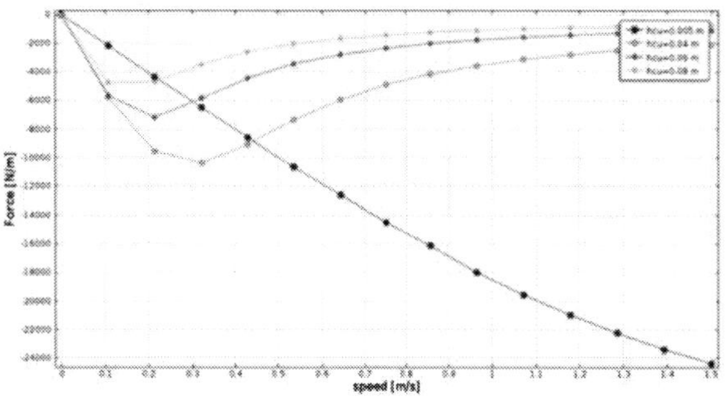

Figure 5. Evolution of the damping force as a function of the excitation speed for the optimized damper for $f = 50\,\mathrm{Hz}$, i.e. an maximum amplitude of speed worth $v_x = 0.16\,\mathrm{m/s}$.

IV. Optimization results

The value of the optimized parameters for an excitation frequency of $100\,\mathrm{Hz}$, and 2 teeth are given in Table II, and the equivalent 2-D representation of the damper is illustrated on Fig. 6. The objective function Eq. (2) is then worth $2.66\ 10^6\,\mathrm{Ns/m^5}$ for an excitation current density of $5\,\mathrm{A/mm^2}$, and the geometric dimensions (Table II) give a damper volume (see Fig. 2) of $0.021\,\mathrm{m}^3$.

Tracing the evolution of the objective function with respect to the teeth proportion, we can see on Fig. 7(a) that the optimal damper does not have teeth. Indeed, given that the damping force expression is $\bar{F} = q(\bar{v} \times \bar{B})$, there is no gain in having teeth which give a lower magnetic flux density inside the conductor.
On Fig. 7(b), the evolution of the objective function with the conductor thickness is shown. There is an optimal thickness, and this thickness is quite high. The fact that the iron reluctance is not negligible probably plays a role in this optimal

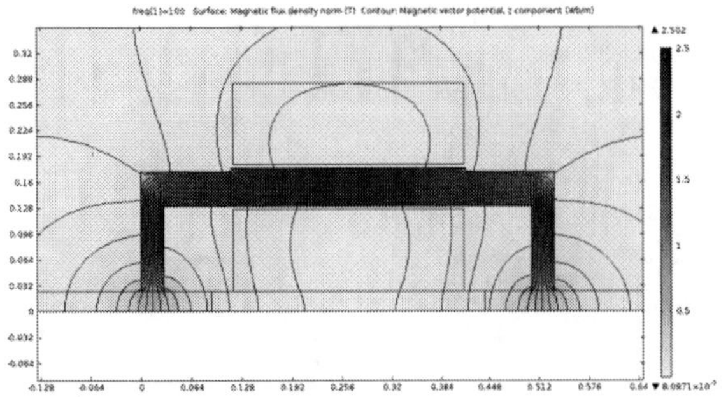

Figure 6. Equivalent 2D representation of the optimized damper, for 2 teeth and an excitation frequency of 100 Hz.

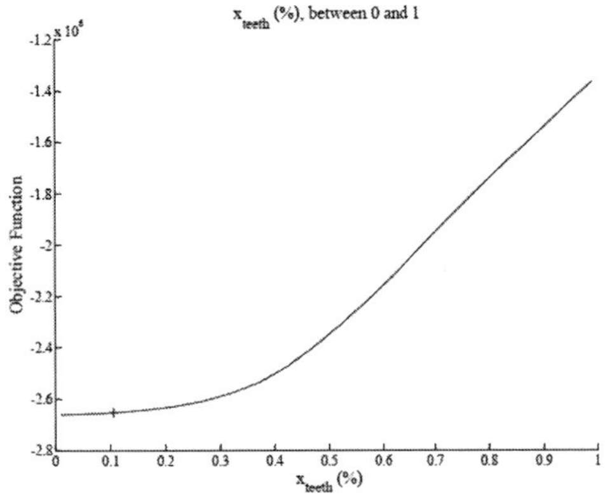

(a)

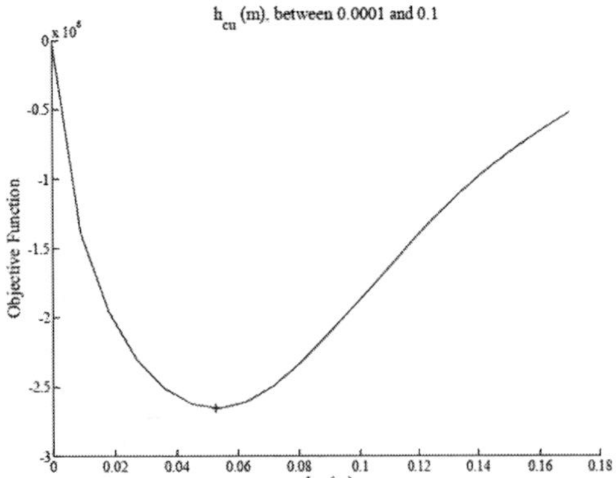

(b)

Figure 7. Evolution of the objective function as a function of the optimization parameters (a) x_{teeth} the teeth proportion (b) h_{cu} the thickness of the conductor .

Table II
OPTIMIZED GEOMETRIC PARAMETERS OF THE DAMPER, FOR $f = 100$ Hz

Parameter	Optimized Value
L_1	173 mm
L_2	138 mm
l_1	29 mm
l_2	43 mm
l_3	95 mm
h_{cu}	51 mm
n_{teeth}	2
x_5	5%

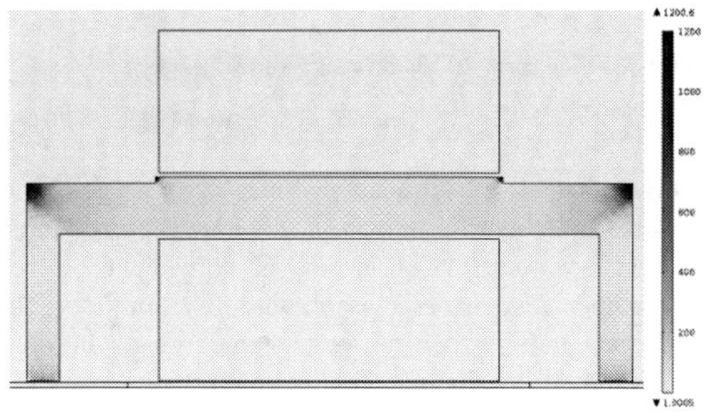

(a)

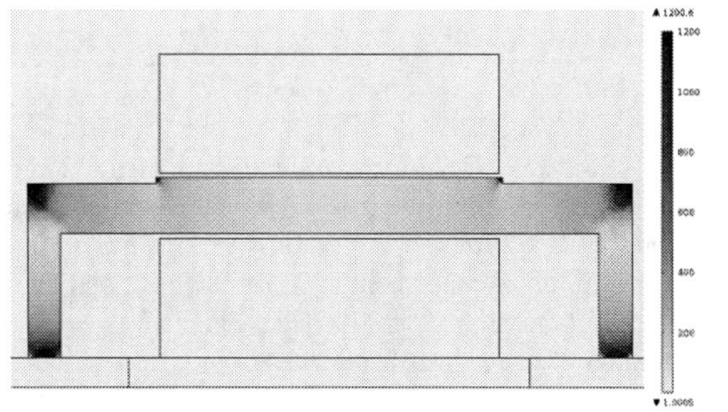

(b)

Figure 8. Representation of the iron relative permeability (a) for $h_{cu} = 10$ mm and (b) for $h_{cu} = 52$ mm

value: when the copper thickness decreases, the iron becomes highly saturated, as can be seen on Fig. 8.

The evolution of the normalized parameters with the excitation frequency is represented on Fig. 9. We can observe that their values are almost constant with frequency. This could be expected, as the force amplitude increases linearly with the frequency (see Fig. 4): the damping coefficient is constant and remains optimal for all the considered frequencies. The small variations of the optimal parameters are also due to the

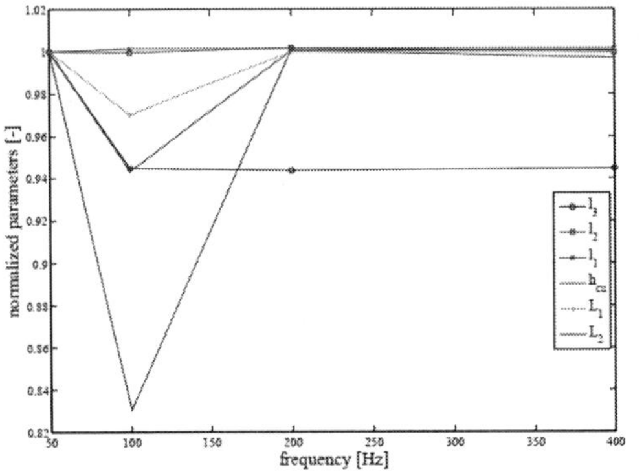

Figure 9. Evolution of the normalized parameters with the excitation frequency.

sensitivity of the objective function which becomes low when approaching the optimum, and they would dissapear if more generations were taken in the genetic algorithm. This can be observed on Fig. 7 and Fig. 10, where we are not exactly on the optimum.

V. Conclusion

Previous works on electrodynamic bearings have shown the importance of adding stabilizing damping to the system. In this paper we have proposed an inherently stable radial electrodynamic damper with a robust design. This damper can easily be used with an electrodynamic bearing or with passive magnetic bearings.

By an equivalent plane 2-D representation and harmonic perturbation FE simulations, an optimization of the geometric dimensions of the damper has been conducted, the objective function being the volumetric damping coefficient.

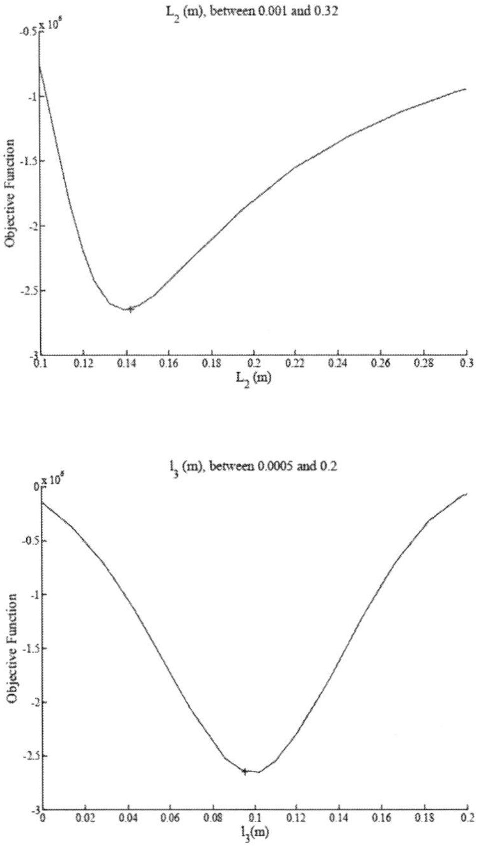

(a)

(b)

Figure 10. Evolution of the objective function as a function of the optimization parameters (a) L_2 the damper external radius (b) l_3 the magnetic circuit central leg diameter .

References

[1] A. Borisavljevic, "Limits, modeling and design of high-speed permanent magnet machines," Ph.D. dissertation, TUDelft, 2011.

[2] A. Looser and J. W. Kolar, "An active magnetic damper concept for stabilization of gas-bearings in high-speed permanent magnet machines," *IEEE Transactions on industial electronics*, vol. 61, no. 6, pp. 3089–3098, June 2014.

[3] F. Jiancheng, L. Yun, S. Jinji, and W. Kun, "Analysis and design of passive magnetic bearing and damping system for high-speed compressor," *IEEE Transactions on Magnetics*, vol. 48, no. 9, pp. 2528–2537, 2012.

[4] V. Kluyskens and B. Dehez, "parameterized electromechanical model for magnetic bearings with induced currents," *Journal of system design and dynamics*, vol. 3, no. 4, pp. 453–461, 2009.

[5] A. Tonoli, N. Amati, F. Impinna, and J. G. Detoni, "A solution for the stabilization of electrodynamic bearings: modeling and experimental validation," *Journal of Vibration and Acoustics*, vol. 133, 2011.

[6] Y. Kligerman and M. S. Grushkevich A.and Darlow, "Analytical and experimental evaluation of instability in rotordynamic system with electromagnetic eddy-current damper," *Journal of Vibration and Acoustics*, vol. 120, pp. 272–278, 1998.

[7] N. Amati, X. D. Lepine, and A. Tonoli, "Modeling of electrodynamic bearings," *Journal of vibrations and acoustics*, vol. 130, pp. 061 007–1-061 007-9, 2008.

[8] V. Kluyskens and B. Dehez, "Dynamical electromechanical model for magnetic bearings subject to eddy currents," *IEEE Transactions on Magnetics*, vol. 49, no. 4, pp. 1444–1452, 2013.

[9] K. E. Graves, D. Toncich, and P. G. Iovenitti, "Theoretical comparison of motional and transformer emf device damping efficiency," *Journal of Sound and Vibration*, vol. 233, no. 3, pp. 441–453, 2000.

[10] T. A. Lembke, "Understanding electrodynamic dampers," in *Proceedings of ISMB10*, 2006.

[11] K. Davey, A. Filatov, and R. Thompson, "Design and analysis of passive homopolar null flux bearings," *IEEE transactions on magnetics*, vol. 41, no. 3, pp. 1169–1175, 2005.

[12] A. Tonoli, "Dynamic characteristics of eddy current dampers and couplers," *Journal of Sound and Vibration*, vol. 301, pp. 576–591, 2007.

[13] N. Amati, A. Tonoli, A. Canova, F. Cavalli, and M. Padovani, "Dynamic behavior of torsional eddy-current dampers: Sensitivity of the design parameters," *IEEE Transactions on Magnetics*, vol. 43, pp. 3266–3277, 2007.

[14] T. A. Lembke, "1-dof bearing arrangement with passive radial bearings and highly efficient integrated electrodynamic dampers, edd," in *International Symposium on Magnetic Bearings*, 2012.

[15] E. Matagne, G. Cividjian, and V. Kluyskens, "Exact expression of corner reluctances in a magnetic circuit of rectangular section," in *XIV International Symposium on electromagnetic fields in mechatronics, electrical and electronic engineering*, 2009.

Active Magnetic Bearings for Rolling Element Bearings Outer Race Incipient Defect Diagnosis

XU Yuanping[a], ZHOU Jin[a], DI Long[b], XU Longxiang[a]

[a] College of Mechanical and Electrical Engineering, Nanjing University of Aeronautics and Astronautics, Yudao Street 29, 210016 Nanjing, China, ypxu@nuaa.edu.cn
[b] Charles L. Brown Department of Electrical and Computer Engineering, University of Virginia, P.O. Box 400743, VA 22904-4743 Charlottesville, U.S.A

***Abstract*—The active health monitoring of rotordynamic systems in the presence of bearing outer race defect is considered in this paper. The shaft is assumed to be supported by conventional mechanical bearings and an active magnetic bearing (AMB) is used in the mid of the shaft or outboard location as an actuator to apply constant force to the system. We present a nonlinear bearing-pedestal system model with the outer race defect under the AMB force. The nonlinear differential equations are integrated using the fourth-order Runge—Kutta algorithm in MATLAB and the simulation result is obtained. The simulation results show that the characteristic outer race defect signal is significantly amplified under the AMB force, which is helpful to improve the diagnosis accuracy of outer race defect.**

I. INTRODUCTION

Rolling element bearing is one of the most common components in the rotating machinery. Its health state directly determines the rotating machinery performance. Many system models [1][2][10] have been created and adopted in the research process of the rolling bearing fault characteristics.

During the bearing incipient fault state, it is difficult to extract the defect characteristic signal because the malfunction itself is not significant and the incipient fault signal is overwhelmed by the random noise and vibration caused by other moving components. For a rolling bearing with zero contact angle, the varying compliance frequency (VC) in the vibration spectrum adversely caused by the time-varying ball position equals to the outer race fault characteristic frequency, which makes it more difficult to detect the incipient small outer race defect.

Active magnetic bearing (AMB) is mainly used to support any rotating element. In addition, AMB can be used as force actuators and sensor [3]. As a force actuator, its amplitude and frequency can be easily altered and it can apply online non-contact excitation force [4]. Nordmann using AMBs as force actuators to identify dynamic characteristics of rotor components [5,6]. Kasarda [7] proposed a method using AMB for the non-destructive evaluation of manufacturing process. Xiangming Lou using AMBs as force actuators to identify the shaft unbalance [4]. Zhu et al. studied the dynamics characteristics of a rotor with crack supporting by AMBs [8]. Sawicki [9] applies a sinusoidal force from the AMB to help identify crack in the rotor. However, the application of AMB as force actuators on the rolling bearing fault diagnosis is relatively rare.

Building on the Feng et al. [11] model, this paper develops a bearing-pedestal model under the AMB force and establishes the outer race defect model. Taking SKF 61901 bearing as an example, we analyze the characteristics of fault signal under the AMB force.

The remainder of the paper is organized as follows. Section 2 presents the model of rolling element bearing with outer race defect and the equation of motion under AMB force. Section 3 describes the simulation results. Conclusions are drawn in Section 4.

II. ROLLING ELEMENT BEARING MODELING

2.1 Contact force

In the rotating state, the rolling number and radial load force of the bearing lead to the stiffness changes periodically, resulting in the varying compliance (VC) vibration. Figure 1 shows the rolling element bearing schematic diagram.

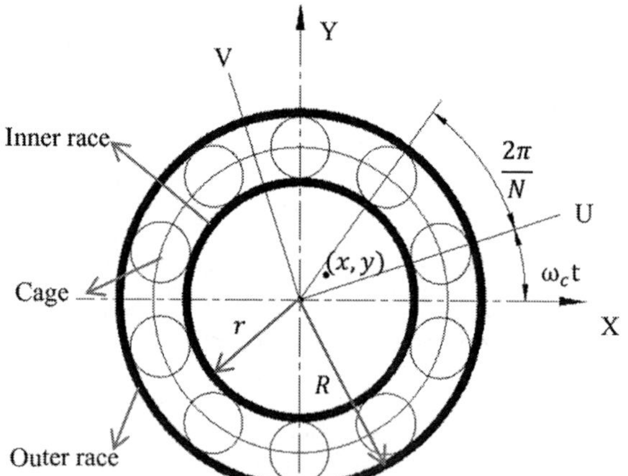

Figure 1. Schematic diagram of a rolling element bearing.

The outer race is fixed in the pedestal and the slippage of rolling elements is ignored. The displacement of the shaft can be divided into x and y and r_0 is the clearance between rolling elements and bearing races. The contact deformation for the ith rolling element δ_i is given by

$$\delta_i = x\cos\theta_i + y\sin\theta_i - r_0. \tag{1}$$

Here the angle position of the rolling element θ_i is a function of the time t, which is given as

$$\theta_i = \omega_c t + \frac{2\pi}{N}(i-1)$$
$$\omega_c = \frac{\omega_s r}{(R+r)}$$

where ω_s is the shaft speed; ω_c is the cage speed(can be calculated from geometry and the shaft speed ω_s assuming no slippage); r and R are the inner and outer race radius respectively; N is the number of rolling elements. The VC frequency is given as

$$\omega_{vc} = \frac{\omega_s r}{(R+r)} N = \omega_c B_N$$

The contact force F of the ball raceway is calculated using traditional Hertz theory (nonlinear stiffness) from

$$F = K\delta^n$$

where K is the contact stiffness depending on the bearing geometry and the elasticity of material; the exponent $n = 3/2$ for ball bearings. Accounting for the fact that F occurs only for positive values of δ, where " + " indicate that. The total contact forces in x and y direction can be calculated as follows:

$$F_{bx} = \sum_{i=1}^{N} K_b(x\cos\theta_i + y\sin\theta_i - r_0)_+^{3/2}\cos\theta_i,$$

$$F_{by} = \sum_{i=1}^{N} K_b(x\cos\theta_i + y\sin\theta_i - r_0)_+^{3/2}\sin\theta_i$$

2.2 Outer race defect modeling

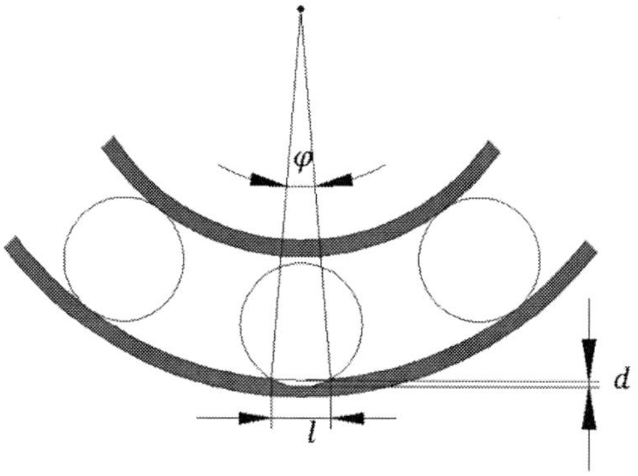

Figure 2. Defect on outer race.

The most failure feature of bearing is the local defect, such as fatigue spall and the corrosive pitting. The outer race defects can be modeled as a slight dent in Figure 2, the length of the dent is l. According to the geometry relationship

$$d = r_b - \sqrt{{r_b}^2 - (l/2)^2}, \quad (8)$$

$$\varphi = 2\arcsin(l/2R), \quad (9)$$

where r_b is the ball radius; φ is the central angle of the outer defect; d is the max increment of the radical clearance. When the ball rolling into the defect area, the radial clearance will increase suddenly, resulting in a decline of the contact force. The outer race defect angular position is assumed as $3\pi/2$ at the bottom of the outer race and the varying clearance Δd caused by the defect is modeled as a half sinusoidal wave, which is give as follows:

$$\Delta d = d\sin\left[\pi\frac{\mathrm{mod}(\theta_i,\ 2\pi) - \left(\frac{3\pi}{2} - \frac{\varphi}{2}\right)}{\varphi}\right]$$

$$\frac{3\pi}{2} - \frac{\varphi}{2} < \mathrm{mod}(\theta_i, 2\pi) < \frac{3\pi}{2} + \frac{\varphi}{2}$$

When the ball is located in the defected area, the total contact force can be calculated by

$$F_{bxd} = \sum_{i=1}^{N} K_b(x\cos\theta_i + y\sin\theta_i - (r_0 + \Delta d))_+^{3/2}\cos\theta_i$$

$$F_{byd} = \sum_{i=1}^{N} K_b(x\cos\theta_i + y\sin\theta_i - (r_0 + \Delta d))_+^{3/2}\sin\theta_i$$

2.3 AMB force modeling

The rotor AMB system contains a PID controller, sensors and power amplifiers. If only power amplifiers are activated and a specific signal is imported, the AMB can be converted to a force actuator. Ignoring the hysteresis loss and edge effect, the AMB force can be calculated by

$$f_{AMB} = \frac{1}{4}\mu_0 n^2 A \frac{i^2}{s^2}\cos\beta$$

Where μ_0 stands for the magnetic permeability of a vacuum; n is the windings of a coil; A is the cross section of the air gap; i is the current value; s is the length of air gap and β is the an angle between forces and magnetic poles.

2.4 Bearing-pedestal modeling

Here we adopt the model of Feng et al. [12] for studying the dynamics of rolling element bearings, which is symbolically represented in Figure 3. The model has four DOF, including two DOF of inner race (x_s, y_s) and two DOF of pedestal (x_p, y_p).

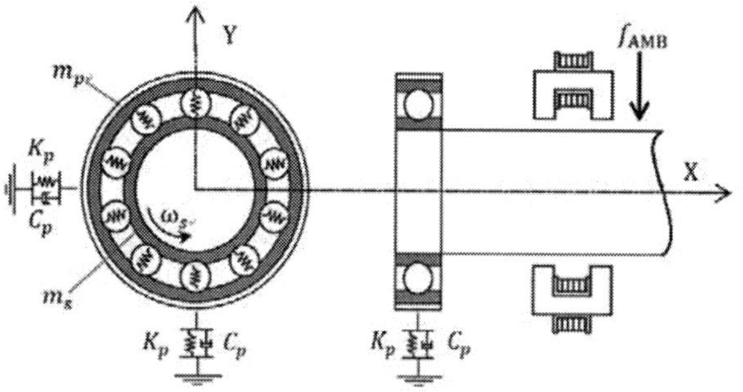

Figure 3. Four-DOF bearing-pedestal model.

If the rolling element is not located in the defect region, the actual radical clearance r_a is given by

$$r_a = r_0$$

If the rolling element is located in the defect region, the actual radical clearance r_a is given by

$$r_a = r_0 + \Delta d. \quad (16)$$

The total contact forces of the model can be calculated as follows:

$$f_{bx} = \sum_{i=1}^{N} K_b[(x_s - x_p)\cos\theta_i + (y_s - y_p)\sin\theta_i - r_a]_+^{3/2} \cos\theta_i$$

$$f_{by} = \sum_{i=1}^{N} K_b[(x_s - x_p)\cos\theta_i + (y_s - y_p)\sin\theta_i - r_a]_+^{3/2} \sin\theta_i$$

Taking into the nonlinear contact force, the governing equations of motion can be written as follows:

$$m_s\ddot{x}_s + c_s\dot{x}_s + f_{bx} = F_u\cos\omega_s t, \quad (19)$$

$$m_s\ddot{y}_s + c_s\dot{y}_s + f_{by} = F_u\sin\omega_s t - m_s g - f_{AMB}, \quad (20)$$

$$m_p\ddot{x}_p + c_p\dot{x}_p + K_p x_p - f_{bx} = 0, \quad (21)$$

$$m_p\ddot{y}_p + c_p\dot{y}_p + K_p y_p - f_{by} = -m_p g, \quad (22)$$

where m_s is the mass of the shaft and the inner race; c_s is the bearing contact damping; m_p is the mass of the pedestal; K_p and c_p are the stiffness and damping of the pedestal. Constant AMB force f_{AMB} is applied in the middle of the shaft or outboard location in the y direction. F_u is the unbalance force which is zero for a balanced rotor.

III. Result

3.1 Inputs for the model

The nonlinear differential equations are evaluated numerically using the fourth-order Runge-Kutta algorithm in MATLAB and the transient response is obtained. In order to get the desired results, the inputs used are shown in Table I.

The dynamic response of a health bearing, fault bearing, health bearing under AMB force and fault bearing under AMB force are calculated respectively. And the response is analyzed between 0.7 and 0.8s.

TABLE I. INPUTS FOR THE MODEL

1. Parameters of SKF 61901 ball bearings	
Outer race radius R	9 mm
Inner race radius r	mm
Number of ball N	
Contact stiffness K_b	7×10^9 N/m$^{3/2}$
Ball diameter d	mm
Pitch diameter D	mm
Contact angle α	0°
Radical clearance r_0	µm
B_N	
2. Other inputs	
Mass of shaft/inner race m_s	1.2kg
Bearing contact damping c_s	3000N · s/m
Mass of pedestal m_p	0.2kg
Pedestal stiffness K_p	1.5×10^7N/m
Damping in pedestal c_p	3000N · s/m
AMB force f_{AMB}	60N

3.2 Outer race defect frequency under AMB force

Actually, even a perfect bearing will generate VC vibrations due to the time-varying distribution of the balls relative to the inner and outer races, which is shown in Figure 4. For a bearing with a stationary outer race, the outer race defect frequency is given by

$$f_o = \frac{1}{2} f_s \left(1 - \frac{d}{D} cos\alpha\right) N$$

For a deep groove ball bearing, $f_o = f_{vc}$ accounting the zero contact angle α, which makes it more difficult to diagnose the outer race defect.

Figure 4 shows the simulation acceleration response of a healthy bearing at 3600 rpm in the y direction. It can be seen that even a perfect bearing will generate VC vibrations due to the time-varying distribution of the balls relative to the inner and outer races. But the AMB force has little influence on the vibration response for a healthy bearing.

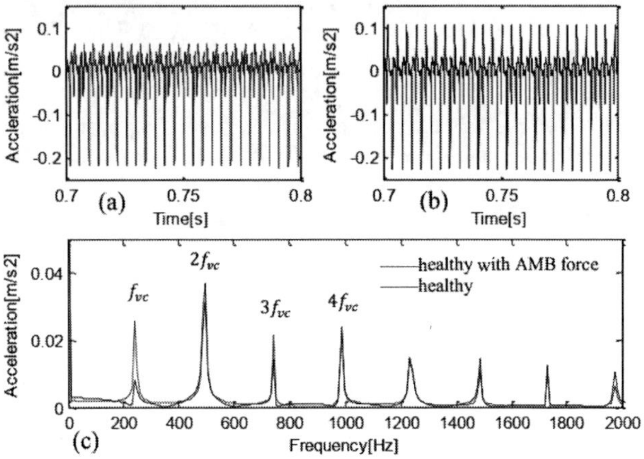

Figure 4. Vibration response of healthy bearing, at 3600 rpm: (a)healthy beaing without AMB force, (b) healthy bearing with AMB force and (c) Frequency spectrum of the envelope between 0.7 and 0.8s.

Figure 5 is the simulation results of a fault bearing with outer race defect at 3600rpm. It can be seen that for an incipient small defect, the amplitudes of defect characteristic frequency f_o and multiples are small in the frequency spectrum. However, there is significant increase in the amplitude at these frequencies under AMB force. The same conclusion can be obtained at 100 Hz (6000 rpm) in Figure 6.

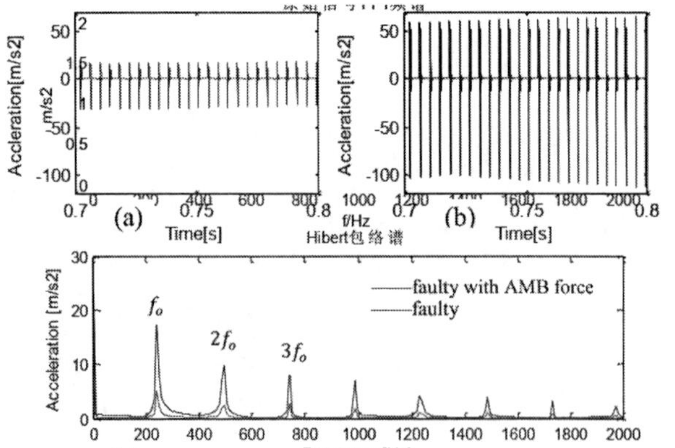

Figure 5. Vibration response of bearing with the outer race small defect, l = 0.2mm, at 3600 rpm: (a)defect bearing without AMB force, (b) defetc

bearing with AMB force and (c) Frequency spectrum of the envelope between 0.7 and 0.8s.

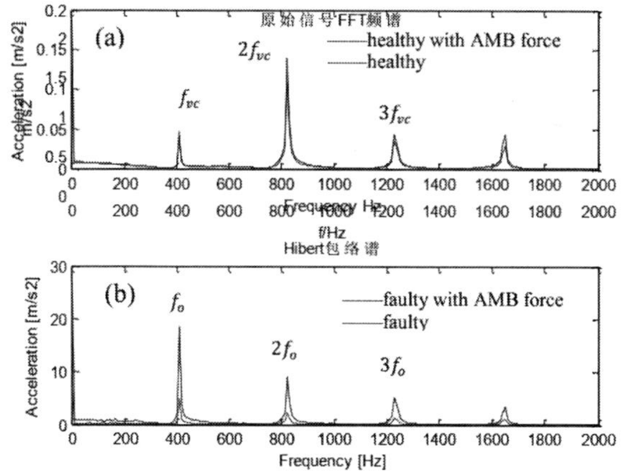

Figure 6. Frequency spectrum of the envelope between 0.7 and 0.8s, at 6000 rpm: (a) healthy bearing and (b) bearing with outer race defect; small defect l = 0.2mm.

3.3 Effect of AMB force on the dynamic response

The amplitude of AMB force has an effect on the fault characteristic frequency. Figure 7 shows the effect of the AMB force amplitude on the dynamic response of a bearing with the outer race defect. Figure 8 shows the defect characteristic frequency f_o trend along with AMB force. It can be seen that the amplitude of the characteristics defect frequency increases with the AMB force rising.

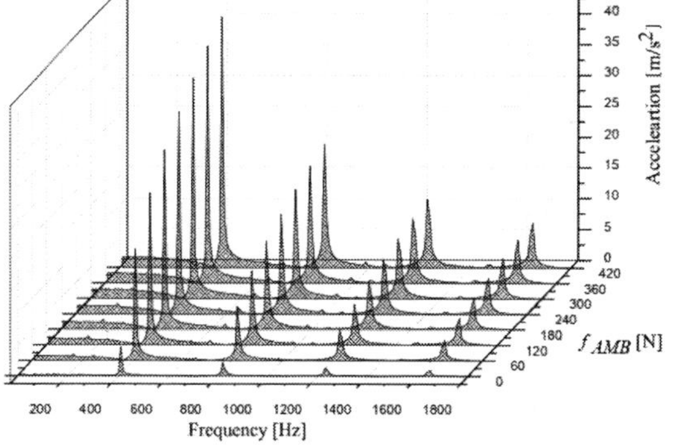

Figure 7. The effect of the AMB force amplitude on the dynamic response of bearing with the outer race defect, at 6000 rpm, small defect l = 0.2mm.

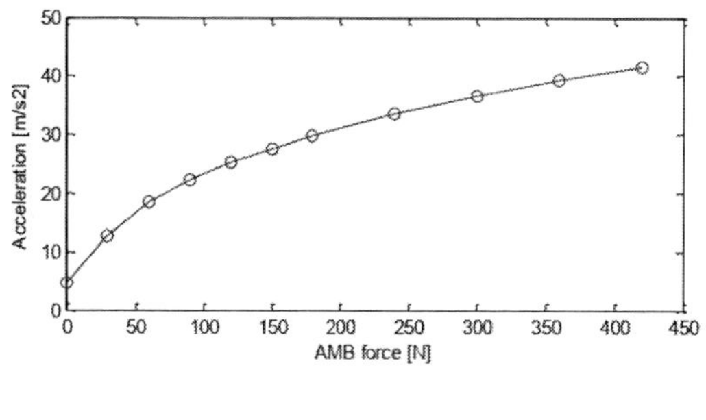

Figure 8. The effect of the AMB force amplitude on the outer race defect amplitude

IV. CONCLUSIONS

The simulation results show that AMBs can be used as an online force actuator to diagnose the bearing incipient outer race small defect. There is significant increase in the characteristic frequency amplitude with the increase in the AMB force, which is very helpful to improve the diagnosis accuracy. Furthermore, the effect of adding an extra force to the system might urge the defect growth, which is obviously a disadvantage. Current efforts are directed toward experimental verification of this work and the application of other bearing and shaft damage.

ACKNOWLEDGMENT

This research has been supported by the Fundamental Research Funds for the Central Universities and Foundation of Graduate Innovation Center in NUAA, Grant No. KFJJ20130215.

REFERENCES

[1] David Brie, ―Modelling of the spalled rolling element bearing vibration signal: an overview and some new results," *Mechanical Systems and Signal Processing*, vol. 14, no. 3, pp. 353–369, 2000.

[2] P. McFadden, J. Smith, ―Model for the vibration produced by a single point defect in a rolling element bearing," *Journal of Sound and Vibration*, vol. 96, no. 1, pp. 69–82, 1984.

[3] J. Marshall, M. Kasarda, J. Imlach, ―A multi-point measurement technique for the enhancement of force measurement with active magnetic bearings," IGTI/TURBO EXPO, New Orleans, LA, 2001 (2001-GT-0246).

[4] Xiangming Lou, ―Study on the identification of unbalance rotor in running condition," Ph.D. Dissertation, ZJU, Hangzhou, China, 2001.

[5] Nordmann, R., Matros, M., Neumer, T., ―Parameter identification in rotating Machinery by Means of Active Magnetic Bearings," IFToMM, 4th Conference on Rotor Dynamics, Chicago, Illinois, 1994.

[6] Knopf E, Nordmann R, ―Active magnetic bearings for the identification of dynamic characteristics of fluid bearings, Sixth International Symposium on Magnetic Bearings," Massachusetts Institute of Technology, Cambridge, USA, 1998.

[7] M. Kasarda, J. Imlach, P.A. Balaji, ―The concurrent use of magnetic bearings for rotor support and force sensing for the nondestructive evaluation of manufacturing processes," in: *SPI Seventh International Symposium on Smart Structures and Materials*, Newport Beach, CA, 2000.

[8] C. Zhu, D.A. Robb, D.J. Ewins, ―The dynamics of a cracked rotor with an active magnetic bearing," *Journal of Sound and Vibration*, vol. 256, no. 3, pp. 469–487, 2003.

[9] J.T. Sawicki, M. Friswell, Z. Kulesza, A. Wroblewski, J.D. Lekki, ―Detecting cracked rotors using auxiliary harmonic excitation," *Journal of Sound and Vibration*, vol. 330, no. 7, pp. 1365–1381, 2011.

[10] N. Sawalhi, R.B. Randall, ―Simulating gear and bearing interactions in the presence of faults, Part I. The combined gear bearing dynamic model and the simulation of localised bearing faults," *Mechanical Systems and Signal Processing*, vol. 22, no. 8, pp. 1924–1951, 2008.

[11] N.S. Feng, E.J. Hahn, R.B. Randall, ―Using transient analysis software to simulate vibration signals due to rolling element bearing defects," *Proceedings of the 3rd Australian Congress on Applied Mechanics*, Sydney, pp. 689–694, 2002.

Application of a Magnetic Actuator as an External Excitation Source in Fault Detection

Ricardo Ugliara Mendes[a], Gregory Bregion Daniel[b], Helio Fiori de Castro[c], Katia Lucchesi Cavalca[d]

[a]rumqld@gmail.com, [b]gbdaniel@fem.unicamp.br, [c]heliofc@fem.unicamp.br, [d]katia@fem.unicamp.br
[abcd] Laboratory of Rotating Machinery, Faculty of Mechanical Engineering, University of Campinas, 200 Mendeleiev Street, 13083-970 Campinas, São Paulo, Brazil

***Abstract*—Rotating machinery covers a broad range of industrial applications. Consequently, unexpected failures may lead to suddenly manufacturing break-off involving high maintenance costs. The most common faults in such machines introduce anisotropy to the system causing changes in its modal parameters and increasing the response backward component. Modal analysis methods can be used to identify changes in the machine behavior and foresee incipient faults. This paper presents the time domain Multiple Output Backward Autoregression (MOBAR) method and discusses its advantages over frequency domain methods on field tests in industry. For the MOBAR method, a tuned non-synchronous excitation called blocking test is applied to the shaft using a magnetic actuator, and when the excitation is turned off, the decay response is used to estimate natural frequencies of the system and the corresponding modal damping. The changing of these parameters can be used to predict system failures. To demonstrate the usability of electromagnetic actuators to apply this method for fault diagnosis and identification, a case study is presented in which the directional coordinates modal parameters are identified for a rotor supported by journal bearings using the MOBAR technique.**

I. Introduction

Heavy rotors, widely used in industries and power plants, have shafts rotating at high speeds that must have high trust levels. In order to obtain safety operating conditions and avoid sudden manufacturing break-off, an efficient maintenance program must be designed, once few hours out of the production line usually means considerable rising of the costs. Therefore, not only the maintenance process must be fast, but also diagnosis tests must not be time consuming.

In rotordynamics, many of the possible failures affect the anisotropy degree of the system; namely journal bearings and seals wear and oil-induced instabilities; and consequently the modal parameters of the system (natural frequencies and modal damping) change. Thereby the identification of the modal parameters change can be used in fault detection, isolation and identification (the three steps in Fault Diagnosis and Identification – **FDI** – according to [1]).

The parameter identification can be performed using one of the many existing modal analysis methods. Such methods are divided into two categories: frequency domain and time domain methods. The first group needs the acquisition of the frequency response function (**FRF**) of the rotor in the entire operational range in order to accurately identify the system modal parameters. The problem is that **FRF** measurements need excitation and response measurements, and it can be very time consuming, e.g. a stepped-sine measurement takes a long time to be made and sine-sweep or white noise techniques require many averages. As time domain methods are always based on transient response, the tests required are much faster; and there is no need to acquire excitation signals.

A remarkable family of time domain methods is the auto-regressive algorithms (**AR**). A backward **AR** method based in a single channel response data was developed in [2]; however methods which take into account multiple response signals achieve better accuracy, with less bias and deviation due to noise ([3], [4]). The method developed in [2] was extended to address multiple outputs in [4] and is referred to as **MOBAR** (Multiple Output Backward Autoregression). Multiple output techniques have better performance due to the richer modal information contained in the data set, presenting also better capability to identify close spaced modes, which are a common characteristic of rotating machinery (e.g. modes crossing due to the gyroscopic effect or a close pair of forward and backward modes); this capability is important in fault diagnosis once the anisotropy inserted by many failures affect both forward and backward modes, that must be identified for correctness of the **FDI** procedure.

Regarding external excitation source for modal analysis of rotating machinery, there is a practical limitation when using shakers or impact hammers, once due to friction, undesired tangential forces and noise can be acting on the system. Moreover, those procedures are hard to be applied on real machines. Hence, the use of an external magnetic actuator is very convenient when working with rotating systems. For this reason, a magnetic actuator has been a promising solution to apply the external excitation in the rotor, as it was done by [3], [5] and [6].

In this context, this article aims to present the application of a magnetic actuator to the **MOBAR** method and analyze its robustness, as well as provide a step-by-step guide to its application on field tests in industry, when short break-off times are mandatory. A case study regarding the identification of directional coordinates modal parameters in a journal bearing supported rotor is presented to demonstrate the potential applicability of this technique in **FDI**.

II. MOBAR

A. Test Procedure

As previously mentioned, time domain techniques are based in transient response measurements. In the **MOBAR** case, the recommended excitation is the blocking test ([3]). In this case, a tuned-sinusoidal excitation is applied to the system and then turned off. The response decay is used for the modal parameter estimation. This sinusoidal excitation is tuned in the frequency of the natural mode to be identified; making the contribution of that mode the dominant component in the system vibration (overcoming the synchronous unbalance and any other signals); i.e. the signal-to-noise ratio (**SNR**) of the mode is maximized and the decay observed in the response is relative to that mode only. It is also important to know the direction of the mode being excited (backward or forward), once that the blocking test is a rotating force that should have the same direction of the mode to guarantee the maximization of the mode **SNR**.

The step-by-step guide to apply the MOBAR technique in a field test is described below (based in [3]).

1. *Determine the natural frequencies*. Before applying the blocking test, the frequencies and directions of the modes to be analyzed should be estimated. This can be done in several ways: based in a model simulation (if there is enough precision) or stepped-sine (needs averaging). Stepped-sine excitation was chosen in this work because it can be done only in the interest frequency range and it is a tuned excitation (presenting better results than white noise measurements), and has the advantage over the sine sweep because it does not have transient effects as excitation frequency changes.
2. *Apply blocking test*. After obtaining modes and frequencies, the blocking test can be applied to the system for each of the natural frequencies to be analyzed. It is recommended to perform more than one blocking test per frequency for averaging purpose (it is observed at this point that the blocking test is a fast procedure, once it takes only enough time to the response due to the excitation achieve permanent regime, and decay after the excitation being turned off). Reference [3] observed that the amplitude of the blocking test should be high to maximize the **SNR** of the mode, but low enough to avoid non-linear effects.
3. *Apply the MOBAR method*. After recording the free decay of the response, the **MOBAR** method can be applied off-line in order to identify the modal parameters of the system. Details of the **MOBAR** method are given in the following section.

One advantage of the **MOBAR** method over the frequency domain methods is that it does not need the **FRF** of the entire operational frequency range of the machine in order do the modal parameters estimation; only the natural frequencies need to be known. Thus the stepped-sine needs to be performed only in a narrow frequency range in the resonance region, significantly reducing the test duration. An indicative of this frequency range can be obtained using a numerical model of the rotor.

However the main advantage of using **MOBAR** is that it can independently identify forward and backward natural modes, even if they are close spaced; which is a known drawback of frequency domain methods or other time domain methods based on one output signal only.

B. MOBAR method

Autoregressive model is a family of algorithms in time domain widely used in system identification. The name of the algorithms are due to how the models are built, where the response for a given time instant is written as a function of the future time instants (backward autoregressive model - **BAR**) as given in (1), or as a function of the previous time instant[1] (forward autoregressive model - **FAR**) as in (2).

$$y_t = \sum_{i=1}^{q} a_i y_{t+i} = a_1 y_{t+1} + a_2 y_{t+2} + \cdots + a_q y_{t+q} \quad (1)$$

$$y_t = \sum_{i=1}^{q} b_i y_{t-i} = b_1 y_{t-1} + b_2 y_{t-2} + \cdots + b_q y_{t-q} \quad (2)$$

q represents the model order, a_i and b_i are the model coefficients and y_i is the measured time series data.

The **MOBAR** uses a backward autoregressive model and the presented development is based in [4]. The method can identify systems with many modes, even if they are close spaced, through the use of multiple measurement channels (which can be signals of different nature as displacement, velocity and acceleration). The terms y_i in (1) are column vectors of length p, where p is the number of sensors, and the coefficients a_i are square matrixes of dimension p. From N measurement points, the matrix equation (3) can be written using (1), and can be solved using least mean square or singular value decomposition ([2], [3], [4], [5]). Once the a_i coefficients were obtained, the backward autoregressive model can be written in the state space form as shown in (4) using the state vector $z_t=[y_t^T \; y_{t+1}^T \ldots y_{t+q-1}^T]^T \in R^{pqx1}$.

$$\begin{bmatrix} y_1^T \\ y_2^T \\ \vdots \\ y_{N-q}^T \end{bmatrix}^T = \begin{bmatrix} a_1^T \\ a_2^T \\ \vdots \\ a_q^T \end{bmatrix}^T \cdot \begin{bmatrix} y_2 & y_3 & \cdots & y_{N-q+1} \\ y_3 & y_4 & \cdots & y_{N-q+2} \\ \vdots & \vdots & & \vdots \\ y_{q+1} & y_{q+2} & \cdots & y_N \end{bmatrix} \quad (3)$$

$$z_t = A_d \cdot z_{t+1}$$

$$z_t = \begin{bmatrix} a_1 & \cdots & a_{q-1} & a_q \\ I_p & \cdots & 0_p & 0_p \\ \vdots & \ddots & \vdots & \vdots \\ 0_p & 0_p & I_p & 0_p \end{bmatrix} \cdot z_{t+1} \quad (4)$$

The **BAR** model (4) can be transformed in a **FAR** model by isolating z_{t+1} as in (5). The eigenvalues of the transformed **FAR** model (**TFAR**) have an interesting property: the

[1] A **FAR** model can be transformed in a **BAR** model by isolating the term y_{t-q} and adding q to the sub-indexes (for notation purpose only); the inverse transformation is analogous.

eigenvalues located inside the unit circle are attributed to the system modes and the others are spurious modes [4]. Therefore, the system modes can be easily identified[2] and the reduced model (A_r) can be converted to a continuous model (A_s) using (6); where dt is the sampling time. The modal information of the physical system (natural frequencies and damping) is obtained from the eigenvalues of A_s.

$$z_{t+1} = A_d^{-1} \cdot z_t$$
$$z_{t+1} = \begin{bmatrix} 0_p & I_p & \cdots & 0_p \\ \vdots & \vdots & \ddots & \vdots \\ 0_p & 0_p & \cdots & I_p \\ a_q^{-1} & -a_q^{-1}a_1 & \cdots & -a_q^{-1}a_{q-1} \end{bmatrix} \cdot z_t \quad (5)$$

$$A_s = \ln(A_r)/dt \quad (6)$$

Overspecify the model order q is an important step to improve accuracy of the algorithm ([2], [4], [5]). Even though the **MOBAR** has the inherent property of separating system modes from spurious modes, when the measurements are subjected to significant levels of noise or the system has high damping – fast decay response, the separation may not be done properly. Therefore, the stabilization diagram may be used to help the system modes identification. The stabilization diagram is widely used in many modal analysis methods and it consists in applying the method for many different orders q and plotting the identified eigenvalues in a diagram as a function of q. In the diagram it is possible to analyze which eigenvalues do not change among the different model orders (in terms of frequency, damping and eigenvector), i.e. identify the system modes.

III. Experimental Setup

A. Test Rig

The test rig setup was defined to reproduce some typical dynamic effects of a journal bearing supported turbomachinery; in this case similar to the real machine analyzed in [6]. Instability due to the presence of the journal bearings was chosen to validate the **FDI** method capability because it is directly related to the anisotropy changing of the bearings as the rotational speed of the rotor increases.

The test rig presented in Fig. 1a consists of a steel SAE 1030 shaft supported by two journal bearings and one rolling bearing. There is also a disc and the electromagnetic actuator journal, both made of steel SAE 1020. As previously mentioned, the actuator is usually inserted at one end of the shaft, however as the test rig shaft is thin, the rolling bearing at the shaft end was used in order to avoid the actuator journal to be overhung and present high amplitude vibrations. A schematic drawing of the shaft with the main dimensions is presented in Fig. 1b and its design parameters are found in Table I.

Displacement sensors measure the vibration amplitude at the journal bearings. Disc vibration, oil temperature and load on the bearings are also measured for monitoring purposes only.

[2] The matrices of the **TFAR** model are different from the common **FAR** model, which has all of its eigenvalues located inside the unit circle.

(a)

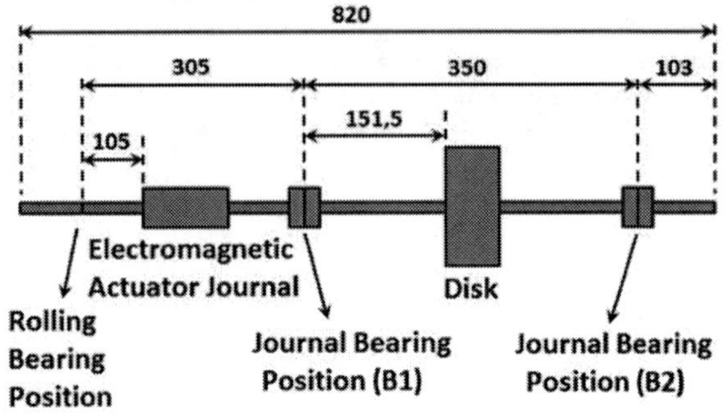

(b)

Figure 1. Experimental setup: (a) test rig; (b) schematic drawing.

TABLE I. Design Parameters of the Test Rig

Parameter	Value
Shaft Length	820 mm
Shaft Diameter	12 mm
Disk Length	47 mm
Disk Diameter	95 mm
Electromagnetic Actuator Journal Length	80 mm
Electromagnetic Actuator Journal Diameter	40 mm
Journal Bearing Length	20 mm
Journal Bearing Diameter	30 mm
Lubricated Oil	ISO VG 32
Journal Bearing Radial Clearance	90 μm
Lubricated Oil Temperature in the Tank	25-27 °C

B. Electromagnetic Actuator

The electromagnetic actuator in Fig. 2a was the external excitation source used to apply the blocking test and the stepped-sine in the shaft during the tests. It was designed and built at the Laboratory of Rotating Machinery (LAMAR) at UNICAMP [7]. Having an homopolar configuration, the actuator is able to apply an excitation of 150 N in a frequency range up to 100 Hz, when operating with an air-gap (radial clearance) of 2.5 mm and a current of 3 A. Therefore, it is capable of exciting all the inertia involved in the test rig throughout the entire operating frequency range of the rotor. The first natural mode of the actuator was evaluated through a finite element model simulation (Fig. 2b) and was found to be at 576 Hz, which is far above the operating frequency of the system, as desired.

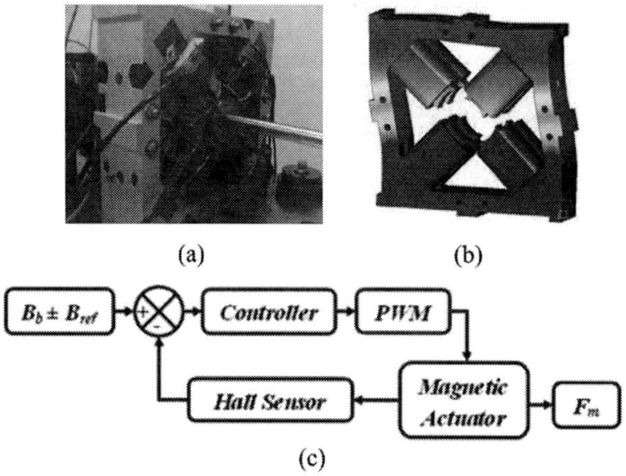

Figure 2. Electromagnetic Actuator: (a) test rig assembly; (b) first natural mode at 576 Hz; (c) actuator system diagram.

The force (F_m) produced by an electromagnetic actuator in equation (7) is proportional to the pole area of the actuator (A_g), the air magnetic permeability (μ_g) and the magnetic flux density (B), which is given by equation (8) and proportional to the current (i), the number of coils (N) and the air-gap length (l_g), according to [8] and [9]. For a differential assembly, the magnetic force, equation (7) becomes equation (9), where B_1 and B_2 are the magnetic field of the opposite coils. Using a bias magnetic field (B_b) added to the reference field (B_{ref}), the linearized force given in equation (10) is obtained [10].

$$F_m = \frac{B^2 \cdot A}{\mu_g} \tag{7}$$

$$B = \frac{\mu_g \cdot N \cdot i}{2 \cdot l_g} \tag{8}$$

$$F_m = \frac{A_g}{\mu_g}\left(B_1^{\ 2} - B_2^{\ 2}\right) \tag{9}$$

$$F_m = \frac{A_g}{\mu_g} 4 B_b B_{ref} \tag{10}$$

Therefore, the actuator force control loop can be designed in terms of magnetic field ([7], [11]) as shown in Fig. 2c, where B_{ref} is the magnetic flux density necessary to obtain the desired reference force. The reference magnetic field is compared to the actual magnetic field, measured by the hall sensor, and the result is the input of the proportional controller. The output voltage of the controller is converted to current by the PWM amplifier (operating in current control mode) and sent to the actuator coils, generating the magnetic force F_m.

IV. Results and Discussion

In order to investigate the robustness of the MOBAR method using an electromagnetic actuator as external excitation force, a series of tests were accomplished. For this paper purposes, the results are shown in two rotational speeds of the rotor: 65 Hz and 70 Hz.

The first step of the test procedure consists of determining the natural frequencies of the rotor to apply the blocking test. Different methods were briefly discussed, however, if the system presents high levels of noise or if the natural frequency appears as a flat surface in the **FRF** instead of a sharp peak, it may be difficult to precisely identify the natural frequency regardless of the method used.

In this work, the two natural frequencies associated to the first forward and the first backward bending modes were identified from the **FRF** obtained using the stepped-sine method, and are presented in Table II. To evaluate the robustness of the method, the blocking tests were applied for different frequencies around the natural frequencies to be identified in a range of ± 3 Hz (from 55 Hz to 61 Hz for the forward frequency and from 57 Hz to 63 Hz for the backward frequency); corresponding to approximately ± 5% of error in the first step of the test procedure, i.e., the natural frequency identification. The accuracy of the method was evaluated by repeating 10 times each blocking test in each of the defined frequencies.

TABLE II. NATURAL FREQUENCIES OF THE FIRST BENDING MODES IDENTIFIED FROM THE FRF USING STEPPED-SINE

Rotational Speed	Forward	Backward
65 Hz	58.1 Hz	60.9 Hz
70 Hz	56.8 Hz	60.2 Hz

It is important to remark that, although the disc vibration is measured, only the measurements in the bearings position will be used with **MOBAR**, once that is the common scenario when dealing with real machines in industry.

The results were obtained using a time interval of 0.17 s of the four signals (horizontal and vertical displacements in the position of both journal bearings) after the blocking test excitation was turned off. Fig. 3 presents a measurement of the vertical displacement of the first journal bearing where it can be seen the blocking test excitation been turned off at the first vertical line in 0.5 s, and the transient decay of the response until 0.67 s, at the second vertical line; the transient response between the vertical lines is the signal used as input for the **MOBAR** technique.

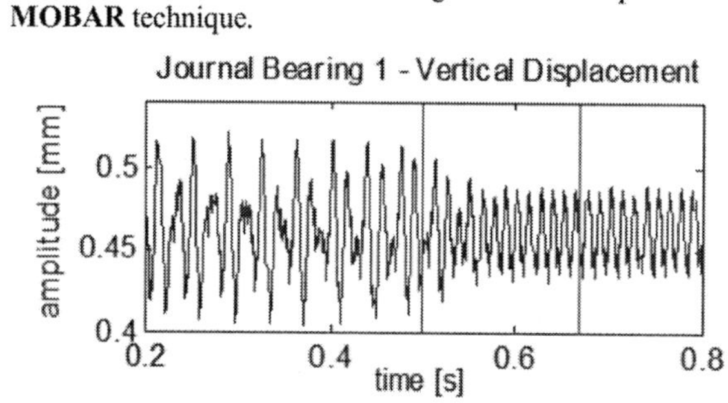

Figure 3. Vertical displacement of the first journal bearing.

After the measurements, as previously mentioned, the **MOBAR** method was applied using different model orders. For this rotor, it was notice that varying the model order from 20 to 50 presented good results. In the most of the measurements, higher orders implied in the method loss of ability to identify the natural frequencies. The authors believe this effect is due to the short time interval used to calculate a high order model, even though the natural frequency was well identified for lower orders, as exemplified in Fig. 4.

In the legend of Fig. 4: asterisk means an unstable pole, triangle is a pole which has its natural frequency close (inside a given range) from a pole of the previous model order, a square pole is the one which frequency and the modal damping have changed (again, inside a given range) from the previous model order, black circle represents a pole whose frequency and eigenvector (modal assurance criterion was used) were kept almost unaltered from the previous model order, and finally, white circle represents a stable pole whose three parameters (natural frequency, modal damping and eigenvector) present small changes from the previous model order. Therefore, in the stabilization diagram analysis, a pole which becomes stable for different model orders is very likely to be a real pole of the system. It should also be noticed that this stability concept is relative to the range defined by the user. In this work, the acceptable change from one model order to other for the pole to be considered stable was 2% for the natural frequency, 10% for the modal damping and 5% for the eigenvector. The modal damping has a higher tolerance once it is the most difficult parameter to be identified.

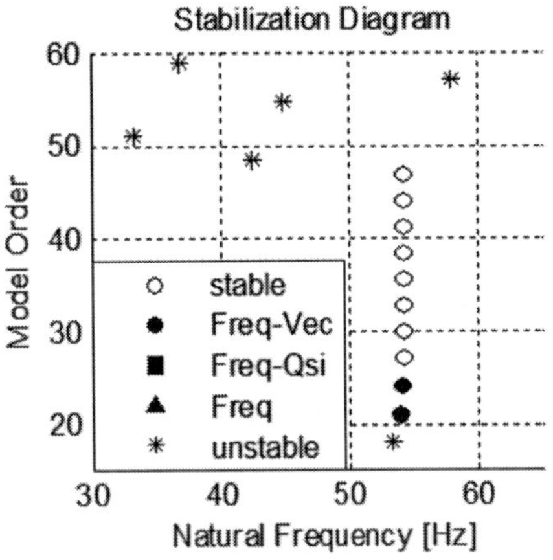

Figure 4. Example of stabilization diagram.

The results of the natural frequency identifications are presented in Figs. 5a and 5b for the rotational speeds of 65 Hz and 70 Hz, respectively. From the ten trials in each excitation frequency, not all of then resulted in a good identification (a stable mode), thus, the black numbers in the figures represent how many of the tests provided good identifications, the vertical bars are ± 1 standard deviation, and the lines connect the mean values. The black line represents forward natural modes and the backward natural modes are represented in grey.

Recalling that the forward natural frequency is around 57 Hz and the backward natural frequency is around 60.5 Hz, it can be noted that in both rotational speeds (Figs. 5a and 5b) analogous results were obtained. In the identification of the backward natural frequency, a higher excitation frequency (60 Hz to 63 Hz) presented better estimates than lower frequencies (smaller changes in the mean value); in both cases, a backward excitation of 58 Hz was not able to identify the backward mode, but the forward only. Otherwise, the forward natural frequency was also identified for lower excitation frequencies (56 to 58 Hz), still it was identified in the entire excitation range (except in the excitation frequency of 59 Hz for the rotational speed of 65 Hz).

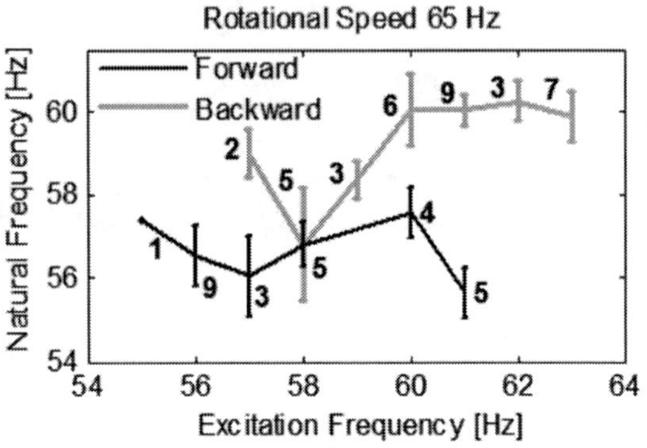

(a)

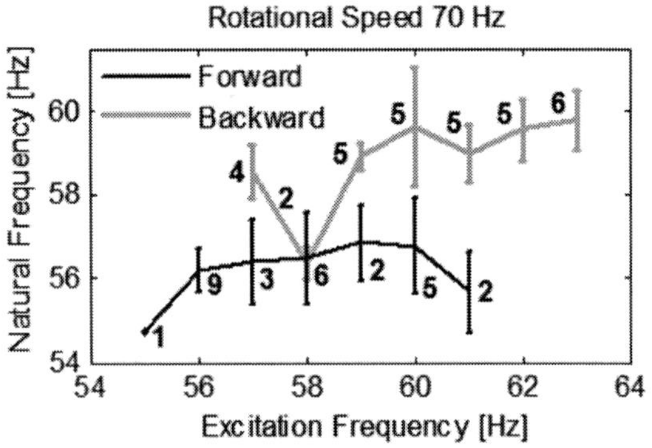

(b)

Figure 5. Natural frequencies identified: (a) rotational speed of 65 Hz; (b) rotational speed of 70 Hz.

The modal damping identification for the rotational speeds of 65 Hz and 70 Hz are presented in Figs. 6a and 6b. The same symbology from Fig. 5 is used in Fig. 6: black numbers represent how many tests (from ten performed for each excitation frequency) resulted in stable poles, vertical bars represent standard deviation of the tests and the lines connect the mean values (black for the forward mode and grey for the backward mode).

In the identified modal damping, the backward mode damping presented smaller standard deviation and smaller changes in the mean value for higher excitation frequencies (60 Hz to 63 Hz).

The identification results of the forward modal damping presented relatively large standard deviations for the rotational speed of 65 Hz. However, it can be noted that the modal damping identified for both the forward and backward mode presented better results in the rotational speed of 70 Hz.

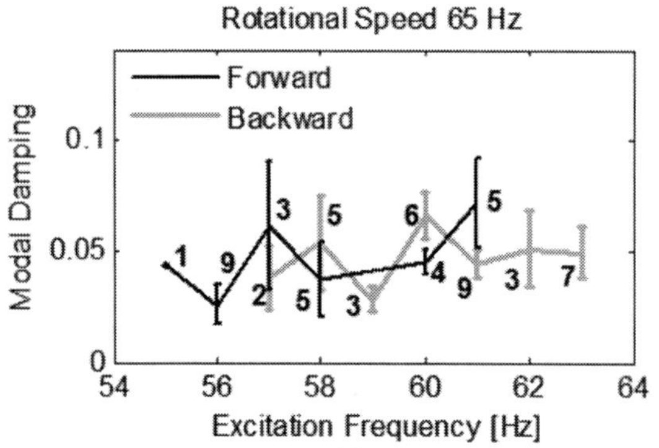

(a)

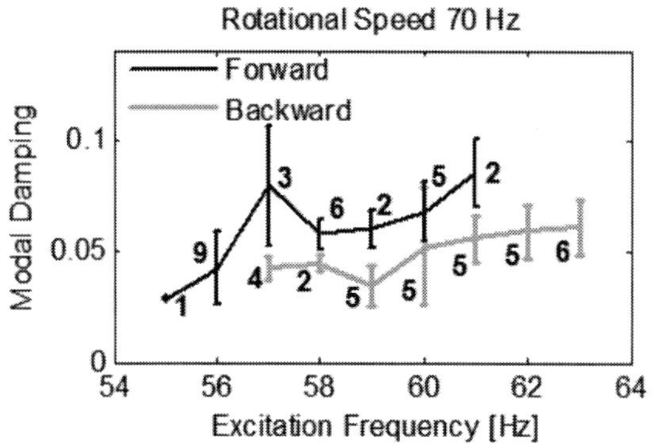

(b)

Figure 6. Modal dampings identified: (a) rotational speed of 65 Hz; (b) rotational speed of 70 Hz.

From all the 280 tests performed, nearly 43.5% produced valid identifications (black numbers in Fig. 5), showing the importance of performing repeated tests when using **MOBAR** to identify modal parameters.

The largest standard deviation presented in the frequencies identification was around 2.7% of the frequency value. The largest standard deviation in the modal damping identification was 70% in the worst case (rotational speed of 65 Hz and excitation of 57 Hz), but only three measurements resulted in stable poles in this set of tests. The excitation frequencies in which 5 or more tests resulted in stable poles presented a standard deviation around 40%. It is important to highlight that the system is highly damped, as shown in Fig. 3, which implies in added difficulty in tests performance and identification procedure.

V. Conclusions

It is concluded from the obtained results that the **MOBAR** method is very accurate and robust in the identification of natural frequencies, but it needs averaging in order to obtain accurate results in the modal damping identification.

The results obtained have shown that the **MOBAR** technique with blocking test excitation using an electromagnetic actuator as external excitation source can be successfully used to identify modal parameters of a rotating system. However it should be remarked that averaging is an important part of the process for highly damped rotors or rotors with a high signal to noise ratio (**SNR**). Conveniently, the blocking test is not time consuming since the duration of each test is about one second, thus one minute under test would provide 60 measurements, producing much better results. Since, modal damping identification is challenging for any identification method, the small time consumed in the blocking test is an important advantage over other methods.

Future works consist of two steps: first, use the MOBAR technique with the magnetic actuator as external excitation source in fault diagnosis and identification (**FDI**) when multiple incipient faults are present in the system; and second, extent the research towards fault tolerant control (**FTC**) using the same magnetic actuators/bearings.

The possible faults to be identified are those which change the characteristics of the system. For example, hydrodynamic bearing wear, cavitation, starvation and oil overheating; all of the mentioned faults change the stiffness and/or damping coefficients of the bearings, influencing the natural modes of the system and even, in more severe cases, making the system unstable. Once one of these conditions are diagnosed and identified (**FDI**), the fault tolerant control (**FTC**) can act through the magnetic actuator itself, allowing the system to safely shut-down for maintenance.

References

[1] J. Chen, R. J. Patton, "Robust Model-Based Fault Diagnosis for Dynamic Systems". *Massachusetts, Kluwer Academic Publishers*, 1999.

[2] R. Kumaresan and D. W. Tufts, "Estimating the parameters of exponentially damped sinusoids and pole-zero modeling in noise". *IEEE Transactions on Acoustics, Speech, and Signal Processing*, vol. ASSP-30, no. 6, pp. 833–840, 1982.

[3] C. H. Cloud, E. H. Maslen and L. E. Barret, "Damping ratio estimation techniques for rotordynamics stability measurements". *Proceedings of ASME Turbo Expo 2008: Power for Land, Sea and Air*, pp. 1-10, 2008.

[4] C. F. Hung and W. J. Ko, "Identification of modal parameter from measured output data using vector backward autoregressive model". *Journal of Sound and Vibration*, vol. 256, no. 2, pp. 249-270, 2002.

[5] C. H. Cloud, "Stability of rotors supported by tilting-pad journal bearings". PhD thesis, University of Virginia, 2007.

[6] B. C. Pettinato, C. H. Cloud and R. S. Campos "Shop acceptance testing of compressor rotordynamic stability and theoretical correlation", *Proceedings of the thirty-ninth turbomachinery symposium*, pp. 31-42, 2010.

[7] R. U. Mendes, L. O. S. Ferreira and K. L. Cavalca, "Analysis of a complete model of rotating machinery excited by magnetic actuator system", *Proc IMechE Part C: J Mechanical Engineering Science*, pp. 1–17, 2012.

[8] A. Chiba, T. Fukao, O. Ishikawa, et al., "Magnetic bearings and bearingless drives", *Oxford: Newnes*, 1st edn., pp. 381, 2005.

[9] G. Schweitzer and E. H. Maslen (eds), "Magnetic bearings – theory design and application to rotating machinery", 1st edn. Berlin: Springer-Verlag, 2009, p.535.

[10] E. H. Maslen, "Magnetic Bearings", *Technical Report*, University of Virginia, 2000.

[11] R. M. Furtado, "A Magnetic Actuator Development for Contactless Excitation in Rotor Systems", Dissertation (Ph.D.) – Faculty of Mechanical Engineering, University of Campinas, Campinas – SP, Brazil, 2008

Energy Cost assessment of the Active Control of a the flexible rotor supported by AMB

Benjamin DEFOY[a], Thomas ALBAN[a], Jarir Mahfoud[b]

[a] GE Oil&Gas, Turbomachinery departement, Le Creusot, France
[b] Université de Lyon, INSA-Lyon, LaMCoS UMR5259, Villeurbanne, France, jarir.mahfoud@insa-lyon.fr

***Abstract*— From a designer point of view, optimizing the energy necessary for the control is an important element that could lead to downsize the control cabinet and to an increase of margins according to power amplifier capabilities. Also, reducing the energy used, generates a less significant environmental impact. Two fuzzy logic based controllers were developed and assessed. First, the "Mechanical" performances are compared with respect to an augmented PID controller. Then, the energy necessary for the control is compared.**

I. Introduction

Active Magnetic Bearings (AMB) are widely utilized in different industrial applications. Several applications have been successfully implemented in the field of Turbomachinery [1,2]. Several studies have been devoted to the elaboration of new controllers that enable better design and performance in operating situations, with acceptable levels of stability and robustness [3-7]. The most successful, for industrial applications, is the augmented PID controller, since it permits designers to master the control process. However, the control of flexible rotors remains difficult due to the unavoidable spillover effects.

Industrial requirements involve several constraints regarding performance, robustness, easiness of implementation and final tuning. These requirements are dictated by international standards such as ISO 14839 and API 617 [8-11]. The specifications of the final users must also be taken into account. The spectrum of applications covered by AMBs is nowadays largely diversified. Each application presents its own operating conditions and excitations.

Fuzzy controllers are well adapted for controlling flexible structures. The advantages of fuzzy control are that it can be used in complex systems such as nonlinear, time-variant and systems including uncertainties [12]. They are less sensitive to variations of system parameters and they allow the utilization of membership functions adapted to the dynamic behavior of the system considered. They also enable the calculation of nuanced actions and take into account several data variations, thereby ensuring robust behavior. Defoy et al. [13] developed a new control strategy that utilize polar coordinate. The originality of the methodology developed is that it manages two significant physical quantities, namely tangential and radial velocities, which are associated to steady state and transient behaviors respectively. The outputs are the forces computed in the polar coordinates and are converted into currents that drive the action lines at the same time.

In this paper the mechanical performances and also the energy necessary for the control of an academic test rig were assessed experimentally. The aim is to evaluate the possibility of optimizing the energy consumption that could lead to downsize of the control cabinet and to an increase of margins according to power amplifier capabilities. In addition, reducing the energy used will be transformed in a less significant environmental impact, which is a current concern nowadays.

First the test rig will be presented, then the different control strategies will be described and finally the results will be discussed.

II. Test Rig

The experiments were performed using an academic test rig (Fig. 1). It is a commercial product manufactured by SKF® and was delivered with a dedicated PID controller. The test rig is equipped with two identical AMB called NDE (Non Drive End) and DE (Drive End) bearings. Each bearing has a maximum static capability of 280N. The action lines are positioned in the configuration load between axes. They are powered in differential driving mode with a bias current of 1A. Current are provided in the range of 0-3A using PWM amplifiers. Two displacement sensors (variable reluctance probes) are integrated in the housing of each bearing and are non-colocalised with actuators. The Input/Output panel gives access to the displacements measured and enables entering current settings for the amplifiers. Each AMB has one back-up bearing with a clearance radius of 0.1mm.

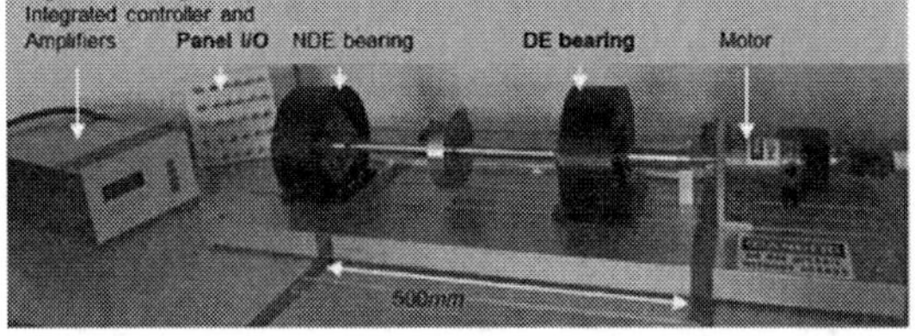

Figure 1. Test rig

The shaft is composed of three parts bolted together. A central part (diameter: 19.05mm; length 344mm) with a disc 120mm in diameter and 25mm long placed at mid-span, together with two shaft ends (35mm of main diameter). The stack of laminated steel sheets is shrunk on each of these two shafts. Due to the different lengths of the latter (190.5mm at DE and 110.5mm at NDE), the rotor is not symmetric in relation to the central disk. The total rotor length is 645mm. The rotor mass is 5.89kg. Three balancing planes are available: one at the central disc and one at each bolted plane. The rotor is driven by a 500W electric motor with a maximum speed of 12,600rpm. Power transmission is provided by a flexible coupling. The operating speed range used in this work is 0 to 10,000rpm, which includes three critical speeds (two forward rigid modes and the first forward flexible mode). The speed of the rotor is monitored by using a speed sensor placed close to the motor.

The frequency bandwidth considered in this study is 0-2kHz. The test rig can be controlled either by using the dedicated device or externally via the input/output panel. The new controller is implemented under Simulink® and dSpace® environments, with a sampling frequency of 20kHz.

The AMB is powered in differential driving mode with a bias current I0 and a control current i (1).

$$F_B = 4\,\xi_B\,\mu_0\,S\,N^2 \left[\frac{(I_0 - i)^2}{\left(2\,g_0 - 2\,x + \frac{L}{\mu_r}\right)^2} - \frac{(I_0 + i)^2}{\left(2\,g_0 + 2\,x + \frac{L}{\mu_r}\right)^2} \right] \quad (1)$$

Where ξ_B is a coefficient depending on the bearing geometry (0.92 in our case), μ_0 the permeability of vacuum space, S the pole area, N the number of turn in one coil, g_0 the nominal air gap, x the rotor position, μ_r the relative permeability of the iron core and L the average length of magnetic flux lines.

III. CONTROL STRATEGIES

Three control strategies were assessed. The first is the one developed by the test rig constructor, a fuzzy controller and a polar fuzzy controller (Fig. 2).

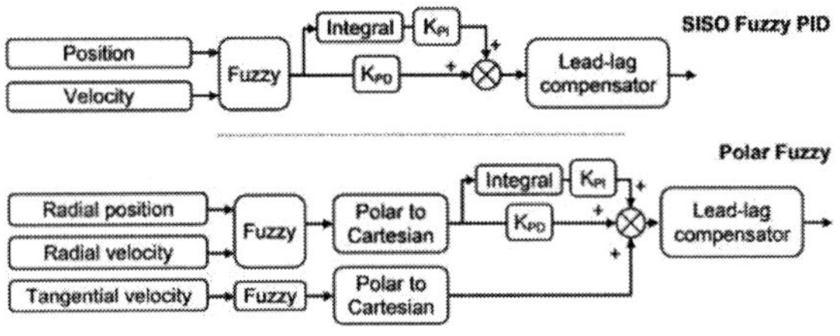

Figure 2. Schemes of the controllers

A. Fuzzy Controller

It consist of the sum of a fuzzy PI with a fuzzy PD controller. The displacement was measured and the velocity calculated by numerical derivation. The two data (displacement and velocity) were used as inputs for the fuzzy controller. Four generalized bell shape membership functions were utilized [14]. These membership functions are associated with four fuzzy sets: positive/negative displacements and positive/negative velocities

The design of a membership function take into account the noise of measurement. The displacements and velocities are measured, and there respective amplitude gives directly the measurement uncertainty. These amplitudes of displacements and velocities should be insufficient to change the prevalent membership function. On the other hand, in order to be efficient small amplitude of displacement and velocity should fully belong to one of the two fuzzy sets (positive or negative). Finally, as nominal behavior is considered and, the displacements and velocities are relatively small compared to noise of measurement, only two membership functions were utilized for each input of the fuzzy controller.

The rules are chosen in order to optimize the energy dissipation and to minimize the kinetic energy of the rotor (the force is always opposite to the velocity). The rules selected were:

- When position and velocity have the same sign, then the proportional gain is activated. Note that the other gains are always activated.
- When the displacement and velocity have opposite signs, the damping gain is conserved but the stiffness is set to zero.

This study considers only the case of nominal behavior with low displacements. Consequently, if the membership functions are tuned too smoothly, a linear controller can approximate the non-linear one. And, if the membership functions are tuned with too sharp transitions, disturbances can appear, and can lead to system instability. In our case, where the ratio of tolerated displacements over the uncertainties due to measurements is small, there is a limited range of variation for the membership functions that enables an interesting benefit of the fuzzy controllers without disturbing the system.

B. Polar Controller

Generally, the response of the system, as acceleration or displacement, is measured by using sensors placed along the structure. Regarding the lateral behavior, two sensors oriented in two perpendicular axes are sufficient to describe the dynamic behavior. Then, the measured signals are processed and analyzed either in time, frequency or time-frequency domains.

We believe that this approach is quite suitable for non-rotating structures, but when dealing with rotating machinery there is a lack of information regarding the rotational speed. This lack could be made up by describing the response measured in polar coordinates. The observation of the measurements in the polar domain can lead to an easier interpretation of the dynamic behavior, where the steady state and the transient behaviors could be directly distinguished, particularly in the case of real time controlled systems.

The transformation from Cartesian to polar representation is obtained classically as indicated in (2). Where x and y are the measured displacements along the x and y direction in the

Cartesian representation, while r and θ are the corresponding polar quantities.

$$\begin{cases} r = \sqrt{x^2 + y^2} \\ \tan\theta = \dfrac{y}{x} \end{cases} \tag{2}$$

Then the radial and the tangential velocities, Vr and Vt respectively, are calculated as follow:

$$\begin{aligned} V_r &= \dot{r} \\ V_t &= r\,\dot{\theta} \end{aligned} \tag{3}$$

Where $\dot{r}$ is obtained by numerical differentiation of the radial position.

In a steady state case and in the presence of synchronous excitation as unbalance, the orbit is circular (symmetric rotor), the radial displacement is constant consequently the radial velocity is nil while the tangential velocity is constant. It is worth mentioning that in the case of system dissymmetry, the displacement orbit is elliptical and the radial speed and position as well as the tangential speed have harmonic variation that corresponds to the second harmonic of the rotating speed.

The actuating in the polar domain has also several advantages by introducing targeted action on stiffness and damping. In this work, the controller was developed in the polar coordinate system, measurements and actuating are still realized in the Cartesian coordinate system. This is due to the fact that technology used is designed to operate in one direction.

The physical quantities introduced by polar transformation enable the controller to distinguish the disturbance produced by the unbalance from the transient disturbance that excites the rotor radially. Indeed, when the rotor is submitted only to unbalance excitations, the rotor orbit is circular and the radius of the whirl orbit is constant or change slowly during run up. The radial velocity is almost nil, the variations are only due to the noise of measurement.

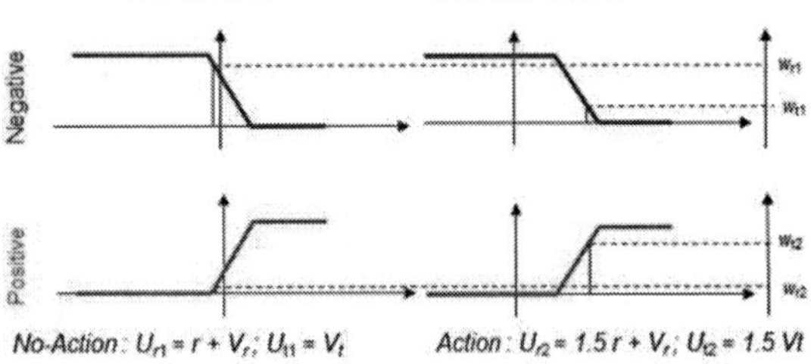

Figure 3. Membership functions of the polar fuzzy controller

Four trapezoidal membership functions were utilized (Fig. 3). The rules selected are:

1. For steady-state behavior, when tangential velocity is high tangential damping is increased by 50%.

2. For transient behavior, when radial velocity is positive, radial stiffness is increased by 50%.

3. If the velocities variation is small or negative, there is no action.

Then the fuzzy controller commands are expressed as actions along x and y direction:

$$\begin{cases} F_x = F_r \cos\theta - F_t \sin\theta \\ F_y = F_r \sin\theta + F_t \cos\theta \end{cases} \tag{4}$$

IV. EXPERIMENTAL RESULTS

The fuzzy controllers were compared with the initial augmented PID controller delivered with the test rig. The unbalance responses during run-up from 0 to 10,000rpm in 100 seconds was assessed.

Results are presented among the two direction of the drive end bearing (V13 & W13) and the non-drive end (V24 & W24).

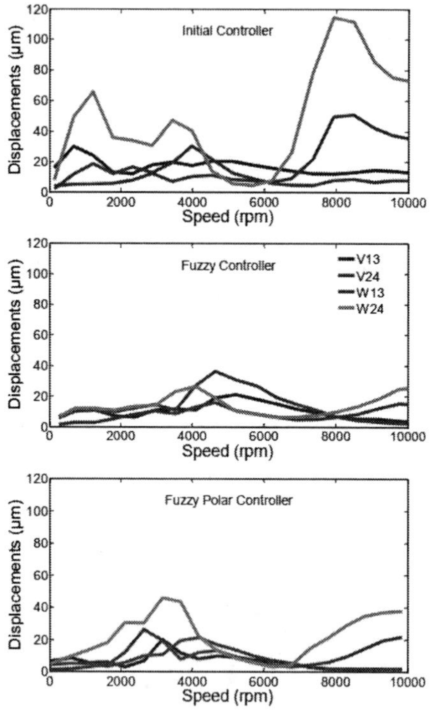

Figure 4. Unbalance responses during run-up

First, the displacement levels resulting from the three controllers assessed are compared in figure 4. The rms value of the displacement was chosen. The unbalance responses exhibited three critical speeds: the first around 1000rpm, that occurred only when using the initial PID controller was identified as foundation mode; the second in the vicinity of 4000rpm, corresponds to the cylindrical mode; and the last, near 9000rpm, was identified as the first flexible critical speed. The conical mode was damped too strongly to appear in the unbalance response.

The three controllers provided performances that match the requested specifications. Form a performance point of view, the fuzzy controller exhibits the best results. Nevertheless, this controller exhibited anisotropic behavior that could be undesirable. This anisotropy is due to the fact that the membership functions were expressed in the Cartesian coordinate system. Consequently, the control forces generated, when reaching displacement thresholds, lead to square orbit shape.

The rms value of the current was calculated (Fig. 5). It is representative of the energy necessary for the control in each case. No calculations, at this stage, were performed to quantify the energy. We estimate that the rms value is properly representative of the energy consumed.

The aim is not to minimize the energy necessary for the control, which is negligible with respect to the energy necessary to perform the run-up of the system. But as the AMB are designed (even partially) as a function of the max current, we believe that the controller with minimum energy could lead to a downsizing of the control cabinet and to an increase of margins according to power amplifier capabilities.

The results obtained show that the fuzzy polar controller represent the optimal behavior performance / consumption.

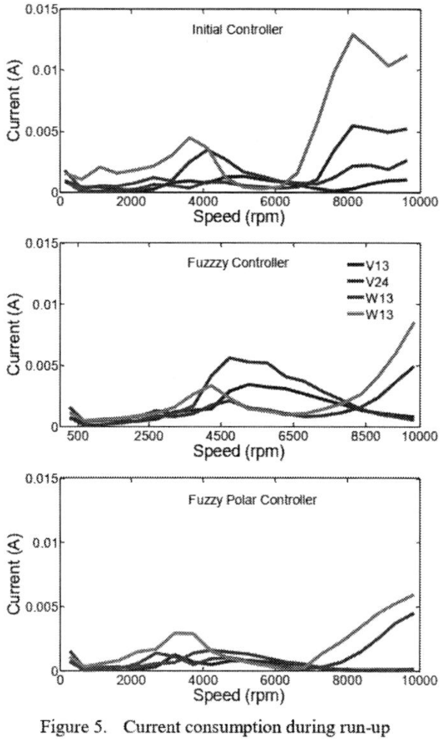

Figure 5. Current consumption during run-up

V. Conclusion

The mechanical performances and the energy necessary for the control of an academic test rig were assessed experimentally. The aim is to evaluate the possibility of optimizing the energy consumption that could lead to downsize of the control cabinet and to an increase of margins according to power amplifier capabilities.

Two fuzzy controllers were presented and compared experimentally to a PID controller. Better performances with fuzzy controllers were obtained. The energy consumption was compared also. It could be noticed that the Fuzzy controllers are significantly less energy consuming.

The idea is to select a controller, with an optimized energy consumption / mechanical behavior, aiming at optimizing the design of AMB. The results obtained are encouraging for the ongoing development.

Only those three controllers were assessed. Effectively, for a completeness of the study more controllers have to be assessed, but at this stage, the aim was to evaluate the fuzzy logic bases strategies for the control of turbomachines.

References

[1] E. H. Maslen, "Smart Machine Advances in Rotating Machinery", *Proceeding 9th International Conference on Vibrations in Rotating Machinery, IMechE*, Exeter, UK, pp.3-14.I., 2008.

[2] M. K. Swann, A. P.Sarichev and E.Tsunoda, "A diffusion model for active magnetic bearing systems in large turbomachinery", *Proceeding 11th ISMB*, Japan, pp. 380-384, 2008.

[3] G. Schweitzer and E. H Maslen, Magnetic Bearings, Theory, Design, and Application to Rotating Machinery, Springer-Verlag, 535p, 2009.

[4] P.-Y. Couzon and J. Der Hagopian, "Neuro-fuzzy active control of rotor suspended on active magnetic bearing". J. Vib. Control, 13(4), pp. 365–384, 2007.

[5] N. M. Sahinkaya, A.-H. G. Abulrub, C. R. Burrows and P. S. Keogh, "A Multiobjective Adaptive Controller for Magnetic Bearing Systems". *J. Eng. Gas Turbines Power*, 132, pp.122503, 2010.

[6] R. L. Fittro and C. R. Knospe, "The μ Approach to Control of Active Magnetic Bearings", *J. Eng. Gas Turbines Power*, Vol 124, No. 3, pp. 566-570, 2002.

[7] S. Lei, and A. B. Palazzolo, "Control of flexible rotor systems with active magnetic bearings", *Journal of Sound and Vibration*, Vol. 314, Issues 1-2, pp.19-38, 2008.

[8] ISO 14839-1. Mechanical Vibration – Vibration of rotating machinery equipped with active magnetic bearings – Part 1: Vocabulary, 2002.

[9] ISO 14839-2. Mechanical Vibration – Vibration of rotating machinery equipped with active magnetic bearings – Part 2: Evaluation of vibration, 2004.

[10] ISO 14839-3. Mechanical Vibration – Vibration of rotating machinery equipped with active magnetic bearings – Part 3: Evaluation of stability margin, 2006.

[11] API 617. Axial and centrifugal compressors and expander-compressors for petroleum, chemical and gas industry service, 7th ed. 2002.

[12] J.Mahfoud and J. Der Hagopian, "Fuzzy Active Control Of Flexible Structures By Using Electromagnetic Actuators", *ASCE's Journal of Aerospace Engineering*, Vol. 24, No. 3, pp. 329-337, 2011.

[13] B. Defoy, T. Alban and J. Mahfoud, "Assessment of the Effectiveness of a Polar Fuzzy Approach for the Control of Centrifugal Compressors", *ASME, Journal of dynamic systems, measurement, and control*. [DOI: 10.1115/1.4026468]

[14] Matlab : The MathWorks – Matlab and Simulink for Technical Computing, www.mathworks.com.

Optimization based AMB Controller Design and Verification for Flexible Rotors

Bin Liu[1,a], Johan Sjöberg[1,b], Antti Laiho[2,c]

[1] ABB Corporate Research, Forskargränd 8, 72178 Västerås, Sweden
[2] ABB Discrete Automation and Motion, Strömbergintie 1B, 00380 Helsinki, Finland

[a] bin.liu@se.abb.com, [b] johan.sjoberg@se.abb.com, [c] antti.laiho@fi.abb.com

***Abstract*—Engineering costs especially for controller design are substantial and obstruct active magnetic bearings (AMBs) for broader industrial applications. An optimization based AMB controller design method is developed to solve this problem. The optimization criteria are selected to describe AMB practical performance. Controller components are chosen considering that parameters can be manually interpreted and modified on-site for commissioning. A multi-objective genetic algorithm (MOGA) toolbox is used to tune these controller component parameters to minimize the design criteria automatically. The method has been verified in a controller design process for an AMB levitated machine. With this method, the engineering cost for controller design can be reduced significantly.**

I. Introduction

Compared to traditional mechanical bearings, AMB offers many benefits for machines, for instance, lubrication free, contamination free, and long life operation, because there is no mechanical contact between the rotor and the stator[1]. In addition, AMB has inherent monitoring, adjustable stiffness and damping features which are really important for high speed and high reliability applications. There is a trend to equip AMBs in high speed machines to meet performance requirements[2]. One of the main barriers today for broader industrial AMB applications is the cost. AMBs can double the price of a large machine. Engineering costs especially controller design are substantial for AMB systems. Another barrier is that AMB requires specific knowledge. Tuning of controller parameters requires engineering experience in AMB area, especially for flexible rotors applications.

Genetic algorithms can solve these problems which have been widely used to industrial design and optimization applications[3]. Skalak introduced genetic algorithm to choose AMB actuator parameters[4]. Wei introduced multi-objective optimization to tune AMB position controller parameters[5][6]. However, the optimization criteria in [5] are not intuitively related to AMB performances which increases commissioning difficulty. A more practical method is proposed in this paper: optimization criteria are chosen to describe AMB system performances, and controller components are selected considering that parameters can be manually interpreted and modified on-site for commissioning. The verification shows that the method works well in a test machine.

This paper is organized as follows: In Section 2, the AMB-rotor system modeling is introduced. In Section 3, the optimization criteria are discussed and controller design is given and analyzed. In Section 4, the experimental verification is given. Finally, a conclusion is given in Section 5.

Figure 1. Test rotor with finite element model.

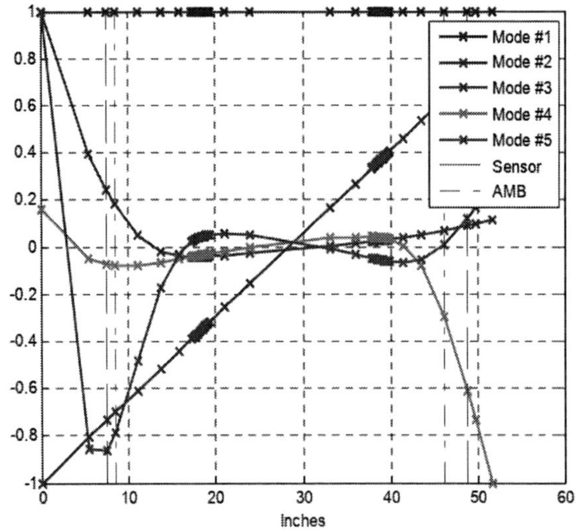

Figure 2. Undamped rotor modal shapes with sensor and actuator locations.

II. AMB-Rotor System Modeling

A. Rotor model

In order to describe rotor flexible modes precisely, a test rotor is modeled in finite element method with 75 nodes, as shown in Figure 1. Figure 2 shows the analyzed undamped rotor modal shapes from the FE model. The equation of motion can be described as follow, in which the matrices are calculated from the FE model:

$$\begin{cases} \dot{X}_r = [A_r + \Omega G_r]X_r + B_{r\_a}F_a + B_{r\_d}F_d \\ Y_a = C_{r\_a}X_r \\ Y_s = C_{r\_s}X_r \end{cases} \quad (1)$$

where X_r are the rotor displacements in modal coordinate, G_r is the gyroscopic matrix, and Ω is the rotational speed. Y_a and Y_s are the rotor displacements at actuator and sensor position, respectively. B_{r_a} and B_{r_u} are the input matrices for actuator force and unbalance disturbance, respectively. F_a and F_d are the actuator force and disturbance force, respectively.

B. *Actuaor model*

Actuator model can be described as follows:

$$\begin{cases} \dot{X}_a = A_a X_a + B_{a\_u}u + B_{a\_y}Y_a \\ F_a = C_a X_a \end{cases} \quad (2)$$

where X_a are the internal states of actuators and power amplifiers, and u are the control signals. B_{a_u} and B_{a_y} are input matrices for control signals and rotor displacement at actuator position.

C. *Complete model*

The complete model can be assembled by combing the rotor model and the actuator as follows:

$$\begin{cases} \begin{bmatrix} \dot{X}_r \\ \dot{X}_a \end{bmatrix} = \begin{bmatrix} A_r + \Omega G_r & B_{r\_a}C_a \\ B_{a\_y}C_{r\_a} & A_a \end{bmatrix} \cdot \begin{bmatrix} X_r \\ X_a \end{bmatrix} + \\ \qquad \begin{bmatrix} B_{r\_d} \\ 0 \end{bmatrix} \cdot F_d + \begin{bmatrix} 0 \\ B_{a\_u} \end{bmatrix} \cdot u \\ Y_s = \begin{bmatrix} C_{r\_s} & 0 \end{bmatrix} \cdot \begin{bmatrix} X_r \\ X_a \end{bmatrix} \end{cases} \quad (3)$$

III. Optimization Based Controller Design

A. *Optimization Criteria*

Guoxin had an insightful description about AMB controller objectives in his PhD thesis[7]:

1) In the low frequency range below 1.0 Hz, the controller was an integrator with a flat static gain.

2) In the rigid body frequency range from 1 Hz to 100 Hz, the controller was essentially a PD controller with stiffness and damping required to stabilize the rigid body modes.

3) In the frequency range above 100 Hz, the controller placed a series of complex poles and zeros (generalized second order filters) to stabilize the flexible rotor and substructure modes. The controller gain rolls off above 500 Hz.

Combined with ISO standards[8][9], the optimization criteria for controller design are given as follows, and the expressions are given to calculate the criteria values during the optimization process.

a) closed loop system is stable

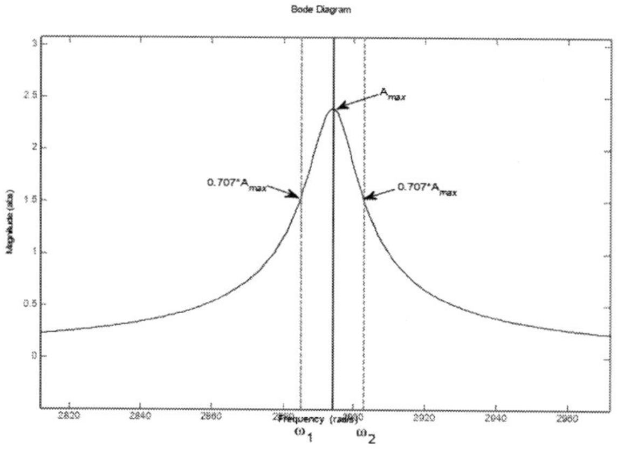

Figure 3. Rotor benidng mode Q factor calculation.

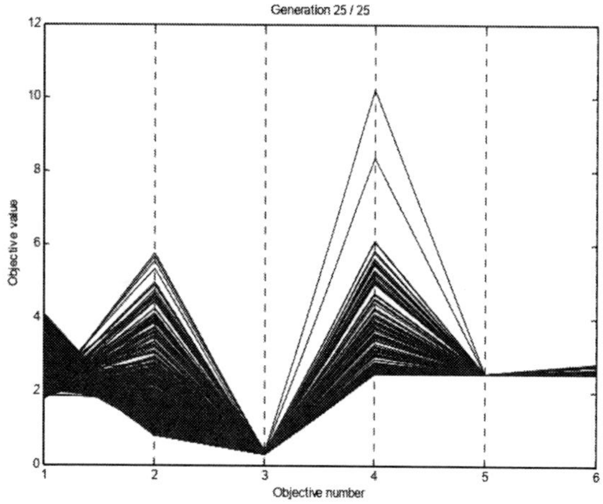

Figure 4. Optimization results for controller design.

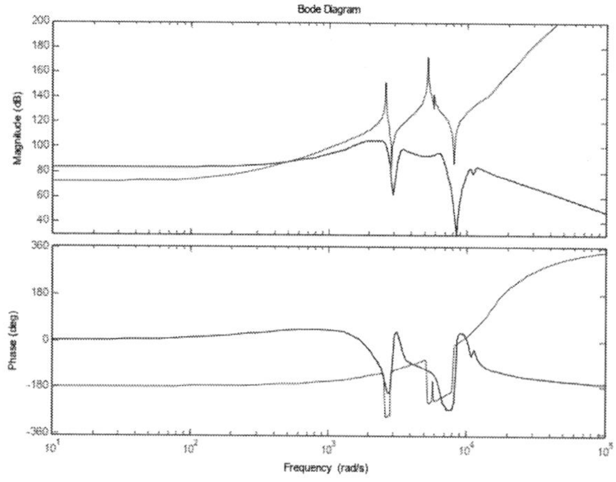

Figure 5. Designed controller(blue) vs inversed plant model(green).

$$G_s(s) = \frac{1}{1 + P(s) \cdot C(s)} \tag{4}$$

b) closed loop system has robust performance (sensitivity function is less than 3)

$$\max\{S(s)\} = \max\left\{\frac{1}{1 + P(s) \cdot C(s)}\right\} < 3 \tag{5}$$

c) closed loop system has stable margin at critical speeds

$$G_Q(s) = \frac{P(s)}{1 + P(s) \cdot C(s)} = P(s) \cdot S(s) \tag{6}$$

$$Q = \frac{A_{\max}(G_Q)}{\omega_2 - \omega_1} \tag{7}$$

d) controller has low gain at high frequency for lower noise sensitivity

$$J = \sum_{f_h}^{f_B} A_c(f) \tag{8}$$

B. *Controller components*

A PID controller is used to stabilize AMB-rotor system rigid modes. A roll-off filter is used to eliminate high frequency noises. Lead-lag compensators and notch filters are added to deal with flexible modes. The equations are shown as follows respectively:

$$G_{PID}(s) = K_p + \frac{K_i}{s} + \frac{K_d s}{T_f s + 1} \tag{9}$$

$$G_R(s) = \frac{\omega_r^2}{s^2 + 2\zeta_r \omega_r s + \omega_r^2} \tag{10}$$

$$G_L(s) = \frac{s^2 + 2\zeta_2 \omega_2 s + \omega_2^2}{s^2 + 2\zeta_1 \omega_1 s + \omega_1^2} \tag{11}$$

$$G_N(s) = \frac{\alpha^2 s^2 + 2\alpha\xi\omega_n s + \omega_n^2}{s^2 + 2 \cdot 0.05 \cdot \omega_n s + \omega_n^2} \tag{12}$$

A multi-objective genetic algorithm(MOGA) toolbox, e.g. Matlab global optimization toolbox, could be used to tune controller parameters automatically. Figure 4 shows the optimization results after 25 generations.

IV. VERIFICATION RESULTS

In order to verify the design method, an experimental verification was executed in a test machine as shown in Figure 6. The test machine has radial bearings in both drive and non-drive end. The levitation process went on really smooth. In the first try the rotor was levitated but with small vibration. By changing only one parameter (controller gain), the test rotor achieved stable levitation.

Figure 6. Photo of the test machine.

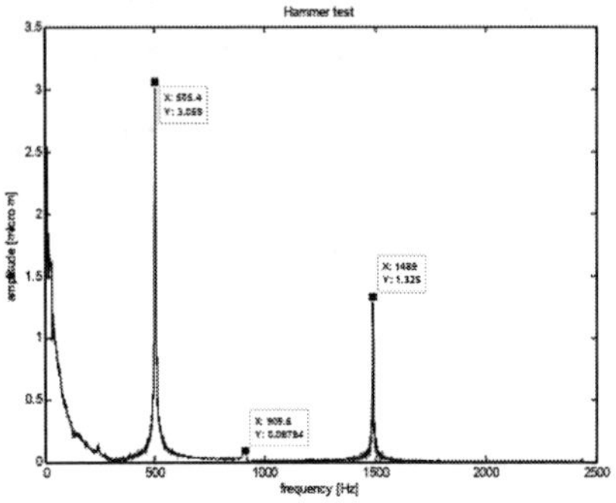

Figure 7. Hammer test result when the rotor was leviated.

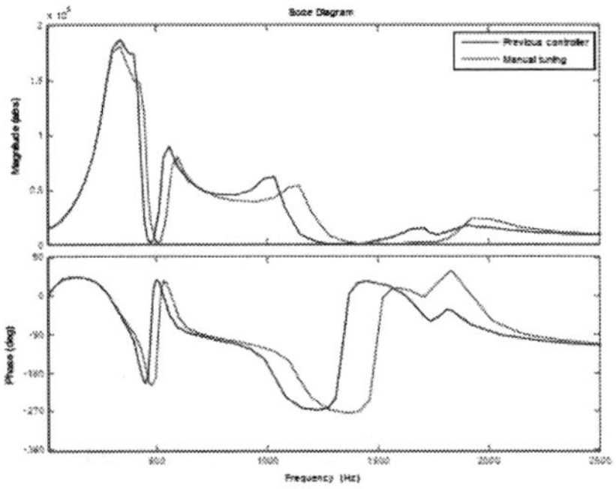

Figure 8. Manually tuned controller result.

After the levitation, a hammer test was carried out to verify the rotordynamic features. The hammer test result is given in Figure 7. The result shows that the first bending mode is 505Hz and the third bending mode is 1489Hz, while in the rotordynamic model they are 473Hz and 1339Hz respectively. With the hammer test result, the notch filter frequencies for the first and third bending modes have been changed manually, as shown in Figure 8.

Without any more parameter changing, the designed controllers levitated the rotor and rotated to the maximum continuous speed 200Hz(12000rpm). Figure 9 shows the maximum displacements of drive and non-drive end during the speed ramp up and down test. Figure 10 shows the rotor orbits measured by the position sensor at 200Hz in 0.5 seconds. The orbits in both ends are less than 3μm, which shows the levitation is in a really stable status.

V. Conclusion

An optimization based AMB controller design method has been described in this paper. The optimization criteria are chosen to describe AMB performances, and controller components are selected considering manually tuning on-site. A multi-objective genetic algorithm toolbox can be used to tune the parameters automatically. The design results have been verified in experiments. During the experimental verification, only 3 of the total 34 parameters have been changed and the machine rotates to the maximum continuous speed (12000rpm) with performances much better than the ISO standard requirement.

For a newly built AMB system, if the objectives and variables are set up properly, the AMB position controller design process can be done in half an hour. Even including the setup of the rotordynamic features and other system parameters such as actuators, sensors, and amplifiers, the AMB controller design process can be finished in several hours. The engineering cost for controller design can be reduced significantly.

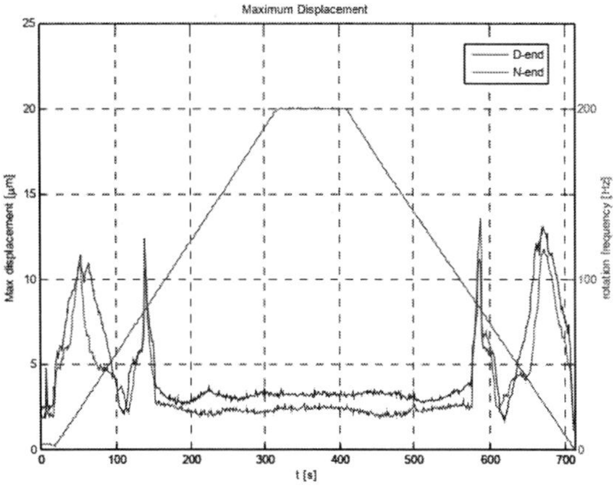

Figure 9. D-end and N-end maximum rotor displacement during a ramp up and down test from 0 to 200Hz.

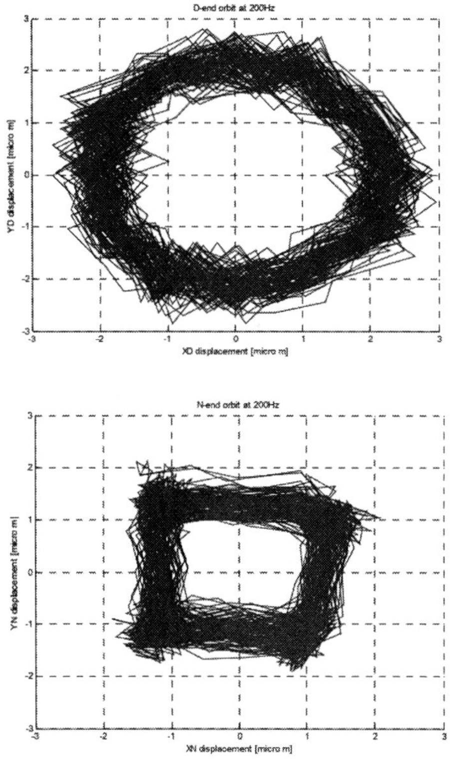

Figure 10. D-end and N-end rotor orbit at 200Hz.

References

[1] G. Schweitzer, E. Maslen, "Magnetic bearings, theory, design, and applications to rotating machinery," Springer, 2009.

[2] A. Masala, G. Vannini, M. Lacour, et al, "Lateral rotordynamic analysis and testing of a vertical high speed 12.5MW motorcompressor levited by active magnetic bearings," *Proceedings on 13th Int. Conf. on Magnetic Bearings*, pp: 8-13, 2010.

[3] P.J. Fleming, R.C. Purshouse, "Evolutionary algorithms in control system engineering: a survey." *Journal of Control Engineering Practice*, vol. 10, 2002.

[4] S. C. Skalak, E. Maslen, Y. Teng, "Magnetic bearing actuator design using genetic algorithms," *Journal of Engineering Design*, pp. 143–164, vol. 10, no. 2, 1999.

[5] C. Wei, D. Söffker, "Improving PID-control of AMB-Rotor system deisgn I: optimization strategy," *Proceedings on 13th Int. Conf. on Magnetic Bearings*, paper ID: 78, 2012.

[6] C. Wei, D. Söffker, "Improving PID-control of AMB-Rotor system deisgn II: experimental results," *Proceedings on 13th Int. Conf. on Magnetic Bearings*, paper ID: 79, 2012.

[7] G. Li, "Robust stabilization of rotor-active magnetic bearing systems," Ph.D dissertation, 2007.

[8] ISO Standard 14839-3, "Mechanical vibration–vibration of rotating machinery equipped with active magneitc bearings–part 3: evaluation of stability margin," 2006.

[9] ISO Standard 14839-4, "Mechanical vibration–vibration of rotating machinery equipped with active magneitc bearings–part 4: technical guidelines," 2012.

Robust control and loop shaping design with the application to flexible rotor levitated with AMBs

Goranka Štimac[a], Sanjin Braut[a], Roberto Žigulić[a], Ante Skoblar[a]

[a] Faculty of Engineering, University of Rijeka, Vukovarska 58, 51000 Rijeka, Croatia, gstimac@riteh.hr

***Abstract*— The aim of this research is to ensure stable operation at a nominal rotation speed (above the first critical speed) of a flexible rotor levitated with active magnetic bearings (AMBs) and to guarantee good performance of the rotor/AMB system. Initially, minimizing the ITAE performance index, PID control system is designed to stabilize the rotor. The obtained transfer function is adopted as a compensator to shape the singular values of the nominal plant. By applying the Glover-McFarlane control method, a robust controller which includes coprime factor uncertainties is designed. The results show that the AMB system satisfies Zone A specifications defined by the ISO 14839 standard for newly commissioned machines with magnetic bearings.**

I. INTRODUCTION

Flexible rotors supported by active magnetic bearings (AMBs) are characterized by multiple lightly damped bending modes below the operating speed or within the controller bandwidth. Consequently, control of flexible rotors presents a challenge since linear design approaches are inherently only approximations of nonlinear dynamics [1]. Therefore, since PID control laws largely ignore model uncertainty and the coupled dynamics, they cannot guarantee robust stability and performance of rotor/AMB systems [2]. However, modern robust and optimal control theories provide a systematic method to optimize stability and performance of rotor/AMB systems [3].

Regarding the H_∞ robust method, two basic approaches can be distinguished, an approach based on the loop shaping and an approach based on the signals. Glover-McFarlane H_∞ method [4] is an open loop shaping method which has found its application in the field of AMBs as well [5, 6]. The main idea behind the loop-shaping in the open loop is that the singular values of the closed loop transfer function matrices can be determined directly from the singular values of the open loop system.

This paper focuses on the robust controller design for a flexible rotor/AMB system using the procedure as presented by Glover and McFarlane. The design procedure combines the loop shaping techniques from classical control with H_∞ synthesis so as to optimize the robustness against coprime factor uncertainty. In the first step, PID controller is designed to stabilize the rotor. In the second step, the singular values of the nominal plant are shaped by implementing the obtained PID controller model as a pre-compensator and the robust controller is designed by applying the Glover-McFarlane control method.

II. MATHEMATICAL MODEL

Fig. 1 presents the considered rotor/AMB system configuration. It consists of the flexible rotor levitated by two radial AMBs with non-collocation. The mass of each AMB rotor part is 8 kg. The length and the diameter of the shaft are 1.1 m and 0.025 m, respectively. Both AMBs operate with 0.3 mm radial air gap. In addition, three discs of the same dimensions (0.22 m in diameter and 0.05 m wide) are mounted on the shaft.

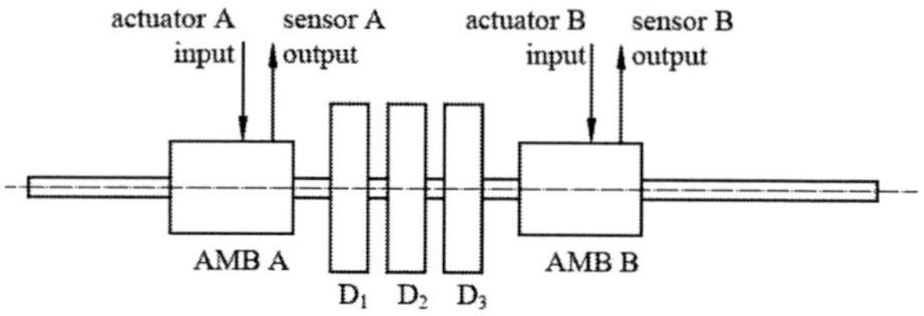

Figure 1. Flexible rotor/AMB system configuration, with three discs D_1, D_2 and D_3

By using the data from the finite element analysis, a structural mathematical model of the flexible rotor/AMB system in the modal coordinates can be constructed [7]. A nominal model **G** is defined as a state-space model described by the matrices **A**, **B**, **C** and **D**, as follows:

$$\mathbf{A}=\begin{bmatrix} \mathbf{0} & \mathbf{I} \\ -\tilde{\mathbf{K}}+\mathbf{\Phi}^{T}\mathbf{K}_s\mathbf{\Phi} & -\tilde{\mathbf{D}} \end{bmatrix}$$
$$\mathbf{B}=\begin{bmatrix} \mathbf{0} \\ \mathbf{\Phi}^{T}\mathbf{K}_i \end{bmatrix} \quad (1)$$
$$\mathbf{C}=\begin{bmatrix} \mathbf{T}_S\mathbf{\Phi} & \mathbf{0} \end{bmatrix}$$
$$\mathbf{D}=\begin{bmatrix} \mathbf{0} \end{bmatrix}$$

with

$$\tilde{\mathbf{D}}=\text{diag}\left(2\zeta_1\omega_1 \quad 2\zeta_2\omega_2 \quad \cdots \quad 2\zeta_i\omega_i\right)$$
$$\tilde{\mathbf{K}}=\text{diag}\left(\omega_1^2 \quad \omega_2^2 \quad \cdots \quad \omega_i^2\right) \quad , \quad (2)$$

where ω_i and ζ_i are the natural frequency and the modal damping ratio of the i-th vibration mode. Furthermore, $\mathbf{\Phi}$ is the modal matrix of the mass normalized eigenvectors, $\mathbf{K}_s$ and $\mathbf{K}_i$ are 4×4 diagonal matrices of the force-displacement coefficients k_s and the force-current coefficients k_i, and $\mathbf{T}_S$ is the Boolean matrix transforming the displacement outputs at the corresponding degrees of freedom. The obtained natural frequencies of the flexible rotor model are: ω_1 = 266 rad/s, ω_2 = 729 rad/s and ω_3 = 1346 rad/s. The corresponding modal damping is 0.02, 0.01 and 0.005, respectively. In order to show rotor behavior at the AMB locations, the mode shapes (1st, 2nd and 3rd) of the system are shown in Fig 2.

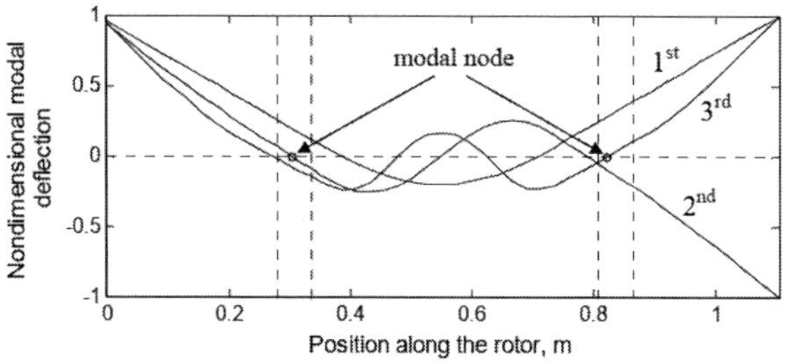

Figure 2. The natural mode shapes (1st, 2nd, 3rd) of the rotor/AMB system

The power amplifiers used to drive the actuator coils are modeled as a second order filter, with the cut-off frequency f_{amp} = 700 Hz and the damping ratio ζ_{amp} = 0.85. The position sensors are modeled as a simple constant K_{sens} = 8000 V/m. Moreover, the total phase delay due to unmodeled dynamics of the sensors and other electronic devices, including sampling time of the digital controller is modeled as a total time delay of 1.3 ms using the Padé approximation of the first order.

Finally, Figs. 3 and 4 show frequency response plots of the AMB A and AMB B from power amplifier inputs to position sensor outputs in y-axis.

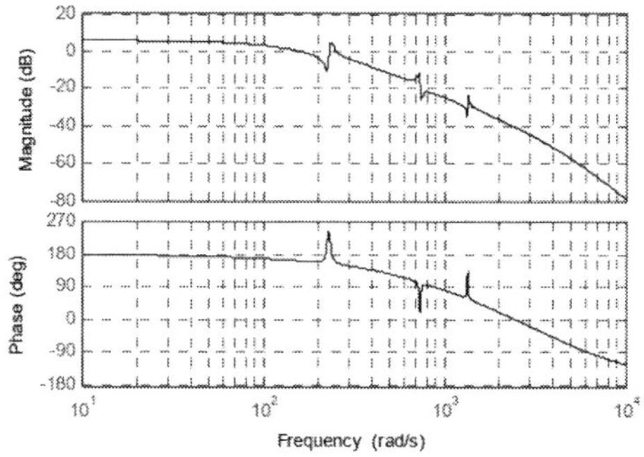

Figure 3. Frequency response plot of AMB A in y-axis

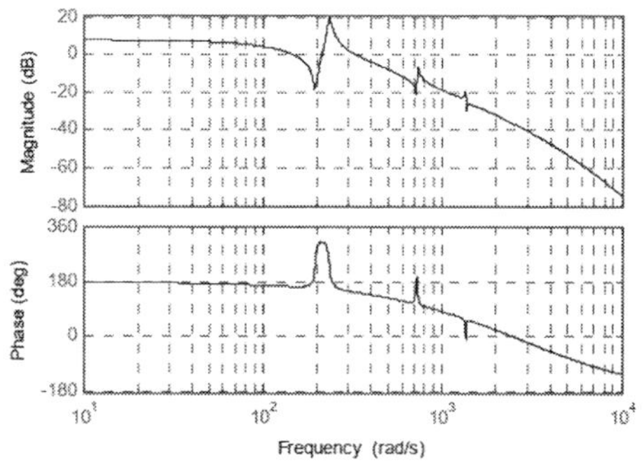

Figure 4. Frequency response plot of AMB B in y-axis

In Fig. 3 (at the 2nd bending mode of AMB A) and in Fig. 4 (at the 3rd bending mode of AMB B) it can be noticed that the pole appears before the corresponding zero. This indicates that the applied force at the near resonant rotor speeds would be 180° out of phase with respect to the measured displacement and special attention should then be paid to the controller design. Such observations are in accordance with the modal nodes position (Fig. 2).

III. CONTROL

A. PID control and gain optimization

A practical PID controller is characterized by the transfer function:

$$G_{pid} = K_p + \frac{K_{int}}{s} + K_d \frac{N_f}{1 + N_f / s}, \tag{3}$$

where K_p is the proportional gain, K_{int} is the integral gain, K_d is the derivative gain and N_f is the filter coefficient. In order to define the set of parameters (K_p, K_{int}, K_d, N_f) the closed loop system structure of the system presented in Fig. 5 is adopted for controller tuning.

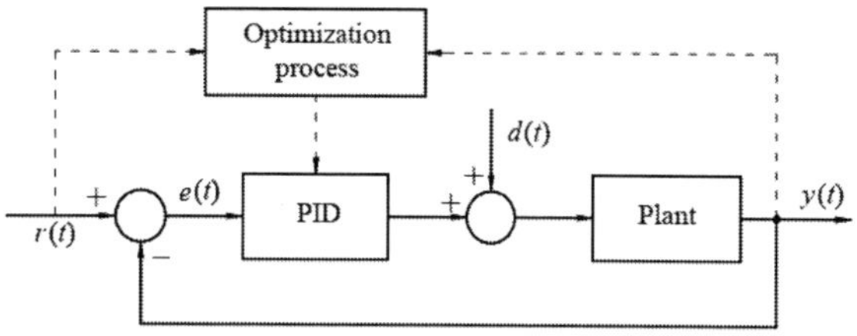

Figure 5. System structure adopted for PID controller tuning

Output response $y(t)$ of the system is compared with the desired response $r(t)$ and the resulting error $e(t) = r(t) - y(t)$ is minimized through an optimization performance criterion to determine the optimal PID gains. Here, the minimization of the integral of time multiplied by the absolute value of error (ITAE) is adopted, which is defined as:

$$J\left(K_p, K_{int}, K_d, N_f\right) = \sum_{i=1}^{m} \int_0^T t\left|e_i(t)\right| dt, \tag{4}$$

where T is the time of integration and $m = 4$ is the total number of control axes (two per one AMB), i.e. PID controllers.

The search of the PID controller design may be stated as a constrained multivariable optimization problem defined as:

$$\min_x J(x) \text{ subject to } x > 0, \text{ where}$$
$$x = \left[K_p, K_{int}, K_d, N_f\right] \tag{5}$$

For the minimization of the performance index (4), with the constraints defined in (5), built-in function *fmincon* of the Matlab Optimization Toolbox is used by applying the interior-point algorithm.

B. Glover–McFarlane H_∞ loop-shaping design procedure

The loop shaping procedure includes two steps. In the first step, the nominal model **G**, given in the form of its left coprime factorization is shaped using the pre-compensator matrix $\mathbf{W}_1$ and the post-compensator matrix $\mathbf{W}_2$ (Fig. 6), so as to obtain the shaped plant:

$$\mathbf{G}_s = \mathbf{W}_2\mathbf{G}\mathbf{W}_1 . \quad (6)$$

In the second step, robust stabilization of the shaped plant with respect of the coprime factor uncertainties is performed by synthesizing a stabilizing controller $\mathbf{K}_s$ [8]. The final controller can be then constructed by combining $\mathbf{K}_s$ with $\mathbf{W}_1$ and $\mathbf{W}_2$, such that

$$\mathbf{K} = \mathbf{W}_1\mathbf{K}_s\mathbf{W}_2 . \quad (7)$$

An important advantage of the presented method is that it does not require explicit definition of uncertainties.

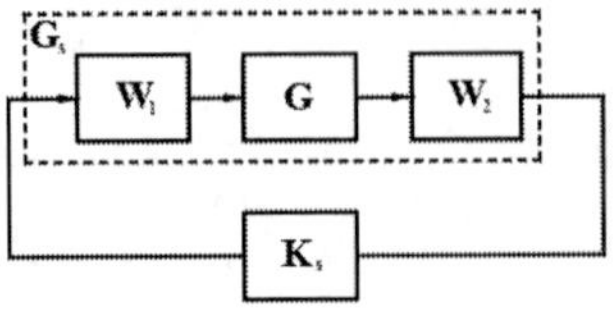

Figure 6. Shaped model $\mathbf{G}_s$ with controller $\mathbf{K}_s$

C. Compensator selection

The diagonal elements of $\mathbf{W}_1$ and $\mathbf{W}_2$ are selected so as to provide the desired shape of the singular values of the open-loop frequency response. However, in many practical applications, including AMBs, systems are unstable and can possess more than one gain crossover frequency. In such cases, the selection of the transfer functions for the design of the open loop process becomes difficult.

In this paper it is shown that $\mathbf{W}_1$ can be efficiently selected as the following diagonal transfer function matrix:

$$\mathbf{W}_1 = \begin{bmatrix} w & 0 & 0 & 0 \\ 0 & w & 0 & 0 \\ 0 & 0 & w & 0 \\ 0 & 0 & 0 & w \end{bmatrix}, \quad w = G_{\mathrm{pid}} G_{\mathrm{notch2}} G_{\mathrm{notch3}} , \quad (8)$$

where G_{pid} is the PID controller transfer function defined in (3) and G_{notch2} and G_{notch3} are the notch filters related to the 2nd and the 3rd bending modes defined as:

$$G_{\mathrm{notch2}} = \frac{s^2 + \omega_{\mathrm{notch2}}^2}{s^2 + 2\zeta_{\mathrm{notch2}}\omega_{\mathrm{notch2}} s + \omega_{\mathrm{notch2}}^2} ,$$
$$G_{\mathrm{notch3}} = \frac{s^2 + \omega_{\mathrm{notch3}}^2}{s^2 + 2\zeta_{\mathrm{notch3}}\omega_{\mathrm{notch3}} s + \omega_{\mathrm{notch3}}^2} \quad (9)$$

The corresponding central frequencies of the notch filters are ω_{notch2} and ω_{notch3} and the corresponding damping ratios are ζ_{notch2} and ζ_{notch3}. Controller parameters (K_p, K_{int}, K_d, N_f) are obtained by minimization of ITAE performance index (4).

Since in the observed AMB system there is no difference between importance of the outputs, $\mathbf{W}_2$ is selected as 4×4 unit diagonal matrix.

IV. SIMULATION RESULTS AND CONCLUSION

This chapter presents an analysis of the closed loop sensitivity functions of the rotor/AMB system, referring to the ISO 14839-3 standard [9]. From the measured sensitivity function $S(j\omega)$ of each control axis (i=1,...,m) in the frequency domain, the index to be evaluated is obtained from the relationship:

$$S_{\max} = \max_i \left[\max \left| S(j\omega) \right| \right], \quad 0 \le \omega \le 2\pi f_{\max} , \quad (10)$$

where f_{max} is the maximum frequency in the observed frequency range. Equation (10) generally states that the overall system rating is determined as the worst rating of any of the transfer functions individually measured for all four control axes.

The Glover–McFarlane H_∞ loop-shaping controller design procedure was carried out for the rotational speed Ω = 30 Hz (subcritical speed). Figs. 7 – 14 show plots of output sensitivity of the rotor/AMB system for the y-axis of AMB A and y-axis of AMB B for both PID and robust control methods. The magnitudes are measured at standstill (Ω = 0 rad/s), at Ω = 30 Hz (188.5 rad/s), at Ω = 50 Hz (314.2 rad/s) and at Ω = 60 Hz (377 rad/s). The presented results are obtained with the following parameters: K_p = 0.6393, K_{int} = 6.271, K_d = 0.004, N_f = 80000, ω_{notch2} = 729 rad/s, ω_{notch3} = 1346 rad/s and ζ_{notch2} = ζ_{notch3} = 0.15.

Since the output sensitivity function provides a means of quantifying the disturbance rejection of the closed loop system, it can be concluded that the AMB system controlled by the proposed algorithm (PID + Glover-Mcfarlane) provides better disturbance rejection property than the system controlled solely by PID. Moreover, peak sensitivity lower than 3 (in absolute terms) satisfies *Zone* A specifications for newly commissioned machines with magnetic bearings defined by ISO 14839-3 and, therefore, indicates very good stability margin (as in Figs. 7 – 11). *Zone* B (peak sensitivity from range 3 – 4) is normally considered acceptable for unrestricted long-term operation (as in Figs. 12 and 13). However, maximal magnitude of the output sensitivity for AMB B at Ω = 60 Hz (as in Fig. 14) is equal to 5.38, indicating very poor stability margin (*Zone* D). This peak corresponds to excitation of the 2nd bending mode of the rotor which introduces non-collocation into the system (pole before zero at the 2nd bending mode of AMB A, Fig. 3).

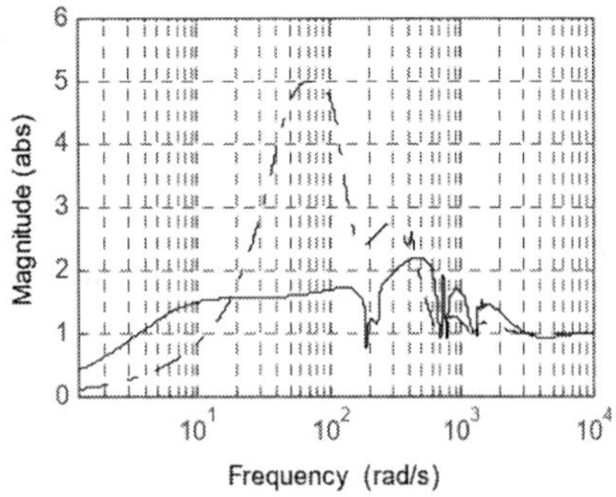

Figure 7. Output sensitivity function of the AMB A (Ω = 0 rad/s); PID(–.), PID+Glover-McFarlane (—)

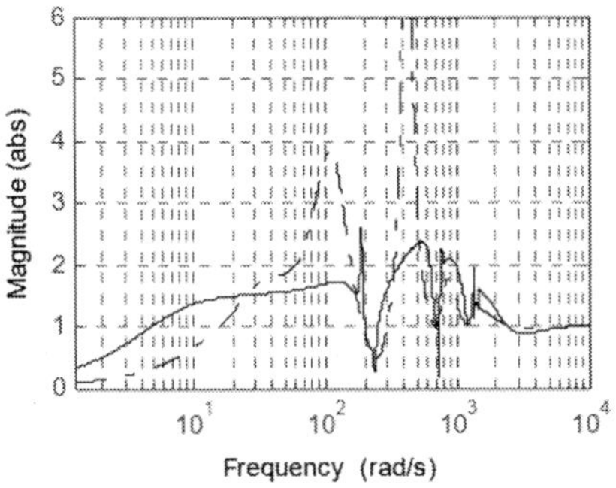

Figure 8. Output sensitivity function of the AMB B (Ω = 0 rad/s); PID(-.), PID+Glover-McFarlane (—)

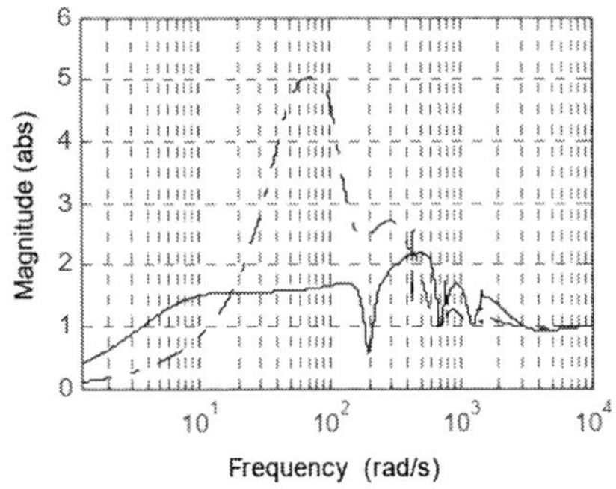

Figure 9. Output sensitivity function of the AMB A (Ω = 188.5 rad/s); PID(-.), PID+Glover-McFarlane (—)

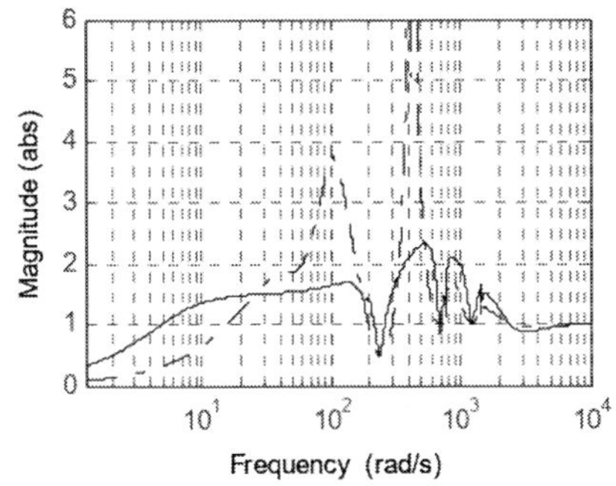

Figure 10. Output sensitivity function of the AMB B (Ω = 188.5 rad/s); PID(-.), PID+Glover-McFarlane (—)

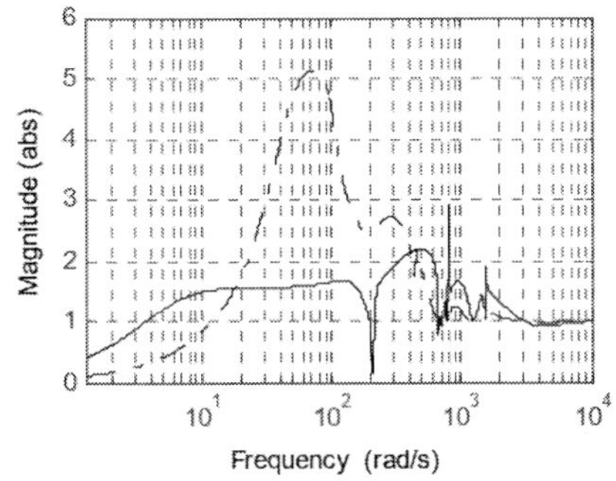

Figure 11. Output sensitivity function of the AMB A (Ω = 314.2 rad/s); PID(-.), PID+Glover-McFarlane (—)

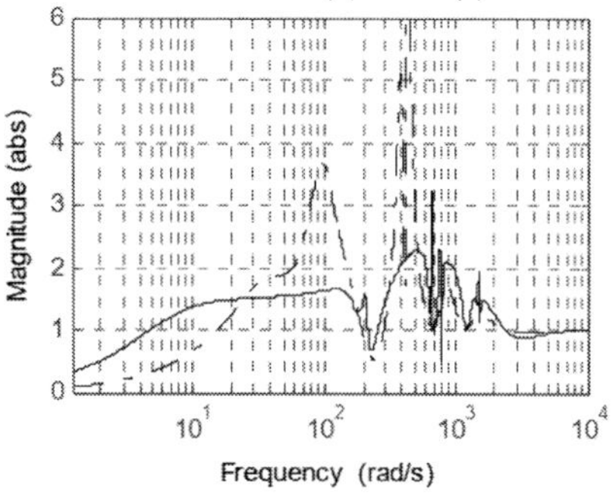

Figure 12. Output sensitivity function of the AMB B (Ω = 314.2 rad/s); PID(-.), PID+Glover-McFarlane (—)

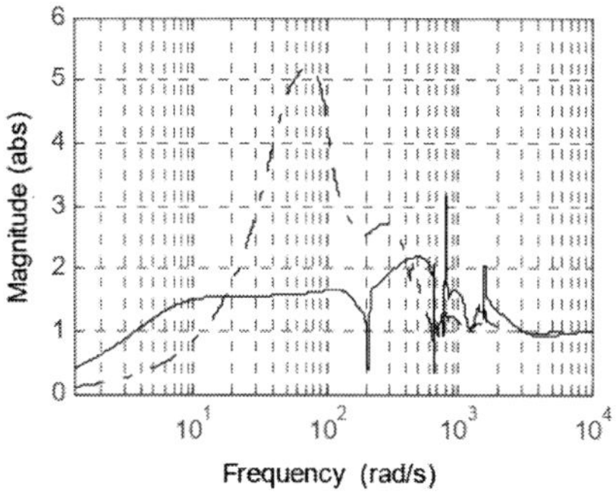

Figure 13. Output sensitivity function of the AMB A (Ω = 377 rad/s); PID(-.), PID+Glover-McFarlane (—)

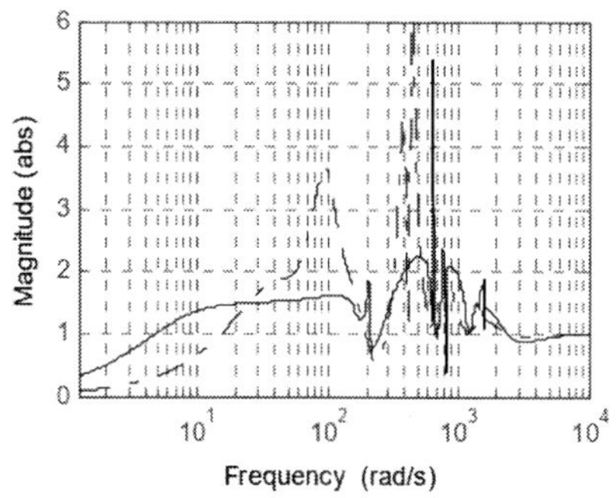

Figure 14. Output sensitivity function of the AMB B (Ω = 377 rad/s); PID(-.), PID+Glover-McFarlane (—)

In conclusion, the proposed procedure does not satisfy the ISO standard in the entire frequency range if the non-collocation problem exists. However, with no collocation problems it provides stable operation of the flexible rotor/AMB system above the first critical speed.

ACKNOWLEDGEMENT

Research reported in this paper was supported by University of Rijeka, Croatia Grants No. 13.09.1.2.11 and No. 13.09.2.2.19.

References

[1] G. Balas, J. Doyle, "Robustness and performance trade-offs in control design for flexible structures," *IEEE Transactions on Control System and Technology*, vol. 2, no. 4, pp. 352-361, 1994.

[2] S. E. Mushi, Z. Lin, P. E. Allaire, "Design, construction and modeling of a flexible rotor active magnetic bearing test rig," *IEEE/ASME Transactions on Mechatronics*, vol. 17, no. 6, pp. 1170-1182, 2012.

[3] K. Zhou, J. Doyle, "Robust and Optimal Control," Prentice-Hall, Upper Sadle River, New Jersey, 1996.

[4] K. Glover, D. McFarlane, "Robust stabilization of normalized coprime factor plant descriptions with H_∞ bounded unceraínty," *IEEE Transations on Automatic Control*, vol. 34, pp. 821–830, 1989.

[5] A. L .F. Driemeyer, H. Bourles, E. R. De Pieri, "A robust nonlinear controller with application to a magnetic bearing system," *Proceedings of the 44th IEEE Conference on Decision and Control*, 2005, pp. 4927–4932, 2005.

[6] E. S. Armstrong, "Robust controller design for flexible structures using normalized coprime factor uncertainty," *NASA Technical Paper 3325*, Langley Research Center, Hampton, Virginia, 1993.

[7] G. Štimac, S. Braut, N. Bulić, R. Žigulić, "Modeling and Experimental Verification of a Flexible Rotor/AMB System," *COMPEL: The International Journal for Computation and Mathematics in Electrical and Electronic Engineering*, vol. 32, pp.1244-1254, 2013.

[8] S. Skogestad, I. Postlethwaite, "Multivariable feedback control," John Wiley & Sons, 1996.

[9] ISO 14839-3:2006, Mechanical vibration — Vibration of rotating machinery equipped with active magnetic bearings, Part 3: Evaluation

A Method of Simple Adaptive Control for MIMO Nonlinear AMB-flywheel Levitation System

Yong He[a], Kenzo Nonami[b], Fumiya Shimizu[c]

[a] Department of Electronics and Mechanical Engineering, 1-33, Yayoi-cho, Inage-ku, Chiba-shi, Japan, heyong83@graduate.chiba-u.jp

[b] Department of Electronics and Mechanical Engineering, 1-33, Yayoi-cho, Inage-ku, Chiba-shi, Japan, nonami@faculty.chiba-u.jp

[b] Department of Electronics and Mechanical Engineering, 1-33, Yayoi-cho, Inage-ku, Chiba-shi, Japan, z9t0632@students.chiba-u.jp

***Abstract*—In this paper, AMB-flywheel is desirable to minimize the energy required for its stabilization control. We propose a design method for performing SAC control for a MIMO AMB-flywheel system with five degrees of freedom; The SAC controller was proposed for application to an AMB-flywheel system and control performance was improved by combining SAC with PID controller. The PID based SAC controller is evaluated via experiments. We present the analysis results of the influence of external disturbances from the EV driving on the road and high-speed rotation experimental results about the AMB-flywheel system. The results demonstrate that the SAC controller is feasible and effective.**

I. Introduction

In recent years, energy consumption certainly increases year by year, but the fossil fuels, such as petroleum, coal and natural gas, is limited, and will dry up in the future. The renewable natural energy resources (Wind, solar, wave, and geothermal energies) are considered replaces the major role of fossil energy. In this new energy field, the current energy storage system is used chemical batteries, which are still not fully satisfying the requirement. The flywheel energy storage system to store electricity is one solution for energy optimal utilization[1], [2].

Most of studies about active magnetic bearing flywheel energy storage system is fixed on the ground, as used in wind-power plant or UPS systems; we developed electrical vehicle with AMB-flywheel as shown in Fig 1. In this paper, we are focusing on the use of AMB-flywheel energy storage system on electric vehicle. An AMB-flywheel system is a replacement of conventional chemical battery for electric vehicle.

II. Modeling of AMB-flywheel

Accurate mathematical description of the physical system is the important part of effective controller design. Building a mathematical mode of the system to represent the actual physical system can be used to verify the performance of system through digital simulation work. It is also providing a structural form for new control algorithms[3]. Stable operation of flywheel, which consists of active magnetic bearings, is achieved through appropriate magnetic forces generated by

Figure 1. Overview of over of AMB-flywheel powered EV

the magnetic bearing actuator. The physical properties of the active magnetic bearing have been presented in Table I This AMB-flywheel was designed as a vertical structure. In this research, it will be installed on the electrical vehicle (EV) as the vertical direction, so we should defined z-axis as the gravity direction in the modeling of it in Fig 2.

As covered in last section, the linearization model could provide satisfactory system response at approximately close to the linearized region, but nonlinear control techniques developed based on nonlinear AMB model can be achieved at wider range of system operations. Let us consider the rotor of the AMB system as a rigid rotor. The fundamental equations of motion of a rigid rotor-active magnetic bearing system in horizontal direction is derived as (1)

$$\begin{cases} m\ddot{x} = F_x + m\delta\omega^2\cos(\omega t) \\ I_r\ddot{\theta}_y - I_z\omega\dot{\theta}_x = N_y + ml_c\delta\omega^2\cos(\omega t) \\ m\ddot{y} = F_y + m\delta\omega^2\sin(\omega t) \\ I_r\ddot{\theta}_x + I_z\omega\dot{\theta}_y = N_x + ml_c\delta\omega^2\sin(\omega t) \end{cases} \quad (1)$$

Where x and y denote the linear displacement of the rotors

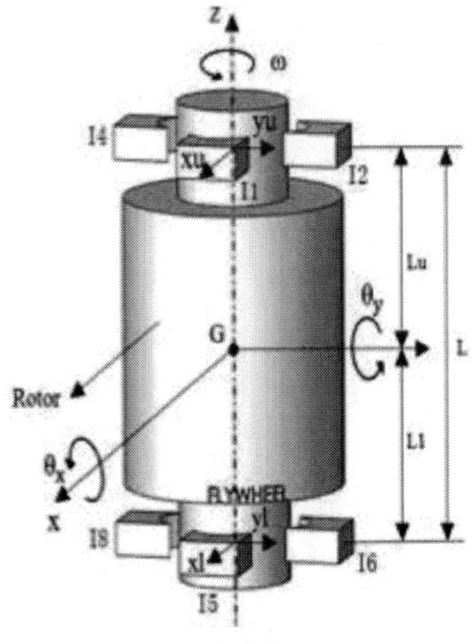

Figure 2. Overview of over of AMB-flywheel powered EV

Table I
PARAMETERS OF THE AMB-FLYWHEEL

Item	Value	Unit
m	Rotor mass	100 kg
I_z	Polar moment of interia	$1.114\,\text{kg.m}^2$
I_r	Other moment of interia	$2.610\,\text{kg.m}^2$
L_u	Distance of upper AMB from center of gravity	$1.815e-4\,\text{m}$
L_l	Distance of lower AMB from center of gravity	$3.086e-4\,\text{m}$
X_0	Nominal air gap in radial direction	$0.5e-3\,\text{m}$
Z_u	Nominal air gap in axial direction	$0.4e-3\,\text{m}$
I	Allowable current	6 A

center of mass along the x and y axes, respectively. Similarly, θ_y and θ_x denote the angular displacements of the rotor around the x and y axes, respectively. F_x and F_y denote the electromagnetic forces acting on the bearing in the x and y directions, respectively. ω denotes the flywheel rotation speed; δ , the eccentricity; m , the mass of the rotor; I_z , the moment of inertia about the z axis; and I_r , the moment of inertia about the x and y axes. N_x and N_y denote the moment of force about the x and y axes, respectively. F_x , F_y , N_x , and N_y can be expressed as(2)

$$\begin{cases} F_x = f_{xu} + f_{xl}, F_y = f_{yu} + f_{yl} \\ N_x = f_{yl}L_l - f_{yu}L_u, N_y = f_{xu}L_u - f_{xl}L_l \end{cases} \tag{2}$$

III. ANALYSIS AND DESIGN SAC FOR AMB-FLYWHEEL

The operating environment of the automotive is complex. The road conditions impacts the AMB-flywheel when vehicle is driving, such as different vibration, acceleration, deceleration, steering and so on. Therefore, it is necessary to design a strong stiffness active magnetic bearing controller for suitable the vehicle dynamic specific. There are two key techniques for development of AMB-flywheel powered Electric Vehicle (EV):
1. AMB stable suspension control under driving disturbances.
2. Reducing the energy consumption.

An electrical vehicle with a flywheel energy storage system using a five-degree-of-freedom magnetic bearing has been carrying out as shown in Fig 1. Simple adaptive control is a simple and robustness algortithm, which could be apply to AMB flywheel system[4], [5]. According to the work condition of controlled object AMB-flywheel powered EV, We apply a PID based simple adaptive control scheme designs combined with zero-bias control rule for the AMB-flywheel energy storage system.

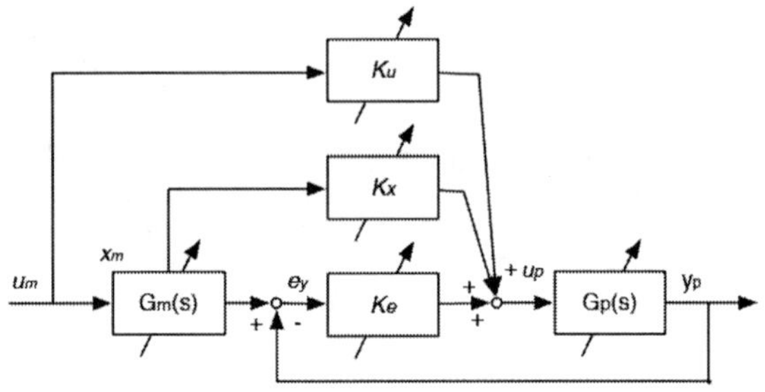

Figure 3. Block diagram of the SAC scheme

The SAC algorithm structure and overview of control object AMB-flywheel are shown as Fig.3. The system was simulated, and the final result shows the control method is feasible and effective. SAC is a simple yet robust algorithm for unmodeled dynamics [6]. In this algorithm, a plant model of order n_p is described as

$$\begin{cases} \dot{x}_p = A_p x_p(t) + B_p u(t) \\ y_p(t) = C_p x_p(t) \end{cases} \tag{3}$$

where $x_p(t)$ is the n_p th-order plant state vector, $u(t) \in R^{nj\times 1}$ is the system input vector,$y_p(t) \in R^{nj\times 1}$is the system output vector, and observable but unknown parameter plant model of order. The plant output $y_p(t)$ is required to asymptotically track the output of the following model, y_m.

$$\begin{cases} \dot{x}_m = A_m x_m(t) + B_m u_m(t) \\ y_m(t) = C_m x_m(t) \end{cases} \tag{4}$$

If we let tracking error$e_y(t) = y_m(t) - y_p(t)$, the SAC controller can be defined as

$$k(t) = k_I(t) + k_p(t) \tag{5}$$

$$\dot{k}_I(t) = \Gamma_I z(t) e_y(t) - \delta(t) k_I(t) \tag{6}$$

$$k_p(t) = \Gamma_p z(t) e_y(t) \tag{7}$$

$$\delta(t) = \frac{\delta_1 e_y(t)^2}{\delta_3 + e_y(t)^2} + \delta_2 \tag{8}$$

where δ_1, δ_2 and δ_3 are small positive constants and Γ_I and Γ_p are constant matrices. $\Gamma_I = \Gamma_I^T > 0$ and $\Gamma_p = \Gamma_p^T > 0$.

We need to tune parameters of Γ_I and Γ_p for SAC controller by ourselves in the simulation and the experiment for AMB-flywheel system.

In order to prove the stability of the closed-loop system, the controlled plant is required to be SPR or ASPR. The ASPR conditions are as follows:

A. The system plant is strictly proper or proper

B. $G_p(s)$ is the minimum-phase

C. $G_p(s)$ has the minimal realization A, B, C, where $CB > 0$.

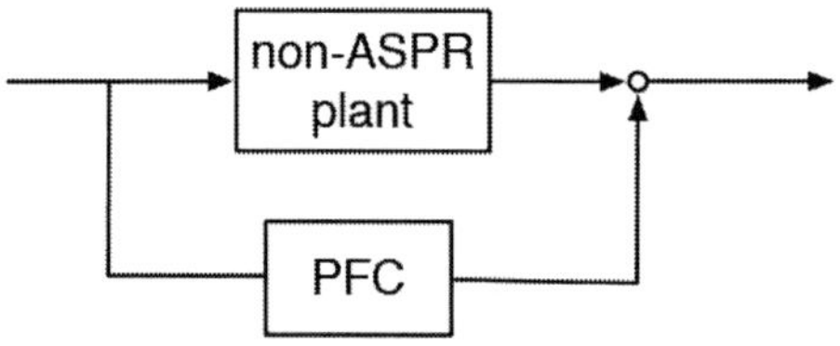

Figure 4. Parallel feedforward compensator for non-ASPR plant

However, since most practical systems do not satisfy the ASPR condition, the ASPR conditions are the fundamental restrictions for practical applications of the output-feedback-based adaptive control.A PFC should be developed to make the augmented system a minimum-phase system and ensure that it has a relative degree one.

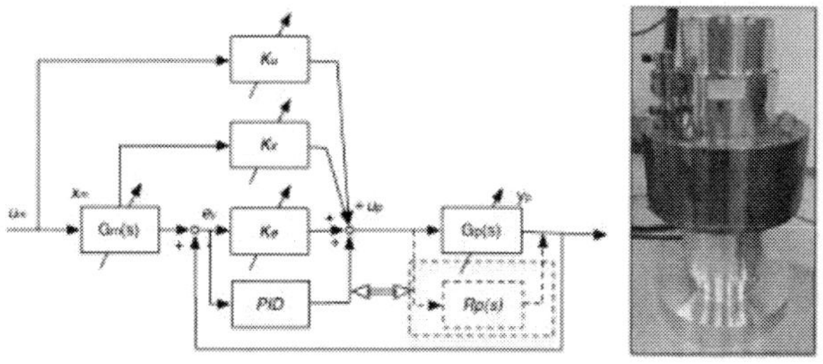

Figure 5. Block diagram of SAC-PID double controller

In this case, the design of a PFC for AMB-flywheel system consists of two steps. We first design the PID Controller that stabilizes the model of the AMB flywheel at the center position. Then inverse transfer function of the stabilizing PID controller is used for the PFC. The PFC renders the augmented system to be almost strictly positive real (ASPR)[7]. We have to design the PFC such that it renders the augmented system to be ASPR. The first step designs a PID controller for without SAC controller. This PID controller can stabilize the flywheel rotor at the operating point. Inverse transfer function of the obtained PID controller is used for the PFC in Fig 2. The air gap is very small between the rotor of flywheel and touchdown bearing, so AMB controller require high response speed and small overshoot. In SAC, auto-adjustment of the gain parameters is originally based on the PI adaptive. Here, for a quicker response, a PID rule is added. This PID-SAC combined with PID and SAC controller, which has advantages of quicker response and stronger adaptability and robustness characteristics for AMB-flywheel system. SAC with PID rule is shown in Fig 5, and the reference model is set as zero. In this double controller system, PID is defined as the PFC for AMB-flywheel plant. Hence, the SAC controller can be defined as achieve $\lim_{t\to\infty} e_y(t)$. Hence, the controller can be defined as

$$u_p(t) = k(t)z(t) \tag{9}$$

$$z(t) = e_y(t) \tag{10}$$

$$k(t) = k_I(t) + k_p(t) + k_{PID} \tag{11}$$

$$k_p(t) = \Gamma_p z(t) e_y(t) \tag{12}$$

$$\dot{k}_I(t) = \Gamma_I z(t) e_y - \sigma(t) k_I(t) \tag{13}$$

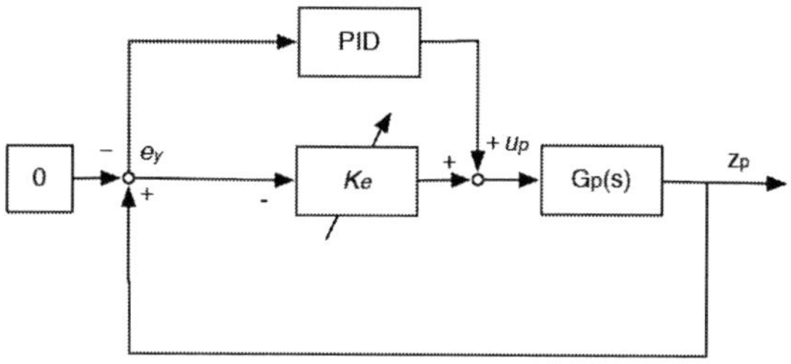

Figure 6. Simplified block diagram of SAC with PID

IV. Zero-Bias Current Magnetic Bearing

Since the AMB-flywheel system is used as a storage device for electrical energy, it is desirable to minimize the energy required for its stabilization control. Zero-bias current magnetic bearings have the potential to reduce power losses because only a single electromagnet of the pair is operational at any time. The switching rule of zero-bias control is defined as

$$f = f_1 + f_2 \tag{14}$$

$$f = \begin{cases} f_1 \quad (f_2 = 0), & if x \geq 0 \\ f_2 \quad (f_1 = 0), & other \end{cases} \tag{15}$$

The SAC control input $U = (f_{xu}, f_{xl}, f_{yu}, f_{yl})$ is provided in the form of an attractive force between the electromagnets, but it should be converted into electric current.The zero-bias modes of other electromagnet pairs follow the same form.

V. Experimental Results

Experiments were performed on the PID based SAC with the zero-bias modes. The orbit of 120Hz CFRP rotor flywheel rotation is shown in Fig.7. The displacement of charging and discharging experiment is shown as Fig.8 from 0Hz to 100Hz and 100Hz to 0Hz.

The control currents in the direction of the upper electromagnets during the rotation experiment are shown in Fig.5. As seen in the Fig.9, the maximum current is less than 1A, when the flywheel rotated from 0HZ to 100Hz.

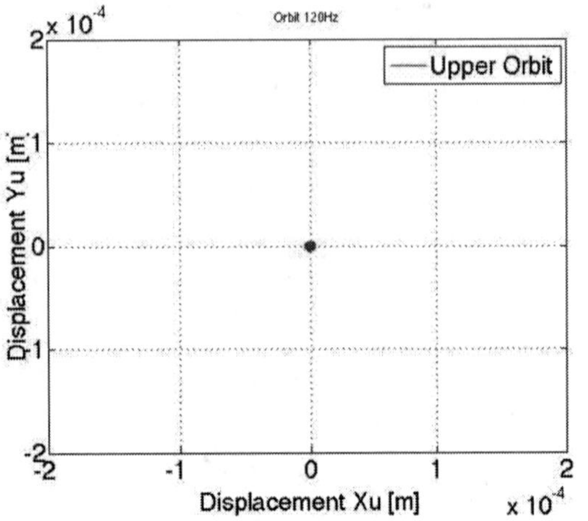

Figure 7. The orbit of 120Hz flywheel rotation

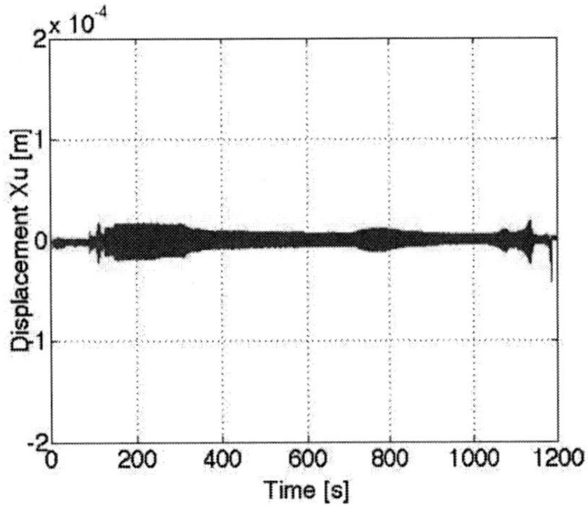

Figure 8. Displacement Xu from 0Hz-100Hz-0Hz rotations

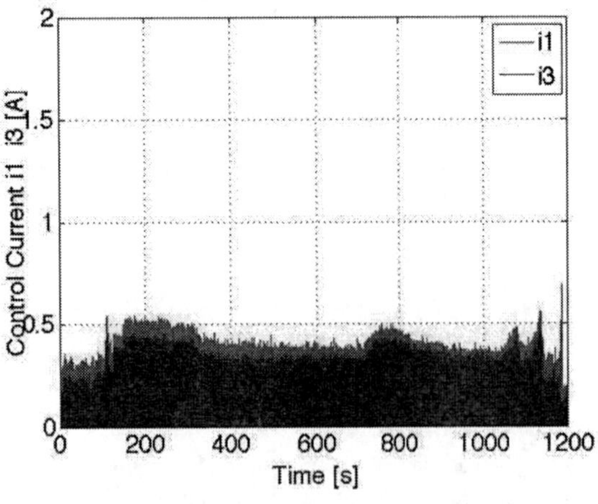

Figure 9. The orbit of 120Hz flywheel rotation

VI. Conclusion

This paper is concerned with the design of a simple adaptive control (SAC) technique and application into AMB-flywheel system, and effort provided an analysis of SAC with a focus on operating with minimal power, and showed that it worked. The results of the analysis of the simple adaptive zero bias control data has been presented. The analytical results were based on nonlinear model, and the experimental results indicate how a magnetic bearing operates using the SAC method. PFC cause the offset problem of levitating of flywheel rotor, it effects the rotation of flywheel rotor on radial direction offset. Then, we modified the basic SAC scheme with PID based SAC. The control performance was improved by combining SAC with PID controller. In the proposed approach, SAC was successfully applied to vertically design five-axis controlled AMB-flywheel systems. The experimental results were found to satisfy the requirements of the AMB-flywheel system. Low-loss operation is very important in AMB system applications, especially in high speed energy flywheel energy storage system. Since eddy current losses are propositional to the flux, zero-bias control scheme for AMB are advantageous for low-loss operation of AMBs. The zero-bias mode did not deteriorate the stability of the control system; further, energy consumption in this mode was low.

The AMB system that means low power and low loss without bias currents has been proposed using SAC design of nonlinear control theory with zero-bias current control for a AMB system. In this study, SAC controller was successfully applied to a vertical designed five-axis AMB-flywheel system. The high-speed rotation experimental results have presented in this shaft paper. Using PID based SAC controller can suspended the flywheel rotor rapidly and stably.

References

[1] S. Sivrioglu, M. Saigo, and K. Nonami, "Adaptive backstepping control design for a flywheel zero-bias amb system," *IEEE Control Applications*, vol. 2, no. 3, pp. 1106–1111, 2003.

[2] S.Sivrioglu, K.Nonami, and M.Saigo, "Low power consumption nonlinear control with h compensator for a zero-bias flywheel amb system10: 1151." *Journal of Vibration and Control*, vol. 10, no. 1151, 2004.

[3] A. Rashid and B. Husain, "Multi objective sliding mode control of active magnetic bearing system," *Faculty of Electrical Engineering University Technology Malaysia*, 2009.

[4] B. Rachmanto and K. Nonami, "A study on amb-flywheel powered electric vehicle," *Journal of System Design and Dynamics*, pp. 659–670, 2009.

[5] S. Shah, Z. Iwai, and I. Mizumoto, "Simple adaptive control of processes with time-delay," *J. Pro.Cont*, vol. 7, no. 6, pp. 439–449, 1997.

[6] H. Kaufman, I. Barkana, and K. Sobel, "Direct adaptive control algorithms theory and applications-2 ed," *Springer Verlag New York*, 1997.

[7] I. Mizumoto, D. Ikeda, T. Hirahata, and Z. iwai., "Design of discrete time adaptive pid control systems with parallel feedforward compensator," *control engineering practice*, 2010.

Benchmark and Verification of Control Algorithm for Flywheel with Active Magnetic Bearing on Electric Vehicle and Proposal of New SAC Algorithm (ε_1 modification and bias variable Γ_p approach)

Fumiya SHIMIZU[a], Kenzo NONAMI[b]

[a] Chiba University Graduate School of Engineering, 1-33, Yayoi-cho, Inage-ku, Chiba-shi, Chiba, Japan, z9t0632@students.chiba-u.jp

Abstract— **This paper presents the requirements which a good rotor controller of flywheel energy storage system (FESS) with Active Magnetic Bearing (AMB) on Electric Vehicle (EV) should satisfy. AMB-FESS is used for energy regeneration. The dynamics of the flywheel rotor on EV is very complex and to compare various features a benchmark is necessary. Controllers are evaluated according to certain criteria as used by many motor companies. Controllers are selected to satisfy the design requirements prescribed by many motor companies. Thus the robust adaptive controllers were chosen and benchmarked using simulations and experiments. The investigation revealed that a controller with high rigidity gain and a reference model can not only suppress disturbance effects but also lower power consumption. Therefore Simple Adaptive Control with epsilon1 modification and bias variable gamma P approach is proposed. The utility of the new controller experimentally was proven.**

Keywords— Flywheel Energy Storage System, Active Magnetic Bearing, Electric Vehicle, epsilon1 modification approach, bias variable gamma P approach

I. INTRODUCTION

Recently, it is becoming popular applying flywheel to vehicles for energy regeneration. Otaki and Kosuda had revealed that gyroscopic precession of flywheel does not affect the drivability and stability of the vehicle in theory and experiments. [1] Saito showed that Flywheel Energy Storage Systems (FESS) are superior to capacitor or Li-ion batteries when used for the regeneration systems and power sources. [2] Yokota reported that hybrid system which consists of hydraulic drive flywheel and gasoline engine is effective for decreasing fuel consumption. [3] Moreover, Jens has investigated the control method for flywheel and CVT. He revealed that the system can increase drivability and decrease power consumption. [4] In industry, flywheels are widely used. TOROTRAK, RICARDO, Optare and Allison Transmission developed flywheel bus. [5]

However, the research used mechanical bearing flywheels. Because of the friction heat, energy efficiency is low. Therefore we propose Active Magnetic Bearing (AMB) - FESS. Recent research has been focused on AMB-FESS mounted on vehicles. The University of Texas, Center for Electro Mechanics (UT-CEM) and Department of Defense has been developing an AMB-FESS combat vehicle. [6] However this system uses permanent magnets and needs big control inputs. Moreover, UT-CEM had developed a flywheel bus. [7], [8], [9], [10], [11] In this research, flywheel design, control design of magnetic bearing, design of gimbals for sustaining flywheel and its efficiency are described.

Though resent research or applications of AMB-FESS has been focused on large vehicles, very few research focused on AMB-FESS for consumer vehicles. It is more difficult to reduce power consumption, miniaturize and make flywheel robust. This system requires a high positioning accuracy. Therefore, it is necessary to compare and develop a good control algorithm for AMB-FESS on Electric Vehicle (EV). The Uppsala University has been focused on the control algorithm for the inverter/converter, analysis of the magnetic bearing, and design of the driveline [12], [13], [14], [15]. However, recent research hasn't been focused on the control algorithm for AMB flywheel rotor on EV.

The objective of this research is to compare several control methods and grasp the features which control algorithm is good for AMB-FESS on EV. We compare several methods by evaluating criteria as used in many motor companies (-difficulty of design, power consumption, and external disturbance suppression) and propose the best controller for AMB-FESS on EV. In this investigation, we focused on robust adaptive control algorithms and its features. The algorithms can tune gains and change controllers adaptively and continuously so designers can design controllers easily and improve their quality. Moreover, these methods are robust. Therefore they can compensate for variable external disturbance. Furthermore, they can lower power consumption because of adaptive gain tuning.

In this investigation, several robust adaptive control algorithms were benchmarked using simulations and experiments. Generalized Minimum Variance Control (GMVC) (One of the Self Tuning Regulators), Model Reference Adaptive Control (MRAC), and Simple Adaptive Control (SAC) are simulated and experimentally evaluated. These controllers use sigma-modification approach or epsilon 1 modification algorithm and/or bias variable gamma P approach. These algorithms were compared with Linear Quadratic Gaussian (LQG) algorithm and PID algorithm which are widely used in industry.

Thorough this research, we found the features which a good performance controller for AMB-FESS on EV should have. The feature is that the controller with high rigidity and a reference model, can suppress disturbance effects, lower power consumption and improve positioning accuracy. Therefore we improved the SAC algorithm and named it SAC with epsilon 1 modification and bias variable gamma P approach.

In this paper, we also examine the performance of a new SAC algorithm by simulation and experiment. We also showed that the new algorithm is the best robust adaptive controller for AMB-FESS on EV.

II. CONTROL SYSTEM DESIGN

A. Equations of Motion

To apply robust adaptive control algorithms, AMB-FESS on EV must be described by equations of motion. Our system satisfies the following 4 assumptions.

- Vehicle cannot move in yaw direction
- Flywheel rotor rotates at constant speed
- Flywheel is fixed in the center of gravity of the vehicle
- Flywheel rotor and vehicle move or incline slightly

Based on these assumptions, AMB-FESS on EV is described by the equations of motion. Coordinate system of AMB-FESS on EV is set as shown in Fig. II. In Fig. II (a), x_u, y_u, x_l and y_l are axes of upper and lower radial direction of flywheel. I_1 to I_8, I_u and I_l are the input currents to each electric magnet. G is the position of center of gravity. X and Y are axes of radial direction of EV. θ_x and θ_y are tilting angles of flywheel. L_u and L_l are the distances between center of gravity and x_u-y_u/x_l-y_l plane. L is the distance between x_u-y_u plane and x_l-y_l plane. ω is rotation speed. In Fig. II (b), x-y-z and X-Y-Z are the coordinate axes of flywheel/ vehicle. Origin is located of the center of gravity. δ is the mass eccentricity, and l_c is the distance between the origin of the coordinate system and the geometric origin. Generally, l_c is unknown. Θ_X and Θ_Y are tilting angles of EV. K is the spring constant of EV and the anti-vibration rubber. Details of the other parameters are as described in Table. II.

Derivation of the equations of motion [16] is very complex, and its summary is described below. Models are described in moving coordinate system. Considering roll-pitch motion, our models derive from law of conservation of momentum and law of conservation of angular momentum. These models are shown as equation (1) and (2). Equation (1) is the model of the EV while equation (2) is the model of the flywheel. This model was calculated in respect to the center of gravity. The model can be transformed to sensor coordinates (X_u, Y_u, X_l, and Y_l) by using simple transforming matrixes.

(a) Flywheel (b) Fly-car

Figure I. Overview of flywheel and Overview of fly-car

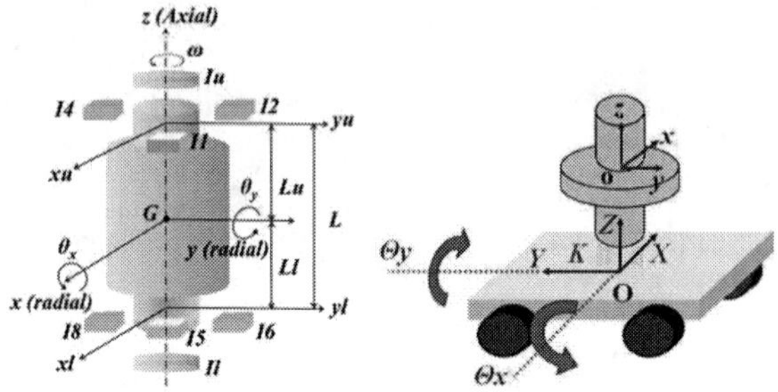

(a) Coordinate system of flywheel (b) Coordinate system of vehicle

Figure II. Coordinate system

TABLE I. SPECIFICATION OF FLYWHEEL SYSTEM

Parameter	Value
Rotor mass	100 Kg
Flywheel diameter	0.4 m
Flywheel thickness	0.04 m
Polar moment of inertia	0.877 Kg · m^2
Tilting moment of inertia	2.438 Kg · m^2
Constant of radial AMB	25.196 N · mm^2/A^2
Constant of axial AMB	70.568 N · mm^2/A^2
Upper distance from center of gravity	0.186 m
Lower distance from center of gravity	0.304 m
Radial nominal air gap	0.5 mm
Axial upper nominal air gap	0.33 mm
Axial lower nominal air gap	0.6 mm
Radial nominal touch-down gap	0.2 mm
Axial nominal touch-down gap	0.4 mm
Allowable current	6 A

$$\begin{cases} I_{VX}\ddot{\Theta}_X = -N_x + N_X \\ I_{VY}\ddot{\Theta}_Y = -N_y + N_Y \\ M\left(\ddot{X} + \ddot{Y}\Theta_X - \ddot{Z}\Theta_Y\right) = -(F_x + mg\theta_Y) + \\ \qquad \{F_X - (F_Z - \mathrm{Mg})\Theta_Y + K_X X - K_Z Z\Theta_Y\} \\ M\left(\ddot{Y} + \ddot{Z}\Theta_X\right) = -\left(F_y - mg\theta_X\right) + \\ \qquad \{F_Y + (F_Z - \mathrm{Mg})\Theta_X + K_Y Y - K_Z Z\Theta_X\} \\ M\left(\ddot{Z} + \ddot{X}\Theta_Y - \ddot{Y}\Theta_X\right) = -(F_z - mg) + \{F_X\Theta_Y \\ \qquad -F_Y\Theta_X + (F_Z - \mathrm{Mg}) + K_X X\Theta_X + K_Z Z\} \end{cases} \quad (1)$$

$$\begin{cases} I_r(\ddot{\theta}_X - \omega\dot{\theta}_Y - \omega\dot{\theta}_y + \ddot{\theta}_x) + I_z\omega(\dot{\theta}_Y + \dot{\theta}_y) \\ \qquad = N_x + ml_c\delta\omega^2 \sin\omega t \\ I_r(\ddot{\theta}_Y + \omega\dot{\theta}_X + \omega\dot{\theta}_x + \ddot{\theta}_y) - I_z\omega(\dot{\theta}_X + \dot{\theta}_x) \\ \qquad = N_y + ml_c\delta\omega^2 \cos\omega t \\ m(\ddot{x} + \ddot{X}) = F_x - mg\theta_Y - K_X X + K_Z Z\theta_Y \\ \qquad +m\delta\omega^2 \cos\omega t \\ m(\ddot{y} + \ddot{Y}) = F_y + mg\theta_X - K_Y Y - K_Z Z\theta_X \\ \qquad +m\delta\omega^2 \sin\omega t \\ m(\ddot{z} + \ddot{Z}) = F_z - mg - K_X X\theta_X - K_Z Z \end{cases} \quad (2)$$

TABLE II. PARAMETERS IN THE EQUATION OF MOTION

Parameter	vehicle	Flywheel
Mass	M	m
Moment of inertia	I_X, I_Y, I_Z	I_r, I_p
Center of gravity	X, Y, Z	x, y, z
Tiliting angle	θ_X, θ_Y	θ_x, θ_y
Disturbance force	F_X, F_Y, F_Z	
Disturbance moment	N_X, N_Y	
Control force		F_x, F_y, F_z
Control moment		N_x, N_y
Spring constant	K_X, K_Y, K_Z	

B. Control algorithms

Adaptive control method can be designed by using implicit or explicit methods. Considering difficulties of design, this research adopts implicit methods.

[Model Reference Adaptive Control System (MRACS)]

MRACS is one of model following control systems. MRACS defines control input as denoted by equation (3) [17].

$$u(t) = \theta(t)^T z(t) \qquad (3)$$

Here, $z(t)$ is the data vector. This term consists of output of the reference model, output of the plant and input to the plant. $\theta(t)$ is the state value vector. This term is defined as follows.

$$\dot{\theta}(t) = -\Gamma(t)z(t)\varepsilon_1(t) \qquad (4)$$

Here, $\Gamma(\Gamma = \Gamma^T > 0)$ is the weight matrix and $\varepsilon_1(t)$ is the augmented error.

[Simple Adaptive Control System (SACS)]

SACS is a simplified MRACS. SACS makes control input as in equation (5).

$$u(t) = K(t)z(t) \qquad (5)$$

Here, $z(t)$ is same as in equation (3). $K(t)$ is the adaptive gain vector. This term is defined in equations (6) to (8). [18], [19]

$$K(t) = K_P(t) + K_I(t) \qquad (6)$$

$$K_P(t) = -\Gamma_P(t)z(t)e(t) \qquad (7)$$

$$\dot{K}_I(t) = -\Gamma_I(t)z(t)e(t) - \sigma_I K_I(t) \qquad (8)$$

Here, weight matrix Γ_P, Γ_I and σ_I have the following properties - $\Gamma_P = \Gamma_P^T > 0, \Gamma_I = \Gamma_I^T > 0, \sigma_I > 0$. The term $e(t)$ is the residual tracking error. Equations (6) to (8) are proportion + integral adjustment rule. σ_I is sigma-modification approach. Details are described below.

[Generalized Minimum Variance Control System (GMVCS)]

GMVCS is the well known self tuning control system. GMVCS uses an evaluation function. The plant is described as follows.

$$A(z^{-1})y(t) = z^{-1}B(z^{-1})u(t) + C(z^{-1})w(t) \qquad (9)$$

Here, $y(t)$ is the output of the plant, $u(t)$ is the input to the plant and $w(t)$ is white noise. A, B and C are some polynomials. Evaluation function is described in equation (10).

$$J = \bar{E}[\{P(z^{-1})y(t+\tau) + Q(z^{-1})u(t) - R(z^{-1})r(t)\}^2] \quad (10)$$

Here, $\bar{E}[\,]$ means expectation (spatial averaging). τ is a constant value. P, Q and R are polynomials and these terms satisfy equation (11) - Diophantine equation.

$$P(z^{-1})C(z^{-1}) = A(z^{-1})E(z^{-1}) + z^{-1}F(z^{-1}) \quad (11)$$

Here, E and F are polynomials. Control input: $u(t)$ is defined equation (12).

$$u(t) = {[C(z^{-1})R(z^{-1})r(t) - F(z^{-1})y(t) + G(z^{-1})u(t)]}\Big/{[b_0 + q_0]} \quad (12)$$

Here, b_0 and q_0 is the first term of B or Q, respectively, and $G(z^{-1})$ satisfy equation (13).

$$G(z^{-1}) = E(z^{-1})B(z^{-1}) + C(z^{-1})Q(z^{-1}) - (b_0 + q_0) \quad (13)$$

[Sigma-Modification Approach]

Sigma-Modification Approach is often used in SACS algorithms. This approach increases robustness of adaptive control [20], [21], [22]. However, this method can't decrease residual tracking error. Our group defines σ as follows.

$$\sigma(t) = \sigma_1 + {\sigma_2 e(t)^2}\Big/{[\sigma_3 + e(t)^2]} \qquad (14)$$

Here, σ_1 to σ_3 are tuning parameters.

[Epsilon1 Modification Approach]

The epsilon1 Modification Approach uses equation (14) instead of equation (15). [23] Using this approach the reference value can be reached.

$$\sigma(t) = \gamma_I \|\varepsilon_1(t)\| \qquad (15)$$

Here, $\gamma(\gamma_I = \gamma_I^T > 0)$ is the weight matrix. This approach and sigma-modification approach can be applied to MRACS, SACS and GMVCS.

[Bias Variable gamma P Approach]

If SACS with epsilon1 modification approach is used, residual tracking error goes to zero, but at the same time, it loses its rigidity. From equation (7), it is obvious that rigidity gain goes to zero when residual tracking error goes to zero. If AMB-FESS doesn't have rigidity, it cannot suppress external disturbance. Therefore, Bias Variable gamma P approach is proposed to avoid this problem.

In this research, Γ_P (in equation (7)) is defined as

$$\Gamma_P = c_1 + \frac{c_2}{\delta + z(t)e(t)} \tag{16}$$

This is the bias variable gamma P method. c_1 and c_2 are constant values. δ is a very small constant value. δ is introduced to prevent division by zero and smoothing of the gain changing. Using equation (16), if residual tracking error goes to zero, the system keeps its rigidity. The method described above revealed to be very useful for the system which needs high rigidity gains.

III. Benchmark

In this investigation, at first, several control algorithms designed for axial direction are benchmarked in simulations. The model of axial direction is very simple. Moreover, the axial direction controller and the radial direction controllers should have same structures to make the design process easier.

Therefore, we simulated and compared controllers using the axial direction model. We selected controllers that gave good simulation results and then investigated their efficiency experimentally.

We compared control algorithms based on 3 criteria common in industry – external disturbance suppression, power consumption, and difficulty of design [24].

External disturbance suppression is evaluated by 2 norm the displacement and infinity norm the displacement. The 2-norm of displacement evaluates controller's robustness against external impulse disturbances. Infinity norm the displacement evaluates a controller's robustness against persistent external disturbances. Power consumption is evaluated by the power consumption evaluation function denoted as follows.

$$J_p \overset{\text{def}}{=} \frac{1}{\tau}\int_0^{\tau} I(t)^2 \, d\tau \tag{17}$$

"I" is the control currents and τ is driving time. Because AMB has magnetic flux leakage, it is impossible to know the real resistance value. However, currents are measured by sensors. Therefore the criterion is introduced. Difficulty of design is evaluated by the number of tuning parameters.

A. Simiation

Simulation model considers white noise and disturbances measured in a driving experiment. The maximum disturbance in axial direction is almost 1[G]. The MRACS's reference model is a 2nd order system. The damping ratio is 0.8 and the setting time is 0.2 second. The SACS'S reference model is the 1st order system and the other terms are the same as MRACS. Moreover, all algorithms use the bias control method. Results are shown in Table. III.

Accordingly to Table. III, MRACS with epsilon1 modification and bias variable gamma P approach is the best controller in respect to 2norm, infinity norm, and power consumption.

TABLE III. Controller Evaluation

Controller	2norm	∞ norm	P.C.E.V.	N. T.P.
MRACS + ε + Γ	7.8×10^{-5}	4.4×10^{-6}	0.89	22
SACS + σ	9.3×10^{-4}	2.7×10^{-5}	1.83	8
SACS + ε + Γ	3.7×10^{-4}	1.4×10^{-5}	1.02	8
GMVCS	Failed to Stabilize			10
LQG	7.1×10^{-3}	3.2×10^{-4}	2.07	8
PID	1.3×10^{-3}	2.2×10^{-5}	1.85	3

*ε+Γ: epsilon1 modification and bias variable gamma P approach
*σ: sigma modification approach
*P.C.E.V.: Power Consumption Evaluation Value
*N.T.P.: the Number of Tuning Parameters

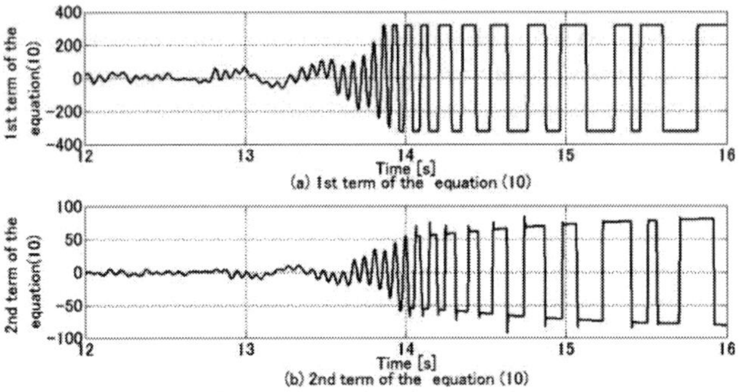

(a) 1st term of the equation (10)
(b) 2nd term of the equation (10)
Figure III. Each term value of equation (10)

However, the controller is not good in terms of difficulty of design. If MRACS is mounted on radial direction controllers, the designer has to tune (at least) 88 parameters. As a whole, SACS with epsilon1 modification and bias variable gamma P approach is a good performance controller.

Additionally, SACS with epsilon1 modification and bias variable gamma P approach is better than SACS with sigma modification approach. Sigma modification cannot decrease a residual tracking error. Therefore, the controller needs much power consumption. Moreover, the norm values are higher than SACS with epsilon1 modification and bias variable gamma P approach

GMVCS cannot suppress disturbances. Fig. III shows each term value in the equation (10). Because reference value is zero, the third term of the equation (10) is omitted. According to the equation (10) and/or Fig. III, the controller provides appropriate input after flywheel moves extremely. Therefore, GMVCS cannot suppress disturbances. Moreover, the PID controller is superior to the LQG controller for the same as GMVCS.

On the other hand, MARCS and SACS controllers can suppress impulse/persistent external disturbance compared with the other controllers. These controllers have the reference models. Controllers force the flywheel to follow the reference model if flywheel moves extremely. Therefore these controllers are robust against large disturbances.

Furthermore, the performances of SACS with sigma modification approach are similar to PID controllers. This result proves that the performance of epsilon1 modification and bias variable gamma P approach is superior to sigma modification approach.

TABLE IV. CONTROLLER EVALUATION

Controller	2norm/ M.D. [G]	∞ norm/ M.D. [G]	P.C.E.V. / A.D. [G]	N.T.P.
SACS $+\varepsilon+\Gamma$	5.1×10^{-4}	2.6×10^{-5}	0.71	8
LQG	6.4×10^{-3}	2.6×10^{-4}	2.23	8
PID	9.1×10^{-4}	1.3×10^{-4}	1.18	3

*M.D.: maximum disturbance
*A.D.: average disturbance
*$\varepsilon+\Gamma$: epsilon1 modification and variable gamma P approach
*P.C.E.V.: Power Consumption Evaluation Value
*N.T.P.: the Number of Tuning Parameters

In this investigation, we also simulated the adaptive controllers without the robust structures or zero bias controllers. However these controllers couldn't suppress disturbances and white noise. We conclude that adaptive controllers need robust structures.

As a matter of course, these results depend on gain tuning. However it can be said with high possibility that the number of tuning parameters and the features which good performance controllers have doesn't change.

B. Experiments

According to simulation results, SACS with epsilon1 modification and bias variable gamma P approach is a good performance controller for AMB-FESS on EV. To verify its performance, we compared it with LQG controller and PID controller by a driving experiment.

These controllers were compared while the flywheel rotor rotated at 100 Hz. The model of AMB-FESS (equation (1) and (2)) doesn't have coupled terms between axial and radial direction. However, the real plant has coupled affects. Therefore, performance of axial direction controllers must be investigated while the flywheel rotor rotates.

Experiment results are shown in Table.IV and Fig. IV to Fig. VII. According to Table.IV, 2 norm and infinity norm of the displacement are divided by the maximum disturbance. Moreover, the power consumption evaluation value is divided by the average disturbance.

The disturbance values are different from each experiment. Therefore, these criteria must be evaluated by the same condition.

According to Table. IV, the features of the controller's performances are similar to Table III. SACS with epsilon1 modification and bias variable gamma P approach is good performance controller for AMB-FESS on EV in respects to external disturbance suppression and power consumption.

Fig.IV to Fig.VI show the each experiment data. According to these data, SACS with epsilon1 modification and bias variable gamma P approach received the largest disturbance (almost 2.5[G]). This controller can suppress external disturbance, in contrast to the LQG and PID controllers which cannot. The flywheel was almost touched-down.

Fig. VII shows the disturbance and gains of SACS with epsilon1 modification and bias variable gamma P approach

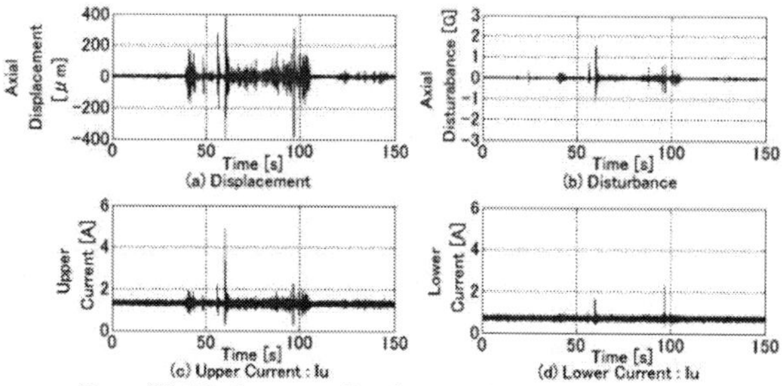

Figure IV. Displacement, disturbance and input current (LQG)

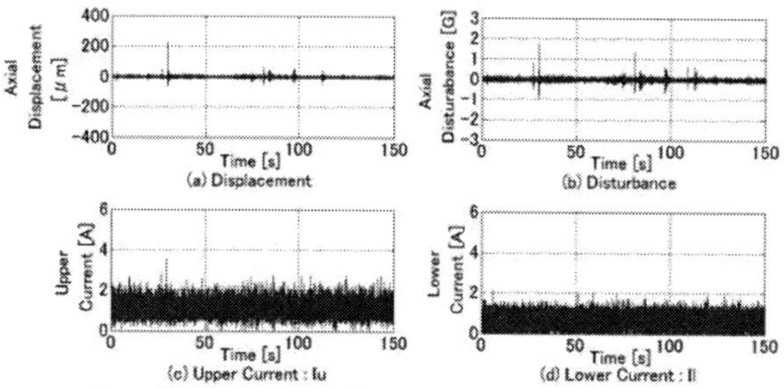

Figure V. Displacement, disturbance and input current (PID)

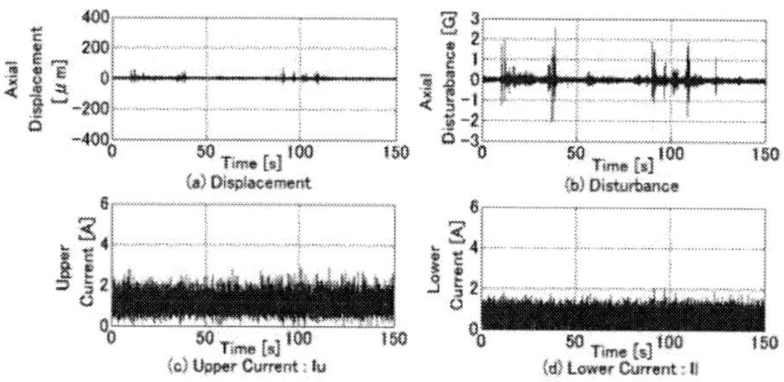

Figure IV. Displacement, disturbance and input current (SACS)

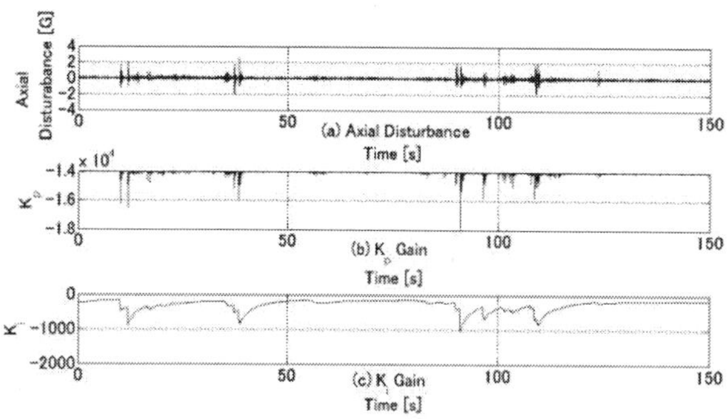

Figure VII. Disturbance and Kp, Ki gain (SACS)

during experiment. When the flywheel rotor vibrates on a large scale, K_p gain increases instantly. A few sampling steps after that, K_i gain increases to follow the reference value. Therefore the controller can suppress external disturbance well.

IV. CONCLUSIONS

In this investigation, several robust adaptive controllers for AMB-FESS on EV were benchmarked. We compared the controllers by 3 criteria – external disturbance suppression, power consumption, and difficulty of design. External disturbance suppression is evaluated by 2 norm and infinity norm of the displacement. Power consumption is evaluated by input currents. Difficulty of design is evaluated by the number of tuning parameters.

Through this investigation, it is clarified that good performance controllers for AMB-FESS on EV have following 7 features.

Good performance controllers ...

- don't evaluate control inputs
- tune gains adaptively
- have a robust structures
- have a reference model.
- have the structure which decrease a residual tracking error
- doesn't have zero rigidity gain at the same time reference values are reached
- have appropriate bias rigidity gains

As a matter of course, the results depend on gain tuning. However, it can be said with high probability that the features which good performance controllers have don't change.

Moreover, in this investigation, we proposed Simple Adaptive Control System with epsilon1 modification and bias variable gamma P approach. This controller is easy to design and shows good performance.

Hereafter, Simple Adaptive Control System with epsilon1 modification and bias variable gamma P approach will be mounted to radial direction's controllers. Moreover, the features which are found in this investigation are evaluated quantitatively. Furthermore, our group will make energy regeneration structure and investigate the utility of AMB-FESS EV.

REFERENCES

[1] H. Otaki, M.Kosuda, 'Some Analysys and Development of Flywheel Powered Vehicle', *JSME Transactions*, Series C, Vol 46, No. 411, pp.1375, 1980

[2] T.Saito, H.Ogasawara, N. Yamada, 'Study of Flywheel Energy Storage System and Application to Electric Vehicle', *JSME Transactions*, Series B, Vol 70, No. 697, pp.2482-2489, 2004

[3] S.Yokota, 'A Flywheel Hybrid Vehicle Making Use of Constant Pressure System : Fabrication of Stationary Test Facility and Experiment of Urban Driving Schedule', *JSME Transactions*, Series C, Vol 68, No. 671, pp.2127-2132, 2002

[4] Jens Schlurmann and Dierk Schroder, 'Compensation of Dynamic Torques and Flywheel Start in a CVT Based Hybrid Powertrain', *Proceeding of the 2006 IEEE International Conference on Control Application*, FrA06.4

[5] TOROTRAK plc. (online), available from < http://www.torotrak.com/>, (accessed on 10th June, 2014)

[6] M.A.Pichot, J.P.Kajs, B.R.Murphy, A.Ouroua, B.M.Rech, R.J.Hayes, J.H.Beno, G.D.Buckner, and A.B.Palazzolo, "Active Magnetic Bearing for Energy Storage System for Combat Vehicles", *IEEE Transactions on magnetics*, Vol 37, No.1, pp.318- 323,2001

[7] B.T.Murphy, D.A.Bresie, H.H.Beno, "BEARING LOADS IN A VEHICULAR FLYWHEEL BATTERY", *SAE Technical Paper* 970213, 1997

[8] Lawrence A.Hawkins, Brian T.Murphy, "APPLICATION OF PERMANENT MAGNET BIAS MAGNETIC BEARINGS TO AN ENERGY FLYWHEEL", *Fifth Symposium on Magnetic Suspension Technology* 1999

[9] Lawrence A.Hawkins, Brian T.Murphy, John Kajs, "ANALYSYS AND TESTING OF A MAGNETIC BEARING ENERGY STORAGE FLYWHEEL WITH GAIN-SCHEDULED MIMO CONTROL", *Proceedings of ASME TURBOEXPO 2000, Munich, Germany*, ASME 2000-GT-405, 2000

[10] Lawrence HAWKINS, Brian MUROHY, Joseph ZIERER, Richard HAYES, "Shock and Vibration Testing of an AMB Supported Energy Storage Flywheel", *JSME international journal* Series C, Vol.46, No.2 pp.429-435, 2003

[11] R.J.Hayes, J.P.Kajs, R.C.Thompson, "J.H.Beno, Design and Testing of a Flywheel Battery for a Transit Bus", *SAE Technical Paper* 1990-01-1159, 1999

[12] Johan Abrahamsson, Magnus Hedlund, Tobias Kamf, Hans Bernhoff, "High-Speed Kinetic Energy Buffer: Optimization of Composite Shell and Magnetic Bearings", *IEEE Transaction*, Vol.61, No.6 pp.3012-3021, 2014

[13] Johan Abrahamsson, Janaina Goncalves de Oliveira, Juan de Santiago, Johan Lundin, Hans Bernhoff, "On the Effeciency of a Two-Power-Level Flywheel-Based All-Electric Driveline", *Energies*, ISSN 1996-1073, pp.2794-2817, 2012

[14] Janaina G. Oliveira, Hans Bernhoff, "Power electronics and control of two-voltage-level flywheel based all-electric driveline", *ISIE IEEE international symposium*, pp.1659-1665, 2011

[15] Janaina Goncalves de Oliveira, Hans Bernhoff, "Battery Recharging Issue for a Two-Power-Level Flywheel System", *Journal of Electric and Computer Engineering*, Vol. 2010, No.37, 2010

[16] Fumiya Shimizu, "Development of Simulator for AMB-Flywheel Energy Storage System", *JSME Transaction MOVIC2013*, B05, 2013

[17] R.V. Monopoli, 'Model Reference Adaptive Control with an Augmented Error Signal', *IEEE Transaction*, AC-19-5, pp.474-484, 1974

[18] J. Broussard and M.J. O'Brien, 'Feedforward Control to Track the Output of a Forced Model', *IEEE Transaction*, AC-25-4, 851-853, 1980

[19] H. Kaufman, I. Bar-Kana, K. Sobel, 'Direct Adaptive Control Algorithms', *Theory and Applications* ,2nd Ed., Springer-Verlag ,1998

[20] P.A.Ioannou and P.V.Kokotovic, "Robust Redesign of Adaptive Control", *IEEE Trans*, AC-29-3, pp.202-211, 1984

[21] I. Mizumoto, Z. Iwai, "Recent trend of SAC", *JSIE transaction*, Vol.40, No.10, pp.723-728, 2001

[22] P.A.Ioannou, 'Robust Adaptive Controller with Zero Residual Tracking Errors', *IEEE Transaction*, AC-31-8, pp.773-776 , 1986

[23] K.S.Narendra and A.M. Annaswamy, "Robust Adaptive Control in the Presence of Bounded Disturbance", *IEEE Trans*, AC-31-4, pp.306-315, 1986

[24] Yuji Yasui, Ikue Kawasumi, Kousuke Higashitani, 'The Adaptive Air-Fuel Ratio Control for High-Efficiency and Low-Emission Engines', *SICE Transaction on Industrial Application*, Vol.9 No.3 pp.11-16 , 2010

Integrated Controller Based on Floating Point Operations in an FPGA for a Magnetically Suspended Flywheel

Li Zhang, Kun Liu, Kai Xiao

College of Aerospace Science and Engineering, National University of Defense Technology, Changsha 410073, China, zhangkaiyue2004@aliyun.com, liukun@nudt.edu.cn, xiaokai@nudt.edu.cn

***Abstract*—In order to reduce the size and power loss of the electronic package for a magnetic suspended flywheel, an integrated controller of the magnetic bearings is proposed. The integrated controller optimizes the power amplifier and the position controller with signal conditioning as a whole. All the necessary control tasks, namely, signal conditioning, rotor position control, generalized notch filter, electromagnetic coil current control and three-level pulse width modulation for the H-bridge are implemented in an FPGA. The multiply-accumulate calculation involved in the above tasks is implemented by reusing the floating point operation IP cores in timepieces. A universal block for typical one order transfer function is proposed to improve the computation efficiency. An automatic calibration method for the rotor position signals is proposed and implemented. Finally, tests results show that the performance of the integrated controller is satisfying.**

Introduction

Reaction flywheels are main actuators of attitude control systems for spacecrafts such as satellites. They are often the main bottleneck of high precision satellites. An effective way to improve the precision of attitude control and extend the lifetime of the system is replacing the traditional ball bearings with magnetic bearings. Magnetic bearings are non-contact actuators which can offer many distinctive advantages such as extremely low friction loss, vibration control, and high rotational speed [1]. For a magnetically suspended flywheel (MSFW), the flywheel is suspended and stabilized within a ring of electromagnets whose coil currents are actively controlled [2].

The electronic package which operates the magnetic bearings and its control strategies is always the focus in the research. Considering the strongly nonlinear relationship between the magnetic force and the coil current combined with hysteresis, flux leakage, and saturation effects [3], many advanced design approaches have been used. These include eigenstructure assignment dynamical compensation [4], filtered- x least mean square algorithm [5], nonlinear observers [6], sliding mode control [7], genetic algorithms [8], time-delay control [9], and μ -synthesis [10]. However, analysis of the trajectories in the current-displacement plane tends towards the zones where linearity is good. As a consequence, the traditional proportional integral differential (PID) control combined with some unbalance compensation is enough for most application.

When the rotor's geometric axis and mass axis are not one, because of the synchronous unbalance force, the rotor will not rotate around the bearing center line. Consequently, the rotational accuracy is reduced. At some critical frequencies, the rotor runout rises dramatically that the rotor may hit the retainer bearing. For the unbalance effects, generalized notch filter [11], Q-parameterization theory [12], iterative learning control [13-14], et al are used. Among these algorithms, the generalized notch filter based on coordinate transform is proved to be suit for digital implement with high computation efficiency [23].

State-of-art platforms implementing the control tasks are usually dSPACE [4-11], [14], digital signal processors (DSPs) [15-16], and field-programmable gate arrays (FPGAs) [17-20]. Among these devices dSPACE is the most powerful, but it is only used in labs. DSPs are chosen for many industrial applications, but the software solution in DSPs is time-consuming in some high bandwidth cases. Recently, FPGAs are becoming more and more attractive because of their flexible performance and time to market and of their quasi-analog characteristics. Dedicated architectures based only on the specific needs of the algorithm can be implemented using FPGAs. One disadvantage of FPGA-based controller is that the bits of the fixed-format of each control variable must be chosen carefully to achieve the best tradeoff between precision and resources utilization [21]. Recently, floating point operations in FPGAs are shown to be feasible and effective [20], [22], and the resources utilization seems to be acceptable [20].

This paper is organized as follows. Section 2 presents the control models of a 4-DOF hybrid magnetically suspended outer-rotor flywheel. Section 3 describes the control strategies of the integrated controller. Section 4 gives the detailed FPGA solution of the control strategies. In this section a universal block for typical one order transfer function and an automatic calibration method for the rotor position signals are proposed. Section 5 presents the test results. Section 6 concludes the paper.

II. Control Models Of Magnetically Suspended Flywheel System

In a 4-DOF hybrid magnetically suspended flywheel, the axial movement of the rotor is stabilized through permanent magnet. The radial movement and rotation are controlled actively through the two radial magnetic bearings. The rotation along the rotor axis is driven by non-contact brushless DC motor.

Fig. 1 shows the rotor dynamic model of the magnetic bearings. O is the center of gravity, Ω is the rotor rotation speed, x_1, f_{x1} represent displacement from the balance position and net force at x direction in the upper radial magnetic bearing, y_1, f_{y1}, x_2, f_{x2}, y_2, f_{y2} analogous. l is the displacement between the two bearing planes.

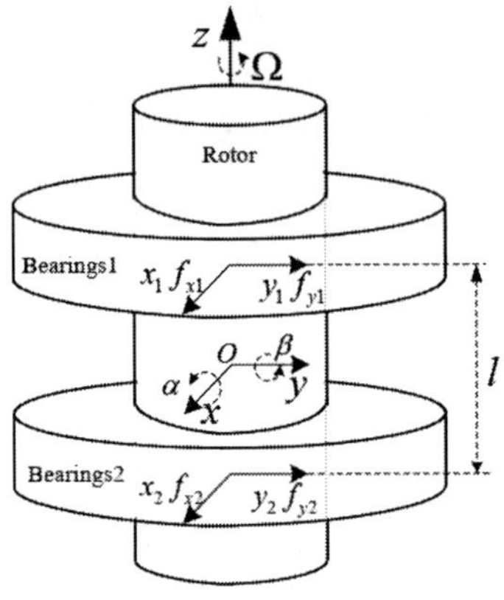

Fig. 1. Model of a 4-DOF MSFW

The general displacement $\mathbf{q}$, and the general force $\mathbf{F}$ are defined as:

$$\mathbf{q}=\begin{bmatrix} x & \beta & y & -\alpha \end{bmatrix}, \mathbf{F}=\begin{bmatrix} f_x & p_y & f_y & -p_x \end{bmatrix} \quad (1)$$

where f_x is the net force in x direction, and p_x is the net torque around x-axis. The rotor dynamic model can be established as:

$$\mathbf{M}\ddot{\mathbf{q}}+\mathbf{G}\dot{\mathbf{q}}=\mathbf{F} \quad (2)$$

where $\mathbf{M}$ is the mass matrix, and $\mathbf{G}$ is the gyro matrix.

$$\mathbf{M}=\begin{bmatrix} m & 0 & 0 & 0 \\ 0 & J_d & 0 & 0 \\ 0 & 0 & m & 0 \\ 0 & 0 & 0 & J_d \end{bmatrix}, \mathbf{G}=\begin{bmatrix} 0 & 0 & 0 & 0 \\ 0 & 0 & 0 & J_z\Omega \\ 0 & 0 & 0 & 0 \\ 0 & -J_z\Omega & 0 & 0 \end{bmatrix} \quad (3)$$

where m is the rotor mass matrix, J_d and J_z are the transverse and polar mass moments of inertia of the rotor.

The bearing force $\mathbf{f}$ and the rotor displacement in bearing coordinate $\mathbf{q}_m$ are:

$$\mathbf{q}_m=\begin{bmatrix} x_1 & x_2 & y_1 & y_2 \end{bmatrix}, \mathbf{f}=\begin{bmatrix} f_{x1} & f_{x2} & f_{y1} & f_{y2} \end{bmatrix} \quad (4)$$

The relationship between $\mathbf{q}_m$, $\mathbf{f}$ and $\mathbf{q}$, $\mathbf{F}$ can be described as

$$\mathbf{q}=\frac{0.5}{l}\begin{bmatrix} l & l & & \\ 2 & -2 & & \\ & & l & l \\ & & 2 & -2 \end{bmatrix}\mathbf{q}_m, \mathbf{F}=\begin{bmatrix} 1 & 1 & & \\ l/2 & -l/2 & & \\ & & 1 & 1 \\ & & l/2 & -l/2 \end{bmatrix}\mathbf{f} \quad (1)$$

By taking the Taylor's series expansions, the linearized magnetic force is given by

$$\mathbf{f}=k_x\mathbf{q}_m+k_i\mathbf{i},\ \mathbf{i}=[i_{x1}, i_{x1}, i_{y1}, i_{y2}]^{\mathrm{T}} \quad (2)$$

where k_x, k_i are the displacement stiffness(in N/m) and the current stiffness(in N/A), $i_{x1}, i_{x1}, i_{y1}, i_{y2}$ are the currents in the corresponding windings. Choosing the state vector as $X^T=[x_1, x_2, y_1, y_2, \dot{x}_1, \dot{x}_2, \dot{y}_1, \dot{y}_2]$ and the control input vector as $U^T=[i_{x1}, i_{x1}, i_{y1}, i_{y2}]$, the state space model of the 4-DOF magnetic bearings control system can be obtained as:

$$\begin{cases} \dot{X}=\mathbf{A_C}X+\mathbf{B_C}U \\ Y=\mathbf{C}X \end{cases} \quad (3)$$

$$\mathbf{A_C}=\begin{bmatrix} 0_{4\times4} & I_4 \\ M_p & M_c \end{bmatrix}, \mathbf{B_C}=\begin{bmatrix} 0_{4\times4} \\ M_f \end{bmatrix}, \mathbf{C}=\begin{bmatrix} I_4 & 0_{4\times4} \end{bmatrix} \quad (4)$$

$$M_p=\begin{bmatrix} m_{p1} & m_{p2} & 0 & 0 \\ m_{p2} & m_{p1} & 0 & 0 \\ 0 & 0 & m_{p1} & m_{p2} \\ 0 & 0 & m_{p2} & m_{p1} \end{bmatrix}, \begin{cases} m_{f1}=\dfrac{k_x(J_d+ml^2)}{mJ_d} \\ m_{f2}=\dfrac{k_x(J_d-ml^2)}{mJ_d} \end{cases} \quad (5)$$

$$M_c=\begin{bmatrix} 0 & 0 & -m_c & m_c \\ 0 & 0 & m_c & -m_c \\ m_c & -m_c & 0 & 0 \\ -m_c & m_c & 0 & 0 \end{bmatrix}, m_c=\frac{J_z\Omega}{2J_d} \quad (6)$$

$$M_f=\begin{bmatrix} m_{f1} & m_{f2} & 0 & 0 \\ m_{f2} & m_{f1} & 0 & 0 \\ 0 & 0 & m_{f1} & m_{f2} \\ 0 & 0 & m_{f2} & m_{f1} \end{bmatrix}, \begin{cases} m_{f1}=k_i\dfrac{J_d+ml^2}{mJ_d} \\ m_{f2}=k_i\dfrac{J_d-ml^2}{mJ_d} \end{cases} \quad (7)$$

III. Control Strategies Of The Integrated Controller

The block diagram of the integrated controller is shown in Fig. 2. There are two loops in the control structure. The inner loop is the coil current loop, which is stabilized with a proportional-integral current controller; the outer is the rotor position loop.

The four-DOF rotor position is stabilized using four proportional-integral-derivative (PID) controllers which are implemented in decentralized control mode. Besides, two "add-on" controller, the cross-feedback controller and the

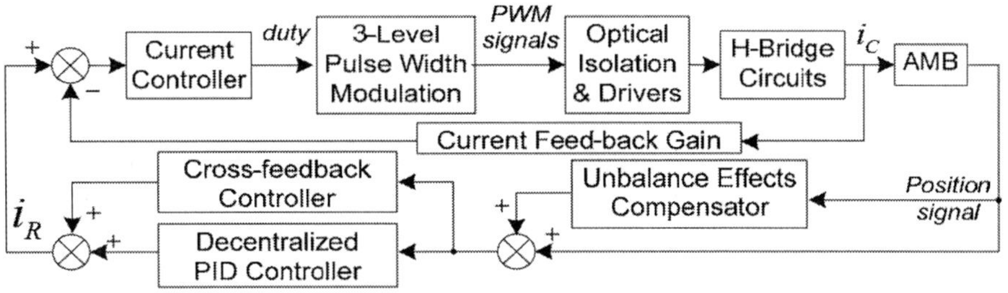

Fig. 2. Diagram of the integrated controller.

unbalance compensator are adopted to reject the gyroscopic effects and unbalance effects.

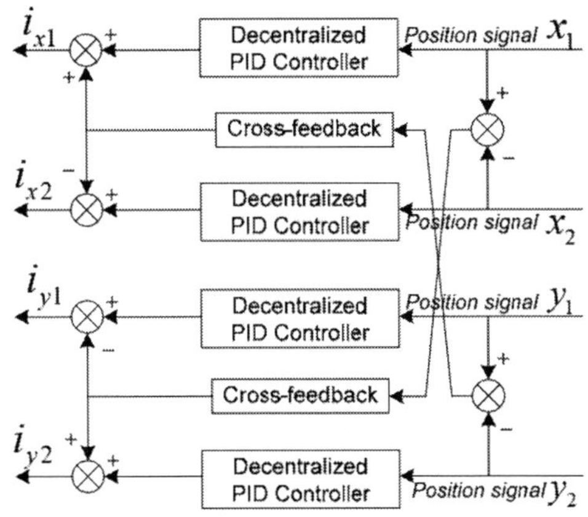

Fig. 3. Structure of the cross-feedback controller.

A. Decentralized PID Controller

To stabilize the rotor's four-DOF planar motion, the four-input-four-output system is divided into four SISO subsystems. All the four SISO subsystems have the same PID controller. The transfer function of the PID controller is

$$G_{PID}(s) = G_{LP}(s)G_{PD1}(s)G_{PD2}(s)G_I(s) \qquad (12)$$

where $G_{LP}(s)$ is a low-pass filter; $G_{PD1}(s)G_{PD2}(s)$ is a second-order phase-lead controller; $G_I(s)$ is a integral controller which is used to cancel the static position error.

Note that the second-order phase-lead controller is composed of two uniform one-order phase-lead controllers, and the total transfer function $G_{PID}(s)$ can be divided into four typical one-order transfer functions. The purpose is maximizing the reuse of the modules.

B. Cross-feedback controller

The structure of the cross-feedback controller is shown in Fig. 3. The transfer function of the cross-feedback controller is also one-order.

C. Unbalance Compensator

The unbalance compensation scheme based on coordinate transformation is adopted in this paper [23]. The scheme is using two coordinate transformations to pick the synchronous rotor runout signal component up from the position signals. The synchronous component is added to the original position signals to achieve unbalance compensation.

The structure of the unbalance compensator is shown in Fig. 4. Note that it is a feed-forward open loop compensator,

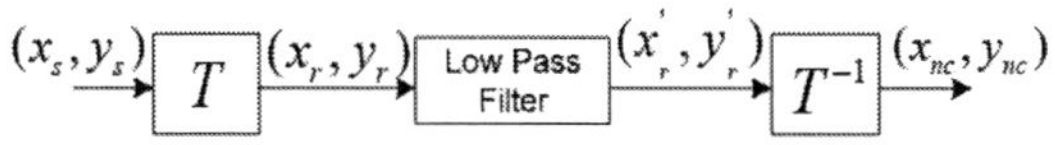

Fig. 4. Structure of the unbalance effects compensator.

and it will not affect the system stability as its notch-filter counterpart does [14]. Another advantage of this compensation scheme is that it can obtain two-DOF synchronous signals in one processing unit, because it uses the characteristic that there is a 90 degree phase difference between the x-direction synchronous signal and the x-direction synchronous signal. Therefore the computation efficiency is high, so it is suit for digital implement.

The steps to get the synchronous component from the rotor position signals are:

1) Use coordinate transformation T to get the position coordinate (x_r, y_r) of C in $X_r - C - Y_r$. Coordinate $X_r - C - Y_r$ is fixed with the rotor, rotating around O at the same speed with the rotor.

$$\begin{cases} x_r = x_s \cos\Omega t + y_s \sin\Omega t \\ y_r = -x_s \sin\Omega t + y_s \cos\Omega t \end{cases} \qquad (13)$$

2) Filter the position signals (x_r, y_r) with a low pass filter $G_{nf}(s)$.

$$G_{nf}(s) = \frac{1}{s/2\pi f + 1}, f = 10 \sim 20Hz \qquad (14)$$

3) Transform the filtered position signals (x_r', y_r') back into $X_s - O - Y_s$ to get the unbalance compensation signals (x_{nc}, y_{nc})

$$\begin{cases} x_{nc} = x_r' \cos\Omega t - y_r' \sin\Omega t \\ y_{nc} = x_r' \sin\Omega t + y_r' \cos\Omega t \end{cases} \qquad (15)$$

D. Current control loop

The current loop in Fig. 2 contains five parts, namely, the current controller, H-bridge electronic package driven by

PWM generator, electromagnetic coil, Hall-effect current sensor, and the current feedback gain. The current feedback gain is a key parameter which determines the current loop's response time.

The current controller is composed of a one-order phase-lead controller and a signal conditioner to transform the output of the phase-lead controller into PWM signals' duration ratio (duty) with polarity. The signal conditioning algorithm is

$$d = \mathrm{sgn}(y)\cdot[0.5+\frac{y}{2y_{\max}}] \quad (16)$$

where y is the output of phase-lead controller, $y_{\max}$ is the saturating value of y . d is the output of the current controller, which is the input of the PWM generator. The PWM generator generates the PWM switch signals base on three-level PWM scheme. The PWM signals are used to drive the H-bridges through optical isolation and gate driving circuits. The H-bridge is composed of metallic oxide semiconductor field effect transistor (MOSFET). The upper leg of the bridge is composed of P-channel MOSFET and the lower the N-channel MOSFET.

IV. FPGA Solution Of The Control Strategies

The VHDL code of the integrated controller is developed, synthesized, and implemented under Xilinx Foundation ISE circumstance. The VHDL entity has two main processes. One of them carries out the floating point control algorithms; the other implements the three-level PWM scheme. The sampling frequency and the controlling frequency are both 10 kHz. The PWM frequency is 20 kHz. The frequencies are chosen to minimize the noise coupled in the circuit.

A. *Process of floating point control algorithms*

1) Time allocation and the floating point operator

In the process of floating point control algorithms, the 100 μs sampling/controlling period is achieved by a counter range from 1 to 5000. The counter counts the rising edge of the system clock signal, whose frequency is 50 MHz. Therefore there are 5000 time pieces in one processing period, each is 20 ns long. The value of the counter can be used as a time flag in the calculation process to allocate time for different tasks.

The four-DOF rotor position signals (differential) and the four-DOF coil current signals are sampled using three four-channel analog to digital (A/D) converter. The A/D conversion time for the four channels' position signals and coil current signals is 12 μs , consuming 600 time pieces. The A/D conversion Output coding is 14-bit signed complement binary. The 14-bit position and current signals are converted to floating point signals using Xilinx ISE floating point operator IP core. The data structure of the floating point signals complies with the IEEE-754 standard. The precision is 24 bits with 16-bit fraction and 8-bit exponent. The latency of the fix-point to floating-point conversion is 6 time pieces.

After the fix-point to floating-point conversion, the entire control algorithm is implemented using only floating point operator IP cores and lookup tables (LUTs). One floating-point multiplication needs 7 time pieces, and one floating-point addition 12 time pieces. Besides, the logic resources used by one floating-point multiplication IP core are 339 LUTs and 401 flip flops (FFs), one floating-point addition 438 LUTs and 430 flip flops (FFs). Considering that there are total 84 multiplications and 56 additions in the four channels' control algorithm and one X3CS1000 FPGA has only 15360 LUTs and Slice FFs, reuse of modules is necessary. In addition, some algorithm refinement must be done in order to implement the calculation of the sine and cosine value in the unbalance compensation using only LUTs and MACs (Multipliers and Accumulators).

2) Modularity and algorithm refinement

The controller transfer functions can be separated into several typical one-order transfer functions, which are chosen to be the design modules in the implementation of the control algorithm. Using bilinear transform, a typical one-order transfer function in Laplace-domain

$$G(s)=\frac{as+b}{cs+d} \quad (17)$$

can be transformed into z -domain as

$$G(z)=\frac{y(z)}{u(z)}=\frac{(2a+bT)+(bT-2a)z^{-1}}{(2c+dT)+(dT-2c)z^{-1}} \quad (18)$$

which can be expressed as

$$y(k)=par1\cdot u(k)+par2\cdot u(k-1)+par3\cdot y(k-1) \quad (19)$$

where the three coefficients are

$$\begin{cases} par1=(2a+bT)/(2c+dT) \\ par2=(bT-2a)/(2c+dT) \\ par3=(2c-dT)/(2c+dT) \end{cases} \quad (20)$$

Equation (20) can be implemented using MACs as shown in Fig. 5.

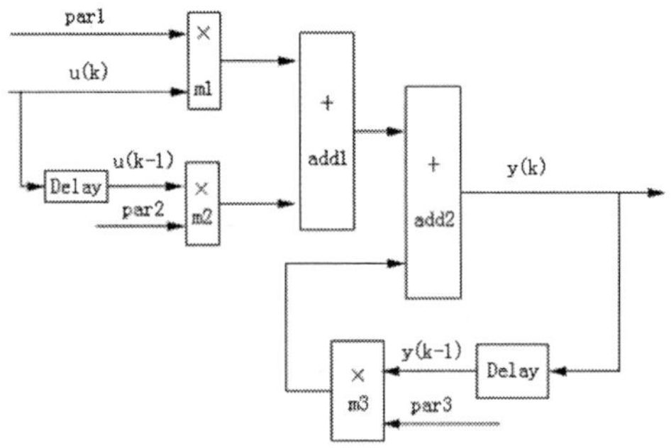

Fig. 5. Hardware module of one-order transfer function

The block module shown in Fig. 5 contains three floating-point multiplication IP cores and two floating-point addition IP cores. The 84 multiplications and 56 additions in the four channels' control algorithm can be done by reusing this single block. The logic resources used are 1893 LUTs and 2063 flip flops (FFs), and the latency of the block is 37 time pieces.

In the process of control algorithm, the unbalance compensator adopted an incremental algorithm when calculating $\sin\Omega t$ and $\cos\Omega t$. The incremental algorithm is

$$\begin{cases}\sin(\Omega t+\Omega T)=\sin(\Omega t)\cos(\Omega T)+\cos(\Omega t)\sin(\Omega T)\\ \cos(\Omega t+\Omega T)=\cos(\Omega t)\cos(\Omega T)-\sin(\Omega t)\sin(\Omega T)\end{cases} \quad (21)$$

where T is the sampling period for A/D conversion on the FPGA board. The values of $\sin\Omega T$ and $\cos\Omega T$ are stored in a lookup table (LUT) indexed by rotational speed Ω. Thus $\sin(\Omega t)$ and $\cos(\Omega t)$ can be calculated using only LUTs and MACs, given initial values $\sin(\Omega t_0)$ and $\cos(\Omega t_0)$.

B. *Process of 3-level PWM*

Three-level PWM scheme uses two control bits to generate the four switching signals for an H-bridge arrangement of power switches: a PWM bit and a Direction bit [24]. The direction control bit determines whether the current is decreasing or increasing. The PWM bit determines whether the

bearing is connected to the power supply or is simply circulating current around the inductive load.

In the integrated controller, the PWM period is 50 μs, which is generated by a counter range from 1 to 2500. Consequently, there are 2500 time pieces, and the resolution of the switching signals' modulation is 1/2500. The value of the counter is defined as time flag k.

If the input of the process of 3-level PWM is d ,which is a duration ration with polarity, then the PWM bit b_p and the Direction bit b_d can be expressed as

$$\begin{cases}b_d=\mathrm{sgn}(d)\\ b_p=\mathrm{sgn}(2500\cdot|d|-k)\end{cases} \quad (22)$$

The H-bridge arrangement of power switches $S_1 \sim S_4$ shown in Fig. 9 are divided into two groups: the positive group S_1, S_3, and the negative group S_2, S_4. When b_d is high, the positive group is the master group while the negative group is the slave group, and the switching signals $q_1 \sim q_4$ can be expressed as

$$\begin{cases}q_1=b_p\\ q_2=q_1\cap\overline{q_3}\\ q_3=q_1\cdot e^{j180^\circ}\\ q_4=\overline{q_1}\cap q_3\end{cases} \quad (23)$$

In Equation (25), $q_3=q_1\cdot e^{j180^\circ}$ means that q_3 is generated by adding a 180° phase delay into q_1. The 180° phase delay is generated by using a shift register in the FPGA. The depth of the shift register is 1250 bits.

If b_d is low, the negative group is the master group while the positive group is the slave group, and the switching signals $q_1 \sim q_4$ can be expressed as

$$\begin{cases}q_1=q_2\cap\overline{q_4}\\ q_2=b_p\\ q_3=\overline{q_2}\cap q_4\\ q_4=q_2\cdot e^{j180^\circ}\end{cases} \quad (24)$$

Therefore the switching signals $q_1 \sim q_4$ can be derived from b_d and b_p.

C. *An automatic calibration method*

The four DOF position signals x_1, x_2, y_1, y_2 shown in Fig. 4 are generated using eight eddy current sensors in differential arrangement. Take x_1 shown in Fig. 5 for example, in ideal situation, $x_1 = x_{s1} - x_{s2}$, x_{s1} and x_{s2} are the position detected by eddy current sensors S_4 and S_2. In practice, x_{s1} and x_{s2} need calibration before use. Calibration of the position signals are traditionally implemented using potentiometers which require additional circuits. The integrated controller adopts a digital calibration method as follows (take x_1 for example).

1) Set the duration ratio of the two X plain H-bridges as +1, wait for 500 ms, calculate the position signal $x_1 = x_{s1} - x_{s2}$, store it as x_{1+}.
2) Set the duration ratio of the two X plain H-bridges as -1, wait for 500 ms, calculate the position signal $x_1 = x_{s1} - x_{s2}$, store it as x_{1-}.
3) Calculate the calibration signal x_{adj}.

$$x_{adj} = -\frac{x_{1+}+x_{1-}}{2} \quad (25)$$

After the calibration steps, the position signal x_1 is replaced with $x_1 + x_{adj}$ before sending to the position controller.

V. Experiment Results

The integrated controller utilizes a Xilinx X3CS1000 FPGA to implement all the above control strategies. Fig. 6 shows the integrated controller board. Fig. 7 shows the experimental setup. Fig. 8 gives the position signals at the starting time.

From Fig. 8 we can notice that after the power-on instruction, the FPGA needs about 400 ms for loading. The process of calibrating x_1, x_2, y_1, y_2 takes about 2 second. After calibration, the rotor goes into stable suspension in 400 ms. The protection gap of the magnetic bearings is 1650 mV. In stable suspension, the rotor displacement is about 100 mV peak-to-peak, which is less than 6% of the total gap. Considering the eddy current displacement sensor's sensitivity is 8 mV/μm, the peak-to-peak rotor displacement is about 12.5 μm in static suspension.

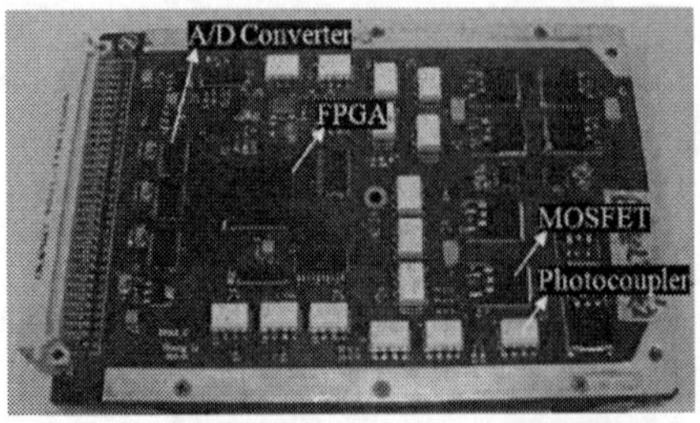

Fig. 6. The integrated controller board

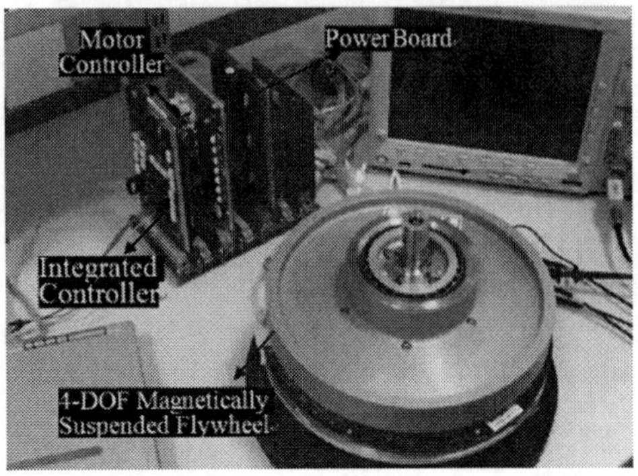

Fig. 7. Experimental setup

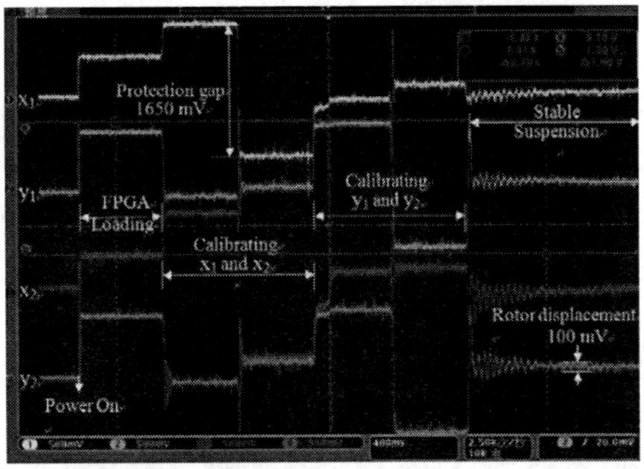

Fig. 8. Auto calibration and static suspension result of the magnetically suspended flywheel

Fig. 9 gives the rotor displacement in x_1 direction before unbalance compensation at the rotation speed of 1800 rpm. The rotor runout is about 30 μm . The value drops to 15 μm after unbalance compensation as shown in Fig. 10. The effectiveness of the unbalance compensation is therefore validated.

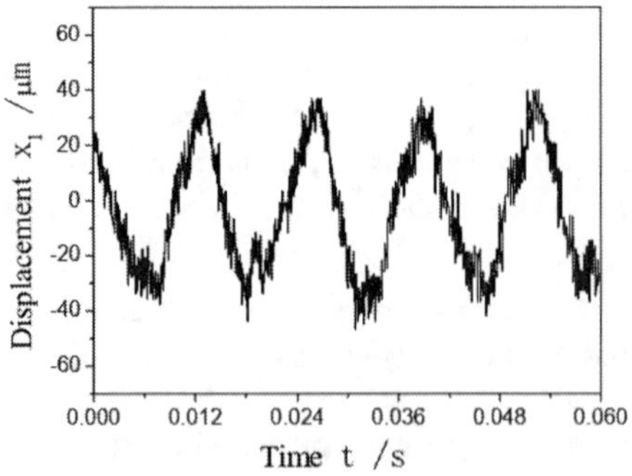

Fig. 9. Rotor displacement in x_1 direction before unbalance compensation at the rotation speed of 1800 rpm

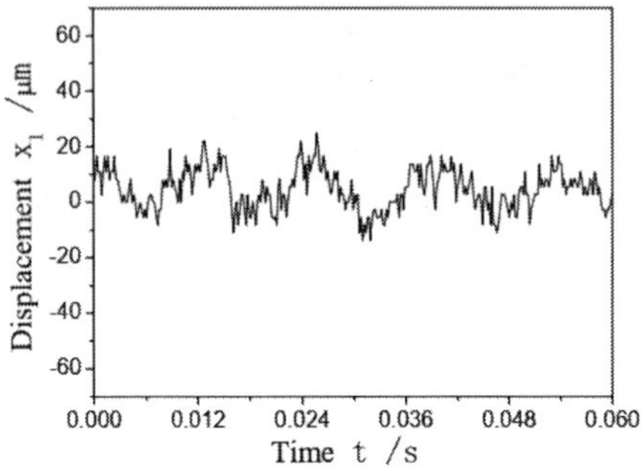

Fig. 10. Rotor displacement after unbalance compensation

TABLE I
RESOURCES SUMMARY

Target Device : xc3s1000		
Number of Slice Flip Flops:	5,642 out of 15,360	36%
Number of 4 input LUTs:	6,646 out of 15,360	43%
Number of occupied Slices:	5,797 out of 7,680	75%
Total Number of 4 input LUTs:	7,122 out of 15,360	46%
Number of bonded IOBs:	27 out of 173	15%
Number of MULT18X18s:	13 out of 24	54%
Number of GCLKs:	2 out of 8	25%
Number of DCMs:	1 out of 4	25%
Total equivalent gate count for design: 163,631		

Table 1 shows the FPGA logic resources used to implement the integrated controller. The total equivalent gate count for design is 163,631.

VI. CONCLUSIONS

This paper proposes an integrated controller to reduce the size and power loss of the electronic package for a 4-DOF magnetic suspended satellite Reaction flywheel. All the necessary control tasks are implemented in one single FPGA. The multiply-accumulate calculation involved in the above tasks is implemented by reusing the floating point operation IP cores in timepieces. Total equivalent gate count for design is about 160,000. Using the integrated controller, the rotor can goes into stable suspension in 400 ms after the power-on instruction, and the rotor displacement is less than 6% of the total protection gap in static suspension. The unbalance compensation reduces the rotor runout dramatically. A universal block for typical one order transfer function based on floating point MACs is proposed to improve the computation efficiency, and is validated. The logic resources used are 1893 LUTs and 2063 flip flops (FFs), and the latency of the block is 37 time pieces. An automatic calibration method for the rotor position signals is proposed and tested to be effective.

REFERENCES

[1] G. Schweitzer, H. Bleuler, A. Traxler. *Active magnetic bearing: basics, properties, and applications*. Zurich, Switzerland: vdf Hochschulverlag, 1994.

[2] R. D. Williams, F. J. Keith, P. E. Allaire. "Digital Control of Active Magnetic Bearings," *IEEE Transactions on Industrial Electronics*, vol. 37, no. 1, pp. 19-27, Feb. 1990.

[3] S. C. Mukhopadhyay, T. Ohji, M. Iwahara. "Modeling and Control of a New Horizontal-Shaft Hybrid-Type Magnetic Bearing," *IEEE Transactions on Industrial Electronics*, vol. 47, no. 1, pp. 100-108, 2000.

[4] G. Duan, D. Howe. "Robust Magnetic Bearing Control via Eigenstructure Assignment Dynamical Compensation," *IEEE Transactions on Control Systems Technology*, vol. 11, no. 2, pp. 204-212, Apr. 2003.

[5] M. S. Kang, W. H. Yoon. "Acceleration Feedforward Control in Active Magnetic Bearing System Subject to Base Motion by Filtered-X LMS Algorithm," *IEEE Transactions on Control Systems Technology*, vol. 14, no. 1, pp. 134-140, Feb. 2006.

[6] Xinkai Chen, Chunyi Su, T. FUKUDA. "A Nonlinear Disturbance Observer for Multivariable Systems and Its Application to Magnetic Bearing Systems," *IEEE Transactions on Control Systems Technology*, vol. 12, no. 4, pp. 569-577, Aug. 2004.

[7] J. H. Lee, P. E. Allaire, Gang Tao, et al. "Experimental Study of Sliding Mode Control for a Benchmark Magnetic Bearing System and Artificial Heart Pump Suspension," *IEEE Transactions on Control Systems Technology*, vol. 11, no. 1, pp. 128-136, Feb. 2003.

[8] P. Schroder, B. Green, P. J. Fleming. "On-line evolution of robust control systems: An industrial active magnetic bearing application," *Contr. Eng. Practice*, vol. 9,no. 1, pp. 37-49, 2001.

[9] K. Youcef-toumi, S. Reddy. "Dynamic analysis and control of high speed and high precision active magnetic bearings," *ASME J. Dyn. Syst.,Meas., Contr.*, vol. 112, no. 1, pp. 133-142, 1990.

[10] K. Nonami, T. Ito. " μ Synthesis of Flexible Rotor-Magnetic Bearing Systems," *IEEE Transactions on Control Systems Technology*, vol. 4, no. 5, pp. 503-512, Sep. 1996.

[11] R. Herzog, P. Buhler, C. Gahler, et al. "Unbalance Compensation using generalized notch filters in the multivariable feedback of magnetic bearings," *IEEE Transactions on Control Systems Technology*, vol. 4, no. 5, pp. 580-586, Aug. 1996.

[12] A. M. Mohamed, Busch-Vishniac. "Imbalance compensation and automation balancing in magnetic bearing systems using the Q-parameterization theory," *IEEE Transactions on Control Systems Technology*, vol. 3, no. 2, pp. 202-211, Apr. 1995.

[13] C. R. Knopse, S. M. Tamer, S. J. Fedigan. "Synthesis of robust gain matrices for adaptive rotor vibration control," *ASME J. Dyn. Syst.,Meas., Contr.*, vol. 120,no. 7, pp. 298-300, 1997.

[14] Bi Chao, Wu Dezheng, Jiang Quan, et al. "Automatic Learning Control for Unbalance Compensation in Active Magnetic Bearings," *IEEE Transactions on Magnetics*, vol. 41, no. 7, pp. 2270-2280, Jul. 2005.

[15] H. Bleuler, C. Gahler, R. Herzog, et al. "Application of Digital Signal Processors for Industrial Magnetic Bearings," *IEEE Transactions on Control Systems Technology*, vol. 2, no. 4, pp. 280-289, Apr. 1994.

[16] C. R. Knospe, S. J. Fedigan, R. W. Hope, et al. "A Multitasking DSP Implementation of Adaptive Magnetic Bearing Control," *IEEE Transactions on Control Systems Technology*, vol. 5, no. 2, pp. 230-238, Apr. 1997.

[17] Zhengwei Fang, J. E. Carletta, R. J. Veillette. "A Methodology for FPGA-Based Control Implementation," *IEEE Transactions on Control Systems Technology*, vol. 13, no. 6, pp. 977-987, Nov. 2005.

[18] F. Krach, B. Frackelton, J. Carletta, et al. "FPGA-Based Implementation of Digital Control for a Magnetic Bearing," in *IEEE Proceedings of the American Control Conference*, pp. 1080-1085, 2003.

[19] R. Jastrzebski, R. Pollanen, O. Pyrhonen. "Analysis of System Architecture of FPGA-based Embedded Controller for Magnetically Suspended Rotor," in *IEEE Proc. Int. Symposium on System-on-chip, Tampere, Finland*, pp. 128-132, 2003.

[20] Li Zhang, Kun Liu, Xiaofei Chen. "FPGA Implementation of a Three-Level Power Amplifier for Magnetic Bearings," in *IEEE Proceedings of 9th International Conference on Electronic Measurement & Instruments*, Beijing, China, vol. 1, pp. 455-460, 2009.

[21] E. Monmasson, M. N. Cirstea. "FPGA design methodology for industrial control systems-A review," *IEEE Transactions on Industrial Electronics*, vol. 54, no. 4, pp. 1926-1936, Aug. 2007.

[22] Z. Salcic, J. Cao, S. K. Nguang. "A Floating-Point FPGA-Based Self-Tuning Regulator," *IEEE Transactions on Industrial Electronics*, vol. 53, no. 2, pp. 693-704, Apr. 2007.

[23] H. Mou. "Research on Unbalance Compensation of Active Magnetic Bearing," M.S. thesis, Nanjing University of Aeronautics and Astronautics, Nanjing, China, 2003.

[24] T. A. Harris. "Current mode switching for tri-state amplifiers in magnetic bearing control systems," U.S. Patent, 6 005 316, December 21, 1999.

Emulation of Energy Storage Flywheels on a Rotor-AMB Test Rig

Xujun Lyu[a,b], Long Di[b], Se Young Yoon[b], Zongli Lin[b], Yefa Hu[a]

[a] School of Mechanical and Electronic Engineering, Wuhan University of Technology, Wuhan, Hubei, China, emmalxj198762@gmail.com, huyefa@whut.edu.cn

[b] Charles L. Brown Department of Electrical and Computer Engineering, and Rotating Machinery and Controls Laboratory (ROMAC), University of Virginia, Charlottesville, VA 22904-4743, USA, {ld4vv, syy5b, zl5y}@virginia.edu

***Abstract*—Because of their many appealing features, such as high specific energy, high specific power, ultra-compactness, short charging time (in minutes), long life span and no pollution, energy storage flywheels suspended on active magnetic bearings (AMBs) have drawn worldwide attention. In this paper, we use an existing rotor-AMB test rig to emulate the operation of energy storage flywheels suspended on AMBs. The test rig contains four active magnetic bearings, two located at the two ends of the rotor, one located at the rotor mid-span and the last located at the rotor quarter span. We will use the two magnetic bearings in the mid and quarter spans to emulate the negative stiffness of the generator and the gyroscopic effect of the flywheel disk. Simulation and experimental results are presented.**

I. Introduction

With the constant development of the economy and the society, the demand for energy storage, which helps to maintain the power system stability, improve the power quality and aid distributed generation, keeps rising [1]. Compared with other energy storage technologies such as super conducting magnetic energy storage, conventional electrochemical batteries and ultra capacitors, flywheels have notable advantages such as high specific energy, high specific power, ultra-compactness, short charging time (in minutes), long life span and no pollution. These benefits of flywheels make them ideal for storing energy in space applications such as satellites, spacecraft and space stations [2]. For hybrid electric vehicles, flywheels can discharge rapidly in order to improve the drivability of vehicles under acceleration or hill-climbing condition. On the other hand, when under deceleration or down-hill condition, flywheels charge quickly in order to improve the efficiency in storing the regenerative energy [3]–[5]. Flywheels are also used for regulating the peak power load of grids, wind power systems and uninterruptible power systems.

Flywheel energy storage systems are typical mechanical batteries. The kinetic energy is stored in a high speed rotating disk of the flywheel ([3], [6]). This mechanical energy is converted back to electric energy by a generator, which is mounted on the same rotor as the flywheel disk. The stored energy for a flywheel system is proportional to the square of the rotating speed Ω, and is given by

$$E = \frac{1}{2} J_{\rm p} \Omega^2, \tag{1}$$

where $J_{\rm p}$ is the polar moment of inertia of the flywheel disk.

Advanced flywheels operate at high rotating speeds to store a large amount of energy, which results in a high demand on the bearing system for stability and performance. The conventional rolling element bearings, sliding bearings and hydraulic bearings are not adequate to meet such a demand. Active magnetic bearings use magnetic forces to suspend the rotor, and the forces they produce can be actively controlled. Moreover, they possess appealing features such as no mechanical contact, no friction losses, no wear, high speed capability and clean operation. As a result, active magnetic bearings are ideal for supporting high-speed flywheel rotors.

Compared with rotors in many other applications, rotors in the flywheel systems are highly complex. First, in an effort to minimize the overall size of the system, the flywheel disk and the generator are usually mounted on the same shaft, causing a coupling effect between the generator and the rotor dynamics [3]. Second, unlike with applications where the ratio of polar-to-transverse moments of inertia is small and the gyroscopic effects can be neglected ([7], [9]), the large flywheel disks generate strong gyroscopic effects that have to be taken into account in the design of the AMB controller.

The complexities of the flywheel systems entail sophisticated AMB controllers. Brown et al. [7] and Dever et al. [8] confirmed that gyroscopic modes can be stabilized by using cross-axis proportional gains and experimentally demonstrated this fact on the NASA Glenn D1 flywheel. Ahrens et al. [9] and Kucera et al. [10] proposed a cross feedback control method to compensate gyroscopic effects of flywheel AMB systems and experimental results showed that it improve system performance. Hawkins et al. [11] used a gain-scheduled MIMO control algorithm to control an energy storage flywheel, with the consideration of gyroscopic effects. Sivrioglu et al. [12] used a nonlinear control approach with an H_∞ compensator for a zero-bias flywheel AMB system, and experimental results showed that it is a reliable method to control the rotor. Sakai et al. [13] designed a new passivity based control without conventional cross-feedback, and both the simulation and experimental results on a flywheel AMB system showed its advantages in terms of low computational costs and strong robustness.

In summary, the control of flywheel AMB systems has been studied extensively in the literature. Despite this abundance of resources in the literature, the experimental validation of the theoretical results presented in many of these papers has always been a difficult task. The reason is that it is very expensive and technically challenging to build a flywheel AMB test rig. Among many difficulties, specialized material and complex manufacturing techniques are required to withstand the stress of high speed rotation. We propose to emulate the rotordynamic characteristics of energy storage flywheels on an existing rotor AMB test rig in the Rotating Machinery and Controls Laboratory (ROMAC) at University of Virginia. The test rig contains four active magnetic bearings, two located at the two ends of the rotor, one at the mid span and the other at the quarter span.

The remainder of this paper is organized as follows. Section II describes the rotor AMB test rig and state the problem to be solved in this paper. Section III analyzes the negative stiffness caused by the generator and the gyroscopic effects of the flywheel disk. Section IV presents both the simulation and experimental results. Section V draws the conclusions to this paper.

II. The Test Rig and Problem Statement

Shown in Fig. 1 is a schematic drawing of the rotor-AMB test rig [14] we are to use to emulate the operation of an energy storage flywheel system. This rotor-AMB test rig is a research platform constructed in our ROMAC laboratory. The original purpose of this test rig was to emulate an industrial size centrifugal gas compressor [15]. In particular, disk 1 and disk 2 emulate the wheels in a compressor. There are four AMBs in the test rig. Two radial support AMBs are located at the two ends of the rotor. One exciter AMB is at the mid-span and the other is at the quarter span of the rotor. This combination of four radial AMBs allows the simulation of different operating conditions of the compressor.

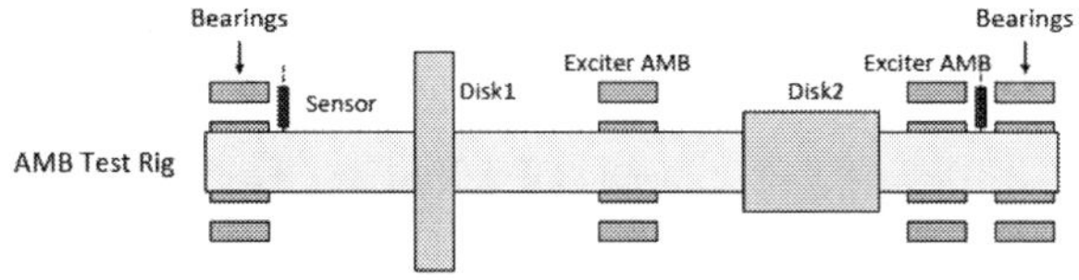

Figure 1: A schematic of the rotor-AMB test rig at University of Virginia.

The rotor in the test rig is 1.23 m long and weights 44.9 kg. Four laminated steel journals are mounted on the shaft respectively for the two radial support AMBs at the non-driven end (NDE) and driven end (DE), and the two radial exciter AMBs at the rotor mid and quarter spans. There are also two auxiliary ball bearings mounted at the support AMB locations to prevent damage to the AMBs in the event of a rotor drop. A 3.7 kW high speed motor with variable frequency drive (VFD), Colombo RS-90/2, drives the rotor in the test rig to speeds up to 18,000 rpm. The entire flexible rotor AMB test rig is shown in Fig. 2.

We use the two AMBs in the mid and quarter spans to emulate the negative stiffness of the generator and the gyroscopic effects of the flywheel disk on the rotor dynamics.

A: Non-driven End Support AMB | B: Gyroscopic Disk 1 | C: Rotor Shaft
D: Mid-span Disturbance AMB | E: Gyroscopic Disk 2 | F: Electric Motor
G: Quarter-span Disturbance AMB | H: Driven End Support AMB | I: Control Station
J: Variable Frequency Drive | K: Amplifiers & Sensor Conditioning Station

Figure 2: An overview of the flexible rotor AMB test rig.

III. Analysis of Flywheels

The generator mounted on the rotor and the gyroscopic effects caused by the flywheel disk both affect the rotor dynamics of the AMB system. The motion equation of the AMB system is given by

$$M\ddot{q} + (C + \Omega G)\dot{q} + Kq = F_{\text{mag}} + F_{\text{ext}}, \quad (2)$$

where,

M: the symmetric rotor mass matrix,
C: the symmetric support damping matrix,
G: the skew-symmetric gyroscopic effect matrix,
K: the symmetric support stiffness matrix,
Ω: the rotating speed,
F_{mag}: the forces provided by support AMBs,
F_{ext}: the external forces acting on the rotor,
q: the generalized displacement vector.

In the following subsections, we will analyze the effects of the generator and the flywheel disk on the rotor dynamics.

A. Negative Stiffness Caused by the Generator

The effect of the generator on the rotor dynamics is in the form of radial forces, which can be represented as a negative stiffness. Kascak et al. [16] calculated the forces on the rotor caused by a generator for different rotor displacements from the center. It was shown that for small displacements the negative stiffness is about 9.76×10^4 N/m. We use the exciter AMB at the quarter span to emulate this negative stiffness effect of the generator.

B. Gyroscopic Effects of the Flywheel Disk

A change in the direction of the rotational axis in a spinning flywheel results in a precession rotation generated by the gyroscopic effect. Consider a rotary disk in Fig. 3 mounted on a rigid and massless shaft, spinning about the z axis with an angular velocity Ω rad/s, and supported by two bearings A and B. A precession of the rotor on the x-z plane generates a gyroscopic couple represented by the coupled forces F_y at

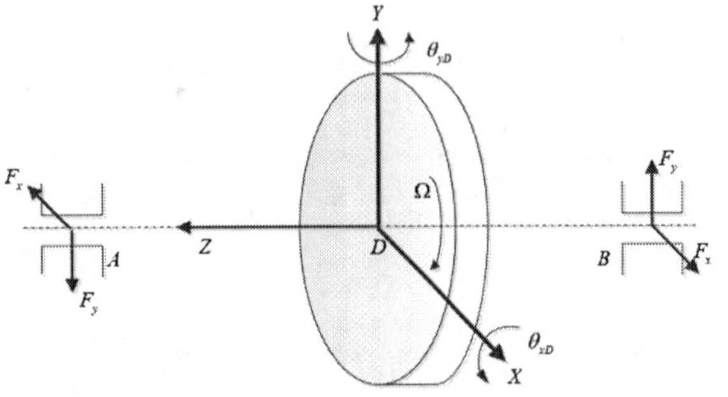

Figure 3: A schematic of gyroscopic effect of the flywheel disk.

bearings A and B, which in turn rotates the orientation of the z axis on the y-z plane. Similarly, F_x represents the gyroscopic couple generated by a precession on the y-z plane.

Fig. 3 illustrates the angular displacement of the disk about its mass center, where $\theta_{x\mathrm{D}}$ and $\theta_{y\mathrm{D}}$ are the mass center angular displacements in the x and y directions, respectively. The total moments about the mass center in the x direction, $M_{x\mathrm{D}}$, and in the y direction, $M_{y\mathrm{D}}$, are given respectively by

$$M_{x\mathrm{D}} = J_{\mathrm{t}}\ddot{\theta}_{x\mathrm{D}} + J_{\mathrm{p}}\Omega\dot{\theta}_{y\mathrm{D}}, \quad (3)$$

$$M_{y\mathrm{D}} = J_{\mathrm{t}}\ddot{\theta}_{y\mathrm{D}} - J_{\mathrm{p}}\Omega\dot{\theta}_{x\mathrm{D}}, \quad (4)$$

where J_{p} is the polar moment of inertia of the disk and J_{t} is the transverse moment of inertia of the disk; $J_{\mathrm{t}}\ddot{\theta}_{x\mathrm{D}}$ and $J_{\mathrm{t}}\ddot{\theta}_{y\mathrm{D}}$ are respectively the inertia in the x and y directions; and $J_{\mathrm{p}}\Omega\dot{\theta}_{y\mathrm{D}}$ and $J_{\mathrm{p}}\Omega\dot{\theta}_{x\mathrm{D}}$ are respectively the gyroscopic moments in the x and y directions.

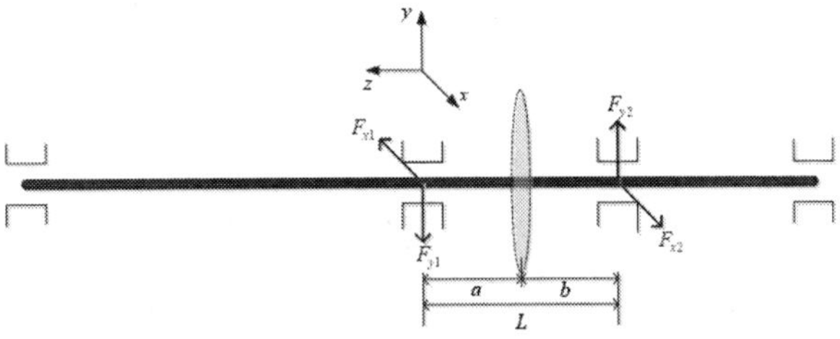

Figure 4: A diagram for the emulation of gyroscopic effects.

For the rotor-AMB test rig shown in Fig. 2, we use the two exciters AMBs at the rotor mid and quarter spans to emulate the gyroscopic moments exerted by the flywheel disk located between the exciter bearings, as shown in Fig. 4. In the figure, F_{x1} and F_{x2} are the external forces generated by the exciter bearings to emulate the gyroscopic couple caused by a precession about the x axis. Similarly, F_{y1} and F_{y2} are the exciter bearing forces emulating the gyroscopic couple caused by a precession about the y axis. We use the exciter AMBs at the rotor mid span to generate the forces F_{x1} and F_{y1} and the exciter AMBs at the rotor quarter span to generate the forces F_{x2} and F_{y2}. The relationships between the gyroscopic moments and corresponding exciter AMB forces are as follows,

$$J_{\mathrm{t}}\ddot{\theta}_{x\mathrm{D}} + J_{\mathrm{p}}\Omega\dot{\theta}_{y\mathrm{D}} = -aF_{x1} + bF_{x2}, \quad (5)$$

$$J_{\mathrm{t}}\ddot{\theta}_{y\mathrm{D}} - J_{\mathrm{p}}\Omega\dot{\theta}_{x\mathrm{D}} = -aF_{y1} + bF_{y2}, \quad (6)$$

$$F_{x1} = -F_{x2}, \quad (7)$$

$$F_{y1} = -F_{y2}, \quad (8)$$

where a and b are the distances between the center of the flywheel disk we want to emulate and the exciter AMBs at the rotor mid and quarter spans, respectively.

For AMBs, the magnetic forces are given by,

$$F_{x1} = K_{\mathrm{i}1}i_{\mathrm{p}x1} - K_{x1}x_1, \quad (9)$$

$$F_{x2} = K_{\mathrm{i}2}i_{\mathrm{p}x2} - K_{x2}x_2, \quad (10)$$

$$F_{y1} = K_{\mathrm{i}1}i_{\mathrm{p}y1} - K_{x1}y_1, \quad (11)$$

$$F_{y2} = K_{\mathrm{i}2}i_{\mathrm{p}y2} - K_{x2}y_2, \quad (12)$$

where,

$K_{\mathrm{i}1}$: current gain of the mid-span AMB,
$K_{\mathrm{i}2}$: current gain of the quarter-span AMB,
K_{x1}: negative stiffness of the mid-span AMB,
K_{x2}: negative stiffness of the quarter-span AMB,
x_1: rotor x axis displacement at the mid span,
x_2: rotor x axis displacement at the quarter span,
y_1: rotor y axis displacement at the mid span,
y_2: rotor y axis displacement at the quarter span,
$i_{\mathrm{p}x1}$: current for the mid-span AMB to generate F_{x1},
$i_{\mathrm{p}x2}$: current for the quarter-span AMB to generate F_{x2},
$i_{\mathrm{p}y1}$: current for the mid-span AMB to generate F_{y1},
$i_{\mathrm{p}y2}$: current for the quarter-span AMB to generate F_{y2}.

The displacements x_1, y_1, x_2 and y_2 are related to the angular displacements $\theta_{x\mathrm{D}}$ and $\theta_{y\mathrm{D}}$ by the following equations,

$$\theta_{x\mathrm{D}} = \frac{y_2 - y_1}{L}, \quad (13)$$

$$\theta_{y\mathrm{D}} = \frac{x_2 - x_1}{L}. \quad (14)$$

In order to emulate the gyroscopic couple using the exciter bearings, we need to find the expressions of the currents. Substituting (9)-(14) into (5)-(8) results in,

$$i_{\mathrm{p}x1} = \frac{K_{x1}}{K_{\mathrm{i}1}}x_1 - \frac{J_{\mathrm{t}}}{K_{\mathrm{i}1}L^2}(\ddot{y}_2 - \ddot{y}_1) - \frac{J_{\mathrm{p}}\Omega}{K_{\mathrm{i}1}L^2}(\dot{x}_2 - \dot{x}_1), \quad (15)$$

$$i_{\mathrm{p}x2} = \frac{K_{x2}}{K_{\mathrm{i}2}}x_2 + \frac{J_{\mathrm{t}}}{K_{\mathrm{i}2}L^2}(\ddot{y}_2 - \ddot{y}_1) + \frac{J_{\mathrm{p}}\Omega}{K_{\mathrm{i}2}L^2}(\dot{x}_2 - \dot{x}_1), \quad (16)$$

$$i_{\mathrm{p}y1} = \frac{K_{x1}}{K_{\mathrm{i}1}}y_1 - \frac{J_{\mathrm{t}}}{K_{\mathrm{i}1}L^2}(\ddot{x}_2 - \ddot{x}_1) + \frac{J_{\mathrm{p}}\Omega}{K_{\mathrm{i}1}L^2}(\dot{y}_2 - \dot{y}_1), \quad (17)$$

$$i_{\mathrm{p}y2} = \frac{K_{x1}}{K_{\mathrm{i}1}}y_2 + \frac{J_{\mathrm{t}}}{K_{\mathrm{i}2}L^2}(\ddot{x}_2 - \ddot{x}_1) - \frac{J_{\mathrm{p}}\Omega}{K_{\mathrm{i}2}L^2}(\dot{y}_2 - \dot{y}_1). \quad (18)$$

We can then carry out the simulation of the gyroscopic effects with the forces generated by the exciter AMBs to make it equal to the gyroscopic forces generated by the flywheel disk.

IV. Simulation and Experimental Results

In order to verify the method presented in Section III, a simulation study is first conducted based on the Simulink model derived from the experimental test rig, which contains all essential components of the rotor-AMB system. The Simulink model is formulated in a state space form consisting of 36 states with 12 inputs and 20 outputs. The 12 inputs include the unbalance forces, external forces from the exciter AMBs and the control voltages for the support AMBs; the 20 outputs include the rotor displacements at the four AMB locations, the corresponding displacement sensor outputs, and the support AMB forces.

Because of the complex dynamics and the uncertainties in the test rig, the AMB system is stabilized by a model based μ-synthesis controller [14], which is formulated in a state space form with 48 states, 4 inputs and 4 outputs. The controller can effectively handle the first and second flexible modes at near 243 Hz and 552 Hz, respectively.

For the flywheel disk, the polar moment of inertia is $J_p = 0.21\,\text{kg·m}^2$ and the transverse moment of inertia is $J_t = 0.105\,\text{kg·m}^2$. The properties of the exciter AMBs are summarized in Table I. The effect on the test rig model caused by the added negative stiffness and the emulated gyroscopic forces are shown and discussed in the following subsections.

Table I: Exciter AMBs Properties

Property	mid/quarter	Units
Bias current, I_b	1	A
Number of poles, n	8	--
Estimated air gap flux density, B	0.27	T
Current gain, K_i	94/91	N/A
Negative stiffness, K_x	165/186	kN/m
Copper turns, N	94	--
Projected pole area, A_p	700	mm^2
Nominal air gap, g_0	0.575/0.491	mm

A. Simulation of Generator Effects

The generator effect is in the form of radial forces, and it is interpreted as a negative stiffness acting on the rotor. The exciter AMB at the quarter span is used to provide the required negative stiffness around 9.76×10^4 N/m. Shown in Figs. 5 and 6 are the rotor displacements without and with the generator negative stiffness at the speed of 6,000 rpm, respectively. We observe that the peak rotor displacement at the quarter-span bearing under the μ-synthesis control is less than 2×10^{-5} m (0.78 mils) for both cases, which corresponds to a peak force from the generator negative stiffness of about 1 N. It is observed that the disturbance force is small and its effect on the rotor dynamics is minimal for even higher rotating speeds up to 15,000 rpm.

B. Simulation of Gyroscopic Effects

The gyroscopic effect is emulated using both the mid-span and quarter-span AMBs. The perturbation currents are generated following Eqs. (15)-(18). In order to verify that the emulated gyroscopic effect is generated properly, the gyroscopic forces emulated by the exciter AMBs are compared to the gyroscopic forces the rotor would experience which are calculated based on the gyroscopic matrix G. For this

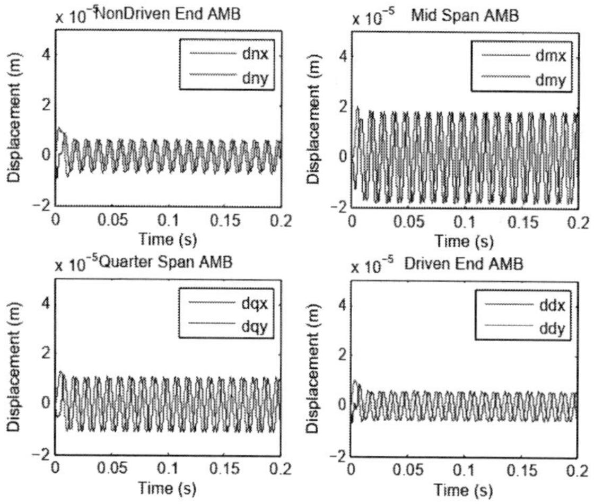

Figure 5: Rotor displacements without the generator negative stiffness.

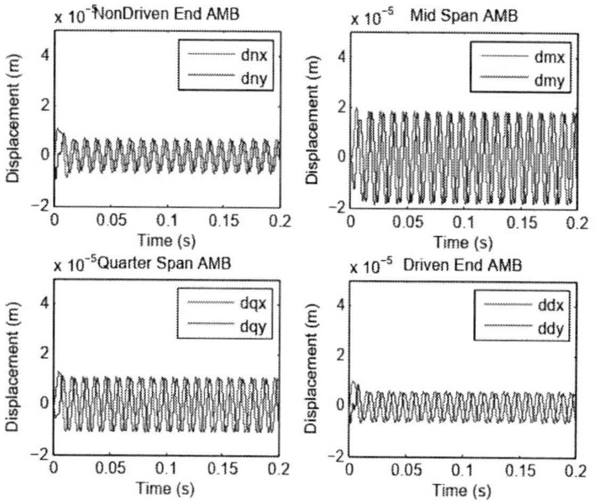

Figure 6: Rotor displacements with the generator negative stiffness.

comparison, we first observe that the gyroscopic matrix G results from Disk 1 on the rotor AMB test rig. Therefore, in order to present a relevant comparison between the gyroscopic effects of similar rotors, with gyroscopic forces represented at opposite sides of the flywheel disk, the forces due to the matrix G are also shown at the locations of the support AMBs. With this observation, the gyroscopic forces in both cases are on opposite sides of the main disk to reveal a similar pattern between the emulated gyroscopic forces and the actual forces caused by the gyroscopic matrix G.

Shown in Fig. 7 is the emulated gyroscopic forces generated by the exciter AMBs at the speed of 6,000 rpm. Figure 8 shows the actual forces caused by the gyroscopic matrix G at the locations of the support AMBs. It can be observed that the pattern of the emulated gyroscopic forces and the actual gyroscopic forces are similar. There is a difference in

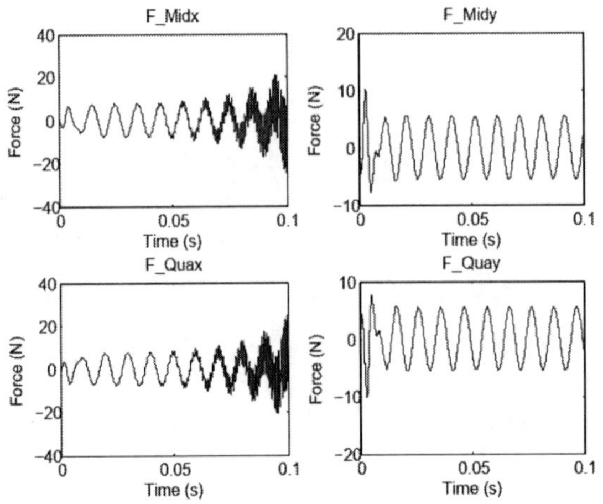

Figure 7: Emulated gyroscopic forces generated by the exciter AMBs.

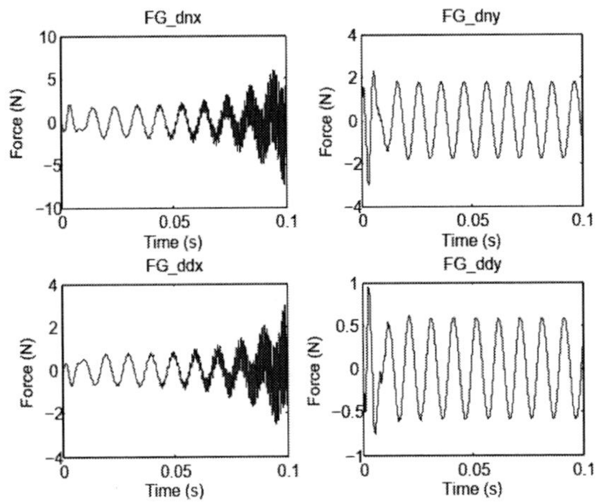

Figure 8: Actual forces caused by the gyroscopic matrix G at the support AMBs.

the magnitude of the forces due to the difference in the size and location of the contributing disks, but the relative phase and frequency of the gyroscopic forces are well aligned.

It is also observed in Figs. 7 and 8 that a high frequency oscillation appears in the simulated forces, which increases in magnitude over time. This phenomenon is due to the excitation of the rotor third flexible mode by the emulated gyroscopic forces. Because of the locations of the mid-span and quarter-span AMBs, the injected forces from the emulation of the gyroscopic effect lie right at the two peaks of the mode shape corresponding to the third flexible mode. The excited mode causes an asynchronous high-frequency vibration on the rotor, with the magnitude that increases at high rotating speeds together with the emulated gyroscopic forces.

In order to further demonstrate the proposed emulation approach and avoid exciting the third bending mode, the original polar moment of inertia J_p from the flywheel disk is reduced by 80%. The emulated gyroscopic forces generated by

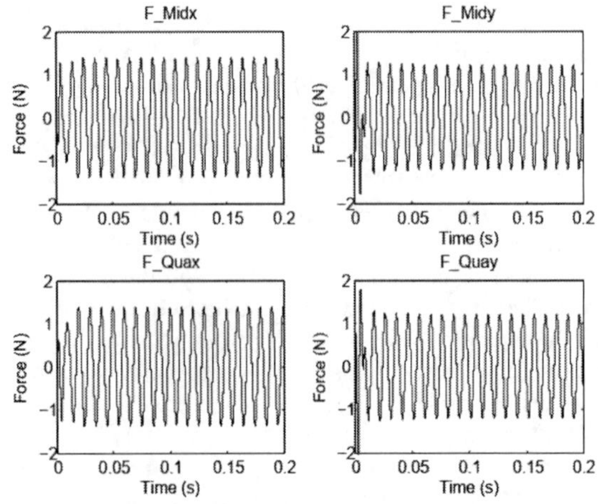

Figure 9: Emulated gyroscopic forces generated by the exciter AMBs with a reduced polar moment of inertia.

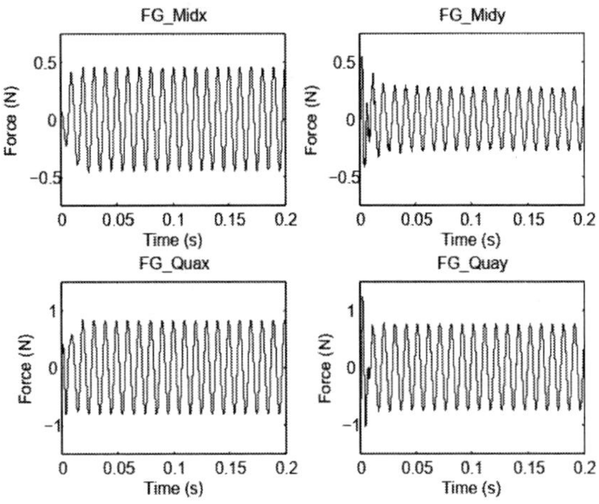

Figure 10: Actual forces caused by the gyroscopic matrix G at the exciter AMBs with a reduced polar moment of inertia.

the exciter AMBs, and the actual gyroscopic forces calculated from matrix G at the locations of the exciter AMBs, are shown respectively in Figs. 9 and 10 at the rotating speed of 6,000 rpm. The emulated and the actual gyroscopic forces are seen to be very similar, which demonstrates that the proposed emulation approach is feasible.

C. Experimental Results

The proposed emulation approach is also experimentally validated on the AMB test rig described in Section II. Since the generator negative stiffness causes a negligible effect on rotor dynamics, here only the gyroscopic effect of the flywheel disk is emulated on the test rig. Magnetic bearings D and G generate the required excitation forces in opposite directions and the rotor displacements are saved at 6,000 rpm. Shown in Fig. 11 are the sensor measurements of the rotor displacement at the DE and NDE locations. The exciter AMBs are switched on at around 13.6 s, and it is observed that the

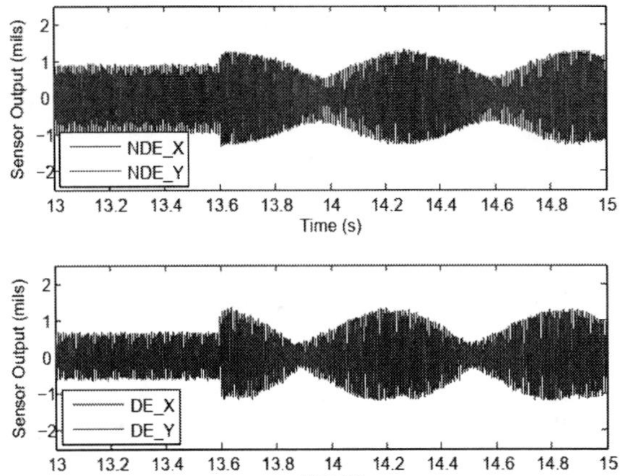

Figure 11: Sensor measurements.

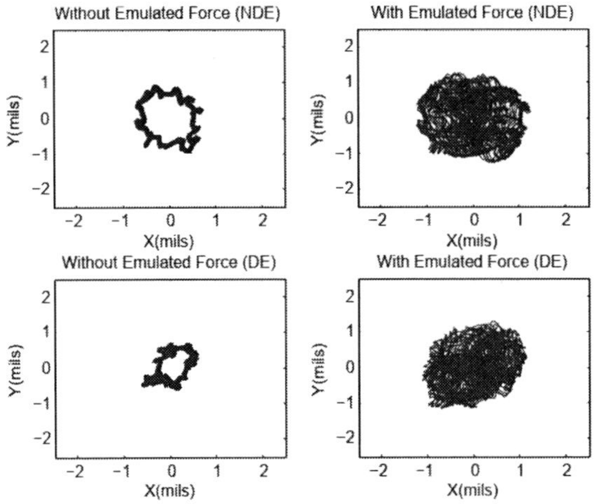

Figure 12: Rotor orbits.

magnitude of the rotor displacement increases in both the x and the y axes. Besides, sinusoidal movements of the rotor also appear under the excitation forces generated by our emulation scheme. Fig. 12 compares the rotor orbits without and with the excitation forces. The figure reveals the increases in the displacement magnitude and the gyroscopic like effect.

V. Conclusions

In this paper, an existing rotor-AMB test rig with exciter AMBs is adopted to emulate the operation of energy storage flywheels. The two AMBs at the mid and quarter spans of the shaft are used to emulate the negative stiffness of the generator and the gyroscopic effect of the flywheel disk. The simulation results have shown the negative stiffness provided by the generator causes a negligible effect on the rotor dynamics. The emulated gyroscopic forces generated by the exciter AMBs are similar to the actual gyroscopic forces produced by the gyroscopic matrix obtained from a previous finite element analysis, which demonstrates that the emulation approach presented in the paper is feasible.

For the future work, we will design a controller to improve the stability and performance of the rotor-AMB test rig in the presence of the emulated gyroscopic forces of the flywheel disk.

VI. Acknowledgements

This work was supported in part by the Key Project of Hubei Provincial Natural Science Foundation (No. 2009CDA028) and the China Scholarship Council.

References

[1] P. F. Ribeiro, B. K. Johnson, M. L. Crow, A. Arsoy and Y. Liu, "Energy storage systems for advanced power applications," *Proceedings of the IEEE*, vol. 89, no.12, pp.1744-1756, 2001.

[2] C. M. Reid, T. B. Miller, M. A. Hoberecht, P. L. Loyselle, L. M. Taylor, S. C. Farmer and R. H. Jansen, "History of electrochemical and energy storage technology development at NASA Glenn Research Center," *Journal of Aerospace Engineering*, vol. 26, no. 2, pp. 361-371, 2013.

[3] R. Hebner, J. Beno, and A. Walls, "Flywheel batteries come around again," *IEEE Spectrum*, vol. 39, no. 4, pp. 46-51, 2002.

[4] S.M. Lukic, J. Cao, R.C. Bansal, F. Rodriguez and A. Emadi, "Energy storage systems for automotive applications," *IEEE Transactions on Industrial Electronics*, vol. 55, no. 6, pp. 2258-2267, 2008.

[5] J. G. Bitterly, "Flywheel technology: Past, present and 21st century projections," *IEEE aerospace and electronic systems magazine*, vol. 13, no. 8, pp. 13-16, 1998.

[6] G. O. Cimuca, C. Saudemont, B. Robyns and M. M. Radulescu, "Control and performance evaluation of a flywheel energy-storage system associated to a variable-speed wind generator," *IEEE Transaction on Industrial Electronics*, vol. 53, no. 4, pp. 1074-1085, 2006.

[7] G. V. Brown, A. Kascak, R. H. Jansen, T. P. Dever and K. P. Duffy, "Stabilizing gyroscopic modes in magnetic-bearing-supported flywheels by using cross-axis proportional gains," in *Proceedings of the AIAA Guidance, Navigation, and Control Conference*, San Francisco, California, 2005, pp. AIAA-2005-5955.

[8] T. P. Dever, G. V. Brown, K. P. Duffy and R. H. Jansen, "Modeling and development of a magnetic bearing controller for a high speed flywheel system," in *Proceedings of the 2nd International Energy Conversion Engineering Conference*, Providence, RI, 2004, pp. AIAA-2004-5626.

[9] M. Ahrens and L. Kucera, "Cross feedback control of a magnetiic bearing system," in *Proceedings of 3rd International Symposium on Magnetic Suspension Technology*, Tallahassee, Florida, 1995, pp. 177-191.

[10] M. Ahrens, L. Kucera and R. Laesonneur, "Performance of a magnetically suspended flywheel energy storage device," *IEEE Transactions on Control Systems Technology*, vol. 4, no. 5, pp. 494-502, 1996.

[11] L. A. Hawkins, B. T. Murphy and J. Kajs, "Analysis andtesting of a magnetic bearing energy storage flywheel with gain-scheduled, MIMO control," in *Proceedings of ASME Turboexpo*, Munich, Germany, 2000, pp. 1-8.

[12] S. Sivrioglu, K. Nonami and M. Saigo, "Low power consumption nonlinear control with H_∞ compensator for a zero-bias flywheel AMB system," *Journal of Vibration and Control*, vol. 10, pp. 1151-1166, 2004.

[13] S. Sakai, K. Kuriyama and K. Nonami, "A novel passivity based control of active magnetic bearing systems without conventional cross feedback," *Journal of System Design and Dynamics*, vol. 3, no. 4, pp. 540-550, 2009.

[14] S.E. Mushi, "Robust Control of Rotordynamic Instability in Rotating Machinery Supported by Active Magnetic Bearings," Ph.D. Dissertation, Uniersity of Virginia, Charlottesville, VA, 2012.

[15] L. Di and Z. Lin, "Control of a flexible rotor active magnetic bearing test rig: a characteristic model based all-coefficient adaptive control approach," *Control Theory and Technology*, vol. 12, no. 1, pp. 1-12, 2013.

[16] P. E. Kascak, T. P. Dever and R. H. Jansen, "Magnetic circuit model of PM motor-generator to predict radial forces," in *Proceedings of 1st International Energy Conversion Engineering Conference*, Cleveland, 2003.

An Energy Storage Flywheel Supported by Hybrid Bearings

Kai Zhang[a], Xingjian Dai[a], Jinping Dong[a]

[a] Department of Engineering Physics, Tsinghua University, Beijing, China, zhangkai@mail.tsinghua.edu.cn

***Abstract*—Energy storage flywheels are important for energy recycling applications such as cranes, subway trains. In a petroleum field, a drilling platform runs with big load variation. A vertical flywheel energy storage system had been tested to stabilize the load fluctuation and proved its effectiveness. To improve bearing life and reliability, a new flywheel bearing system was designed. The key was the use of hybrid bearings including an axial permanent magnetic bearing (PMB), a lower end ball bearing and an upper end active magnetic bearing (AMB). The design effectively combined advantages of the different bearings. The original upper ball bearing was replaced with the AMB to avoid the use of a lubrication system for it and improve the damping ability for the rotor vibration modes. The vibration mode and the harmonic response were provided in the rotor dynamic analysis for the new flywheel rotor. The analysis results proved that the new rotor structure and the AMB design were appropriate.**

I. INTRODUCTION

In an oil drilling platform, the drill is a key instrument and its power is often provided by a diesel power system or a natural gas engine. When a drill operates, its power consume varies a lot with time and ongoing operation. Using diesel or natural engines to operate the drill would pose problems of low efficiency and large amount of emissions. It is attractive to replace such drilling energy sources with a new one which can make the entire drilling process economic and environmentally friendly [1], [2].

One of the advanced technologies is adding a flywheel subsystem to an original diesel engine system. When a drill operates in a low load state, the diesel engine could still operate in a full power mode and the redundant power from the diesel could be used to drive the flywheel rotor through a DC motor. When the rig operates in a high load station, energy shortage for a peak power requirement could be supplied by the flywheel system. The flywheel energy storage system would discharge and supply power to the rig through the DC motor.

A flywheel energy storage system (FESS) is one of options among available renewable energy resources. It has a high output power, a long life and a high response speed [3]. It is economically and practically feasible to modern oil and gas industry.

A large power flywheel (Flywheel I) had been successfully designed and tested in a drill system in Zhongyuan Petroleum Exploration Bureau of China, whose driving power was based on a diesel engine system. Its effects had been proved [1]. In the test rig system, the output time of the power system for a pulse drilling load of 300-500 kW was 2-20 s. During one operation period, the diesel engine had an abundant output time of 80-120 s at 100-150 kW power level. So in its operation mode, the flywheel worked as a motor at 90-130 kW power level for 120 s and then worked as a generator at 300-500 kW power level for 20 s. The flywheel operated as a power amplifier in a mode with lower charge power and higher discharge power.

The structure of the Flywheel I rotor was shown in Fig. 1. It was a vertical flywheel supported by an upper ball bearing, a lower ball bearing and a permanent magnetic bearing (PMB). The flywheel housing was filled with helium to decrease wind loss of the rotor in operation. The ball bearings were used to support radial load and partial axial load of the flywheel rotor whose maximum rotation speed was 3600 rpm. For each of the bearings, there was a plastic ring mounted outside to decrease its radial stiffness and provide necessary damping for the rotor bearing system. The PMB was used to counteract the most rotor weight. It effectively decreased the axial load of the lower ball bearing and was benefit for the bearing life. The lower bearing was immersed in oil and its life could be ensured by a good heat transfer and a good lubrication. For the upper bearing, it was lubricated by grease. The lubrication was not as good as the lower bearing and its life would be influenced.

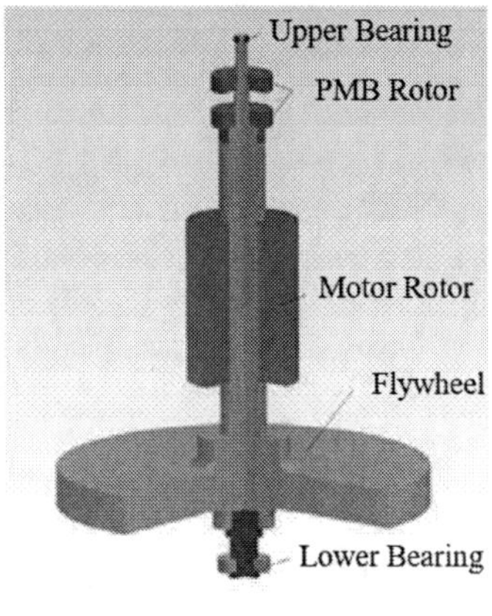

Figure 1. The structure of the Flywheel I rotor.

Active magnetic bearings are attractive for their special advantages compared with traditional ball bearings and fluid film bearings, such as contactless, no wear, no oil, low power consumption, low maintenance cost, and controllability of bearing dynamic characteristics and of rotor unbalance response [4].

To further improve bearing life, especially for the upper bearing, a new hybrid bearing strategy was designed and it would be tested in a new flywheel system (Flywheel II). The hybrid bearings included an axial permanent magnetic bearing (PMB), a lower end ball bearing and an upper end active magnetic bearing (AMB). The key was to replace the original upper ball bearing with the new AMB.

Such a design combined advantages of the different bearings and could improve system reliability. The PMB was cheap and energy-saving. It removed most static load for the vertical rotor. So it was beneficial for improving the ball bearing life and decreasing the system power lost. The ball bearing was cheap and reliable but needed lubrication. Its static load was limited at high rotation speeds and was suitable to be used as the lower end bearing. For the upper end support, a lubrication system for a ball bearing was complex and difficult to maintain. The lubrication problem of the upper bearing could be avoided by using the AMB. The AMB with no need for lubrication was dynamic controllable and suitable for the high speed rotor with a small dynamic load. Though cost more, it could provide a strong ability of restraining the rotor vibration, effectively improve the bearing life and reduce the maintenance cost.

The new flywheel system structure with the new hybrid bearing system would be shown and the important feathers of the hybrid bearing system would be provided.

To prove the effectiveness of the AMB design, the rotor dynamic analysis for the new flywheel rotor would be done. The observability and controllability of the AMB for the vibration modes interested would be discussed, and the load capacity of the AMB would be compared with the harmonic response analysis results of the rotor system.

A. Flywheel I rotor system

The Flywheel I rotor was designed according to the FESS power and energy requirement. The overall energy efficiency of the system in one working period should be larger than 80%. Some important system parameters were listed in Table 1.

TABLE I. SYSTEM PARAMETERS OF THE FLYWHEEL I

Parameter	Value
Recycle speed range	1800-3600 rpm
Total kinetic energy at the highest rotation speed	16.3 MJ
Motor power	300-500 kW
Transverse moment of inertia	230 kg.m^2
Highest line speed of the rotor	226 m/s
Overall high of the rotor	1800 mm
Largest diameter of the rotor	1200 mm
Total weight of the rotor	1600 kgf

The design of the Flywheel I had been proven successful by field experiments. But for a long-term operation, the current lubrication of the upper bearing was not satisfied. It was hard to build a new lubrication system for the upper bearing to increase its reliability. So a new hybrid bearing strategy was designed and would be tested in the Flywheel II rotor system.

B. Flywheel II rotor system

The new flywheel system structure with the new hybrid bearing system was shown as Fig. 2.

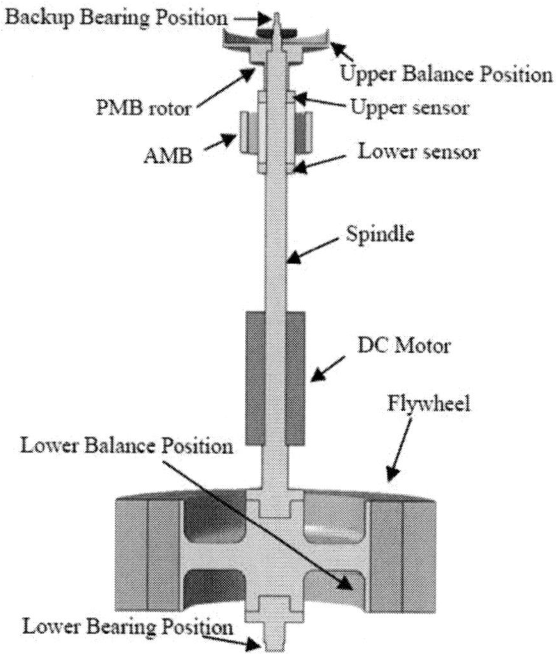

Figure 2. The structure of the Flywheel II rotor.

In the new design, an upper active magnetic bearing was added to the original bearing systems and it was used to replace the upper ball bearing to support the rotor at the upper end. A backup ball bearing was located in the position where the original ball bearing located. But there was a small gap of 0.25 mm between its inner ring and the flywheel rotor. Such a backup bearing was used to protect the AMB rotor from colliding with the AMB stator when the rotor was in a critical condition where the rotor vibrated violently, or when the AMB stopped working.

For the AMB, there were two steel rings fixed on the upper and the lower sides of the AMB rotor. They were both used as detection targets of eddy current sensors. The displacement signal of the AMB center was calculated from the signals of the two sensors.

The flywheel was designed to store a total energy of 2.47 kWh at 18,000 rpm and deliver 40 kW for more than 160 seconds. Some important system parameters were listed in Table II.

The flywheel fitted on a steel hub was composed by two kinds of composite materials. The steel hub was first wrapped with glass fiber composites and then carbon fiber composites were wrapped outside the inner glass fiber.

TABLE II. SYSTEM PARAMETERS OF THE FLYWHEEL II

Parameter	Value
Recycle speed range	9000-18000 rpm
Total kinetic energy at the highest rotation speed	8.9 MJ
Motor power	20-40 kW
Transverse moment of inertia	5 $kg.m^2$
Highest line speed of the rotor	565 m/s
Overall high of the rotor	1150 mm
Largest diameter of the rotor	600 mm
Total weight of the rotor	170 kgf

C. *Active magnetic bearing*

The AMB was placed above the motor and it was used to replace the original upper radial ball bearing. When it worked, it would levitate the rotor from the upper backup bearing (a ball bearing), provide stiffness and damping for the rotor and help the rotor running stably across several critical speeds in a high speed operation. Because the distance from the AMB center to the mass center was much larger compared with the distance from the lower bearing and the rotor was vertically oriented, the static load for the AMB was not large. But its dynamic load capacity should be large enough to restrain the rotor's synchronous critical vibration when the rotor crossed its critical speeds.

The basic parameters of the AMB were shown in Table III.

TABLE III. AMB PARAMETERS OF THE FLYWHEEL II

Parameter	Value
Stator length	70 mm
Bearing load	1000 N
Diameter of the bearing rotor	70 mm
Gap between the bearing rotor and the stator	0.5 mm
Stator pole number	8

D. *Rotor vibration modes*

For the Flywheel II, an important consideration in the rotor design was to assure the effectiveness of the AMB in the rotor vibration control. The key was that the observability and controllability of the AMB should be suitable. So an accurate rotor vibration mode analysis should be carried out.

To obtain accurate dynamic behavior of a rotor, finite element methods are usually used. Many commercial software packages, such as NX Nastran, have provided the ability to construct a model for a rotor with complex meridian cross sections. In such a model, gyroscopic effects and centrifugal softening effects can be effectively taken into account.

For simplicity, 2D Fourier elements were used to model the flywheel rotor. With the 2D Fourier elements, axisymmetrical rotors can be modeled using 2D axisymmetrical shell or volume finite elements whose displacement field is developed in Fourier series [5]. The corresponding model was shown in Fig. 3. There were two springs connecting the ground with the rotor at the positions corresponding to the AMB and the lower bearing. The spring stiffness was set as 1000 N/mm and 10000 N/mm for the two bearings respectively. The damping was set the same as 500 N.s/m for them. The first 5 vibration modes of the flywheel rotor at standstill were shown as Fig. 4. The modes included two rigid modes and 3 flexible modes. Because the eigenfrequencies of the other modes were much higher than the maximum rotation frequency of the rotor, the analysis for them was ignored here.

Figure 3. 2D FEM model of the flywheel rotor.

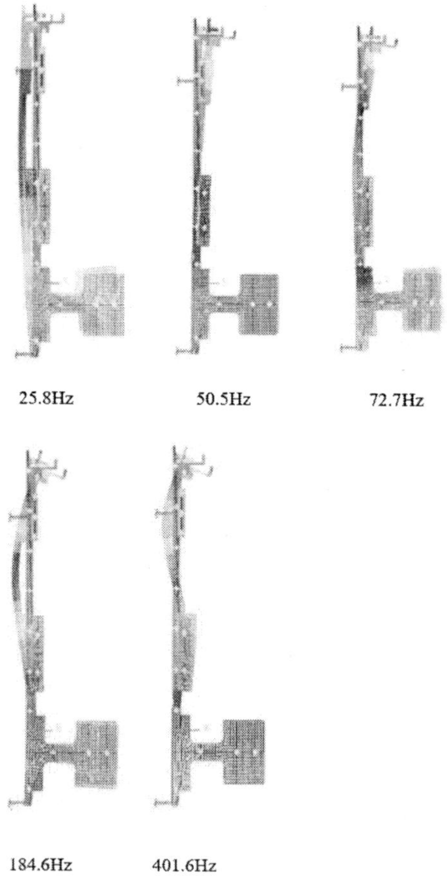

Figure 4. Vibration modes of the flywheel rotor.

When the rotor runs from 0 to 18000 rpm, the vibration of the 5 modes must be well restrained. So their observability and controllability condition at the AMB position should be assured. The FEM calculated results for them were listed in Table IV. The values in the table showed the vibration amplitude of the rotor relative to the maximum amplitude values of the corresponding vibration modes. For the AMB, the mode vibration amplitude of the 5 modes was acceptable and it assured that the AMB force could effectively control the mode vibration. But at the position of the upper sensor, the vibration amplitude of the 4th mode was relatively small, namely, that the observability of the sensor for the 4th mode was not good. A similar case happened at the lower sensor position for the 3rd mode. If the two sensors were combined to provide an average value, the AMB center displacement measurement accuracy for all the modes could be acceptable.

TABLE IV. OBSERVABILITY AND CONTROLLABILITY CONDITION OF THE FIRST 5 MODES

Position	Mode number				
	1st	2nd	3rd	4th	5th
Upper sensor center	-0.86	0.68	0.50	-0.24	-0.50
AMB center	-0.90	0.55	0.29	-0.50	-0.82
Lower sensor center	-0.94	0.43	0.12	-0.80	-0.88

E. *Gyroscopic effects*

Because of the flywheel body on the rotor, its inertia was high and the gyroscopic effects were strong. The gyroscopic effects greatly influenced the rotor dynamic behavior of the rotor at a high rotation speed. The eigenfrequencies at different rotation speeds were obtained from the FEM analysis and the Campbell diagram for the rotor was shown in Fig. 5.

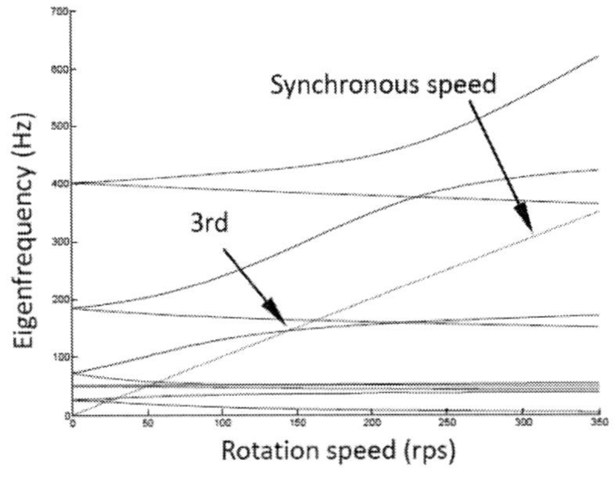

Figure 5. Campbell diagram for the rotor.

From Fig. 5., it could be seen that the rotor's rigid body modes and bending modes were both influenced by the gyroscopic effects obviously. For a full speed operation, the rotor had to pass 3 critical speeds totally. Because the 3rd critical speeds was much higher than the other critical speeds, as shown in the figure, it was the most difficult to be passed. The AMB should have a large enough dynamic load capacity to provide a force to counteract the excitation force from the rotor unbalance at this critical speed. Furthermore, the rotor should be properly balanced before a high speed operation to decrease the load of the bearings.

F. *Frequency response analysis*

To verify that the AMB could effectively restrain the synchronous vibration of the rotor at the 3rd critical speeds, a frequency response analysis for the rotor was necessary.

After a good dynamic balance, if the balance quality grade of the rotor could reach G6.3, the residual unbalance of the rotor would be 536 g.mm. When the residual unbalance was placed at the rotor mass centre and the rotor was suspended by the two bearings as shown in Fig. 3, the harmonic response of the rotor at the upper and the lower bearings was shown in Fig. 6. It could be seen that the reaction force of the upper bearing was much less than its load capacity within the rotation speed range when a damping of 500 N.s/m was provided (such a damping was reasonable for a flywheel AMB [6]). So for an operation above the 1st flexible critical speed, the AMB load capacity would be sufficient.

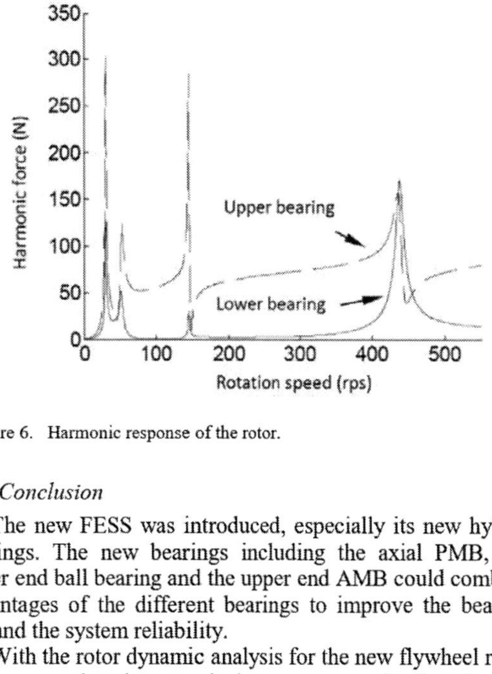

Figure 6. Harmonic response of the rotor.

G. *Conclusion*

The new FESS was introduced, especially its new hybrid bearings. The new bearings including the axial PMB, the lower end ball bearing and the upper end AMB could combine advantages of the different bearings to improve the bearing life and the system reliability.

With the rotor dynamic analysis for the new flywheel rotor, it was seen that the rotor had to run across its first flexible critical speed to reach its maximum rotation speed. The observability and controllability of the AMB for the vibration modes interested were suitable. The harmonic response analysis results showed that the bearing load was enough for such a supercritical operation.

This work is supported by NSFC(11175100) and "The Importation and Development of High-Caliber Talents Project of Beijing Municipal Institutions"(YETP0114)

REFERENCES

[1] L.B. Jin, X.J. Jiang, C.P. Zhang, M.J. Tong, "Research on oil rig based on the flywheel energy storage system, " *Micromotors*, vol. 45, no. 11, pp. 27-30, 2012, (in Chinese)

[2] S. Ahmed, R. Tafreshi, A. Palazzolo, T. Randall, "Drilling rig fuel and emissions reduction through regenerative braking, load leveling and grid drilling," *International Petroleum Technology Conference*, 2009
[3] A. Rojas, "Flywheel energy matrix systems – today's technology, tomorrow's energy storage solution," *IEEE Power Engineering Society General Meeting*, 2004
[4] G. Schweitzer, H. Bleuler, A. Traxler, "Active magnetic bearings basics, properties and application of active magnetic bearings," ETH, Switzerland: Hochschulverlag AG, 1994
[5] D. Combescurea, A. Lazarus, "Refined finite element modelling for the vibration analysis of large rotating machines: Application to the gas turbine modular helium reactor power conversion unit," *Journal of Sound and Vibration*, vol. 318, no. 4-5, pp. 1262-1280, 2008
[6] L. Hawkins, S. Imani, D. Prosser, M. Johnston, "Design and shop testing of a 165kw cryogenic expander/generator on magnetic bearings," *Ninth International Symposium on Magnetic Bearings*, 2004

Flywheel Energy Storage System to be Applied for the Railway System

Jun-Ho Lee[a]

[a] Korea Railroad Research Institute
#176, Cheoldo, Bangmulgwan-ro, Uiwang, GyeonggiDo, Korea, 437-757

***Abstract*— Design and manufacturing of flywheel energy storage system and energy conversion to be applied for the railway system are presented. In general railway trains generate regenerative energy for 10-20 sec when the train brakes and high traction energy is needed for very short moment (10 sec) when the train increases the traction force. Considering such characteristics of the railway system energy storage system for the railway should have very fast response property. Among the various energy storage systems flywheel energy storage system has the fastest response property, which means that flywheel ESS is the most suitable for the railway system.**

I. Introduction

Many researchers have invested their effort to develop flywheel ESS focusing on the development of the high speed flywheel module rather than the system level development including an operational system for the energy conversion.
The representative groups for the flywheel ESS are KTSi(Kinetic Traction System Inc.), Vycon, University of Virginia, KRRI(Korea Railroad Research Institute), Piller Power Systems, etc. Among them KTSi and KRRI have tried to develop flywheel ESS which is applicable to the railway system. KTSi has tested its flywheel ESS in the Bombardier light rail test track in 2012 which is in Kingston, Canada. It has 35,000rpm operational speed, 200kW*2, and completely passive type. On the other hand KEPRI(Korea Electrical Power Research Institute) has tried to develop 100kWh, 350kW/unit super conductor flywheel ESS.
In this paper the system configuration for the flywheel ESS including operational system for the energy conversion and the design and manufacturing of small-scale flywheel energy storage system (five degree of freedoms) supported by magnetic bearings are presented.
The effectiveness of the proposed flywheel energy storage system is shown based on the high speed rotational test and also the applicability of the proposed configurations to the railway system is presented by using the energy conversion of the flywheel energy storage system that indicates the charging of the regenerative energy and discharging for the traction force based on the load test.

II. Mathematical Model and Configuration

Fig. 1 presents a flywheel rotor supported by the magnetic bearings which has four-degrees-of-freedom. For the rotational rotor, left and right electromagnetic bearings support the rotor to make levitation so that the rotor can rotate without any contact with the stator. Displacement sensors are attached to very close to the electromagnetic bearings to make feedback of the displacement deviation which make it possible for the rotor to be controller actively.

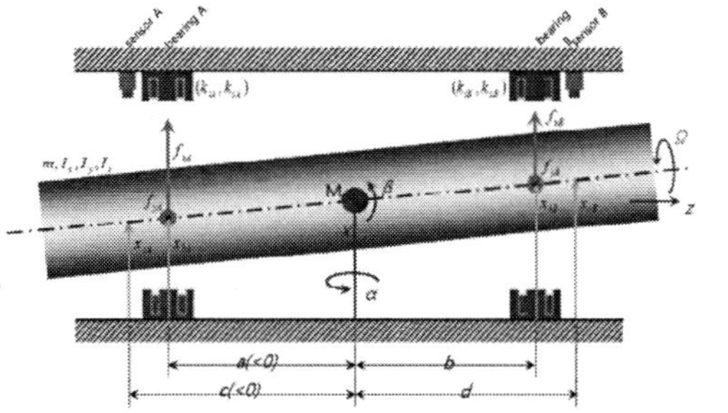

Fig. 1 Flywheel rotor which has four-degrees-of-freedom

Equation of motion of the rotor shown in Fig. 1 is such as:

$$M\ddot{q} + G\dot{q} = {}_{M}T_{B}u_{AMB} \qquad (1)$$
$$y = {}_{S}T_{M}q$$

where

$$M = \begin{bmatrix} I_y & 0 & 0 & 0 \\ 0 & m & 0 & 0 \\ 0 & 0 & I_x & 0 \\ 0 & 0 & 0 & m \end{bmatrix}, G = \begin{bmatrix} 0 & 0 & I_z\Omega & 0 \\ 0 & 0 & 0 & 0 \\ -I_z\Omega & 0 & 0 & 0 \\ 0 & 0 & 0 & 0 \end{bmatrix},$$
$${}_{M}T_{B} = \begin{bmatrix} a & b & 0 & 0 \\ 1 & 1 & 0 & 0 \\ 0 & 0 & a & b \\ 0 & 0 & 1 & 1 \end{bmatrix}, {}_{S}T_{M} = \begin{bmatrix} c & 1 & 0 & 0 \\ d & 1 & 0 & 0 \\ 0 & 0 & c & 1 \\ 0 & 0 & d & 1 \end{bmatrix} \qquad (2)$$

$$q = (\beta, x, -\alpha, y)^T$$
$$u_{AMB} = (f_{xA}, f_{xB}, f_{yA}, f_{yB})^T \quad (3)$$
$$y = (x_{sA}, x_{sB}, y_{sA}, y_{sB})^T$$

M is the mass matrix that is composed of mass and moment of inertia, G is the gyroscope matrix that is 0 when the rotor stops, however the amplitude becomes greater gradually as the rotational speed goes up. ${}_M T_B$ is the transformation matrix to transform the bearing coordinates to mass center coordinates, and ${}_S T_M$ is the transformation matrix to transform mass center coordinates to displacement sensor coordinates. q is the vector that expresses the position of the radial directions and the rotational angles in the mass center coordinates. u_{AMB} the force that is generated from the magnetic bearings, and y is the distance that is measured from the displacement sensor to the rotor.

Fig. 2 shows the overall system configuration that consists of flywheel system, vacuum pump, and levitation controller. The flywheel system has the flywheel rotor and the housing that rotational motor stator, upper, lower, and thrust magnetic bearings are to be installed in it. The vacuum pump is connected with the housing to make vacuum the inside of the hosing, which increases the efficiency of the power conversion. The levitation controller levitates the rotor and controls the vibration of the rotor during the high speed rotation.

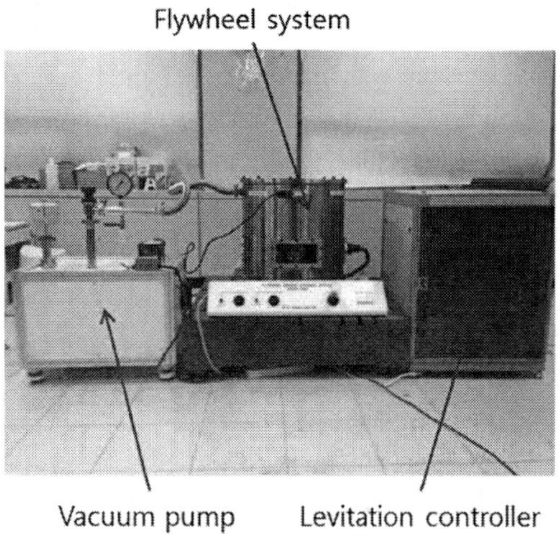

Fig. 2 FESS (Flywheel Energy Storage System) configuration

III. Rotor Design

Fig. 3 shows the shape of the rotational rotor. The rotational rotor is composed of the upper and lower radial bearing to support the rotor in the radial direction, the thrust plate which is used for the levitation of the rotor in the horizontal direction, the flywheel to store the rotational energy, and the PM type motor to make the rotor rotate.

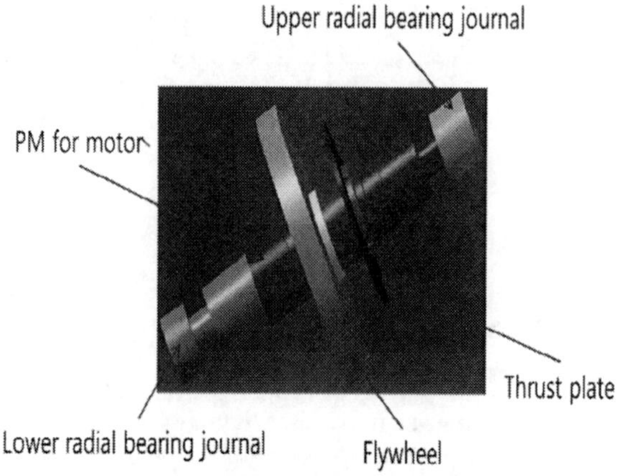

Fig. 3 Shape of the rotational rotor

Fig. 4 shows the FRF(Frequency Response Function) test result that has been produced from the general natural frequency test using the impact hammer. As seen in the figure the first mode is at 1,160Hz ($\cong$ 69,600rpm).

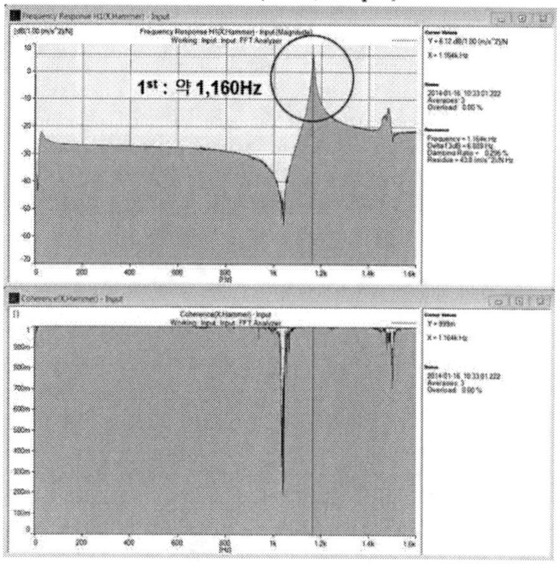

Fig. 4 FRF test result

Fig . 5 presents the measured data to analyze the relations between the rotational speed and the natural mechanical modes that appear in the rotational machine when the rotor rotates in high speed. Fig. 5 indicates that the first natural frequency appears at 1,050Hz~1,100Hz, and the rotational speed in that frequency is 10,000rpm ~ 12,000rpm that meets with the 5th and 6th mechanical modes and makes the amplitude of the rotational vibration at these modes be increased. Based on these analyses the rotational speed is set to 12,000rpm to avoid the mechanical resonance in this paper.

The 1st natural frequency in the FRF test is 1,160Hz and the 1st natural frequency in the rotational mode test is 1,050Hz~1,100Hz. The 60Hz frequency difference is considered as measuring error.

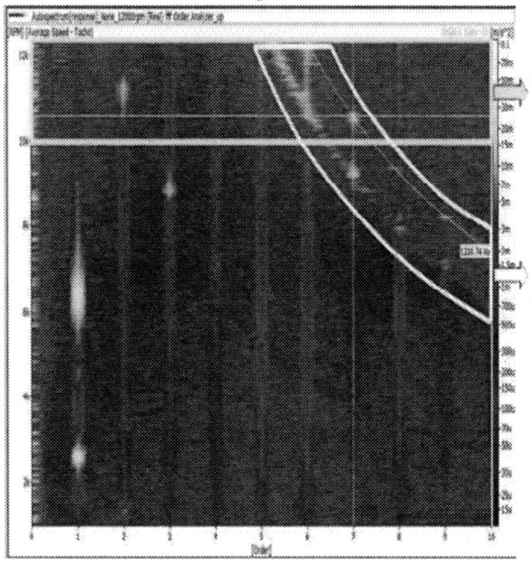

Fig. 5 Rotational modes analysis

IV. MANUFACTURING

In this section each component that consist of the flywheel energy storage system is presented.

4.1 Rotational rotor

Fig. 6 presents rotational rotor that has thrust plate for the rotor levitation and wheel to store the excess electrical energy. The total weight of the rotor is 23[kg].

Fig. 6 Flywheel Rotor

4.2 Magnetic bearings

Fig. 7 shows the radial and thrust magnetic bearings to support the rotor in radial and in axial direction. They have 500[N] and 2,500[N] maximum load capacity, respectively. Table 1and 2 indicate the design requirements for the radial and the axial magnetic bearings.

Table 1. requirements for the radial MB

Variables	Values
Number of poles (n_p)	8
Iron ratio (f_i)	0.6
Unbalance grade	6.3
Saturation flux density (B_{sat})	1.2 T
Mass of the rotor	32 Kg
Max. operating speed	30,000rpm
Nominal air gap (g_Q)	0.6mm
Bearing journal diameter	69.4mm

Table 1. requirements for the axial MB

Variables	Values
Max. current density	6A/mm²
Saturation flux density (B_{sat})	1.2T
Mass of the rotor	32Kg
Max. operating speed	30,000 rpm
Nominal air gap (g_Q)	0.6mm

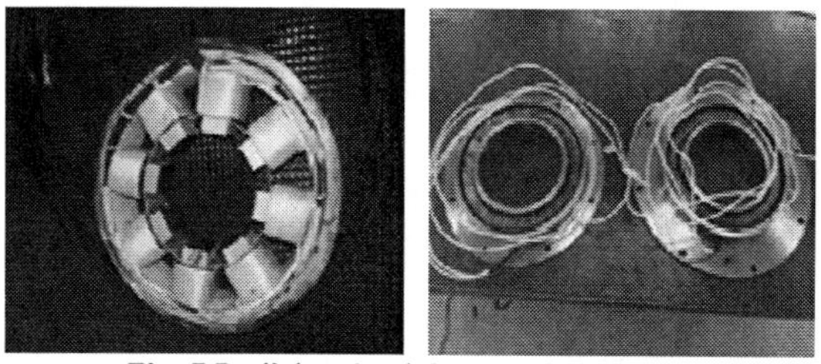

Fig. 7 Radial and axial magnetic bearings

4.3 Housing

Fig. 8 indicates the hosing for the flywheel energy storage system. The height and the diameter of the hosing are 585[mm] and 550[mm], respectively.

Fig. 8 Housing

V. SIMULATIONS

For the simulations of the system that mentioned in the above sections PID controllers (seen in Fig. 9) are installed to control input current of the magnetic bearings which play an important role to maintain stable air gap between magnetic bearings and the rotor. Table 3 is the parameters for the simulations

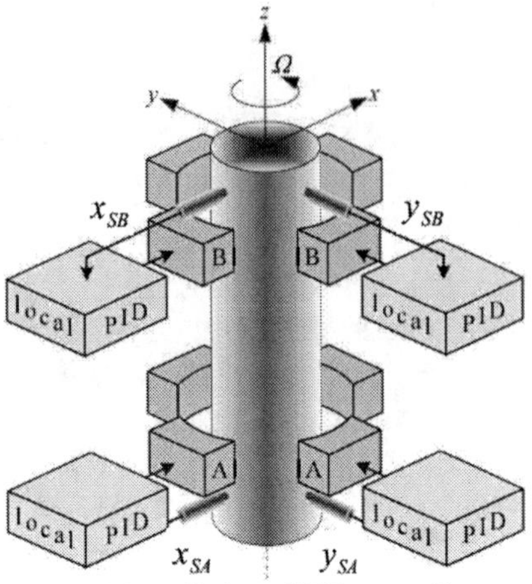

Fig. 9 Installation of PID controllers

Table 1. Parameters for the flywheel energy storage system

Variables	Value	Unit
Coil Inductance : L	50	[mH]
Coil Resistance : R	1	[Ω]
Steady Current : i_0	1	[A]
Cross Sectional Area : A	1.666×10^{-4}	$[m^2]$
Number of turns : N	152	[turns]
Permeability : μ_0	$4\pi\times10^{-7}$	[H/m]

Fig. 10 - 11present simulation result. For the simulations 10[g-mm] and 100[g-mm] unbalance masses added in the rotor and the initial position of the rotor is 0.1[mm] from the orgin. The red line indicates nominal airgap, 0.3[mm]. The rotational speed changes from 0[rpm] to 12,000[rpm] . As shown in the figures the trajectory of the rotor starts at the initial position (0.1[mm]) and reachs to the zero position even if there are unbalance masses (10[g-mm] and 100[g-mm]), and also the air gap deviation is very small even if the rotational speed changes from 0[rpm] to 12,000[rpm].

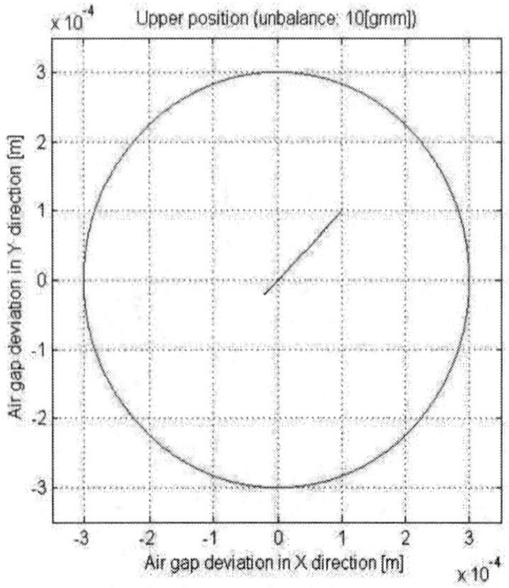

Fig. 10 Plotting in upper XY direction (10[g-mm])

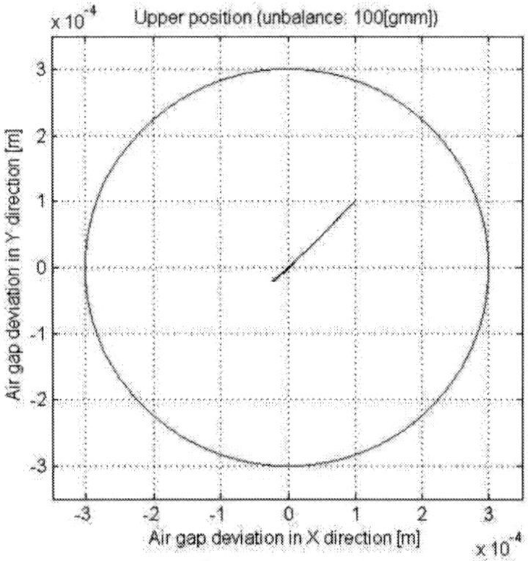

Fig. 11 Plotting in upper XY direction (100[g-mm])

VI. EXPERIMENTAL RESULTS

Fig. 12 shows the run down test results for the vacuum pump when the speed of the rotor has been increased up to the 6,000[rpm]. In the figure the blue line and the green line indicate the non- vacuum and the vacuum, respectively. The blue line reaches to zero speed in 178[sec] faster than that of the green line, which means the rotor rotates longer time in the vacuum.

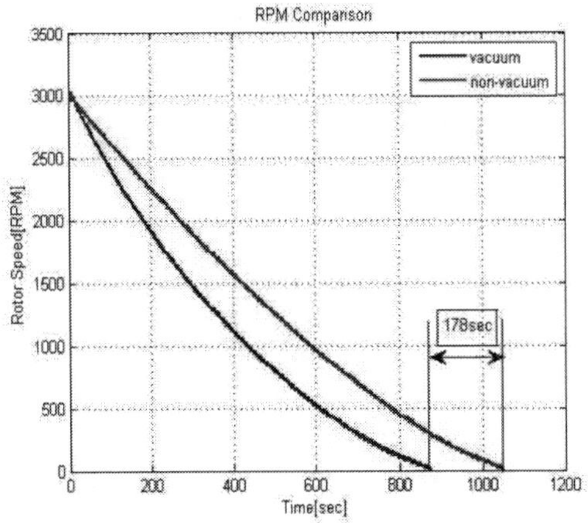

Fig. 12 Run down test for the vacuum pump

Fig 13 and 14 are the rotational test results at 12,000[rpm] that show 0.01[mm] vibration amplitude, which means that the rotational rotor is in very good centering position. Natural frequency at 4,400[rpm] (rigid mode) causes the higher vibration amplitude. In these figures the red line indicates the nominal air gap distance, 0.3[mm].

Fig. 15 presents the output power for the different load level. In case of 4[Ω] the output power at 12,000[rpm] is 3.6[kW] and it is reduced to about 96[W] at the rotational speed 1,769[rpm]. However for the lower load level (6[Ω] or 10[Ω]) than 4[Ω] the rotational speed to reach to about 96[W] (2362[rpm], 2860[rpm]) is higher than that of the 4[Ω] load level, which means that higher load can use more energy.

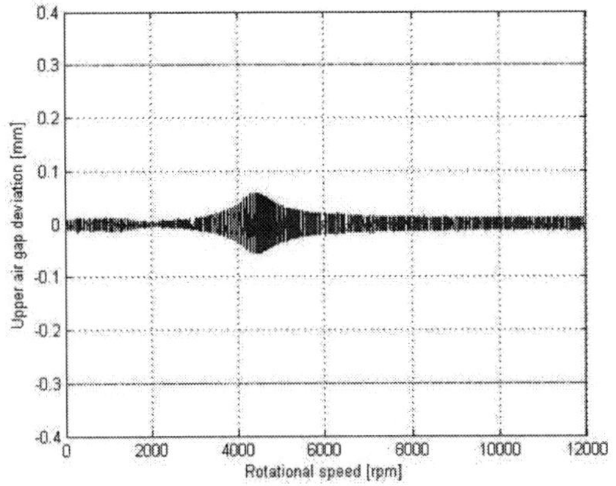

Fig. 13 Rotational speed test from 0 to 12,000rpm

VII. CONCLUSIONS

In this paper flywheel energy storage system to be applied for the railway system is presented. Mathematical model for the flywheel rotor (four-degrees-of-freedom), manufacturing of the FESS, and rotational test results were shown. Output power versus different load level tests were also performed and indicated that higher load can use more energy.

In the future to define the FESS efficiency various experimental tests will be performed.

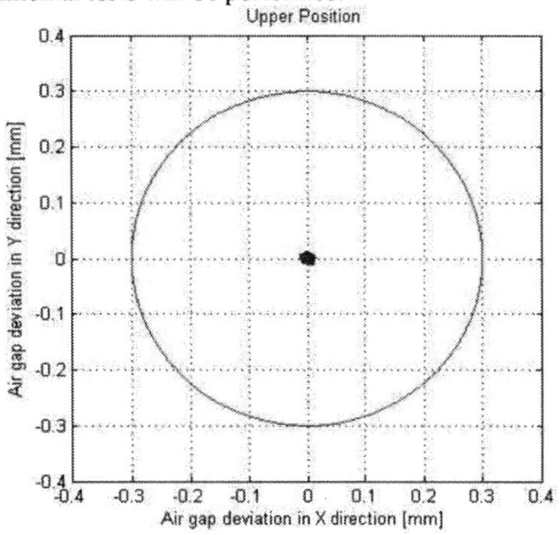

Fig. 14 Rotational speed test from 0 to 12,000rpm

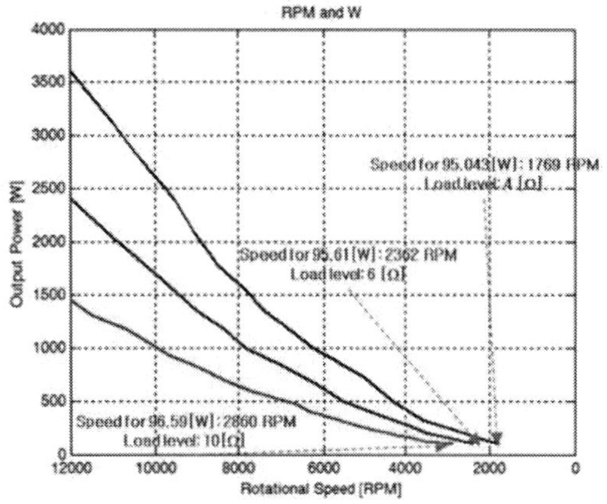

Fig. 15 Output power for the different load level

REFERENCES

[1] Jun-Ho Lee, *etal.* "A Study on the Design Procedure of the Eight Pole Magnetic Bearings for the Inner-rotor and the Outer-rotor type", Journal of Electrical Engineering Technology, vol. 8-6, pp. 1424-1430, 2013.

[2] G. Schweitzer and E. H. Maslen, eds., Magnetic Bearings: Theory, Design, and Application to Rotating Machinery, Springer, 2010.

[3] G. Schweitzer, H. Bleuler, A. Traxler, " Active Magnetic Bearings: Basics, Properties and Applications of Active Magnetic Bearings" , vdf Houchschulverlag AG der ETH Zurich,1994

[4]. A. Ortiz Salazar, W. Dunford, R. Stephan, E. Watanabe, "A Magnetic Bearing System Using Capacitive Sensors for Position Measurement" , IEEE Transactions on Magnetics, Vol. 26, No. 5, September 1990

Losses in an Outer-Rotor-Type Kinetic Energy Storage System in Active Magnetic Bearings

Hendrik Schaede [a], Michael Richter [a], Lukas Quurck [a], Stephan Rinderknecht [a]

[a] IMS - Institute for Mechatronic Systems, Technische Universität Darmstadt,
Otto-Berndt-Straße 2, 64287 Darmstadt, Germany, schaede@ims.tu-darmstadt.de

Abstract **- This paper starts with a general description of the losses and the classification of the loss components in kinetic energy storage systems. Based on the design of an outer-rotor-type kinetic energy storage system an analytical loss model is presented. It considers the conversion losses and the drag losses of the electric drive, the drag losses of the homopolar active magnetic bearings and the gas friction losses, as well as the constant losses such as caused by the bias magnetization of the active magnetic bearings. In detail a new approach for the analytical calculation of the drag losses of the homopolar active magnetic bearings is presented, taking the eddy current and hysteresis losses as well as the penetration depth of the magnetic field into the rotor in account. The paper ends with the presentation of a developed measurement method for the experimental determination of the loss map of kinetic energy storage systems. It enables for the isolation of the drag loss components of the drive system, the magnetic bearings and the gas friction losses.**

TABLE I. USED VARIABLES

Symbol	Name (all units SI)
P_{Iron}	Iron losses
m_{Iron}	Mass of magnetized iron
f	Frequency
B	Flux density
d	Penetration depth
μ	Permeability
κ	Electrical resistivity
ω	Angular frequency
Θ	Polar moment of inertia
E	Energy content
P	Power
Δt	Duration of measurement

I. INTORDUCTION

Losses in energy storage systems can be divided into power-dependent losses, state-of-charge-dependent losses and constant losses. In a kinetic energy storage system the power-dependent losses result from the energy transfer in the drive system. The state-of-charge-dependent losses are caused by drag losses on the rotor, which are mainly the drag losses in the bearings, the drag losses of the motor as well as the gas friction losses. Constant losses result from the control system, the vacuum-system and, in case of active magnetic bearings, from the bias magnetization. A visual representation of the overall loss structure gives the three dimensional loss-map. Figure 1 shows an analytically calculated loss map of a kinetic energy storage system. The loss power is plotted on the vertical axis over the systems state defined by the energy content and the power transfer on horizontal axes.

With a higher energy content, which is quadratic to the rotational speed, the drag losses arise. These loss components are mainly proportional to the quadratic rotational speed, hence the drag losses appear mainly linear with the energy content in the loss map (see Power = 0 kW). Caused by the ohmic losses in the electric drive system, the power-dependent conversion losses rise quadratic to the system power. With lower rotational speeds the efficiency of the electric drive decreases, resulting in a higher loss power when energy is transferred.

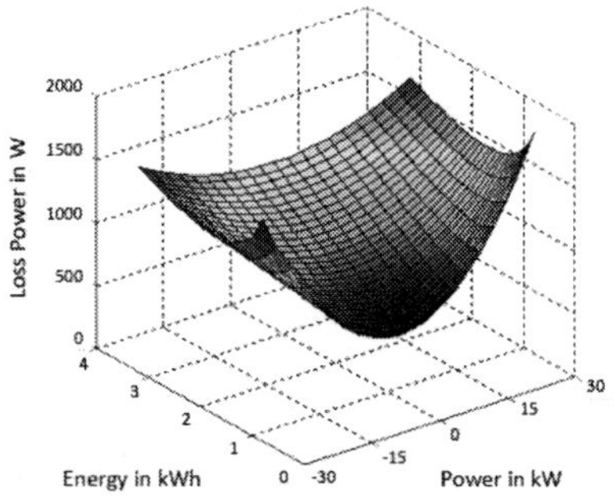

Figure 1. Analytically calculated lossmap of a Kinetic-Energy-Storage-System.

Figure 2 gives the basic structure of an outer-rotor-type kinetic energy storage system. The rotor in form of a hollow-cylinder made of fiber-reinforced-plastics (FRP) rotates around the stator, the rotor components of the magnetic bearings and the drive are integrated on its inner side.

Such systems have already been in discussion in the year 1980 [1]. This rotor geometry provides the maximum polar moment of inertia for the used rotor material, hence it promises the highest achievable energy density resulting in lower costs per capacity of the complete system. Yet no such systems are commercially available. In addition to the normal problems concerning inner-rotor kinetic energy storage systems, like highly gyroscopic effects, new problems arise with this topology. The comparable low stiffness of the rotor made of FRP and the centrifugal forces, caused by the rotation, result in a widening of the rotor of up to 1 % of its inner diameter. Depending on the properties of the overall system, the electric drive and the magnetic bearings must deal with an air-gap enlargement of approx. 0.5 mm to 1.5 mm. With the current state of development of FRP, electric machines and active magnetic bearings (AMB) currently new approaches to develop such outer-rotor-type kinetic energy storage systems are undertaken [2],[3].

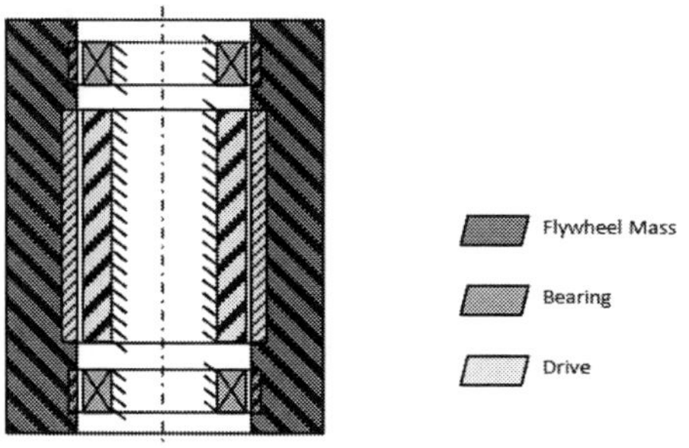

Figure 2. Structure of an outer-rotor-type kinetic energy storage system

Also at the IMS such systems are under theoretical and practical investigation. Figure 3 shows a photo of the set up prototype during the assembly. The rotor inner diameter is 180 mm, its outer diameter 300 mm and the rotor height 500 mm. Resulting in an inertia of the complete rotor of 0.7 kgm². In axial direction a passiv magnetic bearing (PMB) compensates the weight of the vertical rotor of 50 kg. The radial forces are compensated by active magnetic bearings of homopolar topology with active bias magnetization. The system is accelerated up to 20,000 rpm in a high vacuum environment, resulting in an energy content of approx. 0.4 kWh. A synchronous electric drive with Halbacharry-Rotor acts as motor and generator, providing 36 kW at 20.000 rpm. Figure 4 shows a photo of the complete test-rig.

Figure 3. Picture of the rotor on the stator during system assembly

Figure 4. Picture of the test-rig

II. Analytical Loss-Model

With the aim to find an optimal kinetic energy storage system for a given application an analytical model of the complete system is set up, allowing for fast studies over a wide range of parameters. For different parameters like inner-diameter, outer-diameter, height and maximum rotational speed as well as required capacity and power the model calculates the properties of the FRP-rotor, the AMB-system, as well as the drive. For the assessment of the resulting system its operational losses over the applications load profile are calculated via an analytical loss model. This loss model considers the conversion losses, the drag losses as well as the constant losses. While the constant- and the power-dependent losses can be calculated relatively easily, the drag losses of the rotor are more complex.

For most rotor geometries and rotating speeds the gas friction losses in high vacuum result from molecular flow. Because of the little number of left molecules under the low pressure environment, the gas molecules do not interact with each other like in fluids. Interaction takes place between the molecules and surfaces of the rotor, the stator and the containment. The resulting drag is calculated based on the momentum transferred between the molecules and the rotor.

The drag losses of the motor-generator result from the iron losses caused by the rotating rotor field. They can be calculated based on the approach of Steinmetz [4] or the more evolved approach of Bertotti [5]. In both cases the equations consider the mass of the magnetized material (m_{iron}), the magnetization frequency (f) and the magnetic flux density (B). Both approaches consider the hysteresis effects (k_h) and the eddy current effects (k_e) in the iron, the approach of Bertotti additionally considers the influence of abnormal eddy currents in the magnetic domains (k_a).

$$P_{iron} = m_{iron}(k_h \cdot fB^2 + k_e \cdot f^2B^2 + k_a \cdot f^{1,5}B^{1,5}) \quad (1)$$

These approaches take sinusoidal fields into account as they mainly appear in synchronous drives. Figure 5 compares the flux density in the air-gap of heteropolar and homopolar magnetic bearings. In both cases the magnetic field is approximated by a rectangular waveform. Fringing effects and leakage flux smoothen the waveform, so the approach of

Bertotti is often adopted to heteropolar bearings, as for example suggested in [6]. The analytical calculation of the drag losses in homopolar magnetic bearings is more complex. [7] proposes the Fourier transformation of the air-gap field to calculate the drag loss for each harmonic frequency of the signal. The set up model considers leakage flux and fringing effects via a cosine function, see Figure 5. The width of the cosine is estimated as 1.9 times the air-gap, following [9].

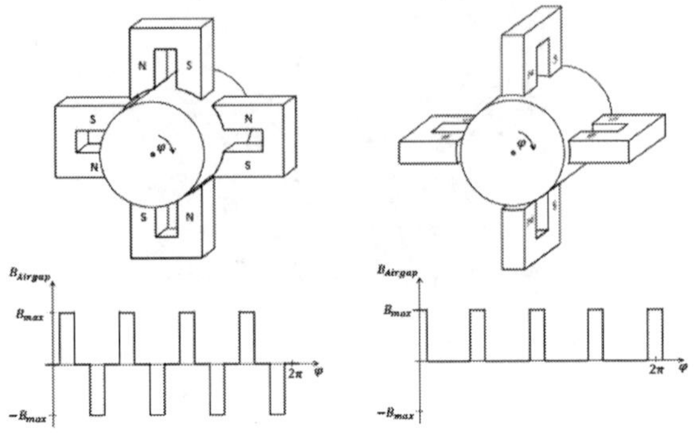

Figure 5. Comparison of the air-gap flux density of heteropolar (left) and homopolar (right) active magnetic bearings

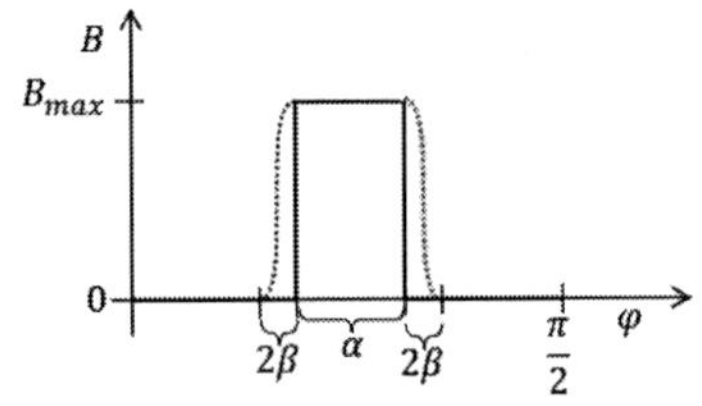

Figure 6. Approximation of the fluxdensity on the rotor surface under one pole of the homopolar bearing

With these assumptions the Fourier coefficients for the field of the homopolar bearing result as:

$$F(\omega) = \frac{a_0}{2} + \sum_{n=1}^{\infty} a_n cos\,(4n\varphi) \qquad (2)$$

With:

$$a_0 = 1 + \frac{4}{\pi}(\beta - \alpha)$$

$$a_n = a_1(n) + a_2(n) + a_3(n)$$

$$a_1(n) = \frac{1}{n\pi}(\sin(4n\alpha) - \sin(4n(\alpha - 2\beta)))$$

$$a_2(n) = -\frac{2}{\pi}\left(\frac{\sin(\pi - 4n\alpha) + \sin(4n(\alpha - 2\beta))}{\left(\frac{\pi}{2\beta} - 4n\right)} + \frac{\sin(\pi + 4n\alpha) - \sin(4n(\alpha - 2\beta))}{\left(\frac{\pi}{2\beta} + 4n\right)}\right)$$

$$a_3(n) = -\frac{2}{n\pi}\sin(4n\alpha).$$

The total drag losses of the homopolar bearing result by summing up the losses of the harmonic components. Especially for high magnetization frequencies, as they occur in this application, eddy currents reduce the depth the magnetic field penetrates the rotor. Hence the magnetized mass m_{iron} is reduced. For a two dimensional field the penetration depth results dependent on the materials electrical resistivity κ and permeability μ and the angular magnetization frequency ω as:

$$\mathrm{d} = \sqrt{\frac{2}{\mu\kappa\omega}} \qquad (3)$$

In the rotor of the homopolar magnetic bearing a three dimensional field is present which cannot be calculated analytically. Figure 6 shows the geometric approximation enabling the calculation.

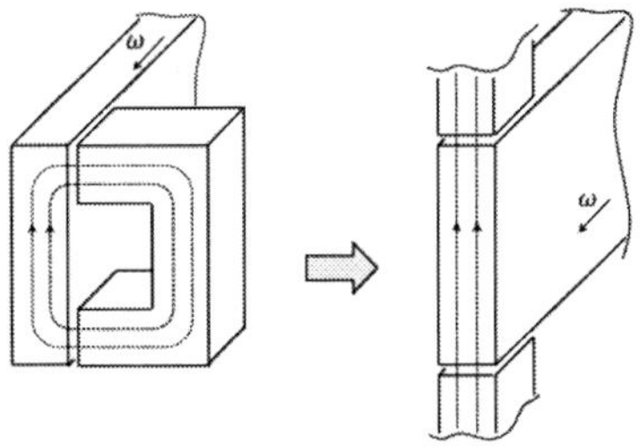

Figure 7. Approximation of the bearing structure for the calculation of the eddy current dept

For each harmonic frequency of the Fourier transformation the magnetized mass is calculated based on the penetration depth (3). With the flux density of the harmonics (2), the drag of each harmonic is calculated via equation (1). Figure 7 gives an example of the calculated losses over the rotation frequency. The consideration of the penetration depth leads to discontinuities in the calculated losses. These occur when the penetration depth of one Fourier coefficient is smaller than the needed depth to penetrate the rotor in total. Such discontinuities do not exist in a real system, the resulting error is considered to fall within the general accuracy of the model.

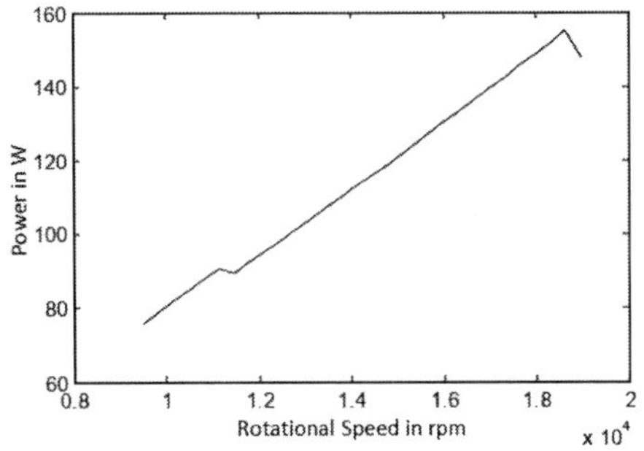

Figure 8. Calculated drag losses of both bearing planes of the homopolar active magnetic bearing

Figure 9 compares the gas friction losses, the drag losses of the AMB as well as the motor-generator unit and the summed up total losses over the rotational frequency. A detailed description of the underlying models is given in [10].

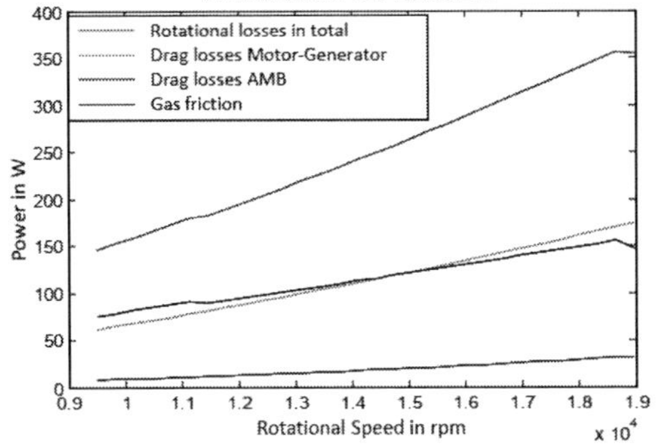

Figure 9. Rotational losses over rotation frequency

III. Intended Procedure for the Experimental Determination of the Losses

The losses for each operational point $P_{loss}(E,P)$ of the loss map can be calculated as the difference between the power transferred to the system and the resulting change in its energy content divided by the duration of the measurement. The energy transferred to the system is measured exactly with a wattmeter based on the drive currents and voltages. With the measured rotational speed and the known inertia θ of the rotor also its current state of energy can be calculated exactly.

$$P_{loss}(E,P) = P_{Drive} - \frac{\frac{1}{2}\theta(\omega_2{}^2 - \omega_1{}^2)}{\Delta t} \quad (4)$$

For the measurement two possibilities arise, conducting the measurement for a constant time period or a constant change of the rotational speed. To achieve a high accuracy the measurement should be conducted for a relevant change in the systems energy content. For operating points with a high system power the conversional losses dominate the drag losses. Also the change of the energy content is linear dependent on the power and the duration of the power transfer. Hence for these operating points the measurement should be conducted for a constant time period dependent on the systems power.

For little or no power transfers the drag losses dominate. Caused by the quadratic dependency between rotational speed and energy content, as well as rotational speed and the drag losses, the measurement should be conducted for a constant change in the rotational speed dependent on the system energy content.

A special case of the loss measurements is the coast down, when no energy is transferred to the rotor ($P_{Drive} = 0$). During this measurement only drag losses slow down the rotor, so it is an adequate measurement to determine the drag losses. Since the different loss components (gas friction, drag of the bearings and drag of the drive) occur simultaneously, they cannot be determined separately. In order to isolate the three different components two system parameters must be varied. The set up prototype allows for the variation of the gas pressure inside the containment and the variation of the bias magnetization, so the two needed parameters are present. An appropriate influence on the system's drag losses is expected for doubling the gas pressure from 0.03 Pa to 0.06 Pa, leading to approx. twice the gas friction losses. The reduction of the bias magnetization will reduce its drag losses, but also the maximum rotational speed of the system, since it reduces the force slew rate of the magnetic bearings. Since the bearings drag losses are proportional to the quadratic flux density, a reduction of the bias magnetization by 25 % will lead to a significant influence on the drag losses.

To speed up the coast down measurements, the measurements will be conducted at different operating points which are approached through a directed charge or discharge of the system.

IV. Discussion and Conclusion

An outer-rotor-type kinetic energy storage system in active magnetic bearings has been developed and setup. The system's current maximal rotating speed is 20,000 rpm, resulting in an energy content of 0.4 kWh. For further developments an analytical loss model is set up. It considers the conversion and the drag losses of the electric drive, the drag losses of the homopolar active magnetic bearings and the gas friction losses as well as the constant losses. For the calculation of the drag losses of the homopolar magnetic bearings a new approach combining different models given in the literature was developed. It enhances the approach of Bertotti for the magnetization waveform of the homopolar magnetic bearing and the field's penetration depth. The model shows small discontinuities over the rotational speed, which are expected to be within the models general uncertainty.

Our next step is the model update via the experimental determination of the losses on the set up prototype system. In order to achieve a high accuracy the presented measurement method takes the system's operating point into account. Special effort is needed for the isolation of the drag loss components of the permanent magnetic synchronous drive, the drag losses of the homopolar magnetic bearings and the gas friction losses.

Acknowledgement

The research and results described in this paper has been co-funded by a LOEWE-Project of the Hessen-Agentur of the federal state of Hesse, Germany.

REFERENCES

[1] *1980 Flywheel Technology Symposium, October* 1980, Scottsdale Arizona

[2] KAILASAN, A.; Jiang W.: Dimond T.: Sheffler D. and Allaire P.: Design and analysis of the magnetic suspension system in an energy storage flywheel. In: *Proceedings of ISMB13*, 2012

[3] TOH, C.S.; CHEN S.L.; JUANG D.P.: Development of a Ring-type Flywheel Battery System for Storage of Renewable Energy. In: *Proceedings of ISMB13*, 2012

[4] STEINMETZ, Charles Proteus: *On the Law of Hysteresis (Part III.), and the Theory of Ferric Inductances.* In: *Transactions of the American Institute of Electrical Engineers* XI (1894), S. 570–616

[5] BERTOTTI, G.; BOGLIETTI, A.; CHIAMPI, M.; CHIARABAGLIO, D.; FIORILLO, F.; LAZZARI, M.: *An improved estimation of iron losses in rotating electrical machines.* In: *IEEE Transactions on Magnetics* 27 (1991), Nr. 6, S. 5007–5009

[6] SCHWEITZER, G.; MASLEN, Eric H. (Eds.): *Magnetic Bearings - Theory, Design, and Application to Rotating Machinery.* URL http://www.springer.com/engineering/mechanical+engineering/book/978-3-642-00496-4 – Überprüfungsdatum 2014-02-02

[7] MATSUMURA, F.; HATAKE, K.: Relation between Magnetic Pole Arrangement and MAgnetic Loss in Magnetic Bearing. In: *Proceedings of the ISMB 3.*

[8] KLOHR, Markus: *Entwicklung und Konstruktion einer umrichtergespeisten magnetgelagerten Permanentmagnet-Synchronmaschine für 40kW/ 40000 /min.* Aachen : Shaker, 2008 (Berichte aus der Elektrotechnik)

[9] TRAXLER, A.: *Eigenschaften und Auslegung von berührungsfreien elektromagnetischen Lagern.* Dissertation. ETH Zürich, 1986

[10] SCHAEDE, H.: *Dezentrale elektrische Energiespeicherung mittels kinetischer Energiespeicher in Außenläufer-Bauform.* Dissertation. Technische Universitaet Darmstadt, 2014

Topology Optimization of a Flywheel Energy Storage Rotor using a Genetic Algorithm

Thomas Hinterdorfer, Alexander Schulz, Harald Sima, Stefan Hartl and Johann Wassermann

Vienna University of Technology, Institute of Mechanics and Mechatronics,
Wiedner Hauptstraße 8-10, A-1040 Vienna, Austria,
{givenname.surname}@tuwien.ac.at

***Abstract*—This paper deals with topology optimization of the rotor of a flywheel energy storage system (FESS). For isotropic materials the constant stress disc (CSD) is the best choice to maximize energy density. Modern FESS are manufactured of fiber reinforced plastics (FRP), due to their high specific strength. Because of the transverse isotropic material description the optimal rotor topology changes. A genetic algorithm (GA) for nonlinear integer problems will be used to find the best rotor structure which can be manufactured within one filament winding process.**

I. Introduction

Flywheel energy storage systems represent an ecologically and economically sustainable technology for decentralized energy storage. Compared to other storage technologies such as e.g. accumulators, they offer longer life cycles without performance degradation over time and usage and need almost no systematic maintenance. To achieve a high energy density FESS rotors are usually manufactured of FRP, which enables high tip speeds. For stable operation and minimal friction loss high-end FESS are supported by magnetic bearings. To minimize drag the rotor operates in a vacuum chamber. An overview and the requirements on the individual components of a FESS are given in [1].

The aim of rotor design is to maximize the energy density κ which is defined as the kinetic energy per unit mass

$$\kappa = \frac{\frac{1}{2} I_z \Omega_{max}^2}{m} \tag{1}$$

with the moment of inertia I_z, the maximum angular velocity Ω_{max} and the rotor mass m. The maximum possible energy density for an isotropic material is equal to the materials specific strength

$$\kappa_{max} = \frac{X}{\rho} \tag{2}$$

where X denotes the ultimate strength and ρ the density. Moreover the reached energy density depends on the geometry. The geometry factor is defined as

$$K = \frac{\kappa}{\kappa_{max}} \tag{3}$$

and is a measure for the cost-effectiveness of the rotor shape, because a higher value of K leads to a higher amount of stored energy at constant mass. The values of K for different rotor geometries are listed in Table I, where a poisson's ratio of $\nu = 0.3$ was assumed.

Table I
Geometry factor K for different rotor cross-sections for isotropic material with $\nu = 0.3$ (cf. [2])

Flywheel geometry	cross section	K
theoretical constant stress disc		1.000
real constant stress disc		0.7-0.98
conical disc	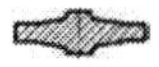	0.7-0.95
constant thickness disc		0.606
thin hollow cylinder		0.5
disc with rim		0.4-0.5
pierced constant thickness disc		0.303

When an isotropic material is assumed, the shape factor K equals one when the equivalent stress reaches its maximum value in every point of the structure with respect to some stress criterion, which corresponds to the constant stress disc. Due to the tri-axial stress state the contour of the CSD cannot be derived analytically. Some suggestions are given in [3], where an analytical function for the CSD is described. In the range of $r \leq \beta_r R_o$ the contour is given as a function of the normalized radius $\chi = r/R_o$

$$H_u(r) = h_c \left\{ e^{\frac{-B\chi^2}{2}} + \overline{\alpha} e^{\frac{-B\beta_r^2}{2}} \left[\frac{1}{2} + \frac{1}{\pi} \text{atan}\left(\frac{\overline{\chi}}{a_r} \right) \right] \right\} \tag{4}$$

for $r > \beta_r R_o$ and $h < h_{max}$

$$H_u(r) = h_c e^{\frac{-B\beta_r^2}{2}} \left[\frac{1+\alpha_r}{2} + \left(\frac{\overline{\alpha}}{\pi a_r} - B\beta_r \right) \overline{\chi} \right] \tag{5}$$

and and in the case of $r > \beta_r R_o$ and $h \geq h_{max}$

$$H_u(r) = h_{max} = \alpha_r h_c e^{\frac{-B\beta_r^2}{2}} \tag{6}$$

with the radius

$$\beta_r = \left\{ \frac{2}{B\alpha_r} \left[\overline{\alpha} + \sqrt{\frac{\alpha_r^2 B(B-2+2\nu)}{(1-\nu)^2} + \overline{\alpha}^2} \right] - \frac{1+\nu}{1-\nu} \right\}^{\frac{1}{2}} \tag{7}$$

and the abbreviations $\overline{\alpha} = \alpha_r - 1$ and $\overline{\chi} = \chi - \beta_r$ and the constants B, α_r, h_c and a_r. The described function will be used as reference solution in chapter IV-A.

As a result of (2) carbon and glass fiber reinforced plastics (CFRP, GFRP) are the material of choice for high performance FESS, because of their high strength and low density. Fig. 1

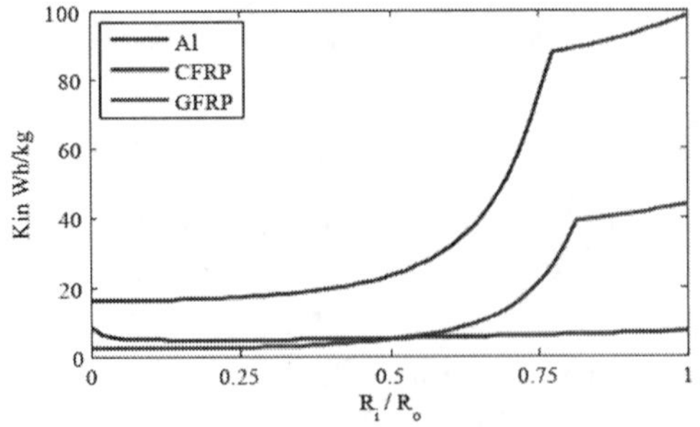

Figure 1. Energy density of a disc as function of ratio inner to outer radius for different materials

presents the energy density of a disc as a function of ratio inner radius R_i to outer radius R_o for different materials and demonstrates the potential of those materials in comparison to aluminium. The calculation was done by using the material described in chapter III-A. The maximum energy density is reached when R_i/R_o is close to 1. Then the radial stress is almost zero and the fibers are fully loaded in circumferential direction. Vice versa, at smaller values the radial stress is rather high and therefore the load is transverse to fiber direction.

Due to easy manufacturing mostly a simple cylinder is used like in [4], where the rotor is assembled of rings with different materials. It concludes that stiffer materials at increased radius leads to a decrease of radial stresses due to the different radial expansion. In [5] also a rectangular rotor shape is used which consists of several rings press-fitted into each other. This results in compressive residual stresses and therefore an increase of energy density. A low cost flywheel for rural energy storage is described in [6]. A kind of constant stress disc is used as rotor shape, which is legitim due to the quasi-isotropic rotor made of short fiber composites which is molded and than milled.

Usually for topology optimization the SIMP-method is used, which is a challenge when inertia loads are applied (cf. [7]). In [8] a turbine disk profile is optimized by metamorphic development. For this purpose the strain energy density is evaluated at the rotor surface, which decides if the structure will grow or degrade in a special position. Here only isotropic materials are considered and the objective function is not useful for flywheel applications. [9] uses an injection island genetic algorithm to optimize the rotor topology and different algorithms are compared to each other. Both papers only consider isotropic materials.

There are some subtle constructions in literature. A thick hollow cylinder is used as inertia in e.g. [10]. This cylinder is supported by a hub which is connected to the shaft. The focus lies on the hub design which is formed to enable an operation between first and second bending frequency.

II. Assumptions and Overview

Material transitions are always a soft spot. So this work will build up a rotor structure that is made of a single material which can be manufactured within one filament winding process. The shaft is planned to be shrink-fitted into the composite rotor. This leads to compressive residual stresses in radial direction. For the sake of simplicity the shaft will be neglected in the following optimization, which is a conservative assumption in strength analysis. Also eigenfrequencies and eigenmodes will not be considered. All calculations where done with a safety margin of 2.

In chapter III the finite element model and the used GA are described. The optimization results for isotropic as well as transverse isotropic materials are presented in chapter IV. Chapter V concludes the paper.

Table II
Material properties of CFRP and Aluminium

		CFRP	Al
E_φ	longitudinal young's modulus, GPa	145.38	70
E_r	transverse young's modulus, GPa	9	70
$G_{\varphi r}$	shear modulus in φr-plane, GPa	4.97	26.9
$\nu_{\varphi r}$	poisson's ratio φr	0.24	0.3
ν_{zr}	poisson's ratio zr	0.27	0.3
ρ	density, kg/m^3	1535.5	2700
X	longitudinal tensile strength, MPa	2179	275
X'	longitudinal compressive strength, MPa	1702	275
Y	transverse tensile strength, MPa	98	275
Y'	transverse compressive strength, MPa	210	275
S	shear strength, MPa	115	275

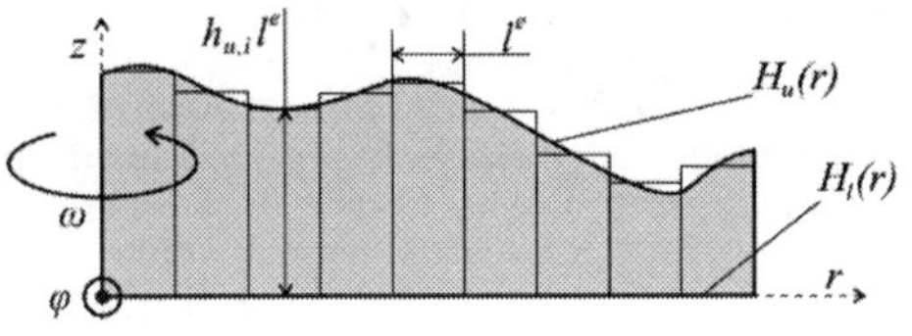

Figure 2. Discretization of the rotor geometry and parameter definitions

III. Optimization Problem

A. Finite Element Model

The properties of the applied aluminium and unidirectional CFRP with fibers oriented in circumferential direction are summerized in Table II. Together with (8) the elasticity tensor can be set up

$$G_{rz} = \frac{E_r}{2(1+\nu_{rz})}, \quad \frac{\nu_{r\varphi}}{E_r} = \frac{\nu_{\varphi r}}{E_\varphi}, \quad E_z = E_r, \tag{8}$$
$$G_{\varphi z} = G_{\varphi r}, \quad \nu_{\varphi r} = \nu_{\varphi z}, \quad \nu_{r\varphi} = \nu_{z\varphi}, \quad \nu_{rz} = \nu_{zr}.$$

For aluminium the von Mises yield criterion and for the CFRP the Puck criterion is used ([11], [12] and [13]).

The stress calculation was performed with an axisymmetric finite element model. Quadriliteral elements with a square cross section of length l^e and linear shape functions were applied. The discretization method can be seen in Fig. 2. The z-axis is the axis of rotation and the rotor is assumed to be symmetrical with respect to the $r\varphi$-plane. The rotor contour consists of the upper $H_u(r)$ and the lower contour $H_l(r)$. The optimization was carried out by the use of the rate of change of the contour as optimization variable

$$H_u(r) = H_{u,0} + \int_0^r \frac{dH_u(\eta)}{d\eta}\, d\eta$$
$$H_l(r) = \int_0^r \frac{dH_l(\eta)}{d\eta}\, d\eta \tag{9}$$

where η is an integration variable. Thereby, bounds can be applied on the optimization variables to avoid large steps. The rotor is discretized with N elements along the radial direction and also the heigth can only change in steps of l^e, which leads to the approximation of the contour

$$H_u\left(i \le \frac{r - R_i}{l^e} < i+1\right) \approx \left[h_{u,0} + \sum_{i=1}^{N-1} \underbrace{(h_{u,i} - h_{u,i-1})}_{\Delta h_{u,i}}\right] l^e$$
$$H_l\left(i \le \frac{r - R_i}{l^e} < i+1\right) \approx \left[\sum_{i=1}^{N-1} \underbrace{(h_{l,i} - h_{l,i-1})}_{\Delta h_{l,i}}\right] l^e \tag{10}$$

where $h_{u,0}$, $\Delta h_{u,i}$ and $\Delta h_{l,i}$ are integer values. $h_{l,0}$ is zero to keep the geometry in the origin.

The vertical displacement of the nodes along the r-axis was set to zero to ensure a symmetric rotor. Also the horizontal displacement along the z-axis was set to zero. The applied inertial load is the volume force due to the angular velocity

$$F_{\text{Vol}} = \int_m r\omega^2\, dm \tag{11}$$

where ω is an arbitrary angular velocity.

The stresses were evaluated in the nodes and afterwards averaged over the whole element to smooth the results and avoid numerical singularities. To evaluate the stress criterion a stress ratio R where introduced

$$\sigma^*_{ij} = \frac{\sigma_{ij}}{R} \tag{12}$$

where σ^*_{ij} denotes the stress on the failure surface of the appropriate criterion and σ_{ij} the actual stress state. The stress state is within the failure surface if $0 \le R < 1$ is fulfilled. Higher values of R means failure. Due to linear analysis the angular velocity, which leads to failure of the structure with a safety margin S can be derived as

$$\Omega_{max} = \omega \sqrt{\frac{1}{S \cdot \max\{R(\omega)\}}} \quad . \tag{13}$$

Together with (1) the maximum energy density of a given structure can be calculated.

B. Optimization Algorithm

Due to the finite element discretization the optimization variables are integer values. This leads to the disadvantage that calculating an optimization step with a gradient based method is not possible. Therefore a genetic algorithm which can handle integer variables is used. The fitness function is to maximize the energy density

$$\min_{\boldsymbol{x}} \{-\kappa\} \tag{14}$$

where $\boldsymbol{x}$ is the vector of optimization variables presented later. In the following the function `ga(...)` of MATLAB R2010b with some adaptions will be used.

As a start a set of unique individuals of the optimization variables $h_{u,0}$, $\Delta h_{u,i}$, $\Delta h_{l,i}$ in (10) is generated by a creation function and the energy density is evaluated. A number of elite individuals are guaranteed to survive into the next generation. A selection function chooses parents which are then combined by a crossover function to childrens for the next generation. Afterwards the set of individuals for the following generation is filled with randomly created individuals by a mutation function. This procedure is done till the algorithm stops, which is the case if no improvement is reached within a choosen number of generations.

The creation function and mutation function are identical and are creating a random unique population, which fulfills the bounds and constraints and do not lead to intersections of the geometry. Thus it is not necessary to solve a constraint optimization problem. The crossover function randomly composes two parent genes. The parent individuals are chosen by a stochastic universal sampling as a selection function.

In the following examples the population size was 400 and ten elite individuals were used. Any other options where left default.

IV. Optimization Results

A. Isotropic Material - Constant Stress Disc

To validate the functionality of the algorithm in a first step the CSD will be calculated. For this purpose the lower contour and the inner radius were set to zero

$$H_l(r) = 0, \qquad R_i = 0 \tag{15}$$

and the upper contour is optimized. The goal is a fully stressed structure. The vector of optimization variables is then

$$\boldsymbol{x} = \left(h_{u,0}, \Delta h_{u,1}, \ldots, \Delta h_{u,N-1}\right)^T \tag{16}$$

with the length N which was chosen to be 100. The optimization bounds were defined as

$$-2 \le \Delta h_{u,i} \le 2 \tag{17}$$

so the maximum vertical variation was allowed to be 2 elements within one radial element step. To avoid intersections with the r-axis the upper contour must fulfill

$$H_u(r) > 0 \quad . \tag{18}$$

The maximum possible energy density with aluminium defined in Table II is

$$\kappa_{max} = \frac{X}{S \cdot \rho} = 14.15\, \frac{Wh}{kg} \tag{19}$$

with a safety margin of 2.

The algorithm stopped at generation 69 and the result is shown in Fig. 3 and compared to the suggestion presented

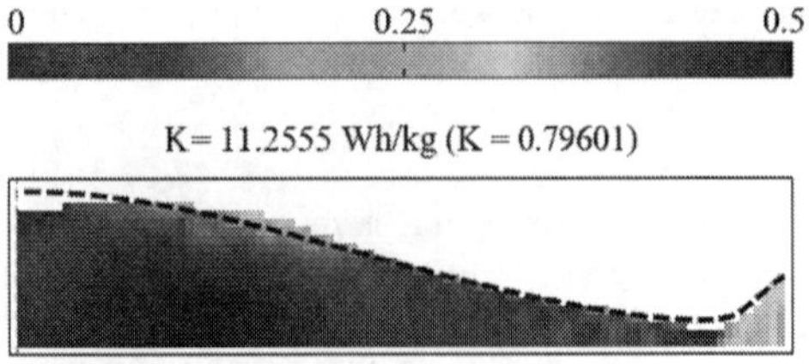

Figure 3. Optimization result with aluminium; Geometry factor $K = 0.80$; Suggestion of [3] for the CSD with $B = 5.25$, $\alpha_r = 4.89$, $h_c = 0.39$ and $a_r = 0.0244$; Contour plot of von Mises yield criterion

in [3] with the parameters $B=5.25$, $\alpha_r=4.89$, $h_c=0.39$ and $a_r=0.0244$. The results are in good agreement. On the axis of rotation the heigth is $H_u(r=0)/R_o=0.18$ and then decreases to a minimum of $H_u(r/R_o=0.9)/R_o=0.02$. At the outer radius a rim is formed with a height of $H_u(r=R_o)/R_o=0.09$. The reached energy density is 11.26 Wh/kg, which corresponds to a geometry factor of $K=0.80$.

B. Transverse Isotropic Material

Now also the lower rotor contour is unconstrained and optimized and the inner radius is no longer zero. The optimal rotor shape in the case of FRP is a thin cylinder with fibers in circumferential direction. Here the radial stress is very small and the whole load is carried by the fibers. For this reason an optimization algorithm will always try to converge to a thin cylinder. Therefore it will be requested that the rotor has a minimum radial thickness in the midplane

$$H_l(r \leq R_m) = 0 \tag{20}$$

with R_m as the minimal outer radius at $z = 0$. To achieve this the appropriate values of $\Delta h_{l,i}$ are removed from the vector of optimization variables, which yields to

$$\boldsymbol{x} = \left(h_{u,0}, \Delta h_{u,1}, \ldots, \Delta h_{u,N-1}, \Delta h_{l,M}, \ldots, \Delta h_{l,N-1}\right)^T \tag{21}$$

with

$$M = \left\lceil N \cdot \left(1 - \frac{R_m - R_i}{R_o - R_i}\right)\right\rceil \tag{22}$$

where $\lceil \ldots \rceil$ is the ceil-operator.

Again 100 elements along the r-axis were used. Due to the manufacturing process the contour should increase monotonically, which is fulfilled by defining the optimization bounds as

$$\begin{aligned} 0 \leq \Delta h_{u,i} \leq 4 \\ 0 \leq \Delta h_{l,i} \leq 4 \quad . \end{aligned} \tag{23}$$

To achieve a reasonable rotor the following constraints where defined

$$\begin{aligned} \frac{R_i}{R_o} = 0.3\,, \qquad \frac{h_{u,0}}{h_{max}} \geq 0.3\,, \\ \frac{R_m - R_i}{R_o - R_i} \geq 0.3\,, \quad H_u(r) > H_l(r) \end{aligned} \tag{24}$$

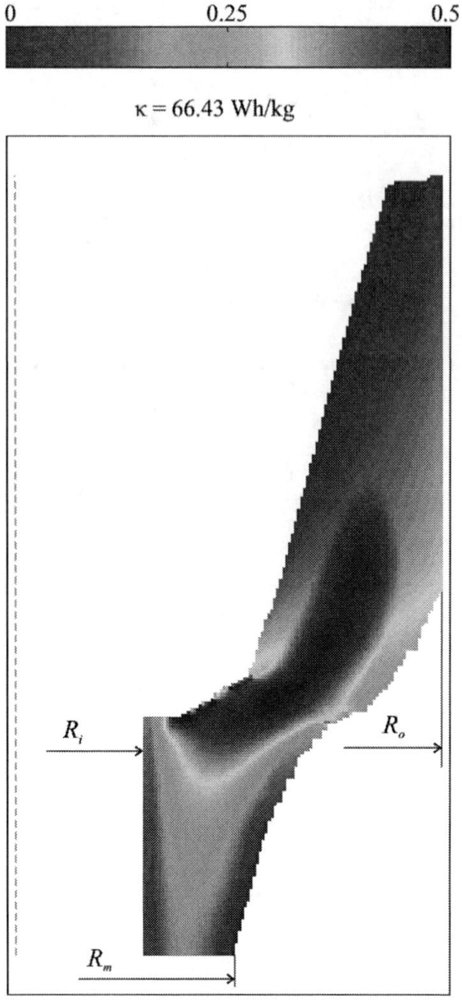

Figure 4. Optimization result after 83 generations with CFRP; $\kappa = 66.43\ Wh/kg$; Contour plot of Puck criterion

with

$$h_{max} = \frac{H_u(r = R_o)}{l^e} \quad . \tag{25}$$

The optimization result after 83 generations is shown in Fig. 4. An energy density of $\kappa = 66.43\ Wh/kg$ is reached. Due to the tendency to converge against a thin hollow cylinder, the constraints in (24) are active. The upper contour is flat starting from R_i, then a radius with a subsequent high slope is formed. Contrary H_l increases very fast and then flattens, therefrom in the shaft region, mass will be removed. In spite of decreasing moment of inertia the radial stress is reduced in this region enabling higher angular velocities. Here the highest specific stress is still in radial direction. The cantilever part is forming a hollow cone and the stresses are shifted in circumferential direction. Furthermore the slope of the upper contour leads to a larger radius of inertia and therefore an increase of performance.

Fig. 5 demonstrates the influence of the parameters listed

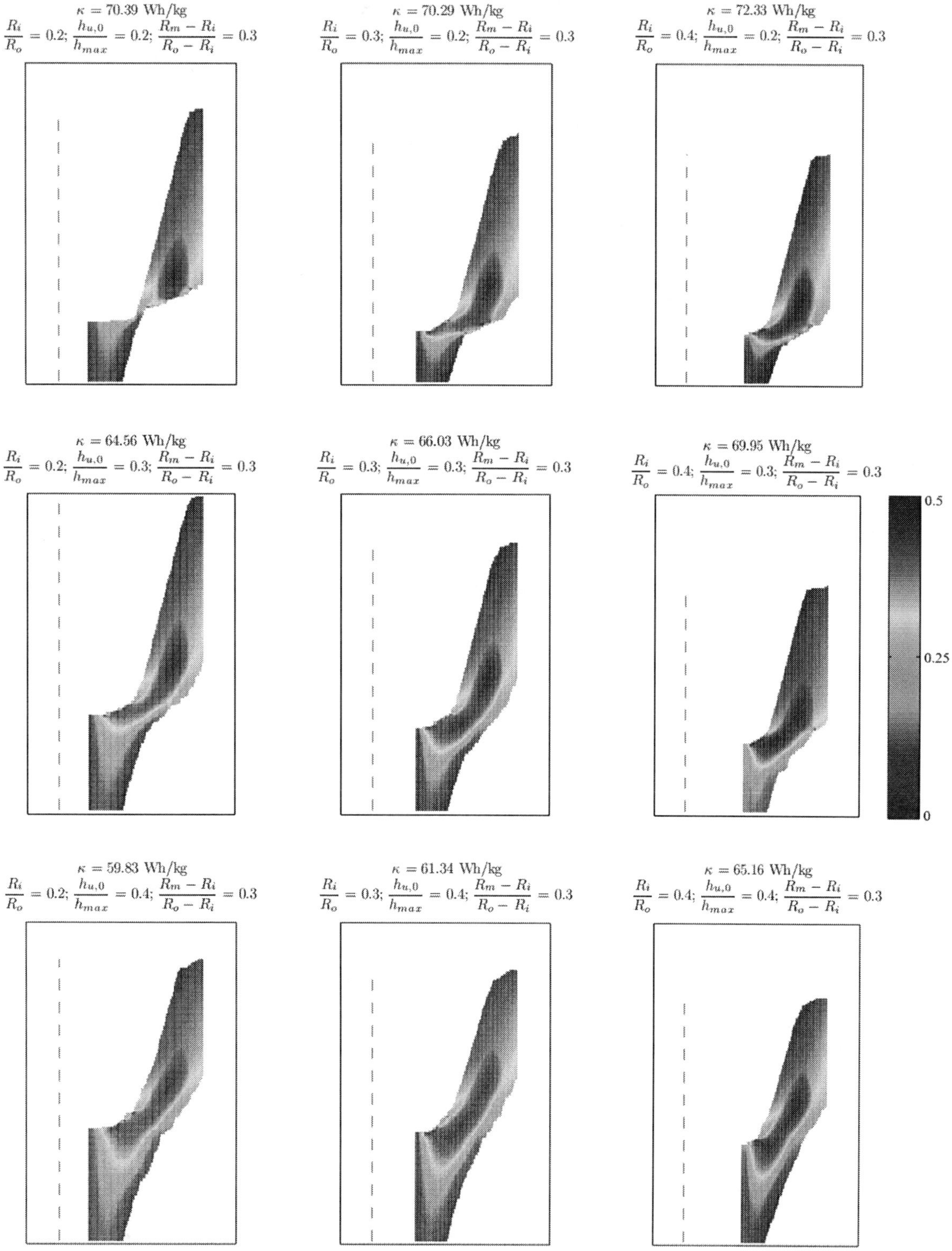

Figure 5. Optimization results for different sets of contraints; $(R_m - R_i)/(R_o - R_i) = 0.3$ and different values for R_i/R_o and $h_{u,0}/h_{max}$

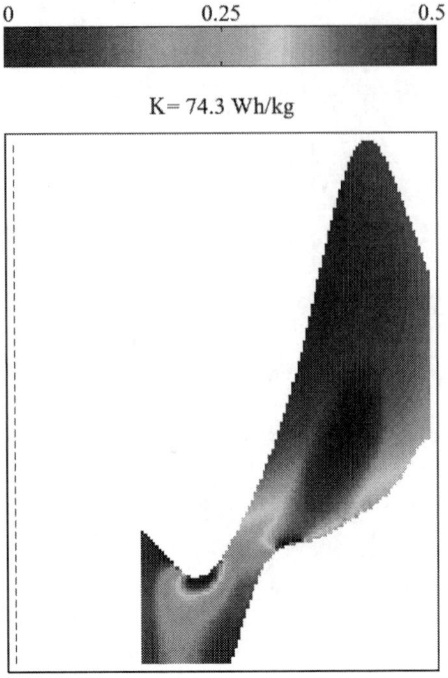

Figure 6. Optimization result after 84 generations with CFRP and non-monotonic rotor shape; $\kappa = 74.3\ Wh/kg$; Contour plot of Puck criterion

in (24). R_i/R_o and $h_{u,0}/h_{max}$ were varied with 0.2, 0.3 and 0.4 while $(R_m - R_i)/(R_o - R_i)$ was held constant at 0.3. The maximum value of the height h_{max} decreases with increasing R_i/R_o. In general the energy density increases with R_i/R_o. κ increases with decreasing values of $(R_m - R_i)/(R_o - R_i)$ and $h_{u,0}/h_{max}$. Low values of the height constraint are leading to a thin structure. The energy density can further be increased when the condition for monotonicity in (23) is dropped. In Fig. 6 the optimization result with

$$\begin{aligned} -4 \leq \Delta h_{u,i} \leq 4 \\ -4 \leq \Delta h_{l,i} \leq 4 \end{aligned} \tag{26}$$

as constraints, leading to a non-monotonic rotor shape, is shown. The upper contour was fitted to a polynomial of 12th order and the lower contour to one with 7th order to get a smoother rotor shape. One can see that the height at the beginning $h_{u,0}$ is rather low and the contour begins decreasing, forming a large radius to the cantilever. Furthermore the maximum height is smaller which leads to a more compact and plump structure. The lower contour is still monotonically increasing while the upper contour formes a tilted "S". Due to the larger radius at the beginning of the cantilever the energy density can be increased.

V. Conclusion

This paper presents the optimization of a FESS rotor by GA. A finite element model was used to find the shape of the CSD and furthermore a CFRP rotor was optimized for high performance FESS.

Improvements can be done by introducing triangular elements to get a smoother contour by avoiding sharp corners in the mesh. Also high numerical stresses can be avoided by advanced stress smoothing methods.

The optimized rotor shape can be manufactured just as a disc shaped rotor. The fibers are immersed in a resin bath and then wound on a mandrel. Care has to be taken in the area of the cantilever. When the inclination is too high, slipping of the fibers from the mandrel can occur. When the resin is cured, the mandrel can be removed. The whole rotor can be manufactured within one process or the outer contour can be filled and then formed by turning.

VI. Acknowledgment

The presented work is part of the research project "Optimum Shape-Flywheel" and is supported by "Klima und Energiefonds" in line with the program "NEUE ENERGIEN 2020", conducted by the Austrian Research Promotion Agency (FFG).

References

[1] B. Bolund, H. Bernhoff, M. Leijon M, "Flywheel energy and power storage systems," *Renewable and Sustainable Energy Reviews* 11 (2007) 235-258.
[2] G. Genta, "Kinetic energy storage, theory and practice of advanced flywheel systems", *Butterworths*, London Boston, 1985.
[3] G. Genta, "Some considerations on the constant stress disc profile", *Meccanica*, December 1989, Vol. 24, Issue 4, pp. 235-248.
[4] S.K. Ha, D.J. Kim and T.H. Sung , "Optimum design of multi-ring composite flywheel rotor using a modified generalized plane strain assumption", *International Journal of Mechanical Sciences*, 43 (2001) 993-1007.
[5] J.D. Kwon, S.J. Kim, S.U. Nasir, S.K. Ha, "Design and Fabrication of Hybrid Composite Flywheel Rotor," *World Academy of Science*, Engineering and Technology 60, 2011.
[6] R. Okou, A.B. Sebitosi, A. Pillay, "Flywheel rotor manufacture for rural energy storage in sub-Saharan Africa", *Energy*, Vol. 36, Issue 10, Oct. 2011, pp. 6138-6145.
[7] M.P. Bendsøe, O. Sigmund, "Topology Optimization, Theory, Methods and Applications", Springer-Verlag Berlin Heidelberg, 2003.
[8] J.S. Liu, G.T. Parks, P.J. Clarkson, "Optimization of Turbine Disk Profiles by Metamorphic Development", *Journal of Mechanical Design*, 2002.
[9] D. Eby, R.C. Averill, W.F. Punch, E.D. Goodman, "Optimal design of Fywheels using an injection island genetic algorithm," *Artificial Intelligence for Engineering Design, Analysis and Manufacturing - AI EDAM*, Vol. 13, pp. 327-340.
[10] D. Hockney, D. Ansbigian, M. Polimeno, W. Spears, A. Palazzolo, S. Pekarek, "Extreme Energy Density Flywheel Energy Storage System for Space Applications", *6th International Energy Conversion Engineering Conference (IECEC)*, Ohio, Juli 2008.
[11] A. Puck, H. Schürmann, "Failure analysis of FRP laminates by means of physically based phenomenological models", *Composites Science and Technology* 62 (2002) 1633-1662.
[12] A. Puck, J. Kopp, M. Knops, "Guidelines for the determination of the parameters in Puck's action plane strength criterion", *Composites Science and Technology* 62 (2002) 371-378.
[13] M.J. Hinton, A.S. Kaddour, P.D. Soden, "A comparison of the predictive capabilities of current failure theories for composite laminates, judged against experimental evidence," *Composites Science and Technology* 62 (2002) 1725-1797.

High Speed Flywheels for Vehicular Applications

Johan Lundin [a], Tobias Kamf [a], Johan Abrahamsson [a], Juan de Santiago [a], Magnus Hedlund [a], Hans Bernhoff [a]

[a] Uppsala University, Division for Electricity, Lägerhyddsvägen 1, Uppsala, Sweden, Johan.Abrahamsson@angstrom.uu.se

***Abstract*—This article analyzes two business cases using flywheels in vehicles. Firstly, a diesel based car ferry, in Gothenburg, performing 102 5 minute long journeys per day. Secondly, a diesel based city bus. A theoretical analysis of the benefits, technical and economical, of electrifying these vehicles using flywheels is performed.**

As basis for the analysis is a constructed high-speed flywheel, designed to store up to 870 W h at a rotational speed of 30 000 rpm. The rotor of the flywheel is constructed from a permanent magnet machine surrounded by a shell of composite material. The flywheel is optimized for low loss as well as low cost.

The economical analysis of the two business cases show great potential for cost reduction by either utilizing a flywheel as temporary energy storage for capturing regenerative braking energy or replacing the diesel generator completely as main energy storage.

The main material cost of the constructed flywheel pertains to the position sensors and the permanent magnet material used in the magnetic bearing.

I. Introduction

One of the big challenges for the electric vehicle is finding a technology able to store large amounts of energy, while at the same time being capable of handle high, bidirectional flows of energy. The total amount of available energy limits the reach of the vehicle. The maximum power not only determines the acceleration of the vehicle, but also how much of the braking energy that can be reused. As none of the methods of energy storage (batteries, fuel cells, petroleum based generators) presently fulfills both these requirements optimally, it would be of benefit to separate the main energy storage from the power capabilities of the driveline. This could be achieved by an energy buffer, capable of high power but at the same time able to store a moderate amount of energy. Flywheels have been identified as a promising technology for such a buffer.

One of the first implementations of a flywheel for vehicular applications was as main energy storage in the *Gyrobus*, first appearing in public in Switzerland in 1950. The steel rotor of the flywheel in the Gyrobus had a weight of 1500 kg and a diameter of 1.6 m. The maximum rotational speed was 3000 rpm corresponding to approximately 6.6 kW h. Energy flux to and from the rotor was achieved through a three-phase, 52 kW, asynchronous machine [1], [2].

An implementation of the concept of the modern flywheel for vehicular applications with composite material and magnetic bearings was realized at the University of Austin, Texas in 1999 [3], [4]. The unit stored 2 kW h of energy and was able to deliver 150 kW of continuous power. Additionally, a theoretical investigation on the concept of low-speed flywheels for a bus was completed, prompted by the high cost of the high-speed prototype [5].

In 2007 the company *Flybrid Systems* developed a completely mechanical system for energy storage in vehicles called *KERS*, kinetic energy recovery system [6]. The unit was made specifically for the purpose of meeting the 2009 Formula One regulations. The KERS was capable of storing up to 111 W h of usable energy, utilizing a 5 kg rotor. The system was rated at 60 kW with a maximum rotational speed of 60 000 rpm. In 2011, Volvo Car Corporation announced their intention of building flywheel hybrid cars for mass production using a similar concept to that of Flybrid Systems [7]. A prototype of such a car was finished in 2013, and is presently being tested.

In 2012, an *Audi R18 e-tron quattro* won the 24 Hours of Le Mans race, the worlds oldest competition in endurance racing, as well as the six hours of Silverstone race. The car was equipped with a flywheel accumulator system from *Williams Hybrid Power*, recently purchased by the British company *GKN*. The system comprised magnetically loaded composite material and ceramic bearings. Two pressure chambers, separated by a turbo-molecular pump, ensured low pressure around the flywheel, while the bearings operated at a slightly higher pressure. The system was able to store 140 W h of usable energy with a power-rating of 150 kW [8]. Later the same year, Williams Hybrid Power announced a joint cooperation with the *Go-Ahead Group* of producing six prototype buses with flywheel systems. These are presently being tested, and initial results show fuel savings with up to 30 %.

II. Studied Applications

Flywheels are very well suited for applications where the number of charge/discharge cycles as well as the peak power is high, but the demand on specific energy storage is low to moderate. *City buses* and *near shore ferries* are two applications that match these specifications.

The purpose of the flywheel may be to absorb high-power charge/discharge peaks, leaving the main energy storage with the task of providing a smooth and low-power current going solely in the discharge direction. This reduces the losses in the main energy storage and – in case it consists of a battery – extends its lifetime. The amount of batteries may thereby be reduced as batteries with high energy density, but low power density, can be used [9].

In addition, such a system may equally well be used for storing regenerative braking energy in a drivline based around an internal combustion engine. Another possible application is as load leveling in drivelines based around fuel cells.

In some cases, the flywheel may even be even be used as the main energy storage itself. These are cases where recharging of the flywheel is readily available, and can be performed frequently enough to limit the required amount of stored energy.

This paper investigates two business cases using flywheels in originally diesel based vehicles, one passenger ferry and one city bus. These two applications have some similarities in that they both mean long working days, a harsh environment and high cost for any unplanned down-time. They are also characterized by frequent starts and stops, making the flywheel's fast-charging capabilities and long cycle life valuable.

In this article two cases are studied for each application; one where the flywheel is inserted between the diesel engine and the traction motor, enabling regenerative braking. And one where the diesel engine is completely replaced by a flywheel system. These two cases are then compared with a reference case using the standard diesel engine alone.

A. City bus

A city bus is characterized by frequent accelerations and decelerations with resulting bidirectional power flows. The demand on the energy storage is although high. It has to withstand numerous short but heavy charge and discharge currents in a harsh environment during long working days. It has to be able to operate within a wide range of ambient temperature. And on top every hour of unplanned down-time is expensive and highly undesirable. ICE buses have been around for a century and their advantages and disadvantages are well known. Hybrid and all-electric buses are beginning to intrude the market, with batteries as the predominant electric energy storage. They show significantly higher energy efficiency but batteries are not well-suited for all the requirements.

However, these requirements make a hybrid or all-electric city bus a good application for a flywheel as a power handling device. A typical city bus weighs 10 t and handles about 150 kW–200 kW of maximum power in both directions. With an average power demand of 30 kW–40 kW the difference between the average and maximum power ranges between four and seven times. Furthermore the average power demand is in only one direction while the instantaneous power naturally goes in both directions.

B. Ferry

Recent interest in electrification of boats and ferries have materialized in several successful projects, see [10] and [11].

Flywheels have several advantages in marine applications with periodic start and stop cycles. They can be rapidly charged each time the ferry reaches shore. Batteries are typically only charged once overnight. The energy storage capacity required for the flywheel is then much smaller than for the battery, opening up for possibly lighter and more economical systems.

One-time charging at night or continuous charging at each stop have been evaluated for a commuter shuttle called *Älvskytteln* in Gothenburg (Sweden) [12]. The shuttle transports vehicles between the villages Lindholmen and Rosenlund, a distance of 900 m. The drive cycle, which can be seen in Figure 1, is repeated 102 times per day. Using a flywheel with high speed on-shore charging requires an energy storage of approximately 1 % of the total energy used in one day.

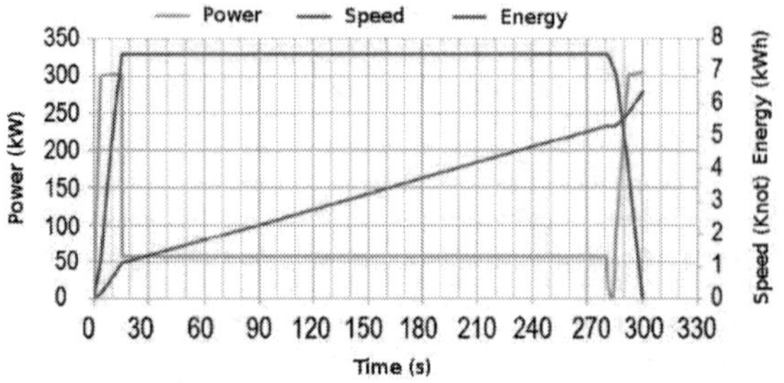

Figure 1. Power, energy and speed of a commuter shuttle in Gothenburg, Sweden, during a day without wind. Picture from [12]

Maintenance is also favorable to the flywheel; battery users have reported problems in off shore applications due to harsh environment. Flywheels design can suppress the gyroscopic effect, but in marine applications the gyroscopic effect can be an advantage. A flywheel system that stabilizes small size boats has been reported [13].

III. Inner and outer losses

Inner losses are all losses occurring inside the drivetrain and thus affected by the vehicle. They consist of losses in the power electronics, in the mechanical parts of the drivetrain, in the energy storage and so on. Outer losses are the losses occurring outside the drivetrain and thus affected by the drive cycle.

Outer losses can be categorized in two sub-groups. The first group, including air drag and rolling resistance, are called *definite losses* since they are not possible to recover in any way. The second group, including acceleration loss and slope resistance loss, is more interesting from an energy saving point of view. They are definitive losses in a traditional ICE drivetrain but not in an electric drivetrain that provides regenerative braking. Therefore, they are called *recoverable losses*.

Practically, not all of the recoverable outer losses can be recovered. First there are cases where the recoverable losses to some or full extent are used to even out the definitive losses to obtain the wanted speed of the vehicle. For example this happens when the vehicle is accelerating and/or going in a downhill.

IV. Simulation set-up

A model of both a city bus in urban traffic, and the ferry Älvskytteln were modeled in an in-house simulation code (Matlab) used to evaluate drivelines.

The city bus in the simulation had a frontal area of 4.8 m^2 and an air resistance coefficient of 0.8. Its maximum acceleration was set to 2.5 m^2. Every trip included 20 stops and took 30 minutes to fulfill. Around five of these minutes consisted of recharging at an end stop.

The bus was assumed to run for 18 hours a day, i.e. 36 trips a day. Ten drive cycles were randomized and simulated, each with 20 stops where the bus stops for 30 s each. The road was built up by 100 parts, each one with the length randomized from a one-sided normal distribution with a mean of 100 m and a standard deviation of 50 m. The gradient of the road for every part was set by a two-sided normal distribution with a mean of 0 degrees and a standard deviation of 0.5°. The speed limit of the road was set to 50 km/h.

The simulation set-up for the ferry application was aimed at the specific case of the commuter shuttle described above. The mass of the ferry was 200 t, the maximum motor power was set to 300 kW and the number of trips per day was 102. For a complete description of the input parameters to the simulation software, see Table I.

Table I
PARAMETERS USED IN SIMULATION

	City bus	Ferry
Mass [t]	10	200
Max motor power [kW]	170	300
Max braking power [kW]	250	300
Energy content in diesel [kW h/L]	9.8	
Average diesel motor efficiency	25 %	
Optimal diesel motor efficiency	30 %	
Electric drivetrain efficiency	80 %	
Regenerative braking roundtrip efficiency	75 %	
Energy stored in flywheel available for use	75 %	
Safety margin for energy storage in flywheel [-]	1.3	
Diesel price (Sweden, excl. VAT) [€/L]	1.22	
Electricity price (Sweden, excl. VAT) [€/kWh]	0.09	
Trips per day [-]	36	102
Running days per year [-]	300	
Max power of flywheel [kW]	170	300
Discount rate [per year]	5 %	

V. RESULTS OF SIMULATION

The maximum size of the flywheel for the two cases is calculated as the kinetic energy of the vehicle at top speed (50 km/h for the bus and 14 km/h for the ferry) plus the potential energy of the vehicle of the total altitude span of the drive cycle.

Further, the flywheel is assumed to have an operating window between 50 % and 100 % of its speed, which means that 75 % of the actual energy stored in the flywheel is available for use. Upon that a safety margin of 1.3 is added. Results of the simulations can be found in Table II.

A. *Flywheel as energy buffer*

An average total flywheel size of 1.1 kW h for the bus and 0.72 kW h for the ferry was found to be required for the cases the flywheel acts as energy buffer. In both cases, significant savings in diesel cost was found.

Table II
RESULTS OF SIMULATION

	City bus	Ferry
Total energy [kW h]	15.4(0.28)	6.37
Total braking energy [kW h]	4.06(0.02)	1.01
Total distance travelled [km]	11.2(0.33)	1.10
Total running time [s]	1534(24)	300
Diesel only	**City bus**	**Ferry**
Diesel consumption per trip [L]	6.27(0.11)	2.60
Diesel cost per year [k€]	82.6(1.50)	97.1
Diesel + flywheel	**City bus**	**Ferry**
Diesel consumption per trip [L]	4.19(0.10)	1.91
Diesel cost per year [k€]	55.2(1.28)	71.3
Yearly savings [k€]	27.4(0.23)	25.8
Required size of flywheel [kW h]	1.10(0.19)	0.72
Present value of savings (20 years) [k€]	341(11.5)	322
Flywheel only	**City bus**	**Ferry**
Electricity consumption per trip [kW h]	15.4(0.36)	7.0
Electricity cost per year [k€]	15.0(0.35)	19.3
Yearly savings [k€]	67.6(1.16)	77.7
Required size of flywheel [kW h]	19.6(0.50)	8.0
Present value of savings (20 years) [k€]	843(14.4)	969

In the case of the city bus, the yearly savings from the reduction of fuel by almost 35 % amounted to 27.4 k€, or 341 k€ in present value over 20 years.

The corresponding numbers for the ferry application was a fuel reduction of 25 % corresponding to 25.8 k€ annually, or 322 k€ in present value.

Assuming a total cost of 100 k€ for power electronics (AC/DC/AC to and from the flywheel), the upper estimate of the cost per stored kW h for the city-bus application becomes 220 k€/kWh and 310 k€/kWh for the ferry.

B. *Flywheel as main energy storage*

An average total flywheel size of 19.6 kW h for the bus and 8.0 kW h for the ferry was found to be required in case the flywheel should act as main energy source.

In the case of the city bus, the yearly savings from replacing the diesel fuel with electricity amounted to 67.6 k€, or an estimated 843 k€ in present value for a system life of 20 years.

The corresponding number for the ferry application was 77.7 k€ annually, or an estimated 969 k€ in present value. These quite large savings stem primarily from the higher efficiency of the electric driveline, and the lower price of electric energy, compared with the diesel based driveline.

Assuming also here a total cost of 100 k€ for power electronics yields an upper estimate on the specific cost of 38 k€/kWh for the city bus, and 110 k€/kWh for the ferry application.

Note that in this calculation, only the fuel savings are included, not the significant cost reduction resulting from removing the now obsolete diesel driveline. Also not included in the estimation is the fact that in some cases (most likely the case of the ferry) the local grid is not strong enough to provide the power required to charge the energy storage. Then, a secondary energy buffer is required at each charging station, or the grid connection has to be improved.

VI. Constructed Flywheel

In order to gain experience with the costs associated with the construction of a flywheel energy storage, a flywheel aimed at functioning as an energy buffer in an electric bus was designed and constructed, see [14]. The design limit of kinetic energy in the system was 0.87 kW h at a rotational speed of 30 000 rpm.

A composite shell was used to obtain the targeted amount of stored energy within a compact form-factor. The shell was constructed from two layers of unidirectional composite material, an inner layer of glass fiber surrounded by an outer layer of carbon fiber. The use of an inner rim of glass fiber, as well as producing the shell in one curing, resulted in a significant cost reduction.

A permanent magnet rotor was enclosed within the composite shell. The magnets were positioned in a cylindrical Halbach structure within a high-strength aluminum cylinder. The flux-path was closed through an inner concentric steel cylinder. The stator was constructed using Litz wire and fixated to a non-conductive plastic structure in order to reduce eddy current losses.

The rotor was suspended with the help of magnetic bearings in order to decrease stand-by losses and vibrations, and increse lifetime. Passive thrust magnets levitated the rotor axially. An active magnetic bearing (AMB) comprising eight actively controlled heteropolar electromagnetic actuators fixated the remaining four degrees of freedom.

A picture of the constructed prototype, as well as some of its main parameters, can be found in Fig. 2.

Parameter	Unit	Value
Total mass	[kg]	73.0
Rotor mass	[kg]	46.6
Total height	[mm]	535
Rotor outer diameter	[mm]	350
Thickness of shell	[mm]	76.4
Moment of inertia	[kg·m^2]	0.636
Max rotational velocity	[rpm]	30000
Max kinetic energy	[W h]	871.8
LP RMS phase-voltage	[V]	116
HP RMS phase-voltage	[V]	269
Continuous LP power	[kW]	2.0
Continuous HP power	[kW]	4.7

Figure 2. A magnetically levitated flywheel with composite shell has been designed and constructed. Picture from [15]

A. Rotor construction

At the core of the constructed rotor there is an aluminum cylinder (22), see Figure 3. A shaft (18) is mounted in the central hole of the aluminum cylinder. Also found on the main shaft is an aluminum cup (20), a steel cylinder (21) and a plastic cup (19) (containing the axial bearing magnets). A similar plastic cup can be found on the opposite side of the structure (24), carrying the other half of the axial bearing magnets.

Inside the aluminum cylinder there is a plastic matrix (26), containing the main rotor magnets of the machine (25). Lastly the whole core is inserted into a thick composite shell (23), which both acts as energy storage by increasing the systems moment of inertia, and as structural support by preventing the core from expanding as the rotational speed is increased.

The composite shell consists of both a glass-fiber inner layer and a carbon fiber outer layer, with a ratio of around 50/50 between the two composites.

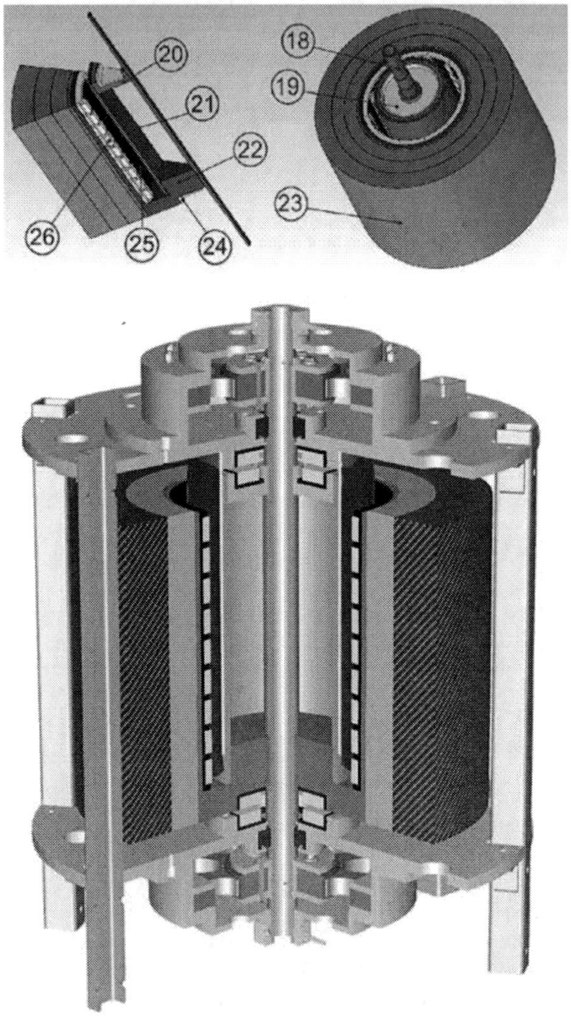

Figure 3. Cut view of CAD design for the constructed flywheel. Top: Rotor assembly. Bottom: Overview of complete flywheel.

B. Magnetic bearing construction

Eight laminated actuators and two targets were constructed from sheets of non-oriented electrical steel, approximately one square meter in size. The legs of the actuators, and the inside of the guiding plates holding the laminates together, were covered with electrically insulating but thermally conducting tape. Each individual electromagnet was thereafter manually wound with 1 mm enamelled copper wire. Each actuator was wound with 400 turns resulting in a filling factor of 45 %.

The control system was centered around a programmable automation controller (PAC) from National Instruments, the CompactRIO (C-RIO). Contactless position sensors were used to measure the position of the rotor shaft accurately and with high bandwidth. The position of the shaft was translated into a required current in an outer control loop in the PAC.

Further, sensors using the Hall effect measured current through each electromagnet. The values of current were fed to an inner control loop in the PAC, which forced the current to the desired value by varying the duty cycle of the MOSFETs.

All circuits were encapsulated in aluminum boxes, and connected with shielded cables to minimize electro-magnetic interference, see Figure 4. For a more detailed description of the test set-up, see [15].

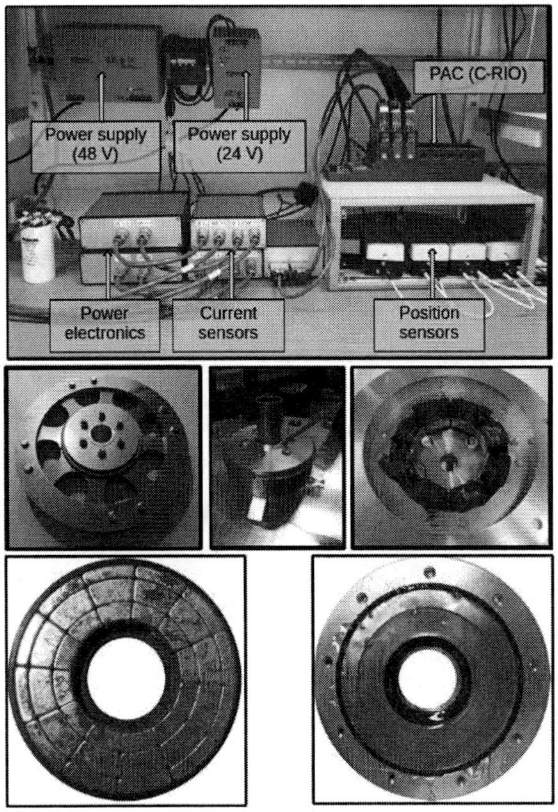

Figure 4. Magnetic bearing. Top: Power electronics with current control for radial active magnetic bearings. Middle: Constructed active magnetic bearing actuators. Bottom: Constructed passive axial thrust magnets.

A first test of the magnetically levitated system can be seen in Figure 5. The test compares the spin-down loss of the rotor, in air, with a previous flywheel of similar size with mechanical bearings. Note that power loss associated with generation of the bias current (approximately 40 W) is not included.

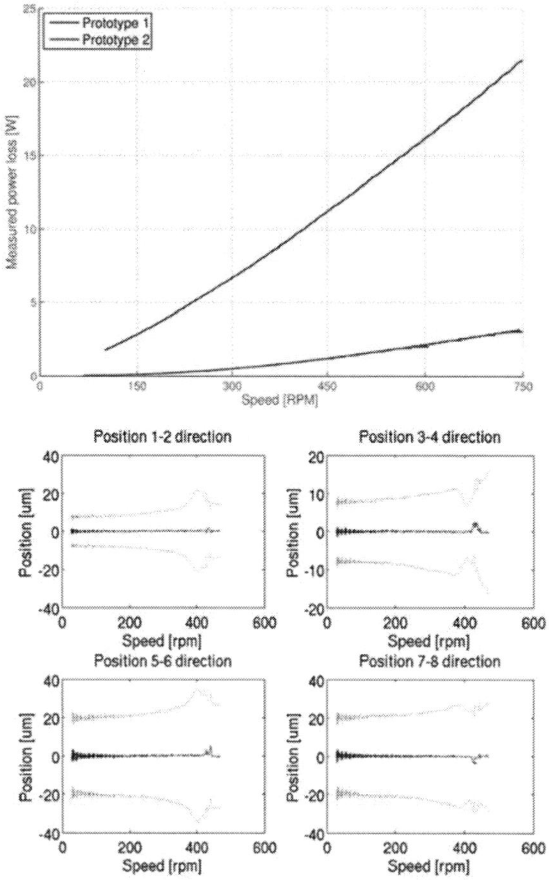

Figure 5. First spin down test of flywheel (note: in air). Top: comparison of the spin-down losses of the magnetically levitated flywheel (Prototype 2) with a previously constructed mechanically levitated flywheel (Prototype 1). Note: the power loss of the active magnetic bearings are not included. None of the two flywheels have been balanced. Bottom: Deviation of measured position from set position for the upper and lower radial active magnetic bearings during spin-down.

The flywheel operated stably until the designated top speed of 8000 rpm. The deviation from the set-point of both upper and lower bearing was always under 40 µm, although the rotor has not yet been balanced.

The difference between the intrinsic losses in a flywheel with mechanical bearings as apposed to one with magnetic bearings is clearly visible although the spin-down test was performed in air. At low speeds, where most of the loss in the mechanical flywheel comes from the friction of the rubber seals in the ball-bearings, the intrinsic loss in the magnetically levitated system is at least one order of magnitude smaller.

However, note that in this plot the power required to drive the bias current in the actuators (around 40 W) is not included.

C. Material cost

Although a magnetically levitated flywheel may seem like an overly complicated and expensive solution to store energy, the actual hardware needed to build one is not that hard to come by, nor is it very expensive. Furthermore, the computational power needed to make the flywheel function is becoming cheaper by the year, thanks to the rapid development within the field of electronics and microcomputers. Today a single FPGA have enough computational power to levitate a fully-fledged flywheel, as demonstrated in the setup presented within this paper.

In the cost breakdown for the prototype, Table III, one can see that the most expensive parts, with good margin, was the inductive position sensors. Followed by the combined cost for all the neodymium magnets. Those two components stand for 8.4 k€ of the total 13 k€. The reason for the high price on the sensors is that they have a very high accuracy (< 4 µm) and high bandwidth (20 kHz). This accuracy is not really needed to get the system to work, but is is advantageous to have in a prototype in order to get precise measurements.

Table III
MATERIAL COST OF FLYWHEEL

Part	Cost [k€/kWh]	Fraction of total cost [%]
Rotor		
Carbon fiber [Toray T700]	0.6	3.1
Glass fiber [Advantex]	0.05	0.2
Additional rotor structure	0.6	3.1
NeFeB magnets	2.15	11.0
Chamber		
Support structure	0.6	3.1
Electromagnets	0.35	1.8
Stator	0.35	1.8
Vaccum System	0.45	2.3
Electronics		
FPGA (CompactRIO)	1.7	8.7
3-phase inverter	1.1	5.6
Current control circuit	0.55	2.8
Position sensors	11.1	56.6
Total	19.6	

D. Rotor Manufacturing

In order to reduce the manufacturing cost of the composite shell, and to allow for shorter manufacturing time, the shell is was wound and cured all in one cycle. The alternative being winding the composite in separate rings and curing them individually and then press-fitting them on top of each other. However, the downside with the single cure approach is that it induces radial tensions in the shell. In order to control these radial tensions and following strains special care has to be taken when winding the composite but also to the geometry of the rest of the machine.

Additionally, the inner parts of the rotor is intentionally made of weaker, heavier materials, like aluminum and pieces of sintered NdFeB magnets. So when system begins to rotate these relatively "weak and heavy" parts will begin pressing against the inner walls of the much stiffer composite, due to the centripetal forces. This in turn creates compressive radial forces in the composite, counteracting the radial tensile stresses left over from the manufacturing process.

The trade-off for these compressive radial stresses is an increase in the hoop stress of the composite. Note though, that in a unidirectional filament wound carbon-composite, which this rotor is, the hoop direction is by far the strongest direction of all, having tensile strengths many times that of even steel. This allows for much higher rotational speeds that those achievable by a pure metal flywheels of similar size.

In this prototype the composite shell has an total thickness of 74 mm. Any thicker and the radial stresses left over from manufacturing would delaminate the material. Although the thickness of the composite shell is maximized, the 50/50 composite ratio is not optimum from the point-of-view of stored energy per Euro. For this a ratio containing much more glass-fiber, up to around 80 %, would be better due to the lower cost of glass-fiber.

The reason for not using this 80/20 design is purely due to manufacturing limitations. When the composite shell is cured as a whole and as a mixture of different fibers, the different thermal expansion coefficients of the materials have to be taken into account as the curing occurs at around 120°. Here the carbon fiber have around 1/10:th the expansion of the glass-fiber. If the glass-fiber layer is too thick, buckling of the fibers occur, see Figure 6.

Figure 6. Manufacturing of composite shells. A layer of glass-fiber is wound unidirectionally on a mandrel, after which a layer of carbon fiber is added. The middle of the three constructed shells (bottom) shows buckling from thermal expansion of the glass fiber during curing.

VII. Discussion

An introduction of flywheels into the drive-train is profitable both for the ferry and for the city bus even though the ferry seems to be the most interesting business case. The ferry also has other advantages such as a higher tolerance to added mass and volume.

However, some investments have to be done. Except for the obvious investment in the flywheel itself also charging facilities have to be set up at the end stops (for bus) and on shore (for ferry). The bus has to be equipped with an electric traction motor. This is not needed on the ferry since it already is diesel-electric – like almost all ferries.

The magnitude of the C-rates required, especially for the hybrid case but also for the all-electric case, indicates that batteries is far from a good choice for this application. They would also very fast be worn-out by the high number of cycles per year.

The main material cost for the flywheel at the moment are the very expensive position sensors. Secondly, the cost of permanent magnetic material is very high. The material cost of the composite material is relatively low, especially since the shell is made partly of glass fiber composite.

The cost of manufacturing is significant, but difficult to estimate. The position sensors used in the set-up requires no additional manufacturing, except assembly. However, the composite shell requires winding and curing in specialized machinery. Depending on where and how this is done, the cost of production of the rotor may increase significantly.

The next step of the development of the flywheel is to balance it, optimize the control algorithm and make accurate models of the different loss mechanisms during operation.

REFERENCES

[1] "The oerlikon electrogyro," tech. rep., Automobile Engineer, 1955.

[2] B. Storsand, "Electric vehicle running between two charging stations without a contact-line," June 1952.

[3] M. M. Flynn, J. J. Zierer, and R. C. Thompson, "Performance testing of a vehicular flywheel energy system," in *Proceedings of the SAE World Congress & Exhibition*, pp. 1–6, 2005.

[4] R. Hayes, J. Kajs, R. Thompson, and J. Beno, "Design and testing of a flywheel battery for a transit bus," tech. rep., Society of Automotive Engineers, SAE, 1999.

[5] C. Hearn, M. Flynn, M. Lewis, R. Thompson, B. Murphy, and R. Longoria, "Low cost flywheel energy storage for a fuel cell powered transit bus," in *Vehicle Power and Propulsion Conference, 2007. VPPC 2007. IEEE*, pp. 829 –836, sept. 2007.

[6] D. Cross and J. Hilton, "High speed flywheel based hybrid systems for low carbon vehicles," in *Hybrid and Eco-Friendly Vehicle Conference, 2008. IET HEVC 2008*, pp. 1 –5, dec. 2008.

[7] Volvo, "Volvo car corporation tests flywheel technology." https://www.media.volvocars.com/global/en-gb/media/pressreleases/38273, May 2011.

[8] C. Tarrant, "Flywheel assembly," 02 2012.

[9] J. Lundin, "Flywheel in an all-electric propulsion system," 2011.

[10] H. Abrahamson, "Laddbar färja får långtidskontrakt." http://www.nyteknik.se/nyheter/fordon_motor/fartyg/article3614754.ece, Jan. 2013.

[11] Trafikverket, "Eldriven färja förbättrar miljön i hamburgsund." http://www.driftunderhallsnyheter.se/2012/10/eldriven-f-rja-f-rb-ttrar-milj-n-i-hamburgsund, 2012.

[12] S. Carlsen and A. Gustavsson, "Elektrifiering av Älvskytteln," Master's thesis.

[13] J. D. Adams, S. W. McKenney, *et al.*, "Gyroscopic roll stabilizer for boats," Dec. 13 2005. US Patent 6,973,847.

[14] J. Abrahamsson, M. Hedlund, T. Kamf, and H. Bernhoff, "High-speed kinetic energy buffer: Optimization of composite shell and magnetic bearings," *Industrial Electronics, IEEE Transactions on*, vol. 61, pp. 3012–3021, June 2014.

[15] J. Abrahamsson, *Kinetic Energy Storage and Magnetic Bearings : for Vehicular Applications*. PhD thesis, Uppsala University, Electricity, 2014.

Increasing bearing life-time of mobile flywheels by using bearing-less drive methods

Manes Recheis[a]

[a] Institute of Electrical Measurement and Measurement Signal Processing, Inffeldgasse 23/II, 8010 Graz, Austria, recheis@tugraz.at

***Abstract*—Fully electric and hybrid electric power trains enable reuse of braking energy. This decreases the amount of lost energy dramatically, especially for discontinuous operation. Good examples are urban traffic, public transportation or delivery services. Mobile flywheels provide an excellent potential for the emerging market of electrified power trains, but they are currently only used in a few applications such as motor sports (Formula 1, Le Mans, GT3 car racing). To enable a wider use in personal or public transportation, the life-span of the flywheel's bearings have to be increased significantly. This paper presents an alternative approach to extend the life-span of the rolling element bearings by using bearing-less drive methods.**

I. Introduction

Energy storage in combination with renewable energies is key to a low carbon society. For mobile power-trains energy storage also enables a more efficient operation by restoring recuperated kinetic and potential energy. This justifies the enormous effort put into the research and development related to batteries, capacitors and flywheels.

A more efficient and sustainable mobility requires energy storage that enables vehicles to take advantage of features such as load averaging and regenerative braking. The energy storage therefore underlies requirements like a high power density, a high cycle life-time and a long service life. Flywheels meet these requirements and provide further advantages such as no capacity fade over time and an extremely wide operating temperature range. Further no limited or environmentally harmful resources need to be used and the state of charge (SOC) is always known exactly.

II. Motivation

To enter the mobile energy storage market successfully, the weight, size and costs of flywheel systems have to be a minimum. With magnetic bearings, the flywheel system has virtually unlimited bearing life, but it would require back-up bearings and a more sophisticated and therefore more costly rotor design. Previous research conducted at Graz University of Technology has shown that the weight, size and costs of a mid-sized car's flywheel energy storage system (FESS) increases dramatically when magnetic bearings are used (project "PowerKERS", FFG 825.553). Additionally the rotor gets decoupled thermally from the stator, which can be cooled easily and therefore increases the risk of overheating due to the eddy current and hysteresis losses of the rotor lamination. Since the flywheel is operated in vacuum to reduce air friction loss, there is no convection cooling inside the vacuum housing. Therefore radiation would be the only way to dissipate power if there are no mechanical bearings used.

These facts are also represented by the current market situation too. All suppliers of mobile flywheel energy storage systems worldwide apply lighter, smaller and cheaper hybrid rolling element bearings. Unfortunately they suffer from a significant life-time shortening by loads which occur in driving operation. With an average car-life of 250.000 to 300.000 kilometers and an assumed average speed of 60 km/h the average car life-time is estimated with 4500 hours. The life-time of rolling element bearings can be estimated by the product of rotational speed and bearing diameter. To reach the 4500 hours of required car operation with an average flywheel rotor speed of about 40.000 RPM the maximum bearing diameter is 18 mm for steel ball bearings and 31 mm for hybrid ceramic ball bearings. The results are based on the extrapolated data given from Schaeffler's FAG product data sheet of deep groove ball bearings as depicted in Fig. 1 [1] and do not include occurring bearing loads.

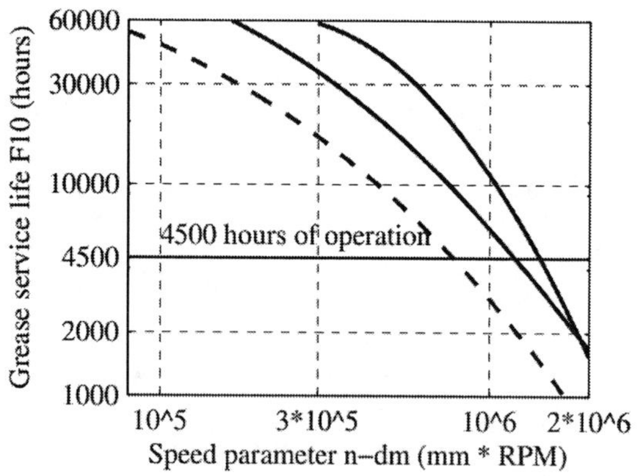

Figure 1. Estimated bearing life-time by grease service life for steel ball bearings (dashed) and hybrid ceramic ball bearings (solid) [1]

The graph shows that ceramic ball bearings can manage the high rotational speeds over the required period of time.

Table I. EVALUATED AND WEIGHTED DESIGN METHODS FOR STANDARD BEARING SYSTEMS AND THE PROPOSED BEARING SYSTEM

Main FESS application		stationary FESS & racing	stationary FESS	proposed method for flywheel hybrid cars
	Weight	Mechanical bearings	Active magnetic bearings	Mechanical bearings with additional brush-less load relief
Bearing life	5	- - -	+++	+
Construction space	3	+++	- - -	-
Variable damping	3	- - -	+++	+++
Rotor dynamics	2	+++	- - -	++
Thermal conductivity	2	+	- - -	+
Costs	2	+++	- - -	- -
Weight	1	+++	- - -	-
Sum of points		2	-6	10

Unfortunately they do not last long enough when the occurring bearing loads are considered. The estimated bearing life time for the ceramic rollers is higher than for steel ball bearings, since they feature a lower density, which reduces centrifugal forces on the outer bearing ring. They also have a much harder surface, which reduces wear. With the hybrid ceramic rolling element bearings it is possible to meet the requirement of 4500 hours of operation. Unfortunately the bearing life also strongly depends on occurring bearing loads, the operating temperature and contamination of lubricant. With the estimated occurring bearing load spectrum that results from the conducted research project "PowerKERS' (FFG 825.553) at Graz University of technology, the bearing life is not sufficient though. The new proposed method combines the advantages of both, simple mechanic bearings and active magnetic suspension without the need of backup bearings. The results are listed in Table I.

III. TECHNICAL OUTLINE

The motivation is clearly to reduce the bearing loads due to driving operation. The bearing loads are compromised of static and dynamic rotor imbalance, linear vehicle acceleration and gyroscopic torque. The share of rotor imbalance depends of course on the balancing quality. Acceleration and deceleration of the vehicle and gyroscopic reaction torque due to vehicle deflection are the main contributors to bearing's loads.

In stationary flywheel energy storage systems with ball bearings a resilient bearing mount is used. The resilient bearing mount is basically a soft mechanical bearing mount with a progressive stiffness, as can be found in elastomers and springs. This bearing mount reduces imbalance loads drastically, especially in overcritical operation. To avoid excessive rotor oscillation caused by gyroscopic reaction, rotor bending or vehicle movement, a certain rate of damping is required. In overcritical operation damping of rotor oscillation leads directly to loss of rotational energy. Therefore a variable rate of damping is desired.

A constant rate of damping can be applied by squeeze film dampers, which are used in turbo machinery. In this case the damping forces are transferred to the rotor through the rolling element bearings, and that results in constant bearing loads in overcritical operation. To avoid this constant bearing loads the damping forces have to be applied to the rotor contact-less. Applying damping to a rotor with no contact is possible by active and passive type magnetic bearings [2].

A. *Which magnetic bearing type to choose?*

Superconducting magnetic bearings feature good stiffness and damping properties but both characteristics are not variable and they require cryogenic cooling, which is not possible with reasonable effort for vehicles. Especially not if the are occasionally in use. Therefore cryogenic cooled superconducting bearings are applied successfully in large scale stationary FESSs (e.g. Beacon Power LLC, Boeing R&T). Passive magnetic bearings, e.g. "Null-Flux" homopolar induction bearings or permanent magnet bearings feature a certain rate of stiffness and damping. They can be used to compensate the axial bearing loads due to the rotor's weight. Their power density is lower than active bearings though and a variable rate of damping is missing.

That leaves active control as best option. A previous investigation at Graz University of Technology has shown, that even a magnetic bearing, that produces just a fraction of the estimated maximum bearing force would increase the rolling element bearing's life-time dramatically [3]. The reason for this is that smaller bearing loads are much more likely in the bearing's load spectrum and therefore a small load relief has still a remarkable effect. The bearing load spectrum is basically an estimation of the probability density function of occurring bearing forces.

Additional active magnetic bearings (AMBs) are equipped with sophisticated design efforts for the flywheel's rotor structure. They also increase the rotor length, reducing bending critical speeds. Even though smaller AMB are able to extend the FESS's service interval, it still requires additional construction space and a more sophisticated rotor design. To just gain bearing life, a much simpler approach is to adapt the flywheel's stator to produce radial force and relief the mechanical bearings. The proposed method is in general used in bearing-less drives and there are three types of electric motors for flywheels. Induction motors, permanent magnet motors, and reluctance motors [4].

B. *Which motor type to choose?*

There are many possible types, which will be reviewed in the following paragraphs from the perspective of vehicular flywheel systems.

1) Induction machines: They can be constructed to produce radial force by means of 2 separate sets of windings as proposed by [5]. IMs with a squirrel cage structure feature for good driving torque performance. Without the squirrel

Table II. EVALUATED AND WEIGHTED BEARING-LESS DRIVE TYPES FOR THE PROPOSED FLYWHEEL BEARING SYSTEM

Motor type	Weight	**Induction motor**	**Permanent magnet motor**	**Switched reluctance motor**	**Synchroneous reluctance motor**
Rotor losses	5	- - -	+++	- - -	+++
Power density	3		+++	+	+
High rotor Temp.	3	- -	- - -	+++	+++
Costs	3		- -	+++	++
Mechanical safety	2	- -	- - -	++	++
No-load torque	2	+++	- - -	+++	+++
Round-trip efficiency	2		+++	-	+
Noise	1	+	+++	- - -	
Reactive Power	1	-	+++	- -	- -
Points		-20	9	9	43

cage, a laminated or massive rotor is coated with a plated copper layer. The copper coated type induction machines are used for high speed applications. The massive rotor design is not suitable for flywheels due to the classical magnetic losses in the solid rotor. Additionally in case of a rotor failure, the released energy of solid rotor fragments is dangerous. In contrast, sheets of soft magnetic lamination do not break simultaneously and therefore less energy is released at the same time. Further the thin sheets are expected to dissipate the kinetic energy faster into deformation.

2) Permanent Magnet Motors: They benefit from a high power density and there are bearing-less types available in a large variety of topologies. Cylindrical PM rotors, inset type PM rotors, buried PM rotors and slice rotors [6]. The permanent magnets no load flux flows through the stator even when the flywheel is in standby and that results in magnetization drag. This affects flywheels that have a relatively high power compared to their stored energy. That applies for vehicular systems. An example is a short term flywheel energy storage for hybrid vehicles. An automotive flywheel would therefore be empty in estimated 15 minutes. Such a high self-discharge rate is still not a problem for mobile FESSs. The rotor drag increase the fuel consumption theoretically by estimated 0.1 liters or less of gasoline per 100 driven kilometers.

The limited and costly rare earth elements of the permanent magnets are a serious drawback for a possible market entry of mobile flywheels. They also limit the rotor temperature to about 120 to 140 °C. High temperature samarium cobalt magnets extend the possible rotor temperature to 200 °C and more. Unfortunately they are more brittle than the neodymium iron boron magnets. To hold the rotor's permanent magnets in place a thin layer of high tensile strength material is needed to preserve a small magnetic gap. Carbon fibre composites feature the actual highest ratio of tensile strength to density and are therefore the best choice. Nevertheless this facts increase the costs drastically and the standard epoxy resin matrix is degraded by rotor temperatures above 110 °C. In terms of safety massive magnets are also critical. In the event of a bursting rotor smaller fractions release less energy at the same time. With circumferential speeds of several hundred meters per second the rotor fractions collide with the crash ring with the same speed as a rifle bullet.

3) Reluctance motors: They are in general robust, cost effective and sustain high rotor temperatures theoretically up to 300 °C. The rotor temperature is actually limited by the rolling element bearings. Equally as induction machines they have no no-load drag and they also require a higher amount of reactive power.
Switched reluctance motors (SRMs) are particularly cheap and robust, since the have at least 3 concentrated windings. As long as the motor keeps spinning, they even produce torque with just one remaining active phase. That is very valuable in case of a single or multiple phase-fault. SRMs suffer from high magnetic losses in the rotor and the stator due to the repeated current excitation of all phases for a single revolution. With an alternating bi-directional current pattern, they magnetization losses can even be reduced by about 20 % [7]. Previous work has shown that the rotor losses of SRMs can be estimated by simulation based on the soft magnetic material properties [8]. For an example lets assume an average power dissipation of 300 W at a rated power of 20 kW with a duty cycle of 50 %. Under these circumstances the rotor of a car flywheel reaches estimated temperature of 200 °C.

The synchronous reluctance machine is the best overall choice for a simple, cost effective, robust and efficient flywheel for hybrid electric or fully electric vehicles as summarized in Tabluar II. The radial forces are produced by combining 2 sets of windings. One 3-phase rotating magnetic flux field for the torque production (flux oriented d-q control) and one set of windings for radial forces [6]. The superposition of both magnetic fluxes enables to relief bearing loads and to gain bearing life.

IV. METHOD

A. Rotor Design

To design an internal motor flywheel (simplest and easiest method) with a desired energy content, the rotor geometry can be simplified by a cylindrical shape. With the known density ρ of the soft magnetic rotor laminations the energy content can be formulated directly by the rotor's diamter d and height h. That leaves the diameter to height ratios as one degree of freedom for the rotor design. A ratio smaller than one ($d/h < 1$) results in a larger area to produce radial force and also in a longer lever to counteract gyroscopic torque. Long and slim rotors ($d/h < 1$) tend to be softer in axial bending than short and wide rotors ($d/h > 1$). The slim rotors also enable to use soft magnetic steel sheets with less tensile strength. To increase tensile strength cobalt alloys are used. With added cobalt costs increase dramatically as well as hysteresis losses due to the increasing grain size. On the contrary the cobalt alloys feature saturation flux densities B_s of about 2.2 Tesla whereas Si-Fe sheets usually saturate at 1.8

to 2.0 Tesla.
Figure 2 shows that a rotor with 90 Wh of stored energy can be built out of standard steel sheets, but it is not recommended due to the resulting long and thin rotor geometry which reduces the axial bending stiffness of the rotor.

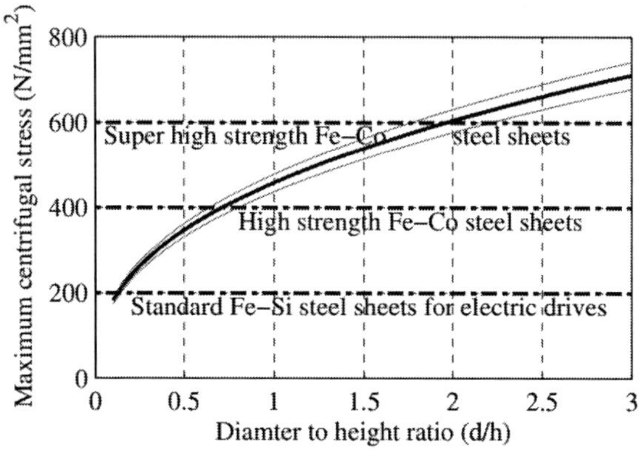

Figure 2. Maximum tensile strength for 90 Wh of stored energy (uncertainty boundaries are ±10 Wh) with a maximum rotational speed of 60.000 RPM and a density of $\rho = 8.2\ g/cm^3$. The material limits (dashed-dotted lines) include a safety factor of 2.

With high strength cobalt alloys d/h ratios of 0.5 to 0.7 are possible resulting in strong radial forces and good torque capability. A good method would be to produce driving torque with flux densities up to 1.8 Tesla and leave a flux density reserve of 1.8 to 2.0 Tesla for radial force production and counteracting gyroscopic torque. The ratio in Figure 3 is calculated with this proposed flux density reserve of 0.2 Tesla and the principle of virtual displacement.

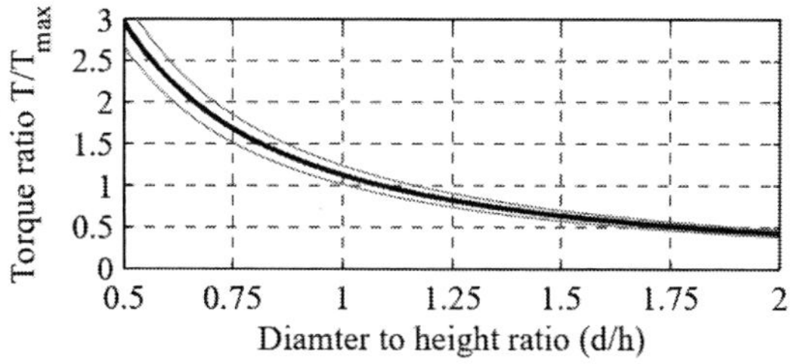

Figure 3. Ratio of highest counteracting torque T to the defined maximum gyroscopic torque ($T_{max} \approx 130$ Nm) occurring at an angular vehicle deflection speed of 1 rad/s which is equivalent to driving over a speed bump at 50 km/h. The uncertainty boundaries are ±10 % of the radial force capability.

Bearing loads due to gyroscopic reaction torque can be counteracted with d/h ratios of up to 1.08. Since the defined maximum gyroscopic torque is mainly depending on the stored energy the ratio gets better for thinner rotors. Another important bearing load source is linear acceleration forces. The analysed measurement data from the research project "PowerKERS' (FFG 825.553) conducted at Graz University of Technology has shown that the highest amplitudes of linear acceleration that do not lead to a complete technical failure of the vehicle occur at fender benders and small crashes. The g-rates are in the range of 10 g (98.1 m/s^2) as long as the impact is absorbed by plastic and thin metal components. That leaves the vehicle by definition in an operational state and such g-rates should therefore be covered by the proposed system. Figure 4 shows that for d/h ratios below 0.84 g-rates can be counteracted for the given energy content of 90 Wh and the maximum rotational speed of 60.000 RPM.

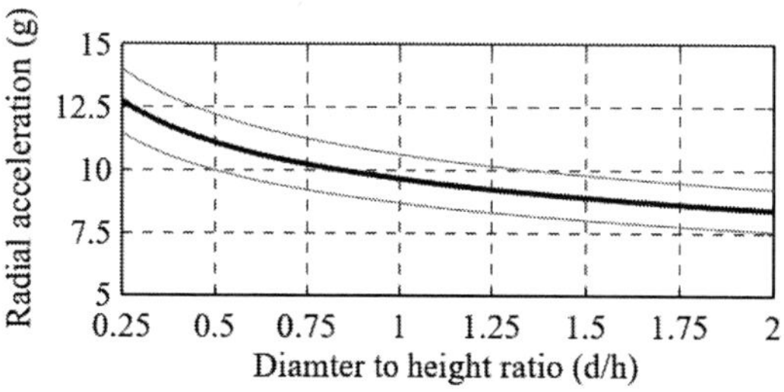

Figure 4. Maximum radial acceleration as a function of the rotor geometry to counteract linear acceleration.

B. Estimating rolling element bearing life-time

As depicted by Figure 5 a test car from the FH Joanneum University of Applied Sciences was equipped with linear acceleration sensors and angular speed sensors in 3 spatial axes. From the recorded measurement the bearing load shares are estimated on the basis of the Euler Equations, the flywheel geometry and the flywheel's suspension. It should be noted, that the test car was not equipped with a flywheel. The feedback from the flywheel's reaction torque to the car is negligibly small which was proven by simulation in [9]. With the estimated bearing force statistics basic life-time estimation formulas from rolling element bearing suppliers can be fed. This step is explained in more details in [3].

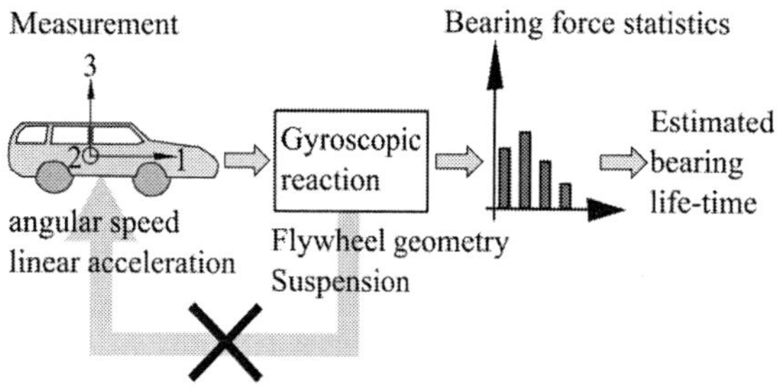

Figure 5. Overview of the rolling element bearing force statistics and life-time estimation.

The following results are calculated with a d/h ratio of 0.7, a stored energy of 90 Wh and a maximum rotational speed of 60.000 RPM. The rotor's diameter results in 0.11 m and the height is 0.157 m. The measurement data was collected on a course with a length of 60 km comprised of single-lane and dual-lane city traffic, stop-and-go traffic, inter-urban roads and a motorway segment. Driving over a speed bump is included as well. To get a one-dimensional bearing force probability density function as depicted in Figure 6 the resulting bearing force components are added geometrically.

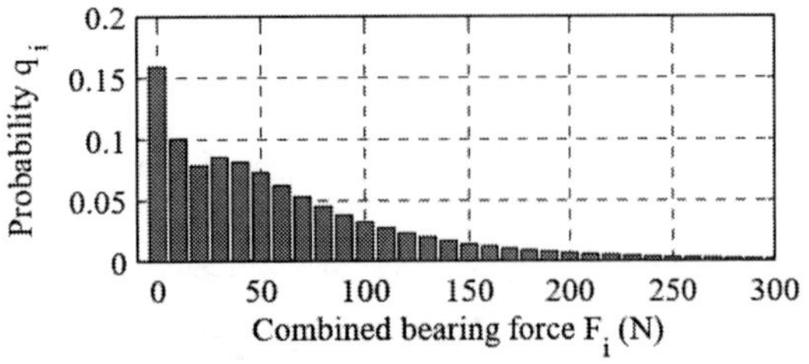

Figure 6. Estimated probability density function of the occurring bearing loads based on measurement data from 4 test-cars (ML320, Opel Movano, VW Passat and Smart Roadster). The bearing forces include linear acceleration loads and gyroscopic torque loads.

From this estimated probability density function the estimated fatigue life L10h can be calculated by Equation 1. L10h is the estimated amount of operational hours that a large number of bearings will sustain with a probability of 90 %. The rotational speed n in RPM is assumed with 44.750 RPM, which corresponds to the flywheel's state of charge (SOC) of 50 %. The corresponding rotational speed at the moment of the occurring bearing load could be considered. This relatively big step in complexity would additionally require to simulate the driving operation with the hybrid drive concept and was therefore left out.

$$L_{10h} = \frac{16666 \cdot C^3}{n \cdot \sum_{i=1}^{N} q_i \cdot F_i^3} \tag{1}$$

C. Lubrication method

The hybrid ceramic bearings can be operated grease-lubricated and oil-lubricated. Oil-lubrication requires a closed oil circuit with a filter system and a small oil pump. The oil-flow has to be controlled depending on the rotational speed to reduce unnecessary viscous losses. Oil-lubrication increases the design effort and therefore mainly the costs. On the contrary it features the possibility of direct cooling and higher rotational speeds for rolling element bearings. Therefore higher dynamic load ratings can be achieved for the same speed range and that enables a longer bearing life. Grease-lubrication is just less sophisticated and less costly. Tabular III compares the estimated fatigue life of a grease-lubricated and an oil-lubricated hybrid ceramic bearing.

D. Results

The data presented in Tabular III shows that even with oil-lubrication 10 % of all flywheels would become inoperative within half of the required life-time of 4500 hours. The average estimated car-life of 4500 hours is reached in public transportation applications within 1 year.

Table III. ESTIMATED FATIGUE LIFE FOR 2 HYBRID CERAMIC BEARINGS FROM UKF [10]

Bearing Type	719 UHC 30 A15	719 UHC 20 A15
Lubrication	oil	grease
Diamter	30 mm	20 mm
Dynamic load rating C	2470 N	1820 N
Axial pretensioning	60 N	40 N
L10h (90 %)	2106 h	894 h
L5h (95 %)	1306 h	554 h
L2h (98 %)	695 h	295 h
Required life-time factor	2.1 - 6.5	5 - 15.2

E. Assessing the bearing-life extension

The chosen geometry enables enough area to produce the required radial forces. Depending on the power electronics intermediate circuit voltage the number of windings of the radial force windings have to be adjusted. The aim is to be able to apply the required control current fast enough. The length of the maximum time-lag depends on the way the flywheel's is allowed to travel. Due to the progressive stiffness of the resilient bearing mount the rotor can move in a small orbit around the center of rotation without producing significant bearing loads. The goal is to keep the rotor inside the no-load orbit. With a given air gap of 600 μm and the diameter to height ratio of 0.7 about 40 % of the air gap (240 μm) have to be reserved for centrifugal growth of the rotor and for temperature dependent growth of the rotor (maximum Temperature assumed T_{max} = 200 °C). This leaves about 300 μm for the resilient bearing mount to get from very soft to hard. This should leave an orbit of about $\pm$100 μm for the proposed brush-less method to counteract the rotor-movement. In the range of $\pm$100 μm to $\pm$300 μm the resilient bearing mount acts as retainer bearing.

With a linear acceleration of 98.1 m/s^2 (10 g), which equals a fender bender, the flywheel rotor leaves the 100 μm orbit in 1.4 ms. Driving over a speed-bump at 50 km/h results in gyroscopic torque in the range of 100 to 130 Nm. With this torque applied to the vertical flywheel axis the rotor leaves the 100 μm orbit in 0.82 to 0.94 ms. The dead time due to unfortunate rotor position lies in between 0.25 and 0.75 ms depending on the flywheel's state of charge (SOC). Therefore the force rise-time has to be below 1 ms.

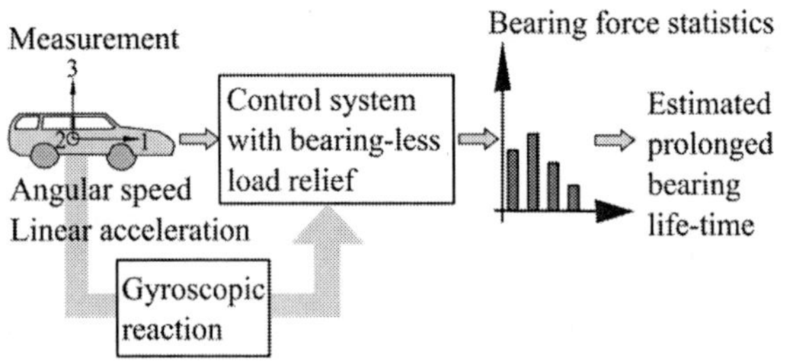

Figure 7. Overview of the system to prolong the rolling elements bearing life

V. RESULTS

By comparing the statistics of the bearing loads an estimated life-time expansion factor of about 10 to 12 is feasible with manageable efforts in power electronics and control. Higher factors are difficult to achieve with just measuring the rotor displacement. One way to improve the system is to measure the vehicles linear acceleration and angular velocity in driving operation and use the measurement signals to calculate a feed forward control as depicted in Fig.7. This possible improvement was not evaluated here and is part of ongoing research activities.

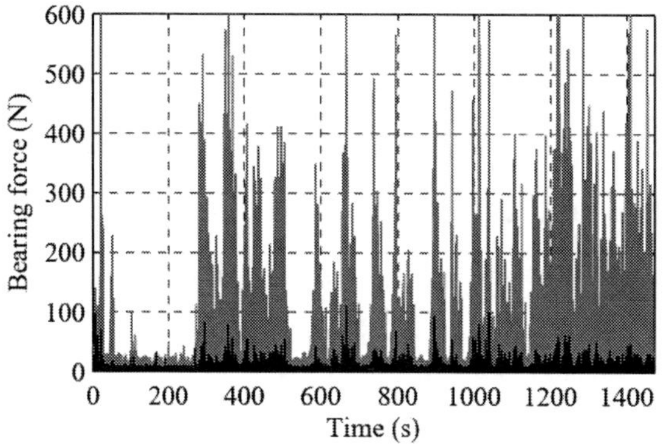

Figure 8. Time dependent combined bearing force (grey) and reduced bearing force (black) from a slice of the test drives (ML320).

With the estimated life-time extension factor of 10 to 12 it could be possible to even use the grease-lubricated bearing and achieve a 5 % drop-out rate in 80 % of the mean car life-time. With the oil-lubricated bearing the life-time could be extended to 15.000 hours which would be suitable for public transportation as well (3 years service interval). Test drive measurements of public transportation systems have not been evaluated yet to estimate the bearing life-time of such a flywheel system.

VI. CONCLUSION AND OUTLOOK

The introduction and motivation brings the reader closer to the special topic of flywheel energy storage systems. The technical outline explains why a brush-less design in combination with a synchronous reluctance machine is the best overall choice for a simple, cost effective and easy to design flywheel system that has the best chances to enter the market of mobile high power energy storage systems. Section IV explains how the simple internal motor topology can be designed to fulfil certain requirements. Further the method of estimating the rolling element bearings fatigue life on test drive measurement data is introduced. Results of the measurement data evaluation is presented by 2 example bearings. The results are shown for two different lubrication methods. The limitations of assessing the bearing life-time is described as well and an estimated achievable bearing life-time extension factor is presented. A possible improvement of the overall system is also mentioned.

REFERENCES

[1] "Schaeffler Group: FAG - Deep grove ball bearings data sheet," 2014, available online at www.schaeffler.com, last visit June 2014.

[2] G. Schweitzer and E. Maslen, Eds., *Magnetic Bearings*. Springer, 2009.

[3] M. Recheis *et al.*, "Improving kinetic energy storage for vehicles through the combination of rolling element and active magnetic bearings," in *39th Annual Conference of IEEE Industrial Electronics Society*, 2013.

[4] A. Rahman *et al.*, "Super high speed electrical machines - summary," in *IEEE Power Engineering Society General Meeting*, 2004.

[5] A. Chiba *et al.*, "Characteristics of a bearingless induction motor," *IEEE Transactions on Magnetics*, vol. 27, no. 6, pp. 5199–5201, 1991.

[6] *Magnetic Bearings and Bearingless Drives*. Elsevier, 2005.

[7] B. Schweighofer, M. Recheis *et al.*, "Rotor losses in a switched reluctance motor - analysis and reduction method," *EPJ Web of Conferences*, vol. 40, no. 17008, 2013.

[8] M. Recheis *et al.*, "Selection of magnetic materials for bearing-less high speed mobile flywheel energy storage systems," in *Proceedings of the 21st Soft Magnetic Material Conference*, 2013.

[9] G. Bischof *et al.*, "Investigation of a passenger car's dynamic response due to a flywheel-based kinetic energy recovery system," *International Journal of Vehicle Mechanics and Mobility: Vehicle System Dynamics*, vol. 52, no. 2, pp. 201–217, 2014.

[10] "UKF hybrid cermaic bearing product datasheets," 2014, available online at www.ukf.de, last visit June 2014.

Kinetic Energy Storage: an Ideal Application for Magnetic Bearings

Giancarlo Genta

Politecnico di Torino, Corso Duca degli Abruzzi 24, 10129 Torino, Italy, Giancarlo.genta@polito.it

Abstract—**Kinetic energy storage systems have a long history, but in the last half a century many studies and projects aimed to make this form of energy storage competitive with other systems were developed. One of the main problems related to flywheel energy storage is linked to the energy dissipations due to aerodynamic and bearing drag: Magnetic bearings, also due to their ability of working in vacuum, are thus intrinsically required for these applications.**

I. INTRODUCTION

Storing energy in the form of kinetic energy is an old idea and its applications in the last hundred twenty years were many. However, it not easy to define what we mean with *kinetic energy storage*: any rotating object stores energy in kinetic form and thus any machine containing rotating elements could be considered as an accumulator of kinetic energy. Going further along this way, any moving object is a kinetic energy accumulator, being immaterial whether its motion is translational or rotational.

It is only slightly better if we define a kinetic energy storage system as a device which contains an element that stores energy by rotating (a flywheel), without requiring the motion of the machine as a whole. Apart from the fact that devices that comply with this definition are very old, and some of them dating back from 6,000 years ago have been found.

Flywheels have been used for millennia to regularize the angular velocity of rotating elements, in particular in connection with the use of cranks, and their use became widespread with the introduction of reciprocating thermal engines, both steam or internal combustion engines[1].

However, the use of flywheels to regularize the rotational motion involves storage of energy for a very short period of time, usually linked with the device's rotational speed. To be more precise, the time the flywheel goes through a complete charge-discharge cycle depends on the time passing between two subsequent 'pushes' the rotating shaft receives from what supplies the force setting it in motion. The latter could be the foot of the potter in the most ancient applications of the potter wheel, or the piston in a reciprocating engine. This time could span from a few seconds in the potter wheel to a small fraction of a second (for instance, in a four strokes cycle single cylinder piston engine running at 3000 rpm it is of 40 ms).

It must also be noted that in many applications a specific flywheel (or rotating mass) was not even required to regularize the motion of a reciprocating engine: the huge translating mass of a steam locomotive was enough to store the kinetic energy required to regularize the motion of its rotating parts (the wheels).

It is thus possible to devise a criterion to distinguish between what is simply motion regularization and true energy storage in the form of kinetic energy: in the latter case the energy must be stored for a much longer time, and in particular it might be advanced the suggestion that this time must not be linked with the period of rotation of the flywheel itself and at any rate must be larger than the latter by some orders of magnitude.

One of the first applications in history which satisfies this requirement is the Howell torpedo (Fig. 1), built in 1888: a flywheel rotating initially at 21,000 rpm gave it a range of 1,500 m at 55 km/h [1]

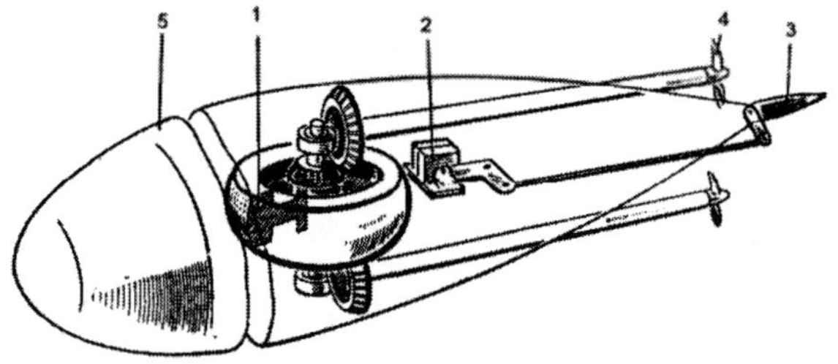

Figure 1. Flywheel torpedo built by Howell in 1888. 1) Flywheel, 2) steering mechanism, 3) rudder, 4) variable pitch propellers, 5) warhead.

The kinetic energy was stored for a time more than 35,000 times greater than the rotational period of the flywheel.

All the old applications of kinetic energy storage (for instance, the energy accumulator for a wind power plant designed by Ufimtsev in 1931 [1] and the Gyrobus built by Oerlikon in the 1960s [2]) shared this characteristics.

II. EFFICIENCY AND DRAG

Kinetic energy storage systems, like any other energy storage systems, are effective only if they are able to give back during the discharge a substantial amount of the energy they stored during the charge.

In the case of kinetic energy storage systems the losses that make it impossible to recover all the stored energy are mainly of two types: the drag torque on the flywheel (aerodynamic drag torque) and the drag torque of the bearings. Other losses, like the losses in the transmission, must be added.

To obtain the efficiency of the charge-discharge cycle the efficiency of the device that converts the energy must be accounted for. In the case of what is usually called a kinetic

[1] Even some of the earliest electric motors worked on the same scheme: a coil operated a sort of 'piston' which set in rotation a shaft through a connecting rod and a crank. They needed a flywheel as well.

battery the energy is exchanged in the form of electric energy, so the efficiency of the electric motor (in the charge phase) and generator (in the discharge) must be accounted for.

Clearly, both the above mentioned losses act for all the time the energy is stored in the system and not only for the time in which the accumulator is charged or discharged, like the efficiency of the mechanical transmission or of the motor and the generator. This is much worse than what is typical of most other energy storage devices.

To reduce flywheel losses (aerodynamic losses) the only way is to have the flywheel to operate in a very low pressure container, although it may also help to use a low viscosity fluid. For this reason, all modern flywheel energy storage systems, in particular if the storage time is not short, operate in high vacuum, with all the difficulties this may involve.

Bearing losses might be even more serious and they affect much the efficiency of any flywheel energy storage system. Actually, the bearings were the main weak points of all old flywheel systems like that in Fig. 1, and, even if at that time rolling element bearings were much less common than today, they were a must in flywheel systems. Rolling element bearings have also a limited duration, which involves the need of bearing replacement as a part of the maintenance cycle.

It is possible to say that all flywheel storage system suffer from a self-discharge problem. A simple (and quite approximate) way to evaluate the seriousness of this problem is to evaluate the energy the flywheel accumulator loses during the standby phase. Assume that a flywheel accumulator is fully charged when the flywheel spins at a speed Ω_{max} and fully discharged when it spins at speed Ω_{min}. Ratio $\alpha = \Omega_{min}/\Omega_{max}$ is the depth of discharge parameter.

Assume that the flywheel and bearing drag moment is $M(\Omega)$, and that the flywheel stores the energy for a time t, the speed loss is:

$$\Delta\Omega = \frac{1}{J\overline{(1/M)}}t \tag{1}$$

where

$$\overline{\left(\frac{1}{M}\right)} = \frac{1}{\Delta\Omega}\int_{\Omega_{max}-\Delta\Omega}^{\Omega_{max}} \frac{1}{M} d\Omega \tag{2}$$

is the average value of the reciprocal of the drag torque during slowing down and J is the moment of inertia of the flywheel plus the other elements which rotate together with it.

By introducing the energy density of the flywheel D, defined as the energy stored referred to the mass of the rotating elements of the system, the energy used to charge the kinetic accumulator (from its fully discharged to the fully charged state) and referred to the energy that can be extracted, is

$$e_{in} = \frac{1}{2}J\left(\Omega_{max}^2 - \Omega_{min}^2\right) = mD\left(1-\alpha^2\right) \tag{3}$$

The energy that can be extracted after time t is

$$e_{out} = \frac{1}{2}J\left[\left(\Omega_{max} - \Delta\Omega\right)^2 - \Omega_{min}^2\right] \tag{4}$$

With simple computations it is possible to compute the efficiency with which the flywheel can store the energy for time t

$$\eta = \frac{e_{out}}{e_{in}} = \frac{\left(1 - t\frac{1}{\overline{(1/M)}\sqrt{2DmJ}}\right)^2 - \alpha^2}{1-\alpha^2} \tag{5}$$

The efficiency reduces to 0, i.e. the flywheel accumulator self-discharges completely, at time t_{sd}

$$t_{sd} = \overline{(1/M)}\sqrt{2DmJ}(1-\alpha) = \overline{(1/M)}m\rho\sqrt{2D}(1-\alpha) \tag{6}$$

where ρ is the radius of inertia of the flywheel

By introducing the nondimensional time $t^* = t/t_{sd}$, the expression for the efficiency reduces to

$$\eta = \left(1-t^*\right)\left(1 - t^*\frac{1-\alpha}{1+\alpha}\right) \tag{7}$$

The efficiency is plotted as a function of the nondimensional time for some values of the depth of discharge in Fig. 2.

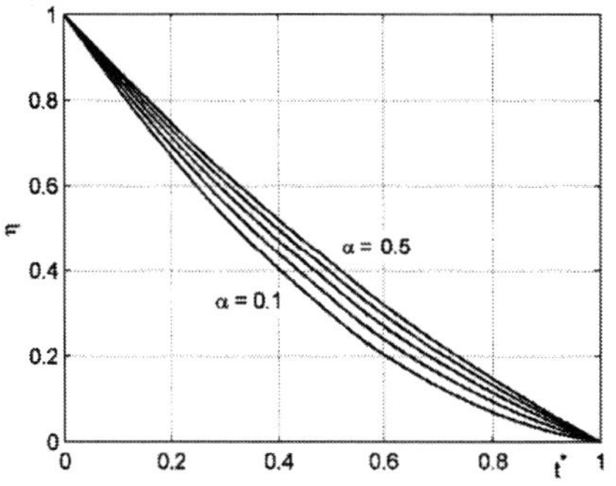

Figure 2. Efficiency of a flywheel system as a function of the nondimensional time t^* for various values of the depth of discharge α

Eq. (6) may be misleading: if the drag moment M is a function of the speed, as it is usually the case, $\overline{(1/M)}$ depends on the speed range and hence on time t. The self-discharge time cannot be computed in closed form and, while Eq. (6) still holds, it cannot be used directly for computing t_{sd}.

Another point which must be stated is that the value of the efficiency so computed is just a storage efficiency: the overall charge-discharge efficiency is lower, since it should include also the efficiency during the charge and the discharge phases. If the system is left at idle between subsequent charge-discharge cycles, the energy losses at idle must be also accounted for and also the possibility that the flywheel speed goes below the minimum speed Ω_{min}, which implies that some energy must be spent to restore the minimum conditions.

The drag moment $M(\Omega)$ is the sum of several contributions. For the aerodynamic drag, an expression can be

$$M_a = \rho_g \Omega^2 r_o^5 C_M \tag{8}$$

where ρ_g is the density of the gas in the container and r_o is the outer radius of the flywheel [3]. The moment coefficient C_M is a function of at least 3 nondimensional parameters, the Reynolds number, the Mach number and the Knudsen number. Since these three numbers depend on the speed, the dependence of the aerodynamic drag moment on the speed is actually not quadratic, and the integral to obtain $\overline{(1/M)}$ must be performed numerically.

The Knudsen number is particularly important in high vacuum applications [4]. An expression for the drag torque in high vacuum (free molecular flow, Knudsen number > 10), valid only for a thin disc, is

$$M_a = \rho_g \Omega r_o^4 \sqrt{\frac{m_m}{2KT}} \quad (9)$$

where K is the Boltzman constant, T is the absolute temperature and m_m is the mass of the molecules, i.e. the molecular mass of the gas divided by the Avogadro number. The drag torque is in this case linear with the speed.

For the bearing drag several empirical formulae exist. For ball and roller bearings a formula that can be used is [3, 5]

$$M_b = C_1 mg + C_2 \Omega^{2/3} \quad (10)$$

where C_1 and C_2 are coefficient depending on the type and size of the bearings, on the direction of the load, the lubrication, and other factors.

For magnetic bearings an empirical formula is [6]

$$M_b = mg(C_1 + C_2\Omega) \quad (11)$$

where C_1 and C_2 are empirical coefficients. This relationship dates back to the early developments of magnetic bearings, but it still can be used, provided that the two empirical coefficients are given updated values. It just approximates the dependence of drag with speed with a linear law.

Another type of drag linked with the bearing system, which is almost always overlooked, is the rotordynamic drag. Simply put, when a rotor spins in the supercritical mode, some bearing damping (nonrotating damping) is required to fight rotordynamic instability. Since the rotor spins in a more or less self-centered condition and cannot be perfectly balanced, the centers of the bearings have an orbital motion, which cause some energy dissipation which is seen as a further drag form. Rotordynamic drag has a strong increase when the rotor operates close to a critical speed, which may be of use when the rotor is free spinning, since it causes it to slow down below the critical speed, preventing it from operating for a long time in high vibration conditions. However, it increases with speed in the whole supercritical range.

In the case of a Jeffcott rotor with eccentricity ε and nonrotating damping coefficient c_n, the rotordynamic drag torque in the high supertitical range is asymptotically [7]

$$M_{rot} = \varepsilon^2 c_n \Omega \quad (12)$$

This drag may be high, particularly at high speed, if supercritical operation is chosen to relax the balancing requirements of the rotor, or in case strict balancing tolerances cannot be reached, as is the case with many composite material flywheels.

III. Critical Design Points

The difficulties to overcome in the design of flywheel storage systems had always been linked with:

- The high mass of the rotor, or better its low energy density
- The high flywheel (or aerodynamic) drag
- The high bearing drag

A further critical point was the need of a variable ratio transmission, in case the output of the storage system was mechanical, or of a mechanical-electric energy conversion when the output was electric.

Kinetic energy storage systems had a revival in the 1970s, after the energy crisis of 1973. With the increase of the price of oil and the forecasts of an impending drying out of the oil reserves, the research in the field of energy storage multiplied. It seemed that the perspectives of kinetic energy systems were bright and many research projects were started.

In particular, the three mentioned difficulties were thought to be easily manageable owing to some technological advancements:

- The use of composite materials and/or new geometries could increase, even substantially, the energy density of the flywheel
- Operation in high vacuum could reduce substantially the flywheel drag
- Magnetic bearings could allow to decrease substantially bearing drag.

The difficulties related with the energy input/output could be also improved, in particular with research in the field of variable ratio transmissions for mechanical solutions, and the striking progress in the field of power electronics could allow managing the power input/output in a completely sealed unit in case of electric solutions.

Some attempts were made to design variable inertia flywheels, with the aim of operating at a fixed speed and exchanging energy by varying the moment of inertia of the rotor instead of varying its speed. It is, however, possible to demonstrate that the mechanism producing the variation of the moment of inertia has a complexity and manages a power that are of the same magnitude than the device (transmission) that causes the speed to vary in conventional applications. As a consequence, variable inertia flywheels remained on the paper.

Some of these provisions reinforce each other, for instance using power electronics for a completely electrical transmission allows to completely seal the housing, without having a shaft connecting the flywheel to the outside world. This makes it much easier to have high vacuum, thus reducing aerodynamic drag.

However, most of them act in opposite ways, for instance to exploit high performance materials to improve the energy density implies running at higher speeds (at equal size of the device), which makes things much worse for the aerodynamic drag and the bearing problems. Increasing vacuum makes lubrication of standard bearings more difficult, and so on.

Above all these technical considerations there are economical considerations: kinetic energy storage systems are competing with energy accumulators of different types, mainly electrochemical batteries, and to be competitive they must not only show a better, or at least a comparable performance, but above all a lower cost. An exception may be those few applications like military or space applications where cost is not so important – or at least was not so important, since nowadays even those applications are much more cost-sensitive than in the past.

All these points were at the core of the revolution in the flywheel field which occurred starting from the beginning of the 1970s. At that time the idea that a large improvement of flywheel energy storage system was possible became quite widespread and it was a common opinion that flywheels would take over in many different applications, from hybrid vehicles (at that time seriously considered after a gap of 70 years from the first realizations of the 1900s), to stationary devices (mostly for renewable energy sources), from uninterruptible power supplies (UPS) for computers to aerospace applications, and so on.

The amount of research done at that time was impressive and most of the theoretical basis in the field date back to the 1970s. Molt of the references of the present paper date from those times, and most of what has been done later is at best implementation of what was then studied.

IV. SUPERFLYWHEELS OF THE 1970S

The confidence that high performance flywheels would change completely the prospects in the energy storage field was so widespread that a new term was introduced to designate these advanced flywheel system: **superflywheels**.

Superflywheels where characterized by

- a composite material structure, possibly with an unconventional shape,
- high vacuum operation,
- magnetic bearings
- and possibly some advanced mechanical or electric transmission

The idea that much better flywheels could be built using a composite structure was substantiated by a relationship yielding the maximum energy density a flywheel can attain:

$$\frac{e}{m} = K\frac{\sigma}{\rho_m} \quad (13)$$

where σ and ρ_m are respectively the stress present when storing the energy e and the material density, and K is a coefficient depending only on the flywheel shape, which can vary between 0.3 to 1.

To keep the costs compatible with the applications, many low-cost, high performance materials were suggested. Flywheels made of wood or even paper were seriously considered at that time, in particular for low cost applications.

However, in spite of all this work, the wide application of flywheel energy storage systems did not materialize, and this can be ascribed to a wide spectrum of reasons.

As a first point, many of these designs were unrealistic, or even basically flawed. Some configurations thus could never be tested or, if tested, didn't reach the performance their designers predicted. This was due to their geometry or design, to the material or both.

In other cases some overoptimistic assumption were made at the beginning of the design phase: the system could work, but never reached the predicted performance and their overall advantages over conventional systems vanished.

In particular, some confusion was made between the energy density of the flywheel and that of the accumulator and between the ultimate stress or the material and the stress that can be reached in operation, including the consideration of fatigue. To account for this a modification of Eq. (13) was proposed [3]:

$$\left(\frac{e}{m}\right)_{overall} = \alpha_1\alpha_2\alpha_3 K\frac{\sigma}{\rho_m} \quad (14)$$

where α_i are 3 coefficients smaller than unity which account for the realistically safe value of the stress, the depth of discharge and the ratio between the mass of the flywheel and that of the whole accumulator. The reduction due to these factors is easily by an order of magnitude or (much) more.

The disappointing outcome of many flywheel system was however linked in the majority of cases with the impossibility of reaching the cost objectives stated in the early design phase, in particular when the goal was that of building very low cost storage systems.

The introduction of new types of batteries, with much higher performance than the lead-acid batteries which were so common in the middle of the twentieth century, in term of energy density but even more of power density, was another good reason that hampered the diffusion of flywheels. Even in cases where this kind of energy storage system seem to have many advantages on other approaches like the KERS (Kinetic Energy Recovery System [8]) used on racing cars, the solution based on flywheels is less widespread than solutions based on batteries.

V. MAGNETIC SUSPENSIONS FOR FLYWHEELS

The problem of decreasing drag (aerodynamic and, above all, bearing) is an essential one for flywheel energy storage systems, in particular for cases where the energy must be stored for a time that is not very short. For this reason most advanced flywheels studied since the 1970s were designed to operate in high vacuum and incorporated magnetic bearings. Although magnetic bearings were not originated from studies aimed to store energy in flywheels, most of their earlier applications were connected with them [9, 10].

In general, flywheel systems are considered an ideal application for magnetic suspensions since they have relaxed specifications for what the suspension stiffness is concerned; this is particularly important since many advanced flywheel designs are characterized by intrinsic difficulties in achieving accurate balancing, compelling to resort to the self-centering occurring in the supercritical range. As already stated, in general this leads to higher rotordynamic losses.

Moreover, in stationary applications the only load applied to the flywheel is its own weight, which allows low stiffness operation without large and varying displacement of the rotor in the container. However, if the motor/generator is suspended together with the flywheel or the transmission system exerts forces on the latter, there are limitations to the bearing compliance which may become quite strict, in particular in case of applications characterized by high power density.

In case of applications on board of vehicles, inertia forces and gyroscopic moments due to the vehicle maneuvering are present and this may well pose strict limitations to the suspension compliance too.

All kind of magnetic bearings have been suggested for flywheel systems and most of them have been used at least on some prototype or demonstrator:

- Full 5-axes active suspensions
- Passive suspensions with at least one active axis
- Superconducting passive suspension

- Full passive suspensions stabilized by rotation (Levitron-style)
- Electrodynamic passive suspensions
- Magnetic/mechanical hybrid suspensions.

Generally speaking, each of them has its own advantages and there is a rationale under all choices. As a general rule, passive suspensions are often regarded as optimal for their low cost but usually are characterized by a low stiffness and low damping. The only possibility of having a fully passive suspension which is stable in the whole speed range is by resorting to superconducting levitation [11], which however needs low temperature operation and has particularly low stiffness and damping. High temperature superconductors seemed to be the solution to many problem, but they anyway require the use of liquid nitrogen (boiling point 77 K) or even lower temperatures to get higher performances.

Even if several designs of superconducting magnetic bearings have been developed and a number of prototypes of flywheel systems based on them have been built, this seems to be a technology limited to some niche applications.

The other fully passive technologies mentioned above require anyway external stabilization at low speed and have also other limitations, like for instance the relatively high bearing drag displayed by electrodynamic bearings. They too may be considered for niche applications.

Finally, hybrid mechanical-magnetic bearings in which the magnetic system acts as a load reliever for the mechanical bearings have several advantages, among which there is low cost and simple design, but they share with the traditional mechanical solutions the need of lubrication and a bearing drag which is higher than that of magnetic bearings (although less than purely mechanical solutions due to the lower mechanical bearings load). Solutions in which the flywheel axis is vertical and the passive magnetic bearing supports the axial load while the mechanical bearings give stability and deal with dynamic loads and unbalance have been used several times.

Active magnetic bearings, or a suspension system with at least some active axis, can be considered as the most suitable solution for what performance is concerned, but in some cases they can lead to costs which are higher than what can be accepted for low cost applications. The number of papers published on this subject is large, and many ideas have been forwarded and tested on demonstrators and full scale prototypes to lower the bearing drag, improve self balancing, increase the stiffness, make it easier to cross the critical speeds, etc.

Active magnetic bearings can be tailored on the application, allowing to solve the technical problems which still face the designer of flywheel systems. Whether the solution is also feasible from the economical viewpoint depends on the particular application.

VI. EXAMPLE 1

Consider a flywheel system based on 'pre-1970' technology: a 1600 mm diameter 1500 kg steel flywheel connected to a motor-generator whose rotor has a mass of 250 kg, storing 3.3×10^7 J = 9.15 kWh at a top speed of 3000 rpm. The minimum speed is 1500 rpm. This system is similar to the one that was installed on the Gyrobus city bus and is representative of 1950-1960 technology. It is thus possible to compute that the moment and radius of inertia are respectively of 668.7 kg m^2 and 668 mm.

The energy density of the flywheel alone is 6.11 Wh/kg but, if the whole accumulator is accounted for, a value of 3 or 4 Wh/kg is a realistic one.

Assuming that the flywheel operates in a hydrogen atmosphere at a pressure of 10 Torr, a practice then quite common to reduce the gas density while allowing a certain cooling to the motor/generator, the Reynolds, Mach and Knudsen numbers are respectively 2.44×10^4, 1.17×10^{-5} and 0.193. The flow around the flywheel is thus laminar, subsonic and not free-molecular.

The aerodynamic drag at top speed is about 0.88 Nm, and reduces to 0.31 Nm at the minimum speed. The self-discharge time due to aerodynamic drag is quite long, namely 55 hours.

In this case the bearing drag is much higher: assuming a pack of two preloaded angular contact ball bearings plus a deep grove bearing, a first approximation evaluation yields a constant bearing drag of 2 Nm. The total discharge time reduces then to about 11 hours. The storage efficiency as a function of the storage time is reported in Fig. 3.

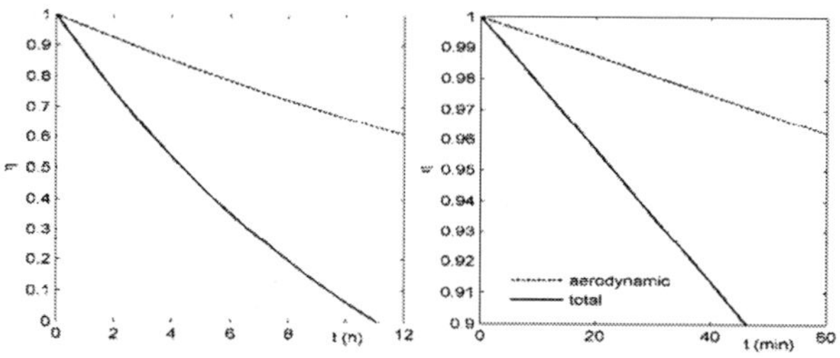

Figure 3. Efficiency of the flywheel system of Example 1 as a function of time; the zone of the plot related to short discharge times is repoorted on the figure on the right. The curves showing the effects of aerodynamic and total drag are reported.

It must be remembered that the actual efficiency is much lower, since also the drag during charging and discharging must be accounted for and other losses, like transmission losses and electric losses at idle must be included in the computation. The example shows however that a flywheel system of this kind can be used only for storing energy for quite a short time, of the order of a few tens of minutes as a maximum. It is also clear that with this technology the bearing drag is dominant, even in the case the flywheel operates in a partial vacuum.

VII. EXAMPLE 2

Consider now a 'modern' flywheel system, able to store the same quantity of energy, but using a CRP unidirectional composite. The cylindrical flywheel has an outer and inner diameter of 609 and 487 mm, a length of 1182 mm, a mass of 198 kg and a moment of inertia of 15.05 kg m^2. At a maximum speed of 20,000 rpm it stores 3.3×10^7 J = 9.15 kWh, with an energy density of 46.3 Wh/kg.

The energy density of the whole accumulator is much lower, and a value of 20 Wh/kg is a realistic one.

Assuming that the flywheel operates in air at very low pressure, namely 10^{-3} Torr, the Reynolds, Mach and Knudsen numbers are respectively 16.9, 1.86 and 0.165. The flow around the flywheel is thus laminar and supersonic. The Knudsen number is already in a region in which some free molecular flow effect start to be felt, but this can be neglected as a first approximation.

The aerodynamic drag at top speed, computed taking into account the flywheel length in an approximate way, is about 0.18 Nm, and reduces to 0.06 Nm at the minimum speed. The self-discharge time due to aerodynamic drag is quite long, namely about 40 hours.

It is quite difficult to state the bearing drag without specifying the details of their design, however using eq. (11) with reasonable values of the coefficient a bearing drag of 0.17 Nm at top speed, slightly smaller than the aerodynamic drag, can be evaluated. At low speed the bearing drag is larger than that due to air drag. The total discharge time is thus of about 18 hours. The storage efficiency as a function of the storage time is reported in Fig. 4.

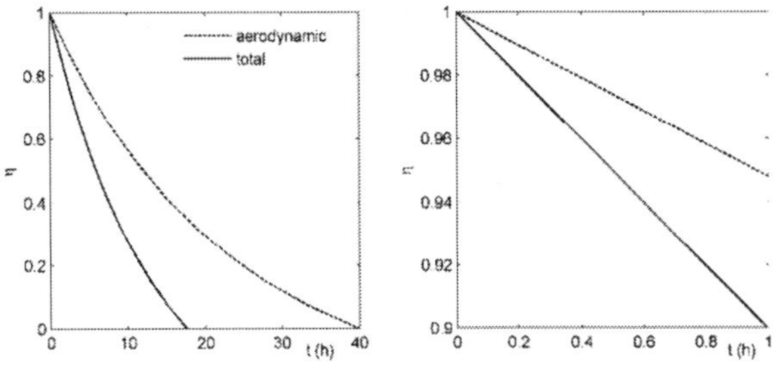

Figure 4. As Fig. 3, but related to Example 2.

VIII. CONCLUSIONS

Almost half a century has passed since when the research in the field of advanced flywheel energy storage systems has started. A large number of studies were performed and many prototypes were built with mixed success, but the wide diffusion of these systems on the market never materialized.

Some of the possible reasons may well be due to psycological factors: kinetic energy storage systems are something going against one of the most established trends of modern technology: the shift from rotating machinery technology to static machinery. An example of the same trend is the substitution of rotating converters (frequency converters and ac/dc converters) with converters based on power electronics. Also, rotating converters had a good flywheel effect, being able to use the energy stored as kinetic energy to smooth load variations.

Owing to the very high speeds (not only angular velocity but above all peripheral velocity), rotating elements compels to use high vacuum and complex magnetic suspension systems. This is perceived as a useless complication, causing also an unnecessary cost increase at least if the performance achieved is not much higher than that achievable with competing systems. When flywheels had to compete with lead acid and NiCd batteries this was considered well justified, while now the high energy and power density of advanced batteries make them less competitive.

Another point that may be labeled more as subjective than objective is safety. High speed rotating machines are perceived as quite dangerous, and the fact that failure modes of advanced, composite material, flywheels may well be more benign than those of solid steel flywheels (or at least they are advertised as such) doesn't change substantially the problem. It must be remembered that when a flywheel fails, there are not only the fragments to contain, but also the angular momentum to dissipate, which is difficult, being immaterial the type of flywheel. And that carbon dust, resulting from the flywheel of an advanced flywheel, ignites very easily once the container fails. Some severe accidents caused by the explosion of an energy storage flywheel [see, for instance, 12] make things worse. It is true that also advanced batteries have their own risks, and fuel tanks perhaps even more, but we are used to the latter and to the safety precautions allowing to use safely fuel. A crash between vehicles carrying a fast spinning flywheel or an advanced battery on board involves dangers we are not used to and thus we fear more. Many things in modern technology are "dangerous", and we have learned to deal with them safely, and there is no doubt that also flywheels may be made safe, but this requires much R&D work.

Magnetic bearing suffer themselves of this problem, as shown by the requirement of installing backup bearings. When the technology will be fully mature it is likely that the safety issue will be dealt with in another way, apart from the fact that it is unlikely that a touchdown bearing can guarantee a safe slowdown of a high speed flywheel in which a large quantity of energy is stored.

Several other technological issues have prevented flywheel energy storage systems from becoming common, one of which being the issues linked with the transmission: by their own nature flywheel energy storage supply a mechanical power output with a speed strictly linked with its state of charge (variable inertia flywheels have little definite advantage in this area). Variable ratio mechanical transmissions able to transfer a large power with a wide speed ratio range and good efficiency didn't materialize and electrical transmission, which on the contrary had striking advancements owing to brushless motors and power electronics, are in direct competition with all electric systems based on advanced electrochemical batteries.

It is thus likely that flywheel energy storage will be restricted to some niche applications, and that their success will be strictly dependent on a suitable set of magnetic bearings, the only technology able to allow storing energy in the mechanical form efficiently for periods of times comparable with those of other energy accumulators.

REFERENCES

[1] N. V. Gulia, *Flywheel Engines* (in Russian), Mashinostroyenye Press, Moscow, 1976.

[2] ----, *The Oerlikon Eletrogyro*, Automobile Engineer, Dec. 1955,

[3] G. Genta, *Kinetic Energy Storage*, Butterworths, London, 1985.

[4] A. Gu, *On the Viscous Drag on a Rotating Disc*, Journ. Of Fluid Eng., June 1974, pp. 184-186.

[5] T.A. Harris, Rolling Bearing Analysis, Wiley, New York, 1966.

[6] H. Haberman, G. Liard, *Les Premieres applications du palier magnetique actif*, La review des roulements, n. 409, pp. 1-13.

[7] G. Genta, *Dynamics of Rotating Systems*, Springer, New York, 2005

[8] http://www.racecar-engineering.com/articles/f1/flywheel-hybrid-systems-kers/

[9] P.C. Poubeau, *High Speed Flywheels Operating on 'One Active Axis' Magnetic Bearings*, 1977 Flywheel Tech. Symp., S. Francisco, Oct. 1977, pp. 229-240.

[10] P.C. Poubeau, *Flywheels Energy Storage Systems Operating on Magnetic Bearings*, 1980 Flywheel Tech. Symp., Scottsdale, Oct. 1980.

[11] F.C. Moon, *Superconducting Levitation*, Wiley, New York, 1994.

[12] https://www.osha.gov/pls/imis/accidentsearch.accident_detail?id=14515597

Modeling and Control of a Hybrid Magnetic Bearing in a Ring-type Flywheel System

Chow-Shing Toh[a] and Shyh-Leh Chen[b]

[a] Department of Mechanical Engineering and Advanced Institute of Manufacturing with High-tech Innovations, National Chung Cheng University, Chiayi 621, Taiwan, dennis_tcs@hotmail.com

[b] Department of Mechanical Engineering and Advanced Institute of Manufacturing with High-tech Innovations, National Chung Cheng University, Chiayi 621, Taiwan, imeslc@ccu.edu.tw

***Abstract*—In this study, a hybrid magnetic bearing (HMB) is proposed for a ring-type flywheel system. The HMB is composed of passive and active units. The rotor weight will be fully supported by the lateral forces of the passive units. The other 4 DOFs will be levitated and controlled by the active units. We will first establish the magnetic force model of the passive and active units of the proposed HMB. With the magnetic force models in hand, a stabilizing levitation controller for the HMB is designed. Moreover, numerical simulation results verify the effectiveness of the controller.**

I. Introduction

Magnetic bearing is an important type of noncontact bearing. It possesses the advantages of no friction, less energy loss, and environmental friendliness [1, 2]. Therefore, they can find wide applications, such as those requiring high speed. In general, there are three types of magnetic bearings: passive, active, and hybrid. Passive magnetic bearing (PMB) has the advantage of no energy consumption, but cannot be used alone for stable levitation. Stable levitation can be achieved by the active magnetic bearing (AMB) that uses the electromagnets to actively control the magnetic forces, with the disadvantage of higher cost and energy consumption. Hybrid magnetic bearing (HMB) is an integration of PMB and AMB. In other words, some degree-of-freedoms (DOF) of the rotor are levitated by PMB and the other DOFs are levitated by AMB.

The objective of this study is to propose an HMB for a ring-type flywheel system [3]. The HMB is composed of passive and active units. The rotor weight will be fully supported by the lateral forces of the passive units. The other 4 DOFs will be levitated and controlled by the active units. We will first establish the magnetic force model of the passive and active units of the proposed HMB. With the magnetic force model, we can then design a stabilizing controller for the HMB for stable levitation of a rigid rotor. For the passive unit, the magnetic force is a function of the air gap and offset between the stator and rotor Halbach arrays 4, 5]. For the active unit, the magnetic force is a function of the air gap and the coil current. With the magnetic force models in hand, a stabilizing levitation controller for the HMB will be designed. Numerical simulation will be conducted to verify the effectiveness of the controller.

II. DESCRIPTION OF HMB

In this study, a novel type of HMB is proposed, which consists of active and passive magnetic units. For the passive unit, both the rotor and stator sides are equipped with Halbach array of permanent magnets, as shown in Fig. 1. For the active unit, the stator is composed of a permanent magnet and two electromagnets and the rotor consists of the Halbach array of permanent magnets, as shown in Fig. 2. The two electromagnets are wound together in a differential way so that the coil currents will be provided with the differential driving. As a result, the stator will work as a Halbach array, but its magnetic field can be actively controlled.

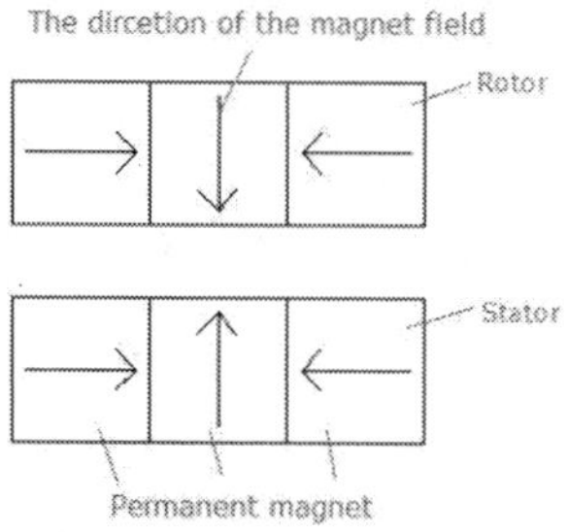

Fig. 1 The passive unit of the HMB

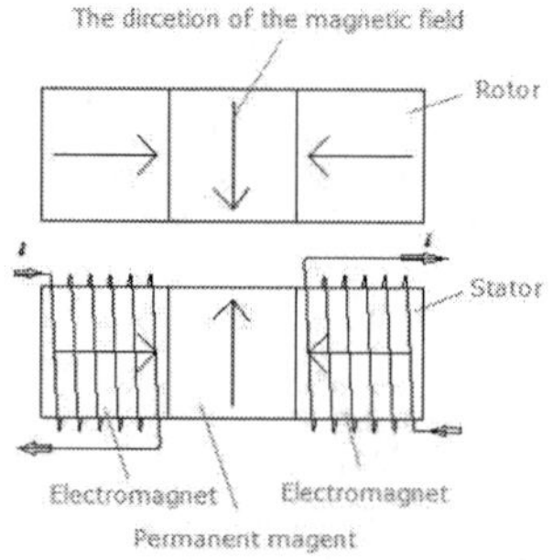

Fig. 2 The active unit of the HMB

III. MAGNETIC FORCE MODEL

The analytical magnetic force models will be obtained experimentally. For each magnetic unit, there exist normal magnetic force and lateral magnetic force. We denote them by f_{pn} and f_{pl}, respectively, for the passive magnetic unit; and f_{an} and f_{al}, respectively, for the active magnetic unit. For the passive magnetic unit, the magnetic forces are functions of air gap (l) and offset (s). Please refer to Fig. 3 for the definition of air gap and offset. For the active magnetic unit, the magnetic forces are functions of air gap and coil current (i). It is assumed that the offset in the active magnetic unit is negligible. In summary, the following four functions will be obtained experimentally

normal force of passive unit: $f_{pn}(l, s)$

lateral force of passive unit: $f_{pl}(l, s)$

normal force of active unit: $f_{an}(l, i)$

lateral force of active unit: $f_{al}(l, i)$

For each pair of air gap, offset (for passive unit only), and coil current (for active unit only), both the normal and lateral magnetic forces are measured using a load cell, as shown in Fig. 3 for the measurement of normal force and Fig. 4 for the measurement of lateral force.

Fig. 5 shows several sets of measurement data for $f_{pn}(l, s)$. With the experimental data, the least square based curve fitting method will be employed to obtain the magnetic force model $f_{pn}(l, s)$. Since the magnetic force model is a function of multiple variables, it is difficult to choose a proper candidate function for curve fitting. It is found that for each air gap, if the magnetic forces are normalized with respect to the maximum force, the magnetic forces with distinct air gaps almost coincide, as shown in Fig. 6. In other words, it is independent of air gap. This implies that the magnetic force model can be expressed as a function of separate variables. That is, it is a multiplication of two single-variable functions, the function of air gap and that of offset. In other words,

$$f_{pn}(l, s) \approx k f_{pn1}(l) f_{pn2}(s)$$

The two functions can be obtained using the technique of curve fitting separately. The polynomial functions of least degree are used for curve fitting.

Similar technique can be applied to obtain the other magnetic force models: $f_{pl}(l, s)$, $f_{an}(l, i)$, and $f_{al}(l, i)$. The details will be presented in the full-length paper.

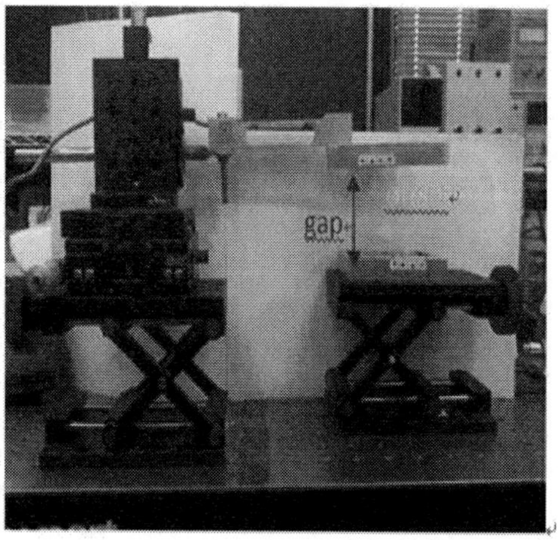

Fig. 3 Measurement set-up for normal force of the passive unit

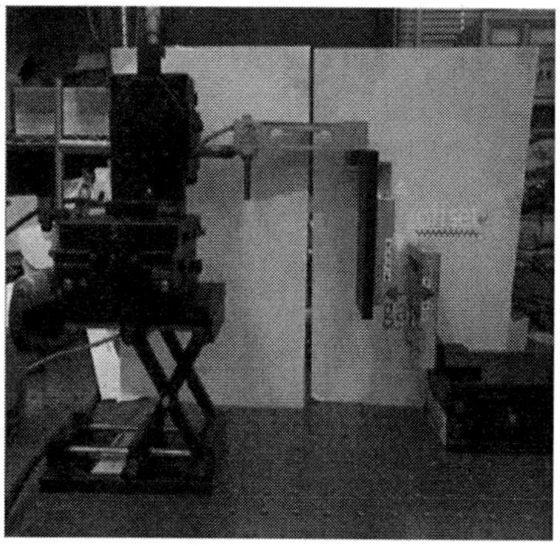

Fig. 4 Measurement set-up for lateral force of the passive unit

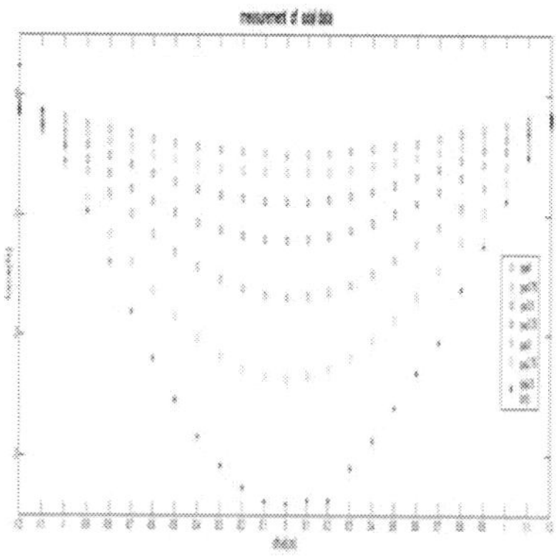

Fig. 5 Measurement of normal force for the passive unit

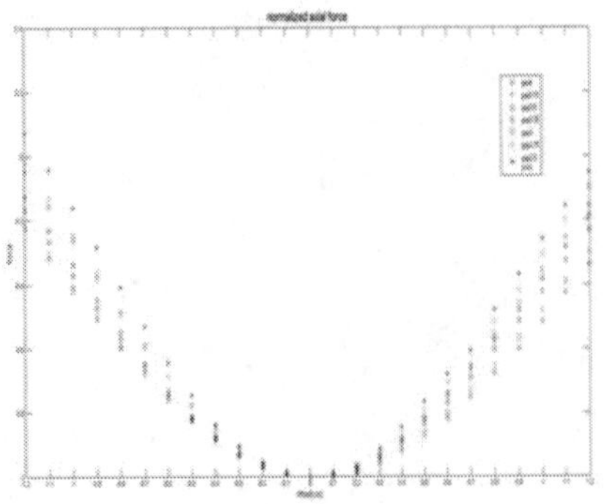

Fig. 6 Normalized normal force for the passive unit

IV. APPLICATION TO A RING-TYPE FLYWHEEL SYSTEM

The proposed HMB is intended to be used in a ring-type flywheel system, as shown in Fig. 7. The rotor weight will be fully supported by the lateral forces of the passive units. The other 4 DOFs will be levitated and controlled by the active units. The magnetic force models established in the previous sections can be generalized to the ring-type HMB. With the magnetic force model, one can then get the dynamic model and design the stabilizing controller accordingly.

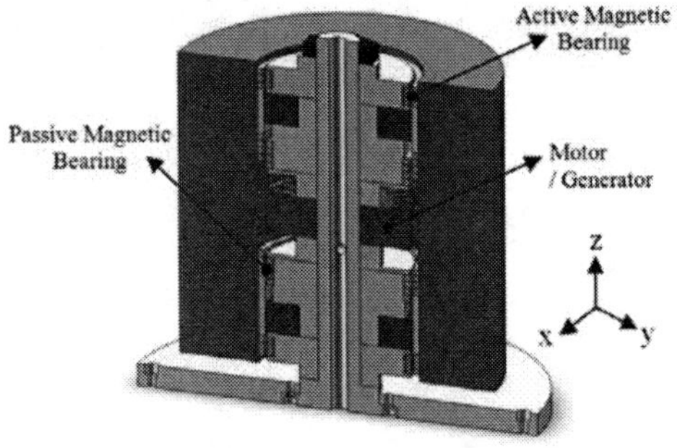

Fig. 7 The ring-type flywheel system

In the present system, an octagon structure with 8 cuboid active magnetic units is designed to form a ring-type magnetic bearing, as shown in Fig. 8. There are 8 coil sets on the stator and the opposite coil sets are wired in a differential way. As a result, there are four independent control currents in this system, and bias currents are not necessary. In other words, the magnetic force for the 4 DOFs can be approximated as

$$f_{atx} \approx k_1 i_{tx} \ , \ f_{aty} \approx k_2 i_{ty} \tag{1}$$

$$f_{abx} \approx k_3 i_{bx} \ , \ f_{aby} \approx k_4 i_{by} \tag{2}$$

where i_{tx}, i_{ty}, i_{bx}, i_{by} are control currents and k_1, k_2, k_3, k_4 are constant. The subscript with "a" denotes AMB and subscript with "t" and "b" denotes top and bottom.

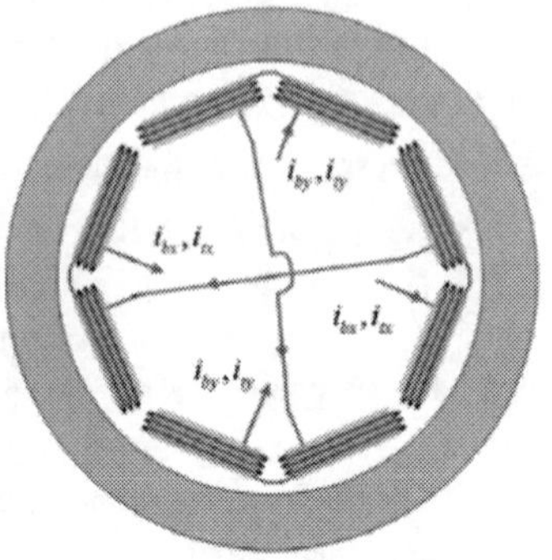

Fig. 8 An octagon structure with 8 cuboid active magnetic units

Next, the passive magnetic unit consisting of ring-type permanent magnet Halbach array is considered. Similar to the active unit, the magnetic force is formed by a deci-hexagon structure with 16 cuboid passive magnetic units, as shown in Fig. 9. When the rotor is running at the steady state, the summation force at the radial directions will be zero. The magnetic force on the passive magnetic unit can be expressed as

$$f_{pl} \approx k_{pl}(z,s) \ , \ f_{pl} \approx k_{pv}(z,s) \tag{3}$$

where z is offset. Note that k_{pl} and k_{pv} is a parameter in terms of air gap and offset. The subscript with "p" denotes PMB and subscript with "l" and "v" denotes normal and vertical.

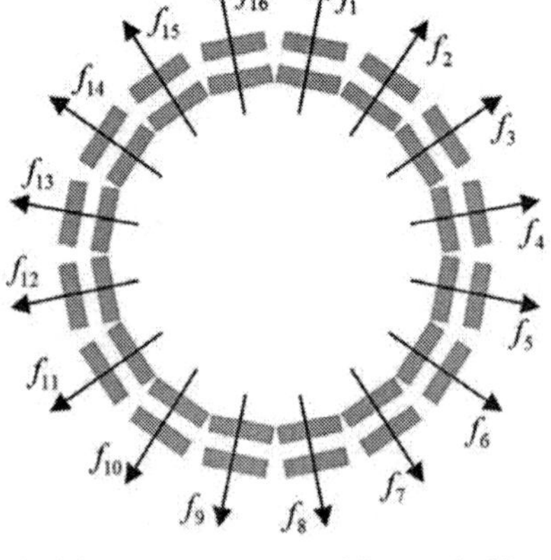

Fig. 9 A deci-hexagon structure with 16 cuboid passive magnetic units

Because the wall of ring-type magnets is quite thin and the diameters are large, the ring-type magnets are difficult to be fabricated and magnetized, so the passive unit consists of segmented arc-shaped permanent magnets.

In order to obtain the equation of motions for this flywheel battery prototype, energy method is applied here. It is assumed that the rotor displacement at z -direction is small. Then, the equations of motion can be obtained and expressed as

$$\sum F_x = m\frac{l_t\ddot{x}_b + l_b\ddot{x}_t}{l_t + l_b} \tag{4}$$

$$\sum F_y = m\frac{l_t\ddot{y}_b + l_b\ddot{y}_t}{l_t + l_b} \tag{5}$$

$$\sum M_x = I\frac{\ddot{y}_b - \ddot{y}_t}{l_b + l_t} + I_z\omega\frac{\dot{x}_t - \dot{x}_b}{l_b + l_t} + I_z\dot{\omega}\frac{x_t - x_b}{l_b + l_t} \tag{6}$$

$$\sum M_y = I\frac{\ddot{x}_t - \ddot{x}_b}{l_b + l_t} - I_z\omega\frac{\dot{y}_b - \dot{y}_t}{l_b + l_t} \tag{7}$$

where

$$\sum F_x = k_1 i_{tx} + k_3 i_{bx} + k_{13} x_t + k_{14} x_b \tag{8}$$

$$\sum F_y = k_2 i_{ty} + k_4 i_{by} + k_{15} y_t + k_{16} y_b \tag{9}$$

$$\sum M_x = k_4 i_{by} l_b - k_2 i_{ty} l_t + k_{17} y_t + k_{18} y_b \tag{10}$$

$$\sum M_y = k_1 i_{tx} l_t - k_3 i_{bx} l_b + k_{19} x_t + k_{20} x_b \tag{11}$$

$$k_{13} = k_5 + k_9 \,,\; k_{14} = k_6 + k_{10} \tag{12}$$

$$k_{15} = k_7 + k_{11} \,,\; k_{16} = k_8 + k_{12} \tag{13}$$

$$k_{17} = k_{11} l_{bp} - k_7 l_{tp} \,,\; k_{18} = k_{12} l_{bp} - k_8 l_{tp} \tag{14}$$

$$k_{19} = k_5 l_{tp} - k_9 l_{bp} \,,\; k_{20} = k_6 l_{tp} - k_{10} l_{bp} \tag{15}$$

The superscript "$\cdot$" denotes differentiation with respect to time. Note that l_b , l_t is distance between center of gravity and active magnetic unit, l_{bp} , l_{tp} is distance between center of gravity and passive magnetic unit and m is mass of rotor. Also, cross-section radial and polar mass moment of inertia are indicated as I and I_z , and ω is rotating speed.

With the mathematical model obtained as above, one can then design a stabilizing levitation controller for the system. Here, the integral sliding mode control (ISMC) is adopted. The detail of the controller design is omitted here for brevity. In this study, all simulations are done by MATLAB/Simulink. The system parameters and geometric sizes are given in TABLE I, and the detailed information of magnetic bearings is listed in TABLE II. Moreover, it is assumed that the rotor begins to rotate at t=0.2s with the torque of 1Nm, and the damping coefficient will be neglected, so there are levitation phase and rotation phase. If the maximum of rotating speed is set to 10000rpm, implying that the battery is full, it requires 183.84 seconds and the total stored energy is 9.626×10^4 J .

The initial states of simulation case are x(0)=[-3.53x10-4m 0 3.53x10-4m 0 3.53x10-4m 0 -3.53x10-4m]T. Note that two additional states are needed for ISMC, which are taken to be xm(0)=[0,0]T. The initial states for ISMC are thus [xm(0), x(0)]T. Moreover, the parameters of ISMC controller are listed on TABLE III, mainly provided by trial and error. The results of numerical simulation in duration of $t = 0 \sim 1$sec are shown in Fig. 10~13, including variation of rotating speed, trajectory and displacements of rotor at AMBs, and the control current i_{bx}, i_{by}, i_{tx}, i_{ty} .

V. CONCLUSIONS

In this paper, a novel type of active magnetic unit is proposed and applied to a ring-type flywheel battery system. The active magnetic unit consists of a conventional Halbach array with additional coil to actively adjust the magnetic force. The measurements of active and passive magnetic units with segmented Halbach array are analyzed. The flywheel energy storage system has been stabilized with ISMC controller. The illustrative example reveals that controller parameters are adequately adjusted to stabilize the unstable rotor. The robustness issue of suppressing the external disturbances and model uncertainties, such as assembly errors, manufacturing errors, mass unbalance and sensor noise, are further work scope. The fabrication and manufacturing of proposed flywheel battery prototype will be done in near future, and the experiment and development of driver of motor/generator are main work scope in next stage.

ACKNOWLEDGMENT

This work is supported in part by the Ministry of Science and Technology of Taiwan, ROC, under the Grant NSC 102-2221-E-194 -026.

REFERENCES

[1] Schweitzer, G. and Eric H. Maslen (Ed.), *Magnetic Bearing*, Springer-Verlag, Berlin Heidelberg, 2009.

[2] Schweitzer, G. Bleuler, H., and Traxler, A., *Active Magnetic Bearings*, Hochschleverlag AG an der ETH Zurich, 1994.

[3] Chow-Shing Toh and Shyh-Leh Chen, "Development of a Ring-type Flywheel Battery System for Storage of Renewable Energy," *The 13th International Symposium on Magnetic Bearing*, 2012, Arlington, Virginia, USA.

[4] S. Jinji, R. Yuan and F. Jiancheng, Passive axial magnetic bearing with Halbach magnetized array in

magnetically suspended control moment gyro application, *Journal of Magnetism and Magnetic Materials 323*, 2011, pp. 2103-2107

[5] Y.M. Choi, M.G. Lee, D.G. Gweon, and J. Jeong, A new magnetic bearing using Halbach magnet arrays for a magnetic levitation stage, *Review of scientific instruments 80*, 045106, 2009

TABLE I. PARAMETERS OF PROPOSED FLYWHEEL BATTERY PROTOTYPE

m(kg)	I(kgm^2)	I_z(kgm^2)
19.62	0.1623	0.1756

l_b(m)	l_t(m)	l_{bp}(m)	l_{tp}(m)
9.794×10^{-2}	9.805×10^{-2}	4.112×10^{-2}	4.523×10^{-2}

TABLE II. PARAMETERS OF ACTIVE AND PASSIVE MAGNETIC BEARING

Nominal air gap of AMB	2×10^{-4}m
Nominal air gap of PMB	2.5×10^{-4}m
Cross section of magnetic pole	1.24×10^{-3} m^2
Number of coil turns	270
Width of permanent magnet, a	5×10^{-3}m
Height of permanent magnet, b	5×10^{-3}m

TABLE III. PARAMETERS OF INTEGRAL SLIDING MODE CONTROLLER

b_1	b_2	ρ	α	k_c	ε
50	150	5	2	0.7	0.5

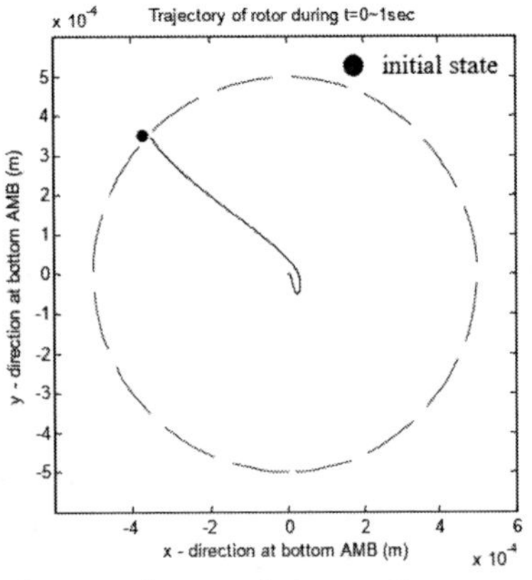

Fig. 10 Trajectory of rotor at bottom active magnetic bearing

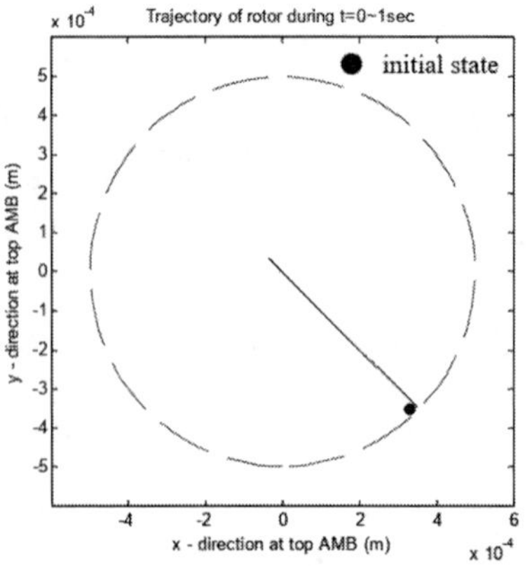

Fig. 11 Trajectory of rotor at top active magnetic bearing

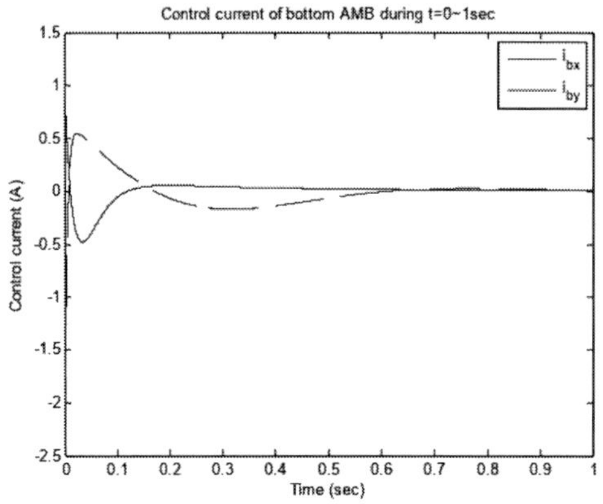

Fig. 12 Control current for i_{bx} and i_{by}

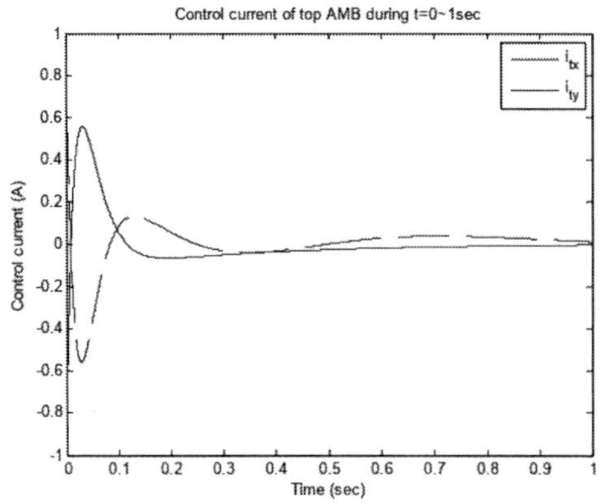

Fig. 13 Control current for i_{tx} and i_{ty}

Superconducting Bearing Design for Outer Rotor Flywheel Using Lumped Parameter Techniques

Clay Hearn[a], Siddharth Pratap[a], Dongmei Chen[b], Raul Longoria[b]

[a] University of Texas at Austin, Center for Electromechanics, Austin, Texas, USA
hearn@cem.utexas.edu
[b] University of Texas at Austin, Department of Mechanical Engineering, Austin, Texas, USA

***Abstract*— This paper describes the application of lumped parameter modeling techniques to designing high temperature superconducting bearings for outer-rotor flywheel energy storage systems. The lumped parameter models decrease computational time by 99% compared to Finite Element Analysis (FEM) without compromising fidelity needed to capture the non-linear and hysteretic force-displacement behavior between a levitated permanent magnet and bulk superconductor. The techniques formulated can be used to quickly evaluate lifting capacity and translational stiffness for a superconducting bearing design. The validity of the modeling approach has been verified by comparing results from FEM studies and experimental tests.**

I. Introduction

Flywheel energy storage has become an attractive storage technology for specific grid utility services where high power density and cycle rates are required. At the moment, potential applications for flywheel energy storage lie in shorter term storage applications below 20 minutes, such as voltage/frequency regulation services and power smoothing [1, 2].

A limitation to increase efficient storage times of flywheels from minutes to hours and days has been due to inherent storage losses, or leakage [3]. This leakage is primarily attributed to aerodynamic windage and losses through the bearings. Windage losses can be minimized by operating under hard vacuums, typically under 1 Torr [4], whereas magnetic bearings are currently employed to minimize bearing losses. Although magnetic bearings significantly reduce losses compare to contact style bearings, these bearings still suffer from losses due to eddy currents and magnetic hysteresis, effects that can lead to loss rates up to 2%-5% per hour [5, 6]. Optimization of bearing designs can drop the loss rate down to 0.5% per hour [7].

High temperature superconducting (HTSC) bearings can lead to further reduce loss rates associated with bearings. Flywheel systems with HTSC bearings have been demonstrated at very low loss rates of 0.1% per hour [5], which translates into a free spin down time constant of 1000 hrs. HTSC bearings are also passive and require little to no additional control structure for stable operation. The basic construction of an HTSC bearing for a flywheel application consists of the stable levitation of a permanent magnet pinned to a bed of bulk high temperature superconductors.

Compared to designing active magnetic bearings, where a magnetic circuit can be defined to estimate air gap fluxes and potential bearing stiffness, there are very few low order methods for describing the non-linear and hysteretic behavior between a levitated permanent magnet pinned to a bulk superconductor. The most simple model for this interaction is the advanced mirror image method [8, 9]. This method models the permanent magnet as a dipole interacting with a frozen image and diamagnetic mirror image. Although this modeling technique can be readily implemented, the model assumes the bulk superconductors are perfect conductors and ignores the highly non-linear losses which limit induced currents to the critical current density. High level finite element methods (FEM) can capture these effects [10] but, while they are the most accurate methods to use, they are computationally intensive. For the initial design stage of a flywheel with high temperature superconducting bearings, a lower order modeling method is desired to estimate lifting capacity and bearing stiffness. This is especially necessary for outer-rotor flywheel designs where many design iterations may be required to balance magnet banding strength and mass loading against the supporting composite bandings.

The authors of this paper have presented methodologies for lumped parameter models in an axisymmetric [11] and 3D frame [12] to describe the dynamic interaction between a levitated permanent magnet pinned to a bulk high temperature superconductor. Both models in their respective publications were confirmed against FEM analysis and experimental tests. Compared to FEM methods, the lumped parameter models reduce computational time by up to 99%, without sacrificing the nonlinear and hysteretic system dynamics. This paper describes how these techniques can be used to assess lifting capacity and bearing stiffness for HTSC bearing designs of outer rotor flywheel systems.

II. Bearing Topology and Model Development

Lumped parameter modeling is a technique that subdivides a system into discrete elements which can be represented by a select, finite, number of states. The state space formulation of these models is represented by a set of ordinary differential equations (ODEs) that can be quickly solved through numerical integration. For modeling bearing systems that

This work was supported by Stanford University through the Global Climate and Energy Project (GCEP).

utilize the stable levitation of permanent magnets over bulk high temperature superconductors, the lumped parameter models sub-divide the superconductors into pre-defined elements to estimate bulk average current flow, which is the local state. For the axisymmetric case [11], the superconducting plate is sub-divided into nested concentric ring elements which are concentrically constrained to the fields produced by the permanent magnet. For mechanical considerations, this model formulation is useful for estimating the bearing lifting capacity and the dynamic axial stiffness. In the 3D formulation [12], the bulk superconducting region is subdivided into overlapping disc elements which allow the model to simulate non-circular surface currents that will be induced due to translational movement of the magnet. A representation of a single permanent magnet levitating over a bulk HTSC disc is shown in Figure 1 for both respective model formulations.

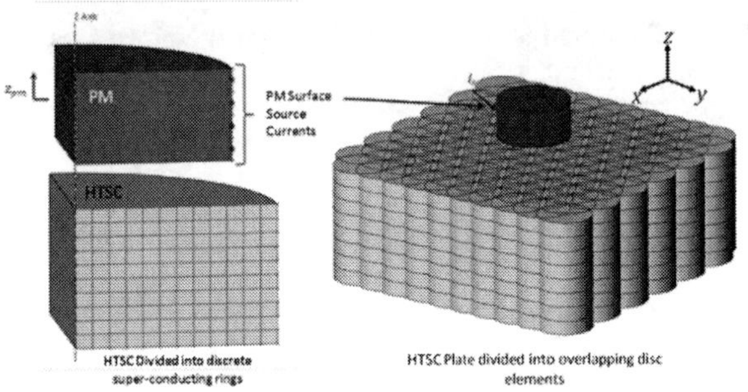

Figure 1: Left: Permanent magnet levitating over a bulk HTSC disc modeled as concentric ring elements for the axisymmetric model. Right: Permanent magnet levitating over a bulk HTSC block modeled as overlapping disc elements for the 3D model.

Utilizing Faraday's law of induction, a set of differential equations (1) can be developed to estimate the induced element currents, I_n, in the respective lumped parameter models. The spatial gradients of the magnetic fields, ϕ, produced by the permanent magnet array link the component velocity, v, of the magnet array to the electromotive force induced on each element. From bond graph terminology [13], these spatial gradients can be identified as gyrator moduli, which represent ideal power conversions between the magnetic and mechanical elements within the system. These moduli are also used to calculate the resultant force on the magnet array due to movement (2). The inductance matrix, **L**, is a symmetric matrix that includes all self and mutual inductance values for the superconducting elements within the system.

$$\begin{bmatrix} \frac{dI_1}{dt} \\ \frac{dI_2}{dt} \\ \vdots \\ \frac{dI_N}{dt} \end{bmatrix} = -L^{-1} \left[\begin{bmatrix} e_1 \\ e_2 \\ \vdots \\ e_N \end{bmatrix} + \begin{bmatrix} \frac{d\phi_1}{dx} \\ \frac{d\phi_2}{dx} \\ \vdots \\ \frac{d\phi_N}{dx} \end{bmatrix} v_x + \begin{bmatrix} \frac{d\phi_1}{dy} \\ \frac{d\phi_2}{dy} \\ \vdots \\ \frac{d\phi_N}{dy} \end{bmatrix} v_y + \begin{bmatrix} \frac{d\phi_1}{dz} \\ \frac{d\phi_2}{dz} \\ \vdots \\ \frac{d\phi_N}{dz} \end{bmatrix} v_z \right] \quad (1)$$

$$\boldsymbol{F} = \left(\sum_{j=1}^{N} -\frac{d\phi_j}{dx} I_j\right)\hat{\imath} + \left(\sum_{j=1}^{N} -\frac{d\phi_j}{dy} I_j\right)\hat{\jmath} + \left(\sum_{j=1}^{N} -\frac{d\phi_j}{dz} I_j\right)\hat{\boldsymbol{k}} \quad (2)$$

For high temperature superconductors, once the upper critical magnetic field is exceeded, the material loses the property of superconductivity and becomes resistive. At this point, the current flow within the material is saturated at the critical current density, J_c. The model formulation in (1) includes power-law based resistance components, e_n, to model the rapid rise in resistivity once the critical current density is exceeded. The power law method has been effectively used in FEM studies to model these loses and saturation effects [10].

To estimate the magnetic flux produced by the permanent magnet that penetrates each ring or disc element of the respective axisymmetric or 3D models, the magnet is modeled by discrete current loops on the exterior surface which produce an equivalent magnetic field strength. For each surface current loop, I_s, the magnetic vector potential, $\boldsymbol{A}$, can be calculated at any point with respect to the loop by using a formulation of the Biot-Savart Law, as presented by Smythe [14]. From Figure 3, Smythe's formulation of the vector potential is shown in (3), where K and E are elliptical integrals of the first and second kind, with modulus k_{si} (4).

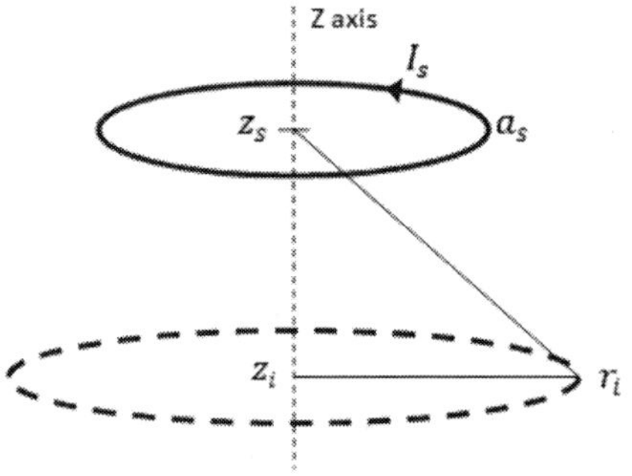

Figure 2: Schematic of fixed source current loop (on magnet surface) with current, I_s, radius, a_s, and axial position z_s, to a passive filament ring element (within superconducting element) with radius, r_i and axial position z_i.

$$\boldsymbol{A} = \frac{\mu_0 I_s}{\pi k_{si}} \left(\frac{a_s}{r_i}\right)^{\frac{1}{2}} \left(\left(1 - \frac{1}{2}k_{si}^2\right) K(k_{si}) - E(k_{si}) \right) \quad (3)$$

$$k_{si} = \sqrt{\frac{4 a_s r_i}{(a_s + r_i)^2 + (z_i - z_s)^2}} \quad (4)$$

From the magnetic vector potentials, Stokes Theorem (5) can be used to calculate the magnetic flux through any surface or closed path. From here, the Leibnitz rule is used to calculate the spatial derivatives (where q_i is x_i, y_i, or z_i respectively) of the magnetic flux (6) through a closed circular loop of a superconducting element to calculate the gyrator moduli used in (1).

$$\phi = \iint \nabla \times \boldsymbol{A} \cdot d\boldsymbol{S} = \oint \boldsymbol{A} \cdot d\boldsymbol{l} \quad (5)$$

$$\frac{d\phi_{ij}}{dq_i} = \int_0^{2\pi} \frac{\partial}{\partial q_i}[\phi_{ij}(q,\theta)]\, d\theta \quad (6)$$

The respective papers on axisymmetric and 3D model development, [11] and [12], give detailed techniques for calculating these moduli based on the specific geometric constraints for each model.

III. Application to Bearing Design

The lumped parameter models may be applied during the initial design phase of a superconducting bearing for an outer rotor flywheel to determine lifting capability and translational stiffness. The bearings should be sized to support the weight of the flywheel, and axial and translational stiffness values must be assessed to ensure no instabilities occur during operation. An outer rotor flywheel consists of a set of high strength composite rings that are levitated and spun to store energy in the kinetic form. Rotor side bearing and motor-generator elements are typically located on the ID (inner diameter) bore of the composite bandings. This design topology maximizes energy stored to mass ratios for the final flywheel system. For the superconducting bearings, the size and layout of the rotor side permanent magnets must be optimized to limit mass loading of the surrounding composite bandings during operation.

A topology for a 25 kWh outer-rotor flywheel with end face HTSC bearings is shown in Figure 3 [6]. This proposed flywheel is a notional design for utility scale load-leveling that was developed to demonstrate the lumped parameter modeling technique for a complete bearing set. An array of permanent magnets sits within the end-faces of the outer rotor composite bandings on either side to stably levitate the flywheel and allow near lossless rotation about the spin axis. Typically, the permanent magnets are arranged in a Halbach or alternating pole pair arrangement to increase field gradients and overall stiffness [15, 16]. The bearing arrangement may also be located on the ID bore of the composite bandings in a journal bearing configuration [17]. Flywheel dimensions, mass, and operating speed are shown in Table I.

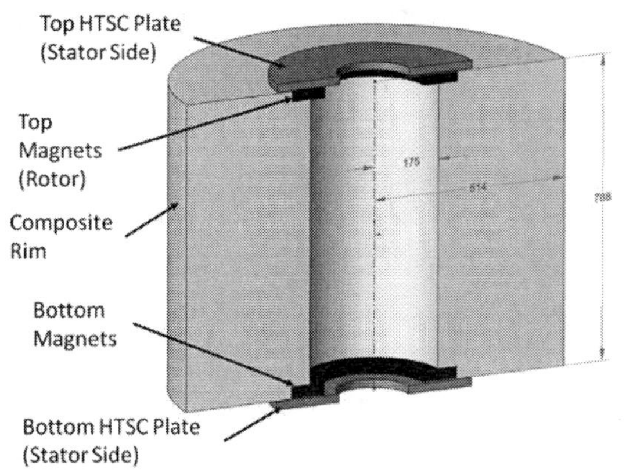

Figure 3. Notional 25 kWh flywheel rotor design using HTSC face bearings for levitation.

Table 1: Flywheel dimensions for bearing analysis

Inner Radius	Outer Radius	Length	Mass	Delivered Energy	Speed
0.175 m	0.514 m	0.768 m	878 kg	25 kWh	13,000 RPM

IV. Axisymmetric Analysis

The initial design analysis utilizes the axisymmetric modeling methodology to determine the bearing lifting capability. The bearings must support the full weight of the flywheel assuming that the flywheel is vertically oriented. The equations for axisymmetric model are solved with a fixed step, 4th order Runge-Kutta solver, which improves computational time and allows optimization studies to be performed. For this analysis a parametric study was performed to optimize pole pair arrangement of the permanent magnets to maximize lifting force and vertical stiffness.

A schematic of one of the superconductor bearing sets is shown in Figure 4. For the axisymmetric model, the HTSC plate is sub-divided into 300 concentric superconducting ring elements assuming a critical current density of 12 kA/cm^2. The plate of bulk superconductors has an inner radius of 105 mm, outer radius of 265 mm, and an axial thickness of 15 mm. The corresponding permanent magnets, which are located on the end faces of the composite bandings, have an inner radius of 145mm, outer radius of 225mm and axial thickness of 30mm. The inner radius of the permanent magnet set, which is smaller than the inner radius of the bandings described in Table 1, would require additional support structure that would need to be considered in a structural analysis. The permanent magnets are assumed to have an equivalent surface current strength of 1030 kA/m. Discrete current loops, I_s, on the cylindrical faces of the permanent magnets are employed to generate the equivalent magnetic fields, as shown in the break-out image in Figure 4. When the flywheel is centered axially, there is a 2mm air gap between the permanent magnets and bulk superconductors on the top and bottom faces.

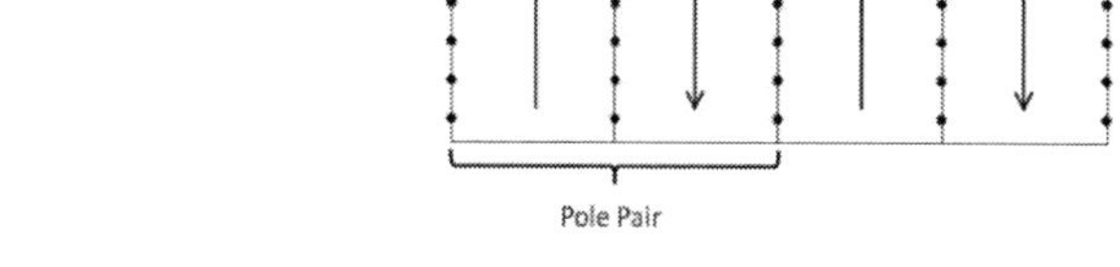

Figure 4: Schematic of pole pair arrangement for HTSC bearing.

An analysis using the axisymmetric modeling method evaluated the impact of pole pair arrangement of the permanent magnets for lifting capacity and axial stiffness. This analysis assumed 1, 2, and 3 alternating north-south pole pairs. Since the inner and outer radii of the permanent magnet region was assumed locked for this study, increasing the pole pair count reduced the radial width of the permanent magnet elements. To determine potential lifting capacity and stiffness, the flywheel rotor is axially swept through available gap space. This movement occurs after cooling the bulk superconductors below critical and is necessary in order to induce currents within the bulk HTSC, that will provide the levitation force. The bearing superconductors are initially cooled with the flywheel positioned at $z_{fw} = +1.8mm$. The flywheel is then swept between $z_{fw} = +1.8mm$ and $z_{fw} = -1.8mm$ for 2 cycles over a time span of 60 seconds to determine the force displacement characteristic. This low rate of axial speed, v_z = 0.24 mm/s, to avoid inducing additional losses within the bulk superconductors.

This analysis generates a force-displacement profile for the HTSC bearing set where lifting capability and axial stiffness can be evaluated. Figure 5 presents the results for the 2 pole pair magnet arrangement, where the top graph shows the forces for the top and bottom bearing plates and the bottom graph shows the net axial bearing force. Through the movement cycle, there are 4 equilibrium points where the net bearing force equals the weight of the flywheel rotor. This analysis reveals a nonlinear hysteresis in the force-displacement plots that is due to the internal losses which result in trapped fields within the bulk superconductors.

Table 2 summarizes the study results and shows that 2 pole pairs is an optimal magnet arrangement for this design. Using 2 pole pairs maximizes the vertical stiffness of the bearing set and produces an equilibrium which is closest to center for the axial position of the flywheel rotor.

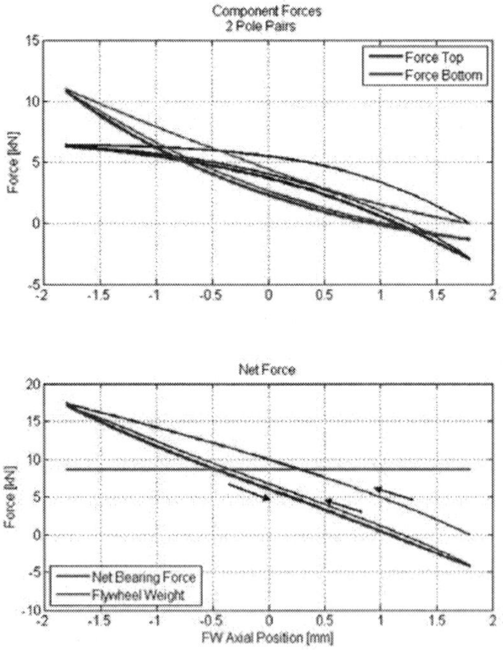

Figure 5: Force displacement curve of flywheel swept two times between +/- 1.8mm with 2 pole pairs.

Table 2: Summary of axisymmetric analysis

Pole Pairs	1	2	3
Magnet Width [mm]	40.0	20.0	13.3
Equilibrium [mm]	-0.912	-0.339	-0.707
Stiffness [kN/m]	3750	5500	5417

V. Translational Analysis

From the results of the axisymmetric analysis, the 3D lumped parameter modeling method [12] is used to estimate the translational bearing stiffness of the 2 pole pair bearing configuration. This method requires sub-dividing the superconducting region into individual, overlapping superconducting disc elements to estimate induced current flow. Compared to the axisymmetric method, the 3D model is more computationally intensive for calculating the field gradients per (6) since symmetry cannot be used. To improve model performance, these field gradients are calculated off-line and then stored in 3-D lookup tables for solving the main body of equations in (1).

In order to achieve accurate analysis results, the meshed overlapping disc elements for the superconducting plates must have diameters less than the interacting pole widths. This meshing requirement can significantly increase model order and calculation time. In order to reduce element count, the HTSC plate thickness is reduced to 5mm, which covers the penetration of induced currents per the axissymmetric analysis. The mesh for the transverse analysis uses 2 layers of overlapping disc elements, each with a diameter of 14mm. The layer closer to the interacting magnet had a thickness of 1 mm, while the second layer had the remaining thickness of 4mm. A total of 6,156 disc elements were used for each HTSC plate, as shown in Figure 6.

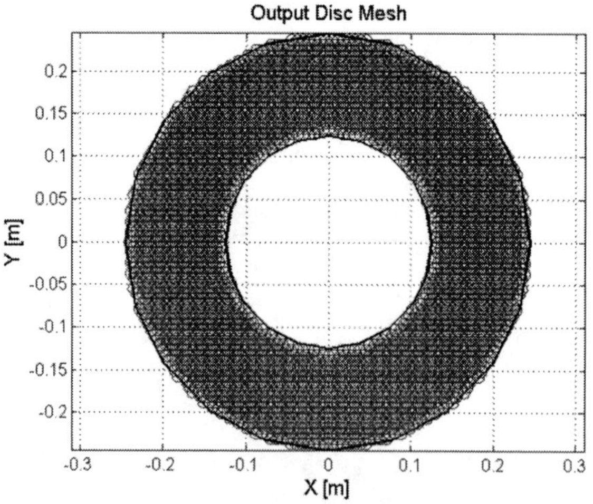

Figure 6: Disc element mesh for transverse analysis

The first step in performing the translational analysis with the 3D model is to re-perform the axial sweep and settle the model at the equilibrium position. This step is performed to estimate the induced currents and flywheel axial position at equilibrium, which will be used as initial conditions for the transverse sweep simulation.

Once the initial conditions at equilibrium are determined, a controlled constant velocity sweep in the x direction is performed, similar to the vertical sweep performed in the axisymmetric analysis. For the translational analysis, the flywheel is swept between +/- 1 mm for 2 cycles over a time span of 30 seconds. Again, this model is solved with a 4th order Runge-Kutta using a time step of 2 ms. The results from the translational sweep analysis are shown in Figure 7. As the bearing is moved off-center, the induced currents within the HTSC plates produce a translational force in the x direction on the bearing to push it back to the original position. In addition to this translational force, an axial force is also generated on the permanent magnets in the z direction. These force components can be clearly identified in (2) where axial flux gradients, $\frac{d\phi_i}{dz}$, couple with induced currents. This repulsive axial force is a summation of a decrease in lifting capacity of the top plate and an increase of lifting capacity on the bottom plate.

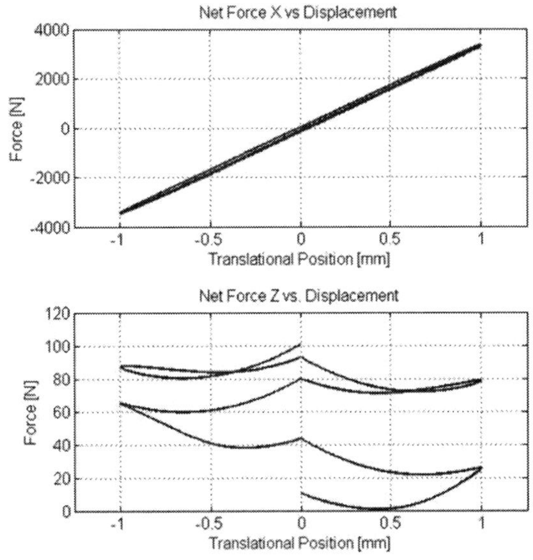

Figure 7: Bearing translational force versus translational movement

VI. Application Towards Rotor Dynamics

The results from this analysis can be applied to rotor dynamic analyses to identify rigid body modes. HTSC bearings typically have low stiffness values and will have to traverse one or two critical modes on initial spin up. Ideally, the flywheel should operate in a speed regime where no modes are present, but in order to reach operating speed it may run through a mode. It is necessary to determine the location of these modes during the design process to assess how they may affect operation. The bearing stiffness values for the top and bottom pairs are shown in Table 3.

Table 3: Axial, translational, and cross coupling bearing stiffness predicted from lumped parameter models.

	Top	Bottom
K_{zz}	1833 kN/m	3667 kN/m
K_{xx}	1770 kN/m	1610 kN/m
K_{xz}	-143 kN/m	158 kN/m

For a first pass analysis, a rigid body assumption can be used to estimate critical modes with respect to operating speeds. Linearized equations, presented by Schweitzer [18], were used with the respective translational stiffness values, K_{xx}, for the top and bottom bearing pairs. In this type of analysis, the gyroscopic coupling is the only variable that affects the resonance frequencies as a function of speed. From this analysis, a Campbell diagram can be produced which shows the critical resonance frequencies as a function of rotor speed, Figure 8.

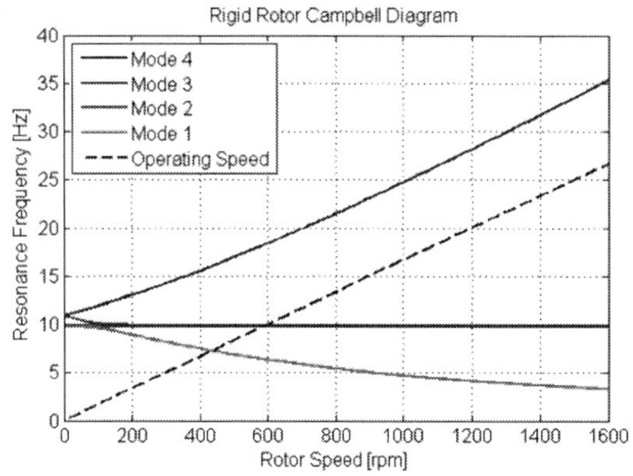

Figure 8: Rigid body vibration modes versus rotor speed and operational load line.

From the rigid body analysis for the 25 kWh outer rotor flywheel design, the flywheel will need to contend with crossing a single critical mode, Mode 2, at a speed of 600 rpm. Modes 1 and 3 are backward whirl modes that should not be excited since they are opposite the spin direction. Mode 2 is a forward whirl mode which can be excited by potential imbalances in the flywheel. Mode 4 is a conical whirl mode, but should not be excited during operation at any point since the polar to transverse inertia ratio is greater than 1 for the flywheel design.

The nominal operating speed of this notional flywheel design is between 6,500 and 13,000 rpm, which is significantly higher than the 2nd mode crossing, but to initially get the flywheel up to speed, this mode will need to be considered. Possible ways to get through this mode include the addition of copper rings to add damping to the system which will attenuate the effects of the mode. This method would induce a constant source of eddy current losses that will increase the overall bearing loss. Another method to transverse this mode is to add a set of active magnetic bearings which can be activated to change the radial stiffness and rigid body dynamics so that the flywheel passes through the mode. These supplemental active magnetic bearings should not be designed using a permanent magnet bias design, as this would constantly induce losses during operation. Since this bearing should only be "activated" during a small portion of time when the flywheel is spinning up to operation speed, a bias field current could be used for bearing control to eliminate losses during nominal operation.

VII. Conclusions

The lumped parameter modeling method significantly improves calculation time over FEM analysis without ignoring internal losses within the bulk superconductors that limit maximum current density and produce the non-linear hysteretic force-displacement behavior. The equations are formulated in state-space as a set of ordinary differential equations that can be efficiently and deterministically solved using fixed step Runge-Kutta methods. The axisymmetric method is used to evaluate the axial stiffness and weight carrying capability for a proposed flywheel design. The 3D method is used to estimate the translational bearing stiffness, which will affect rotordynamics. Both of the methods presented in this paper can be incorporated into larger mechanical models to fully simulate flywheel dynamics.

The analysis presented in this paper is one of the first steps in a bearing design for an outer rotor composite flywheel. Along with bearing performance, the mechanical performance of the loaded composite rotor must be assessed. The addition of permanent magnets on the end faces of the composite rotor will cause additional mass loading of the surrounding composite bandings at speed. Iterative structural analyses will be required to determine the impact of this additional mass loading, and may require a redesign of the proposed magnet arrangement. The quick turn-around time afforded by the techniques presented will greatly aid such an iterative design analysis approach.

References

[1] P. Denholm, E. Ela, B. Kirby and M. Milligan, "The Role of Energy Storage with Renewable Electricity Generation," NREL/TP-6A2-47187, Golden, CO, January 2010.

[2] M. L. Lazarewicz and T. M. Ryan, "Integration of flywheel based energy storage for frequency regulation on deregulated markets," in Power and Energy Society General Meeting, July 2010.

[3] C. S. Hearn, M. C. Lewis, S. B. Pratap, R. E. Hebner, F. M. Uriate, D. Chen and R. G. Longoria, "Utilization of Optimal Control Law to Size Grid-Level Flywheel Energy Storage," IEEE Transactions on Sustainable Energy, vol. 4, no. 3, pp. 611-618, July 2013.

[4] M. M. Flynn, "A Methodology for Evaluating and Reducing Losses, Heating, and Operational Limitations of High-Speed Flywheel Batteries," Dissertation to The University of Texas at Austin, August 2003.

[5] M. Strasik, P. E. Johnson, A. C. Day, J. Mittleider, M. D. Higgins, J. Edwards, J. R. Schindler, K. E. McCrary, C. R. McIver, D. Carlson, J. F. Gonder and J. R. Hull, "Design, Fabrication, and Test of a 5-kWh/100-kW Flywheel Energy Storage Utilizing a High-Temperature Superconducting Bearing," IEEE Transactions on Applied Superconductivity, vol. 17, no. 2, pp. 2133-2137, June 2007.

[6] C.S. Hearn, " Design Methodologies for Advanced Flywheel Energy Storage", Dissertation to The Univerity of Texas at Austin, August 2013.

[7] M. A. Pichot and M. A. Driga, "Loss Reduction Strategies in Design of Magnetic Bearing Actuators for Vehicle Applications," IEEE Transactions on Magnetics, vol. 41, no. 1, pp. 492-496, Jan. 2005.

[8] A. A. Kordyuk, "Magnetic Levitation for Hard Superconductors," Journal of Applied Physics, vol. 83, pp. 610-612, 1998.

[9] J. R. Hull and A. Cansiz, "Vertical and Lateral Forces Between a Permanent Magnet and a High Temperature Superconductor," Journal of Applied Physics, vol. 86, no. 11, pp. 6396-6404, 1999.

[10] F. Grilli, et al. Finite-Element Method Modeling of Superconductors: From 2-D to 3-D. IEEE Trans. Appl. Supercond., vol. 15, no. 1, pp. 17-25, March 2005.

[11] C.S. Hearn, S.B. Pratap, D. Chen, and R.G. Longoria, "Reduced Order Dynamic Model of Permanent Magnet and HTSC Interaction in an Axisymmetric Frame," IEEE Transactions on Mechatronics, in press, Aug. 2013.

[12] C.S. Hearn, S.B. Pratap, D. Chen, and R.G. Longoria, " Lumped Parameter Model to Describe Dynamic Translational Interaction for High Temperature Superconducting Bearings," IEEE Transactions on Applied Superconductivity, in press, Jan. 2014.

[13] Karnopp, Dean C., Margolis, Donald L., and Rosenberg, Ronald C. System Dynamics: Modeling and Simulation of Mechatronic Systems. 3rd Ed. John Wiley and Sons, New York, 2000.

[14] Smythe, William R., Static and Dynamic Electricity, Hemisphere Publishing, New York, 1989, pp. 234, 290-291

[15] M. Ikeda, A. Wongsatanawarid, H. Seki and M. Murakami, "Interaction of bulk superconductors with flywheel rings made of simple permanent magnets," Physica C, vol. 469, pp. 1270-1273, 2009.

[16] G. G. Sotelo, A. C. Ferreira and J. R. de Andrade, "Halbach Array Superconducting Magnetic Bearing for a Flywheel Energy Storage System," IEEE Transactions on Applied Superconductivity, vol. 15, no. 2, pp. 2253 - 2256, June 2005.

[17] F. N. Werfel, U. Floegel-Delor, R. Rothfeld, T. Riedel, B. Goebel, D. Wippich and P. Schirrmeister, Superconductor Bearings, Flywheels and Transportation, Superconductor Science and Technology, vol. 25, 014007 (16pp), 2012.

[18] G. Schweitzer, "Chapter 7: Dynamics of the Rigid Rotor," in *Magnetic Bearings; Theory, Design, and Application to Rotating Machinery*, Berlin, Springer-Verlag, 2009, pp. 167-189.

Energy Storage Flywheel Magnetic Bearing System – Magnetic Linear Circuit vs. 3-D Finite Element Model

Arunvel Kailasan[a], Tim Dimond[b], Paul Allaire[c]

[a] Gardner Denver, Inc., 100 Gardner Park, Peachtree City, GA 30269, USA, arunvel.kailasan@gardnerdenver.com
[b] Rotor Bearing Solutions International, 3277 Arbor Trace, Charlottesville, VA 22911, USA, tim.dimond@rotorsolution.com
[c] Rotor Bearing Solutions International, 3277 Arbor Trace, Charlottesville, VA 22911, USA, paul.allaire@rotorsolution.com

***Abstract*—A 1 kW-hr (3.6 MJ) energy storage flywheel with a design operating speed of 40,000 rpm and an inside-out flywheel has been designed and optimized. It is supported in two thrust and two radial active magnetic bearings. The purpose of this paper is to present the active magnetic bearing system modeling including the magnetic analysis.**

At the design stage, there are two ways to evaluate the magnetic bearing properties: linear circuit model or finite element model. The linear circuit models are simple lumped mass models. The finite element model discretizes the bearing geometry and the surrounding space and requires a great deal more work. There are not many full 3-D finite element magnetic bearing systems that have been fully compared to the equivalent linear circuit model. There are legitimate questions about accuracy of magnetic bearing modeling results.

This paper considers these two approaches to compare the results for a realistic energy storage flywheel design. If the differences are large, it is expected that the 3-D finite element model is more likely to be accurate. This paper documents the differences for the flywheel magnetic bearing designs.

The linear circuit model prediction of axial force capacity was 17% lower and the prediction of coil indcutance was 8% lower than the predictions from 3-D finite element model. The linear circuit model of the radial bearing underestimated the force capacity by 14% and underestimated the inductance by 8.5% when compared to the 3D finite element model. Values were also compared for the open loop stiffness, current gain and force slew rate for both bearing designs.

These relatively large differences in magnetic bearing performance parameters obtained between the linear circuit model and the 3-D finite element model, shows that the linear circuit model is not very accurate. Thus, the concern is that designing and optimizing a complex magnetic bearing supported device requires 3-D finite element modeling at the design stage.

The flywheel has not been constructed and the magnetic bearings are thus not available for testing. It is hoped that this can be future work.

I. Introduction

Active magnetic bearings have advantages over traditional rolling-element or fluid film bearing designs. Active magnetic bearing systems have significantly lower power losses than fluid film bearing systems and allow for effective system damping, unlike rigidly-mounted rolling element bearing systems. As a results, AMBs are an enabling technology for efficient high-speed flywheel design. When combined with a vacuum chamber, parasitic losses resulting from flywheel rotation are minimized. The design of the magnetic bearing system is therefore an integral part of having a good, highly efficient flywheel design.

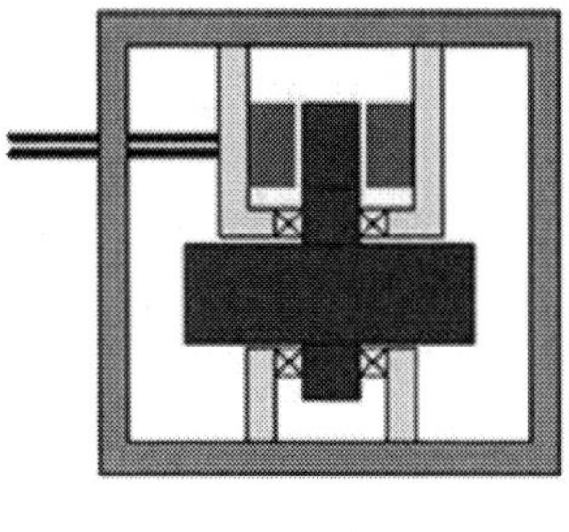

(a) Conventional Schematic

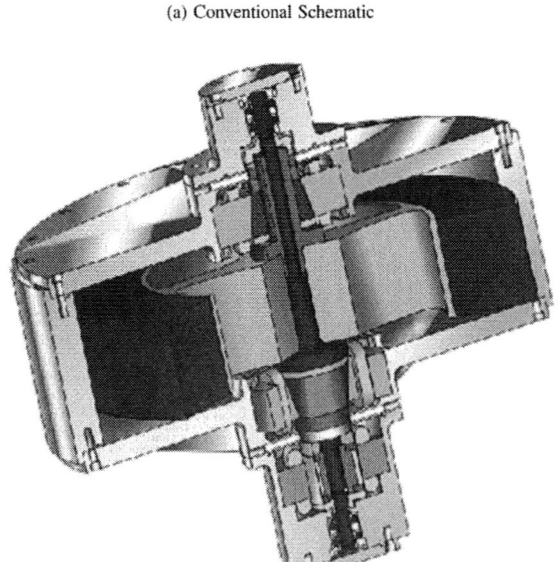

(b) Conventional Example

Figure 1: Conventional Flywheel Design

There are two basic design configurations of flywheel systems commercially available: a conventional design and an integrated design. In the "conventional design," the rotor has a large diameter section, where most of the kinetic energy is stored, attached to a smaller diameter section, which is used by the motor to spin the flywheel. This is the most common design. However, this configuration tends to have a larger housing and containment structure because of the additional rotor length. This results in a very heavy flywheel configuration which can be expensive to manufacture. A schematic of the conventional configuration is shown in Fig. 1a. A generic conventional flywheel design is shown in Fig. 1b.

The second configuration is an "integrated design" in which the electromagnetic and energy storage portions of the rotor are combined. This type of design is found in newer generation flywheels but is not well suited for composite rotor flywheels because of the need for electromagnetic material for torque production. However, in the integrated configuration, the housing and stator of the motor comprise a large portion of the vacuum and burst containment for the flywheel. A typical 1kW-hr integrated flywheel weighs about 200 -250 Kg. A schematic of the integrated design is shown in Fig. 2a and the integrated design from Active Power [1] is shown in Fig. 2b.

A variant of the integrated flywheel is the barrel type flywheel, shown schematically in Fig. 3a. An example barrel configuration flywheel design from PowerTHRU (formerly Pentadyne) [2] is shown in Fig. 3b. This design variant places the electromagnetic parts of the rotor on the inner diameter of the rotor. With this configuration, the outer diametral portion of the rotor can be either steel or composite material for energy storage.

At the design stage, the magnetic bearing properties can be determined either with linear circuit models or with finite element analysis. Jang et al. [3] used linear circuit models to develop initial design parameters for thrust AMBs, including thrust load, coil resistance, and coil inductance. The designs were confirmed with finite element analysis and experiment.

Jang et al. [4] introduced fringing correction factors into a linear circuit model of an active magnetic thrust bearing to improve load predictions. Results were compared to a finite element analysis and experiment. Permeance correction factors based on bearing geometry were developed from analysis. These correction factors brought the circuit model into agreement with the finite element model within 1% and also closely tracked experimental measurements. The uncorrected linear circuit model underpredicted the thrust load both as a function of position and as a function of current.

Zhang and Tian [5] developed an analytical expression for eddy current losses in active magnetic thrust bearings. The magnetic diffusion equations were solved for a non-laminated thrust bearing and compared to a finite element analysis. The two analyses gave identical results for frequencies up to 1 kHz, and were within 10% for frequencies from 1 kHz-10 kHz.

The flywheel design for this study is a barrel-type design, shown schematically in Fig. 4. The 5-axis AMB system for the flywheel is comprised of a double-acting thrust bearing and two radial magnetic bearings. The double acting thrust bearing system is a normal AMB with one coil and two poles in the upper and lower bearings. A linear magnetic circuit for the thrust bearing is used to determine the magnetic flux density, magnetic pole forces, the inductance, the open loop stiffness, the current and the slew rate for a given coil current. The linear magnetic circuit values obtained are compared to a full 3-D analysis of the same geometry thrust bearing.

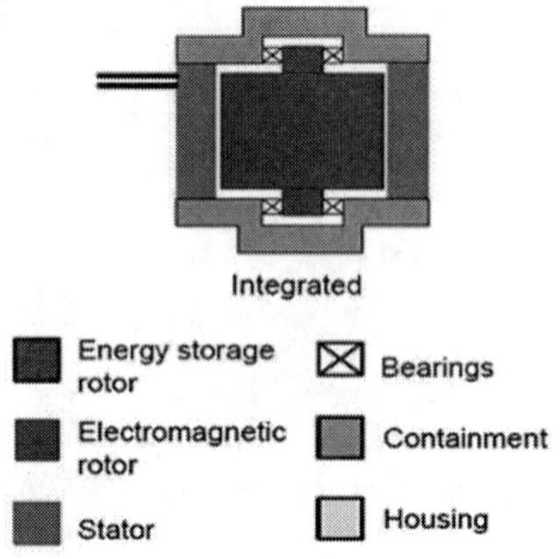

(a) Integrated Schematic

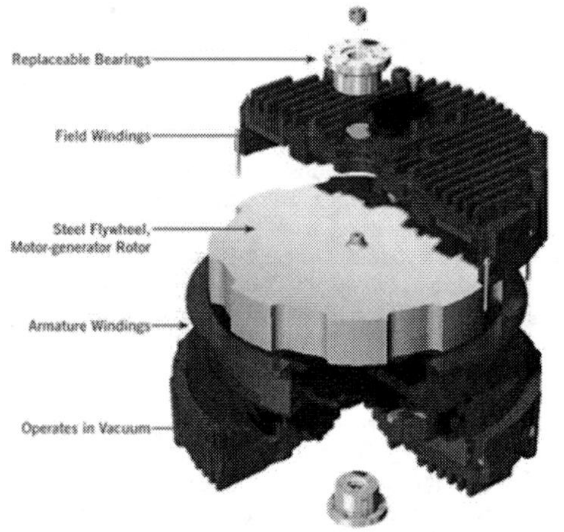

(b) Integrated Example - Active Power Flywheel [1]

Figure 2: Integrated Flywheel Design Example

The radial bearings were designed as inside-out operating bearings with the rotor on the outside and the stator on the inside. They had twelve poles in an E-core design. The radial force capability, inductance, open loop stiffness, current gain and force slew rate were all evaluated using a linear circuit model and a 3-D finite element model using a similar procedure to the thrust bearing analysis.

II. Active Magnetic Bearings

The bearing analysis was conducted in two parts. The initial designs were obtained assuming linear magnetic circuit theory. Peak magnetic fluxes were limited to ensure that the linearity

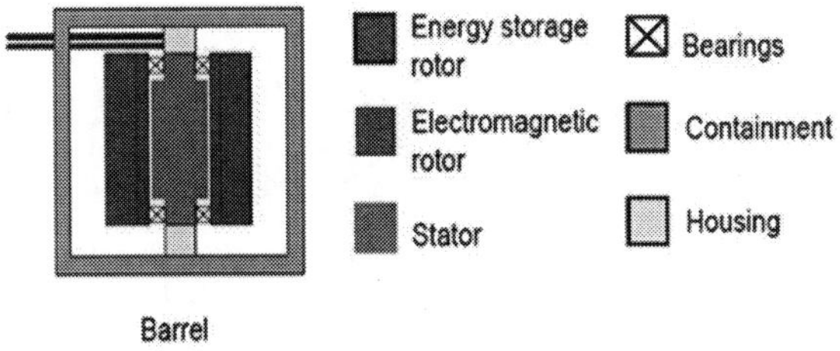

(a) Barrel Schematic

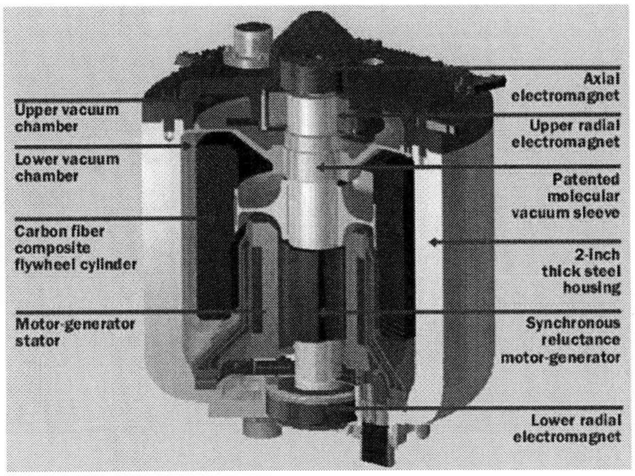

(b) Barrel Example - PowerTHRU Flywheel [2]

Figure 3: Barrel Flywheel Design Example

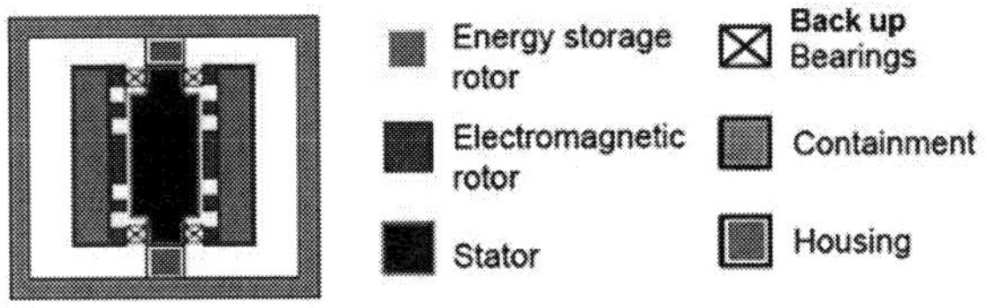

Figure 4: Integrated Flywheel Design, Present Study

assumption remained valid. The design flux density at the knee of the curve was taken as 1.2 T. This is a typical linear B-H flux limit for silicon iron alloys. A packing factor of 0.7 was applied in computing the coil cavity dimensions and the amount of copper wire available to energize the bearing. The packing factor accounts for unused space with round wire winding and wire insulation. The packing factor results in an increased packing volume when calculating the magnetic bearing size. A 14 AWG wire was used for the analysis.

Then, the results of the linear analyses were confirmed using non-linear finite element analysis (FEA) performed with a commercial code. The FEA included saturation effects in the magnetic materials.

A. Linear Circuit Models

The equations for the linear magnetic circuit models presented are available in [3]. The fundamental equations for magnetic circuit analysis are Ampere's loop law and conservation of magnetic flux. Ampere's loop law describes the induced magnetic field due to an applied current in a wire coiled around magnetic material and is given by

$$\sum_{i=1}^{n_i}\left(\frac{Bg}{\mu_0}\right)_i=\sum_{i=1}^{n_i}(NI)_i \tag{1}$$

where B is the magnetic flux density, g is the nominal air gap, μ_0 is the permitivitty of free space, $4\pi\cdot10^{-7}$ H/m, N is the number of winding turns, and I is the input current in A. Conservation of magnetic flux indicates that the total flux Φ_i through a node in the magnetic circuit is zero:

$$\sum_{i=1}^{n_i}\Phi_i=0 \tag{2}$$

The available force $\mathbf{f}$ was obtained in terms of the peak magnetic flux density B_{sat} for a lumped model as:

$$\mathbf{f}=\frac{A_pB_{sat}^2}{2\mu_0}\mathbf{n} \tag{3}$$

The vector $\mathbf{n}$ is a unit normal to the pole face area A_p. The maximum available pole face area was limited by the back iron cross-sectional area and the stator inner and outer diameters. The back iron was sized to saturate at the same input current density as the poles.

The slew rate magnitude was calculated in terms of linearized bearing operation for opposing pairs of the AMB designs in the circuit models as

$$\frac{d\left|\mathbf{f}\right|}{dt}=\frac{\mu_0A_pN^2I_b}{2g}\frac{dI_p}{dt} \tag{4}$$

where I_b is the bias current and I_p is the perturbation current. The rate of change of perturbation current is limited by the open-loop inductance L and the peak amplifier voltage V_c as

$$\frac{dI_p}{dt}=\frac{V_c}{L} \tag{5}$$

To properly design the bearing controller, the open loop characteristics of the bearings are needed to characterize the plant. The open loop stiffness is found as

$$k_x\approx-\left.\frac{\partial\left|\mathbf{f}\right|}{\partial x}\right|_{x=x_0,i_p=i_{p_0}} \tag{6}$$

The open loop stiffness is negative, indicating that AMBs are open-loop unstable.

The current gain for the bearing is found as

$$k_i\approx\left.\frac{\partial\left|\mathbf{f}\right|}{\partial i_p}\right|_{x=x_0,i_p=i_{p_0}} \tag{7}$$

B. Finite Element Analysis

Results for a three-dimensional finite element analysis performed using commercial finite element software were compared to the bearing design parameters obtained from the linear circuit models. The dimensions of the initial bearing size using the linear circuit models described in Section II-A is presented in Kailasan [6]. The magnetic flux paths was determined for both the thrust and radial magnetic bearings, and important parameters such as load capacity, flux density, inductance, magnetic field energy and slew rate were calculated.

The finite elements implement the fundamental constitutive electromagnetic relations in a discretized form. The fundamental relationships are a subset of Maxwell's equations and include magnetomotance, given by

$$\nabla \times \mathbf{H} = \mathbf{J} \tag{8}$$

where **H** is the magnetic field intensity and and **J** is the current density.

The conservation of magnetic flux **B** at a given cross-sectional area in the bearing in differential form is expressed as:

$$\nabla \cdot \mathbf{B} = 0 \tag{9}$$

The applied force from the magnetic fields developed in the AMB and directed through the air gaps is given by

$$\mathbf{f} = \oint \mathbf{T} \cdot d\mathbf{S} \tag{10}$$

The term **T** in (10) is the Maxwell stress tensor, expressed using indicial notation as

$$T_{ij} = \frac{1}{\mu_0}\left(B_i B_j - \frac{1}{2}\delta_{ij} B_{ij}^2\right) \tag{11}$$

where δ_{ij} is the Kronecker delta.

Coil inductance is not directly available from the finite element solution. However, the inductance can be found from the total magnetic field energy produced by the maximum current. The inductance can then be calculated as [7]:

$$L_c = \frac{2W}{I^2} \tag{12}$$

where L_c is the inductance in H, W is the magnetic field energy in J, and I is the current applied to the magnetic bearing coils in A.

C. Thrust Bearing

A double-acting thrust magnetic bearing was designed for the flywheel. The section view of the thrust magnetic bearing is shown in Fig. 5. With a vertical rotor orientation, the main function of the thrust bearing is to support the entire flywheel mass and to resist dynamic thrust loads while in operation. The bearing consists of three major parts: the thrust disk, which is the rotating part of the bearing and is attached to the flywheel, the thrust stator which is attached to the stator back iron, and the coils inside the stator.

A three dimensional finite element analysis using commercial software was conducted to verify the parameters obtained from the linear circuit model of the thrust bearing. A mesh density study was performed to ensure that the results had numerically converged. About 1 million elements were used in the analysis, with a denser mesh in the air gap to improve accuracy. Figure 6 shows the modeled thrust magnetic bearing.

The flux path and the maximum load capacity were calculated with maximum current applied to the coils. Other parameters of interest are the inductance, slew rate and magnetic field energy. The finite element method allows for modeling

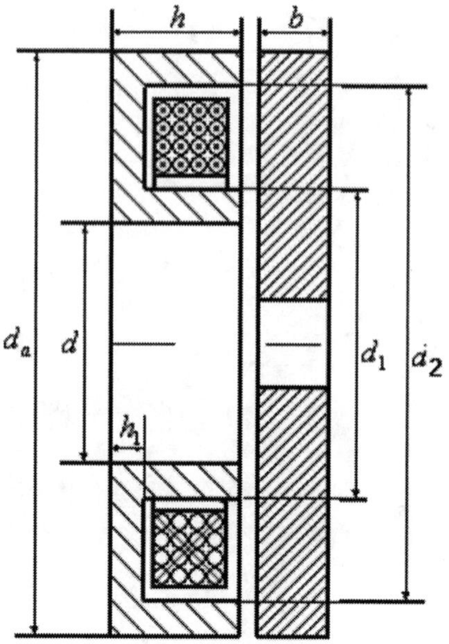

Figure 5: Thrust AMB

of flux leakage and is more accurate than linear circuit models that assume all flux is in the air gap.

To calculate the value of open loop stiffness, the rotor was centered between the opposing thrust magnets while applying equal currents to each coil, resulting in zero net force. The rotor was then moved axially in increments of 0.05 mm, and a force versus displacement curve was calculated based on five increments. The slope of the line gives the open loop stiffness k_x.

The current gain was obtained by varying the value of the perturbation current with the rotor centered axially in the model and plotting net force versus perturbation current. The slope of the line for force versus perturbation current gives the current gain k_i.

D. Radial Bearings

The inside-out rotor design with the stator in the center required an innovative inside-out E-core type radial magnetic bearing design for the flywheel, as shown in Fig. 7. In contrast to traditional magnetic bearing designs, the bearing consists of an inner laminated stator component and an outer rotor lamination stack. The stator poles are pointed radially outwards as shown in Fig. 7. The stator component is attached to the stator core while the laminations are attached to the inner steel spline rotor.

In this design, the rotating part of the bearing is the outer ring. The bearing poles, windings, and the back iron are inside of the rotating ring and are on the stator. The 12-pole E-core arrangement has four quadrants. The main pole in the center of each quadrant has about twice the pole face area as the two auxiliary poles.

As with the case of the thrust bearing, a 3-D FEA was conducted to compare results for the load capacity, flux path and

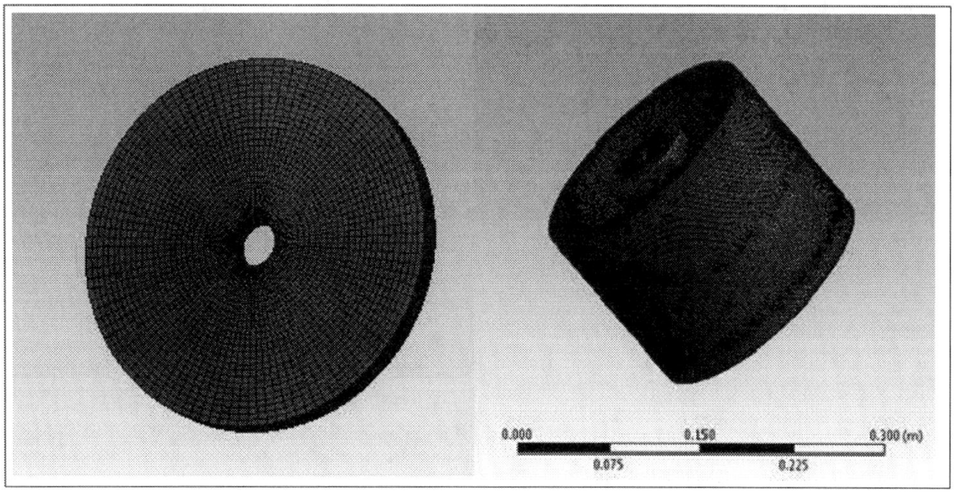

Figure 6: 3D Thrust AMB Finite Element Mesh

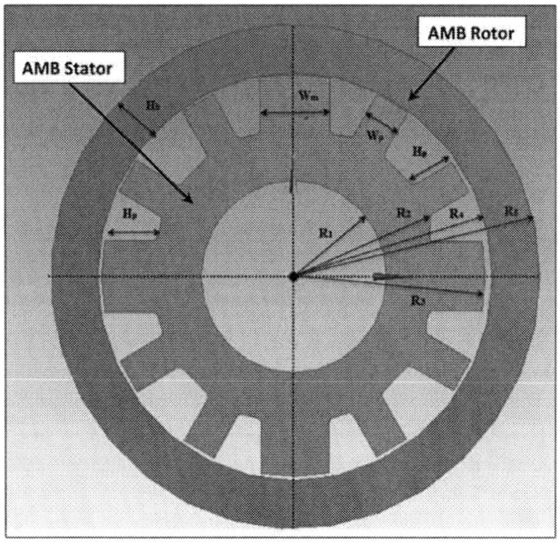

Figure 7: Radial AMB Inverted Design

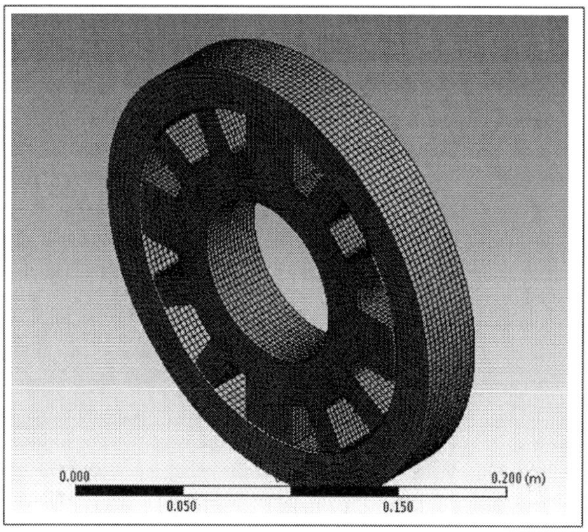

Figure 8: Radial AMB Finite Element Mesh

flux density while calculating parameters such as inductance, magnetic energy and slew rate. The location and extent of magnetic saturation was also evaluated with the finite element analysis. The E-core model is shown in Fig. 8. A similar perturbation procedure to that used for the thrust bearing was used to determine open loop stiffness k_x and current gain k_i. The inductance was computed based on Eq. (12).

III. Results

A. *Thrust Magnetic Bearing*

Table I shows a comparison between the linear circuit model and the 3-D FEA values obtained for the flywheel thrust bearing. A sample thrust bearing flux density finite element solution is shown in Fig. 9. The peak calculated load capacity from the FEA was 1800 N. This was 17% larger than load capacity of 1500 N calculated using the linear circuit model. The inductance predicted by the finite element model was 9% higher than what was predicted by the linear circuit model. This led to a corresponding decrease in force slew rate predicted by the finite element model when compared to the linear circuit model.

Differences were also found in the predictions of actuator open-loop characteristics. When compared to the finite element model, the linear circuit model overpredicted open loop stiffness magnitude by 19% and underpredicted current gain by 4.2%. While there is a large difference in predicted open loop stiffness, this large a difference can be treated as an uncertainty in modern control methods such as H_∞ or μ-synthesis.

The linear circuit model also overpredicted slew rate by 21% when compared to the 3-D FEA. Since slew rate is a linear function of coil inductance, this was to be expected.

Table I: Thrust Bearing Design Values

Property	Value Linear Circuit Model	3D FEA
F_{max}, N	1500	1812
W, J	-	3.82
L_c, mH	68.1	74.5
k_x N/mm	-172.1	-143.9
k_i N/A	286.6	299
Slew, N/s	$3.9 \cdot 10^6$	$3.2 \cdot 10^6$

Table II: Radial Bearing Design Values

Property	Value Linear Circuit Model	FEA
F_{max}, N	850	990
W, J	-	4.68
L_c, mH	92.4	101.2
k_x N/mm	-128.3	-113.4
k_i N/A	42.6	45.3
Slew, N/s	$7.1 \cdot 10^6$	$6.4 \cdot 10^6$

However, it is an indication that the linear circuit model is overly optimistic in estimating the ability of the magnetic bearing to react to rapid changes in force.

The peak magnetic flux obtained from the finite element model was about 1.24 T. Equal and opposite forces were found on the thrust disk and the stator, indicating that the analysis had converged. There was strong agreement between the peak flux in the 3-D FEA and the 1.2 T peak flux assumed in the linear circuit model for design purposes.

B. Radial Magnetic Bearing

As with the thrust bearing, a 3-D finite element model was made for comparison to the results obtained from a linear circuit model. A mesh convergence study was done to performed to ensure that the analysis had numerically converged. About 1 million elements were used for the analysis. An example flux density solution from the radial bearing finite element model is shown in Fig. 10.

A summary of the FEA results compared to the values obtained from linear circuit models is presented in Table II. The maximum load capacity from the 3-D FEA was 990 N. This was 16% higher than the 850 N obtained from the linear circuit model. As with the thrust bearing, the inductance predicted by the radial bearing finite element model was about 9% higher than that predicted by the linear circuit model.

As with the thrust bearing, there were significant differences in the actuator open-loop characteristics between the 3-D FEA and the linear circuit model. When compared to the FEA, the linear circuit model prediction for open loop stiffness magnitude was 13% higher. The linear circuit model prediction for current gain was 6% lower than the FEA prediction.

The slew rate prediction from the linear circuit model was 11% higher than the 3-D FEA prediction for the radial bearing. This is consistent with the difference in prediction for the inductance for this bearing.

The FEA predicted a peak flux of 1.36 T, which is entering the saturation flux region of silicon iron alloys. This may account for some of the difference in load capacity between the FEA and the linear circuit model, since magnetic flux was limited to 1.2 T in the linear circuit model.

IV. Conclusions

In this paper, linear circuit models for an active magnetic thrust bearing and an active magnetic radial bearing were compared to 3D finite element models. A consistent finding from both analyses was that the finite element analysis consistently predicted higher load capacity than was found using linear circuit models. This was surprising, since the linear circuit model treated magnetic components as ideal lumped parameters. It is expected that fringing and leakage effects would have served to reduce the load capacity since some of the magnetic flux is not used to generate force. However, this finding was consistent with the findings of Jang et al. [4]. Another possibility is that the calculated peak flux from the FEA was higher than the assumed peak flux from the linear circuit models. The peak flux in the linear circuit model is set at 1.2T. In the finite element model, calculated peak flux was as high as 1.34 T. This additional flux would result in increased force and may also part of the difference between the linear circuit models and the FEA.

The finite element prediction for current gain was up to 19% higher than the prediction from the linear circuit model for the current gain. While this is a large difference, it should be manageable since there is some model reconciliation that is part of the initial levitation process. Similar comments apply to the difference in open loop stiffness between models. While an underestimate of the open loop stiffness could result in an unstable closed-loop system if the control gain is too low, this should be able to be corrected during initial levitation.

Some of these differences between linear circuit models and finite element analysis can be mitigated by using derating factors on the linear design equations to account for fringing and leakage effects, as shown by Jang et al. [4]. However, these derating factors must be empirically confirmed. The FEA accounts for fringing, leakage, and non-linearity effects and is easier to verify.

Based on these results, linear circuit models are suitable for preliminary sizing of AMB actuators and for quickly iterating on design options. The final design and final actuator size should be based on the finite element analysis.

References

[1] Active Power, "Active Power CleanSource UPS Multi-Module Systems," Downloaded from `http://www.activepower.com/Documents/Datasheets/CSMMS-US-W.pdf` on 15 June 2014.

[2] "Pentadyne's VSSDC," *Broadcast Engineering*, vol. 48, no. 9, pp. 90–92, 2006.

[3] S.-M. Jang, U.-H. Lee, J.-Y. Choi, and J.-P. Hong, "Design and analysis of thrust active magnetic bearing," *Journal of Applied Physics*, vol. 103, no. 7, pp. 07F122–07F122, 2008.

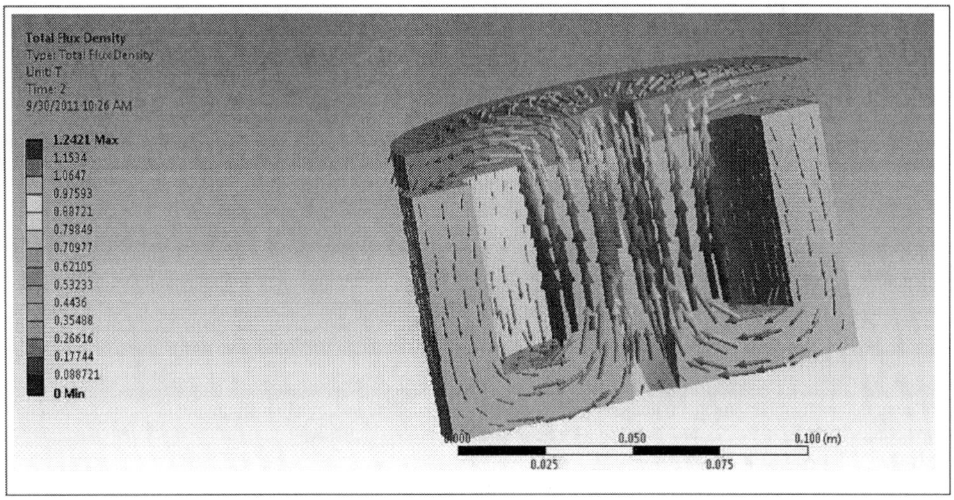

Figure 9: 3D Thrust AMB Finite Element Analysis

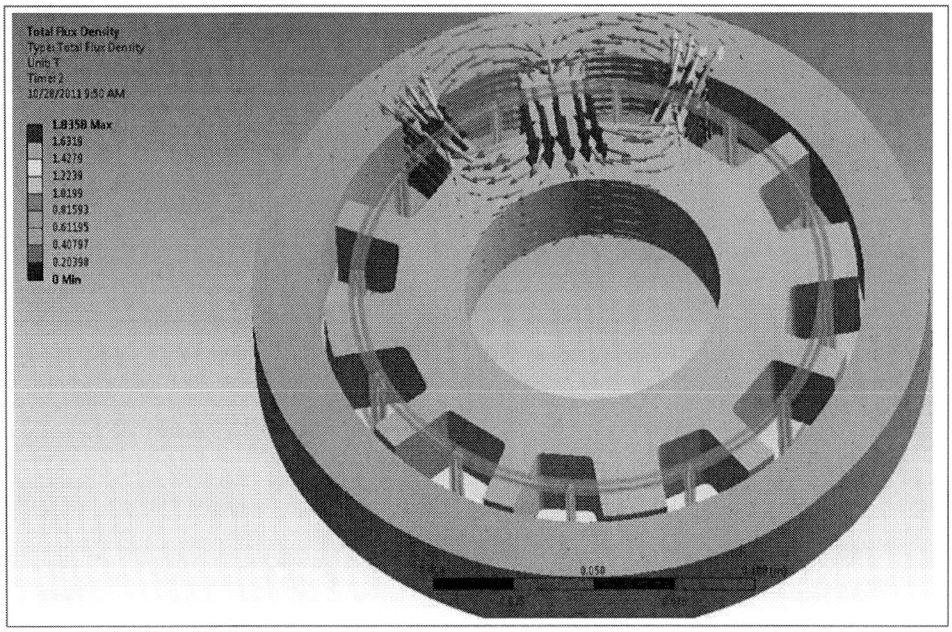

Figure 10: Radial AMB Finite Element Solution

[4] S.-M. Jang, K.-H. Kim, K.-J. Ko, J.-H. Choi, S.-Y. Sung, and Y.-B. Lee, "Improved thrust calculations of active magnetic bearings considering fringing flux," *Journal of Applied Physics*, vol. 111, no. 7, p. 07E726, 2012.

[5] M. Zhang and Y. Tian, "Frequency domain modeling of eddy current loss for magnetic thrust bearings," *Proceedings of the Institution of Mechanical Engineers, Part J: Journal of Engineering Tribology*, vol. 227, no. 7, pp. 718–728, 2013.

[6] A. Kailasan, "Preliminary Design and Analysis of an Energy Storage Flywheel," PhD Thesis, University of Virginia, Charlottesville, VA", 2013.

[7] T. W. Nehl, F. A. Fouad, and N. A. Demerdash, "Determination of saturated values of rotating machinery incremental and apparent inductances by an energy perturbation method," *IEEE Transactions on Power Apparatus and Systems*, no. 12, pp. 4441–4451, 1982.

Multicriteria Design Process for Outer-Rotor Kinetic Energy Storage Systems

Maximilian Schneider[a], Hendrik Schaede[a], Stephan Rinderknecht[a]

[a] Institute for Mechatronic Systems in Mechanical Engineering, Otto-Berndt-Straße 2, 64287 Darmstadt, Germany,
schneider@ims.tu-darmstadt.de

***Abstract*—Due to the global energy transition, energy storage capacities are necessary. Kinetic energy storage systems can be used to provide valuable grid services but have to be redesigned to fulfill this task. A structured process is necessary in order to cope with the complexity of those systems and to maximize performance and profitability. This multicriteria design process is described in detail for an outer-rotor kinetic energy storage system. The approach is based on a detailed system model consisting of the several sub-models. It incorporates the design of the drive system, the magnetic bearing system, as well as the composite rotor. This approach enables the use of optimization algorithms in order to find systems matching the application specific objectives such as minimal operational losses or a maximal energy density. The multicriteria design process is then applied to an exemplary application.**

I. Introduction

In many countries, the government has initiated a transition of the energy system. Conventional power sources such as coal or nuclear power are being replaced by renewable energies such as wind and solar in order to provide a more sustainable energy supply. Most renewables, however, come with a main disadvantage. Their power output is volatile and uncertain resulting in high prognosis errors. In combination with the stochastic demand of electricity a stable and secure energy supply can only be efficiently achieved by installing a certain amount of storage capacity.

When analyzing the characteristics of this storage demand it becomes clear that different storage characteristics are needed. This is especially true for grid stabilization services, where systems with a moderate storage capacity, high electrical power, and a high cycle life are necessary. At first glance, conventional kinetic energy storage systems, which are already commercially available for uninterruptable power supply applications, seem to be able to fulfill those requirements. In-depth analyses show, however, the need for a redesign of the basic technology in order to be competitive with chemical batteries and capacitors within this field of application. Moreover a customization of the storage technology to the final application is crucial to its performance and thus its profitability.

This contribution introduces a detailed technical design process for kinetic energy storage systems in outer-rotor design. Firstly, different topologies are compared by analyzing their energy density and degree of integration. Secondly, the technical development process is put into perspective to the overall development process (SDA methodology). Thirdly, the optimization process is described in detail and advantages as well as challenges are discussed. The last chapter applies the process to a real application.

II. Analysis of Kinetic Energy Storage System Designs

Before the development process can be analyzed in detail, some considerations about the technology itself and different designs are necessary.

A. Overview of Possible System Designs

A kinetic energy storage system (KESS) stores energy in the rotation of a mass with certain inertia. The stored kinetic energy E_{kin} is given by equation (1):

$$E_{kin} = \frac{1}{2}\,\theta\omega^2\,. \tag{1}$$

It is square-depended of the rotating frequency ω and linear-depended from the inertia θ. While the inertia for a cylindrical tube depends linearly on the mass of the rotor m and squarely on the inner and outer radii:

$$\theta = \frac{1}{2}\,m(r_{outer}^2 + r_{inner}^2)\,. \tag{2}$$

A maximum energy density is achieved by high rotating frequencies as well as large radii, requiring high strength materials, such as carbon fibers, in order to be able to reach high tangential velocities [1]. All types of kinetic energy storage systems have certain main parts in common. A schematic drawing is shown in Fig. 1.

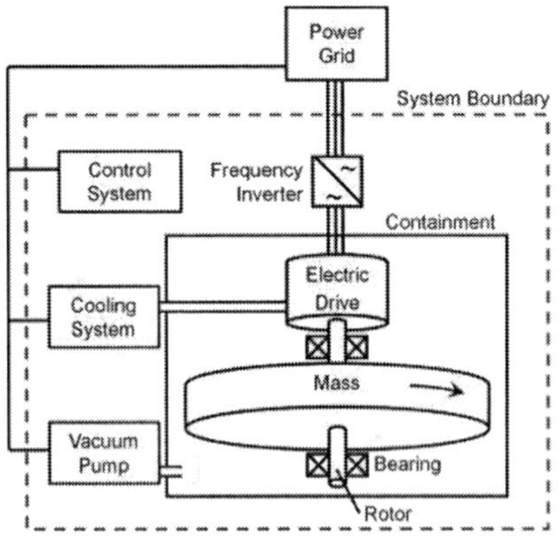

Figure 1. Components of a typical kinetic energy storage system

Via an electric drive the rotor is accelerated and hence its energy content varied. The rotor can be supported by different kinds of bearings. All kinds of magnetic bearings are in discussion; roller bearings can also be found in commercial systems. Depending on the design and the components used, different peripheral subsystems, such as a vacuum pump, a cooling as well as a control system, are needed to efficiently operate the kinetic energy storage system.

As already mentioned before, kinetic energy storage systems have been investigated for a long time, explaining the different topologies available today. Four main types can be distinguished [2], see Fig. 2.

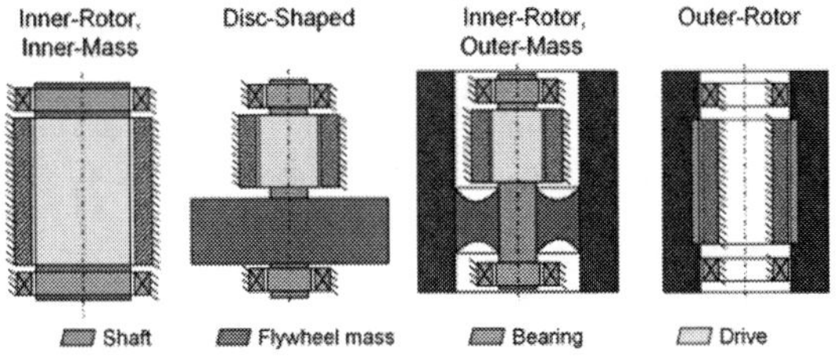

Figure 2. Overview of different flywheel types [3]

B. Degree of Integration

A suitable characteristic to compare different kinetic energy storage system designs is their degree of integration (DoI). A high DoI is necessary to maximize the energy density of a kinetic energy storage system and to minimize the production cost. It can be defined as the extent to which different system components are used to generate a high inertia. The disc-shaped flywheel type can be found in commercially available systems and has the lowest DoI. Only the actual flywheel mass is used to store the energy. The inner–rotor/outer-mass concept has a higher DoI because the flywheel mass rotates around the drive and bearings, leading to a higher inertia. A high DoI can be observed for the inner-rotor/inner-mass concept, because the rotor parts of the electric drive and the bearings largely contribute to the moment of inertia. Since the electric drive and the bearings have to have the diameter of the mass, the overall system size is limited. The highest DoI is reached with the outer-rotor concept because all components of the rotor are placed at the largest possible radius leading to a high inertia of the rotor with respect to the rotors mass. The outer-rotor concept is the only topology that is not commercially available today. One reason for this may be the high development effort that is still necessary for such systems. Only little knowledge is present on outer-rotor magnetic bearings as well as outer-rotor high-speed electric drives. In spite of the straight forward overall system structure, interdependencies between the different components of these highly integrated systems lead to a complex overall system. The remainderof this contribution will focus on the outer-rotor topology because it has the highest potential in terms of energy density and DoI but is not sufficiently developed yet.

III. The Overall Development Process

In [3] a general methodology for the development of energy storage systems is described as the so-called Specification, Design and Assessment Methodology (SDA-Methodology). It enables the efficient and reproducible development of energy storage systems for a specific field of application. Based on this method, requirements for the design process of a certain storage technology can be derived.

A. The SDA Methodology

The methodology consists of the three meta-steps: 'Specification', 'Design' and 'Assessment'. Every meta-step can be divided into several sections. The exact process is displayed in Fig. 3. The specification process starts with the load profile synthesis. Here the necessary storage power profile is derived from the application load profile. Based on this, the energy storage system can be dimensioned. The storage dimensions are forwarded to the design process that starts with a decision for a certain storage technology. Within the block 'Energy Storage' the system is designed and optimized. In order to ensure a correct and efficient operation of the system, a suitable operational strategy has to be developed. The designed storage technology is then forwarded to the step 'Assessment'. Here the technology can be tested and evaluated based on simulations. Necessary iterations and optimization steps can then be initiated.

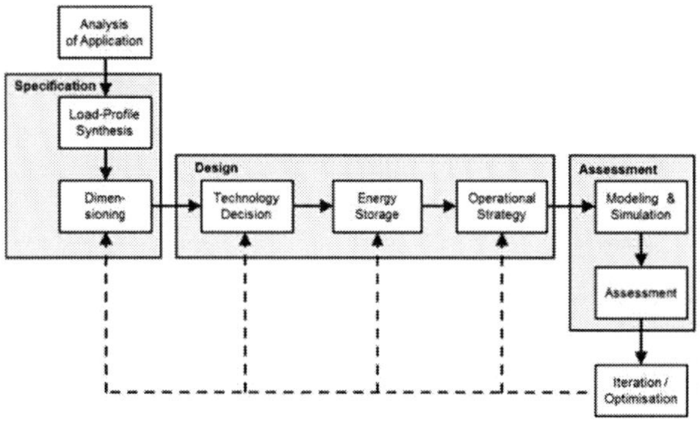

Figure 3. Specification, Design and Assessment methodology [3]

B. Requirements Resulting from Specification

Depending on the field of application of the kinetic energy storage system different requirements can result from the 'Specification'. Indispensable for the design process, however, is the nominal electrical power and the usable storage capacity. Other requirements such as certain efficiencies, cycle life, safety, total costs or energy/power densities can be defined by the specification process and have to be fulfilled by the design process. In the following we assume that in the step 'Technology Decision' the outer-rotor concept has been selected.

IV. Design Process for the Outer-Rotor Conept

This contribution proposes an integrated design approach for an outer-rotor KESS. The optimal design of KESS has been discussed in various publications. Most of the authors, however, focus only on the rotor design (e.g. [4], [5], [6]). The contributions considering the overall system design do not focus on the characteristics of the outer-rotor design (e.g. [7], [8]).

Before the actual design process can be initiated the interdependencies between the different system components must be determined. The insight into those relations leads to a better understanding of the system itself and the optimization problem. Afterwards an overall system model can be developed which is used to find an optimal solution to the specified requirements.

A. Identifying Important Interdependencies

According to Fig. 1, the important parts of every kinetic energy storage system are the bearings, the rotor and the electric drive. Additional properties can be identified that characterize the system such as vacuum quality and rotational speed. Table I gives the results of an analysis of interdependencies between components and properties.

TABLE I. INTERDEPENDENCIES BETWEEN IMPORTANT SYSTEM COMPONENTS AND PROPERTIES OF THE OUTER-ROTOR DESIGN

	Bearing type	**Rotor geometry**	**Electric drive**	**Rotor material**	**Vacuum quality**	**Rotational speed**
Bearing type	/	●	◐	◐	●	●
Rotor geometry		/	●	●	◐	●
Electric drive			/	○	○	●
Rotor material				/	◐	●
Vacuum quality					/	●
Rotational speed						/

●: high, ◐: medium, ○: low

As can be seen in Table I, especially the rotor geometry and the rotational speed have high interdependencies with all other components and properties. This means that a high change in rotational speed, for example, leads to a strong change in rotor material or electric drive. The connection between bearing type and electric drive, for instance, is low because the same electric drive can be operated with different bearing types.

B. Optimization Problem

In the previous chapters the complexity of the optimization problem has been described. It is obvious that with the high DoI of the outer-rotor concept and the resulting large number of interdependencies a sequential selection and design of the different components does not lead to an optimal system design. In addition to the interdependencies several non-linear constraints, such as the maximum stress in the material or the design of the magnetic bearings and the electric drive have to be considered. Depending on the exact field of application several objective functions need to be optimized at the same time. For most systems at least two of those objectives can be conflicting goals, for example the energy density and the operational losses. Combining the requirements imposed on the optimization process a non-linear, multicriteria optimization with a large number of input variables has to be carried out.

C. Developing a System Model

In order to be able to optimize the KESS a detailed system model is necessary. In the following a model for an outer-rotor kinetic energy storage system with active magnetic bearings (AMB) for the radial suspension, a rotor out of fibre-plastic composite material, and a permanent magnet excited synchronous motor is described, see Fig. 4. To reduce the large number of input variables only four important geometric variables (outer-rotor radius, inner-rotor radius, rotor height, radial length of fittings), the number of stator windings of the electric drive and one operational variable (rotational frequency) are used as input parameters.

The first step of the system model is to complete the system parameters so that the requirements (maximal power, storage capacity) are fulfilled using geometrical, material and electrical, operational assumptions. The geometrical assumptions consist of geometrical parameters that are kept constant or within a certain range, such as the air gap between rotor and stator, the balancing quality of the rotor or the number of pole pairs of the electrical drive. The material assumptions are densities, maximal stresses, electrical conductivities, temperatures and pressures. Electrical assumptions are defined as maximal currents and voltages for the electric drive and the AMB as well as maximal flux densities and loss factors. Operational assumptions are important for the next steps as they define operational situations that have to be fulfilled by the components, such as power and force requirements for the electric drive and AMBs.

The next step within the system model is the design of the fiber-composite rotor, the active magnetic bearings and the electric drive. The rotor stresses are calculated using the analytical approach described in [9] for a one-layer, elastic and anisotropic composite rotor using the Euler–Cauchy governing equation. The AMB is designed such that it can provide the maximal forces occurring when exciting the nutation eigenfrequency of the rotor superimposed by the maximal radial coupling force of the permanent magnet axial bearing. The electric drive has to be designed for the maximal torque occurring when the maximal power is demanded at minimal rotational speed.

In the next step the constraints gate has to be passed. Here properties of the system that were calculated before are evaluated. Basic geometrical constraints such as a larger outer than inner diameter and a smaller total height of the fittings than the total rotor height have to be fulfilled. Also the maximum rotor strain at maximum rotational frequency has to be smaller than the critical strain defined by the material. Dynamic constraints that are verified are the eigenfrequencies of the stator or the controllability measured by the inertia ratio. Electrical constraints that are checked are the maximal current and voltage of the electric drive and the AMBs. The KESS is only forwarded to the loss model if all constraints are fulfilled by the system configuration. Otherwise the system configuration is discarded.

D. Optimization Objectives: Loss and Energy Density Model

Up to this point the model is independent from the optimization objectives. The Results are alternative designs fulfilling the constraints. In the following, the properties of each design concerning the optimization objectives are determined. Here, these objectives are minimal operational losses and a maximum energy density. Within the loss model three major loss components of the kinetic energy storage configuration are calculated. The operational losses are caused by air friction losses, losses of the electric drive (conversion losses and drag losses) and drag losses of the AMB as well as its Ohmic losses. The calculation is carried out using stationary analytical loss models. The loss estimations of the electric drive and the AMBs are based on material specific loss parameters that consider frequency dependent hysteresis and eddy current losses as well as Ohmic losses. The energy density of the configuration regarding the rotor is also calculated.

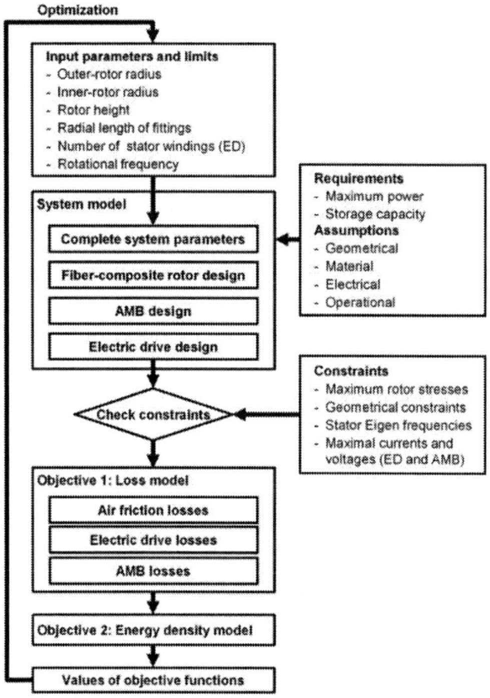

Figure 4. System model for the optimization

V. EXAMPLARY APPLICATION: FACTORY GRID

Kinetic energy storage systems can be used for different applications. Applications with a high number of cycles at high electrical powers compared to the total transferred energy are especially suitable for a KESS. As an example a small factory grid is analyzed in the following and a KESS shall be designed for this application.

A. Specification

According to the SDA methodology the dimensions of the energy storage system (power, energy capacity) have to be specified before it can be designed. Therefore the load profile is analyzed and the storage dimensions are derived in the following paragraphs.

Figure 6 (top) shows the load profile of the factory grid for about 5.5 hours. The resolution of the signal is 1 second resulting in very high power peaks up to 150 kW. The aim of the KESS is to smooth the load profile and at the same time limit the maximum power to 100 kW. The resulting reduced volatility of the profile generates a value for the storage operator because electricity costs depend on the maximum power at the grid connection point of the factory. Moreover it helps to relieve the load of the distribution grid leading to a higher capacity to integrate renewable energies.

In order to generate the desired load profile, the original load profile is filtered. An analogue low-pass filter could be used but would lead to a phase delay of the filtered signal. The field of digital signal processing provides methods to overcome this drawback. One possibility is to use a zero-phase digital filter to generate the generic load profile. Such filters process the input data in both the forward and reverse direction [10]. Since the full load profile has to be known at the time of filtering this method cannot be used for real time applications. For the suggested procedure no online capabilities are necessary because the ‘Specification’ process is only performed offline and before installing the KESS. The filtered load curve was generated using a corner frequency of 0.0105 Hz.

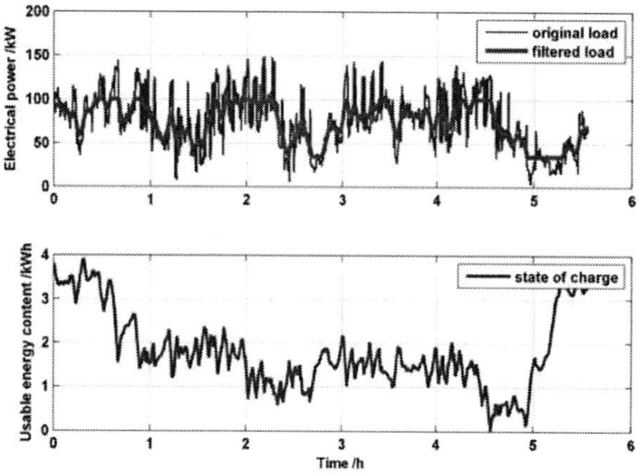

Figure 5. Load curve of the industrial plant (top) and the resulting energy curve of the KESS (bottom)

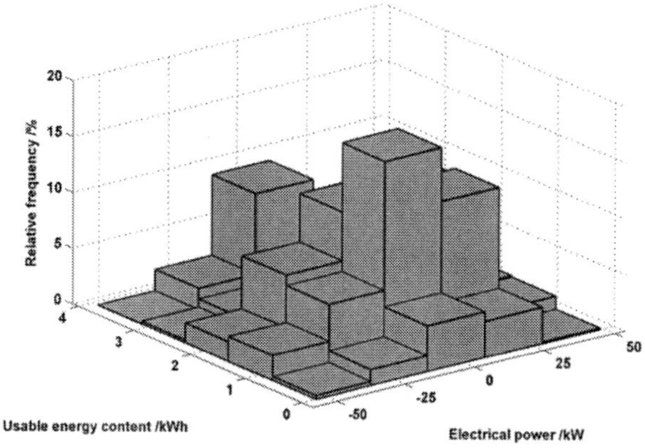

Figure 6. Relative frequencies of operating points

The KESS has to provide the difference between the filtered and the original load profile. A positive power difference equals charging and a negative difference discharging of the KESS. Fig. 6 (bottom) shows the resulting energy curve of the KESS during operation.

In order to analyze the operating points of the KESS, defined by power and energy content, the relative frequencies are determined. As can be seen in Fig. 6, the KESS mostly operates at medium to high energy levels and small powers. The maximum power is only needed a few times. From this analysis it becomes clear that a maximal electrical power of 54.3 kW and a usable energy capacity of 3.9 kWh are necessary to fulfill the task.

B. Design

For the design phase an optimization algorithm is needed. The problem is a non-linear multicriteria optimization problem with non-linear constraints and limited input variables. For such complex problems evolutionary algorithms deliver satisfying results (see e.g. [11]). Thus a genetic algorithm is applied here. In order to thoroughly search the design space the population size is set to 4000 and the number of generations is set to 500.

The information generated during the specification phase is also implemented. The maximal electrical power and the usable energy capacity are given to the storage model as requirements. The relative frequencies are used to calculate combined total losses for different operating points. During the optimization process the total losses, consisting of air friction, electric drive and active magnetic bearing losses, are calculated for every operating point. The optimization algorithm, however, can only process a single value for every optimization criterion. Therefore the different losses are weighted with the relative frequency of their operation point. The combined total losses are then calculated by summing up the weighted losses. In addition the energy density is calculated using the rotor weight and the usable energy content.

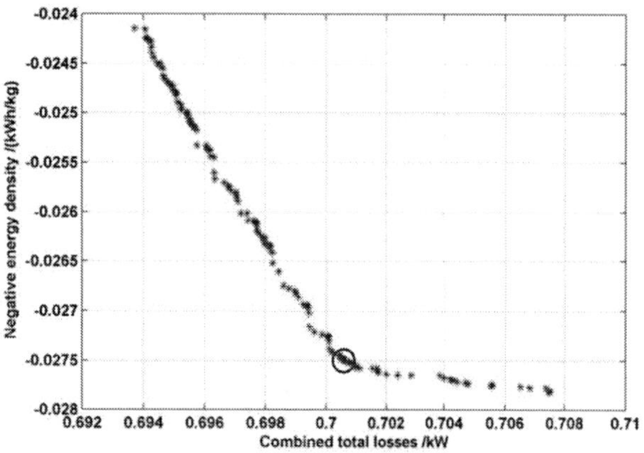

Figure 7. Pareto frontier of the optimization process

The optimization results can be displayed in form of a Pareto frontier of the two optimization criteria, see Fig. 7. Since the energy density is a criterion that is to be maximized, the sign has to be switched in order to get a minimization problem.

The total number of individuals on the Pareto frontier is 135. The optimization was terminated because the maximum number of generations was exceeded after about 22 minutes.

The Pareto frontier can be interpreted as follows. For a given value of combined total losses, there is no individual with a higher energy density than the one on the frontier. As can be seen from its form, there is one group of solutions that is likely to be chosen by a designer with balanced preferences (red circle). From there a further decrease of the losses would mean that a lot of energy density has to be sacrificed (left side of the circle). In order to increase the energy density even further, the losses increase significantly (right side of the circle).

In the following, the KESS in the middle of the red circle is analyzed in detail. The combined total losses of this system lie at 0.705 kW and the energy density at 0.0275 kWh/kg.

Table II shows the input parameters that are necessary to gain the selected KESS.

TABLE II. INPUT PARAMETERS OF THE SELECTED KESS

Parameter	Value
Outer-rotor diameter	234 mm
Inner-rotor diameter	115 mm
Total height	612 mm
Number of stator windings of electric drive	14
Radial length of fittings	8 mm
Scaling factor of rotational frequency	1.5 %

Together with the requirements, assumptions and the storage model, these parameters lead to the complete KESS. The geometrical dimensions of the system are displayed in Fig. 8.

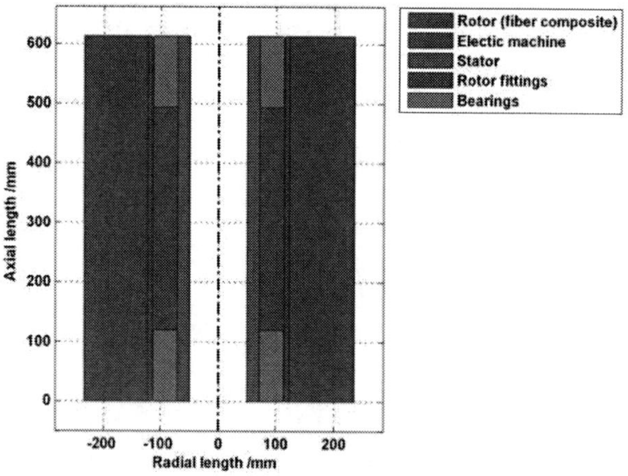

Figure 8. Geometrical dimensions of the selected KESS

TABLE III. OPERATION PARAMETERS OF THE SELECTED KESS

Parameter	Value
Minimal rotational speed	15,666 1/min
Maximal rotational speed	28,830 1/min
Moment of inertia	4.37 kg·m^2
Minimal physical energy content	1.63 kWh
Maximal physical energy content	5.53 kWh

In addition to the geometrical dimensions the operation parameters result from the same model. As can be seen in

Table III the KESS is operated between 15,666 and 28,830 1/min.

With the moment of inertia of 4.37 kg·m^2 this can be converted to a minimal physical energy content of 1.63 kWh where the KESS is considered empty because it cannot provide the full electrical power anymore. The maximal physical energy content lies at 5.53 kWh. Here the stresses within the rotor reach their fatigue limits. The stresses of the fiber-composite rotor at maximal rotational frequency are shown in Table IV.

TABLE IV. MATERIAL PARAMETERS (FIBER-COMPOSITE) OF THE SELECTED KESS

Parameter	Value	Limit
Maximal tangential stress	738.2 N/mm^2	740 N/mm^2
Minimal radial stress	-66.1 N/mm^2	-120 N/mm^2
Maximal radial stress	34.9 N/mm^2	35 N/mm^2

Since the losses of the system are fundamental to the optimization, they are analyzed in depth. Figure 9 shows a loss map of the KESS for different operational points. The losses for electrical power equal to zero are caused by the AMB losses, the air friction losses and the drag losses of the electric drive. All of them increase with increasing rotational speed resulting in an angled map. For positive or negative electric powers the conversion losses of the electric drive occur and contribute to the total losses. This leads to a bent loss map in the direction of electric power.

The total losses reach a maximum of 935 W for maximal rotational speed (28,830 1/min) and maximal negative electrical power (-54.3 kW). The loss minimum lies at 518 W for minimal rotational speed (15,666 1/min) and no electrical power.

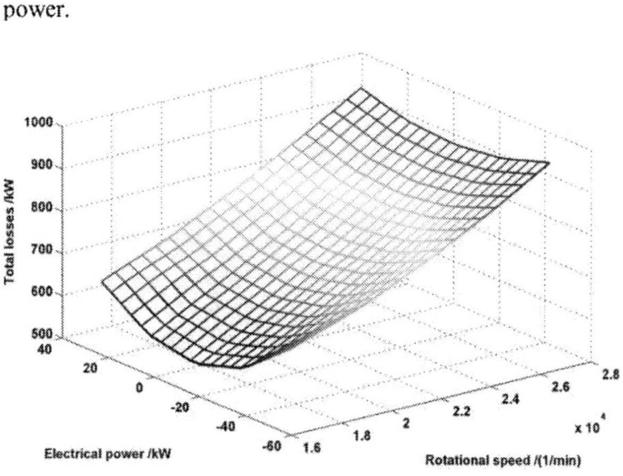

Figure 9. Loss map of the selected KESS

In order to get an idea of the order of magnitude of the different loss components, Table V shows their mean values of all operating points.

The electric drive losses have the highest contribution to the total losses, but also the AMB losses play an important role. The air friction losses are low because vacuum quality is very high (0.005 Pa).

TABLE V. AVERAGE LOSS PARAMETERS OF THE SELECTED KESS

Parameter	Value
Mean AMB losses	119.35 W
Mean electric drive losses	500.04 W
Mean air friction losses	2.43 W

VI. CONCLUSION AND OUTLOOK

Driven by the transition of the energy supply, new and efficient energy storage technologies are necessary. Outer-rotor kinetic energy storage systems have the capability to contribute to a sustainable energy supply. Their design process, however, is complex due to a high degree of integration leading to a large number of interdependencies between system components. A special optimization procedure based on a detailed system model is necessary to generate results suiting the detailed requirements of the application.

In this contribution a multicriteria optimization regarding the total system losses and the energy density is carried out. In order to perform such an optimization not only a detailed system model is needed but also a loss and energy density model. Since the optimization is based on those models the model accuracy directly contributes to the results. Very detailed and thus very exact models, however, dramatically increase the computation time. In order to make decisions concerning the right degree of abstraction, experimental validations of the system and loss models are essential and should be carried out in the next step.

REFERENCES

[1] G. Genta, "Kinetic Energy Storage: Theory and Practice of Advanced Flywheel Systems", Butterworth, London and Boston, 1985.

[2] H. Schaede, M. Schneider, M. Richter, L. Quurck, S. Rinderknecht, "Kinetic Energy Storgae – Energy Efficient and Highly Dynamic for Decentralised Applications", *Proceedings International Electric Drives Production Conference 2013, 2013.*

[3] H. Schaede, A. Von Ahsen, S. Rinderknecht. and D. Schiereck, "Electric Energy Storages – a Method for Specification, Design and Assessment", *Int. J. Agile Systems and Management*, vol. 6, no. 2, pp. 142-163, 2013.

[4] S. K. Ha, S. J. Kim, S. U. Nasir, and S. C. Han, "Design optimization and fabrication of a hybrid composite flywheel rotor," *Composite Structures*, vol. 94, no. 11, pp. 3290–3299, 2012.

[5] A. C. Arvin and C. E. Bakis, "Optimal design of press-fitted filament wound composite flywheel rotors," *Composite Structures*, vol. 72, no. 1, pp. 47–57, 2006.

[6] J. L. Pérez-Aparicio and L. Ripoll, "Exact, integrated and complete solutions for composite flywheels," *Composite Structures*, vol. 93, no. 5, pp. 1404–1415, 2011.

[7] C. Zhang and K. J. Tseng, "Design and control of a novel flywheel energy storage system assisted by hybrid mechanical-magnetic bearings," *Mechatronics*, vol. 23, no. 3, pp. 297–309, 2013.

[8] B. Bolund, H. Bernhoff, and M. Leijon, "Flywheel energy and power storage systems," *Renewable and Sustainable Energy Reviews*, vol. 11, no. 2, pp. 235–258, 2007.

[9] S.M Arnold, A.F Saleeb, N.R Al-Zoubi, "Deformation and life analysis of composite flywheel disk systems", *Composites Part B: Engineering*, vol. 33, no. 6, pp. 433–459, 2002.

[10] A.V. Oppenheim, R.W. Schafer, "Discrete-Time Signal Processing", Prentice-Hall, pp. 284–285, 1989.

[11] M. Gen, Genetic Algorithms and Engineering Design. John Wiley & Sons, 1997.

Novel Cylindrical Magnetic Levitation Stage with High Precision Motion

Jeon Jeong-Woo[a], Lee Chang-Lin[a], Oh Hyeon-Seok[a], Kim Jong-Moon[a]

[a] Korea Electrotechnology Research Institute, Bulmosan Street 10-12, Chang-Won, South Korea, jwjeon@keri.re.kr

***Abstract*—This paper proposed novel cylindrical magnetic levitation stage. This stage can move the cylinder to rotations and translations as well as levitations with high precision in vacuum. This advantage is useful to make a nano patterning on the surface of cylindrical specimen by using electron beam lithography. In this paper, a conceptual design and a detailed design of the novel cylindrical magnetic levitation stage are shown. Firstly, mathematical modeling and simulation results are shown and then experimental results are compared after manufactured.**

I. Introductions

Nowadays, an application of nano patterns is expanding for display panels and solar cell as well as semiconductors. Roller type nanoimprint lithography is one of best solution to make a nano patterned mass production. There are lots of researches which are related with roller type nanoimprint lithography [1-2]. But, most of these researches are to imprint with a nano patterned flexible plate which was rapped on cylinder. A nano patterned area from those researches can be limited by such as a seam. If nano patterns can be directly manufactured on a surface of a cylindrical stamp, it is very useful to make a nano patterned mass production with roller type nanoimprint lithography. The cylindrical stamp can be manufactured by conventional lithography method such as electron beam lithography technology [3-4]. Here, a special bearing is needed to reduce errors from rotating machine. This special bearing should be rotated with nano resolutions in vacuum. One of best solutions is a magnetic-typed bearing [5-6]. The magnetic-typed bearing can be controlled with nano resolutions and operated in vacuum. But, there are two limitations. The one is that the most magnetic bearing cannot be moved along axles of it. The other is that there is interfere to install with electron beam column. In this paper, novel cylindrical magnetic levitation stage which is proposed can overcome those limitations. A concept of this stage was come from planar typed magnetic levitation stage [7-8]. The cylindrical magnetic levitation stage can move rotations and translations as well as levitations without contacts. A conceptual and detailed design was already proposed [9]. The cylindrical mag-lev stage is composed of a mover and a stator

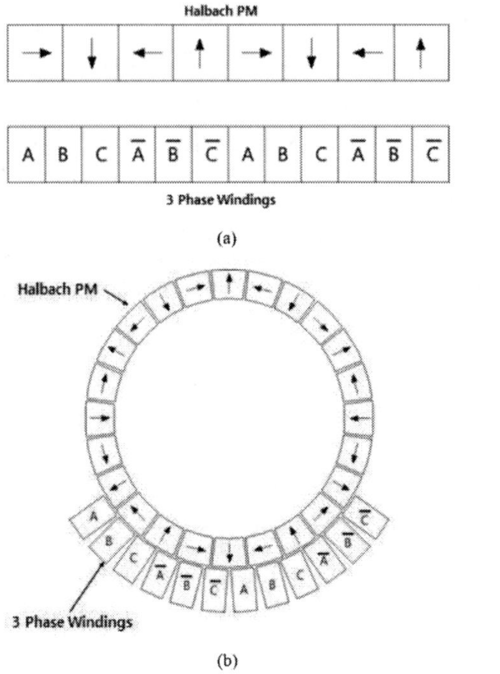

Figure 1. Concepts of MagLev Stage (a) The Plane-Type MagLev Stage (b) The Cylindrical-Type MagLev Stage

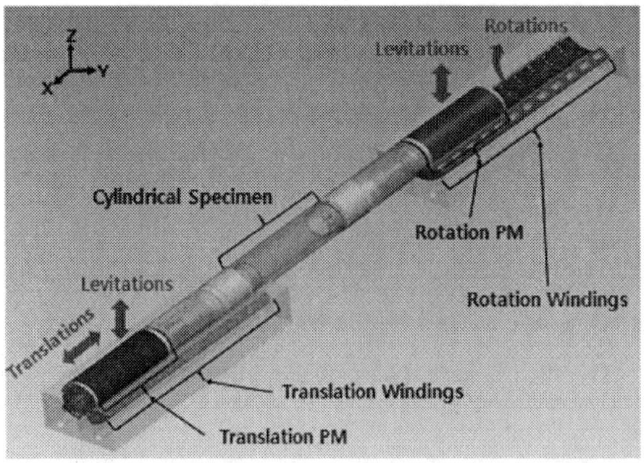

Figure 2. The Novel Cylindrical-Type MagLev Stage (3D Model)

part. The mover part is composed of permanent magnet array and the stator is composed of winding array. This cylindrical mag-lev stage is installed and tested on a test bench and is successfully levitated and maintained with high precisions.

II. Design

A. Conceptual Design

Conventional plane-type MagLev Stage is shown in Fig. 1 a). A magnetic repulsion force is generated between Halbach array and windings by properly induced currents at windings. The magnetic repulsion force can generate a traction force as well as a levitation force simultaneously. The mover can be levitated by the levitation force and then moved horizontally by the traction force. A conceptual design of novel cylindrical maglev stage is shown in Fig. 1 b). The Halbach permanent magnet array can be formed like a cylinder shape. The 3 phase winding can be formed like a hemi-cylinder shape. A magnetic repulsion force between cylindrical shaped permanent magnet array and hemi-cylindrical shaped 3 phase windings can be generated similar with it of plane-type maglev stage. This force can generate a rotation force (instead of a traction force at plane-type maglev stage) as well as a levitation force.The 3D model of the novel cylindrical maglev stage is shown in Fig. 2. It can be levitated and then rotated as well as translations of axial with contactless. The cylindrical maglev stage is composed of a cylindrical shape mover and two hemi-cylindrical shape stators. The mover is composed of cylindrical specimen and two cylindrical-shape Halbach permanent magnet array. These Halbach magnet arrays are assembled on both ends of the cylindrical specimen. Array directions of these Halbach magnet array are perpendicular each other. Two hemi-cylindrical shape stators are composed of 3 phase windings. Array directions of two stators are also perpendicular each other like as two magnet array. The mover is put on two stators.

B. Detailed Design

A detail designed cylindrical maglev stage is shown in Fig. 3. Magnetic repulsion forces are generated between Halbach permanent magnet array and 3 phase windings. It can generate traction forces as well as levitation forces. Here, the traction forces are differently generated due to the array direction of permanent magnet and windings. In Fig. 3 (a), the array directions of permanent magnets and windings are located with rotationally. Rotation forces as well as levitation forces are generated. In Fig. 3 (b), the array directions of permanent magnets and windings are located along the axial. Translation forces as well as levitation forces are generated. Two-type permanent magnet arrays are combined in the mover. All magnetic force which are generated from two-type permanent magnet arrays and 3 phase windings are combined due to a properly induced current at windings. Finally, the mover can be rotated and translated simultaneously as well as levitated.

C. Magnetic Forces

In Fig. 3, the 3 phase windings are assembled with I and J rows. One winding module (which is indicated by RED dotted line) is composed of 6 windings. The magnetic repulsion force (Fn) can be represented by Lorenz force. The Lorenz force is

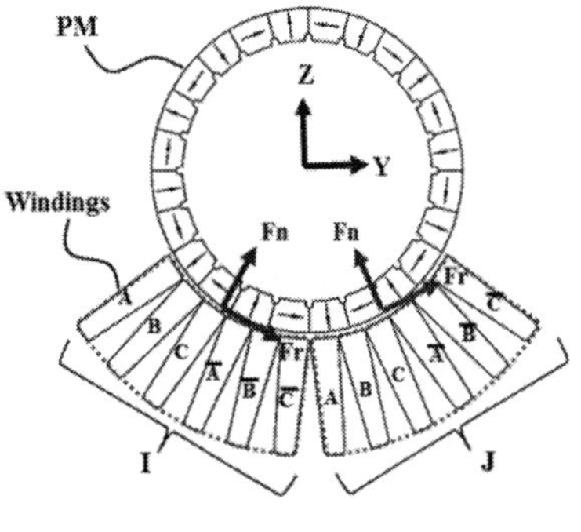

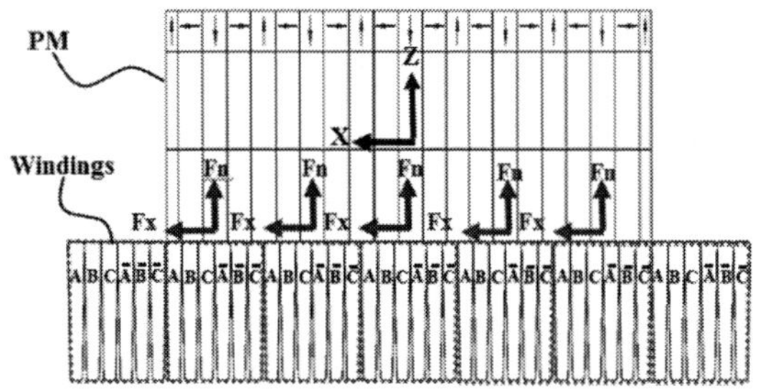

Figure 3. The Detailed Design of Novel Cylindrical-Type MagLev Stage (a) The Front View of the Rotational Part (b) The Side View of The Translation Part

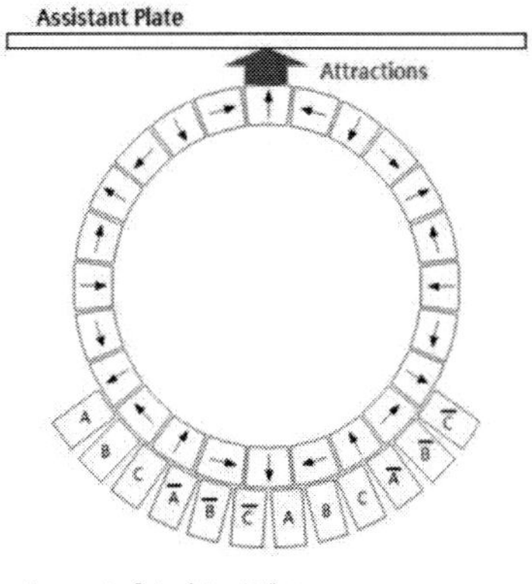

Figure 4. The Concept of Assistant Plate

generated at between 3 phase winding module and permanent magnet array. A direction of the magnetic repulsion force (Fn) is determined by a direction of 3 phase winding module. Values of direction of I and J windings are ±30°. The magnetic repulsion force (Fn) can be divided by the magnetic levitation force (Fz) and the magnetic guidance force (Fy). In Fig. 3 (a), the orthogonal force (Fr) of the magnetic repulsion force (Fn) can be generated due to rotationally arrangements of permanent magnet array and 3 phase winding array. The mover can be rotated. In Fig. 3 (b), the orthogonal force (Fx) of the magnetic repulsion force (Fn) can be generated due to straightly arrangements of permanent magnet array and 3 phase winding array. The mover can be translated.

D. Assistant Force

The magnetic levitation force can be controlled by induced currents at 3 phase winding module. Here, high currents cause high temperature of windings. This is disadvantage in vacuum because a heat cannot transfer through vacuum. In this research, the assistant plate of magnetic levitation is considered to reduce heat problem of windings as shown in Fig. 4. This plate is located upper of the permanent magnet. The attraction force which is generated between plate and permanent magnet can help to levitate the cylindrical mover. This assistance force can be adjusted by distance between plate and permanent magnet. Induced currents can be reduced due to this assistance force. Here, the amount of assistant force is considered with a half of total required levitation forces.

III. Modeling

A. Mathematical Model

The magnetic repulsion forces are generated between the permanent magnets and 3 phase windings for reason of Lorentz Force. These forces are composed of the magnetic repulsive force (Fn) and the orthogonal force (Fr and Fx) which are shown in Fig. 3. The formulas of these forces for one pitches of the magnetic levitation stage are as the following equations [7-9]:

$$\begin{bmatrix} F_r, F_x \\ F_n \end{bmatrix} = \frac{1}{2}\mu_0 M_0 \eta_0 N_m G e^{-\frac{2\pi}{l} z_0} \begin{bmatrix} i_q \\ i_d \end{bmatrix} \quad (1)$$

$$G = \frac{\sqrt{2}\omega l^2}{\pi^2}(1 - e^{-\frac{2\pi}{l}\Gamma})(1 - e^{-\frac{2\pi}{l}\Delta}) \quad (2)$$

Here, $\mu_0 M_0$ which is magnet remanence is 1.465 [T], η_0 which is turn density is 1.6026e^{+6} [Turn/mm^2], N_m which is number of active magnet pitch is 1, ω which is permanent magnet width of one pitch is 52.18 [mm], l which is 3 phase winding width of one pitch is 52.18 [mm], Γ which is PM array thickness is 10 [mm], Δ which is coil array thickness is 14.04 [mm], Z_0 which is Nominal gap is 0.95 [mm]. The repulsive and orthogonal forces of one pitch of our magnetic levitation stage are as the following:

$$\begin{bmatrix} F_r, F_x \\ F_n \end{bmatrix} = 12.1695 \begin{bmatrix} i_q \\ i_d \end{bmatrix} \quad (3)$$

If the values of input current (i_q, i_d) are 1 [A] for each, the values of repulsive force (Fn) and orthogonal force (Fr, Fx) are 12.1695 [N]. The magnetic repulsive force (Fn) is composed of vertical forces (Fz) and horizontal forces (Fy) which are the following equations:

$$\begin{bmatrix} F_z \\ F_y \end{bmatrix} = \begin{bmatrix} F_n \cdot \cos\theta \\ F_n \cdot \sin\theta \end{bmatrix} \quad (4)$$

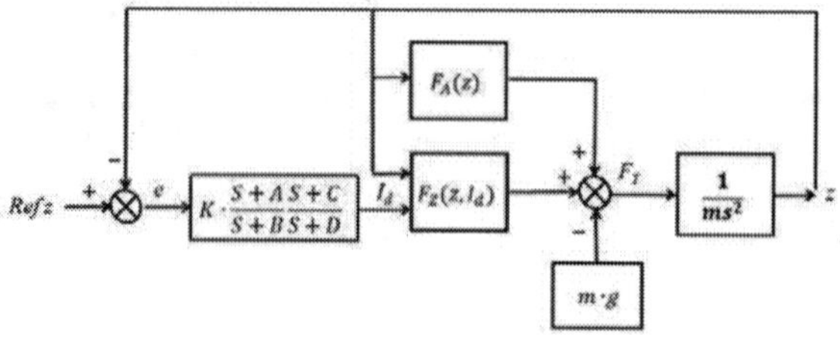

Figure 5. The Dyanmic Model of Simulation

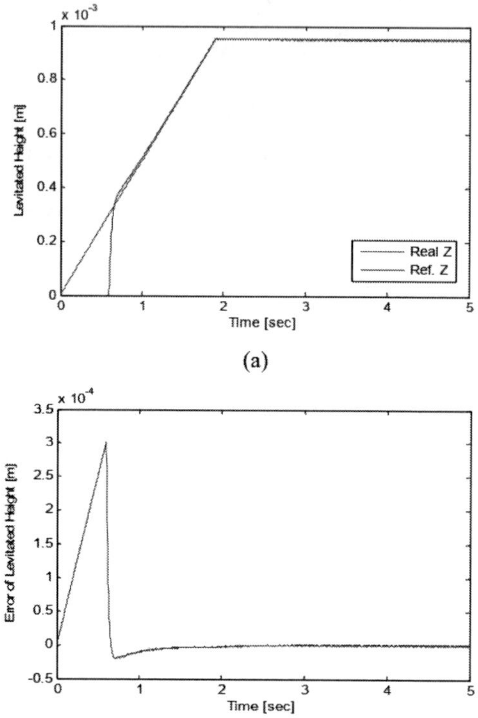

(a)

(b)

Figure 6. The Simulation Results (a) Levitation (b) Error

Here, θ is the angle of the magnetic repulsive force and the value is 30 degree. The values of vertical forces (Fz) and horizontal forces (Fy) are about 10.5391 [N] and about 6.0847 [N]. If the valid ratio of winding shape (L2/L1 = 0.731) is considered, these values is corrected by 7.7041 [N] and 4.4479 [N]. Here, a sum of vertical forces (Fz) is 154.0814 [N] because a number of one pitch winding modules of rotation part and translation part are 10 each other. This calculated magnetic vertical force (Fz) is not much different than results from the above analysis.

B. Dynamic Model

The dynamic model for simulation of magnetic levitation is configured in Fig. 5. The levitation height (z) is feed back into the controller, the assistant force calculation model and the magnetic levitation stage model. The controller output (Id) is feed into the magnetic levitation stage model. The dynamic model of the cylindrical magnetic levitation stage is as the following equations:

$$m \cdot \ddot{z} = F_z(z, I_d) + F_A(z) - m \cdot g \quad (5)$$

Here, m is total mass of the mover and the value is 30 [kg], z is the levitation height, I_d is input currents of 3 phase windings, Fz(z, I_d) which is came from equation (2) is total vertical force of magnetic repulsive forces (Fn), $F_A(z)$ is the attractive force between the assistant plate and permanent magnets, g is gravity accelerations. Total vertical force (Fz(z, I_d)) is sum of each 10 winding modules of the rotation part and the translation part. The assistant force ($F_A(z)$) is calculated by the electromagnetic FEM analysis. The assistant force ($F_A(z)$) is as following equations:

$$F_A(z) = 4{,}948e(-238.1 \cdot z) + 24.02e(11.78 \cdot z) \quad (6)$$

Here, total assistant force is twice time of $F_A(z)$ because these assistant forces are simultaneously generated at the rotation part as well as the translation part. The Lead-Lag compensator is used for magnetic levitation control as following equations:

$$C(s) = K\frac{s+A}{s+B}\frac{s+C}{s+D} \quad (7)$$

Here, K = 1.051 x 10^6, A = 38.7857, B = 1252.3954, C = 2.9483, D = 0.0013. The linearization result of closed loop system with controller at steady state is:

$$G(s) = \frac{6.054\times10^5(s+38.79)(s+2.948)}{(s+41.02)(s+3.061)(s^2+1{,}208s+5.513\times10^5)} \quad (8)$$

Here, crossover frequency is 356.6469 [rad/sec] and phase margin is 134.9171 [deg].

C. Simulations

The simulation results of dynamic model are shown in Fig. 6. A reference levitation height (Ref. Z) is smoothly increased up to 950 [um]. This is to minimize transient state at the beginning of levitations. In Fig. 6 (a), the real levitation height (Real Z) is very well following the reference. In Fig. 6 (b), the error is maintained within small ranges. After levitations, the net force of assistant plates is about 166 [N] and the net force of magnetic levitation is about 128 [N]. The total levitation force which is sum of these forces is about 294[N]. The current of one winding module is about 0.83 [A]. The assistant force is help to reduce required currents of winding modules for levitations.

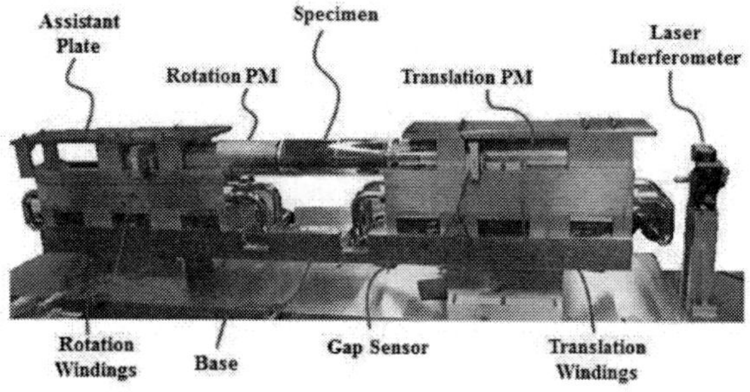

Figure 7. The Test Bench

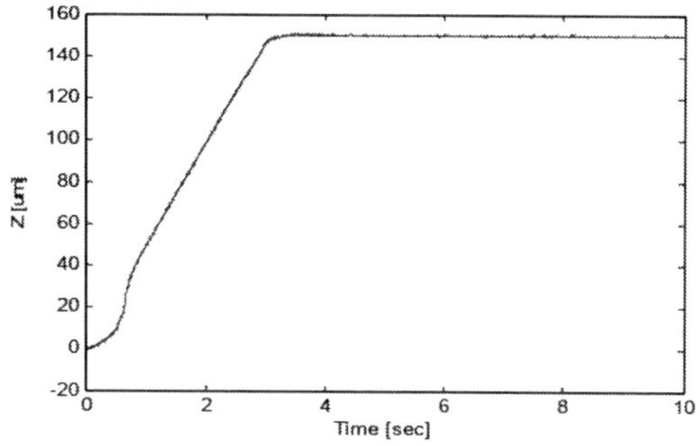

(a)

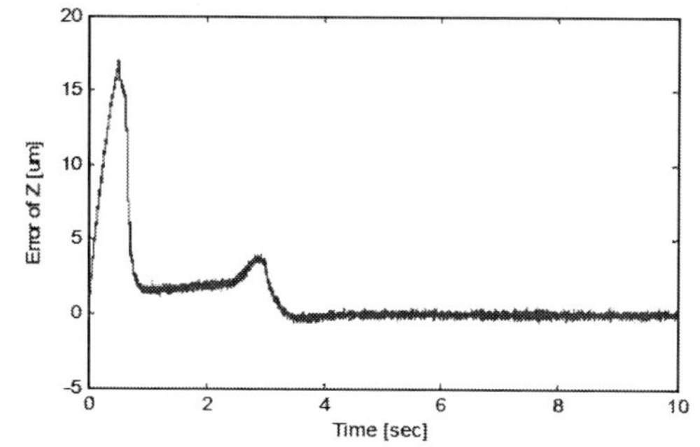

(b)

Figure 8. The Experimental Results (a) Levitation (b) Error

D. Experiments

The test bench of cylindrical maglev stage is shown in Figure 7. This test bench is composed of a base part, mover parts (cylindrical specimen, rotation PM, and translation PM), stator parts (rotation windings and translation windings), assistant plates, cooling pipe, and measurement sensors (one laser interferometer, four capacitive probe sensors, and one optical rotary encoder). The levitation test results of cylindrical maglev stage are shown in Fig. 8. The goal of levitation height is 150 [um]. The beginning of levitation is shown to unstable situations. After take-off, the error is shown to be stable. The errors of Z axis is maintained under 1 [um].

IV. CONCLUSIONS

This paper proposed novel cylindrical magnetic levitation stage. This stage can move the cylinder to rotations and translations as well as levitations with high precision in vacuum. This advantage is useful to make a nano patterning on the surface of cylindrical specimen by using electron beam lithography. The cylindrical specimen with nano patterned is applicable for roller-typed nano imprint lithography for mass productions. The mathematical modeling was used for dynamic simulation with the lead-lag compensator. The assistant plates which were located at top of the permanent magnet array were considered. These plates are useful to reduce required currents of winding modules for levitations and minimize heat problems in vacuum. The simulation and experimental results were shown that the cylindrical stage can be levitated and controlled by the lead-lag compensator. For the future, experimental results of rotations as well as translations will be showed.

REFERENCES

[1] C.M. Sotomayor Torres, S. Zankovych, J. Seekamp, A.P. Kam, C. Clavijo Cedeno, T. Hoffmann, J. Ahopelto, F. Reuther, K. Pfeiffer, G. Bleidiessel, G. Gruetzner, M.V. Maximov, B. Heidari, "Nanoimprint lithography: an alternative nanofabrication approach", Materials Science and Engineering C 23, pp.23-32, 2003.

[2] Shuhuai Lan, Hyejin Lee, Jun Ni, Soohun Lee, Moongu Lee, "Survey on Roller-type Nanoimprint Lithography (RNIL) Process", ICSMA 2008, pp.371-376, April 2008.

[3] Shih Chun Tseng, Wen Yang Peng, Yi Fan Hsieh, Ping Jen Lee, Wen Lang Lai, "Electron Beam Lithography on Cylindrical Roller", Journal Microelectronic Engineering, Vol. 87, Issue 5-8, pp.943-946, May 2010.

[4] Noriyuki Unno, Jun Taniguchi, and Kiyoshi Ishikawa, "Fabrication of a seamless roll mold using inorgainc electron beam resist with postexposure bake", Journal of Vacuum Science & Technology B, Vol. 29, Issue 6, October 2011.

[5] A. Chiba, T. Fukao, O. Ichikawa, M. Oschima, M. Takemoto and D. G. Dorrell, "Magnetic Bearings and Bearingless Drives", Elsevier 2005, ISBN 0-7506-5727-8.

[6] M. G. Noh, S. R. Cho, J. H. Kyung, S. K. Ro, and J. K. Park, "Design and Implementation of a Fault-Tolerant Magnetic Bearing System for Turbo-Molecular Vacuum Pump", IEEE/ASME Transactions on Mechatronics, Vol. 10, No. 6, pp.626-631, 2005.

[7] Won-jong Kim, David L. Trumper, Jeffrey H. Lang, "Modeling and Vector Control of Planar Magnetic Levitator", IEEE Trans. On Industry Applications, Vol. 34, No. 6, pp.1254-1262, 1998.

[8] Jeong-Woo Jeon, Mitica Caraiani, Hyeon-Seok Oh, Sungshin Kim, "Experiments of a Novel Magnetic Levitation Stage for Wide Are Movements", Journal of Electrical Engineering & Technology, Vol. 7, No. 4, pp.558-563, 2012.

[9] Jeong-Woo, Jeon, Mitica Caraiani, Chang-Lin Lee, Yeon-Ho Jeong, Jong-Moon Kim, Hyeon-Seok Oh, Sungshin Kim, "Novel Cylindrical Magnetic Levitation Stage for Rotation as well as Translation along Axles with High Precisions", The Transactions of the Korean Institute of Electrical Engineers, Vol.61, No.12, pp.1828-1835, 2012

Active Reduction of Bearing Forces for Short Stroke Linear Actuators

Florian Poltschak[a]

[a] JKU Hoerbiger Institute for Smart Actuators, Johannes Kepler University, Altenbergerstrasse 69, 4040 Linz, Austria, florian.poltschak@jku.at

***Abstract*—Linear magnetic actuators are used in industry for a wide area of applications. In order to reach high values for the axial thrust force, the mover commonly features a permanent magnet excitation. Thus, in combination with a slotted stator layout, these actuators exhibit a high destabilizing stiffness in the direction perpendicular to the direction of motion which has to be compensated by the linear bearings. This paper introduces a linear short stroke actuator with an integrated active bearing force compensation to minimize the forces acting on the mechanical bearings. The potential and limits of a bearing force compensation is analyzed based on a short stroke linear actuators with an E-shaped stator layout. It is outlined how far bearing forces can be compensated even for simple actuator layouts. Especially for linear drives oscillating with a high speed this compensation can significantly reduce wear and thus improve lifetime and efficiency as well as reduce the size of the mechanical bearings.**

I. Introduction

As soon as the mover of a linear motor is equipped with a permanent magnet (PM) excitation, attractive forces towards the stator exist at least if the mover is not in a perfect central position. With the need for high thrust forces, which can be achieved with slotted stator layouts, these forces are becoming a problem and a limiting factor for the bearing life time. A common measure to reduce or ideally cancel these forces is a symmetric design which can be tubular or planar. For an ideally centric mover no resulting forces will occur, but however, the often required dry running linear bush bearings are a subject to wear what results by the time in an eccentricity. While the wear increases the eccentricity, the bearing forces rise even more because of the destabilizing stiffness of the magnetic forces. This increases the friction and deteriorates the efficiency.

This problem has been investigated in [1] for a pick and place robot. There the attractive forces of the linear motor have been counteracted passively by fixing PMs underneath a soft magnetic plane.

In [2] the control of the rolling motion of a magnetically levitated linear motor is investigated. Small additional electromagnetic actuators are used to stabilize the bearingless linear motor.

Other examples for linear motors with active magnetic bearings are given in [3] and [4]. In these cases the linear motor is integrated in a free piston compressor and features two active magnetic bearings to compensate for the radial motor and gravity forces.

The objective of this paper is to analyze the potential and limits to reduce bearing forces of a short stroke actuator with single phase characteristics, based on E-shaped stator elements. These actuators are intended to be used as oscillating drives integrated into the respective target application which could be a previously mentioned free piston compressor. Thus it is assumed that a dry running linear bearing is available in the system but need not be exposed to the large attractive PM forces resulting from the magnetic actuator. However this bearing allows to overcome axial positions where no or only a weak compensation of the PM attractive forces is possible. Hence, the second objective of the paper is to keep the system as simple as possible.

II. System Layout

The system layout is given in Fig. 1. It features a two-sided planar E-shaped stator with one or alternatively two individual coils on each side. The planar mover is mounted to an axis which features a bearing ring at each end. It holds two permanent magnets (PM) on each side. The system is designed to cancel the lateral forces when it is in a perfect central position. External forces are assumed only to act along the y-axis.

However, as soon as wear or initial eccentricities in the x-direction cause an imbalance of the attractive forces, bearing forces occur. To measure the bearing forces the bearing rings run each in a radial force measurement gauge. Additionally the axial and lateral position of the axis are measured to determine the eccentricity and tilting angle.

The actuator has been designed to show a constant force/current ratio in y-direction for a stroke $s = \pm 5mm$ around the center position. Exceeding the maximal stroke the force/current ratio decreases rapidly. Hence a mechanical stop is required to keep the mover in the desired region of operation.

III. Bearing Forces

From the six degrees of freedom of the rigid body motion one degree of freedom is actively controlled to perform the required axial motion in y-direction. Two degrees of freedom are stabilized passively by choosing the active width of the mover in x-direction larger than the height in z-direction. This refers to the linear motion in z-direction and the tilting around the y-axis. The tilting motion around the x-axis is held by the mechanical bearings. Hence remain the attractive

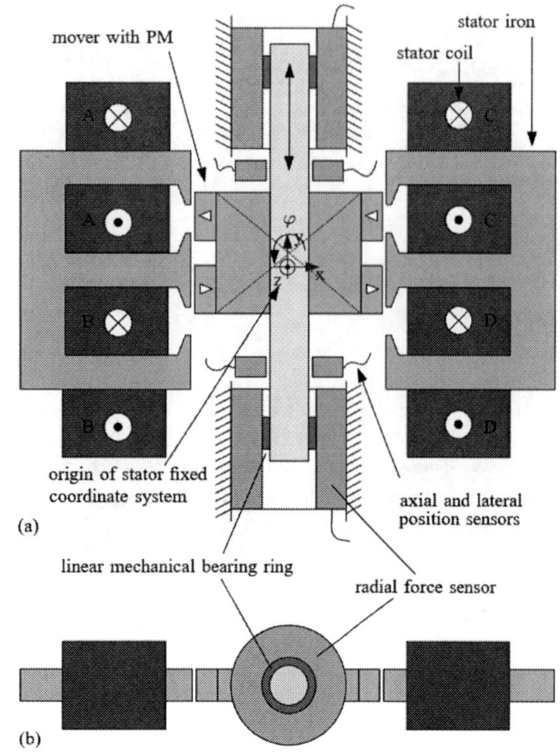

Figure 1. (a) Top view and (b) front view of the investigated linear actuator system with 4 phases labeled with A-D.

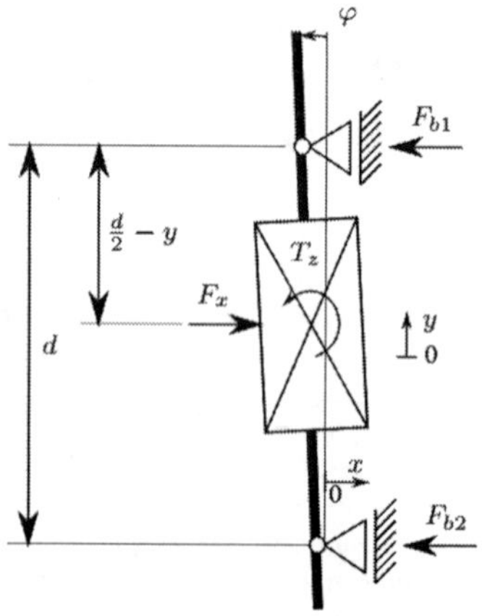

Figure 2. Model of static bearing forces under influcence of eccentricity and tilting.

forces of the PMs in x-direction and the tilting around the z-axis that require to be compensated actively. Remaining forces that cannot be compensated are stabilized by the mechanical bearings.

A. Wear of linear bearings

Compensating bearing forces is of main interest when the actuator requires to be integrated into a system. Here a typical requirement is often a small system size and high system overall efficiency. For oscillating actuators this results in the need for dry running bearings using a durable bearing material. However in the end wear limits the life time of the mechanical bearings. Assuming an oscillating actuator with the mean mover velocity v_m and the mean bearing pressure p_m the differential equation for wear is empirically deduced and given as

$$\dot{w}(t) = K_0 p_m(t) v_m \tag{1}$$

with the parameter K_0 having the unit $\mathrm{ms}^2/\mathrm{kg}$ which is the wear parameter of the selected bearing material.
It is assumed that all forces acting on the bearing are negligible compared to the resulting passive PM forces perpendicular to the direction of motion. Hence, the mean bearing pressure

$$p_m(t) = f(w(t)) \tag{2}$$

is a function of the wear, which results in an eccentricity of the mover.

For the investigated symmetrical electromagnetic actuators with small air-gap width, the forces perpendicular to the direction of motion can be approximated to vary linear with the eccentricity or wear respectively. The linear dependency is defined by the stiffness k_x. This leads to the expression

$$p_m(t) = \frac{k_x w(t)}{n h_b d_b \pi / 3} \tag{3}$$

for the mean bearing pressure. The installed n bearing rings all have the same geometric dimensions of length h_b and diameter d_b. It is assumed that each ring bears only on one third of its circumferential surface. Given an inevitable initial eccentricity of w_0 the wear results in:

$$w(t) = w_0 e^{\frac{3K_0 k_x t v_m}{n d_b h_b \pi}}. \tag{4}$$

From (4) follows, that for fixed geometrical dimensions of the bearing rings the stiffness k_x directly influences the lifetime

$$t_{life} = \frac{n d_b h_b \pi \ln\left(\frac{w_{max}}{w_0}\right)}{3 K_0 k_x v_m}. \tag{5}$$

of the bearing, that is reached as soon as w reaches w_{max}.

B. Calculation of static bearing forces

According to Fig. 2 the equilibrium of the bearing forces F_{b1}, F_{b2} and the force F_x resulting from eccentricities in x-direction

$$F_{b1}(y) + F_{b2}(y) = F_x(y) \tag{6}$$

and the torque equilibrium

$$0 = F_{b2}(y)d + T_z(y) - F_x(y)\left(\frac{d}{2} - y\right) \tag{7}$$

yield

$$F_{b1}(y) = \frac{F_x(y)d + 2T_z(y) + 2F_x(y)y}{2d} \tag{8}$$

$$F_{b2}(y) = \frac{F_x(y)d - 2(T_z(y) + F_x(y)y)}{2d} \tag{9}$$

with a distance d between the two bearings. The torque T_z and the force F_x are both dependent on y and are acting on the center of the mover (Fig. 2).

IV. System Model

A. Force and torque model

The generalized forces

$$\mathbf{Q}_F = \begin{bmatrix} F_x \\ F_y \\ T_z \end{bmatrix} \tag{10}$$

with respect to $\mathbf{q} = [x, y, \varphi]^T$ can be developed into a Taylor series at $\mathbf{i} = \mathbf{0}$ and $x_0 = 0$ what results in the definition of the forces in the form

$$\mathbf{Q}_F = \mathbf{M}_L \cdot \mathbf{i} + \mathbf{M}_C \tag{11}$$

if all terms of higher order with respect to $\mathbf{i}$ are neglected. The terms $\mathbf{M}_L$ holds the coefficients linear to the coil currents and terms $\mathbf{M}_C$ is independent of the currents. The vector $\mathbf{i}$ has the dimension $m \times 1$ with m being the number of phases. The applied superposition requires the model to be linear with respect to the phase currents $\mathbf{i}$. However, a nonlinear behavior can occur in the direction of motion y. According to [5] the linear model can be expressed as

$$\mathbf{Q} = \mathbf{T}_m \mathbf{i} = \mathbf{Q}_F - \mathbf{M}_C. \tag{12}$$

The entries of the matrices $\mathbf{M}_L$ or $\mathbf{T}_M$ and $\mathbf{M}_C$ are calculated using a finite element (FE) analysis of the investigated system given in Fig. 1.

B. Dynamic model

Assuming current control a model based on applied phase currents can be derived. For the given setup the mover of the linear actuator should be operated a the point $x = 0$ and $\varphi = 0$ whereas the motion along the y-axis has no limitation as long as the previous set assumption of linearity with respect to the current holds true.
Thus, the equation of motion

$$\mathbf{M}\ddot{\mathbf{q}} = \mathbf{Q} - \mathbf{Q}_l. \tag{13}$$

with the generalized mass matrix $\mathbf{M}$, and the generalized load force $\mathbf{Q}_l = [0, F_{ly}, 0]^T$ can be developed into a Taylor series. Neglecting all terms with a higher order as one this results in

$$\begin{aligned}\mathbf{M}\ddot{\mathbf{q}} = &\left.\frac{\partial(\mathbf{M}_L\mathbf{i})}{\partial \mathbf{q}}\right|_{x_0,y,\varphi_0,\mathbf{i}_0} \mathbf{q} + \left.\frac{\partial \mathbf{M}_C}{\partial \mathbf{q}}\right|_{x_0,y,\varphi_0,\mathbf{i}_0} \mathbf{q} + \\ &+ \mathbf{M}_L(x_0, y, \varphi_0)\,\mathbf{i} + \left.\frac{\partial \mathbf{M}_C}{\partial \mathbf{i}}\right|_{x_0,y,\varphi_0}.\end{aligned} \tag{14}$$

The term $\mathbf{M}_C$ is per definition independent of the current $\mathbf{i}$ and the term $\mathbf{M}_L\mathbf{i}$ can approximately be seen as independent from x and φ. With the stiffness matrix

$$\mathbf{k}_q(y) = \left.\frac{\partial \mathbf{M}_C}{\partial \mathbf{q}}\right|_{x_0,y,\varphi_0,\mathbf{i}_0} \tag{15}$$

and (12) the equation of motion in (14) can be simplified to

$$\mathbf{M}\ddot{\mathbf{q}} = \mathbf{k}_q(y)\,\mathbf{q} + \mathbf{T}_{m0}(y)\,\mathbf{i}. \tag{16}$$

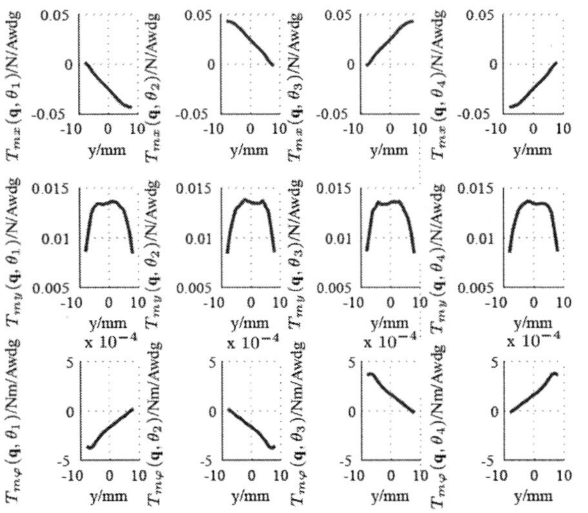

Figure 3. $\mathbf{T}_m(\mathbf{q}, \theta)$ matrix of the actuator with four phases and $\mathbf{q} = [x_0, y, \varphi_0]^T$ as a result from a FE analysis.

Neglecting all minor couplings in the stiffness matrix it results in the structure

$$\mathbf{k}_q(y) = \begin{bmatrix} k_{xx} & 0 & k_{x\varphi} \\ 0 & k_{yy} & 0 \\ k_{\varphi x} & 0 & k_{\varphi\varphi} \end{bmatrix} \tag{17}$$

showing a decoupling of the direction of motion from the other directions. The x- and φ-coordinates are coupled with $k_{x\varphi}$ and $k_{\varphi x}$. In general all entries of $\mathbf{k}_q$ are dependent on y.

V. Bearing Force Reduction

To reduce both bearing forces and to achieve the desired motion in axial direction independently the $\mathbf{T}_m$ matrix has to fulfill the rank condition [5]

$$\mathrm{rank}(\mathbf{T}_m) = 3. \tag{18}$$

Hence the system requires to have at least three phases. The currents in these phases have to be controlled in a way to generate the desired force and torque components $\mathbf{Q}$. Thus (12) requires to be solved for $\mathbf{i}$ what results in

$$\mathbf{i} = \mathrm{inv}(\mathbf{T}_m)(\mathbf{q})\mathbf{Q} = \mathbf{K}_m(\mathbf{q})\mathbf{Q} \tag{19}$$

with $\mathrm{inv}(\mathbf{T}_m)$ being a generalized matrix-inverse of $\mathbf{T}_m$.

A. Four coil actuator

The classic matrix-inverse can with (19) only be implemented for a three phase system. For the investigated case of four phases further constraints can be applied. In [6] this problem is solved to achieve minimum copper losses and the sum of all phase currents being zero. Hence the inverse of $\mathbf{T}_m$ can be calculated as

$$\mathbf{K}_m(y) = \mathbf{A}(\mathbf{T}_m\mathbf{A})^{-1} \tag{20}$$

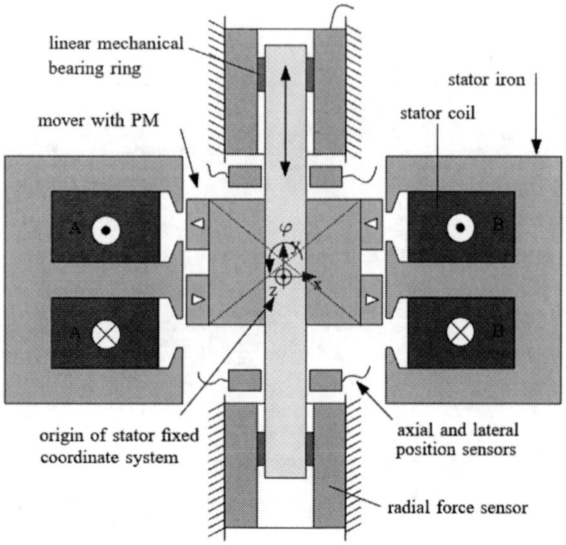

Figure 4. Top view of the actuator system reduced to two coils.

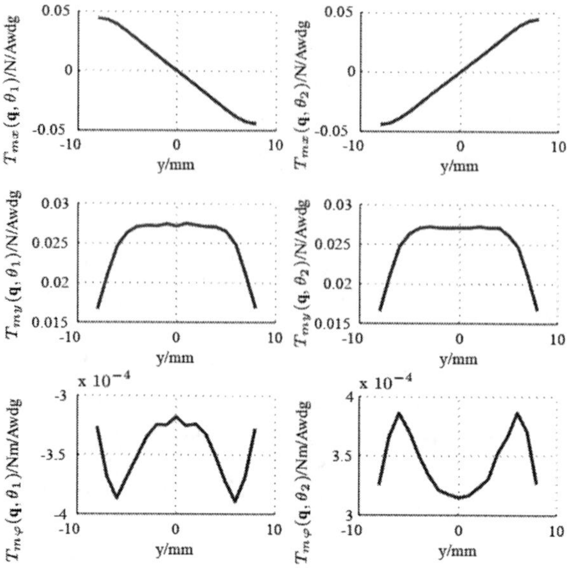

Figure 5. $\mathbf{T}_m(\mathbf{q},\theta)$ matrix of the actuator with two phases and $\mathbf{q} = [x_0, y, \varphi_0]^T$ as a result from a FE analysis.

with

$$\mathbf{A} = \left(\mathbf{T}_m^T - \frac{1}{m}\mathbf{1}\mathbf{1}^T\mathbf{T}_m^T\right). \tag{21}$$

The $\mathbf{T}_m$ matrix for the investigated system is visualized in Fig. 3, fulfills the rank condition of (18) and has the structure

$$\mathbf{T}_m = \begin{bmatrix} T_{mx1} & T_{mx2} & -T_{mx1} & -T_{mx2} \\ T_{my} & T_{my} & T_{my} & T_{my} \\ T_{m\varphi 1} & T_{m\varphi 2} & -T_{m\varphi 1} & -T_{m\varphi 2} \end{bmatrix} \tag{22}$$

with all entries dependent on y. Thus, the system described in (16) can be separated into a subsystem describing the axial motion and an independent subsystem concerning the bearing forces. With this setup it is possible to cancel both bearing forces F_{b1} and F_{b2}. As disadvantage remains the fact that the system needs four separate phases. This requires a realization with a minimum of four coils, what is double the amount as originally required for the y-direction motion.

B. Two coil actuator

Without the need for bearing forces reduction, the actuator system of Fig. 1 would simply be constructed with two coils connected in series, one on each side. Controlling the two coils independently, this system allows a bearing force reduction, too. The system with two phases is shown in Fig. 4. This setup still requires with two full bridges a power electronics of the same size as the star connected four-coil system. However, the mechanical setup remains the same as for an actuator without bearing force reduction.

In this case again a $\mathbf{T}_m$ matrix can be calculated according to (12). As a matter of fact the rank of $\mathbf{T}_m$ cannot exceed 2 in this case. A visualization of the $\mathbf{T}_m$ matrix featuring $\text{rank}(\mathbf{T}_m) = 2$ is given in Fig. 5. Two actuating variables now have to control the motion in the three directions of $\mathbf{q}$. The structure of the $\mathbf{T}_m$ matrix

$$\mathbf{T}_m(y) = \begin{bmatrix} T_{mx} & -T_{mx} \\ T_{my} & T_{my} \\ T_{m\varphi} & -T_{m\varphi} \end{bmatrix} \tag{23}$$

with all entries dependent on y and the stiffness matrix (17) yield that the direction of motion y can again be independently controlled from the bearing forces. To control the bearing forces now only one actuating variable is available. The matrix $\mathbf{K}_m(y)$ is calculated pointwise numerically as the Moore-Penrose pseudo-inverse of $\mathbf{T}_m(y)$ for each position y.
The pseudo-inverse satisfies with the following four conditions

$$\begin{aligned} &\mathbf{T}_m\mathbf{K}_m\mathbf{T}_m = \mathbf{T}_m \\ &\mathbf{K}_m\mathbf{T}_m\mathbf{K}_m = \mathbf{K}_m \\ &\mathbf{T}_m\mathbf{K}_m \text{ is hermitian} \\ &\mathbf{K}_m\mathbf{T}_m \text{ is hermitian} \end{aligned} \tag{24}$$

most of the requirements to an inverse matrix. However, the product

$$\mathbf{T}_m\mathbf{K}_m \tag{25}$$

need not be identical to the unity matrix and indicates which of the variables x and φ can be controlled with two phases. From the visualization of this matrix product (25) in Fig. 6 it can be concluded that the y-position can be controlled all time, whereas the x-position can be controlled for all y-positions except the center, where only the angle φ can be influenced. From (16), (23) and (17) follows that a current in the form

$$\begin{aligned} i_A &= i_y + \Delta i \\ i_B &= i_y - \Delta i \end{aligned} \tag{26}$$

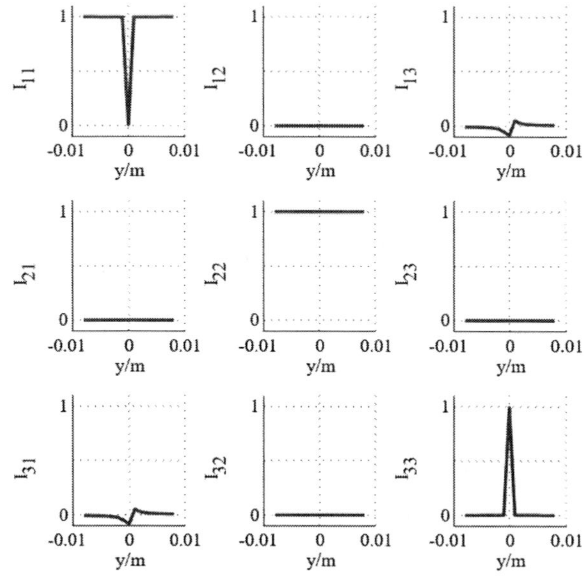

Figure 6. The product of $\mathbf{T}_m\mathbf{K}_m$ has a singularity point at $y = 0$.

allows to control the force in y direction with the current component i_y and the bearing forces with the component Δi. Consequently follows from (9) and (16) with (26) for the static bearing forces

$$\begin{aligned} F_{b1} &= \frac{dk_{xx}+2k_{\varphi x}+2k_{xx}y}{2d}x + \frac{2dT_{mx}+4(T_{m\varphi}+T_{mx}y)}{2d}\Delta i + \\ &+ \frac{dk_{x\varphi}+2(k_{\varphi\varphi}+k_{x\varphi}y)}{2d}\varphi \\ F_{b2} &= \frac{dk_{xx}-2k_{\varphi x}-2k_{xx}y}{2d}x + \frac{2dT_{mx}-4(T_{m\varphi}+T_{mx}y)}{2d}\Delta i + \\ &+ \frac{dk_{x\varphi}-2(k_{\varphi\varphi}+k_{x\varphi}y)}{2d}\varphi. \end{aligned} \tag{27}$$

Defining the movement in x-coordinate as eccentricity e resulting from wear, the max. tilting angle can be expressed as

$$\varphi_{max} = \arctan\left(\frac{d}{e}\right) \approx \frac{d}{e}. \tag{28}$$

Using the maximum tilting angle to compare the terms influenced by φ in (27) shows a negligible influence of the tilting angle on the bearing forces of the investigated actuator. Together with the result of (25) follows that the bearing forces can be reduced controlling the eccentricity in x-direction, while the tilting can be neglected.

VI. Optimal reduction of bearing forces

For the two coil actuator follows from (27) that both bearing forces cannot be canceled completely. Thus the function

$$f(y) = F_{b1}(y)^2 + F_{b2}(y)^2 \tag{29}$$

is introduced to minimize both bearing forces simultaneously. Solving

$$\frac{\mathrm{d}f}{\mathrm{d}\Delta i}(y) = 0 \tag{30}$$

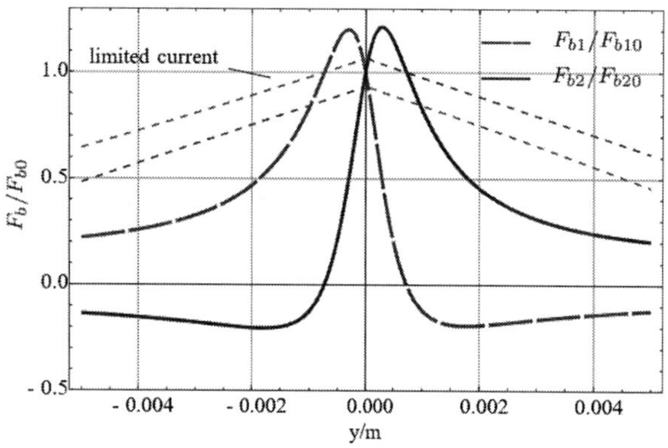

Figure 7. Relative force reduction for the two phase system with minimum bearing forces (thick) and a current limit of 5% of the rated current (dashed).

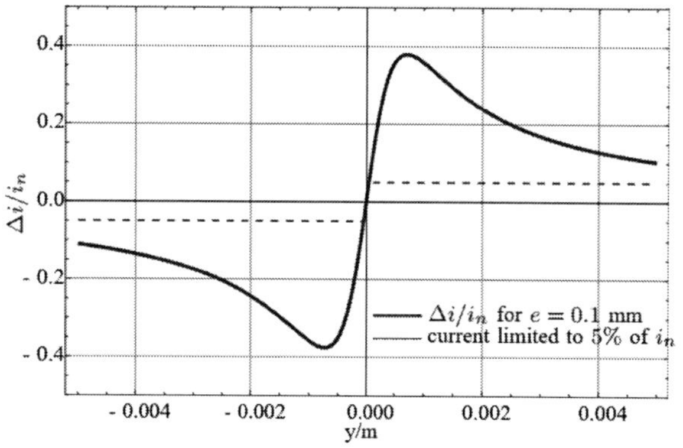

Figure 8. The current $\Delta i/i_n$ required for the two phase system to achieve minimum bearing forces for an eccentricity of 0.1mm due to wear is shown with a thick line and the current limited to 5% of the rated current is shown with a dashed line.

results in

$$\begin{aligned} \Delta i(y) = &-\frac{d^2k_{xx}T_{mx} + 4T_{ma}\left(k_{\varphi x} + k_{xx}y\right)}{2\left(d^2T_{mx}^2 + 4T_{ma}^2\right)}x - \\ &-\frac{d^2k_{x\varphi}T_{mx} + 4T_{ma}\left(k_{\varphi\varphi} + k_{x\varphi}y\right)}{2\left(d^2T_{mx}^2 + 4T_{ma}^2\right)}\varphi \end{aligned} \tag{31}$$

with

$$T_{ma}(y) = T_{m\varphi} + T_{mx}y,$$

where again the term proportional to φ can be neglected for the investigated actuator.
The minimized bearing forces result in

$$\begin{aligned} F_{b1,2}(y) = &\pm \frac{dT_{mx} \mp -2T_{ma}}{d^2T_{mx}^2 + 4T_{ma}^2} \cdot \\ &\cdot \left(\left(k_{\varphi x}T_{mx} - k_{xx}T_{m\varphi}\right)x + \right. \\ &\left. + \left(k_{\varphi\varphi}T_{mx} - k_{x\varphi}T_{m\varphi}\right)\varphi\right) \end{aligned} \tag{32}$$

and will only disappear at selected y positions as long as an eccentricity is present. Figure 7 shows the analytic calculation based on FE data of the resulting minimum bearing forces with respect to the original bearing forces for the investigated

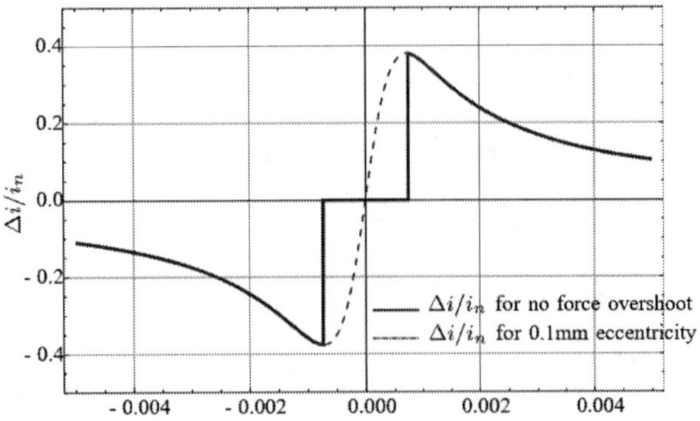

Figure 9. The current $\Delta i/i_n$ required for the two phase system to achieve minimum bearing forces for an eccentricity of 0.1mm due to wear is shown as dashed line and the current to avoid any overshoot of the bearing forces is shown with a thick line.

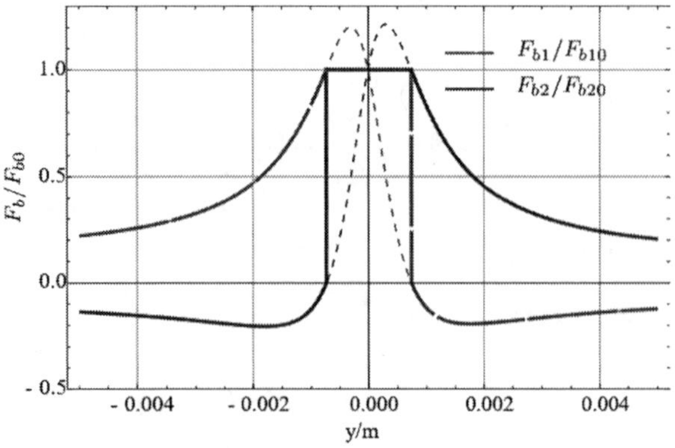

Figure 10. Relative force reduction of two phase system with minimum bearing forces (dashed) and no overshoot of bearing forces (thick).

system. It confirms the result derived from the $\mathbf{T}_m$ matrix, that it is not possible to cancel the bearing forces around the center position $y = 0$. The limit can be found from (32) as

$$y = \pm \left(\frac{d}{2} + \frac{T_{m\varphi}}{T_{mx}} \right). \tag{33}$$

Within this region only one bearing force can be reduced, while the other one increases.

A. Influence of limits

The current Δi that is required to reduce the bearing forces scales linear with the eccentricity. Thus, the maximum current density (here equal to the rated current) defines for a required force F_y, the maximum force reduction. Figure 8 shows the required current in relation to the rated current i_n to achieve the force reduction of Fig. 7 with an eccentricity due to wear of 0.1mm as thick line. The peak amount of roughly 40% of the rated current is considered rather high. As an example the dashed line in Fig. 8 shows a current restricted to only 5% of the rated current, what leaves 95% of the rated current for the actuation in y-direction. Even with this restriction the bearing forces can considerably be reduced as shown in Fig. 7 with dashed lines. The terms F_{b10} and F_{b20} are the bearing forces if no compensation is applied.
To avoid any increase or overshoot of bearing forces in the area around the poorly controllable center position, the current Δi has to be set to zero in the interval given in (33) as shown in Fig. 9. This results in a force reduction given in Fig. 10.

VII. Conclusion and Outlook

In this paper the feasibility of compensating the bearing forces resulting from the PM attractive forces of a linear actuator are investigated based on analytical and numerical models. Though the bearing forces are canceled for an ideally centered position of the mover, wear and an eccentricity will occur by the time. Thus, the possibility is examined to reduce these emerging forces, to reduce the wear and increase the life time of the bearing ring.
The FE analysis shows that the actuator can be linearized with respect to current and magnetic stiffness. The linearized model has been derived and shows that at least four star connected coils are required to cancel both bearing forces for all y positions completely. For the actuator with two phases the forces can no longer be canceled, but a significant reduction can be achieved. The required current and the limits of operation are derived analytically. This results in an approximation of the expected reduction of bearing forces.
The results show the feasibility of a bearing force reduction even for an actuator featuring only two coils.
Currently, a prototype of the system is built to allow an experimental validation of the presented results.

Acknowledgment

This work has been carried out at LCM GmbH as part of a project funded by the Austrian COMET K2 program. The program is supported by the Austrian federal government, the federal state of Upper Austria, the Johannes Kepler University and all of the scientific partners which are part of the COMET K2 Consortium.

References

[1] J. J. H. Paulides, J. L. G. Janssen, and E. Lomonova, "Bearing lifetime of linear pm machines," in *Energy Conversion Congress and Exposition, 2009. ECCE 2009. IEEE*, 2009, pp. 1083–1090.

[2] Woo-Yeon Kim, Jong-Min Lee, and Seung-Jong Kim, "Rolling motion control of a levitated mover in a permanent-magnet-type bearingless linear motor," *Magnetics, IEEE Transactions on*, vol. 46, no. 6, pp. 2482–2485, 2010.

[3] Nan-Chyuan Tsai and Chao-Wen Chiang, "Design and analysis of magnetically-drive actuator applied for linear compressor," *Mechatronics*, vol. 20, no. 5, pp. 596–603, 2010.

[4] ——, "High-frequency linear compressor and lateral position regulation," *Control Systems Technology, IEEE Transactions on*, vol. 20, no. 1, pp. 127–138, 2012.

[5] S. Silber and W. Amrhein, "Force and torque model for bearingless pm motors," *Proceedings of the International Power Electronics Conference (IPEC), Vol. 1, Tokyo*, Apr. 2000.

[6] ——, "Power optimal current control scheme for bearingless pm motors," *7th International Symposium on Magnetic Bearings (ISMB), Zurich, Switzerland*, pp. 401–406, Aug. 2000.

Design of Electromagnetic Levitation Linear Bearing for FPD Glass Delivery Applications

Ki-Chang Lee[a], Jun-Young Baek[b], Ji-Won Kim[a], Min-Hyuk Ahn[a], Tae-Uk Jung[c] and Min-Cheol Lee[b]

[a] Korea Electrotechnology Research Institute, 12 Bulmosan-ro 10beon-gil, Sungsan-gu, Changwon-si, Gyeongsangnam-do, 642-120, Korea, leekc@keri.re.kr

[b] School of Mechanical Engineering, Pusan National University, 2, Busandaehak-ro 63beon-gil, Geumjeong-gu, Busan 609-735, Korea, mclee@pusan.ac.kr

[c] Department of Electrical Engineering, Kyungnam University, 7, Kyungnamdaehak-ro, Masanhappo-gu, Changwon-si, Gyeongsangnam-do, 631-701, Korea, tujung@kyungnam.ac.kr

***Abstract* — In flat panel display (FPD) manufacturing, especially in next-generation organic light-emitting diode (OLED) display manufacturing, a high-weighted glass carrying tray is required to be position servo-controlled precisely in long range of vacuum process line without mechanical contact. So special contactless linear bearing, which is called a electromagnetic levitation linear bearing (ELLB) is suggested in this paper. It can expend its operating distance by adding series of electro-magnets and shifting the electromagnets by section switching algorithms suggested in reference [12]. In order to prove the feasibility of the ELLB, a test setup with specially designed ELLB is made-up and the performance of it is observed. With the proposed ELLBs and proper linear propulsion system, the glass-carrier system in vacuum FPD manufacturing process line is expected to be constructed easily in the near future.**

I. Introduction

The electromagnetic linear bearings (ELB) have been a essential elements in modern industries because of its contactless property. The ELB can be classified by several types based on its operating principles – eddy-current bearing, repulsive linear synchronous motor bearing, electromagnetic suspensions (EMS) and so on. The eddy-current bearings were widely studied in the earlier high speed railways. For example, the Inductrack was a passive, fail-safe electro-dynamic magnetic levitation system, which uses only unpowered loops of wire in the track and permanent magnets on the vehicle to achieve magnetic levitation [1]. The repulsive coreless (air-cored) linear synchronous motor propulsions have been the most popular precision levitation system in semiconductor manufacturing applications where the precision servo-control is most important for the light-weighted tray [2] ~ [6]. The electromagnetic suspension (EMS) can have high load capability and good levitation gap control performance. It is defined as the magnetic levitation of an object achieved by constantly alternating the strength of a magnetic field produced by electromagnets using a feedback loop [7]. With proper design of actuator and controller, the EMS can achieve nanometer position accuracy for high-weighted tray [8], [9].

Meanwhile, in recent manufacturing automation industries, especially in flat panel display (FPD) manufacturing such as organic light-emitting diode (OLED) display in–line fabrication process, precise contactless linear bearings with high load capabilities are required. In conventional semiconductor manufacturing applications, only high precision contactless linear bearings were required, but in upcoming FPD manufacturing applications, heavy-weighted tray is required to move precisely and seamlessly in entire all the process line. So the combination of electromagnetic linear bearing and linear propulsion system is required in the applications [10].

As new type of linear bearing for FPD glass delivery applications, an electromagnetic levitation linear bearing (ELLB) is suggested in this paper. This new bearing is based on the system shown in [9] in structure. The differences are a series of electromagnets fixed on the rail, and these electromagnets can be extended as long as the tray moves. The cut-away view of the suggested transportation system of the FPD glass-carrying tray is shown in Figure 1. In the Figure, 2 ELLBs and one linear active magnetic bearing (LAMB) are used in order to support the tray in five degrees of freedom, and longitudinal position of the tray is controlled by additional linear motor. Because the weight of tray is over hundreds kilograms, the high load capability of the ELLB is essential.

The description of the paper is as follows: In section II, the design strategy of the ELLB is shown and feedback controller design of the ELLB is shown in section III. Experimental setups and results are shown in section IV and V respectively. And final conclusions are shown in section V.

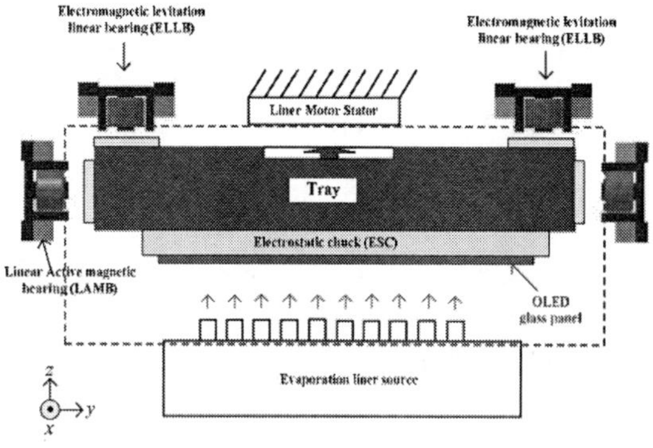

Figure 1. Transportation system of the FPD glass carring tray.

II. Design of a ELLB

A. Design of the ELLB

The ELLB is composed of a series of electromagnets on the rail, which build up attraction forces on the core in the tray against the gravity. A simple transport system of tray with one ELLB is shown in Figure 2, where 5 adjacent electromagnets among a series of electromagnets have a role in attracting the tray against the gravity at the same time. So the number of active electromagnets becomes five. If we can properly define the "Section" as some area among periodic displacement of the electromagnets and the tray can pass through this Section without losing the gap and pitch control performance, it means that long stroke operation of the ELLB is possible. In the Figure, the k-th section is defined as the area occupied by most left one among active five electromagnets. Because the length of tray is the same as distance of 5 electromagnets, the center of mass point of the tray exists above the $(k+2)$-th electromagnet. – Figure 2 shows the situation when the Section, $k = 0$. In the Figure, the vertical gap and pitch angle of the tray should be maintained as constant values by the ELLB, while the longitudinal x position of the tray is servo-controlled by additional linear motor which is not shown in the figure. Other degrees of freedom such as lateral moment, roll and yaw of the tray are neglected for simplicity. So the ELLB has two degree of freedom. A series of gap sensors can be fixed on the rail to calculate the average air-gap at the center of mass position (CMP) and pitch angle of the tray. The average gap and pitch angle are controlled as constant values by controlling attraction forces between core in the tray and electromagnets on the rail. Figure 2 also shows the attraction forces built by 5 active electromagnets on the rail, when the tray moves.

From the geometry, the resultant force f_{net} and the moment M_{net}, which should be built by the control current of the electromagnets, can be calculated as follows:

$$f_{net} = f_k + f_{k+1} + f_{k+2} + f_{k+3} + f_{k+4} \tag{1}$$

$$\begin{aligned} M_{net} &= (2P + \Delta x) f_k + (P + \Delta x) f_{k+1} + (\Delta x) f_{k+2} \\ &\quad - (P - \Delta x) f_{k+3} - (2P - \Delta x) f_{k+4} \end{aligned} \tag{2}$$

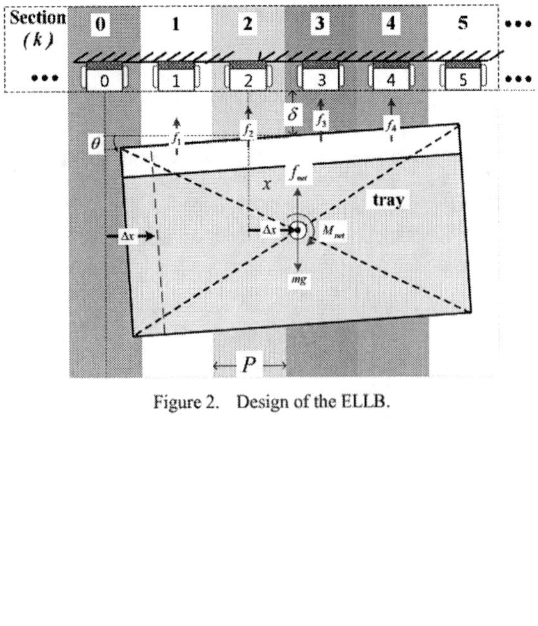

Figure 2. Design of the ELLB.

B. Dynamic model of the tray supported by the ELLB

If longitudinal x axis degree of freedom of the tray is neglected, the dynamic equation of the tray levitated by the electromagnets on the rail becomes as follows where m is the tray's mass and J_{yy} is moment of inertia of the tray for pitch direction and P is interval of electromagnet placement.

$$m\ddot{\delta} = mg - f_{net} \tag{3}$$

$$J_{yy}\ddot{\theta} = M_{net} \tag{4}$$

C. Design of electromagnet and stiffness of the ELLB

If the air-gap between k-th electromagnet and core in the tray is δ_k, the attraction force exerted by the electromagnet which carries control current i_k is as follows:

$$f_k = K_{mag} \frac{i_k^{\,2}}{\delta_k^{\,2}} \tag{5}$$

If the center of mass point of the levitated tray is exactly under (k+2)-th electromagnet and pitch angle of the tray is zero, only three electromagnets among five section magnet, which are (k+1)-th, (k+2)-th, and (k+3)-th electromagnets, levitate the tray with the same control current. At this moment, each electromagnet can be thought as a mechanical spring with open-loop position stiffness of $k_s = 2K_{mag} i_k / \delta_k^{\,3}$ as shown in Figure 3. So linearized open loop bearing stiffness of the ELLB becomes follows:

$$k_{eq} = 3k_s = 6K_{mag} \frac{i_k}{\delta_k^{\,3}} \tag{6}$$

Normally the closed loop bearing stiffness can be designed the same order as the open-loop position stiffness in the design of controller for the levitation control system [11].

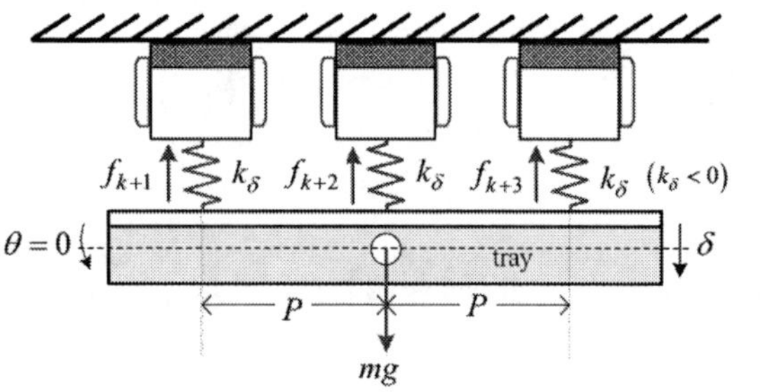

Figure 3. Bearing stiffness model for the ELLB.

III. Closed Loop Control of the ELLB

A. Overall controller design for the ELLB

The overall control block diagram of the ELLB is shown in Figure 4, and parameters for the ELLB and controllers are shown in Table I and Table II respectively. The reference gap δ is 0.7 millimeter and pitch angle is zero for the 18 kilogram weighted tray. A conventional proportional integral and derivative controller and a special section switching algorithm (SSA) is designed, simulated and tested for control of the tray. The precise description about plant modeling, controller design, and simulation results are shown in reference [12].

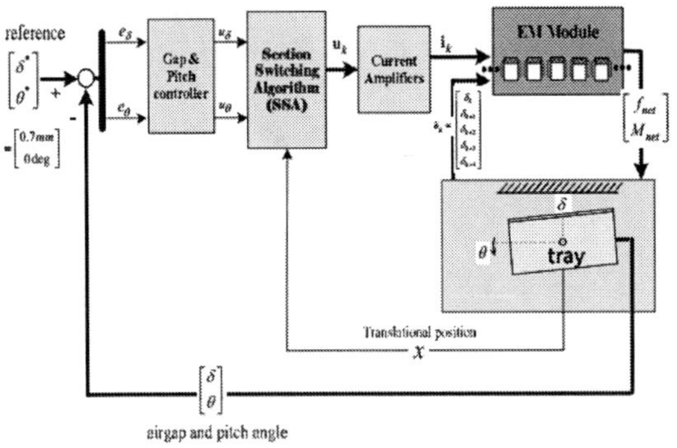

Figure 4. Block diagram of controller for the proposed ELLB.

TABLE I. Control parameters for the ELLB

Description		Abb.	Value	Unit
Gap control	Mass	k_P	23	kg
	Moment	ω_D	87	kgm^2
	length	$k_{I\delta}$	15	m
Picth control	Mass	k_P	18	kg
	Moment	ω_D	1.16	kgm^2
	length	$k_{I\delta}$	20	m

TABLE II. Parameters of ELLB and electromagnets

Description		Abb.	Value	Unit
Pitch of the EMs		P	0.09	m
Electromagnet constant		K_{mag}	5.27×10^{-6}	Nm^2/A^2
Nominal gap		δ_0	0.7	mm
Tray	Mass	m	18	kg
	Moment	J_{yy}	0.2225	kgm^2
	length		0.45	m
	height		0.25	m

IV. Experimental Setups

In order to verify the feasibility of the suggested ELLB, a simple test-setup with the closed loop control of the ELLB is made as shown in Figure 5. A ball-screw linear motor is used to control the longitudinal x displacement of the tray. Total 6 numbers of electromagnets are placed with P meter intervals in order to build up levitation forces and two gap-sensors are located between magnets with $2P$ interval on the rail in order to measure gaps and attitude of the tray. For the control of the ELLB, two proportional-integral-derivative (PID) levitation controllers for the pitch and gap control of the tray and a special SSA for successful Section changes are simulated in Matlab/Simulink software package and implemented with the Real-Time Workshop toolbox from MathWorks Inc. The final control algorithms are implemented with a commercial real-time controller, the DS1104 Power-PC control board. Six numbers of current controllers which are composed of liner power OP-Amps are used to apply currents to each electro-magnet as the control commands.

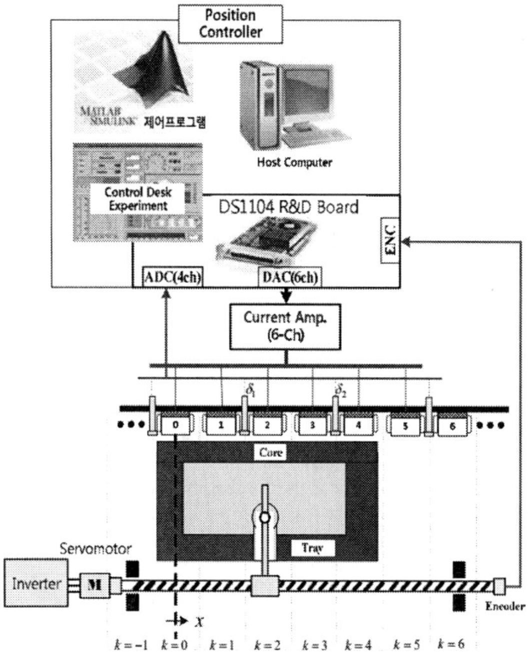

Figure 5. Moving tray model with the proposed ELLB.

The experiments are performed as follows: When the closed loop control of the ELLB is working, the tray' CMP is moved from 2nd magnet position (Section $k=0$) to the 4th magnet position (Section $k=2$) with $2P$ intervals. This means that the Section is changed twice from 0 to 2. Therefore a long distance operation of the ELLB is possible by extension of Sections and switching control of the Sections. This is implemented by just adding another electromagnets and transition control of the Sections.

V. Experimental Results

The experimental result with the test setup is shown in Figure 6 ~ 8. The center gap δ and pitch angle θ of the tray is calculated from the real gap sensor signals, δ_1 and δ_2. When the levitation and pitch control of the tray is working, the tray is moved forward with constant speed of 5 millimeter per second by the ball-screw linear motor driver.

Figure 6 shows the longitudinal x displacement of the tray, and Section and Section deviations as the tray moves. The control commands of the levitation and pitch controllers also shown in the Figure.

Figure 7 shows the current control commands which is calculated from the SSA. It can be inferred that because the tray's parameters are the same at every moment, the sum of the control current should be the same. But small deviations in control current of the electromagnets was monitored when switching of the electromagnets are occurred.

Figure 8 shows the performance of the closed-loop control of the tray. It is also monitored that the center gap δ and pitch θ of the tray is deviated when switching of the electromagnets are occurred. This can be also monitored from the sensor signals δ_1 and δ_2 . These deviations are due to control command changes to each electromagnet when Section changes occur. More observations and precise controls should be done in order to reduce these deviations in near future.

VI. Conclusion

New type of magnetic levitation, which can support heavy-weighted moving tray in the long distance rail very precisely with the help of periodic placement of EMs on a rail, is proposed. For the proof of feasibility of the proposed ELLB and SSA, test setup of a moving tray with one ELLB is made up. With the help of a special SSA suggested in reference [12] and proper conventional levitation control techniques, the suggested linear bearing can support a 18 kg weighted, 450 mm long test tray with gap control performance under ±40 μm in the experiment. This new contactless linear bearing could be very useful for next-generation OLED in-line fabrication process, where huge heavy tray is required to move precisely in vacuum environment.

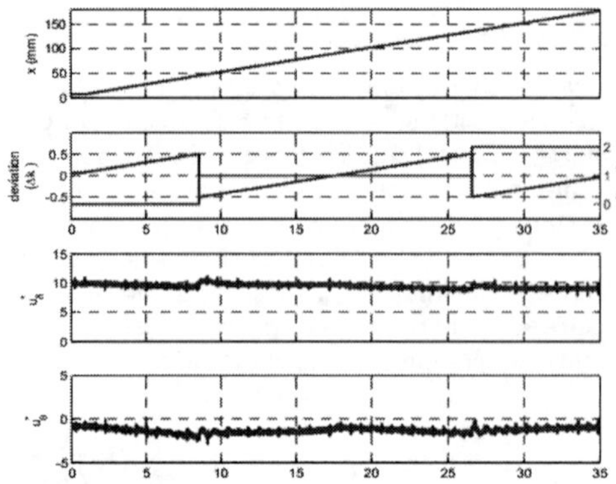

Figure 6. Control commands and Sections.

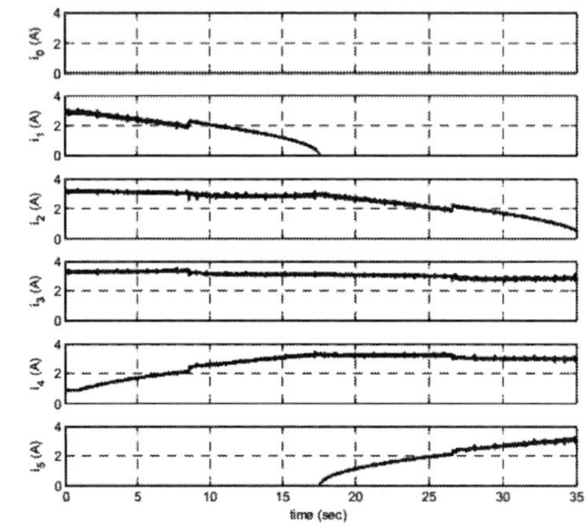

Figure 7. Control currents of the electromagnets.

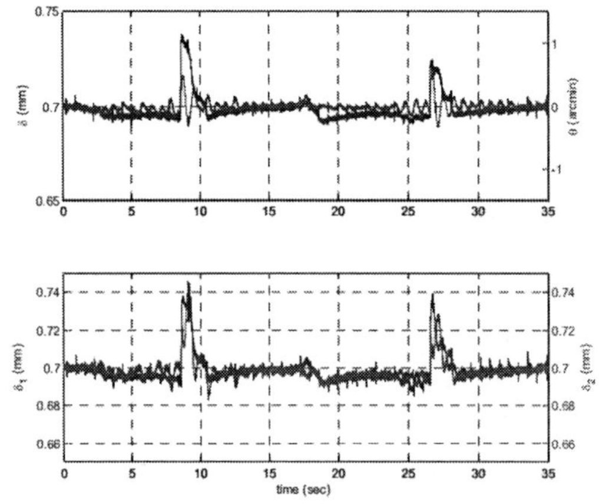

Figure 8. Performance of the ELLB

REFERENCES

[1] Jorgen L. Nikolajsen, "Experimental verification of an eddy-current bearing", http://ntrs.nasa.gov/archive/nasa/casi.ntrs.nasa.gov/19890013542.pdf

[2] W.J. Kim, D.L. Trumper, J.H. Lang, "Modeling and Vector Control of Planar Magnetic Levitator," *IEEE Trans. On Industry Applications*, Vol. 34, No. 6, pp. 1254-1262, Nov.dec. 1998.

[3] C.M.M van Lierop, J. W. Jansen, A. A H Damen, E.A. Lomonova, P. P J Van den Bosch, and A. J A Vandenput, "Model-Based Commutation of a Long-Stroke Magnetically Levitated Linear Actuator," Industry Applications, IEEE Transactions on , vol. 45, No. 6, pp. 1982 ? 1990, Nov.-dec. 2009.

[4] Jeroen de Boeij, E. Lomonova, and Jorge Duarte, "Contactless Planar Actuator With Manipulator: A Motion System Without Cables and Physical Contact Between the Mover and the Fixed World," Industry Applications, IEEE Transactions on, vol.45, no.6, pp.1930-1938, Nov.-dec. 2009. doi: 10.1109/TIA.2009.2031858

[5] J. W. Jansen, C. M M Van Lierop, , E. A. Lomonova, and A. J A Vandenput, "Modeling of Magnetically Levitated Planar Actuators With Moving Magnets," Magnetics, IEEE Transactions on, vol.43, no.1, pp.15-25, Jan. 2007. doi: 10.1109/TMAG.2006.886051

[6] Zhu Yu, Shengguo Zhang, Haihua Mu, Kaiming Yang, and Wensheng Yin, "Augmentation of Propulsion Based on Coil Array Commutation for Magnetically Levitated Stage," Magnetics, IEEE Transactions on , vol.48, no.1, pp.31-37, Jan. 2012. doi: 10.1109/TMAG.2011.2166559

[7] http://en.wikipedia.org/wiki/Electromagnetic_suspension

[8] Trumper, D.L., and Slocum, A. H., "Five-Degree-of-Freedom Control of an Ultra-Precision Magnetically-Suspended Linear Bearing," NASA Workshop on Aerospace Applications of Magnetic Suspension Technology, NASA Langley Research Center, Hampton, VA, Sept. 25-27, 1990.

[9] D. L. Trumper and M.A. Queen, "Precision Magnetic Suspension Linear Bearing", Proceedings of ISMST1, 1991.

[10] Seok-Rak Chang et al. "Apparatus for organic layer deposition and method for manufacturing of organic light emitting display apparatus using the same," US patent 20130009177 10 January 2013.

[11] G. Schweitzer, "Active magnetic bearings - chances and limitations, Proc. 6th Internat. IFToMM Conf. on Rotor Dynamics, Sydney, Sept. 30-Oct. 3, 2002.

[12] Ki-Chang Lee, Seokhwan Moon, Hyunuk Ha, Byoung-Gun Park, Ji-won Kim, Jun-Young Baek, Min-Cheol Lee, "A Novel High Precision Electromagnetic Suspension for Long-Stroke Movement and Its Performance Evaluation", Journal of Electrical Engineering & Technology, vol. 9, no. 2, pp. 514-522. March 2014.

[13] Kichang Lee, "Section Control of the Electromagnetic Levitation Conveyor(EMLC) for a FPD Glass carrying Robot", Ph.D. thesis, Pusan National University, Busan, Korea, 2014. (in Korean)

An Eddy Current Damper for Reaction Force Compensation for a Linear Motor Motion Stage

Hyeong Joon Ahn[a], Yo Han You[a]

[a] Soongsil University, Sangdo-dong, Dongjak-ku, Seoul, Korea, ahj123@ssu.ac.kr

***Abstract*—This paper presents an eddy current damper (ECD) for reaction force compensation for a linear motor motion stage. A reaction force compensation mechanism with a movable magnet track and an ECD (or conductor plate) is developed. Lumped model of the ECD is derived considering sinusoidal magnetic flux density distribution and effective width of the ECD. In addition, effects of both the location and width of the conductor plate are discussed. Multi-physical transient analysis of ECD is analyzed using commercial finite element software, which is compared with the lumped parameter analysis. Finally, the derived lumped model is validated with experiments.**

I. INDTRODUCTION

Vibration of the machine base due to reaction force during rapid acceleration or deceleration motion is one of main causes for reducing life and productivity of the system. A large force during accelerating a payload generates an equal opposite reaction force to the system base. This reaction force causes the system to oscillate with unacceptable amplitude or in long settling time, which has a bad influence on production quality of manufacturing process and vibration sensitive device [1]. In addition, the residual vibration of the system base degrades the system productivity since the system cannot precede the next process without attenuating the vibration.

There are two main reaction force compensation (RFC) methodologies to reduce the residual vibration of the system [2-4]. One is to use an isolated mechanical structure for an independent transfer path of the reaction force [2]. The other is using a dummy motion stage to generate counter force and to cancel out the reaction force [3, 4].

A reaction force compensation mechanism with a movable magnet track and an ECD (or conductor plate) is developed. This mechanism relaxes disadvantages of the spring based reaction force compensation mechanism such as resonance and difficulty of assembly and design due to spring [5]. In addition, design conditions such as force transmission ratio and magnet track displacement can be satisfied by adjusting the shape and location of the conductor plate [6].

II. REACTION FORCE COMPENSATION MECHANISM BASED ON ECD

A. Principle

Reaction force compensation systems of moving magnet type based on spring and ECD are shown in Fig. 1. Both reaction force compensation systems consists of movable magnet track with additional mass and restrictor of magnet track displacement (either spring or ECD). In particular, ECD or conductor plate at both end of magnet track generates easily damping force for the magnet track. Desired performance can be achieved by adjusting additional mass and spring or ECD.

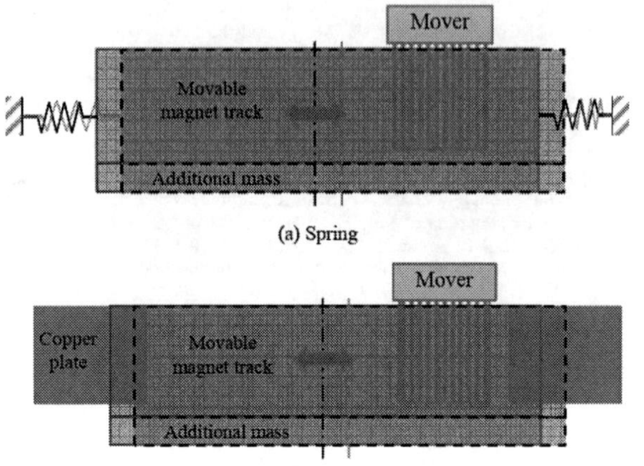

(a) Spring

(b) ECD

Figure 1. Passove reaction force compensation mechanisms

Reaction force compensation device based on ECD has several advantages compared with that based on spring. Although ECD has nonlinear characteristic related to the displacement of magnet track, reaction force system with ECD has no resonance (due to removing spring), high DOF on design, low cost for assembly and manufacturing. However, design and analysis of the ECD is difficult and there may be a homing problem of magnet track due to friction.

B. Modeling

Both reaction force compensation system can be modeled as Eq. (1) by generalizing the restricted force of either spring or ECD for magnet track as F_r. Magnet track (m_{MT}) with additional mass (m_D) can move (x_{MT}) freely with respect to the machine base (m_{Base}) and either spring or ECD generates restricted force ($f_R(x_{MT}, \dot{x}_{MT})$) for the movement of magnet track. While the restricted force of spring for magnet track can simply be expressed with f_R=$k_S\, x_{MT}$ in case of spring, the restricted force of ECD for magnet track has complex form and will be provided later in detail. Part of reaction force or restricted force is transferred to the machine base through the spring or ECD, as shown in Eq. (1).

$$(m_{MT}+m_D)\dot{x}_{MT}+f_R(x_{MT},\dot{x}_{MT})=f_T(t) \tag{1}$$

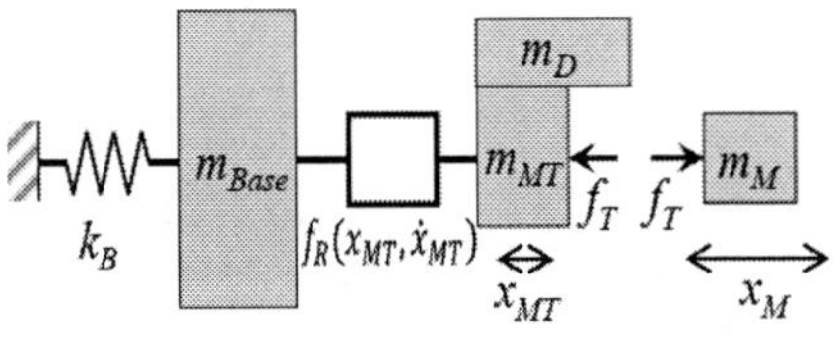

Figure 2. Dynamic model for reaction force compensation mechanism of moving magnet type

C. ECD

Damper force for moving conductor plate in the constant magnetic field can be expressed with Eq. (2) and is proportional to geometric coefficient (α), square of magnetic flux density (B), conductivity (σ), thickness (t), height (l), width (w) and velocity of the conductor (v).

$$F_d=\alpha\sigma t B^2 l w v \tag{2}$$

Effective with of ECD changes according to movement of magnet track, width and location of the conductor plate, as shown in Fig. 1 (b). If the conductor plate is completely inside of magnet track, the effective width is same as the width of conductor plate irrespective of the displacement of magnet track. However, the effective width of the ECD depends on the displacement of the magnet track if a part of conductor plate is not inside of the magnet track.

Flux density is sinusoidal along motor moving direction. Alternating arrangement of south and north poles produces sinusoidal flux density, which generates thrust force with three phase current of coil.

We can calculate ECD force over velocity considering its effective width and sinusoidal magnetic flux density using Eq. (3). The final expression may be different according to displacement of the magnet track. The ECD force over velocity according to the magnet track position can be calculated as shown in Fig. 4.

$$F=\alpha\sigma t l B^2\int \sin^2\left(\frac{\pi}{p}x\right)dx \tag{3}$$

III. LUMPED ANALYSIS OF REACTION FORCE COMPENSATION MECHANISM BASED ON ECD

A. Simulation model.

Simulink model for simulation is shown in Fig. 4. Thrust force is calculated considering mover acceleration and guide friction. Then, response of magnet track is calculated with Eq. (1) considering mathematical model of ECD (Eq. (3)) and effective width of ECD, as shown Fig. 4. The linear motor used for simulation has continuous force of 208N and max. velocity of 5 m/s. The sinusoidal variation of the magnetic flux density of magnet track along moving direction is neglected and the geometric coefficient for ECD is determined considering the rectangular shape of magnet [7]. The motion file has each moving distance of 100mm, max. velocity of 500mm/s and max. acceleration of 10m/s^2.

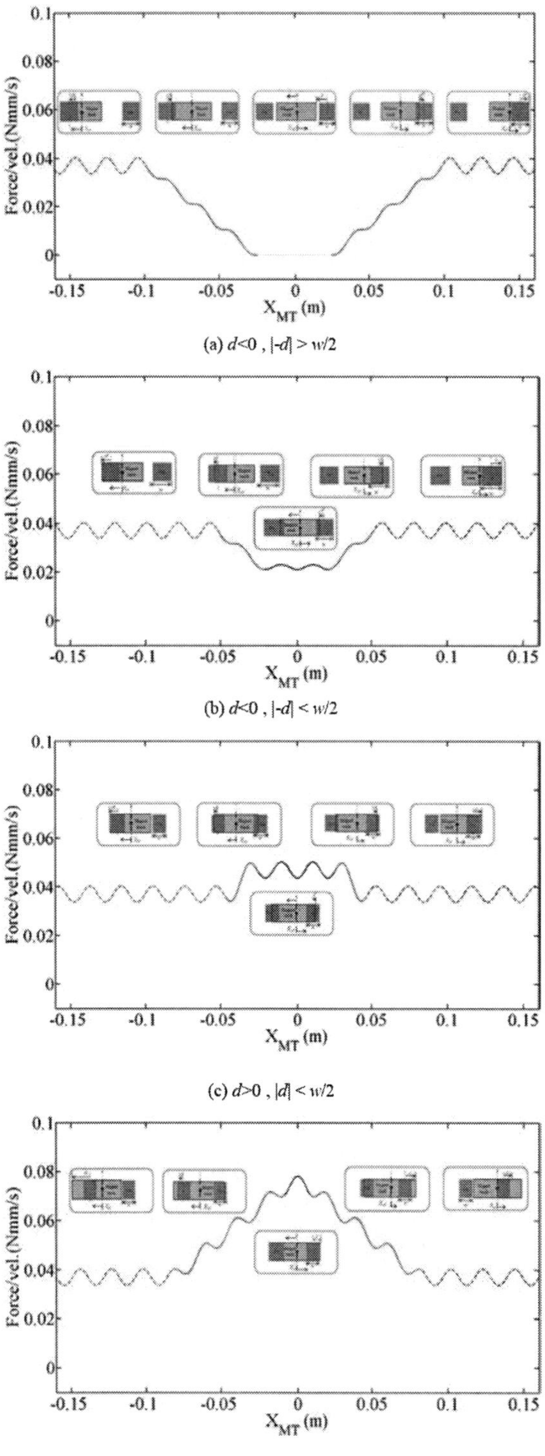

Figure 3. Eddy current force variation according to the magnet track position

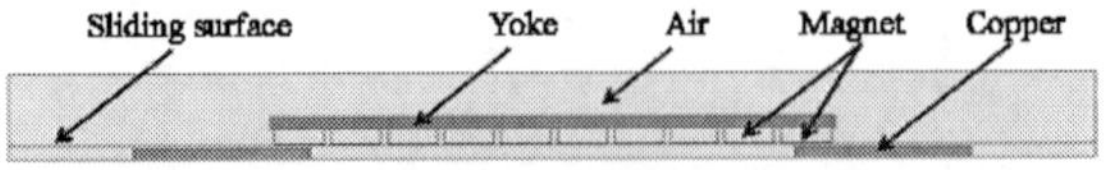

(a) FE model

(b) Magnetic flux density

(c) Current density

Figure 4. FEM analysis of the ECD with moving magnet track

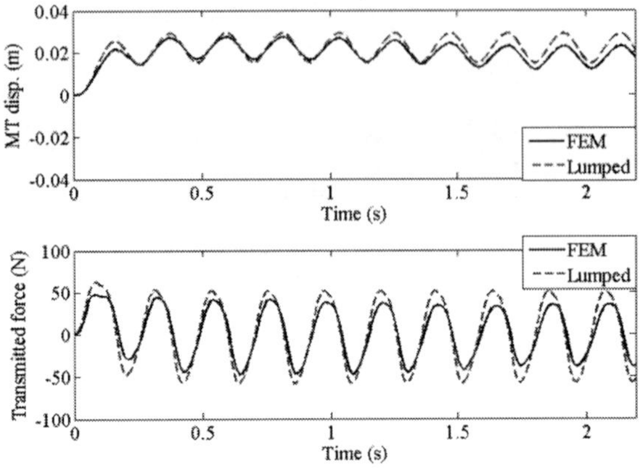

Figure 5. Comparison of Lumped and FEM analyses

B. *Multi-physical analysis of reaction force compensation system*

Finite element model and mesh generation for reaction force compensation system with ECD is shown in Fig. 4 (a). Only half of reaction force compensation system is modeled, which consists of copper plate, permanent magnet, yoke and air as shown in Fig. 4(a). Movement of magnet track is modeled using deformable mesh is used and sliding surface is set as identity pair for fixed and movable geometries. First, static analysis is performed for initial values of transient analysis. Then, magnetic field and motion equation of Eq. (1) are analyzed simultaneously with COMSOL [8]. Static magnetic field analysis is performed at initial position. Then, time-dependent coupled analysis of magnetic field and motion equation of magnet track are performed. Fig. 4 (b) and (c) show magnetic potential distribution and current density of copper plate at 2 s, respectively. The eddy current is induced according to the movement of the magnet track, which produces eddy current damping force interacting with magnetic field.

Magnet track displacement and transmitted force of both lumped and FEM analyses are compared in Fig. 5. The lumped analysis matches quite well with FEM analysis.

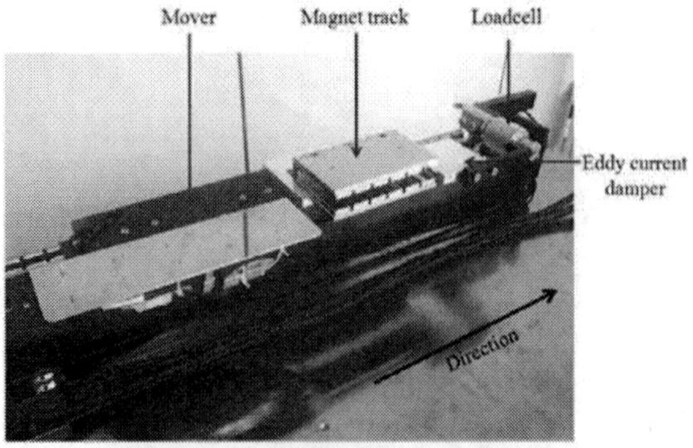

Figure 6. Experimental set-up

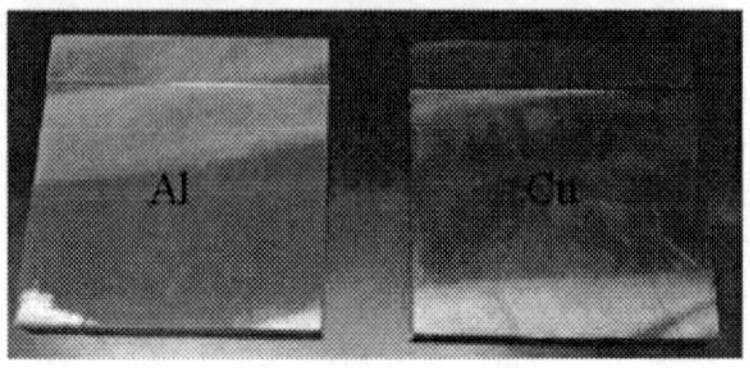

Figure 7. Al and Cu plate for ECD

IV. EXPERIMENTAL VALIDATION

A. *Experimental set-up.*

We built an experimental set-up with a linear motor motion stage as shown in Fig. 6 and its specifications are shown in Table I. Experimental set-up consists of a linear motor stage, load cell, magnet track and ECD plate (Al or Cu plate of Fig. 7). The small magnet track is attached on the mover and the ECD plate is assembled with the load cell. The load cell measures eddy current damper force during mover motion.

TABLE I. SPECIFICATION OF EXPERIMENTAL SET-UP

Product	Name		Features or unit	Value
Umac (controller)	Acc-24E2	PWM command output /digital input	phase voltage commands channel	2
			Encoder channel	2
	Acc-51E	Sinoidal encoders	times	4096
			Input Voltage	1
			Input channel	4
	Acc-36E	ADC	Output voltage	±10
			resolution(bit)	12
JSME-02D (Servo Amp)	Peak Output Current		Arms	15
	Input Frequency		Hz	50/60
	Phase Requirement		Phase	1 or 3
	Maximum PWM Frequency		kHz	15
	Minimum Dead Time		μs	3.5
	resolution		bit	12
Loadcell	Rated output		mV/ V	2.0 ± 0.005
	Repeatability		%	0.01
	Max. capacity		kgf	50

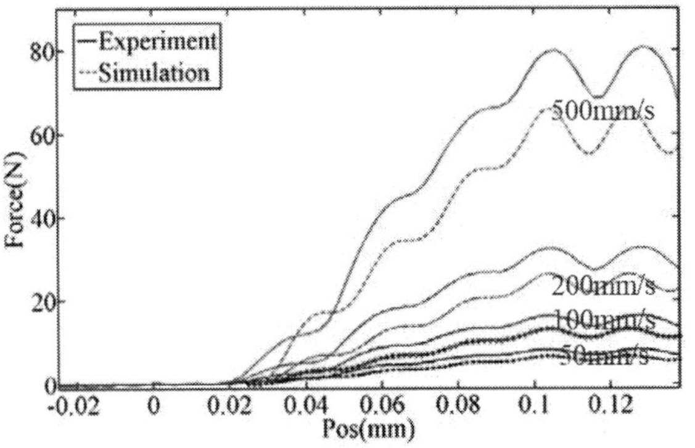

(a) Al plate (9 mm thickness)

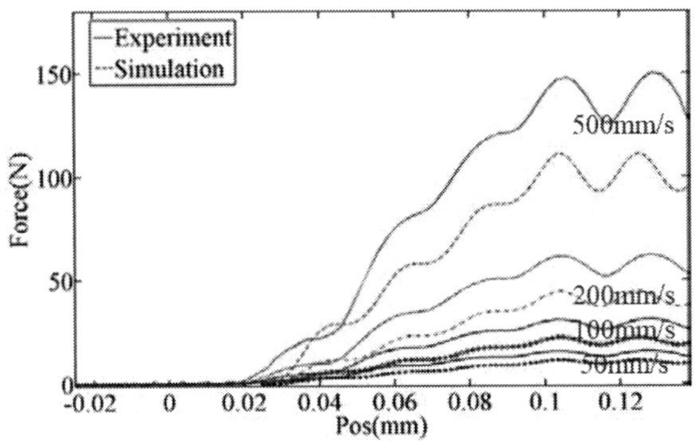

(b) Cu plate (9 mm thickness)

Figure 8. Comparison of meausred and simulated ECD force

B. *Experimental results.*

Experimental and simulation results are compared in Fig. 8. We generate mover motion with different speed (50, 100, 200 and 500 mm/s) and measure the ECD force of Al or Cu plate. Fringe effect makes large error at the beginning of the magnet track. Although there are mismatches due to ignoring harmonic components of the magnetic flux density, simulation results matches well with the experimental results. In particular, experimental results of the Cu plate has substantial deviation from the simulation results since

V. Conclusion

This paper presents an ECD for reaction force compensation for a linear motor motion stage. Lumped model of the ECD is derived considering sinusoidal magnetic flux density and effective width of ECD. In addition, effects of both the location and width of the conductor plate are discussed. Multi-physical transient analysis of ECD is analyzed using commercial finite element software, which is compared with the lumped parameter analysis. Finally, the derived lumped model is validated with experiments.

Acknoledgement

This research was supported by the MSIP (Ministry of Science, ICT & Future Planning), Korea, under the Convergence - ITRC (Convergence Information Technology Research Center) support program (NIPA-2014-H0401-14-1005) supervised by the NIPA and by Basic Science Research Program through the National Research Foundation of Korea (NRF) funded by the Ministry of Education, Science and Technology (NRF-2013R1A1A2010764).

References

[1] Jang, J. W., Park, S. W. and Hong, S. W., 2008, "Command Generation Method for High-speed and Precise Positioning of Positioning Stage," J. of KSPE, 25(10), pp. 122-129.

[2] Greene, P. M., Hero, S., Bitter, D., 2005, "Reaction Force Transfer System," United States Patent, US006844635B2.

[3] Galburt, D. N., 2004, "Method, System, and Apparatus for Management of Reaction Loads in a Lithography System," United States Patent, US006784978B2.

[4] Ka, A., Poon. T., Wai, L., Kho, F., Yang, P.-H., Chang, P.-W., 2005, "Modular Stage with Reaction Force Cancellation," United States Patent, US006917412B2.

[5] Cho, J.K., Choi, D.S., and Ahn, H.J, 2011, "Mechanism and control of a reaction force compensation of XY linear motion stage system," J. of KSME A, 35(6), pp. 599-607

[6] You, Y.H., Ahn, H.J., 2014, "A Passive Reaction Force Compensation (RFC) Mechanism for a Linear Motor Motion Stage," IJPEM, Vol. 15, No. 5, pp 1-5.

[7] Shiau, J.-K., Ma, D.-M. and Jou, M., 2008, ''Analysis and experiments of eddy current brakes with moving magnets,'' Materials Science Forum, Vols. 575–578, pp. 1299–1304, 2008

[8] COMSOL, Multiphysics Modeling, Finite Element Analysis, and Engineering Simulation Software, www.comsol.com

Design of Novel Bearing-less Electromagnetic Actuator in Valve Operation

František Mach, Pavel Karban, Ivo Doležel

University of West Bohemia, Univerzitní 26, 306 14 Pilsen, Czech Republic, {fmach, karban, idolezel}@kte.zcu.cz

***Abstract*—A new bistable electromagnetic actuator is designed and modelled. The actuator with a magnetically moved plunger represents the main part of a valve intended for control of flow of electrically non-conductive liquids. Presented is the basic concept of the valve together with its mathematical model that is solved numerically by the code Agros2D developed by the authors. The aim of the work is to verify the functionality of the device and find its basic operation characteristics.**

I. Introduction

One of the main problems of existing valves is their structural complexity. In many cases they use various drag-bars for the transfer of force effects and other movable elements, which reduces the safety and reliability of their operation. Even more complicated is the situation when the valve should work in a controllable bistable regime (which is secured, for example, by return springs representing further movable parts). In the food, chemical and other industries with extremely high demands on the cleanliness of operation, effective cleaning of such valves is also an issue of principal importance.

The existing solutions to the problem abound, but most of them are technologically rather complicated [1]–[5], see Fig. 1. The goal of this paper is to design a structurally very simple valve for operation in a bistable regime, verify its functionality and find its basic operation characteristics.

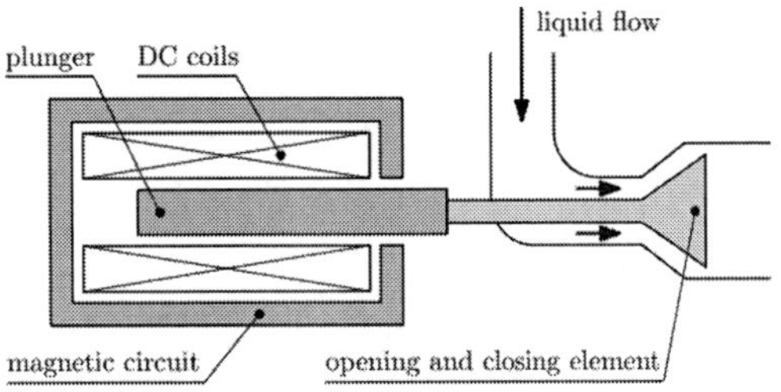

Figure 1. Classical electromagnetic valve system

II. Basic Concept

Figure 2 shows the principal arrangement of the designed valve. The device consists of an electromagnetic actuator, two linear magnetic bearings made of permanent magnets, hollow cylindrical plunger, shell, and element opening or closing the flow of liquid through the valve.

The valve works in two modes, closed and open. Both modes are depicted in Fig. 3. Figure 4 shows the detailed configuration of the system together with the principal dimensions of the most important parts. The permanent magnet bearings help to stabilize the movable plunger (opening and closing elements represent props for the non-magnetic shell in both described modes, open and closed). Thus, they do not support any levitation, but only the stabilisation of the plunger against optional force impacts.

The bistable mode of the valve (the plunger has only one stable position) is secured by the short magnetic circuit with permanent magnet drawn in Fig. 4. The plunger is in the closed position (stable position, see left part in Fig. 3) if the coil is without field current. The magnetic flux produced by DC field current in the long magnetic circuit demagnetizes the adjacent part of the plunger, so that the force caused by the short magnetic circuit is suppressed. The plunger starts moving to the open position (unstable position, see the right part in Fig. 3). The plunger is moved to the stable position again when the field current is switched off.

Advantages of the new concept:

- Bistable regime secured by the permanent magnets.
- The permanent magnets are not demagnetized in the course of the process (local demagnetization of the plunger is used instead), which contributes to the safety and reliability of the valve.
- The valve does not contain any further movable elements such as drag-bars or springs and its cleaning is very easy.
- The valve saves energy due to local demagnetization of the plunger.
- The plunger is stabilized by the magnetic bearing. This measure prevents the valve from leaking at pressure jumps in the connected pipe. It also prevents the fer-

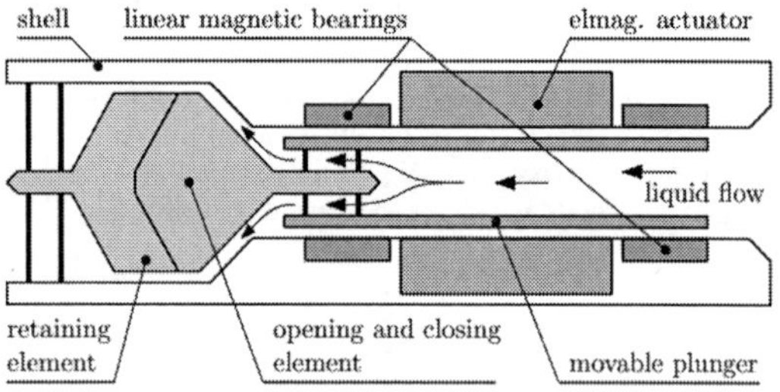

Figure 2. Basic concept of designed valve

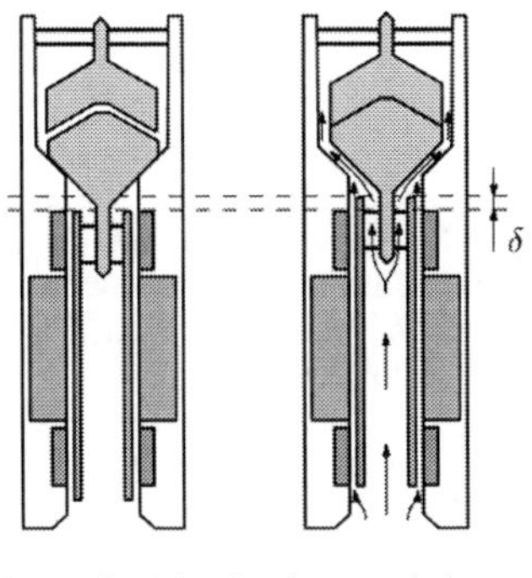

Figure 3. Operation modes (closed and open valve)

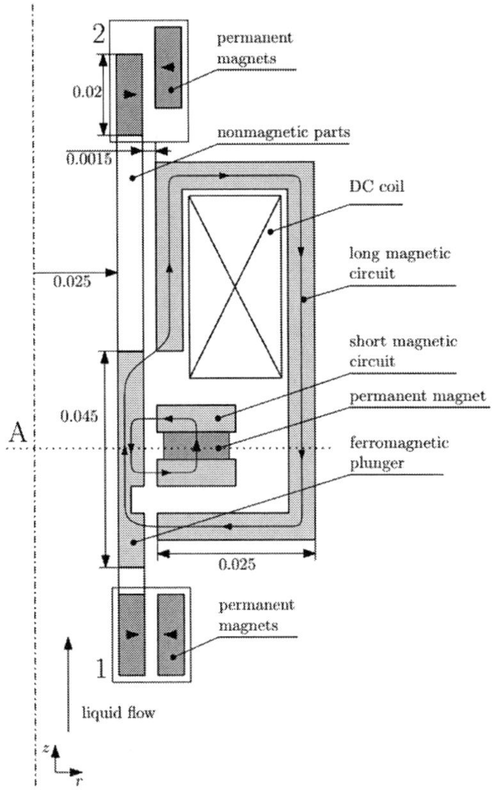

Figure 4. Magnetic circuit of designed valve with main dimensions in m

romagnetic plunger from undesirable approaching to the other magnetic parts.

- The process of opening and closing the valve can be fully controlled by DC field coil current.

III. Mathematical Model

The investigated valve was studied by numerical simulation of magnetic field, incompressible flow field and also dynamics of the movable plunger.

A. Model of magnetic field

The distribution of magnetic field in the system (for any position of the plunger) is described by the equation for magnetic vector potential $\boldsymbol{A}$ in the form

$$\mathrm{curl}\ \left(\mu(\boldsymbol{B})^{-1}\mathrm{curl}\ \boldsymbol{A}-\boldsymbol{H}_{\mathrm{c}}\right)=\boldsymbol{J}_{\mathrm{ext}}\,, \tag{1}$$

where μ denotes the magnetic permeability, $\boldsymbol{B}$ flux density, symbol $\boldsymbol{H}_{\mathrm{c}}$ stands for the coercive force and $\boldsymbol{J}_{\mathrm{ext}}$ denotes the density of the field current in the field coil.

The model is solved as a 2D axi-symmetric problem. The solution area is closed by the Dirichlet boundary condition in the form $\boldsymbol{A}=\boldsymbol{0}$.

Important in the first step of modelling are the static characteristic of the device both when closing and opening the valve, i.e., the dependences of the axial force acting on the plunger on its position in the system. The force can be determined from the actual distribution of the magnetic field and can be expressed using the formula

$$\boldsymbol{F}_{\mathrm{m}}=\oint_{S}\boldsymbol{S}_{\mathrm{m}}\,\mathrm{d}\boldsymbol{S}\,, \tag{2}$$

where $\boldsymbol{S}_{\mathrm{m}}$ is the Maxwell stress tensor and the integration is performed over the complete boundary S of the plunger ($\boldsymbol{S}$ being the outward normal to this boundary).

The Maxwell stress tensor $\boldsymbol{S}_{\mathrm{m}}$ is given by the expression

$$\boldsymbol{S}_{\mathrm{m}}=-\frac{1}{2}(\boldsymbol{H}\cdot\boldsymbol{B})\boldsymbol{I}+\boldsymbol{H}\otimes\boldsymbol{B}\,, \tag{3}$$

where $\boldsymbol{I}$ is the unit diagonal matrix and symbol $\otimes$ denotes the dyadic product.

B. Model of incompressible flow

The incompressible flow in the working area of the valve is described by the Navier-Stokes equation for liquid velocity $\boldsymbol{v}$ in the steady state (partial simplification of the model) and pressure p in the form

$$\rho(\boldsymbol{v}\cdot\mathrm{grad}\ )\boldsymbol{v}=-\mathrm{grad}\ p+\eta\triangle\boldsymbol{v}+\boldsymbol{f}\,, \tag{4}$$

$$\mathrm{div}\ \boldsymbol{v}=0\,, \tag{5}$$

where ρ is the density of the liquid, η represents its dynamic viscosity and finally $\boldsymbol{f}$ denotes the internal forces (for example gravitational force).

C. Dynamics of the plunger

The dynamics of the movable plunger is described by the motion equations

$$m\frac{\mathrm{d}\boldsymbol{v}}{\mathrm{d}t}=\boldsymbol{F}_{\mathrm{m}}+\boldsymbol{F}_{\mathrm{f}}+\boldsymbol{F}_{\mathrm{g}}\,, \tag{6}$$

$$\boldsymbol{v}=\frac{\mathrm{d}\boldsymbol{s}}{\mathrm{d}t}\,, \tag{7}$$

where $\boldsymbol{v}$ is the velocity of the plunger, t is the time, $\boldsymbol{F}_{\mathrm{m}}$ is the force produced by the magnetic field, $\boldsymbol{F}_{\mathrm{f}}$ stands for the pressure and viscous forces, $\boldsymbol{F}_{\mathrm{g}}$ is the gravity force, m denotes the mass of the plunger and, finally, $\boldsymbol{s}$ represents the trajectory of the plunger.

IV. Numerical Analysis

The numerical solution of the problem is realized by a fully adaptive higher-order finite element method whose algorithms are implemented into the codes Agros2D [6] and Hermes [7]. Both codes have been developed in our group for a couple of years. The codes are intended for monolithic numerical solution of systems of generally non-linear and non-stationary second-order partial differential equations and their principal purpose is to model complex physical problems in the hard coupled formulations. Both codes are freely distributable under the GNU General Public License.

Figure 5 shows the convergence of the Newton method for two cases (excited and unexcited DC coil) for the position of plunger $\delta = 0.125\,\mathrm{mm}$. The calculation of the force was realized by the Eggshell method [8].

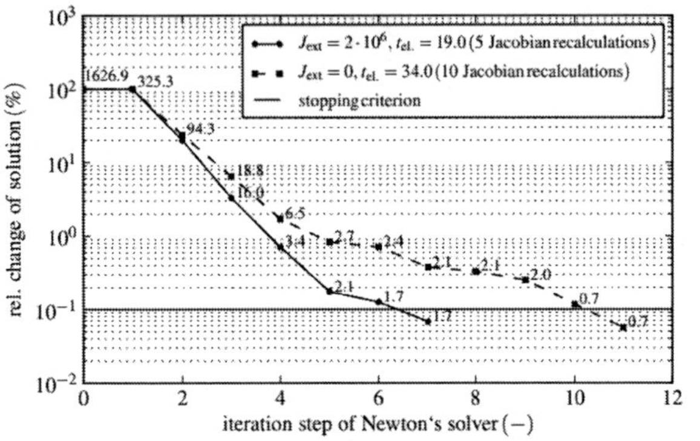

Figure 5. Convergence of the Newton solver (dependence of relative change of solution on number of iteration steps for plunger position $\delta = 0.125\,\mathrm{mm}$). The numbers at point denote residual, $t_{el.}$ is the elapsed time.

Figure 6 shows the convergence of the static characteristic in the regime "opening". The computations are performed for ten uniformly distributed points.

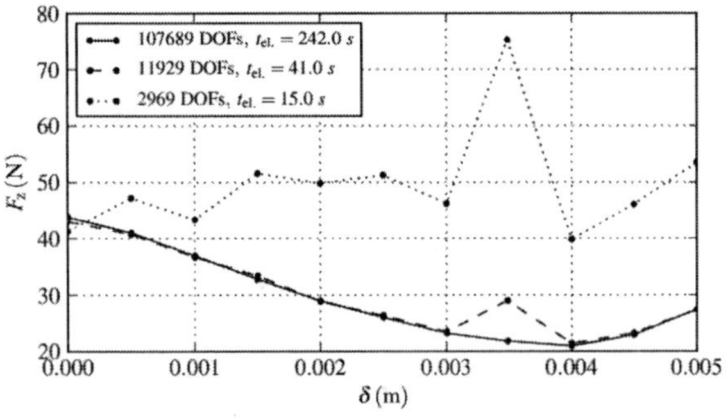

Figure 6. Convergence of static characteristics in regime opening

V. Illustrative Example

A. Input data

The field coil of the actuator contains 600 turns wound by a thin copper conductor. The field current density is $2\cdot10^6\,\mathrm{A\,m^{-2}}$. The ferromagnetic parts of the magnetic circuit are manufactured of iron whose saturation curve is depicted in Fig. 7. The permanent magnets VMM10 in magnetic bearings are of NdFeB type and their manufacturer provides the following parameters: remanence $B_r = 1.45\,\mathrm{T}$, relative permeability in the second quadrant $\mu_r = 1.21$ and maximum allowable temperature is $T_{\max} = 80\,^\circ\mathrm{C}$. For simulation of liquid flow we used water with dynamic viscosity $\eta = 0.001\,\mathrm{Pa\,s}$ and density $\rho = 10^3\,\mathrm{kg\,m^{-3}}$. For the entry part of the valve a permanent velocity of flow was prescribed, that continuously decreases to the wall of the channel.

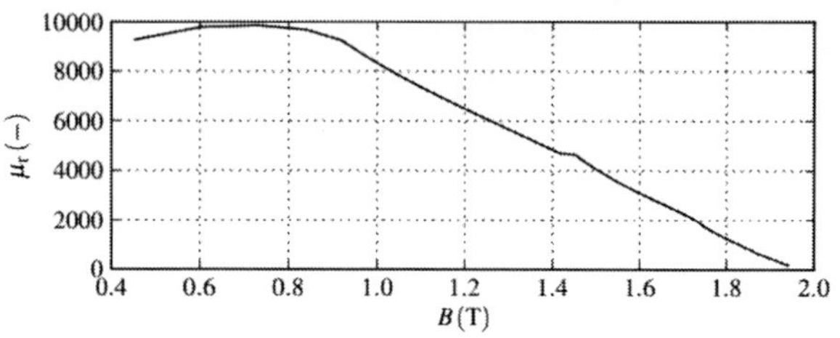

Figure 7. Relative permeability of used iron versus magnetic flux density

B. Selected results

Figures 8 and 9 show the distributions of flux density and force lines at the beginning and end of both processes - opening and closing of the valve.

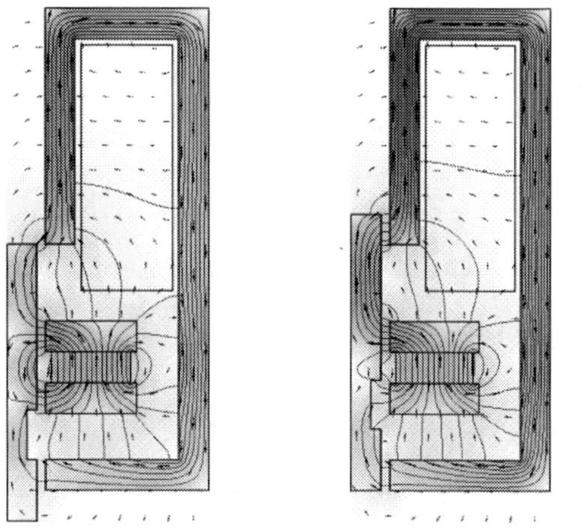

Figure 8. Opening of valve (distribution of flux density and force lines)

Both static characteristics depicted in Fig. 10 are of similar shapes and may be considered "almost" linear. These shapes were reached thanks to the cooperation of the long and short magnetic circuits indicated in Fig. 4. This fact may

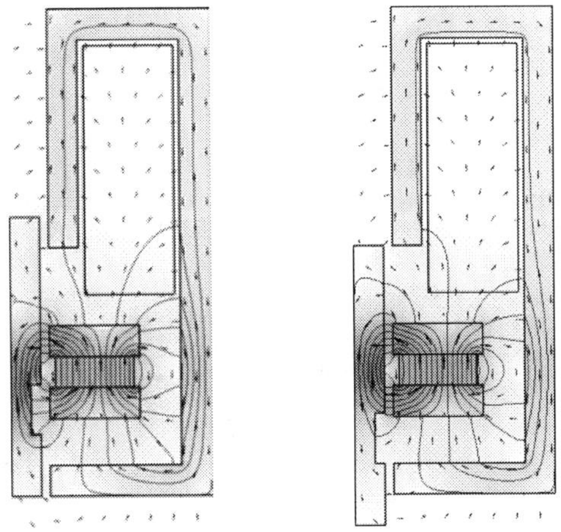

Figure 9. Closing of valve (distribution of flux density and force lines)

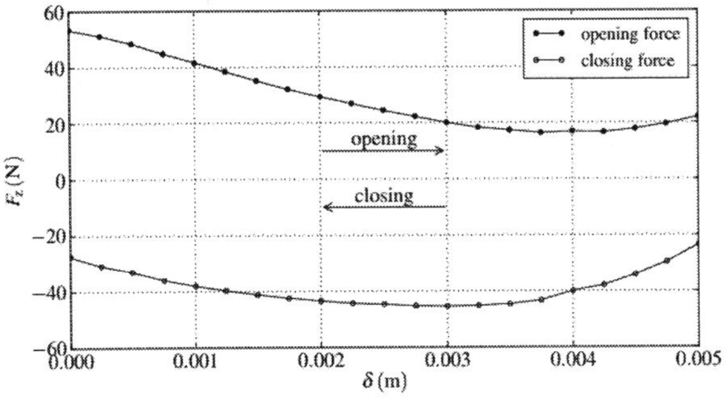

Figure 10. Static characteristics of actuator

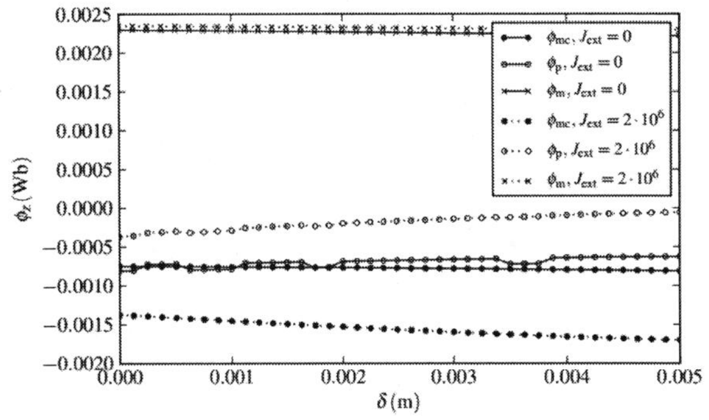

Figure 11. Magnetic fluxes in long magnetic circuit (ϕ_{mc}), plunger (ϕ_p) and permanent magnet of short magnetic circuit (ϕ_m). All fluxes are calculated in cut A (see Fig. 4)

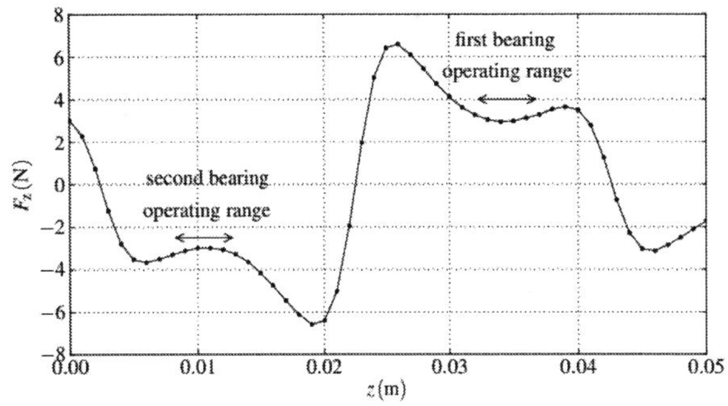

Figure 12. Static characteristic of bearing

conveniently be used for the control of the valve due to uniform dynamic characteristics for its opening and closing.

The short notch in the ferromagnetic part of the plunger allows increasing of the force effect of the short magnetic circuit, which is important for reaching a sufficient downforce of the closed valve and prevention of leakage.

Figure 11 shows the magnetic fluxes in several parts of the magnetic circuit for both excited ($J_{ext} = 0$) and unexcited ($J_{ext} = 2 \cdot 10^6\,\mathrm{A \cdot m^{-2}}$) coils. From the comparison of both states it is clear, that the magnetic flux produced by the DC coil reduces the local magnetization of the plunger and magnetic flux in permanent magnet. This means that the force produced by the short magnetic circuit is reduced and the force produced by the coil makes the plunger mover.

The operation regime of the bearing is defined in the linear part of the static characteristics. This allows stabilizing the plunger in the axial direction because the force effects secure that the plunger keeps the bearing inside the actuator. This can also be seen in Fig. 12 with the indicated ranges of operations of both bearings: the peaks push the bearings to the corresponding local minimums.

Figures 13 show the distributions of the liquid velocity in the valve for two different positions of the plunger. The pressure and viscous forces acting on the opening and closing element are depicted in Fig. 14 together with the static characteristic of the valve (superposition of the magnetic force acting to the plunger and axial force acting on the magnetic bearing). The operation of the valve closing is characterized by a considerable increase of the velocity of flow in the closing channel, which results in the pressure and viscous forces pressing on the valve. This phase is critical for error-less closing of the valve and it is necessary to produce a sufficient force acting on the plunger.

Dynamics of the movable plunger was computed with respecting pressure and viscous forces. The time-dependent characteristic of plunger position δ and plunger velocity v are shown in Fig. 15.

From the above figures it is clear that the valve is able to close the flow despite a considerable growth of the pressure force on the closing-up element at the increasing velocity of flow through the valve. The action is, moreover, very fast and time of closing very low ($t < 11\,\mathrm{ms}$).

VI. Conclusion

The paper presents basic design of the novel valve. The simulations show that the system exhibits very good parameters that allows wide industrial applications. Further work in the

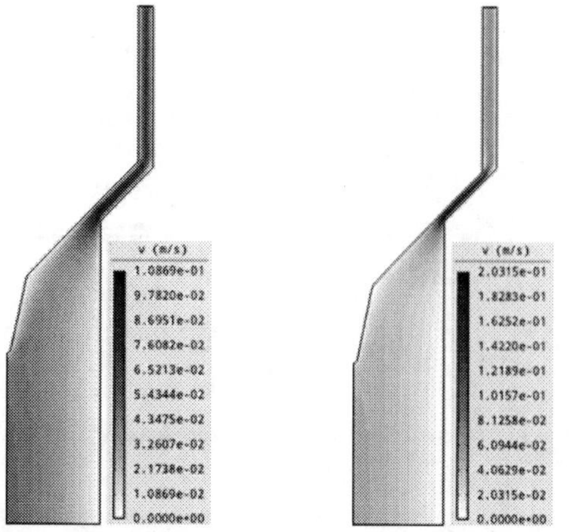

Figure 13. Distribution of liquid velocity for plunger position $\delta = 0.005\,\mathrm{m}$ (open valve) and $\delta = 0.0025\,\mathrm{m}$

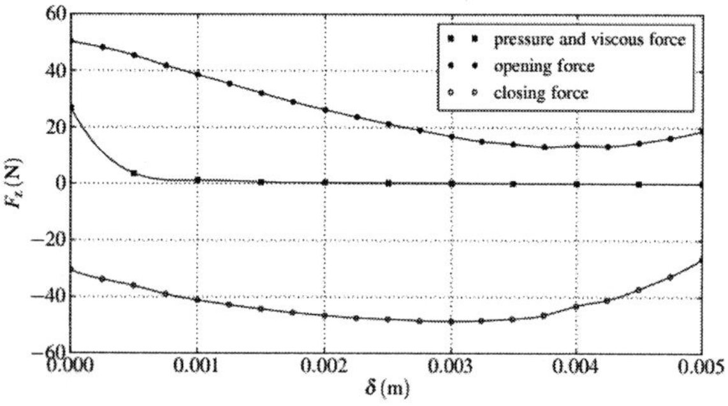

Figure 14. Static characteristics of actuator with magnetic bearing and pressure and viscous force acting to opening and closing element

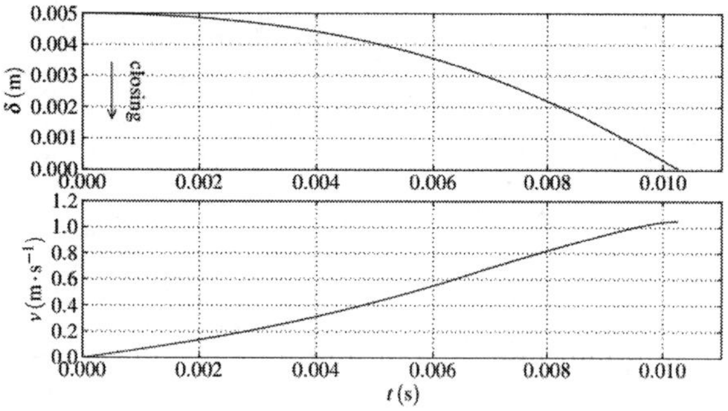

Figure 15. Dynamic characteristics of valve with respecting pressure and viscous force acting to movable plunger

domain will be aimed at optimizing the whole system and also in building the physical model of the device and experimental verification of the theoretical results.

VII. Acknowledgements

This work was supported by the European Regional Development Fund and Ministry of Education, Youth and Sports of the Czech Republic (project No. CZ.1.05/2.1.00/03.0094: Regional Innovation Center for Electrical Engineering - RICE), by the project P102/11/0498 (Grant Agency of the Czech Republic) and project SGS-2012-039.

References

[1] Yee-Pien, Y., Jieng-Jang, L., Da-Hau, Y., Yi-Ruei, C., Pai-Hsiu, L.: Multiobjective Optimal Design and Soft Landing Control of an Electromagnetic Valve Actuator for a Camless Engine, Transactions on Mechatronics, 2013, Volume 18, Issue 3, pp 963– 972

[2] Albert, J., Banucu, R., Hafla, W., Rucker, W.M.: Simulation Based Development of a Valve Actuator for Alternative Drives Using BEM-FEM Code, IEEE Transactions on Magnetics, 2009, Volume 45, Issue 3, pp 1744–1747

[3] Baoping, C., Yonghong, L., Xiaojie, T., Zelu, W., Fei, W., Hang, L., Renjie, J.: Optimization of Submersible Solenoid Valves for Subsea Blowout Preventers, IEEE Transactions on Magnetics, 2010, Volume 47, Issue 2, pp 451–458

[4] Parlikar, T.A., Chang, W.S., Qiu, Y.H., Seeman, M.D., Perreault, D.J., Kassakian, J.G., Keim, T.A.: Design and experimental implementation of an electromagnetic engine valve drive, Transactions on Mechatronics, 2005, Volume 10, Issue 5, pp 482 –494

[5] Parlikar, T.A., Chang, W.S., Qiu, Y.H., Seeman, M.D., Perreault, D.J., Kassakian, J.G., Keim, T.A.: Design and experimental implementation of an electromagnetic engine valve drive, Transactions on Mechatronics, 2005, Volume 10, Issue 5, pp 482–494

[6] Karban, P., Mach, F., Kůs, P., Pánek, D., Doležel, I.: Numerical solution of coupled problems using code Agros2D, Computing, 2013, Volume 95, Issue 1, pp 381–408

[7] Solin, P., Cerveny, J., Dolezel, I.: Arbitrary-Level Hanging Nodes and Automatic Adaptivity in the hp-FEM, Math. Comput. Simul., 2008, Volume 77, pp 117–132

[8] Henrotte, F., Felden, M., Vander, G. M., Hameyer, K.: The eggshell approach for the computation of electromagnetic forces in 2D and 3D, COMPEL, 2004, Volume 23, Issue 4, pp 996–1005

Magnetically suspended motor system for artificial hearts and blood pumps

Toru Masuzawa

Ibaraki University, 4-12-1 Nakanarusawa-cho, Hitachi, Ibaraki, Japan, masuzawa@mx.ibaraki.ac.jp

***Abstract*—Magnetic levitation technology, for magnetic bearings and magnetically suspended motors, is a cutting edge technology to produce artificial hearts and higher performance blood pumps. A wider blood gap and eliminating the contacting parts in the device based on the maglev technology provide better blood compatibility and higher durability of the device. Novel inventions about the maglev mechanism are required to overcome the size and power limitation of artificial hearts. Specialization of the maglev system is also necessary due to different requirements with the spreading application of artificial hearts. The current research topics of the maglev system for artificial hearts and blood pumps will be presented in the mini-symposium.**

I. Introduction

Artificial hearts and mechanical circulation support devices have been studied since 1958. Ventricular assist devices with rotary blood pump are commercially available at the outset of the 21st century [1]. The rotary blood pumps are very common as blood pumps for the mechanical circulatory support right now, but the mechanical bearing mechanism has potential problems at the contacting part suspected to be a cause of hemolysis and thrombogenecity. The magnetic bearing or self-bearing motor are promising technologies to give us long device durability and higher blood compatibility of the blood pumps. The magnetic suspension technologies produce wider blood gaps between the levitated and rotated impeller and pump casing which eliminates mechanical contact and wear. These features provide better blood compatibility to avoid such as blood trauma and thrombogenecity. Active position control of the levitated impeller provides safer operation of the blood pumps and new feature of the devices. I will introduce several types of the magnetic suspension system for artificial heart and blood pumps developed in Ibaraki University.

II. Magnetic suspension system and blood pumps

A. Axial magnetic suspension system for mechanical circulatory support

A levitated impeller is set between a magnetic bearing stator and a motor stator and is axially suspended by the magnetic suspension system to which balances axial forces on the impeller [2] [3]. Fig. 1 shows the extracorporeal blood pump with an axial magnetic suspension system. This pump has been developed under collaborative research with the Department of Therapeutic Strategy for Heart Failure, the University of Tokyo, National Cerebral and Cardiovascular Center Research Institute and Ibaraki University. The pump is developed for acute heart failure patients to enhance recovery of natural heart. The magnetic bearing produces an upward attractive force to balance the downward attractive force produced by the motor stator. There are bias permanent magnets on the impeller and a hybrid magnetic bearing is constructed to produce sufficient attractive force with the miniaturized device. The position and tilting are controlled actively with the magnetic bearing, and radial position is restricted with passive magnetic stability .

The structure of the magnetic bearing is shown in Fig. 2.

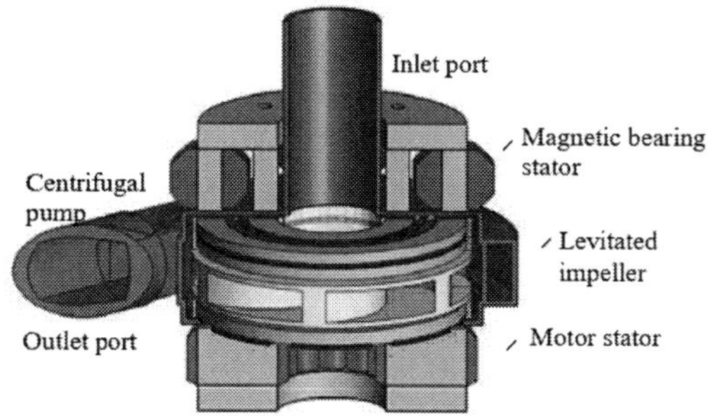

Figure 1. Cross sectional view of maglev blood pump with axial suspension system. A centrifugal pump is set between magnetic bearing and axial motor stator.

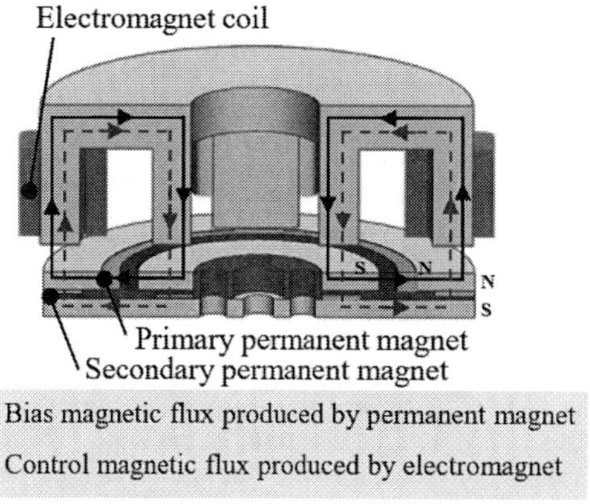

Figure 2. Structure of double bias hybrid magnetic bearing. Magnetic flux produced by the secondary permanent magnet is overlapped with it of the primary permanent magnet.

Figure 3. Exterior view of the extracorporeal maglev blood pump with axial magnetic suspension system.

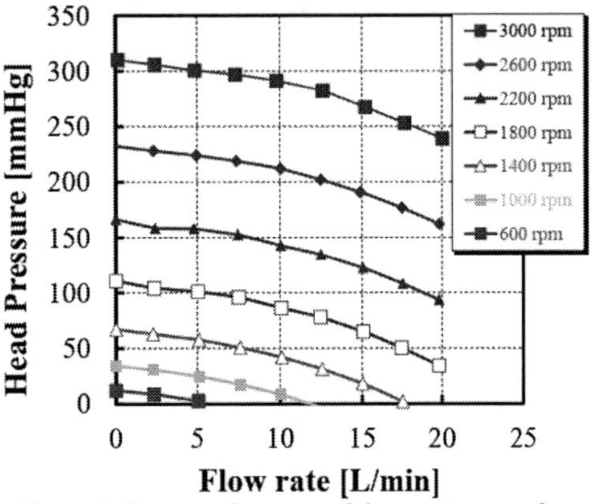

Figure 4. Pump performance of the extracorporeal maglev blood pump with the axial magnetic suspension system.

The double bias hybrid magnetic bearing had been invented to develop a small and powerful magnetic bearing. In the hybrid magnetic bearing, the bias permanent magnet is used to produce bias magnetic flux, but the permanent magnet (PM) inserted in the magnetic circuit is a resistance for the electric magnet. We can enlarge the net of the attractive force but the force factor to the exciting current into the electro magnet is reduced by the PM. We have tried to overcome the problem with deterioration of the force factor by using double bias PMs. The primary PM produces main bias magnetic flux and the secondary PM regulates the magnetic flux path. The primary PM is magnetized in the radial direction and the secondary PM is magnetized in the axial direction. The yokes are set at the both circumferential side of the primary PM and the back of the secondary PM. The PM are set like the same pole of the PMs faces to face. The magnetic flux produced by the primary PM, indicated as solid line in Fig. 2, goes through the yoke, the air gap, and the stator core of the magnetic bearing and restricted its direction with the secondary PM. The magnetic flux produced by the secondary PM is overlapped with that of the primary PM at the air gap and enhances the magnetic attractive force at the air gap. The magnetic path of the magnetic flux produced by the electromagnet of the magnetic bearing, indicated with dashed line in Fig. 2, is constructed with the stator core of the magnetic bearing, the air gap, the secondary PM and the back yoke of the secondary PM. This is the shortest path for the magnetic flux because the primary PM exhibits a higher magnetic circuit resistance. The double bias hybrid magnetic bearing produces higher magnetic flux in the air gap coupled with lower magnetic resistance for the electromagnet which enables a higher attractive force and higher force factor with excitation current. In our maglev pump, the PMs and the yoke behind the secondary PM is set on the impeller.

Fig. 3 and Fig. 4 show the extracorporeal maglev blood pump with the axial magnetic suspension system and the pump performance. The blood pump is designed to produce higher flow rate with extracorporeal connection and to produce pulsatile flow by changing the rotating speed synchronized with the heart beat to possibly enhance the chance for myocardial recovery due to increased coronary artery flow. The developed maglev pump can produce more than 20 L/min against a head pressure of 250 mmHg.

B. Axial magnetic suspension system for total artficial heart

An axial magnetic suspension system is also used for a total artificial heart [3, 4]. The maglev total artificial heart is developed under a collaborative research with BiVACOR Inc., Texas Heart Institute in Houston, TX USA, ICET lab in Brisbane, Australia and Ibaraki University in Hitachi, Japan. There are two types of the total artificial heart with the axial magnetic suspension system. The BiVACOR® device utilizes a similar structure with the axial magnetic suspension system for mechanical circulatory support explained already above [3] whilst another system, called the IBHEART and is explained as follows, has a reverse structure [4]. Fig. 5 shows the structure of the magnetically suspended motor. The stators of the magnetic bearing and the motor are combined at the center of the device back-to-back to form a single levitated motor. The left and right pumps are constructed at both ends of the stator core. Impellers of both pumps are connected with a rod through the center hole of the stator core. Permanent magnets generating the magnetic force toward the stator core are set on the backside surface of the impellers and attractive forces produced by the magnetic bearing and the motor are balanced by the digital PID controller to suspend the impellers. The unique feature of the artificial heart is its preload sensitivity and flow rate balancing function between the left

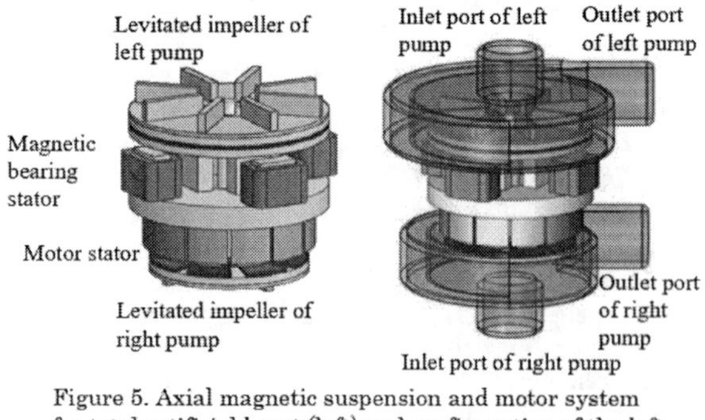

Figure 5. Axial magnetic suspension and motor system for total artificial heart (left) and configuration of the left and right pump with the actuator (right side). Centrifugal pumps are set at both sides of the actuator.

and right blood pump. Preload means an inlet pressure related with the atrial pressure. The pump flow rate must increase when the preload increases to prevent lung congestion because the increasing of the atrial pressure means large amount of the blood is accumulated in the lung. Management ability of the blood volume based on the preload is an essential function for the total artificial heart. But, implantation of the flow rate probes is difficult with current technology and an auto regulation mechanism of the pump flow rate is a key to develop the artificial heart. The auto regulation mechanism of the artificial heart is implemented based on the magnetic suspension mechanism and the zero power control with the hydrodynamic force in our device. The axial impeller position is regulated with a zero power controller. The zero power controller searches force minimal impeller axial position that makes a balance the PM attractive force and the hydraulic force acts on the levitated impeller. For example, the impeller is shifted upward when the left pump preload increases and is shifted downward when the preload decreases. The hydraulic pump efficiency and the flow rate increase when the impeller is shifted upward because the blood gap between the impeller vane tip and the pump casing changes narrower by shifting impeller position and the secondary flow that is a cause of the inefficiency of the pump decreases. The pump efficiency and the flow rate decrease when the impeller is shifted downward. The flow rate of the pump is changed with changing of the preload in our device.

Fig. 6 shows exterior view of the prototype pump of the total artificial heart. The diameter and the width of the IBHEART artificial heart are 76 mm and 62 mm, respectively. Fig. 7 shows pump performance of the prototype pump. Maximum flow rate is 20 L/min and the 10 % of the flow rate can be adjusted by shifting the impeller position by 0.8 mm with same rotating speed of the impeller.

C. Radial magnetic supesion sysstem

The radial magnetic suspension system that suspends the levitated rotor in the radial direction is used for the ventricular assist device to prevent severity of heart disease. Fig. 8 shows a thin cascade blood pump with the radial magnetic suspension system for the mild heart disease patients [5]. This pump has been developed under collaborative research with the Department of Therapeutic Strategy for Heart Failure and Ibaraki University. Ring shaped rotor is set around the maglev stator as an outer rotor structure. Two independent electric coils are wound on the stator poles to construct a self-bearing motor. One coil is used to control radial position of the impeller and the other is to control rotation of the impeller. The levitation coil produces attractive force toward the center of the stator and controls rotor radial position. The rotor position and rotation can be regulated independently with P+/-2 poles theory. Axial movement of the levitated rotor is

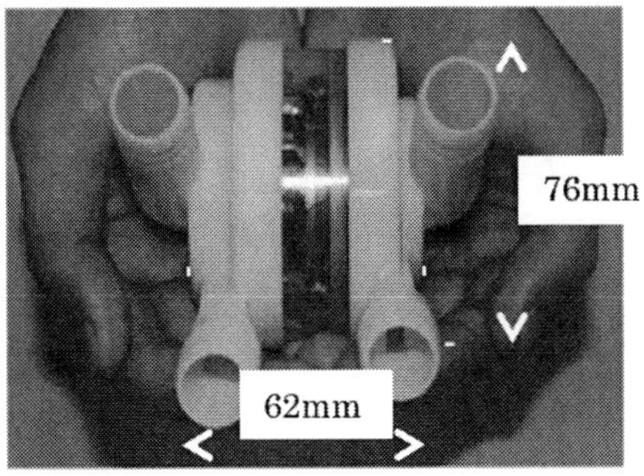

Figure 6. Exterior view of the maglev total artificial heart.

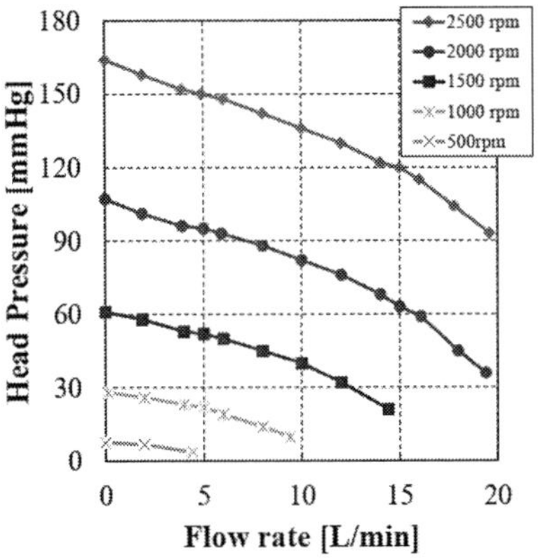

Figure 7. Pump performance of lefty pump of the maglev total artificial heart.

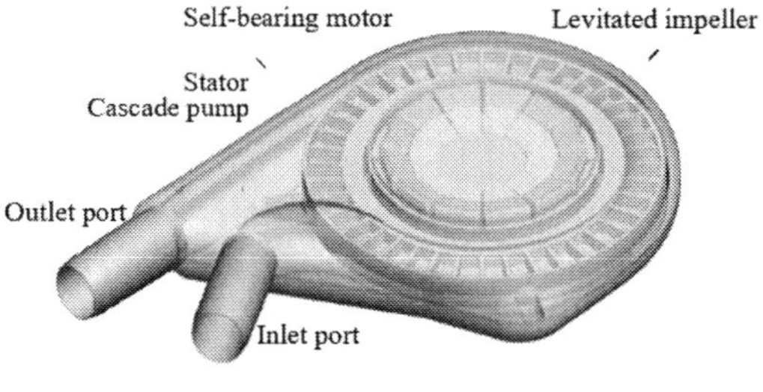

Figure 8. Blood pump with radial magnetic suspension system. A cascade pump is adopted to make the device thin.

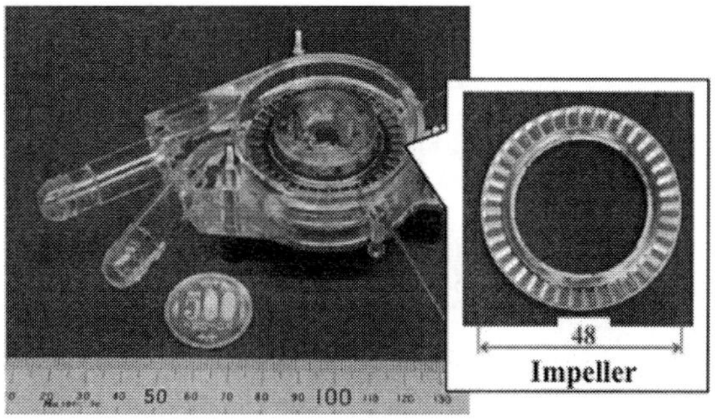

Figure 9. Exterior view of the maglev cascade blood pump.

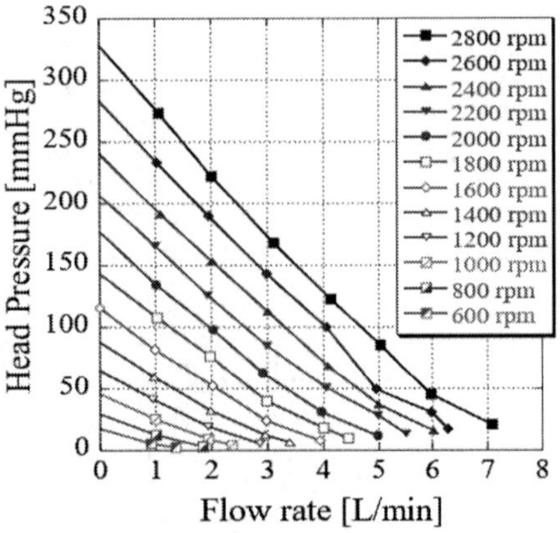

Figure 10. Pump performance of the maglev cascade blood pump.

restricted with passive stability of the magnetic system. The radial magnetic suspension system is good to make thinner device. A cascade pump with an inlet port and an outlet port on same plane is adopted as a blood pump to make the device thin. The pump is designed for implantation under thorax muscle layer to achieve easy pump replacement and for long term circulation assist more than ten years. Exterior view of the prototype pump is shown in Fig. 9. The ring shaped impeller has pockets at its circumference to convey the fluid from the inlet port to outlet port. There is a channel around the impeller to transport the blood with the rotation of the impeller and the channel is closed between the inlet and outlet port to pump out the blood from the outlet port. The outer diameter of the ring shaped impeller is 48 mm and the thickness of the prototype pump is 22 mm. Target flow rate of the pump is 2.0 L/min against a head pressure of 100 mmHg. Fig. 10 shows the pump performance of the maglev cascade pump. Maximum flow rate and maximum head pressure are 7 L/min and 300 mmHg, respectively. The pump indicates sufficient pump performance as a ventricular assist device.

D. Axial suspension sysstem with double stators

Fig. 11 shows five-axis control maglev motor for a pediatric ventricular assist device [6]. This device has been developed under collaborative research with National Cerebral and Cardiovascular Center Research Institute and Ibaraki University. We overcome the deterioration of the motor torque due to miniaturization of the motor by adopting double stator mechanism. The proposed mechanism also permits us to control five-axis of the levitated impeller. The stators set at both sides of the levitated impeller have two electric coils, one is to control axial position and rotation and the other is to control tilting and radial position of the impeller. The axial position and rotation of the levitated impeller are controlled with vector control algorithm. The tilt and radial position of the impeller are controlled with a double stator mechanism and P+/-2 poles theory. Fig. 12 explains generation of the restoring torque and radial force on the levitated impeller. The rotating magnetic field is generated with the P+/-2 poles theory and the torque and the radial force on the levitated impeller are generated simultaneously by both of the stator.

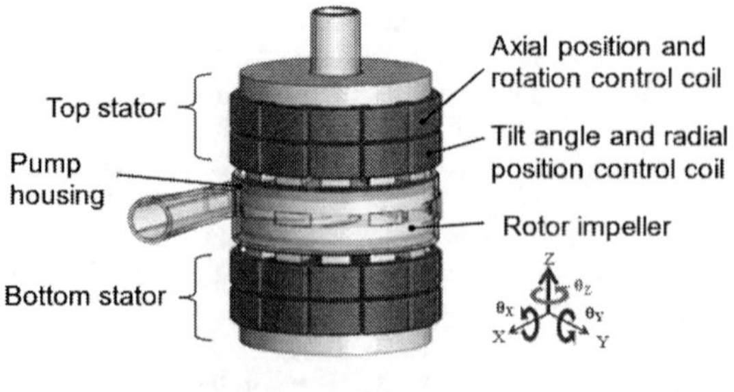

Figure 11. Five-axis controlled double stator levitated motor for pediatric ventricular assist device. A centrifugal pump is set between motor stators.

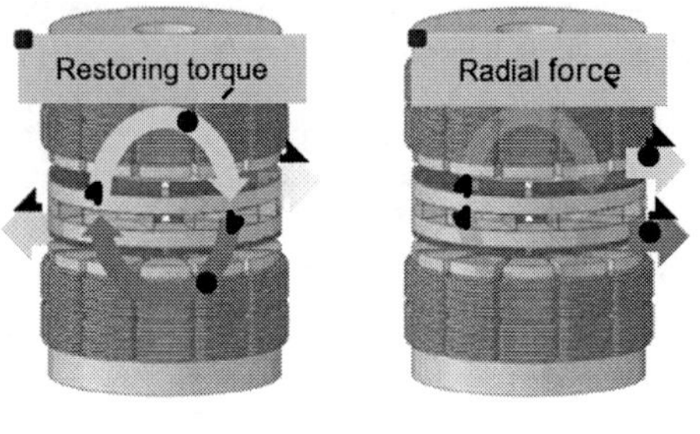

Figure 12. Generation of the restoring torque and radial force with double stators

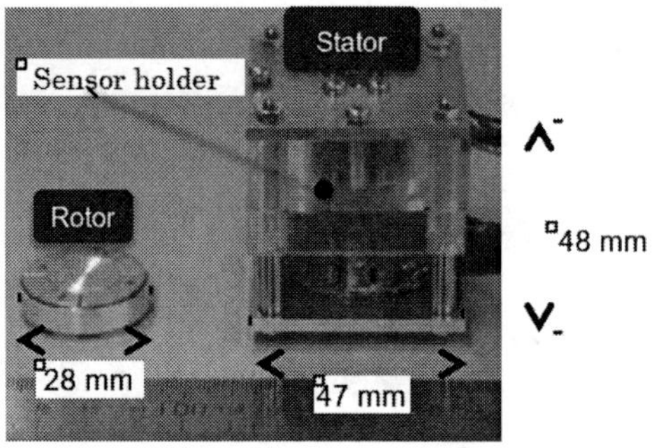

Figure 13. Exterior view of 5-axis controlled double stator levitated motor for pediatric ventricular assist device.

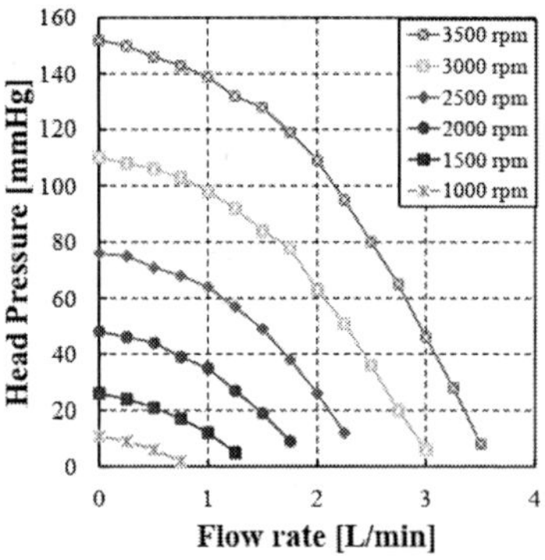

Figure 14. Pump performance of the maglev pediatric ventricular assist device.

The restoring torque only can be extracted by overlapping the forces produced by both stators. Also, only radial force can be extracted with same manner.

The exterior view of the prototype motor is shown in Fig. 13. The developed motor has an outer diameter of 28 mm, height of 41 mm and magnetic air-gap length of 1.5 mm. The impeller can be levitated and rotated up to a rotating speed of 6,500 RPM with an oscillation amplitude of less than 100 micrometers in air. Fig. 14 shows pump performance of a pediatric ventricular assist device with five-axis control maglev motor. Target flow rate of the pump is 1.5 L/min against a head pressure of 100 mmHg, the prototype pump can achieve the target performance with a rotating speed of 3000 RPM.

III. Summary

Several types of magnetic suspension system for artificial heart and mechanical circulatory support devices developed at Ibaraki University are introduced. The roll of mechanical circulatory support and its usage spreads widely. Specialized magnetic suspension system should be developed corresponding to each device application.

Acknowledgement

The author indicates special thanks to all collaborators of these researches, especially to Dr. Daniel L Timms at BiVACOR inc., Prof. Shunei Kyo and Prof. Takashi Nishimura at the University of Tokyo, Dr. Eisuke Tatsumi at National Cerebral and Cardiovascular Center, Japan.

References

[1] David Joyce, et al, "Mechanical Circulatory Support: Principles and Applications," McGraw-Hill Professional, 2011.

[2] M. Kitago, T. Masuzawa, T. Nishimura, S. Kyo, "Axial-type Magnetically suspendxed motor for a therapeutice artificial heart," Journal of the Japan Society of Applied Electromagnetics and Mechanics, vol. 19, no. 2, pp. 280-285, 2011. (in Japanese)

[3] N. A. Greatrex, D. L. Timms, N. Kurita, W. W. Palmer, T. Masuzawa, "Axial Magnetic Bearing Development for the BiVACOR Rotary BiVAD/TAH," IEEE Transactions on Biomedical Engineering, 57(3), 514-721, 2010.

[4] N. Nishimura, T. Masuzawa, D. L. Timms, "Total artificial heart with a single magnetically suspended motor," Journal of the Japan Society of Applied Electromagnetics and Mechanics, vol. 21, no. 2, pp. 152-159, 2013. (in Japanese)

[5] K. Ukita, T. Masuzawa, H. Onuma, T. Nishimura and S. Kyo, "A Radial Type Self-Bearing Motor for Small Maglev Regenerative Blood Pump," Journal of the Japan Society of Applied Electromagnetics and Mechanics, vol. 20, no. 2, pp. 312-318, 2012.

[6] M. Osa, T. Masuzawa, E. Tatsumi, "5-DOF control double stator motor for paediatric ventricular assist device," Proceedings of ISMB13, paper 41, 2012.

Magnetic Bearing System for Left Ventricular Assist Artificial Heart Pump – Combined Active and Passive Configuration

Paul Allaire[a], Wei Jiang[b], Arunvel Kailasan[c], Tim Dimond[d]

[a] Rotor Bearing Solutions International, 3277 Arbor Trace, Charlottesville, VA 22911-7580, USA, paul.allaire@rotorsolution.com
[b] University of Virginia, 122 Engineer's Way, Charlottesville, VA 22904, USA
[c] Gardner Denver, Inc., 100 Gardner Park, Peachtree City, GA 30269, USA, arunvel.kailasan@gardnerdenver.com
[d] Rotor Bearing Solutions International, 3277 Arbor Trace, Charlottesville, VA 22911-7580, USA, tim.dimond@rotorsolution.com

***Abstract*—Magnetic bearings continue to contribute to the development of ventricular assist artificial heart pumps. This paper describes the magnetic suspension for a unique prototype axial flow pump designed for approximately 6 l/min at 100 mm Hg performance with an operating speed of approximately 7,000 rpm. The integrated magnetic suspension design provides a direct non-contact blood flow path through the pump with no obstructions which might create low flow areas and thrombosis (blood clots). The magnetic suspension is a combination of permanent magnets (PMs) and active magnetic bearings (AMBs). There are two radial AMBs which support the four degrees of freedom at the ends of the axial pump impeller and an axial PM thrust bearing. The axial PM bearing supports the direction of the largest fluid force on the impeller. A major objectives of artificial hearts is to have extremely low power consumption. Thus the integrated PM and AMB suspension system has an operating magnetic suspension power of approximately 2 watts. The design, numerical modeling, and testing of the magnetic suspension system to support the fluid loads and the g loads is described in the paper.**

I. Introduction

The problem of heart disease is a high priority medical research It remains a very significant cause of death or limited capability to live a robust life for many individuals. The number of heart transplants in the United States are severely limited because there are only approximately 2,500 donors per year [1]. Another problem is matching the transplant recipient to the donor. Currently, the use of cyclosporin as an anti-coagulant has made survival on a heart transplant much easier than in the years prior to 1980.

Left ventricular devices (LVADs), designed to support the left ventricle, have proven extremely valuable in use as a bridge to transplant in many patients. In 1994, the US Federal Drug Agency (FDA) approved the first LVAD [1]. The current leader in the field is the Heartmate II, made by Thoratec, a finger sized axial flow pump, approved by the FDA in 2008. The Heartmate II weighs 0.28 kg (10 oz) and requires a sling holding a power module and 3.6 kg (8 lbm) of batteries that require recharging after 8 hours [1]. The unit is plugged into the wall power outlet at night. The current Heartmate II has blood lubricated needle bearings which require recipients to take significant anti-coagulation medication. Another popular LVAD is made by Heartware International [1]. In 2012, about 2,000 LVADs were implanted, as compared to 200 in 2006 [1]. As many as 40% of LVADs are now categorized as destination therapy or permanent assist devices. It is noted that the future likely holds magnetic suspension for LVADs, according to Thoratec officials [1].

The first LVAD configuration utilizing an active magnetic suspension in a centrifugal pump, including a 90 day animal implant, was reported in a number of publications in the mid-1990s [2,3]. It used a full electromagnetic suspension but this type of suspension produced high currents and too much heat. The next centrifugal pump utilizing a magnetic suspension employed permanent magnets in combination with the electromagnets, called a hybrid system, greatly reduced the coil currents [4-12, 16]. There is no method of supporting a body solely with permanent magnets so at least one axis must be actively controlled, as proven in Earnshaw's theorem. The axis along the centrifugal impeller axis was actively controlled. This pump configuration eventually was tested in animal and human tests and produced for clinical use by World Heart under the title Levacor. Manufacture of this device ceased in 2011.

The surgeons associated with the development of the Levacor LVAD desired an axial flow pump which would be more suited for implantation. An axial flow LVAD design, called the Lifeflow LVAD, formulated in the late 2000s, was developed with a combined active/permanent magnet system that is reported in this paper. The basis for the permanent magnet design was originally developed in [13,14]. A general paper on sliding mode control of active magnetic bearings related to heart pumps was developed in [15]. An initial paper on this new axial flow suspension, in the Lifeflow LVAD pump, was written in 2007 [17]. The major objective of the suspension was to support the impeller when subject to blood flow forces due to the flow in the pump with a minimum of power consumption, yet be capable of preventing contact between the impeller and LVAD housing when subject to an 8 g dynamic load in any direction. The Lifeflow magnetic suspension, described in this paper, consists of a

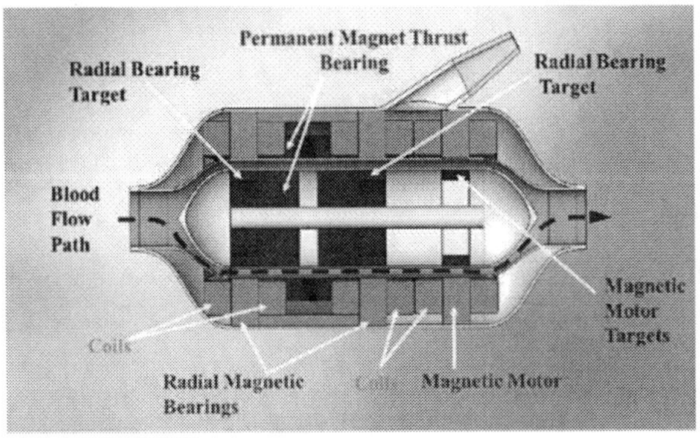

Figure 1. Cutaway View of Lifeflow LVAD Configuration

permanent magnetic axial bearing and two radial active magnetic bearings with permanent magnet bias. The measured power consumption in a suspension test rig was 2 watts when operated at 7,000 rpm.

II. MAGNETIC SUSPENSION SYSTEM

A cutaway view of the LVAD is shown in Fig. 1. The bloodflow path is a direct, unobstructed flow path around the pump axial impeller, which has a series of spiral vanes on the outer diameter. It has a flow rate of 3 to 8 l/min with a design flow of approximately 6 l/min at 7,000 rpm. The hybrid 5 axis magnetic suspension system for the LVAD consists of a combination of PM magnet bearings and active magnetic bearings (AMBs), as illustrated in Fig. 1. Figure 2 shows another, more detailed view of the pump and the magnetic bearing system.

A perspective view of the magnetic suspension stator design for this LVAD is shown in Fig. 3, a stator length of 0.03 m.. The rotor outer diameter is 8 mm and the stator inner diameter is 10 mm with an airgap of 2 mm. Included in the airgap will be the titanium canning material. The highest loads on the pump impeller are due to the pressure difference across the pump. The axial axis PM generates a positive PM stiffness k_z, which places the impeller in the axial center of the clearance space inside the impeller housing. The radial bearings are AMBs, both front and back inside the housing clearance. The AMBs and back iron are made with cobalt iron laminations. The physical magnetic suspension system stator prototype is shown in Fig. 4. The silver components are the permanent magnets while the coils and darker AMB lamination stacks of the two radial bearings are clearly shown. The magnetic suspension test rig for operating the motor and magnetic suspension is shown in Fig. 5. Also, the Lifeflow rotor with magnetic components in the center for both radial AMB lamination stacks and PMs. Steel shaft ends were added for displacement sensors in the testing.

The two radial bearings are each six pole AMBs, as illustrated in Fig. 6, with an equal number of wire turns on each magnetic pole. The pole face area for each pole is 41.4 sq mm. The laminations were constructed of M-19 steel. The backiron allows for magnetic communication between the poles.

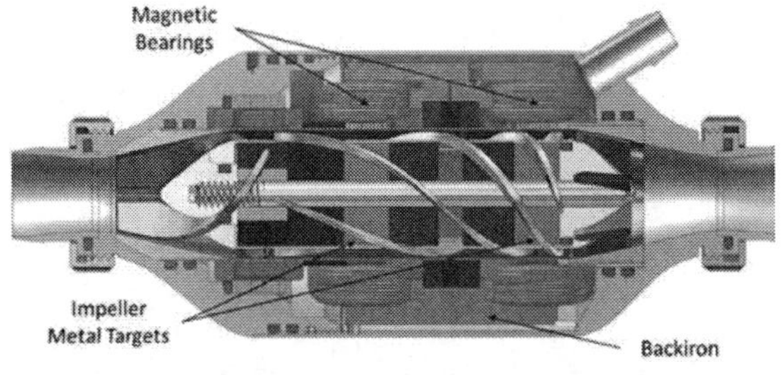

Figure 2. Lifeflow pump schematic showing flow path and magnetic bearings

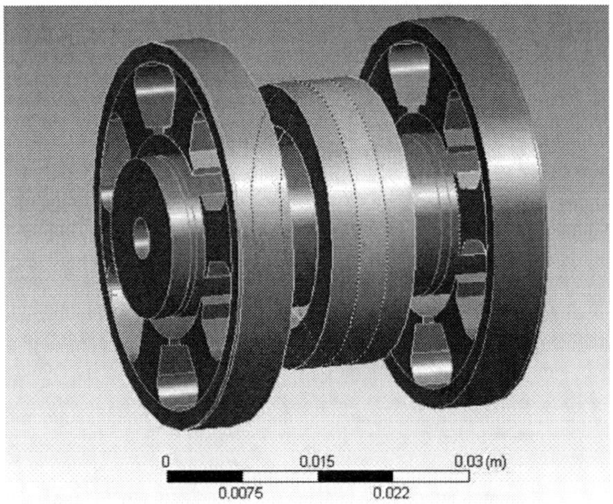

Figure 3. Perspective view of the bearing stator structure

Figure 4. Stator Lifeflow Magnetic Suspension System Prototype

It is well known that the way to reduce the coil power consumption is to supply the bias flux using a permanent magnet flux. The hybrid configuration shown in Fig. 5 uses the PM and AMB flux paths in a way to generate the combined radial flux which add the PM flux in the rotor to the two radial bearing active flux magnitude when they go through the radial bearings. The AMB M-19 iron has a linear operation value near 1.3 T and a saturation flux near

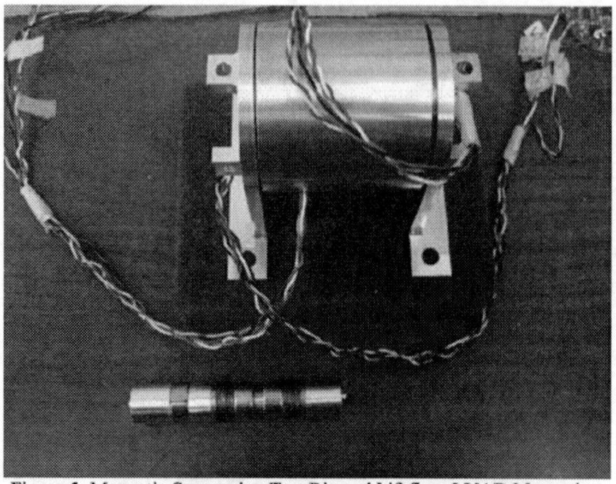

Figure 5. Magnetic Suspension Test Rig and Lifeflow LVAD Magnetic Suspension Rotor Component

2.0 T, so the PM bias flux is set at approximately half of that or 0.6 T. The coils were selected to provide the magneto-motive force NI to attain the remaining airgap flux density near the linear limit of approximately 1.3 T.

The PM flux path requires a high flux level which is not attainable with a simple PM circuit design. The three PM components in the stator (housing) part of the are configured in the reverse polarity way, as shown in Fig. 1, such that the polarities are 1) S-N (shown in the left red block), 2) N-S (shown in the center blue block), and 3) S-N (shown in the right red block). This special configuration gives a much higher flux density level than a simple, single set of PMs. In the rotor, two simple PMs are placed as shown in Fig. 1 (as shown in the two red blocks). These rotor PMs have two purposes: first, supply the PM bias flux for the radial bearings and second, axial stiffness desired to keep the impeller centered when it is subject to the applied fluid load. The PM material was NdFeB 35 EH.

A three dimensional finite element model of the six pole active magnetic bearing structure is shown in Fig. 7. The rotor three dimensional finite element model is shown in Fig. 8.

A 3-D finite element analysis of the combined AMB-PM system without control currents using ANSYS was also formulated. Figure 9 shows an example of the three dimensional flux plot including both active and passive magnetic suspension system. Also, Fig. 10 shows an example side view of the three dimensional plot. This analysis was used to determine the positive open loop axial stiffness with the result of k_z = 8,65 N/mm, as shown in Fig. 11. The negative open loop radial stiffness, also without coil currents was k_x = - 17.65 N/mm. When coil currents are added, the linearized radial current gain was k_i = 0.688 N/A to be used in controlling the magnetic suspension system. Figure 12 shows the plot of force vs. coil current. The finite element flux levels in the radial air gaps were found to be in the range of 0.31 to 0.36 Tesla.

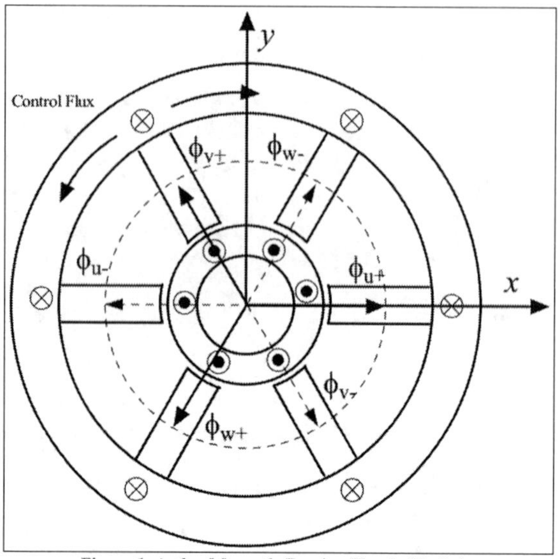

Figure 6. Active Magnetic Bearing Flux in Six Poles

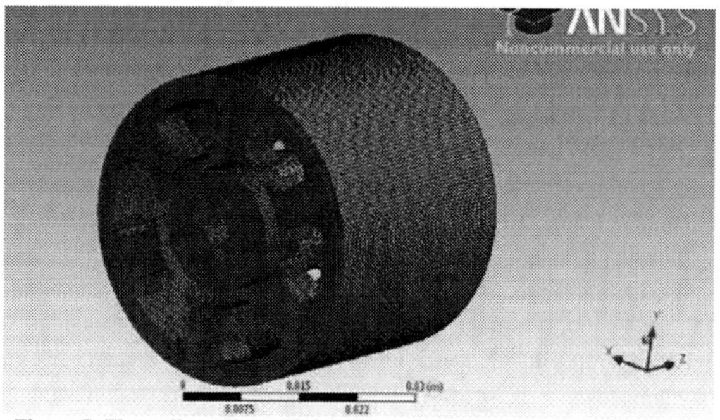

Figure 7. Three dimensional finite element model of the six pole active laminated magnetic bearing structure

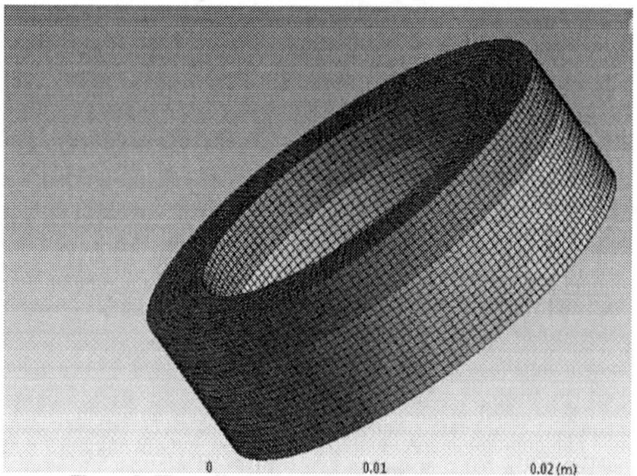

Figure 8. Three dimensional model of the rotor structure in the active magnetic bearing components

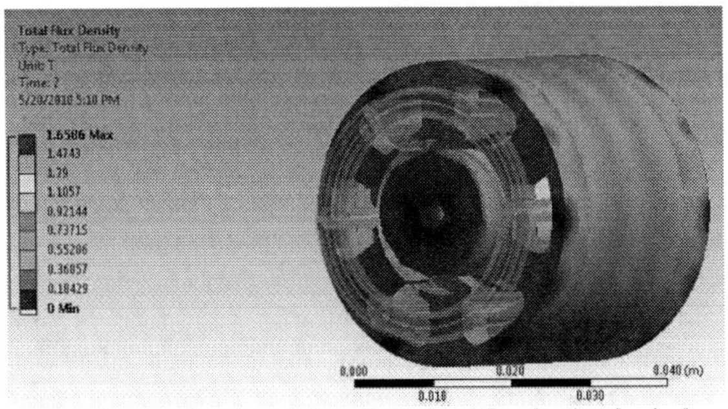

Figure 9. Perspective view of three dimensional flux density plot in the combined active and passive magnetic suspension system.

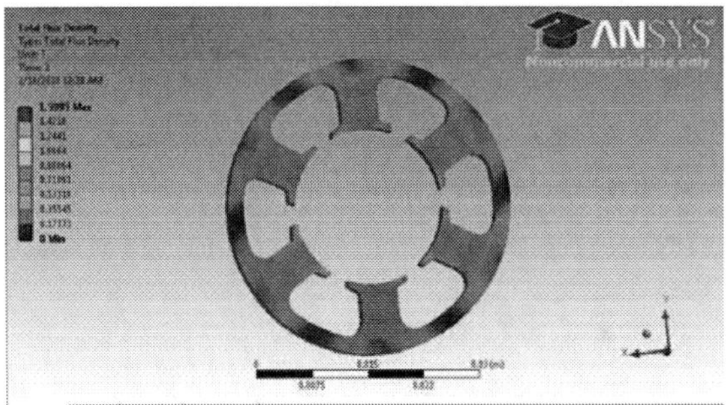

Figure 10. Side view of three dimensional flux density plot in the combined active and passive magnetic suspension system.

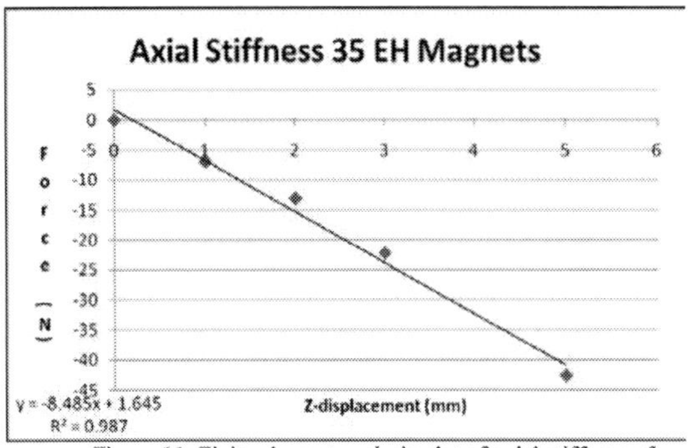

Figure 11. Finite element analysis plot of axial stiffness of combined acive and passive magnetic suspension system

It is also important to evaluate the influence of the blood flow on the rotor. These values were obtained using computational fluid dynamics where the impeller was calculated in the centered position and then moved radially a small distance [18]. The calculation was performed in the rotating coordinate system and then transformed into the fixed coordinate system. The effective radial stiffness is k_{xx} = -5.9 N/mm, k_{xy} = 7.3 N/mm, k_{yx} = -9.1 N/mm, and k_{yy} = -6.5 N/mm at 7,000 rpm. The effective radial damping is c_{xx} = $5.7\cdot10^{-4}$ N-s/mm, c_{xy} = $3.68\cdot10^{-4}$ N-s/mm, c_{yx} = $-3.28\cdot10^{-4}$ N-s/mm, and c_{yy} = $2.2\cdot10^{-4}$ N-s/mm also at 7,000 rpm. The added mass terms are m_{xx} = $0.72\cdot10^{-4}$ N-s^2/mm, m_{xy} = $0.09\cdot10^{-4}$ N-s^2/mm, m_{yx} = $-0.29\cdot10^{-4}$ N-s^2/mm, and m_{yy} = $0.79\cdot10_{-4}$ N-s^2/mm.

III. Magnetic Circuit Model

The control of the radial AMB axes was carried out with a magnetic flux circuit model. The magnetic flux simple circuit model is shown in Fig. 13 with both the AMB and PM circuits included. The magneto-motive force relations for the six pole radial bearings are given by:

$$\begin{aligned} &\phi_{u+}R_{u+} - NI_u = \phi_{u-}R_{u-} + NI_u \\ &= \phi_v + R_{v+} - NI_v = \phi_{v-}R_{v-} + NI_v \\ &= \phi_{w+}R_{w+} - NI_w = \phi_{w-}R_{w-} + NI_w \end{aligned} \quad (1)$$

There are three control current components, after taking the opposite poles as employing paired components of the system. The flux constraint equation is

$$\phi_{u+} + \phi_{u-} + \phi_{v+} + \phi_{v-} + \phi_{w+} + \phi_{w-} = \phi_0 \quad (2)$$

The air gap reluctances for the six poles are given in the following set of equations

$$\begin{aligned} R_{u+} &= \frac{g_0}{\mu_0 A}(1-x) \\ R_{u-} &= \frac{g_0}{\mu_0 A}(1+x) \\ R_{v+} &= \frac{g_0}{\mu_0 A}\left(1+\frac{x}{2}-\frac{\sqrt{3}y}{2}\right) \\ R_{v-} &= \frac{g_0}{\mu_0 A}\left(1-\frac{x}{2}+\frac{\sqrt{3}y}{2}\right) \\ R_{w+} &= \frac{g_0}{\mu_0 A}\left(1+\frac{x}{2}+\frac{\sqrt{3}y}{2}\right) \\ R_{w-} &= \frac{g_0}{\mu_0 A}\left(1-\frac{x}{2}-\frac{\sqrt{3}y}{2}\right) \end{aligned} \quad (3)$$

Assembling all of these equations into matrix form yields the AMB control matrix in the form, involving the pole flux variables, as shown in the Appendix. This is a rather large matrix system of size 6x6 that needs to be reduced for efficient control operation.

A. *System Size Reduction*

The controller must be placed in a small controller space, on single circuit board, in the on-body LVAD controller box. A smaller order control system matrix will be more compact and simplify the circuit board design and construction. We

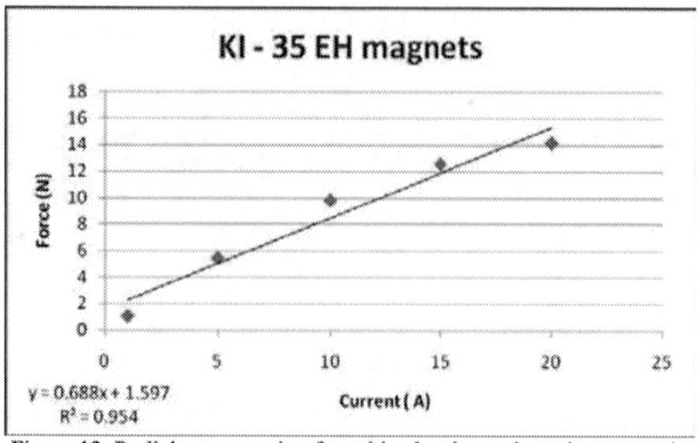

Figure 12. Radial current gain of combined active and passive magnetic bearing system evaluated with three dimensional finite elements.

introduce a combination of the basic flux variables of the form:

$$
\begin{aligned}
\phi_{us} &= \phi_{u+} - \phi_{u-} \\
\phi_{ua} &= \phi_{u+} + \phi_{u-} \\
\phi_{vs} &= \phi_{v+} - \phi_{v-} \\
\phi_{vm} &= \phi_{v+} + \phi_{v-} \\
\phi_{ws} &= \phi_{w+} - \phi_{w-} \\
\phi_{wa} &= \phi_{w+} - \phi_{w-}
\end{aligned} \qquad (4)
$$

We assume that the rotor radial displacements x, y are small compared to the radial clearance between the rotor surface and the housing. Thus, second order terms in the approximation are neglected. Then, the system control matrix equations reduce to the form shown in the Appendix.

The force equations have the form

$$
\begin{aligned}
F_u &= \left(\phi_{u+}^2 - \phi_{u-}^2\right)\frac{1}{2\mu_0 A} = \phi_{us}\phi_{ua}\frac{1}{2\mu_0 A} \\
F_v &= \left(\phi_{v+}^2 - \phi_{v-}^2\right)\frac{1}{2\mu_0 A} = \phi_{vs}\phi_{va}\frac{1}{2\mu_0 A} \\
F_w &= \left(\phi_{w+}^2 - \phi_{w-}^2\right)\frac{1}{2\mu_0 A} = \phi_{ws}\phi_{wa}\frac{1}{2\mu_0 A} \\
F_x &= F_u - \frac{1}{2}F_v - \frac{1}{2}F_w \\
F_y &= \frac{\sqrt{3}}{2}F_v - \frac{\sqrt{3}}{2}F_w
\end{aligned} \qquad (5)
$$

Here the first three relations in Eq. (5) give the net forces in the three primary pole axes in the three pole pair axes. The next two force equations give the force components in the x, y coordinate directions. The combined flux variables in the first three force equations above are related to the control currents via the relations:

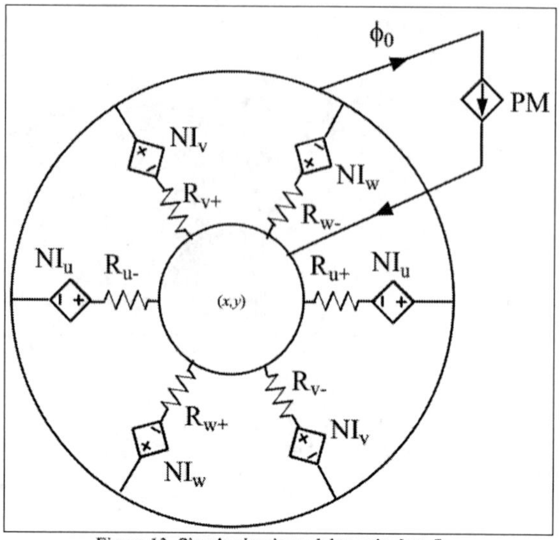

Figure 13. Simple circuit model to calculate flux

$$
\begin{aligned}
\phi_{us} &= 2\frac{\mu_0 NA}{g_0} I_u + x\frac{\phi_0}{3} \\
\phi_{ua} &= \frac{\mu_0 NA}{g_0}\left(\frac{4x}{3} I_u + \frac{x - \sqrt{3}y}{3} I_v + \frac{x + \sqrt{3}y}{3} I_w\right) + \frac{\phi_0}{3}
\end{aligned} \qquad (6)
$$

The values of the AMB and PM fluxes are indicated above to obtain the PM flux level at approximately 0.6 T and the AMB maximum flux level also equal to 0.6 T.

A simple PID controller was used for the control of the radial bearings. Non contact displacement sensors were used in the test rig for levitation testing. Various values for the PID control were tested. Operation above 8,000 rpm was attained. In the final pump, it was intended that a self sensing system, as reported in [17], was used for feedback control.

With this configuration, a sudden load to one side of the impeller is calculated to withstand a g load of 8 based upon the mass of the impeller, the maximum available control current and the reduced order control matrix. It may be noted that a patient with an LVAD falling flat on the ground would produce a load of approximately 2g. Thus, the design for 8g should provide an excellent margin of safety for impact loading on the magnetic suspension.

IV. Conclusions

This paper describes a new hybrid magnetic suspension designed for an axial flow left ventricular assist device. It has a combined active and passive magnetic bearing which minimizes the power consumption of the suspension system with a measured power loss, in the test rig described above, of 2 watts when operating centered in the clearance space at 7,000 rpm. When compared to the calculated motor power of 10 watts when operating with blood flow, this is quite a small value.

A 3-D magnetic finite element model was developed to evaluate the open loop stiffness and current gain values. Also, the effective stiffness, damping and added mass terms were added to the AMB controller model. It should be noted that the blood flow radial stiffness values are of similar magnitude of the magnetic open loop stiffness.

The method of developing the control system matrix based upon the basic pole flux values was developed. However it was rather large in size. A new method of control using a combination of pole flux values was developed. It resulted in a reduced control matrix for easier implementation in a single circuit board controller suitable for a wearable LVAD. The magnetic suspension was constructed and operated in a test rig. Also, the small operating power of the magnetic suspension will likely increase battery life with a single charge. Unfortunately, the complete pump with blood flow and magnetic suspension was never built due to lack of funds.

AUTHOR'S NOTE

The authors are grateful for the funding provided for this project through the ROMAC Industrial Research Program under the Direction of Prof. Paul Allaire.

REFERENCES

[1] A. Allen, "Heart to Heart", *Washington Post,* 2013

[2] P. E. Allaire, H. C. Kim, E. H. Maslen, G. B. Bearnson, and D. B. Olsen, "Design of a Magnetic Bearing Supported Prototype Centrifugal Artificial Heart Pump," *Tribology Transactions*, vol. 39, no. 3, pp. 663–669, 1996.

[3] G. B. Bearnson, E. H. Maslen, D. B. Olsen, P. E. Allaire, P. S. Khanwilkar, J. W. Long, and I. C. Kim, "A proposal and trial on a model of a motion monitor system for a ski jumper using terrestrial magnetism and acceleration sensors," *Journal of American Society of Artificial Internal Organs*, vol. 20, no. 3, pp. 275–281, 1996.

[4] P. Khanwilkar, D. Olsen, G. Bearnson, P. Allaire, E. Maslen, R. Flack, and J. Long, "Using Hybrid Magnetic Bearings to Completely Suspend the Impeller of a Ventricular Assist Device," *Artificial Organs*, vol. 20, no. 6, pp. 597–604, 1996.

[5] E. Hilton, P. Allaire. M. Baloh, E. Maslen, G. Bearnson, P. Khanwilkar, and D. Olsen, "Magnetic suspension controls for continuous flow ventricular assist device," *ASAIO Journal*, vol. 43, no. 5, pp. M598–M602, 1997.

[6] P. Allaire, G. Bearnson, R. Flack, E. Maslen, M. Baloh, E. Hilton, D. Baun, D. Noh, P. Khanwilkar, D. Olsen, A. Goshorn, and T. Waters, "Continuous Flow Magnetically Suspended Centrifugal Flow Pump For Ventricular Assist Device (CFVADIII): Design and Testing," *Artificial Organs*, vol. 22, no. 6, pp. 475–480, 1996.

[7] Y. Nose, T. Tsutsui, K. Butler, R. Jarvik, Y. Takami, C. Nojiri, P. Allaire, O. Van Ram, and J. Dow, "Rotary Pumps – New Developments & Future Perspectives," *ASAIO Journal*, vol. 44, no. 3, pp. 234–237, 1998.

[8] G. Bearnson, D. Olsen, P. Khanwilkar, J. Long, M. Sinnott, A. Kumar, P. Allaire, M. Baloh, and J. Decker, "An Implantable Centrifugal Pump with Hybrid Magnetic Bearings," *ASAIO Journal*, vol. 44, no. 5, pp. M733–M736, 1998.

[9] M. J. Baloh, P. E. Allaire, E. F. Hilton, N. Wei, E. H. Maslen, D. Baun, R. D. Flack, D. B. Olsen, G. B. Bearnson, and P. Khanwilkar, "Magnetic Bearign System for Continuous Flow Ventricular Assist Device," *Artifical Organs Journal*, vol. 45, no. 5, 1998.

[10] M. J. Baloh, P. E. Allaire, E. F. Hilton, N. Wei, D. Baun, R.. Flack, G. B. Bearnson, D. B. Olsen, and P. Khanwilkar, "Characterization of a Magnetic Bearing System and Fluid Properties for a Continuous Flow Ventricualar Assist Device," *Artifical Organs*, vol. 23, no. 8, pp. 792-796, 1999.

[11] P. E. Allaire, J. Decker, M. Baloh, J. Lee, and J. Long, "Magnetic Suspension of an Artificial Heart Pump Using EM/PM Bearings," *Proceedings of the NASA Fifth International Conference on Magnetic Suspension Technology*, 1999.

[12] S. M. Fiser, H. G. Wood, P. E. Allaire, J. M. Long, A. K. Kaza, I. L. Kron, J. A. Kern, and C. G. Tribble, "A Long Term Ventricular Assist Device Utilizing Magnetic Bearing System and Implantable Controller," *Heart Surgery Forum*, vol. 4, no. 1, pp.11-12, 2001.

[13] W. Jiang, P. E. Allaire, M. Baloh, and H. G. Wood, "Magnetic Vector Potential Analysis of Permanent Magnet Suspension System: Forces and Moments," *Proceedings of the Eighth International Symposium on Magnetic Bearings*, pp. 521-525, 2002.

[14] W. Jiang, P. E. Allaire, M. Baloh, and H. G. Wood, "Analysis of Permanent Magnet Suspension System: Stiffness and Stability," *Proceedings of the Eighth International Symposium on Magnetic Bearings*, pp. 527-532, 2002.

[15] J. H. Lee, P. E. Allaire, G. Tao, and J. A. Decker, "Experimental Study of Sliding Mode Control for a Benchmark Magnetic Bearing System and Artifical Heart Pump Suspension," *IEEE Transactions on Control Systems Technology*, vol. 11, no. 1, pp. 128-138, 2003.

[16] X. Song, W. Jiang, H. G. Wood, P. E. Allaire, and D. B. Olsen, "Numerical Analysis for Off-center Performance of Magnetic Bearings," *30th Annual Conference of European Society for Artifical Organs*, 2003.

[17] P. E. Allaire and W. Jiang, "Design and Testing of the Lifeflow Total Magnetic Suspension Left Ventricular Assist Device," *15th Congress of International Society of Rotary Blood Pumps*, 2007.

[18] A. Untariou, H. G. Wood, P. E. Allaire and T. W. Dimond, "Calculation of Dynamic Coeficients for a Magnetically Levitated Artifical Heart Pump Using CFD Approach," *Proceedings of the ASME International Mechanical Engineering Conference*, 2008.

APPENDIX: RADIAL BEARING CONTROL MATRICES

The large control matrix for the radial bearings in terms of the basic pole flux variables and control currents is

$$\begin{bmatrix} 1-x & -1-x & 0 & 0 & 0 & 0 \\ 0 & 1+x & -1-\frac{x}{2}+\frac{\sqrt{3}y}{2} & 0 & 0 & 0 \\ 0 & 0 & 1+\frac{x}{2}-\frac{\sqrt{3}y}{2} & -1+\frac{x}{2}-\frac{\sqrt{3}y}{2} & 0 & 0 \\ 0 & 0 & 0 & 1-\frac{x}{2}+\frac{\sqrt{3}y}{2} & -1-\frac{x}{2}-\frac{\sqrt{3}y}{2} & 0 \\ 0 & 0 & 0 & 0 & 1+\frac{x}{2}+\frac{\sqrt{3}y}{2} & -1+\frac{x}{2}+\frac{\sqrt{3}y}{2} \\ 1 & 1 & 1 & 1 & 1 & 1 \end{bmatrix} \begin{Bmatrix} \phi_{u+} \\ \phi_{u-} \\ \phi_{v+} \\ \phi_{v-} \\ \phi_{w+} \\ \phi_{w-} \end{Bmatrix} = \begin{bmatrix} 2 & 0 & 0 \\ -1 & -1 & 0 \\ 0 & 2 & 0 \\ 0 & -1 & -1 \\ 0 & 0 & 2 \\ 0 & 0 & 0 \end{bmatrix} \frac{\mu_0 NA}{g_0} + \begin{Bmatrix} 0 \\ 0 \\ 0 \\ 0 \\ 0 \\ \phi_0 \end{Bmatrix} \quad (7)$$

in terms of the pole geometry, basic pole flux values, pole axis control currents, and the permanent magnet flux value. The reduced radial bearing control matrix in terms of combined flux variables, control currents, and permanent magnetic flux value is given by

$$\begin{Bmatrix} \phi_{us} \\ \phi_{ua} \\ \phi_{vs} \\ \phi_{va} \\ \phi_{ws} \\ \phi_{wa} \end{Bmatrix} = \frac{\mu_0 NA}{g_0} \begin{bmatrix} 2 & 0 & 0 \\ \frac{4x}{3} & \frac{x-\sqrt{3}y}{3} & \frac{x+\sqrt{3}y}{3} \\ 0 & 2 & 0 \\ -\frac{2x}{3} & \frac{-2x+2\sqrt{3}y}{3} & \frac{x+\sqrt{3}y}{3} \\ 0 & 0 & 2 \\ -\frac{2x}{3} & \frac{x-\sqrt{3}y}{3} & \frac{-2x-2\sqrt{3}y}{3} \end{bmatrix} \begin{Bmatrix} I_u \\ I_v \\ I_w \end{Bmatrix} + \frac{\phi_0}{6} \begin{Bmatrix} 2x \\ 2 \\ -x+\sqrt{3}y \\ 2 \\ -x-\sqrt{3}y \\ 2 \end{Bmatrix} \quad (8)$$

This is much smaller and more suitable for a single circuit board controller.

Development of Extracorporeal Maglev Blood Pumps

Tadahiko Shinshi[a], Wataru Hijikata[a],Taichi Mamiya[a], Setsuo Takatani[b]

[a] Tokyo Institute of Technology, 4259 Nagatsuda-cho, Midori-ku, Yokohama, Japan, shinshi@pi.titech.ac.jp
[b] MedTech Heart Inc., 3-24-8, Asakusabashi, Taito-ku, Tokyo, Japan

Abstract- **In centrifugal blood pumps, contactless support of the impeller and a sufficient fluid gap in the pump head have been identified as features that have the potential to keep blood damage to a minimum and enhance the durability of the blood pump. We have developed three types of cost-effective magnetic bearings and contactless torque transmission mechanisms for disposable centrifugal blood pumps which can be used for extracorporeal circulation. In this paper, we introduce prototype maglev centrifugal blood pumps that utilize these mechanisms and we present both the experimental results and the results of simulations carried out on them.**

INTRODUCTION

For bridge to implant of a donor heart or expensive implantable ventricular assist device, a cost-effective, highly durable extracorporeal blood pump system that does not cause thrombus formation and hemolysis is required. We have developed three types of magnetic bearing and contactless torque transmission mechanisms for centrifugal blood pump systems that have disposable simply structured magnetically-levitated impeller-rotors.

First, in this paper, we introduce these three types of magnetic bearing and contactless torque transmission mechanisms [1]-[5]. Then, the designs and prototypes of three types of disposable maglev centrifugal blood pump that use these mechanisms are described. Finally, the mechanical performance of all three pumps in mock circulatory systems is described. Details of biocompatibility of Types 1 and 2 pump heads evaluated using animal tests [6]-[9] are given. Furthermore, the possibility of blood damage in Types 1 and 3 pump heads is discussed based on CFD simulation results.

MAGNETIC BEARING AND TORQUE TRANSIMISSION MECHANISMS FOR EXRTACORPOREAL MAGLEV PUMPS

Figures 1(1), (2) and (3) show the proposed magnetic bearing and contactless torque transmission mechanisms. The stators in each case are placed on the outer side of the disposable rotors. The radial motion of the rotor is actively controlled by the electromagnets in the stator. The axial and tilt motion of the rotor is passively supported by the magnetic coupling between the rotor and the core of the electromagnet. In Type 1, the rotor consists of a rare earth magnet ring sandwiched between two slotted-iron rings. In Types 2 and 3, in order to reduce the cost of the permanent magnets, the disposable rotor consists of slotted iron parts only.

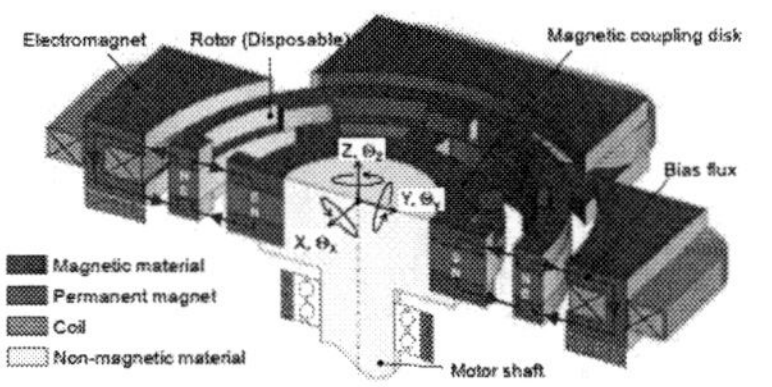

(1) Outer magnetic bearing and inner torque transmission mechanism using a rare earth magnet ring in a disposable rotor (Type 1)

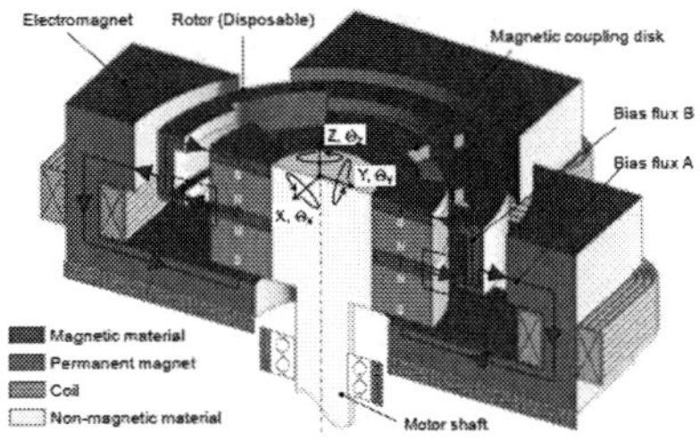

(2) Outer magnetic bearing and inner torque transmission mechanism using no permanent magnet ring in a disposable rotor (Type 2)

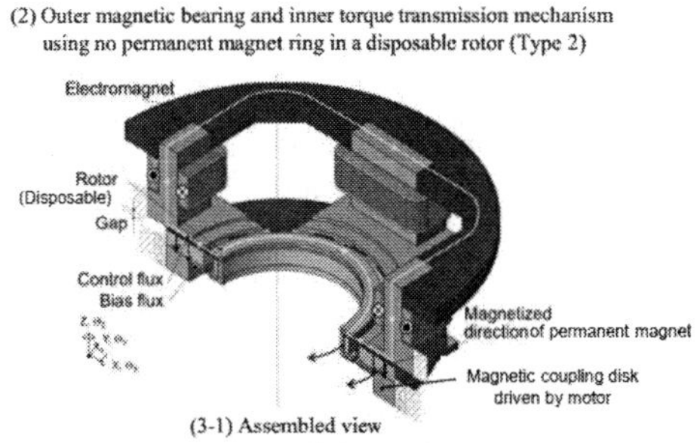

(3-1) Assembled view

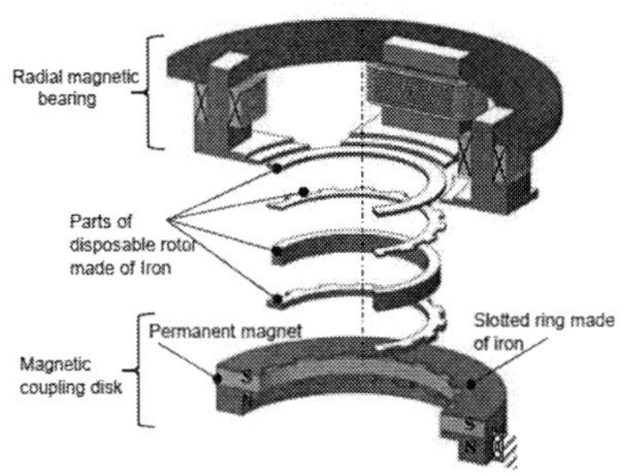

(3-2) Exploded view

(3) Outer magnetic bearing and outer torque transmission mechanism using no permanent magnet ring in a disposable rotor (Type 3)

Figure 1. Proposed magnetic bearing and noncontact torque transmission mechanisms for disposable centrifugal blood pumps.

To rotate the rotor, three types of torque transmission mechanism are proposed. In Types 1 and 2, the magnetic couplings are placed on the inner side of the disposable rotor. The motor torque is transmitted to the rotor using the magnetic coupling between the rotor and a coupling disk, which is directly driven by a motor.

Type 3 also uses magnetic coupling mechanism to transmit the motor torque, but the coupling disk is placed on the outer side of the disposable rotor. Compared with Types 1 and 2, the configuration of Type 3 has high design flexibility for the impeller-rotor and smooth secondary flow can be generated under the impeller-rotor in the pump head as shown in Figure 2. The smooth secondary flow is expected to reduce blood damage.

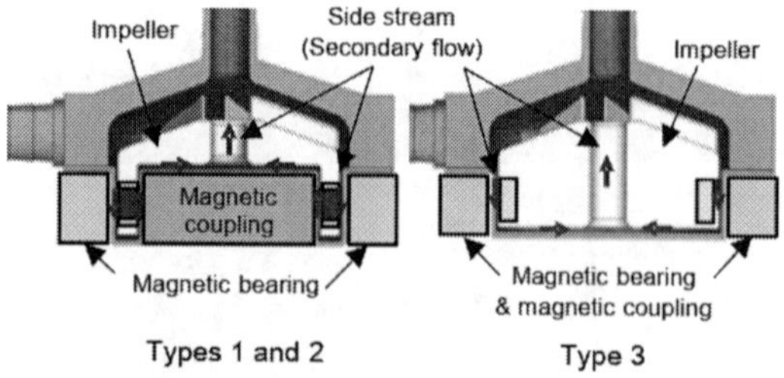

Figure 2. Magnetic bearing and magnetic coupling configurations and the secondary flow channel design in the pump heads.

DESIGN AND FABRICATION OF THE MAGLEV CENTRIFUGAL PUMPS

Three types of maglev centrifugal blood pump were designed and fabricated as shown in Figures 3 (1), (2) and (3). Figure 4 shows a photograph of the preclinical maglev centrifugal blood pump system including the magnetic bearing and motor drivers, the monitor for the Type 1 blood pump and an artificial lung.

All the pumps consist of a pump head, a two-degree-of - freedom controlled magnetic bearing, a magnetic coupling disk and noncontact displacement sensors and a motor. The pump head consists of disposable top and bottom housings and an impeller-rotor. The external dimensions of the impeller used for these pumps are almost the same. The impeller has six straight vanes and its diameter is 50mm. The radial and axial minimum fluid gaps between the rotor and the bottom housing is set to be 300μm and 1mm, respectively.

These magnetic bearings were designed to have the same magnetic gaps and almost the same passive stiffness in the axial and tilt directions. Furthermore, the designed maximum transmittable torque using the magnetic coupling is almost the same.

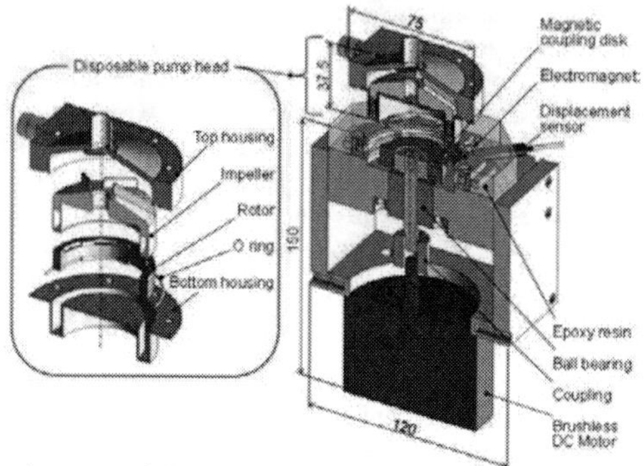

(1) Outer magnetic bearing and inner torque transmission mechanism using a rare earth magnet ring in a disposable rotor (Type 1)

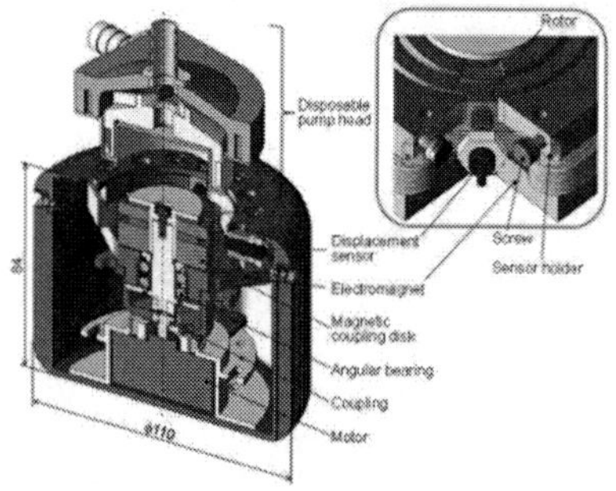

(2) Outer magnetic bearing and inner torque transmission mechanism using no rare earth magnet ring in a disposable rotor (Type 2)

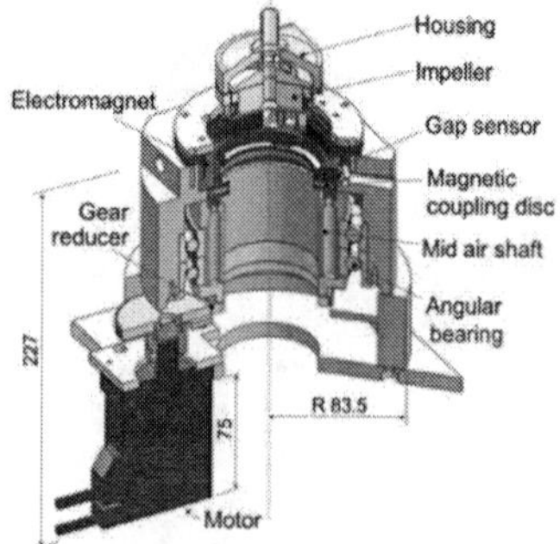

(3) Outer magnetic bearing and outer torque transmission mechanism using no rare earth magnet ring in a disposable rotor (Type 3)

Figure 3. Designed and fabricated maglev centrifugal blood pumps.

EVALUATION OF PUMP PERFORMANCE AND BIOCOMPATIBILITY

In the development of blood pumps, not only pump performance but also evaluation of the biocompatibility is important. All the maglev pumps achieved contactless levitation and rotation at speeds of more than 3,250rpm. The rotational accuracy of the impeller-rotor was measured. The radial vibration of the three types of pump during pumping was approximately within ±50μm which is much smaller than the fluid gap of 300μm.

Figure 5 shows the measured H-Q characteristics of the Type 2 blood pump in a 40 per cent glycerol–water solution. The other H-Q characteristics of Types 1 and 3 blood pumps are the same. This performance satisfies the requirements for both left ventricular assist and heart-lung support.

Hemolytic characteristics of Types 1 and 2 blood pumps were evaluated by in-vitro tests using fresh porcine blood. The normalized index of hemolysis (NIH) is half of that generated by a conventional centrifugal blood pump using a contact bearing (BioPump BPX-80, Medtronic).

In the Tokyo Medical and Dental University, Types 1 and 2 blood pumps were implanted into young Holstein calves for ventricular assist tests for up to two months as shown in Figure 6 and for cardiopulmonary bypass tests for up to three weeks as shown in Figure 7. After the improvements of manufacturing process of the pump heads, no problems occurred during the animal tests and the disposable pump heads did not need to be changed. Both types of blood pump achieved high durability and high biocompatibility.

Figure 4. Preclinical maglev centrifugal blood pump system using Type 1 magnetic bearing and magnetic coupling mechanisms

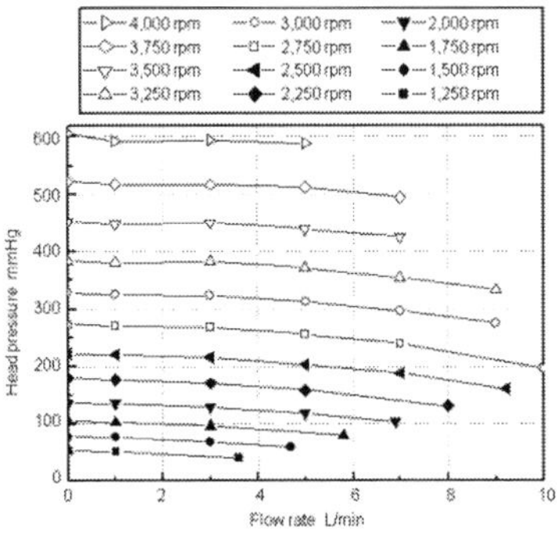

Figure 5. Experimental HQ characteristics in a 40 per cent glycerol-water solution (Type 2 blood pump).

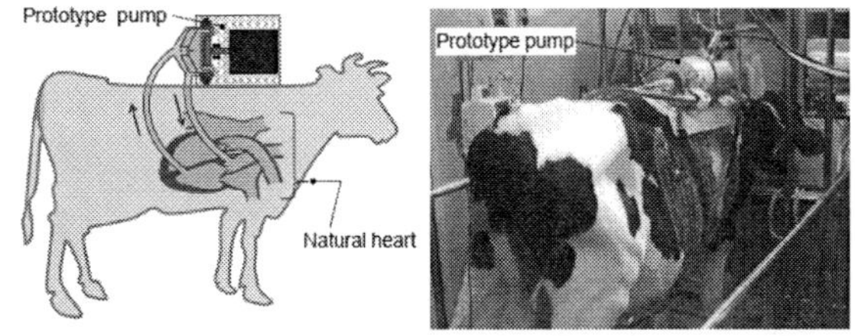

Figure 6. Left ventricular assist test using a centrifugal blood pump of Type 1 blood pump.

In order to evaluate the effectiveness of using a Type 3 pump head on blood damage, streak-lines in the secondary flow in Types 1 and 3 pump heads were simulated and compared using the CFD models shown in Figure 8. The pump head profile of Type 1 is the same as that of Type 2. Assuming a flow rate of 5L/min at 2,000rpm, the transit time and the length of the trajectory of each particle passing through the area between the impeller and bottom housing were calculated and the results are summarized in Figure 9.

The transit time and length of trajectory of the particle passing through the Type 3 pump head are smaller than those for Type 1 pump head. It is concluded that the secondary flow pass in the Type 3 pump head is much smoother and the damage to blood cells is expected to be less compared with Types 1 and 2 pump heads.

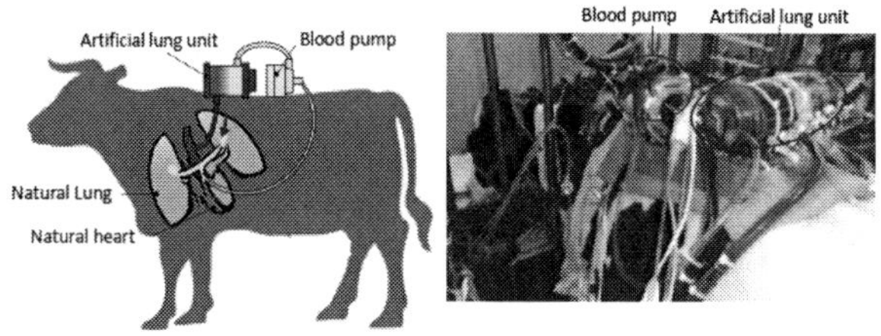

Figure 7. Cardiopulmonary bypass test using a centrifugal blood pump of Type 1.

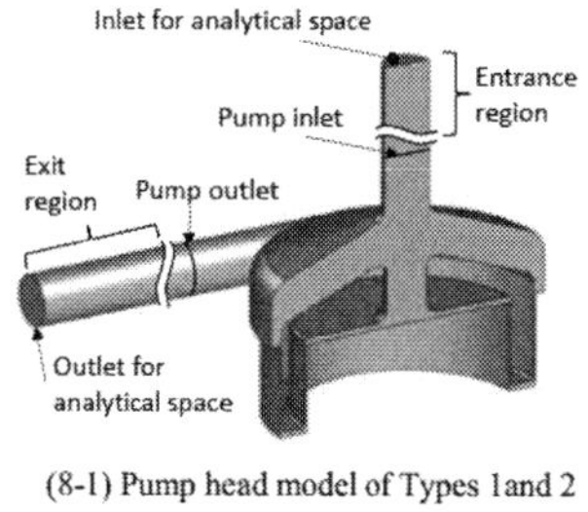

(8-1) Pump head model of Types 1and 2

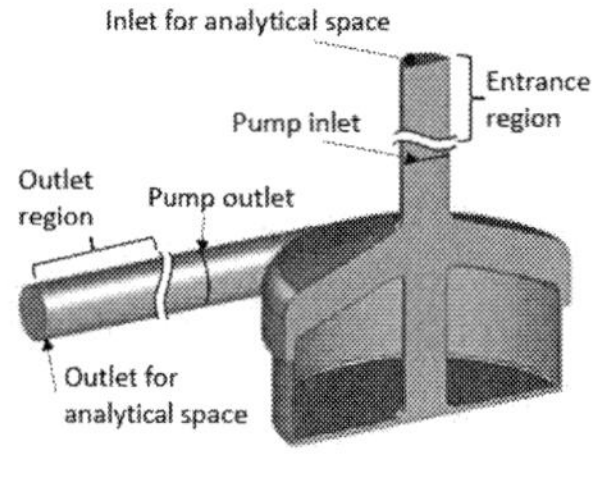

(8-2) Pump head model of Type 3

Figure 8. CFD models of Types 1 and 3 pump heads

CONCLUSION

In this paper, three types of disposable centrifugal blood pump using cost-effective maglev impeller-rotors were introduced and the pump performances and biocompatibilities reported. In future work, we plane to carry out in-vitro and in vivo tests using a Type 3 blood pump and to develop a more compact bearingless motor using disposable rotor-impeller without a permanent magnet.

This work was partly supported by JSPS KAKENHI Grant No. 24360059 and a research grant from the Mitutoyo Association for Science and Technology.

REFERENCES

[1] W. Hijikata, T. Shinshi, J. Asama, L. Li, H. Hoshi, S. Takatani, A. Shimokohbe, " A MagLev Centrifugal Blood Pump with a Simple-Structured Disposable Pump Head, "Artificial Organs, vol. 32, no. 7, pp. 351-540, 2008.

[2] W. Hijikata, H. Sobajima, T. Shinshi, Y. Nagamine, S. Wada, S. Takatani, A. Shimokohbe, "Disposable MagLev Centrifugal Blood Pump Utilizing a Cone-Shaped Impeller," Artificial Organs, vol. 34, no. 8, pp. 669-676, 2010.

[3] W Hijikata, T Mamiya, T Shinshi, S Takatani, "A cost-effective extracorporeal magnetically-levitated centrifugal blood pump employing a disposable magnet-free impeller," Proc. IMechE, J. Engineering in Medicine, vol. 225, pp. 1149-1157, 2011.

[4] T. Mamiya, W. Hijikata, T. Shinshi, "A Disposable Magnetic Bearing System for an Extracorporeal Centrifugal Blood Pump," Proceedings of The 13th International Symposium on Mgnetic Bearings, Washignton, USA, 2012.

[5] K. Momose, T. Mamiya, W. Hijikata, T. Shinshi, "An Extracorporeal Maglev Centrifugal-type Blood Pump Employing a Permanent Magnet-free Disposable Pump Head and an Outer Magnetic Coupling Mechanism, " Journal of the Japan Society for Precision Engineering, vol. 80, no. 2, pp. 81-88. 2014. (in Japanese)

[6] T. Someya, M. Kobayashi, S. Waguri, T. Ushiyama, E. Nagaoka, W. Hijikata, T. Shinshi, H. Arai, S. Takatani, "Development of a Disposable Mag-Lev Centrifugal Blood Pump (MedTech Dispo) Intended for One Month Support in Bridge-to-Bridge Applications - In Vitro and Initial In Vivo Evaluations," Artificial Organs, vol. 33, no.9, pp.704-713, 2009.

[7] E. Nagaoka, T. Someya, T. Kitao, T. Kimura, T. Ushiyama, W. Hijikata, T. Shinshi, H. Arai, S. Takatani, "Development of a Disposable Mgnetically Levitated Centrifugal Blood Pump (MedTech Dispo) Intended for Bridge-to-Bridge Applications Two-Week In Vivo Evaluation s, "Artificial Organs, vol. 34, no. 9, pp. 778-783, 2010.

[8] T. Fujiwara, E. Nagaoka, T. Watanabe, N. Miyagi, T. Kitao, D. Takota, T. Mamiya, T. Shinshi, H. Arai, S. Takatani, " New Generation Extracorporeal Membrane Oxygenation With MedTech Mag-Lev, a Single-Use, Magnetically Levitated, Centrifugal Blood Pump: Preclinical Evaluation in Calves," Artificial Organs, vol. 37, no. 5, pp. 447-456, 2013.

[9] E. Nagaoka, T. Fujiwara, D. Sakota, T. Shinshi, H. Arai, S. Takatani, "MedTech Mag-Lev, Single-use, Extracorporeal Magnetically Levitated Centrifugal Blood Pump for Mid-term Circulatory Suppor, "ASAIO Journal, vol. 59, no.3, pp. 246-252, 2013.

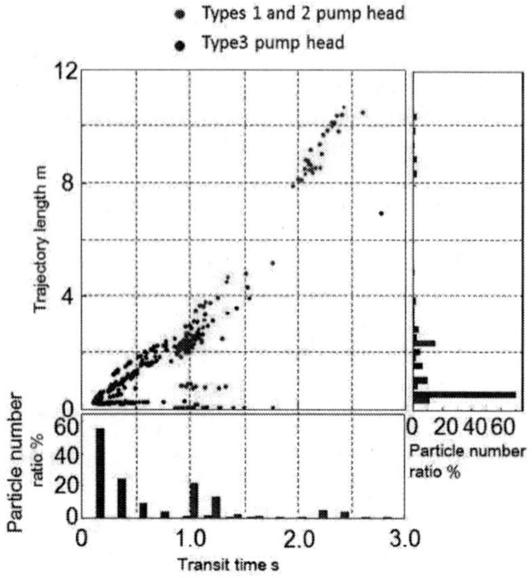

Figure 9. Analysis of streak-lines in the secondary flow of Type 1 and Type 3 pump heads

Optimization design of magnetically suspended system for the BiVACOR total artificial heart

Nobuyuki Kurita[a], Daniel Timms[b], Nicholas Greatrex[b], Matthias Kleinheyer[b], Toru Masuzawa[c]

Gunma University, 1-5-1 Tenjin, Kiryu, Gunma, JAPAN, nkurita@gunma-u.ac.jp
[b] Texas heart Institute, 6770 Bertner Avenue, Houston, Texas, USA, dtimms@texasheart.org
[c] Ibaraki University, 4-12-1 Nakanarusawa, Hitachi, Ibaraki, JAPAN, masuzawa@mx.ibaraki.ac.jp

***Abstract*— The BiVACOR device is a third generation rotary total artificial heart under development to treat end stage cardiovascular disease. Non-contact suspension and rotation of the impeller of the blood pump was achieved by a combination of a permanent magnet hybrid type magnetic bearing and an axial flux brush-less DC motor. Moreover, according to a series of animal trials, functionality of the proposed magnetic system in the total artificial heart application was clarified.**

I. INTRODUCTION

Due to the high rate of heart disease and shortage of donor hearts [1], there has been a longstanding interest to develop mechanical replacement for the heart. Recently, third generation rotary blood pumps of implantable size have shown considerable promise in this application [2]–[6]. These devices utilize magnetic or hydrodynamic bearings support the impeller which eliminates any mechanical contact. Therefore, the operational mechanical lifetime of these devices may exceed ten years.

The BiVACOR, third generation rotary total artificial heart (TAH) is under development to treat end stage cardiovascular disease [7], [8]. It utilizes the magnetic suspension system to not only reduce device wear, but to assist in automatically balancing left and right device outflows by utilizing virtual zero power control algorithms. This paper provides an insight into the TAH's magnetic system components, including the magnetic bearing and axial motor technologies, the resulting flow balancing performance when implemented in a chronic animal trial.

II. EXPERIMENTAL DEVICE

A. Total Artificial Heart

Figure 1 shows an exploded view of the BiVACOR TAH. The outer diameter and height of the device are approx. 60 mm and 70 mm, respectively. The device includes a set of left and right impeller vanes positioned on opposing sides of the rotor core to form double-sided centrifugal impeller. Active axial rotor suspension is achieved with three axial magnetic bearings and rotated via an axial flux motor. Although they are omitted in figure, three eddy current displacement sensors are arranged between the magnetic bearings to measure axial displacement.

Since the left and right impeller blades are inherently coupled to the common rotating rotor core, if rotational speed is slowed, the both of the outflow of the left and right pumps

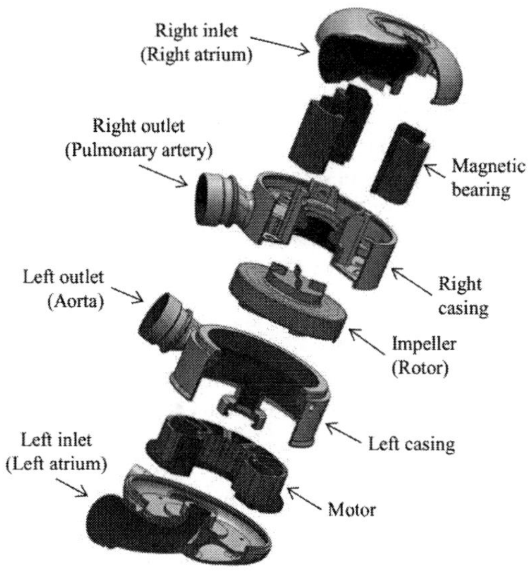

Figure 1. Schematic of developed magnetically levitated TAH

will be reduced simultaneously. However, to account for the required changes in relative left and right outflow, the magnetic bearing system can displace the impeller axially with in the pump cavity. This action simultaneously alters the efficiency of each semi-open impeller vane set in an inverse relationship. For example, a movement toward the left cavity will increase left out flow while reducing right outflow, and vice versa [7].

B. Magnetically Levitation Rotation System

The magnetic bearing is composed of three horseshoe shaped electromagnets in the casing and two ring shape permanent magnets on the rotor to produce bias flux. These magnetic bearings are evenly spaced 120° apart; regulating axial displacement and tilt of the rotor impeller, depending on the feedback from the displacement sensor signals. Figure 2(a) shows the components of the magnetic bearing. The horseshoe shaped stator core has a coil winding on its one leg. Two ring shape permanent magnets (Inner ring PM and outer ring PM) are attached on the rotor disk as the bias PMs which produces bias flux at the airgap between the stator and the rotor. Figure 2 (b) and (c) shows the operating principle of the magnetic bearing for the position control mode. As mentioned above, the bias flux path was generated by the bias PM as shown in the figure by the green dotted line. When the rotor was moved by the disturbance force, the displacement sensors are detects the rotor movements and feedback the displacement signal to the PID controller. The PID controller applies adequate control current to the windings. Figure 2(b) shows operating principle in order to increase magnetic force. The flux density at the airgap is strengthened by the control flux shown by the orange solid arrow. In a contrasting situation, the airgap flux density is weakened by the negative control current in the Figure 2(c).

Figure 2(d) and (e) shows the operating principle for the zero power modes. When the rotor was moved by the disturbance force, the detected displacement signals are feedback the signal to the virtual zero power (VZP) controller which regulates the control current to be zero. As shown in Figure 2(d), when the disturbance force acts on the rotor downward, the rotor moves upward where the disturbance force and the permanent magnet attractive force are balanced. In a contrasting situation, when the disturbance force acts on the rotor upward, the rotor moves downward in order to weaken the airgap flux density as shown in Figure 2(e).

According to the VZP control, the distance of the rotor displacement D_R (m) is depends on the strength of the disturbance force F_D (N) and the stiffness of the magnetic levitation system K_{MLV} (N/m) as shown in Equation (1).

$$D_R = \frac{F_D}{K_{MLV}} \quad (1)$$

In general zero-power control, the stiffness of the magnetic levitation system K_{MLV} is a constant value which is the difference between the magnetic bearing suspension force stiffness K_{MB} and the motor attractive force stiffness K_M. However, the magnetic bearing suspension force stiffness is controllable by the small constant control current of the electromagnets, i.e. ΔK_E. By substituting K_{MB}, K_M and ΔK_E to Equation (1), D_R is expressed as follows.

$$D_R = \frac{F_D}{(K_{MB} + \Delta K_E) - K_M} \quad (2)$$

According to the Equation (2), the distance of the rotor displacement D_R is also controllable in VZP mode. In this paper, however, disturbance force which acts on the rotor is unmeasurable. Only the pressures (inlet and outlet pressure of both of left and right pumps) are measurable. Thus, we don't use disturbance force but disturbance pressure. Because the surface area of the impeller won't change, the force and the pressure essentially have the same implications in this paper.

The axial flux motor is composed of twelve stator poles and eight permanent magnets on the rotor as field magnets. It is operated as a permanent magnetic synchronous motor using back EMF to determine the rotor position. The left impeller is installed in the motor side airgap. Therefore, the airgap width between the surface of the stator pole and permanent magnet is very wide (3.4mm).

C. Automatic Rotor Position Control

Systemic vascular resistance (SVR) and the pulmonary vascular resistance (PVR) are altered by various causes such as changes in posture, sneezing, exercise and so on. Decrease

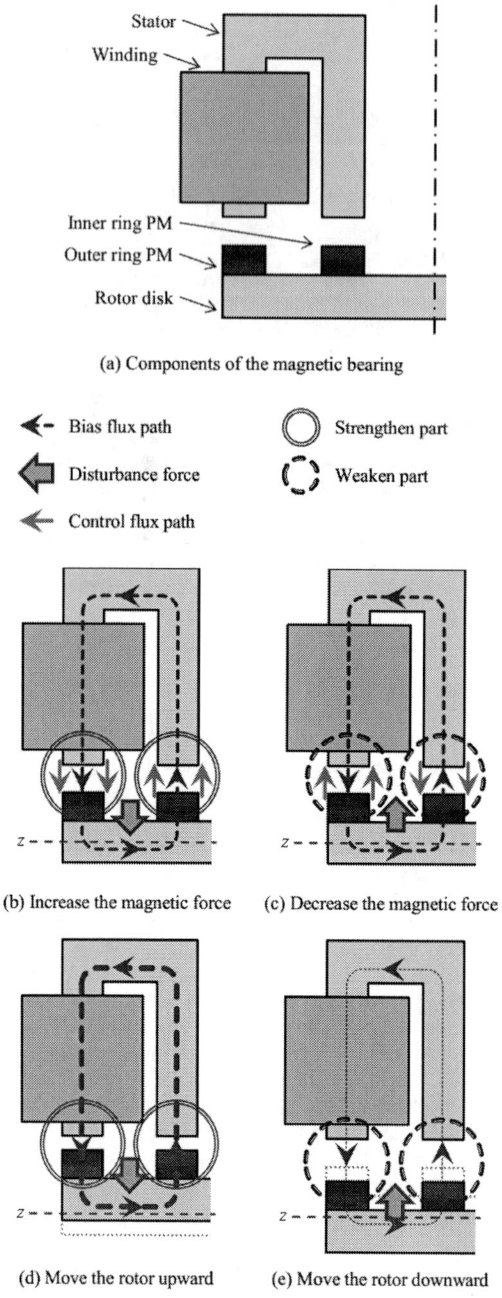

Figure 2. Operating principle of the magnetic bearing

of SVR results in an increase of left sided flow rate of the TAH. This causes a drop in the left atrial pressure (LAP) and the aortic pressure (AOP). If the LAP become below 0 mmHg, a suction event may occur. In order to avoid such an event, the rotor should move toward right side to reduce the pressure differential and thus outflow rate of the left pump. As previously described, to accommodate these required changes in flow rate and pressure, the rotor axial position should be varied by the magnetic bearing. Automatic rotor position compensation is achieved by the VZP. When the AOP and the

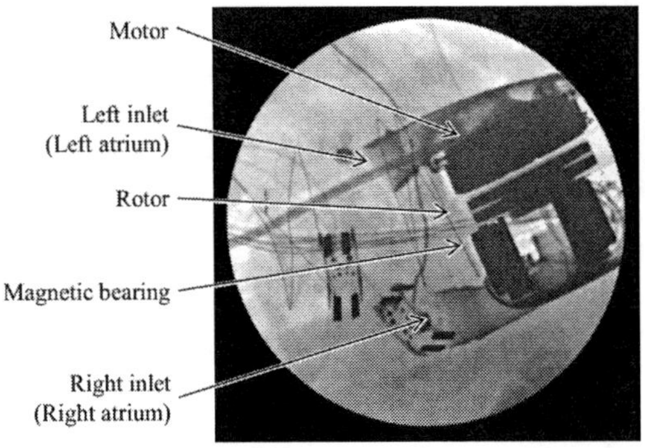

Figure 3. Angiogram of the BiVACORE during animal trial

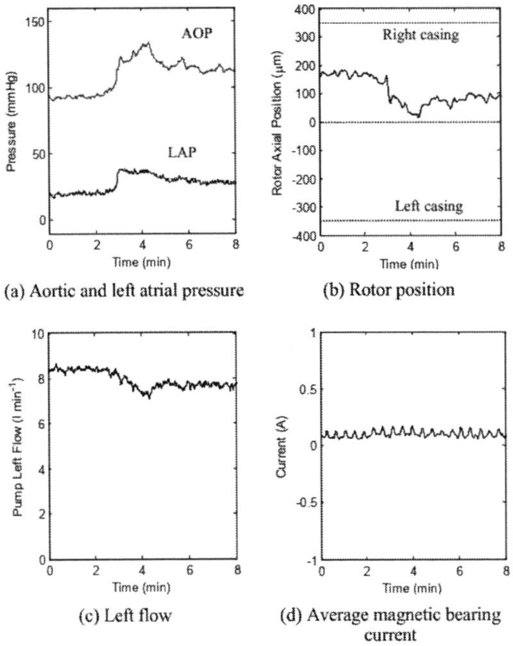

(a) Aortic and left atrial pressure (b) Rotor position

(c) Left flow (d) Average magnetic bearing current

Figure 4. Results of the chronic animal trial

LAP decrease, the hydraulic force from left side to right side also decreases. Consequently VZP moves the rotor toward right side to balance the hydraulic force and the magnetic force of the magnetic system. An increase of SVR causes an increase of the LAP. In order to decrease the LAP and also increase the right atrium pressure (RAP) and pulmonary artery pressure (PAP), the rotor should be moved toward left side. When the LAP increased, the hydraulic force from left side to right side also increases. Then VZP automatically responds to move the rotor toward left side.

Although this controller is designed to normally operate with a target current of 0 A, the set point of the current control can be modified to alter the rotor's steady state position. This can be used by a user to improve the balancing of the TAH in particular patients and scenarios.

III. Experimental Result

In order to verify the functionality of the developed BiVACOR TAH, a recovery trial was completed. An 87.5 kg male calf was supported for more than 2 days. Figure 3 shows angiogram of the BiVACOR TAH during the animal trial. According to the figure, it is clear that the stable magnetic levitation was achieved by magnetic bearings and the axial flux motor. In the position control mode, the axial directional rotor position movement of ±0.3mm was also achieved. Moreover, the stable levitated rotation was also achieved in the objective speed range from 1,800 min^{-1} to 2,800 min^{-1}. In addition, the pulsatile speed control ability was also confirmed, in order to produce washout flow to prevent thrombus formation within the device.

Figure 4 shows the results of an event in which the calf returned to the sitting position after standing for approximately 15 minutes. As shown in Figure 4(a), when the sit-down event occurs, at t = 3 s, the AOP and LAP rise causing the hydraulic force acting on the rotor to increase towards the right casing. The VZP controller acts to resolve this force and moves the rotor 130 μm towards the left of the pump casing increasing the relative efficiency of the left pump (Figure 4 (b)). Approximately 5 seconds after the event occurred the system stabilizes with a steady state rotor position of approximately 70 μm towards the left casing from its pre-event position. The change in posture caused a steady state increase to the SVR resulting in a new flow rate of 7.5 ℓ/m which is 0.5 ℓ/m lower than before the event (Figure 4 (c)). The AOP and LAP increased by 20 mmHg and 7mmHg respectively. As mentioned above, the stiffness of the rotor movement versus disturbance pressure is controllable by the small constant control current of the electromagnets. The average magnetic bearing current remained close to its set point of 0.1 A throughout the event (Figure 4 (d)) with the periodic breathing events clearly visible on the signal.

IV. Discussion

The performance of the magnetic system was sufficient to maintain stable levitation and rotation at speeds required for TAH support. The VZP controls the rotor axial position automatically to compensate flow rate and pressure to avoid suction events.

By the VZP, the rotor moves axially due to the pressure difference of the left and right pump. The pressure acting on the rotor can be calculated as follows.

$$\Delta P = (AOP - LAP) - (PAP - RAP)$$

The pressure difference before and after the stand-up event was $\Delta P = 9$ mmHg. The change of rotor displacement was 70 μm. Therefore, the system stiffness is 7.8 μm/mmHg. This system stiffness was achieved by configuring the magnetic suspension system to have a low magnetic stiffness. The magnetic airgap of the motor and magnetic bearings were chosen to be large to lower the magnetic stiffness while maintaining the location of the magnetic zero point. The resulting system has yielded appropriate system stiffness; however the larger airgaps reduce the efficiency and performance of the motor and magnetic bearings. To improve the power consumption of the device changes to the motor, such as increasing the winding number and pole configuration will be investigated, whilst trying to maintain the desired stiffness demonstrated in this system.

The axial flux motor was driven by sensorless position detection using back EMF. It achieved pulsatile speed control, in order to produce washout flow to prevent potential thrombus formation. However, wide magnetic airgaps and cyclic speed change leads to a reduction of efficiency. The energy consumption of the implantable TAH should be as small as possible. Therefore, efficiency of the axial flux motor should be improved by an increase of the winding number and different configuration of motor poles.

V. CONCLUSION

A magnetic system for the BiVACOR TAH was designed and developed. The magnetic levitation system enables axial movement of the rotor during rotation. According to a chronic animal trial, stable levitated rotation was achieved in the objective speed range of the designed blood pump. The VZP indicated good automatic rotor position control performance in response to variations in pressure, and thus effectively balanced left/right outflows. Moreover, the trial clarified the functionality of the BiVACOR TAH.

REFERENCES

[1] American Heart Association, "Heart disease and stroke statistics – 2009 update", A report from the American Heart Association Statistics Committee and Stroke Statistics Subcommittee (2009)

[2] Peter A. Watterson, John C. Woodard, Victor S. Ramsden and John A. Reizes, "Ventr Assist Hydrodynamically Suspended, Open, Centrifugal Blood Pump", Artificial Organs, Volume 24 Issue 6, pp. 475-477 (2001)

[3] Toru Masuzawa, Shiroh Ezoe, Tsuyoshi Kato and Yohji Okada, "Magnetically Suspended Centrifugal Blood Pump with an Axially Levitated Motor", Artificial Organs, Volume 27 Issue 7, Pages 631-638 (2003)

[4] Hideo Hoshi, Tadahiko Shinshi and Setsuo Takatani, "Third-generation Blood Pumps With Mechanical Noncontact Magnetic Bearings", Artificial Organs, Volume 24 Issue 6, pp. 324-338 (2006)

[5] Hideyuki Fumoto, David J. Horvath, Santosh Rao, Alex L. Massiello, Tetsuya Horai, Tohru Takaseya, Yoko Arakawa, Nicole Mielke, Ji-Feng Chen, Raymond Dessoffy, Kiyotaka Fukamachi and Leonard A. R. Golding, "In vivo acute performance of the Cleveland Clinic self-regulating, continuous-flow total artificial heart", The Journal of Hert and Lung Translplantation, Vol. 29, No. 1, pp. 21-26 (2010)

[6] O.H. Frazier and William E. Cohn, "Continuous-Flow Total Heart Replacement Device -Implanted in a 55-Year-Old Man with End-Stage Heart Failure and Severe Amyloidosis-", Tex Heart Institute Journal Vol. 39, No. 4, pp. 542-546 (2012)

[7] Nobuyuki Kurita, Daniel Timms, Nicholas Greatrex, Toru Masuzawa, "Axial Magnetic Bearing Development for the BiVACOR Rotary BiVAD/TAH", The proceedings of the 11th International Symposium on Magnetic Bearings, pp. 217-224 (2008)

[8] Nicholas A. Greatrex, Daniel L. Timms, Nobuyuki Kurita, Edward W. Palmer and Toru Masuzawa, "Axial Magnetic Bearing Development for the BiVACOR Rotary BiVAD/TAH", IEEE Transactions on Biomedical Engineering, Vol. 57, No. 3, pp 714-721 (2010)

Landing Tests with a 6300rpm, 9t AMB Rotor in Rolling Element Back-up Bearings

Dr. Joachim Denk[a], Dr. Bert-Uwe Köhler[a], Dr. Günther Siegl[a], Peter Siebke[b]

[a] Siemens AG, I DT LD AP MG MF-DW EN 4 1, Nonnendammallee 72, 13629 Berlin, Germany, joachim.denk@siemens.com
[b] Schaeffler Technologies GmbH & Co. KG, Industriestrasse 1-3, 91074 Herzogenaurach, Germany, siebkpte@schaeffler.com

***Abstract*— This article describes a systematic approach to qualify backup bearings for very large rotors levitated by active magnetic bearings (AMB). The approach comprises dynamic simulations of the landing events prior to the testing with good prediction quality as part of the safety concept. The approach was applied for the backup bearing qualification of a 9t rotor with a rotational speed up to 6420rpm. The results of the landing tests are shown and compared with the simulation results.**

I. Introduction

A. Background

As previously described in [1], Siemens AG has developed a new active magnetic bearing system based on SINAMICS standard drive technology. As part of the development, back-up bearings based on ball bearing technology have been developed in cooperation with Schaeffler Technologies GmbH & Co. KG (Schaeffler). It is widely agreed in the AMB community that a rotor in ball back-up bearings is less likely to develop a backward whirl than a rotor in sleeve bearings due to more advantageous friction conditions. To qualify the back-up bearings for large high-speed applications with high rotor weights, several speed-graded landing tests [2] or [3] with an electric 6300rpm synchronous double-end drive prototype motor are performed. To our knowledge, using ball back-up bearing technology in a motor application of this size with a 9t rotor, a maximum speed of 6300rpm and a backup bearing inner diameter of 360mm is new.

B. Objectives of the landing test program

With a specific customer application in mind, the landing test qualification program is to provide evidence that the back-up bearings fulfill the customer, i.e. that they are capable of

- withstanding a minimum of 3 drops from full speed (6300 rpm) with a braking time of 13 seconds realized with the converter actively braking the motor
- handling one full rundown with a braking time as long as 100 seconds without damage to the motor, while back-up bearing damage is allowed; the prolonged braking time simulates a malfunction of the braking system, in which case the process forces of the application define the braking duration

If the condition of the backup bearings still permits it, after the qualification program has been completed, the setup will be used for additional experiments to gain experience on special malfunctions in the magnetic bearing system, such as the failure of only one magnetic bearing axis.

C. Overview of the preparation and execution of the tests

Due to the novelty of the application, considerable efforts were made during the development of the bearings as well as the preparation and execution of the tests in order to ensure safety and to make the tests and thus the qualification of the product a success:

- Extensive simulation studies during development of the back-up bearings [4] performed by Schaeffler, including FEM and transient landing simulations.
- Landing tests in a smaller application with a 1.3t rotor with higher rotor speed (12,700rpm) in order to gain experience with comparable circumferential bearing speeds, which are an important parameter in a ball-bearing application.
- Definition of a qualification program delivering a certain level of confidence that the back-up bearing design will fulfill the customer requirements with a sufficient safety margin.
- Transient landing simulations of the complete qualification program as preparation for the landing tests, using both a simulation environment developed by Schaeffler [4] and a simulation environment developed by Siemens AG [5].
- Safety concept.
- Extensive landing test procedure with well-defined checks after a landing as well as stop/go criteria for execution of the next landing test.

The various steps are explained in more detail in the following.

II. System Setup

A. Prototype Motor Setup

The prototype double-end drive motor has been designed for a maximum speed of 6300rpm and a trip speed of 6420rpm (exceeding the trip speed results in an automatic braking command for the converter). The motor comprises two radial magnetic bearings, see Figure 1. The two radial back-up bearings are installed on the outboard side of the magnetic bearings, see Figure 2. The motor is not equipped with an axial bearing, since axial bearings are part of the driven machine. However, axial bump stops are integrated into the radial back-up bearings.

Figure 1. Radial magnetic bearing of the prototype motor from the SIMOTICS Active Magnetic Bearing series.

Figure 2. Radial back-up bearing of the prototype motor.

The back-up bearings are double-raceway ball bearings with ceramic balls and an elastic element [4]. The back-up bearing gap for this large application is 500μm to account for thermal expansion and assembly tolerances. The back-up bearing gap is about 25% of the AMB air gap, and significantly smaller than the air gap of the motor stator, which prevents, under all circumstances, rotor-stator contact at any points other than the back-up bearings.

In order to simulate the future on-site system configuration as closely as possible during the landing tests, the rotor is held in its axial position by the original EDE (exciter drive end) coupling, which is attached to a spindle-like device with two pre-stressed pairs of ball bearings, see Figure 3. The expected radial deviation of the rotor during the landing tests can be tolerated by the EDE coupling. At the DE side, the actual DE (drive end) coupling is simulated using a dummy coupling to obtain identical rotor dynamics.

Figure 3. Motor setup during the landing tests with the original coupling and coupling spacer mounted on the EDE side to provide axial guidance.

The motor is located in a room with reinforced walls for safety reasons. Due to the restricted space in this room, the motor cover and cooler were removed for the landing tests.

The motor is driven by a converter so that the motor can be braked from top speed within less than 20sec. The rotor position in the magnetic bearing, the bearing current, the bearing voltage and the rotor speed are monitored by the SIMOTICS active magnetic bearing system. A trip signal is generated if certain threshold values are exceeded indicating e.g., that the rotor has left a position window around the setpoint position. In turn, a trip causes the converter to perform an emergency stop and brake the motor.

A drop switch deactivates the enable signals of the AMB axes, causing the rotor to land in the back-up bearings. Depending on the parameterization of the drives, all or specific bearing axes can be deactivated. Reactivating the enable signals with the drop switch allows the rotor to be relevitated, also see section IV.B.

The motor can be braked by the converter at any time by pressing an emergency stop button.

B. Back-up bearing instrumentation

The radial magnetic bearing position sensors are mounted close to the radial back-up bearing. This means that their signals can be used as a measure for the rotor position in the back-up bearings.

Since the back-up bearings are based on ball bearing technology, it can be expected that the temperatures in the backup bearings during run-down remain low. This is verified by measuring outer ring- and inner ring temperatures of one back-up bearing using probes in direct contact (outer ring) or optical temperature probes (inner ring).

III. Landing Test Program

A. Preliminary landing tests

To establish an initial proof of concept that the ball back-up bearing design will be able to cope with the high circumferential bearing speeds during a landing specific to the target application, landing tests with a 1.3t high speed rotor ([1], [7]) have been performed prior to the tests with the 9t rotor. The extensive test program comprised 60 landing tests with full run-down at different speeds and braking times from speeds of up to 12,700rpm on one set of back-up bearings. After this test program was completed, the bearings were still fit for the purpose. In this way, a circumferential back-up bearing speed of 94% of the target application was achieved. In addition to gaining experience, these tests were performed to develop the safety concept and to write a dynamic simulation software that is able to predict the experimental results of back-up bearing tests.

B. *Qualification program*

In order to demonstrate the endurance of the back-up bearings and that customer specifications are complied with as listed in section I.B, the test program has been made more stringent compared to customer specifications:

- API 417 Annex 4F for expander compressors on magnetic bearings suggests in section F7.7 a test for new back-up bearings, which includes a 3s delevitation into the back-up bearings at maximum speed followed by a relevitation. This test simulates a situation that is as close as possible to a real run-down situation in the field with respect to temperature rise, see Note 2 for section F7.7. Instead of this suggestion, braked run-downs to zero speed have been scheduled, which are even closer to the actual field conditions.
- The braking time from trip speed is increased to 19sec compared to the customer requirement of 13sec – with the exception of the 100sec landing
- The braking curve is programmed to be linear. In the future field application the process load of the driven machines will result in a braking torque proportional to the square of speed, i.e. the initial slope in the speed decrease will be steeper than for a linear braking curve. In the field, the rotor will thus leave the critical high speed range faster than in the conservative linear approximation

Furthermore, the number of landings has been increased compared to what the customer specified:

- A landing at 1200rpm, 2400rpm, 3600rpm, 4800rpm and 6000rpm each preceding the landings at trip speed, see also the section about safety features IV.E.
- 5 landings at the trip speed of 6420rpm and 19sec braking time with deactivation of all 4 bearing axes instead of the required 3 landings
- 1 landing at the trip speed of 6420rpm and 100sec braking time

C. *Additional system testing after the qualification program*

If the state of the back-up bearings permits it, after completion of the qualification program, the following optional additional tests have been scheduled. The purpose of these tests is to gain experience with the behavior of the rotor in conjunction with the closed-loop controlled bearing system in exceptional circumstances, which in principle could occur in the AMB system setup:

- deactivation of a DE and an EDE bearing axis, i.e. two bearing axes at trip speed
- deactivation of a DE bearing axis at trip speed
- deactivation of an EDE bearing axis at trip speed

IV. Safety Concept and Test Execution

Besides a loss of rotor integrity during the landing test, a backward whirl [6] is the most critical system behavior that might occur, since the bearing forces that occur during a backward whirl can severely damage the back-up bearing system. Thus, measures need to be taken to minimize the risk that the above events occur. The following main measures have been identified based on a risk assessment.

A. *Design phase and test preparation*

- Experimental modal analyses of the rotor and calibration of the simulation environment
- Validation that the weakest rotor parts, such as the end winding covers and coupling bolts between exciter and main shaft, are able to cope with the simulated forces that occur when a rotor drops
- Verification calculation that the connection between motor base frame and foundation as well as the connection between motor stator and base frame will withstand even the worst case scenario that a rotor locks
- Extensive simulation studies for drops with various initial conditions, worst case unbalance situations and worst case eccentricities of the back-up bearing sleeves on the rotor to gain confidence that the design is robust with respect to a backward whirl, i.e. that a backward whirl does not have to be expected

B. *Test system setup*

- The drop switch can be remotely actuated at a safe distance away from the motor in axial direction
- The rotor behavior can be monitored at the remote location using the HMI system of the SIMOTICS Active Magnetic Bearings
- The drop switch allows the rotor to be relevitated in the event of a potentially critical situation
- The emergency brake button allows the rotor to be braked in the event of a potentially critical situation
- Online monitoring of the inner ring temperature to allow relevitation, should the bearing temperature-gradient deviate from what is expected
- UPS allowing levitation of the rotor for more than 30min, should the power supply in the plant fail; the power supply must be restored within this time so that the motor can be actively braked

C. *Initial checks*

- Check that the automatic braking of the rotor is initiated reliably and instantly when the rotor reaches a trip level, in particular the position trip level
- Check that the converter can brake the rotor from trip speed (6420rpm) in less than 19s
- Check the functionality of the UPS, emergency brake button and inner ring temperature measurement

D. *Initial tests*

- Smoothly place the rotor into the back-up bearing; check the effect on the radial rotor position and calibrate the simulation environment

- Landing at 0 rpm and calibration of simulation environment to rotor behavior
- Ensure that the deviation of the delevitated rotor from the back-up bearing center is within expected values (elasticity effects)

E. Test execution

- Start data recording
- The rotor is accelerated up to the appropriate landing speed; to reduce the probability of a back-up bearing failure at high speed due to a design flaw, the landings are graded according to speed (1200rpm, 2400rpm, 3600rpm, 4800rpm, 6000rpm, 6420 rpm)
- The drop switch is activated and the rotor falls into the back-up bearings
- As soon as the rotor position reaches the trip level, a trip is automatically generated by the AMB system and the converter brakes the rotor to a standstill

F. Test evaluation

- Comparison between the results of the dynamic simulation and experiment in terms of orbits, rotor deflection in the bearings, and whirling tendency; since dynamic simulations are time consuming, all landings of the test program have been simulated in advance; to account for uncertainty in the unbalance situation, the simulations have been executed for an in-phase (static) and out-of-phase (dynamic) unbalance situation; nevertheless, an exact reproduction of measured trajectories is not possible because of the complexity and the number of parameters that influence the initial conditions in a particular landing (e.g. rotor position when the drop switch is actuated); it is therefore more reasonable to compare the major trends of a landing that are independent of the details of the initial contact; should the prediction quality of these trends deviate from expectation, the parameterization of the simulation environment needs to be improved based on the measurement results; otherwise, the test program does not have to be delayed by on-site simulation runs
- Visual inspection of the back-up bearing assembly, of the rolling elements and of the raceways using an endoscope
- Verification that the rotor deviation from the center position of the back-up bearing is small enough and within what can be expected, i.e. that the bearing deflection is within expected tolerances, and that there is no chance of the rotor coming into contact with the stator
- Verification that the result of a clearance check does not show any increase in the bearing air gap, which might result from plastic deformation; for the bearing clearance check the rotor is pushed gradually against the back-up bearing's inner ring in the radial direction at equally spaced angular intervals with an automated procedure; the bearing/rotor contact is detected by changes in the measured current; the coordinates of the bearing/rotor contact points are then used to estimate an average clearance
- Proof by dynamic simulation that the next landing test at higher speed will not result in unwanted dynamics.

If the outcome of all above points is positive, the next landing of the landing test program can be prepared and executed.

V. Test and Simulation Results

A. Qualification program

The eleven landings of the qualification program including the 100sec landing were successfully completed and without any unexpected effects. The rotor exhibited good behavior at all times. The prediction quality of the simulation runs performed prior to the tests was good enough regarding the key features so that the test program could be continued without any need for simulation parameter calibration.

Selected measurement results are shown in the following, together with simulation runs performed by Siemens AG. More details about the simulation environment as well as further simulation and measurement results can be found in [5]. Simulation results from the simulation environment of the Schaeffler company are described in [4].

Measurement results of the first three landings at 6420rpm with 19sec are shown in Figure 4. From the measurement results it can be concluded that the deviation of the rotor from the setpoint position during the landing due to elasticity of the back-up bearing is significantly smaller than the active magnetic bearing air gap of about 2mm. Thus, there is no risk that the rotor comes into contact with the AMB stator. It can also be noted, that the rotor behaved in a similar fashion during the three landings regarding the area in the back-up bearing in which the rotor moves as well as in the number of larger loops in vertical direction.

The corresponding simulation result for the landing at 6420rpm with 19sec is shown in Figure 5. Although the horizontal movement of the rotor in the back-up bearing is wider, key features such as the bouncing height have been correctly predicted. Also, as in reality, the bouncing height at the DE-side is higher than at the EDE-side. Most importantly, neither in the simulation nor in the measurement, was there any indication of undesirable dynamics, such as a forward or a backward whirl.

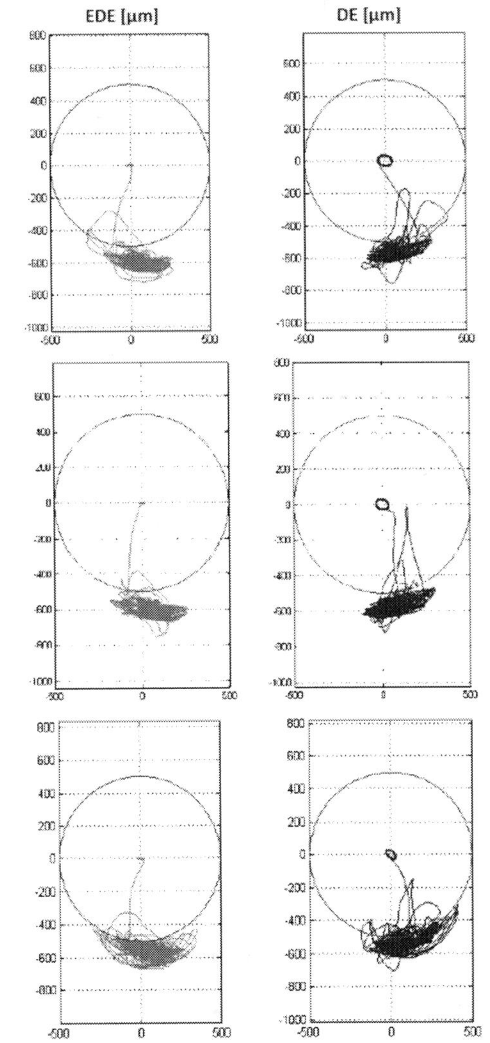

Figure 4. Measurement results of the first three landing tests at 6420rpm with a braking time of 19sec.

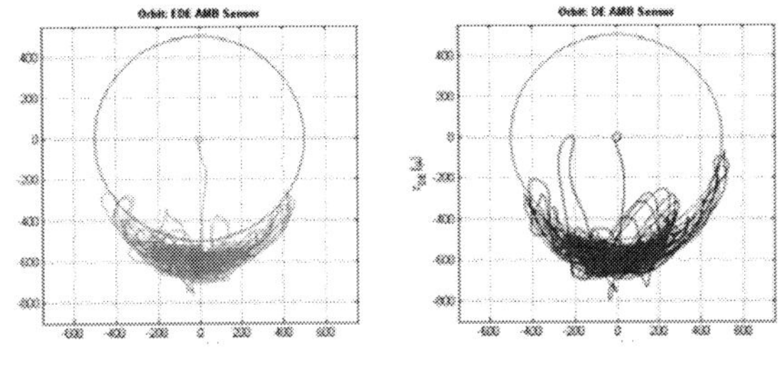

Figure 5. Simulation result of the 6420rpm landing with a braking time of 19sec.

An additional example is the landing test at 6000rpm, which is shown in Figure 6. as measurement and in Figure 7. as simulation result. Also here, the bouncing height has been correctly predicted by the simulation, however, the movement of the rotor in the horizontal direction is larger in simulation than in reality. However, since the measured movement in horizontal direction is more confined – and thus even less critical with respect to the development of unwanted dynamics than in the conservative simulation – it was not necessary to tune the simulation model.

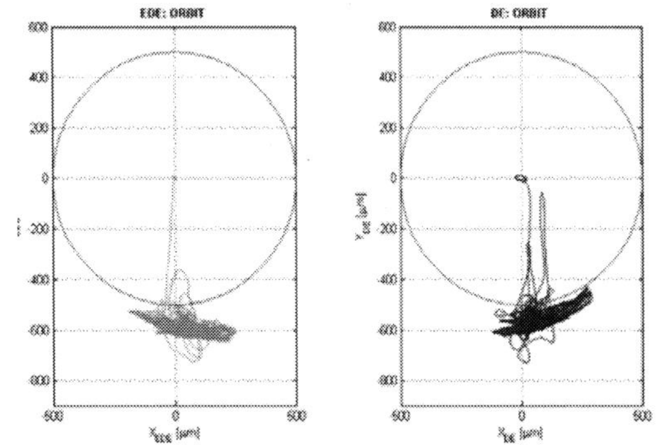

Figure 6. Measurement results of the landing tests at 6000rpm.

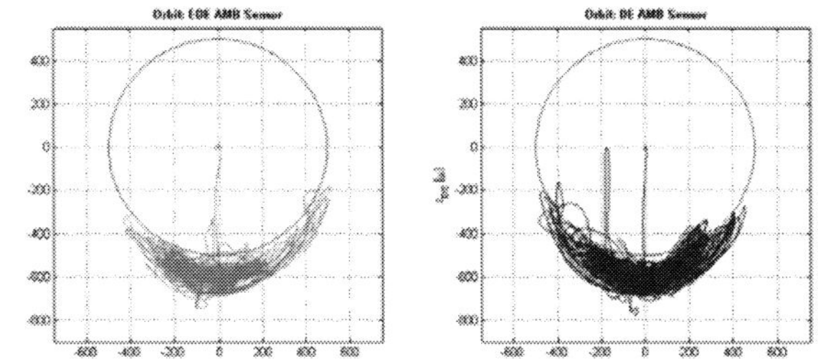

Figure 7. Simulation result of the landing tests at 6000rpm.

Figure 8. shows the result of a clearance check after the 100sec landing at the end of the qualification program at the DE and EDE back-up bearing. The bearing air gap (~550µm), given by the radius in the above diagrams, did not change with respect to the result of the clearance checks before and during the landing program, i.e. there are no indications that the bearing geometry changed.

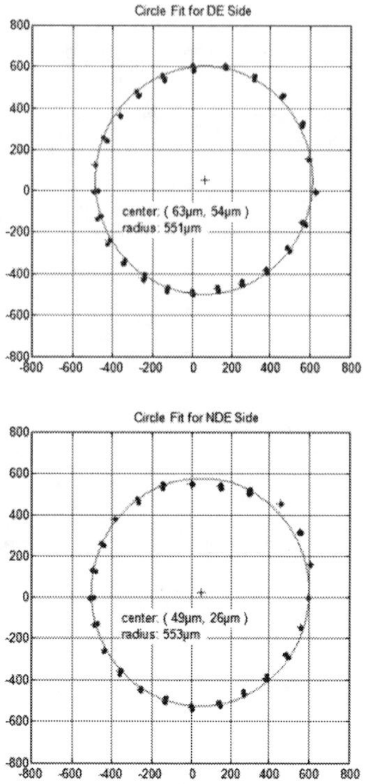

Figure 8. Result of the clearance check at the DE and EDE bearing after the 100sec landing at the end of the qualification program.

The endoscopic examinations of the bearings after the individual drops did not show any problematical amount of wear patterns at the balls or the bearing rings; see the example in Figure 9.

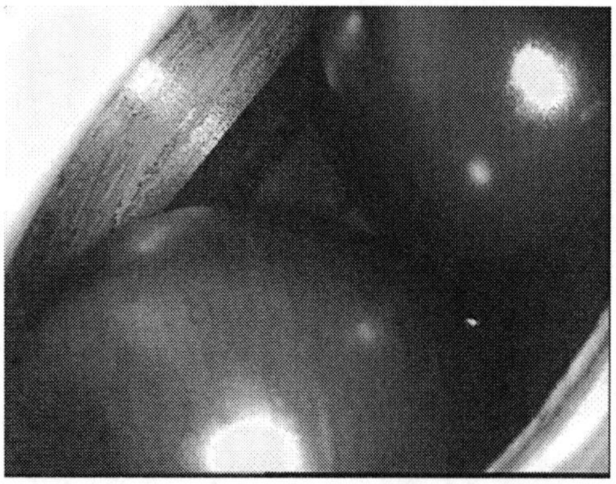

Figure 9. Optical endoscopic examination of the bearings after the 100sec drop at different regions of the bearing did not show any worrying amount of wear.

The temperature measurement at the bearing's inner ring indicated a typical temperature rise of 3 K during the 19sec drops and a temperature rise of 16 K during the 100sec drop. These temperature rises lie significantly below the worst case design values.

B. Gaining further experience

In order to learn more about the behavior of the closed-loop controlled system in case of single-axis drops, e.g. due to an AMB amplifier failure, landing tests in individual bearing axes have been performed. As an example, a landing into only one axis of the EDE bearing is shown in Figure 10. It can be seen that the system response is well behaved, and that the rotor comes to a safe standstill with all but one axis in levitation.

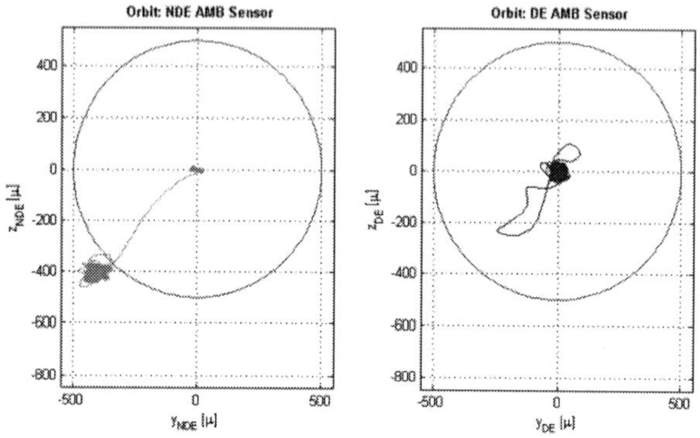

Figure 10. Simulation result of a single axis landing with only one axis of the EDE bearing having been deactivated.

C. Discussion of the results

The presented theoretical and experimental results demonstrate the durability of ball back-up bearings in very large machine applications. The credibility of the methodology comes from the comparison between theoretical and experimental data as well as from structural integrity checks, which were performed on the bearings after each drop. The objective of this test strategy is twofold: i) to show that ball bearings can withstand the load levels associated with challenging landing tests, and ii) to validate the simulation methodology through the correlation between experimental and simulated results. The latter in particular provides confidence in the predicted results of such simulations for development and design purposes. Testing to bearing failure would have been very informative, however, very expensive and unnecessary with respect to ensuring that customer specifications are complied with. Thus, the confidence gained from the test program described above is seen as a good compromise.

VI. Conclusion

The landing test concept described in this article can be used as a guideline to back-up bearing qualification. It allowed the systematic and safe execution of landing tests on a very large AMB machine with a maximum landing speed of 6420rpm and a rotor weight of 9t. The dynamic simulation runs of the landing events performed prior to each landing test as part of the safety concept predicted the rotor behavior measured during the landing with good quality. Accordingly, there was no indication of unwanted dynamics such as backward or forward whirl during the tests. The landing program carried out with a total of 11 landings (5 at 6420rpm and 19sec braking time, one at 6420rpm and 100sec braking time) proves the feasibility of ball back-up bearings for the prototype machine described. To our knowledge, the use of ball back-up bearings for machines of this size and speed is unique.

References

[1] J. Denk, H.-G. Köpken, D. Stoiber, F. Viering, "Active Magnetic Bearing Systems for Turbomachinery Applications based on Standard Drive Technology," *Proceedings of the 13th International Symposium on Magnetic Bearings (ISMB13), Arlington, Virginia, USA, 2012.*

[2] J. Schmied and J. C. Pradetto, "Behaviour of a one ton rotor being dropped into auxiliary bearings," *Proceedings of the 3rd International Symposium on Magnetic Bearings, Alexandria, VA, USA, 1992.*

[3] J. Anders, P. Leslie, L.-E. Stacke, "Transient Rotor Drop Simulations of Touchdown Bearings in Active Magnetic Bearings Applications and Experimental Validation," *Proceedings of the Gas Machinery Conference, Austin, TX, USA, 2012.*

[4] P. Siebke, "A novel rolling element backup bearing for a 9t rotor application," *Proceedings of the 14th International Symposium on Magnetic Bearings (ISMB14), Linz, Austria, 2014 (to appear).*

[5] G. Siegl, T. Tzianetopoulou, "Simulation and Experimental Validation of a 9t AMB Rotor Landing in Rolling Element Backup Bearings," *Proceedings of the 14th International Symposium on Magnetic Bearings (ISMB14), Linz, Austria, 2014 (to appear).*

[6] J. Jiang, Z. Y. Shang, L. Hong "Characteristics of dry friction backward whirl—A self-excited oscillation in rotor-to-stator contact systems," Science China, Technological Sciences, Vol. 53, No.3: 674-683, March 2010

[7] Renwanz, W., Liebich, R., Fehres, R. "Validierung und Parameter-anpassung von Fanglager-Simulationsmodellen," 10. Internationale Tagung Schwingungen in rotierenden Maschinen, SIRM 2013, Berlin, Feb. 2013

Simulation and Experimental Validation of a 9t AMB Rotor Landing in Rolling Element Back-up Bearings

Guenther Siegl[a], Theodora Tzianetopoulou[a], Joachim Denk[a]

[a] Siemens AG, I DT LD AP MG MF-DW EN 4 1, Nonnendammallee 72, 13629 Berlin, Germany, joachim.denk@siemens.com

Abstract—Backup Bearings (BuB) must be designed to withstand the shock loads which occur during impacts with heavy and/or high-speed rotors. Both experimental testing and transient dynamic simulation are necessary for the study of BuB performance. This paper presents the transient dynamic simulation methodology developed within SIEMENS for the study of full-scale rotor landings in rolling element BuBs. The developed models were validated with a substantial landing program led and performed by SIEMENS, whose objective was to qualify the safe operation of hybrid rolling element BuBs for an industrial application utilizing an AMB-levitated 9t supercritical motor. The paper describes the BuB model, summarizes the simulation results and offers a comparison with experimental measurements from the aforementioned successful landing tests program.

I. INTRODUCTION

The implementation of Active Magnetic Bearing (AMB) technology for the support and operation of large rotating machinery has been gaining ground in recent years due to the advantages it introduces in comparison to conventional fluid-film bearing technology, such as frictionless operation and the flexibility to implement active control concepts. A primary concern for rotating machinery OEMs and their customers is the continuously safe operation of equipment under all operating conditions. BuBs are a necessary and central component for the safety of AMB-levitated rotor systems. In the event of an AMB shut-down or failure, the BuBs will support the rotating parts during the landing, absorb the impact loads and will allow the rotor to coast down to standstill.

For large rotor AMB-machines the use of dry friction BuBs has been the preferred practice in the industry. It is common knowledge that friction-BuBs promote the contact conditions necessary to result in the backward whirling of rotors. High rotor vibration amplitudes due to fully developed backward whirl-type rotor motion can cause severe damage or even destroy the rotating machinery. The kinematics of rolling element bearings significantly reduce the possibility of backward whirling during landings. However, technical literature on the subject cites cases where rotating machines can exhibit backward whirl type rotor behavior during landing in rolling element BuBs as well [1].

This paper presents the methodology developed by SIEMENS to simulate numerically the transient dynamic behavior of rotors during landings. The developed numerical framework is based on a detailed kinematic description of the rotor and bearing component movements and of their interactions during a landing event. The methodology has been validated with measurements from an extensive landing program which was performed by SIEMENS Dynamowerk in Berlin to qualify a rolling element BuB-solution for a large 9t motor drive application [2]. To our knowledge, the utilized BuBs are the largest to have been reported for this type of application.

The BuBs are based on a novel cage-less design developed by Schaeffler Technologies GmbH & Co. KG in collaboration with SIEMENS. They consist of two rows of ceramic ball rolling elements, a common inner ring with two separate outer rings, and a housing with asymmetric compliance [3].

The objective of the BuB qualification program was to demonstrate the safety of transient operation, not only for the particular project-specific application, but also to prove that rolling element BuBs can fully withstand the landing of heavy rotating machines without promoting backward whirling. The presented simulation method was an integral part of the safety concept for the 9t rotor landing qualification program. Landing simulations were performed prior to the landing experiments to predict the occurrence of possible mechanical failures during a landing.

The simulation results presented in this paper aim to demonstrate the agreement achieved between experimental measurements and simulation predictions and to show that the existing numerical framework can be used to make reliable predictions for the rotor-dynamic behavior and the mechanical performance of rotating machinery in the future.

II. 9T-ROTOR LANDING TESTS ON ROLLING-ELEMENT BACK-UP BEARINGS: EXPERIMENTS & SIMULATION

A. *Modeling of Transient Rotor Dynamics*

The simulations are performed with proprietary software developed within SIEMENS for transient dynamic calculations.

The rotor modeling is based on the finite element method, and serves as input information for the SIEMENS code. The rotor is discretized with an appropriate number of cylindrical beam elements with the commercial FEM software MADYN, and the developed model is calibrated against a free-free rotor modal test. After modal fitting, the FE description is converted to a standard dynamic system state-space formulation for numerical integration. The stiffness, mass and damping matrices of the tuned model are exported for further use as input to SIEMENS' transient dynamics solver. External system disturbances enter the system of linear algebraic equations through the force vector. Such excitations can be unbalance forces, gravity loads, process loads, braking torques or AMB centering forces. Forces due to rotor-BuB contact interactions enter the system of equations in the same manner through the force vector.

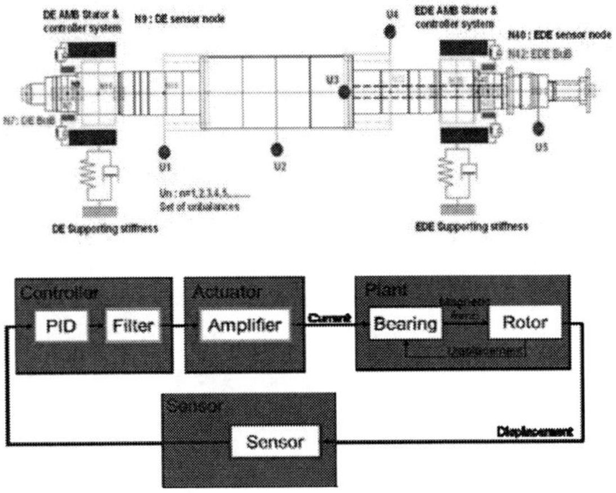

Figure 1 Complete mechatronic simulation model.

The key feature of the Siemens code is the full 3D kinematic modeling of the complete BuB system. The kinematics of each BuB component and the interactions between neighboring bodies are described in detail for all six degrees of freedom (DOFs) with the relevant equations of motion (see Figure 2).

The contact forces which develop during landing between the rotor shaft and the BuB inner ring are based on Hertzian contact theory considering the stiffness properties of the contacting materials and the geometries of the contacting surfaces. Similarly, Hertzian contact laws are used for the calculation of contact forces, which arise from the eventual ball-to-ball contact (due to the cage-less design) as well as for the continuous ball to inner and outer ring contacts. Friction forces are derived from the normal component of the corresponding normal contact forces using static/sliding friction coefficients, which depend on the properties of the contacting materials, the differential speed and the roughness of the surfaces involved.

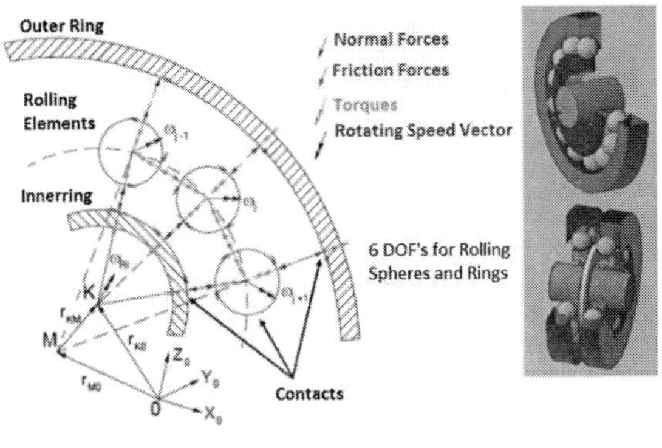

Figure 2 2D schematic simplification of the Siemens BuB model.

B. *Experimental Landing Tests*

A number of landings have been performed with the 9t prototype machine [2]. For safety purposes it was decided that rotor testing should not be initiated with the trip speed of 6420rpm before the system landing behavior was studied at lower landing speeds.

Five preliminary landings took place with gradually increased initial speeds in order to observe any unwanted or unexpected tendencies in the system's dynamic behavior. The rotor was dropped from 1200rpm, 2400rpm, 3600rpm, 4800rpm and 6000rpm prior to the intended qualification drops. Each landing was triggered by simultaneously deactivating both supporting radial AMBs. The rotor was decelerated to zero speed with a constant (speed-independent) electromagnetic braking torque, which caused the rotor speed to decrease linearly. The braking torque was selected to decelerate the rotor from 6420rpm speed to rest in 19sec. The same torque was used in each landing test, thus test duration was proportional to the initial drop speed.

After the successful completion of the lower speed landings, the BuBs were qualified for drops at the trip level speed. The rotor was dropped five times from the bearing's center at 6420rpm speed with a 19sec braking time, again triggered by simultaneous deactivation of all four AMB axes.

Landing from the trip speed with braking only due to aerodynamic resistance (and in absence of any electromechanical braking torque as the case would be in compressor or turbine applications) was simulated with a landing test in which the rotor speed decreased from 6420rpm to zero over 100sec. A lower braking torque was applied in this case to achieve the desired deceleration.

Finally, the prototype's rotor-dynamic behavior was also evaluated for different patterns of AMB failure with additional drop tests from the trip speed. Drop tests were performed with:

- one AMB axis deactivated on both AMBs, (only the x- or only the y-axis active at both AMBs), and
- one AMB axis deactivated at one of the AMBs.

In the industrial application being considered, the rotor is utilized as a double-end drive, and its axial movement on the drive train is constrained by the axial AMB installed on the rotating machine coupled to the rotor's EDE (exciter drive end). The couplings used are compliant in their lateral direction to eliminate radial interactions between coupled drive train components. For the landing tests, the couplings attached to the rotor ends are replaced by dummy weights of appropriate weight and inertial moments. To ensure that the dropped 9t rotor would not axially shift during the landing program, it was axially coupled to a rolling element ball bearing, which restrained axial motion but allowed rotation at all speeds. The coupling used during the tests was radially compliant and did not influence the lateral movements of the rotor during landing and coast-down. In simulations, this condition was applied by eliminating the DOF corresponding to the rotor's longitudinal translation.

C. *Results: Test and Numerical Simulation*

This section presents a representative set of landing test results recorded during qualification program together with the corresponding simulations. As mentioned in the previous section, several drops with a gradually increasing initial speed were performed. The results presented here are from tests with the most challenging landing conditions for the rotor and the BuBs (speed and duration). Two of the five 19sec landings from 6420rpm speed are presented along with the 100sec landing test from the same speed.

For completeness in presenting the system's physical response and the code's predictive capacity, additional results from tests with inherently different initial conditions are also presented: a landing from an intermediate initial speed (3600rpm), a landing from the maximum speed (6420rpm) with single-axis AMB deactivation at the DE and EDE, and a maximum speed drop with one AMB deactivated (at the DE).

The simulated behavior for each of the experiments presented here is compared to the experimental measurements to validate performance of the SIEMENS code.

- Numerical Modeling Preliminaries

The rotor's FE model is based on the available design geometry and material properties. Thus, all mass distributions and moments of inertia are accurately transferred into the model. Further, the rotor model was calibrated with data from modal analysis (hammer) tests to match the measured mode shapes and modal frequencies. Modal damping of 0.5% was assumed as realistic for all simulations.

The AMB controller parameters and magnetic field properties used for the simulations are those of the prototype machine.

To ensure modeling accuracy, the stiffness and damping properties of the BuB pedestal model were calibrated using rotor vibration data recorded by the AMB sensors in tests where the rotor was dropped from different positions within the BuB clearance with zero rotational speed. In all simulations, gravity was acting in the vertical axis, and was applied equally to the rotor, the AMB housing (stator) and to all components of the BuB; namely, the rolling elements, the inner- and outer rings, and the bearing housing.

In contrast to the aforementioned quantities – which are either known or can be directly determined from existing measurements – the friction coefficients, as well as the stiffness and damping values for the contacts between the rotor and the BuB inner ring, as well as between rolling elements, and between rolling elements and bearing raceways can only be estimated. The selected values must be physically reasonable, i.e., based on widely accepted values for contact between metals and/or ceramics, and with rational proportionality to the damping and stiffness values used for the remaining components of the model.

The rotor was balanced prior to the landing tests. However, the exact distribution of the residual unbalance along the rotor was unknown. Thus, for the simulation model, the unbalance weight distribution (location and phase) was varied until an acceptable match was accomplished between simulated and measured orbit amplitudes for rotation inside the AMBs. An in-phase unbalance grade of G1.5 was selected.

Finally, the landing position inside a BuB and the contact conditions during this initial touchdown of the rotor can be a determining factor for the initiation of whirling. Thus, an effort was made in simulations to reproduce not only the measured orbit, but also the rotor's trajectory after the AMBs had been deactivated, such that the initial landing conditions for a particular landing experiment and the corresponding simulation would be as close as possible. This was achieved by matching the position on the orbit at which the bearing was switched off.

Without loss of physical accuracy and as described in the previous section, all simulations were performed with the rotor's longitudinal translation (x-axis translation) fixed and the five remaining DOFs free (y- and z-translation; x-, y-, and z-rotation).

- 19sec landing: 6420rpm initial rotor speed

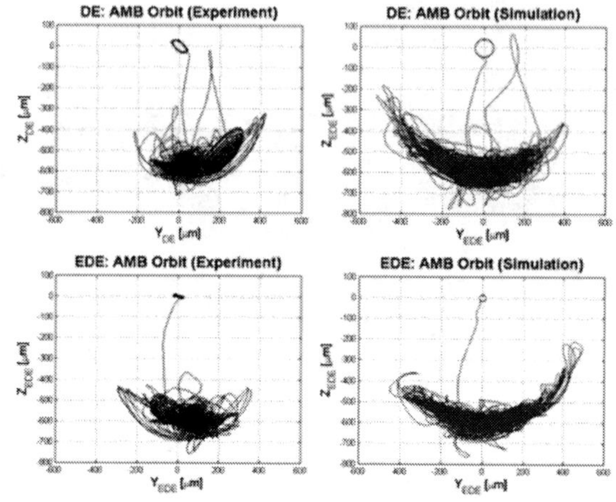

Figure 3 Measured and simulated rotor orbits at the AMB locations. (5th landing from 6420rpm)

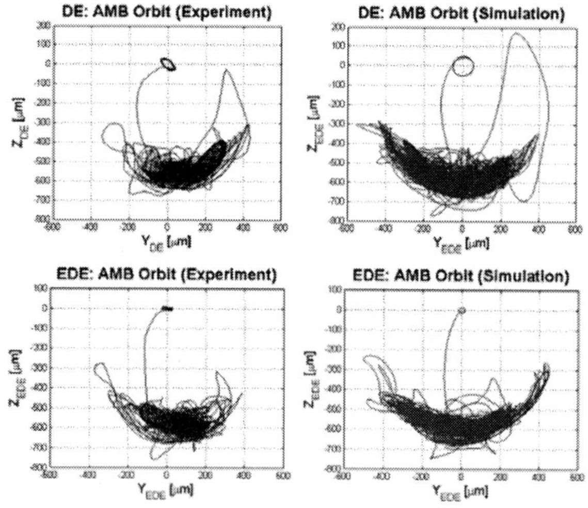

Figure 4 Measured and simulated rotor orbits at the AMB locations. (4th landing from 6420rpm)

The measured orbits for each of the five 19sec braked drops from maximum speed (6420rpm) were practically identical. In addition, for the same drop speed, the similarity in the behavior between the 19sec and 100sec drops is remarkable (when the results are presented in terms of rotating speed). Therefore, it is evident from the measurements that the system's response is highly repeatable and consistent. The orbital characteristics of the rotor as it passes through different critical frequencies during deceleration remain unchanged regardless of the initial drop speed or the duration of the coast-down

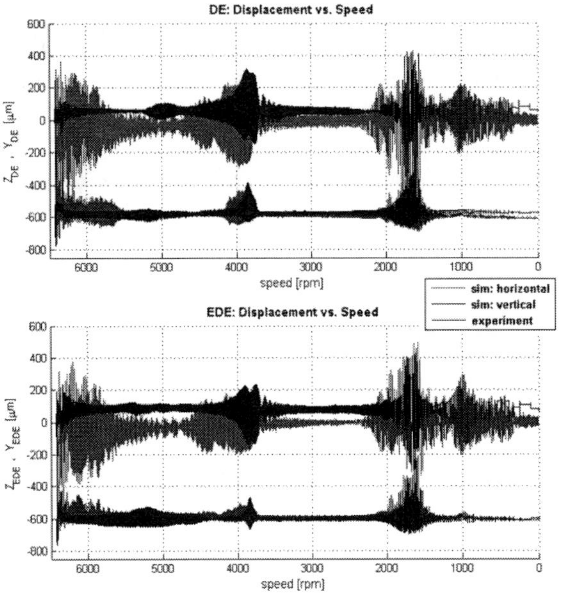

Figure 5 Measured and simulated lateral displacement at AMBs vs. speed. (5th landing from 6420rpm).

- 100sec landing: 6420rpm initial rotor speed

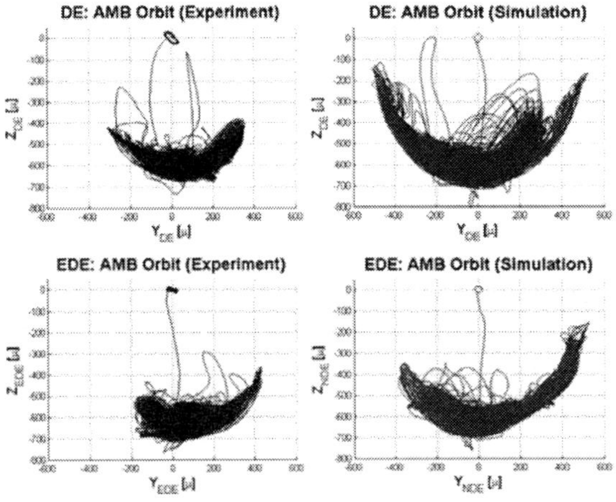

Figure 6 Measured and simulated rotor orbits at the AMBs. (100sec landing from 6420rpm).

The two additional resonances which appear in the 100sec deceleration around 5600rpm and 4800rpm are identified as natural frequencies of the BuB pedestals, which are not excited during the 19sec coast-downs, as the rotor only remains at these critical speeds for a limited time when compared to the 100sec landing scenario.

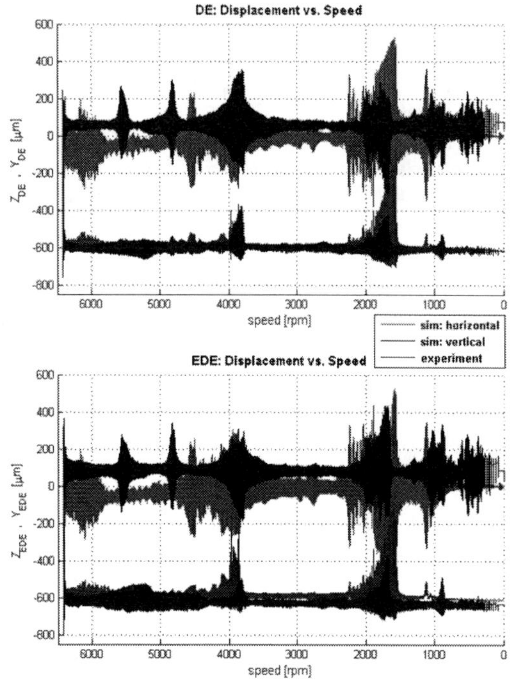

Figure 7 Measured and simulated lateral displacement at AMBs vs. speed. (100sec landing from 6420rpm).

It should be noted that the design feature introduced by Schaeffler Technologies GmbH & Co. KG (see [3]) to add compliance at the lower part of the BuB-housing in an area concentrated at the vicinity of rotor touchdowns has not been accounted for in the model presented here. This feature ``arrests´´ the rotor after a landing and limits its pendulum motion. The discrepancy between the simulated and measured horizontal movement of the rotor (the horizontal component, shown by the red line in Figure 5 , Figure 7 and Figure 9) is attributed to this omission and an improvement is due in this direction.

It should be noted that the measured rotor orbits are not perfectly cyclical during operation in the AMBs, due to asymmetrical properties of the foundation in the testing facility where the landings took place.

Drops from lower speeds exhibited the same characteristics for overlapping speed ranges, indicating that the rotor assumes the same deflection shapes at a given speed. These observations qualified the prototype machine – including the rotor and all bearings – as highly robust, as no alterations or failures were manifested throughout the landing program.

- 12sec landing: 3600rpm initial rotor speed

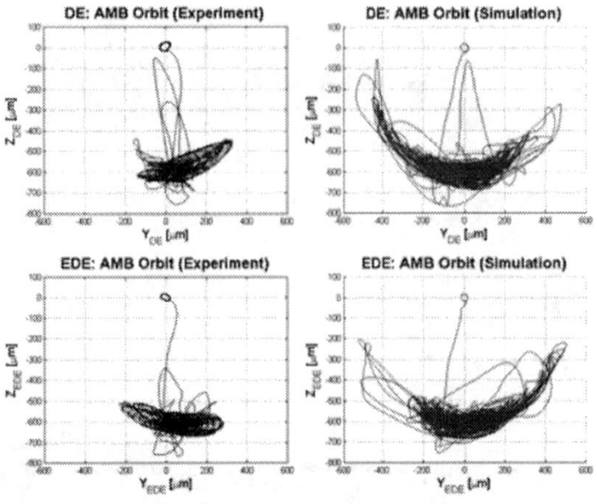

Figure 8 Measured and simulated rotor orbits at the AMBs. (12sec landing from 3600rpm).

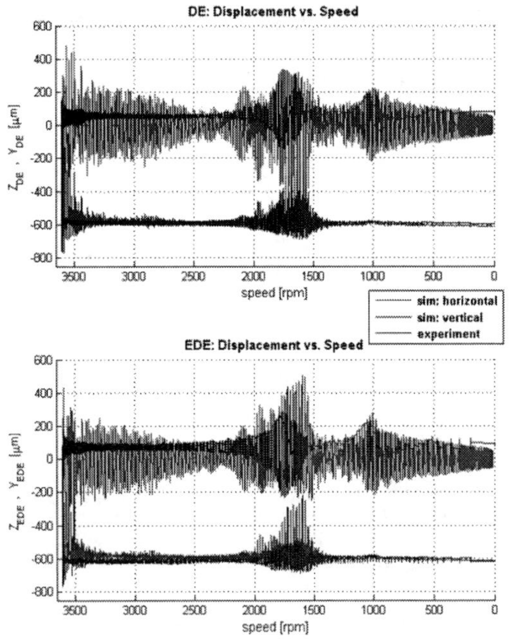

Figure 9 Measured and simulated lateral displacement at AMBs vs. speed (12sec landing from 3600rpm).

- Single AMB single-axis shutdown landing: 6420rpm

Figure 10 presents the measured and simulated orbits for a single axis drop from the trip speed. The x-axis of only the DE AMBs is deactivated. The rotor slides down sideways at the DE without any striking effects. It can be seen that the still fully actively controlled side is disturbed and tends to follow the initial trajectory of the landing side, is however, steered back to the AMB center position. The disturbed orbit inside the fully active AMB is only a result of excitations from the highly contained orbit inside the BuB at the DE.

Figure 11 presents the measured and simulated orbits after deactivation of the x axis at both AMBs. The AMB axes are rotated by 45° to the horizontal. With only the y-axis of the AMBs active, the rotor slides downwards under the influence of gravity, and lands sideways in the BuB at an inclination of 45° (the direction of the deactivated x axis). The tendency of the DE to bounce more after the first touchdown is evident in both measurements and simulations.

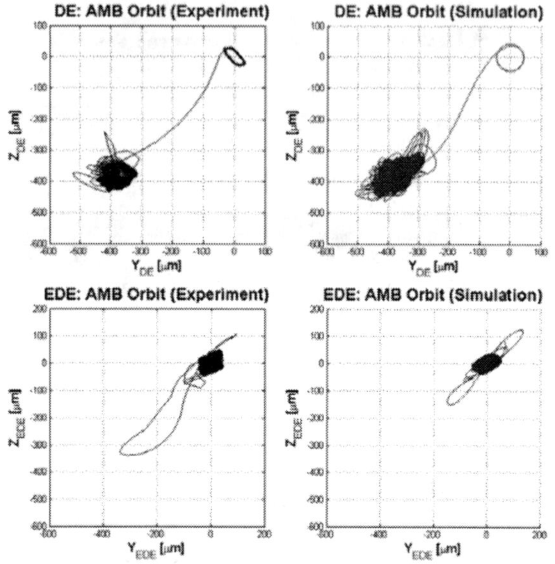

Figure 10 Measured and simulated orbits after deactivation of the X-axis at the DE AMB .

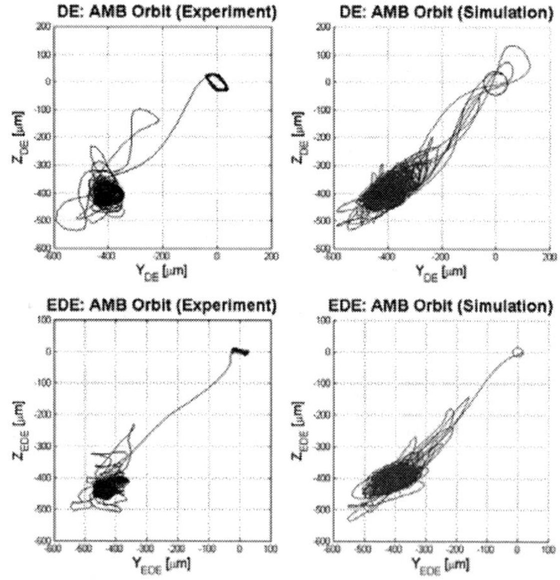

Figure 11 Measured and simulated orbits after deactivation of the X axis at both DE and EDE AMBs.

Effort was again made in the aforementioned simulations to deactivate the AMBs at the appropriate time instance so that the rotor would depart from its circular orbit at the correct position and follow the trajectory observed in the landing tests before touchdown in the BuBs. The quality achieved in the simulated response is visible in the characteristics of the initial transients (e.g., the eight shape in Figure 10).

- Discussion

The software used for transient rotor dynamics within SIEMENS has been validated against the results of the landing experiments. It can be used to predict backup bearing performance in other landing situations, e.g. higher rotor unbalance, without the need for further experiments. Moreover, it allows for design optimization as it permits the numerical monitoring of quantities which otherwise cannot be recorded, but might be essential for analysis. Such quantities can be internal to the BuBs, such as inner ring and rolling element speeds, contact stresses at BuB components, BuB temperature rise due to friction, rotation of components in space or compliant/damping member deformations when the BuB is in use.

Other important quantities are the developing contact forces and the resulting torques acting on the rotor and the BuB inner ring during touch-down and deceleration.

Figure 12 and Figure 13 offer an example of the detail in BuB analysis which can be achieved with the developed software. Such information is crucial for the development of reliable equipment.

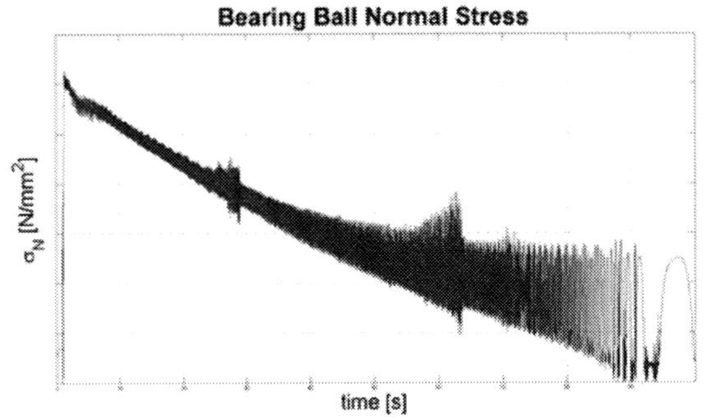

Figure 12 Simulated time evolution of the contact stress on a randomly selected rolling element from the 100sec drop.

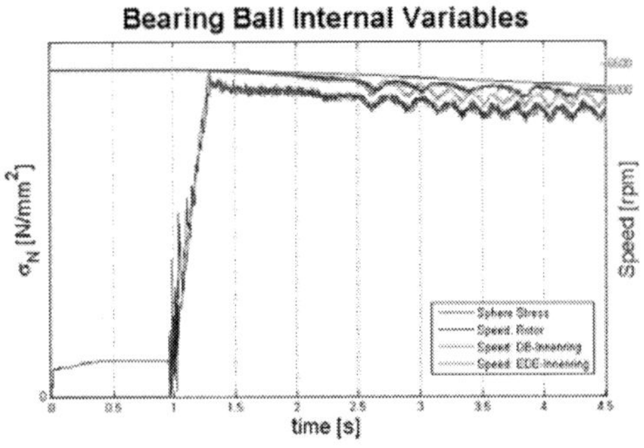

Figure 13 Simulated contact stress on a sphere element and its correlation to inner ring speed directly after rotor touchdown in the BuB. The stress curve is a detailed excerpt from Figure 12 .

III. Conclusions

It has been demonstrated that rolling element bearings can perform as backup bearings for high-speed, heavy rotating machinery. A full-scale demonstration took place at SIEMENS Dynamowerk in Berlin with a 9t heavy rotor system rotating at a speed of 6420rpm. To our knowledge, this has been the most demanding application reported for backup ball bearings.

The performed experimental program included a set of rigorously planned and monitored landing tests, which can serve as a benchmarking quality standard for the qualification of backup bearing concepts which have to satisfy requirements beyond the established experience in the field.

A methodology developed by SIEMENS for transient dynamic simulations was presented. The proprietary software can be used for predictive rotordynamic analysis during emergency landings in BuBs. The code accounts for all bearing subsystems and for their physical interactions with the rotating components. It was validated through several landing experiments and provides a reliable tool for the prediction, analysis and improvement of the performance and reliability of future rotor system designs.

References

[1] J. Schmied, J. Pradetto:. "Behaviour of a One Ton Rotor Being Dropped into Auxiliary Bearings," Proc. of 3rd International Symposium on Magnetic Bearings, Alexandria, Virginia, 1992.

[2] J. Denk , B.-U. Köhler , G. Siegl , P. Siebke: "Landing Tests with a 6300rpm, 9t AMB Rotor in Rolling Element Back-up Bearings", Proceedings of the 14th International Symposium on Magnetic Bearings (ISMB14), Linz, Austria, 2014 (to be published).

[3] P. Siebke: "A novel rolling element back-up bearing for a 9t rotor application", Proceedings of the 14th International Symposium on Magnetic Bearings (ISMB14), Linz, Austria, 2014 (to be published).

A novel rolling element back-up bearing for a 9 t rotor application

Peter Siebke[a], Hermann Golbach[b]

[a] Schaeffler Technologies GmbH & Co. KG, Industriestraße 1-3, 91074 Herzogenaurach, Germany, siebkpte@schaeffler.com
[b] Schaeffler Technologies GmbH & Co. KG, Industriestraße 1-3, 91074 Herzogenaurach, Germany, golbahrm@schaeffler.com

Abstract—**The paper outlines the development of a novel back-up bearing unit which complements Siemens' new magnetic bearings for an electric motor with a 9 t rotor. The back-up bearing utilizes rolling bearing elements. Principles for the rotor supported by the back-up bearings are derived and used to design the novel, so-called "Smiley" design. Dynamic simulations are presented predicting that the system fulfils the requirements. This is proven by comparison with experimental data.**

I. Introduction

A. Background

Active magnetic bearings are a favorable bearing concept for applications with high load, high speed and high efficiency requirements. Correspondingly, Siemens has developed new magnetic bearings based on standard drive technology for its new 23 MW electric motor [1] illustrated in figure 1. For safety reasons, active magnetic bearings require back-up bearings. During landing of a rotor on the back-up bearing regime, the rotor can develop undesired backward whirls, resulting in unacceptable high and hence destructive forces, if the back-up bearing design is not appropriate.

The development of whirls is a result of friction conditions in the system, and it is widely recognized, that a rolling element back-up bearing might provide advantageous friction conditions [2]. Schaeffler Technologies GmbH & Co. KG with its brands INA and FAG has a vast knowledge in rolling bearings for a huge variety of applications, so it was only natural that it would accept the challenge to design a rolling element back-up bearing.

B. Specific operation conditions

The operation of back-up bearings differs from that of normal rolling element bearings, e.g.

- After the active magnetic bearings are switched off, the rotor lands on the back-up bearings, this causes an impact on the back-up bearings. The impact leads to both, high forces and consequently high angular acceleration of the bearing inner rings. This back-up bearing acceleration is larger than normal for rolling element bearing applications.
- Back-up bearings operate for only a short period of time at high speed, therefore the dimensioning of the back-up bearings doesn't rely on the lifetime calculation of standardized rolling element bearings.

Figure 1. Simplified 3D model of an electric 23 MW motor with magnetic levitated shaft.

Other specific conditions are:

- The movement of the rotor in back-up bearings can be chaotic [3], which makes a prediction potentially difficult.
- The rotor might develop a backward whirl in the system, which must be avoided by the bearing design.
- The specific application restricts design possibilities.

This paper focuses on two situations the back-up bearing has to resist:

- impact with high forces and
- dynamics movements during run down of the rotor

by elementary and numerical investigations.

II. Design Principles

As a first stage in the design process, elementary investigations were performed to find principles fulfilling the above requirements, i.e. resistance against impact with high forces and dynamics during run down.

A. Initial landing impact

The impact caused by the landing rotor once it contacts the back-up bearing is the most relevant load condition the back-up bearing has to resist. For the draft design of the bearing, the system is simplified to a linear one. With the assumption of a constant back-up bearing stiffness, the contact forces dependent on the radial deflection are determined. The integral of the contact force over the deflection is the contact energy. The maximum deflection and maximum contact force are determined by the equilibrium of the initially potential energy, which the landing shaft converts to translational kinetic energy, and the contact energy. This is illustrated in figure 2.

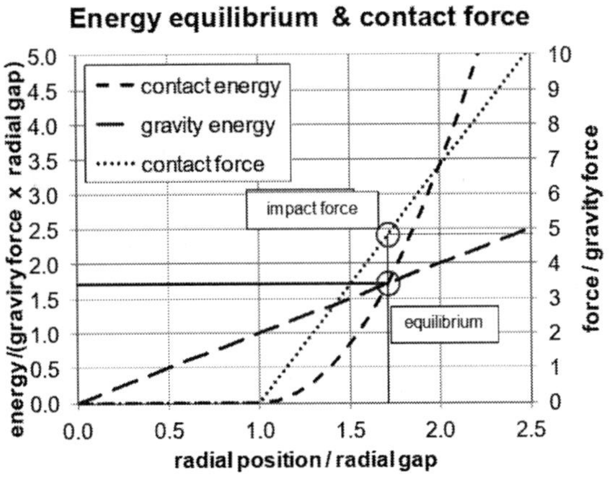

Figure 2. Illustration of the energy equilibrium and the resulting contact force as shock factor. When the gap is closed, the contact force rises linearly with radial position. The contact energy is the integral of the contact force over the radial position. Values normalized by the radial gap between rotor and bearing and by the weight.

So the contact force depends on the initial potential energy of the shaft, and the contact stiffness.

A variation of the contact stiffness influences the contact force: the lower the contact stiffness, the lower the contact force, hence a low stiffness is desirable. On the other hand, the lower the contact stiffness, the higher the deflection. So the specified maximum deflection describes a lower limit to the contact stiffness.

The assumption of a constant stiffness is not valid for back-up bearings with rolling elements, but the procedure can also be used for nonlinear systems.

B. *Run down the back-up bearings*

During run down the desired position of the shaft is as close as possible to the static equilibrium position. A linearization of the stiffness at this equilibrium position can be used to describe the system. The linear contact force in radial direction is described by

$$|F(r)| = \begin{cases} 0 & r < 0 \\ c(r - r_0) & r \geq r_0 \end{cases} \tag{1}$$

c *radial stiffness*
F *radial force*
r *radial position*
r_0 *radial gap.*

The linearization of the system at the equilibrium is

$$F(x,y) \approx \begin{pmatrix} 0 \\ F_G \end{pmatrix} + \left(\left. \frac{\partial F(r)}{\partial x} \right|_{\substack{x=0 \\ y=-r_0-\frac{F_G}{c}}} \quad \left. \frac{\partial F(r)}{\partial y} \right|_{\substack{x=0 \\ y=-r_0-\frac{F_G}{c}}} \right) \begin{pmatrix} x \\ y \end{pmatrix} \tag{2}$$

x *horizontal coordinate direction*
y *vertical coordinate direction*
F_G *gravity force*
others as above.

This leads to a system with an orthotropic support with the equations for the stiffness in horizontal direction

$$c_x = \frac{c\,F_G}{F_G + c\,r_0} \tag{3}$$

c_x *horizontal stiffness*
others as above

and in vertical direction

$$c_y = c \tag{4}$$

c_y *vertical stiffness*
others as above.

With this linearization the critical frequencies of the rotor lying in the back-up bearing can be predicted. Figure 3 shows the natural frequencies of the shaft depending on the stiffness of the back-up bearing.

Depending on the obtained natural frequencies and the operating conditions, the contact stiffness can now be adopted to the application.

Finally, the stability of an overcritical, orthotropic system can be considered as a design criterion according to Gasch [3]. The author introduces an orthotropic parameter μ_L which is the ratio of radial stiffness difference over radial stiffness sum

$$\mu_L = \frac{c^2\,r_0}{2\,c\,F_G + c^2\,r_0} \tag{5}$$

μ_L *orthotropic parameter*
others as above.

The orthotropic parameter showed to be not critical for this system.

III. Novel Back-up Bearing Design

The Schaeffler back-up bearing design is a further development of the published design from Schmied and Pradetto [8], consisting of two angular contact rolling element bearings combined as face-to-face arrangement. The new bearings are fully complemented with ceramic rolling elements.

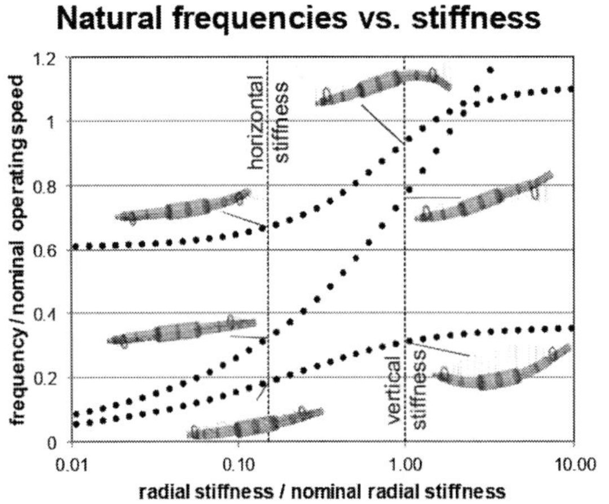

Figure 3. Natural frequencies of the shaft depending on the radial stiffness of the back-up bearing. The horizontal und vertical stiffness calculated is marked. Stiffness normalized by nominal stiffness and frequency normalised by nominal operating speed.

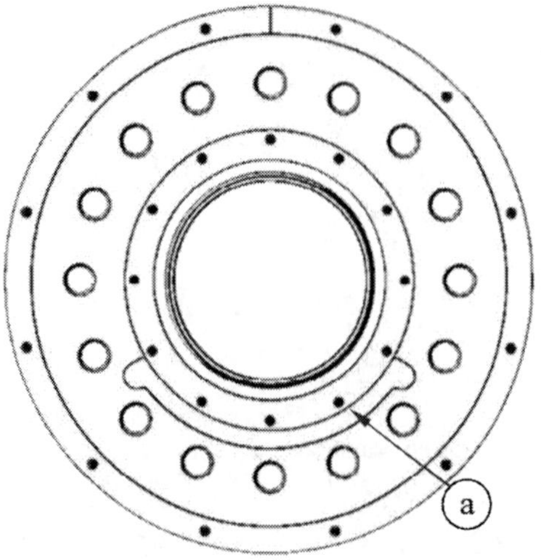

Figure 4. Back-up bearing housing with "Smiley" design at position a) to reduce the stiffness in the direction of the main load and to disturb the symmetry of the back-up bearing.

To resist the higher impact loads the basic design was improved. As shown for the linear system, a stiffness reduction leads to lower impact forces. Additionally, Simon showed [9] for a pendulum type centrifuge that a polygonal back-up bearing reduces maximum contact forces at the back-up bearing during run up. Helfert [6] tried to adopt this effect with two misaligned rolling element bearings. To combine both effects, the reduced stiffness in the main load direction and the non symmetric properties, the housing was modified by introducing a cut-out, which is due to its shap called "Smiley" design (figure 4) which reduces the stiffness in the direction of the gravity.

The new design reduces the impact and overcomes the rotational symmetry of the back-up bearing thus reducing the probability of whirl.

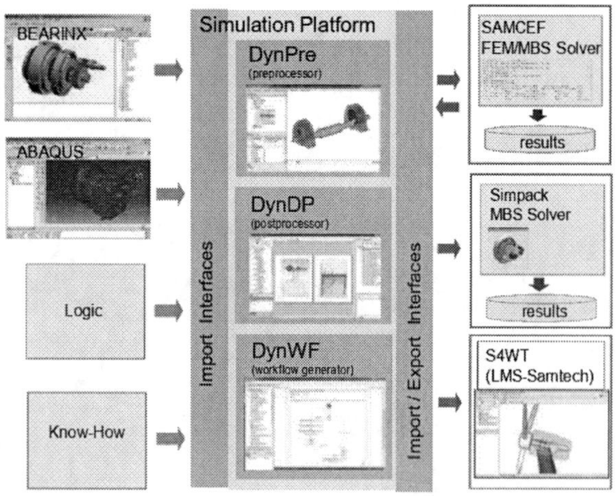

Figure 5. Scheme of the Schaeffler Simulation Platform.

IV. Simulation

A. Motivation

The above principles explain the smiley design for back-up bearing systems but are too coarse to dimension it. For this task the wide range of simulation methods present at Schaeffler for roller bearing development were utilized. These simulation methods enable investigations from the microscopic level of the roller raceway contact to complex system simulations of full drive trains including control units [7].

B. Schaeffler Tools

Finite Element Analysis was performed to analyze the stress and strain due to the impact to the back-up bearings in its pedestal. Thereby, the simulation gained accuracy from the available ABAQUS roller element developed by Schaeffler, enabling exact consideration of each individual roller [4] [5].

To investigate the behavior of the full rotor due to the bearings during run down, dynamic simulation had to be carried out. For this class of problem Schaeffler developed its "Schaeffler Simulation Platform". The platform focuses on modeling quality and efficiency by automation of frequently used process sequences for drive trains. As figure 5 illustrates, the simulation platform collects different model definitions e.g. from BEARINX®, Schaeffler's tool for bearing analysis, or from FEA codes. It even allows the import of customer models in collaborative projects. The collected data are then used to create and run the dynamic simulation models in various commercial solvers, such as the MBS code Simpack and the MBS/FEA code SAMCEF Mecano. The Schaeffler Simulation Platform is completed by features allowing efficient and consistent post-processing and workflow control. Due to all the features it became apparent that the Schaeffler Simulation Platform enables application engineers to access the complex field of dynamic simulation.

C. Run down simulation

For the run down simulation of the rotor in the system as displayed in figure 6, a co-simulation combining BEARINX® with an elastic multibody simulation was performed, with all subsystems verified in the Schaeffler Simulation Platform

The simulation of the back-up bearing behavior was performed in BEARINX® using directly its files.

The results of the FEA for the housing including the smiley design were condensed to a non-linear spring damper

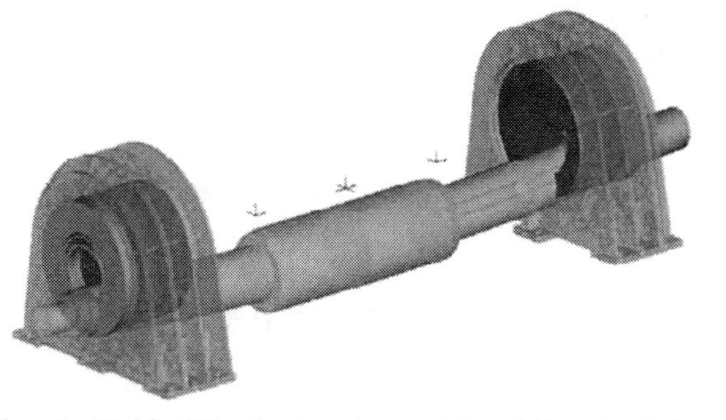

Figure 6. Model of the electric motor consisting of the rotor, back-up bearings and elastic pedestals.The deformation of the shaft and the bearings are scaled 1000 times to visualize the deformations.

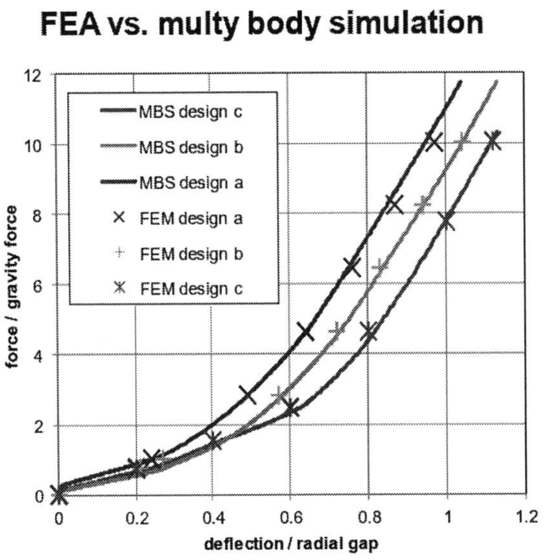

Figure 7. Comparison of defelection in smiley direction between FEM and multi body simulation for different designs with same parameters. The force is normalized by gravity, and the deflection by the radial air gap.

system. The validity of the approach was checked by another investigation in the Schaeffler Simulation Platform: a submodel containing the FEA model of the housing, the back-up bearing and their non-linear linkage was compared to back-up bearings and the spring damper system. Figure 7 illustrates for the vertical direction, that both approaches show identical properties which enable parameter studies with the spring damper model.

The model of the shaft utilizes Siemens' experience in producing and simulating rotors for electric motors. Geometry and material properties of the rotor were imported using the Schaeffler Simulation Platform. The model was validated by a comparison between natural frequencies of the free shaft determined with the Schaeffler Simulation Platform and those determined by Siemens.

The bearing pedestals are modeled with superelements. This modeling incorporates the experience of the influence of elastic structures on the load on bearings.

D. *Sample Results*

Figure 8 and figure 10 show results of the simulation in waterfall diagrams. The frequencies that are excited at run down correspond to the frequencies predicted with the linearization illustrated in figure 3. The comparison between the horizontal and the vertical displacements shows that the resonance frequencies differ e.g. line "A" in figure 8 (approx. 0.2) and figure 10 (0.3). This is due the orthotropic effects of the design and shaft position.

E. *Experiment*

To qualify the application Siemens and Schaeffler performed tests of the motor landings above nominal speed, see [2] for a description of the procedure of the tests.

Figures 9 and 11 show results of the measurements as waterfall diagrams. The frequencies excited during the run down correspond to the frequencies in the simulation. The difference between resonance frequencies in the horizontal and vertical directions differs as predicted in the simulations.

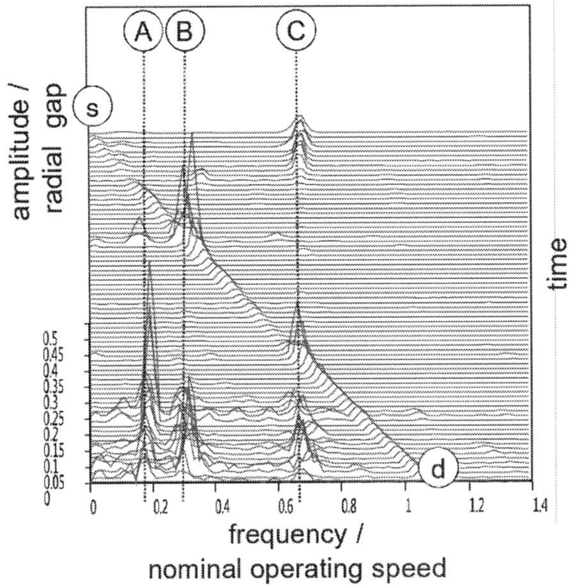

Figure 8. Simulation results: Waterfall diagram of the horizontal rotor displacement in the magnetic bearing. The rotor is dropped at d and stopped at s. A, B and C are the predicted natural frequencies (cf. fig 3). The frequency is normalized by the nominal operating speed, the amplitude by the radial gap.

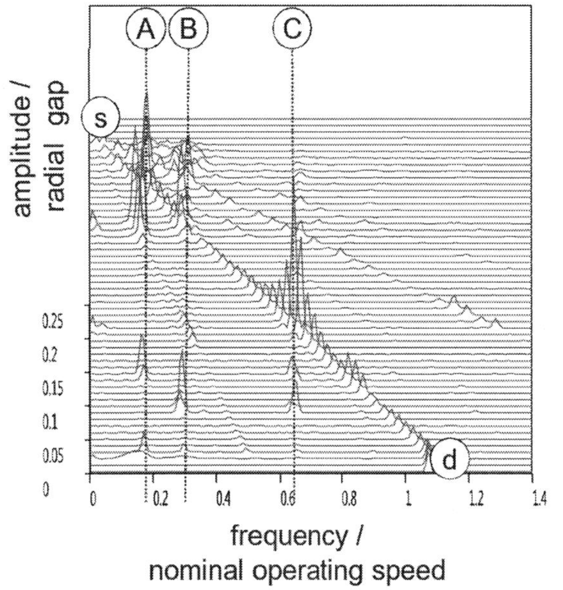

Figure 9. Experimental results: Waterfall diagram of the horizontal rotor displacement in the magnetic bearing. The rotor is dropped at d and stopped at s. A, B and C are the predicted natural frequencies (cf. fig 3). The frequency is normalized by the nominal operating speed, the amplitude by the radial gap.

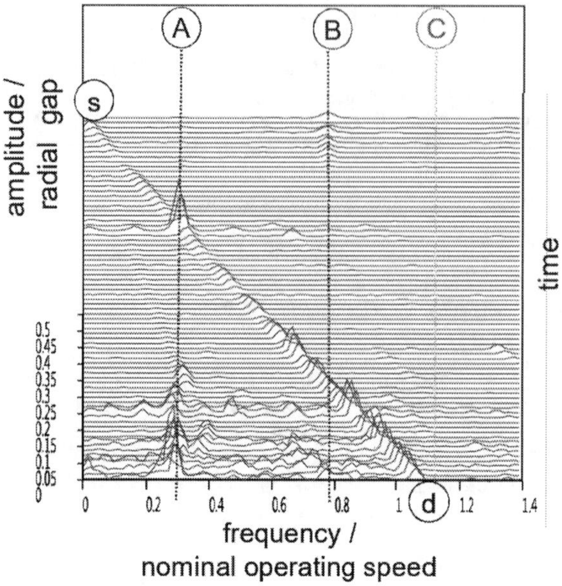

Figure 10. Simulation results: Waterfall diagram of the vertical rotor displacement in the magnetic bearing. The rotor is dropped at d and stopped at s. A, B and C are the predicted natural frequencies (cf. fig 3). The frequency is normalized by the nominal operating speed, the amplitude by the radial gap.

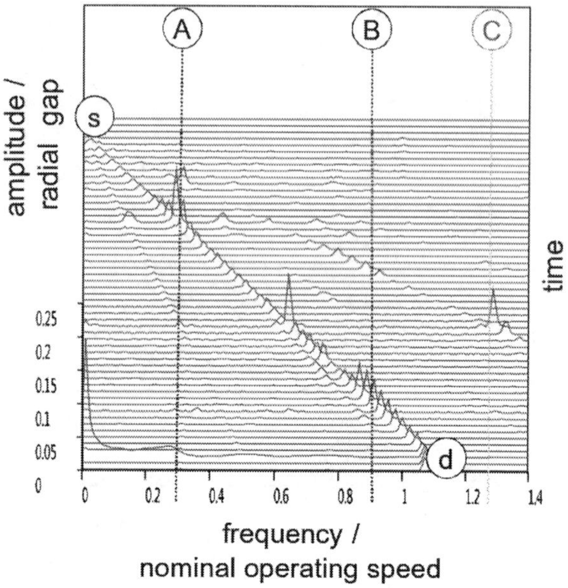

Figure 11. Experimental results: Waterfall diagram of the vertical rotor displacement in the magnetic bearing. The rotor is dropped at d and stopped at s. A, B and C are the predicted natural frequencies (cf. fig 3). The frequency is normalized by the nominal operating speed, the amplitude by the radial gap.

Hence the orthotropic effects are confirmed.

V. Summary and Conclusion

For a novel rolling element back-up bearing used for a rotor in a magnetic bearing application, design principles were presented: Firstly, a method to estimate the required vertical stiffness of the system was derived. Secondly a method to calculate the horizontal stiffness, which affects the natural frequencies for the system of the back-up bearings supporting the shaft during run down was presented.

Furthermore a detailed numerical approach was shown. Comparisons with experimental data showed good agreement, and proved the theoretical approach and the simulations.

VI. Acknowledgement

The author would like to thank Schaeffler Technologies for permission to publish these results. The author is grateful to Dr. Joachim Denk of Siemens and his team. This paper could not be presented without their contributions.

References

[1] J. Denk, H.-G. Köpken, D. Stoiber, F. Viering (2012): "Active Magnetic Bearing Systems for Turbomachinery Applications based on Standard Drive Technology"; Proceedings of the 13th International Symposium on Magnetic Bearings (ISMB13); Arlington, Virginia (USA).

[2] J. Denk, B.-U. Köhler, G. Siegl, P. Siebke (2014): "Landing Tests with a 6300 rpm, 9 t AMB Rotor In Rolling Element Back-up Bearings"; Proceedings of the 14th International Symposium on Magnetic Bearings (ISMB14); Linz (Austria) - to be published.

[3] R. Gasch, R. Nordmann, H. Pfützner (2005): „Rotordynamik"; 2. Auflage, Berlin: Springer Verlag.

[4] H. Golbach (1999): "Integrated Non-Linear FE Module for Rolling Bearing Analysis"; Proceedings of the NAFEMS World Congress; Newport, Rhode Island (USA).

[5] H. Golbach (2004): "The Application of Modern FE Simulation Methods in the Product Development of Transmission Components"; Proceedings of the SAE World Congress; Brasilia (Brazil)

[6] M. Helfert (2009): "Rotorabstürze in Wälzlager: Experimentelle Untersuchung des Rotor-Fanglager-Kontakts"; Suedwestdeutscher Verlag für Hochschulschriften (30. November); o. O.

[7] S. Lenssen, A. Degtiarev (2013): "Durchgängige Werkzeuge zur Systemsimulation – präsentiert am Beispiel einer Windenergieanlage"; Proceedings of the CWD 2013 (Conference for Wind Power Drives); Aachen (Germany).

[8] J. Schmied, J. Pradetto (1992): "Behaviour of a One Ton Rotor Being Dropped into Auxiliary Bearings"; Proceedings of the 3th International Symposium on Magnetic Bearings (ISMB3); Alexandria, Virginia (USA).

[9] U. Simon (2001): "Rotor–Stator–Kontakt in polygonförmigen Fanglagern"; Dissertation; TU Braunschweig. Zugleich Fortschritts-Bericht VDI Reihe 11, Nr. 308; VDI-Verlag.

SIEMENS

Christoph Grund, Dr. Detlef Haje

Qualification of Magnetic Bearings for Industrial Turbines

„14th International Symposium on Magnetic Bearings"
11./14. August 2014, Johannes Kepler University, Linz

SIEMENS

Qualification of Magnetic Bearings for Industrial Steam Turbines

- Vision of an oil-free turbine
- AMB turbine test bed
- Qualification tests
- Prototype application
- Summary

08/2014

Grund, Haje / E P SU R&D IST PFT

SIEMENS

Vision of an “oil-free turbine”

Advantages for steam turbines:

- Improved turbine efficiency by up to 1% by reduced bearing friction
- Reduced fire hazard by avoiding the lube oil system
- Reduced maintenance scope
- Wear-free bearings
- Online Rotor Monitoring by AMB Controller

08/2014

Grund, Haje / E P SU R&D IST PFT

SIEMENS

AMB turbine test bed

Main development targets for the steam turbine test bed at the University of Applied Sciences Zittau/Görlitz:

1. Steam Temperature simulation to verify the thermal effectiveness of the patented cooling system
2. Verify the handling of simulated steam forces by the AMB
3. Verify the function of auxiliary bearings

Qualification of the AMB technology for the first application at Industrial Steam Turbines

SIEMENS

AMB turbine test bed

Test bed key parameter		
Speed	5800	rpm
Elect. Power rating Elect. Brake Power	280 640	kW
Loads: Radial AMB Axial AMB Load Bearing	 25 25/50 35	kN
Rotor Mass ca. Length ca.	 2500 4000	 kg mm
Inductive Heating Power	40	kW

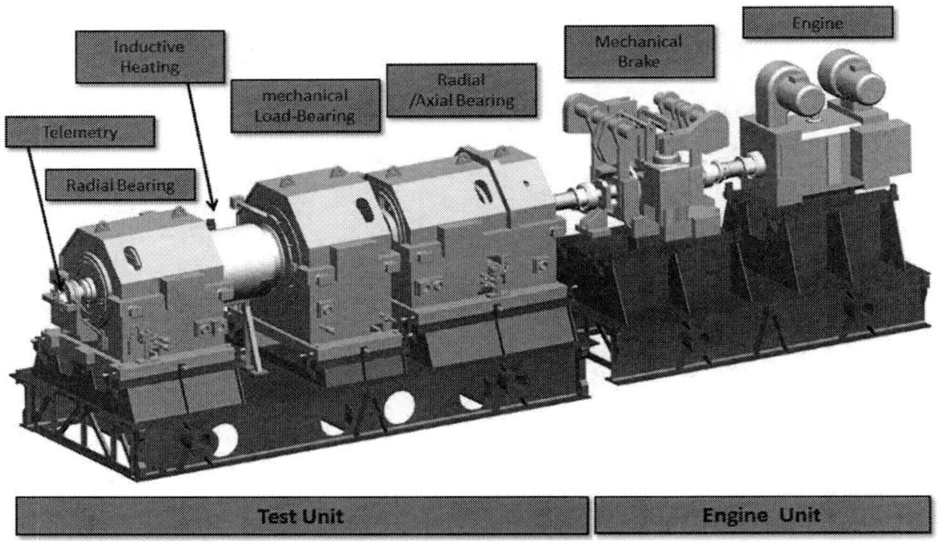

08/2014

Grund, Haje / E P SU R&D IST PFT

SIEMENS

AMB turbine test bed

08/2014

Grund, Haje / E P SU R&D IST PFT

SIEMENS

AMB turbine test bed

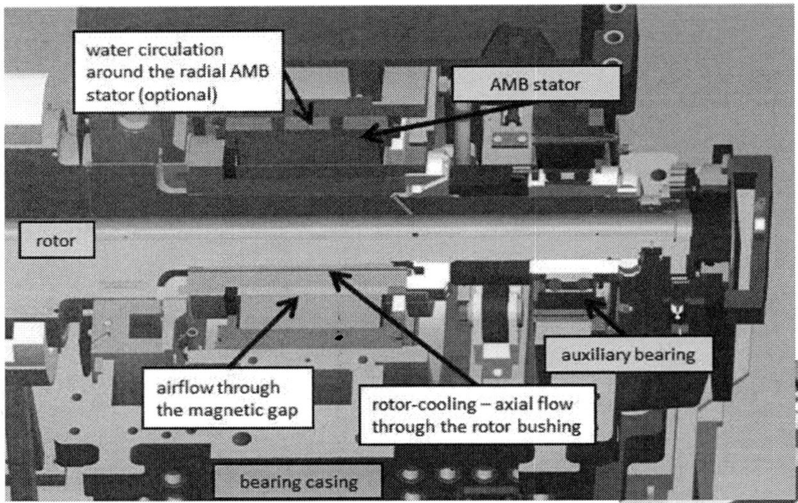

AMB cooling system

Horizontally divided radial AMB

08/2014

Grund, Haje / E P SU R&D IST PFT

SIEMENS

Instrumentation

- Temperature Probes within the rotor (transmitted via telemetry-system)
- Temperature Measurement at the rotor surface with Pyrometers
- Vibration measurement on all bearing casings
- Speed measurement at inner Ring of the auxiliary Bearing
- 6 temperature Probes at the outer Ring of the auxiliary Bearing
- Force Measurement at one auxiliary Bearing
- Condition Monitoring for both auxiliary Bearings
- Volume flow, temperature and pressure measurement for the cooling medias
- AMB Position Sensors

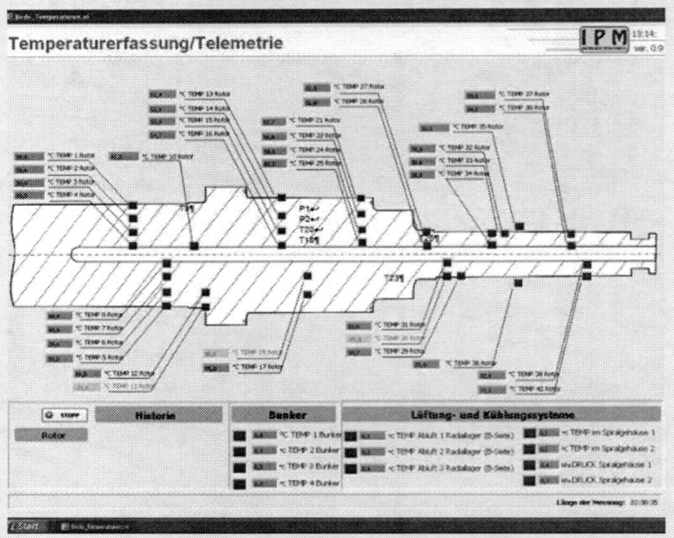

Operation interface screenshot

08/2014

Grund, Haje / E P SU R&D IST PFT

SIEMENS

Qualification tests

Executed Tests

1. Commissioning
2. Flow test AMB rotor cooling
3. Steam temperature tests
4. Auxiliary bearing tests
5. Steam force tests
6. Cooling air parameter optimization

Rotor cooling flow test

SIEMENS

Test target:
Verification of cooling air flow for the whole operation range.

Result:
Verification successful. Ram pressure slightly increasing with rotational speed.

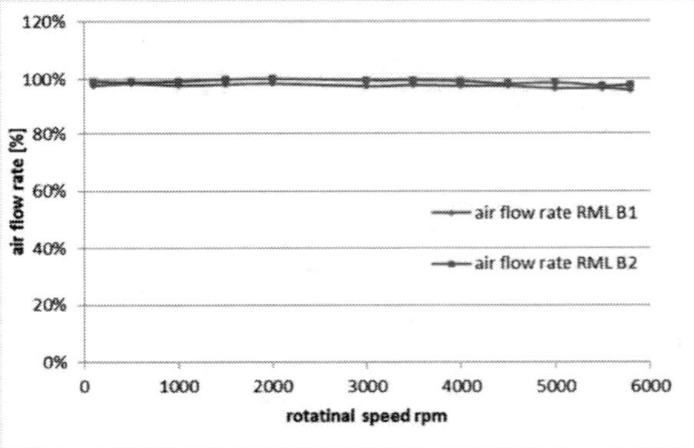

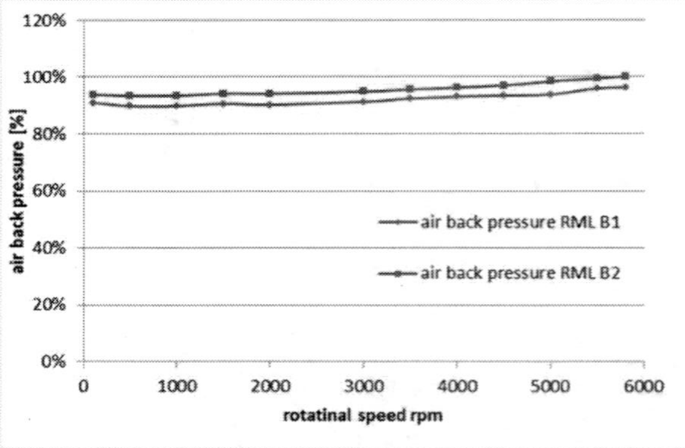

08/2014

Grund, Haje / E P SU R&D IST PFT

SIEMENS

Steam temperature test

Test target:
Verify cooling power to ensure function of AMB

Results:
Stable rotor cooling proven for 510 ° C

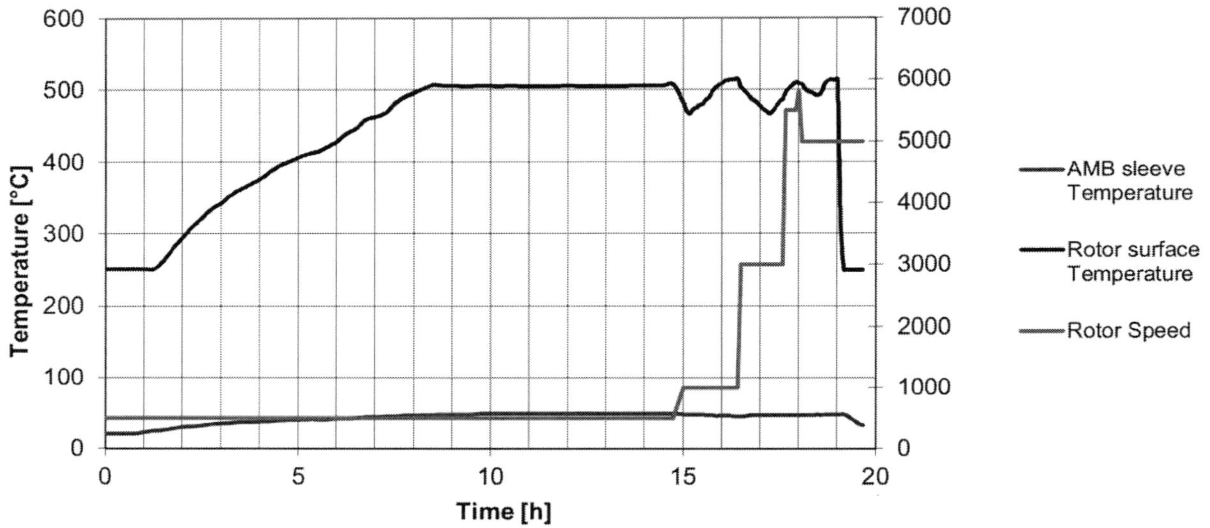

08/2014

Grund, Haje / E P SU R&D IST PFT

SIEMENS

Auxiliary bearing test

Test target:

Verify function of auxiliary bearing

Result:

Completed test series with full speed drop tests

Auxiliary bearing qualified for application

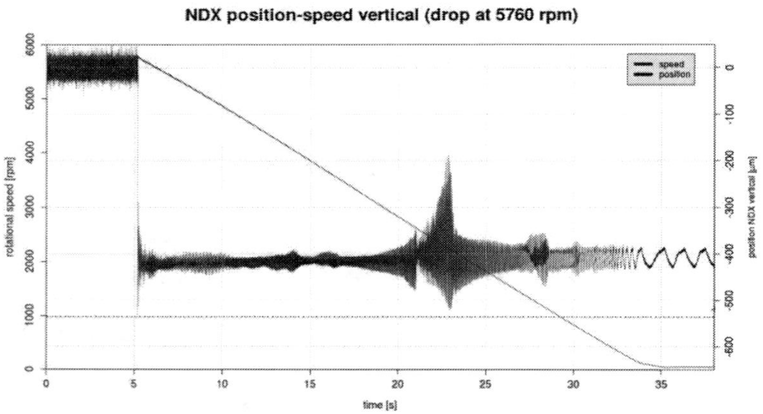

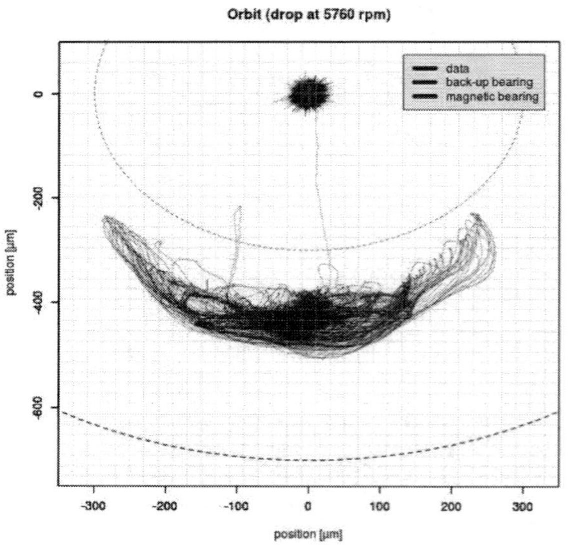

08/2014

Grund, Haje / E P SU R&D IST PFT

SIEMENS

Steam force simulation

Test target:
Verify compensation of a radial process force

Result:
Simulated steam force can be compensated

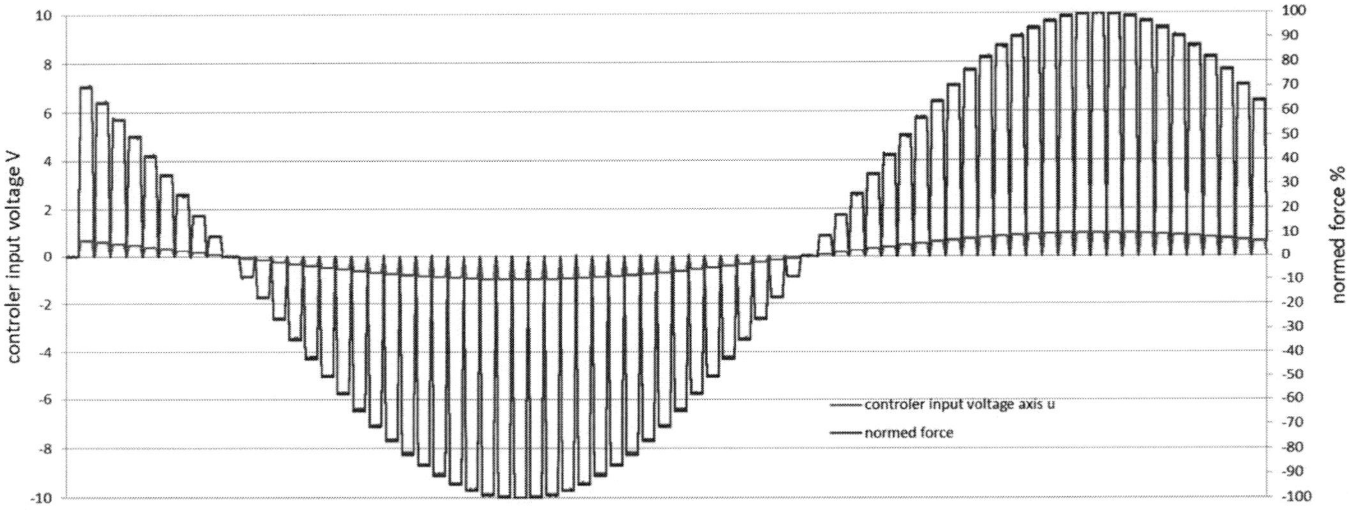

08/2014

Grund, Haje / E P SU R&D IST PFT

SIEMENS

Optimization of cooling air parameter

Test target:
Identification of the influence of the different cooling systems on the rotor temperature

Result:
Optimized air flow rate determined
for AMB operation.

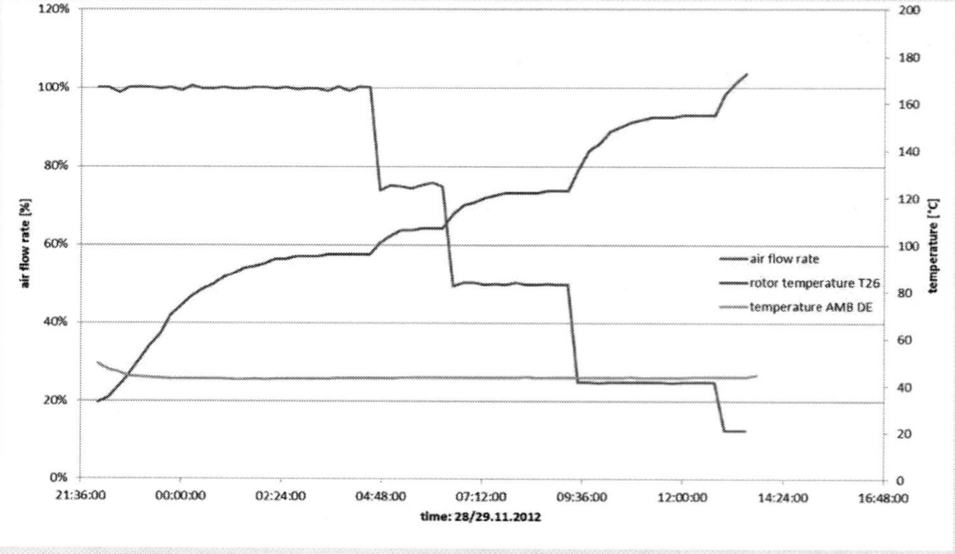

08/2014

Grund, Haje / E P SU R&D IST PFT

SIEMENS

Prototype application

Key Data

- Backpressure turbine
- Rotor weight: approx. 2,5 t
- Turbine Power: approx. 10 MW
- Operational speed: 4500-5400 rpm
- Double End Pump Drive
- Turbine Delivery in August 2014
- Siemens Simotics AMB
- Radial AMB horizontal divided
- FAG Auxiliary Bearings

Prototype steam turbine during manufacturing in Goerlitz

08/2014

Grund, Haje / E P SU R&D IST PFT

SIEMENS

Summary

- ✓ Successful operation of the heated rotor in AMB at 5760 rpm
- ✓ Successful test of the AMB cooling System up to maximum speed
- ✓ Successful test of the AMB system under simulated power plant environment
- ✓ Successful test with simulated steam force
- ✓ Successful test of the auxiliary bearings

AMB technology qualified for first Industrial Steam Turbine Prototype

08/2014

Grund, Haje / E P SU R&D IST PFT

Summary

SIEMENS

08/2014

Grund, Haje / E P SU R&D IST PFT

SIEMENS

Disclaimer

This document contains forward-looking statements and information – that is, statements related to future, not past, events. These statements may be identified either orally or in writing by words as "expects", "anticipates", "intends", "plans", "believes", "seeks", "estimates", "will" or words of similar meaning. Such statements are based on our current expectations and certain assumptions, and are, therefore, subject to certain risks and uncertainties. A variety of factors, many of which are beyond Siemens' control, affect its operations, performance, business strategy and results and could cause the actual results, performance or achievements of Siemens worldwide to be materially different from any future results, performance or achievements that may be expressed or implied by such forward-looking statements. For us, particular uncertainties arise, among others, from changes in general economic and business conditions, changes in currency exchange rates and interest rates, introduction of competing products or technologies by other companies, lack of acceptance of new products or services by customers targeted by Siemens worldwide, changes in business strategy and various other factors. More detailed information about certain of these factors is contained in Siemens' filings with the SEC, which are available on the Siemens website, www.siemens.com and on the SEC's website, www.sec.gov. Should one or more of these risks or uncertainties materialize, or should underlying assumptions prove incorrect, actual results may vary materially from those described in the relevant forward-looking statement as anticipated, believed, estimated, expected, intended, planned or projected. Siemens does not intend or assume any obligation to update or revise these forward-looking statements in light of developments which differ from those anticipated.

Trademarks mentioned in this document are the property of Siemens AG, its affiliates or their respective owners.

08/2014

Grund, Haje / E P SU R&D IST PFT

Test Field for Magnetic Bearing Applications under Extreme Conditions

Stephan Duesterhaupt[a], Holger Neumann[a], Christian Panescu[a], Torsten Rottenbach[a], Frank Worlitz[a]

[a] Institute of Process Technology, Process Automation and Measuring Technology
Hochschule Zittau/Goerlitz - University of Applied Sciences, Theodor-Koerner-Allee 16, 02763 Zittau, Germany
s.duesterhaupt@hszg.de, h.neumann@hszg.de, c.panescu@hszg.de, t.rottenbach@hszg.de, f.worlitz@hszg.de

***Abstract*—Active magnetic bearings (AMB) are an essential key technology in modern power plant concepts. The application of AMBs in turbo machines (steam turbines, cooling pumps and compressors) is characterized by extreme process and environmental conditions like high temperatures, corrosive fluids and gases. Especially the use of organic materials limits the insulation level of electromagnetic actuators. Currently the allowed maximum continuous temperature of coils and laminated cores is 160 °C. Known solutions like ceramic insulation systems allow actuator temperatures up to 450 °C. But their production is complex and the specific actuator size increases compared to conventional insulated actuators.**

The development of new surface technologies and advances in material science allow new basic approaches for insulations and thereby for High Temperature Active Magnetic Bearings (HTAMBs). For instance the use of the physical vapor deposition technology (PVD) allows low cost coatings with thicknesses of only a few micrometers. Furthermore, the fusion of AMBs with catcher bearings (CB) to an integrated unit is advantageous for a compact machine design. It is one objective to place the bearing units close to the process section, if appropriate into the process section. So the CBs also have to fulfil their functions under extreme conditions.

I. Introduction

In November 2011 the University Zittau/Goerlitz received funding for three projects placed in the research field of energy and environment with a volume of 5.75 million euros. Taking the required equity into consideration the total volume is 6.35 million euros. The project is financed within the "program of energy efficiency, renewable energies" by the European Regional Development Fund (ERDF) and the Saxon State Ministry for Economic Affairs, Labour and Transport (SMWA).

The aim within the three scientific projects is to develop and optimize technologies, which can increase the efficiency of energy systems. It will be shown how energy can optimally be used and transformed in subsystems of large and small power plants. This allows to conserve the natural resources. The three projects are:

- Increase in the energy efficiency of turbo machines in power plants through innovative bearing concepts (MFLP)
- Energy efficiency in thermal power plants, thermal energy storage system (THERESA)
- Thermochemical energy storage system (TCV)

Therefore it is planned to construct and build three separate test fields within a new university power plant laboratory called "Zittauer Kraftwerkslabor".

For simulating scaled process conditions – like in power plants – THERESA will provide superheated steam for the MFLP with a maximum controlled flow rate of $0.1\,\mathrm{kg\,s^{-1}}$ and with a maximum controlled temperature of 250 °C.

The steam is necessary for studying the AMB and CB behaviour concerning temperature and corrosive influences. Alternatively the interaction between the two test fields can be used for the simulation of different operation points in a turbomachine. Concerning the steam process this would be startup, shutdown and emergency shutdown. Here the question is how the thermohydraulic process can follow. Another question is for example whether it is possible to store the steam fast enough in a thermal energy storage system.

The demand for

- electric power of 680 kVA,
- steam with a maximum mass flow of $5\,\mathrm{kg\,s^{-1}}$ and with a maximum temperature of 220 °C,
- an area with $300\,\mathrm{m^2}$

requires the reconstruction and expansion of a machine hall in the area of the public utilities company "Stadtwerke Zittau GmbH" (SWZ).

II. Test field MFLP

The main objective within the project "Increase in the energy efficiency of turbo machines in power plants through innovative bearing concepts" is to develop, construct and build a complete test field for longterm tests of AMBs and CBs under extreme process and environmental conditions.

The test field consists of a test rig for magnetic and catcher bearings which is called MFLP and also of different peripheral subsystems. These include energy distribution, electric drive, process control, uninterruptible power supply and steam pipes.

The target within the development process is to achieve a test station providing the possibility to investigate several configurations of AMBs and CBs.

Therefore the following requirements must be fulfilled:

- application of steam specifically to the CBs and separately from the AMBs

- minimum effort for assembly and disassembly
- emergency CBs for testing the regular CBs up to the design limit
- modular setup of the housing to allow installation of several types of bearings.

Test rig – The MFLP is a horizontal machine. Figure 1 shows an overview of the recipient with the components installed (Table I). During the design of the test rig one aim was to get a high modularity and an easy access for assembly and disassembly. As Figure 1 shows the recipient can be divided horizontally. Each component has its own case and can be separated from the recipient with the help of a crane.

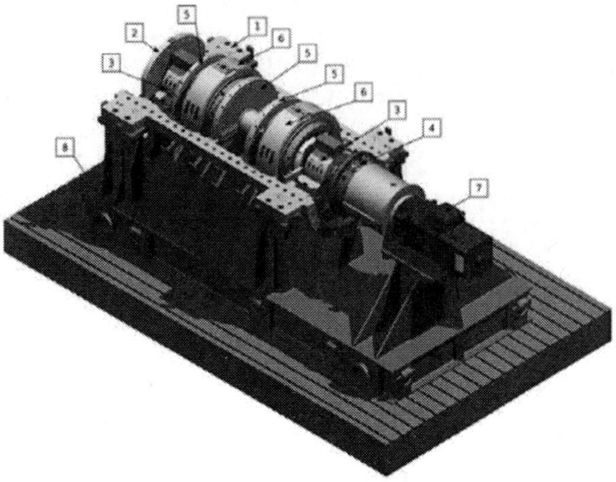

Figure 1. Test rig with installed components

Table I
COMPONENTS OF TEST RIG MFLP

No.	
1	Recipient
2	Emergency CB
3	Radial CB
4	Axial CB
5	Seal
6	AMB
7	Drive
8	Mounting base

The main part of the test rig is a magnetically levitated rotor having a mass of approximately 1 300 kg. On drive end (DE) as well as on non-drive end (NDE) combined axial and radial active magnetic bearings are installed. They are identical. This guarantees optimum position of the center of gravity point. In cases of deactivation or malfunction of AMBs catcher bearings guarantee the rotor stability. On the DE and NDE side it is possible to install ball or plain bearings as radial CBs. Furthermore, on the DE side an axial plain bearing is installed.

The main parameters of the test rig are summarized in Table II.

Table II
MAIN PARAMETERS OF TEST RIG MFLP

Parameter	
Rotor	
Mass	1 300 kg
Length	2 653 mm
Maximum speed	3 600 rpm
Recipient	
Mass	4 500 kg
Length	2 700 mm
Diameter	1 700 mm
Magnetic bearings	
Axial force	20 kN
Standard airgap, axial	500 µm
Radial force	35 kN
Standard airgap, radial	800 µm
Catcher bearings	
Standard airgap	300 µm
Standard airgap, emergency	500 µm

Rotor – The speed of the rotor which is mounted horizontally (Figure 2) will be increased up to maximum rotation of 4 000 rpm. The rotor mainly consists of

- shaft,
- two tension discs (component of axial magnetic bearing),
- two rotor laminations (component of radial magnetic bearing),
- sleeves (component of operation bearing and emergency catcher bearing) and
- clutch (mechanic connection to the electric drive).

Figure 2. Pre-assembled rotor

Emergency catcher bearings – The CBs are supposed to be strained up to the design limit. Due to this, a malfunction is possible. Therefore the test rig has special emergency catcher bearings which are designed as plain bearings. They are positioned at the DE and NDE and do have the longest distance to the center of gravity point of the rotor. In case of a malfunction of the radial CBs the emergency CBs prevent the recipient, rotor and the active magnetic bearings from damage.

Recipient – The whole rotor including the bearings will be placed in a pressure-resistant recipient (Figure 3). It is necessary to simulate different temperature and environmental conditions along the rotor. So the recipient is separated into four sections with a separate steam connection each. The sections are separated by a labyrinth seal system which is typical for turbo machines. In sum there is more than one case to manipulate the conditions. Steam can be applied to the process room, CB or AMB separately. In another case the process room and the CB section can be applied simultaneously for example. In this way several practice-oriented temperature profiles can be adjusted.

Figure 3. Pressure-resistant recipient with welded flanges

Grounding – The pressure-resistant recipient is mounted on a base frame which is also copied for technical solutions like steam turbines in power plants. The whole machine is connected to a mounting plate on a massive grounding. Massive dynamic forces have to be absorbed by the CBs and therefore also by the basement in case of a rotor touchdown. The grounding has a mass of approximately 132 000 kg.

Figure 4 shows the machine basement. It is a solid grounding having a mass of approximately 120 000 kg. The mounting

Figure 4. Massive grounding

plate having a mass of 13 000 kg was mounted onto the rotor basement (Figure 5).

Figure 5. Mounting plate

The resulting machine grounding provides the stiffness, inertia and damping required. This is necessary to deal with the intense reaction forces expected in case of rotor touchdowns.

III. Objectives of Planned Experiments and Tests

Model validation – Design, modelling and simulation of magnetic bearings and catcher bearings form the basis of the experimental work executed at test rig MFLP. Existing algorithms for the design of magnetic bearings and sub-models for simulation of bearing dynamics are applied. The model must be extended to be applicable for new bearing concepts and for investigations on reliability of bearing components under extreme environmental conditions. Thus, non-linearities or temperature-depending drift-offs must be considered more strongly.

Based on the experimental results the methods will be validated with this new test facility.

Testing new catcher bearing concepts – Another focus is to experimentally investigate radial plain catcher bearings. At present, mainly ball CBs are used in large-scale machines. They allow the control of dynamics during contact of rotor and catcher bearing to the greatest possible extent. Extensive operating experience is available that allows industrial application. The constructive design consisting of inner ring, ball element, outer ring and rotor sleeve is more complex in comparison with the plain bearing. It consists only of 2 wearing components – Stator sleeve and rotor sleeve. Enhancing magnetic bearings for use in turbo machines requires departing from traditional construction already in the design process. The increase of operation temperatures allows placing the bearing components in the vicinity of or directly in the process room. This also applies for catcher bearings.

Catcher bearings fulfill safety requirements. The influence of steam or aqueous solutions leads to premature aging of bearing components. Dispersion grease in ball bearings is rinsed out or burned and steel alloy corrodes. Thus, the temperature range for operation is from the start limited in a higher degree than for plain bearings. In switched-off mode the named impacts cause damage. This will cause malfunction and failure particularly for ball bearings.

Derived from the construction principle of hydraulic bearings, the application of split plain bearings is interesting. Figure 6 shows a plain CB with horizontal split. The bearing components (Table III) were designed and constructed in the course of the project.

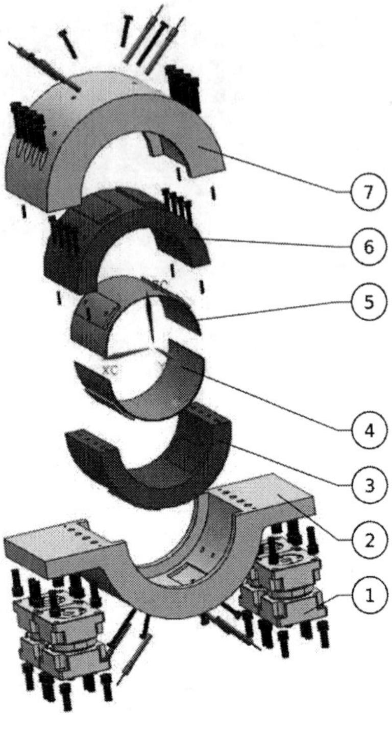

Figure 6. Radial plain CB with horizontal split

Table III
COMPONENTS OF RADIAL PLAIN CB

No.	
1	Load measurement device
2	Bearing housing, lower part
3	Bearing stator, lower part
4	Lower bearing shell
5	Upper bearing shell
6	Bearing stator, upper part
7	Bearing housing, upper part

Subsequently, the bearing construction as a result of strength analysis was evaluated (Finite Element Method) at various load situations. Figure 7 shows an example for mechanical total deformation of the lower bearing shell. Assuming the line compression and a force of 180 kN, the bearing shell will deform max. 85 µm.

This novel construction of plain catcher bearings is intended to be tested in follow-up projects under normal and extreme process conditions. The results of numerical simulation will

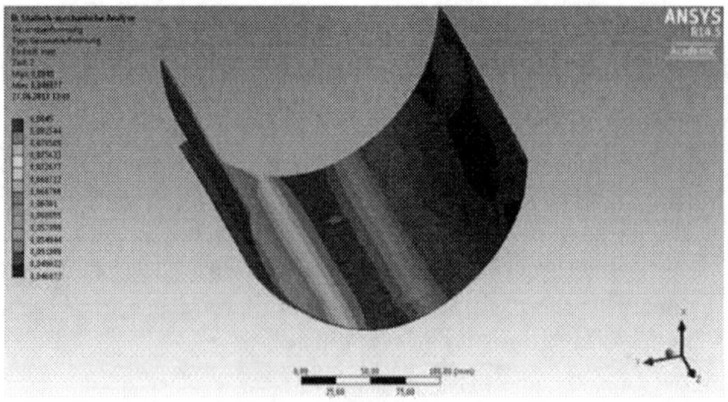

Figure 7. Compression of lines of lower bearing shell

be validated within this framework as well. Moreover, optimizations in terms of geometry and material selection will be derived on the basis of results and operational experiences.

Long-term behavior of magnetic and catcher bearings and its components – Another goal is to extend the range of operation temperature of bearing components. As mentioned before, progress in the field of materials science allows also to increasingly integrate bearings into the process room. Turbo machines can thus be constructed in a more compact way. This leads to considerable improvement of rotor dynamics.

In industrial steam turbines, temperatures of more than 250 °C are expected in the location of bearings. Relating the magnetic bearings, the electrical isolation of lamination and winding limits the operational temperature which is currently at 180 °C. Contrary to hydraulic bearings as they are used in turbo machines, there is no cooling medium in active magnetic bearings. The stator can be cooled effectively with liquid but, the cooling of the rotor lamination, however, can currently be done only with very effortful periphery. The cooling units cause additional effort which is also visible in a reduced degree of efficiency. It is intended to stepwise reduce this cooling demand in future.

The solution is to enhance the electrical isolation. It presently consists of organic materials like varnish and resin. It is intended to use anorganic isolation materials, e.g. ceramic materials in coatings in future. Using state-of-the-art surface technologies (Physical vapor deposition by means of sputtering) it is possible to apply coatings in the µm range.

Besides thermal influence, the long term effect of water vapor and water on the bearing components is going to be investigated. Here, the focus is placed on corrosion susceptibility of ferric and non-ferric metals. The components of magnetic bearing stators and catcher bearing units can be produced of chrome steel either partly or not at all. Here, the lamination and the ball bearing rings can be mentioned in particular. Moreover, also rotor sleeves are susceptible to corrosion. Contrary to that, the ball bearing elements do no further depict a problem. They are made of ceramics.

Long-term investigations will be made to show, how and to which extent it is possible to encapsulate magnetic bearings and catcher bearings in one unit.

IV. Conclusion

The test field is designed for long-term tests of AMBs and CBs under extreme process and environmental conditions. Therefore a modular case and sub-system concept is used. The paper summarizes the needed strategy concerning the recipient, rotor and bearing design.

To reach the experimental goals a test rig rotor has been completely supported horizontally with active magnetic bearings. It is accelerated with an electrical shaft drive. The maximum number of rotations is 4 000 rpm.

The rotor is surrounded by a pressure vessel. The volume of this container is separated by means of inside radial sealing layers. Due to a connected hydraulic cycle it is possible to apply in sections either vapor or compressed air to the rotor. Thus, it is possible to investigate the thermal and chemical behavior of the rotor and the rotor components. Furthermore, the pressure vessel has also the function of a barrier. During selected damage escalations at the machine, the failure of mechanical structures may occur. This is particularly applicable for the investigations planned on plain and ball CBs. Bursting of components thus leads to detaching of fragments. The pressure vessel avoids them entering the space of the test hall.

The test rig is designed in a modular way. Plain CBs and rolling CBs can optionally be mounted radially and axially in the corresponding space. For this reason sufficiently dimensioned mechanical interfaces are available. A key solution here is the additional grading of the circulating air gap in the radial CB which is a plain bearing. The emergency CBs are designed for one-time contact with rotor catcher bearing. They allow secure shut-down of the test rig even in case of burst catcher bearings.

With this test field it is possible to develop and test new and innovative bearing concepts. The investigation will be carried out under scaled power plant conditions.

Backup Bearing Modelling for Turbo Machines with High Axial and Radial Loads

Dr. Jan J. Janse van Rensburg[a], Christian Vanek[b], Prof. Dr.-Ing. Frank Worlitz[b]

[a] School of Mechanical and Nuclear Engineering, North-West University, Potchefstroom 2520, South-Africa, Jan.JanseVanRensburg@nwu.ac.za
[b] Hochschule Zittau/Görlitz IPM, Theodor-Körner-Allee 16, 02763 Zittau, Germany, cvanek@hszg.de; f.worlitz@hszg.de

Abstract— **The use of active magnetic bearings (AMBs) in turbo machines is an active research topic and the implementation in industrial applications is becoming commonplace. This is especially true when investigating the backup bearings (BBs) for high speed rotors with high axial and radial loads. In AMB supported turbo machines there usually are high axial process forces present, and should the axial AMB (AAMB) fail, these forces need to be supported by the axial backup bearings (ABBs). In order to design appropriate BBs the movement of the rotor and the associated forces produced during a rotor delevitation event (RDE) is of critical importance. These forces should be determined using a rotor delevitation model that includes the forces on all the backup bearings and the effect that these forces have on the rotor stability. This paper presents a short overview of various types of ABBs in the literature and the associated models for these BBs. The second part of the paper presents a model for a novel BB currently under construction at the IPM, and how this model is integrated into the rotor delevitation model BBSim [1]. The third part of the paper presents some results obtained using the proposed ABB model. The results investigate the effect that the radial distance where contact occurs and the axial force present on the rotor has on the RDE severity based on the *Vval* values.**

I. Background on ABBs

The process forces in turbo machines cause high forces in the axial direction. During AMB failure these forces on the ABB has a destabilizing effect in the radial direction. This destabilizing force is dependent on the contact between the ABB and the rotor. Schmied and Pradetto identified the phenomena of forward whirling in [2]. Wilkes et al presented the consequences axial contact has on the shaft movement during a rotor drop in the backup bearings [3]. These authors show that a sub-critical frequency can develop destructive whirling if the rotor is in contact with the ABB during a RDE. In machines with axial forces present on the BBs it is important to stabilize the whirl motions in order to achieve safe rotor delevitation. The effect of these axial forces and contact with the ABB on the behavior of the rotor during a RDE should be known and is important in the design of an intrinsically safe machine.

A. *Layout of ABBs in the literature*

Machines with high axial loads are usually designed to have a vertically orientated rotor. In vertical machines with a combined axial/radial BB the gravity force is used to stabilize the rotor [4-6]. A vertical machine was developed for the high temperature reactor; the BB for this machine is shown in Fig. 1. The tapered carbon clutch helps to stabilize the rotor in the radial direction [5].

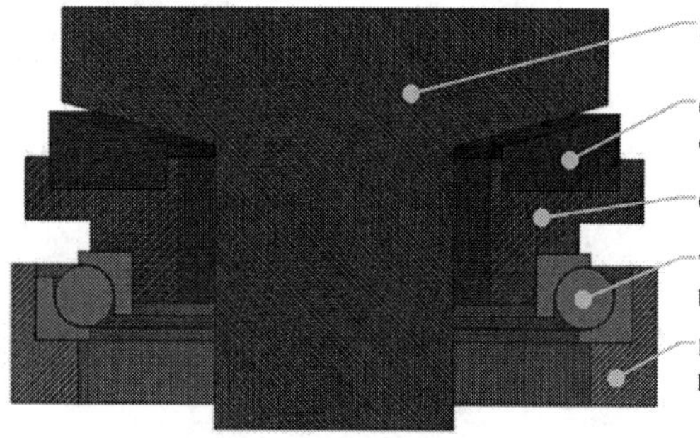

Figure 1. Schematic of the design of an ABB for vertical machines [5].

Since there is no stabilization effect in horizontal machines the layout as shown in Fig. 1 would also be advantages for use in these machines. The most commonly utilized layout for ABBs is a combined axial-radial BB [5, 7-9]. When using a combined axial-radial BB the influence of the radial stiffness and damping has on the axial forces experienced by the BB should be investigated. Some other machines use brass rings that is integrated into the axial AMB as the ABBs [10] These types of ABBs are usually implemented when the axial loads are relatively low and the wear of these rings can be managed. It is important to note that for radial BBs, rolling element bearings are commonly used in vertical machines and plain bearings is more commonly used in horizontal machines.

B. *ABB Modeling*

In 2002, Orth and Nordmann presented a tool to calculate the non-linear dynamics of AMBs [11]. The tool includes a model for the ABBs. Sun et al presented a detailed rolling element bearing model for the radial BBs and a point mass model for the ABB [9]. The schematic of the dynamic ABB model is shown in Fig. 2

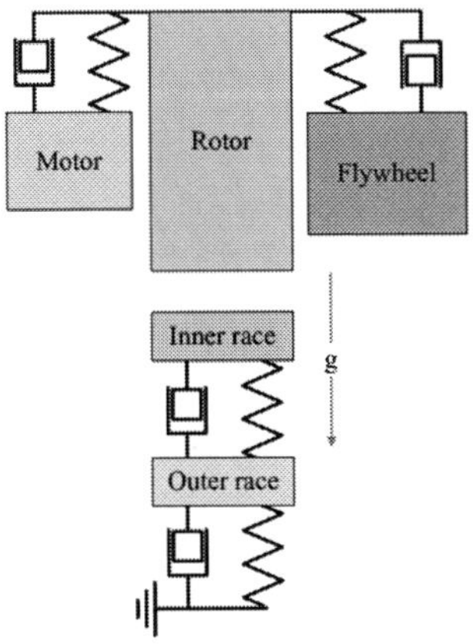

Figure 2. Schematic of the dynamic model of an ABB developed by Sun et al. [9]

II. AXIAL BACKUP BEARING (ABB)

A. Mechanical design of the ABB

The University of applied sciences Zittau/Görlitz is developing a test bed for AMBs and BBs the MFLP [12]. The goal of this test bed is research in the field of AMB applications in extreme conditions. Plain BBs were chosen for this test bed since these are commonly used in extreme conditions to prevent contamination of the process fluid, especially in energy applications.

The test bed allows for the testing of both rolling element BBs and plain BBs. This enables the testing of various types of BB combinations and examining the influence that the ABB has on the rotor behavior and forces on the radial BBs. The design of the ABB for this test bed is shown in Fig. 3. The test bed BB includes the possibility to measure the following parameters:

- force on each of the bearing pads, as shown in Fig. 3
- displacement of the shaft at the bearing location
- temperature on each of the bearing pads, as shown in Fig. 3

Figure 3. Illustration of the axial Backup bearing for the MFLP Test-Bed

The bearing is designed so that various materials can be used for the pads in order to manipulate the friction factor, thus enabling an investigation into effect that the friction factor has on the behavior of the rotor. Furthermore, various damping elements can also be used in order to test various configurations.

B. Modelling principle for the MFLP axial backup bearing using BBSim

The concept of the BBSim model is the division into simpler sub-models of the complex interaction between the rotor, the AMBs, the BBs and the ABB. The main sub-models include a rotor-model, AMB-models and BB/stator-models in the translational direction and the rotational direction.

The model for the radial BBs, radial AMBs and rotor is described in previously published work [1, 13-15]. Similarly, to the previous published works on BBSim the axial backup bearing (ABB) model is divided into a linear and rotational model.

In order to determine the axial position of the rotor a simple mass-damper-spring system is used as shown in Fig. 4. The airgap size is a variable depending on the angular displacement of the shaft as shown in Fig. 5. As the angular displacement of the rotor increases the airgap size decreases.

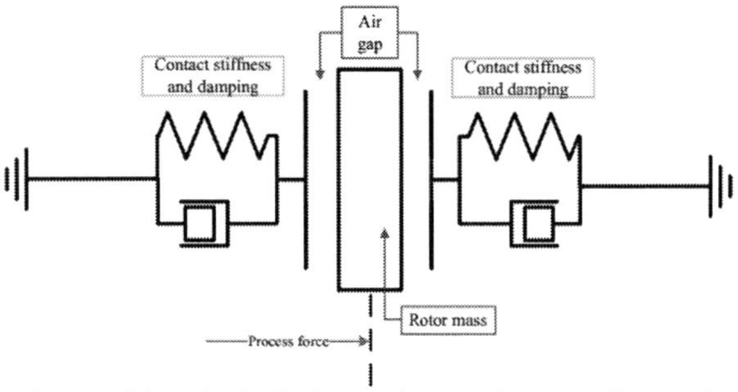

Figure 4. Schematic of a simple mass-damper-spring system for use with the ABB model

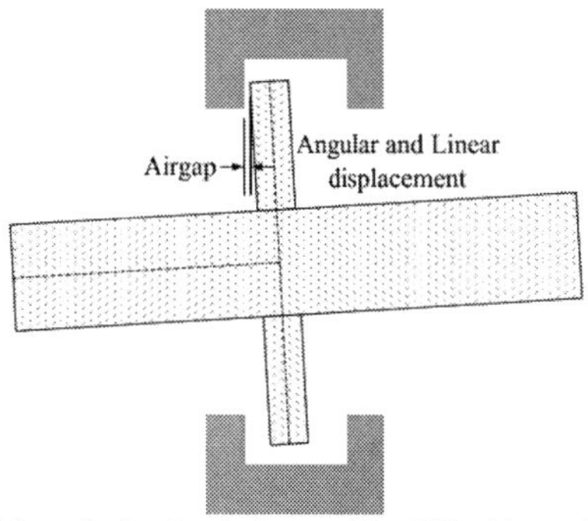

Figure 5. Schematic showing the layout of the ABB with angular and linear displacement of the rotor

The amount that the airgap in effect decreases is expressed by (1).

$$\Delta Z_{ag} = R_{ABB} \cdot (\sin \phi) \quad (1)$$

where ΔZ_{ag} is the amount that the airgap is reduced by the angular deflection of the shaft, R_{ABB} is the radius where contact occurs in the ABB and ϕ is the angular deflection of the shaft at the ABB location. Contact normal force and contact detection is based on (2).

$$F_{N_{ABB}} = \begin{cases} 0 & if \quad Z_r < Z_{eff} \\ K_{ABB} \cdot (Z_r - Z_{eff}) + \\ C_{ABB} \cdot \dfrac{d(Z_r - Z_{eff})}{dt} & if \quad Z_r \geq Z_{eff} \end{cases} \quad (2)$$

$$\text{with} \quad Z_{eff} = (R_{ag} - \Delta Z_{ag})$$

where $F_{N_{ABB}}$ is the normal contact force, K_{ABB} and C_{ABB} is the ABB stiffness and damping in the axial direction, Z_r is the linear axial position of the rotor as determined by the model shown in Fig. 4, R_{ag} is the size of the nominal airgap of the ABB.

Because the rotor is rotating during contact there is friction forces present. The friction and normal forces present on the rotor can be transformed into force-moment pairs as shown in Fig. 6. The friction force is defined using normal Coulomb friction as given in (3) with μ_{ABB} the dynamic friction factor between the contacting surfaces in the ABB.

$$F_{\mu_{ABB}} = F_{N_{ABB}} \cdot \mu_{ABB} \quad (3)$$

The magnitude of the breaking moment is dependent on the radius where contact occurs and the friction force caused by contact, as shown in (4). This breaking torque is then added to the breaking torque of the rotational rotor model as given in [1].

$$M_{break} = F_{\mu_{ABB}} \cdot R_{ABB} \quad (4)$$

The magnitude of the bending moment on the rotor is dependent on the normal force caused by contact and the radius where the contact occurs as shown in (5). The bending moment is added to the translational rotor model as discussed in [1].

$$M_{ABB} = F_{N_{ABB}} \cdot R_{ABB} \quad (5)$$

Using the described model together with the translational and rotational models in BBSim provides a complete solution for RDEs where the AMBs, the rotor, the ABB and the BBs are modelled.

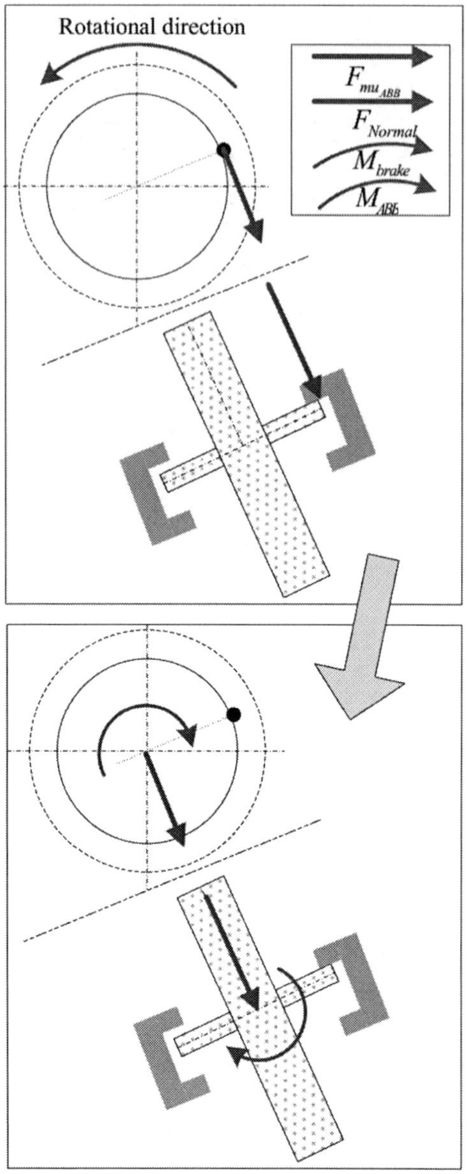

Figure 6. Schematic showing the force-moment pairs created by the normal and friction forces on the rotor

III. Results obtained using the ABB model in BBSim

For the purpose of this paper an investigation was done using BBSim and the proposed ABB model to determine the influence that the ABB has on the behavior of the rotor during a RDE. In order to determine the magnitude of the influence the RDEs need to be quantified.

A. *Rotor delevitation quantification (Vval)*

A way to measure the severity of a RDE is to use a nondimensionalised distance over time, to produce a non-dimensionalised speed. The rationale behind the nondimensionalised velocity can be found in the well-known formulas for impulse (6) and translational kinetic energy (7).

$$I = \overline{F} \cdot \Delta t = m \cdot \Delta v \tag{6}$$

$$E_k = \frac{1}{2} \cdot m \cdot v^2 \tag{7}$$

From the above equations, it is clear that the translational velocity is an important variable in both cases. The rationale is to obtain an indication of the amount of energy transformed into transverse motion. The transverse motion is an indication of the energy that the backup bearings dissipate or transform during a certain RDE and this is an indication of the degradation of the bearing-quality

The distance is non-dimensionalised by dividing with the airgap radius in order to produce a value that represents the number of times that the rotor travelled the entire distance of the airgap radius. The calculation of the non-dimensionalised distance is shown in (8), with n the window size (number of samples), i the index number, r_{airgap} the radius of the backup bearing clearance, x the position of the rotor center in the x-direction, y the position of the rotor center in the y-direction, k the index number for $Dval$ and n_{total} the total number of samples in the RDE data [1,15].

$$Dval_k = \frac{\sum_{i=((k \cdot n)-n)+1}^{k \cdot n} \sqrt{(x_i - x_{i-1})^2 + (y_i - y_{i-1})^2}}{r_{airgap}} \tag{8}$$

$$\text{with } k = \left[1, \ldots, \frac{n_{total}}{n}\right]$$

This non-dimensionalised distance is divided by the time over which it was calculated to produce an average non-dimensionalised speed with the unit of per second. If the chosen window size is the same as the total number of samples in the RDE's data, there is only one $Dval$ value. When the window size and total number of samples are the same (9) is used to determine the $Vval$ value, with f_s the sampling frequency and n the window size [1].

$$Vval = Dval_1 \cdot \left(\frac{f_s}{n}\right) \tag{9}$$

It is important to note that the maximum value for $Vval$ does not necessarily occur directly after the rotor delevitation. The dynamics of the system could cause a critical frequency being traversed at a lower rotational speed than the delevitation rotational speed. When a critical frequency of the system is traversed the behavior of the rotor in the backup bearings can become much more violent [1].

B. Rotor model

The rotor model used to determine the results presented in this paper is shown in Fig. 7. The rotor weight is approximately 2.4 tons and the rotor has a length of approximately 3.8 m.

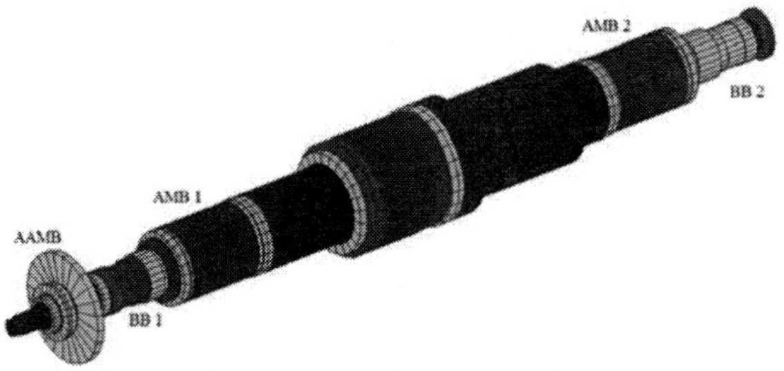

Figure 7. Rotor model used to produce results presented

The rotor is modelled in RotFE and consists of 118 nodes, with the locations for the AAMB, BB1, AMB1, AMB2 and BB2 as shown in Fig. 7.

C. BBSim ABB results

The proposed model was used to determine the effect that two parameters has on the rotor behavior during a delevitation event these parameters are the radial distance where contact occurs within the ABB (shown in Fig. 8) and the axial force present on the rotor (shown in Fig. 9).

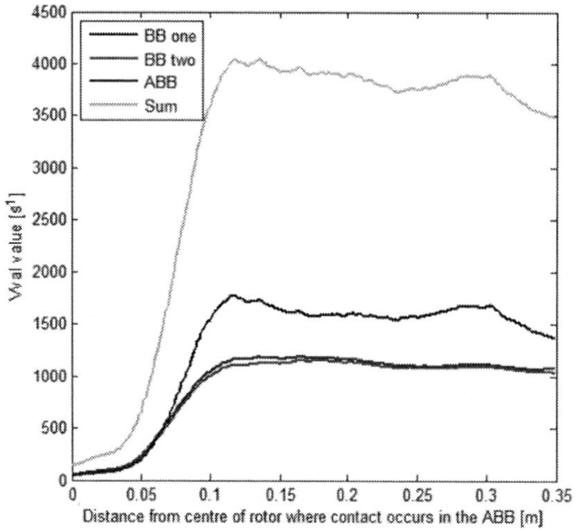

Figure 8. Results showing RDEs' Vval values for increasing radial distance where the ABB makes contact.

As can be seen in Fig. 8 the radial distance where contact occurs has a large influence on the rotor delevitation severity up until a radial distance of 0.12 it would seem that any further increase in the radial distance does not adversely affect the rotor behavior. It is important to note that as the distance increases the magnitude of $F_{\mu_{ABB}}$ and M_{break} also increases. The increase in the friction force would force the rotor to move radially and perpendicular to the contact location, this movement could possibly lead to forward whirling or forward whip as stated in [3]. The increased breaking torque would cause the rotor to decelerate quicker thus causing the rotor to have a less violent behavior since the rotor speeds through any critical frequencies faster compared to a smaller breaking torque.

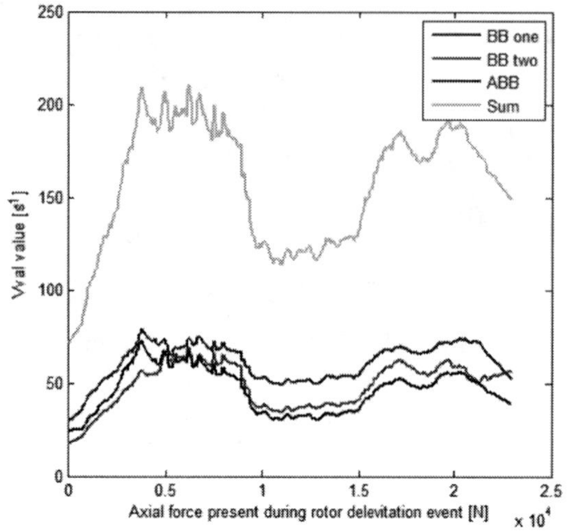

Figure 9. Results showing rotor delevitation behavior for increasing axial load.

As shown in Fig. 9 initially the axial force increases the *Vval* value up until 5 000 N where after the *Vval* value reaches a plateau until about 10 000 N where there is a sharp decrease in the *Vval* value, this pattern appears to be repeated with increments of 5 000 N, although the trend of the curves is slightly upward.

The upward trend of this curve would indicate that the rotor behavior is dependent on the axial force present; the fluctuations of the *Vval* values might be caused by axial critical frequencies cross-coupled with the radial critical frequencies of the system. The excitement of these frequencies is dependent on the magnitude of the axial force and the damping and stiffness present in both the radial and axial BB mounts.

CONCLUSION AND DISCUSSION

The model is able to calculate the influence that the forces present in the axial backup bearing has on the shaft motion during a RDE. The new test bed in Zittau will be used to generate results to validate the model, and to compare the severity of RDEs when using various sliding pad materials.

The BBSim model enables a total simulation solution when investigating AMB systems. The BBSim model includes the capability to simulate any number of AMBs and BBs (Sliding and rolling element) including the rotor dynamics of the rotor.

Although some conclusions can be made about the influence of the magnitude of the axial forces present and the radial distance where contact occurs, these conclusions require further investigation.

REFERENCES

[1] J.J. Janse van Rensburg, "Delevitation modelling of an active magnetic bearing supported rotor," North West University, Potchefstroom, South Africa, PhD Thesis 2014.

[2] J. Schmied and J. C. Pradetto, "Behaviour of a One Ton Rotor Being Dropped Into Auxiliary Bearings," , 1992.

[3] J. Wilkes, J. Moore, D. Ransom, and G. Vannini, "An improved catcher bearing model and an explanation of the forward whirl/whip phenomenon observed in active magnetic bearing transient drop experiment," in *Proceedings of ASME Turbo Expo 2013: Turbine Technical Conference and Exposition*, San Antonio, TX, USA, 2013.

[4] D. Ransom, A. Masala, J. Moore, G. Vannini, and Massimo Camatti, "Numerical and Experimental Simulation of a Vertical High Speed Motorcompressor Rotor Drop onto Catcher Bearings," in *11th International Symposium on Magnetic Bearings (ISMB11)*, Nara, Japan, 2008, pp. 136-143.

[5] S.R. Penfield and E. Rodwell, "Auxiliary Bearing Design Considerations for Gas-Cooled Reactors," International Atomic Energy Agency, Palo Alto, USA, Report of a technical committee meeting (TCM) on Gas Turbine Power Conversion Systems for Modular HTGRs 2001.

[6] M.T. Caprio, B.T. Murphy, and J.D. Herbst, "Spin Commissioning and Drop Tests of a 130 kW-Hr Composite Flywheel," , 2004.

[7] E.E. Swanson, J.F. Walton, and H. Heshmat, "A 35,000 rpm Test Rig for Magnetic, Hybrid and Back-Up Bearings," *Transactions of the ASME*, vol. Paper No. 99-GT-180, pp. 1-7, 1999.

[8] E.N. Cuesta, L.U. Medina, V.R. Rastelli, N.I. Montbrun, and S.E. Diaz, "A Simple Kinematic Model for the Behaviour of a Magnetically Levitated Rotor Operating in Overload Regime," , 2003.

[9] G. Sun, A.B. Palazzolo, A. Provenza, and G. Montague, "Detailed Ball Bearing Model for Magnetic Suspension Auxiliary Service," *Journal of Sound & Vibration*, vol. 269, no. 3, pp. 933-963, 2004.

[10] T. Rottenbach and F. Worlitz, "Long-term operation of an AMB-supported coolant pump in a lignite-fired power station - results and operational experience," in *11th International Symposium on Magnetic Bearings*, 2008.

[11] M. Orth and R. Nordmann, "ANEAS, a Modelling Tool for Nonlinear Analysis of Active Magnetic Bearing Systems," in *Proceedings of the IFAC Conference of Mechatronic Systems*, Berkeley, CA, USA, 2002.

[12] H. Neumann, C. Panescu, S. Düsterhaupt, and F. Worlitz, "MFLP - magnet- und Fanglagererprüfstand des Zittauer Kraftwerkslabores," in *9 Workshop Mechatronische Systeme*, 2013.

[13] J.J. Janse van Rensburg, G. van Schoor, and P.A. van Vuuren, "Delevitation modelling of an active magnetic bearing supported rotor," in *Proceedings of the 12th International Symposium on Magnetic Bearings (ISMB12)*, Wuhan, China, 2010.

[14] C. Vanek, F. Worlitz, J.J. Janse van Rensburg, and G. van Schoor, "Experimentelle und theoretische Untersuchungen zum Rotorabsturz in Wälzlager," in *Workshop Magnetlagertechnik Zittau-Chemnitz*, Zittau, 2011, pp. 97-101.

[15] J.J. Janse van Rensburg, G. van Schoor, and P. van Vuuren, "The characterization of the severity of rotor delevitation events - A parametric study," in *Proceedings of the 13th International Symposium on Magnetic Bearings (ISMB13)*, Arlington, VA, USA, USA, 2012.

Fabrication of Non-contact Carrier System using Solar Magnetic Suspension

Yuji ISHINO[a], Masaya TAKASAKI[a], Takeshi MIZUNO[a]

[a] Saitama University, Shimo okubo 255, Sakura-ku, Saitama, JAPAN, yishino@mech.saitama-u.ac.jp

***Abstract*— A solar-powered magnetic suspension carrier system is fabricated which achieves noncontact power supply to the carrier. There are two conventional methods of supplying power to a carrier. One is through wires or contacts such as pantograph and the other is a battery in the carrier. However, the noncontact characteristics are lost in the former, and stations for charging the battery are necessary in the latter. These problems are overcome by the proposed carrier system using solar magnetic suspension that can achieve continuous levitation. In rooms like in factory, the generation of power by solar cells is low so that the magnetic suspension needs power saving. A zero power control system is applied to the suspension of the floator. In addition, low-power peripheral devices are also fabricated.**

I. Introduction

Solar power generation systems need no mechanical motion, conventional fuel consumption and operating cost. These systems are widely used in houses recently because it is a countermeasure against global warming. Consequently, solar cells are becoming a popular and cost-saving green power generation source year by year[1]. Solar power systems are suitable for small outdoor standalone apparatus where electric power transmission infrastructure is not available. However, the power generated by a solar cell is sensitive to solar conditions because the generated solar power fluctuates due to illuminance. Hence, standalone systems powered by solar energy need buffered capacitors or batteries to operate even in low-illuminance environments including night and cloudy weathers while they are charged during daylight. Such buffered techniques regarding energy storage have been also progressive and lower in cost recently.

Magnetic suspension systems [2] are widely used in non-contact and friction-free suspension. Magnetic suspension has various fields of applications such as ultra high rotational speed bearings [3], rapid-transit carrier [4], guide for high-speed elevator [5], carrier systems in clean room [6], rotating gyroscopic sensors [7], space instruments [8] and blood pumps [9]. To control them, various control methods are proposed and applied. Among them, the zero-power control [10], [11] has been used in suspension system with combined electromagnet and permanent magnet (hybrid magnet). The zero-power control achieves the steady states in which the attractive force produced by the permanent magnets balances the weight of the suspended object (floator) the control current converges to zero. Because there is no steady energy consumption at the stable levitation, this method is used in energy saving devices.

A single-degree-of-freedom solar magnetic suspension system has been developed [12]. The system is combined with solar power generation and zero-power magnetic suspension technique. Dedicated power-saving peripheral devices are newly fabricated for solar magnetic suspension system. The system achieved the stable suspension of a 90-gram floator even under a fluorescent lamp of 5[klx].

This paper proposes to apply solar magnetic suspension to a carrier system. The conventional magnetic suspension carrier systems can be categorized into two types: guide rail with magnetic coils and magnetic coils on board the car. The former does not need supply to the car for suspension. However, this system needs a lot of magnetic coils along the rail. Therefore, this system is disadvantageous in installing cost. On the other hand, latter type needs a power supply to the carrier. There are two conventional methods of supplying power to the carrier. One is through wires or contacts such as pantograph and the other is a battery in the carrier. However, the non-contact characteristics are lost in the former, and stations for charging the battery are necessary in the latter. These problems are overcome by the proposed carrier system using solar magnetic suspension that can achieve continuous levitation without any external power supply. Therefore, this system keeps the non-contact property permanently.

In this paper, the magnetic suspension carrier system is assumed to be used in an indoor condition. Usually the illuminance of indoor is darker than sunlight. The solar cells produce small generation power. Therefore, it is necessary for the magnetic suspension system to become extremely power saving.

II. Solar Magnetic Suspension Carrier

A. Outline

Figure 1 shows the outline of the solar magnetic suspension carrier including the power supply flow. In this figure, solid and dashed lines represent signal and power flows, respectively. The unit of the magnetic suspension consists of an electromagnet with a permanent magnet, a displacement sensor, a power amplifier and a controller.

The carrier has three magnetic suspension units to control three degrees of freedom of motion. The power supply for the magnetic suspension units is common. The power supply has solar cells, a buffered storage device and a constant voltage circuit for the peripheral devices. The electric power generated by the solar cells are charged to the storage devices. The constant voltage power supply is for the displacement sensor and the controller.

B. Experimental apparatus

Figure 2 shows a photograph of the fabricated magnetic suspension carrier. The size of the apparatus is 340×200×160[mm]. The loading platform has six solar cells in three side two in each. The solar cell is amorphous silicon type, the shape is rectangular and the size and thickness are 167.5×148 [mm] and 0.4[mm], respectively. Table 1 shows the ratings of the solar cell [13]. The three magnetic suspension units are installed on the top plate of the carrier. They are used to control the three motions of the carrier: like perpendicular direction (z), rotation about the horizontal axis (*pitch*) and rotation about the rail direction (*roll*). The units are independently controlled. Horizontal direction (x) and rotation about the z axis (*yaw*) are passively stabilized by the edge effect of the magnetic fluxes.

Figure 3 shows a photograph of the magnetic suspension unit. The size of the unit is 60×50×55[mm]. It has a hybrid magnet, a pickup coil for displacement sensor and four touch down bearings. Figure 4 shows a flow diagram of magnetic suspension unit. Each magnetic suspension unit has a peripheral electric circuit.

An inductive-type displacement sensor was fabricated for power saving. A typical commercially available eddy-current displacement sensor consists of an oscillating circuit, a resonator, a rectifier, a smoothing circuit, a linearizer and a stabilized power supply. The fabricated displacement sensor only uses an oscillator IC, a resonator, a diode rectifier and a RC smoothing circuit. The oscillator produces a rectangular-wave with a frequency of 1[MHz] and an amplitude of 3.3[V] (Fig. 4 (A)). The resonator consists a single FET, a pickup coil and a capacitor . The inductance of the pickup coil decreases as the gap between the pickup coil and the rail increases. The gap becomes larger when the carrier moves up. The resonance frequency increases as the inductance of the pickup coil decreases. The resonance frequency set to be lower than the oscillation frequency at any distance. The resonant frequency approaches the carrier frequency 1[MHz] when the gap increases. Thus, the output of the resonator decreases as the gap decreases (Fig. 4 (B)). The output is transmitted to the controller through a rectifier circuit (Fig. 4 (C)) and a low path filter (Fig. 4 (D)). This principle of rectifier and smoothing circuit is same as that of AM demodulating circuit.

A peripheral interface controller (PIC) is used as a controller. PICs are popular to industrial developers, hobby users and students due to their low consumed power, wide availability and low cost. The pin function of PIC is selectable by programing. The PIC for magnetic suspension controller uses a single-channel A/D converter, a single-channel enabled signal from the power supply circuit for active mode, two-channel direction signal of the coil current via amplifier and a single-channel PWM output. It implements the zero-power control. The zero-power control is realized by feeding back the integral of the control voltage. The controlled period is 1[msec]. The supplied voltage for the sensor circuit and PIC are 3.3[V].

Table 1 Nominal ratings of solar photovoltaic cell [13].

Parameter	Ratings
Typical open circuit voltage	4.7[V]
Typical open circuit current	175[mA]
Reference maximum power	517[mW]

Solar simulator 50 [kLx] nominal condition.

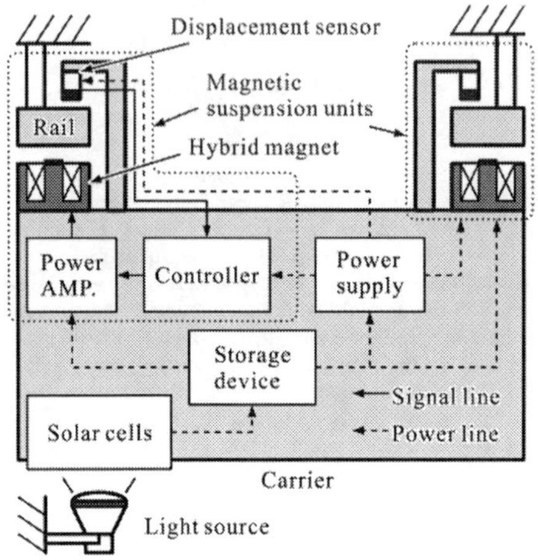

Fig. 1 Outline of solar magnetic carrier with signal and power flow.

Fig. 2 Photograph of a magnetic suspension carrier.

Fig. 3 Photograph of a magnetic suspension unit.

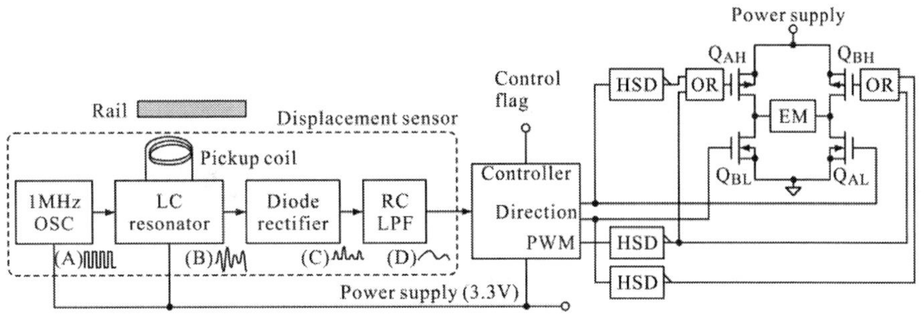

Fig. 4 Flow diagram of magnetic suspension unit.

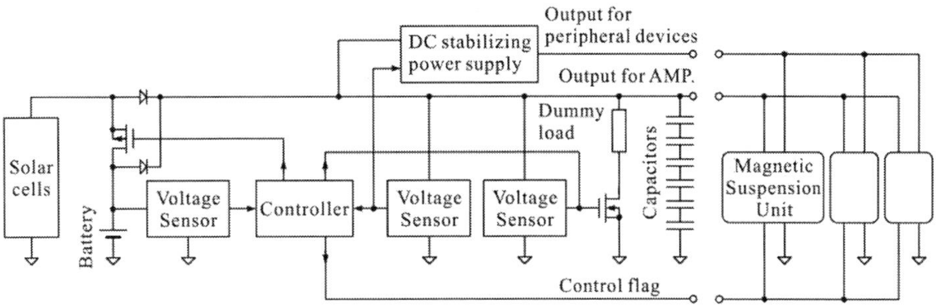

Fig. 5 Flow diagram of power supply circuit.

A fabricated amplifier is a voltage-controlled H-bridge constructed from two low side N-ch FETs (Q_{AL}, Q_{BL}), two high side P-ch FETs (Q_{AH}, Q_{BH}), three pre-high-side-driver FET (HSD in the figure) and two OR gates. The direction signal has 3 states of a positive, minus and zero. The high side FETs are driven by the compared signals. The signals compared direction signals with the PWM signal. When the direction signal is positive, Q_{AH} in PWM operation, Q_{AL} is turned on, and other side FETs (Q_{BH} , Q_{BL}) are kept turned off, and vice versa. When the control value is zero, all the transistors are turned off.

Figure 5 shows a flow diagram of power supply circuit. A power storage device is needed to a standalone active system using solely solar energy. The fabricated system uses electric-double-layer-capacitor and a nickel-hydrogen secondary battery for power storage. The energy density of typical electric-double-layer capacitors is hundreds to thousands times greater than that of a conventional electrolytic capacitor of the same dimensions [14]. The storage circuit of the fabricated system has six series-connected capacitors with parallel-connected balanced registers individually. The leakage currents are different among the capacitors. The voltage of the series-connected capacitors vary individually according to the leakage currents. The balancing registers balance the voltage for each capacitor. The capacity of the secondary battery is 200[mAh]. As shown later, magnetic suspension system consumes 350[mW]. Therefore, the expected levitation period is 4.8hours without solar power supply. The secondary battery has a constant voltage characteristic. Therefore, the output voltage of the power supply for the amplifier is approximately constant. When the generation power of the solar cells exceeds consumption power, the output voltage is raised and the current is decreased. If output voltage exceeds a threshold voltage, surplus power is consumed by a dummy load. When overvoltage or undervoltage, power supply circuit stops suspension control and charging for the battery.

III. EXPERIMENTAL RESULT

A. Consumption power of peripheral devices

The consumption power of the suspension unit was measured. The power of the H-bridge amplifier is 2.7[mW] with an applied voltage of 8.4[V] and no load. The power is consumed mainly in charging the gate-source capacitor in the FET and high-side-driver (HSD). The consumption power in the amplifier becomes zero when the coil current of the electromagnet is zero. The consumption power of PIC is 1.8[mW]. The average consumption power of displacement sensor is 0.7[mW]. Commercially available displacement sensors with similar dimensions consume electric power of more than 200[mW] [15]. The consumption power of peripheral device in the unit is 5.2[mW].

The consumption power of the power supply circuit was measured. The power loss due to leakage current of the electro-double-layer-capacitor circuit is 1[mW] in the case of 10[V].

When the charging voltage is high, the leakage current is large. The consumption power in the power supply circuit depends on the output voltage. The power consumption is 1.9[mW] at the other than the zero. The power of peripheral devices for floating is less than 20[mW] in the suspension system.

B. Consumption power at the floating

The consumed power of the experimental apparatus was measured. In this measurement, an 8.4-V power supply is substituted for the solar cells. The current is measured through a sensing resistor inserted into the circuit. Figure 6 shows an experimental result of the consumption power for hybrid magnet in the steady suspension. The average consumption power is approximately 330[mW] for the suspension. The total of suspension system consumption power is 350[mW]. Illuminance of approximately 15[klx] is necessary for the magnetic suspension system.

IV. CONCLUSIONS

A non-contacted carrier system using solar magnetic suspension was fabricated. The carrier was controlled vertically passively stabilized horizontally. Dedicated power-saving peripheral devices were newly fabricated. The consumed power of the fabricated system was measured. The total consumed power was approximately 350[mW] in the steady-state suspension.

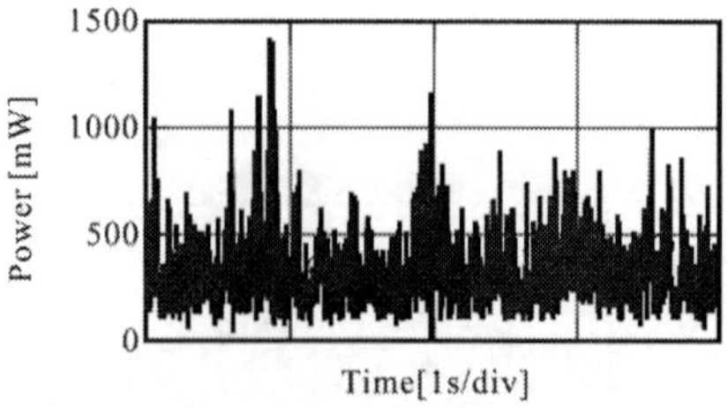

Fig. 6 Consumption power characteristic for electromagnet at the steady state floating.

REFERENCES

[1] Martin A. Green, Keith Emery, Yoshihiro Hishikawa, Wilhelm Warta and Ewan D. Dunlop, Solar cell efficiency tables (version 39) ,Progress in photovoltaics research and applications 2012, 20, pp.12-20.

[2] Frazier,R.H.,Gilinson Jr, P.J. amd Oberbeck, G.A., Magnetic and electric suspensions, (1974), pp.314-318, USA: The MIT press.

[3] J.W. Beams, Magnetic Suspension for Small Rotors, Review of Scientific Instruments, Vol. 21, (1950), pp.182-184

[4] H. Weh, Linear Synchronous motor development for urban and rapid transit systems, IEEE Transaction on magnetics, vol. MAG-15, No.6, (1979), pp.1422-1427.

[5] Morishita, M., Seki, Y., Zero Power Control Applied to Electromagnetic Non-contact Guide System for Elevator Cars, SICE The 1st Control Division Conference, (2001), pp.225-228.

[6] Morishita, M., Azukizawa, T., Kanda, S., Tamura, N. and Yokoyama, T., A New Maglev System for Magnetically Levitated Carrier System, IEEE Trans, Vehicular Technology, Vol.38, No.4 , (1989), 230-236.

[7] Maruyama, Y., Mizuno, T., Takasaki, M., Ishino, Y., Ishigami, T. and Kameno, H., An Application of Active Magnetic Bearing to Gyroscopic and Inertial Sensors, Journal of System Design and Dynamics, Vol.2, No.1, (2008), pp.155-164.

[8] Henrikson, C. H., Lyman, J. and Studer, P. A., Magnetically Suspended Momentum Whells for Spacecraft Stabilization, AIAA paper (1974), pp. 74-128.

[9] Masuzawa T, Onuma H, Kim SJ, et al: Magnetically suspended centrifugal blood pump with a self bearing motor. ASAIO J 48:437-42, 2002.

[10] Morishita, M., Azukizawa, T., Zero power Control Method for Electromagnetic Levitation System, Trans IEE Japan (1988), 108-D, pp. 447-454.

[11] Mizuno, T., Takemori, Y., A Transfer-Function Approach to the Analysis and Design of Zero-Power Controllers for Mag-netic Suspension System, Trans IEE Japan (2001), 121-D, No.9, 933-940.

[12] Ishino, Y., Takasaki, M., Mizuno, T., Development of a Solar Photovoltaic Magnetic Suspension System, Proceedings of the 13th International Symposium on Magnetic Bearings (ISMB13), paper_35, (2012.8.7)

[13] SANYO Semiconductor Co. Ltd., Data sheet AT-7666 No.A0062, (2008).

[14] Nichicon corporation, Electric double layer capacitors "EVerCAP " CAT.1500G, (2011).

[15] Baumer Electric Ltd., IWRM 18U9501 Data sheet, (2012).

Proposal of AC Magnetic Suspension Using Magnetic Resonant Coupling

Takeshi MIZUNO[a], Kei TAKAHASHI[a], Yuji ISHINO[a], Masaya TAKASAKI[a]

[a] Department of Mechanical Engineering, Saitama University, Shimo-Okubo 255, Sakura-ku, Saitama 338-8570, JAPAN
mizar@mech.saitama-u.ac.jp

***Abstract*— A novel AC magnetic suspension using magnetic resonant coupling is proposed and studied both theoretically and experimentally. The conventional AC magnetic suspension with energy transfer function has a problem that energy transfer efficiency is very low because it has a gap. In contrast, even if it has a gap, the energy transfer method using magnetic resonant coupling has high efficiency. In this paper, this method is combined with AC magnetic suspension. An experimental apparatus is fabricated for basic study. The floator was successfully suspended without control, and the self-stabilization characteristic is confirmed.**

I. Introduction

There are several methods of suspending a body without contact by using magnetic forces [1], [2]. One of them uses AC electromagnet instead of controlled DC electromagnet. The tuned LCR circuit levitation was investigated extensively in some period [3]-[6] mainly because it can be stable without a control loop. However, this characteristic has lost significance recently because powerful controllers are available at rather low cost [2].

Other forms of suspension using AC electromagnet have been developed to achieve energy transfer to the suspended object (floator) simultaneously [7]-[9]. When energy is consumed in the floator, either wire connection between the suspended object and ground facilities or the installation of a battery on the object [10] is necessary in suspension systems using controlled-DC electromagnet. However, the former breaks the noncontact property. In the latter, exchanging or recharging a battery is unavoidable. Such problems can be solved by using AC electromagnet not only for suspension but also for energy transfer [7]-[9].

However, the conventional AC magnetic suspension with energy transfer function has a problem that energy transfer efficiency is rather low because of a rather wide gap. To overcome this problem, a novel AC magnetic suspension using magnetic resonant coupling is proposed. It is to be noted that the energy transfer method using magnetic resonant coupling has high efficiency in spite of wide gap [11].

In this work, magnetic resonant coupling is introduced into AC magnetic suspension for improving the energy transfer performances. It is also demonstrated that the proposed suspension system has a self-stabilization characteristic so that stable suspension is achieved without any active control. An experimental apparatus is fabricated for basic study. Several experiments carried out with the apparatus are presented to demonstrate the fundamental properties of the proposed magnetic suspension system.

II. Principles

A. Force Generation

The principle of suspension is illustrated in Fig.1. The stator electromagnet is a part of a series-resonant circuit fed by an alternating-voltage source (primary circuit). An electromagnet is attached to the floator and is also a part of another series-resonant circuit (secondary circuit). The two series-resonant circuits are adjusted to have a common resonant angular frequency ω_r. The two electromagnets face each other across a gap. Even if the gap is rather wide, a high-efficiency energy transfer is expected from the stator to the

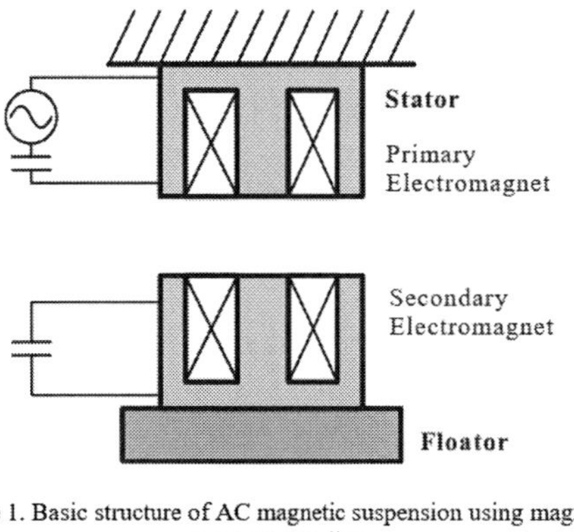

Figure 1. Basic structure of AC magnetic suspension using magnetic resonant coupling

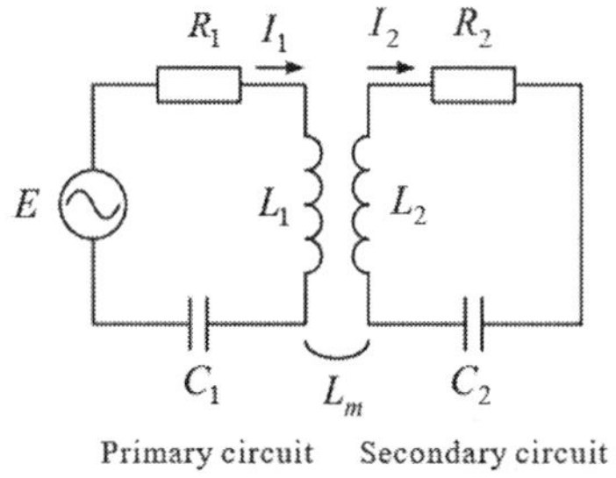

Figure 2. Equivalent circuit

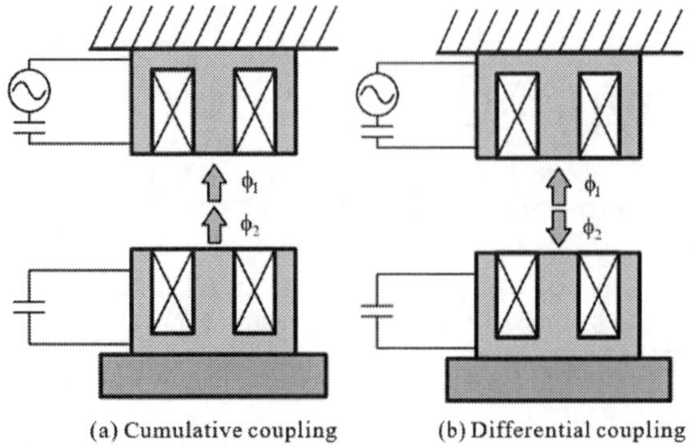

Figure 3 Direction of magnetic fluxes

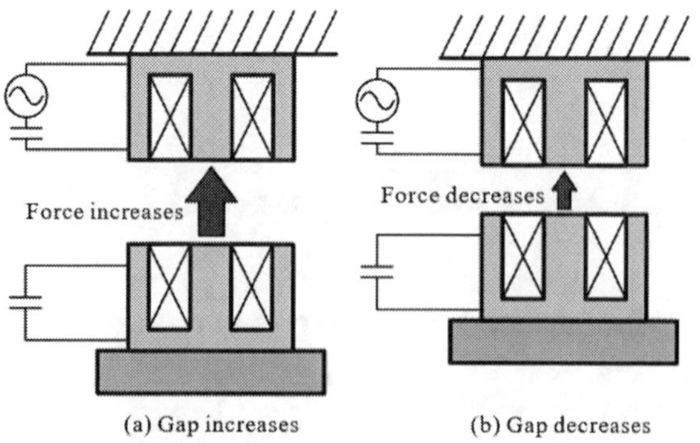

Figure 4. Self-stabilization characteristic

floator because of magnetic resonant coupling.

Figure 2 shows an equivalent circuit. It is assumed for simplicity that the primary and secondary circuits have a same self-inductance L and a same capacitor C with no loss ($R_1=R_2=0$). The mutual inductance L_m is given by kL where k is the coupling coefficient. The circuit equations are given by

$$j\left(\omega L-\frac{1}{\omega C}\right)I_1 - j\omega kLI_2 = E\ , \tag{1}$$

$$j\left(\omega L-\frac{1}{\omega C}\right)I_2 - j\omega kLI_1 = 0\ , \tag{2}$$

where ω: voltage-source angular frequency, I_1: amplitude of current flowing through the stator electromagnet, I_2: amplitude of current flowing through the floator electromagnet, and E: amplitude of voltage source. From (1) and (2), each amplitude of current is given by

$$I_1 = -j\frac{\left(\omega L-\dfrac{1}{\omega C}\right)}{\left(\omega L-\dfrac{1}{\omega C}\right)^2-\omega^2k^2L^2}E\ , \tag{3}$$

$$I_2 = -j\frac{\omega kL}{\left(\omega L-\dfrac{1}{\omega C}\right)^2-\omega^2k^2L^2}E\ . \tag{4}$$

From (3) and (4), we find that

$$I_1I_2 < 0 \text{ for } \omega < \omega_r \quad \textit{Cumulative coupling,} \tag{5}$$

$$I_1I_2 > 0 \text{ for } \omega > \omega_r \quad \textit{Differential coupling.} \tag{6}$$

Figure 3 illustrates the direction of force between the electromagnets. In the cumulative coupling mode, the flux generated by the stator electromagnet is reinforced with the flux generated by the floator electromagnet so that the force between them is attractive as shown in Fig.3(a). In contrast, the force can be repulsive in the differential coupling mode as shown in Fig.3(b).

Figure 5. Photograph of experimental apparatus

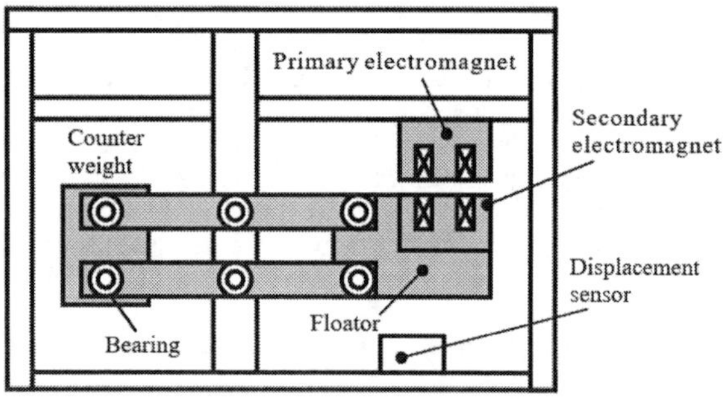

Figure 6. Schematic drawing of experimental apparatus

B. *Self-Stabilization*

It is assumed for simplicity that only the coupling coefficient k is a function of the gap between the stator and the floator. Generally, the coefficient increases as the gap decreases. The resonant condition of the whole circuit shown in Fig.2 is given by

TABLE I. CIRCUIT PARAMETERS

	Parameter	Value
L_1	Self-inductance of stator electromagnet	9.30 mH
C_1	Capacitance of primary circuit	1.03 μF
L_2	Self-inductance of floator electromagnet	9.37 mH
C_2	Capacitance of secondary circuit	1.04 μF

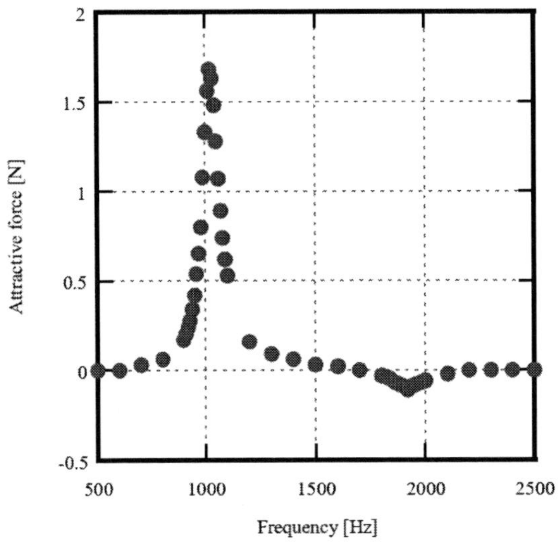

Figure 7. Force between the electromagnets as a function of frequency ($E = 15\text{V}$, Gap: 1mm, $R_L = 1\Omega$)

$$k = 1 - \frac{1}{\omega^2 LC}. \tag{7}$$

For a fixed given ω, this value is designated by k_r. Assume that $k_r < k$ and the coupling is cumulative. In these conditions, as the gap increases, k becomes smaller and approaches to k_r so that the amplitude of current and the attractive force increase as shown in Fig.4(a). Inversely, as the gap decreases, the attractive force decreases as shown in Fig.4(b). It indicates that this suspension system has a self-stabilization characteristic that was proper to the conventional AC magnetic suspension using a tuned LCR circuit [3]-[6].

III. EXPERIMENT

A. Apparatus

Figures 5 and 6 show a photograph and a schematic drawing of the fabricated apparatus for experimental study. It has a seesaw-type floator. An electromagnet for suspension is fixed at one end of the floator while a counter weight is attached to the other end for the adjustment of static unbalance. The stator electromagnet is fixed above the floator electromagnet. Under the floator, an eddy-current gap sensor is installed to measure the displacement of the floator. The parameters are listed in Table I.

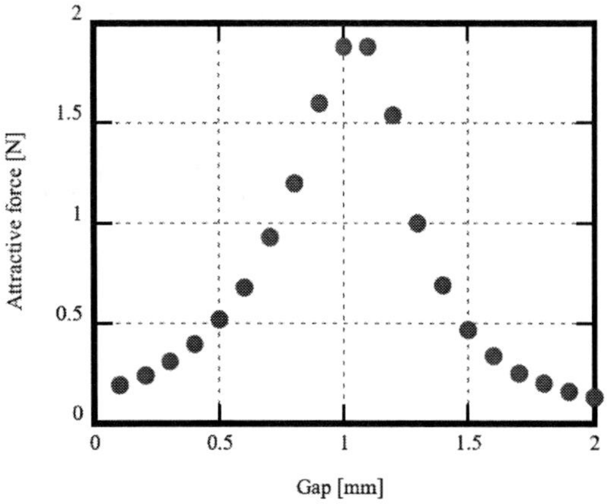

Figure 8. Force as a function of gap ($E = 15\text{V}$, $\omega = 6.28\times10^3$rad/s, $R_L = 1\Omega$).

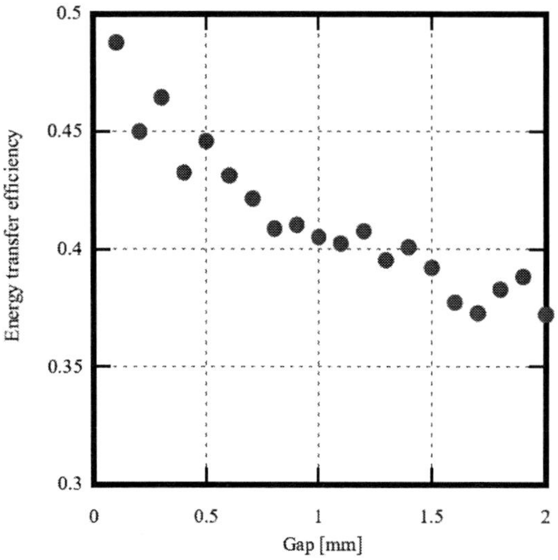

Figure 9. Energy transfer efficiency ($E = 15\text{V}$, $\omega = 6.28\times10^3$rad/s, $R_L = 10\Omega$).

B. Force characteristics

Figure 7 shows a measured force-frequency characteristic. During the measurement, the floator is fixed to a load cell installed below the floator and the gap is kept to be 1mm. A load resistance R_L is inserted to short the secondary circuit; it is set to be 1 Ω. The amplitude of the applied voltage E is 15 V. It is found that the maximum attractive and repulsive forces are obtained at 1020 Hz and 1910 Hz, respectively. Not only attractive force but also repulsive force were generated as predicted in Fig.3.

Figure 8 shows a measured force-gap characteristic with E=15 V and $\omega = 6.28\times10^3$ rad/s (correponding to 1100 Hz). It is found that the force increases as the gap increases from 0

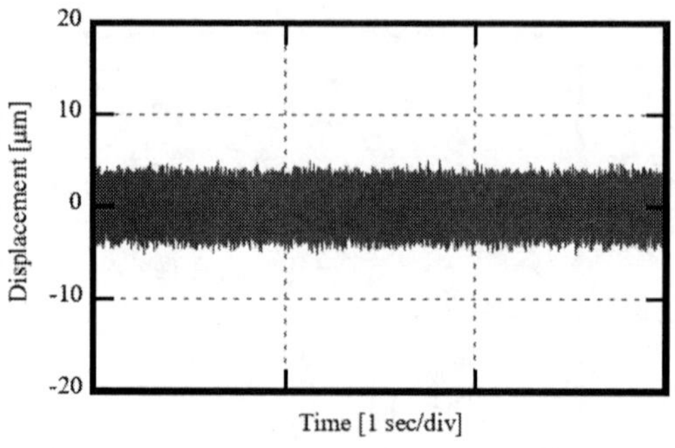

Figure 10. Displacement of floator during levitation ($E = 15\text{V}$, $\omega = 6.28\times10^3$rad/s, $R_L = 1\Omega$).

to 1 mm. This result indicates the possibility of self-stabilization.

C. *Energy transfer characteristics*

Figure 9 shows a measured energy transfer efficiency as a function of the gap under the conditions where $E = 15\text{V}$, $\omega = 6.28\times10^3$ rad/s (1000 Hz) and $R_L = 10\Omega$. It is found that the efficiency decreases as the gap increases but the ratio of decrease is not so large as that in the conventional AC magnetic suspension system [9].

D. *Suspension without control*

Self-stabilization suspension was achieved with E=20 V and $\omega = 6.28\times10^3$ rad/s (1000 Hz). The gap was 0.8 mm approximately. The displacement of the floator during suspension is shown in Fig.10. The floator is kept at a position without any active control.

IV. Conclusions

The AC magnetic suspension using magnetic resonant coupling was proposed. The fundamental characteristics of the proposed suspension system were studied both analytically and experimentally. It is demonstrated that the proposed system has the same characteristic of self-stabilization as the conventional tuned LCR circuit levitation. The self-stabilization suspension was achieved successfully in the fabricated apparatus.

References

[1] B. V. Jayawant, "Electromagnetic Levitation and Suspension Technique," pp.1-31., Rdward Arnold, London, 1981.

[2] G. Schweitzer and E. H. Maslen (Ed.), "Magnetic Bearings," pp.10-15, Springer, 2009 .

[3] B. V. Jayawant and D. P. Rea, "New electromagnetic suspension and its stabilisation," Proc. IEE, Vol.115, No.4, pp.549-554, 1968.

[4] B. V. Jayawant and B. Z. Kaplan, "The development and stabilization of a new electromagnetic suspension," Journal of Physics E: Scientific Instruments, Vol.4, pp.301-306, 1971.

[5] B. Z. Kaplan and D. Regev, "Dynamic stabilization of tune-circuit levitatiors," IEEE Trans. on Magnetics, Vol.12, No.5, pp.556-559, 1976.

[6] S. Hagihara, "Performance and stability of a magnetic suspension device using a tuned LCR circuit," Proc. IEE, Vol.125, No.2, pp.153-156, 1978.

[7] O. Tsukamoto, K. Yasuda and J. Z. Chen, "A new levitation system with AC magnets," IEEE Trans. on Magnetics, Vol.24, No.2, pp.1497-1500, 1988.

[8] T. Hirata, T. Hikihara and Y. Hirane, "Susepension characteristics of magnetic suspension system bylinear induction motor," Trans. IEE Japan, Vol.110-D, No.10, pp.1091-1099, 1990.

[9] Mizuno, T., Kishita, D., Takasaki, M. and Ishino, Y., "PM-biased AC magnetic suspension having energy transfer function," Proc. 9th International Symposium on Magnetic Bearings, 70, 2004.

[10] M. Morishita, T. Azukizawa, S. Kanda, N. Tamura and T. Yokoyama, "A new Maglev system for magnetically levitated carrier system," IEEE Trans. on Vehicular Technology, 38, No.4, pp.230-236, 1989.

[11] André Kurs, et al., "Wireless Power Transfer via Strongly Coupled Magnetic Resonances," Science Express, Vol. 317, No.5834, pp. 83-86, 2007.

Simulation Analysis of Dust-free Transit System Using Permanent Magnetic Suspension

Junjie Jin [a], Feng Sun[a], Yingyuan Lu [a], Koichi Oka[b]

[a] Shenyang University of Technology, No.111, Shenliao West Road, Economic & Technological Development Zone, Shenyang,110870, P.R.CHINA, sunfeng@sut.edu.cn
[b] Kochi University of Technology, 185 Miyanokuchi, Tosayamada, Kami city, Kochi 782--8502, Japan

***Abstract*—This paper proposes a dust-free transit system using permanent magnetic suspension. In this system, a transit vehicle hangs below two guide rails; four non-contact active suspension supporting devices are arranged symmetrically in two flanks of the transit vehicle. An active suspension supporting device mainly consists of an AC servo motor, a disk radial magnetized permanent magnet and magnetizer. The magnet provides magnetic force, the motor drives disc-magnet rotate to change the magnetic flux path flowing through the guide rail. The mechanical dynamics model of permanent magnetic suspension system is set up. The suspension characteristic of the system is simulated by LQR method. The results show that this system can be suspended stably, theoretically.**

I. INTRODUCTION

At present, in the dust-free transit workplace, such as the production line of processing electronic components, biology laboratory and so on, the unmanned vehicles and FOUP&MAC are widely used at home and abroad. The metal powder and the lubrication oil are inevitable to exist in traditional transit equipments. Since the characteristics of non-contact, frictionless and lubrication free [1], magnetic levitation technology is used in dust-free transit vehicles, which can meet the high requirements of air purity.

Dr. Song designed a magnetic suspension feed mechanism with the linear motor, which could realize non-contact driving and satisfy the clean environment of the integrated circuit processing [2]. Pro. Morishita developed a hung transit device using the electromagnetic levitation technology, which had been applied in an electronic components production line [3].

Recently, with development of the rare earth permanent magnet materials, the attraction force magnet is increased sharply. Therefore, advantages of compact structure and strong attractive force, make the permanent magnetic suspension technology applied widely. The author presents a permanent magnet suspension system with suspension force control method of variable magnetic circuit[4]. The suspension force is real-time adjusted by changing the angle of the disc radial permanent magnet.

II. STRUCTURE AND PRINCIPLE

A. System Structure

The structure of dust-free transit system is shown in Figure 1. In the system, a transit vehicle hangs below two guide rails, four non-contact active suspension supporting devices are arranged symmetrically in two flanks of the transit vehicle, the sensors and limit wheels (not shown) are arranged nearby the suspension support. There are disc radial magnetized permanent magnets, magnetizers and servo motors (directly connect to magnet). The material of magnet is NdFeB30; the magnetizer material is permalloy 1J85, the rail material is Q235 steel.

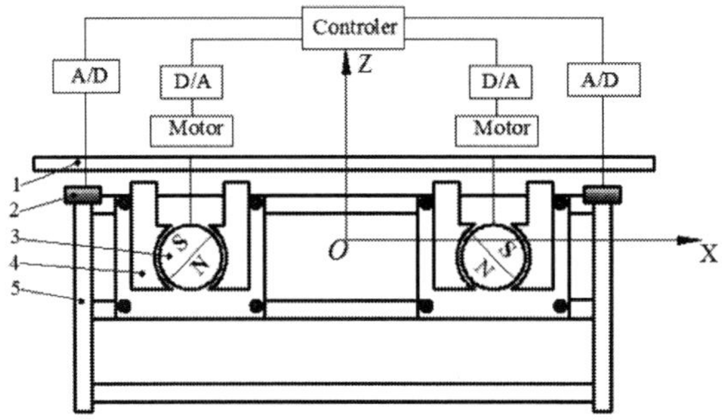

1-Guide rail, 2-Sensor, 3-Permanent magnet, 4- Magnetizer, 5- Frame

Figure 1. Structure diagram of dust-free transmission device with permanent magnetic suspension

B. Principle

The system suspension principle is adjusting the magnet rotation angle to realize the suspension force of four suspension supports real-time control, balance the device gravity, achieve the movement on the Z axis direction (vertical direction), the rotation around the X axis and Y axis.

Four suspension supports working principle are identical, so take one of them for example. It adopts the principle of flux path control, the motor rotates the disc magnet to change the magnetic flux through the guide rail, control the suspension force. Three working states of magnet is shown in Figure 2. The initial state of the magnet rotating to 0° is shown in Figure 2.(a). The magnetic flux being from N pole goes back to S pole only through the magnetizer, and hardly any flux flows through the guide rail, there is no attractive force between the magnetizer and the guide rail. Figure 2.(b) shows the state of the magnet rotates to θ. A part of the magnetic flux flows S through the guide rail, the suspension force (F) is generated between the magnetizer and the guide rail. As shown Figure 2.(c), when the magnet being 90° state, the magnetic flux is almost entirely flowing through the guide rail, so the suspension force is largest. While the magnet rotation angle being 90° to 360° like in Figure 2.(a).

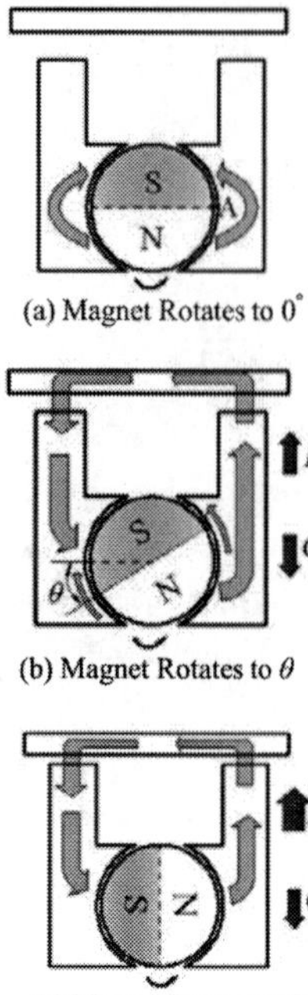

(a) Magnet Rotates to 0°

(b) Magnet Rotates to θ

(c) Magnet Rotates to 90°

Figure 2. Variable Magnetic Circuit Control Principle

III. SYSTEM MODEL

Defining the system balance position in stable suspension state: all the air gap length between the suspension supports and the guide rail are equal, the sum of suspension force balance to the system gravity. The system force model is shown in Figure 3. Table 1 shows the symbols and value of system parameters in the equilibrium position. Table 2 shows the system variables relative to the equilibrium position.

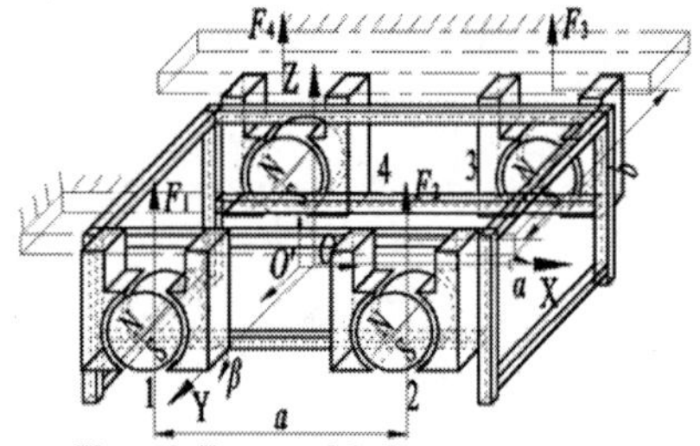

Figure 3. Force model of permanent magnet suspension system

TABLE I. SYMBOLS AND VALUE OF SYSTEM PHYSICAL PARAMETERS

	Parameter	Value
z_0	Air-gap length under equilibrium position	3mm
m	Device mass	17.18kg
J_α	Rotational inertia of device around X axis	0.4723kg·m2
J_β	Rotational inertia of device around Y axis	0.5048kg·m2
K_d	Electromagnetic torque coefficient of motor	0.512N·m/A
J	Rotational inertia of motor and magnet	1.492×10^{-4}kgm^2
a	Distance between suspension support 1 and 2	282mm
b	Distance between suspension support 2 and 3	506mm

TABLE II. VARIABLE QUANTITY SYMBOL

	Variable quantity
F_j (j=1,2,3,4)	Micro suspension force in suspension supports
z_j (j =1,2,3,4)	Micro variable of suspension air-gap
θ_j (j =1,2,3,4)	Micro variable of motor rotation angle
i_j (j =1,2,3,4)	Control input current of motor
z	Displacement in Z axis direction
α	Rotation quantity around X axis
β	Rotation quantity around Y axis

A. Device's Dynamic Model

Defining vertical up is Z axis's positive direction, the clockwise around X and Y axis is α and β positive direction. The device's barycenter may deviate from the equilibrium position to O' point while the device being movement. Setting the generalized coordinate of O' point is $\boldsymbol{q}=[z \quad \alpha \quad \beta]^T$.

Assumed that the suspension system is a rigid body, defining the generalized coordinates of system suspension supports is $\boldsymbol{q}_b=[z_1 \quad z_2 \quad z_3 \quad z_4]^T$.

The coordinate transformation between system barycenter O' point and air-gap variable of suspension supports is

$$\begin{cases} z=\frac{1}{4}(z_1+z_2+z_3+z_4) \\ \alpha=\frac{1}{2b}(z_3+z_4-z_1-z_2) \\ \beta=\frac{1}{2a}(z_1+z_4-z_2-z_3) \end{cases} \tag{3}$$

Where, a is the center distance between suspension support 1 and 2 in Figure 3, b is the center distance between suspension support 1 and 4.

The device's kinetic energy Ek can be expressed as

$$E_k=\frac{1}{2}m\dot{z}^2+\frac{1}{2}J_\alpha\dot{\alpha}^2+\frac{1}{2}J_\beta\dot{\beta}^2 \tag{4}$$

According to the Lagrange equation, ignoring the system potential energy, the system dynamic equation is expressed as

$$\boldsymbol{M}\ddot{\boldsymbol{q}}_b=\boldsymbol{F}_j \quad (j=1,2,3,4) \tag{5}$$

Where,

$$\boldsymbol{M}=\begin{bmatrix} k_1 & k_2 & k_3 & k_4 \\ k_2 & k_1 & k_4 & k_3 \\ k_3 & k_4 & k_1 & k_2 \\ k_4 & k_3 & k_2 & k_1 \end{bmatrix}, \quad \ddot{\boldsymbol{q}}_b=\begin{bmatrix} \ddot{z}_1 \\ \ddot{z}_2 \\ \ddot{z}_3 \\ \ddot{z}_4 \end{bmatrix}, \quad \boldsymbol{F}_j=\begin{bmatrix} F_1 \\ F_2 \\ F_3 \\ F_4 \end{bmatrix},$$

$$k_1=\frac{m}{16}+\frac{J_\alpha}{4b^2}+\frac{J_\beta}{4a^2}, \quad k_2=\frac{m}{16}+\frac{J_\alpha}{4b^2}-\frac{J_\beta}{4a^2},$$

$$k_3=\frac{m}{16}-\frac{J_\alpha}{4b^2}-\frac{J_\beta}{4a^2}, \quad k_4=\frac{m}{16}-\frac{J_\alpha}{4b^2}+\frac{J_\beta}{4a^2}$$

B. Suspension Support Mathematical Model

Every suspension support is a small system, its model as shown Figure 4, the suspension force f and load torque T_L of permanent magnet expressions are

$$f=K_m\frac{\sin^2\theta}{(z+d_m)^2} \quad T_L=K_T\cdot\frac{\sin 2\theta}{z+d_m} \tag{6}$$

Where K_m is suspension force coefficient, K_T is torque coefficient, θ is the permanent magnet rotation angle, z is air gap length between magnetizer and the rail, d_m is air gap compensating coefficient.

In Figure 4, defining vertical upward (air gap length decreases) is positive direction, counterclockwise of magnet rotation angle is positive direction, the system has micro motion of deviating from the equilibrium position, the system micro suspension force F_j is

$$F_j = k_{z1} z_j + k_{\theta 1}\theta_j \quad (j\text{=}1,2,3,4) \tag{7}$$

Where, k_{z1} is the open loop force-displacement rigidity of suspension support, $k_{\theta 1}$ is the open loop force-angle rigidity of suspension support, and

$$k_{z1} = -\frac{2K_m \sin^2\theta_0}{(z_0 + d_m)^3} \qquad k_{\theta 1} = \frac{K_m \sin 2\theta_0}{(z_0 + d_m)^2}$$

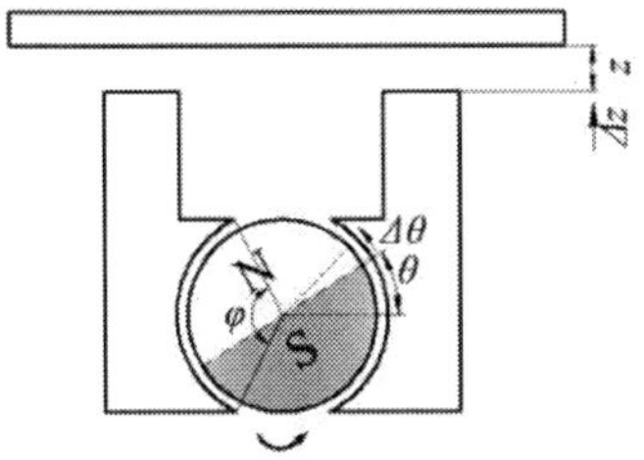

Figure 4. Suspension support model

The rotation differential equation of suspension support is

$$J\ddot{\theta}_j = K_d i_j - k_{z2} z_j - k_{\theta 2}\theta_j \quad (j\text{=}1,2,3,4) \tag{8}$$

Where, k_{z2} is open loop torque-displacement rigidity of permanent magnet, $k_{\theta 2}$ is open loop torque-angle rigidity of permanent magnet, and

$$k_{z2} = \frac{2K_T \sin 2\theta_0}{(\Delta d + z_0)^2} \qquad k_{\theta 2} = \frac{2K_T \cos 2\theta_0}{\Delta d + z_0}$$

The dynamic model of suspension system is established by Eq.(5) and Eq. (8)

$$\begin{cases} \boldsymbol{M}\ddot{\boldsymbol{q}}_b = \boldsymbol{F}_j \\ J\ddot{\boldsymbol{\Theta}} = K_d \boldsymbol{I} + k_{z2}\boldsymbol{q}_b + k_{\theta 2}\boldsymbol{\Theta} - c_2\dot{\boldsymbol{\Theta}} \end{cases} \tag{9}$$

Where,

$\boldsymbol{I} = [i_1 \quad i_2 \quad i_3 \quad i_4]^T$ $\boldsymbol{\Theta} = [\theta_1 \quad \theta_2 \quad \theta_3 \quad \theta_4]^T$

From Eq.(9), it can be concluded that, considering the state variable θ, the system state space equation order will be double, what makes system performance analysis and controller design much more difficult.

The variable θ can be expressed by input i and state variables z, according to Eq.(8), Eq.(7) can be expressed as

$$F_j = K_d \frac{k_{\theta 1}}{k_{\theta 2}} i_j + \left(k_{z1} - \frac{k_{\theta 1}k_{z2}}{k_{\theta 2}}\right) z_j \quad (j\text{=}1,2,3,4) \tag{10}$$

C. System Optimization

There are four suspension supports determining the Z axis direction displacement in the suspension system. Because three points can determine a plane, the redundant control quantity of Z axis displacement can be resolved by the following equation

$$z_1 + z_3 = z_2 + z_4 \tag{11}$$

The system dynamic differential equation, Eq.(9), can be expressed by the suspension supports 1, 2, 3(in Figure 3)

$$\boldsymbol{N}\ddot{\boldsymbol{q}}_{b1} = \boldsymbol{F}_j \quad (j\text{=}1,2,3) \tag{12}$$

Where, $\ddot{\boldsymbol{q}}_{b1} = [\ddot{z}_1 \quad \ddot{z}_2 \quad \ddot{z}_3]^T$

$$\boldsymbol{N} = \begin{bmatrix} \frac{m}{4} + \frac{J_\beta}{a^2} & -\frac{J_\beta}{a^2} & \frac{m}{4} \\ -\frac{J_\beta}{a^2} & \frac{J_\alpha}{b^2} + \frac{J_\beta}{a^2} & -\frac{J_\alpha}{b^2} \\ \frac{m}{4} & -\frac{J_\alpha}{b^2} & \frac{m}{4} + \frac{J_\alpha}{b^2} \end{bmatrix} \quad \boldsymbol{N}^{-1} = \begin{bmatrix} t_{11} & t_{12} & t_{13} \\ t_{12} & t_{11} & t_{23} \\ t_{13} & t_{23} & t_{11} \end{bmatrix}$$

$\boldsymbol{N}^{1}$ is inverse matrix of the matrix of $\boldsymbol{N}$

IV. SUSPENSION CHARACTERISTICS ANALYSIS

A. Control Characteristic

In order to judge the system suspension characteristic, the system space state equation is

$$\begin{cases} \dot{\boldsymbol{X}} = \begin{bmatrix} \boldsymbol{O} & \boldsymbol{I} \\ \boldsymbol{A}_1 & \boldsymbol{O} \end{bmatrix} \boldsymbol{X} + \begin{bmatrix} \boldsymbol{O} \\ \boldsymbol{B}_1 \end{bmatrix} \boldsymbol{u} \\ \boldsymbol{Y} = \boldsymbol{C}\boldsymbol{X} \end{cases} \tag{13}$$

Where, $\boldsymbol{O}$ is three order zero matrix, $\boldsymbol{I}$ is three order unit matrix,

$$\boldsymbol{A}_1 = \begin{bmatrix} \left(k_{z1} - \frac{k_{\theta 1}k_{z2}}{k_{\theta 2}}\right)t_{11} & \left(k_{z1} - \frac{k_{\theta 1}k_{z2}}{k_{\theta 2}}\right)t_{12} & \left(k_{z1} - \frac{k_{\theta 1}k_{z2}}{k_{\theta 2}}\right)t_{13} \\ \left(k_{z1} - \frac{k_{\theta 1}k_{z2}}{k_{\theta 2}}\right)t_{12} & \left(k_{z1} - \frac{k_{\theta 1}k_{z2}}{k_{\theta 2}}\right)t_{22} & \left(k_{z1} - \frac{k_{\theta 1}k_{z2}}{k_{\theta 2}}\right)t_{23} \\ \left(k_{z1} - \frac{k_{\theta 1}k_{z2}}{k_{\theta 2}}\right)t_{13} & \left(k_{z1} - \frac{k_{\theta 1}k_{z2}}{k_{\theta 2}}\right)t_{23} & \left(k_{z1} - \frac{k_{\theta 1}k_{z2}}{k_{\theta 2}}\right)t_{33} \end{bmatrix}$$

$$\boldsymbol{B}_1 = \begin{bmatrix} \frac{k_{\theta 1}}{k_{\theta 2}} K_d t_{11} & \frac{k_{\theta 1}}{k_{\theta 2}} K_d t_{12} & \frac{k_{\theta 1}}{k_{\theta 2}} K_d t_{13} \\ \frac{k_{\theta 1}}{k_{\theta 2}} K_d t_{12} & \frac{k_{\theta 1}}{k_{\theta 2}} K_d t_{11} & \frac{k_{\theta 1}}{k_{\theta 2}} K_d t_{23} \\ \frac{k_{\theta 1}}{k_{\theta 2}} K_d t_{13} & \frac{k_{\theta 1}}{k_{\theta 2}} K_d t_{23} & \frac{k_{\theta 1}}{k_{\theta 2}} K_d t_{11} \end{bmatrix}$$

$\boldsymbol{X} = [z_1 \quad z_2 \quad z_3 \quad \dot{z}_1 \quad \dot{z}_2 \quad \dot{z}_3]^T$ $\boldsymbol{u} = [i_1 \quad i_2 \quad i_3]^T$

$\boldsymbol{Y} = [z_1 \quad z_2 \quad z_3]^T$

The controllability matrix ($\boldsymbol{P}_C$) and the observability matrix ($\boldsymbol{P}_O$) are

$$\boldsymbol{P}_C = [\boldsymbol{B} \quad \boldsymbol{AB} \quad \boldsymbol{A}^2\boldsymbol{B} \quad \boldsymbol{A}^3\boldsymbol{B} \quad \boldsymbol{A}^4\boldsymbol{B} \quad \boldsymbol{A}^5\boldsymbol{B}]$$

$$\boldsymbol{P}_O = [\boldsymbol{C} \quad \boldsymbol{CA} \quad \boldsymbol{CA}^2 \quad \boldsymbol{CA}^3 \quad \boldsymbol{CA}^4 \quad \boldsymbol{CA}^5]^T$$

Matrix $\boldsymbol{B}_1$ and $\boldsymbol{I}$ are all non-zero matrix, so Matrix $\boldsymbol{P}_C$ is column full rank, Matrix $\boldsymbol{P}_O$ is row full rank, then the system is controllable and observable.

B. LQR Simulation

In order to verify the system response characteristic, on the base of Table 1, the system is simulated in LQR method. The weight parameters of the state quantity and input as following according to the desired performance index

$$Q=\begin{bmatrix}10&0&0&0&0&0\\0&1&0&0&0&0\\0&0&1&0&0&0\\0&0&0&1&0&0\\0&0&0&0&1&0\\0&0&0&0&0&1\end{bmatrix}\qquad R=\begin{bmatrix}1&0&0\\0&1&0\\0&0&1\end{bmatrix}$$

The optimal control matrix (feedback matrix) $\boldsymbol{K}$ is

$$K=\begin{bmatrix}380.4&14.73&-21.2&4.088&-1.316&0.873\\18.85&366.8&50.69&-1.083&3.559&-0.119\\-24.31&51.68&369.5&0.573&-0.019&3.228\end{bmatrix}$$

When the system is in the steady suspension state, a step input of 0.1mm is given to the suspension support 1, but no to the suspension support 2 and 3. The system response result is shown in Figure 5.

It is concluded that, the input being to one suspension support, three motor current and air-gap length will change. Another, the control current of motor 1 is bigger than and motor 2 and 3, displacement change of suspension support 1 is more obvious than suspension support 2 and 3. Because the input to suspension support 1, mainly causes motor 1 current and the displacement of suspension support 1 change, but the system itself is rigid, so also cause motor 2 and 3 control current and the displacement of suspension support 2 and 3 have smaller change.

What is more, when the up input of air gap length

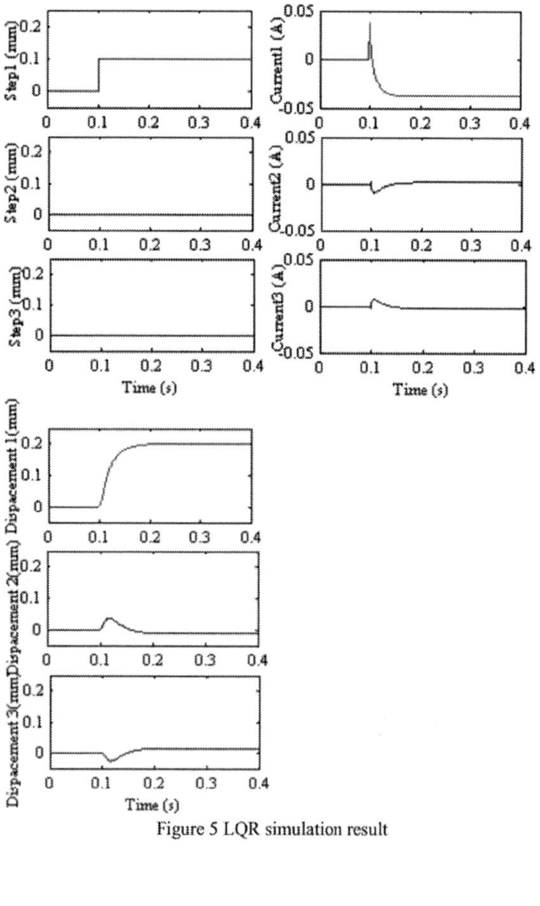

Figure 5 LQR simulation result

decreasing is given to suspension support, the motor control current increases rapidly, the magnet slightly rotates to the direction of angle increasing from the initial equilibrium, the suspension force becomes bigger than the gravity, so the air gap length decreases. Subsequently, in order to achieve a new equilibrium position, the motor current decreases rapidly, the magnet rotates to the direction of angle decreasing correspondingly because of reaction torque, the suspension force is decreased to balance the gravity. In the new equilibrium state, the motor current is a litter smaller, the device has micro up movement, the air gap length is reduced. The system is rigid, the variation trend of motor 2 current is opposite to current 1, the variation trend of motor 3 current is same to current 1, but they are smaller. The displacement variation of suspension on support 2 is opposite to suspension support 1 trend, the displacement variation of suspension support 3 is same to suspension support 1 trend.

When the step input to suspension support 2 (or 3), the system response is similar to above, not repeat.

V. SUMMARY

A permanent magnetic suspension system has been proposed, which is applied in the dust-free transit field. The system realizes real time control of magnetic force, can reach zero suspension force.

The mechanical dynamics model of dust-free transit system using permanent magnetic suspension has been built. The system can realize stable suspension after the step input by simulation of LQR method.

The future works are carrying out the suspension experiment of this system.

ACKNOWLEDGEMENTS

This work was financially supported by the National Natural Science Foundation of China (Grant No. 51105257 and 51310105025) and the Doctoral Found of Liaoning Province of China (Grant No. 20111042).

REFERENCES

[1] Junichi Asama, Ryo Nakamura, Hiroya Sugimoto ,et al. "Evaluation of Magnetic Suspension Performance in a Multi-Consequent-Pole Bearingless Motor," *IEEE Transactions on Magnetics*, vol. 47, no. 10, pp.4262-4265, 2011

[2] Wenrong Song, Yang Chen, Xiaodong Song, et al. "Magnetic force coupling in maglev feeding mechanism," *Optics and Precision Eng ineering*, vol. 12, no. 4, pp. 411–414, 2004.

[3] Morishita Mimpei. "An adaptive-controlled self-gap-detecting electromagnetic suspension system," *IEEE International Symposium on Industrial Electronics*, vol. 1, no. 7, pp. 2562–2565,2006.

[4] Sun Feng, Oka Koichi. "Development of Noncontact Suspension Mechanism Using Flux Path Control Disk Magnet Rotation," *Transactions of the Japan Society of Mechanical Engineers*, vol. 76, no. 771, pp.2916-2922, 2010.

6D Planar Magnetic Levitation System - Mag6D

Christoph Schaeffel[a]; Michael Katzschmann[a], Hans-Ulrich Mohr[a], Rainer Gloess[b], Christian Rudolf[b], Christopher Mock[b], Carolin Walenda[b]

[a] IMMS, Institut fuer Mikroelektronik- und Mechatronik-Systeme gemeinnuetzige GmbH, Ehrenbergstr. 27, 98693 Ilmenau, Germany
christoph.schaeffel@imms.de
[b] Physik Instrumente (PI) GmbH & Co. KG, Auf der Roemerstr. 1, 76228 Karlsruhe, Germany

***Abstract* – The paper describes the structure, the performance and a set of model based design steps of a novel 6 DOF planar magnetic drive system with nanometer positioning accuracy. The movable platform of the system is free of any wiring connection and the upper surface is freely accessible.**

I. Introduction

The paper describes the structure of a novel 6 DOF planar magnetic drive system with nanometer accuracy. Some details of the model based design are described and a few qualification results are given.

A main advantage of the system is that it uses only six, i.e. the minimum number of sensors and coils to control all six DOFs what keeps the system simple and makes it very effective. Furthermore the measurement system and the actors are completely located below the slider, hence the platform is freely accessible from above.

II. Structure of the System

The new developed, magnetically levitated motion system with six degrees of freedom allows a motion of the slider-platform of 100 x 100 mm^2 in x-y-direction and 100 µm in z-direction. Rotation around z-axis is limited by the measurement system to ±0.25°, rotation around x- and y-axis is limited by the geometry of the system to ±100 µrad.

The system is actuated by 3 coil pairs mounted at the stator which interact with 3 Halbach arrays in the moving platform. Using only 3 arrays instead of 4 (see e.g. [4]) minimizes the required components of the system. Furthermore a system with only 3 arrays is smaller and lighter what decreases the power losses and hence the thermal load of the system. The more elaborate commutation can easily be done by the controller. The use of efficient Halbach arrays gives high magnetic field components below the slider reducing the power consumption of the coils and hence the heat generation in the system. Temperature stability is an important requirement for high accurate positioning technology.

The measurement system consist of a compact 6D-measuring head, placed between the coil pairs and a moving grid plate mounted on the bottom side of the slider platform. The planar degrees of freedom (x, y, r_z) are measured by 3 optical sensors which scan the grid. The vertical degrees of freedom (z, r_x, r_y) are determined by 3 pairs of capacitive sensors, which measure the vertical distance to the grid plate. The sensor resolution is in the nanometer and tenth of micro radiant range. A main advantage is the grid plate inside the movable platform. Therefore the sensor system allows the access to all sides of the platform, a requirement in inspection systems. Furthermore the moving platform is completely passive. No cables or other connections between slider and stator which can introduce disturbance forces are needed.

To avoid disturbances on the capacitive sensor values the frequencies of the PWM amplifiers and the measurement frequency of the sensors have to be synchronized.

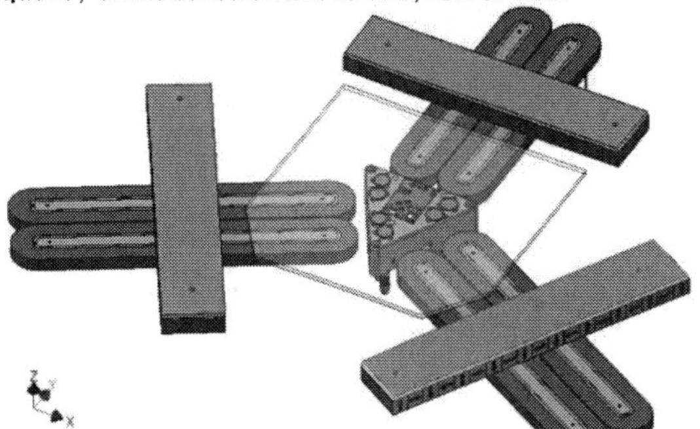

Figure 1. Schematic structure of the Prototype System

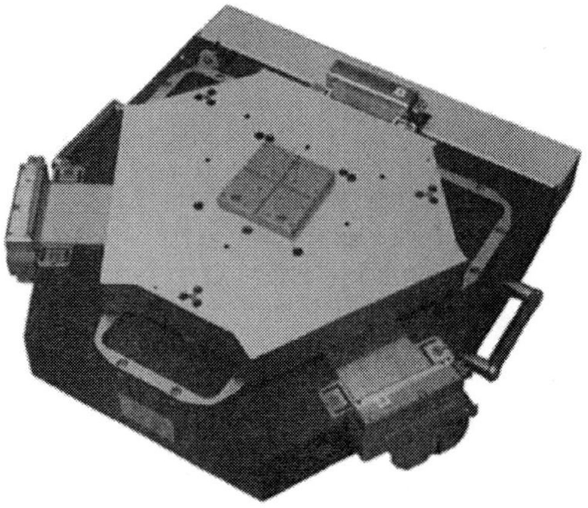

Figure 2. Setup of the Prototype System

III. Halbach Array Configuration and Commutation

Halbach array configurations are known for their high flux density on one side and small stray fields on the opposite side. The high field components below the magnets minimize the power consumption of the coils. Due to that characteristic they were chosen for the setup. The Sm_2Co_{17}-magnets with medium magnetization grade generate a flux density of about 0.5 T below the array. This material is also ultra-high vacuum compatible and can be used in applications that require a strong hydrogen atmosphere. Hence these magnets were used in a first prototype.

Figure 3 shows a schematic of an array with a FEM result. Depending on the coil position with respect to the magnets coil currents generate horizontal and vertical forces. Additional a torsional momentum around an axis perpendicular to the image plane of Figure 3 is generated.

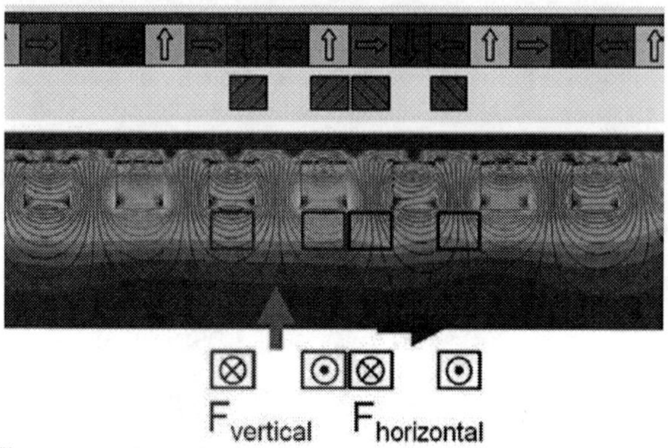

Figure 3. Model of a Halbach array (above) and flux distribution calculated by FEM (below), the sketch shows, how vertical and horizontal forces can be generated by a coil pair

Because no iron material is used in the actuator system no nonlinear effects like hysteresis or saturation occur. Hence the generated forces and momentums depend linear on the coil currents and superposition can be assumed. Figure 4 shows the almost sin/cos-like proportionality factors calculated by FEM.

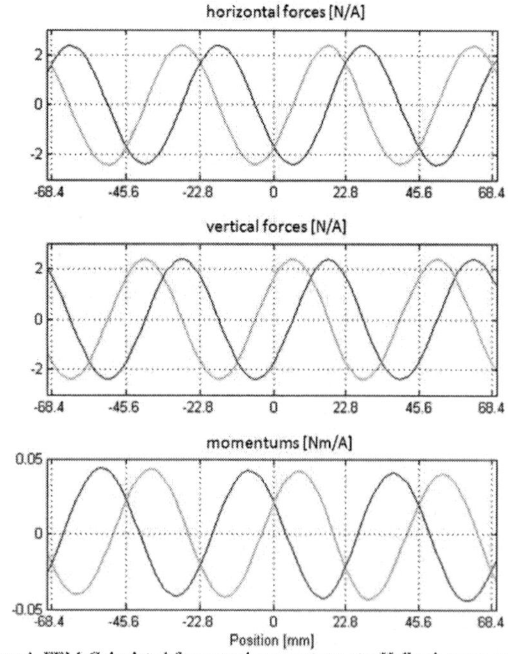

Figure 4. FEM-Calculated forces and momentums at a Halbach array, caused by the left (blue curves) and the right (green curves) coil

Depending on the slider platform position this generated forces and momentums can be transferred into a platform fixed Cartesian coordinate system. Thus the position depending relation K(x,y) between the 6 coil currents $i_1 ... i_6$ and the forces F_x, F_y, F_z as well as the momentums M_x, M_y, M_z can be determined.

$$\begin{pmatrix} F_x \\ F_y \\ F_z \\ M_x \\ M_y \\ M_z \end{pmatrix} = K(x,y) \cdot \begin{pmatrix} i_1 \\ i_2 \\ i_3 \\ i_4 \\ i_5 \\ i_6 \end{pmatrix} \quad (1)$$

The (6x6)-Matrix K(x,y) is invertible in the whole travel range and thus forms the basis for the commutation of the coils. With the inverse Matrix $K(x,y)^{-1}$ the force set points, calculated by a controller, can be converted into the set points for coil currents.

These simulation based results were checked by measuring of the magnetic flux density below the Halbach arrays. The horizontal flux component, which generates vertical forces, is shown in Figure 5 and Figure 6. The measured flux densities match well with the simulation results. Close to the Halbach array the horizontal component of the magnetic field shows harmonics of 5th and 9th order. But the coil geometry that is exposed to the magnetic field works like a filter function. Finally the forces are almost sinusoidal as shown in Figure 4.

The final test for simulation accuracy was done by measuring the required coil currents in closed loop mode when the slider levitates at a fixed position. The maximum deviation from the predicted coil currents was less than 2.5 percent. This can be explained by simulation errors, variations of the magnetic properties and unexpected variations of the geometry or mounting of the Halbach arrays. The blue curves in Figure 10 show the errors, caused by these deviations when moving with constant velocity. A system specific correction of the matrix $K(x,y)^{-1}$ can be derived from the deviation between predicted and measured coil currents which minimizes the position errors. The red curves in Figure 10 show the result.

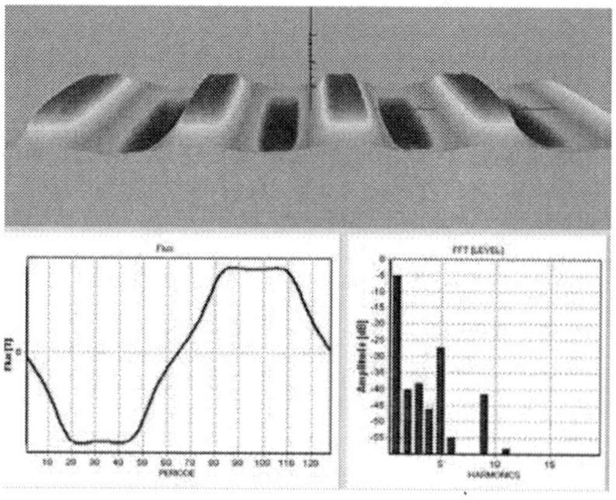
Figure 5. Measured horizontal flux component X, 1 mm below the Halbach array, harmonic distortion of 5th and 9th order

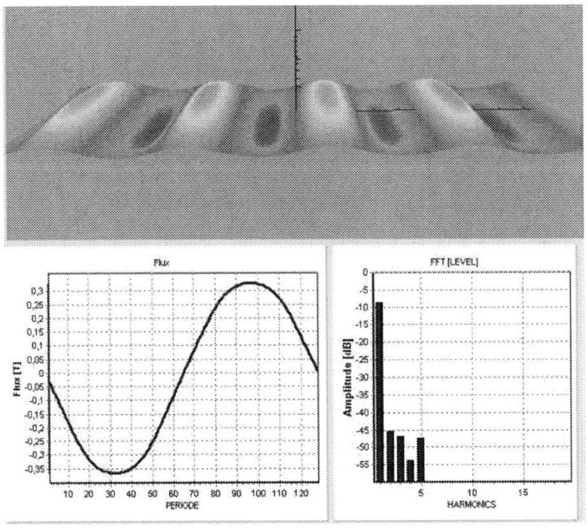

Figure 6. Measured horizontal flux component X, small distortions in 4 mm distance

IV. Thermal Management

Applying forces and momentums generated by the slider platform weight to equation (1) the power losses caused by gravitation can be calculated from the necessary coil currents. Figure 7 shows the result for the first prototype.

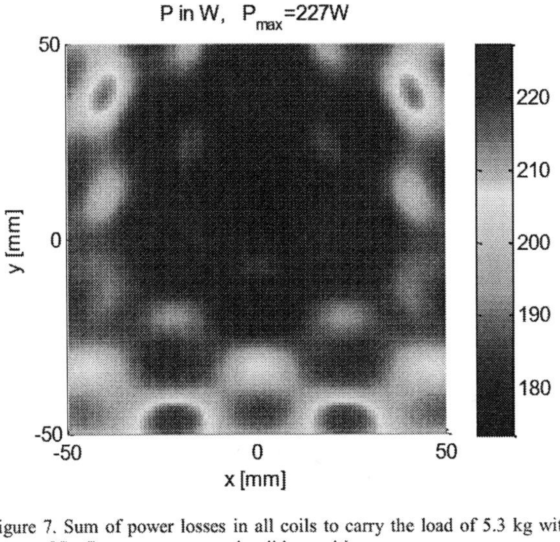

Figure 7. Sum of power losses in all coils to carry the load of 5.3 kg with arrays of SmCo-magnets versus the slider position

For a second prototype with a lighter slider platform and with NdFeB-magnets Figure 8 shows, that the power losses can be considerably reduced to less than 100 W.

Because temperature stability is an important issue for highly accurate positioning technology the coils are equipped with very flat cooling sandwiches based on efficient liquid pipe technology. Thereby the temperature rise on the top of the coil housing remains less than 1.5 K for the first prototype and less than 0.3 K for the second prototype.

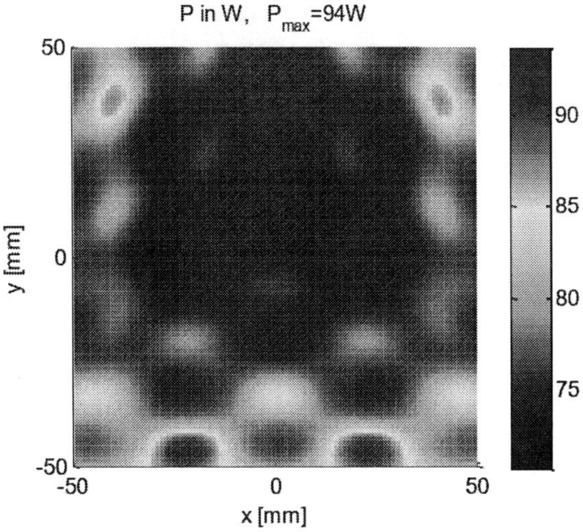

Figure 8. Sum of power losses in all coils to carry the load of 4.2 kg with arrays of NdFeB-magnets versus the slider position

V. Digital Controller

The control for all DOF of the platform is based on PI's digital industrial modular controller concept. The position controller uses six independent servo loops. The position inputs of the servo loops are calculated by a coordinate transformation which transforms the measured sensor values to the DOF-positions. The current set points for the six amplifiers are calculated from the output of the servo loops, using the position dependent commutation matrix $K(x,y)^{-1}$. These set points are input of six PWM current amplifiers.

The real time operating system of the CPU is synchronized by a FPGA interrupt mechanism. The controller runs with a PC-ETX processing module with a servo cycle time of 200 µs. Trajectory generation, simple to use command set and PI's real time interface is implemented. The necessary initialization and shutdown procedures are developed to drive the system in a well-defined starting position or switch it off at the parking position.

An algorithm, which measures the necessary data and calculates the correction for the commutating matrix $K(x,y)^{-1}$ is developed. Hence the correction can be adjusted specifically to every system.

VI. Results

The prototypes show position stability in the resolution of the sensor system. Also for fast dynamic motion sequences tracking errors are in the nm range when using correction matrices to compensate the difference between the FEM based commutation matrix $K(x,y)^{-1}$ and the true relation between forces/momentums and coil currents.

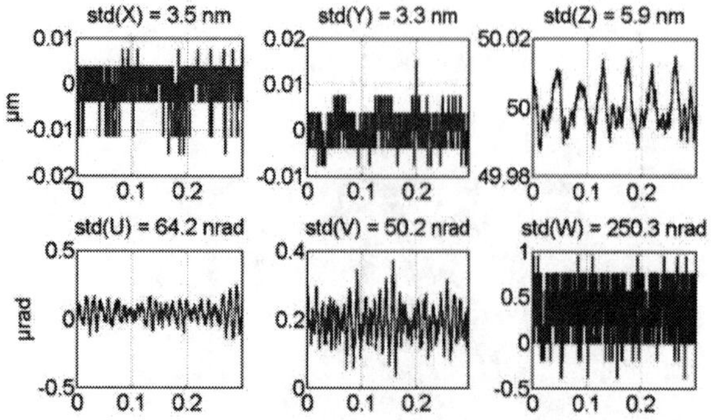

Figure 9. Position stability at position x = y = 0, z = 50 µm; u = v = w = 0

Horizontal movements with 50 mm/s and 2 m/s² are done with a vertical deviation (z) of less than 0.25 µm (see Figure 10 red curves). The difference between the blue and the red curves shows the strong influence of the correction matrices.

The z-curve in Figure 10 still shows a systematic error motion which will be compensated in an updated controller version.

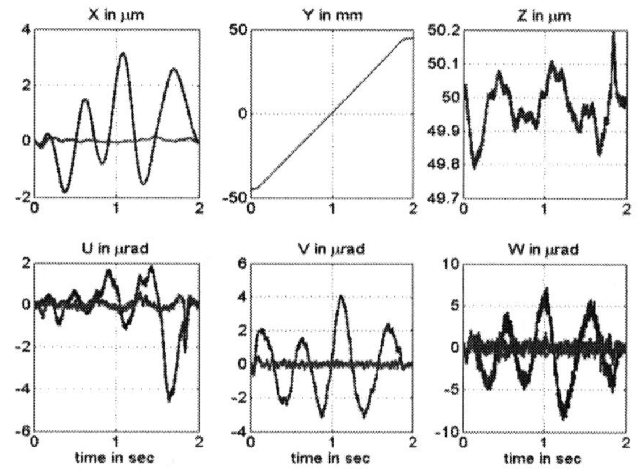

Figure 10 Horizontal movement with 50 mm/s without (blue curves) and with correction matrices (red curves)

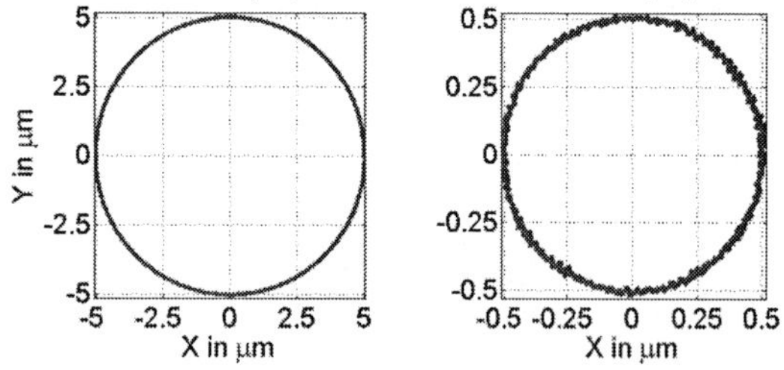

Figure 11. Circular motion in x-y-plane with diameters of d = 10 µm and d = 1 µm, blue curve: measured positions, red curve: low passed positions (f_{Pass} = 50 Hz, f_{Stop} = 100 Hz, Stopband: -40 dB)

Figure 11 shows circular motions of 2 different diameters. There are no hysteresis effects observed. The accuracy is in nanometer level. Due to the z-travel of 100 µm also Helix movements can be performed with some nm-resolution (Figure 12).

The controller setup in the prototype limits the maximum velocity to 100 mm/s at the moment but there are no physical limits caused by the structure.

The high eigenfrequency of the platform assembly of more than 500 Hz allows high control bandwidth. The small noise level of all axes shown in Figure 9 opens the way for further improvements with new high resolution sensor systems in the servo loops.

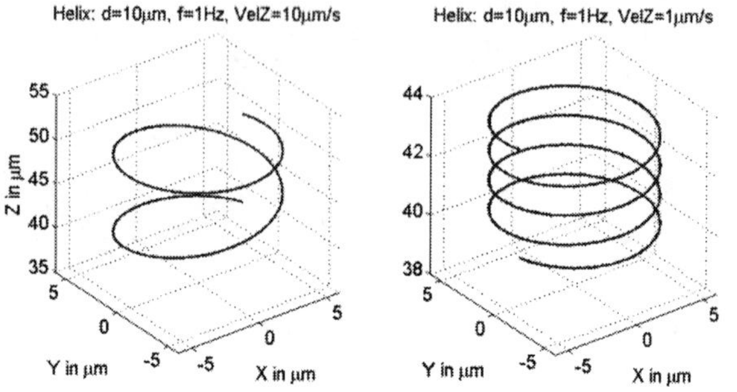

Figure 12. Helix movements with a diameter of 10 µm and a z-velocity of 10 µm/s and 1 µm/s

VII. Conclusion

A prototype of a 6 DOF planar magnetic levitation system with nanometer positioning capability as shown in Figure 2 has been constructed. It comprises an integrated compact 6D sensor and is controlled by PI's modular E-712 digital nanopositioning controller, using precalculated commutation functions for the three coil pairs realizing a smooth path control.

ACKNOWLEDGEMENT

This work was supported by the Federal Ministry of Economics and Technology of Germany in the projects KF2361602BN9 & KF2534501BN9.

References

[1] C. Rudolf, C. Mock, C. Walenda, R. Gloess, R. Liang, C. Schaeffel, M. Katzschmann, H.-U. Mohr, "6D Magnetic Levitation Positioning System with Compact Integrated 6D Sensor"; *Proc. of 12th euspen International Conference*, Stockholm, June 2012

[2] R. Gloess, C. Mock, C. Rudolf, C. Walenda, C. Schaeffel, M. Katzschmann, H.-U. Mohr, "Magnetic Levitation in 6DOF with Halbach Array Configuration", *Proc. of Actuator 2012, International Conference*, Bremen, June 2012

[3] C. Schaeffel, M. Katzschmann, H.-U. Mohr, R. Gloess, C. Mock, C. Rudolf, C.Walenda, "Planar Magnetic 6D Levitating System with compact Sensor"; *Proc. of ASPE 2012*, International Conference, San Diego, October 2012

[4] S. D. Ruben, T.-C. Tsao, T. Lemmons, G. Caskey, R. Hocken, "Intrinsic Damping of Linear Halbach Brushless Motors"; *Proc. of ASPE 2011*, International Conference, Denver, November 2011

[5] Won-jong Kim, D. Trumper, "High- precision magnetic levitation stage for photolithography"; *Precision Engineering 22*:66-77 1998

Nd-Fe-B permanent magnets for passive and hybrid magnetic bearings with extended application temperature range

Michael Weickhmann, Matthias Katter, Bernd Schleede
VACUUMSCHMELZE GmbH & Co. KG, Hanau, Germany

***Index Terms*—VACODYM®,VACOMAX® magnetic Bearing, Surface charge model, amperien current density, coulombian representation, FEA**

I. CATEGORIES

Digest Categories: Magnetic bearing

***Abstract*—Passive magnetic bearings or hybrid passive magnetic bearings are commonly built up with Sm-Co or with Nd-Fe-B magnets. One of the most important item of those magnets is it's homogeneity in flux density distribution and the precise angularity. A miss-match would cause an undesired displacement. It may therefore be very important to measure these homogeneity's and angular deviation very precisely. Some application call for a higher temperature, thus a need of higher remanence B_r is inevitable and in some cases the temperature coefficient T_k of remanence B_r is also of importance. This may be achieved by applying the Grain Boundary Diffusion GBD. In the following paper we will give an overview of the current development to ensure this future demand. A variation of parameter studies will give an overview of the main influencing factors which act on the bearing forces and stiffness.**

II. INTRODUCTION

Turbo molecular pumps for example are equipped with passive magnetic bearings on the high vacuum side. The passive magnetic bearing are to stabilise the turbine shaft radially, while the axial stabilisation is achieved by an active (electric) magnetic bearing or a mechanical bearing. Whenever a parametric analysis of magnetic bearings is conducted, the analytical approach by simulating the magnet with surface charge modules is very convenient, since its results may be very rapidly obtained. The programs used in this field are commonly written in Matlab®/Scilab® or in Mathematica®. Of interest are the forces F_z obtained in a bearing and its stiffness K_z versus displacement. In this paper, we will give a brief introduction of these analytical methods and we will compare those results with Finite Element Analysis FEA. At the same time we will try to show the realistic influences of the variation of dimensions, remanence and the magnetic orientation onto the forces and the stiffness of a magnetic bearing.

III. ANALYTICAL APPROACH OF AXIALLY ORIENTED MAGNET RINGS FOR MAGNETIC BEARINGS

In general let us first try to simulate a magnetic ring by using two surface charged coils which may substitute a ring magnet refer to eqn.(1) and eqn.(2) as used in [1] and [2].

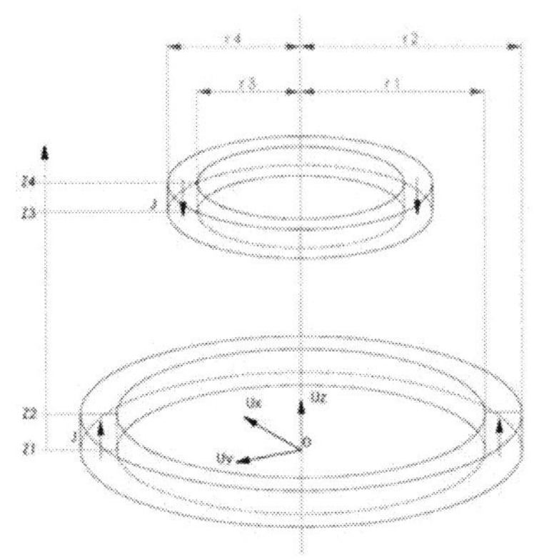

Figure 1. Two axially oriented magnet rings as an axial bearing.

$$F_z = \frac{J_1 J_2}{2\mu_0} \sum_{i=1}^{2} \sum_{j=3}^{4} (-1)^{1+i+j} \{f_z(r_i, r_j)\} \quad (1)$$

$$f_z(r_i, r_j) = r_i r_j \int_{z_3}^{z_4} \int_{z_1}^{z_2} \int_{0}^{2\pi} \frac{(\tilde{Z} - Z)cos(\Theta)dzdzd\Theta}{(r_i^2 + r_j^2 - 2r_i r_j cos(\Theta) + (z-z)^2)^{\frac{3}{2}}} \quad (2)$$

$$F_z = \frac{J_1 J_2}{2\mu_0} \sum_{i,k=1}^{2} \sum_{j,l=3}^{4} (-1)^{1+i+j+k+l} F_{i,j,k,l} \quad (3)$$

with eqn.(2) inserted into eqn.(1)

$$F_{i,j,k,l} = r_i r_j g(z_k - z_l, r_i^2 + r_j^2 + (z_k - z_l)^2, -2r_i r_j) \quad (4)$$

$$g(a, b, c) = A + S \quad (5)$$

With A and S being:

$$A = \frac{a^2 - b}{c}\pi + \frac{\sqrt{c^2 - (a^2 - b)^2}}{c}(a_1 + a_2) \quad (6)$$

with

$$a_1 = \log[\frac{-16c^2}{(c^2 - (a^2 - b)^2)^{\frac{3}{2}}}] \quad (7)$$

$$a_2 = \log[\frac{c^2}{(c^2 - (a^2 - b)^2)^{\frac{3}{2}}}] \quad (8)$$

and

$$S = S_1 + S_2 + S_3 \quad (9)$$

$$S_1 = \frac{2a}{c\sqrt{b+c}}((b+c)E[\Phi_1, m_1] - cF[\Phi_1, m_1]) \quad (10)$$

In eqn.(10) $\Phi_1 = arcsin[\sqrt{\frac{b+c}{b-c}}]$
In eqn.(10) $m_1 = arcsin[\frac{b-c}{b+c}]$

$$S_2 = \frac{2a}{c\sqrt{b+c}\sqrt{\epsilon}}(\frac{c}{\sqrt{\mu}}E[m_2] - c\sqrt{\mu}K[m_2]) \quad (11)$$

In eqn.(11) $m_2 = \frac{b-c}{b+c}$
In eqn.(11) $\epsilon = \frac{c}{c-b}$
In eqn.(11) $\mu = \frac{c}{b+c}$

$$S_3 = \sqrt{\epsilon\beta^{-1}}((b-a^2)K[2\mu] + (a^2 - b + c)\Pi[\frac{2c}{b+c-a^2}, 2\mu]) \quad (12)$$

In eqn.(12) $\mu = \frac{c}{b+c}$, $\beta = \frac{b+c}{b-c}$ and $\epsilon = \frac{c}{c-b}$
These analytical calculation and formulae of $g(a, b, c)$ in eqn. (10 to 12) contain elliptical integrals $K[m]$, $F[\Phi, m]$, $E[\Phi, m]$, $E[m]$ and $\Pi[m, n]$, $\Pi[n, \Phi, m]$ as described in the following equations (13) to (18):

$$K[m] = \int_0^{\frac{\pi}{2}} \frac{1}{\sqrt{1 - msin(\Phi)^2}} d\Phi \quad (13)$$

Where m in eqn.(13) is valid for $<= 1$.

$$F[\Phi, m] = \int_0^{\Phi} \frac{1}{\sqrt{1 - msin(\Phi)^2}} d\Phi \quad (14)$$

$$E[\Phi, m] = \int_0^{\Phi} \sqrt{1 - msin(\Phi)^2} d\Phi \quad (15)$$

$$E[m] = \int_0^{\frac{\pi}{2}} \sqrt{1 - msin(\Phi)^2} d\Phi \quad (16)$$

$$\Pi[n, m] = \Pi[n, \frac{\pi}{2}, m] \quad (17)$$

$$\Pi[n, \Phi, m] = \int_0^{\Phi} \frac{1}{\sqrt{1 - nsin(\Phi)^2}} \frac{1}{\sqrt{1 - msin(\Phi)^2}} d\Phi \quad (18)$$

The exact analytical formulation of K_z as in [2] can be expressed in the following equations.

$$K_z = \frac{J_1 J_2}{2\mu_0} \sum_{i,k=1}^{2} \sum_{k,l=3}^{4} (-1)^{1+i+j+k+l} C_{i,j,k,l} \quad (19)$$

where

$$C_{i,j,k,l} = 2\sqrt{\alpha}E[m_4] - 2\frac{r_i^2 + r_j^2 + (z_k z_l)^2}{\sqrt{\alpha}} K[m_4] \quad (20)$$

with $\alpha = (r_i - r_j)^2 + (z_k - z_l)^2$ and $m_4 = \frac{-4r_i r_j}{\alpha}$.

And m in eqn.(14) to (18) and n eqn.(17) and eqn.(18) are valid for $m, n <= 1$ respectively. As the magnet ring is substituted by two surface coils on the inner and outer circumference, the magnetic orientation is fixed to the axial direction, thus any deviation of orientation may not be calculated. This calls for a different approach, which may be found in [3]. In this paper the focus is on a dipole method and gives a 2D representation of permanent magnet bearings. The results differ slightly from the fully analytical approach given by [2] as it is an abbreviation.

$$F_z = p\frac{J_1 J_2}{2\pi\mu_0}\frac{2}{R^3} S_1 S_2 sin(\beta_1 + \beta_2 - 3\Theta) \quad (21)$$

For the axial forces versus axial displacement F_z is calculated and simulated in eqn. (21).

$$K_z = -2K_r = p\frac{J_1 J_2}{2\pi\mu_0}\frac{6}{R^4} S_1 S_2 cos(\beta_1 + \beta_2 - 4\Theta) \quad (22)$$

In eqn. 22 one may observe, that the radial stiffness is half the axial stiffness, i.e. $K_r = -\frac{K_z}{2}$. This is the reason, why in this paper we concentrate on calculating the axial stiffness K_z only.

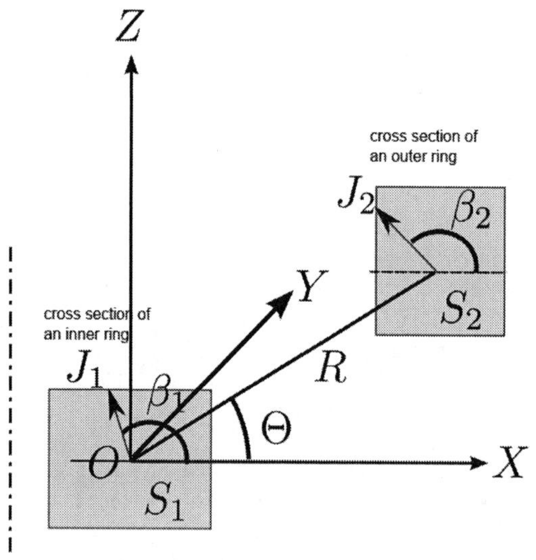

Figure 2. 2D Rings

The bearing stiffness K_z versus the axial displacement is calculated in eqn. (22). The advantage of this approach is, that it allows a deviation of orientation, but only to the extent, that the orientation is uniform over the entire ring. This uniform orientation may give a diverging or converging flux density above or below the rings pole surface, which is also know as "hot"- Side / "cold" - Side effect [4]. As stated in the paper of Bekinal et. al. [3] there will be a slight difference in results compared to the Yonnet et. al. publications [5] and

[2], thus a comparison with 2D FEA [1] calculations is therefore useful. A single angular deviation in one single direction super positioned with the normal axial direction, e.g. X-direction or Y-direction is not possible to simulate using these analytical formulae. This may be done by super positioning a diametrical ring. An analytical approach needs to be elaborated but is not the scope of this paper, thus to obtain any results for this kind of constellation, it would be necessary to calculate those bearings with 3D FEA. [2]

IV. Parameter study of influencing factor on bearing forces and stiffness

All analytical calculations as done by e.g. [5] [2] [6] [7] [8] [9] [10] [11] [12] are a very fast and convenient approach to render results, but do not reflect processing variances of remanence, dimension and orientation deviation. We therefore have undertaken a parameter study to find the most significant influencing factors which act on the bearing forces and stiffness. As a reference we may consider a passive magnetic bearing with the following dimension and magnetic properties, homogeneity and orientation deviation (by angular deviation causing a so called "Hot" and "Cold Side of each magnetic ring. If one would vary the outer and inner ring magnet of an axial bearing by a parameter study, one would need to vary the dimension of the inner ring denoted with suffix $_1$ D_{outer_1}, D_{inner_1},$Height_1$, Br_1 and β_1 and the same for the outer ring denoted by suffix $_2$ D_{outer_2}, D_{inner_2},$Height_2$, Br_2 and β_2 accordingly and keeping each of the five parameter on three levels, i.e. minimal, nominal and maximal. This would need $3^{10} = 59049$ combinations. To run a calculation of so many variations would be a rather endless undertaking. We therefore try to reduce the variation by combining the rings' volume and remanence to its magnetic moment, i.e. $M_{mag_1} = Vol_1 Br_1$ for the inner ring and $M_{mag_2} = Vol_2 Br_2$ for the outer ring. By doing so we now reduce the number of combinations, i.e. M_{mag_1} and β_1 for the inner ring and M_{mag_2} and β_2 for the outer ring, to $3^4 = 81$ combinations.

	Nominal	LSL	USL
Outer Ring outer Dia. [mm]	64.00	63.98	64.02
Outer Ring inner Dia. [mm]	44.00	43.98	44.02
Inner Ring outer Dia. [mm]	40.00	39.98	40.02
Inner Ring inner Dia. [mm]	20.00	9.98	10.02
Height [mm]	10.0	9.99	10.01
Remanence Br [T]	1.00	0.97	1.03
Angle of Orientation [deg]	90	88.5	91.5

Table I
DIMENSION AND PROPERTIES OF INNER RING MAGNET BEARING

Tab. I gives an overview of the parametric variation of dimension, magnetic properties and deviation of orientation, which then is reduced to the magnetic moment and the deviation of orientation as shown in Tab. II . This approach simplifies the number of variation significantly.

[1] http://www.femm.info/wiki/HomePage

[2] http://www.ansys.com/Products/Simulation+Technology/Systems+&+Multiphysics/Multiphysics+Enabled+Products/ANSYS+Maxwell

	Nominal	LSL	USL
Magnetic Moment M_{mag_1} inner Ring IR [nVsm]	9.424	8.924	10.001
Magnetic Moment M_{mag_2} outer Ring OR [nVsm]	16.965	16.231	17.714
Angle of Orientation β_1 of IR [deg]	88.5	90	91.5
Angle of Orientation β_2 of OR [deg]	88.5	90	91.5

Table II
DIMENSION AND PROPERTIES OF OUTER RING MAGNET BEARING

V. Result of parameter study

The analytical result were compared with 2D FEA results to see how good these calculation fit.

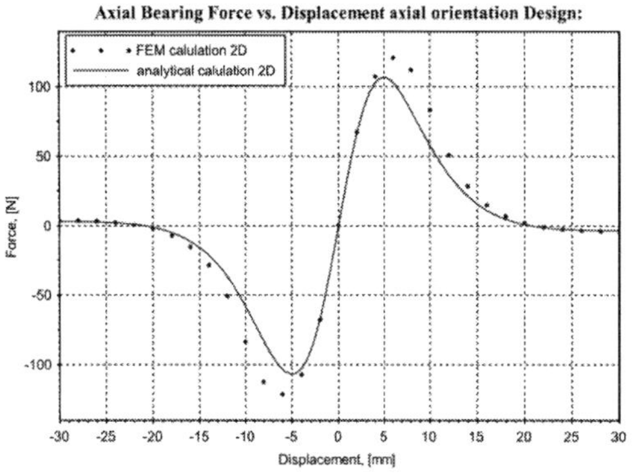

Figure 3. Force Fz vs. axial displacement 2D FEA and analytical calculation

It can be seen in fig. 3 the analytical results (red line) fit well. Only the maximum and the minimum differ slightly, the 2D FEA results (blue dots) show a higher maximum and lower minimum respectively.

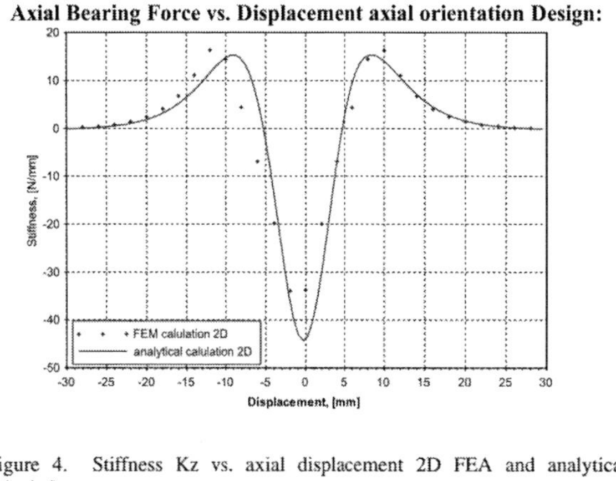

Figure 4. Stiffness Kz vs. axial displacement 2D FEA and analytical calculation

As far as the stiffness is concerned, the 2D FEA results differ slightly from the analytical results, especially with the maxima's fig. 4 as described by Bekinal [3].

First let us look at the influence of varying the magnetic moment and keeping the orientation constant at $90[deg]$.

The forces are symmetrical over the displacement, which can be seen in fig. 5, the maximum in magnetic moment

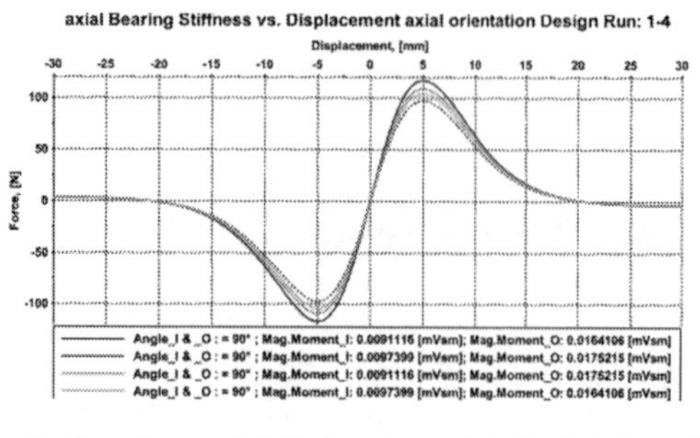

Figure 5. Force Fz vs. axial displacement analytical calculation with orientation 90 [deg] for various magnetic moments

produces the highest force where the lowest force is at $98[N]$ (black line) and the highest $118[N]$ (dark blue line), thus producing a difference of $20[N]$. It also shows, that the smallest inner magnetic moment combined with the highest outer magnetic moment (green line) renders a higher force maximum than the highest inner magnetic moment combined with the lowest outer magnetic moment (light blue line). All in all the maximum force depend very strongly on the magnetic moment only, so the higher the magnetic moment of both rings is, the higher the maximum force.

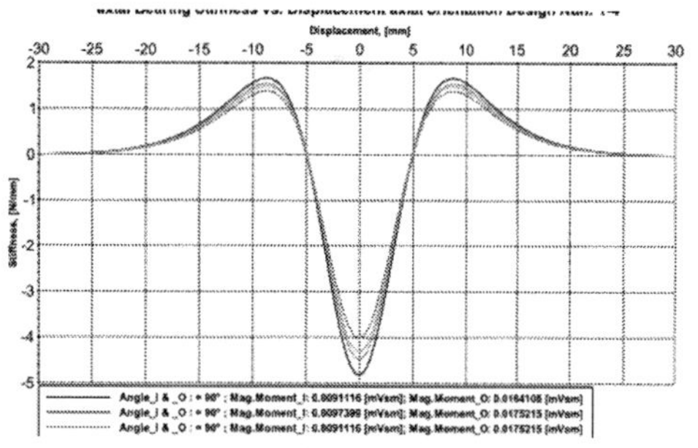

Figure 6. Stiffness Kz vs. axial displacement analytical calculation with orientation 90 [deg] for various magnetic moments

Considering the same with the stiffness Plot fig. 6 a dislocation of the inner magnet ring from the symmetry either in upward or downward displacement with varying magnetic moment renders a difference from $-4.0[\frac{N}{mm}]$ to $-4.7[\frac{N}{mm}]$ in stiffness. Since, according to eqn. 22, the radial stiffness $K_r = -\frac{K_z}{2}$ with the variation of the magnetic moments by a total of appr. $21[\%]$ yields a variation of the radial bearing stiffness by appr. $18[\%]$. The lowest magnetic moment in both rings render the lowest stiffness Kz (black line). The highest magnetic moment renders the highest stiffness Kz (dark blue line). The combination of the lowest inner magnetic moment with the highest outer magnetic moment (green line) renders a higher stiffness than the combination of the highest inner magnetic moment with the lowest outer magnetic moment (light blue line). Again the Stiffness is only correlated with the variance of magnetic moment.

It is important to say, that the respective maxima are symmetrical, thus the forces and stiffness's are the same regardless the displacement in the positive or negative direction from the line of symmetry, where the inner ring is within the outer ring.

Now considering the same analysis by varying the angle of orientation, thus, the so called "hot"- side "cold"- side effect by varying the angle from $90[deg]$ + / - $1.5[deg]$ symmetrical keeping the magnetic moment constant at the nominal value similar plots can be analysed.

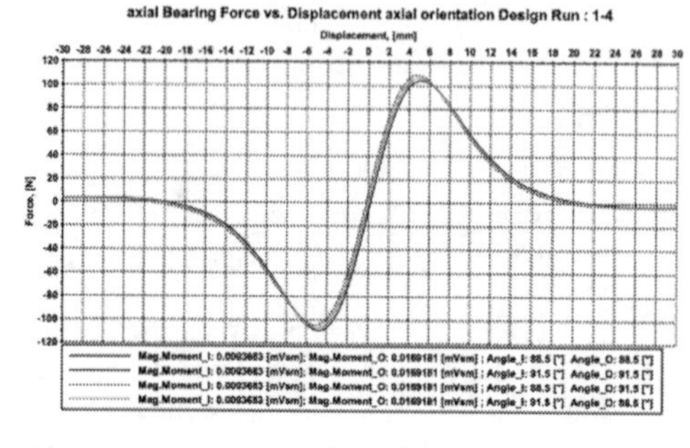

Figure 7. Force Fz vs. axial displacement analytical calculation with varying angle of orientation

When the angle of orientation of the inner and the outer ring are at 88.5 deg (both rings have a diverging field density distribution), the forces shown in fig. 7 are not symmetrical, thus moving the inner ring from the negative to the positive displacement in z-direction on the negative side the forces are less than on the positive side (black line). When the angles of orientation of both rings are at 91.5 deg, thus converging field distribution, the forces are also not symmetrical and on the negative side are higher than on the positive side of the displacement in z-direction (dark blue line). Combining a diverging inner ring (angle 88.5 deg) with a converging outer ring (angle 91.5 deg) renders a symmetrical force (green line). Where as combining a converging inner ring (angle 91.5 deg) with a diverging outer ring (angle 88.5 deg) also renders a symmetrical force and both maxima are on the same level (light blue line). The light blue line covers identically the green line.

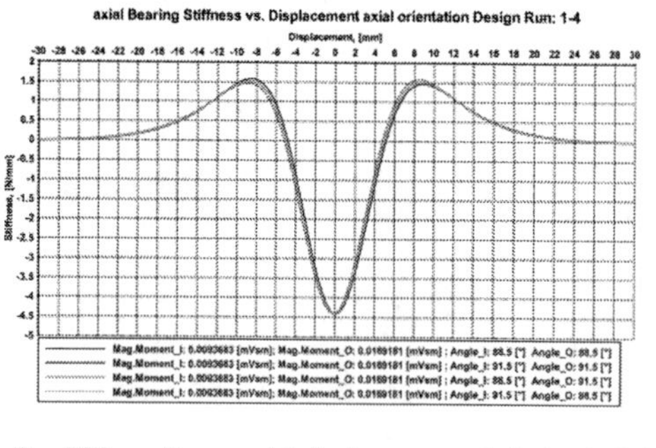

Figure 8. Stiffness Kz vs. axial displacement analytical calculation with varying angle of orientation

As expected from the forces, the stiffness refer to fig. 8 show the same, if both are diverging rings (i.e. angle 88.5 deg) they render on the negative side of the displacement a lower and on the positive side of the displacement a higher stiffness (black line). If both rings are converging (angle 91.5 deg) then the stiffness on the negative side is higher than on the positive side (dark blue line). If the inner ring is diverging and the outer ring converging, then the stiffness is again symmetrical (green line). If the inner inner ring is converging and the outer ring diverging, then the stiffness is on the same level as later.

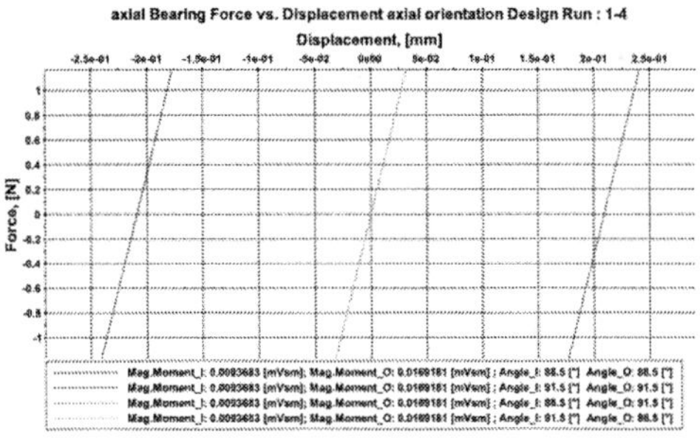

Figure 9. Force Fz vs. axial displacement analytical calculation with varying angle of orientation

The above described influences can be analysed in more detail by magnifying fig. 9 in the vicinity of the origin. In general, one may deduce, that a unsymmetrical force- or stiffness- distribution may be neutralised by combining a "hot side" inner ring with a "cold side" outer ring or vice verse. When two diverging or two converging magnetic rings are combined within a bearing, a so called "parasitic axial force" at $z = 0.0$ is $+/-8[N]$ which can be observed by extrapolating to $z = 0.0$. These parasitc forces may be neutralised by shifting one of the rings in axial direction by $+/-0.21[mm]$ refer 9. So it is important to measure these effect very precisely. This phenomenon was also patented by VACUUMSCHMELZE GmbH & Co. KG (VAC) [3].

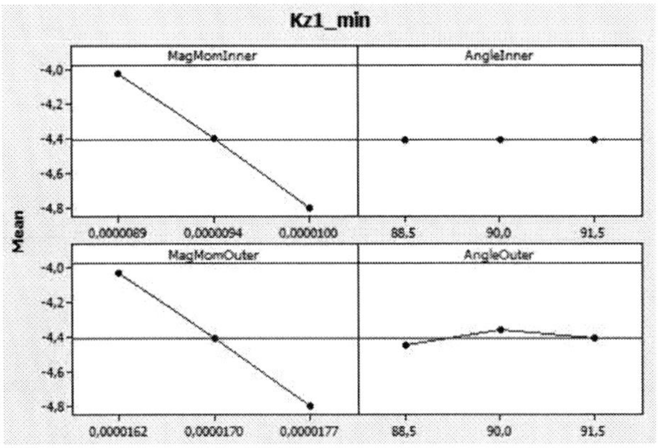

Figure 10. Main Effect magnetic moment acting on stiffness Kzmin. Mean taken over all 81 parameter sets

Analysing the 81 combinations possible, one may deduce, that the magnetic moment is a so called main effect, which acts on the maximum and minimum Force and Stiffness refer to fig. 10.

Where as the symmetrical or unsymmetrical distribution of forces or Stiffness cause a so called "parasitic axial force" and only depends on the angle of orientation fig. 11 or "Hot side"

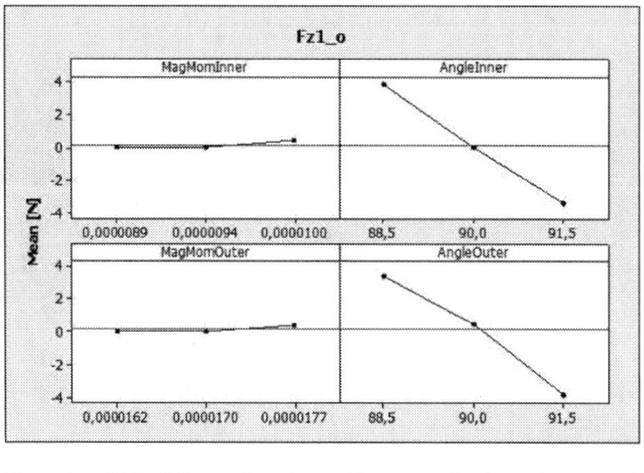

Figure 11. Main Effect angle acting on Fz at z = 0.0. Mean taken over all 81 parameter sets

and "Cold Side" of the ring magnets and it may be neutralised by a sorted and paired combination of rings.

VI. Outlook

Most magnetic baerings are currently equipped with Sm-Co magnets. But resently even Nd-Fe-B magnets are being considered as an alternative solution. With a new family of grades and technologies, one may be able to increase the share of Nd-Fe-B magnets in magnetic bearings, when ever a higher environmental temperature, thus coercivity is needed.

A proposal to meet these demands is to use Grain Boundary Diffusion (GBD) magnets as in fig. 12. By implementing this technology the coercivity may be increased by $> 400\frac{kA}{m}$ and keeping the remanence constant. As shown in fig 12 by the blue line, which represents a typical demagnetisation curve of transversal pressed VACODYM 131 TP, while the black lines shows the increase of coercivity by applying Dy/Tb Grain Boundary Diffusion. By developing new magnet grades using this advanced processing step as described, VACUUMSCHMELZE (VAC) has not only met and fulfilled the demand from the market to reduce the heavy RE content of Nd-Fe-B magnets, but also significantly improved the magnetic properties. Thus, by reducing the heavy RE content, the security of supply of RE elements may be enhanced. These additional process steps of applying RE elements and exposing the magnet to 800 $[degC]$ - 1000 $[degC]$ do have an influence on the production costs. The overall result may not necessarily be cost neutral, depending of course on the level of the raw material prices. However, such a technique demonstrates that new grades can be tailored to meet specific customer and application demands, either in the form of higher H_{cJ} or B_r values, enabling precise optimisation of the magnet grade to the working conditions of the customers application.

VII. Conclusion

In this paper, we tried to find an analytical method, with which we are able to find the main influencing factors which act on a axial magnetic bearing forces and stiffness. By carrying out a parametric variation, we were able to find

[3]Patentschrift DE 103 21 925 B4 2008.10.02

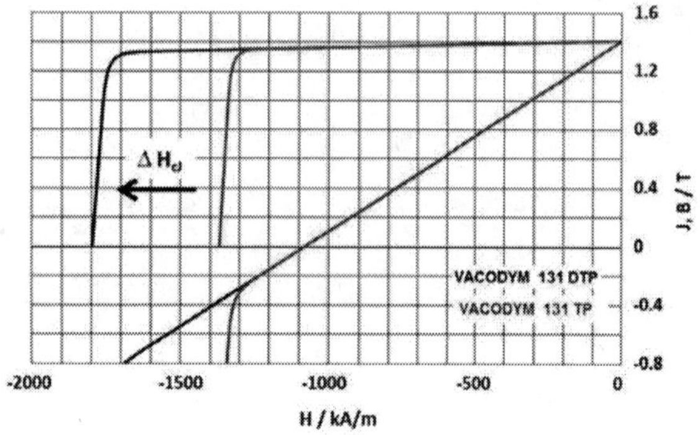

Figure 12. The increase in magnetic properties with diffusion treatment.

that the forces and stiffness not only depend on the mutual magnetic moments of the inner and outer magnetic ring, but also on their mutual so called" "Hot Side" or "Cold Side" effect, which was represented here by the angular orientation. The approaches by Yonnet et. al. [5] [2] through [11] [12] could not be used, because the analytical calculation does not take into account, that a diverging or converging orientation of a ring is possible. Only with the work of Bekinal et. al. [3] it is possible to consider this effect.

We also were able to show that a paired combination of diverging and converging rings may reduce the unsymmetrical force and stiffness distribution of a magnetic bearing.

REFERENCES

[1] V.Lemarquand G.Lemarquand. Passive permanent magnet bearings for rotating shaft : Analytical calculation. *Intech,croatia*, ISBN 978-953-307-148-0:132, 2010.

[2] V. Lemarquand R.Ravaud, G.Lemarquand. Force and stiffness of passive magnetic bearings using permanent magnets. part 1 : Axial magnetization. *IEEE Transactions of Magnetics*, 45,7:2996, 2009.

[3] S. Jana M.I. Peerzade S.I. Bekinal, T.R. Anil. Optimization of halbach magnetized permanent magnet axial bearing. *IJEIT*, 2,1:5, 2012.

[4] M. Weickhmann. Nutzung des Nord- / Sd- Effektes bei permanenterregten Maschinen. *FEMAG Anwendertreffen, Offenbach*, 2011.

[5] J.P. Yonnet. Permanent magnet bearing and couplings. *IEEE Transaction on Magnetics*, 17:1169–1173, 1981.

[6] R. Ravaud G.Lemarquand. Halbach structure for permanent magnets bearing. *Progress In Electromagnetics Research M*, 14:263–277, 2010.

[7] R. Ravaud G.Lemarquand. Magnetic field produc ed by a parallelepipedic magnet of various and uniform polarization. *Progress In Electromagnetics Research PIER*, 98:207–219, 2009.

[8] G.Lemarquand R.Ravaud. Halbach structures for permanent magnets bearings. *Progress In Electromagnetcs Research M*, 14:263–277, 2010.

[9] V.Lemarqund R.Ravaud, G.Lemarquand. Force and stiffness of passive magnetic bearings using permanent magnets. part 2: Radial magnetization. *IEEETransactions on Magnetics*, 45,9:3334–3342, 2009.

[10] V.Lemarqund R.Ravaud, G.Lemarquand. Force and stiffness of passive magnetic bearings using permanent magnets. part 2: Radial magnetization. *IEEETransactions on Magnetics*, 45,9:3334–3342, 2009.

[11] G.Lemarquand V. Lemarquand. Passive permanent magnet bearing for rotating shaft : Analytical calculation. In *Magnetic Bearings, Theory and Application*, 2010.

[12] Anthony Zander W. Robertson, Ben Cazzolato. A simplified force equation for coaxial cylindrical magnets and thin coils. *School of Mechanical Engineering, University of Adelaide, SA 5005, Australia*, pages 1–5.

High strength iron-cobalt materials for magnetic bearings

14th International Symposium on Magnetic Bearings
August 11- 14, 2014, Linz, Austria

Frederik Fohr, Niklas Volbers, Joachim Gerster

VACUUMSCHMELZE

ADVANCED MATERIALS – THE KEY TO PROGRESS

Content

- General properties of 49%Fe 49%Co 2%V iron- cobalt alloys
- High strength iron- cobalt alloys for high speed rotors
- Micrometer rolled laminations for smallest losses at high frequencies

Content

- General properties of 49%Fe 49%Co 2%V iron- cobalt alloys
- High strength iron- cobalt alloys for high speed rotors
- Micrometer rolled laminations for smallest losses at high frequencies

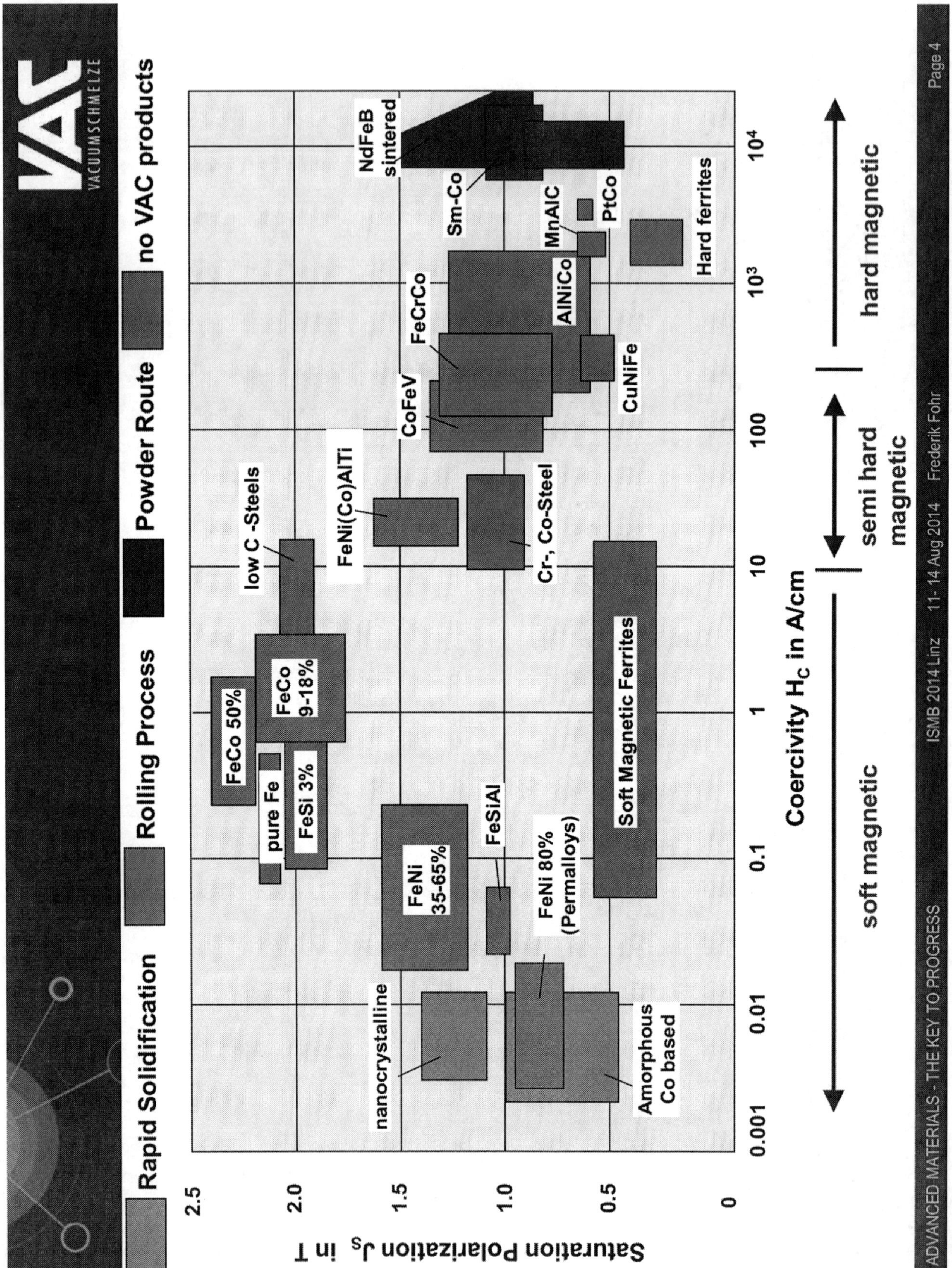
VAC
VACUUMSCHMELZE
Rapid Solidification
Rolling Process
Powder Route
no VAC products
Saturation Polarization J_S in T
2.5
2.0
1.5
1.0
0.5
0
0.001
0.01
0.1
1
10
100
10^3
10^4
Coercivity H_C in A/cm
nanocrystalline
Amorphous Co based
FeNi 35-65%
FeNi 80% (Permalloys)
FeSiAl
pure Fe
FeSi 3%
FeCo 50%
FeCo 9-18%
low C -Steels
FeNi(Co)AlTi
Cr-, Co-Steel
Soft Magnetic Ferrites
CoFeV
FeCrCo
AlNiCo
CuNiFe
MnAlC
PtCo
Sm-Co
NdFeB sintered
Hard ferrites
soft magnetic
semi hard magnetic
hard magnetic
ADVANCED MATERIALS - THE KEY TO PROGRESS
ISMB 2014 Linz
11- 14 Aug 2014
Frederik Fohr
Page 4

Comparison of inductions of different soft magnetic alloys

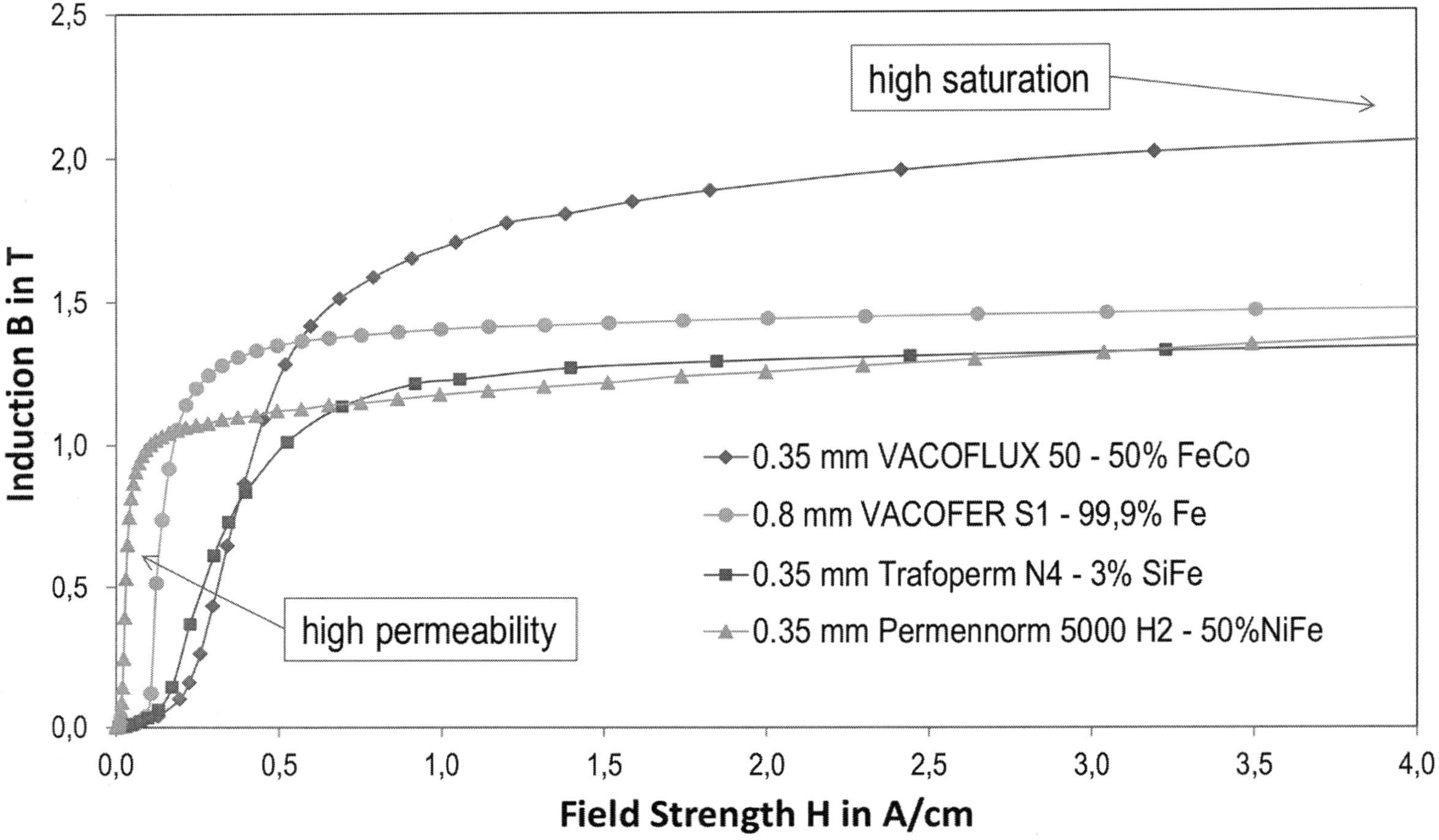

VACOFLUX© 50
$B_{sat} = 2{,}35$ T

Electrical 3%SiFe steel
$B_{sat} = 2{,}03$ T

VACOFLUX© 50
$B_{sat} = 2{,}35$ T

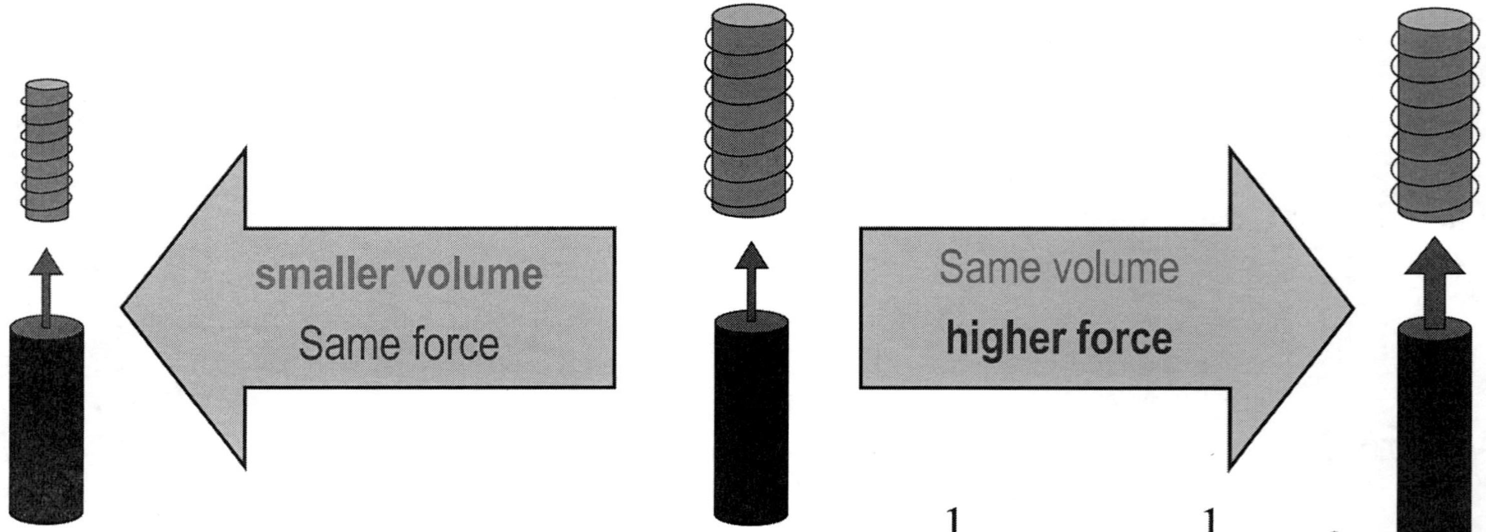

$$w = \frac{1}{2} H \cdot B = \frac{1}{2\mu_r \mu_0} B^2$$

$$\xrightarrow{H \to \infty} \frac{1}{2\mu_0} B_{sat}^2$$

$$\Rightarrow +34\% \; energy\ density$$

Energy density w for high fields depends only on the satuartion level

Comparison of losses of 50% FeCo and 3% SiFe

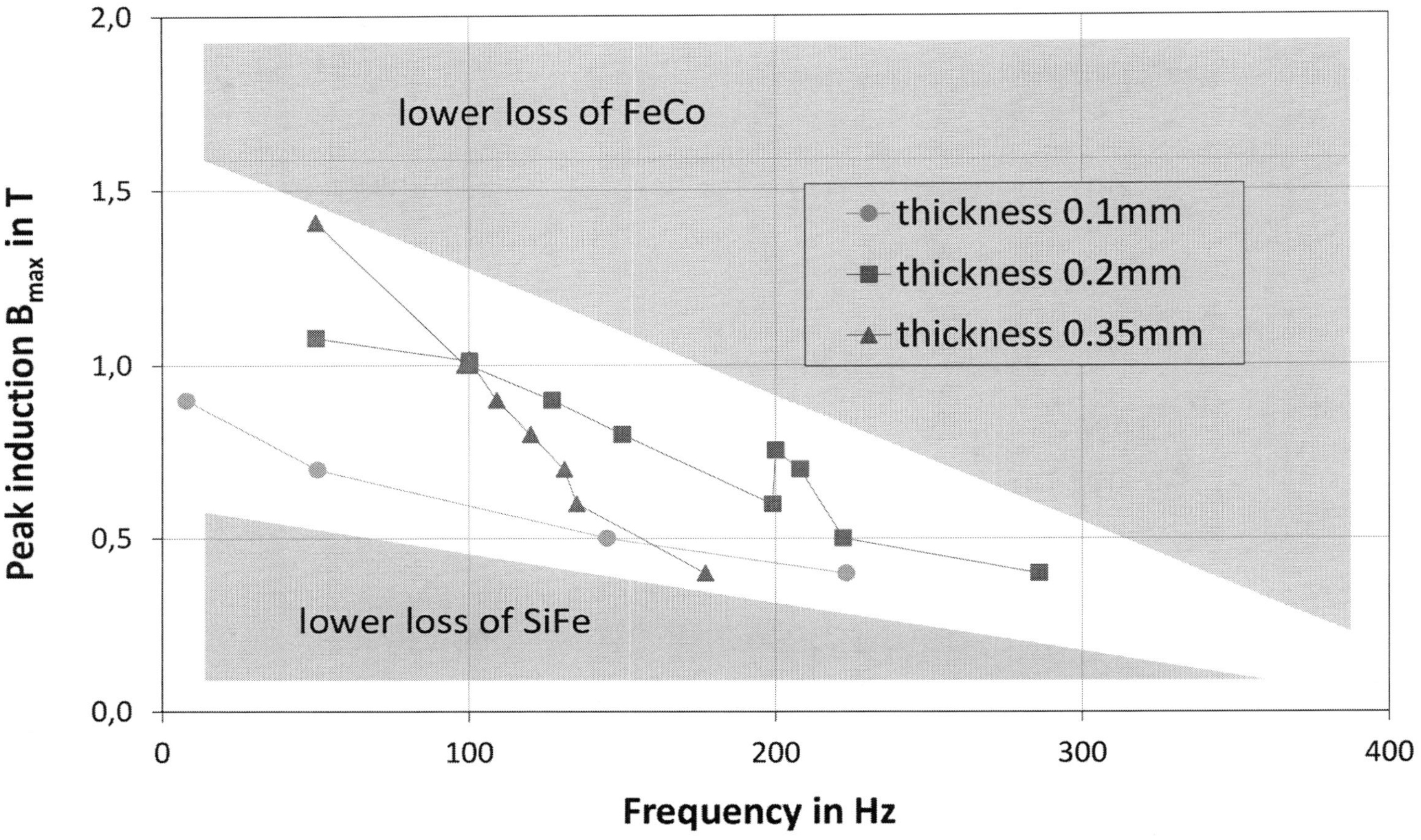

Ref.: W. Pieper, J. Gerster, JAP, Vol. 109, No. 7, April 2011

Comparison of soft magnetic alloys for magnetic bearings

	Material	Saturation in T	Curie Temp. T_c in °C	Yield Strength in MPa	Electrical Resistivity in μΩm
cold rolling	FeCo 50wt%	2.35	950	200 - 800	0.40
cold rolling	Pure Iron	2.15	770	100	0.10
cold rolling	SiFe 3wt%	2.03	740	300 - 400	0.40
cold rolling	FeNi 50wt%	1.55	440	140 - 200	0.45
rapid solidification	Amorphous FeSiB	1.56	420	≈ 3000	1.30

Content

- General properties of 49%Fe 49%Co 2%V iron- cobalt alloys
- High strength iron- cobalt alloys for high speed rotors
- Micrometer rolled laminations for smallest losses at high frequencies

VACODUR® – high strength FeCo alloys

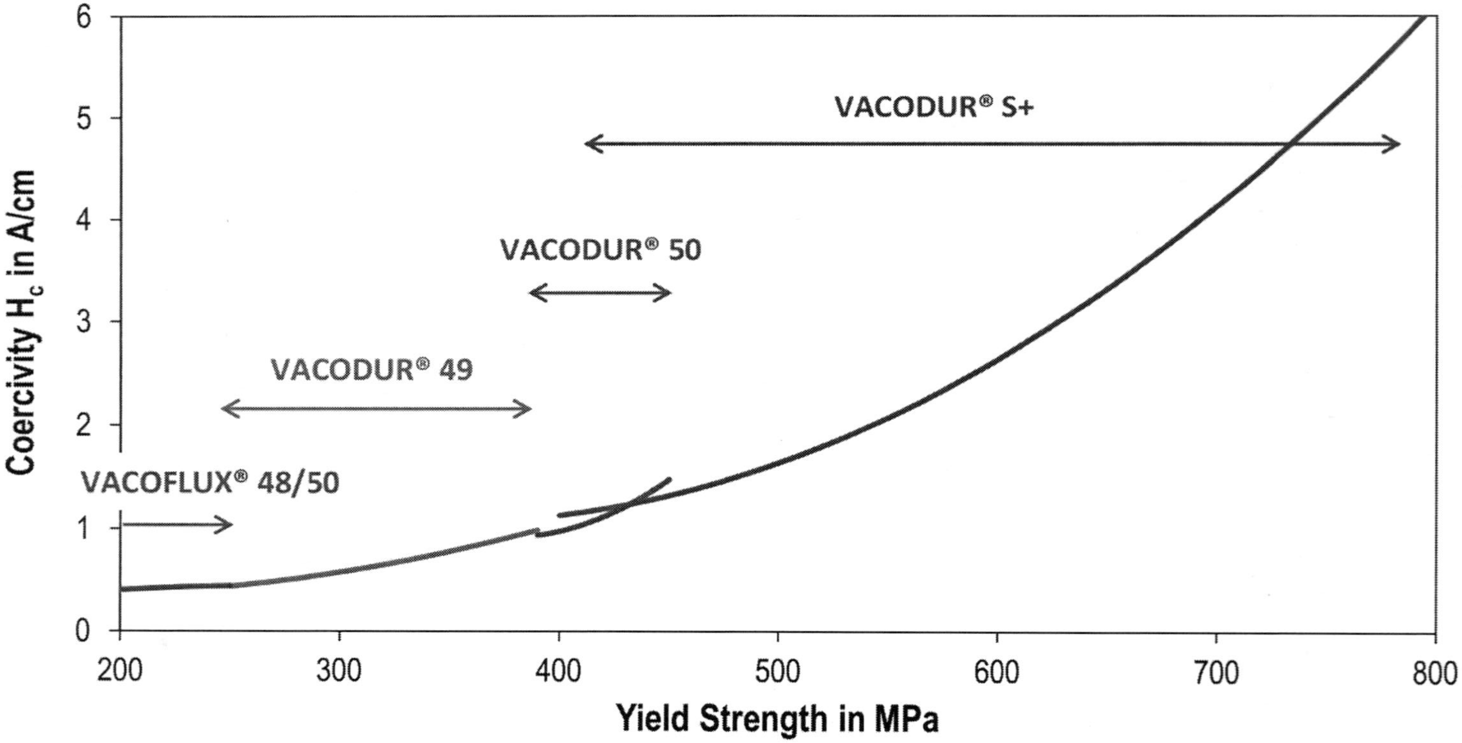

VAC covers the complete yield strength range from 200 to 800 MPa

Adjustment of the yield strength by grain size

- **Coercivity:** Law by Mager $H_c = H_0 + \frac{m}{d}$

- **Yield Strength** (Hall-Petch relation) $R_{p0,2} = \sigma_0 + \frac{k}{\sqrt{d}}$

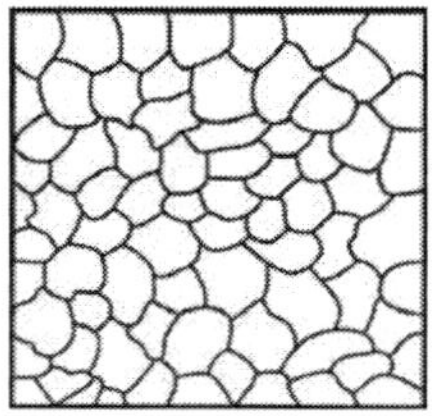

small grain
high Hc, high strength

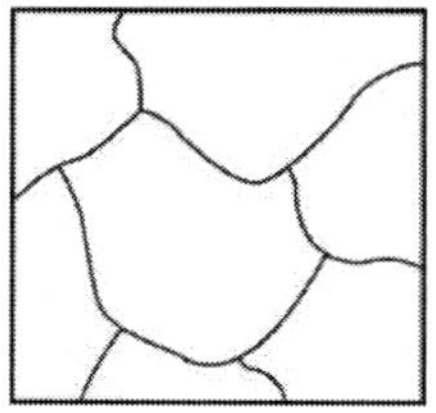

coarse grain
small Hc, low strength

$$H_c = H_0 + \frac{m}{k^2}(R_{p0.2} - \sigma_0)^2$$

Ref: E.O. Hall, Proc. Phys. Soc. **64** (1951). 742.747 / N.J.Petch: J. Iron Steel Inst. **174** (1963), 25-28 / A. Mager. Ann. Phys. 11 (1952), 11-12.

Adjustment of the grain size by final annealing

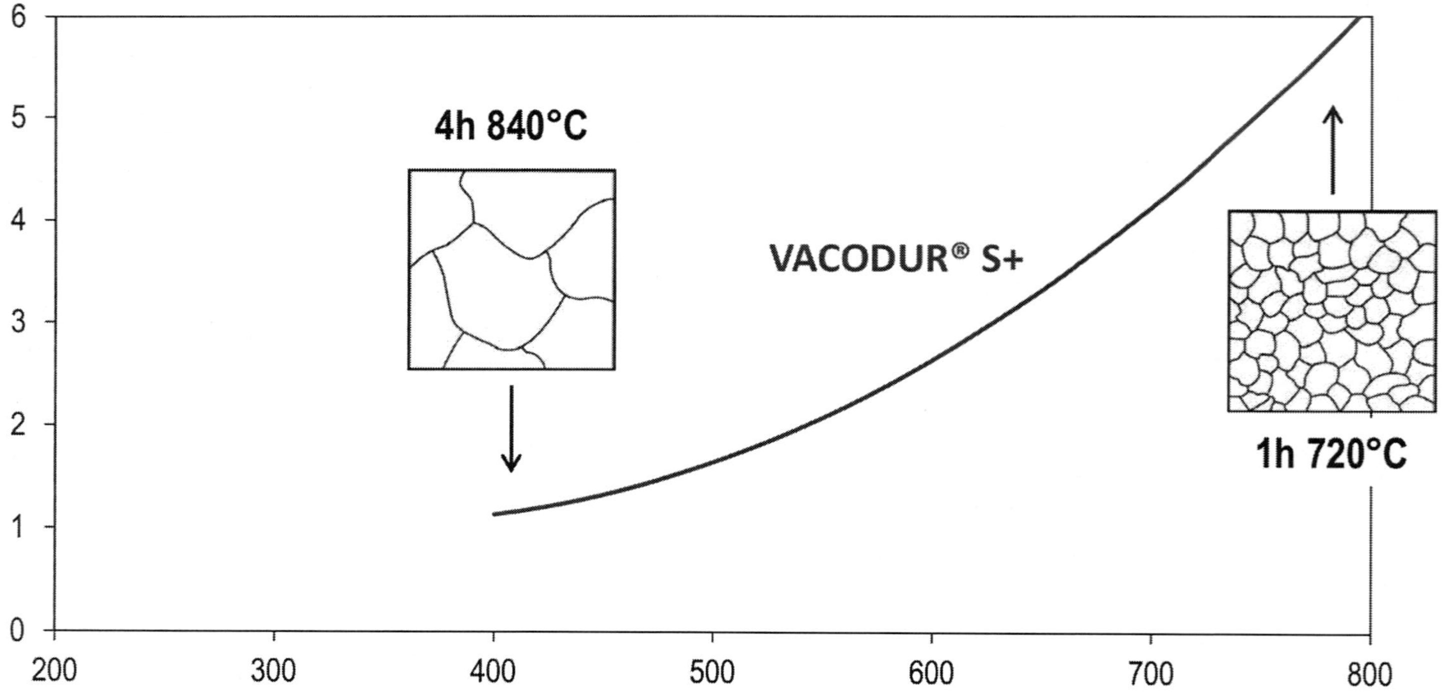

Grain size determines mechanical and magnetic properties

Adjustment of the grain size by final annealing

- Dry hydrogen atmosphere prevents oxidation and removes impurities
- Recovery from mechanical stress due to mechanical processing
- Grain size adjustment by variation of temperature and holding time
- Recrystallisation of the microstructure

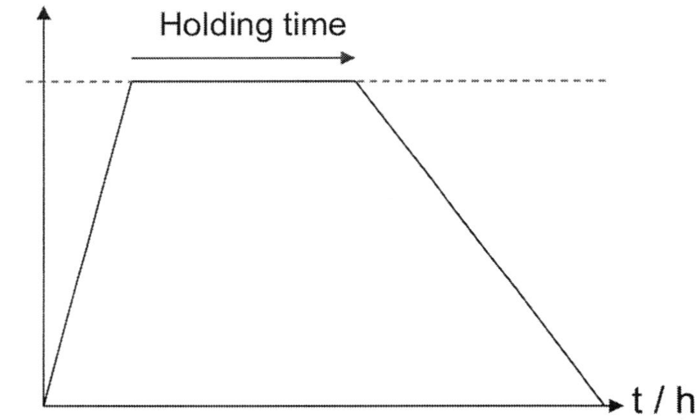

Typical temperature profile

Hood type furnace

VACUUMSCHMELZE

Content

- General properties of 49%Fe 49%Co 2%V iron- cobalt alloys
- High strength iron- cobalt alloys for high speed rotors
- Micrometer rolled laminations for smallest losses at high frequencies

Cold rolling mill

- Mill- power is supplied by two small working rolls
- Small working rolls permit thin strip thicknesses
- The working rolls are backed by larger idling rolls
- Idling rolls permit the application of high pressure

Arrangement of rolls

20-high Rohn mill

cold rolled coils

Core Loss reduction in thin strip material

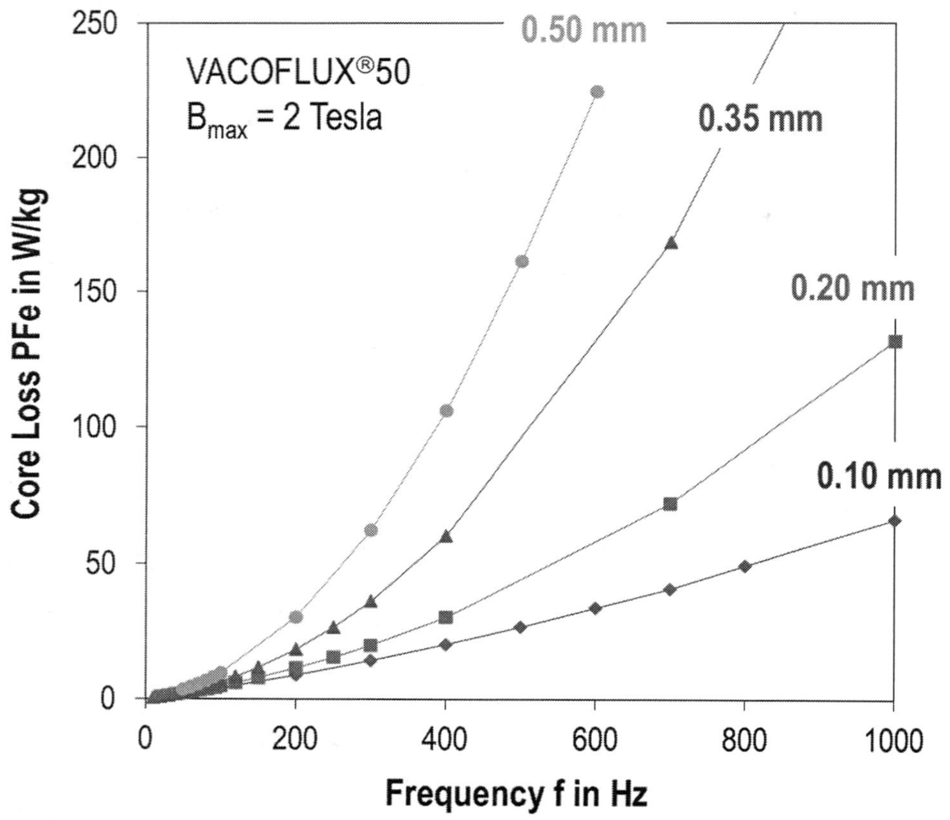

- Eddy currents decrease quadratically with smaller strip thickness

$$P_{ed} = \frac{\pi^2 \sigma\, d^2 B_{\max}{}^2}{6\rho} f^2$$

- Laminated FeCo- stacks with strip thickness down to 50 µm are in VAC series production (VACSTACK®)

VACSTACK®- VAC's laminated FeCo- Stacks

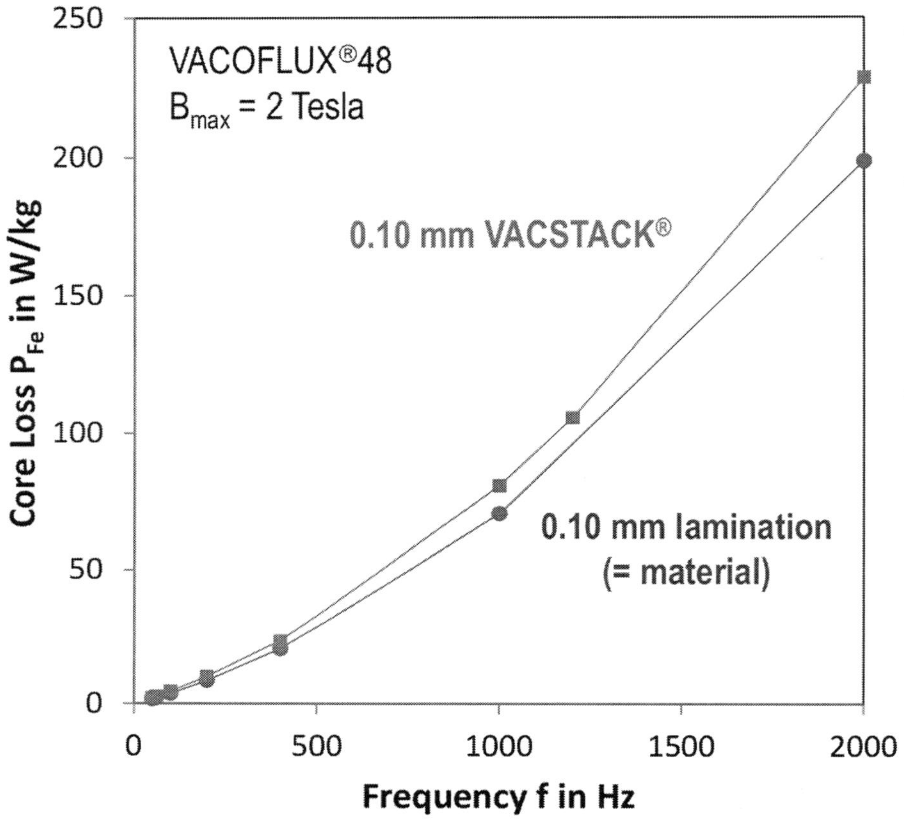

- VACSTACK® losses are close to the material losses
- VACSTACK® procedure ensures minimal deterioration of the magnetic properties
- High- precision packaging and tight tolerances

Comparison VACSTACK® in lamination thicknesses 0.10 mm and 0.05 mm

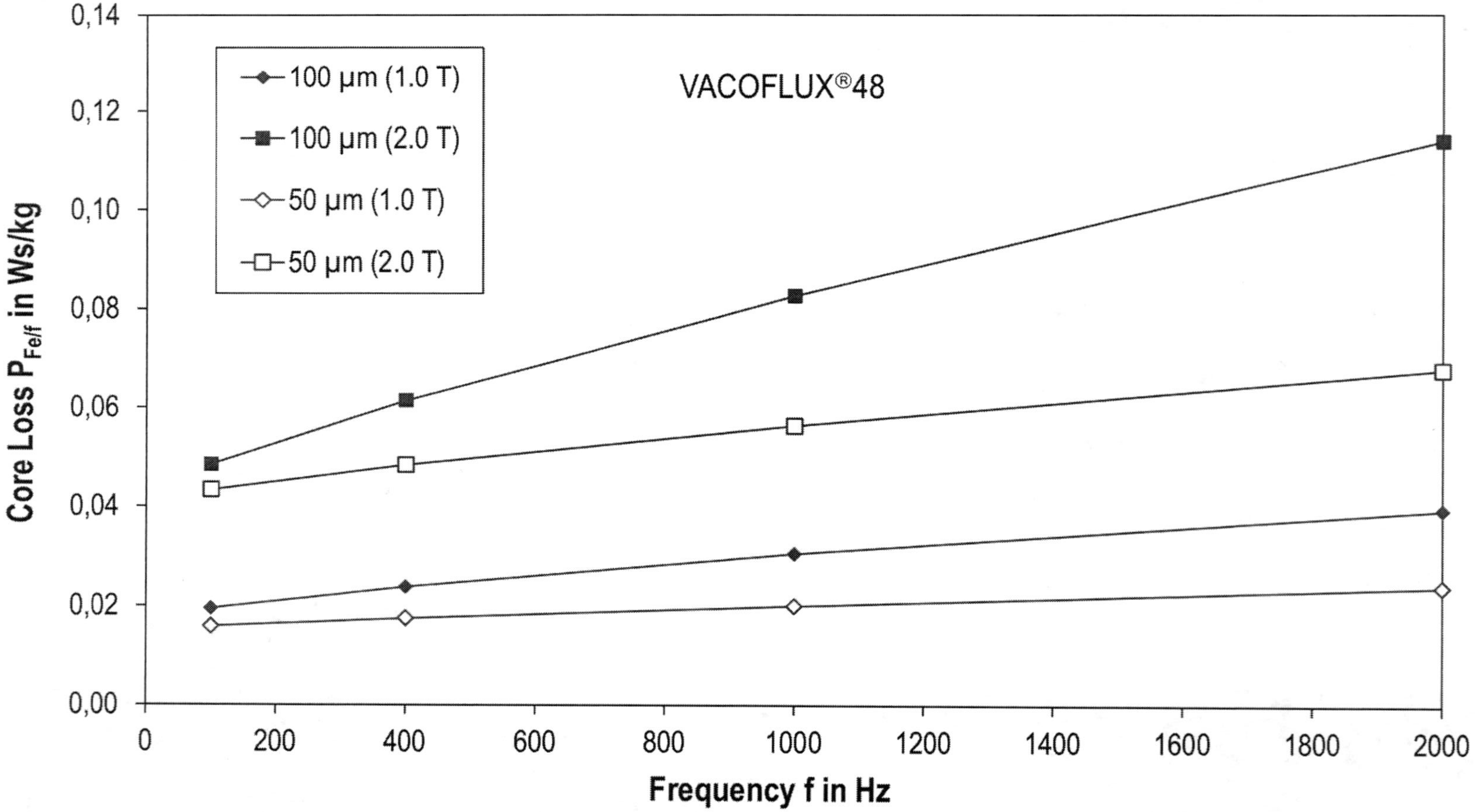

VACSTACK® filling factor

- Filling factor determines the total magnetic volume
- Typical VACSTACK® filling factors of 96- 98%
- Filling factor decreases with laminations thickness

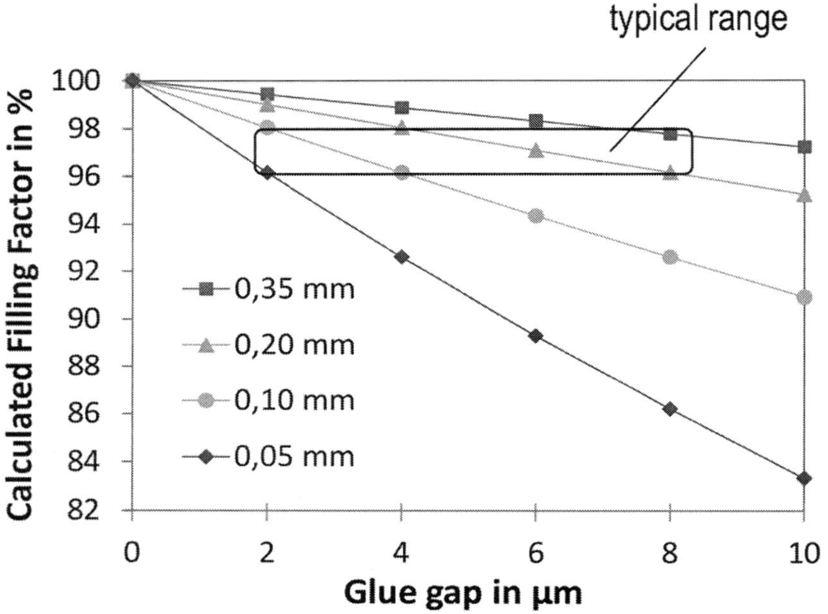

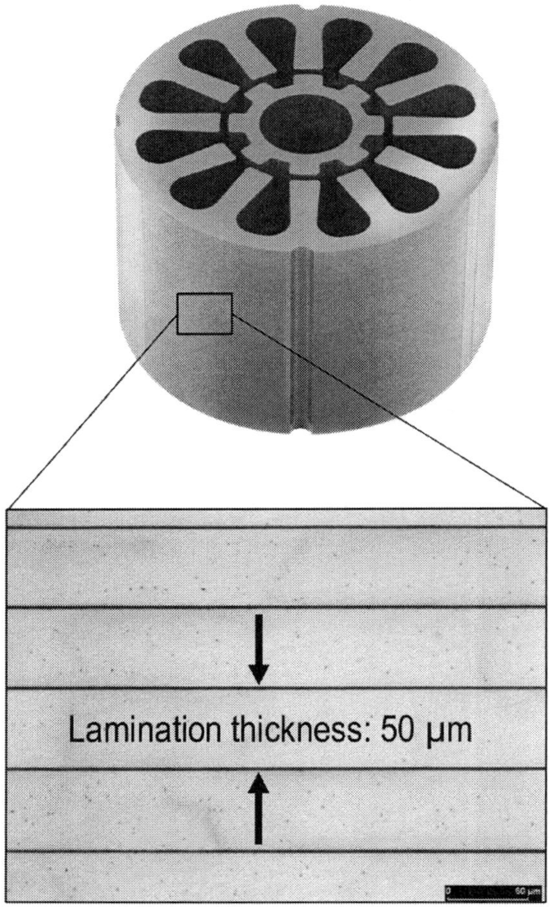

Summary

VAC's iron- cobalt materials supply:

- Highest magnetic saturation for weight saving and/or highest forces – **VACOFLUX®**
- Improved tensile properties for high speed rotors – **VACODUR®**
- High performance stacks for motors & generators – **VACSTACK®**

Thank you for your attention

contact: frederik.fohr@vacuumschmelze.com

http://www.vacuumschmelze.com/en/downloads/ht-brochures/soft-magnetic-materials.html

Basics of Superconducting Levitation and its Use in the Transport System SupraTrans II

Ludwig Schultz[a,b,c], Oliver de Haas[a], Dietmar Berger[b], Lars Kühn[a,b,c], Anne Berger[b,c], Tilo Espenhahn[b,c]

[a] evico GmbH Dresden, Germany , [b] IFW Dresden, Helmholtzstrasse 20, D-01069 Dresden, Germany,,
[c] TU Dresden, Germany, l.schultz@ifw-dresden.de

New means of urban transportation and logistics will become realistic with superconducting magnetic bearings using nanostructured bulk high temperature superconductors [1,2]. Superconducting magnetic levitation works passively stable without any electronic control but with attracting and repelling forces to suspend a vehicle pendant or standing upright from zero to high speed - perfect conditions for the idea of rail-bound individual transport with cabins for 4 - 5 passengers called on demand only. They will levitate noiseless over the track made of RE permanent magnets to the chosen destination saving energy and travel time [3]. A big step forward to this vision can be tested at Dresden. The world largest research and test facility for transport systems using HTS bulk material in the levitation and guidance system in combination with a permanent magnet track was put into operation. A vehicle for 2 passengers, equipped with linear drive propulsion, non-contact energy supply, second braking system and various test and measurement systems is running on an 80 m long oval driveway.

In the presentation the superconducting materials as well as the principle of superconducting levitation by flux pinning in high temperature superconductors will be described. An overview of the SupraTrans II research facility and future directions of superconductivity-based magnetic levitation and bearing for automation technology and transportation will be given.

References:

1. G. Fuchs, P. Schaetzle, G. Krabbes, S. Gruss, P. Verges, K.-H. Mueller, J. Fink, L. Schultz, *Trapped magnetic fields larger than 14 T in bulk YBa2Cu3O7-x*, Appl. Phys. Letters 76 (2000) 2107-2109.
2. S. Gruss, G. Fuchs, G. Krabbes, P. Verges, G. Stoever, K.-H. Mueller, J. Fink, L. Schultz, *Superconducting bulk magnets: Very high trapped fields and cracking*, Appl. Phys. Lett. 79 (2001) 3131-3133.
3. L. Schultz, O. de Haas, P. Verges, C. Beyer, S. Roehlig, H. Olsen, L. Kuehn, D. Berger, U. Noteboom, U. Funk, *Superconductively levitated transport system - The SupraTrans project*, IEEE Trans. Applied Superconductivity 15 (2005) 2301-2305.

Basics of Superconducting Levitation and its Use in the Transport System SupraTrans II

Ludwig Schultz[a,b,c], Oliver de Haas[a], Dietmar Berger[b], Lars Kühn[a,b,c], Anne Berger[b,c] and Tilo Espenhahn[b,c]
[a] evico GmbH Dresden, Germany
[b] IFW Dresden, Helmholtzstrasse 20, D-01069 Dresden, Germany
[c] TU Dresden, Germany
l.schultz@ifw-dresden.de

- Suspension by magnetic fields
- High temperature superconductors
- The SupraTrans project
- Energy-efficient applications

Levitation why?

- Fascinating
- No friction, energy efficient
- Soft bearing
  - Incredibly comfortable
  - High velocity
  - Endless durability
  - New traffic concepts

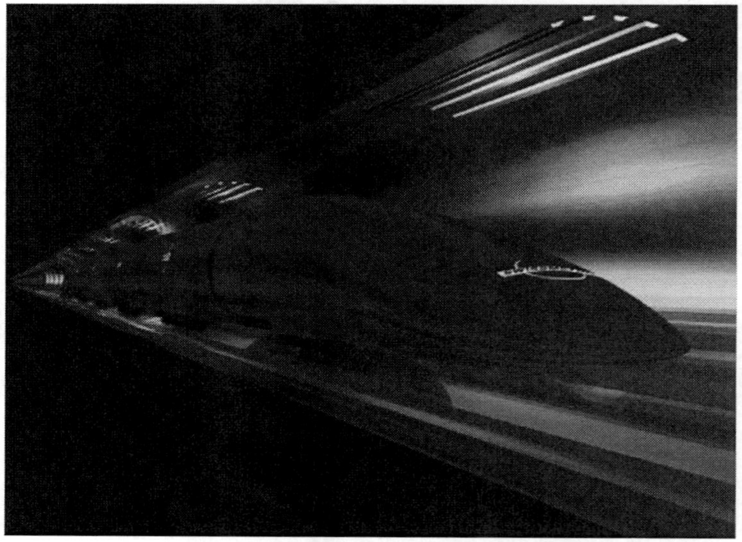

Superconductivity

Two properties:

1) Zero resistance R = 0
for $T < T_c$

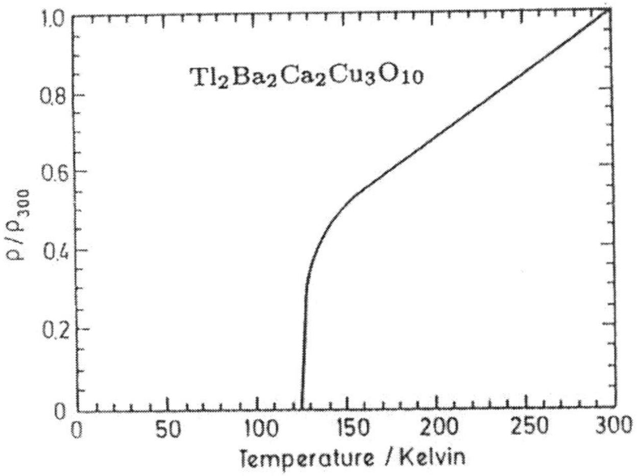

2) Magnetic field is pushed out of the superconductor

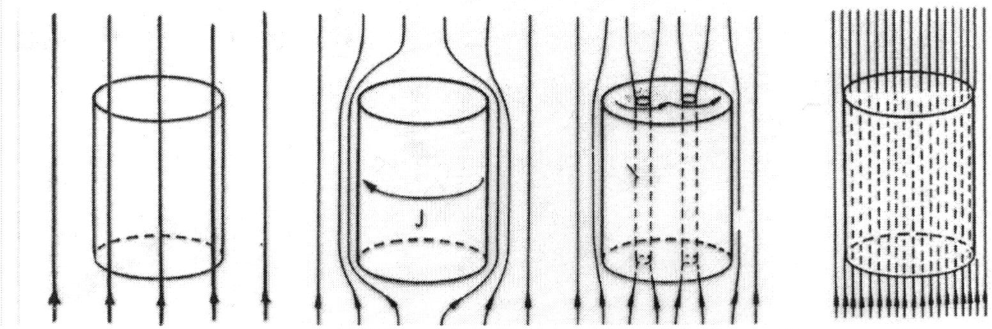

normal state

$B < B_{c1}$ Meissner state

$B_{c1} < B < B_{c2}$ mixed state

$B > B_{c2}$ normal state

Development of the critical temperature T_c

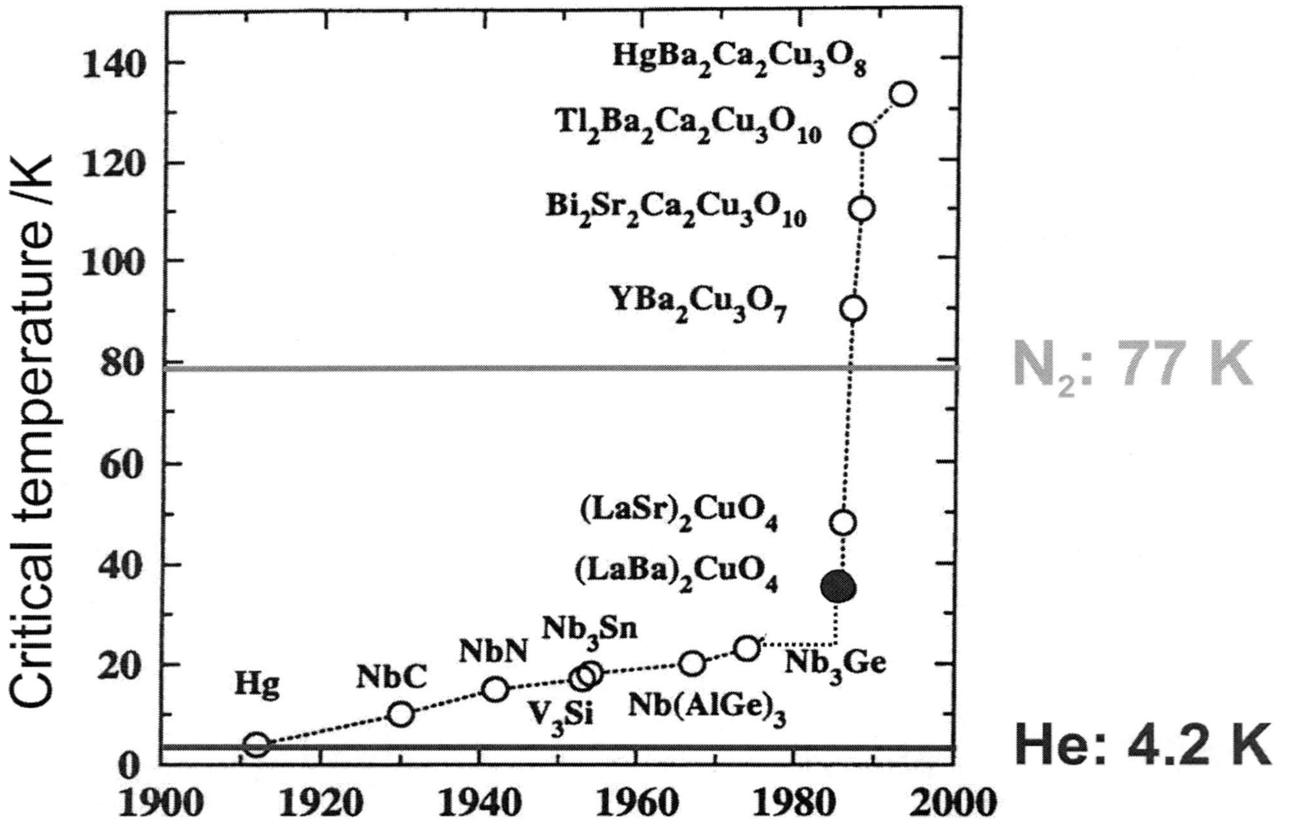

Superconductor in a magnetic field

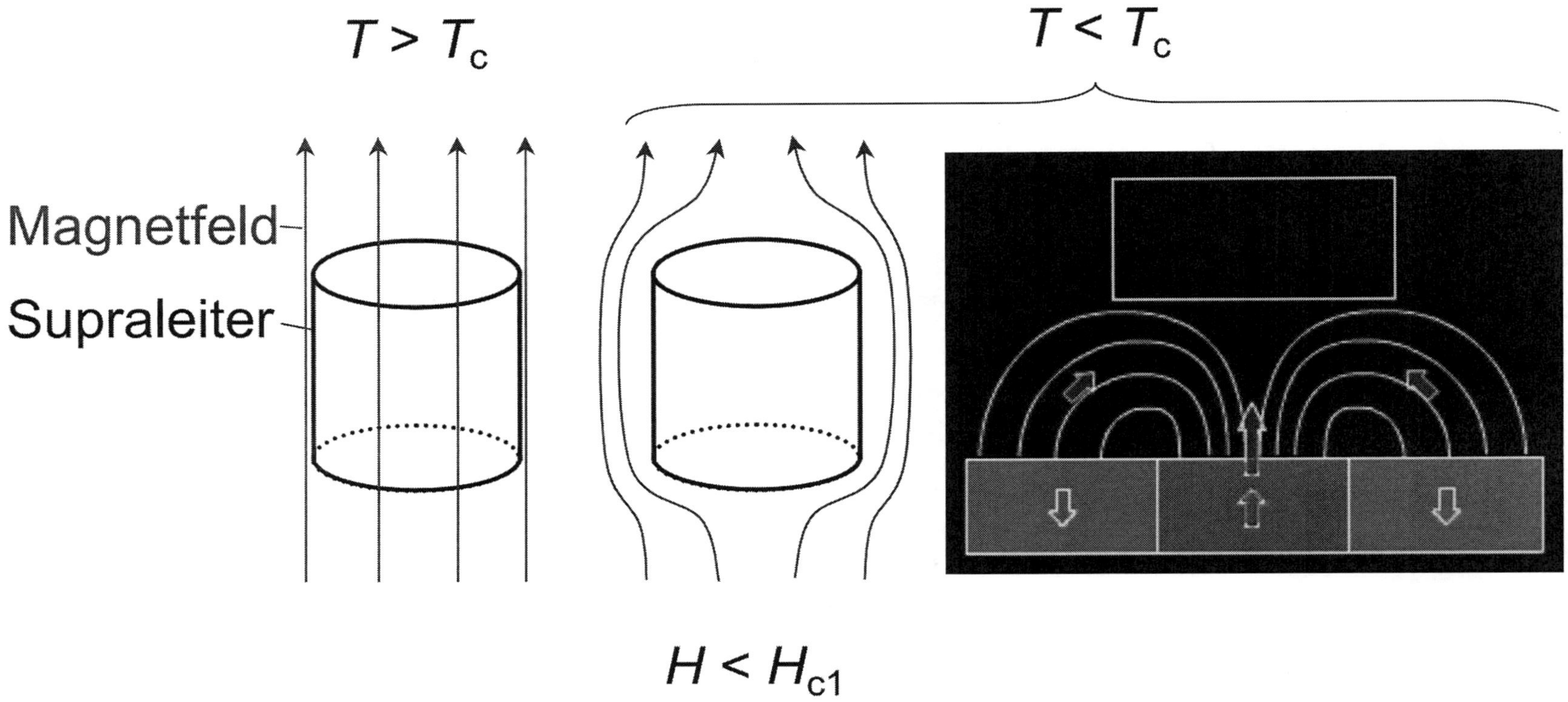

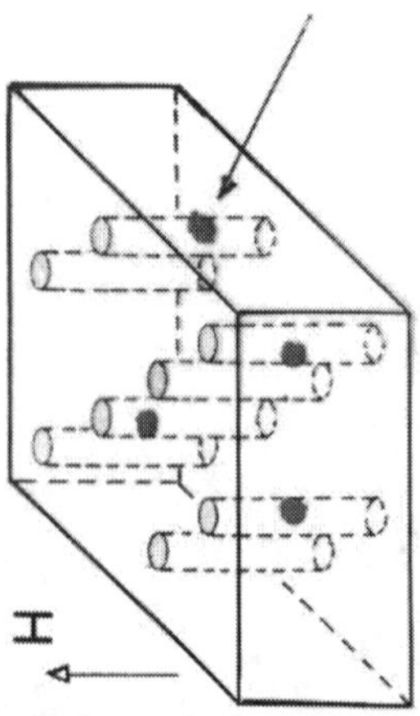
H

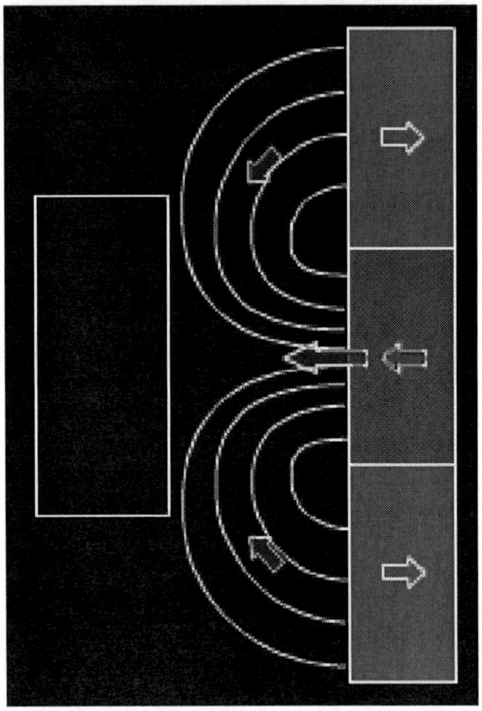

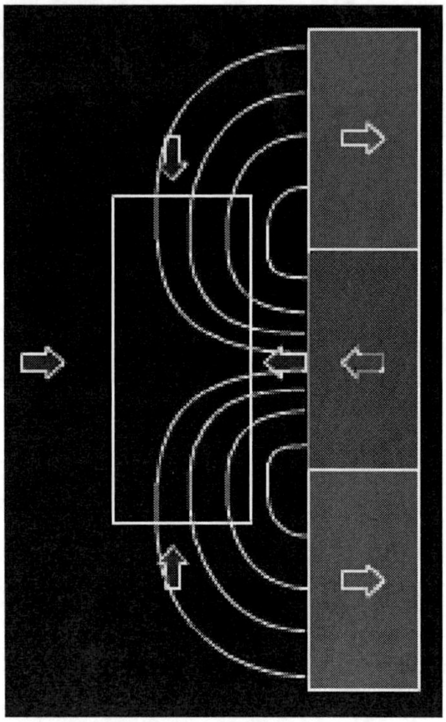

Project SUPRATRANS

Goals:

- qualification of the superconducting suspension technology for application in transport systems
- increasing size, increasing levitation forces
- construction of a demonstrator
- system integration for propulsion, safety, logistics, power transfer, etc.

SupraTrans II

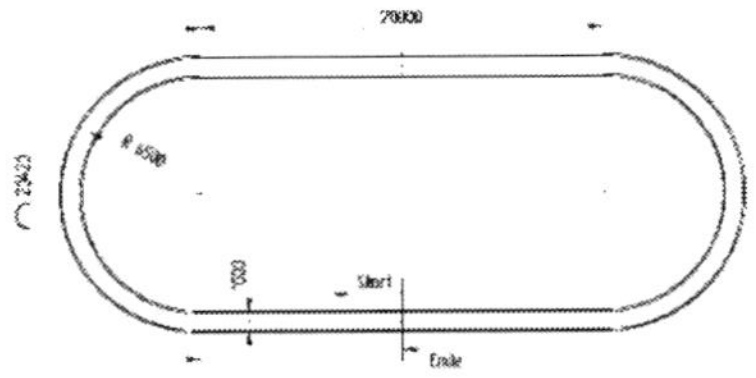

Permanent magnetic rail

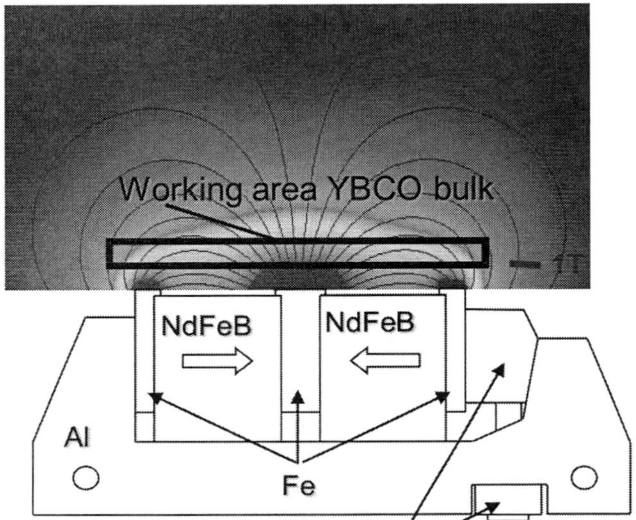

Magnetic design	
Architecture	3 pole - flux collector
Pole distance	48 mm (centre-centre)
Perm. magnets	NdFeB N40
B_r	*1276,8 mT*
Dimensions (mm)	*50 x 50 x 40*
Weight	*750 g* } 30 kg/rail
Magnets / rail	*40* } 30 kg/rail
Flux collector	Steel - softmagnetic
B_{surf}	*1.1 T*

vehicle

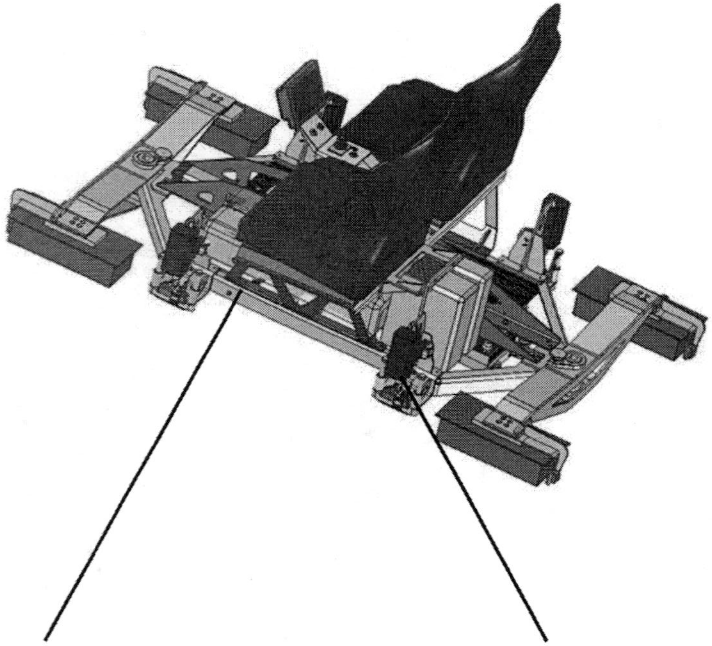

Spezification

- Total weight = 400 kg
- Air gap = 10 mm
- 2 braking system
- a = $1 m/s^2$
- v = 20 km/h

4 cryostats
- 24 YBCO bulks/cryostat
- YBCO area = 500 cm^2/cryostat
- Force density = 4 – 4.5 N/cm^2 @ 8 – 10 mm
- Levitation force = 8 – 9 kN @ 8 – 10 mm

4 x lift-/ brake system
- liftsystem brings vehicle into cooling position
- auxiliary brake

Propulsion system

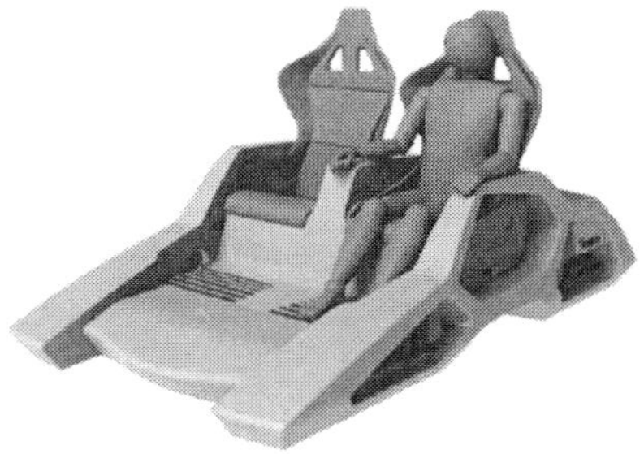

Features

U_n = 380 V
I_n = 45 A
f = 0 – 80 Hz
F_{lin} = 1000 N
η = 30 %
m_{Stator} = 43 kg

Asynchronous short stator linear motor

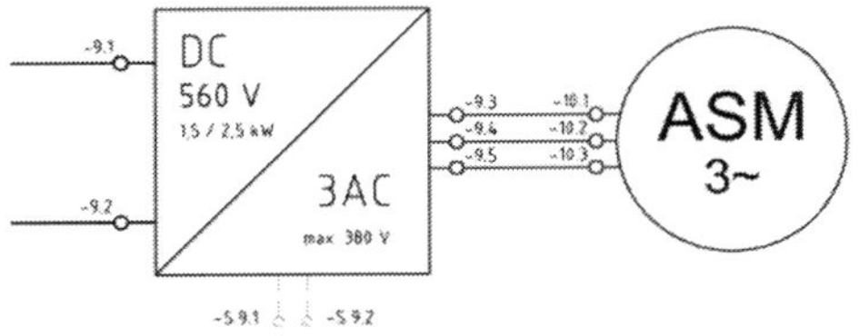

Track

Short circuit "rotor" (5 mm Cu + iron yoke)

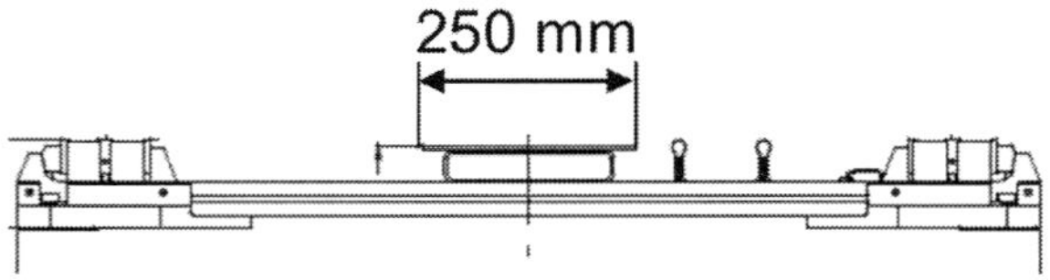

Vehicle

Short stator (ASM 3-phase winding)
(by Oswald GmbH)

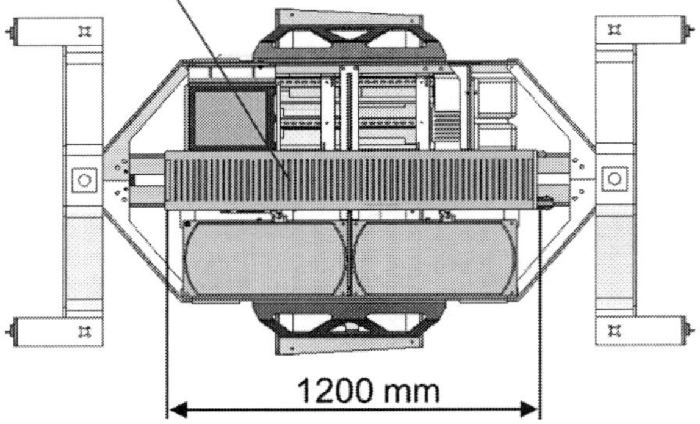

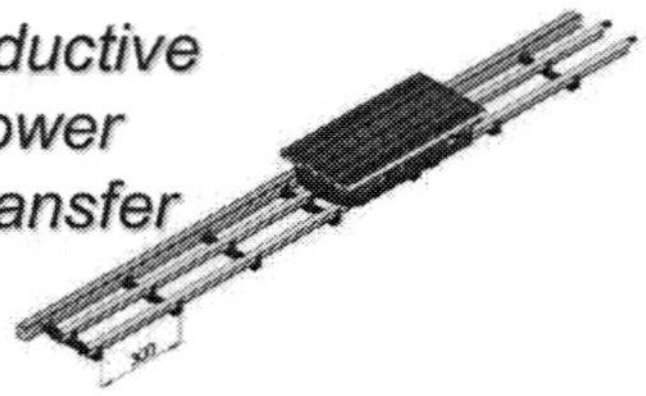

Energy supply

Inductive power transmission (non-contact)
→ transformer principle

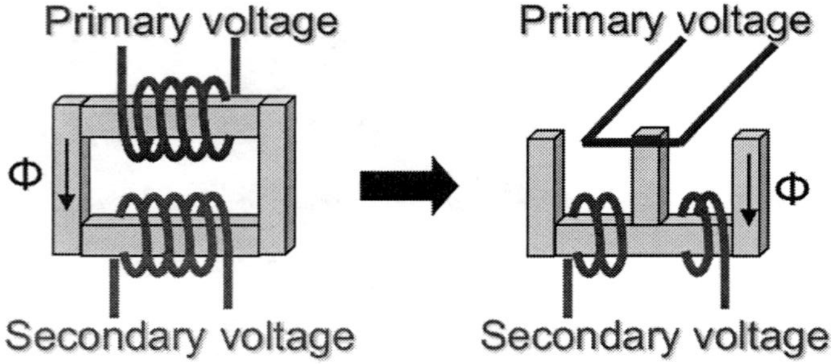

$U_P \sim f * N_P * \Phi$

$U_S \sim - f * N_S * \Phi$

<u>16 kW Track Supply Module</u>
- track current 125 A
- frequency 20 kHz

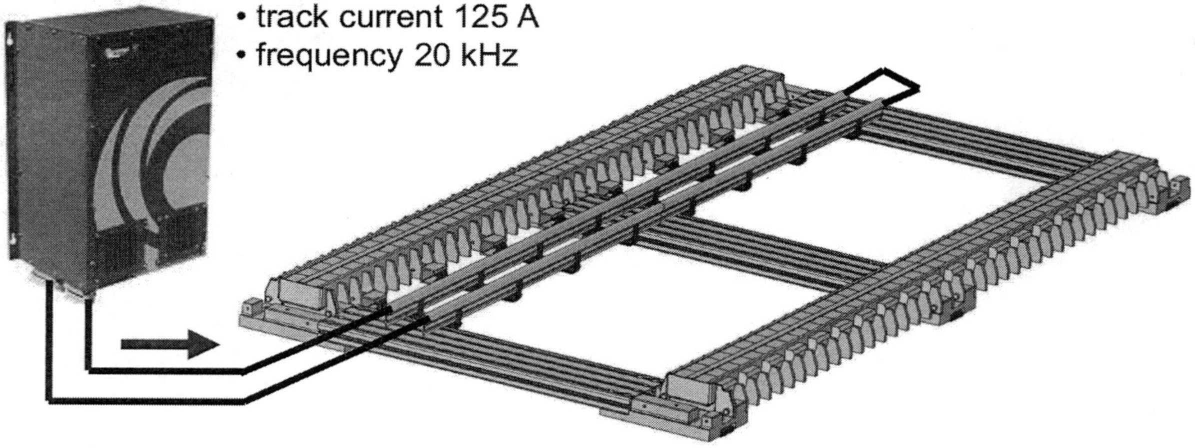

Preliminary conclusions

- Stable superconductive suspension on a magnetic track is possible
- High energy efficiency due to zero friction
- Noiseless and extremely soft motion

- Remaining issues:
  - Optimization of the permanent magnet track in relation to the shape of the superconductor:
    - reduction of size, weight, costs
    - cheaper magnets
  - Electromagnetic track, crossings, turnout switches
  - Non-liquid cryocooler

- Next demonstration projects:
  - people transportation
  - automation technology (cooperation with Festo)
  - hyper gravity – human centrifuge

Track-bound individual urban transportation

- **Requirements:**
  - magnetic railway
  - cabins for 4 or 5 persons
  - central control
  - central power supply (for the drive)
  - low energy consumption
  - fast turnout switches

- **Advantages:**
  - individual mobility
  - no parking space problem
  - effective use of vehicles
  - fast (compared to mass transportation)
  - service on demand, no fixed schedule
  - continuous night service
  - personal security

SupraTrans II

Superconducting magnetic levitation

Advantages for the construction of transport systems:

- ✓ Passively stable levitation
- ✓ Environmentally friendly, no CO_2 emission
- ✓ Absolutely noiseless
- ✓ 24 h use with just one cryostate filling
- ✓ Ice on the track: no problem

evico

STAATSMINISTERIUM FÜR WISSENSCHAFT UND KUNST

Freistaat SACHSEN

Large Human Centrifuge: System Considerations

- Gravity: 1 g vertically
- Rotation: 1 g horizontally (centrifugal force)
  rim speed must be proportional to SQR (radius)
  (i.e.: about 90 km/h at 150 m diameter)
- Resulting gravity: 1.4 g under 45 degrees
- Main problem for human beings: Coriolis force,
  proportional to SQR (1/radius) at constant
  centrifugal force
- Our proposal: mobile home on a circular magnetic track
  (diameter: 150 m)
  with superconducting suspension
  (soft bearing, comfortable, noiseless)

Thank you for your attention

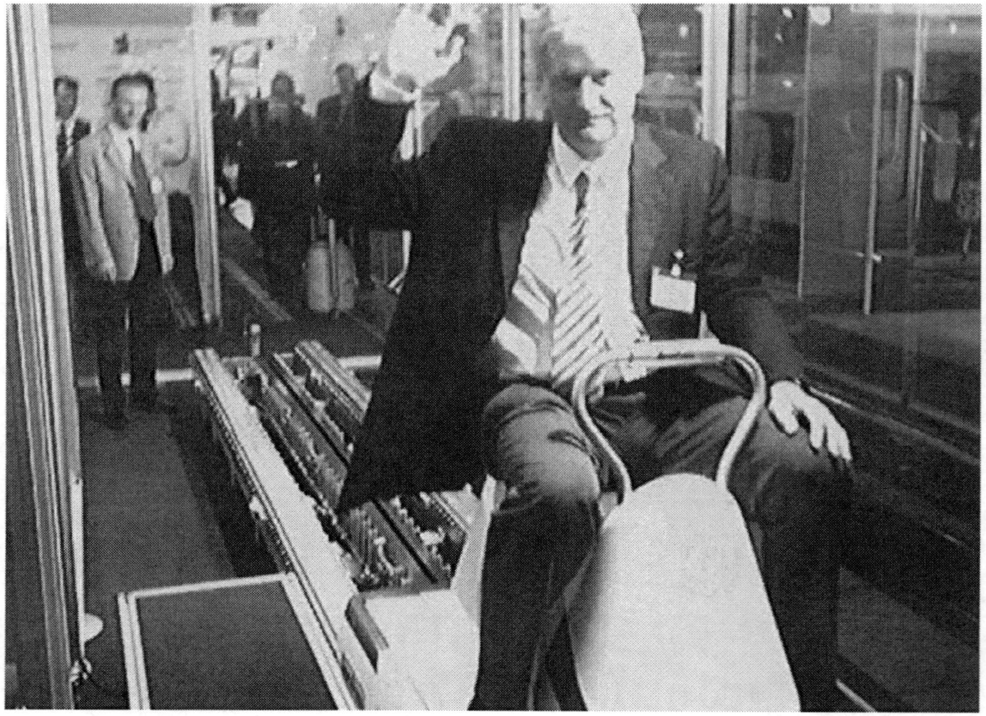

Acknowledgements

DFG, SMWK, SMWA, SAB, BMBF

G. Fuchs, K. Nenkov, B. Holzapfel, G. Krabbes

evico

Superconducting magnetic bearings for industrial applications

Dr. Oliver de Haas

Superconductivity

- Superconductivity (SL)
  - \+ Zero electric resistivity
  - \+ Zero losses
  - \- Sophisticated cooling
- High-temperature - superconductivity (HTSC)
  - \+ Zero electric resistivity
  - \+ Zero losses
  - \- Easy cooling

3 x smaller **3 x lighter** **efficient** **CO_2 reduction**

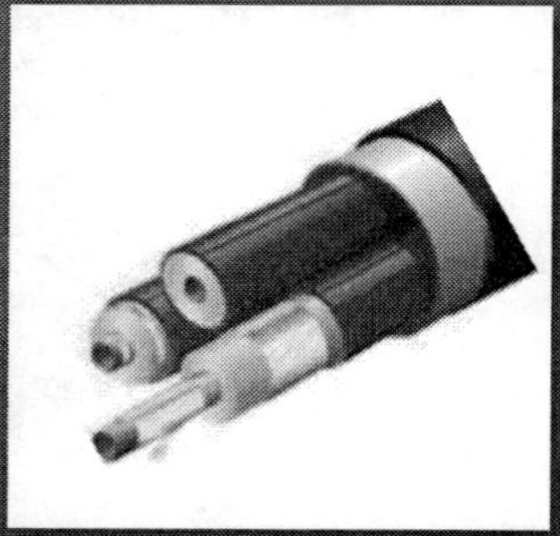

Cable

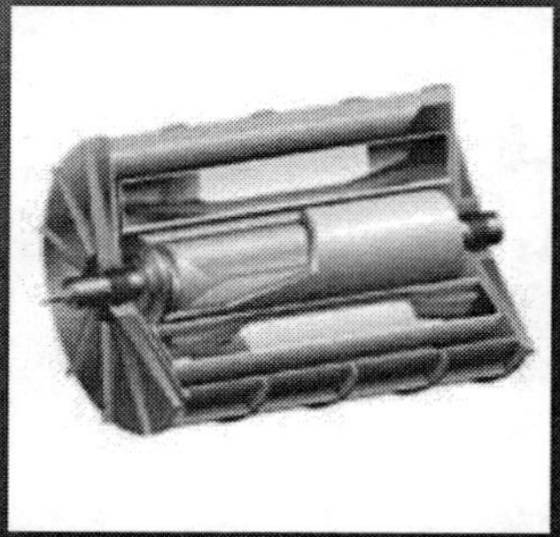

Motor

Trafo

Generator

Technische Anwendung und Markt

evico

Superconductor in a magnetic field

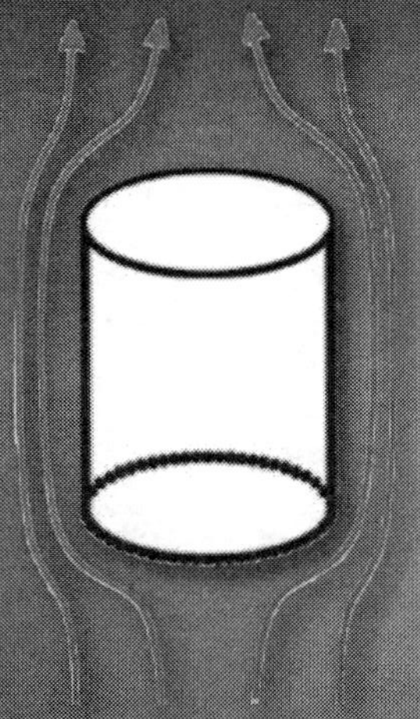

- Superconductor displaces magnetic field
- Meißner-Ochsenfeld-Effect or
- Ideal diamagnet
- **Magnetic field pushes out the Superconductor!**

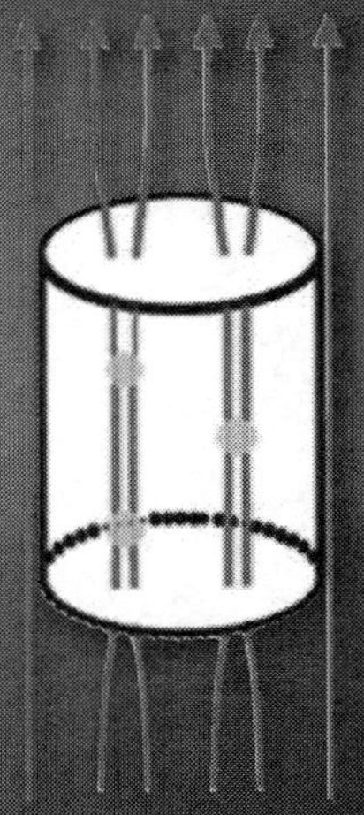

- Typ-II-Superconductor
- Magnetic field can penetrate the SC - Fluxlines
- Pinningeffect
- **Superconductor memorizes the field**

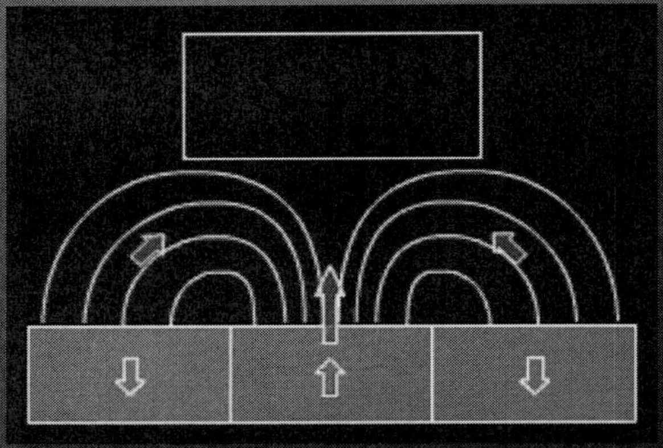

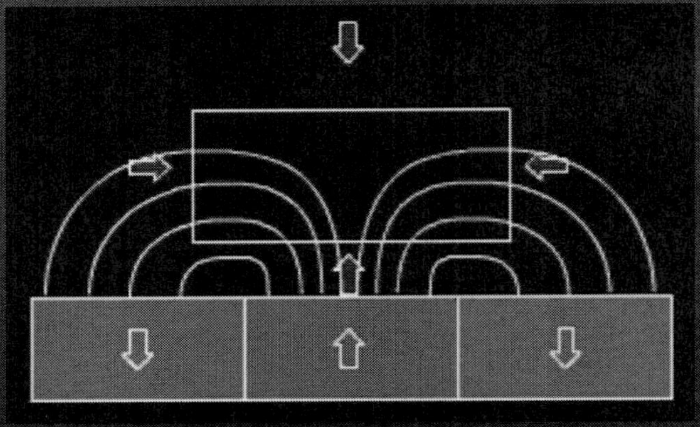

Field displacement and Fluxpinning

evico

Supraleiter im Magnetfeld

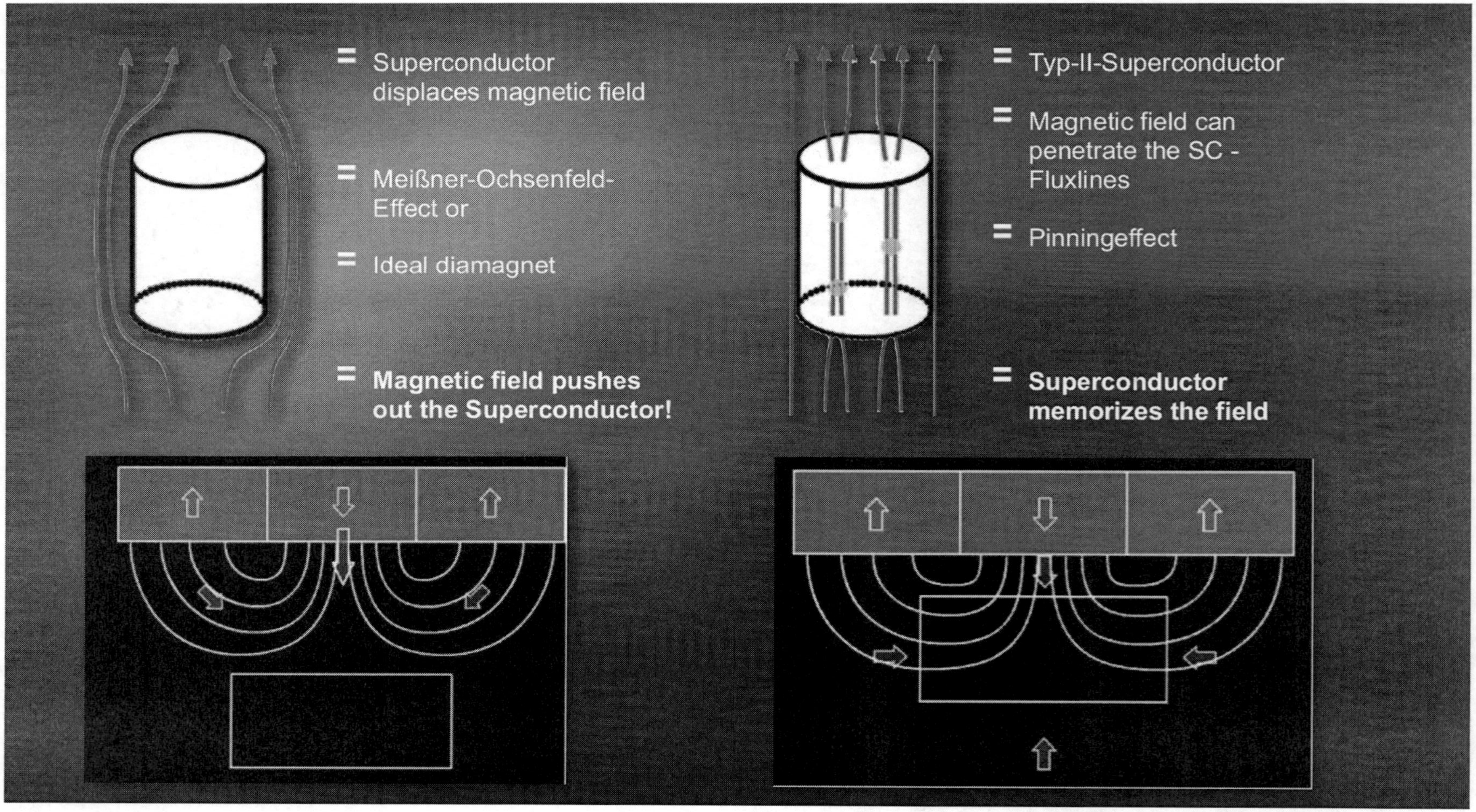

Field displacement and Fluxpinning

evico

Bearing types

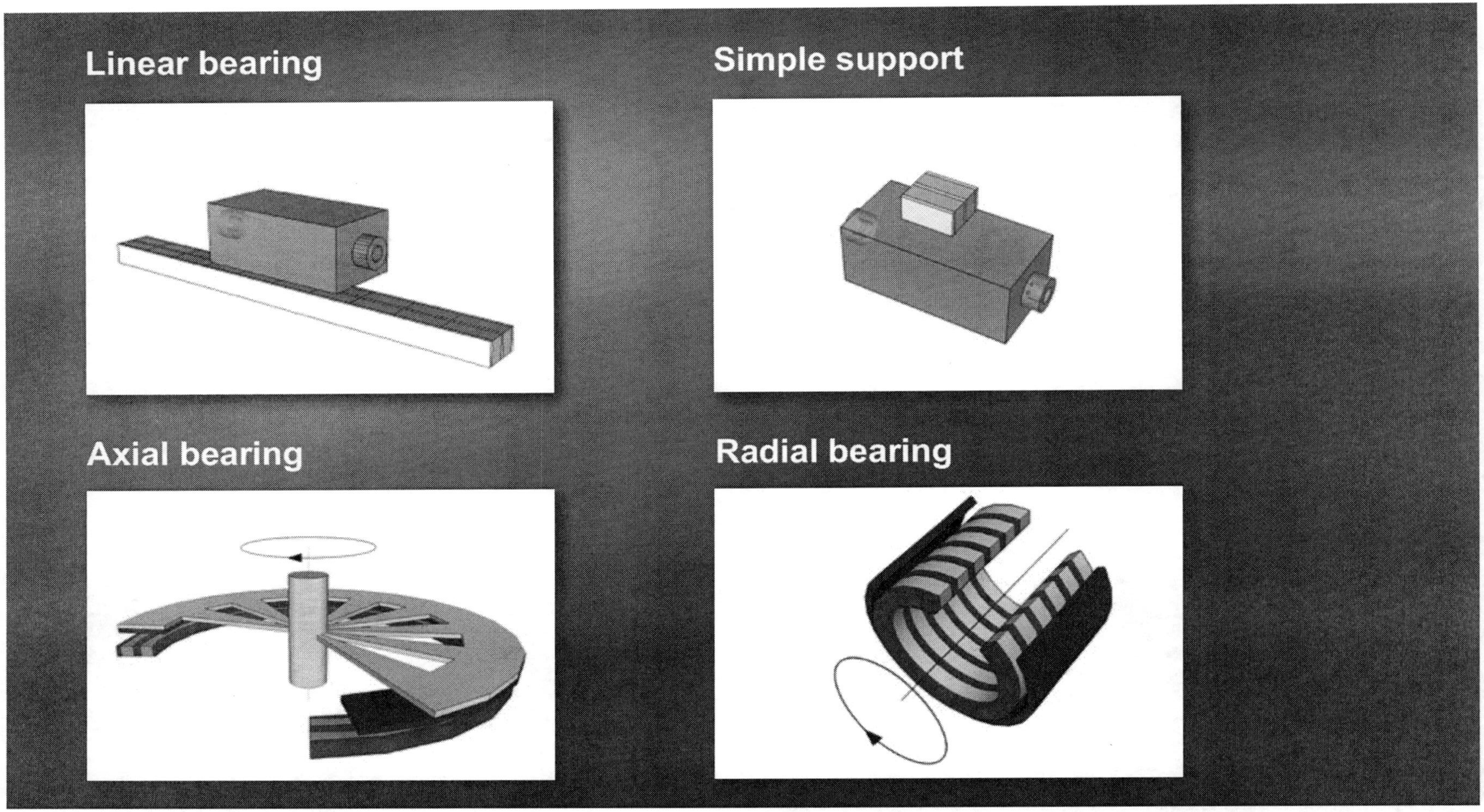

evico

Levitation Characteristicis

Linear Bearing

SC area: 120 x 30 mm

SC temperature: 77 K

Magnetic guidway width: 120

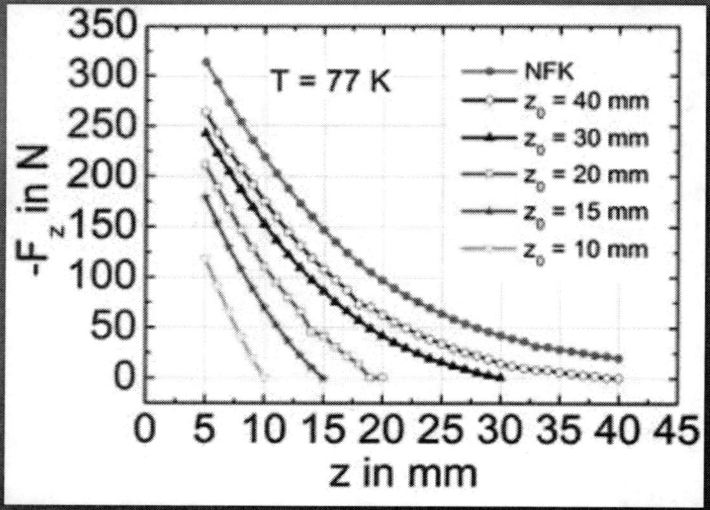

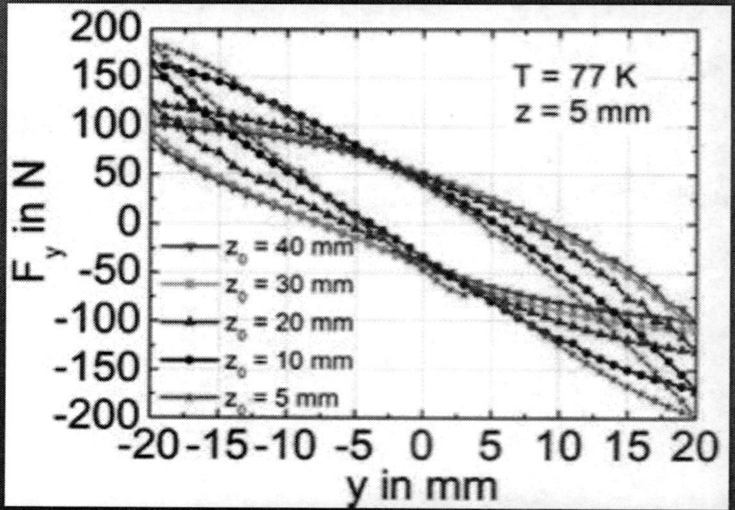

evico

SupraTrans II

SupraTrans

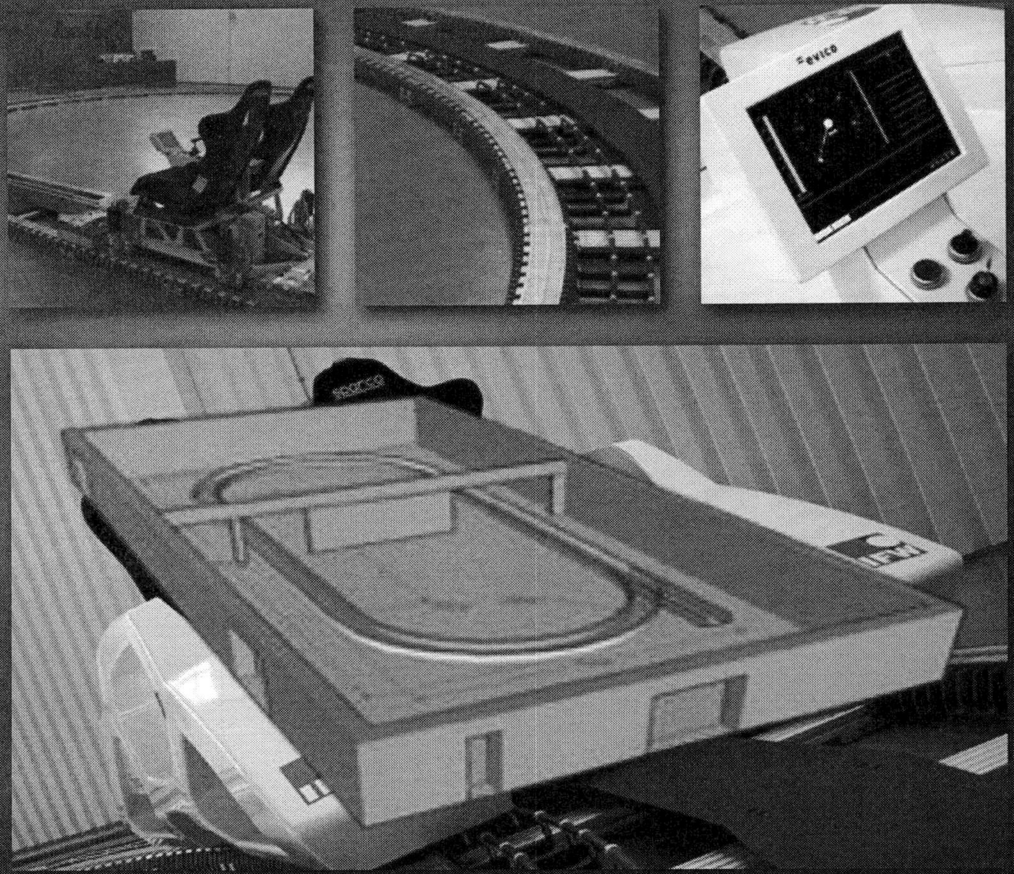

SupraTrans II

- 80 m closed loop
- Guideway switch
- Wireless energy transmission
- 200 kg pay-load
- 0,1 l /h LN2 consumption
- 1 m/s^2 acceleration
- 2 independent braking systems
- On bord / remote vehicle control

Design: Airport Gate-to-Gate Shuttle

Wir bauen Zukunft

Leibniz-Institut für Festkörper- und Werkstoffforschung Dresden

evico

Supraleiterlager + Manipulatoren

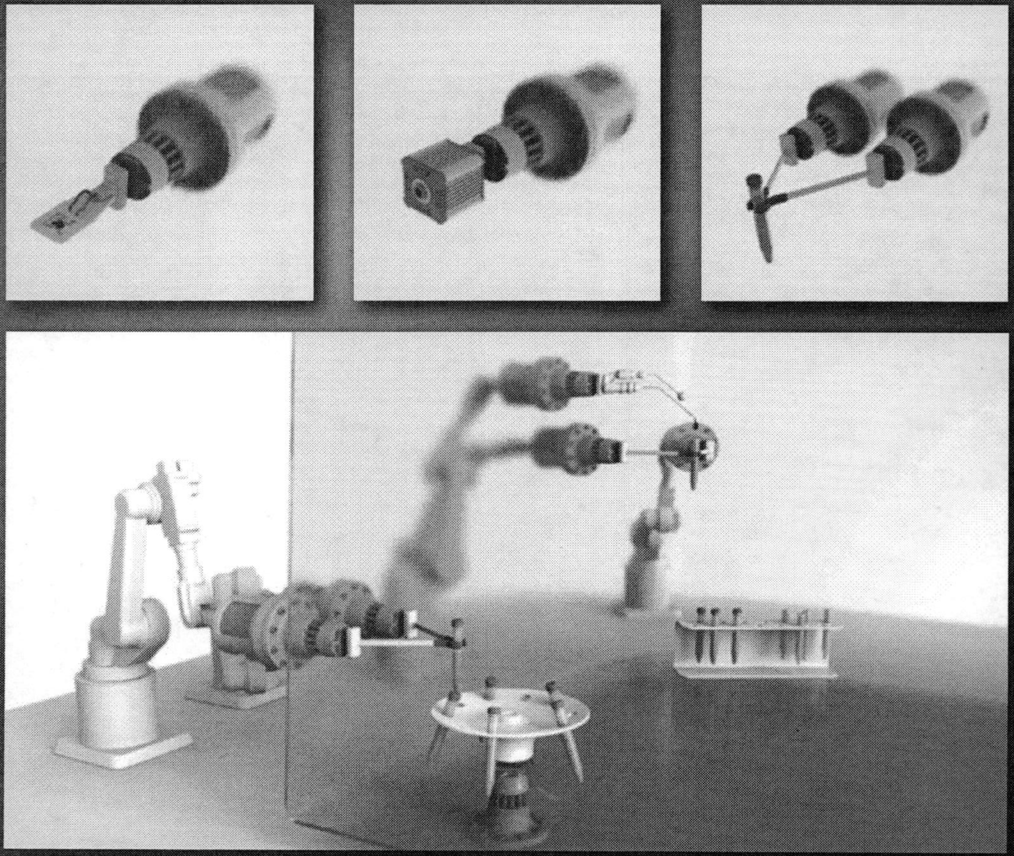

- Manipulation of samples
- Process control

10 Jahre evico

evico

Supraleiter-Kryostat

Working gap	5 mm	3 mm
Vertical stiffness	12 N/mm	21 N/mm
Lateral stiffness*	4 N/mm	6.3 N/mm
Superconductor area	261 mm²	
Cooling power at 67 K	1.6 W	
Minimum temperature	65 K	
Input power	60 W	
Voltage	28 V	
Dimensions (Diameter, Height)	ø 194 x 225 mm	
Weight cryosystem	5.2 kg	
Weight magnetic puck	0.67 kg	
Ambient temperature range	- 40 ... + 40 °C	

- compact
- plug & play
- efficient !

- Livetime to be improved

10 Jahre evico

evico

FESTO SupraMotion

- Search for scenarios of the application of superconducting bearings in industrial automation.
- SupraLinearMotion
- SupraHandling
- SupraPicker

10 Jahre evico

evico

FESTO SupraMotion 2.0

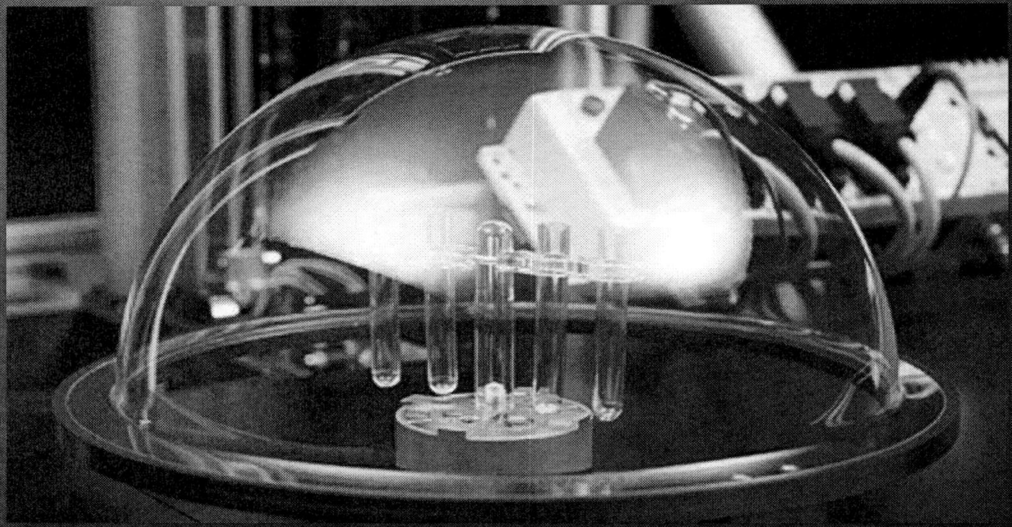

- Search for scenarios of the application of superconducting bearings in industrial automation.
- SupraHandling 2.0
- SupraShuttle
- SupraChanger

10 Jahre evico

evico

evico
Thank you
for your attention

Recent Development of Soft Magnetic Composite Materials and its Application

Dr. Lars-Olov Pennander

Höganäs AB, Bruksgatan, SE-26383 Höganäs, Sweden, lars-olov.pennander@hoganas.com

Abstract—**This paper explores recent material development for Soft Magnetic Composites (SMC) which are based on pure iron metal powders combined with inorganic resistive coatings. SMC components are manufactured by compaction in a shaped tool followed by heat-treatment of the component for stress relaxation and to reach optimal magnetic and mechanical performance. The recently developed Somaloy® 5P SMC materials are presented and compared to former technology. The unique combination of material properties and manufacturing process is here concluded to enable possibilities to realize compact and efficient high-frequency electromagnets, a technique that can be an interesting opportunity for large volume, cost efficient magnetic bearings and bearingless applications.**

I. Introduction

Soft magnetic components are used in electromagnets to provide a well-defined, efficient magnetic flux path that enables high performance of the application. Iron, cobalt and nickel are the main ferromagnetic base elements. Iron is the most common and also the least expensive of these three. The magnetic properties are focused to the base elements and the performance is to a large extent ruled by the purity and cleanness of the final engineering material.

The total losses connected to the magnetization are commonly subdivided into hysteresis-, eddy current- and the remaining anomalous losses. Eddy current losses are frequency dependent and proportional to the materials electric resistivity but depends also on the cross-sectional dimensions of the component perpendicular to the magnetic flux path. Electrically isolated steel-sheet lamination is the predominant solution to the eddy current problem.

The limitations of the lamination technique is the two-dimensional structure that results in sharp edge rectangular cross-sections of the soft magnetic core that also rules the process of winding and do often require quite thick bobbins to avoid short-circuits between the winding and the core. Another restriction is the cost for making thin laminations as the rolling process will be less productive and technically difficult when the lamina thickness decreases. Punching of thin laminas is also an area of extended challenges.

The idea to use metal particles as soft magnetic material for electromagnetic cores is not new; already in 1886 Fritts [1] applied for a patent of a generator using e.g. gutta-percha filled with iron particles as core material. A 'newborn' interest in SMC materials for machines started in the 1990s followed by intense material development to improve the competiveness of SMC. This development is driven by development of switch mode power supplies and demands on higher frequencies still providing commercial cost efficiency in volume applications. Typical development areas are high efficient traction drives for electric vehicles, other automotive applications forced by extended electrification mainly driven by efficiency demands, home appliances like HVAC pump drives and many other applications.

The introduction of SMC is, in the majority of cases, connected to new development since many traditional solutions based on rather low-cost laminations are difficult to challenge with the more engineered and also somewhat more costly SMC technology when just compared core-to-core in a direct replacement situation.

The SMC components will show a unique property profile and a successful SMC application should of course then make the best use of these special values. It can be of great help to focus on some simple rules to more efficiently judge the applicability of SMC to a specific application. At least the following five points are recommended to be considered when approaching SMC as core material for an electromagnetic application:

- The application works at higher frequencies and demands lowest losses
- The aim is efficient volume production
- Compact design is rewarded
- There is a value in savings of other components
- When it may be beneficial to integrate functionality by other geometrical features direct in the core body

The list implies that SMCs are not simply just new materials but also to a large extent add values that are uniquely connected to the forming and production process of components and systems. These aspects do also indicate the requirement of special competences that should be present to ensure a successful development work with SMC. The electric application engineer is advised to gain some understanding of the forming and production process of SMC powder components since this actually can be a critical point of the application development.

II. Modern Soft Magnetic Composite Materials

The first 'modern' SMC materials were also based on an organic matrix that bonds the particles and also provide the resistive coating. The resistive coating creates the bulk resistivity of the material. The organic matrix can positively supply quite a high mechanical strength and resistance to eddy currents in the SMC components. However, the disadvantage is the limited heat-treatment temperatures that are restricted to about 200-300°C. This temperature level is not enough to bring full stress relaxation of the deformed powder particles and will unconditionally result in relatively high hysteresis losses then taking out a part of the effect of the typically low eddy current losses.

A more attractive alternative is to use an inorganic particle surface coating system that can stand high temperatures and that allows for optimal relaxation of the particle for minimal hysteresis losses.

Another important condition is the lubrication during the press and compaction process. There are mainly two methods available; die wall lubrication (DWL) adding lubricant direct to the walls of the tool and/or internal lubrication by mixing a lubricant with the magnetic powder recognized as a premix.

The main benefits of using a premix is the much higher productivity and less variation in final properties of the component compared to DWL. It is also very important that the internal lubricant will add to preserve the resistivity of the particle coating else the eddy current protection will be lost and losses will increase.

Saturation induction of SMCs is quite high and the internal in-particle eddy current losses are significantly lower due to the smaller particle size, typically 5-400μm compared to the thickness of the steel sheet laminations of normally 100-1000μm. As a result of the isolating coating layer, the SMCs have distributed air gaps leading to lower permeability shown in Fig.1.

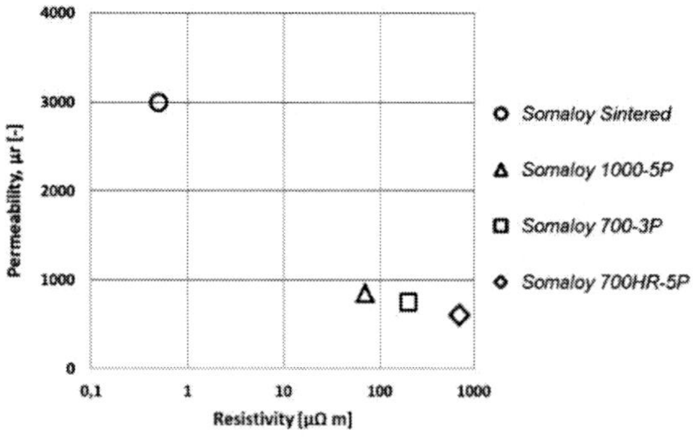

Fig.1 Relation of permeability to resistivity for a selection of Somaloy SMC grades with similar particle size.

The high pressures applied to the powder in the press-tool will deform the particles and will give high density. The softer the particle is the less effort is required to achieve high density. Alloyed or badly conditioned as well as small particles will be more difficult to deform into high densities. High quality SMC powder products have very good compressibility. Typical forming pressures for SMCs are in the range of 600MPa to 800MPa resulting in SMC component densities typically in the range of 7.3 g/cm^3 to 7.5 g/cm^3.

III. Recent Development of Somaloy® 5P Materials

Höganäs AB have since the beginning of the 1990s developed and produced the Somaloy® SMC grades based on pure iron particles combined with inorganic coatings. The magnetic properties have been improved significantly since

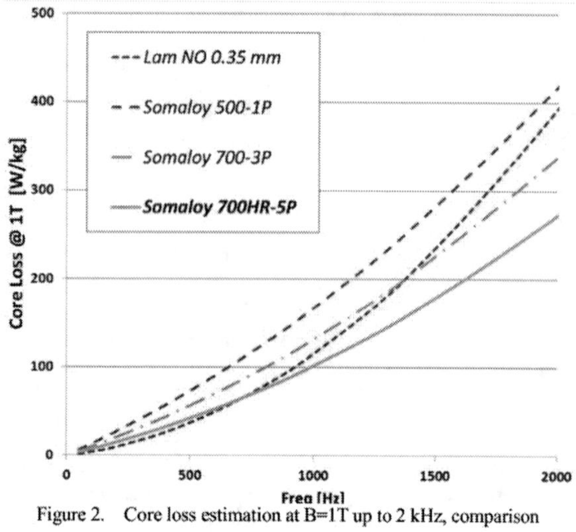

Figure 2. Core loss estimation at B=1T up to 2 kHz, comparison

the first introduction of Somaloy 500-1P in 1996, thanks to the focused development work done with the vision to supply state-of-art SMC products to the electromagnetic application market as shown in Table 1-3 and Fig.2.

The recently developed Somaloy® 5P technology [2] is based on a new inorganic particle coating that can be heat-treated at up to 650°C for optimal stress relief of the compacted particles. These high temperatures will drastically decrease hysteresis losses and thereby H_c from about 220 A/m for Somaloy 700-3P down to a level of 120 A/m for Somaloy 700HR-5P.

TABLE I. Overview of Somaloy® Grades up to 2KHz

Material	Strength TRS [MPa]	Perm. μr_max [-]	Resistiv. ρ [μΩ m]	Induction B@10kA/m [T]
Somaloy 500-1P	50	500	70	1.51
Somaloy 700-3P	125	750	200	1.61
Somaloy 700HR-5P	**60**	**600**	**700**	**1.57**
Somaloy 1000-5P*	**60**	**850**	**70**	**1.63**

*application range <400 Hz

TABLE II. Overview of Somaloy® High Frequency Grades

Material	Strength TRS [MPa]	Perm. μr_max [-]	Resistiv. ρ [μΩ m]	Induction B@10kA/m [T]
Somaloy 130i-1P	30	290	8000	1.40
Somaloy 130i-5P	**45**	**330**	**>20000**	**1.46**

TABLE III. Typical Loss Somaloy® High Frequency Grades

Material	Density P [kg/m^3]	Loss 1T/100Hz [W/kg]	Loss 0.1T/10kHz [W/kg]	Loss 0.1T/30kHz [W/kg]
Somaloy 130i-1P	7.35	12	27	110
Somaloy 130i-5P	**7.43**	**8**	**22**	**96**

The possibility to heat-treat the SMC component at high temperatures will also be mirrored in the maximum temperature of operation in the application. Somaloy components can quite easily stand constant temperatures up to about 200°C with no degrading over time. The particle coating will supply a basic resistance to corrosion similar to typical lamination steel sheets. In corrosive environment it is recommended to apply external corrosion protection.

Mechanical properties of SMC are different from laminated steel sheets. The strength is lower and the materials show a more brittle behavior that must be considered when handling the SMC components, performing secondary operations or assembling a structure. Machining of SMC components should be avoided due to the hard and brittle performances that may negatively influence the components properties. The best recommendation for assembly is to use gentle clamping and glueing.

IV. SMC Component Manufacturing and Design

SMC components of Somaloy® are formed and compacted into net-shape in a press with a specific tool set. The component design and how it will be pressed will set the specification of the press regarding complexity of tooling and maximum press force. The component design will have a direct impact on the component cost. Design-for-production aspects are recommended to be implemented as early as possible in the design phase.

The here described SMC materials are typically compacted at pressure levels of about 800 MPa or higher even up to levels of 2000 MPa in extreme cases. At these conditions the lubrication both internally in-between the particles and to the walls of the tool is very critical. Some lubrication systems like the Somaloy 3P will also require controlled temperature of the die to deliver optimal performance. The lubricant occupies some volume of the powder mix and high lubricant content will reduce the final density of the component.

It is popular to describe SMC materials as 3-dimensional but in fact there are two features covered in this designation. First is the almost isotropic material property that gives similar magnetization conditions in all directions. Secondly is the feature of 3-d forming of components that is connected to the production process. This means in practice that a SMC component with simple shape can utilize an advanced magnetic flux path. A too complicated 3-d shaped component may require quite complex and more costly tooling and press process.

When opening for 3-d flux path designs that are possible with the almost isotropic SMC materials will bring new design alternatives and new possibilities to optimize not only the magnetic core but also the winding.

A popular SMC feature is the 3-d 'field concentration' as shown in Fig. 3, which can allow adaption between a large permanent magnet area and a smaller cross-section area of the core in the coil. This feature combined with a more circular core shape can reduce the copper winding length significantly and the net result is decreased winding losses and also reduced winding cost. Flux concentration can open up for Ferrite magnets as a low-cost alternative.

Some machine topologies are designed for 3-d flux path's and thereby specially aimed to utilize SMC components.

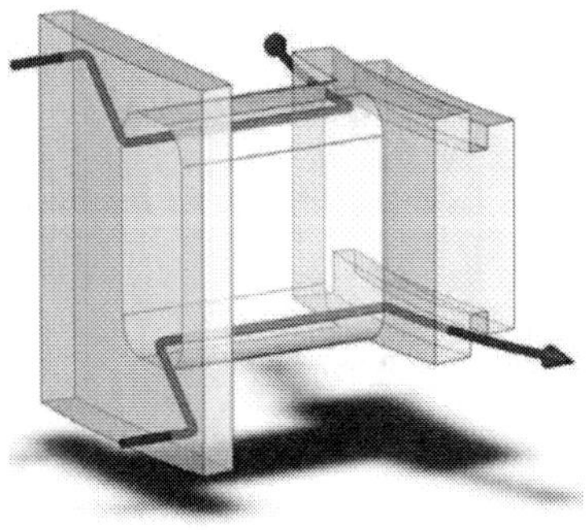

Figure 3. 3-dimensional flux concentration in SMC component..

Figure 4. Clawpole TFM topology

Examples is the transverse flux machine (TFM) of clawpole type [3] shown in Figure 3 with modulated poles characterized by a single hoop winding and a large number of poles resulting in high torque densities.

V. SMC Potential For Magnetic Bearings or Bearingless Applications

The unique combination of 3-dimensional resistance to high frequency eddy-currents can bring potential benefits using SMC compared to laminated steel-sheet cores; the low eddy-currents at higher frequencies can potentially increase bandwidth for a magnetic bearing system as described by Hijikata et al. [4]. Utilization of 3-d flux patterns can result in space efficient solutions like the combined radial and axial magnetic bearing described by Kim and Kim [5].

VI. Conclusions

Recent improvements of inorganic coating systems for Somaloy® 5P SMC materials have resulted in significantly reduced coercive force and hysteresis losses. Bulk resistivity remains at high level and the result is a non-alloy cost-efficient iron SMC material with losses competing <0.35 mm steel-sheet laminations.

For the application of these materials it is found important to consider both material and component forming processes early in the design work. It is also observed that a major part of the component properties are set in the manufacturing process, at compaction and heat-treatment that also must be seriously considered in the component development. Further, some examples are identified showing potential improvement regarding bandwidth and compact, space efficient designs beneficially applicable for magnetic bearing systems.

REFERENCES

[1] C. E. Fritts, "Apparatus or means for the Production, Transmission and Distribution of Electric Currents," *Pat. Appl. US278.866*, pp. 11–12, 1886.

[2] Z. Ye, M. Lenberg and C. Pompermaier, "A New Generation of SMC Materials with Low Core Loss," *Proceedings of Powder Metallurgy World Congress PM2012, Yokohama, 2012.*

[3] P.G. Dickinson, A.G. Jack and B.C. Mecrow, "Improved Permanent Magnet Machines with Clawpole Armatures", *Proceedings of ICEM 2002*, 2002.

[4] W. Hijikata et al, "Cost-effective Magnetic Bearing System using a Single-Use Magnet-Less Rotor for an Extracorporal Centrifugal Blood Pump," *Proceedings of The 12th Int. Symp. On Magnetic Bearings ISMB12*, pp. 654-659, 2010.

[5] H.Y. Kim and S.J. Kim, "An Effective Way to Combine Radial and Axial Magnetic Bearings in a Unit," *Proceedings of The 10th Int. Symp. On Magnetic Bearings ISMB10*, 2006.

Radial force modeling for bearingless motors based on analyze of field origins

Blaise Lapôtre[a], Noureddine Takorabet[a], Farid Meilbody-Tabar[a], Ramdane Lateb[b], Joaquim Da Silva[b]

[a] Université de Lorraine, GREEN, 2 av. de la forêt de Haye, 54516 Vandœuvre-lès-Nancy, France, Blaise.lapotre@univ-lorraine.fr
[b] SKF Magnetic Mechatronics 2 rue des champs, 27250 Saint Marcel, France

Abstract— **This paper proposes a global harmonic analysis of a multiphase bearingless PM-motor topology to determine the radial force by considering normal and tangential field components. This allows establishing the semi-analytical expressions of the radial force considering the location of field origin. The accuracy of the model is verified by full FEM model using a statistic approach. The semi-analytic model allows predicting the force waveforms according t the stator-rotor topology. The analysis of two specific topologies is performed and highlights the origin of force ripples. The fastness of this semi-analytical model makes it suitable to implement in a design optimization process.**

A. Introduction

In the last decade, high speed motors and active magnetic bearings have been widely used in high pure or wear-free applications. One challenging solution is to use bearingless motors, providing both functions, torque as well as levitation force generation with the same machine.

The principle of bearingless motors has been adapted for different motor topologies: PM motor [1], reluctance machine [2] or induction motor [3]. The main idea consists in creating levitation forces by using two consecutive field space harmonics of ranks p and p + 1 [4]. The field is either created by windings or magnets according to the adopted motor topology.

Several techniques have been developed to determine the radial forces in electric machines. In the particular case of bearingless motors many authors have proposed interesting contributions for different topologies. For toothless machines the Lorentz force can be considered [5]. Another way consists in using magnetic co-energy method to evaluate the forces [2]. An alternate approach is to use the Maxwell stress tensor method in the air gap [6].

The analytical calculation of radial forces and torque in bearingless machines is quite complex when field harmonics should be taken into account. In some cases, it is possible to assume some assumptions to make the calculation easier. Some authors have proposed to neglect the tangential component of the field in the air-gap [7]. This approach is appropriate for small air-gap machines and when normal component of the field is dominant. This is the case of induction machines or variable reluctance machines [2].

Other authors prefer eliminate directly the harmonic by winding configuration or topology but without consideration presumed parasitic force norm or direction of [8-9].

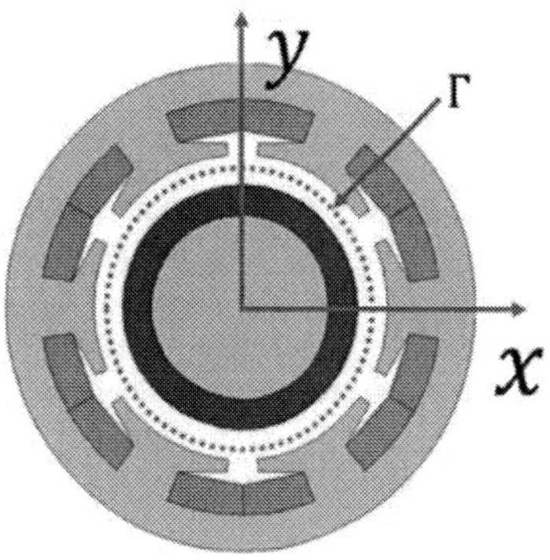

Figure 1 : Six slot bearingless motor geometries

Under other conditions we can establish an analytical model with a large number of adjustable parameters [10] or with utilization of force calculated by Maxwell stress tensor and Lorenz force [11].

In this paper, the authors propose an analytical model based on the spectral decomposition of the fields created by different sources (stator winding, rotor magnets, ...). The calculation of the forces is performed by using the Maxwell stress tensor expressed in terms of different harmonics of the normal and tangential components of the flux density in the air-gap. In order to simplify the expressions and reduce the number of parameters, the magnetic vector potential formulation has been introduced. This approach allows evaluating the contribution of each couple of consecutive harmonics to the total levitation force. Thus, the design process of these machines is made more straightforward and easier.

B. Radial force formulation

As mentioned above [4], the radial force components of bearingless motor are resulting from the interaction of magnetic consecutive field space harmonics of ranks p and p + 1. Using Maxwell tensor method on contour Γ figure 1, the projection of the force components in two axes (x, y) fixe relative to the stator are given in equation (1):

$$F_x = \frac{r.L_z}{2\mu_0}\int_0^{2\pi}[(B_n^2 - B_t^2)\cos\theta_s - 2B_tB_n\sin\theta_s]d\theta_s$$
$$F_y = \frac{r.L_z}{2\mu_0}\int_0^{2\pi}[(B_n^2 - B_t^2)\sin\theta_s + 2B_tB_n\cos\theta_s]d\theta_s \qquad (1)$$

B_n and B_t are respectively the normal and tangential components of the air gap flux density. With L_z , r respectively, the iron length and the radius in the middle of the air-gap, θ_s is the angular coordinate of an arbitrary point on the considered contour and μ_0 is the permeability of vacuum.

Using Fourier transformation, normal and tangential components of the flux density can be expressed as follows:

$$\begin{cases} B_{n\varrho}(\theta_{n\varrho},\theta_s) = \frac{1}{2}\sum_k |Bn_{\varrho,k}| e^{j\theta_{n\varrho,k}} e^{jk\theta_s} + |Bn_{\varrho,k}| e^{-j\theta_{n\varrho,k}} e^{-jk\theta_s} \\ B_{t\varrho}(\theta_{t\varrho},\theta_s) = \frac{1}{2}\sum_k |Bt_{\varrho,k}| e^{j\theta_{t\varrho,k}} e^{jk\theta_s} + |Bt_{\varrho,k}| e^{-j\theta_{t\varrho,k}} e^{-jk\theta_s} \end{cases} \quad (2)$$

The subscript ϱ indicates the nature of the considered field source which can be either permanent magnets or the current distributions.

$|Bn_{\varrho,k}|$ and $|Bt_{\varrho,k}|$ are the amplitudes of the kth harmonics of the normal and tangential components of the air-gap flux density. $\theta_{n\varrho,k}$ and $\theta_{t\varrho,k}$ are their phases.

By neglecting the saturation of the iron core, the superposition principle can be applied. The combination of equations (1) and (2) allows determining the components F_x and F_y of the force which can then be expressed by:

$$\begin{cases} F_x = \frac{r.L_z\pi}{2\mu_0}\sum_{\varrho,\varsigma,k} FX_{k,\varrho,\varsigma} \\ F_y = \frac{r.L_z\pi}{2\mu_0}\sum_{\rho,c,k} FY_{k,\varrho,\varsigma} \end{cases} \quad (3)$$

The components $FX_{k,\varrho,\varsigma}$ and $FY_{k,\varrho,\varsigma}$ are the elementary forces generated by the interaction of two successive harmonics of rank k and k+1 of the flux density given in (2) due to the same source or different ones ϱ,ς.

$$\begin{aligned} FX_{k,\varrho\varsigma} = & |B_{n\varrho,i}B_{n\varsigma,i+1}| cos(\theta_{n\varrho,i} - \theta_{n\varsigma,i+1}) \\ & - |B_{t\varrho,i}B_{t\varsigma,i+1}| cos(\theta_{t\varrho,i} - \theta_{t\varsigma,i+1}) \\ & - |B_{n\varrho,i}B_{t\varsigma,i+1}| sin(\theta_{n\varrho,i} - \theta_{t\varsigma,i+1}) \\ & - |B_{t\varrho,i}B_{n\varsigma,i+1}| sin(\theta_{t\varrho,i} - \theta_{n\varsigma,i+1}) \end{aligned} \quad (4)$$

$$\begin{aligned} FY_{k,\varrho\varsigma} = & |B_{n\varrho,i}B_{n\varsigma,i+1}| \sin(\theta_{n\varrho,i} - \theta_{n\varsigma,i+1}) \\ & - |B_{t\varrho,i}B_{t\varsigma,i+1}| \sin(\theta_{t\varrho,i} - \theta_{t\varsigma,i+1}) \\ & + |B_{n\varrho,i}B_{t\varsigma,i+1}| \cos(\theta_{n\varrho,i} - \theta_{t\varsigma,i+1}) \\ & + |B_{t\varrho,i}B_{n\varsigma,i+1}| \cos(\theta_{t\varrho,i} - \theta_{n\varsigma,i+1}) \end{aligned} \quad (5)$$

The subscripts ϱ and ς denote the nature of the considered field sources which can be either permanent magnets or the current source.

The main idea is to reduce the number of parameter in equations (4) and (5). By considering the magnetic vector potential MVP A on the contour Γ, the Fourier decomposition of this potential is:

$$A_\varrho = \frac{1}{2}\sum_k A_{\varrho,k}e^{ik\theta_s} + A^*_{\varrho,k}e^{-ik\theta_s} \quad (6)$$

where

$$A_{\varrho,k} = |A_{\varrho,k}|e^{i\theta_{\alpha,k}} \quad (7)$$

By the derivation of potential the tangential and normal component are given by:

$$\begin{aligned} B_{n\varrho} = -\frac{1}{r}\frac{\partial A}{\partial \theta_s} = \frac{1}{2}\frac{1}{r}\sum_k & k|A_{\varrho,k}|e^{i(\theta_{\varrho,k}+\frac{\pi}{2})}e^{ik\theta_s} \\ & + k|A_{\varrho,k}|e^{-i(\theta_{\varrho,k}+\frac{\pi}{2})}e^{-ik\theta_s} \end{aligned} \quad (8)$$

and

$$\begin{aligned} B_{t\varrho} = -\frac{\partial A}{\partial r} = \sum_k & -\frac{\partial|A_{\varrho,k}|}{\partial r}e^{i\theta_{\varrho,k}}e^{ik\theta_s} \\ & -\frac{\partial|A_{\varrho,k}|}{\partial r}e^{-i\theta_{\varrho,k}}e^{-ik\theta_s} \end{aligned} \quad (9)$$

By identification is possible two establish a relationship between tangential and normal field phase. Depending on the sign of $\partial|A_{\varrho,k}|/\partial r$ two cases are possible:

- $\theta_{\varrho t,k} = \theta_{\varrho n,k} - \frac{\pi}{2}$ when $\partial|A_{\varrho,k}|/\partial r$ are positive, the field source is located outside the air gaps contour
- $\theta_{\varrho t,k} = \theta_{\varrho n,k} + \frac{\pi}{2}$ when $\partial|A_{\varrho,k}|/\partial r$ are negative, in this case the field source is located inside the air gaps contour.

Introducing the phase relationship given above in both equations (4) and (5), $FX_{i,\varrho\varsigma}$ and $FY_{i,\varrho\varsigma}$ can be simplified as follows:

$$\begin{cases} FX_{i,\varrho\varsigma} = K\,cos(\theta_{n\varrho,i} - \theta_{n\varsigma,i+1}) \\ FY_{i,\varrho\varsigma} = K\,sin(\theta_{n\varrho,i} - \theta_{n\varsigma,i+1}) \end{cases} \quad (10)$$

The expression of the coefficient K depends on the location of the field sources with respect to the contour: inside the contour or outside the contour. The four possible cases are summarized in Table I.

The expressions given in Table I show the impact of different field space harmonics on the mean value or the harmonics of the radial force. This can be used for improving the design of bearingless motor. Illustrative examples will be given in the following sections.

TABLE I COEFFIENT (K) EXPRESSIONS

	$B_{i,\varrho external}$	$B_{i,\varrho internal}$
$B_{i+1,\varsigma external}$	$(\lvert B_{n\varsigma,i+1}\rvert + \lvert B_{t\varsigma,i+1}\rvert).(\lvert B_{n\varrho,i}\rvert - \lvert B_{t\varrho,i}\rvert)$	$(\lvert B_{n\varsigma,i+1}\rvert + \lvert B_{t\varsigma,i+1}\rvert).(\lvert B_{n\varrho,i}\rvert + \lvert B_{t\varrho,i}\rvert)$
$B_{i+1,\varsigma internal}$	$(\lvert B_{n\varsigma,i+1}\rvert - \lvert B_{t\varsigma,i+1}\rvert).(\lvert B_{n\varrho,i}\rvert - \lvert B_{t\varrho,i}\rvert)$	$(\lvert B_{n\varsigma,i+1}\rvert - \lvert B_{t\varsigma,i+1}\rvert).(\lvert B_{n\varrho,i}\rvert + \lvert B_{t\varrho,i}\rvert)$

C. determination of harmonic evolution

N-phase, 2-pole Permanent Magnet motor is considered. The stator winding is composed of n concentrated coils, one coil for one tooth.

The stator phase currents (i_j; $j=1$ to n) is the superposition of two n-phase current systems (I1 and I2) defined by.

$$\begin{cases} i_{j,1} = I_1 \cos\left(\omega_1 t + \phi_1 + \dfrac{(j-1)\pi}{n}\right) \\ i_{j,2} = I_2 \cos\left(\omega_2 t + \phi_2 + \dfrac{2(j-1)\pi}{n}\right) \end{cases} \quad (11)$$

In the general case, I_1, I_2, ω_1, ω_2, ϕ_1, and ϕ_2 represent the amplitudes, frequencies and the phases of the two systems respectively.

The current system I1 generates a 2-pole magnetomotive force (mmf) and the current system I2 generates a 4-pole mmf.

According to the multiphase winding theory, the two n-phase systems I1 and I2 generate forward harmonics of the air-gap flux density of ranks $nm+1$ and $nm+2$ and backward harmonics of ranks $nm-1$ and $nm-2$.

The phase angles of these space harmonic depend directly of current phase for forward harmonics the evolution are in same way while the evolution are reverse for backward harmonic

As well as the norm of every harmonic have a linear dependence to the current norm.

$$\begin{cases} |Bn_{\varrho,k}| = K_{n,k}.I_\varrho & \theta_{n\varrho,k} = \theta_{0n\varrho,k} \pm \omega_\varrho t \\ |Bt_{\varrho,k}| = K_{t,k}.I_\varrho & \theta_{t\varrho,k} = \theta_{0t\varrho,k} \pm \omega_\varrho t \end{cases} \quad (12)$$

The magnets create a radial flux density which harmonic spectrum depends on the size of the magnets and the nature of the magnetization. If the slot effect is neglected the amplitudes of the flux density harmonic are constant while their phases increase linearly with the speed of the rotor and the harmonic rank (13).

$$\begin{cases} |Bn_{m,k}| = Cst & \theta_{nm,k} = \theta_{0nm,k} + k\omega t \\ |Bt_{m,k}| = Cst & \theta_{tm,k} = \theta_{0tm,k} + k\omega t \end{cases} \quad (13)$$

These considerations can estimate the flux density and determine the force for any position of rotor and current configuration by performing only one FE calculation for each field sources.

D. validation of model by statistic analyses

In order to validate the developed model, it is necessary to test it on a large number of machines whose geometrical parameters Figure 2 vary in a 5-D space corresponding to geometrical parameters and for different scenarios of operation.

The 5-D space of parameters is presented in Table II which gives the upper and lower bounds of each geometrical parameter.

The different scenarios of operation are summarized in Table III. They concern mainly the values of the currents, their phases and the rotor position.

The number of geometric and electrical parameters leads to a 10-D space. Therefore a statistical approach comparing FE and semi-analytical results is adopted. It consists in randomizing a machine and an operating point and computing their performances with both models. The comparison is focused on the radial forces determined by Maxwell stress tensor in the air-gap. This process is performed on 1000 machines randomly chosen in the 10-D space.

TABLE II INTERVAL OF TOPOLOGY VARIABLES

Parameter	Interval
Slot and phase number	5 or 6
Stator inner radius R_a	10 at 40 mm
Air gap e	1 at 3mm
Thickness of magnet a	1 at 5 mm
Slot opening λ	0% to 33 %

TABLE III INTERVAL OF ELECTRICAL VARIABLES

Parameter	Interval
Rotor position	0 to 2π rad
Norm of current system I1 : I_1	0 to 400 At
Phase of current system I1 : Φ_1	0 to 2π rad
Norm of current system I2 : I_2	50 to 550 At
Phase of current system I2 : Φ_2	0 to 2π rad

The relative errors between forces computed by the two methods are calculated in terms of norm and direction. Figure 3 shows the corresponding histogram of the errors which have Gaussian distributions.

For the direction of the force, the errors do not exceed ±1.5°. The evaluation of normal distribution gives a mean value of -0.002° with a variance of 0.011°. For the norm of the force, the errors is less than 5% and the normal distribution is characterized by a mean value of 0.4% and variance of 0.062%.

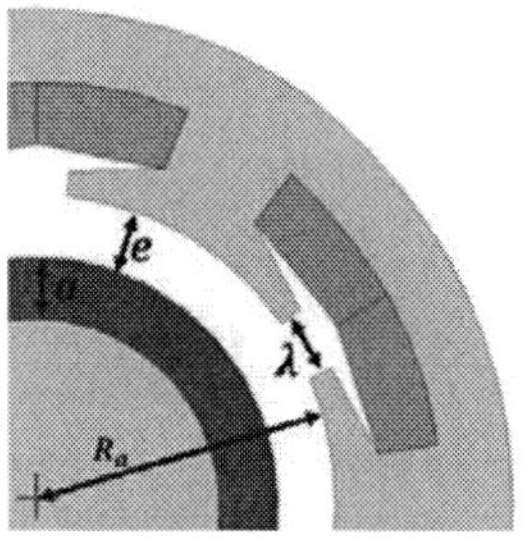

Figure 2 : Representation of topology variable

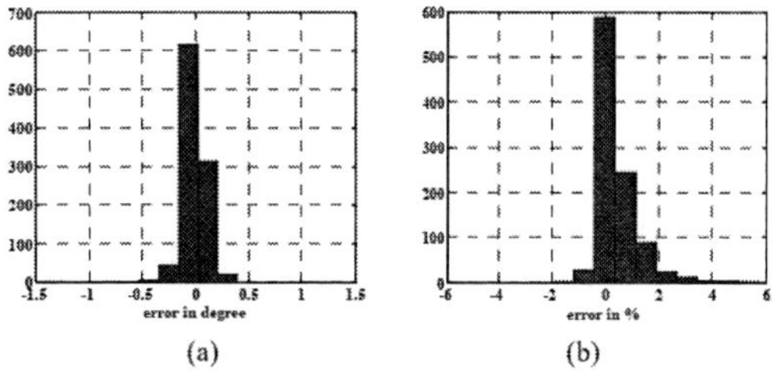

Figure 3 : (a) histogram of angle error in direction of force between semi analytical model and FEM (b)histogram of relative norm error of force between semi analytical model and FEM

These two results allow validating this statistical approach in terms of size of population (1000). Consequently, the proposed semi-analytical model is validated by means of full FE calculation.

E. Example of a bearingless 6 phase PM-motor

We consider a 6-phase, 2-pole permanent magnet motor. The stator winding is composed of 6 concentrated coils considered like external sources of field. The rotor is composing of one magnet ring with radial or diametrical magnetization Figure 4.

Their configuration create harmonic of flux density in air-gap depending on their sources (magnet , system current I1 and I2) according to equation (12) and (13) these existing harmonics and their frequencies are summarized in Table IV and V. In this tables we find the different pairs of harmonics which interact and generate radial force components. They can be gathered in three groups highlighted with the three colors (red, blue, green) shown in Table IV and V. They correspond also to the different cases listed in Table I:

- red: force component which norm is the product of two sums,
- blue: forces component which norm is the product of two subtractions,
- green: forces which norm is product of subtraction and sum.

<u>Case of parallel magnetized rotor :</u>

The Table IV allows highlighting the interactions between the different field harmonics. The general form of the elementary force due to a couple of field harmonics of generate by currents is given as bellow:

$$K\cos(\omega_1 t - \omega_2 t + \phi) \qquad (14)$$

where ϕ is the shift phase between the two field harmonics.

The interaction between the fields generated by the magnets and the currents can be resumed in one interaction of 2 first harmonic with equation:

$$K\cos(\omega t - \omega_2 t + \phi) \qquad (15)$$

It should be reminded that to have a constant torque, the electric pulsation of the system I1 and the mechanical pulsation must be equal $\omega = \omega_1$ (the rotor has one pole pair,

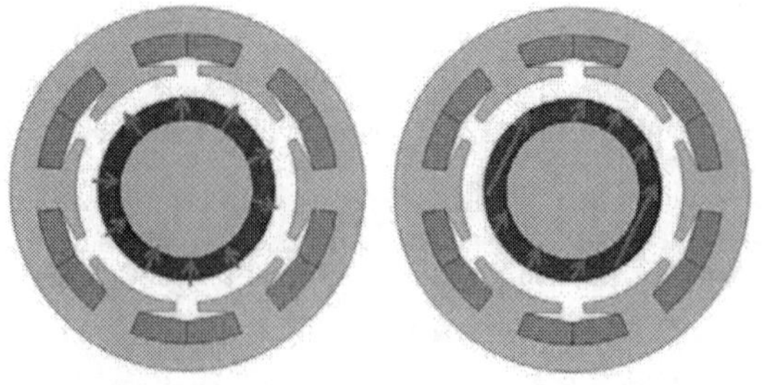

(a) Radial (b) Parallel

Figure 4 : Magnetization topologies of the rotor

Table IV). In addition, to obtain a constant levitation force it is necessary to have $\omega_1 = \omega_2$ (equ. 13). Figure 5 Shows the force components F_x and F_y obtained by the proposed semi-analytical model with the 16 first harmonics and by a full FEM computation. There is an excellent accordance between the two models; then the semi-analytical model is perfectly validated for parallel magnetized rotors.

<u>Case of radial magnetized rotor :</u>

As mentioned above Radially magnetized rotor generates flux density in the air-gap with odd space harmonics of frequencies $(2m + 1)\omega$. These field harmonics interact with the field harmonics generated by the current system I2 which are shown in red and blue in Table V. The general form of the elementary forces due to a couple of field harmonics is in the following form:

$$K\cos(\mp(2m + 1)\omega t \mp \omega_2 t + \phi) \qquad (16)$$

On Figure 6, we show the F_x and F_y component of the levitation force obtained in the case of radial magnetized rotor As in the case of parallel magnetized rotor, the mechanical pulsation and the electrical pulsation of the current system I_2 must be equal to $\omega = \omega_1 = \omega_2$. In this condition, contrary to the case of parallel magnetized rotor, the levitation force has not only a constant component, it contains also the harmonics as of frequency 2mω.

TABLE IV HARMONIC PULSATIONS FOR 6 PHASE AND RADIAL (IN READ) OR PARALLEL MAGNTIZATION

Root field	Space harmonium								
	1	2	3	4	5	6	7	8	9
I1	ω_1				$-\omega_1$		ω_1		
I2		ω_2		$-\omega_2$				ω_2	
Magnet	ω								

TABLE V HARMONIC PULSATIONS FOR 6 PHASE AND RADIAL (IN READ) OR PARALLEL MAGNTIZATION

Root field	Space harmonium								
	1	2	3	4	5	6	7	8	9
I1	ω_1				$-\omega_1$		ω_1		
I2		ω_2		$-\omega_2$				ω_2	
Magnet	ω		3ω		5ω		7ω		9ω

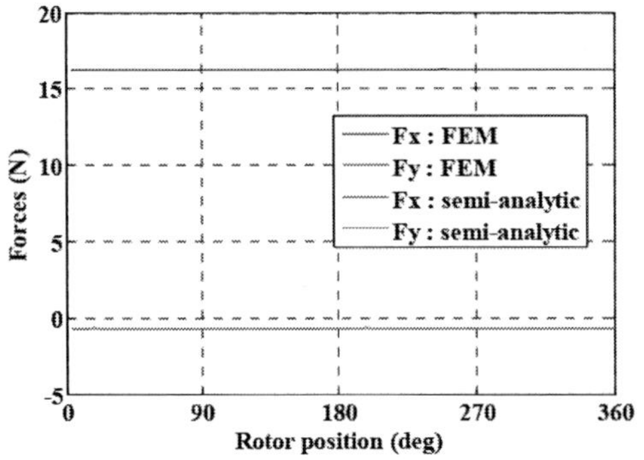

Figure 5 : Force x-y components of the studied six-phase motor with parallel magnetization

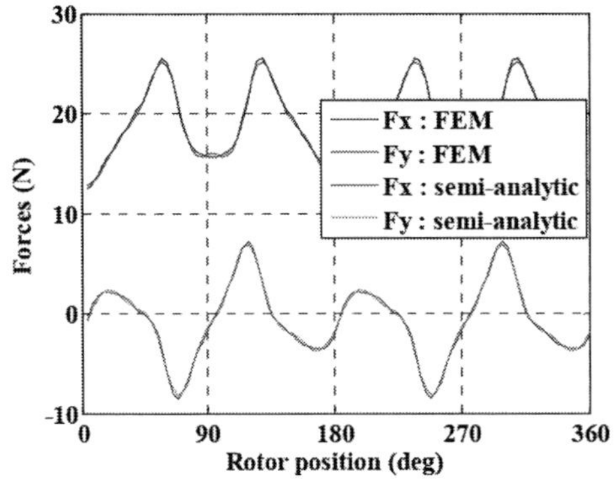

Figure 6 : Force x-y components of the studied six-phase motor with radial magnetization

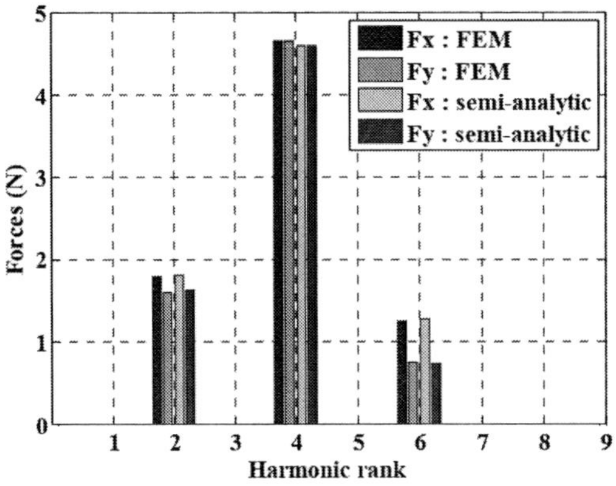

Figure 7 : Force harmonic analysis for radial magnetization

by using both analytical model with 16 first harmonics and full FEM computations. In this case also there is an excellent agreement between the results obtained by these two models.

The spectral analysis of the forces waveforms is shown on Figure 7 for both F_x and $F_y$components. This spectral analysis allows the identification of the interaction between the different harmonics of the flux density. For example, the 2nd harmonic of the force is produced by the interaction of the 2nd field harmonic generated by the current system I2 and the 3rd field harmonic generated by the magnets. In order to indicate the mentioned field harmonics interaction we use henceforth the following notation $(2_{I2}, 3_{mag} \Rightarrow 2_{force})$.

The interaction $(4_{I2}, 3_{mag} \Rightarrow 4_{force})$ corresponds to the red group characterized by the product of two sums (Table I) while the interaction $(2_{I2}, 3_{mag} \Rightarrow 2_{force})$ corresponds to the blue group characterized by the product of two subtractions (Table I). That's why the amplitude of the 4th harmonic of the force is higher than the amplitude of its 2nd harmonic.

Furthermore, it can be seen that the force harmonics of higher ranks may have several origins. For example the interactions $(4_{I2}, 5_{mag} \Rightarrow 6_{force})$ and $(8_{I2}, 7_{mag} \Rightarrow 6_{force})$ lead to the 6th harmonic of force. These harmonics are weak because they are generated by high rank field harmonics which have low amplitudes.

F. CPU time considerations

In the proposed semi-analytic model, the determination of radial force is performed in two steps:

- The first step consists in calculating the air-gap flux density and its Fourier decomposition for each source (magnet and current systems I1 and I2). This is performed once and requires only 3 FE calculations. This step needs fixed time called initialization.
- The second step consists to compute the force for a given instant (time) t_k. Using equations (12) and (13), the air-gap flux density is reconstituted for the considered instant and the force components are calculate by equations (10) and Table I. This is performed for each instant and does not require FE calculations. This step needs variable time depending on the number of considered instants.

For example in the machine considered in the previous section, the number of nodes us fixed to 7000 node and the calculations are performed with a processor Intel core i7-3740 (2.7GHz). The first step takes 9.3 seconds and the calculation of the forces one point in the second step need 1 millisecond. For 100 points over a mechanical period the proposed semi analytical model takes less than 10 seconds while the full FE method takes 235seconds.

The CPU time is strongly reduced with the proposed model which can be used in optimization processes or in coupled models of control on Matlab-Simulink platforms.

G. Conclusion

In this paper, the authors present a new approach for modeling of radial forces in bearingless machines. The model is based on the spectral decomposition of the flux density distribution in the air-gap, generated by different sources (magnets in rotor and currents in stator). The spectral formulation of the radial force is performed with Maxwell stress tensor. It allows the calculation of the different interaction between the field harmonics and predicts the harmonic content of the force waveforms. Such model is useful for the designer which may specify the undesirable field harmonics before starting the design.

This model needs few FEM calculations and consequently it is very fast compared to full FEM computations.

REFERENCES

[1] M. Oshima, S. Miyazawa, T. Deido, A. Chiba, F. Nakamura, T. Fukao, "Charactreristics of a permanent magnet type bearingless motor", *IEEE trasactions on industry application*, vol. 32, no. 2, pp. 363–370, 1996.

[2] A. Chiba, M. Azizur Rahman, T. Fukao, "Radial force in a bearingless reluctance motor" *IEEE transactions on magnetics*,vol 27,no 2, pp. 786–790, 1991.

[3] M. Kang, J. Huang, J. Yang, H. Jiang, "Analysis and experiment of a 6 phase bearingless induction motor", *Proceedings on Electrical Machines and Systems. ICEMS*, pp. 990–994, 2008.

[4] Y. Okada, S. Miyamoto, T. Ohishi, "Levitation and torque control of internal permanent magnet type bearingless motor," *IEEE transaction on control systems technology*, vol. 4, no. 5, pp. 565-571, 1996.

[5] D. Steinert, T. Nussbaumer, J. Kolar "Concept of a 150 krpm bearingless slotless disc drive with combined windings", *Proceedings on Electric Machines & Drives Conference, IEMDC,* pp. 311–318, 2013

[6] G. Munteanu, A. Binder, T. Schneider, "Loss measurement of a 40 kw high-speed bearingless PM synchronous motors", *Proceedings on Energy Conversion Congress and Exposition, ECCE,*, pp. 722–729, 2011,

[7] S. Serri, A. Tani, G. Serra, "Analytical model of radial forces considering mutual effects between torque and levitation current space vectors in 5-phase PM bearingless motors", *Proceedings on Industrial Electronics Society, IECON*, pp. 5142–5147, 2013

[8] T. Schneider, J. Petersen, G. Serra, "Influence of pole pair combinations on high-speed bearingless permanent magnet motor performance", *Proceedings on Power Electronics, Machines and Drives, PEMD.*, pp. 707–711, 2008

[9] B. Li, J. Huang, W. Kong, L. Zhao, "Analysis of Half-coiled Short-pitch Windings with Different Phase Belt for Multiphase Bearingless Motor", *Journal of Electrical Engineering & Technology, vol 9 No1* pp. 162–168, 2014

[10] G. Munteanu, A. Binder, T Schneider, B. Funieru, "No-load tests of a 40 kW high-speed bearingless permanent magnet synchronous motor", *Proceedings of* International Symposium on *Power Electronics Electrical Drives Automation and Motion,* SPEEDAM, Pisa, Italy, pp 1460-1465, 2010

[11] F. Zürcher, T. Nussbaumer, J.W. Kolar, "Principles of magnetic levitation force and motor torque generation by superposition of harmonics in bearingless brushless motors", *Proceedings on Industrial Electronics,. IECON '09,* pp. 1246–1251, 2009

Active Magnetic Bearing Interdisciplinary Design Modelling and Control Tool

Adam Krzysztof Piłat[a]

[a] AGH University of Science and Technology, Al. A. Mickiewicza 30, 30-059 Krakow, Poland, ap@agh.edu.pl

Abstract— **This elaboration presents a modern approach in the field of active magnetic bearing design based on computations supported by partial differential equations (PDEs) and ordinary differential equations (ODEs). The magnetic field problems are solved using the finite element method (FEM), while the levitating rotor motion is calculated using ODEs. The novel Active Magnetic Bearing Interdisciplinary Design Modelling and Control (AMBIDMC) study tool was developed customising COMSOL Multiphysics and MATLAB software. This paper summarises the author's recent work in the field of integrated active magnetic levitation (AML) systems design and modelling including control and motion dynamics. A discussion of the features and limitations of the field of control systems is provided in this paper, showing how the developed model can be useful for controller studies. The discussion is illustrated by simulation data collected from working virtual prototypes.**

I. INTRODUCTION

An overview of the literature [29, 32, 35] focussed on the design and control of active magnetic levitation systems: suspensions and bearings shows different approaches and solutions related to AML system design, modelling, control studies and real-time control.

A control study for the AMB is a key point in AMB design. The controller is designed at the end of the prototyping stage, when the device is manufactured and identified. Subsequently, the modelling stage and controller study are activated. The open question is whether the identification supplies sufficient information for the controller study. It is rather difficult to investigate the rotor-AMB system in terms of all possible scenarios of control and measurement signals versus rotor displacement. With a complex model, an investigative experiment can be planned. It seems very practical to use a such model to identify cross couplings in the rotor-AMB system and to use them in the controller structure for optimal rotor control.

Nowadays, rapid prototyping methods and mechatronic devices with complex designs are becoming a reality; thus, the traditional approach of serial design needs to be modified.

A novel concept (see Fig. 1) of design, modelling and control study was proposed [19] in the field of control and active magnetic levitation. A simultaneous design, optimisation and control study is a desirable approach to prototyping new devices. Therefore, controller design and performance should be analysed at the design stage. A number of available software tools enable the design of so-called virtual prototypes (VPs), or computer models of the target device. The most advanced VPs are characterised by embedded dynamics. In the case of mechatronic systems, the controller is embedded to carry out dedicated tasks. Working with this type of VP enables verification of the concept, optimisation of design, and improvement of control quality.

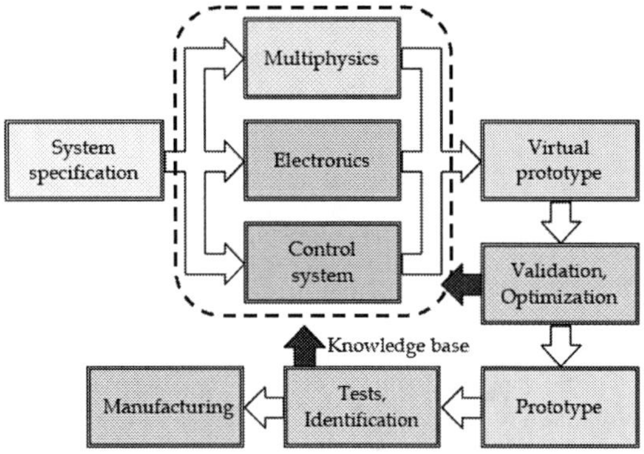

Figure 1. The proposed interdisciplinary design approach according to which the virtual prototype is being developed and studied

Typically, the active magnetic bearing control is designed on the basis of a mathematical model based in turn on ordinary differential equations (ODEs). The modelling stage is supported by the calculation of the electromagnetic foce characteristics on the baseis of the finite element methods (FEM). This method is very useful for the detailed analysis of magnetic circuits, mechanical features, thermal effects, etc. Finite element analysis can accurately account for the magnetic field which lies outside of the iron and thereby properly account for its effect on the actuator forces and electrical properties [15]. This method was succesfully applied to magnetic flux analysis [6, 7, 14, 17, 18, 34]. The finite element method was also applied to the analysis of eddy current losses in the thin surface layer of a laminated core [16]. The FEM calculations are very useful for the prototyping stage, optimization [5, 12, 16, 30]. The FEM method is used to calculate the shaft bending modes [2, 11, 12]. To perform a FEM analysis a wide range of software tools is used. One can find modelling supported by ANSYS [3, 14] , COMSOL [1, 9, 10, 13, 25, 26], COSMOS [12]. Recently, these calculations have been supported by the electromagnetic force characteristics obtained from FEM modelling [8,10, 31, 33, 35]. Very often, this stage is accomplished following a set of practical identification experiments. Typically, 2D analysis is used due to the computational effort and complexity of 3D modelling. However, the current trend is towards 3D modelling, in order to achieve a complex model of a rotor-AMB system.

One can find a number of different control strategies [32, 35]: linear, non-linear, and intelligent. The last-mentioned uses linear or non-linear models developed in a local (single electromagnet) or global case (complete AMB). Additionally, complete machine dynamics are modelled and simulated.

The aim of this research was to build complex dynamical models of the Active Magnetic Bearing. In a such model we wish to simulate: shaft levitation, controller action and electromagnetic forces acting on the shaft in a time domain.

In this research, which is focussed on a steered active magnetic bearing virtual prototype with truly levitating rotor motion solved in the time domain, the following questions can be addressed:

- *What are the main benefits of interdisciplinary design of AMB modelling, simulation and control?*
- *Is it possible to propose identification and control algorithms and test them with a virtual prototype?*
- *Is this kind of model/virtual prototype suitable for ODE-based modelling and controller studies?*

II. Active Magnetic Bearing Interdisciplinary Design Modelling and Control study (AMBIDMC) Tool

This research was motivated by the manual prototyping of active magnetic suspensions and bearings, [21] magnetic field analysis and geometric optimisation, [20] and a controller study [29] supported by a set of experimental studies including hard real-time control [23, 24, 29].

The main idea was to enable a complex study in the field of AML-based devices. Therefore, dedicated software had to be developed. To obtain a professional solution, the COMSOL and MATLAB packages were combined to take advantage of their best features. Software integration and customisation enabled a custom solution. This tool especially supports AMBs. It uses the COMSOL Multiphysics software as the core of the PDE and ODE calculations. Additionally, the MATLAB software is used to customise the design and carry out optimisation tasks, simulation, analysis and controller synthesis. The first application of the developed tool was an AMB shape generator [21]. Subsequently, the complexity was increased. Currently, the 2D modelling stage has been completed, and the completely functioning AMB-rotor VP is being simulated. The most important component, a controller, can be embedded either into the COMSOL model as internal mathematical formulas, or, alternatively, into m-coded functions and linked to the model.

The model is processed in the time domain in a parallel way, solving the magnetic field with the support of PDEs, computing rotor motion dynamics using ODEs, and solving controller equations. The applied method for the deformed mesh in time enables the rotor to be placed in true levitation mode in the AMB plane.

Using this tool, the user proceeds with the following steps:

- decision/concept stage: configuration, e.g. 3 or 4 electromagnets with a C-shaped core, solid or ring-type rotor, smooth or sharp poles [28],
- geometry design - dimensions of particular components,
- materials assignment: materials used in the model, such as air, copper, steel with or without B-H characteristics,
- definition of physics: assignment of magnetic field phenomena, coil regions, current- or voltage-driven coils with an RL equation,
- motion dynamics: formulas for rotor motion in the AMB plane,
- free-moving components: automatic assignment of the rotor properties available for motion in the AMB plane,
- time-dependent solver: time step and simulation time,
- analysis and post-processing: predefined post-processing quantities such as a 2D plot of magnetic flux density, rotor position, and control, coil current time diagrams, and a film representing rotor motion.

This tool is capable of performing a wide range of optimisation experiments, including: rotor unbalance, external vibrations and loads, material properties, heating, parameter perturbation, linear and non-linear control, forces, torque, mechanical stress, heat losses, eddy currents, etc. The range of analysis depends on the physical phenomena included in the model.

III. Rotor levitation in AMBs with Four and Three Electromagnets

Using the AMBIDMC tool, two AMB dynamical models were developed. Both operate with heteropolar configurations of 3 and 4 C-shape electromagnets. Both models were inspired by partly realised calculations and manufactured prototypes (see Figs. 2a and 5a).

A. Slim type AMB with four C shaped electromagnets and rotor stalibisation mode

Fig. 2a presents a slim-type AMB (10mm thick) called AMB4EM and ring-type rotor prototype; Fig. 2b is a schematic diagram with a smooth stator design [21].

A summary of the AMB4EM parameters is given in Table I with symbols corresponding to the AMB standard [4]. Due to the ring type rotor, the following properties were introduced: internal diameter ϕd_{int}, outer stator diameter D_{ext}, and W_{ext} representing the extended pole width.

TABLE I. AMB4EM Parameters

	Parameter	Value
D_{ext}	Outer diameter of radial stator core	90.0 mm
D	Inner diameter of radial stator core	50.0 mm
d	Outer diameter of radial rotor core	48.0 mm
d_{int}	Inner diameter of radial rotor core	40.0 mm
L	Effective radial bearing length	10.0 mm
W	Width of a magnetic pole	10.0 mm
W_{ext}	Width of a magnetic pole	12.0 mm

The numerical model was realised as current-driven. Rotor motion equations were added to translate the rotor in the AMB plane. The magnetic flux density at every pole was examined and the *XY* components of the electromagnetic force were calculated. The simulation results presented in Figs. 3 and 4 show the start-up and levitation of the rotor without the presence of gravity and without bias current. At the start-up, the rotor was located at the position of (150μm, -200μm), and other states to zero. The summary of AMB states calculated at the first time step is given in Table II. Two PD controllers operating in differential mode were applied for stabilisation, i.e. one for the *X*-axis and the other for the *Y*-axis.

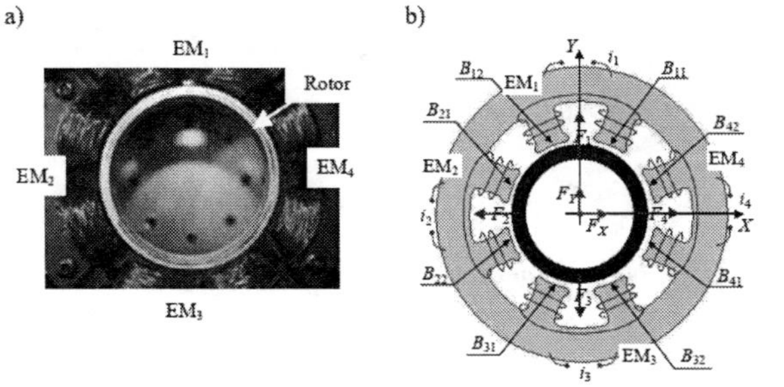

Figure 2. AMB with four actuators, smooth poles, and a ring-type rotor in the slim configuration: AMB and rotor thickness is 10mm

The simulation carried out with rotor planar motion represents control generated by the PD controllers attracting the rotor towards the bearing centre. The most interesting results are revealed in the magnetic flux density *Bij* where *i*-corresponds to the electromagnet *i*={1, 2,3, 4}, *j* represents the pole *j*={1, 2}. Thanks to this modelling approach, the differences in *B* values for every electromagnet are easily visible, which is crucial to the controller and observer studies. Moreover, the rotor displacement is visible in these signals indirectly, what allows to provide research towards self-sensing control.

TABLE II. AMB4EM STATES CALCULATED AT THE FIRST TIME STEP

	Parameter	Value
x, y	rotor X, Y position	150.0 µm,-200.0 µm
i_1, i_2, i_3, i_4,	EM 1÷4 coil current	0.00 A, 0.00 A, 1.59 A, 1.59 A,
Fx	*Electromagnetic force in X, Y axis*	2.73 N, -5.03 N
B_{11}, B_{12}	Magnetic inductance at EM 1	0.72 mT, 0.23mT
B_{21}, B_{22}	Magnetic inductance at EM 2	0.23 mT, 0.94 mT
B_{31}, B_{32}	Magnetic inductance at EM 3	77.00 mT, 83.97 mT
B_{41}, B_{42}	Magnetic inductance at EM 4	62.12 mT, 55.44mT

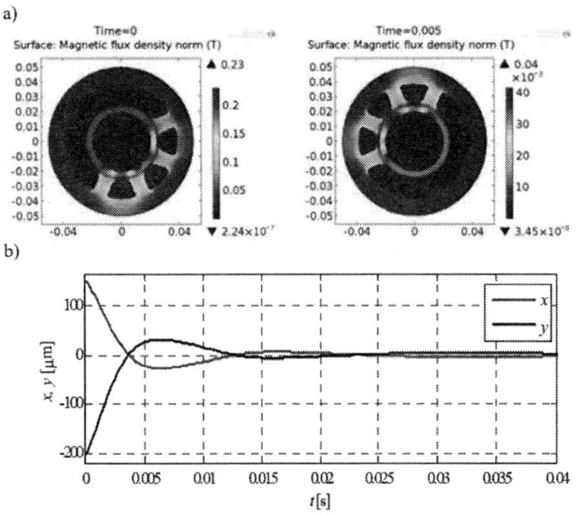

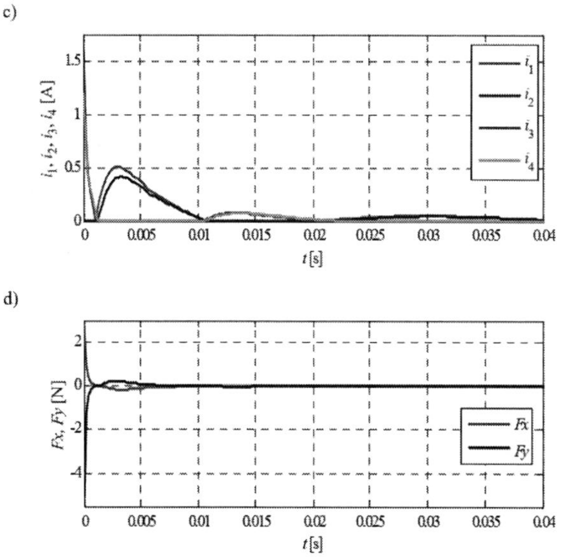

Figure 3. The simulation results of the stabilisation task - distribution of magnetic flux density (a); rotor displacement (b); coil curerents (c) and electromagnetic force components (d).

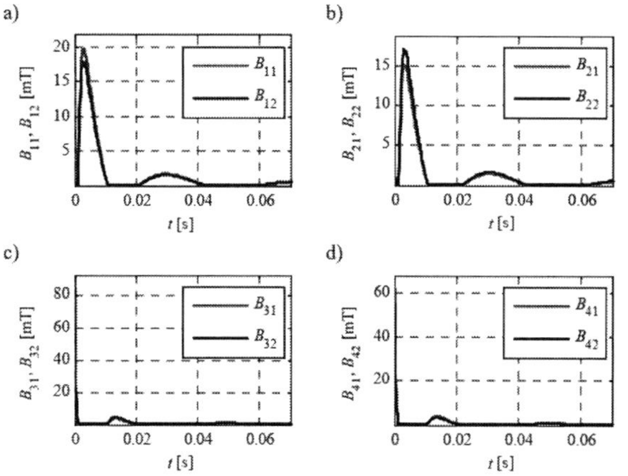

Figure 4. Magnetic flux desity measured at every AMB pole (refer to Fig. 2) in the stabilisation task (Fig. 3).

B. AMB with three C shaped electromagnets: rotor stabilisation and rotations

The second configuration to be considered is an AMB with three electromagnets (AMB3EM) [26]. A summary of the AMB3EM parameters is given in Table III. Symbols correspond to the AMB standard [4] extended with a few new properties due to the ring type rotor: internal diameter ϕd_{int}, outer stator diameter D_{ext}, and W_{ext} representing the extended pole width. The irregular displacement of electromagnets and smooth stator poles are the main features differentiating this configuration from standard ones. For this configuration, the motion dynamics in the AMB plane are modelled by a set of two differential equations [22, 26].

TABLE III. AMB3EM PARAMETERS

	Parameter	Value
D_{ext}	Outer diameter of radial stator core	95.0 mm
D	Inner diameter of radial stator core	51.8 mm
d	Outer diameter of radial rotor core	48.8 mm
d_{int}	Inner diameter of radial rotor core	42.0 mm
L	Effective radial bearing length	25.0 mm
W	Width of a magnetic pole	10.0 mm
W_{ext}	Width of a magnetic pole	13.0 mm

The shaft stiffness and damping coefficients were set to zero and, instead of standard electromagnetic force formulas, the electromagnetic force *X* and *Y* components were applied to both equations appropriately. The rotor imbalance was set to zero, but the rotations were enforced to drive the rotor via the transformation of the external coordinates. In this case the rotations operate as a disturbance for the stabilisation task. The rotation profile was set as a desired speed-up curve. Three local PD controllers were applied to stabilise the rotor at the bearing centre. These controllers operated on the basis of local distances between the rotor and electromagnet axis, calculated on the basis of geometrical rotor position. The bias current value was set at 0.1A. The summary of AMB states calculated at the first time step is given in Table IV. These simulations show that the basic PD controller is able to stabilise the rotor over a wide range of rotational speeds, although it is not robust for variable speed. However, using the proposed modelling approach, a number of modifications (including gain scheduling, adaptive or robust) to control methods can be proposed and validated.

TABLE IV. AMB3EM STATES CALCULATED AT THE FIRST TIME STEP

	Parameter	Value
x	rotor *X* position	-150.0 μm
y	rotor *Y* position	-300.0 μm
i_1, i_2, i_3	EM 1÷3 coil current	659.8mA, 0.0 A, 140.2mA
F_X, F_Y	*Electromagnetic force in X, Y axis*	-83.1 mN, -52.7 mN
B_{11}, B_{12}	Magnetic inductance at EM 1	6.24 mT, 8.24 mT
B_{21}, B_{22}	Magnetic inductance at EM 2	0.28 mT, 0.29 mT
B_{31}, B_{32}	Magnetic inductance at EM 3	28.29 mT, 25.74 mT

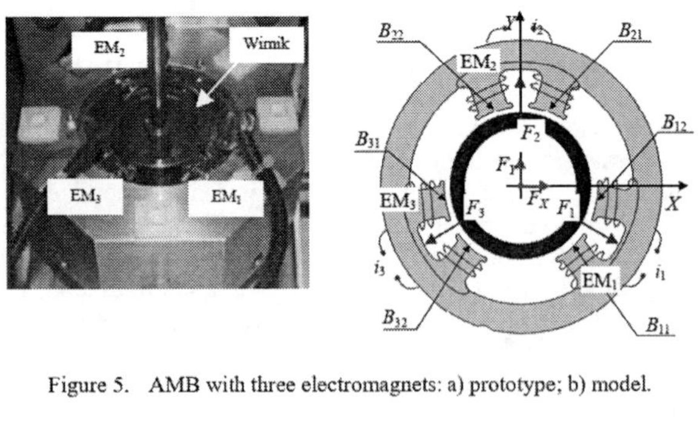

Figure 5. AMB with three electromagnets: a) prototype; b) model.

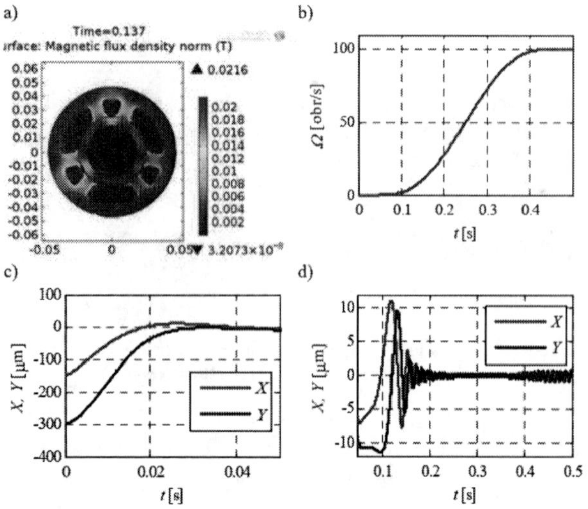

Figure 6. Dstribution of magnetic flux density (a); rotor velocity profile (b) and rotor displacement (c) time diagrams: start-up (d) and stabilisation phase at rotations (e).

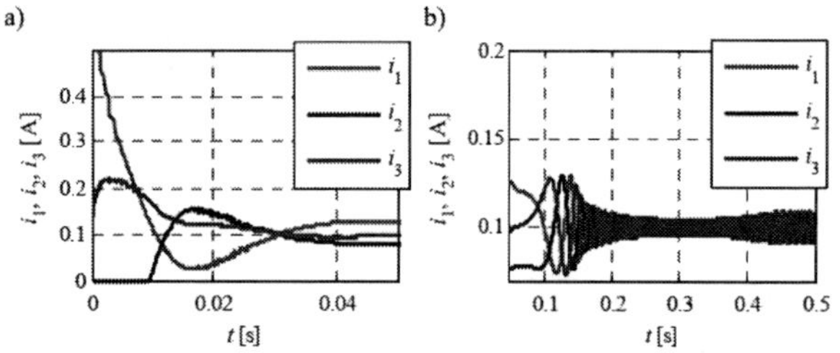

Figure 7. Control signals (coil currents) applied at the start-up and stabilisation phases under external rotary excitations

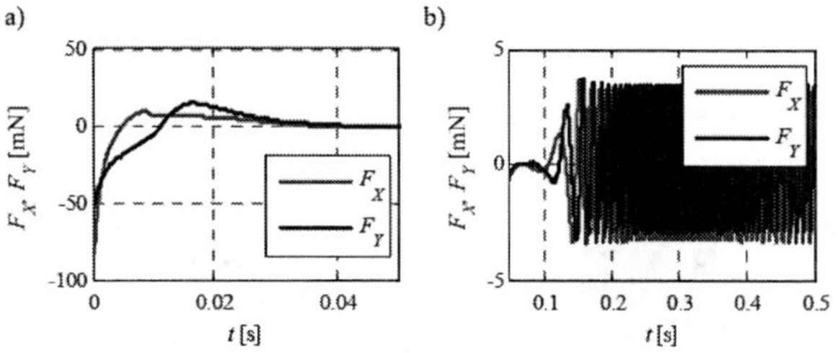

Figure 8. Caclulated electromagnetic forces obtained at the start-up (a) and stabilisation (b) phases under external rotary excitations

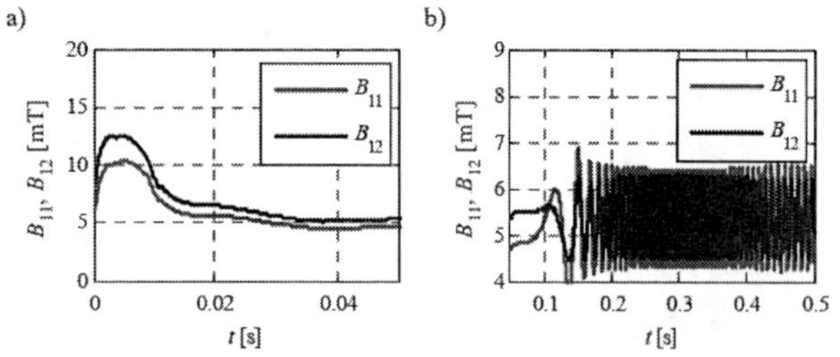

Figure 9. Measured magnetic flux density at EM1 poles, collected at the start-up (a) and stabilisation (b) phases under external rotary excitations

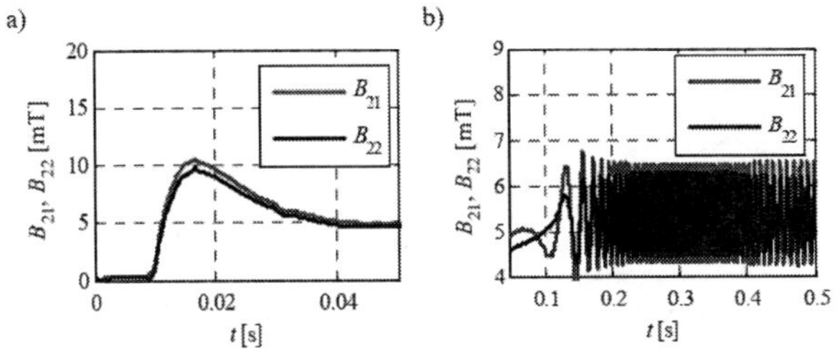

Figure 10. Measured magnetic flux density at EM2 poles - collected at the start-up (a) and stabilisation (b) phases under external rotary excitations.

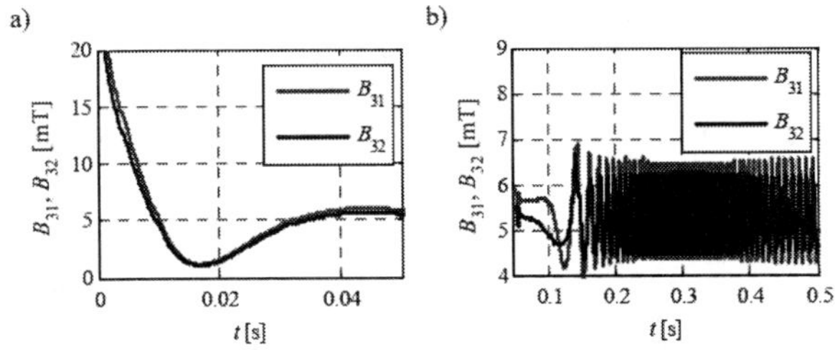

Figure 11. Measured magnetic flux density at EM3 poles - collected at the start-up (a) and stabilisation (b) phases under external rotary excitations.

Again, for the AMB3EM the magnetic flux density measured at the front of poles, allows to estimate the rotor position. Thus, the developed model is well suitable for a nonlinear controller synthesis, observers design and research towards self-sensing control. All these aspects will be considered in further research.

C. *Discussion on simulation results*

The completed simulations (see also [25, 26, 27, 28]) show that the electromagnetic force F_{Em} calculated on the basis of the standard formula (1) is a simplification and does not cover geometrical and magnetic properties; nor is it truly rotor-dependent.

$$F_{\mathrm{Em}}(a,i) = -\frac{K_{\mathrm{Em}}}{4}\frac{i^2}{a^2}, \text{ or } , F_{\mathrm{Em}}(B) = -\frac{A}{\mu_0}B^2 \qquad (1)$$

where: K_{Em} - electromagnet constant, i - coil current, A - area of electromagnet cross-section, a - distance between the levitated object and electromagnet pole, B - magnetic flux density. The magnetic flux density measured at both poles of the electromagnets varies; therefore it must be considered in the identification and magnetic flux density based controllers and observers. The implemented rotor motion in the AMB plane enables testing of the controller. The structural rotor imbalance can be computed on the basis of rotor geometry and used in the simulation model.

Simulated models correspond to existing prototypes, but the comparison and convergence discussion on the simulation and experimental results is planned to be executed. This requires the proposal and identification of scenarios in which both variants can be realised.

IV. Modelling aspects towards controller study

The developed AMB model using the AMBIDMC tool makes it possible to conduct a set of studies:

- This type of synergistic model enables an unlimited number of experiments and dependencies existing in levitation systems. As the most impressive results, the author cites the complete dynamics solved in the time domain, including rotor displacement and its influence.
- Identification of electromagnetic force. Thanks to the complexity of the model, the rotor position and the coil current can be adjusted statically or dynamically, thus making it possible to achieve static and dynamic characteristics. Direct identification in the AMB plane is a difficult task and requires the use of a special apparatus, but gives a overview of the true force characteristics.
- Identification of the actuator properties. Using the AMBIDMC tool, magnetic flux density vs coil current and rotor position can be analysed. Additionally, coil current vs supply control signal can be analysed when the model is extended with a RL circuit representing the electromagnet winding.
- Admissible sets can be achieved for a study of control signals and rotor displacements in the region of levitation. This enables the study of the static and dynamic properties of the complete solution with embedded control.
- The magnetic flux density is a valuable quantity for magnetic-flux-based feedback controllers and rotor position observers.
- All quantities can be analysed statically, in the time and frequency domains.
- Other quantities corresponding to mechanical and thermal properties are available for analysis. Mechanical strain can be achieved under the rotor load of the electromagnetic forces. For the electromagnet windings, core and surrounding elements, thermal analysis can be accomplished on the basis of resistive heating of coils.
- The study, focussed on the influence of material properties on AMB dynamics and operation, can easily be used for the definition of constants or characteristics.
- The linear and non-linear controller formulas can be embedded in the form of equations, including integration or derivative calculations. The controller embedded in the model and its parameters can be easily tuned using the optimisation stage for defined control quality criteria.
- For those wishing to calculate the controller analytically, the developed model can be a source of data for ODE model tuning. In this case there is no need to build the prototype initially and perform a number of expensive tests, some of which cannot be carried out due to specific apparatus requirements, e.g. magnetic flux density, electromagnetic force value for every rotor position and any control signal. Certain electromagnetic force investigations and thermal analyses were conducted [29].

With the proposed model, carrying out identification schemes is much more flexible. Please note that simulation models offer much more than just experiments. However, it is obvious that the models (including material properties, physical phenomena, and real controller implementation) ought to be verified on the basis of experimental results. Thus, the AMBIDMC tool can be used to verify the identification concepts behind practical experiments.

V. Colnclusions

The synergistic dynamical models of the active magnetic bearings were realised and allow to provide time domain simulations. The developed AMBIDMC tool can be helpful in the analysis of existing AMB configurations as well for new designs. It helps designers to check and validate the device concept at the simulation stage, thanks to the full functionality of the virtual prototype. Model verification is available and convergence analysis is required on the basis of experimental investigations. It appears that the complex model gives a better overview of physical quantities than mathematical approximations based only on identification data. The complex model enables the collection of much more information and many more dependencies due to an unlimited number of possible identification scenarios. Finally, such a complex numerical model is well suited for controller design, tuning and tests. The tool is conducive to the addition of new features and will be expanded in the direction of 3D modelling and complex machine design.

REFERENCES

[1] Ajith K.V., Mashuq-un-Nabi, Modeling and Analysis of a Feedback-Controlled Active Magnetic Levitation System Using COMSOL Multiphysics Finite Element Software, COMSOL Conference, Bangalore, 2009

[2] AL-Khazali H.A., Askari M.R., Accurate Identification of Performance for Rotor-Bearing Systems Using the Modified Modelling Under Gyroscopic Effect, International Journal of Modern Engineering Research (IJMER), Vol.1, Issue.2, ISSN: 2249-6645, pp 527-537, 2011

[3] Biswas P.K., Banerjee S., ANSYS Based FEM Analysis for Three and Four Coil Active Magnetic Bearing-a Comparative Study, International Journal of Applied Science and Engineering, 11, 3, pp. 277-292, 2013

[4] British Standard, ISO 14839-1:2002, Mechanical vibration — Vibration of rotating machinery equipped with active magnetic bearings — Part 1: Vocabulary

[5] Cheng S., Olles M.W., Burger A.F., Day S.W., Optimization of a hybrid magnetic bearing for a magnetically levitated blood pump via 3-D FEA, Mechatronics, Vol. 21, pp. 1163–1169, 2011

[6] Chiba, A., Onoya, S., Kikuchi, T., Ooshima, M., Miyazawa, S., Nakamura, F. and Fukao, T., An Analysis of a Prototype Permanent-Magnet Bearingless Motor using Finite Element Method, ISMB5, 1996

[7] Eryong H., Kun L., Finite Element Analysis on Magnetic and Mechanical Characteristics for Hybrid Magnetic Bearing, Proceedings of 2012 International Conference on Mechanical Engineering and Material Science (MEMS 2012), Atlantis Press, pp. 225-230, 2012

[8] Eugen O. Ranft, An improved model for self-sensing heteropolar active magnetic bearings, Ph.D. Thesis, Promoter: G. van Schoor, North-West University, 2007

[9] Falkowski K., Simulation of Passive Magnetic Bearing using COMSOL Multiphysics, COMSOL Conference, Stuttgart, 2011

[10] Ferfecki P., Frantisek O., Comparison of methods for computation of electromagnetic forces in the radial active magnetic bearing, Transactions of the VŠB – Technical University of Ostrava, Mechanical Series, No. 3, vol. LV, pp. 51-58, 2009

[11] Hynynen, K., Jastrzebski, R. and Pöllänen, R. Identification and Modeling of AMB Rotor System Using Complex Lead and Lag Compensator Structures, ISMB11, 2008

[12] Kimman M.H., Langen H.H., Schmidt R.H., A miniature milling spindle with Active Magnetic Bearings, Mechatronics, vol. 20, pp. 224–235, 2010

[13] Kluyskens V., Dehez B., Computation of the Forces Acting on a Magnetic Bearing Due to Eddy Currents, COMSOL Conference, Grenoble, 2007

[14] Liao P., Ma S., Wu G, Jingfeng Mao J., Jiang A., The Analysis of Radial Magnetic Bearing's Magnetic Field in Active Magnetic Bearing Electric Spindle Unit, Advanced Materials Research Vols. 291-294, pp. 1593-1599, 2011

[15] Maslen E., Magnetic Bearings, University of Virginia, Department of Mechanical, Aerospace, and Nuclear Engineering, Charlottesville, Virginia, 2000

[16] Muramatsu K., Shimizu T., Kameari A., Yanagisawa I., Tokura S., Saito O., et al., Analysis of eddy currents in surface layer of laminated core in magnetic bearing system using leaf edge elements, IEEE Trans. Magn., vol. 42, no. 4, pp. 883–886, 2006

[17] Pichot, M.A., Driga, M.D., Magnetic Circuit Analysis Of Homopolar Magnetic Bearing Actuators, ISMB9, 2004

[18] Pilat A., FEMLab software applied to active magnetic bearing analysis, International Journal of Applied Mathematics and Computer Science, vol. 14, no 4, pp 497-501, 2004

[19] Pilat A., Active magnetic suspension and bearing. Modelling and simulation. Vienna: InTech Education and Publishing. ISBN 978-3-902613-25-7, pp. 453–470, 2008

[20] Pilat A., AMB Construction Optimization Tool. Solid State Phenomena. Trans Tech Publications. Vol. 147-149, pp. 393–398, 2009

[21] Pilat A., Analytical modeling of active magnetic bearing geometry. Applied Mathematical Modelling. ISSN 0307-904X, Vol. 34. No 12, pp. 3805–3816, 2010

[22] Pilat A., Testing of the programmable AMB spring and damping forces in a flexible rotor test-rig mode. Proceedings of the twelfth International Symposium on Magnetic Bearings: August 22–25, Wuhan, China. Wuhan University of Technology Press, ISBN 978-7-900503-06-0, pp. 336–343, 2010

[23] Pilat A., Control toolbox for industrial programmable analog controller embedding state feedback controller. IEEE 17th Conference on Emerging Technologies Factory Automation (ETFA), pp. 1–4, 2012

[24] Pilat A., The programmable analog controller: static and dynamic configuration, as exemplified for active magnetic levitation". Przegląd Elektrotechniczny. ISSN 0033-2097. R. 88 nr 4b, pp. 282–287, 2012

[25] Pilat A., Modeling, Simulation and Control of Dual Electromagnet Active Magnetic Levitation. COMSOL Conf., Milan, pp.1–4, 2012

[26] Pilat A., An synergistic dynamic model of an active magnetic bearing with three electromagnets, International Symposium on Electrodynamic and Mechatronic System (SELM), pp. 23–24, 2013

[27] Pilat A., Electromagnetic force components analysis in an active magnetic bearing numerical model, International Symposium on Electrodynamic and Mechatronic System (SELM), pp. 21–22, 2013

[28] Pilat A., Modelling, investigation, simulation, and pid current control of active magnetic levitation fem model," 18th International Conference on Methods and Models in Automation and Robotics (MMAR), pp. 299–304, 2013

[29] Pilat A., Active magnetic levitation systems, ISSN 0867-6631, ISBN 978-83-7464-621-5, AGH Press, Krakow, 2013

[30] Sakagami, M., Chiba, A., Ooshima, M. and Fukao, T., Torque Improvement In A Buried-Type IPM Bearingless Motor With A Novel Rotor Structure, ISMB 9, 2004

[31] Schmidt, E., Platter, T. and Springer, H., Force and Stiffness Calculations in Magnetic Bearings - Comparison between Finite Element Method and Network Theory, ISMB5, 1996

[32] Schweitzer G., Maslen E. H., Bleuler H., Cole M., Keogh P., Larsonneur R., Nordmann R., Okada Y., Traxler A. Magnetic Bearings: Theory, Design, and Application to Rotating Machinery. Springer, 2009

[33] Sugimoto, H. and Chiba, A., Precise Measurement of Suspension Force in Two-Axis Actively-Positioned Bearingless Motor for a Comparison to 3D-FEM Analysis, ISMB13, 2012

[34] Zimon J., Tomczyk B., Wejnert D., Eddy Current Influence on the Parameters of Active Magnetic Bearing, ISMB12, 2012

[35] Magnetic bearings repository: http://magneticbearings.org

Nitsche-type Mortaring for Simulation of Electrical Machines

Armin Fohler[a], Walter Zulehner[b]

[a] Linz Center of Mechatronics, Altenberger Straße 69, 4040 Linz, Austria, armin.fohler@lcm.at
[b] Johannes Kepler University Linz, Altenberger Straße 69, 4040 Linz, Austria

***Abstract*—We will present a finite element program implementing the idea of Nitsche-type mortaring in the framework of simulating electrical machines. Using this method it is possible to split the domain into subdomains only connected via a function on the interface. From this approach we gain the advantage that adjacent meshes (e.g. rotor and air gap) do not have to fit together nodewise. The mortaring will be used on the interface between rotor and air gap as well as on the symmetry edges of the motor, since only a part of the machine is considered in the simulation due to motor symmetry.**

I. Introduction

The program at issue was designed to deal with the magnetostatic case of Maxwell's equations in 2D, as it occurs as a simplification in the simulation of electrical machines. Starting with:

$$\begin{aligned} \operatorname{curl}\mathbf{E} &= -\frac{\partial \mathbf{B}}{\partial t}; \\ \operatorname{curl}\mathbf{H} &= \mathbf{J} + \frac{\partial \mathbf{D}}{\partial t}; \\ \operatorname{div}\mathbf{D} &= \rho; \\ \operatorname{div}\mathbf{B} &= 0; \end{aligned}$$

with material laws:

$$\begin{aligned} \mathbf{B} &= \mu(|\mathbf{H}|)\mathbf{H} + \mathbf{M}; \\ \mathbf{D} &= \epsilon\mathbf{E} + \mathbf{P}; \\ \mathbf{J} &= \sigma(\mathbf{E} + \mathbf{v}\times\mathbf{B}). \end{aligned}$$

we can derive the magnetostatic case. Here $\mathbf{E}$ denotes the electric field intensity, $\mathbf{P}$ the polarization. $\frac{1}{\mu} = \nu$, with μ the permeability and ν the reluctivity, ρ stands for the charge density. The electric conductivity is represented by σ, v the velocity and ϵ is the permitivity. Under the assumptions, that the magnetic flux density $\mathbf{B}$, the magnetic field intensity $\mathbf{H}$ and the electric current density $\mathbf{J}$ are time-independent and $\frac{\partial \mathbf{D}}{\partial t} \ll \mathbf{J}$, with $\mathbf{D}$ denoting the electric flux density we arrive at the magnetostatic formulation.

Those assumptions are reasonable in the framework of electrical machines since we are dealing with slowly varying magnetic fields.

On a single computational domain with $\partial\Omega$ symbolizing its boundary we get the following boundary value problem:

Find $u : \overline{\Omega} \to \mathbb{R}$ such that

$$\begin{aligned} -\operatorname{div}(\nu(|\nabla u|)\nabla u) &= J_3 - \frac{\partial M_1}{\partial x_2} + \frac{\partial M_2}{\partial x_1} && \text{in } \Omega \\ u &= 0 && \text{on } \partial\Omega \end{aligned}$$

If we consider a computational domain divided into subdomains Ω_i with interfaces Γ_{ij} between the domains Ω_i and Ω_j we have to impose additional conditions on the interfaces. In our case those subdomains will be the rotor, the stator and the air gap. The adapted boundary value problem is he following:

Find $u : \overline{\Omega} \to \mathbb{R}$ such that

$$\begin{aligned} -\operatorname{div}(\nu(|\nabla u_i|)\nabla u_i) &= J_{3,i} - \frac{\partial M_{1,i}}{\partial x_2} + \frac{\partial M_{2,i}}{\partial x_1} && \text{in } \Omega_i \\ u_i &= u_j && \text{on } \Gamma_{ij} \\ \nu(|\nabla u_i|)\nabla u_i\cdot n_i - M_i\cdot\tau_i & \\ = \nu(|\nabla u_j|)\nabla u_j\cdot n_j - M_j\cdot\tau_j & && \text{on } \Gamma_{ij} \\ u &= 0 && \text{on } \partial\Omega \end{aligned}$$

In this system of equations M denotes the magnetization if the domain is a permanent magnet, J_3 is the only nonzero component of the current density vector and n_i stands for the outer normal vector and τ_i for the tangential vector of the subdomain Ω_i. The reluctivity ν depends nonlinearily on u. The second condition states that the solution u does not have jumps on the interface between two domains. The third condition characterizes the flux in normal direction.

II. Nonlinear materials

During computation we deal with the dependency of ν on u in nonlinear materials by applying a Newton method. For the termination criterion we measure the L^2-norm of the residual

$$F(.) := b - A(\nu(|\nabla u_{n-1}|))\,u_n.$$

Where b denotes the right-hand-side of the multidomain formulation and A the matrix representing the linearized system. The index $n, n-1$ refers to the solution of the Newton method in the n and $n-1$ step respectively. If the residual of the last computed u_n for the current rotor-stator-constellation is smaller than ϵ times the residual of the initial guess u_{00} in the L^2-norm:

$$F(u_n) \le \epsilon F(u_{00})$$

we accept the last computed u_n as solution. Here the tolerance parameter ϵ is chosen 10^{-6}.

III. Domain decomposition techniques

We want to approximate a full rotation of the motor without remeshing the whole domain in every discrete step. One possibility to do this is the moving band technique [1]. The idea is to fix the meshes in two adjacent domains and only generate a new mesh in a small region between those. If we can impose an equidistant mesh on the two adjacent parts we can even use the same mesh for each calculation. Using this approach we have a conforming mesh in every rotation step. The downside of this technique is that we cannot use nonconforming meshes and – as mentioned in [2] – in the 3D case the remeshing for each calculation is impractical.
Focusing on domain decomposition techniques which allow us to generate one mesh for each subdomain independently for all rotational displacements we find three major approaches:

- the mortar element method
- the discontinuous Galerkin method
- the Nitsche-type mortaring.

In the framework of the mortar element methods on the interface between two domains one domain is defined as the masterside and one as the slave. To glue both domains together Lagrange-multipliers are used. In the standard form this leads to a saddle point problem. By choosing special shape functions [3] for the Lagrange multipliers this saddle point form can be avoided [2].
The classical discontinuous Galerkin approach is formulated for each triangle of the mesh, see also [4]. The ansatz functions are discontinuous over the element interfaces. Stability is guaranteed by adding penalisation terms for the jumps of the solution between to adjacent triangles in the bilinear form. This can be easily generalized to the domain level.

IV. Nitsche-type mortaring

Originally the idea of Nitsche was to incorporate Dirichlet boundary data directly into the variational formulation and not – as usual – into the function space [5].
The same principle can be applied on an interface between two subdomains. In the literature we can find slightly different approaches all named Nitsche(-type) mortaring (see [6], [7]), we will follow the formulation proposed in [8]. The idea is similar to the incorporation of boundary conditions: The function which was given in the case of incorporating Dirichlet boundary condition is now an unknown function λ on the interface between the two domains. This unknown function represents the solution on the interface $u\big|_{\Gamma}$.
To get a continuous solution on the whole domain we have to ensure that there are no jumps between the solutions on adjacent subdomains. Here the function on the interface comes into play as we enforce the solution on each of these subdomains to match λ.
To show the difference between the standard variational formulation (1) in comparison to Nitsche-type mortaring (2) both are given below. The variational form of the boundary value problem for the single computational domain writes as:

Find $u : \overline{\Omega} \to \mathbb{R}$ such that

$$\int_{\Omega} \left(\nu\left(|\nabla u|\right)\nabla u - (M_1, M_2)_i^T\right) \cdot \nabla v dx = \int_{\Omega} J_3 v dx \quad (1)$$

$\forall v : \overline{\Omega} \to \mathbb{R}, v\big|_{\partial\Omega} = 0;$

whereas the Nitsche-type formulation based on the multidomain approach with the notation $u_i := u\big|_{\Omega_i}$ is given by:

Find $u : \overline{\Omega} \to \mathbb{R}$, $\lambda_{ij} : \Gamma_{ij} \to \mathbb{R}$ such that

$$\begin{aligned}
&\sum_i \Bigg(\int_{\Omega_i} \left(\nu\left(|\nabla u_i|\right)\nabla u_i - (M_1, M_2)_i^T\right) \cdot \nabla v_i dx \\
&\quad - \underbrace{\sum_j \int_{\Gamma_{ij}} \left(\nu\left(|\nabla u_i|\right)\nabla u_i \cdot n_i - (M_1, M_2)_i^T \cdot \tau_i\right)(v_i - \phi_{ij})\, dS}_{(i)} \\
&\quad + \beta \underbrace{\sum_j \int_{\Gamma_{ij}} \left(\nu\left(|\nabla u_i|\right)\nabla v_i \cdot n_i\right)(u_i - \lambda_{ij})\, dS}_{(ii)} \\
&\quad + \alpha \underbrace{\sum_j \int_{\Gamma_{ij}} \nu\left(|\nabla u_i|\right)(u_i - \lambda_{ij})(v_i - \phi_{ij})\, dS}_{(iii)} \Bigg) \\
&= \sum_i \int_{\Omega_i} J_3\ v_i dx \qquad (2)
\end{aligned}$$

$\forall v : \overline{\Omega} \to \mathbb{R}, v\big|_{\partial\Omega} = 0$ and $\forall \phi_{ij} : \Gamma_{ij} \to \mathbb{R}$

The functions ϕ appearing in the formula are the test functions on the boundary.
This formulation with interior interfaces Γ_{ij} can be easily adjusted to the case of symmetry edges.
We observe, that if the exact solution $\overline{u}$ of (1) is smooth enough, then it also solves the formulation of the Nitsche-type mortaring (2). This holds since the term (i) vanishes due to $\frac{\partial \overline{u}_i}{\partial n_i}\big|_{\Gamma_{ij}} = -\frac{\partial \overline{u}_j}{\partial n_j}\big|_{\Gamma_{ij}}$ on the interface between the domain Ω_i and Ω_j.
The terms (ii) as well as (iii) are equal to zero in case of such a solution $\overline{u}$ because $\overline{u}_i - \lambda_{ij} = 0$, considering $\overline{u}_i = \overline{u}_j = \lambda_{ij}$ on Γ_{ij}.
Therefore the formulation (2) is consistent.
Due to the fact that we solve the problem on a mesh we only get a discrete approximation of $\overline{u}$. Therefore the term (iii) will act as a penalisation term.
Via the weight factors α and β we can control the properties of the system. For example by assigning $\beta = -1$ we get a symmetric system. To ensure solvability we have to assign α in a proper way. For instance if $\beta = -1$, α has to be suitably large, whereas in the case $\beta = 1$ there are no restrictions for α.

The right hand side of the resulting system of linear equations can be represented by a matrix of the following form:

$$\begin{pmatrix} A_{\Omega_1} & & \alpha\tilde{C}_{\Gamma_1} + \beta B_{\Gamma_1} \\ & A_{\Omega_2} & \alpha\tilde{C}_{\Gamma_2} + \beta B_{\Gamma_2} \\ \alpha\tilde{C}_{\Gamma_1} - B_{\Gamma_1} & \alpha\tilde{C}_{\Gamma_2} - B_{\Gamma_2} & \alpha\hat{C}_{\Gamma_1} + \alpha\hat{C}_{\Gamma_2} \end{pmatrix} \begin{pmatrix} u_{\Omega_1} \\ u_{\Omega_2} \\ \lambda \end{pmatrix}$$

And the loadvector has the form:

$$\begin{pmatrix} \bar{c}_1(M_1, M_2)^T_{\Omega_1} + \bar{c}_2 J_{\Omega_1,3} - \bar{c}_3(M_1, M_2)^T_{\Omega_1} \cdot \tau_{\Omega_1} \\ \hat{c}_1(M_1, M_2)^T_{\Omega_2} + \hat{c}_2 J_{\Omega_2,3} - \hat{c}_3(M_1, M_2)^T_{\Omega_2} \cdot \tau_{\Omega_2} \\ 0 \end{pmatrix}$$

One of the advantages of this approach is the fact that due to the separation of the meshes by the interface function we do not have to provide matching meshes at the interface. This is of particular interest when simulating electrical machines, since it allows for an arbitrary rotational displacement between the motor parts without remeshing.

V. Model Problem

The motor we want to simulate is depicted in Fig. 1. Material in grey is iron, yellow materials are magnets. Their magnetisation is parallel in the direction as indicated in the picture. There are 12 slots with 4 pole-pairs in the stator – the double layer winding is defined according to the scheme in the figure.

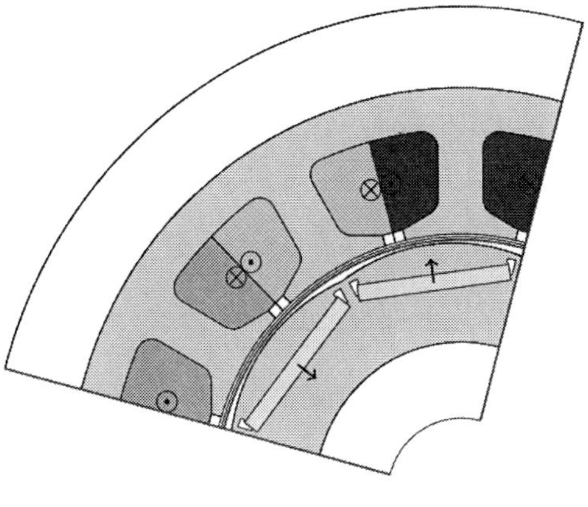

Figure 1. Motor sector (mortar edges depicted in red)

In case of electrical machines we use 3 different meshes; one for the rotor, one for the stator and one for the air gap domain. The rotor and the stator meshes are generated by NETGEN [9] resulting in an unstructured mesh, in the air gap we provide a structured mesh.

VI. Results

To compare the results computed by the finite element program at issue, the motor with the same specification was simulated with FEMAG [10].

FEMAG used a motor mesh with a total of 10552 nodes whereas the mesh for our program consited of 7799 nodes including 181 on the interface and 118 on the symmetry edges. The parameters in the interface integrals were chosen as $\alpha = 100$ and $\beta = 1$.

On the interface between rotor and air gap we get three different solution for the vector potential coming from the mesh on the rotor, one from the air gap mesh and additionaly the function λ on the interface. The differences of these solutions are shown in the following Fig. 2:

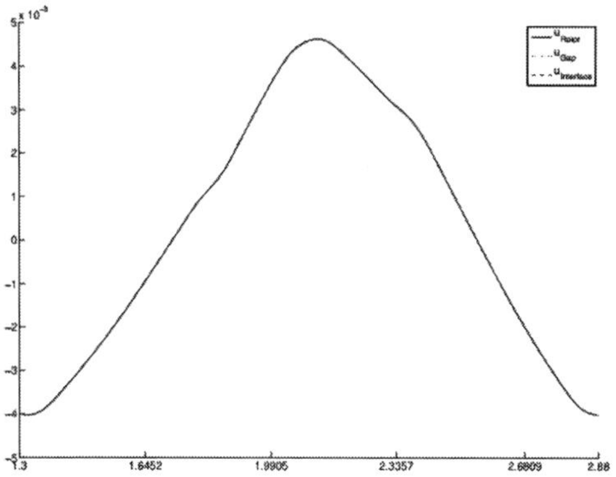

Figure 2. Comparison of the vectorpotential on the interface

In a postprocessing step the torque and linked fluxes are calculated. A comparison of these quantities with those estimated by FEMAG is depicted in Fig. 3 and 4 respectively.

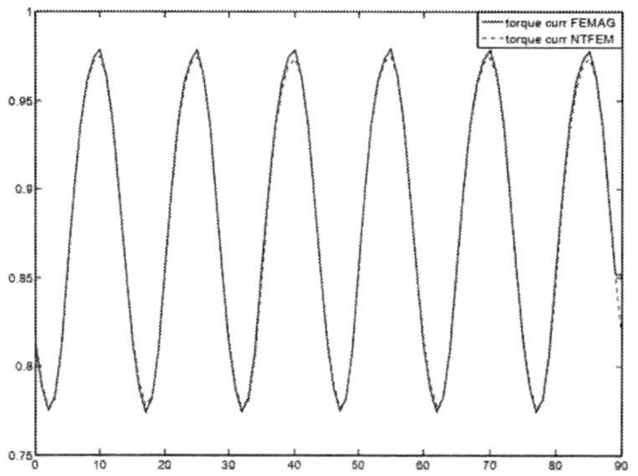

Figure 3. Torque comparision for the motor with currents

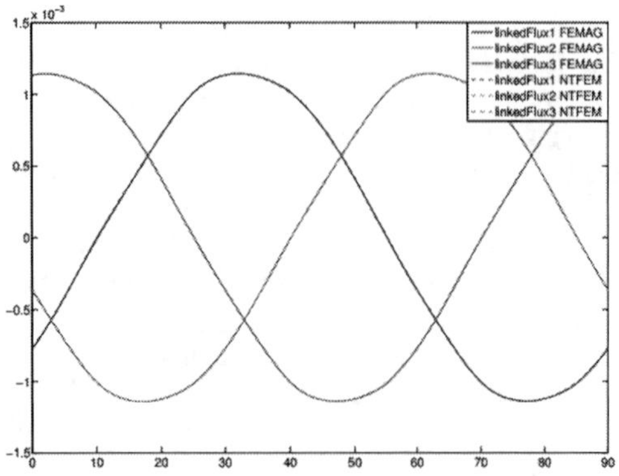

Figure 4. Linked flux comparision for the motor with currents

The differences in the curves of the torque vanish if we refine the mesh sufficiently.

VII. Future Work

Using any sort of domain decomposition technique enables us to use not only nonconforming meshes between the domains but as a result allows an arbitrary rotational displacement between rotor and stator. We will use this advantage in future work to construct more efficient path-following algorithms with respect to ϕ.

Acknowledgement

This work has been supported by the Austrian COMET-K2 program of the Linz Center of Mechatronics (LCM), and was funded by the Austrian federal government and the federal state of Upper Austria.

References

[1] B. Davat, Z. Ren, and M. Lajoie-Mazenc, "The movement in field modeling," *Magnetics, IEEE Transactions on*, vol. 21, no. 6, pp. 2296–2298, Nov 1985.

[2] E. Lange, F. Henrotte, and K. Hameyer, "A variational formulation for nonconforming sliding interfaces in finite element analysis of electric machines," *Magnetics, IEEE Transactions on*, vol. 46, no. 8, pp. 2755–2758, Aug 2010.

[3] B. Wohlmuth, "A mortar finite element method using dual spaces for the lagrange multiplier," *SIAM Journal on Numerical Analysis*, vol. 38, no. 3, pp. 989–1012, 2000. [Online]. Available: http://dx.doi.org/10.1137/S0036142999350929

[4] B. Rivière, *Discontinuous Galerkin Methods for Solving Elliptic and Parabolic Equations*. Society for Industrial and Applied Mathematics, 2008. [Online]. Available: http://epubs.siam.org/doi/abs/10.1137/1.9780898717440

[5] J. Nitsche, "Über ein Variationsprinzip zur Lösung von Dirichlet-Problemen bei Verwendung von Teilräumen, die keinen Randbedingungen unterworfen sind," *Abhandlungen aus dem Mathematischen Seminar der Universität Hamburg*, vol. 36, no. 1, pp. 9–15, 1971.

[6] R. Stenberg, "Mortaring by a method of J.A. Nitsche," *Computational mechanics*, 1998.

[7] K. Pönitz, "Finite-Elemente-Mortaring nach einer Methode von J.A. Nitsche für elliptische Randwertaufgaben," Ph.D. dissertation, Chemnitz University of Technology.

[8] K. Hollaus, D. Feldengut, J. Schöberl, M. Wabro, and D. Omeragic, "Nitsche-type Mortaring for Maxwell's Equations," *Progress In Electromagnetics Research Symposium Proceedings*.

[9] Netgen. [Online]. Available: http://www.hpfem.jku.at/netgen/

[10] Femag. [Online]. Available: http://people.ee.ethz.ch/ femag/

Experimental Verification of a Rotor Unbalance Response Based Approach to the Identification of Magnetic Bearing Support Parameters

Long Di[a], Jin Zhou[b], Changli Cheng[b], Zongli Lin[a]

[a] Charles L. Brown Department of Electrical and Computer Engineering, and Rotating Machinery and Control Laboratory (ROMAC), University of Virginia, Charlottesville, VA 22904-4743, USA, {ld4vv, zl5y}@virginia.edu

[b] College of Mechanical and Electrical Engineering, Nanjing University of Aeronautics and Astronautics, nuaazhoujin@gmail.com, 970053197@qq.com

Abstract—**The support parameters, the stiffness and damping coefficients, of active magnetic bearings (AMBs) have direct influence on the dynamic response and stability of a rotor bearing system. The unbalance response method is a simple but effective way to identify the AMB support parameters. In this paper, we present experimental verification of this identification method. We first briefly introduce the unbalance response method and the error response surface approach to rectifying the identification error induced by the rotor flexibility. The final AMB stiffness and damping coefficients are determined as the sum of the nominal values identified by the unbalance response method and the identification errors derived from the error response surfaces. We will then present the experimental verification of this combined unbalance response and error response surface identification method on a rotor AMB test rig. In order to verify the identification method, the parameters thus identified based on the experimental data are integrated with the rotor finite element model to form the rotor-AMB model. The model unbalance response is obtained through simulation and compared with experimentally measured unbalance response. The close agreement between the model unbalance response and the experimentally measured response demonstrates that the proposed method is effective in identifying the AMB support parameters.**

I. Introduction

Active magnetic bearings (AMBs) find a wide range of applications in energy storage, machine tool operation, artificial heart pumps and hybrid vehicles [1], because of their lubrication free and non-contact operation. The stiffness and damping coefficients of a rotor AMB system have substantial effect on the system dynamic response and stability, and many methods have been reported on their determination. Humphris et al. derived equivalent stiffness and damping coefficients from the AMB system transfer function [2], Yu [3] and Hu et al. [4] analyzed the theoretical AMB static and dynamic characteristics, Wang and Jiang [5] proposed a method to estimate the bearing stiffness using the control current ratio. These methods are based on a simplified model and the support parameters of AMBs are identified only through the theoretical analysis, which leads to inaccurate results. In addition, it is difficult to identify support parameters based on theoretical analysis methods when complex AMB control algorithms are employed.

Compared to the theoretical analysis approach, which requires simplified model of the rotor AMB system, experimental identification approach is an easier but effective way to identify the AMB stiffness and damping coefficients. Wang [6] and Schweitzer et al. [7] confirmed that the AMB stiffness and damping coefficients are closely related to the controller frequency response, with the controller bandwidth having the strongest impact. Yang et al. [8] used the AMB as an exciter for the stiffness identification. Since the excitation force is obtained through the electrical current conversion, magnetic flux leakage causes the actual force to be inconsistent with the theoretical value. Wu [9] and Wang and Gao [10] used the external load method to identify the static stiffness, but they did not consider the dynamic characteristics of magnetic bearings. Zhou and Lin [11] used the hammer impact method to estimate the stiffness and damping of a single degree of freedom (SDOF) AMB system, and then verified the results through the sine sweep frequency response. According to the amplitude-frequency characteristic curve, the estimated system stiffness is close to the value obtained by the hammer impact method, but the study was only limited to an SDOF system. Shen and Yu [12] used the multiple frequency excitation approach, in combination with the least squares method, to identify the AMB support parameters, but a rigid rotor motion equation was used as the rotor model. The drawback for these existing experimental identification approaches is that they mostly do not consider rotor flexibility, which leads to inconsistent results at different rotating speeds.

In summary, there are basically two main approaches to the identification of the AMB support parameters. One is based on the frequency characteristics of the control system, which has been studied through theoretical analysis but lacks experimental verification. The other one is to analyze response to the excitation. The second approach is mostly based on a rigid rotor motion and does not take the rotor flexibility into consideration. Treating the unbalance force an excitation, Zhou et al. [16] developed an unbalance response based method to identify AMB support parameters, which takes the rotor flexibility into consideration during the identification process. The unbalance response based support parameter identification method has already been widely used in mechanical bearings, but few works have been documented for applying this method to AMBs. Because the unbalance force is the simplest form of external excitation, it does not require additional devices, which makes experimental studies easier to conduct [13]–[15].

In this paper, we present experimental verification of the unbalance response based method that was developed in [16] on a flexible rotor AMB test rig. Two sets of balancing weights are attached to the rotor disks to obtain the unbalanced rotor displacements, from which the nominal stiffness and damping coefficients are derived. We then rectify the identification error caused by the rotor flexibility with an error response surface approach. The final parameters thus identified are integrated with the rotor finite element model to form the rotor-AMB model. The model unbalance response is obtained through simulation and compared with experimentally measured unbalance response. The close agreement between the model unbalance response and the experimentally measured response demonstrates that the proposed method is effective in identifying AMB support parameters.

The remainder of this paper is organized as follows. Section II describes the experimental setup and the unbalance response based method we are to verify. Section III presents the results. Section IV draws a conclusion to the paper.

II. The Identification Method

This section contains three subsections. Subsection II-A describes the flexible rotor AMB test rig where the proposed identification method is applied. The identification results are also verified based on this test rig in Sec. III. Subsection II-B introduces a rigid rotor model based identification method to calculate the nominal values of the AMB stiffness and damping coefficients from the measured unbalance response. Subsection II-C adopts a finite element model of the rotor to construct error response surfaces for rectifying the identification errors induced by the rigid rotor modeling of the flexible rotor.

A. Experimental Setup

Shown in Fig. 1 is the experiment setup that we use to verify the identification method described in Section II. The setup contains a five DOF rotor AMB test rig and two sets of balancing weights that are added to the rotor disks. This test rig was built to simulate an industrial high speed centrifugal gas compressor and two radial AMBs (H and A) are located on the driven (DE) and nondriven end (NDE) of the rotor. The nominal bias current is around 3.8 A so the support AMBs' load capacity is around 1336 N. Each AMB is equipped with four Model 422 PWM amplifiers from Copley Controls and each amplifier can generate up to 10 A current and 3.2 kW peak power output to provide the maximum required force slew rate. The rotor movement is monitored by a 10 channel Kaman eddy current sensor system. For each AMB control axis, there is one pair of sensor probes in a differential arrangement to measure the rotor displacement in the direction of the AMB control quadrants. To avoid damage when the rotor drops, two auxiliary bearings are also mounted in the casing with 10 mils clearance. The steel rotor has a length of 1.23 m and weighs around 44.9 kg. The first rigid body mode is around 2500 RPM (41.67 Hz) and the first bending mode is around 14,800 RPM (246.67 Hz). A 3.7 kW Colombo RS-90/2 high speed motor with variable frequency drive (VFD) is used to move the rotor up to 18,000 RPM.

There are two discs attached to the rotor and they serve as the balance planes. For disc 1, the distance between the mounting holes and center is 0.11265 m while for disc 2, it is 0.03799 m. The first set of balancing weights are $m_1 = 1.262$ g and $m_2 = 3.926$ g. The second set are $m_1 = 1.415$ g and $m_2 = 4.103$ g. We are going to identify the supporting stiffness and damping coefficients of the two magnetic bearings A and H.

The digital control system is based on the Innovative Integration M6713 PCI board and a TI C6713B 32-bit floating point digital signal processing (DSP) chip is used for the implementation of the digital control algorithm with an updating frequency around 12 kHz. Besides the DSP, a Labview data acquisition console is also added to interface with the computer for additional data logging. The data logging function is written in Labview 2010 and the sensor measurements and control outputs of all four control axes are saved at 6.4 KHz, which are much higher than the logging function provided by the DSP graphic user interface. They are also converted from time domain response to frequency domain response using the FFT block provided by Labview so it is possible to precisely identify at which frequency the maximum vibration happens. The rotating speed is also saved at 6.4 KHz, so these information can be used to identify the unbalance weight and estimate the stiffness and damping coefficients. Identification of AMB support parameters requires the steady state rotor displacement amplitude and phase information at AMB locations. The displacement amplitudes are obtained through eddy current sensors and the phase information is obtained through fiber optic sensors.

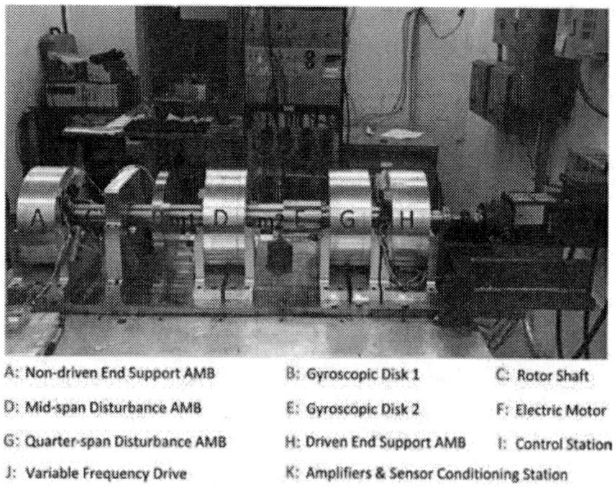

Figure 1: A Five DOF rotor AMB test rig with balancing weights m_1 and m_2.

B. Unbalance Response Based Identification

Shown in Fig. 2 is the schematic of the rotor AMB test rig we are to verify the unbalance response based identification method on.

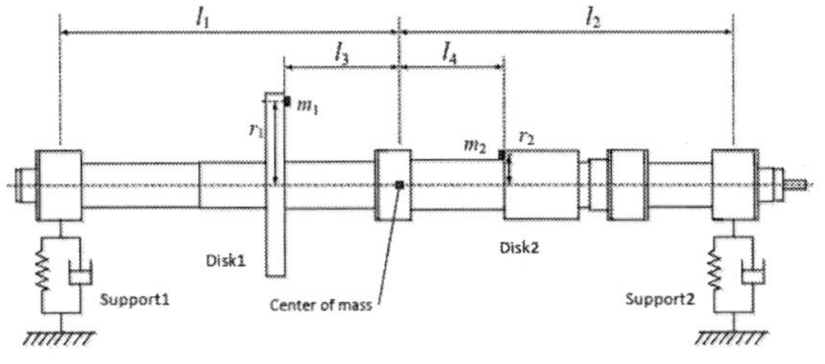

Figure 2: The schematic of the rotor AMB test rig.

In the Fig 2, l_1 and l_2 respectively represent the distances from the AMBs at both ends of the rotor to the rotor centroid, l_3 and l_4 are respectively the distances from disk 1 and disk 2 to the rotor centroid, m_1 and m_2 are the balancing weights attached respectively to the left and right disks, r_1 and r_2 are the distances between the disk centers and locations of the balancing weights.

We will model the AMBs with a combination of springs and dampers. The effects of unbalance force on the rotor center of mass and the effect of AMB force are equivalent to each other. The following equations are derived respectively from the balance of force and the balance of torque,

$$m\ddot{x}_c = f_{ux1} + f_{ux2} + f_{x1} + f_{x2}, \quad (1)$$

$$m\ddot{y}_c = f_{uy1} + f_{uy2} + f_{y1} + f_{y2}, \quad (2)$$

$$I_t\ddot{\theta}_y = -f_{ux1}l_3 + f_{ux2}l_4 - f_{x1}l_1 + f_{x2}l_2 - I_p\Omega\dot{\theta}_x, \quad (3)$$

$$I_t\ddot{\theta}_x = -f_{uy1}l_3 + f_{uy2}l_4 - f_{y1}l_1 + f_{y2}l_2 + I_p\Omega\dot{\theta}_y, \quad (4)$$

where x_c and y_c are the movements of the rotor center of mass in the x and y directions, respectively, m is the rotor mass, f_{ux1}, f_{ux2}, f_{uy1} and f_{uy2} are the unbalance forces generated by the balancing weights, f_{x1}, f_{x2}, f_{y1} and f_{y2} are the supporting forces from the AMBs, θ_x and θ_y are the rotor tilt angles in the x and y axes, I_t and I_p are respectively the transverse and polar moments of inertias, and Ω is the rotor rotating speed.

The AMB supporting forces can be expressed as

$$\begin{aligned} f_{x1} &= -\left(k_{x1}(\Omega)x_1 + c_{x1}(\Omega)\dot{x}_1\right), \\ f_{y1} &= -\left(k_{y1}(\Omega)y_1 + c_{y1}(\Omega)\dot{y}_1\right), \\ f_{x2} &= -\left(k_{x2}(\Omega)x_2 + c_{x2}(\Omega)\dot{x}_2\right), \\ f_{y2} &= -\left(k_{y2}(\Omega)y_2 + c_{y2}(\Omega)\dot{y}_2\right), \end{aligned}$$

where $k_{xi}(\Omega)$, $c_{xi}(\Omega)$, $k_{yi}(\Omega)$ and $c_{yi}(\Omega)$, $i = 1, 2$, are the stiffness and damping coefficients of AMBs in the x and y directions, with $i = 1$ denoting the AMB at the left end and $i = 2$ denoting the AMB at the right end. For notational brevity, in what follows we will often drop the dependence on Ω of these coefficients, and simply write k_{xi}, c_{xi}, k_{yi} and c_{yi}.

Based on Eqs. (1)-(4), the motion equation of the rotor can be further derived in the following complex form [16]

$$M\ddot{q} + (C + \Omega G)\dot{q} + Kq = F_0 e^{j\Omega t}, \quad (5)$$

where M is the rotor mass matrix, C is the support damping matrix, G is the gyroscopic effect matrix, K is the support stiffness matrix, $F_0 e^{j\Omega t}$ is the rotor unbalance force vector, and $q = q_0 e^{j\Omega t}$ are the rotor displacements vector with $q_0 = [x_1\ x_2\ y_1\ y_2]^{\mathrm{T}}$. Matrices M, K, G and C are defined as follows,

$$M = \frac{1}{L}\begin{bmatrix} ml_2 & ml_1 & 0 & 0 \\ -I_t & I_t & 0 & 0 \\ 0 & 0 & ml_2 & ml_1 \\ 0 & 0 & -I_t & I_t \end{bmatrix},$$

$$K = \begin{bmatrix} k_{x1} & k_{x2} & 0 & 0 \\ -k_{x1}l_1 & k_{x2}l_2 & 0 & 0 \\ 0 & 0 & k_{y1} & k_{y2} \\ 0 & 0 & -k_{y1}l_1 & k_{y2}l_2 \end{bmatrix},$$

$$G = \frac{1}{L}\begin{bmatrix} 0 & 0 & 0 & 0 \\ 0 & 0 & -I_p & I_p \\ 0 & 0 & 0 & 0 \\ I_p & -I_p & 0 & 0 \end{bmatrix},$$

$$C = \begin{bmatrix} c_{x1} & c_{x2} & 0 & 0 \\ -c_{x1}l_1 & c_{x2}l_2 & 0 & 0 \\ 0 & 0 & c_{y1} & c_{y2} \\ 0 & 0 & -c_{y1}l_1 & c_{y2}l_2 \end{bmatrix},$$

where L is the rotor span.

The total unbalance excitation vector F_0 is given by

$$F_0 = \begin{bmatrix} f_1 + f_2 \\ -f_1l_3 + f_2l_4 \\ -j(f_1 + f_2) \\ -j(f_1l_3 - f_2l_4) \end{bmatrix}, \quad (6)$$

where

$$\begin{aligned} f_1 &= m_1 r_1 \Omega^2 e^{-j\phi_1}, \\ f_2 &= m_2 r_2 \Omega^2 e^{-j\phi_2}, \end{aligned}$$

with ϕ_1 and ϕ_2 being the phase angles for the two balancing weights.

Substitution of $q = q_0 e^{j\Omega t}$ into (5) results in

$$j\Omega C q_0 + K q_0 = F_0 + (\Omega^2 M - j\Omega^2 G)q_0. \quad (7)$$

The values of the AMB support parameters can be solved from expanded Eq. (7), which can be reorganized into the following two equations,

$$Q_x P_x = F_x, \quad (8)$$

$$Q_y P_y = F_y, \quad (9)$$

where

$$Q_x = \begin{bmatrix} x_1 & x_2 \\ -l_1x_1 & l_2x_2 \end{bmatrix},$$

$$P_x = \begin{bmatrix} k_{x1} + j\Omega c_{x1} \\ k_{x2} + j\Omega c_{x2} \end{bmatrix},$$

$$Q_y = \begin{bmatrix} y_1 & y_2 \\ -l_1 y_1 & l_2 y_2 \end{bmatrix},$$

$$P_y = \begin{bmatrix} k_{y1} + j\Omega c_{y1} \\ k_{y2} + j\Omega c_{y2} \end{bmatrix},$$

$$F_x = \begin{bmatrix} f_1 + f_2 \\ -f_1 l_3 + f_2 l_4 \end{bmatrix} + \frac{\Omega^2}{L} \begin{bmatrix} ml_2 & ml_1 & 0 & 0 \\ -I_t & I_t & -jI_p & jI_p \end{bmatrix} \begin{bmatrix} x_1 \\ x_2 \\ y_1 \\ y_2 \end{bmatrix},$$

$$F_y = \begin{bmatrix} -j(f_1 + f_2) \\ -j(f_1 l_3 - f_2 l_4) \end{bmatrix} + \frac{\Omega^2}{L} \begin{bmatrix} 0 & 0 & ml_2 & ml_1 \\ jI_p & -jI_p & -I_t & I_t \end{bmatrix} \begin{bmatrix} x_1 \\ x_2 \\ y_1 \\ y_2 \end{bmatrix}.$$

Then we can solve P_x and P_y from Eqs. (8) and (9) to get the nominal values of the support parameters k_{xi}, c_{xi}, k_{yi} and c_{yi}, $i = 1, 2$,

$$P_x = Q_x^{-1} F_x, \quad P_y = Q_y^{-1} F_y.$$

We denote the nominal values as $k_{xi}^0(\Omega)$, $c_{xi}^0(\Omega)$, $k_{yi}^0(\Omega)$ and $c_{yi}^0(\Omega)$, $i = 1, 2$.

C. *Parameter Rectification Based on Response Surface*

In the unbalance response identification procedure described in the previous subsection, the rigid rotor model is adopted, which is expected to result in substantial identification errors. We will adopt the response surface approach to improve the precision. The response surface approach relies on the finite element model of the flexible rotor.

Because the coupling between the x and y axes of a radial AMB is small, we will establish an error response surface based on four support parameters in the x and y axes independently. Four control quadrants are assigned to each AMB and each control channel covers the two opposing quadrants. The top two quadrants both have 45° offset from the vertical so the two control channels can evenly carry the rotor weight, which makes the dynamics for both x and y axes almost identical. Therefore, the error response surfaces in the x and y axes can be established similarly. The error response surfaces in the x axes are formulated as follows,

$$\Delta k_{x1} = F_1(k_{x1}^0, k_{x2}^0, c_{x1}^0, c_{x2}^0),$$
$$\Delta k_{x2} = F_2(k_{x1}^0, k_{x2}^0, c_{x1}^0, c_{x2}^0),$$
$$\Delta c_{x1} = F_3(k_{x1}^0, k_{x2}^0, c_{x1}^0, c_{x2}^0),$$
$$\Delta c_{x2} = F_4(k_{x1}^0, k_{x2}^0, c_{x1}^0, c_{x2}^0),$$

where $F_i(k_{x1}^0, k_{x2}^0, c_{x1}^0, c_{x2}^0)$, $i = 1, 2, 3, 4$, are polynomials in terms of the four nominal AMB support parameters k_{x1}^0, k_{x2}^0, c_{x1}^0 and c_{x2}^0, as identified by the unbalance response method from the experimental measurements.

We approximate these error response surfaces with polynomials of fifth degree in the normalized nominal stiffness and damping coefficients $\tilde{k}_{x1}$, $\tilde{k}_{x2}$, $\tilde{c}_{x1}$, $\tilde{c}_{x2}$. Take Δk_{x1} at speed Ω as an example, we have

$$\begin{aligned} \Delta k_{x1} = {} & \beta_0(\Omega) + \beta_1(\Omega)\tilde{k}_{x1} + \beta_2(\Omega)\tilde{k}_{x2} + \beta_3(\Omega)\tilde{c}_{x1} \\ & + \beta_4(\Omega)\tilde{c}_{x2} + \cdots + \beta_{17}(\Omega)(\tilde{k}_{x1})^5 + \beta_{18}(\Omega)(\tilde{k}_{x2})^5 \\ & + \beta_{19}(\Omega)(\tilde{c}_{x1})^5 + \beta_{20}(\Omega)(\tilde{c}_{x2})^5 + \beta_{21}(\Omega)\tilde{k}_{x1}\tilde{k}_{x2} \\ & + \beta_{22}(\Omega)\tilde{k}_{x1}\tilde{c}_{x1} + \cdots + \beta_{26}(\Omega)\tilde{c}_{x1}\tilde{c}_{x2}. \end{aligned}$$

The procedure for determining the coefficients $\beta_i(\Omega)$, $i = 0, 1, \cdots, 26$, is described in detail in [16] and is briefly summarized in Fig. 3.

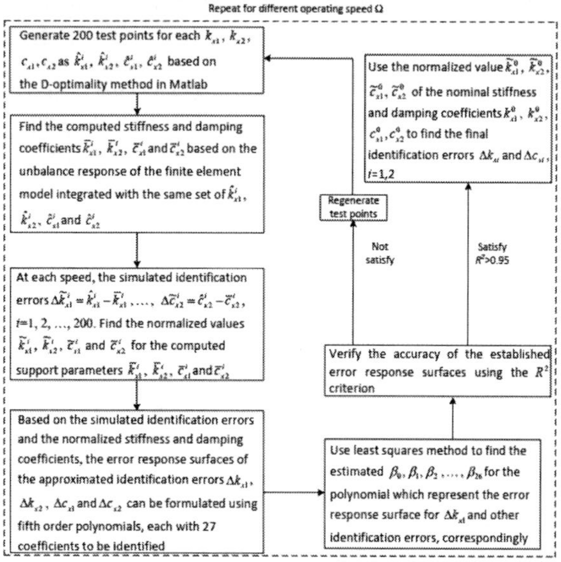

Figure 3: The procedure for determining the coefficients of the error response surfaces.

With these coefficients determined, the identification error can be calculated as

$$\begin{aligned} \Delta k_{x1}(\Omega) = {} & \beta_0(\Omega) + \beta_1(\Omega)\tilde{k}_{x1}^0(\Omega) + \beta_2(\Omega)\tilde{k}_{x2}^0(\Omega) \\ & + \beta_3(\Omega)\tilde{c}_{x1}^0(\Omega) + \beta_4(\Omega)\tilde{c}_{x2}^0(\Omega) + \cdots \\ & + \beta_{17}(\Omega)(\tilde{k}_{x1}^0(\Omega))^5 + \beta_{18}(\Omega)(\tilde{k}_{x2}^0(\Omega))^5 \\ & + \beta_{19}(\Omega)(\tilde{c}_{x1}^0(\Omega))^5 + \beta_{20}(\Omega)(\tilde{c}_{x2}^0(\Omega))^5 \\ & + \beta_{21}(\Omega)\tilde{k}_{x1}^0(\Omega)\tilde{k}_{x2}^0(\Omega) + \cdots \\ & + \beta_{26}(\Omega)\tilde{c}_{x1}^0(\Omega)\tilde{c}_{x2}^0(\Omega), \end{aligned}$$

where $\tilde{k}_{x1}^0(\Omega)$, $\tilde{k}_{x2}^0(\Omega)$, $\tilde{c}_{x1}^0(\Omega)$, $\tilde{c}_{x2}^0(\Omega)$ are normalized values of $k_{x1}^0(\Omega)$, $k_{x2}^0(\Omega)$, $c_{x1}^0(\Omega)$ and $c_{x2}^0(\Omega)$.

The other error response surfaces on the x axis, $\Delta k_{x2}(\Omega)$, $\Delta c_{x1}(\Omega)$ and $\Delta c_{x2}(\Omega)$, can be formulated and obtained in the same procedure.

The final identified AMB stiffness and damping coefficients on the x axis can be obtained as,

$$k_{x1}(\Omega) = k_{x1}^0(\Omega) + \Delta k_{x1}(\Omega),$$

$$k_{x2}(\Omega) = k_{x2}^0(\Omega) + \Delta k_{x2}(\Omega),$$
$$c_{x1}(\Omega) = c_{x1}^0(\Omega) + \Delta c_{x1}(\Omega),$$
$$c_{x2}(\Omega) = c_{x2}^0(\Omega) + \Delta c_{x2}(\Omega).$$

III. Experimental Verification

The experimental identification process for the stiffness and damping coefficients is summarized in Fig. 4.

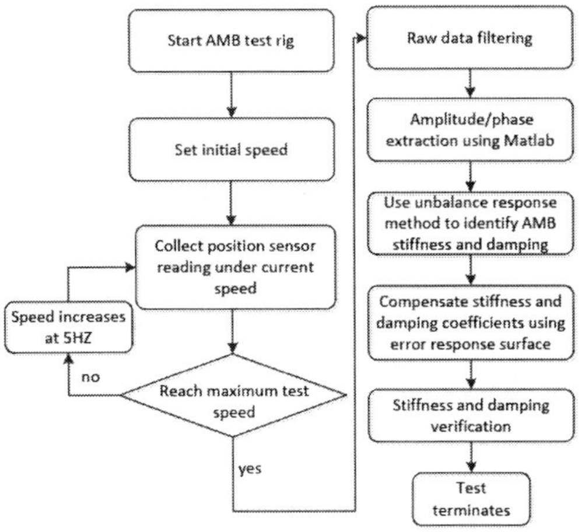

Figure 4: Experimental identification process.

A. Identification Results

The data logging starts from 600 rpm to 6000 rpm. The rotor displacements and rotating speed are saved every 300 rpm for around 10 seconds. The measured displacements go through unbalance signal filtering using Kaiser bandpass filters and zero phase digital filters to extract the steady state amplitude and phase information. Based on the experimental measurements, the nominal values of the stiffness and damping coefficients can be identified using the unbalance response method described in Subsection II-B.

To construct the error response surfaces as described in Subsection II-C, we create a finite element model of the test rig in MSC.Patran. The stiffness and damping coefficients of AMBs are closely related to the feedback control gains. Generally, the control gains for each control channel are slightly different, so the stiffness and damping coefficients for each axis are different. Therefore, we used different test points in constructing the error response surfaces for the support parameters. Two virtual balancing weights are attached to the model discs and then we analyze the unbalance response through the finite element model from 600 rpm to 6000 rpm with an interval of 300 rpm. Based on the rectification method proposed in Subsection II-C, we can develop the error response surfaces for different operating speeds.

The identified stiffness and damping coefficients for the two sets of balancing weights are shown in Figs. 5 and 6. It is observed that the AMB stiffness coefficients on both axes increase with the operating speed because the frequency of the unbalance force is directly related to the operating speed.

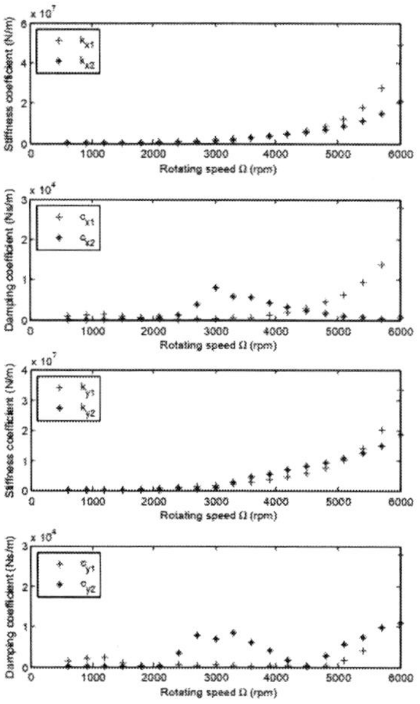

Figure 5: Identified stiffness and damping coefficients for Set 1 of balancing weights. The upper two plots: k_x and c_x; The lower two plots: k_y and c_y.

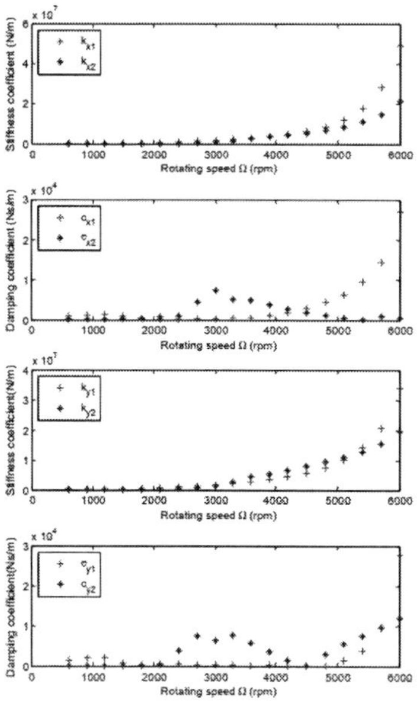

Figure 6: Identified stiffness and damping coefficients for Set 2 of balancing weights. The upper two plots: k_x and c_x; The lower two plots: k_y and c_y.

B. Verification of the Identification Results

To verify the accuracy of the identification results, we integrate identified stiffness and damping coefficients with the finite element model of the rotor to form a full model of the rotor-AMB test rig. Based on this model, we obtain the model unbalance response and compare them with the experimental measurements. The comparison results are shown in Figs. 7 and 8.

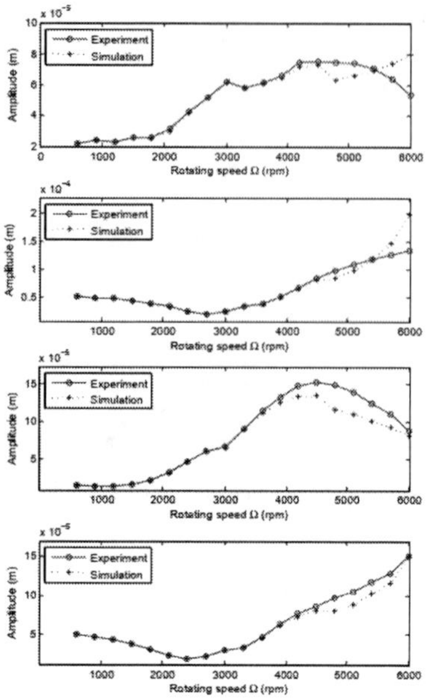

Figure 7: Comparison of displacements between experiments and simulation under Set 1 of balancing weights. The upper two plots: x_1 and x_2; The lower two plots: y_1 and y_2.

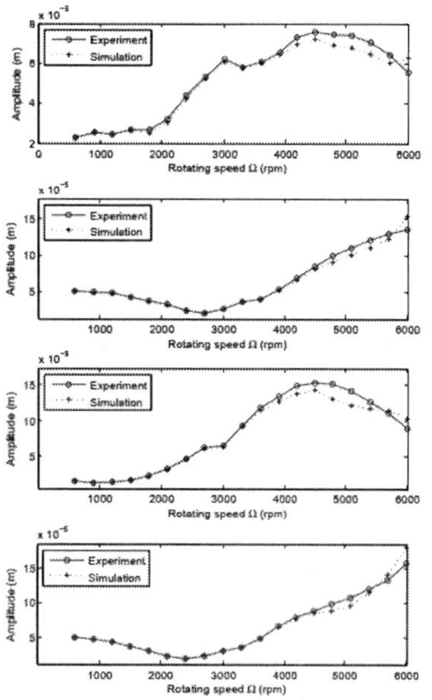

Figure 8: Comparison of displacements between experiments and simulation under Set 2 of balancing weights. The upper two plots: x_1 and x_2; The lower two plots: y_1 and y_2.

It can be observed that the unbalance response from the finite element model closely agrees with the experimental measurements at speeds ranging from 600 rpm to 6000 rpm. This verifies the effectiveness of the identification method.

IV. Conclusions

This paper presented the experimental verification of an unbalance response based identification method we recently proposed to identify AMB stiffness and damping coefficients. Considering the rotor flexibility, the final AMB support parameters are determined as the sum of the nominal values calculated from the experimental unbalance response of a flexible rotor AMB test rig and the identification errors derived from the error response surfaces. The identified support parameters are combined with the finite element model of the test rig to generate unbalance response and they are compared with experimental measurements. Their close agreements indicate that the proposed method is effective in identifying the AMB stiffness and damping coefficients.

V. Acknowledgements

This work was in part supported by National Natural Science Foundation of China (No. 51075200) and the Jiangsu provincial Natural Science Foundation (No. BK2011070).

References

[1] G. Schweitzer and E. H. Maslen, *Magnetic Bearings*. Springer, 2009.
[2] R. Humphris, R. D. Kelm, and D. W. Lewis, "Effect of control algorithms on magnetic journal bearing properties," *Journal of Engineering for Gas Turbines and Power, Transactions of the ASME*, vol. 108, pp. 624–632, 1986.
[3] L. Yu, *Controllable Magnetic Levitationrotor System*. Beijing: Science Press, 2003.
[4] Y. Hu, Z. Zhou, and Z. Jiang, *Foundation Theory and Application of Active Magnetic Bearing*. Beijing: China Machine Press, 2006.
[5] H. Wang and Z. Jiang, "Research on the electromagnetic force and the stiffness of the active magnetic bearing," *Journal of Tianjin University*, vol. 10(4), pp. 15–20, 1992.
[6] X. Wang, "Parametric Design and Application Investigation of Electromagnetic Bearing System," Ph.D. Dissertation, Xi'an Jiaotong University, Xi'an, 1994.
[7] G. Schweitzer, R. Siegwart, and R. Herzog., "Stiffness of magnetic bearings," in *Proc. of 4th Int. Symposium on Magnetic Bearings*, Switzerland, 1994, pp. 251–256.
[8] Z. Yang, L. Zhao, and H. Zhao, "Automatic measurement of dynamic stiffness for active magnetic bearings," *Chinese Journal of Mechanical Engineering*, vol. 37(3), pp. 25–29, 2001.
[9] H. Wu, "Study on dynamic characteristics for AMB suspension rotor," Ph.D. Dissertation, Wuhan University of Techonology, Wuhan, 2005.
[10] X. Wang and C. Gao, "Measurement methodof stiffnessfor magnetic bearings," *China Mechanical Engineering*, vol. 21(8), pp. 889–891, 2010.
[11] J. Zhou and J. Lin, "The theoretical analysis and experimental research on the supporting characteristics in axial magnetic bearing," *Machine Design and Research*, vol. 26(5), pp. 71–73, 2010.
[12] Y. Shen and L. Yu, "Identification of dynamic coefficients of active magnetic bearings," *Machine Tool and Hydraulics*, vol. 6, pp. 8–11, 2001.
[13] O. D. Santiago and L. S. Andres, "Field methods for identification of bearing support parameters-part i: Identification from transient rotor dynamic response due to impacts," *Journal of Engineering for Gas Turbines and Power*, vol. 129, pp. 205–212, 2007.
[14] Z. Li, J. Zhang, and C. Lu, "Study on identification of dynamic characteristics of flexible rotor system from the rotor's disequilibrium response," *Lubrication Engineering*, vol. 5, pp. 23–25, 1999.
[15] O. D. Santiago and L. S. Andres, "Field methods for identification of bearing support parameters-part ii: Identification from rotor dynamic response due to imbalances," *Journal of Engineering for Gas Turbines and Power*, vol. 129, pp. 213–219, 2007.
[16] J. Zhou, L. Di, C. Chen, Y. Xu, and Z. Lin, "A rotor unbalance response based approach to the identification of the stiffness and damping coefficients of active magnetic bearings," submitted for publication.

Modeling and Control of Magnetic Bearings with Nonlinear Magnetization

Ali Gerami, Paul Allaire, Roger Fittro

Ph.D. Candidate Dept. of Mechanical and Aerospace Eng., University of Virginia Charlottesville, VA 22903, ag4nt@virginia.edu
Chief Technical Officer, Rotor Bearing Solutions International, Charlottesville, VA 22903, paul.allaire41@gmail.com
Associate Director, Romac laboratory, Dept. of Mechanical and Aerospace Engineering University of Virginia, Charlottesville, VA 22903, rlf9w@virginia.edu,

***Abstract*- In this paper a nonlinear modeling approach is proposed to operate magnetic bearings above the linear range and near the material saturation point. The goal is to get higher load capacity from an existing AMB system by just changing the software. In order to do this, the system should be operated in the nonlinear region of the magnetization curve and therefore should be modeled nonlinearly. Simulation results show approximately twice the force capability. The controller designed using this approach has a much better transient response and domain of attraction.**

***Keywords*- Nonlinear Modeling; Magnetic Bearing; Generalized Lur'e Method; Linear Matrix Inequalities; Balance Beam**

I. INTRODUCTION

In this paper we introduce a new nonlinear modeling and control method to increase the load capacity of magnetic bearings. To increase the load, the electric current must be increased. This extra electric current and therefore extra flux density makes the system highly nonlinear. Therefore there is a need for a nonlinear model that can accurately model these higher operating conditions. Since it is a common industrial practice to include amplifiers with more capacity than needed for AMB systems as a safety factor, our proposed method can potentially increase the load capacity by just a change in the software. Therefore the proposed method can offer a low cost solution for occasional harsh situations with high transient loads that need to be handled by the AMB. In this study, a nonlinear model for magnetic journal bearings is developed considering the core material nonlinear behavior. This nonlinear behavior is shown in Fig. 1.

Extensive research has been reported to take into account AMB nonlinearities. The force created by an electromagnet is proportional to the square of the electric current and inversely proportional to the square of the air-gap between the rotor and the stator. The AMB force is thus nonlinear, and leads to some complexities in control synthesis. Much research has been devoted to the synthesis of such control laws. The majority of the research in nonlinear AMB's has considered current and airgap nonlinearity. The research done by Yin [1], Inoue [2], Abdelfatah [3], Hong [4], Smith [5] are just a few examples. The force, displacement, and current relationship have usually been studied in single-DOF systems. Walsh [6] considered the geometric coupling of a 2-DOF AMB and studied the changes in stiffness and the resulting bifurcation. Abdelfatah [3] studied the nonlinear oscillations caused by the gyroscopic effect. He [7] and Huang [8] modeled and controlled 5 and 6-DOF systems. Other researches have considered geometric nonlinearities by Hu [9], amplifier nonlinearities by Inoue [2], hysteresis by Wang [10] and control system delays by Tsuyoshi [2,11] and Zheng [12,13]. More recently, self-sensing magnetic bearings by Noh [14] and contact between rotor and the auxiliary bearing by Foiles [15] have been studied. Steinschaden [16] and some others considered the magnetic flux saturation by using a simple bilinear model. Widger [17] and Rivas [18] incorporated the material saturation by curve fitting. Even though magnetization nonlinearity is modeled in some of the previous work, these models are not used for controller development. In this paper, the magnetization nonlinearity is modeled precisely, using the Lur'e method. This method makes the model more suitable for control design, which in turn enables the design of controllers that significantly improve the load capacity of the system.

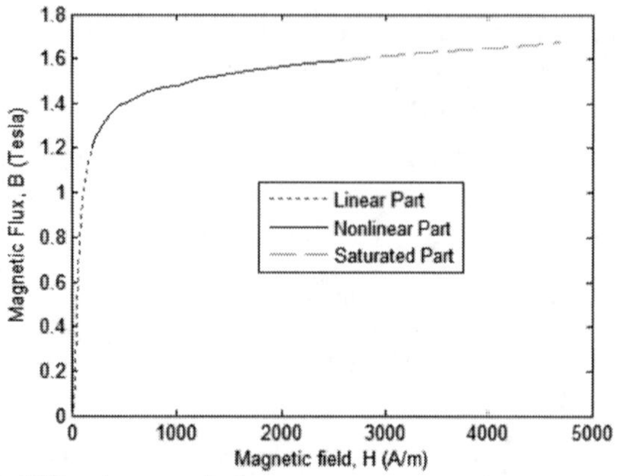

Fig. 1 Silicon iron magnetization curve

II. Dynamic Model and Constraints

A. Introduction

In this paper a balance beam is used for dynamic modeling and simulation. The mechanical part is basically a beam free to rotate on a pivot. The angle of the beam with the horizon is controlled by the two electromagnets on both ends of the beam. This system emulates the dynamics of a single DOF thrust magnetic bearing. The equation of motion can be written as follows:

$$J\ddot{\theta} = -D\dot{\theta} + L_a\left(F_2 - F_1\right) \tag{1}$$

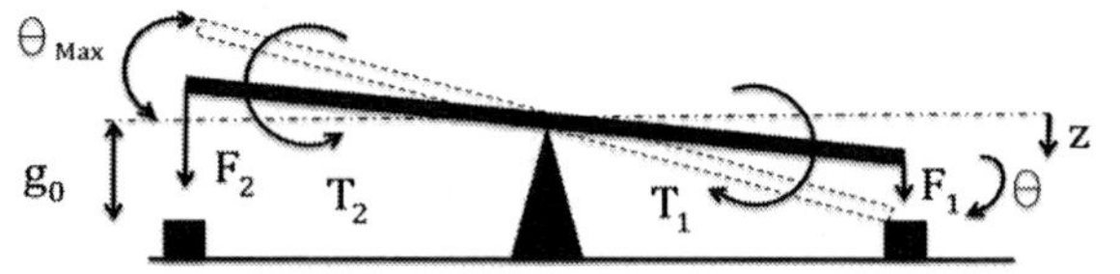

Fig. 2 The balance beam

Here J is the mass moment of inertia of the beam, θ is the angle between the beam and the horizontal direction. D is the system damping due to the pivot and air friction. F_1 and F_2 are the left and right electromagnetic forces and L_a is the distance between electromagnets and pivot (see Fig. 2).

B. Nonlinear Force Model

Since the proposed model works in the nonlinear B-H region, the flux density is very high and the magnetic reluctance of the core material cannot be neglected. Therefore the flux density in the electromagnets can be calculated as follows:

$$B_{1,2} = \frac{NI_{1,2}}{\frac{2\left(g_0 \pm z\right)}{\mu_0} + \frac{L_s}{\mu_r}} \tag{2}$$

Here B_1 and B_2 are the flux density of the electro magnets, L_s is the length of silicon iron in the magnetic circuit, N is the number of turns in each coil, μ_0 is the permeability of free space, μ_r is the relative permeability of silicon iron, g_0 is the airgap, $I_1 = I_b + I_c$ and $I_2 = I_b - I_c$ are the electric currents, I_c is the control current, I_b is the bias current, L_s / μ_r is the magnetic reluctance of the silicon iron, and $2g_0 / \mu_0$ is the magnetic reluctance of the airgap.

In the nonlinear region of the B-H curve the relative permeability of the silicon iron can be modeled as follows:

$$\mu_r = aB + b \tag{3}$$

Here a and b are two constants that were found by curve fitting on the nonlinear region of the B-H curve. The actuator force is related to the flux density by the following equation:

$$F = \frac{A}{\mu_0} B^2 \tag{4}$$

By using a combination of Eqs. 2,3, and 4, the force of each electromagnet can be calculated using the following equations:

$$F_1 = \frac{1}{g_0^2\left(1 - \frac{2z}{g_0}\right)}\begin{bmatrix} c_1 I_1^2 - c_2 z + c_3 \\ +\left(c_4 I_1 - c_5 z + c_6\right)\sqrt{c_7 I_1^2 + c_8 I_1 z - c_{10} z + c_{11}} \end{bmatrix} \tag{5}$$

$$F_2 = \frac{1}{g_0^2\left(1 + \frac{2z}{g_0}\right)}\begin{bmatrix} c_1 I_2^2 + c_2 z + c_3 \\ +\left(c_4 I_2 + c_5 z + c_6\right)\sqrt{c_7 I_2^2 + c_8 I_1 z + c_{10} z + c_{11}} \end{bmatrix} \tag{6}$$

Here $c_1, c_2, \ldots, c_{11}$ are constants based on system parameters and they can be found in the appendix. z is the displacement of the end of the beam. These force equations are valid for the nonlinear part of the B-H curve. For the linear part of the B-H curve, the conventional equations are used. Figure 3 shows that the proposed nonlinear model closely matched the manufacturer material data. The balance beam dimensions and other physical parameters are listed in Table 1.

Table 1-Balance beam parameters

	Parameter	Value
L_a	Electromagnet distance from the pivot	0.1667 m
D	Pivot friction coefficient	0 Ns/m
J	Moment of inertia	16.32 gm^2
N	Number of turns per coil	110 Turns
g_0	Airgap	0.397 mm
L_s	Distance the flux travels inside the core	0.178 m
I_b	Bias current	1.55 A
K_s	Stiffness	1236540 N/m
I_{max}	The maximum electric current	9.61 A
θ_{max}	The maximum angle of rotation	0.0024 Rad

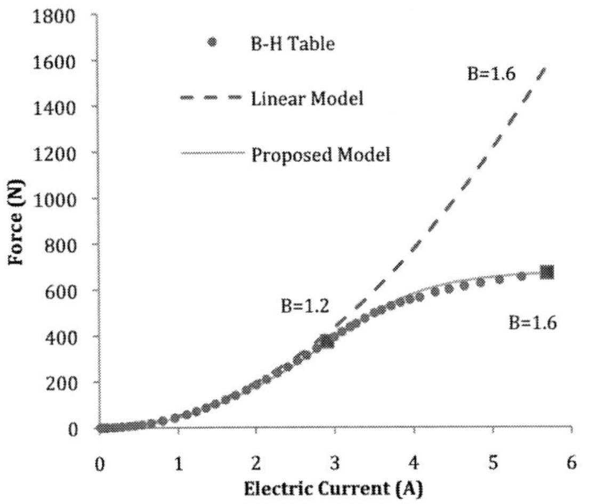

Fig. 3 Each electromagnet's force vs., coil current

III CONTROL

The Lur'e method was used for nonlinear modeling. The control problems described by the Lur'e method have a forward path that is linear and time-invariant, and a feedback path that contains a memoryless, possibly time-varying, nonlinearity (see Fig. 4). In the figure, A is the state matrix, B_u is the input matrix, B_w is the disturbance input matrix, and C_y is the output matrix. This method breaks the nonlinear model into linear and nonlinear parts, which simplifies the control synthesis process.

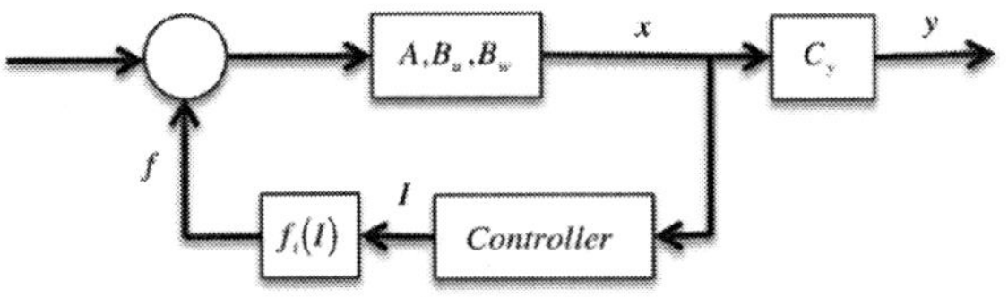

Fig. 4 Balance beam in terms of Lur'e problem

The total balance beam force can be calculated as follows:

$$F = K_s z + f_i(I_c) \tag{7}$$

$$z = L_a \theta \tag{8}$$

Here, F is the net actuator force ($F = F_1 - F_2$). K_s can be obtained by linearizing the force around its equilibrium point, and $f_i(I_c)$ is the nonlinear force due to the electric current assuming no displacement. The balance beam can be modeled into the Lur'e system formulation as follow,

$$\begin{bmatrix} \dot{\theta} \\ \ddot{\theta} \end{bmatrix} = \begin{bmatrix} 0 & 1 \\ \frac{L_a^2 K_s}{J} & -\frac{D}{J} \end{bmatrix} \begin{bmatrix} \theta \\ \dot{\theta} \end{bmatrix} + \begin{bmatrix} 0 \\ \frac{L_a}{J} \end{bmatrix} f_i \tag{9}$$

In terms of the Lur'e problem, $f_i(I_c)$ is the nonlinear input and should therefore satisfy the sector condition. The sector condition constraints the system within two sector bounds which normally are two straight lines. In addition to the regular sector condition, the generalized sector condition that is developed by Lin et al. [19] is used in this work (see Fig. 5). The generalized condition includes using two segmented lines instead of two straight lines, which enables the resulting control design to be less conservative.

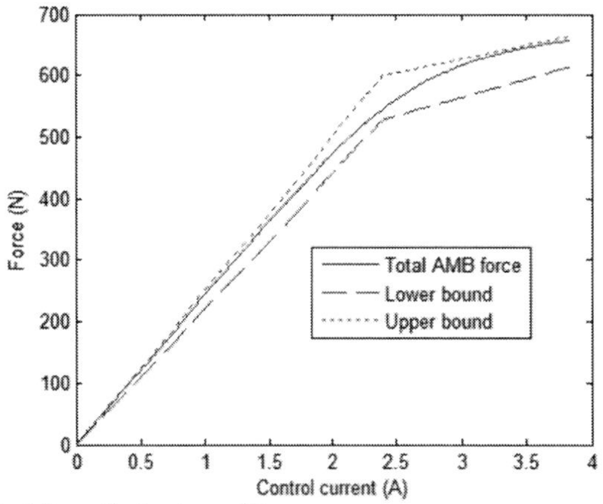

Fig. 5 Generalized sector condition

In addition to the sector condition, there are other physical constraints that need to be satisfied to make sure 1) the beam will not contact the electromagnets, 2) the coils will not overheat, 3) and the total current in each electromagnet is positive. These conditions are expressed mathematically as follows:

$$|I_c| \le I_{\max} \tag{10}$$

$$I_1, I_2 > 0 \tag{11}$$

$$|\theta| \le \theta_{\max} \tag{12}$$

Here $I_{\max}$ is the maximum tolerable current without overheating the coil and $\theta_{\max}$ is the beam's maximum allowable rotation angle.

IV Results

By using linear matrix inequalities and linear programming, an optimized control for the system of Equations (9) is designed that satisfies the physical constraints. The conventional linear model is compared to the nonlinear model with both normal and general sector conditions. The first objective was to obtain the largest possible domain of attraction (invariant ellipsoid). In other words, a controller is designed that can provide the largest initial angular displacement which also ensures system stability. The following optimization problem is solved to achieve this objective:

$$\begin{aligned}
&\sup_{P>0,F} \alpha \\
&(a)\ \alpha x_i \in \varepsilon(P), i = 1,2,\cdots,l \ \ and \ \ x_i \in X_0 \\
&(b)\ (A + k_1 BF)^T P + P(A + k_1 BF) \le 0 \\
&\quad\ (A + k_2 BF)^T P + P(A + k_2 BF) \le 0 \qquad (13) \\
&(c)\ \varepsilon(P) \subset L(F) \\
&(d)\ \varepsilon(P) \subset L(G)
\end{aligned}$$

Here, P is a positive definite variable, $\varepsilon(P)$ is the invariant ellipsoid, and G is the state constraint matrix which in here can be defined as $G = [1/\theta_{\max}\ \ 0]$. In Part (a) of the optimization problem of Equation (12), the x_i's are chosen points that determine the shape of the domain of attraction (the ratio of the major and minor axis of the invariant ellipsoid). The optimization problem maximizes α which corresponds to the size of the invariant ellipsoid. Part (b) guarantees the stability of the system on the normal sector bounds. k_1 and k_2 are the slopes of the two sector bounds. For generalized sector bounds the following constraint should be used instead of constraint (b):

$$\begin{aligned}
&b1)\ (A + k_{0m} BF)^T P + P(A + k_{0m} BF) < 0 \\
&b2)\ (A + BH_{im})^T P + P(A + BH_{im}) < 0,\ i \in I[1,N] \qquad (14) \\
&b3)\ \varepsilon(P) \subset \bigcap_{i=1}^{N} L\left(\frac{H_{im} - k_{im} F}{c_{im}}\right)
\end{aligned}$$

Here, H_{im} is a new variable in the optimization problem, k_{im} is the slope of different segments of the sector bound and c_{im} is the y-intercept of these segments. For example k_{31} is the slope of the fourth segment of the first sector bound and c_{12} is the y-intercept of second segment of the second sector bound. For the first segment $i = 0$ and since it passes through the origin, the y-intercept is zero.

Part (c) and part (d) define the amplifier and state constraints respectively. The domain of attraction and the constraints are depicted in Fig. 6. Three different control designs are compared. "Nonlinear Regular" is the domain of attraction of the system under a feedback control that is designed using the nonlinear model with regular sector conditions; "Nonlinear Generalized" is the domain of attraction of the system under the controller designed using the nonlinear model with generalized sector conditions, and "Jacobean" marks the domain of attraction obtained by a controller designed based on the linear Jacobean model.

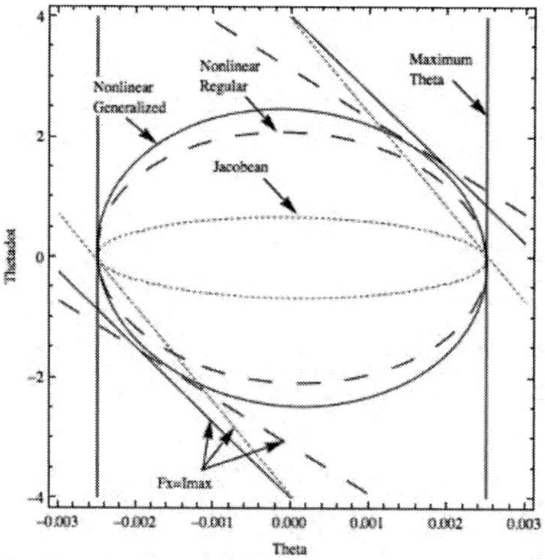

Fig. 6 Domain of attraction for linear and nonlinear models

The sloped lines represent the constraint imposed by the maximum possible electric current. By using the nonlinear model, these constraints are less restrictive and allow larger initial velocities.

The second objective was to obtain the fastest transient response. To achieve this objective the following optimization problem is solved:

$$\sup_{P>0,F} \beta$$

$$(a)\ x_i \in \varepsilon(P), i = 1,2,\cdots,l \ \ and \ \ x_i \in X_0$$

$$(b)\ (A + k_1BF)^T P + P(A + k_1BF) \le -\beta P$$

$$(A + k_2BF)^T P + P(A + k_2BF) \le -\beta P \quad (15)$$

$$(c)\ \varepsilon(P) \subset L(F)$$

$$(d)\ \varepsilon(P) \subset L(G)$$

The various parts of this optimization problem represent the same constraints as the previous optimization problem, but here a larger β represents a better transient response. Therefore β is optimized. As for the previous optimization problem, to use the generalized sector condition, Equation (14) should replace condition (b) in Equation (15). The results show that using the nonlinear model can significantly improve the transient response of the system (see Fig. 7). Systems responses to a step input and zero initial conditions are shown in Figure 7. As can be seen, the two controllers that are designed using the nonlinear model have significantly better transient responses. These control designs result in approximately half of the settling time compared to the linear model, which is a significant improvement.

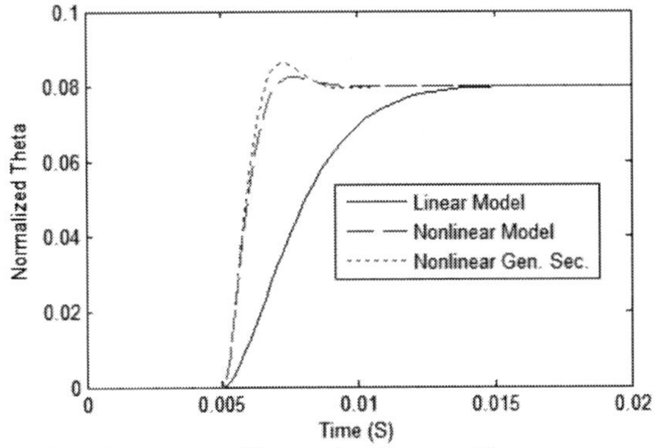

Fig. 7 Transient response of linear and nonlinear models

V CONCLUSIONS

In this paper a new nonlinear model using modeling material magnetization in AMBs was proposed for the first time. State and input constraints were also considered in modeling. LMI optimization and the Lur'e method were utilized for control synthesis. Simulation results illustrate a larger domain of attraction. Also the proposed model enabled controllers to be designed with significantly better transient response. In general, the proposed model provides the system with extra force when it is needed. More importantly, these improvements can be achieved by only changing the control algorithm and no change in the hardware is necessary. This extra capability can potentially be used to design smaller, lighter, and more reliable magnetic bearings.

APPENDIX

Force constants:

$$c_1 = \frac{A\mu_0}{8N^2}, \quad c_2 = \frac{a(L_1 + L_2)A + 2a^2 A g_0}{2\mu_0 b^2}$$

$$c_3 = \frac{A(L_1 + L_2)^2 + 4a(L_1 + L_2)A g_0 + 4a^2 A g_0}{8\mu_0 b^2}$$

$$c_4 = \frac{AN}{2}, c_5 = \frac{aA}{b\mu_0}, c_6 = \frac{A(L_1 + L_2) + 4Aag_0}{4b\mu_0}$$

$$c_7 = \frac{\mu_0^2}{16N^2}, \quad c_8 = \frac{\mu_0 N(2ag_0 - L_1 - L_2)}{8b}$$

$$c_9 = \frac{\mu_0 Na}{4b}, \quad c_{10} = \frac{a(L_1 + L_2) + 2a^2 g_0}{4b^2}$$

$$c_{11} = \frac{(L_1 + L_2)^2 + 4a(L_1 + L_2)g_0 + 4a^2 g_0^2}{16b^2}$$

REFERENCES

[1] Yin, Lingbing, and Lei Zhao. "Nonlinear Control for a Large Air-gap Magnetic Bearing System." Proc. of SMiRT 19, Toronto. N.p.: n.p., n.d. N. pag. Web. Aug. 2007.

[2] Inoue, Tsuyoshi, Yukio Ishida, and Shin Murakami. "Theoretical Analysis and Experiments of the Nonlinear Vibration in a Vertical Rigid Rotor Supported by the Magnetic Bearing System." *Ournal of System Design and Dynamics* 1.2 (n.d.): 295-306. Web. 2007.

[3] Abdelfatah, Mohamed M., and Emad P. Fawzi. "Nonlinear Oscillations in Magnetic Bearing Systems." *Ieee* 8th ser. 38 (1993): 1242-245. Print.

[4] Hong, Sung K., and Reza Langari. "Fuzzy Modeling and Control of a Nonlinear Magnetic Bearing System." *Journal of Intelligent & Fuzzy Systems* 7.4 (1999): 335-46. Print.

[5] Smith, R. D., and W. F. Weldon. "Nonlinear Control of a Rigid Rotor Magnetic Bearing System: Modeling and Simulation with Full State Feedback." *Ieee* 31.2 (1995): 973-80. Print.

[6] Walsh, Thomas F. *Nonlinear Dynamic Analysis of a Magnetic Bearing System with Flux Control: The Effects of Coordinate Coupling*. N.p.: DukeUniversity, 1993. Print.

[7] He, Rong, Kang Z. Liu, and Osami Saito. "Nonlinear Output Feedback Control of a 5DOF AMB System." Proc. of 16th IFAC World Congress, Czech Republic. Vol. 16. N.p.: n.p., 2005. N. pag. Print. Ser. 1.

[8] Huang, Shi J., and Lih C. Lin. "Stable Fuzzy Control with Adaptive Rotor Imbalance Compensation For nonlinear Magnetic Bearingsystems." *Journal of the Chinese Institute of Engineers* 4th ser. 28 (2005): 589-603. Print.

[9] Hu, Tingshu, and Zongli Lin. *Control Systems with Actuator Saturation: Analysis and Design*. Boston: Birkhäuser, 2001. Print.

[10] Wang, Yong, James D. Paduano, and Richard M. Murray. "BIFURCATION CONTROL: THEORIES, METHODS, AND APPLICATIONS." *IEEE* (1999): 730-36. Print.

[11] Inoue, Tsuyoshi, Motoki Sugiyama, Yasuhiko Sugawara, and Yukio Ishida. "Nonlinear Vibration in a Vertical Rigid Rotor Supported by the Magnetic Bearing: Case Considering the Delays of Both Electric Current and Magnetic Flux." *21st Biennial Conference on Mechanical Vibration and Noise / Modeling, Vibration and Control of Rotor*

Systems. Proc. of ASME, Nevada, USA, Las Vegas. Vol. 1. N.p.: n.p., 2007. 1195-203. Print.

[12] Zheng, Kai, Heng Liu, and Lie Yu. "Robust Fuzzy Control of A nonlinear Magnetic Bearing system with Computing Time Delay." *IEEE/ASME* (2008): 839-43. Print.

[13] Zheng, Kai, and Lie Yu. "Effects of Time Delay on Resonance of a Nonlinear Magnetic Bearing System with Delayed Feedback." *International Journal of Applied Electromagnetics and Mechanics* 3-4 33 (2010): 1547-554. Print.

[14] Noh, Myounggyu D., and Eric H. Maslen. "Self-Sensing Magnetic Bearings Using Parameter Estimation." *IEEE TRANSACTIONS ON INSTRUMENTATION AND MEASUREMENT* 46.1 (1997): 45-50. Print.

[15] Foiles, W. C., and P. E. Allaire. "Nonlinear Transient Modeling of Active Magnetic Bearing Rotors during Rotor Drop on Auxiliary Bearing." *Industrial Conference and Exhibition on Magnetic Bearings*. Proc. of MAG '97, Radisson Plaza Hotel at Mark Center, Alexandria, Virginia. N.p.: n.p., 1997. 154-63. Print.

[16] N. Steinschaden "Nonlinear Stability Analysis of Active Magnetic Bearings" *in fifth Int. Symp. Magnetic Suspension technology*, 1999, pp. 411-427.

[17] J. Rivas, J. M. Zamarro, E. Martin, and C. Pereira, "Simple approximation for magnetization curves and hysteresis loops," *IEEE Trans. Magn.*, vol. MAG-17, no. 4, pp. 1498–1502, Jul. 1981.

[18] G. F. T. Widger, "Representation of magnetisation curves over extensive range by rational-fraction approximations," *Proc. Inst. Elect. Eng.*, vol. 116, pp. 156–160, 1969.

[19] Huang, Bin, and Zongli Lin. "Absolute Stability With a Generalized Sector Condition." Automatic Control, IEEE 49.4 (2004): 535-48. Print.

Stability analysis of a new Passive PMs Bearing

Ernesto Tripodi [a], Antonino Musolino [a], Rocco Rizzo [a], Dante Casini [a]

[a] Department of Energy and System Engineering, University of Pisa, Largo Lucio Lazzarino, I-56126 Pisa, Italy,
ernesto.tripodi@dsea.unipi.it

***Abstract*—In this paper a new Permanent Magnets (PMs) bearing is presented. The device is composed of a rotor capable to levitate at a short distance from a dedicated stator. The magnetic suspension is achieved by proper configurations of PMs arranged on both the stator and the rotor. The device is characterized by intrinsic instability and passive stabilization is attempted exploiting eddy currents on a conducting sheet surrounding the stator magnets. The system has been simulated by means of a dedicated numerical tool capable to take into account the effects of the magneto-mechanical coupling. In particular, the coupled problem has been integrated by means of a prediction-correction nested scheme. The simulation activity has produced some interesting results, that are here extensively discussed. More specifically, it has been shown that the stability with respect to the center of mass translation can be passively obtained, if the rotations are actively prevented. An idea of an active stabilizing system for the rotations, exploiting the results of the simulation is finally briefly described.**

I. Introduction

The modern development of Magnetic Levitation systems (known as MAGLEV) started in the late 1960s, when the idea of using magnetic force to levitate vehicles became bearable, mainly due to some recent discoveries: the development of low-temperature superconducting wire, the transistor and chip based electronic control technology [1], [2]. MAGLEV provides high-speed motion, safety, reliability, low environmental impact and minimum maintenance [3]. There are two basic options to obtain magnetic levitation: electromagnetic system [4], [5], [6] working in attraction mode with forces generated by electromagnets and electrodynamic system [7], [8] working in repulsive mode with forces generated by superconductive coils. Both the solutions are characterized by unstable behavior. In particular the first one is unstable in the levitating direction (typically vertical): in fact as the two parts of the systems approach one the other the attractive force increases. The electrodynamic system is unstable in the direction transverse to levitation and in the motion directions. The nowadays availability of rare heart PMs (e.g. NdFeB) characterized by high values of remnant field has made possible to conceive a new class of MAGLEV systems where the suspension is assured by the repulsion of properly shaped PMs [9], [10]. As known, stability of levitation systems based on PMs is prevented by Earnshaw's theorem [11]. This theorem states that a set of steady charges, magnetizations, or currents cannot stay in stable equilibrium under the action of stady electric and magnetic field alone. Several applications of MAGLEV or magnetic bearing devices must be fail-safe, and this poses severe constraints on the design and operation of the stabilization systems. In this context a great effort is devoted to the design of passive and more reliable stabilization devices. There are some circumstances under which electric and magnetic systems can avoid the consequences of the Earnshaw's theorem: time varying fields (e.g., eddy currents, alternating gradient), active feedback, ferrofluids, superconductors and diamagnetic systems. In this paper, the use of eddy current stabilization for the reduction or the elimination of the intrinsic instability of the bearing is investigated. More precisely, if some magnetized parts of a system are in motion near conductive materials, eddy currents are induced and the system is not under the action of steady magnetic fields alone. The hypothesis of the Earnshaw's theorem, that is a direct consequence of the Laplace equation, is not valid in this case since the system is now governed by the diffusion equation. In this paper some preliminary results of the coupled electromechanical analysis of a PMs bearing are discussed, showing how the presence of motional induced eddy currents have a positive effect on the dynamic of the bearing device, so reducing the complexity of the control system. In particular, Section II introduces the proposed device, Section III briefly describes the numerical tool used for the analysis of the device, while Section IV discusses the obtained results. In Section V the results obtained are summarised and a simple control system is proposed to achieve the overall stability.

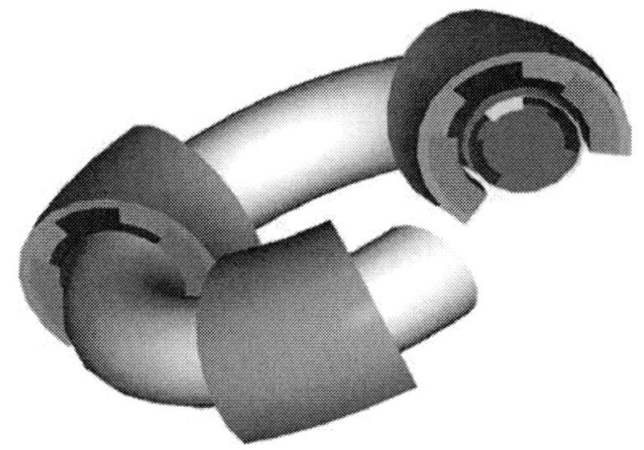

Figure 1. A 3D view of the analysed device.

II. Proposed Device

The proposed system, exploiting the induced eddy currents to contrast the instability related to the PMs arrangement is shown in figures 1 and 2. It is composed of a toroidal stator and a segmented rotor made of (at least) three blocks equally spaced along the circumference. The stator is fixed while the rotor can move with 6 degrees of freedom (DoFs).

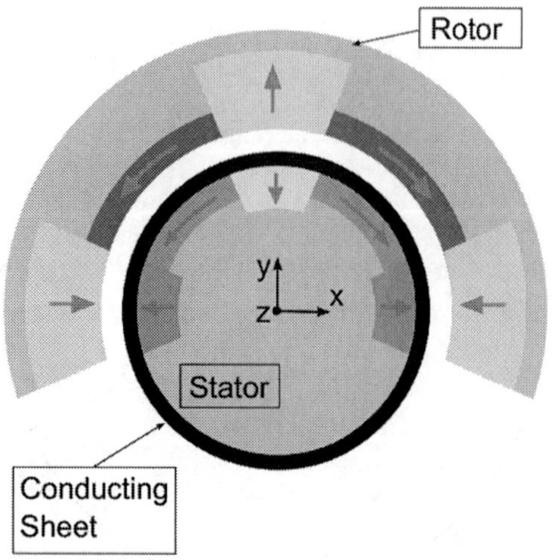

Figure 2. Cross section of the bearing.

The PMs in both the stator and the rotor are arranged in Halbach array configurations [12] focusing the field lines in the airgap between them. A sheet of conductive material surrounds the stator (black in 2. Levitation (along z-direction) is achieved by the repulsion of oppositely magnetized PMs. In order to describe the dynamics of the proposed device let us assume that the moving part rotates around a vertical shaft (directed as the z-axis) coincident with the symmetry axis of the stator. A system of motion induced eddy currents flows on the stator sheet; as a consequence a levitation force in the z-axis direction (which sums with the levitation force of the PMs) and a magnetic drag torque are observed. In fact, the motion induced eddy currents interact with the PMs of the rotor in order to reduce the cause that produces the eddy currents themselves. This is achieved by a velocity reduction (a drag force on the rotor) but also by moving the source of the inducing field (the PMs on the rotor) away from the conductive region. The system is stable in the levitation direction since the levitation force is a decreasing function of the distance. In correspondence to the symmetric configuration radial forces cancel. If the symmetry condition does not hold (e.g. because of a radial shift of the rotation shaft) a net force in the radial direction is expected to appear. This force is the resultant of the forces between the two PMs systems and those between the motion induced eddy currents and the PMs on the rotor. As known (Earnshaw's theorem) the forces between the PMs are destabilizing and as a consequence the rotor would touch the stator after a short time. Referring to figure 2, if we consider a displacement of the rotor in the y-axis direction we see that the currents induced on the conductive sheet are stronger on the side characterized by a reduced distance with the rotor, while are weaker on the side where the distance is greater. The net resulting force is then directed along the negative y-axis direction so performing a stabilizing action. The described stabilizing contribution of the forces between the induced currents and the PMs is present in the case of angular displacement of the rotation shaft with respect to the vertical (z-axis) direction. Considering that the rotor is segmented in at least three sectors, as a results of this angular displacement some of its sectors are closer to the stator, others more distant. The induced currents on the regions of the stator which correspond to the nearest sectors of the rotor are more intense and consequently the forces exerted on the PMs in these sectors of the rotor are strongest. On the contrary the forces on the rotor sectors that are more distant from the stator are decreased. The resultant effect is a torque restoring the vertical position of the rotation shaft.

III. The Numerical Electromechanical Formulation

In order to investigate the performance of this system it is necessary to use a numerical model. The equations describing the rotor dynamics with six DoFs are inherently nonlinear because of the dependence of the force on the position of the rotor itself. Moreover the dynamic problem is coupled with the diffusion equation of the magnetic field. The solution of the electromagnetic problem has been carried out by an integral formulation that reduces the diffusion equation to an equivalent network with time varying parameters. The values of the parameters in the electrical equations are function of the position of the rotor. The details of the adopted formulation are reported in [13], [14] and the development of a C code exploiting the GPGPU Nvidia CUDA libraries is extensively described in [15].
Under the hypothesis of linear magnetisable materials the equations of the problem produced by the equivalent network formulation coupled with the Newton-Euler equations of motion, can be written as:

$$\begin{aligned} &\mathbf{L}(\mathbf{C}(t))\frac{d\mathbf{i}}{dt} + \left[\mathbf{R}(\mathbf{C}(t)) + \mathbf{K}\left(\mathbf{C}(t), \dot{C}(t)\right)\right]\mathbf{i} = \mathbf{e}(t) \\ &\mathbf{F}(\mathbf{C}(t)) = m\ddot{\mathbf{q}} \\ &\mathbf{M}(\mathbf{C}(t)) = \mathbf{I}_{\theta\theta}\dot{\boldsymbol{\omega}} + \boldsymbol{\omega} \times \mathbf{I}_{\theta\theta}\boldsymbol{\omega} \end{aligned} \quad (1)$$

where $\mathbf{e}(t)$ represents the vector of the applied voltage generators and $\mathbf{i}$ is the vector of the currents in the elementary volumes, included the equivalent magnetization currents. All the coefficients matrices are function of which represents the system configuration at the instant t. $\mathbf{C}(t)$ is defined as the set of the positions and orientations of all the elementary volumes in which the device is discretized:

$$\mathbf{C}(t) = (\mathbf{x}(t), \mathbf{y}(t), \mathbf{z}(t), \boldsymbol{\phi}(t), \boldsymbol{\theta}(t), \boldsymbol{\psi}(t))$$

$\mathbf{L}(\mathbf{C}(t))$ denotes the inductance matrix; $\mathbf{R}(\mathbf{C}(t))$ is the resistance matrix and $\mathbf{K}\left(\mathbf{C}(t), \dot{\mathbf{C}}(t)\right)$ takes into account the electromotive force due to the motional effects. In particular $\dot{\mathbf{C}}(t)$, termed as the derivatives of the system configuration at the instant t, describes the velocity of every elementary volume in the hypothesis of rigid body. The $\dot{\mathbf{C}}_i(t)$ corresponding to the i-th elementary volume is constituted by the three components of the translation velocity, the three components of the angular velocity, and the three coordinates of the center of rotation.
Equation (1) is solved by a prediction correction nested scheme. The rationale behind it is the search for an approximation of the time dependence of the coefficients in

electrical and mechanical equations. The predictor-corrector approach is used to obtain an approximate behavior of the named quantities by a linear interpolation between the known values at the previous time step and the predicted values at the next time step. Inserting this knowledge in the equations has the effect of considering updated values of the coefficients, allowing a coupling between the equations which is stronger than the one in a simply staggered scheme and comparable with a monolithic approach. The integration algorithm can be described as follows ($\Delta t = t_{n+1} - t_n$):

- assuming $\mathbf{C}(t)$ constant in the interval Δt an estimate of the currents at t_n is obtained by a trapezoidal rule applied on (1a);
- an estimate of $\mathbf{C}(t_{n+1}) = \mathbf{C}_{n+1}$ is obtained by applying forward Euler integration to (1b) and (1c);
- a piecewise linear approximation is assumed for $\mathbf{L}$ in Δt, similarly for $\mathbf{R}$ and $\mathbf{K}$:

$$\dot{\mathbf{L}}(t_n) \simeq \frac{\mathbf{L}\left(\tilde{\mathbf{C}}(t_{n+1})\right) - \mathbf{L}\left(\tilde{\mathbf{C}}(t_n)\right)}{\Delta t}$$

and so

$$\mathbf{L}(t) \simeq \mathbf{L}(t_n) + \dot{\mathbf{L}}(t_n) \cdot (t - t_n) = \mathbf{L}_n + \dot{\mathbf{L}}_n \cdot (t - t_n);$$

- The expressions are introduced in (1a):

$$\left(\mathbf{L}_n + \dot{\mathbf{L}}_n \cdot (t - t_n)\right) \frac{d\mathbf{i}}{dt} + \left[\left(\mathbf{R}_n + \dot{\mathbf{R}}_n \cdot (t - t_n)\right) + \left(\mathbf{K}_n + \dot{\mathbf{K}}_n \cdot (t - t_n)\right)\right] \mathbf{i} = \mathbf{e}(t) \quad t \in [t_n, t_{n+1}] \tag{2}$$

- integrating with a trapezoidal-like rule we obtain the corrected values of the currents at the instant t_{n+1}. We write:

$$\int_{t_n}^{t_{n+1}} \left(\mathbf{L}_n + \dot{\mathbf{L}}_n (t - t_n)\right) \frac{d\mathbf{i}}{dt} dt + \int_{t_n}^{t_{n+1}} \left[\left(\mathbf{R}_n + \dot{\mathbf{R}}_n (t - t_n)\right) + \left(\mathbf{K}_n + \dot{\mathbf{K}}_n (t - t_n)\right)\right] \mathbf{i} dt = \int_{t_n}^{t_{n+1}} \mathbf{e}(t) dt \tag{3}$$

- and after the numerical integration:

$$\mathbf{L}_n (\mathbf{i}_{n+1} - \mathbf{i}_n) + \frac{\Delta t \dot{\mathbf{L}}_n}{2} (\mathbf{i}_{n+1} + \mathbf{i}_n) + \frac{\Delta t (\mathbf{R}_n + \mathbf{K}_n)}{2} (\mathbf{i}_{n+1} + \mathbf{i}_n) + \frac{\Delta^2 t \left(\dot{\mathbf{R}}_n + \dot{\mathbf{K}}_n\right)}{2} \mathbf{i}_{n+1} = \frac{\Delta t (\mathbf{e}_{n+1} + \mathbf{e}_n)}{2} \tag{4}$$

- Collecting terms finally gives a linear system where the unknowns are the corrected currents at the instant t_{n+1}:

$$\left(\mathbf{L}_n + \frac{\Delta t \dot{\mathbf{L}}_n}{2} + \frac{\Delta t (\mathbf{R}_n + \mathbf{K}_n)}{2} + \frac{\Delta^2 t \left(\dot{\mathbf{R}}_n + \dot{\mathbf{K}}_n\right)}{2}\right) \mathbf{i}_{n+1} = \frac{\Delta t (\mathbf{e}_{n+1} + \mathbf{e}_n)}{2} + \left(\mathbf{L}_n - \frac{\Delta t \dot{\mathbf{L}}_n}{2} - \frac{\Delta t (\mathbf{R}_n + \mathbf{K}_n)}{2}\right) \mathbf{i}_n \tag{5}$$

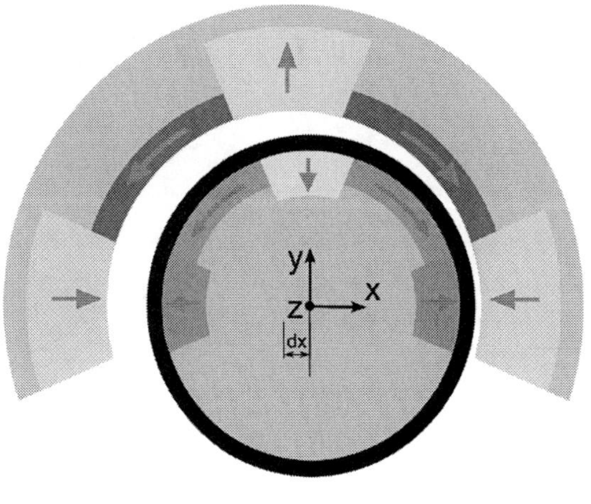

Figure 3. Configuration with the rotor shifted along the x-axis.

- Once (5) is solved, force and torque are evaluated again by using the force coefficient in the predicted configuration $\tilde{\mathbf{\Phi}}(t_{n+1})$ with the corrected values of the currents $\mathbf{i}(t_{n+1})$ just obtained. The integration of the mechanical equations yields the corrected position of the moving body.

The integration method has been validated and the results exhibit a very good accuracy respect to the experimental data found in literature, as [16].

IV. Simulation Results

As a case study we considered a device with an average radius of 8 cm. The tube constituting the stator has a radius of 2.2 cm and the clearance between rotor and stator is 7 mm. We made a set of simulations driving the rotor at different rotational speeds. We noticed that, at low speeds (less than 3000 rpm), the stabilizing effect was not enough (i.e. the system behavior was really close to the static case, without enough levitation force), while, increasing the speed over a certain value (more than 4000 rpm) have the effect of reducing the magnetic drag force, as expected, but without significative variations in the stabilizing effect. We report here the results obtained with a rotational speed of 4800 rpm.

The levitation force at rest is 450 N and increases to 470 N at motion. The radial magnetized red and blue sectors have an angle of 67.50°, the green and yellow an angle of 45° while the azimuth magnetized purple and pink sectors have an angle of 33.75°. The radial width is 0.5 cm for the red and yellow sectors, 1 cm for the green and blue, and 0.3 cm for the purple and pink ones. The thickness of the conductive (aluminum) sheet is 2mm. The numerical formulation in Section III has been used to perform the analysis of some configurations obtained by the described device.

We started considering the device with the conductive sheet removed with an initial position characterized by the rotation shaft shifted of 1.5 mm with respect to the symmetry axis as shown in figure 3. The results of simulation are shown in figures 4 and 5. As expected the system is unstable with

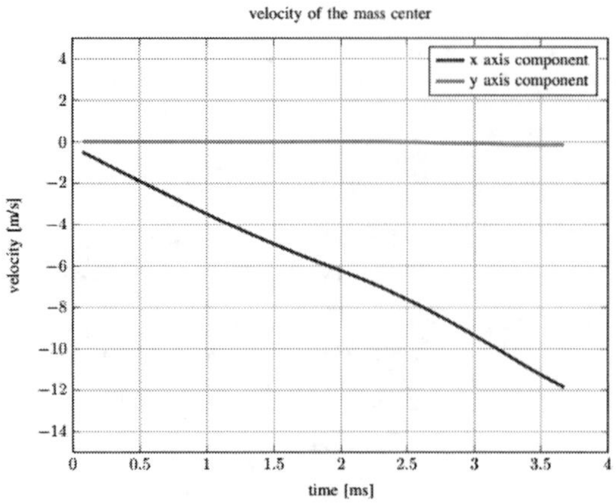

Figure 4. Velocity of the center of mass of the rotor.

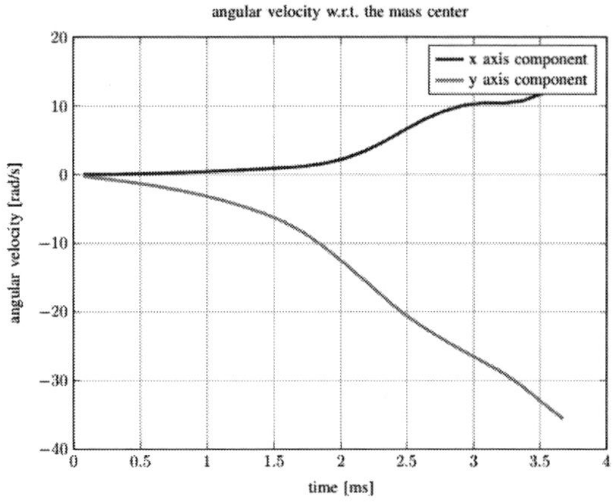

Figure 5. Components of the angular velocity.

respect to the radial direction and with respect the rotation around the x and the y axes. The contact between the rotor and the stator happens after about 3 ms. Figures 6 and 7 refer to the configuration where the conductive sheet has been introduced. Also in this case the rotor touches the stator as a consequence of the unstable behavior of the system. We can observe that the presence of the conductive sheet (and of the induced currents) is able to slower the unstable dynamics; the rotor takes a longer time (about 17 ms) to touch the stator.

Despite the eddy currents on the sheet are not able to stabilize the device, they can be used to reduce the complexity of the control system since a lower dynamic requires a slower control action that is easier to be designed. To further investigate the effects of the currents on the conductive sheet we performed simulations of the device with a reduced number of DoFs preventing rotations respect to x and y axes assuming the presence of two of the mentioned bearings arranged in a vertical configuration. Figure 8 shows the velocity waveforms while figure 9 shows the force waveforms; the simulation has been done with the same lateral displacement $dx = 1.5$ mm.

According to the results of the simulation, the dynamic of the system is stable. This is actually a good result if we think

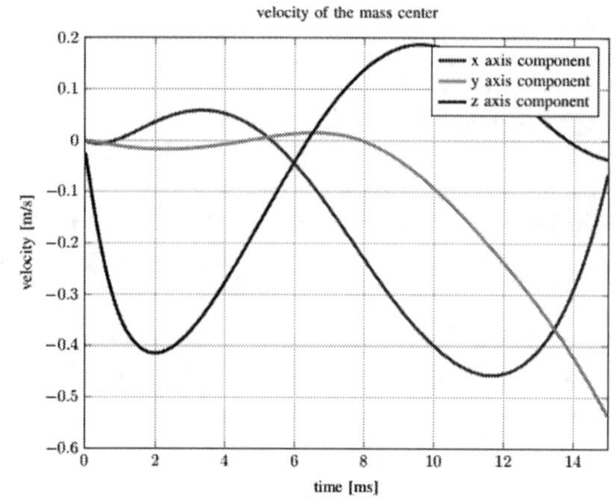

Figure 6. Velocity of the center of mass of the rotor.

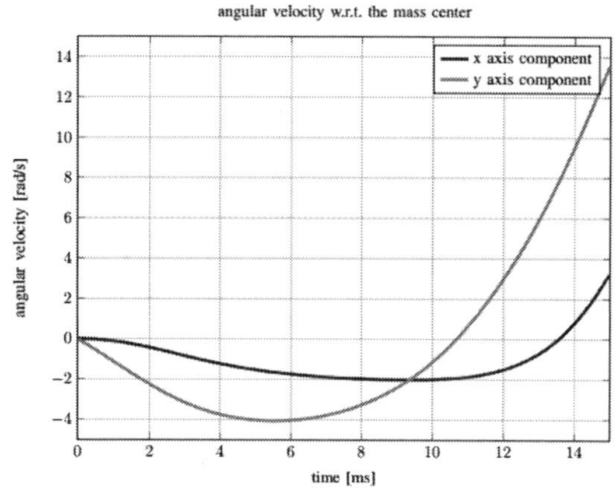

Figure 7. Components of the angular velocity.

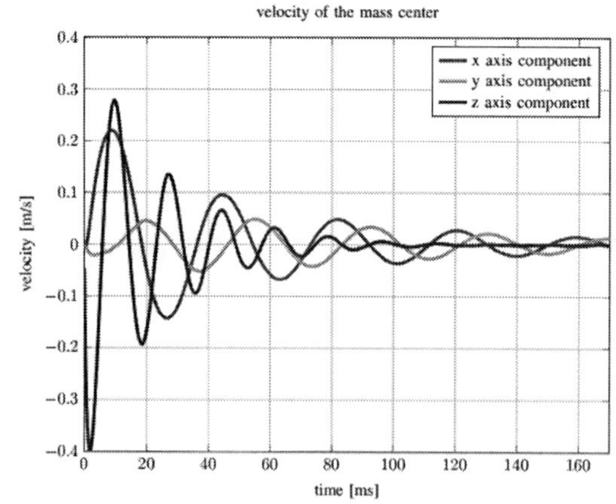

Figure 8. Velocity of the center of mass of the rotor with 3 DoFs.

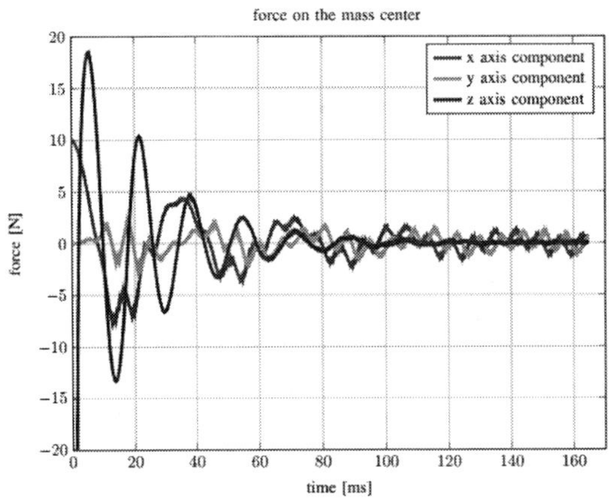

Figure 9. Force the center of mass of the rotor with 3 DoFs.

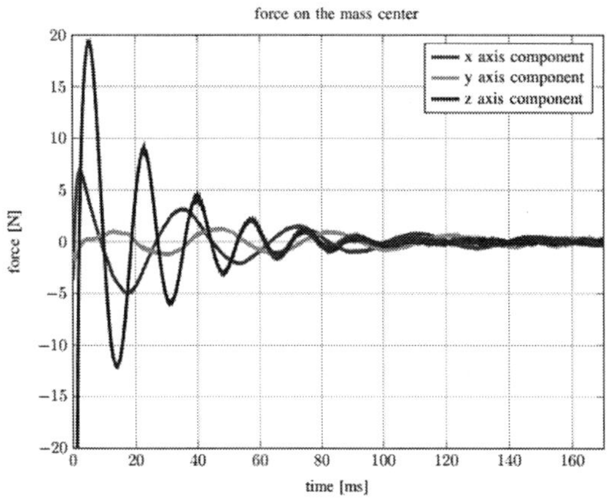

Figure 11. Forces on the center of mass of the rotor with 3 DoFs.

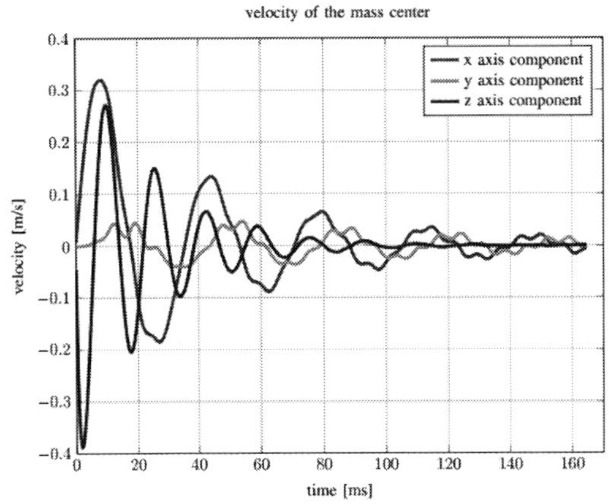

Figure 10. Velocity of the center of mass of the rotor with 3 DoFs.

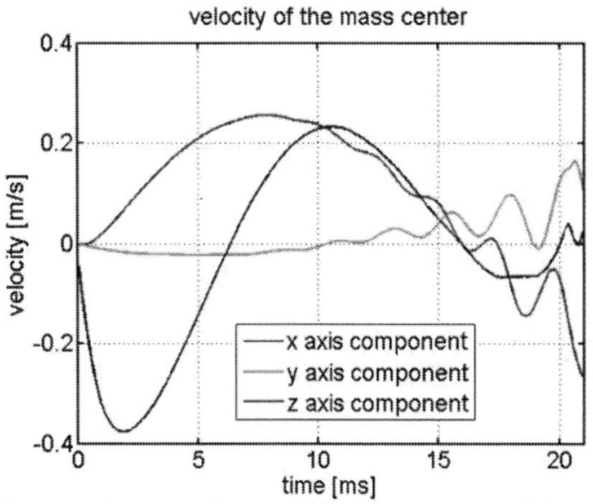

Figure 12. Velocity of the center of mass of the rotor.

that in this kind of devices, stability has to be discussed in the contest of dynamics. This means that the robustness requirements previously mentioned, involve concepts of dynamic stability in presence of modeling error due to uncertainties (modern non linear dynamics). This theory, usually, not only requires the knowledge of how the forces and torques change with the position and orientation but also of how they changes with both linear and angular velocities. The control systems are consequently usually really complex: this result is then really interesting because it permits a simpler synthesis of the active controller and so reducing the cost. Another similar simulation has been done applying a lateral force of 10 N to the rotor. The results are shown in figures 10 and 11; the system is able to compensate the lateral force as well as for the lateral displacement.

We finally performed a simulation (6 DoFs) where only the vertically magnetized PMs (the yellow and the green ones in figure 2) are retained in the stator. The results are reported in figs. 12 and 13 respectively for the velocities, while in fig.14 for what concerns the forces on the center of mass. The system remains unstable, but its response has changed showing an oscillatory behavior. The contact occurs after 22 ms. The levitation force has about halved.

V. Conclusions

An exhaustive simulation activity has been performed on a PMs bearing based on Halbach array configurations. The conductive sheet is not sufficient to stabilize to system, but it makes slower the unstable dynamics, actually simplifying the control systems action. Since the actuators basically consist of coils, a slower dynamics will need slower control actions, and then smaller voltages. The main result of the our analysis relies in the intrinsic stability with respect to the mass center translation, once the rotations with respect x and y axes are prevented. This means that a stabilization system is needed only to maintain the direction of the rotation axis parallel to the z-axis, while the system is able to stabilize the position of the rotation axis. Equivalently, if the stabilizing action results in a net force beside the needed torque, the system is able to compensate this force by adjusting it position. A possible setup for a stabilization device is shown in figure 15, where four electromagnets are positioned in each side of the rotating shaft where some ferromagnetic material has been inserted.

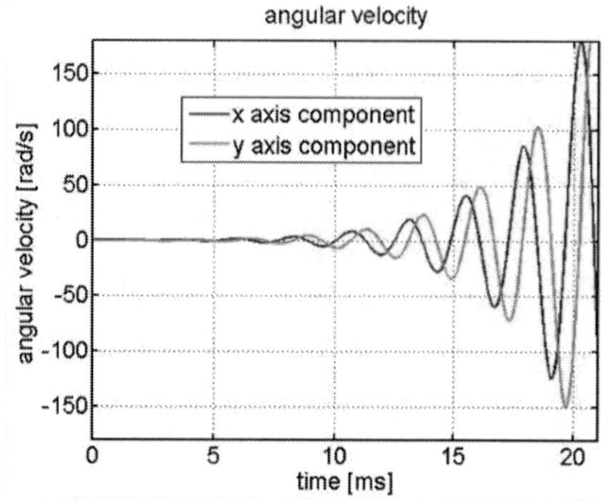

Figure 13. Components of the angular velocity.

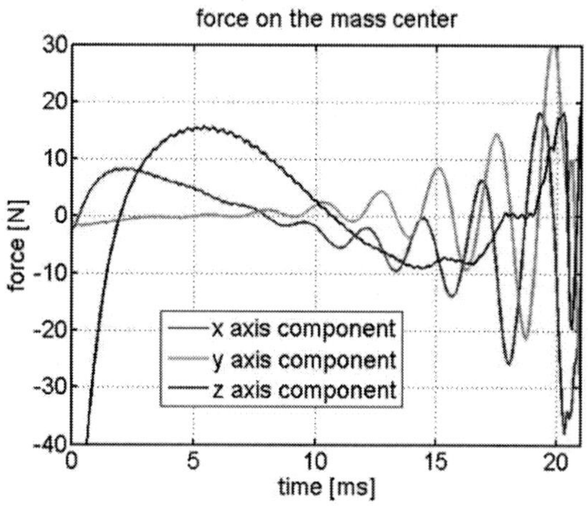

Figure 14. Forces on the center of mass of the rotor with 3 DoFs.

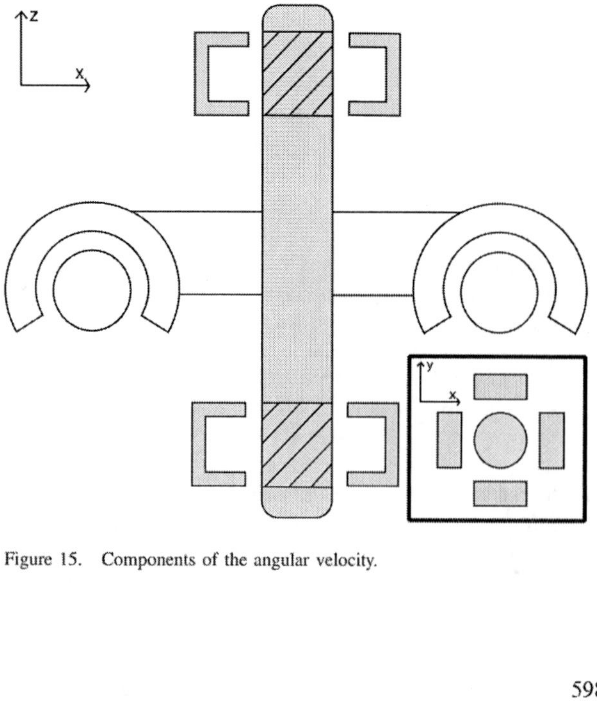

Figure 15. Components of the angular velocity.

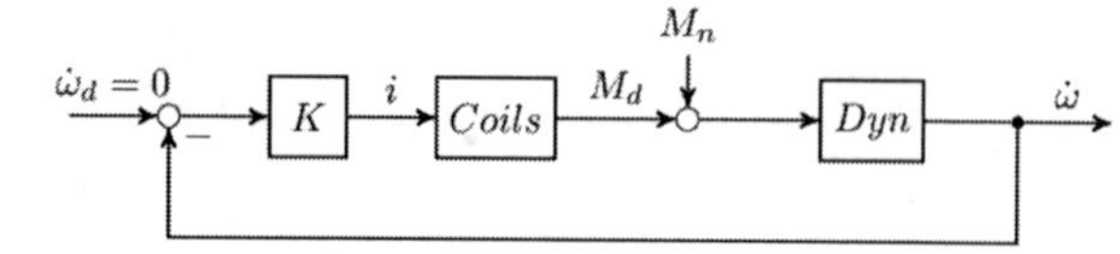

Figure 16. Control scheme.

The current in the coils of the electromagnets can be controlled by the value of the angular acceleration, as in the simple control scheme shown in figure 16.

REFERENCES

[1] V. Kozorez and O. Cheborin, "On stability of equilibrium in a system of two ideal current rings," in *Dokl. Akad. Nauk UkrSSR, Ser. A*, vol. 4, 1988, pp. 80–81.

[2] F. C. Moon, *Chaotic and Fractal Dynamics: Introduction for Applied Scientists and Engineers*. John Wiley & Sons, 2008.

[3] H.-W. Lee, K.-C. Kim, and J. Lee, "Review of maglev train technologies," *Magnetics, IEEE Transactions on*, vol. 42, no. 7, pp. 1917–1925, 2006.

[4] J. Meins, L. Miller, and W. Mayer, "The high speed maglev transport system transrapid," *Magnetics, IEEE Transactions on*, vol. 24, no. 2, pp. 808–811, 1988.

[5] N. Amati, A. Tonoli, F. Impinna, and J. Girardello Detoni, "A solution for the stabilization of electrodynamic bearings: Modeling and experimental validation," *Journal of Vibration and Acoustics*, vol. 133, pp. 021 004–1, 2011.

[6] A. Tonoli, N. Amati, F. Impinna, J. G. Detoni, H. Bleuler, and J. Sandtner, "Dynamic modeling and experimental validation of axial electrodynamic bearings," in *Proceedings of the 12th International Symposium on Magnetic Bearings*, 2011, p. 68.

[7] M. Ono, S. Koga, and H. Ohtsuki, "Japan's superconducting maglev train," *Instrumentation & Measurement Magazine, IEEE*, vol. 5, no. 1, pp. 9–15, 2002.

[8] E. Diez-Jimenez, J.-L. Perez-Diaz, and J. C. Garcia-Prada, "Local model for magnet–superconductor mechanical interaction: Experimental verification," *Journal of Applied Physics*, vol. 109, no. 6, pp. 063 901–063 901, 2011.

[9] N. Paudel, J. Z. Bird, S. Paul, and D. Bobba, "Modeling the dynamic suspension behavior of an eddy current device," in *Energy Conversion Congress and Exposition (ECCE), 2011 IEEE*. IEEE, 2011, pp. 1692–1699.

[10] A. Musolino, R. Rizzo, M. Tucci, and V. M. Matrosov, "A new passive maglev system based on eddy current stabilization," *Magnetics, IEEE Transactions on*, vol. 45, no. 3, pp. 984–987, 2009.

[11] S. Earnshaw, "{On the nature of the molecular forces which regulate the constitution of the luminiferous ether}," *Trans. Camb. Phil. Soc*, vol. 7, pp. 97–112, 1842.

[12] K. Halbach, "Application of permanent magnets in accelerators and electron storage rings," *Journal of Applied Physics*, vol. 57, no. 8, pp. 3605–3608, 1985.

[13] A. Musolino, R. Rizzo, M. Toni, and E. Tripodi, "Acceleration of electromagnetic launchers modeling by using graphic processing units," in *Electromagnetic Launch Technology (EML), 2012 16th International Symposium on*. IEEE, 2012, pp. 1–6.

[14] A. Musolino and R. Rizzo, "Numerical modeling of helical launchers," *Plasma Science, IEEE Transactions on*, vol. 39, no. 3, pp. 935–940, 2011.

[15] A. Musolino, R. Rizzo, E. Tripodi, and M. Toni, "Modeling of electromechanical devices by gpu-accelerated integral formulation," *International Journal of Numerical Modelling: Electronic Networks, Devices and Fields*, vol. 26, no. 4, pp. 376–396, 2013.

[16] H. Karl, J. Fetzer, S. Kurz, G. Lehner, and W. M. Rucker, "Description of team workshop problem 28: an electrodynamic levitation device," in *Proceedings of the TEAM Workshop in the Sixth Round*. Citeseer, 1997, pp. 48–51.

Realization of a Multi-Pole Permanent Magnetic Bearing with Rotating Magnetization

Edmund Marth[a], Gerald Jungmayr[a], Jörg Kobleder[a], Martin Panholzer[a] and Wolfgang Amrhein[a]

[a] Department of Electrical Drives and Power Electronics, Johannes Kepler University, Altenbergerstrasse 69, 4040 Linz, Austria, edmund.marth@jku.at

***Abstract*—The optimization of stiffness of a permanent magnetic single ring bearing is presented in this paper. This is achieved by only changing the magnetization pattern. More precisely, a homogeneous magnetization is replaced by a rotating magnetization with two magnetic poles, leading to a 3.7-fold improvement in stiffness. Usually, such a rotating magnetization is realized by stacking several homogeneously magnetized rings with different magnetization direction, which is referred to as Halbach stacking. Our approach was to realize a continuously rotating magnetization pattern in one ring. In order to achieve the maximum possible stiffness for the given bearing dimensions, the optimization of the magnetization process is crucial and also presented in the paper.**

I. Introduction

For the design of permanent magnetic ring bearings guidelines are available which help to obtain a bearing with a high functional value, i.e., stiffness or force, with respect to the volume of the permanent magnetic rings. This objective is crucial since the total volume of the magnetic material is directly linked to the costs of the final bearing. The possible design variables are the geometrical dimensions, the number of magnetic poles, and the shape of the magnetization. How these parameters influence the stiffness or force value of the resulting bearing is described in several publications, e.g., [1]–[5]. In summary, a "good" bearing should have cross section dimensions (per magnetic pole) in the size of the air gap, at least three magnetic poles, and a rotating magnetization. As shown in [5] the optimal cross section dimensions, in terms of maximal stiffness per magnet volume, are often unrealistic. Either the necessary geometric dimensions are too small to be fabricated or the absolute stiffness value is too low (see [5] for detailed information). Concerning the rotating magnetization there are basically two methods for realization: (i) several permanent magnet rings with homogeneous, but in its direction rotating magnetization can be stacked together or (ii) one permanent magnet ring is magnetized to reach a continuously rotating magnetization, cf. [1], [4]. Such magnetization patterns are also referred to as Halbach stacking or continuous Halbach magnetization respectively. What we present in this paper is the realization of such a multi-pole bearing with continuously rotating magnetization. As the bearing should replace a standard bearing used in an existing permanent magnetic levitated system, the geometric dimensions of the bearing are predefined. Thus, the paper basically deals with the maximization of the radial stiffness by means of an optimized magnetization pattern. This also includes the optimization of the necessary magnetization coil as well as the magnetization process.

Basic configurations of permanent magnetic bearings with continuously rotating magnetization are shown in Fig. 1(a) and (b) in attractive and repulsive configuration respectively. The single ring bearing with homogeneous magnetization shown in Fig. 2 is the starting point for optimization. The corresponding parameters are given in Table I. The permanent magnetic material used for the realization is an isotropic ferrite. The characteristic parameters are summarized in Table II.

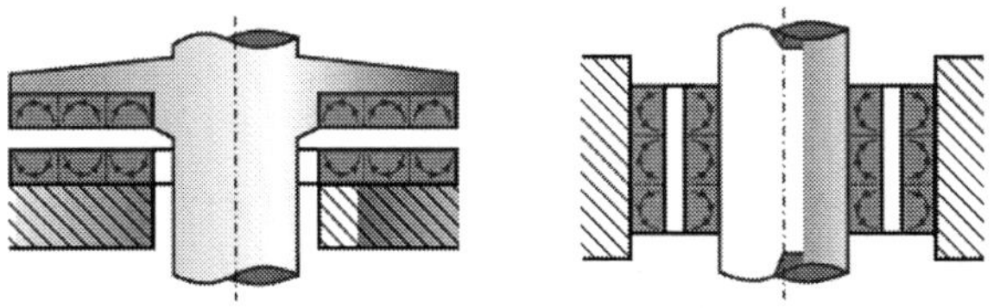

(a) attractive (b) repulsive

Figure 1: Exemplary configurations of radially stable bearings with three magnetic poles and continuous Halbach magnetization.

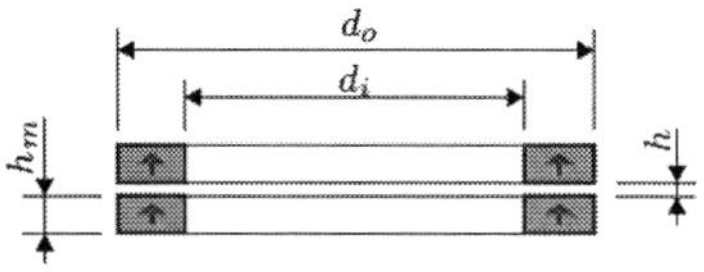

Figure 2: Configuration which is to be optimized with respect to maximum stiffness.

Table I: Geometric parameters of the attractive bearing to be realized

Parameter	Value	Description
d_i	21 mm	inner diameter
d_o	29 mm	outer diameter
h_m	2.2 mm	axial height of one magnet
h	0.7 mm	axial air gap of the bearing

Table II: Permanent magnetic material data, measured at 24°C

Parameter	Value	Parameter	Value
B_r	0.208 T	$H_{c,B}$	-147 kA/m
$H_{c,J}$	-279 kA/m	$(BH)_{max}$	7.8 kJ/m³
μ_r	1.13		

II. Magnetization of the Annular Rings

A. Magnetization Coil

To obtain a continuous rotating magnetization for bearings in attractive configuration the necessary arrangement of the magnetization coils with the proper current direction is shown in Fig. 3. That the number of coils n_c can be higher than the number of magnetic poles p is an additional degree of freedom for optimization. This might be advantageous for minimizing edge effects. To ensure full magnetization of the permanent

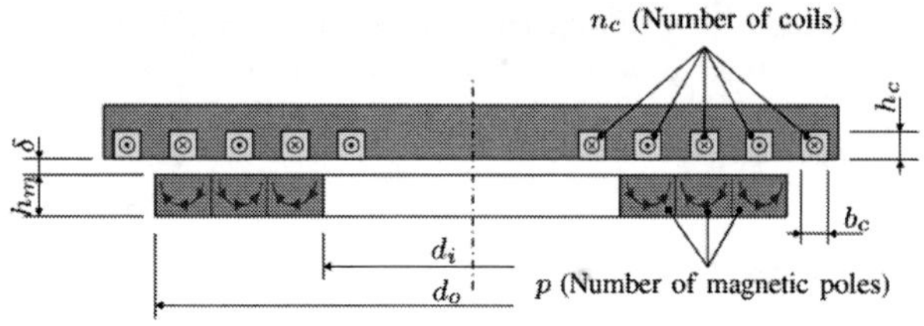

Figure 3: Parametric model of the magnetization coil.

magnet, the magnetic field generated by the coils has to be roughly the triple coercitivity of the polarization $H_{c,J}$, about 840 kA/m in our case. Since a sinusoidal magnetization is desired, cf. [4], [5], no iron can be used for flux guiding. As a consequence, very high currents are necessary to obtain this flux density. Hence, the temperature of the coils will rise significantly, which is the limiting factor in the design of the magnetization coil.

B. Magnetization Current and Thermal Behavior

The equivalent circuit for the simulation of the magnetization current is shown in Fig. 4 and includes the magnetization coil and the magnetizer used to energize the coil. The magnetizer is an impulse magnetizer manufactured by the

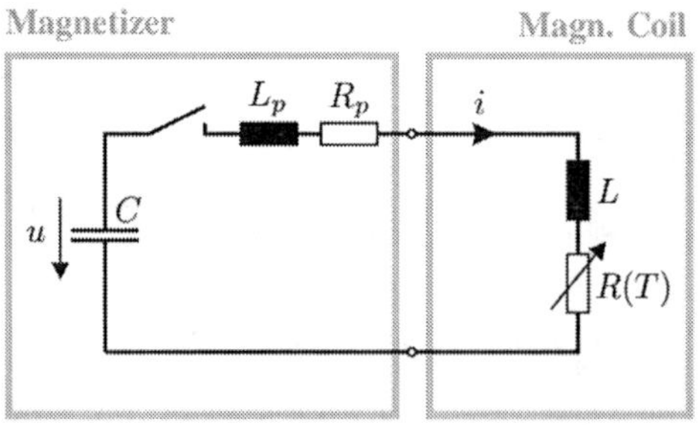

Figure 4: Equivalent circuit for the magnetization process, including magnetization coil and magnetization stage.

mPulse company[1]. The important ratings are given in Table III. The values of the parasitic parameters R_p and L_p where identified from prior measurements. Since the magnetization

Table III: Magnetizer ratings

Parameter	Value	Comment
$U_{C,0}$	up to 3000 V	continuously adjustable; $U_{C,0} = u_{(t=0)}$
C	fix: 540 µF 2 × 810 µF 4 × 1080 µF	arbitrary combination possible; $C = 540\,\mu$F up to 6480 µF
i_{max}	50 kA	maximum output current
W_{max}	29, 16 kJ	maximum magnetization energy
R_p	14 mΩ	identified parasitic resistance
L_p	2.5 µH	identified parasitic inductivity

current i can not be arbitrarily big due to thermal reasons, the coil temperature T has to be modeled by defining the ohmic resistance of the coil as function of the temperature:

$$R(T) = R_{20} \cdot (1 + \alpha \cdot \Delta T) \ . \tag{1}$$

Here, ΔT is $T - 20°$C, R_{20} denotes the resistance at room temperature (20°C), and α denotes the temperature coefficient of copper. The ohmic resistance of the coil at room temperature can be calculated by

$$R_{20} = \rho \cdot \frac{l}{A} \tag{2}$$

with the specific resistance of copper ρ, the total length l of all n_c coils, and the cross section area per single coil $A = b_c \cdot h_c$.

The thermal energy E_{th} stored in the copper of the coil is

$$E_{th} = c_p \cdot m \cdot T_K \ ,$$

where m is the mass of the coil, c_p the specific heat capacity of copper and T_K the absolute temperature of the coil in Kelvin. Assuming, that the total thermal energy E_{th} of one single magnetizing event is stored in the copper, i.e., no energy is emitted by heat transfer, radiation or convection, one can calculate the increase of the coil temperature via

$$\mathrm{d}E_{th} = R(T) \cdot i^2 \mathrm{d}t \ ,$$

$$\frac{\mathrm{d}E_{th}}{\mathrm{d}t} = c_p \cdot m \frac{\mathrm{d}T_K}{\mathrm{d}t} = R(T) \cdot i^2$$

and finally

$$\dot{T} = \frac{R(T) \cdot i^2}{c_p \cdot m} \ . \tag{3}$$

Since only a temperature gradient is calculated in (3), the absolute temperature T_K is substituted by the temperature in degrees Celsius, T. Although equation (3) gives a worst case estimation, pratical results are expected to be close due to the short period while the coil is energized, cf. Fig. 5. The final state equations describing the dynamic behavior of the magnetization current i and the Temperature of the copper are

$$\dot{u} = \frac{i}{C} \tag{4}$$

$$\dot{i} = \frac{u - i \cdot (R_{20} \cdot (1 + \alpha_{20} \cdot \Delta T) + R_p)}{L + L_p} \tag{5}$$

$$\dot{T} = \frac{i^2 \cdot R_{20}(1 + \alpha_{20} \cdot \Delta T)}{c_p \cdot m} \ . \tag{6}$$

[1]www.m-pulse.biz

All parameters used in equations (1) to (6) are summarized in Table IV. The coil inductivity L was calculated with the

Table IV: Parameters of the magnetization-current simulation

Parameter	Value/Unit	Comment
α	$3.9 \cdot 10^{-3}\,\mathrm{K}^{-1}$	temperature coefficient of resistance of copper
c_p	$384\,\frac{\mathrm{J}}{\mathrm{kgK}}$	specific heat capacity of copper
ρ	$17.8 \cdot 10^{-9}\,\Omega\mathrm{m}$	specific resistance at 20°C
$i(t)$	A	magnetization current, state variable
$u(t)$	V	voltage drop over C, state variable
$T(t)$	°C	coil temperature, state variable
$R(T)$	Ω	coil resistance, function of the geometric design parameters
R_{20}	Ω	coil resistance at 20°C, function of the geometric design parameters
L	H	coil inductivity, function of the geometric design parameters

FEMM finite element software[2], based on the given geometric parameters and the number of coils n_c. As an additional parameter, each coil can have N windings. The resistance was also calculated as function of the geometric parameters, n_c, and N. Furthermore, an estimated value of the resistance representing the supply wires and the interconnections of the single coils was added.

The simulated magnetization current and the simulated coil temperature are shown in Fig. 5 together with the measured magnetization current. The measurement was done with the configuration built up and which is described in detail in the realization section III-C. Fig. 5 shows that the derived model is well suited to predict the magnetization current.

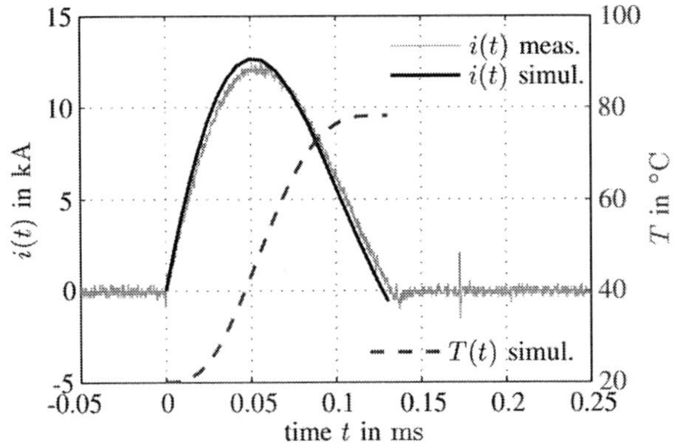

Figure 5: Magnetization current and coil temperature during one energizing process, simulation and measurement.

III. Optimization of the Magnetizing Process

A. Optimization Problem Formulation

As mentioned, the main goal of this optimization is to maximize the radial stiffness of a bearing with given cross section dimensions by adapting the magnetization pattern. The bearing we want to optimize is pictured to scale in Fig. 2 and its dimensions are given in Table I. Due to the available magnetizer and the thermal constraints, the whole magnetization process, as presented in the previous section, has to be considered in such an optimization. The available design parameters with the allowed range of values are defined in Table V. Objectives and constraints of the optimization

Table V: Design variables for the optimization

Parameter	Min.	Max.	Unit	Comment
δ	0.1	2	mm	
h_c	0.5	5	mm	
p	1	4		$p \in \mathbb{N}$
x_{b_c}	0.1	0.8		$b_c = x_{b_c} \cdot \frac{d_o - d_i}{2p}$
n_c	p	$p+8$		$n_c \in \mathbb{N}$
N	1	10		$N \in \mathbb{N}$
$U_{C,0}$	100	3000	V	
C	540	6480	μF	$C \in \{540, 1350, 2160, 3240, 4320, 5400, 6480\}\mu$F

are given in Table VI. For the temperature T it is clear that it should be minimal. In order to reduce the solution space to a feasible region, an upper constraint of $T = 250\,°\mathrm{C}$ was defined. The peak value of the magnetization current $\hat{i}$ is no objective, but again an upper limit is defined due to the limitations of the magnetizer. Instead of maximizing the radial stiffness of the bearing the negative axial stiffness $-k_z$ is maximized. The advantage of this objective is, that the axial stiffness can be calculated with a two dimensional finite element program due to the rotational symmetry, whereas the calculation of the radial stiffness represents a three dimensional problem which would lead to a highly increased time used for optimization. This approach is valid, because the axial stiffness is directly linked to the radial stiffness by the so called Earnshaw Theorem, ([6], [7]). Earnshaws Theorem states that if no magnetically conducting material is involved and $\mu_r = 1$ also for the magnets,

$$k_z = -2 \cdot k_r \tag{7}$$

holds true. Even if Earnshaw's assumptions are not perfectly fulfilled, ($\mu_r = 1.13$!), maximization of $-k_z$ will lead to high radial stiffness.

Table VI: Objectives and constraints

		Constraints		
Parameter	Objective	Min.	Max.	Unit
$-k_z$	maximized			N/mm
T	minimized	–	250	°C
$\hat{i}$	no Obj.	–	50	kA

B. Optimizing using MagOpt

For the optimization of this multi objective problem the software tool MagOpt (short for Magnetic Optimization[3]) is

[2]www.femm.info

[3]It should be noted, that the tool has a very general structure and thus is not limited to magnetic problems.

used [8]. It is developed at the Department of Electrical Drives and Power Electronics and the Linz Center of Mechatronics and is designed to serve as a kind of communication- or interface manager. In MagOpt, which is implemented in the Java programming language, the whole optimization project can be set up in the form of a tree structure, allowing to link different simulation-, calculation-, and CAD-software with each other. MagOpt is capable of checking all dependencies within the project and consequently executing the whole tree in the proper order. Furthermore, MagOpt has a built in optimization tool using genetic algorithms, which was used for this optimization. Let us look at our problem, also to clarify MagOpt's working principle.

In order to find the optimal solution, we first have to generate an initial parameter set (cf. Table V). When using genetic algorithms, such a set of design parameters is called an individual. Typically hundreds of such individuals with different parameter sets form a generation and are preferably processed in parallel. The following description is also visualized in Fig. 6 and explains the processing of one individual.

The design parameters are defined as so called "parameter fields" within the MagOpt tree[4]. These parameters can now act as an input for any function defined in the tree. For example, solving our problem requires the calculation of the magnetization coils' inductivity. This is done, as mentioned earlier, by involving the finite element software FEMM, which receives all relevant parameters from MagOpt. After calculation is finished, MagOpt collects the results from the FEMM output and makes them available within the MagOpt tree for further processing. As simple algebraic equations can be solved directly within MagOpt, the total resistance of the magnetization coil can be determined. Thus, all necessary parameters to simulate the magnetization current and the temperature of the coil can be committed to Matlab/Simulink, where equations (4) to (6) are implemented. The results of this simulation, namely the peak current $\hat{i}$ and the coil temperature T, are again collected by MagOpt. By applying $\hat{i}$ to the magnetization coil, the resulting magnetization pattern can be determined. Therefore, again in FEMM, the magnet is sectioned in lots of small elements. Each element is magnetized according to the strength and orientation of the magnetic field produced by the coil. To determine the strength of the permanent magnetic magnetization the initial magnetization curve pictured in Fig. 7 was used. To determine the axial stiffness of the whole bearing, this single ring is mirrored with reversed direction of magnetization and the axial force as function of axial displacement is calculated. The final magnetization pattern with the corresponding field lines and nominal air gap is shown in Fig. 8. The stiffness value is again collected by MagOpt.

Now the optimizer evaluates the results of all individuals, namely the objective values, and generates new datasets by combining the best individuals. A new generation is born and ready to be evaluated. All the steps and functionality performed by external programs such as FEMM are controlled mainly by script files, e.g., LUA-script for FEMM, which are also defined in MagOpt.

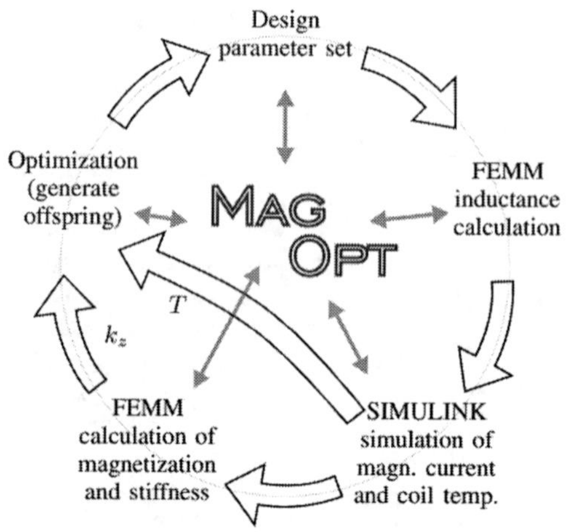

Figure 6: Optimization path

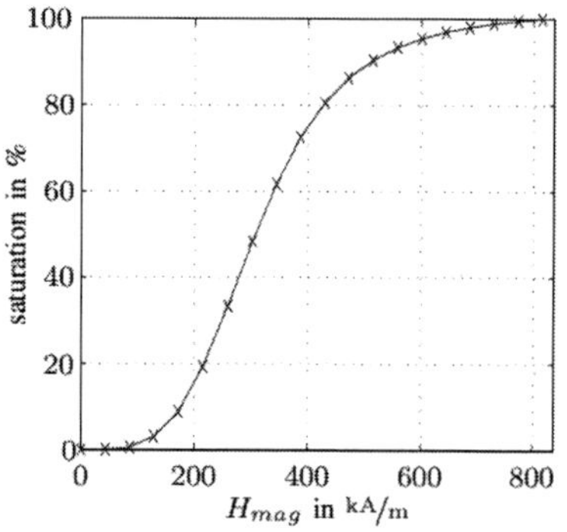

Figure 7: Assumed initial curve of the polarization's saturation used to determine the polarization in each section of the permanent magnet. Full magnetization is achieved for field strength $H_{mag} \geq 3 \cdot H_{c,J}$.

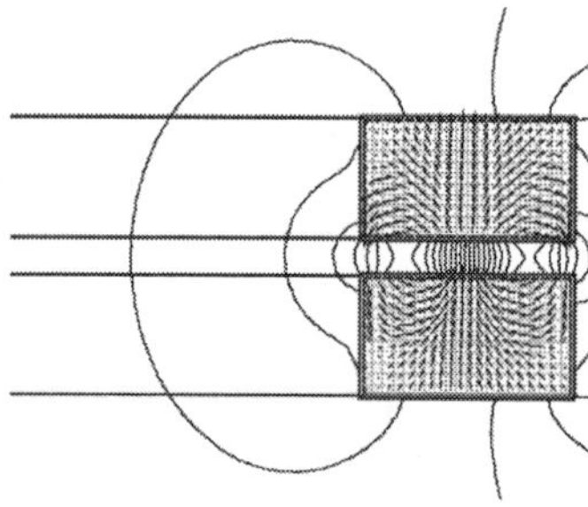

Figure 8: Calculated magnetization pattern and magnetic field lines of the realized bearing.

C. Pareto– optimal Solution and Realization of the Magnetization Coil

The outcome of the multi-objective optimization is the Pareto front shown in Fig. 9. The parameters of the chosen configuration are summarized in Table VII. The slight deviation from the Pareto optimal solution is due to the required discretization of the coils cross section area to fit available wire dimensions. In our case a $0.6 \times 1.3\,\mathrm{mm}$ wire is used.

[4]This tree has the same structure for all individuals, but different parameter values.

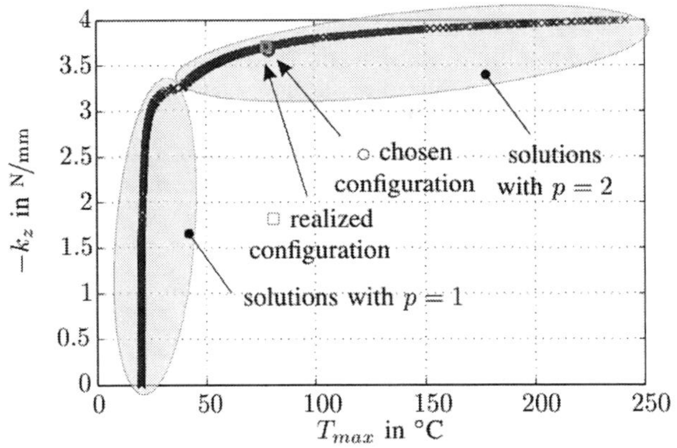

Figure 9: Pareto front showing axial stiffness vs. magnetization coil temperature.

Table VII: Selected configuration

Parameter	Value	Parameter	Value
DESIGN PARAMETERS			
n_c	6	N	2
b_c	1.6 mm	h_c	1.3 mm
$U_{C,0}$	1200 V	C	540 μF
p	2	δ	0.1 mm
OBJECTIVE VALUES AND SUBSIDIARY PARAMETERS			
$-k_z$	3.66 N/mm	T_{max}	79°C
L	319 nH	R_{20}	20.8 mΩ
i_{max}	11.8 kA	m	8.6 g

The realized magnetization coil is shown in Fig. 10. From the real setup, a slightly different resistance and mass was measured, leading to a necessary voltage of the magnetizer of $U_{C,0} = 1330V$ in order to reach a similar peak current and maximum coil temperature as gained from the optimization. The real parameters are the corresponding simulation results are listed in table VIII.

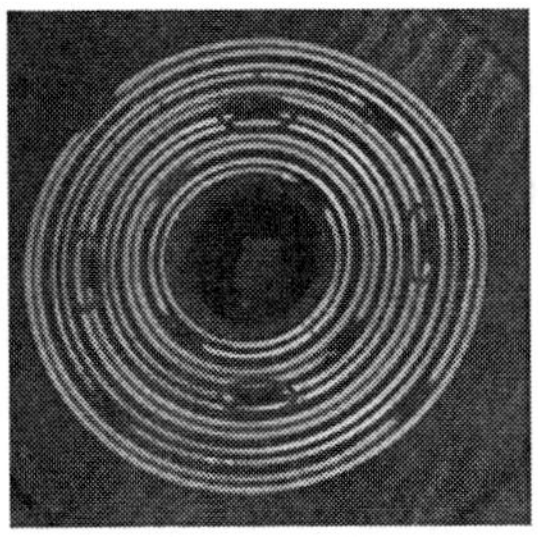

(a)

(b)

Figure 10: Realized magnetization coil. (a) Detailed view on the realized magnetization coil; every second single coil is shaded for better distinction; observe the symmetric positioning of the threwholes – indicated by the small ovals – in order to minimize cogging effects. (b) Complete setup.

Table VIII: Deviating parameters of the realized configuration with respect to table VII

Parameter	Value	Parameter	Value
R_{20}	25 mΩ	m	12 g
$U_{C,0}$	1330 V	i_{max}	12.6 kA
$-k_z$	3.7 N/mm	T_{max}	78°C

The simulated magnetization current and the measured one has already been shown in Fig. 5. The data used for simulation corresponds to table VIII and the magnetizer was also initialized with $U_{C,0} = 1330\,V$.

IV. Force Measurements

For comparison and validation of the calculation methods used for the optimization, the force of the single ring bearing with axial magnetization was measured. The setup for the force measurements is pictured in Fig. 11. Since magnetization

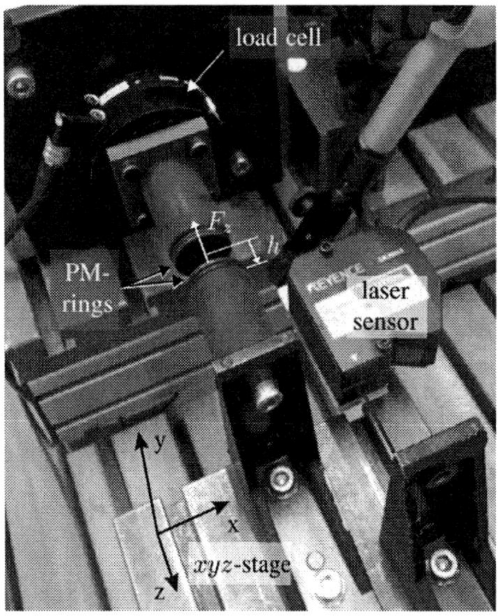

Figure 11: Setup to measure the axial force of permanent magnetic bearings. Force is measured by a load cell, the air gap distance is captured by a laser sensor.

in axial direction is easy to realize, full magnetization can be assumed within the whole volume of the magnet. The measured axial force as function of the air gap dimension h and the corresponding finite element calculation done with FEMM for the axially magnetized bearing are pictured in Fig. 12, showing a very good agreement. Furthermore, the force measurement is fitted by a polynomial and the stiffness is calculated therefrom by $k_{ax} = \frac{dF_z}{dh}$. Positive values of h an F_z point in the directions as indicated in Fig. 11.

Similar data is given for the realized bearing with rotating magnetization in Fig. 13. These results show a certain deviation between measured and calculated data. Possible reasons for this difference are: (i) the material is not isotropic, as assumed (the data from table II are measured only in axial

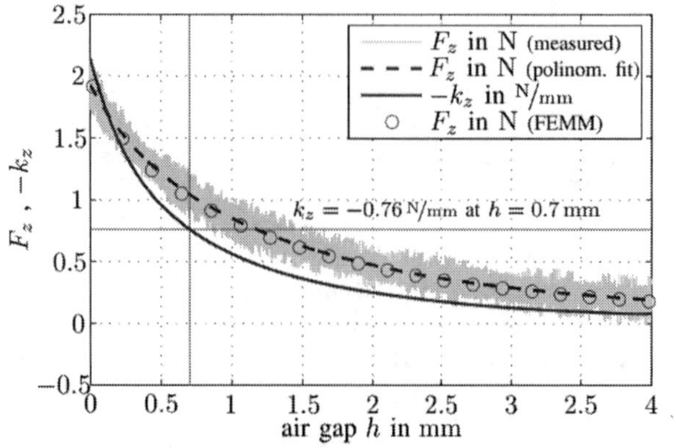

Figure 12: Axial force and stiffness value of the bearing with axial magnetization (cf. Fig. 2); the stiffness value is gained from the polynomial fit of the axial force.

direction) and/or (ii) the initial curve of the material, which we did not have at our disposal, was badly assumed (cf. Fig. 7). Unfortunately, these influences are hard to validate due to the lack of material data. That the (theoretically) desired peak current was reached during the magnetization process can be seen from Fig. 5. In this context it might be worth to be mentioned, that even bearings with peak currents during magnetization of up to 15 kA did not lead to improved results.

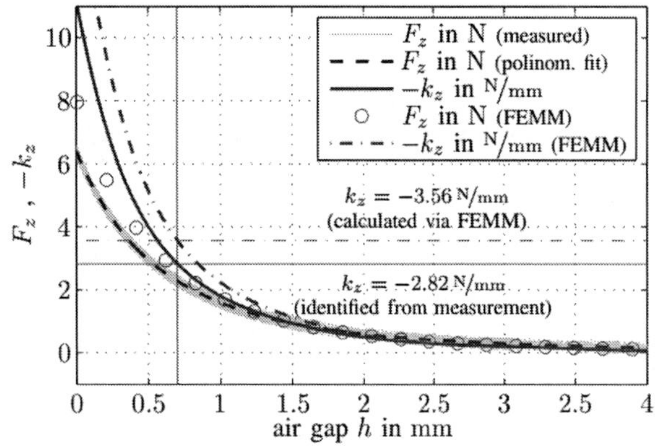

Figure 13: Axial force and stiffness value of the bearing with rotating magnetization (cf. Fig. 8); stiffness values are gained from polynomial fits of the axial forces (fit of FEMM-calculated force values are not shown).

Nevertheless, comparing the axially homogeneous magnetized bearing with the optimized one, a 3.7- fold improvement in the stiffness value and a 2.7- fold improvement in force value was achieved just by optimizing the magnetization pattern.

Concerning the stiffness calculation, the different values for k_z given in table VIII and Fig. 13 originate from a slightly different calculation method. During optimization the stiffness is calculated from the force values via the central differential quotient, while in Fig. 13 the stiffness is calculated from a polynomial fit of the calculated force values. This also points out the general problem of stiffness calculation via the finite element method.

Referring to [4], [5], the geometric parameters of the single ring bearing are very bad. But even compared to a standard double ring bearing as shown in Fig. 14, where a stiffness of $k_z = -1.8$ N/mm and a force of $F_z = 1.4$ N were calculated via FEMM, the improvement is still notable.

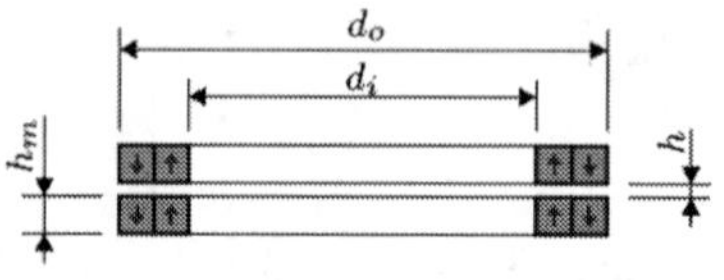

Figure 14: Bearing in double ring configuration; geometric parameters as given in table I.

Another interesting aspect can be seen from the magnetization pattern, Fig. 8. Due to the geometric specifications it was not possible to reach a rotating magnetization within the whole cross section of the magnets. In the outer regions the pattern resembles rather the initial axial magnetization than the rotating one. Thus, if the bearing is thought as a combination of two bearings as pictured in Fig. 15, the outer bearing has bad magnetization pattern and a big air gap. The contribution of the outer bearing to the total stiffness is therefore low compared to the magnet volume. For the inner bearing, the stiffness calculates to 77% of the value given in table VIII with only 50% of magnet volume.

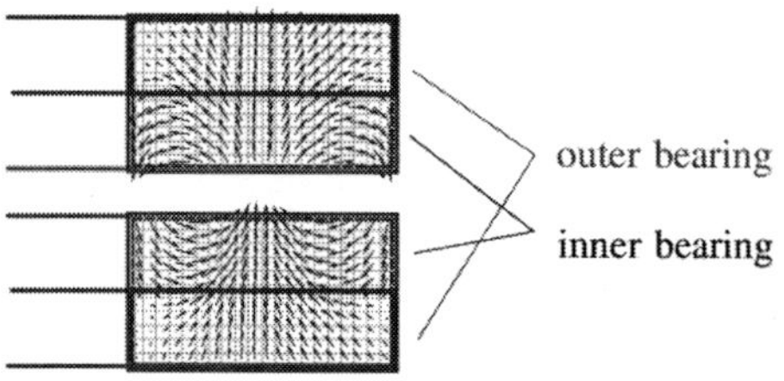

Figure 15: Conceptional configuration with two separate bearings.

The latter example of conceptionally splitting up the bearing is intended to show another possible issue of an optimized design and the importance of the proper geometric dimensions. In particular the strong correlation between the possible magnetization depth (in h_m direction) and the pole width $^{d_o - d_i}/_{2p}$ is a realization specific aspect to be considered. An advantage of the continuous rotating magnetization as presented in this paper is, that the number of poles not necessarily has to be an integer value. Looking at the Pareto front, Fig. 9, a rational number of poles leads to an improved Pareto front in the region between solutions with $p = 1$ and $p = 2$, i.e., in the region between $25°C < T_{max} < 75°C$.

V. Further Work

Concerning the deviation of calculation and measurement of the continuous Halbach bearing further investigations are necessary. First of all, the anisotropic initial magnetization curves of the material are important in order to eliminate the possible errors mentioned.

VI. Conclusion

The presented work shows the possible improvement of stiffness and force values due to an optimized magnetization pattern. Also the importance of including the magnetization process itself becomes obvious.

Including the magnetization coil and the magnetization process in the bearing optimization is a novel approach for permanent magnetic bearing design. It represents the whole physical magnetization mechanism to reach maximum stiffness: the simulation of the magnetization current and coil temperature, the alignment of the elementary magnetic dipoles and the calculation of the stiffness of the final bearing.

Highly efficient designs can be expected if also the geometric parameters are adjusted during optimization.

VII. Acknowledgment

Parts of this work were supported by the Linz Center of Mechatronics (LCM) GmbH, a K2-center of the COMET program of the Austrian Government. The authors thank the Austrian and Upper Austrian Government for their support.

References

[1] J.-P. Yonnet, G. Lemarquand, S. Hemmerlin, and E. Olivier-Rulliere, "Stacked structures of passive magnetic bearings," *Journal of Applied Physics*, vol. 70, no. 10, pp. 6633–6635, November 1991.

[2] M. Lang and J. K. Fremerey, "Optimization of permanent-magnet bearings," in *6th International Symposium on Magnetic Suspension Technology*. Turin, Italy: Politecnico di Turino, October 2001. [Online]. Available: www.magneticbearings.org

[3] R. Moser, J. Sandtner, and H. Bleuler, "Optimization of repulsive passive magnetic bearings," *Magnetics, IEEE Transactions on*, vol. 42, no. 8, pp. 2038–2042, 2006.

[4] E. Marth, G. Jungmayr, M. Panholzer, and W. Amrhein, "Optimization of stiffness per magnet volume ratio of discrete and continuous Halbach type permanent magnetic bearings," in *ISMB13, 13th International Symposium on Magnetic Bearings*, 2012. [Online]. Available: www.magneticbearings.org

[5] E. Marth, G. Jungmayr, and W. Amrhein, "A 2-d-based analytical method for calculating permanent magnetic ring bearings with arbitrary magnetization and its application to optimal bearing design," *Magnetics, IEEE Transactions on*, vol. 50, no. 5, pp. 1–8, May 2014.

[6] S. Earnshaw, "On the nature of molecular forces wich regulate the constitution of luminiferous ether," *Transactions of the Cambridge Philosophical Society*, vol. 7, no. 1, pp. 97–112, 1842.

[7] W. Braunbeck, "Freischwebende Körper im elektrischen und magnetischen Feld," *Zeitschrift Physik*, vol. 112, pp. 753–763, 1939.

[8] S. Silber, W. Koppelstätter, G. Weidenholzer, and G. Bramerdorfer, "MagOpt – optimization tool for mechatronic components," in *Proceedings of the 14th International Symposium on Magnetic Bearings*, 2014. [Online]. Available: www.magneticbearings.org

"Whirl Imposer"--- Proposal for a novel passive magnetic rotor bearing system

Qingwen Cui [a], Jan Sandtner[b], Hannes Bleuler[a]

[a] Laboratory of robotic systems (LSRO), Ecole Polytechnique Fédérale de Lausanne (EPFL),CH- 1015 Lausanne, Switzerland,
qingwen.cui@epfl.ch, hannes.bleuler@epfl.ch
[b] Silphenix GmbH, Sonnenweg 10, CH-4436 Oberdorf, Switzerland

***Abstract*—Passive magnetic bearing systems are enticing in applications where low stiffness is acceptable, but where cost, life-time, power consumption and reliability are critical.**

This article investigates a new proposal for stabilization of permanent magnet (i.e. passive) bearing systems based on rotor precession. There will be a minimal and a maximal rotational speed for stable contact-free levitation.

The feasibility of this new stabilization scheme is investigated with the help of realistic numerical data, a demonstration system is currently in planning.

I. Introduction

Passive magnetic bearing systems [1], i.e. systems with complete contact-free levitation, but without any active control, are very appealing for certain cost-critical applications where low stiffness may be tolerated, but where simplicity, reliability and low power consumption are important.

Earnshaw's theorem [2] states that for a static system perfect (i.e., in all degrees of freedom) passive magnetic levitation is impossible. That is to say, at least one degree of freedom has to be controlled with other methods [3]-[6].

However levitation using solely permanent magnets can be achieved under certain additional conditions. Gyroscopic effects, eddy currents or superconduction can stabilize contact-free levitation. The "Levitron" toy is a proof. It is known that precession can provide a stabilizing mechanism [7].

The design proposed here consists of two passive radial bearings, for which the axial (thrust) stiffness is negative by definition. Then, negative axial stiffness compensation is added at the cost of lowering the radial stiffness. The essential part Whirl Imposer, which induces positive axial stiffness and precession, is introduced as well. A proposal will now be presented for achieving overall stable levitation in a certain range of rotational speed.

II. Concept

A. Rotating system with radial bearings

According to the Earnshaw's theorem

$$2k_r + k_z = 0 \tag{1}$$

where k_r... radial stiffness
k_z...axial stiffness

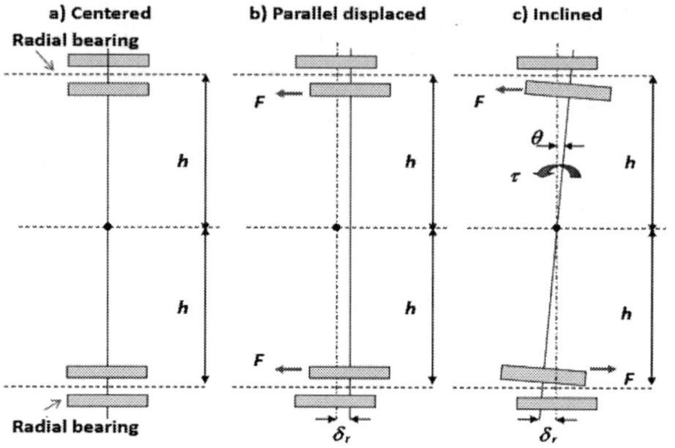

Figure. 1 Radial and conical stiffness of radial PM bearings.

Radial bearings with a positive radial stiffness therefore have a negative axial stiffness:

$$k_z = -2k_r \tag{2}$$

In order to be simple, we will now directly present a realistic numerical example of a potential test system. Consider the situation shown in Fig. 1a:

Both upper and lower bearing is composed of two identical axially magnetized permanent magnet rings in an attractive mode with stiffnesses

$$k_r = 12N/mm, k_z = -24N/mm \tag{3}$$

Let the vertical distance between radial bearings be

$$2h = 2 \times 150mm = 300mm \tag{4}$$

The total stiffness of both bearings together for the cylindrical mode will then be

$$k_{rtot} = 24N/mm, k_{ztot} = -48N/mm \tag{5}$$

If the shaft is radially displaced (i.e. parallel to the vertical axis) by a distance δr = 1 mm (Fig. 1b), then the restoring force will be F = 24 N.

If the shaft is inclined so that the upper end is displaced to the right and the lower end to the left by a distance δr = 3 mm (Fig. 1c), which corresponds to an angle

$$tg\theta = \frac{\delta_r}{h} = \frac{3}{150} = 0.02 \Rightarrow \theta = 0.02rad \tag{6}$$

the restoring force at shaft ends will be

$$F_\theta = k_r \times \delta_r = 12N/mm \times 3mm = 36N \quad (7)$$

Forces at both shaft ends produce a restoring torque

$$\tau = F_\theta \times 2h = 36N \times 300mm = 10.8Nm \quad (8)$$

Thus the conical stiffness will be

$$k_\theta = \frac{\tau}{\theta} = \frac{10.8}{0.02} = 540Nm/rad \quad (9)$$

B. Principle of negative axial stiffness compensation

Consider now what happens, when a compensation for the negative axial stiffness will be introduced as shown in (Fig. 2a).

The negative axial stiffness compensation system consists of two coaxial axially magnetized permanent magnet cylinders in an attractive mode. The system is located exactly at the middle between radial bearings. The bearing-rotor system's properties with respect to parallel displacement are chosen to be the same as above (Fig. 2b).

The radial stiffness will be increased, say, by a factor 5 (arbitrary chosen), i.e. from

$k_r = 12N/mm$ to $k_r' = 60N/mm$

The new axial stiffness will be

$$k_z' = -120N/mm$$

Total stiffnesses of both radial bearings together

$$k_r' = 120N/mm, k_z' = -240N/mm$$

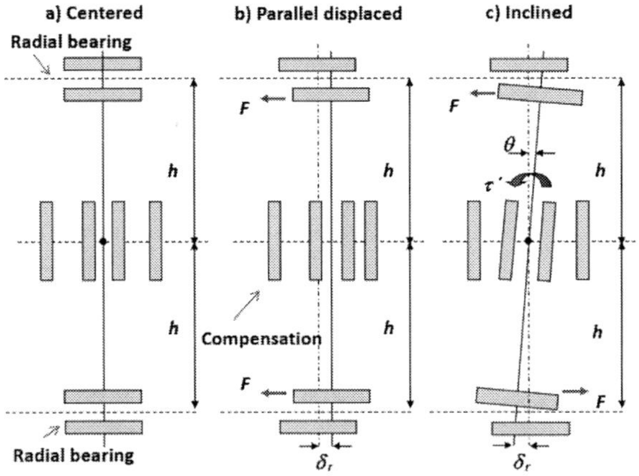

Figure 2. Bearing behavior with negative stiffness compensation.

In order to obtain the same properties as above, the compensating system has to possess the following stiffnesses

$$k_{cr}' = -96\,\text{N/mm},\ k_{cz}' = 192N/mm$$

(Note that, this structure introduces a negative radial and positive axial stiffness). When all corresponding stiffnesses are added together

$$k'_{rtot} = 2 \times 60 - 96 = 24N/mm,$$
$$k'_{ztot} = -2 \times 120 + 192 = -48N/mm \quad (10)$$

which is the same as in the first case.

However, concerning the conical stiffness, with the same inclination, i.e. also with the same displacement distance δr = 3 mm as above (Fig. 2c), the forces at both shaft ends will be

$$F'_\theta = k'_r \times \delta_r = 60N/mm \times 3mm = 180N \quad (11)$$

These forces produce a stabilizing torque

$$\tau' = F'_\theta \times 2h = 180N \times 300mm = 54Nm \quad (12)$$

Hence the conical stiffness will be

$$k'_\theta = \frac{\tau'}{\theta} = \frac{54}{0.02} = 2700Nm/rad \quad (13)$$

which is five times more than without the compensation.

By increasing radial bearing stiffness of the radial bearings x-times, then with the negative axial stiffness compensation the resulting conical stiffness will also be increased by the same factor, on condition that the radial and axial stiffness of the system remain the same.

Note that when the rotor is working in conical mode, there will be much increased restoring force and stabilizing torque, due to the increased radial stiffness of the outer radial bearings. This is why precession (i.e. the conical mode of the rotor) could contribute to stabilizing.

C. Whirl Imposer

Whirl imposer consists of a rotor and a stator, which are coaxial discs. There are magnets distributed equally in circumferential direction on both the rotor and the stator. The magnets are oriented so that the forces between the stator and rotor are attractive. Thus the whirl imposer introduces positive axial stiffness. A sample is shown in Fig.3.

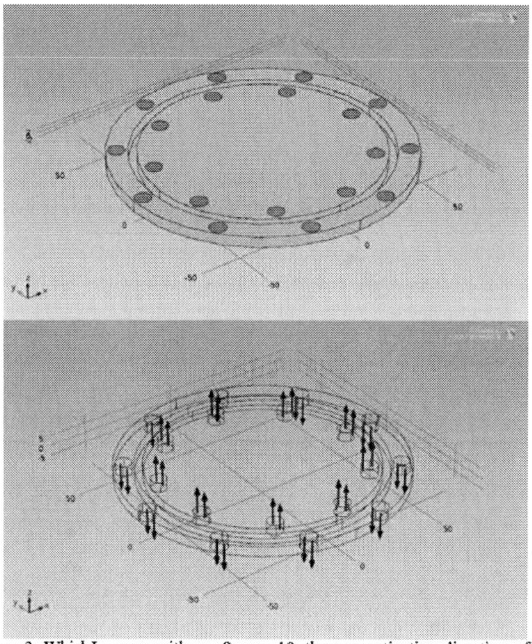

Figure 3. Whirl Imposer with $n_R = 9$, $n_S = 10$, the magnetization direction of magnets are shown.

Moreover, due to the disc shape of the rotor, gyroscopic stiffness will be high, which does contribute to stability, similar to the gyroscopic effect in the Levitron toy [8].

In principle, precession can easily be introduced in a rotating system by means of a weak unbalance with two additional masses at appropriate rotor locations. This is called torque-free precession, which means there is no external torque applied to the rotor [9]. The forces depend strongly on the rotation speed and rotate synchronously with the rotor. This property may not be desirable for our purpose

However, when the principle of whirl imposer is used, precession will be induced as well, but the forces don't depend on the rotational speed and rotate either in the same or in the opposite sense than the rotor angular speed. In this case, the precession is induced by external forces (magnetic force on the rotor of whirl imposers, F_R). The angular speed of these forces can be made different from the rotor angular speed, depending on numbers of permanent magnet poles on stator and rotor.

The multiplication coefficient κ between the angular speed of the radial force F_R and rotor angular speed is given by the following expression:

$$\kappa = \frac{n_R - n_S}{n_R} \qquad (14)$$

where

n_R ...number of PMs within the rotor

n_S ...number of PMs within the stator

It can be seen that this factor is a positive or negative rational number, hence both forward and backward whirls may be achieved.

Precession angular speed will be inversely proportional to the rotor angular speed in this case, which will induce an upper limit of the rotational speed in order not to obtain a too low precession speed. It is expected that a stable behaviour will only be achieved in a limited range of rotational speeds. This range of rotational speeds will depend on the properties of the whole system.

The PMs may be oriented either radially or axially. The operation of both variants is similar but not exactly the same (PMs are used more efficiently in the second option, as shown in Fig.3). In order to obtain some preliminary feeling about whirl imposer behaviour, both variants have been manufactured (but not yet the whole system).

The number combination of PMs on the whirl imposer can be chosen such as to minimize torque ripple. Essentially, the whirl imposer introduces a positive axial stiffness to the system. It can thus compensate for the residual axial negative stiffness. For a given azimuth angle the Earnshaw theorem (equation (1)) is not satisfied; however, if the torques are integrated over all azimuth angles then the theorem holds as well for the whirl imposer.

The magnitudes of axial stiffness and radial forces depend on the number and size of magnet poles, and on the air gap between rotor and stator. These properties may be chosen for optimal overall system behaviour. The precession producing torque depends on the whirl imposer axial locations with respect to the axial centre of the rotor.

The FEM code COMSOL 4.4 has been used to simulate a realistic model of whirl imposer with n_R= 9 n_S= 10 as shown in Fig.3. Our interest is mainly with the radial force on the rotor. Here κ is expected to be -9 according to equation (14).

TABLE I. PARAMETERS OF A WHIRL IMPOSER

	Parameter	Value
d_r	Outer rotor diameter	112 mm
ds	Outer stator diameter	140 mm
dsi	Inner stator diameter	116 mm
t	Thickness of magnets	5 mm
rm	Radius of magnets	4 mm
mB	Magnetic flux density	1.22 T

Parameters of this whirl imposer are listed in Table I while some results of obtained forces on the rotor are shown in Table II. In this case, the rotor is in the nominal position $(x, y, z)=(0, 0, 0)$ and rotates in the CCW direction.

For azimuth angle 4° and its multiples, two rotor and stator magnets are located at the same radius vector. This is repeated periodically.

TABLE II. RESULTS OF FORCES ON THE ROTOR AT NOMINAL POSITION

φ [°]	F_x [N]	F_y [N]	F_R [N]	φ_F [°]
0	2.27	0.01	2.27	0
1	2.19	-0.38	2.22	-9.79
2	2.13	-0.69	2.24	-17.95
3	2.03	-1.06	2.29	-27.46
4	1.89	-1.30	2.29	-34.52

where:

φ... rotor position azimuth in degrees.

F_x... x-component of rotor force in N.

F_y... y-component of rotor force in N.

φ_F...azimuth of radial force in degrees.

Results show that the force vector magnitude remains always the same and it rotates with a higher speed than the rotor in the CW direction ($\kappa = -9$ proved here).

When the rotor is displaced off the nominal position, the force vector magnitude will vary according to rotor position azimuth angle. Then the period 4° will not hold, and turns to 40° depending only on the symmetry of the rotor magnets distribution.

For the radial displacement d_x=2mm, a simulation has also been made in COMSOL.

Forces on the rotor are listed in Table III. The azimuth of the rotor is in the range (0,20°).

TABLE III. RESULTS OF FORCES ON THE ROTOR WHEN DX=2MM

φ[°]	F_x [N]	F_y [N]	F_R [N]	φ_F [°]
0	4.031	0.11	4.032	1.563
2	3.82	-1.116	3.98	-16.29
4	2.96	-2.047	3.60	-34.67
5	2.512	-2.286	3.40	-42.30
6	2.129	-2.487	3.27	-49.44
8	1.054	-2.602	2.81	-67.95
10	0.294	-2.30	2.315	-82.72
10.5	0.051	-2.134	2.135	-88.66
11	-0.066	-2.195	2.196	-91.72

φ[°]	F_x [N]	F_y [N]	F_R [N]	φ_F [°]
11.5	-0.342	-2.013	2.041	-99.65
12	-0.353	-1.87	1.903	-100.69
14	-0.89	-1.484	1.73	-120.95
15	-1.102	-1.178	1.612	-133.09
16	-1.086	-1.00	1.475	-137.39
18	-1.266	-0.443	1.34	-160.70
20	-1.292	-0.02	1.292	-179.12

With the Curve Fitting tool in Matlab, the function of the force vector magnitude could be approximately written as:

$$F_R = 1.433\cos(0.151\varphi + 0.346) + 2.75 \qquad (15)$$

while the azimuth of the radial force could be written as a function of the rotor azimuth angle:

$$\varphi_F = -8.94\varphi + 4 \qquad (16)$$

Figures of F_R and φ_F are plotted according to the above functions.

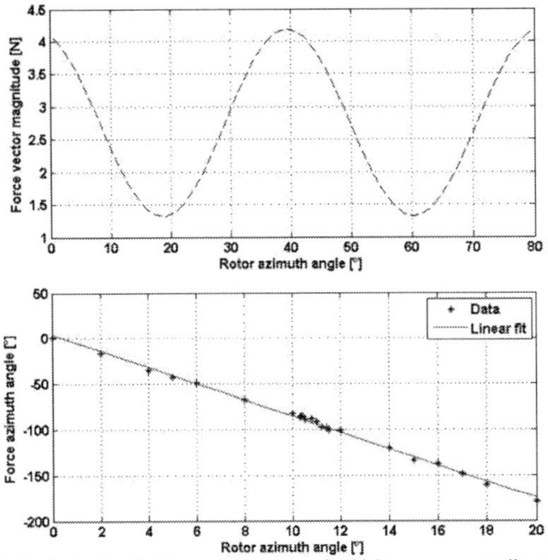

Figure 4. Analysis of radial force on the rotor of Whirl Imposer according to rotor azimuth angle.

For azimuth of the rotor φ=0°, simulation has been made as well. Magnitude of F_R is plotted in Figure 5 for radial displacement varying from 0 to 4.5mm.

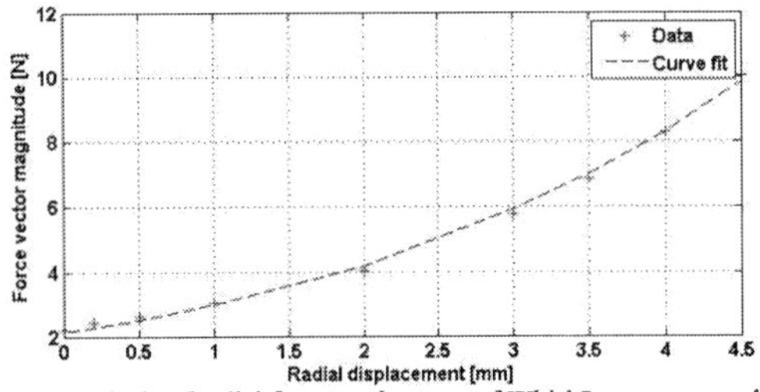

Figure 5. Analysis of radial force on the rotor of Whirl Imposer according to radial displacement.

Function of F_R in this case could be written as following with Curve Fitting:

$$F_R = 2.113e^{0.343 d_x} \qquad (17)$$

The results obtained with FEM calculations show that the magnitude of force vector on the rotor is no longer a fixed value if the rotor moves off the nominal position, instead, it varies as a sinusoid wave according to the azimuth of the rotor. It varies as an exponential function according to the radial displacement. The azimuth of the radial force is proportional to the azimuth of the rotor with a coefficient κ close to −9 (the value from equation (14)), which shows the relationship between the angular speed of the radial force and the rotor angular speed.

D. *A passive magnetic rotor bearing system*

Fig.6 shows the bearing system with combination of the negative axial stiffness compensation and two whirl imposers. The whirl imposers compensate the negative axial stiffness and introduce negative radial stiffness, the overall radial stiffness is thus negative.

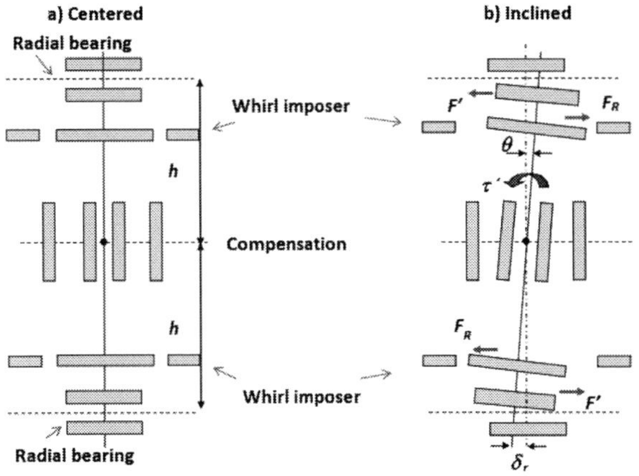

Figure 6. A bearing system combined with the negative axial stiffness compensation and two whirl imposers.

The whirl imposers could be located between the two radial bearings or outside of them. In order to achieve a torque onto the system axis, the azimuthal orientation of whirl imposers will be such that the rotating force of the upper imposer will be shifted with respect to the lower one by 180 degrees. These forces should have the same magnitude but opposite orientation (Fig. 6b). Thus the torque will be exerted onto the shaft about the central point of shaft symmetry (i.e. at the middle of the shaft). The torque produces precession, which always changes its orientation in accordance with the torque vector. Thus an unremitting **conical mode** arises. For a given rotational speed the cone angle (with respect to the vertical axis) remains constant.

If the shaft is inclined (Fig. 6b) in conical mode, then there will be the attractive force from the whirl imposers as well as restoring force due to the radial bearings. Even a small torque due to the whirl imposer is enough for modulating the restoring torque due to the radial bearings, thus the conical mode will be ensured.

In the rotating system, the spin angular momentum L is along the rotation axis, but the torque due to F_R about the centre of the shaft is in a direction perpendicular to the angular momentum. The torque produces a change in L which is perpendicular to L. Such a change causes a change in direction of L but not a change in its size. This motion is the so called torque-induced precession.

For a spinning top (or a gyroscope), the torque is induced by gravity and it rotates at the same speed as precession. In our case, the torque τ_p is induced by F_R. Thus the rotating speed of τ_p is basically proportional to the spinning speed of the rotor with a coefficient κ. Besides, the magnitude of τ_p varies periodically due to spinning of the rotor and varies exponentially with the radial displacement.

Analysis of the rotordynamics of this system is done with the FEM code DYNROT. Some realistic parameters are chosen for a numerical simulation. The radial bearings, the whirl imposer and the negative axial stiffness compensation are as described above.

Assume that the system is axially stable, which means enough axial stiffness has been provided. Gravity can easily be compensated for a vertical rotor, so it is not considered at this moment.

The rotor of the system is modelled with 1-D beam elements, while the rotating parts of the two whirl imposers are modelled with rigid mass elements.

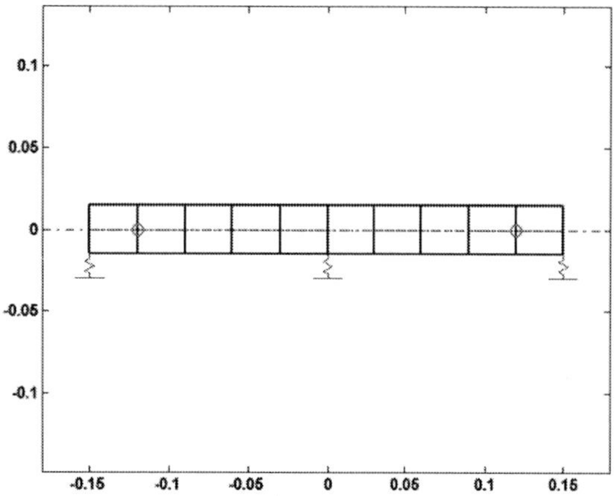

Figure 7. Modelling of the rotor system in DYNROT.

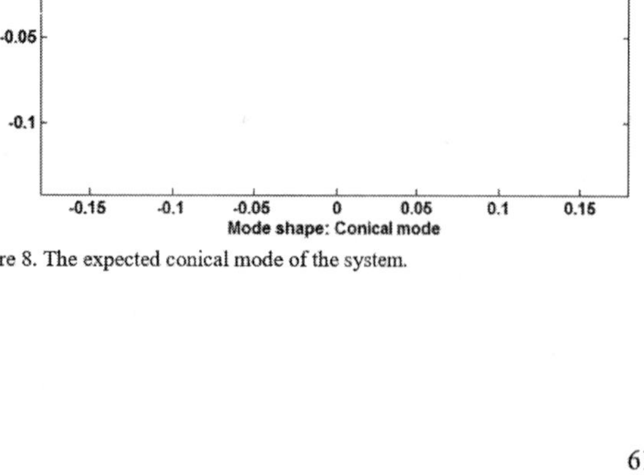

Figure 8. The expected conical mode of the system.

Figure 8 shows the desired conical mode.

The radial forces due to whirl imposers vary not only with radial displacement, but also with the azimuth of the rotor. This kind of force is not easy to handle in commercial FEM codes which assume rotational symmetry.

As shown in Figure 4, F_R varies in a function of the radial displacement. Thus, for simplicity, the maximum and minimum values are used in a first analysis for prediction of approximate dynamic behaviour. A more complete simulation is underway.

III. Conclusion

This paper introduces a novel design of a passive magnetic bearing system using solely permanent magnets. On the base of a rotating system equipped with radial bearings, two subsystems have been designed: Negative axial stiffness compensation and Whirl Imposer. Conical stiffness can thus be increased without modifying radial and axial stiffness of the system. The new component proposed here is the "whirl imposer", consisting of a number of stator and rotor permanent magnets, which induce precession and at the same time compensate for the residual axial negative stiffness.

Although the system is radially unstable at standstill, a conical mode is expected to stabilize the system in a certain working speed range.

References

[1] J.P.Yonnet, "Passive magnetic bearings with permanent magnets,"*IEEE Trans.Magn.*,vol.MAG-14,no.5,pp.803-805,Sep.1978

[2] S. Earnshaw, "On the nature of the molecular forces which regulate the constitution of the lumiferous ether", *Trans. Cambridge. Phil. Soc.*7, pp. 97-112(1842).

[3] J. Sandtner, J.L. Bermudez, and H.Bleuler, "High speed passive magnetic bearing with increased load supporting capacities," in *Proc. Seventh Int. Symp. Magnetic Bearings*, Aug. 2000, pp. 159-164

[4] G. Schweitzer, H. Bleuler, and H. Traxler, *Active Magnetic Bearings: Basic, Properties and Applications.* Heidelberg, Germany: Springer, 1993

[5] J. Sandtner and H. Bleuler, "Electrodynamic passive magnetic bearing using reluctant forces," in *Proc. 17th Int. Conf. Magnetically Levitated System and Linear Drives (MAGLEV)*, Lausanne, Switzerland,2002

[6] R. Moser and H. Bleuler, "Precise positioning using electrostatic glass motor with diamagnetically suspended rotor," *IEEE Trans. Appl. Supercond.*, vol.12, no. 1, pp. 937-939, Mar.2002

[7] Martin D. Simon, Lee O.Heflinger, and S.L.Ridgway, "Spin stabilized magnetic levitation," *Am. J. Phys.*, Vol.65, No.4, April 1997.

[8] Genta, G., Del Prete, C., Rondano, D.(1999)," Gyroscopic Stabilization of Passive Magnetic Levitation", *Meccanica* 34,4-11-424

[9] http://en.wikipedia.org/wiki/Precession

Analysis of passive magnetic bearings for kinetic energy storage systems

Elkin Rodriguez[a], Juan de Santiago[b], J. José Pérez-Loya[b], Felipe S. Costa[a], Guilherme G. Sotelo[c], Janaína G. Oliveira[d], Richard M. Stephan[a]

[a] Universidade Federal do Rio de Janeiro, Technology Center I-148, CEP 21941-909, Rio de Janeiro – RJ, Brazil, elkinrodvel@msn.com
[b] Uppsala University, Ångström laboratory, 751 21, Uppsala, Sweden
[c] Fluminense Federal University, Passo da pátria Street 156 – room 431-E, CEP 24210-240 Niterói – RJ, Brazil
[d] Universidade Federal de Juiz de Fora, Cidade Universitária – Plataforma 5 Martelos, CEP 36036-900, Juiz de fora – MG, Brazil

***Abstract*— This paper describes a passive magnetic bearing system intended to be used in a flywheel. The upper bearing consists of two rings of permanent magnets operating as a radial bearing. The lower bearing consists of high temperature superconductors in the stator and a permanent magnet cylindrical disk in the rotor, which contributes to the radial and axial stability of the system. Those bearings have been tested in an experimental rig specially designed for it. Measurements of the radial and axial forces, both static and dynamic, for different gaps are presented in this paper**

I. Introduction

Flywheel Energy Storage Systems (FESS) are globally gaining attention. They raised in popularity when a FESS was introduced in Formula 1 in 2009 [1][2]. Other applications include electric grid voltage and frequency regulation [3], energy storage and load leveling for integration of renewable energy sources, as solar and wind power [4].

Energy storage systems are classified according to their energy capacity, power rate, life time, availability factor, etc. Flywheels are an environmentally friendly storage solution since no hazardous materials are used in their construction, resulting in a small environmental impact. But, on the other hand, high performance flywheels can only be achieved with magnetic bearings.

Flywheels store energy in form of rotational energy as presented in Eq. 1.

$$E = \frac{1}{2} I\omega^2 \tag{1}$$

High rotational speed increases energy density but also increases bearing losses. Magnetic bearings are therefore an interesting alternative to traditional rolling bearings. Magnetic Bearings can be classified into two major types: Active Magnetic Bearing (AMB) and Passive Magnetic Bearing (PMB). AMB requires accurate control of the current through the windings of electromagnets. PMB uses the repulsion forces without active control systems. The main advantages of PMB are the low energy consumption.

The authors are investigating the bearing set up presented in Fig. 1. The upper bearing is a permanent magnet bearing, and it consists of two rings of axially oriented permanent magnets. The lower bearing is a Superconducting Magnetic Bearing (SMB), and it is composed by melt textured $Y_1Ba_2Cu_3O_{7-\delta}$ superconductor (YBCO) in the stator and a rotor made with permanent magnets and steel arranged in a flux collector configuration.

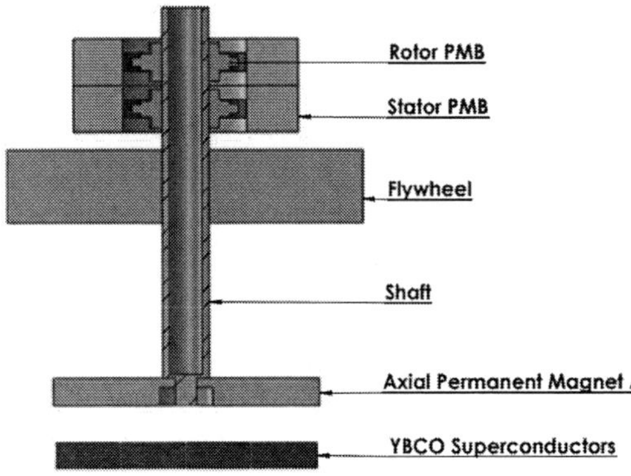

Figure 1. Schematic representation of the pasive magnetic bearing system.

The upper radial magnetic bearing properties were presented in [5]. The results of the static and dynamic tests to characterize the lower passive magnetic bearing are reported in the present paper.

The SMBs present different properties than rolling bearings and AMBs. They have lower force density and require broader air gaps. SMBs present limitations concerning the weight of the flywheel and the air gap required to motor/generator.

II. Magnetic bearings structure

The magnetic bearing system is presented schematically in Fig. 1. It consists of two PMB: one radial permanent magnet bearing on the top and one SMB in the bottom. This section presents details of the bearings' designs.

A. *Axial superconducting magnetic bearing*

The SMB is made up of 8 YBCO bulks enclosed in a static cryostat. The dimensions of each bulk are 67x32x14 mm^3, approximately. Fig. 2 shows two individual YBCO bulks and the complete array mounted inside the cryostat.

A disc made with Nd-Fe-B permanent magnets arranged and a soft ferromagnetic material arranged in an axial flux collector topology is mounted on the rotor [6][7]. The layout and dimensions of this setup are shown in Fig. 3.

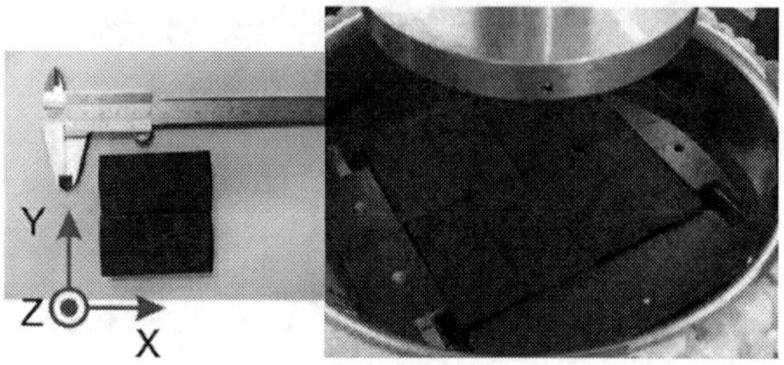

Figure 2. Superconductor blocks.

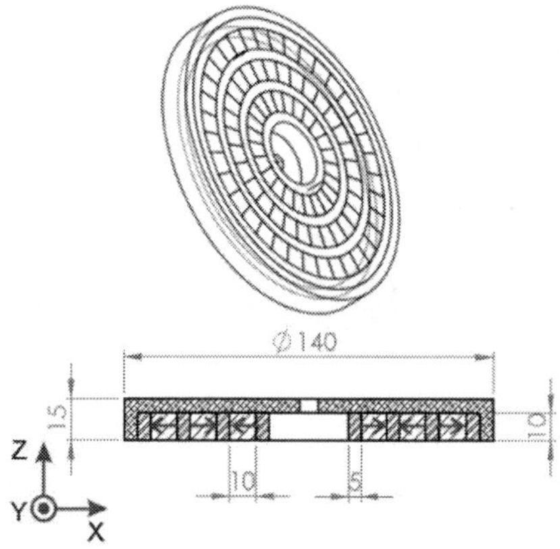

Figure 3. Axial flux concentrator permanent magnet array for the rotor of the supermconductive magnetic bearing.

B. *Radial permanent magnetic bearing (RPMB)*

This bearing is constructed by two concentric permanent magnet rings. This structure was previously investigated in [5], [8] and [9]. The main parameters of the upper radial magnetic bearing are summarized in Table I and Fig. 4.

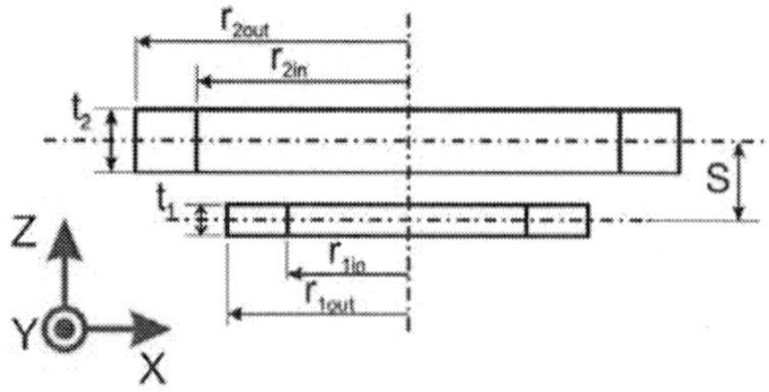

Figure 4. Parameterization of ring magnets used for the construction of the proposed Passive Magnetic Bearings.

TABLE I. RADIAL PERMANENT MAGNETIC BEARING

Parameters	Description	Length (mm)
r_{1in}	Inner radius of inner ring	20
r_{1out}	Outer radius of inner ring	30
t_1	Thickness of inner ring	5
r_{2in}	Inner radius of outer ring	35
r_{2out}	Outer radius of outer ring	45
t_2	Thickness of outer ring	10
S	Axial distance from geometric center of each magnet	--

III. PMB TEST RESULTS

The measurements were performed in a test bench that can be moved in the X and Z axes while a load cell is fixed over it, as shown in Fig. 5. This structure has two linear actuators, a six-axis (it measures X, Y and Z forces and X, Y and Z torques) ATI load cell and the National Instruments PCI-6040E board.

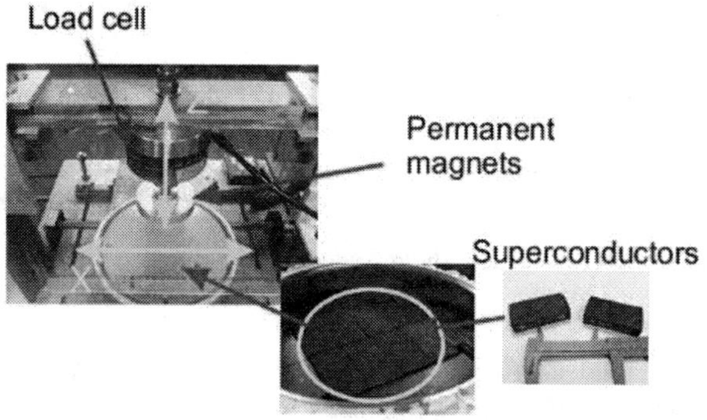

Figure 5. Measurement system.

A. *Radial permanent magnetic bearing (RPMB)*

The rotor has no physical contact with the stator. Therefore, the rotor can move in both radial and axial directions. The stator of the radial bearing was fixed in the test bench; the rotor was displaced from the stator center and the resulting force registered. This test was performed both for the axial and radial direction. Fig. 6 shows that the force measured suffers small variations also due to displacements in orthogonal directions. The radial direction has a positive stiffness, while the axial direction has a negative stiffness; the RPMB is therefore unstable in the axial direction and requires either active control or a stabilizing lower bearing.

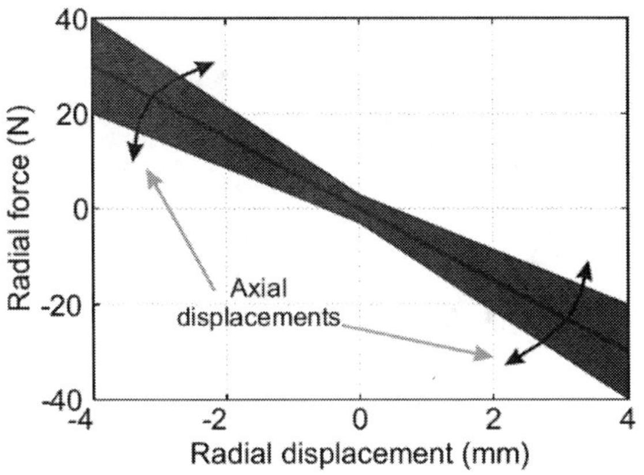

(a)

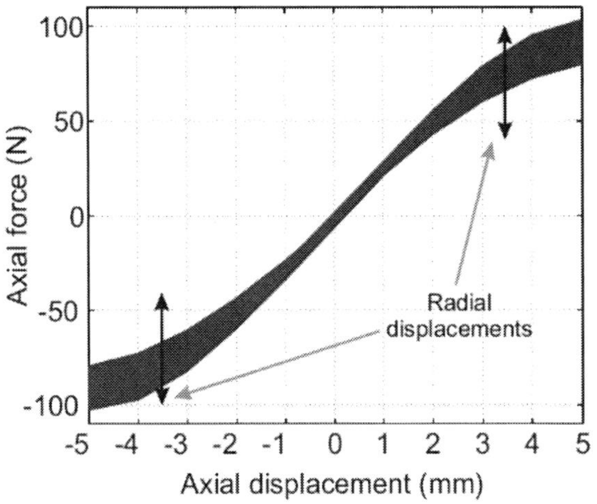

(b)

Figure 6. Behavior of the axial and radial forces for small axial and radial displacements.

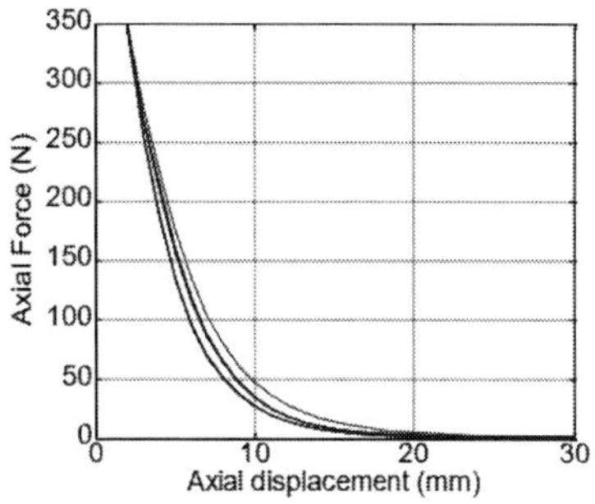

Figure 7. Axial force for different air gaps. ZFC test.

2) FC static

The superconductive material was cooled down below its critical temperature in the presence of a magnetic field. The rotor was lifted a certain distance while the stator was cooled down. The static tests have been performed in two steps after FC: the magnetic flux pumping and the displacement from the operation point. The process is explained in Fig. 8. The experiment was repeated for 4 different distances of FC so that we had different behaviors of magnetic field strength. The closer the rotor is to the superconductive element, the higher the magnetic field strength.

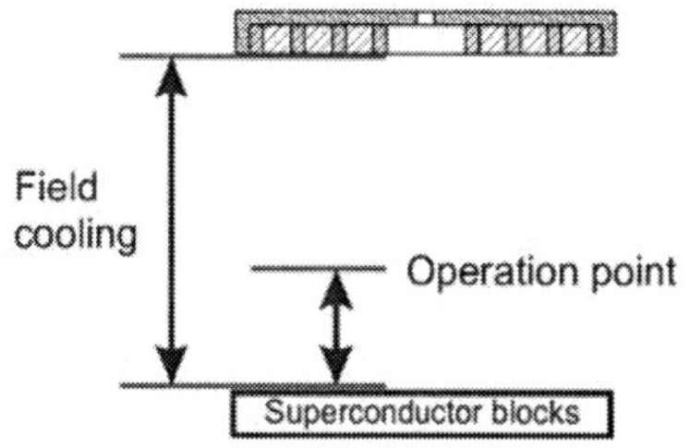

Figure 8. Schematic representation of the static tests of the superconducting magnetic bearing.

Figure 9 shows the magnetic flux pumping accomplished for each FC point, 12, 17, 22 and 27 mm between the rotor and the YBCO bulks.

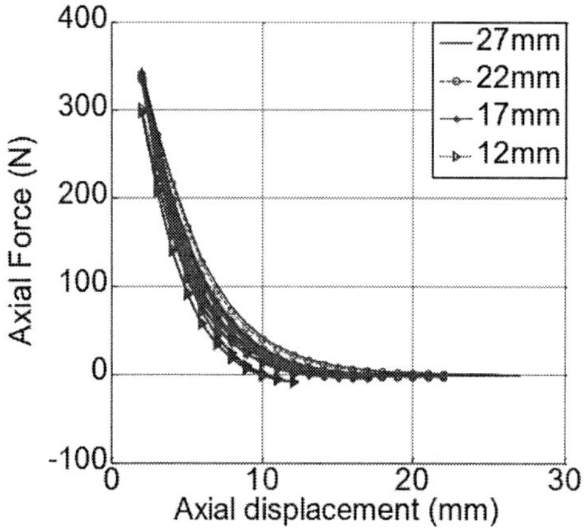

Figure 9. Axial force for different FC points.

B. Axial superconducting magnetic bearing

Superconductors present their diamagnetic and zero resistance effects only when they are cooled below a certain critical temperature. Liquid nitrogen was used in order to cool down the YBCO bulks in these experiments. When the superconductor is cooled in the presence of a magnetic field, the process is referred as field-cooling (FC). Zero field cooling (ZFC) means that the permanent magnet rotor is removed from the test set up when the superconductors were cooled down. Both FC and ZFC tests have been performed.

1) ZFC static

Rotor and stator were taken apart 130 mm before the superconductive material was cooled down below its critical temperature. After the cool down, the magnetic rotor was moved in a quasi-static movement, in steps of 1 mm. After each movement, the system remains static during one second in order to the flux can relax inside the superconductor. Fig. 7 shows the measured axial forces when the air gap was reduced and increased, forming a hysteresis loop.

The second test led the rotor to the operation point. This is located at 7 mm from the upper face of the superconductor blocks. Two different displacements were performed: the radial and axial displacements.

Smaller displacements in axial direction were accomplished to assess the behavior of the axial force. Figure 10 shows the axial force to each FC.

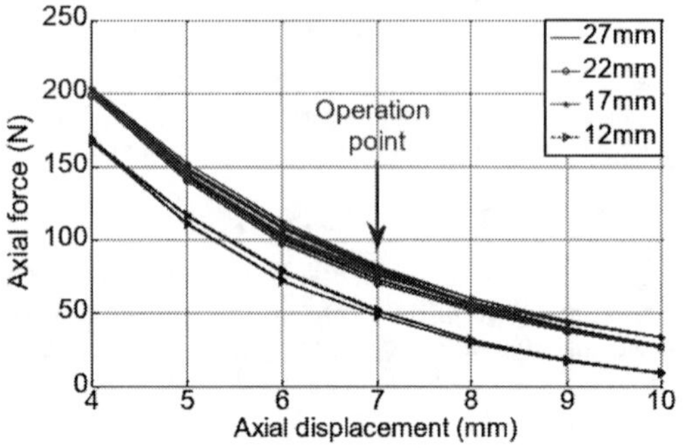

Figure 10. Axial force to small axial displacements in the operation point.

The radial stability has also been studied. The rotor was misaligned from the center of stator in X direction. Radial displacements with an amplitude of 5 mm were performed. Figure 11 shows the behavior of the radial force for these displacements for each FC height.

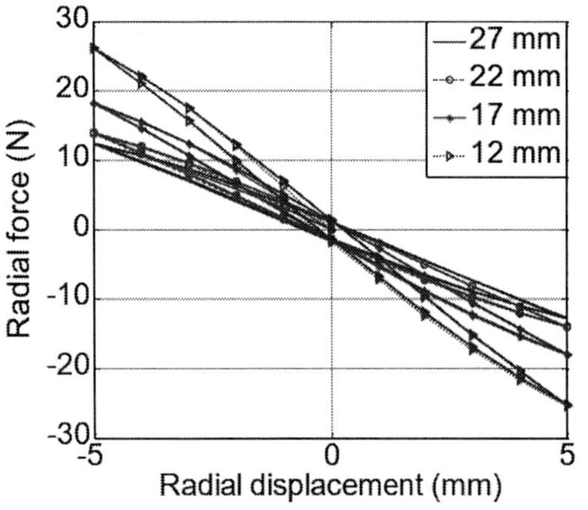

Figure 11. Radial force produced by each FC in the operation point.

3) Dynamic test

Dynamic tests were performed to assess the behavior of the SMB without any upper bearing. The test bench used to measure the force is shown in Fig. 12.

The FC height was 12 mm. After FC, the rotor was accelerated to 2650 rpm with the external help of a drill machine. At this speed, the rotor was released free. Figure 13 shows the speed decrease vs. time.

The radial and axial forces are shown in Figs. 14 and 15.

The passage through resonance frequencies can be clearly observed.

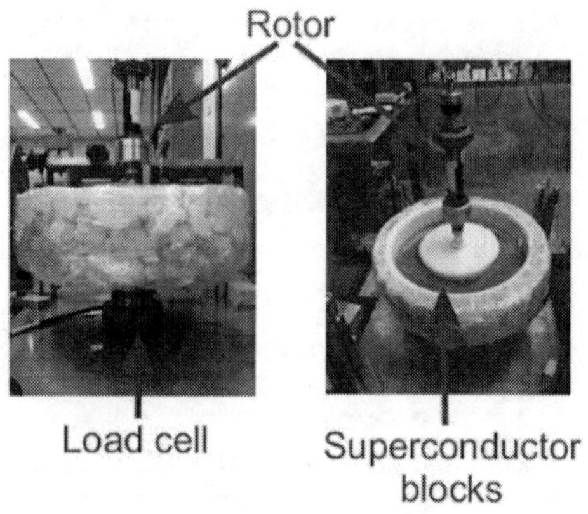

Figure 12. Superconducting magnetic bearing system to measure forces.

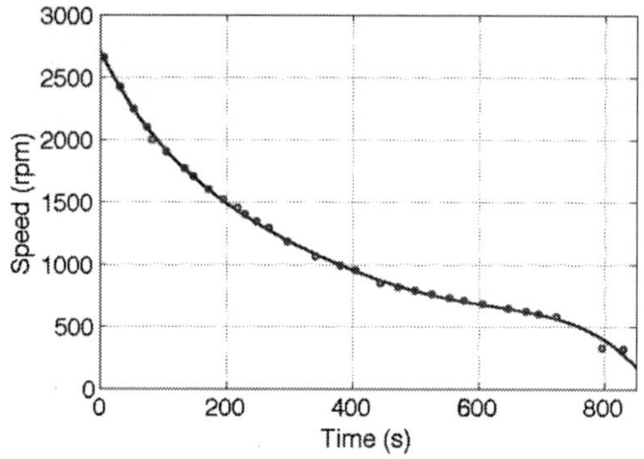

Figure 13. Speed decrease vs. time

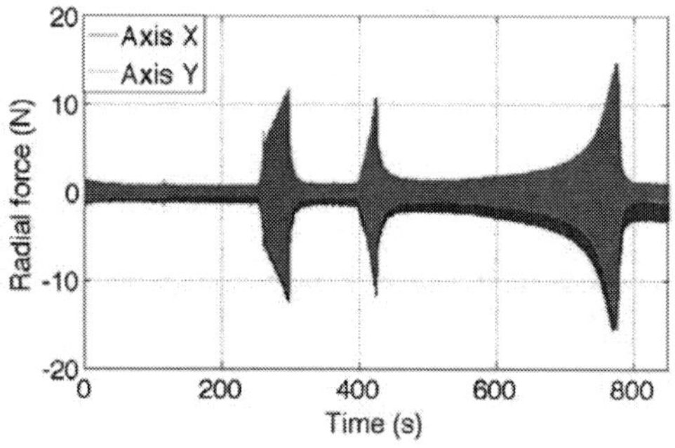

Figure 14. Radial forces vs. time.

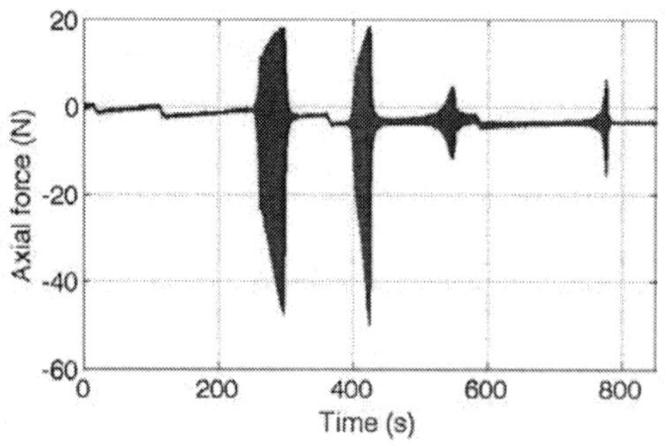

Figure 15. Axial force vs. time.

IV. Discussion

Another configuration with 2 superconducting bulks has also been tested. Static results are presented in Figs. 16 and 17 to show the behavior of axial and radial forces interaction due to radial and axial displacements. These results were obtained with FC 20 mm. Axial force is measured for different air gaps and different radial displacements, and the influence of the radial displacement is not significant and only observed for big radial displacements.

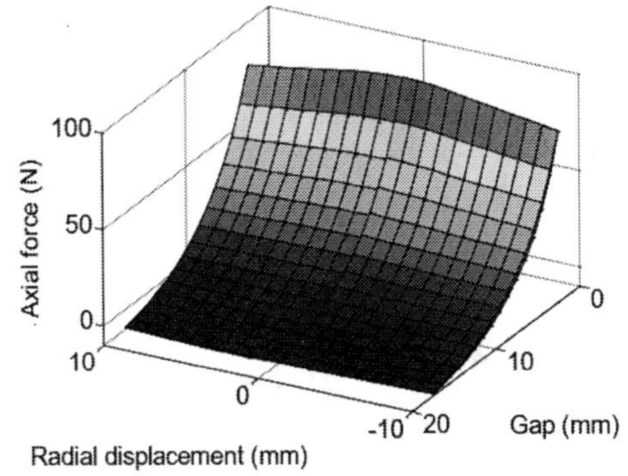

Figure 16. Axial force meassured due to axial displacement for different radial displacements. The test was performed with only 2 superconductive bulks in the SMB stator.

The radial force was also measured for different air gaps and the results are presented in Fig. 13. The SMB presents a non-linear region for small air gaps, but the radial force is independent from the air gap when a value of 7 mm or higher was chosen. This non-linear region is probably due to the small size of the superconductive array.

V. Conclusion

A passive magnetic bearing set up with a radial magnetic bearing on the top and a superconducting magnetic bearing on the bottom was presented. The bearings have been tested independently and the results were presented.

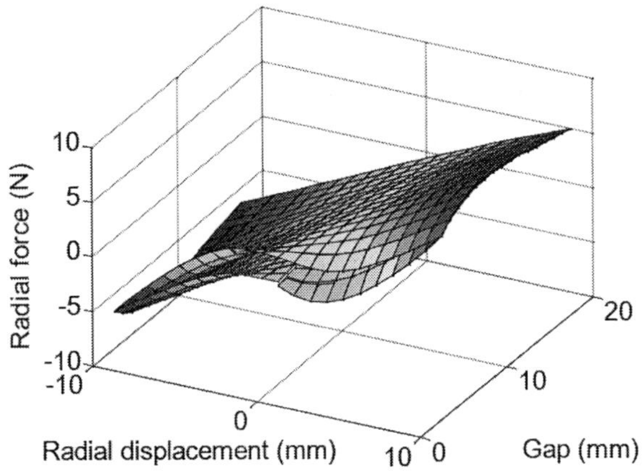

Figure 17. Radial force meassured for different axial and radial displacements. The test was performed with only 2 superconductive blocks in the SMB stator.

Radial permanent magnet bearings are intrinsically unstable, as the force increases proportionally to the displacement in the axial direction. They require some element to give stability to the set up, such as active control, a rolling bearing in the other side of the shaft or a superconducting magnetic bearing as applied in our set up.

The superconducting magnetic bearing constructed consists of a stator with an array of 8 pieces of high temperature superconductors (YBCO) and an axially-oriented permanent magnet flux concentrator array rotor. The test includes axial and radial force measurements with zero field cooling and field cooling in static and dynamic conditions.

The results have show that the SMB radial force with 12 mm FC (fig. 11) is similar to the results that obtained for RPMB (Fig. 6). The radial force increases for lower FC heights (Fig.11), however, the axial force decreases (Fig. 9 and 10).

For the complete implementation with passive magnetic bearings, the SMB with 12 mm FC is a good option, since it presents a similar radial force behavior to that of the RPMB. Moreover, the axial force is reduced but its rate of increase is bigger that with RPMB. Tests are under course and will be reported in the near future.

Acknowledgements

The authors would like to thank the financial support received from FAPERJ, Swedish Research Council, The Swedish Centre for Smart Grids and Energy Storage (SweGRIDS), STandUP for Energy and CONACYT.

References

[1] J. Cibulka, "Kinetic energy recovery system by means of flywheel energy storage", Advanced Engineering, vol 3, n.1, p. 27-38, 2009.

[2] A. A. Boretti, "Improvements of vehicle fuel economy using mechanical regenerative braking", International Journal of Vehicle Design", vol. 55, n. 1, p. 35-48, 2011.

[3] R. de Andrade, *et al.* "Flywheel energy storage system description and tests", IEEE Transactions on applied superconductivity , vol. 17, n. 2, p. 2154-2157, 2007.

[4] J. M. Carrasco, *et al.*, "Power-electronic systems for the grid integration of renewable energy sources: A survey". IEEE Transactions on Industrial Electronics, vol. 53, n.4, p. 1002-1016, 2006.

[5] E. Rodriguez and R. Stephan, "Study of Passive Magnetic Bearings for Vertical-Axis Wind Turbines", 13 th International symposium on magnetic bearings, Arlington – USA, no. 32, august 7-9, 2012.

[6] G. G. Sotelo, A. C. Ferreira ans R. de Andrade, "Halbach Array Superconducting Magnetic Bearing for a Flywheel Energy Storage System", IEEE Transactions on applied superconductivity, vol. 15, no. 2, June 2005.

[7] G. G. Sotelo, R. de Andrade and A. C. Ferreira, "Magnetic Bearing sets for a Flywheel System", IEEE Transactions on applied superconductivity, vol. 17, no. 2, June 2007.

[8] E. Rodriguez and R. Stephan, "Radial and axial forces of a passive magnetic bearing", 1st Brazilian workshop on magnetic bearings, Rio de janeiro – Brazil, no. 22, october 25-26, 2013.

[9] J. J. Pérez-Loya, et. al, "Magnetic Modeling and measurement of forces between permanent magnet rings used as passive magnetic bearings", 1st Brazilian workshop on magnetic bearings, Rio de janeiro – Brazil, no. 25, october 25-26, 2013.

Innovative Magnetic Levitation System by Using Persistent Current in Superconducting Coil

Mochimitsu Komori, Singo Takase, Kaoru Nemoto, Nobuo Sakai, Ken-ichi Asami
Dept. of Applied Science for Integrated System Engineering, Kyushu Institute of Technology
1-1 Sensui, Tobata, Kitakyushu, Fukuoka 804-8550, Japan
E-mail: komori_mk@yahoo.co.jp

***Abstract*---Superconducting technique is applied to the levitation system. Persistent current in superconducting coil and control current in copper coil are used for levitating object and controlling object, respectively. The system is composed of a superconducting coil, a copper coil, a levitated object, a photo sensor, a PD controller, and power amplifiers. In this paper, basic study on superconducting coil and solenoid coil, and the dynamic characteristics of levitated object are performed. This may be the first trial that superconducting coil is used for magnetic levitation.**

I. INTRODUCTION

Magnetic levitation technique is a very promising technology because of no mechanical contact [1], [2]. In general, magnetic levitation technique uses electromagnets (EMs) for controlling the positions of levitated objects. When EMs are applied to levitation systems, much electrical power is necessary for levitating them. Wind channel test equipment with an airplane model is shown in Fig. 1(a). The airplane model (or other models) is supported by a mechanical support to keep it in the center of the wind channel. However, the wind channel needs nothing around the model, because the mechanical support disturbs the environment around the airplane model (or other models). Then, in order not to disturb the environment, magnetic levitation technique is very promising for wind channel test equipments. By using magnetic levitation technology, a wind channel test equipment is proposed as shown in Fig. 1(b). The test equipment is composed of an airplane model, some EMs for levitation and some sensors for detecting model position. In order to get larger distance between EMs and airplane model (levitated object), much larger ampere-turns are necessary. However, in order to get larger ampere-turns, there may be a technical limit for conventional EMs systems.

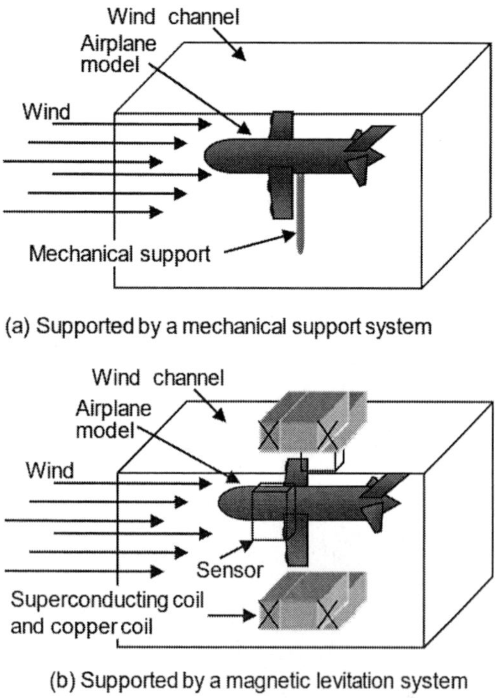

Figure 1. Wind channel test equipment with an airplane supported by (a) a mechanical support system and by (b) a magnetic levitation system

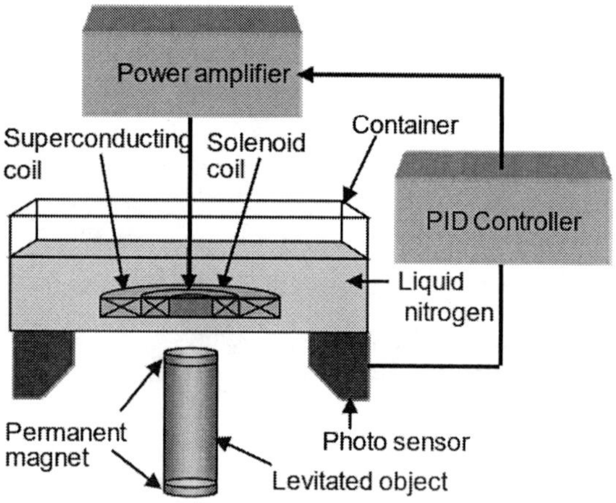

Figure 2. Wind channel test equipment with an object supported by a magnetic levitation system

Table 1 Specifications of supercon-ducting coil

	Superconducting coil
Material	BSCCO
Outside diameter	103.4mm
Inside diameter	60mm
Thickness	10mm
Turn	99.5turn

Figure 3. Superconducting coil

Then, in this paper superconducting technique is applied to a levitation system such as a wind channel test equipment [3], [4], [5]. Persistent current in superconducting coil and control current in copper coil are used for levitating object and controlling object, respectively. In this case, there is no electrical power loss in superconducting coil because of no electric resistance of superconducting coil. In this paper, basic studies on superconducting coil and solenoid coil are performed, and dynamic characteristics of levitated object are studied. This may be the first trial that superconducting coils are used for magnetic levitation.

II. Levitation Method and Experimental Setup

Fig. 2 shows a schematic illustration of a wind channel test equipment with a levitated object supported by a magnetic levitation system. This system is just a levitation system for studying a wind channel test equipment using superconducting technique. The system is composed of a superconducting coil, a copper coil, a levitated object, a photo sensor, a PD controller and a power amplifier. The copper coil (Φ 0.3 mm $\times$500 turns) measures 40 mm in outer diameter, 10 mm in inner diameter and 11 mm in thickness, which is set inside the superconducting coil. Since these coils have no core inside themselves, it is not necessary to worry about core saturation. The superconducting coil and the copper coil are immersed in liquid nitrogen inside the container, and are used for levitating object and controlling object, respectively. Each coil is connected to a power supply, respectively. In the experiment, persistent current is applied to the superconducting coil and control current is applied to the copper coil. The photo sensor is used for detecting the distance of levitated object.

Fig. 3 shows the superconducting coil (BiSCCO) for studying persistent current. Table 1 shows the specifications of the superconducting coil, which measures 103.4 mm in outer diameter, 60 mm in inner diameter and 10 mm in thickness. The maximum permitted current of the superconducting coil is 160 A. In the experiment, a persistent current 11.1 A corresponding to the levitated object weight 68.0 g is applied to the superconducting coil.

Persistent current mode test is performed by using the superconducting coil shown in Fig. 3. In order to make a persistent current, constant current is applied to the superconducting coil by using a DC power supply. After a heater switch is turn ON, the persistent current mode is continued to perform. In the experiment, magnetic flux density on the coil surface corresponding to persistent current is measured. Fig. 4 shows one of the experimental results for persistent current mode test, showing the relationship between magnetic flux density and time. After the heater switch is turned OFF at a time $\approx$ 90 s, the applied current by power supply gradually decreases to zero. Then, the persistent current mode starts at a time $\approx$ 120 s. The persistent current is 20 A, and the persistent current continues for several hundred seconds as shown in Fig. 4. The magnetic flux density decreases to $\cong$ 0.022 T at a time 600 s. This is caused by the energy loss at the solder joint of superconducting coil. If our technique for solder joint is so

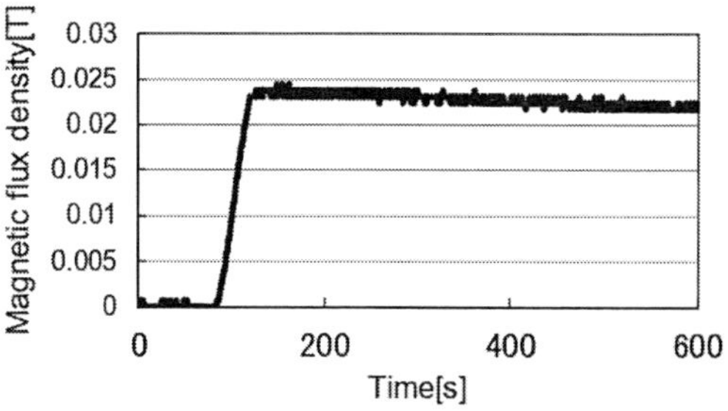

Figure 4. Persistent current mode test, showing the relationship between magnetic flux density and time

satisfactory, the decrease in persistent current will be very small and neglected.

III. STATIC CHARACTERISTICS

A. Attraction force

Attractive forces between superconducting coil and permanent magnet (PM) are measured. In the experiment, the current from 0 to 20 A is applied to the superconducting coil. Fig. 5 shows the relationship between attractive force and driving current for various distances of 10, 15 and 20 mm. The distance means the gap between superconducting coil surface and PM surface. The figure shows the linear relationships between force and current. The attractive forces at a distance of 20 mm are more than ≅ 0.7 N. The attractive forces are enough in comparison with the levitated object weight 68.0 g.

Attractive forces between superconducting coil and PM

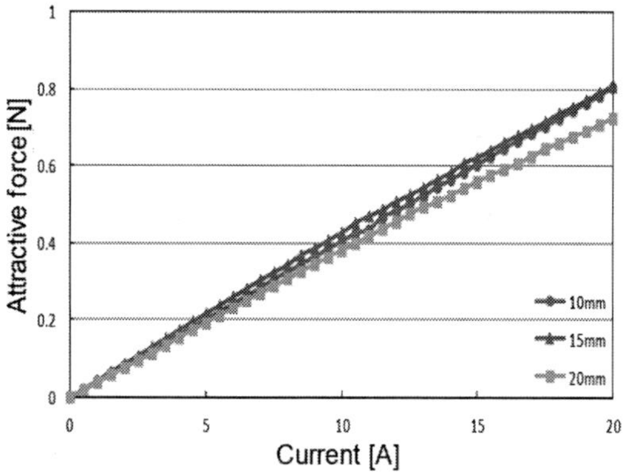

Figure 5. Relationship between attraction force and current

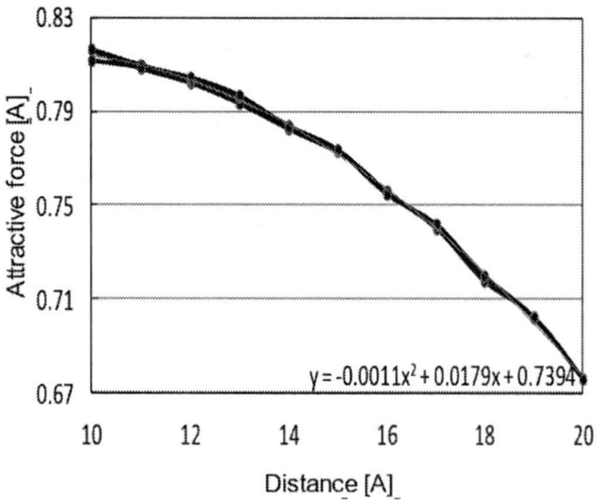

Figure 6. Relationship between attraction force and distance

are measured in the distance range of 10 mm – 20 mm. In the experiment, the applied current is kept constant at 20 A. The initial distance is set at 15 mm, then the distance is changed as 10 mm→20 mm→10 mm. The experimental result is shown in Fig. 6. With increasing distance, the attractive force becomes smaller from ≅ 0.81 N to ≅ 0.68 N. The change in attractive force is not so large with increasing distance. There is no hysteresis loop in the figure.

B. Influence between superconducting coil and solenoid coil

In order to study the influence between superconducting coil and copper coil, magnetic flux densities generated by the superconducting coil and the copper coil are measured. Fig. 7 shows an experimental setup for measuring magnetic flux densities. The experimental setup is composed of the superconducting coil, the copper coil and a Hall sensor. During the experiment, the superconducting coil and copper coil are immersed in liquid nitrogen. The magnetic flux density is measured at a distance 15 mm from the superconducting coil surface by using the Hall sensor (184 mV/T).

Fig. 8 shows the relationship between magnetic flux density and applied current to copper coil. Fig. 8(a) shows the result of magnetic flux density by using copper coil without superconducting coil. The magnetic flux density increases with increasing applied current as shown in Fig. 8(a). The relationship is almost linear over a wide current range. The maximum magnetic flux density at a current 2.0 A is ≅ 5.7×10^{-3}T. Fig. 8(b) shows the result of magnetic flux density by using copper coil inside the superconducting coil. In the experiment, the superconducting coil is not excited. The magnetic flux density increases with increasing applied current. The relationship is almost linear over a wide current range. The relationship is almost the same as that in Fig. 8(a). The maximum magnetic flux density at a current 2.0 A is ≅ 4.9×10^{-3}T, which is a little smaller than the flux density ≅ 5.7×10^{-3}T in Fig. 8(a). This might be caused by the

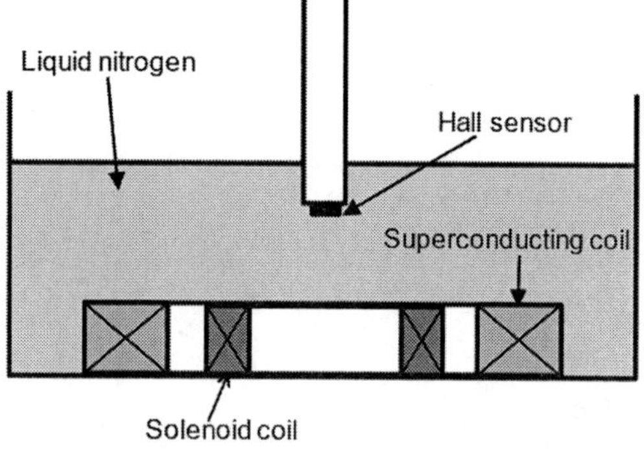

Figure 7. Experimental setup for studying influences between copper coil and superconducting coil

induced current in the superconducting coil, because each end of the superconducting coil is connected.

Fig. 9 shows the relationship between magnetic flux density and applied current to the copper coil. In the experiment, a persistent current 15.2 A (amplifier display: 21 A) is applied to the superconducting coil. The difference between 15.2 A and 21 A is caused by the electric leakage in the driving circuit. The current is applied to the copper coil in the same direction as the persistent current as shown in Fig. 9(a). In the experiment, the applied current is changed from 0 A to 2.0 A. The magnetic flux density increases with increasing applied current as shown in Fig. 9(a). The relationship is almost linear over a wide current range. The minimum and maximum magnetic flux densities at a currents 0 A and 2.0 A are $\cong 16\times10^{-3}$ T and $\cong 21\times10^{-3}$ T, respectively. Thus, the difference between them is $\cong 5.0\times10^{-3}$ T, which is almost equal to the flux density $\cong 4.9\times10^{-3}$ T (at a current 2.0 A) in Fig. 8(b). A small hysteresis is observed in Fig. 9(a).

The current is applied to the copper coil in the opposite direction against the persistent current as shown in Fig. 9(b). In the experiment, the applied current is changed from 0 A to 2.0 A. The magnetic flux density decreases with decreasing applied current. The relationship is almost linear over a wide current range. The minimum and maximum magnetic flux densities at a currents 0 A and 2.0 A are $\cong 16\times10^{-3}$ T and $\cong 11\times10^{-3}$ T, respectively. Thus, the difference between them is $\cong 5.0\times10^{-3}$ T, which is almost equal to the flux density $\cong 4.9\times10^{-3}$ T (at a current 2.0 A) in Fig. 8(b). A small hysteresis is observed in Fig. 9(b). As a result of these experiments in Figs. 8 and 9, it is found that the superconducting coil a little influences the magnetic flux density by copper coil. However, since the influence is not so large, it is neglected in the experiments hereafter.

C. Sensor characteristics

Fig. 10 shows the experimental setup of distance detecting system by using photo sensor. The system is composed of a photo sensor and a LED light (1.94 W, 3.5 V). Between them a levitated object is set, that is, the object is set just under the superconducting coil center. In the figure, LED light is represented by the red lines and the levitated object disturbs the red light. The photo sensor detects the LED light proportional to the distance of levitated object as shown in Fig. 10. In the experiment, the levitated object is moved upward like a thick arrow as shown in Fig. 10. That is, the LED light on the photo sensor decreases with moving object upward. During the motion upward, the sensor output voltage is measured.

The experimental result is shown in Fig. 11, showing the relationship between sensor output voltage and distance from the container bottom surface. The result shows that the sensor output voltage in the distance range from 8 mm to 12 mm is

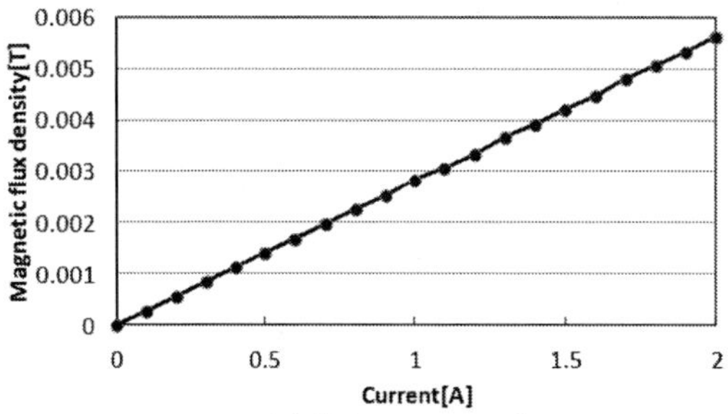

(a) Only copper coil

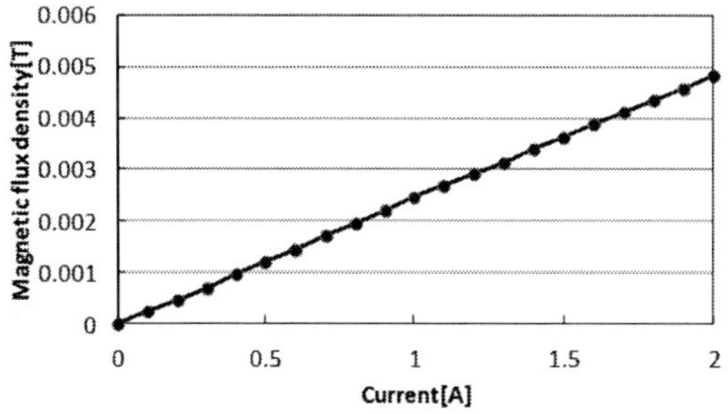

(b) Copper coil inside the superconducting coil

Figure 8 Relationship between magnetic flux density and applied current to copper coil

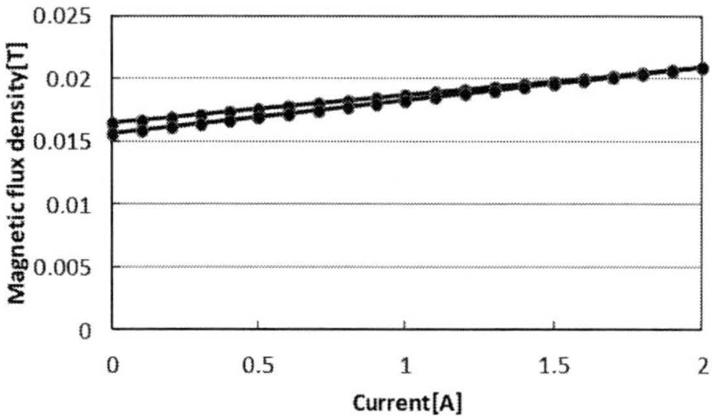

(a) Same direction

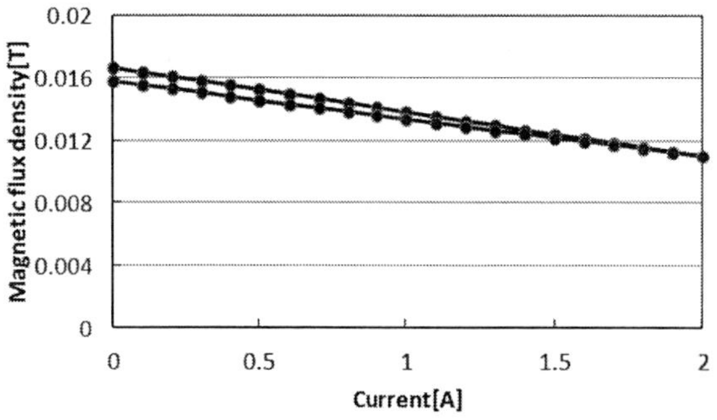

(a) Opposite direction

Figure 9. Relationship between magnetic flux density and applied current to copper coil

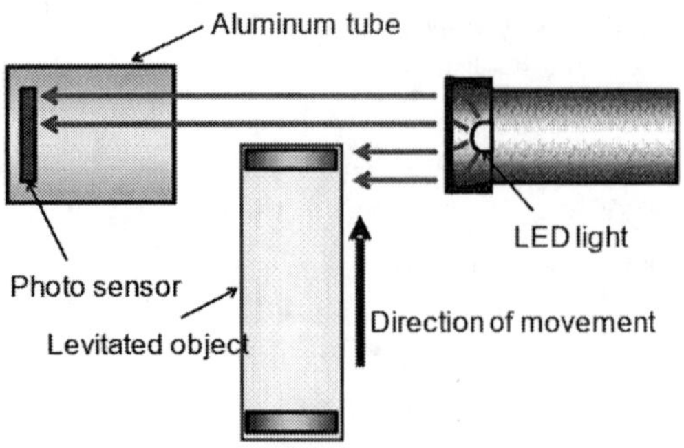

Figure 10. Distance detecting system by using photo sensor

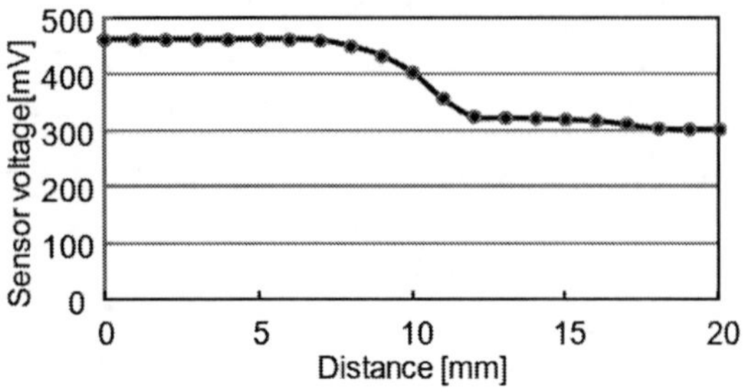

Figure 11. Relationship between sensor output voltage and displacement

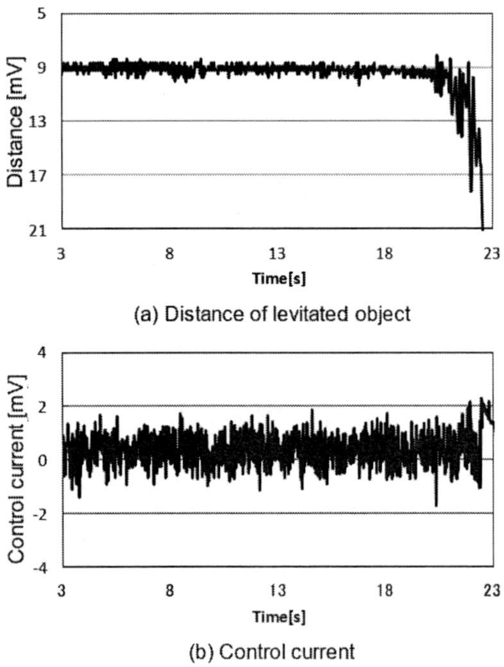

Figure 12. Distance of levitated object and control current

sensitive and useful for sensing the distance. In the experiment hereafter, the distance range from 8 mm to 12 mm is used for the levitation.

IV. Dynamic Characteristics

A levitated object (Φ 28.0 $\times$100 mm, 68.0 g) is made of white plastic, "Duracon". Doughnut-shaped NdFeB PMs are attached to both ends of the levitated object. The PM measures 24 mm (outer diameter) ×8 mm (inner diameter) ×4.5 mm (thickness) in size. The magnetic flux density on the PM surface is 0.209 T. In the experiment, PD control is applied to the levitating system. The experimental result is shown in Fig. 12, representing (a) the distance of levitated object and (b) control current. The levitated object continues to levitate at a distance 9.0 mm for more than $\approx$15 s as shown in Fig. 12(a). The distance 9.0 mm is the gap between the levitated object top surface and the container bottom surface, which is equal to a distance 15 mm between the levitated object top surface and superconducting coil bottom surface. The displacement amplitude is between -0.5 mm and 0.5 mm, which is not so small. This is because the photo sensor has some noises and the sensor is affected by a lot of frost. For the levitation, the control current applied to the copper coil is shown in Fig. 12(b) and the persistent current is 14.4 A. The levitated object drops at a time $\approx$20 s as shown in Fig. 12(a). This is because the photo sensor has some noises and the sensor is affected by a lot of frost.

V. Summary

Superconducting technique is applied to the levitation system. Persistent current in superconducting coil and control current in copper coil are used for levitating object and controlling object, respectively. The basic study on superconducting coil and solenoid coil are performed. It is found that the superconducting coil a little influences the magnetic flux density by copper coil. However, the influence is not so large. The photo sensor output voltage in the distance range from 8 mm to 12 mm is sensitive and useful for sensing the distance. As a result, it is found that the levitated object continues to levitate at a distance 9.0 mm for more than $\approx$15 s.

Anyway, the object is succeeded to levitate by using the persistent current in the superconducting coil. It may be the first trial that persistent current in superconducting coil is applied to the levitation system.

References

[1] Mizuno, T., Sekine, D., Ishino, Y. and Takasaki, M., Proposal of Force Measurement Using Series Magnetic Suspension, ASME 2012 5th Annual Dynamic Systems and Control Conference joint with the JSME 2012 11th Motion and Vibration Conference (2012), pp. 55-60

[2] Kawamura, Y. and Mizota, T., Wind Tunnel Experiment of Bluff Body Aerodynamic Models Using a New Type of Magnetic Suspension and Balance System, J. Fluids Eng., Vol. 135, No. 10 (2013), pp.1-5.
[3] Hull, J. R., Mulcahy, T. M., Uherka, K. L., Abboud, R. G., Low rotational drag in high-temperature superconducting bearings. IEEE Trans. Applied Superconductivity, Vol.5, No. 2 (1995), pp.626-629.
[4] Komori, M., Kobayashi, H. and Kumamoto, M., Hybrid high Tc superconducting magnetic bearing (SMB) system with no bias currents, 6th International symposium on magnetic bearings (1998), pp. 214-223.
[5] Maruo, E. and Komori, M., A trial of superconducting magnetic levitation (SML) with superconducting coil, Journal of the Japan Society of Applied Electromagnetics and Mechanics, Vol. 19, No. 3 (2011), pp. 79-82.

Experimental Tests and Simulations to the Design of an Electrodynamic Bearing

Marcelo de Almeida Lopes[a], Guilherme G. Sotelo[b], Elkin Rodriguez[a] and Richard M. Stephan[a]

[a] Federal University of Rio de Janeiro, Technology Center I-148, Rio de Janeiro, Brazil, marcelolopes@ufrj.br
[b] Fluminense Federal University, Passo da Pátria Street 156, Niterói, Brazil

***Abstract*—This paper describes an electrodynamic suspension system built with permanent magnets and aluminum plates. Initially, a simple case was studied to demonstrate the behavior of the levitation and drag forces for different conditions, such as air gap, angular velocity and thickness of the aluminum target. In the second part, numerical and experimental tests were performed to design an homopolar electrodynamic bearing (EDB). Forces obtained with eccentric operation as well as different rotor radial velocities are presented.**

I. Introduction

A. Overview

Active magnetic bearings, based on electromagnetic attraction forces to establish the conditions for dynamic positioning, occupy much of the technical literature and industrial applications [1]. However, they require sensors, power electronics and controllers that are not necessary for the operation of electrodynamic bearings (EDBs). This magnetic levitation technology, based on repulsive forces, uses the same principle applied in the Yamanashi JR-MagLev test line [2]. A disadvantage of EDB is the force dependence with the operational speed, which must be carefully considered [3-6].

B. Basic Principle of the Electrodynamic System

When a constant magnetic field crosses a moving surface made of conductive material, induced currents are produced and flow in the conductor. The current amplitude is a function of the velocity, eccentricity, air gap, electric conductivity as well as the magnetic field distribution.

The interaction of the magnetic induction (*B*) and the current density (*J*) is described by the Lorentz force law, which is written as:

$$\vec{f} = \vec{J} \times \vec{B} \quad (1)$$

The force density, *f*, has two components that are known as drag and lift forces. The first one acts against the rotational movement while the second one collaborates for suspension of the system. The force components should be optimized according to the objective. For example, drag force must be maximized in a brake system, while lift force is improved for a suspension system.

A levitation system can be designed by analytical calculus [7] or numerical modeling followed by experimental tests. The second method is chosen in this paper due to the facility given by finite elements simulation (FEM) to achieve an optimized system.

II. Disk-PM Case

A. Bench Characteristcs

A simple experimental system was constructed to illustrate the characteristics of the electrodynamic levitation as shown in Figure 1. The experiment has provided results for different conditions. Most of them are related to forces as function of the air gap, thickness, and speed.

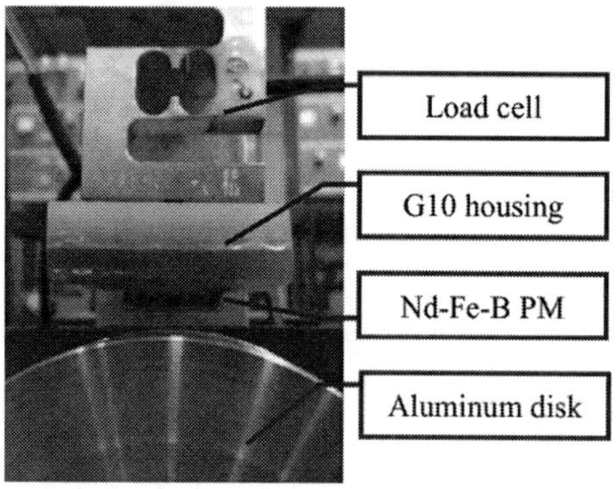

Figure 1. Experimental rig used for levitation and drag force measurements.

Table I presents the main parameters of the disk-PM bench, in which PM's magnetic flux was in the vertical direction. These values were also used to make numerical models, specifically, bi- and tri-dimensional models.

TABLE I. Disk-PM Parameters

Parameter	Value
PM Dimensions (mm)	25.4 x 25.4 x 12.7
Disk Dimensions (mm)	Φ150 x 40
PM Material	Nd-Fe-B N35
Disk Material	Aluminum

All simulations and experiments have considered different values for air gap, angular velocity and disk thickness (5 mm, 10 mm, and bulky disk).

B. Results

Experimental results have been used to validate bi- and tri-dimensional simulations. Figure 2 shows simulations as well as experimental results considering a 2 mm air gap, bulky disk and 30°C of temperature.

The curves in Figure 2 demonstrate that both numerical results tend to converge for high velocities, showing that bi-dimensional models are satisfactory in this case. Although tri-dimensional models seems to be more accurate, bi-dimensional represents less computer processing time. In the case of low velocities, tri-dimensional models seem to be mandatory.

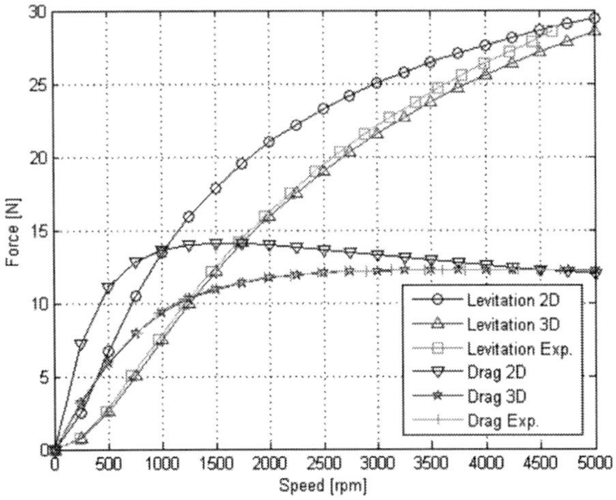

Figure 2. Levitation and drag forces for a 2mm air gap and bulky disk.

The system design must consider heat transfer, since temperature variations can reduce the system performance, and, in the worst case, damage the PMs.

In Figure 3 the drag and levitation forces are compared for temperature variation. The conclusion for this specific simulated case is that the levitation force is more sensible than the drag one. The levitation force decreases 78%, when compared initial (0 °C) and final (150 °C) temperatures, while drag force decays 40%. Therefore, the design of electrodynamic system must consider cooling factors. High temperatures affect the aluminum resistivity as well as the demagnetization of the permanent magnets.

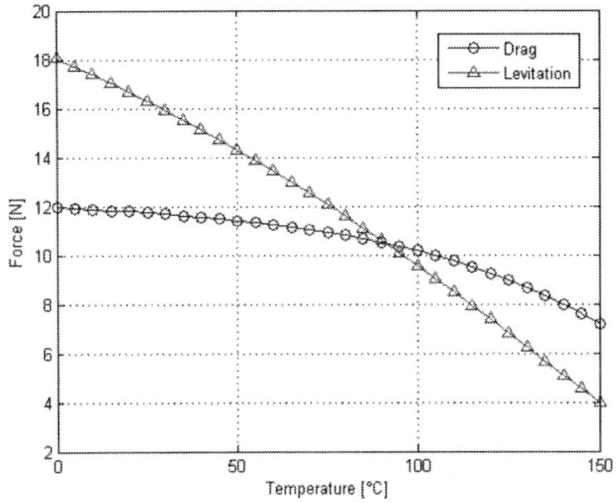

Figure 3. Simulated forces behavior with temperature for 2 mm air gap, 2,000 rpm and bulky disk.

The electrodynamic system is subjected to induced currents, and its amplitude is a function of the conductor parameters. When thickness is under analysis, it is possible to see a saturation value. In other words, the forces in a levitation system vary as a function of thickness while the skin depth (ζ) is not reached. If the thickness is higher than ζ, the force saturates and assumes a practically constant value.

The behavior of the electrodynamic forces with thickness shows that there is a specific thickness value, which can be chosen to optimize levitation force as well as conductor mass. Figure 4 shows simulated results of levitation force as function of thickness for different rotational velocities. In all four curves the saturation value can be seen, and it differs because the skin depth is dependent of the frequency. Therefore, the conductor thickness must be chosen for the required operation condition in order to guarantee a saturated operation. In the other hand, when the design is based on a brake system, drag force should be maximized, and then, thickness could be less than the skin depth.

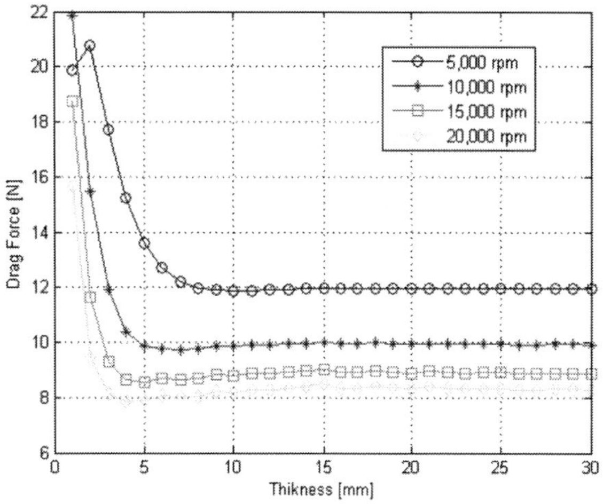

Figure 4. Simulated drag forces for a 2 mm air gap as function of thickness.

Figure 5 shows simulations of the levitation force curves for different disk thickness, from 1 to 30 mm. In contrary to drag force, levitation force increases for high thickness. Thus, an electrodynamic levitation system must have at least the skin depth thickness. In Figure 5, for example, the levitation force for a 20,000 rpm operation can be optimized with a 2.5 mm of thickness.

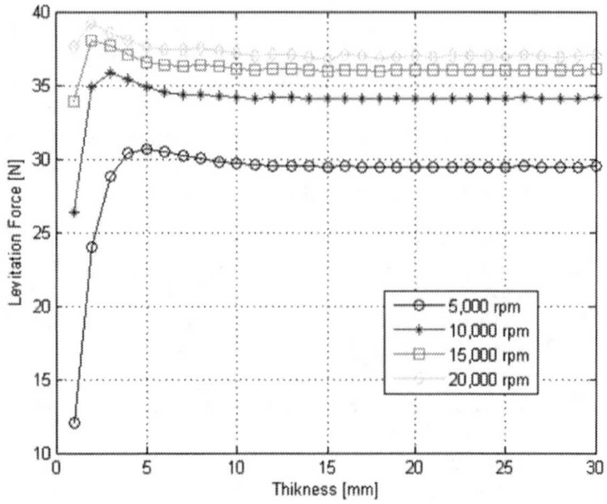

Figure 5. Simulated levitation forces for a 2 mm air gap as function of thickness.

III. HOMOPOLAR EDB

A. General Issues

The greatest advantage of radial homopolar EDB [4] is the low eddy current losses for centralized operation, since ideally no currents are induced in the rotor. In practice, permanent magnets (PM) are not perfectly magnetized and small eccentricities exist, departing from this ideal condition. Moreover, radially magnetized Nd-Fe-B rings are more expensive and difficult to produce than axially magnetized ones. Iron flux shapers can minimize these problems and are employed in this experiment, as shown in Figure 6. The PMs are axially magnetized and the flux shaper acts directing a radial magnetic flux. These flux shapers can be designed in order to optimize the radial magnitude of the magnetic field and maximize the stiffness and damping. Nevertheless, PMs with similar characteristics must be suitably selected.

The second challenge is to choose the ideal air gap that maximizes the levitation force and offers a secure operational condition.

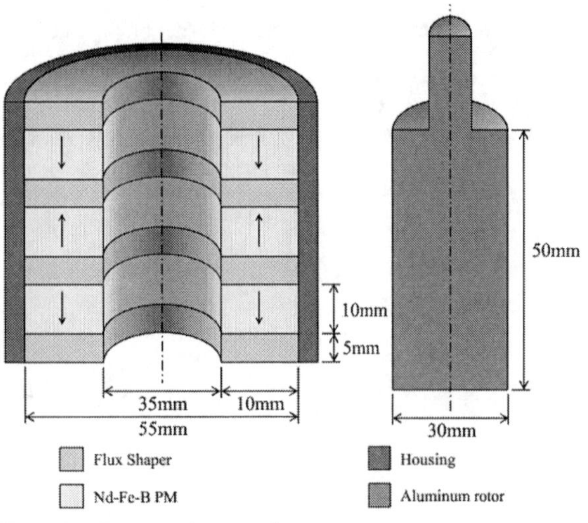

Figure 6. EDB stator and rotor section.

B. Experimental Bench

A second experimental bench composed by a CNC lathe, a force/torque sensor, and a data acquisition system was designed to measure the bearing forces as show Figure 7. The motor used is limited to 18,000 rpm. The positioning of the EDB rotor could be performed by the CNC lathe, which has tridimensional movement. Results have been obtained by sample average for each eccentricity and rotational velocity. By this way, it was possible to filter the noise.

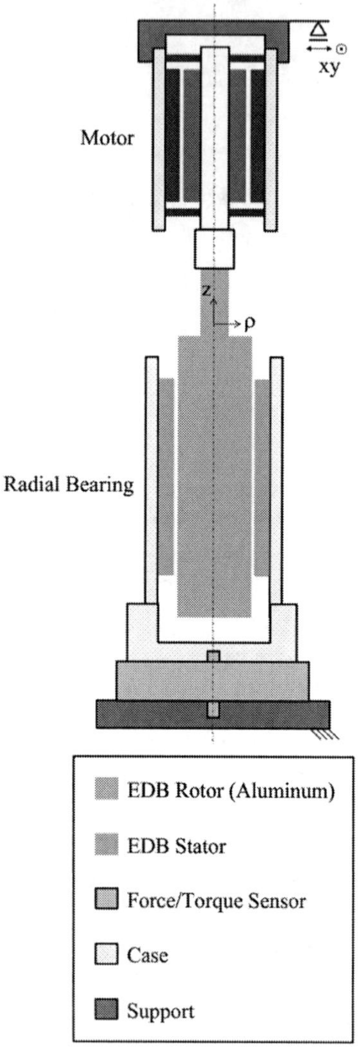

Figure 7. Representation of the experimental bench of the EDB.

Figure 8 shows the experimental bench during the tests. The EDB was built according to the parameters of Figure 6. An important characteristic of the homopolar bearing is low loss for centralized operation. This advantage can be noted in the linear force curves obtained for different eccentricities. The small deviations could be associated to losses. As shown in Figure 3, the forces decrease as temperature gets higher. Thus, as the system induces more current, greater is the loss and smaller are the forces.

Figure 8. Experimental bench with EDB.

C. Results

The force values have been obtained varying eccentricity and angular velocity. The first by 0.1 mm steps, and the second by 2,000 rpm steps.

Figure 9 presents numerical and experimental results. In addition, the graphic shows the linear relationship with radial displacement revealing a constant stiffness for each rotation.

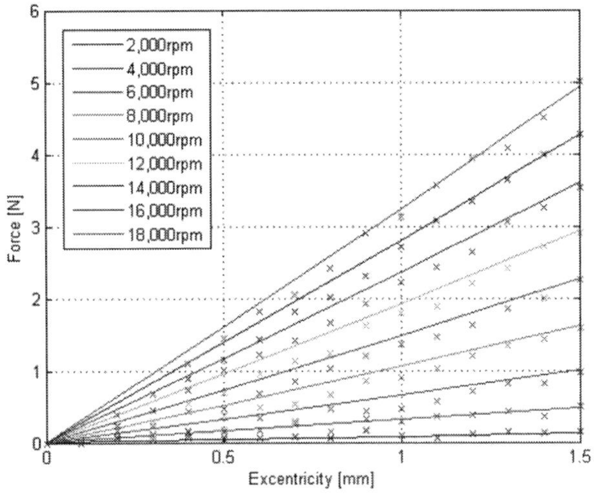

Figure 9. Levitation force in the EDB rotor due to excentricity (- numerical | x experimental).

Unfortunately, the force does not act only in opposition to the displacement. There is another component, which is perpendicular to it. This one attributed to inductance of the system that delays the induced current, changing the resultant force angle. Therefore, the EDB has a singular characteristic. It has a levitation and a drag stiffness.

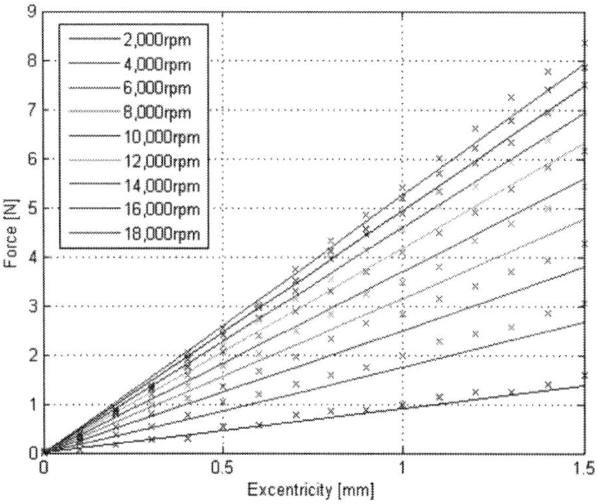

Figure 10. Drag force in the EDB rotor due to excentricity (- numerical | x experimental).

Linear force behavior is also achieved when radial velocity occurs. Therefore, the mechanical parameters of the bearing are related to both conditions: eccentricity as well as radial movement, being stiffness a function of the first and damping to the second. Figure 11 shows lift force dependence with radial velocity.

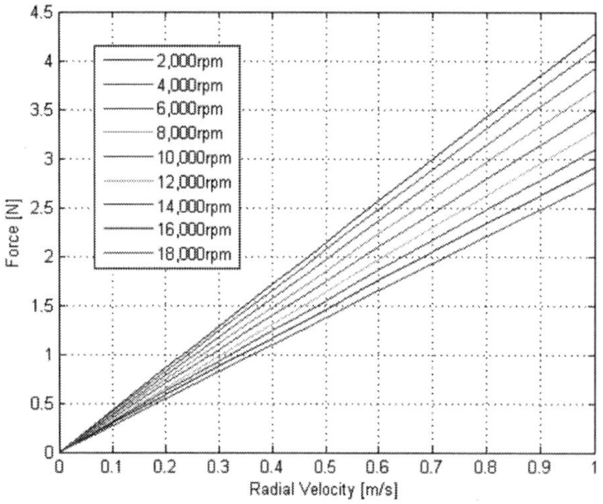

Figure 11. Simulated levitation force in the EDB rotor due to radial velocity.

Comparing levitation and drag forces associated with radial velocity (Figure 11 and Figure 12), it is noted that the last one increases with velocity while the first decreases. Therefore, the levitation damping of the EDB proposed decreases as speed gets high. In other way, drag damping increases with speed.

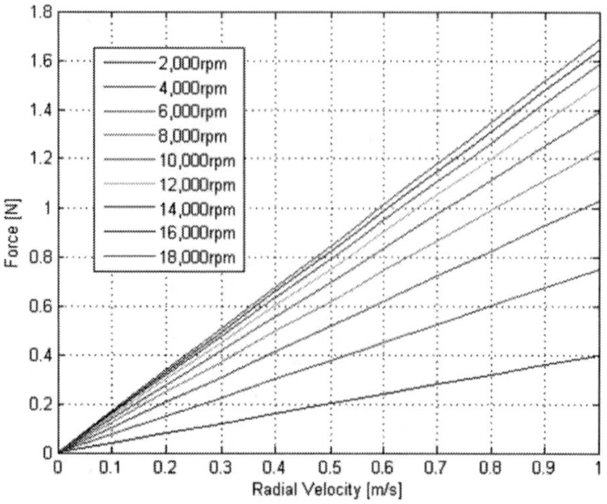

Figure 12. Simulated drag force in the EDB rotor due to radial velocity.

D. *Optimization*

Figure 13 shows numerical results for two EDB's. One of them is based on the Figure 6 values and is called "Non-Optimized" system. The other one is called "Optimized" system. In this case, flux concentrators were dimensioned in order to maximize radial flux.

The flux concentrators for the optimized system have 6 mm width, and the rotor thickness adopted was 9 mm. This value could minimize rotor mass and has not affected forces amplitude.

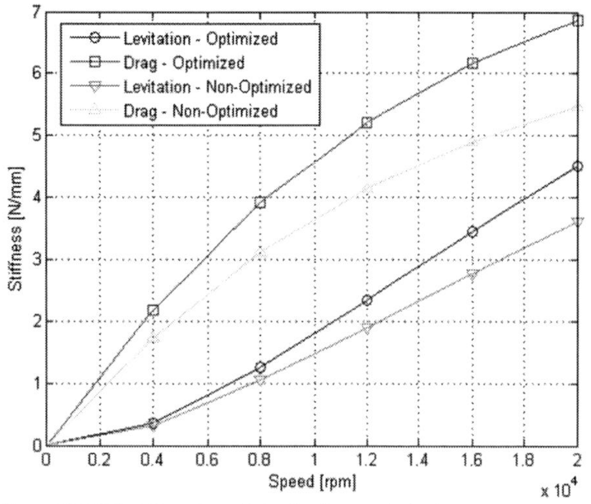

Figure 13. EDB stiffness for optimized and non-optimized system.

IV. CONCLUSIONS

The first part of this paper showed the possibilities and limitations offered by finite elements simulations to predict levitation and drag forces presented in electrodynamic magnetic bearings. Based on this knowledge, an electrodynamic bearing has been designed, constructed and tested. The advantages of an homopolar configuration were pointed out. Further investigation must be carried out to improve the forces magnitude.

V. ACKNOWLEDGMENTS

The authors thank BNDES, CNPq, FAPERJ and CAPES for the financial support.

REFERENCES

[1] G. Schweitzer, H. Bleuler, and A. Traxler. "*Active Magnetic Bearings*," Springer, 2009.

[2] J. L. He, D. M. Rote, and H. T. Coffey. "*Study of Japonese Electrodynamic – Suspension Maglev Systems*," Center for Transportation Research Argonne National Laboratory, 1994.

[3] I. Fabrizio. "*Electrodynamic Bearings – Modeling and Design*," Ph.D. Thesis, Politecnico di Turino, Italy, 2010.

[4] T. A. Lembke. "*Design and Analysis of a Novel Low Loss Homopolar Electrodynamic Bearing*," Ph.D. Thesis, KTH Electrical Engineering, Sweden, 2005.

[5] M. T. Tompson. "Eddy Current Magnetic Levitation – Models and Experiments," *IEEE Potentials*, vol. 19, pp. 40-44, 2000.

[6] M. T. Tompson. "Electrodynamic Magnetic Suspension – Models, Scaling Laws, and Experimental Results," *IEEE Transactions on Education*, vol. 43, no. 3, pp. 336-342, 2000.

[7] SINHA, P. K. *Electromagnetic Suspension: Dynamics & Control*. France: Peter Peregrinus, v. 30, 1987.

Design Synthesis and Experimental Validation of an AMB Controller with a Power Dense Amplifier for Retrofit Applications

Praveen Kumar[a], Richard Jayawant

[a] Waukesha Magnetic Bearings, Worthing, UK. Email pkumar@waukbearing.com

Abstract—**This paper describes the design, development and validation of a high power amplifier used to up-rate the axial power amplifiers of a standard AMB controller. The VA rating of the amplifiers was doubled whilst maintaining the original footprint of the amplifier.**
The thermal management and other constructional challenges presented by such up-rating and high power density are discussed, together with some of the techniques used to overcome them.

This up-rated system is suitable for retrofit applications where the controller is coupled to pre-existing AMB hardware. Preliminary results from such an installation are presented.

I. INTRODUCTION

A. Retrofit requirements

When considering selection of an AMB controller for an application in which the controller is replacing an existing controller, there are a number of constraints which apply, which would not necessarily apply in the case of a complete new bearing system design. By the very nature of the exercise, the controller which is being replaced is likely to be quite old and will almost certainly have been developed using outdated power semi-conductors. This may well mean that:

- The DC link voltage of the amplifier system is lower than would be selected for a modern design of controller.
- In order to achieve a required dynamic force capability the controller will have a minimum reactive power (VA) requirement and this may lead to higher currents than would be applicable with a more modern controller.

A simple solution to the problem of matching the VA requirements of the bearing to the voltage and current capabilities of the new controller would be to re-wind the bearing with a suitable number of turn and wire gauge together with replacement of the field cables. However, such rewind of the bearing may not be practicable due to such items as:

- Timescales for the controller exchange
- Not wanting to dismantle the machinery
- Existing pressure penetrator voltage ratings and wire gauges.
- Existing field cable runs being buried or otherwise inaccessible.
- Added cost

Where the DC link voltage of the selected controller is fixed and a bearing re-wind is ruled out, the cable and bearing insulation systems must be re-qualified for the higher DC link voltage, and then the only subsequent challenge is to select a controller of the correct current rating. This is likely to give a system with an increased VA rating and consequently better dynamic force capability – leading to operational performance benefits.

This re-qualification of the windings, cables and pressure penetrators would be either by review of the documented ratings of the components (they may have been originally selected with a margin which permits operation at the higher voltage), or by performing a insulation test at a voltage significantly higher than the proposed operational voltage. Such an insulation test carries significant risk, since in the event of a test failure the winding (or cable) would no longer be suitable for continued operation either with the new controller or the existing controller.

Where the DC link voltage of the selected controller can be modified to match the existing controller then the challenge is again to select a controller of the correct current rating. This is likely to give a system with unused VA rating within the controller, due to operation at reduced DC link voltage.

B. Power electronic systems in magnetic bearings

In magnetic bearing applications, power electronic systems are usually designed with high power-semiconductor switches and diodes in discrete form. The preferred method of achieving weight and cost reduction of these systems is through the use of higher switching frequencies and devices with low on state resistance. Though semiconductors have been driving force for power electronic system developments, passive components used in conjunction with switching devices have different form factors and poor thermal properties. The assembly of components on a single PCB results in poor volume utilization and low power densities [1]. Thermal management is done at component level and is based on single sided heat removal with bulky and heavy heat sinks. The passive components often operate at high temperatures even when exhibiting relatively low dissipation levels due to low thermal conductivity of passive materials. The production cost escalates due to the complex assemblies associated with discrete parts and labor intensive assembly processes. The design cycle time also increases since standardization of such circuits and parts becomes difficult.

C. *Impact of power electronic system design on reliability of magnetic bearings*

The demand for reliable magnetic bearing systems has resulted in development of amplifiers with novel thermal management techniques. It is worth noting that ambient air temperatures in AMB controllers have a direct impact on lifetime of these systems. Low operating temperatures improve reliability; whereas, higher temperatures offer more efficient material usage due to a higher electromagnetic excitation capability. Furthermore, the convection rate of heat to the ambient from any component surface relies heavily on the difference between surface and ambient temperatures. A uniform temperature distribution on the component surface will therefore result in more effective convection of heat to the ambient, due to a higher average temperature difference between surface and ambient allowing for a higher convection rate. This then results in overall lower component surface temperatures.

D. *Thermal management of power electronic systems in magnetic bearings*

Components inside an amplifier dissipate power losses resulting in temperature rise of the component material as well as that of the neighbouring components. Depending on the properties of materials, they exhibit optimal and maximum temperatures. The optimal operation temperature of a component is a compromise between the lifetime or reliability of the component and the level of electromagnetic excitation of the component.

To reduce size and cost of amplifiers, research into new system integration and packing technologies has gathered momentum. To increase the power densities by improving material usage, a design technique is required that provides a good interaction between thermal management and integration technologies implemented towards reducing size of amplifiers. In most electrical designs, one strives to achieve the maximum functionality with the minimum material and lowest possible parts count. The same principle should be applied to the thermal-management system i.e. it should be as simple as possible, should not use many parts, and should use as little material as possible.

Amplifiers generally use multiple discrete semiconductor devices, which are mounted on a heat sink to dissipate heat. In order to provide isolation between heat sink and device and prevent short circuit, an interface pad made of epoxy or a polymer, filled with ceramic particles such as boron nitride or alumina is used. The ceramic particles are added enhance thermal properties of the interface pad which also acts as a heat conduction path. The interface pad being a conformable material also helps to reduce the contact resistance between the metallic surfaces of the heat sink and the device. However, the interface pad is a source of high thermal resistance in the overall conduction path between device and heat sink.

II. DESIGN TECHNIQUES

A. *Parameters affecting amplifier volume*

Amplifier size and construction is dependent on the following factors:

- Sizes of passive components such as transformers and chokes. These are affected by inter-winding isolation material in inductor and transformer structures.
- Electrically conductive paths. They include cable routes, electrical package pins and tabs, cable cross sectional areas and material for wire bonds.
- Thermal cladding, thermal interface and potting material.
- Thermal layers on PCBs dedicated to cooling.
- Capacitor housings or brackets to increase rate of cooling.

B. *TMLD and TDR*

In this paper, thermal management effectiveness is evaluated by two figures of merit:

Thermal Management Loss Density (TMLD) and Thermal Design Rating (TDR)

TMLD addressees the *effectiveness* of the implemented material and parts used for thermal management by *volume*, while TDR addresses the *thermal performance* of implemented material and parts [2].

$$TMLD = \frac{P_{losses}}{V_{tm}} \quad (1)$$

P_{losses}: *total power dissipation.*
V_{tm}: *volume of all material contributing toward thermal management of the converter.*

A high TMLD value is preferred as it gives greater power density.

$$TDR = e^{-1/2\frac{(T-T_{optimal})^2}{2T_{dev}^2}} \quad (2)$$

$$T_{dev} = \frac{T_{max}-T_{optimal}}{4}\ ;\ T_{dev} > 0$$

T_{max} : *component maximum temperature ;*
$T_{optimal}$: *component optimal operating temperature*
T_{dev}: *Temperature deviation*

Components with similar TDR values can be grouped together. High power density results in higher component operating temperatures. The flipside is that reliability is reduced. In order to achieve both higher power density and higher component reliability, TDR values should be taken into account along with values for TMLD [2].

C. High current PCBs

Often overlooked but an important building block in amplifiers is the power PCB itself. Most of the PCBs used for low power applications are manufactured with copper traces made of copper weights ranging from ½ oz/Ft² to 3 oz/Ft². For high power applications, boards can be manufactured with copper weights between 4 oz/Ft² to 20 oz/Ft².

With higher copper weights, designers can avail advantages [3] such as:

- Higher endurance to thermal strains and increased mechanical strength at connector sites.
- Increased amperage capacity.
- Overall reduction in size of amplifier by incorporating multiple copper weights on same layer of circuitry.
- Reduced cable routing.

By manufacturing PCBs with higher copper weights, high current circuits can be combined with thicker copper plating in the vias. Plated through holes will eliminate the need to add duplicate layers in parallel, thereby eliminating any concern of load sharing among multiple layers. Measuring temperature rise in boards becomes simplified and failures due to thermal stresses are minimized, the overall advantage being a cooler running and more reliable PCB [5].

Approximate current for a given track dimension in external layers is

$$I_{ext} = 0.048 \times dt \times 0.044 \times w \times th \times 0.725 \quad (3)$$

While for internal layers is

$$I_{int} = 0.024 \times dt \times 0.044 \times w \times th \times 0.725 \quad (4)$$

I_{ext} *and* I_{int} *are currents* (A); *dt: temperature rise* $(°C)$; *w : width of trace* (mm); *th: thickness of trace* (mm)

Commonly used dielectric material is FR4 with maximum operating temperature of 130°C. Higher temperature polyamides are also available with 250°C rating.

Electrical vias can substantially reduce the thermal resistance of a PCB. They are commonly used in medium and high power density boards. In principle each via is a copper sleeve through the board or through a part of the board (blind or buried via). In addition to its electrical function, a via can also be used as a thermal short.

D. Gate drive design

Gate drive design: A good gate drive design revolves around selecting a device that satisfies ideal conditions as much as possible. The design can then be optimised to realize the best amplifier performance.

In general four parameters V_{gg+}, R_g, I_g and drive layout are the parameters that should be considered very carefully. V_{gg+}(Positive gate bias voltage) is the voltage across gate and emitter with IGBT conducting and it affects on state loss and switching speed. R_g is the value of gate turn on resistor and affects switching performance of the device.

$V_{CE(sat)}$ is inversely proportional to V_{gg+} for a fixed value of collector current I_c. For a gate threshold voltage $V_{ge(th)}$, gate drive current

$$I_g = \frac{V_{gg+} - V_{ge(th)}}{R_g} \quad (5)$$

An increase in V_{gg+} reduces the switching time of the device and switching losses become smaller. This also increases I_g which charges gate emitter capacitance C_{ge} faster which leads to quicker increase in collector current I_c. This in turn leads to an increase in $\frac{dI_c}{dt}$. Large values of $\frac{dI_c}{dt}$ and $\frac{dV_c}{dt}$ indicate lower switching losses and switching speed which is an advantage but from an EMI point of view, it is desirable to limit them. A device with good built in diode recovery characteristics can limit rates of rise of collector current and voltage [4].

To achieve a good balance between switching losses and EMI, a commonly used technique is to select smaller value of R_g and reduce switching losses by increasing I_g. This method also reduces $\frac{dV}{dt}$ and in turn surge voltages caused by miller effect. However, minimum R_g values are limited by characteristics of freewheeling diode of IGBT on opposite side in multi-pack IGBT's. So a turn off gate resistor $R_{g(off)}$ and a turn on resistor $R_{g(on)}$ can be used. They can be accordingly adjusted for optimum performance. It is preferred to not share stray inductance of IGBT leads and PCB tracks between common emitter terminal and power emitter terminal of device since they can have undesirable effect on switching speed of the device [3].

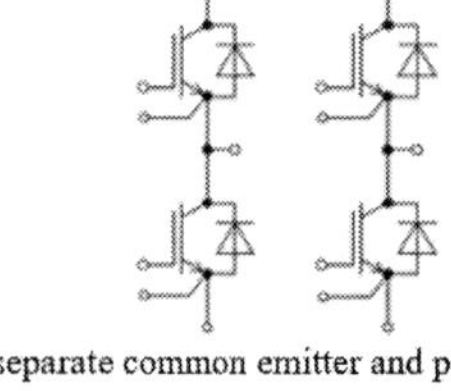

Figure 1. IGBT with separate common emitter and power emitter.

III. RETROFIT EXERCISE

A. Requirements summary

Waukesha Magnetic Bearings worked with a leading rotating equipment manufacturer with a retrofit solution to replace an aging controller with more sophisticated AMB controller to realize higher dynamic force and computational capabilities, and exploit benefits offered by modern power electronic devices. A low cost AMB controller 'Zephyr' was chosen for the retrofit since the product offered significant advantages such as increased reliability, ease of assembly, small lead times and less space requirements.

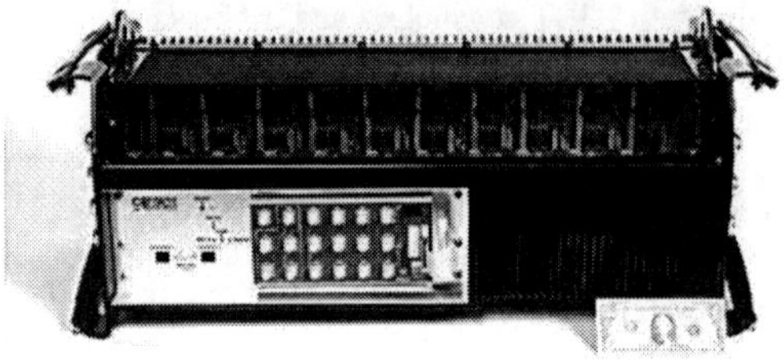

Figure 2. Zephyr Controller.

In order to match the VA requirements of the bearing to the voltage and current capabilities of Zephyr an upgrade of axial amplifier was deemed necessary due to constraints imposed by existing mechanical setup. This included:

- The requirement that the existing windings be retained.
- The maximum static load (and the consequent maximum static current) when operating at full gas density and full power.

TABLE I. VA RATINGS OF THE 2 CONTROLLERS

Parameter	Existing controller	Zephyr controller
Radial Bearing Current (peak)	30 A	27A
Radial Bearing Volts	160 V	390 V
Axial Bearing Current (peak)	60 A	27A
Axial Bearing Volts	160 V	390 V

It can be seen from the Table I that the peak VA requirement of Zephyr on the axial bearing could only be matched with an amplifier rated for 60A i.e. more than 2 times that offered by standard amplifier rated for 27A. The redesign methodology used to up-rate axial amplifier along with the field test results are discussed in the sections that follow.

B. Amplifier design

It has been discussed in earlier sections that reliability increases with lower operating temperatures while higher temperatures allow more efficient thermal usage due to increased electromagnetic excitation capability. It has also been pointed that, a design technique that provides a good interaction between thermal management and integration technologies can increase the power densities of amplifiers and significantly reduce cost per Kilo-watt.

One of the amplifier re-design constraints was that the retrofit amplifier had to fit in the same area envelope as the existing one i.e. 140mm ×205mm (W×L). The current density of the amplifier PCB would then have to increase from 0.1883A/cm² (27A current) to 0.419A/cm² (60A current). This is a manifold increase in current density that can only be achieved by novel thermal management techniques and low component count.

Figure 3. Existing Zephyr Amplifier PCB.

In order to evaluate TMLD of IGBTs for this exercise, different packages were evaluated for their suitability.

TABLE II. TMLD RATINGS FOR DIFFERENT PACKAGE TYPES FOR 60A CURRENT

Package type	TO-3P	TO-247	EasyPack	SOT-227
P_{losses}(W)	90	90	90	90
V_{tm}(mm³)	3095	3488	12478	20473
TMLD (W/cm³)	29.07	25.80	7.212	4.39
T_{rise} above ambient (°C)	112	101	40	31.7

To achieve a balance between TMLD and temperature and thereby obtain better reliability figures, the EasyPack module and SOT-227 modules were considered for further study. Both modules do not require thermal interface pads for isolation which can be a barrier to efficient cooling.

The EasyPack module offered significant advantages over SOT-227:

- Ease of assembly onto a PCB.
- Separate common emitter and power emitter terminals which reduce effects of stray inductance on device switching speed.
- Modular package with multiple devices as opposed to a single device in SOT package.

The next challenge was to design a PCB capable of carrying 60A(peak) current. In order maintain a balance between device temperature and favourable track temperature (40°C above ambient) for 60A, the track width was determined to be at least 3.06mm for external layers and 7.96mm for internal layer. This was calculated based on IPC-2221 formula with 12 oz/Ft² copper weight. The copper weight was a trade-off between high current and optimal temperature rise for reliable operation of the amplifier. A further design challenge was presented with surface mount components associated with gate drive and current sensor circuits. The footprints could not be included on the 12oz PCB due to the practical difficulties associated with etching small pads and tracks with this thickness of copper. The solution was to split the card into 2 separate sections, one 12 oz/Ft^2 cu and a second 2oz/Ft^2 which carried all the SMD components. The SMD components have high TDR values and were mounted on a daughter board which was connected to power board through a network of copper vias and edge plates.

The gate drive circuit was designed with few components using methods described in section II. A turn-on and a turn-off gate resistor were used to achieve minimum values of gate resistance. This resulted in an increase in I_g which charges gate emitter capacitance C_{ge} faster leading to quicker increase in collector current I_c . Achieving an optimum value of V_{gg+} for I_g was easier since the IGBT device from Infineon itself had small $V_{ge(th)}$. Further a compatible gate drive IC from Infineon was chosen which matched device input impedances well and thereby helped in maximum transfer of power from source to the device.

The use of a daughter board aided the layout process and gate-emitter loop could be kept as small as possible. Moreover with separate common emitter and power emitter terminals on this device, parasitic inductances were reduced significantly and SMD components could be placed close to the device which reduced EMI. The power PCB was designed to be 4 layer, with wider track lengths on inner layers to enable dissipation of heat efficiently. The daughter board was designed with 2 layers.

Figure 4. Up rated Zephyr Amplifier PCB.

A well designed layout should also cater to reducing electrical stress on gate low voltage control ICs. Electrical stresses can be caused by switching transients which are a result of mismatch between source and load impedance and high slew rates of control IC's [6]. EMI filters are necessary to reduce ringing on output lines and it is important that they do not overheat. Heating alters the properties of inductor cores and performance degradation can occur at high temperatures. PCB mounted versions of common mode filters with current ratings up to 50A (convection cooled) and 80A (forced cooled) have recently become available in market. They are dense packages with nano-crystalline cores which allow greater inductance values and higher current capacities, which was not possible until a few years ago. However these cores are brittle and age quickly at elevated temperatures. Forced convection cooling was necessary to obtain necessary performance at 60A (peak) currents. Similarly a parallel combination of PCB mounted differential mode filters was used to increase current handling capacity to 60A.

IV. TEST RESULTS

The amplifiers were assembled onto a common heat sink. A cross flow blower with an air flow rate of 217.77cu.ft/min was used to cool the main heat sink and two axial fans with a flow rate of 100cu.ft/min were used for reasons discussed above. The cross flow blower lowered the main heat sink temperature by about 7.5°C, while the axial fans lowered the temperatures of passive components by 16°C. The Boost-brake module which provides the 390V DC link voltage to the amplifiers retained its power rating of 3kW. To cope with the increased power requirement of the axial bearing (and the associated increase in DC link ripple current), the reservoir capacitance associated with this axis was doubled.

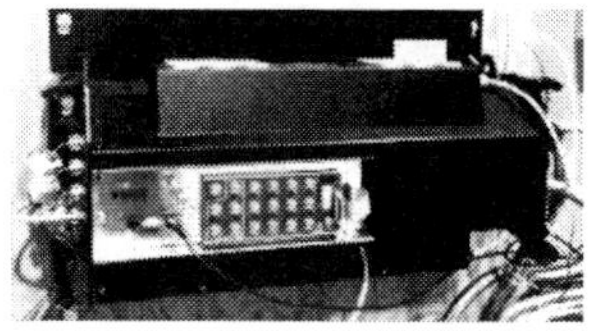

Figure 5. Up rated Zephyr Controller

A thermal profile of major components was plotted during in-house testing and is shown in Figures 6 and 7. The current magnitudes in amplifiers for the duration of this test are given in Table III. It was seen that the up rated axial amplifier was running almost at same temperature as the radial amplifiers for 2.5 times increase in current. This can be attributed to choice of the device at the design stage, layout of the board and the uniform heat distribution provided by thicker PCB tracks.

TABLE III. IGBT TEMPERATURES ON EXISTING AMPLIFIER AND UP-RATED AMPLIFIER AFTER 2 HOUR CONTINUOUS RUN

Axis	IGBT temperature(°C)	Bias Current(A)
J1P	69	20
J1M	70	20
J2P	68	20
J2M	68	20
J3P	67	20
J3M	67	20
J4P	67	20
J4M	67	20
ThP	66	50
ThM	64	50

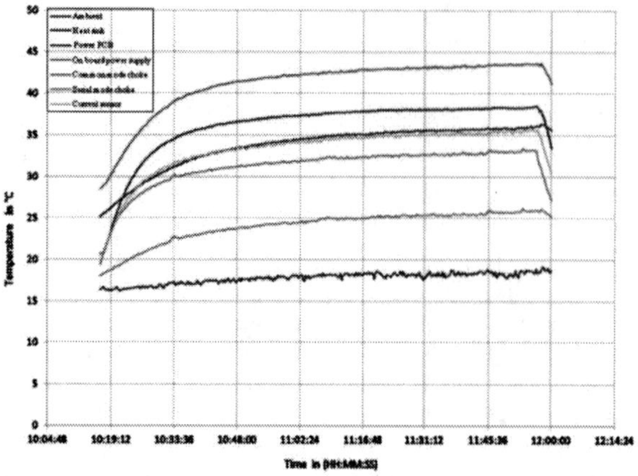

Figure 6. Thermal profile of retrofit axial amplifier

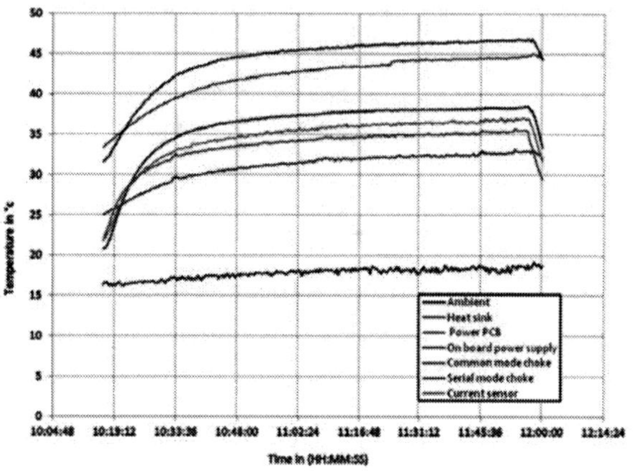

Figure 7. Thermal profile of existing radial amplifier

From the thermal profiles, it can be seen that the retrofitted amplifier PCB is cooler by 8°C. The passive components (common mode and serial mode chokes) can be seen to be running cooler by 6°C. The ambient temperature internal to Zephyr was maintained around 40°C catering to good reliability, high power densities and a nice thermal balance.

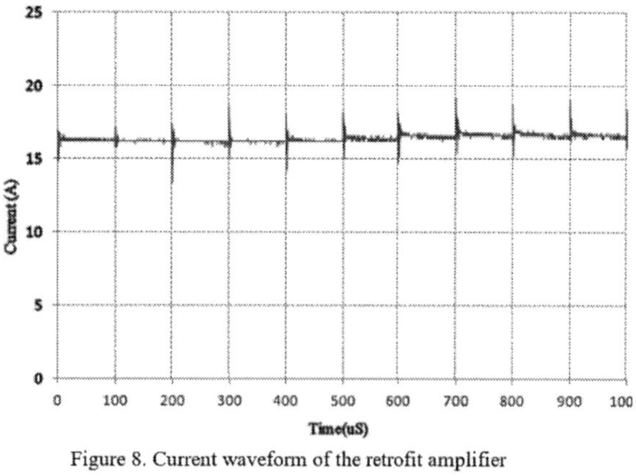

Figure 8. Current waveform of the retrofit amplifier

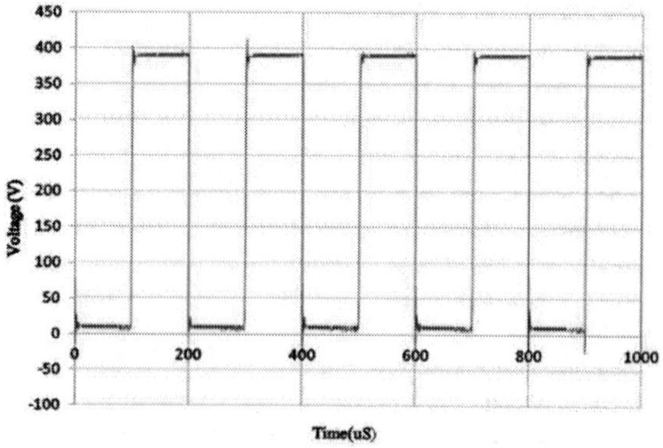

Figure 9. Voltage waveform of the retrofit amplifier

Figures 8 and 9 show plots of load current and voltage from the actual installation on site. It can be seen that the ringing on lines is only marginal. The switching transient at turn-on is small. The axial current needed for levitation was 16A. The turn-off transient also reduced due to use of a dedicated 'gate turn-off resistor'. The layout was kept simple with use of a low component count on the power card. This minimized track routing complexity and possibility of earth loops which are a pre-dominant cause of EMI.

V. CONCLUSIONS

Thermal management of amplifiers in magnetic bearings is crucial for systems requiring high reliability and power density. This paper demonstrates the optimization of design can be done by using high current PCBs; devices with good TMLD and TDR values and a gate drive which provides a balance between switching losses and EMI. The techniques discussed in this paper have been demonstrated (through field experience) to produce good results in the design of amplifiers for a retrofit application.

REFERENCES

[1] I.Josefovic, A.Popovic-Gerber. J.A Ferriera, "Thermal Management of Compact SMT Multilayer Converters," *IEEE Energy Conversion Congress and Exposition (ECCE)*,2011

[2] Erik C.W.de Jong, J. A Ferriera and Pavol Bauer, "Design Techniques for Thermal Management in Switch Mode Converters," *IEEE Transactions On Industry Applications, VOL.42, NO.6*, November/December 2006

[3] Dave Basista, "Using Heavy Copper and Extreme Copper in PCB Design and Fabrication for Maximum Reliability," *EPEC engineered technologies.*

[4] K.J Um, "IGBT Basics II, Application note 9020," Fairchild Semiconductor, April 2002.

[5] Martin März, "Thermal Management in High-Density Power Converters," Fraunhofer Institute of Integrated Systems and device technology, *IEEE International conference on industrial technology December 2003.*

[6] Alexandre Boyer, Sonia Ben Dhia, Binhong Li, Nestor Berbel and Raul Fernandez-Garcia, "Experimental Investigations into the Effetcs of Electrical Stress on Electromagnetic Emission from Integrated Circuits," *IEEE Transactions on Electromagnetic compatibility* VOL. 56, NO.1, February 2014.

Switch-Mode Current Amplifier for Active Magnetic Bearings

Sergei Basovich, Tomer Ben-Moha, Mor Mordechai Peretz, Shai Arogeti, and Ziv Brand

Ben-Gurion University of the Negev, P.O.B 653 Beer-Sheva 84105, Israel, basovich@post.bgu.ac.il

***Abstract*—This paper presents the concept and the practical implementation of a switch-mode current amplifier for the use in Active Magnetic Bearings (AMB). The switch-mode amplifier consists of a full-bridge capable of operating in 2-level, 3-level or phase-shifted PWM mode. An optimal digital current-programmed mode controller lays out the foundations to a proximate-optimal, robust, large-signal dynamic response of the system under extreme cases of loading and parameters variations. In particular, the current amplifier developed is capable of following the desired reference for large inductance variations (due to displacement). The power stage combines a newly developed, efficient, passive-based current sensor that is capable of sensing bi-directional and near-zero currents, while being robust to external disturbances and maintaining simple and efficient structure. The presented current amplifier concept has been verified experimentally on a single-axis AMB platform using a dual channel amplifier prototype, controlled by a single 16 bit low-cost microcontroller.**

I. Introduction

The key factor that is required for reliable operation for Active Magnetic Bearings (AMB) is a well-regulated current-sourcing current amplifier with low quiescent noise to assure steady-state stability [1]. Another requirement is that the current amplifier is capable of sustaining large-signal load variations, and follow a desired reference with guaranteed dynamic performance under wide changes in the system parameters. The latter is due to the fact that the effective inductance of the actuator significantly varies with the displacement of the rotor [2],[3].

Two general types of current amplifiers exist in the literature in the context of AMB. One approach is based on the classic linear amplifiers, and it is inherently associated with very poor efficiency [4]. The more preferred approach to drive magnetic actuators is by switch-mode converters technology which has a proven superiority in terms of efficiency and reduced size at the cost of a larger, yet still acceptable, quiescent noise [5]. To further reduce the steady-state noise, a three-level PWM mode of operation is commonly used [2]. However, the controller implementation may become complex in particular when multiple actuators per control-axis are required.

Following the recent growth in popularity of digital control, adaptation of control concepts from motor drives and switch-mode power supplies (SMPS) applications that were applied to AMB systems, have demonstrated conceptual operation with good dynamic performance [6]. However, as in the initial steps in digital control of SMPS [7],[8] similar to analog control, operation is obtained using high-performance microprocessors which makes this type of solution economically-prohibitive, in particular for lower power AMB systems.

Another important element in the design of current-sourcing converter for AMB is the current sensor. Although the subject of current sensing has not been widely addressed hitherto, it is a key factor to assure the desired performance, especially for the case of switching systems. In [9], the actuator current is sensed via a series resistor, therefore introducing additional losses. These can be eliminated using a hall-effect device [10],[11], at the cost of sensitivity to external magnetic fields and more complex design. However, both methods exhibit poor SNR at lower current levels. Since the nonlinear AMB plants are usually controlled using the feedback linearization method [12], which involves the on-off switching of the opposite electromagnets, the capability of the proper near-zero currents sensing is crucial to the reliable AMB system operation. An alternative approach to current sensing derived from SMPS technology that is pursued in this study is using a current transformer. Here, the current of the switching element is passed via a pulse transformer and the full behavior of the signal is restored using additional circuitry. The main limitation of such design is associated with a reset of the pulse transformer [13].

The objective of this study is thus to introduce an efficient current amplifier for AMB applications with fully-integrated digital current-programmed mode controller and a newly developed digital-control-oriented current sensing method using a simple pulse transformer. The current amplifier for a 1-degree-of-freedom (DOF) AMB system (Fig.1) consists of two H-bridge converters, synchronously operating in a three-level PWM mode, governed and controlled by a single 16 bit, fixed-point, microcontroller. The amplifier was validated on a real AMB system, it exhibits robust response for changes in the reference, loading conditions and system parameters.

The paper is organized as follows; the current amplifier structure and principle of operation, as well as the description of the new current sensing method are given in Section II. The discrete optimal controller design is described in Section III. Section IV presents the experimental results, and Section V concludes the paper.

II. Current Amplifier Overview

A. *Structure and Operating Principle*

The switch-mode current amplifier (Fig.1) consists of eight MOSFETs, each four of them in H-bridge configuration, and

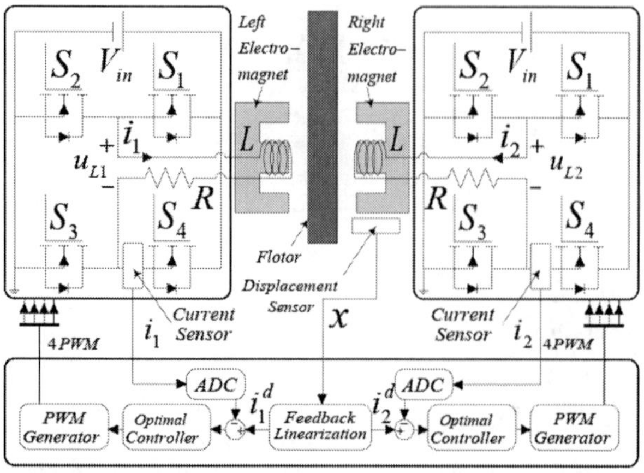

Figure 1: Structure of the two channel current amplifier for a 1-DOF AMB system

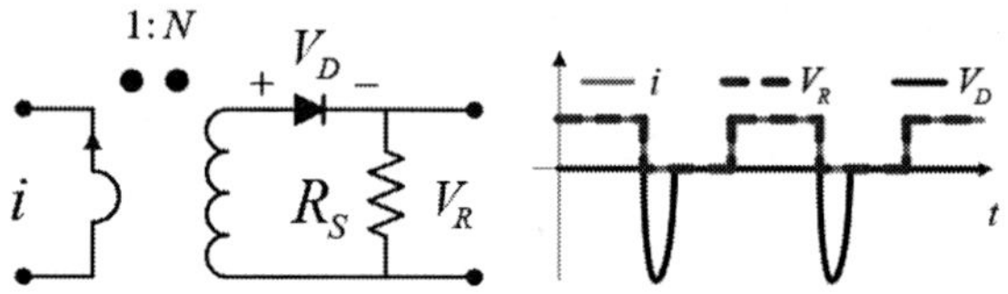

(a) Sensing circuit (b) Typical waveform

Figure 2: DC current transformer sensing

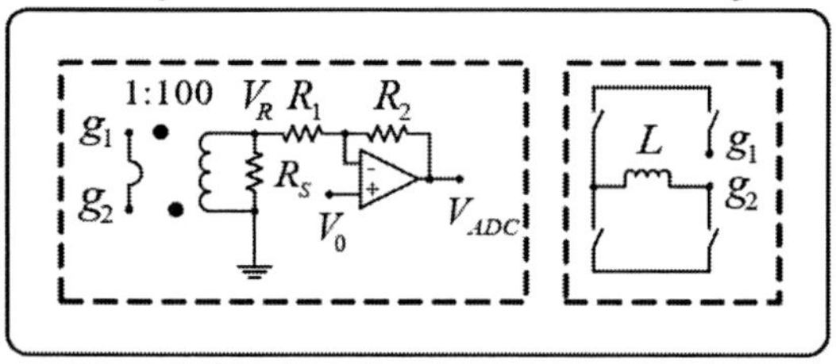

Figure 3: The introduced current sensor scheme

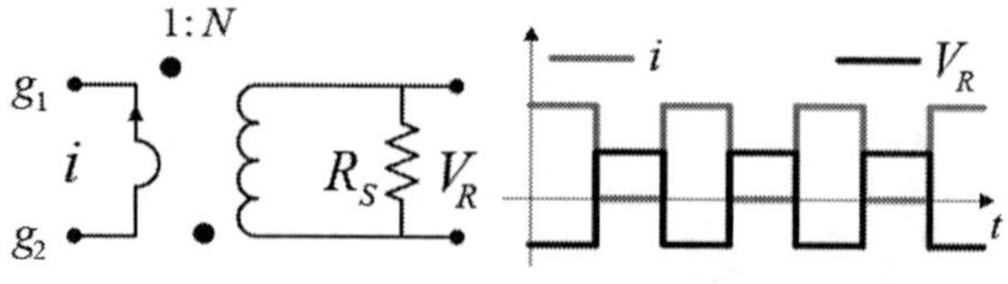

(a) Sensing circuit (b) Typical waveform

Figure 4: Introduced current transformer sensing

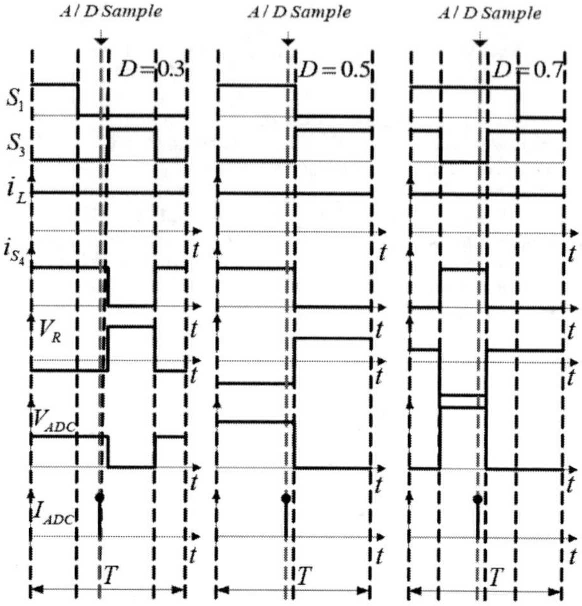

Figure 5: Current sensor operating principle for a variety of PWM signals (T is the sampling period, S_2 and S_4 are complimentary to S_1 and S_3 respectively)

a microcontroller capable of sampling voltage using A/D and generating 8 PWM signals.

According to schematic representation in Fig.1 a PC executing the feedback linearization based control law for an AMB receives the flotor displacement and calculates reference currents i_1^d, i_2^d for the amplifier control loop. The ADC of the microcontroller which is running the amplifier control loop samples the inductor current and reference current signals at a predefined rate, which in this study chosen to be at the switching frequency (upper bound of the switching system) to allow rapid response of the controller. The calculated error between the measured currents i_1, i_2 and i_1^d, i_2^d is then serves as an input to the optimal current-control algorithm developed to produce the required correction signal such to regulate the average actuator current PWM signals (S_1, S_2, S_3, S_4 where S_2,S_4 are complimentary to S_1,S_3 respectively) at the switching frequency of 100[kHz]. The PWM signals for two H-bridges are produced simultaneously using a dedicated per-converter PWM generator. The notation L, R, V_{in} is used to represent the actuator inductance, parasitic resistance and input voltage, respectively. The voltage between the actuator terminals in both H-bridges is denoted by u_{L1} and u_{L2}. As can be observed from Fig.1, the current regulation depends on the capability of accurately sensing actuator's current. To this end, a simple and efficient current sensor based on a pulse transformer, has been developed and is detailed in the following section.

B. Current Sensing Method

Accurate current measurement is essential to assure the desired closed-loop performance of the current amplifier. Similarly to inductor current sensing in SMPS, the main challenge in sensing the actuator's current, is the need to measure the DC component, which cannot be applied using simple current transformers. to this end, numerous current sensing methods for switch-mode applications have been covered in the literature, among them sense resistor [14], Hall-effect [15] and current transformers paired with signal reconstruction circuits [13]. However, in the context of AMB applications, the realization of current sensors has been conventionally applied by either resistive or Hall-effect concepts due to their capability to obtain both the AC and DC components of the signal. The main drawbacks of resistive measurement are the introduction of additional losses in the conduction path, poor signal-to-noise ratio at lower currents, and that they require instrumentation with relatively high common-mode rejection ratio. Simple Hall-effect devices are prone drifts and are sensitive to external magnetic fields and nearby currents [15].

Current transformers are limited to AC measurements due to the need for flux balance (i.e. reset) to avoid saturation of the magnetic element. DC measurement via a transformer conventionally applies a reset circuitry, similar to the operation of a forward converter [13], i.e. the DC current is passed through the transformer in intervals sufficiently spaced to allow balancing of the flux. One possible implementation, widely used in SMPS applications (Fig.2) is to locate the primary winding of current transformer in series to a switching element and a diode in series to the measurement (on a resistor) at the secondary. In this way, the peak value of the pulsating current is accurately obtained during the conduction time of the switch and the reset of the transformer is facilitated during its off interval, however, the voltage developed at the secondary during the reset interval may be significantly high. By reversing the reset concept, another possible solution is the transformer-based current sense circuit that has been presented in [14] and further advanced in [16] for compatibility with digital control operation. There, DC current of the primary winding is obtained using an AC equivalent of the secondary winding. An external excitation in the secondary alternates the current, creating a pulsed waveform with peak current that is proportional to primary DC value. Assuming negligibly small ripple, a snapshot of the signal per sampling interval is sufficient data to reconstruct the actual current waveform. In addition to isolation, this method offers the possibility to directly measure the DC current through the actuator, however, it requires additional timing and control circuitry. Another drawback of both solutions in the context of AMB is the capability of measuring both positive and negative currents with the same fixture.

In this study, a modified approach, that combines the benefits of the above-mentioned current transformer based methods with the benefits of a digital control, has been developed. The description of the sensor and its principle of operation is assisted by Fig.3, Fig.4 and Fig.5.

As shown in Fig.4a, a conventional AC current transformer is realized to obtain the pulsating current shape of the switch current. The primary winding is connected in series to the switch and the current value is obtained on a resistor at the secondary. Since the original signal has a DC component, which is not passed through the transformer (Fig.4b), the relationship between the primary and the secondary peak currents depends on the duty ratio $(1-D)$ of the switch, and is expressed as:

$$i_{sec} = \frac{i_{avg} - i}{N} = \frac{-D\,i}{N}, \tag{1}$$

where i_{avg} is the DC component, N is the winding ratio of the transformer and D is the duty cycle of S_1 and S_3.

The operation of an AMB involves both positive and negative current through the actuator, depending on the duty ratio of the H-bridge. To allow acquisition of both negative and positive current values using an ADC, a fixed DC value, V_0, is added to the signal as can be observed in Fig.3. The sensed signal by the ADC, V_{ADC} can be expressed as:

$$V_{ADC} = \left(\frac{R_1+R_2}{R_1}\right)V_0 + \left(\frac{R_2}{R_1}\right)\frac{R_S}{N}D\,i \tag{2}$$

where R_S is the resistor on which the secondary current convert to voltage, R_1 and R_2 are the resistors that sets the gain of the amplifier.

Since the entire controller operation is facilitated digitally, the current duty ratio of all switches is available at all times. Given D, the real DC value of the actuators current can be obtained by:

$$i_{ADC} = \left[V_{ADC} - \left(\frac{R_1+R_2}{R_1}\right)V_0\right]\frac{R_1}{R_2}\frac{N}{R_S}\frac{1}{D} \tag{3}$$

which can be obtained using the ADC within the entire conduction time frame of the switch.

Fig.5 illustrates a detailed timing diagram and key waveforms of the introduced current sensor for various operating points of the current amplifier. Due to V_0 offset, the actuator current can be successfully restored for both negative (V_R>0) and positive (V_R<0) values. Using the current sensing method described here, the following advantages are obtained: no additional losses in the conduction path, galvanic isolation, simple realization without special instrumentation, EMI immunity, and lowered reset voltage compared to other solutions.

III. Control Design

The following section presents the control scheme for an 1DOF AMB actuator. The controller consists of two layers: the inner layer which regulates the AMB current, and the outer layer which controls the AMB flotor displacement.

A. *Optimal Controller for Current Amplifier*

This subsection is devoted to the optimal discrete controller design. In view of [17], the controller is designed for the regulation case. Following [18], the relation between the inductor voltage and current can be described by

$$\frac{di}{dt} + \frac{R}{L_n}i = \frac{u_L}{L_n} \tag{4}$$

where L_n is the nominal actuator inductance. Define the control signal u, as follows

$$u = 2D - 1, \quad -1 < u < 1 \tag{5}$$

Neglecting the switches on resistance and substituting (5) in (4) yields

$$\frac{di}{dt} + \frac{R}{L_n}i = \frac{V_{in}}{L_n}u \tag{6}$$

The following state variables

$$x_1 = \int_0^t i(\tau)d\tau, \quad x_2 = i \tag{7}$$

are used to obtain the state space representation of (6)

$$\dot{x} = Ax + Bu, \quad y = Cx \tag{8}$$

where $x \triangleq \{x_1, x_2\}^T$, and

$$A \triangleq \begin{bmatrix} 0 & 1 \\ 0 & -\frac{R}{L} \end{bmatrix}, B \triangleq \begin{bmatrix} 0 \\ -\frac{V_{in}}{L_n} \end{bmatrix}, C \triangleq \begin{bmatrix} 0 & 1 \end{bmatrix} \tag{9}$$

Assuming that the sampling period T is small with respect to the system bandwidth, the first order approximation can be used to represent the system (8)-(9) in discrete state space by

$$x(k+1) = A_d x(k) + B_d u(k), \quad y(k) = Cx(k) \quad (10)$$

$$A_d \triangleq I + AT, \quad B_d \triangleq BT \quad (11)$$

where I is identity matrix. Applying the discrete linear quadratic regulation (LQR) methodology yields the control law of the form

$$u(k) = -Kx, \quad K = (B_d^T S B_d + R)^{-1} B^T S A_d \\ S, R > 0 \quad (12)$$

which minimizes the following cost function

$$J(u) = \sum_{n=1}^{\infty} (x(k)^T Q x(k) + u(k)^T R u(k)) \quad (13)$$

when $S > 0$ solves the algebraic Riccati equation

$$A_d^T S A_d + Q - S - A_d^T S B_d (B_d^T S B_d + R)^{-1} B^T S A_d = 0 \quad (14)$$

It can be verified that the following holds

$$rank\{B_d, A_d B_d\} = 2 \quad (15)$$

therefore a couple $\{A_d, B_d\}$ is controllable. Since a couple $\{A_d, C\}$ is not observable, in order to assure stability of (10)-(11) with (12) the weight matrix Q must be chosen in such a way that a couple $\{A_d, Q\}$ will be detectable [19]. In other words Q have to satisfy [20]

$$\exists P > 0 \quad | \quad A_d^T P A_d - P - Q^T Q < 0 \quad (16)$$

Thus, solving (14) for

$$Q = \begin{bmatrix} 2.3575e8 & 0 \\ 0 & 37 \end{bmatrix}, \quad R = 0.1 \quad (17)$$

one can obtain the following control gain

$$K = \begin{bmatrix} 3599.2 & 18 \end{bmatrix} \quad (18)$$

B. Feedback-linearization Controller for AMB

A 1-DOF AMB plant (Fig.6) was used in order to validate the presented power module. This AMB system consists of two electromagnets, a moving part, a static part and a proximity sensor. The moving part is connected to the static platform by a flexible link, the displacement of the moving part is measured by the proximity sensor. The moving part motion in the x direction can be described by a following equation

$$m\ddot{x} + kx = F \quad (19)$$

where m is the moving part equivalent mass, k is the moving part stiffness, and F is a force applied by electromagnets. This force can be modeled as [12]

$$F = F_1 - F_2 = c\frac{i_1^2}{(g_0 - x + x_f)^2} - c\frac{i_2^2}{(g_0 + x + x_f)^2} \quad (20)$$

where F_1 and F_2 are the forces applied by a left and right electromagnets respectively, c is the electromagnetic coefficient, g_0 is a nominal (corresponding to x=0) air gap, x_f is an additional length above the air gap due to a final permeability of the electromagnets core material, i_1 and i_2 are the control currents supplied to the left and right electromagnet respectively. The values of parameters for the employed in this study AMB system are given in Tab.I.

From (20) it can be observed that the force F is nonlinear in the moving part position x. To control the moving part

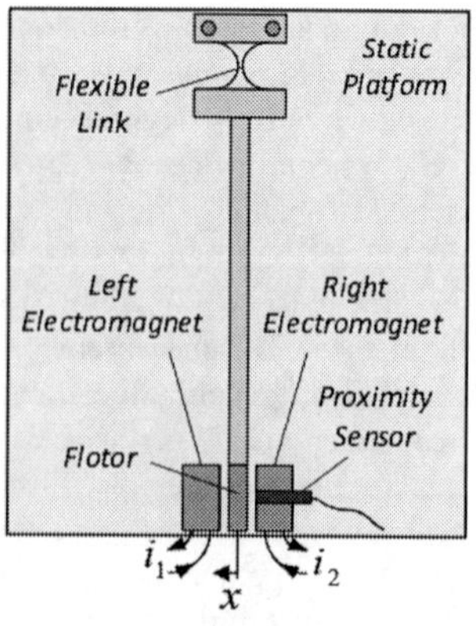

Figure 6: 1-DOF AMB system

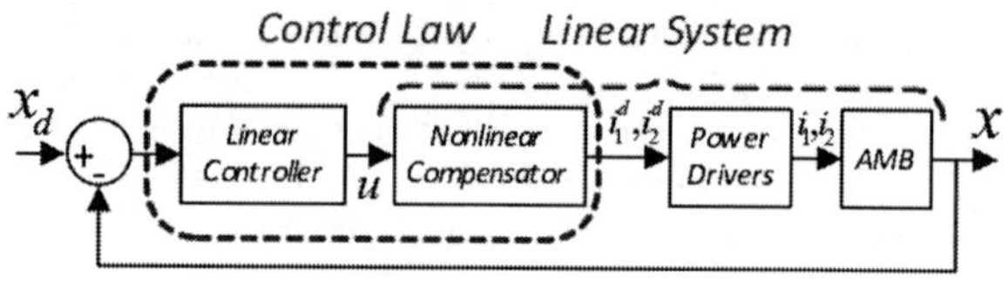

Figure 7: Feedback-linearization scheme

motion we use the feedback linearization technique. In general, this technique is based on a transformation of a nonlinear system to the linear one via feedback and a transformation of the state vector, and in this work we will use it to obtain a linearized expression for F (Fig.7). To this end, let us define the following transformation

$$i_1 = \begin{cases} (g_0 - x + x_f)\sqrt{um/c}, & u > 0 \\ 0, & otherwise \end{cases} \\ i_2 = \begin{cases} (g_0 + x + x_f)\sqrt{-um/c}, & u < 0 \\ 0, & otherwise \end{cases} \quad (21)$$

Transformation (21) means that the position of the moving part is controlled by switching between i_1 and i_2, i.e., a single control signal u is sufficient to carry out the regulation task. Substitution of (21) into (19)-(20) yields

$$\ddot{x} + \frac{k}{m}x = u \quad (22)$$

while for u we select the following PD control

$$u = -c_1(x - x_d) - c_2(\dot{x} - \dot{x}_d) \quad (23)$$

The Bode plot of the transference between $X_d(s)$ and $X(s)$ for

$$c_1 = 70000, \quad c_2 = 480 \quad (24)$$

is shown in Fig.8.

Table I: AMB Parameters

Parameter	*Description*	*Value*
m	Moving part equivalent mass	$0.48[kg]$
k	Flexible link stiffness	$1272.8[N/m]$
c	Electromagnetic coefficient	$3.73e-6[Nm^2/A^2]$
g_0	Nominal air gap	$313[\mu m]$
x_f	Add. length due to final permeab.	$1.8634[\mu m]$

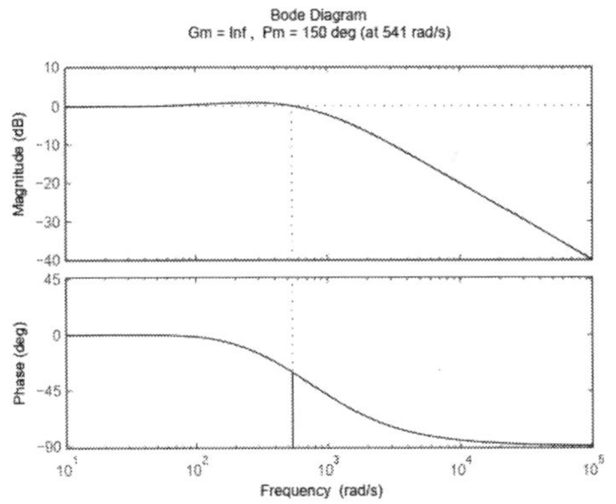

Figure 8: Transference for the closed loop system (22)-(23)

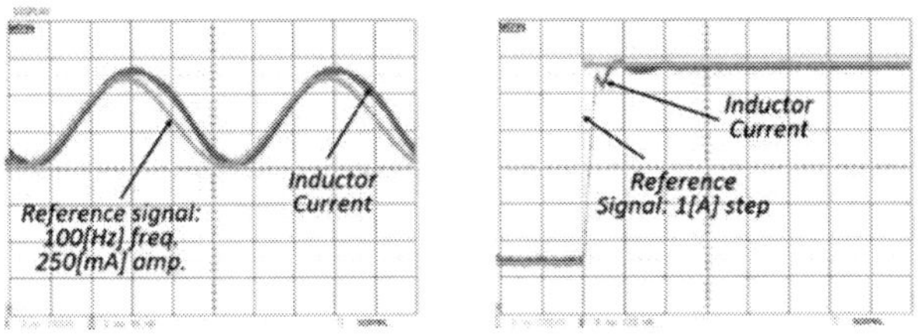

(a) Responce to sinusoidal signal (b) Step response

Figure 9: Tracking performance

IV. Experimental Results

The following section presents the experimental results conducted on the current amplifier. The first part presents the dynamic performance of the current amplifier prototype, when the second presents the dynamic performance of the 1 DOF AMB while driven by the current amplifier prototype.

A. Current Amplifier

The presented current amplifier concept has been experimentally validated by a prototype as described in Fig.1, operating at $100\,[\mathrm{kHz}]$. The system parameters were: $L = 17 - 45\,[\mathrm{mH}]$, $R = 1.6\,[\Omega]$, $V_{in} = 25\,[\mathrm{V}]$. The digital controller has been implemented with a single 16-bit, fixed-point, microcontroller that comprises a 10-bit ADC, and eight PWM outputs.

The current amplifier performance was validated for both set-point and sinusoidal wave tracking. The results of the experiments are shown in Fig.9a and Fig.9b. An experimental demonstration of the current sensing method is shown in Fig.10.

In order to evaluate the dynamic performance of the current amplifier (Fig.11), it should be noted that the relevant performance measure of the current amplifier should be obtained observing the response to a large-signal perturbation. Thus, the information available from the small-signal bode plot can lead to deceptive conclusions. On the other hand, the large-signal dynamic characteristics of the current amplifier can be estimated using the rise time of the response in Fig.9b as follows: $\omega_{BW} = 2/t_r$, where ω_{BW} $[\mathrm{rad/sec}]$ is the equivalent bandwidth, and $t_r = 1.5 \times 10^{-3}$ $[\mathrm{sec}]$ is the rise time.

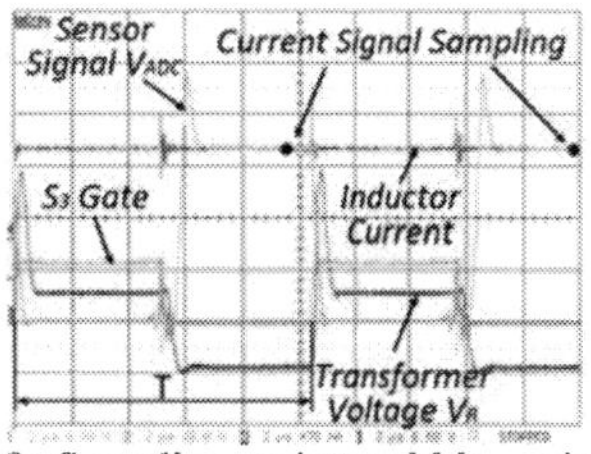

Figure 10: Sampling points within a single cycle

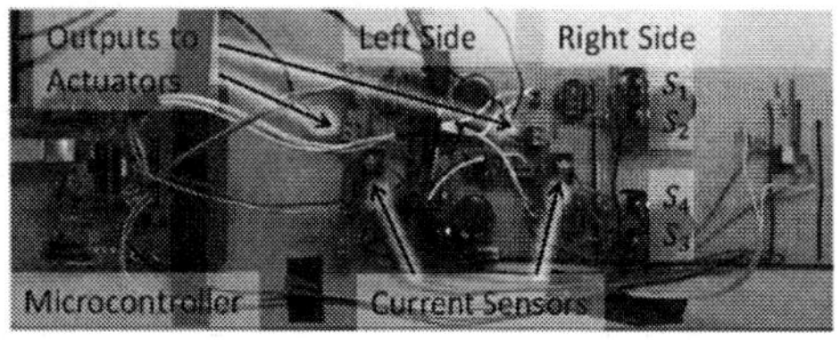

Figure 11: Power Amplifier

B. AMB System

The feasibility of the presented current amplifier was demonstrated using a real AMB system. (Fig.12). The outer control loop (AMB position regulation) was implemented using the MATLAB real-time toolbox and an acquisition card. The performed experiments include tracking the square wave reference signal and attenuation of $12[Hz]$ frequency vibration. In order to generate the vibration a DC motor with eccentric load (Fig.12) was attached to the moving part of the AMB system. The results of experiments for square wave tracking are shown in Fig.13, and the results for vibration attenuation are shown in Fig.14.

V. Conclusion

The AMB current amplifier with an effective current sensor was presented and implemented in this paper. The key idea behind the presented current sensing method is the sampling of a signal that caries information about the inductor current, instead of its full reconstruction. The experimental results indicates a satisfactory compliance of the current amplifier prototype with an AMB actuator.

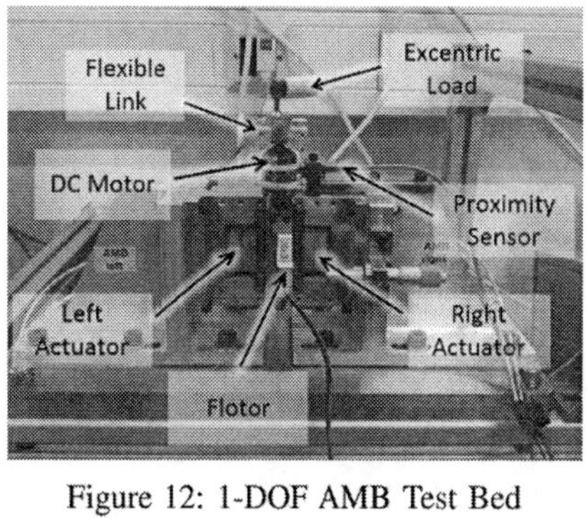

Figure 12: 1-DOF AMB Test Bed

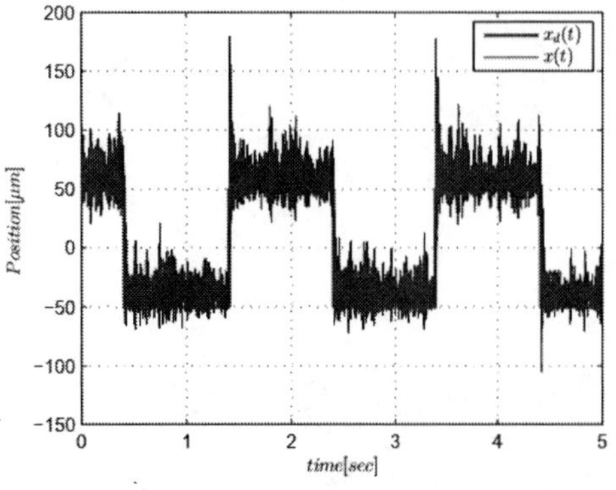

(a) AMB Position

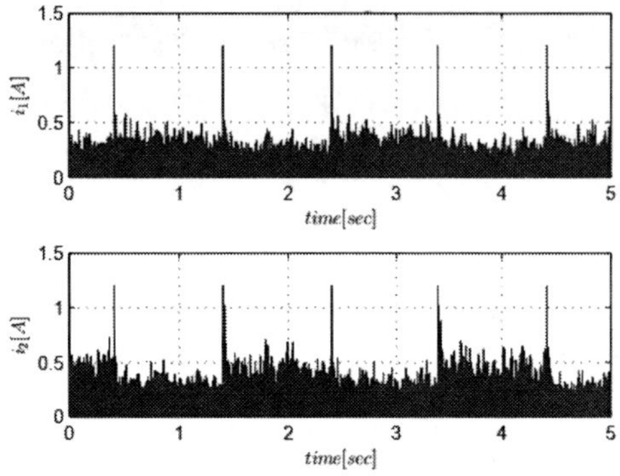

(b) Control Currents

Figure 13: Square wave tracking

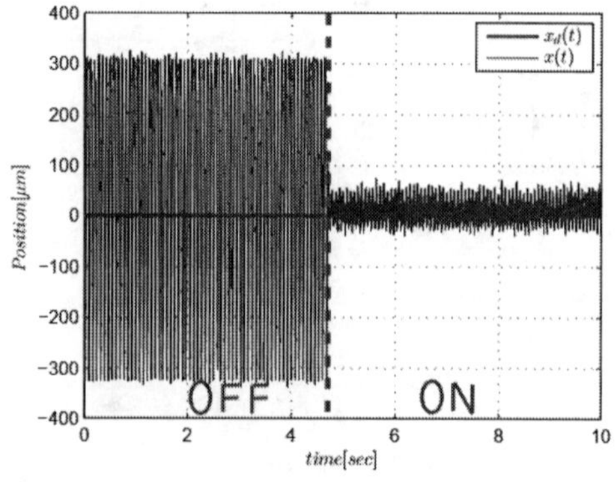

(a) AMB Position

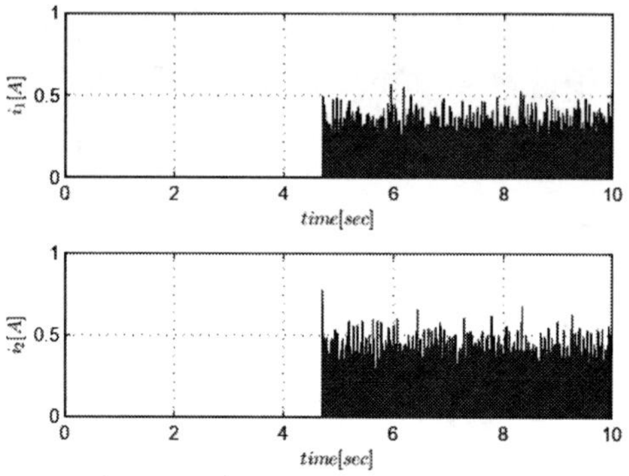

(b) Control Currents

Figure 14: Attenuation of 12[Hz] frequency vibration

REFERENCES

[1] G. Schweitzer and E. H. Maslen, *Magnetic Bearings Theory, Design and Applications to Rotating Machinery*. Springer-Verlag, 2009.

[2] J. Zhang and N. Karrer, "Igbt power amplifiers for active magnetic bearings of high speed milling spindles," in *Industrial Electronics, Control, and Instrumentation, 1995., Proceedings of the 1995 IEEE IECON 21st International Conference on*, vol. 1, 1995, pp. 596–601 vol.1.

[3] J. Y. Hung, N. G. Albritton, and F. Xia, "Nonlinear control of a magnetic bearing system," *Mechatronics*, vol. 13, no. 6, pp. 621 – 637, 2003. [Online]. Available: http://www.sciencedirect.com/science/article/pii/S095741580200034X

[4] J. Zhang and J. O. Schulze, "Synchronous three-level pwm power amplifier for magnetic bearings," in *International Symposium on Magnetic Bearings*, 1996.

[5] Z. Changsheng and M. Zhiwei, "A pwm based switching power amplifier for active magnetic bearings," in *Electrical Machines and Systems, 2005. ICEMS 2005. Proceedings of the Eighth International Conference on*, vol. 2, 2005, pp. 1563–1568.

[6] M. Peretz and S. Ben-Yaakov, "Time-domain design of digital compensators for pwm dc-dc converters," *Power Electronics, IEEE Transactions on*, vol. 27, no. 1, pp. 284–293, 2012.

[7] S. Lei and A. Palazzolo, "Real time digital control of magnetic bearings with microprocessors," in *Innovative Computing, Information and Control, 2006. ICICIC '06. First International Conference on*, vol. 2, 2006, pp. 154–157.

[8] W. Zhang, Y. Dun, Y. Sun, L. Yu, W. Shi, W. Zhang, and M. Shi, "High productivity reconfigurable digital control system and its application on magnetic bearing," in *Industrial Electronics and Applications, 2009. ICIEA 2009. 4th IEEE Conference on*, 2009, pp. 1844–1850.

[9] L. Zhang and J. Fang, "Pwm power amplifier with pd correction for magnetic suspending flywheel," vol. 5253, 2003, pp. 741–746. [Online]. Available: http://dx.doi.org/10.1117/12.521984

[10] L. Zhang, K. Liu, and X. Chen, "Fpga implementation of a three-level power amplifier for magnetic bearings," in *Electronic Measurement Instruments, 2009. ICEMI '09. 9th International Conference on*, 2009, pp. 1–455–1–460.

[11] A. Bonfitto, G. Botto, M. Chiaberge, L. Suarez, and A. Tonoli, "A multi-purpose control and power electronic architecture for active magnetic actuators," in *Power Electronics and Motion Control Conference (EPE/PEMC), 2012 15th International*, 2012, pp. DS2b.9–1–DS2b.9–5.

[12] D. Trumper, S. Olson, and P. Subrahmanyan, "Linearizing control of magnetic suspension systems," *Control Systems Technology, IEEE Transactions on*, vol. 5, no. 4, pp. 427 –438, jul 1997.

[13] R. W. Erickson and D. Maksimovic, *Fundamentals of Power Electronics (Second Edition)*, 2nd ed. Springer, 2001. [Online]. Available: http://www.worldcat.org/isbn/0792372700

[14] B. Mammano, "Current sensing solutions for power supply designers," Texas Instruments Incorparated, Tech. Rep., 2001.

[15] P. Ripka and A. Tipek, *Modern Sensors Handbook*. Wiley-ISTE, 2007.

[16] S. Ziegler, L. Borle, and H. H. C. Lu, "Transformer based dc current sensor for digitally controlled power supplies," in *AUPEC*, 2007.

[17] K. Ogata, *Modern Control Engineering (4th edition)*. Prentice-Hall, Inc. Upper Saddle River, New Jersey, 2002.

[18] J. Wang and L. Xu, "System model of three-level switching power amplifier for magnetic bearing," in *Measuring Technology and Mechatronics Automation, 2009. ICMTMA '09. International Conference on*, vol. 2, 2009, pp. 708–711.

[19] A. Locatelli, *Optimal Control An Introduction*. Birkhauser Basel, 2001.

[20] J. P. Hespanha, *Linear systems theory / Joao P. Hespanha*. Princeton University Press, 2009, formerly CIP.

Analysis of Nonlinear Behavior in Power Amplifier of AMB System

Yin Zhang, Jiancheng Fang, Yinxiao Jiang, Jinxiang Zhou, Qi Chen

School of Instrumentation Science and Opto-Electronics Engineering, Beijing University of Aeronautics and Astronautics, Beijing, 100191 P. R. China

***Abstract*—In order to make the AMB system satisfied the high energy-efficiency demands, the nonlinear behavior of H-bridge inverter, which is a power amplifier for AMB system, is studied in this paper. The discrete mapping model of the H-bridge inverter is established with some improvement using stroboscopic method. Based on this model, the dynamic evolution process of the inverter from stable state to chaos is directly exhibited, and the stable variation ranges of feedback gain and coil inductances are determined. Furthermore, a new simulation nonlinear model for an AMB system based on nonlinear H-Bridge inverter amplifier is set up, the influence of H-bridge inverter' s nonlinear behavior on the entire AMB system with two alternate change parameters is investigated. The simulation and experimental results demonstrate the existence of chaos phenomenon in AMB system and provide a stable range for AMB coil and control system design. The new AMB system model may have some benefits to the study of chaos control method for avoiding bifurcation and chaos phenomenon in AMB system.**

***Index Terms*—H-bridge inverter, power amplifier, discrete mapping model, chaos phenomenon, stable variation range, nonlinear model, chaos control method**

I. INTRODUCTION

Nowadays, in order to acquire high energy-efficient drives, the development trend of rotating machinery is high speed, which makes active magnetic bearings (AMB) a crucial element because of negligible friction resulting from non-contact suspension capability and the ability of unbalance vibration control. AMB system has been used in many industrial fields such as vacuum and cleanroom systems, medical devices and turbo-machinery [1]. However, new problems emerge as the AMB system applied in certain applications such as compressor and air blower etc. Sometimes the rotor's volume and weight of the compressor or air blower are very considerable, which means the rotor exists large unbalance vibration while rotating. In order to suppress large unbalance vibration, AMB system demands high-power electromagnetic actuator, which needs large and rapidly changing current. As a result the power amplifier of the AMB system operates in the nonlinear section more frequently. The nonlinear problems of AMB system have attracted more and more attention in the last decade. Much work has been done in several aspects such as analyzing nonlinear phenomena in the AMB system [2-3], establishing nonlinear models of the AMB system [4-5] and designing control method for nonlinear model [6-8], which brought great improvement to AMB system. Recently, the study of nonlinear problems of AMB system is focusing on the instability of the AMB system caused by time delays of digital controllers. Zheng Kai and Yu Lie proposed a robust fuzzy logic-base control scheme for a nonlinear magnetic bearing system that is subject to time delays in feedback loop [9]. Yuan Ren and Jiancheng Fang present an ingenious design, proportional-differential current-sensing resistor networks, to solve the common problems associated with the digital control delay [10].

With the rapid development of the study of nonlinear circuits, another nonlinear issue in AMB system attracts our attention. H-bridge inverter, as the power amplifier for active magnetic bearing system, is considered as a proportional component for most of time. However, as a kind of typical piecewise smooth circuit [11], the H-bridge inverter has many nonlinear features, which is ignored by most of AMB models. The study of the nonlinear behavior of H-Bridge inverter starts from 1997. Robert et al. introduced the bifurcation and chaos in the proportion controlled circuit for the first time [12-16], and they prove that the time-delayed feedback control (TDFC) method is an effective method to stabilize chaotic H-Bridge inverter. Hiroyuki Asahara continues studying on the chaotic evolution process of the H - bridge inverter with a width coefficient range [17-18]. Wang Xue-Mei points out that the fundamental cause of chaos in H-bridge inverter is the boundedness of duty cycle [19]. The anterior researchers have given a quiet accurate description and study method of the H-bridge inverter's nonlinear behavior. However, all the work above considered H-bridge inverter as an independent system rather than a component of the entire AMB system.

In this paper, we establish a discrete mapping model of the H-bridge inverter with some improvements, which makes the model more suitable for a genuine AMB power amplifier circuit. Furthermore, we analyze the H-bridge inverter's nonlinear behavior, not only as an independent system, but also as a component of the entire AMB system.

The brief is organized as follows. In Section II, we establish the discrete mapping model of the H-bridge inverter. The stable variation range of feedback gain and coil inductance is analyzed in Section III. In Section IV, we set up a new nonlinear model for the AMB system based on nonlinear H-bridge inverter amplifier and analyze nonlinear behavior with two alternate change parameters. Finally, we provide an experimental result of static state suspension with appropriate feedback gain and unsuitable feedback gain.

II. MODELING OF H-BRIDGE INVERTER

The idea circuit of the H-bridge inverter and its load is shown in Fig. 1. The circuit consists of a voltage source (E), four switch transistors ($VT1 \sim VT4$), an inductor (L) and a resistor (R). It is running in a fixed switching period (T). The circuit working principle can be described as follows: At the beginning of each period the current is sampled and compared to the reference current to give a pulse width modulated (PWM) signal, which controls the duty cycle of that period. The reference current is given by the AMB system controller, which usually contains displacement of the rotor loop and current of the coil loop in a genuine AMB system. In this section, we consider H-bridge inverter as an independent system; the reference current is given in two cases, direct current and sinusoidal wave, corresponding to static state suspension and rotary state suspension.

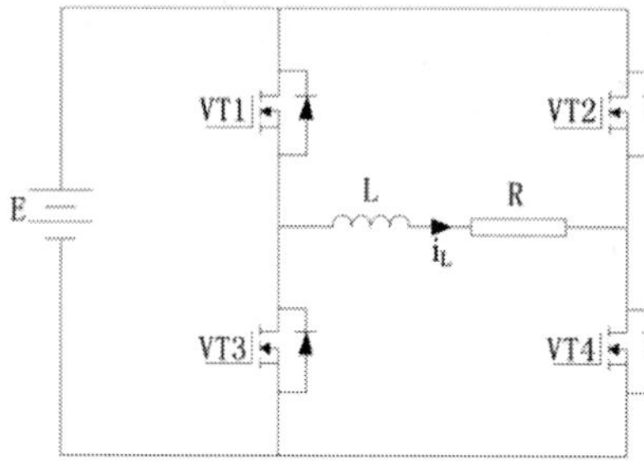

Figure 1. Ideal H-bridge inverter and its load

Depending on the state of switch transistors, the H-bridge inverter has two working states, denoted as S_1 and S_2. S_1 represents current increasing state, while S_2 represents current decreasing state. When the circuit is working on S_1, PWM signal is "high" and two switch transistors ($VT1$ and $VT4$) are turned on. When the circuit is working in S_2, PWM signal is "low", and the other two switch transistors ($VT2$ and $VT3$) are turned on. In both states, $VT1$ and $VT4$ are always in complementary states to $VT2$ and $VT3$. The state variable is the inductor current i_L. The relationship of $VT1$ and i_L is shown in Fig. 2. (As $VT4$ is operating same as $VT1$, $VT2$ and $VT3$ are operating opposite to $VT1$, the relationships between $VT2$, $VT3$, $VT4$ and i_L are no longer need to exhibit.) During one period T the H-bridge inverter involves a sequence of three switch states $S_1-S_2-S_1$. ($t1$

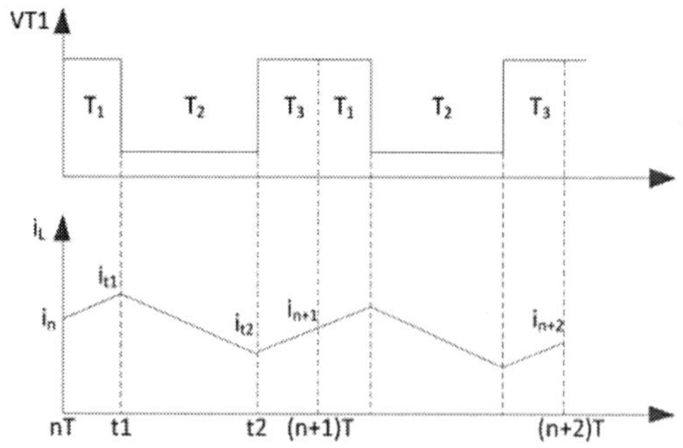

Figure 2. The relationship of $VT1$ and i_L

denotes the first switching time, $t2$ denotes the second switching time.) The duty cycle d_n is the proportion of time duration for S_1 to the switching period. It is obvious that d_n must take value between zero and one. (The subscript n denotes the values at the beginning of the nth switching period.) The inductor current's dynamics can be described by

$$S_1: \frac{di_L}{dt} = -\frac{Ri_L}{L} + \frac{E}{L} \tag{1}$$

$$S_2: \frac{di_L}{dt} = -\frac{Ri_L}{L} - \frac{E}{L} \tag{2}$$

The inductor current can be described by the following equations according to Fig. 2, Eq. (1) and Eq. (2).

For $nT < t < t1$, state S_1

$$i_{t1} = (i_{t2} - \frac{E}{R})e^{-\frac{T1}{\tau}} + \frac{E}{R} \tag{3}$$

$$T1 = t1 - nT \tag{4}$$

For $t1 < t < t2$, state S_2

$$i_{t2} = (i_{t1} + \frac{E}{R})e^{-\frac{T2}{\tau}} - \frac{E}{R} \tag{5}$$

$$T2 = t2 - t1 \tag{6}$$

For $t2 < t < (n+1)T$, state S_1

$$i_{n+1} = (i_{t2} - \frac{E}{R})e^{-\frac{T3}{\tau}} + \frac{E}{R} \tag{7}$$

$$T3 = T - T1 - T2 = T1 \tag{8}$$

$$d_n = \frac{T1 + T3}{T} \tag{9}$$

By solving the corresponding equation, i_{n+1} can be expressed in terms of i_n and d_n by

$$i_{n+1} = i_n e^{-\frac{T}{\tau}} + 2\frac{E}{R}e^{\frac{-T}{2\tau}}[2\sinh(\frac{d_n T}{2\tau}) - \sinh(\frac{T}{2\tau})] \tag{10}$$

where $\tau = \frac{L}{R}$. Then, we have to find out the feedback relationship that connects the duty cycle to the state variable. As the duty cycle is computed by the controller of AMB system, we need to know something about the AMB system's features. A typical single channel schematic diagram of the decentralized controlled AMB system is shown in Fig. 3. A contact-less position sensor measures the displacement x between rotor and AMB, and then feeds this information into the controller. The controller sends out a desired current signal i_{ref} to the electromagnetic actuator which transforms this signal in the electromagnetic coil to generate the desired magnet force. As a result the AMB system contains two feedback loops, the displacement loop and the current loop. In this section we focus on the current loop. The current loop contains power amplifier (H-bridge inverter), electromagnetic

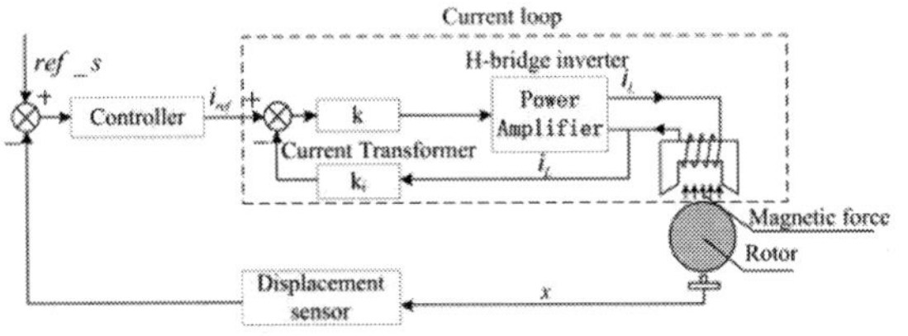

Figure 3. Single channel schematic diagram of decentralized controlled AMB system

coil, current transformer and feedback gain k . The duty cycle derived from Fig. 3 is written as

$$d_n = \frac{1}{2} + k(i_{ref} - k_i i_n) \tag{11}$$

where k_i is the proportional coefficient of current transformer, k is the feedback gain of current loop.

III. SIMULATION OF CURRENT LOOP

The simulations of current loop are based on the equations in Section II. Firstly, for each set of parameter values, time-domain cycle-by-cycle waveforms are generated by solving Eq. (10) and Eq. (11) in a sub-interval of time. Secondly, we capture the steady-state time-domain waveforms of the state variable (inductor current) under each set of parameter values to obtain bifurcation diagram. The primary circuit parameters used in simulations are shown in TABLE I. There are two notes of caution here. Firstly, duty cycle d_n is strained between 0.1 and 0.9, which is necessary in a real AMB system. Secondly, the carrier frequency F is the three times of the reciprocal of switching period, which is applied in the practical system.

TABLE I. PARAMETERS USED IN SIMULATIONS OF THE CURRENT LOOP

	Parameter	Value
T	Switching Period	0.15 ms
E	Input Voltage	90 V
L	Coil Inductance	10 mH
R	Coil Resistance	2 Ω
F	Carrier frequency	20KHz
d_n	Duty cycle	$0.1< d_n<0.9$
k	Feedback gain	0.8
k_i	Proportional coefficient	0.9

A. Bifurcations with amplification coefficient k

Considering the feedback gain k as the bifurcation parameter, we study the dynamics of the state variable i_L . The obtained bifurcation diagram is shown in Fig. 4 and Fig. 5, corresponding to static state suspension and rotary state suspension respectively. The diagrams summarize the values of i_L as k increases in the two suspension conditions. From the diagrams, we observe that the current loop subsystem exhibits a T periodic orbit when k is relatively small. This orbit exists whatever k is, but it is stable when $k < 1.48$. When $k > 1.48$, a $2T$ periodic orbit appears and then chaos is observed. The change of dynamical behavior is due to a nonstandard bifurcation known as border-collision bifurcation, as studied in [12] and [18]. As a result, in order to stabilize the current loop we need to choose feedback gain in the stable range from 0 to 1.48.

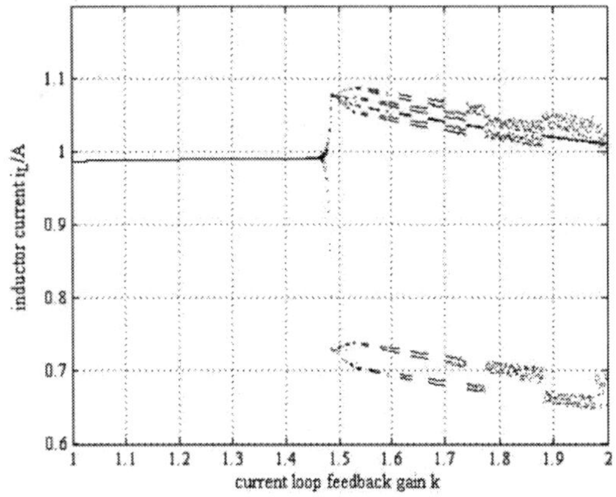

Figure 4. Bifurcation diagram of the coil current and feedback gain with direct current reference

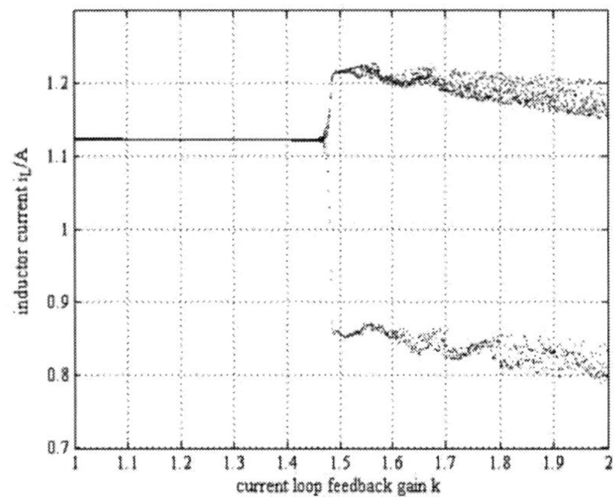

Figure 5. Bifurcation diagram of the coil current and feedback gain with sinusoidal wave current reference

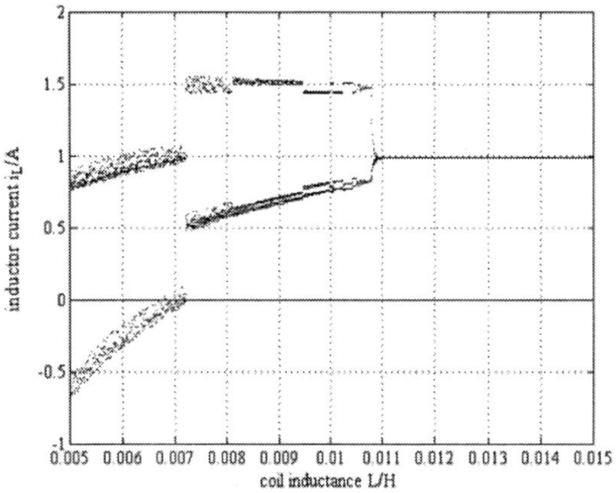

Figure 6. Bifurcation diagram of the coil current and coil inductance with direct current reference

B. Bifurcations with coil inductance L

Considering coil inductance L as the bifurcation parameter, we obtain bifurcation diagrams which are shown in Fig. 6 and Fig. 7. The diagrams show that a relatively small coil inductance L will cause bifurcation. The bifurcation is observed when L is about 11mH with parameters in Table I. Similarly, we have the stable range for L, which is bigger than 11mH.

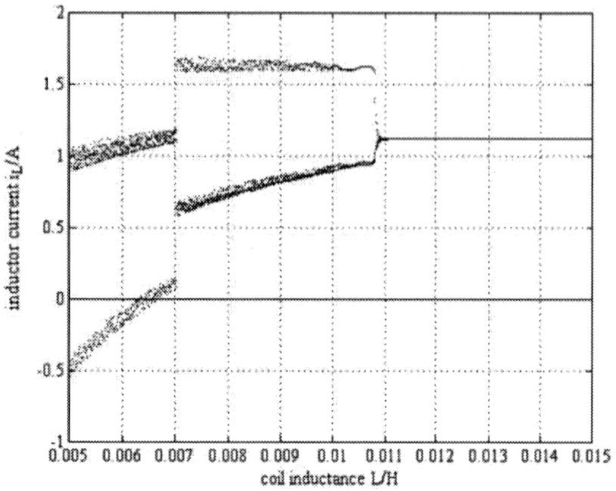

Figure 7. Bifurcation diagram of the coil current and coil inductance with sinusoidal wave current reference

In this section, we obtained bifurcation diagrams of i_L with k and L, and parameter stable range. The simulation result may be helpful for designing AMB's control and physical parameters. But all the simulations and analyses above consider the current loop as an independent module. In the next section we analyze the H-bridge inverter as a component of the entire AMB system.

IV. Simulations of AMB System

The simulations of AMB system are based on the Simulink Module of Matlab. Firstly, we established AMB system model based on nonlinear H-bridge inverter amplifier as shown in Fig. 8. Secondly, with each set of parameter values, time-domain cycle-by-cycle waveforms are generated by running the AMB system nonlinear model. Thirdly, we capture the steady-state time-domain waveforms of the state variable (inductor current) under each set of parameter values to obtain a bifurcation diagram.

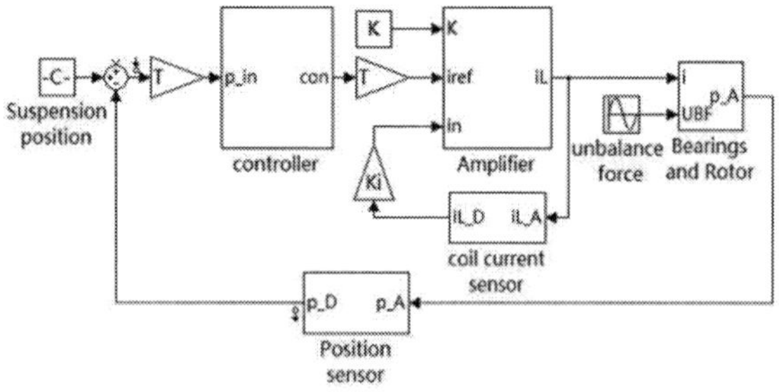

Figure 8. AMB system nonlinear model

Once the AMB system is assembled, the carrier frequency, bus voltage and coil inductance are already determined, we can't adjust these parameters while the AMB system is operating. Therefore, we pay more attentions to the adjustable parameters while the system is running, which is the proportional coefficient of current transformer k_i and the feedback gain of the current loop k. Fig. 9 is the bifurcation diagram of the coil current with proportional coefficient and current transformer adjusting at the same time while the AMB system is working on the rotating state. The primary parameters used in our simulations are same as TABLE I. The simulation result shows that the bifurcation point of i_L with changing k decreases while k_i increases. The stable zone for feedback gain k and current transformer k_i is achieved. Fig. 10 is the contrast of coil current with stable parameters and unstable parameters. We can notice that unstable parameters caused oscillation of coil current is obvious in Fig. 10(b). The current oscillation caused by the nonlinearity of H-bridge inverter is the performance of bifurcation behavior in the time-domain.

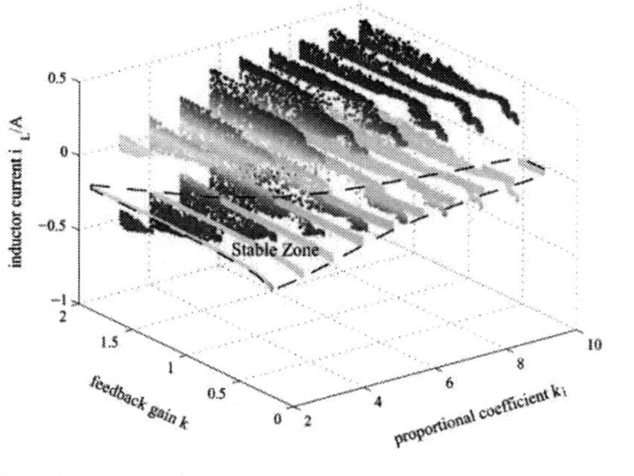

Figure 9. Bifurcation diagram of the coil current, proportional coefficient and current transformer while rotary state suspension

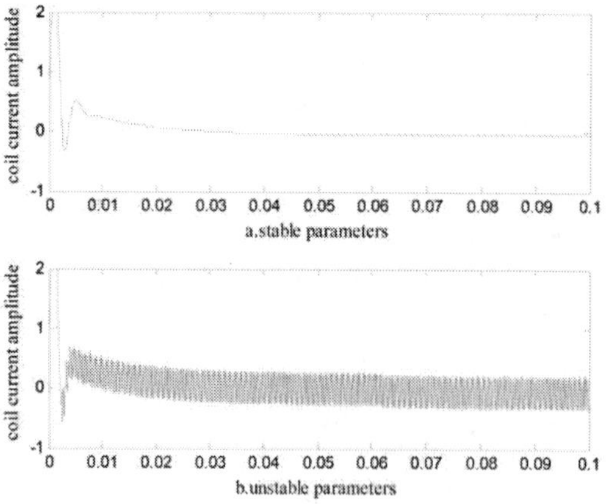

Figure 10. Coil current with stable parameters and unstable parameters

V. Experiment of Static State Suspension

The static state suspension experimental investigations with different proportional coefficient of current transformer k_i and the feedback gain of current loop k are presented in this section. The AMB system is applied to a 100Kw high speed permanent magnet motor as an experimental system which is shown in Fig 11.

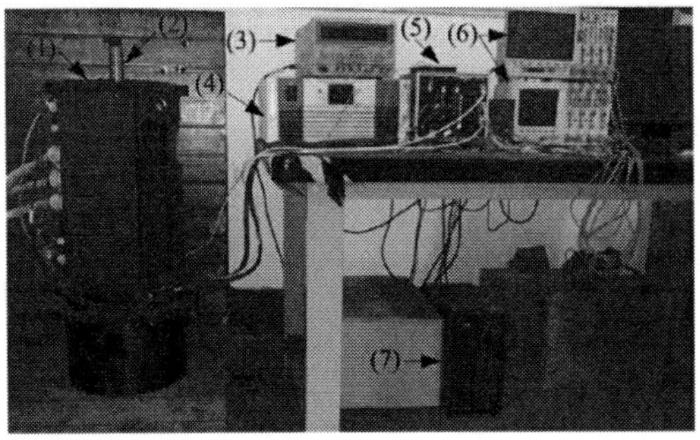

Figure 11. The experimental system. (1) Magnetically suspended motor. (2) Rotor. (3) Controller power supply 48V. (4) Amplifier power supply 90V. (5) Control system and Power amplifier. (6) Oscilloscope. (7) UPS.

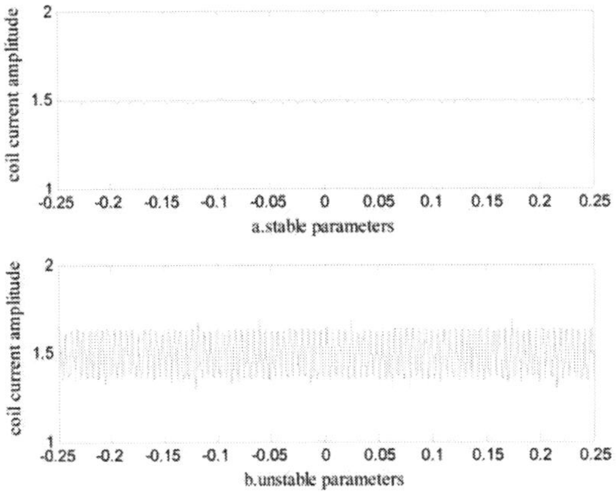

Figure 12. Coil current with stable parameters and unstable parameters

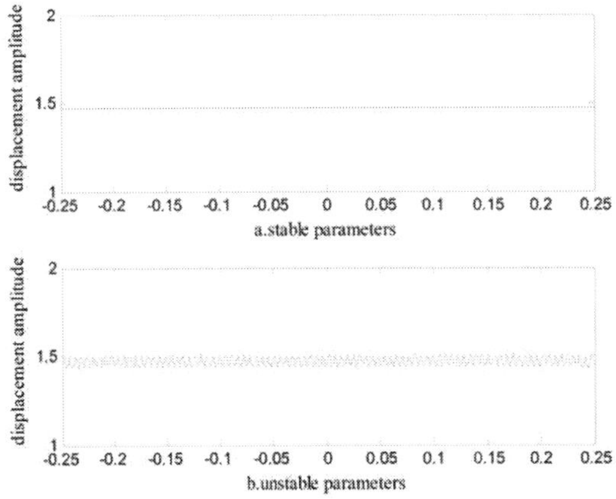

Figure 13. Rotor displacement with stable parameters and unstable parameters

Fig. 12 and Fig. 13 are the contrast of coil current and rotor displacement with stable parameters and unstable parameters of one radial suspension channel. The parameters of Fig. 12(a) and Fig. 13(a) are $k = 0.32$ and $k_i = 4.3$, the parameters of Fig. 12(b) and Fig. 13(b) are $k = 0.75$ and $k_i = 8.2$. The oscillation of rotor displacement and coil current are clearly observed in Fig. 12(b) and Fig. 13(b) respectively. The current oscillation is caused by the nonlinearity of the H - bridge inverter; the rotor displacement oscillation is caused by current oscillation which leads to oscillations of magnetic force.

VI. Conclusion

The simulation and experimental results demonstrate the existence of chaos phenomenon in power amplifier of AMB system, which will lead to oscillation of rotor displacement and affect the stability of the system. The simulation of current loop provides a stable range for feedback gain and coil inductance which may have some help with AMB coil's parameters design. The simulation of AMB system based on nonlinear power amplifier model achieves a stable zone for feedback gain k and current transformer k_i. The new AMB system model may have some benefits to the study of chaos control method for avoiding bifurcation and chaos phenomenon in AMB system.

References

[1] G. Schweitzer., E. H. Maslen, H. Bleuler, M. Cole, P. Keogh, R. Larsonneur, R. Nordmann, Y. Okada, A. Traxler "Magnetic Bearings: Theory, Design, and Application to Rotating Machinery," Springer (2009).

[2] J.C. Ji, L. Yu and A.Y.T. Leung, "Bifurcation behavior of a rotor supported by active magnetic bearings", *Journal of Sound and Vibration*, vol. 235(1), pp. 133-151, 2000.

[3] J.C. Ji, "Stability and Hopf bifurcation of a magnetic bearing system with point delays", *Journal of Sound and Vibration*, vol. 259(4), pp. 845-856, 2003.

[4] J. Jugo, I. Lizarraga, I. Arredondo, "Nonlinear analysis of an AMB system using harmonic domain LTV models", *Proceedings of the 2006 IEEE International Conference on Control Applications*, Munich, Germany, October 4-6, 2006.

[5] Martin Ruskowski, Karl Popp, "Nonlinear modeling of a magnetically guided machine tool axis", *Seventh International Symposium on Magnetic Bearings*, August 23-25 , 2000, ETH Zurich.

[6] Selim Sivrioglu, Kenzo Noami, Atsushi Kubo, Ryouichi Takahata, "Nonlinear adaptive control for a flywheel rotor-AMB system with unknown parameter", *Eighth International Symposium on Magnetic Bearing*, August 26-28, 2002, Mito, Japan.

[7] Christian Huettner, "Nonlinear state control of a left ventricular assist device (LAVD)", *Eighth International Symposium on Magnetic Bearing*, August 26-28, 2002, Mito, Japan.

[8] Herbert Arabner, Martin Reisinger ,Siegfried Silber, Wolfgang Amrhein, Christian Redemann, Peter Jenckel, "Nonlinear Feedback Control of a Five Axes Active Magnetic Bearing", *The Twelfth International Symposium on Magnetic Bearings*, Wuhan, China, August 22-25, 2010.

[9] Zheng Kai, Liu Heng, Yu Lie, "Fuzzy Modeling and Output Feedback Stabilization of a Nonlinear Magnetic Bearing with Delayed Feedback", *The Twelfth International Symposium on Magnetic Bearings*, Wuhan, China, August 22-25, 2010.

[10] Yuan Ren, Jiancheng Fang, "Current-Sensing Resistor Design to Include Current Derivative in PWM H-Bridge Unipolar Switching Power Amplifiers for Magnetic Bearings", *IEEE Transactions of industrial electronics*, vol. 59, No. 12, december 2012.

[11] Faqiang Wang, "Analysis and control of Multi-scale Bifurcation and Chaos in the Piece-wise Smooth Circuit" [D], *Xi'an Jiaotong University*, 2009.

[12] D. Carton, B. Robert, M. C. Pera, and C. Goeldel, "The use of methods studying chaos to analyze a current mode pulse width modulation H-bridge" *Proc. Eur. Conf. on Power Electronics and Applications*, 1997, pp. 3272–3277.

[13] B. Robert and C. Robert, "Border collision bifurcations in a one-dimension piecewise smooth map for a PWM current-programmed H-bridge inverter," *Int. J. Contr.*, vol. 75, no. 16/17, pp. 1356–1367, 2002.

[14] ROBERT B, LU H H C, FEKI M. "Adaptive time-delayed feedback for chaos control in a PWM single phase inverter" *Journal of Circuits, Systems, and Computers*, Computers, 2004, 13 (3): 519-534.

[15] FEKI M, ROBERT B, LU H H C, "A proportional plus extended time-delayed feedback controller for a PWM inverter" *35th Annual IEEE Power Electronics Specialists Conference,* June 20-25, 2005, Aachen, Germany. 2004: 3317-3320.

[16] ROBERT B,FEKI M." Control of a PWM inverter using proportional plus extended time-delayed feedback", *International Journal of Bifurcation and Chaos,* 2006, 16(1): 113-128.

[17] ASAHARA H, KOUSAKA T. "Bifurcation analysis in a PWM current-controlled H-bridge inverter" *International Journal of Bifurcation and Chaos,* 2011, 21(3) : 985- 996.

[18] AKATSU S, TORIKAI H, SAITO T." Zero-cross instantaneous state setting for control of a bifurcating H-bridge inverter" *International Journal of Bifurcation and Chaos*, 2007, 17 (10):3571-3575.

[19] Wang Xue-Mei, Zhang Bo. "The fast and slow-scale stabilities and chaotic motion of H-bridge sine inverter" *Acta Physica Sinica,* 2009, 58(4):2448-2454.

Active Magnetic Bearings as a Tool for Parameter Identification of Journal Bearings

Patrick Felscher[a], Katrin Baumann[b], Oliver Alber[a]

[a] Inst. of Structural Dynamics, TU Darmstadt, Otto-Berndt-Str. 2, 64287 Darmstadt, Germany, felscher@sdy.tu-darmstadt.de
[b] ICS Engineering GmbH, Am Lachengraben 5, 63303 Dreieich, Germany, katrin.baumann@ics-engineering.com

Abstract—**In this overview, it is presented how Active Magnetic Bearings (AMBs) can successfully be used for the measurement of equilibrium positions and the identification of stiffness and damping coefficients of journal bearings. Therefore, the theory of force measurements in AMBs by using either a theoretical or an experimentally derived relation is provided. Different operation modes of AMBs are considered, including the operation as a magnetic actuator. The successful parameter identification of journal bearing properties using AMBs is demonstrated by measurements.**

I. Introduction

Active Magnetic Bearings (AMBs) are a powerful tool for investigating rotating and floating machine parts such as journal bearings. The challenge of testing journal bearing dynamics is the changing shaft position within the bearing depending on the rotational speed of the rotor. The changing shaft position affects possible excitation signals as well as measurement tools and methods. AMBs can handle these varying conditions quite well as they operate contact-free. They also allow for very high rotational speeds. Furthermore, AMBs do not only support the rotor, but they can simultaneously be used for the application of defined additional loads to the rotor as well as for measuring the system's reaction at the same position.

These capabilities of AMBs can advantageously be used for the identification of journal bearing properties. The most interesting parameters of journal bearings with respect to their dynamic behavior are

- the equilibrium position of the shaft in the bearing depending on the operation conditions and
- the linearized, speed dependent stiffness and damping coefficients.

Experimental identification of these parameters usually involves the measurement of the resulting journal bearing force for a defined position of the shaft. The journal bearing force can be measured either by conventional means such as piezoelectric force sensors or strain gauges at the journal bearing base or alternatively by using additional AMBs. In this work, two methods of AMB force measurements are described in Section II. An alternative approach for experimental identification of journal bearing properties is the measurement of the shaft position for a defined additional force on the rotor. Therefore different operation modes for AMB control are considered in Section III.

A. Equilibrium Position of the Shaft

To identify the equilibrium position of the shaft, Glienicke, [2], applied a static force on the bearing housing using air pressure bellows and measured the distance between bearing and shaft with capacitive position sensors. Knopf, [3], utilized AMBs for setting the rotor to a defined radial position and measured the journal bearing force by Hall sensors inside the AMBs. This method is adopted to force measurements with the AMBs by Baumann, [1], and applied to the test rig shown in Figure 1 (presented in Section IV).

B. Stiffness and Damping Coefficients of Journal Bearings

The identification of journal bearing stiffness and damping coefficients usually requires a small additional rotor excitation (either deflection or force) at the specific equilibrium position and the measurement of the resulting journal bearing forces respectively the resulting rotor deflections. Knopf, [3], and others applied a sinusoidal motion on the shaft and measured the AMB force using Hall sensors. Someya, [4], and Glienicke, [2], applied a small sinusoidal force on the bearing using a vibration exciter. Hagg and Sankey, [5], used an unbalance force for the excitation of the shaft. Nordmann, [6], applied an impact force to the shaft and performed parameter identification in the frequency domain. This method generally lacks repeatability of the applied force. Baumann, [1] and [7], introduces a method using AMBs simultaneously for excitation and measurement, which is presented in Section V.

Section VI comments on the possile sources of errors for the presented measurement techniques.

Compared to previous approaches using AMBs for journal bearing parameter identification, both excitation and measurement can simultaneously be executed in a wide frequency domain by using only one device as actuator and sensor at the same time. Once the test setup is arranged, identification can be performed very quick and with a high degree of repeatability. Advantage can be taken of the already existing position sensors and magnetic actuators in the AMB.

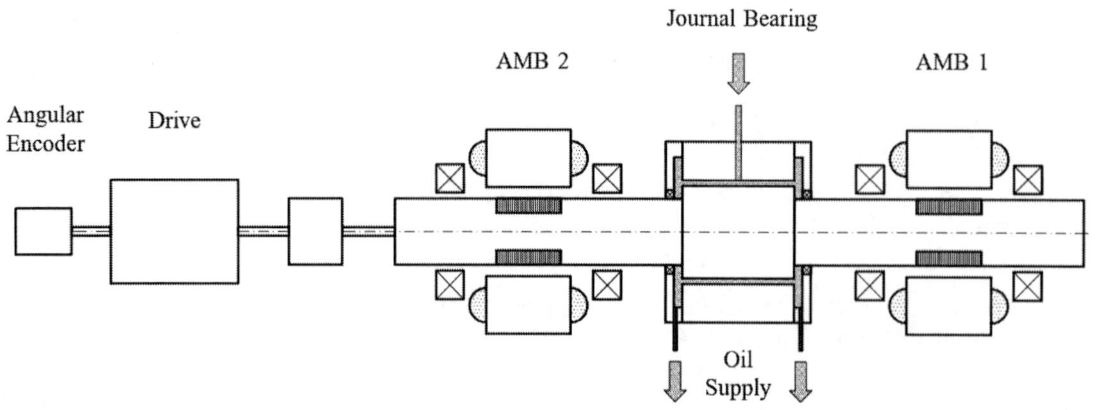

Figure 1. Test rig for parameter identification of a journal bearing using AMBs [1].

II. AMB FORCE MEASUREMENTS

The AMB force measurement is generally based on the relation between the rotor position s, the AMB control current i and the resulting magnetic force F acting on the rotor. A first, straight-forward approach is to use the theoretical s-i-F-relation as shown for example by Hoffmann, [8]. But as AMBs are influenced by several nonlinearities, a second, experimental approach is used.

A. Theoretical s-i-F-Relation

Usually, there are at least four coils at the circumference of a radial AMB for applying magnetic forces on the rotor in four directions. A sketch of a typical AMB assembly is shown in Figure 2, [8]. The magnetic force in one direction composes from the two coils located on opposite sides,

$$F_{js} = F_{js+} - F_{js-} = k_{ML}\left[\left(\frac{i_V + i_{js}}{s_0 + s_j}\right)^2 - \left(\frac{i_V - i_{js}}{s_0 - s_j}\right)^2\right], \quad (1)$$

where s is either the z or y direction, $F_{js} = F_{jz}$ or F_{jy} respectively and $j = 1, 2$ (describing AMB 1 or 2 in the test rig, see Figure 1). Herein, s_0 is the length of the initial air gap in the magnetic bearing, i is the current in the coils, generally consisting of a control current i_{js} and a constant premagnetization current i_V. s_j is the distance of the shaft from the center position of the AMB and k_{ML} is a characteristic constant of the magnetic bearing resulting from material properties and geometry.

B. Interpolation of an Experimentally Derived s-i-F-Map

Unfortunately, AMBs are influenced by several nonlinearities which cannot be considered in equation (1). Therefore, an experimental approach for determining the AMB force was used by Hoffmann, [8], and Baumann, [1]. The idea is to map the magnetic force for numerous combinations of rotor position and control current within the reasonable operating range of the AMB. Based on this data base, the magnetic force can then be inter- or even extrapolated for arbitrary rotor positions and control currents, see Figure 3.

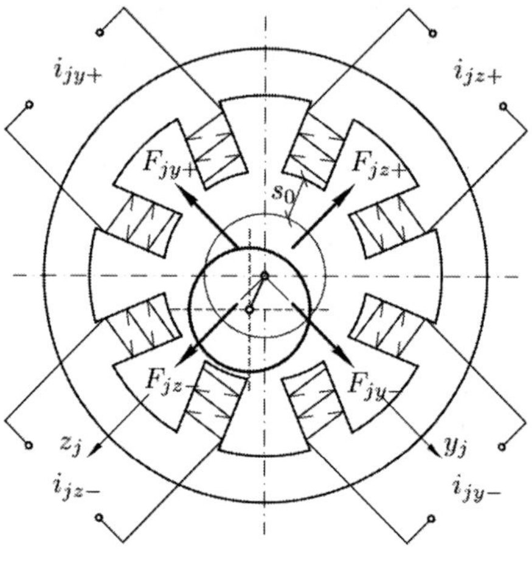

Figure 2. Assembly of an Active Magnetic Bearing [8].

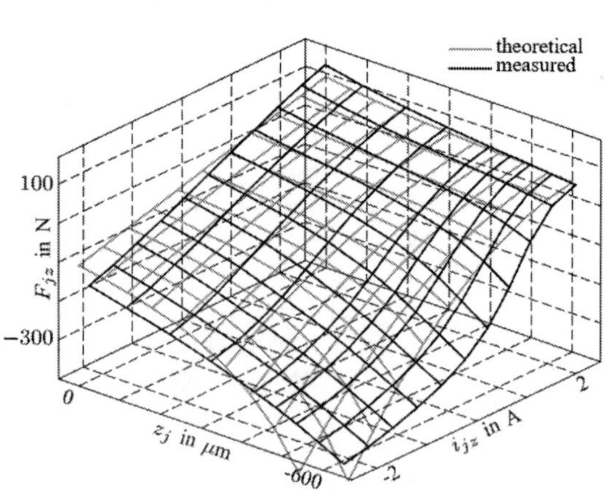

Figure 3. Comparison between the theoretical and the measured s-i-F-relation for the AMB used in [1] (with $i_{jy}=0$, $y_j=0$ and $i_V=2.5\,\text{A}$).

For the force mapping, Hoffmann and Heiland, [8] and [9], developed a test rig in which a non-rotating shaft was supported by thin legs carrying strain gauges. The shaft was then fixed in an AMB at the desired positions and a defined control current was applied on the AMB magnets. The magnetic force acting on the shaft was then measured by strain gauges and mapped. For mapping and for interpolation, it was taken advantage of the symmetry of the magnetic bearing.

III. Alternative Operation Modes of AMBs

Concerning the necessary control of AMBs and as well their application as actuators, different operation modes have to be considered:

A. Position Controlled AMBs

AMBs are generally used as contactless bearings which shall support a rotor at a specified position. In this case, the rotor shaft is forced to a defined position s_{contr} by adjusting the magnets' current i_{js} and consequently the magnetic force F_{js} by a PID control as shown in Figure 4. The possibility to change the rotor position in the AMBs allows the identification of journal bearing parameters at any position in the fluid film bearing independently from the rotor speed. Also, misalignment effects can be considered.

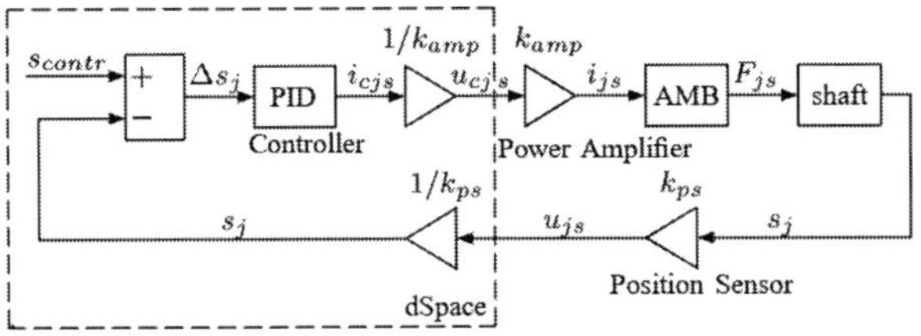

Figure 4. Position controlled Active Magnetic Bearing.

B. Force Controlled AMBs

If the AMB is primarily used as an actuator, it is more suitable to control the magnetic force of the AMB instead of the position, see Figure 5. In [10], the following approach is presented: First the magnetic force F_{js} of the magnetic actuator is estimated (F_{est}) using the AMB control current i_{cjs} and the position measurement of the shaft s_j. For this, different estimation functions F_{est} can be used: either the theoretical s-i-F-function or the experimentally derived s-i-F-map. In a second step, the estimated magnetic force of the actuator is controlled by a PID control, delivering the final control current i_{cjs} for the AMB. In [10], the control of the AMB is realized by a dSpace system and the magnetic force is estimated using the theoretical s-i-F-function (equation 1).

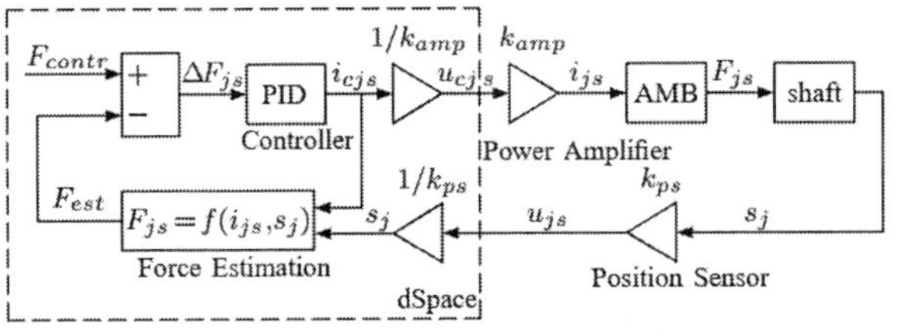

Figure 5. Force controlled Active Magnetic Bearing.

C. Force and Position Controlled AMBs

If the AMB is only used as an actuator as presented in Section B, additional bearings are needed to carry the rotor weight and to ensure the position of the shaft. At the test rig shown in Figure 1, the AMBs are primarily controlling the correct rotor position in the fluid film bearing. In order to run the journal bearing with different additional external loads applied by the AMBs, a combination of force and position control is necessary. In addition to the force control, Roy [11] controls the inclination angle of the rotor system to zero with forces acting in opposite directions in the two AMBs. This allows the shaft to move in lateral direction in the journal bearing depending on rotational speed and external load. Applying additional loads to the journal bearing using the AMBs gives the opportunity to change the operating conditions (Sommerfeld number, eccentricity, stiffness and damping coefficients) of the journal bearing at constant speed. Also, time dependent loads can be applied by the AMBs based on the s-i-F-map, simulating real load characteristics. In a test

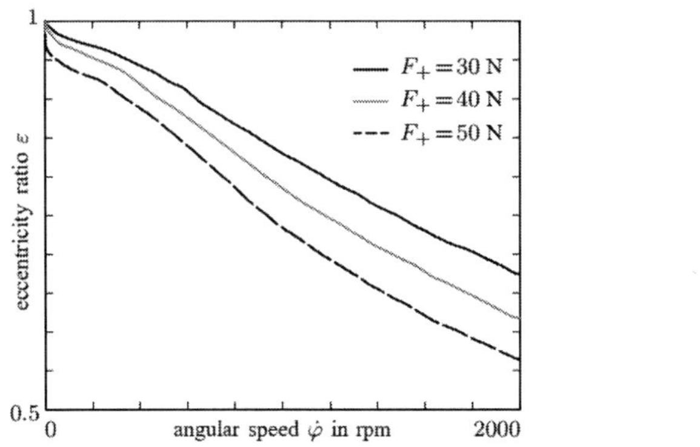

Figure 6. Eccentricity ratios ε depending on angular speed $\dot{\varphi}$ and three load conditions F_+, [11]

three different additional static forces F_+ acting in opposite direction of the gravity are applied. The resulting eccentricity ratios ε of the equilibrium positions in the journal bearing depending on the angular speed $\dot{\varphi}$ and the load conditions F_+ are shown in Figure 6.

IV. Measurement of the Journal Bearing Equilibrium Position with AMBs

The position control proposed in section III is used within the method for measuring the equilibrium rotor position in a journal bearing introduced by Knopf, [3], and revised by Baumann, [1]: For defined radial rotor positions, the resulting journal bearing force is measured with the help of the AMBs. As the resulting force in a journal bearing acts in the same direction as the applied load, the attitude angle γ_0 is known from the direction of the resulting force.

In detail, the rotor is placed at several defined radial positions along one of the magnetic axes of the AMBs by controling the AMB position. Then the rotor position, the control currents of the AMBs, and the rotational speed are measured in a first run at constant speed (with journal bearing operation, subscript n) and in a second run at zero rotational

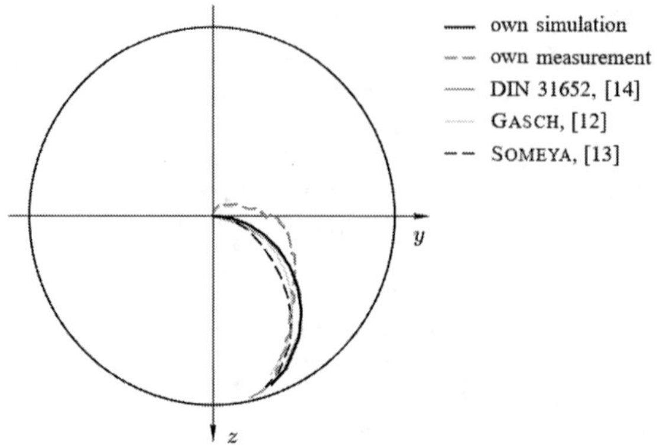

Figure 7. Equilibrium rotor position of the tested journal bearing, [1]

speed (without journal bearing operation, subscript 0). In the next step, the AMB force components F_{1z}, F_{2z}, F_{1y} and F_{2y} for both magnetic bearings 1 and 2 are either calculated using the theoretical approach (1) or interpolated using the experimentally derived s-i-F-map. The journal bearing force F_G then results from the difference between the magnetic forces in both runs (with and without journal bearing operation),

$$\begin{aligned} F_{Gz} &= F_{1z_0} + F_{2z_0} - F_{1z_n} - F_{2z_n} \\ F_{Gy} &= F_{1y_0} + F_{2y_0} - F_{1y_n} - F_{2y_n} \,. \end{aligned} \tag{2}$$

Finally, the attitude angle γ_0 is calculated from

$$\gamma_0 = \arctan \frac{F_{Gz}}{F_{Gy}} \,. \tag{3}$$

Figure 7 presents the experimental and some numerical results for the tested journal bearing in comparison to several reference data from GASCH, NORDMANN and PFÜTZNER, [12, Tab. 12.1], from SOMEYA, [13, Calc. No. 2], and from DIN 31652, [14, Tab. 18]. The experimental results agree with the reference data in an accaptable manner. Only for small rotor deflections ($\varepsilon < 0.5$), some moderate differences have to be noticed.

V. IDENTIFICATION OF JOURNAL BEARING STIFFNESS AND DAMPING COEFFICIENTS WITH AMBS

A. General Measurement Method

In general, the stiffness and damping coefficients of journal bearings are linearized at the equilibrium position for small deflections, [12]. Therefore, they can be identified by applying a force on the rotor and measuring the rotor's deflection or vice versa, [2] or [4]. As the equilibrium position of the shaft is speed dependent, the stiffness and damping coefficients are speed dependent as well.

AMBs are quite useful for the identification as they allow to place the rotor at a desired (equilibrium) position and as they can apply small additional magnetic forces on the rotor simultaneously. The resulting small rotor deflections around the equilibrium position can also be measured by the AMB displacement sensors.

Baumann, [1], proposes the following method for the identification of stiffness and damping coefficients of journal bearings using AMBs:

For the measurement, the rotor is placed either at the equilibrium position depending on the rotational speed or at the corresponding radial position at one of the AMB axes. In the latter case, the final stiffness and damping coefficients have to be converted by a geometric transform connecting the equilibrium position with the chosen AMB axis. Now, small additional excitation currents i_{ejs} are defined and added to the control currents i_{cjs} for applying additional excitation forces on the rotor by the AMB. Then, the AMB currents with and without the additional excitation currents, the rotor position, and the rotational speed are measured. For each operating point, several samples are recorded.

Afterwards, the additional rotor deflections $\Delta\mathbf{r} = [\Delta y, \Delta z]^T$ around the equilibrium position are calculated by subtracting the equilibrium position from the measured rotor position. The additional excitation force $\mathbf{F} = [F_y, F_z]^T$ is obtained from the difference between the total magnetic force (resulting from the total AMB current) and the magnetic control force (only due to the control current). Because of the AMB's nonlinearities, it is not advisable to estimate the excitation forces directly from the excitation current. For describing the forces, the theoretical approach or preferably the experimental s-i-F-map can be used.

Now, the speed dependent journal bearing stiffness and damping coefficients c_{kl} und b_{kl} with $k, l = y, z$ can be determined from the complex coefficient matrix

$$\mathbf{K}(\overline{\Omega}) = \begin{bmatrix} c_{yy} + i\overline{\Omega} b_{yy} + \overline{\Omega}^2 m_{yy} & c_{yz} + i\overline{\Omega} b_{yz} + \overline{\Omega}^2 m_{yz} \\ c_{zy} + i\overline{\Omega} b_{zy} + \overline{\Omega}^2 m_{zy} & c_{zz} + i\overline{\Omega} b_{zz} + \overline{\Omega}^2 m_{zz} \end{bmatrix} \tag{4}$$

derived from the linearized equation of motion

$$\hat{\mathbf{F}}(\overline{\Omega}) = \mathbf{K}(\overline{\Omega})\, \Delta\hat{\mathbf{r}}(\overline{\Omega}) \,. \tag{5}$$

The complex coefficient matrix can be calculated from the cross power spectral density matrix $\widetilde{\mathbf{S}}_{Fr}$ between the excitation $\mathbf{F}$ and the response $\Delta\mathbf{r}$, and from the auto spectral density matrix $\widetilde{\mathbf{S}}_{rr}$ of the responses $\Delta\mathbf{r}$ by (for details see [1])

$$\mathbf{K}^*(\overline{\Omega}) = \widetilde{\mathbf{S}}_{Fr}(\overline{\Omega})\, \widetilde{\mathbf{S}}_{rr}^{-1}(\overline{\Omega}) \,, \tag{6}$$

where $\mathbf{K}^*(\overline{\Omega})$ is the complex conjugate stiffness matrix $\mathbf{K}(\overline{\Omega})$. At last, the stiffness and damping coefficients c_{kl} and b_{kl} as well as the mass parameters m_{kl} are determined by splitting the complex coefficient matrix $\mathbf{K}(\overline{\Omega})$ in its real and imaginary part,

$$\mathrm{Re}\{k_{kl}\} = c_{kl} + \overline{\Omega}^2 m_{kl} \quad \text{and} \quad \mathrm{Im}\{k_{kl}\} = \overline{\Omega}\, b_{kl} \,, \tag{7}$$

and a final least square fitting.

B. Types of Excitation

The use of AMBs for exciting rotating systems allows a fast and reproducible identification of journal bearing parameters due to the short measurement period for each sample also at different operation conditions. Depending on the measurement focus, several types of excitation can be realized:

- Noise excitation: Providing a broadband white noise signal, all relevant frequencies are excited.
- Sine sweep excitation: Providing a sine sweep as excitation current, a specific frequency range is excited at an increased energy level compared to the noise excitation.
- Impulse excitation: A short force impulse excites all relevant frequencies of the system. The excitation time is minimal.

The proposed identification method was already tested for a simoultaneous, but uncorrelated noise excitation in two independent directions at the test rig in Figure 1. The identification method was not only applied at constant rotational speeds but also at rotor run-up processes in a short speed interval around the nominal rotational speed, [1]. Some results of these tests are shown in Figures 8 and 9. The tests with sine sweep and impulse execution at the same test rig are still going on.

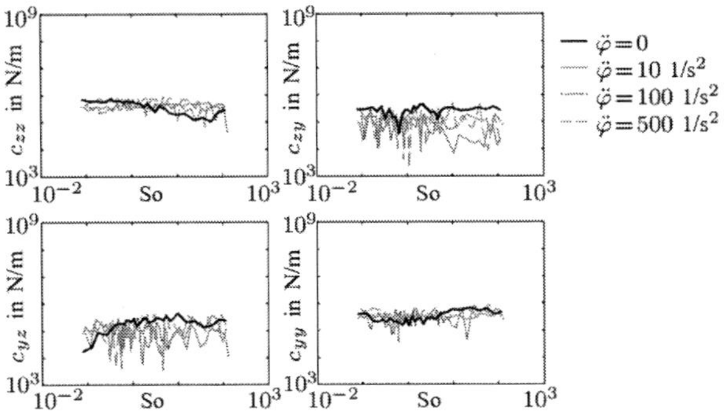

Figure 8. Measured journal bearing stiffness coefficients c_{kl} at stationary operation and during run-up processes depending on the Sommerfeld number So, [1]

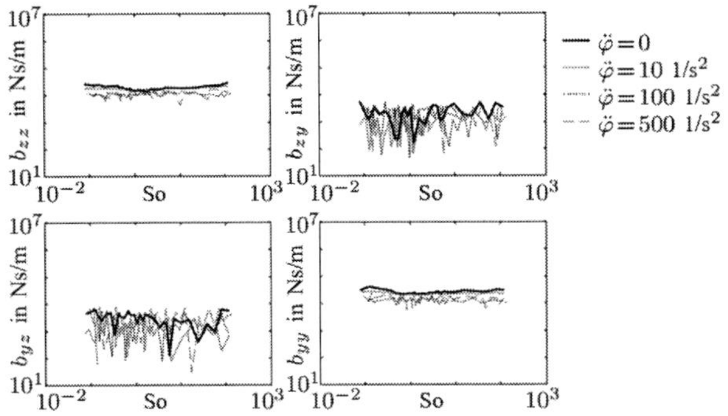

Figure 9. Measured journal bearing damping coefficients b_{kl} at stationary operation and during run-up processes depending on the Sommerfeld number So, [1]

VI. Possible Sources of Errors

The correct identification of journal bearing characteristics strongly depends on the exact measurement of the journal bearing force. As the journal bearing force is determined from the difference of the AMB forces in two separate runs respectively with/without the small excitation current, situations with almost identical or very small AMB forces are critical to the exactness. These critical situations are at the one hand especially measurements with small journal bearing forces as they occur for small Sommerfeld numbers or small rotor eccentricities in the journal bearing, and at the other hand measurements with small AMB forces which occur for large Sommerfeld numbers respectively for large rotor eccentricities in the journal bearing.

Additionally, the measurement accuracy of the AMB rotor position and the accuracy of the AMB control chain have to be kept in mind. Furthermore, the s-i-F-map has been derived experimentally by using the same technical devices and additional strain gauges.

For the used test rig, the s-i-F-map itself has a measurement inaccuracy of 6,4 %, and the force measurement by the s-i-F-map of the AMBs has an inaccuracy of 12,2 %. Summed up, the determination of the attitude angle and the journal bearing rotor position have a measurement inaccuracy of 24.4 % respectively 5 %. This means that the above experimental results are quite acceptable.

Besides the measurement inaccuracies, the influence of oil grooves and other form deviations of the journal bearing have to be considered, and the exact positioning of particularily the AMB sensors has to be ensured.

VII. Summary

Active Magnetic Bearings are a powerful tool for the parameter identification of journal bearings. The position of the shaft in the bearing and the loading of the shaft can be set independently in a certain range. Additionally, it is possible to apply a reproducible excitation on the bearing at the same time easily.

The equilibrium position and the stiffness and damping coefficients of a journal bearing have been measured successfully by using AMBs as support, sensors and actuatos simultaneously. For this, it was necessary to determine the magnetic forces in the AMB. It has been shown that the presented idealized nonlinear equation is not sufficient, but the use of an experimentally derived s-i-F-map delivers good results. This s-i-F-map can be measured with a certain effort.

The described methods using AMBs for parameter identification in rotating systems can be adapted to other applications, for example squeeze film dampers, tilting pad journal bearings or components with elastomers.

References

[1] K. Baumann, *Dynamische Eigenschaften von Gleitlagern in An- und Auslaufvorgängen.* Dissertation TU Darmstadt, 2010.

[2] J. Glienicke, *Feder- und Dämpfungskonstanten von Gleitlagern für Turbomaschinen und deren Einfluss auf das Schwingungsverhalten eines einfachen Rotors.* Dissertation TH Karlsruhe, 1966.

[3] E. Knopf, *Identifikation der Dynamik turbulenter Gleitlager mit aktiven Magnetlagern.* Dissertation TU Darmstadt, 2001.

[4] T. Someya, *Stabilität einer in zylindrischen Gleitlagern laufenden, unwuchtfreien Welle – Beitrag zur Theorie des instationär belasteten Gleitlagers.* Dissertation TH Karlsruhe, 1962.

[5] A. C. Hagg and G. O. Sankey, "Some dynamic properties of oil-film journal bearings with reference to the unbalance vibration of rotors," *Journal of Applied Mechanics*, pp. 302–306, 1956.
[6] R. Nordmann and K. Schöllhorn, "Ermittlung der Feder- und Dämpfungskonstanten von Gleitlagern durch parametrische Identifikation," *VDI-Berichte*, vol. 381, 1980.
[7] K. Baumann, F. Dohnal, and R. Markert, "Active magnetic bearings as actuators and sensors in journal bearing measurements," in *Proceedings on The Twelfth International Smposium on Magnetic Bearings (ISMB12)*, 22.-25. August 2010, pp. 686–693.
[8] K.-J. Hoffmann, *Integrierte aktive Magnetlager*. Dissertation TU Darmstadt, 1999.
[9] R. Heiland, *Konstruktion einer Vorrichtung zur Messung von Magnetlagerkräften*. Studienarbeit TU Darmstadt, 1995.
[10] G. B. Daniel, K. L. Cavalca, and O. Alber, "Application of active magnetic actuator for excitation of rotating systems supported by tilting pad journal bearing," in *22nd International Congress of Mechanical Engineering (COBEM 2013)*, 2013.
[11] H. Roy, *Force Control of Magnetic Bearing*. Master-Thesis TU Darmstadt/IIT Delhi, 2013.
[12] R. Gasch, R. Nordmann, and H. Pfützner, *Rotordynamik*, 2nd ed. Berlin: Springer-Verlag, 2006.
[13] T. Someya, *Journal-Bearing Databook*. Berlin: Springer-Verlag, 1989.
[14] "Gleitlager – hydrodynamische radial-gleitlager im stationären betrieb: Berechnung von kreiszylinderlagern." Berlin, 1996.

Magnetic Bearing Rotor Vibration Analysis Based on the Harmonic Wavelet Package Algorithm

Li Deguang, Liu Shuqin*

School of electrical engineering, Shandong university, Jinan, Shandong ,China, 250061
* Corresponding author: Liu shuqin

Abstract—**Analysis of the magnetic bearing rotor vibration is the base of the optimizing design, supervise and diagnosis of the magnetic bearing. Harmonic wavelet package was used for the analysis of the vibration signal, the 1 dimension time domain signal was mapped to 2 dimension time-frequency plane, and the 3 dimension time-frequency domain energy graph was constructed, then the analysis of the rotor vibration became intuitionistic and convenient. Via analysis of the time-frequency graph, the vibration in each time and each frequency was obtained. After the feature extraction from the time-frequency graph, the supervise and the diagnosis of the magnetic bearing can be realized.**

I. Introduction

The real-time analysis of magnetic bearing rotor operating conditions is the basis and premise of the monitoring and fault diagnosis of the magnetic bearing. Many scholars have conducted research on this issue, the methods used are mostly based on frequency domain analysis, usually through the rotor vibration signal FFT transform to analyze the operating conditions of the rotor. Because Fourier analysis is a kind of overall transformation, either in the time domain completely, or in the frequency domain completely, it can't organically combine signal's nature in two domains. This is because the signal waveform in the time domain does not contain any information of frequency domain, and the Fourier spectrum has no function of localized signal analysis, don't have any information of time domain. Thus signal analysis faced with a pair of fundamental contradiction: the localization of time domain and frequency domain. In the actual signal processing, especially for non-stationary signal such as vibration signals, the frequency domain characteristics of the signal at any time points are important. This prompts us to adopt time-frequency analysis method, so time and frequency domains can be combined to describe the joint characteristic of the two domains, and the time-frequency spectrum can be constituted. Wavelet analysis is an ideal time-frequency domain analysis, it is a kind of time-scale(time-frequency) analysis method. It has the characteristic of multi-resolution analysis, and has the ability to characterize the local features of time and frequency domain. It is a kind of time and frequency localized analysis method with variable time window and frequency window. I.e., in the low frequency portion it has higher frequency resolution and lower time resolution, while in the high frequency portion it has higher time resolution and a low frequency resolution, known as microscope of signal analysis. It is because of this feature, that wavelet analysis is self-adaptive. In engineering applications, particularly in signal processing, image processing, pattern recognition, seismic exploration, fault diagnosis and other monitoring field, wavelet transform is considered to be a major breakthrough on the tools and methods in recent years. It has become another sharp essential mathematical tools for technology workers. Magnetic bearing rotor vibration is a typical random signal, many of its important features are reflected in the details of some transient small-signals. As a time-frequency domain analysis method, wavelet transform can decompose the signal in time and frequency domain simultaneously, can easily capture the transient characteristics of the signal, particularly suitable for the analysis of random signal of the magnetic bearing rotor vibration. Although ordinary wavelet function is compactly supported in frequency domain, but its frequency domain characteristic is not ideal box-shaped, when multi-scale analyzing of signal, spectrum between adjacent scale aliased to each other, resulting in specific band signal extraction and analysis difficult. In this paper, harmonic wavelet with ideal frequency characteristics was introduced, and improved algorithm was given and was used to the analysis of the rotor vibration signal, the problems mentioned above were resolved. Experimental results show that the harmonic wavelet packet analysis is very suitable for the magnetic bearing rotor vibration analysis.

II. Improved Harmonic Wavelet Analysis

Harmonic wavelet is a kind of complex wavelet raised by Cambridge university professor Newland[4], it has a desired box-shape property in frequency domain, Which avoids the spectrum leak and aliasing between adjacent subspace, maintains their respective independence, and greatly facilitates the analysis of signal characteristics.

Generating function of the harmonic wavelet is:

$$\psi(t) = \frac{\exp(i4\pi t) - \exp(i2\pi t)}{i2\pi t} \quad (1)$$

Its box-shape spectrum was shown in Figure 1 .

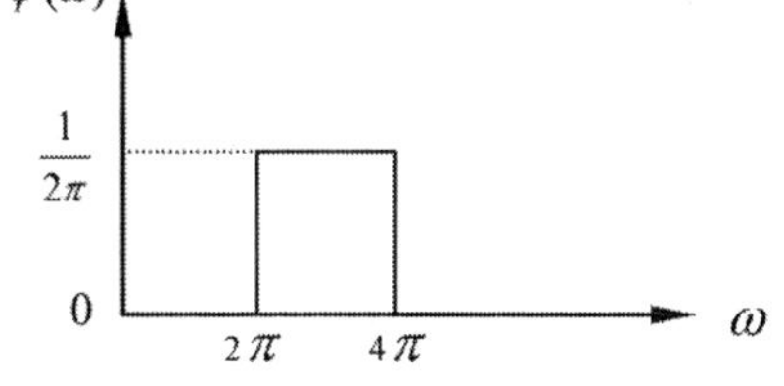

Figure1. spectrum of the harmonic wavelet generating function

Define harmonic wavelet's scale $m,n \in R^{+}$ ($m<n$) and translation k, then harmonic wavelet function is:

$$\psi(m,n,k)=\frac{\exp\left[i2n\pi(t-\frac{k}{n-m})\right]-\exp\left[i2m\pi(t-\frac{k}{n-m})\right]}{i2\pi(n-m)(t-\frac{k}{n-m})} \quad (2)$$

For the time-series signal $f(r), r=1,2,\cdots,N-1$,the discrete harmonic wavelet transform expression is:

$$W_d(m,n,k)=\frac{n-m}{N}\sum_{r=0}^{N-1} f(r)\bar{\psi}_{m,n}(r-\frac{k}{n-m}) \quad (3)$$

The traditional harmonic wavelet packet algorithm was divided by progressive one-half octave band split [4]. In the case of a certain sampling frequency, the frequency band division mode was fixed. In order to decompose the signal to either fine scale required, harmonic wavelet packet algorithm must be improved .

Define Hilbert space $\varphi_1,\varphi_2,\cdots\varphi_\infty$,where φ_j is generated by the translation series $\psi(m_j,n_j,k)$ ($1\le j<\infty$),and $n_j=m_{j+1}$. Let $n_j-m_j=n_{j+1}-m_{j+1}$,then $\varphi_1,\varphi_2,\cdots\varphi_\infty$ split the entire band evenly, each function space occupied band width $2\pi(n_j-m_j)$ .Can be proved:

1)The translation functions in a certain scale are mutually orthogonal.

2)Function $\psi(m_{j1},n_{j1},k)$ and $\psi(m_{j2},n_{j2},k)$ ($j1\neq j2$) belong to different scales are mutually orthogonal.

Therefore, when $b=n_j-m_j$ changing from small to large, decomposition of an time discrete signal by Formula(3) constitutes a different resolution harmonic wavelet packet analysis, this wavelet packet analysis can decompose the signal to any fine scale needed.

In this paper, Newland fast algorithm was used to fulfill the harmonic wavelet packet decomposition. The implementation process is as follows :

1)Determining the frequency resolution and time resolution according to the need, determining the sampling sequence length according to the sampling frequency.

2)Fulfill FFT to the sampling sequence, obtain frequency domain sequence having the equal length of time domain sequence.

3)According to the frequency and time resolution split the frequency domain sequence to segments with equal length .

4)Fulfill IFFT to each segment separately.

III. THE CONSTRUCTION OF THE HARMONIC WAVELET PACKET TIME-FREQUENCY GRAPH

Complex Fourier coefficients are obtained through the Fourier decomposition to signal, and the results can be visually displayed using spectrum graph. Similarly, the complex wavelet coefficients obtained through harmonic wavelet packet transform can be visually represented using wavelet packet time-frequency graph. Wavelet time-frequency graph is a diagram of time frequency grid, it is the three-dimensional graphic representation of the energy distribution of the signal in time-frequency domain. The construction approach of harmonic wavelet packet time-frequency graph is: Using wavelet coefficient number k (corresponding to the time t) as the horizontal axis, and the wavelet decomposition level j (corresponding frequency) as vertical axis to construct a plane, a surface is composed with the plane as reference plane. The height $H_{j,k}$ (expressed energy) of each point on the surface based on the plane is $H_{j,k}=2|a_{j,k}|^2$,where $a_{j,k}$ is the harmonic wavelet package coefficient. This makes the wavelet time-frequency grid graph. $H_{j,k}$ is chose in such a way so as to make the volume enclosed between the surface and the reference plane is exactly equal to the energy of the discrete signal $f(r)$, which meet the Parseval theorem.

Time-frequency energy distribution graph obtained by above method is dramatic ups and downs. In this paper , the log of the $H_{j,k}$ was took , so the grid graph is not so obtrusive, which is more conducive to observe less energy peaks , namely $H_{j,k}=\log(2|a_{j,k}|^2)$. When we are more concerned about the overall characteristics of the signal, the general form of the wavelet time-frequency graph is more appropriate, it can clearly shows the real energy distribution in each band of the signal. When we are more concerned with the details of local signals with weak energy , using the logarithmic form of wavelet time-frequency graph is more appropriate, it can prominently display the details of the high and low frequency component of the signal.

Select a number of rotor vibration signal segment from eddy current sensor output signal, and apply spectral analysis to them, a section of the spectrum was shown in Figure 2.

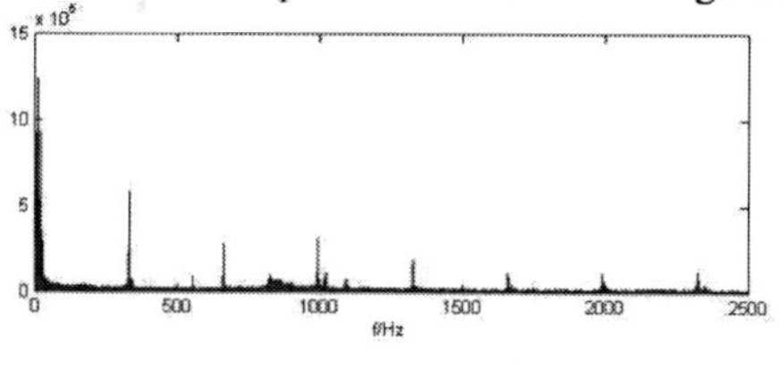

Figure2. Rotor vibration signal spectrum

As can be seen from Figure 2, at the frequency corresponding the rotor speed frequency (333Hz) and its high harmonic frequencies , the spectrum has high narrow peak , which is mainly caused by the active magnetic bearing rotor's imbalance. Besides these peaks, the spectrum of the rotor vibration is mainly concentrated within the low frequency of 100Hz, in addition, there are several peaks at high frequencies, mainly within 1500Hz.

Figure 3 shows the harmonic wavelet packet decomposition time-frequency graph of several rotor vibration signal segment. Since the effective band of the signal is mainly within 1500Hz, the frequency range in graph is 0-1500Hz.

Low-frequency part of the time-frequency graph, i.e. waveform 0Hz nearby along the time axis provides an overview of the signal analyzed, the trends of the rotor vibration can be clearly seen from this section. High-frequency part of the time-frequency graph represents the vibration intensity of the signal in different time and frequency point on the basis of the signal profile. Therefore, harmonic wavelet package analysis has very good positioning capability in both time domain and frequency domain, the transient characteristics of the signal can be vividly demonstrated.

IV. ELECTROMAGNETIC BEARING ROTOR VIBRATION ANALYSIS BASED ON HARMONIC WAVELET PACKET FREQUENCY GRAPH

In order to finely analyze the time-frequency characteristics of the rotor vibration signal, frequency resolution and time resolution should be as high as possible, but the two resolutions are mutually constrained. In a certain sampling frequency, the two resolutions can be increased simultaneously by increasing the sampling length of digital series. But in order to ensure real-time analysis, the length of the sample sequence must be limited.

Considering the rotor vibration signal is mainly concentrated in the low frequency band , the frequency resolution is determined as 5Hz, in addition , restricted by the real-time constraints and the processor processing speed, the time resolution is determined as 0.2s. In the case of systematic sampling frequency of 5000Hz , a complete characteristics analysis cycle is 1s. Magnetic bearing rotor vibration signal wavelet packet frequency time-frequency graph of several different work conditions are shown in Figure 3.

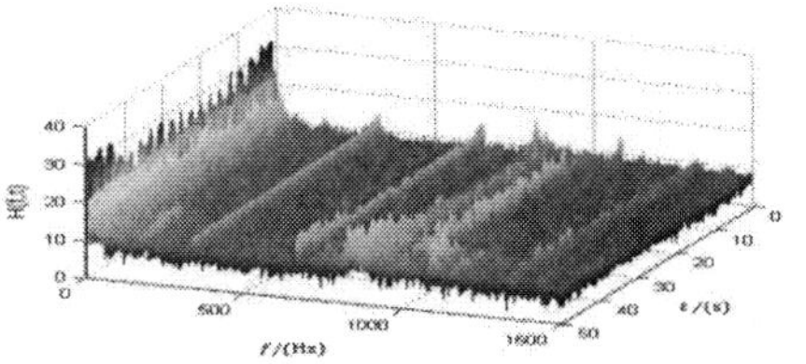

(a)

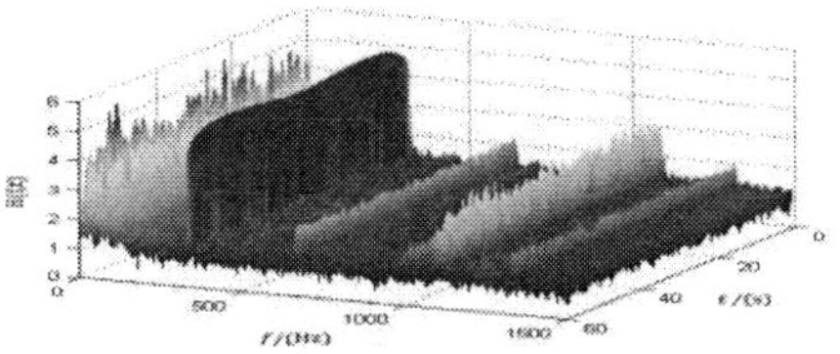

(b)

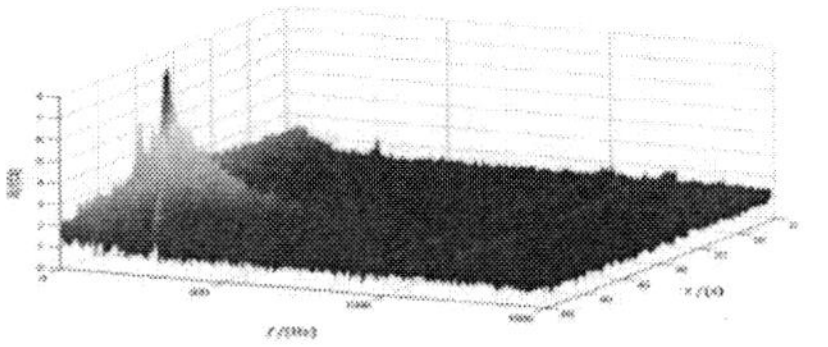

(c)

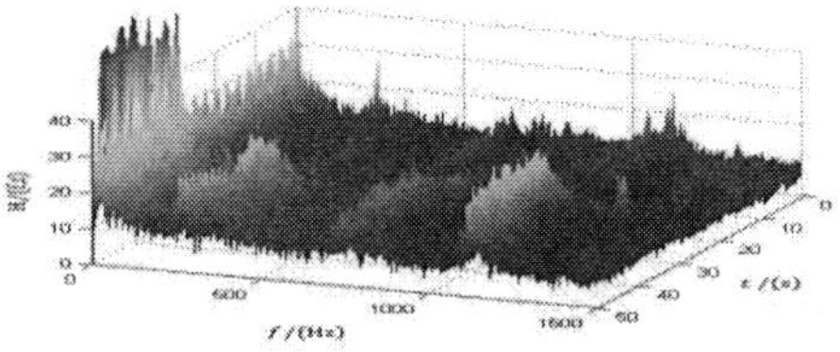

(d)

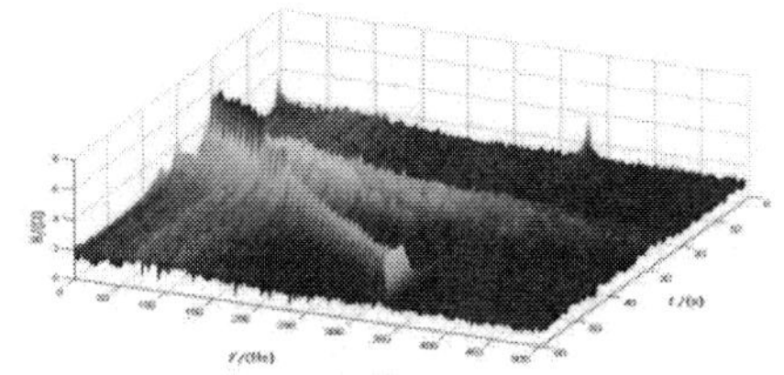

(e)

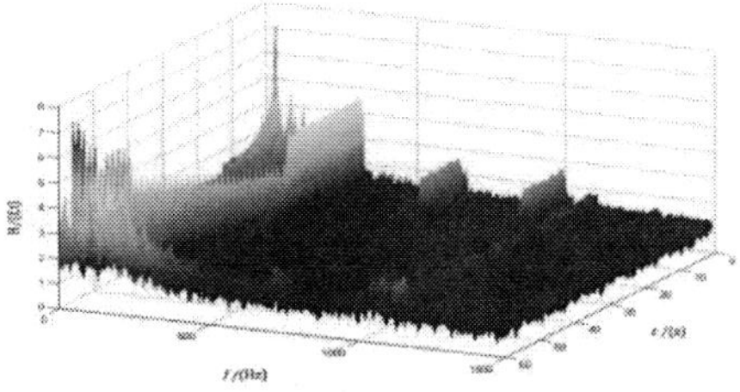

(f)

Figure3. Harmonic wavelet package time-frequency graph of rotor vibration signal in several different work conditions

Figure 3 (a) shows harmonic wavelet packet time-frequency graph of a normally rotating electromagnetic bearing. As can be seen from the figure, the vibration of the rotor are mainly concentrated in low frequency, rotor imbalance vibration in rotor rotating frequency (333Hz) and its 2,3,4 order harmonic are obvious distributed parallel to the time axis , but the magnitude is not very high, this shows the working condition of the rotor is well. In Figure (b) , the vibration amplitude with rotating frequency becomes large significantly, the unbalance vibration became the dominant vibration of the rotor, this means poor balancing of the rotor, the rotor still works stable, although there is a larger vibration amplitude. In Figure (c), the amplitude of the time-frequency graph mutated at about 35 seconds, and soon returned to normal levels, indicating that the intermittent collision failure occurs between rotor and auxiliary bearing. The failure vibration frequency range was 0-500Hz, and the peak was at around 100Hz. In Figure (d) , the amplitude mutated from about 35 seconds in each different frequency bands, the amplitude became significant large, and continued along the time axis , indicating the occurrence of a falling accident of the rotor. Figure (e) is the time-frequency graph during the startup of motor. It could be seen from the figure, from the startup of the motor at about 20 second, the rotating speed was straight up in the first 50 seconds until reached predetermined speed of 20000rpm (333Hz), and maintain that speed along the time axis. Within the first 10 seconds after starting of the motor, the rotor amplitude is large, but the amplitude becomes smaller along the time axis. Throughout the startup process, the vibration of 2,3,4 order harmonics were very obvious. Figure (f) is the time-frequency graph of the process of the motor was stopped, it can be seen from the figure, the rotor began linear deceleration from 20 second, and the rotor was completely stopped at 56 second. After the motor power was turned off, the vibration of the 2,3 order harmonics disappeared immediately. When approaching the rotor's stop, there was significantly large vibration amplitude, and the vibration frequency was mainly concentrated in the 0-250Hz.

V. CONCLUSION

Harmonic wavelet has Box-shape frequency domain characteristic, and it has fast algorithm based on FFT, this makes it gets successful applications in the vibration signal characteristic analysis of magnetic bearing rotor.

The grid time-frequency graph constructed based on harmonic wavelet packet fast algorithm can reflect the intuitive and specific information in each time-frequency point of the vibration signal, indicating the harmonic wavelets time-frequency graph has great value in engineering.

On the basis of harmonic wavelet packet decomposition of magnetic bearing rotor vibration signal, extracting rotor vibration characteristics from time-frequency expansion coefficients can further achieve real-time monitoring and fault diagnosis of the rotor operating conditions .

Acknowledgment

This work was supported by Excellent Young Scientists Award Fund of Shandong Province BS2011ZZ009, Foundation of Shangdong University (2012JC011).

REFERENCES

[1] Varghese B,Parhare S,Gao R,Guo C,Malkin S.Development of a sensor -integrated intelligent grinding wheel for in-process monitoring [J],Annals of the CIRP, 2000, 49(1) :231-234.

[2] Pathare S,Gao R,Varghese B,Guo C,Malkin S.A DSP based telemetric data acquisition system for in-process monitoring of grinding operation [J],IEEE Instrumentation and measurement conference,St, Paul,1998 MN,1:191-196.

[3] Stephane LaChance,Andrew Warkentin,Robert Baure. evelopment of an automated system for measuring grinding wheel wear flats.Journal of Manufactruing Systems. Vol.22/No.2,2003.pp:130-135.

[4] D.E Newland,Harmonic wavelet nanlysis,Proceedings of the Royal Society of London,Vol 448,pp.203-225,Oct.1993.

[5] Sebastien Auchet, Pierre Cherrier, Michel Lacour, Paul Lipinski. A new method of cutting force measurement based on command voltages of active electro-magnetic bearings[J].International Journal of Machine Tools & Manufacture 44(2004).1441-1449.

[6] Mani, D.D.Quinn, M.Kasarda. Active health monitoring in a rotating cracked shaft using active magnetic bearings as force actuators[J]. Journal of Sound and Vibration 294(2006) 454-465.

[7] Aenis,E.Knopf,R.Nordmann. Active magnetic bearings for the identification and fault diagnosis in turbomachinery[J]. Mechatronics 12 (2002).pp:1011-1021.

[8] Process monitoring for a machine tool spindle with magnetic bearings[C]. Seventh International Symp.on Magnetic bearings, August 23-25.2000.195-199.

[9] Feng,B.S.Kim,A.Shih,J.Ni.Tool wear monitoring for micro-end grinding of ceramic materials[J].Journal of Materials Processing Technology 209 (2009) 5110-5116.

[10] E.Kasarda,J,Marshall,R.Prins. Active magnetic bearing based force measurement using the multi-point technique[J]. Mechanics Research Communications 34(2007) 44-53.

A Novel Twin-Shaft Rotor with Active Magnetic Couplings for Vibration Control

Chris Lusty[a], Necip Sahinkaya[a,b], Patrick Keogh[a]

[a] University of Bath, Claverton Down, Bath, UK, C.Lusty@bath.ac.uk
[b] School of Mechanical Engineering, Kingston University, London, UK

***Abstract*—A novel rotor topology is proposed with the goal of reducing vibration in high speed rotor systems. Primarily, reduction in synchronous vibration at critical speeds is targeted. In essence, the novel topology consists of a hollow tube rotor coupled to a secondary non-rotating shaft through a number of actively managed magnetic couplings. An outline of the new topology is provided, along with a finite element analysis comparing the out-of-balance vibration response of the proposed rotor to a comparable simple rotor. Designs for a test rig based on this topology are presented, together with an outline of anticipated testing and potential future development of the idea.**

I. Introduction

In order to alter the vibration characteristics of a rotor, it is proposed to use a novel arrangement of components coupled with active control. The key principle of the concept is to include, in addition to the primary rotor, a secondary "damping shaft". This second shaft runs concentrically through the primary rotor (which must naturally be hollow) and does not rotate. The two shafts are independently supported - the primary rotor on some form of bearing (e.g. rolling element), and the secondary shaft simply clamped.

Mounted along the length of the secondary shaft are one or more magnetic bearings. These magnetic bearings are "inside-out" compared to standard magnetic bearings, in that the stator is on the inside, and the rotor on the outside. In ordinary low speed operation of the rotor (i.e. when vibration is not a problem), these magnetic bearings are not activated, and thus there is no link whatever between the two shafts of the system. However, as critical speeds are approached by the rotor, the magnetic bearings can be activated, coupling the two shafts and changing the vibration behaviour of the rotor.

By altering the axial position and/or number of magnetic bearings activated at any given time, the overall system characteristics can be altered, and thus a variety of frequencies of vibration mitigated. Via active control, it is therefore anticipated that such a system will offer a powerful and flexible way to control excessive vibration.

II. Prior Work in Similar Areas

A good overview of the mechanisms leading to synchronous rotor vibration is provided by Nelson [1]. Reducing vibration amplitudes theoretically permits higher speed applications, and thus smaller, higher power density machines. From a research point of view, active rotor vibration reduction techniques may be considered to split into three distinct groups:

1. Those concerned with active balancing, i.e. redistributing mass around the rotor system to remove the driving force of synchronous vibration. These methods tend to be adaptations of off-line balancing methods. Such techniques have been a topic of interest since Van de Vegte's paper [2] proposing the design of an active balancing head. Various developments in both control logic and physical design have been presented since [3]–[8]. A good review of this field is provided by Zhou and Shi [9].

2. Those concerned with directly applying forces to the rotor; generally this is by means of lateral forces either at the supporting bearings or mid-span along the shaft. Direct force application methods are primarily the domain of the magnetic bearing. Much work has been done on optimising the control strategies for the bearings to minimise vibration, but they are necessarily limited in their performance by constraints of physical size and load capacity. The function and application of magnetic bearings is comprehensively considered by Maslen and Schweitzer [10], with a recent review paper of research topics in the field provided by Schweitzer [11]. Other types of active bearing have also been considered, including squeeze film dampers, gas bearings and laterally actuated passive bearings [12]–[15].

In terms of applying forces at mid-span (i.e. non-bearing) locations, a significant amount of research has been done surround the concept of the Active Magnetic Damper (AMD). An early example is provided by Nikolajsen *et al.* [16], who model a marine propulsion shaft, one of whose members rotates super-critically. By applying an AMD to this shaft, Nikolajsen demonstrates substantial reduction of synchronous vibration amplitude. Nikolajsen's work was extended by Kasarda *et al.* [17] with the addition of disks to an otherwise simple rotor. Kasarda proves that not only does an AMD have significant scope to reduce the amplitude of the vibration, but also that the exact placement of the AMD is not critical to its successful performance. Particular comparison is drawn between locating the AMD near mid-span (i.e. point of maximum deflection for first mode vibration), near a disk not at mid span, and close to one of the supporting bearings. In a later paper, Kasarda *et al.* [18] focus on the potential for AMDs to be used in the control of sub-synchronous vibrations cause by external influences.

The common theme throughout all of the research presented on AMDs is that they are external-stator devices, and they are rigidly fixed to the ground/machine base. This differs from the work presented in this paper, which uses internal-stator bearings mounted on a flexible support shaft.

3. Those that are less clearly defined in nature. Techniques here generally involve changing some physical property of the system (other than mass redistribution), for example stiffness or geometry, and thus alter vibration response.

The technique proposed in this paper falls somewhere between this category and the "direct force application" category (2). Here the aim is to alter the definition of the rotor system in such a way as to bypass vibration peaks, which involves an element of lateral force application to the rotor. A previous attempt at geometry-alteration method is presented by Ortega *et al.* [19], where it is proposed to axially move one of the support bearings in order to change the rotor's effective length. Conversely this paper advocates activating *additional* bearings at other axial locations.

III. The Proposed Topology

Figure 1 shows a simplified illustration of the proposed rotor layout. The rotor is a hollow tube of soft magnetic metal (e.g. steel) supported on external rolling element, or other passive bearings. The damping shaft is supported entirely separately by non-rotating supports, either pin-jointed or clamped. The material for the non-rotating shaft is chosen on a basis of its stiffness qualities, and it need not have appreciable magnetic permeability. The fact that this shaft does not rotate gives the advantage that any imbalance present in the member does not add to or complicate the dynamics of the primary rotor when the two shafts are coupled. The active couplings are achieved by internal-stator magnetic bearings. A cross sectional view of one of these is provided in Figure 2.

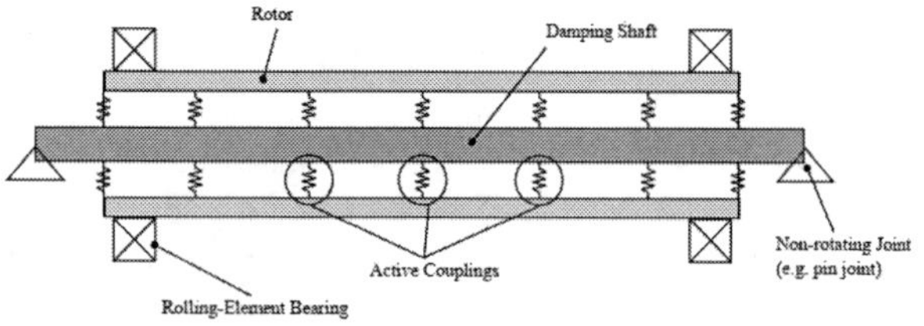

Figure 1. Schematic view of proposed novel rotor topology. Note that the number/position of "active couplings" is arbitrary.

IV. Magnetic Bearing Considerations

In addition to the new ideas regarding rotor topology presented here, significant consideration is given to the magnetic bearings that are to form the active coupling element between the shafts. In the vast majority of magnetic bearing systems used today, the stator of the bearing is outside of the rotor (external stator systems). This arrangement is chosen because it removes size constraints from the design of the stator allowing more powerful bearings to be designed. It is clearly also infeasible to use internal stator bearings on either solid shafts or very thin shafts. However, the external stator does have disadvantages. It takes up significant space along the shaft's length that may otherwise be either used for mounting other components. Furthermore, in certain applications there may not be space in the vicinity of the shaft to use an external stator bearing.

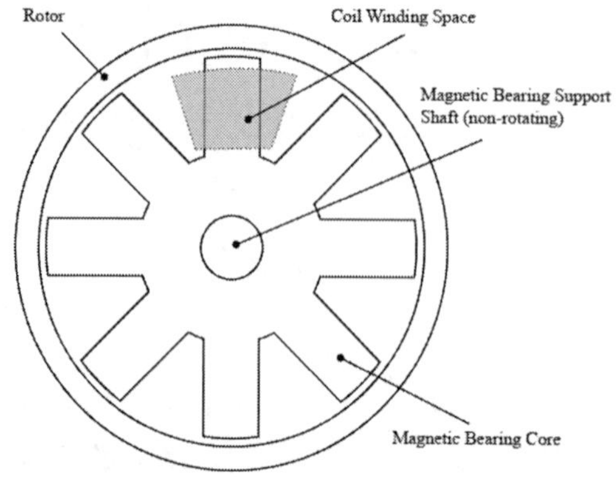

Figure 2. Schematic cross-section of internal magnetic bearing

The design presented here uses custom designed internal stator magnetic bearings. In addition to addressing the space issues presented by by external stator bearings, the circumstance of the internal bearings being mounted on a flexible shaft is central to the design and exploits possibilities relating to the overall system dynamics not considered by other work on active magnetic damping.

A. Stator Core Material

It is almost universal practice in magnetic bearing design to construct the cores out of laminated electrical steel. This provides a high level of magnetic permeability and good saturation strength, while reducing energy losses due to eddy current build up. However, design geometry possibilities are considerably limited when working with laminated steel. In particular, to facilitate manufacture it is desirable for the design to be prismatic in the stacking direction (to enable every individual lamination to be identical). It is well documented that this requirement makes the manufacture of homopolar magnetic bearings more complex and costly than heteropolar designs. Indeed despite the fact that homopolar designs incur lower eddy current losses than heteropolar arrangements, the difficulty of manufacture leads to the majority of magnetic bearings using the less efficient, but easier to manufacture, heteropolar layouts.

To overcome this trade off, it is proposed that laminated steel is replaced with a powder metal based Soft Magnetic Composite (SMC) as the core material for a magnetic bearing. SMC offers several clear advantages over laminated steel:

- It has effectively isotropic lamination and is thus capable of supporting alternating magnetic flux in any direction without eddy current build up
- It has the ability to be formed into complex 3D shapes

For this project, the adoption of SMC allows the magnetic bearings to use a compact, single piece homopolar design similar to that shown in Figure 3. Fabricating this design directly from laminated steel would incur greater eddy current losses than an SMC core as not all directions are insulated in the steel. Fabricating a comparable equally compact homopolar core in a fully insulated way from a multi-part laminated steel design would be substantially more complex.

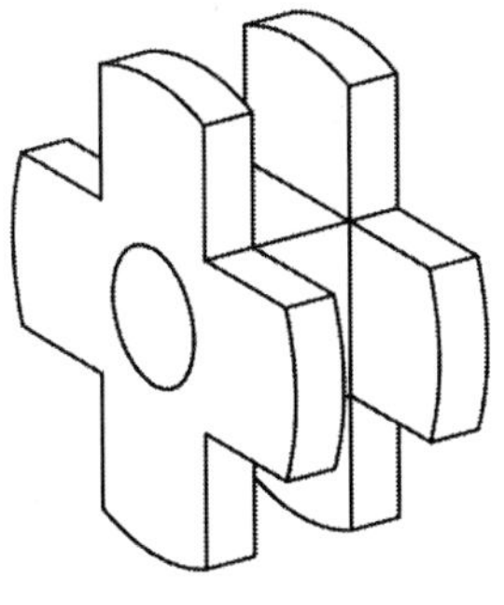

Figure 3. An internal-stator homopolar bearing core

In general, SMC is fabricated from iron powder, chemically treated to leave each individual grain with a thin layer of electrically insulating material around the outside. The powder is then compressed in a mold to form the final shape of the component required. Post forming heat treatments are also applied to relieve stress from the compaction process.

Of course, the compaction-forming of powder-metal components is generally only financially viable on production scale runs, not for small/one-off jobs. For this reason, prefabricated discs of a version of the SMC material specially treated to have good machinability properties are used. It is noteworthy that in altering the material to give good machinability, the magnetic properties of the SMC are slightly adversely affected.

Of course, one still has the issues of space limitation when designing an internal-stator magnetic bearing. In general this means one is limited by the load capacity of the bearing. This is, however, by no means a prohibitive difficulty in this design, owing to the fact that the bearings are not required to support the mass of the rotor (which is dealt with by external rolling element bearings).

V. Numerical Modelling

A. Finite Element Model

The following derivation is based on standard Finite Element rotor modelling, described by Nelson [20]. The basic unit of the finite element model used is an eight degree-of-freedom rotor section having degrees-of-freedom collected into a single vector:

$$\mathbf{q}(t) = (x_1(t), y_1(t), \theta_1(t), \phi_1(t), x_2(t), y_2(t), \theta_2(t), \phi_2(t))^T$$

where (x_i, y_i) are node lateral displacement, and (θ_i, ϕ_i) are node angular displacements.

The element equation of motion has the form:

$$\mathbf{m}\ddot{\mathbf{q}} - \Omega\mathbf{g}\dot{\mathbf{q}} + \mathbf{k}\mathbf{q} = \mathbf{f} \qquad (1)$$

Here $\mathbf{m}$, $\mathbf{k}$ and $\mathbf{g}$ are mass, stiffness and gyroscopic matrices respectively. $\mathbf{f}$ is a column vector of forces applied to each of the degrees of freedom in $\mathbf{q}$. Ω is the speed of shaft rotation (rad/s).

The complete equation of motion for a multi-element rotor is:

$$\mathbf{M}\ddot{\mathbf{Q}} - \Omega\mathbf{G}\dot{\mathbf{Q}} + \mathbf{K}\mathbf{Q} = \mathbf{F} \qquad (2)$$

where $\mathbf{M}$, $\mathbf{K}$, $\mathbf{G}$, $\mathbf{F}$ and $\mathbf{Q}$ are the global equivalents of the elemental terms $\mathbf{m}$, $\mathbf{k}$, $\mathbf{g}$, $\mathbf{f}$ and $\mathbf{q}$, respectively.

This is fairly standard analysis in rotor dynamics. From this point, modelling of the coupled shaft system is undertaken as follows. Firstly, the $\mathbf{Q}$ vector is extended to include degrees of freedom for both the rotor ($\mathbf{Q_R}$) and the secondary shaft ($\mathbf{Q_S}$):

$$\mathbf{Q} = \begin{bmatrix}\mathbf{Q_R} & \mathbf{Q_S}\end{bmatrix}^T$$

Using this definition, an expanded form of Equation (2) is presented, which demonstrates where coupling terms are applied between the two shafts:

$$\begin{bmatrix}\mathbf{M_R} & 0 \\ 0 & \mathbf{M_S}\end{bmatrix}\begin{bmatrix}\ddot{\mathbf{Q}}_\mathbf{R} \\ \ddot{\mathbf{Q}}_\mathbf{S}\end{bmatrix} + \begin{bmatrix}\Omega\mathbf{G} & 0 \\ 0 & 0\end{bmatrix}\begin{bmatrix}\dot{\mathbf{Q}}_\mathbf{R} \\ \dot{\mathbf{Q}}_\mathbf{S}\end{bmatrix} + \dots$$
$$\dots\begin{bmatrix}\mathbf{K_R}+\mathbf{K_C} & -\mathbf{K_C} \\ -\mathbf{K_C} & \mathbf{K_S}+\mathbf{K_C}\end{bmatrix}\begin{bmatrix}\mathbf{Q_R} \\ \mathbf{Q_S}\end{bmatrix} = \begin{bmatrix}\mathbf{F_R} \\ \mathbf{F_S}\end{bmatrix} \qquad (3)$$

Note that due to its non-rotation, there is no gyroscopic term for the secondary shaft. The $\mathbf{K_C}$ terms in Equation (3) represent the stiffness of the magnetic coupling between the two shafts.

Equation (3) is representative of the kind of coupling shown in Figure 1, i.e. simple spring couplings, and this is indeed the kind of coupling used in the results presented in this paper. However, due to the actively controlled nature of the magnetic bearings, it is possible to implement other forms of coupling. For instance, damping can be added by applying a control force proportional to the relative velocity, rather than the relative displacement, of the two shafts. This may be modelled by modifying the second matrix term in Equation (3) to:

$$\begin{bmatrix}\Omega\mathbf{G}+\mathbf{C_C} & -\mathbf{C_C} \\ -\mathbf{C_C} & \mathbf{C_C}\end{bmatrix}$$

where $\mathbf{C_C}$ includes a damping rate. For the present case however, the model is left as in Equation (3).

To achieve a frequency analysis of a rotor subject to any given forcing, one can take the Laplace transform of Equation (3), set $s = j\omega$, and rearrange into the form of:

$$\mathbf{Q}(j\omega) = (-\mathbf{M}\Omega^2 - \Omega^2\mathbf{G}j + \mathbf{K})^{-1}\mathbf{F}(j\omega) \qquad (4)$$

It can be seen that for a specific rotor with a specific out of balance mass, all the terms on the right hand side of Equation (4) are known, and thus the response in the various degrees of freedom may be calculated.

B. Case Study

The application of this model to a particular system is now presented to illustrate the effectiveness of the internal shaft/magnetic bearing system. The rotor modelled is topologically as shown in Figure 1, with the difference that just a single magnetic bearing at the axial mid-point is used. The rotor is modelled as a 1 m long steel tube, of outside diameter 80 mm and wall thickness 2 mm. It is acknowledged that a 2 mm thickness would be insufficient to carry the full flux of the magnetic bearing at moderate to high field strengths, and thus an external steel collar at the bearing location would be needed in practice.

The inner (non-rotating) shaft used in the model is a 1.2 m long solid aluminium bar with a diameter of 20 mm. The supports for both rotor and secondary shaft are modelled as pin joints. The magnetic bearing is modelled to have a 1 mm clearance between its outer diameter and the inner diameter of the rotor giving a maximum force capability of 500 N and leading to a maximum available stiffness of 5×10^5 N/m.

A frequency response was calculated for this rotor with an arbitrary out-of-balance mass at a point along its length. For the first simulation, the rotor was entirely decoupled from the secondary shaft (magnetic bearing switched off). The simulation was then repeated with the magnetic bearing activated, and thus the two shafts coupled. Figure 4 shows both frequency responses. It is observed that the plot of the response of the uncoupled rotor includes two critical speeds at around 110 Hz and 440 Hz. These can easily be verified analytically using:

$$\omega_n = a_n \sqrt{\frac{EI}{\mu l^4}} \tag{5}$$

For a simply supported beam, a_n takes values of $(i\pi)^2$ for ω_i.

It is clear that the peak at $\simeq$ 110 Hz in the simple rotor response is absent in the coupled system response. Thus if the magnetic bearing is activated at, say 85 Hz, and then deactivated at around 165 Hz, that entire vibration peak is avoided. Note that the frequency responses are RMS averages of the responses at all nodes.

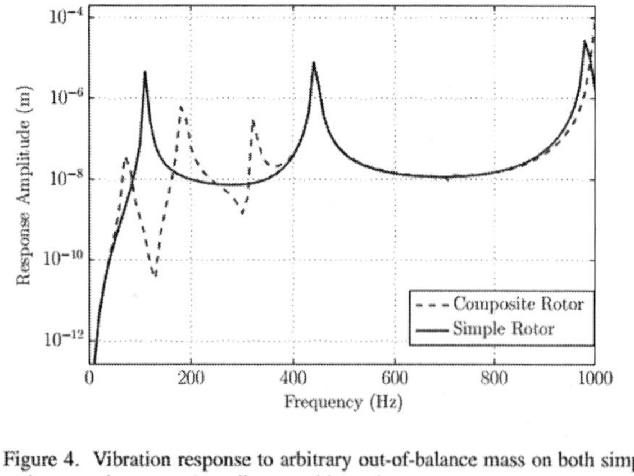

Figure 4. Vibration response to arbitrary out-of-balance mass on both simple and composite rotors (coupling at mid-span)

It is of interest to examine exactly what vibration, in the sense of mode shapes and amplitudes, the rotor (and secondary shaft, where relevant) is experiencing at key frequencies. For instance, if the magnetic bearing is activated at 90 Hz to circumvent the rotor's first critical speed as suggested by Figure 4, the response seen in Figure 5 is obtained. The amplitude of the rotor's vibration is much the same both before and after bearing activation, and it remains in a classic "first-mode" shape in both conditions. This agrees with this frequency being approximately the intersection of the two vibration responses shown in Figure 4. It is observed that the secondary shaft takes a shape having a combination of first and third modes. This may be explained by examining the natural frequencies of the shaft itself, calculated analytically via Equation (5). The shaft's second natural frequency (ignoring the effect of the magnetic bearing which will tend to reduce the value) is around 110 Hz. Of course the shaft is prevented from exhibiting any second mode vibration shapes by the fact that the excitation is applied at the centre - a node in all even mode shapes. Thus the combination is of first and third shapes.

At the first critical speed of the uncoupled rotor, 110 Hz, the mode shape diagram in Figure 6 reinforces just how much of an effect coupling to the second shaft has in terms of amplitude reduction. Indeed to see the shapes of the members when coupling is active, it is necessary to plot a separate zoomed-in figure of those responses, such is the difference in their relative scales. This is presented in Figure 7. It can be seen that the secondary shaft is once more vibrating with a combination of first and third modes, with increasing emphasis on the third mode as the exiting frequency nears the shaft's own third natural frequency.

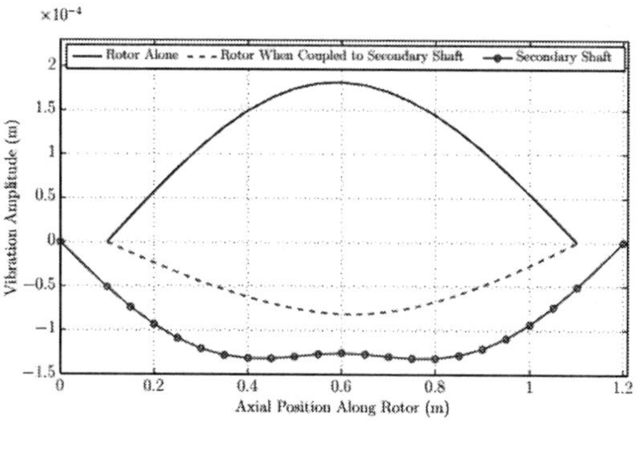

Figure 5. Response shapes at 90 Hz

It is noted that the arrangement for which Figure 4 shows the vibration response does not offer any avoidance of the second critical speed ($\simeq$ 440 Hz) as it does for the first. This is expected given that the mode shape of the second natural frequency or the rotor will have a node at central span, and thus a coupling located here is of no consequence whatever. This is the reason for proposing either multiple or mobile (or both) couplings. Figure 8 shows the frequency response of the exact same compound rotor, but this time with magnetic bearings activated near the $^1/_4$ and $^3/_4$ axial points rather than the centre. This arrangement allows for successful avoidance of the second resonant peak in the simple rotor response. In this

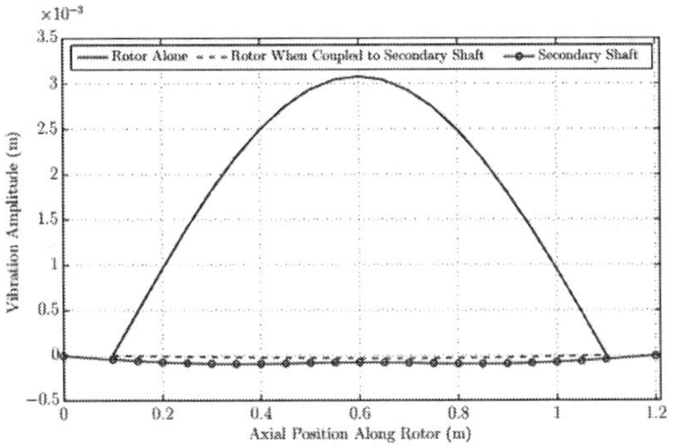

Figure 6. Response shapes at 110 Hz

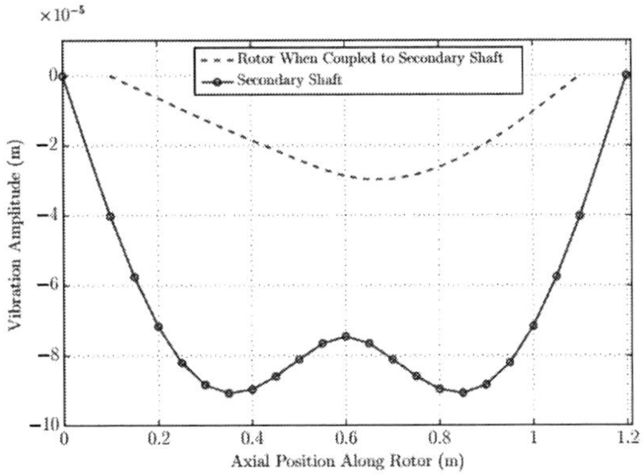

Figure 7. Response shapes at 110 Hz - magnification to show coupled member shapes

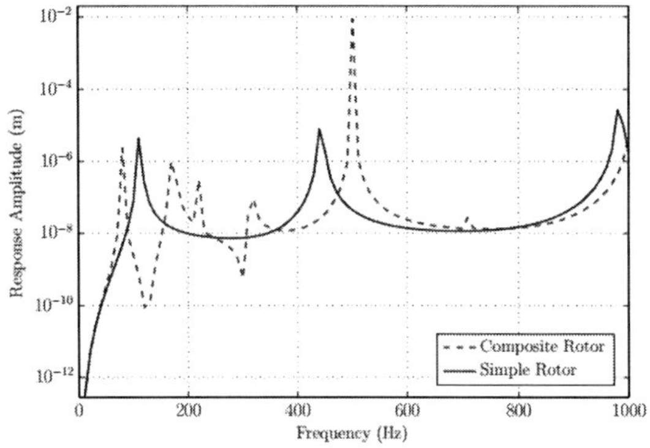

Figure 8. Vibration response to arbitrary out-of-balance mass on both simple and composite rotors (coupling around 1/4 and 3/4 axial points)

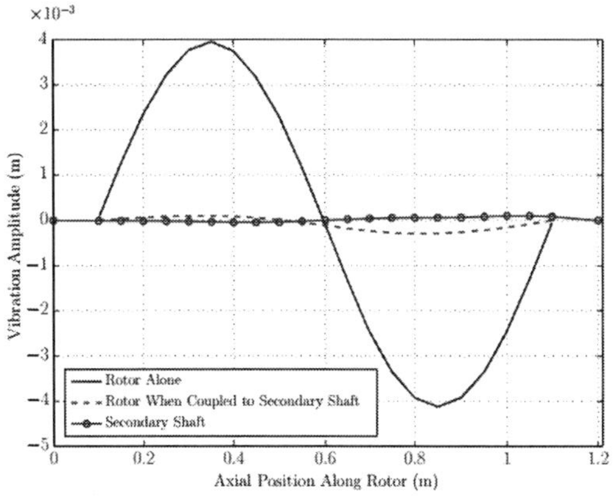

Figure 9. Mode Shapes at 440 Hz with 1/4 and 3/4 point coupling

model, it is apparent that such coupling also allows avoidance of the first critical speed. A mode shape plot is provided for the system at 440 Hz in Figure 9. One clearly sees the second mode shape of the rotor vastly decreased in amplitude by the coupling. Note that the slight asymmetry observable in the mode shape plots can be attributed to the fact that the out of balance mass being simulated is applied to an off-centre node in the finite element model (to allow symmetric mode shape responses to be observed).

C. Motion of Inner Shaft

A question arises regarding the stiffness and/or motion of the inner shaft. Namely, it is vital to ensure that no contact will occur between the rotor and the magnetic bearing. It is noted that providing the displacement of the central shaft relative to the displacement of the rotor does not cause a collision between the magnetic bearing and the rotor, it is permissible for this shaft to vibrate.

To address the collision concern, an out-of-balance mass is simulated on the rotor, and the amplitude of the displacement of the magnetic bearing relative to the corresponding (adjacent) location of the rotor is plotted. When considering the resultant plot shown in Figure 10, two facts are borne in mind. Firstly, at rest there is designed to be a 1 mm gap between the AMB and the rotor, and secondly the magnetic bearing is only to be engaged at certain frequencies. Specifically, the magnetic bearing is only to be switched on at frequencies where the amplitude of rotor vibration is reduced. For the rotor model referred to in this paper, with a magnetic bearing located at mid span, avoiding excessive vibration at the first critical speed requires engaging the magnetic bearing between around 85 and 160 Hz. The corresponding speed range on Figure 10 is highlighted. It can then be observed that at no point in the range where the bearing is actually active does the relative displacement come even close to the 1 mm boundary. Thus there is no collision.

VI. Practical Testing

A test rig has been designed to test the validity of the concept presented - schematic overviews are provided in Figures 11 and 12. This is currently under construction; when completed it should provide results to validate the numerical simulation. Initial testing is anticipated to consist of impulse testing of both simple and composite rotor arrangements to prove the difference in fundamental frequency. After that it is hoped to complete supercritical rotation tests to conclusively demonstrate the merit of the new topology. The rotor of the

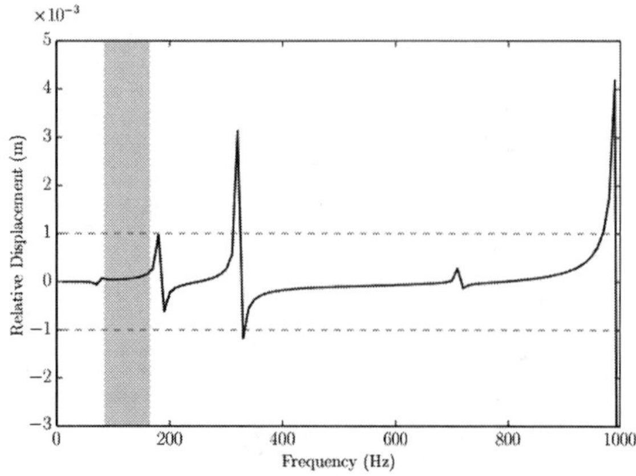

Figure 10. Displacement of magnetic bearing relative to corresponding location of rotor at various frequencies

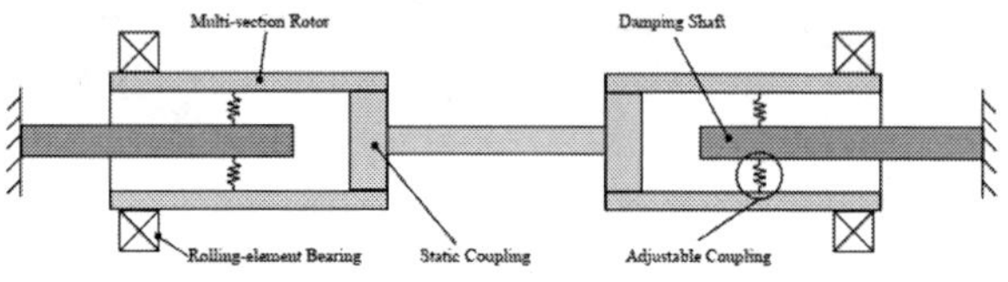

Figure 11. Schematic of novel rotor test rig

rig has been specially designed to allow supercritical rotation at relatively low speeds by building the rotor out of several sections, including a comparatively flexible middle section.

VII. Conclusions

A novel concept of coupling a rotor to a secondary shaft by way of internal-stator Active Magnetic Bearings has been proposed. Finite element modelling has been performed for such a rotor, and responses to an out-of-balance mass are plotted. It is clearly illustrated that it is possible to design a system whereby the critical speeds of the primary rotor are substantially altered by coupling it so a secondary shaft. Therefore by actively engaging and disengaging the magnetic couplings, it is possible to avoid the excessive vibration associated with the rotor's critical speeds. Furthermore, mode shape plots at rotor critical speeds are presented illustrating the difference in amplitude between the simple and coupled rotors, with maximum vibration amplitude reductions of around an order of magnitude observed. A design concept for a practicable laboratory-scale rig to confirm the finite element results is presented.

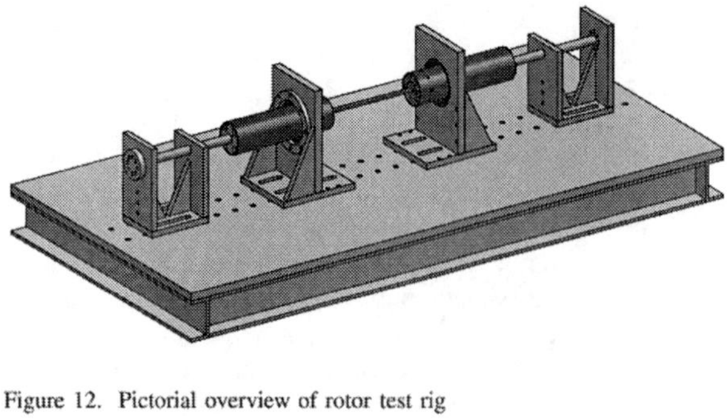

Figure 12. Pictorial overview of rotor test rig

VIII. Acknowledgements

The authors are grateful for the award of a James Dyson Foundation Studentship to support this research.

References

[1] FC Nelson. Rotor dynamics without equations. *International Journal of COMADEM*, 10(3):2, 2007.

[2] J Van De Vegte. Continuous automatic balancing of rotating systems. *IMechE Journal of Mechanical Engineering Science*, 6(3):264–269, 1964.

[3] AS Sekhar and Sarangi Debraj. Online balancing of rotors. In *National Conference on Mechanisms and Machines*, 2003.

[4] J Van de Vegte and RT Lake. Balancing of rotating systems during operation. *Journal of Sound and Vibration*, 57(2):225–235, 1978.

[5] J Van De Vegte. Balancing of flexible rotors during operation. *IMechE, Part C, Journal of Mechanical Engineering Science*, 23(5):257–261, 1981.

[6] RED Bishop and GML Gladwell. The vibration and balancing of an unbalanced flexible rotor. *IMechE Journal of Mechanical Engineering Science*, 1(1):66–77, 1959.

[7] RED Bishop. On the possibility of balancing rotating flexible shafts. *IMechE, Part C, Journal of Mechanical Engineering Science*, 24(4):215–220, 1982.

[8] Stephen W Dyer and Jun Ni. Adaptive influence coefficient control of single-plane active balancing systems for rotating machinery. *ASME Journal of Manufacturing Science and Engineering*, 123(2):291–298, 2001.

[9] Shiyu Zhou and Jianjun Shi. Active balancing and vibration control of rotating machinery: a survey. *Shock and Vibration Digest*, 33(4):361–371, 2001.

[10] Eric H Maslen and Gerhard Schweitzer. *Magnetic Bearings: Theory, Design, and Application to Rotating Machinery*. Springer, 2009.

[11] G. Schweitzer. Applications and research topics for active magnetic bearings. In *IUTAM Symposium on Emerging Trends in Rotor Dynamics*, pages 263–273. Springer, 2011.

[12] CR Burrows, MN Sahinkaya, and OS Turkay. An adaptive squeeze-film bearing. *ASME Journal of Tribology*, 106(1):145–151, 1984.

[13] Ilmar F Santos. Trends in controllable oil film bearings. In *IUTAM Symposium on Emerging Trends in Rotor Dynamics*, pages 185–199. Springer, 2011.

[14] Jinhao Qiu, Junji Tani, and Taekyu Kwon. Control of self-excited vibration of a rotor system with active gas bearings. *ASME Journal of Vibration and Acoustics*, 125(3):328–334, 2003.

[15] Ricardo C Simões, Valder Steffen, Johan Der Hagopian, and Jarir Mahfoud. Modal active vibration control of a rotor using piezoelectric stack actuators. *Journal of Vibration and Control*, 13(1):45–64, 2007.

[16] JL Nikolajsen, R Holmes, and V Gonhalekar. Investigation of an electromagnetic damper for vibration control of a transmission shaft. In *Institution of Mechanical Engineers Proceedings*, volume 193, 1979.

[17] MEF Kasarda, PE Allaire, RR Humphris, and LE Barrett. A magnetic damper for first-mode vibration reduction in multimass flexible rotors. *ASME Journal of Engineering for Gas Turbines and Power*, 112(4):463–469, 1990.

[18] MEF Kasarda, H Mendoza, RG Kirk, and A Wicks. Reduction of subsynchronous vibrations in a single-disk rotor using an active magnetic damper. *Mechanics Research Communications*, 31(6):689–695, 2004.

[19] Andrés Blanco Ortega, Francisco Beltrán Carbajal, Gerardo Silva, and Marco Antonio Oliver Salazar Navarro. Active vibration control of a rotor-bearing system based on dynamic stiffness control activo de vibraciones en un sistema rotor-chumaceras basado en la rigidez dinámica. *Revista Facultad de Ingeniería Universidad de Antioquia*, (55):125–133, 2010.

[20] HD Nelson and JM McVaugh. The dynamics of rotor-bearing systems using finite elements. *ASME Journal of Engineering for Industry*, 98:593, 1976.

Experimental modal analysis of a gyroscopic rotor in active magnetic bearings

Gudrun Mikota [a], Andreas Josef Pröll [b], Siegfried Silber [c]

[a] Johannes Kepler University Linz, Altenbergerstrasse 69, 4040 Linz, Austria, gudrun.mikota@jku.at
[b] Johannes Kepler University Linz, Altenbergerstrasse 69, 4040 Linz, Austria
[c] Linz Center of Mechatronics GmbH, Altenbergerstrasse 69, 4040 Linz, Austria

Abstract—**For a gyroscopic rotor in active magnetic bearings, the rigid body natural frequencies and mode shapes are investigated. An impulsive excitation force is generated by two rectangular pulses in the current control reference signal. With vibration measurements available in two perpendicular directions for each bearing, experimental modal analyses are performed at rotational speeds up to 100000 rpm. Identification results demonstrate the gyroscopic split of natural frequencies.**

I. Introduction

The dynamic behavior of rotors differs from the behavior of non-rotating structures. The multi-degrees-of-freedom description of a rotor is characterized by non-symmetric system matrices arising from gyroscopic and circulatory effects. An axially symmetric rotor exhibits double natural frequencies at rest, which are separated by the gyroscopic split from the onset of rotation. Such effects may be modeled and predicted in theory, but the results will depend on the accuracy of material parameters and the underlying properties of the bearings. As an alternative, natural frequencies and mode shapes can be identified by experimental modal analysis, which has been applied to rotors since first experiments described in [1]. The identification requires measured frequency response functions between excitation force and vibration response. Compared to non-rotating structures, rotors are difficult to excite; the hammer blow [1], shaker excitation with an auxiliary bearing [2], rotating unbalance [3], and electromagnetic exciters [4] have all been used with different effort, influence on the dynamic behavior of the rotor, and accuracy of the measured excitation force.

Active magnetic bearings offer another way to apply controlled excitation forces. There have been some attempts to use them for experimental modal analysis. In [5], sophisticated a test stand is described, and it is suggested to determine the magnetic force from measurements of the magnetic flux; however, the resulting modal parameters are missing. Lee, Ha, and Kim [6] described an experimental modal analysis of a shaft-type rotor in active magnetic bearings. The magnetic forces are measured by piezoelectric force transducers; together with voltage and displacement signals, they are used to identify the dynamic current to voltage relationship without directly measuring the control current. An experimental modal analysis of the closed loop system is performed with bandlimited random noise or sinusoidal excitation signals. At a rotational speed of 6800 rpm, four natural frequencies are

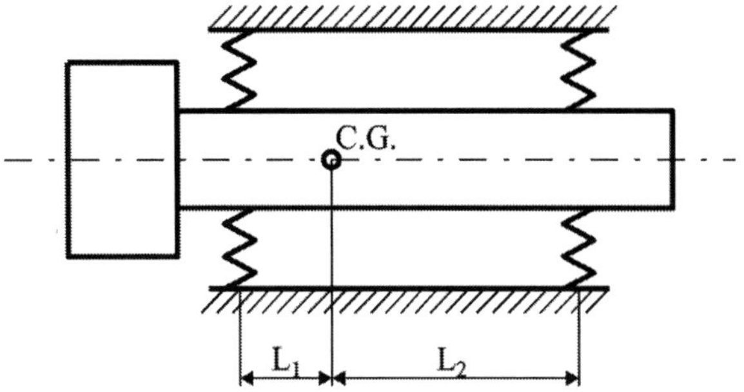

Figure 1. Schematic of rotor supported by active magnetic bearings.

identified together with the corresponding damping ratios. Backward and forward frequencies only differ by a few per cent and are extracted through complex modal testing [7].

In this paper, active magnetic bearings are used to determine the four rigid body modes of a gyroscopic rotor depending on the rotational speed. The magnetic bearing system includes an underlying current controller. An impulsive excitation force is generated by two rectangular pulses in the reference current; the resulting impulse in the actual current is measured in open loop and then applied to the closed loop system. The rotor is tested at rotational speeds up to 100000 rpm to obtain an experimental record of the gyroscopic effect.

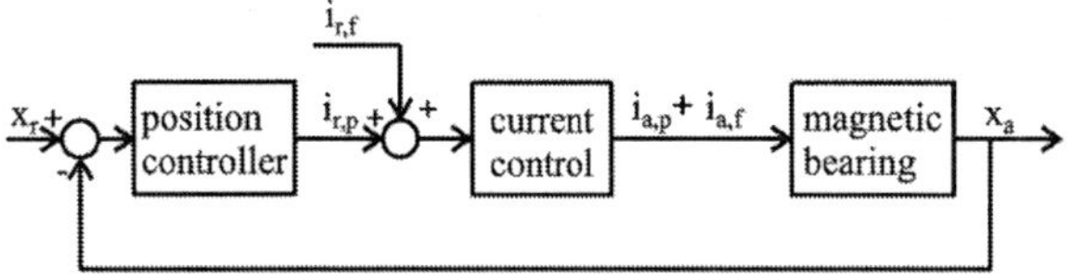

Figure 2. Block diagram of position control circuit.

II. Experimental Setup

A schematic of the experimental setup is shown in Fig. 1. The rotor consists of a shaft that is supported by two active magnetic bearings and carries a cantilevered disc at one of its ends. Position sensors are situated immediately next to the bearings and measure the position of the rotor in two perpendicular directions. The rotor's weight is supported by permanent magnets. The position of the rotor is stabilized by the active magnetic bearing position control in two perpendicular directions for each bearing. It is assumed that in the immediate vicinity of the equilibrium position, the magnetic force from the bearing coils can be approximated by a linear function of coil current and position.

The system under investigation is constituted by the rotor in its position controlled active magnetic bearings. The same active magnetic bearings are used to generate the excitation force, which is possible by adding a defined impulse to the position control current. Figure 2 shows the respective block diagram. A suitable reference signal $i_{r,f}$ is added to the reference current $i_{r,p}$ from the position controller; assuming linear behavior of the current control, the resulting impulse $i_{a,f}$ will be added to the actual position control current $i_{a,p}$. To determine the required reference signal $i_{r,f}$, the dynamic behavior of the current control is tested separately with the position control switched off for a few milliseconds.

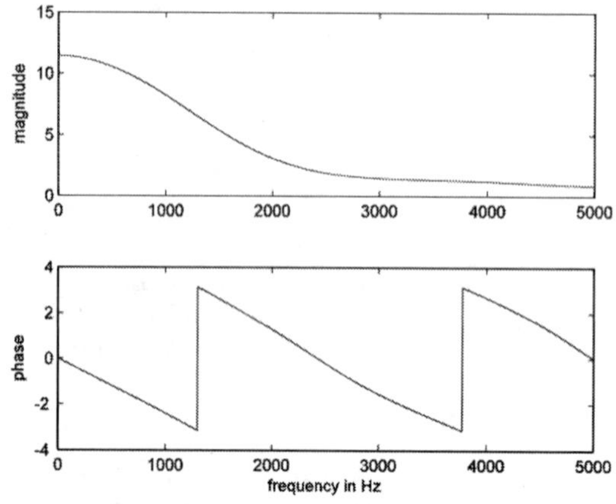

Figure 5. Frequency spectrum of actual current for excitation.

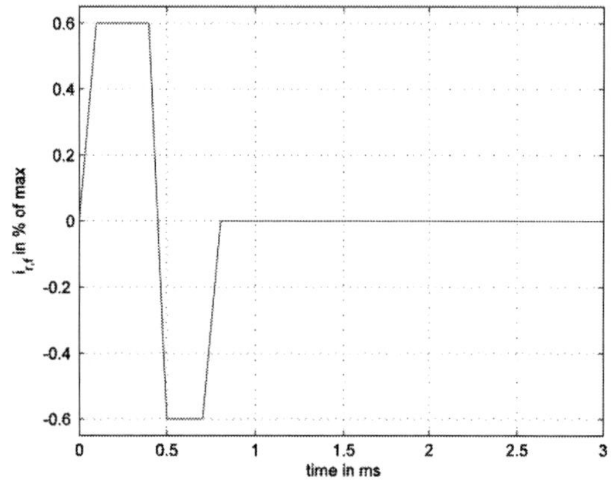

Figure 3. Reference current for excitation.

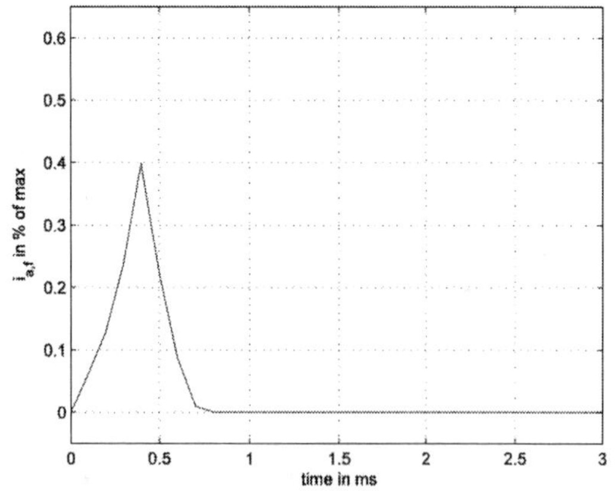

Figure 4. Actual current for excitation.

III. Excitation Signal

To obtain an excitation that is easy to implement in the existing software, an impulsive force signal is sought, which should be broadband enough to cover the frequency range of interest. Figure 3 shows the selected reference signal $i_{r,f}$; it consists of two rectangular pulses of different signs and lengths. The resulting impulse $i_{a,f}$ after the current control is depicted in Fig. 4. The shape of the impulse is nearly triangular, and its frequency spectrum is shown in Fig. 5. The impulse $i_{a,f}$ in the bearing coil current is assumed to generate a proportional magnetic force.

The excitation can be applied in two perpendicular directions per bearing, which is done in separate tests. Since measurements are made for a number of rotational speeds, stepped sine testing would be a tedious procedure. With random noise, there would be no simple control over the maximum amplitude of the response. Compared to chirp, impulsive excitation is much easier to implement and is therefore used although moderate accuracy must be expected.

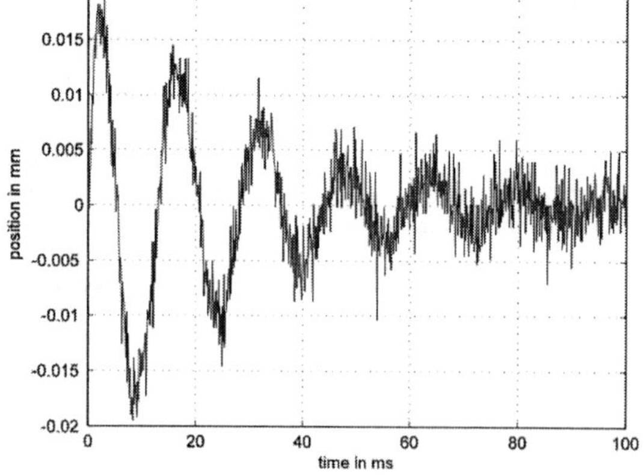

Figure 6. Vibration response at bearing 1, direction x, rotor at rest.

IV. VIBRATION RESPONSES

To determine the natural frequencies and mode shapes of the rotor, the impulsive force is applied in one degree of freedom, while vibration responses are measured in all degrees of freedom. The active magnetic bearings provide four position sensors so that a rigid body model with four degrees of freedom can be studied. Figure 6 shows the vibration response at bearing 1, direction x, of the rotor at rest with the impulsive force applied in same degree of freedom; to extract the relevant transient process, an exponential window is applied to the signal, the result of which is displayed in Fig. 7. The application of an exponential window introduces additional damping into the system; as natural frequencies and mode shapes are hardly changed, this is common practice to avoid leakage errors with impulsive excitation [2].

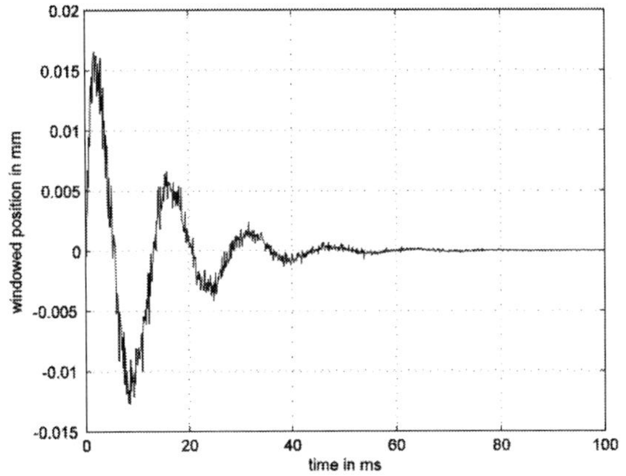

Figure 7. Vibration response at bearing 1, direction x, rotor at rest, after application of an exponential window.

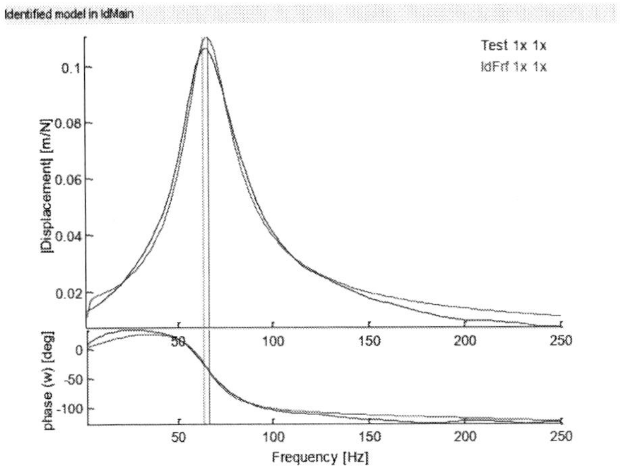

Figure 8. Measured and estimated point frequency response function, rotor at rest.

V. MODAL PARAMETERS

Natural frequencies and mode shapes of the rotor are determined at rest and at rotational speeds of 20000, 40000, 60000, 80000, and 100000 rpm. In each case, a single test provides four position signals which contain the vibration responses. With these signals windowed, the frequency response functions with respect to the impulsive force signal are calculated by Discrete Fourier transform, averaged over four measurements, and passed on to the Matlab Structural Dynamics Toolbox for modal parameter extraction [8].

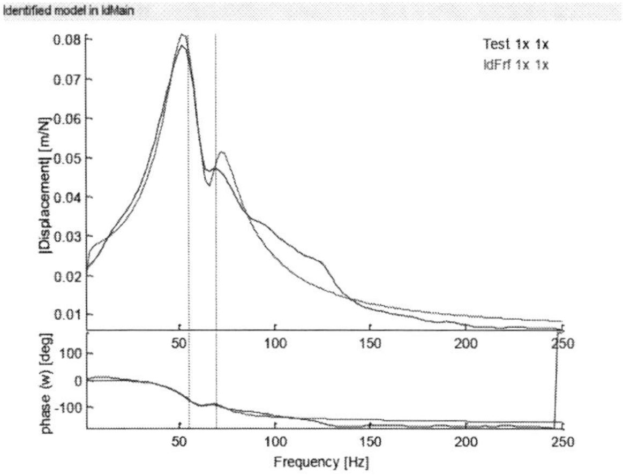

Figure 9. Measured and estimated point frequency response function at 60000 rpm.

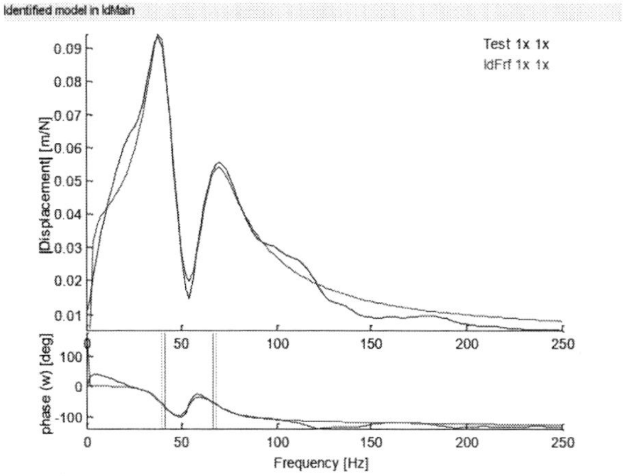

Figure 10. Measured and estimated point frequency response function at 100000 rpm.

First tests are done with the excitation applied at bearing 1, direction x. Figure 8 shows a comparison between the measured and estimated point frequency response function for the rotor at rest. The first resonance at 66 Hz is clearly reproduced by the estimate. In Fig. 9, measured and estimated point frequency response functions are depicted for a rotational speed of 60000 rpm. Due to the gyroscopic effect, the first resonance has split into one at 55 Hz and another at

71 Hz. This goes further at 100000 rpm, where these two resonances appear at 40 Hz and 70 Hz as can be seen in Fig. 10. For both 60000 and 100000 rpm, the mode shape animation shows that the lowest natural frequency belongs to a backward mode; the next mode shape, which should be forward, comes out less clearly.

From Figs. 8, 9, and 10, it is not clear what becomes of the second rigid body resonance of the rotor at rest. With the excitation applied at bearing 1, this resonance was excited near a node. All tests are therefore repeated with the excitation applied at bearing 2, direction x. For the second pair of natural frequencies, backward and forward modes can be distinguished from 40000 rpm upwards.

Experimental modal analyses are also performed with the excitation applied at bearing 1, direction y, and at bearing 2, direction y.

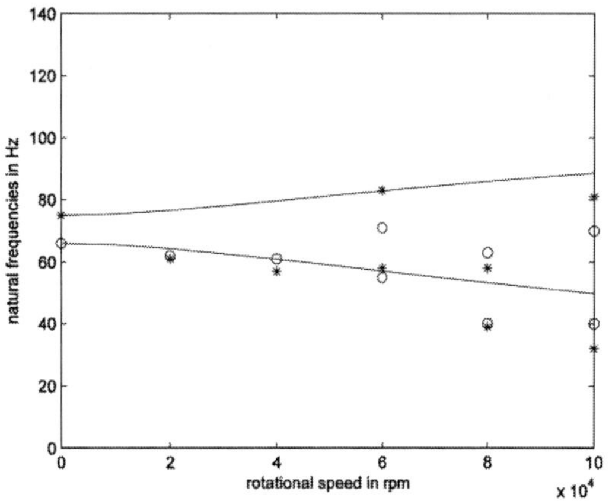

Figure 11. Calculated (-) and measured natural frequencies, excitation at bearing 1, directions x (o) and y(*), depending on the rotational speed.

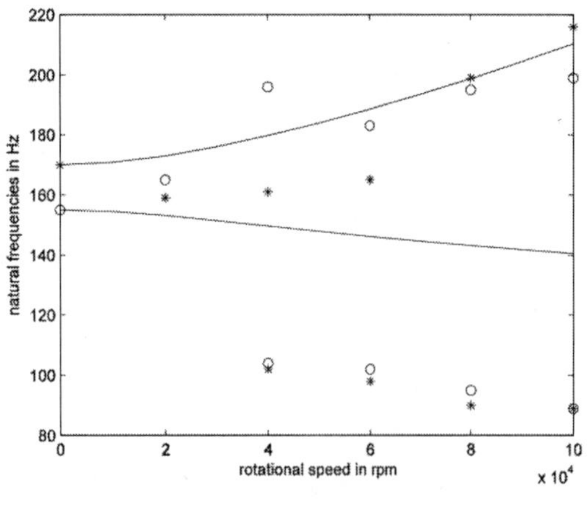

Figure 12. Calculated (-) and measured natural frequencies, excitation at bearing 2, directions x (o) and y(*), depending on the rotational speed.

VI. Stiffness Identification

All four bearing stiffness values are identified using the mass geometry of the rotor and the first and second natural frequencies in directions x and y at rest. With the identified stiffness values, the relationship between the rotational speed and the first pair of natural frequencies is depicted in Fig. 11. This figure also shows the measured natural frequencies with the excitation applied at bearing 1 in directions x and y. From 60000 rpm upwards, two natural frequencies can be distinguished for each excitation direction. In theory, these should be independent of the excitation direction and follow the calculated values. The remaining discrepancies may be explained by the poor frequency resolution resulting from the shortness of the windowed vibration responses.

Calculated and measured values for the second pair of natural frequencies are given in Fig. 12. In this case two branches are measured from 40000 rpm upwards. It is remarkable that for the lower branch of natural frequencies, measured results show a strong deviation from calculated ones although they agree fairly well among both excitation directions. This could mean that the rotor did not behave like a rigid body and may be explained by the type of connection between the disc and the shaft.

VII. Conclusions

The rigid body natural frequencies and mode shapes of a gyroscopic rotor in active magnetic bearings have been investigated. The active magnetic bearings have been used to generate an impulsive force signal, which offered a straightforward way to perform an experimental modal analysis. For both pairs of natural frequencies, identification results demonstrate the gyroscopic split. However, advanced excitation methods will be required to improve the accuracy of the results.

References

[1] R. Nordmann, "Identification of the modal parameters of an elastic rotor with oil film bearings," *Transactions of the ASME - Journal of Vibration, Acoustics, Stress, and Reliability in Design*, vol. 106, pp. 107–112, 1984.

[2] D. J. Ewins, "Modal Testing: Theory, Practice and Application," *second ed. Research Studies Press Ltd., Baldock, Hertfordshire, England*, 2000.

[3] A. Muszynska, "Modal testing of rotors with fluid interaction," *International Journal of Rotating Machinery*, vol. 1, no. 2, pp. 83–116, 1995.

[4] H. Irretier, "Experimentelle Modalanalyse in der Rotordynamik," *VDI Berichte*, Nr. 1550, 2000.

[5] P. Förch, A. Reister, C. Gähler, and R. Nordmann, "Modale Analyse an rotierenden Maschinen mittels Magnetlager," *Proceedings of Schwingungen in rotierenden Maschinen III*, pp. 245–253, Universität Kaiserslautern, 1995.

[6] C.W. Lee, Y.H. Ha, and C.S. Kim, "Identification of active magnetic bearing system using magnetic force measurement," *Proceedings of Fourth International Symposium on Magnetic Bearings*, pp. 305–309, ETH Zurich, 1994.

[7] C.W. Lee, "A complex modal testing theory for rotating machinery," *Mechanical Systems and Signal Processing*, vol. 5, no. 2, pp. 119-137, 1991.

[8] E. Balmès, J.M. Leclère, "Structural Dynamics Toolbox User's Guide," *SDTools, 52 rue Vergniaud, 75013 Paris, France*, 1991-2003.

Identification of Stiffness and Damping via SPHS Excitation and Multi-DOF Active Magnetic Bearing-Rotor Model

Jiang KeJian[a, b] Zhu ChangSheng[b]

a. College of Informatics and Electronics, Zhejiang Sci-Tech University, Hangzhou, China, 310016.

b. College of Electrical Engineering, Zhejiang University, Hangzhou, China, 310027.

Abstract: **The equivalent stiffness and the equivalent damping are the important and commonly used parameters for representing the support characteristic of the Active Magnetic Bearing(AMB). A method for measuring the equivalent stiffness and the equivalent damping is proposed with parameter identification via the multi-frequency excitation. The parameter identification is based on the multi-degree of freedom(DOF) rotor model, not the single DOF model, which can not be suitably applied to the multi-DOF AMB-rotor system. Additionally, the Schroeder Phased Harmonic Sequences(SPHS) are applied to generate the signal for the multi-frequency excitation. SPHS can achieve the lowest peak value by means of appropriate selection for the relative phasing of each frequency component, so that the possibility of the rotor vibration exceeding clearances of AMB and the magnetic force reaching saturation is minimized. Finally, the experiments indicate that the proposed method can efficiently reduce the peak value for the superimposed multi-frequency excitation and correctly identify the equivalent stiffness and equivalent damping of the AMB-rotor system.**

Keywords: **active magnetic bearing, multi-frequency excitation, Schroeder phased harmonic sequences, equivalent stiffness, equivalent damping**

1. Introduction

The active magnetic bearing(AMB) can not only provide the levitating support for the rotor without any mechanical friction, but also apply the real-time magnetic force to suppress the rotor vibration. For the performance advantages of the AMB technology, more and more attentions have been focused on the AMB application.

For either the traditional mechanical bearing or the active magnetic bearing, the support characteristic is the foundation for the rotor dynamics analysis. In the traditional mechanical bearing, the commonly used parametric representation for the support characteristic are the stiffness and the damping, which are also used for the active magnetic bearing, namely, the equivalent stiffness and the equivalent damping.

In recent years, there have been a great number of investigations into measuring and analyzing the AMB support characteristic. Tsai et al.[1] applied the wavelet transform algorithm to identify the magnetic damping and magnetic stiffness coefficients of the AMB system and analyzed their nonlinear order. Lim et al.[2] identified the equivalent stiffness and the equivalent damping for the AMB with the PID control strategy and analyzed the change rule of the support characteristic under the different PID parameters. Kim et al.[3] achieved to measure the force-current factor and the force-displacement factor by converging of adaptive filter. Mehmet et al.[4] and Lim et al.[5] presented the stiffness identification with multi-frequency excitation, which adopted the SPHS to avoid the superimposing peak of multi-frequency component. Bauomy et al.[6] investigated the response of the AMB-rotor system subjecting to a periodically time-varying stiffness. The stability of system near the simultaneous combined and sub-harmonic resonance was analyzed by the 4th Rung Kutta method. Sayyad et al. [7] presented the use of a variable stiffness type magnetic vibration absorber to control the vibration of a beam structure. Its principle is similar to the rotor vibration of AMB with variable stiffness in single DOF. Baloh et al. [8] proposed a method to measure the force-current factor and the force-displacement factor of AMB by adaptive estimation with least square algorithm.

2. Problem Formulation

In this study, the attention would be confined to solve two problems for measuring the stiffness and the damping of the AMB-rotor system.

1: Most aforementioned methods proposed for measuring the stiffness and the damping of AMB are based on the levitation model in single degree of freedom(DOF), which can not be suitably applied to the multi-DOF rotor systems. A multi-DOF modeled parameter identification method is needed.

2: The multi-frequency excitation is a normal method for the parameter identification in AMB. It is desirable that each frequency component of the multi-frequency excitation should be strong enough so that the vibration responses of every component are vivid. However, the vibration peak due to superimposing all the frequency component may result in unacceptably large rotor vibration or cause force saturating of the AMB.

In this paper, a multi-DOF rotor model based method is proposed for measuring the stiffness and the damping via multi-frequency excitation. Schroeder Phased Harmonic Sequences(SPHS) is adopted to achieve the lowest peak value of the multi-frequency excitation by means of appropriate selection for the relative phasing of each components, so that the possibility of the vibration displacement exceeding clearances or the bearing force reaching saturation is minimized.

3. Identification of Stiffness and Damping with Multi-DOF Rotor model

If the AMB control system adopts the decentralized control for each DOF, namely, the AMB• s control current in a certain direction is decided only by the rotor displacement in its own direction, the AMB support can be equivalent to two mutually perpendicular spring-damping structures as shown in Fig.1, which is the so-called equivalent stiffness and equivalent damping. It is similar to the concept of the traditional mechanical bearing.

Hereby, provided a rotor is supported by two AMBs, A and B, shown in Fig.1. The support of two AMBs are equivalent to 4 spring-damping structures in 4 directions respectively. The equivalent stiffness or the equivalent damping in a certain direction is decided only by the rotor displacement in its own direction respectively, namely, the spring-damping structures of every direction are independent each other.

In Fig.1, O is the mass center of the rotor, the distance between the rotor mass center and two AMBs respectively are l_a and l_b. The axis z is the rotating axis of the rotor.

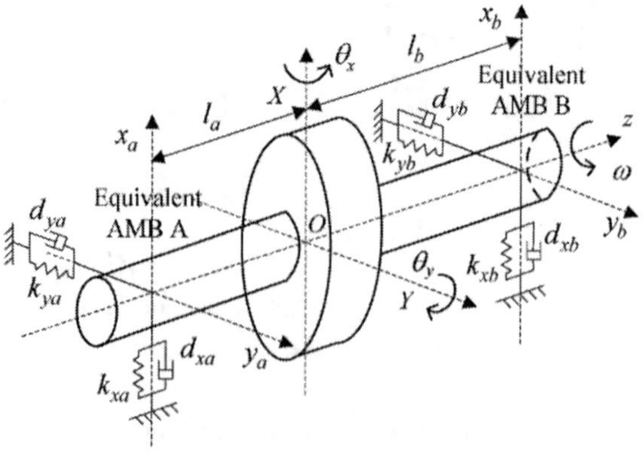

Fig.1 equivalent model forAMB-rotor system

The rotor attitude vector can be defined as $[X,\theta_y,Y,\theta_x]^T$, where, X,Y are the displacements of the rotor mass center in x and y directions, θ_y and θ_x are the angular displacements of rotor rotating around axis y and axis x. The positive directions of every motion are shown in Fig. 1.

Otherwise, the rotor attitude vector also can be defined by the rotor location in the place of two AMBs as $[x_a,x_b,y_a,y_b]^T$, where, x_a,y_a is the rotor!s location coordinates in the place of the AMB A, and x_b,y_b is that of AMB B, as shows in Fig.1.

The transfer relation between two types of the rotor attitude vectors can be given as follows

$$\begin{bmatrix} X \\ \theta_y \\ Y \\ \theta_x \end{bmatrix} = \frac{1}{l} \begin{bmatrix} l_b & l_a & 0 & 0 \\ 1 & -1 & 0 & 0 \\ 0 & 0 & l_b & l_a \\ 0 & 0 & 1 & -1 \end{bmatrix} \begin{bmatrix} x_a \\ x_b \\ y_a \\ y_b \end{bmatrix} \tag{1}$$

Firstly, the equations of rotor motion can be written in the form of stiffness and damping, (In this paper, only the radial supports of rotor are considered and the gyroscopic effect is neglectable)

$$\begin{aligned} m\ddot{X} &= -(d_{xa}\dot{x}_a + k_{xa}x_a) - (d_{xb}\dot{x}_b + k_{xb}x_b) + F_x \\ J\ddot{\theta}_y &= -(d_{xa}\dot{x}_a + k_{xa}x_a)l_a + (d_{xb}\dot{x}_b + k_{xb}x_b)l_b + M_x \\ m\ddot{Y} &= -(d_{ya}\dot{y}_a + k_{ya}y_a) - (d_{yb}\dot{y}_b + k_{yb}y_b) + F_y \\ J\ddot{\theta}_x &= -(d_{ya}\dot{y}_a + k_{ya}y_a)l_a + (d_{yb}\dot{y}_b + k_{yb}y_b)l_b + M_y \end{aligned} \tag{2}$$

Where, m is the mass of rotor, J is the transverse moment of inertia. F_x,F_y are the exciting forces respectively in direction x and y, and M_x,M_y are the exciting moments around the x axis and y axis respectively. The equivalent stiffness of every directions are denoted respectively by $k_{xa},k_{xb},k_{ya},k_{yb}$, and that for the equivalent damping are $d_{xa},d_{xb},d_{ya},d_{yb}$.

Eq.(2) is rebuilt by substitution of Eq.(1). The

following matrix equations can be obtained

$$\frac{1}{l}\begin{bmatrix} ml_b & ml_a & 0 & 0 \\ J & -J & 0 & 0 \\ 0 & 0 & ml_b & ml_a \\ 0 & 0 & J & -J \end{bmatrix}\begin{bmatrix} \ddot{x}_a \\ \ddot{x}_b \\ \ddot{y}_a \\ \ddot{y}_b \end{bmatrix} + \begin{bmatrix} d_{xa} & d_{xb} & 0 & 0 \\ d_{xa}l_a & -d_{xb}l_b & 0 & 0 \\ 0 & 0 & d_{ya} & d_{yb} \\ 0 & 0 & d_{ya}l_a & -d_{yb}l_b \end{bmatrix}\begin{bmatrix} \dot{x}_a \\ \dot{x}_b \\ \dot{y}_a \\ \dot{y}_b \end{bmatrix} \quad (3)$$

$$+\begin{bmatrix} k_{xa} & k_{xb} & 0 & 0 \\ k_{xa}l_a & -k_{xb}l_b & 0 & 0 \\ 0 & 0 & k_{ya} & k_{yb} \\ 0 & 0 & k_{ya}l_a & -k_{yb}l_b \end{bmatrix}\begin{bmatrix} x_a \\ x_b \\ y_a \\ y_b \end{bmatrix} = \begin{bmatrix} F_x \\ M_x \\ F_y \\ M_y \end{bmatrix}$$

By the Laplace transform, the frequency domain equations of Eq.(3) is

$$\begin{bmatrix} -m\omega^2\frac{l_b}{l}+k_{xa}+j\omega d_{xa} & -m\omega^2\frac{l_a}{l}+k_{xb}+j\omega d_{xb} & 0 & 0 \\ -J\frac{\omega^2}{l}+k_{xa}l_a+j\omega d_{xa}l_a & J\frac{\omega^2}{l}-k_{xb}l_b-j\omega d_{xb}l_b & 0 & 0 \\ 0 & 0 & -m\omega^2\frac{l_b}{l}+k_{ya}+j\omega d_{ya} & -m\omega^2\frac{l_a}{l}+k_{yb}+j\omega d_{yb} \\ 0 & 0 & -J\frac{\omega^2}{l}+k_{ya}l_a+j\omega d_{ya}l_a & J\frac{\omega^2}{l}-k_{yb}l_b-j\omega d_{yb}l_b \end{bmatrix}$$

$$=\begin{bmatrix} F_x(j\omega) \\ M_x(j\omega) \\ F_y(j\omega) \\ M_y(j\omega) \end{bmatrix}\begin{bmatrix} x_a(j\omega) \\ x_b(j\omega) \\ y_a(j\omega) \\ y_b(j\omega) \end{bmatrix}^{-1} \quad (4)$$

To simplify the expression, let

$$\begin{aligned} A_{xa} &= k_{xa} + j\omega d_{xa} \\ A_{xb} &= k_{xb} + j\omega d_{xb} \\ A_{ya} &= k_{ya} + j\omega d_{ya} \\ A_{yb} &= k_{yb} + j\omega d_{yb} \end{aligned} \quad (5)$$

Eq.(4) is rebuilt by substitution of Eq.(5). It can be written as

$$\begin{bmatrix} -m\omega^2\frac{l_b}{l}+A_{xa} & -m\omega^2\frac{l_a}{l}+A_{xb} & 0 & 0 \\ -J\frac{\omega^2}{l}+A_{xa}l_a & J\frac{\omega^2}{l}-A_{xb}l_b & 0 & 0 \\ 0 & 0 & -m\omega^2\frac{l_b}{l}+A & -m\omega^2\frac{l_a}{l}+A_{yb} \\ 0 & 0 & -J\frac{\omega^2}{l}+A_{ya}l_a & J\frac{\omega^2}{l}-A_{yb}l_b \end{bmatrix} \quad (6)$$

$$\begin{bmatrix} x_a(j\omega) \\ x_b(j\omega) \\ y_a(j\omega) \\ y_b(j\omega) \end{bmatrix} = \begin{bmatrix} F_x(j\omega) \\ M_x(j\omega) \\ F_y(j\omega) \\ M_y(j\omega) \end{bmatrix}$$

Obviously, Eq.(6) is the linear equations of complex number with four unknowns, which are $A_{xa}, A_{xb}, A_{ya}, A_{yb}$. In the parameter identification, $[F_x(j\omega), M_x(j\omega), F_y(j\omega), M_y(j\omega)]^T$ are the excitation signals and $[x_a(j\omega), x_b(j\omega), y_a(j\omega), y_b(j\omega)]^T$ are the displacement responses. They all are known quantities. Therefore, $A_{xa}, A_{xb}, A_{ya}, A_{yb}$ can be solved as follows

$$A_{xa} = \frac{F_x(j\omega)l_b + M_x(j\omega) + \left(m\omega^2\frac{l_al_b}{l} - J\frac{\omega^2}{l}\right)x_b(j\omega)}{x_a(j\omega)l} + m\omega^2\frac{l_b^2}{l^2} + J\frac{\omega^2}{l^2}$$

$$A_{xb} = \frac{F_x(j\omega)l_a - M_x(j\omega) + \left(m\omega^2\frac{l_al_b}{l} - J\frac{\omega^2}{l}\right)x_a(j\omega)}{x_b(j\omega)l} + m\omega^2\frac{l_a^2}{l^2} + J\frac{\omega^2}{l^2}$$

$$A_{ya} = \frac{F_y(j\omega)l_b + M_y(j\omega) + \left(m\omega^2\frac{l_al_b}{l} - J\frac{\omega^2}{l}\right)y_b(j\omega)}{y_a(j\omega)l} + m\omega^2\frac{l_b^2}{l^2} + J\frac{\omega^2}{l^2}$$

$$A_{yb} = \frac{F_y(j\omega)l_a - M_y(j\omega) + \left(m\omega^2\frac{l_al_b}{l} - J\frac{\omega^2}{l}\right)y_a(j\omega)}{y_b(j\omega)l} + m\omega^2\frac{l_a^2}{l^2} + J\frac{\omega^2}{l^2}$$

(7)

Finally, the equivalent stiffness and equivalent damping can be deduced as

$$\begin{aligned} k_{xa} &= \mathrm{Re}\{A_{xa}\} \quad ; \quad d_{xa} = \mathrm{Im}\{A_{xa}\}/\omega \\ k_{xb} &= \mathrm{Re}\{A_{xb}\} \quad ; \quad d_{xb} = \mathrm{Im}\{A_{xb}\}/\omega \\ k_{ya} &= \mathrm{Re}\{A_{ya}\} \quad ; \quad d_{ya} = \mathrm{Im}\{A_{ya}\}/\omega \\ k_{yb} &= \mathrm{Re}\{A_{yb}\} \quad ; \quad d_{yb} = \mathrm{Im}\{A_{yb}\}/\omega \end{aligned} \quad (8)$$

Where, $\mathrm{Re}\{\ \}$ and $\mathrm{Im}\{\ \}$ respectively is to obtain the real part and imaginary part of a complex number.

4. Schroeder Phased Harmonic Sequences

The multi-frequency excitation is the normal method for parameter identification of AMB. For the AMB system, however, if all the frequency components of multi-frequency excitation are arbitrarily added together, the resulting response may produce unacceptably large rotor vibration or cause force saturating of the magnetic bearing, which may sharply increase the undesired nonlinearity of measurement. It is not a good solution to decrease the excitation amplitude of every frequency, because this will result in weak perturbation in the frequency domain and consequently increase the identification errors.

Schroeder Phased Harmonic Sequences(SPHS)[11] can achieve the lowest possible peak value signal by means of superimposing these components with appropriate selection of the relative phasing, so that the possibility of the vibration displacement exceeding clearance or the bearing force reaching saturation is minimized.

4.1 Composing of SPHS

SPHS reduces the peak value by re-organizing the relative phasing of each harmonic. The Composing of SPHS is very simple and its expression is given by

$$x(t) = \sum_{k=1}^{N}\sqrt{2P_k}\cos(2k\pi t/T + \theta_k) \quad (9)$$

Where, N is the number of harmonics, T is the period of the fundamental-harmonic, P_k is the power of the kth harmonic.(if signal is $A\cos(2\pi t/T)$, its power is $A^2/2$, so $\sqrt{2P_k}$ is the amplitude of the kth harmonic, θ_k is the phase angle of the kth harmonic.)

In order to obtain the lowest peak value of harmonics superimposing, SPHS designates the phase angle of each harmonic as follows

$$\theta_k = \theta_1 - 2\pi\sum_{i=1}^{k-1}(k-i)P_i \qquad (10)$$

Where, θ_1 is the phase angle of the fundamental-harmonic, which can be assigned at will. In most applications of SPHS, θ_1 will be assigned as 0 or π. Then, Eq.(10) can be modified as

$$\theta_k = \pi[[\sum_{i=1}^{k-1}(k-i)P_i]] \qquad (11)$$

Where, [[ ]] computes the largest integer not larger than their contents, namely, rounding operation.

Eq.(11) can be further simplified if all of harmonics have the same amplitude, namely, it!s the flat-spectrum signal, $P_i = P/N(i=1,2,\cdots,N)$, then

$$\theta_k = \pi[[k^2/2N]] \qquad (12)$$

Eq(12) is the most simplified SPHS representation, which is widely used in practical applications.

4.2 Illustrative Examples of SPHS

In this section, the illustrative examples comparing the SPHS signal with the non-SPHS signal will be given.

Let the frequency of the fundamental harmonic is 1Hz, the amplitude of all harmonics are 1 unit, the number of harmonics are respectively given as N=25• 50• 100. For the non-SPHS signal, the phase angles of all harmonics are always defined as $\theta_i = 0\ (i=1,2,\cdots,N)$, which is called zero phase angle signal in this paper. The illustrative examples for comparing are shown from Fig.2 to Fig.7.

It can be observed that the peak value of the zero phase angle signal sharply increases with the harmonics number growing. The peak value are about 20,40,80 for N=25• 50• 100 respectively. Contrastively, the peak value of SPHS increases gently in spite of the number of harmonics being doubled and redoubled. The peak value are about 9,12,15 for N=25• 50• 100 respectively. By all appearances, if the zero phase angle signal is adopted, the possibility of the vibrationt exceeding clearance or the magnetic force reaching saturation will be greater.

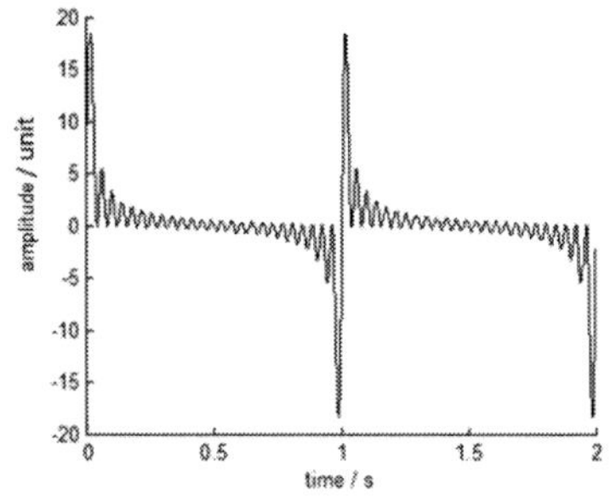

Fig.2 N=25 zero phase angle signal

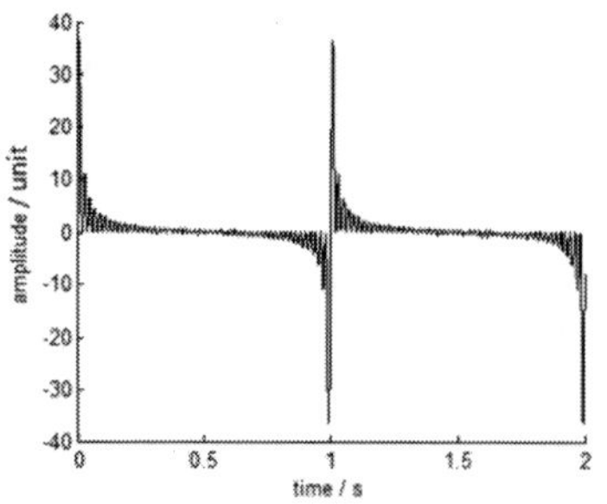

Fig.3 N=50 zero phase angle signal

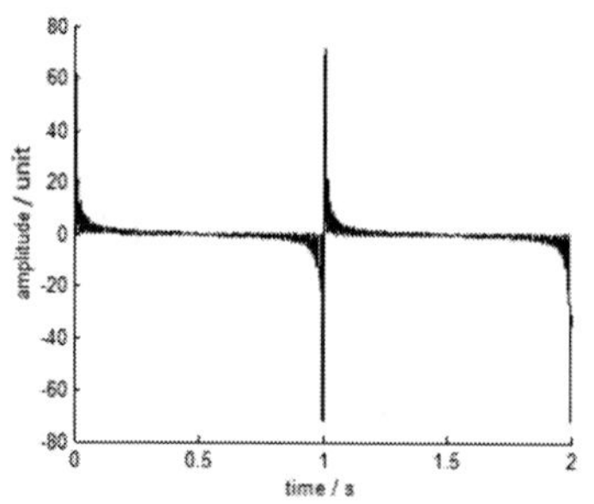

Fig.3 N=100 zero phase angle signal

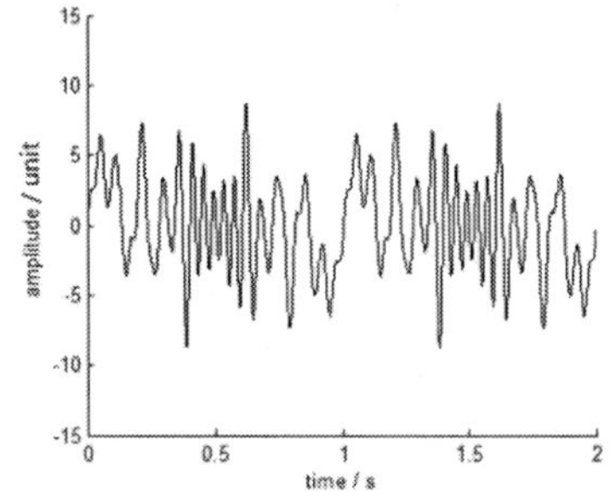

Fig.5 N=25 SPHS signal

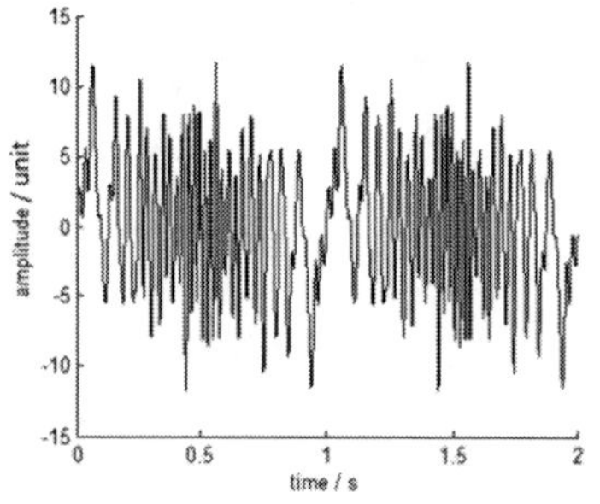

Fig.6 N=50 SPHS signal

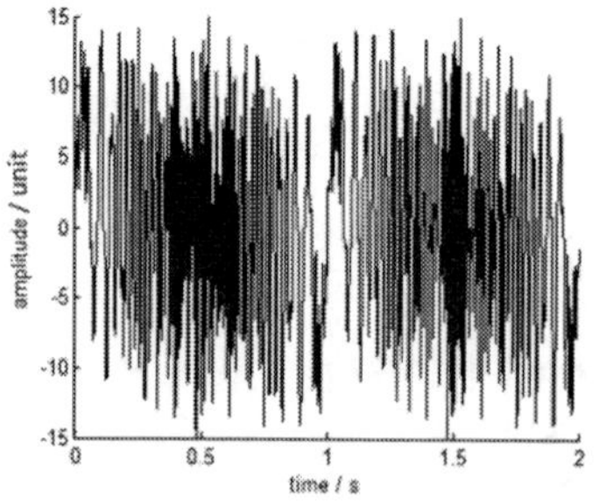

Fig.7 N=100 SPHS signal

5. Experiment

5.1 AMB Rig

To evaluate the effectiveness of the proposed method, The experiments were carried out on a rig of AMB flywheel rotor system with vertical shaft structure as shown in Fig. 8. The axial support for the rotor is provided by a permanent magnetic bearing and the radial support by two radial AMBs, upper AMB and lower AMB. The equivalent stiffness and the equivalent damping of two radial AMBs need measuring in expectation. Fig.9 is the drawing of rotor structure, whose parameters are that the rotor mass including flywheel is m=58.32kg, the transverse moment of inertia is J=1.878 $\mathrm{kg} \times \mathrm{m}^2$, the clearance of touch-down bearing is 0.2mm, $l_a = 305\mathrm{mm}$, $l_b = 222\mathrm{mm}$. The force-current factor of the upper AMB is 326N/A, and that of the lower is 111.32N/A. The experiments were operated on the d-SPACE DS1103 with the decentralized PID control for rotor levitating.

Firstly, for checking the vibration condition, the amplitude-frequency characteristics of the rotor vibration are carefully measured. When the excitation is imposed on the upper AMB, the amplitude-frequency characteristics of 4 directions in two radial AMB are shown in Fig.10 and Fig.11. It can be observed that there are some peaks of rotor vibration in the range of 20Hz-60Hz. Among them, the vibration peaks in 26Hz and 54Hz are the two highest peaks, which may be, by estimate, the critical frequency(resonance frequency) of cylindrical mode and conical mode of rotor motion. Because the critical frequency values of every directions are different and close each other, so that their vibration peaks are close one another and crowd in the range of 20Hz-60Hz. When the frequency is beyond 80Hz, the amplitude-frequency characteristic sharply descends close to zero.

Fig.8 rig of AMB flywheel rotor system

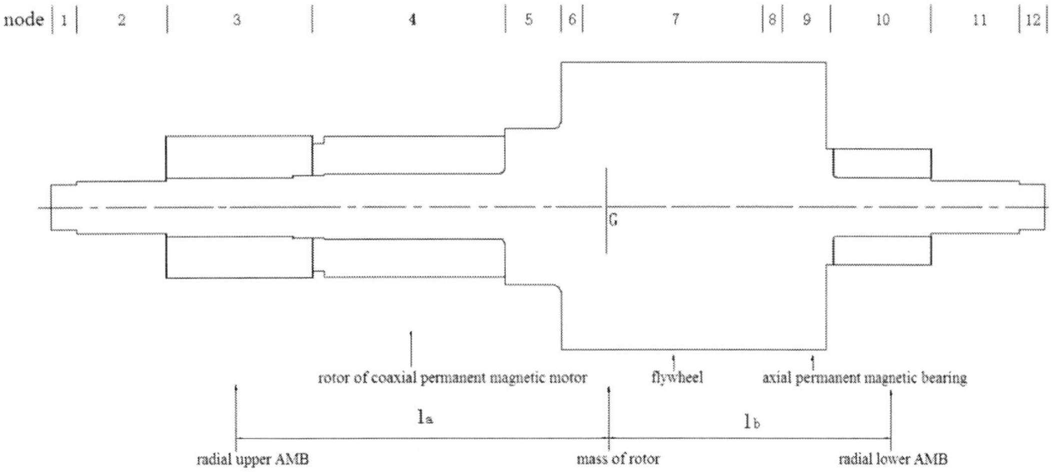

Fig.9 rotor structure drawing

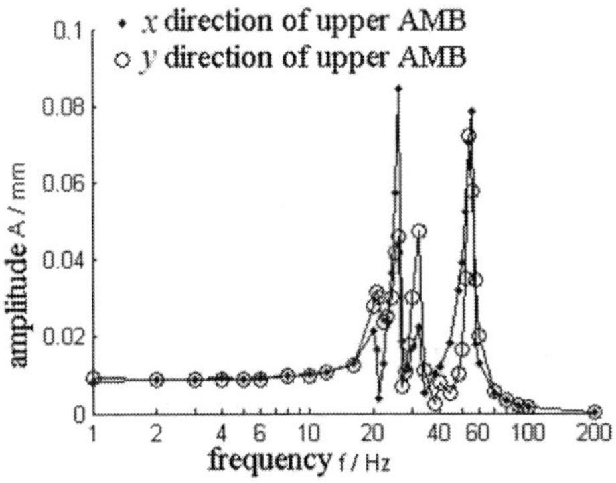

Fig.10 amplitude -frequency characteristic of upper AMB

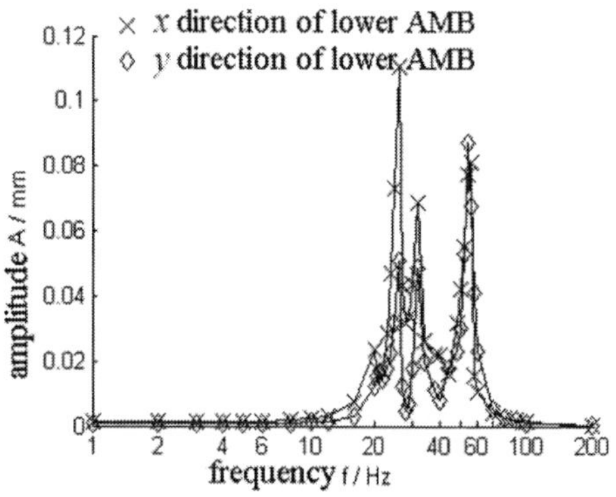

Fig.11 amplitude -frequency characteristic of lower AMB

5.2 Test for SPHS Effectiveness

In this test, the contrasts of the practical effectiveness between SPHS and zero phase angle signal will be tested in expectation.

A rotating vector-like signal is injected into the control current of the upper AMB, that is, all the sine harmonics are injected into x direction and all the cosine harmonics are injected into y direction. Therefore, the magnetic exciting force in every harmonics frequency are all the rotating vector. The adopted multi-frequency excitation is a flat-spectrum signal in the range of 1-200Hz. The fundamental-harmonic is 1Hz and N=200. In test, the amplitudes of harmonics are respectively set up as 0.4N, 0.8N, 1.2N, 1.6N.

The experiments will record the practical rotor vibration orbit under the different amplitude of excitation. Fig.12 and Fig.13 are the response orbit for the excitation harmonics amplitude of 0.4N. Fig.14 and Fig.15 are that of 0.8N. It is clearly that the greater the excitation harmonic amplitude is, the more the vibration response orbit enlarges. However, the rotor orbit enlarging of zero phase angle signal is much more than that of SPHS.

It can be observed in Fig.14 that, when the excitation amplitude reaches to 0.8N for zero phase angle signal, the rotor orbit in lower AMB is close 0.2mm and it meams that the rotor just comes into contact with the touch down bearing. It implies that the excitation amplitude can not be increased any more in the case of zero phase angle signal.

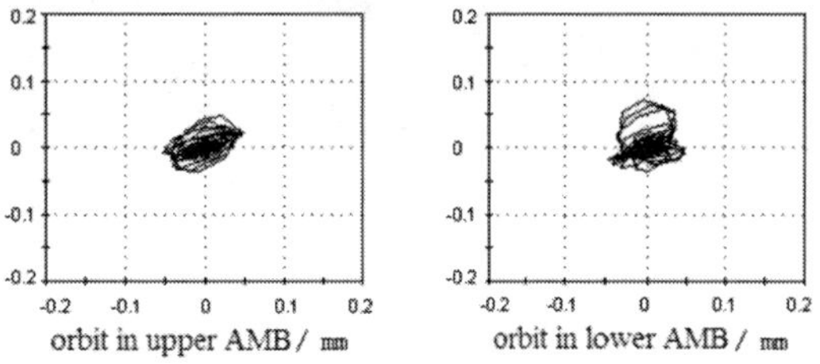

Fig.12 response orbit of 0.4N with zero phase angle

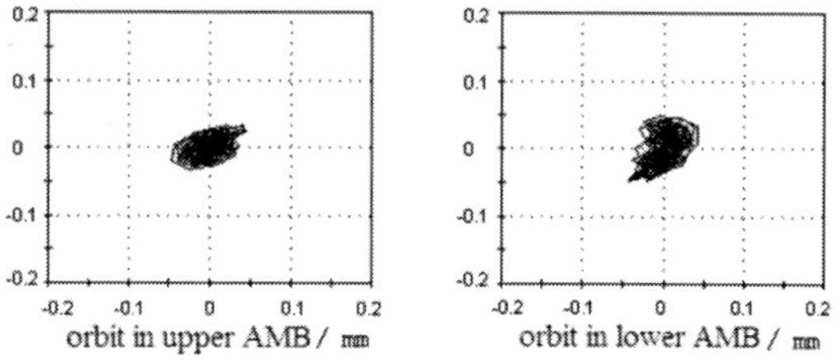

Fig.13 response orbit of 0.4N with SPHS

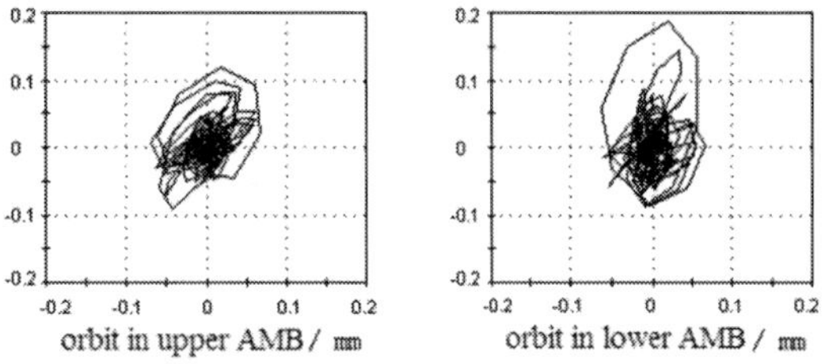

Fig.14 response orbit of 0.8N with zero phase angle

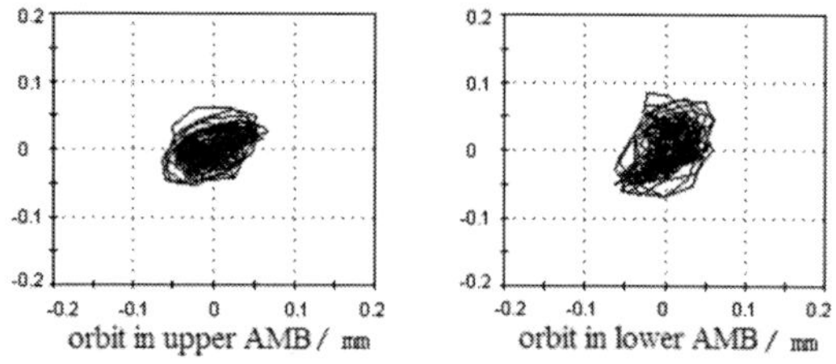

Fig.15 response orbit of 0.8N with SPHS

Then, the excitation amplitude continues to be increased for SPHS. Fig.16 is the rotor orbit for the excitation amplitude increasing to 1.2N with SPHS and Fig.17 is for 1.6N. The vibration orbit enlarges with the excitation increasing. For SPHS, when the excitation amplitude reaches to 1.7N, the rotor evidently contacts the touch down bearing.

It is concluded that, the upper limit of excitation harmonics amplitude is 1.6N for SPHS, and 0.8N for the zero phase angle signal. It proves that SPHS can efficiently minimize the peak value for the superimposed multi-frequency signal.

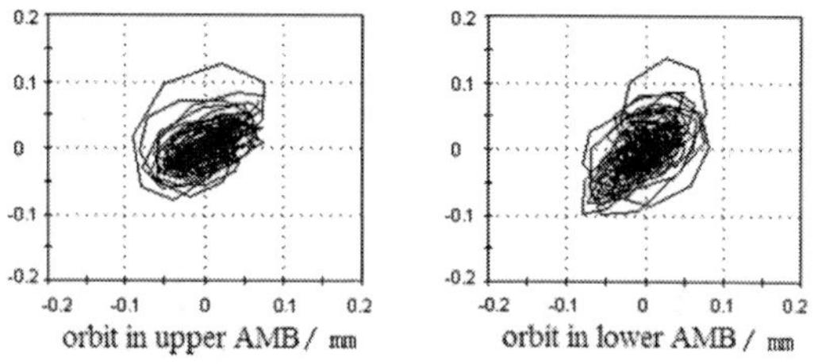

Fig.16 response orbit of 1.2N with SPHS

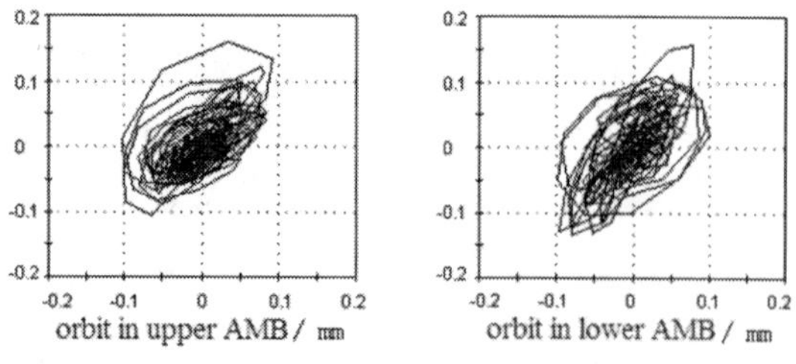

Fig.17 response orbit of 1.6N with SPHS

5.3 Measuring Stiffness and Damping with multi-frequency excitation

According to Eq.(7)and (8), the equivalent stiffness and the equivalent damping in 4 directions of the AMB rig will be measured by the proposed method.

Fig.18 is the measurement results of the equivalent stiffness under the excitation harmonics amplitude of 1.6N. Because the equivalent stiffness in 4 directions is similar to each other, the corresponding direction of each diagram line for the equivalent stiffness is not denoted in Fig.18. The equivalent stiffness of AMB increases gradually along with the frequency rising, which is the same conclusion drawn in other academic literatures[5,9,10].

Fig.18 shows that the measuring values have a visible drop in the band of critical frequency (resonance frequency) in the range of 20Hz-60Hz. The reason is that the rotor vibration is much great in the band of critical frequency. The greater the rotor vibration displacement is, the larger control current is needed. The AMB of the experiment rig adopts the differential mechanism. When the control current is less than the bias current, the magnetic force would be twice produced by both two sides AMB poles. Correspondingly, when the control current is more than the bias current, the magnetic force would only be produced by one pole of single side in the AMB. Consequently, the equivalent stiffness would drop, when the control current is more than the bias current. The same phenomenon have been observed in the reference[10].

Above 80Hz, the diagram lines of the equivalent stiffness is in disorder for identification errors due to the rotor weak vibrating.

The measurement results of the equivalent damping are shown in Fig.19 with the excitation harmonics amplitude of 1.6N. It is visible that the equivalent damping values are very little in most frequency, and also is disordered for identification errors beyond 80Hz.

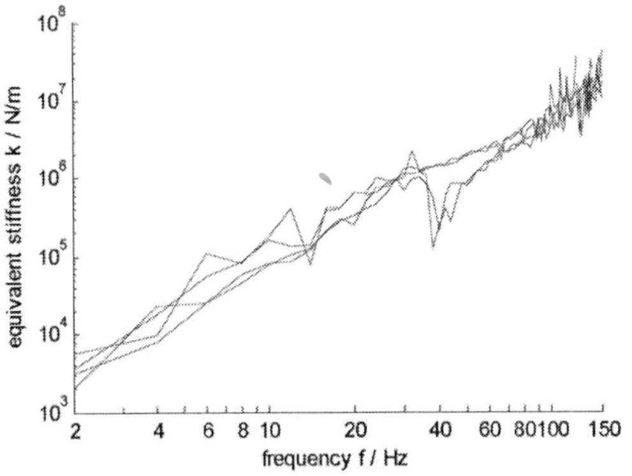

Fig.18 equivalent stiffness in 4 directions under 1.6N excitation

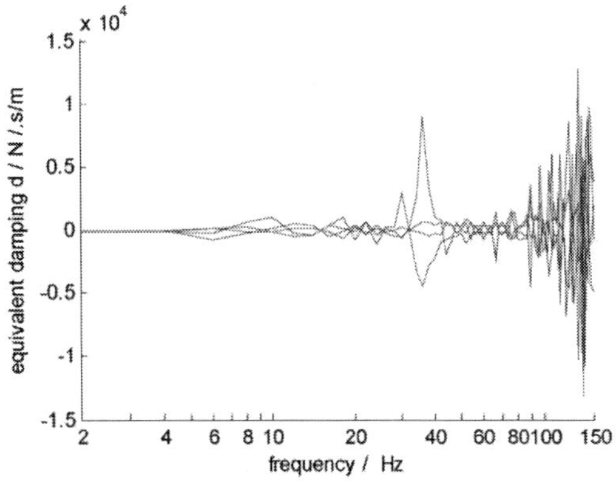

Fig.19 equivalent damping in 4 directions under 1.6N excitation

6. Conclusion

In this study, a method for measuring the equivalent stiffness and the equivalent damping is proposed based on the multi-DOF rotor model. SPHS is adopted in the multi-frequency excitation to achieve the lowest possible peak value signal by means of appropriate selection for the relative phasing of each component without decreasing the amplitude of every components, so that the possibility of the vibration displacement exceeding clearances or the bearing force reaching saturation is minimized.

Finally, the experiments indicate that the proposed method can correctly identify the equivalent stiffness and the equivalent damping of the AMB-rotor system. It really has the better practical utility for the dynamic analysis of AMB.

Acknowledgement

The authors would like to acknowledge the National Natural Science Foundation of China(11172261, 11272288), the Postdoctoral Science Foundation (2013M531452), the Zhejiang Provincial Natural Science Foundation(LZ13E070001, LY12E05027), the Science Foundation of Zhejiang Sci-Tech University (1104827-Y).

References

[1] N. Tsai, H. Li, C. Lin, et al. Identification of rod dynamics under influence of active magnetic bearing[J]. Mechatronics. 2011, 21(4): 1013-1024.

[2] T. Lim and S. Cheng. Parameter estimation and statistical analysis on frequency dependent active control forces[J]. Mechanical Systems and Signal Processing. 2007, 21(5):2112-2124.

[3] S. Kim and C. Lee. On-line identification of current and position stiffness by LMS algorithm in active magnetic bearing system

equipped with force transducers[J]. Mechanical Systems and Signal Processing. 1999, 13(5): 681-690.

[4] N. Mehmet, O. Matthew, T. Cole, et al. On the use of Schroeder phased harmonic sequences in multi-frequency vibration control of flexible rotor/magnetic bearing systems[C]. Proceedings of 8th International Symposium on Magnetic Bearings. Mito, Japan, 2002, pp:218-222.

[5] T. Lim, S. Cheng and C. Poh. Parameter estimation of one-axis magnetically suspended system with a digital PID controller[C]. Proceedings of 1st International Conference on Sensing Technology. New Zealand, 2005, pp:419-424.

[6] H. Bauomy. Stability analysis of a rotor-AMB system with time varying stiffness[J]. Journal of the Franklin Institute. 2012, 349(2): 1871 1890.

[7] F. Sayyad and N. Gadhave. Variable stiffness type magnetic vibration absorber to control the vibration of beam structure[J]. Journal of Vibration and Control. 2013, 7, published online:1-7.

[8] M. Baloh, G. Tao and P. Allaire. Adaptive estimation of magnetic bearing parameters[C]. Proceedings of the IEEE International Conference on Control Applications. Hawaii, USA, 1999, pp:1193-1198.

[9]Yang Zuoxing, Zhao Lei, Zhao Hongbi. Automatic measurement of dynamic stiffness for active magnetic bearings[J]. Chinese Journal of Mechanical Engineering. 2001, 37(3):25-29. (in Chinese)

[10]Jiang Kejian, Zhu Changsheng. On-line Measurement of Supporting Characteristics of an Active Magnetic Levitations with Unknown Transfer Function[J]. China Mechanical Engineering. 2010, 21(8): 883-888. (in Chinese)

[11]M. Schroeder. Synthesis of Low-Peak-Factor Signals and Binary Sequences with Low Autocorrelation [J]. IEEE Transaction on Information Theory. 1970, 16(1): 85-89.

Position Estimation for Self-Sensing Magnetic Bearings Based on Double Detection of Current Slopes

Jinou Wang, Andreas Binder

Technische Universität Darmstadt, Institute of Electrical Energy Conversion, Landgraf-Georg-Str. 4, 64283 Darmstadt, Germany, jwang@ew.tu-darmstadt.de, abinder@ew.tu-darmstadt.de

***Abstract*—This paper introduces a new approach to estimate the rotor position of self-sensing magnetic bearings, which works on detecting the current slopes during both positive and negative voltage switching of PWM amplifiers. The position is estimated using the current slopes straightforward. The consideration of double current slopes shows advantages concerning the reduction of the influences of the coil resistance and the induced voltage due to the rotor movement. The current slope is calculated by multiple current sampling, therefore, no additional hardware is needed for the current slope measurement. Experimental results verified the proposed approach.**

I. Introduction

Self-sensing magnetic bearing technology has been researched intensively since the last decades. Although commercial applications already exist [1], [2], there are still technical challenges concerning the improvements of dynamic and linearity of the position estimation. These two factors are very important for a fast and accurate position control.

The modern self-sensing techniques all seem to use a kind of high-frequency interrogation signal for position estimation [3-6]. To avoid the hardware signal-injection unit, the switching ripple, which is a nature of PWM power amplifiers, is most applied. The reason is that the switched voltage as an interrogation signal has the highest frequency and amplitude, which should deliver the best estimation performance. But due to the position control, the duty cycle of the switched voltage varies, which leads to considerable decrease of the achievable performance. Many experts have worked on this issue in the past years [3, 5].

The above approaches model the triangle current ripple waveform with a set of sinusoidal harmonics and consider the position estimation in the linear periodic point of view. Due to the nature of the inductance, high frequency harmonics have more amplitude deviations with the same inductance change caused by the rotor movement, and therefore make much more contribution to the SNR (signal-to-noise ratio) and to the dynamic of the position estimation. But the estimator based on the above modelling cannot handle all frequencies, especially in the high frequency region. Therefore, the current ripple is theoretically not fully utilized for the position estimation.

To fully exploit the current ripple, it seems better to model it directly with triangle signals. Then, the current slopes will be used, as first introduced in [7]. This perspective looks directly at the nature of the current response of electromagnets

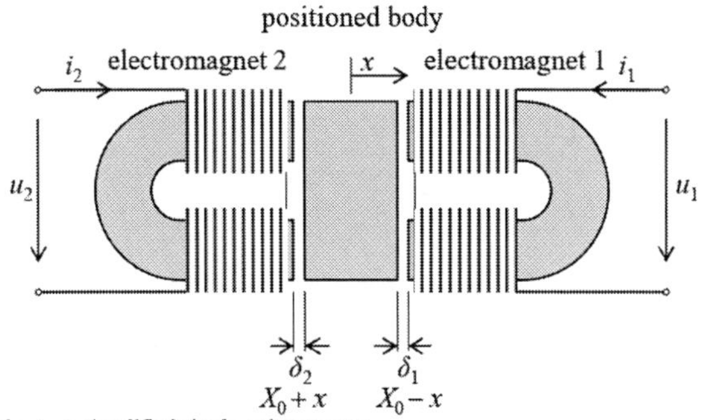

Fig. 1: A simplified single-axis actuator

driven by switched voltages. In [7], the rotor position is sensorless controlled, but there is no explicit position output, therefore, the position control and the position estimation are not decoupled. Independent position estimation makes the design of the position controller more flexible, so in [8], a slope based estimation approach was proposed.

Till now, the influences of the coil resistance and the induced voltage due to the rotor movement are neglected in the position estimation. The proposed approach in this paper takes them also into account and eliminates them. Thus, it should deliver better estimation performance. Details about the approach will be given in the following sections. Beginning with the simplified modelling in Section II, the basic idea of the proposed approach will be explained. Then, mechanisms to realize the idea will be given in Section III. This self-sensing technique will be applied to a real magnetic bearing, which has coupled control fluxes and permanent magnet excited bias fluxes. The modelling is quite different and has to be adapted. This will be explained in Section IV. Experimental results will be discussed in Section V. Conclusions and future work will be given in the last section.

II. Modelling of a Single-Axis Actuator

To explain the proposed approach, the model of a single-axis actuator is shown in Fig. 1. We point out, that the proposed method is also valid for higher number of axes, as long as the magnetic paths of each electromagnet are independent from the other, which is the case for most hetero-polar magnetic bearings.

We define the origin of the x-axis in the middle between the two electromagnets and distinguish the electric and geometric quantities of different electromagnets with subscripts $k = 1{,}2$. For simplification, we neglect eddy currents, flux leakage and edge flux effects and assume a sufficient high permeability $\mu_{\mathrm{Fe}} \to \infty$ of the laminated iron cores. Then, according to Faraday's law and to Ampere's law, we get the following basic equations:

$$\dot{\psi}_k + R_k i_k = u_k\,,\ \ k = 1{,}2. \tag{1}$$

$$\psi_k = \frac{2\mu_0 A N^2 i_k}{\delta_k}\,,\ \ k = 1{,}2. \tag{2}$$

In the above equations, ψ_k is the magnetic flux linkage of the electromagnet coil, R_k is the total coil resistance of each electromagnet, i_k is the coil current, u_k is the coil terminal voltage of the electromagnet, μ_0 is the permeability of free space, A is the area of each pole face, and N is the number of turns of one coil. So each electromagnet has a total number of turns of $2N$, and δ_k is the individual air gap width of each electromagnet. Additionally, we have the geometric constraints,

$$\delta_1 = X_0 - x\,, \tag{3}$$

$$\delta_2 = X_0 + x\,, \tag{4}$$

where X_0 is the nominal air gap width, and x is the position deviation. We apply the time derivative to both sides of (2) and get

$$\dot{\psi}_k = \frac{2\mu_0 A N^2}{\delta_k}\cdot\left(\dot{i}_k - \frac{i_k}{\delta_k}\dot{\delta}_k\right). \tag{5}$$

Both (1) and (5) are valid in the time domain. We put (1) into (5) and get

$$u_k - R_k i_k = \frac{2\mu_0 A N^2}{\delta_k}\cdot\left(\dot{i}_k - \frac{i_k}{\delta_k}\dot{\delta}_k\right). \tag{6}$$

If we use a two-level PWM (Pulse Width Modulation) amplifier, then the coil voltage u_k changes between U_{dc} and $-U_{\mathrm{dc}}$, where U_{dc} is the DC-link voltage of the PWM amplifier. Within one switching period, the following conditions,

$$u_k = -U_{\mathrm{dc}}\,, \tag{7}$$

during the negative switching interval, and

$$u_k = U_{\mathrm{dc}}\,, \tag{8}$$

during the positive switching interval, are given. In a digital control system, we can choose the control frequency equal to the PWM switching frequency f_{T}, which means that the fastest possible control is applied. Further, we apply two samples per PWM cycle. At these sample points, (7) and (8) are fulfilled respectively, as shown in Fig. 2. We consider the nth PWM cycle and use the subscript α to represent the time interval T_α, when condition (7) is fulfilled, and β to represent the time interval T_β, when condition (8) is fulfilled: $T_\alpha + T_\beta = 1/f_{\mathrm{T}} = T_0$. Then we get two equations (9), (10), which are the discrete form of (6) during the negative and positive switching, respectively.

$$-U_{\mathrm{dc}} - R_k i_{k,n,\alpha} = \frac{2\mu_0 A N^2}{\delta_{k,n,\alpha}}\cdot\left(\dot{i}_{k,n,\alpha} - \frac{i_{k,n,\alpha}}{\delta_{k,n,\alpha}}\dot{\delta}_{k,n,\alpha}\right), \tag{9}$$

$$U_{\mathrm{dc}} - R_k i_{k,n,\beta} = \frac{2\mu_0 A N^2}{\delta_{k,n,\beta}}\cdot\left(\dot{i}_{k,n,\beta} - \frac{i_{k,n,\beta}}{\delta_{k,n,\beta}}\dot{\delta}_{k,n,\beta}\right). \tag{10}$$

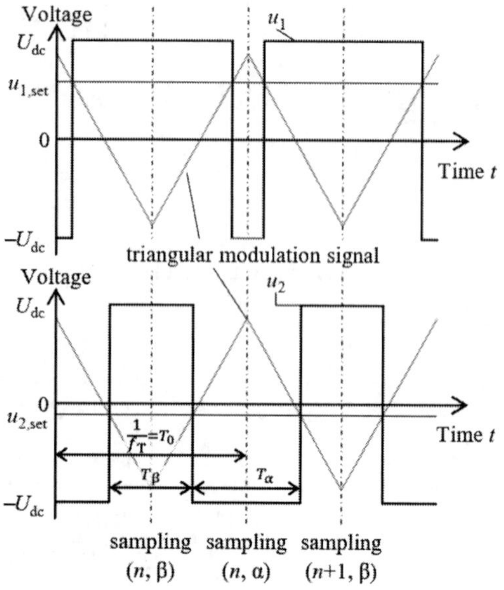

Fig. 2: Two-level PWM with double sampling per period

The time interval between the two samples (n, α) and (n, β) is half of the PWM period $1/(2f_{\mathrm{T}})$, which is sufficient small, so we could neglect the changes of the air gap width and the velocity of the positioned body. Then we have

$$\delta_{k,n,\alpha} \cong \delta_{k,n,\beta} = \delta_{k,n}\,, \tag{11}$$

$$\dot{\delta}_{k,n,\alpha} \cong \dot{\delta}_{k,n,\beta} = \dot{\delta}_{k,n}. \tag{12}$$

If the duty cycle of the nth PWM cycle according to $T_\alpha = T_\beta$ is 50%, we can further assume that

$$i_{k,n,\alpha} \cong i_{k,n,\beta} = i_{k,n}\,. \tag{13}$$

With the simplifications of (11)-(13), we derive the following equation from (9) and (10),

$$\delta_{k,n} = -\frac{\mu_0 A N^2}{U_{\mathrm{dc}}}\dot{i}^{*}_{k,n}\,, \tag{14}$$

where

$$\dot{i}^{*}_{k,n} = \dot{i}_{k,n,\alpha} - \dot{i}_{k,n,\beta}\,. \tag{15}$$

Equation (14) estimates the air gap width of the corresponding electromagnet. In the modeling we do not neglect the influences of the coil resistance and the induced voltage due to the body movement, and we eliminate these influences with the double detection of the current slopes. Thus, (14) will deliver a more accurate position estimation. Further, if we exploit the geometric constraints of (3) and (4), then we can also eliminate the position offsets, consisting of the mechanical nominal air gap width and the additional equivalent air gap width due to the limited iron permeability $\mu_{\mathrm{Fe}} < \infty$! So from (3) and (4) we get

$$x_n = -\frac{\delta_{1,n} - \delta_{2,n}}{2}. \tag{16}$$

Finally, considering (14) and (16) we derive the final expression of the estimated position during the nth cycle:

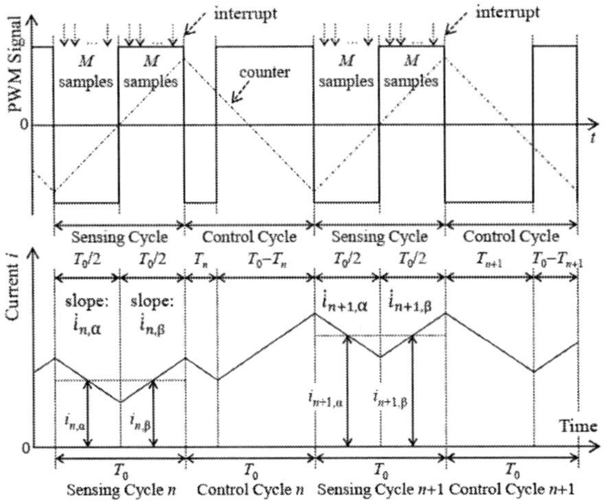

Fig. 3: PWM pattern with split sensing and control cycles (T_α=T_β=T_0/2)

$$x_n = \frac{\mu_0 A N^2}{2U_{\text{dc}}} \cdot \left(\dot{i}_{1,n}^* - \dot{i}_{2,n}^*\right). \tag{17}$$

In the next section, implementation details will be explained, concerning the measurement of the current slopes and the mechanism to ensure the 50% PWM duty cycle for the position estimation.

III. Implementation of the Approach

The easiest way to measure the current slope is to use voltage transformers. With the current flowing through the primary coil of the transformer, the open terminal voltage of the secondary coil is proportional to the current slope, as described in [7]. This approach demands space for installing transformers, and it also increases the cost. So a modern way is to use high-speed A/D (Analogue to Digital) converters to sample several points within one switching state and calculate the current slope with the method of least squares [9-11].

As shown in the preliminary work in [12], the current slopes are calculated via software. The number of multiple samples, which determines the accuracy of the slope calculation, is therefore limited by the system data transfer rate. Considerable improvement was made in [13], in which the calculation is performed in the FPGA (Field Programmable Gate Array). Further, the introduction of splitting of the PWM pattern into a sensing cycle, which has a fixed duty cycle of 50%, and a control cycle, which has the variable duty cycle (T_0 $-T_n$)/T_0, (T_0 $-T_{n+1}$)/T_0, … , for the current control, fits the requirement of (13) perfectly (Fig. 3). The duty cycle 50% gives zero change of the average current and does therefore not disturb the current control.

As shown in Fig. 3, we apply multiple sampling with M sample points twice during the sensing cycle. We show here only the slope calculation for negative switching as an example. For the nth PWM cycle, we obtain with the M measured current values $i_{n,\alpha,0}$; … ; $i_{n,\alpha,M-1}$, measured at the time instants $t_{n,\alpha,0}$; … ; $t_{n,\alpha,M-1}$, the following current vector

$$\boldsymbol{i}_{n,\alpha} = \left[i_{n,\alpha,0}\ i_{n,\alpha,1}\ i_{n,\alpha,2}\ \cdots\ i_{n,\alpha,M-1}\right]^T, \tag{18}$$

with the corresponding time vector,

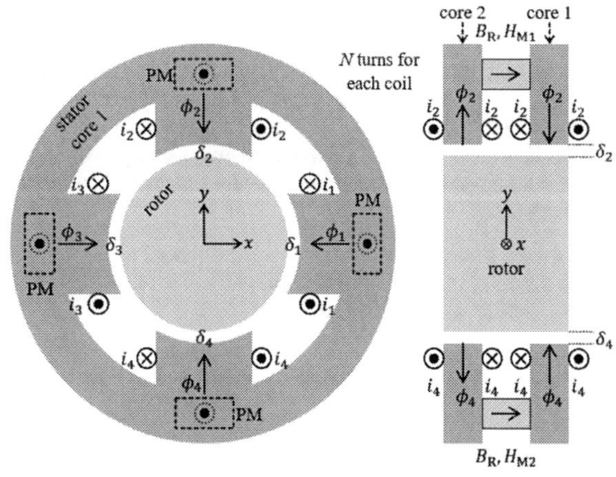

Fig. 4: Schematic representation of the used industrial magnetic bearing

$$\boldsymbol{t}_{n,\alpha} = \left[t_{n,\alpha,0}\ t_{n,\alpha,1}\ t_{n,\alpha,2}\ \cdots\ t_{n,\alpha,M-1}\right]^T. \tag{19}$$

Then the current slope can be calculated using the method of least squares with

$$\dot{i}_{n,\alpha} = \frac{\sum_{j=0}^{M-1}\left(i_{n,\alpha,j} - \bar{i}_{n,\alpha}\right)\cdot\left(t_{n,\alpha,j} - \bar{t}_{n,\alpha}\right)}{\sum_{j=0}^{M-1}\left(t_{n,\alpha,j} - \bar{t}_{n,\alpha}\right)^2}, \tag{20}$$

where $\bar{i}_{n,\alpha}$ and $\bar{t}_{n,\alpha}$ are the mean value of the M components of the vectors (18) and (19). To simplify the calculation in FPGA, we choose a fixed time interval $T_{\text{m.s.}}$ between neighbouring samples. Then the components of the time vector in (19) can be described by

$$t_{n,\alpha,j} = t_{n,\alpha,0} + j \cdot T_{\text{m.s.}},\ \ j = 0, 1, 2, \cdots, M-1. \tag{21}$$

Considering (21), equation (20) becomes

$$\dot{i}_{n,\alpha} = \frac{\sum_{j=0}^{M-1}(2j - M + 1)\cdot i_{n,\alpha,j}}{\frac{1}{6}M \cdot (M^2 - 1)\cdot T_{\text{m.s.}}}. \tag{22}$$

In the same way, the current slope of positive switching is given as

$$\dot{i}_{n,\beta} = \frac{\sum_{j=0}^{M-1}(2j - M + 1)\cdot i_{n,\beta,j}}{\frac{1}{6}M \cdot (M^2 - 1)\cdot T_{\text{m.s.}}}. \tag{23}$$

With (22), (23) and (17), we determine the estimated position. The advantage is the simplicity of the system structure. From the hardware point of view, it is the same as a pure current control. Fewer hardware components mean increased reliability and more cost saving.

A compromise is made between the accuracy of the position estimation and the dynamic of the current control. Due to the additional sensing cycle, whose average voltage is zero but which lasts for T_0 like the control cycle (Fig. 3), the effective DC-link voltage for the current control is reduced to its half. Therefore, the current change rate, caused by sensing and control cycle, is slower than with a normal PWM pattern without additional sensing cycles. An approach to achieve the fastest possible current control under this limitation was also given in [13].

IV. APPLICATION TO A REAL MAGNETIC BEARING

The proposed self-sensing technique is applied to an industrial magnetic bearing, which has bias magnetic fluxes in axial direction, excited by permanent magnets, and coupled control fluxes in radial direction, excited by coils, as shown in Fig. 4. Therefore, due to the coupled magnetic paths of the two orthogonal x and y axes, the modelling is quite different to the one described in Section II.

We take the same assumption for the magnetic properties as in Section II ($\mu_{\mathrm{Fe}} \to \infty$) and define the quantities, shown in Fig. 4, as follows: For the subscripts $k = 1,2,3,4$, δ_k is the individual air gap width; ϕ_k is the total magnetic flux through the poles of the stator iron cores, which has opposite direction in stator core1 and core 2; i_k is the controlled coil current. Each coil has N turns, and the coils, which are located in the same position on stator core 1 and 2, are connected in series as a coil group. The voltage applied to each coil group is u_k; $H_{\mathrm{M},k}$ is the magnetic field strength in each permanent magnet in axial direction. Each magnet has the remanence B_{R}, the length in axial direction l_{M}, and the cross section area in the radial plane A_{M}. The cross section area of the poles is A_{P}. The permeabilities of the air and of the magnets are μ_0 and μ_{M}, respectively.

According to Ampere's law and Gauss's law for magnetism, we obtain 8 equations to describe the electromagnetic properties of the given magnetic bearing. These equations are written in matrix form

$$\boldsymbol{A} \cdot \boldsymbol{s} = \boldsymbol{b} \tag{24}$$

with the matrix $\boldsymbol{A}$ defined as

$$\begin{bmatrix} a_0 & 0 & 0 & 0 & \delta_1 & 0 & 0 & 0 \\ 0 & a_0 & 0 & 0 & 0 & \delta_2 & 0 & 0 \\ 0 & 0 & a_0 & 0 & 0 & 0 & \delta_3 & 0 \\ 0 & 0 & 0 & a_0 & 0 & 0 & 0 & \delta_4 \\ 0 & 0 & 0 & 0 & -\delta_1 & \delta_2 & 0 & 0 \\ 0 & 0 & 0 & 0 & 0 & -\delta_2 & \delta_3 & 0 \\ 0 & 0 & 0 & 0 & 0 & 0 & -\delta_3 & \delta_4 \\ -a_1 & -a_1 & -a_1 & -a_1 & 1 & 1 & 1 & 1 \end{bmatrix}, \tag{25}$$

where $a_0 = 0.5\mu_0 A_{\mathrm{P}} l_{\mathrm{M}}$ and $a_1 = \mu_{\mathrm{M}} A_{\mathrm{M}}$. The vector of unknowns $\boldsymbol{s}$ is defined as

$$[H_{\mathrm{M},1} \quad H_{\mathrm{M},2} \quad H_{\mathrm{M},3} \quad H_{\mathrm{M},4} \quad \phi_1 \quad \phi_2 \quad \phi_3 \quad \phi_4]^{\mathrm{T}}. \tag{26}$$

The free term vector $\boldsymbol{b}$ is defined as

$$\mu_0 A_{\mathrm{P}} N \cdot \left[i_1 \quad i_2 \quad -i_3 \quad -i_4 \quad -i_1 + i_2 \quad -i_2 - i_3 \quad i_3 - i_4 \quad \frac{4B_{\mathrm{R}} A_{\mathrm{M}}}{\mu_0 A_{\mathrm{P}} N} \right]^{\mathrm{T}} \tag{27}$$

with the positive direction of current flow as shown in Fig. 4.

We apply the time derivative to both sides of (24) and get

$$\dot{\boldsymbol{A}} \cdot \boldsymbol{s} + \boldsymbol{A} \cdot \dot{\boldsymbol{s}} = \dot{\boldsymbol{b}}. \tag{28}$$

Using the double slope detecting technique for the current signals, for the given n^{th} sensing cycle, we have

$$\dot{\boldsymbol{A}}_{n,\alpha} \cdot \boldsymbol{s}_{n,\alpha} + \boldsymbol{A}_{n,\alpha} \cdot \dot{\boldsymbol{s}}_{n,\alpha} = \dot{\boldsymbol{b}}_{n,\alpha}, \tag{29}$$

and

$$\dot{\boldsymbol{A}}_{n,\beta} \cdot \boldsymbol{s}_{n,\beta} + \boldsymbol{A}_{n,\beta} \cdot \dot{\boldsymbol{s}}_{n,\beta} = \dot{\boldsymbol{b}}_{n,\beta}. \tag{30}$$

The assumptions (11)-(13) are still valid, and we further assume due to the high switching frequency

$$H_{\mathrm{M},k,n,\alpha} \cong H_{\mathrm{M},k,n,\beta} = H_{\mathrm{M},k,n} \tag{31}$$

and

$$\phi_{k,n,\alpha} \cong \phi_{k,n,\beta} = \phi_{k,n}. \tag{32}$$

Then, (29) and (30) can be united to (33)

$$\boldsymbol{A}_n \cdot \left(\dot{\boldsymbol{s}}_{n,\alpha} - \dot{\boldsymbol{s}}_{n,\beta}\right) = \dot{\boldsymbol{b}}_{n,\alpha} - \dot{\boldsymbol{b}}_{n,\beta}. \tag{33}$$

We further define

$$\dot{H}^*_{\mathrm{M},k,n} = \dot{H}_{\mathrm{M},k,n,\alpha} - \dot{H}_{\mathrm{M},k,n,\beta} \tag{34}$$

and

$$\dot{\phi}^*_{k,n} = \dot{\phi}_{k,n,\alpha} - \dot{\phi}_{k,n,\beta}, \tag{35}$$

so we get

$$\boldsymbol{A}_n \cdot \dot{\boldsymbol{s}}^*_n = \dot{\boldsymbol{b}}^*_n. \tag{36}$$

According to Faraday's law, we have with the given arrow system of Fig. 4,

$$R_k i_k + 2N\frac{\mathrm{d}\phi_k}{\mathrm{d}t} = u_k, \quad \text{for } k = 1,2, \tag{37}$$

and

$$R_k i_k - 2N\frac{\mathrm{d}\phi_k}{\mathrm{d}t} = u_k, \quad \text{for } k = 3,4. \tag{38}$$

The coil terminal voltages during the α and β interval of the nth sensing cycle are

$$u_{k,n,\alpha} = -U_{\mathrm{dc}} \text{ and } u_{k,n,\beta} = U_{\mathrm{dc}} \text{ for } k = 1,2,3,4. \tag{39}$$

Therefore, we get with (35), (37), (38) and (39)

$$\dot{\phi}^*_{k,n} = -\frac{U_{\mathrm{dc}}}{N}, \quad k = 1,2, \tag{40}$$

and

$$\dot{\phi}^*_{k,n} = \frac{U_{\mathrm{dc}}}{N}, \quad k = 3,4. \tag{41}$$

Now we construct a modified equation system, which takes the air gap widths $\delta_{k,n}$ as unknown variables instead of $\dot{\phi}^*_{k,n}$.

$$\widetilde{\boldsymbol{A}} \cdot \tilde{\boldsymbol{s}}_n = \widetilde{\boldsymbol{b}}_n \tag{42}$$

In (42) the matrix $\widetilde{\boldsymbol{A}}$ with constant coefficient is

$$\begin{bmatrix} a_0 & 0 & 0 & 0 & -a_2 & 0 & 0 & 0 \\ 0 & a_0 & 0 & 0 & 0 & -a_2 & 0 & 0 \\ 0 & 0 & a_0 & 0 & 0 & 0 & a_2 & 0 \\ 0 & 0 & 0 & a_0 & 0 & 0 & 0 & a_2 \\ 0 & 0 & 0 & 0 & a_2 & -a_2 & 0 & 0 \\ 0 & 0 & 0 & 0 & 0 & a_2 & a_2 & 0 \\ 0 & 0 & 0 & 0 & 0 & 0 & -a_2 & a_2 \\ a_1 & a_1 & a_1 & a_1 & 0 & 0 & 0 & 0 \end{bmatrix}, \tag{43}$$

where $a_2 = \frac{U_{\mathrm{dc}}}{N}$. The modified vector of unknowns $\tilde{\boldsymbol{s}}_n$ is

$$[\dot{H}^*_{\mathrm{M},1,n} \quad \dot{H}^*_{\mathrm{M},2,n} \quad \dot{H}^*_{\mathrm{M},3,n} \quad \dot{H}^*_{\mathrm{M},4,n} \quad \delta_{1,n} \quad \delta_{2,n} \quad \delta_{3,n} \quad \delta_{4,n}]^{\mathrm{T}} \tag{44}$$

and the modified free term vector $\widetilde{\boldsymbol{b}}_n$ is

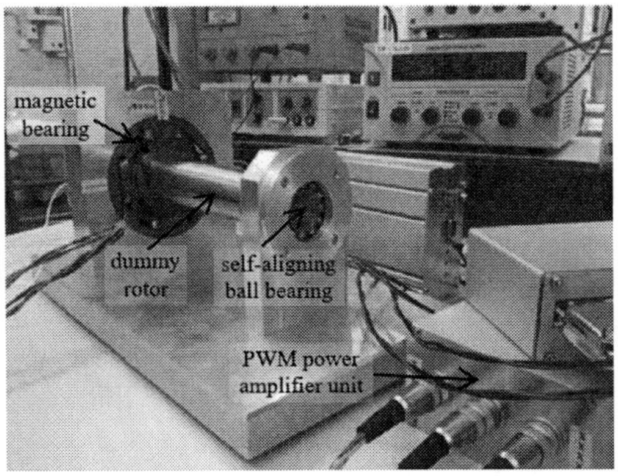

Fig. 5: Test bench of the self-sensing magnetic bearing

$$\begin{gathered}\mu_0 A_P N \cdot \left[\dot{i}^*_{1,n} \quad \dot{i}^*_{2,n} \quad -\dot{i}^*_{3,n} \quad -\dot{i}^*_{4,n} \right. \\ \left. -\dot{i}^*_{1,n}+\dot{i}^*_{2,n} \quad -\dot{i}^*_{2,n}-\dot{i}^*_{3,n} \quad \dot{i}^*_{3,n}-\dot{i}^*_{4,n} \quad 0\right]^T .\end{gathered} \tag{45}$$

We solve the linear equation system (42) for the air gap widths and get the four solutions

$$\delta_{k,n} = -\frac{\mu_0 A_P N^2}{U_{dc}} \dot{i}^*_{k,n} , \tag{46}$$

which have a similar expression as (14). Again, we apply the geometric constrains given as

$$x = -\frac{\delta_1 - \delta_3}{2} , \tag{47}$$

and

$$y = -\frac{\delta_2 - \delta_4}{2} . \tag{48}$$

Thus we finally obtain the expression of the position estimation as

$$x_n = \frac{\mu_0 A_P N^2}{2U_{dc}} \cdot (\dot{i}^*_{1,n} - \dot{i}^*_{3,n}) \tag{49}$$

and

$$y_n = \frac{\mu_0 A_P N^2}{2U_{dc}} \cdot (\dot{i}^*_{2,n} - \dot{i}^*_{4,n}) . \tag{50}$$

Equations (49) and (50) imply that only the axial fluxes, which are magnetically decoupled, are relevant for the position estimation. The coupled radial control fluxes do not seem to affect the algorithm. But we point out that due to the existing relative big leakage fluxes in axial direction, which are not included in the modelling, the position estimation is largely influenced by the coupled magnetic paths in the radial plane. The coupling trends to deform the estimated orbit in the radial plane into an ellipse form, as shown later in Section V.

V. Experimental results

We built a test bench to validate the proposed self-sensing technique. As shown in Fig. 5, the dummy rotor is supported by the magnetic bearing on the left side and by a self-aligning ball bearing on the right side. The used industrial magnetic bearing was equipped originally with a differential control winding, which is not suitable for the proposed self-sensing technique, because the current slopes of the individual coil groups are not accessible. Therefore, we spilt the connection between the coil groups to drive the four coil groups separately.

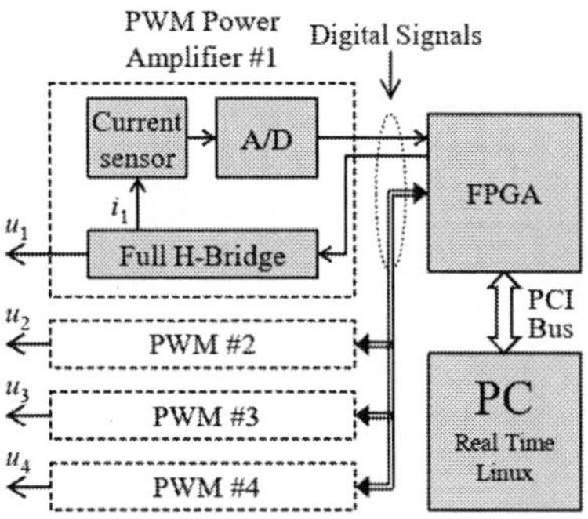

Fig. 6: Block diagram of the control system

Fig. 6 illustrates the structure of the control system. The PWM power amplifier unit is our own design, which has on-board current sensors and A/D converters, so that the currents are measured and digitized locally and transmitted digitally, which reduces the distortion of signal transmission, caused by electromagnetic interference of the excited coil groups. The full-bridge circuit and the A/D converter are driven by a FPGA board, which generates the PWM signals and synchronizes the multiple current sampling according to the timing illustrated in Fig. 3. The FPGA calculates additionally the current slope with (22) and (23). Thus, the FPGA serves as an interface between the magnetic bearing system and the control software, running on a PC under real time Linux.

The PWM switching frequency is $f_T = 20\,\text{kHz}$. The digital control works therefore on half of that frequency $f_T/2 = 10\,\text{kHz}$ due to the timing with $2T_0$ of sensing and control cycle, as shown in Fig. 3. The present FPGA firmware features $M = 8$ as multiple current samples with $T_{m.s.} = 2$ µs, which results in a 500 kHz sampling rate for the A/D converters. The used 12-bit A/D converter can be driven up to 1 MHz. So further improvements concerning the multiple sampling are possible. We also mounted position sensors on the magnetic bearing to get a reference signal for the position in comparison to the estimated position. A first evaluation concerning the static and transient performance of the position estimation was made with the enabled closed loop position control, using the measured position.

Fig. 7 shows the measured quasi-static performance of the position estimation. We control the rotor position so that it follows a circular orbit with a radius of about 75 µm around the central axis. It moves with a period of 5 s. We compare the estimated position with the measured position. The direct estimation output, using (49) and (50), is plotted in Fig. 7a and 7c. We obtained a different phase shift in x- and y-signals beside the general gain error for both axes: the estimated position signal of x-axis lags the measured position signal, while the estimated position signal of y-axis leads the measured position signal. This phase shift results in an elliptical deformation of the estimated orbit. As mentioned in Section IV, this deformation is caused by the coupled radial fluxes. This is a nature of the magnetic bearing, which means it could be corrected by an appropriate model. We choose the simple ellipse transformation for the correction and obtain the corrected position $(\tilde{x}_n, \tilde{y}_n)$ with the following equation

$$\begin{bmatrix}\tilde{x}_n \\ \tilde{y}_n\end{bmatrix} = \begin{bmatrix} g_1 & g_1 g_2 \\ g_1 g_2 & g_1 \end{bmatrix} \cdot \begin{bmatrix} x_n \\ y_n \end{bmatrix} , \tag{51}$$

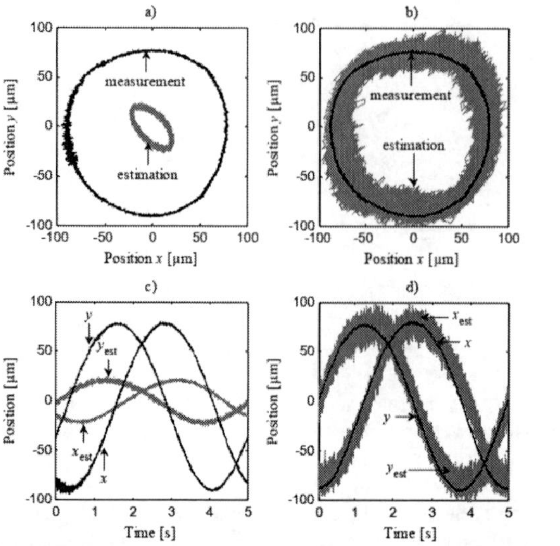

Fig. 7: Original and corrected position estimation vs. position measurement: a) Orbit of original position estimation, c) corresponding position curve of a), b) corected position estimation, d) corresponding position curve of b).

where g_1 is the common gain, which corrects the general gain error; and g_2 is the cross gain, which compensates the different phase shift of the position on the two orthogonal axes. After this correction (51), the estimated orbit now matches the measured orbit, as shown in Fig. 7b. The estimation and measurement are also in phase now, as shown in Fig. 7d. Due to the gain scaling, the noise band is generally amplified, which requires a further improvement concerning noise cancelling.

Fig. 8 shows the transient performance of the position estimation. The *x*-position of the rotor is kept at zero in the center. We give the *y*-axis position controller a step position command of 175 μm and record the measured and estimated position. The step response is given in Fig. 8c. A zoomed curve is plotted in Fig. 8d. During the step change of the real position on the *y*-axis, there is no noticeable delay of the estimated position, which means a very good dynamic performance. The estimated position on the *x*-axis remains unchanged, therefore, the cross coupling due to the coupled radial fluxes is completely eliminated.

VI. Conclusions and Future Work

A position estimation technique for self-sensing magnetic bearings is proposed in this paper. The derivation of the algorithm and the technical implementation is explained in detail. Experimental results show that the proposed technique also applies to existing industrial active magnetic bearings with coupled radial control fluxes and permanent magnet excited bias fluxes.

The future work will be focused on the improvement of signal processing to further reduce the noise level without loss of estimation dynamic. The position controller will be adapted for the estimated position as control feedback.

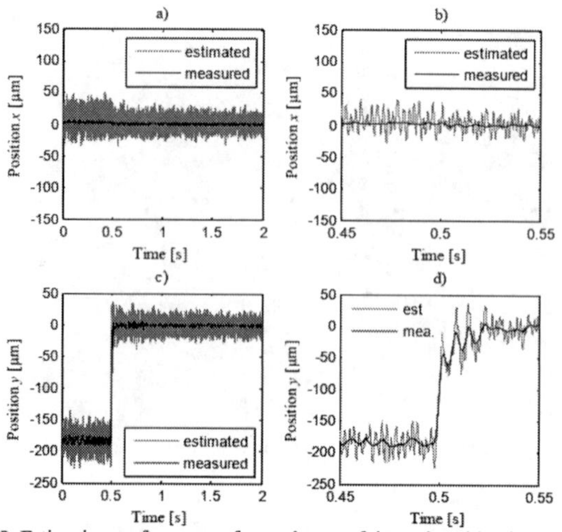

Fig.8: Estimation performance of step change of the real position in *y*- axis: a) Position of *x*-axis during step change of the position in *y*-axis, b) zoom of a), c) step change of the real position in *y*- axis, d) zoom of c).

References

[1] M. Brunet, U. Schroeder, Y. Tremaudant, "Active magnetic bearing with automatic detection of the position thereof," Patent WO2005103517 A1, 3 Nov., 2005.

[2] M. Morishita, H. Itoh, "The self-gap-detecting electromagnetic suspension system with robust stability against variation of levitation mass," in Proceedings of International Symposium on Power Electronics, Electrical Drives, Automation and Motion (SPEEDAM), 2006, pp. 1178-1183.

[3] K. Morita, T. Yoshida, K. Ohniwa, "Improvement of Sensing Characteristics of Self-Sensing Active Magnetic Bearings," Electrical Engineering in Japan, vol. 166, pp. 70-77, Jan. 2009.

[4] M. D. Noh, E. H. Maslen, "Self-sensing magnetic bearings using parameter estimation," IEEE Transactions on Instrumentation and Measurement, vol. 46, pp. 45-50, 1997.

[5] A. Schammass, R. Herzog, P. Bühler, et al., "New results for self-sensing active magnetic bearings using modulation approach," IEEE Transactions on Control Systems Technology, vol. 13, pp. 509-516, 2005.

[6] Y. Jung-Sik, K. Jang-Hwan, S. Seung-Ki, et al., "Sensorless position control of active magnetic bearings based on high frequency signal injection method," in Proceedings of Applied Power Electronics Conference and Exposition, Miami, USA, 2003, vol. 1, pp. 83-88.

[7] L. Lichuan, T. Shinshi, A. Shimokohbe, "State feedback control for active magnetic bearings based on current change rate alone," IEEE Transactions on Magnetics, vol. 40, pp. 3512-3517, 2004.

[8] A. C. Niemann, "Self-sensing Algorithms for Active Magnetic Bearings," Ph.D. Thesis, North-West University, 2008.

[9] J. Weigold, M. Braun, "Predictive Current Control Using Identification of Current Ripple," IEEE Transactions on Industrial Electronics, vol. 55, pp. 4346-4353, 2008.

[10] F. Becker, H. Ennadifi, M. Braun, "Straightforward current control - One step controller based on current slope detection," in Proceedings of 14th European Conference on Power Electronics and Applications (EPE), Birmingham, UK, 2011, 10 pages.

[11] A. Liske, P. Hofmeier, M. Braun, "Extended straightforward current control for permanent magnet synchronous machines," in Proceedings of 15th European Conference on Power Electronics and Applications (EPE), Lille, France, 2013, paper no. 0841, 10 pages.

[12] J. Wang, A. Binder, "Self-sensing magnetic bearings using multiple sampling of currents alone," in Proceedings of 15th European Conference on Power Electronics and Applications (EPE), Lille, France, 2013, paper no. 0210, 10 pages.

[13] J. Wang, A. Binder, "Current slope calculation in FPGA for sensorless control technique and associated slope based predictive precise current control," in Proceedings of IEEE International Symposium on Sensorless Control for Electrical Drives and Predictive Control of Electrical Drives and Power Electronics (SLED/PRECEDE) Munich, Germany, 2013, paper no. S3P1, 8 pages.

Improved Sensorless Control of a Modular Three Phase Radial Active Magnetic Bearing

Matthias Hofer, Markus Hutterer, Thomas Nenning, Manfred Schrödl

Institute of Energy Systems and Electric Drives, Vienna University of Technology,
Gusshausstraße 25, 1040 Vienna, Austria,
matthias.hofer@tuwien.ac.at

***Abstract*—This paper discusses new results and improvements on the sensorless control of a permanent magnet biased three phase radial active magnetic bearing (AMB), which are also known as hybrid magnetic bearings (HMB). Self-sensing methods offer cost potentials because of a more simple bearing system architecture, but show lower performance in accuracy and bandwidth compared to external position sensors. Thus, making self-sensing technologies competitive to systems with position sensors is a broad field of research and as well a goal of this paper. The proposed sensorless method is based on inductance measurement by voltage pulse injection. Applying an oversampling strategy for current slope measurement reduces the noise of the sensorless obtained rotor position significantly. Timing limitations and operational bandwidth can be improved by utilizing alternative voltage pulse sequences. Further, variations of the permanent magnet bias flux of the HMB prototype are investigated to show the influences on the sensorless rotor position detection method and limitations for applications.**

I. Introduction

Active magnetic bearings are used for levitation of rotating shafts. Usually they are used in applications where classical technologies are not useful, especially at high operational speeds and/or under high environmental requirements [1]. Here the advantages of a frictionless and a wearless mechanical support with low maintenance and additional functionality like active balancing are utilized. Definitely, compared to classical bearing types the AMB system is more complex and has higher costs. One potential for cost reduction is found in the self-sensing technology which reduces the number of components and increase utilization of of already used components. In general, sensorless means, that the electromagnetic actuator itself is used as position sensor too and rotor position information is evaluated by electrical quantities, namely currents and voltages. Thus, external position sensors are avoided. At AMB applications the position sensing system takes over a major functionality. First because of the unstable positioning behavior at ferromagnetic levitation systems. And second, the sensor parameters affect the overall AMB system and its operation boundaries like bandwidth, noise, accuracy, etc. Thus, self-sensing or respectively sensorless technologies for AMBs are field of research since many years [2] with the target to reach parameters as sensor systems do have. The sensorless control approach used in this work is based on the so called INFORM (Indirect Flux detection by Online Reactance Measuremnt) method which is known from sensorless control of electrical drive applications. At AMBs this method allows the rotor position determination during voltage injection pulses by using a three phase bearing architecture (Fig. 1) together with a three phase voltage source inverter similar to electric drives [3], [4]. Hence, using common inverter technology offers a further cost reduction potential. A short introduction in the INFORM method at AMB is given at the beginning of this work. First, alternative voltage injection sequences and their position sensing accuracy are discussed. Limitation factors for self-sensing performance are already known from literature [5], [6]. Especially nonlinearities like saturation have high impact on self-sensing technologies. Therefore one goal of this work is to identify the impact of high flux levels or even saturation on the sensorless obtained rotor position by the INFORM method.

Figure 1. Stator lamination with various permanent magnet configurations - left: N_M=24 ferrite magnets, right: N_M=36 NdFeB magnets

II. Active Magnetic Bearing Design

In this work a three phase permanent magnet biased homopolar AMB with a modular approach is used. By utilizing exchangeable permanent magnets according to Fig. 1 instead of a permanent magnet ring , e.g. as shown in [7], the bias flux level can be adjusted. A wide bias flux range is reached by varying the number of magnets N_M and even the magnet types according to Tab. I. Combined with different coil configurations a variable and modular bearing setup is given which enables the identification of parameter dependencies and possible operating limits of the sensorless control method.

Table I

BEARING SETUP FOR DIFFERENT BIAS FLUX - ANALYTICAL MODEL

N_M	Magnet type	Bias flux density B_0
12	Ferrite	0.079 T
24	Ferrite	0.182 T
36	Ferrite	0.269 T
12	NdFeB	0.319 T
24	NdFeB	0.649 T
36	NdFeB	0.949 T

The bias flux density B_0 affects the bearing force characteristic of the AMB, e.g the force behavior in x direction (vertical) is shown in Fig. 2. As known, at increasing bias flux density B_0 the force generation per phase current increases. Consequently the force current stiffness factor K_i gets higher for permanent magnet biased bearings and result in a lower current consumption and higher bearing efficiency.

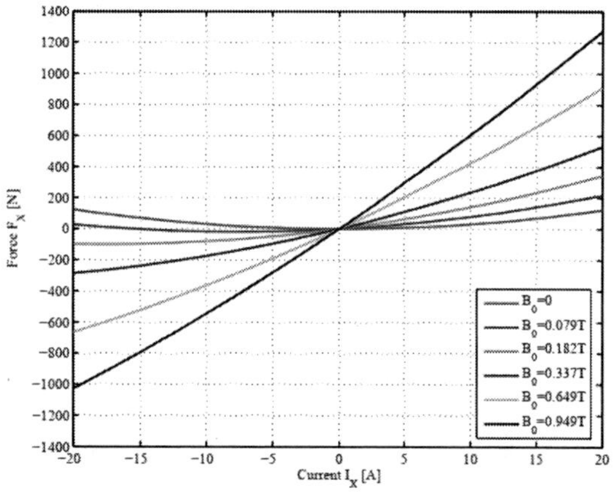

Figure 2. Modeled bearing forces F_X at various permanent magnet configurations for a coil with N=20 turns

In Fig. 3 the phase current for several bias flux densities B_0 is modeled. Additionally measurement results for the AMB prototype setup with N_M=24 Ferrite magnets show a bias current of I_{X0}=7.5A. For another setup with higher premagnetization by using a combination of 12 ferrite and 12 NdFeB magnets per stator stack the bias current consumption is reduced to I_{X0}=4.65A. These measurements confirm the principle behavior of the analytical model, although the measured current consumption is slightly higher due to model uncertainties.

III. SENSORLESS CONTROL METHOD

A. Sensorless position detection

For the capability of a sensorless rotor position detection the AMB stator inductance dependency on the rotor position has

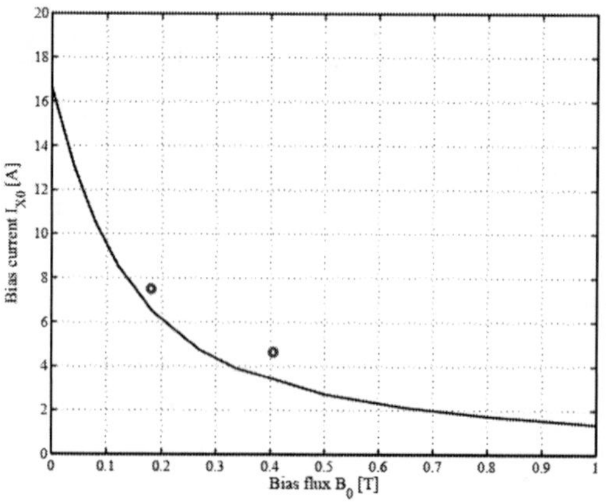

Figure 3. Bias Current I_{X0} for several bias flux densities B_0 at N=20 turns for a rotor weight m = 8.64kg- analytical model (black) and measurement results (red).

to be considered. Using a two axis notation the flux linkages ψ_X and ψ_Y derive as

$$\begin{aligned} \psi_X &= L_{XX} \cdot I_X + L_{XY} \cdot I_Y \\ \psi_Y &= L_{YX} \cdot I_X + L_{YY} \cdot I_Y \end{aligned} \quad (1)$$

with the self inductances L_{XX}, L_{YY} and the cross coupling inductances $L_{XY} = L_{YX}$. Herein, the complex stator current space phasor is represented by the components I_X and I_Y. For measuring the inductance characteristic and its position dependency the inverse inductance is helpful. Using the nominal inductance L_0 at centered rotor the inverse of the rated self inductance is build as

$$\frac{1}{l_{XX}|_{y=0}} = \frac{L_0}{L_{XX}} \quad (2)$$

An analysis of the analytic electromagnetic AMB model show a dependency on the rotor position x according to Fig. 4. For an ideal permeable iron ($\mu_{r,Fe} \to \infty$) a simple characteristic is given at a nominal airgap δ_0 as

$$\frac{1}{l_{XX}|_{y=0}} = 1 - \frac{x}{2\delta_0} \quad (3)$$

Changing the relative permeability of the iron path from an ideal permeable material to a real parameter value of e.g. $\mu_{r,Fe}$ = 1000 a reduction of the position sensitivity is identified. Thus, to reach high accuracy in self-sensing the relative permeability should be as high as possible. To evaluate the rotor displacement out of this fundamental characteristic a linear combination of three current slope measurements in each phase direction U, V and W is used.

B. Symmetric voltage pulses

A typical current response during an applied symmetric voltage injection pulse sequence is shown in Fig. 5. For a measurement in x direction (aligned with phase direction U) a voltage space phasor pattern $\underline{u}_{u-} \to \underline{u}_{u+} \to \underline{u}_{u-} \to \underline{u}_{u+}$ is applied. This classical pulse sequence result in a symmetric current slope detection around the operation point.

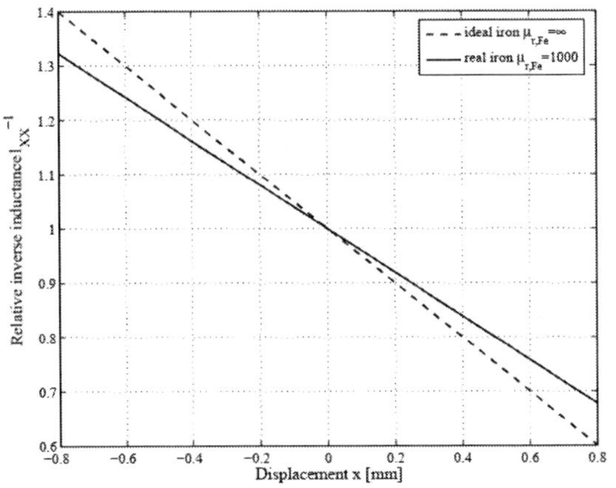

Figure 4. Relative inverse inductance $l_{XX}{}^{-1}$ analytically derived. Dependency on the rotor position $x|_{y=0}$ for ideal permeable material and $\mu_{r,Fe} = 1000$

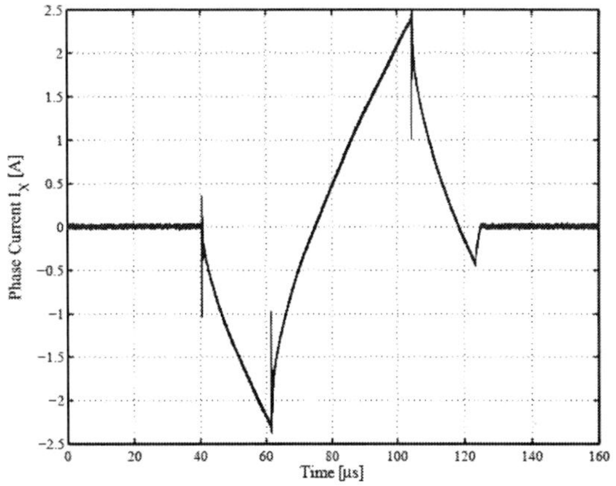

Figure 6. Inductance measurement of $L_{XX}|_{y=0}$ by asymmetric voltage pulse injections at position $x = -0.4mm$ at U_{DC}=120 V. Measured current reaction I_X.

The measured current response show eddy current effects immediately after each voltage switching. Therefore current slope measurement is only performed in the "linear" range of the current response, but this reduces the sensitivity compared to the maximum theoretical value without eddy currents.

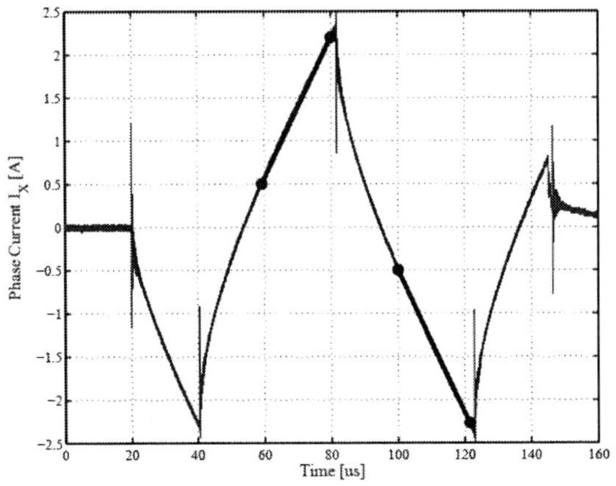

Figure 5. Inductance measurement of $L_{XX}|_{y=0}$ by symmetric voltage pulse injections at position $x = 0.8mm$, U_{DC}=120 V. Measured current reaction I_X and linear current slope approximation by two single samples per slope (black)

C. Asymmetric voltage pulses

Alternative to a symmetric pulse pattern a shortened asymmetric pulse sequence can be used as depicted in Fig. 6. Here the current slope is not measured symmetrically around the magnetic operation point due to eddy current effects. By shortening the pulse sequence duration the position measurement frequency can be increased for improving self-sensing bandwidth but the signal content for slope detection is only half compared to the symmetric voltage pulses.

D. Three active voltage pulses

The three active sequence applies voltage space phasors in all three phase directions within one sequence. E.g using negative space phasors $\underline{u}_{u-} \rightarrow \underline{u}_{v-} \rightarrow \underline{u}_{w-}$ the current slopes Δi_U, Δi_V and Δi_W are available immediately and result in a higher self sensing bandwith at the same pulse rate. Using pre-phasors and post-phasors according to Fig. 7 the slope measurements are symmetric around the operation point. The measured current reaction is depicted in Fig. 8.

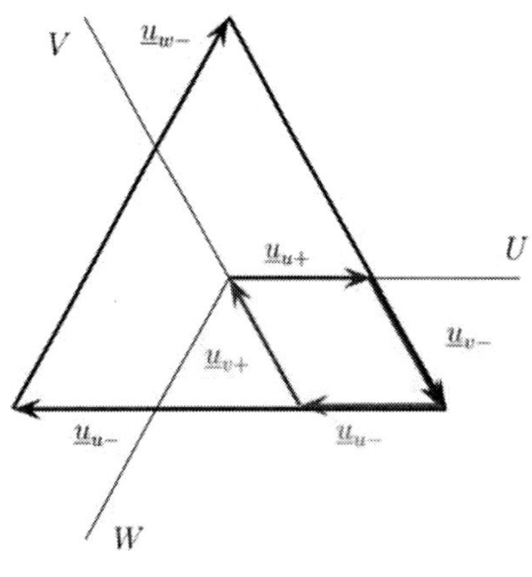

Figure 7. Voltage space phasors for three active voltage pulse injection including pre-phasors (blue) and post-phasors (red) achieving symmetric operation points

E. Noise reduction by oversampling

New digital signal processors (DSPs) offer fast analog to digital conversions (ADC) with high resolution. Therefore increasing the number of samples during the voltage pulse sequence is feasible. This oversampling strategy reduces the noise of the sensorless determined rotor position for same pulse pattern as shown in Tab. II. Here statistical methods

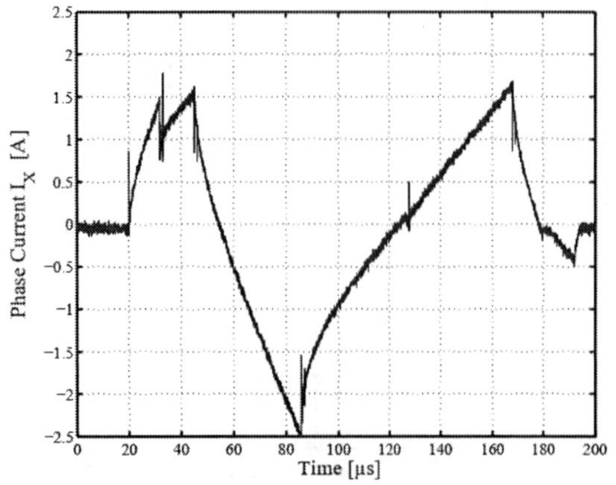

Figure 8. Inductance measurement $L_{XX}|_{y=0}$ by three active voltage pulse injections at position $x = -0.4mm$ at U_{DC}=120 V. Measured current reaction I_X.

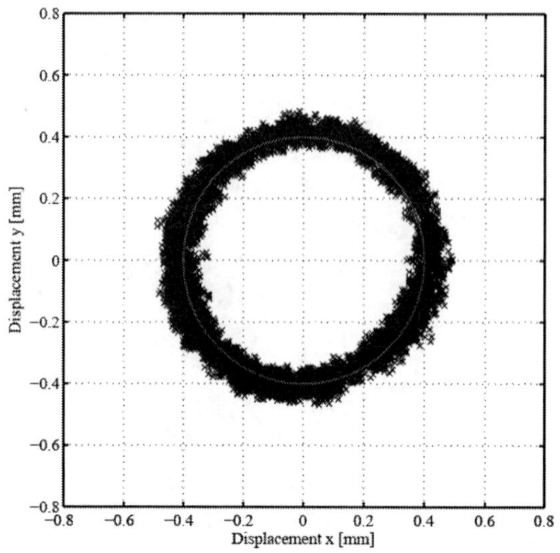

Figure 9. Trajectory of the sensorless position measurements with double sampling and symmetric voltage pulses at radial rotor displacements r=0.4mm for a setup with N_M=24 ferrite magnets (black). Rotor displacement (red).

are used to evaluate the accuracy of the sensorless rotor position according [8]. A standard deviation of σ=25 μm can be reduced to σ=15 μm by changing from a double sampling strategy to quad sampling for symmetric pulses. Depending on electronic hardware capabilities (computation power of the DSP and the ADC) current sampling even up to a continuous oversampling of the current signal during the injected voltage pulse can offer further improvements.

Table II
STANDARD DEVIATION OF THE ROTOR POSITION FOR DIFFERENT INFORM PULSES

Pulse type	Sampling	Std. dev. σ	Bearing setup
symmetric	2	25 μm	24 ferrites
symmetric	4	15 μm	24 ferrites
symmetric	4	12 μm	12 ferrites, 12 NdFeB
symmetric	8	11 μm	12 ferrites, 12 NdFeB
asymmetric	4	19 μm	24 ferrites
asymmetric	4	19 μm	12 ferrites, 12 NdFeB
asymmetric	8	19 μm	12 ferrites, 12 NdFeB
three active	16	8 μm	12 ferrites, 12 NdFeB

The sensorless position signal accuracy can also be evaluated considering mechanically defined rotor positions. Measurement results in Tab. II show the position noise at one mechanical fixed rotor position. Alternatively trajectories at certain radial displacements are used to discuss the self sensing positon accuracy. Such trajectories for a radial displacement of r=0.4 mm (40 % of the airgap defined by touchdown bearings) are depicted in Fig. 9, Fig. 10 and Fig. 11 for different voltage pulse sequences and number of current slope samples.

For the sensorless control of the permanent magnet biased AMB prototype a three phase voltage source inverter with digital control is used. The current measurement is implemented with shunt resistors at the low side switches according to Fig. 12. This current sensing architecture is advantageous for current slope measurement during negative voltage phasors $\underline{u}_{u-}$, $\underline{u}_{v-}$ and $\underline{u}_{w-}$ because here the phase currents can be measured directly. At positive voltage space phasors the phase currents have to be evaluated indirectly out of the other two phase currents. For this reason the position measurement by the three active sequence allows higher number of current samples of the sequential ADC and result in a higher position accuracy.

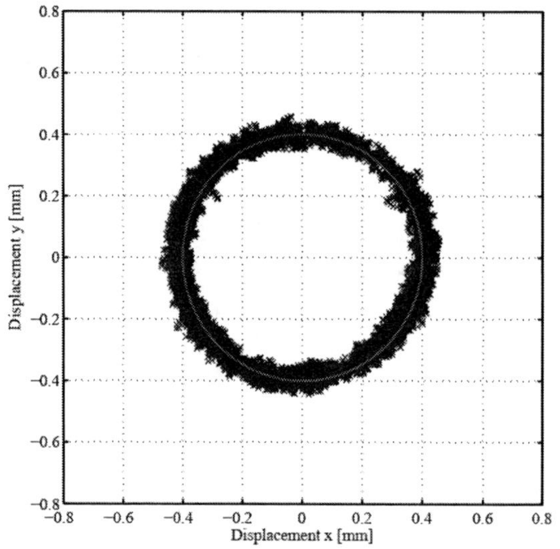

Figure 10. Trajectory of the sensorless position measurements with quad sampling and asymmetric voltage pulses at radial rotor displacements r=0.4mm for a setup with N_M=24 ferrite magnets (black). Rotor displacement (red).

IV. Bias Flux Influence on the Sensorless Rotor Position

Permanent magnet biased AMBs can achieve high load forces and and a high current force coefficient K_i at relatively

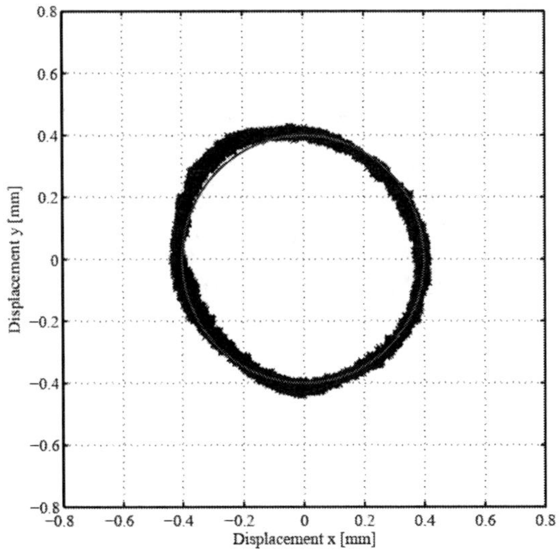

Figure 11. Trajectoriy of the sensorless position measurements with 16 samples at three active voltage pulses at radial rotor displacements r=0.4mm for a setup with 12 ferrite magnets (black). Rotor displacement (red).

small bearing dimensions. Thus a high utilization of the magnetic materials with high flux density levels is given and can lead to saturation. As saturation of the iron path influences self sensing performance [5], [6] this behavior and possible limitations for the applied sensorless method are a main interest in this work too. According to Tab. II a comparison of two different bias flux levels is given for several voltage pulse sequences. The increased bias flux level do not affect the self sensing position noise in the open loop, but the trajectory shape is slightly deformed at the higher bias flux as shown in Fig. 11.

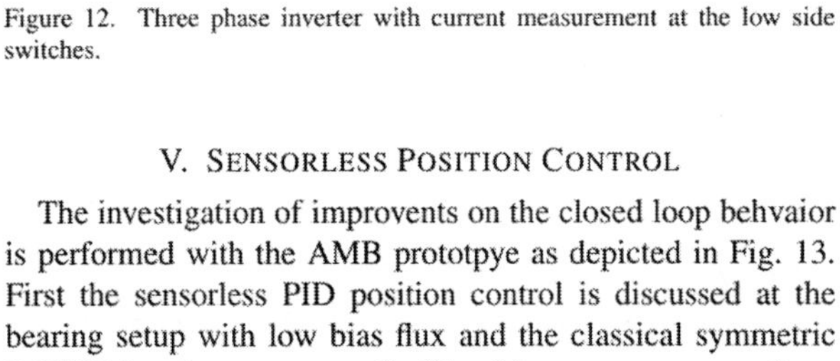

Figure 12. Three phase inverter with current measurement at the low side switches.

V. Sensorless Position Control

The investigation of improvents on the closed loop behvaior is performed with the AMB prototpye as depicted in Fig. 13. First the sensorless PID position control is discussed at the bearing setup with low bias flux and the classical symmetric INFORM pulse sequence. In Fig. 14 a step response from $\pm 0.2mm$ (20% of the airgap) is depicted. The position signal does only contain position sensing noise which is not much higher than in the open loop and no clear oscillations are identified. Despite the low bias flux and low gain of the position controller the load capacity is low.

The sensorless control with asymmetric INFORM pulses at

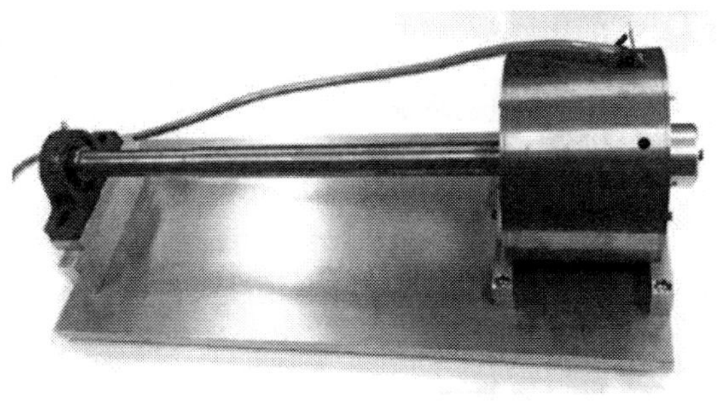

Figure 13. Laboratory setup of the homopolar three phase AMB.

centered rotor is presented in Fig. 15. The position detection rate is increased compared to the symmetric pulses, but according to Tab. II the position noise is much higher even with 8 current samples. This measurement is done at higher bias flux, where the position noise result in high oscillations at levitation. Definitely the maximum bearing load is increased, but the control behavior looks nervous.

The sensorless control behavior with three active INFORM

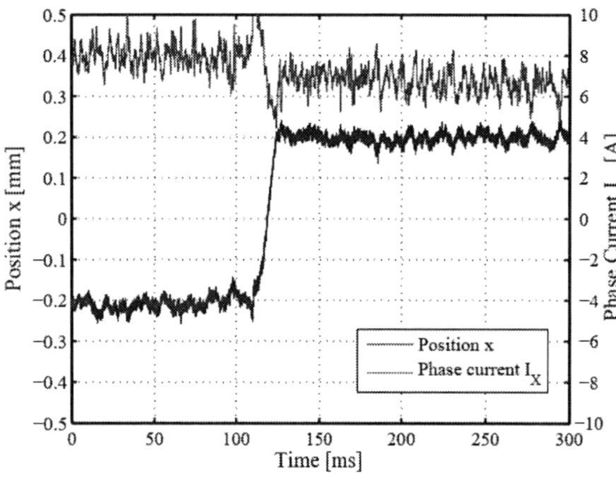

Figure 14. Sensorless movement from x=-0.2mm to x=-0.2mm and Phase current I_X, setup with $N_M = 24$ ferrite magnets at $U_{DC} = 120\,\text{V}$

at centered rotor position is shown in Fig. 16. Compared to the single pulse strategy from the symmetric and the asymmetric sequence the three active INFORM results directly in a complete position information of all three coils after one INFORM pulse. Thus the information evaluation rate for one coil is three times higher compared to the symmetric and asymmetric INFORM sequences. Thus an increased sensing bandwidth at a low position noise is achieved. Here even with the low position sensing noise oscillations in the rotor position are identified.

Considering the overall control behavior of the sensorless three phase HMB system for different pre magnetization and INFORM sequences, a design compromise between bias current consumption and self sensing accuracy is identified in this

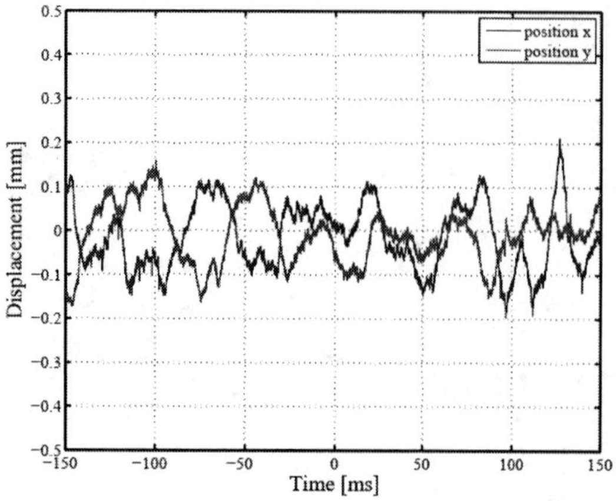

Figure 15. Sensorless operation with asymmetric INFORM pulses with 8 samples at centered rotor, setup with a combination of 12 ferrite and 12 NdFeB magnets at $U_{DC} = 120\,\mathrm{V}$

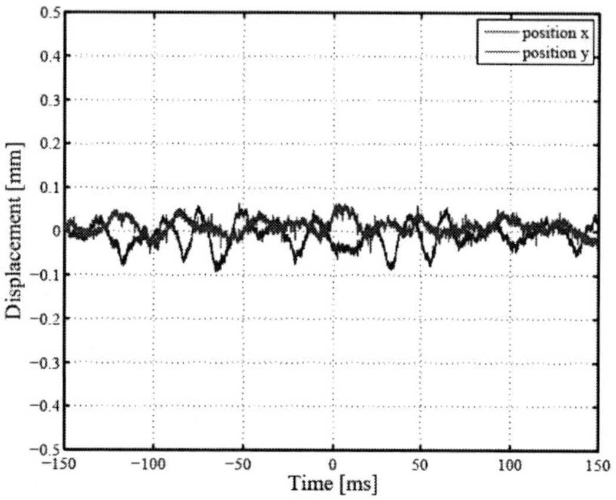

Figure 16. Sensorless operation with three active INFORM pulses with 16 samples at centered rotor, setup with a combination of 12 ferrite magnets and 12 NdFeB magnets at $U_{DC} = 120\,\mathrm{V}$

work. The reason is, that a higher bias flux is recommended to reduce the current consumption in operation, but this negatively affects the sensorless accuracy in an indirect way. Although the position detection noise does not change due to a higher bias flux, but bearing parameters K_x and K_i change. Thus, with the required increase of the controller gain increase the noise in the closed loop system is amplified. To overcome this trade off a further reduction of self sensing position noise is required. A first improvement can be the integration of the three active INFORM pulse sequence directly into the PWM voltage pattern of the current control as described in [9]. The self-sensing bandwidth gets increased and control behavior is expected to be improved. Because the operational modes of current control and position measurement do not work sequentially by interrupts and they are merged and distortions of the current control get reduced. Additionally, an eddy current reduction by utilization of enhanced lamination materials improves the position signal noise by increasing the "linear" range of the current slope measurement.

VI. Conclusions

In this paper new results and improvements regarding the sensorless position detection and sensorless position control of a three phase permanent magnet biased radial AMB are presented. Following concluding statements are identified:

- Utilizing oversampling strategies in the DSP increases the accuracy of the current slope measurement and consequently the sensorless rotor position noise is reduced and sensing accuracy is improved.
- Alternative INFORM pulse sequences allow an increase of position sensing frequency and result in a higher possible control bandwidth. At the three active sequence a rotor position is immediately available after one interruption. Additionally this sequence show the lowest position noise.
- A dependency of the sensorless rotor position detection method from the bias flux level was not identified in the open loop, but changed bearing parameters and increased control gain affect the overall behavior. A high sensibility of the position noise is given which result in oscillations at levitated operation.

Although this AMB prototype setup show limitations for self sensing especially at high bias flux for reaching low current consumption, two possible improvements are identified. First, reducing eddy currents for increased signal utilization and second, integrating the three active INFORM pulse sequence in a three active PWM control pattern. These topics will be investigated in future research work.

VII. Acknowledgment

This work is conducted within the project P21631-N22 of the Vienna University of Technology with the Austrian Science Fund (FWF). The authors thank the Austrian Science Fund for supporting this project.

References

[1] G. Schweitzer, *Magnetic Bearings, Theory, Design, and Application to Rotating Machinery.* Springer Berlin, 2009.

[2] E. Malsen, "Selfsensing for active magnetic bearings: overview and status," in *Proceedings of the Tenth International Symposium on Magnetic Bearings (ISMB10)*, 2006.

[3] M. Hofer, W. Staffler, and M. Schrödl, "Sensorless control of a three phase radial active magnetic bearing," in *Proceedings of the Twelfth International Symposium on Magnetic Bearings (ISMB12)*, 2010, pp. 680–685.

[4] M. Hofer, "Design and sensorless position control of a permanent magnet biased radial active magnetic bearing," Ph.D. dissertation, Vienna University of Technology, June 2013.

[5] D. Montie, "Performance limitations and self-sensing magnetic bearings," Ph.D. dissertation, University of Virginia, 2003.

[6] M. D. Noh and E. H. Maslen, "Self-sensing magnetic bearings (part ii) - effects on saturation," in *Proceedings of the Fifth International Symposium on Magnetic Bearings (ISMB5)*, Aug. 1996, pp. 113–118.

[7] H. Zhu, H. Chen, Z. Xie, and Y. Zhou, "Configuration and control for ac-dc three degrees of freedom hybrid magnetic bearings," in *Proceedings of the Tenth International Symposium on Magnetic Bearings (ISMB10)*, 2006.

[8] M. Hofer and M. Schrödl, "Statistic properties of a sensorless control method for a three phase permanent magnet biased radial active magnetic bearing," in *Proceedings of the 13th European Conference on Power Electronics and Applications (EPE 2009)*, 2009.

[9] T. Nenning, M. Hofer, M. Hutterer, and M. Schrödl, "Setup with two self-sensing magneitc bearings using differential 3-active inform," in *Proceedings of the Fourteenth International Symposium on Magnetic Bearings (ISMB14)*, 2014.

Statistic Errors of Different INFORM Evaluation Methods applied to Magnetic Bearings

Thomas Nenning, Matthias Hofer, Markus Hutterer, Manfred Schrödl

Institute of Energy Systems and Electrical Drives, Vienna University of Technology, Gußhausstraße 25-27, 1040 Vienna, Austria, thomas.nenning@tuwien.ac.at

Abstract—**Active magnetic bearings offer many advantages compared to conventional bearings. But due to economic reasons their field of application is relatively small. Self-sensing magnetic bearings can contribute to a significant decrease of the system costs, of course at limited accuracy in position detection compared to systems with sensors.**
The later on described INFORM method is one approach to self-sensing magnetic bearings. The position depending reluctance is measured through high frequent voltage injections and the resulting current slope. This paper discusses how different evaluation methods capable of calculating the current slope from the raw measurement data, differ in sensitivity and noise rejection. Therefore first the theoretical background of the used self-sensing method is explained. Next the error propagation of different evaluation methods is analysed and finally compared to measurements.

I. Introduction

Major field of research for self-sensing magnetic bearings is to improve the position detection to facilitate more accurate and stable position control algorithms. This paper focuses on how different evaluation methods for the INFORM-measurement affect the position detection accuracy.
The INFORM method is well known from self-sensing synchronous electrical drives [1]. Basic idea is to measure the inductance of a phase coil by injection of high frequent voltage impulses and measuring the resulting slope of the phase current. If the reactance of the iron path and leakage fluxes are neglected, then the coil inductance solely dependents on the air-gap between rotor and stator. Hence the INFORM method can be used for self-sensing magnetic bearings as proved in several studies [2][3].
A first study on the statistical behaviour of INFORM position detection [4] showed that the resolution was approximately one tenth of the air gap. This means that the quality of self-sensing position measurements has to be improved if used in highly unstable system like a active magnetic bearing with bias flux.

II. The INFORM method

The so called INFORM method (Indirect Flux Detection by Online Reactance Measurement) was originally developed for the rotor position detection of synchronous drives [1] and is well known for the sensorless control of synchronous drive applications. It relies on measuring the position by high frequency voltage injection pulses. If the iron path and leakage fluxes are neglected, then the coil inductance is solely dependent on the air-gap between rotor and stator. Assuming a three pole stator arranged in Y-shape, then the coil inductance can be calculated from:

$$L(x)|_{y=0} = \frac{\mu_0 N^2 A}{\frac{3}{2}\left(l_0 - \frac{x}{2}\right)} \tag{1}$$

with the number of coil turns N , the air gap area A, the nominal air gap length l_0 and the displacement x at $y = 0$. If the inductances of the three coils are known, the 2D radial displacement (x,y) can be calculated. In the following paragraph it is described how these inductances are obtained.

A. 3-Active Working principle

In the used prototype, the so called 3-Active INFORM is implemented. It is a enhanced variant of the classical INFORM. The INFORM pulse sequence is produced according to [5], which is a combination of three active test voltage space phasors, approximating arbitrary voltage space phasors within a certain limit. Hence, INFORM measurement information and desired space phasors according to current control can be realized at the same time.
The zero-voltage space vector u_0 is generated if all three ON-times are equal within a PWM cycle. The corresponding voltage and current curves are shown in figure 1.
Using equation 2, the coil inductance can be calculated out of the known dc link voltage and the measured current slope ($r_s = 0\Omega$).

$$u = L \cdot \frac{di(t)}{dt} \rightarrow L = u \cdot \frac{\Delta\tau}{\Delta i} \tag{2}$$

When the coil inductances are known the position can be calculated using calculation rules basing upon the mathematical model described in [5].

III. Theory of error propagation

IV. Obtaining the inclination

In order to obtain a high Signal to noise ratio (SNR) for the current slope measurement, the current curve is oversampled several times. Result is a measurement vector with a sample series of the phase current curves. In a first approach the

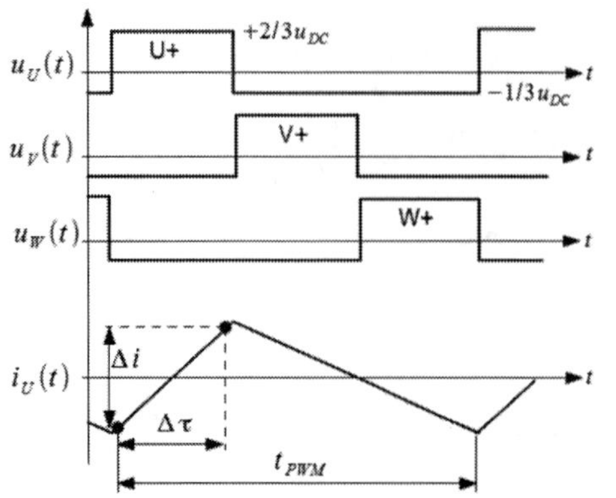

Figure 1. Voltage and current curves in 3-Active PWM modus

single measurement points can be seen as not correlative with equidistant time stamps. The following chapters will describe four different methods to obtain the position dependent inclination of the current slope.

A. Mean-difference method

The first is to average all the points of the first half of the n measurement points to the mean value $\bar{x_1}$, and all points of the second half to $\bar{x_2}$ as shown in figure 2. Goal is to fit these two

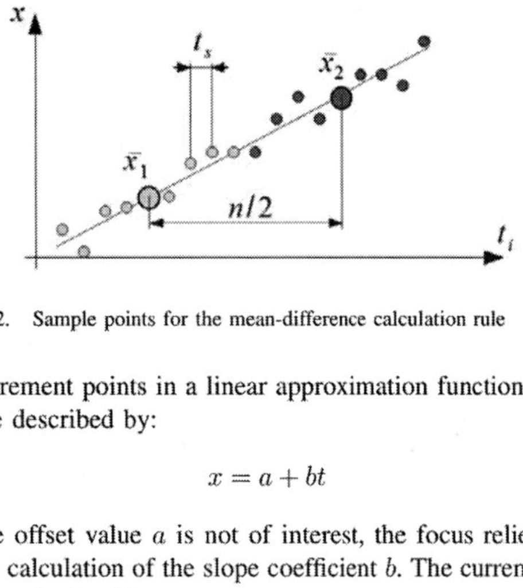

Figure 2. Sample points for the mean-difference calculation rule

measurement points in a linear approximation function which can be described by:

$$x = a + bt \tag{3}$$

As the offset value a is not of interest, the focus relies only on the calculation of the slope coefficient b. The current slope can be calculated by:

$$b = \frac{\bar{x_2} - \bar{x_1}}{\frac{n}{2} \cdot t_s} \tag{4}$$

with the number of samples n and the sampling rate time t_s.

B. Linear regression method

Second considered method is a linear regression. A linear curve is fitted into the measurement vector by minimizing the sum of the squares of the errors. The slope coefficient b can be calculated by

$$b = \frac{SS_{tx}}{SS_{tt}} = \frac{\sum_{i=1}^{n}(t_i - \bar{t})(x_i - \bar{x})}{\sum_{i=1}^{n}(t_i - \bar{t})^2} \tag{5}$$

SS_{tx} is the empirical covariance of t and x and SS_{tt} the empirical variance of t. For the sake of simplicity the mean time $\bar{t}$ is set to zero (see figure 3).

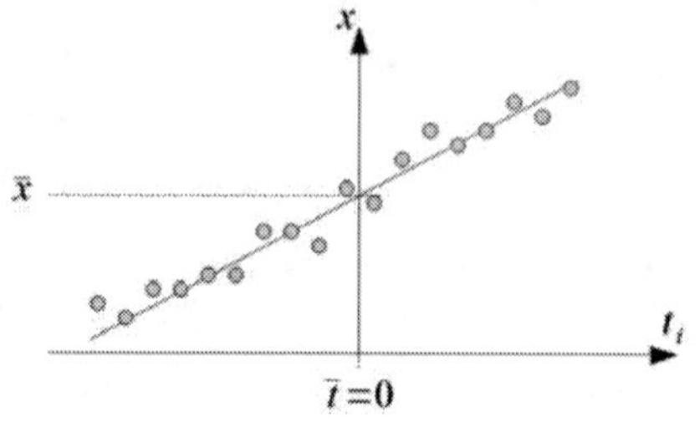

Figure 3. Sample points for the linear regression calculation rule

This simplifies equation (5) to

$$b = \frac{\sum_{i=1}^{n} t_i x_i}{\sum_{i=1}^{n} t_i^2} \tag{6}$$

C. Reverse mean-difference method

The working principle of the calculation rules before was to first take samples from the mid point and let the observation horizon grow from "inside to outside" with an increasing number n. The reverse approach is to first take the most outer sample points and grow from "outside to inside" with an increasing number n. Hence the calculation rule changes to

$$b = \frac{\bar{x_2} - \bar{x_1}}{(n_{max} - n/2 + 1)t_s} \tag{7}$$

with the number of totally sampled points n_{max}

D. Reverse linear regression method

The reverse linear regression is similar to the above described reverse mean-difference method, where first the outer sample points are utilized growing from "outside to inside" with an increasing number n. The resulting calculation rules for b is very similar and just defers in the indices of the sum operator:

$$b = \frac{\sum_{i=n_{max}-n}^{n_{max}} t_i x_i}{\sum_{i=n_{max}-n}^{n_{max}} t_i^2} \tag{8}$$

V. Error propagation with multiple inputs

A. Basic considerations

Assuming small errors $|\Delta x| << |x|$, a common approach is an approximation to a first-order Taylor series of the function b(x) [6]. For a single input system the resulting equations would be

$$b(x + \Delta x) = b(x) + \frac{db(x)}{dx}\Delta x \tag{9}$$

$$b(x + \Delta x) - b(x) = \Delta b = \frac{db(x)}{dx}\Delta x \tag{10}$$

with the input error Δx and the output error Δb.
If every measurement point is considered as independent and mutually uncorrelated input variable, then the output error yields to

$$\Delta b = \frac{\partial b(x)}{\partial x_1}\Delta x_1 + \frac{\partial b(x)}{\partial x_2}\Delta x_2 + ... + \frac{\partial b(x)}{\partial x_n}\Delta x_n \quad (11)$$

Equation (11) describes how to calculate the resulting absolute error. To evaluate how measurement uncertainties affect the result's uncertainty, the very similar Gaussian error propagation law offers an approach [6]:

$$S_b = \sqrt{\sum_{i=1}^{n} \left(\frac{\partial b(x)}{\partial x_i}\right)^2 (S_{x_i})^2} \quad (12)$$

where S_{x_i} is the absolute standard deviation of each measurement x_i, and S_b the absolute standard deviation of the slope coefficient b. For this law, the measurement noise needs to be a Gaussian distribution. Figure 4 illustrates a histogram of a typical current sample measurement. It can be seen that this assumption is roughly fulfilled.

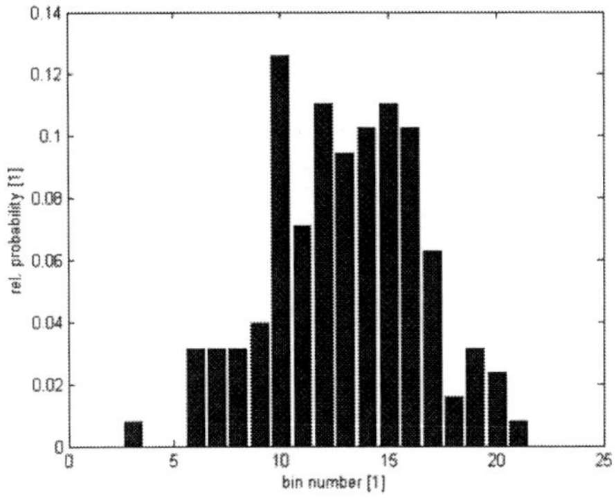

Figure 4. Histogramm of a measured current signal point in 3-Active PWM mode

This was also already shown in a previous study [4].

B. Error propagation of the mean-difference method

To calculate how different numbers of samples used by the mean-difference method affect the result, the Gaussian error propagation law (12) has to be applied to the calculation rule (4). This yields to

$$S_b = \sqrt{\sum_{i=1}^{2} \left(\frac{\partial}{\partial x_i}\frac{\bar{x}_2 - \bar{x}_1}{\frac{n}{2}\cdot t_s}\right)^2 (S_{\bar{x}_{1,2}})^2} \quad (13)$$

and further to:

$$S_b = \frac{2}{n \cdot t_s}\sqrt{2} S_{\bar{x}_{1,2}} \quad (14)$$

with the number of utilized sample points n, the sampling period time t_s and the standard deviation of the two average points $S_{\bar{x}_{1,2}}$. Thereby it was assumed that the standard deviation for all measurement points is equal. The standard deviation of the two points x_1 and x_2 yields from averaging each with the half number of total sampling points. It can be calculated with:

$$S_{\bar{x}_{1,2}} = \frac{1}{\sqrt{n/2}} S_x \quad (15)$$

with the standard deviation of a single measurement point S_x. If equation 15 is applied to equation 14, the overall calculation rule leads to:

$$S_b = \frac{4}{\sqrt{n^3} \cdot t_s} S_x \quad (16)$$

C. Error propagation of the linear regression method

Equivalent to the mean-difference method the Gaussian error propagation law (12) is applied to the calculation rule (6). This yields to

$$S_b = \sqrt{\sum_{j=1}^{n} \left(\frac{\partial}{\partial x_j}\frac{\sum_{i=1}^{n} t_j x_j}{\sum_{i=1}^{n} t_i^2}\right)^2 (S_{x_i})^2} \quad (17)$$

and further to:

$$S_b = \sqrt{\frac{1}{\sum_{i=1}^{n} t_i^2}} S_x \quad (18)$$

Again under the assumption the standard deviation for all sample points is equal.

D. Error propagation of the reverse mean-difference method

When the Gaussian error propagation low is applied to the calculation rule 7, then the standard deviation of the reverse mean-difference method consequently yields to:

$$S_b = \frac{2}{\sqrt{n}(n_{max} - n/2 + 1)t_s} S_x \quad (19)$$

with the number of sample points n used for the calculation, and the total number of sampling points n_{max}.

E. Error propagation of the reverse linear regression method

When the Gaussian error propagation low is applied to the calculation rule 7, then the standard deviation of the reverse mean-difference method consequently yields to:

$$S_b = \sqrt{\frac{1}{\sum_{i=n_{max}-n}^{n_{max}} t_i^2}} S_x \quad (20)$$

with the number of sample points n used for the calculation, and the total number of sampling points n_{max}.

VI. Experimental Results

To verify the theory, a set of slope gradient measurements was recorded, where the inclination was calculated by the micro controller using the mean-difference, the linear regression and their reverse variants. The signal to noise ratio SNR is defined as

$$\text{SNR} = 10 \cdot log\left(\frac{\bar{x}^2}{Var(x)}\right) \quad (21)$$

and used to compare the results of the theoretical model with the measurements. The results for the inner to outer calculation rules are shown in figure 5 the results for the reverse outer to inner in figure 6.

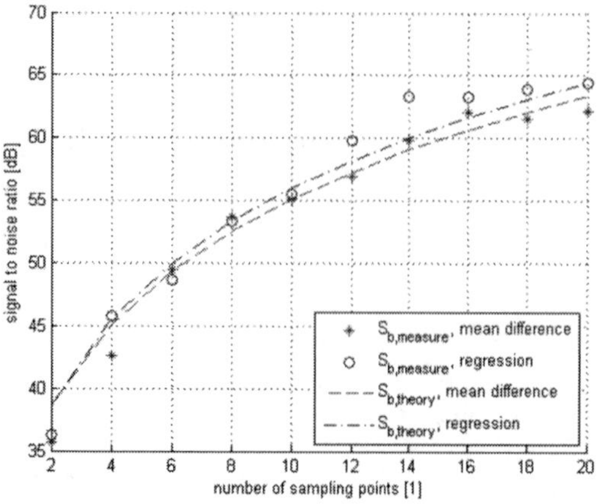

Figure 5. Comparison of the mean-difference / regression methods theoretical and measured SNR

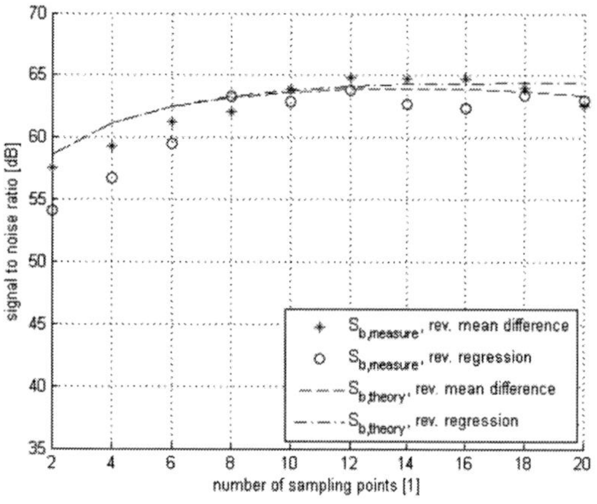

Figure 6. Comparison of the reverse mean-difference / regression methods theoretical and measured SNR ($n_{max} = 20$)

VII. Conclusion

The curves show a good match between theory and measurement. Two main conclusions can thus be drawn. First the difference between the linear regression and the mean difference method is neglectable. The linear regression shows a little higher SNR with an increasing number of samples, but this does not stand against the drastically increased computation time due to the necessary multiplications. Second it can clearly be seen that the reverse approach is advantageous as the consideration of only a few sample points is enough to nearly reach the maximum SNR. Thus a lot of computation time can be saved. Figure 7 summarizes the results

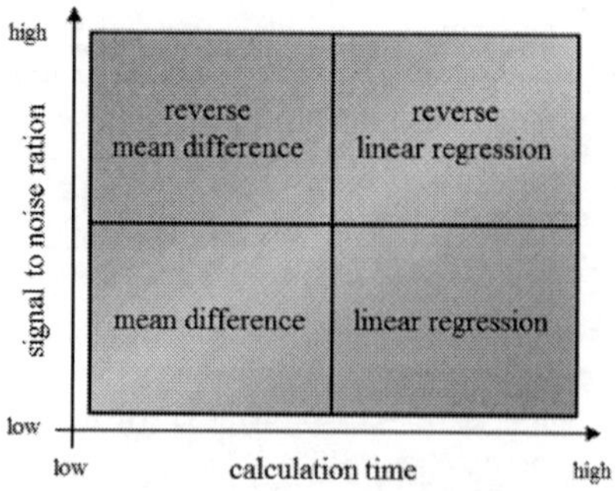

Figure 7. Comparison of the different calculation rules

References

[1] M. Schrödl, "Sensorless control of permanent-magnet synchronous machines at arbitrary operating points using a modified inform 'flux' model," *European Transac-tions on Electrical Power Engineering, ETEP, Vol.3, Nr.4*, 1993.

[2] E. Schmidt and M. Hofer, "Inductances and forces of a three phase permanent magnet biased radial active magnetic bearing in dependence on the rotor eccentricity," *Electric Machines and Drives Conference, 2009. IEMDC '09. IEEE International*, pp. 1–7, 2009.

[3] M. Hofer and M. Schrödl, "Test bench for three phase permanent magnet biased radial AMBs with a sensorless control strategy based on inform method," *Proceedings of 11th International Symposium on Magnetic bearings*, pp. 1–5, 2008.

[4] M. Hofer and M. Schrödl, "Statistic properties of a sensorless control method for a three phase permanent magnet biased radial active magnetic bearing," *Power Electronics and Applications, 2009. EPE '09. 13th European Conference*, 2009.

[5] G. Kalteis, "Method and circuit configuration for approximate adjustment of a voltage space vector," 2006, eP Patent App. EP20,060,009,591. [Online]. Available: http://www.google.com/patents/EP1722470A2?cl=en

[6] P. Fornasini, "The uncertainty in physical measurements: An introduction to data analysis in the physics laboratory," *Electric Machines and Drives Conference, 2009. IEMDC '09. IEEE International*, p. p. 161 ff., 2008.

Setup with two Self-Sensing Magnetic Bearings using Differential 3-Active INFORM

Thomas Nenning, Matthias Hofer, Markus Hutterer, Manfred Schrödl

Institute of Energy Systems and Electrical Drives, Vienna University of Technology, Gußhausstraße 25-27, 1040 Vienna, Austria, thomas.nenning@tuwien.ac.at

Abstract—**This paper describes how a rotor with two radial self-sensing Hybrid Magnetic Bearings (HMB) with a novel self-sensing method is set up.**
Replacing position sensors with low cost alternatives currently is a broad field of research as it allows to decrease costs significantly. First the adapted version of the INFORM method in use is described in detail, explaining the advantages compared to the classic INFORM. Next the results of a 3D electro-magnetic finite element simulation are shown. The simulation was used to find an optimal geometry concerning the mutually exclusive goals of maximising bearing force and sensitivity / linearity of the INFORM method.
Finally the simulation results are compared to measurements on the prototype.

I. Introduction

The Department of Electrical Drives and Machines at the Technical University of Vienna has already tested and proofed the applicability of the so called INFORM method for self-sensing magnetic bearings in previous prototypes [1][2]. The new prototype introduced in this paper uses a differental measuring approach, which inreases the signal to noise ratio significantly compared to the classical INFORM approach. Further it is the first prototype set up with two radial self-sensing HMB based on INFORM.

II. The INFORM method

A. Operating principle

The so called INFORM method (Indirect Flux Detection by Online Reactance Measurement) was originally developed for the rotor position detection of synchronous drives [3] and is well known for the sensorless control of synchronous drive applications. It relies on measuring the position by high frequency voltage injection pulses. If the iron path and leakage fluxes are neglected, then the coil inductance is solely dependent on the air-gap between rotor and stator. Assuming a three pole stator arranged in Y-shape, then the coil inductance can be calculated from:

$$L(x)|_{y=0} = \frac{\mu_0 N^2 A}{\frac{3}{2}\left(l_0 - \frac{x}{2}\right)} \tag{1}$$

with the number of coil turns N , the air gap area A, the nominal air gap length l_0 and the displacement x at $y = 0$. If the inductances of the three coils are known, the 2D radial displacement (x,y) can be calculated. The following paragraph describes how these inductances are obtained.

B. 3-Active INFORM

The classic INFORM measure sequence stops the current-control-loop in determined time steps to inject the needed high frequency voltage pulses [4]. The cycles in between this voltage pulses is used by the current controller to adjust the winding current using a classical pulse width modulation. This intercepting behaviour limits the position sampling rate and consequently causes a low bandwidth of the position controller. This is especially critical in an originally unstable system like a magnetic bearing with a constant bias flux. Therefore the so called 3-Active INFORM was introduced to increase the INFORM sampling rate.
The pulse sequence is produced according to [5], which is a combination of three active test voltage space phasors, approximating arbitrary voltage space phasors within a certain limit. Hence, INFORM measurement information and desired space phasors according to current control can be realized at the same time. Figure 1 illustrates two example voltage space vectors. The zero-voltage space vector u_0 is generated if all

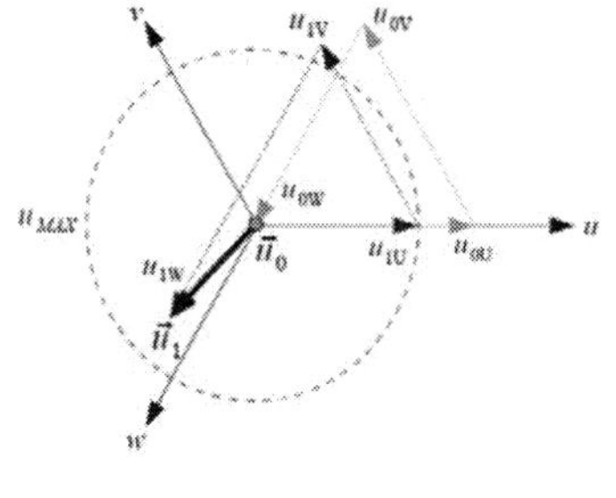

Figure 1. Two voltage space phasors examples $\vec{u_0}$ and $\vec{u_1}$ in 3-Active PWM modus

three ON-times are equal. If the ON-times differ from each other, a voltage space vector is generated, in this example u_1. Due to the sequence pattern, a positive voltage in U,V or W direction is applied on the bearings coils at every moment which allows the evaluate the position after each current control cycle. The corresponding voltage and current curves for one cycle are shown in figure 2.

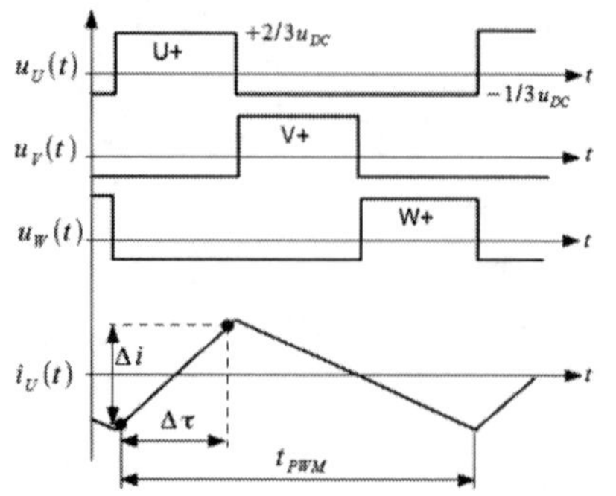

Figure 2. Voltage and current curves in 3-Active PWM modus

Using equation 2, the coil inductance can be calculated out of the known dc link voltage and the measured current slope ($r_s = 0\Omega$).

$$u = L \cdot \frac{di(t)}{dt} \rightarrow L = u \cdot \frac{\Delta\tau}{\Delta i} \quad (2)$$

When the coil inductances are known the position can be calculated using calculation rules basing upon the mathematical model described in [5].

Downside of the 3-Active PWM pattern is a loss in the applicable voltage, as there is a minimum ON-time required for each phase. This typically limits the voltage space phasor magnitude to a maximum of 25% of the direct current link voltage.

C. *Differential 3-Active INFORM*

The 3-Active INFORM increases the sample rate a lot compared to the classical INFORM, as the position can be detected in every current control cycle. Beside the smaller available voltage space phasor magnitude, another downside is the small amplitude of the injected current ripple compared to the nominal current of the magnetic bearing. If the current slope now is measured with the ADC-input for the working current, the resolution is considerably low, as the ripple magnitude is usually abut one tenth of the nominal current.

The solution is a differential approach which has been introduced by [6]. The in this prototype used variant consists of a stator with six poles, where the differential winding current of two opposite poles is measured. The differential current is proportional to the inductance difference and thereby also to the displacement. As the rotor is displaced in y-direction, the inductance of the upper coil is bigger. The corresponding currents for one 3-Active PWM cycle are shown in figure 3 The differential current is measured with a transformer where the two coil-wires are wound in different directions on the primary side. The secondary side is terminated with a small resistance. The resulting voltage signal is converted in an separate ADC input of the IC. This allows to measure the current ripple independently from the working current with full available resolution. Figure 4 shows the schematic circuit for one pole-pair. The

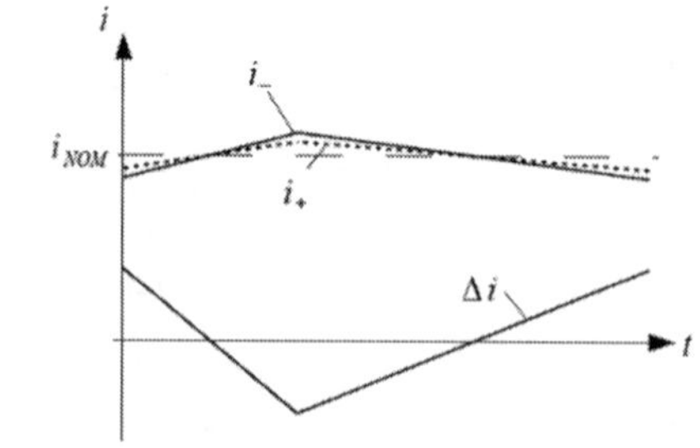

Figure 3. Current slopes in two opposite coils at displacement

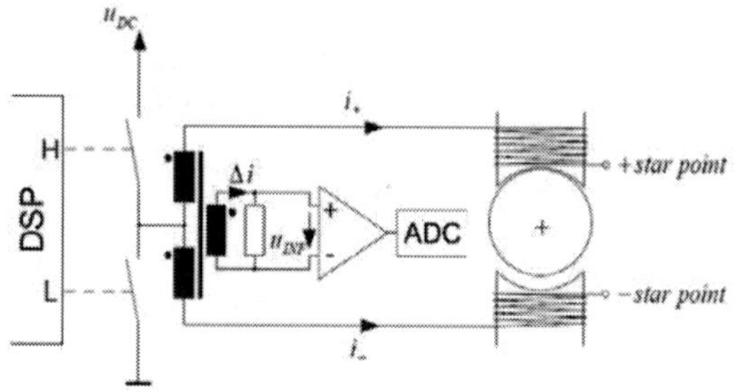

Figure 4. Schemtic measurement circuit for the differental INFORM method

differential approach increases the signal to noise ration (SNR) significantly. Figure 5 shows the position detection in three different positions. The corresponding standard

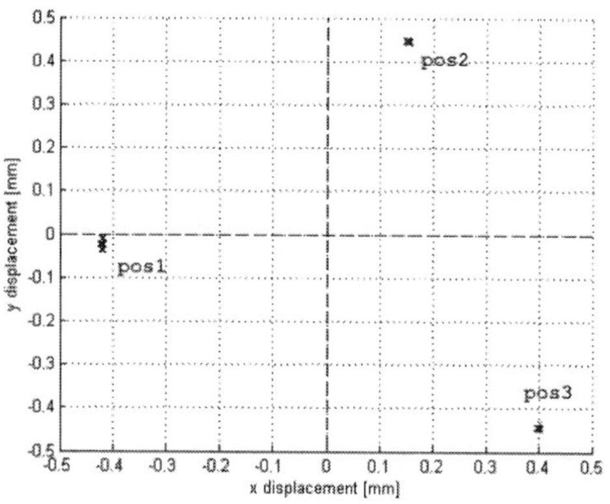

Figure 5. Position detection with differential 3-Active INFROM measurement

deviations of the position measurement are given in table I.

Table I
STANDARD DEVIATION OF THE POSITION MEASUREMENT WITHIN THE THREE POSITIONS (RADIAL AIR GAP = 0.8MM)

position	x-direction	y-direction
pos1	$0.812\,\mu m$	$1.992\,\mu m$
pos2	$1.350\,\mu m$	$0.7937\,\mu m$
pos3	$1.276\,\mu m$	$0.9097\,\mu m$

A noise of approx. $1\mu m$ at an air gap of $0.8mm$ is at

the class of accuracy of sensors and facilitates high dynamic control algorithms. At this resolution, the noise of the ADC chain becomes a considerable factor, making it necessary to also focus on electromagnetic compatibility and print design. Figure 6 shows the measured position and velocity estimated by a Kalman filter at a step of the target position (PD position controller). The curves prove the high SNR. This allows a stiff

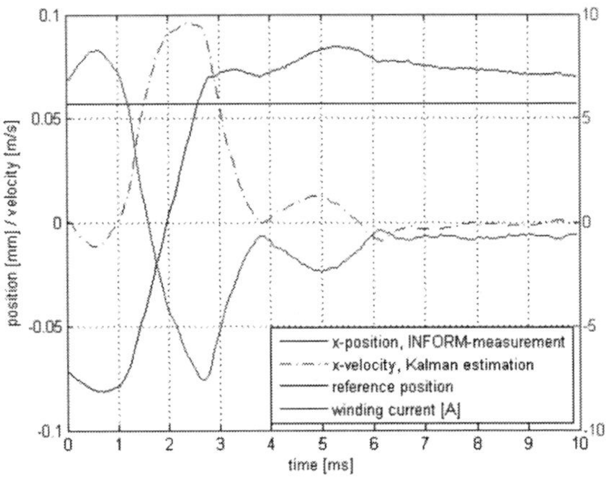

Figure 6. Step response of a change in x-direction

control and thus a quick change of position. Interestingly the position first changes to the opposite direction for a bit. One possible explanation could be the influence of the winding current on the position measurement. Further investigations will be done on that effect.

III. Design and Construction

A. General requirements

The basic parameters like outer diameter and nominal voltage were chosen by the given electrical drive and voltage inverter. In order to be able to use three phase inverter from the field of electrical drives and therefore reduce costs, the number of phases was chosen with three. It was shown in [7] that magnetic bearings can be realized through three poles arranged in Y-shape. As this prototype uses a differential approach, the number of poles was chosen to be six. Further operating costs can be sunk when using permanent magnets for bias flux as shown in [8]. So this prototypes uses HMBs with permanent magnets to create a homo-polar bias flux.
The air gap was chosen as a compromise between available bearing force and the available accuracy of production. The general parameters are listed in table II.

B. Bearing force

To find the optimal distribution of copper and iron for the given outer diameter, a static-electromagnetic 3D finite element model was created. Optimization goals were maximising the bearing force while ensuring a good sensitivity and linearity for the INFORM method. Hence primary bottlenecks for the magnetic flux have to be avoided

Table II
Basic Parameters of the Prototype

Parameter	Value
Outer bearing diameter	120 mm
Air gap	0.8 mm
Mass of rotor	0.8 kg
Bearing force	20 N
Number of poles	6

and a good balance between space for iron for the guidance of the magnetic flux and space for copper to conduct the current has to be found.

The simulation of the final configuration (figure 7) shows that the centering force decreases with an increasing displacement. This is due to leakage fluxes that can not be influenced by the control current and the negative stiffness due to the bias flux. Compared to figure 7, the measured bearing

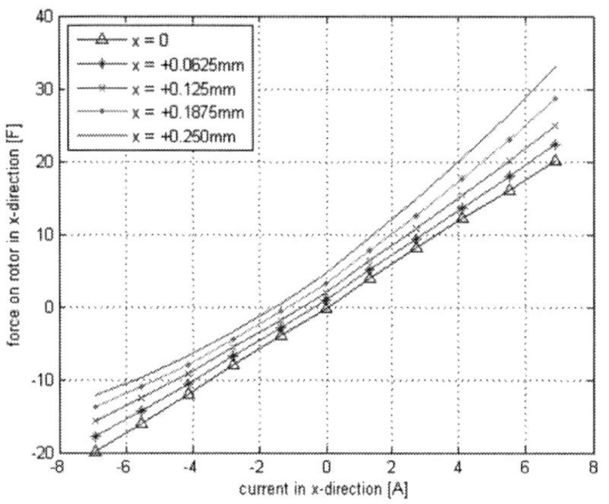

Figure 7. Simulated bearing force over winding current and y-displacement

force is shown in figure 8 Possible cause for the difference between the simulation and the measurement are not exact model parameters which are especially hard to obtain for the laminated iron.
Second reason could be the ripple current of the 3-active INFORM measurement. As the reluctance force is a quadratic function of the magnetic flux, the ripple could cause a higher force. Further investigations will be done on that topic.

C. The prototype

Figure 9 shows the current state of the prototype. It consists of a rotor supported by two independent HMB in radial direction. In axial direction the rotor is supported passive, as the energetic minimum of the magnetic field is in the centred position. The inverter was originally designed for a conventional electrical drive application. The only adaptation necessary was the adjustment of the voltage and current measurement range, and the gain of the additionally ADC

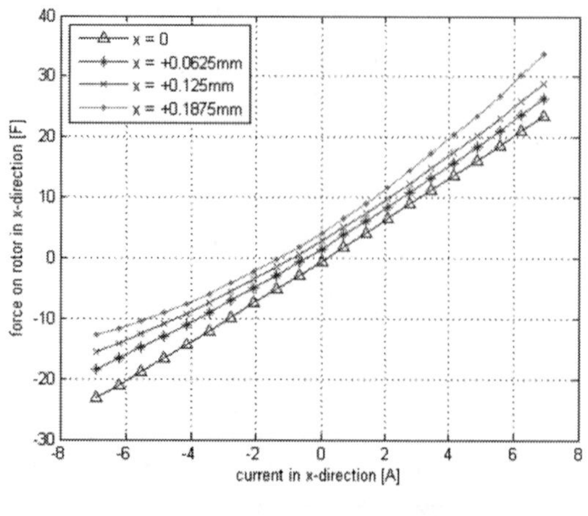

Figure 8. Measured bearing force over winding current and y-displacement

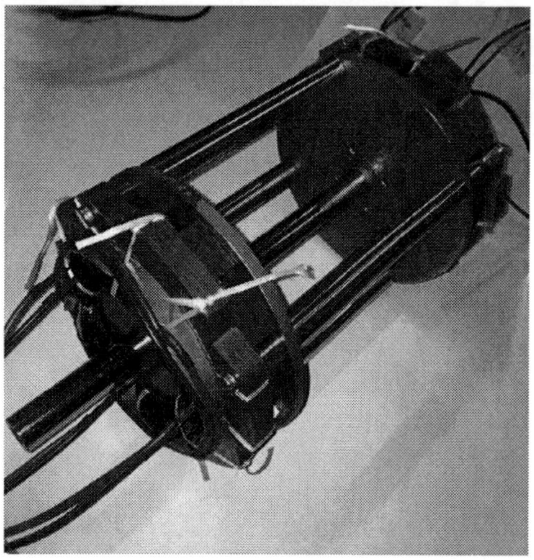

Figure 9. Prototype with two self-sensing HMBs

inputs for the differential current signal. The rest remained the same, which proves the ability of the presented self-sensing method for being realized with conventional 3-phase inverters.

IV. Conclusion

Throughout different measurements, the differential 3-Active INFORM proved to oppose a big improvement compared to the classical INFORM method [4]. Downside of the presented method is the influence of saturation in the iron path on the position detection. This limits the available bearing force, as the control-currents need to be limited. More intense investigations on methods to compensate these influence on the accuracy of the implemented self-sensing method should be done.

Upside is the good and resolution and the high bandwidth. At this level of accuracy the noise of the whole ADC-chain becomes a considerable factor. Especially the short distance between the measure transformer and the ADC input of the inverters controller, help reducing the ADC-chain noise.

Summarized it can be said that this first results show that this self sensing method is promising way and worth to be further investigated.

References

[1] E. Schmidt and M. Hofer, "Inductances and forces of a three phase permanent magnet biased radial active magnetic bearing in dependence on the rotor eccentricity," *Electric Machines and Drives Conference, 2009. IEMDC '09. IEEE International*, pp. 1–7, 2009.

[2] M. Hofer and M. Schrödl, "Test bench for three phase permanent magnet biased radial ambs with a sensorless control strategy based on inform method," *Proceedings of 11th International Symposium on Magnetic bearings*, pp. 1–5, 2008.

[3] M. Schrödl, "Sensorless control of permanent-magnet synchronous machines at arbitrary operating points using a modified inform 'flux' model," *European Transac-tions on Electrical Power Engineering, ETEP, Vol.3, Nr.4*, 1993.

[4] M. Hofer, "Design and sensorless position control of a permanent magnet biased radial active magnetic bearing," *Dissertation, Vienna University of Technology*, 2013.

[5] G. Kalteis, "Method and circuit configuration for approximate adjustment of a voltage space vector," Nov. 15 2006, eP Patent App. EP20,060,009,591. [Online]. Available: http://www.google.com/patents/EP1722470A2?cl=en

[6] Kenichi Matsuda and Yohji Okada, "Self-sensing magnetic bearing using the principle of differential transformer," *Proceedings of 5th International Symposium on Magnetic bearings*, pp. 1–6, 1996.

[7] S.-L. Chen and C.-T. Hsu, "Optimal design of a three-pole active magnetic bearing," *IEEE Transactions on Magnetics*, vol. 38, no. 5, pp. 3458 – 3466, 2002.

[8] M. Reisinger, W. Amrhein, S. Silber, C. Redemann, and P. Jenckel, "Development of a low cost permanent magnet bias bearing," *Proceedings of 9th International Symposium on Magnetic Bearings*, pp. 1–6, 2004.

Investigations on the "Direct Digital Inductance Estimation"-concept for self-sensing AMBs under influence of eddy currents

Michael Richter[a], Hendrik Schaede, Stephan Rinderknecht

[a] Institute for Mechatronic Systems in Mechanical Engineering TU Darmstadt, Otto-Berndt-Str. 2, 64287 Darmstadt, Germany, richter@ims.tu-darmstadt.de

Abstract—**This article states the challenges connected with eddy-currents by use of the direct digital inductance estimation (DDIE) concept for self-sensing active magnetic bearings. Starting with a short introduction to the estimation algorithm, the negative influences of eddy currents on the position estimation are described step by step. Such influences are an increasing position noise as well as an interval and duty cycle dependent position error. The article closes with an outlook on additional investigations to further separate, quantify, and reduce the associated position error in order to extend the performance of the DDIE-concept.**

I. MOTIVATION

Based on the broad advantages of self-sensing active magnetic bearings (above all higher degree of integration, lower investment costs and installation space) a variety of different self-sensing concepts can be found in literature. Following the classification of [1], two main groups can be identified:

The "State Estimation Approach" accounts for the fact to be able to stabilize the active magnetic bearing (AMB) by measuring the coil current alone. Unfortunately, this concept group is inherently less robust compared to a sensor driven configuration [2].

A generally higher grade of robustness can be obtained with the "Modulation Approach". If the AMB is driven by a switching power amplifier, the resulting high frequency current ripple is inherently modulated by the changing inductance due to a changing magnetic air gap. By demodulating this interrogation signal, the air gap can be obtained. An overview of different demodulation algorithms can be found in [3]. Most of them use analog low- and band-pass filters which add a phase lag to the estimated position and thus reducing the robustness.

In order to overcome the limitations of the referred self-sensing schemes, a new and promising DDIE-concept group was identified. To rate their possible higher degree of performance, the DDIE-concept developed in [4] was successfully implemented on a uniaxial test rig (Fig.1). In [5] a detailed description of this test rig as well as the essential enhancements and adaptations of the estimation algorithm are given. A first insight during the implementation phase implies a non-negligible impact of eddy currents (EC) on the estimated position. According to that, the impact of eddy currents on the estimated position are investigated, their different effects are quantified, and possible improvement measures are drawn.

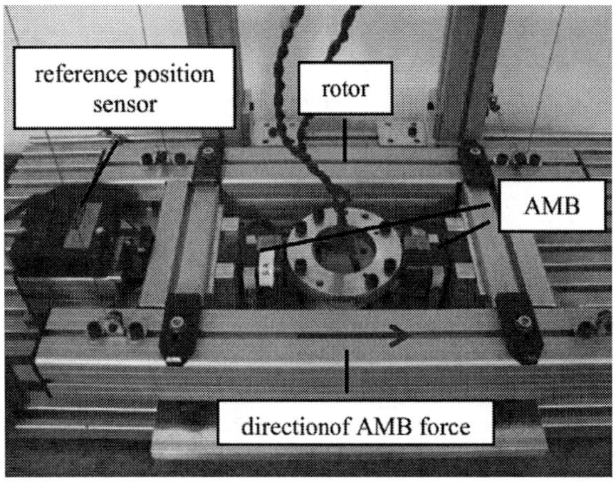

Figure 1. Uniaxial test rig with differential inner actuators and a horizontal lifted outer rotor

II. DDIE-CONCEPT

The basic idea behind the DDIE-concepts is to estimate the inductance and therefore the position of the rotor by means of the current ripple resulting from a two-stage switching power amplifier. Every current ripple is evaluated digitally in the time domain without additional low- and band-pass filters. This causes the estimated position to be of minimum phase lag contrary to other self-sensing concepts [1], [3]. Furthermore, the utilized DDIE-concept also accounts for a time dependent coil resistance, the rotor velocity and measurement noise [4], [5].

The sequence of the position estimation algorithm is shown in Fig. 2. The electromagnetic core equation (a) serves as the basis for the estimation. Here u is the coil voltage, i the coil current, s the air gap, $\hat{R}$ the estimated coil resistance, ΔR the estimation error of the coil resistance, Ψ the magnetic flux linkage and L_d the differential inductance of the magnetic bearing. Since (a) cannot be directly evaluated due to the unknown rotor velocity ds/dt and ΔR, both are initially neglected. In case of a sampling rate over 100 times higher than the switching frequency f_{PWM}, the truncated equation (b) can now be evaluated by means of a least square estimation. If

(b) is evaluated for two different intervals of the ripple e.g. its rising and $\hat{L}'_d$ falling edge $\hat{L}''_d$ (c) the resulting estimation error due to the truncation of (a) can be compensated (d). The compensation holds true, if the unknown rotor velocity ds/dt and ΔR are equal for both consecutive intervals.

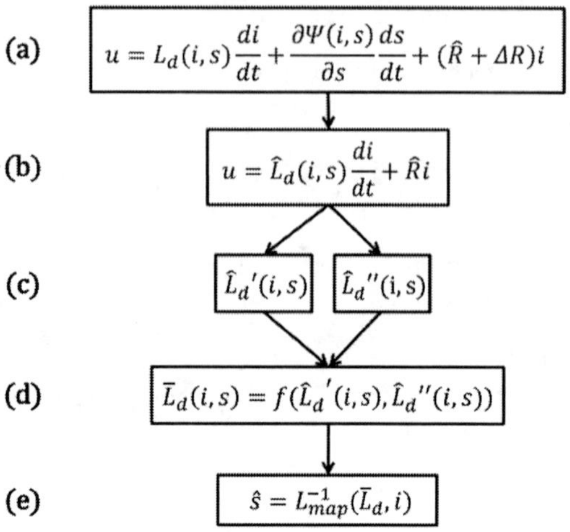

Figure 2. Sequence of the utilized DDIE – concept

By mapping the estimated average differential inductance $\overline{L_d}$ with means of the inverted inductance map L_{map}^{-1} the position of the rotor i.e. the estimated air gap $\hat{s} = L_{map}^{-1}(\overline{L_d}, i)$ can be obtained (e). The inverted inductance map can be understood as a calibration of the magnetic bearing as a sensor. Fig. 3 shows the measured inductance map of one electromagnet of the bearing. The inductance is estimated by the algorithm for different mean coil currents and for different constant air gaps (measured by a reference position sensor).

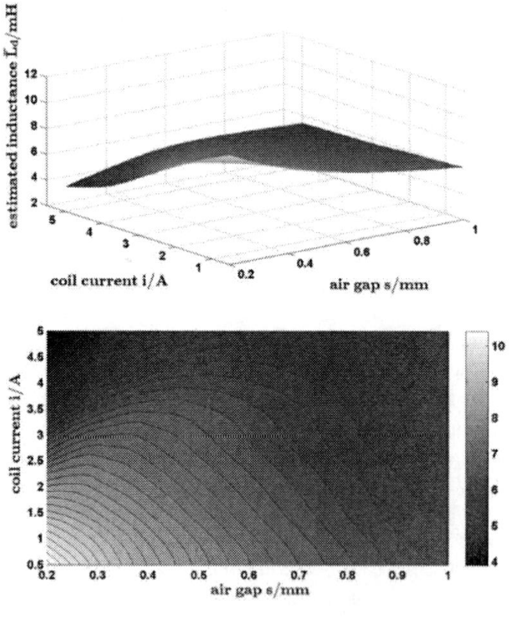

Figure 3. Inductance map of one electromagnet of the bearing

III. Influences of Eddy-Currents

Eddy-currents arise out of the switching of the power amplifier itself (PWM-EC) as well as out of the changing duty cycle (CDC-EC) i.e. the changing control current. They both effect the inductance estimation in several ways.

A. Increasing position noise

In case of a switching power amplifier, the presence of eddy currents can be seen equally by a varying current gradient or by a varying inductance. With increasing eddy currents, the inductance tends towards zero and the current gradient raises or respectively drops depending on the sign of the coil voltage. This effect can be seen right after each switching of the power amplifier (Fig. 4).

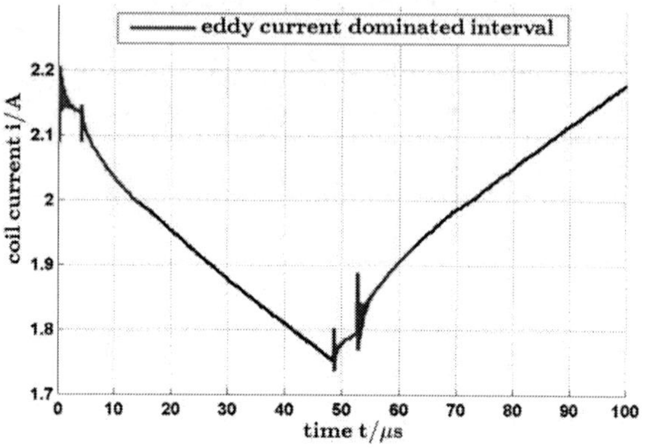

Figure 4. Current ripple with eddy current dominated intervals (air gap s = 0.5 mm, f_{PWM} =10 kHz)

Within the mentioned intervals, the current ripple is mainly dominated by eddy currents and not by the air gap. Thus, causing the position estimation to fail [5]. As a consequence, the most influenced interval of the current ripple right after each switching has to be omitted for the inductance estimation. The truncation reduces the remaining interval width Δt_{est} (i.e. the number of current samples N_{cs} when considering a fixed sampling period T_{sample}) for the inductance estimation. Since the current noise sensitivity of the estimated inductance rises with a reduced number of current measurements, the unwanted noise of the estimated position will increase too. (cf. Fig. 5 and Eq. 1). As a consequence, as many current samples as possible should be used for the inductance estimation.

$$\Delta s_{noise} = \left|\frac{\partial s}{\partial L_d}\right| \Delta L_{noise} = \left|\frac{\partial s}{\partial L_d}\right| \frac{L_d^2}{|u|} \frac{\Delta i_{noise}}{T_{sample}\sqrt{\frac{N_{cs}^3 - N_{cs}}{12}}} \quad (1)$$

Equation (1) expresses the mentioned dependency between the noise amplitude of the measured current (Δi_{noise}) and the resulting noise amplitude of the estimated air gap (Δs_{noise}). It considers the least square estimation but neglects the coil current and rotor velocity. Equation (1) as well as Fig. 5 also show a dependence of Δs_{noise} on the inductance of the bearing. Near the saturation of the magnetic material, the noise amplitude increases and becomes maximum which can be traced back to a near zero gradient of the inductance with respect to the air gap within this range ($\partial L_d/\partial s \approx 0$).

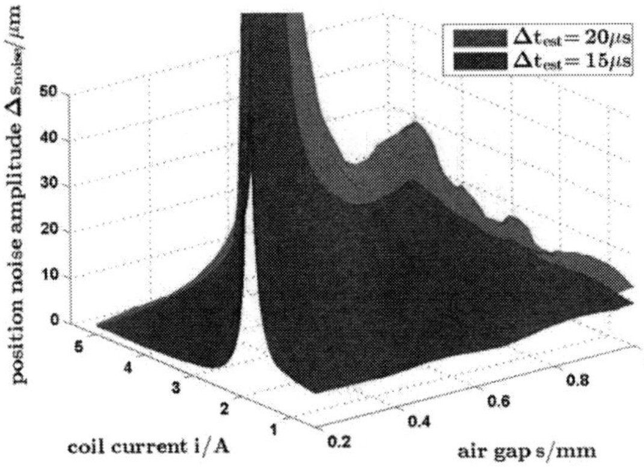

Figure 5. Position noise amplitude for different interval widths Δt_{est} used for the estimation of the inductance

B. *Interval dependend position error*

Eddy-currents induced by the switching of the amplifier (PWM-EC) drop in time, meaning not only the foremost samples of the ripple are influenced by them but almost the whole ripple. As a result, the modulus of the gradient of the ripple drops gradually until the next switching of the amplifier takes place. Since eddy currents are not considered by the estimation algorithm, the estimated inductance and therefore the position varies with the interval taken for the estimation. If the PWM-EC do not change from one estimation to another their influence can be compensated if the same interval of the ripple is used for the estimation as well as for the inverted inductance map.

TABLE I. MEASUREMENT CONFIGURATION OF FIG. 6

	Parameter	Value
$\bar{i}$	mean current	2 A
s	air gap	0.5 mm
Δt_{cut}	truncated time range after switching	14 µs
Δt_{map}	width of interval used for inverted inductance map	15 µs
Δt_{est}	width of interval used for estimation	10, 15 and 20 µs
T_{PWM}	modulation period	100 µs

In Fig. 6 the measured position error (reference sensor position minus estimated position) over 35 PWM periods for a constant air gap, a constant coil current (i.e. constant pulse width) and different interval widths is shown (see Table I). The data for $\Delta t_{est} = \Delta t_{map} = 15\mu s$ confirms the assumption to be able to compensate the influence of the PWM-EC in this special case. The remaining position error solely results from the current noise (III.A).

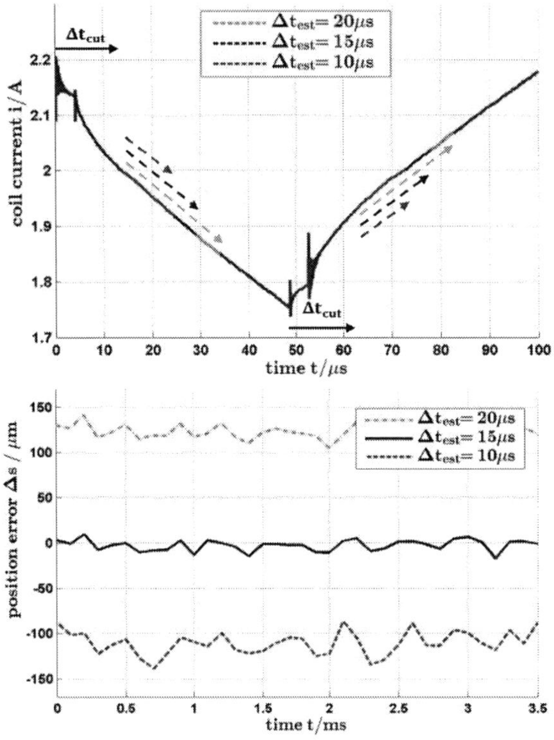

Figure 6. Influence of different sample intervals on the estimated position

The measurement further reveals the need of a compensation scheme since already less diverging intervals cause a high position error Δs. Unfortunately, a fixed interval after the switching for the compensation conflicts with a low current noise sensitivity for pulse widths differing from 0.5 (50% duty cycle). The maximum interval width and therefore N_{cs} is limited by the minimum pulse width for the rising edge i.e. maximum pulse width for the falling edge of the current ripple. Nevertheless, a low noise position estimation can be achieved if the inverted inductance map also depends on the pulse width. The solution has the benefit of an easy implementation in form of a three dimensional lookup-table but can only be obtained by a demanding and complex measurement (the inductance has to be measured for different air gaps, at different mean coil currents and for different duty cycles). For the following investigations only a fixed interval after the switching is used.

C. *Duty cycle dependent position error*

Eddy currents due to a changing duty cycle (CDC-EC) are also influencing the gradient of the current. Unfortunately, they cannot be compensated only by means of an inverted inductance map since the main frequency of the changing duty cycle and the amplitude of the oscillation are changing in time. Thus the inverted inductance map has to be frequency and amplitude dependent if their impact on the current ripple is dominant.

Nearly the same holds if the PWM-EC do not decay until the next switching takes place. An estimation of the inductance for a fixed interval width Δt_{est} of 15µs with a varying truncated time range Δt_{cut} from 9µs to 29µs for the current ripple of Fig. 6 is shown in Fig. 7. Thus the measurement shows the estimated inductance along the current ripple.

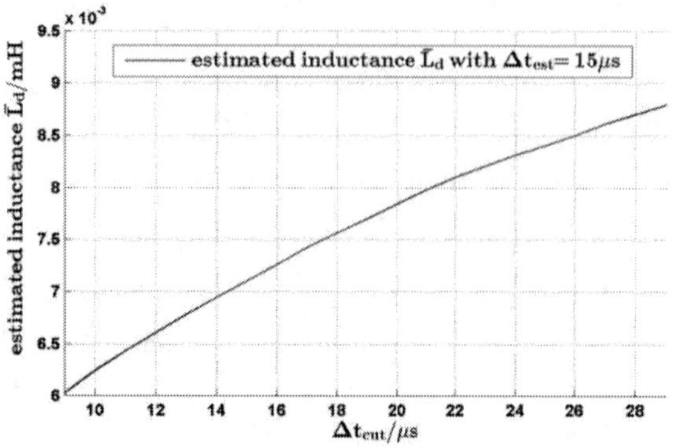

Figure 7. Estimated inductance along curent ripple with $\bar{i}$ = 2 A and s = 0.5 mm

The change of inductance and therefore the change of di/dt does not reach a constant level. As a result the PWM-EC are not decayed until the next switching takes place. Thus they do not only affect the ripple interval right after the switching but also the interval after the subsequent switching. A superposition of PWM-EC resulting from both switches in an estimation interval occurs. Unfortunately, the superposition varies with the duty cycle as the time between two switches varies accordingly. Until now the inverted inductance map is evaluated for duty cycles near 0.5, which causes it to be unable to compensate the influence of the PWM-EC in case of duty cycles differing from 0.5. A duty cycle dependent position error occurs.

In order to assess the resulting position errors due to the CDC and PWM-EC, a comparison measurement between a constant reference position and the estimated position for a sinusoidal changing coil current is done. The parameters of the measurement are summarized in Table II.

TABLE II. MEASUREMENT CONFIGURATION OF FIG. 8 & 9

	Parameter	Value
$\bar{i}$	mean current	1.8 A
Δi	current amplitude	0.9 A
f	frequency	125 Hz, 250 Hz
s	constant air gap	0.5 mm
Δt_{cut}	truncated time range after switching	14 µs
Δt_{map}	width of interval used for inverted inductance map	15 µs
Δt_{est}	width of interval used for estimation	15 µs
T_{PWM}	modulation period	100 µs

The measurement for $f = 250\ Hz$ (Fig. 8) reveals a high position error which varies almost chaotically in the range of low and high duty cycles (i.e. high coil current gradients). For a pulse-width near 0.5 (cf. t=2ms and t=6ms) the error is comparatively low. The observation indicates a duty cycle dependent position error which can be mainly assigned to a slow decay of the PWM-EC if the impact of the CDC-EC is low.

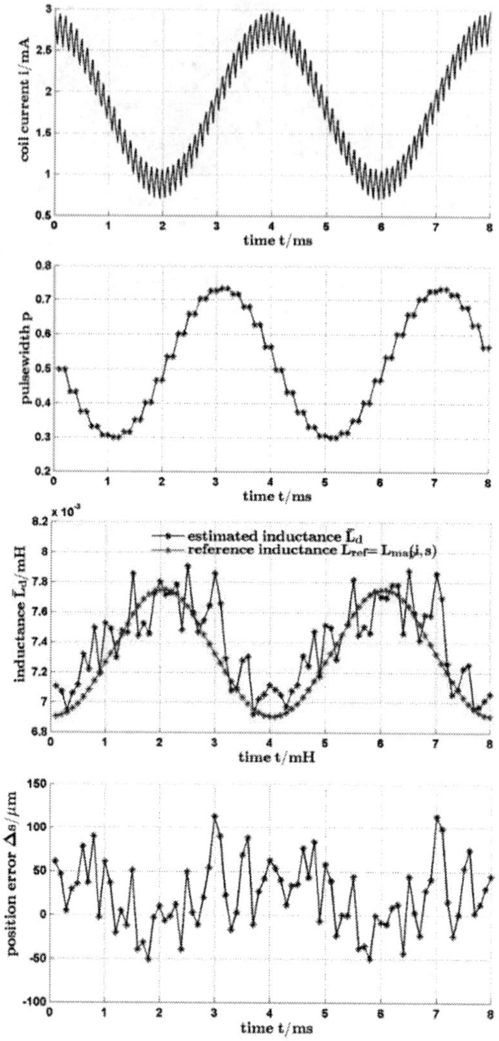

Figure 8. Position error for a fast changing coil current f = 250 Hz

A second measurement (Fig. 9) with a halved main frequency of 125 Hz confirms this assumption. Even though the theoretical eddy currents losses

$$P_{ec} \sim f^2 \tag{3}$$

of the CDC-EC should be divided by four compared to those of the previous measurement with 250 Hz, the resulting position error does not behave in the same way. The position error stays high, thus the impact of CDC-EC on the position error is comparatively low.

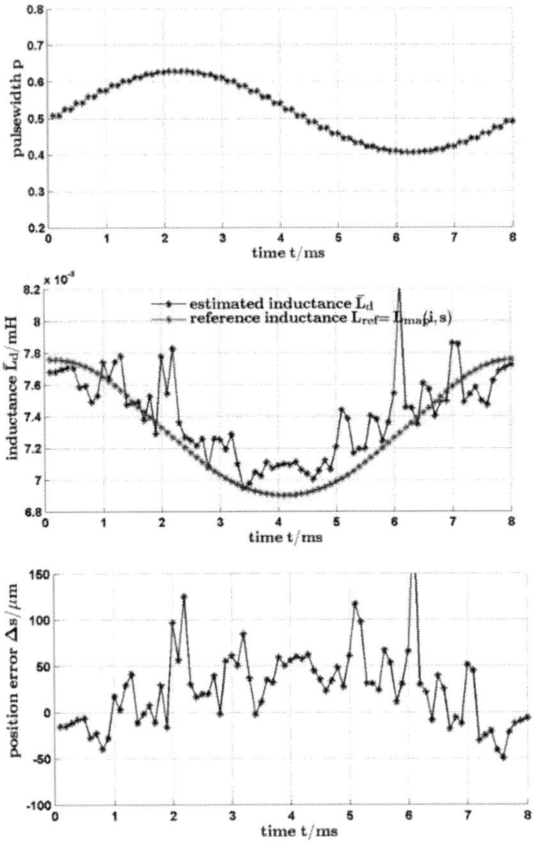

Figure 9. Position error for a fast changing coil current f = 125 Hz

The measurements also show an influence of the hysteresis. The position error distinguishes between the falling and rising edge of the fundamental oscillation of the control current i.e. it depends on the historical background of the magnetic material.

The measurements support the expected negative effect of eddy currents on the position estimation via the DDIE-concept. Nevertheless, a distinct quantitative assignment of the eddy currents arising out of the switching of the power amplifier (PWM-EC), out of the changing duty cycle (CDC-EC), or the hysteresis of the material to the resulting position error cannot be drawn yet. The following possible improvement measures are served to further support or disprove the shown dependencies:

- If the superposition of the PWM-EC influences the position error as proposed, a reduced PWM Frequency should decrease the superposition and as a result the duty cycle dependency of the position error. Thus the measurements will be repeated for a PWM Frequency of 5000 Hz instead of 10.000 Hz.
- A second procedure to further investigate the duty cycle dependent superposition of the PWM-EC and as a consequence of the position error can be an extension of the inverted inductance map by different duty cycles. Thus, the inverted inductance map will be able to account for this effect, and if the assumption holds true the resulting position error will decrease.

IV. Discussion and Conclusion

The suspected non-negligible impact of eddy currents on the estimation algorithm gave the impulse to this detailed investigation. Indeed, their negative impact is confirmed.

Eddy currents rising from the switching of the power amplifier cause an unacceptably high position error (~ 20% of the initial air gap of s=0.5) if the interval of the current ripple taken for the estimation does not equal the interval taken for the inverted inductance map. Both intervals can be kept equal even for a varying duty cycle with the drawback of an increasing position noise. With an extension of the inverted inductance map with respect to the duty cycle the trade-off can be reduced.

Until now the position error induced by a fast varying duty cycle represents the major challenge. The resulting position error of nearly 20% of the initial air gap is unacceptable. A quantitative assignment to the PWM and CDC-EC as well as to the hysteresis of the material is hard to draw. Nevertheless the measurements reveal a high impact of the superposition of PWM-EC on the position error. Thus, two improvement measures will be investigated in the future. First, the PWM frequency will be reduced. The PWM-EC have more time to decay until the next switching takes place. The resulting superposition and therefore the position error should decrease. Unfortunately, a reduced PWM frequency increases the phase lag of the position estimation as well as of the overall control of the AMB. Thus, it is a first workaround and serves mainly to support or disprove the shown dependencies. A second improvement measure is an extension of the inverted inductance map with respect to the duty cycle. Thus the varying superposition of the PWM-EC are considered by the overall estimation and the position error should decrease.

The soft magnetic material under consideration is an electrical steel of 0.35 mm thickness (M165-35S). Since eddy currents are frequency dependent, a magnetic material with lower core losses, especially in the high frequency domain, could also further reduce the negative effects of the PWM-EC on the estimated position. An electromagnet made out of soft magnetic composite (SMC) with the same dimensions as the laminated one is already built (Fig. 10) and will be investigated in the future.

Figure 10. Electromagnet made out of SMC

REFERENCES

[1] A. Schammass, R. Herzog, P. Buhler, und H. Bleuler, „New results for self-sensing active magnetic bearings using modulation approach“, *Control Syst. Technol. IEEE Trans.*, vol. 13, no. 4, pp. 509–516, 2005.

[2] N. Morse, R. Smith, and B. P. J. Antaki, "Position sensed and selfsensing magnetic bearing configurations and associated robustness limitations," in Proc. 37th IEEE Conf. Decision Control, Tampa, FL, Dec. 1998, pp. 2599–2604.

[3] A. C. Niemann, „Self-sensing algorithms for active magnetic bearings“, Thesis, 2008.

[4] T. Glück, W. Kemmetmüller, C. Tump, und A. Kugi, „A novel robust position estimator for self-sensing magnetic levitation systems based on least squares identification“, *Control Eng. Pr.*, vol. 19, no. 2, pp. 146–157, 2011.

[5] M. Richter, H. Schaede, L. Quurck, S. Rinderknecht, "Untersuchung und Erhöhung der Praxistauglichkeit sensorloser Magnetlager nach dem "Direct Digital Inductance Estimation"-Konzept", *Proceedings of Workshop on Magnetic Bearing Technology 9*,pp. 59-65, 2013.

[6] G. Schweitzer und E. H. Maslen, *Magnetic Bearings*. Heidelberg: Springer, 2013.

The Comparison of excitation source for sensors of the AMB system in HTGR

Xiaohan Liu[a], Ni Mo[b], Yan Zhou[b],

[a] Institute of Nuclear and New Energy Technology, Tsinghua University, The Key Laboratory of Advanced Reactor Engineering and Safety, Ministry of Education, Beijing, 100084, China, xiaohan_liu@tsinghua.edu.cn
[b] Institute of Nuclear and New Energy Technology, Tsinghua University, The Key Laboratory of Advanced Reactor Engineering and Safety, Ministry of Education, Beijing, 100084, China

***Abstract*—Nowadays the High Temperature Gas-Cooled Reactor (HTGR) is developed widely over the world. A helium turbine of a 10 MW HTGR test module reactor with the core made of spherical fuel elements is under research at the Institute of Nuclear and New Energy Technology of Tsinghua University in China. In HTGR, the magnetic bearings are widely applied, where the rotating machineries are running under highly pure helium environment. Compared with conventional bearings, Active Magnetic Bearings (AMBs) possess several attractive advantages, such as contact-free, no-lubricating and active damping vibration. The sensors are the key components of the AMBs system and are responsible for measuring the position and movement of the rotor, the operating temperature and so on. In order to operate the sensors successfully, a stable and efficient excitation source is needed. The paper will present the design, simulation and test of Class AB and Class D type power amplifier excitation sources for the sensors of the AMBs system.**

I. Background

A. High Power Sinewave Signal Excitation Source

A high power sinewave signal excitation source [1] is always used to drive the sensor in the AMBs system [2-6], which can be obtained by a basic sinewave signal after amplification by a power amplifier device or circuit. Several methods could be used to generate the basic sinewave signal, such as Wien-bridge circuit [7] and direct digital frequency synthesizer (DDFS) [8]. The Wien-bridge circuit is analog circuit and the DDFS is digital circuit that often needs a CPU. Power amplifier (PA) circuits are often classified as Class A, B, AB, D and so on. Class A designs are the simplest but with the lowest efficiency. Class B amplifiers can achieve 78.5% efficiency theoretically but introducing crossover distortion. Class AB is a good compromise with a higher efficiency than Class A and a lower distortion than Class B, so it is widely used for audio power amplifiers. Currently both Class AB and Class D type PAs for the excitation source of sensors are developed. The power of Class AB type is about 100~500W and could not reach a high value because its power dissipation results in a high temperature condition that will introduce much more noise and distortion. A cooling fan helps to improve the condition but contributes to a more complex circuit. However, the class D type can achieve up to 90~95% efficiency and therefore reduces the power cost, temperature, and noise so that the power could be in the range of 100~1000W even without the cooling fan. Though there are many Class AB and Class D high power PA products for concert audio speakers and other audio field, the frequency of excitation source, in the range of 10kHz ~ 22 kHz, is much higher than that of audio speakers within 0 ~ 4 kHz and that makes the designs are different.

II. Class AB Power Amplifier

A. Basic Class AB Topology

Fig. 1 shows a simplified Class AB amplifier structure. The upper transistor is NPN or NMOS and the lower one is PNP or PMOS. In this operation, each transistor amplifies half of the input waveform, but also conducts a small amount on the other half. As a result the region where both transistors simultaneously are nearly off is reduced. The output waveform is combined by two amplified half and the crossover is minimized greatly.

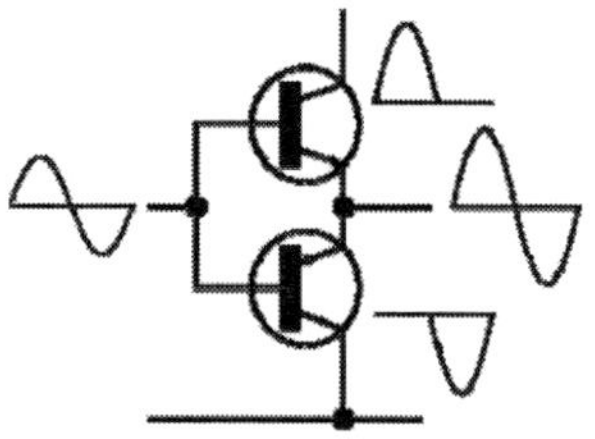

Figure 1. A basic Class AB amplifier.

The Class AB integrated circuit, TDA8547, is used to implement the amplified part for the excitation source of the sensors, as the Fig. 2 illustrated. The frequency of this IC is limited to 20 kHz because it is designed for audio field originally and then the amplitude attenuation would exist so that an extra amplitude tuner is needed.

B. Basic Sinewave Signal Generator

To simplify the sinewave generation, an integrated circuit, MAX038, is utilized instead of discrete component. The

MAX038 is a high-frequency function generator that

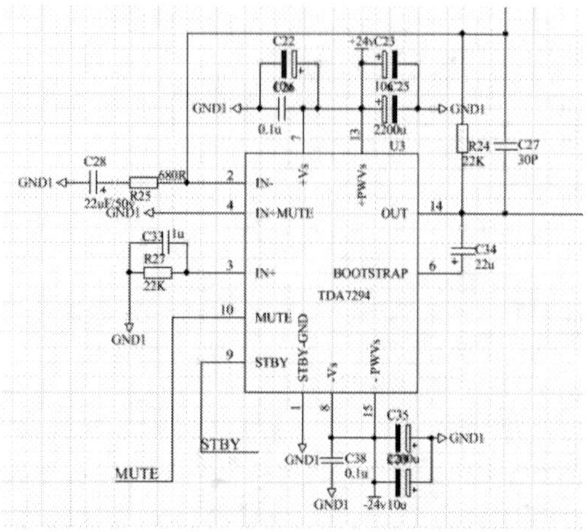

Figure 2. Class AB part: TDA8547 IC.

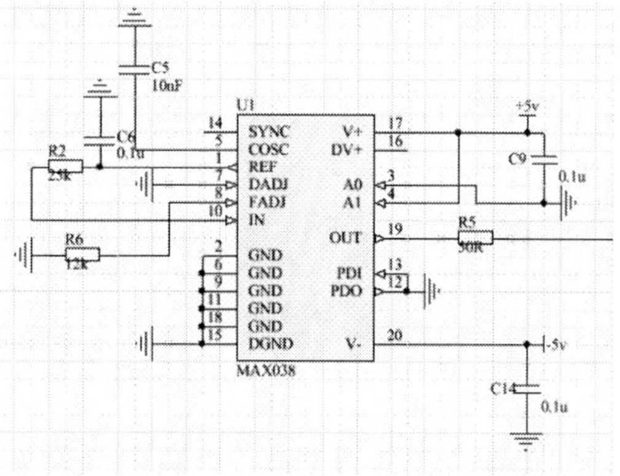

Figure 3. sinewave generator: MAX038 IC.

produces low-distortion sine, triangle, sawtooth, or square waveforms at frequencies from less than 1Hz to 20MHz or more, using a minimum of external components. Fig. 3 shows the implementation of this IC. The frequency of sinewave is calculated by Eq. 1.

$$f = 2 \times 2.5 / (R_2 C_5) \quad (1)$$

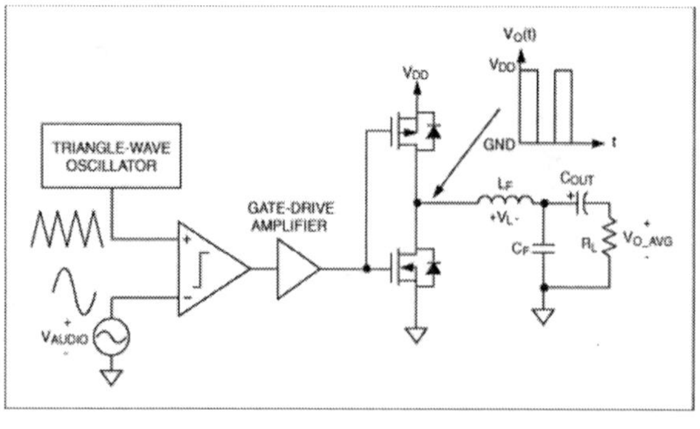

Figure 4. A basic class D amplifier.

III. Class D Power Amplifier

A. Basic Class D Topology

Fig. 4 shows a simplified functional block diagram illustrating a basic half-bridge class D amplifier. It consists of four parts: pulse width modulation (PWM) wave generator, gate driver, half-bridge or full-bridge power MOSFET and LC low-pass filter. The PWM wave is generated by using a sinewave sampled by a triangle-wave with a dozens of sample rates. A higher rate helps to achieve a less distorted final amplified sinewave but requires a gate-driver with a higher switching frequency. The gate driver produces a couple of switching signals for a half bridge or a full bridge with double amplitude using inverted input. Note that a full bridge consists of two half bridges. All NMOS devices are always preferred because NMOS has several advantages over PMOS, such as smaller on-resistance, higher efficiency, and lower temperature. A bootstrap circuit is needed for the bridge driver to drive the upper NMOS in the bridge circuit. The LC filter is for reducing the harmonics of the original sinewave and the sampling triangle wave.

B. PWM Wave Generator

The PWM wave generator consists of three parts: the original sinewave, the sampling triangle wave and the comparator. Many methods can be used to generate the sinewave using analog circuits such as the Wien-bridge but with poor ability to tune the frequency. A simple STM32F108 MCU using the SPI port can help to produce an easily tunable PWM wave with a high sample frequency. Fig. 5 shows a 20 kHz sinewave and a 450 kHz triangle wave and the generated PWM wave using a 9 MHz SPI clock.

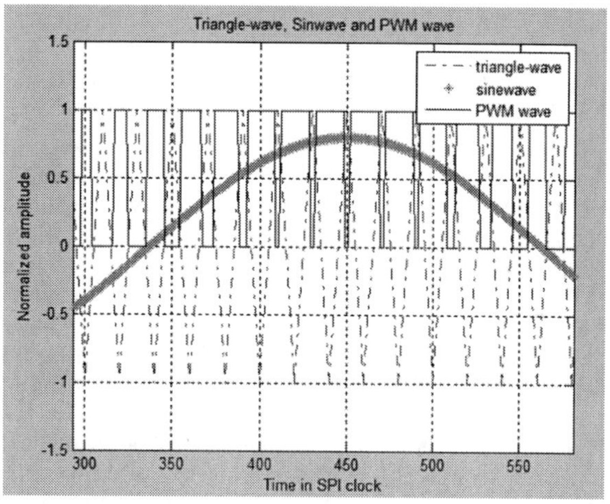

Figure 5. The PWM wave generated by a triangle-wave and a sinewave comparation.

C. Half-bridge Gate Driver

The main difference between the high power concert audio power amplifier and this excitation source is that the audio frequency range is normally about 0~4 kHz while the excitation source is in 10 kHz~22 kHz range. The much higher sample rate requires a higher switching frequency gate driver. Because the gate driver needs a dead time to control the half bridge to reduce distortion and power dissipation, the swing rate of the half bridge should be high and the dead time should be small. For example, an IRS2104S half-bridge driver with about 500 ns dead time at either the rising and falling edges will occupy about 50% of the duty cycle at the peak and trough of the sinewave in Fig. 2. This increases the distortion and improvement is required.

D. *Half-bridge and Full-bridge Comparison*

Class D amplifiers can be categorized into the half-bridge and full-bridge configurations. A half bridge using two MOSFETs is simpler while a full bridge using four MOSFETs is better in performance because the differential output structures of the two half bridges cancels the even-order harmonic distortions and DC offsets and generates two times the output swing. Table I shows a simple comparison. The excitation source for the sensor needs 40-100 V output swing and a full-bridge circuit has more choices in power devices and is easier to design.

TABLE I. A COMPARISON OF HALF-BRIDGE AND FULL-BRIDGE

	Half-bridge	Full-bridge
Output swing	1	2
MOSFET	2	4
Gate driver	1	2

E. *LC Low-pass Filter*

The amplitude-frequency response of the PWM wave is shown in Fig. 6. It is a 20 kHz sinewave modulated by a 450 kHz triangle wave. The signal above 20 kHz is of no use and should be removed by an LC low-pass filter (LPF) which behaves like a Butterworth LPF. The LC component value is determined by the LPF stop band frequency (f_{stop}) and the characteristic impedance of the load (Z_c), shown in Eq. 2. Setting f_{stop} to 50 kHz and for a load with a 2 Ω resistance in series with a 0.1 mH inductance, L is 10 μH and C is 3.3 μF approximately.

$$L = \sqrt{2} / f_{stop} \times Z_c$$
$$C = \sqrt{2} / f_{stop} / Z_c \quad (2)$$

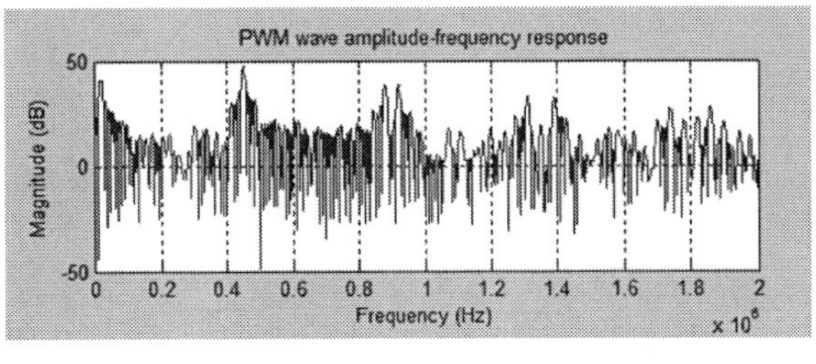

Figure 6. The amplitude-frequency response of PWM wave.

IV. Simulation And Test Result

A. *Circuit Schematic And Simulation Result*

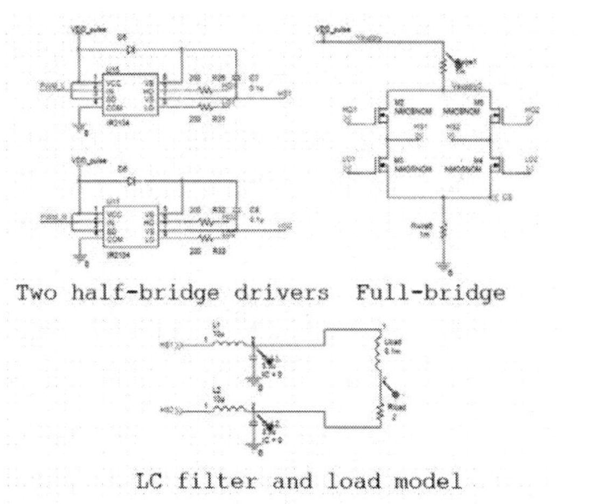

Figure 7. The main part of excitation source.

Fig. 7 shows the main part of the Class D excitation source. It consists three parts: two half-bridge drivers, a full-bridge and an LC LPF with the load model. The PWM_L in Fig. 7 is the PWM wave as shown in Fig. 5 and the PWM_R is the inverted one. The half-bridge driver is an IRS2104S and the power MOSFET is an IRFH5053. Two inductances of 10 μH bearing up to 6-10A maximum current are needed for the LC filter with a 2Ω load in series of 0.1 mH inductance. The simulation result using Cadence Pspice is shown in Fig. 8. The output waveform is indeed an enlarged sinewave and drives the load with less then 5A.

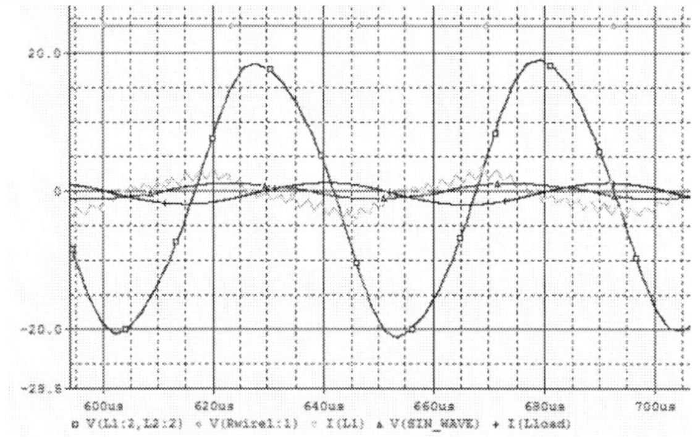

Figure 8. The simulation result using Cadence Pspice software.

B. *Test Result*

The printed circuit boards (PCBs) are manufactured to verify the class AB and class D power amplifier excitation sources. The result of the PWM wave using an STM32F108 MCU with the algorithm of Fig. 5 is shown in Fig. 9. The output sinewave of the excitation source is shown in Fig. 10. These results illustrate that the PWM wave generator works well and the two half-bridge drivers, the full bridge and the LC filter are suitable for this source. The Class AB type also works well with a lower distortion but with a higher current and needs a cooling fan. The waveforms are the same and so here is not shown.

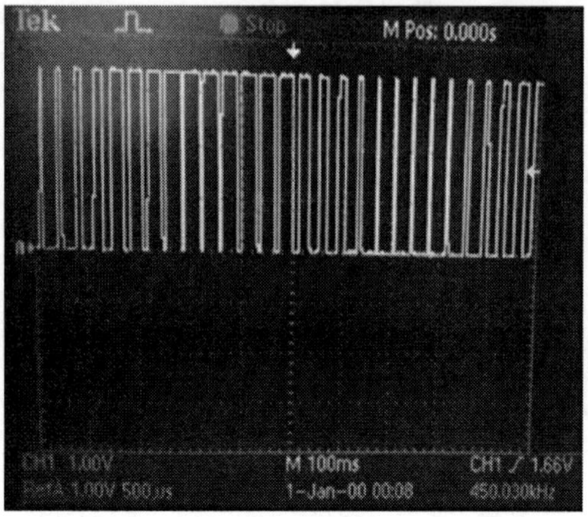

Figure 9. The PWM wave generated by an STM32F108 MCU.

V. Conclusion

A. Improvement and Future Work

Two types of excitation source for the sensor of the AMBs system have been verified and especially the proposed Class D power amplifier sinewave excitation source is shown to work by simulation and verified by experimental tests. It increases the efficiency and reduces an extra cooling fan versus the Class AB type source and this helps the sensor work stably. Future work will be carried out to reduce the distortion for the high frequency switch of the half-bridge driver.

References

[1] Athanasios Chasalevris, Fadi Dohnal, Ioannis Chatzisavvas, "Experimental detection of additional harmonics due to wear in journal bearings using excitation from a magnetic bearing," *Tribology International*, vol. 71, pp. 158–167, 2014.

[2] Lei Shi, Suyuan Yu, Guojun Yang, Zhengang Shi, Yang Xu, "Technical design and principle test of active magnetic bearings for the turbine compressor of HTR-10GT," *Nuclear Engineering and Design*, vol. 251, pp. 38–46, 2012.

[3] Martin Gronek, Torsten Rottenbach, Frank Worlitz, "A contribution on the investigation of the dynamic behavior of rotating shafts with a Hybrid Magnetic Bearing Concept (HMBC) for blower application," *Nuclear Engineering and Design*, vol. 240, pp. 2436–2442, 2010.

[4] Xu Yang, Shi Zhengang, Yang Guojun, Zhao Lei, Yu Suyuan, "Design aspects and achievements of active magnetic bearing research for HTR-10GT," *Nuclear Engineering and Design*, vol. 238, pp. 1121–1128, 2008.

[5] S. E. Mushi, Zongli Lin, P. E. Allaire, "Design, Construction, and Modeling of a Flexible Rotor Active Magnetic Bearing Test Rig," *IEEE/ASME Transactions on Mechatronics*, vol. 17, pp. 1170–1182, 2012.

[6] Kejian Jiang, Changsheng Zhu, "Multi-frequency periodic vibration suppressing in active magnetic bearing-rotor systems via response matching in frequency domain," *Mechanical Systems and Signal Processing*, vol. 25, pp. 1417–1429, 2011.

[7] Yongan Li, "A new single MCCCDTA based Wien-bridge oscillator with AGC," *AEU - International Journal of Electronics and Communications*, vol. 66, pp. 153–156, 2012.

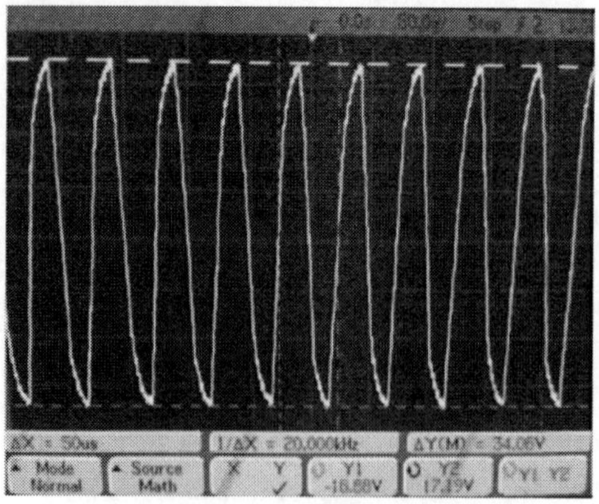

Figure 10. The output sinewave of the excitation source.

[8] B. K. Vasan, S. K. Sudani, D. J. Chen, R. L. Geiger, "Low-Distortion Sine Wave Generation Using a Novel Harmonic Cancellation Technique," *IEEE Trans. Circuits and Systems I*, vol. 60, issue. 5, pp. 1122–1134, 2013.

On the Direct Use of Electromagnetic Field Information for Magnetic Rotor Position Control

Hubert Mitterhofer[a], Mikhail Rakov[b]

[a] Johannes Kepler University Linz, Altenberger Strasse 69, 4040 Linz, Austria, hubert.mitterhofer@jku.at
[b] RLR Consulting, 641 B Por La Mar Circle, Santa Barbara, CA 93103, USA, mikhrak@aol.com

Abstract—**This article describes a simple principle for obtaining information about the radial rotor position and later on shows how this information can be used to stabilize the rotor deflection in a magnetically levitated system. It is important to notice that the considered demonstration system was designed in a way the discussed effects are clearly noticeable. It is not meant to be a design proposal for an actual motor.**

I. Introduction

There are not many inventions that have had such a profound impact on human technology as the invention of the wheel. One of these rare exceptions is the invention of a rotating electromagnetic field, that significantly shaped the industrial world as we know it today. Unlike the case of the wheel, the process of creating the concept and finding a practical application of the rotating electromagnetic field took a relatively short period of about one decade, from 1882 to 1891, and is connected to a small group of brilliant scientists and inventors. The first in this group is the Serbian inventor Nikola Tesla, who in February of 1882 had a vision of a rotating electromagnetic field as a result of the interaction of poly-phase alternating currents, and later developed many of its important practical applications. The other two were the Italian Galileo Ferraris, and the Russian/Polish engineer and inventor Mikhail Dolivo-Dobrovolski who in 1891 at the International Electrotechnical Exhibition triumphantly demonstrated the possibilities of generating, transmitting and practically using poly-phase electrical currents. Not surprisingly, the main interest and research efforts for newly discovered theoretical concepts were focused on the areas of energetics in the final years of the 19th century. Essentially, human society was still in the very beginning of its information stage, and poly-phase signals, as well as the resulting rotating electromagnetic field, were considered predominantly as a means of efficient creation, distribution and consumption of energy. However, as we can see it now, they have much more intrinsic properties, important specifically for the exchange of information signals. These properties with their basic principles of work and a potential use for the area of magnetic levitation are the subject of the present work.

II. Rotating electromagnetic field as a source of information

While the primary purpose of rotating electromagnetic fields is the brushless delivery of energy to and from rotors of electric motors and generators, this field can perform one more very important function – to serve as an instrument of information exchange between the parts of electromechanical devices. By its very nature, a rotating electromagnetic field presents a radar-like, permanent monitor of the position of the rotor and its relative position with regard to other components - the stator, its poles and windings, etc. And, while a perfect poly-phase system is completely symmetrical and balanced, any violation of this symmetry will result in a disbalanced signal containing information about the source and nature of this disbalance. This basic fact has been used for detecting the rotor position in several publications. The authors of [1], [2], [3] inject additional high frequency voltage and current components for detecting the current rise as a function of the coil inductances and thus, of the rotor position. With sophisticated signal treatment, this so called sensorless principle was used to levitate the rotor of a magnetic bearing. Another example can be found in [4] and later by different authors in a very similar publication [5], where the axial position of a magnetically levitated blood pump was detected using Hall sensors which measured the electromagnetic field, distorted by the rotor deflection. Also most existing eddy current sensors, for which countless examples exist in literature and on the market, could be put into the group of devices which actively generate an electromagnetic field which is altered due to the rotor position. This change is detected by the passive sensor part and is then used in order to obtain the rotor position.

All the examples mentioned above have in common that the electromagnetic field variation is captured and fed into a signal treatment process of varying complexity. Looking at a very basic field configuration, we can, however, notice an extremely natural form of phase and amplitude of the harmonic signals as we will shown in the following sections. The principles are based on the findings described in [6] and [7]. The following paragraphs will demonstrate the method when applied to a demo structure for a 3D-FE simulation.

A. Demonstration system

Before diving off into the description of an actual prototype, we want to point out that the present structure is not meant to constitute an actual motor or bearing design but rather to allow a 3D-FE simulation, demonstrating the very simple and natural way of obtaining radial rotor position information from the electromagnetic field. For this simulation, the following

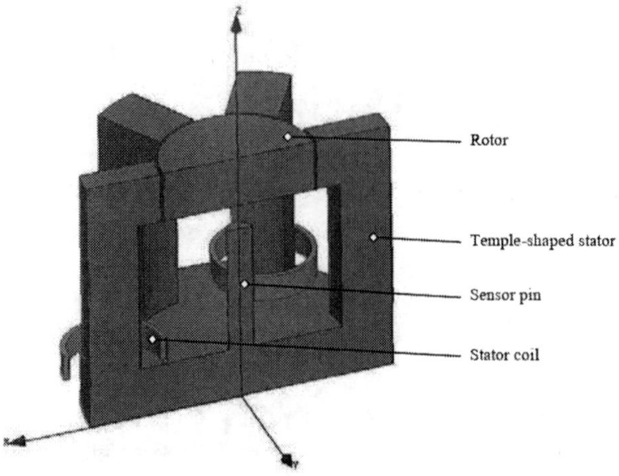

Figure 1. Simulated structure for showing the rotor position dependent electromagnetic field

assumptions were made:

- All magnetically conducting parts in both, the rotor and the stator of the used structure are considered as ideal magnetic conductors - saturation effects are not being taken into account.
- The axial rotor stabilization is not considered, the focus of this investigation is on the radial position information.
- All dimensions are chosen arbitrarily, they do not follow any design for an actual drive but shall be suitable for the 3D-FE simulation, demonstrating the mentioned effects.
- The resulting absolute values for flux densities, forces or other physical characteristics are not meant for designing an actual drive but have to be seen as relative indicators for the described phenomena.

Let us suppose a rotating, symmetric 3-phase, star-connected electrical system where each phase drives its current through one coil, being wound as a concentrated coil around one single stator tooth. The 6 teeth of which only 3 carry windings are located at a mechanical angle of 60° and are formed in a temple motor-style in order to provide sufficient space for all components. One of the special features of the demonstration system is a sensor pin located below the rotor. This pin has a large air gap to the rotor in order to keep the axial force influence to a minimum. A small but proportional part of the sum flux in the rotor will flow through this sensor pin. Together with an iron rotor disk, the motor topology of the system shown in Figure 1 resembles an induction machine principle.

B. *Flux distribution*

For the most simple case, let us assume sinusoidal currents $I_1(t)$, $I_2(t)$ and $I_3(t)$ of identical amplitude in the three phases, having an electrical phase shift of 120° to each other. The coils are placed such that coil 1 has its center at the x-axis of the Cartesian coordinate system marked in Figure 1. The remaining two coils are positioned at 120° and 240° in mathematically positive sense. The flux generated by the phase currents passes through the stator tooth, the rotor, the remaining teeth and the stator base plate functioning as a yoke. In the case of a completely centered rotor disk, no flux is conducted in the thin sensor pin which is located below the rotor disk since the sum of the coil flux components

$$\Phi_1(I_1) + \Phi_2(I_2) + \Phi_3(I_3) = 0 \tag{1}$$

is equal to zero. This fully symmetric situation can now be disturbed by a radial shift of the rotor which shall be investigated in three different scenarios.

1) *Centered position* (S0)
2) *Eccentric axis* (S1): The rotor is displaced radially and keeps on rotating about its (now displaced) axis of geometry. With the arbitrarily chosen values, the rotor is e.g. shifted towards coil 1 for half the air gap width δ. This means that the displacement vector in cylinder coordinates r, φ and z is

$$S1_{shift} = \begin{bmatrix} \frac{\delta}{2} \\ 0 \\ 0 \end{bmatrix}. \tag{2}$$

3) *Axis orbit* (S2): The axis of rotation is displaced from the center of geometry which means that the axis of geometry will describe a circular orbit as the rotor spins about its axis of rotation. For the simulation, the rotating displacement vector has a length of $\frac{\delta}{2}$ and was chosen to rotate at the same rate as the electrical phase vector but with a mechanical phase of $\varphi_{mech} = \varphi_{el} - 90°$, making the displacement vector

$$S2_{shift} = \begin{bmatrix} \frac{\delta}{2} \\ \varphi_{el} - 90° \\ 0 \end{bmatrix}. \tag{3}$$

This means that the rotor would come as close as $\frac{\delta}{2}$ to the stator tooth 1 (carrying coil 1) at the moment when the electrical phase angle is 90°, creating an identical situation point for both eccentricity scenarios.

The cross section through the entire structure at half the height of the sensor pin is given in Figure 2. Due to the shift of the rotor in x-direction in S1, the inductance value of coil 1 rises, the sum of all flux components is no longer equal to zero according to (1) and a proportional flux component can be found flowing in the sensor pin[1]. Figure 3 shows the mean flux density values in z-direction over the cross section of the sensor pin at half of its the vertical height for all scenarios S0, S1 and S2. As it could be expected, there is no sum flux for S0, there is sinusoidal sum flux for S1 and there is a sum flux showing a constant offset value and three times increased frequency for S2.

[1]At this point we want to recall that the dimensions and the air gap of the sensor pin was chosen arbitrarily. Any kind of real application would have to design this flux path in accordance to the flux sensor topology and its flux sensing range and resolution.

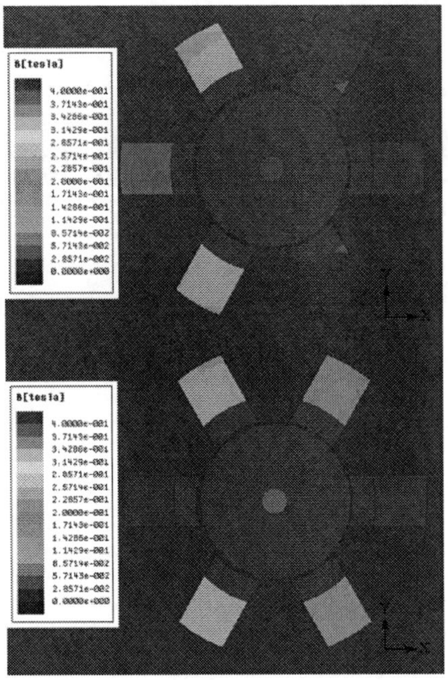

Figure 2. Flux distribution in a horizontal cross section at half the height of the sensor pin for a centered rotor S0 (top) and an eccentric rotor in S1 (bottom). The bottom figure is generally true for S1 and can additionally be interpreted as S2 at an electrical phase angle of 90°.

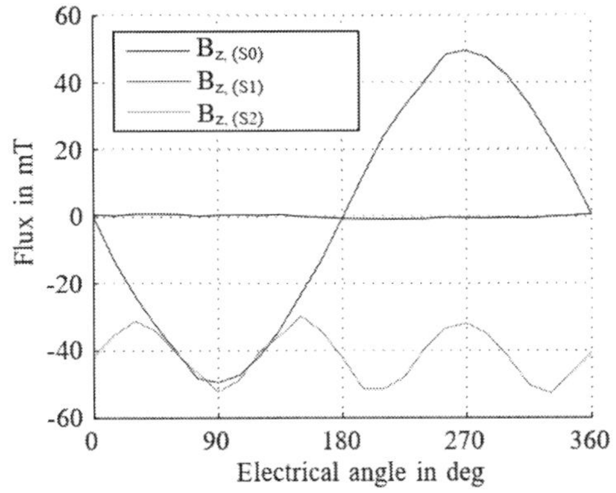

Figure 3. Mean flux density in positive z-direction in the central sensor pin over the electrical angle for different eccentricity situations

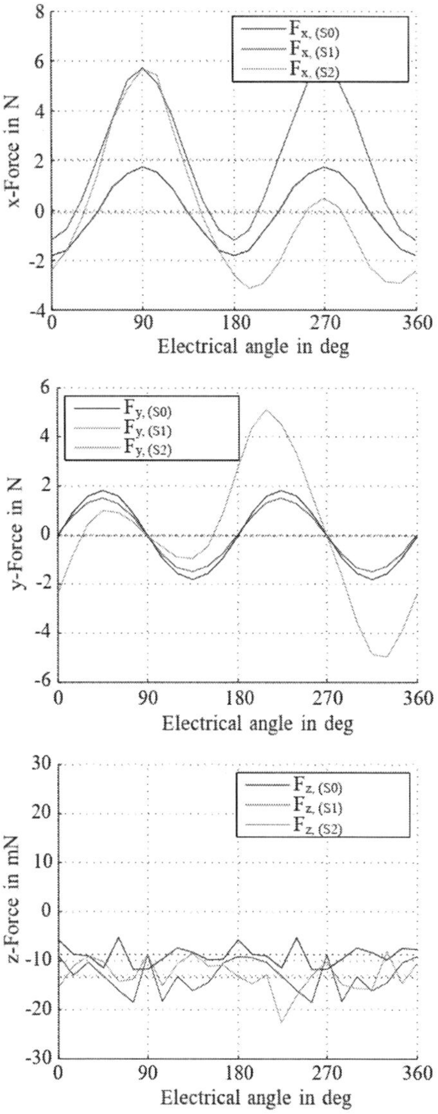

Figure 4. Forces distributions and mean force values acting on the rotor, given in Cartesian stator coordinates. Each component (x-forces on top, y-forces in the middle, z-forces on the bottom) was evaluated for all scenarios S0, S1 and S2.

C. Resulting reluctance force

The radial reluctance forces between the stator teeth and the rotor are identical in amplitude for all three air gap regions between the rotor and one of the coil-carrying stator teeth in case of the centered rotor. This is, of course, due to equal air gap flux density and equal geometric dimensions in the air gaps. The forces also have the same phase shift as the coil currents, leading to a rotating force vector acting on the rotor disk.

Shifting the rotor by applying the two eccentricity scenarios S1 and S2 will result in a changed force vector. Figure 4 shows the force distribution and the mean force values over one electric period. As the rotor was moved in positive x-axis direction (equivalent to the mechanical angle of $\varphi_{mech} = 0°$), the x-force orbit in S1 is also shifted to more positive values in this direction. The mean value reflects this simple correspondence clearly. For S2 with its orbiting axis, the force vector changes its amplitude over the electric angle but, as for the centered rotor, has a mean value of zero. The mentioned identical situation for the two eccentricity scenarios S1 and S2 can be seen at the identical curve height at an electrical angle of 90°.

Concerning the y-forces on the rotor, the force in S1 is almost identical to the case with centered rotor position except for the slightly smaller amplitude. Even if this is not the case for S2, the mean y-force for all scenarios is equally zero.

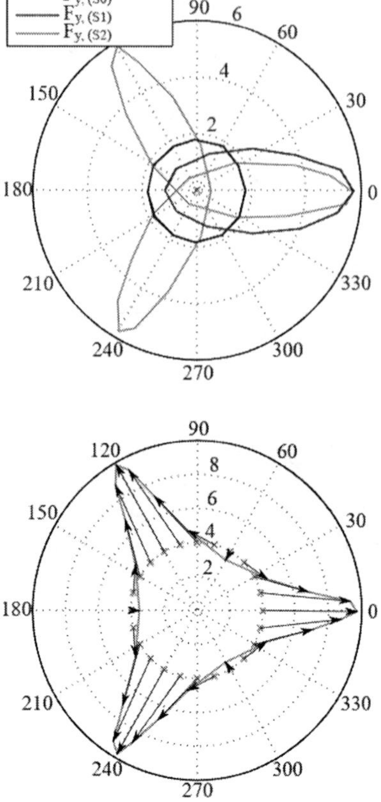

Figure 5. The cylindrical coordinate system offers a more natural way of seeing the occurring force orbits in newtons with their mechanical angle in degrees as the electric angle rotates for one period (top). Scenario S2 is shown again (bottom) with an arbitrarily chosen offset in order to reflect the changing rotor center position marked by the crosses and the corresponding force vector.

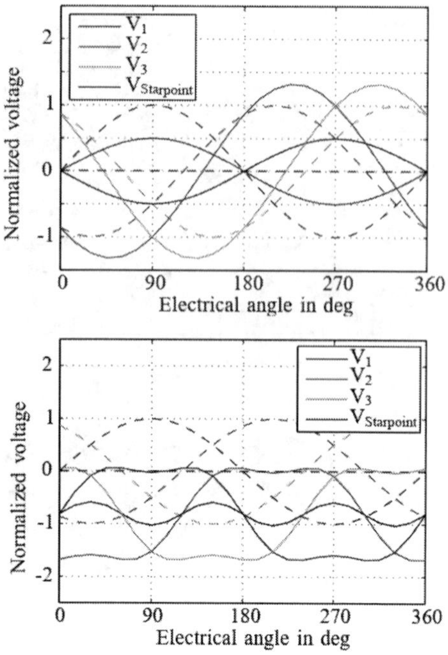

Figure 6. Symmetric 3-phase system (dotted lines) with star-point modulated according to flux in eccentricity situation S1 (top) and S2 (bottom) and resulting phase voltages (solid lines). All values are referenced to the peak value of the symmetric voltages.

The z-force which is also displayed in Figure 4 is slightly negative due to the attractive reluctance force between rotor and sensor pin but the values are by far below the x- and y-forces and close to the simulation resolution, making them negligible for the moment.

Using the cylindrical coordinate system as shown in Figure 5 offers a direct view on the presented circumstances. The given angle scale refers to the mechanical force angle φ_{mech}. A vector pointing in direction of $\varphi_{mech} = 0$ is aligned with the positive x-axis as defined in Figure 2. The fact that the electrical angle φ_{el} is not directly visible may be irritating but must be kept in mind.

While the depicted force orbit for S0 is a circle and the one for S1 has elliptical shape, the curve of the force orbit in S2 is not very intuitive. Only when looked at with a certain offset shown in the bottom part of Figure 5, the characteristic becomes more clear. This offset is added for helping to relate the rotor center position (marked with 'x' tags) to the forces occurring in that position. Its value was chosen arbitrarily but was maintained constant for both, Figure 5 and Figure 8.

III. Application of the Gained Information

In the sensorless motor control techniques referenced in the beginning of section II which are also applied in most eddy current sensors, certain parameters of the altered magnetic field are captured and treated in a signal processing part in order to extract the values of interest. Here, we want to show a much more straightforward approach by directly using the phase and amplitude information of the field through the central sensor pin, measured by one sensor (different sensor systems may be applied) in order to control the star point of the electrical system. Figure 6 shows the voltage traces of an original symmetric 3-phase system in dashed lines. Additionally, the resulting unsymmetrical phase and star point voltages are drawn in solid lines when the star-point of the system is modulated as shown in the black curve. In this case, the height of the star-point voltage is proportional to the flux value shown in Figure 3 for the eccentricity situation S1 and S2 with an arbitrarily chosen but constant ratio of amplification. The resulting currents in the stator coils increase or decrease the radial forces on the rotor.

As shown in Figure 7, these forces tend to correct the rotor eccentricity in case S1 by decreasing the current amplitude in the coil which the rotor is shifted towards. At the angle of 90° for instance, there is still a slightly attractive force (positive x-value) which would further deflect the rotor from its central position but the mean x-force over one electrical period is negative, thus stabilizing the rotor. For S2, the mean force is still zero which is necessary in order not to add a constant offset like in S1 to the rotating displacement vector.

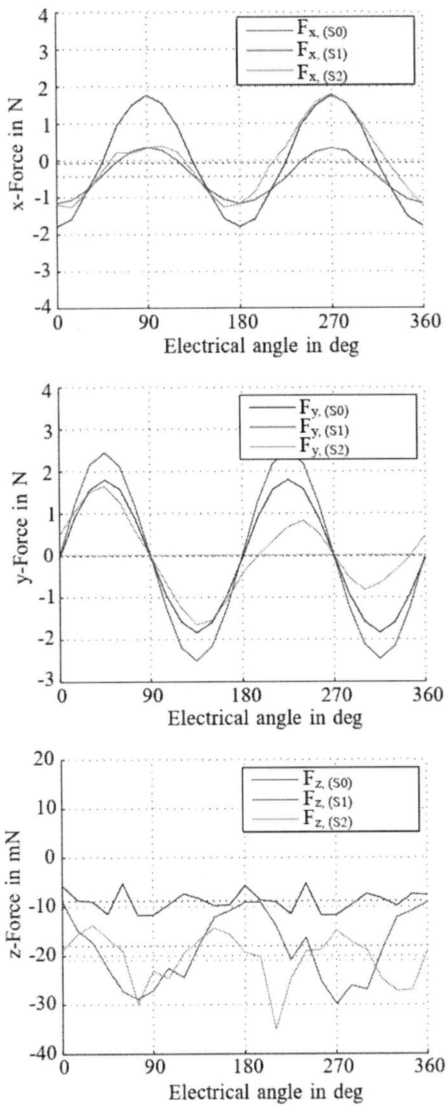

Figure 7. Forces distributions and mean force values acting on the rotor, when phase voltages are shaped by star-point modulation according to the measured sensor pin flux. Each component (x-forces on top, y-forces in the middle, z-forces on the bottom) was evaluated for all scenarios S0, S1 and S2.

Generally, the occurring z-forces shown in Figure 7 have increased slightly relative to Figure 4 but still range in a negligible magnitude.

Finally, let us again take a look at the picture in cylindrical coordinates, shown in Figure 8. In the image on top, the force ellipse of scenario S1 has deformed significantly in comparison to the constant star-point situation of Figure 5. The orbit clearly shows how the forces on the rotor would correct the rotor offset by acting towards a mechanical angle of 180° instead of pulling the rotor more towards 0°. As above, the force orbit of scenario S2 is difficult to interpret which is why it is drawn again with the same offset value as in Figure 5.

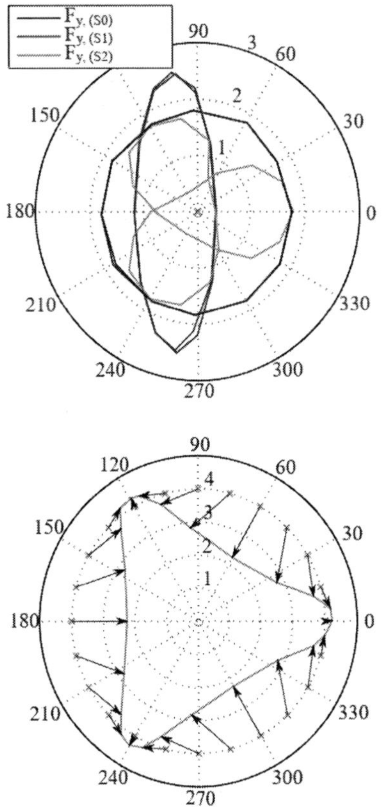

Figure 8. The cylindrical coordinate system with forces in newtons over the mechanical angle in degrees again offers a different view, showing clearly the stabilizing forces in cases S1 and S2. As for the Figure 5, scenario S2 (bottom) is shown with the arbitrarily chosen, but constant offset shown above in order to reflect the changing rotor center position marked by the crosses and the corresponding force vector.

The resulting force orbit and especially, the sketched force vectors, show how the rotor is moved back towards the center in almost every position except for the moments when the rotor axis is closest to one of the coil carrying stator teeth. This force amplitude, however, is only a result of the chosen amplification of the measured flux for modulating the star-point. It can basically be increased to much higher levels, thus providing much more aggressive stabilization forces.

IV. Discussion

When thinking of using the presented principles, several more or less severe limitations for practical use come up.

- Flux measurement is a delicate operation and e.g. Hall sensors which are able to measure flux with no or very little phase delay are unfortunately not very stable over temperature. Even if a homogeneous temperature distribution over the entire motor is assumed, the changing temperature and thus, the changing flux sensitivity would have to be compensated in order not to directly influence the actively created radial force amplitude. Another problem for the flux measurement is a possible axial deflection. As this would change the axial air gap, any axial movement would have severe influence on the measured sensor pin flux level in the demonstration system. A more sophisticated system of measuring this component could resolve the issue.

- Irregularities in the magnetic circuit in one of the teeth, e.g. due to rolling, punching or other manufacturing processes or material characteristics can easily result in a permanent shift of the central target position for the presented rotor position control.
- The flux sensor information was only used to influence the motor voltage proportionally, here. For an actual radial levitation control, of course also the derivative value would have to be taken into account in order to provide damping.
- The stability of the axial degree of freedom needs to be investigated thoroughly in order not to create unstable behavior due to the attractive reluctance forces between rotor and sensor pin.
- There are no permanent magnets in the presented demonstration system. This means that for the resting rotor or rather, for the case of the electric system being shut down, there will be no rotor position detection due to the missing sensor flux and also no stabilizing force of any kind on the rotor. Even if the electric system is up and running, the necessary field frequency would have to be well above the radial resonance frequency of a magnetically stabilized system so the rotor would not be severely deflected synchronously to the rotating flux vector.

V. Conclusion

The presented paper shows, how the electromagnetic field is influenced by the radial rotor position. For this reason, a disk shaped demonstration system was designed and simulated, using idealized materials and conditions. The gained information from the deformed field can now be used to modulate the starpoint of a three phase system, driving the motor. It has been demonstrated that this modulation changes the phase currents in a way to produce stabilizing radial forces acting on the rotor in the two essential eccentricity situations with constant and rotating deflection vector, respectively. The proposed method is essentially different from most eddy current sensors and sensorless techniques as it uses no digital signal treatment but merely amplifies the measured flux signal and, thus, retains its valuable phase and proportional amplitude information. As this work focuses on the demonstration of the principle, several more or less severe constraints for bearingless motor operation have not been considered here, however, the mentioned obstacles could probably be overcome in a sophisticated drive design. One may expect that the general considerations presented in this paper will result in significant improvement of the characteristics of the devices and systems mentioned in [6], [7].

References

[1] M. Hofer, E. Schmidt, and M. Schroedl, "Comparison of a solid and a laminated rotor for sensorless control of radial active magnetic bearings," *Proc. International Electric Machines and Drives Conference*, pp. 625–630, 2009.

[2] M. Hofer and M. Schroedl, "Statistic properties of a sensorless control method for a three phase permanent magnet biased radial active magnetic bearing," *Proc. European Conference on Power Electronics and Applications*, pp. 1–8, 2009.

[3] M. Hofer, E. Schmidt, and M. Schroedl, "Design of a three phase permanent magnet biased radial active magnetic bearing regarding a position sensorless control," *Proc. Applied Power Electronics Conference and Exposition*, pp. 1716–1721, 2009.

[4] S.-M. Yang and Y.-C. Chang, "Axial and radial position sensing for a magnetically levitated rotor using hall sensors," *Proc. 33rd IEEE Industrial Electronics Society Conference*, pp. 2225–2229, 2007.

[5] Y. Zhang, S. Liu, Y. Guan, H. Li, and Y. Fan, "The axial position sensing and signal processing in maglev artificial heart pump," *Proc. 5th Int. Conference on Theories and Applications*, pp. 396–400, 2010.

[6] M. Rakov, "Method and systems for operating magnetic bearings and bearingless drives," US Patent 8,115,358B1, 2012.

[7] M. Rakov and T. Shinano, "Reactance sensors of radial position for magnetic bearings and bearingless drives," US Patent US 8,299,781B2, 2012.

Eddy current sensors for magnetic bearings of the textile spinning machines

Miroslav Štusák

Rieter CZ s.r.o., Ústí nad Orlicí, Czech Republic, E-mail: miroslav.stusak@rieter.com

***Abstract*—The position sensors for magnetic bearings are very important parts of such a system. It is possible to say they are one of the most exacting part there because lot of conflicting demands are posed on them. This presentation describes the sensors for textile spinning machines which must have not only usual technical features but, in addition, they should fulfill the specific requirements of this industry, especially functionality in very dusty environment, resistance against foreign magnetic fields and last not least also potential for economical mass production without individual hardware setting and without selecting of specific components. Simply told, the prices at this market segment are forced down but the quality and reliability must be kept high.**

I. Introduction

A. Open End Spinning Process

One of the textile principles how to produce yarn is the open end spinning process, based on coiling of individual textile fibers in the spinning rotor with rotating speed often significantly over 100.000rpm. Conventional ball bearings for such speeds are expensive and they have limited lifetime. In addition, power consumption is high because of mechanical friction losses. Due to these facts, the alternatives were investigated and one of them was the magnetic bearing principle presented in [1]. Just at the beginning of the project, it was clear that a suitable position sensor will be one of the most significant components of such a system.

B. A History of Development

At first, a sample of the magnetic bearing without special sensors was tested, it had been assumed to sense the current of the bearing coils for a sensorless rotor position detection as proposed by [2]. This idea was left soon because of very low sensitivity and accuracy. As a next step, an eddy current sensor working on 500kHz with a synchronous rectifier has been investigated which provided good results but which was not suitable for mass production because individual hardware adjustment was needed and handling of the special thin wired coils was complicated. Also the influence of the magnetic field, generated by the motor driver, was visible because the working frequency of the sensor was not far off the used sensing frequency. We have also tested a capacitive principle which seemed to be very promising up to the point of high speed rotation when the noise on the floating electrode, which is the rotor, was too high, probably caused by static electricity. Optical methods were not investigated because of expected heavy dust deposit in the system. Because of these facts, we have concentrated on the eddy current sensor principle demonstrated in [3], trying to eliminate or to reduce all known disadvantages described above.

Figure 1. A sample of a HF transformer made of a PCB

C. A General Approach to the Problem

In the beginning there was a requirement to design a suitable sensor for a magnetic bearing within a relatively short time, using the known facts from the area of common general purpose electronics. It had been fulfilled by optimizing such methods under a lot of measurements and practical evaluations up to the moment when the results were good enough. No extremely high theories and no special expensive equipment was used but only step by step development on the right way to get a fit system for practical use.

II. Eddy Current Principle on a Higher Frequency

A. Sensor Design

From the principles that had already been investigated it seems that the eddy current principle could fulfill all the demands since it is enough reliable and accurate (see [4]). When a high frequency around 50MHz is used, then the sensing coils can be made as simple, cheap and easy to install PCB transformers when some general design rules as presented in e.g. [5] are followed. The design shown in Figure 1 will allow repeatable mass production with acceptable costs. In addition, the strong magnetic field with the frequency in the area of several kHz, which should be expected inside the system because of the motor drive signal, could be significantly suppressed, thanks to the very low inductance of the sensing coils. In this case, the coils had 5 windings what means about 200nH of inductance.

As an alternate way, ferrite coils with one winding were put up here but their features was not acceptable because certain of a non-linearity under the influence of strong magnetic field in the respective surroundings, see [6]. Therefore, only the

Figure 2. HF transformers in two orthogonal axes

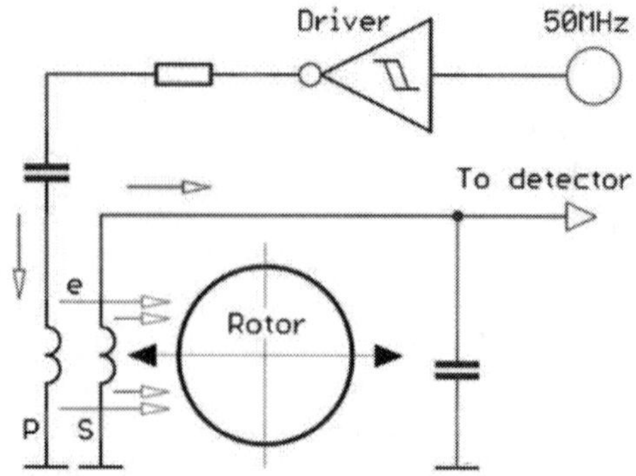

Figure 3. A principal single axis sensor configuration

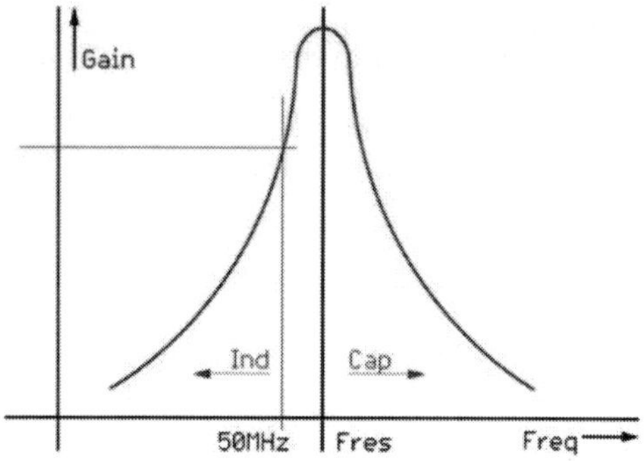

Figure 4. Sensor working point over frequency

air core PCB transformers were developed and used. These sensors were intended to be used for sensing a rotating part with diameter about 32mm, what means circumferential speed of approximately 250m/s at a rotor speed of 150.000rpm. This speed had no significant influence to the sensor features, no problematic behavior like for example missing target effect have been observed. The target part of the rotating element can be made of steel or aluminum alloy. The influence of these materials was measured as unessential. It is caused by skin depth which is given by

$$\delta = \sqrt{\frac{2\rho}{\omega\mu}} \tag{1}$$

where ρ is the specific resistivity, ω stands for the angular frequency and where μ denotes the permeability (cf. [6]).

B. Signal Detection

Now, the only remaining problem which has to be solved is a precise detection of the output signal level without any ambient influences, especially temperature and manufacturing tolerances. Finally, the suitable solution was found in the technique known from radio engineering, see Figure 7 and Figure 8.

III. High Frequency Signal Detection

A. Signal Generation

There are always two single sensors in one axis of the rotor position detecting system. Figure 2 shows the actual sensor PCB.

The real position of the rotor is evaluated from the differential signal between them. These sensors are made as PCB transformers with their primary coils powered from one common source of HF signal. The induced voltage in the secondary coils reflects the rotor distance to this transformer because it is influenced by the eddy currents induced in the rotor's conductive surface, see Figure 3. To increase the basic sensitivity, the primary and secondary windings are designed as coupled parallel resonant circuits. Their resulting resonant frequency must be slightly higher than the frequency that the sensor is powered with, in order to keep the inductive character of those circuits as a monotonic function of the output signal, even if the rotor moves back and forth, see Figure 4.

As a rectifier, the depletion mode field effect transistor (FET) is used because it does not show a voltage drop as in diodes and because of its nonlinear transfer characteristic shown in Figure 5 and Figure 6with a certain range of temperature independence. After the detection of the HF signals in the operational amplifier circuit shown in Figure 7, there are two DC signals from both sensors in one axis available. These two signals are of the same value in case that the rotor is in the middle position. A difference between these two DC signals shows the deflection of the actual rotor position from the middle one. It means that such a signal could be used as an input for the stabilization of the rotor position in the active magnetic bearing system as described in [7].

B. Sensor Signal Evaluation Circuit

Such four sensors described above can be used directly inside the textile spinning unit in two orthogonal axes to sense the position of the spinning rotor. The output DC signals are evaluated using a differential amplifier for each axis, shown in Figure 9. The mechanical and electrical tolerances are then compensated by digital potentiometers with internal memory where the initial correction is stored after the assembly and initial calibration of each magnetic bearing. Together with the control electronics and with the power stabilization coils, this generates a compact system that was also documented in [7].

Table I
SKIN DEPTH VERSUS FREQUENCY RELATION

Material	Resistivity in $\frac{\Omega mm^2}{m}$	Relative permeability	Skin depth in mm						
			50Hz	500Hz	5kHz	50kHz	500kHz	5MHz	50MHz
Iron lamination	0.19	1000*	1	0.3	0.1	0.03	0.01	0.003	0.001
Aluminum	0.027	1	11.8	4	1.18	0.4	0.12	0.04	0.012
Copper	0.017	1	9.4	2.9	.94	0.29	0.09	0.029	0.009

* The actual relative permeability of iron strongly depends on the operation point in its B-H curve, this is an exemplary value.

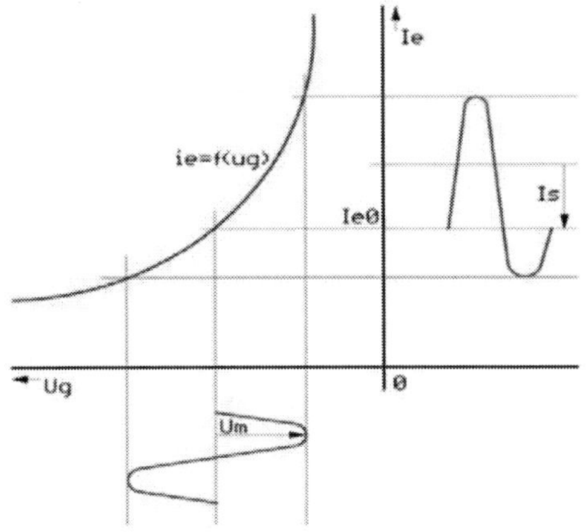

Figure 5. Typical transfer characteristic of depletion mode FET

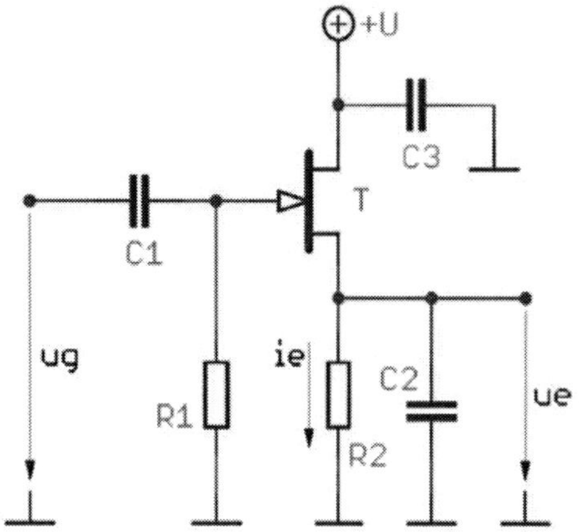

Figure 7. Sample of detector circuit

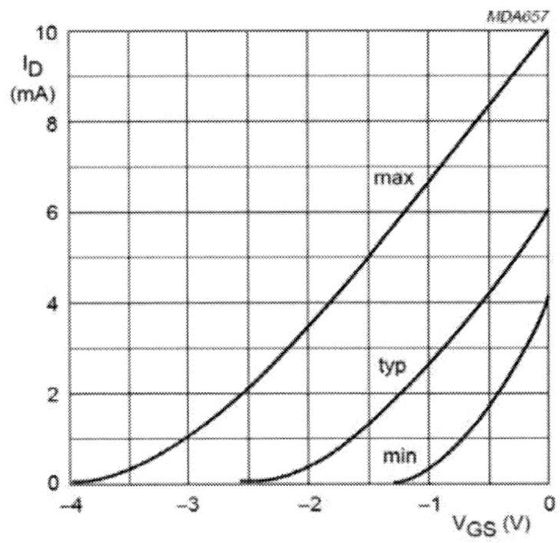

Figure 6. Real characteristic of depletion mode FET BFR31

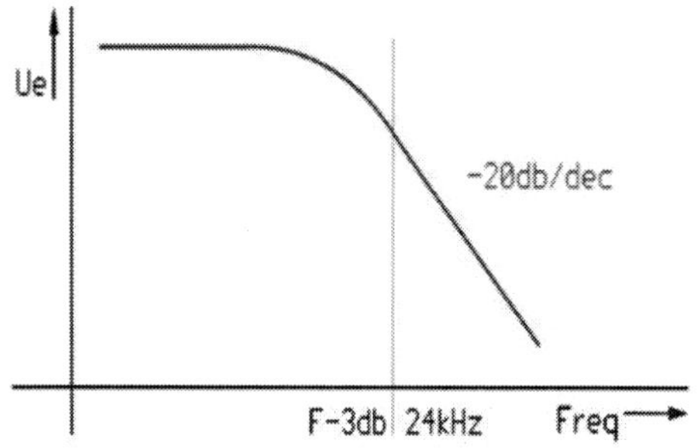

Figure 8. Frequency of characteristic low pass filter

IV. MEASURED PARAMETERS

The FET detector output is represented by the source voltage Ue which depends on the position of the rotor because of the eddy current effect. This voltage, of course, goes down when the rotor is leaving its center position and is approaching the secondary coil of the sensing transformer as shown in the right part of Figure 10. When the rotor shifts away from the transformer, the influence of the eddy currents decreases and the source voltage Ue goes to the upper limit value, see the left part of the Figure 10. Because of the above facts, the course

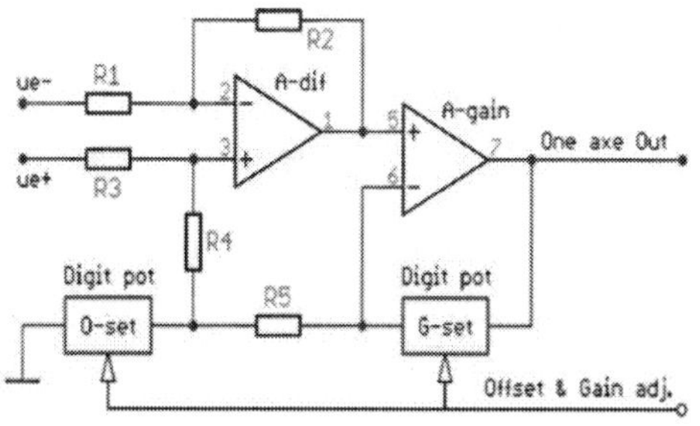

Figure 9. Principle of the differential amplifier for one axis with digital setting of the offset and gain

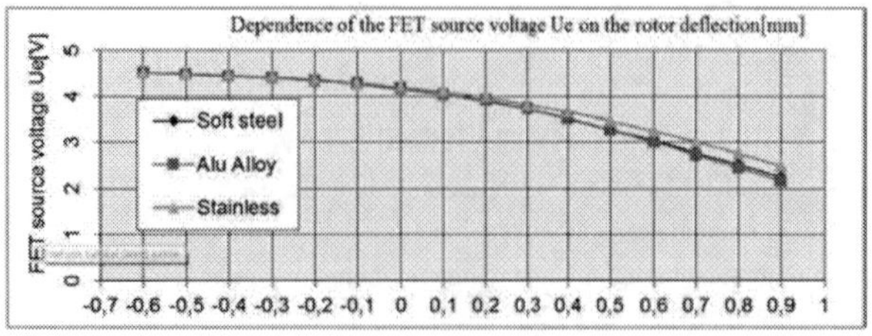

Figure 10. Relation between FET source voltage Ue and the rotor deflection from the center for various materials

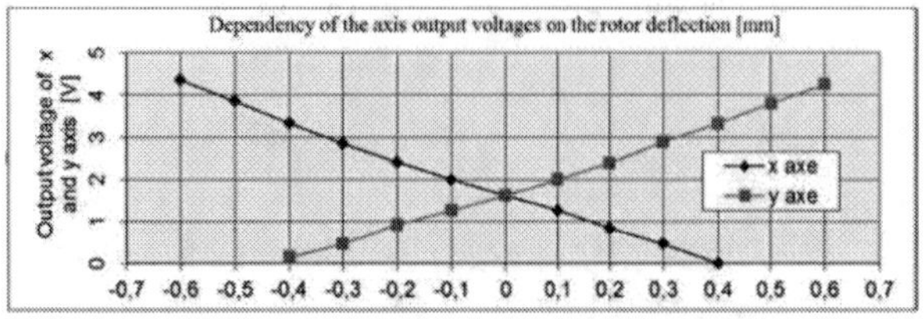

Figure 11. Resulting characteristics of the sensor system for both, x and y axes

of the dependency is not linear. What is interesting there, is that the curves are almost the same for rotors made of soft steel, special stainless submarine steel or aluminum alloy.

Using the differential amplifier of Figure 9 for subtracting the Ue voltages of the two FET detectors per axis results in an almost linear dependency between the rotor position and the differential output voltage. These curves are the same for both of x and y axes. The proportion of Ue as a function of the rotor shift, $U_e = f(shift)$, is approximately $1\frac{V}{mm}$ for one sensor around the middle position. After differential evaluation the final sensitivity is $2\frac{V}{mm}$. It is sufficient to use an amplifier with Gain of 2 to get the required ratio of $4\frac{V}{mm}$. The final response after those operations is displayed in Figure 11.

The resulting features of a complete one axis sensor are given Table II. These achieved parameters meet the required ones with a certain reserve. In accordance to Figure 9, the system allows to set the gain and to correct the tolerance offset independently and to store the correcting values into the internal memories of the digital potentiometers. It is remarkable that neither of components needs to be specifically selected or manually adjusted. One of the other big advantages is the fact that the correcting values are stored directly on the sensor board which allows to replace one sensor unit by another one without any necessary adjustment.

Table II
MEASURED VALUES FOR EDDY CURRENT SENSORS AS TWO-AXIS SYSTEM

Sensitivity in all range of the rotor position	$\frac{2V}{0.5mm}$
Normalized sensitivity	$4\frac{V}{mm}$
Output noise level on one axis output	$4mV_{pp}$
Theoretical accuracy	1μm
Practically usable accuracy	$10\mu m$
Measuring time for one position value	$50\mu s$
Number of measured values at 150.000rpm	8/rev

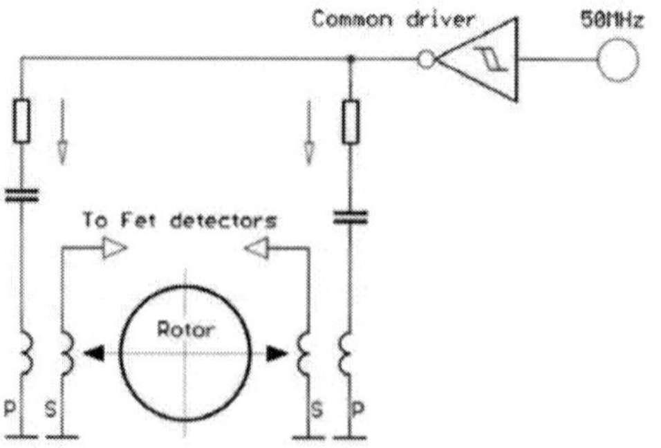

Figure 12. Principle of common power source for both primary coils of one axis

V. Temperature Stability

The next very important area of the sensor behavior in general is its temperature stability. There are lot of possibilities how the temperature could affect the sensors, since they are working on the eddy current principle. Practically, all the parts of the sensing evaluation chain have their certain temperature dependency. The most efficient tool for eliminating those effects is the differential principle of the signal handling from the AC source up to the final evaluation for each axis. Of course, the temperature stability of each individual part should be as high as possible. Keeping the principle above, the power source of the HF signal for the primary coils of the sensing transformers must be only one common for each axis as displayed in Figure 12. This is equally true for the DC powering of all differential circuits in one axis and for the referential voltages for the operational amplifiers.

The next important idea in this method is to use the same components with identical configuration for both sensors in one axis, believing that their temperature influences are also identical and they will be eliminated by the differential evaluation of the sensor signals. That should be right concerning all the passive and active components, an exception could be FET detector, only. To prevent such trouble, its working point must be adjusted to a temperature independent area by setting the optimal Id current, shown in Figure 13. The final temperature behavior of the sensing system which has been designed and made in accordance to the requirements above, is shown in Figure 14. In this experiment, the rotor was levitated magnetically and observed by the proposed sensor while the temperature was changed from 25°C at the start of the measurement to 60°C at the end of the 30 minute measuring duration. It is visible in this picture that the temperature influence on the differential output signal is below an absolute variation of +/- 10mV, representing a mechanical inaccuracy about $5\mu m$.

VI. Production and Operational Experience

The goal was to develop a position sensor for a magnetic bearing, suitable to be used in a textile spinning process and having specific functional features, such as accuracy of the position, reliability and resistance against the ambient influences which are especially temperature, dust and vibration. From

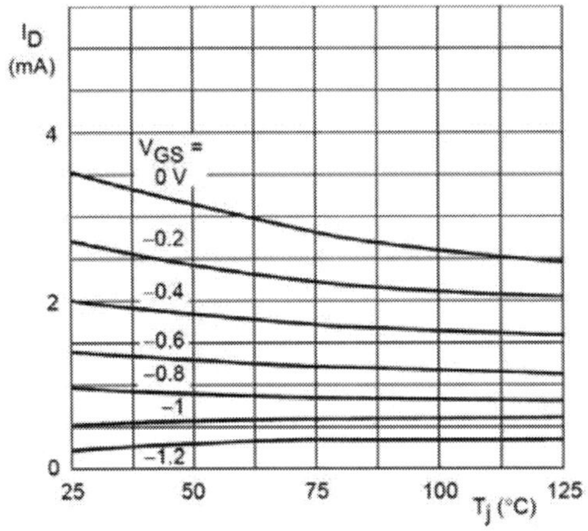

Figure 13. Real temperature behavior of depletion type FET

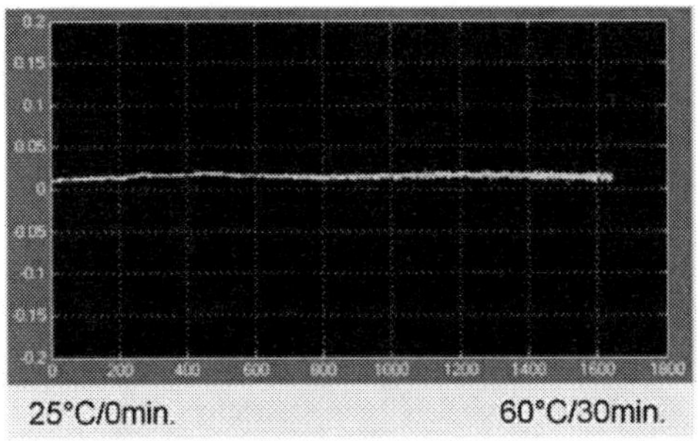

Figure 14. Measured temperature behavior of the one axis sensor system from 25°C to 60°C

the production point of view, there is no need for specially selected components and the individual adjustment is very easy and can be done automatically just via software as a last step of the final assembly. There are already some good practical experiences available, showing that there is a potential for some future industrial application of such a sensor concept. Two photos of the front and back side of the actual sensor board are visible in Figure 15. The complete schematic is given in Figure 16.

VII. Conclusion

This presentation describes the sensors of a high speed magnetic bearing for the textile industry which have been designed using known principles and keeping the known rules of such a design. A sensor unit was successfully designed and tested which is easy to produce at acceptable costs. Because of that, there is a good chance to use such a sensor also in other fields of magnetic bearing application with similar requirements.

Figure 15. HF transformers in two orthogonal axes

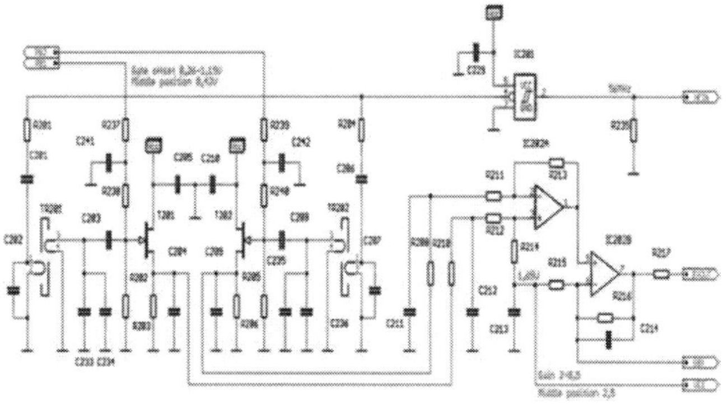

Figure 16. Complete sensor schematic

References

[1] S. Silber, P. Dirnbereger, W. Amrhein, J. Sloupensky, M. Moravec, and M. Reisinger, "Integrated rotor spinning drive system," *The 13th Mechatronics forum*, 2012.

[2] S. C. Mukhopadhyay, "Do we really need sensors? A sensorless magnetic bearing perspective," *1st International Conference on Sensing Technology*, 2005.

[3] P. Buehler, "Device for contactless measurement of distances in multiple directions," *EP 1 422 492 A1*, 2004.

[4] B. Syrovatka, "Zpusob stabilizace levitujiciho rotujiciho elementu a zarizeni pro stabilizaci levitujiciho rotujiciho elementu," *PVCZ2009 0568A3*, 2009.

[5] Y. Lee, "Antenna circuit design for RFID," *Microchip Inc., Application Notes, ww1.microchip.com/downloads/en/AppNotes/00710c.pdf, status July 2014*, 2003.

[6] B. Rall, H. Zenkner, A. Gerfer, and T. Brander, "Trilogy of magnetics, 4th edition," *Wuerth Electronics Midcom*, 2010.

[7] M. Stusak, "Device for detecting position of rotating working means in active magnetic bearing," *EP 1415 8472*, 2013.

Researches on Substrates of High Temperature Eddy Current Sensor

Chao Han[a], Longxiang Xu[b], Yingzhe Lin[c]

[a] College of Mechanical and Electrical Engineering, Nanjing University of Aeronautics and Astronautics, Yudao St 29, B221 Nanjing, China, cold-wave777@163.com
[b] College of Mechanical and Electrical Engineering, Nanjing University of Aeronautics and Astronautics, Yudao St 29, B315 Nanjing, China, fqp@nuaa.edu.cn
[c] Nanjing CIGU Limited Corporation, Jiuzhu Rd 100, Nanjing, China, lyz@cigu.org.cn

Abstract*—In this paper, substrates of eddy current sensor were composed by silver films and glass ceramic disks, all of these can resist high temperature. The thin silver films were deposited by radio frequency magnetron sputtering on machinable glass ceramic at room temperature. Variations of sputtering power, bios voltage and power density were carried out for each deposition, then parts of as-deposited samples were subjected to annealing at 600℃ within a vacuum chamber. Structural properties were studied by X-ray diffraction (XRD) and scanning electron microscope (SEM). It was shown that structural properties had a strong dependency on sputtering power and annealing temperature. Electrical contact resistance measured by a four point probe instrument was directly affected by the thickness of films. It was also found that the film conductivity, especially in thinner films, was improved by the increasing grain size. Further more, the film adhesion was observed by scratch tests experiment. Finally, the subsequent processes of high temperature eddy current sensor were introduced.

1 Introduction

There is a general agreement in the literature that metal films are widely used as coatings in the field of electronics, meanly because their low resistivity, high-thermal conductivity and excellent ductility [1,2]. And their properties have been extensively studied, but few researches have been offered to the application of metal films in high temperature. Traditionally, some active metals (Cr, Cu, Ni) are used for coating films. But they are easily contaminated by oxygen in high temperature. On the contrary, some inert metals (Au, Pt) which show strong stability in high temperature are difficult to obtain [3]. Finally, considering about the performance in inertness and cost-effective, silver attracts specific interest of researchers in coating films.

Different technologies have been reported to produce silver films with adequate performances for application in electronic devices: physical vapor deposition (PVD), vacuum evaporation, vacuum winding coating, DC and radio frequency (RF) magnetron sputtering [4]. In this work, silver films are prepared by RF magnetron sputtering for its some advantages: high deposition rate, high power efficiency, low temperature and good performances in smoothness, uniformity and binding force.

The use of glass ceramic disks as substrates for deposition has several advantages: machinability makes it can be machined to be any shape, high temperature resistance makes it can be worked in high temperature environment and high thermal expansion coefficient (8.7×10^{-6} /℃) generates small thermal stress between silver films and substrates under large temperature difference[5,6].

2 Experimental details

Magnetron sputtering deposition were done with the sputtering target of a 60mm diameter 99.99% purity pure silver disk to the substrates of the 3mm thickness and 10 mm diameter glass ceramic disks. Considering about the bond strength between silver films and substrates, the glass ceramic disks were slightly polished to the roughness Ra under 1 μm .

All the substrates were cleaned in an ultrasonic bath with acetone and dried before inserted into vacuum chamber. The distance between target and substrate was 50 mm. The working pressure in vacuum chamber could reach approximately 0.001 Pa due to the utilization of a turbo molecular-diaphragm pump combination.

The sputtering gas was filled with high-purity argon (99.99%) and the pressure was stably about 4 Pa. A gas flow controller was used to maintain the argon flow in the chamber at 10 SCCM.

Before film deposition, the target was pre-sputtered on the shield about 30 minutes to remove the contaminations in the target surface and stabilize the magnetron parameters. In the deposition, the performances of the silver films were determined by the magnetron parameters, such as the sputtering power, bias voltage, power density, deposition rate, deposition time, deposition thickness and annealing condition et al. therefore, the eight samples were distributed into four groups based on various magnetron parameters (as shown in Table 1).

TABLE I. SAMPLES WITH DIFFERENT PROCESSING PARAMETERS

Sample No.	Sputtering power (W)	Bias voltage (V)	Power density (W/cm2)	Deposition rate (nm/min)	Deposition time (min)	Deposition thickness (μm)	Annealing condition
a1	40	340	1.42	37	40	1.5	
a2							annealed
b1	100	525	3.54	50		2	
b2							annealed
c1	160	650	5.66	63		2.5	
c2							annealed
d1	250	820	8.85	125		5	
d2							annealed

The samples (a2,b2,c2,d2) we chose from each group were placed into a vacuum chamber which was then pumped to a pressure of 1.3 × 10-3 Pa and annealed stepwise at temperature of 600 °C to anneal with 30 minutes. After definite time intervals the annealing was interrupted, and after cooling to room temperature the chemical composition and crystal structure of all the films was analyzed by X-ray diffraction (XRD) D8 ADVANCE with Cu K α radiation and parallel beam geometry with a 2° incidence angle. After that, the samples were cut apart and inlaid. Then, the inlaid samples were polished in order to observe the cross-section surface profile.

Moreover, the cross-section morphologies were observed by SEM. Considering about the non-conductivity of cross-section, the inlaid samples ware coated with 20nm gold in 4 minutes by sputter coater K550X EMITECH in order to add their conductivity.

Electrical resistivity of silver films was acquired at room temperature by four point probe instrument RG-7C from NAPSON Corporation. The thickness of films was also input into the four point probe instrument to ensure the accuracy. The contact force between probes and samples was detected and adjusted during measurement to avoid that the probes puncture thin films. This analysis delivered ρ values within an error margin of~1%.

Finally, the film adhesion was measured by a micro scratch tester WS-2005. It is equipped with a standard diamond indenter with 120° taper angle and 0.2mm radius. During the measurement, a loading rate of 25N/min was used with scratches rate of 1mm/min. The critical load was detected by a force sensor within the micro scratch tester in rupture position of films to calculate the adhesive stress by equation (6), thereafter the scratches images were observed by SEM.

3 Results

3.1 X-ray diffraction

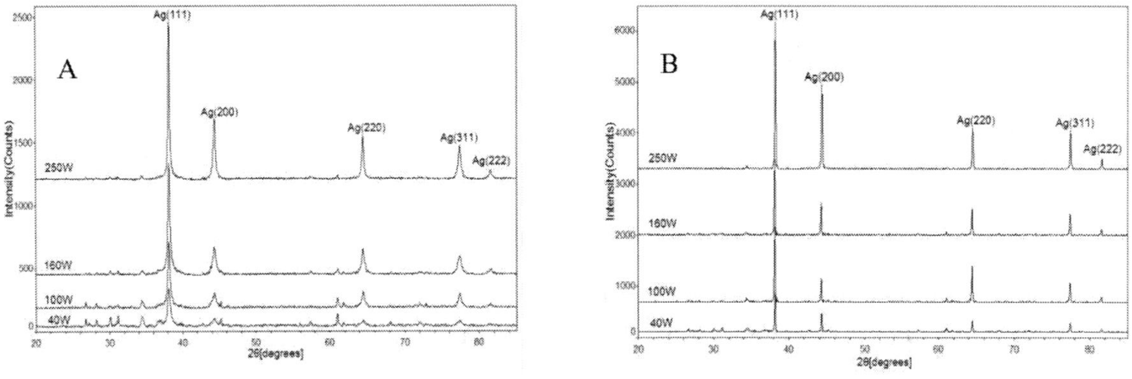

Figure. 1. XRD patterns for: A) as-deposited silver films with different sputtering power, B) annealed silver films with different sputtering power

Fig.1 shows the XRD spectrums of the silver films for the various different sputtering powers and annealing conditions. In order to minimize the errors of the experiment, prior to the analysis of the XRD spectrums, the back ground was subtracted. After this process, the diffraction peaks are completely symmetric and without any background. As shown in fig.1, the peaks indicate the polycrystalline nature of the silver films. And the strongest diffraction in each sample is the Ag (111), which can be observed in each sample. In addition, the peak becomes sharper with the increasing of sputtering power, meanwhile the splitting of diffraction peaks decrease or even disappear.

From the FWHM of diffraction peak (111), the average grain diameter of silver films can be estimated by the Scherrer formula as [7]

$$d = \frac{k\lambda}{\beta \cos\theta} \tag{1}$$

where d is the grain size in the silver films, the Scherrer constant k is taken equal to 0.9, λ (0.15406nm) is the wavelength of the X-ray, β is the FWHM of peak (111) can be indicated from

Table 2, and θ (2 θ =38.2°) is the diffraction angle of peak (111).

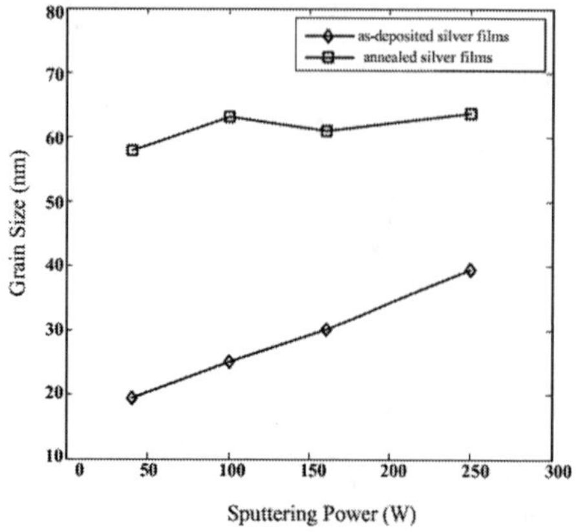

Figure.2. Grain size of silver films with different sputtering power and different annealing condition

Fig.2 shows that the grain size of films with the range form 19.4nm to 39.5nm for various sputtering power. And they steadily increase during annealing to about 60nm. With a higher growth rate, the initial silver crystals enlarge quickly before the nuclei are uniformly formed. And the grain sizes become larger after annealing due to the merging processes.

3.2 Microstructure

Fig.3 shows cross-sectional SEM images of two silver films with same sputtering power but different annealing condition. These two images show that the annealing process induce some changes in the morphological features of silver films. From the cross-section images with higher magnification (right column), one can clearly observe that silver aggregates have a significant change in size and shape, with diameters of grain varying from a few nanometers (as-deposited sample) to about 60nm (annealed sample).

The reason to believe that this suggestion is true is the result of work by K. Heinemann and H. Poppa [8] in which direct evidence is reported for the simultaneous occurrence of Ostwald ripening and short distance cluster mobility during annealing of silver films onto MGC surfaces in the temperature about 600°C. And this is confirmed by the variation of grain size detected by XRD shown in Fig.2.

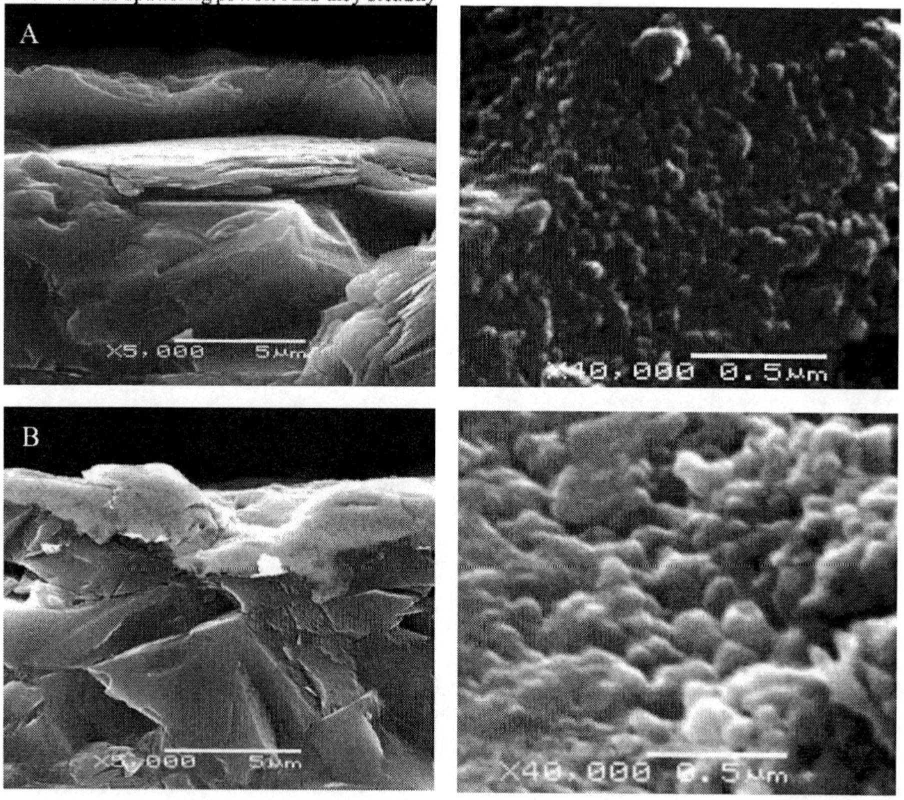

Figure 3. Fractured cross-section SEM images of two silver films with different annealing condition: (A) 160 W, (B) 160 W and annealed. The images in the right column are the same films taken at a higher magnification.

3.3 Electrical properties

The electrical resistivity of the different films at room temperature was measured by four probe measurements. The numerical value ρ_0 showed in instrument is calculated by the following equation [9],

$$\rho_0 = 2\pi S \frac{V}{I} \quad (2)$$

Where S is the distance between two adjacent probes, V is the potential difference between two inside probes, and I is the current flowing thought two outside probes. ρ_0 can be considered to be electrical resistivity ρ theoretically, if the distance between test point and border of films is greater than 4S. In this experiment, the condition of distance is not met. So the modifying factor B_0 should be taken in consideration, and electrical resistivity ρ should be written as

$$\rho = \frac{\rho_0}{B_0} = 2\pi S \frac{V}{I} \cdot \frac{1}{B_0} \quad (3)$$

The electrical resistivity ρ of silver films with different thickness and annealing condition after correction is showed in Fig. 4.

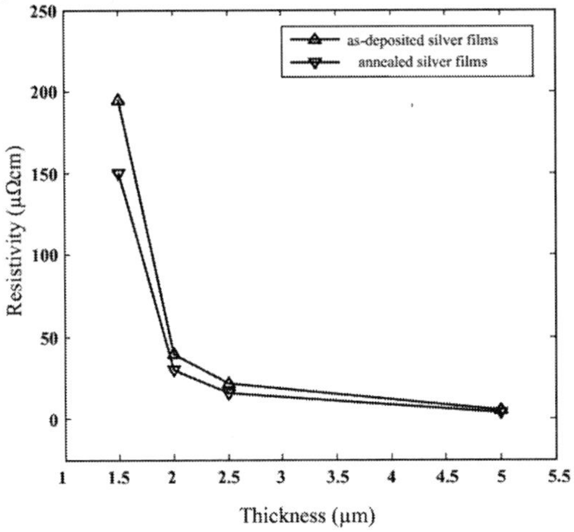

Fig.4. Electrical resistivity as a function of the thickness of silver films with different annealing condition

From Fig.4 one may notice that silver films at thickness of 1.5 μm exhibit higher resistance values (about 200 $\mu\Omega cm$) than that of others and this can explain the fact that the silver films at the lowest sputtering power exhibit darker images which showed in Fig3. When the film thickness increases to 2 μm, the ρ value decreases dramatically to about 40 $\mu\Omega cm$. When the film thickness further increases about 5 μm, the ρ value reaches about 3 $\mu\Omega cm$, being almost the same as that of bulk Ag (1.6 $\mu\Omega cm$). So a conclusion can be made that the electrical resistivity is directly affected by the thickness of films.

One may also notice that, the ρ value of silver films at thickness of 1.5 μm is lower after the annealing treatment than they have been before it, falling from 196 $\mu\Omega cm$ to 150 $\mu\Omega cm$. This is due to the fact that the increasing crystal grain size caused by annealing treatment decreases the grain boundary scattering of electrons [10,11].

3.4 Scratch tests

Scratch tests of silver films were carried out for with a loading rate of 25N/min. Fig.5A shows the total scratch track of film for an increasing load from 0 to 100N in 4 minutes. It is obvious that the width of scratch track is broadened with the increasing of load [12]. Fig.5B shows the scratch track in the rupture position. Where, the contact force added up to the critical load and the delaminating between film and substrate occurred.

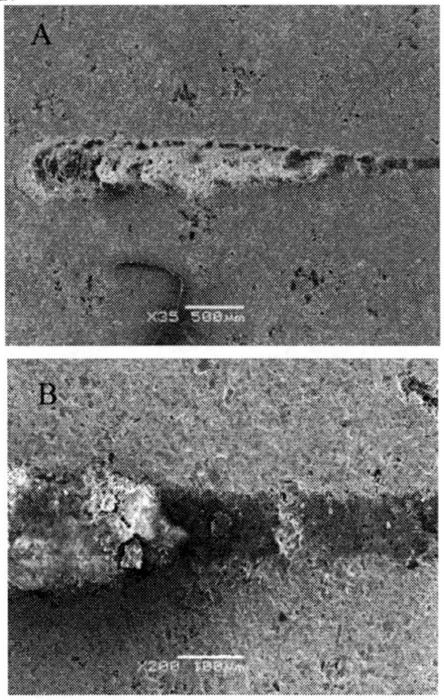

Figure 5. SEM images of scratch tests for silver films: (A) scratch track of film with an increasing load from 0 to 100N, (B) scratch track of film in rupture position

The critical load for all films shows the similar value about 30N. It can be deduced that what directly affects the critical load are not the film thickness and annealing condition but the property of substrate and coating technique.

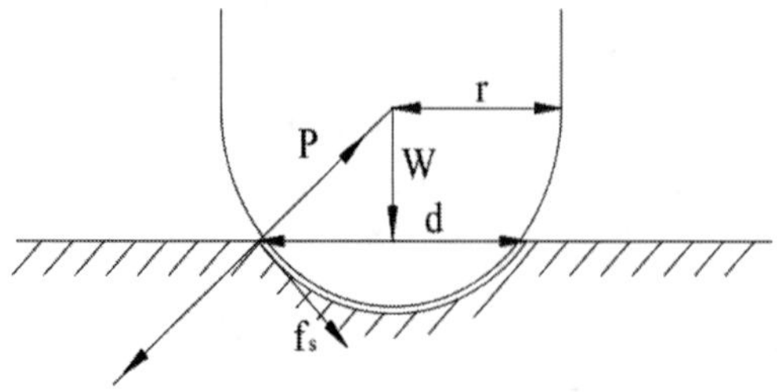

Figure.6. Schematic diagram of relationship between force and deformation of film

Fig.6 shows the relationship between force and deformation of film in the scratch tests. Where r is the radius of indenter with 0.2mm, W is the contact force given by indenter, d is the width of scratch track which can be measured from Fig.5, P is the resistance given by substrate, which can be expressed as [13]

$$P = \frac{W}{\pi d^2} \tag{4}$$

And f_S is shear press of film which can be calculated by the following equation,

$$f_S = \sqrt{\frac{W}{\pi r^2 P - W}} \cdot P \tag{5}$$

When contact force (W) equals to the critical load (W_C), the critical value of f_S can be considered as the adhesive stress and it should be equal to

$$f_{SC} = \sqrt{\frac{W_C}{\pi r^2 P - W_C}} \cdot P \tag{6}$$

According to equation (6), the range of critical values (f_{SC}) is calculated to be 245 N/mm^2. Comparing with the traditional methods such as electroplating and directing bonding (80 N/mm^2), the films deposited by RF magnetron sputtering shows a better performance due to the excellent penetrating power of high kinetic energy silver atoms.

4 Subsequent processes

The silver films made by radio frequency magnetron sputtering have good performance in structure, conductivity and adhesion. They are ideal material for metallizing of glass ceramic. Because of the conductivity of continuous silver films, the films cannot be used for micro electroforming directly. So lithograph and thin film etching is necessary. The process contains for parts: depositing of positive photoresist, exposure, development and etching of silver film. After that, the silver films after etching have ideal pattern and the second lithograph will be performed for micro electroforming. The process contains depositing of negative photoresist, exposure and development. After that, the silver films after etching expose for conductivity layer and micro electroforming can be performed.

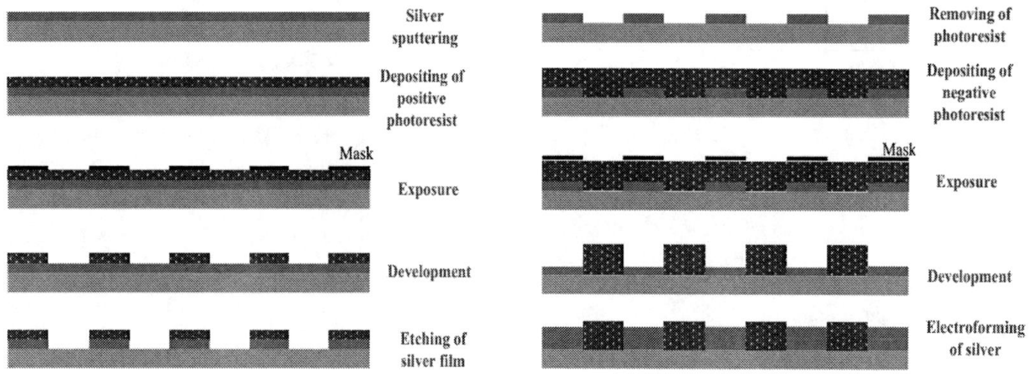

Figure 7. The subsequent processes of high temperatureeddy current sensor

5 Conclusions

In this paper, thin silver films were deposited on the glass ceramic by RF magnetron sputtering at room temperature. As a result, the deposition rate rises from 50nm/min to 125nm/min with the increasing of the sputtering power.

After that, annealing treatments at 600℃ were carried out on parts of as-deposited samples. And the microstructure, the topography, the electrical properties and the adhesive stress of films were investigated by X-ray diffraction, scanning electron microscope, four point probe instrument and micro scratch tester respectively.

The experimental results show that the increase of sputtering power will increase the grain size as well as the surface roughness. In addition, due to the annealing treatments, the grain size is increased further, whereas, the surface roughness is dropped.

For thicker films (5 μm), deposition at 250W led to a low value of the resistivity (3 $\mu\Omega cm$) compared to that of bulk Ag (1.6 $\mu\Omega cm$). Annealing treatment reduced the resistivity due to the increasing of crystal grain size. But, the observed changes of conductivity were meanly attributed to differences in the film thickness and the surface topography. And, the adhesive stress deposited by RF magnetron sputtering reached 245 N/mm^2, which shows a significant improvement than traditional methods.

Finally, the subsequent processes were introduced briefly. They contained two lithography, thin film etching and micro electroforming. These processes ensured the manufacture of silver coils on glass ceramic.

REFERENCES

[1] T. Suzuki, Y. Abe, M. Kawamura, et al. Optical and electrical properties of pure Ag and Ag-based alloy thin films prepared by RF magnetron sputtering [J]. Vacuum, 2002, 66: 501-504.

[2] K. Sarakinos, J. Wördenweber, F. Uslu, et al. The effect of the microstructure and the surface topography on the electrical properties of thin Ag films deposited by high power pulsed magnetron sputtering [J]. Surf. Coat. Technol, 2008, 202: 2323-2327.

[3] R.C. Adochite, D. Munteanu, M. Torrell, et al. The influence of annealing treaments on the properties of Ag:TiO2 nanocomposite films prepared by magnetron sputtering [J]. Surf. Sci, 2012, 258: 4028-4034.

[4] P.G. Kelly, R.D. Arnell. Magnetron sputtering: a review of recent developments and applications [J]. Vacuum, 2000, 56:159-172.

[5] S. Voss, S. Gandikota, R. Tao, et al. Adhesion studies of CVD copper metallization [J]. Microelectronic Eng, 2000, 50:501-508.

[6] N. Iwase, A. Tsuge, Y. Sugiura. Development of a thermal conductive AIN ceramic substrate technology [J]. Hybrid Microelectron, 1984, 7:49-53.

[7] L. Duan, X. Yu, L. Ni, Z. Wang. ZnO:Ag film growth on Si substrate with ZnO buffer layer by rf sputtering [J]. Appl. Surf. Sci, 2011, 257: 3463-3467.

[8] L.F. Senna, C.A. Achete, T. Hirsch, et al. Structural, chemical, mechanical and corrosion resistance characterization of TiCN coatings prepared by magnetron sputtering [J]. Surf. Coat. Technol, 1997, 94-95: 390-397.

[9] H.B. Wang, Z.Q. Wu. Advocating the physical experiment method [M]. Higher Education Press, 1990.

[10] D.M. Sun, Z.Q. Sun. Metal ceramic film and its applications in photoelectron technology [M]. Science Press, Beijing, 2004.

[11] N. Nedfors, O. Tengstrand, E. Lewin, et al. Structural, mechanical and electrical-contact properties of nanocrystalline-NbC/amorphous-C coatings deposited by magnetron sputtering [J]. Surf. Coat. Technol, 2011, 206: 354-359.

[12] M. Jing, R.L. Fu, H. He, et al. Direct bonded metal substrates and bonding methods [J]. Aerospace Materials and Technology, 2008, 3:1-7.

[13] D.Q. Xiao, J.G. Zhu, J.L. Zhu, et al. Film physics and device [M]. National Defense Industry Press, Beijing, 2011.

Giant Magnetostrictive Material-Fiber Bragg Grating Magnetic Field Sensors used in Magnetic Bearings

Jiayi Liu[a], Guoping Ding*[b], Bin Gao[b], Biyun Zhang[b]

[a]School of Information Engineering, Wuhan University of Technology, Luoshi Road, 430070 Wuhan, Hubei, China
[b]School of Electromechanical Engineering, Wuhan University of Technology, Luoshi Road, 430070 Wuhan, Hubei, China
dgp7910@163.com

***Abstract*—A novel FBG-GMM sensor was proposed to measure magnetic flux density, in which the FBG was stuck perpendicular to the GMM's principal magnetostriction orientation to make a thin-slice sensor. The FBG-GMM magnetic field sensor's properties were tested on a U-shape electromagnet test setup, the measured data were consistent with the theoretical prediction and the sensor' linear range is from 0.2T to 0.6T. The FBG-GMM magnetic field sensor was introduced to measure a radial magnetic bearing's air-gap flux density. Static flux density measurement was performed through two FBG-GMM sensors and one temperature compensation FBG, and the measured data were consistent with the sensor properties obtained from the U-shape electromagnet test, which validate the FBG-GMM magnetic field sensor's feasibility.**

I. Introduction

A magnetic bearing utilizes magnetic forces to levitate a rotating shaft in the bearing stator that includes actuator control coils, and a narrow air-gap separates the bearing stator and the rotor. In analyzing the behavior and performance of a magnetic bearing, the primary objective is to determine the forces generated by the actuator in response to currents of the coils and motion of the rotor. The magnetic field is physical medium of magnetic levitation and motion control, which essentially describes the coil/geometry–forces relationship. Due to the essential open-loop instability of the magnetic bearings system, the magnetic field is of remarkable dynamic and nonlinear features.

Normally magnetic circuit or finite element methods are used to predict magnetic field in magnetic bearings. However these theoretical analyses have inevitable limitations that their assumptions, models or boundary conditions cannot accurately describe magnetic field features, especially when the magnetic bearings operate at nonlinear region. So it is necessary to perform magnetic field online measurement to accumulate data for future design calculations and even online control.

This paper proposed a fiber Bragg grating (FBG) based Giant Magnetostrictive Material magnetic field sensors for magnetic bearings. Fiber Bragg grating sensors offer several significant advantages over conventional electrical sensors, such as immunity to electromagnetic noise and radio frequency interference, easy multiplexibility, small size, light weight, corrosion resistance and higher temperature capability. FBG sensors have become an important class of sensing device for long-distance and distributed sensing of various types of physical parameters such as strain, pressure, temperature, etc. in a broad domain of industrial fields. However, FBG sensors fall short of any magnetic field sensing because of their inherently weak magneto-optical Faraday effect. Recently, it has been demonstrated that a FBG can be stuck onto a Giant Magnetostrictive Material(GMM) to form a GMM–FBG sensor for magnetic field or electric current sensing[1-5]. *TbDyFe*, whose saturation magnetostriction is approximate 2000ppm, is one of the most promising giant magnetostrictive materials. It will expand when being put in magnetic field, which has a linear and reverse response to the strength of the magnetic field. If these magnetostrictive materials are bonded with an FBG fiber, the elongation of the magnetostrictive film due to magnetic field will result in a change in grating period in FBG fiber, and therefore generate a shift of the FBG central wavelength. In this way the drift of the FBG wavelength to the magnetic field response is correlated. Compared to conventional non-optical magnetic field or electric current sensors such as Hall-effect sensors and reluctance coils, this type of GMM–FBG sensor features a high level of immunity to electromagnetic interference, a great potential for large-scale multiplexing and a large capability for self-reference.

This paper proposed a novel FBG-GMM sensor to measure magnetic flux density, in which the FBG was stuck perpendicular to the GMM's principal magnetostriction orientation to make a thin-slice sensor to be used in magnetic bearings. Firstly, a U-shape electromagnet test setup was constructed and the GMM–FBG magnetic field sensor's static properties were tested. Then the GMM–FBG magnetic field sensor was introduced to magnetic bearing's air-gap flux density measurement. A radial magnetic bearing test setup was constructed and the air-gap flux density was measured through two FBG-GMM sensors and one temperature compensation FBG in static conditions

II. FBG-GMM MAGNETIC FIELD SENSOR

A. Giant Magnetostrictive Material (GMM)

A magnetostrictive material develops large mechanical deformations when subjected to an external magnetic field as shown in Fig.1. This phenomenon is attributed to the rotations of small magnetic domains in the material, which are randomly oriented when the material is not exposed to a magnetic field. The orientation of these small domains by the imposition of the magnetic field creates a strain field. As the intensity of the magnetic field is increased, more and more

* Corresponding author: dgp7910@163.com; Tel:0086-18071110860; Fax:0086-27-87651793

magnetic domains orientate themselves so that their principal axes of anisotropy are collinear with the magnetic field in each region and finally saturation is achieved.

An equation about Giant Magnetostrictive Material should be taken into consideration:

$$\delta = \varepsilon_{GMM} = \frac{\Delta l}{l} \qquad (1)$$

In this equation, δ is the magnetostrictive coefficient, ε_{GMM}, l and Δl is the strain, original length and elongation of the magnetostrictive material respectively.

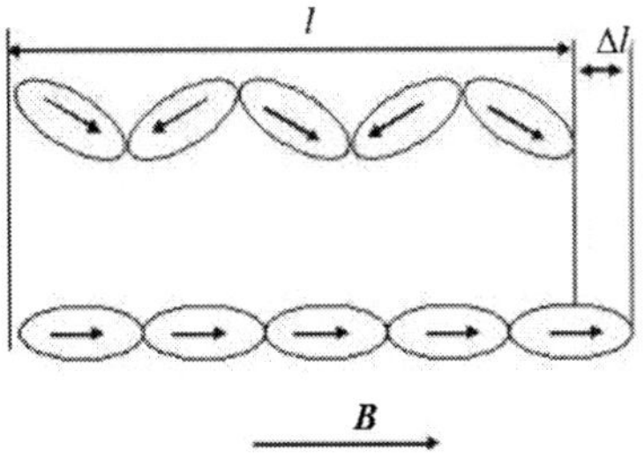

Fig.1 Magnetostriction

Some magnetostrictive material, such as *RFe*, *TbDy* and *TbDyFe*, has much lager magnetostrictive coefficient, higher energy density and response speed than normal ones and is called Giant Magnetostrictive Material (GMM).GMM is widely used in many research fields such as the measurement of magnetic field, sonar system, shock mitigation system. In this paper, we used *TbDyFe* as sensitive material whose magnetostriction curve is shown in Fig. 2.

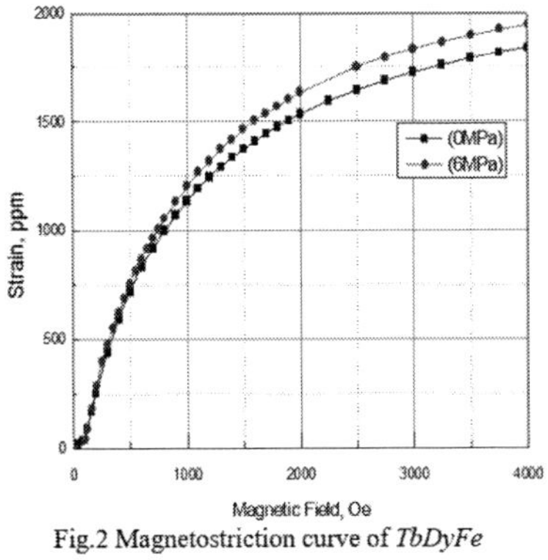

Fig.2 Magnetostriction curve of *TbDyFe*

B. *Sensing Principle of Fiber Bragg Grating(FBG)*

The fiber Bragg grating (FBG) is photo-inscribed into an optical fibre using a high power UV laser beam and is associated with a small periodic refractive index variation in the fibre core. Fig. 3 shows the basic sensing principle of an FBG. When the light of a broadband spectrum is guided through the optical fibre to the FBG, a narrow band component as indicated by the spike in the reflection spectrum is reflected back at the Bragg resonance wavelength, given by:

$$\lambda_B = 2n\Lambda \qquad (2)$$

Where Λ is the grating pitch and n is the fibre refractive index of the core. In Eq. (2) it can be seen that the Bragg wavelength is changed with a change in the grating pitch or the effective refractive index.

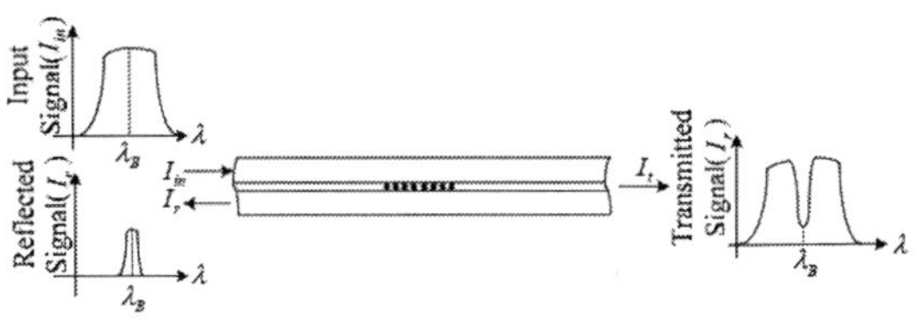

Fig. 3 Sensing principle of FBG

The Bragg wavelength is strain-dependent and temperature-dependent. The strain response arises due to both the physical elongation of the sensor (and corresponding fractional change in grating pitch), and the change in fiber index due to photo-elastic effects, whereas the thermal response arises due to the inherent thermal expansion of the fiber material and the temperature dependence of the refractive index. The shift in Bragg wavelength with strain and temperature can be expressed using

$$\Delta\lambda_B = 2n\Lambda\left(\left\{1-\left(\frac{n^2}{2}\right)[P_{12}-\nu(P_{11}+P_{12})]\right\}\varepsilon+\left[\alpha+\frac{(\frac{dn}{dT})}{n}\right]\Delta T\right) \qquad (3)$$

where ε is the applied strain, $P_{i,j}$ coefficients are the Pockel's (piezo) coefficients of the stress-optic tensor, ν is Poisson's ratio, and α is the coefficient of thermal expansion of the fiber material (e.g., silica), and ΔT is the temperature change. The factor $\left(\frac{n^2}{2}\right)[P_{12}-\nu(P_{11}+P_{12})]$ has a numerical value of ≈ 0.22. The measured strain response at constant temperature is

$$\frac{1}{\lambda_B}\frac{\delta\lambda_B}{\delta\varepsilon} = 0.78\times10^{-6}\mu\varepsilon^{-1} \qquad (4)$$

This responsivity gives a "rule-of-thumb" measure of the grating shift with strain of 1 nm per 1000 $\mu\varepsilon$ at 1300nm. In silica fibers, the thermal response is dominated by the dn/dT effect, which accounts for 95% of the observed shift. The normalized thermal responsivity at constant strain is

$$\frac{1}{\lambda_B}\frac{\delta\lambda_B}{\delta_T} = 6.67\times10^{-6} {}^{\circ}C^{-1} \qquad (5)$$

In this paper, we used FBGs of about 1300nm wavelength, that a wavelength resolution of 1 pm (0.001 nm) is required to resolve a temperature change of 0.1℃, or a strain change of 1 μ strain.

C. *FBG-GMM Magnetic Field sensor*

The FBG-GMM magnetic sensor's principle is that the FBG stuck onto the GMM detects the GMM's stain induced by magnetostriction when it is exposed to magnetic field. Normally the FBG was stuck onto the GMM along its principal magnetostriction orientation to constitute a FBG-GMM magnetic field sensor as shown in Fig.4 (a).

However this configuration was not feasible for magnetic bearing's air-gap flux density measurement. The main reason was that the magnetic bearing's air-gap was too narrow to accommodate the FBG as it was stuck along the flux density orientation, which was also the GMM's principal magnetostriction orientation.

Therefore, we customized a 1.5mm-thick *TbDyFe* slice whose principal magnetostriction is along thickness orientation. When the *TbDyFe* slice was stuck onto the magnet pole's end surface, the flux density passing through the *TbDyFe* slice produced magnetostriction along thickness orientation. Since the FBG's length (almost 10mm) is much larger than the air-gap's thickness, the FBG was stuck onto side of the *TbDyFe* slice along its length orientation, which is perpendicular to the principal magnetostriction orientation as shown in Fig.4 (b). Fortunately, *TbDyFe* is of quasi-linear magnetostriction that means magnetostriction along thickness orientation causes relevant deformation along length and width orientation, and volume of *TbDyFe* slice almost maintains the same. Therefore the above adaptation was feasible in principle and it was necessary to test this novel FBG-GMM sensor's properties before its practical application.

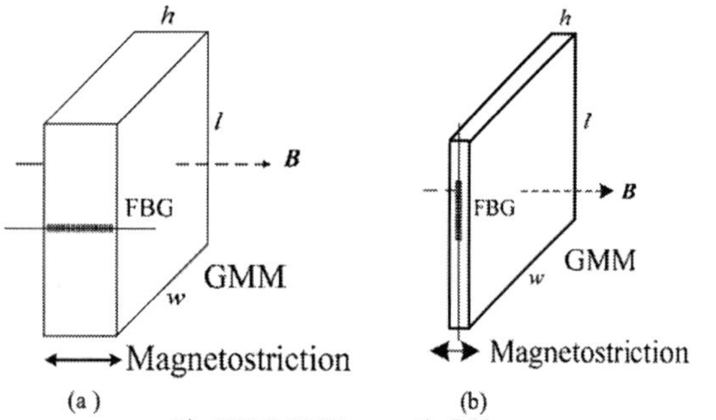

Fig.4 FBG-GMM magnetic field sensor
(a)FBG was stuck along principal magnetostriction orientation
(b)FBG was stuck perpendicular to the principal magnetostriction orientation

III. TEST OF FBG-GMM MAGNETIC FIELD SENSOR PROPERTIES

Based on the basic principle of the FBG-GMM magnetic sensor proposed in Section II, a set-up was constructed to test the FBG-GMM magnetic field sensor's properties including linearity, sensitivity, and range.

Fig.5 shows the schematic diagram of the test set-up, which was composed of a pair of U-shape electromagnets powered by two DC supplies, a FBG-GMM magnetic sensor and a FBG wavelength demodulator connected with a computer.

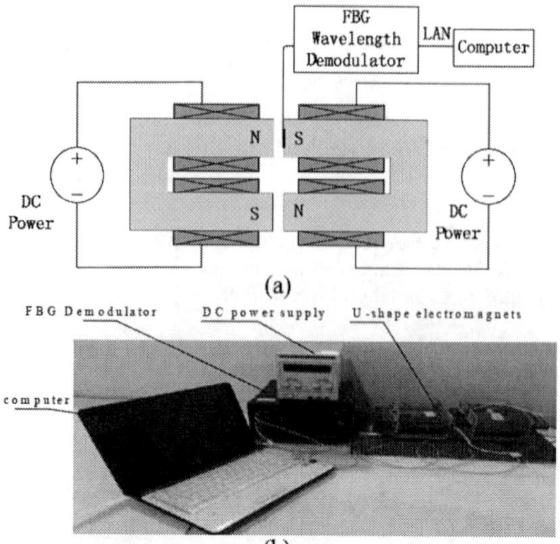

Fig.5 A test set-up of the FBG-GMM sensor's properties
(a) schematic diagram (b) actual system

Each U-shape electromagnet was made of a 30-piece silicon steel lamination stack wound with an 800-turn coil. The air-gap between the two electromagnets was fixed at 5mm, and a FBG-GMM magnetic sensor was stuck onto the surface of one electromagnet's pole to measure the flux density in the air-gap. FBG-GMM magnetic sensor's wavelength shift was demodulated and sent to the computer.

In the static properties test, DC currents from 0.1A to 2.2A with an interval of 0.1A were applied on the coils. A basic magnetic circuit equation shown in (6) was used to calculate the air-gap flux density according to the applied DC currents.

$$B_0 = \frac{\mu_0 N * I}{x_0} \quad (6)$$

where, $\boldsymbol{B}_0$ is the air-gap flux density, μ_0 is the magnetic permeability of a vacuum, N is number of the coils, I is the current applied on the coils and x_0 is the length of the air-gap.

It should be mentioned that since the air-gap between the electromagnet pair was large enough to dissipate the heat and the applied current was relatively small, we ignored the temperature effort on the sensor. Moreover, a Gauss-meter was also used to measurement the air-gap flux density to validate the calculated flux density

The FBG-GMM sensor's wavelength shift relevant with air gap flux density is illustrated in Fig.6

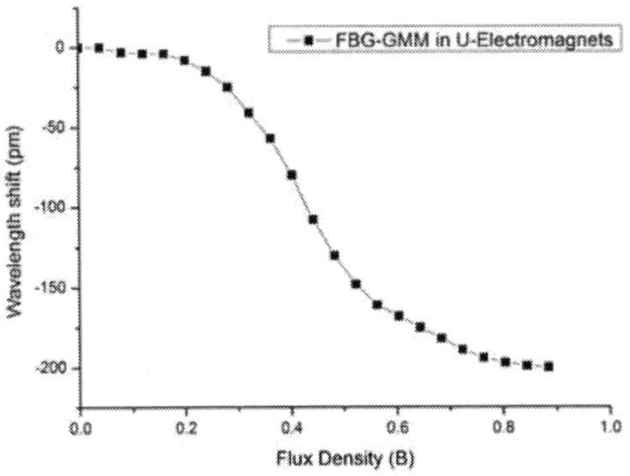

Fig.6 FBG-GMM sensor's wavelength shift relevant with air-gap flux density (DC currents, U-Electromagntes)

In Fig.6, the FBG's wavelength shift was negative with flux density from 0 to 0.9T, and it decreased with increasing flux density. It was consistent with the theoretical prediction that when the flux density increased, the GMM slice's thickness increased while its length and width decrease to remain the same volume. Shrink of the length caused decrease of FBG wavelength shift.

Moreover, the FBG-GMM sensor's wavelength shift curve could be divided to three parts. The first part, from 0 to 0.2T, it seemed like a "dead zone" where the wavelength shift varied very slightly; the second part, from 0.2T to 0.6T, the wavelength shift decreased almost linearly with increasing flux density; and the last part, from 0.6T to 0.9T, the wavelength shift curve slowed down to saturation. Although the total wavelength shift was about 200pm, amount to $200u\varepsilon$ with a 1300um-center wavelength, which was almost a tenth of the strain produced by *TbDyFe* principal magnetostriction, it was large enough for the air-gap flux density measurement.

IV. FBG-GMM MAGNETIC FIELD SENSOR USED IN MAGNETIC BERINGS

Based on the tests of the FBG-GMM magnetic sensor's properties in Section III, the sensor was introduced to magnetic bearings for the air-gap flux density measurement. Fig. 7 depicts the sensors detail and their mounted positions in a radial magnetic bearing. Two FBG-GMM magnetic sensors were stuck onto the surface of a magnet pole (pole 7), especially, since the bearing's air-gap was too narrow to dissipate the heat and the FBG was sensitive to temperature variation, an additional FBG was stuck onto the surface of the same magnet pole to measure the temperature variation and to be used to compensate the temperature influence on the FBG-GMM magnetic sensors.

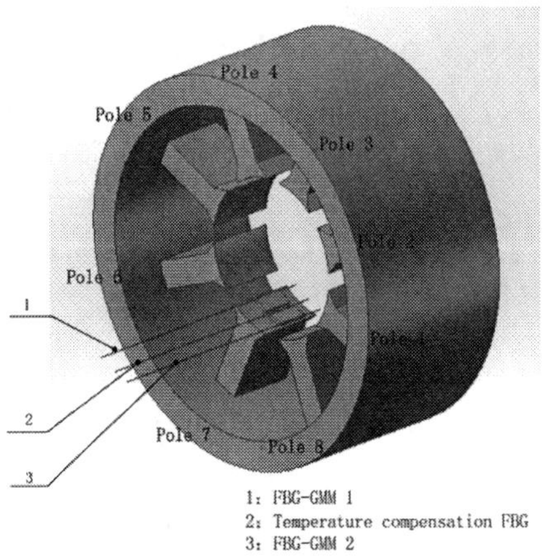

(a)

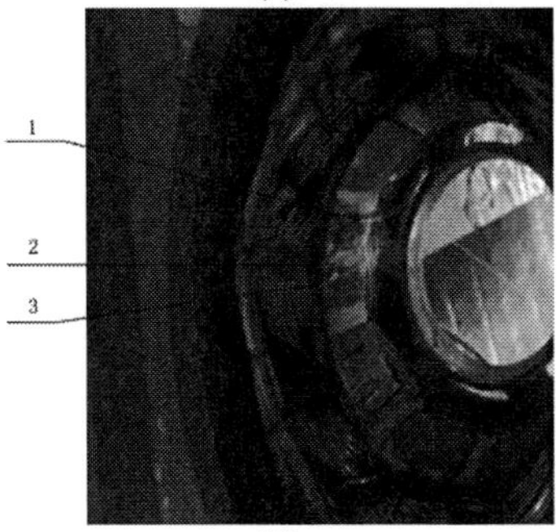

(b)

Fig.7 FBG-GMM sensors mounted in a radial magnetic bearing
(a) schematic diagram (b) actual device

Fig.8 shows the measurement system constructed to perform the magnetic bearing's air-gap flux density measurement. In the system, a motor with a frequency converter drove the rotor through a coupling, and a DC power supply provided currents to the coils wound on the magnet poles of the radial bearing, and the FBG-GMM magnetic sensor's wavelength shift was demodulated and sent to the computer.

The radial magnetic bearing was of an 8-pole structure with NNSS configuration. The rotor's inner diameter and outer diameter were respectively 45mm and 66mm. The stator's inner diameter and outer diameter were respectively 71mm and 144mm. The air-gap between the rotor and the stator of this radial bearing was fixed at 2.5mm. Each magnet pole-pair had a 334-turn coil.

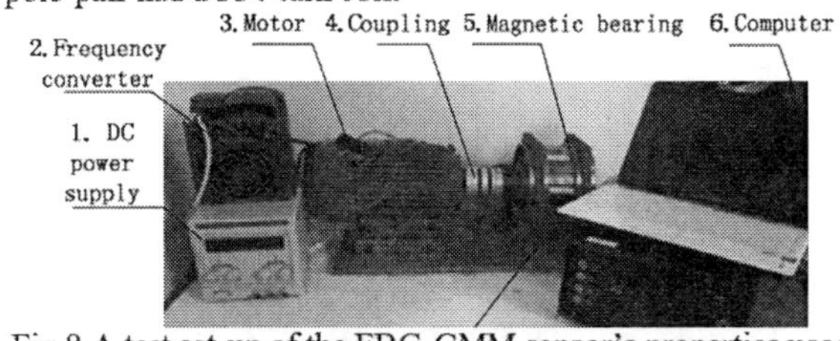

Fig.8 A test set-up of the FBG-GMM sensor's properties used in magnetic bearings
(a) schematic diagram (b) actual system

In this paper we focused on the feasibility and basic properties of the sensor, so the air-gap flux density was measured in static condition that was static rotor with DC currents excitation. DC currents from 0A to 4.8A with an interval of 0.2A were applied on the coils of one pole pair (pole 6 and pole 7). The air-gap flux density was calculated based on the magnetic circuit equation shown in (6).

Fig. 9 shows the wavelength shift of the two FBG-GMM sensors and the temperature compensation FBG.

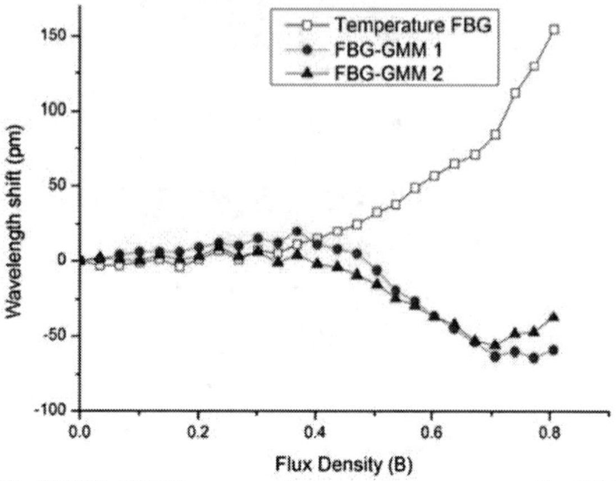

Fig.9 FBG-GMM sensor and temperature compensation FBG wavelength shift

In Fig. 9 the maximum wavelength shift of the temperature compensation FBG reached about 150pm, according to the temperature sensibility illustrated in Section II, the temperature increment was about 15℃ as the currents reached 4.8A, which meant the temperature effect on the FBG-GMM sensors could not be ignored.

The temperature FBG's wavelength shift was subtracted from the FBG-GMM sensors' wavelength shift to get the temperature compensation. Fig.10 shows the compensated wavelength shift of the two FBG-GMM sensors.

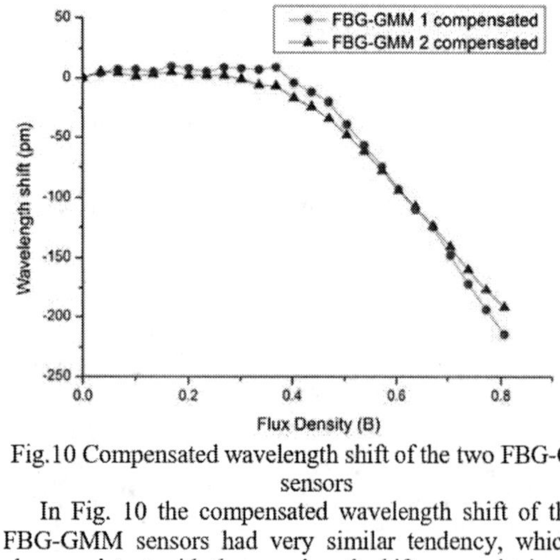

Fig.10 Compensated wavelength shift of the two FBG-GMM sensors

In Fig. 10 the compensated wavelength shift of the two FBG-GMM sensors had very similar tendency, which was also consistent with the wavelength shift curve obtained from U-electromagnets experiment shown in Fig.6. It can be concluded that the FBG-GMM sensor was feasible to be used to measure the air-gap flux density of magnetic bearings.

CONCLUSION

A Giant magnetostrictive material-fiber Bragg grating magnetic field sensors used in magnetic Bearings was investigated experimentally. Based on this study, the following conclusions can be drawn:

1. A novel FBG-GMM sensor was proposed to measure magnetic flux density, especially in small gap.
2. The FBG-GMM magnetic field sensor's properties were tested on a U-shape electromagnet test setup, the measured data were consistent with the theoretical prediction and the sensor' linear range was from 0.2T to 0.6T.
3. The FBG-GMM magnetic field sensor was introduced to measure a radial magnetic bearing's air-gap flux density. Static flux density measurement was performed through two FBG-GMM sensors and one temperature compensation FBG, and the measured data were consistent with the sensor properties obtained from the U-shape electromagnet test, which validated the FBG-GMM magnetic field sensor's feasibility.

ACKNOWLEDGEMENTS

This work was supported by the National Science Foundation of China (Project No. 51105285).

REFERENCES

[1] J.L. Cruz, A. Diez, M.V. Andres, A. Segura, B. Ortega, L. Dong, Fiber Bragg gratings tuned and chirped using magnetic fields, Electron. Lett. 33 (1997) 235–236.

[2] K.S. Chiang, R. Kancheti, V. Rastogi, Temperature-compensated fiber-Bragggrating-based magnetostrictive sensor for dc and ac currents, Opt. Eng. 42(2003) 1906–1909.

[3] D. Satpathi, J.A. Moore, M.G. Ennis, Design of a Terfenol-D based fiber-optic current transducer, IEEE Sens. J. 5 (2005) 1057–1065.

[4] C. Ambrosino, S. Campopiano, A. Cutolo, A. Cusano, Sensitivity tuning in Terfenol-D based fiber Bragg grating magnetic sensors, IEEE Sens. J. 8 (2008)1519–1520.

[5] A.A. Moghadas, M. Shadaram, Fiber Bragg grating sensor for fault detection in radial and network transmission lines, Sensors 10 (2010) 9407–9423.

Leibniz Institute for Solid State and Materials Research Dresden

TECHNISCHE UNIVERSITÄT DRESDEN

Ultra-thin and flexible Hall effect sensorics

GEFÖRDERT VOM Bundesministerium für Bildung und Forschung

Deutsche Forschungsgemeinschaft DFG

Ingolf Mönch, Michael Melzer, Falk Bahr, Denys Makarov,
Wilfried Hofmann und Oliver G. Schmidt

IFW DRESDEN INSTITUTE FOR MAGNA INTEGRATIVE NANOSCIENCES

Application potential

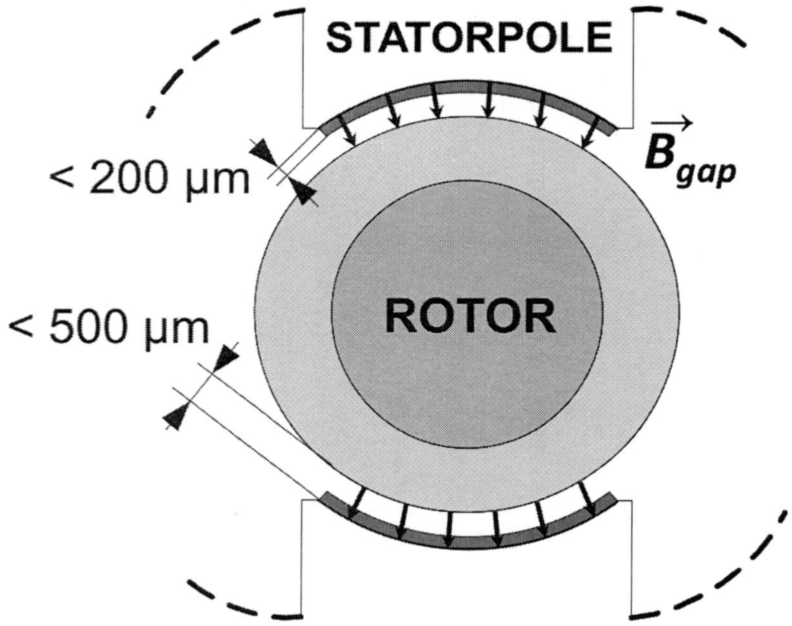

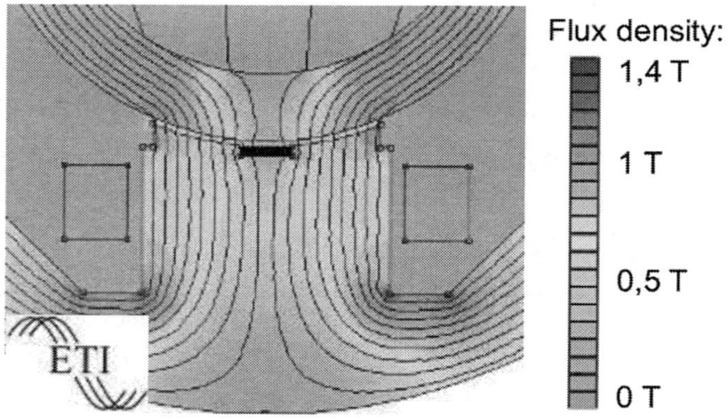

- Flux density measurement in the curved and narrow air gaps of electrical machines and drives (e.g. for flux based control of AMBs)

With commercial Hall sensors

→ Element height: ≈ 0,6 mm

- Increasing gap width results in reduced flux density
  → lower magnetic force
- Inhomogeneous field distribution for sensor countersunk in statorpole

Our approach: Integration of ultra-thin and flexible Hall sensors onto the statorpole‘s curved surface

Flexible Hall sensorics

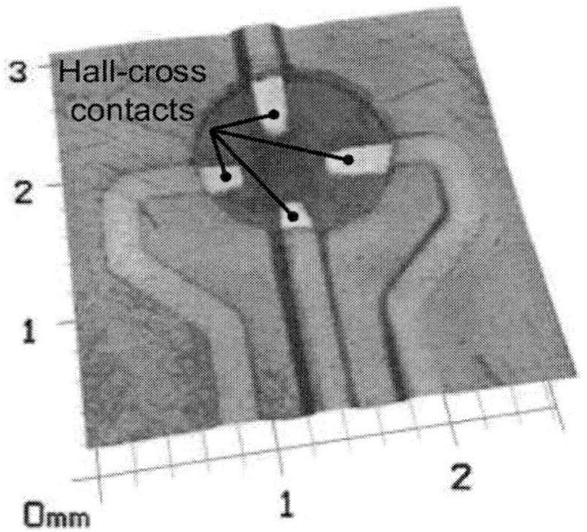

FPC with exposed contacts

Deposited Bi thin film

Sensor on a stator of a brushless motor

Combination of polymeric supports (FPC) and metallic thin films (Bismuth)

- Hall-cross geometry pre-defined by exposed FPC electrodes
- Bismuth deposition by magnetron sputtering

→ Total height of flexible element: 150 μm including wiring

Flexible Hall sensorics

Signal change under bending

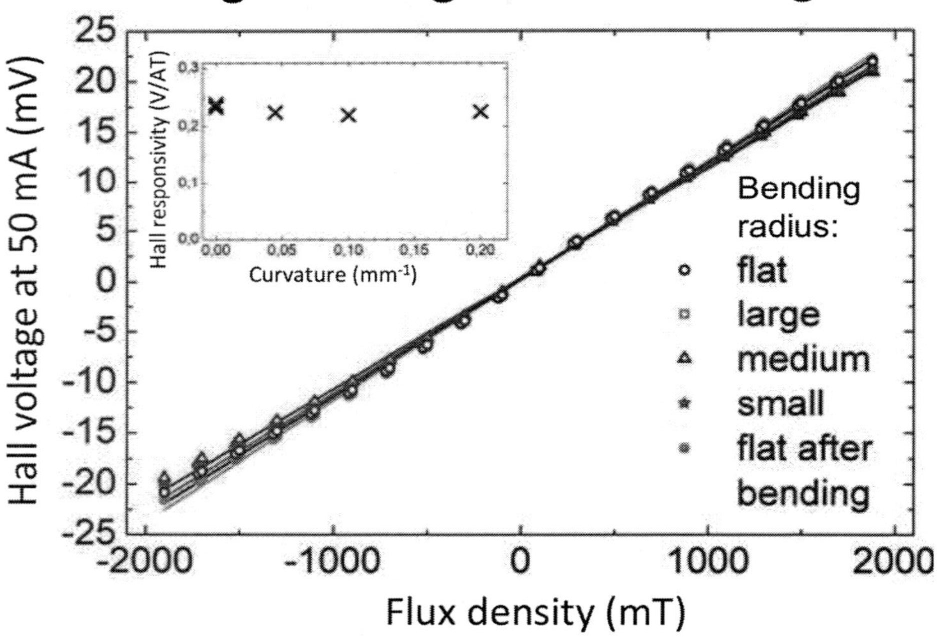

- Stable Hall signal under bending down to 5 mm radius of curvature
- 16 channel sensor arrays in 1D and 2D configurations

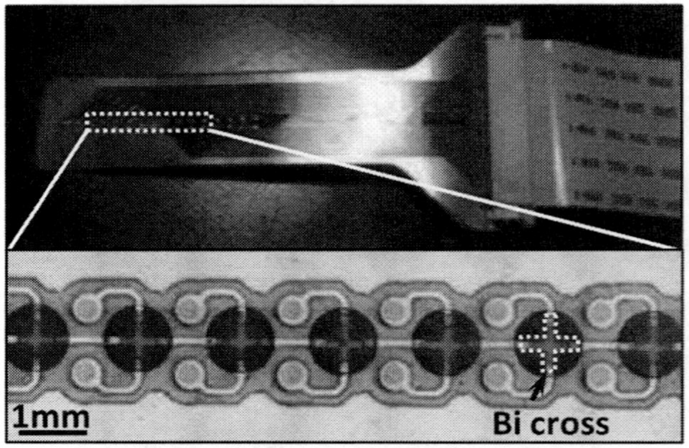

Linear positioning sensor

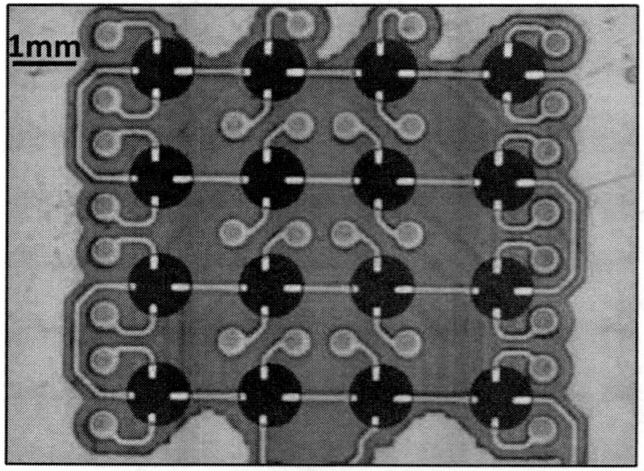

2D magnetic mapping

IFW DRESDEN INSTITUTE FOR MAGNA INTEGRATIVE NANOSCIENCES

Thermal post-processing

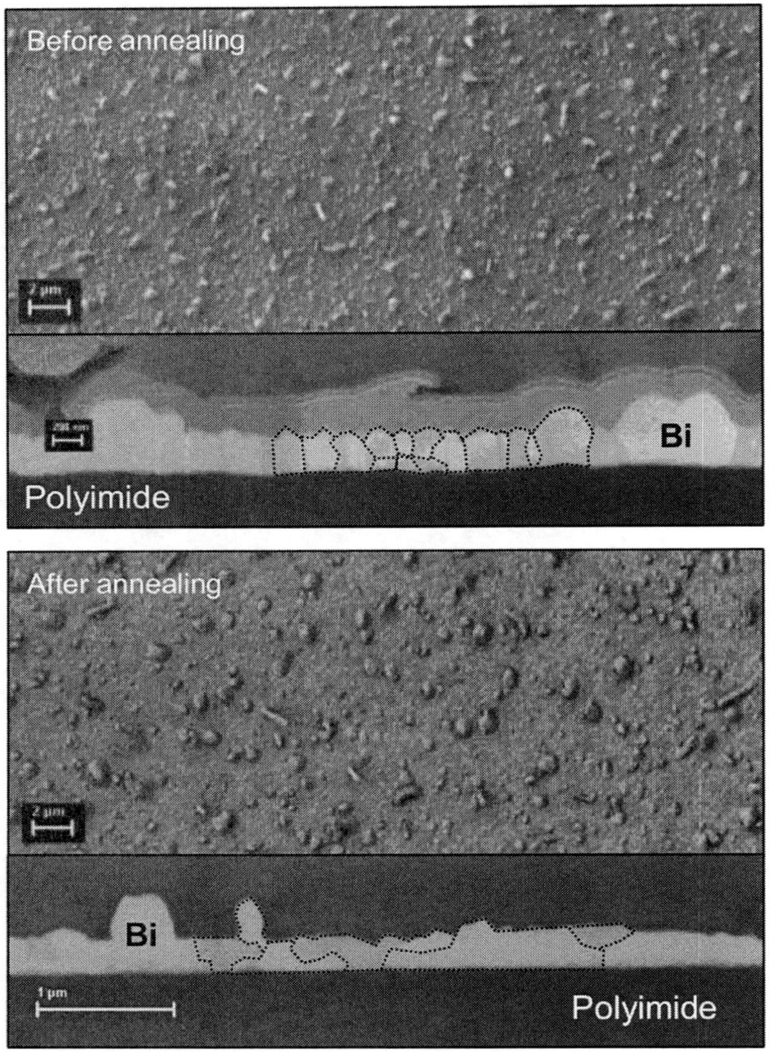

230 nm Bismuth Film

→ Granular morphology

- Post annealing at 250°C for 3h

→ Grain size: 250 nm → 750 nm

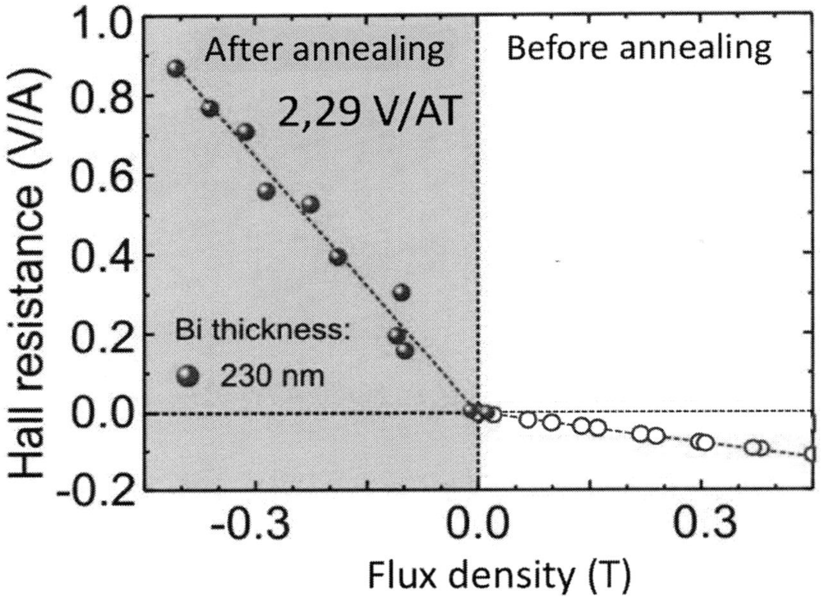

IFW DRESDEN INSTITUTE FOR MAGNA INTEGRATIVE NANOSCIENCES

Optimization of contact pads

Cu contact pad on FPC (28 µm height)

→ Contacting 230 nm Bismuth film

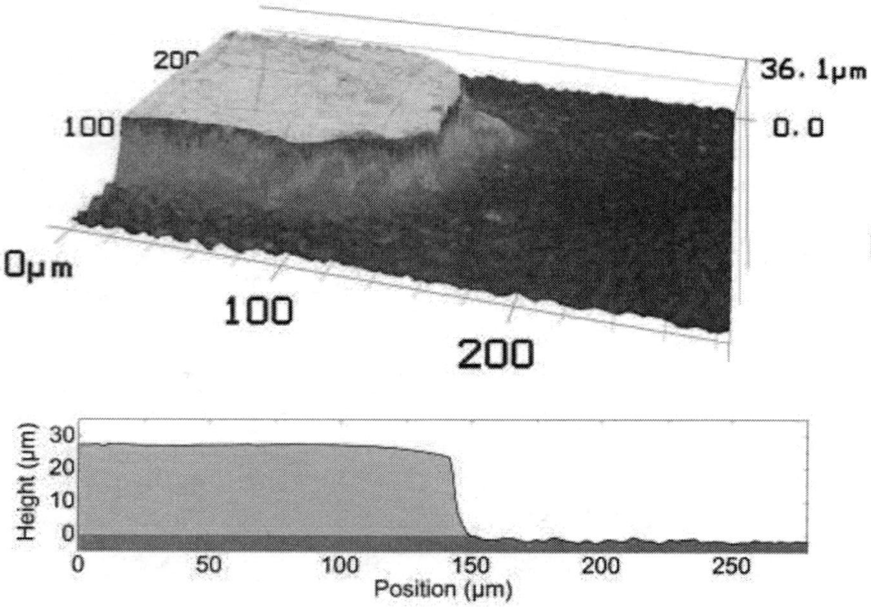

Local electrochemical etching

→ Reduction of contact height

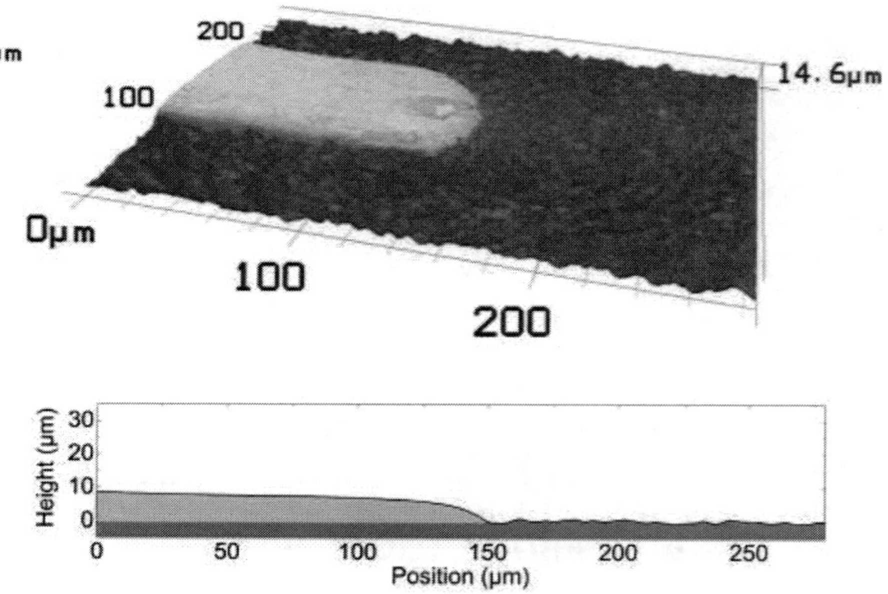

→ Enhanced contact reliability due to smoother pad edge

IFW DRESDEN INSTITUTE FOR MAGNA INTEGRATIVE NANOSCIENCES

Outlook: Integration into AMB

TECHNISCHE UNIVERSITÄT DRESDEN

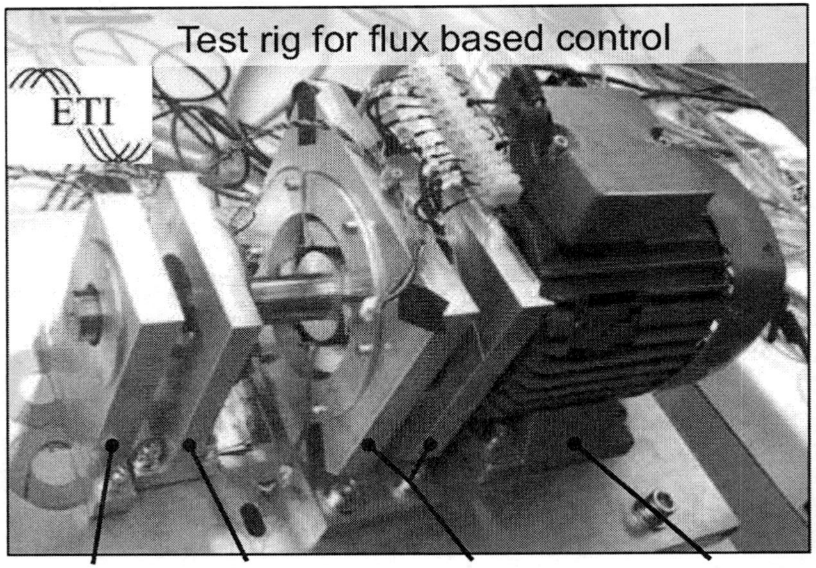

Savety bearing | Position gauge | **Magnetic bearing** | Electric motor

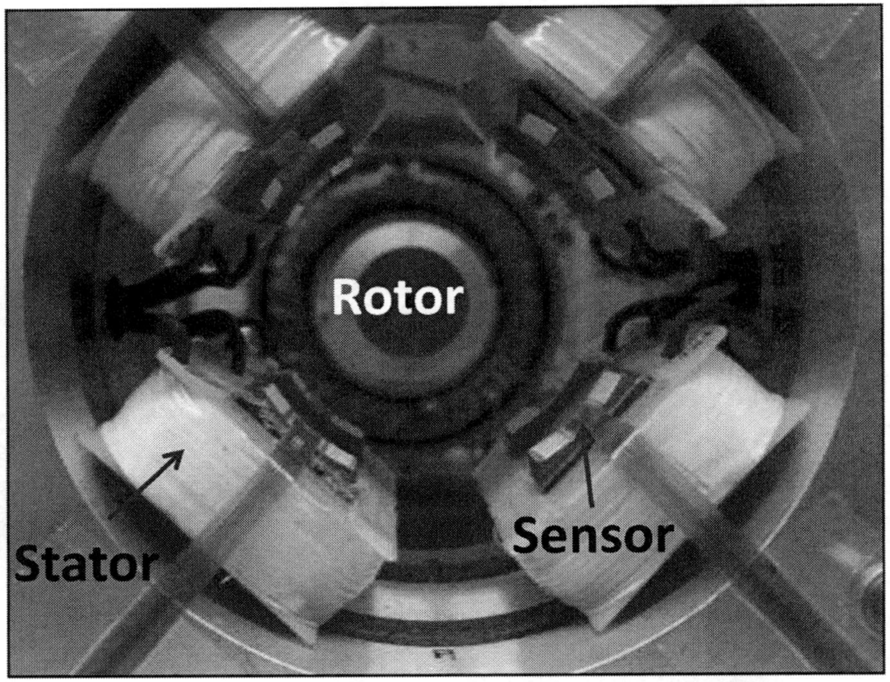

- Radial AMB with permanent pre-magnetization
- Capacitive position gauge
- Integration of 8 individual flexible Bi Hall sensors in the air gap with differential operation

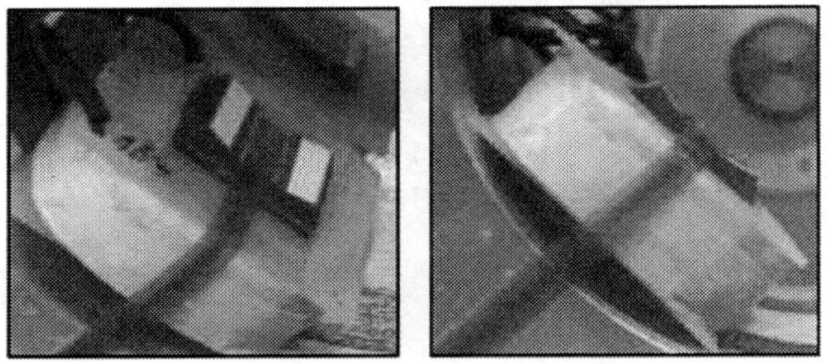

Bi Hall sensor fixed to the AMB stator pole

IFW DRESDEN INSTITUTE FOR MAGNA INTEGRATIVE NANOSCIENCES

The MAGNA Group at IFW Dresden

Collaborations:

ETI
Falk Bahr, Wilfried Hofmann

AVT
Daniel Ernst, Thomas Zerna

JKU Johannes Kepler Universität Linz
Martin Kaltenbrunner, Ingrid Graz, Siegfried Bauer

The University of Tokyo
Takao Someya, Tsuyoshi Sekitani

Funding:

Gefördert vom Bundesministerium für Bildung und Forschung

nanett nano system integration network of excellence

erc

Deutsche Forschungsgemeinschaft DFG

IFW Dresden Institute for MAGNA Integrative Nanosciences

Effect of Eddy Currents on the Stray Flux Based Measurement System for Magnetic Bearing

Johannes Rudolph[a], Prof. Dr.-Ing. Ralf Werner[b]

[a] Technische Universität Chemnitz, Faculty of Electrical Engineering and Information Technology, Electrical Energy Conversion Systems and Drives, Reichenhainerstrasse 70, 09126 Chemnitz, Germany, johannes.rudolph@etit.tu-chemnitz.de
[b] Technische Universität Chemnitz, Faculty of Electrical Engineering and Information Technology, Electrical Energy Conversion Systems and Drives, Reichenhainerstrasse 70, 09126 Chemnitz, Germany, ralf.werner@hrz.tu-chemnitz.de

***Abstract*—This paper concerns the behaviour and special characteristics of an alternative measurement system for magnetic bearing based on stray flux which depends on the rotor position. The impact of eddy currents on the dynamic properties will be analysed and methods for compensation will be presented. Furthermore the effect of electromagnetic disturbance will be a matter and how it can be reduced by special sensor design.**

I. Basics

A. Introduction

The measurement of the rotor position using flux sensors is a rarely used method. The integration of flux sensors into the air gap is associated with some problems. Most gaps are to small for using standard field sensors. A good possibility is the integration of very thin magnetic sensitive elements [1]. An other option is given by application of additional magnetic circuits which contain flux sensors like Hall elements [2]. A new and alternative opportunity is a measurement system which is based on the often neglected stray flux which also depends on the rotor position [3], [4]. Low cost standard Hall sensors are integrated in the space between adjacent pairs of poles. The magnetic stray flux is also a function of the air gap-length and thus of the rotor position. In previous considerations it has been shown that the direction of the flux is an important fact for positioning flux sensors with only one magnetic sensitive direction. But there is a further correlation between the stray flux and the coil current which has to be considered when the position of the rotor is derived from the stray flux.

B. Problem

While testing the stray flux based measurement system in passive mode a particular behaviour occurred when a step response was recorded. The rotor position is changed stepwise in the way the air gap reduces. Normally the measured flux should increase in the same manner. But like shown in figure 1 there is a short negative peak at the beginning. Caused by the fast rise of the coil current eddy currents appear in the laminated core, which leads to a damped increase of the main flux. Thereby the influence of the coil current is over compensated by the controller for a short time. Just in case of highly dynamic operation modes this effect leads to adverse effects which can compromise a stable operation of the magnetic bearing.

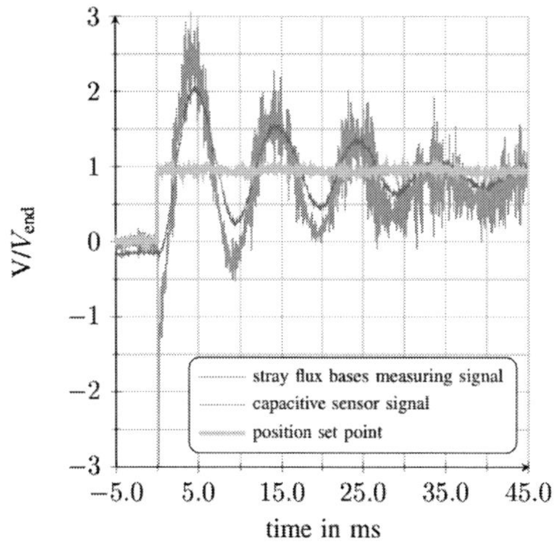

Figure 1. Step response while running the stray flux based measurement system in passive mode. The bearing operates with auxiliary capacitive sensors and PID controller

C. Delay of the magnetic field rise caused by eddy currents and overcompensation of the stray flux based position signal

When a current is applied to an air-core coil the magnetic field around the coil is rising in the same speed as the coil current is rising. Adding a ferromagnetic core to the coil causes a delay in the rise of the magnetic field. The reason for this can be shown with a simplified example. If a monolithic ferromagnetic core is applied the single coil will become a transformer. The secondary coil is formed by the lateral area of the core which represents a single short circuited winding. If the current in the primal coil rises its magnetic field will induce a voltage in the secondary single winding. Because it's short circuited a current begins to flow. Hence the magnetic field counteracts its cause. The rise of the primal magnetic field is damped. If the monolithic core is substituted by a laminated one the effect will be less pronounced. But if the current rises in a steep way there will be still a delay of the magnetic field like shown in figure 2. The control current and the Hall-signal are recorded at the test bearing after a voltage step. It is obvious that the field has a small delay, especially in the beginning when the current is steeply rising. Remembering that the current value is measured and used to calculate a current-compensated Hall-signal it is easy to see that a steeply rising current leads to over compensation.

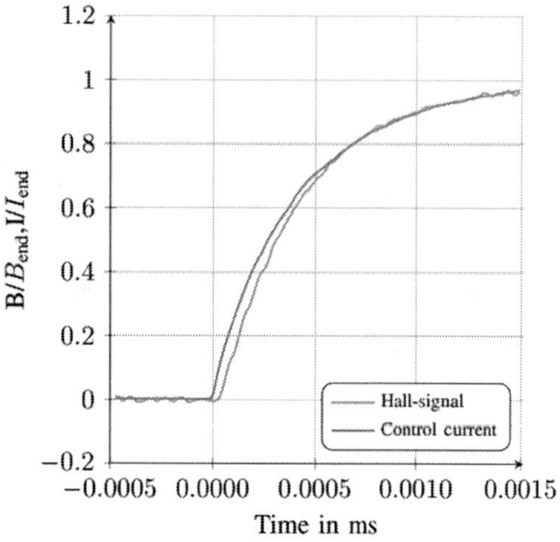

Figure 2. Delay of the magnetic stray flux relative to the control current

In case of suddenly occurring disturbance forces the controller reacts normally with corresponding current to counteract.Especially in this case of steeply rising currents there is a high risk that overcompensation (particular if it is acting contrary) compromise a stable and save operation of the magnetic bearing.

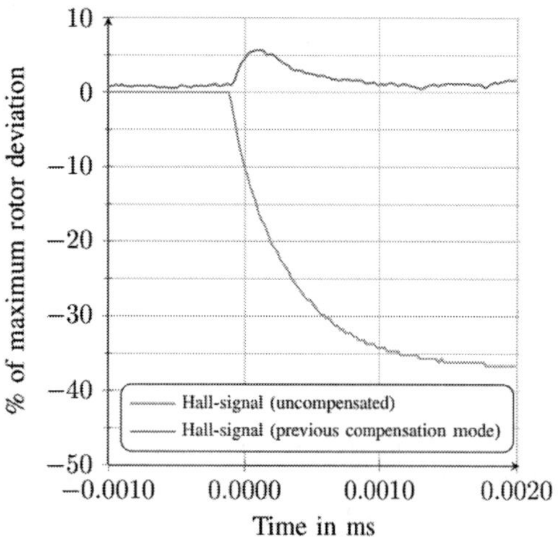

Figure 3. Uncompensated Hall-signal during current step and Hall-signal with previous compensation mode.

The previous current compensation is shown in figure 3 related to the uncompensated Hall-signal after a current step. It is clearly to see that an contrary overcompensation occurs in the beginning when the current rise is steep. The figure shows the difference of the two opposing Hall-signals at bottom and

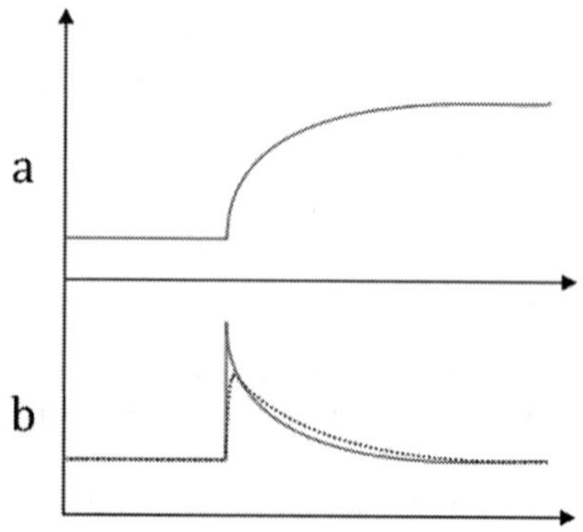

Figure 4. Idealised current raising after a voltage step (a) and its derivation (b-continuous) as well as the filtered derivation (b-dotted).

top position ([4] and [3]) in percent of the maximum rotor deviation. Neither the current compensated stray flux based measuring signal is inaccurate and pretends an rotor deviation of more than 5%. In this experiment the voltage was limited to 15V. In normal operation mode of the test bearing the voltage is set to 325V. Thus the current can rise more steeply and the overcompensation can lead to a bigger position error.

D. Advanced current compensation

To minimize the error of the stray flux based measurement system according to fast and wide variations in the control current it is necessary to ascertain not only the momentary value of the control current. There is also a need to know its increase. This can be done by numerical derivation of the measured current signal. But it is well known that the derivation of real and noisy signals is not as simple as the derivation of mathematical function. Every small peek is gained multiple. The result is a signal with a grade of disturbance which is magnitudes higher than the measured current signal. A popular way to reduce this effect is the application of a simple filter.

An idealised course of the current after a voltage step and its derivation is shown in figure 4. The figure also shows the filtered derived current signal. It is clear to see that filtering leads to a less pronounced peak and slower decreasing of the signal which does not reflect the current rising accurately. Based on this filtered derivation only an incomplete current compensation can be realised. A better result can be achieved when the peak is increased and the decreasing is gained in a way that the results look like the course of the unfiltered signal. A feasible way to achieve this is to scale the derived signal with a coefficient which depends also on the derivation. With this method the signal is gained when the raising is steep because the coefficient is big. When the signal decreases the coefficient decreases too which leads to narrower peak. To improve the previous current compensation the current scaling factor is now multiplied by the filtered and modified derivation of the current signal like shown in figure 5. The output is

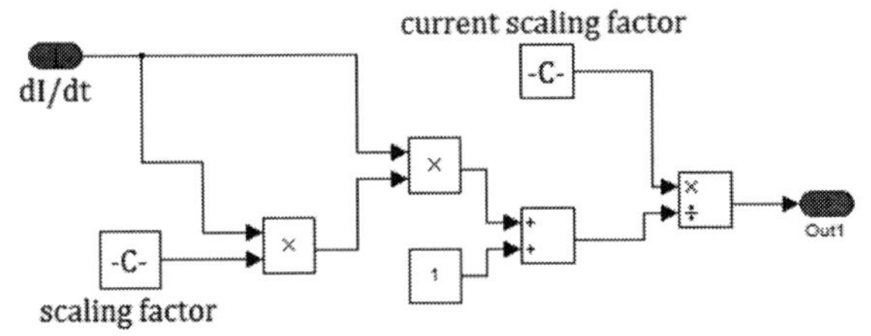

Figure 5. Model of the improved current compensation.

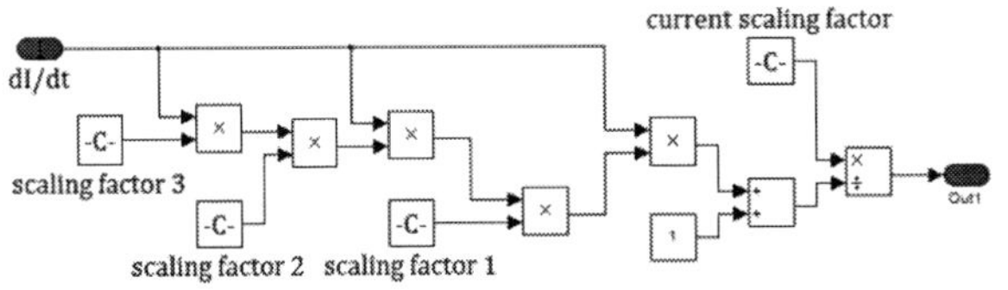

Figure 6. Model of the multi stage current compensation.

added or subtracted from the Hall-signal depending if x- or y-axis control current is treated.

Due to the fact that the scaling factor also depends on the filtered derivation the result of the improved current compensation is better than the previous method but still imperfect. For a further improvement it has been found that a multi staged scaling can be useful. For better understanding the block diagram is shown in figure 6. It is obvious that the first scaling factor is varied by a second factor. This factor also depends on the derived current signal and is varied by a third factor in turn. The scaling can be cascaded more than three times like the following equation shows.

$$c = \frac{c_{\text{static}}}{\left(\frac{dI}{dt}\right)^n \cdot c_1 \cdot c_2 .. c_n + 1} \tag{1}$$

If the current rises in an steep way the left part of the numerator increases and the compensation factor c becomes smaller to suppress overcompensation. If the current changes slowly or has a constant value the left part of the numerator becomes near zero so that the compensation factor c is dominated by the static current scaling factor c_{static}. At the test rig a real time system manages the current compensation. To achieve short cycle times the number of stages is set to three. More stages would improve the result of the current compensation but more time for the calculation is needed. It's a compromise between speed and accuracy.

The application of this improved current compensation method leads to a more suppressed influence of the control current on the stray flux based measurement system. The way the current scaling factor is changing during a step response is shown in figure 7. If the current is not or only slowly changing the scaling factor c is set to 90. It is reduced to almost 40 when a steep current rise appears. These values where experimentally found at the test bearing. By this the stray flux based measuring signal changes rarely like shown in figure 8 when the rotor is fixed in the test bearing and the current raises fast. The improved current compensated stray flux based position signal is shown in figure 8 related to the old compensation method. The level of the peak could be reduced from almost 6% of the maximal rotor deviation to about 2.5% and the time range of overcompensation is shortened clearly.

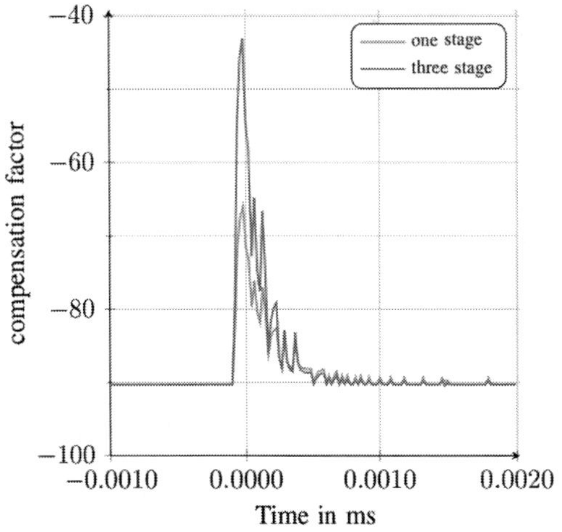

Figure 7. Variation of the current scaling factor with one stage and three stage compensation mode

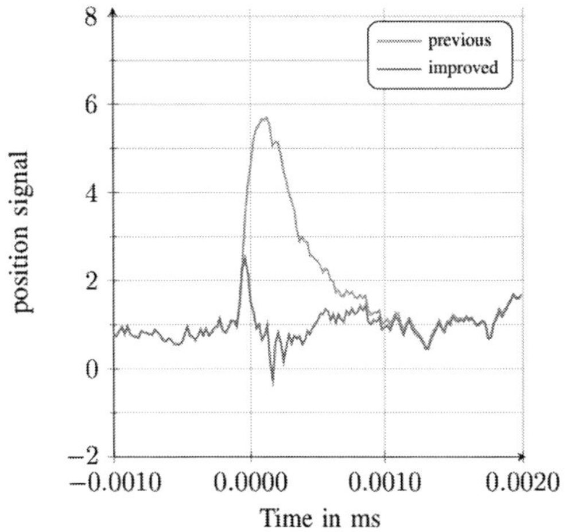

Figure 8. Stray flux based measuring signal previously compensated and improved compensated mode. Rotor is fixed in the test bearing while step response was recorded.

E. Sensor design

In previous experiments with former sensor designs it became clear that electromagnetic disturbance caused by the control coils has a significant influence on the signals from the Hall-sensors. Since the sensors are directly placed near to the coils they are susceptible for such disturbance. In former versions of the sensor there was only one way of signal transmission given and ground areas have been modelled imperfect. The newest Sensor provides direct transmission of

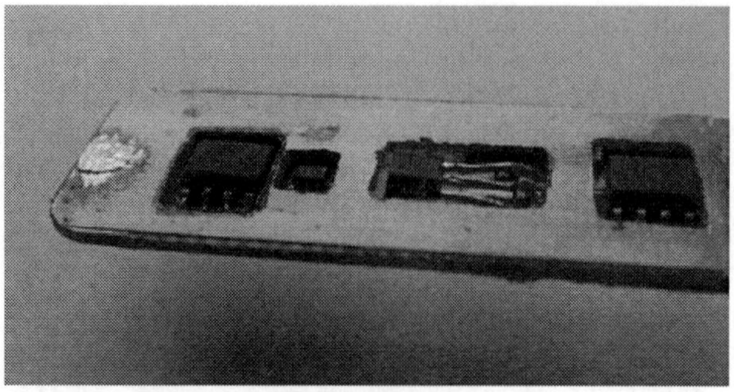

Figure 9. Stray flux sensor V3 (newes version)

Figure 10. Sensors mounted at the test bearing.The lower sensor is the new version.

the Hall-sensor output voltage and two more possibilities like a current based and a differential transmission. The new sensor consists also of four-layer pcb[1] formed by two dual-layer pcb which have been glued together. Now every layer carries a ground area. This is the first step to reduce the influence of electromagnetic disturbance at the place where the Hall-signal is generated. A second step is the implementation of methods for a signal transmission which are more insensitive against electromagnetic distortions like a highly resistive transmission by voltage like it is provided from the Hall-element directly. So the new sensor includes a voltage to current converter and differential signal generator. The signal transmission by current is a well known industrial proved method and the second method originates from audio techniques. As shown in figure 9 the components are placed and soldered at the lower inner layer to minimize the sensor hight. A mounted sensor can be seen at figure 10.

Because only one prototype has been mounted until now extensive measurements and experiments could not accomplished yet but first test has been auspicious.

F. Conclusion

In this paper the delay between current and magnetic field rise caused by eddy currents and their impact on the current compensation of the stray flux based measurement system for the rotor position in magnetic bearing has been discussed. A method has been presented to reduce over compensation significantly by using the derivation of the measured control current. It has been shown how the derived signal should be treated to achieve good results with arguable computing time. Furthermore a new prototype of a stray flux sensor has been presented which permits less noisy measuring signals. Looking ahead all stray flux sensors have to be replaced by the new version and the stray flux based measurement system should switch from passive to active mode and work as input for the loop controller.

[1]pcb - printed circuit board

References

[1] F. Bahr, M. Melzer, D. Karnaushenko, *et al.*, "Erfassung der luftspaltinduktion aktiver magnetlager auf basis ultradünner, flexibler hall-sensoren", *Tagungsband 9. Workshop Magnetlagertechnik Zittau/Chemnitz*, p. 96, Oct. 2013.

[2] W. H. Marco Schramm, "Novel magnetic displacement sensor for mechatronical systems", in *The 33rd Annual Conference of the IEEE Industrial Electronics Society*, IEEE, Nov. 2007.

[3] J. Rudolph, R. Werner, and I. Maximow, "Measurement of the rotor position of an active magnetic bearing using interpolar stray flux", in *1. Brazilian Workshop on Magnetic Bearings*, 2013.

[4] ——, "Messung magnetischer streuflüsse zur bestimmung der wellenposition im magnetlager", in *9. Workshop Magnetlagertechnik Zittau-Chemnitz*, 2013.

Hysteresis self-bearing motor

Mohammad Imani-Nejad[a], David L. Trumper[b]

[a] Massachusetts Institute of Technology , 77 Massachusetts ave, Cambridge, USA, m_imani@alum.mit.edu
[b] Massachusetts Institute of Technology , 77 Massachusetts ave, Cambridge, USA, trumper@mit.edu

***Abstract*—This paper presents a hysteresis self-bearing motor along with its nonlinear analysis. In this work we introduced a new type of self-bearing motor with segmented stator. The major advantages of this stator are: easy and low cost manufacturing, higher filling factor and shorter magnetic path for the stator. We tested the self-bearing concept successfully with this new configuration. Hysteresis motors are suitable for high speed application since their rotor has neither permanent magnets nor any winding or mechanical features. We used FEA analysis to check the feasibility of proposed idea. Then we used Chua's hysteresis model to simulate the nonlinear behavior of this system.**

***Index Terms*—Self-bearing motor, Axial flux, Hysteresis, Chua, Segmented.**

I. Introduction

Many types of self-bearing motors have been presented in the literature. Permanent magnet self-bearing motors are widely proposed for a variety of applications; however they have some disadvantages for high speed and high temperature systems. Induction self-bearing motors have been proposed since at least early 1990 [1] and [2]. For an output of more than a few kilowatts, induction machines are competitive to permanent magnet motors. They are also less expensive and more robust compared to PM machines.

Hysteresis motors are a synchronous induction machine. Steinmetz apparently introduced the first hysteresis motor in 1917, as stated in [3]. This type of motor has a simple solid rotor of material with large magnetic hysteresis. They also typically have a polyphase stator winding similar to an induction machine. Like induction machines, a slotted stator having distributed windings is used to produce as nearly as possible a sinusoidal magnetic flux distributed spatially. The rotating magnetic field creates magnetic poles in the rotor by means of induction. These poles follow the stator magnetic field. Because of hysteresis, the rotor poles lag behind the inducing magneto motive force wave by the hysteretic lag angle. Before the rotor reaches synchronous speed, a given region of the rotor experiences a cyclic hysteresis loop at the slip frequency. Under this circumstance, the lag angle is only a function of the rotor material, and it remains at its maximum value. When the rotor reaches synchronous speed with an initial load, due to the rotor inertia interacting with magnetic stiffness, the lag angle starts to oscillate about its equilibrium point until the hysteresis torque is sufficient to carry the load. Typically the period of these oscillations is about 1-5 seconds. This phenomena is called hunting.

On the other hand different types of multiple winding structure have been presented in the literature. This type of machines have suspension windings in addition to motor windings. Typically, by controlling current through a combination of P pole and P$\pm$2 pole windings the rotor is levitated and rotated. One of the main disadvantages of such a system, especially for axial type motors is the low filling factor for the stator. This is a practical issue because winding of such a stator is not an easy task. On the contrary, U core actuators have a very good filling factor. Moreover, these type of actuators are much easier to manufacture and less expensive compared to conventional axial stators. Another advantage of a segmented stator is the short magnetic path in the stator. In this paper we present a segmented stator design for hysteresis self-bearing motor along with non-linear analysis.

II. Principle of Operation

A. Configuration

The primary goal of designing this system is introducing a new self-bearing motor for high speed flywheels. High speed flywheel rotors experience large expansion due to high centrifugal force. This causes large gap variation in radial flux motors. Thus we chose an axial flux type self-bearing motor where the expansion occurs in the non-sensitive direction.

Figure 1 shows the fabricated prototype of segmented stator and hysteretic rotor. The complete setup has two stators: one on the top and one on the bottom for torque control. In our experimental setup we used 12 U cores of laminated silicon iron to build the segmented stator. The rotor is fabricated from D2 tool steel. This configuration enables us to implement either a two or a four pole motor simply by changing the current pattern.

B. Magnetic flux distribution

It is very important to explain how the magnetic path is formed in the rotor in the first place. Even though a simple solid ring would work in this design, instead we propose a composite ring such as shown in figure 2.

As we can see in this figure, the gap between the two rings causes a circumferential magnetic field in both rings. We used epoxy to fill this gap, however for real application this gap can be filled with composite material such as carbon fibre. This is an advantage for high speed systems, where we can use high hysteresis material for the rings and over wrap it with a strong material such as carbon fiber. In this case, steels with large density provide the mass whereas the carbon fibre gives

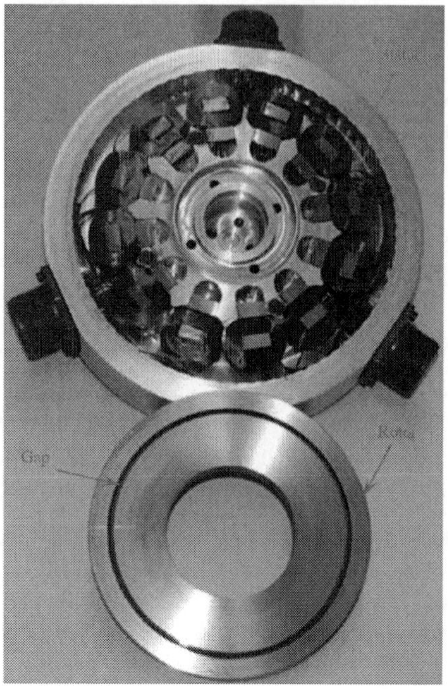

Fig. 1. Hysteresis self-bearing motor. The rotor ring is levitated and rotated by a hysteresis motor structure.

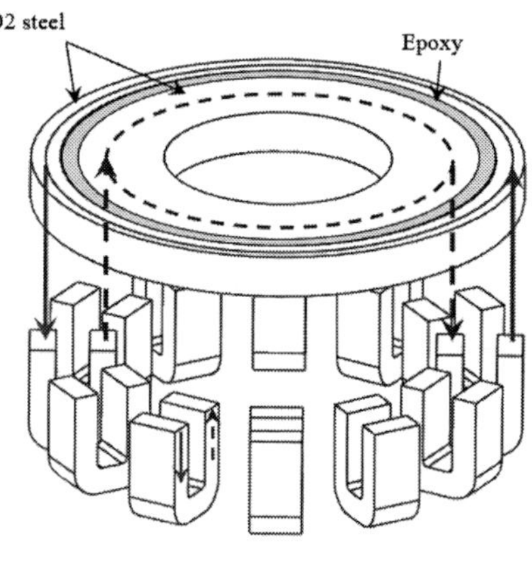

Fig. 2. Segmented stator-rotor magnetic path in composite ring rotor.

the strength to the rotor structure. The other advantage of the segmented stator configuration which can be seen in figure 2 is that the magnetic path in the stator is relatively short.

Figure 3 shows an FEA simulation for the two pole motor configuration. The FEA analysis shows that magnetic in the rotor is similar to that generated in the regular stator.

C. Control

Figure 4 shows the control block diagram for our system.

As shown in this figure there are three control loops for levitation control ,i.e., the x and the two tip tilt θ_z and θ_y axes. The commutated signals are added and subtracted to the upper and lower stator respectively. The torque is controlled

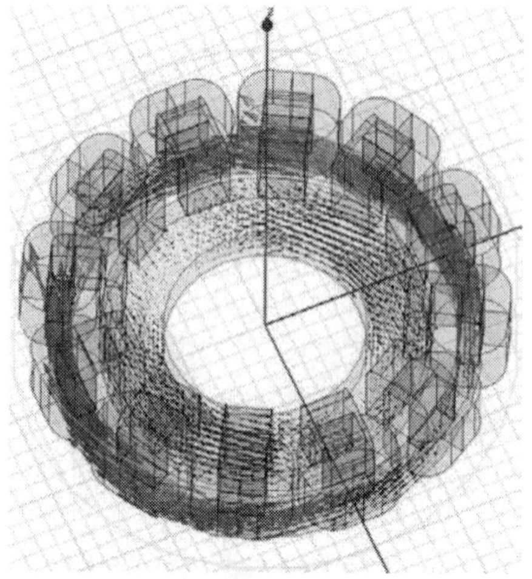

Fig. 3. FEA simulation for segmented stator in two pole configuration.

by a bias current that is shown by T. This bias current is added differentially to top and lower stator; thus it enables us to vary the shear magnetic field. Since the hysteresis motor can run synchronous, we can simply use a $\frac{V}{f}$ control strategy for speed regulation. However we can also run the system with a closed loop velocity control as needed.

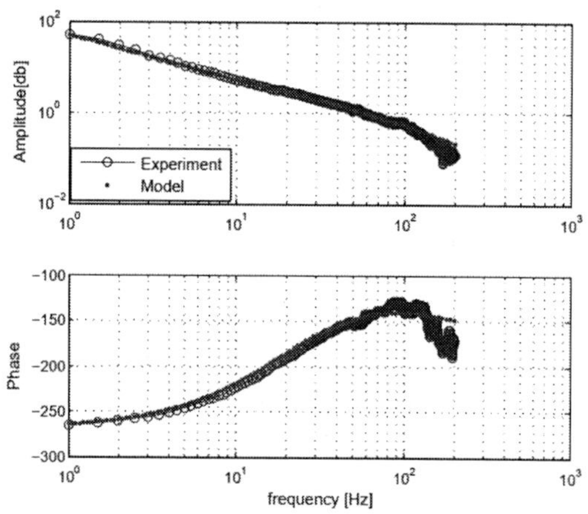

Fig. 5. Loop transfer function frequency response for axial direction. Cross over frequency is $f_c = 60$Hz with phase margin $\phi_m = 40°$

A lead-lag controller is designed for a cross over frequency of $f_c = 60$Hz and phase margin of $\phi_m = 40°$ which provides approximately $\zeta = 0.4$ damping. The loop transfer function frequency responses for both vertical and tilting axes are given in Figures 5 and 6.

As we can see in figure 6, the tip tilt phase differs significantly from theory at low frequencies. We believe that this is due to lateral mode coupling. Since the rotor is passively stable in lateral axes thus it can couple with a lateral mode. We can also see a resonance at 200Hz which is a structural resonance.

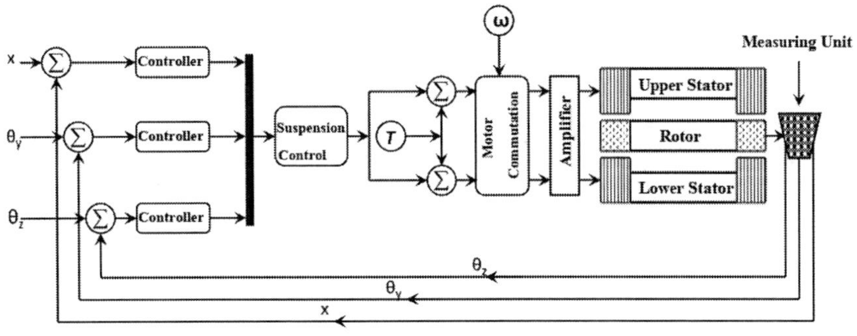

Fig. 4. Single winding self-bearing control block diagram.

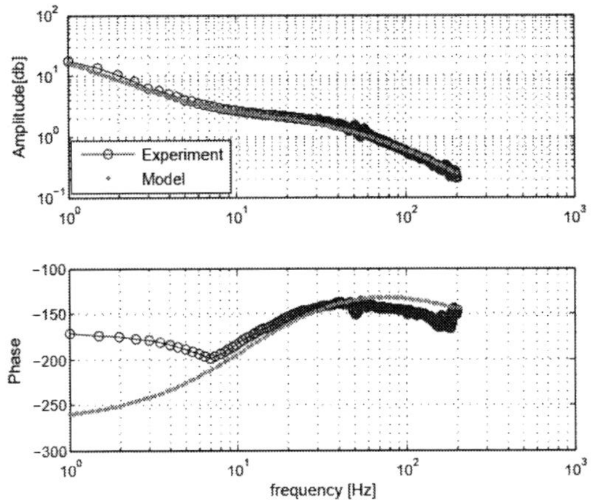

Fig. 6. Loop transfer function frequency response for tip/tilt direction (θ).

III. Non-linear Analysis

A. Chua Hysteresis Model

In this analysis we use a Chua nonlinear hysteresis model [4]. The Chua hysteresis model is given by

$$\frac{dy(t)}{dt} = W\left(\frac{dx(t)}{dt}\right) \times h\left(y(t)\right) \times g\left[x(t) - f(y(t))\right]. \quad (1)$$

In this equation $x(t)$ is the input and $y(t)$ is the output. The model is specified by two monotonically increasing functions: f the *restoring function* which describes the energy storage, and g, the *dissipation function* which describes the energy dissipation. The W and h functions provide the loop widening and weighting functions respectively. These four functions should be tuned such that the desired hysteresis loop is attained. Chua and Bass have illustrated the hysteresis loop for two different types of material in [5]. In this paper they also present several possible functions for loop widening.

As an example, functions f and g were replaced by the logarithmic functions given below to simulate the D2 steel hysteresis loop shown in figure 7. We chose absolute function for W and unity gain for h.

$$f(y) = \frac{\text{sign(y)}}{-0.005} \ln\left[1 - \frac{1}{0.7}|y|\right]$$
$$g(u) = -50 \ln\left[1 - \frac{1}{1500}|u|\right] \quad (2)$$

The B-H curve used in this analysis is the Chua's model fitted for D2 steel shown in figure 8.

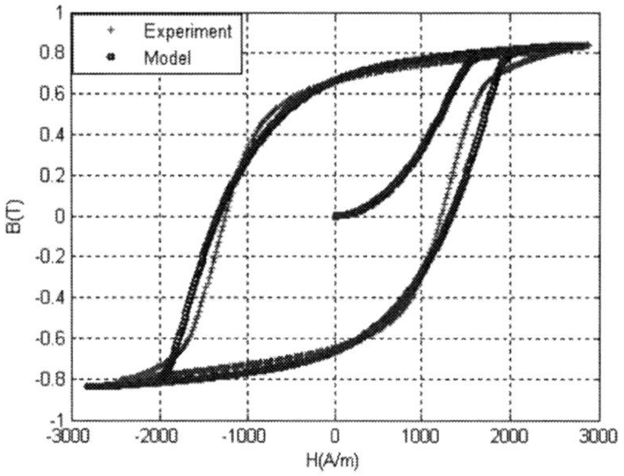

Fig. 8. Hysteresis model of Chua's configuration as fitted to D2 steel experimental data.

B. Suspension Forces and Moments

In this section we present the analysis of suspension forces and moments for hysteresis self-bearing motors.

We derive the governing equations and present the SIMULINK block diagram for axial force. A similar analysis can be easily used for inclination moments. For this purpose we only need to replace the force, displacement and mass by moment, inclination, and moment of inertia, respectively.

Figure 9 shows the magnetic path for a U core actuator and a segment of the rotor. In case of homogenous magnetic field the magnetic normal stress σ is given:

$$\sigma = \frac{B^2}{2\mu_0}. \quad (3)$$

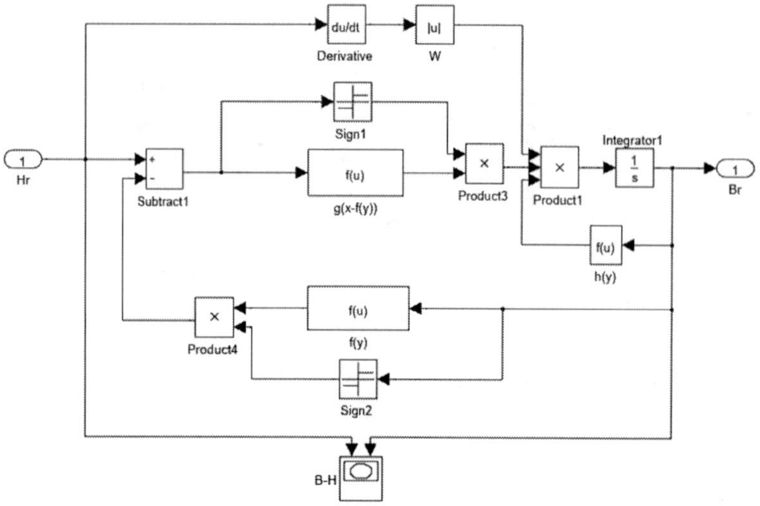

Fig. 7. Chua model Simulink block diagram.

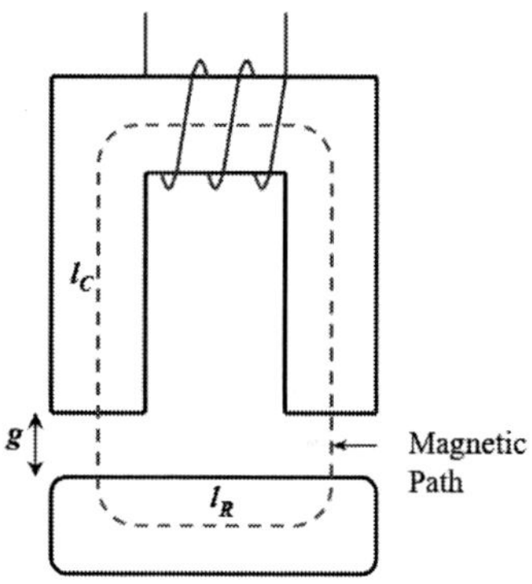

Fig. 9. Magnetic path for U core actuator.

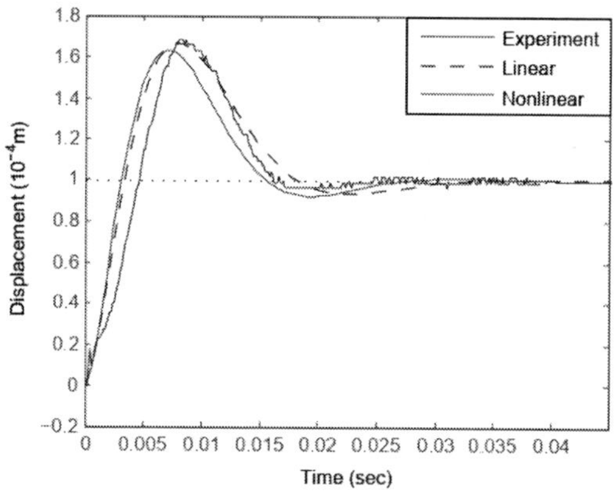

Fig. 12. Step response. Zero reference corresponds to a gap of 0.5mm.

Therefore the force is given by:

$$f = A\frac{B^2}{2\mu_0}; \tag{4}$$

where A is the active surface area of the electromagnet. Meanwhile we can write Ampere's law along the magnetic path

$$Ni = H_R l_R + H_g l_g + H_c l_c, \tag{5}$$

where l_R, l_g and l_c are the length of magnetic paths in rotor, gap and stator respectively. In this analysis we assume that the stator permeability is much larger than the air and the rotor. Thus we can write:

$$H_R = \frac{(Ni - H_g l_g)}{l_R}. \tag{6}$$

Figure 10 shows the block diagram of force generation. The block is constructed based on the equations presented above. The block has two inputs: current and displacement (inclination). The magnetic field intensity H_R for the rotor is calculated by using (6). The magnetic flux density is then computed through Chua's model. For homogenous magnetic flux density and ignoring fringes, the magnetic flux density is constant i.e. $B_g = B_R$. Thus we can calculate the magnetic field intensity in gap by:

$$H_g = \frac{B_g}{\mu_0}. \tag{7}$$

Once we have the magnetic field, force is obtained by using (4).

C. Results

Figure 12 shows the step response for the vertical direction, motion of rotor. In this figure we present both fitted linear and nonlinear models versus experimental data. As we can see the simulation results are quite acceptable.

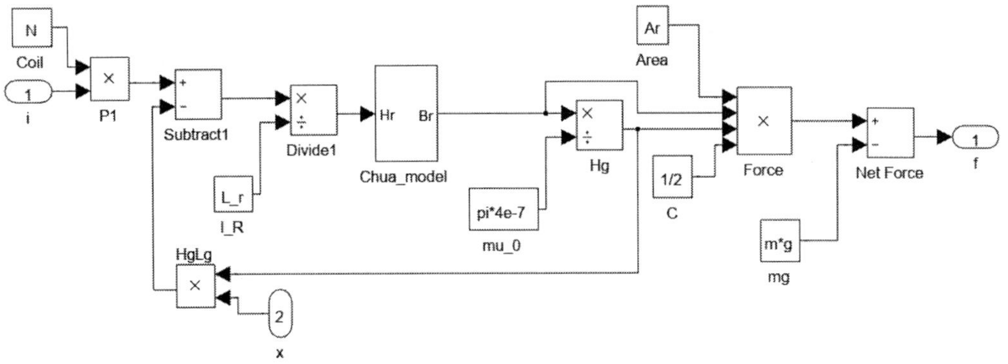

Fig. 10. Force generation in nonlinear bearing model including Chua hysteresis model.

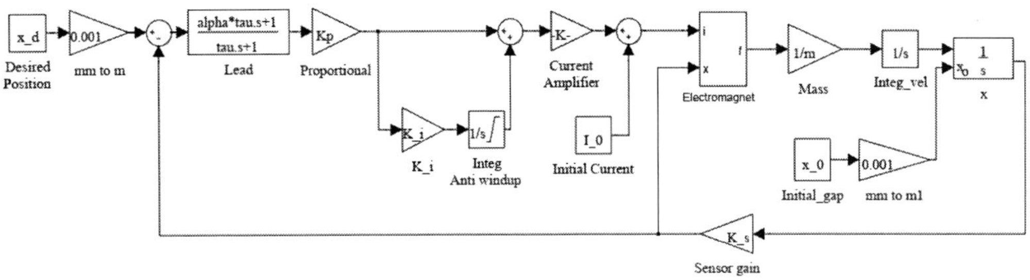

Fig. 11. Hysteresis self-bearing suspension control block diagram. Nonlinear force model is enclosed in block labeled Electromagnet.

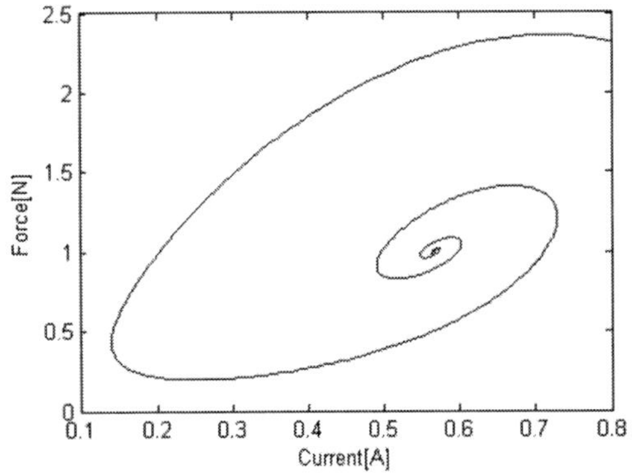

Fig. 13. Force v.s. Current during transition of an 0.1mm step input from initial 0.5mm gap.

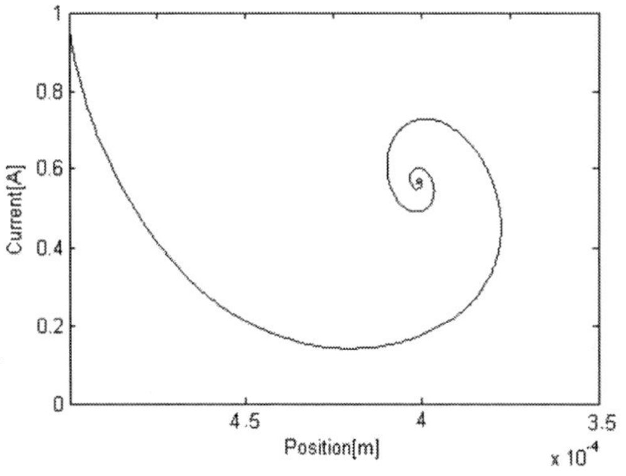

Fig. 14. Current v.s. Displacement during transition of an 0.1mm step input from initial 0.5mm gap.

Figures 13 and 14 shows the force versus current and current versus displacement during the step transition. These simulations are run for a gap=0.5mm and a step of 0.1mm.

Figure 15 shows the B-H trajectory in the rotor during a step transition.

The motor torque analysis is also calculated and presented in [6]. Figure 16 shows the rotor speed versus time for both model and experimental results. Figure 17 shows the hunting phenomenon that was measured experimentally at 2100 rpm. As we can see, it takes about 1 minute for the oscillation to die. This is because we didn't actively control the speed; however if hunting is an issue for a particular application it can be easily resolved by speed control.

IV. Conclusion

In this paper, a novel self-bearing motor has been proposed. A segmented stator hysteresis self-bearing motor was designed and tested successfully. FEA simulation was done to check the

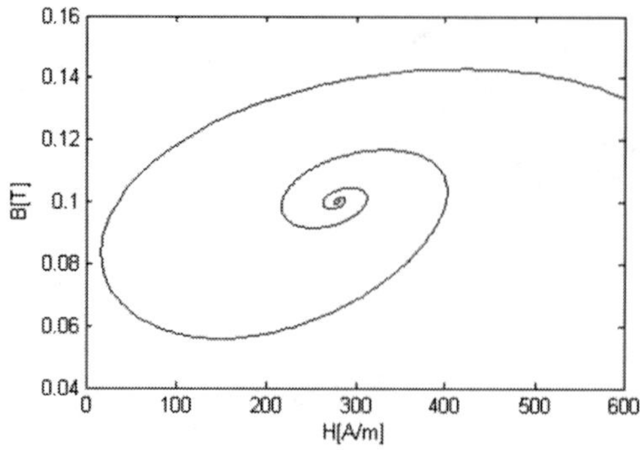

Fig. 15. Plant B-H during transition of an 0.1mm step input from an initial 0.5mm gap.

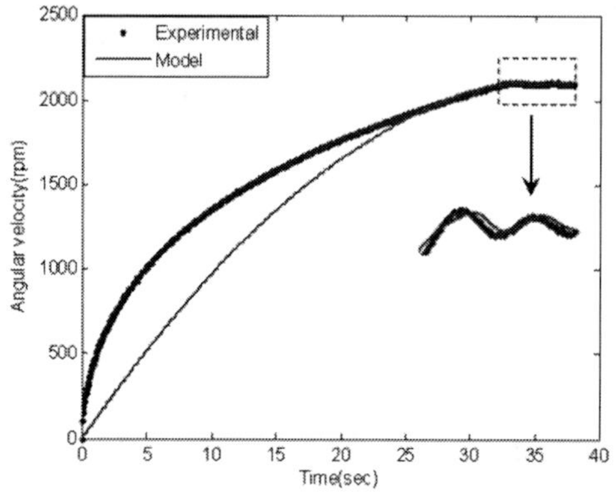

Fig. 16. Speed vs time under 0.5 A rms excitation. Blue curve is experiment and red curve is model.

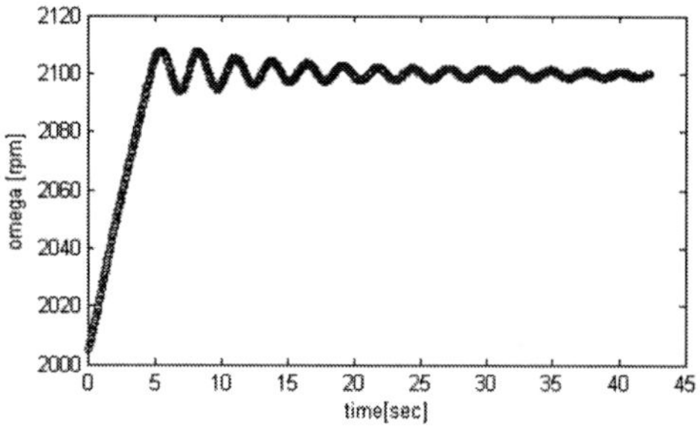

Fig. 17. Measured speed hunting at 2100 rpm.

feasibility of such a system. Later a nonlinear analysis based on Chua model is provided to simulate the system behavior.

V. Acknowledgment

The authors would like to thank the MIT Energy Initiative for funding this project.

References

[1] A. Chiba, D. T. Power and M. A. Rahman, "Charachteristic of a bearingless induction motor," *IEEE Trans. Magn.*, Vol.27, no.6, pp 5199-5201,Nov. 1991.

[2] R. Schoeb and J. Bischel, "Vector control of the bearingless motor," in *Proc. 4th ISMB*, Zurich, Switzerland, pp 327-332,1994.

[3] J. J. Nitao, E. T. Scharlemann and B. A. Kirkendal, "Equivalent circuit modeling of hysteresis motors," in *Techincal report, Lawrence Livermore National Laboratory*,2009.

[4] L. O. Chua and K. A.. Stromsmoe, "Lumped circuit model for nonlinear inductors exhibiting hysteresis loops," *IEEE Trans. Circuit Therory.*,.1970.

[5] L. O. Chua and S. C.. Bass, "A generalized hysteresis model," *IEEE Trans. Circuit Therory.*, Vol.19, no.1, pp 36-48,Nov. 1972.

[6] M. Imani Nejad and D. L.. Trumper, "Self-bearing motor design and control," *PhD Thesis.*, Mechanical Engineering, MIT,. Cambridge. 2013.

*** Formatting Issue - Best Available Paper/Graphic ***

Magnetically Suspended Reaction Sphere with One-axis Hysteresis Drive

Lei Zhou, Mohammad Imani Nejad, David L. Trumper

Dept. of Mechanical Engineering, Massachusetts Institute of Technology
77 Massachusetts Avenue, Cambridge, MA, USA

{leizhou, m_imani, trumper} @mit.edu

Abstract—**We have designed, built and tested a magnetically suspended reaction sphere with one-axis hysteresis drive (1D-MSRS) in the Precision Motion Control Lab at MIT. This project is aiming at two goals: (a) exploring the possible design of a magnetically suspended reaction sphere for small spacecraft's attitude control; and (b) evaluating the performance of a hysteresis motor for the reaction wheel/sphere application. The 1D-MSRS is a motor with spherical rotor, which is magnetically suspended in all translational directions, and is driven by hysteresis motor about vertical axis. For the suspension of the sphere, single dimensional magnetic levitation in vertical direction and bearingless motor around the rotor are combined. An equivalent circuit model of the hysteresis motor is used to analyze the motor's dynamic behavior. Good agreements between computed and measured motor performance validates the model, which provides a useful simulation tool for further MSRS modeling and control. Test results of the 1D-MSRS show that a starting torque of 8.15 mNm is achieved at 0.7 A zero-peak excitation current, and with this excitation the sphere can run up to 12,000 rpm synchronously in lab with the existence of air drag.**

I. Introduction

In a spacecraft's flight control, the Attitude and Orbit Control System (AOCS) of the spacecraft is responsible for the high precision of control in orbit. When the spacecraft is maneuvering, it requires an external force, or torque, which is often provided by a minimum of three reaction wheels (often 4-5 wheels are used for optimization and redundancy [1]). By accelerating the appropriate wheels, the system can produce a zero-mean torque about any axis to the spacecraft, and momentum can be stored as well [2]. Such wheels are often used for both spacecraft attitude control [3] and large angle slewing maneuvers [4]. Other applications include vibration compensation and orientation control of solar arrays [5], as well as energy storage.

As an alternative to a combination of several reaction wheels, we propose to use of a magnetically levitated reaction sphere (MSRS) for satellite's attitude control. Our alternate version is that the sphere can be angularly accelerated about any axis by a three dimensional (3D) motor, making the attitude of the spacecraft in all axes controllable by a single device. Due to its symmetry, a sphere can always gives the same inertia, independent of its rotational axis. Thus using reaction sphere can reach smaller size and mass, which is an essential consideration for space device design. Also, magnetic suspension eliminates imbalance, so the vibrations emitted by the reaction wheel that degrade the performance of precision instruments in space are suppressed. Mechanical friction is also eliminated by the magnetic bearing, and thus low steady-state power consumption may be possible. This also enables it work without lubrication and can be expected to have longer maintain-free lifespan. Furthermore, the rotation of a sphere has no gyroscopic coupling between the rotations about the three axes, which potentially enables better control possibilities.

The idea of a reaction sphere was proposed more than two decades ago [6], but to our knowledge none have reached a ready-for-commercialization technology. To our understanding, this is mainly due to the difficulties of 3D angular position sensing for spherical rotor and the complexity due to coupling of fields by stator segments and between bearing and motor functions. In recent years several spherical motors, mainly for robotics application, have been developed [7] [8] [9] [10] [11] [12]. These design often include a mechanical shaft coupling that induces a rotational singularity, which prevents them from directly being used as an all degree-of-freedom actuator for attitude control. A permanent-magnet-motor-based reaction sphere is presented in [13]. This design successfully solved the problem of sensing and field coupling, which makes it a good actuator about all axis. But the complexity of its magnetic structure may prevent it from being suitable for small satellites application, and the strength of the rotor limits its maximum rotational speed, which may limit the performance of the actuator.

Among many motor driving principles, the hysteresis motor is receiving increasing attention due to its advantages of simple structure, vibration-free operation, self-starting and constant torque production in starting. Another distinct feature of this motor is that its rotor can be made out of a single piece of hard and strong steel, which allows the rotor to stand large stresses and makes this motor concept attractive for high-speed applications. Reference [14] reported the performance of several hysteresis motors of different scales, and the results make us believe that this kind of motor concept may demonstrate good performance in reaction wheels. However, to the best of our knowledge, the hysteresis motor has not been introduced to the development of reaction wheels until now.

Aiming at the dual goal of exploring the design for a magnetically suspended reaction sphere and evaluating hysteresis motor performance for reaction wheels application, we decide to focus this project on the development of a magnetically suspended reaction sphere with one-axis hysteresis drive (1D-MSRS). The hardware demonstrates a motor with a spherical rotor driven by hysteresis motor, and is magnetically suspended in all translation directions. For the suspension of the sphere, single dimensional magnetic levitation in vertical direction and bearingless motor around the rotor are combined.

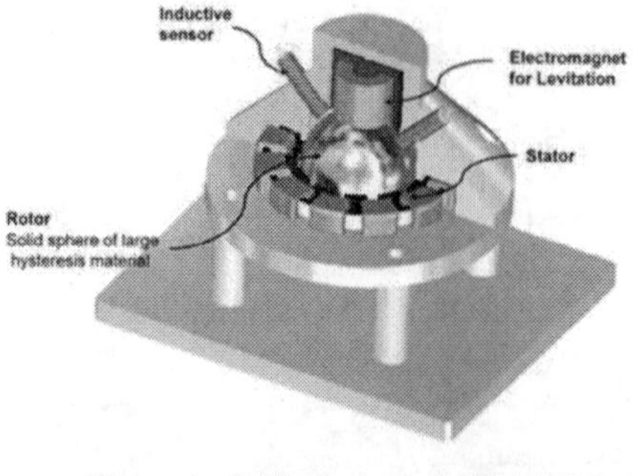

Figure 1: CAD design of 1D-MSRS.

(a) (b)

Figure 2: Photograph of the 1D-MSRS.
(a) Structure; (b) stator and rotor.

We believe that the bearingless motor is a potential way to solve the field coupling between motor and bearing field for the 3D spherical motor design.

This paper focuses on the design characteristics of the suspension and driving system of the 1D-MSRS. What follows are the simulation and experimental results that demonstrate the performance of the 1D-MSRS. A more detailed analysis of the design, modeling, and control of the system is presented in [15].

II. Single DOF Reaction Sphere Design and Integration

The 1D-MSRS is a motor with a magnetically levitated spherical rotor that can rotate and stores momentum about the vertical axis.

A. System Overview

Figure 1 shows the CAD design of 1D-MSRS. The rotor is a 54 mm diameter sphere of hardened D2 steel. Four induction position sensors are placed around the rotor to measure the sphere's position in three translational degrees of freedom. The rotor sphere is magnetically levitated in the vertical direction by a reluctance actuator placed at the north pole. The stator is arranged around the sphere's equator line, and serves both for levitating the sphere in the horizontal plane and for generating torque about z axis simultaneously by a bearingless motor configuration. A reflective optical tachometer is used for speed detection of the reaction sphere. The structure of the device is presented in Figure 2.

(a) (b)

Figure 3: Stator for the 1D-MSRS without and with winding.

B. Position Sensors

The spherical rotor displacements in all three translational degrees of freedom are monitored by position sensors, which are of induction type. The four sensors are arranged 45 degrees to the vertical axis, and pointing to the center of the sphere. In this configuration the translation of the sphere in 3 directions (two horizontal directions and one vertical direction, also x, y and z) are sensed and noise is mitigated by taking the average of the signals.

C. Rotor

The rotor presents in the 1D-MSRS is a 54 mm diameter solid sphere of magnetically hard material. According to the analysis of hysteresis motor torque, the torque is proportional to rotor material's hysteresis loop area [16]. Although materials with better hysteresis properties exists, we selected D2 steel for the rotor of 1D-MSRS for the proof of our design, as this material was readily available. D2 steel is a kind of high Carbon, high Chromium type tool steel. It contains 11 to 15 percent Chromium, which makes it a deep hardening, highly wear resistant and magnetically hard alloy. These properties of the D2 steel makes it a good candidate for the rotor material of a hysteresis motor.

D. Stator

The stator for the 1D-MSRS is a custom-made stator with 24 slots and a length of 9.50 mm, which is stacking 12 layers of motor stator laminations cut from non-oriented electrical steel sheets of gauge 24. Since the stator needs to work as a magnetic bearing and motor stator simultaneously, the multiple winding approach of a bearingless motor is used, with its 4-pole winding for rotation and 2-pole winding for stabilization. Figure 3 shows the stator for the reaction sphere, where (a) shows the stator laminations without winding, and (b) shows the wound stator.

E. Vertical levitation actuator

An electromagnet actuator is used for the position control of the reaction sphere in the vertical direction. Figure 4 shows the rotor sphere being levitated in the vertical direction by this actuator.

To make the system more energy efficient, a piece of permanent magnet is embedded in the magnetic path of the levitation actuator. With this design, most of the weight is carried by the permanent magnet flux and the required excitation DC coil current can be greatly reduced.

*** Formatting Issue - Best Available Paper/Graphic ***

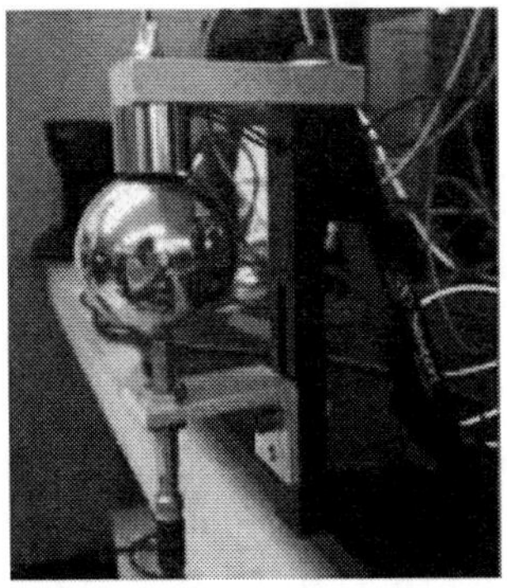

Figure 4: Single DOF levitation of sphere by reluctance actuator with permanent magnet.

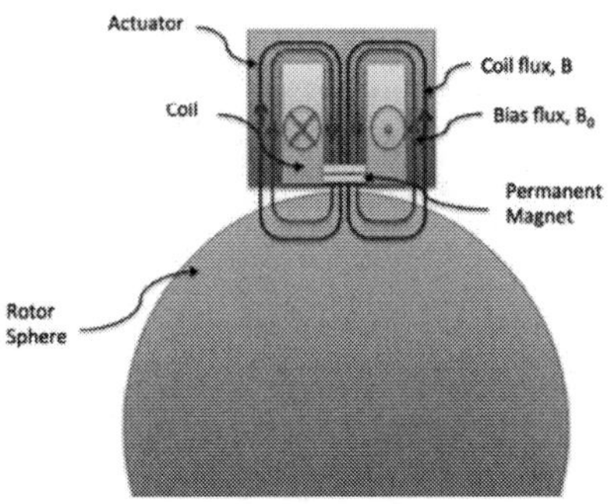

Figure 5: A cross section of the biased levitation diagram

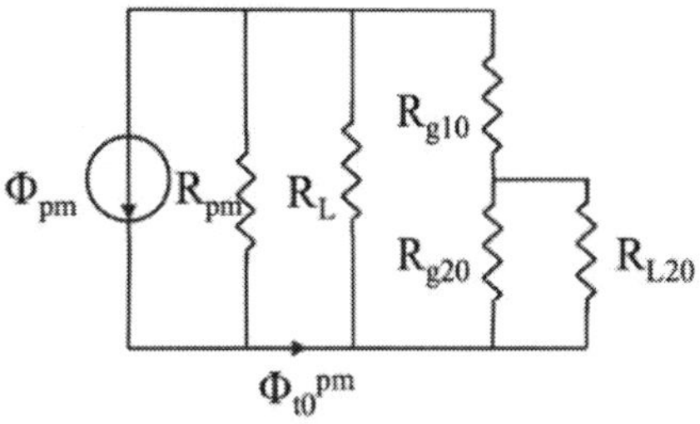

Figure 6: Equivalent magnetic circuit model for flux distribution in flux-biased levitation actuator.

F. Controller

The digital controller is selected as high speed digital signal processing computer, with the capability of hardware real-time interrupt processing. In the development of the 1D-MSRS prototype, an NI PXI chassis with real-time controller is used and is running at a sampling and processing frequency of 10 kHz. Five A/D and seven D/A converter channels with 14 bits precision are used. The controller's functions are:

- Receive signals for displacement from inductive probes and motor rotational speed from tachometer;
- Receive control commands from the host computer to change parameters of the levitation and self-bearing motor system;
- Generate and send current control signals according to the control algorithms;
- Generate acceleration trajectory for the sphere's rotation;
- Diagnose the state of the elements of the self-bearing motor system and transmit messages to the host computer.

G. Power amplifier

In order to directly drive the current in the actuators, seven separated relatively large bandwidth current control amplifiers are used. Though the driving efficiency of switching amplifier is higher, linear amplifiers are chosen in this project to avoid high frequency switching noise. The linear amplifier chip PA12 from APEX is selected, which has built-in fly-back diodes that protect the output from over voltage due to the fly-back effect. By building the current feedback loop around this amplifier for current control, a closed-loop bandwidth of 20 kHz is reached, and the amplifier can provide up to 5 A continuous current with output voltages up to ±45 V. The design for the current control amplifiers are described in [14] in detail.

III. Single DOF Reaction Sphere Modeling and Control

In this section modeling and control of the 1D-MSRS's suspension and rotation is presented.

A. Vertical Suspension

In this section the levitation and vertical position control for the rotor sphere is introduced. Figure 5 shows a simplified diagram of the flux biased magnetic levitation actuator, which contains both bias flux from a permanent magnet and stabilizing flux generated by a coil in the slot.

First the levitation force generated by the permanent magnet is analyzed. Since the permeability of the sphere and actuator core is much larger than that of air, we assume that the magnetic reluctances are concentrated in the air gaps and in the permanent magnet. A magnetic circuit model is used to analyze the flux distribution, which is presented in Figure 6. In this model, Φ_{pm} is the remanence flux of the permanent magnet. Φ_{t0}^{pm} is the total flux that flows out from the permanent

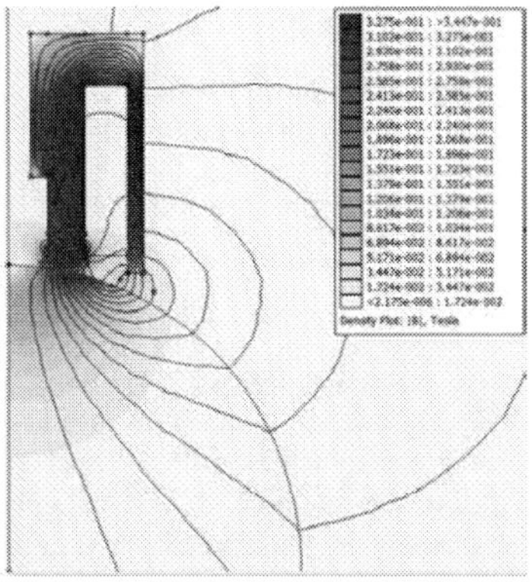

Figure 7: FEA solution of DC field in biased flux levitation actuator using typical design parameters. The permeability of the actuator core and sphere materials is assumed as $5000\mu_0$.

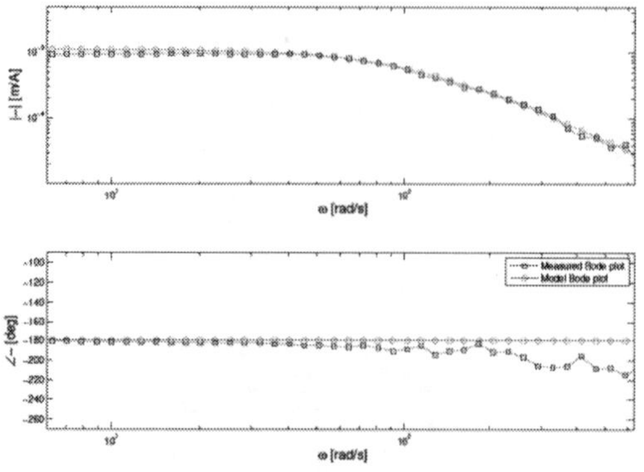

Figure 8: Plant Bode plot of the magnetic suspension system from coil current input [A] to sphere position output [m]. Blue: experimental data; green: model.

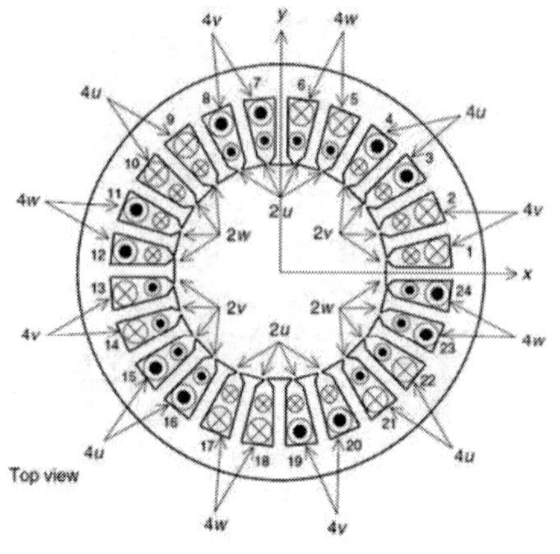

Figure 9: Multiple winding diagram for 1D-MSRS.

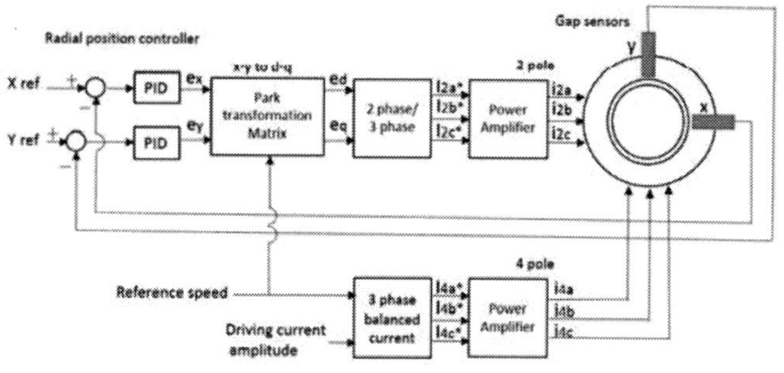

Figure 10: Bearingless motor principle diagram.

magnet. R_{g10} and R_{g20} are the magnetic reluctances of air gap in the center and in the periphery, respectively. R_L and R_{L20} are the reluctances for leakage paths. Here the leakage reluctance R_L is difficult to calculate using lumped parameter model, but it can be derived from numerical calculation. Figure 7 shows a flux plot calculated by the finite element method under the assumption of cylindrical symmetry. The ratio of flux leakage can be achieved from this numerical result. Then the air gap flux generated by the permanent magnet can be calculated.

In additional to the permanent magnet in the magnetic circuit, the coil in slots is excited with controlling current to levitate the sphere. The total coil winding $N \cdot i$ Ampere-turns generates controlling flux density in the air gap. By superposition of both the permanent magnet flux and the coil flux, we can calculate the total flux in the air gap, and further the reluctance force for magnetic levitation can be calculated by either an energy method or via the Maxwell stress tensor.

The frequency response from the current to the position of the rotor is measured to identify the system, and the obtained Bode plot is given in Figure 8 (Blue). An analytical modal of the magnetic suspension plant is developed, and the modeled Bode plot is shows also in Figure 8 (green). Good agreement between the model and the measured plant dynamics verified our model.

To stabilize the system, the sphere's position signal is fed back, and a PID controller is implemented in the real-time processor. A cross-over frequency of 300 $\mathrm{rad/s}$ is reached, with a phase margin of 40°.

B. Bearingless motor

In order to drive the sphere's spinning and suspension simultaneously, a bearingless motor is adopted in the 1D-MSRS. Bearingless motor is achieved by using two sets of 3 phase winding on a single stator. By correctly configure and control current in these motor windings, the machine can generate radial force for suspension as well as torque for spinning with only one stator [17]. Figure 9 shows the winding pattern of the 1D-MSRS's stator for driving and suspension in horizontal plane.

Figure 10 presents the working principle of bearingless motor. By exciting the motor winding (here the 4-pole windings) with rotational three-phase current, a torque can be created on the rotor. The air gap lengths in stationary x and y coordinates are measured by displacement sensors. Controllers are designed for lateral levitation of the sphere, and the control effort signals are transferred to rotational three-phase coordinate fixed to stator coordinate by means of Park and Clarke transformation. Then the control signals are fed into amplifiers and energizes the radial position control windings (here the 2-pole windings). Figure 10 shows the block diagram of the 1D-MSRS's bearingless motor system control.

The frequency response of the lateral position control loop is measured under different excitation current amplitude in the driving winding (here the 4 pole motor winding). The measured Bode plots is depicted in Figure 11.

Plotting the break frequencies of the plant Bode plots under different driving current amplitudes, we get the data shown in Figure 12. We fitted these data by linear function: with Eq. 1.

$$f_{break}[Hz] = 3.1429 \times I[Amp] + 3.1905. \qquad (1)$$

This changing of the break frequency of the lateral levitation plant of bearingless motor system shows that as the amplitude of the excitation strength decreases, the radial force from stator is getting smaller, thus the 'negative stiffness' in this radial suspension system decreases. This implies that bearingless motor system requires a minimum driving current amplitude in the motor windings, even when no driving torque is needed.

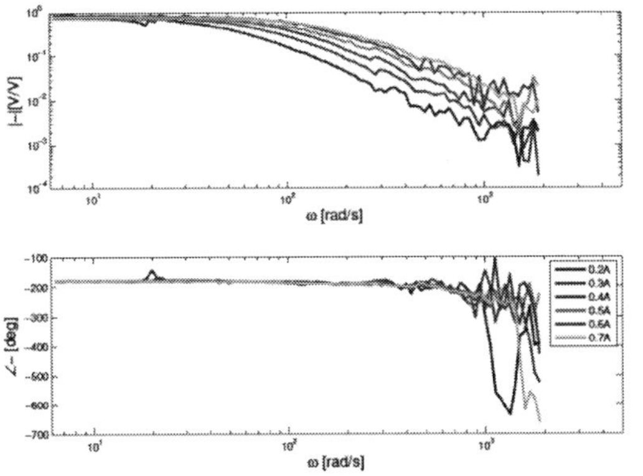

Figure 11: Experimentally measured plant frequency response for X direction sphere suspension with different excitation amplitude (zero-to-peak).

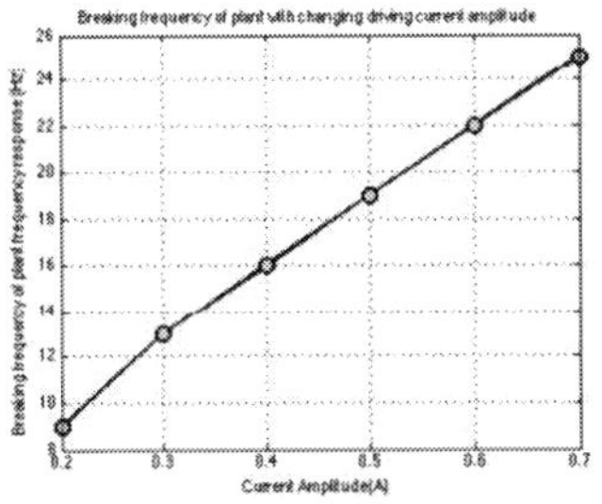

Figure 12: Break frequency of plant with changing driving current amplitude.

This change of the system's frequency response also requires an adjustment in the controller to keep the system stable under all driving conditions. To address it, PID controllers are designed to deal with this nonlinearity by adaptively adjusting its parameters according to Eq. 1. With the designed controllers, the loop reached a bandwidth varying with the excitation current amplitude and a fixed phase margin of 45 degree. The bandwidth under 0.2 A excitation current is 170 rad/s and with 0.7 A excitation the bandwidth is 471 rad/s. In this way the radial position of the sphere is successfully regulated at the center of the stator under all excitation amplitudes.

C. *Hysteresis Motor*

Modeling of the hysteresis motor is difficult due to the complexity of hysteresis material properties. Researchers have developed several different kinds of hysteresis motor analysis models. Reference [18] presented an analytical study of hysteresis motors and introduced the idea of using the fundamental harmonic response to the magnetizing field by simplifying the hysteresis loop into a parallelogram shape. Reference [19] introduced an elliptical hysteresis loop model and derived an equivalent circuit model for the hysteresis motor under this assumption. Several other papers [20] [21] presented different set of hysteresis motor models using the analogies to other types of motors.

In our work, an elliptical approximation of hysteresis loop is used to analyze the dynamic behavior of the 1D-MSRS, and the corresponding equivalent circuit of hysteresis motor was developed by Miyairi and T. Kataoka [19] and further presented in [22] and [14]. This model is used to analyze both the steady state running and starting up of the reaction sphere, and we believe it is reasonable for several reasons. First, we are using the self-bearing motor for the lateral suspension of the 1-D MSRS, thus the sphere is always first levitated in radial directions and then being driven by rotating magnetic field. In this way, under DC excitation conditions or rotating under the excitation of rotating magnetic field, the magnetization pattern and pole distribution of the sphere surface is the same. Therefore we do not need to model the virgin magnetization curve (B-H curve starting from zero magnetization) as most hysteresis motor transient simulation does. Second, the reaction sphere is excited below saturation. In this circumstance the real hysteresis loop has less higher order harmonics, which makes it more similar to a phase lag between B field and H field. Third, due to the effect of eddy current, the hysteresis property is showing dependency to excitation frequency. The loop widening effect makes the hysteresis loop get more close to an elliptical model. Last, this model has an assumption that the hysteresis lag angle is held constant and is equal to it's maximum when slip is large, which makes the motor hysteresis starting torque maximum and does not depend on slip during starting up. This assumption is close to what is really happening when a hysteresis motor is starting spinning, and makes this equivalent circuit model valid for motor transient behavior simulation, although usually they are not accurate when slip is large.

In the hysteresis motor's equivalent circuit model, a balanced single-frequency, three-phase source is assumed. The B-H curve of the rotor material is approximated by a ellipse:

$$B = B_m cos(\theta) \tag{2a}$$

$$H = \frac{B_m}{\mu} cos(\theta + \delta). \tag{2b}$$

Here δ is the lag angle between H and B. Thus the permeability μ is the ratio of the maximum value of B and H is $\mu = \frac{B_m}{H_m}$. Higher order harmonics of the flux density are neglected. Based on these assumptions, the motor equations can be derived, and Figure 13 shows the motor equivalent circuit. The values of the circuit elements derived in [19] are:

$$L_g = \frac{2mK_w{}^2N^2\mu_0 lRg}{p^2\pi g} \tag{3a}$$

$$R_r = \omega_b \frac{mK_w{}^2N^2V_r\mu}{\pi^2 r_r{}^2} sin\delta \tag{3b}$$

$$L_r = \omega_b \frac{mK_w{}^2N^2V_r\mu}{\pi^2 r_r{}^2} cos\delta \tag{3c}$$

Figure 13: Equivalent circuit model of hysteresis motor.

Here δ is the lag angle; K_w is the winding factor; m is the number of phase;p is the number of poles; r_r is mean radius of length of the magnetic field path within rotor; r_g is the radius of the air gap; ω_b is the input angular frequency; N is the number of winding per phase; g is the air gap length; and μ is the permeability of the rotor material. and the circuit parameters are calculated by:

The detail derivation of the circuit parameters are shown in [19]. After solving the circuit for apparent currents, we can calculate the electrical torque of the motor by:

$$T_e = \frac{m}{2}\frac{P}{2}L_g I_r I_g sin\delta \quad (4)$$

Then the motor's mechanical speed is calculated by the rotor's dynamic equation:

$$T_e - T_d = \frac{J}{p}\frac{dw_r}{dt} \quad (5)$$

The lag angle between electrical angular position and mechanical angular position is given by:

$$\frac{\omega_b - \omega_r}{p} = \frac{d\delta}{dt} \quad (6)$$

Simulation of the reaction sphere's dynamics can be conducted by selecting initial conditions and the above calculations. Matlab/Simulink is used for the simulation of the motor system, and the results are reported in the later section.

IV. Experimental and Simulation Results

Table I shows the design parameters for the 1D-MSRS.

Table I: Parameters for equivalent circuit model simulation

parameter	value
Sphere radius	27 mm
Stator height	9.5 mm
Stator back iron length	12mm
Stator pole width	3 mm
Air gap length between stator and sphere (each side)	0.4 mm
Stator number of slots	24
Number of poles for driving winding	4
Number of wires per slot for driving winding	80
Number of poles for suspension winding	2
Number of wires per slot for suspension winding	40

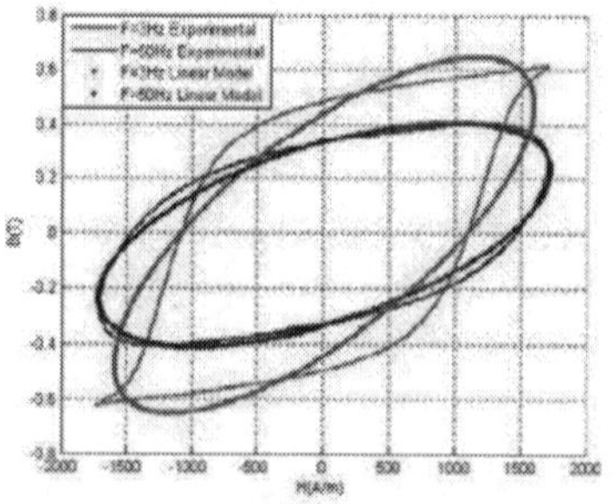

Figure 14: D2 steel linear model approximation from [14].

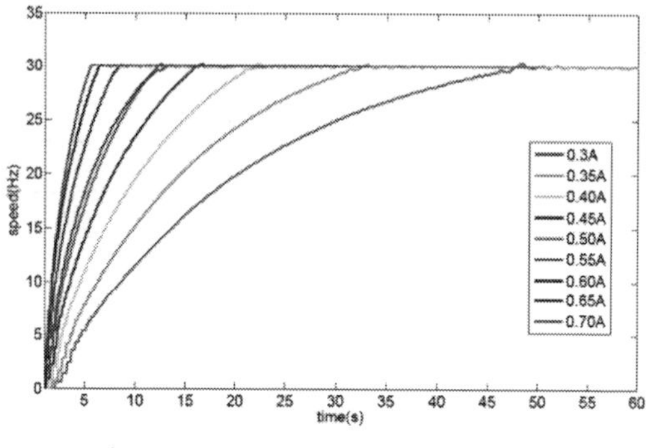

Figure 15: Acceleration curves of 1D-MSRS under different excitation current amplitude.

The B-H characteristics of the sphere material are required for the simulation of the reaction sphere. As we discussed before, the hysteresis loop has a dependency on the excitation frequency. Figure 14 presents the B-H curve of the rotor material (D2 steel) under different excitation frequencies, and their linearized elliptical models. The original data and the measuring process are given in [14]. We chose the appropriate permeability μ and lag angle δ for simulation according to the measurement under certain frequency with taking the loop widening effect into account.

Figure 15 presents the acceleration curves of the 1D-MSRS under different amplitudes of excitation current. We can see

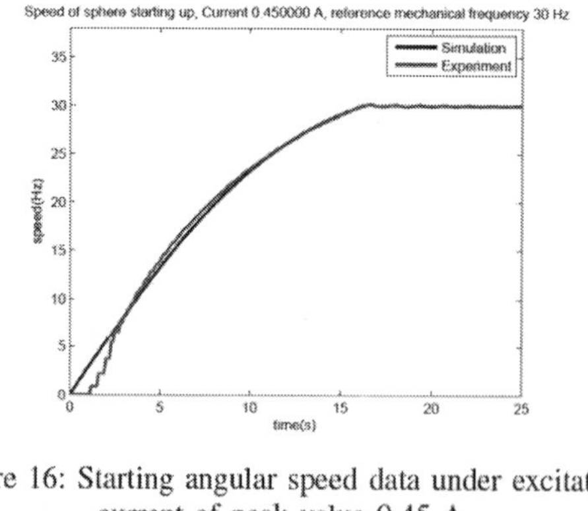

Figure 16: Starting angular speed data under excitation current of peak value 0.45 A.

that all curves locks into the reference speed of 30 Hz (1800 rpm) after acceleration, and a $3-5$ Hz speed oscillation occurs when the speed reaches synchronous. THis is due to the hunting dynamics of the hysteresis motor. Data shows that with 0.7 A excitation current, the sphere can reach the synchronous speed of 30 Hz (1800 rpm) within 6 seconds. Also, a starting torque of $8.15 \times 10^{-3} N.m$ is demonstrated under 0.7 A exciting current.

The maximum achievable speed for the 1D-MSRS is also tested.When running in the lab with the existence of air drag, the sphere can run up to 200 Hz (12,000 rpm) synchronous speed. We believe this specification can be improved when the sphere is running in vacuum.

Figure 16 shows the measured and simulated starting angular speed of the 1D-MSRS plotting together. Good agreement between simulation and experimental data confirms the validity of the equivalent circuit model for analysis of the 1D-MSRS.

V. Conclusion and Future Work

The development of a magnetically suspended reaction sphere based on a single degree-of-freedom hysteresis drive (1D-MSRS) is presented in this paper. Magnetic suspension, bearingless drive and hysteresis motor principles are used in the design for 1D-MSRS. An equivalent circuit model for hysteresis motor is used to analyze the dynamic behavior of the 1D-MSRS. A starting torque of 8.15 mNm is achieved under 0.7 A of excitation current. The sphere can run up to 200 Hz (12,000 rpm) synchronous speed in lab.

Further research on speed hunting suppression, power efficiency optimization and motor performance tests in closed-loop speed control for the 1D-MSRS is ongoing in the Precision Motion Control Lab at MIT.

With these results in hand, the next steps to take in the future would be the design of a three dimensional reaction sphere. This study shows that hysteresis motor is a promising motor concept for space application, especially when it is assisted with magnetic bearings. It enables vibration-free operations and a very high rotational speed. The performance of the 1D-MSRS demonstrates that bearingless motor is a promising method to solve the field coupling between motor and bearing function for a 3D magnetic levitated motor. Through the tests of the 1D-MSRS also shows that with its advantages of high-speed, quiet operation and self-starting, hysteresis motor is a promising motor concept for reaction wheel applications.

Acknowledgment

The authors would like to thank MIT Lincoln Laboratory for supporting this project via the Advanced Concept Committee. We also thank National Instruments for equipment donations for the project.

References

[1] I. A. Budianto and J. R. Olds, "Design and deployment of a satellite constellation using collaborative optimization," *Journal of spacecraft and rockets*, vol. 41, no. 6, pp. 956 963, 2004.

[2] B. Bialke, "High fidelity mathematical modeling of reaction wheel performance," *Guidance and control 1998*, pp. 483–496, 1998.

[3] T. Marshall, T. Gunderman, and F. Mobley, "Reaction wheel control of the msx satellite," in *Guidance and Control 1991*, vol. 1, 1991, pp. 119–138.

[4] L. P. Davis, J. Wilson, R. Jewell, and J. Roden, "Hubble space telescope reaction wheel assembly vibration isolation system," *NASA Marshall Space Flight Center, Huntsville, Alabama*, vol. 9, 1986.

[5] T. Fukada, H. Hosokai, and N. Yajima, "Flexibility control of solar battery arrays: 3rd report vibration and attitude control with consideration of the dynamics of a reaction wheel as an actuator," *Bulletin of JSME*, vol. 29, no. 255, pp. 3121–3125, 1986.

[6] W. H. Isely, "Magnetically supported and torqued momentum reaction sphere," Sep. 16 1986, uS Patent 4,611,863.

[7] K. Kaneko, I. Yamada, and K. Itao, "A spherical dc servo motor with three degrees of freedom," *Journal of dynamic systems, measurement, and control*, vol. 111, pp. 398–402, 1989.

[8] G. J. Vachtsevanos and K. R. Davey, "Spherical motor particularly adapted for robotics," Apr. 19 1988, uS Patent 4,739,241.

[9] C. Xia, H. Li, and T. Shi, "3-d magnetic field and torque analysis of a novel halbach array permanent-magnet spherical motor," *Magnetics, IEEE Transactions on*, vol. 44, no. 8, pp. 2016–2020, 2008.

[10] K.-M. Lee, G. Vachtsevanos, and C. Kwan, "Development of a spherical stepper wrist motor," *Journal of Intelligent and Robotic Systems*, vol. 1, no. 3, pp. 225–242, 1988.

[11] K.-M. Lee, H. Son, J. Joni *et al.*, "Concept development and design of a spherical wheel motor (swm)," in *IEEE International Conference on Robotics and Automation*, vol. 4. IEEE; 1999, 2005, p. 3652.

[12] W. Wang, J. Wang, G. Jewell, and D. Howe, "Design and control of a novel spherical permanent magnet actuator with three degrees of freedom," *Mechatronics, IEEE/ASME Transactions on*, vol. 8, no. 4, pp. 457–468, 2003.

[13] E. Onillon, O. Chételat, L. Rossini, L. Lisowski, S. Droz, and J. Moerschell, "Reaction sphere for attitude control," in *Proc. 13th European Space Mechanisms and Tribology Symposium*, 2009.

[14] M. Imani Nejad, "Self-bearing motor design & control," Ph.D. dissertation, Massachusetts Institute of Technology, 2013.

[15] L. Zhou, "Magnatically suspended reaction sphere with one-axis hysteresis drive," Master's thesis, Massachusetts Institute of Technology, 2014.

[16] B. R. Teare, "Theory of hysteresis-motor torque," *Electrical Engineering*, vol. 59, no. 12, pp. 907–912, 1940.

[17] A. Chiba, T. Fukao, O. Ichikawa, M. Oshima, M. Takemoto, and D. G. Dorrell, *Magnetic bearings and bearingless drives*. Access Online via Elsevier, 2005.

[18] M. Copeland and G. Slemon, "An analysis of the hysteresis motor i-analysis of the idealized machine," *Power Apparatus and Systems, IEEE Transactions on*, vol. 82, no. 65, pp. 34–42, 1963.

[19] S. Miyairi and T. Kataoka, "A basic equivalent circuit of the hysteresis motor," *Elect. Engng. Japan (USA)*, vol. 85, pp. 41–50, 1965.

[20] M. A. Rahman, "Analytical models for polyphase hysteresis motor," *Power Apparatus and Systems, IEEE Transactions on*, no. 1, pp. 237–242, 1973.

[21] D. O'Kelly, "Theory and performance of solid-rotor induction and hysteresis machines," in *Proceedings of the Institution of Electrical Engineers*, vol. 123, no. 5. IET, 1976, pp. 421–428.

[22] J. Nitao, E. Scharlemann, and B. Kirkendall, "Equivalent circuit modeling of hysteresis motors," Lawrence Livermore National Laboratory (LLNL), Livermore, CA, Tech. Rep., 2009.

Magnetic Performance of Halbach Array Branching Mechanism Proto-model using Cylinder Shaped Permanent Magnets

Mikael J. Bragge[a], Atsushi Ito[b], Haruhiko Suzuki[b]

[a] Helsinki Metropolia University of Applied Sciences, Kalevankatu 43, 00180 Helsinki, Finland, Mikael.Bragge@metropolia.fi
[b] Fukushima National College of Technology, Nagao 30, Kami-arakawa, Taira, Iwaki, Fukushima 970-0834, Japan

***Abstract*—Basic Halbach arrays that are used to move objects, such as superconductors or pyrolytic graphite plates on them, lack the ability to switch the object from one array to another one in a simple manner. This requires a special branching point for the array.**

This paper provides a proposition of a model for such branching point. Built from cylinder shaped permanent magnets and causing them to rotate in the same direction, allows the object to move between different arrays. Simulations and actual measurements done on a purpose-built testing equipment provide the details on the performance of said model.

I. INTRODUCTION

Passive levitation is used in various fields for industry. Being an important part of many systems, Halbach arrays have gained a lot of use. These arrays require special turning points, as explored in [1] and [2], for moving objects from one array to another. This is one proposition of such turning point.

The idea is to create a Halbach array arrangement, normally done with rectangular magnets, by using cylinder shaped magnets with similar magnetization direction as the rectangular ones used normally. This allows the rotation of these cylinder shaped magnets along their main axis. By rotating all magnets into the same direction and by same amount, it is assumed to cause the wave-like flux pattern of the Halbach array to move along the array. One point of observation of this pattern is marked with dashed line in Figure 1 for clarity. This makes it possible to create a junction to move an object from one array to another, as the object pinned to the flux is assumed to move with the constantly changing field. Other assumed possibility is that the pinned object resists the movement of the flux, making rotating of the magnets nearly impossible. This is our main point of interest.

II. SIMULATION

A simulation was created with MagNet to verify the field lines movement as the magnets were rotated 10 degrees at a time for one full circle, 36 different steps. Figure 2 shows simulated fields for 0°, 45° and 90° of rotation of magnets. A previous work [3] using cylinder shaped permanent magnets

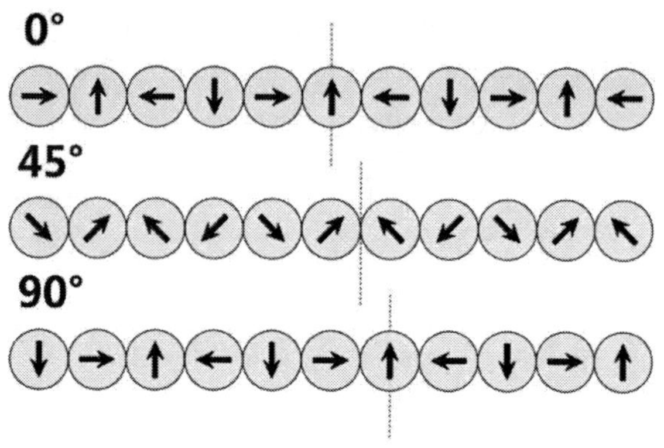

Figure 1. Magnet arrangement and 3 phases of rotation. Rotating all magnets clockwise.

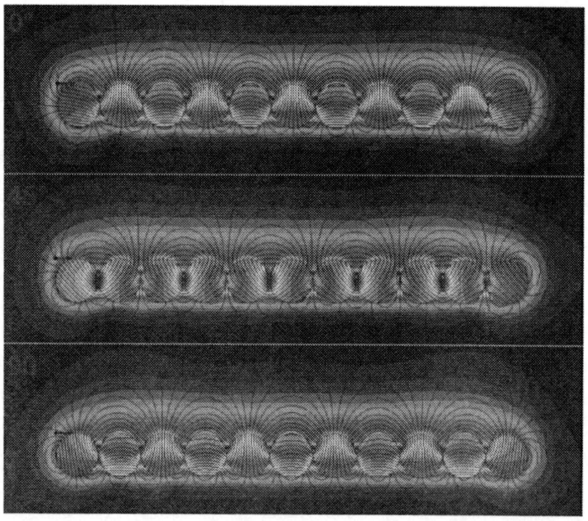

Figure 2. Simulations on 3 different phases of rotation, 0°, 45° and 90°

to build a Halbach array was found, and it was made clear that construction of an array like this was possible in terms of high enough flux density for some practical uses, thus decision of moving into measurements was made. The rotating direction of magnets in this previous work was every other magnet clockwise and every other counterclockwise and without a flux pinned object on the field. This means that it is still unknown if it will be possible to do rotation with an object pinned to the flux, as is desired in our case.

III. SETUP FOR MEASUREMENTS

A custom measuring stand and holders for magnets, as shown in Figures 3 and 4, were designed with 3D CAD software and manufactured. These were used to hold the magnets in place, so that measurement could be taken with the magnets aligned into Halbach array configuration and few intermediate stages of the rotation while the flux pattern was moved a length of the distance between magnets center points, corresponding to total of 90 degrees of rotation on the magnets. This covers the cases sown in Figure 1 and Figure 2. Measurements were initially taken with 0mm air-gap between magnets, but was adjusted later to 2mm for future measurements.

The magnets used are cylinder shaped neodymium alloy based, measuring 5mm in diameter and 30mm in length. The measuring stands are all made of aluminum alloy.

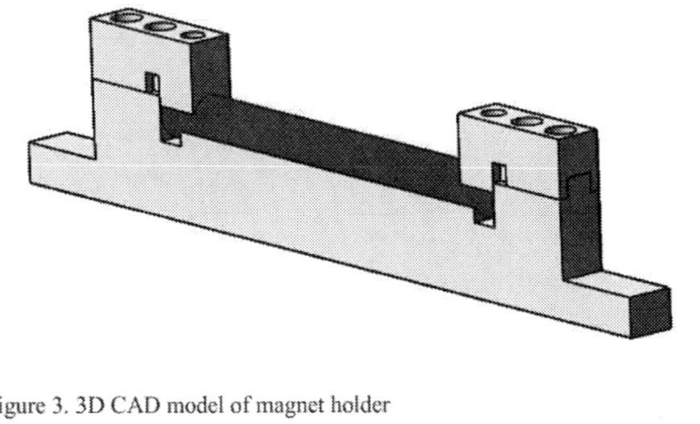

Figure 3. 3D CAD model of magnet holder

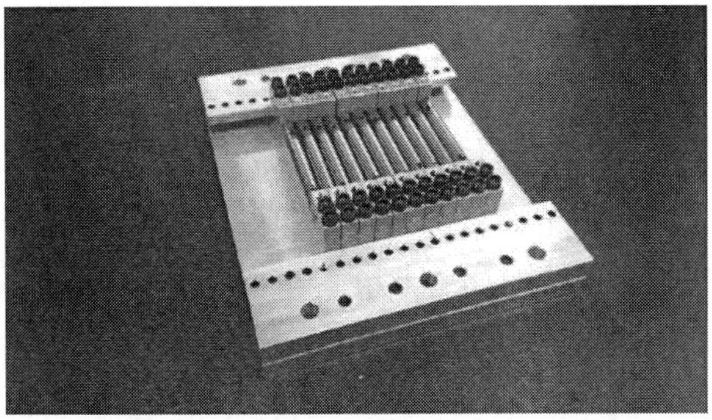

Figure 4. Actual measuring stand and 11 holders with magnets inserted

IV. MEASUREMENTS

Measurements for the top and bottom sides of the Halbach array have been taken for 0° of rotation, 1mm above the array with air-gap of 0mm. The measurements have been visualized in Figures 5 and 6. Additional measurements in Figures 7 and 8 are for 45° degrees of rotation.

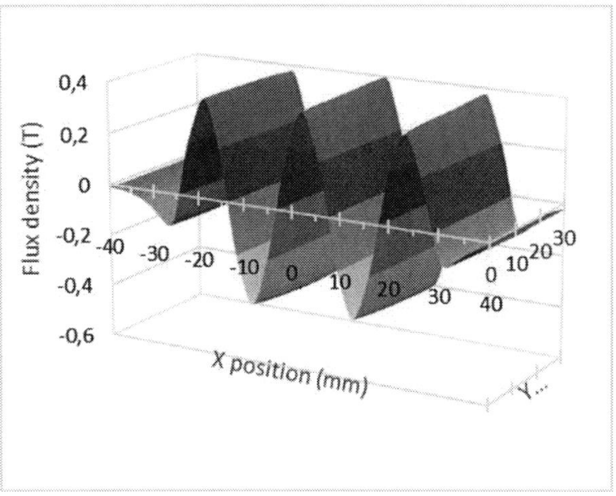

Figure 5. Cylinder Halbach array top side measurements 0° of rotation and 0mm air-gap, sensor head 1mm away

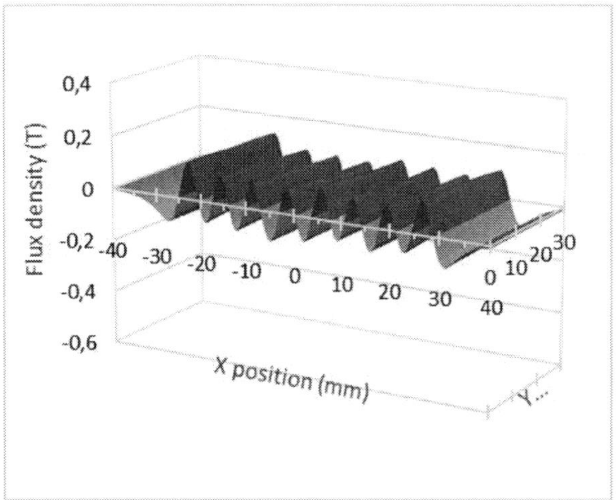

Figure 6. Cylinder Halbach array bottom side measurements 0° of rotation and 0mm air-gap, sensor head 1mm away

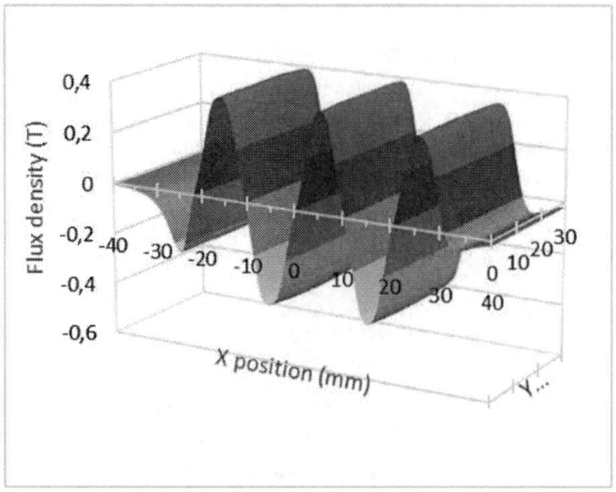

Figure 7. Cylinder Halbach array top side measurements 45° of rotation and 0mm air-gap, sensor head 1mm away

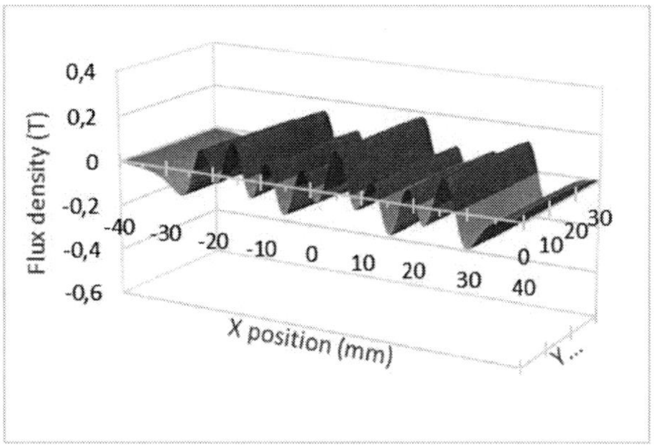

Figure 8. Cylinder Halbach array bottom side measurements 45° of rotation and 0mm air-gap, sensor head 1mm away

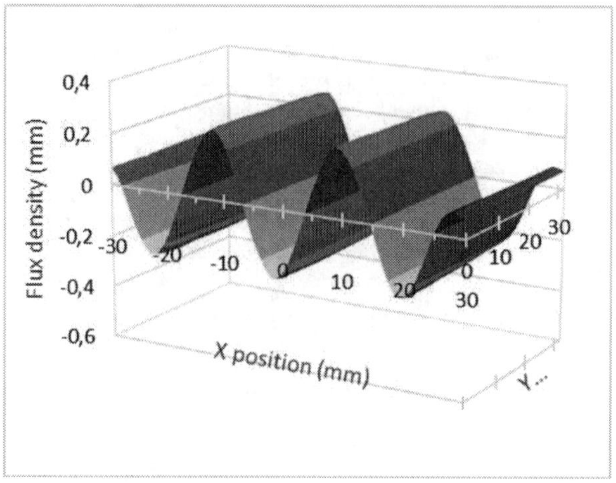

Figure 9. Cylinder Halbach array top side measurements 0° of rotation and 2mm air-gap, sensor head 1mm away

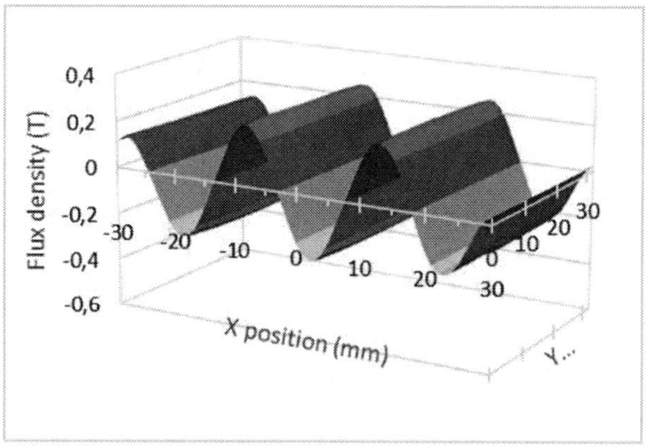

Figure 10. Cylinder Halbach array top side measurements 45° of rotation and 2mm air-gap, sensor head 1mm away

This presents a few problems with practical implementation. With 0mm of air-gap between magnets, rotating them or manufacturing other equipment to mount the magnets into rotatable configuration is rather complex. For this purpose, another measurements with the same stands and magnets were made, while increasing the air-gap into 2mm. These results can be found in Figures 9 and 10. These results give us confidence, that a simple rotating model can be built to verify this ideas plausibility.

V. PRACTICAL APPLICATION

For practical use of this system, an arrangement is suggested as in Figure 11. In this system, gears are used to hold the magnets in place and distributing applied torque while keeping the magnets in correct orientation for halbach array to be formed. This arrangement can be utilized as turning point in a linear track application as pictured in Figures 12 and 13.

Figure 11. Cylinder Halbach array, rotatable arrangement 3D CAD suggestion

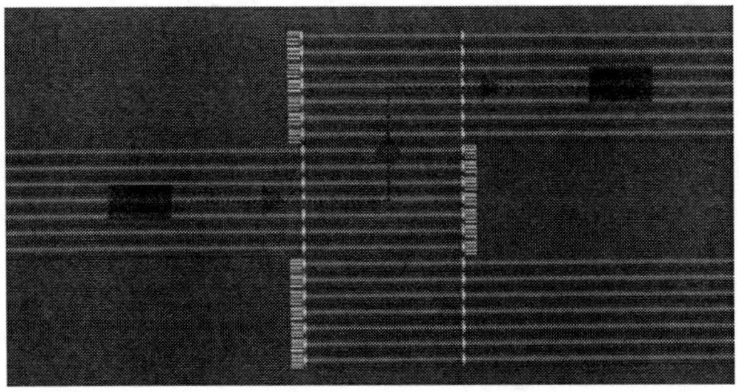

Figure 12. Cylinder Halbach array practical arrangement, moving object with zero velocity relative to tracks direction

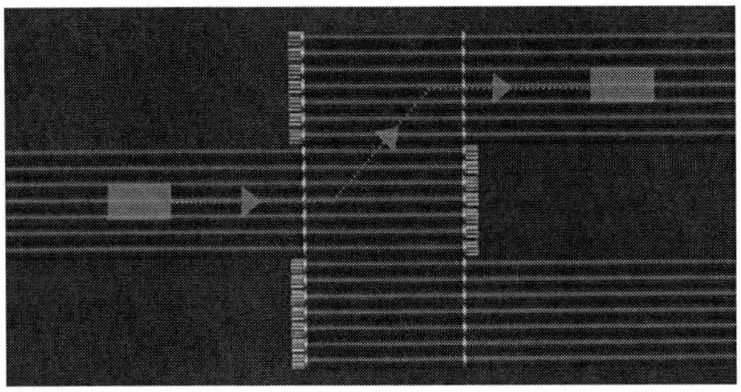

Figure 13. Cylinder Halbach array practical arrangement, moving object with non-zero velocity relative to tracks direction

In an ideal situation, where magnets have exactly the same flux pattern and gears holding the magnets in correct orientation have no backlash, the total magnetic force creating momentum for rotating magnets in the arrangement is assumed to equal zero. This can be easily seen in Figure 1, at 45° phase, where equal number of magnetic pole pairs with attraction and repulsion forces exist.

VI. RESULTS

The results confirm, that it is possible to build a Halbach array out of cylinder shaped magnets and pin an object to it. Building a rotating array and verifying that it is possible to move a pinned object on the array by rotating the magnets is left for a future project. For practical applications, where a large amount of non-ideal factors have to be accounted for, more simulations with applied forces from magnets in the arrangements and torque required to rotate the arrangement are to accompany this project.

REFERENCES

[1] H.Suzuki, K.Kusano, K.Yogiashi, M.Takashige, "A Model Vehicle with Magnetic Flux Pinning Effects of High-Tc Oxide Superconductors", The 4th International Symposium on Magnetic Bearings, ISMB-4, Aug. 1994, ETH Zurich, pp.423-428.

[2] C.Beyer, L.Kuhn, O.D.Haas, L.Schults, "A turnout switch for the superconductively levitated transport system SupraTrans", The 19th International Conference on Magnetically Levitated Systems and Linear Drives, MAGLEV2006, Sep. 2006, Dresden, pp.257-261.

[3] J.E. Hilton, S.M. McMurry, An adjustable linear Halbach array, Journal of Magnetism and Magnetic Materials 324 (2012) 2051–2056.

Controller Structure for Magnetic Bearing for Achieving both Unbalance Suppression and Stabilization

Yuichi Ariga[a], Yohei Saitoh[b], Keiji Watanabe[a]

[a] Faculty of Engineering, Yamagata University, 4-3-16 Jonan, Yonezawa, Japan, y_ariga@yz.yamagata-u.ac.jp
[b] Matsue College of Technology, 14-4 Nishiikuma-cho, Matsue, Japan

***Abstract*— This study proposes a control system in which the controller structure is divided in consideration of the unstable zeros of the controller. This controller can both inhibit a sine wave disturbance intended to produce unbalanced vibrations and stabilize the system with desired bearing stiffness. The effectiveness of the proposed method is verified through a simulation.**

INTRODUCTION

One of the features of active magnetic bearings is the ability to control unbalances. Already, many studies have focused on unbalance compensation methods [1]. Because "generalized notch filters" [2] can be used within rigid body critical speeds, these are very effective methods. However, they require the measurement of the closed-loop sensitivity function after designing the stabilized controller.

At the same time, many control methods have also been proposed to solve various control specifications, such as H_∞ control [3][4] and linear matrix inequality (LMI) [5]. However, because a determination of a weight function which satisfies simultaneously various control specification is difficult, more advanced techniques are required to design a good controller. To solve this problem, model bridge control (MBC), which is based on internal model control [6], is proposed [7][8]. MBC is a modification of the generalized stabilizer. It has a control structure in which adjustable models bridge over the gaps between the external signals and the desired outputs. MBC has been used to configure an unbalance control system for a magnetic bearing, and this was verified through a simulation [9]. The simulation results show that the proposed control method makes it possible to pass the rigid body critical speed safely, control housing vibrations, and avoid amplifier saturation. However, it is necessary to design a high supported stiffness controller for inhibiting a large overshoot of displacement for levitating. This is because the zeros of the unbalanced disturbance compensator closer to the origin cause a large overshoot.

This study proposes a controller structure that is divided in consideration of the unstable zeros of the controller. By this method, without changing the performance of the sine wave disturbance suppression, a response for levitating and good support desired rigidity are obtained.

CONTROLLED OBJECT

The configuration of a vertical-type rotor system supported by magnetic bearings is shown in Figure 1. The nomenclature used in this paper is shown in Table 1.

A controlled object is assumed to be a one-degree-of-freedom system under decentralized control. This system is unstable when a constant attractive force produced by a bias current is greater than the spring force of a shaft. The coil current is considered a control input and mass displacement, the output. The state equation is as given below.

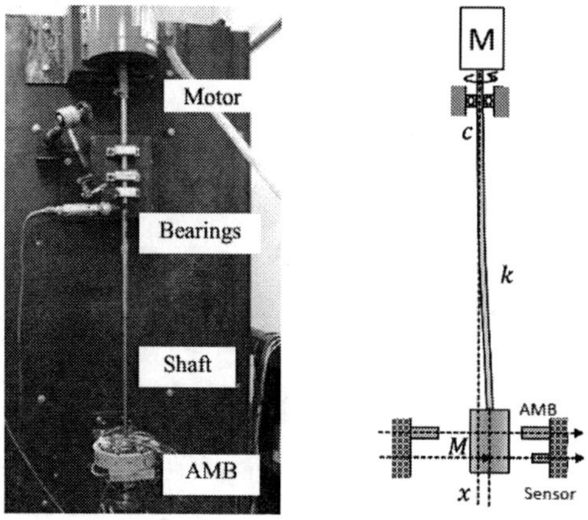

(a) Experimental equipment (b) Model

Figure 1. Controlled object

TABLE I. PARAMETER DEFINITION

	Parameter	Value
M	Equivalent mass	1.47 kg
F_0	Constant attractive force	3.5 N
X_0	Air gap	2.0 mm
I_0	Bias current	0.35 A
k	Spring constant	1.28 kN/m
d_c	Damping constant	0.016 μNs/m

$$\dot{X}_g = A_g X_g + B_g u$$
$$= \begin{bmatrix} 0 & 1 \\ \left(4\frac{F_0}{X_0} - k\right)\Big/M & d_c\big/M \end{bmatrix} \begin{bmatrix} x \\ \dot{x} \end{bmatrix} + \begin{bmatrix} 0 \\ 2\frac{F_0}{I_0}\frac{1}{M} \end{bmatrix} u, \quad (1)$$

$$y = C_g X_g = [1 \quad 0] X_g \quad (2)$$

The transfer function of a controlled system is expressed as follows.

$$G(s) = C_g(sI - A_g)B_g = \frac{k}{(s-\alpha)(s+\beta)} \quad (3)$$

where α is an unstable pole and β, a stable pole.

STABILIZING CONTROLLER, AND ITS STRUCTURE

A. Basic-type controller

A feedback control system using a controller $G_c(s)$ is formed as shown in Figure 2. $y(s)$ is the output, $r(s) = 1/s$ is a step reference input, and $d(s)$ is a sine wave disturbance expressed as follows.

$$d(s) = \omega/(s^2 + \omega^2) \quad (4)$$

A controller for stabilizing the system of Figure 2 is generally represented by the following internal model parameterization (IMP).

$$G_c(s) = \frac{P(s)}{1 - G_c(s)P(s)} \quad (5)$$

Here, $P(s)$ is a free parameter that is expressed as follows.

$$P(s) = \frac{(s-\alpha)(s+\beta)}{k(1+\tau s)^2} \frac{a + bs + cs + ds^2}{(1+\tau s)^3} \quad (6)$$

τ is the time constant of the closed loop system, and the response of this system is determined by its value. If its value is small, quick levitation response and high bearing stiffness can be obtained. Contrarily, if a larger value is set, slow levitation response and low bearing stiffness are obtained. The coefficients a, b, c and d are calculated by a simultaneous equation based on the following conditions of an internal stabilization (See APPENDIX).

$$1 - G(s)P(s)\big|_{s=\alpha} \quad (7)$$

$$1 - G(s)P(s)\big|_{s=0} \quad (8)$$

$$1 - G(s)P(s)\big|_{s=\omega j} \quad (9)$$

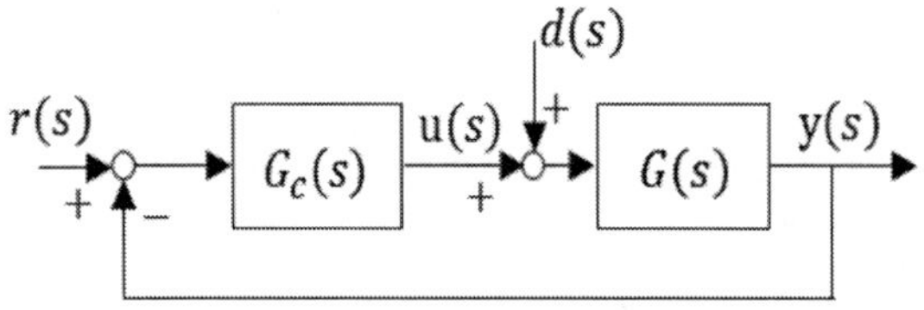

Figure 2. Feedback control system

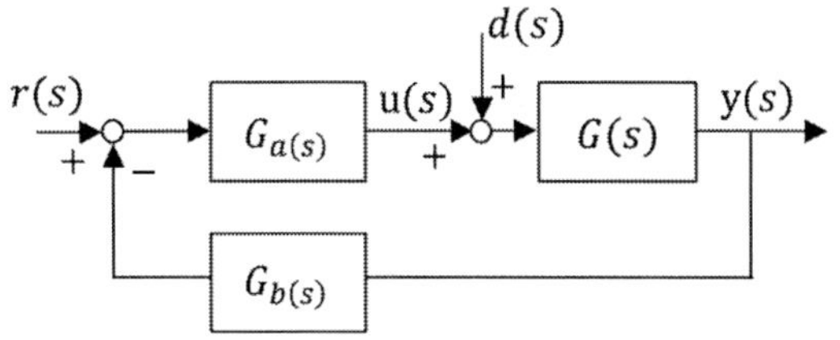

Figure 3. Separated-type feedback control system

$P(s)$ as calculated by the above method has zeros owing to a sine wave disturbance. If these zeros are close to the origin, they worsen a step response. Therefore, it had to be set to a small value of τ for inhibiting a large overshoot of displacement levitating. This constraint causes that cannot be set only high bearing stiffness.

B. Separated-type controller

The structure of the controller $G_c(s)$ obtained in the previous section can be changed to a distributed structure, as shown in Figure 3. $P(s)$ for this structure is defined as follows:

$$P(s) = P_a(s)P_b(s)$$
$$P_a(s) = \frac{(s-\alpha)(s+\beta)}{k(1+\tau s)^2} \quad (10)$$
$$P_b(s) = \frac{a + bs + cs + ds^2}{(1+\tau s)^3}$$

$P_b(s)$ contains unstable zeros of $G_c(s)$. Based on equation (10), the controller $G_c(s)$ is divided into two elements.

$$G_c(s) = G_a(s)G_b(s)$$
$$G_a(s) = \frac{P_a(s)}{1 - G_c(s)P(s)}, \quad G_b(s) = P_b(s) \quad (11)$$

Thus, the step response can be improved without changing the disturbance suppression performance.

SIMULATION RESULTS

The controllers are designed to suppress a sine wave disturbance of 5Hz, and their control performance is verified through a simulation.

Basic-type controllers were designed for two conditions—a low bearing stiffness of $\tau = 0.05$ and a high bearing stiffness of $\tau = 0.02$. Figure 4 shows a comparison of the results for the levitation response. A large overshoot occurs if τ is large. This causes sustained vibrations between the touchdown bearing in the experiment.

A separated-type controller is configured to separate the basic-type controller designed based on Equations 10 and 11. Figure 5 shows a comparison of the results for the levitation response. Regardless of the value of τ, a separated-type controller shows smooth levitation without overshoot.

Figure 6 shows the verification results of disturbance suppression performance of each type of controller. The controllers are designed with $\tau = 0.05$. A sine wave disturbance is input from 4.0 [s] at the place that has levitated stably. Both controllers suppress the disturbance with identical performance.

CONCLUSION

In this study, the effectiveness of the controller structure divided in consideration of the unstable zeros of the controller was shown. The future task is to experimentally verify the effectiveness of the proposed method in a two-degree-of-freedom system.

REFERENCES

[1] G. Schweiter and E. H. Maslen. Magnetic Bearings. Springer, 2009.

[2] H. Herzog, P. Bühler, C. Gähler, and R. Larsonneur. Unbalance compensation using generalized notch filters in the multivariable feedback of magnetic bearings. IEEE Trans. On Control Systems Technology, vol.4 no.5 pp.580-586, 1996.

[3] J. Doyle, K. Glover, P. P. Khargonekar and B. A. Francis. State space solution to standard and control problems. IEEE Trans. On Autom. Contr., vol.34, no.8, pp.831-847, 1989.

[4] K. Zhou, J. Doyle and L. Glover. Robust and Optimal Control. Prentice Hall, 1996.

[5] S. Boyd and C. H. Barratt. Linear Controller Design: Limits of Performance. Prentice Hall, 1991.

[6] M. Morari and E. Zafiriou, Robust Process Control. Prentice Hall, 1989.

[7] R. Wang, K. Watanabe, E. Muramatsu, Y. Ariga and S. Endo. Robust control via internal model principle. In 17th International Symposium on MTNS, pp.2587-2589, 2006.

[8] R. Wang and K. Watanabe, Three-degree of freedom internal model control for unstable process. In SICE Annual Conf. pp.1631-1634, 2007.

[9] Y. Ariga and K. Watanabe,Unbalance control of magnetic bearings by multi-degree of freedom internal model control, ISMB13 paper,-46, 2012.

APPENDIX: CONTROLLER DESIGN METHOD

From Equations 5 and 6

$$1 - G(s)P(s) = \frac{(1-a)+(5\tau-b)s+(10\tau^2-c)s^2+(10\tau^3-c)s^3+5\tau^4 s^4+\tau^5 s^5}{(1+\tau s)^5} \quad \text{(A1)}$$

Than a condition of Equation 8

$$1 - a = 0 \quad \therefore a = 1 \quad \text{(A2)}$$

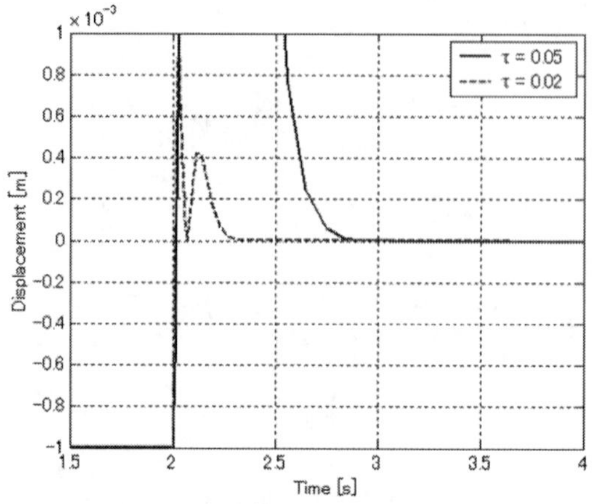

Figure 6. Levitation responses (basic-type controller)

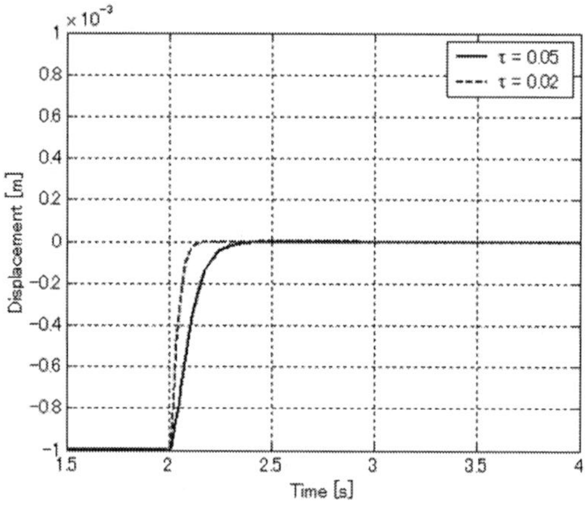

Figure 4. Levitation responses (separated-type contrller)

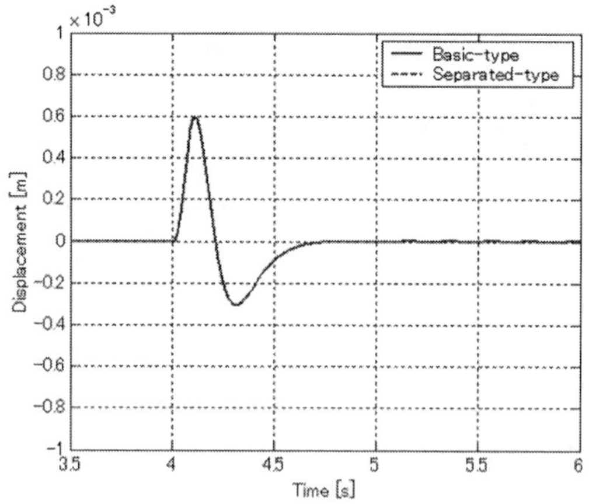

Figure 5. Responses of sine wave disturbance ($\tau = 0.05$)

From Equation 7, which is a condition for the unstable pole α

$$1-G(s)P(s)\Big|_{s=\alpha}$$
$$=\frac{(5\tau-b)\alpha+(10\tau^2-c)\alpha^2+(10\tau^3-c)\alpha^3+5\tau^4\alpha^4+\tau^5\alpha^5}{(1+\tau\alpha)^5} \tag{A3}$$

For the above equation becomes zero

$$(5\tau-b)\alpha+(10\tau^2-c)\alpha^2+(10\tau^3-d)\alpha^3+5\tau^4\alpha^4+\tau^5\alpha^5=0$$

Thus

$$b+c\alpha+d\alpha^2=5\tau+10\tau^2\alpha+10\tau^3\alpha^2+5\tau^4\alpha^3+\tau^5\alpha^4 \tag{A4}$$

Similarly, for an angular frequency of the agitation,

$$1-G(s)P(s)\Big|_{s=\omega j}$$
$$=\frac{(5\tau-b)\omega j+(10\tau^2-c)\omega^2+(10\tau^3-c)\omega^3 j+5\tau^4\omega^4+\tau^5\omega^5 j}{(1+\tau\omega j)^5} \tag{A5}$$

From the condition that Equation A5 becomes zero, the following equations are provided.

$$-(10\tau^2-c)\omega^2+5\tau^4\omega^4=0 \tag{A6}$$

$$(5\tau-b)\omega-(10\tau^3-d)\omega^3+\tau^5\omega^5=0 \tag{A7}$$

From Equations A6

$$c=10\tau^2-5\tau^4\omega^2 \tag{A8}$$

The following equation is obtained and rearranging the equation A7.

$$b-\omega^2 d=5\tau-10\tau^3\omega^2+\tau^5\omega^4 \tag{A9}$$

Substituting Equation A8 in Equation A4

$$b+\alpha^2 d=5\tau+10\tau^3\alpha^2+5\tau^4\alpha(\omega^2+\alpha^2)+\tau^5\alpha^4 \tag{A10}$$

Subtracting Equation A9 from Equation A10.

$$(\omega^2+\alpha^2)d=10\tau^3(\omega^2+\alpha^2)+5\tau^4\alpha(\omega^2+\alpha^2)-\tau^5(\omega^2+\alpha^2)(\omega^2-\alpha^2)$$

$$\therefore d=10\tau^3+5\tau^4\alpha-\tau^5(\omega^2-\alpha^2) \tag{A11}$$

Substituting equation A11 in equation A9

$$b-\omega^2\left\{10\tau^3+5\tau^4\alpha-\tau^5(\omega^2-\alpha^2)\right\}=5\tau-10\tau^3\omega^2+\tau^5\omega^4$$

$$\therefore b=5\tau+5\tau^4\alpha\omega^2+\tau^5\alpha^2\omega^2 \tag{A12}$$

In the above, all coefficients of the free parameter are determined. The controller G_c is obtained as the following form in the end.

$$G_c(s)=\frac{P(s)}{1-G_c(s)P(s)}$$
$$=\frac{(s-\alpha)(s+\beta)(a+bs+cs^2+ds^3)}{k\left\{(5\tau-b)s+(10\tau^2-c)s^2+(10\tau^3-d)s^3+5\tau^4s^4+\tau^5s^5\right\}} \tag{A13}$$

The denominator becomes the following equation.

$$k\left\{(5\tau-b)s+(10\tau^2-c)s^2+(10\tau^3-d)s^3+5\tau^4s^4+\tau^5s^5\right\}$$
$$=k\left[-\tau^4\omega^2\alpha(5+\tau\alpha)s+5\tau^4\omega^2s^2-5\tau^4\alpha s^3-\tau^5(\alpha^2-\omega^2)s^3+5\tau^4s^4+\tau^5s^5\right]$$
$$=k\tau^4s\left[-\omega^2\alpha(5+\tau\alpha)+5\omega^2s-5\alpha s^2-\tau(\alpha^2-\omega^2)s^2+5s^3+\tau s^4\right]$$
$$=k\tau^4s\left[5\omega^2(s-\alpha)+5s^2(s-\alpha)+\tau\left\{s^4+(\omega^2-\alpha^2)s^2-\omega^2\alpha^2\right\}\right]$$
$$=k\tau^4s\left[5\omega^2(s-\alpha)+5s^2(s-\alpha)+\tau\left\{(s^2-\alpha^2)(s^2+\omega^2)\right\}\right]$$
$$=k\tau^4s(s-\alpha)\left[5(s^2+\omega^2)+\tau(s+\alpha)(s^2+\omega^2)\right] \tag{A14}$$

Rearranging Equation A13 using Equation A14, the following controller G_c is obtained.

$$G_c(s)=\frac{(s-\alpha)(s+\beta)(a+bs+cs^2+ds^3)}{k\tau^4s(s^2+\omega^2)\left\{\tau(s+\alpha)+5\right\}} \tag{A15}$$

Influence on Unbalanced Magnetic Pull

Henning Kasten [a], Christian Redemann [b]

[a] Levitec GmbH, 35633 Lahnau, Germany, henning.kasten@levitec.de
[b] Levitec GmbH, 35633 Lahnau, Germany, christian.redemann@levitec.de

Abstract— **This article describes the effects of the unbalanced magnetic pull and the opportunities to suppress asymmetries of the air gap field.**

I. Introduction

Today, high speed drives are often realized with magnetic bearings. One of the main challenges of developing and building-up such machines is the rotor dynamics. Unbalanced magnetic pull, UMP, and eccentric air gap fields occur when the rotor is not perfectly symmetrically built or centered in the stator [1]. The disadvantages of this phenomenon are:

- Reducing of the rotor eigen frequency
- Additional noise and vibrations
- Higher currents in the coils of the magnetic bearings to compensate the additional forces.

It is possible to use specific windings, which suppress eccentric fields and undesired parasitic forces. Such windings are not new [2, 3]. Many large machines have been constructed with these windings. The strands of such windings are divided in many parallel connected winding parts. Unsymmetrically fields cause balancing currents, which reduce the eccentric fields but also increase the additional winding losses. This paper shows ways to reduce the UMP with specific windings and shows FEM aided forces and loss calculations. With the results the characteristics of these specific windings with parallel connected paths can be evaluated.

II. Types of Air Gap Eccentricities

Figure 1 above shows an ideal cylindrically symmetrical rotor, which is not perfectly centered. This circumstance will create a static eccentricity. The most minimal air gap length is at a position that will not change when the machine works. At this location the smallest magnetic reluctance and the highest values of the magnetic induction appear. Therefore, the deformed air gap field causes a constant force in the direction to the smallest air gap length. Production tolerances cause the static eccentricity. Furthermore, some applications have eccentric rotors to reach better grinding surfaces [4]. Unsymmetrical end windings and anisotropic stacks cause also static eccentricities but their magnitudes are very small. Figure 1 bottom shows another case. It is called dynamic eccentricity. The axis of the shaft is exactly placed in the stator hole. However, the rotor surface turns eccentrically to the shaft. As a result the minimal air gap length will turn around. Such rough rotors are not manufactured. But magnet material variations also cause dynamic eccentricity. To reach small eccentricity fields the rotor must be very exactly mounted. That causes high effort. In some applications this cannot be reached. Furthermore the rotor magnets must not have tolerances. In general, these tolerances produce the highest amount of eccentric fields.

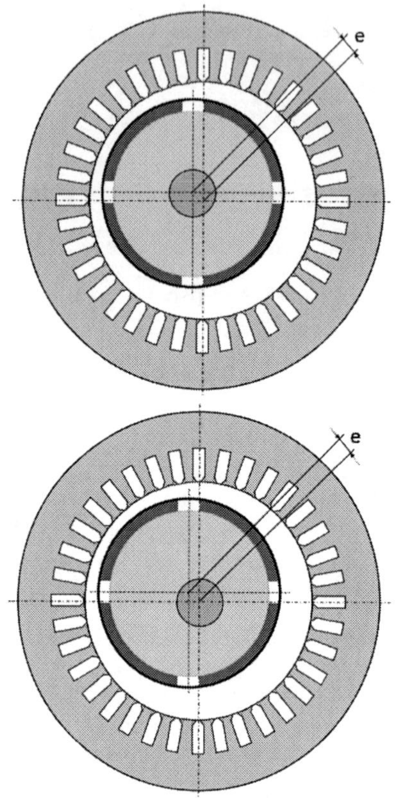

Figure 1. Static eccentricity (top) and dynamic eccentricity (bottom)

III. Rotor Measurement

Figure 2 shows magnetic fields of electric machines with very different magnet heights. In spite of the high difference, the 2-pole machine has a symmetrical air gap field. The nature of the magnetic flux $\left(\operatorname{div}\vec{B} = 0\right)$ explains this phenomenon. The fluxes of both poles must be equal. Therefore, 2-pole machines are insensitive to magnet asymmetries. The shown 4-pole machine has more degrees of freedom. The smallest rotor pole has the smallest flux and the highest rotor magnet causes the highest flux. This effect will also appear in motors with higher number of poles. To reach symmetrical fluxes it is necessary to use magnets with very small tolerances or to measure and adjust the finished rotor. Figure 3 shows an instrument for rotors with surface mounted permanent magnets. If the rotor turns, the field can be measured along the circumference in 6 positions.

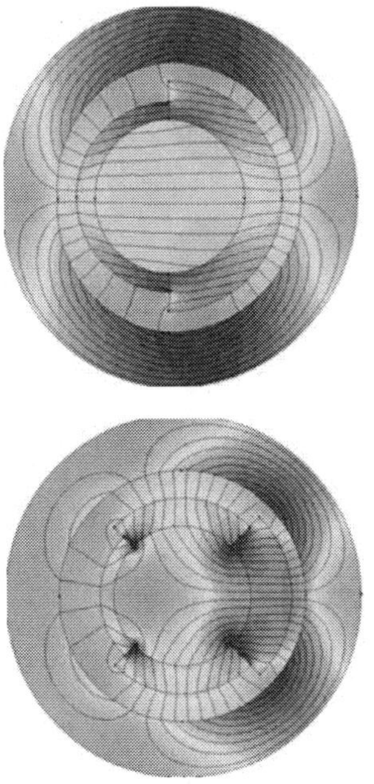

Figure 2. Magnetic fields of a 2-pole machine (top) and a 4-pole machine (bottom)

If the rotor length is greater than 50 mm, the rotor magnets will be divided in many segments to reduce the eddy current losses. Such rotors are built up with 50 or more magnet segments. However, to measure and select high numbers of magnets needs extra time and therefore is more expensive.

IV. HARMONICS OF THE ECCENTRIC FIELDS

Every real air gap field can be described by the theory of harmonics. Every single harmonic with an integer order number cannot create eccentric fields or resultant forces between stator and rotor. Eccentric field will be created by many harmonics with different order numbers. Figure 4 shows an eccentric field that is caused by two harmonics. The difference of both order numbers is only one. This small difference is typical for all eccentric fields. The most significant harmonics in machines with p pole pairs have the order numbers:

$$\nu = p \pm 1 . \qquad (1)$$

They are also called eccentric harmonics. Both harmonics have the same magnitudes.

An exception are machins with only 2 poles ($p = 1$). In this case (1) is not to be used. The eccentric field of these rotors is created by harmonics with the order numbers $\nu = p = 1$ and $\nu = p + 1 = 2$.

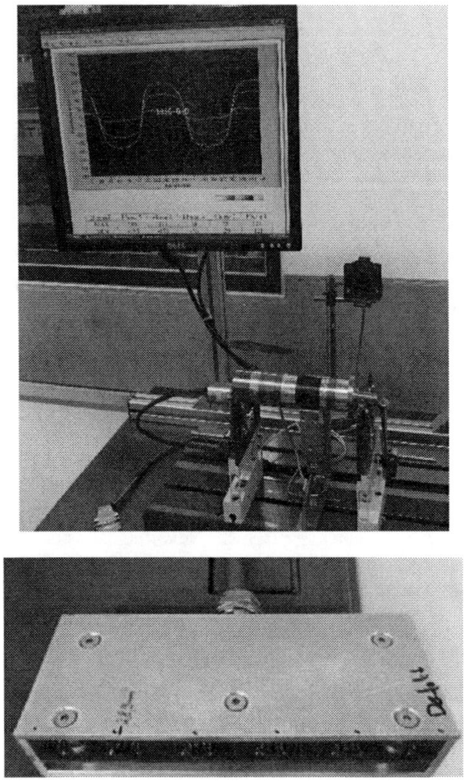

Figure 3. Automatic measuring machine for rotor magnets and an additional sensor for larger machines (bottom)

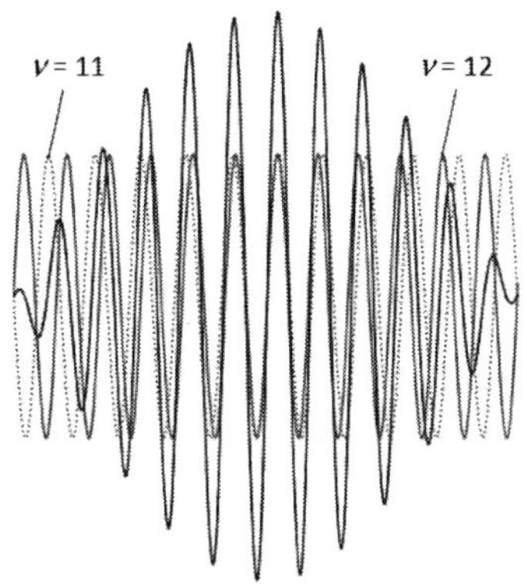

Figure 4. Superposition of two harmonics

V. WINDINGS WITH PARALLEL PATHS

Windings with parallel paths can help to reduce the asymmetries of the air gap field. On the other hand, additional circulating currents cause higher winding losses.

A. Parallel Paths with diametric arranged coils

At first only one phase of a 4-pole winding is analyzed. The number of slots per pole and phase is only one (Figure 5). The voltage u of one a current-carrying coil amounts:

$$u_c = Ri + w_c \frac{\mathrm{d}\Phi_c}{dt}. \qquad (2)$$

In electric machines, especially in those, which work with high speeds, the ohmic voltage drop can be neglected. For that reason, same magnetic fluxes Φ_c must pass in both parallel connected coils. The symmetry of both fluxes is achieved by an additional circulating current. This current increases the flux in one of both coils and decreases the flux in the other coil. Nevertheless, because of the stray flux Φ_σ the similar coil fluxes are different to the fluxes in the air gap Φ_δ. These three fluxes must fulfill the following equation:

$$\Phi_c = \Phi_\delta + \Phi_\sigma. \qquad (3)$$

Accordingly small stray fluxes are essential to reach an air gap field with small eccentricity. Instead of the in (3) used fluxes it is also possible to handle this equation with inductances. With small stray inductivities a very symmetrical air gap field can be reached. Nevertheless, because of the additional circulating currents higher winding currents and losses will also occur.

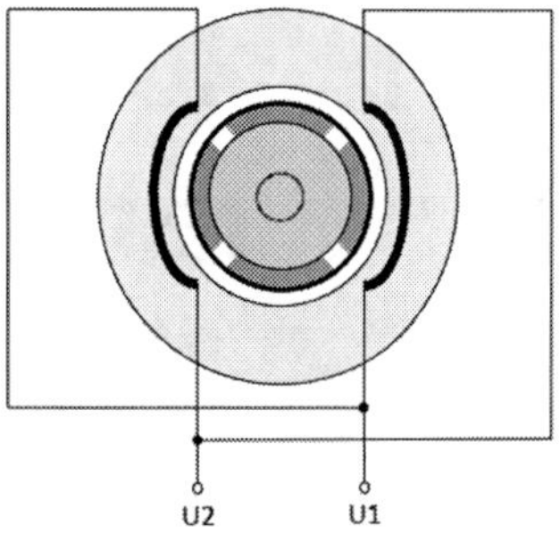

Figure 5. A single phase of a 4-pole Machine with parallel paths

A good suppression of eccentricity fields will be achieved by connecting all diametric placed coils parallel. This claim can be fulfilled with all two-shift windings and one-shift windings that have an even number of pole pairs.

B. One-shift windings with an uneven number of pole pairs

Figure 6 gives an example for this case. It can be seen an illustration of one phase of a 6-pole one-shift winding. In this case the suppression of the eccentricity is more obscured. The explanation will be given with the eccentric harmonics. The harmonics with the 2nd and 4th order numbers cause mainly the eccentric field. The three parallel connected coils induce voltages that are $\nu\,120°$ phase shifted to one another. The main field ($\nu = 3$) causes voltages that are in phase, wherefore the circulating current does not occur. But the eccentric harmonics generate phase shifted voltages by 120° degrees. The voltage that drives circulating currents averages:

$$U_2 = \mathrm{U}_{c,2}(1 - \mathrm{e}^{\mathrm{j}240°}) = 1.732 U_{c,2}$$
$$U_4 = \mathrm{U}_{c,4}(1 - \mathrm{e}^{\mathrm{j}120°}) = 1.732 U_{c,4}\,. \qquad (4)$$

It is obvious, that this harmonics cause a high current in the shorted circuit. This current suppresses the eccentric harmonics and thereby the asymmetries in the air gap field.

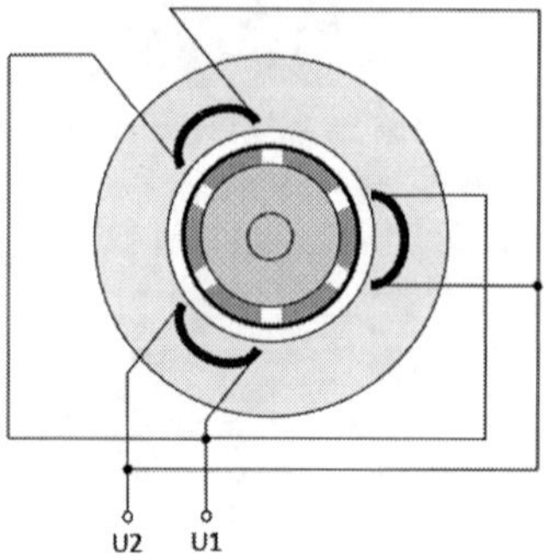

Figure 6. A single phase of a 6-pole one-shift winding

C. Windings with a higher number of slots per pole and phase

Figure 7 left shows two paths with many in series connected coils in a group. This connection causes only symmetrical fluxes of both groups. The coil fluxes do not have same values. Therefore the suppression of the eccentric fields is weaker than when only on coil per group is used. The suppression can be strengthened by using additional junctions, which are shown in figure 7 mid and right. Because only the additional circulating currents will pass through the junctions, the cross sections may be dimensioned smaller.

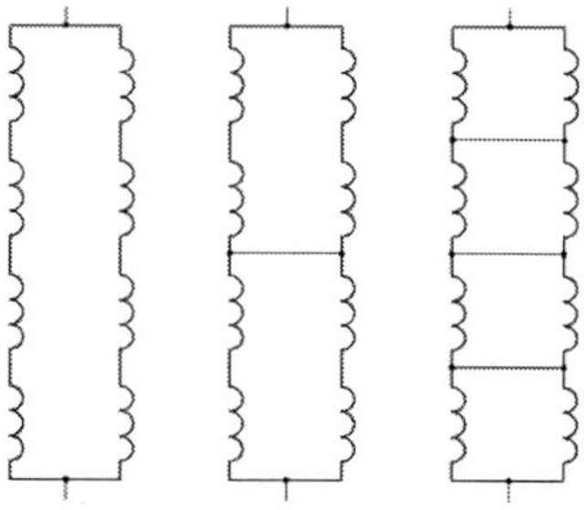
Figure 7. Phase with two parallel paths: without any additional junctions and additional junctions

VI. Favorable Number of Slots in Cage Rotors

With asynchronous machines it is possible to reach peripheral speeds greater than 200 m/s [3, 5]. The used rotors are very symmetrical. However, the rotor losses and the high harmonic content of such rotors are detrimental. A rotor with an even number of slots produces harmonics with uneven order numbers. Three phase windings show the same behavior. A superposition with two waves shown in figure 4 is not possible. Therefore rotors with an even number of slots cannot generate eccentric fields by the harmonic content. Rotors with an uneven number of rotor slots create harmonics with even order numbers. For that reason, such rotors can produce eccentric fields. But the advantage of this slot numbers is avoiding oscillating torques. By using uneven slot numbers, the rules in (5) should be adhered to.

$$N_2 \neq N_1 \pm 1 \cap N_2 \neq N_1 \pm (2p \pm 1). \qquad (5)$$

N_1 stands for the number of stator slots and N_2 for the number of rotor slots.

VII. Calculation of Forces and Currents

The force calculation is important to know the bearing forces and to estimate the decrease of the rotor eigen frequency. Furthermore, suppressing of eccentric fields of particular windings can be determined. A description of an analytical way is too complex for this paper. In brief, an FEM-aided calculation should be given. At first, the machine will be examined without any winding currents. Abnormalities can be observed by changing the material properties and displacing the rotor position. Figure 8 shows a 2D-simulation with an eccentric placed rotor. The most of FEM software programs are able to determine the force between stator and rotor. [4] describes the effects of such forces and the influence of the critical speed. To determine the magnitudes of the eccentric harmonics, one rotor pole should be placed symmetrical to the smallest air gap length. The radial components of the air gap field at the smallest air gap length and in the opposite can be ascertained with the functions of the FEM program. Both calculated flux densities are local maximums of the fundamental wave. The difference is caused by the eccentricity. Therefore, the determined inductions are named with $\hat{B}_{\min}$ and $\hat{B}_{\max}$. The magnitudes of the most important eccentric harmonics can be calculated:

$$\hat{B}_{p\pm 1} = \frac{\hat{B}_{max} - \hat{B}_{min}}{4}. \qquad (6)$$

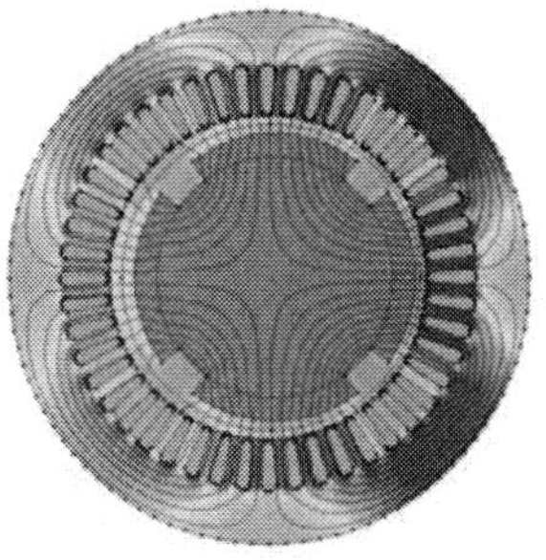

Figure 8. Magnetic field calculation with static eccentricity

In the next step the effect of windings with parallel paths can be calculated iteratively. Figure 9 shows the FEM calculation with the additional current. In this position and moment the circulating currents of the other phases are zero. Figure 10 shows the difference of the flux linkages of both groups and the resultant force in addition to the circulating current. The force is 8.2 kN/m without the circulating current. By using parallel phases the flux difference will be almost zero and the force is reduced by 4.4 kN/m.

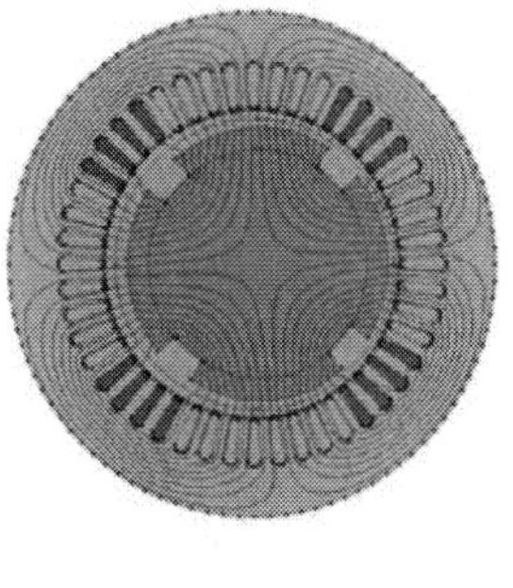

Figure 9. Field of the rotor and of the circulating current

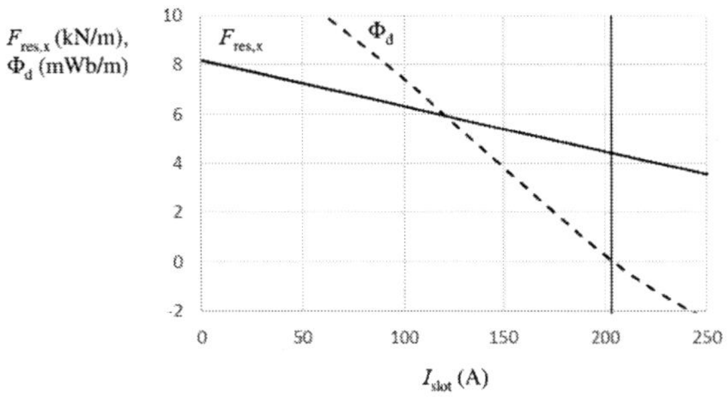

Figure 10. Resultant forces and flux difference as a function of circulating current

VIII. Conclusions

The paper has shown the disadvantages of unsymmetrical manufactured high speed machines with magnetic bearings. The eccentric air gap field was described with two harmonics, which can be suppressed with specific shown windings. Furthermore, ways have been shown to calculate with 2D-FEM forces and currents that are caused by the eccentricity.

References

[1] M. Wallin, "Measurement and modelling of unbalanced magnetic pull in hydropower generators", Uppsala University, 2013

[2] K.-H, Ketteler, "Über die Wirkung von parallelen Ankerzweigen auf das Oberfeldverhalten von Käfigläufern", Hannover University, 1983

[3] G. Müller, K. Vogt, B. Ponick, "Berechnung elektrischer Maschinen", Weinheim: Wiley-VCH, 2008

[4] SKF, "Von der Produktneuheit zur bewährten Lösung", EVOLUTION, 2003

[5] Huppunen, J, "High-speed solid-rotor induction machine – electromagnetic calculation and design" PhD thesis, Lappeenranta University of Technology, Finland, 2004

Comparison of a Model-based and a Model-free Unbalance Control Methods in Magnetic Bearing System

Ying He[a,b], Zhe Sun[a,b], Jingjing Zhao[a,b], Zhengang Shi[a,b], Lei Shi[a,b]

[a] Institute of Nuclear and New Energy Technology, Tsinghua University, BeiJing, 100084, PR China, heying11@mails.tsinghua.edu.cn

[b] The key laboratory of advanced reactor engineering and safety, Ministry of Education, Beijing, 100084, PR China

***Abstract*—Synchronous vibration caused by rotor imbalance is an obvious vibration problem in rotating machinery. Since active magnetic bearing has been proved to be an effective approach to solve this problem, many researchers have proposed various methods to reduce synchronous vibration and those methods can be classified into two categories: model-based method and model-free method. In the paper, we use a model-free method, iterative learning control (ILC) and a model-based adaptive forced balancing methods (MBAFB) to attenuate synchronous vibration respectively and then compare their performances in several aspects. The results of this paper can provide references for the users to choose an appropriate unbalance control algorithm.**

I. Introduction

Compared with traditional bearings, magnetic bearings possess several advantages, such as, contact free, no lubrication, and the unique ability of active control. Those advantages promote the developments and applications of magnetic bearings. At present, they are widely used in vacuum and clean room technology, turbo machinery and centrifuges [1].

Rotor imbalance results from materials or manufacture errors is inevitable. For a rotating system, the imbalance may generate a centrifugal force which would cause undesirable noise, rotor run-out and housing vibration. More specially, the centrifugal force is proportional to the square of rotating speed. Therefore, when the rotor operates at high speed, the synchronous vibration caused by rotor imbalance is obvious and even has an effect on the stability of the system. Many mechanical measures have been presented and applied to balancing the rotor. However, those measures are always costly and time consuming. Take dynamic balancing for example, rotor imbalance is reduced by adding or removing a small mass from the rotor. During the balancing process, the system need to start first then stop for several times and the rotor need to be re-balanced when the operation speed changes.

Since the active magnetic bearing has been proved to be an effective approach to solve this problem, many researchers have proposed several effective control methods to reduce synchronous vibration and those methods can be classified into two categories: model-based method and model-free method [2], [3], [4], [5]. In the paper, first of all, we build the model of the whole system. It is worth mentioning that a flexible rotor model is built using lumped mass method which promote the accuracy of objective system model. Then, synchronous vibration control of the system is achieved using a model-free method and a model-based method respectively. Finally, we compare the performances of those unbalance control methods.

II. Model of the Objective System

An AMB system is proposed of several components: rotor, magnetic bearings, sensors, controller, amplifier. In this section, we build the model of each components and the details are described below

A. Model of the Flexible Rotor

It is well-known that motion of the rotor can be described by following equation:

$$M\ddot{X} + \Omega G\dot{X} + KX = F \tag{1}$$

Where M, G, K are respectively mass matrix, gyroscopic matrix, stiffness matrix. Here, X is the rotor displacements vector, Ω is rotating speed and F is the magnetic bearing forces vector. If we choose $Z = \begin{bmatrix} X^T & \dot{X}^T \end{bmatrix}^T$ as state variables, then the rotor model can be rewritten in the state space form:

$$\begin{aligned} \dot{Z} &= \begin{bmatrix} 0 & I \\ -M^{-1}K & -\Omega M^{-1}G \end{bmatrix} Z + \begin{bmatrix} 0 \\ I \end{bmatrix} F \\ Y &= \begin{bmatrix} 0 & I \end{bmatrix} Z \end{aligned} \tag{2}$$

Where, Y is the output vector and I is identity matrix.

Since, magnetic bearings are widely applied in high speed rotating machinery. The rotor, in generally, operates beyond the rigid modes and several flexible modes. In this case, the simple rigid model of the rotor is ill-suited and a flexible rotor model is required. In the paper, parameters of rotor model is obtained and rotor model is established using lumped mass method. The basic principle of this method is to simplify rotor as several discs with mass and the moment of inertia lumped and massless elastic rods and the details about this method can be seen in literatures [6], [7]. To obtain a flexible rotor model, several flexible modes are maintained during the model reduction process.

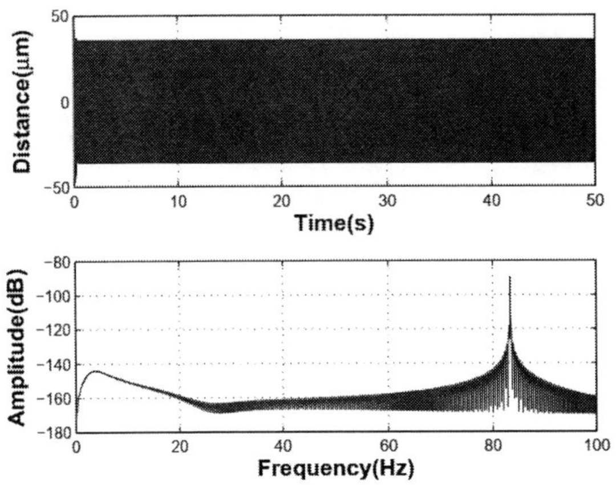

Figure 1. Rotor runout curve at axis X1 without unbalance control at $5000rpm$.

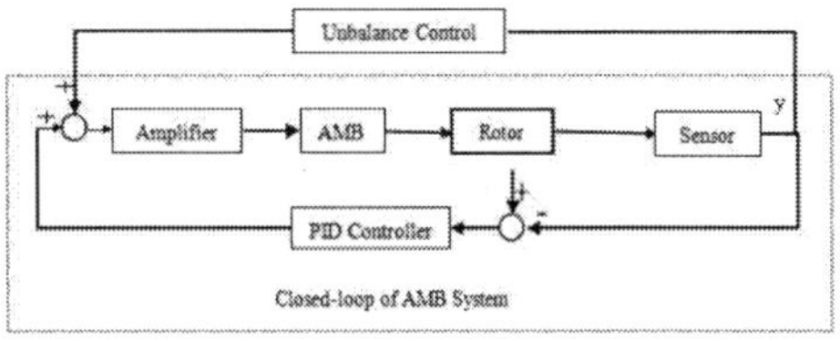

Figure 2. Block diagram of closed-loop control AMB system with unbalance control.

B. Models of Other Parts

The models of other parts of the objective system is given below:

Sensors and amplifiers are simplified into proportional components and represented by K_S, K_A respectively. Magnetic bearings are described by the classical linear form as following:

$$F_M = -K_X X + K_I I_C \tag{3}$$

Where F_M is magnetic bearing forces matrix, K_X is the force-displacement factor matrix, and K_I is the force-current factor matrix. Here, I_C is control current vector. As we all know that, the open loop of magnetic bearing system is inherently unstable. Therefore, a PID controller is introduced as negative feedback control to realize the stable of the system.

III. Unbalance Analysis and Compensation

A. Unbalance Analysis

When rotor rotates at speed ω, there will be a centrifugal force F_C results from the misalignment between the mass center and the geometric center of the rotor. Assume that the rotor imbalance is m and the distance between the mass center and the geometric center of the rotor is e. Then this centrifugal force can be described as:

$$F_C = me\omega^2 \sin(\omega t + \varphi) \tag{4}$$

However, this periodical centrifugal force acting on the rotor could induce sinusoidal disturbance to the displacement of the rotor if only the PID controller applied. The simulation result in Fig.1 also confirms this phenomenon. The sinusoidal disturbance can be written as:

$$d(t) = \alpha_d(t)\sin(\omega t) + \beta_d(t)\cos(\omega t) \tag{5}$$

Here, $\alpha_d(t)$, $\beta_d(t)$ are the fourier coefficients of the sinusoidal disturbance.

In generally, the sinusoidal disturbance is the mainly part of rotor vibration. It obstructs the applications of AMBs in high-lever rotating accuracy conditions. In these cases, unbalance compensation techniques can be applied to reduce the rotor runout. Among various of methods, open loop feed-forward control is most widely used. These methods add additional control signals to the controller outputs according to the displacements of the rotor. The structure of AMB system with unbalance compensation is illustrated in Fig.2. In this paper, the additional control signals are generated based on ILC and MBAFB respectively.

B. Iteration learning control

The original idea of iterative learning control was first raised by Uchiyama in 1978 [8] and further proposed by Arimoto and others. Iterative learning control is a model-free system synthesis method which yields the control signal u_{k+1} for the next cycle using the current control signal u_k and an error correction item, the product of learning gain and e_k, the difference between desired output y_d and the current output y_k, or the derivative of e_k. There are a series of ILC control methods, P-type, D-type, PID-type and so on. In the paper, we will use P-type ILC law which yields u_{k+1} by adding u_k and error correction item, the product of learning gain and e_k only. Therefore, this method has significant advantages: physically easy to implement and insensitive to measurement noises. The open loop P-style learning law in time domain can be described as:

$$\begin{aligned} u_{k+1}(n) &= u_k(n) + Le_k(n) \\ n &\in 0, 1, \cdots, n_f - 1 \end{aligned} \tag{6}$$

Where, $n_f = 2\pi/(T_s\omega)$, T_s is the sample time. $u_k(n)$ is the current control input, and $u_{k+1}(n)$ is the control input of next cycle at sample point n. Here, L is the iterative learning gain and convergence condition can be found in [9]. The error defined as:

$$e_k(n) = y_d(n) - y_k(n) \tag{7}$$

Here, $y_d(n)$ is the desired output.

C. Model Based Adaptive Forced Balancing

The simplified AMB system with MBAFB control is illustrated in Fig.3. The block of plant denotes the close loop of AMB system. And, it's transfer function $G(s)$ can be derived according to Fig.2. Here, we assume that $G(j\omega) = Ae^{j\theta}$ and the synchronous compensation signal has the following time domain representation:

$$r(t) = \alpha_r(t)\sin(\omega t) + \beta_r(t)\cos(\omega t) \tag{8}$$

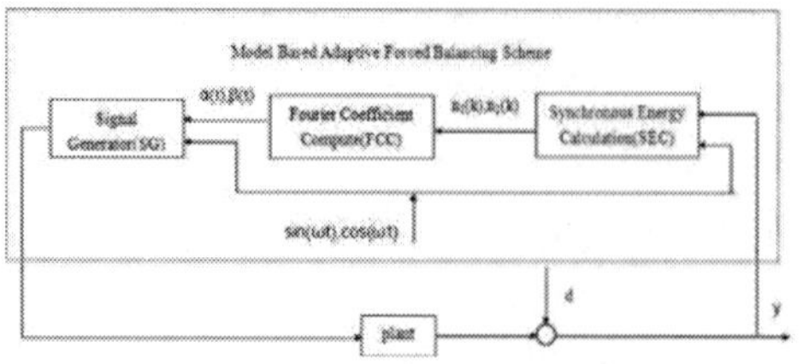

Figure 3. Block diagram of closed-loop AMB system with AFB.

Then, the output of the system can be derived as:

$$\begin{aligned} y(t) &= \alpha_d(t)\sin(\omega t) + \beta_d(t)\cos(\omega t) \\ &+ A\alpha_r(t)\sin(\omega t + \theta) + A\beta_r(t)\cos(\omega t + \theta) \end{aligned} \tag{9}$$

As shown in Fig.3, the MBAFB control algorithm consists of three parts: synchronous energy calculation (SBC), fourier coefficient computer (FCC) and signal generater (SG) [10]. In the first part, the so-called synchronous energy signals are calculated according to the system output signal $y(t)$. In order to obtain $n_1(t)$, $n_2(t)$, $y(t)$ is demodulated and then filtered using low pass filter to filter out the non-DC components. With the assumption that high frequency components are canceled out, $n_1(t)$, $n_2(t)$ can be derived and then be used to generate the fourier coefficients of the synchronous compensation signal in the FCC part. FCC plays an important role in the adaptive algorithm, in this part, the fourier coefficients are calculated online and updated every sample interval. The adaptive laws of $\alpha(k)$, $\beta(k)$ can be written as follows

$$\alpha(k+1) = \alpha(k) - (cos(\theta)n_1(k) + sin(\theta)n_2(k))/A \tag{10}$$

$$\beta(k+1) = \beta(k) - (cos(\theta)n_2(k) - sin(\theta)n_1(k))/A \tag{11}$$

Where $\alpha(k) = \alpha(t)$, $\beta(k) = \beta(t)$ for $t = kTs$, $k = 0, 1, \cdots$. Finally, the synchronous compensation signal can be constructed using the following equation

$$r(t) = \alpha_r(t)\sin(\omega t) + \beta_r(t)\cos(\omega t) \tag{12}$$

IV. Simulation and Comparison of the Methods

A. Attenuation of Synchronous Vibration

To validate the algorithms introduced above, simulations with ILC and MBAFB control algorithm are carried out at $5000rpm$. The simulation results in Fig.4 and Fig.5 clearly demonstrated that either the ILC algorithm or the MBAFB control algorithm can attenuate the synchronous vibration to some degree. As shown in Fig.1, without unbalance compensation control, the maximum rotor runout is about $35\mu m$. With ILC control, the maximum rotor runout attenuate gradually and significantly with the increase of repetition number (see Fig.5). The steady-state of rotor runout can be achieved by setting stop condition of ILC algorithm. With MBAFB control, the steady-state rotor runout can be obtained after several times of repetitions and maximum rotor runout is about $30\mu m$. Compared with ILC algorithm, the MBAFB algorithm applied is less efficient (see Fig.4).

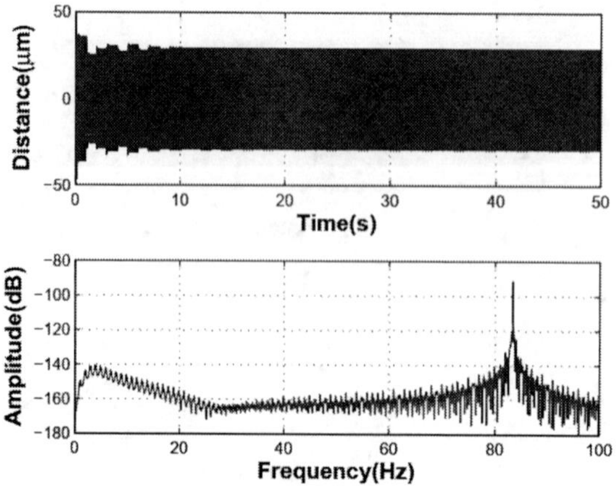

Figure 4. Rotor runout curve at axis X1 with MBAFB at $5000rpm$.

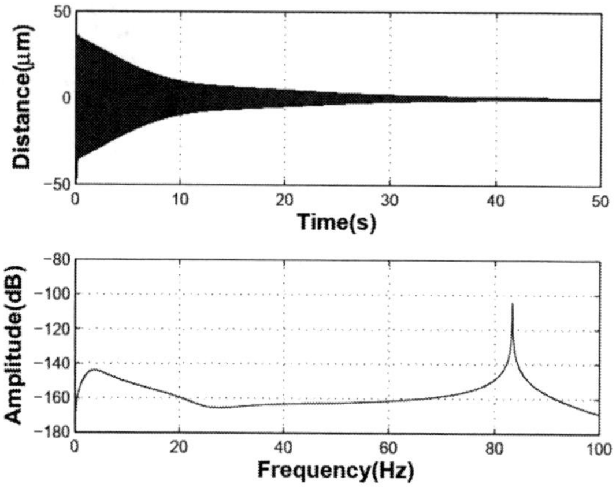

Figure 5. Rotor runout curve at axis X1 with ILC at $5000rpm$.

B. Tuning Lever of Controller Parameters

P-style ILC algorithm applied in this paper has a controller parameter L. The learning gain L is required to satisfy convergence condition below

$$\|I - LCB\| < 1 \tag{13}$$

Here, C and B are parameters of discrete-time close-loop system. However, the system parameters can be neglected when tuning the learning gain. Satisfied learning gain could be obtained by online adjusting[2]. In addition, according to [11], the range of learning gain L depend on the sample time as discrete-time close-loop system parameters come from continuous-time close-loop system. For MBAFB algorithm, there are no additional control parameters except the magnitude and phase of the close-loop system.

C. Sensitivity to Rotating Speed

In the simulation above, we assume that rotating speed of the rotor is known precisely. However, in practice, rotor

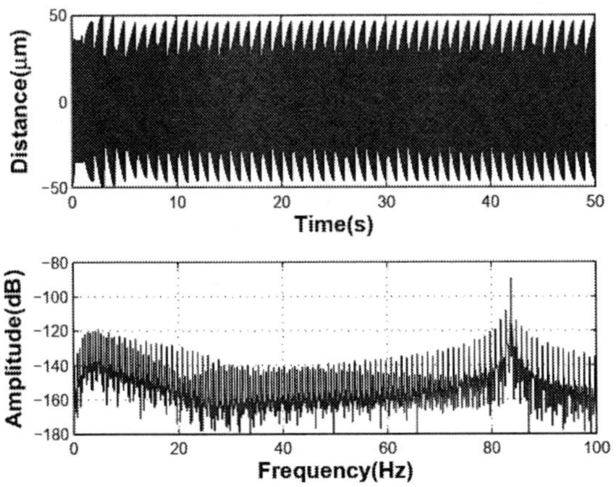

Figure 6. Rotor runout curve at axis X1 with MBAFB, 0.5% speed error

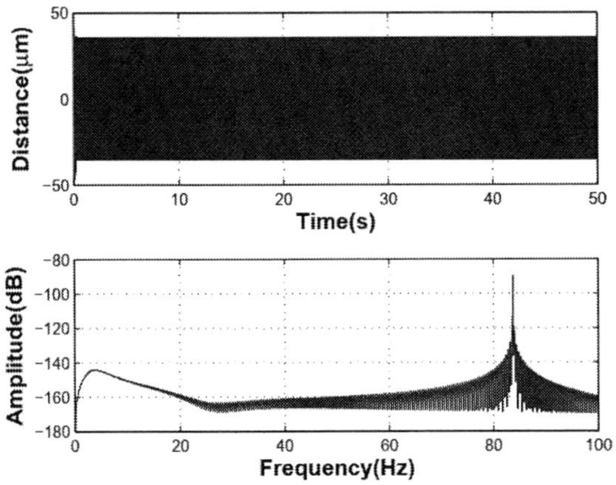

Figure 7. Rotor runout curve at axis X1 with ILC, 0.5% speed error

speed varies in a certain range. The performance of MBAFB algorithm with 0.5% speed error is illustrated in Fig.6. we find that, with speed error, MBAFB become invalid and even worsen the rotor runout. Fig.7 shows the performance of ILC algorithm with 0.5% speed error. ILC become invalid similarly. That's because synchronous vibration detection component is involved in the ILC algorithm to cancel out the affection of noises in practice[2]. From analysis above, we can see that both ILC and MBAFB algorithms are sensitive to the speed error. The algorithms should be developed to overcome this defect.

V. Conclusion

Synchronous vibration is a major vibration problem associated with rotating machinery. For a few years, researchers have devoted themselves to study unbalance compensation control algorithms using active magnetic bearing through active control. The algorithms presented can be divided into two categories: the model-free method and model-based method. In this paper, we establish the AMB system model and then use a model-free method, iterative learning control and a model-based adaptive forced balancing control to attenuate the unbalance vibration respectively. The performances of these unbalance control methods are compared from several aspects. The results of this paper reflect some properties of model-free and model-based unbalance compensation methods and can provide references for the users to choose an appropriate unbalance control algorithm.

VI. Acknowledgement

Thanks go to Zhe Sun for his useful suggestions and constant guidance . I also would like to extend my gratitude to my adviser Lei Shi for his instructive advice and Professor Yu, who led me into this area. This paper is financially supported by the National Science and Technology Major Project of China (2011ZX069) and Project 61305065 supported by NSFC.

References

[1] G. Schweitzer, H. Bleuler, and A. Traxler, "Active magnetic bearings-basics, properties and application of active magnetic bearings," *ETH, Switzerland*, 1994.

[2] C. Bi, D. Wu, Q. Jiang, and Z. Liu, "Automatic learning control for unbalance compensation in active magnetic bearings," *Magnetics, IEEE Transactions on*, vol. 41, no. 7, pp. 2270–2280, 2005.

[3] K. W. Eure, "Adaptive predictive feedback techniques for vibration control," Ph.D. dissertation, Citeseer, 1998.

[4] W.-J. Lee, S. S. Oh, and D. Cheong, "Rotor unbalance compensation without angular position sensor for active magnetic bearing," in *Power Electronics and ECCE Asia (ICPE & ECCE), 2011 IEEE 8th International Conference on.* IEEE, 2011, pp. 2446–2449.

[5] Z.-h. Liu, K. Nonami, and Y. Ariga, "Adaptive unbalanced vibration control of magnetic bearing systems with rotational synchronizing and asynchronizing harmonic disturbance," *JSME International Journal Series C*, vol. 45, no. 1, pp. 142–149, 2002.

[6] H. D. GU, "Investigations of Active Magnetic Bearing-Flexible Rotor System," PhD Thesis, Tsinghua University, October 2005.

[7] Y. E. ZHONG, Y. Z. HE, Z. WANG, and F. Z. LI, Eds., *rotor dynamics*, 1st ed., ser. Relationship Between Critical Speed of the Rotor System and Certain Factors. Beijing: Tsinghua University Press, 1987, Chap. 4, pp. 69–73.

[8] M. Uchiyama, "Formation of high-speed motion pattern of a mechanical arm by tial," *Transactions of the Society of Instrumentation and Control Engnineers*, vol. 14, no. 6, pp. 706–712, 1978.

[9] Z. Bien and J.-X. Xu, *Iterative learning control: analysis, design, integration and applications.* Kluwer Academic Publishers, 1998.

[10] E. Vahedforough, B. Shafai, and S. Beale, "Estimation and rejection of unknown sinusoidal disturbances using a generalized adaptive forced balancing method," in *American Control Conference, 2007. ACC'07.* IEEE, 2007, pp. 3529–3534.

[11] H. Elci, R. W. Longman, M. Q. Phan, J.-N. Juang, and R. Ugoletti, "Simple learning control made practical by zero-phase filtering: Applications to robotics," *Circuits and Systems I: Fundamental Theory and Applications, IEEE Transactions on*, vol. 49, no. 6, pp. 753–767, 2002.

Compensation of Unbalance Forces of a Switched Reluctance Machine with Combined Current and Fluxlinkage Control

Thomas Hinterdorfer, Alexander Schulz, Harald Sima, Stefan Hartl and Johann Wassermann

Vienna University of Technology, Institute of Mechanics and Mechatronics,
Wiedner Hauptstraße 8-10, A-1040 Vienna, Austria,
{givenname.surname}@tuwien.ac.at

***Abstract*—The presented work handles the compensation of unbalance forces for a 6/4-Switched Reluctance Machine (SRM), which generate significant radial magnetic forces in an eccentric motor shaft position. When the rotor is operating e.g. in its principal axis of inertia, these forces must be compensated by the bearings. This is of major interest for long term storage Flyweels where bearing efficiency is a key factor for the overall system performance.**

Due to the back electromotive force (BEMF) in high speed operation, precise current control is a major challenge. Hence a combination of current and fluxlinkage control is used to minimize the generated radial forces. It can be used for low and high speed applications.

For the observation of fluxlinkage two different methods are described. The functionallity is shown by means of simulations and measurement.

I. Introduction

Flywheel Energy Storage Systems (FESS) represent an ecologically and economically sustainable technology for decentralized energy storage. Compared to other storage technologies such as e.g. accumulators, they offer longer life cycles without performance degradation over time and usage and need almost no systematic maintenance. To achieve a high energy density FESS rotors are usually manufactured of fiber reinforced plastics, which enables tip speeds up to 800 m/s.

For long term storage Flywheel applications magnetic bearings for rotor support are used and the rotor is operated in an evacuated containment to minimize drag. Due to losses in coils, iron, electronics and air drag the self-discharge is usually rather high. The bearings are optimized with respect to efficiency to maximize storage time.

The stator and the rotor of a SRM have salient poles, see Fig. 1, where the geometry of a 6/4-SRM is shown. Hence, the airgap changes with the rotor angle θ. When a phase, consisting of two opposing coils, is excited, the rotor attempts to align with the stator to minimize reluctance. Thus, by sequentially exciting the machines phases, a continuous rotation can be achieved. The SRM has no permanent magnets in its assembly and therefore minimal holding torque. Furthermore, the rotor mass and its moment of inertia is rather small refered to its torque. These are essential advantages for long term storage Flywheels. The machines disadvantages are

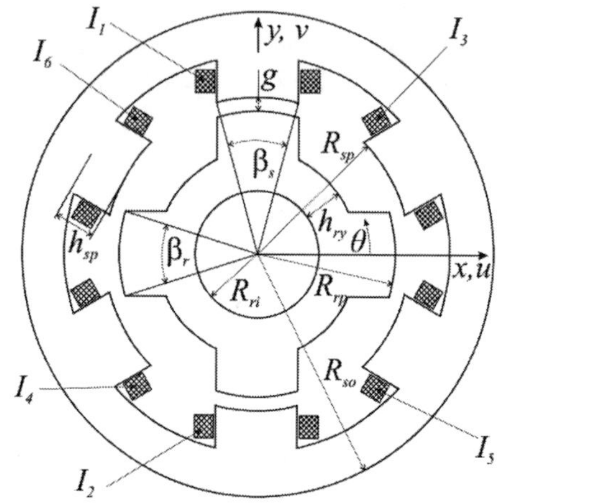

Figure 1. Definitions and geometry of the examined machine in aligned position with the rotor angle $\theta = 0°$

high torque ripple and a complex control due to the nonlinear behaviour. Furthermore the SRM requires an electronic power converter to run.

To maximize efficiency of the radial magnetic bearings the rotor is held in its principle axis of inertia by unbalance control, to cancel out speed synchronous currents and consequently provide a minimum of energy for radial support. Hence, the machine's operating point will be in an eccentric position. Moreover, the high tip speed causes large rotor deformations leading to an unbalance which depends on the angular velocity. When the rotor of an SRM is in an eccentric position radial forces are generated due to different reluctances of the two opposing airgaps. This forces must be compensated again by the magnetic bearings and therefore decreasing there efficiency. Thereby the concept presented in this paper is to control the two coils of a SRM-phase separatly, with the goal of generating no resulting forces and thereby minimizing the influence of the machine on the magnetic bearings. Furthermore the presented concept also enables a bearingless operation of the machine, which increases the reliability of a FESS and allows a smaller design of the bearings.

Compensation methods for unbalanced synchronous- and

induction machines are described e.g. in [1], where connecting the phase coils in parallel is leading to a balanced fluxlinkage. An analogous method is described in [2] and [3], where the unbalance forces of a three phase 12/8-SRM are compensated. In that paper only machine topologies with four coils or more in one phase are investigated. It was shown that two parallel branches with neighboring coils in series and opposing coils in parallel and an equalizer between them is leading to the best compensation.

[4] discusses the effect of short circuited coils on the rotor which avoids an asymmetric flux distribution. The advantage of this method is, that no additional power converters are necessary, but it results in an increased constructional effort.

In [5] the compensation is done by adjusting the ampere turns of the phase. The number of turns of the stator coils can be increased or decreased by switches. The change of torque is adjusted by the amplitude of the current. The disadvantage of this method is that the number of turns can only be changed in discrete steps. The required correction is dependent on the actual eccentricity and can be detected by mutual coupling which is described in [6].

[7] uses an analytical function to determine the reference current for force compensation, which are dependent on the rotor angle and the displacement. A 12/8-SRM is described and the compensation method is to control the current of opposing coils separatly.

II. Overview

SRMs generate high magnetic forces in an eccentric run. Therefore in the following a comepensation method will be presented. Chapter III deals with a current controlled machine to achieve the required power in combination with a fluxlinkage controller to compensate unbalance forces. For this purpose two different fluxlinkage observers will be discussed. In chapter IV the used model will be presented which then will be compared to measurement data discussed in chapter VI done at the test rig presented in chapter V. Finally chapter VII concludes the paper.

The data of the examined machine with a nominal power of 5 kW and an operating range of 650 rad/s to 2000 rad/s are summarized in Table I and the definitions are shown in Fig. 1.

In the following mutual coupling of the phases is neglected and only one phase with coils 1 and 2 is considered.

III. Compensation of Unbalance Forces

To compensate unbalance forces and enable a bearingless operation, the two opposing coils of one phase are controlled separately. A block diagram of the control scheme is shown in Fig. 2. To achieve the needed torque or power a superior controller generates firing angles and a reference current I_{cmd} for a current controller, which calculates the needed normalized voltage as an input for a pulse-width-modulator

$$d_{cmd,1,2} = \frac{1}{U_{DC}} \left(K_{P,I} L(\theta) \left(I_{cmd} - I_{1,2} \right) + \underbrace{\frac{dL(\theta)}{d\theta} I_{1,2}\omega}_{\text{BEMF}} \right) \quad (1)$$

Table I
Data of the examined SRM

number stator poles	P_s	6
number rotor poles	P_r	4
rotor pole angle, °	β_r	38
stator pole angle, °	β_s	37
shaft radius, mm	R_{ri}	45.88
rotor pole radius, mm	R_{rp}	79
rotor yoke thickness, mm	h_{ry}	25.38
stator pole height, mm	h_{sp}	24.13
stator pole radius, mm	R_{sp}	80
stator outer radius, mm	R_{so}	139.7
stack length, mm	L_s	67.99
airgap length, mm	g	1
turns per coil	N	80

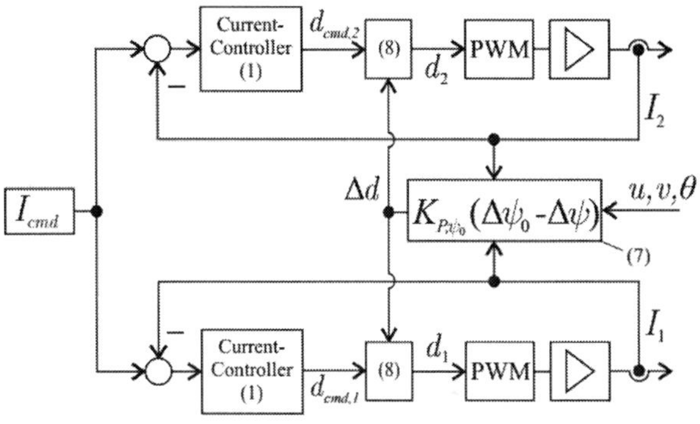

Figure 2. Block diagram of the control scheme

with the link voltage U_{DC}, the overall inductance in centered position L, the actual current $I_{1,2}$ in coil 1 or 2 and the angular velocity ω. A proportional controller with $K_{P,I}$ as gain and feedforward control of the BEMF is used. $K_{P,I}$ can be a function of ω to enable more flexibility.

A. Control Scheme

In case of an eccentric run, the concept of the presented controller is to use the current controller in (1) for the coil which has the larger airgap. In the coil with the smaller airgap the normalized voltage is decreased to adapt the fluxlinkage to achieve vanishing forces. The acting forces and the fluxlinkages can be determined by a magnetic circuit model or a finite element model as e.g. in [8]. When a positive displacement v in y-direction is assumed, the current in coil 2 can be controlled to the reference value and the current in coil 1 must be decreased. The difference of the two fluxlinkages of coil 1 and 2 can be derived by solving for vanishing force in y-direction at constant rotor angle θ, excitation current I_2 as well as translatory displacements u and v

$$F_y(\theta, I_1 = I_{cmd} - \Delta I_0, I_2 = I_{cmd}, u, v) = 0 \quad . \quad (2)$$

This results in a current difference ΔI_0 which can be substituted into the fluxlinkages, leading to the reference value of

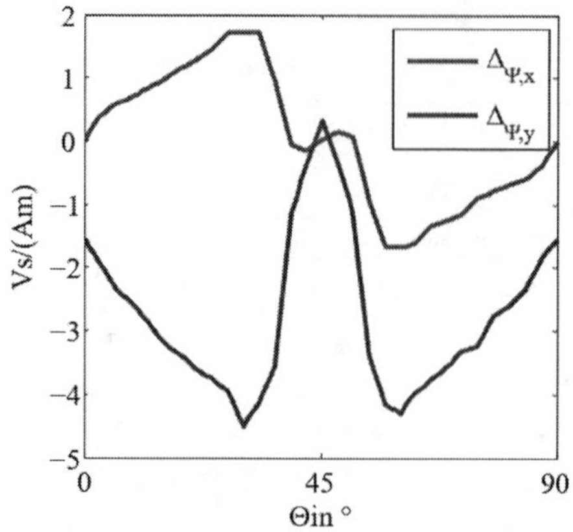

Figure 3. Fluxlinkage correction due to displacement u and v

the fluxlinkage difference

$$\Delta\psi_0 = \psi_1 - \psi_2 \tag{3}$$

which is also a funtion of the rotor position and the excitation currents. When linear material behaviour is assumed, $\Delta\psi_0$ variies linear with the current. Due to increasing BEMF with higher angular velocity the actual current won't match the reference so the actual current is used to form the reference of the fluxlinkage difference

$$\Delta\psi_0 = \begin{cases} \Psi_0(\theta, u, v)\, I_1 & \text{for} \quad \Psi_0 > 0 \\ \Psi_0(\theta, u, v)\, I_2 & \text{for} \quad \Psi_0 \leq 0 \quad . \end{cases} \tag{4}$$

Furthermore the superposition principle can be used and Ψ_0 can be written as a linear combination of u and v

$$\Psi_0(\theta, u, v) = \Delta_{\psi,x}(\theta)\, u + \Delta_{\psi,y}(\theta)\, v \tag{5}$$

with the fluxlinkage correction $\Delta_{\psi,x}$ and $\Delta_{\psi,y}$, which can be seen in Fig. 3, due to displacements u and v, respectively. Due to Faraday's law of induction

$$d_{1,2} = \frac{U_{1,2}}{U_{DC}} = \frac{1}{U_{DC}} \frac{d\psi_{1,2}}{dt} \tag{6}$$

the fluxlinkage can be adjusted by a voltage correction Δd which can be done by a proportional controller

$$\Delta d = K_{P,\psi_0}(\Delta\psi_0 - \Delta\psi) \tag{7}$$

with d_1 and d_2 as the normalized voltages of the two coils, the applied voltages U_1 and U_2, the controller gain K_{P,ψ_0} and the actual fluxlinkage $\Delta\psi$.

To prevent overcurrent a distinction of cases is advantageous

$$\left.\begin{aligned} d_2 &= d_{cmd,2} \\ d_1 &= d_2 - \Delta d \end{aligned}\right\} \quad \text{for} \quad \Delta\psi_0 \leq 0$$
$$\left.\begin{aligned} d_1 &= d_{cmd,1} \\ d_2 &= d_1 + \Delta d \end{aligned}\right\} \quad \text{for} \quad \Delta\psi_0 > 0 \quad . \tag{8}$$

In case of negative $\Delta\psi_0$, which arises for a positive displacement v the voltage correction is subtracted from d_1, and for a positive value it is added to d_2. This procedure should prevent overcurrent and is not sensitive to noise in comparison to the usage of Δd for the distinction in (8).

B. Fluxlinkage Observer

In (7) the actual fluxlinkage difference is needed which can hardly be measured, hence an observer will be used. One method is to directly use the differential equation

$$\begin{aligned} \Delta\psi &= \psi_1 - \psi_2 \\ &= \int_0^t \Big[U_{DC}(d_1 - d_2) - R(I_2 - I_1)\Big]\, dt \end{aligned} \tag{9}$$

with the ohmic resistance R. Transformation into discrete time domain leads to

$$\Delta\psi = \Delta\psi z^{-1} + T_s \Big[U_{DC}(d_1 - d_2) + R(I_2 - I_1)\Big] \tag{10}$$

with the sampling time T_s. This kind of observer is very accurate at high speeds and when the current drops to zero after excitation. Otherwise a measurement or sampling error will be summed up due to the integral behaviour. Especially at low speeds this leads to instabilities due to the large excitation time span and therefore to high radial forces. For this reason this observer won't be considered in chapter VI.

Another method is to use the definition of secant inductance

$$\psi_1 = L_{11}(\theta, u, v) I_1 + L_{12}(\theta, u, v) I_2 \tag{11}$$

with the self-inductance L_{11} and the mutual inductance L_{12} which are both functions of the rotor position. For small displacements u and v, i.e. in the linear part of the magnetization curve, the fluxlinkage of coil 1 can be linearized around the geometric center as follows

$$\begin{aligned} \psi_1 \approx &\left[L_{11}(\theta)\big|_0 + \frac{\partial L_{11}(\theta)}{\partial x}\bigg|_0 u + \frac{\partial L_{11}(\theta)}{\partial y}\bigg|_0 v\right] I_1 + ... \\ &\left[L_{12}(\theta)\big|_0 + \frac{\partial L_{12}(\theta)}{\partial x}\bigg|_0 u + \frac{\partial L_{12}(\theta)}{\partial y}\bigg|_0 v\right] I_2 \quad . \end{aligned} \tag{12}$$

Due to symmetry the relations for ψ_2 are similar

$$\begin{aligned} L_{22}(\theta)\big|_0 &= L_{11}(\theta)\big|_0 &&= L_{11,0} \\ L_{21}(\theta)\big|_0 &= L_{12}(\theta)\big|_0 &&= L_{12,0} \\ \frac{\partial L_{11}(\theta)}{\partial x}\bigg|_0 &= -\frac{\partial L_{22}(\theta)}{\partial x}\bigg|_0 &&= L_{11,x0} \\ \frac{\partial L_{12}(\theta)}{\partial x}\bigg|_0 &= -\frac{\partial L_{21}(\theta)}{\partial x}\bigg|_0 &&= L_{12,x0} \\ \frac{\partial L_{11}(\theta)}{\partial y}\bigg|_0 &= -\frac{\partial L_{22}(\theta)}{\partial y}\bigg|_0 &&= L_{11,y0} \\ \frac{\partial L_{12}(\theta)}{\partial y}\bigg|_0 &= -\frac{\partial L_{21}(\theta)}{\partial y}\bigg|_0 &&= L_{12,y0} \end{aligned} \tag{13}$$

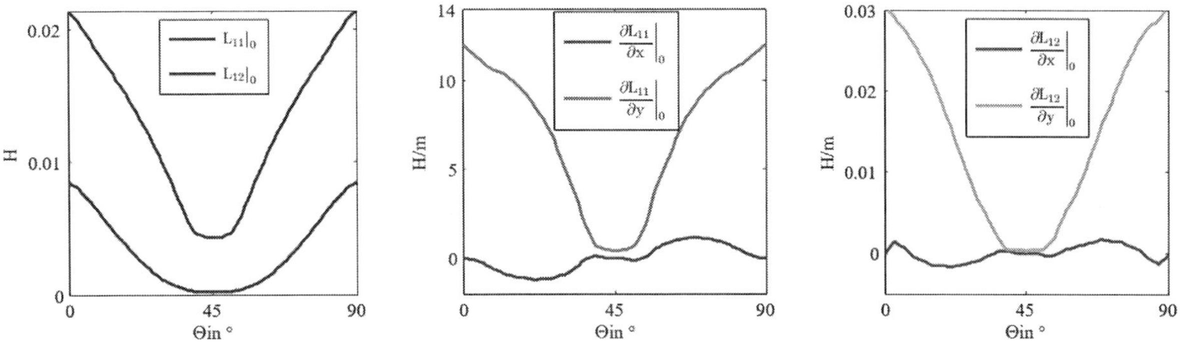

Figure 4. Self-inductance and mutual inductance (left); Derivative of self-inductance with respect to x and y (middle); Derivative of mutual inductance with respect to x and y (right)

and the fluxlinkage difference is derived as

$$\Delta\psi = (L_{11,0} - L_{12,0})(I_1 - I_2) + \dots$$
$$\Big[(L_{11,x0} + L_{12,x0})u + \dots \tag{14}$$
$$(L_{11,y0} + L_{12,y0})v\Big](I_1 + I_2)$$

which now has low computational effort. The self- and mutual inductance as well as their derivatives are shown in Fig. 4. The fluxlinkages can be determined by three look-up tables which are only a function of the rotor angle and therefor can be calculated online with low requirement on memory.

IV. Simulation Model

For an accurate analysis of the machines behaviour, a detailed model was built. It contains the machines characteristics in look-up-tables determined by a 2D finite element analysis containing a nonlinear material description, described in [8], therefore end-effects were neglected. The characterestics of torque, forces and the fluxlinkages depend on the rotor position and the excitation currents. The fluxlinkages are unique functions of the currents, therefore they can be converted into each other and the integral form of (6) can be used for simulation

$$\psi_1 = \int_0^t \Big[U_1 - RI_1(\theta, \psi_1, I_2, u, v)\Big]\, dt$$
$$\psi_2 = \int_0^t \Big[U_2 - RI_2(\theta, I_1, \psi_2, u, v)\Big]\, dt \quad . \tag{15}$$

The model also includes noise of the current sensor, which was assumed to be triangular with an amplitude of 0.4 A at a frequency of 475 kHz. Before converting the signals into the digital domain by analog-to-digital converters (ADC) the current signal is fed through a Bessel filter of sixth order and cutoff frequency of 62.5 Hz. The ADCs resolution is 12 bit and quantization noise was assumed to have a variance of 10 quantization steps and average 0. The sampling frequency was 800 kHz and the last sampled current value in the past controller period was used for control, to avoid calculation time in the real-time-implementation. The controller frequency was 80 kHz. The power converters were modeled as ideal switches. All calculated controller parameters, as e.g. $L_{11,0}$ etc., were approximated with steps of 1°.

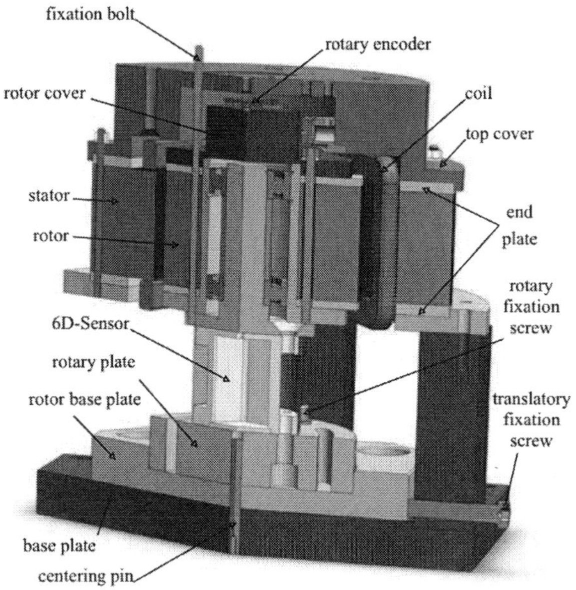

Figure 5. 3D plot of the test rig construction

V. Test Rig

The test rig was constructed to enable static as well as dynamic measurements. In special the acting magnetic forces were of interest. In Fig. 5 a 3D-plot of the test rig is shown.

The rotor base plate can be moved with respect to the base plate to reach every rotor displacement u and v. The rotary plate carries a 6D-force sensor, which is connected to an axle with two ball bearings supporting the machines rotor. The permanent magnet for the rotary encoder is mounted on the rotor cover. The coils are mounted on the machines stator, which carries a top cover to ensure a safety operation and holding the stator part of the encoder.

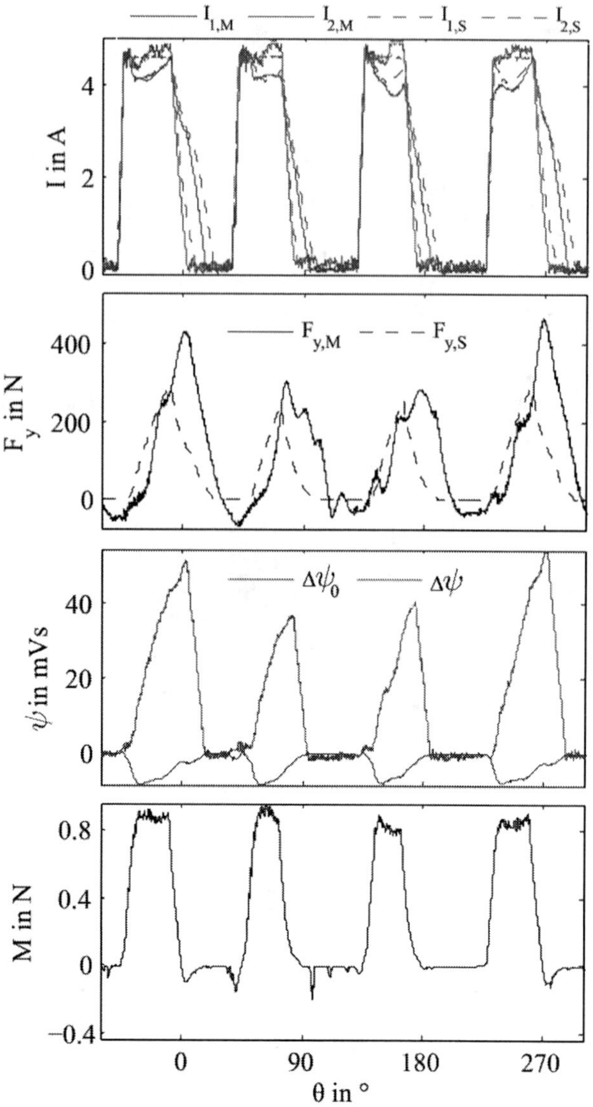

Figure 6. Comparison of measurement and simulation results in normal operation at 90 rad/s with a reference current of 4.5 A

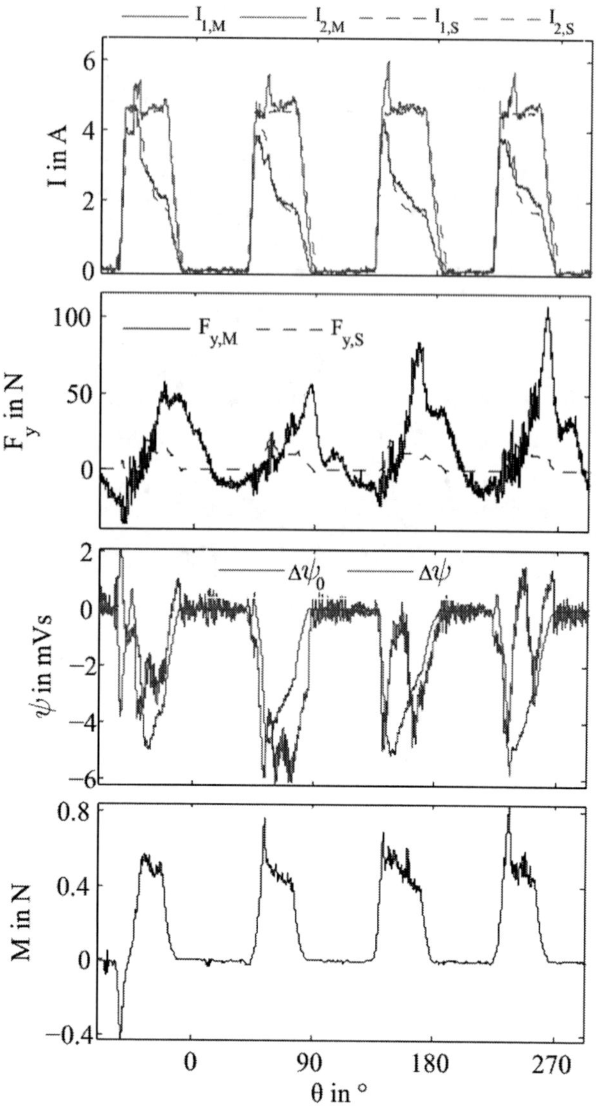

Figure 7. Comparison of measurement and simulation results with unbalance force compensation at 90 rad/s with a reference current of 4.5 A

The controller was implemented on an ARM-Cortex-M4 DSP. The rotary encoder was operated in PWM-Mode with a frequency of 976.56 Hz. For that reason an angular observer was implemented to improve the measurement results. For this the angular verlocity was approximated with a least-squares based derivative estimator of third order and the actual angle approximated with the difference quotient.

The current was measured with a fluxgate sensor for the controller and a current probe for the measurement. The eight analog signals and the digital angle signal were measured with two mixed signal oscilloscopes, which were externally triggered by a PWM-output which was set to high when the reference angular velocity was reached.

VI. Measurement Results

To demonstrate the functionallity and to validate the simulation model, in this chapter a comparison to measurements is given. For this a constant angular velocity is assumed and the power supply is regulated to a link voltage of 30 V. The rotor is fixed at a translatory displacement of $u = 0$ mm and $v = 0.46$ mm. Furthermore only one phase will be considered, subsequent mutual coupling is eliminated. The measurement and the simulation was done once in normal operation with both currents having the same reference and once with the control scheme illustrated before.

The first bending mode of the rotor structure appears at 136 Hz and therefore the measured force was filtered with a second order IIR-notch filter with cutoff frequency 136 Hz.

Fig. 6 shows the comparison of measurement and simulation

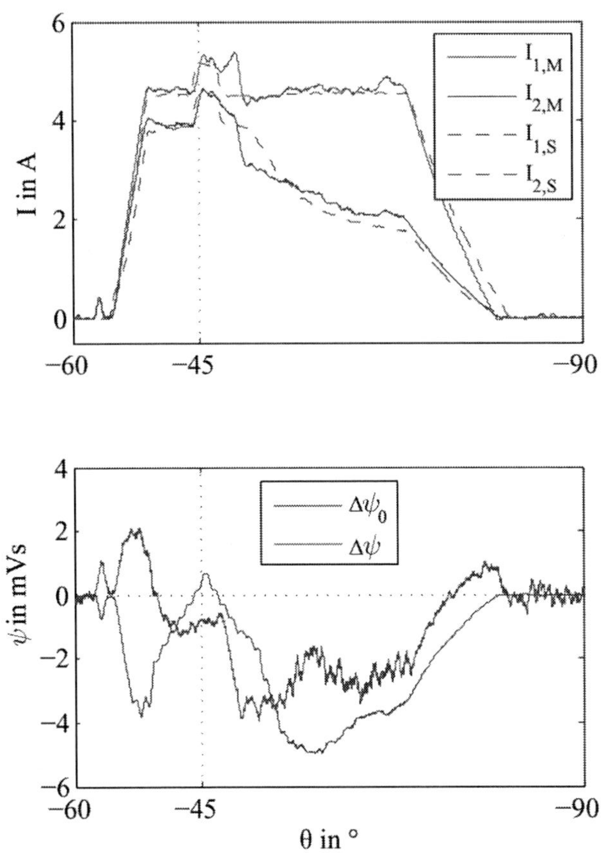

Figure 8. Detailed plot of measurement and simulation results with unbalance force compensation at 90 rad/s with a reference current of 4.5 A

in normal operation at 90 rad/s with a reference current of 4.5 A. The turn-on and turn-off angles for the simulation were adapted to the measurement. The measured currents in the two coils are denoted as $I_{1,M}$ and $I_{2,M}$ and the simulated ones as $I_{1,S}$ and $I_{2,S}$. The same was done with unbalance force compensation, which is shown in Fig. 7. In both figures, the fluxlinkage difference was determined with (14) with simulated inductances but measured currents. The ploted torque is interpolated to the measured currents from the simulated characteristics.

In normal operation the maximum force is above 400 N while in Fig. 7 the force can be compensated to values about 100 N. Due to the large displacement the BEMF is very large for coil 1. Therefore in normal operation I_1 is below the reference current.

Due to the decreased current I_1 in Fig. 7 compared to normal operation the generated torque also decreases. This would be compensated by a superior controller to ensure to commanded torque.

In Fig. 8 the results of Fig. 7 are shown in more detail. One can see the good agreement of the simulated data to the measurements results. In the range of $\theta = 45°$ the current I_2 begins to rise above the reference value and I_1 reaches the reference current. This is due to the positive value of the reference fluxdifference $\Delta\psi_0$ in the range of $\theta = 45°$ where the case distinction in (8) changes.

In contrast to the normal operation, the difference of the two fluxlinkages ist much smaller with force compensation. Nevertheless the result can be approved by using a PD-controller to minimize the control error in (7). Its worth noting, that also in the real-time-implementation machine characteristics from a 2D finite element analysis were used. Further improvement can be reached by using measured data.

VII. Conclusion and Outlook

In this paper a control method to compensate the unbalance forces of a SRM is presented. The key of this method is the separate control of the phase coils and adjust the fluxlinkages to achieve vanishing forces. Furthermore two fluxlinkage observers where discussed. Simulation results demonstrated the effectiveness of the controller in contrast to the method in [8], where major problems arised due to BEMF and thereby imprecise current control.

The validation was done at a special manufactured test rig. The results were in good agreement to the simulation results.

The real-time-implementation was done with simulated data only. Further optimization and better compensation can be reached by usage of measured values.

VIII. Acknowledgment

The presented work is part of the research projects "SEE-Flywheel" and "LTS-Flywheel" and is supported by "Klima und Energiefonds" in line with the program "NEUE ENERGIEN 2020" and "Building of the Future Plus", respectively, both conducted by the Austrian Research Promotion Agency (FFG).

References

[1] O.W. Andersen, "Compensation of Unbalanced Magnetic Forces by Distributed Parallel Circuits," *International Conference on Electrical Machines ICEM*, Krakow, Aug. 2004.

[2] J. Li, D. Choi and Y.H. Cho, "Compensation of Unbalanced Magnetic Force in Switched Reluctance Motor with Airgap Nonuniformity," *EPE Chapter Electric Drives Joint Symposium*, France, Jul. 2009.

[3] J. Li, D. Choi and Y.H. Cho, "Analysis of Rotor Eccentricity in Switched Reluctance Motor With Parallel Winding Using FEM," *IEEE Transactions on Magnetics*, Jul. 2009, Vol. 45, No. 6, pp. 2851-2854.

[4] H. Mansir, "Switched Reluctance Machine Having Unbalance Forces Compensation Coils," U.S. Patent 5 422 525, June 6, 1995.

[5] M. R. Tavakoli, H. Torkaman and E. Afjei, "Eccentricity faults compensation in SRMs by counterbalancing ampere turns in facing poles," *Power Electronics, Drive Systems and Technologies Conference (PEDSTC)*, 2013, pp. 66-71.

[6] H. Torkaman, E. Afjei and P. Yadegari, "Static, Dynamic, and Mixed Eccentricity Faults Diagnosis in Switched Reluctance Motors Using Transient Finite Element Method and Experiments," *IEEE Transactions on Magnetics*, 2012, Vol. 48, No. 8, pp. 2254-2264.

[7] J. Ji, Y. Sun, W. Zhao, X. Zhu and C. Song, "Calculation and remedial strategy of radial force of switched reluctance motors with eccentricity," *International Conference on Electrical Machines and Systems ICEMS*, 2008, pp. 3932-3935.

[8] T. Hinterdorfer, A. Schulz and H. Sima, "Force Prediction and Radial Force Compensation of a Switched Reluctance Motor," *39th Annual Conference on Industrial Electronics, IECON 2013*, pp. 2872-2877.

Zero-Power Control for Magnetic Bearings in Artificial Heart Pumps

Shuqin Liu[a], Zhongguo Bian[a], Youpeng Fan[b], Yunpeng Zhang[a]

[a] School of electrical engineering, Shandong university, Jinan, China, lshuqin@sdu.edu.cn bzhongguo@sdu.edu.cn
[b] State grid of China technology college, Jinan, China

Abstract—Artificial heart pump is a new approach to the treatment of heart failure, and the main representative of the third generation of pump is the application of the magnetic suspension or other suspension technology. Energy saving operation of artificial heart pump is very important. This paper studies the application of zero power control in the artificial heart pump. The model of zero power controller and operation theory is presented, and the stiffness characteristic of rotor under zero power control is analyzed. The control method to change equivalent stiffness of rotor system is given according to suspension position, verifying the results.

I INTRODUCTION

According to the survey of WHO, the cardiovascular disease and heart disease accounted for about 30% of all diseases. Various types of heart disease turns to heart failure at last, and about 1/5 of people will eventually become the heart failure patients according to statistics. The rapid developments of drugs and interventional techniques have saved a lot of acute myocardial infarction patients' life. However, the current treatment of heart failure is mainly depends on drugs , whose effects are limited. Effective therapy of heart failure disease is heart transplant. Due to the impact of donor and cost, the treatment is not widely applied. The majority of heart failure patients are dead in waiting for donor heart!

In recent years, ventricular assist device (VAD), artificial heart pump shows its unique advantage in clinical treatment, and has been increasingly applied in clinical so that it plays an important role in medicine and surgery of heart. VAD can compensate or replace heart function, and decrease the after-loading at the same time, in favor of myocardial metabolism and the recovery and balance of nutritional. Also, VAD can promote the recovery of cardiac function. For those heart that can't recovery heart function completely, VAD can replace the natural heart and exert physiological blood pumping function in a long term, which can extend the life of patients and improve their quality of life. Artificial heart has good application prospect and it's a new way for the treatment of heart failure. The countries all over the world have invested a lot of manpower and material resources for its development. Artificial heart pump has been developed to the third generation, and the main symbol is the application of the magnetic suspension technology or other suspension technology. More and more evidence shows that the third generation of the blood pump is better than the first two generation products in performance. The third generation of the blood pump uses the suspension technology, without the problem of bearing wear, improving the using durability of the blood pump. As a result, it's more suitable for long-term circulatory support. The impeller is suspended in the pump body, having no friction and extrusion and decreasing hemolysis significantly, so that the incidence rate of thrombosis reduced significantly. Because of avoiding energy consumption and temperature increase caused by mechanical friction, the third generation of the heart pump has higher energy efficiency ratio and it's more conducive to long-term assistance. The third generation of the heart pump attracts more research and development personnel.

Energy saving operation of artificial heart pump is very important! How to realize stable suspension in the rotor and have low power consumption is a key problem in maglev artificial heart pump. At first, zero power control mainly uses permanent magnet to provide levitation force, but it doesn't always save energy with the permanent magnet. In order to improve the utilization rate of permanent magnets, we need to search how to set quiescent operation point to make the system have stronger robustness and larger stability range. In view of the above problems, this paper mainly studies the zero power control system of Maglev artificial heart pump.

II THE PRINCIPLE OF MAGNETIC BEARING ZERO POWER CONTROLLER

A. The principle of the zero power control

Zero power control strategy makes the electromagnetic coil current is near to zero when rotor is suspended stably. When subjected to static force within the allowable range, the system can always keep the coil current in small vibration around zero by adjusting the suspension gap, therefore it can reduce power consumption. But zero power controller has low stiffness and when subjected to strong disturbance, it is easy to make the collision between rotor and stator lose stability. Therefore, it's particularly important to research on high stiffness control of zero power control. Zero power control also should have strong adaptability to interference. In order to solve the anti-interference problem of zero power control, the mirror-image current integral feedback is added to achieve zero power control.

B. Mirror-image current integral feedback

The zero power control block diagram of mirror-image current integral feedback is shown in Figure 1.

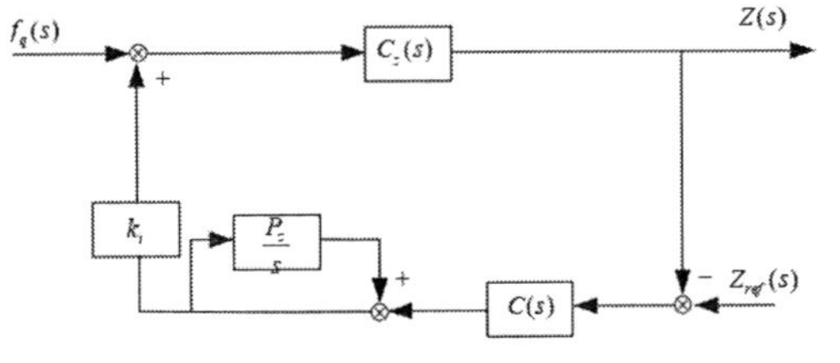

Figure1. Zero power control block diagram

Where $C_z(s)=\frac{1}{ms^2+k_x}$, $C(s)=P_d+P_v s$.

Taking the disturbance force as input, the coil current as output, the transfer function of the system is

$$I(s)=\frac{-\left(P_v s^2+P_d s\right)}{ms^3+\left(k_i P_v-P_z\right)s^2+\left(k_i P_d+k_x\right)s-k_x P_z}F_q(s) \tag{1}$$

When the disturbance force is a step force, that is $F_q(s)=\frac{F_0}{s}$. According to final value theorem

$$\lim_{t\to\infty} i(t)=\lim_{s\to 0} sI(s)=0 \tag{2}$$

Equation (2) shows when the magnetic bearing is in the steady state, the coil current is 0. The rotor relies on the balance between axial force generated by radial bearing and axial magnetic force by balance to achieve the rotor suspension, indicating that zero power control can be achieved.

When rotor is suspended stably, axial displacement can be expressed as

$$\lim_{z\to\infty} z(t)=\lim_{s\to 0} sZ(s)=\frac{F_0}{k_x} \tag{3}$$

Equation (3) shows that when the system is in a steady state, the rotor will deviate from the equilibrium position. The degree of deviation is related to disturbance force and force displacement stiffness coefficient, at the same time it shows that stiffness parameter is very important for zero power control.

III RESEARCH ONSTIFFNESSOFZEROPOWERCONTROLLER

Equivalent stiffness and equivalent damping of rotor have the same meaning as ordinary magnetic suspension system. Equivalent stiffness and equivalent damping of rotor shown as:

$$\begin{cases} K=k_x-\mathrm{Re}\left(\frac{K_i(k_d s^2+k_p s)}{P_z-s}\right)=k_x+\frac{K_i\omega^2(P_z k_d+k_p)}{P_z^2+\omega^2} \\ d=\frac{k_i\bullet\mathrm{Im}\left(-\frac{K_i(k_d s^2+k_p s)}{P_z-s}\right)}{\omega}=-k_i\frac{k_p P_z-k_d\omega^2}{P_z^2+\omega^2} \end{cases} \tag{4}$$

Dynamic stiffness characteristics of the rotor is

$$\frac{F_q(s)}{Z(s)} = K_{dy} = \left|\frac{(ms^2+k_x)(P_z-s)-K_i(k_d s^2+k_p s)}{P_z-s}\right| \tag{5}$$

Equation (4) shows that the displacement of the rotor static offset rotor is related to the axial open-loop stiffness and the amplitude of disturbance force.

The rotor's dynamic stiffness and damping ratio under zero power control of changes with the integral parameters as shown in figure 2. In Figure 2, the black line shows stiffness damping characteristics when the mirror-image integral feedback coefficient is 0, and it's the common PD control at this time. Green line, blue line and red line respectively correspond to pz=50, pz=100, pz=150. In general, in the low frequency band, compared to common PD control, the value of dynamic stiffness of zero power control decreases, indicating that the sensitivity of the rotor to static load is increased. That is, when the rotor is suffered by external interference, the balance position of the rotor will be changed, while the damping is decreased. At high frequency band, zero power control is the same with PD control. For different mirror-image integral feedback coefficient, the changes of dynamic stiffness and damping ratio is also different. The smaller mirror-image integral feedback coefficient is, the more quickly dynamic stiffness and damping ratio increases with the frequency change.

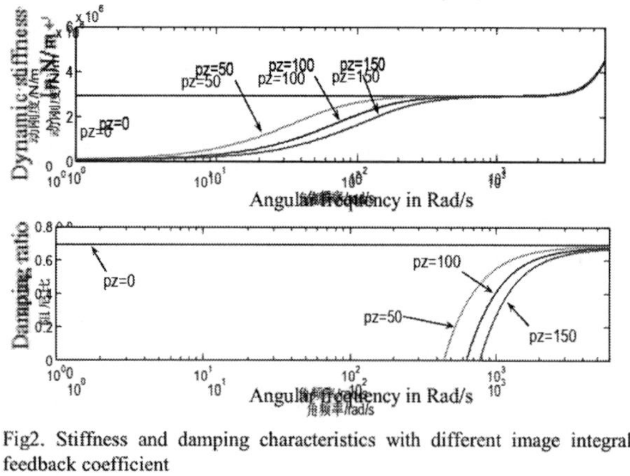

Fig2. Stiffness and damping characteristics with different image integral feedback coefficient

Ⅳ IMPROVED ZERO POWER CONTROLSTRATERY

In order to improve the stiffness of zero power control, we can introduce a small displacement feedback to the current terminal to increase the axial stiffness. The structure diagram is shown in figure 3.

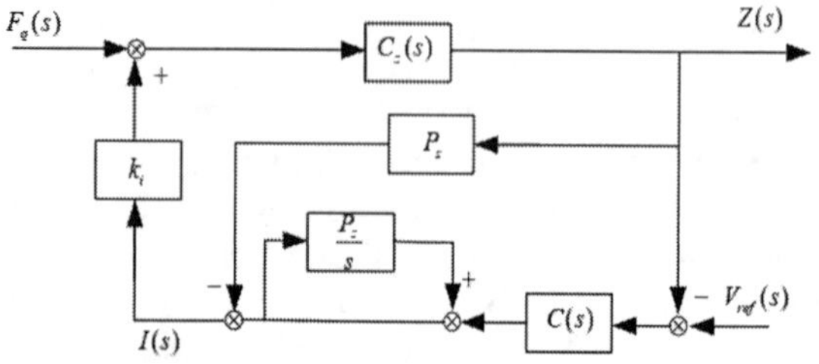

Figure 3. Diagram of improved zero power control

We can list the transfer function of disturbance force and output current by same method, which is shown in equation (6)

$$I(s) = \frac{-[k_d s^2+(k_p+P_s)s-P_sP_z]}{ms^3+(k_ik_d-mP_z)s^2+(k_ik_p+k_iP_s+k_x)s-P_z(k_iP_s+k_x)}F_q(s) \tag{6}$$

When the disturbance force is a step force, that is, $F_q(s)=\frac{F_0}{s}$, According to final value theorem

$$\lim_{t\to\infty} i(t) = \lim_{s\to 0} sI(s) = \frac{P_sF_0}{k_iP_s+k_x} \tag{7}$$

At this time when the current is in steady state, the coil current is not zero, but the constant., whose numerical value is related with the amplitude of interference force, force current stiffness coefficient and $\frac{P_s}{k_x}$. And the current of rotor is

$$Z(s) = \frac{-(P_z-s)}{ms^3+(k_ik_d-mP_z)s^2+(k_ik_p+k_iP_s+k_x)s-P_z(k_iP_s+k_x)}F_q(s) \tag{8}$$

By simplification

$$Z(s) = \frac{P_z-s}{(ms^2+k_x)(P_z-s)-K_i[k_d s^2+(k_p+P_s)s-P_zP_s]}F_q(s) \tag{9}$$

The axial displacement after rotor is suspended steadily

$$\lim_{z\to\infty} z(t) = \lim_{s\to 0} sZ(s) = \frac{F_0}{k_x+k_iP_s} \tag{10}$$

Compared to equation (3), $k_i P_s$ is added to the denominator, that is, we can increase axial stuffiness by adjusting P_s.

V EXPERIMENTALRESULTS

Based on the above analysis and research, we design and fabricate a maglev artificial heart pump prototype, as shown in figure 4.

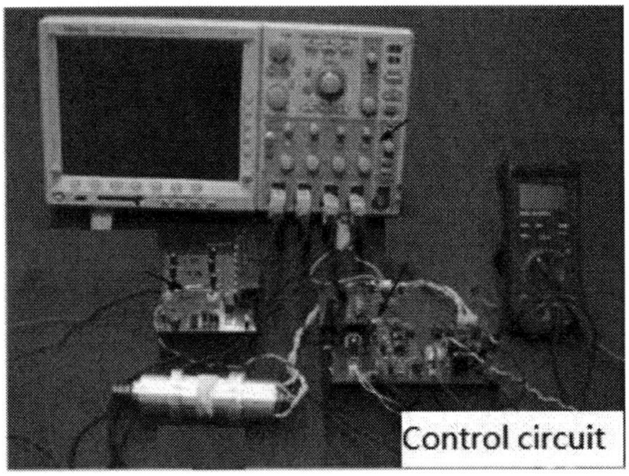

Figure 4. Maglev artificial heart pump prototype

Among them, Hall sensor UGN3503 is used to measure the rotor displacement. The sensor gain becomes 20V/mm and the amplifier gain becomes 0.03A/V after the signal conditioning circuit. The closed-loop control of coil current can be achieved by sampling resistor. By changing the position of the prototype, the load is changed. Figure 5 is the current curve after the change of load.

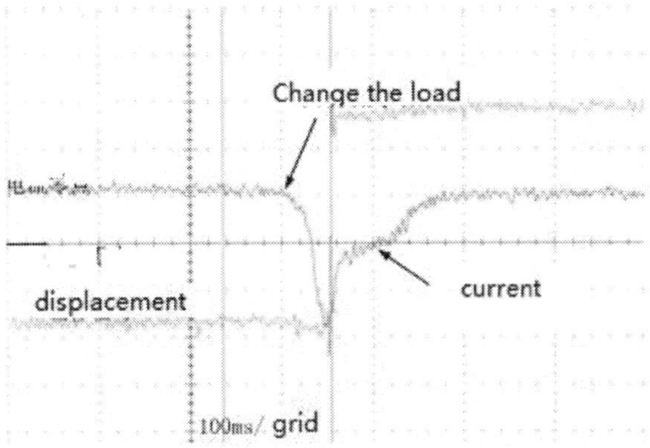

Figure 5. Zero power control curve when load changes

From Figure 5 we can know that the coil current is zero under the initial state, and the displacement of rotor deviates from the zero position because permanent magnetic force is not equal to radial force. When the load is changed, the coil current becomes reverse 0.3V, and then gradually decays to zero. The displacement of rotor turns to another equilibrium state, achieving zero power control.

Acknowledgment

This work was supported by National Natural Science Foundation of China (No. 51075236)

REFERENCES

[1] Choi H S, Lee J H, Cho B H, et al. Analysis and design considerations of zero-voltage and zero-current-switching (ZVZCS) full-bridge PWM converters[C]. Power Electronics Specialists Conference, 2002. pesc 02. 2002 IEEE 33rd Annual. IEEE, 2002, 4: 1835-1840.

[2] Ren Y, Fang J. Current-sensing resistor design to include current derivative in PWM H-bridge unipolar switching power amplifiers for magnetic bearings[J]. Industrial Electronics, IEEE Transactions on, 2012, 59(12): 4590-4600.

[3] Chen J, He Z F, Qi X. A new control method for MIMO first order time delay non-square systems[J]. Journal of Process Control, 2011, 21(4): 538-546..

[4] Deane J H B, Hamill D C. Instability, subharmonics, and chaos in power electronic systems[J]. Power Electronics, IEEE Transactions on, 1990, 5(3): 260-268.

[5] Akatsu S, Torikai H, Saito T. Zero-cross instantaneous state setting for control of a bifurcating H-bridge inverter[J]. International Journal of Bifurcation and Chaos, 2007, 17(10): 3571-3575.

[6] [6]Fan Youpeng,Research on Magnetic Bearing Power Amplifier and Low Power Control,ph D Dissertation of Shandong University,2013

Force Free Control in Active Magnetic Bearing Systems for the Magnetically Suspended Compound Molecule Pump

Pingfan Liu[a], Lei Zhao[a], Ying He[a], Kai Zhang[b]

[a] Institute of Nuclear and New Energy Technology
[a] The key laboratory of advanced reactor engineering and safety, Ministry of Education
[b] Department of Engineering Physics
Tsinghua University
Beijing, 100084
PR China
Email: lpfhit@163.com

***Abstract*—A force free control method is studied based on the active magnetic bearings (AMBs) system of the magnetically suspended compound molecule pump. This method uses the feed-forward strategy. The output signals generated with the least mean square (LMS) algorithm are the same as synchronous disturbances, which are caused by unbalance forces. So through this approaches, the controller need not respond to synchronous disturbances. This method can let the rotor rotate around its inertia axis and eliminate the synchronous disturbances of the currents in the electromagnet windings to significantly attenuate the mechanical vibration of the shell. The experimental results verify that the proposed approach is significantly minimized the amplitude of the shell vibration of the magnetically suspended compound molecule pump, which will provide good experience and design references for the future application of the AMBs in the molecule pump.**

I. INTRODUCTION

It is well known that the active magnetic bearings (AMBs) system have many advantages including no contact, no lubrication, low vibration and efficient operation at high speed. So it has been widely used in high-speed turbines such as the magnetically suspended compound molecule pump[1], [2]. Because of the existence of unbalanced masses, unbalance vibrations are generated when the shafts are rotating. They are also referred to as synchronous disturbances. With the rotating speed increasing, synchronous disturbances are more and more significant. Therefore, how to attenuate synchronous disturbances and increase system stability is of extreme importance in the AMBs design[3], [4]. To attenuate synchronous disturbances in AMBs, a number of techniques have been applied[5], [6], [7], including both feed-forward and feedback techniques. A force free control method can be used to reduce the unbalance vibrations by reducing the synchronous disturbances of the currents, which is physically equivalent to the shaft rotating about its principal axis of inertia. In this paper the force free control technique is applied in the magnetically suspended compound molecule pump. This method uses the feed-forward strategy. The output signals generated with the least mean square (LMS) algorithm are the same as synchronous disturbances[8], which are caused by unbalanced masses. So through this approaches, the controller need not respond to synchronous disturbances. This method can let the rotor rotate around its inertia axis and eliminate the synchronous disturbances of the currents in the electromagnet windings to significantly attenuate the mechanical vibration of the shell. The experimental results verify that the proposed approach is significantly minimized the amplitude of the shell vibration of the magnetically suspended compound molecule pump, which will provide good experience and design references for the future application of the AMBs in the molecule pump.

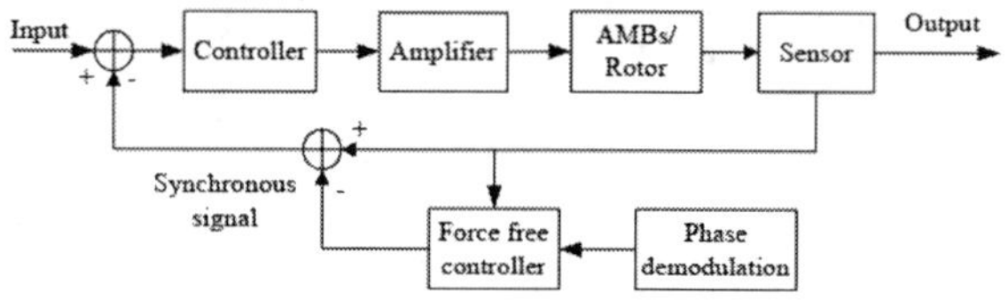

Figure 1. Block diagram of the force free control method

II. FORCE FREE CONTROL

A. General Description

In order to attenuate the mechanical vibration of the shell of the magnetically suspended compound molecule pump, a force free control method is to prevent the controller from saturate by subtracting the synchronous component from input of controller. The block diagram is shown in Fig. 1.

Figure 1 shows the proposed the force free controller block for the minimization of unbalance vibrations in AMBs. A controller is used for stabilizing AMBs, while the force free controller is used to generate compensating signals to minimize synchronous disturbances. The output of the force free controller is injected at the summing junction of the magnetic bearing feedback control loop.

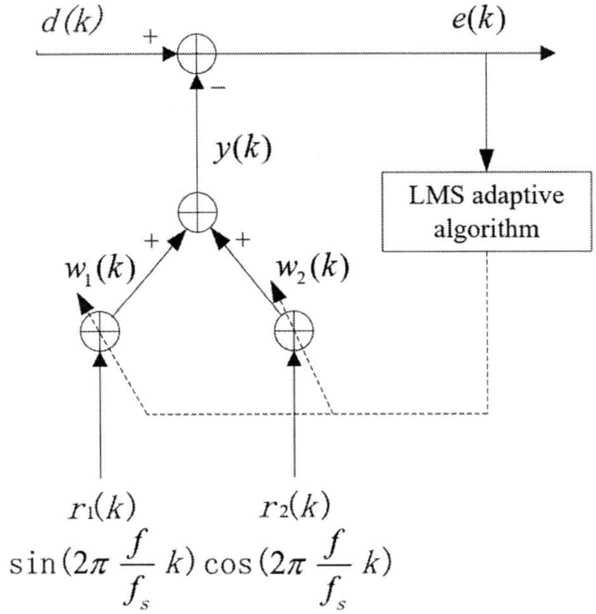

Figure 2. Structure of the force free controller

The basic idea of force free control is to generate a compensate signal with the same amplitude and phase of the synchronous disturbances signal, which is used for offsetting rotor vibration signals of the rotor displacements. And then controller need not respond to synchronous disturbances. So the core issue of force free control is how to get the compensation signal. In other words, the problem is how to extract the synchronous disturbances from the vibration signals.

B. LMS Algorithm

A detailed structure of the force free controller is shown in Fig. 2. Where f represents the rotational frequency in Hz, f_s represents the sampling frequency, and T is the sampling period, ie. $f_s = 1/T$. $d(k)$ is the input displacement signal, $y(k)$ is the compensation signal, and $e(k)$ is the error signal after compensation. Two reference signals $r_1(k)$ and $r_2(k)$ are adjusted by two weighting functions $w_1(k)$ and $w_2(k)$ to generate an estimation of the synchronous disturbance signal in $d(k)$.

These weighting functions are to be updated using the LMS algorithm derived below. LMS algorithm has a very simple structure, it modifies the gain parameter every sample time. It uses a momentary gradient method, and has the advantage of simple structure of algorithm. Assume that the input displacement signal to the force free controller is:

$$d(k) = sin(2\pi \frac{f}{f_s} k + \varphi) \tag{1}$$

Assume that the vector form of the two reference signals is:

$$X(k) = \begin{bmatrix} r_1(k) \\ r_2(k) \end{bmatrix} = \begin{bmatrix} \sin(2\pi \frac{f}{f_s} k) \\ \cos(2\pi \frac{f}{f_s} k) \end{bmatrix} \tag{2}$$

And the vector form of the two weighting functions also is:

$$W(k) = \begin{bmatrix} w_1(k) \\ w_2(k) \end{bmatrix} \tag{3}$$

So the compensation signal from the force free controller is:

$$y(k) = W'(k)X(k) \tag{4}$$

The objective of the adaptation algorithm is to seek parameters vector $y(k)$ so as to minimize mean squared vibration performance cost function:

$$J(W) = e^2(k) \tag{5}$$

where the error signal is:

$$e(k) = d(k) - y(k) = d(k) - W'(k)X(k) \tag{6}$$

And now the LMS algorithm is used to update the weighting functions. The basis for the LMS algorithm is the momentary quadratic error. The momentary gradient is calculated by:

$$\nabla_W \{J(W)\} = -2e(k)X(k) \tag{7}$$

The weight coefficient is updated by using the momentary gradient:

$$W(k+1) = W(k) - c\nabla_W \{J(W)\} = W(k) + 2ce(k)X(k) \tag{8}$$

where c is the adjustment factor. When c is small, the adaptive process is not fast and not sensitive to the change of the signal amplitude. But when c is too large the process is sensitive to noise and tends to unstable. The force free control method requires a relatively small c for it uses a momentary gradient.

III. EXPERIMENTS RESULTS

Experiments have been carried out in the AMBs system of the magnetically suspended compound molecule pump. The force free control method has been tested and investigated. Because the performances of the upper and lower radial AMBs are similar, so only the lower radial AMB is discussed, which is represented by AMB1.The force free controller is injected at the summing junction of the feedback control loop at the rotating speed of 200 Hz.

The measured actuator current in the x-direction of the AMB1 is represented by i_x . Fig. 3 shows i_x time response with and without the force free controller. Fig. 4 and Fig. 5 show the frequency response. It will be noted that the amplitude of the actuator current is significantly reduced when implementing the force free controller, and even no synchronous disturbances are observed. As a consequence the unbalance forces at the synchronous speed are negligible. Therefore the amount of improvement achieved in reduction of the shell vibration of the magnetically suspended compound molecule pump is more than 15dB.

Furthermore, the effects of the force free controller on the displacement of the AMB1 are also analyzed. Fig. 6 shows the shaft centerline orbits at the AMB1 without the force free

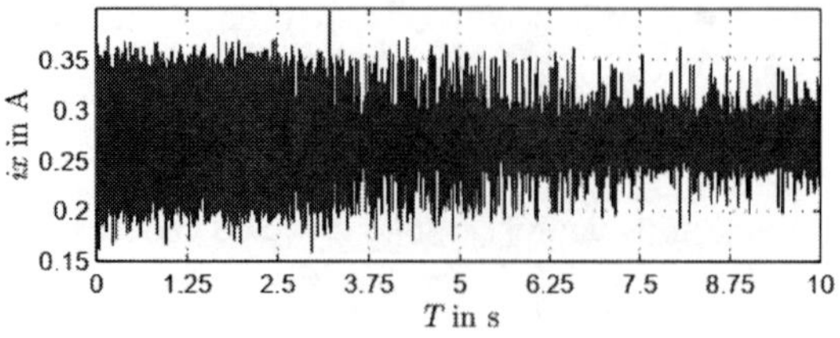

Figure 3. i_x time response with and without the force free controller

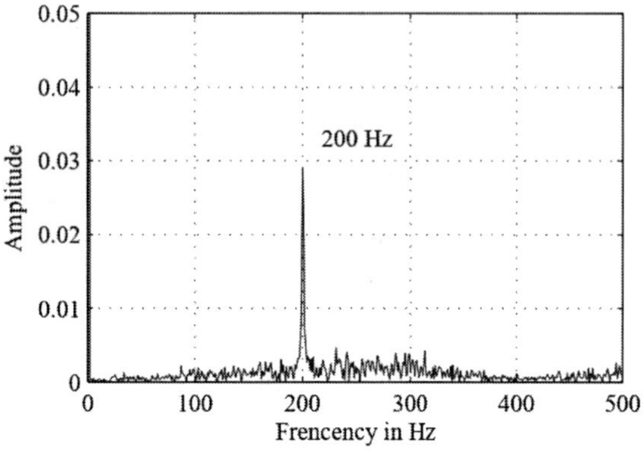

Figure 4. i_x frequency response without the force free controller

controller. The shaft centerline orbits at the AMB1 with the force free controller are shown in Fig. 7. Fig. 8 and Fig. 9 show the frequency response of the displacement in the x-direction of the AMB1. The comparison of them indicates that the size of the orbits is slightly big with the use of the force free controller at the rotating speed of 200 Hz. The amplitude of the rotor displacement is increased from 16 μm to 20 μm. But it is also much smaller than the protection gap of the magnetically suspended compound molecule pump. So the performance of the magnetically suspended compound molecule pump is not affected.

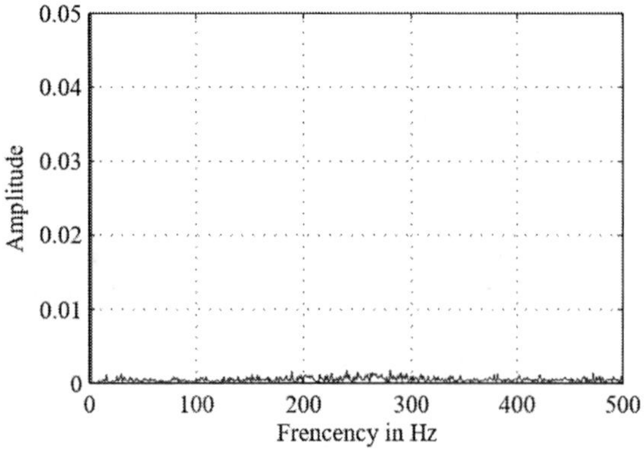

Figure 5. i_x frequency response with the force free controller

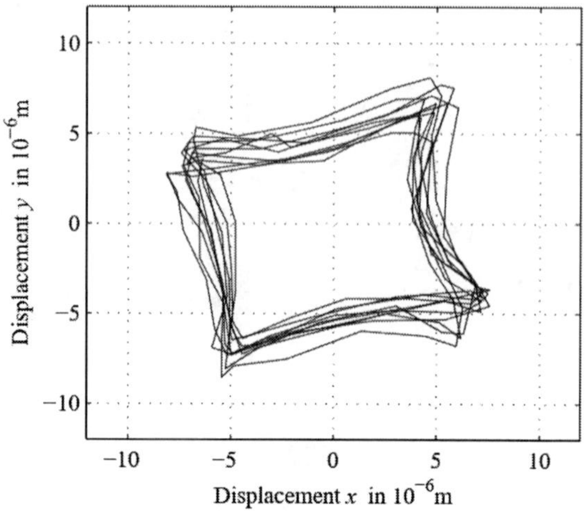

Figure 6. Shaft centerline orbits at the AMB1 without the force free controller

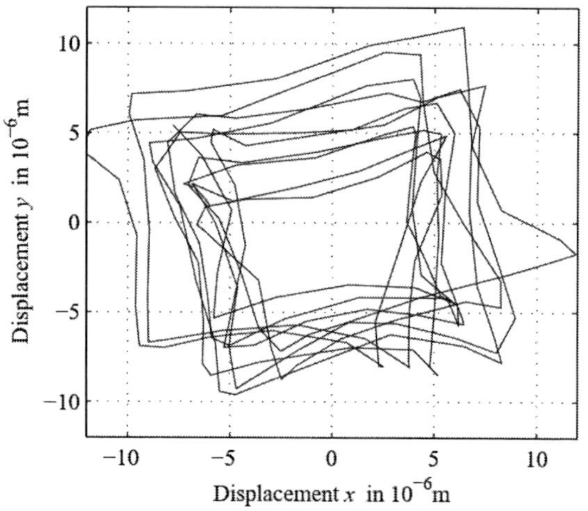

Figure 7. Shaft centerline orbits at the AMB1 with the force free controller

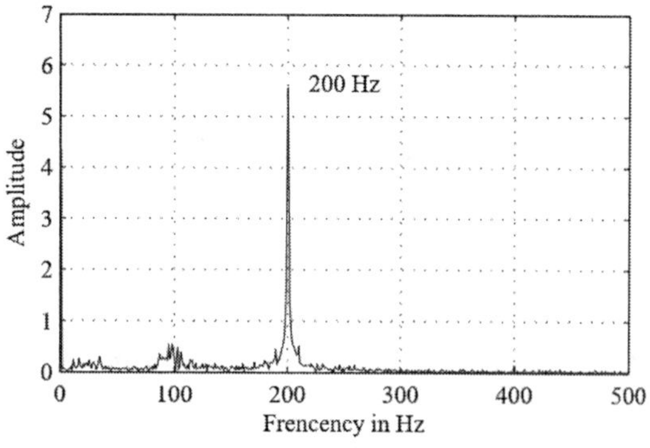

Figure 8. Displacement x frequency response without the force free controller

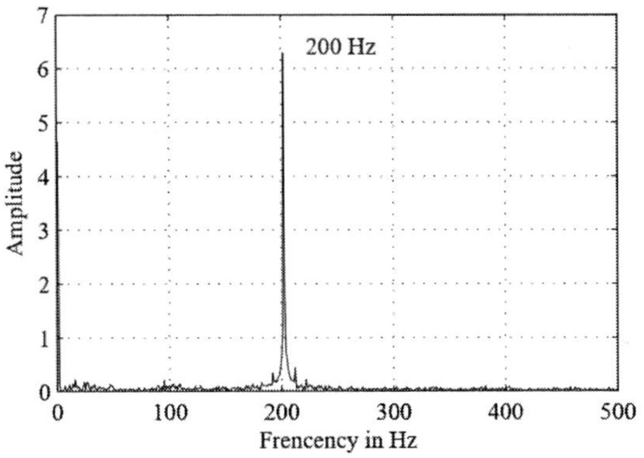

Figure 9. Displacement x frequency response with the force free controller

IV. CONCLUSIONS

In this paper, we have investigated the force free control method in the AMBs system of the magnetically suspended compound molecule pump. The LMS algorithm is proposed to get the compensation signal. The synchronous disturbances of the currents in the electromagnet windings have been eliminated. So the rotor has rotated around its inertia axis, and finally the machine vibration has been restrained. The force free control method has been successfully implemented in real time operation using the magnetically suspended compound molecule pump at Tsinghua University and this technique prove to be significantly practical and it achieved good results.

V. ACKNOWLEDGMENT

Thanks go to Lei Zhao for his constant encouragement and guidance. I also would like to express my thanks to Kai Zhang who have instructed and helped me a lot. This paper is financially supported by the National Science and Technology Major Project of China (2011ZX069).

REFERENCES

[1] G. Schweitzer, H. Bleuler, and A. Traxler, "Active magnetic bearings-basics, properties and application of active magnetic bearings," *ETH, Switzerland*, 1994.

[2] K. Wada, T. Inohara, M. Yoshida, Y. Yamato, T. Nakayasu, M. Iguchi, Y. Hikichi, N. Ogiwara, and K. Mio, "Development of the radiation-hardened magnetically suspended compound molecular pump," *Vacuum*, vol. 84, no. 5, pp. 699–704, 2009.

[3] K. Zhang and X. Zhang, "A review of unbalance control technology of active magnetic bearings," *China Mechanical Engineering*, vol. 21, no. 8, pp. 897–903, 2010.

[4] C. R. Knospe, R. W. Hope, S. J. Fedigan, and R. D. Williams, "Experiments in the control of unbalance response using magnetic bearings," *Mechatronics*, vol. 5, no. 4, pp. 385–400, 1995.

[5] K. Zhang, J. Dong, X. Dai, and X. Zhang, "Vibration control of a turbo molecular pump suspended by active magnetic bearings," in *Proceedings of ASME Turbo Expo 2011*. ASME, 2011.

[6] J. Shi, R. Zmood, and L. Qin, "The indirect adaptive feed-forward control in magnetic bearing systems for minimizing selected vibration performance measures," in *Proceedings of the Eighth International Symposium on Magnetic Bearings*, 2002.

[7] C. Hui, L. Shi, J. Wang, and S. Yu, "Adaptive unbalance vibration control of active magnetic bearing systems for the htr-10gt," in *Proceedings of the 18th International Conference on Nuclear Engineering*, 2010, pp. 793–802.

[8] Z. Dekui, J. Wei, and Z. Hongbin, "Unbalance vibration control methods for active magnetic bearings system," *Journal of Tsinghua University*, vol. 10, p. 008, 2000.

Structures and Characteristics of Magnetic Bearing with Zero Power Controlled Electromagnetic Suspension

Yoshitake Kamijo[a], Hiroaki Ito[a], Yutaka Maruyama[a]

[a] Toshiba Corporation, Toshiba-Cho 1, Fuchu-Shi, Tokyo, Japan, yoshitake.kamijo@toshiba.co.jp

***Abstract*— This paper proposes a new magnetic bearing in order to decrease rotor eccentricity and power consumption. The proposed bearing system consists of moving frames, suspensions, permanent magnets, electromagnets and Zero Power controller. The system enables to avoid rotor touchdowns. A clearance between a rotor and our proposed bearing may be increased compared to conventional electromagnetic bearings. The paper shows the experimental results to verify testing system.**

I. Introduction

There is Zero Power Control as a means to realize magnetic levitation by power saving. Zero Power Control enables to converge electromagnetic current to zero, using magnet units which consisted of permanent magnets and electromagnets, keeping a levitating object stable [1]. As application of this Zero Power Control, we aim at development of magnetic bearings with Zero Power Controlled electromagnetic suspension paying attention to the magnetic bearing which carries out non-contact support of a rotor. Power saving magnetic bearings can be constituted by applying Zero Power Control. Furthermore, since the gap between magnet units and a rotor can be set up greatly, the risk of touchdown can be reduced and improvement in reliability can be aimed at. However, if Zero Power Control is applied to magnetic bearings, a severe problem emerges. When a levitating rotor is subjected to external steady force, to keep equilibrium between external force and magnetic force, a position of levitating rotor may be changed. Therefore, when Zero Power Control is applied to the magnetic bearing, there is a problem that rotor eccentricity increases by external force. To overcome this difficulty, this paper proposes a new magnetic bearing structure with moving frames and suspensions. The structure enables to decrease a rotor eccentricity, and a risk of contacting touchdown bearings with fixed frames is reduced. The paper shows the experimental results to verify the system.

II. Magnetic Bearing with Zero Power Controlled Electromagnetic Suspension

A. Structure of Magnetic Bearing

Fig. 1 shows the fundamental composition of magnetic bearings to propose. Magnet units are placed on opposite sides of moving frames, and a levitating rotor is placed between magnet units. Moving frames are supported by means of springs, and enable to displace according to the external force acting on the levitating rotor. Touchdown bearings are placed on fixed frames, and not interlocked with a motion of moving frames. The levitating rotor is biaxially controlled by placing magnetic support equipment in a vertical and horizontal direction of the levitating rotor.

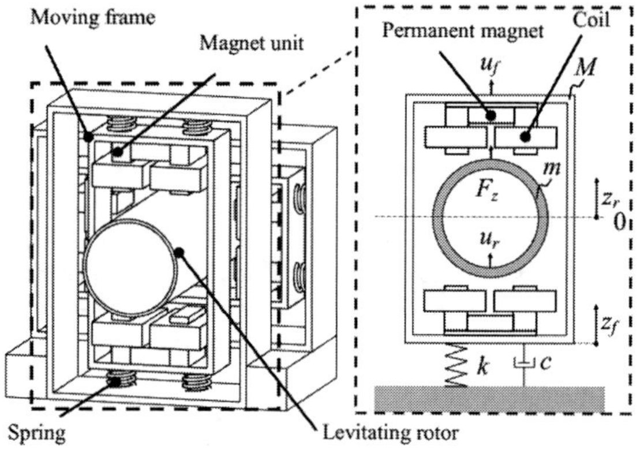

Fig. 1. Structure of Magnetic Bearing

B. Principle of fixed levitating position

As shown in Fig. 1, when the coupled model of magnetic levitation systems and spring systems is defined, motion equation for a levitating rotor and moving frames is given by equation (1) and (2).

$$m(\Delta\ddot{z}_r + \Delta\ddot{z}_f) = \frac{\partial F_z}{\partial z_r}\Delta z_r + \frac{\partial F_z}{\partial i_z}\Delta i_z + u_r \tag{1}$$

$$M\Delta\ddot{z}_f = -\frac{\partial F_z}{\partial z_r}\Delta z_r - \frac{\partial F_z}{\partial i_z}\Delta i_z - k\Delta z_f - c\Delta\dot{z}_f + u_f \tag{2}$$

where z_r is a gap between levitating rotor and magnetic units, z_f is the absolute position of moving frames, m is the mass of the levitating rotor, M is the mass of the moving frame, F_z is the total of the electromagnetic force received from the vertical or horizontal magnet units in a static state, i_z is exciting current, u_r is the external force acting on the levitating rotor, u_f is the external force acting on the moving frame, k is a spring constant, and c is a damping coefficient.

As $u_f = 0$, when the external force u_r acts on the levitating rotor, variation on the absolute position of the levitating rotor is represented by $\Delta z_r + \Delta z_f$. If Zero Power Control is applied, it is set to $i_z = 0$ in a static state. Therefore, when the time-derivative term is zero in equation (1), static displacement of the levitating rotor and the moving frame is given by equation (3).

$$\Delta z_r = -\frac{u_r}{\partial F_z / \partial z_r}, \ \Delta z_f = \frac{u_r}{k} \tag{3}$$

Therefore, $\Delta z_r + \Delta z_f$ is given by equation (4).

$$\Delta z_r + \Delta z_f = -\frac{u_r}{\partial F_z / \partial z_r} + \frac{u_r}{k} \tag{4}$$

Therefore, if spring constant k equals to magnetic spring constant $|\partial F_z / \partial z_r|$, the absolute position of the levitating rotor is not be theoretically changed to static external force. There is previous work used this principle [2]. However, there is no example of application to a levitating rotor. By placing magnetic units in vertical or horizontal axis, High linearity area exists broadly. However, since it is not strict linearity, some variation arises (Fig. 2).

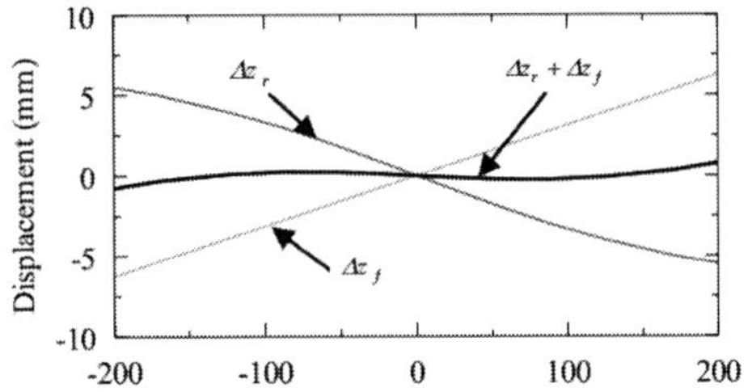

Fig. 2. Suspension property

C. Zero Power Controller

Fig. 3 shows the composition of Zero Power Controller. The control input u is calculated from the gain compensation loop which multiples gap and exciting current detected by displacement and current sensor, the change speed of the gap and external force acting on the levitating rotor by the gain F_d and the current integration loop which multiplies the integration value of exciting current by the gain K.

It is made to stabilize by designing each gain suitably. In addition, change speed and external force of the gap were presumed using the minimum dimension observer.

Voltage equation is given by equation (5). where e_z is exciting voltage, R is coil resistance, N is coil turns, ϕ_z is interlinkage magnetic flux, and L is inductance.

$$e_z = R\Delta i_z + N\frac{\partial \phi_z}{\partial z_r}\Delta \dot{z}_r + L\Delta \dot{i}_z \tag{5}$$

State equation is given by equation (6) derived from equation (1), (2), and (5).

$$\begin{cases} \dot{x} = Ax + bu + df \\ y = cx \end{cases} \tag{6}$$

where

$$x = \begin{pmatrix} \Delta z_r \\ \Delta z_f \\ \Delta \dot{z}_r \\ \Delta \dot{z}_f \\ \Delta i_z \end{pmatrix}, \ A = \begin{pmatrix} 0 & 0 & 1 & 0 & 0 \\ 0 & 0 & 0 & 1 & 0 \\ a_{31} & a_{32} & 0 & a_{34} & a_{35} \\ a_{41} & a_{42} & 0 & a_{44} & a_{45} \\ 0 & 0 & a_{53} & 0 & a_{55} \end{pmatrix}, \ b = \begin{pmatrix} 0 \\ 0 \\ 0 \\ 0 \\ b_{51} \end{pmatrix},$$

$$c = \begin{pmatrix} 1 & 0 & 0 & 0 & 0 \\ 0 & 1 & 0 & 0 & 0 \\ 0 & 0 & 0 & 0 & 1 \end{pmatrix}, \ d = \begin{pmatrix} 0 & 0 \\ 0 & 0 \\ d_{31} & d_{32} \\ 0 & d_{42} \\ 0 & 0 \end{pmatrix}, \ f = \begin{pmatrix} u_r \\ u_f \end{pmatrix}$$

$$a_{31} = \left(\frac{1}{M} + \frac{1}{m}\right)\frac{\partial F_z}{\partial z_r}, \ a_{32} = \frac{k}{M}, \ a_{34} = \frac{c}{M},$$

$$a_{35} = \left(\frac{1}{M} + \frac{1}{m}\right)\frac{\partial F_z}{\partial i_z}, \ a_{41} = -\frac{1}{M}\frac{\partial F_z}{\partial z_r}, \ a_{42} = -\frac{k}{M},$$

$$a_{44} = -\frac{c}{M}, \ a_{45} = -\frac{1}{M}\frac{\partial F_z}{\partial i_z}, \ a_{53} = -\frac{N}{L}\frac{\partial \phi_z}{\partial z_r},$$

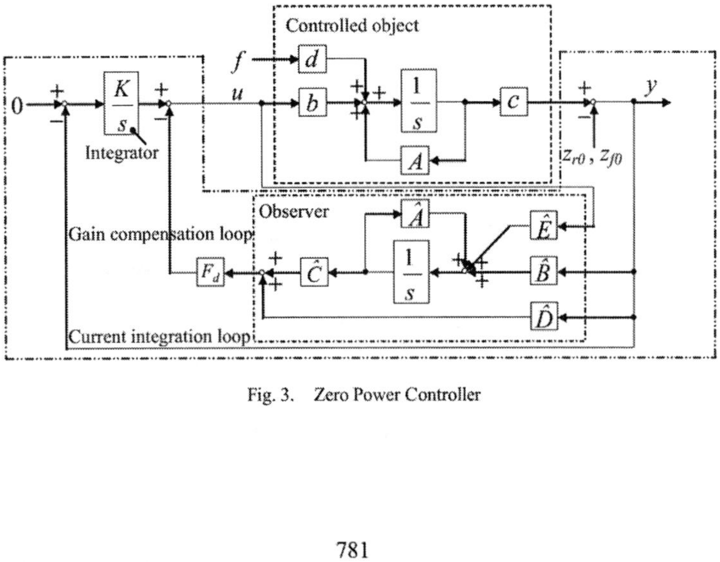

Fig. 3. Zero Power Controller

$$a_{55} = -\frac{R}{L},\ b_{51} = \frac{1}{L},\ d_{31} = \frac{1}{m},\ d_{32} = -\frac{1}{M},\ d_{42} = \frac{1}{M}$$

Moreover, u is a control input (exciting voltage e_z).

At this time, a control rule of Fig. 3 is given by equation (7).

$$u = -F_d \hat{x} - K\int y dt \quad (7)$$

where

$$F_d = \begin{pmatrix} F_1 & F_2 & F_3 & F_4 & F_5 & F_6 & F_7 \end{pmatrix},\ \hat{x} = \begin{pmatrix} \Delta z_r \\ \Delta z_f \\ \Delta \dot{z}_{rh} \\ \Delta \dot{z}_{fh} \\ \Delta i_z \\ u_{rh} \\ u_{fh} \end{pmatrix},$$

$$K = \begin{pmatrix} 0 & 0 & K_3 \end{pmatrix}$$

In addition, $\hat{x}$ is an output of the minimum dimension observer and is defined by equation (8).

$$\begin{cases} \dot{z}_{ob} = \hat{A} z_{ob} + \hat{B} y + \hat{E} u \\ \hat{x} = \hat{C} z_{ob} + \hat{D} y \end{cases} \quad (8)$$

where

$$z_{ob} = \begin{pmatrix} \Delta \dot{z}_{roh} \\ \Delta \dot{z}_{foh} \\ u_{roh} \\ u_{foh} \end{pmatrix},\ \hat{A} = \begin{pmatrix} -\alpha_1 & a_{34} & d_{31} & d_{32} \\ 0 & a_{44} - \alpha_2 & 0 & d_{42} \\ -\alpha_3 & 0 & 0 & 0 \\ 0 & -\alpha_4 & 0 & 0 \end{pmatrix},$$

$$\hat{B} = \begin{pmatrix} a_{31} - \alpha_1^2 + d_{31}\alpha_3 & a_{32} + a_{34}\alpha_2 + d_{32}\alpha_4 & a_{35} \\ a_{41} & a_{42} + (a_{44} - \alpha_2)\alpha_2 + d_{42}\alpha_4 & a_{45} \\ -\alpha_1\alpha_3 & 0 & 0 \\ 0 & -\alpha_2\alpha_4 & 0 \end{pmatrix},$$

$$\hat{C} = \begin{pmatrix} 0 & 0 & 0 & 0 \\ 0 & 0 & 0 & 0 \\ 1 & 0 & 0 & 0 \\ 0 & 1 & 0 & 0 \\ 0 & 0 & 0 & 0 \\ 0 & 0 & 1 & 0 \\ 0 & 0 & 0 & 1 \end{pmatrix},\ \hat{D} = \begin{pmatrix} 1 & 0 & 0 \\ 0 & 1 & 0 \\ \alpha_1 & 0 & 0 \\ 0 & \alpha_2 & 0 \\ 0 & 0 & 1 \\ \alpha_3 & 0 & 0 \\ 0 & \alpha_4 & 0 \end{pmatrix},\ \hat{E} = \begin{pmatrix} 0 \\ 0 \\ 0 \\ 0 \end{pmatrix}$$

α_1, α_2, α_3, α_4 are parameters which determine the poles of the observer. Laplace transform of equation (6), (7), and (8) is respectively carried out, and solved for state variables. As a result, it is given by equation (9).

$$X(s) = \left(sI - A + bF_d\hat{C}\left(sI - \hat{A}\right)^{-1}\hat{B}c + b\left(F_d\hat{D} + \frac{K}{s}\right)c \right)^{-1} \left(x_0 - bF_d\hat{C}(sI - A)^{-1}z_0 + dF(s)\right) \quad (9)$$

$\Phi(s)$ is defined by equation (10).

$$\Phi(s) \overset{def}{\equiv} \left(sI - A + bF_d\hat{C}\left(sI - \hat{A}\right)^{-1}\hat{B}c + b\left(F_d\hat{D} + \frac{K}{s}\right)c \right)^{-1} \quad (10)$$

$\Phi(s)$ is given by equation (11).

$$\Phi(s) = \frac{1}{\Gamma(s)\Gamma_{rob}(s)\Gamma_{fob}(s)} \begin{pmatrix} \phi_{11} & \phi_{12} & \phi_{13} & \phi_{14} & \phi_{15} \\ \phi_{21} & \phi_{22} & \phi_{23} & \phi_{24} & \phi_{25} \\ \phi_{31} & \phi_{32} & \phi_{33} & \phi_{34} & \phi_{35} \\ \phi_{41} & \phi_{42} & \phi_{43} & \phi_{44} & \phi_{45} \\ \phi_{51} & \phi_{52} & \phi_{53} & \phi_{54} & \phi_{55} \end{pmatrix} \quad (11)$$

The characteristic equation of proposed system is given by equation (12).

$$\Gamma(s)\Gamma_{rob}(s)\Gamma_{fob}(s) = 0 \quad (12)$$

where $\Gamma(s) = 0$ is a characteristic equation of a magnetic levitation control drive. $\Gamma_{rob}(s) = 0$ is a characteristic equation of the observer on the levitating rotor, and $\Gamma_{fob}(s) = 0$ is a characteristic equation of the observer on the moving frames. $\phi_{mn}(m,n = 1\sim5)$ is a polynomial of s. Stability of a system is aimed at by determining α_1, α_2, α_3, α_4, and feedback gains $F_1 \sim F_7$ that the characteristic root obtained from equation (11) is in left half area of s plane.

III. EXPERIMENT VERIFICATION

A appearance of verification magnetic bearing is shown in Fig . 4, and design specifications are given by Table 1. Non-contact support of the levitating rotor is carried out as a magnetic bearing by 2 axis control of vertical and horizontal sides, it is necessary to place the magnetic unit and moving frame of each axis so that it may operate independently. In that case, installation positions are shifted so that mutual operation may not interfere, and equipment is enlarged axially. Moreover, it is necessary to set up the control parameters independently since the operating point of the magnetic force which acts on the rotor differs. As a result, complication of control system design and decreased stability may be caused. Then, in 2 axis control, the vertical and horizontal frames are placed on the same plane as U-shaped frames. By this composition, the operating point of the magnetic force which acts on the rotor becomes the same, and it can make the same vertical and horizontal control drive designs.

The verification experiment of the rotational characteristic was conducted using the magnetic bearing of this composition. The action on the relative displacement of a rotor and a moving frame is shown in Fig. 5. The action on the moving frame is shown in Fig. 6. The action on the absolute position of the rotor is shown in Fig. 7. The response of exciting current is shown in Fig. 8. Although the relative position of a rotor and a moving frame, and the position of a moving frame are changed respectively, absolute position change of the rotor determined by the balance is a maximum of 0.4 (mm). Since the nominal gap of this equipment is set as 4 (mm), the rotor eccentricity rate showing the rate of the absolute position change to a nominal gap is set to 10 (%).

As a feature of an action, it turns out that absolute position change of a rotor is increasing in the 700 (rpm) neighborhood. This is a peak by the natural frequency of a magnetic levitation system. The natural frequency of a magnetic levitation system is determined by the response of control. If a control gain is set up highly, a response will go up and natural frequency will become high. This control drive designed so that natural frequency might be set to about 11 (Hz) in consideration of a response and stability. Therefore, it is the characteristic as a design mostly. Unlike the general magnetic bearing, this bearing has the double structure where a moving frame is arranged in a fixed frame. Although a rotor position change can be controlled, without increasing bearing stiffness theoretically, resonance of a moving frame exists according to double structure. The natural frequency of a moving frame is from 800 (rpm) to 900 (rpm), and if this is exceeded, exciting current increases to a maximum of 2 (A).

Characteristic of moving frames are shown in Fig .9. Spring has vibration free range and vibration isolation range. In the vibration free range, a rotor and a moving frame serve as a controlled object. However, since it is hard coming to move a moving frame in the vibration isolation range, it may be considered that a mass point is only a rotor. It is important for the characteristic to switch two control. The result of having switched control according to the characteristic of moving frames is shown in Fig. 10 and Fig. 11. This system balances rotor eccentricity rate and power consumption.

Fig. 4. Verification magnetic bearing (2-axis)

Table 1. Specification (2-axis)

Magnet unit	Outside dimensions	87 (mm)×87 (mm)×40 (mm)
	Magnet material	Nd-Fe-B
	Coil resistance	1.3 (Ω)
	Coil turns	250 (turn)
	Nominal gap	4 (mm)
Mechanical parts	Constant of spring	39.8 (N/mm)
	Levitation mass	5.8 (kg)

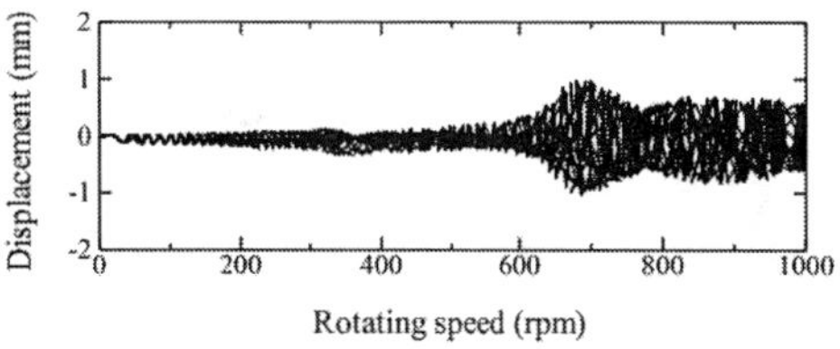

Fig. 5. Displacement between rotor and moving frame at y axis

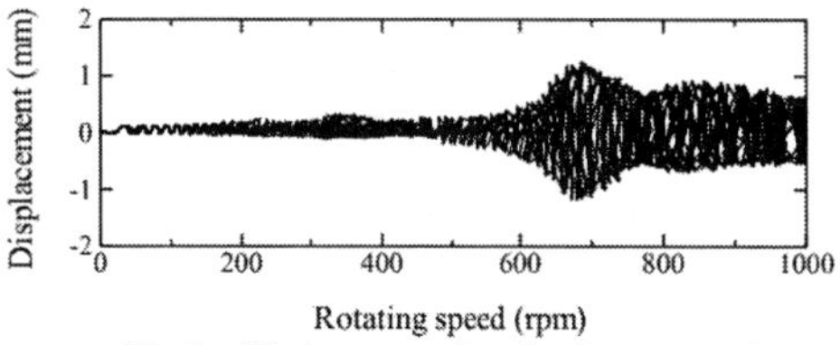

Fig. 6. Displacement of moving frame at y axis

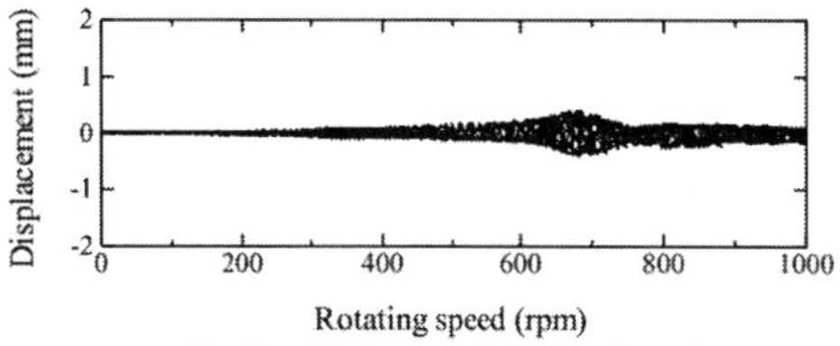

Fig. 7. Displacement of rotor at y axis

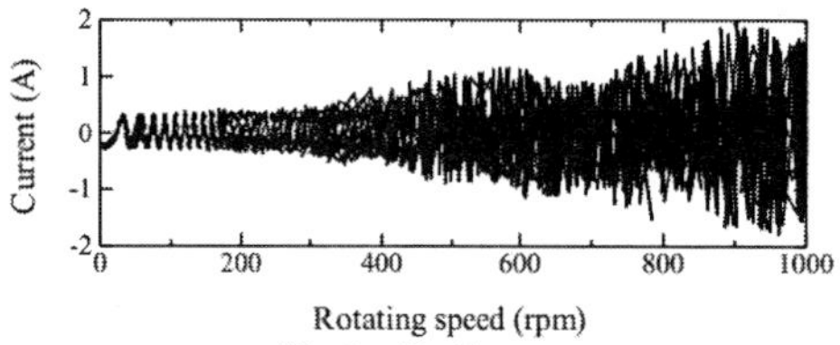

Fig. 8. Exciting current

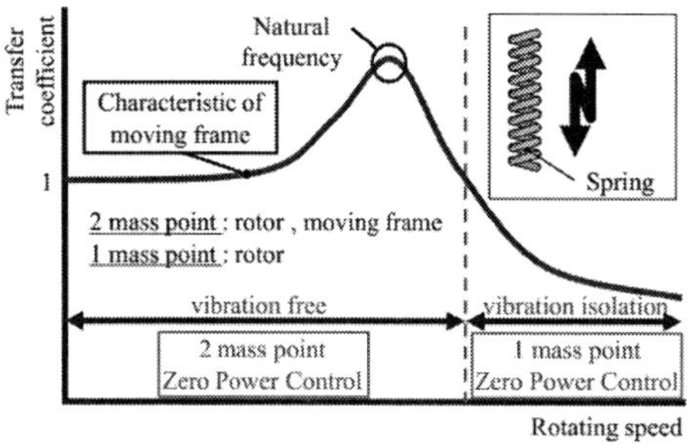

Fig. 9. Characteristic of moving frame

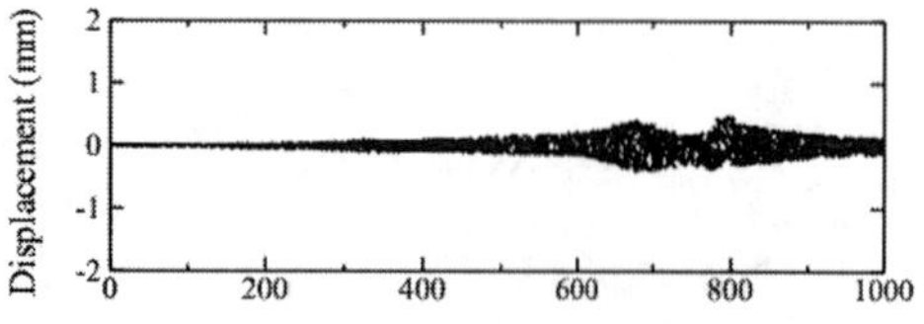

Fig. 10. Displacement of rotor with switching control

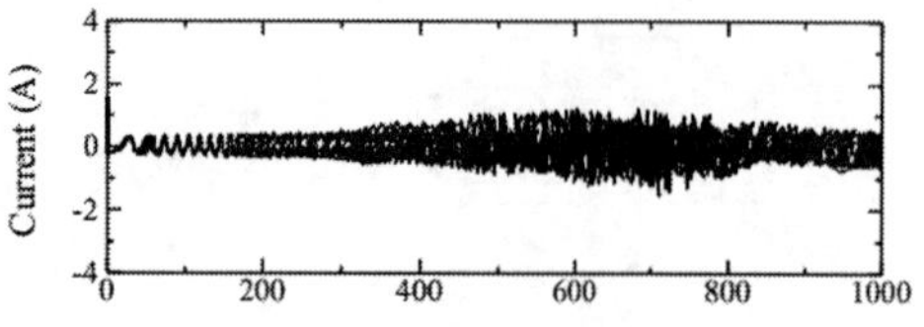

Fig. 11. Exciting current with switching control

IV. CONCLUSION

This paper proposes a new magnetic bearing in order to decrease rotor eccentricity and power consumption. Proposed bearing system consists of moving frames, suspensions, permanent magnets, electromagnets and Zero Power controller. The rate of the absolute position change to a nominal gap is set to 10 (%). The paper shows the experimental results to verify testing system.

REFERENCES

[1] M.Morishita, T.Azukizawa, "Zero Power Control Method for Electromagnetic Levitation System", Trans. IEE Japan, Vol.108-D, No.5, pp. 447-454, 1988.

[2] T.Mizuno, R.Yoshitomi, "Vibration Isolation System Using Zero-Power Magnetic Suspension", Trans. JSME, Vol.68-C, No.673, pp. 2599-2604, 2002.

Control of a Flexible Rotor AMB Compressor Test Rig with Aerodynamic Cross Coupling

Simon Mushi[a,b], Paul Allaire[a,c]

[a] Rotor Bearing Solutions International, LLC, 3277 Arbor Trace, Charlottesville, Virginia 22911, USA
[b] simon.mushi@rotorsolution.com
[c] paul.allaire@rotorsolution.com

Abstract—**Many industrial companies are adopting magnetic bearings for high speed compressors at this time. A flexible rotor test rig of approximately 1.2 m in length was developed with four active magnetic radial bearings. It was employed to evaluate the control of destabilizing aerodynamic cross coupling stiffness Q in the center of the machine. The two end AMBs were used for support and one center AMB used to apply the cross coupled stiffness corresponding to a typical location of gas seal cross coupled stiffness sources. The fourth AMB was employed as a system exciter. Unfortunately for the compressor manufacturer and the end user, often the magnitude of the cross coupled stiffness changes with various compressor operating conditions. The test rig shaft operated up to about 18,000 rpm, through the first bending critical speed which was at approximately 14,000 rpm.**

μ-synthesis control was used to achieve highly effective control and low vibration response with the selection of an averaged value of cross coupled stiffness Q and a set of useful system uncertainty weighting functions when the controller does not have exact advance notice of the cross coupling stiffness amplitude. Details of the application of the mu synthesis controller implementation are presented. A physical approach to modeling the Q uncertainty involved selecting an expected nominal value of the real part of the first rotor system eigenvalue for the rotor first bending mode with an expected range of uncertainly above and below that value. Then suitable weighting functions were developed for the mu synthesis controller design. The measured sensitivity function was below 3 at 7,000 rpm when the rotor was subjected to high cross coupled stiffness at the center.

(a) C16 Compressor Rotor

(b) C16 Compressor

Figure 1: Solar Turbines C16 Pipeliner Compressor

I. Introduction

One of the more pressing problems in the rotating machinery industry today concerns that of high pressure compressors with gas seal/liquid aerodynamic cross coupling forces near the center of a flexible rotor [1]. Applications include compressors in the oil, gas and chemical industries. A typical compressor rotor is shown in Fig. 1a and a compressor train in Fig. 1b. An interior photograph of a multistage compressor compressor is shown in Fig. 2.

Magnetic bearings are increasingly used for the bearing system and always used in subsea applications. Often, the value of the cross coupled stiffness is not well known. Also, the seal location is near the center of the rotor but the source of damping, whether the radial bearings are fluid film or magnetic, is located at the bearings which are placed at the ends of a long flexible rotor usually operating above the first bending critical speed [2], [3]. The result of high cross coupled stiffness is the excitation of rotor whirl at the compressor first bending critical speed, as illustrated in Fig. 3. An important issue with magnetic bearing systems in such cases is the treatment of uncertainty in the magnetic bearing system.

The frequency spectrum for a compressor undergoing such instability is shown in Fig. 4. The subsynchronous vibrations (SSV) are at 2520 cpm when the operating speed is 16,000 rpm. As the compressor designs continue to evolve with higher operating speeds and pressures in magnetic bearing supported machine, the problem will require a lot of improved analysis and control technology.

The concept of a destabilizing aero cross coupled stiffness acting on a flexible rotor is illustrated by Ehrich [1] as shown in Fig. 5. The destabilizing cross coupled force F_{xc} acts in the direction of positive whirl. The pattern of poorly damped linear unstable rotor vibrations is shown in Fig. 6a where the

Figure 2: A multistage centrifugal compressor with upper casing removed

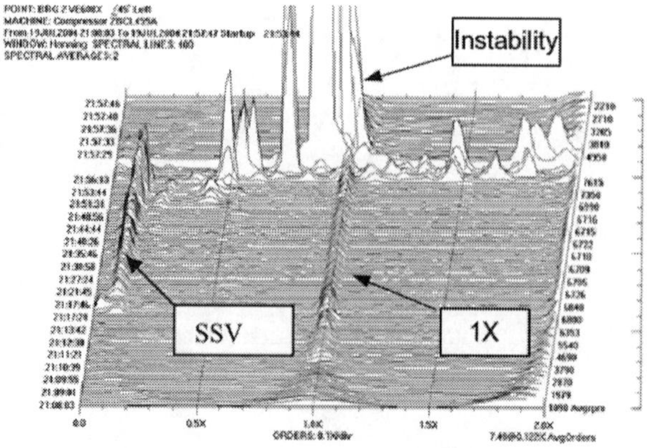

Figure 3: High Pressure Centrifugal Compressor Waterfall Plot Showing Rotordynamic Instability

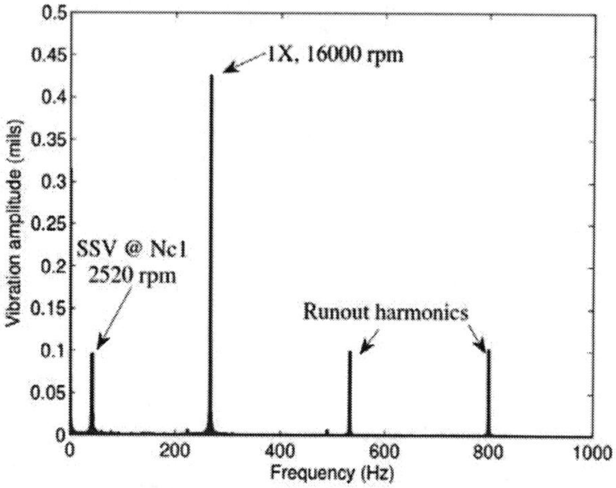

Figure 4: Spectrum Plot Showing Subsynchronous Vibration

vibrations are predicted to go off to infinity. In reality, the shaft eventually hits something, such as the auxiliary bearings and the nonlinear amplitude plot, as illustrated in Fig. 6b goes into a limit cycle. However, severe damage to the auxiliary bearings or other machine close clearance components may occur if the rotor instability is not robustly controlled.

Some formulas are used to estimate cross coupled stiffness Q such as the modified Alford equation [3] as given in Eq. (1). However, the numerical value of Q is often not well known, as noted above. This model uncertainty presents the opportunity for the application of robust modern control, which is the main subject of this paper.

$$Q = \frac{16 \times P \times MW}{D \times h \times f} \times \frac{\rho_D}{\rho_S} \tag{1}$$

The primary objective of this work is to develop a state space model of a flexible rotor-AMB system which models a compressor undergoing cross coupled excitation. The model includes uncertainty features: plant disturbances, sensor noise, actuator and system error signal [4]. A four block mixed sensitivity problem is formulated and diagonal weighting functions developed [5]. DK iteration was employed to optimize the system performance [6]. The controller was then implemented experimentally while operating well above the first bending critical speed. The system sensitivity function performance was evaluated and system damping measured using a single channel backward autoregression approach [7].

II. Flexible Rotor - AMB Test Rig with Exciters

The test rig is shown in Fig. 7. The shaft has length 1.23 m, shaft diameter 100 mm, and weight 440 N. Two disks are placed on the shaft as shown in Fig. 8. There are four magnetic bearings – two end bearings for support of the shaft, one at the center for simulation of the aero cross coupled stiffness, and the fourth one for external excitation to evaluate the system damping properties, as illustrated in Fig. 9. The direct drive motor is capable of a maximum operating speed of 18,000 rpm. A full description and characterization of the system is given in Mushi, et al. [8]. It has auxiliary bearings with radial clearance of 0.38 mm. Table I gives the details of the test rig design.

The rotor has eigenvalues (undamped natural frequencies) which change with speed due to gyroscopic forces, as illustrated in a Campbell diagram of rotor natural frequency vs. operating speed shown in Fig. 10. The red lines in Fig. 10 represent 1X and 2X times operating speed. The shaft free-free eigenvectors are shown in Fig. 11.

The plant model must be accurately determined for the control process to work as intended. The rotor characteristic transfer functions for the model nondrive end (NDEX) and drive end bearings (DEX) are shown in Fig. 12 and compared to experimental results. The agreement is good.

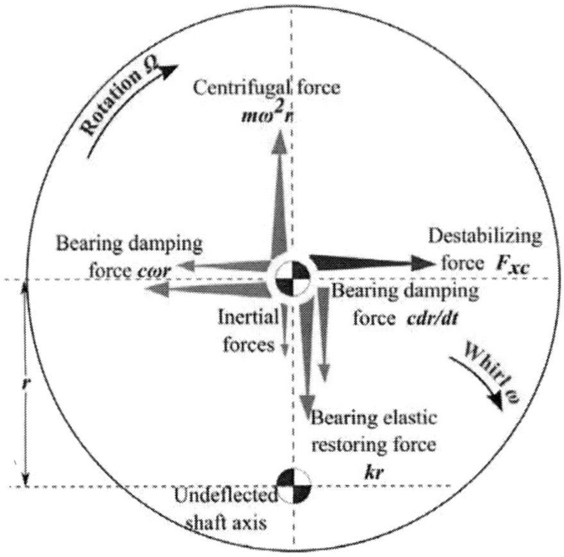

Figure 5: Cross-coupled forces

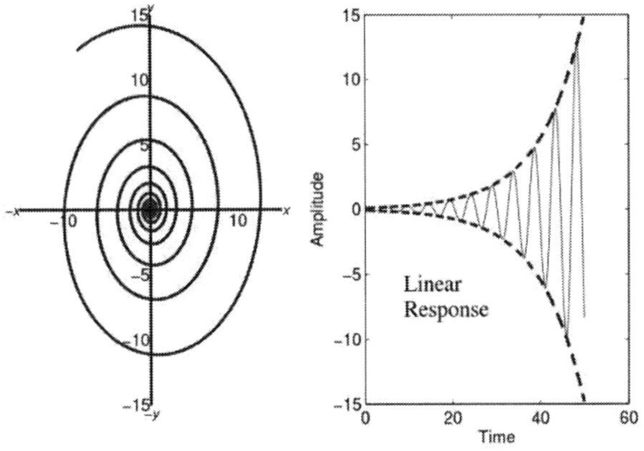

(a) Linear Response

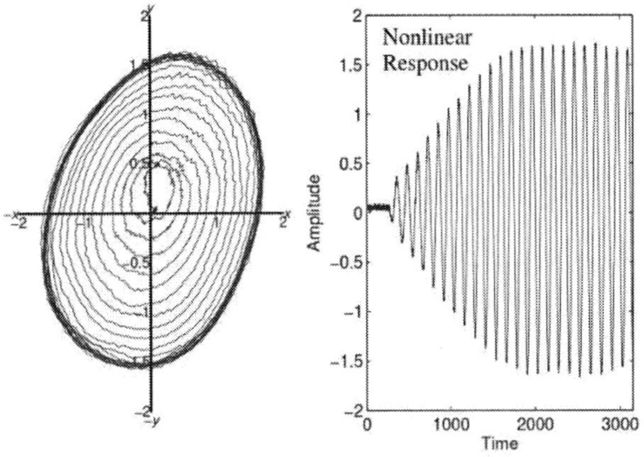

(b) Non-linear Response

Figure 6: Rotordynamic Instability Response

Figure 7: Active Magnetic Bearing Test Rig

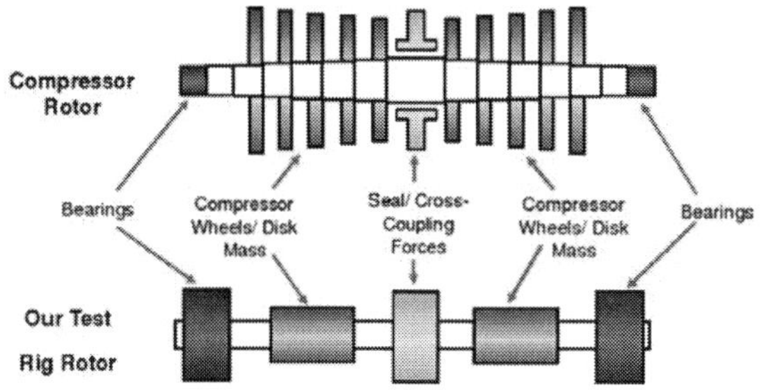

Figure 8: AMB Test Rig Approximation of a Compressor

Table I: Test Rig Properties

Property	Value	Units
Total rotor weight	440	N
Rotor length	1068	mm
Bearing span	918	mm
Average shaft diameter	57.7	mm
Area moment of inertia	5.43×10^{-7}	m^4
Mass moment of inertia	0.0880	$kg{\cdot}m^2$
1st free-free lateral bending mode	13,200	rpm
Maximum speed	18,000	rpm
Motor power	3.7	kW
AMB radial clearance (average)	550	μm
Backup bearing clearance	250	μm
Support AMB stator length	43.6	mm
Support AMB bias current	4.0	A
Support AMB negative stiffness	0.6 - 0.9	MN/m
Support AMB current gain	90-100	N/A
Power amplifier DC bus	150	V
Controller sampling rate	12	kHz

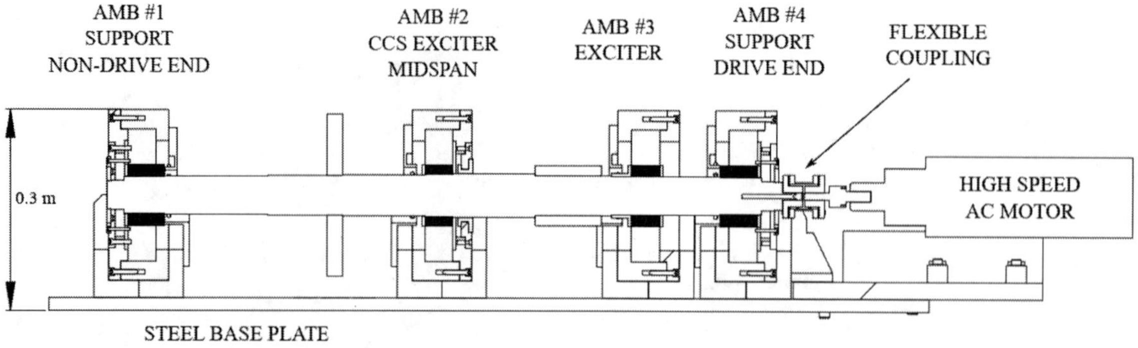

Figure 9: AMB Test Rig Elevation View

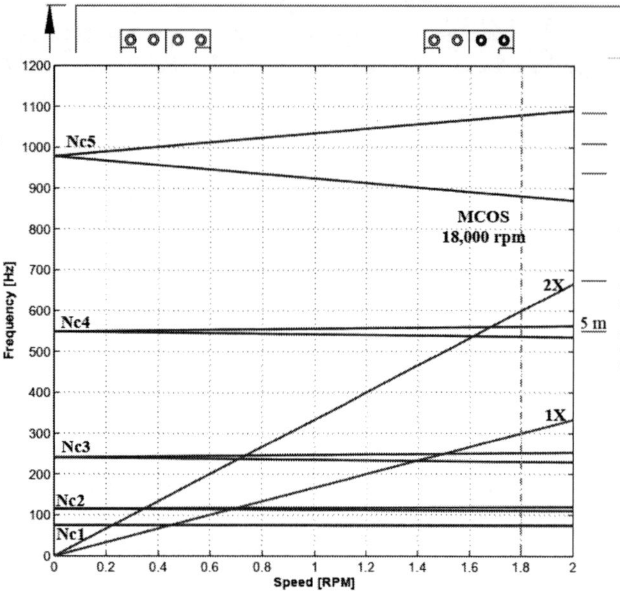

Figure 10: Rotor Campbell Diagram

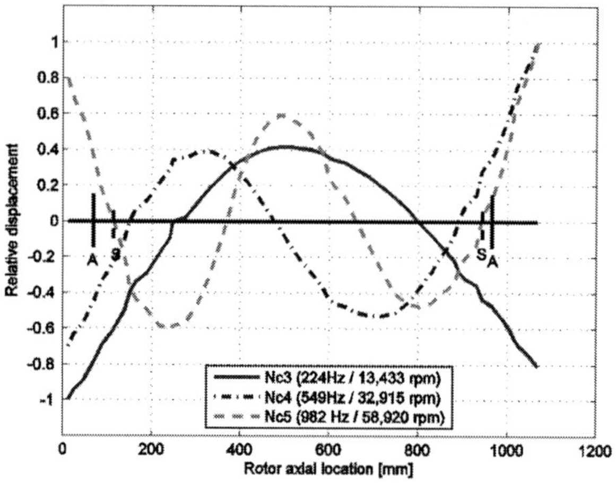

Figure 11: Rotor Undamped Mode Shapes

III. A CONTROL SYSTEMS PERSPECTIVE: MODEL BASED SYNTHESIS AND ANALYSIS TOOLS

A. Mu-synthesis and analysis

The μ-synthesis technique is the only multivariable control design approach that directly addresses the robust performance problem, i.e., the design of a stabilizing control law that guarantees a performance specification for all plant model perturbations within a defined set [9]. The theory was introduced in the 1980s, and steadily improving commercial software tools have been available in the last decade. A few notable prototype industrial rotor-AMB systems utilizing μ-synthesis have been documented in open literature [10], [11], but there is scant evidence the application of μ-analysis techniques. These facts highlight the significant challenges which remain to be addressed in order for the full potential of these modern techniques to be realized.

The most prominent challenge is the relatively high design complexity involved with these techniques and concerns regarding practical implementation and field troubleshooting. A state space block diagram of the flexible rotor-AMB plant for this test rig is shown in Fig. 13.

The last issue, field troubleshooting, is helped by the advent of remote diagnostic and communications capabilities of current AMB control systems that allow online system identification and auto-tuning [12], [13]. The second concern is less of an issue nowadays as the available computing power for executing real-time control algorithms continues to increase.

The first issue mentioned looms large as the translation of engineering performance specifications into formal constraints in the form of weighting functions, uncertainty models, and other design criteria remain a largely heuristic process [14]. These issues are not unique to μ-techniques, but symptomatic of the limited penetration of advanced multivariable control into industrial systems. One may argue why bother with an investment in μ-synthesis if a hand-tuned compensator designed by a control practitioner has the potential to deliver similar performance (this is particularly relevant in the industrial control system community dominated by hand tuned PID-like compensators). Maslen and Sawicki [15] answer this question by suggesting that investments in μ-techniques result in an improvement in the engineering process. The complexity and performance demanded of modern precision control systems is such that the notion when provided with a model and performance specification one can "turn a crank" and deliver a control law is naive [16]. Assuming that an accurate nominal plant model is available (a nontrivial task in many cases [17]), the remaining challenge is the development and continuous refinement of performance specifications and uncertainty models that yield controllers approaching the theoretical maximum achievable performance.

B. Mixed sensitivity weighting function selection

Performance specifications are characterized in the frequency domain as typical rotor-AMB systems lack specific time domain specifications, e.g., overshoot and settling time and these do not fit directly into the frequency based robust H_∞ and μ-synthesis frameworks [14]. The loop-shaping approach is preferred over the signal based approach as it does not require implementation specific information on the magnitudes of signals in the closed-loop system [18]. Sensitivity functions serve as useful measures of control performance in the frequency domain. Considering a unity feedback connection of a plant $G(s)$, controller $K(s)$, with exogenous disturbance forces acting on the plant input and output d_i and d_o respectively, sensor noise n and control effort u, controlled variable output y and controller reference input signal r. A block diagram of the flexible rotor AMB system is given in Fig. 14.

$$y(s) = T_i(s)r(s) + S_o(s)d_o(s) + GS_o(s)d_i(s) - T_o(s)n(s), \quad (2)$$

$$u(s) = KS_o(s)r(s) - LS_o(s)d_o(s) - S_i(s)d_i(s) - KS_o(s)n(s) \quad (3)$$

where several sensitivity functions can be defined [19]:

- Input Sensitivity $S_i(s) \triangleq (I + K(s)G(s)) - 1$: a measure of system rejection of disturbances at plant inputs;
- Complementary Input Sensitivity $Ti(s) \triangleq K(s)S_o(s)G(s)$: a measure of the effect of disturbance at the reference input on the control signal;
- Output Sensitivity $S_o(s) \triangleq (I + G(s)K(s)) - 1$: a measure of the noise rejection and closed-loop command tracking;
- Complementary Output Sensitivity $T_o(s) \triangleq GS_i(s)K(s)$: a measure of the effect of noise on the control signal;
- Process Sensitivity $GS_i(s)$: a measure of the closed loop mechanical compliance (reciprocal of stiffness) of the rotor-bearing;
- Control Sensitivity $KS_o(s)$: a measure of the control effort.

The ISO/API input sensitivity function limit for class A (new machines) is less than 3.0. An output sensitivity function for the test rig NDEX and DEX sensors with collocated PID control is given in Fig. 15. It exceeds the class A limit.

A bound on a single sensitivity function, such as the output sensitivity takes the form of $||W_s(s)S(s)||_\infty < 1$, where $W_s(s)$ is a stable, proper transfer function. If the inequality is satisfied, the inverse weighting function $W_s^{-1}(s)$ represents the upper bound on $S(s)$. A typical inverse weighting function to achieve low frequency disturbance attenuation, for example, is shown in Fig. 16.

For high performance motion control systems typified by rotor-AMB applications, a single sensitivity provides insufficient degrees of freedom to simultaneously specify disturbance rejection region while minimizing controller bandwidth and as well as reference tracking performance [4]. By combining several weighted sensitivity functions, i.e., $||W_s(s)S(s)||_\infty < 1$, $||W_u(s)KS(s)||_\infty < 1$, and $||W_r(s)T(s)||_\infty < 1$, into a stacked objective:

$$\min_K \left\| \begin{matrix} W_s(s)S(s) \\ W_u(s)KS(s) \\ W_r(s)T(s) \end{matrix} \right\|_\infty < \gamma \quad (4)$$

and letting the $H\infty$ synthesis process handle the trade-offs between competing goals so that maximum performance may be attained.

Experimental measurements of the sensitivity function for the x and y axes with cross coupled stiffness applied at the center AMB with standard PID control are presented here. The first one is given in Fig. 17 for a cross coupled stiffness value of 2.6MN/m and a peak sensitivity function of 2.5. With this level of cross coupling, the rotor satisfies the criteria of class A operation. With a higher value of cross coupled stiffness of 3.15 MN/m, the peak sensitivity function shown in Fig. 18 is 4.6 MN/m. Now the rotor does not qualify as a class A operation.

The four-block problem in Eq. (4) provides a means for any three non-complimentary functions to be weighted individually. A diagram of the four block system is shown in Fig. 18. The advantages of the four-block scheme over the two or three block problems are its well-posedness and avoidance of pole-zero cancellation by the synthesized controller [20], [21].

Figure 19 illustrates the feedback connection of the weighting functions to the plant and controller blocks. The reference input r and disturbance input d are new signals introduced. Shaping the control sensitivity function $KS(s)$ is necessary to effect a controller gain roll-off at high frequencies to avoid exciting unmodeled dynamics or amplifying sensor noise. Shaping the process sensitivity function $GS(s)$ affects the closed-loop damping, while the shaping the output sensitivity $S(s)$ translates directly to modifying the disturbance rejection properties. The complimentary sensitivity $T(s)$ cannot be shaped independently of the aforementioned functions as a result of the complimentary nature of the four sensitivity functions.

The system gain from disturbance inputs to performance outputs is

$$\begin{bmatrix} z_1 \\ z_2 \end{bmatrix} = T_{zw} \begin{bmatrix} d \\ r \end{bmatrix} \quad (5)$$

and to guarantee the desired nominal performance specified by the above sensitivity functions, the cost γ to be minimized by the controller K must be less than unity.

This results in upper bounds on **S**, **GS**, **KS** and **T**, are $(\mathbf{W_p W_r})^{-1}$, $(\mathbf{W}_p \mathbf{W}_d)^{-1}$, $(\mathbf{W}_u \mathbf{W}_r)^{-1}$, and $(\mathbf{W}_u \mathbf{W}_d)^{-1}$, respectively. The cost function represented by Eq. (6) solves the

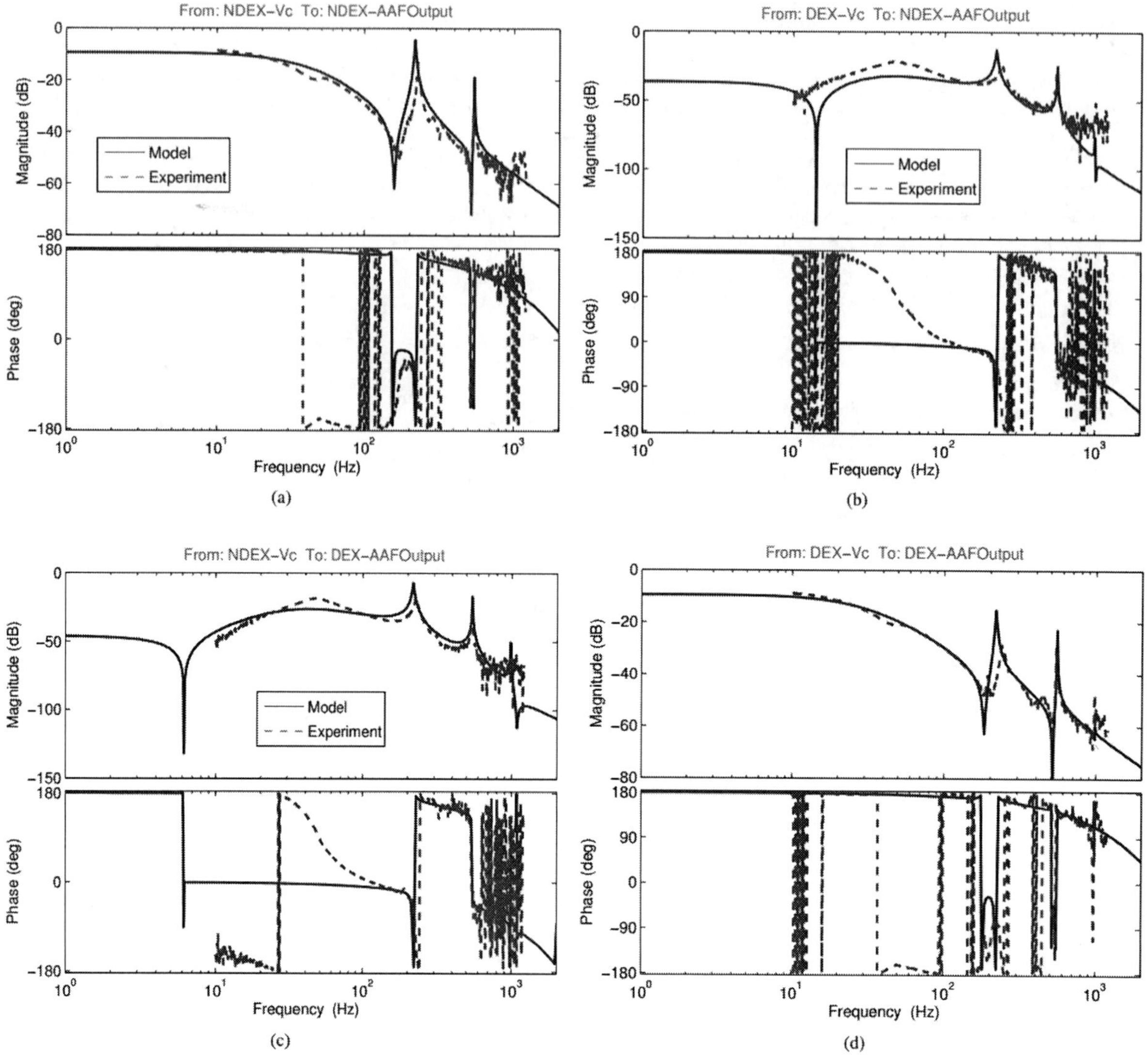

Figure 12: Rotor Transfer Functions

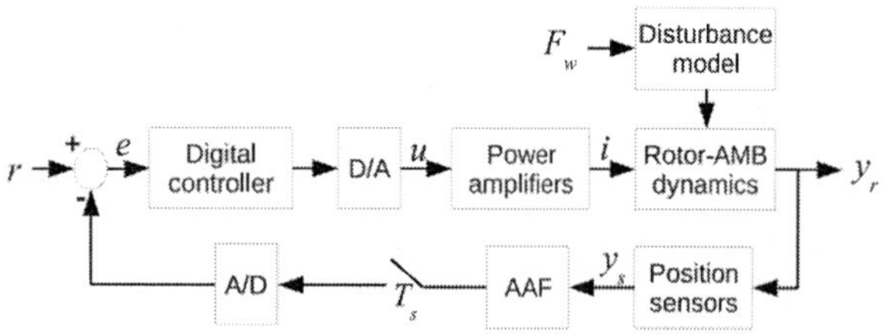

Figure 13: Control System Block Diagram

nominal performance problem. Since Δ has been defined in block diagonal form, the straightforward extension of Eq. (6) to solve the robust performance problem is carried out by defining the structured uncertainty set Δ and finding the controller that minimizes $\mu_\Delta(\mathbf{M})$ using D-K iteration [22].

$$\min_K \|T_{zw}\|_\infty < \gamma$$

$$\left\| \begin{bmatrix} \mathbf{W}_p(\mathbf{I}+\mathbf{GK})^{-1}\mathbf{W}_r & \mathbf{W}_p\mathbf{G}(\mathbf{I}+\mathbf{GK})^{-1}\mathbf{W}_d \\ \mathbf{W}_u\mathbf{K}(\mathbf{I}+\mathbf{GK})^{-1}\mathbf{W}_r & \mathbf{W}_u\mathbf{GK}(\mathbf{I}+\mathbf{GK})^{-1}\mathbf{W}_d \end{bmatrix} \right\|_\infty < \gamma$$

$$\left\| \begin{bmatrix} \mathbf{W}_p\mathbf{S}\mathbf{W}_r & \mathbf{W}_p\mathbf{G}\mathbf{S}\mathbf{W}_d \\ \mathbf{W}_u\mathbf{K}\mathbf{S}\mathbf{W}_r & \mathbf{W}_u\mathbf{T}\mathbf{W}_d \end{bmatrix} \right\|_\infty < \gamma \tag{6}$$

IV. Modeling Uncertainty

Modeling of the impact of the cross coupled stiffness (CCS) or Q on the open-loop plant dynamics can be carried out directly from first principles or phenomenologically. In the case of the former, an uncertain CCS is modeled directly

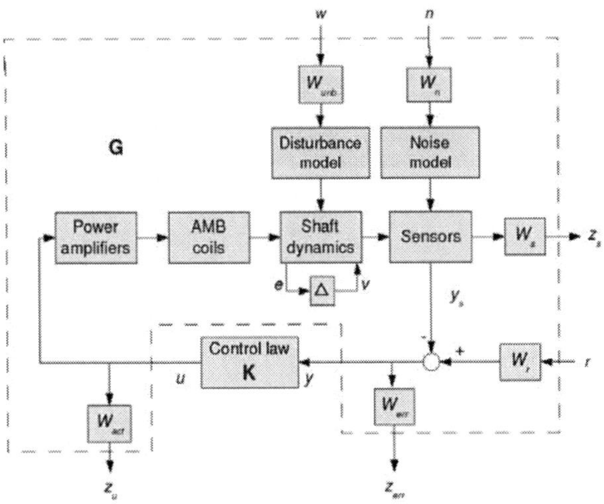

Figure 14: AMB System Block Diagram

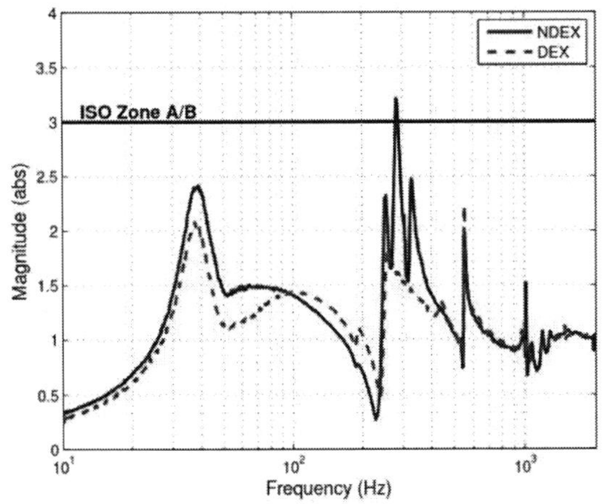

Figure 15: Output sensitivity of NDEX and DEX channels of collocated PID controller.

as a real-valued stiffness varying between 0 and 4 MN/m acting at the rotor midspan. Therefore, the associated rotor displacements as shown in Eq. (7) are the shaft finite element locations x_{23} and y_{23}. The system schematic is shown in Fig. 8.

$$\mathbf{M}\ddot{\mathbf{q}} + (\mathbf{D}+\Omega\mathbf{G})\dot{\mathbf{q}} + \mathbf{K}_s\mathbf{q} = \mathbf{B}_{\text{mag}}\mathbf{F}_{\text{mag}} + \mathbf{B}_w\mathbf{F}_w \quad (7)$$

$$\mathbf{y}_r = \mathbf{C}\mathbf{q} \quad (8)$$

Assuming the CCS (Q) is the sole exogenous disturbance in the system:

$$\mathbf{B}_w\mathbf{F}_w = Q\begin{bmatrix} 0 & 1 \\ -1 & 0 \end{bmatrix}\begin{Bmatrix} q_{x_{23}} \\ q_{y_{23}} \end{Bmatrix} \quad (9)$$

where the stiffness Q can be modeled in two ways:

1) As a real parameter varying from 0 to 1.8 MN/m with a nominal value of 0.9 MN/m denoted CCS Model 1, or

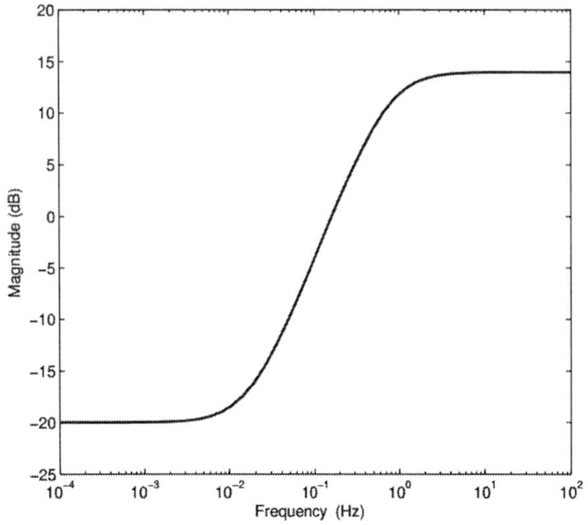

Figure 16: Typical inverse performance weighting function $W_p^{-1}(s)$

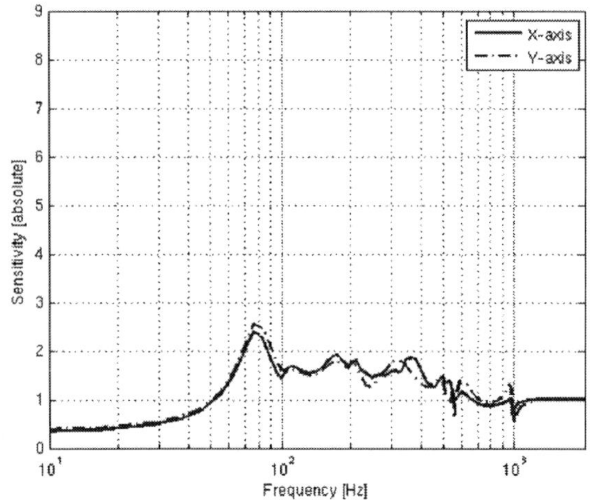

Figure 17: Sensitivity Function, Q = 2.6 MN/m

2) As a real parameter varying from 0 to 2.4 MN/m with a nominal value of 1.2 MN/m denoted CCS Model 2.

The resulting model shows, as expected, increased coupling between the x and y lateral rotor displacements at all rotating speeds. Figure 20 demonstrates the coupling with four Bode plots and a singular value plot of the nominal plant augmented with the uncertainty defined by CCS Model 1.

While the demonstrated coupling between the axes is mathematically correct, the additional dynamics complicate the model and present an open-loop plant that is more of a challenge to validate and control. Increased conservatism arises as Q affects several parameters within the plant model. While the effect on the Nc1 rotor pole is important and the most desired to be captured by this uncertainty definition,

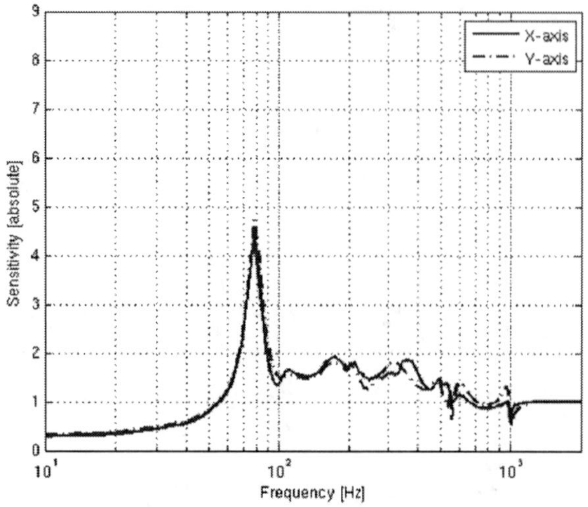

Figure 18: Sensitivity Function, Q = 3.15 MN/m

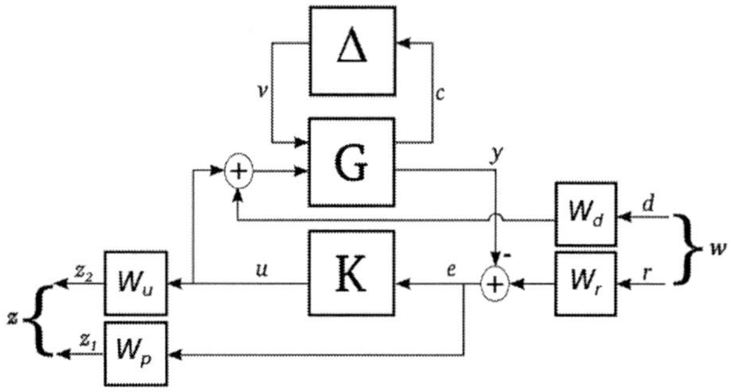

Figure 19: Interconnection of performance weights with plant, uncertainty description and controller

several collateral effects occur such as the perturbation of Nc3 and Nc4 poles and zeros are not desired.

Phenomenological modeling of the uncertainty attempts to model the empirical effect of the CCS on the plant rather than the intrinsic dynamics of the rotor-AMB system which have been shown to be highly conservative [22]. By far the most significant effect of CCS on the plant is the trajectory of the RHP pole of Nc1 as CCS is varied (see Fig. 21 and Table II). Increasing the CCS has the effect of splitting the real RHP pole due to Nc1 into a complex pole that moves further to the right of the complex plane. Assuming the Nc1 complex eigenvalue (an RHP pole) of the form $\sigma \pm j\omega$, the effect of CCS on σ can be modeled either:

1) as a real-valued uncertainty varying from 200 to 240 rad/s with a nominal value of 220 rad/s denoted CCS Model 3, or
2) as a real-valued uncertainty varying from 200 to 260 rad/s with a nominal value of 230 rad/s denoted CCS Model 4.

An eigenvalue perturbation description was used to represent the variations above, and combine into the state-space description of the rotor-AMB system. The advantages of

Table II: The effect of varying CCS magnitude on the location of Nc1 eigenvalue

Q, N/m	Eigenvalue	f_n, Hz	ζ
0	195	31	-1
$5 \cdot 10^5$	$199 \pm 55.7j$	32.9	-0.963
$1 \cdot 10^6$	$199 \pm 55.7j$	32.9	-0.963
$1.5 \cdot 10^6$	$207 \pm 80.7j$	35.5	-0.928
$2 \cdot 10^6$	$217 \pm 104j$	38.2	-0.904
$2.5 \cdot 10^6$	$227 \pm 125j$	39.8	-0.908
$3 \cdot 10^6$	$237 \pm 104j$	44.1	-0.855
$4 \cdot 10^6$	$256 \pm 104j$	44.9	-0.818

the phenomenological approach are that x,y coupling (off-diagonal gain) is minimized, allowing the nominal model developed. Effects of the uncertain CCS modeled through a single eigenvalue perturbation remain largely in the low frequency band around Nc1. The uncertainty set described by CCS Model 3 and 4 is thus compact for the equivalent physical responses. The gap metrics of CCS Model 3 and CCS Model 4 are identically 0.624 indicating a not unreasonable distance from the nominal plant description.

V. Controller Evaluation

The optimum benchmark controller was developed after several iterations of the weighting function modification and control synthesis. Details of the optimization process are given in Mushi [22]. The benchmark serves as the root from which other controllers were developed. Key properties of the benchmark controller are evaluated below to highlight the details of its construction. The analysis largely focuses on the control response, and the Bode plots and the singular value plots are used to show the magnitude and phase, MIMO response, and closed-loop actuator stiffness of the controller.

Convergence of the DK iteration of the μ-upper bound occurred within 4 iterations. A zeroth order D-scaling matrix was generated during each iteration. This results in the minimum achievable controller order of 48. A discrete-time implementation of this control law sampled at 12 kHz executes within 52 us on the DSP hardware. This represents 62% of the available interrupt [23]. The evolution of the μ-upper bound as a function of frequency during the D-K iterations and the minimum peak value of μ is 1.02.

Analyzing the pole and zero locations of the controller is important to examine whether pole-zero cancellation has occurred between the controller and plant. Pole-zero cancellation is at best highly undesirable due to the absence of robust stabilization, and at worst likely to produce unstable controllers if the plant has RHP zeros. The poles and zeros of the benchmark controller are shown in Fig. **??**, where it is confirmed no cancellation with plant poles or zeros occurs. This is a byproduct of using the four-block mixed sensitivity framework for specifying the controller performance. Interestingly, the controller contained four pairs of complex RHP zeros at $205 \pm 3449j$, $189 \pm 3431j$, $81 \pm 1180j$, and $80 \pm 1175j$. These non minimum-phase zeros introduce phase lag in the vicinity of the bending modes and serve to augment their damping [24].

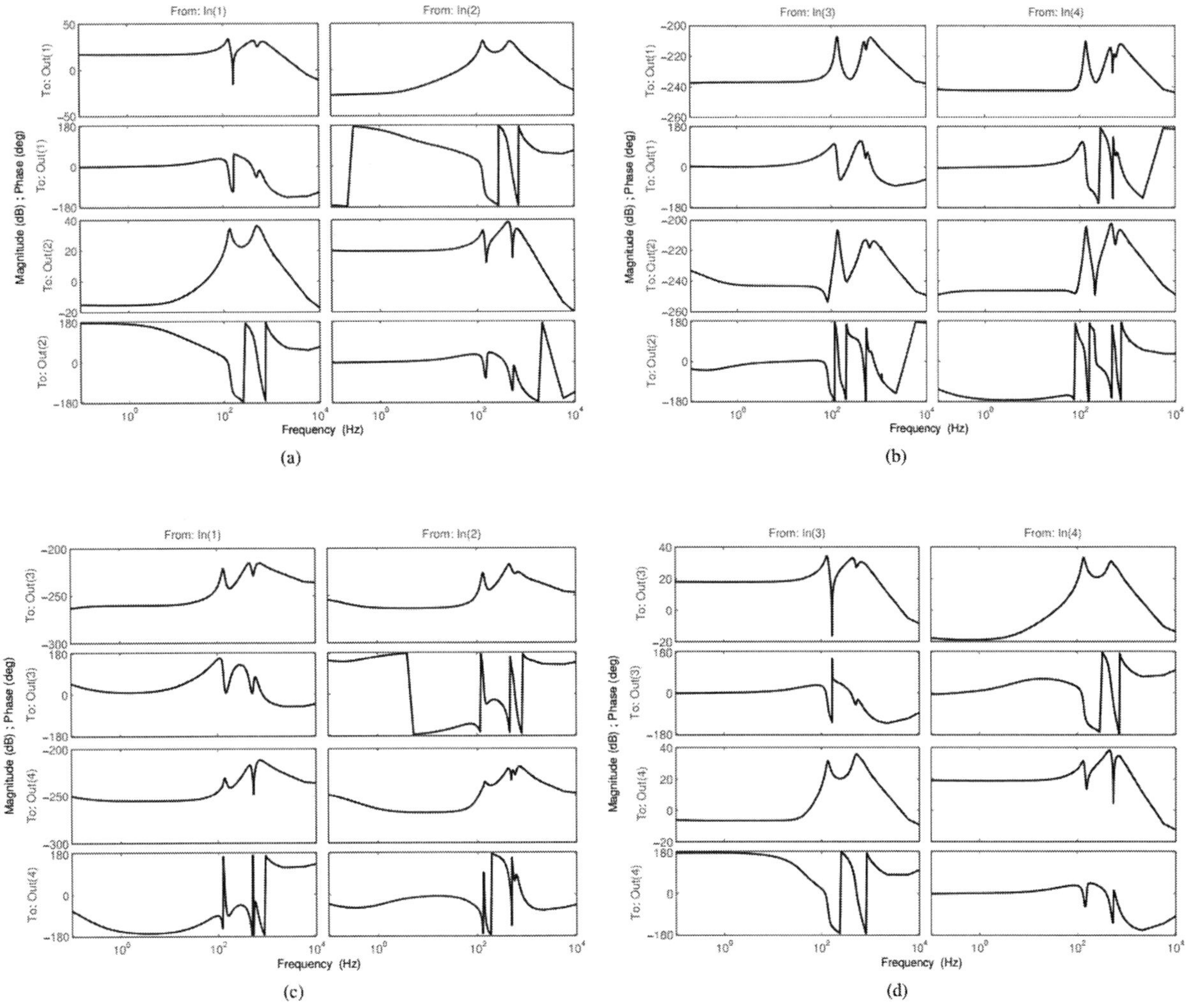

Figure 20: Bode plots of response of Benchmark I controller across all four channels. In and Out refer to input and output directions, while the indices 1, 2, 3 and 4 refer to NDE-X, DE-X, NDE_Y and DE-Y, respectively.

The equivalent closed-loop mechanical stiffness of AMB actuator as a function of frequency [25] is:

$$K_{eq_{ij}}(j\omega) = K_i G_s G_{as} \mathrm{Re}\big(K_{ij}(j\omega)\big) \tag{10}$$

where $\mathbf{G}_s$ is the sensor gain, $\mathbf{G}_{as}$ is the amplifier gain, K_i is the AMB current gain and $K_{ij}(j\omega)$ is the frequency response of the i, j controller channel. This equivalent stiffness is plotted in Fig. 22 can be compared to similar plots for mechanical bearings. The plots show that a similar stiffness is apparent along the x and y axes of each bearing, while the driven-end bearing has an overall higher low-frequency stiffness than the non driven-end bearing.

VI. μ-Analysis of Controllers with Cross-Coupled Stiffness Uncertainty

The aim of this section is to examine the changes introduced by the addition of a cross coupled stiffness Q uncertainty description to a plant model already augmented with gyroscopic model uncertainty. Several different models with CCS uncertainties – CCS models 1, 2, 3, and 4 – are considered in this section.

The two controllers synthesized with block-diagonal performance weight $W_{p,4}$ and either CCS Model 3 or CCS Model 4 yielded a smaller increase in μ from the Benchmark cases as compared to the controller designed with 70% support Figs. 23, 24, the RS bounds have a sharp peak close to 200 Hz seen with CCS Model 3. Breaking down the nominal performance of these two controllers again reveals similar system gain behavior with frequency to Benchmark Ic3 [22]. A lower bias current was necessary for block diagonal $\mathbf{W}_{p,4}$ Ic2 controller to achieve stable suspension of the rotor. This is likely to contribute to slightly reduced performance with respect to CCS since a higher bias current is associated with

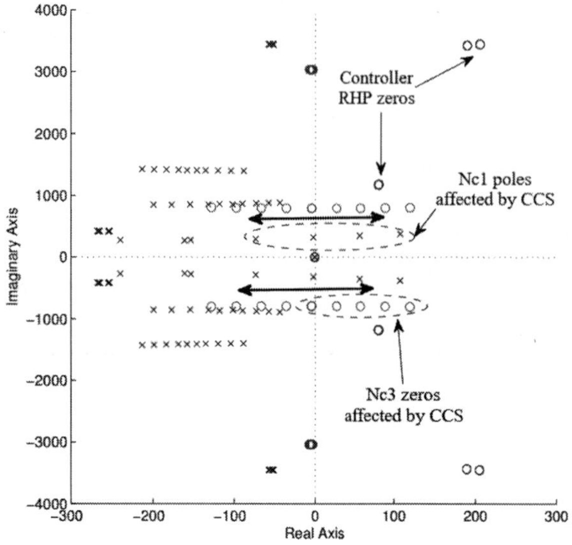

Figure 21: Pole-zero trajectory of closed loop rotor-AMB system at 0 rpm with CCS(0 to 2400 N/mm)

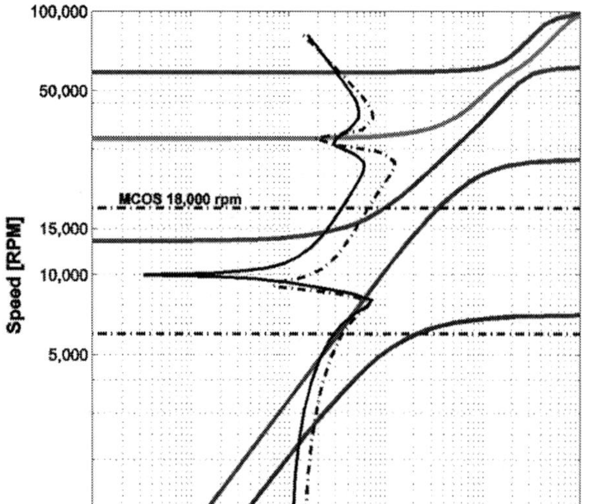

Figure 22: Undamped critical speed map showing eigenvalues as a function of support stiffness. The closed-loop actuator stiffness for an exemplary controller is overlaid.

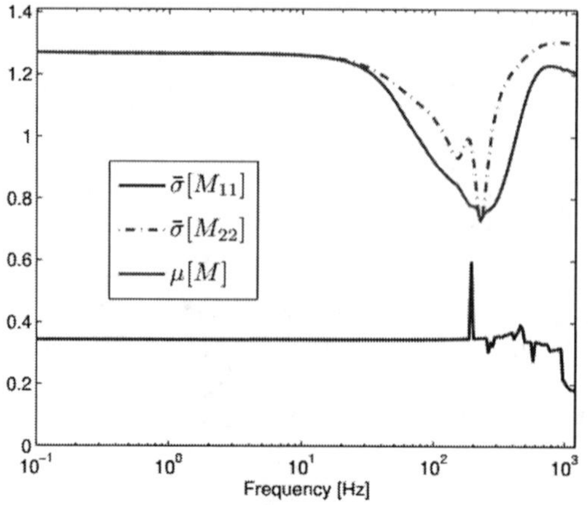

(a) NS, NP, RP bounds.

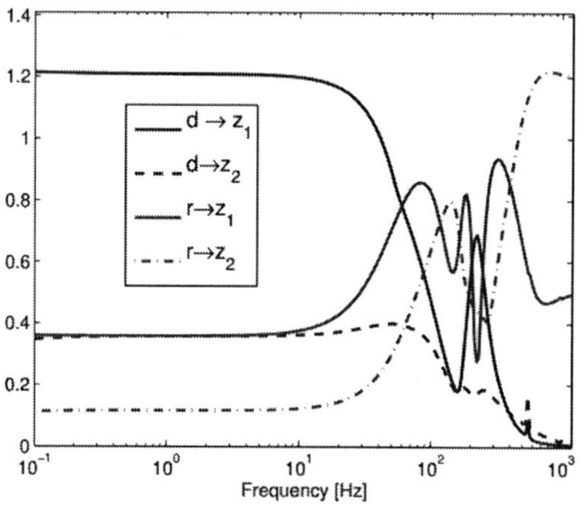

(b) Nominal performance breakdown

Figure 23: μ-Analysis of nondiagonal performance weight $W_{p,4}$ controller designed with plant uncertainty including *Speed Model* 2 and *CCS Model* 3.

higher values of the AMB current gain K_i.

VII. Experimental Results

The benchmark controller was tested in the flexible rotor-AMB test rig. Bode plots and phase angles for 8 of the 16 total loop transfer functions taken at 0 rpm are shown in Fig. 25.

The measured output sensitivity functions measured at 5,000 rpm and 10,000 rpm are shown Bode plots in Fig. 26. The measured amplitude values for the inverse weighting function performance bound is less than the ISO/API class A/B limit indicating high levels of disturbance rejection. The peaks values occur in the range of the rigid body modes suggesting a variation of the actuator gain properties K_i, K_x, from their expected values. There are no peaks near bending modes which is desired for robust performance. The peak sensitivity was measured to be 9.17 dB (2.87) over all of the control channels. Other sensitivity functions were also measured with

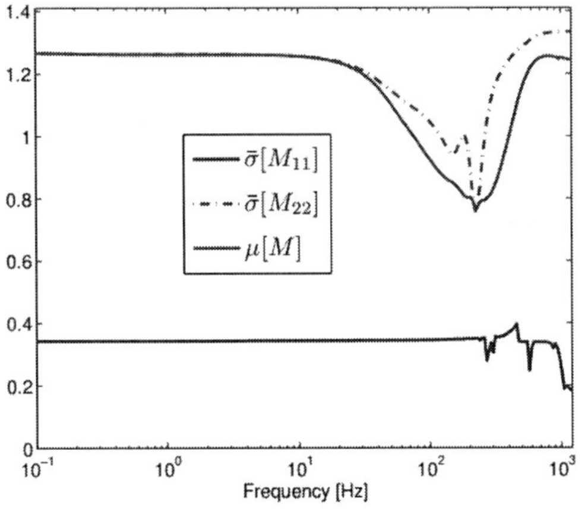

(a) NS, NP, RP bounds.

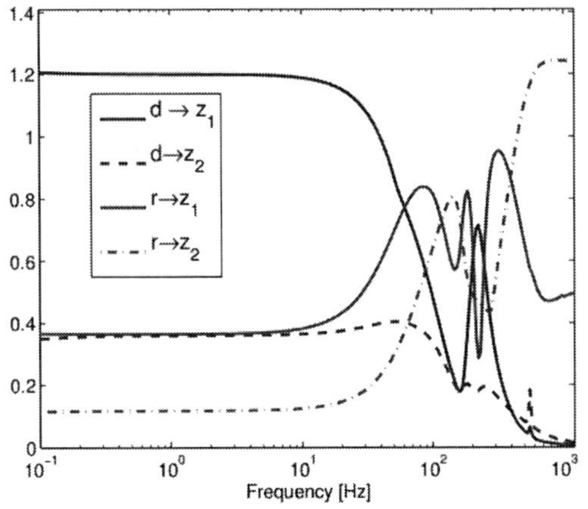

(b) Nominal performance breakdown

Figure 24: μ-Analysis of nondiagonal performance weight $W_{p,4}$ controller designed with plant uncertainty including *Speed Model 2* and *CCS Model 4*.

similar robust results. The complementary sensitivity function results for sensor noise are shown in Fig. 27.

VIII. Forced Response

The rotor was able to operate above the first critical speed, at approximately 14,000 rpm as shown in Fig. 28. A high speed balancing run was required to get through the first critical speed and operate up to 18,000 rpm.

IX. Cross Coupled Stiffness Stability Threshold

The cross coupled stiffness Q destabilizing forces were applied to the rotor using the center AMB actuator. The rotor rigid body modes are easily stabilized [26]. The stability of the first rotor bending mode Nc1 is strongly affected by the cross coupled stiffness. The maximum value of the cross coupled stiffness Q prior to the onset of instability is denoted as the stability threshold. An equivalent statement is that when the system effective damping is zero, the linear stability threshold has been reached. The cross coupled stiffness reduces the damping of the forward whirl mode until it reaches instability.

The two best system models of cross coupled stiffness Q effects assumed that the complex eigenvalue of the excited first rotor bending mode had certain characteristics. In the first Q controller model, labeled No. 3, a real valued uncertainty of the first bending mode eigenvalue from 200 rad/sec to 240 rad/sec was modeled with the nominal value of 220 rad/sec. In the second Q controller, labeled No. 4, a real valued uncertainty of the first bending mode eigenvalue from 200 rad/sec to 260 rad/sec was modeled with the nominal value of 230 rad/sec. This specification of uncertainty is a physical approach to the Q stiffness effects.

To experimentally determine the stability threshold, successively higher cross coupled stiffness values were imposed on the rotor. The system damping ratio was measured by using a blocking approach [27]. Circular rotor excitation was applied using the quarter span AMB an the response measured, as shown in Fig. 29.

Values of Q used in the testing were $Q = 0$, 600, 1200, 1800 N/mm operating at 7,000 rpm. The experimental stability plots of system log decrement vs. Q are given in Fig. 30. The first measurement is without the Q uncertainty model, as shown in Fig. 30a for several benchmark controllers. The second measurement is with the Q uncertainty model, as shown in Fig. 30b. The Q controller model 3 with block diagonal weighting matrices performed the best in Fig. 30b.

X. Conclusions

The issue of robust control related to destabilizing seal aero cross coupled stiffness in high pressure industrial compressors using AMB supports does not have a simple solution. However, the research reported in this paper shows that modern control with the mu synthesis approach provides one suitable method of solution. Usually these compressors are operated supercritical and with the issue of subsynchronous vibration a problem. While a few prototype industrial machines with μ-synthesis control have been reported in the literature, there has not been very much literature reporting on the problem of cross coupled stiffness with the level of detail found in this paper.

A systematic treatment of mu synthesis modeling and analysis has been presented to illustrate the modeling procedure. An extensive treatment of system uncertainties, plant input disturbances, plant output disturbances, sensor noise, control effort and controller input reference signal, has been presented with a four block uncertainty model. A total of 14 controllers were investigated with the focus here on two of the most successful ones. It has been shown that when an accurate model of the flexible rotor-AMB system including all important features – rotor, AMB, sensors, amplifiers, gyroscopic effects, cross coupled stiffness estimates – and the uncertainty models with suitable weighting functions, the controller is successful in achieving robust control. DK iteration optimization has been shown to achieve the desired mu control level in only a few iterations.

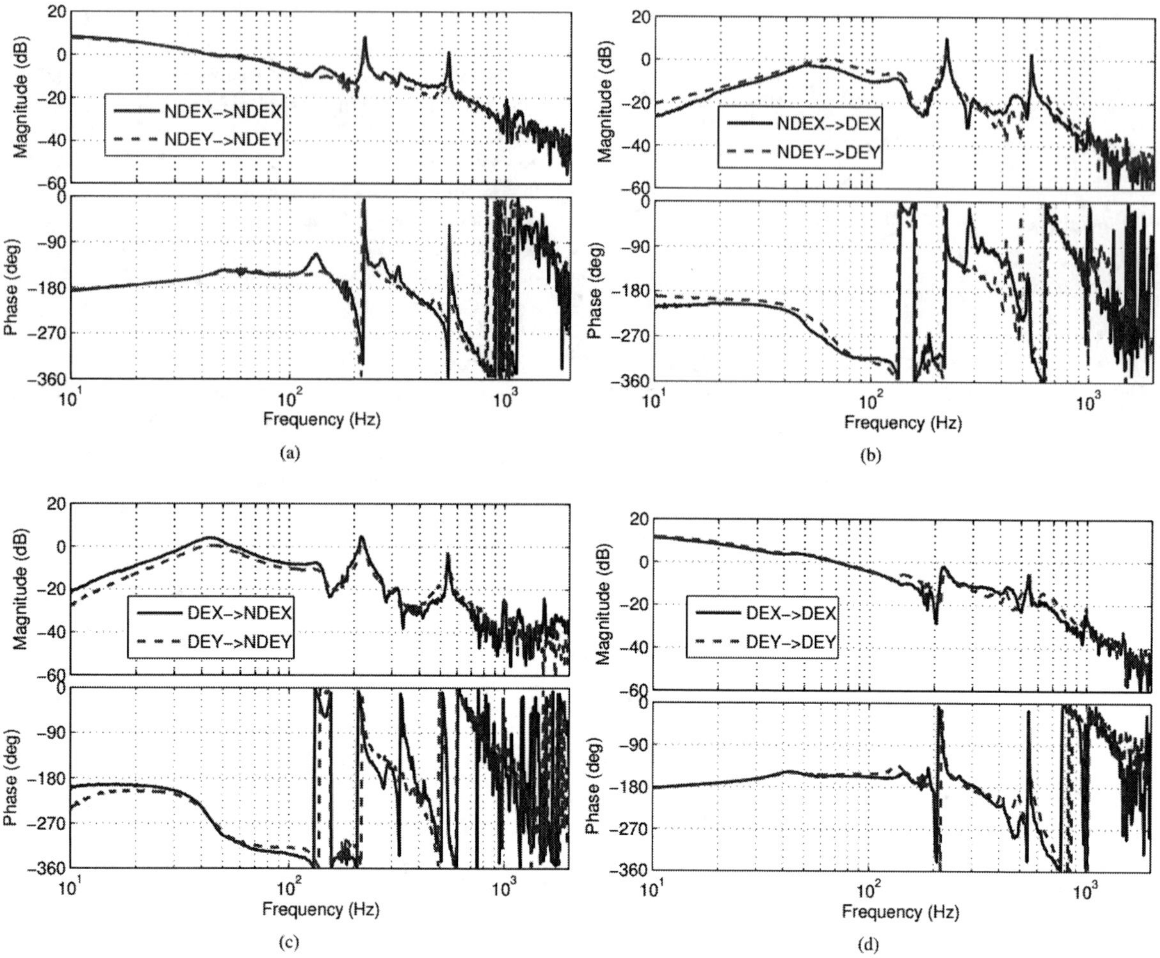

Figure 25: Bode plots of measured loop transfer functions.

The method of assessing the system performance through sensitivity functions was an approach jointly adopted relatively recently by ISO/API and employed in this work. The experimental results show that the best controllers developed in this research were able to qualify the test rig performance as ISO/API class A even with high levels of aero cross coupling. This result has not been reported in the literature prior to this paper.

Both of the best Q controllers were successful in stabilizing the first rotor bending mode using a physical approach. This approach shows that the key to controlling aero cross coupled stiffness due to seals at the center of an industrial compressor is treating the real part of the Nc1 eigenvalue with a nominal value of expected cross coupled stiffness and a range of values treated as uncertainty in the rotor AMB system. A suitable uncertainty weighting function structure must be included to properly stabilize the system.

There were also some unexpected results. For example, the controllers with the block diagonal cross coupled stiffness uncertainty performance weighting did well in providing robust control of the system for the stability threshold. However, another controller without that performance weighting function provided a stability threshold that was as good and slightly better.

It seems likely that the methods developed in this paper have a useful purpose in industrial high pressure compressor with AMB supports. It is hoped that this paper assists in that effort.

Authors' Note

The authors would like to thank the organizers of the ISMB 14 Symposium for allowing such a long keynote paper. While this paper provides a lot of detail regarding this research topic, it is noted here that many aspects of this research project could not be included in this paper even with all of the material which was included. The authors recommend readers who want to know more to consult Mushi [22].

References

[1] F. F. Ehrich, *Handbook of Rotordynamics*. Krieger Publishing Company, 2004.

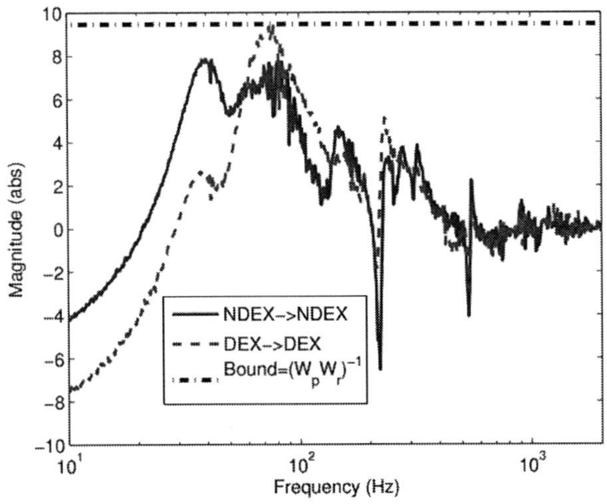

(a) 5,000 rpm

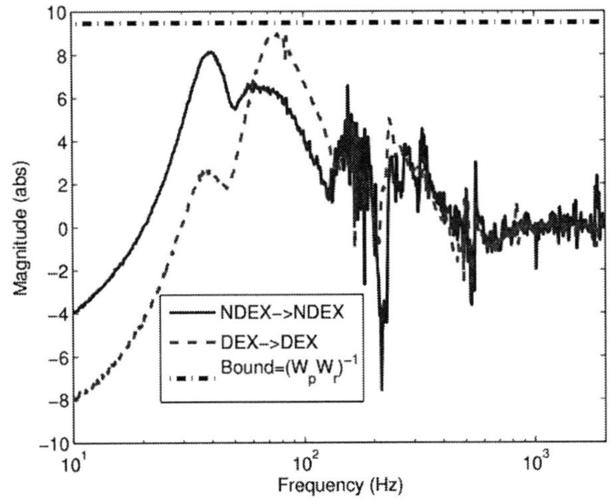

(b) 10,000 rpm

Figure 26: Bode plots of output sensitivity functions measured at various speeds.

[2] L. Barrett, "Stability and Nonlinear Response of Rotor-bearing Systems with Squeeze Film Bearings," Ph.D. dissertation, University of Virginia, 1979.

[3] J. Alford, "Protecting Turbomachinery From Self-Excited Rotor Whirl," *ASME J. Eng. for Power*, vol. 87, no. 4, pp. 333–344, October 1965.

[4] U. Schonhoff, "Practical Robust Control of Mechatronic Systems with Structural Flexibilities," Ph.D. Dissertation, Darmstadt Technical University, 2002.

[5] S. Skogestad and I. Postlethwaite, *Multivariable Feedback Control: Analysis and Design*. John Wiley & Sons, 2005.

[6] J. Doyle, "Analysis of feedback systems with structured uncertainty," *IEE Proc. Part D (Control Theory and Applications)*, vol. 129, pp. 242–250, 1982.

[7] R. Kumaresan and D. Tufts, "Estimating the parameters of exponentially damped sinusoids and pole-zero modeling in noise," *IEEE Trans. Acoust. Spch. Signal Proc.*, vol. 30, no. 6, pp. 833–840, 1982.

[8] S. Mushi, Z. Lin, and P. Allaire, "Design, construction and modeling of a flexible rotor active magnetic bearing Test Rig," in *Proc. ASME Turbo Expo.*, no. GT2010-23619, Glasgow, UK, 2010, pp. 443–452.

[9] Honeywell and Lockheed-Martin, "Application of multivariable control theory to aircraft control laws; Final Report : Multivariable control design guidelines," Wright Laboratory, US Airforce Materiel Command, Tech. Rep. WL-TR-96-3099, 1996.

[10] F. Losch, C. Gahler, and R. Herzog, "Mu-Synthesis controller design for a 3MW pump running in AMBs," in *Proc. 6th Int. Symp. Magn. Bearings*, 1998, pp. 415–428.

[11] R. Fittro and C. Knospe, "Rotor compliance minimization via mu-control of active magnetic bearings," *IEEE Trans. Contr. Syst. Technol.*, vol. 10, no. 2, pp. 238–249, 2002.

[12] F. Losch, "Identification and Automated Controller Design for Active Magnetic Bearing Systems," Ph.D. dissertation, ETH Zurich, 2002.

[13] M. Swann, "Ramping up magnetic bearing use," *Turbomachinery International*, vol. 50, no. 7, Novemer/December 2009.

[14] M. Franchek, "Selecting the Performance Weights for the μ and H∞ Synthesis Methods for SISO Regulating Systems," *ASME J. Dyn. Sys. Meas. Contr.*, vol. 118, pp. 126–131, 1996.

[15] E. Maslen and J. T. Sawicki, "Mu-Synthesis for magnetic bearings: Why use such a complicated tool?" in *Proc. ASME Int. Mech. Engr. Congr. Expo IMECE 2007*, no. 43033, 2007, pp. 1103–1112. [Online]. Available: http://link.aip.org/link/abstract/ASMECP/v2007/i43033/p1103/s1

[16] S. Garg, "Implementation Challenges for Multivariable Control: What You Did Not Learn in School!" NASA Glenn Research Center, Tech. Rep. NASA/TM-2008-215027, 2008.

[17] B. Ogunnaike, "A contemporary industrial perspective on process control theory and practice," *Ann. Rev. Control*, vol. 20, pp. 1–8, 1996. [Online]. Available: http://www.sciencedirect.com/science/article/B6V0H-43MCMVR-2/2/10194ba37549cf255efcacb4695e696c

[18] M. van de Wal, G. van Baars, F. Sperling, and O. Bosgra, "Multivariable H∞/μ feedback control design for high-precision wafer stage motion," *Control Engineering Practice*, vol. 10, pp. 739–755, 2002.

[19] R. Larsonneur, *Magnetic Bearings: Theory, Design, and Application to Rotating Machinery*. Springer, 2009, ch. 2: Principle of Active Magnetic Suspension, pp. 27–68.

[20] M. Englehart and M. Smith, "A four-block problem for H_∞ design: Properties and applications," in *Proc. 29th IEEE Conf. Decision Control*, vol. 4, 1990, pp. 2401–2406.

[21] M. Smith, "Well-Posedness of H^∞ Optimal Control Problems," *SIAM J. of Control and Optim.*, vol. 28, no. 2, pp. 342–358, 1990.

[22] S. E. Mushi, "Robust Control of Rotordynamic Instability in Rotating Machinery Supported by Active Magnetic Bearings," Ph.D. dissertation, University of Virginia, Charlottesville, Virginia, 2012.

[23] S. Mushi, Z. Lin, and P. Allaire, "Robust control design for a supercritical rotor supported by AMBs," *IEEE Amer. Contr. Conf.*, 2012.

[24] G. Li, E. Maslen, and P. Allaire, "A note on ISO AMB stability margin," in *Proc. 10th Int. Symp. Magn. Brgs. ISMB10*, 2006, pp. 124–129.

[25] R. Williams, F. Keith, and P. Allaire, "Digital control of active magnetic bearings," *IEEE Tras. Ind. Electron.*, vol. 37, no. 1, pp. 19–27, 1990.

[26] R. Larsonneur, *Magnetic Bearings: Theory, Design, and Application to Rotating Machinery*. Springer, 2009, ch. 8: Control of the Rigid Rotor in AMBs, pp. 181–228.

[27] C. Cloud, E. Maslen, and L. Barrett, "Damping ratio estimation techniques for rotordynamic stability measurments," *ASME J. Eng. Gas Turb. Power*, vol. 131, 2009.

[28] M. Boerlage, "Rejection of disturbances in multivariable motion systems," Ph.D. Dissertation, Eindhoven University of Tehnology, 2008.

APPENDIX A
STATE SPACE DESCRIPTION

A complete state-space description for the uncertain plant with including the dynamic contribution of performance weighting functions and the modeled uncertainty from the bending modes is

$$P(s) := \begin{cases} \begin{bmatrix} \dot{x}(t) \\ c(t) \\ z(t) \\ y(t) \end{bmatrix} = \begin{bmatrix} A & B_\Delta & B_1 & B_2 \\ C_\Delta & 0 & 0 & 0 \\ C_1 & 0 & D_{11} & D_{12} \\ C_2 & 0 & 0 & 0 \end{bmatrix} \begin{bmatrix} x(t) \\ v(t) \\ w(t) \\ u(t) \end{bmatrix}, \\ v(t) = \Delta c(t) \end{cases} \tag{11}$$

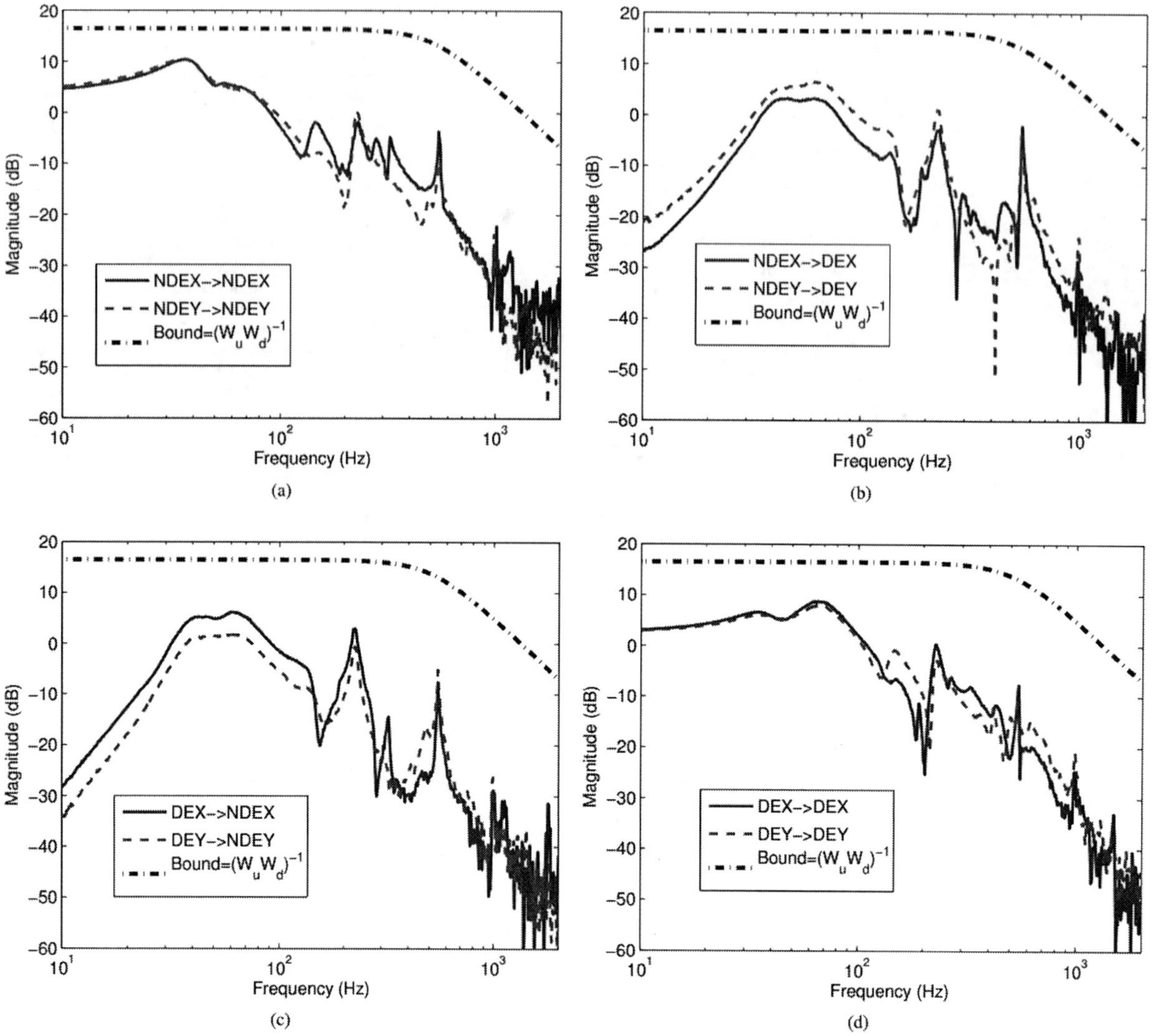

Figure 27: Bode plots of measured complementary sensitivity function or closed-loop response.

where

$$A = \begin{bmatrix} \hat{A}_r & 0 & 0 & 0 \\ B_{d,u}C_{r,s} & A_d & 0 & 0 \\ -B_{w,p}D_dC_{r,s} & -B_{w,p}C_d & A_{w,p} & 0 \\ 0 & 0 & 0 & A_{w,u} \end{bmatrix}$$

$$B_\Delta = \begin{bmatrix} 0 & I_2 & 0 & I_2 & 0 \end{bmatrix}^\top$$

$$c_\Delta = \begin{bmatrix} 0 & -2\omega_3^2 & 0 & -2\zeta\omega_3 & 0 \\ 0 & -2\omega_4^2 & 0 & -2\zeta\omega_4 & 0 \end{bmatrix}$$

$$C_\Delta = \begin{bmatrix} c_\Delta & c_\Delta \end{bmatrix}$$

$$B_1 = \begin{bmatrix} \hat{B}_rW_d & 0 & 0 & 0 \\ 0 & 0 & B_{w,p}W_r & 0 \end{bmatrix}^\top$$

$$B_2 = \begin{bmatrix} \hat{B}_r & 0 & 0 & B_{w,u} \end{bmatrix}^\mathrm{T}$$

$$C_1 = \begin{bmatrix} -D_{w,p}D_dC_{r,s} & -D_{w,p}C_d & C_{w,p} & 0 \\ 0 & 0 & 0 & C_{w,u} \end{bmatrix} \quad (12)$$

$$C_2 = \begin{bmatrix} D_dC_{r,s} & C_d & 0 & 0 \end{bmatrix}$$

$$D_{11} = \begin{bmatrix} 0 & W_rD_{w,p} \\ 0 & 0 \end{bmatrix}$$

$$D_{12} = \begin{bmatrix} 0 \\ D_{w,u} \end{bmatrix}$$

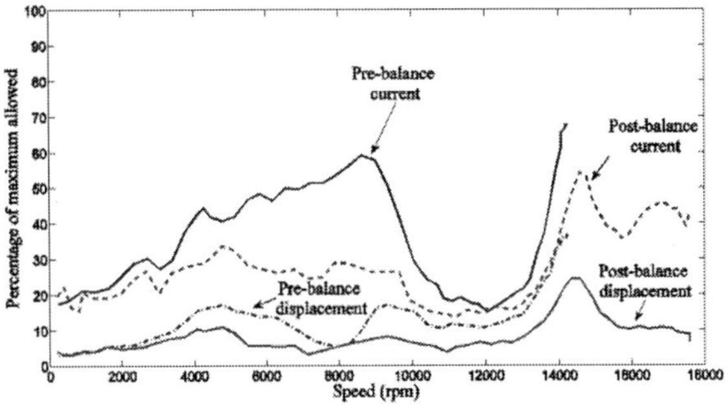

Figure 28: Rotor displacement (0-pk) and perturbation current at the NDE AMB as a percentage of the minimum clearance and bias current, respectively during a run-up to 18,000 rpm. Measurements are taken both before and after dynamic balancing.

and where ω_3 and ω_4 are the nominal natural frequencies of the first two bending modes Nc3 and Nc4, and $\zeta = 0.2\%$ is the modal damping. The uncertain state space model P contains 48 states, a contribution of 36 states from rotor-AMB model G, and 12 states from the dynamics of $W_p(s)$ and $W_u(s)$ repeated across four control channels. Balancing using a diagonal similarity transformation was used to improve the numerical conditioning of state-space representation of P.

Appendix B
Nondiagonal Performance Weights

Nondiagonal performance weights show promise as means of maximally exploiting the available degrees of freedom during control synthesis [28]. The MBTRI plant in the presence of significant cross-coupling the plant becomes less diagonally dominant. This suggests that a controller synthesized with nondiagonal weighting functions may have additional degrees of freedom to optimize the overall closed-loop sensitivity leading to performance improvements. Four prototype nondiagonal scalings were developed to systematically evaluate the effects of coupling and directionality between multiple control channels:

1) Full block performance weight

$$W_{p,1}(s) = \begin{bmatrix} 1 & 1 & 1 & 1 \\ 1 & 1 & 1 & 1 \\ 1 & 1 & 1 & 1 \\ 1 & 1 & 1 & 1 \end{bmatrix} \times \text{blkdiag}\left[W_{p,NDE}, W_{p,DE}, W_{p,NDE}, W_{p,DE}\right] \quad (13)$$

2. Cross-coupled performance weight

$$W_{p,2}(s) = \begin{bmatrix} 1 & 0 & 0.5 & 0 \\ 0 & 1 & 0 & 0.5 \\ 0.5 & 0 & 1 & 0 \\ 0 & 0.5 & 0 & 1 \end{bmatrix} \times \text{blkdiag}\left[W_{p,NDE}, W_{p,DE}, W_{p,NDE}, W_{p,DE}\right], \quad (14)$$

3. Simple block-diagonal

$$W_{p,3}(s) = \begin{bmatrix} 1 & 1 & 0 & 0 \\ 1 & 1 & 0 & 0 \\ 0 & 0 & 1 & 1 \\ 0 & 0 & 1 & 1 \end{bmatrix} \times \text{blkdiag}\left[W_{p,NDE}, W_{p,DE}, W_{p,NDE}, W_{p,DE}\right], \quad (15)$$

4. Block-diagonal with scaled off-diagonal terms

$$W_{p,4}(s) = \begin{bmatrix} 1 & 0.5 & 0 & 0 \\ 0.5 & 1 & 0 & 0 \\ 0 & 0 & 1 & 0.5 \\ 0 & 0 & 0.5 & 1 \end{bmatrix} \times \text{blkdiag}\left[W_{p,NDE}, W_{p,DE}, W_{p,NDE}, W_{p,DE}\right], \quad (16)$$

5. Block-diagonal with negative scaled off-diagonal terms

$$W_{p,5}(s) = \begin{bmatrix} 1 & -0.5 & 0 & 0 \\ -0.5 & 1 & 0 & 0 \\ 0 & 0 & 1 & -0.5 \\ 0 & 0 & -0.5 & 1 \end{bmatrix} \times \text{blkdiag}\left[W_{p,NDE}, W_{p,DE}, W_{p,NDE}, W_{p,DE}\right]. \quad (17)$$

The five controllers synthesized kept the same parameters for W_u, W_d and W_r as the benchmark case.

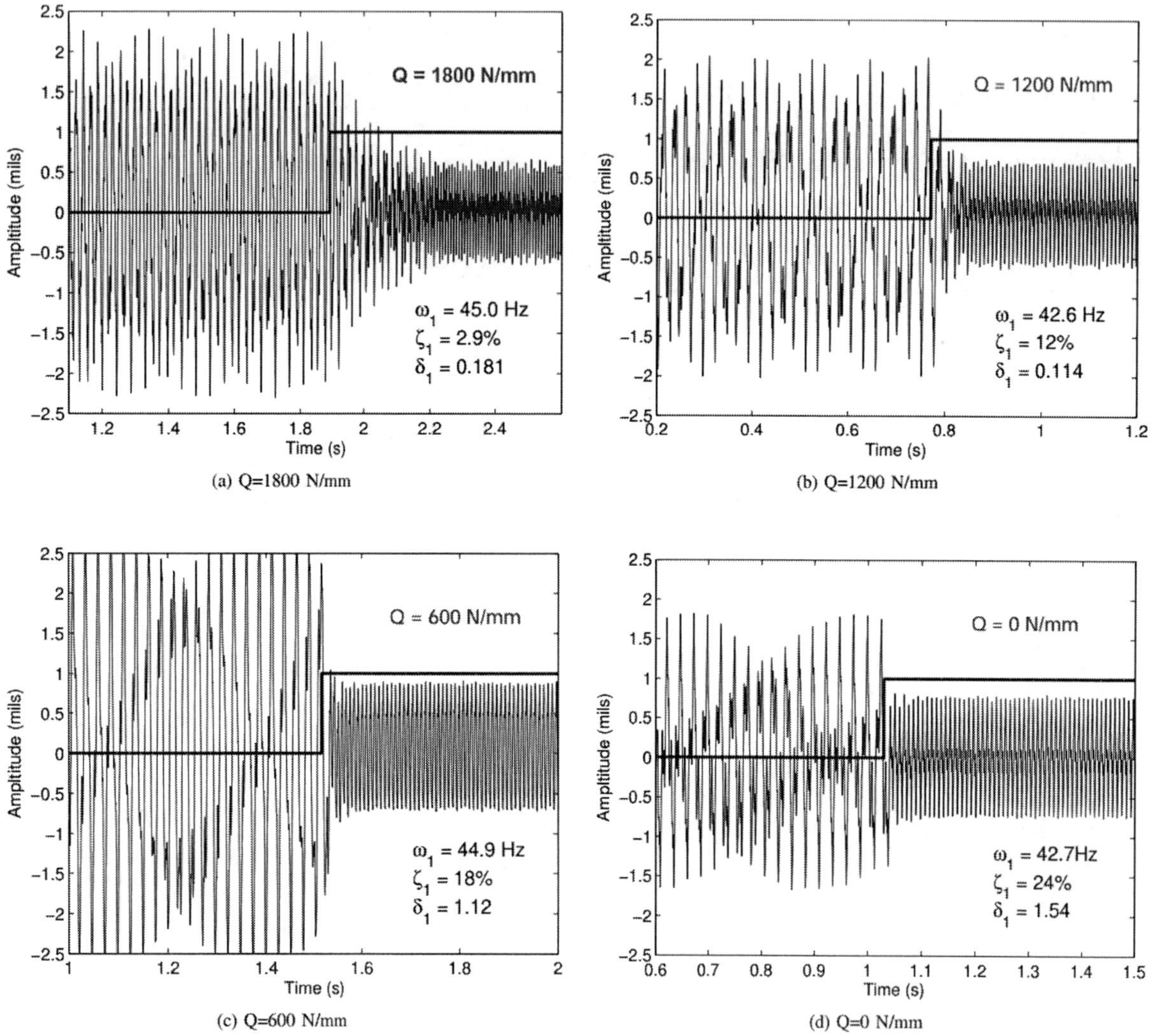

(a) Q=1800 N/mm

(b) Q=1200 N/mm

(c) Q=600 N/mm

(d) Q=0 N/mm

Figure 29: Rotor displacement response at 7,000 rpm on termination of blocking (indicated by the rising edge trigger) under different magnitudes of CCS excitation. Benchmark I controller is used to suspend the rotor and the stability threshold for this controller is 2000 N/mm.

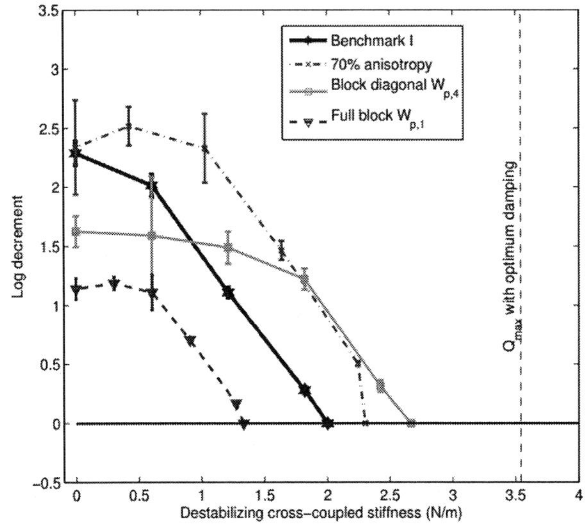

(a) Without CCS uncertainty model.

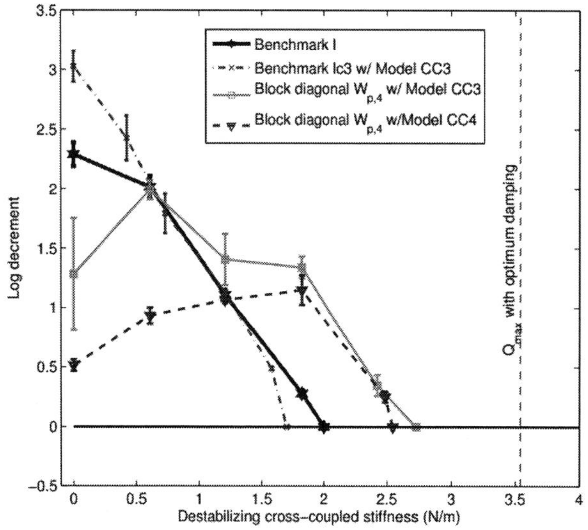

(b) With CCS uncertainty model.

Figure 30: Experimental stability sensitivity plot of log decrement of Nc1 versus destabilizing cross-coupled stiffness Q for several controllers compared with the Benchmark I.

Overview on Various Types of AMBs & their Respective Potential for Applications

Hannes Bleuler [a]

[a] EPFL Lausanne, LSRO-IMT, CH 1015, Lausanne, Switzerland, hannes.bleuler@epfl.ch

Abstract—**Many different types of magnetic bearings are known since a long time: "Active" and "passive", "electrodynamic", "electromagnetic", "hybrid" etc. Certain types have found many applications and others none. A recent successful realization of a complex system (a momentum reaction sphere) based on a Lorentz-force type of bearing will be presented in more detail and the potential of as yet unexploited types of magnetic bearings will be briefly discussed.**

I. Basic Types of Magnetic Bearings

A. *Introduction*

Since Earnshaw's 1842 investigation [1], it is known that magnetic forces alone cannot provide contact-free levitation in all degrees of freedom. It is only in the 1930'ies that practical ways to overcome this fundamental limit are beginning to be reported [2], essentially through the application of stabilizing feedback control. After 1945, especially through the activities of Jesse Beams [3], and, later, of many others, solutions to the problem of magnetic levitation are realized, although some of them have been kept confidential for evident military reasons. Nevertheless, the complete picture of available solutions is well known since many years. These solutions are listed below. However, many potentially interesting types of magnetic levitations have still not found practical applications. This is the issue to be addressed in this contribution.

B. *"Physical" Classification*

The classification of magnetic bearing types follows naturally from the physical mechanism of achieving magnetic force. A first distinction concerns the two basically different mechanisms types of magnetic forces, i.e. forces on moving charge (Lorentz force) and forces on magnetic material (related to energy of the magnetic field). At a fundamental level, both effects are identical, namely the consequence of Einstein's special theory of relativity on electrical charge. From a practical point of view, the two effects are however significantly different (the effects of moving charges of magnetic materials being reduced to material properties).

So at the beginning of the classification, there are the two groups of Lorentz force devices and reluctance force devices.

Lorentz force is perpendicular to the magnetic flux and to the current, depending linearly of the current. Reluctance force is in direction of the magnetic flux and essentially proportional to the square of the current. In each of the two main families, finer distinctions lead naturally to an exhaustive list of the fundamental magnetic levitation types. This is illustrated in Figure 1. In the **first group, Lorentz Force devices**, the energy in the magnetic field remains constant and force is linear with the current, as it results from the vector cross-product of magnetic flux and charge velocity. No magnetic material is involved (except for field enhancement).

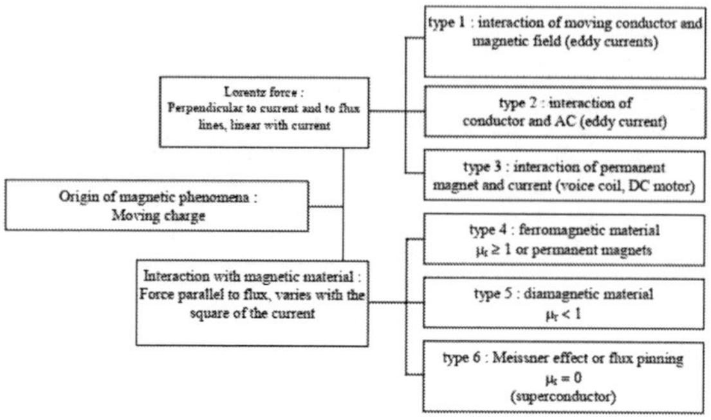

Figure 1. Fundamental types of magnetic levitation. While interactions of type 1 to 3 are all used for electric motors, they are (yet) seldom to be found in magnetic bearings. Technical realizations of magnetic bearings are almost exclusively of type 4

In the **second group, electromagnetic devices or devices involving magnetic material**, force is calculated not as a cross product of field vector and current vector, but from an energy integral over the volume filled by magnetic flux. Force is normal to the surface of magnetic material and (in case currents are involved and in first approximation) proportional to the square of electric current. Permanent magnet devices and superconducting devices also fall into this broad category.

C. *"Engineering" Classification, Current applications*

Parallel to this "physical" classification, technical criteria allow an overview based on more application oriented considerations. This will include criteria such as "active" or "passive" bearings, sensorless (or "self-sensing") bearings, permanent magnet bearings, bearingless motors, gyroscopically stabilized bearings etc. We will not attempt a classification along such criteria in this paper, we just would like to draw the attention to the fact that there exists a rich body of publications, not always available on internet, on the most various types of technical solutions to the magnetic levitation challenge. As a representative example, the full body of the 14 ISMB symposia offers quite a complete picture of the work done in the last 26 years.

From the six fundamental magnetic bearing types in Fig. 1, present realizations almost exclusively concentrate on type four (electromagnetic) bearings. In the next section, a recent successful realization of a type three (Lorentz force) bearing will therefore be briefly presented.

II. Example: A reaction sphere for sattelite attitude control

We would like to present here a challenging project of an unusual type of magnetic bearing which has very recently achieved levitation. It concerns a momentum sphere for satellite attitude control, and involves several partners (essentially CSEM, EPFL, ESA, Maxon Motors). After several years of work (and a number of publications), the project is far from over, but an important milestone has been reached very recently (too late for inclusion as a full paper in this Symposium). Nevertheless, references can be found under [5].

The realization of a truly spherical momentum rotor, controllable around all spatial axes of rotation, is a serious technical challenge. The problem has been solved with an ironless stator of 20 coils arranged on the vertices of a dodecahedron. The rotor is a complex assembly of (Nd-Fe) permanent magnets with 4 pole-pairs in a cubic arrangement.

The stator (diameter 20 cm) is shown open in fig. 2.

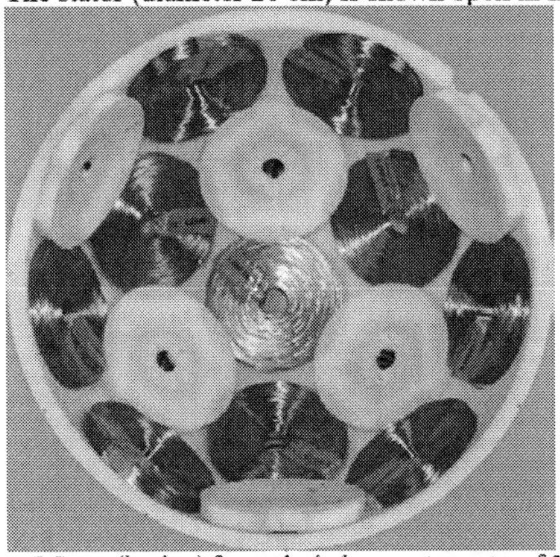

Figure 2 Stator (ironless) for a spherical momentum rotor of 20 cm diameter. Half of the 20 stator coils, arranged at the vertices of a regular dodecahedron, can be seen. The rotor consists of a sphere with Nd-Fe-B magnet shells in a 4 pole-pair arrangement, oriented as the vertices of a cube.

This system has recently been levitated successfully and rotated up to 12 rpm, work continues after this first operation. It is described in the publications and the PhD thesis of Leopoldo Rossini, of EPFL [7,8]

III. Application Potential of other types of magnetic bearings

As illustrated with this application example, magnetic bearing types other than controlled electromagnetic bearings still hold a promising potential for future applications. Among these I would like to mention three bearing types.

1 Superconducting rotor bearings (e.g. [6]). At the 10th ISMB Symposium, engineers from Siemens have presented a magnet an industrial magnetic bearing prototype based on high temperature superconductors. The fundamental technical hurdles seem to have been overcome and I could imaginefurther applications for this type of bearings, perhaps in situations where a high degree of reliability is essential, or in connection with high TC generators.

2 Passive Bearings with eddy current stabilization. This type of bearing has been presented at ISMB 9 (2004) in Lexington, KY [9]. It has the potential of energy-free levitation at room temperature. Losses are only proportional to deviations from rotational symmetry and of course to air friction. This makes it suitable for stationary flywheel energy storage. The problems to be overcome seem concentrated on the mechanical design and the behavior of high-velocity flywheels.

3 Diamagnetic bearings. As to now, a few exotic experiments for flow meter or vacuum gauge have been reported, but there seems to be a potential for clean-room handling and satellite attitude control for small satellites, an emerging activity in space technology.

IV. Conclusions

A classification of basic magnetic bearing types shows that among six basically different principles of magnetic levitation, only one type has found widespread application up to now. Although some obvious facts explain partially this situation, it is believed that some of the unusual types of magnetic bearings could have interesting practical applications, thus opening new perspectives for magnetic bearing engineering recent example of a reaction sphere has been presented in some detail, other new application domains might be accessible by exploiting the potential of all magnetic levitation types.

References

[1] Earnshaw S. "On the nature of Molecular Forces which Regulate the Constitution of the Lumiferous Ether" Trans. Cambridge Philosophical Soc. Vol 7, 1842, pp 97-112

[2] Braunbeck W.: "Frei schwebende Körper im elektrischen und magnetischen Feld", Zeitschrift für Physik, 112 (1939), pp 753-763

[3] Beams, Younf, Moore: "On the Production of High Centrifugal Fields", Journal of Appl. Physics, 1946, pp 886-890

[4] Bleuler, Cole, Keogh, Larsonneur, Malsen, Nordmann, Okada, Schweitzer, Traxler : Magnetic Bearings: Theory, Design & Application to Rotating Machinery, Schweitzer & Malsen, Eds., Springer, 2011.

[5] Chételat O. "Study of a reaction sphere for attitude control (sphere) – reaction sphere design," CSEM CH-2002 Neuch.tel, Tech. Rep. SPH.CSE.TN2.iss6, 2009.

[6] Miyagawa, Kameno, Takahata, Ueyama "A 0.5 kWh flywheel energy storage system using a high-T superconductig magnetic bearing," *IEEE Journal of Applied superconductivity*, vol. 9, no. 2, pp. 996–999, 1999

[7] Rossini Leopoldo: "Electromagnetic modeling and control aspects of a reaction sphere for satellite attitude control", PhD thesis EPFL 2014 (to be published)

[8] L. Rossini, O. Chételat, E. Onillon, and Y. Perriard, "Force and torque analytical models of a reaction sphere actuator based on spherical harmonic rotation and decomposition," Mechatronics, IEEE/ASME Transactions on, vol. 18, no. 3, pp. 1006–1018, 2013

[9] Sandtner, J., Bleuler, H. : « Electrodynamic passive magnetic bearing with planar Halbach arrays », 9th ISMB, Lexington, KY, 2004, also available at http://www.silphenix.ch/

Electrical drive systems with active magnetic bearings in industrial applications - possibilities and challenges

Matthias Kroll

LEViTEC GmbH, Georg-Ohm-Str. 11, 35633 Lahnau, Germany, Matthias.kroll@levitec.de

***Abstract*—This article describes the use of active magnetic bearings in an electrical drive system and how performance, reliability and availability can be increased significantly based on an application example. It is important to know about the strong interaction between the different system components to get the best technical results, as well as considering production related aspects.**

I. Introduction

The demands on machine tools regarding productivity, reliability and life cycle costs (TCO) are increasing permanently. Therefore, it is essential to constantly optimize and improve the entire machine, single components, the production processes, as well as machining parameters and a continuous monitoring of the machining processes is necessary. Machining spindles are key components in machine tools. They have major influence on productivity, output quality and availability. The following statements refer to spindles in grinding applications. Due to the increasing demands that spindles are facing, it is crucial to further examine alternative bearing techniques, in addition to ball- or rolling bearings. Special interest is put into active magnetic bearings, as they provide additional functionality. The past has seen various studies and investigations about the use and benefits of magnetic bearings in grinding spindles [1], [2], [3]. Beside the advantages like no friction and no wear, magnetic bearings i.a. are able to enhance the rotating speed, as well as bearing parameters can be adjusted and a continuous process monitoring and diagnosis is feasible. By using magnetic bearings in spindles, they become intelligent system components of a machine or an entire machining centre. The behavior of the magnetic bearing spindle "LeviSpin 700" has been investigated. Work results, additional functionalities and the value for the customer will be shown.

I. Design and Function of a Magnetic Bearing Spindle

A. Technical Data "LeviSpin 700"

One of the major guidelines to design a magnetic bearing spindle was to observe the size like a conventional spindle. The following table shows the data of the magnetic bearing spindle used in the example [4].

Nominal Power	15 kW
Nominal Speed	42.000 rpm
Nominal Current	32 Amp
Typ of Motor	Permanent magnet machine
Radial Force	800 N
Axial Force	800 N
Tool Taper	HSK 50E
Clamping state monitoring	Yes
Seals	Labyrinth seal with active sealing air
Cooling	Water, closed system

Table I: Technical Data LeviSpin 700

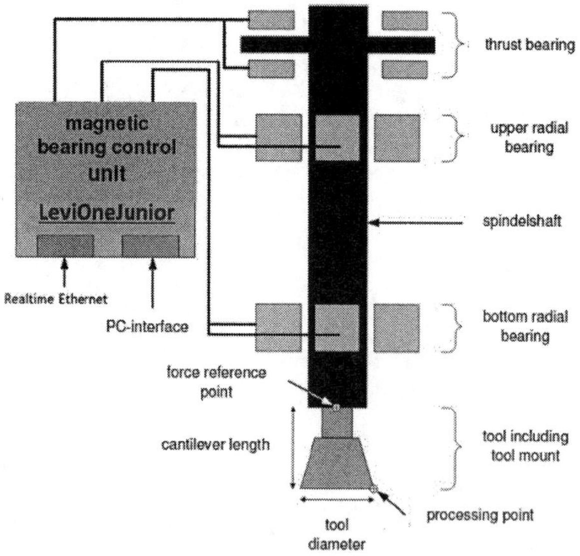

Figure 1. Magnetic bearing spindle basic configuration

In comparison to conventional spindles with ball-or rolling bearings, the shaft in a magnetic bearing spindle is levitated. Besides the rotation, there are additional degrees of freedom, which can be used but also need to be stabilized by the magnetic bearings. These degrees of freedom can be used by moving the shaft from its zero point or through oscillation in axial or radial direction.

Figure 2. Magnetic bearing spindle "LeviSpin 700" and magnetic bearing control unit "LeviOneJunior"

B. Magnetic Bearings and Control Unit

The control of the bearing is exclusively digital, using a high performance digital signal processor (Figure 2). The analogous position- and current signals are digitized and transferred to the processor. The digital control allows the user to access multiple bearing parameters and actual values which can be processed in a higher level control system, for instance in a CNC. The position of the rotor can be adjusted, as well as the stiffness of the bearings. Furthermore, it is possible to allow a force controlled machining, as well as continuous process monitoring [3], [5]. Therefore, a real time ethernet communication to the CNC of the machine is inevitable. It ensures the fast, real time data transfer between the electronic of the magnetic bearing and the CNC. As a consequence the five axis of the magnetic bearing system will be integrated in the CNC-program.

The force is proportional to the distance x between shaft and magnetic bearing stator and the needed current I, and can be calculated according to equation (1). kx and ki are absolute terms which are depending on the particular bearing.

$$F \sim kx * x + ki * I \qquad (1)$$

Position and current signals are reflecting an exact picture of the machining process. These signals, in combination with the cantilever length and the diameter of the tool (Figure 1) allow conclusions to the machining process, as well as operating with defined force like implemented in [3].

Due to the advantages for this application, the spindle in the example operates with unipolar bearings. Combining electromagnets with permanent magnets allows the use of much smaller electromagnets. This leads to a better dynamic, due to smaller inductivity, meeting increasing demands. This fact is used to oscillate the shaft or to fast line-out external disturbances. As already mentioned, using magnetic bearings allows minor adjustments of the rotating shaft. It can be moved several micrometers within the air gap. This fact can be used for tool compensation or superfinishing [10]. Besides, those bearings need less electrical power. They are short and compact, because of the high energy density of the permanent magnets. This has a positive impact on the rotor dynamics and the thermal situation.

C. Motor and Drive

The motor principle has been determined by refering to the induction machine as well as to the permanent magnet machine. The induction machine is commonly used in spindles. Reasons are the easy, solid and cost efficient rotor design and a wide field weakening range, which allows constant power over a wide speed range. On the other hand there are considerable advantages when using the permanent magnet machine. The following facts underline why the permanent magnet machine has been finally used for the above mentioned spindle:

- less rotor losses (rotor temperature)
- less volume / length (rotor dynamics)
- larger shaft diameter possible (rotor dynamics)
- large power efficiency over a wide speed range (energy efficiency)
- larger air gap possible (shaft displacement)
- shorter acceleration time (tool change time)

After setting the motor principle, defining the maximum size, torque and the rotational speed, the design of the motor can be done. Due to the compact construction, the high energy density and thermal utilization of the motor, a water cooling was installed to dissipate the heat loss effectively. For rotor dynamic reasons a four pole motor design was chosen to shorten the winding overhang. According to equation (2) a nominal speed of 42.000 rpm equals an electric frequency of 1.400 Hz. The used drive has to be capable to supply this frequency to this high speed permanent magnet machine.

$$f = \frac{n * p}{60} \qquad (2)$$

TABLE II: DATA OVERVIEW

	Parameter	Value
f	frequency	1.400 Hz
p	Number of pole pairs	2
n	Nominal speed	42.000 rpm

To minimize effects like dimensional deviation caused through thermal growth of the spindle shaft, it is inevitable to minimize losses in the spindle rotor. As shown in [7], the drive is mainly responsible. In order to reduce rotor losses of the motor, caused by current harmonics, the motor needs to supply by sinusoidal quantities. Therefore, the inverter is run with high PWM-frequency. If necessary a sine-wave filter or a choke is installed between the inverter and motor [6]. Figure 3 shows the direct impact of the PWM-frequency on the rotor temperature. The investigation has been executed by the PTW-Institute of the Technical University of Darmstadt. The temperature behavior of a 15 kW permanent magnet motor running at 27.600 rpm with different PWM-frequencies is shown in [8]. For the above mentioned reasons a drive Type "Sinus 4000" was chosen. This drive is a three level inverter with a PWM-frequency of maximum 64 kHz, designed to run high speed permanent magnet motors.

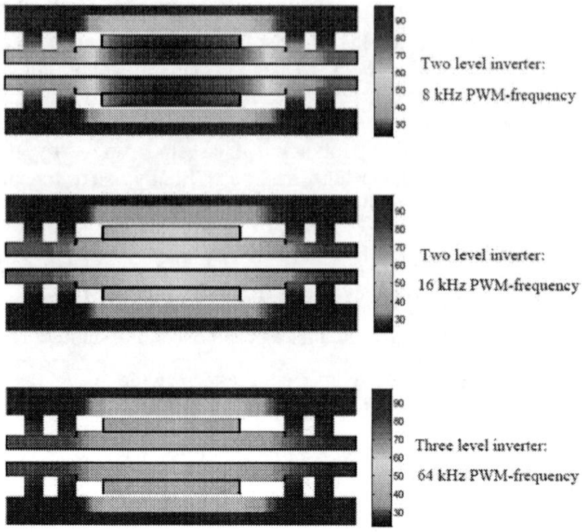

Figure 3. Impact of the PWM-frequency on the rotor temperature in °C

II. MACHINING RESULTS

The following examples refer to grinding tests with different materials at the Furtwangen University. Such as carbide metal, ceramic and steel. The tests were executed with a three-axis CNC-machine [2], [9]. The influence of axial oscillation and specific shaft displacement during the grinding operation on the machining results were of major interest. Additional investigations like measuring the grinding forces by the magnetic bearings, as well as monitoring the grinding process by current and position values were carried out. Figure 4 shows the test arrangement. FT is the tangential, FN the normal force in N, acting on the grinding tool.

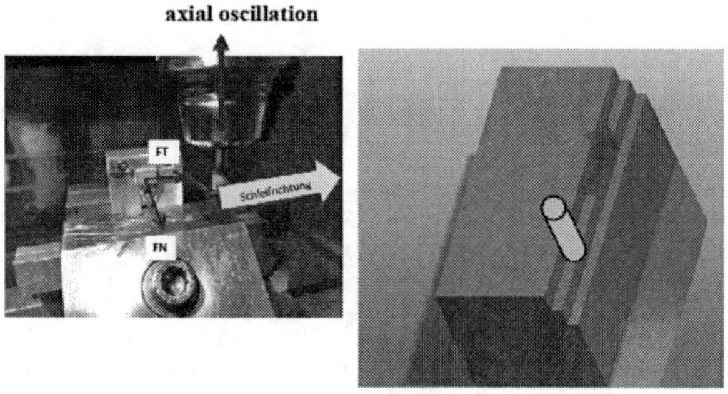

Figure 4. Test arrangement

As already mentioned the spindle shaft can be defined oscillated within the air gap of the touchdown bearings. In comparison to spindles with ultrasonic excitation, magnetic bearings spindles allow low frequency oscillation. Due to electrical and mechanical time constants of the system, the reachable amplitude is decreasing with increasing frequency. While the frequencies below 75 Hz amplitudes up to 120 µm are possible, using 700 Hz only allows 0.9 µm. Figure 5 shows the dependence of amplitude and frequency.

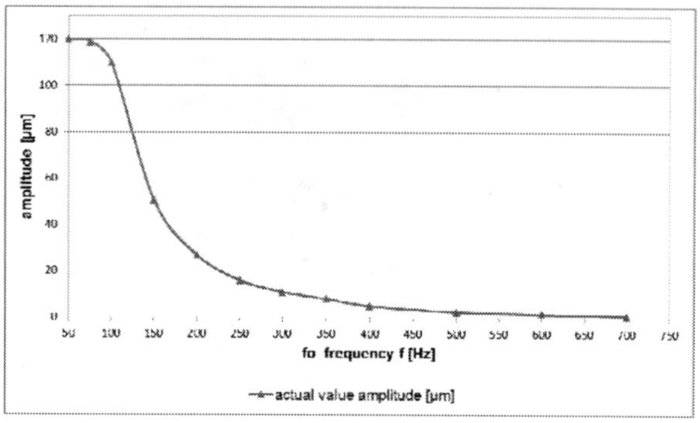

Figure 5. Dependence of amplitude and frequency

Magnetic bearings also permit to limit the amplitude even if larger amplitudes are possible by the system. Figure 6 shows the influence of the oscillation frequency on the grinding forces in tangential direction. The tangential force has been normalized, 100% equals no oscillation. The amplitude has been limited on various values. The optimum, regarding minimize the force, appeared at s~70/90 µm and 80 Hz.

TABLE III: CONSTRAINTS AND MACHINING PARAMETERS

	Parameter	Unit
Process	Peripheral grinding	
Material	Hard metal SMX/P25	
Tool	Grinding pin D91, Ø 10 mm	
v_c	Cutting speed	8,37 m/s
v_f	Feed speed	52 mm/min
a_e	Cutting depth	0,05 mm
FT	Tangential force	N
FN	Normal force	N
Rz	Average roughness	µm
s	Shaft deflection/Oscillation	µm

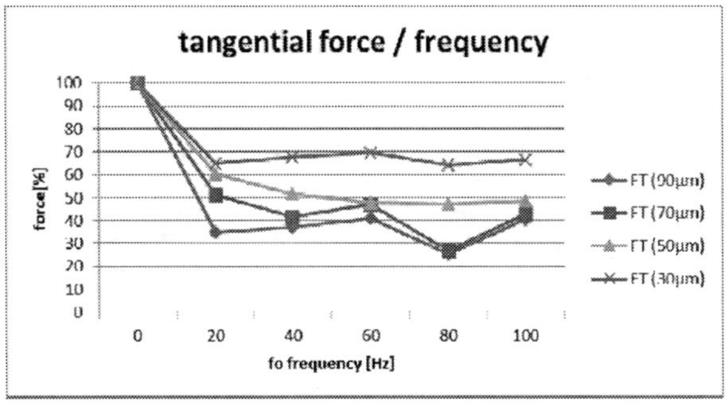

Figure 6. Dependence of amplitude and frequency

Similar results are shown in Figure 7, concerning the surface quality.

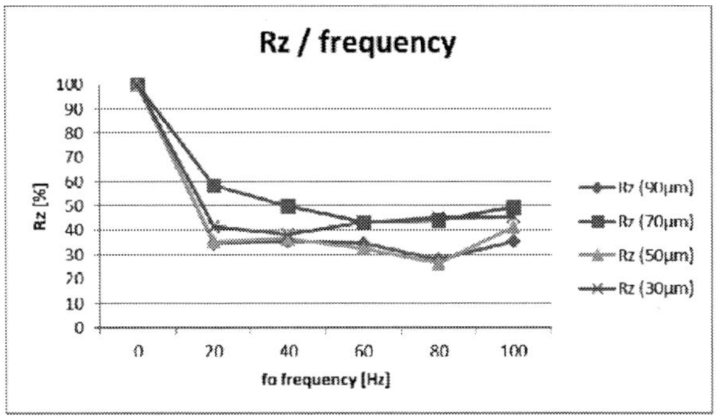

Figure 7. Dependence of average roughness and frequency

III. Customer Value

The use of magnetic bearings in grinding spindles leads to additional functionalities and features. As already shown, the axial oscillation helps to reduce forces during the grinding process. Besides, the surface of the material is of higher quality because of les roughness. By reducing grinding forces, overall productivity is increased through shorter machining time. With other words, this enables higher feed speed and therefore higher metal removal rate. Less forces on the tool extend the operating life. Combined with this energy efficient motor and the maintenance-free magnetic bearings, it will reduce the lifecycle costs significantly. By using the integrated sensors in the magnetic bearings, a continuous process monitoring and diagnosis is feasible. During the process generated data can be used directly and further processed. Abnormalities or faults can be detected early. This improves the availability of the machine.

IV. Conclusion

The behavior of a magnetic bearing spindle during grinding process has been investigated. Furthermore, additional functionality and usefulness were considered. Only if all system components are well aligned and interacting perfectly, the best performance and the highest costumer value can be achieved. In summary it can be said that:

- Magnetic bearings in grinding spindles enable performance increase and additional functionality
- Low-frequency axial oscillation reducing grinding forces and improving the surface quality, in other words increased productivity at higher quality
- conditional monitoring of the machining process is possible, as well as an improvement of the process reliability
- no further sensors needed to control and monitor grinding forces

References

[1] Hörsemann, W.: Hochgeschwindigkeitsschleifen mit aktiv magnetgelagerten Spindeln. Dissertation, TU-Braunschweig, 1992

[2] Grünefeld, S.: Untersuchungen zur Wirkungsweise magnetgelagerter Spindeln auf Schleifprozesse, Bachelorarbeit, Kompetenzzentrum für Schleiftechnologie und Feinstbearbeitung der Hochschule Furtwangen University, 2013

[3] InnoNet Förderprojekt: Präzisionsschleifmaschine mit Magnetlagerspindel zur integrierten Prozessanalyse (PreciGind), Aachen 2007

[4] LEViTEC GmbH: Produktflyer LeviSpin 700, 02/2010

[5] Kroll, M.: Energieeffizienz durch schnelldrehende magnetgelagerte Antriebe, ETG-Kongress, November 2011

[6] Beineke, S., Hebing L., Bünte, A.: High-Speed Drive with Three-Level Inverter for Vacuum Pumps, Laser Cooling and High-Speed Cutting, EPE 2003 Toulouse

[7] Rothenbücher, S., Schiffler, A., Bauer, J.: Die Speisung macht's, Werkstatt & Betrieb, Ausgabe 07/2009

[8] Arbeitskreismotorspindeln VI, Berichte, PTW der TU-Darmstadt 2008

[9] Grünefeld, S., Lohner, R: Interne Untersuchungsberichte, KSF der Hochschule Furtwangen University, 2013

[10] Redemann, C.: Zusätzlicher Kundennutzen durch den Einsatz von Magnetlagern in Bearbeitungsspindeln, Gestaltung von Spindel Lager Systemen, 2006 Aachen

Acknowledgment

The presented research work has been supported by the Linz Center of Mechatronics (LCM GmbH), which is part of the COMET/K2 program of the Federal Ministry of Transport, Innovation and Technology and the Federal Ministry of Economics and Labour, Austria. The author would like to thank the Austrian Government, the Upper Austrian Government and the Johannes Kepler University for their support in making this work possible.

RIETER

Magnetic Bearings: Application to High Speed Textile Processing

- ISMB14 Linz 2014

- Jiří Sloupenský, Vicepresident Product Research Rieter

Spinning = yarn production

From hand made yarn …

Today industrial solutions

Clasical (Ring)

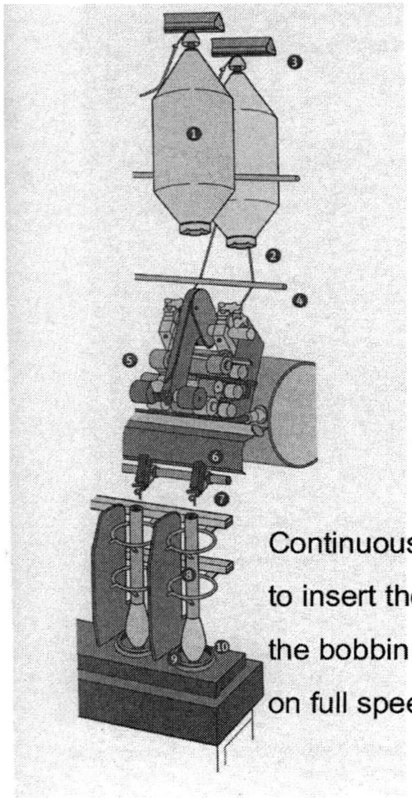

Continuous fiber stream –
to insert the twist
the bobbin has to rotate
on full speed → only 150 g possible

" Open End" (Rotor)

Bobbin rotates just on winding speed
→ weight up to 5 kg possible

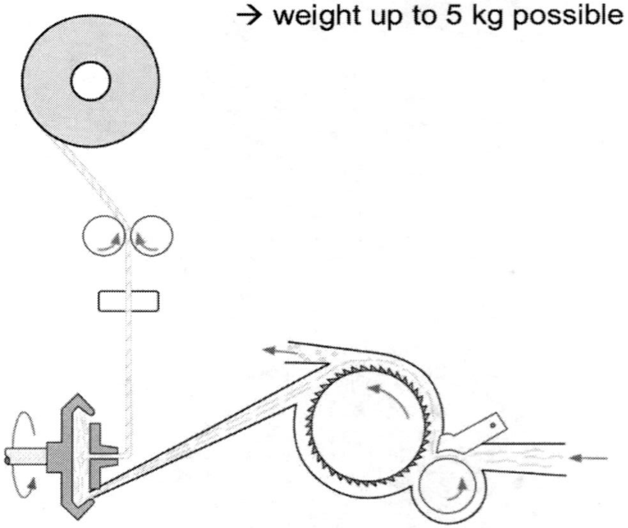

single fibers in the channel
→ Open End of the yarn

Open End (Rotor) spinning history - milestones

RIETER

1937	Berthelsen developed first open end
1950	Meinberg in Germany makes trial with different Open End spinning systems
1965	Czechoslovak KS200 rotor spinning machine was introduced to public runnig at 30 000 rotor rpm.
1967	Improved BD200 was presented with first spinning mill of OE coming under production
1971-1975	Many machine manufacturers started with OE , improved versions of machines were launched with speed up to 100 000 rpm .
1975	Schlafhorst (DE) introduced fully automated Autocoro machine
1999	Rieter introduced "Semi automated machine " for low cost countries

1937 Berthelsen developed first open end

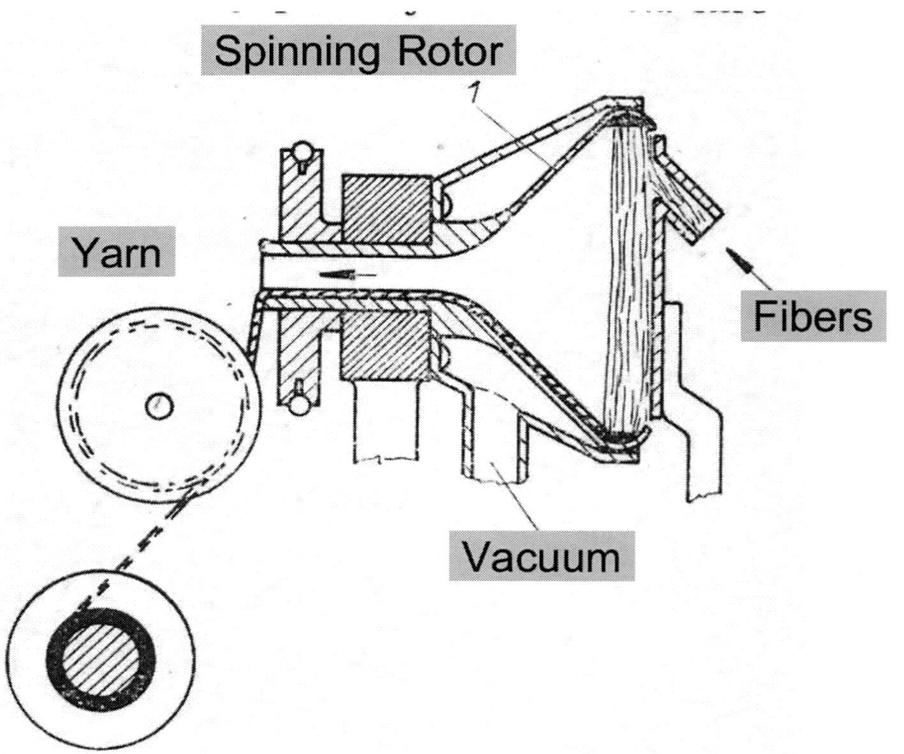

Important : Yarn is taken out opposite to the fiber input

Industrialisation of OE spinning in Czechoslovakia

1965 – first industrial OE machine introduced to public: KS 200

Still yarn is taken out
opposite to the fibber input !!

State of the art - fully automated Rotor machine

Rotor speed – way to productivity

Productivity of Rotor spinning machine is given by the rotor speed

→ high requirements for drive and bearing , but high cost pressure

All historical units had a yarn take up to the opposite direction than fiber input, but it was too complicated for drive and bearing

→ The new concept of Rotor spinning was developed in 60th yarn take up to the same side as fiber input

Direct rotor drive – the most simple solution

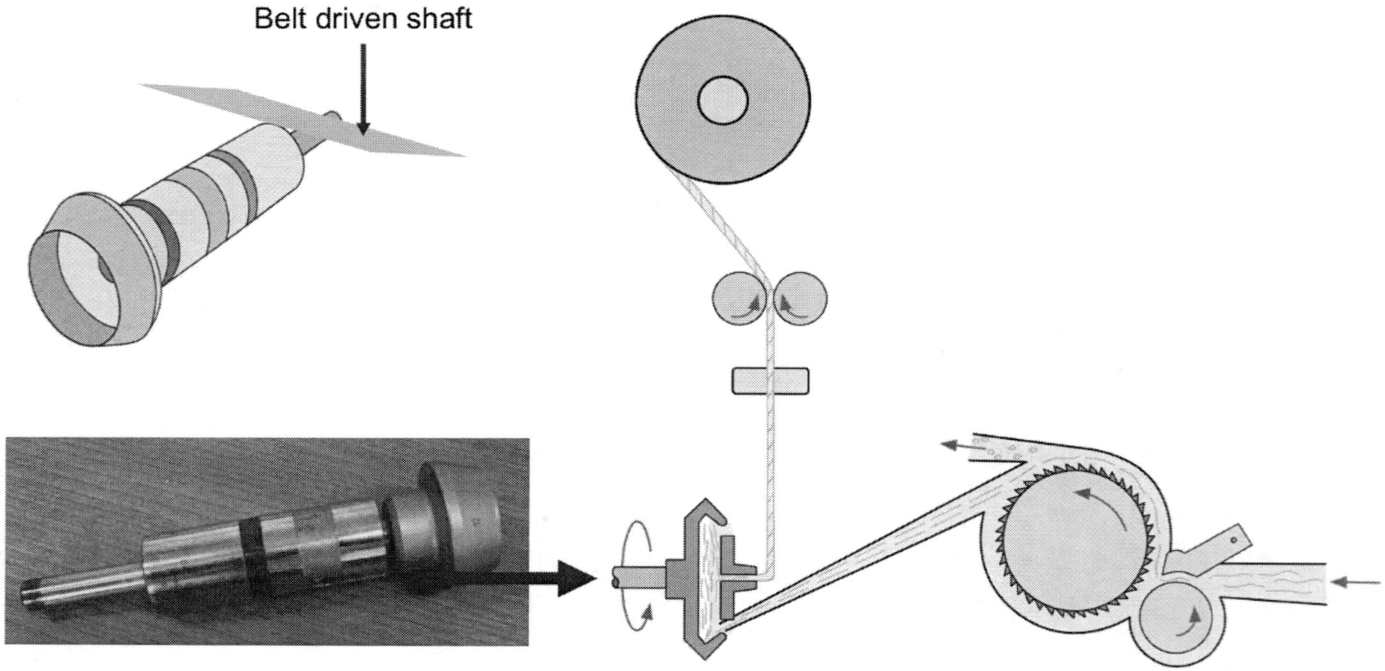

Speed limit about 100 000 r.p.m.

Indirect drive – speed multiplier

Speed up to 160 000 r.p.m possible,
but high friction , high abrasion

→ part wearing
and high energy consumption

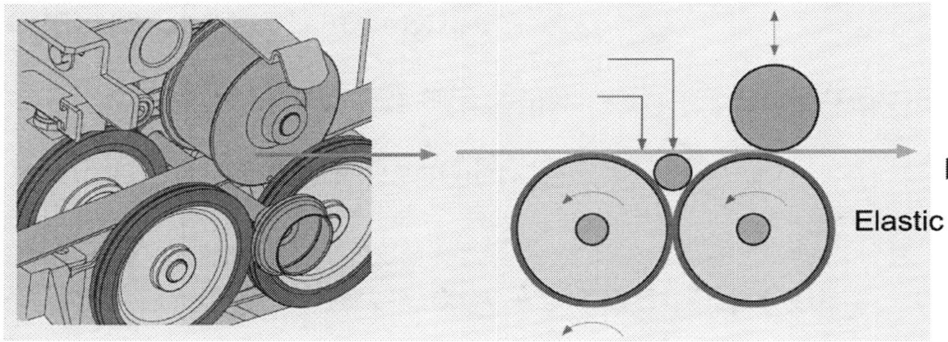

Why Magnetic Bearing ? - Expectations

RIETER

- Reducing energy consumption
- Reducing mechanical abrasion → higher life time
- Possible higher rotor speed ??
- Recovering of the idea of opposite yarn take up
  → different yarn properties ?
  → less dust deposit ?
- Utilization of newly available technologies and components

Project History

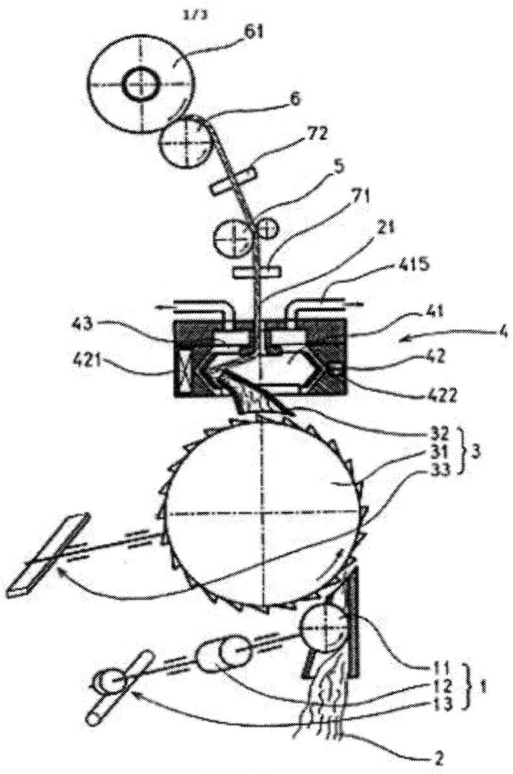

Fig. 1

1999 – first patent applied

2000 .. 2004 – our own (home) experiments

(not successful- only 30 000 r.p.m. reached)

2005 - collaboration with JKU/LCM starts

2006 - first running unit – 85 000 r.p.m.

2007 .. 2013 – many details solved,
step by step development

2012 - 150 000 r.p.m. reached

High-speed motor

Requirements

High efficiency

Low destabilizing stiffness in radial direction.

Short axial length to minimize air drag.

Mechanical protection of the permanent magnet to withstand high centrifugal forces (tensile strength of permanent magnet materials is limited to approx. 100MPa).

Back iron ring is part of the bearing too.

→ Mutual influence should be avoided!

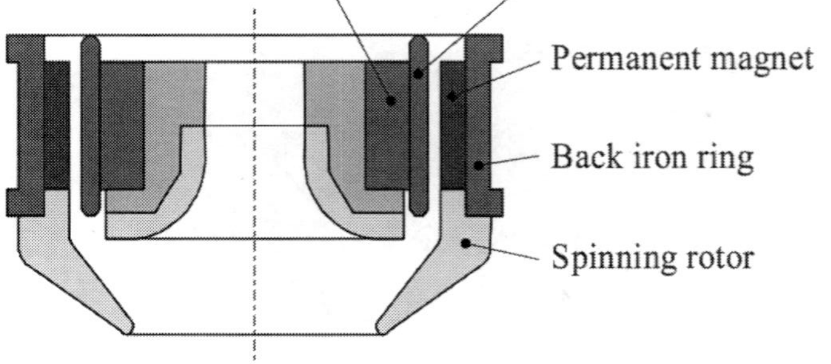

Reality

Rotor

Stator

Magnetic bearing

Combination of active and passive magnetic bearing

Radial DOFs are stabilized actively.

Axial and tilting DOFs are stabilized passively.

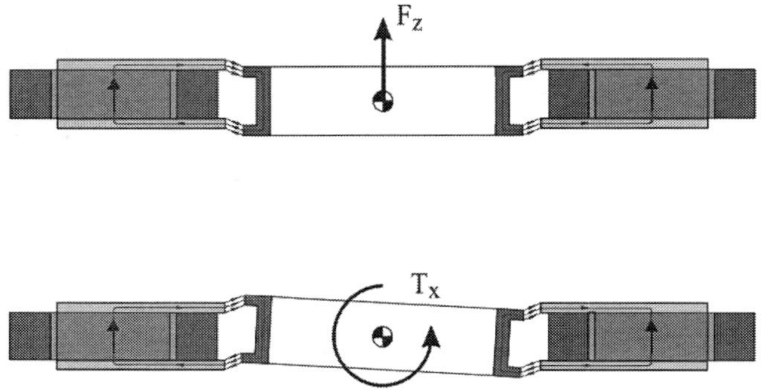

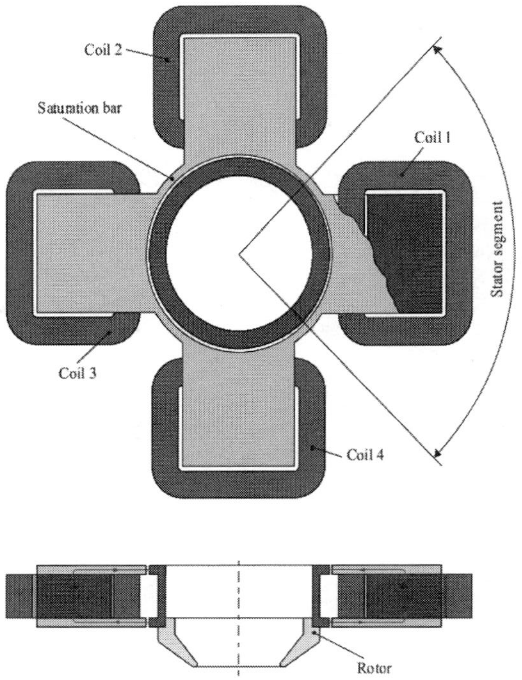

First experimental units

September 2006 : first runnig unit with magnetic bearing (85 000 r.p.m.)

One of the following prototypes (up to 130 000 r.p.m.)

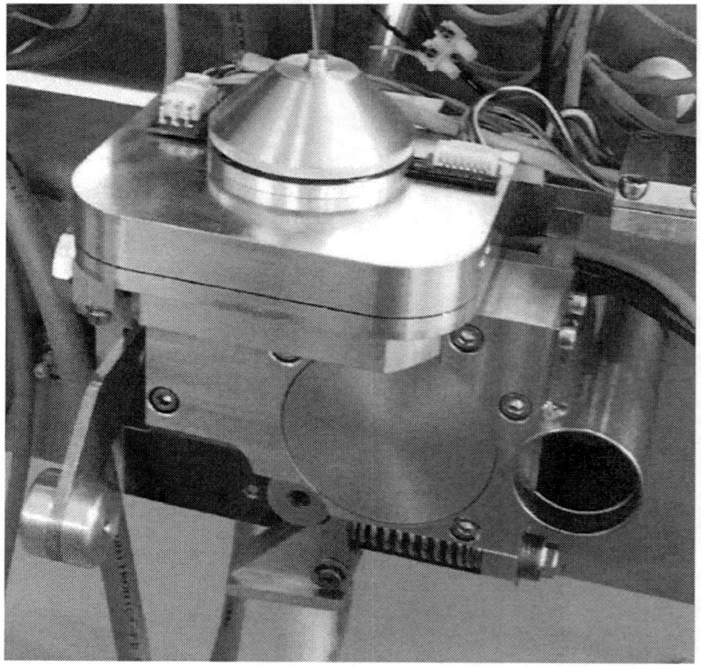

MAGNUM – electronics development

First experiments with LCM modular electronics

Step by step down sizing

and cost reduction

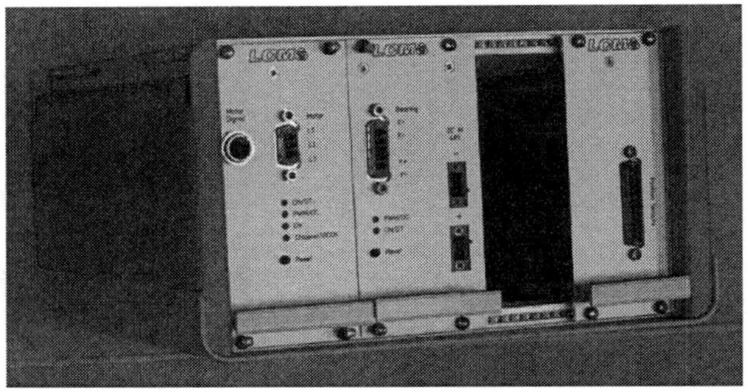

Rieter own development
with JKU/LCM support

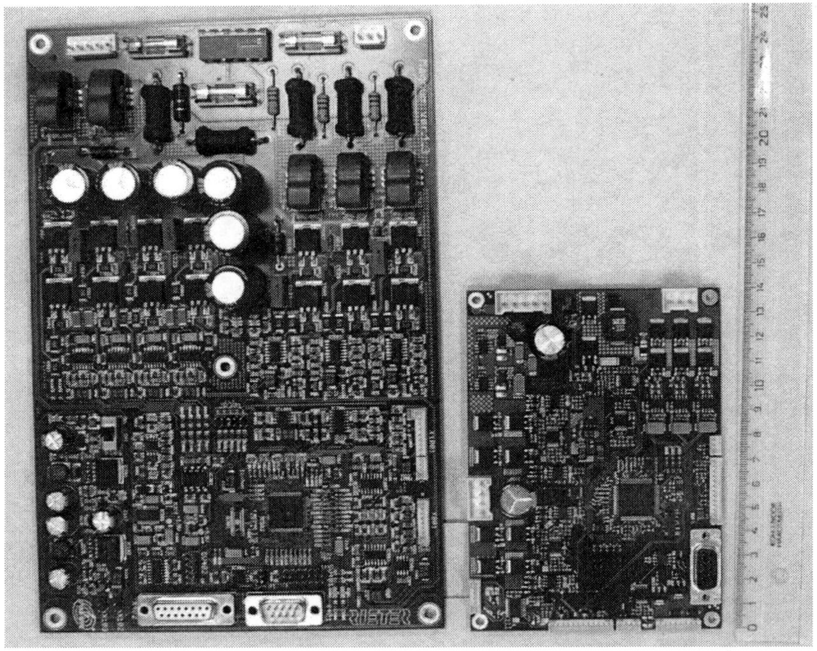

Experimental results - design

Prototype

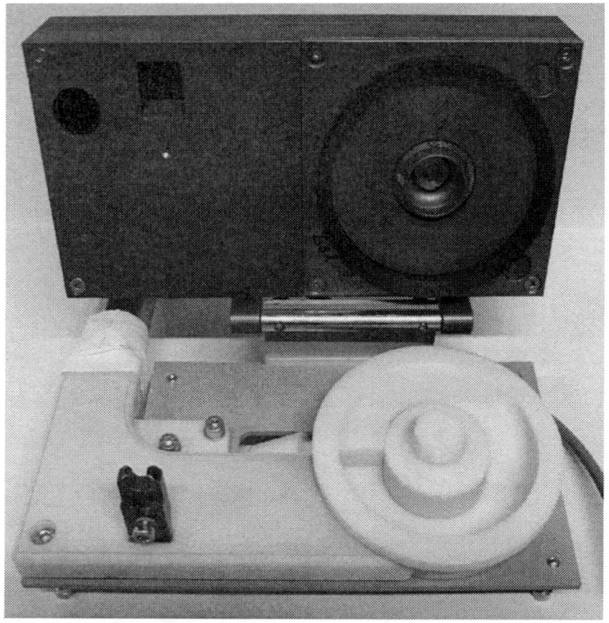

Integrated spinning unit

High speed motor

Magnetic bearing

Experimental results – textile technology, energy

Yarn propertiess – fully comparable with comercial machines

Dust deposit in the rotor groove - at least 10 x less

After 8 hours of spinning

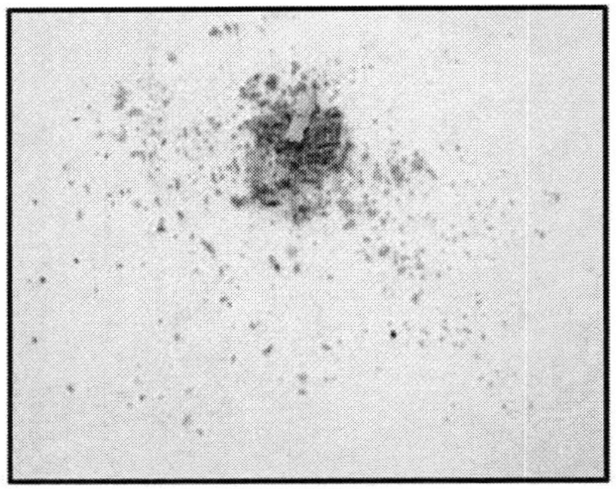

Conventional unit

unit with magnetic bearing

Energy consumption : reduced at least 30 % on 100 000 r.p.m.
(110 W → 75 W)

(The energy for spinning and air friction remain the same)

Success because of close collaboration between university and industry

RIETER

End

Thanks for your attention

Industrialisation Trends for Active Magnetic Bearings in the Turbomachinery Industry

ISMB 14
Linz, Austria
August 2014

WAUKESHA MAGNETIC BEARINGS
A Waukesha Bearings Business

EXPERT ENGINEERING. PROVEN RESULTS.

INTRODUCTION

EXPERT ENGINEERING. PROVEN RESULTS.

Introduction

Thirty Years and Hundreds of Heavy Industrial Applications

EXPERT ENGINEERING. PROVEN RESULTS.

Introduction

- Oil & Gas with trends both upstream & downstream from midstream
- Power Gen (notably, gas cooled nuclear)
- Size: 1-50 MW with most below 25 MW

Introduction

- Three generations of technology w/ 3rd now being introduced
  - 1st Generation: mid 1980's until mid 1990's
  - 2nd Generation: mid 1990's to present day

EXPERT ENGINEERING. PROVEN RESULTS.

GLOBALIZATION OF OPERATIONS

EXPERT ENGINEERING. PROVEN RESULTS.

Global Footprint to Support Growing Markets

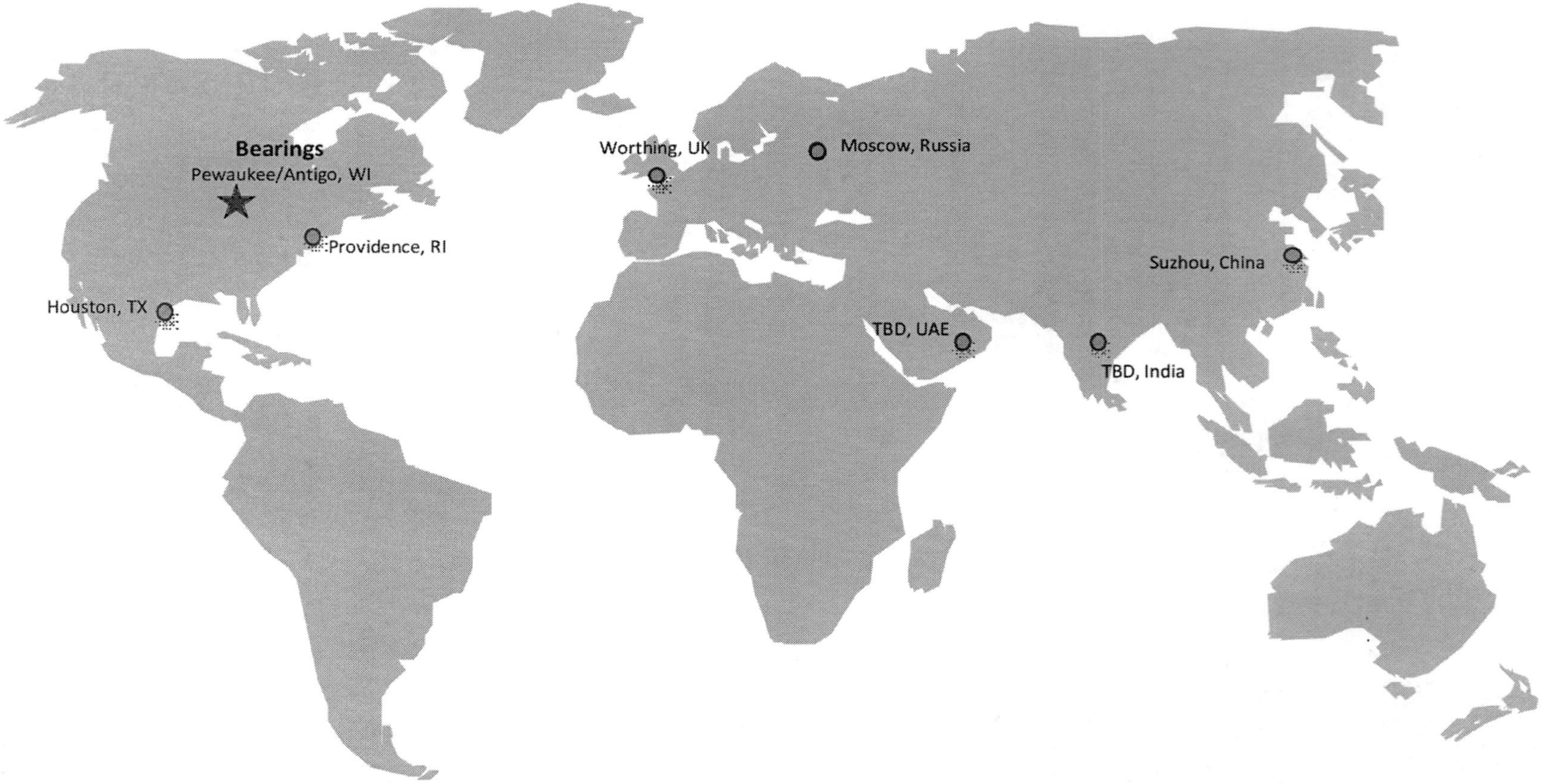

EXPERT ENGINEERING. PROVEN RESULTS.

Globalization of Operations

- Sales
- Manufacturing in Three Locations (China, UK, US)
  - IS0 9001 & 14001 in all locations
- Project engineering
- Product development
- Commissioning & Aftermarket Service
  - Including training programs in China, Japan, Russia, Saudi Arabia, US
  - Formal Obsolescence Policy

Global Operations Footprint

Antigo / Pewaukee

- 180 Employees
- 64,000 sq. ft.

Worthing, UK

- 50 Employees
- 30,000 sq. ft.

Suzhou, China

- 13 Employees
- 12,000 sq. ft.

Russia

- 7 Employees
- 1,000 sq. ft.

West Greenwich, RI

- 53 Employees
- 30,000 sq. ft.

DIFFERENTIATION FROM OTHER INDUSTRIAL PRODUCT

EXPERT ENGINEERING. PROVEN RESULTS.

Differentiation from Other Industrial Product

- Highly engineered
- Project based order volume
- Wide spectrum of types for sourcing & processing
  - Mechanical
  - Electrical
  - Electronic
  - Software
- Relatively low volume in comparison to other industrial or consumer products
- Long time spans between generations

PRODUCT STANDARDISATION TRENDS

EXPERT ENGINEERING. PROVEN RESULTS.

Product Standardisation Trends

- Centred on controls, not bearings

2 versions

4 versions

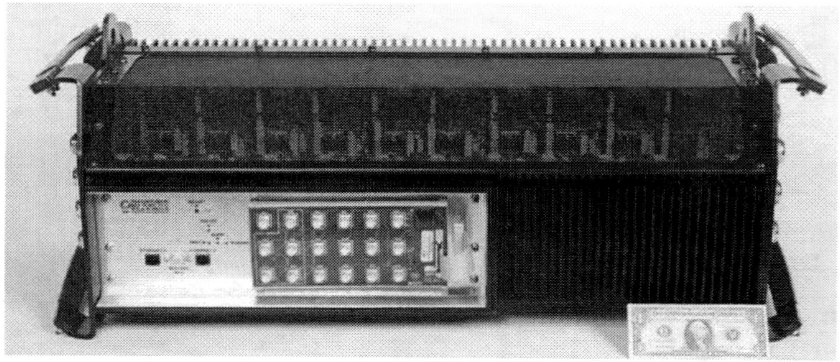

2 versions

EXPERT ENGINEERING. PROVEN RESULTS.

Product Standardisation Trends

- Suggestion for sharing of electronics across AMB suppliers not supported
  - Seen as inhibitor to innovation
  - Costs reductions would also be impeded
  - Contrary to ‘make or buy’ trend

GLOBAL TREND TOWARD 'MAKE' FROM 'BUY'

EXPERT ENGINEERING. PROVEN RESULTS.

Global Trend toward 'Make' from 'Buy'

- Examples D-R, GE, MDT, Siemens
  - Other OEMs have AMB technology in house e.g. KHI, MHI - there are undoubtedly others
- Rationale: Leveraging resources and related technologies that already exist

Global Trend toward 'Make' from 'Buy'

But this cycle was seen in fluid film bearing industry too

Waukesha, as a large worldwide bearing supplier, sees this trend reversing again

- Sharper focus by specialist suppliers
  - Cost and lead times
  - Aftermarket support
  - Product development

EXPERT ENGINEERING. PROVEN RESULTS.

LEAN ENTERPRISE ACTIVITY

EXPERT ENGINEERING. PROVEN RESULTS.

Lean Enterprise Activity

- Alignment and streamlining of supply chain with goals & values
- Six Sigma & Lean Manufacturing leading to
  - Cost reductions
  - Lead time reductions
  - While maintaining highest quality
- Continuous Improvement Program

EXPERT ENGINEERING. PROVEN RESULTS.

Bearing Lead Time Reduction Program

Past State

- Multiple small outside suppliers
  - Varying capabilities for quality and delivery performance
  - Most in high cost countries
- Sequential design and manufacture meaning that WMB is always in the protracted critical path
- Excess time and expense for intra transport

EXPERT ENGINEERING. PROVEN RESULTS.

Bearing Lead Time Reduction Program

<u>Future State</u>

- Cross site project manager
  - Quality gates across all sites
  - Daily constraint meetings
  - First Past Occurrence to capture issues and prevent future occurrences
- Key processes brought in house
  - e.g. coil winding
- Concurrent design & manufacturing
  - e.g. staged machining

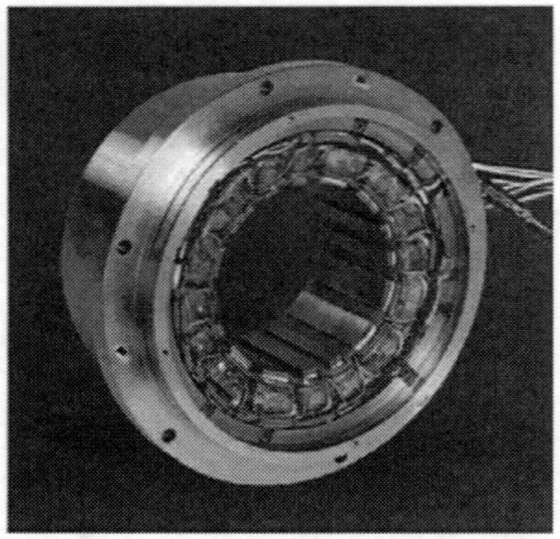

EXPERT ENGINEERING. PROVEN RESULTS.

Bearing Lead Time Reduction Program

PAST:

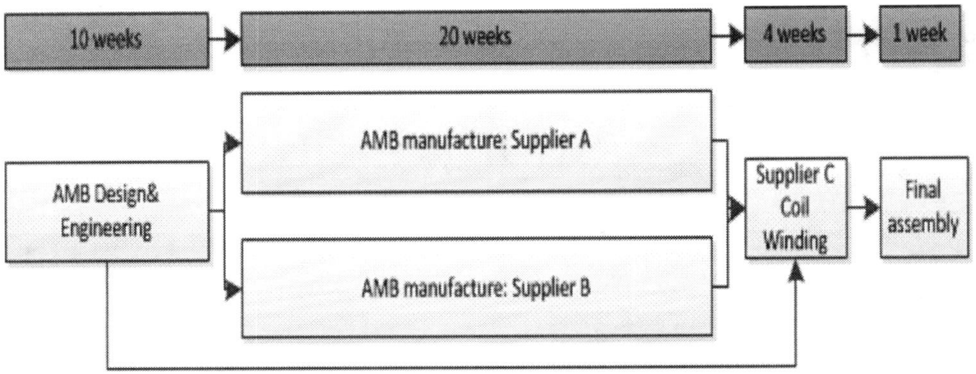

FUTURE:

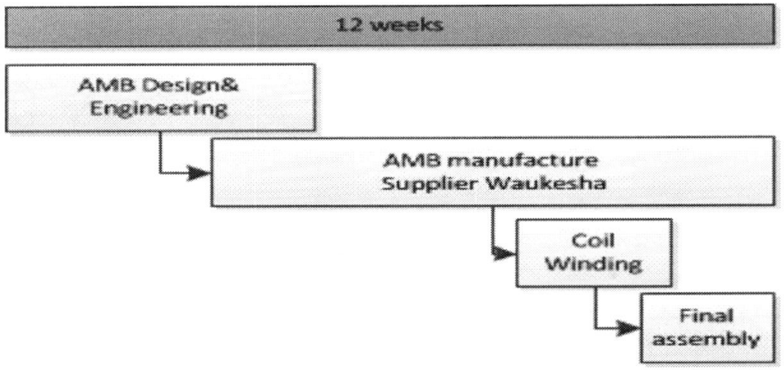

EXPERT ENGINEERING. PROVEN RESULTS.

Bearing Lead Time Reduction Program

<u>Benefits Summary</u>

- Consistent, shorter lead times
- Increased Quality
- One Value Stream
- Engineering and Manufacturing in Synch
  - Design for Manufacture
  - Schedules aligned

EXPERT ENGINEERING. PROVEN RESULTS.

Controller Lead Time Reduction Program

Past State

- Poor visibility on schedule
- High usage of contractors
- No "Demand Pull"
  - Daily stock outs of long lead items
  - No replenishment system
- Cells are not functional with little cross training, no skills matrix
- High internal return rate
- Hand corrected drawings don't get documented

EXPERT ENGINEERING. PROVEN RESULTS.

Controller Lead Time Reduction Program

Future State

- Single piece flow
- Consolidated supply base
- Demand Pulls
  - Replenishment system for consumables
  - High value items on supplier 'call offs'
- Standardized cell layouts
- Quality gates
- First Past Occurrence system

Controller Lead Time Reduction Program

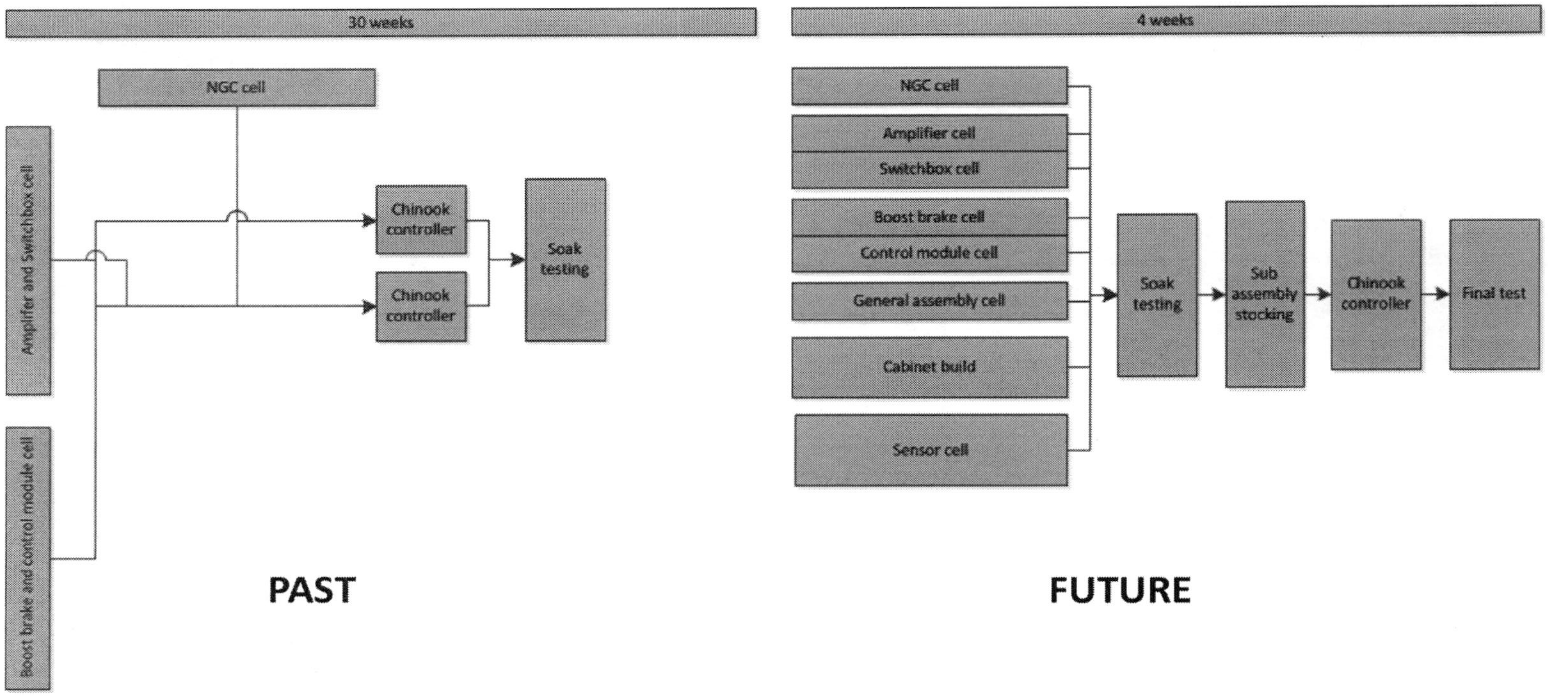

EXPERT ENGINEERING. PROVEN RESULTS.

Controller Lead Time Reduction Program

Benefits Summary

- Lead times: consistent, shorter and visible
- Build to stock via Demand Pull
  - Zero stock outs
- Elimination of contractors
- Increased Quality
  - 100% validation testing
  - Reduced infant mortalities
  - 5S daily audits
  - Consistent build areas, variations visible

EXPERT ENGINEERING. PROVEN RESULTS.

IDEATION

EXPERT ENGINEERING. PROVEN RESULTS.

Formal Ideation System

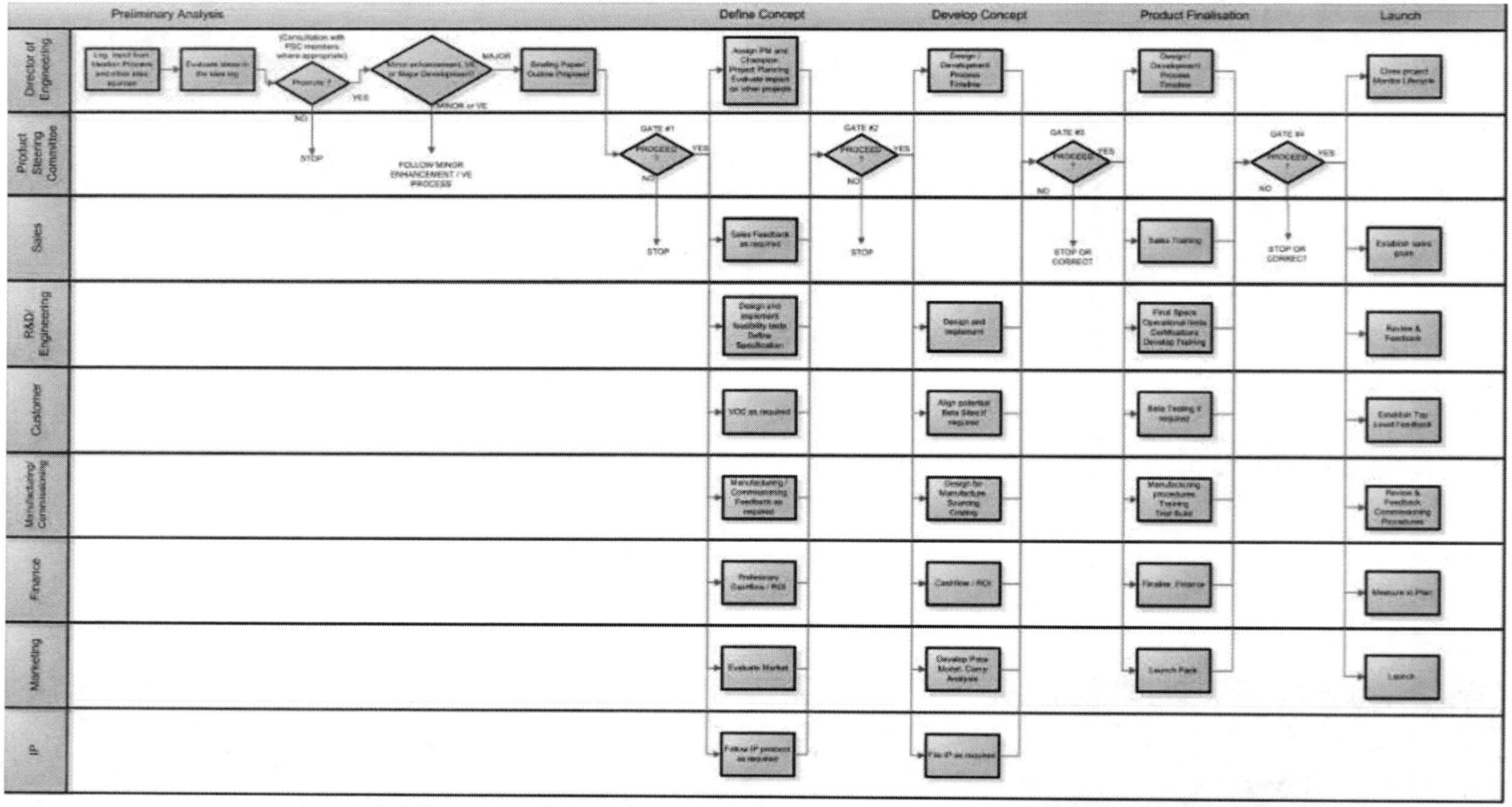

EXPERT ENGINEERING. PROVEN RESULTS.

ISMB 2014 Author Index

A

B

C

D

E

F

G

H

I

J

K

L

M

N

O

P

Q

R

S

T

U

V

W

X

Y

Z